Advanced Dynamics

Advanced Dynamics
Rigid Body, Multibody, and Aerospace Applications

Reza N. Jazar
School of Aerospace, Mechanical, and Manufacturing Engineering
RMIT University
Melbourne, Australia

WILEY

JOHN WILEY & SONS, INC.

This book is printed on acid-free paper.♾

Published by John Wiley & Sons, Inc., Hoboken, New Jersey

Published simultaneously in Canada

Library of Congress Cataloging-in-Publication Data:

Jazar, Reza N.
Advanced dynamics : rigid body, multibody, and aerospace applications / Reza N. Jazar.
p. cm.
Includes index.
ISBN 978-0-470-39835-7 (hardback); ISBN 978-0-470-89211-4 (ebk); ISBN 978-0-470-89212-1 (ebk); ISBN 978-0-470-89213-8 (ebk); ISBN 978-0-470-95002-9 (ebk); ISBN 978-0-470-95159-0 (ebk); ISBN 978-0-470-95176-7 (ebk)
1. Dynamics. I. Title.
TA352.J39 2011
620.1′04—dc22

2010039778

Printed in the United States of America

10 9 8 7 6 5 4 3 2 1

The answer is waiting for the right question.

To my daughter
Vazan,
my son
Kavosh,
and my wife,
Mojgan

Contents

Note: A star (★) indicates a more advanced subject or example that is not designed for undergraduate teaching and can be dropped in the first reading.

Preface

This book is arranged in such a way, and covers those materials, that I would have liked to have had available as a student: straightforward, right to the point, analyzing a subject from different viewpoints, showing practical aspects and application of every subject, considering physical meaning and sense, with interesting and clear examples. This book was written for graduate students who want to learn every aspect of dynamics and its application. It is based on two decades of research and teaching courses in advanced dynamics, attitude dynamics, vehicle dynamics, classical mechanics, multibody dynamics, and robotics.

I know that the best way to learn dynamics is repeat and practice, repeat and practice. So, you are going to see some repeating and much practicing in this book. I begin with fundamental subjects in dynamics and end with advanced materials. I introduce the fundamental knowledge used in particle and rigid-body dynamics. This knowledge can be used to develop computer programs for analyzing the kinematics, dynamics, and control of dynamic systems.

The subject of rigid body has been at the heart of dynamics since the 1600s and remains alive with modern developments of applications. Classical kinematics and dynamics have their roots in the work of great scientists of the past four centuries who established the methodology and understanding of the behavior of dynamic systems. The development of dynamic science, since the beginning of the twentieth century, has moved toward analysis of controllable man-made autonomous systems.

LEVEL OF THE BOOK

More than half of the material is in common with courses in advanced dynamics, classical mechanics, multibody dynamics, and spacecraft dynamics. Graduate students in mechanical and aerospace engineering have the potential to work on projects that are related to either of these engineering disciplines. However, students have not seen enough applications in all areas. Although their textbooks introduce rigid-body dynamics, mechanical engineering students only work on engineering applications while aerospace engineering students only see spacecraft applications and attitude dynamics. The reader of this text will have no problem in analyzing a dynamic system in any of these areas. This book bridges the gap between rigid-body, classical, multibody, and spacecraft dynamics for graduate students and specialists in mechanical and aerospace engineering. Engineers and graduate students who read this book will be able to apply their knowledge to a wide range of engineering disciplines.

This book is aimed primarily at graduate students in engineering, physics, and mathematics. It is especially useful for courses in the dynamics of rigid bodies such as advanced dynamics, classical mechanics, attitude dynamics, spacecraft dynamics, and multibody dynamics. It provides both fundamental and advanced topics on the kinematics and dynamics of rigid bodies. The whole book can be covered in two successive courses; however, it is possible to jump over some sections and cover the book in one course.

The contents of the book have been kept at a fairly theoretical–practical level. Many concepts are deeply explained and their use emphasized, and most of the related theory and formal proofs have been explained. Throughout the book, a strong emphasis is put on the physical meaning of the concepts introduced. Topics that have been selected are of high interest in the field. An attempt has been made to expose the students to a broad range of topics and approaches.

ORGANIZATION OF THE BOOK

The book begins with a review of coordinate systems and particle dynamics. This introduction will teach students the importance of coordinate frames. Transformation and rotation theory along with differentiation theory in different coordinate frames will provide the required background to learn rigid-body dynamics based on Newton–Euler principles. The method will show its applications in rigid-body and multibody dynamics. The Newton equations of motion will be transformed to Lagrangian equations as a bridge to analytical dynamics. The methods of Lagrange will be applied on particles and rigid bodies.

Through its examination of specialist applications highlighting the many different aspects of dynamics, this text provides an excellent insight into advanced systems without restricting itself to a particular discipline. The result is essential reading for all those requiring a general understanding of the more advanced aspects of rigid-body dynamics.

The text is organized such that it can be used for teaching or for self-study. Part I "Fundamentals," contains general preliminaries and provides a deep review of the kinematics and dynamics. A new classification of vectors is the highlight of Part I.

Part II, "Geometric Kinematics," presents the mathematics of the displacement of rigid bodies using the matrix method. The order-free transformation theory, classification of industrial links, kinematics of spherical wrists, and mechanical surgery of multibodies are the highlights of Part II.

Part III, "Derivative Kinematics," presents the mathematics of velocity and acceleration of rigid bodies. The time derivatives of vectors in different coordinate frames, Rāzī acceleration, integrals of motion, and methods of dynamics are the highlights of Part III.

Part IV, "Dynamics," presents a detailed discussion of rigid-body and Lagrangian dynamics. Rigid-body dynamics is studied from different viewpoints to provide different classes of solutions. Lagrangian mechanics is reviewed in detail from an applied viewpoint. Multibody dynamics and Lagrangian mechanics in generalized coordinates are the highlights of Part IV.

METHOD OF PRESENTATION

The structure of the presentation is in a *fact–reason–application* fashion. The "fact" is the main subject we introduce in each section. Then the "reason" is given as a proof.

Finally the "application" of the fact is examined in some examples. The examples are a very important part of the book because they show how to implement the knowledge introduced in the facts. They also cover some other material needed to expand the subject.

PREREQUISITES

The book is written for graduate students, so the assumption is that users are familiar with the fundamentals of kinematics and dynamics as well as basic knowledge of linear algebra, differential equations, and the numerical method.

UNIT SYSTEM

The system of units adopted in this book is, unless otherwise stated, the International System of Units (SI). The units of degree (deg) and radian (rad) are utilized for variables representing angular quantities.

SYMBOLS

- Lowercase bold letters indicate a vector. Vectors may be expressed in an n-dimensional Euclidean space:

$$\begin{array}{cccccc} \mathbf{r}, & \mathbf{s}, & \mathbf{d}, & \mathbf{a}, & \mathbf{b}, & \mathbf{c} \\ \mathbf{p}, & \mathbf{q}, & \mathbf{v}, & \mathbf{w}, & \mathbf{y}, & \mathbf{z} \\ \boldsymbol{\omega}, & \boldsymbol{\alpha}, & \boldsymbol{\epsilon}, & \boldsymbol{\theta}, & \boldsymbol{\delta}, & \boldsymbol{\phi} \end{array}$$

- Uppercase bold letters indicate a dynamic vector or a dynamic matrix:

$$\mathbf{F}, \ \mathbf{M}, \ \mathbf{I}, \ \mathbf{L}$$

- Lowercase letters with a hat indicate a unit vector. Unit vectors are not bolded:

$$\begin{array}{c} \hat{\imath}, \ \hat{\jmath}, \ \hat{k}, \ \hat{e}, \ \hat{u}, \ \hat{n} \\ \hat{I}, \ \hat{J}, \ \hat{K}, \ \hat{e}_\theta, \ \hat{e}_\varphi, \ \hat{e}_\psi \end{array}$$

- Lowercase letters with a tilde indicate a 3×3 skew symmetric matrix associated to a vector:

$$\tilde{a} = \begin{bmatrix} 0 & -a_3 & a_2 \\ a_3 & 0 & -a_1 \\ -a_2 & a_1 & 0 \end{bmatrix} \qquad \mathbf{a} = \begin{bmatrix} a_1 \\ a_2 \\ a_3 \end{bmatrix}$$

- An arrow above two uppercase letters indicates the start and end points of a position vector:

$$\overrightarrow{ON} = \text{a position vector from point } O \text{ to point } N$$

- A double arrow above a lowercase letter indicates a 4×4 matrix associated to a quaternion:

$$\overleftrightarrow{q} = \begin{bmatrix} q_0 & -q_1 & -q_2 & -q_3 \\ q_1 & q_0 & -q_3 & q_2 \\ q_2 & q_3 & q_0 & -q_1 \\ q_3 & -q_2 & q_1 & q_0 \end{bmatrix}$$

$$q = q_0 + q_1 i + q_2 j + q_3 k$$

- The length of a vector is indicated by a nonbold lowercase letter:

$$r = |\mathbf{r}| \qquad a = |\mathbf{a}| \qquad b = |\mathbf{b}| \qquad s = |\mathbf{s}|$$

- Capital letters A, Q, R, and T indicate rotation or transformation matrices:

$$Q_{Z,\alpha} = \begin{bmatrix} \cos\alpha & -\sin\alpha & 0 \\ \sin\alpha & \cos\alpha & 0 \\ 0 & 0 & 1 \end{bmatrix} \qquad {}^{G}T_{B} = \begin{bmatrix} c\alpha & 0 & -s\alpha & -1 \\ 0 & 1 & 0 & 0.5 \\ s\alpha & 0 & c\alpha & 0.2 \\ 0 & 0 & 0 & 1 \end{bmatrix}$$

- Capital letter B is utilized to denote a body coordinate frame:

$$B(oxyz), \quad B(Oxyz), \quad B_1(o_1 x_1 y_1 z_1)$$

- Capital letter G is utilized to denote a global, inertial, or fixed coordinate frame:

$$G, \quad G(XYZ), \quad G(OXYZ)$$

- Right subscript on a transformation matrix indicates the *departure* frames:

$$T_B = \text{transformation matrix from frame } B(oxyz)$$

- Left superscript on a transformation matrix indicates the *destination* frame:

$${}^{G}T_{B} = \text{transformation matrix from frame } B(oxyz) \text{ to frame } G(OXYZ)$$

- Whenever there is no subscript or superscript, the matrices are shown in brackets:

$$[T] = \begin{bmatrix} c\alpha & 0 & -s\alpha & -1 \\ 0 & 1 & 0 & 0.5 \\ s\alpha & 0 & c\alpha & 0.2 \\ 0 & 0 & 0 & 1 \end{bmatrix}$$

- Left superscript on a vector denotes the frame in which the vector is expressed. That superscript indicates the frame that the vector belongs to, so the vector is expressed using the unit vectors of that frame:

$${}^{G}\mathbf{r} = \text{position vector expressed in frame } G(OXYZ)$$

- Right subscript on a vector denotes the tip point to which the vector is referred:

$$^{G}\mathbf{r}_P = \text{position vector of point } P$$
$$\text{expressed in coordinate frame } G(OXYZ)$$

- Right subscript on an angular velocity vector indicates the frame to which the angular vector is referred:

$$\omega_B = \text{angular velocity of the body coordinate frame } B(oxyz)$$

- Left subscript on an angular velocity vector indicates the frame with respect to which the angular vector is measured:

$$_G\omega_B = \text{angular velocity of the body coordinateframe } B(oxyz)$$
$$\text{with respect to the global coordinate frame } G(OXYZ)$$

- Left superscript on an angular velocity vector denotes the frame in which the angular velocity is expressed:

$$^{B_2}_{G}\omega_{B_1} = \text{angular velocity of the bodycoordinate frame } B_1$$
$$\text{with respect to the global coordinate frame } G$$
$$\text{and expressed in body coordinate frame } B_2$$

  Whenever the subscript and superscript of an angular velocity are the same, we usually drop the left superscript:

$$_G\omega_B \equiv {}^{G}_{G}\omega_B$$

  Also for position, velocity, and acceleration vectors, we drop the left subscripts if it is the same as the left superscript:

$$^{B}_{B}\mathbf{v}_P \equiv {}^{B}\mathbf{v}_P$$

- If the right subscript on a force vector is a number, it indicates the number of coordinate frames in a serial robot. Coordinate frame B_i is set up at joint $i+1$:

$$\mathbf{F}_i = \text{force vector at joint } i+1 \text{ measured at the originof } B_i(oxyz)$$

  At joint i there is always an action force $\mathbf{F}_i$ that link (i) applies on link $(i+1)$ and a reaction force $-\mathbf{F}_i$ that link $(i+1)$ applies on link (i). On link (i) there is always an action force $\mathbf{F}_{i-1}$ coming from link $(i-1)$ and a reaction force $-\mathbf{F}_i$ coming from link $(i+1)$. The action force is called the *driving force* , and the reaction force is called the *driven force*.
- If the right subscript on a moment vector is a number, it indicates the number of coordinate frames in a serial robot. Coordinate frame B_i is set up at joint $i+1$:

$$\mathbf{M}_i = \text{moment vector at joint } i+1 \text{ measured at theorigin of } B_i(oxyz)$$

  At joint i there is always an action moment $\mathbf{M}_i$ that link (i) applies on link $(i+1)$, and a reaction moment $-\mathbf{M}_i$ that link $(i+1)$ applies on link (i). On link (i) there is always an action moment $\mathbf{M}_{i-1}$ coming from link $(i-1)$ and a reaction moment $-\mathbf{M}_i$ coming from link $(i+1)$. The action moment is called the *driving moment*, and the reaction moment is called the *driven moment*.

- Left superscript on derivative operators indicates the frame in which the derivative of a variable is taken:

$$\frac{{}^G d}{dt}x, \quad \frac{{}^G d}{dt}\,{}^B\mathbf{r}_P, \quad \frac{{}^B d}{dt}\,{}^G_B\mathbf{r}_P$$

If the variable is a vector function and the frame in which the vector is defined is the same as the frame in which a time derivative is taken, we may use the short notation

$$\frac{{}^G d}{dt}\,{}^G\mathbf{r}_P = {}^G\dot{\mathbf{r}}_P, \qquad \frac{{}^B d}{dt}\,{}^B_o\mathbf{r}_P = {}^B_o\dot{\mathbf{r}}_P$$

and write equations simpler. For example,

$${}^G\mathbf{v} = \frac{{}^G d}{dt}\,{}^G\mathbf{r}(t) = {}^G\dot{\mathbf{r}}$$

- If followed by angles, lowercase c and s denote cos and sin functions in mathematical equations:

$$c\alpha = \cos\alpha \qquad s\varphi = \sin\varphi$$

- Capital bold letter $\mathbf{I}$ indicates a unit matrix, which, depending on the dimension of the matrix equation, could be a 3×3 or a 4×4 unit matrix. $\mathbf{I}_3$ or $\mathbf{I}_4$ are also being used to clarify the dimension of $\mathbf{I}$. For example,

$$\mathbf{I} = \mathbf{I}_3 = \begin{bmatrix} 1 & 0 & 0 \\ 0 & 1 & 0 \\ 0 & 0 & 1 \end{bmatrix}$$

- Two parallel joint axes are indicated by a parallel sign ($\|$).
- Two orthogonal joint axes are indicated by an orthogonal sign ($\vdash$). Two orthogonal joint axes are intersecting at a right angle.
- Two perpendicular joint axes are indicated by a perpendicular sign ($\perp$). Two perpendicular joint axes are at a right angle with respect to their common normal.

Advanced Dynamics

Part I

Fundamentals

The required fundamentals of kinematics and dynamics are reviewed in this part. It should prepare us for the more advanced parts.

1

Fundamentals of Kinematics

Vectors and coordinate frames are human-made tools to study the motion of particles and rigid bodies. We introduce them in this chapter to review the fundamentals of kinematics.

1.1 COORDINATE FRAME AND POSITION VECTOR

To indicate the position of a point P relative to another point O in a three-dimensional (3D) space, we need to establish a coordinate frame and provide three relative coordinates. The three coordinates are scalar functions and can be used to define a position vector and derive other kinematic characteristics.

1.1.1 Triad

Take four non-coplanar points O, A, B, C and make three lines OA, OB, OC. The *triad* $OABC$ is defined by taking the lines OA, OB, OC as a rigid body. The position of A is arbitrary provided it stays on the same side of O. The positions of B and C are similarly selected. Now rotate OB about O in the plane OAB so that the angle AOB becomes 90 deg. Next, rotate OC about the line in AOB to which it is perpendicular until it becomes perpendicular to the plane AOB. The new triad $OABC$ is called an *orthogonal triad*.

Having an orthogonal triad $OABC$, another triad $OA'BC$ may be derived by moving A to the other side of O to make the *opposite triad* $OA'BC$. All orthogonal triads can be superposed either on the triad $OABC$ or on its opposite $OA'BC$.

One of the two triads $OABC$ and $OA'BC$ can be defined as being a *positive triad* and used as a *standard*. The other is then defined as a *negative triad*. It is immaterial which one is chosen as positive; however, usually the *right-handed convention* is chosen as positive. The right-handed convention states that the direction of rotation from OA to OB propels a *right-handed screw* in the direction OC. A right-handed or positive orthogonal triad cannot be superposed to a left-handed or negative triad. Therefore, there are only two essentially distinct types of triad. This is a property of 3D space.

We use an orthogonal triad $OABC$ with scaled lines OA, OB, OC to locate a point in 3D space. When the three lines OA, OB, OC have scales, then such a triad is called a *coordinate frame*.

Every moving body is carrying a *moving* or *body frame* that is attached to the body and moves with the body. A body frame accepts every motion of the body and may also be called a *local frame*. The position and orientation of a body with respect to other frames is expressed by the position and orientation of its local coordinate frame.

When there are several relatively moving coordinate frames, we choose one of them as a *reference frame* in which we express motions and measure kinematic information. The motion of a body may be observed and measured in different reference frames; however, we usually compare the motion of different bodies in the *global reference frame*. A global reference frame is assumed to be motionless and attached to the ground.

Example 1 Cyclic Interchange of Letters In any orthogonal triad $OABC$, cyclic interchanging of the letters ABC produce another orthogonal triad superposable on the original triad. Cyclic interchanging means relabeling A as B, B as C, and C as A or picking any three consecutive letters from $ABCABCABC\dots$.

Example 2 ★ Independent Orthogonal Coordinate Frames Having only two types of orthogonal triads in 3D space is associated with the fact that a plane has just two sides. In other words, there are two opposite normal directions to a plane. This may also be interpreted as: we may arrange the letters A, B, and C in just two orders when cyclic interchange is allowed:

$$ABC, \quad ACB$$

In a 4D space, there are six cyclic orders for four letters A, B, C, and D:

$$ABCD, \quad ABDC, \quad ACBD, \quad ACDB, \quad ADBC, \quad ADCB$$

So, there are six different tetrads in a 4D space.

In an nD space there are $(n-1)!$ cyclic orders for n letters, so there are $(n-1)!$ different coordinate frames in an nD space.

Example 3 Right-Hand Rule A right-handed triad can be identified by a right-hand rule that states: When we indicate the OC axis of an orthogonal triad by the thumb of the right hand, the other fingers should turn from OA to OB to close our fist.

The right-hand rule also shows the rotation of Earth when the thumb of the right hand indicates the north pole.

Push your right thumb to the center of a clock, then the other fingers simulate the rotation of the clock's hands.

Point your index finger of the right hand in the direction of an electric current. Then point your middle finger in the direction of the magnetic field. Your thumb now points in the direction of the magnetic force.

If the thumb, index finger, and middle finger of the right hand are held so that they form three right angles, then the thumb indicates the Z-axis when the index finger indicates the X-axis and the middle finger the Y-axis.

1.1.2 Coordinate Frame and Position Vector

Consider a positive orthogonal triad $OABC$ as is shown in Figure 1.1. We select a *unit length* and define a *directed line* $\hat{\imath}$ on OA with a unit length. A point P_1 on OA is at a distance x from O such that the directed line $\overrightarrow{OP_1}$ from O to P_1 is $\overrightarrow{OP_1} = x\hat{\imath}$. The

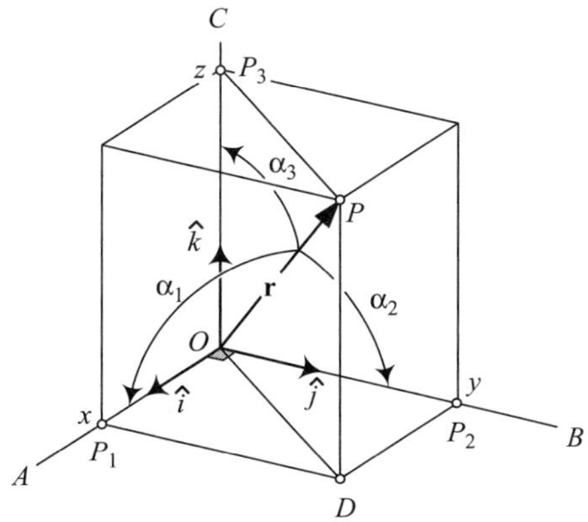

Figure 1.1 A positive orthogonal triad $OABC$, unit vectors $\hat{\imath}$, $\hat{\jmath}$, $\hat{k}$, and a position vector **r** with components x, y, z.

directed line $\hat{\imath}$ is called a *unit vector* on OA, the unit length is called the *scale*, point O is called the *origin*, and the real number x is called the $\hat{\imath}$*-coordinate* of P_1. The distance x may also be called the $\hat{\imath}$ *measure number* of $\overrightarrow{OP_1}$. Similarly, we define the unit vectors $\hat{\jmath}$ and $\hat{k}$ on OB and OC and use y and z as their coordinates, respectively. Although it is not necessary, we usually use the same scale for $\hat{\imath}$, $\hat{\jmath}$, $\hat{k}$ and refer to OA, OB, OC by $\hat{\imath}$, $\hat{\jmath}$, $\hat{k}$ and also by x, y, z.

The scalar coordinates x, y, z are respectively the length of projections of P on OA, OB, and OC and may be called the *components* of **r**. The components x, y, z are independent and we may vary any of them while keeping the others unchanged.

A scaled positive orthogonal triad with unit vectors $\hat{\imath}$, $\hat{\jmath}$, $\hat{k}$ is called an *orthogonal coordinate frame*. The position of a point P with respect to O is defined by three coordinates x, y, z and is shown by a *position vector* $\mathbf{r} = \mathbf{r}_P$:

$$\mathbf{r} = \mathbf{r}_P = x\hat{\imath} + y\hat{\jmath} + z\hat{k} \tag{1.1}$$

To work with multiple coordinate frames, we indicate coordinate frames by a capital letter, such as G and B, to clarify the coordinate frame in which the vector **r** is expressed. We show the name of the frame as a left superscript to the vector:

$$^{B}\mathbf{r} = x\hat{\imath} + y\hat{\jmath} + z\hat{k} \tag{1.2}$$

A vector **r** is expressed in a coordinate frame B only if its unit vectors $\hat{\imath}$, $\hat{\jmath}$, $\hat{k}$ belong to the axes of B. If necessary, we use a left superscript B and show the unit vectors as $^{B}\hat{\imath}$, $^{B}\hat{\jmath}$, $^{B}\hat{k}$ to indicate that $\hat{\imath}$, $\hat{\jmath}$, $\hat{k}$ belong to B:

$$^{B}\mathbf{r} = x\ ^{B}\hat{\imath} + y\ ^{B}\hat{\jmath} + z\ ^{B}\hat{k} \tag{1.3}$$

We may drop the superscript B as long as we have just one coordinate frame.

The distance between O and P is a scalar number r that is called the *length*, *magnitude*, *modulus*, *norm*, or *absolute value* of the vector **r**:

$$r = |\mathbf{r}| = \sqrt{x^2 + y^2 + z^2} \tag{1.4}$$

We may define a new unit vector $\hat{u}_r$ on $\mathbf{r}$ and show $\mathbf{r}$ by

$$\mathbf{r} = r\hat{u}_r \tag{1.5}$$

The equation $\mathbf{r} = r\hat{u}_r$ is called the *natural expression* of $\mathbf{r}$, while the equation $\mathbf{r} = x\hat{\imath} + y\hat{\jmath} + z\hat{k}$ is called the *decomposition* or *decomposed expression* of $\mathbf{r}$ over the axes $\hat{\imath}$, $\hat{\jmath}$, $\hat{k}$. Equating (1.1) and (1.5) shows that

$$\begin{aligned}\hat{u}_r &= \frac{x\hat{\imath} + y\hat{\jmath} + z\hat{k}}{r} = \frac{x\hat{\imath} + y\hat{\jmath} + z\hat{k}}{\sqrt{x^2 + y^2 + z^2}} \\ &= \frac{x}{\sqrt{x^2 + y^2 + z^2}}\hat{\imath} + \frac{y}{\sqrt{x^2 + y^2 + z^2}}\hat{\jmath} + \frac{z}{\sqrt{x^2 + y^2 + z^2}}\hat{k} \end{aligned} \tag{1.6}$$

Because the length of $\hat{u}_r$ is unity, the components of $\hat{u}_r$ are the cosines of the angles α_1, α_2, α_3 between $\hat{u}_r$ and $\hat{\imath}$, $\hat{\jmath}$, $\hat{k}$, respectively:

$$\cos\alpha_1 = \frac{x}{r} = \frac{x}{\sqrt{x^2 + y^2 + z^2}} \tag{1.7}$$

$$\cos\alpha_2 = \frac{y}{r} = \frac{y}{\sqrt{x^2 + y^2 + z^2}} \tag{1.8}$$

$$\cos\alpha_3 = \frac{z}{r} = \frac{z}{\sqrt{x^2 + y^2 + z^2}} \tag{1.9}$$

The cosines of the angles α_1, α_2, α_3 are called the *directional cosines* of $\hat{u}_r$, which, as is shown in Figure 1.1, are the same as the directional cosines of any other vector on the same axis as $\hat{u}_r$, including $\mathbf{r}$.

Equations (1.7)–(1.9) indicate that the three directional cosines are related by the equation

$$\cos^2\alpha_1 + \cos^2\alpha_2 + \cos^3\alpha_3 = 1 \tag{1.10}$$

Example 4 Position Vector of a Point P Consider a point P with coordinates $x = 3$, $y = 2$, $z = 4$. The position vector of P is

$$\mathbf{r} = 3\hat{\imath} + 2\hat{\jmath} + 4\hat{k} \tag{1.11}$$

The distance between O and P is

$$r = |\mathbf{r}| = \sqrt{3^2 + 2^2 + 4^2} = 5.3852 \tag{1.12}$$

and the unit vector $\hat{u}_r$ on $\mathbf{r}$ is

$$\begin{aligned}\hat{u}_r &= \frac{x}{r}\hat{\imath} + \frac{y}{r}\hat{\jmath} + \frac{z}{r}\hat{k} = \frac{3}{5.3852}\hat{\imath} + \frac{2}{5.3852}\hat{\jmath} + \frac{4}{5.3852}\hat{k} \\ &= 0.55708\hat{\imath} + 0.37139\hat{\jmath} + 0.74278\hat{k}\end{aligned} \tag{1.13}$$

The directional cosines of $\hat{u}_r$ are

$$\cos\alpha_1 = \frac{x}{r} = 0.55708$$
$$\cos\alpha_2 = \frac{y}{r} = 0.37139 \tag{1.14}$$
$$\cos\alpha_3 = \frac{z}{r} = 0.74278$$

and therefore the angles between $\mathbf{r}$ and the x-, y-, z-axes are

$$\alpha_1 = \cos^{-1}\frac{x}{r} = \cos^{-1}0.55708 = 0.97993\,\text{rad} \approx 56.146\,\text{deg}$$
$$\alpha_2 = \cos^{-1}\frac{y}{r} = \cos^{-1}0.37139 = 1.1903\,\text{rad} \approx 68.199\,\text{deg} \tag{1.15}$$
$$\alpha_3 = \cos^{-1}\frac{z}{r} = \cos^{-1}0.74278 = 0.73358\,\text{rad} \approx 42.031\,\text{deg}$$

Example 5 Determination of Position Figure 1.2 illustrates a point P in a scaled triad $OABC$. We determine the position of the point P with respect to O by:

1. Drawing a line PD parallel OC to meet the plane AOB at D
2. Drawing DP_1 parallel to OB to meet OA at P_1

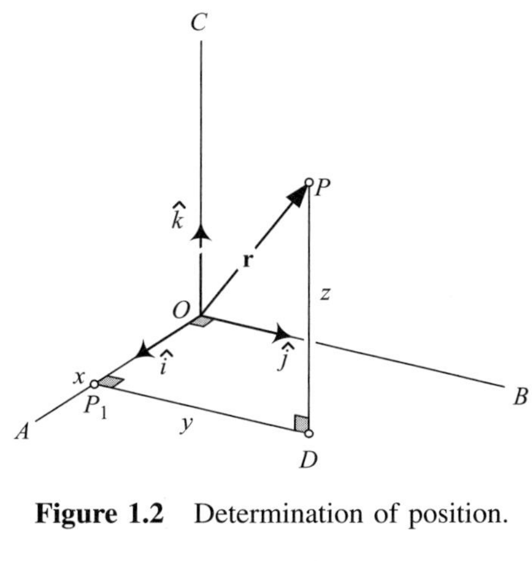

Figure 1.2 Determination of position.

The lengths OP_1, P_1D, DP are the coordinates of P and determine its position in triad $OABC$. The line segment OP is a diagonal of a parallelepiped with OP_1, P_1D, DP as three edges. The position of P is therefore determined by means of a parallelepiped whose edges are parallel to the legs of the triad and one of its diagonal is the line joining the origin to the point.

Example 6 Vectors in Different Coordinate Frames Figure 1.3 illustrates a globally fixed coordinate frame G at the center of a rotating disc O. Another smaller rotating disc with a coordinate frame B is attached to the first disc at a position ${}^G\mathbf{d}_O$. Point P is on the periphery of the small disc.

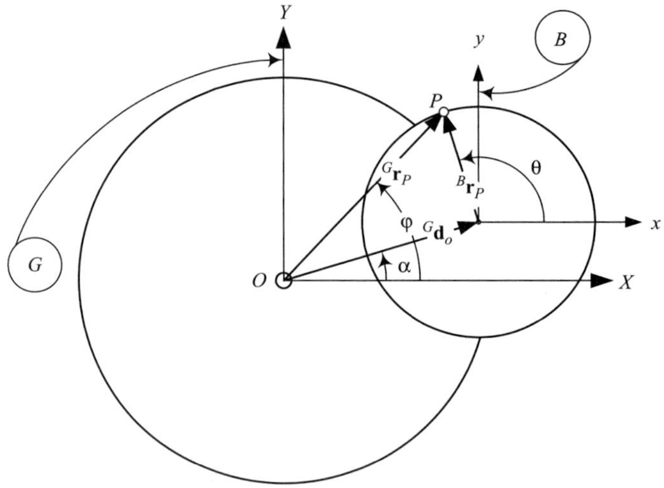

Figure 1.3 A globally fixed frame G at the center of a rotating disc O and a coordinate frame B at the center of a moving disc.

If the coordinate frame $G(OXYZ)$ is fixed and $B(oxyz)$ is always parallel to G, the position vectors of P in different coordinate frames are expressed by

$$ {}^G\mathbf{r}_P = X\hat{\imath} + Y\hat{\jmath} + Z\hat{k} = {}^G r_P \left(\cos\varphi\,\hat{\imath} + \sin\varphi\,\hat{\jmath}\right) \tag{1.16} $$

$$ {}^B\mathbf{r}_P = x\hat{\imath} + y\hat{\jmath} + z\hat{k} = {}^B r_P \left(\cos\theta\,\hat{\imath} + \sin\theta\,\hat{\jmath}\right) \tag{1.17} $$

The coordinate frame B in G may be indicated by a position vector ${}^G\mathbf{d}_o$:

$$ {}^G\mathbf{d}_o = d_o\left(\cos\alpha\,\hat{\imath} + \sin\alpha\,\hat{\jmath}\right) \tag{1.18} $$

Example 7 Variable Vectors There are two ways that a vector can vary: length and direction. A variable-length vector is a vector in the natural expression where its magnitude is variable, such as

$$ \mathbf{r} = r(t)\,\hat{u}_r \tag{1.19} $$

The axis of a variable-length vector is fixed.

A variable-direction vector is a vector in its natural expression where the axis of its unit vector varies. To show such a variable vector, we use the decomposed expression of the unit vector and show that its directional cosines are variable:

$$ \mathbf{r} = r\,\hat{u}_r(t) = r\left(u_1(t)\hat{\imath} + u_2(t)\hat{\jmath} + u_3(t)\hat{k}\right) \tag{1.20} $$

$$ \sqrt{u_1^2 + u_2^2 + u_3^2} = 1 \tag{1.21} $$

The axis and direction characteristics are not fixed for a variable-direction vector, while its magnitude remains constant. The end point of a variable-direction vector slides on a sphere with a center at the starting point.

A variable vector may have both the length and direction variables. Such a vector is shown in its decomposed expression with variable components:

$$\mathbf{r} = x(t)\hat{\imath} + y(t)\hat{\jmath} + z(t)\hat{k} \tag{1.22}$$

It can also be shown in its natural expression with variable length and direction:

$$\mathbf{r} = r(t)\,\hat{u}_r(t) \tag{1.23}$$

Example 8 Parallel and Perpendicular Decomposition of a Vector Consider a line l and a vector $\mathbf{r}$ intersecting at the origin of a coordinate frame such as shown is in Figure 1.4. The line l and vector $\mathbf{r}$ indicate a plane $(l, \mathbf{r})$. We define the unit vectors $\hat{u}_{\parallel}$ parallel to l and $\hat{u}_{\perp}$ perpendicular to l in the $(l, \mathbf{r})$-plane. If the angle between $\mathbf{r}$ and l is α, then the component of $\mathbf{r}$ parallel to l is

$$\mathbf{r}_{\parallel} = r\cos\alpha \tag{1.24}$$

and the component of $\mathbf{r}$ perpendicular to l is

$$\mathbf{r}_{\perp} = r\sin\alpha \tag{1.25}$$

These components indicate that we can decompose a vector $\mathbf{r}$ to its parallel and perpendicular components with respect to a line l by introducing the parallel and perpendicular unit vectors $\hat{u}_{\parallel}$ and $\hat{u}_{\perp}$:

$$\mathbf{r} = \mathbf{r}_{\parallel}\hat{u}_{\parallel} + \mathbf{r}_{\perp}\hat{u}_{\perp} = r\cos\alpha\,\hat{u}_{\parallel} + r\sin\alpha\,\hat{u}_{\perp} \tag{1.26}$$

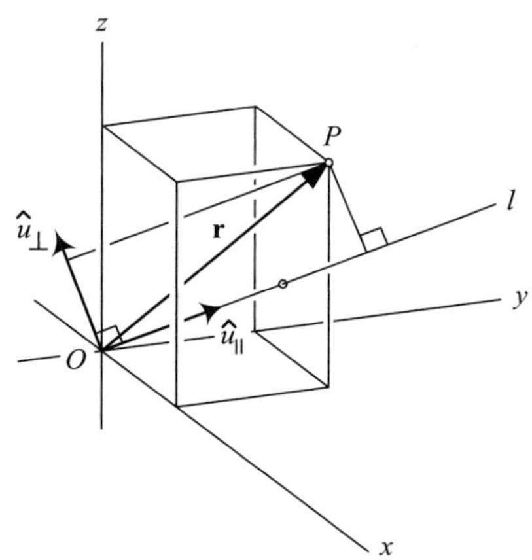

Figure 1.4 Decomposition of a vector **r** with respect to a line l into parallel and perpendicular components.

1.1.3 ★ Vector Definition

By a vector we mean any physical quantity that can be represented by a directed section of a line with a start point, such as O, and an end point, such as P. We may show a vector by an ordered pair of points with an arrow, such as $\overrightarrow{OP}$. The sign $\overrightarrow{PP}$ indicates a zero vector at point P.

Length and direction are necessary to have a vector; however, a vector may have five characteristics:

1. *Length*. The length of section OP corresponds to the magnitude of the physical quantity that the vector is representing.
2. *Axis*. A straight line that indicates the line on which the vector is. The vector axis is also called the *line of action*.
3. *End point*. A start or an end point indicates the point at which the vector is applied. Such a point is called the *affecting point*.
4. *Direction*. The direction indicates at what direction on the axis the vector is pointing.
5. *Physical quantity*. Any vector represents a physical quantity. If a physical quantity can be represented by a vector, it is called a *vectorial physical quantity*. The value of the quantity is proportional to the length of the vector. Having a vector that represents no physical quantity is meaningless, although a vector may be dimensionless.

Depending on the physical quantity and application, there are seven types of vectors:

1. *Vecpoint*. When all of the vector characteristics—length, axis, end point, direction, and physical quantity—are specified, the vector is called a *bounded vector*, *point vector*, or *vecpoint*. Such a vector is fixed at a point with no movability.
2. *Vecline*. If the start and end points of a vector are not fixed on the vector axis, the vector is called a *sliding vector*, *line vector*, or *vecline*. A sliding vector is free to slide on its axis.
3. *Vecface*. When the affecting point of a vector can move on a surface while the vector displaces parallel to itself, the vector is called a *surface vector* or *vecface*. If the surface is a plane, then the vector is a *plane vector* or *veclane*.
4. *Vecfree*. If the axis of a vector is not fixed, the vector is called a *free vector*, *direction vector*, or *vecfree*. Such a vector can move to any point of a specified space while it remains parallel to itself and keeps its direction.
5. *Vecpoline*. If the start point of a vector is fixed while the end point can slide on a line, the vector is a *point-line vector* or *vecpoline*. Such a vector has a constraint variable length and orientation. However, if the start and end points of a vecpoline are on the sliding line, its orientation is constant.
6. *Vecpoface*. If the start point of a vector is fixed while the end point can slide on a surface, the vector is a *point-surface vector* or *vecpoface*. Such a vector has a constraint variable length and orientation. The start and end points of a vecpoface may both be on the sliding surface. If the surface is a plane, the vector is called a *point-plane vector* or *vecpolane*.

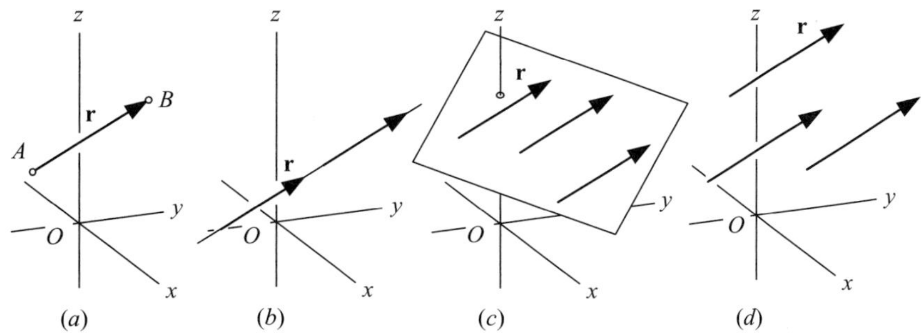

Figure 1.5 (*a*) A vecpoint, (*b*) a vecline, (*c*) a vecface, and (*d*) a vecfree.

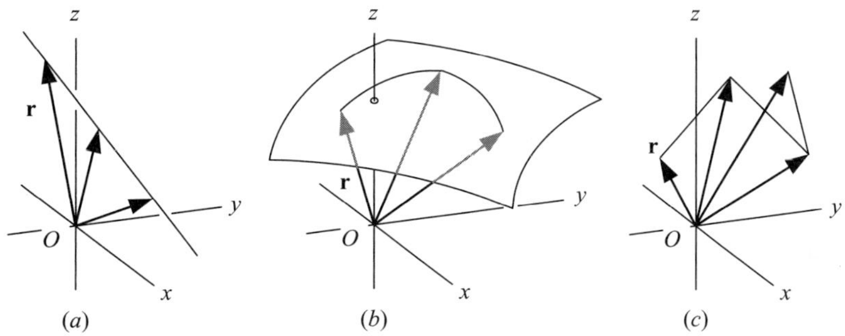

Figure 1.6 (*a*) a vecpoline, (*b*) vecpoface, (*c*) vecporee.

7. *Vecporee*. When the start point of a vector is fixed and the end point can move anywhere in a specified space, the vector is called a *point-free vector* or *vecporee*. Such a vector has a variable length and orientation.

Figure 1.5 illustrates a vecpoint, a vecline, vecface, and a vecfree and Figure 1.6 illustrates a vecpoline, a vecpoface, and a vecporee.

We may compare two vectors only if they represent the same physical quantity and are expressed in the same coordinate frame. Two vectors are equal if they are comparable and are the same type and have the same characteristics. Two vectors are equivalent if they are comparable and the same type and can be substituted with each other.

In summary, any physical quantity that can be represented by a directed section of a line with a start and an end point is a vector quantity. A vector may have five characteristics: length, axis, end point, direction, and physical quantity. The length and direction are necessary. There are seven types of vectors: vecpoint, vecline, vecface, vecfree, vecpoline, vecpoface, and vecporee. Vectors can be added when they are coaxial. In case the vectors are not coaxial, the decomposed expression of vectors must be used to add the vectors.

Example 9 Examples of Vector Types *Displacement* is a vecpoint. Moving from a point A to a point B is called the displacement. Displacement is equal to the difference of two position vectors. A *position vector* starts from the origin of a coordinate frame and ends as a point in the frame. If point A is at $\mathbf{r}_A$ and point B at $\mathbf{r}_B$, then displacement from A to B is

$$\mathbf{r}_{A/B} = {}_B\mathbf{r}_A = \mathbf{r}_A - \mathbf{r}_B \tag{1.27}$$

Force is a vecline. In Newtonian mechanics, a force can be applied on a body at any point of its axis and provides the same motion.

Torque is an example of vecfree. In Newtonian mechanics, a moment can be applied on a body at any point parallel to itself and provides the same motion.

A space curve is expressed by a vecpoline, a surface is expressed by a vecpoface, and a field is expressed by a vecporee.

Example 10 Scalars Physical quantities which can be specified by only a number are called *scalars*. If a physical quantity can be represented by a scalar, it is called a *scalaric physical quantity*. We may compare two scalars only if they represent the same physical quantity. Temperature, density, and work are some examples of scalaric physical quantities.

Two scalars are equal if they represent the same scalaric physical quantity and they have the same number in the same system of units. Two scalars are equivalent if we can substitute one with the other. Scalars must be equal to be equivalent.

1.2 VECTOR ALGEBRA

Most of the physical quantities in dynamics can be represented by vectors. Vector addition, multiplication, and differentiation are essential for the development of dynamics. We can combine vectors only if they are representing the same physical quantity, they are the same type, and they are expressed in the same coordinate frame.

1.2.1 Vector Addition

Two vectors can be *added* when they are *coaxial*. The result is another vector on the same axis with a component equal to the sum of the components of the two vectors. Consider two coaxial vectors $\mathbf{r}_1$ and $\mathbf{r}_2$ in natural expressions:

$$\mathbf{r}_1 = r_1\hat{u}_r \qquad \mathbf{r}_2 = r_2\hat{u}_r \tag{1.28}$$

Their addition would be a new vector $\mathbf{r}_3 = r_3\hat{u}_r$ that is equal to

$$\mathbf{r}_3 = \mathbf{r}_1 + \mathbf{r}_2 = (r_1 + r_2)\hat{u}_r = r_3\hat{u}_r \tag{1.29}$$

Because r_1 and r_2 are scalars, we have $r_1 + r_2 = r_1 + r_2$, and therefore, coaxial vector addition is *commutative*,

$$\mathbf{r}_1 + \mathbf{r}_2 = \mathbf{r}_2 + \mathbf{r}_1 \tag{1.30}$$

and also *associative*,

$$\mathbf{r}_1 + (\mathbf{r}_2 + \mathbf{r}_3) = (\mathbf{r}_1 + \mathbf{r}_2) + \mathbf{r}_3 \tag{1.31}$$

When two vectors $\mathbf{r}_1$ and $\mathbf{r}_2$ are not coaxial, we use their decomposed expressions

$$\mathbf{r}_1 = x_1\hat{\imath} + y_1\hat{\jmath} + z_1\hat{k} \qquad \mathbf{r}_2 = x_2\hat{\imath} + y_2\hat{\jmath} + z_2\hat{k} \tag{1.32}$$

and add the coaxial vectors $x_1\hat{\imath}$ by $x_2\hat{\imath}$, $y_1\hat{\jmath}$ by $y_2\hat{\jmath}$, and $z_1\hat{k}$ by $z_2\hat{k}$ to write the result as the decomposed expression of $\mathbf{r}_3 = \mathbf{r}_1 + \mathbf{r}_2$:

$$\begin{aligned}
\mathbf{r}_3 &= \mathbf{r}_1 + \mathbf{r}_2 \\
&= \left(x_1\hat{\imath} + y_1\hat{\jmath} + z_1\hat{k}\right) + \left(x_2\hat{\imath} + y_2\hat{\jmath} + z_2\hat{k}\right) \\
&= \left(x_1\hat{\imath} + x_2\hat{\imath}\right) + \left(y_1\hat{\jmath} + y_2\hat{\jmath}\right) + \left(z_1\hat{k} + z_2\hat{k}\right) \\
&= (x_1 + x_2)\,\hat{\imath} + (y_1 + y_2)\,\hat{\jmath} + (z_1 + z_2)\,\hat{k} \\
&= x_3\hat{\imath} + y_3\hat{\jmath} + z_3\hat{k}
\end{aligned} \tag{1.33}$$

So, the sum of two vectors $\mathbf{r}_1$ and $\mathbf{r}_2$ is defined as a vector $\mathbf{r}_3$ where its components are equal to the sum of the associated components of $\mathbf{r}_1$ and $\mathbf{r}_2$. Figure 1.7 illustrates vector addition $\mathbf{r}_3 = \mathbf{r}_1 + \mathbf{r}_2$ of two vecpoints $\mathbf{r}_1$ and $\mathbf{r}_2$.

Subtraction of two vectors consists of adding to the minuend the subtrahend with the opposite sense:

$$\mathbf{r}_1 - \mathbf{r}_2 = \mathbf{r}_1 + (-\mathbf{r}_2) \tag{1.34}$$

The vectors $-\mathbf{r}_2$ and $\mathbf{r}_2$ have the same axis and length and differ only in having opposite direction.

If the coordinate frame is known, the decomposed expression of vectors may also be shown by column matrices to simplify calculations:

$$\mathbf{r}_1 = x_1\hat{\imath} + y_1\hat{\jmath} + z_1\hat{k} = \begin{bmatrix} x_1 \\ y_1 \\ z_1 \end{bmatrix} \tag{1.35}$$

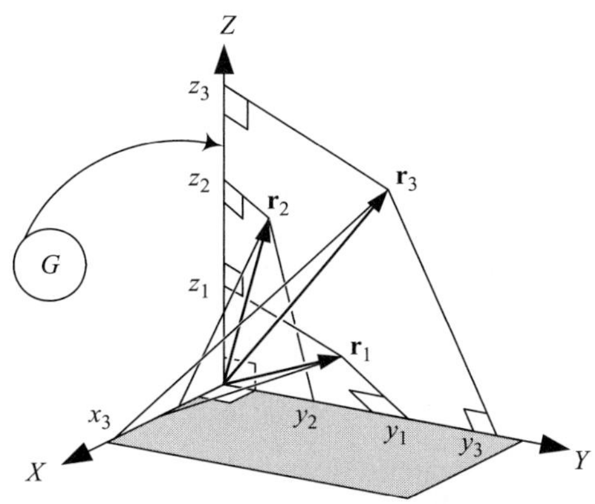

Figure 1.7 Vector addition of two vecpoints $\mathbf{r}_1$ and $\mathbf{r}_2$.

$$\mathbf{r}_2 = x_2\hat{\imath} + y_2\hat{\jmath} + z_2\hat{k} = \begin{bmatrix} x_2 \\ y_2 \\ z_2 \end{bmatrix} \tag{1.36}$$

$$\mathbf{r}_3 = \mathbf{r}_1 + \mathbf{r}_2 = \begin{bmatrix} x_1 \\ y_1 \\ z_1 \end{bmatrix} + \begin{bmatrix} x_2 \\ y_2 \\ z_2 \end{bmatrix} = \begin{bmatrix} x_1 + x_2 \\ y_1 + y_2 \\ z_1 + z_2 \end{bmatrix} \tag{1.37}$$

Vectors can be added only when they are expressed in the same frame. Thus, a vector equation such as

$$\mathbf{r}_3 = \mathbf{r}_1 + \mathbf{r}_2 \tag{1.38}$$

is meaningless without indicating that all of them are expressed in the same frame, such that

$$^{B}\mathbf{r}_3 = {}^{B}\mathbf{r}_1 + {}^{B}\mathbf{r}_2 \tag{1.39}$$

The three vectors $\mathbf{r}_1$, $\mathbf{r}_2$, and $\mathbf{r}_3$ are coplanar, and $\mathbf{r}_3$ may be considered as the diagonal of a parallelogram that is made by $\mathbf{r}_1$, $\mathbf{r}_2$.

Example 11 Displacement of a Point Point P moves from the origin of a global coordinate frame G to a point at $(1, 2, 0)$ and then moves to $(4, 3, 0)$. If we express the first displacement by a vector $\mathbf{r}_1$ and its final position by $\mathbf{r}_3$, the second displacement is $\mathbf{r}_2$, where

$$\mathbf{r}_2 = \mathbf{r}_3 - \mathbf{r}_1 = \begin{bmatrix} 4 \\ 3 \\ 0 \end{bmatrix} - \begin{bmatrix} 1 \\ 2 \\ 0 \end{bmatrix} = \begin{bmatrix} 3 \\ 1 \\ 0 \end{bmatrix} \tag{1.40}$$

Example 12 Vector Interpolation Problem Having two digits n_1 and n_2 as the start and the final interpolants, we may define a controlled digit n with a variable q such that

$$n = \begin{cases} n_1 & q = 0 \\ n_2 & q = 1 \end{cases} \qquad 0 \le q \le 1 \tag{1.41}$$

Defining or determining such a controlled digit is called the interpolation problem. There are many functions to be used for solving the interpolation problem. Linear interpolation is the simplest and is widely used in engineering design, computer graphics, numerical analysis, and optimization:

$$n = n_1(1 - q) + n_2 q \tag{1.42}$$

The control parameter q determines the weight of each interpolants n_1 and n_2 in the interpolated n. In a linear interpolation, the weight factors are proportional to the distance of q from 1 and 0.

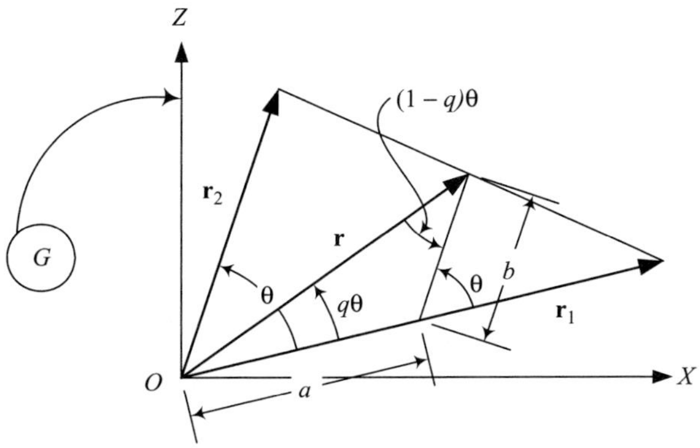

Figure 1.8 Vector linear interpolation.

Employing the linear interpolation technique, we may define a vector $\mathbf{r} = \mathbf{r}(q)$ to interpolate between the interpolant vectors $\mathbf{r}_1$ and $\mathbf{r}_2$:

$$\mathbf{r} = (1-q)\mathbf{r}_1 + q\mathbf{r}_2 = \begin{bmatrix} x_1(1-q) + qx_2 \\ y_1(1-q) + qy_2 \\ z_1(1-q) + qz_2 \end{bmatrix} \tag{1.43}$$

In this interpolation, we assumed that equal steps in q results in equal steps in $\mathbf{r}$ between $\mathbf{r}_1$ and $\mathbf{r}_2$. The tip point of $\mathbf{r}$ will move on a line connecting the tip points of $\mathbf{r}_1$ and $\mathbf{r}_2$, as is shown in Figure 1.8.

We may interpolate the vectors $\mathbf{r}_1$ and $\mathbf{r}_2$ by interpolating the angular distance θ between $\mathbf{r}_1$ and $\mathbf{r}_2$:

$$\mathbf{r} = \frac{\sin\left[(1-q)\theta\right]}{\sin\theta}\mathbf{r}_1 + \frac{\sin(q\theta)}{\sin\theta}\mathbf{r}_2 \tag{1.44}$$

To derive Equation (1.44), we may start with

$$\mathbf{r} = a\mathbf{r}_1 + b\mathbf{r}_2 \tag{1.45}$$

and find a and b from the following trigonometric equations:

$$a\sin(q\theta) - b\sin\left[(1-q)\theta\right] = 0 \tag{1.46}$$

$$a\cos(q\theta) + b\cos\left[(1-q)\theta\right] = 1 \tag{1.47}$$

Example 13 Vector Addition and Linear Space Vectors and adding operation make a *linear space* because for any vectors $\mathbf{r}_1$, $\mathbf{r}_2$ we have the following properties:

1. Commutative:

$$\mathbf{r}_1 + \mathbf{r}_2 = \mathbf{r}_2 + \mathbf{r}_1 \tag{1.48}$$

2. Associative:

$$\mathbf{r}_1 + (\mathbf{r}_2 + \mathbf{r}_3) = (\mathbf{r}_1 + \mathbf{r}_2) + \mathbf{r}_3 \tag{1.49}$$

3. Null element:

$$\mathbf{0} + \mathbf{r} = \mathbf{r} \tag{1.50}$$

4. Inverse element:

$$\mathbf{r} + (-\mathbf{r}) = \mathbf{0} \tag{1.51}$$

Example 14 Linear Dependence and Independence The n vectors $\mathbf{r}_1, \mathbf{r}_2, \mathbf{r}_3, \ldots, \mathbf{r}_n$ are *linearly dependent* if there exist n scalars $c_1, c_2, c_3, \ldots, c_n$ not all equal to zero such that a linear combination of the vectors equals zero:

$$c_1\mathbf{r}_1 + c_2\mathbf{r}_2 + c_3\mathbf{r}_3 + \cdots + c_n\mathbf{r}_n = 0 \tag{1.52}$$

The vectors $\mathbf{r}_1, \mathbf{r}_2, \mathbf{r}_3, \ldots, \mathbf{r}_n$ are *linearly independent* if they are not linearly dependent, and it means the n scalars $c_1, c_2, c_3, \ldots, c_n$ must all be zero to have Equation (1.52):

$$c_1 = c_2 = c_3 = \cdots = c_n = 0 \tag{1.53}$$

Example 15 Two Linearly Dependent Vectors Are Colinear Consider two linearly dependent vectors $\mathbf{r}_1$ and $\mathbf{r}_2$:

$$c_1\mathbf{r}_1 + c_2\mathbf{r}_2 = 0 \tag{1.54}$$

If $c_1 \neq 0$, we have

$$\mathbf{r}_1 = -\frac{c_2}{c_1}\mathbf{r}_2 \tag{1.55}$$

and if $c_2 \neq 0$, we have

$$\mathbf{r}_2 = -\frac{c_1}{c_2}\mathbf{r}_1 \tag{1.56}$$

which shows $\mathbf{r}_1$ and $\mathbf{r}_2$ are colinear.

Example 16 Three Linearly Dependent Vectors Are Coplanar Consider three linearly dependent vectors $\mathbf{r}_1$, $\mathbf{r}_2$, and $\mathbf{r}_3$,

$$c_1\mathbf{r}_1 + c_2\mathbf{r}_2 + c_3\mathbf{r}_3 = 0 \tag{1.57}$$

where at least one of the scalars c_1, c_2, c_3, say c_3, is not zero; then

$$\mathbf{r}_3 = -\frac{c_1}{c_3}\mathbf{r}_1 - \frac{c_2}{c_3}\mathbf{r}_2 \tag{1.58}$$

which shows $\mathbf{r}_3$ is in the same plane as $\mathbf{r}_1$ and $\mathbf{r}_2$.

1.2.2 Vector Multiplication

There are three types of vector multiplications for two vectors $\mathbf{r}_1$ and $\mathbf{r}_2$:

1. *Dot, Inner, or Scalar Product*

$$\mathbf{r}_1 \cdot \mathbf{r}_2 = \begin{bmatrix} x_1 \\ y_1 \\ z_1 \end{bmatrix} \cdot \begin{bmatrix} x_2 \\ y_2 \\ z_2 \end{bmatrix} = x_1 x_2 + y_1 y_2 + z_1 z_2$$
$$= r_1 r_2 \cos\alpha \tag{1.59}$$

The inner product of two vectors produces a scalar that is equal to the product of the length of individual vectors and the cosine of the angle between them. The vector inner product is *commutative* in orthogonal coordinate frames,

$$\mathbf{r}_1 \cdot \mathbf{r}_2 = \mathbf{r}_2 \cdot \mathbf{r}_1 \tag{1.60}$$

The inner product is dimension free and can be calculated in n-dimensional spaces. The inner product can also be performed in nonorthogonal coordinate systems.

2. *Cross, Outer, or Vector Product*

$$\mathbf{r}_3 = \mathbf{r}_1 \times \mathbf{r}_2 = \begin{bmatrix} x_1 \\ y_1 \\ z_1 \end{bmatrix} \times \begin{bmatrix} x_2 \\ y_2 \\ z_2 \end{bmatrix} = \begin{bmatrix} y_1 z_2 - y_2 z_1 \\ x_2 z_1 - x_1 z_2 \\ x_1 y_2 - x_2 y_1 \end{bmatrix}$$
$$= (r_1 r_2 \sin\alpha)\, \hat{u}_{r_3} = r_3 \hat{u}_{r_3} \tag{1.61}$$

$$\hat{u}_{r_3} = \hat{u}_{r_1} \times \hat{u}_{r_2} \tag{1.62}$$

The outer product of two vectors $\mathbf{r}_1$ and $\mathbf{r}_2$ produces another vector $\mathbf{r}_3$ that is perpendicular to the plane of $\mathbf{r}_1$, $\mathbf{r}_2$ such that the cycle $\mathbf{r}_1\mathbf{r}_2\mathbf{r}_3$ makes a right-handed triad. The length of $\mathbf{r}_3$ is equal to the product of the length of individual vectors multiplied by the sine of the angle between them. Hence r_3 is numerically equal to the area of the parallelogram made up of $\mathbf{r}_1$ and $\mathbf{r}_2$. The vector inner product is *skew commutative* or *anticommutative*:

$$\mathbf{r}_1 \times \mathbf{r}_2 = -\mathbf{r}_2 \times \mathbf{r}_1 \tag{1.63}$$

The outer product is defined and applied only in 3D space. There is no outer product in lower or higher dimensions than 3. If any vector of $\mathbf{r}_1$and $\mathbf{r}_2$ is in a lower dimension than 3D, we must make it a 3D vector by adding zero components for missing dimensions to be able to perform their outer product.

3. *Quaternion Product*

$$\mathbf{r}_1\mathbf{r}_2 = \mathbf{r}_1 \times \mathbf{r}_2 - \mathbf{r}_1 \cdot \mathbf{r}_2 \tag{1.64}$$

We will talk about the quaternion product in Section 5.3.

In summary, there are three types of vector multiplication: inner, outer, and quaternion products, of which the inner product is the only one with commutative property.

Example 17 Geometric Expression of Inner Products Consider a line l and a vector $\mathbf{r}$ intersecting at the origin of a coordinate frame as is shown in Figure 1.9. If the angle between $\mathbf{r}$ and l is α, the parallel component of $\mathbf{r}$ to l is

$$\mathbf{r}_{\|} = \overline{OA} = r\cos\alpha \tag{1.65}$$

This is the length of the projection of $\mathbf{r}$ on l. If we define a unit vector $\hat{u}_l$ on l by its direction cosines β_1, β_2, β_3,

$$\hat{u}_l = u_1\hat{\imath} + u_2\hat{\jmath} + u_3\hat{k} = \begin{bmatrix} u_1 \\ u_2 \\ u_3 \end{bmatrix} = \begin{bmatrix} \cos\beta_1 \\ \cos\beta_2 \\ \cos\beta_3 \end{bmatrix} \tag{1.66}$$

then the inner product of $\mathbf{r}$ and $\hat{u}_l$ is

$$\mathbf{r}\cdot\hat{u}_l = \mathbf{r}_{\|} = r\cos\alpha \tag{1.67}$$

We may show $\mathbf{r}$ by using its direction cosines α_1, α_2, α_3,

$$\mathbf{r} = r\hat{u}_r = x\hat{\imath} + y\hat{\jmath} + z\hat{k} = r\begin{bmatrix} x/r \\ y/r \\ z/r \end{bmatrix} = r\begin{bmatrix} \cos\alpha_1 \\ \cos\alpha_2 \\ \cos\alpha_3 \end{bmatrix} \tag{1.68}$$

Then, we may use the result of the inner product of $\mathbf{r}$ and $\hat{u}_l$,

$$\begin{aligned} \mathbf{r}\cdot\hat{u}_l &= r\begin{bmatrix} \cos\alpha_1 \\ \cos\alpha_2 \\ \cos\alpha_3 \end{bmatrix} \cdot \begin{bmatrix} \cos\beta_1 \\ \cos\beta_2 \\ \cos\beta_3 \end{bmatrix} \\ &= r\left(\cos\beta_1\cos\alpha_1 + \cos\beta_2\cos\alpha_2 + \cos\beta_3\cos\alpha_3\right) \end{aligned} \tag{1.69}$$

to calculate the angle α between $\mathbf{r}$ and l based on their directional cosines:

$$\cos\alpha = \cos\beta_1\cos\alpha_1 + \cos\beta_2\cos\alpha_2 + \cos\beta_3\cos\alpha_3 \tag{1.70}$$

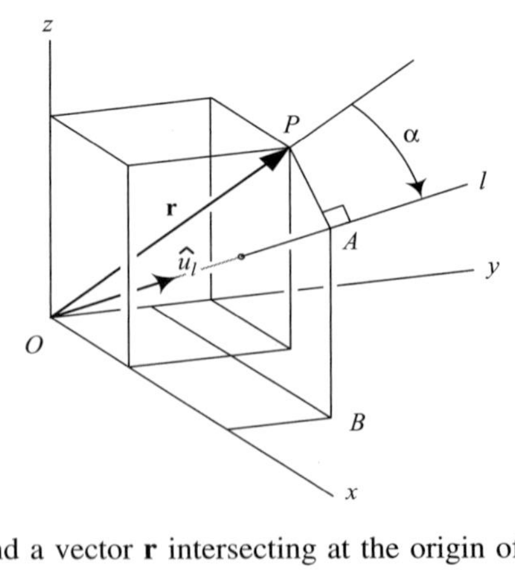

Figure 1.9 A line l and a vector $\mathbf{r}$ intersecting at the origin of a coordinate frame.

So, the inner product can be used to find the projection of a vector on a given line. It is also possible to use the inner product to determine the angle α between two given vectors $\mathbf{r}_1$ and $\mathbf{r}_2$ as

$$\cos\alpha = \frac{\mathbf{r}_1 \cdot \mathbf{r}_2}{r_1 r_2} = \frac{\mathbf{r}_1 \cdot \mathbf{r}_2}{\sqrt{\mathbf{r}_1 \cdot \mathbf{r}_1}\sqrt{\mathbf{r}_2 \cdot \mathbf{r}_2}} \tag{1.71}$$

Example 18 Power 2 of a Vector By writing a vector $\mathbf{r}$ to a power 2, we mean the inner product of $\mathbf{r}$ to itself:

$$\mathbf{r}^2 = \mathbf{r} \cdot \mathbf{r} = \begin{bmatrix} x \\ y \\ z \end{bmatrix} \cdot \begin{bmatrix} x \\ y \\ z \end{bmatrix} = x^2 + y^2 + z^2 = r^2 \tag{1.72}$$

Using this definition we can write

$$(\mathbf{r}_1 + \mathbf{r}_2)^2 = (\mathbf{r}_1 + \mathbf{r}_2) \cdot (\mathbf{r}_1 + \mathbf{r}_2) = \mathbf{r}_1^2 + 2\mathbf{r}_1 \cdot \mathbf{r}_2 + \mathbf{r}_2^2 \tag{1.73}$$

$$(\mathbf{r}_1 - \mathbf{r}_2) \cdot (\mathbf{r}_1 + \mathbf{r}_2) = \mathbf{r}_1^2 - \mathbf{r}_2^2 \tag{1.74}$$

There is no meaning for a vector with a negative or positive odd exponent.

Example 19 Unit Vectors and Inner and Outer Products Using the set of unit vectors $\hat{\imath}, \hat{\jmath}, \hat{k}$ of a positive orthogonal triad and the definition of inner product, we conclude that

$$\hat{\imath}^2 = 1 \qquad \hat{\jmath}^2 = 1 \qquad \hat{k}^2 = 1 \tag{1.75}$$

Furthermore, by definition of the vector product we have

$$\hat{\imath} \times \hat{\jmath} = -\left(\hat{\jmath} \times \hat{\imath}\right) = \hat{k} \tag{1.76}$$

$$\hat{\jmath} \times \hat{k} = -\left(\hat{k} \times \hat{\jmath}\right) = \hat{\imath} \tag{1.77}$$

$$\hat{k} \times \hat{\imath} = -\left(\hat{\imath} \times \hat{k}\right) = \hat{\jmath} \tag{1.78}$$

It might also be useful if we have these equalities:

$$\hat{\imath} \cdot \hat{\jmath} = 0 \qquad \hat{\jmath} \cdot \hat{k} = 0 \qquad \hat{k} \cdot \hat{\imath} = 0 \tag{1.79}$$

$$\hat{\imath} \times \hat{\imath} = 0 \qquad \hat{\jmath} \times \hat{\jmath} = 0 \qquad \hat{k} \times \hat{k} = 0 \tag{1.80}$$

Example 20 Vanishing Dot Product If the inner product of two vectors $\mathbf{a}$ and $\mathbf{b}$ is zero,

$$\mathbf{a} \cdot \mathbf{b} = 0 \tag{1.81}$$

then either $\mathbf{a} = 0$ or $\mathbf{b} = 0$, or $\mathbf{a}$ and $\mathbf{b}$ are perpendicular.

Example 21 Vector Equations Assume **x** is an unknown vector, k is a scalar, and **a**, **b**, and **c** are three constant vectors in the following vector equation:

$$k\mathbf{x} + (\mathbf{b} \cdot \mathbf{x})\, \mathbf{a} = \mathbf{c} \tag{1.82}$$

To solve the equation for **x**, we dot product both sides of (1.82) by **b**:

$$k\mathbf{x} \cdot \mathbf{b} + (\mathbf{x} \cdot \mathbf{b})\, (\mathbf{a} \cdot \mathbf{b}) = \mathbf{c} \cdot \mathbf{b} \tag{1.83}$$

This is a linear equation for $\mathbf{x} \cdot \mathbf{b}$ with the solution

$$\mathbf{x} \cdot \mathbf{b} = \frac{\mathbf{c} \cdot \mathbf{b}}{k + \mathbf{a} \cdot \mathbf{b}} \tag{1.84}$$

provided

$$k + \mathbf{a} \cdot \mathbf{b} \neq 0 \tag{1.85}$$

Substituting (1.84) in (1.82) provides the solution **x**:

$$\mathbf{x} = \frac{1}{k}\mathbf{c} - \frac{\mathbf{c} \cdot \mathbf{b}}{k\,(k + \mathbf{a} \cdot \mathbf{b})}\mathbf{a} \tag{1.86}$$

An alternative method is decomposition of the vector equation along the axes $\hat{\imath}$, $\hat{\jmath}$, $\hat{k}$ of the coordinate frame and solving a set of three scalar equations to find the components of the unknown vector.

Assume the decomposed expression of the vectors **x**, **a**, **b**, and **c** are

$$\mathbf{x} = \begin{bmatrix} x \\ y \\ z \end{bmatrix} \quad \mathbf{a} = \begin{bmatrix} a_1 \\ a_2 \\ a_3 \end{bmatrix} \quad \mathbf{b} = \begin{bmatrix} b_1 \\ b_2 \\ b_3 \end{bmatrix} \quad \mathbf{c} = \begin{bmatrix} c_1 \\ c_2 \\ c_3 \end{bmatrix} \tag{1.87}$$

Substituting these expressions in Equation (1.82),

$$k \begin{bmatrix} x \\ y \\ z \end{bmatrix} + \left(\begin{bmatrix} b_1 \\ b_2 \\ b_3 \end{bmatrix} \cdot \begin{bmatrix} x \\ y \\ z \end{bmatrix} \right) \begin{bmatrix} a_1 \\ a_2 \\ a_3 \end{bmatrix} = \begin{bmatrix} c_1 \\ c_2 \\ c_3 \end{bmatrix} \tag{1.88}$$

provides a set of three scalar equations

$$\begin{bmatrix} k + a_1 b_1 & a_1 b_2 & a_1 b_3 \\ a_2 b_1 & k + a_2 b_2 & a_2 b_3 \\ a_3 b_1 & a_3 b_2 & k + a_3 b_3 \end{bmatrix} \begin{bmatrix} x \\ y \\ z \end{bmatrix} = \begin{bmatrix} c_1 \\ c_2 \\ c_3 \end{bmatrix} \tag{1.89}$$

that can be solved by matrix inversion:

$$\begin{aligned} \begin{bmatrix} x \\ y \\ z \end{bmatrix} &= \begin{bmatrix} k + a_1 b_1 & a_1 b_2 & a_1 b_3 \\ a_2 b_1 & k + a_2 b_2 & a_2 b_3 \\ a_3 b_1 & a_3 b_2 & k + a_3 b_3 \end{bmatrix}^{-1} \begin{bmatrix} c_1 \\ c_2 \\ c_3 \end{bmatrix} \\ &= \begin{bmatrix} \dfrac{kc_1 - a_1 b_2 c_2 + a_2 b_2 c_1 - a_1 b_3 c_3 + a_3 b_3 c_1}{k\,(k + a_1 b_1 + a_2 b_2 + a_3 b_3)} \\ \dfrac{kc_2 + a_1 b_1 c_2 - a_2 b_1 c_1 - a_2 b_3 c_3 + a_3 b_3 c_2}{k\,(k + a_1 b_1 + a_2 b_2 + a_3 b_3)} \\ \dfrac{kc_3 + a_1 b_1 c_3 - a_3 b_1 c_1 + a_2 b_2 c_3 - a_3 b_2 c_2}{k\,(k + a_1 b_1 + a_2 b_2 + a_3 b_3)} \end{bmatrix} \end{aligned} \tag{1.90}$$

Solution (1.90) is compatible with solution (1.86).

Example 22 Vector Addition, Scalar Multiplication, and Linear Space Vector addition and scalar multiplication make a linear space, because

$$k_1 (k_2\mathbf{r}) = (k_1k_2)\,\mathbf{r} \tag{1.91}$$

$$(k_1 + k_2)\,\mathbf{r} = k_1\mathbf{r} + k_2\mathbf{r} \tag{1.92}$$

$$k\,(\mathbf{r}_1 + \mathbf{r}_2) = k\mathbf{r}_1 + k\mathbf{r}_2 \tag{1.93}$$

$$1 \cdot \mathbf{r} = \mathbf{r} \tag{1.94}$$

$$(-1) \cdot \mathbf{r} = -\mathbf{r} \tag{1.95}$$

$$0 \cdot \mathbf{r} = \mathbf{0} \tag{1.96}$$

$$k \cdot \mathbf{0} = \mathbf{0} \tag{1.97}$$

Example 23 Vanishing Condition of a Vector Inner Product Consider three non-coplanar constant vectors **a**, **b**, **c** and an arbitrary vector **r**. If

$$\mathbf{a} \cdot \mathbf{r} = 0 \qquad \mathbf{b} \cdot \mathbf{r} = 0 \qquad \mathbf{c} \cdot \mathbf{r} = 0 \tag{1.98}$$

then

$$\mathbf{r} = 0 \tag{1.99}$$

Example 24 Vector Product Expansion We may prove the result of the inner and outer products of two vectors by using decomposed expression and expansion:

$$\begin{aligned}
\mathbf{r}_1 \cdot \mathbf{r}_2 &= \left(x_1\hat{\imath} + y_1\hat{\jmath} + z_1\hat{k}\right) \cdot \left(x_2\hat{\imath} + y_2\hat{\jmath} + z_2\hat{k}\right) \\
&= x_1x_2\hat{\imath} \cdot \hat{\imath} + x_1y_2\hat{\imath} \cdot \hat{\jmath} + x_1z_2\hat{\imath} \cdot \hat{k} \\
&\quad + y_1x_2\hat{\jmath} \cdot \hat{\imath} + y_1y_2\hat{\jmath} \cdot \hat{\jmath} + y_1z_2\hat{\jmath} \cdot \hat{k} \\
&\quad + z_1x_2\hat{k} \cdot \hat{\imath} + z_1y_2\hat{k} \cdot \hat{\jmath} + z_1z_2\hat{k} \cdot \hat{k} \\
&= x_1x_2 + y_1y_2 + z_1z_2
\end{aligned} \tag{1.100}$$

$$\begin{aligned}
\mathbf{r}_1 \times \mathbf{r}_2 &= \left(x_1\hat{\imath} + y_1\hat{\jmath} + z_1\hat{k}\right) \times \left(x_2\hat{\imath} + y_2\hat{\jmath} + z_2\hat{k}\right) \\
&= x_1x_2\hat{\imath} \times \hat{\imath} + x_1y_2\hat{\imath} \times \hat{\jmath} + x_1z_2\hat{\imath} \times \hat{k} \\
&\quad + y_1x_2\hat{\jmath} \times \hat{\imath} + y_1y_2\hat{\jmath} \times \hat{\jmath} + y_1z_2\hat{\jmath} \times \hat{k} \\
&\quad + z_1x_2\hat{k} \times \hat{\imath} + z_1y_2\hat{k} \times \hat{\jmath} + z_1z_2\hat{k} \times \hat{k} \\
&= (y_1z_2 - y_2z_1)\,\hat{\imath} + (x_2z_1 - x_1z_2)\,\hat{\jmath} + (x_1y_2 - x_2y_1)\,\hat{k}
\end{aligned} \tag{1.101}$$

We may also find the outer product of two vectors by expanding a determinant and derive the same result as Equation (1.101):

$$\mathbf{r}_1 \times \mathbf{r}_2 = \begin{vmatrix} \hat{\imath} & \hat{\jmath} & \hat{k} \\ x_1 & y_1 & z_1 \\ x_2 & y_2 & z_2 \end{vmatrix} \tag{1.102}$$

Example 25 bac–cab Rule If $\mathbf{a}, \mathbf{b}, \mathbf{c}$ are three vectors, we may expand their triple cross product and show that

$$\mathbf{a} \times (\mathbf{b} \times \mathbf{c}) = \mathbf{b}\,(\mathbf{a} \cdot \mathbf{c}) - \mathbf{c}\,(\mathbf{a} \cdot \mathbf{b}) \tag{1.103}$$

because

$$\begin{aligned}
&\begin{bmatrix} a_1 \\ a_2 \\ a_3 \end{bmatrix} \times \left(\begin{bmatrix} b_1 \\ b_2 \\ b_3 \end{bmatrix} \times \begin{bmatrix} c_1 \\ c_2 \\ c_3 \end{bmatrix} \right) \\
&= \begin{bmatrix} a_2\,(b_1c_2 - b_2c_1) + a_3\,(b_1c_3 - b_3c_1) \\ a_3\,(b_2c_3 - b_3c_2) - a_1\,(b_1c_2 - b_2c_1) \\ -a_1\,(b_1c_3 - b_3c_1) - a_2\,(b_2c_3 - b_3c_2) \end{bmatrix} \\
&= \begin{bmatrix} b_1\,(a_1c_1 + a_2c_2 + a_3c_3) - c_1\,(a_1b_1 + a_2b_2 + a_3b_3) \\ b_2\,(a_1c_1 + a_2c_2 + a_3c_3) - c_2\,(a_1b_1 + a_2b_2 + a_3b_3) \\ b_3\,(a_1c_1 + a_2c_2 + a_3c_3) - c_3\,(a_1b_1 + a_2b_2 + a_3b_3) \end{bmatrix}
\end{aligned} \tag{1.104}$$

Equation (1.103) may be referred to as the *bac–cab rule*, which makes it easy to remember. The bac–cab rule is the most important in 3D vector algebra. It is the key to prove a great number of other theorems.

Example 26 Geometric Expression of Outer Products Consider the free vectors $\mathbf{r}_1$ from A to B and $\mathbf{r}_2$ from A to C, as are shown in Figure 1.10:

$$\mathbf{r}_1 = \begin{bmatrix} -1 \\ 3 \\ 0 \end{bmatrix} = \sqrt{10} \begin{bmatrix} -0.31623 \\ 0.94868 \\ 0 \end{bmatrix} \tag{1.105}$$

$$\mathbf{r}_2 = \begin{bmatrix} -1 \\ 0 \\ 2.5 \end{bmatrix} = 2.6926 \begin{bmatrix} -0.37139 \\ 0 \\ 0.92847 \end{bmatrix} \tag{1.106}$$

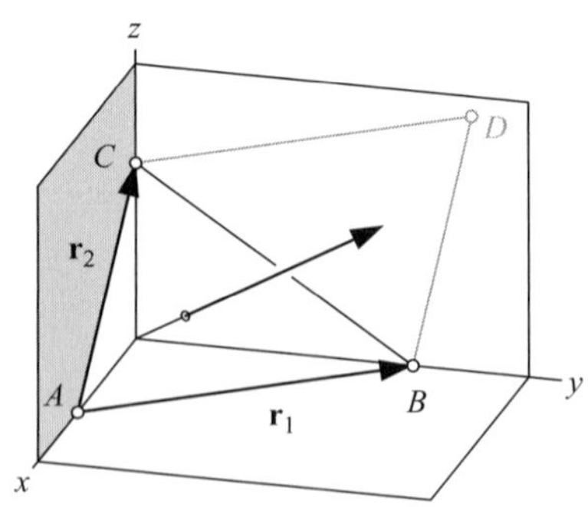

Figure 1.10 The cross product of the two free vectors $\mathbf{r}_1$ and $\mathbf{r}_2$ and the resultant $\mathbf{r}_3$.

The cross product of the two vectors is $\mathbf{r}_3$:

$$\mathbf{r}_3 = \mathbf{r}_1 \times \mathbf{r}_2 = \begin{bmatrix} 7.5 \\ 2.5 \\ 3 \end{bmatrix} = 8.4558 \begin{bmatrix} 0.88697 \\ 0.29566 \\ 0.35479 \end{bmatrix}$$

$$= r_3 \hat{u}_{r_3} = (r_1 r_2 \sin\alpha)\, \hat{u}_{r_3} \tag{1.107}$$

$$\hat{u}_{r_3} = \hat{u}_{r_1} \times \hat{u}_{r_2} = \begin{bmatrix} 0.88697 \\ 0.29566 \\ 0.35479 \end{bmatrix} \tag{1.108}$$

where $r_3 = 8.4558$ is numerically equivalent to the area A of the parallelogram $ABCD$ made by the sides AB and AC:

$$A_{ABCD} = |\mathbf{r}_1 \times \mathbf{r}_2| = 8.4558 \tag{1.109}$$

The area of the triangle ABC is $A/2$. The vector $\mathbf{r}_3$ is perpendicular to this plane and, hence, its unit vector $\hat{u}_{r_3}$ can be used to indicate the plane $ABCD$.

Example 27 Scalar Triple Product The dot product of a vector $\mathbf{r}_1$ with the cross product of two vectors $\mathbf{r}_2$ and $\mathbf{r}_3$ is called the *scalar triple product* of $\mathbf{r}_1$, $\mathbf{r}_2$, and $\mathbf{r}_3$. The scalar triple product can be shown and calculated by a determinant:

$$\mathbf{r}_1 \cdot (\mathbf{r}_2 \times \mathbf{r}_3) = \mathbf{r}_1 \cdot \mathbf{r}_2 \times \mathbf{r}_3 = \begin{vmatrix} x_1 & y_1 & z_1 \\ x_2 & y_2 & z_2 \\ x_3 & y_3 & z_3 \end{vmatrix} \tag{1.110}$$

Interchanging two rows (or columns) of a matrix changes the sign of its determinant. So, we may conclude that the scalar triple product of three vectors $\mathbf{r}_1$, $\mathbf{r}_2$, $\mathbf{r}_3$ is also equal to

$$\begin{aligned} \mathbf{r}_1 \cdot \mathbf{r}_2 \times \mathbf{r}_3 &= \mathbf{r}_2 \cdot \mathbf{r}_3 \times \mathbf{r}_1 = \mathbf{r}_3 \cdot \mathbf{r}_1 \times \mathbf{r}_2 \\ &= \mathbf{r}_1 \times \mathbf{r}_2 \cdot \mathbf{r}_3 = \mathbf{r}_2 \times \mathbf{r}_3 \cdot \mathbf{r}_1 = \mathbf{r}_3 \times \mathbf{r}_1 \cdot \mathbf{r}_2 \\ &= -\mathbf{r}_1 \cdot \mathbf{r}_3 \times \mathbf{r}_2 = -\mathbf{r}_2 \cdot \mathbf{r}_1 \times \mathbf{r}_3 = -\mathbf{r}_3 \cdot \mathbf{r}_2 \times \mathbf{r}_1 \\ &= -\mathbf{r}_1 \times \mathbf{r}_3 \cdot \mathbf{r}_2 = -\mathbf{r}_2 \times \mathbf{r}_1 \cdot \mathbf{r}_3 = -\mathbf{r}_3 \times \mathbf{r}_2 \cdot \mathbf{r}_1 \end{aligned} \tag{1.111}$$

Because of Equation (1.111), the scalar triple product of the vectors $\mathbf{r}_1$, $\mathbf{r}_2$, $\mathbf{r}_3$ can be shown by the short notation $[\mathbf{r}_1\mathbf{r}_2\mathbf{r}_3]$:

$$[\mathbf{r}_1\mathbf{r}_2\mathbf{r}_3] = \mathbf{r}_1 \cdot \mathbf{r}_2 \times \mathbf{r}_3 \tag{1.112}$$

This notation gives us the freedom to set the position of the dot and cross product signs as required.

If the three vectors $\mathbf{r}_1$, $\mathbf{r}_2$, $\mathbf{r}_3$ are position vectors, then their scalar triple product geometrically represents the volume of the parallelepiped formed by the three vectors. Figure 1.11 illustrates such a parallelepiped for three vectors $\mathbf{r}_1$, $\mathbf{r}_2$, $\mathbf{r}_3$.

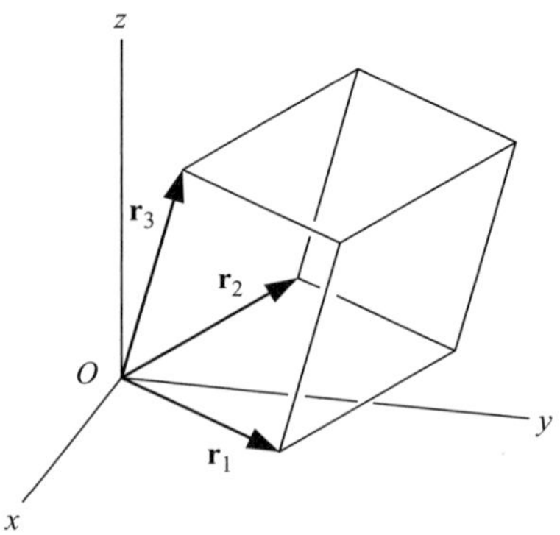

Figure 1.11 The parallelepiped made by three vectors $\mathbf{r}_1$, $\mathbf{r}_2$, $\mathbf{r}_3$.

Example 28 Vector Triple Product The cross product of a vector $\mathbf{r}_1$ with the cross product of two vectors $\mathbf{r}_2$ and $\mathbf{r}_3$ is called the *vector triple product* of $\mathbf{r}_1$, $\mathbf{r}_2$, and $\mathbf{r}_3$. The *bac–cab* rule is always used to simplify a vector triple product:

$$\mathbf{r}_1 \times (\mathbf{r}_2 \times \mathbf{r}_3) = \mathbf{r}_2 (\mathbf{r}_1 \cdot \mathbf{r}_3) - \mathbf{r}_3 (\mathbf{r}_1 \cdot \mathbf{r}_2) \tag{1.113}$$

Example 29 ★ Norm and Vector Space Assume $\mathbf{r}$, $\mathbf{r}_1$, $\mathbf{r}_2$, $\mathbf{r}_3$ are arbitrary vectors and c, c_1, c_3 are scalars. The *norm* of a vector $\|\mathbf{r}\|$ is defined as a real-valued function on a vector space v such that for all $\{\mathbf{r}_1, \mathbf{r}_2\} \in V$ and all $c \in \mathbb{R}$ we have:

1. Positive definition: $\|\mathbf{r}\| > 0$ if $\mathbf{r} \neq 0$ and $\|\mathbf{r}\| = 0$ if $\mathbf{r} = 0$.
2. Homogeneity: $\|c\mathbf{r}\| = \|c\| \, \|\mathbf{r}\|$.
3. Triangle inequality: $\|\mathbf{r}_1 + \mathbf{r}_2\| = \|\mathbf{r}_1\| + \|\mathbf{r}_2\|$.

The definition of norm is up to the investigator and may vary depending on the application. The most common definition of the norm of a vector is the length:

$$\|\mathbf{r}\| = |\mathbf{r}| = \sqrt{r_1^2 + r_2^2 + r_3^2} \tag{1.114}$$

The set v with vector elements is called a *vector space* if the following conditions are fulfilled:

1. Addition: If $\{\mathbf{r}_1, \mathbf{r}_2\} \in V$ and $\mathbf{r}_1 + \mathbf{r}_2 = \mathbf{r}$, then $\mathbf{r} \in V$.
2. Commutativity: $\mathbf{r}_1 + \mathbf{r}_2 = \mathbf{r}_2 + \mathbf{r}_1$.
3. Associativity: $\mathbf{r}_1 + (\mathbf{r}_2 + \mathbf{r}_3) = (\mathbf{r}_1 + \mathbf{r}_2) + \mathbf{r}_3$ and $c_1 (c_2\mathbf{r}) = (c_1 c_2) \mathbf{r}$.
4. Distributivity: $c (\mathbf{r}_1 + \mathbf{r}_2) = c\mathbf{r}_1 + c\mathbf{r}_2$ and $(c_1 + c_2) \mathbf{r} = c_1\mathbf{r} + c_2\mathbf{r}$.
5. Identity element: $\mathbf{r} + \mathbf{0} = \mathbf{r}$, $1\mathbf{r} = \mathbf{r}$, and $\mathbf{r} - \mathbf{r} = \mathbf{r} + (-1)\,\mathbf{r} = \mathbf{0}$.

Example 30 ★ Nonorthogonal Coordinate Frame It is possible to define a coordinate frame in which the three scaled lines OA, OB, OC are nonorthogonal. Defining three unit vectors $\hat{b}_1$, $\hat{b}_2$, and $\hat{b}_3$ along the nonorthogonal non-coplanar axes OA, OB, OC, respectively, we can express any vector $\mathbf{r}$ by a linear combination of the three non-coplanar unit vectors $\hat{b}_1$, $\hat{b}_2$, and $\hat{b}_3$ as

$$\mathbf{r} = r_1\hat{b}_1 + r_2\hat{b}_2 + r_3\hat{b}_3 \tag{1.115}$$

where, r_1, r_2, and r_3 are constant.

Expression of the unit vectors $\hat{b}_1$, $\hat{b}_2$, $\hat{b}_3$ and vector $\mathbf{r}$ in a Cartesian coordinate frame is

$$\mathbf{r} = x\hat{\imath} + y\hat{\jmath} + z\hat{k} \tag{1.116}$$

$$\hat{b}_1 = b_{11}\hat{\imath} + b_{12}\hat{\jmath} + b_{13}\hat{k} \tag{1.117}$$

$$\hat{b}_2 = b_{21}\hat{\imath} + b_{22}\hat{\jmath} + b_{23}\hat{k} \tag{1.118}$$

$$\hat{b}_3 = b_{31}\hat{\imath} + b_{32}\hat{\jmath} + b_{33}\hat{k} \tag{1.119}$$

Substituting (1.117)–(1.119) in (1.115) and comparing with (1.116) show that

$$\begin{bmatrix} x \\ y \\ z \end{bmatrix} = \begin{bmatrix} b_{11} & b_{12} & b_{13} \\ b_{21} & b_{22} & b_{23} \\ b_{31} & b_{32} & b_{33} \end{bmatrix} \begin{bmatrix} r_1 \\ r_2 \\ r_3 \end{bmatrix} \tag{1.120}$$

The set of equations (1.120) may be solved for the components r_1, r_2, and r_3:

$$\begin{bmatrix} r_1 \\ r_2 \\ r_3 \end{bmatrix} = \begin{bmatrix} b_{11} & b_{12} & b_{13} \\ b_{21} & b_{22} & b_{23} \\ b_{31} & b_{32} & b_{33} \end{bmatrix}^{-1} \begin{bmatrix} x \\ y \\ z \end{bmatrix} \tag{1.121}$$

We may also express them by vector scalar triple product:

$$r_1 = \frac{1}{\begin{vmatrix} b_{11} & b_{12} & b_{13} \\ b_{21} & b_{22} & b_{23} \\ b_{31} & b_{32} & b_{33} \end{vmatrix}} \begin{vmatrix} x & y & z \\ b_{21} & b_{22} & b_{23} \\ b_{31} & b_{32} & b_{33} \end{vmatrix} = \frac{\mathbf{r} \cdot \hat{b}_2 \times \hat{b}_3}{\hat{b}_1 \cdot \hat{b}_2 \times \hat{b}_3} \tag{1.122}$$

$$r_2 = \frac{1}{\begin{vmatrix} b_{11} & b_{12} & b_{13} \\ b_{21} & b_{22} & b_{23} \\ b_{31} & b_{32} & b_{33} \end{vmatrix}} \begin{vmatrix} b_{11} & b_{12} & b_{13} \\ x & y & z \\ b_{31} & b_{32} & b_{33} \end{vmatrix} = \frac{\mathbf{r} \cdot \hat{b}_3 \times \hat{b}_1}{\hat{b}_1 \cdot \hat{b}_2 \times \hat{b}_3} \tag{1.123}$$

$$r_3 = \frac{1}{\begin{vmatrix} b_{11} & b_{12} & b_{13} \\ b_{21} & b_{22} & b_{23} \\ b_{31} & b_{32} & b_{33} \end{vmatrix}} \begin{vmatrix} b_{11} & b_{12} & b_{13} \\ b_{21} & b_{22} & b_{23} \\ x & y & z \end{vmatrix} = \frac{\mathbf{r} \cdot \hat{b}_1 \times \hat{b}_2}{\hat{b}_1 \cdot \hat{b}_2 \times \hat{b}_3} \tag{1.124}$$

The set of equations (1.120) is solvable provided $\hat{b}_1 \cdot \hat{b}_2 \times \hat{b}_3 \neq 0$, which means $\hat{b}_1$, $\hat{b}_2$, $\hat{b}_3$ are not coplanar.

1.2.3 ★ Index Notation

Whenever the components of a vector or a vector equation are structurally similar, we may employ the summation sign, $\sum$, and show only one component with an index to be changed from 1 to 2 and 3 to indicate the first, second, and third components. The axes and their unit vectors of the coordinate frame may also be shown by x_1, x_2, x_3 and $\hat{u}_1, \hat{u}_2, \hat{u}_3$ instead of x, y, z and $\hat{\imath}, \hat{\jmath}, \hat{k}$. This is called *index notation* and may simplify vector calculations.

There are two symbols that may be used to make the equations even more concise:

1. *Kronecker delta* δ_{ij}:

$$\delta_{ij} = \begin{Bmatrix} 1 & i = j \\ 0 & i \neq j \end{Bmatrix} = \delta_{ji} \tag{1.125}$$

It states that $\delta_{jk} = 1$ if $j = k$ and $\delta_{jk} = 0$ if $j \neq k$.

2. *Levi-Civita symbol* ϵ_{ijk}:

$$\epsilon_{ijk} = \tfrac{1}{2}(i-j)(j-k)(k-i) \qquad i, j, k = 1, 2, 3 \tag{1.126}$$

It states that $\epsilon_{ijk} = 1$ if i, j, k is a cyclic permutation of 1, 2, 3, $\epsilon_{ijk} = -1$ if i, j, k is a cyclic permutation of 3, 2, 1, and $\epsilon_{ijk} = 0$ if at least two of i, j, k are equal. The Levi-Civita symbol is also called the *permutation symbol*.

The Levi-Civita symbol ϵ_{ijk} can be expanded by the Kronecker delta δ_{ij}:

$$\sum_{k=1}^{3} \epsilon_{ijk}\epsilon_{mnk} = \delta_{im}\delta_{jn} - \delta_{in}\delta_{jm} \tag{1.127}$$

This relation between ϵ and δ is known as the *e–delta* or *ϵ–delta* identity.

Using index notation, the vectors **a** and **b** can be shown as

$$\mathbf{a} = a_1\hat{\imath} + a_2\hat{\jmath} + a_3\hat{k} = \sum_{i=1}^{3} a_i\hat{u}_i \tag{1.128}$$

$$\mathbf{b} = b_1\hat{\imath} + b_2\hat{\jmath} + b_3\hat{k} = \sum_{i=1}^{3} b_i\hat{u}_i \tag{1.129}$$

and the inner and outer products of the unit vectors of the coordinate system as

$$\hat{u}_j \cdot \hat{u}_k = \delta_{jk} \tag{1.130}$$

$$\hat{u}_j \times \hat{u}_k = \epsilon_{ijk}\hat{u}_i \tag{1.131}$$

Example 31 Fundamental Vector Operations and Index Notation Index notation simplifies the vector equations. By index notation, we show the elements r_i, $i = 1, 2, 3$ instead of indicating the vector **r**. The fundamental vector operations by index notation are:

1. Decomposition of a vector **r**:

$$\mathbf{r} = \sum_{i=1}^{3} r_i \hat{u}_i \tag{1.132}$$

2. Orthogonality of unit vectors:

$$\hat{u}_i \cdot \hat{u}_j = \delta_{ij} \qquad \hat{u}_i \times \hat{u}_j = \epsilon_{ijk} \hat{u}_k \tag{1.133}$$

3. Projection of a vector **r** on $\hat{u}_i$:

$$\mathbf{r} \cdot \hat{u}_j = \sum_{i=1}^{3} r_i \hat{u}_i \cdot \hat{u}_j = \sum_{i=1}^{3} r_i \delta_{ij} = r_j \tag{1.134}$$

4. Scalar, dot, or inner product of vectors **a** and **b**:

$$\begin{aligned}\mathbf{a} \cdot \mathbf{b} &= \sum_{i=1}^{3} a_i \hat{u}_i \cdot \sum_{j=1}^{3} b_j \hat{u}_j = \sum_{j=1}^{3} \sum_{i=1}^{3} a_i b_j \left(\hat{u}_i \cdot \hat{u}_j\right) = \sum_{j=1}^{3} \sum_{i=1}^{3} a_i b_j \delta_{ij} \\ &= \sum_{i=1}^{3} a_i b_i \end{aligned} \tag{1.135}$$

5. Vector, cross, or outer product of vectors **a** and **b**:

$$\mathbf{a} \times \mathbf{b} = \sum_{j=1}^{3} \sum_{k=1}^{3} \epsilon_{ijk} \hat{u}_i a_j b_k \tag{1.136}$$

6. Scalar triple product of vectors **a**, **b**, and **c**:

$$\mathbf{a} \cdot \mathbf{b} \times \mathbf{c} = [\mathbf{abc}] = \sum_{k=1}^{3} \sum_{j=1}^{3} \sum_{i=1}^{3} \epsilon_{ijk} a_j b_j c_k \tag{1.137}$$

Example 32 Levi-Civita Density and Unit Vectors The *Levi-Civita symbol* ϵ_{ijk}, also called the "*e*" tensor, *Levi-Civita density*, and *permutation tensor* and may be defined by the clearer expression

$$\epsilon_{ijk} = \begin{cases} 1 & ijk = 123, 231, 312 \\ 0 & i = j \ or \ j = k \ or \ k = 1 \\ -1 & ijk = 321, 213, 132 \end{cases} \tag{1.138}$$

can be shown by the scalar triple product of the unit vectors of the coordinate system,

$$\epsilon_{ijk} = \left[\hat{u}_i \hat{u}_j \hat{u}_k\right] = \hat{u}_i \cdot \hat{u}_j \times \hat{u}_k \tag{1.139}$$

and therefore,

$$\epsilon_{ijk} = \epsilon_{jki} = \epsilon_{kij} = -\epsilon_{kji} = -\epsilon_{jik} = -\epsilon_{ikj} \tag{1.140}$$

The product of two Levi-Civita densities is

$$\epsilon_{ijk}\epsilon_{lmn} = \begin{vmatrix} \delta_{il} & \delta_{im} & \delta_{in} \\ \delta_{jl} & \delta_{jm} & \delta_{jn} \\ \delta_{kl} & \delta_{km} & \delta_{kn} \end{vmatrix} \qquad i, j, k, l, m, n = 1, 2, 3 \tag{1.141}$$

If $k = l$, we have

$$\sum_{k=1}^{3} \epsilon_{ijk}\epsilon_{mnk} = \begin{vmatrix} \delta_{im} & \delta_{in} \\ \delta_{jm} & \delta_{jn} \end{vmatrix} = \delta_{im}\delta_{jn} - \delta_{in}\delta_{jm} \tag{1.142}$$

and if also $j = n$, then

$$\sum_{k=1}^{3}\sum_{j=1}^{3} \epsilon_{ijk}\epsilon_{mjk} = 2\delta_{im} \tag{1.143}$$

and finally, if also $i = m$, we have

$$\sum_{k=1}^{3}\sum_{j=1}^{3}\sum_{i=1}^{3} \epsilon_{ijk}\epsilon_{ijk} = 6 \tag{1.144}$$

Employing the permutation symbol ϵ_{ijk}, we can show the vector scalar triple product as

$$\mathbf{a} \cdot \mathbf{b} \times \mathbf{c} = \sum_{i=1}^{3}\sum_{j=1}^{3}\sum_{k=1}^{3} \epsilon_{ijk} a_i b_j c_k = \sum_{i,j,k=1}^{3} \epsilon_{ijk} a_i b_j c_k \tag{1.145}$$

Example 33 ★ Einstein Summation Convention The *Einstein summation convention* implies that we may not show the summation symbol if we agree that there is a hidden summation symbol for every repeated index over all possible values for that index. In applied kinematics and dynamics, we usually work in a 3D space, so the range of summation symbols are from 1 to 3. Therefore, Equations (1.135) and (1.136) may be shown more simply as

$$d = a_i b_i \tag{1.146}$$

$$c_i = \epsilon_{ijk} a_j b_k \tag{1.147}$$

and the result of $\mathbf{a} \cdot \mathbf{b} \times \mathbf{c}$ as

$$\begin{aligned} \mathbf{a} \cdot \mathbf{b} \times \mathbf{c} &= \sum_{i=1}^{3} a_i \sum_{j=1}^{3}\sum_{k=1}^{3} \epsilon_{ijk} b_j c_k = \sum_{i=1}^{3}\sum_{j=1}^{3}\sum_{k=1}^{3} \epsilon_{ijk} a_i b_j c_k \\ &= \epsilon_{ijk} a_i b_j c_k \end{aligned} \tag{1.148}$$

The repeated index in a term must appear only twice to define a summation rule. Such an index is called a *dummy index* because it is immaterial what character is used for it. As an example, we have

$$a_i b_i = a_m b_m = a_1 b_1 + a_2 b_2 + a_3 b_3 \tag{1.149}$$

Example 34 ★ A Vector Identity We may use the index notation and verify vector identities such as

$$(\mathbf{a} \times \mathbf{b}) \times (\mathbf{c} \times \mathbf{d}) = \mathbf{c}\,(\mathbf{d} \cdot \mathbf{a} \times \mathbf{b}) - \mathbf{d}\,(\mathbf{c} \cdot \mathbf{a} \times \mathbf{b}) \tag{1.150}$$

Let us assume that

$$\mathbf{a} \times \mathbf{b} = \mathbf{p} = p_i \hat{u}_i \tag{1.151}$$

$$\mathbf{c} \times \mathbf{d} = \mathbf{q} = q_i \hat{u}_i \tag{1.152}$$

The components of these vectors are

$$p_i = \epsilon_{ijk} a_j b_k \tag{1.153}$$

$$q_i = \epsilon_{ijk} c_j d_k \tag{1.154}$$

and therefore the components of $\mathbf{p} \times \mathbf{q}$ are

$$\mathbf{r} = \mathbf{p} \times \mathbf{q} = r_i \hat{u}_i \tag{1.155}$$

$$\begin{aligned} r_i &= \epsilon_{ijk} p_j q_k = \epsilon_{ijk} \epsilon_{jmn} \epsilon_{krs} a_m b_n c_r d_s \\ &= \epsilon_{ijk} \epsilon_{rsk} \epsilon_{jmn} a_m b_n c_r d_s \\ &= \left(\delta_{ir}\delta_{js} - \delta_{is}\delta_{jr}\right) \epsilon_{jmn} a_m b_n c_r d_s \\ &= \epsilon_{jmn} \left(\left(c_r \delta_{ir}\right)\left(d_s \delta_{js}\right) a_m b_n - \left(c_r \delta_{jr}\right)\left(d_s \delta_{is}\right) a_m b_n\right) \\ &= \epsilon_{jmn} \left(a_m b_n c_i d_j - a_m b_n c_j d_i\right) \\ &= c_i \left(\epsilon_{jmn} d_j a_m b_n\right) - d_i \left(\epsilon_{jmn} c_j a_m b_n\right) \end{aligned} \tag{1.156}$$

so we have

$$\mathbf{r} = \mathbf{c}\,(\mathbf{d} \cdot \mathbf{a} \times \mathbf{b}) - \mathbf{d}\,(\mathbf{c} \cdot \mathbf{a} \times \mathbf{b}) \tag{1.157}$$

Example 35 ★ bac–cab Rule and ϵ–Delta Identity Employing the ϵ–delta identity (1.127), we can prove the *bac–cab* rule (1.103):

$$\begin{aligned} \mathbf{a} \times (\mathbf{b} \times \mathbf{c}) &= \epsilon_{ijk} a_i b_k c_m \epsilon_{njm} \hat{u}_n = \epsilon_{ijk} \epsilon_{jmn} a_i b_k c_m \hat{u}_n \\ &= \left(\delta_{im}\delta_{kn} - \delta_{in}\delta_{km}\right) a_i b_k c_m \hat{u}_n \\ &= a_m b_n c_m \hat{u}_n - a_n b_m c_m \hat{u}_n \\ &= a_m c_m \mathbf{b} - b_m c_m \mathbf{c} = \mathbf{b}\,(\mathbf{a} \cdot \mathbf{c}) - \mathbf{c}\,(\mathbf{a} \cdot \mathbf{b}) \end{aligned} \tag{1.158}$$

Example 36 ★ Series Solution for Three-Body Problem Consider three point masses m_1, m_2, and m_3 each subjected to Newtonian gravitational attraction from the other two particles. Let us indicate them by position vectors $\mathbf{X}_1$, $\mathbf{X}_2$, and $\mathbf{X}_3$ with respect to their mass center C. If their position and velocity vectors are given at a time t_0, how will the particles move? This is called the *three-body problem*.

This is one of the most celebrated unsolved problems in dynamics. The three-body problem is interesting and challenging because it is the smallest n-body problem that cannot be solved mathematically. Here we present a series solution and employ index notation to provide concise equations. We present the expanded form of the equations in Example 177.

The equations of motion of m_1, m_2, and m_3 are

$$\ddot{\mathbf{X}}_i = -G\sum_{j=1}^{3} m_j \frac{\mathbf{X}_i - \mathbf{X}_j}{\left|\mathbf{X}_{ji}\right|^3} \qquad i = 1, 2, 3 \tag{1.159}$$

$$\mathbf{X}_{ij} = \mathbf{X}_j - \mathbf{X}_i \tag{1.160}$$

Using the mass center as the origin implies

$$\sum_{i=1}^{3} G_i \mathbf{X}_i = 0 \qquad G_i = Gm_i \qquad i = 1, 2, 3 \tag{1.161}$$

$$G = 6.67259 \times 10^{-11}\ \mathrm{m^3\,kg^{-1}\,s^{-2}} \tag{1.162}$$

Following Belgium-American mathematician Roger Broucke (1932–2005), we use the relative position vectors $\mathbf{x}_1$, $\mathbf{x}_2$, $\mathbf{x}_3$ to derive the most symmetric form of the three-body equations of motion:

$$\mathbf{x}_i = \epsilon_{ijk}\left(\mathbf{X}_k - \mathbf{X}_j\right) \qquad i = 1, 2, 3 \tag{1.163}$$

Using $\mathbf{x}_i$, the kinematic constraint (1.161) reduces to

$$\sum_{i=1}^{3} \mathbf{x}_i = 0 \tag{1.164}$$

The absolute position vectors in terms of the relative positions are

$$m\mathbf{X}_i = \epsilon_{ijk}\left(m_k \mathbf{x}_{jj} - m_j \mathbf{x}_k\right) \qquad i = 1, 2, 3 \tag{1.165}$$

$$m = m_1 + m_2 + m_3 \tag{1.166}$$

Substituting Equation (1.165) in (1.161), we have

$$\ddot{\mathbf{x}}_i = -Gm\frac{\mathbf{x}_i}{\left|\mathbf{x}_i\right|^3} + G_i \sum_{j=1}^{3} \frac{\mathbf{x}_j}{\left|\mathbf{x}_j\right|^3} \qquad i = 1, 2, 3 \tag{1.167}$$

We are looking for a series solution of Equations (1.167) in the following form:

$$\mathbf{x}_i(t) = \mathbf{x}_{i_0} + \dot{\mathbf{x}}_{i_0}(t - t_0) + \ddot{\mathbf{x}}_{i_0}\frac{(t - t_0)^2}{2!} + \dddot{\mathbf{x}}_{i_0}\frac{(t - t_0)^3}{3!} + \cdots \tag{1.168}$$

$$\mathbf{x}_{i_0} = \mathbf{x}_i(t_0) \qquad \dot{\mathbf{x}}_{i_0} = \dot{\mathbf{x}}_i(t_0) \qquad i = 1, 2, 3 \tag{1.169}$$

Let us define $\mu = Gm$ along with an ε-set of parameters

$$\mu = Gm \qquad \varepsilon_i = \frac{1}{\left|\mathbf{x}_i\right|^3} \qquad i = 1, 2, 3 \tag{1.170}$$

to rewrite Equations (1.167) as

$$\ddot{\mathbf{x}}_i = -\mu\varepsilon_i\mathbf{x}_i + G_i\sum_{j=1}^{3}\varepsilon_j\mathbf{x}_j \qquad i = 1, 2, 3 \tag{1.171}$$

We also define three new sets of parameters

$$a_{ijk} = \frac{\mathbf{x}_i \cdot \mathbf{x}_j}{|\mathbf{x}_k|^2} \qquad b_{ijk} = \frac{\dot{\mathbf{x}}_i \cdot \mathbf{x}_j}{|\mathbf{x}_k|^2} \qquad c_{ijk} = \frac{\dot{\mathbf{x}}_i \cdot \dot{\mathbf{x}}_j}{|\mathbf{x}_k|^2} \tag{1.172}$$

where

$$a_{iii} = 1 \qquad a_{ijk} = a_{jik} \qquad c_{ijk} = c_{jik} \tag{1.173}$$

The time derivatives of the ε-set, a-set, b-set, and c-set are

$$\dot{\varepsilon}_i = -3b_{iii}\varepsilon_i \tag{1.174}$$

$$\dot{a}_{ijk} = -2b_{kkk}a_{ijk} + b_{ijk} + b_{jik} \qquad \dot{a}_{iii} = 0 \tag{1.175}$$

$$\dot{b}_{ijk} = -2b_{kkk}b_{ijk} + c_{ijk} - \mu\varepsilon_i a_{ijk} + G_i\sum_{r=1}^{3}\varepsilon_r a_{rjk} \tag{1.176}$$

$$\begin{aligned}\dot{c}_{ijk} &= -2b_{kkk}c_{ijk} - \mu\left(\varepsilon_i b_{jik} + \varepsilon_j b_{ijk}\right) \\ &\quad + G_i\sum_{r=1}^{3}\varepsilon_r b_{jrk} + G_i\sum_{s=1}^{3}\varepsilon_s a_{isk}\end{aligned} \tag{1.177}$$

The ε-set, a-set, b-set, and c-set make 84 fundamental parameters that are independent of coordinate systems. Their time derivatives are expressed only by themselves. Therefore, we are able to find the coefficients of series (1.168) to develop the series solution of the three-body problem.

1.3 ORTHOGONAL COORDINATE FRAMES

Orthogonal coordinate frames are the most important type of coordinates. It is compatible to our everyday life and our sense of dimensions. There is an orthogonality condition that is the principal equation to express any vector in an orthogonal coordinate frame.

1.3.1 Orthogonality Condition

Consider a coordinate system $(Ouvw)$ with unit vectors $\hat{u}_u$, $\hat{u}_v$, $\hat{u}_w$. The condition for the coordinate system $(Ouvw)$ to be orthogonal is that $\hat{u}_u$, $\hat{u}_v$, $\hat{u}_w$ are mutually perpendicular and hence

$$\begin{aligned}\hat{u}_u \cdot \hat{u}_v &= 0 \\ \hat{u}_v \cdot \hat{u}_w &= 0 \\ \hat{u}_w \cdot \hat{u}_u &= 0\end{aligned} \tag{1.178}$$

In an orthogonal coordinate system, every vector $\mathbf{r}$ can be shown in its decomposed description as

$$\mathbf{r} = (\mathbf{r} \cdot \hat{u}_u)\hat{u}_u + (\mathbf{r} \cdot \hat{u}_v)\hat{u}_v + (\mathbf{r} \cdot \hat{u}_w)\hat{u}_w \tag{1.179}$$

We call Equation (1.179) the *orthogonality condition* of the coordinate system $(Ouvw)$. The orthogonality condition for a Cartesian coordinate system reduces to

$$\mathbf{r} = (\mathbf{r} \cdot \hat{\imath})\hat{\imath} + (\mathbf{r} \cdot \hat{\jmath})\hat{\jmath} + (\mathbf{r} \cdot \hat{k})\hat{k} \tag{1.180}$$

Proof: Assume that the coordinate system $(Ouvw)$ is an orthogonal frame. Using the unit vectors $\hat{u}_u$, $\hat{u}_v$, $\hat{u}_w$ and the components u, v, and w, we can show any vector $\mathbf{r}$ in the coordinate system $(Ouvw)$ as

$$\mathbf{r} = u\,\hat{u}_u + v\,\hat{u}_v + w\,\hat{u}_w \tag{1.181}$$

Because of orthogonality, we have

$$\hat{u}_u \cdot \hat{u}_v = 0 \qquad \hat{u}_v \cdot \hat{u}_w = 0 \qquad \hat{u}_w \cdot \hat{u}_u = 0 \tag{1.182}$$

Therefore, the inner product of $\mathbf{r}$ by $\hat{u}_u$, $\hat{u}_v$, $\hat{u}_w$ would be equal to

$$\begin{aligned}
\mathbf{r} \cdot \hat{u}_u &= \left(u\,\hat{u}_u + v\,\hat{u}_v + w\,\hat{u}_w\right) \cdot \left(1\hat{u}_u + 0\hat{u}_v + 0\hat{u}_w\right) = u \\
\mathbf{r} \cdot \hat{u}_v &= \left(u\,\hat{u}_u + v\,\hat{u}_v + w\,\hat{u}_w\right) \cdot \left(0\hat{u}_u + 1\hat{u}_v + 0\hat{u}_w\right) = v \\
\mathbf{r} \cdot \hat{u}_v &= \left(u\,\hat{u}_u + v\,\hat{u}_v + w\,\hat{u}_w\right) \cdot \left(0\hat{u}_u + 0\hat{u}_v + 1\hat{u}_w\right) = w
\end{aligned} \tag{1.183}$$

Substituting for the components u, v, and w in Equation (1.181), we may show the vector $\mathbf{r}$ as

$$\mathbf{r} = (\mathbf{r} \cdot \hat{u}_u)\hat{u}_u + (\mathbf{r} \cdot \hat{u}_v)\hat{u}_v + (\mathbf{r} \cdot \hat{u}_w)\hat{u}_w \tag{1.184}$$

If vector $\mathbf{r}$ is expressed in a Cartesian coordinate system, then $\hat{u}_u = \hat{\imath}$, $\hat{u}_v = \hat{\jmath}$, $\hat{u}_w = \hat{k}$, and therefore,

$$\mathbf{r} = (\mathbf{r} \cdot \hat{\imath})\hat{\imath} + (\mathbf{r} \cdot \hat{\jmath})\hat{\jmath} + (\mathbf{r} \cdot \hat{k})\hat{k} \tag{1.185}$$

The orthogonality condition is the most important reason for defining a coordinate system $(Ouvw)$ orthogonal. ■

Example 37 ★ Decomposition of a Vector in a Nonorthogonal Frame Let $\mathbf{a}$, $\mathbf{b}$, and $\mathbf{c}$ be any three non-coplanar, nonvanishing vectors; then any other vector $\mathbf{r}$ can be expressed in terms of $\mathbf{a}$, $\mathbf{b}$, and $\mathbf{c}$,

$$\mathbf{r} = u\mathbf{a} + v\mathbf{b} + w\mathbf{c} \tag{1.186}$$

provided u, v, and w are properly chosen numbers. If the coordinate system $(\mathbf{a}, \mathbf{b}, \mathbf{c})$ is a Cartesian system $(\hat{I}, \hat{J}, \hat{K})$, then

$$\mathbf{r} = (\mathbf{r} \cdot \hat{I})\hat{I} + (\mathbf{r} \cdot \hat{J})\hat{J} + (\mathbf{r} \cdot \hat{K})\hat{K} \tag{1.187}$$

To find u, v, and w, we dot multiply Equation (1.186) by $\mathbf{b} \times \mathbf{c}$:

$$\mathbf{r} \cdot (\mathbf{b} \times \mathbf{c}) = u\mathbf{a} \cdot (\mathbf{b} \times \mathbf{c}) + v\mathbf{b} \cdot (\mathbf{b} \times \mathbf{c}) + w\mathbf{c} \cdot (\mathbf{b} \times \mathbf{c}) \tag{1.188}$$

Knowing that $\mathbf{b} \times \mathbf{c}$ is perpendicular to both $\mathbf{b}$ and $\mathbf{c}$, we find

$$\mathbf{r} \cdot (\mathbf{b} \times \mathbf{c}) = u\mathbf{a} \cdot (\mathbf{b} \times \mathbf{c}) \tag{1.189}$$

and therefore,

$$u = \frac{[\mathbf{rbc}]}{[\mathbf{abc}]} \tag{1.190}$$

where $[\mathbf{abc}]$ is a shorthand notation for the scalar triple product

$$[\mathbf{abc}] = \mathbf{a} \cdot (\mathbf{b} \times \mathbf{c}) = \begin{vmatrix} a_1 & b_1 & c_1 \\ a_2 & b_2 & c_2 \\ a_3 & b_3 & c_3 \end{vmatrix} \tag{1.191}$$

Similarly, v and w would be

$$v = \frac{[\mathbf{rca}]}{[\mathbf{abc}]} \qquad w = \frac{[\mathbf{rab}]}{[\mathbf{abc}]} \tag{1.192}$$

Hence,

$$\mathbf{r} = \frac{[\mathbf{rbc}]}{[\mathbf{abc}]}\mathbf{a} + \frac{[\mathbf{rca}]}{[\mathbf{abc}]}\mathbf{b} + \frac{[\mathbf{rab}]}{[\mathbf{abc}]}\mathbf{c} \tag{1.193}$$

which can also be written as

$$\mathbf{r} = \left(\mathbf{r} \cdot \frac{\mathbf{b} \times \mathbf{c}}{[\mathbf{abc}]}\right)\mathbf{a} + \left(\mathbf{r} \cdot \frac{\mathbf{c} \times \mathbf{a}}{[\mathbf{abc}]}\right)\mathbf{b} + \left(\mathbf{r} \cdot \frac{\mathbf{a} \times \mathbf{b}}{[\mathbf{abc}]}\right)\mathbf{c} \tag{1.194}$$

Multiplying (1.194) by $[\mathbf{abc}]$ gives the symmetric equation

$$[\mathbf{abc}]\,\mathbf{r} - [\mathbf{bcr}]\,\mathbf{a} + [\mathbf{cra}]\,\mathbf{b} - [\mathbf{rab}]\,\mathbf{c} = 0 \tag{1.195}$$

If the coordinate system $(\mathbf{a}, \mathbf{b}, \mathbf{c})$ is a Cartesian system $(\hat{I}, \hat{J}, \hat{K})$, then

$$\left[\hat{I}\hat{J}\hat{K}\right] = 1 \tag{1.196}$$

$$\hat{I} \times \hat{J} = \hat{K} \qquad \hat{J} \times \hat{I} = \hat{K} \qquad \hat{K} \times \hat{I} = \hat{J} \tag{1.197}$$

and Equation (1.194) becomes

$$\mathbf{r} = \left(\mathbf{r}\cdot\hat{I}\right)\hat{I} + \left(\mathbf{r}\cdot\hat{J}\right)\hat{J} + \left(\mathbf{r}\cdot\hat{K}\right)\hat{K} \tag{1.198}$$

This example may considered as a general case of Example 30.

1.3.2 Unit Vector

Consider an orthogonal coordinate system $(Oq_1q_2q_3)$. Using the orthogonality condition (1.179), we can show the position vector of a point P in this frame by

$$\mathbf{r} = (\mathbf{r} \cdot \hat{u}_1)\hat{u}_1 + (\mathbf{r} \cdot \hat{u}_2)\hat{u}_2 + (\mathbf{r} \cdot \hat{u}_3)\hat{u}_3 \tag{1.199}$$

where q_1, q_2, q_3 are the coordinates of P and $\hat{u}_1, \hat{u}_2, \hat{u}_3$ are the unit vectors along q_1, q_2, q_3 axes, respectively. Because the unit vectors $\hat{u}_1, \hat{u}_2, \hat{u}_3$ are orthogonal and independent, they respectively show the direction of change in $\mathbf{r}$ when q_1, q_2, q_3 are positively varied. Therefore, we may define the unit vectors $\hat{u}_1, \hat{u}_2, \hat{u}_3$ by

$$\hat{u}_1 = \frac{\partial\mathbf{r}/\partial q_1}{|\partial\mathbf{r}/\partial q_1|} \qquad \hat{u}_2 = \frac{\partial\mathbf{r}/\partial q_2}{|\partial\mathbf{r}/\partial q_2|} \qquad \hat{u}_3 = \frac{\partial\mathbf{r}/\partial q_3}{|\partial\mathbf{r}/\partial q_3|} \tag{1.200}$$

Example 38 Unit Vector of Cartesian Coordinate Frames If a vector $\mathbf{r}$ given as

$$\mathbf{r} = q_1\hat{u}_1 + q_2\hat{u}_2 + q_3\hat{u}_3 \tag{1.201}$$

is expressed in a Cartesian coordinate frame, then

$$q_1 = x \qquad q_2 = y \qquad q_3 = z \tag{1.202}$$

and the unit vectors would be

$$\begin{aligned} \hat{u}_1 = \hat{u}_x &= \frac{\partial\mathbf{r}/\partial x}{|\partial\mathbf{r}/\partial x|} = \frac{\hat{\imath}}{1} = \hat{\imath} \\ \hat{u}_2 = \hat{u}_y &= \frac{\partial\mathbf{r}/\partial y}{|\partial\mathbf{r}/\partial y|} = \frac{\hat{\jmath}}{1} = \hat{\jmath} \\ \hat{u}_3 = \hat{u}_z &= \frac{\partial\mathbf{r}/\partial z}{|\partial\mathbf{r}/\partial z|} = \frac{\hat{k}}{1} = \hat{k} \end{aligned} \tag{1.203}$$

Substituting $\mathbf{r}$ and the unit vectors in (1.199) regenerates the orthogonality condition in Cartesian frames:

$$\mathbf{r} = (\mathbf{r} \cdot \hat{\imath})\hat{\imath} + (\mathbf{r} \cdot \hat{\jmath})\hat{\jmath} + (\mathbf{r} \cdot \hat{k})\hat{k} \tag{1.204}$$

Example 39 Unit Vectors of a Spherical Coordinate System Figure 1.12 illustrates an option for spherical coordinate system. The angle φ may be measured from the equatorial plane or from the Z-axis. Measuring φ from the equator is used in geography and positioning a point on Earth, while measuring φ from the Z-axis is an applied method in geometry. Using the latter option, the spherical coordinates r, θ, φ are related to the Cartesian system by

$$x = r\cos\theta\sin\varphi \qquad y = r\sin\theta\sin\varphi \qquad z = r\cos\varphi \tag{1.205}$$

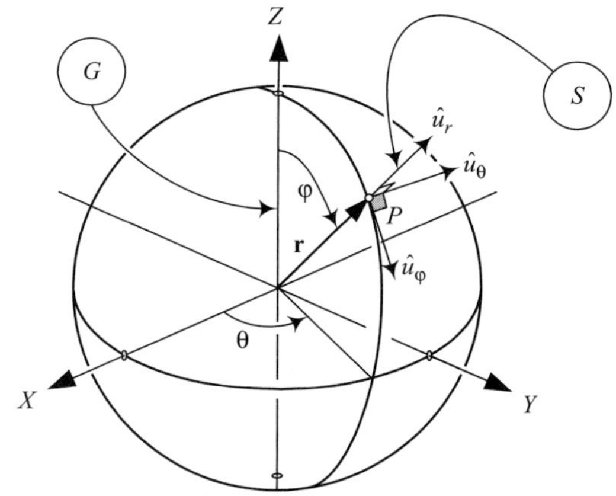

Figure 1.12 An optional spherical coordinate system.

To find the unit vectors $\hat{u}_r$, $\hat{u}_\theta$, $\hat{u}_\varphi$ associated with the coordinates r, θ, φ, we substitute the coordinate equations (1.205) in the Cartesian position vector,

$$\begin{aligned}\mathbf{r} &= x\hat{\imath} + y\hat{\jmath} + z\hat{k} \\ &= (r\cos\theta\sin\varphi)\,\hat{\imath} + (r\sin\theta\sin\varphi)\,\hat{\jmath} + (r\cos\varphi)\,\hat{k}\end{aligned} \tag{1.206}$$

and apply the unit vector equation (1.203):

$$\begin{aligned}\hat{u}_r &= \frac{\partial\mathbf{r}/\partial r}{|\partial\mathbf{r}/\partial r|} = \frac{(\cos\theta\sin\varphi)\,\hat{\imath} + (\sin\theta\sin\varphi)\,\hat{\jmath} + (\cos\varphi)\,\hat{k}}{1} \\ &= \cos\theta\sin\varphi\hat{\imath} + \sin\theta\sin\varphi\hat{\jmath} + \cos\varphi\hat{k}\end{aligned} \tag{1.207}$$

$$\begin{aligned}\hat{u}_\theta &= \frac{\partial\mathbf{r}/\partial\theta}{|\partial\mathbf{r}/\partial\theta|} = \frac{(-r\sin\theta\sin\varphi)\,\hat{\imath} + (r\cos\theta\sin\varphi)\,\hat{\jmath}}{r\sin\varphi} \\ &= -\sin\theta\hat{\imath} + \cos\theta\,\hat{\jmath}\end{aligned} \tag{1.208}$$

$$\begin{aligned}\hat{u}_\varphi &= \frac{\partial\mathbf{r}/\partial\varphi}{|\partial\mathbf{r}/\partial\varphi|} = \frac{(r\cos\theta\cos\varphi)\,\hat{\imath} + (r\sin\theta\cos\varphi)\,\hat{\jmath} + (-r\sin\varphi)\,\hat{k}}{r} \\ &= \cos\theta\cos\varphi\hat{\imath} + \sin\theta\cos\varphi\hat{\jmath} - \sin\varphi\hat{k}\end{aligned} \tag{1.209}$$

where $\hat{u}_r$, $\hat{u}_\theta$, $\hat{u}_\varphi$ are the unit vectors of the spherical system expressed in the Cartesian coordinate system.

Example 40 Cartesian Unit Vectors in Spherical System The unit vectors of an orthogonal coordinate system are always a linear combination of Cartesian unit vectors and therefore can be expressed by a matrix transformation. Having unit vectors of an orthogonal coordinate system B_1 in another orthogonal system B_2 is enough to find the unit vectors of B_2 in B_1.

Based on Example 39, the unit vectors of the spherical system shown in Figure 1.12 can be expressed as

$$\begin{bmatrix} \hat{u}_r \\ \hat{u}_\theta \\ \hat{u}_\varphi \end{bmatrix} = \begin{bmatrix} \cos\theta \sin\varphi & \sin\theta \sin\varphi & \cos\varphi \\ -\sin\theta & \cos\theta & 0 \\ \cos\theta \cos\varphi & \sin\theta \cos\varphi & -\sin\varphi \end{bmatrix} \begin{bmatrix} \hat{\imath} \\ \hat{\jmath} \\ \hat{k} \end{bmatrix} \tag{1.210}$$

So, the Cartesian unit vectors in the spherical system are

$$\begin{aligned} \begin{bmatrix} \hat{\imath} \\ \hat{\jmath} \\ \hat{k} \end{bmatrix} &= \begin{bmatrix} \cos\theta \sin\varphi & \sin\theta \sin\varphi & \cos\varphi \\ -\sin\theta & \cos\theta & 0 \\ \cos\theta \cos\varphi & \sin\theta \cos\varphi & -\sin\varphi \end{bmatrix}^{-1} \begin{bmatrix} \hat{u}_r \\ \hat{u}_\theta \\ \hat{u}_\varphi \end{bmatrix} \\ &= \begin{bmatrix} \cos\theta \sin\varphi & -\sin\theta & \cos\theta \cos\varphi \\ \sin\theta \sin\varphi & \cos\theta & \cos\varphi \sin\theta \\ \cos\varphi & 0 & -\sin\varphi \end{bmatrix} \begin{bmatrix} \hat{u}_r \\ \hat{u}_\theta \\ \hat{u}_\varphi \end{bmatrix} \end{aligned} \tag{1.211}$$

1.3.3 Direction of Unit Vectors

Consider a moving point P with the position vector $\mathbf{r}$ in a coordinate system $(Oq_1q_2q_3)$. The unit vectors $\hat{u}_1$, $\hat{u}_2$, $\hat{u}_3$ associated with q_1, q_2, q_3 are tangent to the curve traced by $\mathbf{r}$ when the associated coordinate varies.

Proof: Consider a coordinate system $\left(Oq_1q_2q_3\right)$ that has the following relations with Cartesian coordinates:

$$\begin{aligned} x &= f\left(q_1, q_2, q_3\right) \\ y &= g\left(q_1, q_2, q_3\right) \\ z &= h\left(q_1, q_2, q_3\right) \end{aligned} \tag{1.212}$$

The unit vector $\hat{u}_1$ given as

$$\hat{u}_1 = \frac{\partial \mathbf{r}/\partial q_1}{\left|\partial \mathbf{r}/\partial q_1\right|} \tag{1.213}$$

associated with q_1 at a point $P\left(x_0, y_0, z_0\right)$ can be found by fixing q_2, q_3 to q_{2_0}, q_{3_0} and varying q_1. At the point, the equations

$$\begin{aligned} x &= f\left(q_1, q_{2_0}, q_{3_0}\right) \\ y &= g\left(q_1, q_{2_0}, q_{3_0}\right) \\ z &= h\left(q_1, q_{2_0}, q_{3_0}\right) \end{aligned} \tag{1.214}$$

provide the parametric equations of a space curve passing through (x_0, y_0, z_0). From (1.228) and (1.358), the tangent line to the curve at point P is

$$\frac{x - x_0}{dx/dq_1} = \frac{y - y_0}{dy/dq_1} = \frac{z - z_0}{dz/dq_1} \tag{1.215}$$

and the unit vector on the tangent line is

$$\hat{u}_1 = \frac{dx}{dq_1}\hat{\imath} + \frac{dy}{dq_1}\hat{\jmath} + \frac{dz}{dq_1}\hat{k} \tag{1.216}$$

$$\left(\frac{dx}{dq}\right)^2 + \left(\frac{dy}{dq}\right)^2 + \left(\frac{dz}{dq}\right)^2 = 1 \tag{1.217}$$

This shows that the unit vector $\hat{u}_1$ (1.213) associated with q_1 is tangent to the space curve generated by varying q_1. When q_1 is varied positively, the direction of $\hat{u}_1$ is called positive and vice versa.

Similarly, the unit vectors $\hat{u}_2$ and $\hat{u}_3$ given as

$$\hat{u}_2 = \frac{\partial \mathbf{r}/\partial q_2}{|\partial \mathbf{r}/\partial q_2|} \qquad \hat{u}_3 = \frac{\partial \mathbf{r}/\partial q_3}{|\partial \mathbf{r}/\partial q_3|} \tag{1.218}$$

associated with q_2 and q_3 are tangent to the space curve generated by varying q_2 and q_3, respectively. ■

Example 41 Tangent Unit Vector to a Helix Consider a helix

$$x = a\cos\varphi \qquad y = a\sin\varphi \qquad z = k\varphi \tag{1.219}$$

where a and k are constant and φ is an angular variable. The position vector of a moving point P on the helix

$$\mathbf{r} = a\cos\varphi\,\hat{\imath} + a\sin\varphi\,\hat{\jmath} + k\varphi\,\hat{k} \tag{1.220}$$

may be used to find the unit vector $\hat{u}_\varphi$:

$$\begin{aligned} \hat{u}_\varphi &= \frac{\partial \mathbf{r}/\partial q_1}{|\partial \mathbf{r}/\partial q_1|} = \frac{-a\sin\varphi\,\hat{\imath} + a\cos\varphi\,\hat{\jmath} + k\,\hat{k}}{\sqrt{(-a\sin\varphi)^2 + (a\cos\varphi)^2 + (k)^2}} \\ &= -\frac{a\sin\varphi}{\sqrt{a^2+k^2}}\hat{\imath} + \frac{a\cos\varphi}{\sqrt{a^2+k^2}}\hat{\jmath} + \frac{k}{\sqrt{a^2+k^2}}\hat{k} \end{aligned} \tag{1.221}$$

The unit vector $\hat{u}_\varphi$ at $\varphi = \pi/4$ given as

$$\hat{u}_\varphi = -\frac{\sqrt{2}a}{2\sqrt{a^2+k^2}}\hat{\imath} + \frac{\sqrt{2}a}{2\sqrt{a^2+k^2}}\hat{\jmath} + \frac{k}{\sqrt{a^2+k^2}}\hat{k} \tag{1.222}$$

is on the tangent line (1.255).

1.4 DIFFERENTIAL GEOMETRY

Geometry is the world in which we express kinematics. The path of the motion of a particle is a curve in space. The analytic equation of the space curve is used to determine the vectorial expression of kinematics of the moving point.

1.4.1 Space Curve

If the position vector ${}^G\mathbf{r}_P$ of a moving point P is such that each component is a function of a variable q,

$$ {}^G\mathbf{r} = {}^G\mathbf{r}(q) = x(q)\,\hat{\imath} + y(q)\,\hat{\jmath} + z(q)\,\hat{k} \tag{1.223} $$

then the end point of the position vector indicates a curve C in G, as is shown in Figure 1.13. The curve ${}^G\mathbf{r} = {}^G\mathbf{r}(q)$ reduces to a point on C if we fix the parameter q. The functions

$$ x = x(q) \qquad y = y(q) \qquad z = z(q) \tag{1.224} $$

are the parametric equations of the curve.

When the parameter q is the arc length s, the infinitesimal arc distance ds on the curve is

$$ ds^2 = d\mathbf{r} \cdot d\mathbf{r} \tag{1.225} $$

The arc length of a curve is defined as the limit of the diagonal of a rectangular box as the length of the sides uniformly approach zero.

When the space curve is a straight line that passes through point $P(x_0, y_0, z_0)$ where $x_0 = x(q_0)$, $y_0 = y(q_0)$, $z_0 = z(q_0)$, its equation can be shown by

$$ \frac{x - x_0}{\alpha} = \frac{y - y_0}{\beta} = \frac{z - z_0}{\gamma} \tag{1.226} $$

$$ \alpha^2 + \beta^2 + \gamma^2 = 1 \tag{1.227} $$

where α, β, and γ are the directional cosines of the line.

The equation of the tangent line to the space curve (1.224) at a point $P(x_0, y_0, z_0)$ is

$$ \frac{x - x_0}{dx/dq} = \frac{y - y_0}{dy/dq} = \frac{z - z_0}{dz/dq} \tag{1.228} $$

$$ \left(\frac{dx}{dq}\right)^2 + \left(\frac{dy}{dq}\right)^2 + \left(\frac{dz}{dq}\right)^2 = 1 \tag{1.229} $$

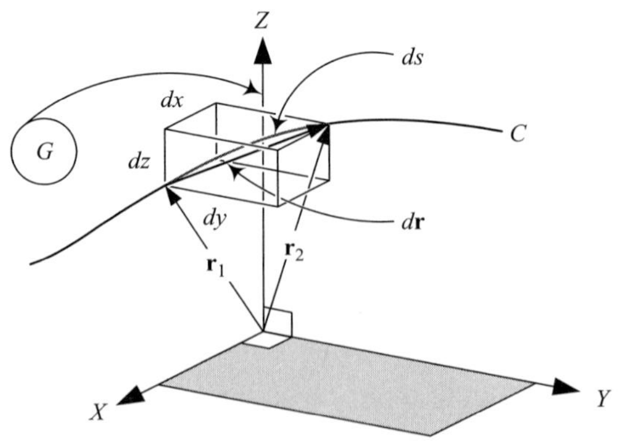

Figure 1.13 A space curve and increment arc length ds

Proof: Consider a position vector ${}^{G}\mathbf{r} = {}^{G}\mathbf{r}(s)$ that describes a space curve using the length parameter s:

$$
{}^{G}\mathbf{r} = {}^{G}\mathbf{r}(s) = x(s)\,\hat{\imath} + y(s)\,\hat{\jmath} + z(s)\,\hat{k} \tag{1.230}
$$

The arc length s is measured from a fixed point on the curve. By a very small change ds, the position vector will move to a very close point such that the increment in the position vector would be

$$
d\mathbf{r} = dx(s)\,\hat{\imath} + dy(s)\,\hat{\jmath} + dz(s)\,\hat{k} \tag{1.231}
$$

The length of $d\mathbf{r}$ and ds are equal for infinitesimal displacement:

$$
ds = \sqrt{dx^2 + dy^2 + dz^2} \tag{1.232}
$$

The arc length has a better expression in the square form:

$$
ds^2 = dx^2 + dy^2 + dz^2 = d\mathbf{r} \cdot d\mathbf{r} \tag{1.233}
$$

If the parameter of the space curve is q instead of s, the increment arc length would be

$$
\left(\frac{ds}{dq}\right)^2 = \frac{d\mathbf{r}}{dq} \cdot \frac{d\mathbf{r}}{dq} \tag{1.234}
$$

Therefore, the arc length between two points on the curve can be found by integration:

$$
s = \int_{q_1}^{q_2} \sqrt{\frac{d\mathbf{r}}{dq} \cdot \frac{d\mathbf{r}}{dq}}\, dq \tag{1.235}
$$

$$
= \int_{q_1}^{q_2} \sqrt{\left(\frac{dx}{dq}\right)^2 + \left(\frac{dy}{dq}\right)^2 + \left(\frac{dz}{dq}\right)^2}\, dq \tag{1.236}
$$

Let us expand the parametric equations of the curve (1.224) at a point $P(x_0, y_0, z_0)$,

$$
\begin{aligned}
x &= x_0 + \frac{dx}{dq}\Delta q + \frac{1}{2}\frac{d^2x}{dq^2}\Delta q^2 + \cdots \\
y &= y_0 + \frac{dy}{dq}\Delta q + \frac{1}{2}\frac{d^2y}{dq^2}\Delta q^2 + \cdots \\
z &= z_0 + \frac{dz}{dq}\Delta q + \frac{1}{2}\frac{d^2z}{dq^2}\Delta q^2 + \cdots
\end{aligned} \tag{1.237}
$$

and ignore the nonlinear terms to find the tangent line to the curve at P:

$$
\frac{x - x_0}{dx/dq} = \frac{y - y_0}{dy/dq} = \frac{z - z_0}{dz/dq} = \Delta q \tag{1.238}
$$

■

Example 42 Arc Length of a Planar Curve A planar curve in the (x, y)-plane

$$y = f(x) \tag{1.239}$$

can be expressed vectorially by

$$\mathbf{r} = x\hat{\imath} + y(x)\,\hat{\jmath} \tag{1.240}$$

The displacement element on the curve

$$\frac{d\mathbf{r}}{dx} = \hat{\imath} + \frac{dy}{dx}\hat{\jmath} \tag{1.241}$$

provides

$$\left(\frac{ds}{dx}\right)^2 = \frac{d\mathbf{r}}{dx} \cdot \frac{d\mathbf{r}}{dx} = 1 + \left(\frac{dy}{dx}\right)^2 \tag{1.242}$$

Therefore, the arc length of the curve between $x = x_1$ and $x = x_2$ is

$$s = \int_{x_1}^{x_2} \sqrt{1 + \left(\frac{dy}{dx}\right)^2}\, dx \tag{1.243}$$

In case the curve is given parametrically,

$$x = x(q) \qquad y = y(q) \tag{1.244}$$

we have

$$\left(\frac{ds}{dq}\right)^2 = \frac{d\mathbf{r}}{dq} \cdot \frac{d\mathbf{r}}{dq} = \left(\frac{dx}{dq}\right)^2 + \left(\frac{dy}{dq}\right)^2 \tag{1.245}$$

and hence,

$$s = \int_{q_1}^{q_2} \left|\frac{d\mathbf{r}}{dq}\right| = \int_{q_1}^{q_2} \sqrt{\left(\frac{dx}{dq}\right)^2 + \left(\frac{dy}{dq}\right)^2}\, dq \tag{1.246}$$

As an example, we may show a circle with radius R by its polar expression using the angle θ as a parameter:

$$x = R\cos\theta \qquad y = R\sin\theta \tag{1.247}$$

The circle is made when the parameter θ varies by 2π. The arc length between $\theta = 0$ and $\theta = \pi/2$ would then be one-fourth the perimeter of the circle. The equation for calculating the perimeter of a circle with radius R is

$$\begin{aligned} s &= 4\int_0^{\pi/2} \sqrt{\left(\frac{dx}{d\theta}\right)^2 + \left(\frac{dy}{d\theta}\right)^2}\, d\theta = R\int_0^{\pi/2} \sqrt{\sin^2\theta + \cos^2\theta}\, d\theta \\ &= 4R\int_0^{\pi/2} d\theta = 2\pi R \end{aligned} \tag{1.248}$$

Example 43 Alternative Space Curve Expressions We can represent a space curve by functions

$$y = y(x) \qquad z = z(x) \tag{1.249}$$

or vector

$$\mathbf{r}(q) = x\hat{\imath} + y(x)\hat{\jmath} + z(x)\hat{k} \tag{1.250}$$

We may also show a space curve by two relationships between x, y, and z,

$$f(x, y, z) = 0 \qquad g(x, y, z) = 0 \tag{1.251}$$

where $f(x, y, z) = 0$ and $g(x, y, z) = 0$ represent two surfaces. The space curve would then be indicated by intersecting the surfaces.

Example 44 Tangent Line to a Helix Consider a point P that is moving on a helix with equation

$$x = a\cos\varphi \qquad y = a\sin\varphi \qquad z = k\varphi \tag{1.252}$$

where a and k are constant and φ is an angular variable. To find the tangent line to the helix at $\varphi = \pi/4$,

$$x_0 = \frac{\sqrt{2}}{2}a \qquad y_0 = \frac{\sqrt{2}}{2}a \qquad z_0 = k\frac{\pi}{4} \tag{1.253}$$

we calculate the required derivatives:

$$\begin{aligned}
\frac{dx}{d\varphi} &= -a\sin\varphi = -\frac{\sqrt{2}}{2}a \\
\frac{dy}{d\varphi} &= a\cos\varphi = \frac{\sqrt{2}}{2}a \\
\frac{dz}{d\varphi} &= k
\end{aligned} \tag{1.254}$$

So, the equation of the tangent line is

$$-\frac{\sqrt{2}}{a}\left(x - \frac{1}{2}\sqrt{2}a\right) = \frac{\sqrt{2}}{a}\left(y - \frac{1}{2}\sqrt{2}a\right) = \frac{1}{k}\left(z - \frac{1}{4}\pi k\right) \tag{1.255}$$

Example 45 Parametric Form of a Line The equation of a line that connects two points $P_1(x_1, y_1, z_1)$ and $P_2(x_2, y_2, z_3)$ is

$$\frac{x - x_1}{x_2 - x_1} = \frac{y - y_1}{y_2 - y_1} = \frac{z - z_1}{z_2 - z_1} \tag{1.256}$$

This line may also be expressed by the following parametric equations:

$$
\begin{aligned}
x &= x_1 + (x_2 - x_1)\, t \\
y &= y_1 + (y_2 - y_1)\, t \\
z &= z_1 + (z_2 - z_1)\, t
\end{aligned}
\tag{1.257}
$$

Example 46 Length of a Roller Coaster Consider the roller coaster illustrated later in Figure 1.22 with the following parametric equations:

$$
\begin{aligned}
x &= (a + b\sin\theta)\cos\theta \\
y &= (a + b\sin\theta)\sin\theta \\
z &= b + b\cos\theta
\end{aligned}
\tag{1.258}
$$

for

$$
a = 200\,\mathrm{m} \qquad b = 150\,\mathrm{m} \tag{1.259}
$$

The total length of the roller coaster can be found by the integral of ds for θ from 0 to 2π:

$$
\begin{aligned}
s &= \int_{\theta_1}^{\theta_2} \sqrt{\frac{d\mathbf{r}}{d\theta} \cdot \frac{d\mathbf{r}}{d\theta}}\, d\theta = \int_{\theta_1}^{\theta_2} \sqrt{\left(\frac{\partial x}{\partial \theta}\right)^2 + \left(\frac{\partial x}{\partial \theta}\right)^2 + \left(\frac{\partial x}{\partial \theta}\right)^2}\, d\theta \\
&= \int_0^{2\pi} \frac{\sqrt{2}}{2}\sqrt{2a^2 + 3b^2 - b^2\cos 2\theta + 4ab\sin\theta d\theta} \\
&= 1629.367\,\mathrm{m}
\end{aligned}
\tag{1.260}
$$

Example 47 Two Points Indicate a Line Consider two points A and B with position vectors $\mathbf{a}$ and $\mathbf{b}$ in a coordinate frame. The condition for a point P with position vector $\mathbf{r}$ to lie on the line AB is that r $-$ $\mathbf{a}$ and $\mathbf{b} - \mathbf{a}$ be parallel. So,

$$
\mathbf{r} - \mathbf{a} = c\,(\mathbf{b} - \mathbf{a}) \tag{1.261}
$$

where c is a parameter. The outer product of Equation (1.261) by $\mathbf{b} - \mathbf{a}$ provides

$$
(\mathbf{r} - \mathbf{a}) \times (\mathbf{b} - \mathbf{a}) = 0 \tag{1.262}
$$

which is the equation of the line AB.

Example 48 Line through a Point and Parallel to a Given Line Consider a point A with position vector $\mathbf{a}$ and a line l that is indicated by a unit vector $\hat{u}_l$. To determine the equation of the parallel line to $\hat{u}_l$ that goes over A, we employ the condition that $\mathbf{r} - \mathbf{a}$ and $\hat{u}_l$ must be parallel:

$$
\mathbf{r} = \mathbf{a} + c\hat{u}_l \tag{1.263}
$$

We can eliminate the parameter c by the outer product of both sides with $\hat{u}_l$:

$$\mathbf{r} \times \hat{u}_l = \mathbf{a} \times \hat{u}_l \tag{1.264}$$

1.4.2 Surface and Plane

A plane is the locus of the tip point of a position vector

$$\mathbf{r} = x\hat{\imath} + y\hat{\jmath} + z\hat{k} \tag{1.265}$$

such that the coordinates satisfy a linear equation

$$Ax + By + Cz + D = 0 \tag{1.266}$$

A space surface is the locus of the tip point of the position vector (1.265) such that its coordinates satisfy a nonlinear equation:

$$f(x, y, z) = 0 \tag{1.267}$$

Proof: The points P_1, P_2, and P_3 at $\mathbf{r}_1$, $\mathbf{r}_2$, and $\mathbf{r}_3$,

$$\mathbf{r}_1 = \begin{bmatrix} -\frac{D}{A} \\ 0 \\ 0 \end{bmatrix} \qquad \mathbf{r}_2 = \begin{bmatrix} 0 \\ -\frac{D}{B} \\ 0 \end{bmatrix} \qquad \mathbf{r}_3 = \begin{bmatrix} 0 \\ 0 \\ -\frac{D}{C} \end{bmatrix} \tag{1.268}$$

satisfy the equations of the plane (1.266). The position of P_2 and P_3 with respect to P_1 are shown by ${}_1\mathbf{r}_2$ and ${}_1\mathbf{r}_3$ or $\mathbf{r}_{2/1}$ and $\mathbf{r}_{3/1}$:

$${}_1\mathbf{r}_2 = \mathbf{r}_2 - \mathbf{r}_1 = \begin{bmatrix} \frac{D}{A} \\ -\frac{D}{B} \\ 0 \end{bmatrix} \qquad {}_1\mathbf{r}_3 = \mathbf{r}_3 - \mathbf{r}_1 = \begin{bmatrix} \frac{D}{A} \\ 0 \\ -\frac{D}{C} \end{bmatrix} \tag{1.269}$$

The cross product of ${}_1\mathbf{r}_2$ and ${}_1\mathbf{r}_3$ is a normal vector to the plane:

$${}_1\mathbf{r}_2 \times {}_1\mathbf{r}_3 = \begin{bmatrix} \frac{D}{A} \\ -\frac{D}{B} \\ 0 \end{bmatrix} \times \begin{bmatrix} \frac{D}{A} \\ 0 \\ -\frac{D}{C} \end{bmatrix} = \begin{bmatrix} \frac{D^2}{BC} \\ \frac{D^2}{AC} \\ \frac{D^2}{AB} \end{bmatrix} \tag{1.270}$$

The equation of the plane is the locus of any point P,

$$\mathbf{r}_P = \begin{bmatrix} x \\ y \\ z \end{bmatrix} \tag{1.271}$$

where its position with respect to P_1,

$$ {}_1\mathbf{r}_P = \mathbf{r}_P - \mathbf{r}_1 = \begin{bmatrix} x + \dfrac{D}{A} \\ y \\ z \end{bmatrix} \tag{1.272} $$

is perpendicular to the normal vector:

$$ {}_1\mathbf{r}_P \cdot ({}_1\mathbf{r}_2 \times {}_1\mathbf{r}_3) = D + Ax + By + Cz = 0 \tag{1.273} $$

■

Example 49 Plane through Three Points Every three points indicate a plane. Assume that (x_1, y_1, z_1), (x_2, y_2, z_2), and (x_3, y_3, z_3) are the coordinates of three points P_1, P_2, and P_3. The plane made by the points can be found by

$$ \begin{vmatrix} x & y & z & 1 \\ x_1 & y_1 & z_1 & 1 \\ x_2 & y_2 & z_2 & 1 \\ x_3 & y_3 & z_3 & 1 \end{vmatrix} = 0 \tag{1.274} $$

The points P_1, P_2, and P_3 satisfy the equation of the plane

$$ \begin{aligned} Ax_1 + By_1 + Cz_1 + D &= 0 \\ Ax_2 + By_2 + Cz_2 + D &= 0 \\ Ax_3 + By_3 + Cz_3 + D &= 0 \end{aligned} \tag{1.275} $$

and if P with coordinates (x, y, z) is a general point on the surface,

$$ Ax + By + Cz + D = 0 \tag{1.276} $$

then there are four equations to determine A, B, C, and D:

$$ \begin{bmatrix} x & y & z & 1 \\ x_1 & y_1 & z_1 & 1 \\ x_2 & y_2 & z_2 & 1 \\ x_3 & y_3 & z_3 & 1 \end{bmatrix} \begin{bmatrix} A \\ B \\ C \\ D \end{bmatrix} = \begin{bmatrix} 0 \\ 0 \\ 0 \\ 0 \end{bmatrix} \tag{1.277} $$

The determinant of the equations must be zero, which determines the equation of the plane.

Example 50 Normal Vector to a Plane A plane may be expressed by the linear equation

$$ Ax + By + Cz + D = 0 \tag{1.278} $$

or by its intercept form

$$\frac{x}{a} + \frac{y}{b} + \frac{z}{c} = 1 \tag{1.279}$$

$$a = -\frac{D}{A} \qquad b = -\frac{D}{B} \qquad c = -\frac{D}{C} \tag{1.280}$$

In either case, the vector

$$\mathbf{n}_1 = A\hat{\imath} + B\hat{\jmath} + C\hat{k} \tag{1.281}$$

or

$$\mathbf{n}_2 = a\hat{\imath} + b\hat{\jmath} + c\hat{k} \tag{1.282}$$

is normal to the plane and may be used to represent the plane.

Example 51 Quadratic Surfaces A quadratic relation between x, y, z is called the quadratic form and is an equation containing only terms of degree 0, 1, and 2 in the variables x, y, z. Quadratic surfaces have special names:

$$\frac{x^2}{a^2} + \frac{y^2}{b^2} + \frac{z^2}{c^2} = 1 \qquad \textit{Ellipsoid} \tag{1.283}$$

$$\frac{x^2}{a^2} + \frac{y^2}{b^2} - \frac{z^2}{c^2} = 1 \qquad \textit{Hyperboloid of one sheet} \tag{1.284}$$

$$\frac{x^2}{a^2} - \frac{y^2}{b^2} - \frac{z^2}{c^2} = 1 \qquad \textit{Hyperboloid of two sheets} \tag{1.285}$$

$$\frac{x^2}{a^2} + \frac{y^2}{b^2} + \frac{z^2}{c^2} = -1 \qquad \textit{Imaginary ellipsoid} \tag{1.286}$$

$$\frac{x^2}{a^2} + \frac{y^2}{b^2} = 2nz \qquad \textit{Elliptic paraboloid} \tag{1.287}$$

$$\frac{x^2}{a^2} - \frac{y^2}{b^2} = 2nz \qquad \textit{Hyperbolic paraboloid} \tag{1.288}$$

$$\frac{x^2}{a^2} + \frac{y^2}{b^2} - \frac{z^2}{c^2} = 0 \qquad \textit{Real quadratic cone} \tag{1.289}$$

$$\frac{x^2}{a^2} + \frac{y^2}{b^2} + \frac{z^2}{c^2} = 0 \qquad \textit{Real imaginary cone} \tag{1.290}$$

$$\frac{x^2}{a^2} \pm \frac{y^2}{b^2} = \pm 1 \quad y^2 = 2px \qquad \textit{Quadratic cylinders} \tag{1.291}$$

1.5 MOTION PATH KINEMATICS

The derivative of vector functions is based on the derivative of scalar functions. To find the derivative of a vector, we take the derivative of its components in a decomposed Cartesian expression.

1.5.1 Vector Function and Derivative

The derivative of a vector is possible only when the vector is expressed in a Cartesian coordinate frame. Its derivative can be found by taking the derivative of its components. The Cartesian unit vectors are invariant and have zero derivative with respect to any parameter.

A vector $\mathbf{r} = \mathbf{r}(t)$ is called a *vector function* of the *scalar variable* t if there is a definite vector for every value of t from a certain set $T = [\tau_1, \tau_2]$. In a Cartesian coordinate frame G, the specification of the vector function $\mathbf{r}(t)$ is equivalent to the specification of three scalar functions $x(t)$, $y(t)$, $z(t)$:

$$^G\mathbf{r}(t) = x(t)\,\hat{\imath} + y(t)\,\hat{\jmath} + z(t)\,\hat{k} \tag{1.292}$$

If the vector $\mathbf{r}$ is expressed in Cartesian decomposition form, then the derivative $d\mathbf{r}/dt$ is

$$\frac{^Gd}{dt}\,{}^G\mathbf{r} = \frac{dx(t)}{dt}\hat{\imath} + \frac{dy(t)}{dt}\hat{\jmath} + \frac{dz(t)}{dt}\hat{k} \tag{1.293}$$

and if $\mathbf{r}$ is expressed in its natural form

$$^G\mathbf{r} = r\hat{u}_r = r(t)\left[u_1(t)\,\hat{\imath} + u_2(t)\,\hat{\jmath} + u_3(t)\,\hat{k}\right] \tag{1.294}$$

then, using the chain rule, the derivative $d\mathbf{r}/dt$ is

$$\begin{aligned}\frac{^Gd}{dt}\,{}^G\mathbf{r} &= \frac{dr}{dt}\hat{u}_r + r\frac{d}{dt}\hat{u}_r \\ &= \frac{dr}{dt}\left(u_1\hat{\imath} + u_2\hat{\jmath} + u_3\hat{k}\right) + r\left(\frac{du_1}{dt}\hat{\imath} + \frac{du_2}{dt}\hat{\jmath} + \frac{du_3}{dt}\hat{k}\right) \\ &= \left(\frac{dr}{dt}u_1 + r\frac{du_1}{dt}\right)\hat{\imath} + \left(\frac{dr}{dt}u_2 + r\frac{du_2}{dt}\right)\hat{\jmath} + \left(\frac{dr}{dt}u_3 + r\frac{du_3}{dt}\right)\hat{k}\end{aligned} \tag{1.295}$$

When the independent variable t is time, an overdot $\dot{\mathbf{r}}(t)$ is used as a shorthand notation to indicate the time derivative.

Consider a moving point P with a continuously varying position vector $\mathbf{r} = \mathbf{r}(t)$. When the starting point of $\mathbf{r}$ is fixed at the origin of G, its end point traces a continuous curve C as is shown in Figure 1.14. The curve C is called a *configuration path* that describes the motion of P, and the vector function $\mathbf{r}(t)$ is its vector representation. At each point of the continuously smooth curve $C = \{\mathbf{r}(t), t \in [\tau_1, \tau_2]\}$ there exists a tangent line and a derivative vector $d\mathbf{r}(t)/dt$ that is directed along the tangent line and directed toward increasing the parameter t. If the parameter is the arc length s of

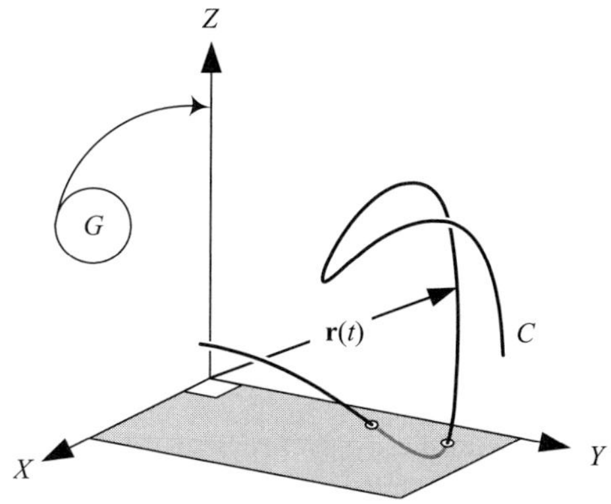

Figure 1.14 A space curve is the trace point of a single variable position vector.

the curve that is measured from a convenient point on the curve, the derivative of ${}^G\mathbf{r}$ with respect to s is the tangential unit vector $\hat{u}_t$ to the curve at ${}^G\mathbf{r}$:

$$\frac{{}^G d}{ds}\,{}^G\mathbf{r} = \hat{u}_t \tag{1.296}$$

Proof: The position vector ${}^G\mathbf{r}$ in its decomposed expression

$${}^G\mathbf{r}\,(t) = x\,(t)\,\hat{\imath} + y\,(t)\,\hat{\jmath} + z\,(t)\,\hat{k} \tag{1.297}$$

is a combination of three variable-length vectors $x\,(t)\hat{\imath}$, $y\,(t)\,\hat{\jmath}$, and $z\,(t)\,\hat{k}$. Consider the first one that is a multiple of a scalar function $x\,(t)$ and a constant unit vector $\hat{\imath}$. If the variable is time, then the time derivative of this variable-length vector in the same frame in which the vector is expressed is

$$\frac{{}^G d}{dt}\left(x\,(t)\,\hat{\imath}\right) = \dot{x}\,(t)\,\hat{\imath} + x\,(t)\,\frac{{}^G d}{dt}\hat{\imath} = \dot{x}\,(t)\,\hat{\imath} \tag{1.298}$$

Similarly, the time derivatives of $y\,(t)\,\hat{\jmath}$ and $z\,(t)\,\hat{k}$ are $\dot{y}\,(t)\,\hat{\jmath}$ and $\dot{z}\,(t)\,\hat{k}$, and therefore, the time derivative of the vector ${}^G\mathbf{r}\,(t)$ can be found by taking the derivative of its components

$${}^G\mathbf{v} = \frac{{}^G d}{dt}\,{}^G\mathbf{r}\,(t) = \dot{x}\,(t)\,\hat{\imath} + \dot{y}\,(t)\,\hat{\jmath} + \dot{z}\,(t)\,\hat{k} \tag{1.299}$$

If a variable vector ${}^G\mathbf{r}$ is expressed in a natural form

$${}^G\mathbf{r} = r\,(t)\,\hat{u}_r\,(t) \tag{1.300}$$

we express the unit vector $\hat{u}_r\,(t)$ in its decomposed form

$$\begin{aligned}{}^G\mathbf{r} &= r\,(t)\,\hat{u}_r\,(t) \\ &= r\,(t)\left[u_1\,(t)\,\hat{\imath} + u_2\,(t)\,\hat{\jmath} + u_3\,(t)\,\hat{k}\right]\end{aligned} \tag{1.301}$$

and take the derivative using the chain rule and variable-length vector derivative:

$$
\begin{aligned}
{}^{G}\mathbf{v} &= \frac{{}^{G}d}{dt}\,{}^{G}\mathbf{r} = \dot{r}\hat{u}_r + r\frac{{}^{G}d}{dt}\hat{u}_r \\
&= \dot{r}\left(u_1\hat{\imath} + u_2\hat{\jmath} + u_3\hat{k}\right) + r\left(\dot{u}_1\hat{\imath} + \dot{u}_2\hat{\jmath} + \dot{u}_3\hat{k}\right) \\
&= \left(\dot{r}u_1 + r\dot{u}_1\right)\hat{\imath} + \left(\dot{r}u_2 + r\dot{u}_2\right)\hat{\jmath} + \left(\dot{r}u_3 + r\dot{u}_3\right)\hat{k}
\end{aligned}
\tag{1.302}
$$

■

Example 52 Geometric Expression of Vector Derivative Figure 1.15 depicts a configuration path C that is the trace of a position vector $\mathbf{r}(t)$ when t varies. If $\Delta t > 0$, then the vector $\Delta\mathbf{r}$ is directed along the secant AB of the curve C toward increasing values of the parameter t. The derivative vector $d\mathbf{r}(t)/dt$ is the limit of $\Delta\mathbf{r}$ when $\Delta t \to 0$:

$$
\frac{d}{ds}\mathbf{r}(t) = \lim_{\Delta t \to 0}\frac{\Delta\mathbf{r}}{\Delta t}
\tag{1.303}
$$

where $d\mathbf{r}(t)/dt$ is directed along the tangent line to C.

Let us show the unit vectors along $\Delta\mathbf{r}$ and $d\mathbf{r}(t)/dt$ by $\Delta\mathbf{r}/\Delta r$ and $\hat{u}_t$ to get

$$
\hat{u}_t = \lim_{\Delta r \to 0}\frac{\Delta\mathbf{r}}{\Delta r} = \lim_{\Delta t \to 0}\frac{\Delta\mathbf{r}/\Delta t}{\Delta r/\Delta t} = \frac{d\mathbf{r}(t)/dt}{dr/dt}
\tag{1.304}
$$

The tangent unit vector $\hat{u}_t$ to the curve C is called the *orientation* of the curve C. When $\Delta t \to 0$, the length of $\Delta\mathbf{r}$ approaches the arc length Δs between A and B. So, Equation (1.304) can also be written as

$$
\hat{u}_t = \lim_{\Delta s \to 0}\frac{\Delta\mathbf{r}}{\Delta s} = \lim_{\Delta t \to 0}\frac{\Delta\mathbf{r}/\Delta t}{\Delta s/\Delta t} = \frac{d\mathbf{r}(t)/dt}{ds(t)/dt}
\tag{1.305}
$$

If $\Delta t < 0$, then the vector $\Delta\mathbf{r}$ is directed toward decreasing values of t.

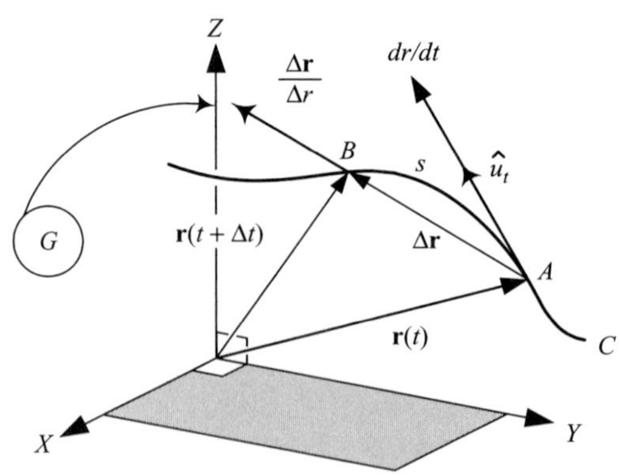

Figure 1.15 The increment vector $\Delta\mathbf{r}$ for $\Delta t > 0$ of a position vector $\mathbf{r}(t)$ is directed along the increasing secant AB of the curve configuration path C.

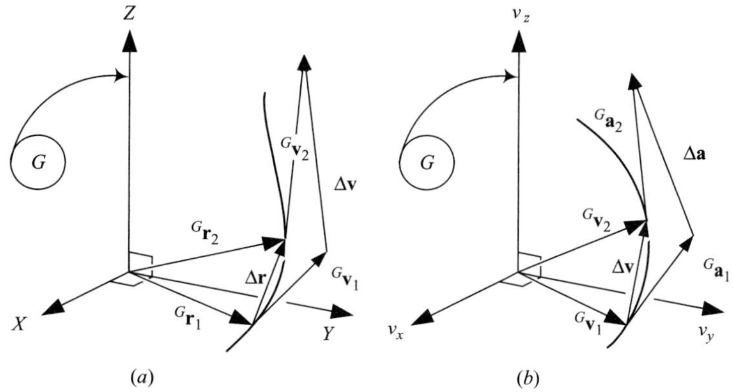

Figure 1.16 (*a*) Path of a position vector $\mathbf{r}$. (*b*) Path of the velocity vector $\mathbf{v} = d\mathbf{r}/dt$.

Consider a moving point P in a coordinate frame $G(x, y, z)$, with a continuously varying position vector $\mathbf{r} = \mathbf{r}(t)$ from a fixed origin, as is shown in Figure 1.16(*a*). The end point of the vector describes a path C when time t varies. Assume that $\mathbf{r} = {}^G\mathbf{r}_1$ is the position vector at a time $t = t_1$ and $\mathbf{r} = {}^G\mathbf{r}_2$ is the position vector at a time $t = t_2$. The difference vector

$$^G\Delta\mathbf{r} = {}^G\mathbf{r}_2 - {}^G\mathbf{r}_1 \tag{1.306}$$

becomes smaller by shortening the time duration:

$$\Delta t = t_2 - t_1 \tag{1.307}$$

The quotient $\Delta\mathbf{r}/\Delta t$ is the average rate of change of $\mathbf{r}$ in the interval Δt. Following the method of calculus, the limit of this quotient when $\Delta t \to 0$ by moving t_2 toward t_1 is the derivative of $\mathbf{r}$ at t_1:

$$\lim_{\Delta t \to 0} \frac{^G\Delta\mathbf{r}}{\Delta t} = \frac{^Gd}{dt}\,{}^G\mathbf{r} = {}^G\mathbf{v} \tag{1.308}$$

where $^G\mathbf{v}$ is a tangent vector to the path C at the position $^G\mathbf{r}_1$ and is called the *velocity* of P.

We may express the velocity vector in a new orthogonal coordinate frame $G(v_x, v_y, v_z)$. The tip point of the velocity vector traces a path in the velocity coordinate frame called a *velocity hodograph*. Employing the same method, we can define the velocity $\mathbf{v} = {}^G\mathbf{v}_1$ at time $t = t_1$ and the velocity $\mathbf{v} = {}^G\mathbf{v}_2$ at time $t = t_2$. The difference vector

$$^G\Delta\mathbf{v} = {}^G\mathbf{v}_2 - {}^G\mathbf{v}_1 \tag{1.309}$$

becomes smaller by shortening the time duration

$$\Delta t = t_2 - t_1 \tag{1.310}$$

The quotient $\Delta\mathbf{v}/\Delta t$ is the average rate of change of $\mathbf{v}$ in the interval Δt. The limit of this quotient is the derivative of $\mathbf{v}$ that makes the acceleration of P:

$$\lim_{\Delta t \to 0} \frac{^G\Delta\mathbf{v}}{\Delta t} = \frac{^Gd}{dt}\,{}^G\mathbf{v} = {}^G\mathbf{a} \tag{1.311}$$

Example 53 A Moving Point on a Helix Consider the point P in Figure 1.17 with position vector ${}^G\mathbf{r}(\varphi)$,

$$ {}^G\mathbf{r}(\varphi) = a\cos\varphi\,\hat{\imath} + a\sin\varphi\,\hat{\jmath} + k\varphi\,\hat{k} \tag{1.312} $$

that is moving on a helix with equation

$$ x = a\cos\varphi \qquad y = a\sin\varphi \qquad z = k\varphi \tag{1.313} $$

where a and k are constant and φ is an angular variable. The first, second, and third derivatives of ${}^G\mathbf{r}(\varphi)$ with respect to φ are

$$ \frac{{}^Gd}{d\varphi}\mathbf{r}(\varphi) = \mathbf{r}'(\varphi) = -a\sin\varphi\,\hat{\imath} + a\cos\varphi\,\hat{\jmath} + k\,\hat{k} \tag{1.314} $$

$$ \frac{{}^Gd^2}{d\varphi^2}\mathbf{r}(\varphi) = \mathbf{r}''(\varphi) = -a\cos\varphi\,\hat{\imath} - a\sin\varphi\,\hat{\jmath} \tag{1.315} $$

$$ \frac{{}^Gd^3}{d\varphi^3}\mathbf{r}(\varphi) = \mathbf{r}'''(\varphi) = a\sin\varphi\,\hat{\imath} - a\cos\varphi\,\hat{\jmath} \tag{1.316} $$

If the angle φ is a function of time t, then the first, second, and third derivatives of ${}^G\mathbf{r}(\varphi)$ with respect to t are

$$ \frac{{}^Gd}{dt}\mathbf{r}(t) = -a\dot{\varphi}\sin\varphi\,\hat{\imath} + a\dot{\varphi}\cos\varphi\,\hat{\jmath} + k\dot{\varphi}\,\hat{k} \tag{1.317} $$

$$ \begin{aligned} \frac{{}^Gd^2}{dt^2}\mathbf{r}(t) &= \left(-a\ddot{\varphi}\sin\varphi - a\dot{\varphi}^2\cos\varphi\right)\hat{\imath} \\ &\quad + \left(a\ddot{\varphi}\cos\varphi - a\dot{\varphi}^2\sin\varphi\right)\hat{\jmath} + k\ddot{\varphi}\,\hat{k} \end{aligned} \tag{1.318} $$

$$ \begin{aligned} \frac{{}^Gd^3}{dt^3}\mathbf{r}(t) &= \left(-a\dddot{\varphi}\sin\varphi - 3a\dot{\varphi}\ddot{\varphi}\cos\varphi + a\dot{\varphi}^3\sin\varphi\right)\hat{\imath} \\ &\quad + \left(a\dddot{\varphi}\cos\varphi - 3a\dot{\varphi}\ddot{\varphi}\sin\varphi - a\dot{\varphi}^3\cos\varphi\right)\hat{\jmath} + k\dddot{\varphi}\,\hat{k} \end{aligned} \tag{1.319} $$

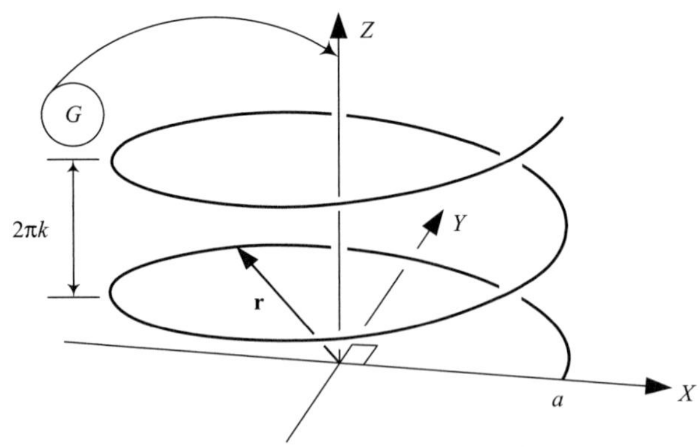

Figure 1.17 Helical path of a moving point.

Example 54 Vector Function If the magnitude of a vector $\mathbf{r}$ and/or direction of $\mathbf{r}$ in a reference frame B depends on a scalar variable, say q, then $\mathbf{r}$ is called a *vector function* of q in B. A vector may be a function of a variable in one coordinate frame but be independent of this variable in another coordinate frame.

1.5.2 Velocity and Acceleration

If the vector $\mathbf{r} = {}^G\mathbf{r}(t)$ is a position vector in a coordinate frame G, then its time derivative is a *velocity* vector ${}^G\mathbf{v}$. It shows the speed and the direction of motion of the tip point of ${}^G\mathbf{r}$:

$$ {}^G\mathbf{v} = \frac{{}^Gd}{dt}\,{}^G\mathbf{r}(t) = \dot{x}(t)\,\hat{\imath} + \dot{y}(t)\,\hat{\jmath} + \dot{z}(t)\,\hat{k} \tag{1.320} $$

The time derivative of a velocity vector ${}^G\mathbf{v}$ is called the *acceleration* ${}^G\mathbf{a}$,

$$ {}^G\mathbf{a} = \frac{{}^Gd}{dt}\,{}^G\mathbf{v}(t) = \ddot{x}(t)\,\hat{\imath} + \ddot{y}(t)\,\hat{\jmath} + \ddot{z}(t)\,\hat{k} \tag{1.321} $$

and the time derivative of an acceleration vector ${}^G\mathbf{a}$ is called the *jerk* ${}^G\mathbf{j}$,

$$ {}^G\mathbf{j} = \frac{{}^Gd}{dt}\,{}^G\mathbf{a}(t) = \dddot{x}(t)\,\hat{\imath} + \dddot{y}(t)\,\hat{\jmath} + \dddot{z}(t)\,\hat{k} \tag{1.322} $$

Example 55 Velocity, Acceleration, and Jerk of a Moving Point on a Helix Consider a moving point P with position vector in a coordinate frame G as

$$ {}^G\mathbf{r}(t) = \cos(\omega t)\,\hat{\imath} + \sin(\omega t)\,\hat{\jmath} + 2t\,\hat{k} \tag{1.323} $$

Such a path is called a *helix* or *screw*. The helix is uniformly turning on a circle in the (x, y)-plane while the circle is moving with a constant speed in the z-direction.

Taking the derivative shows that the velocity, acceleration, and jerk of the point P are

$$ {}^G\mathbf{v}(t) = -\omega\sin(\omega t)\,\hat{\imath} + \omega\cos(\omega t)\,\hat{\jmath} + 2\,\hat{k} \tag{1.324} $$

$$ {}^G\mathbf{a}(t) = -\omega^2\cos(\omega t)\,\hat{\imath} - \omega^2\sin(\omega t)\,\hat{\jmath} \tag{1.325} $$

$$ {}^G\mathbf{j}(t) = \omega^3\sin(\omega t)\,\hat{\imath} - \omega^3\cos(\omega t)\,\hat{\jmath} \tag{1.326} $$

Example 56 ★ Flight of a Bug Consider two cars A and B that are initially 15 km apart. The cars begin moving toward each other. The speeds of cars A and B are 10 and 5 km/h, respectively. The instant they started a bug on the bumper of car A starts flying with speed 12 km/h straight toward car B. As soon as it reaches the other car it turns and flies back. The bug flies back and forth from one car to the other until the two cars meet. The total length that the bug flies would be 12 km.

To calculate the total length of the bug's motion, let us show the velocities of the cars by $\mathbf{v}_A$ and $\mathbf{v}_B$ and the velocity of the bug by $\mathbf{v}_F$. Figure 1.18 illustrates the position of the cars and the bug at a time $t > 0$. Their positions are

$$X_A = v_A t \qquad X_B = l - v_B t \qquad X_F = v_F t \tag{1.327}$$

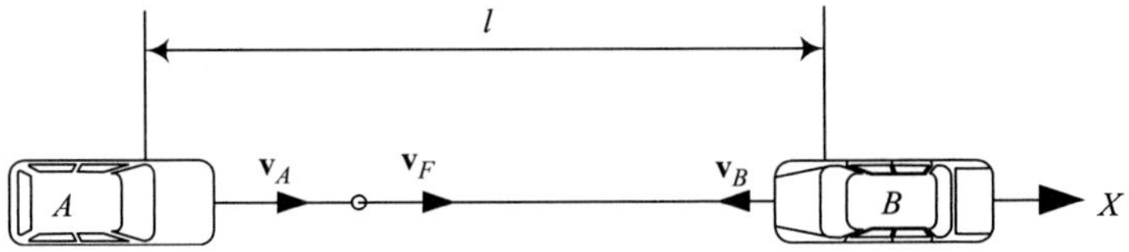

Figure 1.18 Two cars A and B moving toward each other and a bug F flying from one car to the other.

The bug reaches B at time t_1 after flying the distance d_1:

$$t_1 = \frac{l}{v_B + v_F} \qquad d_1 = v_F t_1 = \frac{v_F}{v_B + v_F} l \tag{1.328}$$

At this time, cars A, B and the bug are at

$$X_{A_1} = \frac{v_A}{v_B + v_F} l \tag{1.329}$$

$$X_{B_1} = \left(1 - \frac{v_B}{v_B + v_F}\right) l \tag{1.330}$$

$$X_{F_1} = X_{B_1} \tag{1.331}$$

so their positions when the bug is flying back are

$$X_A = X_{A_1} + v_A t = \frac{v_A}{v_B + v_F} l + v_A t \tag{1.332}$$

$$X_B = X_{B_1} - v_B t = \left(1 - \frac{v_B}{v_B + v_F}\right) l - v_B t \tag{1.333}$$

$$X_F = X_{F_1} - v_F t = \left(1 - \frac{v_B}{v_B + v_F}\right) l - v_F t \tag{1.334}$$

The bug reaches A at time t_2 after flying the distance d_2:

$$t_2 = \frac{l}{v_B + v_F} \frac{v_F - v_A}{v_A + v_F} \qquad d_2 = \frac{v_F}{v_B + v_F} \frac{v_A}{v_A + v_F} l \tag{1.335}$$

At this time cars A, B and the bug are at

$$X_{A_2} = 2 \frac{v_F}{v_B + v_F} \frac{v_A}{v_A + v_F} l \tag{1.336}$$

$$X_{B_2} = \left(1 - \frac{v_B}{v_B + v_F} + \frac{v_B}{v_B + v_F} \frac{v_F - v_A}{v_A + v_F}\right) l \tag{1.337}$$

$$X_{F_2} = X_{A_2} \tag{1.338}$$

so their positions when the bug is flying forward are

$$X_A = X_{A_2} + v_A t = 2\frac{v_A}{v_B + v_F}\frac{v_F}{v_A + v_F}l + v_A t \tag{1.339}$$

$$X_B = X_{B_2} - v_B t = l - \frac{v_B}{v_B + v_F}l\left(1 - \frac{v_F - v_A}{v_A + v_F}\right) - v_B t \tag{1.340}$$

$$X_F = X_{F_2} + v_F t = 2\frac{v_A}{v_B + v_F}\frac{v_F}{v_A + v_F}l - v_F t \tag{1.341}$$

By repeating this procedure, we can find the next times and distances and determine the total time t and distance d as

$$t = t_1 + t_2 + t_3 + \cdots \tag{1.342}$$

$$d = d_1 + d_2 + d_3 + \cdots \tag{1.343}$$

However, there is a simpler method to analyze this problem. The total time t at which the cars meet is

$$t = \frac{l}{v_A + v_B} \tag{1.344}$$

At this time, the bug can fly a distance d:

$$d = v_F t = \frac{v_F}{v_A + v_B}l \tag{1.345}$$

Therefore, if the speeds of the cars are $v_A = 10\,\mathrm{km/h}$ and $v_B = 5\,\mathrm{km/h}$ and their distance is $d = 15\,\mathrm{km}$, it takes an hour for the cars to meet. The bug with a speed of $v_F = 12\,\mathrm{km/h}$ can fly $d = 12\,\mathrm{km}$ in an hour.

Example 57 ★ Jerk, Snap, and Other Derivatives The derivative of acceleration or the third time derivative of the position vector **r** is called the *jerk* **j**; in England the word *jolt* is used instead of jerk. The third derivative may also wrongly be called pulse, impulse, bounce, surge, shock, or superacceleration.

In engineering, jerk is important for evaluating the destructive effects of motion on a moving object. For instance, high jerk is a reason for the discomfort of passengers in a vehicle. Jerk is the reason for liquid splashing from an open container. The movement of fragile objects, such as eggs, needs to be kept within specified limits of jerk to avoid damage. It is required that engineers keep the jerk of public transportation vehicles less than $2\,\mathrm{m/s^3}$ for passenger comfort. There is an instrument in the aerospace industry called a *jerkmeter* that measures jerk.

There are no universally accepted names for the fourth and higher derivatives of a position vector **r**. However, the terms *snap* **s** and *jounce* **s** have been used for derivatives of jerk. The fifth derivative of **r** is *crackle* **c**, the sixth derivative is *pop* **þ**, the seventh derivative is *larz* **z**, the eight derivative is *bong* **b**, the ninth derivative is *jeeq* **q**, and the tenth derivative is *sooz* **u**.

1.5.3 ★ Natural Coordinate Frame

Consider a space curve

$$x = x(s) \qquad y = y(s) \qquad z = z(s) \tag{1.346}$$

where s is the arc length of the curve from a fixed point on the curve. At the point there are three important planes: the *perpendicular plane* to the curve,

$$(x - x_0)\frac{dx}{ds} + (y - y_0)\frac{dy}{ds} + (z - z_0)\frac{dz}{ds} = 0 \tag{1.347}$$

the *osculating plane*,

$$\begin{aligned}\left(\frac{dy}{ds}\frac{d^2z}{ds^2} - \frac{dz}{ds}\frac{d^2y}{ds^2}\right)(x - x_0) + \left(\frac{dz}{ds}\frac{d^2x}{ds^2} - \frac{dx}{ds}\frac{d^2z}{ds^2}\right)(y - y_0)\\ + \left(\frac{dx}{ds}\frac{d^2y}{ds^2} - \frac{dy}{ds}\frac{d^2x}{ds^2}\right)(z - z_0) = 0\end{aligned} \tag{1.348}$$

and the *rectifying plane*,

$$(x - x_0)\frac{d^2x}{ds^2} + (y - y_0)\frac{d^2y}{ds^2} + (z - z_0)\frac{d^2z}{ds^2} = 0 \tag{1.349}$$

The osculating plane is the plane that includes the tangent line and the curvature center of the curve at P. The rectifying plane is perpendicular to both the osculating and normal planes.

The curvature of the curve at P is

$$\kappa = \sqrt{\left(\frac{d^2x}{ds^2}\right)^2 + \left(\frac{d^2y}{ds^2}\right)^2 + \left(\frac{d^2z}{ds^2}\right)^2} \tag{1.350}$$

and the radius of curvature is

$$\rho = \frac{1}{\kappa} \tag{1.351}$$

The radius of curvature indicates the center of curvature in the osculating plane. Figure 1.19 illustrates a space curve and the three planes at a point P. The unit vectors $\hat{u}_t$, $\hat{u}_n$, and $\hat{u}_b$ are indicators of the rectifying, perpendicular, and osculating planes and make an orthogonal triad. This triad can be used to express the velocity and acceleration of the moving point P along the space curve C:

$$\mathbf{v} = \dot{s}\hat{u}_t \tag{1.352}$$

$$\mathbf{a} = \ddot{s}\hat{u}_t + \frac{\dot{s}^2}{\rho}\hat{u}_n \tag{1.353}$$

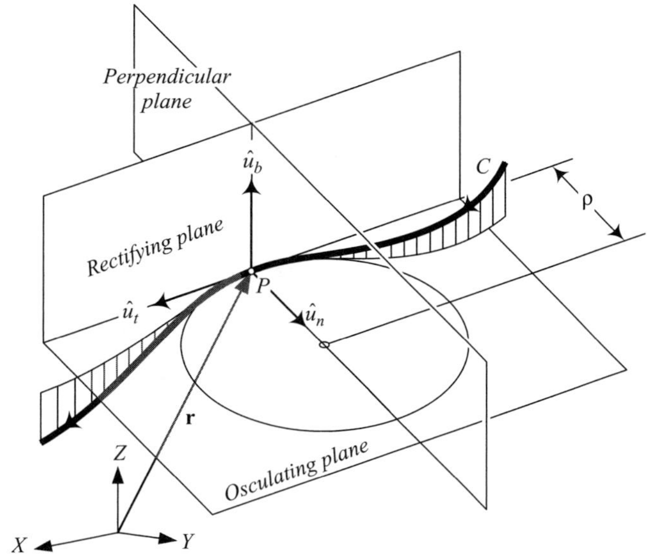

Figure 1.19 A space curve C and the three associated planes to the natural coordinates at a point P.

The orthogonal triad $\hat{u}_t$, $\hat{u}_n$, $\hat{u}_b$ is called the *natural triad* or *natural coordinate frame*:

$$\hat{u}_t = \frac{d\mathbf{r}}{ds} \tag{1.354}$$

$$\hat{u}_n = \frac{1}{\left|d^2\mathbf{r}/ds^2\right|}\frac{d^2\mathbf{r}}{ds^2} \tag{1.355}$$

$$\hat{u}_b = \frac{1}{\left|(d\mathbf{r}/ds) \times (d^2\mathbf{r}/ds^2)\right|}\left(\frac{d\mathbf{r}}{ds} \times \frac{d^2\mathbf{r}}{ds^2}\right) \tag{1.356}$$

Proof: Consider the tangent line (1.228) to the space curve (1.346) at point $P\,(x_0, y_0, z_0)$:

$$\frac{x - x_0}{dx/ds} = \frac{y - y_0}{dy/ds} = \frac{z - z_0}{dz/ds} \tag{1.357}$$

The unit vector along the tangent line l_t is

$$\hat{u}_t = \frac{dx}{ds}\hat{\imath} + \frac{dy}{ds}\hat{\jmath} + \frac{dz}{ds}\hat{k} = \frac{d\mathbf{r}}{ds} \tag{1.358}$$

because dx/ds, dy/ds, dz/ds are the directional cosines of the tangent line. A perpendicular plane to this vector is

$$\frac{dx}{ds}x + \frac{dy}{ds}y + \frac{dz}{ds}z = c \tag{1.359}$$

where c is a constant. The coordinates of $P\ (x_0, y_0, z_0)$ must satisfy the equation of the plane

$$\frac{dx}{ds}x_0 + \frac{dy}{ds}y_0 + \frac{dz}{ds}z_0 = c \tag{1.360}$$

and the perpendicular plane to the space curve at $P\ (x_0, y_0, z_0)$ is

$$(x - x_0)\frac{dx}{ds} + (y - y_0)\frac{dy}{ds} + (z - z_0)\frac{dz}{ds} = 0 \tag{1.361}$$

The equation of any plane that includes $P\ (x_0, y_0, z_0)$ is

$$A\,(x - x_0) + B\,(y - y_0) + C\,(z - z_0) = 0 \tag{1.362}$$

It also includes the tangent line (1.357) if

$$A\frac{dx}{ds} + B\frac{dy}{ds} + C\frac{dz}{ds} = 0 \tag{1.363}$$

and includes the space curve up to Δs^2 if

$$A\frac{d^2x}{ds^2} + B\frac{d^2y}{ds^2} + C\frac{d^2z}{ds^2} = 0 \tag{1.364}$$

Eliminating A, B, and C provides

$$\begin{vmatrix} x - x_0 & y - y_0 & z - z_0 \\ \dfrac{dx}{ds} & \dfrac{dy}{ds} & \dfrac{dz}{ds} \\ \dfrac{d^2x}{ds^2} & \dfrac{d^2y}{ds^2} & \dfrac{d^2z}{ds^2} \end{vmatrix} = 0 \tag{1.365}$$

which is the equation of the osculating plane (1.348). The osculating plane can be identified by its unit vector $\hat{u}_b$, called the *bivector*:

$$\begin{aligned}\hat{u}_b = {} & \frac{1}{u_b}\left(\frac{dy}{ds}\frac{d^2z}{ds^2} - \frac{dz}{ds}\frac{d^2y}{ds^2}\right)\hat{\imath} + \frac{1}{u_b}\left(\frac{dz}{ds}\frac{d^2x}{ds^2} - \frac{dx}{ds}\frac{d^2z}{ds^2}\right)\hat{\jmath} \\ & + \frac{1}{u_b}\left(\frac{dx}{ds}\frac{d^2y}{ds^2} - \frac{dy}{ds}\frac{d^2x}{ds^2}\right)\hat{k} = \frac{1}{u_b}\left(\frac{d\mathbf{r}}{ds} \times \frac{d^2\mathbf{r}}{ds^2}\right)\end{aligned} \tag{1.366}$$

$$\begin{aligned}u_b^2 = {} & \left(\frac{dy}{ds}\frac{d^2z}{ds^2} - \frac{dz}{ds}\frac{d^2y}{ds^2}\right)^2 + \left(\frac{dz}{ds}\frac{d^2x}{ds^2} - \frac{dx}{ds}\frac{d^2z}{ds^2}\right)^2 \\ & + \left(\frac{dx}{ds}\frac{d^2y}{ds^2} - \frac{dy}{ds}\frac{d^2x}{ds^2}\right)^2 = \left(\frac{d\mathbf{r}}{ds} \times \frac{d^2\mathbf{r}}{ds^2}\right)^2\end{aligned} \tag{1.367}$$

The line of intersection of the osculating plane (1.348) and the perpendicular plane (1.361) is called the *principal normal line* to the curve at P. From (1.361) and (1.348) the equation of the principal normal is

$$\frac{x - x_0}{d^2x/ds^2} = \frac{y - y_0}{d^2y/ds^2} = \frac{z - z_0}{d^2z/ds^2} \tag{1.368}$$

The plane through P and perpendicular to the principal normal is called the *rectifying* or *tangent plane*, which has the equation

$$(x - x_0)\frac{d^2x}{ds^2} + (y - y_0)\frac{d^2y}{ds^2} + (z - z_0)\frac{d^2z}{ds^2} = 0 \tag{1.369}$$

The intersection of the rectifying plane and the perpendicular plane is a line that is called the *binormal* line:

$$\frac{x - x_0}{\dfrac{dy}{ds}\dfrac{d^2z}{ds^2} - \dfrac{dz}{ds}\dfrac{d^2y}{ds^2}} = \frac{y - y_0}{\dfrac{dz}{ds}\dfrac{d^2x}{ds^2} - \dfrac{dx}{ds}\dfrac{d^2z}{ds^2}} = \frac{z - z_0}{\dfrac{dx}{ds}\dfrac{d^2y}{ds^2} - \dfrac{dy}{ds}\dfrac{d^2x}{ds^2}} \tag{1.370}$$

The bivector (1.366) is along the binormal line (1.370). The unit vector perpendicular to the rectifying plane is called the normal vector $\hat{u}_n$, which is in the osculating plane and in the direction of the center of curvature of the curve at P:

$$\hat{u}_n = \frac{1}{u_n}\frac{d^2x}{ds^2}\hat{\imath} + \frac{1}{u_n}\frac{d^2y}{ds^2}\hat{\jmath} + \frac{1}{u_n}\frac{d^2z}{ds^2}\hat{k} = \frac{1}{u_n}\frac{d^2\mathbf{r}}{ds^2} \tag{1.371}$$

$$u_n^2 = \left(\frac{d^2x}{ds^2}\right)^2 + \left(\frac{d^2y}{ds^2}\right)^2 + \left(\frac{d^2z}{ds^2}\right)^2 \tag{1.372}$$

The unit vectors $\hat{u}_t$, $\hat{u}_n$, and $\hat{u}_b$ make an orthogonal triad that is called the natural coordinate frame:

$$\hat{u}_t \times \hat{u}_n = \hat{u}_b \tag{1.373}$$

The *curvature* κ of a space curve is defined as the limit of the ratio of the angle $\Delta\theta$ between two tangents to the arc length Δs of the curve between the tangents as the arc length approaches zero:

$$\kappa = \lim_{\Delta s \to 0} \frac{\Delta\theta}{\Delta s} \tag{1.374}$$

The directional cosines of the tangent line are dx/ds, dy/ds, dz/ds at point $P_1\,(x_1, y_1, z_1)$ and $dx/ds + (d^2x)/(ds^2)\Delta s$, $dy/ds + (d^2y)/(ds^2)\Delta s$, $dz/ds + (d^2z)/(ds^2)\Delta s$ at

$$P_2\,(x_2, y_2, z_2) = P_2\left(x_1 + \tfrac{dx}{ds}\Delta s, y_1 + \tfrac{dy}{ds}\Delta s, z_1 + \tfrac{dz}{ds}\Delta s\right)$$

Using the cross product of the unit vectors along the two tangent lines, we have

$$\sin^2\Delta\theta = \left[\left(\frac{dy}{ds}\frac{d^2z}{ds^2} - \frac{dz}{ds}\frac{d^2y}{ds^2}\right)^2 + \left(\frac{dz}{ds}\frac{d^2x}{ds^2} - \frac{dx}{ds}\frac{d^2z}{ds^2}\right)^2 + \left(\frac{dx}{ds}\frac{d^2y}{ds^2} - \frac{dy}{ds}\frac{d^2x}{ds^2}\right)^2\right](\Delta s)^2 \tag{1.375}$$

Because of the constraint among the directional cosines and

$$\lim_{\Delta\theta \to 0} \frac{\sin \Delta\theta}{\Delta\theta} = 1 \tag{1.376}$$

the coefficient of $(\Delta s)^2$ reduces to $(d^2x/ds^2)^2 + (d^2y/ds^2)^2 + (d^2z/ds^2)^2$ and we can calculate the curvature of the curve as

$$\kappa = \frac{d\theta}{ds} = \sqrt{\left(\frac{d^2x}{ds^2}\right)^2 + \left(\frac{d^2y}{ds^2}\right)^2 + \left(\frac{d^2z}{ds^2}\right)^2} \tag{1.377}$$

Consider a circle with $x = \rho \cos\varphi$, $y = \rho \sin\varphi$, $z = 0$. The curvature of the circle would be $\kappa = 1/\rho$ because $ds = \rho\, d\varphi$. Equating the curvature of the curve with the curvature of the circle provides the radius of curvature of the curve:

$$\rho = \frac{1}{\kappa} \tag{1.378}$$

Using the radius of curvature, we may simplify the unit normal vector $\hat{u}_n$ to

$$\hat{u}_n = \rho \left(\frac{d^2x}{ds^2}\hat{\imath} + \frac{d^2y}{ds^2}\hat{\jmath} + \frac{d^2z}{ds^2}\hat{k} \right) = \rho \frac{d^2\mathbf{r}}{ds^2} \tag{1.379}$$

Because the unit vector $\hat{u}_t$ in (1.358) is tangent to the space curve in the direction of increasing curve length s, the velocity vector $\mathbf{v}$ must be tangent to the curve in the direction of increasing time t. Therefore, $\mathbf{v}$ is proportional to $\hat{u}_t$ where the proportionality factor is the speed $\dot{s}$ of P:

$$\mathbf{v} = \dot{s}\hat{u}_t = \dot{s}\left(\frac{dx}{ds}\hat{\imath} + \frac{dy}{ds}\hat{\jmath} + \frac{dz}{ds}\hat{k} \right) = \dot{s}\frac{d\mathbf{r}}{ds} \tag{1.380}$$

$$v = \dot{s} \tag{1.381}$$

The acceleration of P would be

$$\mathbf{a} = \ddot{s}\hat{u}_t + \dot{s}\frac{d}{dt}\hat{u}_t \tag{1.382}$$

However,

$$\frac{d}{dt}\hat{u}_t = \dot{s}\frac{d^2x}{ds^2}\hat{\imath} + \dot{s}\frac{d^2y}{ds^2}\hat{\jmath} + \dot{s}\frac{d^2z}{ds^2}\hat{k} = \frac{\dot{s}}{\rho}\hat{u}_n \tag{1.383}$$

which shows that

$$\mathbf{a} = \ddot{s}\hat{u}_t + \frac{\dot{s}^2}{\rho}\hat{u}_n \tag{1.384}$$

$$a = \sqrt{\ddot{s}^2 + \frac{\dot{s}^4}{\rho^2}} \tag{1.385}$$

The natural coordinate frame $\hat{u}_t$, $\hat{u}_n$, and $\hat{u}_b$ may also be called the *Frenet frame*, *Frenet trihedron*, *repère mobile frame*, *moving frame*, or *path frame*. ■

Example 58 Osculating Plane to a Helix A point P is moving on a helix with equation

$$x = a\cos\varphi \qquad y = a\sin\varphi \qquad z = k\varphi \tag{1.386}$$

where a and k are constant and φ is an angular variable. The tangent line (1.357) to the helix at $\varphi = \pi/4$ is

$$-\frac{\sqrt{2}}{a}\left(x - \frac{1}{2}\sqrt{2}a\right) = \frac{\sqrt{2}}{a}\left(y - \frac{1}{2}\sqrt{2}a\right) = \frac{1}{k}\left(z - \frac{1}{4}\pi k\right) \tag{1.387}$$

Using

$$x_0 = \frac{\sqrt{2}}{2}a \qquad y_0 = \frac{\sqrt{2}}{2}a \qquad z_0 = k\frac{\pi}{4} \tag{1.388}$$

and

$$\frac{dx}{d\varphi} = -a\sin\varphi = -\frac{\sqrt{2}}{2}a \qquad \frac{dy}{d\varphi} = a\cos\varphi = \frac{\sqrt{2}}{2}a \tag{1.389}$$

$$\frac{dz}{d\varphi} = k \tag{1.390}$$

we can find the perpendicular plane (1.347) to the helix at $\varphi = \pi/4$:

$$-\sqrt{2}ax + \sqrt{2}ay + 2zk = \tfrac{1}{2}\pi k^2 \tag{1.391}$$

To find the osculating and rectifying planes, we need to calculate the second derivatives of the curve at $\varphi = \pi/4$,

$$\frac{d^2x}{d\varphi^2} = -a\cos\varphi = -\frac{\sqrt{2}}{2}a \qquad \frac{d^2y}{d\varphi^2} = -a\sin\varphi = -\frac{\sqrt{2}}{2}a \qquad \frac{d^2z}{d\varphi^2} = 0 \tag{1.392}$$

substitute in Equation (1.369) for the osculating plane,

$$\sqrt{2}x - \sqrt{2}ky + 2az = \frac{1}{2}\pi ak \tag{1.393}$$

and substitute in Equation (1.392) for the rectifying plane,

$$\sqrt{2}x + \sqrt{2}y = 2a \tag{1.394}$$

Because of (1.392), the curvature of the helix at $\varphi = \pi/4$ is

$$\kappa = a \tag{1.395}$$

and therefore the curvature radius of the helix at that point is

$$\rho = \frac{1}{\kappa} = \frac{1}{a} \tag{1.396}$$

Having the equations of the three planes and the curvature radius ρ, we are able to identify the unit vectors $\hat{u}_t$, $\hat{u}_n$, and $\hat{u}_b$:

$$\hat{u}_t = \frac{1}{\sqrt{a^2+k^2}}\left(-\frac{\sqrt{2}}{2}a\hat{\imath} + \frac{\sqrt{2}}{2}a\hat{\jmath} + k\hat{k}\right) \tag{1.397}$$

$$\hat{u}_n = -\frac{\sqrt{2}}{2}\hat{\imath} - \frac{\sqrt{2}}{2}\hat{\jmath} \tag{1.398}$$

$$\hat{u}_b = \frac{1}{\sqrt{a^2+k^2}}\left(\frac{1}{2}\sqrt{2}k\hat{\imath} - \frac{1}{2}\sqrt{2}k\hat{\jmath} + a\hat{k}\right) \tag{1.399}$$

We can check and see that

$$\hat{u}_t \times \hat{u}_n = \hat{u}_b \tag{1.400}$$

A helix is a category of space curves with a constant curvature–torsion ratio:

$$\frac{\kappa}{\tau} = \text{const} \tag{1.401}$$

The circular helix is only a special case of the general helix curves.

Example 59 Uniform Motion on a Circle Consider a particle P that is moving on a circle with radius R around the origin of the coordinate frame at a constant speed v. The equation of the circle is

$$\mathbf{r}\cdot\mathbf{r} = r^2 \tag{1.402}$$

where r is the constant length of $\mathbf{r}$. Differentiating (1.402) with respect to time results in

$$\mathbf{r}\cdot\mathbf{v} = 0 \tag{1.403}$$

which shows that $\mathbf{r}$ and $\mathbf{v}$ are perpendicular when $\mathbf{r}$ has a constant length. If the speed of the particle is constant, then

$$\mathbf{v}\cdot\mathbf{v} = v^2 \tag{1.404}$$

which shows that

$$\mathbf{v}\cdot\mathbf{a} = 0 \tag{1.405}$$

Now differentiating (1.403) with respect to time results in

$$\mathbf{r}\cdot\mathbf{a} = -v^2 \tag{1.406}$$

It indicates that $\mathbf{r}$ and $\mathbf{a}$ are collinear and oppositely directed. So, the value of their product must be

$$\mathbf{r}\cdot\mathbf{a} = -ra \tag{1.407}$$

which determines the length of the acceleration vector a on a uniformly circular motion:

$$a = -\frac{v^2}{r} \tag{1.408}$$

Example 60 Curvature of a Plane Curve Let us consider a curve C in the (x, y)-plane as is shown in Figure 1.20, which is defined time parametrically as

$$x = x(t) \qquad y = y(t) \tag{1.409}$$

The curve increment ds is

$$ds^2 = d\mathbf{r} \cdot d\mathbf{r} = dx^2 + dy^2 \tag{1.410}$$

which after dividing by dt would be

$$\dot{s}^2 = \dot{x}^2 + \dot{y}^2 \tag{1.411}$$

Differentiating from the slop of the curve θ,

$$\tan\theta = \frac{dy}{dx} = \frac{\dot{y}}{\dot{x}} \tag{1.412}$$

we have

$$\dot{\theta}\left(1 + \tan^2\theta\right) = \frac{\dot{x}\ddot{y} - \dot{y}\ddot{x}}{\dot{x}^2} \tag{1.413}$$

and therefore

$$\dot{\theta} = \frac{\dot{x}\ddot{y} - \dot{y}\ddot{x}}{\dot{x}^2 + \dot{y}^2} \tag{1.414}$$

However, because of

$$\dot{\theta} = \frac{d\theta}{dt} = \frac{d\theta}{ds}\frac{ds}{dt} = \dot{s}\frac{d\theta}{ds} = \frac{\dot{s}}{\rho} = \frac{\sqrt{\dot{x}^2 + \dot{y}^2}}{\rho} \tag{1.415}$$

we get

$$\kappa = \frac{1}{\rho} = \frac{\dot{x}\ddot{y} - \dot{y}\ddot{x}}{\left(\dot{x}^2 + \dot{y}^2\right)^{3/2}} \tag{1.416}$$

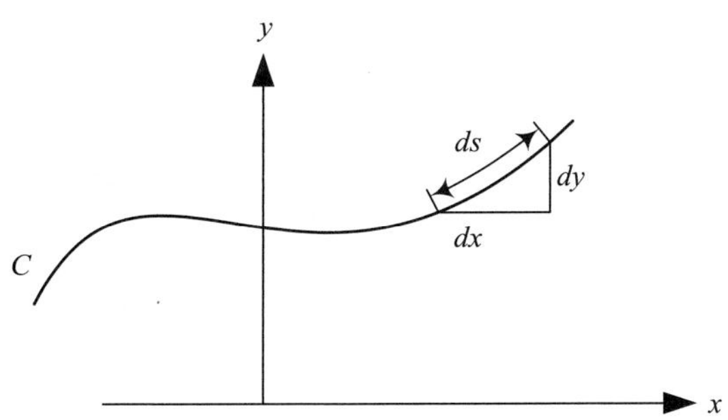

Figure 1.20 A curve C in the (x, y)-plane.

Whenever, instead of (1.409), we have the equation of the plane curve as

$$y = y(x) \tag{1.417}$$

then the curvature equation would simplify to

$$\kappa = \frac{1}{\rho} = \left| \frac{d^2y/dx^2}{\left(1 + (dy/dx)^2\right)^{3/2}} \right| \tag{1.418}$$

As an example, consider a plane curve given by the parametric equations

$$x = t \qquad y = 2t^2 \tag{1.419}$$

The curvature at $t = 3\,\mathrm{s}$ is $2.2909 \times 10^{-3}\,\mathrm{m}^{-1}$ because

$$\kappa = \frac{1}{\rho} = \frac{\dot{x}\ddot{y} - \dot{y}\ddot{x}}{\left(\dot{x}^2 + \dot{y}^2\right)^{3/2}} = \frac{4 - 0}{\left(1 + 16t^2\right)^{3/2}} = 2.2909 \times 10^{-3}\,\mathrm{m}^{-1} \tag{1.420}$$

The same curve can be expressed by

$$y = 2x^2 \tag{1.421}$$

which has the same radius of curvature $\rho = 1/\kappa = 2.2909 \times 10^{-3}\,\mathrm{m}^{-1} = 436.5\,\mathrm{m}$ at $x = 3\,\mathrm{m}$ because $dy/dx = 4x = 12$ and $d^2y/dx^2 = 4$:

$$\kappa = \frac{1}{\rho} = \left| \frac{d^2y/dx^2}{\left[1 + (dy/dx)^2\right]^{3/2}} \right| = \left| \frac{4}{\left[1 + (12)^2\right]^{3/2}} \right|$$
$$= 2.2909 \times 10^{-3}\,\mathrm{m}^{-1} \tag{1.422}$$

Example 61 Natural Coordinate Frame Is Orthogonal To show that the natural coordinate frame $\hat{u}_t$, $\hat{u}_n$, $\hat{u}_b$ in Equations (1.354)–(1.356) is orthogonal, we may differentiate the relation

$$\hat{u}_t \cdot \hat{u}_t = 1 \tag{1.423}$$

with respect to s and get

$$2\frac{d\mathbf{r}}{ds} \cdot \frac{d^2\mathbf{r}}{ds^2} = 0 \tag{1.424}$$

It indicates that $\hat{u}_t$ is orthogonal to $\hat{u}_n$. Equation (1.356) also shows that $\hat{u}_b$ is orthogonal to both $\hat{u}_t$ and $\hat{u}_n$.

Example 62 Vectorial Expression of Curvature Assume that the position vector of a moving point on a space curve is given by

$$\mathbf{r} = \mathbf{r}(s) \tag{1.425}$$

where s is the arc length on the curve measured from a fixed point on the curve. Then,

$$\mathbf{v} = \dot{s}\hat{u}_t \tag{1.426}$$

$$\hat{u}_t = \frac{d\mathbf{r}}{ds} \tag{1.427}$$

$$\frac{d^2\mathbf{r}}{ds^2} = \frac{d}{ds}\hat{u}_t = \frac{1}{\rho}\hat{u}_n = \kappa\hat{u}_n \tag{1.428}$$

and therefore,

$$\kappa = \frac{1}{\rho} = \left|\frac{d^2\mathbf{r}}{ds^2}\right| \tag{1.429}$$

We may also employ the velocity and acceleration vectors of the moving point and determine the curvature of the curve. Because the outer product of $\mathbf{v}$ and $\mathbf{a}$ is

$$\mathbf{v} \times \mathbf{a} = \left(\dot{s}\hat{u}_t\right) \times \left(\ddot{s}\hat{u}_t + \frac{\dot{s}^2}{\rho}\hat{u}_n\right) = \mathbf{v} \times \mathbf{a}_n \tag{1.430}$$

$$|\mathbf{v} \times \mathbf{a}| = v a_n \tag{1.431}$$

we have

$$a_n = \frac{\dot{s}^2}{\rho} = \frac{v^2}{\rho} = \frac{|\mathbf{v} \times \mathbf{a}|}{v} \tag{1.432}$$

and therefore,

$$\kappa = \frac{1}{\rho} = \frac{|\mathbf{v} \times \mathbf{a}|}{v^3} = \frac{|\mathbf{v} \times \mathbf{a}|}{|\mathbf{v}|^3} \tag{1.433}$$

As an example, consider a moving point at

$$\mathbf{r} = \begin{bmatrix} t \\ 2t^2 \end{bmatrix} \tag{1.434}$$

Its velocity and acceleration are

$$\mathbf{v} = \begin{bmatrix} 1 \\ 4t \end{bmatrix} \qquad \mathbf{a} = \begin{bmatrix} 0 \\ 4 \end{bmatrix} \tag{1.435}$$

and therefore the curvature of the motion is

$$\kappa = \frac{1}{\rho} = \frac{|\mathbf{v} \times \mathbf{a}|}{|\mathbf{v}|^3} = \frac{4}{\left(\sqrt{16t^2 + 1}\right)^3} \tag{1.436}$$

The curvature at $t = 3\,\text{s}$ is $\kappa = 2.2909 \times 10^{-3}\,\text{m}^{-1}$.

Example 63 ★ Curvature Vector κ Using the definition of tangential unit vector $\hat{u}_t$,

$$\hat{u}_t = \frac{d\mathbf{r}}{ds} \tag{1.437}$$

and taking a curve length derivative we can define a curvature vector $\boldsymbol{\kappa}$ as

$$\boldsymbol{\kappa} = \frac{d\hat{u}_t}{ds} = \frac{d^2\mathbf{r}}{ds^2} = \kappa\hat{u}_n \tag{1.438}$$

that has a length κ and indicates the curvature center of the curve. So, the curvature vector $\boldsymbol{\kappa}$ points in the direction in which $\hat{u}_t$ is turning, orthogonal to $\hat{u}_t$. The length $\kappa = |\boldsymbol{\kappa}|$ gives the rate of turning. It can be found from

$$\kappa^2 = \frac{d^2\mathbf{r}}{ds^2} \cdot \frac{d^2\mathbf{r}}{ds^2} \tag{1.439}$$

Furthermore, because

$$\hat{u}_t = \frac{\mathbf{v}}{\dot{s}} \tag{1.440}$$

we may also define the curvature vector $\boldsymbol{\kappa}$ as

$$\boldsymbol{\kappa} = \frac{d}{ds}\frac{\mathbf{v}}{\dot{s}} = \frac{1}{\dot{s}}\frac{d}{dt}\frac{\mathbf{v}}{\dot{s}} = \frac{\mathbf{a}\dot{s} - \mathbf{v}\ddot{s}}{\dot{s}^3} \tag{1.441}$$

Example 64* ★ *Frenet–Serret Formulas When the position vector of a moving point on a space curve is given as a function of the arc length s,

$$\mathbf{r} = \mathbf{r}(s) \tag{1.442}$$

we define the unit vectors $\hat{u}_t$, $\hat{u}_n$, and $\hat{u}_b$ and an orthogonal coordinate frame

$$\hat{u}_t \times \hat{u}_n = \hat{u}_b \tag{1.443}$$

that is carried by the point. Because s is the variable that indicates the point, it is useful to determine the derivatives of the unit vectors with respect to s.

Using Equation (1.383), we can find the s-derivative of the tangent unit vector $\hat{u}_t$:

$$\frac{d\hat{u}_t}{ds} = \frac{d\hat{u}_t}{dt}\frac{dt}{ds} = \frac{d\hat{u}_t}{dt}\frac{1}{\dot{s}} = \frac{1}{\rho}\hat{u}_n = \kappa\hat{u}_n \tag{1.444}$$

$$\left|\frac{d\hat{u}_t}{ds}\right| = \frac{1}{\rho} = \sqrt{\left(\frac{d^2x}{ds^2}\right)^2 + \left(\frac{d^2y}{ds^2}\right)^2 + \left(\frac{d^2z}{ds^2}\right)^2} \tag{1.445}$$

To find $d\hat{u}_b/ds$, we may take a derivative from (1.443):

$$\frac{d\hat{u}_b}{ds} = \frac{d}{ds}\left(\hat{u}_t \times \hat{u}_n\right) = \frac{d\hat{u}_t}{ds} \times \hat{u}_n + \hat{u}_t \times \frac{d\hat{u}_n}{ds} = \hat{u}_t \times \frac{d\hat{u}_n}{ds} \tag{1.446}$$

Because $\hat{u}_b$ is a constant-length vector, $d\hat{u}_b/ds$ is perpendicular to $\hat{u}_b$. It must also be perpendicular to $\hat{u}_t$. So, $d\hat{u}_b/ds$ is parallel to $\hat{u}_n$:

$$\frac{d\hat{u}_b}{ds} = -\tau\hat{u}_n = -\frac{1}{\sigma}\hat{u}_n \tag{1.447}$$

The coefficient τ is called the *torsion of the curve*, while $\sigma = 1/\tau$ is called the *radius of torsion*. The torsion at a point of the curve indicates that the osculating plane rotates about the tangent to the curve as the point moves along the curve. The torsion is considered positive if the osculating plane rotates about $\hat{u}_t$ and negative if it rotates about $-\hat{u}_t$. A curve with $\kappa \neq 0$ is planar if and only if $\tau = 0$.

The derivative of the normal unit vector $d\hat{u}_n/ds$ may be calculated from

$$\begin{aligned}\frac{d\hat{u}_n}{ds} &= \frac{d}{ds}\left(\hat{u}_b \times \hat{u}_t\right) = \frac{d\hat{u}_b}{ds} \times \hat{u}_t + \hat{u}_b \times \frac{d\hat{u}_t}{ds} \\ &= -\frac{1}{\sigma}\left(\hat{u}_n \times \hat{u}_t\right) + \frac{1}{\rho}\left(\hat{u}_b \times \hat{u}_n\right) = \frac{1}{\sigma}\hat{u}_b - \frac{1}{\rho}\hat{u}_t\end{aligned} \tag{1.448}$$

Equations (1.444), (1.447), and (1.448) are called the Frenet–Serret formulas. The Frenet–Serret formulas may be summarized in a matrix form:

$$\begin{bmatrix}\dfrac{d\hat{u}_t}{ds} \\ \dfrac{d\hat{u}_n}{ds} \\ \dfrac{d\hat{u}_b}{ds}\end{bmatrix} = \begin{bmatrix} 0 & \kappa & 0 \\ -\kappa & 0 & \tau \\ 0 & -\tau & 0 \end{bmatrix}\begin{bmatrix}\hat{u}_t \\ \hat{u}_n \\ \hat{u}_b\end{bmatrix} \tag{1.449}$$

It shows that the derivative of the natural coordinate unit vectors can be found by multiplying a skew-symmetric matrix and the coordinate unit vectors.

Having the Frenet–Serret formulas, we are able to calculate the kinematics of a moving point on the space curve:

$$\mathbf{v} = \frac{d\mathbf{r}}{dt} = \frac{d\mathbf{r}}{ds}\dot{s} = \dot{s}\hat{u}_t \tag{1.450}$$

$$\mathbf{a} = \frac{d\mathbf{v}}{dt} = \ddot{s}\hat{u}_t + \dot{s}\frac{d\hat{u}_t}{dt} = \ddot{s}\hat{u}_t + \dot{s}^2\frac{d\hat{u}_t}{ds} = \ddot{s}\hat{u}_t + \dot{s}^2\frac{1}{\rho}\hat{u}_n \tag{1.451}$$

$$\mathbf{j} = \frac{d\mathbf{a}}{dt} = \left(\dddot{s} - \frac{\dot{s}^3}{\rho^2}\right)\hat{u}_t + \frac{1}{\rho}\left(3\dot{s}\ddot{s} + \frac{\dot{s}^2}{\rho}\dot{\rho}\right)\hat{u}_n + \frac{\dot{s}^3}{\rho\sigma}\hat{u}_b \tag{1.452}$$

Frenet (1816–1900) and Serret (1819–1885) were two French mathematicians.

Example 65 Characteristics of a Space Curve Consider a space curve C with the parametric equation

$$\mathbf{r} = \mathbf{r}\left(t\right) \tag{1.453}$$

The natural coordinate frame and curve characteristics are

$$\hat{u}_t = \frac{\dot{\mathbf{r}}}{|\dot{\mathbf{r}}|} \tag{1.454}$$

$$\hat{u}_b = \frac{\dot{\mathbf{r}} \times \ddot{\mathbf{r}}}{|\dot{\mathbf{r}} \times \ddot{\mathbf{r}}|} \tag{1.455}$$

$$\hat{u}_n = \hat{u}_b \times \hat{u}_t \tag{1.456}$$

$$\kappa = \frac{\dot{\mathbf{r}} \times \ddot{\mathbf{r}}}{|\dot{\mathbf{r}}|^3} \tag{1.457}$$

$$\tau = \frac{(\dot{\mathbf{r}} \times \ddot{\mathbf{r}}) \cdot \dddot{\mathbf{r}}}{|\dot{\mathbf{r}} \times \ddot{\mathbf{r}}|^2} \tag{1.458}$$

Employing these equations, the Frenet–Serret formulas (1.449) can be determined in time derivatives:

$$\frac{d\hat{u}_t}{dt} = \kappa \, |\dot{\mathbf{r}}| \, \hat{u}_n \tag{1.459}$$

$$\frac{d\hat{u}_n}{dt} = -\kappa \, |\dot{\mathbf{r}}| \, \hat{u}_t + \tau \, |\dot{\mathbf{r}}| \, \hat{u}_b \tag{1.460}$$

$$\frac{d\hat{u}_b}{dt} = -\tau \, |\dot{\mathbf{r}}| \, \hat{u}_b \tag{1.461}$$

Example 66 ★ Osculating Sphere The sphere that has a contact of third order with a space curve at a point $P(x, y, z)$ is called the osculating sphere of the curve at P. If the center of the sphere is denoted by $C\,(x_C, y_C, z_C)$, then the equation of the osculating sphere is

$$(x - x_C)^2 + (y - y_C)^2 + (z - z_C)^2 = R \tag{1.462}$$

where R is the radius of the sphere. Taking three derivatives from (1.462) provides a set of four equations to determine x_C, y_C, z_C and R. To set up the equations, we show the equation of the sphere as

$$(\mathbf{r}_C - \mathbf{r})^2 = R^2$$

where $\mathbf{r}_C - \mathbf{r}$ indicates the position of the center of the sphere from point P. Taking derivatives with respect to the arc length s provides

$$(\mathbf{r}_C - \mathbf{r}) \cdot \frac{d\mathbf{r}}{ds} = 0 \tag{1.463}$$

$$-1 + (\mathbf{r}_C - \mathbf{r}) \cdot \frac{d^2\mathbf{r}}{ds^2} = 0 \tag{1.464}$$

$$-\frac{d\mathbf{r}}{ds} \cdot \frac{d^2\mathbf{r}}{ds^2} + (\mathbf{r}_C - \mathbf{r}) \cdot \frac{d^3\mathbf{r}}{ds^3} = 0 \tag{1.465}$$

Employing the curvature τ and torsion κ, we can rewrite these equations:

$$(\mathbf{r}_C - \mathbf{r}) \cdot \hat{u}_t = 0 \tag{1.466}$$

$$-1 + (\mathbf{r}_C - \mathbf{r}) \cdot \kappa \hat{u}_n = 0 \tag{1.467}$$

$$(\mathbf{r}_C - \mathbf{r}) \cdot \left(\frac{d\kappa}{ds} \hat{u}_n + \kappa \left(\tau \hat{u}_b - \kappa \hat{u}_t \right) \right) = 0 \tag{1.468}$$

Expanding (1.468) yields

$$\frac{d\kappa}{ds} (\mathbf{r}_C - \mathbf{r}) \cdot \hat{u}_n + \kappa \tau (\mathbf{r}_C - \mathbf{r}) \cdot \hat{u}_b - \kappa^2 (\mathbf{r}_C - \mathbf{r}) \cdot \hat{u}_t = 0 \tag{1.469}$$

and using Equations (1.466) and (1.467), we find

$$\frac{1}{\kappa} \frac{d\kappa}{ds} + \kappa \tau (\mathbf{r}_C - \mathbf{r}) \cdot \hat{u}_b = 0 \tag{1.470}$$

Knowing that

$$\frac{d\rho}{ds} = \frac{d}{ds} \left(\frac{1}{\kappa} \right) = -\frac{1}{\kappa^2} \frac{d\kappa}{ds} \tag{1.471}$$

we can simplify Equation (1.470):

$$(\mathbf{r}_C - \mathbf{r}) \cdot \hat{u}_b = \sigma \frac{d\rho}{ds} \tag{1.472}$$

From Equations (1.466), (1.467), and (1.472), we have

$$(\mathbf{r}_C - \mathbf{r}) \cdot \hat{u}_t = 0 \tag{1.473}$$

$$(\mathbf{r}_C - \mathbf{r}) \cdot \hat{u}_n = \rho \tag{1.474}$$

$$(\mathbf{r}_C - \mathbf{r}) \cdot \hat{u}_b = \sigma \frac{d\rho}{ds} \tag{1.475}$$

that indicates $\mathbf{r}_C - \mathbf{r}$ lies in a perpendicular plane. The components of $\mathbf{r}_C - \mathbf{r}$ are ρ along $\hat{u}_n$ and $\sigma\,(d\rho/ds)$ along $\hat{u}_b$:

$$\mathbf{r}_C - \mathbf{r} = \rho \hat{u}_n + \sigma \frac{d\rho}{ds} \hat{u}_b \tag{1.476}$$

Therefore, the position vector of the center of the osculating sphere is at

$$\mathbf{r}_C = \mathbf{r} + \rho \hat{u}_n + \sigma \frac{d\rho}{ds} \hat{u}_b \tag{1.477}$$

and the radius of the osculating sphere is

$$R = |\mathbf{r}_C - \mathbf{r}| = \sqrt{\rho^2 + \sigma^2 \left(\frac{d\rho}{ds} \right)^2} \tag{1.478}$$

Example 67 ★ Taylor Series Expansion of a Space Curve Consider a point P that is moving on a space curve that is parametrically expressed as $\mathbf{r} = \mathbf{r}(s)$. If at $s = 0$ we have the position and velocity of P, it is possible to express the curve by a Taylor expansion:

$$\mathbf{r}(s) = \mathbf{r}(0) + \frac{d\mathbf{r}(0)}{ds}s + \frac{d^2\mathbf{r}(0)}{ds^2}\frac{s^2}{2!} + \frac{d^3\mathbf{r}(0)}{ds^3}\frac{s^3}{3!} + \cdots \tag{1.479}$$

Using the natural coordinate system, we have

$$\frac{d\mathbf{r}}{ds} = \hat{u}_t \tag{1.480}$$

$$\frac{d^2\mathbf{r}}{ds^2} = \kappa\,\hat{u}_n \tag{1.481}$$

$$\frac{d^3\mathbf{r}}{ds^3} = \frac{d}{ds}\left(\kappa\,\hat{u}_n\right) = \frac{d\kappa}{ds}\hat{u}_n + \kappa\left(-\kappa\hat{u}_t + \tau\hat{u}_b\right) \tag{1.482}$$

$$\begin{aligned}\frac{d^4\mathbf{r}}{ds^4} &= \frac{d^2\kappa}{ds^2}\hat{u}_n + \frac{d\kappa}{ds}\left(-\kappa\hat{u}_t + \tau\hat{u}_b\right) + \frac{d\kappa}{ds}\left(-\kappa\hat{u}_t + \tau\hat{u}_b\right)\\ &\quad + \kappa\left(-\frac{d\kappa}{ds}\hat{u}_t - \kappa^2\hat{u}_n + \frac{d\tau}{ds}\hat{u}_b - \tau^2\hat{u}_n\right)\\ &= -3\kappa\frac{d\kappa}{ds}\hat{u}_t + \left(\frac{d^2\kappa}{ds^2} - \kappa^3 - \kappa\tau^2\right)\hat{u}_n + \left(2\tau\frac{d\kappa}{ds} + \kappa\frac{d\tau}{ds}\right)\hat{u}_b\end{aligned} \tag{1.483}$$

and therefore,

$$\frac{d\mathbf{r}(0)}{ds} = \hat{u}_t(0) = \hat{u}_{t_0} \tag{1.484}$$

$$\frac{d^2\mathbf{r}(0)}{ds^2} = \kappa(0)\,\hat{u}_n(0) = \kappa_0\hat{u}_{n_0} \tag{1.485}$$

$$\frac{d^3\mathbf{r}(0)}{ds^3} = -\kappa_0^2\hat{u}_{t_0} + \frac{d\kappa_0}{ds}\hat{u}_{n_0} + \kappa_0\tau_0\hat{u}_{b_0} \tag{1.486}$$

$$\begin{aligned}\frac{d^4\mathbf{r}(0)}{ds^4} &= -3\kappa_0\frac{d\kappa_0}{ds}\hat{u}_{t_0} + \left(\frac{d^2\kappa_0}{ds^2} - \kappa_0^3 - \kappa_0\tau_0^2\right)\hat{u}_{n_0}\\ &\quad + \left(2\tau_0\frac{d\kappa_0}{ds} + \kappa_0\frac{d\tau_0}{ds}\right)\hat{u}_{b_0}\end{aligned} \tag{1.487}$$

Substituting these results in Equation (1.479) shows that

$$\begin{aligned}\mathbf{r} &= \mathbf{r}_0 + s\hat{u}_{t_0} + \frac{1}{2}\kappa_0 s^2\hat{u}_{n_0} + \frac{s^3}{6}\left(-\kappa_0^2\hat{u}_{t_0} + \frac{d\kappa_0}{ds}\hat{u}_{n_0} + \kappa_0\tau_0\hat{u}_{b_0}\right)\\ &\quad + \frac{s^4}{24}\left(-3\kappa_0\frac{d\kappa_0}{ds}\right)\hat{u}_{t_0} + \frac{s^4}{24}\left(\frac{d^2\kappa_0}{ds^2} - \kappa_0^3 - \kappa_0\tau_0^2\right)\hat{u}_{n_0}\\ &\quad + \frac{s^4}{24}\left(2\tau_0\frac{d\kappa_0}{ds} + \kappa_0\frac{d\tau_0}{ds}\right)\hat{u}_{b_0} + \cdots\end{aligned} \tag{1.488}$$

Let us rearrange the equation to determine the natural components of $\mathbf{r} - \mathbf{r}_0$:

$$\mathbf{r} - \mathbf{r}_0 = \left(s - \frac{\kappa_0^2}{6} s^3 - \frac{\kappa_0}{8} \frac{d\kappa_0}{ds} s^4 + \cdots \right) \hat{u}_{t_0}$$
$$+ \left[\frac{1}{2} \kappa_0 s^2 + \frac{1}{6} \frac{d\kappa_0}{ds} s^3 + \frac{1}{24} \left(\frac{d^2\kappa_0}{ds^2} - \kappa_0^3 - \kappa_0 \tau_0^2 \right) s^4 + \cdots \right] \hat{u}_{n_0}$$
$$+ \left[\frac{1}{6} \kappa_0 \tau_0 s^3 + \frac{1}{24} \left(2\tau_0 \frac{d\kappa_0}{ds} + \kappa_0 \frac{d\tau_0}{ds} \right) s^4 + \cdots \right] \hat{u}_{b_0} \tag{1.489}$$

It follows from these equations that in the neighborhood of a point at which $\kappa = 0$ the curve approximates a straight line. Furthermore, if $\tau = 0$ at a point, the curve remains on a plane. Accepting only the first term of each series, we may approximate a curve as

$$\mathbf{r}(s) - \mathbf{r}_0 \approx s\hat{u}_{t_0} + \tfrac{1}{2}\kappa_0 s^2 \hat{u}_{n_0} + \tfrac{1}{6}\kappa_0 \tau_0 s^3 \hat{u}_{b_0} \tag{1.490}$$

Now assume that the position vector of the point P is expressed as a function of time $\mathbf{r} = \mathbf{r}(t)$. If at $t = t_0$ we have the position and velocity of P, it is possible to express the path of motion by a Taylor expansion:

$$\mathbf{r}(t) = \mathbf{r}_0 + (t - t_0)\,\dot{\mathbf{r}}_0 + \frac{(t - t_0)^2}{2!}\ddot{\mathbf{r}}_0 + \frac{(t - t_0)^3}{3!}\dddot{\mathbf{r}}_0 + \cdots \tag{1.491}$$

Using the natural coordinate system (1.454)–(1.461) and defining $\dot{s} = |\dot{\mathbf{r}}|$, we have

$$\dot{\mathbf{r}} = |\dot{\mathbf{r}}|\,\hat{u}_t = \dot{s}\hat{u}_t \tag{1.492}$$
$$\ddot{\mathbf{r}} = \ddot{s}\hat{u}_t + \kappa\dot{s}^2\hat{u}_n \tag{1.493}$$
$$\dddot{\mathbf{r}} = \left(\dddot{s} - \kappa^2\dot{s}^3\right)\hat{u}_t + \kappa\left(3\dot{s}\ddot{s} + \dot{\kappa}\dot{s}^2\right)\hat{u}_n + \kappa\tau\dot{s}^3\hat{u}_b \tag{1.494}$$

and therefore,

$$\mathbf{r}(t) = \mathbf{r}_0 + (t - t_0)\,\dot{s}\hat{u}_t + \frac{(t - t_0)^2}{2!}\left(\ddot{s}\hat{u}_t + \kappa\dot{s}^2\hat{u}_n\right)$$
$$+ \frac{(t - t_0)^3}{3!}\left[\left(\dddot{s} - \kappa^2\dot{s}^3\right)\hat{u}_t + \kappa\left(3\dot{s}\ddot{s} + \dot{\kappa}\dot{s}^2\right)\hat{u}_n + \kappa\tau\dot{s}^3\hat{u}_b\right] + \cdots$$
$$= \mathbf{r}_0 + \left((t - t_0)\,\dot{s} + \frac{(t - t_0)^2}{2!}\ddot{s} + \frac{(t - t_0)^3}{3!}\left(\dddot{s} - \kappa^2\dot{s}^3\right) + \cdots \right)\hat{u}_t$$
$$+ \left(\frac{(t - t_0)^2}{2!}\kappa\dot{s}^2 + \frac{(t - t_0)^3}{3!}\kappa\left(3\dot{s}\ddot{s} + \dot{\kappa}\dot{s}^2\right) + \cdots \right)\hat{u}_n$$
$$+ \left(\frac{(t - t_0)^3}{3!}\kappa\tau\dot{s}^3 + \cdots \right)\hat{u}_b \tag{1.495}$$

Example 68 ★ Torsion of a Space Curve We may use (1.447) to determine the torsion of a curve analytically. Let us start with

$$\tau = -\hat{u}_n \cdot \frac{d\hat{u}_b}{ds} \tag{1.496}$$

and employ

$$\hat{u}_b = \hat{u}_t \times \hat{u}_n \qquad \hat{u}_t = \frac{d\mathbf{r}}{ds} \qquad \hat{u}_n = \rho \frac{d^2\mathbf{r}}{ds^2} \tag{1.497}$$

to get

$$\hat{u}_b = \rho \left(\frac{d\mathbf{r}}{ds} \times \frac{d^2\mathbf{r}}{ds^2} \right) \tag{1.498}$$

and hence

$$\tau = -\rho^2 \frac{d^2\mathbf{r}}{ds^2} \cdot \frac{d}{ds} \left(\frac{d\mathbf{r}}{ds} \times \frac{d^2\mathbf{r}}{ds^2} \right) = \rho^2 \frac{d\mathbf{r}}{ds} \cdot \frac{d^2\mathbf{r}}{ds^2} \times \frac{d^3\mathbf{r}}{ds^3} \tag{1.499}$$

So, the scalar triple product of velocity, acceleration, and jerk $[\mathbf{v}, \mathbf{a}, \mathbf{j}]$ is

$$\left[\frac{d\mathbf{r}}{ds} \frac{d^2\mathbf{r}}{ds^2} \frac{d^3\mathbf{r}}{ds^3} \right] = \frac{d\mathbf{r}}{ds} \cdot \frac{d^2\mathbf{r}}{ds^2} \times \frac{d^3\mathbf{r}}{ds^3} = \tau\kappa^2 \tag{1.500}$$

Example 69 ★ Darboux Vector By defining a vector $\mathbf{u}$ as

$$\mathbf{u} = \frac{1}{\rho} \hat{u}_b + \frac{1}{\sigma} \hat{u}_t \tag{1.501}$$

the Frenet–Serret formulas simplify to

$$\frac{d\hat{u}_t}{ds} = \mathbf{u} \times \hat{u}_t \qquad \frac{d\hat{u}_n}{ds} = \mathbf{u} \times \hat{u}_n \qquad \frac{d\hat{u}_b}{ds} = \mathbf{u} \times \hat{u}_b \tag{1.502}$$

The vector $\mathbf{u}$ is called the *Darboux vector*. Darboux (1842–1917) was a French mathematician.

Example 70 ★ Curvature as the Change of a Deformed Curve Curvature determines how the length of a curve changes as the curve is deformed. Consider an infinitesimal arc ds of a planar curve, as is shown in Figure 1.21. The arc length ds lies to second order on a circle of radius $\rho = 1/\kappa$. Let us push ds a distance dr in the direction of the curvature vector $\boldsymbol{\kappa}$. The arc length ds will change to $(1 - \kappa dr)\, ds$ that is on a new circle of radius $1/\kappa - dr = (1/\kappa)(1 - \kappa dr)$. In general, the displacement is not necessarily in direction $\boldsymbol{\kappa}$ and may be indicated by a vector $d\mathbf{r}$. In this case the change of the arc length is $1 - \boldsymbol{\kappa} \cdot d\mathbf{r}$ and hence, the rate of change of the curve length is $-\int \boldsymbol{\kappa} \cdot d\mathbf{v}\, ds$, where $\mathbf{v} = d\mathbf{r}/dt$.

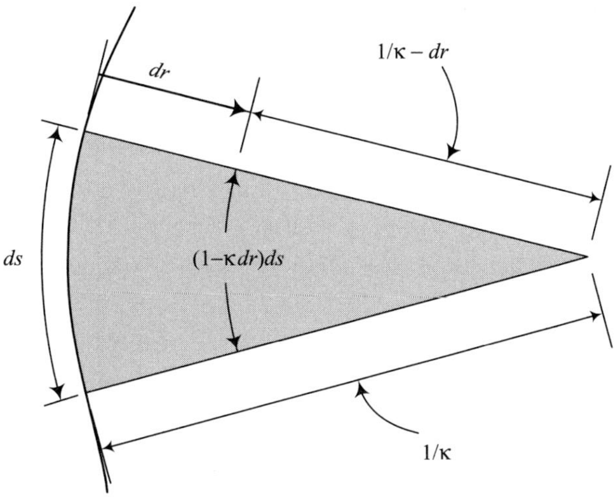

Figure 1.21 An infinitesimal arc ds of a planar curve.

Example 71 ★ Jerk in Natural Coordinate Frame $\hat{u}_t$, $\hat{u}_n$, $\hat{u}_b$ Employing Equation (1.353) and using the derivatives of the unit vectors of the natural coordinate frame,

$$\frac{d\hat{u}_t}{dt} = \frac{d\hat{u}_t}{dt} = \frac{\dot{s}}{\rho}\hat{u}_n \tag{1.503}$$

$$\frac{d\hat{u}_n}{dt} = \frac{\dot{s}}{\sigma}\hat{u}_b - \frac{\dot{s}}{\rho}\hat{u}_t \tag{1.504}$$

$$\frac{d\hat{u}_b}{dt} = -\frac{\dot{s}}{\sigma}\hat{u}_n \tag{1.505}$$

we can determine the jerk vector of a moving point in the natural coordinate frame:

$$\begin{aligned}
\mathbf{j} &= \frac{d}{dt}\mathbf{a} = \frac{d}{dt}\left(\ddot{s}\hat{u}_t + \frac{\dot{s}^2}{\rho}\hat{u}_n\right) \\
&= \dddot{s}\hat{u}_t + \ddot{s}\frac{d}{dt}\hat{u}_t + \frac{2\rho\dot{s}\ddot{s} - \dot{\rho}\dot{s}^2}{\rho^2}\hat{u}_n + \frac{\dot{s}^2}{\rho}\frac{d}{dt}\hat{u}_n \\
&= \dddot{s}\hat{u}_t + \ddot{s}\frac{\dot{s}}{\rho}\hat{u}_n + \frac{2\rho\dot{s}\ddot{s} - \dot{\rho}\dot{s}^2}{\rho^2}\hat{u}_n + \frac{\dot{s}^3}{\rho}\left(\frac{1}{\sigma}\hat{u}_b - \frac{1}{\rho}\hat{u}_t\right) \\
&= \left(\dddot{s} - \frac{\dot{s}^3}{\rho^2}\right)\hat{u}_t + \left(3\frac{\ddot{s}\dot{s}}{\rho} - \frac{\dot{\rho}\dot{s}^2}{\rho^2}\right)\hat{u}_n + \left(\frac{\dot{s}^3}{\rho\sigma}\right)\hat{u}_b
\end{aligned} \tag{1.506}$$

Example 72 ★ A Roller Coaster Figure 1.22 illustrates a roller coaster with the parametric equations

$$\begin{aligned} x &= (a + b\sin\theta)\cos\theta \\ y &= (a + b\sin\theta)\sin\theta \\ z &= b + b\cos\theta \end{aligned} \tag{1.507}$$

for

$$a = 200\,\mathrm{m} \qquad b = 150\,\mathrm{m} \tag{1.508}$$

Such a space curve is on the surface shown in Figure 1.23. The parametric equations of the surface are

$$\begin{aligned} x &= (a + b\sin\theta)\cos\varphi \\ y &= (a + b\sin\theta)\sin\varphi \\ z &= b + b\cos\theta \end{aligned} \tag{1.509}$$

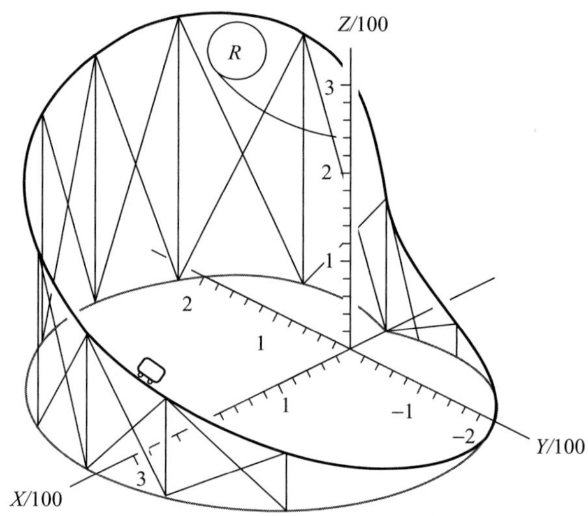

Figure 1.22 A roller coaster.

Let us assume that the car is a particle that moves on the roller coaster when the parameter θ is a function of time. The velocity and acceleration of the particle are

$$\mathbf{v} = \frac{d}{dt}\mathbf{r} = \begin{bmatrix} b\dot{\theta}\cos 2\theta - a\dot{\theta}\sin\theta \\ a\dot{\theta}\cos\theta + b\dot{\theta}\sin 2\theta \\ -b\dot{\theta}\sin\theta \end{bmatrix} \tag{1.510}$$

$$\mathbf{a} = \frac{d}{dt}\mathbf{v} = \begin{bmatrix} (b\cos 2\theta - a\sin\theta)\,\ddot{\theta} - (a\cos\theta + 2b\sin 2\theta)\,\dot{\theta}^2 \\ (a\cos\theta + b\sin 2\theta)\,\ddot{\theta} + (2b\cos 2\theta - a\sin\theta)\,\dot{\theta}^2 \\ -b\ddot{\theta}\sin\theta - b\dot{\theta}^2\cos\theta \end{bmatrix} \tag{1.511}$$

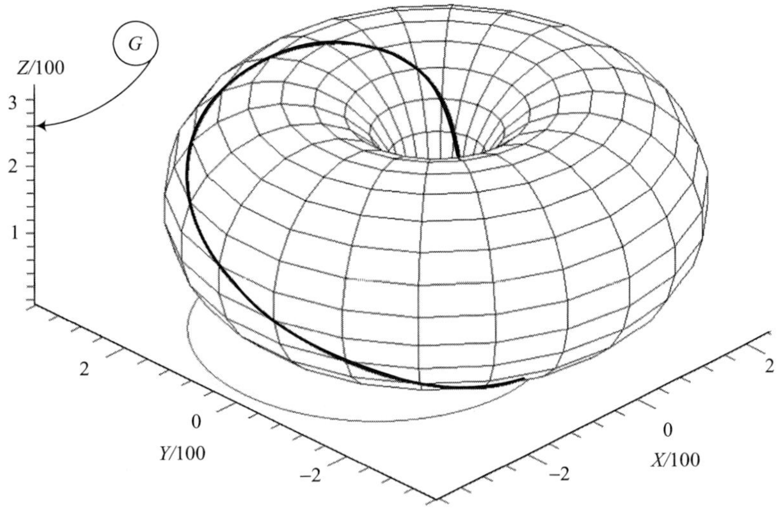

Figure 1.23 The path of the roller coaster is a space curve on the torus.

The equation of the tangent line (1.228) to the space curve is

$$\frac{x - x_0}{b\cos 2\theta - a\sin\theta} = \frac{y - y_0}{b\sin 2\theta + a\cos\theta} = \frac{z - z_0}{-b\sin\theta} \tag{1.512}$$

where

$$\begin{aligned} x_0 &= (a + b\sin\theta_0)\cos\theta_0 \\ y_0 &= (a + b\sin\theta_0)\sin\theta_0 \\ z_0 &= b + b\cos\theta_0 \end{aligned} \tag{1.513}$$

and

$$\begin{aligned} \frac{dx}{d\theta} &= b\cos 2\theta - a\sin\theta \\ \frac{dy}{d\theta} &= b\sin 2\theta + a\cos\theta \\ \frac{dz}{d\theta} &= -b\sin\theta \end{aligned} \tag{1.514}$$

As an example the tangent line at $\theta = \pi/4$ is

$$\frac{x - 216.42}{-141.42} = \frac{y - 216.42}{291.42} = \frac{z - 256.07}{-106.07} \tag{1.515}$$

because

$$\begin{aligned} x_0 &= \left(a + b\sin\frac{\pi}{4}\right)\cos\frac{\pi}{4} = 216.42 \text{ m} \\ y_0 &= \left(a + b\sin\frac{\pi}{4}\right)\sin\frac{\pi}{4} = 216.42 \text{ m} \\ z_0 &= b + b\cos\frac{\pi}{4} = 256.07 \text{ m} \end{aligned} \tag{1.516}$$

$$\frac{dx}{d\theta} = b\cos 2\frac{\pi}{4} - a\sin\frac{\pi}{4} = -141.42 \text{ m/rad}$$
$$\frac{dy}{d\theta} = b\sin 2\frac{\pi}{4} + a\cos\frac{\pi}{4} = 291.42 \text{ m/rad} \tag{1.517}$$
$$\frac{dz}{d\theta} = -b\sin\frac{\pi}{4} = -106.07 \text{ m/rad}$$

The arc length element ds of the space curve is

$$ds = \sqrt{\frac{d\mathbf{r}}{d\theta}\cdot\frac{d\mathbf{r}}{d\theta}}d\theta = \sqrt{\left(\frac{\partial x}{\partial\theta}\right)^2 + \left(\frac{\partial x}{\partial\theta}\right)^2 + \left(\frac{\partial x}{\partial\theta}\right)^2}d\theta$$
$$= \frac{\sqrt{2}}{2}\sqrt{2a^2 + 3b^2 - b^2\cos 2\theta + 4ab\sin\theta d\theta} \tag{1.518}$$

The perpendicular plane (1.347) to the roller coaster curve is

$$(x - x_0)\frac{dx}{d\theta}\frac{d\theta}{ds} + (y - y_0)\frac{dy}{d\theta}\frac{d\theta}{ds} + (z - z_0)\frac{dz}{d\theta}\frac{d\theta}{ds} = 0 \tag{1.519}$$
$$(b\cos 2\theta - a\sin\theta)(x - x_0)$$
$$+ (b\sin 2\theta + a\cos\theta)(y - y_0) - b\sin\theta\,(z - z_0) = 0 \tag{1.520}$$

This perpendicular plane at $\theta = \pi/4$ is

$$-141.42x + 42y - 106.07z - 5302.7 = 0 \tag{1.521}$$

To find the osculating and rectifying planes, we also need to calculate the second derivatives of the curve:

$$\frac{d^2x}{d\theta^2} = -a\cos\theta - 2b\sin 2\theta$$
$$\frac{d^2y}{d\theta^2} = 2b\cos 2\theta - a\sin\theta \tag{1.522}$$
$$\frac{d^2z}{d\theta^2} = -b\cos\theta$$

The osculating plane (1.348) to the roller coaster curve can be found by the derivative with respect to the arc length ds:

$$\left(\frac{dy}{ds}\frac{d^2z}{ds^2} - \frac{dz}{ds}\frac{d^2y}{ds^2}\right)(x - x_0)$$
$$+ \left(\frac{dz}{ds}\frac{d^2x}{ds^2} - \frac{dx}{ds}\frac{d^2z}{ds^2}\right)(y - y_0)$$
$$+ \left(\frac{dx}{ds}\frac{d^2y}{ds^2} - \frac{dy}{ds}\frac{d^2x}{ds^2}\right)(z - z_0) = 0 \tag{1.523}$$

The arc length is a function of θ, so we must transform (1.523) for the derivative with respect to θ. Consider d^2x/ds^2, which we may transform to a function of θ using (1.518):

$$\frac{d^2x}{ds^2} = \frac{d}{ds}\frac{dx}{ds} = \frac{d}{ds}\left(\frac{dx}{d\theta}\frac{d\theta}{ds}\right) = \frac{d^2x}{d\theta^2}\left(\frac{d\theta}{ds}\right)^2 + \frac{dx}{d\theta}\frac{d\theta}{ds}\frac{d}{d\theta}\left(\frac{d\theta}{ds}\right)$$
$$= \frac{4(-2b\sin 2\theta - a\cos\theta)}{6b^2 + 8ab\sin\theta + 4a^2 - 2b^2\cos 2\theta} - \frac{2(b\cos 2\theta - a\sin\theta)(8ab\cos\theta + 4b^2\sin 2\theta)}{\left(6b^2 + 8ab\sin\theta + 4a^2 - 2b^2\cos 2\theta\right)^2} \tag{1.524}$$

Following the same method, Equation (1.523) becomes

$$-\frac{b(a + 2b\sin\theta - 2b\sin\theta\cos^2\theta)}{a^2 + 2b^2 + 2ab\sin\theta - b^2\cos^2\theta}(x - x_0) + \frac{b^2\cos\theta(2\cos^2\theta - 3)}{a^2 + 2b^2 + 2ab\sin\theta - b^2\cos^2\theta}(y - y_0) + \frac{a^2 + 2b^2 + 3ab\sin\theta}{a^2 + 2b^2 + 2ab\sin\theta - b^2\cos^2\theta}(z - z_0) = 0 \tag{1.525}$$

This osculating plane at $\theta = \pi/4$ is

$$-0.395174x + 0.273892y + 1.27943z - 301.37061 = 0 \tag{1.526}$$

The rectifying plane (1.369) is

$$-3.45942x - 1.91823y - .65786z + 1332.29646 = 0 \tag{1.527}$$

Figure 1.24 shows the space curve and the three planes—perpendicular, osculating, and rectifying—at $\theta = \pi/4$.

The curvature κ of the space curve (1.507) from (1.377) and (1.518) is

$$\kappa = \frac{d\theta}{ds} = \frac{2}{\sqrt{4a^2 + 6b^2 - 2b^2\cos 2\theta + 8ab\sin\theta}} \tag{1.528}$$

and therefore the curvature radius of the helix at that point is

$$\rho = \frac{1}{\kappa} = \frac{1}{2}\sqrt{4a^2 + 6b^2 - 2b^2\cos 2\theta + 8ab\sin\theta} \tag{1.529}$$

The equations of the three planes and the curvature κ enable us to identify the unit vectors $\hat{u}_t$, $\hat{u}_n$, and $\hat{u}_b$. The tangent unit vector $\hat{u}_t$ is given as

$$\hat{u}_t = \frac{d\mathbf{r}}{ds} = \begin{bmatrix} \frac{dx}{ds} \\ \frac{dy}{ds} \\ \frac{dz}{ds} \end{bmatrix} = \begin{bmatrix} \frac{dx}{d\theta}\frac{d\theta}{ds} \\ \frac{dy}{d\theta}\frac{d\theta}{ds} \\ \frac{dz}{d\theta}\frac{d\theta}{ds} \end{bmatrix} = \kappa \begin{bmatrix} \frac{dx}{d\theta} \\ \frac{dy}{d\theta} \\ \frac{dz}{d\theta} \end{bmatrix}$$
$$= \kappa \begin{bmatrix} b\cos 2\theta - a\sin\theta \\ b\sin 2\theta + a\cos\theta \\ -b\sin\theta \end{bmatrix} \tag{1.530}$$

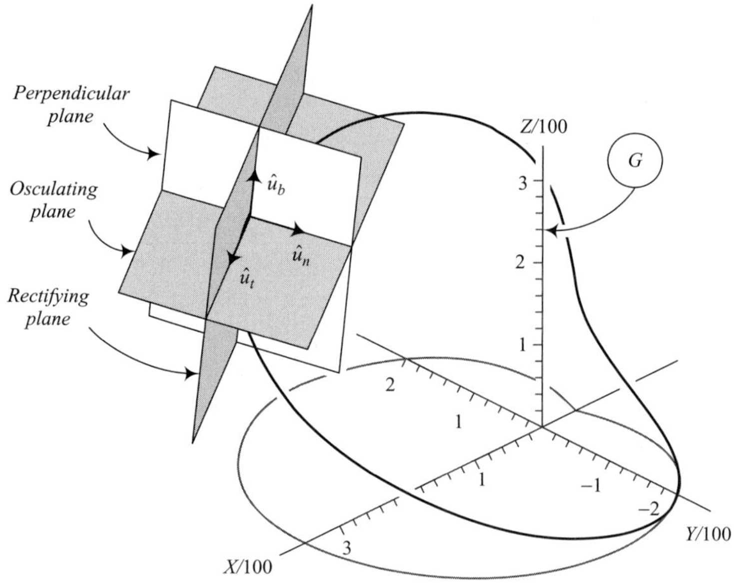

Figure 1.24 The space curve of a roller coaster and the three planes—perpendicular, osculating, and rectifying—at a specific point.

and the normal unit vector $\hat{u}_n$ as

$$\hat{u}_n = \rho \frac{d^2\mathbf{r}}{ds^2} = \rho \begin{bmatrix} \dfrac{d^2x}{ds^2} \\ \dfrac{d^2y}{ds^2} \\ \dfrac{d^2z}{ds^2} \end{bmatrix} = \begin{bmatrix} \kappa \dfrac{d^2x}{d\theta^2} + \dfrac{dx}{d\theta}\dfrac{d\kappa}{d\theta} \\ \kappa \dfrac{d^2y}{d\theta^2} + \dfrac{dy}{d\theta}\dfrac{d\kappa}{d\theta} \\ \kappa \dfrac{d^2z}{d\theta^2} + \dfrac{dz}{d\theta}\dfrac{d\kappa}{d\theta} \end{bmatrix} \tag{1.531}$$

where

$$\frac{d\kappa}{d\theta} = -\frac{b\,(a + b\sin\theta)\cos\theta}{\left(a^2 + 2b^2 - b^2\cos^2\theta + 2ab\sin\theta\right)^{3/2}} \tag{1.532}$$

and the other terms come from Equations (1.528), (1.522), and (1.514).

The bivector unit vector $\hat{u}_b$ from (1.366) and (1.525) is then

$$\hat{u}_b = \frac{\dfrac{d\mathbf{r}}{ds} \times \dfrac{d^2\mathbf{r}}{ds^2}}{\left|\dfrac{d\mathbf{r}}{ds} \times \dfrac{d^2\mathbf{r}}{ds^2}\right|} = \frac{1}{u_b} \begin{bmatrix} \dfrac{dy}{ds}\dfrac{d^2z}{ds^2} - \dfrac{dz}{ds}\dfrac{d^2y}{ds^2} \\ \dfrac{dz}{ds}\dfrac{d^2x}{ds^2} - \dfrac{dx}{ds}\dfrac{d^2z}{ds^2} \\ \dfrac{dx}{ds}\dfrac{d^2y}{ds^2} - \dfrac{dy}{ds}\dfrac{d^2x}{ds^2} \end{bmatrix}$$

$$= \frac{2}{\sqrt{Z}} \begin{bmatrix} -b(a + 2b\sin\theta - 2b\sin\theta\cos^2\theta) \\ b^2\cos\theta\,(2\cos^2\theta - 3) \\ a^2 + 2b^2 + 3ab\sin\theta \end{bmatrix} \tag{1.533}$$

$$Z = 4a^4 + 26b^4 + 38a^2b^2 + 4ab(6a^2 + 15b^2)\sin\theta - 6b^2(3a^2 + b^2)\cos 2\theta - 4ab^3 \sin 3\theta \tag{1.534}$$

Example 73 ★ Curvature Center of a Roller Coaster The position of the center of curvature of a space curve can be shown by a vector $\mathbf{r}_c$, where

$$\mathbf{r}_c = \rho\hat{u}_n \tag{1.535}$$

The radius of curvature and the normal unit vector of the roller coaster space curve (1.507) are give in Equations (1.529) and (1.531). Therefore, the position of the curvature center of the roller coaster is

$$\mathbf{r} + \mathbf{r}_c = \begin{bmatrix} (a + b\sin\theta)\cos\theta \\ (a + b\sin\theta)\sin\theta \\ b + b\cos\theta \end{bmatrix} + \begin{bmatrix} \frac{d^2x}{d\theta^2} + \rho\frac{dx}{d\theta}\frac{d\kappa}{d\theta} \\ \frac{d^2y}{d\theta^2} + \rho\frac{dy}{d\theta}\frac{d\kappa}{d\theta} \\ \frac{d^2z}{d\theta^2} + \rho\frac{dz}{d\theta}\frac{d\kappa}{d\theta} \end{bmatrix} \tag{1.536}$$

$$\rho = \tfrac{1}{2}\sqrt{4a^2 + 6b^2 - 2b^2\cos 2\theta + 8ab\sin\theta} \tag{1.537}$$

Figure 1.25 illustrates the path of motion and the path of curvature center. The initial positions at $\theta = 0$ are indicated by two small circles and the direction of motion by increasing θ is shown by two small arrows.

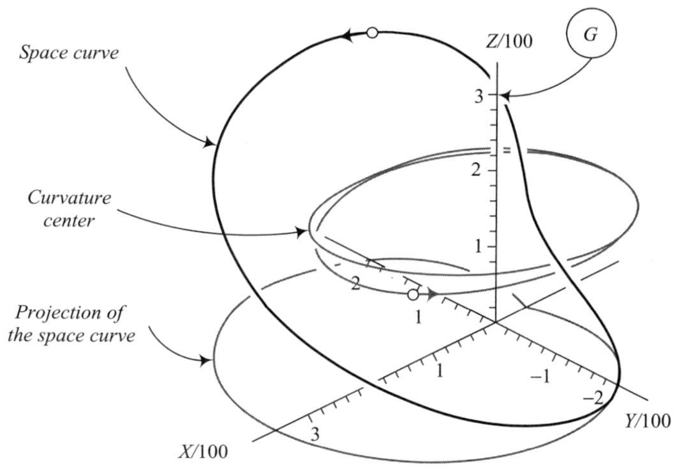

Figure 1.25 The path of motion of a roller coaster and the path of its curvature center.

1.6 FIELDS

A field is a domain of space in which there is a physical quantity associated with every point of the space. If the physical quantity is scalaric, the field is called a *scalar field*, and if the physical quantity is vectorial, the field is a *vector field*. Furthermore, a field is called *stationary* or *time invariant* if it is independent of time. A field that changes with time is a *nonstationary* or *time-variant* field.

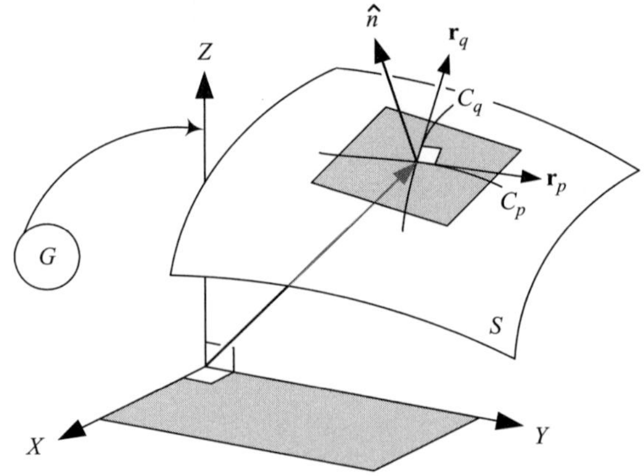

Figure 1.26 A surface ${}^G\mathbf{r} = {}^G\mathbf{r}(p, q)$ and partial derivatives $\mathbf{r}_p$ and $\mathbf{r}_q$.

1.6.1 Surface and Orthogonal Mesh

If the position vector ${}^G\mathbf{r}_P$ of a moving point P is such that each component is a function of two variables p and q,

$$ {}^G\mathbf{r} = {}^G\mathbf{r}(p, q) = x(p, q)\,\hat{\imath} + y(p, q)\,\hat{\jmath} + z(p, q)\,\hat{k} \tag{1.538} $$

then the end point of the vector indicates a surface S in G, as is shown in Figure 1.26. The surface ${}^G\mathbf{r} = {}^G\mathbf{r}(p, q)$ reduces to a curve on S if we fix one of the parameters q or p. The curves C_p and C_q on S at (p_0, q_0) are indicated by single-variable vectors ${}^G\mathbf{r}(p, q_0)$ and ${}^G\mathbf{r}(p_0, q)$, respectively.

At any specific point ${}^G\mathbf{r} = {}^G\mathbf{r}(p_0, q_0)$ there is a tangent plane to the surface that is indicated by a normal unit vector $\hat{n}$,

$$ \hat{n} = \hat{n}(p_0, q_0) = \frac{\mathbf{r}_p \times \mathbf{r}_q}{\left|\mathbf{r}_p \times \mathbf{r}_q\right|} \tag{1.539} $$

where $\mathbf{r}_p$ and $\mathbf{r}_q$ are partial derivatives of ${}^G\mathbf{r}$:

$$ \mathbf{r}_p = \frac{\partial\,\mathbf{r}(p, q_0)}{\partial p} \tag{1.540} $$

$$ \mathbf{r}_q = \frac{\partial\,\mathbf{r}(p_0, q)}{\partial q} \tag{1.541} $$

Varying p and q provides a set of curves C_p and C_q that make a mesh of S. The mesh is called *orthogonal* if we have

$$ \mathbf{r}_p \cdot \mathbf{r}_q = 0 \tag{1.542} $$

Proof: By fixing one of the variables, say $p = p_0$, we can make a single-variable vector function ${}^G\mathbf{r} = {}^G\mathbf{r}(p_0, q)$ to define a curve C_q lying on the surface S. Similarly, we may fix $q = q_0$ to define another single-variable vector function ${}^G\mathbf{r} = {}^G\mathbf{r}(p, q_0)$ and curve C_p. So, there are two curves C_p and C_q that pass through the point (p_0, q_0).

The vectors

$$\mathbf{r}_p = \frac{\partial \mathbf{r}(p, q_0)}{\partial p} = \frac{\partial x(p, q_0)}{\partial p}\hat{\imath} + \frac{\partial y(p, q_0)}{\partial p}\hat{\jmath} + \frac{\partial z(p, q_0)}{\partial p}\hat{k} \tag{1.543}$$

$$\mathbf{r}_q = \frac{\partial \mathbf{r}(p_0, q)}{\partial q} = \frac{\partial x(p_0, q)}{\partial q}\hat{\imath} + \frac{\partial y(p_0, q)}{\partial q}\hat{\jmath} + \frac{\partial z(p_0, q)}{\partial q}\hat{k} \tag{1.544}$$

that are tangent to the curves C_p and C_q are called the *partial derivatives* of ${}^G\mathbf{r}(p, q)$. The vectors $\mathbf{r}_p$ and $\mathbf{r}_q$ define a *tangent plane*. The tangent plane may be indicated by a unit *normal vector* $\hat{n}$:

$$\hat{n} = \hat{n}(p_0, q_0) = \frac{\mathbf{r}_p \times \mathbf{r}_q}{\left|\mathbf{r}_p \times \mathbf{r}_q\right|} \tag{1.545}$$

A surface for which a normal vector $\hat{n}$ can be constructed at any point is called *orientable*. An orientable surface has inner and outer sides. At each point $x(p_0, q_0)$, $y(p_0, q_0)$, $z(p_0, q_0)$ of an orientable surface S there exists a normal axis on which we can choose two directions $\hat{n}_0$, $-\hat{n}_0$. The positive normal vector $\hat{n}_0$ cannot be coincident with $-\hat{n}_0$ by a continuous displacement. The normal unit vector on the convex side is considered positive and the normal to the concave side negative.

If $\mathbf{r}_p \cdot \mathbf{r}_q = 0$ at any point on the surface S, the mesh that is formed by curves C_p and C_q is called an *orthogonal mesh*. The set of unit vectors of an orthogonal mesh,

$$\hat{u}_p = \frac{\mathbf{r}_p}{\left|\mathbf{r}_p\right|} \tag{1.546}$$

$$\hat{u}_q = \frac{\mathbf{r}_q}{\left|\mathbf{r}_q\right|} \tag{1.547}$$

$$\hat{n} = \hat{u}_p \times \hat{u}_q \tag{1.548}$$

defines an orthogonal coordinate system. These definitions are consistent with the definition of unit vectors in Equation (1.200).

★ We assume that the functions $x(p, q)$, $y(p, q)$, and $z(p, q)$ in the parametric expression of a surface in Equation (1.538) have continuous derivatives with respect to the variables q and p. For such a surface, we can define a Jacobian matrix $[J]$ using partial derivatives of the functions x, y, and z:

$$[J] = \begin{bmatrix} x_p & x_q \\ y_p & y_q \\ z_p & z_q \end{bmatrix} \tag{1.549}$$

The surface at a point $P(p_0, q_0)$ is called *regular* if and only if the rank of $[J]$ is not less than 2. A point P at which $[J]$ has rank 1 is called a *singular* point. At a regular point, we have

$$\mathbf{r}_p \times \mathbf{r}_q \neq 0 \tag{1.550}$$

Therefore, we can determine the tangent plane unit-normal vector $\hat{n}$ for every regular point. At a singular point, the rank of $[J]$ is 1 and we have

$$\mathbf{r}_p \times \mathbf{r}_q = 0 \tag{1.551}$$

which indicates $\mathbf{r}_p$ and $\mathbf{r}_q$ are parallel. There is not a unique tangent plane at a singular point.

A surface that has no singularity is called an *immersed surface*. ■

Example 74 Sphere and Orthogonal Mesh A sphere is defined as the position of all points (x, y, z) that have the same distance R from the center (x_0, y_0, z_0):

$$(x - x_0)^2 + (y - y_0)^2 + (z - z_0)^2 = R^2 \tag{1.552}$$

Consider a moving point P on a sphere with a center at the origin. The position vector of P is

$$\begin{aligned}\mathbf{r} &= x\hat{\imath} + y\hat{\jmath} + z\hat{k} \\ &= (R\cos\theta\sin\varphi)\,\hat{\imath} + (R\sin\theta\sin\varphi)\,\hat{\jmath} + (R\cos\varphi)\,\hat{k}\end{aligned} \tag{1.553}$$

Eliminating θ and φ between x, y, and z generates the surface equation:

$$z = \pm\sqrt{R^2 - x^2 - y^2} \tag{1.554}$$

As a sample, when $(\theta, \varphi) = (\pi/3, \pi/4)$, point P is at $(x, y, z) = (0.35355R, 0.61237R, 0.70711R)$ and we may define two curves C_θ and C_φ as

$$C_\theta = \begin{cases} x = R\cos\dfrac{\pi}{3}\sin\varphi \\ y = R\sin\dfrac{\pi}{3}\sin\varphi \\ z = R\cos\varphi \end{cases} = \begin{cases} x = 0.5\,R\sin\varphi \\ y = 0.866\,03\,R\sin\varphi \\ z = R\cos\varphi \end{cases} \tag{1.555}$$

$$C_\varphi = \begin{cases} x = R\cos\theta\sin\dfrac{\pi}{4} \\ y = R\sin\theta\sin\dfrac{\pi}{4} \\ z = R\cos\dfrac{\pi}{4} \end{cases} = \begin{cases} x = 0.707\,11\,R\cos\theta \\ y = 0.707\,11\,R\sin\theta \\ z = 0.707\,11\,R \end{cases} \tag{1.556}$$

The tangent vectors to C_θ and C_φ at arbitrary θ and φ can be found by partial derivatives:

$$\mathbf{r}_\theta = \frac{\partial\,\mathbf{r}(\theta, \varphi)}{\partial\theta} = -R\sin\theta\sin\varphi\,\hat{\imath} + R\cos\theta\sin\varphi\,\hat{\jmath} \tag{1.557}$$

$$\mathbf{r}_\varphi = \frac{\partial\,\mathbf{r}(\theta, \varphi)}{\partial\varphi} = R\cos\theta\cos\varphi\,\hat{\imath} + R\sin\theta\cos\varphi\,\hat{\jmath} - R\sin\varphi\,\hat{k} \tag{1.558}$$

These tangent vectors at the point P reduce to

$$\mathbf{r}_\varphi = \frac{\partial\,\mathbf{r}(\pi/3, \varphi)}{\partial\varphi} = \begin{bmatrix} 0.5R\cos\varphi \\ 0.86603R\cos\varphi \\ -R\sin\varphi \end{bmatrix} \tag{1.559}$$

$$\mathbf{r}_\theta = \frac{\partial\,\mathbf{r}(\theta, \pi/4)}{\partial\theta} = \begin{bmatrix} -0.70711R\sin\theta \\ 0.70711R\cos\theta \\ 0 \end{bmatrix} \tag{1.560}$$

We can check the orthogonality of the curves C_θ and C_φ for different θ and φ by examining the inner product of $\mathbf{r}_\theta$ and $\mathbf{r}_\varphi$ from (1.557) and (1.558):

$$\mathbf{r}_\theta \cdot \mathbf{r}_\varphi = \begin{bmatrix} -R\sin\theta\sin\varphi \\ R\cos\theta\sin\varphi \\ 0 \end{bmatrix} \cdot \begin{bmatrix} R\cos\theta\cos\varphi \\ R\sin\theta\cos\varphi \\ -R\sin\varphi \end{bmatrix} = 0 \tag{1.561}$$

The tangent vectors $\mathbf{r}_\theta$ and $\mathbf{r}_\varphi$ define a tangent plane with a unit-normal vector $\hat{n}$:

$$\begin{aligned} \hat{n} = \hat{n}\left(\frac{\pi}{3}, \frac{\pi}{4}\right) &= \frac{\mathbf{r}_\theta \times \mathbf{r}_\varphi}{\left|\mathbf{r}_\theta \times \mathbf{r}_\varphi\right|} \\ &= \frac{1}{0.70711R^2}\begin{bmatrix} 0.25R^2 \\ 0.43301R^2 \\ 0.5R^2 \end{bmatrix} = \begin{bmatrix} 0.35355 \\ 0.61237 \\ 0.7071 \end{bmatrix} \end{aligned} \tag{1.562}$$

Therefore, we may establish an orthogonal coordinate system at $P(\theta, \varphi, r) = (\pi/3, \pi/4, R)$ with the following unit vectors:

$$\hat{u}_\theta = \frac{\mathbf{r}_\theta}{|\mathbf{r}_\theta|} = \frac{\partial\mathbf{r}/\partial\theta}{|\partial\mathbf{r}/\partial\theta|} = -0.86602\hat{\imath} + 0.5\hat{\jmath} \tag{1.563}$$

$$\hat{u}_\varphi = \frac{\mathbf{r}_\varphi}{\left|\mathbf{r}_\varphi\right|} = \frac{\partial\mathbf{r}/\partial\varphi}{|\partial\mathbf{r}/\partial\varphi|} = 0.35355\hat{\imath} + 0.61238\hat{\jmath} - 0.70711\hat{k} \tag{1.564}$$

$$\hat{n} = \frac{\mathbf{r}_\theta \times \mathbf{r}_\varphi}{\left|\mathbf{r}_\theta \times \mathbf{r}_\varphi\right|} = -0.35356\hat{\imath} - 0.61237\hat{\jmath} - 0.70711\hat{k} \tag{1.565}$$

Example 75 Surface-Analytic Expressions There are several methods to express a surface. Three of them are the most applied forms: parametric, *Monge*, and implicit.

The parametric expression of a surface is when the x-, y-, and z-components of a position vector are functions of two parameters:

$${}^G\mathbf{r} = {}^G\mathbf{r}(p, q) = x(p, q)\,\hat{\imath} + y(p, q)\,\hat{\jmath} + z(p, q)\,\hat{k} \tag{1.566}$$

The Monge expression of a surface is when we eliminate the parameters p and q from x, y, z and define z as a function of x and y:

$${}^G\mathbf{r}(x, y) = x\hat{\imath} + y\hat{\jmath} + z(x, y)\,\hat{k} \tag{1.567}$$

The implicit form of a surface is a nonlinear equation f of x, y, z:

$$f(x, y, z) = 0 \tag{1.568}$$

Example 76 Directional Cosines of Unit-Normal Vector $\hat{n}$ We are able to solve the first two equations of the parametric expression of a surface,

$$x = x(p, q) \qquad y = y(p, q) \qquad z = z(p, q) \tag{1.569}$$

for p and q, and define the surface by a function

$$z = z(x, y) = g(x, y) \tag{1.570}$$

and write the vector representation of the surface by the Monge expression

$$^{G}\mathbf{r}(x, y) = x\hat{\imath} + y\hat{\jmath} + g(x, y)\hat{k} \tag{1.571}$$

The partial derivatives and the equation of the two curves C_x and C_y would be

$$\mathbf{r}_x = \frac{\partial \mathbf{r}}{\partial x} = \hat{\imath} + \frac{\partial g(x, y)}{\partial x}\hat{k} \tag{1.572}$$

$$\mathbf{r}_y = \frac{\partial \mathbf{r}}{\partial y} = \hat{\jmath} + \frac{\partial g(x, y)}{\partial y}\hat{k} \tag{1.573}$$

The cross product of $\mathbf{r}_x$ and $\mathbf{r}_y$ is

$$\mathbf{r}_x \times \mathbf{r}_y = -\frac{\partial g(x, y)}{\partial x}\hat{\imath} - \frac{\partial g(x, y)}{\partial y}\hat{\jmath} + \hat{k} \tag{1.574}$$

and hence the unit-normal vector $\hat{n}$ is

$$\hat{n} = \frac{\mathbf{r}_x \times \mathbf{r}_y}{\left|\mathbf{r}_x \times \mathbf{r}_y\right|} = \frac{-\frac{\partial z}{\partial x}\hat{\imath} - \frac{\partial z}{\partial y}\hat{\jmath} + \hat{k}}{\sqrt{\left(\frac{\partial z}{\partial x}\right)^2 + \left(\frac{\partial z}{\partial y}\right)^2 + 1}} \tag{1.575}$$

The normal vector (1.575) can never be horizontal.

As an example, the position vector of a moving point on the northern hemisphere of the sphere,

$$z = +\sqrt{R^2 - x^2 - y^2} \tag{1.576}$$

is

$$^{G}\mathbf{r}(x, y) = x\hat{\imath} + y\hat{\jmath} + \sqrt{R^2 - x^2 - y^2}\hat{k} \tag{1.577}$$

The partial derivatives $\mathbf{r}_x$ and $\mathbf{r}_y$ and the unit-normal vector $\hat{n}$ are

$$\mathbf{r}_x = \frac{\partial \mathbf{r}}{\partial x} = \hat{\imath} - \frac{x}{\sqrt{R^2 - x^2 - y^2}}\hat{k} \tag{1.578}$$

$$\mathbf{r}_y = \frac{\partial \mathbf{r}}{\partial y} = \hat{\jmath} - \frac{y}{\sqrt{R^2 - x^2 - y^2}}\hat{k} \tag{1.579}$$

$$\hat{n} = \frac{\mathbf{r}_x \times \mathbf{r}_y}{\left|\mathbf{r}_x \times \mathbf{r}_y\right|} = \frac{1}{R}\begin{bmatrix} x \\ y \\ \sqrt{R^2 - x^2 - y^2} \end{bmatrix} \tag{1.580}$$

It shows that the normal vector to a sphere is always in the direction of the position vector $\mathbf{r}$ and away from the center.

These vectors may be used to make an orthogonal coordinate system. At a point such as $(x, y, z) = (0.35355R, 0.61237R, 0.70711R)$, we have

$$\hat{u}_x = \frac{\mathbf{r}_x}{|\mathbf{r}_x|} = \frac{\hat{\imath} - 0.49999\hat{k}}{1.118} = 0.89445\hat{\imath} - 0.44722\hat{k}$$

$$\hat{u}_y = \frac{\mathbf{r}_y}{|\mathbf{r}_y|} = \frac{\hat{\jmath} - 0.86601\hat{k}}{1.3229} = 0.75592\hat{\jmath} - 0.65463\hat{k} \qquad (1.581)$$

$$\hat{n} = \hat{u}_x \times \hat{u}_y = 0.33806\hat{\imath} + 0.58553\hat{\jmath} + 0.67613\hat{k}$$

Example 77 Equation of a Tangent Plane Consider a vector $\mathbf{n}$,

$$\mathbf{n} = a\hat{\imath} + b\hat{\jmath} + c\hat{k} \qquad (1.582)$$

that is perpendicular to a plane at a point (x_0, y_0, z_0). The analytic equation of the plane that includes the point (x_0, y_0, z_0) is indicated by position vector ${}^G\mathbf{r} = x\hat{\imath} + y\hat{\jmath} + z\hat{k}$ such that the vector ${}^G\mathbf{r} - {}^G\mathbf{r}_0$ is perpendicular to $\mathbf{n}$,

$$\left({}^G\mathbf{r} - {}^G\mathbf{r}_0\right) \cdot \mathbf{n} = 0 \qquad (1.583)$$

which reduces to

$$a(x - x_0) + b(y - y_0) + c(z - z_0) = 0 \qquad (1.584)$$

Because the normal vector to a surface $z = g(x, y)$ is

$$\mathbf{n} = -\frac{\partial g(x, y)}{\partial x}\hat{\imath} - \frac{\partial g(x, y)}{\partial y}\hat{\jmath} + \hat{k} \qquad (1.585)$$

the equation of the tangent plane to the surface at a point (x_0, y_0, z_0) is

$$z - z_0 = \frac{\partial g(x_0, y_0)}{\partial x}(x - x_0) + \frac{\partial g(x_0, y_0)}{\partial y}(y - y_0) \qquad (1.586)$$

As an example consider a surface $z = 10 - x^2 - y^2$ and a point P at $(x_0, y_0, z_0) = (1, 2, 5)$. The normal vector at P is

$$\mathbf{n} = 2\hat{\imath} + 4\hat{\jmath} + \hat{k} \qquad (1.587)$$

and the tangent plane at P is

$$z - 5 = -2(x - 1) - 4(y - 2) \qquad (1.588)$$

Example 78 Normal Vector to a Surface Let us eliminate the parameters p and q from the equations of a surface,

$$x = x(p, q) \qquad y = y(p, q) \qquad z = z(p, q) \qquad (1.589)$$

and define the surface by a function

$$z = z(x, y) = g(x, y) \qquad (1.590)$$

or alternatively by

$$f = f(x, y, z) \tag{1.591}$$

So, we theoretically have

$$f(x, y, z) = z - g(x, y) \tag{1.592}$$

The normal vector to surface (1.590) is

$$\mathbf{n} = -\frac{\partial z}{\partial x}\hat{\imath} - \frac{\partial z}{\partial y}\hat{\jmath} + \hat{k} \tag{1.593}$$

However, we may use expression (1.592) and substitute the partial derivatives

$$\frac{\partial f}{\partial x} = -\frac{\partial z}{\partial x} \qquad \frac{\partial f}{\partial y} = -\frac{\partial z}{\partial y} \qquad \frac{\partial f}{\partial z} = 1 \tag{1.594}$$

to define the normal vector to the surface (1.592) by

$$\mathbf{n} = \frac{\partial f}{\partial x}\hat{\imath} + \frac{\partial f}{\partial y}\hat{\jmath} + \frac{\partial f}{\partial z}\hat{k} \tag{1.595}$$

Such an expression of a normal vector to a surface is denoted by $\mathbf{n} = \nabla f$ and is called the *gradient* of f:

$$\nabla f = \frac{\partial f}{\partial x}\hat{\imath} + \frac{\partial f}{\partial y}\hat{\jmath} + \frac{\partial f}{\partial z}\hat{k} \tag{1.596}$$

Example 79 ★ Curvature of a Surface Consider a point P on a surface z with a continuous second derivative, as is shown in Figure 1.27:

$$z = f(x, y) \tag{1.597}$$

To determine the curvature of the surface at P, we find the unit-normal vector $\hat{u}_n$ to the surface at P,

$$\hat{u}_n = \frac{\mathbf{n}}{|\mathbf{n}|} \tag{1.598}$$

$$\mathbf{n} = \left(-\frac{\partial g(x, y)}{\partial x}\hat{\imath} - \frac{\partial g(x, y)}{\partial y}\hat{\jmath} + \hat{k}\right) \tag{1.599}$$

and slice the surface by planes containing $\hat{u}_n$ to consider the curvature vector $\boldsymbol{\kappa}$ of the intersection curve. The curvature vector at P on any intersecting curve will be

$$\boldsymbol{\kappa} = \kappa \hat{u}_n \tag{1.600}$$

The value of κ will change by turning the plane around $\hat{u}_n$. The minimum and maximum values of κ are indicated by κ_1 and κ_2 and are called the *principal curvatures*, where κ_1 and κ_2 occur in orthogonal directions . They may be used to determine the curvature in any other directions.

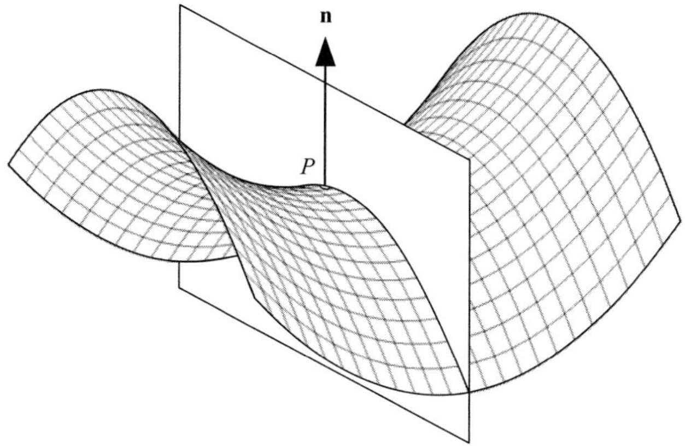

Figure 1.27 A point P on a surface $z = f(x, y)$ with a continuous second derivative.

If the unit vector tangent to the curve at P is shown by $\hat{u}_t$, the intersection curve is in the plane spanned by $\hat{u}_t$ and $\hat{u}_n$. The curvature of the curve κ is called the directional curvature at P in direction $\hat{u}_t$ and defined by

$$\kappa = \hat{u}_t^{\mathrm{T}} \, [D_2 z] \, \hat{u}_t = \hat{u}_t^{\mathrm{T}} \begin{bmatrix} \dfrac{\partial^2 z}{\partial x^2} & \dfrac{\partial z}{\partial x \partial y} \\ \dfrac{\partial z}{\partial y \partial x} & \dfrac{\partial^2 z}{\partial y^2} \end{bmatrix} \hat{u}_t \tag{1.601}$$

The matrix $[D_2 z]$ and vector $\hat{u}_t$ should be determined at point P.

The principal curvatures κ_1 and κ_2 are the eigenvalues of $[D_2 z]$ and their associated directions are called the principal directions of surface z at P. If the coordinate frame (x, y, z) is set up such that z is on $\hat{u}_n$ and x, y are in the principal directions, then the frame is called the principal coordinate frame. The second-derivative matrix $[D_2 z]$ in a principal coordinate frame would be

$$[D_2 z] = \begin{bmatrix} \kappa_1 & 0 \\ 0 & \kappa_2 \end{bmatrix} \tag{1.602}$$

1.6.2 Scalar Field and Derivative

Consider a scalar function f of a vector variable $\mathbf{r}$,

$$f = f(\mathbf{r}) = f(x, y, z) \tag{1.603}$$

such that it provides a number f at a point $P(x, y, z)$. Having such a function is equivalent to associating a numeric value to every point of the space. The space that $f(x, y, z)$ is defined in is called a *scalar field*, and the function f is called the *scalar field function*. The field function is assumed to be smooth and differentiable. A smooth field has no singularity, jump, sink, or source.

Setting f equal to a specific value f_0 defines a surface

$$f(x, y, z) = f_0 \tag{1.604}$$

that is the loci of all points for which f takes the fixed value f_0. The surface $f(x, y, z) = f_0$ is called an *isosurface* and the associated field value is called the *isovalue* f_0.

The space derivative of f for an infinitesimal displacement $d\mathbf{r}$ is a vector:

$$\frac{df(\mathbf{r})}{d\mathbf{r}} = \frac{\partial f}{\partial x}\hat{\imath} + \frac{\partial f}{\partial y}\hat{\jmath} + \frac{\partial f}{\partial z}\hat{k} = \nabla f \tag{1.605}$$

It can also be shown by

$$df = \nabla f \cdot d\mathbf{r} \tag{1.606}$$

where at any point ${}^{G}\mathbf{r} = {}^{G}\mathbf{r}(x, y, z)$ there exists a vector ∇f that indicates the value and direction of the maximum change in f for an infinitesimal change $d\mathbf{r}$ in position.

Figure 1.28 illustrates an isosurface f_0 and the vector ∇f at a point on the isosurface,

$$\nabla f = \nabla f(x, y, z) = \frac{\partial f}{\partial x}\hat{\imath} + \frac{\partial f}{\partial y}\hat{\jmath} + \frac{\partial f}{\partial z}\hat{k} \tag{1.607}$$

The vector ∇f is called the *gradient* of the scalar field . The gradient (1.607) can be expressed by a vectorial derivative operator ∇,

$$\nabla = \frac{\partial}{\partial x}\hat{\imath} + \frac{\partial}{\partial y}\hat{\jmath} + \frac{\partial}{\partial z}\hat{k} = \begin{bmatrix} \dfrac{\partial}{\partial x} \\ \dfrac{\partial}{\partial y} \\ \dfrac{\partial}{\partial z} \end{bmatrix} \tag{1.608}$$

that operates on the scalar field f. The *gradient operator* ∇ is also called the *grad*, *del*, or *nabla* operator.

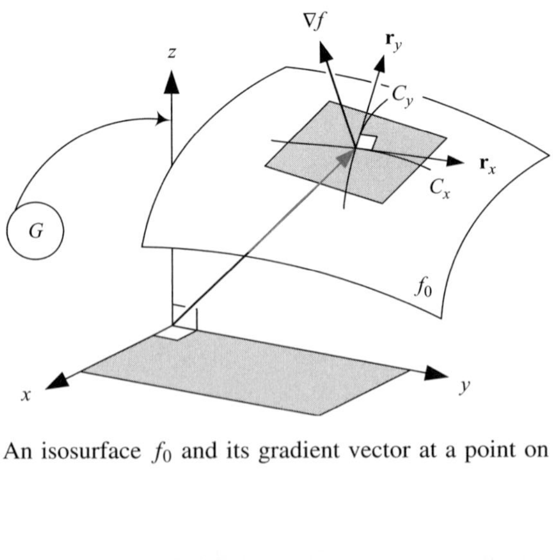

Figure 1.28 An isosurface f_0 and its gradient vector at a point on the isosurface.

Proof: By assigning various values to f, we obtain a family of isosurfaces of the scalar field $f = f(\mathbf{r}) = f(x, y, z)$ as is shown in Figure 1.29. These surfaces serve to geometrically visualize the field's characteristics.

An isosurface $f_0 = f(x, y, z)$ can be expressed by a position vector ${}^G\mathbf{r}$,

$$ {}^G\mathbf{r} = x\hat{\imath} + y\hat{\jmath} + z\hat{k} \tag{1.609}$$

where its components x, y, z are constrained by the isosurface equation (1.604). Let us consider a point P at $\mathbf{r} = \mathbf{r}(x, y, z)$ on an isosurface $f(x, y, z) = f$. Any infinitesimal change

$$d\mathbf{r} = dx\ \hat{\imath} + dy\ \hat{\jmath} + dz\ \hat{k} \tag{1.610}$$

in the position of P will move the point to a new isosurface with a field value $f + df$, where

$$\begin{aligned} df &= f\,(x + dx, y + dy, z + dz) - f(x, y, z) \\ &= \frac{\partial f}{\partial x}dx + \frac{\partial f}{\partial y}dy + \frac{\partial f}{\partial z}dz \\ &= \left(\frac{\partial f}{\partial x}\hat{\imath} + \frac{\partial f}{\partial y}\hat{\jmath} + \frac{\partial f}{\partial z}\hat{k}\right) \cdot \left(dx\ \hat{\imath} + dy\hat{\jmath} + dz\ \hat{k}\right) = \nabla f \cdot d\mathbf{r} \end{aligned} \tag{1.611}$$

So *df* can be interpreted as an inner product between two vectors ∇f and $d\mathbf{r}$. The first vector, denoted by

$$\nabla f = \frac{\partial f}{\partial x}\hat{\imath} + \frac{\partial f}{\partial y}\hat{\jmath} + \frac{\partial f}{\partial z}\hat{k} \tag{1.612}$$

is a Cartesian expression of the *gradient* of the scalar function f, and the second vector $d\mathbf{r}$ is the displacement vector of the point. If the two nearby points lie on the same isosurface, then $df = 0$, $d\mathbf{r}$ would be a tangent vector to this isosurface, and

$$df = \nabla f \cdot d\mathbf{r} = 0 \tag{1.613}$$

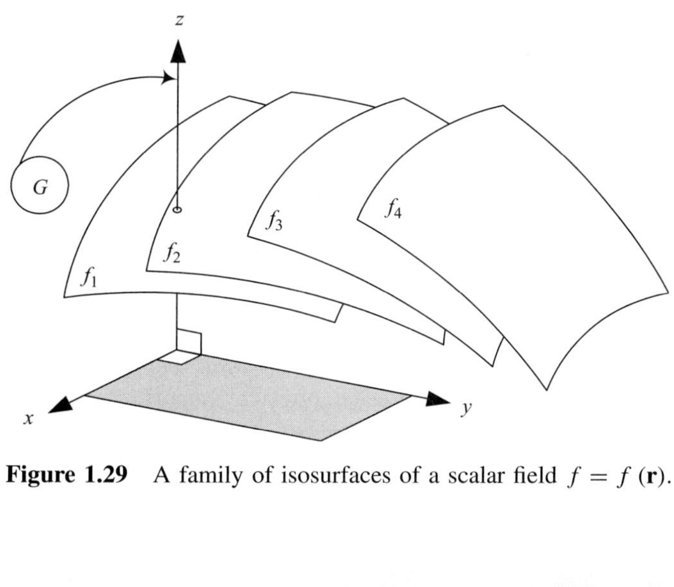

Figure 1.29 A family of isosurfaces of a scalar field $f = f\,(\mathbf{r})$.

Therefore, ∇f is perpendicular to $d\mathbf{r}$ and hence normal to the isosurface f. The gradient of a scalar field is a coordinate-independent property.

We examine a nonstationary field at the interested specific instant of time. ■

Example 80 Derivative of Scalar Function with Vector Variable If $\mathbf{c}$ is a constant vector and $f = \mathbf{c} \cdot \mathbf{r}$ is a scalar field, then

$$\text{grad}\, f = \text{grad}\, (\mathbf{c} \cdot \mathbf{r}) = \mathbf{c} \tag{1.614}$$

If $f = \mathbf{r}^2$, then

$$\text{grad}\, f = \text{grad}\, \mathbf{r}^2 = 2\mathbf{r} \tag{1.615}$$

If $f = |\mathbf{r}|$ and $g = \mathbf{r}^2$, then $f = g^{1/2}$, and therefore,

$$\text{grad}\, f = \frac{1}{2} g^{-1/2}\, \text{grad}\, g = \frac{\mathbf{r}}{|\mathbf{r}|} \tag{1.616}$$

Example 81 Gradient of Scalar Field Consider a scalar field

$$f(x, y, z) = x + x^2 y + y^3 + y^2 x + z^2 = C \tag{1.617}$$

The gradient of the field is

$$\nabla f = \frac{\partial f}{\partial x}\hat{\imath} + \frac{\partial f}{\partial y}\hat{\jmath} + \frac{\partial f}{\partial z}\hat{k} = \begin{bmatrix} y^2 + 2xy + 1 \\ x^2 + 2xy + 3y^2 \\ 2z \end{bmatrix} \tag{1.618}$$

Now assume that the gradient (1.618) is given. To find the field function, we should integrate the components of the gradient:

$$\begin{aligned} f &= \int \left(y^2 + 2xy + 1\right) dx = x^2 y + xy^2 + x + g_1(y, z) \\ &= \int \left(x^2 + 2xy + 3y^2\right) dy = x^2 y + xy^2 + y^3 + g_2(x, z) \\ &= \int (2z)\, dz = z^2 + g_3(x, y) \end{aligned} \tag{1.619}$$

Comparison shows that

$$f(x, y, z) = x + x^2 y + y^3 + y^2 x + z^2 = C \tag{1.620}$$

Example 82 Examples of Scalar and Vector Fields A field is another useful man-made concept to describe physical quantities. We call a function $f = f(x, y, z)$ a scalar field function if it assigns a numeric value to any point $P(x, y, z)$ of space. We call a function $\mathbf{f} = \mathbf{f}(x, y, z)$ a vector field function if it assigns a vector to any point $P(x, y, z)$ of space.

Temperature, density, and humidity are a few examples of scalar fields, and electric, magnetic, and velocity are a few examples of vector fields.

Example 83 Time Derivative of Scalar Field Consider a time-varying scalar field of a vector variable

$$f = f(\mathbf{r}(t)) \tag{1.621}$$

The time derivative of f is

$$\frac{df}{dt} = \frac{df}{d\mathbf{r}} \cdot \frac{d\mathbf{r}}{dt} = \frac{\partial f}{\partial x}\frac{\partial x}{\partial t} + \frac{\partial f}{\partial y}\frac{\partial y}{\partial t} + \frac{\partial f}{\partial z}\frac{\partial z}{\partial t} = \nabla f \cdot \mathbf{v} \tag{1.622}$$

where $\mathbf{v} = d\mathbf{r}/dt$ is called the velocity of the position vector $\mathbf{r}$:

$$\mathbf{v} = \frac{d\mathbf{r}}{dt} = \frac{dx}{dt}\hat{\imath} + \frac{dy}{dt}\hat{\jmath} + \frac{dz}{dt}\hat{k} \tag{1.623}$$

Following the derivative rule of a scalar field function (1.611), we may confirm that the time derivative of a scalar field is

$$\frac{df}{dt} = \nabla f \cdot \frac{d\mathbf{r}}{dt} = \nabla f \cdot \mathbf{v} \tag{1.624}$$

Example 84 Alternative Definition of Gradient Consider the scalar field function

$$f = f(\mathbf{r}) = f(x, y, z) \tag{1.625}$$

When the position vector moves from a point at $\mathbf{r} = \mathbf{r}(x, y, z)$ to a close point at $\mathbf{r} + d\mathbf{r}$, the field function changes from $f(\mathbf{r}) = f$ to $f(\mathbf{r} + d\mathbf{r}) = f + df$:

$$\begin{aligned} f(\mathbf{r} + d\mathbf{r}) &= f(x + dx, y + dy, z + dz) \\ &\approx f(x, y, z) + \frac{\partial f}{\partial x}dx + \frac{\partial f}{\partial y}dy + \frac{\partial f}{\partial z}dz + \cdots \end{aligned} \tag{1.626}$$

Therefore, a change in the field due to an infinitesimal change in position is given as

$$\begin{aligned} df &= f(\mathbf{r} + d\mathbf{r}) - f(\mathbf{r}) \\ &= \frac{\partial f}{\partial x}dx + \frac{\partial f}{\partial y}dy + \frac{\partial f}{\partial z}dz = \nabla f \cdot d\mathbf{r} \end{aligned} \tag{1.627}$$

where df is called the *total derivative* of f.

Example 85 Directional Derivative An isosurface $f(x, y, z) = f$ can be expressed by the position vector ${}^G\mathbf{r}$

$${}^G\mathbf{r} = x\hat{\imath} + y\hat{\jmath} + z\hat{k} \tag{1.628}$$

where its coordinates (x, y, z) are constrained by the isosurface equation (1.604). So, ${}^{G}\mathbf{r}$ is a two-variable vector function where its end point indicates a surface in G. To show this, let us consider a point P at $\mathbf{r} = \mathbf{r}(x, y, z)$ on an isosurface $f(x, y, z) = f$. Any change $d\mathbf{r}$ in the position of P will move the point to a new isosurface with a field value $f + df$:

$$\begin{aligned} df &= f(x + dx, y + dy, z + dz) - f(x, y, z) \\ &= \frac{\partial f}{\partial x}dx + \frac{\partial f}{\partial y}dy + \frac{\partial f}{\partial z}dz \end{aligned} \tag{1.629}$$

Let us move P on a space curve $\mathbf{r} = \mathbf{r}(q)$,

$$\mathbf{r} = x(q)\hat{\imath} + y(q)\hat{\jmath} + z(q)\hat{k} \tag{1.630}$$

The unit vector tangent to the curve at P is

$$\hat{u}_q = \frac{\partial \mathbf{r}/\partial q}{|\partial \mathbf{r}/\partial q|} = \frac{(\partial x/\partial q)\hat{\imath} + (\partial y/\partial q)\hat{\jmath} + (\partial z/\partial q)\hat{k}}{\sqrt{(dx/dq)^2 + (dy/dq)^2 + (dz/dq)^2}} \tag{1.631}$$

For an infinitesimal motion on the curve, we have

$$\frac{df}{dq} = \frac{\partial f}{\partial x}\frac{\partial x}{\partial q} + \frac{\partial f}{\partial y}\frac{\partial y}{\partial q} + \frac{\partial f}{\partial z}\frac{\partial z}{\partial q} \tag{1.632}$$

which can be interpreted as a dot product between two vectors:

$$\frac{df}{dq} = \left(\frac{\partial f}{\partial x}\hat{\imath} + \frac{\partial f}{\partial y}\hat{\jmath} + \frac{\partial f}{\partial z}\hat{k}\right) \cdot \left(\frac{\partial x}{\partial q}\hat{\imath} + \frac{\partial y}{\partial q}\hat{\jmath} + \frac{\partial z}{\partial q}\hat{k}\right) \tag{1.633}$$

$$= \nabla f \cdot \frac{d\mathbf{r}}{dq} = \nabla f \cdot \hat{u}_q \left|\frac{d\mathbf{r}}{dq}\right| \tag{1.634}$$

The first vector, denoted by

$$\nabla f = \frac{\partial f}{\partial x}\hat{\imath} + \frac{\partial f}{\partial y}\hat{\jmath} + \frac{\partial f}{\partial z}\hat{k}$$

is the gradient of the scalar function f expressed in terms of Cartesian coordinates, and the second vector,

$$\frac{d\mathbf{r}}{dq} = \frac{\partial x}{\partial q}\hat{\imath} + \frac{\partial y}{\partial q}\hat{\jmath} + \frac{\partial z}{\partial q}\hat{k} \tag{1.635}$$

is a tangent vector to the space curve (1.630) in the direction of increasing q. The dot product $\nabla f \cdot \hat{u}_q$ calculates the projection of ∇f on the tangent line to the space curve at P. To maximize this product, the angle between ∇f and $\hat{u}_q$ must be zero. It happens when ∇f and $d\mathbf{r}/dq$ are parallel:

$$\frac{\partial f/\partial x}{\partial x/\partial q} = \frac{\partial f/\partial y}{\partial y/\partial q} = \frac{\partial f/\partial z}{\partial z/\partial q} \tag{1.636}$$

A space curve (1.630) with condition (1.636) is perpendicular to the surface (1.628) and is called the *normal* or *flow curve*. Flow curves are perpendicular to isosurfaces of a scalar field f and show the lines of maximum change in field f.

The gradient of the scalar field indicates the direction to move for maximum change in the field, and its magnitude indicates the change in the field for a unit-length move. The product $\nabla f \cdot \hat{u}_q$, which determines the change in the field for a unit-length move in direction $\hat{u}_q$, is called the *directional derivative*.

Example 86 Direction of Maximum Rate of Increase Consider the scalar field

$$\varphi = 10 + xyz \tag{1.637}$$

A point $P(0.5, 0.4, z)$ on an isosurface will have the following z-component:

$$z = \frac{\varphi - 10}{xy} = \frac{\varphi - 10}{0.5 \times 0.4} = 5\varphi - 50 \tag{1.638}$$

The gradient vector $\nabla\varphi$ at point $P(0.5, 0.4, -50)$ on the isosurface $\varphi = 0$ is

$$\nabla\varphi = \begin{bmatrix} \frac{\partial\varphi}{\partial x} \\ \frac{\partial\varphi}{\partial y} \\ \frac{\partial\varphi}{\partial z} \end{bmatrix} = \begin{bmatrix} yz \\ xz \\ xy \end{bmatrix} = \begin{bmatrix} -20 \\ -25 \\ 0.2 \end{bmatrix} \tag{1.639}$$

Example 87 Directional Derivative of a Field at a Point Consider the scalar field

$$f = xy^2 + yz^4 \tag{1.640}$$

Its rate of change at point $P(2,1,1)$ in the direction $\mathbf{r} = \hat{\imath} + 2\hat{\jmath} + \hat{k}$ is found by the inner product of its gradient at P,

$$\nabla f = \begin{bmatrix} \frac{\partial f}{\partial x} \\ \frac{\partial f}{\partial y} \\ \frac{\partial f}{\partial z} \end{bmatrix} = \begin{bmatrix} y^2 \\ z^4 + 2xy \\ 4yz^3 \end{bmatrix} = \begin{bmatrix} 1 \\ 5 \\ 4 \end{bmatrix} \tag{1.641}$$

and $\hat{u}_{\mathbf{r}} = \mathbf{r}/\left|\mathbf{r}\right|$,

$$df = \nabla f \cdot \frac{\mathbf{r}}{|\mathbf{r}|} = \begin{bmatrix} 1 \\ 5 \\ 4 \end{bmatrix} \cdot \begin{bmatrix} \frac{1}{\sqrt{6}} \\ \frac{2}{\sqrt{6}} \\ \frac{1}{\sqrt{6}} \end{bmatrix} = 6.1237 \tag{1.642}$$

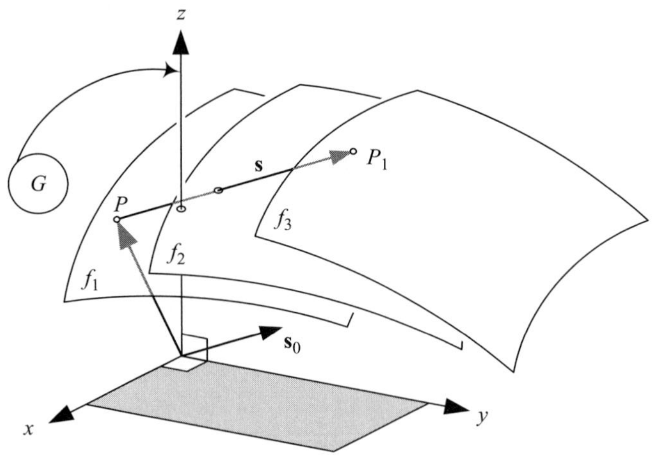

Figure 1.30 Directional derivative of a scalar field $f(\mathbf{r})$ in direction $\mathbf{s}_0$ at a point P is defined by $df/d\mathbf{s}$.

Example 88 ★ Isosurfaces Have No Common Point Consider a field $f = f(\mathbf{r})$ that is defined over a domain Ω of space. The isosurfaces corresponding to different $f = c$ fill the entire of Ω, and no two surfaces $f(\mathbf{r}) = c_1$ and $f(\mathbf{r}) = c_2$, $c_1 \neq c_2$ have common points. The isosurfaces, also called level surfaces, enable us to qualitatively judge the rate of change of the scalar field $f(\mathbf{r})$ in a give direction.

Consider a point P at $\mathbf{r}_P$ in a scalar field $f(\mathbf{r})$ and a fixed direction $\mathbf{s}_0$ as are shown in Figure 1.30. We draw a straight line $\mathbf{s}$ through P parallel to $\mathbf{s}_0$ and pick a point P_1 to define the directional derivative of $f(\mathbf{r})$:

$$\frac{df}{d\mathbf{s}} = \lim_{P_1 \to P} \frac{f(\mathbf{r}_P) - f(\mathbf{r}_{P_1})}{PP_1} \tag{1.643}$$

Such a limit, if it exists, is called the directional derivative of the scalar field $f(\mathbf{r})$ in direction $\mathbf{s}_0$ at point P. Using Equation (1.606), we may show that

$$\frac{df}{d\mathbf{s}} = \nabla f \cdot \hat{u}_{\mathbf{s}} = \frac{\partial f}{\partial x}\cos\alpha + \frac{\partial f}{\partial y}\cos\beta + \frac{\partial f}{\partial z}\cos\gamma \tag{1.644}$$

where α, β, and γ are the directional cosines of $\mathbf{s}$.

1.6.3 Vector Field and Derivative

Consider a vector function $\mathbf{f}$ of a vector variable $\mathbf{r}$,

$$\mathbf{f} = \mathbf{f}(\mathbf{r}) = \mathbf{f}(x, y, z) = f_x(\mathbf{r})\,\hat{\imath} + f_y(\mathbf{r})\,\hat{\jmath} + f_z(\mathbf{r})\,\hat{k} \tag{1.645}$$

so it provides a vector **f** at a point $P(x, y, z)$. Having such a function is equivalent to associating a vector to every point of the space. The space in which there exist an $\mathbf{f}(x, y, z)$ is called a *vector field*, and the function **f** is called the *vector field function*.

The space derivative of $\mathbf{f}(\mathbf{r})$ is a quaternion product of ∇ and **f**,

$$\frac{d\mathbf{f}(\mathbf{r})}{d\mathbf{r}} = \nabla\mathbf{f}(\mathbf{r}) = \nabla \times \mathbf{f} - \nabla \cdot \mathbf{f} = \operatorname{curl}\mathbf{f} - \operatorname{div}\mathbf{f} \tag{1.646}$$

where

$$\nabla \times \mathbf{f} = \operatorname{curl}\mathbf{f} = \begin{vmatrix} \hat{\imath} & \hat{\jmath} & \hat{k} \\ \dfrac{\partial}{\partial x} & \dfrac{\partial}{\partial y} & \dfrac{\partial}{\partial z} \\ f_x & f_y & f_z \end{vmatrix}$$

$$= \left(\frac{\partial f_z}{\partial y} - \frac{\partial f_y}{\partial z}\right)\hat{\imath} + \left(\frac{\partial f_x}{\partial z} - \frac{\partial f_z}{\partial x}\right)\hat{\jmath} + \left(\frac{\partial f_y}{\partial x} - \frac{\partial f_x}{\partial y}\right)\hat{k} \tag{1.647}$$

and

$$\nabla \cdot \mathbf{f} = \operatorname{div}\mathbf{f} = \left(\frac{\partial}{\partial x}\hat{\imath} + \frac{\partial}{\partial y}\hat{\jmath} + \frac{\partial}{\partial z}\hat{k}\right) \cdot \left(f_x\hat{\imath} + f_y\hat{\jmath} + f_z\hat{k}\right)$$

$$= \frac{\partial f_x}{\partial x} + \frac{\partial f_y}{\partial y} + \frac{\partial f_z}{\partial z} \tag{1.648}$$

The first term, $\nabla \times \mathbf{f}$, is a vector and is called the *curl* of the vector field. The second term, $\nabla \cdot \mathbf{f}$, is a scalar and is called the *divergence* of the vector field. The curl of **f** indicates the change in direction and the divergence of **f** indicates the change in the magnitude of **f**.

Proof: If $\mathbf{f}(\mathbf{r})$ is the function of a vector field, then each component of **f** is a scalar function of a vector variable,

$$\mathbf{f}(\mathbf{r}) = f_x(\mathbf{r})\,\hat{\imath} + f_y(\mathbf{r})\,\hat{\jmath} + f_z(\mathbf{r})\,\hat{k} \tag{1.649}$$

So, the differential of the vector field function $\mathbf{f}(\mathbf{r})$ with respect to a change in the position **r** is equal to the gradient of each component of $\mathbf{f}(\mathbf{r})$:

$$\frac{d\mathbf{f}(\mathbf{r})}{d\mathbf{r}} = \frac{d}{d\mathbf{r}}\left(f_x(\mathbf{r})\,\hat{\imath} + f_y(\mathbf{r})\,\hat{\jmath} + f_z(\mathbf{r})\,\hat{k}\right) = \nabla f_x\hat{\imath} + \nabla f_y\hat{\jmath} + \nabla f_z\hat{k}$$

$$= \left(\frac{\partial f_x}{\partial x}\hat{\imath} + \frac{\partial f_x}{\partial y}\hat{\jmath} + \frac{\partial f_x}{\partial z}\hat{k}\right)\hat{\imath} + \left(\frac{\partial f_y}{\partial x}\hat{\imath} + \frac{\partial f_y}{\partial y}\hat{\jmath} + \frac{\partial f_y}{\partial z}\hat{k}\right)\hat{\jmath}$$

$$+ \left(\frac{\partial f_z}{\partial x}\hat{\imath} + \frac{\partial f_z}{\partial y}\hat{\jmath} + \frac{\partial f_z}{\partial z}\hat{k}\right)\hat{k} \tag{1.650}$$

Knowing that

$$\hat{\imath}^2 = \hat{\jmath}^2 = \hat{k}^2 = \hat{\imath}\hat{\jmath}\hat{k} = -1 \tag{1.651}$$

$$\hat{\imath}\hat{\jmath} = -\hat{\jmath}\hat{\imath} = \hat{k} \qquad \hat{\jmath}\hat{k} = -\hat{k}\hat{\jmath} = \hat{\imath} \qquad \hat{k}\hat{\imath} = -\hat{\imath}\hat{k} = \hat{\jmath} \tag{1.652}$$

we can simplify Equation (1.650) to

$$\begin{aligned}
\frac{d\mathbf{f}(\mathbf{r})}{d\mathbf{r}} &= \left(-\frac{\partial f_x}{\partial x} - \frac{\partial f_x}{\partial y}\hat{k} + \frac{\partial f_x}{\partial z}\hat{j}\right) + \left(\frac{\partial f_y}{\partial x}\hat{k} - \frac{\partial f_y}{\partial y} - \frac{\partial f_y}{\partial z}\hat{\imath}\right) \\
&\quad + \left(-\frac{\partial f_z}{\partial x}\hat{j} + \frac{\partial f_z}{\partial y}\hat{\imath} - \frac{\partial f_z}{\partial z}\right) \\
&= \left(\frac{\partial f_z}{\partial y} - \frac{\partial f_y}{\partial z}\right)\hat{\imath} + \left(\frac{\partial f_x}{\partial z} - \frac{\partial f_z}{\partial x}\right)\hat{j} + \left(\frac{\partial f_y}{\partial x} - \frac{\partial f_x}{\partial y}\right)\hat{k} \\
&\quad - \frac{\partial f_x}{\partial x} - \frac{\partial f_y}{\partial y} - \frac{\partial f_z}{\partial z}
\end{aligned} \tag{1.653}$$

which is equal to

$$\frac{d\mathbf{f}(\mathbf{r})}{d\mathbf{r}} = \nabla\mathbf{f}(\mathbf{r}) = \nabla \times \mathbf{f} - \nabla \cdot \mathbf{f} \tag{1.654}$$

The divergence of the gradient of a scalar field f is a fundamental partial differential equation in potential theory called the **Laplacian** of f. The Laplacian of f is shown by $\nabla^2 f$ and is equal to:

$$\begin{aligned}
\nabla^2 f &= \operatorname{div}\operatorname{grad} f = \nabla \cdot \nabla f \\
&= \left(\frac{\partial}{\partial x}\hat{\imath} + \frac{\partial}{\partial y}\hat{j} + \frac{\partial}{\partial z}\hat{k}\right) \cdot \left(\frac{\partial f}{\partial x} + \frac{\partial f}{\partial y} + \frac{\partial f}{\partial z}\right) \\
&= \frac{\partial^2 f}{\partial x^2} + \frac{\partial^2 f}{\partial y^2} + \frac{\partial^2 f}{\partial z^2}
\end{aligned} \tag{1.655}$$

■

Example 89 Derivative of a Vector Function with a Vector Variable If **c** is a constant vector and $\mathbf{f} = \mathbf{c} \times \mathbf{r}$ is a vector field, then

$$\frac{d\mathbf{f}(\mathbf{r})}{d\mathbf{r}} = \operatorname{grad}\mathbf{f} = \operatorname{grad}(\mathbf{c} \times \mathbf{r}) = 3\mathbf{c} \tag{1.656}$$

because $\mathbf{r} = x\hat{\imath} + y\hat{j} + z\hat{k}$ and

$$\begin{aligned}
\frac{d\mathbf{f}(\mathbf{r})}{d\mathbf{r}} &= \nabla \times \mathbf{f} - \nabla \cdot \mathbf{f} = \nabla \times (\mathbf{c} \times \mathbf{r}) - \nabla \cdot (\mathbf{c} \times \mathbf{r}) \\
&= (\nabla \cdot \mathbf{r})\,\mathbf{c} - (\nabla \cdot \mathbf{c})\,\mathbf{r} - (\nabla \times \mathbf{c}) \cdot \mathbf{r} \\
&= \left(\frac{\partial x}{\partial x} + \frac{\partial y}{\partial y} + \frac{\partial z}{\partial z}\right)\mathbf{c} = 3\mathbf{c}
\end{aligned} \tag{1.657}$$

However, if $\mathbf{f} = c\mathbf{r}$, where c is a constant scalar, then its space derivative is a scalar,

$$\frac{d\mathbf{f}(\mathbf{r})}{d\mathbf{r}} = \operatorname{grad}\mathbf{f} = \operatorname{grad} c\mathbf{r} = 3c \tag{1.658}$$

because

$$\frac{d\mathbf{f}(\mathbf{r})}{d\mathbf{r}} = \nabla\times\mathbf{f} - \nabla\cdot\mathbf{f} = \nabla\times c\mathbf{r} - \nabla\cdot c\mathbf{r}$$

$$= \begin{vmatrix} \hat{\imath} & \hat{\jmath} & \hat{k} \\ \dfrac{\partial}{\partial x} & \dfrac{\partial}{\partial y} & \dfrac{\partial}{\partial z} \\ x & y & z \end{vmatrix} - c\left(\frac{\partial x}{\partial x} + \frac{\partial y}{\partial y} + \frac{\partial z}{\partial z}\right) = 3c \tag{1.659}$$

Example 90 Matrix Form of Vector Field Derivative We may arrange the derivative of a vector field $\nabla\mathbf{f}(\mathbf{r})$,

$$\frac{d\mathbf{f}(\mathbf{r})}{d\mathbf{r}} = \nabla\mathbf{f}(\mathbf{r}) = \nabla f_x\hat{\imath} + \nabla f_y\hat{\jmath} + \nabla f_z\hat{k} \tag{1.660}$$

in matrix form:

$$\frac{d\mathbf{f}(\mathbf{r})}{d\mathbf{r}} = \left[\frac{\partial f_i}{\partial x_j}\hat{\imath}_j\hat{\imath}_i\right] = \begin{bmatrix} \dfrac{\partial f_x}{\partial x}\hat{\imath}\hat{\imath} & \dfrac{\partial f_y}{\partial x}\hat{\imath}\hat{\jmath} & \dfrac{\partial f_z}{\partial x}\hat{\imath}\hat{k} \\ \dfrac{\partial f_x}{\partial y}\hat{\jmath}\hat{\imath} & \dfrac{\partial f_y}{\partial y}\hat{\jmath}\hat{\jmath} & \dfrac{\partial f_z}{\partial y}\hat{\jmath}\hat{k} \\ \dfrac{\partial f_x}{\partial z}\hat{k}\hat{\imath} & \dfrac{\partial f_y}{\partial z}\hat{k}\hat{\jmath} & \dfrac{\partial f_z}{\partial z}\hat{k}\hat{k} \end{bmatrix}$$

$$= \begin{bmatrix} -\dfrac{\partial f_x}{\partial x} & \dfrac{\partial f_y}{\partial x}\hat{k} & -\dfrac{\partial f_z}{\partial x}\hat{\jmath} \\ -\dfrac{\partial f_x}{\partial y}\hat{k} & -\dfrac{\partial f_y}{\partial y} & \dfrac{\partial f_z}{\partial y}\hat{\imath} \\ \dfrac{\partial f_x}{\partial z}\hat{\jmath} & -\dfrac{\partial f_y}{\partial z}\hat{\imath} & -\dfrac{\partial f_z}{\partial z} \end{bmatrix} \tag{1.661}$$

The trace of the matrix indicates the divergence of $\mathbf{f}$:

$$\mathrm{tr}\left[\frac{\partial f_i}{\partial x_j}\hat{\imath}_j\hat{\imath}_i\right] = \nabla\cdot\mathbf{f} = \frac{\partial f_x}{\partial x} + \frac{\partial f_y}{\partial y} + \frac{\partial f_z}{\partial z} \tag{1.662}$$

Example 91 Symmetric and Skew-Symmetric Derivative Matrix Recalling that every matrix $[A]$ can be decomposed into a symmetric plus a skew-symmetric matrix,

$$[A] = \tfrac{1}{2}\left[[A] + [A]^{\mathrm{T}}\right] + \tfrac{1}{2}\left[[A] - [A]^{\mathrm{T}}\right] \tag{1.663}$$

we may determine the symmetric and skew-symmetric matrices of the derivative matrix:

$$\left[\frac{\partial f_i}{\partial x_j}\hat{\imath}_j\hat{\imath}_i\right]^{\mathrm{T}} = \begin{bmatrix} -\frac{\partial f_x}{\partial x} & -\frac{\partial f_x}{\partial y}\hat{k} & \frac{\partial f_x}{\partial z}\hat{\jmath} \\ \frac{\partial f_y}{\partial x}\hat{k} & -\frac{\partial f_y}{\partial y} & -\frac{\partial f_y}{\partial z}\hat{\imath} \\ -\frac{\partial f_z}{\partial x}\hat{\jmath} & \frac{\partial f_z}{\partial y}\hat{\imath} & -\frac{\partial f_z}{\partial z} \end{bmatrix} \tag{1.664}$$

$$\left[\frac{\partial f_i}{\partial x_j}\hat{\imath}_j\hat{\imath}_i\right] + \left[\frac{\partial f_i}{\partial x_j}\hat{\imath}_j\hat{\imath}_i\right]^{\mathrm{T}} = \begin{bmatrix} -2\frac{\partial f_x}{\partial x} & \left(\frac{\partial f_y}{\partial x} - \frac{\partial f_x}{\partial y}\right)\hat{k} & \left(\frac{\partial f_x}{\partial z} - \frac{\partial f_z}{\partial x}\right)\hat{\jmath} \\ \left(\frac{\partial f_y}{\partial x} - \frac{\partial f_x}{\partial y}\right)\hat{k} & -2\frac{\partial f_y}{\partial y} & \left(\frac{\partial f_z}{\partial y} - \frac{\partial f_y}{\partial z}\right)\hat{\imath} \\ \left(\frac{\partial f_x}{\partial z} - \frac{\partial f_z}{\partial x}\right)\hat{\jmath} & \left(\frac{\partial f_z}{\partial y} - \frac{\partial f_y}{\partial z}\right)\hat{\imath} & -2\frac{\partial f_z}{\partial z} \end{bmatrix} \tag{1.665}$$

$$\left[\frac{\partial f_i}{\partial x_j}\hat{\imath}_j\hat{\imath}_i\right] - \left[\frac{\partial f_i}{\partial x_j}\hat{\imath}_j\hat{\imath}_i\right]^{\mathrm{T}} = \begin{bmatrix} 0 & \left(\frac{\partial f_y}{\partial x} + \frac{\partial f_x}{\partial y}\right)\hat{k} & \left(-\frac{\partial f_z}{\partial x} - \frac{\partial f_x}{\partial z}\right)\hat{\jmath} \\ \left(-\frac{\partial f_x}{\partial y} - \frac{\partial f_y}{\partial x}\right)\hat{k} & 0 & \left(\frac{\partial f_z}{\partial y} + \frac{\partial f_y}{\partial z}\right)\hat{\imath} \\ \left(\frac{\partial f_x}{\partial z} + \frac{\partial f_z}{\partial x}\right)\hat{\jmath} & \left(-\frac{\partial f_y}{\partial z} - \frac{\partial f_z}{\partial y}\right)\hat{\imath} & 0 \end{bmatrix} \tag{1.666}$$

The skew-symmetric matrix is an equivalent form for $-\nabla \times \mathbf{f}$:

$$\nabla \times \mathbf{f} = \left[\frac{\partial f_i}{\partial x_j}\hat{\imath}_j\hat{\imath}_i\right]^{\mathrm{T}} - \left[\frac{\partial f_i}{\partial x_j}\hat{\imath}_j\hat{\imath}_i\right] \tag{1.667}$$

***Example 92* div r = 3 *and* grad $f(r) \cdot \mathbf{r} = r\, \partial f/\partial r$** Direct calculation shows that if

$$\mathbf{f} = \mathbf{r} \tag{1.668}$$

then

$$\operatorname{div}\mathbf{r} = \nabla \cdot \mathbf{r} = \left(\frac{\partial}{\partial x}\hat{\imath} + \frac{\partial}{\partial y}\hat{\jmath} + \frac{\partial}{\partial z}\hat{k}\right) \cdot \left(x\,\hat{\imath} + y\,\hat{\jmath} + z\,\hat{k}\right)$$
$$= \frac{\partial x}{\partial x} + \frac{\partial y}{\partial y} + \frac{\partial z}{\partial z} = 3 \tag{1.669}$$

To calculate grad $f(r) \cdot \mathbf{r}$ we use

$$r = \sqrt{x^2 + y^2 + z^2} \tag{1.670}$$

$$\frac{\partial r}{\partial x} = \frac{x}{r} \qquad \frac{\partial r}{\partial y} = \frac{y}{r} \qquad \frac{\partial r}{\partial z} = \frac{z}{r} \tag{1.671}$$

and show that

$$\operatorname{grad} f(r)\cdot\mathbf{r} = \nabla f(r)\cdot\mathbf{r} = \left(\frac{\partial f}{\partial x}\hat{\imath} + \frac{\partial f}{\partial y}\hat{\jmath} + \frac{\partial f}{\partial z}\hat{k}\right)\cdot\left(x\,\hat{\imath} + y\,\hat{\jmath} + z\,\hat{k}\right)$$
$$= x\frac{\partial f}{\partial x} + y\frac{\partial f}{\partial y} + z\frac{\partial f}{\partial z} = x\frac{\partial f}{\partial r}\frac{\partial r}{\partial x} + y\frac{\partial f}{\partial r}\frac{\partial r}{\partial y} + z\frac{\partial f}{\partial r}\frac{\partial r}{\partial z}$$
$$= \frac{x^2}{r}\frac{\partial f}{\partial r} + \frac{y^2}{r}\frac{\partial f}{\partial r} + \frac{z^2}{r}\frac{\partial f}{\partial r} = r\frac{\partial f}{\partial r} \tag{1.672}$$

As an application, consider a vector function field $\mathbf{f}$ that generates a vector $f(r)\mathbf{r}$ at every point of space,

$$\mathbf{f} = f(r)\mathbf{r} \tag{1.673}$$

Divergence of $\mathbf{f}$ would then be

$$\operatorname{div}\mathbf{f} = \operatorname{div}(f(r)\mathbf{r}) = \nabla\cdot f(r)\mathbf{r}$$
$$= \left(\frac{\partial}{\partial x}\hat{\imath} + \frac{\partial}{\partial y}\hat{\jmath} + \frac{\partial}{\partial z}\hat{k}\right)\cdot\left[xf(r)\,\hat{\imath} + yf(r)\,\hat{\jmath} + zf(r)\,\hat{k}\right]$$
$$= \frac{\partial}{\partial x}xf(r) + \frac{\partial}{\partial y}yf(r) + \frac{\partial}{\partial z}zf(r)$$
$$= f(r)\left(\frac{\partial x}{\partial x} + \frac{\partial y}{\partial y} + \frac{\partial z}{\partial z}\right) + x\frac{\partial f(r)}{\partial x} + y\frac{\partial f(r)}{\partial y} + z\frac{\partial f(r)}{\partial z}$$
$$= f(r)\nabla\cdot\mathbf{r} + \nabla f\cdot\mathbf{r} = 3f(r) + r\frac{\partial f(r)}{\partial r} \tag{1.674}$$

Example 93 Second Derivative of a Scalar Field Function The first space derivative of a scalar field function $f = f(\mathbf{r})$ is the gradient of f:

$$\frac{df}{d\mathbf{r}} = \nabla f \tag{1.675}$$

The second space derivative of $f = f(\mathbf{r})$ is

$$\frac{d^2 f}{d\mathbf{r}^2} = \frac{d}{d\mathbf{r}}\left(\frac{df}{d\mathbf{r}}\right) = \nabla(\nabla f) = \nabla\times\nabla f - \nabla\cdot\nabla f = -\nabla^2 f$$
$$= -\left(\frac{\partial^2 f}{\partial x^2} + \frac{\partial^2 f}{\partial y^2} + \frac{\partial^2 f}{\partial z^2}\right) \tag{1.676}$$

Example 94 Trajectory of a Vector Field A space curve $\mathbf{r}(s)$ whose tangent at every point has the same direction as a vector field $\mathbf{v}(\mathbf{r})$ is called a *trajectory* of the field.

The trajectories of the vector field $\mathbf{f} = \nabla\varphi$ are the orthogonal curves to the isosurfaces $\varphi = \mathit{const}$ at every point of space. Therefore, the trajectories are the lines of most rapid change of the function $\varphi = \varphi(t)$.

Consider a stationary velocity field $\mathbf{v}(\mathbf{r})$ of a moving fluid:

$$\mathbf{v} = \mathbf{v}(\mathbf{r}) = \mathbf{v}(x, y, z) \tag{1.677}$$

The trajectory of a velocity vector field is called the *streamline* and shows the path of motion of fluid particles. So, the trajectory of a fluid particle is a space curve $\mathbf{r} = \mathbf{r}(s)$ such that

$$d\mathbf{r} \times \mathbf{v}(\mathbf{r}) = 0 \tag{1.678}$$

or equivalently

$$\frac{dx}{v_x(x, y, z)} = \frac{dy}{v_y(x, y, z)} = \frac{dz}{v_z(x, y, z)} \tag{1.679}$$

Equation (1.678) is the vectorial differential equation of the trajectories of the vector field $\mathbf{v}(\mathbf{r})$. Integration of the differential equation provides the family of trajectories of the field. If the vector field is nonstationary, $\mathbf{v} = \mathbf{v}(\mathbf{r}, t)$, the streamlines change with time and do not necessarily coincide with the actual path of particles at a specific time. So, Equation (1.678) will become

$$d\mathbf{r} \times \mathbf{v}(\mathbf{r}, t) = 0 \tag{1.680}$$

or

$$\frac{dx}{v_x(x, y, z, t)} = \frac{dy}{v_y(x, y, z, t)} = \frac{dz}{v_z(x, y, z, t)} \tag{1.681}$$

If $\mathbf{v}(\mathbf{r}) = 0$ at a point P, Equation (1.678) would be indeterminate. Such a point is called a *singular* point.

Example 95 Time Derivative of Vector Field Consider a time-varying vector field of a vector variable:

$$\mathbf{f} = \mathbf{f}(\mathbf{r}(t)) \tag{1.682}$$

The time derivative of $\mathbf{f}$ is

$$\begin{aligned}\frac{d\mathbf{f}}{dt} &= \frac{df_x}{dt}\hat{\imath} + \frac{df_y}{dt}\hat{\jmath} + \frac{df_z}{dt}\hat{k} \\ &= (\nabla f_x \cdot \mathbf{v})\hat{\imath} + \left(\nabla f_y \cdot \mathbf{v}\right)\hat{\jmath} + (\nabla f_z \cdot \mathbf{v})\hat{k}\end{aligned} \tag{1.683}$$

where $\mathbf{v} = d\mathbf{r}/dt$ is the velocity of position vector $\mathbf{r}$.

Example 96 ★ Laplacian of $\varphi = 1/|\mathbf{r}|$ Consider a scalar field φ that is proportional to the distance from a fixed point. If we set up a Cartesian coordinate frame at the point, then

$$\varphi = \frac{k}{|\mathbf{r}|} \tag{1.684}$$

This is an acceptable model for gravitational and electrostatic fields. The Laplacian of such a field is zero,

$$\nabla^2 \varphi = \nabla^2 \frac{k}{|\mathbf{r}|} = 0 \tag{1.685}$$

because

$$\operatorname{grad} \frac{k}{|\mathbf{r}|} = -\frac{k}{|\mathbf{r}|^2} \operatorname{grad} |\mathbf{r}| = -\frac{k\mathbf{r}}{|\mathbf{r}|^3} \tag{1.686}$$

and therefore,

$$\begin{aligned} \operatorname{div} \operatorname{grad} \frac{k}{|\mathbf{r}|} &= \nabla \cdot \frac{-k\mathbf{r}}{|\mathbf{r}|^3} = -\frac{k}{|\mathbf{r}|^3} \nabla \cdot \mathbf{r} - k\mathbf{r} \cdot \nabla \frac{1}{|\mathbf{r}|^3} \\ &= -3\frac{k}{|\mathbf{r}|^3} + 3\frac{k\mathbf{r}}{|\mathbf{r}|^4} \cdot \frac{\mathbf{r}}{|\mathbf{r}|} \end{aligned} \tag{1.687}$$

Example 97 ★ Tensor Fields Recalling that tensor is a general name for any type of physical quantity, such that a tensor of rank 1 is a scalar, rank 2 is a vector, and rank 3 is a 3×3 matrix, we can define a tensor field as a mathematical rule to assign a unique value of a tensor to each point of a certain domain of space. Traditionally tensor is used to indicate a tensor of rank 2 only.

Stress σ and strain ϵ are examples of the fundamental tensors in solid mechanics. A stress field is defined by

$$\left[\sigma_{ij}(\mathbf{r})\right] = \begin{bmatrix} \sigma_x(\mathbf{r}) & \tau_{xy}(\mathbf{r}) & \tau_{xz}(\mathbf{r}) \\ \tau_{yx}(\mathbf{r}) & \sigma_y(\mathbf{r}) & \tau_{yz}(\mathbf{r}) \\ \tau_{zx}(\mathbf{r}) & \tau_{zy}(\mathbf{r}) & \sigma_z(\mathbf{r}) \end{bmatrix} \tag{1.688}$$

A tensor field may be nonstationary if it is a function of space and time. So, for a nonstationary stress field $\sigma_{ij}(\mathbf{r}, t)$, we may define a stress tensor at a specific instant of time.

Example 98 Gradient of a Scalar Field Makes a Vector Field Consider a scalar field $f = f(\mathbf{r})$. The gradient of f assigns a vector ∇f at any position $\mathbf{r}$, and hence, $\mathbf{f} = \nabla f$ defines a vector field in the same definition domain of $f(\mathbf{r})$.

Example 99 Index Notation and Vector Analysis We may show a function $f = f(x, y, z)$ by $f = f(x_1, x_2, x_3)$ or in general by $f = f(q_1, q_2, q_3)$ to make it proper for index notation. If we show the partial derivative of a scalar field function $f = f(q_1, q_2, q_3)$ with respect to q_i by a comma,

$$\frac{\partial f}{\partial q_i} = f_{,i} \tag{1.689}$$

then it is possible to write the vector analysis operations by index notation:

1. Gradient of a scalar field $f = f(q_1, q_2, q_3)$:

$$\nabla f = \operatorname{grad} f = \sum_{i=1}^{3} f_{,i} \hat{u}_i \tag{1.690}$$

2. Laplacian of a scalar function $f = f(q_1, q_2, q_3)$:

$$\nabla^2 f = \sum_{i=1}^{3} f_{,ii} \tag{1.691}$$

3. Divergence of a vector field $\mathbf{r} = \mathbf{r}(q_1, q_2, q_3)$:

$$\nabla \cdot \mathbf{r} = \operatorname{div} \mathbf{r} = \sum_{i=1}^{3} r_{i,i} \tag{1.692}$$

4. Curl of a vector field $\mathbf{r} = \mathbf{r}(q_1, q_2, q_3)$:

$$\nabla \times \mathbf{r} = \operatorname{curl} \mathbf{r} = \sum_{j=1}^{3}\sum_{k=1}^{3} \epsilon_{ijk} \hat{u}_i r_{j,k} \tag{1.693}$$

Example 100 ★ Nabla Identities If $\mathbf{x}$ and $\mathbf{y}$ are two vector functions of $\mathbf{r}$ and φ is a scalar function of $\mathbf{r}$, then we can verify the following identities:

$$\nabla(\mathbf{x} \cdot \mathbf{y}) = (\nabla \mathbf{x}) \cdot \mathbf{y} + \mathbf{x} \cdot (\nabla \mathbf{y}) \tag{1.694}$$

$$\nabla(\mathbf{x} \times \mathbf{y}) = (\nabla \mathbf{x}) \times \mathbf{y} + \mathbf{x} \times (\nabla \mathbf{y}) \tag{1.695}$$

$$\nabla \cdot \varphi \mathbf{x} = \nabla \varphi \cdot \mathbf{x} + \varphi (\nabla \cdot \mathbf{x}) \tag{1.696}$$

$$\nabla \times \varphi \mathbf{x} = \nabla \varphi \times \mathbf{x} + \varphi \nabla \times \mathbf{x} \tag{1.697}$$

$$\nabla \times \nabla \varphi = 0 \tag{1.698}$$

$$\nabla \cdot (\mathbf{x} \times \mathbf{y}) = (\nabla \times \mathbf{x}) \cdot \mathbf{y} + \mathbf{x} \cdot (\nabla \times \mathbf{y}) \tag{1.699}$$

$$\nabla \cdot (\nabla \times \mathbf{x}) = 0 \tag{1.700}$$

$$\nabla \times (\mathbf{x} \times \mathbf{y}) = (\mathbf{y} \cdot \nabla)\mathbf{x} - (\mathbf{x} \cdot \nabla)\mathbf{y} + \mathbf{x}(\nabla \cdot \mathbf{y}) - \mathbf{y}(\nabla \cdot \mathbf{x}) \tag{1.701}$$

$$\mathbf{x} \times (\nabla \times \mathbf{y}) = \nabla \mathbf{y} \cdot \mathbf{x} - \mathbf{x} \cdot \nabla \mathbf{y} \tag{1.702}$$

$$\nabla \times (\nabla \times \mathbf{x}) = \nabla(\nabla \cdot \mathbf{x}) - \nabla^2 \mathbf{x} \tag{1.703}$$

KEY SYMBOLS

$\mathbf{0}$	zero vector
$a, \ddot{x}, \mathbf{a}, \dot{\mathbf{v}}$	acceleration
a_{ijk}	inner product constant of $\mathbf{x}_i$
$\mathbf{a}, \mathbf{b}, \mathbf{c}, \mathbf{p}, \mathbf{q}$	vectors, constant vectors
$[\mathbf{abc}]$	scalar triple product $\mathbf{a} \cdot (\mathbf{b} \times \mathbf{c})$
A, B	points
A, B, C	axes of triad, constant parameters
A, B, C, D	axes of tetrad, coefficient of a plane equation
$\hat{b}_1, \hat{b}_2, \hat{b}_3$	nonorthogonal unit vectors

$\mathbf{b} = \dot{\mathbf{z}}$	bong
b_{ijk}	inner product of $\mathbf{x}_i$, $\dot{\mathbf{x}}_j$
$B(oxyz)$, B_1, B_2	body coordinate frames
c	constant coefficient
c_i	weight factors of vector addition
c_{ijk}	inner product of $\dot{\mathbf{x}}_i$, $\dot{\mathbf{x}}_j$
$\mathbf{c} = \dot{\mathbf{s}}$	crackle
C	space curve
C_p, C_q	space curves on S for constant q and p at (p_0, q_0)
df	total derivative of f
$d\mathbf{r}$	infinitesimal displacement
ds	arc length element
${}^G\mathbf{d}_o$	position vector of B in G
$[D_2z]$	second-derivative matrix
$f = f(\mathbf{r})$	scalar field function
f, g, h	functions of x, y, z of q_1, q_2, q_3
f_0	isovalue
$f(x, y, z)$	equation of a surface
$f(x, y, z) = f_0$	isosurface of the scalar field $f(\mathbf{r})$ for f_0
$\mathbf{f} = \mathbf{f}(\mathbf{r})$	vector field function
G	gravitational constant
$G(OXYZ)$	global coordinate frame
G_i	kinematic constants of three bodies
$\hat{\imath}$, $\hat{\jmath}$, $\hat{k}$	unit vectors of a Cartesian coordinate frame
$\hat{I}$, $\hat{J}$, $\hat{K}$	unit vectors of a global Cartesian system G
$\mathbf{j}$, $\dot{\mathbf{a}}$, $\ddot{\mathbf{v}}$, $\dddot{\mathbf{r}}$	jerk
$[J]$	Jacobian matrix
k	scalar coefficient
l	a line
n	number of dimensions of an nD space, controlled digit for vector interpolation
$\mathbf{n}$	perpendicular vector to a surface $z = g(x, y)$
$\hat{n}$	perpendicular unit vector
O	origin of a triad, origin of a coordinate frame
$OABC$	a triad with axes A, B, C
$(Ouvw)$	an orthogonal coordinate frame
$(Oq_1q_2q_3)$	an orthogonal coordinate system
P	point, particle
q, p	parameters, variables
$\mathbf{q} = \dot{\mathbf{b}}$	jeeq
$r = \lvert\mathbf{r}\rvert$	length of $\mathbf{r}$
$\mathbf{r}$	position vector
$\mathbf{r}_c$	position vector of curvature center of a space curve
${}_B\mathbf{r}_A$	position vector of A relative to B
$\mathbf{r}_p$, $\mathbf{r}_q$	partial derivatives of ${}^G\mathbf{r}$
$\mathbf{r}_\parallel$, $\mathbf{r}_\perp$	parallel and perpendicular components of $\mathbf{r}$ on l
R	radius
s	arc length parameter
$\mathbf{s} = d\mathbf{j}/dt$	snap, jounce
S	surface
t	time

$T = [\tau_1, \tau_2]$	the set in which a vector function is defined
u, v, w	components of a vector $\mathbf{r}$ in $(Ouvw)$
$\mathbf{u}$	Darboux vector
$\mathbf{u} = \dot{\mathbf{q}}$	sooz
u_1, u_2, u_3	components of $\hat{u}_r$
$\hat{u}^{\mathrm{T}}$	transpose of $\hat{u}$
$\hat{u}_l$	unit vector on a line l
$\hat{u}_r$	a unit vector on $\mathbf{r}$
$\hat{u}_1, \hat{u}_2, \hat{u}_3$	unit vectors along the axes q_1, q_2, q_3
$\hat{u}_r, \hat{u}_\theta, \hat{u}_\varphi$	unit vectors of a spherical coordinate system
$\hat{u}_t, \hat{u}_n, \hat{u}_b$	unit vectors of natural coordinate frame
$\hat{u}_u, \hat{u}_v, \hat{u}_w$	unit vectors of $(Ouvw)$
$\hat{u}_\parallel, \hat{u}_\perp$	parallel and perpendicular unit vectors of l
v	speed
$v, \dot{x}, \mathbf{v}$	velocity
$\mathbf{v}(\mathbf{r})$	velocity field
$\mathbf{v}(\mathbf{r}) = 0$	singular points equation
v	vector space
x, y, z	axes of an orthogonal Cartesian coordinate frame
x_0, y_0, z_0	coordinates of an interested point P
$\mathbf{x}, \mathbf{y}$	vector functions
$\mathbf{x}_i$	relative position vectors of three bodies
X, Y, Z	global coordinate axes
$\mathbf{X}_i$	global position vectors of three bodies
$\mathbf{z} = \ddot{\mathbf{p}}$	larz
Z	short notation symbol

Greek

α	angle between two vectors, angle between $\mathbf{r}$ and l
α, β, γ	directional cosines of a line
$\alpha_1, \alpha_2, \alpha_3$	directional cosines of $\mathbf{r}$ and $\hat{u}_r$
δ_{ij}	Kronecker delta
ϵ	strain
$\epsilon_i = 1/\lvert\mathbf{x}_i\rvert^3$	relative position constant of three bodies
ϵ_{ijk}	Levi-Civita symbol
θ	angle, angular coordinate, angular parameter
κ	curvature
$\boldsymbol{\kappa} = \kappa\hat{u}_n$	curvature vector
ρ	curvature radius
σ	stress tensor, normal stress
$[\sigma_{ij}(\mathbf{r})]$	stress field
τ	curvature torsion, shear stress
$\varphi = \varphi(\mathbf{r})$	scalar field function
ω	angular speed

Symbol

$\cdot$	inner product of two vectors
D	dimension
$\times$	outer product of two vectors
∇	gradient operator
$\nabla\mathbf{f}(\mathbf{r})$	gradient of $\mathbf{f}$

$\nabla \times \mathbf{f} = \text{curl}\,\mathbf{f}$	curl of $\mathbf{f}$
$\nabla \cdot \mathbf{f} = \text{div}\,\mathbf{f}$	divergence of $\mathbf{f}$
$\nabla^2 f$	Laplacian of f
$\nabla f = \text{grad}\, f$	gradient of f
$\mathbf{P} = \dot{\mathbf{c}}$	pop
$\triangle$	difference symbol
$\mathbb{R}$	set of real numbers
$\parallel$	parallel
$\perp$	perpendicular

EXERCISES

1. **Position Vector Characteristics** Three position vectors $\mathbf{r}_1 = OA$, $\mathbf{r}_2 = DB$, and $\mathbf{r}_3 = EB$ are illustrated in Figure 1.31.

   (a) Determine the length of OA, DB, and EB.
   (b) Determine the directional cosines of OA, DB, EB, and AO, BD, and BE.
   (c) Determine the angle between OA and DB.
   (d) Determine a vector to be perpendicular to both OA and DB.
   (e) Determine the surface area of the box by using the vectors OA, DB, and EB.
   (f) Determine the volume of the box by using the vectors OA, DB, and EB.
   (g) Determine the equation of the perpendicular plane to OA, DB, and EB.
   (h) ★ Determine the area of the triangle that is made up by the intersection of the planes in (g) if the plane of OA includes point O, the plane of DB includes point D, and the plane of EB includes point E.

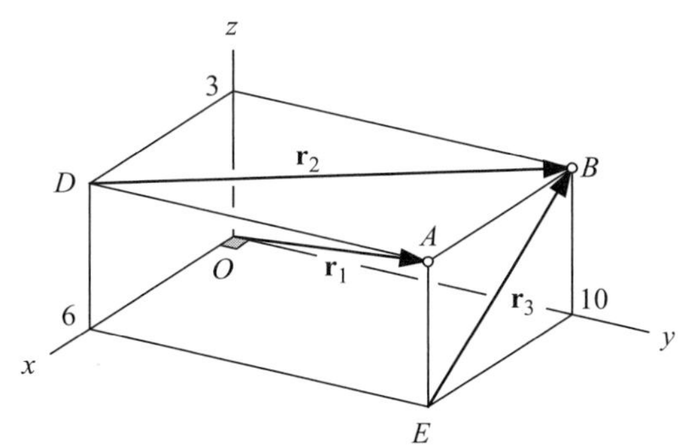

Figure 1.31 Three position vectors OA, DB, and EB.

2. ★ **Independent Orthogonal Coordinate Frames in Euclidean Spaces** In 3D Euclidean space, we need a triad to locate a point. There are two independent and nonsuperposable triads. How many different nonsuperposable Cartesian coordinate systems can be imagined in 4D Euclidean space? How many Cartesian coordinate systems do we have in an nD Euclidean space?

3. **Vector Algebra** Using

$$\mathbf{a} = 2\hat{\imath} - \hat{k} \qquad \mathbf{b} = 2\hat{\imath} - \hat{\jmath} + 2\hat{k} \qquad \mathbf{c} = 2\hat{\imath} - 3\hat{\jmath} + \hat{k}$$

determine

(a) $(\mathbf{a} + \mathbf{b}) \times (\mathbf{a} - \mathbf{b})$ (b) $\mathbf{b} \cdot \mathbf{c} \times \mathbf{a}$

(c) $\mathbf{a} \cdot \mathbf{b} \times \mathbf{c}$ (d) $\mathbf{a} \times \mathbf{b} \times \mathbf{c}$

(e) A unit vector perpendicular to **b** and **c**

4. **Bisector** Assume that in $\triangle OAB$ of Figure 1.32 we have $\angle AOC = \angle BOC$. Show that the vector **c** divides the side AB such that

$$\frac{AC}{CB} = \frac{|\mathbf{a}|}{|\mathbf{b}|}$$

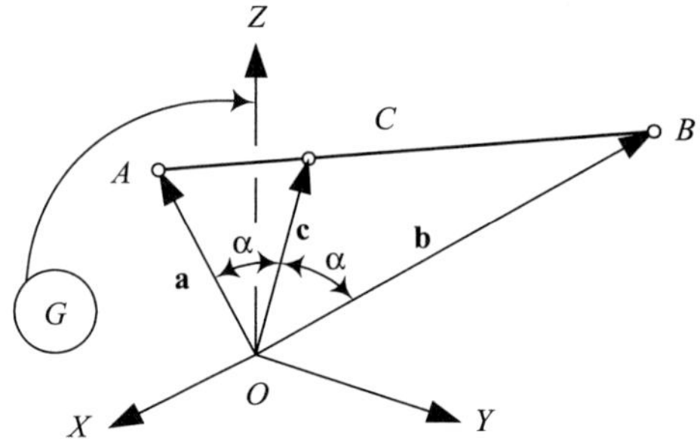

Figure 1.32 The bisector **c** divides the side AB such that $AC/CB = |\mathbf{a}| / |\mathbf{b}|$.

5. **Vector Interpolation** Determine a vector $\mathbf{r} = \mathbf{r}(q)$, $0 \le q \le 1$, to interpolate between two position vectors with the tip points A and B:

(a) $A(1,0,0)$, $B(0,1,0)$ (b) $A(1,1,0)$, $B(-1,1,0)$

(c) ★$A(1,0,0)$, $B(-1,1,0)$

6. **Vectorial Equation** Solve for **x**:

$$a\mathbf{x} + \mathbf{x} \times \mathbf{b} = \mathbf{c}$$

7. **Loci of Tip Point of a Vector** Find the locus of points (x, y, z) such that a vector from point $(2, -1, 4)$ to point (x, y, z) will always be perpendicular to the vector from $(2, -1, 4)$ to $(3, 3, 2)$.

8. **Rotating Triangle** The triangle in Figure 1.33 remains equilateral while point A is moving on an ellipse with a center at O. Assume a corner of the triangle is fixed at O.

(a) What is the path of point B?

(b) What is the area of the triangle?

(c) ★ If the side OA is turning with a constant angular velocity ω, then what is the area of the triangle as a function of time t?

(d) ★ If point A is moving with a constant speed v, then what is the area of the triangle as a function of time t?

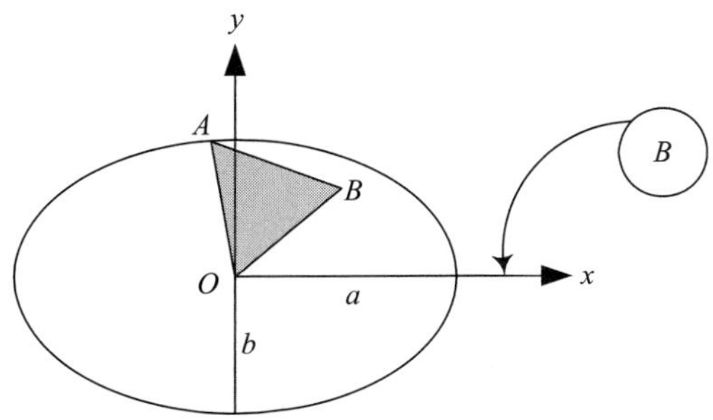

Figure 1.33 A rotating triangle.

9. **Components of an Unknown Vector** Consider a given vector $\mathbf{a}$:

$$\mathbf{a} = \frac{b}{2}\sqrt{x^2 + y^2}\,(\hat{\imath} - \hat{\jmath})$$

Solve the following equations for the components of the vector $\mathbf{r} = x\hat{\imath} + y\hat{\jmath} + z\hat{k}$:

$$\mathbf{r} \times \mathbf{a} = \frac{b}{2}\sqrt{x^2 + y^2}\sinh cy\,\hat{k} \qquad \mathbf{r} \cdot \mathbf{a} = 0$$

10. **Cosine Law** Consider a triangle ABC where its sides are expressed by vectors as

$$\overrightarrow{AB} = \mathbf{c} \qquad \overrightarrow{AC} = \mathbf{b} \qquad \overrightarrow{CB} = \mathbf{a} \qquad \mathbf{c} = \mathbf{a} + \mathbf{b}$$

Use vector algebra and prove the cosine law,

$$c^2 = a^2 + b^2 - 2ab\cos\alpha$$

where

$$\alpha = \angle ACB$$

11. **Trigonometric Equation** Use two planar vectors $\mathbf{a}$ and $\mathbf{b}$ which respectively make angles α and β with the x-axis and prove the following trigonometric equation:

$$\cos(\alpha - \beta) = \cos\alpha\cos\beta + \sin\alpha\sin\beta$$

12. **Spherical Trigonometric Equations** Use vectors to prove the following spherical trigonometric equations in a spherical triangle $\triangle ABC$ with sides a, b, c and angle α, β, γ:

$$\cos a = \cos b\cos c + \sin b\sin c\cos\alpha$$

$$\cos b = \cos c\cos a + \sin c\sin a\cos\beta$$

$$\cos c = \cos a\cos b + \sin a\sin b\cos\gamma$$

13. **Three Colinear Points** Consider three points A, B, and C at $\mathbf{a}$, $\mathbf{b}$, and $\mathbf{c}$. If the points are colinear, then

$$\frac{c_x - a_x}{b_x - a_x} = \frac{c_y - a_y}{b_y - a_y} = \frac{c_z - a_z}{b_z - a_z}$$

Show that this condition can be expressed as

$$(\mathbf{a} \times \mathbf{b}) + (\mathbf{b} \times \mathbf{c}) + (\mathbf{c} \times \mathbf{a}) = 0$$

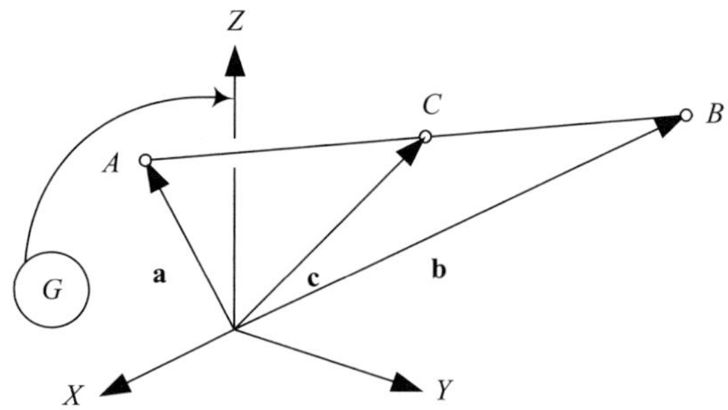

Figure 1.34 Dividing a line in a given ratio.

14. Dividing a Line in a Given Ratio The points A and B are at positions **a** and **b** as shown in Figure 1.34.

(a) Find the position vector **c** of point C that divides the line AB in the ratio of x/y:

$$\frac{AC}{CB} = \frac{x}{y}$$

(b) Show that the equation of a line is

$$\mathbf{r} - \mathbf{a} = k(\mathbf{b} - \mathbf{a})$$

(c) Show that the equation of a plane going through A and B and parallel to a vector **u** is

$$[(\mathbf{r} - \mathbf{a}) \times (\mathbf{b} - \mathbf{a})] \cdot \mathbf{u}$$

(d) Find the equation of a line going through a point A and parallel to a given vector **u**.

15. Volume of a Parallelepiped Consider three points A, B, C and determine the volume of the parallelepiped made by the vectors OA, OB, OC.

(a) $A\,(1, 0, 0)$, $B\,(0, 1, 0)$, $C\,(0, 0, 1)$

(b) $A\,(1, 0, 0)$, $B\,(0, 1, 0)$, C is the center of the parallelepiped in part (a)

(c) ★ $A\,(1, 0, 0)$, $B\,(0, 1, 0)$, C is at a point that makes the volume of the parallelepiped equal to 2. Determine and discuss the possible loci of C.

16. Moving on x-Axis The displacement of a particle moving along the x-axis is given by

$$x = 0.01t^4 - t^3 + 4.5t^2 - 10 \qquad t \geq 0$$

(a) Determine t_1 at which x becomes positive.

(b) For how long does x remain positive after $t = t_1$?

(c) How long does it take for x to become positive for the second time?

(d) When and where does the particle reach its maximum acceleration?

(e) Derive an equation to calculate its acceleration when its speed is given.

17. **Moving on a Cycloid** A particle is moving on a planar curve with the following parametric expression:

$$x = r\,(\omega t - \sin \omega t) \qquad y = r\,(1 - \cos \omega t)$$

(a) Determine the speed of the particle at time t.

(b) Show that the magnitude of acceleration of the particle is constant.

(c) Determine the tangential and normal accelerations of the particle.

(d) Using $ds = v\,dt$, determine the length of the path that the particle travels up to time t.

(e) Check if the magnitude of acceleration of the particle is constant for the following path:

$$x = a\,(\omega t - \sin \omega t) \qquad y = b\,(1 - \cos \omega t)$$

18. **Areal Velocity** Point A in Figure 1.35 is moving on the following circle such that its position vector $\mathbf{r}$ sweeps out with a constant areal velocity h:

$$x^2 - 2Rx + y^2 = 0$$

Determine the velocity and acceleration of the point.

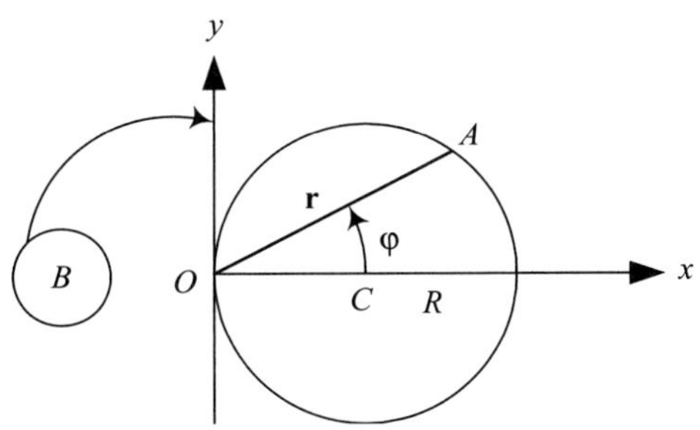

Figure 1.35 A moving point on a circle with constant areal velocity.

19. **Velocity v as a Function of Position x** Determine the acceleration of a particle that is moving according to the following equations:

(a) $v^2 = 2\,(x \sin x + \cos x)$

(b) $v^2 = 2\,(x \sinh x + \cosh x)$

(c) $v^2 = 4x - x^2$

20. **Relative Frequency** Consider a body B that is moving along the x-axis with a constant velocity u and every T seconds emits small particles which move with a constant velocity c along the x-axis. If f denotes the frequency and λ the distance between two successively emitted particles, then we have

$$f = \frac{1}{T} = \frac{c-u}{\lambda}$$

Now suppose that an observer moves along the x-axis with velocity v. Let us show the number of particles per second that the observer meets by the relative frequency f' and the time between meeting the two successive particles by the relative period T', where

$$f' = \frac{c-v}{\lambda}$$

Show that

$$f' \approx f\left(1 - \frac{v-u}{c}\right)$$

21. ★ **A Velocity–Acceleration–Jerk Equation** Show that if the path of motion of a moving particle,

$$\mathbf{r} = \mathbf{r}(t)$$

is such that the scalar triple product of its velocity–acceleration–jerk is zero,

$$\mathbf{v}\cdot(\mathbf{a}\times\mathbf{j}) = 0 \qquad \mathbf{v} = \frac{d\mathbf{r}}{dt} \qquad \mathbf{a} = \frac{d^2\mathbf{r}}{dt^2} \qquad \mathbf{j} = \frac{d^3\mathbf{r}}{dt^3}$$

then $\mathbf{r}(t)$ is a planar curve.

22. **Velocity of End Point of a Stick** Point A of the stick in Figure 1.36 has a constant velocity $\mathbf{v}_A = v\hat{\imath}$ on the x-axis. What is the velocity of point B?

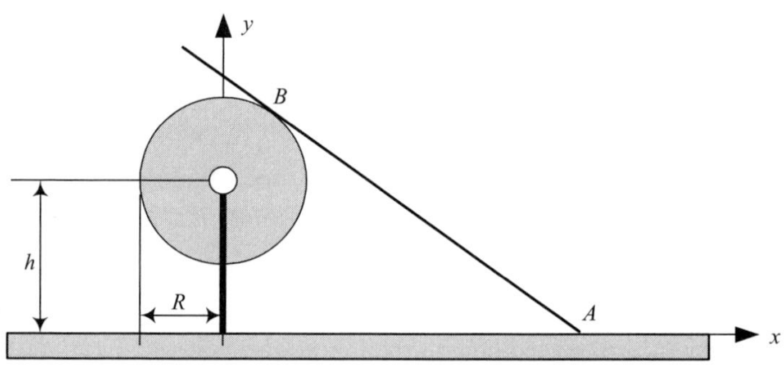

Figure 1.36 A sliding stick.

23. ★ **Disadvantages of a Nonorthogonal Triad** Why do we use an orthogonal triad to define a Cartesian space? Can we define a 3D space with nonorthogonal triads?

24. ★ **Usefulness of an Orthogonal Triad** Orthogonality is the common property of all useful coordinate systems, such as Cartesian, cylindrical, spherical, parabolic, and ellipsoidal coordinate systems. Why do we only define and use orthogonal coordinate systems? Do you think the ability to define a vector based on the inner product and unit vectors of the coordinate system, such as

$$\mathbf{r} = (\mathbf{r}\cdot\hat{\imath})\hat{\imath} + (\mathbf{r}\cdot\hat{\jmath})\hat{\jmath} + (\mathbf{r}\cdot\hat{k})\hat{k}$$

is the main reason for defining the orthogonal coordinate systems?

25. **★ Three Coplanar Vectors** Show that if $\mathbf{a} \times \mathbf{b} \cdot \mathbf{c} = 0$, then $\mathbf{a}$, $\mathbf{b}$, $\mathbf{c}$ are coplanar.

26. **A Derivative Identity** If $\mathbf{a} = \mathbf{a}(t)$ and $\mathbf{b}$ is a constant vector, show that

$$\frac{d}{dt}[\mathbf{a} \cdot (\dot{\mathbf{a}} \times \mathbf{b})] = \mathbf{a} \cdot (\ddot{\mathbf{a}} \times \mathbf{b})$$

27. **Lagrange and Jacobi Identities**

(a) Show that for any four vectors $\mathbf{a}$, $\mathbf{b}$, $\mathbf{c}$, $\mathbf{d}$ the Lagrange identity is correct:

$$(\mathbf{a} \times \mathbf{b}) \cdot (\mathbf{c} \times \mathbf{d}) = (\mathbf{a} \cdot \mathbf{c})(\mathbf{b} \cdot \mathbf{d}) - (\mathbf{a} \cdot \mathbf{d})(\mathbf{b} \cdot \mathbf{c})$$

(b) Show that for any four vectors $\mathbf{a}$, $\mathbf{b}$, $\mathbf{c}$, $\mathbf{d}$ the following identities are correct:

$$(\mathbf{a} \times \mathbf{b}) \times (\mathbf{c} \times \mathbf{d}) = [\mathbf{abd}]\,\mathbf{c} - [\mathbf{abc}]\,\mathbf{d}$$

$$[\mathbf{abc}]\,\mathbf{d} = [\mathbf{dbc}]\,\mathbf{a} + [\mathbf{dca}]\,\mathbf{b} + [\mathbf{dab}]\,\mathbf{c}$$

(c) Show that for any four vectors $\mathbf{a}$, $\mathbf{b}$, $\mathbf{c}$, $\mathbf{d}$ the Jacobi identity is correct:

$$\mathbf{a} \times (\mathbf{b} \times \mathbf{c}) + \mathbf{c} \times (\mathbf{a} \times \mathbf{b}) + \mathbf{b} \times (\mathbf{c} \times \mathbf{a}) = 0$$

28. **★ Flight and Local Time** Figure 1.37 illustrates Earth of radius R with its local coordinate frame E that is turning about the Z-axis of a global coordinate frame G with a constant angular velocity ω. Consider an airplane that is flying at a height h above the spherical Earth. The local time of the airplane is the time of the associated point on Earth right below the airplane. So, the local time of the airplane is determined by its global coordinates. The speed of the airplane $\mathbf{v}$ can be indicated by an angle α with respect to the local constant latitude circle.

(a) An airplane is flying from Tokyo, Japan (35°41′6″N/139°45′5″E), to Tehran, Iran (35°40′19″N/51°25′27″E). What would be the velocity of the plane to have a constant local time. For simplicity assume that both cities are at 35°41′ N.

(b) An airplane is flying from Tehran, Iran (35°40′19″N/51°25′27″E), to Oklahoma City, Oklahoma (35°28′3″N/97°30′58″W). What would be the velocity of the plane to have a constant local time. For simplicity assume that both cities are at 35°40′N.

(c) An airplane is flying from Tehran, Iran (35°40′19″N/51°25′27″E), to Toronto, Canada (43°40′0″N/79°25′0″W). What would be the velocity of the plane to have a constant local time.

(d) An airplane flies from Toronto, Canada (43°40′0″N/79°25′0″W), to Tehran, Iran (35°40′19″N/51°25′27″E). What would be the local time at Tehran if the plane flies with a constant average velocity of part (c) and begins its flight at 1 AM.

(e) An airplane flies from Melbourne, Australia (37°49′0″S/144°58′0″E), to Dubai by the Persian Gulf (25°15′8″N/55°16′48″E). What would be the velocity of the airplane to have a constant local time.

(f) An airplane flies from Melbourne, Australia (37°49′0″S/144°58′0″E), to Dubai by the Persian Gulf (25°15′8″N/55°16′48″E) and returns to Melbourne with no stop. What would be the local time at Melbourne when the airplane is back. Assume the velocity of the airplane on the way to Dubai is such that its local time remains constant and the airplane keeps the same velocity profile on the way back.

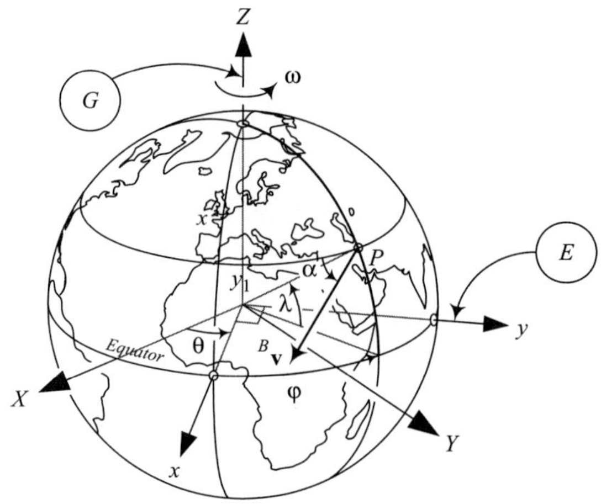

Figure 1.37 Flight and local time on Earth.

29. ★ **Vector Function and Vector Variable** A vector function is defined as a dependent vectorial variable that relates to a scalar independent variable:

$$\mathbf{r} = \mathbf{r}(t)$$

Describe the meaning and define an example for a vector function of a vector variable

$$\mathbf{a} = \mathbf{a}(\mathbf{b})$$

and a scalar function of a vector variable

$$f = f(\mathbf{b})$$

30. ★ **Index Notation** Expand the mass moment of n particles $m_1, m_2, \ldots, m_n$ about a line $\hat{u}$,

$$I_{\hat{u}} = \sum_{i=1}^{n} m_i \left(\mathbf{r}_i \times \hat{u}\right)^2$$

and express $I_{\hat{u}}$ by an index equation.

31. ★ **Frame Dependent and Frame Independent** A vector function of scalar variables is a frame-dependent quantity. Is a vector function of vector variables frame dependent? What about a scalar function of vector variables?

32. ★ **Coordinate Frame and Vector Function** Explain the meaning of ${}^{B}\mathbf{v}_P({}^{G}\mathbf{r}_P)$ if $\mathbf{r}$ is a position vector, $\mathbf{v}$ is a velocity vector, and $\mathbf{v}(\mathbf{r})$ means $\mathbf{v}$ is a function of $\mathbf{r}$.

33. **A Vector Product Identity** Show that for any three vectors $\mathbf{a}$, $\mathbf{b}$, $\mathbf{c}$ in a Cartesian coordinate frame, we have

$$\mathbf{a} \times (\mathbf{b} \times \mathbf{c}) + \mathbf{b} \times (\mathbf{c} \times \mathbf{a}) + \mathbf{c} \times (\mathbf{a} \times \mathbf{b}) = 0$$

34. **Expansion of a Vector with Respect to Two Vectors** Consider two linearly independent vectors $\mathbf{r}_1$ and $\mathbf{r}_2$. Show that every vector $\mathbf{r}_3$ coplanar with $\mathbf{r}_1$ and $\mathbf{r}_2$ has a unique expansion

$$\mathbf{r}_3 = -c_1\mathbf{r}_1 - c_2\mathbf{r}_2$$

35. ★ **Natural Coordinate System of a Parametric Path** Assume the path s can be expressed by a one-parameter function $s = s(\alpha)$, where α is the parameter. Show that

$$\textbf{(a)}\quad \mathbf{r}''s' = s''s'\hat{u}_t + \frac{(s')^3}{\rho}\hat{u}_n$$

$$\textbf{(b)}\quad \hat{u}_n = \frac{\rho}{(s')^3}\left(\mathbf{r}''s' - s''\mathbf{r}'\right)$$

$$\textbf{(c)}\quad \frac{1}{\rho} = \frac{1}{(s')^3}\sqrt{(\mathbf{r}''\cdot\mathbf{r}'')(s')^2 - (\mathbf{r}'\cdot\mathbf{r}'')^2}$$

36. ★ **Natural Coordinate System of a Planar Path** Show that if a planar path is given by a set of equations of a parameter α which is not necessarily the path length

$$x = x(\alpha) \qquad y = y(\alpha)$$

then the natural tangential unit vector and derivatives of the path are

$$\hat{u}_t = \frac{1}{\sqrt{x'^2 + y'^2}}\begin{bmatrix} x' \\ y' \end{bmatrix} \qquad \frac{d\hat{u}_t}{d\alpha} = \frac{x'y'' - x''y'}{x'^2 + y'^2}$$

$$\frac{ds}{d\alpha} = \sqrt{x'^2 + y'^2} \qquad \frac{1}{R} = \frac{\left|x'y'' - x''y'\right|}{\left(x'^2 + y'^2\right)^{3/2}}$$

Use the equations and show that the radius of curvature of the parabola $y = x^2/(4a)$ is

$$R = \frac{4a^2}{\left(4a^2 + x^2\right)^{3/2}}$$

37. ★ **Natural Coordinate System and Important Planes** Consider the space curve

$$x = (10 + 2\sin\theta)\cos\theta \qquad y = (10 + 2\sin\theta)\sin\theta \qquad z = 2 + 2\cos\theta$$

(a) Find the equations of osculating, perpendicular, and rectifying planes and determine them at $\theta = 45^\circ$.

(b) Find the radius and coordinates of the center of curvature of the curve.

38. **Moving on a Given Curve** A particle is moving on a curve $y = f(x)$ such that the x-component of the velocity of the particle remains constant. Determine the acceleration and jerk of the particle.

(a) $y = x^2$

(b) $y = x^3$

(c) $y = e^x$

(d) Determine the angle between velocity vectors of curves (a) and (b) at their intersection.

(e) Determine the exponent n of $y = x^n$ such that the angle between velocity vectors of this curve and curve (a) at their intersection is 45 deg.

39. ★ **A Wounding Cable** Figure 1.38 illustrates a turning cone and wounding cable that supports a hanging box. If the cone is turning with angular velocity ω, determine:

(a) Velocity, acceleration, and jerk of the box

(b) The angular velocity ω such that the velocity of the box remains constant

(c) The angular velocity ω such that the acceleration of the box remains constant

(d) The angular velocity ω such that the jerk of the box remains constant

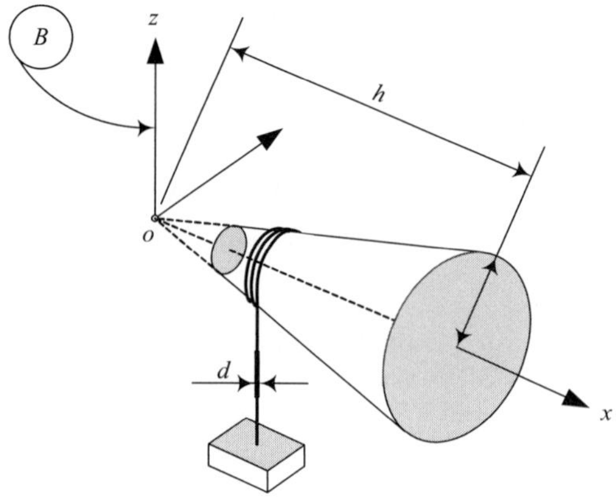

Figure 1.38 A wounding cable on a cone.

40. ★ **Natural Coordinate Unit Vectors** A particle is moving on the following curves. Determine the tangential $\hat{u}_t$, normal $\hat{u}_n$, and binormal $\hat{u}_b$ unit vectors.

(a) $x = \sin\alpha \qquad y = 8x^2$
(b) $x = \cos\alpha \qquad y = -x^2$
(c) $x = \sin^2 t \qquad y = -\sqrt{x}$

41. ★ **Cylindrical Coordinate System and a Helix** A particle is moving on a helix of radius R and pitch a at a constant speed, where

$$z = \frac{a\theta}{2\pi}$$

(a) Express the position, velocity, and acceleration of the particle in the cylindrical coordinate system.

(b) Determine the unit vectors in the cylindrical coordinate system.

(c) Determine the radius of curvature.

42. ★ **Torsion and Curvature of a Helix** A point P is moving with arc length parameter s on a space curve,

$$\mathbf{r}(s) = 10\cos\frac{\sqrt{2}s}{20}\hat{\imath} + 10\sin\frac{\sqrt{2}s}{20}\hat{\jmath} + 10\sqrt{2}s\hat{k} = \begin{bmatrix} 10\cos\dfrac{\sqrt{2}s}{20} \\ 10\sin\dfrac{\sqrt{2}s}{20} \\ 10\sqrt{2}s \end{bmatrix}$$

Determine the curvature κ and torsion τ.

43. **Arc Length Element** Find the square of the element of arc length ds in cylindrical and spherical coordinate systems.

44. **Plane through Three Points** Show that the equation of a plane that includes the three points

$$P_1\ (0, 1, 2) \qquad P_2\ (-3, 2, 1) \qquad P_3\ (1, 0, -1)$$

is

$$4x + 10y - 2z = 6$$

45. Vectorial Operation of Scalar Fields Consider a scalar field $\phi(x, y, z)$,

$$\phi(x, y, z) = \tfrac{1}{2}ax^2 + \tfrac{1}{2}by^2 + c^z$$

(a) Determine grad $\phi(x, y, z) = \nabla\phi(x, y, z)$.

(b) Determine curl grad $\phi(x, y, z) = \nabla \times \nabla\phi(x, y, z)$.

(c) Show that $\nabla \times \nabla\phi = 0$ regardless of the form of ϕ.

(d) Show that $\nabla \cdot (\nabla \times \mathbf{a}) = 0$ regardless of the form of **a**.

(e) Show that $\nabla \cdot (\phi\mathbf{a}) = \nabla\phi \cdot \mathbf{a} + \phi\,(\nabla \cdot \mathbf{a})$.

2

Fundamentals of Dynamics

Laws of motion are experiment-based observations of nature that along with the human-made concept of force are used to model the motion of particles and rigid bodies. In this chapter, we introduce the Newton equation of motion and its application in predicating dynamic phenomena and reviewing the fundamentals of dynamics.

2.1 LAWS OF MOTION

Newton's first law of motion states: Every body persists in a state of rest or of uniform motion in a straight line with a constant velocity unless a force acting upon it changes its state. The first law, also referred to as the *law of force*, defines *force* as an action which tends to change the motion of a body.

Newton's second law of motion states: In a stationary global coordinate frame G, the time rate of change of the momentum of a body is equal to the acting force on the body:

$$\text{Force} = \text{time rate of momentum}$$

The second law, also referred to as the *law of motion*, relates the motion of a body to the causes. The law of motion for a body with constant mass simplifies to

$$\text{Force} = \text{mass} \times \text{acceleration}$$

Newtons law of motion is the foundation for the dynamics of particles and rigid bodies. It is also used to define the *mass* of a body as the ratio of the force acting on the body to the resulting acceleration.

Newton's third law of motion states: When two bodies exert forces on each other, these interacting forces are equal in opposite directions. The third law, also referred to as the *law of reaction*, is important in multibody dynamics.

Example 101 Some Comments on the First Law of Motion The first law defines an inertial frame as the frame in which the first law is correct, and we measure the velocities with respect to the inertial frame. Therefore, every frame in which a force-free body cannot keep its velocity constant is not an inertial frame.

The first law also defines the zero force that is the situation of a force-free body in an inertial frame. However, it does not distinguish between force-free and force-balanced situations.

According to the first law, everything that can change the state of motion is a force. However, it does not provide a clear definition of force independent of motion.

Example 102 Sir Arthur Eddington (1882–1944) Statement According to Eddington's comment, the first law of motion says nothing but: "Every body persists in a state of rest or of uniform motion in a straight line with a constant velocity, unless it does not."

Example 103 Mass, Particle, and Impenetrability We assume a *particle* P is a particular point in three-dimensional Euclidean space where we are interested in its position. The particle permanently carries with it a label m called the *mass* of P. We may also call such a particle a point mass.

Every particle occupies one and only one position in space at any given instant of time t. Particles have the *impenetrability* property, which states: If there exists a single time $t = t_0$ at which any two particles P_1 and P_2 have different positions $\mathbf{r}_1$ and $\mathbf{r}_2$, then their positions never coincide, that is, $\mathbf{r}_1 \neq \mathbf{r}_2\ \forall t$. If there exists a single time $t = t_0$ at which any two particles P_1 and P_2 have the same positions $\mathbf{r}_1 = \mathbf{r}_2$, then their positions coincide permanently, that is, $\mathbf{r}_1 = \mathbf{r}_2\ \forall t$.

Example 104 Some Comments on the Second Law of Motion To discover the second law,

$$\mathbf{F} = \frac{d}{dt}\mathbf{p} \tag{2.1}$$

we should recall that *momentum* is defined as

$$\mathbf{p} = m\mathbf{v} \tag{2.2}$$

When the mass is constant, which is applied to most dynamic problems, the second law reduces to

$$\mathbf{F} = m\mathbf{a} \tag{2.3}$$

where $\mathbf{a} = d\mathbf{v}/dt = d^2\mathbf{r}/dt^2$. It says that the force vector $\mathbf{F}$ and the acceleration vector $\mathbf{a}$ are colinear and proportional. The ratio of $\mathbf{F}/\mathbf{a}$ is called the *mass* m. So, it provides a tool to measure masses because it states that, when the same force $\mathbf{F}$ applies on two particles with masses m_1 and m_2,

$$\frac{a_1}{a_2} = \frac{m_2}{m_1} \tag{2.4}$$

We may use this relation to determine the mass of a particle with respect to a standard mass by comparing their accelerations under the same force.

Being a vectorial equation helps us to decompose the second law in three orthogonal directions along the three axes of the inertial frame that have three scalar equations instead.

The second law also says that the applied force on a mass is proportional to the second (not the first or third) time derivative of its position vector. This fact is the main reason why in the kinematic analysis of dynamic systems we usually talk only about position, velocity, and acceleration.

Although it is not mentioned in the expression of the second law, it is correct only in an inertial frame. In other words, the coordinate frame in which the second law works is an inertial frame.

Example 105 Dynamic Problems A dynamic problem may start from the second law with a given acceleration to search for the position or the problem may start from a given position and ask for the required forces.

In the first type, the resultant applied force on a particle is specified, and the motion is questioned.

In the second type, it is desired to have a particle move in a specified manner, and the required forces are questioned.

Example 106 ★ Limits of the Second Law The second law is the relation between the force $\mathbf{F}$, mass m, position $\mathbf{r}$, and time t. It works as long as the order of magnitude of these parameters is neither very small nor very large relative to human senses. For instance, the second law provides little information for very small and very large masses.

When m is very small, we need quantum mechanics, and when m is too large, we need general relativity. Both of these sciences were discovered and developed when scientists observed poor predications based on the second law.

Newton's second law is the law of motion of massive bodies in Euclidean space along with these assumption that time flows smoothly and independently. However, mass is the thing that makes the space non-Euclidean, and time is dependent on the speed of the mass. So, the second law is not correct in general but it is a very good approximation from an engineering viewpoint.

Example 107 Some Comments on the Third Law of Motion Newton was very brave in stating that we apply the same amount of force on Earth as Earth applies on us.

Today, we can define two types of the third law: weak and strong, or general and constrained. The weak type of the third law states that the forces that the two particles apply on each other are equal and opposite, but not necessarily on the connecting line of the particles. The strong type of the third law states that the forces that two particles apply on each other are equal, opposite, and on the connecting line of the particles. An example of the weak and strong forms of the third law are shown in Figure 2.1.

The third law may also be interpreted as the conservation law of momentum. Consider an isolated system of two masses m_1 and m_2. An *isolated system* does not interact with anything out of the system. The rate of the momentum of the system is

$$\mathbf{F}_{12} + \mathbf{F}_{21} = \frac{d\mathbf{p}}{dt} = \frac{d\left(m_1\mathbf{v}_1 + m_2\mathbf{v}_2\right)}{dt} \tag{2.5}$$

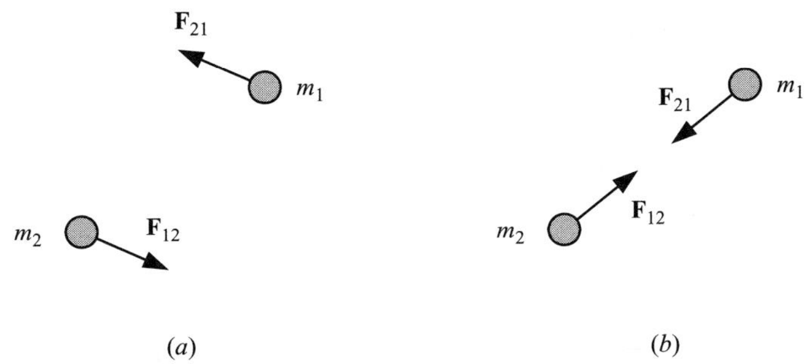

Figure 2.1 The third law of motion: *(a)* weak form; *(b)* strong form.

where $\mathbf{F}_{12}$ is the force that m_1 applies on m_2 and $\mathbf{F}_{21}$ is the force that m_2 applies on m_1. By the third law we have $\mathbf{F}_{12} + \mathbf{F}_{21} = 0$, and therefore, the momentum of the system, $\mathbf{p}$, conserves:

$$m_1\mathbf{v}_1 + m_2\mathbf{v}_2 = \mathbf{c} \tag{2.6}$$

A conserved physical quantity will not change with a change in time.

Example 108 ★ The Practical Newton Laws In modern texts, Newton's laws of motion have been translated and modified to match today's language of science. These laws were originally in the following forms:

> Law 1: *Corpus omne perseverate in statu suo quiscendi vel movendi uniformiter in directum, nisi quantenus illud a viribus impressis cogitur statum suum mutare*. "Any body preserves its state of rest or of uniform rectilinear motion if it is not constrained by induced forces to change its state."
>
> Law 2: *Mutationem motus proportionalem esse vi vel motrici impresse et fieri secundum lineam rectam, qua vis illa imprimitur*. "The variation of motion is proportional to the induced moving force and is directed along the straight line, which is the support of this induced force."
>
> Law 3: *Actioni contrariam semper et aequalem esse reactionem, sive corporum duorum actiones in se muto semper esse aequales at in partes contrarias dirigi*. "The reaction is always opposite and equal to the action or the reciprocal actions of two bodies are always equal and directed in contrary directions."

Example 109 ★ Force Function in Equation of Motion The first and second laws provide no predication or expectation of the force function and its arguments. Qualitatively, force is whatever changes the motion, and quantitatively, force is whatever is equal to mass times acceleration. Mathematically, the equation of motion provides a vectorial second-order differential equation:

$$m\ddot{\mathbf{r}} = \mathbf{F}(\dot{\mathbf{r}}, \mathbf{r}, t) \tag{2.7}$$

We assume that the force function may generally be a function of time t, position $\mathbf{r}$, and velocity $\dot{\mathbf{r}}$. In other words, the Newton equation of motion is correct as long as we can show that force is only a function of $\dot{\mathbf{r}}, \mathbf{r}, t$.

If there is a force that depends on the acceleration, jerk, or other variables that cannot be reduced to $\dot{\mathbf{r}}, \mathbf{r}, t$, we do not know the equation of motion because

$$\mathbf{F}\left(\mathbf{r}, \dot{\mathbf{r}}, \ddot{\mathbf{r}}, \dddot{\mathbf{r}}, \ldots, t\right) \neq m\ddot{\mathbf{r}} \tag{2.8}$$

So, in Newtonian mechanics, we assume that force can only be a function of $\dot{\mathbf{r}}, \mathbf{r}, t$ and nothing else. In the real world, force may be a function of everything; however, we always ignore any other variables than $\dot{\mathbf{r}}, \mathbf{r}, t$, or make some approximations accordingly.

Because Equation (2.7) is a linear equation of force $\mathbf{F}$, it accepts the superposition principle. When a mass m is affected by several forces $\mathbf{F}_1, \mathbf{F}_2, \mathbf{F}_3, \ldots$, we may calculate their summation vectorially,

$$\mathbf{F} = \mathbf{F}_1 + \mathbf{F}_2 + \mathbf{F}_3 + \cdots \tag{2.9}$$

and apply the resultant force on m. So, if a force $\mathbf{F}_1$ provides acceleration $\ddot{\mathbf{r}}_1$ and $\mathbf{F}_2$ provides $\ddot{\mathbf{r}}_2$,

$$m\ddot{\mathbf{r}}_1 = \mathbf{F}_1 \tag{2.10}$$

$$m\ddot{\mathbf{r}}_2 = \mathbf{F}_2 \tag{2.11}$$

then the resultant force $\mathbf{F}_3 = \mathbf{F}_1 + \mathbf{F}_2$ provides the acceleration $\ddot{\mathbf{r}}_3$ such that

$$\ddot{\mathbf{r}}_3 = \ddot{\mathbf{r}}_1 + \ddot{\mathbf{r}}_2 \tag{2.12}$$

To see that the Newton equation of motion is not correct when the force is not only a function of $\dot{\mathbf{r}}, \mathbf{r}, t$, let us assume that a particle with mass m is under two acceleration-dependent forces $F_1(\ddot{x})$ and $F_2(\ddot{x})$ on the x-axis:

$$m\ddot{x}_1 = F_1(\ddot{x}_1) \tag{2.13}$$

$$m\ddot{x}_2 = F_2(\ddot{x}_2) \tag{2.14}$$

The acceleration of m under the action of both forces would be $\ddot{x}_3$,

$$m\ddot{x}_3 = F_1(\ddot{x}_3) + F_2(\ddot{x}_3) \tag{2.15}$$

however, we must have

$$\ddot{x}_3 = \ddot{x}_1 + \ddot{x}_2 \tag{2.16}$$

but we have

$$m\left(\ddot{x}_1 + \ddot{x}_2\right) = F_1(\ddot{x}_1 + \ddot{x}_2) + F_2(\ddot{x}_1 + \ddot{x}_2) \neq F_1(\ddot{x}_1) + F_2(\ddot{x}_2) \tag{2.17}$$

Example 110 ★ Employing Newton's First Law Consider a one-dimensional motion of a particle with $m = 1$ and the equation of motion

$$F = 6x^2 = \ddot{x} \tag{2.18}$$

starting with the initial conditions

$$x_0 = x(0) = 0 \qquad \dot{x}_0 = \dot{x}(0) = 0. \tag{2.19}$$

This equation may have three solutions,

$$x = 0 \qquad x = t^{-2} \qquad x = -t^{-2} \tag{2.20}$$

that satisfy both the equation of motion and the initial conditions. However, according to the first law of motion, only $x = 0$ is an actual solution because at $t = 0$ we have $x = 0$ and therefore no force will apply on the particle. It must remain in its initial position for any time $t > 0$.

Any particle with equation of motion

$$\ddot{x} = an(n-1)\left(\frac{x}{a}\right)^{(n-2)/n} \tag{2.21}$$

may have three solutions:

$$x = 0 \qquad x = at^n \qquad x = -t^{-2} \tag{2.22}$$

The actual solution will be determined by initial conditions.

Example 111 ★ Integral Form of Equation of Motion The law of motion (2.7) is a differential equation. However, we may also express the law of motion with an integral equation

$$m\mathbf{v} = \int_{t_0}^{t} \mathbf{F}(\mathbf{v}, \mathbf{r}, t)\, dt + m\mathbf{v}(t_0) \tag{2.23}$$

where $\mathbf{v}(t_0) = \mathbf{v}_0$ is the initial velocity vector. The integral is called the *impulse* of $\mathbf{F}$ in that time interval. This equation states that the change of momentum $\Delta\mathbf{p} = m\mathbf{v} - m\mathbf{v}_0$ of a particle during the time interval $[t_0, t]$ is equal to the impulse of the resultant force over the same interval.

The differential form of the second law of motion (2.7), or force equals the time rate of momentum, has some disadvantages over the integral form (2.23), or impulse equals momentum change. The position of a particle must be continuous; however, its velocity might have discontinuities. At a velocity discontinuity, the acceleration is not defined, so, at such instants, the differential form of the second law is not valid; however the integral form (2.23) is valid at these instants.

2.2 EQUATION OF MOTION

To find the equation of motion

$$\mathbf{F}(\mathbf{v}, \mathbf{r}, t) = m\mathbf{a} \tag{2.24}$$

we only need to determine the applied force $\mathbf{F}$ and set up the differential equation. Having the Newton equation (2.24), our goal would be to solve the equation of motion and find the position $\mathbf{r}$ as a function of time $\mathbf{r} = \mathbf{r}(t)$.

2.2.1 Force and Moment

In Newtonian mechanics, the acting forces on a system of bodies can be divided into *internal* and *external forces*. Internal forces are acting between bodies of the system, and external forces are acting from outside of the system. Forces can also be divided into *contact* and *body forces*.

External forces and moments are called *load*. The acting forces and moments on a rigid body are called a *force system*. The *resultant* or *total force* **F** is the sum of all the external forces acting on the body, and the *resultant* or *total moment* **M** is the sum of all the moments of the external forces about a point, such as the origin of a coordinate frame:

$$\mathbf{F} = \sum_i \mathbf{F}_i \tag{2.25}$$

$$\mathbf{M} = \sum_i \mathbf{M}_i \tag{2.26}$$

The moment **M** of a force **F** acting at a point P with position vector $\mathbf{r}_P$ about a point Q at $\mathbf{r}_Q$ is

$$\mathbf{M}_Q = \left(\mathbf{r}_P - \mathbf{r}_Q\right) \times \mathbf{F} \tag{2.27}$$

and therefore, the moment of **F** about the origin is

$$\mathbf{M} = \mathbf{r}_P \times \mathbf{F} \tag{2.28}$$

Consider a directional line l and a force **F** acting on a point P at $\mathbf{r}_P$. The line is passing through the origin and its direction is indicated by a unit vector $\hat{u}$. The *moment of the force* **F** about the line l is

$$\mathbf{M}_l = l\hat{u} \cdot (\mathbf{r}_P \times \mathbf{F}) \tag{2.29}$$

The moment of a force may also be called *moment* for simplicity.

The effect of a force system acting on a rigid body is equivalent to the effect of the resultant force and resultant moment of the force system. Any two force systems are equivalent if their resultant forces and resultant moments are equal respectively. If the resultant force of a force system is zero, then the resultant moment of the force system is independent of the origin of the coordinate frame. Such a resultant moment is called *torque*.

Forces are veclines and they can slide on their line of action. So, pulling a body from one side or pushing the body from the other side produces the same external motion. Moments are vecfree and can move parallel to themselves. The magnitude of moment of a force is dependent on the distance between the origin of the coordinate frame and the line of action of the force.

When a force system is reduced to a resultant $\mathbf{F}_P$ and $\mathbf{M}_P$ with respect to a reference point P, we may change the reference point to another point Q and find the new resultants as

$$\mathbf{F}_Q = \mathbf{F}_P \tag{2.30}$$

$$\mathbf{M}_Q = \mathbf{M}_P + \left(\mathbf{r}_P - \mathbf{r}_Q\right) \times \mathbf{F}_P = \mathbf{M}_P + {}_Q\mathbf{r}_P \times \mathbf{F}_P \tag{2.31}$$

where ${}_Q\mathbf{r}_P$ is the position vector of point P with respect to point Q.

Example 112 Every Force System Is Equivalent to a Wrench Poinsot (1777–1859) proved a theorem that says: Every force system is equivalent to a single force plus a moment parallel to the force. A force and a moment about the force axis is called a *wrench*. Poinsot's theorem is similar to Chasles's (1793–1880) theorem: Every rigid-body motion is equivalent to a screw, which is a translation plus a rotation about the axis of translation.

Let $\mathbf{F}$ and $\mathbf{M}$ be the resultant force and moment of a force system. We may decompose the moment into parallel and perpendicular components, $\mathbf{M}_{\|}$ and $\mathbf{M}_{\perp}$, to the force axis. The perpendicular component $\mathbf{M}_{\perp}$ will change when the axis of force $\mathbf{F}$ changes. If Q is a point on the axis of $\mathbf{F}$, by translating the axis of $\mathbf{F}$ to point P such that ${}_Q\mathbf{r}_P$ satisfies

$$\mathbf{M}_{\perp} = {}_Q\mathbf{r}_P \times \mathbf{F} \tag{2.32}$$

we make $\mathbf{M}_{\perp} = 0$. So, the force system of $\mathbf{M}$ and $\mathbf{F}$ at Q can be replaced by force system $\mathbf{M}_{\|}$ and $\mathbf{F}$ at P, and therefore, the force system is reduced to a force $\mathbf{F}$ and a moment $\mathbf{M}_{\|}$ parallel to each other.

Example 113 Newton Law of Gravitation There exists an attractive force between every two massive particles m_1 and m_2. The attraction is called *gravitation* and its associated force is called *gravitational force*. Newton presented a mathematical equation to model the gravitational force of m_1 and m_2 on each other:

$$\mathbf{F}_{21} = -Gm_1m_2\frac{\mathbf{r}_1 - \mathbf{r}_2}{|\mathbf{r}_1 - \mathbf{r}_2|^3} \tag{2.33}$$

$$\mathbf{F}_{12} = -Gm_1m_2\frac{\mathbf{r}_2 - \mathbf{r}_1}{|\mathbf{r}_2 - \mathbf{r}_1|^3} \tag{2.34}$$

where $G = 6.67259 \times 10^{-11}\ \mathrm{m^3\,kg^{-1}s^{-2}}$ is called the *universal gravitational constant*, $\mathbf{F}_{ij}$ is the force of i on j and indicates the force that m_i applies on m_j, and $\mathbf{r}_1$, $\mathbf{r}_2$ are the position vectors of the masses in a global coordinate frame G as are shown in Figure 2.2.

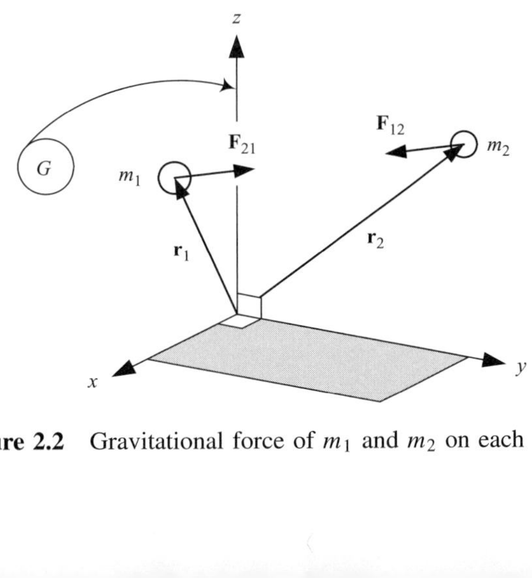

Figure 2.2 Gravitational force of m_1 and m_2 on each other.

Because of the third law of motion, we have

$$\mathbf{F}_{21} + \mathbf{F}_{12} = 0 \tag{2.35}$$

So, the resultant force on m_1 and m_2 is zero, although each feels a force acting upon it.

Following the gravitational law (2.33), we can calculate the resultant force $\mathbf{F}_1$ of n particles m_i, $i = 1, \ldots, n$, on m_1:

$$\mathbf{F}_1 = -Gm_1 \sum_{i=1}^{n} m_i \frac{\mathbf{r}_1 - \mathbf{r}_i}{|\mathbf{r}_1 - \mathbf{r}_i|^3} \tag{2.36}$$

Example 114 Position of Center of Mass The position of the mass center C of a rigid body or a system of particles in a coordinate frame is indicated by $\mathbf{r}_C$. The mass center C is usually measured in a local coordinate frame.

Consider n particles with masses m_i at positions $\mathbf{r}_i$, $i = 1, \ldots, n$. The mass center is a point at $\mathbf{r}_C$ defined by

$$\mathbf{r}_C = \frac{1}{m_C} \sum_{i=1}^{n} m_i \mathbf{r}_i \tag{2.37}$$

$$\begin{bmatrix} x_C \\ y_C \\ z_C \end{bmatrix} = \begin{bmatrix} \frac{1}{m_C} \sum_{i=1}^{n} m_i x_i \\ \frac{1}{m_C} \sum_{i=1}^{n} m_i y_i \\ \frac{1}{m_C} \sum_{i=1}^{n} m_i z_i \end{bmatrix} \tag{2.38}$$

where $m_C = \sum_{i=1}^{n} m_i$ is the total mass of the particles and $\sum_{i=1}^{n} m_i x_i$, $\sum_{i=1}^{n} m_i y_i$, and $\sum_{i=1}^{n} m_i z_i$ are the sums of the moments of m_i about the (y, z)-, (z, x)-, and (x, y)-planes, respectively. In a rigid body, the summation will become integral over the whole rigid body B:

$$\mathbf{r}_C = \frac{1}{m_C} \int_B \mathbf{r} \, dm \tag{2.39}$$

$$\begin{bmatrix} x_C \\ y_C \\ z_C \end{bmatrix} = \begin{bmatrix} \frac{1}{m_C} \int_B x \, dm \\ \frac{1}{m_C} \int_B y \, dm \\ \frac{1}{m_C} \int_B z \, dm \end{bmatrix} \tag{2.40}$$

Example 115 ★ Position of Center of Gravity The gravity center $\mathbf{r}_G$ of a body with mass m is the point at which we must put the whole mass m to be attracted by the same gravitational force of point mass m_0 as the distributed mass of the body:

$$\mathbf{F} = -Gm_0m\frac{\mathbf{r}_G}{r_G^3} = -Gm_0\int_B \frac{\mathbf{r}}{r^3}\,dm \tag{2.41}$$

$$r = |\mathbf{r}| \qquad r_G = |\mathbf{r}_G| \tag{2.42}$$

To determine the location of the gravity center, we have to calculate the result of a spatial integral over the whole body B:

$$\frac{\mathbf{r}_G}{r_G^3} = \frac{1}{m}\int_B \frac{\mathbf{r}}{r^3}\,dm \tag{2.43}$$

This integral cannot be calculated explicitly for a general body, although it can be solved for some special cases.

Consider a uniform rod satellite with mass m as shown in Figure 2.3. Point C indicates the mass center, and point G indicates the gravity center of the rod. Because the rod satellite is a one-dimensional rigid body, Equation (2.43) becomes

$$\frac{1}{r_G^2} = \frac{1}{m}\int_B \frac{1}{r^2}\,dm = \frac{1}{l}\int_{r_C-l/2}^{r_C+l/2} \frac{1}{r^2}\,dr = \frac{4}{4r_C^2 - l^2} \tag{2.44}$$

and the position of the gravity center is at

$$r_G = r_C\sqrt{1 - \frac{l^2}{4r_C^2}} \tag{2.45}$$

The gravity center r_G is closer than the mass center $\mathbf{r}_C$ to the point planet m_0.

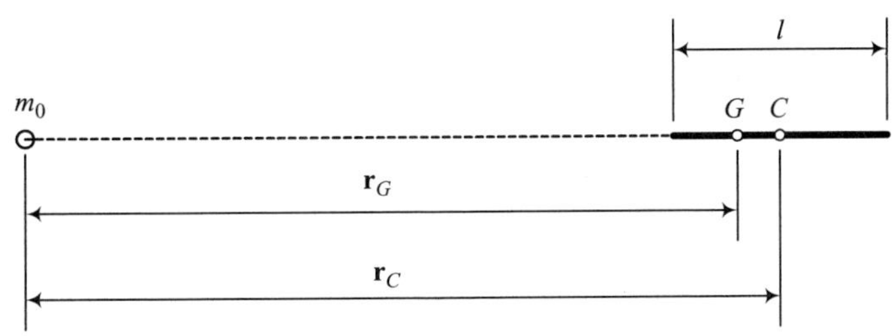

Figure 2.3 A uniform rod satellite with mass m attracteed by a point mass m_0.

The position of the gravity center is not a constant point in the body. It moves based on the orientation of the body and its distance from the source of attraction. Figure 2.4 illustrates a rod satellite at an angle θ with respect to the position vector of its mass center from the planet m_0. To determine the gravity center $\mathbf{r}_G$, we may decompose Equation (2.43) in the coordinate frame (X, Y):

$$\frac{X_G\hat{\imath} + Y_G\hat{\jmath}}{r_G^3} = \frac{1}{m}\int_B \frac{x\hat{\imath} + y\hat{\jmath}}{r^3}\,dm \tag{2.46}$$

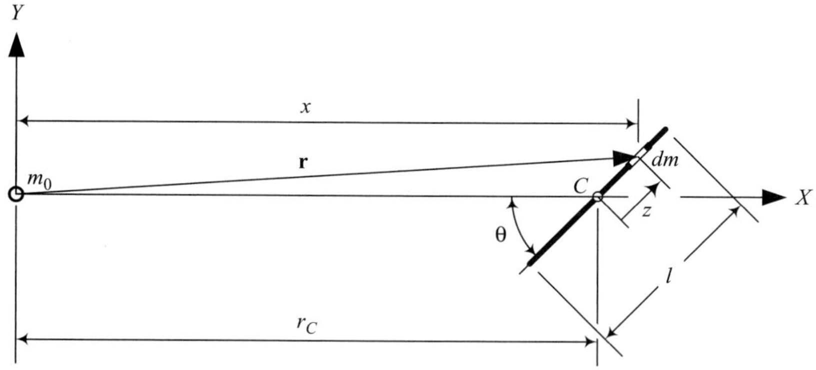

Figure 2.4 A rod satellite at an angle θ with the position vector of mass center.

Using $dm = (m/l)\,dz$ and

$$x = r_C + z\cos\theta \qquad y = z\sin\theta \tag{2.47}$$

$$r = \sqrt{r_C^2 + 2r_C z\cos\theta + z^2} \tag{2.48}$$

we can find the X- and Y-components of $\mathbf{r}_G$:

$$\begin{aligned}\frac{X_G}{r_G^3} &= \frac{1}{l}\int_{-l/2}^{l/2} \frac{r_C + z\cos\theta}{\left(r_C^2 + 2r_C z\cos\theta + z^2\right)^{3/2}}\,dz \\ &= \frac{1}{r_C^2}\left(\frac{1}{\sqrt{\dfrac{l^2}{r_C^2} + \dfrac{4l}{r_C}\cos\theta + 4}} + \frac{1}{\sqrt{\dfrac{l^2}{r_C^2} - \dfrac{4l}{r_C}\cos\theta + 4}}\right)\end{aligned} \tag{2.49}$$

$$\begin{aligned}\frac{Y_G}{r_G^3} &= \frac{1}{l}\int_{-l/2}^{l/2} \frac{z\sin\theta}{\left(r_C^2 + 2r_C z\cos\theta + z^2\right)^{3/2}}\,dz \\ &= -\frac{1}{l r_C\sin\theta}\left(\frac{2 + \dfrac{l}{r_C}\cos\theta}{\sqrt{\dfrac{l^2}{r_C^2} + \dfrac{4l}{r_C}\cos\theta + 4}} - \frac{2 - \dfrac{l}{r_C}\cos\theta}{\sqrt{\dfrac{l^2}{r_C^2} - \dfrac{4l}{r_C}\cos\theta + 4}}\right)\end{aligned} \tag{2.50}$$

Let us define the dimensionless parameters α and β as

$$\alpha = \frac{l}{r_C} \qquad \beta = \frac{r_G}{r_C} \tag{2.51}$$

and rewrite Equations (2.49) and (2.50) as

$$\beta^2 \frac{X_G}{r_G} = \frac{1}{\sqrt{\alpha^2 + 4\alpha\cos\theta + 4}} + \frac{1}{\sqrt{\alpha^2 - 4\alpha\cos\theta + 4}} \tag{2.52}$$

$$\beta^2 \frac{Y_G}{r_G} = -\frac{1}{\alpha\sin\theta}\left(\frac{2+\alpha\cos\theta}{\sqrt{\alpha^2 + 4\alpha\cos\theta + 4}} - \frac{2-\alpha\cos\theta}{\sqrt{\alpha^2 - 4\alpha\cos\theta + 4}}\right) \tag{2.53}$$

or combine them as

$$\beta^4 = \frac{1}{\alpha^2\sin^2\theta}\left(\frac{2+\alpha\cos\theta}{\sqrt{\alpha^2 + 4\alpha\cos\theta + 4}} - \frac{2-\alpha\cos\theta}{\sqrt{\alpha^2 - 4\alpha\cos\theta + 4}}\right)^2 + \left(\frac{1}{\sqrt{\alpha^2 + 4\alpha\cos\theta + 4}} + \frac{1}{\sqrt{\alpha^2 - 4\alpha\cos\theta + 4}}\right)^2 \tag{2.54}$$

Figure 2.5 depicts the function $\beta = \beta(\alpha, \theta)$ for $0 < \alpha < 0.5$ and $-\pi/2\,\text{rad} < \alpha < \pi/2\,\text{rad}$. The difference between r_G and r_C fluctuates more for longer rod satellites when the orientation of the rod satellite changes.

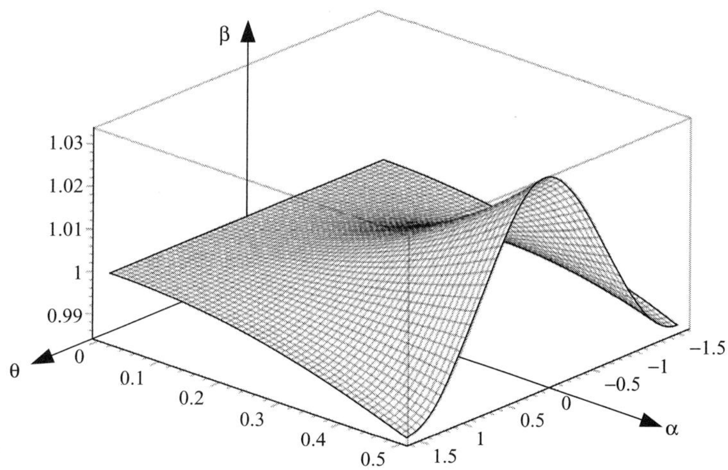

Figure 2.5 The function $\beta = r_G/r_C = \beta(\alpha = l/r_C, \theta)$ for $0 < \alpha < 0.5$ and $-\pi/2\,\text{rad} < \alpha < \pi/2\,\text{rad}$.

2.2.2 Motion Equation

The *momentum* of a body is a vector quantity proportional to the total mass of the body times the translational velocity of the mass center C of the body:

$$\mathbf{p} = m\mathbf{v} \tag{2.55}$$

The momentum is also called *linear momentum* or *translational momentum*. The *moment of momentum* is given as

$$\mathbf{L} = \mathbf{r}_C \times \mathbf{p} \tag{2.56}$$

Consider a directional line l that is passing through the origin. The moment of momentum about l is

$$\mathbf{L}_l = l\hat{u} \cdot (\mathbf{r}_C \times \mathbf{p}) \tag{2.57}$$

where $\hat{u}$ is a unit vector indicating the direction of the line, $\mathbf{r}_C$ is the position vector of the mass center C in the global coordinate frame G, and $\mathbf{L} = \mathbf{r}_C \times \mathbf{p}$ is the moment of momentum about the origin of G. The moment of momentum may also be called *angular momentum*.

The application of a force system is emphasized by the *Newton equation of motion* which states that the global rate of change of the *translational momentum* is proportional to the global *applied force*:

$${}^G\mathbf{F} = \frac{{}^Gd}{dt}\,{}^G\mathbf{p} = \frac{{}^Gd}{dt}\left(m\,{}^G\mathbf{v}\right) \tag{2.58}$$

The law of motion can be expanded to include rotational motion. Hence, the second law of motion also states that the global rate of change of the *angular momentum* is proportional to the global *applied moment*:

$${}^G\mathbf{M} = \frac{{}^Gd}{dt}\,{}^G\mathbf{L} = \frac{{}^Gd}{dt}\,{}^G(\mathbf{r}_C \times \mathbf{p}) \tag{2.59}$$

Proof: Differentiating from the angular momentum shows that

$$\begin{aligned}\frac{{}^Gd}{dt}\,{}^G\mathbf{L} &= \frac{{}^Gd}{dt}\,{}^G(\mathbf{r}_C \times \mathbf{p}) = \left(\frac{{}^Gd\mathbf{r}_C}{dt} \times \mathbf{p} + \mathbf{r}_C \times \frac{{}^Gd\mathbf{p}}{dt}\right) \\ &= {}^G\mathbf{r}_C \times \frac{{}^Gd\mathbf{p}}{dt} = {}^G\mathbf{r}_C \times {}^G\mathbf{F} = {}^G\mathbf{M}\end{aligned} \tag{2.60}$$

The formulation of dynamics developed over several centuries through the work of many scientists. However, because of their critical contributions, the names of Newton (1642–1727) and Euler (1707–1783) have been used to refer to the equations of motion.

Although there is no difference between Equations (2.58) and (2.59) analytically, we call the equation of motion of an applied force (2.58) the *Newton equation* and the equation of motion of an applied moment (2.59) the *Euler equation*. ■

Example 116 Moving on a Parabolic Curve Consider a particle with mass m that moves along a frictionless parabolic path $y = cx^2$ in the (x, y)-plane such that its x-component of velocity is constant:

$$\dot{x} = v_x = \text{const} \tag{2.61}$$

To determine the required force $\mathbf{F}$ to move m, we need to calculate its acceleration $\mathbf{a}$:

$$\mathbf{r} = x\hat{\imath} + cx^2\hat{\jmath} \tag{2.62}$$

$$\mathbf{v} = v_x\hat{\imath} + 2v_xcx\hat{\jmath} \tag{2.63}$$

$$\mathbf{a} = 2v_x^2c\hat{\jmath} \tag{2.64}$$

Therefore, the force on m should be in the y-direction with a constant magnitude proportional to v_x^2:

$$\mathbf{F} = m\mathbf{a} = 2mv_x^2 c\hat{\jmath} \tag{2.65}$$

Now assume that the same particle is moving such that its x-component of acceleration is constant:

$$\ddot{x} = a_x = \text{const} \tag{2.66}$$

To determine the required force $\mathbf{F}$ to move m, we need to calculate its acceleration vector $\mathbf{a}$:

$$\mathbf{r} = x\hat{\imath} + cx^2\hat{\jmath} \tag{2.67}$$

$$\mathbf{v} = \dot{x}\hat{\imath} + 2cx\dot{x}\hat{\jmath} \tag{2.68}$$

$$\mathbf{a} = a_x\hat{\imath} + 2c\left(\dot{x}^2 + a_x x\right)\hat{\jmath} \tag{2.69}$$

Therefore, the required force on m should be

$$\mathbf{F} = m\mathbf{a} = ma_x\hat{\imath} + 2mc\left(\dot{x}^2 + a_x x\right)\hat{\jmath} \tag{2.70}$$

which has a constant x-component and a position- and velocity-dependent y-component.

Example 117 Motion with Constant Moment of Momentum If a particle moves in such a way that its moment of momentum $\mathbf{L}$ is constant,

$$\mathbf{L} = \mathbf{r} \times m\,\mathbf{v} = \mathbf{c} \tag{2.71}$$

We may differentiate the equation

$$\mathbf{v} \times m\,\mathbf{v} + \mathbf{r} \times m\,\mathbf{a} = \mathbf{r} \times m\,\mathbf{a} = \mathbf{r} \times \mathbf{F} = 0 \tag{2.72}$$

to show that there must be no moment applied on the particle. This result also proves that the force $\mathbf{F}$ acting on the particle is always colinear with $\mathbf{r}$. Such a force is called the central force.

Example 118 Motion Equation of a System of Particles Consider a group of n particles $m_i, i = 1, 2, 3, \ldots, n$, with position vectors $\mathbf{r}_i$ in a global coordinate frame G. The position vector of the mass center C of the particles is at

$$\mathbf{r}_C = \frac{1}{m_C}\sum_{i=1}^{n} m_i\mathbf{r}_i \qquad m_C = \sum_{i=1}^{n} m_i \tag{2.73}$$

where m_C is the total mass of the system.

The force acting on each particle m_i can be decomposed into an external force $\mathbf{F}_i$ and an internal force $\sum_{j=1}^{n}\mathbf{f}_{ij}$. The internal force $\mathbf{f}_{ij}$, with the condition $\mathbf{f}_{ii} = 0$, is the force that particle m_j applies on m_i. The motion equation of the particle m_i would be

$$m_i\frac{d^2\mathbf{r}_i}{dt^2} = \mathbf{F}_i + \sum_{j=1}^{n}\mathbf{f}_{ij} \qquad i = 1, 2, 3, \ldots, n \tag{2.74}$$

By adding the n motion equations of all particles, we have

$$\sum_{i=1}^{n} m_i \frac{d^2\mathbf{r}_i}{dt^2} = \sum_{i=1}^{n} \mathbf{F}_i + \sum_{i=1}^{n}\sum_{j=1}^{n} \mathbf{f}_{ij} \tag{2.75}$$

Because of the third law of Newton,

$$\mathbf{f}_{ij} = -\mathbf{f}_{ji} \tag{2.76}$$

the summation of the internal forces is zero:

$$\sum_{i=1}^{n}\sum_{j=1}^{n} \mathbf{f}_{ij} = 0 \tag{2.77}$$

Therefore, the motion equation of all particles reduces to

$$m_C \frac{d^2\mathbf{r}_0}{dt^2} = \mathbf{F}_C \tag{2.78}$$

where

$$\mathbf{F}_C = \sum_{i=1}^{n} \mathbf{F}_i \tag{2.79}$$

is the resultant of all the external forces.

Equation (2.78) states that the motion of the mass center C of a system of particles is the same as if all the masses m were concentrated at that point and were acted upon by the resultant of all the external forces $\mathbf{F}_C$.

Example 119 Angular Momentum of a System of Particles Consider a group of n particles m_i, $i = 1, 2, 3, \ldots, n$, $m_C = \sum_{i=1}^{n} m_i$, with mass center C at $\mathbf{r}_C$ and position vectors $\mathbf{r}_i$ in a global coordinate frame G. Let us show the angular momentum of m_i by $\mathbf{L}_i$:

$$\mathbf{L}_i = \mathbf{r}_i \times m_i \dot{\mathbf{r}}_i \tag{2.80}$$

The summation of the angular momentum of all particles is $\mathbf{L}_O$:

$$\mathbf{L}_O = \sum_{i=1}^{n} \mathbf{L}_i = \sum_{i=1}^{n} (\mathbf{r}_i \times m_i \dot{\mathbf{r}}_i) \tag{2.81}$$

The position vector $\mathbf{r}_i$ of m_i can be shown by an addition $\mathbf{r}_i'$ to the position vector of mass center $\mathbf{r}_C$:

$$\mathbf{r}_i = \mathbf{r}_C + \mathbf{r}_i' \tag{2.82}$$

$$\sum_{i=1}^{n} \mathbf{r}_i' = 0 \tag{2.83}$$

Therefore,

$$\begin{aligned}\mathbf{L}_O &= \sum_{i=1}^{n} \left(m_i \left(\mathbf{r}_C + \mathbf{r}_i'\right) \times \left(\dot{\mathbf{r}}_C + \dot{\mathbf{r}}_i'\right)\right) \\ &= \sum_{i=1}^{n} m_i \left(\mathbf{r}_C \times \dot{\mathbf{r}}_C + \mathbf{r}_i' \times \dot{\mathbf{r}}_C + \mathbf{r}_C \times \dot{\mathbf{r}}_i' + \mathbf{r}_i' \times \dot{\mathbf{r}}_i'\right) \\ &= \mathbf{r}_C \times m_C \dot{\mathbf{r}}_C + \mathbf{L}_C\end{aligned} \tag{2.84}$$

because

$$\sum_{i=1}^{n} m_i \mathbf{r}_i' = 0 \tag{2.85}$$

$$\sum_{i=1}^{n} m_i \dot{\mathbf{r}}_i' = 0 \tag{2.86}$$

$$\sum_{i=1}^{n} \mathbf{r}_i' \times m_i \dot{\mathbf{r}}_i' = \mathbf{L}_C \tag{2.87}$$

where $\mathbf{L}_C$ is the angular momentum of the system about the mass center C.

Equation (2.84) states that the angular momentum $\mathbf{L}_O$ of a system of particles m_i, $i = 1, 2, 3, \ldots, n$, about a fixed point O is equal to the angular momentum $\mathbf{L}_C$ of a particle $m_C = \sum_{i=1}^{n} m_i$ at the mass center $\mathbf{r}_C$ plus the angular momentum $\mathbf{L}_C$ of the particles about the mass center C.

Example 120 Rotational Motion of a System of Particles Consider a group of n particles $m_i, i = 1, 2, 3, \ldots, n$, $m_C = \sum_{i=1}^{n} m_i$, with mass center C at $\mathbf{r}_C$ and position vectors $\mathbf{r}_i$ in a global coordinate frame G. The angular momentum of the system of particles about a fixed point O is equal to the angular momentum of a particle with the total mass at the mass center $\mathbf{r}_C$ plus the angular momentum of the particles about the mass center C:

$$\mathbf{L}_O = \mathbf{r}_C \times m_C \dot{\mathbf{r}}_C + \mathbf{L}_C \tag{2.88}$$

Taking a time derivative of (2.88) in G yields

$$\frac{d\mathbf{L}_O}{dt} = \dot{\mathbf{r}}_C \times m_C \dot{\mathbf{r}}_C + \dot{\mathbf{r}}_C \times m_C \ddot{\mathbf{r}}_C + \frac{d\mathbf{L}_C}{dt} \tag{2.89}$$

and using

$$\dot{\mathbf{r}}_C \times m_C \dot{\mathbf{r}}_C = 0 \tag{2.90}$$

$$m_C \ddot{\mathbf{r}}_C = \sum_{i=1}^{n} \mathbf{F}_i \tag{2.91}$$

$$\mathbf{r}_C = \mathbf{r}_i - \mathbf{r}_i' \tag{2.92}$$

we can rewrite (2.89) as

$$\frac{d\mathbf{L}_O}{dt} - \sum_{i=1}^{n} \mathbf{r}_i \times \mathbf{F}_i = \frac{d\mathbf{L}_C}{dt} - \sum_{i=1}^{n} \mathbf{r}'_i \times \mathbf{F}_i \tag{2.93}$$

or as

$$\frac{d\mathbf{L}_O}{dt} - \mathbf{M}_O = \frac{d\mathbf{L}_C}{dt} - \mathbf{M}_C \tag{2.94}$$

where $\mathbf{M}_O$ and $\mathbf{M}_C$ are the moments of the external forces about O and C, respectively. Because of the Euler equation (2.59), the left-hand side of (2.94) is zero and therefore

$$\mathbf{M}_C = \frac{d\mathbf{L}_C}{dt} = \frac{d}{dt}\left(\mathbf{r}'_i \times m_i \dot{\mathbf{r}}'_i\right) \tag{2.95}$$

Equation (2.95) states that the time derivative of the angular momentum of a system of particles $m_i, i = 1, 2, 3, \ldots, n$, about the mass center C is equal to the moment of external forces about C. This is true even if the mass center has a translational motion relative to the globally coordinate frame.

We can use this fact in the development of rigid-body motion. In studying rotational motion, we may ignore the translational motion of the mass center of the body provided we refer all moments and angular momenta to the mass center.

Example 121 ★ Definition of Dynamics Dynamics is the modeling of motion in nature. We develop some definitions and rules and model the nature's behavior with a set of mathematical equations. Solution of the set of equations is used to predict the behavior of the phenomenon. Being able to predict the behavior of a phenomenon allows us to adjust the parameters of a man-made device to achieve a desired behavior.

Vector calculus is an example of the rules that we made, and the Newton equation of motion is a model of what happens in nature. Both the mathematical rules and our model of nature are subject to change and improvement. Both are approximate or applicable only in a certain domain of the physical world. The model of nature is usually based on experiment or experience. They are applied as long as they are consistent with other models or as long as there is no better model.

Example 122 ★ The Newton–Laplace Principle of Determinacy Having the position $\mathbf{r}$ and velocity $\mathbf{v} = d\mathbf{r}/dt$ of a dynamic system at a moment of time is enough to uniquely determine its future and past motion. Suppose we know $\mathbf{r}_0 = \mathbf{r}(t_0)$ and $\mathbf{v}_0 = \mathbf{v}(t_0)$ at time $t = t_0$. Then the *principle of determinacy* guarantees that we can theoretically determine $\mathbf{r} = \mathbf{r}(t)$ $\forall t \in \mathbb{R}$ by solving the Newton differential equation of motion:

$${}^G\mathbf{F} = \frac{{}^Gd}{dt}\left(m\, {}^G\mathbf{v}\right) = \frac{{}^Gd}{dt}\left(m \frac{{}^Gd}{dt}\, {}^G\mathbf{r}\right) \tag{2.96}$$

So, to determine the past and future of a dynamic system, we only need to have $\mathbf{F}$ and initial conditions at an instant of time.

Example 123 ★ Force Derivative The equation of motion is the connecting point between kinematics and kinetics. Consider a constant mass body for which we can say: The momentum is equal to mass times velocity:

$$ {}^{G}\mathbf{p} = m\, {}^{G}\mathbf{v} = m \frac{{}^{G}d}{dt}\, {}^{G}\mathbf{r} \tag{2.97} $$

The equation of motion says that the applied force is equal to the derivative of the momentum:

$$ {}^{G}\mathbf{F} = \frac{{}^{G}d}{dt}\, {}^{G}\mathbf{p} = m\, {}^{G}\mathbf{a} = m \frac{{}^{G}d^2}{dt^2}\, {}^{G}\mathbf{r} \tag{2.98} $$

Taking a derivative of both sides states that the first derivative of force is equal to mass times jerk $\mathbf{j}$:

$$ \mathbf{Y} = \frac{{}^{G}d}{dt}\, {}^{G}\mathbf{F} = m\, {}^{G}\mathbf{j} = m \frac{{}^{G}d^3}{dt^3}\, {}^{G}\mathbf{r} \tag{2.99} $$

The derivative of force is called the *yank* $\mathbf{Y}$.

The next derivative states that the second derivative of force is equal to mass times snap $\mathbf{s}$:

$$ \mathbf{T} = \frac{{}^{G}d}{dt}\, {}^{G}\mathbf{Y} = \frac{{}^{G}d^2}{dt^2}\, {}^{G}\mathbf{F} = m\, {}^{G}\mathbf{s} = m \frac{{}^{G}d^4}{dt^4}\, {}^{G}\mathbf{r} \tag{2.100} $$

The derivative of yank is called the *tug* $\mathbf{T}$.

The next derivative states that the third derivative of force is equal to mass times crackle $\mathbf{c}$:

$$ \mathbf{S} = \frac{{}^{G}d}{dt}\, {}^{G}\mathbf{T} = \frac{{}^{G}d^3}{dt^3}\, {}^{G}\mathbf{F} = m\, {}^{G}\mathbf{c} = m \frac{{}^{G}d^5}{dt^5}\, {}^{G}\mathbf{r} \tag{2.101} $$

The derivative of tug is called the *snatch* $\mathbf{S}$.

The next derivative states that the fourth derivative of force is equal to mass times pop **þ**:

$$ \mathbf{Ş} = \frac{{}^{G}d}{dt}\, {}^{G}\mathbf{S} = \frac{{}^{G}d^4}{dt^4}\, {}^{G}\mathbf{F} = m\, {}^{G}\mathbf{þ} = m \frac{{}^{G}d^6}{dt^6}\, {}^{G}\mathbf{r} \tag{2.102} $$

The derivative of snatch is called the *shake* **Ş**.

The next derivative states that the fifth derivative of force is equal to mass times larz $\mathbf{z}$:

$$ \mathbf{Z} = \frac{{}^{G}d}{dt}\, {}^{G}\mathbf{Ş} = \frac{{}^{G}d^5}{dt^5}\, {}^{G}\mathbf{F} = m\, {}^{G}\mathbf{z} = m \frac{{}^{G}d^7}{dt^7}\, {}^{G}\mathbf{r} \tag{2.103} $$

The derivative of shake is called the *zoor* $\mathbf{Z}$. The derivative of zoor is called the *setorg* $\mathbf{N}$, the derivative of setorg is called the *gorz* $\mathbf{G}$, and the derivative of gorz is called the *sharang* $\mathbf{H}$. None of these names are standard, mainly because derivatives of the equation of motion are not much applied.

2.3 SPECIAL SOLUTIONS

Special cases of the equation of motion happen when the force $\mathbf{F}$ is only a function of time t, position $\mathbf{r}$, or velocity $\mathbf{v}$. To show these special cases, we consider the motion of a particle in one dimension.

2.3.1 Force Is a Function of Time, $F = F(t)$

The equation of motion in this case is

$$m\frac{dv}{dt} = F(t) \tag{2.104}$$

which can be integrated by separation of variables,

$$m\int_{v_0}^{v} dv = \int_{t_0}^{t} F(t)\,dt \tag{2.105}$$

The result of this integral would be a time-dependent velocity function $v = v(t)$ and can be integrated as

$$\int_{x_0}^{x} dx = \int_{t_0}^{t} v(t)\,dt \tag{2.106}$$

to provide the position x as a function of time $x = x(t)$

Example 124 Exponential Decaying Force Consider a point mass m that is under an exponentially decaying force $F(t) = ce^{-t}$, where c is a constant:

$$m\frac{dv}{dt} = ce^{-t} \tag{2.107}$$

The velocity of the mass is

$$m\int_{v_0}^{v} dv = \int_{t_0}^{t} ce^{-t}dt$$

$$v = v_0 + \frac{c}{m}\left(e^{-t_0} - e^{-t}\right) \tag{2.108}$$

which can be used to find the position:

$$\int_{x_0}^{x} dx = \int_{t_0}^{t} \left(v_0 + \frac{c}{m}\left(e^{-t_0} - e^{-t}\right)\right) dt$$

$$x = x_0 - \frac{c}{m}\left(1 + t_0 - t\right) e^{-t_0} + v_0\left(t - t_0\right) + \frac{c}{m}e^{-t} \tag{2.109}$$

If the initial time t_0 is assumed to be zero, then the position and velocity of the mass are simplified to

$$x = x_0 - \frac{c}{m}\left(1 - t\right) + v_0 t + \frac{c}{m}e^{-t} \tag{2.110}$$

$$v = v_0 + \frac{c}{m}\left(1 - e^{-t}\right) \tag{2.111}$$

Figure 2.6 illustrates the force, velocity, and position of m for

$$m = 1\,\text{kg} \qquad c = 1\,\text{N} \tag{2.112}$$

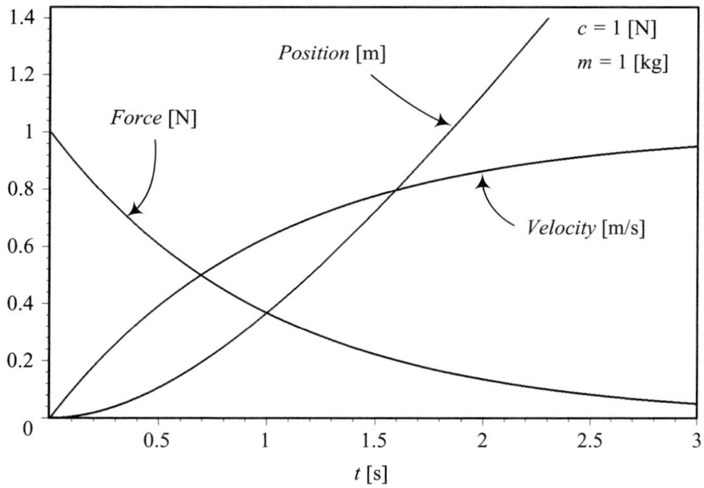

Figure 2.6 Position x and velocity v for force $F = ce^{-t}$.

and initial condition

$$x_0 = 0 \qquad v_0 = 0 \tag{2.113}$$

When the force is only a function of time, then F is proportional to acceleration a and hence to the derivative of velocity v. So, if the force function is complicated to have a closed-form integral, geometric differentiation and integration can provide the solution.

Example 125 Constant Force When the applied force is constant, $F = const$, we may include the problem in the category of $F = F(t)$. In both cases, the problem reduces to differential equations with initial conditions. As an example, consider the two masses m_1 and m_2 linked by a rigid rod as shown in Figure 2.7.

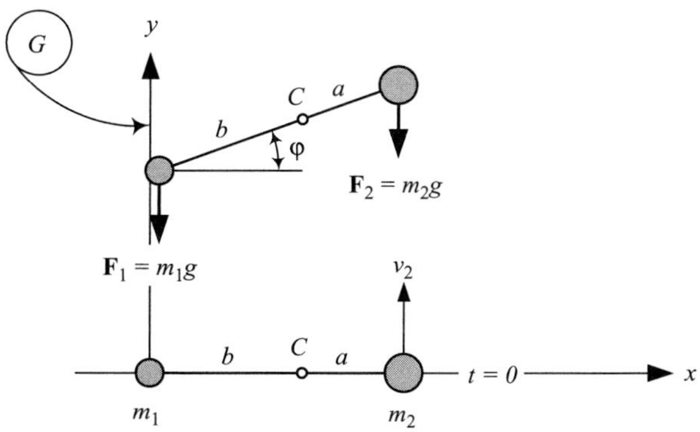

Figure 2.7 Two masses m_1 and m_2 linked by a rigid rod and asymmetric initial velocity.

At $t = 0$, the mass m_1 is at the origin and m_2 is on the x-axis at $x_2 = b + a$. The mass m_2 has an initial velocity $\dot{y}_2 = v_2$. The point C indicates the mass center of the system with $m = m_1 + m_2$. The equations of motion of the mass center and its initial conditions are

$$m\ddot{x}_C = 0 \qquad m\ddot{y}_C = -mg \tag{2.114}$$

$$x_C = b \qquad y_C = 0 \tag{2.115}$$

$$\dot{x}_C = 0 \qquad \dot{y}_C = b\dot{\varphi}_0 \qquad \dot{\varphi}_0 = \frac{v_2}{a + b} \tag{2.116}$$

The solution of this initial-value problem is

$$x_C(t) = b \qquad \dot{y}_C(t) = -\tfrac{1}{2}gt^2 + b\dot{\varphi}_0 t \tag{2.117}$$

It shows that the mass center of the system is moving on a vertical line, while the bar is turning about C with constant angular speed $\dot{\varphi}$.

Example 126 The Projectile Problem When a body is under a constant force $F = F_0$, the above method for $F = F(t)$ can be used successfully:

$$m\frac{dv}{dt} = F \tag{2.118}$$

$$v = v_0 + \frac{F}{m}(t - t_0) \tag{2.119}$$

$$x = x_0 + v_0 t + \frac{F}{2m}(t - t_0)^2 \tag{2.120}$$

Consider a projectile with mass m that is shot with an initial velocity $\mathbf{v}_0$ from the origin of the coordinate frame. The initial conditions of the problem are

$$t_0 = 0 \qquad \mathbf{r}(0) = 0 \qquad \mathbf{v}(0) = \mathbf{v}_0 \tag{2.121}$$

If the velocity of the mass is small, the mass remains near the ground, and we may assume a flat ground with a uniform gravitational attraction $\mathbf{g}$ and no air. So, the weight is the only applied force on the mass and its equation of motion is

$$m\frac{d\mathbf{v}}{dt} = m\mathbf{g} \tag{2.122}$$

$$\mathbf{g} = -g\hat{k} \tag{2.123}$$

$$g = 9.80665 \text{ m/s}^2 \approx 9.81 \text{ m/s}^2 \tag{2.124}$$

This is a classical problem in mechanics and is called the *projectile problem*.

An integral of the equation of motion determines the velocity $\mathbf{v}$ of the projectile:

$$\mathbf{v} = -gt\hat{k} + \mathbf{v}_0 \tag{2.125}$$

Substituting $\mathbf{v} = d\mathbf{r}/dt$ and integrating provide the position of the projectile as a function of time:

$$\mathbf{r} = -\tfrac{1}{2}gt^2\hat{k} + \mathbf{v}_0 t \tag{2.126}$$

The initial velocity of the projectile has a magnitude v_0 and is at an angle θ with respect to the x-axis:

$$\mathbf{v}_0 = v_0 \begin{bmatrix} \cos\theta \\ 0 \\ \sin\theta \end{bmatrix} \tag{2.127}$$

Substituting $\mathbf{v}_0$ in the kinematic equations of the projectile, we find

$$\mathbf{v} = \begin{bmatrix} v_0 \cos\theta \\ 0 \\ v_0 \sin\theta - gt \end{bmatrix} \tag{2.128}$$

$$\mathbf{r} = \begin{bmatrix} v_0 t \cos\theta \\ 0 \\ v_0 t \sin\theta - \tfrac{1}{2}gt^2 \end{bmatrix} \tag{2.129}$$

Equation (2.129) is the projectile's path of motion using t as a parameter. Eliminating t will provide the path of motion in the (x, z)-plane, which is a parabola:

$$z = -\frac{1}{2}g\frac{x^2}{v_0^2\cos^2\theta} + x\tan\theta \tag{2.130}$$

The range R of the projectile on a flat ground can be found by setting $z = 0$ and solving Equation (2.130) for x:

$$R = \frac{v_0^2}{g}\sin 2\theta \tag{2.131}$$

The range becomes a maximum at the optimal angle $\theta = 45\,\text{deg}$:

$$R_M = \frac{v_0^2}{g} \tag{2.132}$$

The projectile reaches the range R at time t_R, which can be found from the z-component of $\mathbf{r}$:

$$t_R = 2\frac{v_0}{g}\sin\theta \tag{2.133}$$

Because of the symmetry of z in Equation (2.130), the projectile will reach its highest point H at

$$t_H = \frac{1}{2}t_R = \frac{v_0}{g}\sin\theta \tag{2.134}$$

which shows that

$$H = \frac{v_0^2}{2g}\sin^2\theta \tag{2.135}$$

These results are independent of the mass of the projectile.

Example 127 ★ Forbidden Umbrella Figure 2.8 illustrates the paths of motion of a projectile, given in Equation (2.130), with a constant initial speed v_0 at different angles θ. It can be seen that, regardless of the shooting angle, the projectile cannot reach out of a closed area. The reachable boundary of space is indicated by a curve that is tangent to all paths. Such a curve that is tangent to all members of a curve family is called the *envelope*.

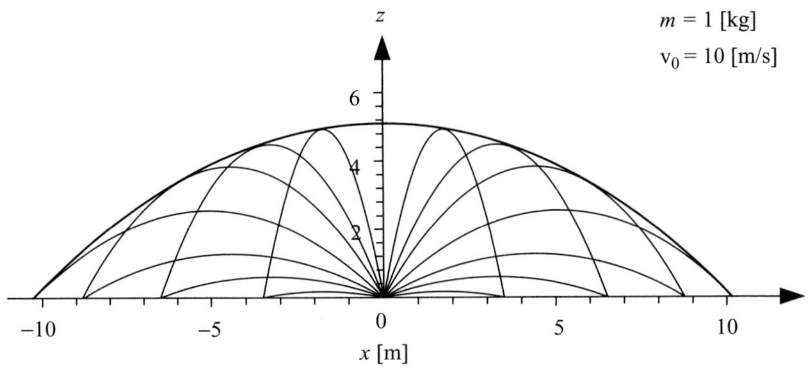

Figure 2.8 The path of motion of a projectile with a constant initial speed v_0 and different angles θ.

Members of curve family are related by changing only one or a set of parameters. To find the envelope of a family, we should eliminate the parameter between the equation of the family and the equation made up of its derivative with respect to the parameter.

The family of a projectile with a constant initial speed is made by changing the angle of the initial velocity. The derivative of z with respect to θ is

$$\frac{dz}{d\theta} = -\frac{1}{v_0^2 \cos^2 \theta} \left(g x^2 \tan \theta - x v_0^2 \right) = 0 \tag{2.136}$$

which yields

$$\tan \theta = \frac{v_0^2}{g x} \tag{2.137}$$

$$\frac{1}{\cos^2 \theta} = 1 + \left(\frac{v_0^2}{g x} \right)^2 \tag{2.138}$$

Substituting (2.137) and (2.133) in (2.130), we find the equation of the envelope:

$$z = \frac{1}{2} \left(\frac{v_0^2}{g} - \frac{g}{v_0^2} x^2 \right) \tag{2.139}$$

When the shooting device is similar to an antiaircraft gun that can turn about the z-axis, the envelope of the reachable space is a circular paraboloid called a *projectile umbrella*:

$$z = \frac{v_0^2}{2g} - \frac{g}{2v_0^2}\left(x^2 + y^2\right) \tag{2.140}$$

Such a paraboloid, illustrated in Figure 2.9, is a *forbidden umbrella* for military aircraft. The reachable space under the umbrella is

$$0 \le z \le \frac{v_0^2}{2g} - \frac{g}{2v_0^2}\left(x^2 + y^2\right) \tag{2.141}$$

These results are independent of the mass of the projectile.

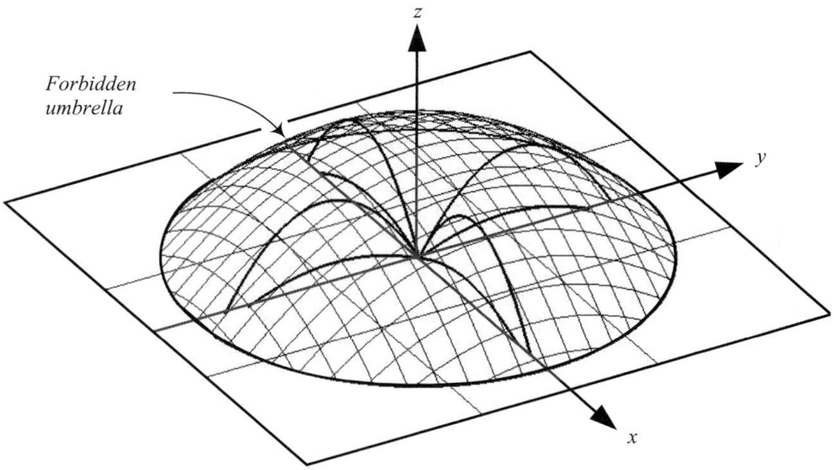

Figure 2.9 Forbidden umbrella.

Example 128 ★ Reach a Point in (x, z)-Plane Equation (2.130) provides the path of motion of a projectile in the no-air and flat-ground conditions when it is thrown from the origin of the (x, z)-plane. Figure 2.8 illustrates the projectile umbrella and reachable area of the projectile in the (x, z)-plane when it is thrown with a constant initial speed v_0 at different angles θ.

Every point under the projectile umbrella can be reached at two different angles and hence at two different times. To show this, we may solve Equation (2.130) for θ:

$$\theta = \tan^{-1}\left(\frac{v_0^2}{gx} + \frac{1}{x}\sqrt{\frac{v_0^4}{g^2} - 2z\frac{v_0^2}{g} - x^2}\right) \tag{2.142}$$

The two values of θ are equal when

$$v_0^4 - 2v_0^2 g z - g^2 x^2 = 0 \tag{2.143}$$

which indicates the same envelope parabola as in Equation (2.139). Therefore, any point under the projectile envelope can be reached at two angles, and any point on the envelope can be reached at only one angle:

$$\theta = \tan^{-1} \frac{v_0^2}{gx} \tag{2.144}$$

If $z = 0$, then $x = v_0^2/g = R_M$ and $\theta = 45$ deg, which confirms that the angle reaches the maximum range.

To reach a point on the ground, we set $z = 0$ and simplify Equation (2.142):

$$\theta = \tan^{-1} \left(\frac{gx}{v_0^2 \pm \sqrt{v_0^4 - g^2 x^2}} \right) \tag{2.145}$$

Figure 2.10 illustrates the shooting angle θ as a function of v_0 for a different horizontal range x. It shows the variety of possible θ and v_0 to reach an x. The two values of θ to reach an x at a fixed v_0 are symmetric with respect to $\theta = 45$ deg. For every v_0, the maximum x is reached at $\theta = 45$ deg.

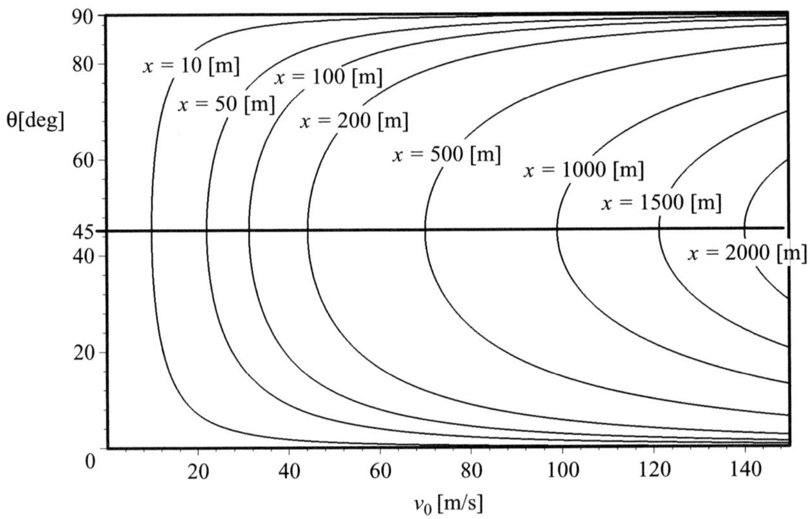

Figure 2.10 Shooting angle θ as a function of v_0 for different values of horizontal range x.

To have a design chart for reaching a point under the projectile umbrella, we write Equations (2.130) and (2.142) as

$$w = -\tfrac{1}{2} u \left(1 + \tan^2 \theta\right) + \tan\theta \tag{2.146}$$

$$\theta = \tan^{-1} \left(\frac{1}{u} + \sqrt{\frac{1}{u^2} - 2\frac{w}{u} - 1} \right) \tag{2.147}$$

where

$$u = \frac{x}{R_M} \tag{2.148}$$

$$R_M = \frac{v_0^2}{g} \tag{2.149}$$

$$w = \frac{z}{x} \tag{2.150}$$

$$0 \le u \le 1 \tag{2.151}$$

Figure 2.11 illustrates w as a function of u for different θ. We have a linear relation between w and u for a given θ. Having v_0 determines R_M, and therefore u would be proportional to x. Now, for any x, the value of u is set and the line uniquely indicates the associated w. The envelope of these curves is

$$w = \frac{1 - u^2}{2u} \tag{2.152}$$

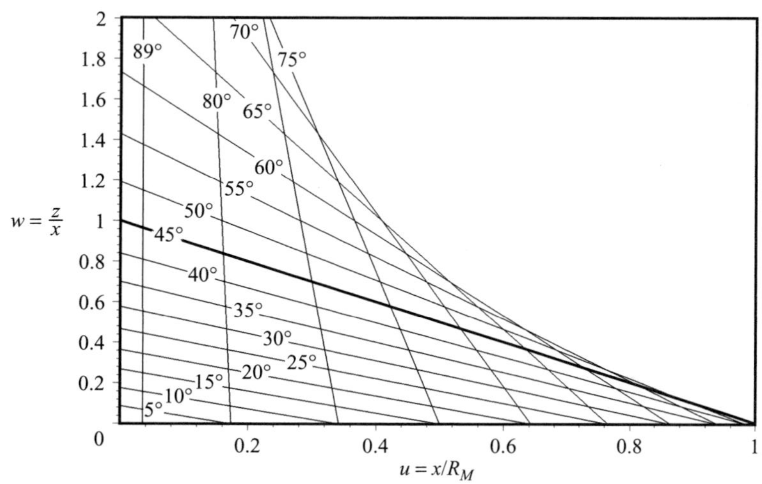

Figure 2.11 The vertical coordinate w as a function of the horizontal coordinate u of a projectile for different shooting angle θ.

As an example, let us set

$$v_0 = 10\,\text{m/s} \qquad g = 9.81\,\text{m/s}^2 \qquad \theta = 30\,\text{deg} \tag{2.153}$$

Then

$$R_M = \frac{v_0^2}{g} = \frac{100}{9.81} = 10.194\,\text{m} \tag{2.154}$$

At any $x \le 10.194$ m, say $x = 5$ m, we have

$$u = \frac{x}{R_M} = \frac{5}{10.194} = 0.49048 \tag{2.155}$$

and hence,

$$w = -\tfrac{1}{2}u\left(1 + \tan^2\theta\right) + \tan\theta = 0.25036\,\text{m} \tag{2.156}$$

which shows that

$$z = xw = 5 \times 0.25036 = 1.2518\,\text{m} \tag{2.157}$$

Figure 2.12 illustrates θ as a function of u for different w. This chart may be used to determine the required v_0 and θ to pass through a desired point under the projectile umbrella. For a given x and z, we determine the value of w and indicate the associated design curve in the chart. If the initial speed v_0 is also given, u is fixed and therefore a vertical line can show the required shooting angles to reach the desired point (x, z). If instead of v_0 we may shoot the projectile at any speed v_0, usually less than a maximum value v_M, there are an infinite number of pairs (v_0, θ) that go through the desired point. To choose a set, we may include a condition such as the minimum time or minimum initial speed.

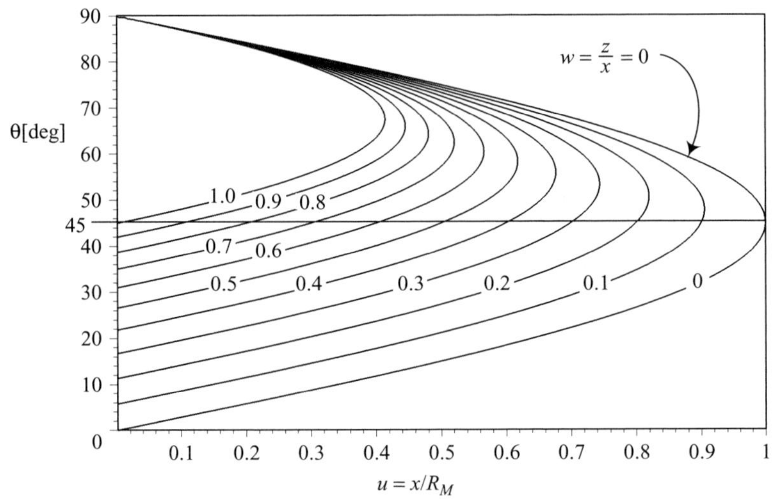

Figure 2.12 Shooting angle θ as a function of horizontal coordinate u for different values of vertical coordinate w.

As an example, let us determine (v_0, θ) to pass through $(x = 5\,\text{m},\ z = 1\,\text{m})$. The associated design curve is indicated by

$$w = \frac{z}{x} = \frac{1}{5} = 0.2 \tag{2.158}$$

Choosing an initial velocity, say $v_0 = 10\,\text{m/s}$, determines a vertical line

$$u = \frac{x}{v_0^2/g} = \frac{5}{100/9.81} = 0.4905 \tag{2.159}$$

that hits the design curve at

$$\theta = 26.963\,\text{deg} \qquad \theta = 74.347\,\text{deg} \tag{2.160}$$

However, if we are free to have any initial velocity within a range, then there is a relationship between v_0 and θ to pass through the point:

$$\tan\theta = \sqrt{4.1564 \times 10^{-4} v_0^4 - 8.1549 \times 10^{-3} v_0^2 - 1} + 2.0387 \times 10^{-2} v_0^2 \quad (2.161)$$

$$v_0 = \frac{11.074}{\cos\theta\sqrt{5\tan\theta - 1}} \quad (2.162)$$

Figure 2.13 illustrates Equation (2.162) graphically. It shows that there is a minimum v_0 to reach the point ($x = 5\,\text{m}$, $z = 1\,\text{m}$). Minimum speed usually means the minimum required effort that is the minimum required explosive material. The minimum point happens at

$$v_0 = 7.735342076\,\text{m/s} \quad (2.163)$$

$$\theta = 50.65496622\,\text{deg} \quad (2.164)$$

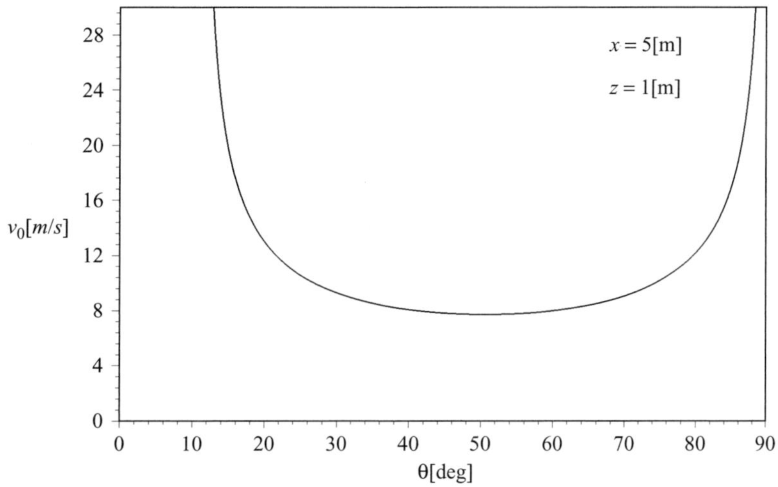

Figure 2.13 The relationship between the initial velocity v_0 and shooting angle θ to reach a specific point.

2.3.2 Force Is a Function of Position, $F = F(x)$

We can write the equation of motion in this case as

$$mv\frac{dv}{dx} = F(x) \quad (2.165)$$

because

$$a = \frac{dv}{dt} = \frac{dx}{dt}\frac{dv}{dx} = v\frac{dv}{dx} \quad (2.166)$$

Equation (2.165) may be integrated by separation of variables,

$$m \int_{v_0}^{v} v\, dv = \int_{x_0}^{x} F(x)\, dx \tag{2.167}$$

to find a position-dependent velocity function $v = v(x)$. This function can be integrated as

$$\int_{x_0}^{x} \frac{dx}{v(x)} = \int_{t_0}^{t} dt \tag{2.168}$$

to provide the position x as a function of time, $x = x(t)$.

Example 129 Harmonic Oscillator Consider a point mass m and an attraction force toward a fixed origin O on the line of the x-axis. The magnitude of attraction is proportional to the distance from O:

$$F = -kx \tag{2.169}$$

The equation of motion of the point

$$\ddot{x} + \omega^2 x = 0 \qquad \omega^2 = \frac{k}{m} \tag{2.170}$$

is a linear equation with constant coefficients. This dynamic system is called a *harmonic oscillator*. Following the $F = F(x)$ method, we may rewrite the equation of motion as

$$\int \dot{x}\, d\dot{x} = -\int \omega^2 x\, dx \tag{2.171}$$

and find

$$\dot{x}^2 + \omega^2 x^2 = \omega^2 C^2 \tag{2.172}$$

$$C^2 = \frac{\dot{x}_0^2}{\omega^2} + x_0^2 \tag{2.173}$$

This motion may be described by a moving point on an ellipse in the $(x, \dot{x})$-plane with semiaxes ωC on the $\dot{x}$-axis and C on the x axis. Such a motion is a *libration* motion between $x = \pm C$.

Integration of Equation (2.172),

$$\int \frac{dx}{\omega\sqrt{C^2 - x^2}} = \int dt \tag{2.174}$$

provides the solution of motion:

$$x = C \sin(\omega t - \varphi) \tag{2.175}$$

$$\varphi = -\arcsin \frac{x_0}{C} \tag{2.176}$$

Example 130 Two-Dimensional Harmonic Motion A force that is a function of position,

$$\mathbf{F} = -k\mathbf{r} \tag{2.177}$$

is applied on a particle with mass m that is moving on the (x, y)-plane. The equation of motion of the particle is

$$m\ddot{\mathbf{r}} = -k\mathbf{r} \tag{2.178}$$

The equation of motion can be decomposed into x- and y-directions as

$$\ddot{x} + \omega^2 x = 0 \tag{2.179}$$

$$\ddot{y} + \omega^2 y = 0 \tag{2.180}$$

where

$$\omega = \sqrt{\frac{k}{m}} \tag{2.181}$$

The solution of Equations (2.179) and (2.180) are

$$x(t) = A_1 \cos(\omega t) + A_2 \sin(\omega t) = A \sin(\omega t - \alpha) \tag{2.182}$$

$$y(t) = B_1 \cos(\omega t) + B_2 \sin(\omega t) = B \sin(\omega t - \beta) \tag{2.183}$$

where α, β, A, and B are related to the initial conditions $x_0 = x(0)$, $y_0 = y(0)$, $\dot{x}_0 = \dot{x}(0)$, and $\dot{y}_0 = \dot{y}(0)$:

$$x_0 = -A \sin\alpha \qquad y_0 = -B \sin\beta \tag{2.184}$$

$$\dot{x}_0 = A\omega \cos\alpha \qquad \dot{y}_0 = B\omega \cos\beta \tag{2.185}$$

To find the path of motion in the (x, y)-plane, we should eliminate time t between x and y. Let us define

$$\gamma = \alpha - \beta \tag{2.186}$$

to expand y,

$$\begin{aligned} y(t) &= B \sin[(\omega t - \alpha) + (\alpha - \beta)] \\ &= B \sin(\omega t - \alpha) \cos(\alpha - \beta) + B \cos(\omega t - \alpha) \sin(\alpha - \beta) \\ &= B \sin(\omega t - \alpha) \cos\gamma + B \cos(\omega t - \alpha) \sin\gamma \end{aligned} \tag{2.187}$$

and substitute x,

$$y = \frac{B}{A} x \cos\gamma + B\sqrt{1 - \left(\frac{x}{A}\right)^2} \sin\gamma \tag{2.188}$$

The path (2.188) can be rearranged to

$$A^2 y^2 - 2ABxy \cos\gamma + B^2 x^2 = A^2 B^2 \sin^2\gamma \tag{2.189}$$

A special case of the path happens when $A = B$ and $\gamma = \pm\pi/2$, which indicates a circular motion:

$$x^2 + y^2 = A^2 \tag{2.190}$$

If $A \neq B$, the path is an ellipse:

$$\frac{x^2}{A^2} + \frac{y^2}{B^2} = 1 \tag{2.191}$$

If $\gamma = 0$ or $\gamma = \pm\pi$, the path reduces to a straight line:

$$y = \frac{B}{A}x \tag{2.192}$$

or

$$y = -\frac{B}{A}x \tag{2.193}$$

Figure 2.14 illustrates the path of motion for $A = B = 1$, $\omega = 1$, and different γ.

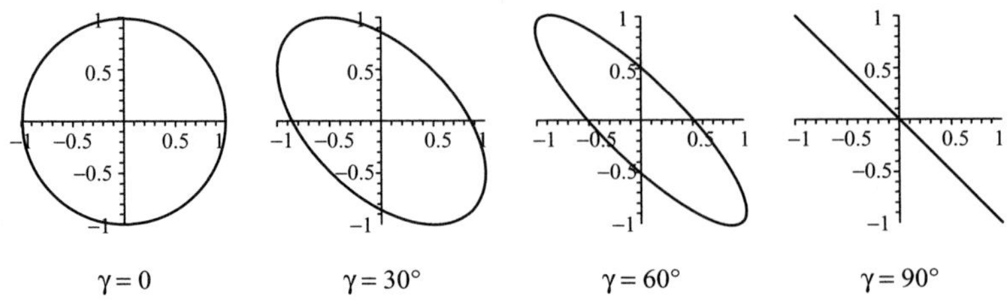

Figure 2.14 Path of the harmonic motion of a particle in the (x, y)-plane.

Example 131 ★ Lissajous Curves Consider a particle with mass m that is moving on the (x, y)-plane under the action of the force $\mathbf{F}$:

$$\mathbf{F} = -k_1 x\,\hat{\imath} - k_2 y\,\hat{\jmath} \tag{2.194}$$

The equation of motion of the particle is

$$m\ddot{\mathbf{r}} = -k\mathbf{r} \tag{2.195}$$

or

$$\ddot{x} + \omega_1 x = 0 \tag{2.196}$$

$$\ddot{y} + \omega_2 y = 0 \tag{2.197}$$

where

$$\omega_1 = \sqrt{\frac{k_1}{m}} \qquad \omega_2 = \sqrt{\frac{k_2}{m}} \tag{2.198}$$

The solutions of Equations (2.196) and (2.197) are

$$x(t) = A_1 \cos(\omega_1 t) + A_2 \sin(\omega_1 t) = A \cos(\omega_1 t - \alpha) \tag{2.199}$$

$$y(t) = B_1 \cos(\omega_2 t) + B_2 \sin(\omega_2 t) = B \cos(\omega_2 t - \beta) \tag{2.200}$$

These equations indicate the parametric path of motion of the particle. If the frequencies ω_1 and ω_2 are commensurable, the path of motion will be closed. The frequencies ω_1 and ω_2 are commensurable when their ratio is a rational fraction, $\omega_1/\omega_2 = m/n$, $\{m, n \in N\}$. The path of motion in this case is called the *Lissajous* curve.

If the frequencies ω_1 and ω_2 are not commensurable, the path of motion will be open, which means the moving particle will never pass twice through the same point with the same velocity.

The Lissajous curves can be described better if we assume $A = B$ and write the parametric equations in the forms

$$x(t) = \cos(r\tau + \delta) \tag{2.201}$$

$$y(t) = \cos(\tau) \tag{2.202}$$

where

$$r = \frac{\omega_1}{\omega_2} \qquad \tau = \omega_2 t - \beta \qquad \delta = \beta r - \alpha \tag{2.203}$$

So, x is a 2π-periodic and y is a $(2\pi/r)$-periodic function of τ.

Figure 2.15 depicts some Lissajous curves and Figure 2.16 shows a few two-dimensional harmonic curves for incommensurable cases.

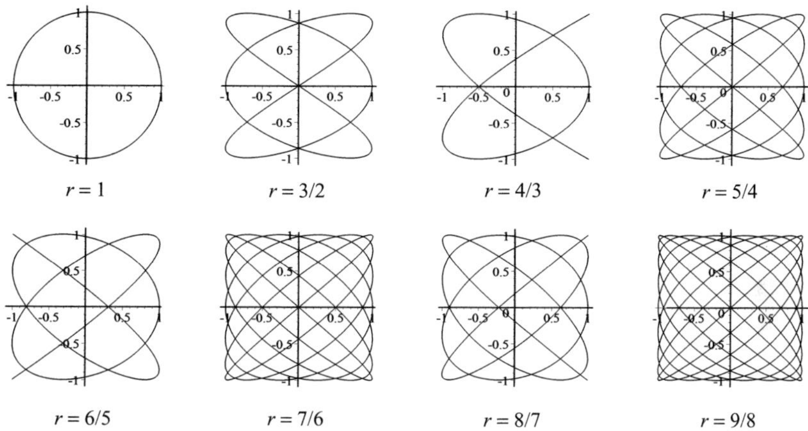

Figure 2.15 Some Lissajous curves for the rational fraction ω_1/ω_2.

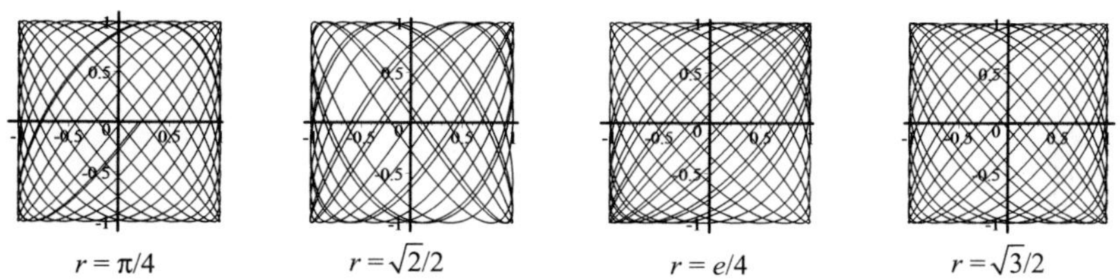

Figure 2.16 Some two-dimensional harmonic curves for incommensurable ω_1 and ω_2.

Example 132 ★ Motion of a Simple Pendulum The Equation of motion of a *simple pendulum*, shown in Figure 2.17, is

$$ml^2\ddot{\theta} + mg\sin\theta = 0 \tag{2.204}$$

By a simple pendulum, we mean a point mass m is attached to the end of a massless bar with length l that is pin joined to the wall. The pendulum is swinging in a uniform gravitational field $\mathbf{g} = -g\hat{k}$. The equation of motion can be simplified to

$$\ddot{\theta} + \frac{g}{l}\sin\theta = 0 \tag{2.205}$$

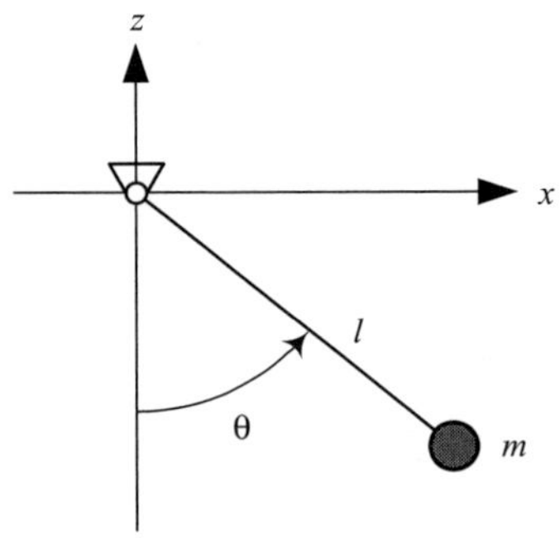

Figure 2.17 A simple pendulum.

Multiplying the equation by $\dot{\theta}$ and integrating yield

$$\frac{1}{2}\dot{\theta}^2 = -\int \frac{g}{l}\sin\theta\, d\theta = \frac{g}{l}\cos\theta + C_1 \tag{2.206}$$

Assuming the initial conditions

$$\theta(0) = \theta_0 \qquad \dot{\theta}(0) = 0 \tag{2.207}$$

we find

$$C_1 = -\frac{g}{l}\cos\theta_0 \tag{2.208}$$

and therefore,

$$\dot{\theta}^2 = 2\frac{g}{l}\left(\cos\theta - \cos\theta_0\right) \tag{2.209}$$

This indicates that, to determine the motion of the pendulum, we need to calculate the equation

$$t = \sqrt{\frac{l}{g}}\int_0^{\theta_0} \frac{d\theta}{\sqrt{2\left(\cos\theta - \cos\theta_0\right)}} \tag{2.210}$$

To calculate the integral, we change the variable from θ to φ such that

$$\sin\frac{\theta}{2} = k\sin\varphi \tag{2.211}$$

$$k = \sin\frac{\theta_0}{2} \tag{2.212}$$

Then we have

$$d\theta = \frac{2k\cos\varphi}{\sqrt{1 - k^2\sin^2\varphi}}d\varphi \tag{2.213}$$

When $\theta = 0$, $\varphi = 0$, and when $\theta = \theta_0$, $\varphi = \pi/2$. Using the new variable, we also have

$$\frac{1}{\sqrt{2\left(\cos\theta - \cos\theta_0\right)}} = \frac{1}{2k\cos\varphi} \tag{2.214}$$

and therefore the integral (2.210) reduces to the complete elliptic integral of the first kind, $F(\pi/2, k)$, with modulus k and amplitude $\pi/2$:

$$\sqrt{\frac{g}{l}}t = \int_0^{\pi/2} \frac{d\varphi}{\sqrt{1-k^2\sin^2\varphi}} = F\left(\frac{\pi}{2}, k\right) \tag{2.215}$$

The time t from the lowest position of the pendulum to any position $\theta < \theta_0$ or $\varphi < \pi/2$ can be found by

$$\sqrt{\frac{g}{l}}t = \int_0^{\varphi} \frac{d\varphi}{\sqrt{1-k^2\sin^2\varphi}} = F(\varphi, k) \tag{2.216}$$

where $F(\varphi, k)$ is the elliptic integral of the first kind with modulus k and amplitude φ. The time for any interval angle from $\varphi = \varphi_1$ at $t = t_1$ to $\varphi = \varphi_2$ at $t = t_2$ is then given by

$$\sqrt{\frac{g}{l}}\,(t_2 - t_1) = F(\varphi_2, k) - F(\varphi_1, k) \tag{2.217}$$

If $k = \sin(\theta_0/2)$ is very small, then the time of swing between $\theta = 0$ and $\theta = \theta_0$ reduces to

$$t = \frac{1}{2}\pi\sqrt{\frac{l}{g}} \tag{2.218}$$

which shows that the linearized period T of the pendulum is

$$T_0 = 2\pi\sqrt{\frac{l}{g}} \tag{2.219}$$

Figure 2.18 illustrates Equation (2.216) for different $k = \sin(\theta_0/2)$. At low values of k the period of oscillation is proportional to the angle θ_0. However, the period grows faster when θ_0 is higher.

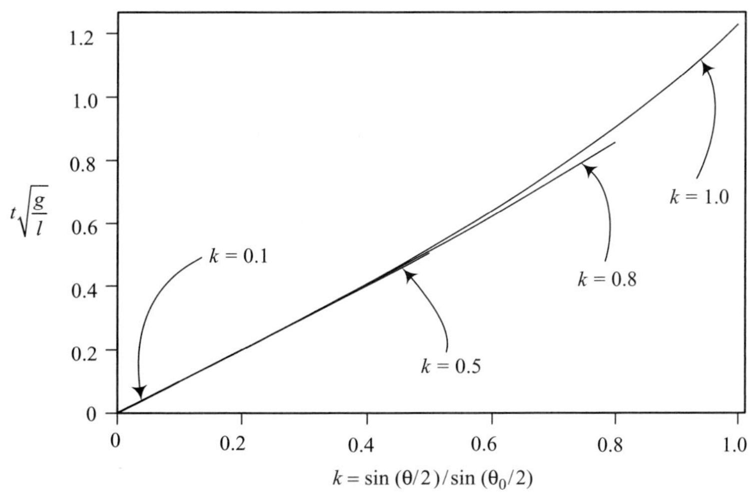

Figure 2.18 Period of oscillation of a simple pendulum for different initial release angle.

Figure 2.19 shows the period of oscillation T/T_0 for different initial angles θ_0.

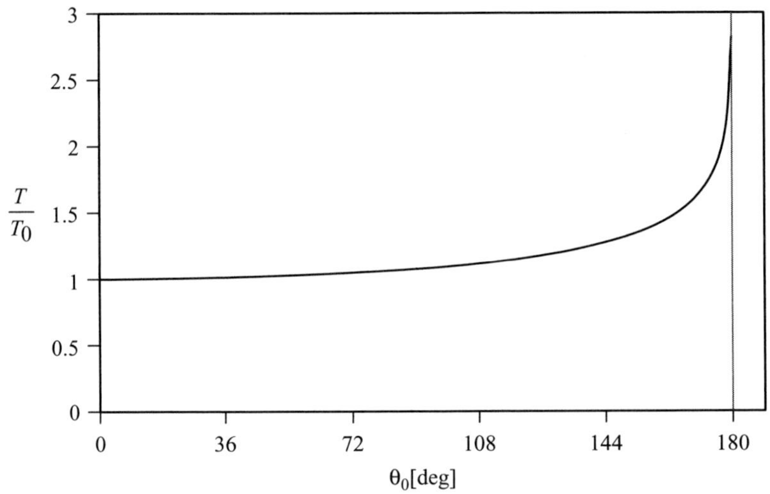

Figure 2.19 Period of oscillation T/T_0 of a simple pendulum as a function of initial angle θ_0.

2.3.3 ★ Elliptic Functions

If $0 \le k \le 1$ and $0 < \varphi \le \pi/2$, then the *elliptic integral of the first kind* is defined as

$$u = F(x, k) = \int_0^x \frac{dy}{\sqrt{(1 - y^2)(1 - k^2 y^2)}} \qquad x \in [-1, 1] \tag{2.220}$$

or

$$F(\varphi, k) = \int_0^\varphi \frac{d\theta}{\sqrt{1 - k^2 \sin^2 \theta}} \qquad \varphi \in [-1, 1] \tag{2.221}$$

$$x = \sin \varphi \tag{2.222}$$

If $x = 1$ or $\varphi = \pi/2$, these definite integrals are called *complete elliptic integrals of the first kind* and denoted by $K(k)$:

$$K(k) = \int_0^1 \frac{dy}{\sqrt{(1 - y^2)(1 - k^2 y^2)}} = \int_0^{\pi/2} \frac{d\theta}{\sqrt{1 - k^2 \sin^2 \theta}} \tag{2.223}$$

The *elliptic integral of the second kind* is defined by

$$E(x, k) = \int_0^x \sqrt{\frac{(1 - k^2 y^2)}{(1 - y^2)}}\, dy \qquad x \in [-1, 1] \tag{2.224}$$

or

$$E(\varphi, k) = \int_0^{\varphi} \sqrt{1 - k^2 \sin^2 \theta} d\theta \qquad \varphi \in [-1, 1] \tag{2.225}$$

$$x = \sin \varphi \tag{2.226}$$

If $x = 1$ or $\varphi = \pi/2$, these integrals are called *complete elliptic integrals of the second kind* and denoted by $E(k)$:

$$E(k) = \int_0^1 \sqrt{\frac{1 - k^2 y^2}{1 - y^2}} dy = \int_0^{\pi/2} \sqrt{1 - k^2 \sin^2 \theta}\, d\theta \tag{2.227}$$

The *elliptic integral of the third kind* is defined by

$$\Pi(x, k, n) = \int_0^x \frac{dy}{\left(1 + n^2 y^2\right) \sqrt{\left(1 - y^2\right)\left(1 - k^2 y^2\right)}} \qquad x \in [-1, 1] \tag{2.228}$$

or

$$\Pi(\varphi, k, n) = \int_0^{\varphi} \frac{d\theta}{\left(1 + n^2 \sin^2 \theta\right) \sqrt{1 - k^2 \sin^2 \theta}} \qquad \varphi \in [-1, 1] \tag{2.229}$$

$$x = \sin \varphi \tag{2.230}$$

If $x = 1$ or $\varphi = \pi/2$, these integrals are called *complete elliptic integrals of the third kind* and denoted by $\Pi(k, n)$:

$$\begin{aligned} \Pi(k, n) &= \int_0^1 \frac{dy}{\left(1 + n^2 y^2\right) \sqrt{\left(1 - y^2\right)\left(1 - k^2 y^2\right)}} \\ &= \int_0^{\pi/2} \frac{d\theta}{\left(1 + n^2 \sin^2 \theta\right) \sqrt{1 - k^2 \sin^2 \theta}} \end{aligned} \tag{2.231}$$

Employing the inverse of the elliptic integral (2.220), we define the following *Jacobi elliptic function*:

$$\text{sn}(u, k) = \sin \varphi = x \tag{2.232}$$

$$\text{cn}(u, k) = \cos \varphi = \sqrt{1 - x^2} \tag{2.233}$$

$$\text{dn}(u, k) = \sqrt{1 - k^2 \sin^2 \varphi} = \sqrt{1 - k^2 x^2} \tag{2.234}$$

When a dynamic problem reduces to an elliptic integral, the problem is considered solved. The behavior of elliptic integrals is well-defined.

The theory of elliptic functions was independently developed by Abel (1802–1829) and Jacobi (1804–1851) in the nineteenth century. Although elliptic functions only enable us to solve a relatively small class of equations of the form $\ddot{x} = F(x)$, some important problems, such as pendulum and torque-free rigid-body motion, belong to this class.

Proof: The elliptic functions are inverses of the elliptic integrals. There are two standard forms of these functions, known as Jacobi elliptic functions and Weierstrass elliptic functions. Jacobi elliptic functions appear as solutions to differential equations of the form

$$\frac{d^2x}{dt^2} = a_0 + a_1 x + a_2 x^2 + a_3 x^3 \tag{2.235}$$

and Weierstrass elliptic functions appear as solutions to differential equations of the form

$$\frac{d^2x}{dt^2} = a_0 + a_1 x + a_2 x^2 \tag{2.236}$$

There are many cases in which $F = F(x)$ in the equation of motion

$$mv\frac{dv}{dx} = F(x) \tag{2.237}$$

is a polynomial or can be expanded as a polynomial,

$$F(x) = a_0 + a_1 x + a_2 x^2 + a_3 x^3 + \cdots \tag{2.238}$$

Consider the motion of a particle with mass $m = 1$ and with the equation of motion of the form

$$\frac{d^2x}{dt^2} = F(x) = a_0 + a_1 x + a_2 x^2 + a_3 x^3 \tag{2.239}$$

which describes the motion of a particle moving under a force function expanded up to third order in displacement x. The solution of this type of equation can be expressed in terms of Jacobi elliptic functions.

Multiplying (2.239) by $\dot{x}$ leads to the first-order differential equation

$$\frac{1}{2}\dot{x}^2 - \left(a_0 x + \frac{a_1}{2}x^2 + \frac{a_2}{3}x^3 + \frac{a_3}{4}x^4\right) = E_1 \tag{2.240}$$

where

$$E_1 = \frac{1}{2}\dot{x}_0^2 - \left(a_0 x_0 + \frac{a_1}{2}x_0^2 + \frac{a_2}{3}x_0^3 + \frac{a_3}{4}x_0^4\right) \tag{2.241}$$

where E_1 is a constant of motion. We may write Equation (2.240) in the form

$$\dot{x}^2 = b_0 + b_1 x + b_2 x^2 + b_3 x^3 + b_4 x^4 \tag{2.242}$$

or

$$\dot{x}^2 = b_4 (x - \alpha)(x - \beta)(x - \gamma)(x - \delta) \tag{2.243}$$

Legendre (1752–1833) could transform Equation (2.243) to

$$\dot{y}^2 = \left(1 - y^2\right)\left(1 - k^2 y^2\right) \tag{2.244}$$

where

$$y^2 = \frac{(\beta - \delta)(x - \alpha)}{(\alpha - \delta)(x - \beta)} \tag{2.245}$$

$$k^2 = \frac{(\beta - \gamma)(\alpha - \delta)}{(\alpha - \delta)(\beta - \delta)} \tag{2.246}$$

The solution of Equation (2.244) is

$$\int dt = \int \frac{dy}{\sqrt{\left(1-y^2\right)\left(1-k^2y^2\right)}} \tag{2.247}$$

Assuming $k^2 < 1$, Legendre transformed the integral

$$F(x,k) = \int_0^x \frac{dy}{\sqrt{\left(1-y^2\right)\left(1-k^2y^2\right)}} \qquad -1 \leqslant x \leqslant 1 \tag{2.248}$$

to

$$F(\varphi,k) = \int_0^\varphi \frac{d\varphi}{\sqrt{1-k^2\sin^2\varphi}} \qquad x = \sin\varphi \tag{2.249}$$

which is called the elliptic integral of the first kind.

The inverse function of the first kind of elliptic integral,

$$u = F(x,k) = F(\sin\varphi,k) = \int_0^\varphi \frac{d\varphi}{\sqrt{1-k^2\sin^2\varphi}} \tag{2.250}$$

is called the Jacobi elliptic function $x = x(u,k)$ and is shown by $\mathrm{sn}(u,k)$:

$$x = \mathrm{sn}(u,k) = \sin\varphi \tag{2.251}$$

So,

$$\mathrm{sn}^{-1}(u,k) = \int_0^\varphi \frac{d\varphi}{\sqrt{1-k^2\sin^2\varphi}} = \int_0^x \frac{dy}{\sqrt{\left(1-y^2\right)\left(1-k^2y^2\right)}} \tag{2.252}$$

The angular variable φ is called the amplitude of u and is denoted by $\mathrm{am}(u)$:

$$\varphi = \mathrm{am}(u) = \sin^{-1}\left[\mathrm{sn}(u,k)\right] \tag{2.253}$$

If $x = 1$ or $\varphi = \pi/2$, the definite integrals (2.248) and (2.249) are called complete elliptic integrals and are denoted by

$$K(k) = \int_0^1 \frac{dy}{\sqrt{\left(1-y^2\right)\left(1-k^2y^2\right)}} = \int_0^{\pi/2} \frac{d\theta}{\sqrt{1-k^2\sin^2\theta}} \tag{2.254}$$

Two other Jacobi elliptic functions are defined by

$$\mathrm{cn}(u,k) = \cos\varphi = \sqrt{1-x^2} \tag{2.255}$$

and

$$\mathrm{dn}(u,k) = \sqrt{1-k^2x^2} = \sqrt{1-k^2\mathrm{sn}^2(u,k)} \tag{2.256}$$

which are the inverse functions of the following integrals respectively:

$$u = \int_1^{\mathrm{cn}(u,k)} \frac{dy}{\sqrt{\left(1-y^2\right)\left(q^2-k^2y^2\right)}} \tag{2.257}$$

$$u = \int_1^{\mathrm{dn}(u,k)} \frac{dy}{\sqrt{\left(1-y^2\right)\left(y^2-q^2\right)}} \tag{2.258}$$

So, the Jacobi elliptic functions are related by the identities

$$\mathrm{sn}^2(u,k) + \mathrm{cn}^2(u,k) = 1 \tag{2.259}$$

$$\mathrm{dn}^2(u,k) + k^2\,\mathrm{sn}^2(u,k) = 1 \tag{2.260}$$

$$\mathrm{dn}^2(u,k) - k^2\,\mathrm{cn}^2(u,k) = 1 - k^2 \tag{2.261}$$

and their derivatives are related by the identities

$$\frac{d}{du}\mathrm{sn}(u,k) = \mathrm{cn}(u,k)\mathrm{dn}(u,k) \tag{2.262}$$

$$\frac{d}{du}\mathrm{cn}(u,k) = -\mathrm{sn}(u,k)\mathrm{dn}(u,k) \tag{2.263}$$

$$\frac{d}{du}\mathrm{dn}(u,k) = -k^2\,\mathrm{sn}(u,k)\mathrm{cn}(u,k) \tag{2.264}$$

■

Example 133 Why $0 \le k \le 1$? Consider the equation of motion of a pendulum,

$$ml^2\ddot{\theta} + mgl\sin\theta = 0 \tag{2.265}$$

which can be reduced to a first-order equation using the energy integral of motion $K + V = E$:

$$\tfrac{1}{2}ml^2\dot{\theta}^2 + mg\,(1 - \cos\theta) = E \tag{2.266}$$

The energy required to raise the pendulum from the lowest position at $\theta = 0$ to the highest position at $\theta = \pi$ is $2mgl$. So, we may write the energy of the system as

$$E = k^2\,(2mgl) \qquad k > 0 \tag{2.267}$$

where the value of k can be calculated by the initial conditions:

$$k = \sqrt{\frac{E_0}{2mgl}} = \frac{1}{2}\sqrt{\frac{1}{gl}\left(2g + l^2\dot{\theta}_0^2 - 2g\cos\theta_0\right)} \tag{2.268}$$

The condition $0 \le k \le 1$ is equivalent to the oscillatory motion of the pendulum.

Example 134 ★ Hypergeometric Functions We can expand the integrand of $K(k)$ by a binomial series:

$$\frac{1}{\sqrt{1 - k^2\sin^2(\theta)}} = 1 + \frac{1}{2}k^2\sin^2\theta - \frac{3}{8}k^4\sin^4\theta + \cdots$$
$$= \sum_{n=0}^{\infty}\frac{(2n-1)!!}{(2n)!!} \tag{2.269}$$

The double factorials are defined as

$$(2n)!! = 2^n n! = 2 \times 4 \times 6 \times \cdots \times (2n) \tag{2.270}$$

$$(2n-1)!! = \frac{(2n-1)!}{2^n n!} = 1 \times 3 \times 5 \times \cdots \times (2n-1) \tag{2.271}$$

For any closed interval $[0, k_{max}^2]$, $k_{max}^2 < 1$, the series is uniformly convergent and may be integrated term by term:

$$K(k) = \frac{\pi}{2}\left[1 + \sum_{n=0}^{\infty}\left(\frac{(2n-1)!!}{(2n)!!}\right)^2 k^{2n}\right] \tag{2.272}$$

Similarly, we find

$$E(k) = \frac{\pi}{2}\left[1 - \sum_{n=0}^{\infty}\left(\frac{(2n-1)!!}{(2n)!!}\right)^2 \frac{k^{2n}}{2n-1}\right] \tag{2.273}$$

These series are called *hypergeometric functions*.

Example 135 ★ Limiting Values of Elliptic Integral Using the series expansions or from the defining integrals of elliptic integrals, we can show that

$$\lim_{k\to 0} K(k) = \frac{\pi}{2} \qquad \lim_{k\to 0} E(k) = \frac{\pi}{2} \tag{2.274}$$

$$\lim_{k\to 1} K(k) = \infty \qquad \lim_{k\to 1} E(k) = 1 \tag{2.275}$$

Example 136 ★ Period of Elliptic Functions Complete elliptic functions are periodic. Let

$$K = \int_0^{\pi/2} \frac{d\theta}{\sqrt{1 - k^2 \sin^2\theta}} \tag{2.276}$$

Then the period of $sn(u,k)$ is $4K$, the period of $cn(u,k)$ is $4K$, and the period of $dn(u,k)$ is $2K$:

$$sn(u, 4K) = snu \qquad cn(u, 4K) = cnu \qquad dn(u, 2K) = \mathrm{dn}u \tag{2.277}$$

Example 137 Circular and Hyperbolic Integrals Consider an integral of the form

$$y = \int \frac{dx}{\sqrt{f(x)}} \tag{2.278}$$

where $f(x)$ is a second-degree polynomial. It can always be possible to reduce $f(x)$ by multiplication by a positive constant either to the form

$$f(x) = a + 2bx + x^2 \tag{2.279}$$

or to the form

$$f(x) = a + 2bx - x^2 \tag{2.280}$$

where a and b are real constants. The function $f(x)$ is assumed to be positive between the limits of integration. Using a change of variable from x to z,

$$x = p + z \tag{2.281}$$

the integrals become

$$y_1 = \int \frac{dz}{\sqrt{(a + 2bp + p^2) + 2(b + p)z + z^2}} \tag{2.282}$$

or

$$y_2 = \int \frac{dz}{\sqrt{(a + 2bp - p^2) + 2(b - p)z - z^2}} \tag{2.283}$$

By choosing $p = -b$ or $p = b$, we eliminate the linear term. In the case of y_1, where $p = -b$, we have $a + 2bp + p^2 = a - b^2$. Then, depending on the value of $a - b^2$, we have three categories:

1. If $a - b^2 > 0$, then with

$$u = \frac{z}{\sqrt{a - b^2}} \tag{2.284}$$

we have

$$y_1 = \int \frac{du}{\sqrt{1 + u^2}} \tag{2.285}$$

2. If $a - b^2 < 0$, then with

$$u = \frac{z}{\sqrt{b^2 - a}} \tag{2.286}$$

we have

$$y_1 = \int \frac{du}{\sqrt{u^2 - 1}} \tag{2.287}$$

3. If $a - b^2 = 0$, then

$$y_1 = \int \frac{dz}{z} = \ln z + C \tag{2.288}$$

In the case of y_2, where $p = b$, we have $a + 2bp - p^2 = a + b^2 > 0$. Then, we have a fourth category:

4. Using

$$u = \frac{z}{\sqrt{a + b^2}} \tag{2.289}$$

we have

$$y_2 = \int \frac{du}{\sqrt{1 - u^2}} \tag{2.290}$$

The integrals obtained above are hyperbolic and circular integrals. They define the inverses of hyperbolic and circular functions:

$$\int_0^w \frac{du}{\sqrt{1+u^2}} = \sinh^{-1} w \tag{2.291}$$

$$\int_0^w \frac{du}{\sqrt{u^2-1}} = \cosh^{-1} w \tag{2.292}$$

$$\int_0^w \frac{du}{\sqrt{1-u^2}} = \sin^{-1} w \tag{2.293}$$

$$\int_w^1 \frac{du}{\sqrt{1-u^2}} = \cos^{-1} w \tag{2.294}$$

This is because

$$\int_0^1 \frac{du}{\sqrt{1-u^2}} = \frac{\pi}{2} \tag{2.295}$$

and we have

$$\cos^{-1} w = \frac{\pi}{2} - \sin^{-1} w \tag{2.296}$$

The definite integral (2.295) is called the complete circular integral.

Example 138 ★ Arc Length of an Ellipse The parametric equations of an ellipse are

$$x = a\cos\theta \qquad y = b\sin\theta \tag{2.297}$$

The differential of the arc length ds is

$$\begin{aligned} ds^2 &= dx^2 + dy^2 = \left(a^2\sin^2\theta + b^2\cos^2\theta\right) d\theta^2 \\ &= \left(a^2 - \left(a^2 - b^2\right)\cos^2\theta\right)\, d\theta^2 \end{aligned} \tag{2.298}$$

Because of

$$a^2 - b^2 = a^2 e^2 \tag{2.299}$$

where e is the eccentricity of the ellipse, we have

$$ds = a\sqrt{1 - e^2\cos^2\theta}\, d\theta \tag{2.300}$$

If we define $\varphi = \pi/2 - \theta$, then

$$ds = a\sqrt{1 - e^2\sin^2\varphi}\, d\varphi \tag{2.301}$$

and therefore the perimeter P of the ellipse is

$$P = 4a\int_0^{\pi/2} \sqrt{1 - e^2\sin^2\varphi}\, d\varphi = 4aE\left(e\right) \tag{2.302}$$

The perimeter of an ellipse is a complete elliptic integral of the second kind. It is a reason that the terminology "elliptic" has been used to describe elliptic integrals.

2.3.4 Force Is a Function of Velocity, $F = F(v)$

The equation of motion in this case is

$$m\frac{dv}{dt} = F(v) \tag{2.303}$$

and can be integrated by separation of variables:

$$m\int_{v_0}^{v}\frac{dv}{F(v)} = \int_{t_0}^{t} dt \tag{2.304}$$

The result of this integral is a time-dependent velocity function $v = v(t)$ and can be integrated as

$$\int_{x_0}^{x} dx = \int_{v_0}^{v} v(t)\,dt \tag{2.305}$$

to provide the position x as a function of time, $x = x(t)$.

Example 139 Projectile in Air Consider a projectile with mass m that is thrown with an initial velocity $\mathbf{v}_0$ from the origin of the coordinate frame. The air provides a resistance force $-c\mathbf{v}$ proportional to the instantaneous velocity $\mathbf{v}$. The *free-body diagram*, a diagram that shows the isolated body with all the external forces acting upon it, is shown in Figure 2.20.

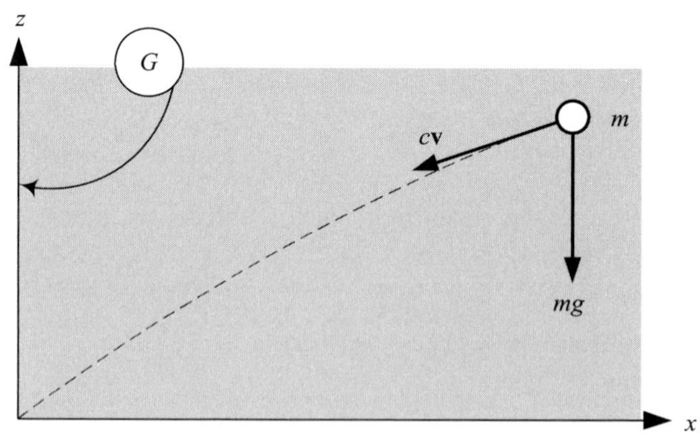

Figure 2.20 Free-body diagram of a projectile in air.

Assuming a flat ground with a uniform gravitational attraction $\mathbf{g}$, the equation of motion of the projectile is

$$m\frac{d\mathbf{v}}{dt} = -mg\hat{k} - c\mathbf{v} \tag{2.306}$$

$$\mathbf{g} = -g\hat{k} \tag{2.307}$$

$$g = 9.80665\,\mathrm{m/s^2} \approx 9.81\,\mathrm{m/s^2} \tag{2.308}$$

To solve the equation of motion (2.306), we may use $\mathbf{v} = d\mathbf{r}/dt$ and integrate the equation to get

$$\mathbf{v} = -gt\hat{k} - \frac{c}{m}\mathbf{r} + \mathbf{v}_0 \tag{2.309}$$

Multiplying Equation (2.309) by $e^{-ct/m}$, we have

$$\frac{d}{dt}\left(\mathbf{r}e^{ct/m}\right) = \mathbf{v}_0 e^{ct/m} - gte^{ct/m}\hat{k} \tag{2.310}$$

and therefore,

$$\mathbf{r}e^{ct/m} = \frac{m}{c}\mathbf{v}_0 e^{ct/m} + \frac{m}{c}ge^{ct/m}\left(\frac{m}{c} - t\right)\hat{k} - \frac{m}{c}\mathbf{v}_0 - \frac{m^2}{c^2}g\hat{k} \tag{2.311}$$

We have found the constant of integration such that $\mathbf{r} = 0$ at $t = 0$. The position vector of the projectile can be simplified to

$$\mathbf{r} = \frac{m}{c}\mathbf{v}_0\left(1 - e^{-ct/m}\right) + \frac{m^2}{c^2}g\left(1 - e^{-ct/m} - \frac{c}{m}t\right)\hat{k} \tag{2.312}$$

and therefore the velocity vector of the projectile is

$$\mathbf{v} = \frac{d\mathbf{r}}{dt} = \mathbf{v}_0 e^{-ct/m} - \frac{m}{c}g\left(1 - e^{-ct/m}\right)\hat{k} \tag{2.313}$$

Equation (2.312) is the solution of the problem. Having $\mathbf{r} = \mathbf{r}(t)$, we are able to determine every kinematic information of the projectile.

At a time $t = t_H$, the projectile is at the maximum height, at which $\mathbf{v}\cdot\hat{k} = 0$, and therefore the *height time* t_H is the solution of

$$\left(\mathbf{v}_0\cdot\hat{k}\right)e^{-ct_H/m} = \frac{m}{c}g\left(1 - e^{-ct_H/m}\right) \tag{2.314}$$

that is,

$$t_H = \frac{m}{c}\ln\left(1 + \frac{c}{m}\frac{1}{g}\mathbf{v}_0\cdot\hat{k}\right) \tag{2.315}$$

The maximum *height* of the projectile is $H = z_M = \mathbf{r}\cdot\hat{k}$ at $t = t_H$:

$$\begin{aligned} H = z_M = \mathbf{r}\cdot\hat{k} &= \frac{m}{c}\left(\mathbf{v}_0\cdot\hat{k} - gt_H\right) \\ &= \frac{m}{c}\left[\mathbf{v}_0\cdot\hat{k} - \frac{m}{c}g\ln\left(1 + \frac{c}{m}\frac{1}{g}\mathbf{v}_0\cdot\hat{k}\right)\right] \end{aligned} \tag{2.316}$$

The *range* of the projectile is $R = |\mathbf{r}|$ when $\mathbf{r}\cdot\hat{k} = 0$ at the *range time* $t = t_R$:

$$\left(1 - e^{-ct/m}\right)\left(\mathbf{v}_0\cdot\hat{k}\right) + \frac{m}{c}g\left(1 - e^{-ct/m} - \frac{c}{m}t\right) = 0 \tag{2.317}$$

The time at which the projectile hits the ground is the solution of a transcendental equation:

$$\left(\mathbf{v}_0 \cdot \hat{k} + \frac{m}{c} g\right)\left(1 - e^{-ct_R/m}\right) - g t_R = 0 \tag{2.318}$$

At $t = t_R$, $\mathbf{r} \cdot \hat{k} = 0$, and therefore $R = \mathbf{r} \cdot \hat{\imath}$:

$$R = \mathbf{r} \cdot \hat{\imath} = \frac{m}{c}\left(1 - e^{-ct_R/m}\right)\left(\mathbf{v}_0 \cdot \hat{\imath}\right) \tag{2.319}$$

If the projectile is thrown in the (x, z)-plane with speed v_0 at an angle θ with respect to the x-axis, the initial velocity is

$$\mathbf{v}_0 = v_0 \begin{bmatrix} \cos\theta \\ 0 \\ \sin\theta \end{bmatrix} \tag{2.320}$$

Substituting $\mathbf{v}_0$ in kinematic equations of the projectile, we find

$$\mathbf{r} = \begin{bmatrix} -\frac{m}{c} v_0 \left(e^{-ct/m} - 1\right) \cos\theta \\ 0 \\ -\frac{m^2}{c^2} g \left(e^{-ct/m} + \frac{c}{m} t - 1\right) - \frac{m}{c} v_0 \left(e^{-ct/m} - 1\right) \sin\theta \end{bmatrix} \tag{2.321}$$

$$\mathbf{v} = \begin{bmatrix} v_0 e^{-ct/m} \cos\theta \\ 0 \\ v_0 e^{-ct/m} \sin\theta - \frac{m}{c} g \left(1 - e^{-ct/m}\right) \end{bmatrix} \tag{2.322}$$

Equation (2.321) is the projectile's path of motion using t as a parameter. We can theoretically find the projectile's path of motion in the (x, z)-plane by eliminating t between the x- and z-components of $\mathbf{r}$. Figure 2.21 illustrates the path of motion of a projectile for different values of the air friction coefficient c.

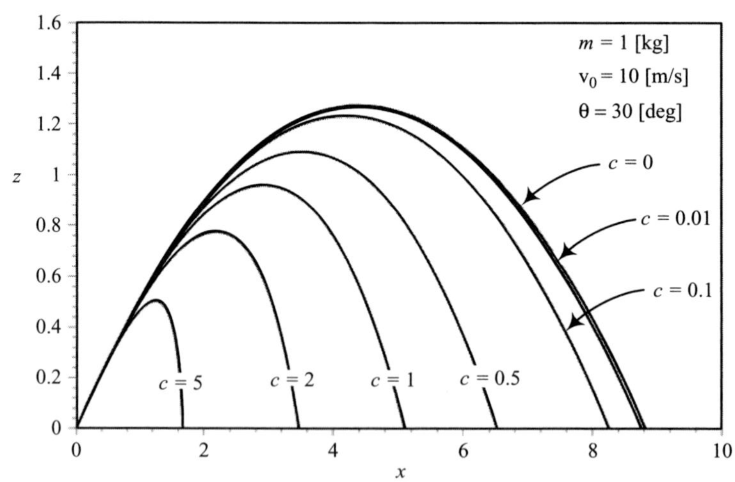

Figure 2.21 The path of motion of a projectile for different values of air friction coefficient c.

The path would be parabolic if there is no air and $c = 0$. To find the path in the no-air condition, we expand $\mathbf{r}$ in a time series

$$\mathbf{r} = \begin{bmatrix} \left(t v_0 + \frac{1}{2}\frac{c}{m}t^2 v_0\right)\cos\theta + O\left(t\left(\frac{c}{m}t\right)^2\right) \\ 0 \\ t v_0 \sin\theta - t^2\left(\frac{1}{2}g + \frac{1}{2}\frac{c}{m}v_0\sin\theta\right) + O\left(t\left(\frac{c}{m}t\right)^2\right) \end{bmatrix} \tag{2.323}$$

and simplify the equations by $c \to 0$ to get

$$\mathbf{r} = \begin{bmatrix} v_0 t\cos\theta \\ 0 \\ v_0 t\sin\theta - \frac{1}{2}g t^2 \end{bmatrix} \tag{2.324}$$

Eliminating t between x and z provides the path of motion:

$$z = -\frac{1}{2}g\frac{x^2}{v_0^2\cos^2\theta} + x\tan\theta \tag{2.325}$$

Example 140 ★ Limit Velocity of a Projectile in Air Let us use $d\mathbf{v}/dt = \mathbf{a}$ and differentiate the equation of motion of a projectile in air (2.306) to find

$$m\frac{d\mathbf{a}}{dt} = -c\mathbf{a} \tag{2.326}$$

This equation can be integrated by separation of variables as

$$m\int_{\mathbf{a}_0}^{\mathbf{a}}\frac{d\mathbf{a}}{\mathbf{a}} = -c\int_0^t dt \tag{2.327}$$

$$\mathbf{a} = \mathbf{a}_0 e^{-ct/m} \tag{2.328}$$

where $\mathbf{a}_0$ is a constant vector. Equation (2.328) indicates that the acceleration vector of the projectile $\mathbf{a}$ is always parallel to a constant vector $\mathbf{a}_0$ with an exponential decreasing magnitude. So, $\mathbf{a} \to 0$ as $t \to \infty$, and therefore $d\mathbf{v}/dt \to 0$ which yields

$$\lim_{t\longrightarrow\infty}\mathbf{v} = -\frac{mg}{c}\hat{k} \tag{2.329}$$

The velocity $\mathbf{v}_l = -(mg/c)\,\hat{k}$ is called the *limit velocity*. There is no limit velocity in the no-air condition.

If the initial velocity vector $\mathbf{v}_0$ is in the (z, x)-plane, the projectile remains in this plane because the applied forces have no component out of this plane. We may show the position vector of the projectile as

$$\mathbf{r} = x\hat{\imath} + z\hat{k} \tag{2.330}$$

The x-component of $\mathbf{r}$ can be found by the dot product of (2.309) and $\hat{\imath}$:

$$x = \mathbf{r}\cdot\hat{\imath} = \frac{m}{c}\left(\mathbf{v}_0 - \mathbf{v}\right)\cdot\hat{\imath} \tag{2.331}$$

Because $\mathbf{v} \to -(mg/c)\,\hat{k}$ as $t \to \infty$, we have

$$x_M = \lim_{t\to\infty} x = \frac{m}{c}\mathbf{v}_0 \cdot \hat{\imath} \tag{2.332}$$

which indicates there is a limiting horizontal range.

Example 141 ★ Kinematic Characteristics of a Projectile in Air Consider a projectile that is thrown in the xz-plane with speed v_0 at an angle θ with respect to the x-axis. The initial velocity of the projectile is given by Equation (2.320):

$$\mathbf{v} = v_0 \cos\theta\,\hat{\imath} + v_0 \sin\theta\,\hat{k} \tag{2.333}$$

The height time t_H and the maximum height of the projectile H are given in Equations (2.315) and (2.316), respectively, which after substituting for $\mathbf{v}_0$ are

$$t_H = \frac{m}{c}\ln\left(1 + \frac{cv_0}{mg}\sin\theta\right) \tag{2.334}$$

$$H = -\frac{m^2}{c^2}g\left(\ln\left(1 + \frac{cv_0}{mg}\sin\theta\right) - cv_0\sin\theta\right) \tag{2.335}$$

and H and t_H may be combined as

$$H = -\frac{m}{c}g\left(t_H - mv_0\sin\theta\right) \tag{2.336}$$

Figure 2.22 illustrates the behavior of t_H and H for different values of the air friction coefficients c. In the no-air condition,

$$t_H = \frac{1}{g}v_0\sin\theta \tag{2.337}$$

$$H = \frac{1}{2g}v_0^2\sin^2\theta \tag{2.338}$$

where H is maximum when c is minimum and $\theta = 90\,\text{deg}$.

The range time t_R is the solution of

$$\left(v_0\sin\theta + \frac{m}{c}g\right)\left(1 - e^{-ct_R/m}\right) - gt_R = 0 \tag{2.339}$$

The solution is not achievable by elementary mathematical functions. However, it is possible to solve the equation numerically and plot the result to have a visual sense of the solution. The range of the projectile is given as

$$R = \mathbf{r}\cdot\hat{\imath} = \frac{m}{c}\left(1 - e^{-ct_R/m}\right)v_0\cos\theta \tag{2.340}$$

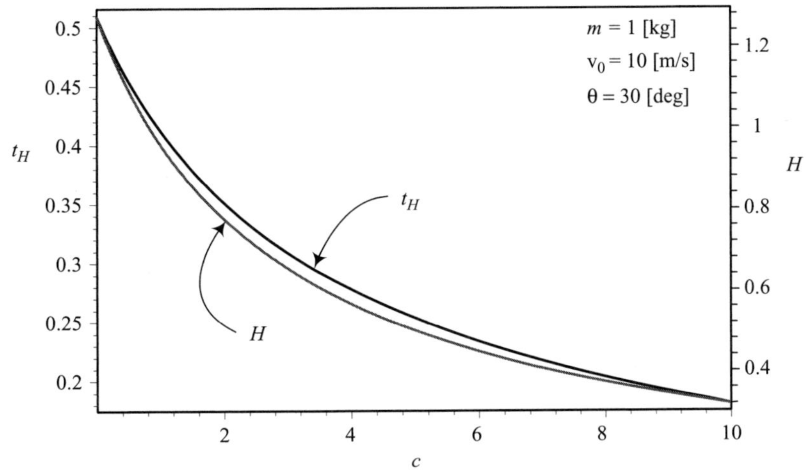

Figure 2.22 The height time t_H and the maximum height H of a projectile in air.

Figure 2.23 illustrates the range time t_R and the range of the projectile R as a function of c. In the no-air condition, t_R and R reduce to

$$t_R = \frac{2}{g} v_0 \sin\theta \tag{2.341}$$

$$R = \frac{1}{g} v_0^2 \sin 2\theta \tag{2.342}$$

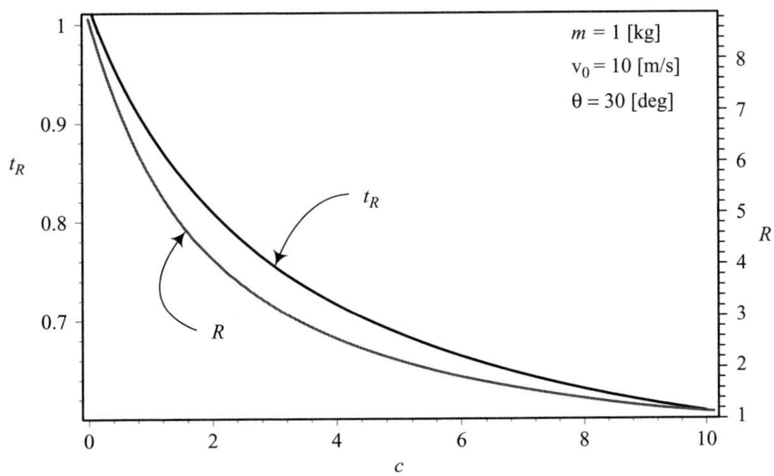

Figure 2.23 The range time t_R and the range R of a projectile in air.

The optimal angle is the shooting angle that maximizes the range R. Figure 2.24 depicts the value of optimal angle for different air friction coefficients c. At the no-air condition, $\theta = 45$ deg is optimal; however, the optimal angle decreases by increasing c.

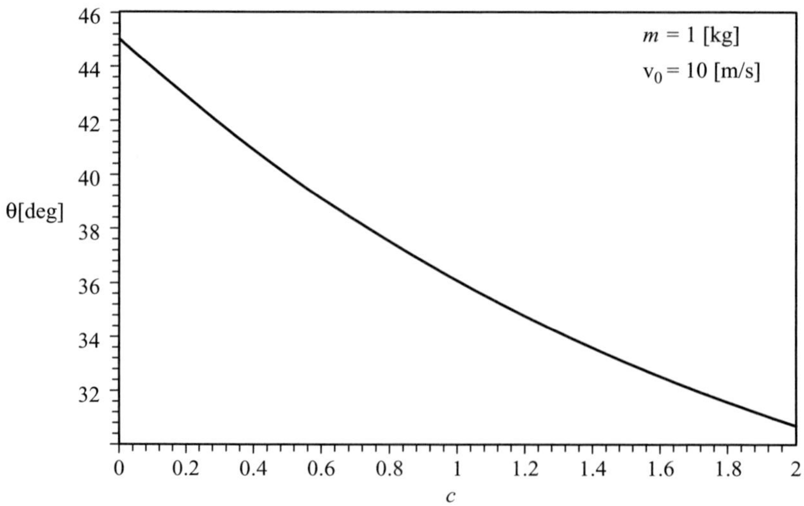

Figure 2.24 The optimal angle θ to maximize the range R.

To determine the optimal angle to maximize the range R approximately, let us expand Equations (2.339) and (2.340) as

$$-g + \left(\frac{c}{m} - \frac{1}{2}\frac{c^2}{m^2}t_R\right)\left(v_0 \sin\theta + \frac{1}{c}gm\right) + O(t_R^2) = 0 \tag{2.343}$$

and

$$R = v_0 t_R \cos\theta + O\left(t_R^2\right) \tag{2.344}$$

respectively, and solve (2.343) for t_R:

$$t_R = 2mv_0 \frac{\sin\theta}{gm + cv_0 \sin\theta} \tag{2.345}$$

So, the range R is approximately equal to

$$R = mv_0^2 \frac{\sin 2\theta}{gm + cv_0 \sin\theta} \tag{2.346}$$

Taking the derivative

$$\frac{d}{d\theta}R = \frac{cmv_0^3 \sin 3\theta + 4gm^2 v_0^2 \cos 2\theta - 3cmv_0^3 \sin\theta}{c^2 v_0^2 + 2g^2 m^2 - c^2 v_0^2 \cos 2\theta + 4cgmv_0 \sin\theta} \tag{2.347}$$

shows that the maximum range happens if the projectile is shot at an angle θ that is the root of

$$-3cv_0 \sin\theta + 4gm \cos 2\theta + cv_0 \sin 3\theta = 0 \tag{2.348}$$

Figure 2.25 illustrates the required shooting angle θ to maximize the range R for different values of the air friction coefficient c, based on Equation (2.348). By increasing the air resistance c, we must decrease θ to achieve a maximum range.

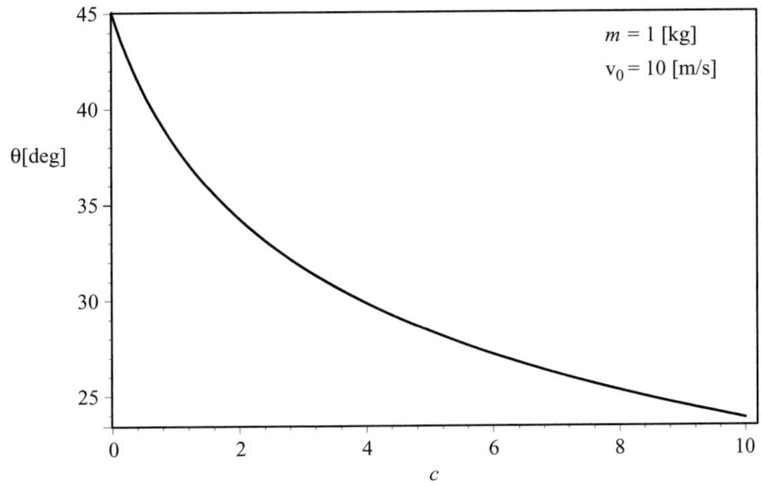

Figure 2.25 The approximate optimal angle θ to maximize the range R.

Expanding Equations (2.339) and (2.340) to a higher degree provides a better approximation.

Example 142 ★ Air Resistance with v^n Let z denote the upward distance from a point at which a particle P with mass $m = 1$ is projected upward with speed v:

$$v = \frac{dz}{dt} \tag{2.349}$$

There is a resistance force proportional to v^n. If the ground is assumed flat with a uniform gravitational attraction, then the equation of motion of the point is

$$\frac{dv}{dt} = -g - kv^n \tag{2.350}$$

To calculate $v(t)$, we may integrate from

$$\frac{dv}{-g - kv^n} = dt \tag{2.351}$$

and to calculate the height z, we can rewrite the equation of motion as

$$v\frac{dv}{dz} = -g - kv^n \tag{2.352}$$

and integrate from

$$\frac{v\,dv}{-g - kv^n} = dz \tag{2.353}$$

If the initial conditions are

$$z(0) = 0 \qquad v(0) = v_0 \tag{2.354}$$

and $n = 1$, then we have

$$\int_{v_0}^{v} \frac{dv}{-g - kv} = t \tag{2.355}$$

$$v = -\frac{g}{k} + \frac{g}{k}\left(1 + \frac{k}{g} v_0 e^{-kt}\right) \tag{2.356}$$

and

$$\begin{aligned} z &= \int_{v_0}^{v} \frac{v\,dv}{-g - kv} = -\frac{v - v_0}{k} + \frac{g}{k^2} \ln \frac{g + kv}{g + kv_0} \\ &= -\frac{g}{k^2}\left(\ln \frac{g + kv_0}{g + kv_0 e^{-kt}} - \frac{k}{g} v_0 \left(1 - e^{-kt}\right)\right) \end{aligned} \tag{2.357}$$

If $n = 2$, then

$$\int_{v_0}^{v} \frac{dv}{-g - kv^2} = t \tag{2.358}$$

$$v = -\sqrt{\frac{g}{k}} \tan\left(\sqrt{gk}t - \tan^{-1}\sqrt{\frac{k}{g}} v_0\right) \tag{2.359}$$

$$z = \int_{v_0}^{v} \frac{v\,dv}{-g - kv} = \sqrt{\frac{1}{k}\left(e^{-2kz} - \frac{g}{k}\right)} \tag{2.360}$$

Example 143 Force Proportional to Velocity Assume that the applied force $\mathbf{F}$ on a particle is proportional to $\mathbf{v}$,

$$\mathbf{F} = k\mathbf{v} \tag{2.361}$$

where k is a constant. Therefore, the equation of motion

$$\mathbf{F} = m\mathbf{a} = k\mathbf{v} \tag{2.362}$$

simplifies to

$$\frac{d^2\mathbf{r}}{dt^2} = \frac{k}{m}\frac{d\mathbf{r}}{dt} \tag{2.363}$$

Integrating (2.363) shows that

$$\frac{d\mathbf{r}}{dt} - \frac{k}{m}\mathbf{r} = \mathbf{c}_1 \tag{2.364}$$

where $\mathbf{c}_1$ is a constant vector. This is a first-order differential equation with the solution

$$\mathbf{r} = \mathbf{c}_2 e^{(k/m)t} - \frac{m}{k}\mathbf{c}_1 \tag{2.365}$$

where $\mathbf{c}_2$ is another constant vector. The path of motion (2.365) shows that, when the applied force is proportional to velocity, the particle will move in a straight line.

2.4 SPATIAL AND TEMPORAL INTEGRALS

Integration of the equation of motion is not possible in the general case; however, there are two general integrals of motion: time and space integrals. The space integral of motion generates the principle of work and energy, and the time integral of motion generates the principle of impulse and momentum.

2.4.1 Spatial Integral: Work and Energy

The Spatial integral of the Newton equation of motion,

$$\int_1^2 {}^G\mathbf{F} \cdot d\mathbf{r} = m \int_1^2 {}^G\mathbf{a} \cdot d\mathbf{r} \tag{2.366}$$

reduces to the *principle of work and energy*,

$${}_1W_2 = K_2 - K_1 \tag{2.367}$$

where K is the *kinetic energy*,

$$K = \tfrac{1}{2} m \, {}^G\mathbf{v}^2 \tag{2.368}$$

and ${}_1W_2$ is the *work* done by the force during the *displacement* $\mathbf{r}_2 - \mathbf{r}_1$:

$${}_1W_2 = \int_1^2 {}^G\mathbf{F} \cdot d\mathbf{r} \tag{2.369}$$

If there is a scalar potential field function $V = V(x, y, z)$ such that

$$\mathbf{F} = -\nabla V = -\frac{dV}{d\mathbf{r}} = -\left(\frac{\partial V}{\partial x}\hat{\imath} + \frac{\partial V}{\partial y}\hat{\jmath} + \frac{\partial V}{\partial z}\hat{k} \right) \tag{2.370}$$

then the principle of work and energy (2.367) simplifies to the *principle of conservation of energy*:

$$K_1 + V_1 = K_2 + V_2 \tag{2.371}$$

Proof: We can simplify the right-hand side of the spatial integral (2.366) by a change of variable:

$$\begin{aligned} \int_{\mathbf{r}_1}^{\mathbf{r}_2} {}^G\mathbf{F} \cdot d\mathbf{r} &= m \int_{\mathbf{r}_1}^{\mathbf{r}_2} {}^G\mathbf{a} \cdot d\mathbf{r} = m \int_{t_1}^{t_2} \frac{d\mathbf{v}}{dt} \cdot \mathbf{v}\, dt \\ &= m \int_{\mathbf{v}_1}^{\mathbf{v}_2} \mathbf{v} \cdot d\mathbf{v} = \frac{1}{2} m \left(\mathbf{v}_2^2 - \mathbf{v}_1^2 \right) \end{aligned} \tag{2.372}$$

The *kinetic energy* of a point P with mass m that is at a position pointed by ${}^G\mathbf{r}$ and having a velocity ${}^G\mathbf{v}$ is defined by (2.368), and the *work* done by the applied force ${}^G\mathbf{F}$ on m in going from point $\mathbf{r}_1$ to $\mathbf{r}_2$ is defined by (2.369). Hence the spatial integral of the equation of motion (2.366) reduces to the principle of work and energy (2.367),

$${}_1W_2 = K_2 - K_1 \tag{2.373}$$

which says that the work ${}_1W_2$ done by the applied force ${}^G\mathbf{F}$ on m during the displacement $\mathbf{r}_2 - \mathbf{r}_1$ is equal to the difference of the kinetic energy of m.

If the force $\mathbf{F}$ is the gradient of a potential function V,

$$\mathbf{F} = -\nabla V \tag{2.374}$$

then $\mathbf{F} \cdot d\mathbf{r}$ in Equation (2.369) is an exact differential, and hence

$$\int_1^2 {}^G\mathbf{F} \cdot d\mathbf{r} = \int_1^2 dV = -(V_2 - V_1) \tag{2.375}$$

$$E = K_1 + V_1 = K_2 + V_2 \tag{2.376}$$

In this case the work done by the force is independent of the path of motion between $\mathbf{r}_1$ and $\mathbf{r}_2$ and depends only upon the value of the potential V at the start and end points of the path. The function V is called the *potential energy*, Equation (2.371) is called the *principle of conservation of energy*, and the force $\mathbf{F} = -\nabla V$ is called a *potential* or a *conservative force*. The kinetic plus potential energy of the dynamic system is called the *mechanical energy* of the system and is shown by $E = K + V$, where E is a constant of motion if all of the applied forces are conservative.

A force $\mathbf{F}$ is conservative only if it is the gradient of a stationary scalar function. The components of a conservative force will only be functions of space coordinates:

$$\mathbf{F} = F_x(x, y, z)\,\hat{\imath} + F_y(x, y, z)\,\hat{\jmath} + F_z(x, y, z)\,\hat{k} \tag{2.377}$$

■

Example 144 Motion on a Planar Curve If the mechanical energy E of a particle is constant, then $\dot{E} = 0$ would give the equation of motion:

$$E = K(x, \dot{x}, t) + V(x) = \text{const} \tag{2.378}$$

$$\dot{E} = \frac{\partial K}{\partial x}\dot{x} + \frac{\partial K}{\partial \dot{x}}\ddot{x} + \frac{\partial K}{\partial t} + \frac{\partial V}{\partial x} = 0 \tag{2.379}$$

Consider a point mass m that slides frictionless on a given curve in the vertical plane (x, z), as is shown in Figure 2.26:

$$z = f(x) \tag{2.380}$$

The kinetic and potential energies of m are

$$K = \frac{1}{2}mv^2 = \frac{1}{2}m\left(\dot{x}^2 + \dot{z}^2\right) = \frac{1}{2}m\dot{x}^2\left(1 + \left(\frac{df}{dx}\right)^2\right) \tag{2.381}$$

$$V = mgz = mgf \tag{2.382}$$

The mechanical energy of m is constant,

$$E = K + V = \frac{1}{2}m\dot{x}^2\left(1 + \left(\frac{df}{dx}\right)^2\right) + mgf = \text{const} \tag{2.383}$$

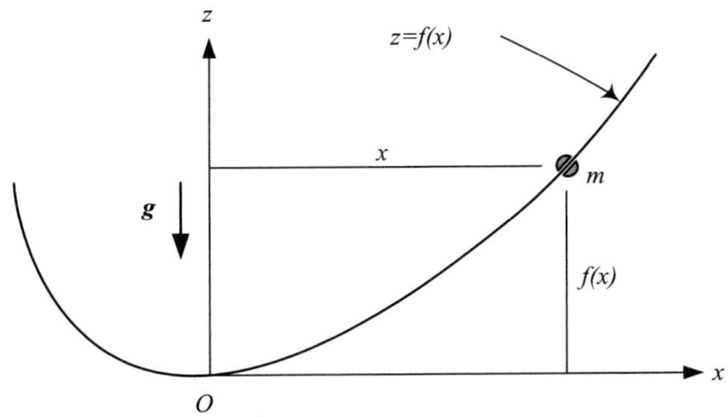

Figure 2.26 A point mass m that slides frictionless on a given curve in the vertical plane (x, z).

and therefore its equation of motion is

$$\dot{E} = m\dot{x}\ddot{x}\left(1 + \left(\frac{df}{dx}\right)^2\right) + m\dot{x}^3 \frac{df}{dx}\frac{d^2 f}{dx^2} + mg\dot{x}\frac{df}{dx} = 0 \tag{2.384}$$

or, after simplification,

$$\ddot{x}\left(1 + \left(\frac{df}{dx}\right)^2\right) + \dot{x}^2 \frac{df}{dx}\frac{d^2 f}{dx^2} + g\frac{df}{dx} = 0 \tag{2.385}$$

As an example, let us try a slope

$$z = kx \tag{2.386}$$

Then

$$\frac{df}{dx} = k \qquad \frac{d^2 f}{dx^2} = 0 \tag{2.387}$$

and therefore the equation of motion is

$$\ddot{x} = -g\frac{k}{1+k^2} \tag{2.388}$$

It is the equation of a free fall under a reduced gravitational acceleration $-kg/(1+k^2)$.

Example 145 A Falling Object on a Spring An object with mass m falls from a height h on a linear spring with stiffness k, as is shown in Figure 2.27. We can determine the maximum compression of the spring using the work–energy principle.

The gravity force mg and the spring force $-kx$ are the acting forces on m. If x_M is the maximum compression of the spring, then $K_2 = K_1 = 0$ and we have

$${}_1W_2 = mg(h + x_M) - \int_0^{x_M} kx\, dx = 0 \tag{2.389}$$

$$x_M = \frac{mg}{k} + \sqrt{\left(\frac{mg}{k}\right)^2 + \frac{2mgh}{k}} \tag{2.390}$$

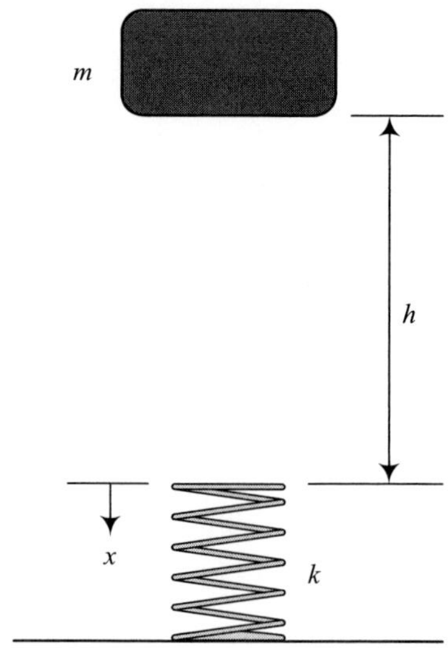

Figure 2.27 A falling object on a spring.

If we put m on the spring, it will deflect statically to

$$x_0 = \frac{mg}{k} \tag{2.391}$$

So we may compare x_M to x_0 and write x_M as

$$x_M = x_0 \left(1 + \sqrt{1 + \frac{2h}{x_0}}\right) \tag{2.392}$$

Consider

$$h = 1\,\text{m} \qquad m = 1\,\text{kg} \qquad k = 1000\,\text{N/m} \tag{2.393}$$

Then if $g = 9.81\,\text{m/s}^2$, we have

$$x_0 = \frac{mg}{k} = \frac{9.81}{1000} = 0.00981\,\text{m} = 9.81\,\text{mm} \tag{2.394}$$

$$x_M = x_0 \left(1 + \sqrt{1 + \frac{2h}{x_0}}\right) = 0.15022\,\text{m} = 150.22\,\text{mm} \tag{2.395}$$

However, if this experiment is performed on the moon surface, then $g = 1.6\,\text{m/s}^2$, and we have

$$x_0 = \frac{mg}{k} = \frac{1.6}{1000} = 0.0016\,\text{m} = 1.6\,\text{mm} \tag{2.396}$$

$$x_M = x_0 \left(1 + \sqrt{1 + \frac{2h}{x_0}}\right) = 0.058191\,\text{m} = 58.191\,\text{mm} \tag{2.397}$$

It is theoretically possible to measure g based on such an experiment:

$$g = \frac{2hk}{m\left[(x_M/x_0 - 1)^2 - 1\right]} \tag{2.398}$$

The mass of the spring and friction are the main sources of error.

Example 146 Variable Potential Energy Figure 2.28 illustrates a uniform rope with mass m and length l lying on a frictionless table with height H. As soon as a short length of the rope hangs over, the rope starts falling.

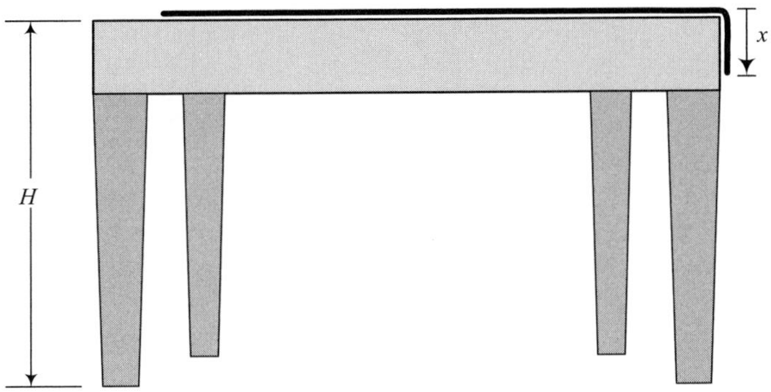

Figure 2.28 A uniform rope with mass m and length l lying on a frictionless table.

To determine the velocity of the rope when its tail leaves the table, we use the energy conservation principle. All particles of the rope are moving together with velocity v, so the kinetic energy of the rope is

$$K = \tfrac{1}{2}mv^2 \tag{2.399}$$

However, the potential energy of the rope belongs to the hanging length x:

$$V = -\frac{m}{l}gx\frac{x}{2} \tag{2.400}$$

Therefore, the energy conservation applies in the form

$$E = K + V = \frac{1}{2}mv^2 - \frac{1}{2}\frac{m}{l}gx^2 = 0 \tag{2.401}$$

which provides the speed of the rope as a function of x:

$$v = \sqrt{\frac{g}{l}}x \qquad 0 \le x \le l \tag{2.402}$$

The speed of the rope when $x = l$ is

$$v_0 = \sqrt{gl} \qquad x = l \tag{2.403}$$

and is independent of the rope's length density.

In the case $H > l$, the rope will have a free fall for a distance $h = H - l$ and its head will touch the ground at speed v_1, which can be found from the work–energy principle:

$$ {}_0W_1 = K_2 - K_1 \tag{2.404} $$

$$ mgh = \tfrac{1}{2}mv_1^2 - \tfrac{1}{2}mv_0^2 \tag{2.405} $$

$$ v_1 = \sqrt{v_0^2 + 2gh} \qquad h = H - l \tag{2.406} $$

In the case $H < l$, the rope's head will touch the ground before its tail leaves the table. If $h = l - H$, then the rope's head will touch the ground at speed v_2:

$$ v_2 = \sqrt{\frac{g}{l}}h \qquad h = l - H \tag{2.407} $$

Example 147 Work of a Planar Force on a Planar Curve Consider a planar force **F**,

$$ \mathbf{F} = 2xy\hat{\imath} + 3x^2\hat{\jmath} \qquad \mathrm{N} \tag{2.408} $$

that moves a mass m on a planar curve

$$ y = x^2 \tag{2.409} $$

from (0, 0) to (3, 1) m.

Using

$$ dy = 2x\,dx \tag{2.410} $$

we can calculate the work done by the force:

$$ \begin{aligned} {}_1W_2 &= \int_{P_1}^{P_2} {}^G\mathbf{F} \cdot d\mathbf{r} = \int_{(0,0)}^{(3,1)} \left(2xy\,dx + 3x^2\,dy\right) \\ &= \int_0^3 \left(2x^3\,dx + 6x^3\,dx\right) = \int_0^3 8x^3\,dx = 162\,\mathrm{Nm} \end{aligned} \tag{2.411} $$

Example 148 Work in a Nonpotential Force Field Consider a force field **F**,

$$ \mathbf{F} = 2xy\hat{\imath} + 3x^3\hat{\jmath} + z\hat{k} \qquad \mathrm{N} \tag{2.412} $$

The work done by the force when it moves from $P_1(0,0,0)$ to $P_2(1,3,5)$ m on a line

$$ \begin{aligned} x &= t \\ y &= \frac{\Delta y}{\Delta x}(x - x_1) = 3t \\ z &= \frac{\Delta z}{\Delta x}(x - x_1) = 5t \end{aligned} \tag{2.413} $$

is

$$
\begin{aligned}
{}_1W_2 &= \int_{P_1}^{P_2} {}^G\mathbf{F} \cdot d\mathbf{r} = \int_{P_1}^{P_2} {}^G\mathbf{F} \cdot \frac{d\mathbf{r}}{dt}\, dt \\
&= \int_{P_1}^{P_2} \begin{bmatrix} 2xy \\ 3x^3 \\ z \end{bmatrix} \cdot \begin{bmatrix} dx \\ dy \\ dz \end{bmatrix} = \int_0^1 \begin{bmatrix} 2xy \\ 3x^3 \\ z \end{bmatrix} \cdot \begin{bmatrix} dt \\ 3\, dt \\ 5\, dt \end{bmatrix} \\
&= \int_0^1 \left(5z + 2xy + 9x^3\right) dt \\
&= \int_0^1 \left(25t + 6t^2 + 9t^3\right) dt = 16.75\,\text{Nm}
\end{aligned}
\tag{2.414}
$$

Let us change the path from a straight line to a curve,

$$
x = t \qquad y = 3t^2 \qquad z = 5t^3 \tag{2.415}
$$

and calculate the work ${}_1W_2$ again:

$$
\begin{aligned}
{}_1W_2 &= \int_{P_1}^{P_2} {}^G\mathbf{F} \cdot d\mathbf{r} = \int_0^1 \begin{bmatrix} 2xy \\ 3x^3 \\ z \end{bmatrix} \cdot \begin{bmatrix} dt \\ 6t\, dt \\ 15t^2\, dt \end{bmatrix} \\
&= \int_0^1 \left(18tx^3 + 15t^2 z + 2xy\right) dt \\
&= \int_0^1 \left(75t^5 + 18t^4 + 6t^3\right) dt = 17.6\,\text{Nm}
\end{aligned}
\tag{2.416}
$$

Because ${}_1W_2$ depends on the path between P_1 and P_2, the force field (2.412) is not a potential field.

Example 149 Work in a Potential Force Field Consider a force field $\mathbf{F}$,

$$
\mathbf{F} = 2xy\hat{\imath} + x^2\hat{\jmath} + 2z\hat{k} \qquad \text{N} \tag{2.417}
$$

The work done by the force when it moves from $P_1(0, 0, 0)$ to $P_2(1, 3, 5)$ m on a line

$$
x = t \qquad y = 3t \qquad z = 5t \tag{2.418}
$$

is

$$
\begin{aligned}
{}_1W_2 &= \int_{P_1}^{P_2} {}^G\mathbf{F} \cdot d\mathbf{r} = \int_0^1 \left(10z + 2xy + 3x^2\right) dt \\
&= \int_0^1 \left(50t + 6t^2 + 3t^2\right) dt = 28\,\text{Nm}
\end{aligned}
\tag{2.419}
$$

We may change the path from a straight line to a curve,

$$
x = t \qquad y = 3t^2 \qquad z = 5t^3 \tag{2.420}
$$

and calculate the work ${}_1W_2$ again:

$$
\begin{aligned}
{}_1W_2 &= \int_{P_1}^{P_2} {}^G\mathbf{F}\cdot d\mathbf{r} = \int_0^1 \begin{bmatrix} 2xy \\ x^2 \\ 2z \end{bmatrix} \cdot \begin{bmatrix} dt \\ 6\,t\,dt \\ 15t^2\,dt \end{bmatrix} \\
&= \int_0^1 2\left(3tx^2 + 15t^2 z + xy\right) dt \\
&= 2\int_0^1 \left(75t^5 + 6t^3\right) dt = 28\,\mathrm{Nm}
\end{aligned}
\tag{2.421}
$$

Because ${}_1W_2$ is independent of the path from P_1 to P_2, the force field (2.417) might be a potential field. To check, we may calculate the curl of the field:

$$
\nabla\times\mathbf{F} = \begin{bmatrix} \partial/\partial x \\ \partial/\partial y \\ \partial/\partial z \end{bmatrix} \times \begin{bmatrix} 2xy \\ x^2 \\ 2z \end{bmatrix} = \begin{bmatrix} 0 \\ 0 \\ 0 \end{bmatrix}
\tag{2.422}
$$

So, the force field is a potential.

Example 150 Moving Electric Charge in an Electromagnetic Field To move an electric charge e with velocity $\mathbf{v}$ in a magnetic field $\mathbf{H}$, we need to provide a force $\mathbf{F}_H$,

$$
\mathbf{F}_H = \frac{e}{c}\left(\mathbf{v}\times\mathbf{H}\right)
\tag{2.423}
$$

where c is the speed of light. A force $\mathbf{F}_E$ will be applied on e when it is in an electric field $\mathbf{E}$:

$$
\mathbf{F}_E = e\mathbf{E}
\tag{2.424}
$$

Hence, the total force on an electric charge e that is moving with velocity $\mathbf{v}$ in an electromagnetic field $\mathbf{E}$ and $\mathbf{H}$ is

$$
\mathbf{F} = e\mathbf{E} + \frac{e}{c}\left(\mathbf{v}\times\mathbf{H}\right)
\tag{2.425}
$$

If the charge moves an incremental distance $d\mathbf{r}$ in time dt, then the work dW done by the force is

$$
\begin{aligned}
dW &= \mathbf{F}\cdot d\mathbf{r} = \mathbf{F}\cdot\frac{d\mathbf{r}}{dt}dt = \mathbf{F}\cdot\mathbf{v}\,dt \\
&= \left(e\mathbf{E} + \frac{e}{c}\left(\mathbf{v}\times\mathbf{H}\right)\right)\cdot\mathbf{v}\,dt = e\mathbf{E}\cdot\mathbf{v}\,dt
\end{aligned}
\tag{2.426}
$$

which shows that the work done by the electromagnetic field is only due to the electric field. We can calculate the rate of work as

$$
P = \frac{dW}{dt} = e\mathbf{E}\cdot\mathbf{v}
\tag{2.427}
$$

to determine the required power to move e at velocity $\mathbf{v}$.

Example 151 Central-Force Motion Consider two particles with masses m_1 and m_2 that interact according to Newton's law of gravitation. The equation of motion of the mass m_1 with respect to m_2 is

$$m_1 \frac{d^2\mathbf{r}}{dt^2} = -G\frac{m_1 m_2}{r^3}\mathbf{r} \tag{2.428}$$

The vector product $\mathbf{L} = \mathbf{r} \times d\mathbf{r}/dt$ is an integral of this motion.

To show this, we cross multiply $\mathbf{r}$ by the equation of motion,

$$\mathbf{r} \times m_1 \frac{d^2\mathbf{r}}{dt^2} = -G\frac{m_1 m_2}{r^3}\mathbf{r} \times \mathbf{r} \tag{2.429}$$

and find

$$\mathbf{r} \times \frac{d^2\mathbf{r}}{dt^2} = 0 \tag{2.430}$$

because $\mathbf{r} \times \mathbf{r} = 0$. Therefore,

$$\frac{d}{dt}\left(\mathbf{r} \times \frac{d\mathbf{r}}{dt}\right) = \mathbf{r} \times \frac{d^2\mathbf{r}}{dt^2} + \frac{d\mathbf{r}}{dt} \times \frac{d\mathbf{r}}{dt} = 0 \tag{2.431}$$

which reduces to

$$\mathbf{r} \times \frac{d\mathbf{r}}{dt} = \mathbf{L} \tag{2.432}$$

where $\mathbf{L}$ is a constant vector.

During the time interval dt, the position vector $\mathbf{r}$ sweeps an area dA,

$$\frac{dA}{dt} = \frac{1}{2}\left|\mathbf{r} \times \frac{d\mathbf{r}}{dt}\right| = \frac{1}{2}\left|\mathbf{L}\right| \tag{2.433}$$

so the magnitude of the integral of motion $\mathbf{L}$ is twice the rate of the swept area by the position vector. Therefore, the position vector $\mathbf{r}$ will sweep equal areas in every time interval independent of the initial position for measurement.

Example 152 Curl of a Potential Force Is Zero: $\nabla \times F = 0$ If $V(x, y, z)$ is a potential energy function with continuous first and second partial derivatives in a given region, then

$$\begin{aligned} -\frac{\partial^2 V}{\partial y\,\partial x} &= \frac{F_x}{\partial y} = \frac{F_y}{\partial x} \\ -\frac{\partial^2 V}{\partial z\,\partial y} &= \frac{F_y}{\partial z} = \frac{F_z}{\partial y} \\ -\frac{\partial^2 V}{\partial x\,\partial z} &= \frac{F_z}{\partial x} = \frac{F_x}{\partial z} \end{aligned} \tag{2.434}$$

To have a function $V(x, y, z)$ such that

$$\mathbf{F} = F_x(x, y, z)\,\hat{\imath} + F_y(x, y, z)\,\hat{\jmath} + F_z(x, y, z)\,\hat{k} \tag{2.435}$$

we must have

$$\frac{F_x}{\partial y} - \frac{F_y}{\partial x} = 0 \qquad \frac{F_y}{\partial z} - \frac{F_z}{\partial y} = 0 \qquad \frac{F_z}{\partial x} - \frac{F_x}{\partial z} = 0 \tag{2.436}$$

These are both necessary and sufficient conditions for the existence of the function $V(x, y, z)$. These conditions may be equivalently expressed by a vector equation:

$$\text{curl}\mathbf{F} = \nabla \times \mathbf{F} = 0 \tag{2.437}$$

Example 153 Tests for a Conservative Force There are four equivalent tests to determine if a force $\mathbf{F}$ is conservative.

1. The force is the gradient of a potential function:

$$\mathbf{F} = -\nabla V \tag{2.438}$$

2. The curl of the force is zero:

$$\text{curl}\mathbf{F} = \nabla \times \mathbf{F} = 0 \tag{2.439}$$

3. The work of the force on every closed path is zero:

$${}_1W_1 = \int_{P_1}^{P_1} {}^G\mathbf{F} \cdot d\mathbf{r} = \oint {}^G\mathbf{F} \cdot d\mathbf{r} = 0 \tag{2.440}$$

4. The work of the force between two points is not path dependent:

$${}_1W_2 = \int_1^2 {}^G\mathbf{F} \cdot d\mathbf{r} = \int_1^2 dV = -(V_2 - V_1) \tag{2.441}$$

Example 154 Kinetic Energy of a System of Particles Consider n particles with masses m_i, $i = 1, 2, 3, \ldots, n$, that are moving with velocities $\mathbf{v}_i = \dot{\mathbf{r}}_i$. The kinetic energy of these particles is

$$K = \frac{1}{2}\sum_{i=1}^{n} m_i \mathbf{v}_i \cdot \mathbf{v}_i = \frac{1}{2}\sum_{i=1}^{n} m_i v_i^2 \tag{2.442}$$

However, the position vector $\mathbf{r}_i$ of the m_i can be expressed by adding $\mathbf{r}_i'$ to the position vector of the mass center $\mathbf{r}_C$:

$$\mathbf{r}_i = \mathbf{r}_C + \mathbf{r}_i' \qquad \sum_{i=1}^{n} \mathbf{r}_i' = 0 \tag{2.443}$$

Therefore, the kinetic energy of the system is

$$\begin{aligned} K &= \frac{1}{2}\sum_{i=1}^{n} m_i \left(\dot{\mathbf{r}}_C + \dot{\mathbf{r}}_i'\right) \cdot \left(\dot{\mathbf{r}}_C + \dot{\mathbf{r}}_i'\right) \\ &= \frac{1}{2}\sum_{i=1}^{n} m_i \left(\dot{\mathbf{r}}_C \cdot \dot{\mathbf{r}}_C\right) + \frac{1}{2}\sum_{i=1}^{n} m_i \left(\dot{\mathbf{r}}_C \cdot \dot{\mathbf{r}}_i'\right) \end{aligned} \tag{2.444}$$

The first term is the kinetic energy of an equivalent mass $m = \sum_{i=1}^{n} m_i$ at the mass center, and the second term is the kinetic energy of particles in their relative motion about the mass center.

Example 155 Work on a System of Particles Consider n particles m_i, $i = 1, 2, 3, \ldots, n$. The force acting on each particle m_i is the external force $\mathbf{F}_i$ plus the resultant of the internal force $\sum_{j=1}^{n} \mathbf{f}_{ij}$ from the particles m_j, $j = 1, 2, 3, \ldots, n$.

When the particles m_i have a displacement $d\mathbf{r}_i$, the work done on it by the external force is

$$dW_i = \mathbf{F}_i \cdot d\mathbf{r}_i \tag{2.445}$$

Employing the motion equation of the particles m_i,

$$m_i \ddot{\mathbf{r}}_i = \mathbf{F}_i + \sum_{j=1}^{n} \mathbf{f}_{ij} \tag{2.446}$$

we can determine the differential work done on the system as

$$dW = \sum_{i=1}^{n} m_i \ddot{\mathbf{r}}_i \cdot d\mathbf{r}_i - \sum_{i=1}^{n} \sum_{j=1}^{n} \mathbf{f}_{ij} \cdot d\mathbf{r}_i \tag{2.447}$$

We can simplify this equation and develop the work–energy principle of a system of particles. Using the equation

$$d\mathbf{r}_i = \dot{\mathbf{r}}_i \, dt \tag{2.448}$$

we rewrite the first term as the differential of kinetic energy:

$$\frac{d}{dt}\left(\frac{1}{2}\sum_{i=1}^{n} m_i \dot{\mathbf{r}}_i^2\right) = dK \tag{2.449}$$

The second term of (2.447) can be written as

$$\begin{aligned} \sum_{i=1}^{n} \sum_{j=1}^{n} \mathbf{f}_{ij} \cdot d\mathbf{r}_i &= \frac{1}{2} \sum_{i=1}^{n} \sum_{j=1}^{n} \mathbf{f}_{ij} \cdot \left(d\mathbf{r}_i - d\mathbf{r}_j\right) \\ &= \frac{1}{2} \sum_{i=1}^{n} \sum_{j=1}^{n} \mathbf{f}_{ij} \cdot d\mathbf{r}_{ij} \end{aligned} \tag{2.450}$$

$$d\mathbf{r}_{ij} = d\mathbf{r}_i - d\mathbf{r}_j \tag{2.451}$$

Assuming that the internal forces are potential, this is the internal potential energy of the system and we may show it by dV_i:

$$dV_i = \frac{1}{2} \sum_{i=1}^{n} \sum_{j=1}^{n} \mathbf{f}_{ij} \cdot d\mathbf{r}_{ij} \tag{2.452}$$

Therefore,

$$dW = dK + dV_i \tag{2.453}$$

and if the external forces are also conservative, then

$$dW = -dV_e \tag{2.454}$$

$$\nabla V_e = -\mathbf{F} \tag{2.455}$$

where V_e is the external potential energy.

Using these results, we may write Equation (2.447) as

$$dT + dV_e + dV_i = 0 \tag{2.456}$$

or

$$T + V_e + V_i = \text{const} \tag{2.457}$$

This is the law of conservation of energy for the system of particles.

The differential of the internal potential of a rigid body is zero,

$$dV_i = 0 \tag{2.458}$$

because the particles are relatively fixed. Therefore, the principle of conservation of energy for a rigid body is similar to that of a particle (2.371):

$$K_1 + V_1 = K_2 + V_2 \tag{2.459}$$

2.4.2 Temporal Integral: Impulse and Momentum

The temporal integral of the Newton equation of motion

$$\int_1^2 {}^G\mathbf{F}\,dt = m\int_1^2 {}^G\mathbf{a}\,dt \tag{2.460}$$

reduces to the *principle of impulse and momentum*

$${}_1\mathbf{I}_2 = \mathbf{p}_2 - \mathbf{p}_1 \tag{2.461}$$

where $\mathbf{p}$ is called the *momentum*,

$$\mathbf{p} = \tfrac{1}{2}m\,{}^G\mathbf{v} \tag{2.462}$$

and ${}_1\mathbf{I}_2$ is the *impulse* of the force during the *time interval* $t_2 - t_1$:

$${}_1\mathbf{I}_2 = \int_1^2 {}^G\mathbf{F}\,dt \tag{2.463}$$

If there is no external force

$$\mathbf{F} = 0 \tag{2.464}$$

then the principle of impulse and momentum (2.461) simplifies to the *principle of conservation of momentum*:

$$\mathbf{p}_1 = \mathbf{p}_2 \tag{2.465}$$

In the case of Euler equation of motion (2.59), we will have the *principle of angular impulse and moment of momentum*,

$$ {}_1\mathbf{H}_2 = \mathbf{L}_2 - \mathbf{L}_1 \tag{2.466} $$

and the *principle of conservation of moment of momentum*,

$$ \mathbf{L}_1 = \mathbf{L}_2 \tag{2.467} $$

Proof: We can simplify the right-hand side of the temporal integral (2.460) by a change of variable:

$$ \begin{aligned} \int_{t_1}^{t_2} {}^G\mathbf{F}\, dt &= m \int_{t_1}^{t_2} {}^G\mathbf{a}\, dt = m \int_{t_1}^{t_2} \frac{d\mathbf{v}}{dt}\, dt \\ &= m \int_{\mathbf{v}_1}^{\mathbf{v}_2} d\mathbf{v} = \frac{1}{2} m \left(\mathbf{v}_2 - \mathbf{v}_1\right) \end{aligned} \tag{2.468} $$

The *momentum* of a point P with mass m at a position pointed to by ${}^G\mathbf{r}$ and having a velocity ${}^G\mathbf{v}$ is defined by (2.462), and the *impulse* of the applied force ${}^G\mathbf{F}$ on m from time t_1 to t_2 is defined by (2.463). Hence the temporal integral of the equation of motion (2.460) reduces to the principle of impulse and momentum (2.461),

$$ {}_1\mathbf{I}_2 = \mathbf{p}_2 - \mathbf{p}_1 \tag{2.469} $$

which says that the impulse ${}_1\mathbf{I}_2$ of the applied force ${}^G\mathbf{F}$ on m during the time interval $t_2 - t_1$ is equal to the difference of the momentum of m.

The temporal integral of the moment equation (2.460),

$$ \int_{t_1}^{t_2} {}^G\mathbf{r} \times {}^G\mathbf{F}\, dt = \int_{t_1}^{t_2} \left(\mathbf{r} \times \frac{d}{dt}\mathbf{p}\right) dt \tag{2.470} $$

provides the principle of angular impulse and moment of momentum. The left-hand side is the time integral of moment $\mathbf{M}$ of a force $\mathbf{F}$ and is shown by the angular impulse $\mathbf{H}$:

$$ \int_{t_1}^{t_2} {}^G\mathbf{r} \times {}^G\mathbf{F}\, dt = \int_{t_1}^{t_2} {}^G\mathbf{M}\, dt = {}_1\mathbf{H}_2 \tag{2.471} $$

The right-hand side of (2.470) is the moment of impulse $\mathbf{I}$ and is shown by the angular momentum $\mathbf{L}$:

$$ \int_{t_1}^{t_2} \left(\mathbf{r} \times \frac{d}{dt}\mathbf{p}\right) dt = \int_{t_1}^{t_2} \frac{d}{dt}\left(\mathbf{r} \times \mathbf{p}\right) dt = \int_{\mathbf{L}_1}^{\mathbf{L}_2} d\mathbf{L} = \mathbf{L}_2 - \mathbf{L}_1 \tag{2.472} $$

■

Example 156 Rectilinear Elastic Collision Consider two particles with masses m_1 and m_2 that are moving on the same axis with speeds v_1 and $v_2 < v_1$, as is shown in Figure 2.29(a). The faster particle hits the slower one, and because of collision, their speed will change to v_1' and $v_2' > v_1'$, as is shown in Figure 2.29(b).

Figure 2.29 Velocity of two particles will change because of collision.

If there is no external force on the particles, the internal force F during the collision would be the only force that affects their motion. The equations of motion of the particles are

$$m_1\ddot{x}_1 = F \tag{2.473}$$

$$m_2\ddot{x}_2 = -F \tag{2.474}$$

Adding these equations yields

$$m_1\ddot{x}_1 + m_2\ddot{x}_2 = 0 \tag{2.475}$$

which shows that the total momentum of the particles is conserved:

$$m_1 v_1 + m_2 v_2 = p \tag{2.476}$$

Therefore, p remains the same before and after the collision:

$$m_1 v_1 + m_2 v_2 = m_1 v_1' + m_2 v_2' \tag{2.477}$$

If the force F is a function of only the relative distance of the particles,

$$F = F(x) \tag{2.478}$$

$$x = x_1 - x_2 \tag{2.479}$$

then there exists a potential energy function $V = V(x)$. The spatial integral of Equation (2.475) in this case indicates that the energy of the particles is conserved too:

$$K + V = E \tag{2.480}$$

Let us consider the potential energy function to be zero when the particles are far from each other. So, V is zero before and after the collision, and therefore, the kinetic energy of the particles is conserved:

$$\tfrac{1}{2}m_1 v_1^2 + \tfrac{1}{2}m_2 v_2^2 = \tfrac{1}{2}m_1 v_1'^2 + \tfrac{1}{2}m_2 v_2'^2 \tag{2.481}$$

Collision of this type, in which there is no loss in kinetic energy, is called the *elastic* collision. It is applied when the particles are hard, such as billiard ball impact.

To determine the velocity of the particles after impact, we may write the momentum and energy conservation equations as

$$m_1 v_1' - m_1 v_1 = m_2 v_2 - m_2 v_2' \tag{2.482}$$

$$m_1 v_1'^2 - m_1 v_1^2 = m_2 v_2^2 - m_2 v_2'^2 \tag{2.483}$$

and find

$$v_2' - v_1' = v_1 - v_2 \tag{2.484}$$

This equation along with (2.482) can be used to determine the final velocities in terms of the initial velocities:

$$v_1' = \frac{m_1 - m_2}{m_1 + m_2} v_1 + \frac{2m_2}{m_1 + m_2} v_2 \tag{2.485}$$

$$v_2' = \frac{2m_1}{m_1 + m_2} v_1 - \frac{m_1 - m_2}{m_1 + m_2} v_2 \tag{2.486}$$

The conditions $v_2 = 0$ and $m_1 = m_2$ may be viewed as special applied cases:

1. Special case $v_2 = 0$. When $v_2 = 0$, the final velocities (2.485) and (2.486) simplify to

$$v_1' = \frac{m_1 - m_2}{m_1 + m_2} v_1 \tag{2.487}$$

$$v_2' = \frac{2m_1}{m_1 + m_2} v_1 \tag{2.488}$$

We may use these equations for an acceptable estimate in different applications. In most applications with $v_2 = 0$, we are interested in the final speed of the stationary particle v_2'. To have a simpler equation, let us define a mass ratio

$$\epsilon = \frac{m_1}{m_2} \tag{2.489}$$

and write Equation (2.488) as

$$v_2' = \frac{2\epsilon}{\epsilon + 1} v_1 \tag{2.490}$$

As an example consider a football player who kicks a stationary ball, as is shown in Figure 2.30. The weight of the ball at the start of the game should not be more than 0.45 kg or less than 0.41 kg. It must be 69 cm $\approx$ 27 in. to 71 cm $\approx$ 28 in. in circumference. A soccer ball of mass 0.43 kg may leave the foot of the very good kicker with an initial speed of 40 m/s. If we assume that the mass of the player is 85 kg and each leg is 10% of the total mass of the body, then

$$\epsilon = \frac{m_1}{m_2} = \frac{0.10 \times 85}{0.43} = 19.767 \tag{2.491}$$

Therefore, the player must kick the ball with at least $v_1 \approx 21$ m/s:

$$v_1 = \frac{\epsilon + 1}{2\epsilon} v_2' = \frac{1 + 19.767}{2 \times 19.767} \times 40 = 21.012 \text{ m/s} \tag{2.492}$$

There are some sources of error. Kicking a soccer ball is not an elastic collision, energy is not conserved during the kicking, and a leg cannot be assumed as a particle.

As a better example of elastic collision, we examine the billiard game. There are 15 balls and 1 cue ball in billiard. The cue ball is used to hit the balls into the six pockets around the billiard table. On average, the balls are 2.25 in. $\approx$ 571.5 mm in diameter and all the balls weigh 5.5 $oz \approx 0.16$ kg except for the cue, which weighs 6 $oz \approx 0.17$ kg. First we hit the cue ball with the cue stick, which has a mass of about 20 $oz \approx 0.6$ kg.

Figure 2.30 A football player kicks a ball.

If the speed of the cue stick reaches 10 m/s, right before the hit, the cue ball will move with $v_2' \approx 15$ m/s:

$$v_2' = \frac{2\epsilon}{\epsilon+1} v_1 = 15.385 \text{ m/s} \tag{2.493}$$

$$\epsilon = \frac{m_1}{m_2} = \frac{20}{6} = 3.3333 \tag{2.494}$$

The cue ball may shoot a billiard ball with a maximum speed $v_2'\prime \approx 16$ m/s; if it collides with a ball instantly,

$$v_2' = \frac{2\epsilon}{\epsilon+1} v_1 = 16.364 \text{ m/s} \tag{2.495}$$

$$\epsilon = \frac{m_1}{m_2} = \frac{6}{5} = 1.2 \tag{2.496}$$

2. Special case $m_1 = m_2$. If $m_1 = m_2$, then the final velocities (2.485) and (2.486) reduce to

$$v_1' = v_2 \tag{2.497}$$

$$v_2' = v_1 \tag{2.498}$$

and the particles trade their speeds.

To examine the effect of the mass ratio in particle collision, let us use ϵ and write Equations (2.485) and (2.486) as

$$v_1' = \frac{\epsilon-1}{\epsilon+1} v_1 + \frac{2}{\epsilon+1} v_2 \tag{2.499}$$

$$v_2' = \frac{2\epsilon}{\epsilon+1} v_1 - \frac{\epsilon-1}{\epsilon+1} v_2 \tag{2.500}$$

If $\epsilon \to 0$ and we hit a very massive particle m_2 with a very light particle m_1, then

$$v_1' = \lim_{\epsilon \to 0} \left(\frac{\epsilon - 1}{\epsilon + 1} v_1 + \frac{2}{\epsilon + 1} v_2 \right) = -v_1 + 2v_2 \tag{2.501}$$

$$v_2' = \lim_{\epsilon \to 0} \left(\frac{2\epsilon}{\epsilon + 1} v_1 - \frac{\epsilon - 1}{\epsilon + 1} v_2 \right) = v_2 \tag{2.502}$$

The heavy particle m_2 will not change its speed, while the light particle m_1 will rebound with its initial speed minus double of the speed of m_2. In the case of $v_2 = 0$, the heavy particle will not move, and the light particle will rebound with its initial speed. It happens when we kick a ball and hit a wall.

If $\epsilon \to \infty$ and we hit a very light particle m_2 with a very massive particle m_1, then

$$v_1' = \lim_{\epsilon \to \infty} \left(\frac{\epsilon - 1}{\epsilon + 1} v_1 + \frac{2}{\epsilon + 1} v_2 \right) = v_1 \tag{2.503}$$

$$v_2' = \lim_{\epsilon \to \infty} \left(\frac{2\epsilon}{\epsilon + 1} v_1 - \frac{\epsilon - 1}{\epsilon + 1} v_2 \right) = 2v_1 - v_2 \tag{2.504}$$

The heavy particle m_1 will not change its speed, while the light particle m_2 will move with double the speed of m_1 minus its own speed. In the case $v_2 = 0$, the heavy particle continues its motion uniformly, and the light particle will gain double the speed of the heavy particle. It happens when we hit a ball with our car.

Example 157 Rectilinear Inelastic Collision In reality, there is always some loss of energy in a collision in the form of heat. This is because the internal force F during the collision is a function of the relative distance and velocity of the particles:

$$F = F(x, \dot{x}) \tag{2.505}$$

$$x = x_1 - x_2 \tag{2.506}$$

$$\dot{x} = \dot{x}_1 - \dot{x}_2 \tag{2.507}$$

So, there is no potential energy function V and the energy of the particles is not conserved. Therefore,

$$\tfrac{1}{2} m_1 v_1^2 + \tfrac{1}{2} m_2 v_2^2 \neq \tfrac{1}{2} m_1 v_1'^2 + \tfrac{1}{2} m_2 v_2'^2 \tag{2.508}$$

However, if there is no external force, we still may add the equation of motion of the particles,

$$m_1 \ddot{x}_1 = F \tag{2.509}$$

$$m_2 \ddot{x}_2 = -F \tag{2.510}$$

and show that the momentum of the particles is conserved,

$$m_1 v_1 + m_2 v_2 = m_1 v_1' + m_2 v_2' \tag{2.511}$$

Collision of this type, in which the kinetic energy is not conserved, is called *inelastic* collision. To analyze inelastic collisions, we define a *restitution coefficient* e by

$$v_2' - v_1' = e\,(v_1 - v_2) \tag{2.512}$$

$$0 \le e \le 1 \tag{2.513}$$

This equation along with the momentum equation (2.511) can be used to determine the final velocities in terms of the initial velocities:

$$v_1' = \frac{m_1 - em_2}{m_1 + m_2} v_1 + \frac{m_2\,(1+e)}{m_1 + m_2} v_2 \tag{2.514}$$

$$v_2' = \frac{m_1\,(1+e)}{m_1 + m_2} v_1 - \frac{em_1 - m_2}{m_1 + m_2} v_2 \tag{2.515}$$

The case $e = 1$ indicates an elastic collision and simplifies Equations (2.514) and (2.515) to (2.485) and (2.486). The case $e = 0$ is called the *plastic collision* for which we have

$$v_1' = v_2' = \frac{m_1}{m_1 + m_2} v_1 + \frac{m_2}{m_1 + m_2} v_2 \tag{2.516}$$

In a plastic collision, m_1 and m_2 will stick together and move with the same speed.

Example 158 Bouncing Ball Consider a ball falls from a height h_0 and hits the ground with restitution coefficient e. It hits the ground with velocity $v_0 = \sqrt{2gh_0}$ at time $t_0 = \sqrt{2h_0/g}$. The rebound velocity of the ball would be $v_1 = ev_0$, which makes the ball to rise to a height $h_1 = e^2 h_0$ at time $t_1 = et_0$. Then, for a second time, it will fall and bounce again, this time to a lesser height $h_2 = e^2 h_1$ at time $t_2 = et_0$. This fall and bounce will continually happen and never stop.

The total distance traveled by the ball is

$$x = h_0 + 2h_1 + 2h_2 + \cdots = -h_0 + 2h_0\left(1 + e^2 + e^4 + \cdots\right) \tag{2.517}$$

and the time taken is

$$t = t_0 + 2t_1 + 2t_2 + \cdots = -t_0 + 2t_0\left(1 + e + e^2 + \cdots\right) \tag{2.518}$$

Recalling that the sum S_n of the first n terms of a geometric series

$$S_n = a + ar + ar^2 + ar^3 + ar^4 + \cdots + ar^{n-1} \qquad (r \neq 1) \tag{2.519}$$

is given by

$$S_n = a\frac{1 - r^n}{1 - r} \qquad (r \neq 1) \tag{2.520}$$

$$\lim_{n\to\infty} S_n = \frac{a}{1 - r} \qquad (r < 1) \tag{2.521}$$

we find the total distance x and time t as

$$x = -h_0 + 2h_0 \frac{1}{1 - e^2} = h_0 \frac{1 + e^2}{1 - e^2} \tag{2.522}$$

$$t = -t_0 + 2t_0 \frac{1}{1 - e} = t_0 \frac{1 + e}{1 - e} \tag{2.523}$$

As an example, assume

$$h_0 = 1\,\text{m} \qquad e = 0.8 \qquad g = 9.81\,\text{m/s}^2 \tag{2.524}$$

Then the ball comes to rest after an infinite number of bounces, when it travels a distance x at time t:

$$x = h_0 \frac{1 + e^2}{1 - e^2} = \frac{1 + 0.8^2}{1 - 0.8^2} = 4.5556\,\text{m} \tag{2.525}$$

$$t = t_0 \frac{1 + e}{1 - e} = \sqrt{\frac{2h_0}{g}} \frac{1 + e}{1 - e} = 4.0637\,\text{s} \tag{2.526}$$

Example 159 Rocket Motion The principle of impulse and momentum is the best method to derive the equation of motion of a variable-mass object such as a rockets.

Consider a rocket with mass m that is moving in free space at velocity $\mathbf{v}$ under no external force, as is shown in Figure 2.31. During the time interval dt, a positive mass dm is ejected from the rocket with a velocity $-\mathbf{v}_m$ with respect to the rocket in the B-frame or the absolute velocity $\mathbf{v} - \mathbf{v}_m$ in the G-frame. After dm is left, the rocket has a mass $m - dm$ and a velocity $\mathbf{v} + d\mathbf{v}$. Therefore, the initial and final momenta in the global coordinate frame are

$$\mathbf{p}_1 = m\mathbf{v} \tag{2.527}$$

$$\mathbf{p}_2 = (m - dm)(\mathbf{v} + d\mathbf{v}) + dm\,(\mathbf{v} - \mathbf{v}_m) \tag{2.528}$$

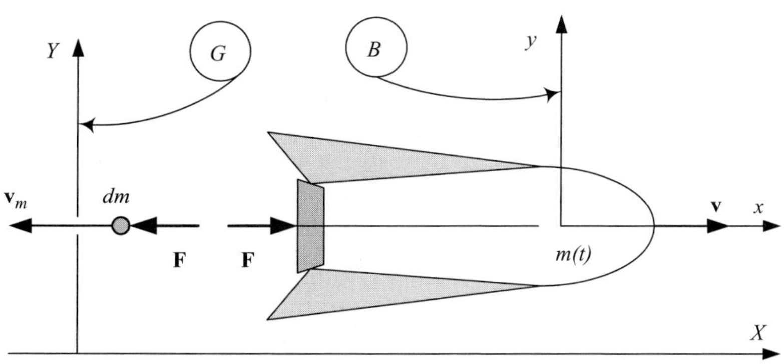

Figure 2.31 A variable mass rocket.

Because there is no external force or the impulse of the external force is zero during dt, we must have

$$\mathbf{p}_1 = \mathbf{p}_2 \tag{2.529}$$

and therefore,

$$m\, d\mathbf{v} = dm\, \mathbf{v}_m \tag{2.530}$$

This equation may be used to determine the trust of the rocket $\mathbf{F}$ that is equal to the mass rate of fuel consumption $\dot{m}$ times the ejection velocity of the burnt gas $\mathbf{v}_m$:

$$\mathbf{F} = m\mathbf{a} = \dot{m}\, \mathbf{v}_m \tag{2.531}$$

We may assume an average density ρ for the exhaust gas and substitute $\dot{m}$ with $\rho A \mathbf{v}_m$ to have

$$F = ma = \rho A\, v_m^2 \tag{2.532}$$

Example 160 Final Velocity of a Rocket The dynamics of a variable-mass object such as a rocket obeys Equation (2.530). If the initial mass and velocity of the rocket are m_0 and $\mathbf{v}_0$, and also the relative velocity $\mathbf{v}_m$ of the ejecting mass dm is assumed to be constant, then we may integrate Equation (2.530) to determine the instantaneous velocity $\mathbf{v}$ of the rocket by having its mass m:

$$\int_{\mathbf{v}_0}^{\mathbf{v}} d\mathbf{v} = \mathbf{v}_m \int_{m_0}^{m} \frac{dm}{m} \tag{2.533}$$

$$\mathbf{v} = \mathbf{v}_0 + \mathbf{v}_m \ln \frac{m_0}{m} \tag{2.534}$$

Having the final mass of the rocket as $m_f = m_0 - m_{fu}$ when there remains no more fuel, the final velocity of the rocket would be

$$\begin{aligned} \mathbf{v}_f &= \mathbf{v}_0 + \mathbf{v}_m \ln \frac{m_0}{m_f} = \mathbf{v}_0 + \mathbf{v}_m \ln \frac{m_0}{m_0 - m_{fu}} \\ &= \mathbf{v}_0 + \mathbf{v}_m \ln \frac{1}{1 - f} \end{aligned} \tag{2.535}$$

$$f = \frac{m_{fu}}{m_0} \tag{2.536}$$

If the rocket starts from rest, its final velocity $\mathbf{v}_f = \mathbf{v}_m \ln(m_0/m_f)$ can be higher than the exhaust velocity $\mathbf{v}_m$ if $m_0/m_f > e\,(= 2.718\,3)$. To maximize the final velocity, we should maximize the exhaust velocity $\mathbf{v}_m$ and the ratio $f = m_{fu}/m_0$. Practically, $f = 0.99$ is a desirable goal, at which the speed change $\Delta\mathbf{v} = \mathbf{v}_f - \mathbf{v}_0$ will be

$$\Delta\mathbf{v}_{\text{Max}} = \mathbf{v}_f - \mathbf{v}_0 = \mathbf{v}_m \ln \frac{1}{1 - 0.99} = 4.6052\mathbf{v}_m \tag{2.537}$$

Assuming $\mathbf{v}_m = 10{,}000\,\text{m/s}$ we may have a rocket with maximum speed

$$\Delta\mathbf{v}_{\text{Max}} = \mathbf{v}_f - \mathbf{v}_0 = 4.6052\mathbf{v}_m = 46{,}052\,\text{m/s} \tag{2.538}$$

Theoretically, $f = 1$ is the maximum value of f at which the speed change $\Delta \mathbf{v} = \mathbf{v}_f - \mathbf{v}_0$ approaches infinity.

The minimum mass of a rocket m_f is a function of the size of the rocket and structural material.

Example 161 ★ Multistage Rocket To minimize the final mass of a rocket, m_f, we may make the rocket multistage. So, we may release the useless stage after burning its fuel. This technique increases the final velocity of the rocket, which is limited by the ratio m_0/m_f.

To clarify the advantage of having a multistage rocket, let us consider a two-stage rocket that goes on a straight line with the following mass parameters:

$m_0 = m_1 + m_2$	total initial mass of the rocket
$m_1 = m_a + m_b$	mass of the first stage
$m_2 = m_c + m_d$	mass of the second stage
m_a	mass of fuel in the first stage
m_b	mass of the first stage shell
m_c	mass of fuel in the second stage
m_d	mass of the second stage shell

When the fuel of the first stage finishes, the rocket is at the first final velocity v_1:

$$v_1 = v_0 + v_m \ln \frac{m_0}{m_2 + m_b} = v_0 + v_m \ln \frac{1}{m_2/m_0 + m_b/m_0} \tag{2.539}$$

At this moment, the first stage will be released to drop the mass m_b, and the second stage ignites. Assuming the same ejection velocity v_m, we reach the final velocity v_2 after burning the fuel in the second stage:

$$v_2 = v_1 + v_m \ln \frac{m_2}{m_d} \tag{2.540}$$

Substituting (2.539) in (2.540) provides the final velocity of the rocket:

$$\begin{aligned} v_2 &= v_0 + v_m \ln \frac{m_0 m_2}{m_d (m_2 + m_b)} = v_0 + v_m \ln \frac{m_0 (m_c + m_d)}{m_d (m_c + m_d + m_b)} \\ &= v_0 + v_m \ln \frac{m_c/m_0 + m_d/m_0}{(m_d/m_0)(m_c/m_0 + m_d/m_0 + m_b/m_0)} \end{aligned} \tag{2.541}$$

We can make the ratio $m_0 m_2/[m_d(m_2 + m_b)]$ greater than $m_0/(m_0 - m_a - m_c)$.

Example 162 ★ Daytime of a Shrinking Earth Assume that the radius of Earth, R, is decreasing by a linear function of time t. This will increase the angular speed of Earth and, hence shortening the length of a day:

$$R = R_0 (1 - kt) \tag{2.542}$$

There is no external moment on Earth and therefore the angular momentum of Earth, $\mathbf{L} = L\hat{K}$, remains constant. The angular momentum of a spherical Earth is

$$L_0 = I_0\omega_0 = \tfrac{2}{3}mR_0^2\omega_0 \tag{2.543}$$

at $t = 0$ and

$$L = I\omega = \tfrac{2}{3}mR_0^2\left(1 - kt\right)^2\omega \tag{2.544}$$

at any other time $t > 0$, where $I = \frac{2}{3}mR^2$ is Earth's mass moment of inertia. Because $L = L_0$, we have

$$\omega = \frac{\omega_0}{(1 - kt)^2} \tag{2.545}$$

If k is very small, this equation simplifies to

$$\omega \approx \omega_0\left(1 + 2kt\right) \tag{2.546}$$

The angular speed of such a shrinking Earth will be increasing with a constant rate:

$$\frac{d\omega}{dt} \approx 2k\omega_0 \tag{2.547}$$

As an example, let us assume that the radius of Earth is shortening by a rate of

$$R_0k = 1\,\text{m/day} \tag{2.548}$$

Then, knowing Earth's information,

$$R_0 \approx 6.3677 \times 10^6\,\text{m} \tag{2.549}$$

$$\omega_0 \approx 2\pi\left(1 + \frac{1}{366}\right)\text{rad/day} = 6.300352481\,\text{rad/day} \tag{2.550}$$

$$k = \frac{1}{R_0}\text{d}^{-1} \approx 1.570425742 \times 10^{-7}\text{day}^{-1} \tag{2.551}$$

we can calculate the angular velocity and its rate of change:

$$\begin{aligned}\frac{d\omega}{dt} &\approx 2k\omega_0 = 2\left(1.570425742 \times 10^{-7}\right)6.300352481 \\ &= 1.9788471442279 \times 10^{-6}\,\text{rad/day}^2\end{aligned} \tag{2.552}$$

$$\begin{aligned}\omega &\approx \omega_0 + \frac{d\omega}{dt}t \\ &= 6.300354460 + 1.9788471442279 \times 10^{-6}t\,\text{rad/day}\end{aligned} \tag{2.553}$$

The angular velocity will increase by $\Delta\omega = \omega - \omega_0$ each day,

$$\Delta\omega = \omega - \omega_0 = 1.9788471442279 \times 10^{-6}\,\text{rad/day} \tag{2.554}$$

and hence each day will be shorter by about 0.027s:

$$\Delta\omega\frac{24 \times 3600}{\omega_0} = 2.7137 \times 10^{-2}\,\text{s} \tag{2.555}$$

Example 163 ★ Motion of a Particle with $(d\mathbf{r}/dt)\cdot\left[(d^2\mathbf{r}/dt^2)\times(d^3\mathbf{r}/dt^3)\right]=\mathbf{0}$
Consider a particle where its motion is under the condition

$$\frac{d\mathbf{r}}{dt}\cdot\left(\frac{d^2\mathbf{r}}{dt^2}\times\frac{d^3\mathbf{r}}{dt^3}\right)=0 \tag{2.556}$$

Because we can interchange the dot and cross products in a scalar triple product, the jerk vector $d^3\mathbf{r}/dt^3$ must be in the plane of velocity $d\mathbf{r}/dt$, and acceleration $d^2\mathbf{r}/dt^2$. Therefore,

$$\frac{d^3\mathbf{r}}{dt^3}=c_1\frac{d^2\mathbf{r}}{dt^2}+c_2\frac{d\mathbf{r}}{dt} \tag{2.557}$$

where c_1 and c_2 are constant scalars. Integrating (2.557) yields

$$\frac{d^2\mathbf{r}}{dt^2}=c_1\frac{d\mathbf{r}}{dt}+c_2\mathbf{r}+\mathbf{c}_3 \tag{2.558}$$

where $\mathbf{c}_3$ is a constant vector. Solution of this equation gives

$$\mathbf{r}=\mathbf{c}_3e^{\lambda_1 t}+\mathbf{c}_4e^{\lambda_2 t}-\frac{1}{c_2}\mathbf{c}_3 \tag{2.559}$$

where $\mathbf{c}_3$ and $\mathbf{c}_4$ are two constant vectors and λ_1, λ_2 are solutions of the characteristic equation

$$s^2-c_1s-c_2=0 \tag{2.560}$$

Solution (2.559) indicates that the motion of a particle under condition (2.556) is a planar motion.

Example 164 ★ Oblique Collision Consider two particles with masses m_1 and m_2 that are moving in the same plane with velocities $\mathbf{v}_1$ and $\mathbf{v}_2$, as is shown in Figure 2.32(*a*). If the particles collide, their velocities will change to $\mathbf{v}_1'$ and $\mathbf{v}_2'$, as is shown in Figure 2.32(*b*).

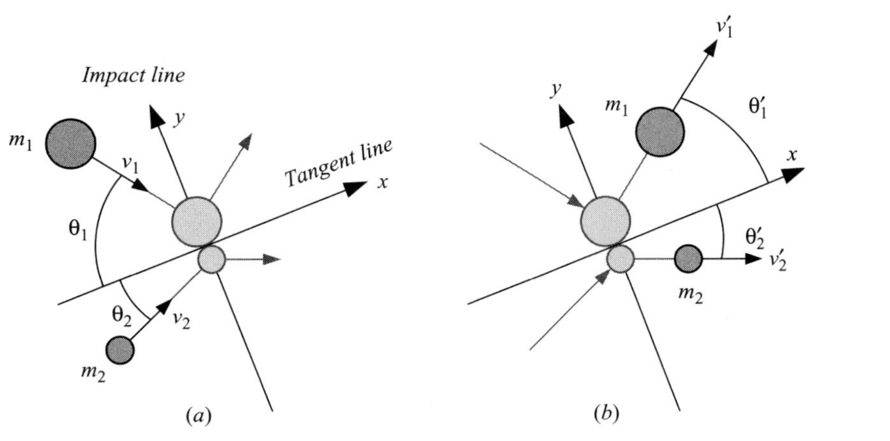

Figure 2.32 Velocity vector of two particles will change because of oblique collision.

At the instant of collision, we define a collision coordinate frame with the x-axis to be tangent to the particles and the y-axis to be the line of centers. The impact happens on the centerline. Assume the velocity vectors of the particles are given before impact. We must be able to determine their velocity vectors after impact by employing the temporal integral of their motions.

If there is no external force on the particles, the internal force **F** during the collision would be the only force that affects their motion. The equations of motion of the particles are

$$m_1\dot{\mathbf{v}}_1 = \mathbf{F} \tag{2.561}$$

$$m_2\dot{\mathbf{v}}_2 = -\mathbf{F} \tag{2.562}$$

Adding these equations,

$$m_1\dot{\mathbf{v}}_1 + m_2\dot{\mathbf{v}}_2 = 0 \tag{2.563}$$

shows that the total momentum of the particles is conserved:

$$m_1\mathbf{v}_1 + m_2\mathbf{v}_2 = \mathbf{p} \tag{2.564}$$

Therefore, **p** remains the same before and after the collision. Let us decompose the conservation of momentum in the collision coordinate frame:

$$m_1 v_1 \cos\theta_1 + m_2 v_2 \cos\theta_2 = m_1 v_1' \cos\theta_1' + m_2 v_2' \cos\theta_2' \tag{2.565}$$

$$m_1 v_1 \sin\theta_1 + m_2 v_2 \sin\theta_2 = m_1 v_1' \sin\theta_1' + m_2 v_2' \sin\theta_2' \tag{2.566}$$

The particles do not collide on the x-axis, so their individual momentum is conserved on the tangent line. It means that Equation (2.565) is a summation of two equations:

$$v_1 \cos\theta_1 = v_1' \cos\theta_1' \tag{2.567}$$

$$v_2 \cos\theta_2 = v_2' \cos\theta_2' \tag{2.568}$$

Another equation for the restitution coefficient e is based on the impact line:

$$v_2' \sin\theta_2' - v_1' \sin\theta_1' = e\left(v_1 \sin\theta_1 - v_2 \sin\theta_2\right) \tag{2.569}$$

Equations (2.566)–(2.569) are the four equations to be solved for the four unknown parameters after the collision.

2.5 ★ APPLICATION OF DYNAMICS

We learn dynamics to achieve four goals:

1. Model dynamic phenomena
2. Determine the equations of motion
3. Determine the behavior of dynamic phenomena
4. Adjust the parameters such that a phenomenon happens in a desired manner

2.5.1 ★ Modeling

Modeling is the way to observe a dynamic phenomenon. Modeling is a compromising job. It must be as simple as possible to be able to develop the equations of motion, and it must be as complete as possible to include the effective parameters.

Example 165 ★ Ocean Tides Assume that Earth is a sphere, its surface is completely covered by water, and there is no other celestial body except Earth and the moon. Figure 2.33 illustrates the positions of Earth E and the moon M in a global coordinate frame G. We consider the attraction of both Earth and the moon on a small mass m placed on the surface of Earth. The position vectors of E, m, and M in G are denoted by $\mathbf{r}_E$, $\mathbf{r}_m$, and $\mathbf{r}_M$. The position vectors of m and E from the moon are $\mathbf{d}$ and $\mathbf{d}_E$, while the position of m is shown by $\mathbf{r}$ from Earth's center.

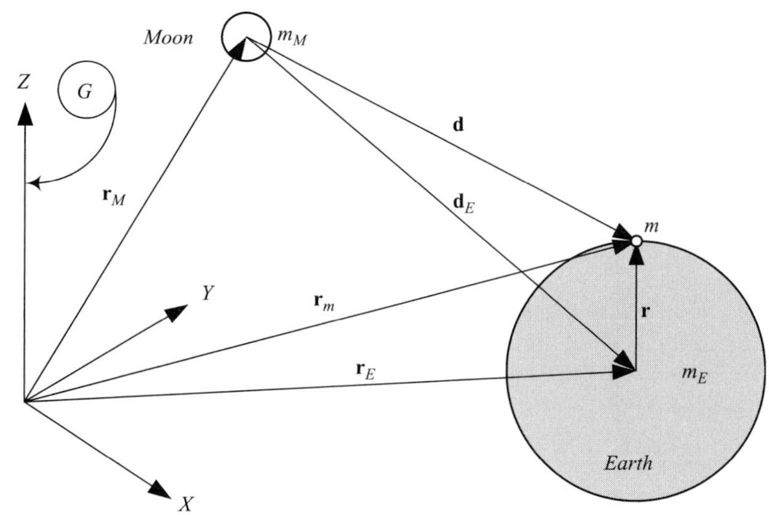

Figure 2.33 The positions of Earth E and the moon M in a global coordinate frame G to examine the effect of their attraction on a small mass m on the durface of Earth.

The equation of motion of m is

$$m\ddot{\mathbf{r}}_m = -G\frac{mm_E}{r^2}\hat{u}_{\mathbf{r}} - G\frac{mm_M}{d^2}\hat{u}_{\mathbf{d}} \tag{2.570}$$

and the equation of motion of Earth due to the attraction of the moon is

$$m_E\ddot{\mathbf{d}}_E = -G\frac{m_M m_E}{d_E^2}\hat{u}_{\mathbf{d}_E} \tag{2.571}$$

Subtracting these equations determines the acceleration of m with respect to Earth's center:

$$\begin{aligned}\ddot{\mathbf{r}} = \ddot{\mathbf{r}}_m - \ddot{\mathbf{d}}_E &= -G\frac{m_E}{r^2}\hat{u}_{\mathbf{r}} - G\frac{m_M}{d^2}\hat{u}_{\mathbf{d}} + G\frac{m_M}{d_E^2}\hat{u}_{\mathbf{d}_E} \\ &= -G\frac{m_E}{r^2}\hat{u}_{\mathbf{r}} - Gm_M\left(\frac{\hat{u}_{\mathbf{d}}}{d^2} - \frac{\hat{u}_{\mathbf{d}_E}}{d_E^2}\right) \end{aligned} \tag{2.572}$$

The first term of (2.572) is the acceleration of m because of Earth and is directed to Earth's center for every point on Earth's surface. The second term is the acceleration due to the difference in the moon's attraction force at Earth's center and at the surface of Earth. The direction of $\hat{u}_{\mathbf{d}_E}$ is always on the Earth–moon centerline however, the direction of $\hat{u}_{\mathbf{d}}$ depends on the position of m on Earth. The second term is called the *tidal acceleration*, which generates the ocean tides.

To visualize the tidal force on different points on Earth, we look at Earth from a polar view as shown in Figure 2.34. The local x-axis is the moon–Earth centerline indicated by the unit vector $\hat{u}\mathbf{d}_E$. The tidal force $\mathbf{F}_T$ on the small mass m on the surface of Earth is

$$\mathbf{F}_T = -Gmm_M \left(\frac{\hat{u}_{\mathbf{d}}}{d^2} - \frac{\hat{u}_{\mathbf{d}_E}}{d_E^2} \right) \tag{2.573}$$

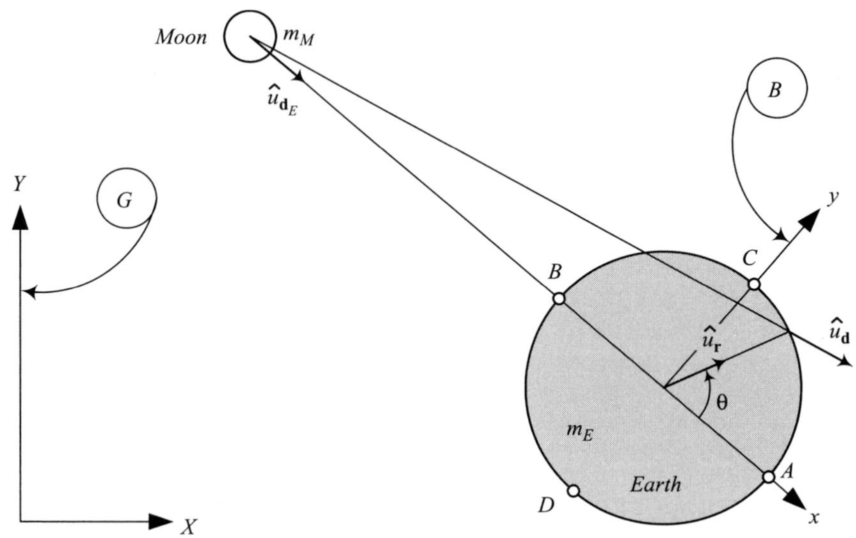

Figure 2.34 Earth–moon model from a polar axis view.

Point A in the figure is the farthest point on Earth from the moon at which $\hat{u}_{\mathbf{d}}$ is in the same direction as $\hat{u}_{\mathbf{d}_E}$. At point A, we have $\mathbf{d} > \mathbf{d}_E$, and therefore $\mathbf{F}_T$ is in the $\mathbf{d}_E$- or x-direction; however, at point B, we have $\mathbf{d} < \mathbf{d}_E$, and therefore $\mathbf{F}_T$ is in the $-\mathbf{d}_E$- or $-x$-direction. Defining a polar coordinate (r, θ) in a local frame B at Earth's center, we may define

$$\hat{u}_{\mathbf{d}_E} = \hat{\imath} \tag{2.574}$$

$$\hat{u}_{\mathbf{r}} = \cos\theta\,\hat{\imath} + \sin\theta\,\hat{\jmath} \tag{2.575}$$

$$d\hat{u}_{\mathbf{d}} = d_E\hat{u}_{\mathbf{d}_E} + r\hat{u}_{\mathbf{r}} = (d_E + r\cos\theta)\,\hat{\imath} + r\sin\theta\,\hat{\jmath} \tag{2.576}$$

$$d^2 = d_E^2 + r^2 + 2rd_E\cos\theta \tag{2.577}$$

to calculate the tidal force:

$$\mathbf{F}_T = -Gmm_M \left(\frac{\hat{u}_{\mathbf{d}}}{d^2} - \frac{\hat{u}_{\mathbf{d}_E}}{d_E^2} \right)$$
$$= -Gmm_M \left[\left(\frac{d_E + r\cos\theta}{d^3} - \frac{1}{d_E^2} \right) \hat{\imath} + \frac{r\sin\theta}{d^3} \hat{\jmath} \right] \tag{2.578}$$

Let us use the data

$$m_M \approx 7.349 \times 10^{22}\,\mathrm{kg}$$
$$d_E \approx 384{,}400\,\mathrm{km}$$
$$m_E \approx 5.9736 \times 10^{24}\,\mathrm{kg}$$
$$r \approx 6371\,\mathrm{km}$$
$$G \approx 6.67259 \times 10^{-11}\,\mathrm{m^3kg^{-1}s^{-2}}$$
$$m = 1\,\mathrm{kg} \tag{2.579}$$

to evaluate the components of the tidal force $\mathbf{F}_T = F_x\hat{\imath} + F_y\hat{\jmath}$ for $0 \le \theta \le 360\,\mathrm{deg}$, as are shown in Figure 2.35. The y-component of $\mathbf{F}_T$ is always toward the center of Earth with zero value at points A and B and maximum values at points C and D. The x-component of $\mathbf{F}_T$ is always outward from the center of Earth with zero value at points C and D and maximum values at points A and B. These components must be added to

$$-G\frac{m_E}{r^2}\hat{u}_{\mathbf{r}} = -9.820087746\left(\cos\theta\hat{\imath} + \sin\theta\hat{\jmath}\right) \tag{2.580}$$

to determine the net force on m. Figure 2.36 illustrates the distribution of tidal force around Earth. An exaggerated model of the level of the oceans is shown in Figure 2.37

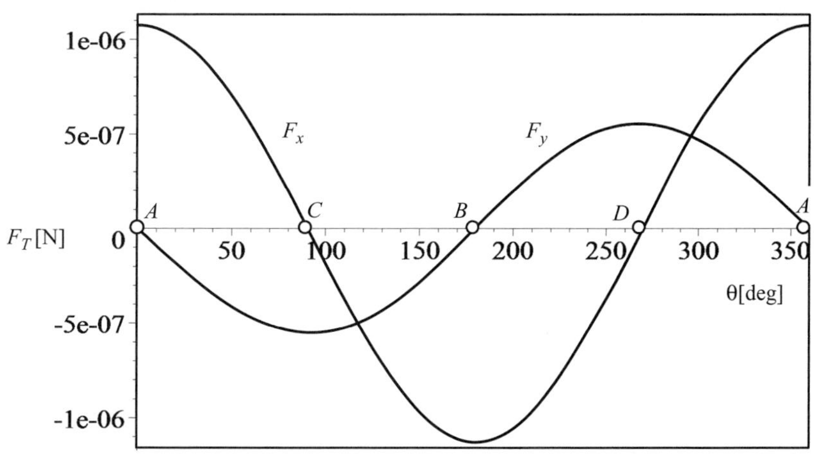

Figure 2.35 The components of the tidal force $\mathbf{F}_T = F_x\hat{\imath} + F_y\hat{\jmath}$.

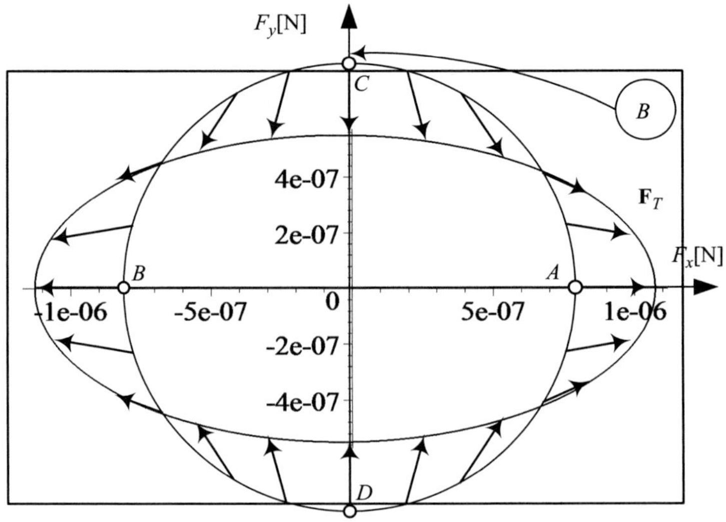

Figure 2.36 Distribution of tidal forces around Earth.

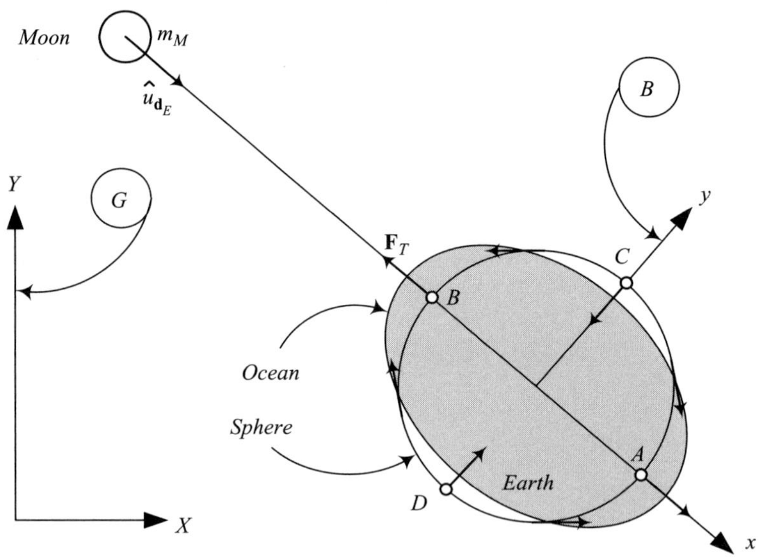

Figure 2.37 An exaggerated model of the level of the oceans because of the tidal effect of the moon.

The Earth's mean radius is determined as the average distance from the physical center to the surface. Figure 2.38 illustrates the maximum, mean, and minimum distances of the moon from Earth. When the moon is at apogee, it is about 11% farther from Earth than it is at perigee.

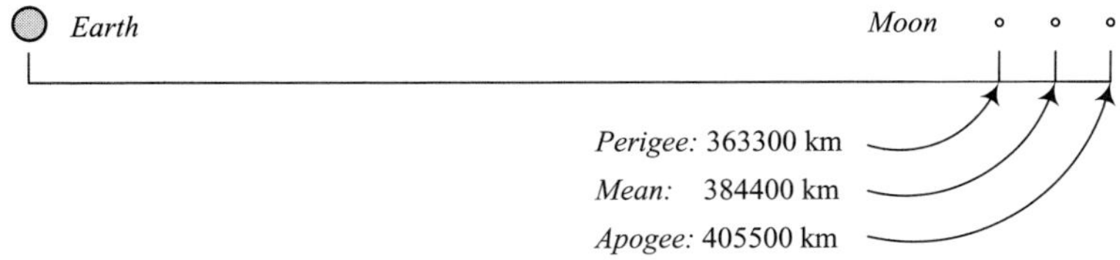

Figure 2.38 In-scale distances of moon from Earth.

The tidal interaction on Earth caused by the moon transfers kinetic energy from Earth to the moon. It slows the Earth's rotation and raises the moon's orbit, currently at a rate of about 3.8 cm per year.

To make a better model of tides, we should add the effect of the sun. When the sun, moon, and Earth are aligned, the gravitational forces of the sun and moon will be combined, causing very high and very low tides. Such a tide, called the *spring tide*, is not related to the season. It happens when the moon is full or new. When the sun and moon are not aligned, the gravitational forces cancel each other out, and the tides are not as high or as low as spring tides. This weak tide is called the *neap tide*. We also need to consider the rotation and oblateness of Earth, the nonzero angle between Earth's polar axis and the orbital plane of the moon, and variations in the moon–Earth distance.

Example 166 Book-stacking Problem Let us model the book-stacking problem and determine how far x the edge of a book can be overhung when stacking books on a table.

Consider a set of n similar books of length l. To balance the first book on a table, we need to adjust the mass center of of the book somewhere over the table. The maximum overhang happens when the mass center is just over the table's edge, $x = l/2$, such as shown in Figure 2.39(a). Now let slide a second book below the first one. To maximize the overhang of two books, the mass center of the stack of two

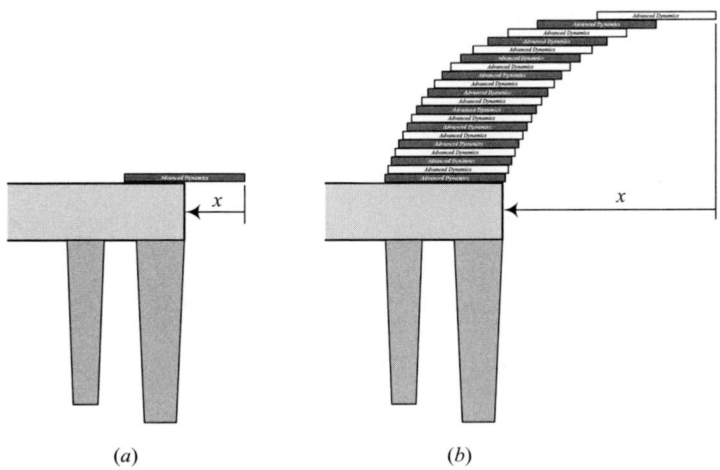

Figure 2.39 The book-stacking problem.

books should be at the edge of the table while the mass center of the first should be directly over the edge of the second. The mass center of the stack of two books is at the midpoint of the books' overlap, so $x = (l + l/2)/2 = 3l/4$. Sliding the third book under the two and adjusting the mass center of the stack of the three books at the edge of the table yield $x = (l + l/2 + l/3)/2 = 11l/6$. By following the same procedure, we find x of n books to be

$$x = \frac{1}{2}\left(1 + \frac{1}{2} + \frac{1}{3} + \cdots + \frac{1}{n}\right) l \tag{2.581}$$

The coefficient of l is called the harmonic number series. There is no limit for the harmonic series, so theoretically, we can move the top book as far as we wish, because

$$\lim_{m\to\infty} \sum_{n=1}^{m} \frac{1}{n} = \infty \tag{2.582}$$

The coefficient would be more than 1 with only four books, and the top book is overhanging farther than the edge of the table:

$$\frac{1}{2}\left(1 + \frac{1}{2} + \frac{1}{3} + \frac{1}{4}\right) = \frac{25}{24} = 1.0417 \tag{2.583}$$

With 10 books we will have $x = 1.4645l$, with 100 books we will have $x = 2.5937l$, and with 1000 books we will have $x = 3.7427l$. Figure 2.39(b) illustrates the solution for $n = 20$.

Example 167 ★ The* n*-Body Problem The n-body problem states: Consider n point masses $m_1, m_2, \ldots, m_n$ in a three-dimensional Euclidean space. Suppose that the force of attraction experienced between each pair of particles is Newtonian. If the initial positions and velocities are specified for every particle at some present time t_0, determine the position of each particle at every future (or past) moment of time. The equations of motion for n bodies using G for the universal constant of gravitation and $\mathbf{X}_i$ to indicate the position vector of the point mass m_i are

$$m_i \ddot{\mathbf{X}}_i = -G \sum_{j=1}^{2} m_i m_j \frac{\mathbf{X}_i - \mathbf{X}_j}{\mathbf{X}_{ij}} \qquad i = 1, 2, 3, \ldots, n \tag{2.584}$$

$$\mathbf{X}_{ij} = \mathbf{X}_i - \mathbf{X}_j \tag{2.585}$$

For $n = 2$ this is called the *two-body problem*, or the *Kepler problem*. Johann Bernoulli first gave the complete solution of the Kepler problem in 1710. The two-body problem has proven to be the only easy and solvable one among the n-body problems. The problems for $n = 2$ and $n > 2$ differ not only qualitatively but also quantitatively.

Example 168 ★ Motion of a Charged Particle in a Magnetic Field To move an electric charge e with velocity $\mathbf{v}$ in a magnetic field $\mathbf{H}$ and an electric field $\mathbf{E}$, we need to provide a force

$$\mathbf{F} = e\mathbf{E} + \frac{e}{c}\,(\mathbf{v} \times \mathbf{H}) \tag{2.586}$$

where c is the speed of light. When $\mathbf{E} = \mathbf{0}$, the particle is moving in a magnetic field and the required force simplifies to

$$\mathbf{F} = \frac{e}{c}\,(\mathbf{v} \times \mathbf{H}) \tag{2.587}$$

Such a force is called the *Lorentz force*. The force is perpendicular to $\mathbf{v}$ and hence the magnetic force does no work on the particle:

$$W = \int \mathbf{F} \cdot d\mathbf{r} = \int \mathbf{F} \cdot \mathbf{v}\,dt = \frac{e}{c} \int (\mathbf{v} \times \mathbf{H}) \cdot \mathbf{v}\,dt = 0 \tag{2.588}$$

Having $W = 0$ is equivalent to having a constant kinetic energy. The equation of motion of the particle is then

$$m\mathbf{a} = \frac{e}{c}\,(\mathbf{v} \times \mathbf{H}) \tag{2.589}$$

Let us assume a uniform magnetic field in the x-direction and substitute for $\mathbf{a}$, $\mathbf{v}$, and $\mathbf{H}$ by

$$\mathbf{v} = \dot{x}\hat{\imath} + \dot{y}\hat{\jmath} + \dot{z}\hat{k} \tag{2.590}$$

$$\mathbf{a} = \ddot{x}\hat{\imath} + \ddot{y}\hat{\jmath} + \ddot{z}\hat{k} \tag{2.591}$$

$$\mathbf{H} = H\hat{k} \tag{2.592}$$

to get

$$m \begin{bmatrix} \ddot{x} \\ \ddot{y} \\ \ddot{z} \end{bmatrix} = \frac{e}{c} \left(\begin{bmatrix} \dot{x} \\ \dot{y} \\ \dot{z} \end{bmatrix} \times \begin{bmatrix} 0 \\ 0 \\ H \end{bmatrix} \right) = \begin{bmatrix} H\dot{y} \\ -H\dot{x} \\ 0 \end{bmatrix} \tag{2.593}$$

The third equation,

$$\ddot{z} = 0 \tag{2.594}$$

can be integrated to find the z-component of motion of the particle:

$$z = \dot{z}_0 t + z_0 \tag{2.595}$$

So, The z-component has a uniform motion at a constant speed. To solve the first and second equations

$$\ddot{x} = k\dot{y} \tag{2.596}$$

$$\ddot{y} = -k\dot{x} \tag{2.597}$$

$$k = \frac{eH}{mc} \tag{2.598}$$

we take a time derivative and get

$$\dddot{x} = k\ddot{y} = -k^2\dot{x} \tag{2.599}$$

$$\dddot{y} = -k\ddot{x} = -k^2\dot{y} \tag{2.600}$$

The solutions of these equations are

$$x - x_0 = A\cos kt + B\sin kt$$
$$y - y_0 = C\cos kt + D\sin kt$$

However, because x and y must satisfy Equations (2.596) and (2.597) at all times, we have

$$-A = D \qquad B = C \tag{2.601}$$

and therefore the components of the position vectors of the particle are

$$x - x_0 = A\cos kt + B\sin kt \tag{2.602}$$
$$y - y_0 = B\cos kt - A\sin kt \tag{2.603}$$
$$z - z_0 = \dot{z}_0 t \tag{2.604}$$

Assuming $\dot{x}_0 = 0$, we find the constants A and B as

$$B = 0 \qquad A = -\frac{1}{k}\dot{y}_0 \tag{2.605}$$

and therefore

$$x - x_0 = -\frac{1}{k}\dot{y}_0 \sin kt \tag{2.606}$$
$$y - y_0 = -\frac{1}{k}\dot{y}_0 \cos kt \tag{2.607}$$
$$z - z_0 = \dot{x}_0 t \tag{2.608}$$

These are the parametric equations of a circular helix with an axis parallel to the z-axis, radius R, and pitch $2\pi/\dot{y}_0$, as is shown in Figure 2.40:

$$R = \frac{1}{k}\dot{y}_0 = \frac{mc}{eH}\dot{y}_0 \tag{2.609}$$

The radius would be larger for larger mass m or higher initial lateral speed $\dot{y}_0$.

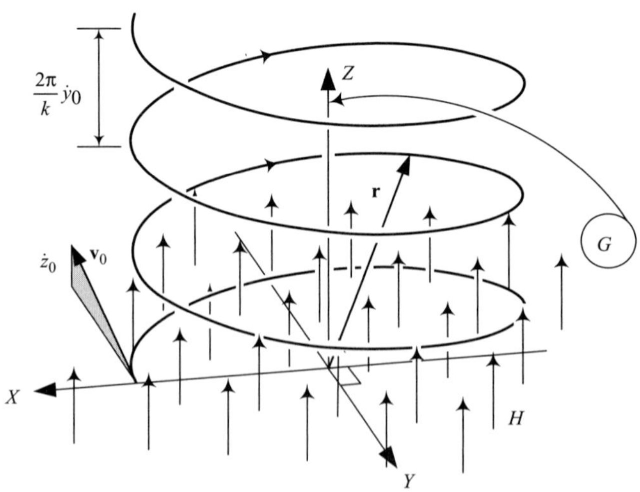

Figure 2.40 Circular helix path of motion of a charged particle in a uniform magnetic field.

The magnetic force on the particle is always perpendicular to its lateral velocity and makes the particle turn on a circular path while moving with a constant speed along the magnetic field. This mechanism will capture charged particles and keep them in the field.

This is the base model used to explain the Van Allen belt around Earth because of Earth's magnetic field and also a magnetic bottle that keeps hot plasma surrounded.

2.5.2 ★ Equations of Motion

Determination of the equations of motion is the most important part of dynamics. Most dynamic problems will end up with a finite set of differential equations along with a set of initial conditions. Having the equations of motion converts a dynamic problem into a mathematical problem. There are few methods to develop the equations of motion of a dynamic phenomenon. The Newton–Euler method is based on an external force and change of momentum. The Lagrange method is based on energy and generalized coordinates. The Hamilton method is based on the Lagrangian, Legendre transformation, and generalized momentum. Although conceptually different, these methods are mathematically equivalent. All three well-developed methods will result in the same equations of motion or equations that may be converted into each other.

Besides the Newton–Euler, Lagrangian, and Hamilton methods, there are other methods of dynamics that generate the equations of motion but follow different concepts. The Gauss principle of minimum constraint of motion, D'Alembert's method of virtual work, and Appell's method of generalized acceleration and force are a few examples.

Example 169 A Body Sliding on an Inclined Plane Consider a body of mass m_1 sliding without friction on the inclined surface of a wedge with mass m_2. The wedge can slide on a horizontal frictionless surface as is shown in Figure 2.41.

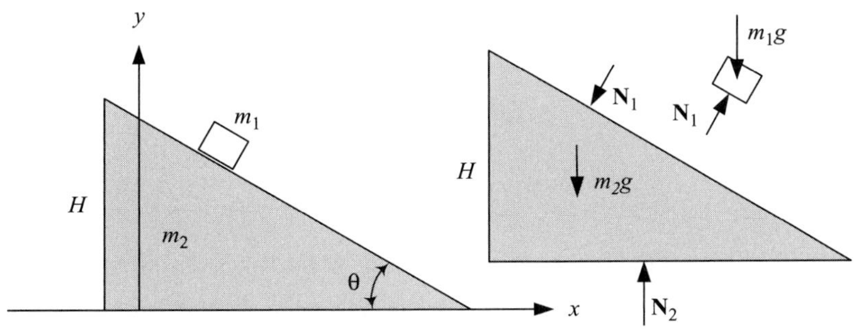

Figure 2.41 A body of mass m_1 sliding without friction down an inclined wedge of mass m_2.

The reaction force $\mathbf{N}_1$ from m_2 to m_1 and the reaction force $\mathbf{N}_2$ from the horizontal surface to m_2 are given as

$$\mathbf{N}_1 = N_1 \sin\theta \hat{\imath} + N_1 \cos\theta \hat{\jmath} \tag{2.610}$$

$$\mathbf{N}_2 = N_2 \hat{\jmath} \tag{2.611}$$

Now we can write the equations of motion of m_1 and m_2:

$$m_1\ddot{x}_1 = N_1 \sin\theta \tag{2.612}$$

$$m_1\ddot{y}_1 = N_1 \cos\theta - m_1 g \tag{2.613}$$

$$m_2\ddot{x}_2 = -N_1 \sin\theta \tag{2.614}$$

$$m_2\ddot{y}_2 = N_2 - N_1 \cos\theta - m_1 g \tag{2.615}$$

Adding the first and third equations and integrating over time provide the conservation of momentum on the x-axis:

$$m_1\dot{x}_1 + m_2\dot{x}_2 = C_1 \tag{2.616}$$

The vertical position of m_2 is constant and hence $\ddot{y}_2 = 0$:

$$N_2 = N_1 \sin\theta + m_1 g \tag{2.617}$$

Let us assume that the initial position of m_2 and m_1 is such that the edge H and m_1 are on the y-axis. So, if m_1 is at x_1 and m_2 is at x_2, then m_1 must be at $y_1 = H - (x_1 - x_2)\tan\theta$, and therefore

$$\dot{y}_1 = -(\dot{x}_1 - \dot{x}_2)\tan\theta \tag{2.618}$$

$$\ddot{y}_1 = -(\ddot{x}_1 - \ddot{x}_2)\tan\theta \tag{2.619}$$

Substituting for $\ddot{x}_1$ and $\ddot{x}_2$ in (2.619) yields

$$\ddot{y}_1 = -\left(\frac{N_1 \sin\theta}{m_1} - \frac{-N_1 \sin\theta}{m_2}\right)\tan\theta \tag{2.620}$$

and employing (2.613) provides

$$N_1 = m_1 g \frac{\cos\theta}{1 + (m_1/m_2)\sin^2\theta} \tag{2.621}$$

So, the acceleration of the wedge and vertical acceleration of m_1 are

$$\ddot{x}_2 = -\frac{m_1}{m_2}\frac{\sin\theta\cos\theta}{1 + (m_1/m_2)\sin^2\theta} g \tag{2.622}$$

$$\ddot{y}_1 = -\left(1 - \frac{\cos^2\theta}{1 + (m_1/m_2)\sin^2\theta}\right) g \tag{2.623}$$

Example 170 ★ Projectile and Variable Gravitational Acceleration When a projectile is always close to the ground, it is reasonable to write its equation of motion as

$$m\frac{d\mathbf{v}}{dt} = m\mathbf{g} \tag{2.624}$$

$$\mathbf{g} = -g\hat{k} \tag{2.625}$$

$$g_0 = 9.80665 \text{ m/s}^2 \approx 9.81 \text{ m/s}^2 \tag{2.626}$$

This is the situation of the flat-ground and no-air conditions examined in Example 126.

Because the gravitational attraction varies with distance from the center of Earth, the gravitational acceleration at a height z is

$$\mathbf{g} = -g_0 \frac{R_0^2}{(R_0 + z)^2} \hat{k} \tag{2.627}$$

if $\mathbf{g}_0 = -g_0 \hat{k}$ at the surface of Earth,

$$R_0 = 6{,}371{,}230 \text{ m} \tag{2.628}$$

The equation of motion of a projectile, where its height is high enough to change the value of g is

$$m \frac{d\mathbf{v}}{dt} = -m g_0 \frac{R_0^2}{(R_0 + z)^2} \hat{k} \tag{2.629}$$

If such a projectile is moving in air with a resistance proportional to the square of its velocity $\mathbf{v}^2$, then its equation of motion would be

$$m \frac{d\mathbf{v}}{dt} = -m g_0 \frac{R_0^2}{(R_0 + z)^2} \hat{k} - c\mathbf{v}^2 \tag{2.630}$$

Now assume that the projectile is a rocket that generates a trust $\mathbf{F}$,

$$\mathbf{F} = -\dot{m}\, \mathbf{v}_m \tag{2.631}$$

where $\dot{m}$ is the mass rate of burning fuel and $\mathbf{v}_m$ is the velocity of the escaping burnt fuel with respect to the rocket. The equation of motion for the projectile rocket in air is

$$m \frac{d\mathbf{v}}{dt} = -\dot{m}\, \mathbf{v}_m - m g_0 \frac{R_0^2}{(R_0 + z)^2} \hat{k} - c\mathbf{v}^2 \tag{2.632}$$

The Earth is not flat, so we should rewrite the equation of motion for a curved ground. If Earth is assumed to be a sphere and $\mathbf{v}_m$ is on the same axis as $\mathbf{v}$,

$$\mathbf{v}_m = -v_m \frac{\mathbf{v}}{|\mathbf{v}|} \qquad v_m > 0 \tag{2.633}$$

then the projectile will have a planar motion, say in the (x, z)-plane:

$$m \frac{d\mathbf{v}}{dt} = \dot{m}\, v_m \frac{\mathbf{v}}{|\mathbf{v}|} - \frac{m g_0 R_0^2}{(R_0 + z)^2} \hat{k} - c v^2 \frac{\mathbf{v}}{|\mathbf{v}|} \tag{2.634}$$

$$\begin{aligned} m\left(\ddot{x}\hat{\imath} + \ddot{z}\hat{k}\right) &= \dot{m} v_m \frac{\dot{x}\hat{\imath} + \dot{z}\hat{k}}{\sqrt{\dot{x}^2 + \dot{z}^2}} - \frac{m g_0 R_0^2}{(R_0 + z)^2} \hat{k} \\ &\quad - c\sqrt{\dot{x}^2 + \dot{z}^2}\left(\dot{x}\hat{\imath} + \dot{z}\hat{k}\right) \end{aligned} \tag{2.635}$$

The equation of motion of such a projectile may be written in polar coordinates (ρ, θ). The gravity of the spherical Earth is

$$\mathbf{g} = -g_0 \frac{R_0^2}{\rho^2} \hat{u}_\rho \tag{2.636}$$

and therefore the equation of motion of the projectile is

$$m\left(\ddot{\rho}-\rho\dot{\theta}^{2}\right)\hat{u}_{\rho}+m\left(2\dot{\rho}\dot{\theta}+\rho\ddot{\theta}\right)\hat{u}_{\theta}$$
$$=\dot{m}v_{m}\frac{\dot{\rho}\hat{u}_{\rho}+\rho\dot{\theta}\hat{u}_{\theta}}{\sqrt{\dot{\rho}^{2}+\rho^{2}\dot{\theta}^{2}}}-mg_{0}\frac{R_{0}^{2}}{\rho^{2}}\hat{u}_{\rho}-c\sqrt{\dot{\rho}^{2}+\rho^{2}\dot{\theta}^{2}}\left(\dot{\rho}\hat{u}_{\rho}+\rho\dot{\theta}\hat{u}_{\theta}\right) \tag{2.637}$$

2.5.3 ★ Dynamic Behavior and Methods of Solution

A dynamic problem is composed of a dynamic phenomenon and its equations of motion based on a proper model. Mathematical tools along with an understanding and observations are used to solve the dynamic problem and determine its dynamic behavior. Today there are three basic methods to solve equations of motion: *analytic*, *numerical*, and *approximate* solutions. The method of *transformation* in real or complex domains can be used to change the coordinates and transform the equation of motion to a simpler form for application of one of the three methods.

Analytic solutions are the best, but they are limited to equations with available closed-form solutions. An analytic solution is the best way to determine every aspect of the behavior of a dynamic phenomenon. Furthermore, the analytic solution is the most challenging method that employs every possible mathematical tools. Many of the mathematical theorems and methods were discovered and invented when scientists tried to solve a dynamic problem analytically.

Approximate solutions, such as *series solutions* or *perturbation methods*, are employed when there is no analytic solution for the equations of motion but we need an equation to analyze the relative effect of the involved parameters. Most of the time, approximate methods provide the second best solutions.

Numerical methods are generally employed when we need to predict the behavior of a given dynamic system starting from an initial condition. They are the only choice when no analytic solution exists, and approximate methods are too complicated.

Example 171 Analytic Solution of First-Order Linear Equations All first-order dynamic systems are in the form

$$\dot{x}+f_{1}(t)\,x=f_{2}(t) \tag{2.638}$$

The first-order equation is a total differential

$$\frac{\partial f}{\partial x}dx+\frac{\partial f}{\partial t}dt=0 \tag{2.639}$$

if it is a derivative of a function $f(x,t)$.

$$f(x,t)=c \tag{2.640}$$

Multiplying the equation by

$$g(t)=\exp\int f_{1}(t)\,dt \tag{2.641}$$

will transform it into

$$\dot{x}g(t) + xf_1(t)\,g(t) = \frac{d}{dt}(xg(t)) = f_2(t)\,g(t) \tag{2.642}$$

which is always solvable:

$$x = \frac{1}{g(t)} \int f_2(t)\,g(t)\,dt \tag{2.643}$$

As an example, consider the dynamics of radioactive matter in a live creature. The molecules of the matter x decompose into smaller molecules and it is natural to assume that the rate of decomposition $\dot{x}$ is proportional to the number of present radioactive molecules x. Furthermore, there would be growth because of food, water, air, and sun:

$$\frac{dx}{dt} = -k_1 x + k_2 x = (k_2 - k_1)\,x \tag{2.644}$$

The solution of the equation would be

$$x = Ce^{(k_2 - k_1)t} \tag{2.645}$$

At the steady-state condition, $k_1 = k_2$, and the level of the number of molecules remains constant, $C = x_0$. To determine the age of a fossil, we may measure x, evaluate the level of $C = x_0$ in a similar live species, and set $k_2 = 0$ to find t:

$$t = -\frac{1}{k_1} \ln \frac{x}{x_0} \tag{2.646}$$

Example 172 ★ No Position in the Equation of Motion If there is no position variable $\mathbf{r}$ in the equation of motion explicitly, the equation has the form

$$f(\ddot{\mathbf{r}}, \dot{\mathbf{r}}, t) = 0 \tag{2.647}$$

In this case we may introduce a new dependent variable $\mathbf{p}$:

$$\mathbf{p} = \dot{\mathbf{r}} \qquad \dot{\mathbf{p}} = \ddot{\mathbf{r}} \tag{2.648}$$

This substitution transforms Equation (2.647) into a first-order equation:

$$f(\dot{\mathbf{p}}, \mathbf{p}, t) = 0 \tag{2.649}$$

If we find a solution for (2.649), we can replace $\mathbf{p} = d\mathbf{r}/dt$ in the solution and solve another first-order equation for the position $\mathbf{r} = \mathbf{r}(t)$. Therefore, if there is no position variable in the equation of motion, we may solve two first-order differential equations instead of one second-order equation.

As an example, consider a particle with $m = 1$ that is under a time-decaying and velocity-dependent force with the equation of motion

$$\ddot{x} = -\dot{x} + e^{-t} \tag{2.650}$$

We may transform the equation to

$$\frac{dp}{dt} + p = e^{-t} \tag{2.651}$$

and find the solution:

$$p = e^{-t} \int e^{-t} e^{t} \, dt = (t + C_1) \, e^{-t} \tag{2.652}$$

Assume general initial conditions as

$$x(0) = x_0 \qquad \dot{x}(0) = p(0) = \dot{x}_0 = p_0 \tag{2.653}$$

Then

$$C_1 = p_0 \tag{2.654}$$

and we have

$$p = (t + p_0) \, e^{-t} \tag{2.655}$$

Substituting back into $\dot{x}$,

$$\frac{dx}{dt} = (t + p_0) \, e^{-t} \tag{2.656}$$

provides the solution

$$x = -(1 + \dot{x}_0 + t)e^{-t} + C_2 \tag{2.657}$$

$$C_2 = 1 + \dot{x}_0 + x_0 \tag{2.658}$$

Example 173 ★ No Time in the Equation of Motion If there is no time t in the equation of motion explicitly, the equation has the form

$$f\left(\ddot{\mathbf{r}}, \dot{\mathbf{r}}, \mathbf{r}\right) = 0 \tag{2.659}$$

In this case we may introduce a new dependent variable **p**,

$$\mathbf{p} = \dot{\mathbf{r}} \qquad \dot{\mathbf{p}} = \ddot{\mathbf{r}} \tag{2.660}$$

and express $\ddot{\mathbf{r}}$ with respect to **r**:

$$\ddot{\mathbf{r}} = \dot{\mathbf{p}} = \frac{d\mathbf{p}}{d\mathbf{r}} \frac{d\mathbf{r}}{dt} = \dot{\mathbf{p}} \frac{d\mathbf{p}}{d\mathbf{r}} \tag{2.661}$$

This substitution transforms the equation of motion into a first-order equation:

$$f\left(\mathbf{p}\frac{d\mathbf{p}}{d\mathbf{r}}, \mathbf{p}, \mathbf{r}\right) = 0 \tag{2.662}$$

If we find a solution for (2.662), we can replace $\mathbf{p} = d\mathbf{r}/dt$ in the solution and solve another first-order equation for the position $\mathbf{r} = \mathbf{r}(t)$. Therefore, if there is no time in the equation of motion, we may solve two first-order differential equations instead of one second-order equation.

As an example, consider a falling bungee jumper with

$$z(0) = z_0 = 0$$
$$\dot{z}(0) = \dot{z}_0 = 0 \tag{2.663}$$

who is under the force of gravity and an elastic rope with stiffness k and a free length l:

$$\ddot{z} = \begin{cases} g & z < l \\ g - \frac{k}{m}(z - l) & z > l \end{cases} \tag{2.664}$$

We use $\dot{z} = p$ and transform the equation to

$$p\frac{dp}{dz} = \begin{cases} g & z < l \\ g - \frac{k}{m}(z - l) & z > l \end{cases} \tag{2.665}$$

and find the solution:

$$p = \begin{cases} \sqrt{2gz + C_1} & z < l \\ \frac{1}{m}\sqrt{2gm^2 z + C_2 m^2 - kmz^2 + 2klmz} & z > l \end{cases} \tag{2.666}$$

Using the initial condition $\dot{z}_0 = p(0) = 0$ and compatibility of the solutions at $z = l$, we find

$$C_1 = 0 \qquad C_2 = -\frac{kl^2}{m} \tag{2.667}$$

and therefore,

$$p = \begin{cases} \sqrt{2gz} & z < l \\ \frac{1}{m}\sqrt{-kl^2 m + 2klmz + 2gm^2 z - kmz^2} & z > l \end{cases} \tag{2.668}$$

Substituting back to $\dot{z} = p$ provides the solution

$$t = \begin{cases} \frac{\sqrt{2z}}{g} + C_3 & z < l \\ \sqrt{\frac{m}{k}}\tan^{-1}\left(\frac{gm + kl - kz}{-k}\sqrt{-\frac{k}{k(z-l)^2 - 2mgz}}\right) + C_4 & z > l \end{cases} \tag{2.669}$$

Because $z(0) = 0$, we have

$$C_3 = 0 \tag{2.670}$$

and because of compatibility of the solutions at $z = l$, we have

$$C_4 = -\sqrt{\frac{2l}{g}} + \sqrt{\frac{m}{k}}\tan^{-1}\left(\sqrt{\frac{mg}{2kl}}\right) \tag{2.671}$$

Let us examine the dynamic of the jumper for the following data:

$$m = 100\,\text{kg} \qquad k = 200\,\text{N/m} \qquad l = 10\,\text{m} \tag{2.672}$$

At $z = l$ we will have

$$t = 1.428\,\text{s} \tag{2.673}$$

After this time elapsed, the solution switches to the case $z > l$ that ends when $\dot{z} = 0$. Solving Equation (2.668) for the maximum stretch of the elastic rope z_M shows that

$$z_M = 25.9575\,\text{m} \tag{2.674}$$

Figure 2.42 illustrates the time history of the falling height of the jumper from $z = 0$ to $z = z_M$. The maximum stretch of the elastic rope could be more easily found from the conservation of energy equation:

$$mgz_M = \frac{1}{2}k\left(z_M - l\right)^2 \tag{2.675}$$

$$z_M = \frac{1}{k}\left(gm + kl + \sqrt{gm\left(gm + 2kl\right)}\right) = 25.958\,\text{m} \tag{2.676}$$

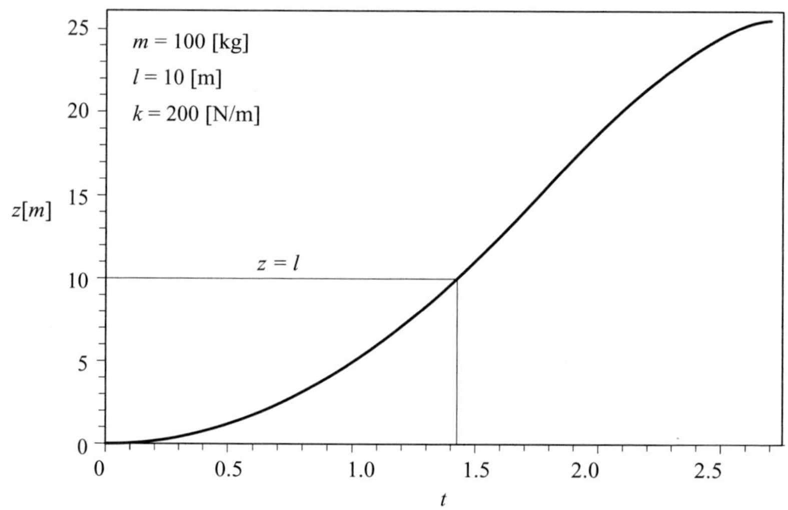

Figure 2.42 Time history of the falling height of a bungee jumper from $z = 0$ to $z = z_M$.

Example 174 ★ Kinematics of a Dragging Point Mass A point mass P is attached to a moving car C at the origin with an extensible cable. The car is moving on the y-axis at a constant speed v and drags P that was initially at $(l, 0)$. Figure 2.43 illustrates the positions of P and C when P is at (x, y).

The equation of the path of P is the solution of a first-order differential equation

$$\frac{dy}{dx} = -\frac{\sqrt{l^2 - x^2}}{x} \tag{2.677}$$

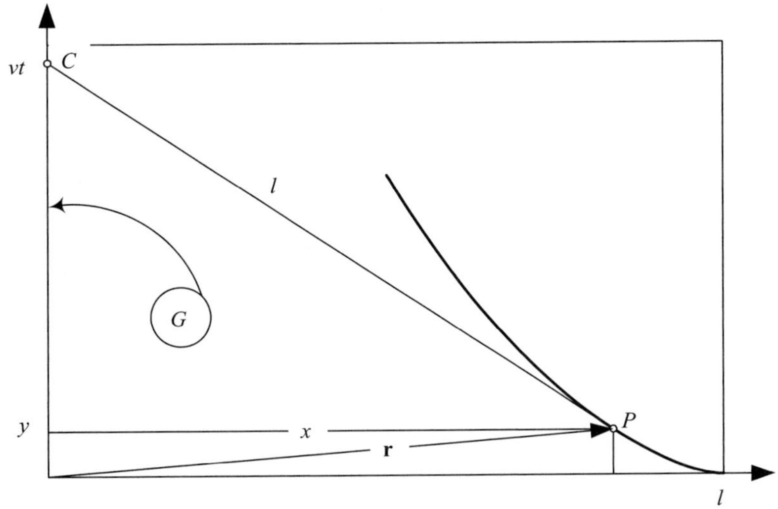

Figure 2.43 The path of motion of a dragged point P by a moving car C.

Using the separation of variables and the initial conditions of P and C, we find

$$y = l \ln\left(\frac{l + \sqrt{l^2 - x^2}}{x}\right) - \sqrt{l^2 - x^2} \tag{2.678}$$

So the position, velocity, and acceleration vectors of P are

$$\mathbf{r}_P = x\hat{\imath} + y\hat{\jmath} \qquad \mathbf{v}_P = \dot{x}\hat{\imath} + \dot{y}\hat{\jmath} \qquad \mathbf{a}_P = \ddot{x}\hat{\imath} + \ddot{y}\hat{\jmath} \tag{2.679}$$

where y is calculated based on (2.678) and the coordinates x and y are related to the position of the car C by

$$x^2 + (y - vt)^2 = l^2 \tag{2.680}$$

Substituting (2.678) in (2.680) provides the time t as a function of the x-component to complete the kinematics of the moving point:

$$t = \frac{1}{v}\left(y + \sqrt{l^2 - x^2}\right) = \frac{l}{v} \ln \frac{l + \sqrt{l^2 - x^2}}{x} \tag{2.681}$$

Using Equation (2.681), we are able to determine x at a time t, and find the y-component from (2.678). Taking the derivative of (2.681) provides $\dot{x}$ as a function of x,

$$\dot{x} = -\frac{vx}{l^2}\sqrt{l^2 - x^2} \tag{2.682}$$

and (2.677) provides $\dot{y}$,

$$\dot{y} = -\frac{\sqrt{l^2 - x^2}}{x}\dot{x} = \frac{v}{l^2}\left(l^2 - x^2\right) \tag{2.683}$$

To determine the acceleration, we may take a derivative from (2.683) and (2.677) to find

$$\ddot{x} = \frac{2x^2 - l^2}{l^2\sqrt{l^2 - x^2}} v\dot{x} = \frac{2x^2 - l^2}{l^4} v^2 x \tag{2.684}$$

$$\ddot{y} = -\frac{2xv}{l^2}\dot{x} = \frac{2\sqrt{l^2 - x^2}}{l^4} v^2 x^2 \tag{2.685}$$

Figure 2.44 depicts the time history of the x- and y-components of the $\mathbf{r}_P$, and Figure 2.45 shows the components of the velocity vector $\mathbf{v}_P$. The components of the acceleration vector $\mathbf{a}_P$ are shown in Figure 2.46. If the mass of point P is m, then the

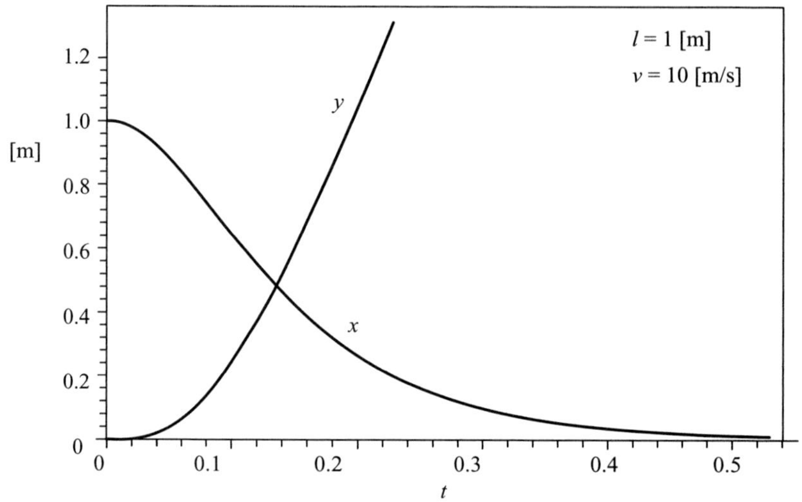

Figure 2.44 Components of position vector $\mathbf{r}_P$ versus time.

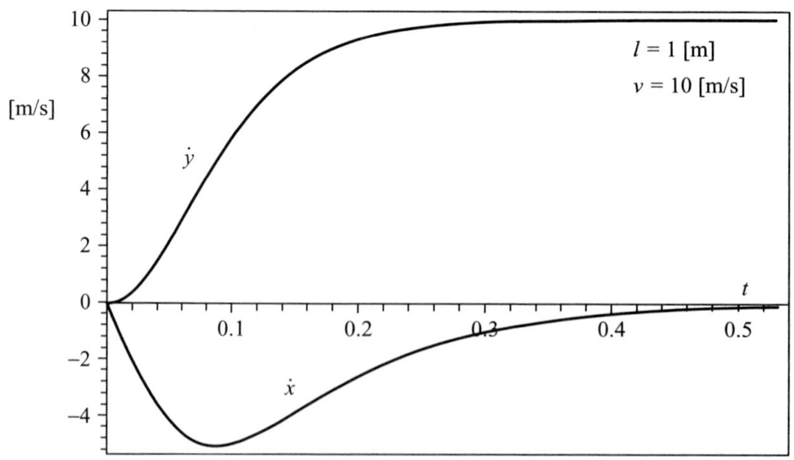

Figure 2.45 The components of velocity vector $\mathbf{v}_P$ versus time.

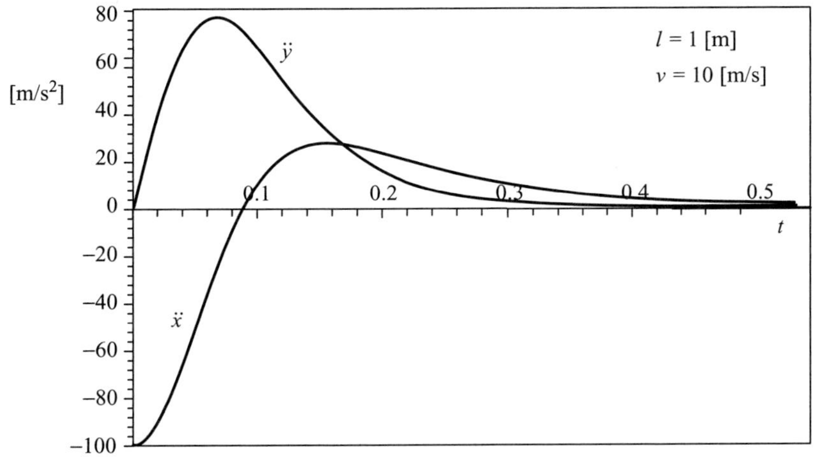

Figure 2.46 Components of acceleration vector $\mathbf{a}_P$ versus time.

required force **F** to move P on the path will be

$$\mathbf{F} = m\mathbf{a}_P = m\left(\ddot{x}\hat{\imath} + \ddot{y}\hat{\jmath}\right) \tag{2.686}$$

Figure 2.47 illustrates the surface of revolution of (2.678) about the x-axis. This surface is called a pseudosphere because it has a constant negative curvature as opposed to a sphere with a constant positive curvature. It is also the mathematical shape of a *Lobachevsky* space on which the sum of the angles of any triangle is less than 180 deg.

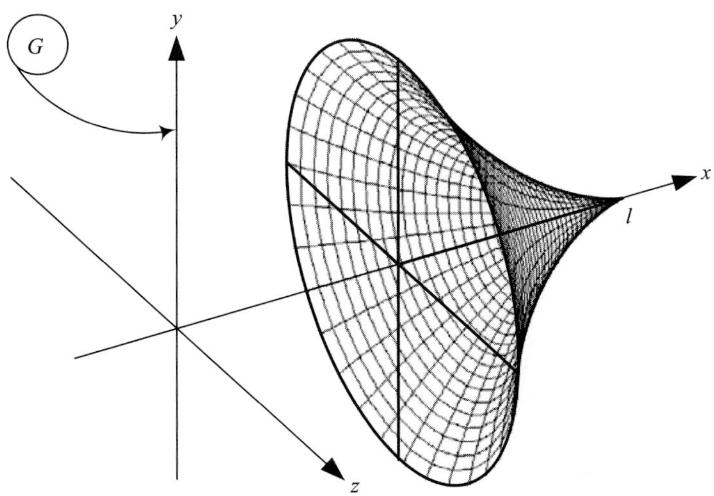

Figure 2.47 The surface of revolution of the motion path of a dragged point.

Example 175 ★ Antiaircraft Rocket An aircraft C is moving on the y-axis with a constant speed v_C. When C is at the origin, a rocket R with speed $v_R = \dot{s}$ is fired from $(a, 0)$ to hit the aircraft. The velocity vector of R is always toward the aircraft and therefore the tangent to the path of the rocket must be

$$\frac{dy}{dx} = -\frac{v_C t - y}{x} \tag{2.687}$$

Figure 2.48 illustrates the velocity vectors of the aircraft C and the rocket R at a time t.

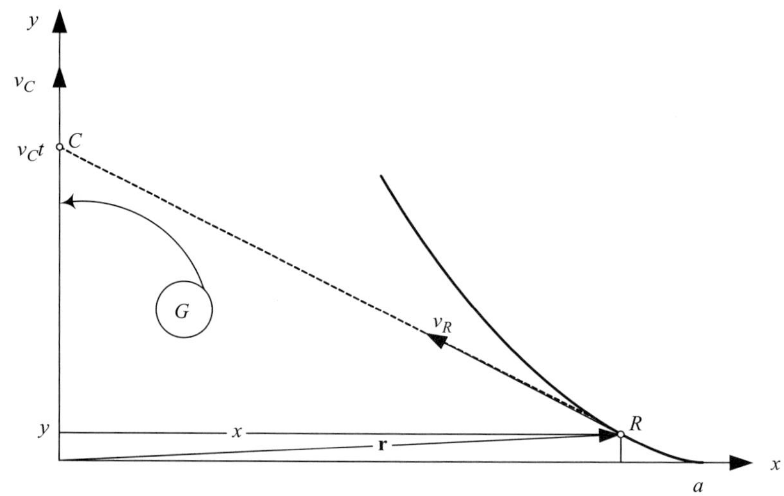

Figure 2.48 The velocity vector of the antiaircraft rocket R is always toward the aircraft C.

To determine the path of R, we should eliminate x, y, or t from (2.687), so we take a time derivative of (2.687):

$$\frac{dt}{dx} = -\frac{x}{v_C}\frac{d^2y}{dx^2} \tag{2.688}$$

Let us eliminate $\dot{s} = v_R$ from dt/dx:

$$\frac{dt}{dx} = \frac{dt}{ds}\frac{ds}{dx} = -\frac{1}{v_R}\sqrt{1 + \left(\frac{dy}{dx}\right)^2} \tag{2.689}$$

where the negative sign appears because x decreases when t increases. Combining (2.688) and (2.689) provides the differential equation of the path:

$$x\frac{d^2y}{dx^2} = k\sqrt{1 + \left(\frac{dy}{dx}\right)^2} \qquad k = \frac{v_C}{v_R} \tag{2.690}$$

We may use $y' = p$ and $y'' = p'$ to transform the equation to

$$x\frac{dp}{dx} = k\sqrt{1 + p^2} \tag{2.691}$$

and find the solution

$$\ln\left(p+\sqrt{1+p^2}\right)=\ln\left(\frac{x}{a}\right)^k \tag{2.692}$$

Solving for p yields

$$2p=2\frac{dy}{dx}=\left(\frac{x}{a}\right)^k-\left(\frac{a}{x}\right)^k \tag{2.693}$$

and integrating determines the path of motion of the rocket:

$$y=\frac{x}{2}\left(\frac{(x/a)^k}{k+1}+\frac{(a/x)^k}{k-1}\right)-\frac{ak}{k^2-1} \tag{2.694}$$

The constant of the integral comes from the initial condition $y(a)=0$. Substituting (2.693) and (2.694) in (2.687), we can find t as a function of x:

$$t=\frac{kx}{2v_C}\left(\frac{(x/a)^k}{k+1}-\frac{(a/x)^k}{k-1}\right)-\frac{ak}{v_C\left(k^2-1\right)} \tag{2.695}$$

Figure 2.49 depicts the coordinates of the rocket as functions of time for the following data:

$$k=0.5 \qquad a=1 \qquad v_C=1 \tag{2.696}$$

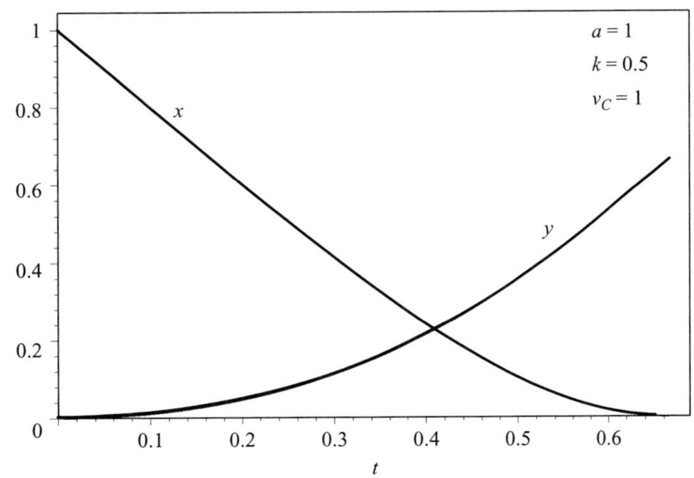

Figure 2.49 Components of position vector of the rocket versus time.

To determine the time t_0 at which the rocket hits the aircraft, we need to solve (2.695) for $x=0$. However, it is not possible to determine $t_0=\lim_{x\to 0} t$ in a general case. The rocket can reach the aircraft for any $0<k<1$, and it is possible to evaluate t_0 for any proper value of k. As an example, we may set $k=0.5$ and evaluate t_0:

$$t_0=\lim_{x\to 0} t=0.6666666667\frac{a}{v_C} \tag{2.697}$$

Example 176* ★ *Series Solution for Two-Body Problem Consider two point masses m_1 and m_2 as are shown in Figure 2.50 at positions $\mathbf{X}_1$ and $\mathbf{X}_2$ that attract each other by a Newtonian gravitational force. If the space is assumed Euclidean and there is no other mass in the space, the equations of motion of m_1 and m_2 are

$$\ddot{\mathbf{X}}_1 = -G_2 \frac{\mathbf{X}_1 - \mathbf{X}_2}{|\mathbf{X}_1 - \mathbf{X}_2|^3} \qquad \ddot{\mathbf{X}}_2 = -G_1 \frac{\mathbf{X}_2 - \mathbf{X}_1}{|\mathbf{X}_2 - \mathbf{X}_1|^3} \tag{2.698}$$

$$G_i = Gm_i \qquad i = 1, 2 \tag{2.699}$$

$$G = 6.67259 \times 10^{-11}\ \mathrm{m^3 kg^{-1} s^{-2}} \tag{2.700}$$

The two-body problem, similar to other initial-value problems in dynamics, is to determine $\mathbf{X}_1(t)$ and $\mathbf{X}_2(t)$ by having the conditions of $\mathbf{X}_1(t_0)$, $\mathbf{X}_2(t_0)$, $\dot{\mathbf{X}}_1(t_0)$, and $\dot{\mathbf{X}}_2(t_0)$ at time t_0.

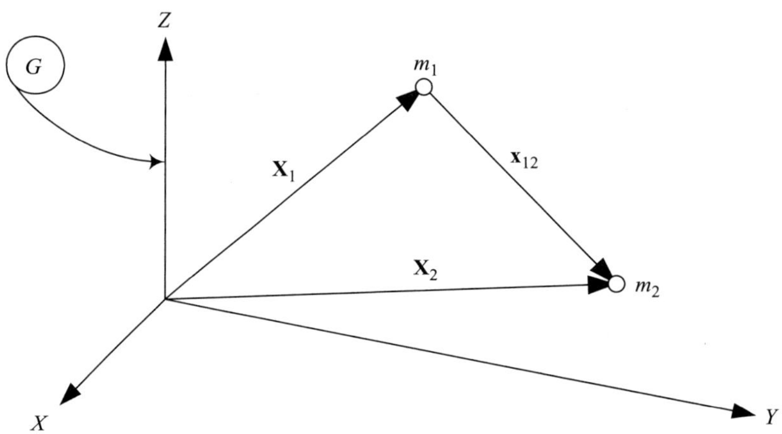

Figure 2.50 Two bodies in Euclidean space.

Subtracting Equations (2.698), we get the fundamental equation of the two-body problem in terms of relative position vector $\mathbf{x}$:

$$\ddot{\mathbf{x}} + \mu \frac{\mathbf{x}}{|\mathbf{x}|^3} = 0 \qquad \mathbf{x} = \mathbf{X}_2 - \mathbf{X}_1 \tag{2.701}$$

$$\mu = G_1 + G_2 \tag{2.702}$$

To solve (2.701), let us search for a series solution in the form

$$\mathbf{x}(t) = \mathbf{x}_0 + \dot{\mathbf{x}}_0 (t - t_0) + \ddot{\mathbf{x}}_0 \frac{(t - t_0)^2}{2!} + \dddot{\mathbf{x}}_0 \frac{(t - t_0)^3}{3!} + \cdots \tag{2.703}$$

$$\mathbf{x}_0 = \mathbf{x}(t_0) \qquad \dot{\mathbf{x}}_0 = \dot{\mathbf{x}}(t_0) \tag{2.704}$$

Because Equation (2.701) is singular only at $\mathbf{x} = 0$, the series (2.703) converges for all nonzero $\mathbf{x}$. To determine the series (2.703), we need to show that the coefficients $\mathbf{x}_0^{(n)}$ can be determined for any desired number of n.

Defining a new parameter ε

$$\varepsilon = \frac{\mu}{|\mathbf{x}|^3} \tag{2.705}$$

we can rewrite Equation (2.701) as

$$\ddot{\mathbf{x}}_1 + \varepsilon \mathbf{x} = 0 \tag{2.706}$$

Recalling the time derivative of the absolute value of a vector,

$$\frac{d}{dt}|\mathbf{x}| = \frac{\mathbf{x}\cdot\dot{\mathbf{x}}}{|\mathbf{x}|} \tag{2.707}$$

and introducing a parameter λ, we find the time derivative of (2.705):

$$\dot{\varepsilon} = -3\mu\frac{\mathbf{x}\cdot\dot{\mathbf{x}}}{|\mathbf{x}|^5} = -3\varepsilon\frac{\mathbf{x}\cdot\dot{\mathbf{x}}}{|\mathbf{x}|^2} = -3\varepsilon\lambda \tag{2.708}$$

$$\lambda = \frac{\mathbf{x}\cdot\dot{\mathbf{x}}}{|\mathbf{x}|^2} \tag{2.709}$$

The time derivative of λ introduces another parameter ψ:

$$\begin{aligned}\dot{\lambda} &= \frac{\dot{\mathbf{x}}\cdot\dot{\mathbf{x}} + \mathbf{x}\cdot\ddot{\mathbf{x}}}{|\mathbf{x}|^2} - 2\frac{(\mathbf{x}\cdot\dot{\mathbf{x}})^2}{|\mathbf{x}|^4} = \frac{\dot{\mathbf{x}}\cdot\dot{\mathbf{x}}}{|\mathbf{x}|^2} - \varepsilon - 2\lambda^2 \\ &= \psi - \varepsilon - 2\lambda^2\end{aligned} \tag{2.710}$$

The time derivative of ψ is a function of only ε, λ, and ψ:

$$\begin{aligned}\dot{\psi} &= 2\frac{\dot{\mathbf{x}}\cdot\ddot{\mathbf{x}}}{|\mathbf{x}|^2} - 2\frac{(\mathbf{x}\cdot\dot{\mathbf{x}})\,(\dot{\mathbf{x}}\cdot\dot{\mathbf{x}})}{|\mathbf{x}|^4} = -2\varepsilon\lambda - 2\lambda\psi \\ &= -2\lambda\,(\varepsilon + \psi)\end{aligned} \tag{2.711}$$

The parameters ε, λ, and ψ are called *fundamental invariants*. They are independent of the coordinate system and form a closed set of time derivatives:

$$\dot{\varepsilon} = -3\varepsilon\lambda \qquad \dot{\lambda} = \psi - \varepsilon - 2\lambda^2 \qquad \dot{\psi} = -2\lambda\,(\varepsilon + \psi) \tag{2.712}$$

Equation (2.712) guarantees the existence of coefficients of the series solution (2.703). The first eight derivatives of $\mathbf{x}$ are

$$\frac{d\mathbf{x}}{dt} = \dot{\mathbf{x}} \tag{2.713}$$

$$\frac{d^2\mathbf{x}}{dt^2} = -\varepsilon\mathbf{x} \tag{2.714}$$

$$\frac{d^3\mathbf{x}}{dt^3} = 3\varepsilon\lambda\mathbf{x} - \varepsilon\dot{\mathbf{x}} \tag{2.715}$$

$$\frac{d^4\mathbf{x}}{dt^4} = \left(-15\varepsilon\lambda^2 + 3\varepsilon\psi - 2\varepsilon^2\right)\mathbf{x} + 6\varepsilon\lambda\dot{\mathbf{x}} \tag{2.716}$$

$$\frac{d^5\mathbf{x}}{dt^5} = \left(105\varepsilon\lambda^3 - 45\varepsilon\lambda\psi + 30\varepsilon^2\lambda\right)\mathbf{x} + \left(-45\varepsilon\lambda^2 + 9\varepsilon\psi - 8\varepsilon^2\right)\dot{\mathbf{x}} \tag{2.717}$$

$$\frac{d^6\mathbf{x}}{dt^6} = \varepsilon\left[\lambda^2\left(-945\lambda^2 - 420\varepsilon + 630\psi\right) + 66\varepsilon\psi - 22\varepsilon^2 - 45\psi^2\right]\mathbf{x} + \varepsilon\lambda\left(420\lambda^2 + 150\varepsilon - 180\psi\right)\dot{\mathbf{x}} \tag{2.718}$$

$$\begin{aligned}\frac{d^7\mathbf{x}}{dt^7} &= \varepsilon\lambda\left[\lambda^2\left(10395\lambda^2 + 6300\varepsilon - 9450\psi\right)\right]\mathbf{x} \\ &\quad + \varepsilon\lambda\left(-2268\varepsilon\psi + 756\varepsilon^2 + 1575\psi^2\right)\mathbf{x} \\ &\quad + \varepsilon\left[\lambda^2\left(-4725\lambda^2 - 2520\varepsilon + 3150\psi\right)\right]\dot{\mathbf{x}} \\ &\quad + \varepsilon\left(+396\varepsilon\psi - 172\varepsilon^2 - 225\psi^2\right)\dot{\mathbf{x}}\end{aligned} \tag{2.719}$$

$$\begin{aligned}\frac{d^8\mathbf{x}}{dt^8} &= \varepsilon\left[\lambda^2\left(-135135\lambda^4 + 155925\lambda^2\psi - 103950\varepsilon\lambda^2\right)\right]\mathbf{x} \\ &\quad + \varepsilon\left[\lambda^2\left(-20160\varepsilon^2 + 60480\varepsilon\psi - 42525\psi^2\right)\right]\mathbf{x} \\ &\quad + \varepsilon\left(\psi^2\left[1575\psi - 3618\varepsilon\right) + \varepsilon^2\left(2628\psi - 584\varepsilon\right)\right]\mathbf{x} \\ &\quad + \varepsilon\lambda\left(62370\lambda^4 + 44100\varepsilon\lambda^2 - 56700\lambda^2\psi\right)\dot{\mathbf{x}} \\ &\quad + \varepsilon\lambda\left(-15876\varepsilon\psi + 9450\psi^2 + 6552\varepsilon^2\right)\dot{\mathbf{x}}\end{aligned} \tag{2.720}$$

Substituting the derivatives of $\mathbf{x}$ into (2.703) and rearranging yield

$$\mathbf{x}(t) = P(t)\,\mathbf{x}(t_0) + Q(t)\,\dot{\mathbf{x}}_0(t_0) = f(\mathbf{x}_0, \dot{\mathbf{x}}_0, t) \tag{2.721}$$

where

$$P(t) = \sum_{i=0}^{\infty} P_i\,(t - t_0)^i \qquad Q(t) = \sum_{i=0}^{\infty} Q_i\,(t - t_0)^i \tag{2.722}$$

The functions P and Q that are power series of ε, λ, and ψ are called Lagrangian coefficients. The first six terms of P and Q are

$$P_0 = 1 \tag{2.723}$$

$$P_1 = 0 \tag{2.724}$$

$$P_2 = -\tfrac{1}{2}\varepsilon_0 \tag{2.725}$$

$$P_3 = \tfrac{1}{2}\varepsilon_0\lambda_0 \tag{2.726}$$

$$P_4 = -\tfrac{1}{12}\varepsilon_0^2 - \tfrac{5}{8}\varepsilon_0\lambda_0^2 + \tfrac{1}{8}\varepsilon_0\psi_0 \tag{2.727}$$

$$P_5 = \tfrac{1}{4}\varepsilon_0^2\lambda_0 + \tfrac{7}{8}\varepsilon_0\lambda_0^3 - \tfrac{3}{8}\varepsilon_0\lambda_0\psi_0 \tag{2.728}$$

$$\begin{aligned}P_6 &= -\tfrac{11}{360}\varepsilon_0^3 + \left(-\tfrac{7}{12}\lambda_0^2 + \tfrac{11}{120}\psi_0\right)\varepsilon_0^2 \\ &\quad + \left(-\tfrac{1}{16}\psi_0^2 + \tfrac{7}{8}\lambda_0^2\psi_0 - \tfrac{21}{16}\lambda_0^4\right)\varepsilon_0\end{aligned} \tag{2.729}$$

$$Q_0 = 0 \tag{2.730}$$

$$Q_1 = 1 \tag{2.731}$$

$$Q_2 = 0 \tag{2.732}$$

$$Q_3 = -\tfrac{1}{6}\varepsilon_0 \tag{2.733}$$

$$Q_4 = \tfrac{1}{4}\varepsilon_0\lambda_0 \tag{2.734}$$

$$Q_5 = -\tfrac{1}{15}\varepsilon_0^2 - \tfrac{3}{8}\varepsilon_0\lambda_0^2 + \tfrac{3}{40}\varepsilon_0\psi_0 \tag{2.735}$$

$$Q_6 = 5/24\varepsilon_0^2\lambda_0 + (\tfrac{7}{12}\lambda_0^3 - \tfrac{1}{4}\lambda_0\psi_0)\varepsilon_0 \tag{2.736}$$

The series solution of the two-body problem is developed by Lagrange (1736–1813). Although this series theoretically converges for all t, it will not provide a suitable approximate solution except for a limited interval of time.

Key to solving the two-body problem, including Lagrange's method, is that the problem can be reduced to one second-order equation.

Example 177 ★ Series Solution for Three-Body Problem Consider three point masses m_1, m_2, and m_3 as are shown in Figure 2.51. Their position vectors with respect to their mass center C are $\mathbf{X}_1$, $\mathbf{X}_2$, and $\mathbf{X}_3$. They attract each other by the Newtonian gravitational force. If the space is assumed Euclidean and there is no other mass in the space, the equations of motion of m_1, m_2, and m_3 are

$$\begin{aligned}
\ddot{\mathbf{X}}_1 &= -G_2\frac{\mathbf{X}_1 - \mathbf{X}_2}{|\mathbf{X}_{21}|^3} - G_3\frac{\mathbf{X}_1 - \mathbf{X}_3}{|\mathbf{X}_{31}|^3} \\
\ddot{\mathbf{X}}_2 &= -G_3\frac{\mathbf{X}_2 - \mathbf{X}_3}{|\mathbf{X}_{32}|^3} - G_1\frac{\mathbf{X}_2 - \mathbf{X}_1}{|\mathbf{X}_{12}|^3} \\
\ddot{\mathbf{X}}_3 &= -G_1\frac{\mathbf{X}_3 - \mathbf{X}_1}{|\mathbf{X}_{13}|^3} - G_2\frac{\mathbf{X}_3 - \mathbf{X}_2}{|\mathbf{X}_{23}|^3}
\end{aligned} \tag{2.737}$$

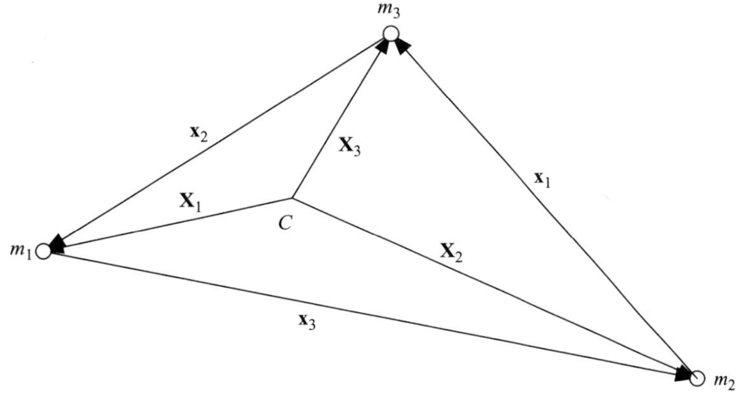

Figure 2.51 Position vectors for three-body problem.

where

$$\mathbf{X}_{ij} = \mathbf{X}_j - \mathbf{X}_i \tag{2.738}$$

$$G_i = Gm_i \tag{2.739}$$

$$G = 6.67259 \times 10^{-11}\,\mathrm{m^3kg^{-1}s^{-2}} \tag{2.740}$$

Using the mass center as the origin implies that

$$G_1\mathbf{X}_1 + G_2\mathbf{X}_2 + G_3\mathbf{X}_3 = 0 \tag{2.741}$$

The equations of motion of three bodies have their most symmetric form when expressed in terms of relative position vectors $\mathbf{x}_1$, $\mathbf{x}_2$, $\mathbf{x}_3$:

$$\begin{aligned} \mathbf{x}_1 &= \mathbf{X}_3 - \mathbf{X}_2 \\ \mathbf{x}_2 &= \mathbf{X}_1 - \mathbf{X}_3 \\ \mathbf{x}_3 &= \mathbf{X}_2 - \mathbf{X}_1 \end{aligned} \tag{2.742}$$

Now the kinematic constraint (2.741) reduces to

$$\mathbf{x}_1 + \mathbf{x}_2 + \mathbf{x}_3 = 0 \tag{2.743}$$

and the absolute position vectors in terms of the relative positions are

$$\begin{aligned} m\mathbf{X}_1 &= m_3\mathbf{x}_2 - m_2\mathbf{x}_3 \\ m\mathbf{X}_2 &= m_1\mathbf{x}_3 - m_3\mathbf{x}_1 \\ m\mathbf{X}_3 &= m_2\mathbf{x}_1 - m_1\mathbf{x}_2 \end{aligned} \tag{2.744}$$

where

$$m = m_1 + m_2 + m_3.$$

Substituting Equations (2.744) in (2.737), we get symmetric forms of the equations of motion:

$$\begin{aligned} \ddot{\mathbf{x}}_1 &= -Gm\frac{\mathbf{x}_1}{|\mathbf{x}_1|^3} + G_1\left(\frac{\mathbf{x}_1}{|\mathbf{x}_1|^3} + \frac{\mathbf{x}_2}{|\mathbf{x}_2|^3} + \frac{\mathbf{x}_3}{|\mathbf{x}_3|^3}\right) \\ \ddot{\mathbf{x}}_2 &= -Gm\frac{\mathbf{x}_2}{|\mathbf{x}_2|^3} + G_2\left(\frac{\mathbf{x}_1}{|\mathbf{x}_1|^3} + \frac{\mathbf{x}_2}{|\mathbf{x}_2|^3} + \frac{\mathbf{x}_3}{|\mathbf{x}_3|^3}\right) \\ \ddot{\mathbf{x}}_3 &= -Gm\frac{\mathbf{x}_3}{|\mathbf{x}_3|^3} + G_3\left(\frac{\mathbf{x}_1}{|\mathbf{x}_1|^3} + \frac{\mathbf{x}_2}{|\mathbf{x}_2|^3} + \frac{\mathbf{x}_3}{|\mathbf{x}_3|^3}\right) \end{aligned} \tag{2.745}$$

We are looking for a series solution for Equations (2.745) in the form

$$\mathbf{x}_i\left(t\right) = \mathbf{x}_{i_0} + \dot{\mathbf{x}}_{i_0}\left(t - t_0\right) + \ddot{\mathbf{x}}_{i_0}\frac{\left(t - t_0\right)^2}{2!} + \dddot{\mathbf{x}}_{i_0}\frac{\left(t - t_0\right)^3}{3!} + \cdots \tag{2.746}$$

$$\mathbf{x}_{i_0} = \mathbf{x}_i\left(t_0\right) \qquad \dot{\mathbf{x}}_{i_0} = \dot{\mathbf{x}}_i\left(t_0\right) \qquad i = 1, 2, 3 \tag{2.747}$$

For simplicity let us define $\mu = Gm$ along with the first set of parameters

$$\mu = Gm \tag{2.748}$$

$$\varepsilon_1 = \frac{1}{|\mathbf{x}_1|^3} \qquad \varepsilon_2 = \frac{1}{|\mathbf{x}_2|^3} \qquad \varepsilon_3 = \frac{1}{|\mathbf{x}_3|^3} \tag{2.749}$$

to rewrite Equations (2.745) as

$$\begin{aligned}
\ddot{\mathbf{x}}_1 &= -\mu\varepsilon_1\mathbf{x}_1 + G_1\left(\varepsilon_1\mathbf{x}_1 + \varepsilon_2\mathbf{x}_2 + \varepsilon_3\mathbf{x}_3\right) \\
\ddot{\mathbf{x}}_2 &= -\mu\varepsilon_2\mathbf{x}_2 + G_2\left(\varepsilon_1\mathbf{x}_1 + \varepsilon_2\mathbf{x}_2 + \varepsilon_3\mathbf{x}_3\right) \\
\ddot{\mathbf{x}}_3 &= -\mu\varepsilon_3\mathbf{x}_3 + G_3\left(\varepsilon_1\mathbf{x}_1 + \varepsilon_2\mathbf{x}_2 + \varepsilon_3\mathbf{x}_3\right)
\end{aligned} \tag{2.750}$$

We also define the following three sets of parameters,

$$\begin{aligned}
a_{111} &= \frac{\mathbf{x}_1\cdot\mathbf{x}_1}{|\mathbf{x}_1|^2} = 1 & a_{112} &= \frac{\mathbf{x}_1\cdot\mathbf{x}_1}{|\mathbf{x}_2|^2} & a_{113} &= \frac{\mathbf{x}_1\cdot\mathbf{x}_1}{|\mathbf{x}_3|^2} \\
a_{121} &= \frac{\mathbf{x}_1\cdot\mathbf{x}_2}{|\mathbf{x}_1|^2} & a_{122} &= \frac{\mathbf{x}_1\cdot\mathbf{x}_2}{|\mathbf{x}_2|^2} & a_{123} &= \frac{\mathbf{x}_1\cdot\mathbf{x}_2}{|\mathbf{x}_3|^2} \\
a_{131} &= \frac{\mathbf{x}_1\cdot\mathbf{x}_3}{|\mathbf{x}_1|^2} & a_{132} &= \frac{\mathbf{x}_1\cdot\mathbf{x}_3}{|\mathbf{x}_2|^2} & a_{133} &= \frac{\mathbf{x}_1\cdot\mathbf{x}_3}{|\mathbf{x}_3|^2} \\
a_{221} &= \frac{\mathbf{x}_2\cdot\mathbf{x}_2}{|\mathbf{x}_1|^2} & a_{222} &= \frac{\mathbf{x}_2\cdot\mathbf{x}_2}{|\mathbf{x}_2|^2} = 1 & a_{223} &= \frac{\mathbf{x}_2\cdot\mathbf{x}_2}{|\mathbf{x}_3|^2} \\
a_{231} &= \frac{\mathbf{x}_2\cdot\mathbf{x}_3}{|\mathbf{x}_1|^2} & a_{232} &= \frac{\mathbf{x}_2\cdot\mathbf{x}_3}{|\mathbf{x}_2|^2} & a_{233} &= \frac{\mathbf{x}_2\cdot\mathbf{x}_3}{|\mathbf{x}_3|^2} \\
a_{331} &= \frac{\mathbf{x}_3\cdot\mathbf{x}_3}{|\mathbf{x}_1|^2} & a_{332} &= \frac{\mathbf{x}_3\cdot\mathbf{x}_3}{|\mathbf{x}_2|^2} & a_{333} &= \frac{\mathbf{x}_3\cdot\mathbf{x}_3}{|\mathbf{x}_3|^2} = 1
\end{aligned} \tag{2.751}$$

$$\begin{aligned}
b_{111} &= \frac{\dot{\mathbf{x}}_1\cdot\mathbf{x}_1}{|\mathbf{x}_1|^2} & b_{112} &= \frac{\dot{\mathbf{x}}_1\cdot\mathbf{x}_1}{|\mathbf{x}_2|^2} & b_{113} &= \frac{\dot{\mathbf{x}}_1\cdot\mathbf{x}_1}{|\mathbf{x}_3|^2} \\
b_{121} &= \frac{\dot{\mathbf{x}}_1\cdot\mathbf{x}_2}{|\mathbf{x}_1|^2} & b_{122} &= \frac{\dot{\mathbf{x}}_1\cdot\mathbf{x}_2}{|\mathbf{x}_2|^2} & b_{123} &= \frac{\dot{\mathbf{x}}_1\cdot\mathbf{x}_2}{|\mathbf{x}_3|^2} \\
b_{131} &= \frac{\dot{\mathbf{x}}_1\cdot\mathbf{x}_3}{|\mathbf{x}_1|^2} & b_{132} &= \frac{\dot{\mathbf{x}}_1\cdot\mathbf{x}_3}{|\mathbf{x}_2|^2} & b_{133} &= \frac{\dot{\mathbf{x}}_1\cdot\mathbf{x}_3}{|\mathbf{x}_3|^2} \\
b_{211} &= \frac{\dot{\mathbf{x}}_2\cdot\mathbf{x}_1}{|\mathbf{x}_1|^2} & b_{212} &= \frac{\dot{\mathbf{x}}_2\cdot\mathbf{x}_1}{|\mathbf{x}_2|^2} & b_{213} &= \frac{\dot{\mathbf{x}}_2\cdot\mathbf{x}_1}{|\mathbf{x}_3|^2} \\
b_{221} &= \frac{\dot{\mathbf{x}}_2\cdot\mathbf{x}_2}{|\mathbf{x}_1|^2} & b_{222} &= \frac{\dot{\mathbf{x}}_2\cdot\mathbf{x}_2}{|\mathbf{x}_2|^2} & b_{223} &= \frac{\dot{\mathbf{x}}_2\cdot\mathbf{x}_2}{|\mathbf{x}_3|^2} \\
b_{231} &= \frac{\dot{\mathbf{x}}_2\cdot\mathbf{x}_3}{|\mathbf{x}_1|^2} & b_{232} &= \frac{\dot{\mathbf{x}}_2\cdot\mathbf{x}_3}{|\mathbf{x}_2|^2} & b_{233} &= \frac{\dot{\mathbf{x}}_2\cdot\mathbf{x}_3}{|\mathbf{x}_3|^2} \\
b_{311} &= \frac{\dot{\mathbf{x}}_3\cdot\mathbf{x}_1}{|\mathbf{x}_1|^2} & b_{312} &= \frac{\dot{\mathbf{x}}_3\cdot\mathbf{x}_1}{|\mathbf{x}_2|^2} & b_{313} &= \frac{\dot{\mathbf{x}}_3\cdot\mathbf{x}_1}{|\mathbf{x}_3|^2} \\
b_{321} &= \frac{\dot{\mathbf{x}}_3\cdot\mathbf{x}_2}{|\mathbf{x}_1|^2} & b_{322} &= \frac{\dot{\mathbf{x}}_3\cdot\mathbf{x}_2}{|\mathbf{x}_2|^2} & b_{323} &= \frac{\dot{\mathbf{x}}_3\cdot\mathbf{x}_2}{|\mathbf{x}_3|^2} \\
b_{331} &= \frac{\dot{\mathbf{x}}_3\cdot\mathbf{x}_3}{|\mathbf{x}_1|^2} & b_{332} &= \frac{\dot{\mathbf{x}}_3\cdot\mathbf{x}_3}{|\mathbf{x}_2|^2} & b_{333} &= \frac{\dot{\mathbf{x}}_3\cdot\mathbf{x}_3}{|\mathbf{x}_3|^2}
\end{aligned} \tag{2.752}$$

$$
\begin{aligned}
c_{111} &= \frac{\dot{\mathbf{x}}_1 \cdot \dot{\mathbf{x}}_1}{|\mathbf{x}_1|^2} & c_{112} &= \frac{\dot{\mathbf{x}}_1 \cdot \dot{\mathbf{x}}_1}{|\mathbf{x}_2|^2} & c_{113} &= \frac{\dot{\mathbf{x}}_1 \cdot \dot{\mathbf{x}}_1}{|\mathbf{x}_3|^2} \\
c_{121} &= \frac{\dot{\mathbf{x}}_1 \cdot \dot{\mathbf{x}}_2}{|\mathbf{x}_1|^2} & c_{122} &= \frac{\dot{\mathbf{x}}_1 \cdot \dot{\mathbf{x}}_2}{|\mathbf{x}_2|^2} & c_{123} &= \frac{\dot{\mathbf{x}}_1 \cdot \dot{\mathbf{x}}_2}{|\mathbf{x}_3|^2} \\
c_{131} &= \frac{\dot{\mathbf{x}}_1 \cdot \dot{\mathbf{x}}_3}{|\mathbf{x}_1|^2} & c_{132} &= \frac{\dot{\mathbf{x}}_1 \cdot \dot{\mathbf{x}}_3}{|\mathbf{x}_2|^2} & c_{133} &= \frac{\dot{\mathbf{x}}_1 \cdot \dot{\mathbf{x}}_3}{|\mathbf{x}_3|^2} \\
c_{221} &= \frac{\dot{\mathbf{x}}_2 \cdot \dot{\mathbf{x}}_2}{|\mathbf{x}_1|^2} & c_{222} &= \frac{\dot{\mathbf{x}}_2 \cdot \dot{\mathbf{x}}_2}{|\mathbf{x}_2|^2} & c_{223} &= \frac{\dot{\mathbf{x}}_2 \cdot \dot{\mathbf{x}}_2}{|\mathbf{x}_3|^2} \\
c_{231} &= \frac{\dot{\mathbf{x}}_2 \cdot \dot{\mathbf{x}}_3}{|\mathbf{x}_1|^2} & c_{232} &= \frac{\dot{\mathbf{x}}_2 \cdot \dot{\mathbf{x}}_3}{|\mathbf{x}_2|^2} & c_{233} &= \frac{\dot{\mathbf{x}}_2 \cdot \dot{\mathbf{x}}_3}{|\mathbf{x}_3|^2} \\
c_{331} &= \frac{\dot{\mathbf{x}}_3 \cdot \dot{\mathbf{x}}_3}{|\mathbf{x}_1|^2} & c_{332} &= \frac{\dot{\mathbf{x}}_3 \cdot \dot{\mathbf{x}}_3}{|\mathbf{x}_2|^2} & c_{333} &= \frac{\dot{\mathbf{x}}_3 \cdot \dot{\mathbf{x}}_3}{|\mathbf{x}_3|^2}
\end{aligned}
\tag{2.753}
$$

The time derivatives of the ε-set are

$$
\dot{\varepsilon}_1 = -3b_{111}\varepsilon_1 \qquad \dot{\varepsilon}_2 = -3b_{222}\varepsilon_2 \qquad \dot{\varepsilon}_3 = -3b_{333}\varepsilon_3 \tag{2.754}
$$

and the time derivatives of the a-set, b-set, and c-set are

$$
\begin{aligned}
\dot{a}_{111} &= 0 \\
\dot{a}_{112} &= -2b_{222}a_{112} + 2b_{112} \\
\dot{a}_{113} &= -2b_{333}a_{113} + 2b_{113}
\end{aligned}
\tag{2.755}
$$

$$
\begin{aligned}
\dot{a}_{121} &= -2b_{111}a_{121} + b_{121} + b_{211} \\
\dot{a}_{122} &= -2b_{222}a_{122} + b_{122} + b_{212} \\
\dot{a}_{123} &= -2b_{333}a_{123} + b_{123} + b_{213}
\end{aligned}
\tag{2.756}
$$

$$
\begin{aligned}
\dot{a}_{131} &= -2b_{111}a_{131} + b_{131} + b_{311} \\
\dot{a}_{132} &= -2b_{222}a_{132} + b_{132} + b_{312} \\
\dot{a}_{133} &= -2b_{333}a_{133} + b_{133} + b_{313}
\end{aligned}
\tag{2.757}
$$

$$
\begin{aligned}
\dot{a}_{221} &= -2b_{111}a_{221} + 2b_{221} \\
\dot{a}_{222} &= 0 \\
\dot{a}_{223} &= -2b_{333}a_{233} + 2b_{223}
\end{aligned}
\tag{2.758}
$$

$$
\begin{aligned}
\dot{a}_{231} &= -2b_{111}a_{231} + b_{231} + b_{321} \\
\dot{a}_{232} &= -2b_{222}a_{232} + b_{232} + b_{322} \\
\dot{a}_{233} &= -2b_{333}a_{233} + b_{233} + b_{323}
\end{aligned}
\tag{2.759}
$$

$$\begin{aligned}
\dot{a}_{331} &= -2b_{111}a_{331} + 2b_{331} \\
\dot{a}_{332} &= -2b_{222}a_{332} + 2b_{332} \\
\dot{a}_{333} &= 0
\end{aligned} \tag{2.760}$$

$$\begin{aligned}
\dot{b}_{111} &= -2b_{111}^2 + c_{111} - \mu\varepsilon_1 + G_1\left(\varepsilon_1 + \varepsilon_2 a_{211} + \varepsilon_3 a_{311}\right) \\
\dot{b}_{112} &= -2b_{222}b_{112} + c_{112} - \mu\varepsilon_1 a_{112} + G_1\left(\varepsilon_1 a_{112} + \varepsilon_2 a_{212} + \varepsilon_3 a_{312}\right) \\
\dot{b}_{113} &= -2b_{333}b_{113} + c_{113} - \mu\varepsilon_1 a_{113} + G_1\left(\varepsilon_1 a_{113} + \varepsilon_2 a_{213} + \varepsilon_3 a_{313}\right)
\end{aligned} \tag{2.761}$$

$$\begin{aligned}
\dot{b}_{121} &= -2b_{111}b_{121} + c_{121} - \mu\varepsilon_1 a_{121} + G_1\left(\varepsilon_1 a_{121} + \varepsilon_2 a_{221} + \varepsilon_3 a_{321}\right) \\
\dot{b}_{122} &= -2b_{222}b_{122} + c_{122} - \mu\varepsilon_1 a_{122} + G_1\left(\varepsilon_1 a_{122} + \varepsilon_2 + \varepsilon_3 a_{322}\right) \\
\dot{b}_{123} &= -2b_{333}b_{123} + c_{123} - \mu\varepsilon_1 a_{123} + G_1\left(\varepsilon_1 a_{123} + \varepsilon_2 a_{223} + \varepsilon_3 a_{323}\right)
\end{aligned} \tag{2.762}$$

$$\begin{aligned}
\dot{b}_{131} &= -2b_{111}b_{131} + c_{131} - \mu\varepsilon_1 a_{131} + G_1\left(\varepsilon_1 a_{131} + \varepsilon_2 a_{231} + \varepsilon_3 a_{331}\right) \\
\dot{b}_{132} &= -2b_{222}b_{132} + c_{132} - \mu\varepsilon_1 a_{132} + G_1\left(\varepsilon_1 a_{132} + \varepsilon_2 a_{232} + \varepsilon_3 a_{332}\right) \\
\dot{b}_{133} &= -2b_{333}b_{133} + c_{133} - \mu\varepsilon_1 a_{133} + G_1\left(\varepsilon_1 a_{133} + \varepsilon_2 a_{233} + \varepsilon_3\right)
\end{aligned} \tag{2.763}$$

$$\begin{aligned}
\dot{b}_{211} &= -2b_{111}b_{211} + c_{211} - \mu\varepsilon_2 a_{211} + G_2\left(\varepsilon_1 + \varepsilon_2 a_{211} + \varepsilon_3 a_{311}\right) \\
\dot{b}_{212} &= -2b_{222}b_{212} + c_{212} - \mu\varepsilon_2 a_{212} + G_2\left(\varepsilon_1 a_{112} + \varepsilon_2 a_{212} + \varepsilon_3 a_{312}\right) \\
\dot{b}_{133} &= -2b_{333}b_{213} + c_{213} - \mu\varepsilon_2 a_{213} + G_2\left(\varepsilon_1 a_{113} + \varepsilon_2 a_{213} + \varepsilon_3 a_{313}\right)
\end{aligned} \tag{2.764}$$

$$\begin{aligned}
\dot{b}_{221} &= -2b_{111}b_{221} + c_{221} - \mu\varepsilon_2 a_{221} + G_2\left(\varepsilon_1 a_{121} + \varepsilon_2 a_{221} + \varepsilon_3 a_{321}\right) \\
\dot{b}_{222} &= -2b_{222}^2 + c_{222} - \mu\varepsilon_2 + G_2\left(\varepsilon_1 a_{122} + \varepsilon_2 + \varepsilon_3 a_{322}\right) \\
\dot{b}_{223} &= -2b_{333}b_{223} + c_{223} - \mu\varepsilon_2 a_{223} + G_2\left(\varepsilon_1 a_{123} + \varepsilon_2 a_{223} + \varepsilon_3 a_{323}\right)
\end{aligned} \tag{2.765}$$

$$\begin{aligned}
\dot{b}_{231} &= -2b_{111}b_{231} + c_{231} - \mu\varepsilon_2 a_{231} + G_2\left(\varepsilon_1 a_{131} + \varepsilon_2 a_{231} + \varepsilon_3 a_{331}\right) \\
\dot{b}_{232} &= -2b_{222}b_{232} + c_{232} - \mu\varepsilon_2 a_{232} + G_2\left(\varepsilon_1 a_{132} + \varepsilon_2 a_{232} + \varepsilon_3 a_{332}\right) \\
\dot{b}_{233} &= -2b_{333}b_{233} + c_{233} - \mu\varepsilon_2 a_{233} + G_2\left(\varepsilon_1 a_{133} + \varepsilon_2 a_{233} + \varepsilon_3\right)
\end{aligned} \tag{2.766}$$

$$\begin{aligned}
\dot{b}_{311} &= -2b_{111}b_{311} + c_{311} - \mu\varepsilon_3 a_{311} + G_3\left(\varepsilon_1 + \varepsilon_2 a_{211} + \varepsilon_3 a_{311}\right) \\
\dot{b}_{312} &= -2b_{222}b_{312} + c_{312} - \mu\varepsilon_3 a_{312} + G_3\left(\varepsilon_1 a_{112} + \varepsilon_2 a_{212} + \varepsilon_3 a_{312}\right) \\
\dot{b}_{313} &= -2b_{333}b_{313} + c_{313} - \mu\varepsilon_3 a_{313} + G_3\left(\varepsilon_1 a_{113} + \varepsilon_2 a_{213} + \varepsilon_3 a_{313}\right)
\end{aligned} \tag{2.767}$$

$$\begin{aligned}
\dot{b}_{321} &= -2b_{111}b_{321} + c_{321} - \mu\varepsilon_3 a_{321} + G_3\left(\varepsilon_1 a_{121} + \varepsilon_2 a_{221} + \varepsilon_3 a_{321}\right) \\
\dot{b}_{322} &= -2b_{222}b_{322} + c_{322} - \mu\varepsilon_3 a_{322} + G_3\left(\varepsilon_1 a_{122} + \varepsilon_2 + \varepsilon_3 a_{322}\right) \\
\dot{b}_{323} &= -2b_{333}b_{323} + c_{323} - \mu\varepsilon_3 a_{323} + G_3\left(\varepsilon_1 a_{123} + \varepsilon_2 a_{223} + \varepsilon_3 a_{323}\right)
\end{aligned} \tag{2.768}$$

$$\begin{aligned}
\dot{b}_{331} &= -2b_{111}b_{331} + c_{331} - \mu\varepsilon_3 a_{331} + G_3\left(\varepsilon_1 a_{131} + \varepsilon_2 a_{231} + \varepsilon_3 a_{331}\right) \\
\dot{b}_{332} &= -2b_{222}b_{332} + c_{332} - \mu\varepsilon_3 a_{332} + G_3\left(\varepsilon_1 a_{132} + \varepsilon_2 a_{232} + \varepsilon_3 a_{332}\right) \\
\dot{b}_{333} &= -2b_{333}b_{333} + c_{333} - \mu\varepsilon_3 + G_3\left(\varepsilon_1 a_{133} + \varepsilon_2 a_{233} + \varepsilon_3\right)
\end{aligned} \tag{2.769}$$

$$\begin{aligned}
\dot{c}_{111} &= -2b_{111}c_{111} - 2\mu\varepsilon_1 b_{111} + 2G_1\left(\varepsilon_1 b_{111} + \varepsilon_2 b_{121} + \varepsilon_3 b_{131}\right) \\
\dot{c}_{112} &= -2b_{222}c_{112} - 2\mu\varepsilon_1 b_{112} + 2G_1\left(\varepsilon_1 b_{112} + \varepsilon_2 b_{122} + \varepsilon_3 b_{132}\right) \\
\dot{c}_{113} &= -2b_{333}c_{113} - 2\mu\varepsilon_1 b_{113} + 2G_1\left(\varepsilon_1 b_{113} + \varepsilon_2 b_{123} + \varepsilon_3 b_{133}\right)
\end{aligned} \tag{2.770}$$

$$\begin{aligned}
\dot{c}_{121} &= -2b_{111}c_{121} - \mu\left(\varepsilon_1 b_{111} + \varepsilon_2 b_{121}\right) + G_1\left(\varepsilon_1 b_{211} + \varepsilon_2 b_{221} + \varepsilon_3 b_{231}\right) \\
&\quad + G_2\left(\varepsilon_1 b_{111} + \varepsilon_2 b_{121} + \varepsilon_3 b_{131}\right) \\
\dot{c}_{122} &= -2b_{222}c_{122} - \mu\left(\varepsilon_1 b_{212} + \varepsilon_2 b_{122}\right) + G_1\left(\varepsilon_1 b_{212} + \varepsilon_2 b_{222} + \varepsilon_3 b_{232}\right) \\
&\quad + G_2\left(\varepsilon_1 b_{112} + \varepsilon_2 b_{122} + \varepsilon_3 b_{132}\right) \\
\dot{c}_{123} &= -2b_{333}c_{123} - \mu\left(\varepsilon_1 b_{213} + \varepsilon_2 b_{123}\right) + G_1\left(\varepsilon_1 b_{213} + \varepsilon_2 b_{223} + \varepsilon_3 b_{233}\right) \\
&\quad + G_2\left(\varepsilon_1 b_{113} + \varepsilon_2 b_{123} + \varepsilon_3 b_{133}\right)
\end{aligned} \tag{2.771}$$

$$\begin{aligned}
\dot{c}_{131} &= -2b_{111}c_{131} - \mu\left(\varepsilon_1 b_{311} + \varepsilon_3 b_{131}\right) + G_1\left(\varepsilon_1 b_{311} + \varepsilon_2 b_{321} + \varepsilon_3 b_{331}\right) \\
&\quad + G_3\left(\varepsilon_1 b_{111} + \varepsilon_2 b_{121} + \varepsilon_3 b_{131}\right) \\
\dot{c}_{132} &= -2b_{222}c_{132} - \mu\left(\varepsilon_1 b_{312} + \varepsilon_3 b_{132}\right) + G_1\left(\varepsilon_1 b_{312} + \varepsilon_2 b_{322} + \varepsilon_3 b_{332}\right) \\
&\quad + G_3\left(\varepsilon_1 b_{112} + \varepsilon_2 b_{122} + \varepsilon_3 b_{132}\right) \\
\dot{c}_{133} &= -2b_{333}c_{133} - \mu\left(\varepsilon_1 b_{313} + \varepsilon_3 b_{133}\right) + G_1\left(\varepsilon_1 b_{313} + \varepsilon_2 b_{323} + \varepsilon_3 b_{333}\right) \\
&\quad + G_3\left(\varepsilon_1 b_{113} + \varepsilon_2 b_{123} + \varepsilon_3 b_{133}\right)
\end{aligned} \tag{2.772}$$

$$\begin{aligned}
\dot{c}_{221} &= -2b_{111}c_{221} - 2\mu\varepsilon_2 b_{221} + 2G_2\left(\varepsilon_1 b_{211} + \varepsilon_2 b_{221} + \varepsilon_3 b_{231}\right) \\
\dot{c}_{222} &= -2b_{222}c_{222} - 2\mu\varepsilon_2 b_{222} + 2G_2\left(\varepsilon_1 b_{212} + \varepsilon_2 b_{222} + \varepsilon_3 b_{232}\right) \\
\dot{c}_{223} &= -2b_{333}c_{223} - 2\mu\varepsilon_2 b_{223} + 2G_2\left(\varepsilon_1 b_{213} + \varepsilon_2 b_{223} + \varepsilon_3 b_{233}\right)
\end{aligned} \tag{2.773}$$

$$
\begin{aligned}
\dot{c}_{231} &= -2b_{111}c_{231} - \mu\left(\varepsilon_1 b_{321} + \varepsilon_2 b_{231}\right) + G_2\left(\varepsilon_1 b_{311} + \varepsilon_2 b_{321} + \varepsilon_3 b_{331}\right) \\
&\quad + G_3\left(\varepsilon_1 b_{211} + \varepsilon_2 b_{221} + \varepsilon_3 b_{231}\right) \\
\dot{c}_{232} &= -2b_{222}c_{232} - \mu\left(\varepsilon_1 b_{322} + \varepsilon_2 b_{232}\right) + G_2\left(\varepsilon_1 b_{312} + \varepsilon_2 b_{322} + \varepsilon_3 b_{332}\right) \\
&\quad + G_3\left(\varepsilon_1 b_{212} + \varepsilon_2 b_{222} + \varepsilon_3 b_{232}\right) \qquad (2.774) \\
\dot{c}_{233} &= -2b_{333}c_{233} - \mu\left(\varepsilon_1 b_{323} + \varepsilon_2 b_{233}\right) + G_2\left(\varepsilon_1 b_{313} + \varepsilon_2 b_{323} + \varepsilon_3 b_{333}\right) \\
&\quad + G_3\left(\varepsilon_1 b_{213} + \varepsilon_2 b_{223} + \varepsilon_3 b_{233}\right)
\end{aligned}
$$

$$
\begin{aligned}
\dot{c}_{331} &= -2b_{111}c_{331} - 2\mu\varepsilon_3 b_{331} + 2G_3\left(\varepsilon_1 b_{311} + \varepsilon_2 b_{321} + \varepsilon_3 b_{331}\right) \\
\dot{c}_{332} &= -2b_{222}c_{332} - 2\mu\varepsilon_3 b_{332} + 2G_3\left(\varepsilon_1 b_{312} + \varepsilon_2 b_{322} + \varepsilon_3 b_{332}\right) \qquad (2.775) \\
\dot{c}_{333} &= -2b_{333}c_{333} - 2\mu\varepsilon_3 b_{333} + 2G_3\left(\varepsilon_1 b_{313} + \varepsilon_2 b_{323} + \varepsilon_3 b_{333}\right)
\end{aligned}
$$

The concise forms of these equations are presented in Example 36. We have defined 84 fundamental parameters and showed that their time evolutions are independent of the selected coordinate system, making a closed set under the operation of time differentiation. Therefore, employing Equations (2.748)–(2.775) we are able to find the coefficients of series (2.746) to develop the series solution for the three-body problem. The first four coefficients of the series are derived below:

$$\frac{d\mathbf{x}_i}{dt} = \dot{\mathbf{x}}_i \qquad (2.776)$$

$$\frac{d^2\mathbf{x}_i}{dt^2} = -\mu\varepsilon_i\mathbf{x}_i + G_i\sum_{j=1}^{3}\varepsilon_j\mathbf{x}_j \qquad (2.777)$$

$$\frac{d^3\mathbf{x}_i}{dt^3} = \mu\varepsilon_i\left(3b_{iii}\mathbf{x}_i - \dot{\mathbf{x}}_i\right) - G_i\sum_{j=1}^{3}\varepsilon_j\left(3b_{jjj}\mathbf{x}_j - \dot{\mathbf{x}}_j\right) \qquad (2.778)$$

$$\frac{d^4\mathbf{x}_i}{dt^4} = -3\mu b_{iii}\varepsilon_i\left(3b_{iii}\mathbf{x}_i - \dot{\mathbf{x}}_i\right) + \mu\varepsilon_i Z_i - G_i\sum_{j=1}^{3}S_j \qquad (2.779)$$

where

$$
\begin{aligned}
Z_i = \Bigg[& 3b_{iii}\dot{\mathbf{x}}_i + 3\mathbf{x}_i\left(-2b_{iii}^2 + c_{iii} - \mu\varepsilon_i a_{iii} + G_i\sum_{k=1}^{3}\varepsilon_k a_{kii}\right) \\
& + \mu\varepsilon_i\dot{\mathbf{x}}_i - G_i\sum_{k=1}^{3}\varepsilon_k\mathbf{x}_k\Bigg] \qquad (2.780)
\end{aligned}
$$

$$S_j = -3b_{jjj}\varepsilon_i^2\left(3b_{jjj}\mathbf{x}_j - \dot{\mathbf{x}}_j\right)Z_j \qquad (2.781)$$

The complexity of the coefficients grows rapidly, indicating that the use of the series solution is not very practical.

2.5.4 ★ Parameter Adjustment

Adjustment of the parameters in a dynamic problem to control the behavior of the phenomenon such that a desired behavior appears is the most important task in dynamics.

Example 178 ★ Antiprojectile Gun A projectile is fired with initial speed v_1 at angle θ_1 in the (x, z)-plane to hit a target at R_1:

$$R_1 = \begin{bmatrix} 0 \\ 0 \\ \frac{v_1^2}{g} \sin 2\theta_1 \end{bmatrix} \tag{2.782}$$

The position vector and the path of the projectile are

$$\mathbf{r} = \begin{bmatrix} v_1 t \cos\theta_1 \\ 0 \\ v_1 t \sin\theta_1 - \frac{1}{2} g t^2 \end{bmatrix} \tag{2.783}$$

$$z = -\frac{1}{2} g \frac{x^2}{v_0^2 \cos^2\theta} + x \tan\theta \tag{2.784}$$

This projectile can hit any point under the projectile umbrella,

$$z = \frac{v_0^2}{2g} - \frac{g}{2v_0^2}\left(x^2 + y^2\right) \tag{2.785}$$

If the target is at the same level as the shooting point, then the projectile will reach the target at time t_1:

$$t_1 = 2\frac{v_1}{g} \sin\theta_1 \tag{2.786}$$

At time $t_p < t_1$, the projectile is at $\mathbf{r}_p$:

$$\mathbf{r}_p = \begin{bmatrix} v_1 t_p \cos\theta_1 \\ 0 \\ v_1 t_p \sin\theta_1 - \frac{1}{2} g t_p^2 \end{bmatrix} \tag{2.787}$$

Assume that there is an antiprojectile gun at

$$\mathbf{r}_0 = \begin{bmatrix} x_0 \\ y_0 \\ z_0 \end{bmatrix} \tag{2.788}$$

that can shoot a bullet with an initial speed v_0, that is, n times faster than the projectile's initial speed:

$$v_0 = n v_1 \qquad 0 < n \qquad n \in \mathbb{R} \tag{2.789}$$

The antiprojectile gun can reach any point (x, y, z) under its bullet umbrella:

$$z = z_0 + \frac{v_0^2}{2g} - \frac{g}{2v_0^2}\left[(x - x_0)^2 + (y - y_0)^2\right] \tag{2.790}$$

The bullet can hit the projectile if the bullet umbrella covers the target and hence overlaps the projectile umbrella, as is shown in Figure 2.52.

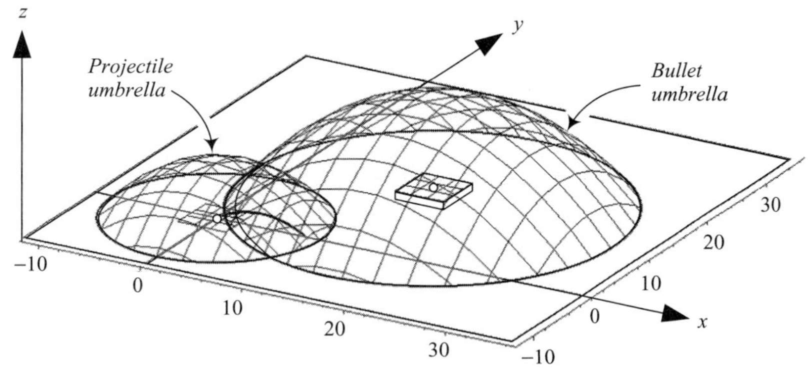

Figure 2.52 Overlap of bullet and target umbrellas.

If the antiprojectile bullet shoots at time $t = t_p$ when the projectile is at (2.787), the bullet must hit the projectile within a time $t_p < t < 2v_1 \sin\theta_1/g$. The path of motion of the bullet is similar to the path of the projectile with a time lag, a displacement in the origin, and a rotation with respect to the coordinate frame. The time lag is t_p and the displacement is $\mathbf{r}_0$. The angular position of the base of the antiprojectile is

$$\alpha_0 = \pi + \tan^{-1}\frac{y_0}{x_0} \tag{2.791}$$

however, the angle of the osculating plane of the bullet can make an arbitrary angle α with respect to the x-axis. Therefore, the path of motion of the bullet is given as

$$\mathbf{r}_b = \begin{bmatrix} x_0 + \left(t - t_p\right) v_0 \cos\theta_0 \cos\alpha \\ y_0 + \left(t - t_p\right) v_0 \cos\theta_0 \sin\alpha \\ z_0 + \left(t - t_p\right) v_0 \sin\theta_0 - \frac{1}{2} g \left(t - t_p\right)^2 \end{bmatrix} \qquad t > t_p \tag{2.792}$$

The bullet can hit the projectile if its four coordinates x, y, z, t are the same as the projectile's. Equating (2.792) and (2.783) provides three equations that must be solved for the unknown parameters α, v_0, t_h, θ_0,

$$x_0 + \left(t_h - t_p\right) v_0 \cos\theta_0 \cos\alpha = v_1 t_h \cos\theta_1 \tag{2.793}$$

$$y_0 + \left(t_h - t_p\right) v_0 \cos\theta_0 \sin\alpha = 0 \tag{2.794}$$

$$z_0 + \left(t_h - t_p\right) v_0 \sin\theta_0 - \frac{1}{2} g \left(t_h - t_p\right)^2 = v_1 t_h \sin\theta_1 - \frac{1}{2} g t_h^2 \tag{2.795}$$

where t_h is the time of impact, $t_p < t_h < 2v_1 \sin\theta_1/g$.

We usually like to hit the projectile as soon as possible. So, we set the antiprojectile gun to its maximum speed. Assume that the maximum value of v_0 is

$$v_0 = 2v_1 \tag{2.796}$$

and solve the equations for α, t_h, θ_0.

We may find $\sin\alpha$ and $\cos\alpha$ from Equations (2.793) and (2.794),

$$2v_1\left(t_h - t_p\right)\sin\alpha = -\frac{y_0}{\cos\theta_0} \tag{2.797}$$

$$2v_1\left(t_h - t_p\right)\cos\alpha = \frac{-x_0 + v_1 t_h \cos\theta_1}{\cos\theta_0} \tag{2.798}$$

and calculate α as a function of t_h:

$$\alpha = \tan^{-1}\left(\frac{y_0}{x_0 - v_1 t_h \cos\theta_1}\right) \tag{2.799}$$

Having $\sin\alpha$ from (2.797), we can calculate $\cos\alpha$ as

$$\cos\alpha = \sqrt{1-\sin^2\alpha} = \frac{1}{2}\sqrt{4 - \frac{y_0^2}{v_1^2\left(t_h - t_p\right)^2\cos^2\theta_0}} \tag{2.800}$$

substitute it in (2.798), and find

$$\frac{-x_0 + v_1 t_h \cos\theta_1}{2v_1\left(t_h - t_p\right)\cos\theta_0} = \frac{1}{2}\sqrt{4 - \frac{y_0^2}{v_1^2\left(t_h - t_p\right)^2\cos^2\theta_0}} \tag{2.801}$$

Eliminating θ_1 between (2.801) and (2.795) gives a transcendental equation to calculate the impact time t_h:

$$\begin{aligned} z_0 &+ \sqrt{4\left(t_h - t_p\right)^2 v_1^2 - y_0^2 - \left(-x_0 + v_1 t_h \cos\theta_1\right)^2} \\ &= v_1 t_h \sin\theta_1 + \tfrac{1}{2} g t_p \left(t_p - 2t_h\right) \end{aligned} \tag{2.802}$$

Calculating t_h from (2.802), we can find α from (2.799) and θ_0 from (2.801).

As an example consider a problem with the following data:

$$\begin{gathered} g = 9.81\,\mathrm{m/s^2} \\ v_1 = 10\,\mathrm{m/s} \qquad v_0 = 20\,\mathrm{m/s} \\ \theta_1 = \pi/6\,\mathrm{rad} \\ x_0 = 10\,\mathrm{m} \qquad y_0 = 11\,\mathrm{m} \qquad z_0 = 0\,\mathrm{m} \end{gathered} \tag{2.803}$$

The projectile will reach the target at time t_1:

$$t_1 = 2\frac{v_1}{g}\sin\theta_1 = 2\times\frac{10}{9.81}\sin\frac{\pi}{6} = 1.019367992\,\mathrm{s} \tag{2.804}$$

Let us assume that the antiprojectile is fired at a time t_p:

$$t_p = 0.1\,\mathrm{s} \tag{2.805}$$

Equation (2.802) provides

$$t_h = 0.7013308212\,\mathrm{s} \tag{2.806}$$

and Equations (2.799) and (2.801) give

$$\theta_0 = 0.2407653799\,\text{rad} \approx 13.795\,\text{deg} \tag{2.807}$$

$$\alpha = -1.9136\,\text{rad} \approx -109.64\,\text{deg} \tag{2.808}$$

Figure 2.53 illustrates the projectile and the antiprojectile gun that can hit the projectile at t_h when both are at the following position:

$$x = 6.073703078$$
$$y = 0 \tag{2.809}$$
$$z = 1.094056669$$

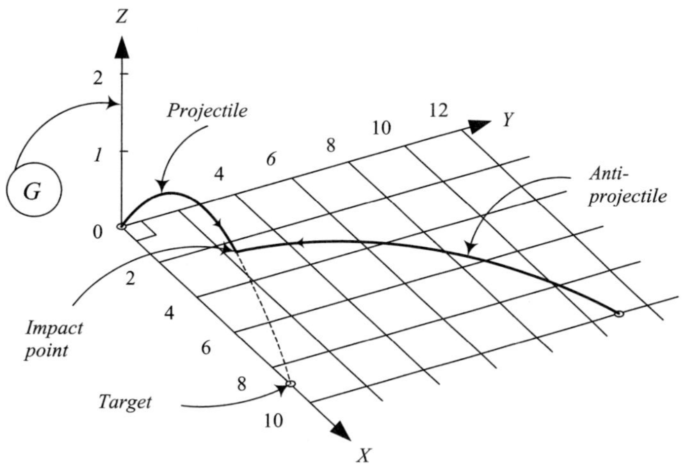

Figure 2.53 Projectile and antiprojectile gun that can hit the projectile at t_h.

KEY SYMBOLS

$\mathbf{0}$	zero vector
$a, \ddot{x}, \mathbf{a}, \dot{\mathbf{v}}, \ddot{r}$	acceleration
$\mathbf{a}, \mathbf{b}, \mathbf{c}, \mathbf{p}, \mathbf{q}$	vectors, constant vectors
A	area
A, B	points
A, B, C	constant parameters
$\mathbf{b} = \dot{\mathbf{z}}$	bong
B, B_1, B_2	body, body coordinate frames
c_i	weight factors of vector addition
$\mathbf{c}$	crackle, constant vector
C	space curve, mass center
d	distance
$d\mathbf{r}$	infinitesimal displacement
ds	arc length element
${}^G\mathbf{d}_o$	position vector of B in G

$[D_2z]$	second-derivative matrix
e	restitution coefficient
E	mechanical energy
$E(k)$	complete elliptic integral of the second kind
$E(\varphi, k)$	elliptic integral of the second kind
$\mathbf{E}$	electric field
$f = f(\mathbf{r})$	scalar field function
f_0	isovalue
$f(x, y, z)$	equation of a surface
$f(x, y, z) = f_0$	isosurface of the scalar field $f(\mathbf{r})$ for f_0
$F(\varphi, k)$	elliptic integral of the first kind
$\mathbf{F}$	force
$g, \mathbf{g}$	gravitational acceleration
$G, G(OXYZ)$	global coordinate frame,gravitational constant
$G_i = Gm_i$	equivalent mass
$\mathbf{G}$	gorz
H	height
$\mathbf{H}$	magnetic field, sharang
$\hat{\imath}, \hat{\jmath}, \hat{k}$	unit vectors of a Cartesian coordinate frame
$\hat{I}, \hat{J}, \hat{K}$	unit vectors of a global Cartesian system G
$\mathbf{I}$	impulse
$\mathbf{j}, \dot{\mathbf{a}}, \ddot{\mathbf{v}}, \dddot{\mathbf{r}}$	jerk
k	scalar coefficient
K	kinetic energy
$K(k)$	complete elliptic integral of the first kind
l	a line
$L, \mathbf{L}$	moment of momentum
m	mass
M	maximum
$M, \mathbf{M}$	moment
$\mathbf{M}_Q$	moment about point Q
$\mathbf{M}_l$	moment about line l
n	number of particles
$\mathbf{n}$	perpendicular vector to a surface $z = g(x, y)$
$\mathbf{N}$	reaction force, setorg
O	origin of a triad, origin of a coordinate frame
$(Ouvw)$	orthogonal coordinate frame
$\mathbf{p}$	momentum
P	point, particle, perimeter
Q	point
q, p	parameters, variables
$\mathbf{q} = \dot{\mathbf{b}}$	jeeq
$r = \lvert\mathbf{r}\rvert$	length of $\mathbf{r}$
$\mathbf{r}$	position vector
R	radius, range
s	arc length
$\mathbf{s} = d\mathbf{j}/dt$	snap, jounce
S	surface
$\mathbf{S}$	snatch
t	time
T	period

$T = [\tau_1, \tau_2]$	set in which a vector function is defined
$\mathbf{T}$	tug
u	dimensionless x
u, v, w	components a vector $\mathbf{r}$ in $(Ouvw)$
$\mathbf{u}$	Darboux vector
$\mathbf{u} = \dot{\mathbf{q}}$	sooz
$\hat{u}_r$	a unit vector on $\mathbf{r}$
$\hat{u}_1, \hat{u}_2, \hat{u}_3$	unit vectors along the axes q_1, q_2, q_3, components of $\hat{u}_r$
$\hat{u}_r, \hat{u}_\theta, \hat{u}_\varphi$	unit vectors of a spherical coordinate system
$\hat{u}_t, \hat{u}_n, \hat{u}_b$	unit vectors of natural coordinate frame
$\hat{u}_u, \hat{u}_v, \hat{u}_w$	unit vectors of $(Ouvw)$
$\hat{u}_\parallel, \hat{u}_\perp$	parallel and perpendicular unit vectors of l
v	speed
$v, \dot{x}, \mathbf{v}, \dot{\mathbf{r}}$	velocity
$\mathbf{v}(\mathbf{r})$	velocity field
V	volume
w	dimensionless z
W	work
x, y, z	axes of an orthogonal Cartesian coordinate frame
x_0, y_0, z_0	coordinates of an interested point P
$\mathbf{x}, \mathbf{y}$	vector functions, relative position vectors
X, Y, Z	global coordinate axes
$\mathbf{X}$	absolute position vectors
$\mathbf{Y}$	yank
$\mathbf{z} = \dot{\mathbf{þ}}$	larz
Z, S	short notation symbol
$\mathbf{Z}$	zoor

Greek

α	angle between two vectors, angle between $\mathbf{r}$ and l
α, β, γ	directional cosines of a line
$\alpha_1, \alpha_2, \alpha_3$	directional cosines of $\mathbf{r}$ and $\hat{u}_r$
ϵ	strain
$\varepsilon, \lambda, \psi$	Lagrange parameters of series solutions
ε, a, b, c	Jazar parameters of series solutions
θ	angle, angular coordinate, angular parameter
κ	curvature
$\kappa = \kappa \hat{u}_n$	curvature vector
$\mu = G_1 + G_2$	equivalen mass in two-body problem
ρ	curvature radius, density
σ	stress tensor, normal stress
$[\sigma_{ij}(\mathbf{r})]$	stress field
τ	curvature torsion, shear stress
$\varphi = \varphi(\mathbf{r})$	scalar field function
ω, ω	angular velocity
$\Pi(\varphi, k, n)$	elliptic integral of the third kind
$\Pi(k, n)$	complete elliptic integral of the third kind

Symbol

$\cdot$	inner product of two vectors
$\times$	outer product of two vectors

∇	gradient operator
$\nabla\mathbf{f}(\mathbf{r})$	gradient of **f**
$\nabla\times\mathbf{f} = \text{curl}\,\mathbf{f}$	curl of **f**
$\nabla\cdot\mathbf{f} = \text{div}\,\mathbf{f}$	divergence of **f**
$\nabla^2 f$	Laplacian of f
$\nabla f = \text{grad}\, f$	gradient of f
∞	infinity
$\parallel$	parallel
$\perp$	perpendicular
þ= ċ	pop
$\text{cn}(u, k)$	Jacobi elliptic function
$\text{sn}(u, k)$	Jacobi elliptic function
$\text{dn}(u, k)$	Jacobi elliptic function
Ş	shake

EXERCISES

1. **Upward Shot in Air** Assume we shoot a particle with mass m upward with initial velocity v_0. Air exerts a drag force $-mkv$, where $k > 0$ is a scalar.

   (a) Determine the maximum height achieved by the particle.

   (b) Determine the time to reach the maximum height.

   (c) Calculate the terminal velocity of the particle.

2. **Free Fall in Air** A particle with mass m falls along the $-z$-axis in air and exerts a resistance force kmv^2, where $k > 0$ is a scalar. Show that the velocity of m is

$$v = \left(1 - \frac{2}{e^{2kv_\infty t} + 1}\right) v_\infty$$

3. **First Guldin Theorem** The area of a surface generated by the revolution of a plane curve about an axis lying in the plane of this curve and not cutting the curve is equal to the product of the length of the curve and the length of the path described by the mass center of the curve. This is the first theorem of Guldin (1577–1643).

   (a) Use the Guldin theorem to determine the mass center of the semicircle arc of Figure 2.54 (a).

   (b) Use the result of part (a) to determine the surface area of the object that is made by the revolution of the semicircle arc of Figure 2.54(b) about the Y-axis.

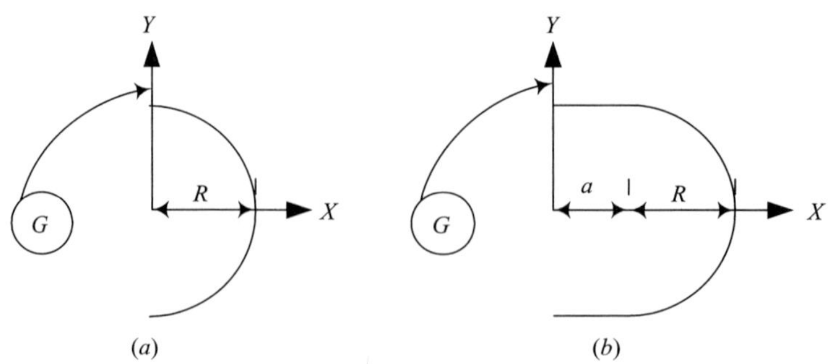

Figure 2.54 A semicircle arc.

4. **Motion along a Vertical Curve** A particle m slides on a curve C with equation $z = f(x)$ in the (x, z)-plane. If m is at (x_0, z_0) when $t = 0$, show that the required time t to be at a point (x, z) is

$$t = \int_{x_0}^{x} \sqrt{\frac{1 + f'^2(x)}{2g(f(x_0) - f(x))}} dx \qquad f' = \frac{df}{dx}$$

5. **Second Guldin Theorem** The volume of a solid generated by the revolution of a plane region about an axis lying in the plane of this region and not cutting the region is equal to the product of the area of the region and the length of the path described by the mass center of the region. This is the second theorem of Guldin (1577–1643).

   (a) Use the Guldin theorem to determine the mass center of the semicircle area of Figure 2.55(a).

   (b) Use the result of part (a) to determine the volume of the solid object that is made by the revolution of the semicircle area of Figure 2.55(b) about the Y-axis.

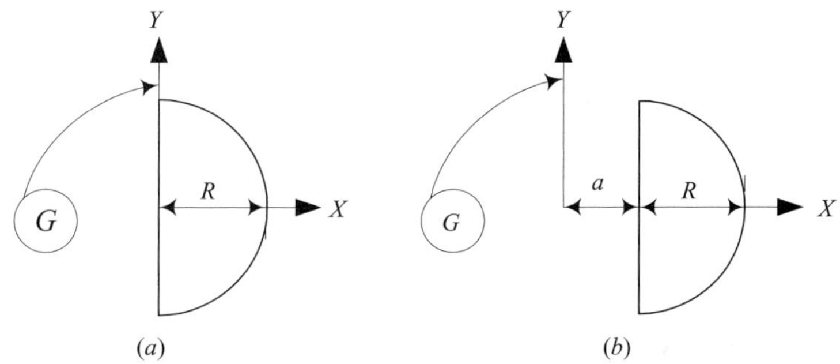

Figure 2.55 A semicircle area.

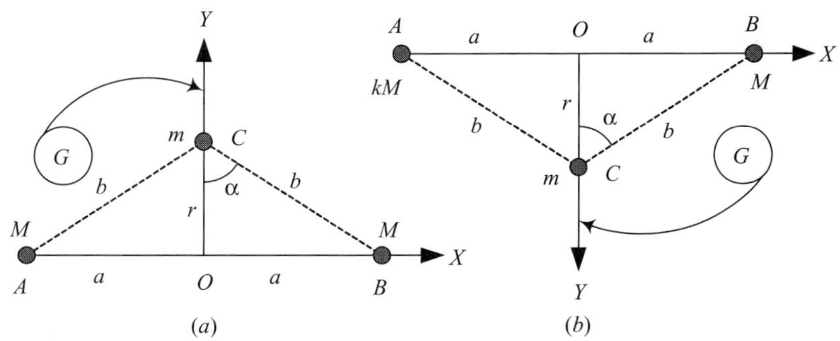

Figure 2.56 A particle C with mass m is attracted by two particles at A and B.

6. **Attraction of Two Point Masses**

   (a) Determine the gravitational force on particle C with mass m that is attracted by two particles A and B with mass M. The particles are shown in Figure 2.56(a).

   (b) Let us assume that the mass of the particle at A is kM, $k > 0$. Determine the gravitational force on particle C with mass m that is attracted by particles A and B, as shown in Figure 2.56(b).

7. **Attraction of a Uniform Bar**

   (a) Determine the gravitational force on particle C with mass m that is symmetrically attracted by a uniform bar with mass M, as shown in Figure 2.57(a).

   (b) Determine the gravitational force on particle C with mass m that is asymmetrically attracted by a uniform bar with mass M, as shown in Figure 2.57(b).

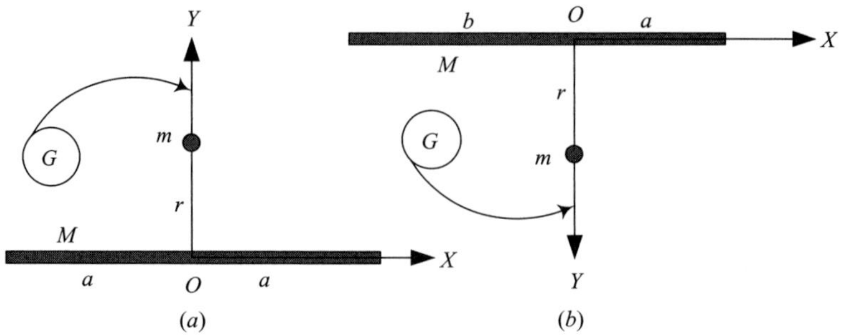

Figure 2.57 A particle C with mass m is attracted by a uniform bar with mass M.

8. **Attraction of a Uniform Disc**

   (a) Determine the attraction force on a particle with mass m at a distance r on the axis of symmetry of a disc with mass M, as shown in Figure 2.58(a).

   (b) ★ Determine the attraction force on a particle with mass m at a distance r from a uniform disc and far from the axis of symmetry of a disc with mass M, as shown in Figure 2.58(b).

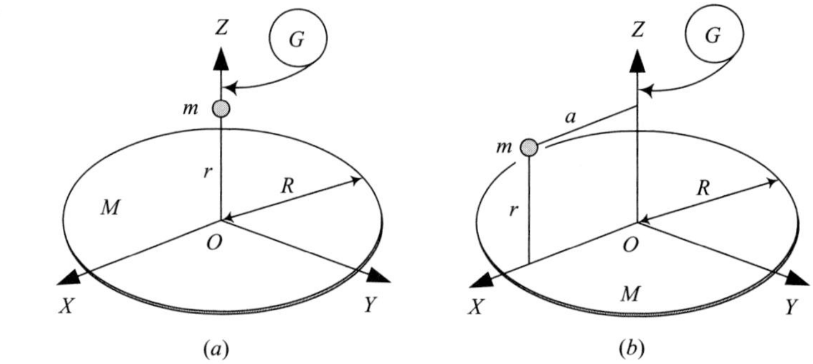

Figure 2.58 Attraction of a particle with mass m by a uniform disc with mass M.

9. **Force for a Given Motion**

   (a) Prove that if the force on a moving particle is always along the tangent to the trajectory, then the trajectory is a straight line.

   (b) Prove that if the trajectory of a moving particle is such that

$$\dot{\mathbf{r}} \cdot (\ddot{\mathbf{r}} \times \dddot{\mathbf{r}}) = 0$$

   then $\mathbf{r} = \mathbf{r}(t)$ is a planar curve.

   (c) Prove that the path of motion of a particle moving under the gravitational attraction is a conic section.

10. ★ **Attraction of a Spherical Shell** Determine the gravitational potential of a spherical shell with inner radius R_1 and outer radius R_2 for (a) inside the shell and (b) outside the shell. (c) Determine the external attraction of a solid sphere by $R_2 \to 0$.

11. **Attraction of a Sphere** Consider a uniform sphere with radius R and mass M, as shown in Figure 2.59. Determine the gravitational attraction force of M on a particle m at a distance a from the center of M.

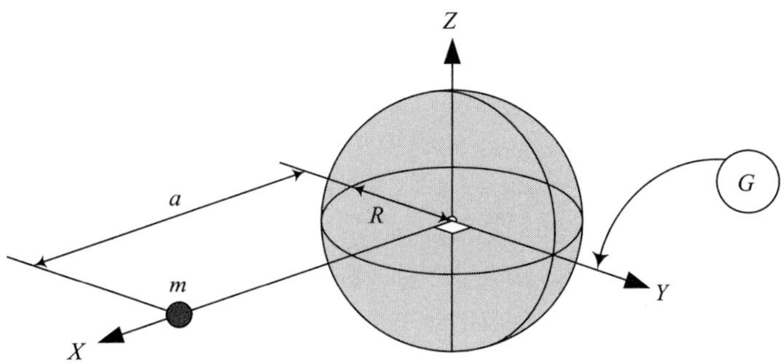

Figure 2.59 Attraction of a uniform sphere with radius R and mass M on a particle with mass m.

12. **Escape from the Moon** A bullet with mass m is fired upward from the surface of the moon with initial speed v_0. Show that the bullet cannot escape from the moon if $v_0^2 < 2MG/R$, where M is the mass of the moon and R is the radius of the moon.

13. **Angular Momentum of a Particle** A particle P with mass m is moving such that its position vector has the components $x = a\theta^3$, $y = 2a\theta$, $z = 0$, where $\theta = \theta(t)$. Determine the angular momentum of the particle about the point $B(0, a, 0)$.

14. **Integrals of a Spherical Pendulum** A spherical pendulum with mass m and length l is shown in Figure 2.60. The angles φ and θ may be used to describe the coordinates of the system.

    (a) Show that the angular momentum of the pendulum about the Z-axis is conserved.
    (b) Determine the mechanical energy $E = K + V$ of the pendulum.
    (c) Use the energy and angular momentum equations to determine $\dot{\varphi}$ and $\dot{\theta}$.

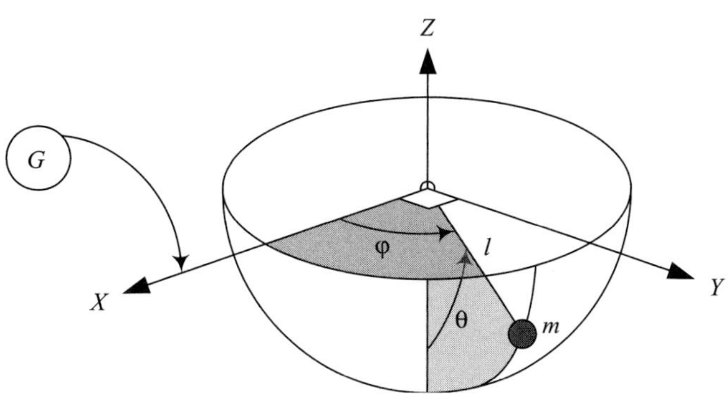

Figure 2.60 A spherical pendulum.

15. **Cartesian-to-Cylindrical System** A force $\mathbf{F} = 10\hat{\imath} + 9\hat{\jmath} - 5\hat{k}$ is acting on a point P at $\mathbf{r} = 3\hat{\imath} + \hat{\jmath} - \hat{k}$. Express $\mathbf{F}$ in (a) cylindrical and (b) spherical coordinate systems.

16. **Total Moment of Momentum** The total moment of momentum $\mathbf{L}$ of a system of particles is the sum of all the individual moments of momentum Show that

$$\begin{aligned}\frac{d\mathbf{L}}{dt} &= \sum_{i=1}^{n} \mathbf{r}_i \times \mathbf{F}_i + \sum_{i=1}^{n}\sum_{j=1}^{n} \mathbf{r}_i \times \mathbf{F}_{ij} \\ &= \sum_{i=1}^{n} \mathbf{r}_i \times \mathbf{F}_i + 0 = \sum_{i=1}^{n} \mathbf{M}_i = \mathbf{M}\end{aligned}$$

17. **A Moving Load on a Barge** A load $m = 2000\,\text{kg}$ is on the smooth deck of a barge of mass $M = 8000\,\text{kg}$. The load is pulled with a constant velocity of $0.1\,\text{m/s}$ toward a winch, as shown in Figure 2.61. Initially both the barge and the load are at rest with respect to the water.

    (a) Determine the speed of the barge and the load with respect to the shore.
    (b) What is the position of the barge after 10 s?
    (c) ★ Is it possible to move the load and give a velocity to the barge so it moves on water with constant velocity?

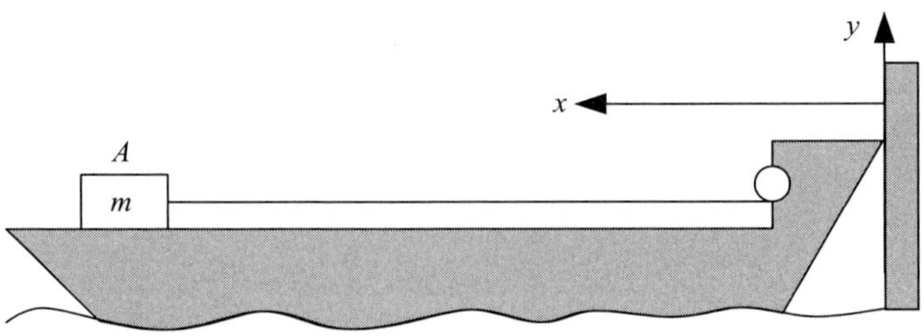

Figure 2.61 A moving load on a barge.

18. **A Fuel Consumption Equation** A rocket has an initial mass m_0. After it fires, the mass of the rocket would be

$$m = \tfrac{1}{2} m_0 \left(1 + e^{-kt}\right)$$

where k is a constant. If the velocity of the rocket is $v = 100\,t^2\,\text{m/s}$ in a straight line, what would be the thrust force after 1 s?

19. **A Falling Chain** A chain with length density $\rho = m/l$ and a mass center at C is hung at rest as shown in Figure 2.62 (a). We release the attachment at point B, as shown in Figure 2.62(b).

    (a) Determine the equation of motion using x as the variable.
    (b) Determine the tension of the chain at A as a function of x.

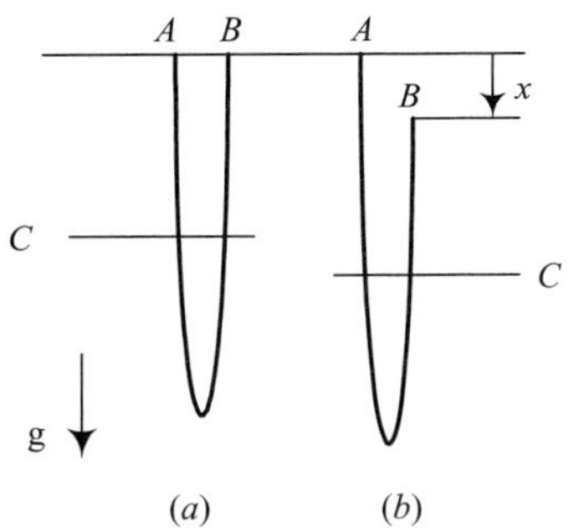

Figure 2.62 A hanging chain.

20. Newton Equation in Coordinate System Determine the Newton equations of motion of a particle of mass m in

(a) a cylindrical coordinate system and

(b) a spherical coordinate system.

(c) Assume the particle m is free to slide from $r = R$, $z = 0$ on a smooth helical wire with cylindrical equation $r = R$, $z = c\theta$. If the gravitational acceleration is $\mathbf{g} = g\hat{k}$. Determine the force $\mathbf{F}$ that the wire exerts on m.

21. Nonstraight Path and Travel Time A particle of mass m has an initial velocity $\mathbf{v} = v_0\hat{\imath}$ and is moving on a frictionless surface as shown in Figure 2.63.

(a) Prove that the particle in Figure 2.63(a) will travel the distance l faster than the particle in part (b).

(b) Show the effect of the function $y = f(x)$ on the total time t to travel the distance l if h is kept constant.

(c) ★ Is there an optimal function $y = f(x)$ to minimize t?

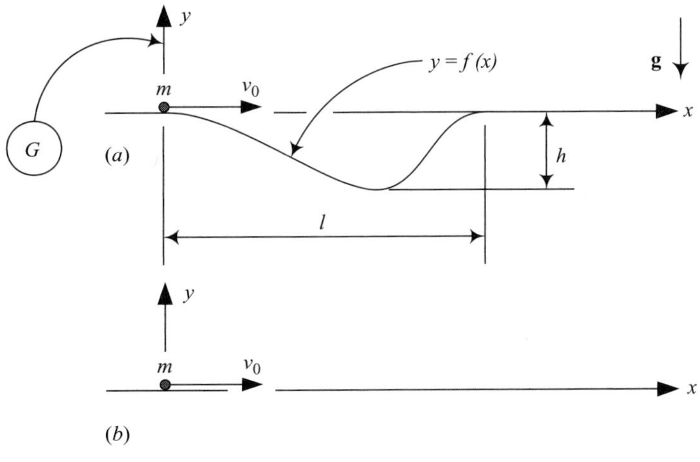

Figure 2.63 A particle will move faster if there is a frictionless dip on its path.

22. Derivative of Kinetic Energy Consider a particle with mass m that under a force $\mathbf{F}$ is moving with velocity $\mathbf{v}$. Show that

$$\text{(a)} \quad \frac{dK}{dt} = \mathbf{F} \cdot \mathbf{v} \qquad \text{(b)} \quad \frac{d(mK)}{dt} = \mathbf{F} \cdot \mathbf{p}$$

23. Double Pendulum

(a) Using a free-body diagram and the Newton method, determine the equations of motion of the double pendulum shown in Figure 2.64.

(b) Determine the kinetic energy of the pendulum.

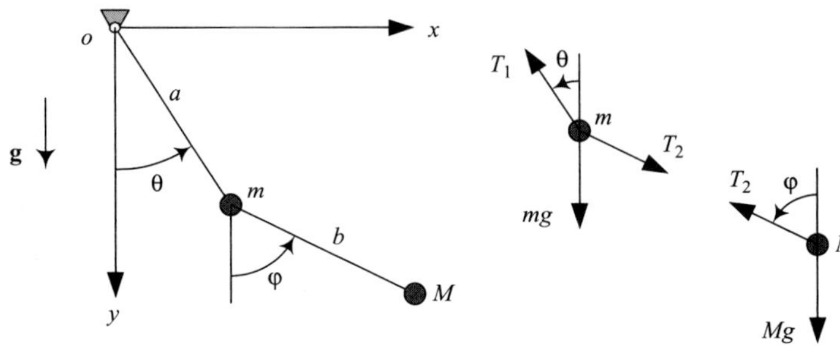

Figure 2.64 A double pendulum.

24. Conservative Force Check Determine whether each of the following force fields $\mathbf{F}$ is conservative and determine the potential of the field relative to the origin for those forces that are conservative:

$$\text{(a)} \quad \mathbf{F} = e^y \hat{\imath} + (z + xe^y)\,\hat{\jmath} + (1 + y)\,\hat{k}$$

$$\text{(b)} \quad \mathbf{F} = \left(y/\sqrt{x^2 + y^2}\right)\hat{\imath} + \left(x/\sqrt{x^2 + y^2}\right)\hat{\jmath}$$

$$\text{(c)} \quad \mathbf{F} = \left(y/\left(x^2 + y^2\right)\right)\hat{\imath} + \left(x/\left(x^2 + y^2\right)\right)\hat{\jmath}$$

$$\text{(d)} \quad \mathbf{F} = 2xy^3z^4\hat{\imath} + 3x^2y^2z^4\hat{\jmath} + 4x^2y^3z^3\hat{k}$$

25. A Particle in a Cone A particle of mass m slides without friction inside an upside-down conical shell of semivertical angle α, as shown in Figure 2.65.

(a) Use cylindrical coordinates and the Euler equation $\mathbf{M} = (d/dt)\mathbf{L}$, to determine the equations of motion of the particle.

(b) Show that it is possible for the particle to move such that it is at a constant R with the cone axis.

(c) Determine the angular speed of the particle for a uniform motion of part (b).

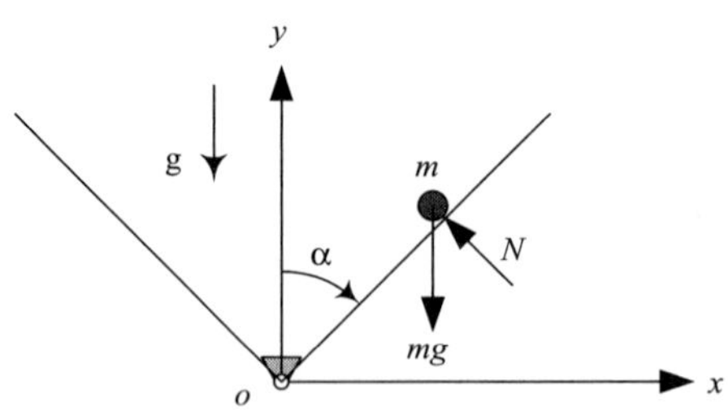

Figure 2.65 A particle of mass m slides inside a conical shell.

26. ★ Elliptic Integrals

(a) Show that

$$\int_{\alpha_1}^{\alpha_2} \sqrt{1 - k^2 \sin^2 \theta}\, d\theta = E(k, \alpha_2) - E(k, \alpha_1)$$

(b) Using $c^2 \cos\theta \, d\theta = \cos\varphi \, d\varphi$, transform the following integral into the standard form:

$$\int_0^{\alpha} \sqrt{1 - c^2 \sin^2 \theta}\, d\theta \qquad c^2 > 1$$

(c) Transform the following integral into an elliptic integral:

$$\int_0^{\alpha} \frac{d\theta}{\sqrt{\cos 2\theta}}$$

27. A Particle on a Curved Surface Draw FBD of the particle in Figures 2.66 (a) and (b) for $a = cR$, $c < 1$, and determine their equation of motion. The spring is linear and applies a tangential force on m.

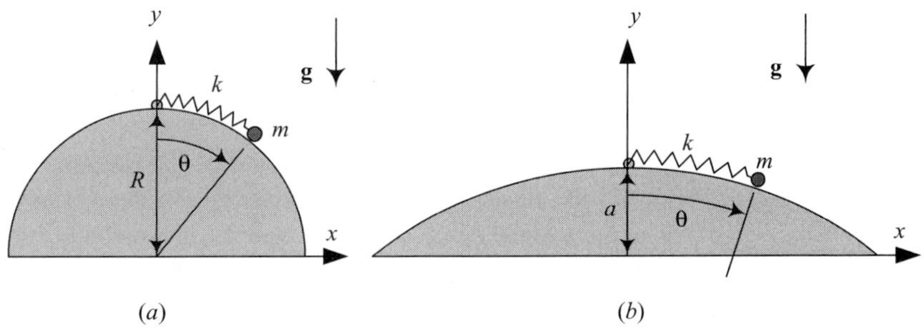

Figure 2.66 A particle on a curved surface.

28. Work on a Curved Path A particle of mass m is moving on a circular path given by

$${}^G\mathbf{r}_P = \cos\theta\, \hat{I} + \sin\theta\, \hat{J} + 4\hat{K}$$

Calculate the work done by a force ${}^G\mathbf{F}$ when the particle moves from $\theta = 0$ to $\theta = \pi/2$:

$$\text{(a)} \qquad {}^G\mathbf{F} = \frac{z^2 - y^2}{(x+y)^2}\hat{I} + \frac{y^2 - x^2}{(x+y)^2}\hat{J} + \frac{x^2 - y^2}{(x+z)^2}\hat{K}$$

$$\text{(b)} \qquad {}^G\mathbf{F} = \frac{z^2 - y^2}{(x+y)^2}\hat{I} + \frac{2y}{x+y}\hat{J} + \frac{x^2 - y^2}{(x+z)^2}\hat{K}$$

29. Moment of Momentum Determine the moment of momentum of m in Figure 2.67. The mass m is attached to a massless rod with length l. The rod is pivoted to a rotating vertical bar that is turning with angular speed ω.

(a) Determine the moment of momentum of the system.

(b) If $m = 2\,\text{kg}$, $l = 1.2\,\text{m}$, and $\omega = 10\,\text{rpm}$ when $\theta = 30\,\text{deg}$, determine ω when $\theta = 45\,\text{deg}$.

(c) Draw a graph to show ω versus θ.

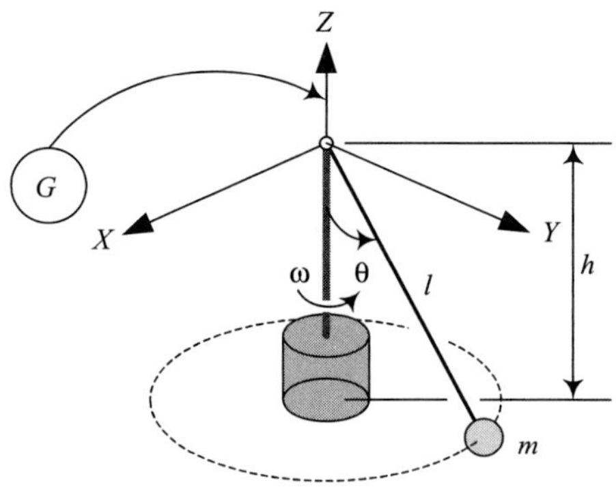

Figure 2.67 A mass m is attached to a massless rod which is turning with angular speed ω.

30. **Impact and Motion** Figure 2.68 illustrates a carriage with mass M that has an oblique surface. The carriage can move frictionless on the ground. We drop a small ball of mass m on the surface when the carriage is at rest.

    (a) Assume the ball falls from a height h above the impact point on the oblique surface. Determine the velocity of M as a function of θ if the restitution coefficient is e.

    (b) Is there any optimal value for θ to provide a maximum speed of M?

    (c) ★ Assume $h >> l$ and the first ball hits at the top point of the surface. If we drop balls every t_1 seconds, then what would be the maximum speed of M when the final ball hits the carriage? How many balls will hit the carriage?

    (d) ★ Assume h and l are comparable and the first ball hits at the top point of the surface. If we drop balls every t_1 seconds, then what would be the maximum speed of M when the final ball hits the carriage? How many balls will hit the carriage?

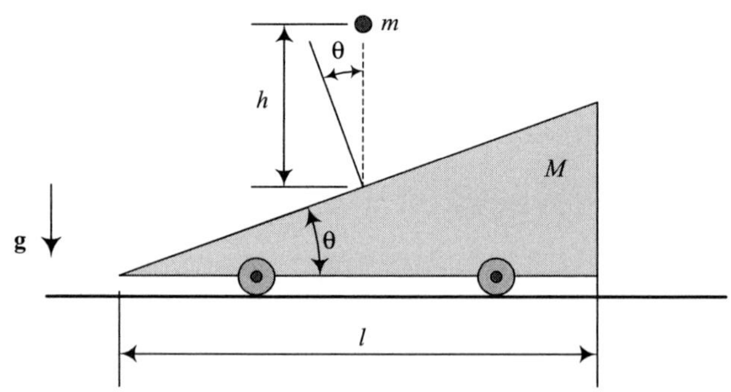

Figure 2.68 A carriage with mass M and an oblique surface can be moved by the impact of dropping balls.

31. **Impact on a Rotating Plate** Figure 2.69 illustrates a rotating plate with a constant angular velocity ω. A series of equal particles of mass m are released from a height h above the center of rotation. Ignore the thickness of the table and assume $e = 1$.

(a) Determine x at which m hits the x-axis if $\theta = 0$ at $t = 0$.

(b) Determine x at which the second m hits the x-axis if it is released when the first m hits the table and $\theta = 0$ at $t = 0$.

(c) Determine x at which the second m hits the x-axis if it is released when the first m hits the x-axis and $\theta = 0$ at $t = 0$.

(d) Assume $\omega = 0.1\ \text{rad/s}$, $h = 1\ \text{m}$ and determine the values of $x_i, i = 1, 2, \ldots, 10$, for the first 10 masses if we release m_{i+1} when m_i hits the table.

(e) Assume $\omega = 0.1\ \text{rad/s}$, $h = 1\ \text{m}$ and determine the values of $x_i, i = 1, 2, \ldots, 10$, for the first 10 masses if we release m_{i+1} when m_i hits the x-axis.

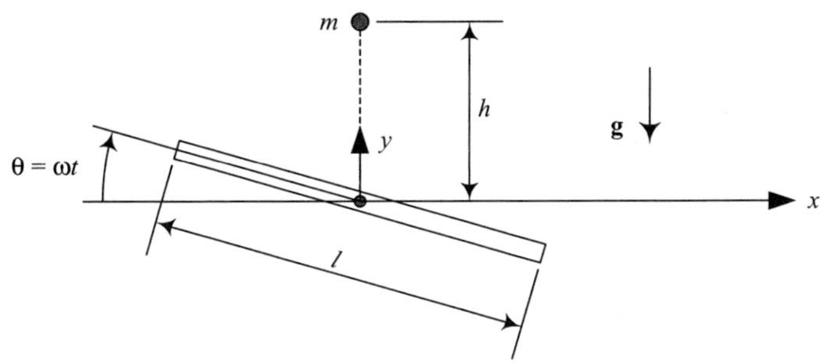

Figure 2.69 Falling particles on a rotating plate.

32. Offset Impact on a Rotating Plate Substitute Figure 2.70 with Figure 2.69 and solve Exercise 32 again.

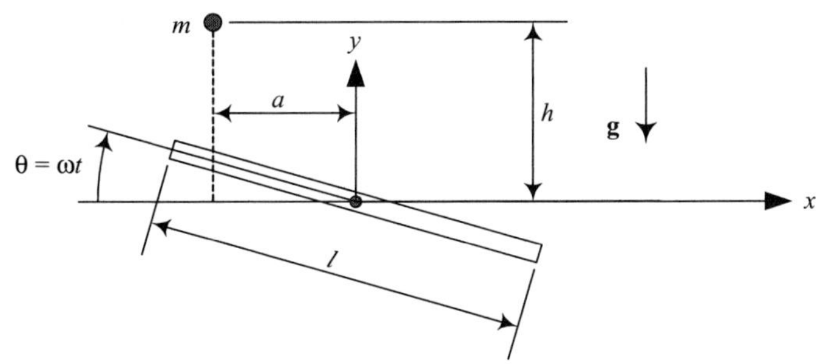

Figure 2.70 Offset falling particles on a rotating plate.

33. Falling Particle on a Semicircle A particle m is falling from a height h on a semicircle solid object with radius R as shown in Figure 2.71. Assume $e = 1$.

(a) Determine a for a given value of m, h, R, l.

(b) Determine a to maximize l.

(c) Determine the central angle of the impact point on the semicircle for l_{Max}. Is the angle 45 deg?

(d) Determine H when $l = l_{\text{Max}}$.

(e) Determine a to have $H = h$.

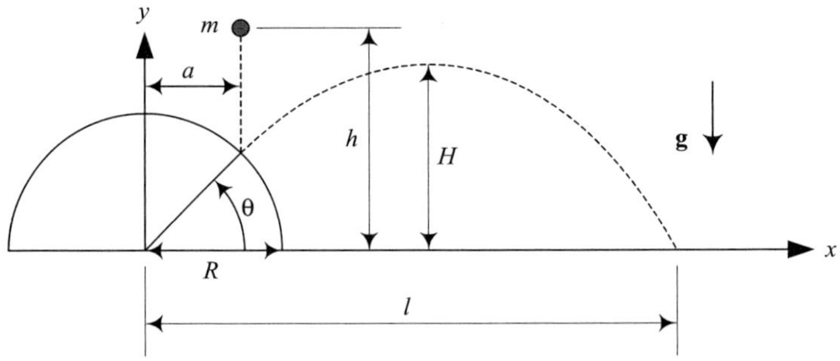

Figure 2.71 Falling particle on a semi-circle.

34. **Impact of a Particle in a Circle** A particle m is fired with initial velocity v_0 from the intersection of a circle with radius R and the x-axis. Assume $e = 1$ and determine the coordinates of the third impact point of the particle with a wall in Figures 2.72(a) and (b).

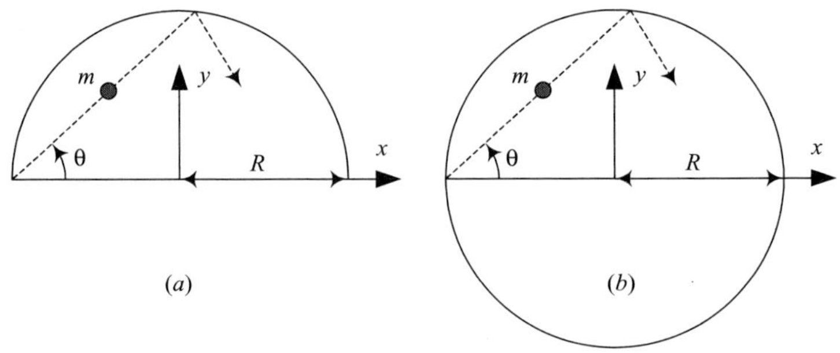

Figure 2.72 Impact of a particle in a semicircle and a circle.

35. **A Rolling Disc and Projectile** Consider the uniform disc of mass m and radius r in Figure 2.73. The disc will begins falling from point $(-l, h)$. Assume $R = ch$, $c < h$.

    (a) Determine the point at which the disc hits the ground if the disc is in a pure roll while it is moving on the ground.

    (b) What is the angular velocity of the disc when it leaves the ground.

    (c) Determine the point at which the disc hits the ground if the disc slides on the ground.

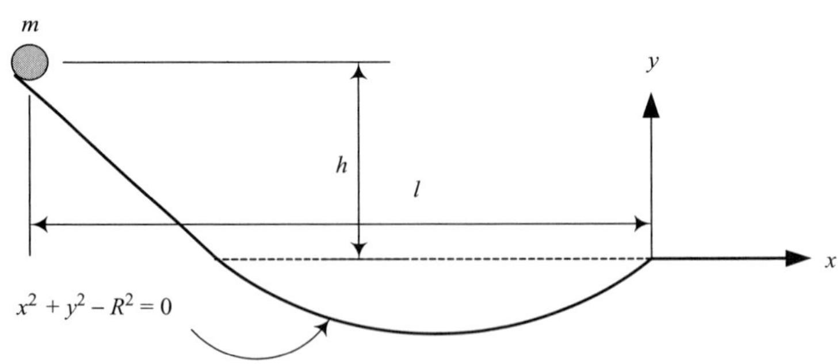

Figure 2.73 A rolling disc acts as a projectile.

36. **Projectile Impact** In Figure 2.74 the particle B is released from rest at time $t = 0$. The particle A is thrown after $t = 1$s with velocity $c\mathbf{v}$, where $\mathbf{v}$ is the velocity of B at $t = 1$ s.

(a) Determine α such that A hits B if $c = 5$ and $l = 100$ m.

(b) Is there a limit of c for a constant l and α to have an impact?

(c) What is the relationship of c and α for a constant l to have an impact?

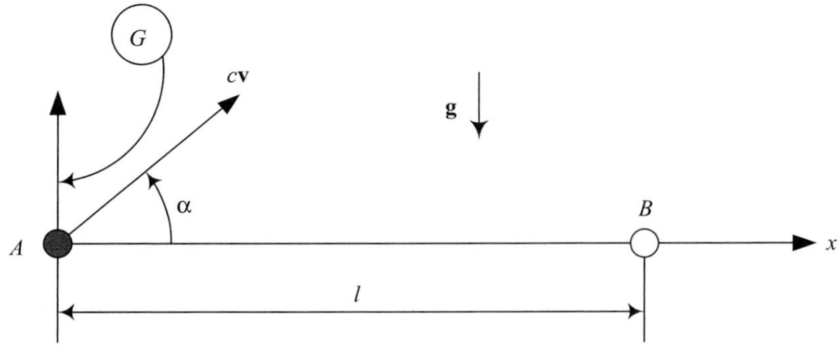

Figure 2.74 A projectile impact to hit a falling particle.

37. **Falling on a Spring** A solid mass m falls on a spring as shown in Figure 2.75(a) or with a spring as shown Figure 2.75(b). The spring exerts a stiffness force F_s. Determine the maximum compression x_{Max} of the springs if:

(a) $e = 0$ and $F_s = kx$

(b) $e = 0$ and $F_s = kx^3$

(c) $e = 1$ and $F_s = kx^3$

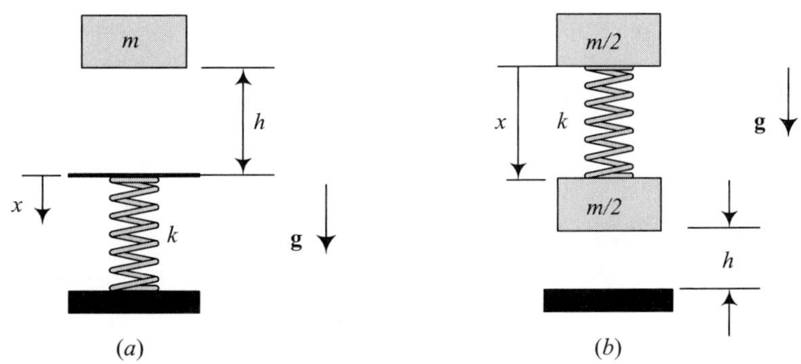

Figure 2.75 A solid mass m falls on or with a spring.

38. Motion in a Fluid When a heavy body B of density ρ_B and volume V is immersed in a fluid of density ρ_F, the weight W of the body would be

$$W = (\rho_B - \rho_F)\, V g$$

(a) Show that we may define the gravitational acceleration as $g' = g\,(1 - \rho_F/\rho_B)$.

(b) Derive the equation of the path of a projectile in the fluid.

(c) ★ Assume $\rho_F = \rho e^{-z}$ and derive the equation of the path of a projectile in the fluid.

39. Hammer Impact in a Circle As is illustrated in Figure 2.76, a hammer with a head of mass m_2 is released from a horizontal position and hits a particle of mass m_1 sitting motionless at the bottom of a circular cage.

(a) Determine the height h that m_1 achieves for a restitution factor e. Examine the answer for extreme real values of $e = 0$ and $e = 1$.

(b) Assume $0 < e < 1$ and determine the position where m_1 hits the circle after separation.

(c) Assume $e = 1.2$ and determine the position where m_1 hits m_2 for the second time. Ignore the size of the head.

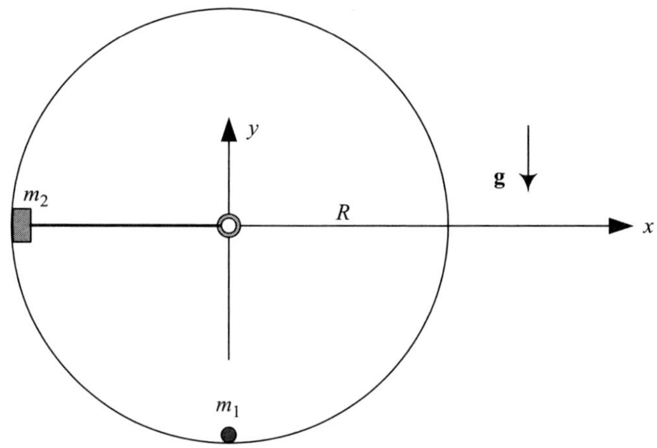

Figure 2.76 Hammer impact in a circle.

40. Free Fall on Earth from a Great Distance If a particle of mass m is released from rest at a distance r from the center of Earth, then its equation of motion is

$$m\frac{d^2r}{dt^2} = -G\frac{mM}{r^2}$$

where G is the gravitational constant, R is the radius of Earth, and M is the mass of Earth.

(a) Show that m hits the surface of Earth with velocity $v = \sqrt{2gR}$ when it is released from $r = \infty$.

(b) ★ Determine the velocity of impact if m is released from $r = R + h$, $h << R$.

41. **A Sliding Bid on a Turning Bar** Figure 2.77 illustrates a bar in the horizontal position at rest. There is a bid on the bar which can slide on the bar with no friction.

(a) Assume the bar is massless and has a constant angular velocity $\dot{\theta}$. When the bar is horizontal, m is at $u = a$. Determine the position u of the bid m when the bar passes $\theta = 90\,\text{deg}$.

(b) Assume the bar is massless and release the bar from the horizontal position. Determine the position u of the bid m when the bar passes $\theta = 90\,\text{deg}$.

(c) Determine the maximum value of u for $0 < \theta < 360\,\text{deg}$ in part (b).

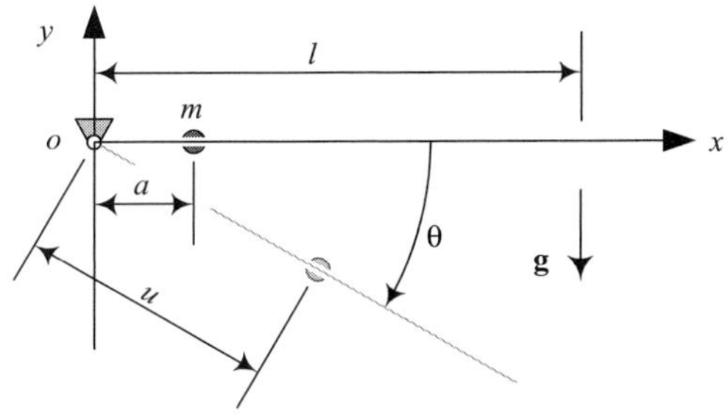

Figure 2.77 A sliding bid on a turning bar.

42. **Force as a Function of Velocity** Consider a particle that is moving according to the equation of motion

$$m\ddot{x} = -C_1 - C_2\dot{x}^2 = -C_1 - C_2 v^2$$

and show that

$$x_{\text{Max}} = \frac{m}{2C_2} \ln\left(1 + \frac{C_2}{C_1} v_0^2\right)$$

43. **Projectile in Air** Consider a projectile in air with a resistance force proportional to v^2 on a flat ground. Show that the equtions to determine the path of motion are

$$m\ddot{x} = -c\dot{x}^2\sqrt{1 + \frac{\dot{y}^2}{\dot{x}^2}} \qquad m\ddot{y} = -c\dot{x}\dot{y}\sqrt{1 + \frac{\dot{y}^2}{\dot{x}^2}} - mg$$

44. **Vertical Projectile in a Fluid** A particle is projected vertically upward with an initial velocity v_0 in a medium with a resistance force $F = cv^2$. Determine $z = z(t)$ from $t = 0$ to the final time $t = t_f$ when the particle gets back to $z = 0$.

45. **Change of Variable** If the applied force on a particle is

$$F = f(x)\,\dot{x}^n + g(x)\,\dot{x}^2$$

show that the equation of motion becomes linear by $\dot{x}^{2-n} = y$.

46. **Projectile Motion** Two particles are projected from the same point with equal velocities. If t and t' are the times taken to reach the other common point of their path and T and T' are the times to the highest point, show that $tT + t'T'$ is independent of the directions of projectiles.

Part II

Geometric Kinematics

Kinematics, the English version of the French word *ciné matique* from the Greek $\kappa\acute{\iota}\nu\eta\mu\alpha$ ("movement"), is a branch of science that studies geometry in motion. By *motion* we mean any type of displacement, which includes changes in position and orientation. Therefore, *displacement*, and its successive derivatives with respect to time, *velocity, acceleration*, and *jerk*, all combine into kinematics. In kinematics, we do not pay attention to what causes the motion.

Geometric kinematics explains how the position vectors are calculated and how they are expressed in different coordinate frames. The orthogonality condition

$$\mathbf{r} = (\mathbf{r} \cdot \hat{\imath})\hat{\imath} + (\mathbf{r} \cdot \hat{\jmath})\hat{\jmath} + (\mathbf{r} \cdot \hat{k})\hat{k}$$

and the kinematic or geometric transformation

$$^{G}\mathbf{r} = {}^{G}R_{B}\,{}^{B}\mathbf{r}$$

are the keys of geometric kinematics.

3

Coordinate Systems

The principal unit vectors of the Cartesian coordinate system are independent of position; however, in a general coordinate system, vectors are position dependent and will be defined at the interested position. Using the Cartesian coordinate system $G(x, y, z)$ as a base, we can transform the position, velocity, and acceleration of a moving particle to other coordinate systems $Q(q_1, q_2, q_3)$. The coordinate orthogonality condition in Cartesian or other orthogonal coordinate systems,

$$\mathbf{r} = (\mathbf{r} \cdot \hat{\imath})\hat{\imath} + (\mathbf{r} \cdot \hat{\jmath})\hat{\jmath} + (\mathbf{r} \cdot \hat{k})\hat{k} \tag{3.1}$$

$$= (\mathbf{r} \cdot \hat{u}_1)\hat{u}_1 + (\mathbf{r} \cdot \hat{u}_2)\hat{u}_2 + (\mathbf{r} \cdot \hat{u}_3)\hat{u}_3 \tag{3.2}$$

is the main tool for coordinate system transformation. The kinematic information of a particle may be expressed in any coordinate system.

3.1 CARTESIAN COORDINATE SYSTEM

To set up a Cartesian coordinate system, we use three sets of parallel and mutually perpendicular planes. One plane of each set is assigned as the *principal* or *zero* plane. The intersection of each pair of the zero planes makes an axis, and the three intersecting axes make an orthogonal triad G. Each axis is perpendicular to a plane that carries the name of the axis. So, the x-axis is perpendicular to the x-plane that is the (y, z)-plane. Similarly, y- and z-axes indicate the y- and z-planes.

The position of a point in space is indicated by three intersecting planes each one parallel to one of the principal planes. The point P in Figure 3.1 is at the intersection of three planes, one parallel to (y, z) at a distance $x = 4$, one parallel to (z, x) at a distance $y = 5$, and one parallel to (x, y) at a distance $z = 3$. We show the coordinates of P by $P(x, y, z) = P(4, 5, 3)$.

The position, velocity, and acceleration vectors of a moving point P in a Cartesian coordinate frame $G(x, y, z)$ are

$${}^G\mathbf{r} = x\hat{\imath} + y\hat{\jmath} + z\hat{k} \tag{3.3}$$

$${}^G\mathbf{v} = \frac{{}^Gd}{dt}\,{}^G\mathbf{r}(t) = \dot{x}\hat{\imath} + \dot{y}\hat{\jmath} + \dot{z}\hat{k} \tag{3.4}$$

$${}^G\mathbf{a} = \frac{{}^Gd}{dt}\,{}^G\mathbf{v}(t) = \ddot{x}\hat{\imath} + \ddot{y}\hat{\jmath} + \ddot{z}\hat{k} \tag{3.5}$$

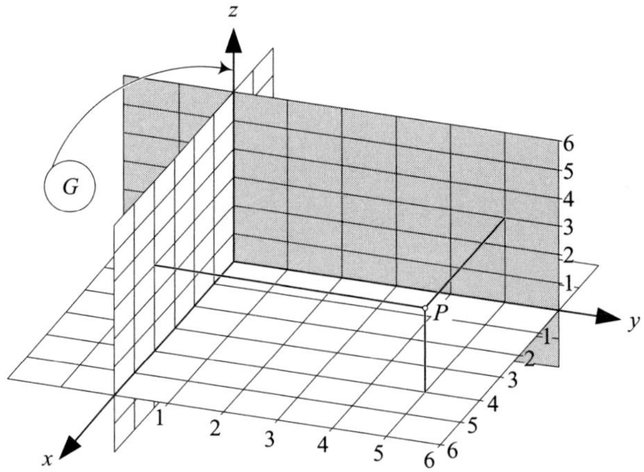

Figure 3.1 Cartesian coordinate system.

Proof: The Cartesian coordinate is the only system that has three invariant unit vectors $\left(\hat{\imath}, \hat{\jmath}, \hat{k}\right)$. An *invariant vector* is a free vector that has a constant length and a constant direction. Applying the unit vector definition (1.200) on the Cartesian expression of a position vector $\mathbf{r}$,

$$^{G}\mathbf{r} = x\hat{\imath} + y\hat{\jmath} + z\hat{k} \tag{3.6}$$

and noting that the unit vectors $\hat{\imath}$, $\hat{\jmath}$, and $\hat{k}$ have constant length and direction, we find

$$\hat{u}_x = \frac{\partial \mathbf{r}/\partial x}{|\partial \mathbf{r}/\partial x|} = \frac{\hat{\imath}}{1} = \hat{\imath} \tag{3.7}$$

$$\hat{u}_y = \frac{\partial \mathbf{r}/\partial y}{|\partial \mathbf{r}/\partial y|} = \frac{\hat{\jmath}}{1} = \hat{\jmath} \tag{3.8}$$

$$\hat{u}_z = \frac{\partial \mathbf{r}/\partial z}{|\partial \mathbf{r}/\partial z|} = \frac{\hat{k}}{1} = \hat{k} \tag{3.9}$$

Therefore, any derivative of the position vector (3.6) can be found by taking the derivative of its components. Showing the time derivative of components x, y, z by an overdot, $dx/dt = \dot{x}$, $dy/dt = \dot{y}$, $dz/dt = \dot{z}$, we can find the velocity, acceleration, and jerk of a moving point by

$$^{G}\mathbf{v} = \frac{^{G}d}{dt}\, {}^{G}\mathbf{r}(t) = \dot{x}\hat{\imath} + \dot{y}\hat{\jmath} + \dot{z}\hat{k} \tag{3.10}$$

$$^{G}\mathbf{a} = \frac{^{G}d}{dt}\, {}^{G}\mathbf{v}(t) = \ddot{x}\hat{\imath} + \ddot{y}\hat{\jmath} + \ddot{z}\hat{k} \tag{3.11}$$

$$^{G}\mathbf{j} = \frac{^{G}d}{dt}\, {}^{G}\mathbf{a}(t) = \dddot{x}\hat{\imath} + \dddot{y}\hat{\jmath} + \dddot{z}\hat{k} \tag{3.12}$$

■

Example 179 Cartesian Kinematic Vectors Consider a point that moves with the position vector $\mathbf{r}$,

$$\mathbf{r} = ae^{-t^2} \sin \omega t \, \hat{\imath} + \tan^{-1}\left(\frac{1}{t^3}\right) \hat{\jmath} + t \cosh \omega t \, \hat{k} \tag{3.13}$$

where a and ω are constants. The velocity, acceleration, and jerk of the point are

$$\begin{aligned} \mathbf{v} &= \frac{d}{dt}\mathbf{r} = ae^{-t^2} (\omega \cos \omega t - 2t \sin \omega t) \, \hat{\imath} - 3\frac{t^2}{t^6+1} \hat{\jmath} \\ &\quad + (\cosh \omega t + \omega t \sinh \omega t) \, \hat{k} \end{aligned} \tag{3.14}$$

$$\begin{aligned} \mathbf{a} &= \frac{d}{dt}\mathbf{v} = -ae^{-t^2} \left(2 \sin \omega t - 4t^2 \sin \omega t + \omega^2 \sin \omega t + 4\omega t \cos \omega t\right) \hat{\imath} \\ &\quad + 6\frac{t\left(2t^6 - 1\right)}{\left(t^6+1\right)^2} \hat{\jmath} + \left(2\omega \sinh \omega t + \omega^2 t \cosh \omega t\right) \hat{k} \end{aligned} \tag{3.15}$$

$$\begin{aligned} \mathbf{j} &= ae^{-t^2} \left(\left(6\omega^2 t - 8t^3 + 12t\right) \sin \omega t - \left(\omega^3 - 12\omega t^2 + 6\omega\right) \cos \omega t\right) \hat{\imath} \\ &\quad - 6\frac{10t^{12} - 25t^6 + 1}{\left(t^6+1\right)^3} \hat{\jmath} + \left(3\omega^2 \cosh \omega t + \omega^3 t \sinh \omega t\right) \hat{k} \end{aligned} \tag{3.16}$$

Example 180 A Planar Spiral Motion A point P is moving on a curve with the following parametric equations:

$$\begin{aligned} x &= a\left(1 - e^{-t}\right) \sin\left(bt^2\right) \\ y &= a\left(1 - e^{-t}\right) \cos\left(bt^2\right) \\ z &= 0 \end{aligned} \tag{3.17}$$

Remembering that $x = a - a \sin(bt)$, $y = a - a \cos(bt)$ indicates a uniform circular motion with angular velocity $\omega = b$ and radius $R = a$ about the center $C(a, a)$, we may say that Equation (3.17) indicates a circular path about $C(a, a)$ with a shrinking radius $R = a(1 - e^{-t})$ and an increasing angular velocity $\omega = bt$. Figures 3.2 illustrates the

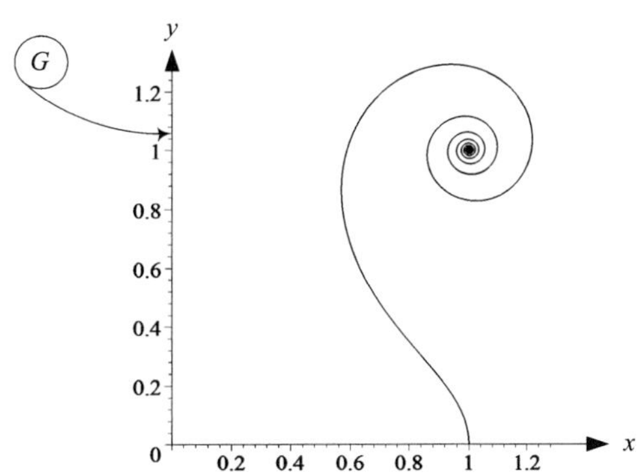

Figure 3.2 The path of the spiral motion in the (x, y)-plane.

path of motion in the (x, y)-plane for

$$a = 1 \qquad b = 2 \tag{3.18}$$

The velocity and acceleration components of the point are

$$\dot{x} = ae^{-t}\sin\left(bt^2\right) + 2abt\left(1 - e^{-t}\right)\cos\left(bt^2\right) \tag{3.19}$$

$$\dot{y} = ae^{-t}\cos\left(bt^2\right) - 2abt\left(1 - e^{-t}\right)\sin\left(bt^2\right) \tag{3.20}$$

$$\ddot{x} = 2ab\left[1 - (1 - 2t)\,e^{-t}\right]\cos bt^2 + a\left[4b^2t^2\left(e^{-t} - 1\right) - e^{-t}\right]\sin bt^2 \tag{3.21}$$

$$\ddot{y} = a\left[4b^2t^2\left(e^{-t} - 1\right) - e^{-t}\right]\cos bt^2 - 2ab\left[1 - (1 - 2t)\,e^{-t}\right]\sin bt^2 \tag{3.22}$$

Figures 3.3–3.5 depict the velocity, acceleration, and jerk of the motion. The final destination $C(x, y) = C(1, 1)$ is the equilibrium point at which the point approaches asymptotically and stays there because the velocity and acceleration curves are approaching zero while P approaches the equilibrium point.

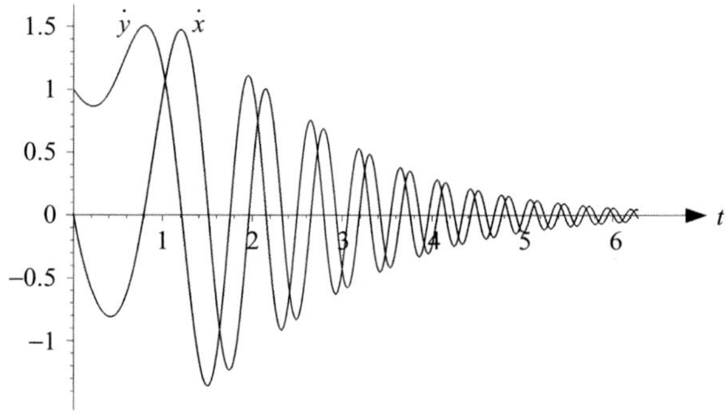

Figure 3.3 Velocity components of moving point P on the spiral motion.

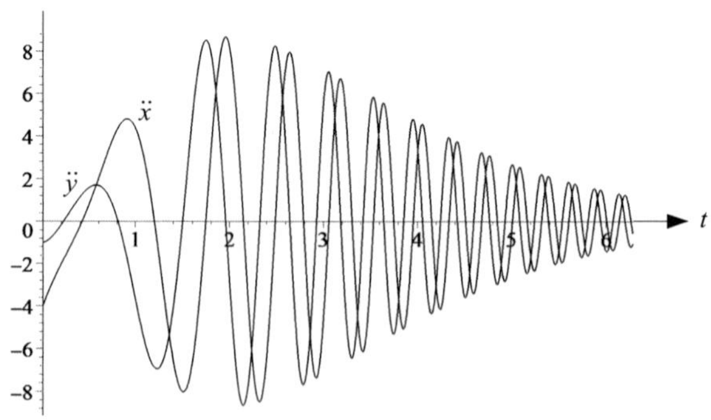

Figure 3.4 Acceleration components of moving point P on the spiral motion.

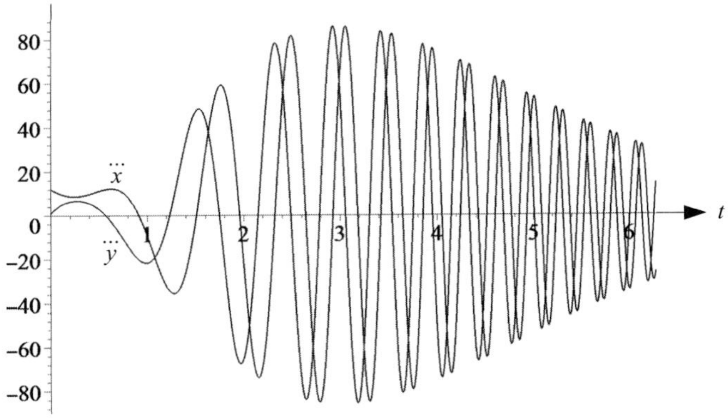

Figure 3.5 Jerk components of moving point P on the spiral motion.

Example 181 Hypertrochoid When a circle is rolling on a straight line, any point of the circle moves on a *cycloid* path. If the radius of the rolling circle on the x-axis is b and a point P is attached to the circle at a distance c, then the parametric equations of the path of P are

$$x = bt - c\sin t \qquad y = b - c\cos t \tag{3.23}$$

The curve that is traced by P is called *ordinary cycloid* if $c = b$, *prolate cycloid* if $c > b$, and *curtate cycloid* if $c < b$. Figure 3.6 illustrates a rolling circle with radius $b = 1$ and three cycloids for $c = 0.4$, $c = 1$, and $c = 1.6$.

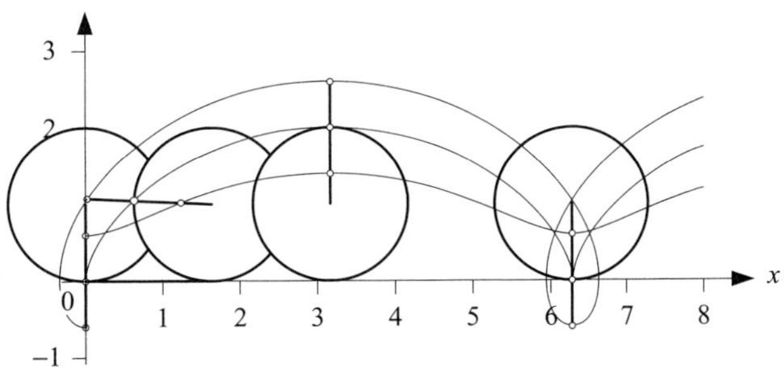

Figure 3.6 A rolling circle with radius $b = 1$ and three cycloids for $c = 1$, $c = 0.4$, and $c = 1.6$.

The *trochoid* is the circular equivalence of the cycloid for which the moving circle rolls on another fixed circle. When the moving circle rolls around inside the fixed circle, the path of the points of the rolling circle is called the *hypotrochoid*, and when the moving circle rolls around outside the circle, the path of the points of the rolling circle is called the *epitrochoid*. Consider a fixed circle with radius a, a rolling circle with radius b, and a fixed arm extending from the center of the rolling circle with length c.

The parametric equations of the path of the tip point of the arm are

$$x = (a - b)\cos t + c\cos\left(\frac{a-b}{b}t\right) \tag{3.24}$$

$$y = (a - b)\sin t - c\sin\left(\frac{a-b}{b}t\right) \tag{3.25}$$

Figure 3.7 illustrates a rolling circle with radius $b = 1$ inside a fixed circle with radius $a = 2$ and four trochoids for $c = 0$, $c = 0.4$, $c = 1$, and $c = 1.6$.

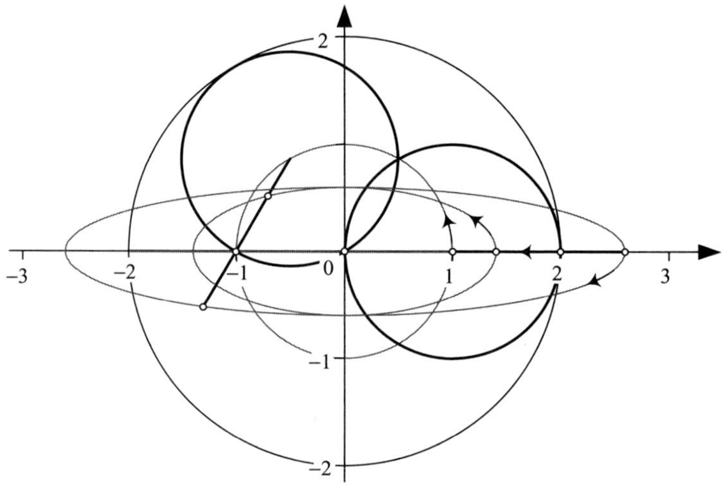

Figure 3.7 A rolling circle with radius $b = 1$ inside a fixed circle with radius $a = 2$ and four trochoids for $c = 0$, $c = 0.4$, $c = 1$, and $c = 1.6$.

The velocity and acceleration components of the point are

$$\dot{x} = -\frac{a-b}{b}\left[c\sin\left(\frac{a-b}{b}t\right) + b\sin t\right] \tag{3.26}$$

$$\dot{y} = -\frac{a-b}{b}\left[c\cos\left(\frac{a-b}{b}t\right) - b\cos t\right] \tag{3.27}$$

$$\ddot{x} = -\frac{a-b}{b^2}\left[b^2\cos t + c\,(a-b)\cos\left(\frac{a-b}{b}t\right)\right] \tag{3.28}$$

$$\ddot{y} = -\frac{a-b}{b^2}\left[b^2\sin t - c\,(a-b)\sin\left(\frac{a-b}{b}t\right)\right] \tag{3.29}$$

In the case

$$a = 2b = 2c \tag{3.30}$$

the equations simplify to

$$x = b\cos t \qquad y = 0 \tag{3.31}$$

which show that a point on the periphery of the rolling circle has a harmonic rectilinear motion. Such a motion may be used to design a device with a required pure harmonic motion. Figure 3.8 illustrates the idea.

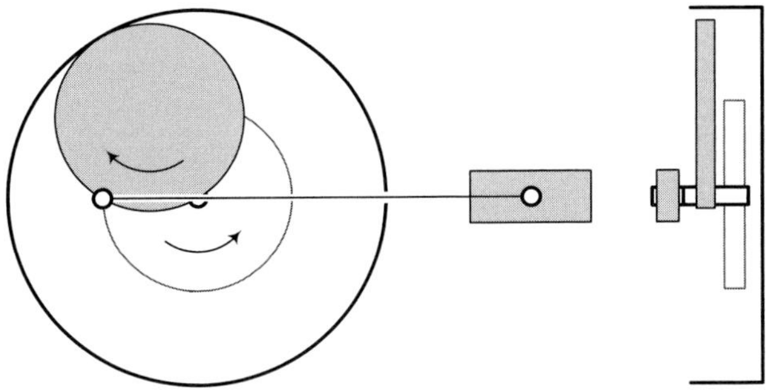

Figure 3.8 A rolling circle with radius b inside a fixed circle with radius $a = 2b$ provides a rectilinear harmonic motion.

Example 182 Ferris Wheel A *Ferris wheel* is made up of an upright wheel with passenger gondolas attached to the rim and hanging under gravity. A Ferris wheel is also known as an observation wheel and a big wheel. Figure 3.9 illustrates a Ferris wheel that has a rim with radius a and hanging gondolas with length b. The parametric equations of the path of a nonswinging passenger are

$$x = a\cos\theta \qquad y = a\sin\theta - b \tag{3.32}$$

which show a passenger will move on a circle with a center at $C(0, -b)$:

$$x^2 + (y + b)^2 = a^2 \tag{3.33}$$

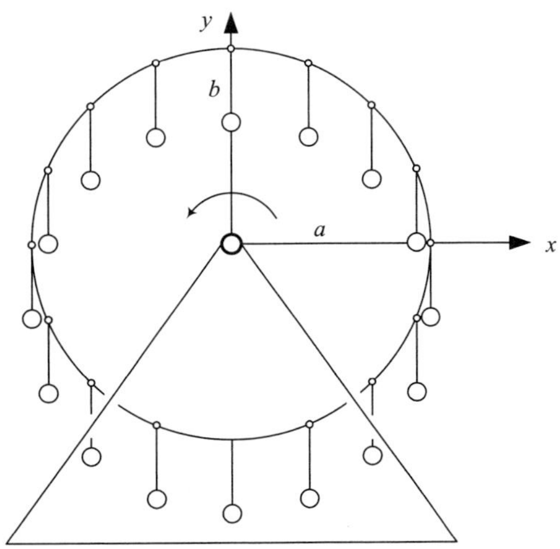

Figure 3.9 A Ferris wheel that has a rim with radius a and hanging gondolas with length b.

If we assume the passenger as a point mass at its mass center, its path of motion would be circular with an angular velocity $\dot{\theta}$. The passenger as a rigid body will not have any angular velocity about its mass center.

3.2 CYLINDRICAL COORDINATE SYSTEM

In a cylindrical coordinate system, the coordinate of a point is at the intersection of two planes and a cylinder, as shown in Figure 3.10. The coordinates of cylindrical and Cartesian coordinate systems are related by

$$x = \rho\cos\theta \qquad y = \rho\sin\theta \qquad z = z \tag{3.34}$$

or inversely by

$$\rho^2 = x^2 + y^2 \qquad \theta = \tan^{-1}\frac{y}{x} \qquad z = z \tag{3.35}$$

The position, velocity, and acceleration of a moving point in a cylindrical coordinate system are

$$\mathbf{r} = \rho\hat{u}_\rho + \dot{z}\hat{u}_z \tag{3.36}$$

$$\mathbf{v} = \dot{\rho}\hat{u}_\rho + \rho\dot{\theta}\hat{u}_\theta + \dot{z}\hat{u}_z \tag{3.37}$$

$$\mathbf{a} = \left(\ddot{\rho} - \rho\dot{\theta}^2\right)\hat{u}_\rho + \left(2\dot{\rho}\dot{\theta} + \rho\ddot{\theta}\right)\hat{u}_\theta + \ddot{z}\hat{u}_z \tag{3.38}$$

Proof: Starting with a Cartesian position vector

$$\mathbf{r} = x\hat{\imath} + y\hat{\jmath} + z\hat{k} \tag{3.39}$$

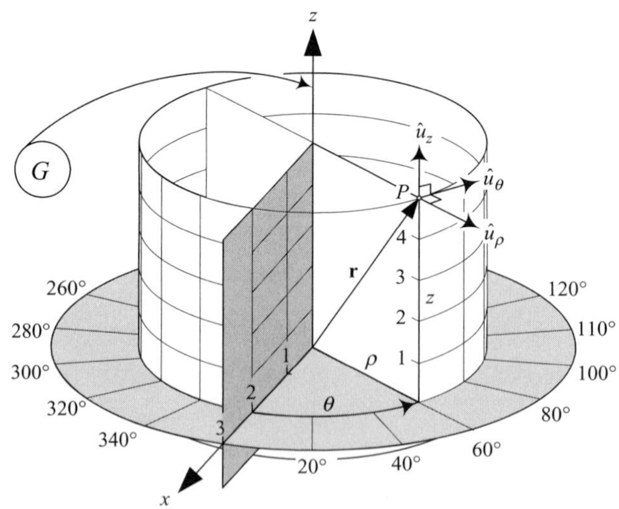

Figure 3.10 Cylindrical coordinate system.

we substitute the Cartesian–cylindrical relations (3.34) to find the Cartesian position vector with cylindrical components

$$\mathbf{r} = \rho\cos\theta\,\hat{\imath} + \rho\sin\theta\,\hat{\jmath} + z\,\hat{k} = \begin{bmatrix} \rho\cos\theta \\ \rho\sin\theta \\ z \end{bmatrix} \tag{3.40}$$

Employing the definition of unit vectors in (1.200) we find the unit vectors of the cylindrical system expressed in the Cartesian system,

$$\hat{u}_\rho = \frac{\partial\mathbf{r}/\partial\rho}{|\partial\mathbf{r}/\partial\rho|} = \frac{\cos\theta\,\hat{\imath} + \sin\theta\,\hat{\jmath}}{1} = \cos\theta\,\hat{\imath} + \sin\theta\,\hat{\jmath} = \begin{bmatrix} \cos\theta \\ \sin\theta \\ 0 \end{bmatrix} \tag{3.41}$$

$$\hat{u}_\theta = \frac{\partial\mathbf{r}/\partial\theta}{|\partial\mathbf{r}/\partial\theta|} = \frac{-\rho\sin\theta\,\hat{\imath} + \rho\cos\theta\,\hat{\jmath}}{\rho} = -\sin\theta\,\hat{\imath} + \cos\theta\,\hat{\jmath} = \begin{bmatrix} -\sin\theta \\ \cos\theta \\ 0 \end{bmatrix} \tag{3.42}$$

$$\hat{u}_z = \frac{\partial\mathbf{r}/\partial z}{|\partial\mathbf{r}/\partial z|} = \frac{\hat{k}}{1} = \hat{k} = \begin{bmatrix} 0 \\ 0 \\ 1 \end{bmatrix} \tag{3.43}$$

Using the coordinate orthogonality condition (3.2) we can transform the position vector from a Cartesian expression to a cylindrical expression,

$$\begin{aligned} \mathbf{r} &= (\mathbf{r}\cdot\hat{u}_\rho)\hat{u}_\rho + (\mathbf{r}\cdot\hat{u}_\theta)\hat{u}_\theta + (\mathbf{r}\cdot\hat{u}_z)\hat{u}_z \\ &= \rho\hat{u}_\rho + z\hat{k} \end{aligned} \tag{3.44}$$

where

$$\mathbf{r}\cdot\hat{u}_\rho = \begin{bmatrix} \rho\cos\theta \\ \rho\sin\theta \\ z \end{bmatrix} \cdot \begin{bmatrix} \cos\theta \\ \sin\theta \\ 0 \end{bmatrix} = \rho \tag{3.45}$$

$$\mathbf{r}\cdot\hat{u}_\theta = \begin{bmatrix} \rho\cos\theta \\ \rho\sin\theta \\ z \end{bmatrix} \cdot \begin{bmatrix} -\sin\theta \\ \cos\theta \\ 0 \end{bmatrix} = 0 \tag{3.46}$$

$$\mathbf{r}\cdot\hat{u}_z = \begin{bmatrix} \rho\cos\theta \\ \rho\sin\theta \\ z \end{bmatrix} \cdot \begin{bmatrix} 0 \\ 0 \\ 1 \end{bmatrix} = z \tag{3.47}$$

To find the velocity vector $\mathbf{v} = d\mathbf{r}/dt$, we take a derivative from the Cartesian expression of the position vector (3.40),

$$\begin{aligned} \mathbf{v} &= \left(\dot{\rho}\cos\theta - \rho\dot{\theta}\sin\theta\right)\hat{\imath} + \left(\dot{\rho}\sin\theta + \rho\dot{\theta}\cos\theta\right)\hat{\jmath} + \dot{z}\hat{k} \\ &= \begin{bmatrix} \dot{\rho}\cos\theta - \rho\dot{\theta}\sin\theta \\ \dot{\rho}\sin\theta + \rho\dot{\theta}\cos\theta \\ \dot{z} \end{bmatrix} \end{aligned} \tag{3.48}$$

and employ the coordinate orthogonality condition (3.2) to find the velocity vector in a cylindrical system,

$$\begin{aligned}\mathbf{v} &= (\mathbf{v}\cdot\hat{u}_\rho)\hat{u}_\rho + (\mathbf{v}\cdot\hat{u}_\theta)\hat{u}_\theta + (\mathbf{v}\cdot\hat{u}_z)\hat{u}_z \\ &= \dot{\rho}\hat{u}_\rho + \rho\dot{\theta}\hat{u}_\theta + \dot{z}\hat{k}\end{aligned} \tag{3.49}$$

where

$$\mathbf{v}\cdot\hat{u}_\rho = \begin{bmatrix}\dot{\rho}\cos\theta - \rho\dot{\theta}\sin\theta \\ \dot{\rho}\sin\theta + \rho\dot{\theta}\cos\theta \\ \dot{z}\end{bmatrix}\cdot\begin{bmatrix}\cos\theta \\ \sin\theta \\ 0\end{bmatrix} = \dot{\rho} \tag{3.50}$$

$$\mathbf{v}\cdot\hat{u}_\theta = \begin{bmatrix}\dot{\rho}\cos\theta - \rho\dot{\theta}\sin\theta \\ \dot{\rho}\sin\theta + \rho\dot{\theta}\cos\theta \\ \dot{z}\end{bmatrix}\cdot\begin{bmatrix}-\sin\theta \\ \cos\theta \\ 0\end{bmatrix} = \rho\dot{\theta} \tag{3.51}$$

$$\mathbf{v}\cdot\hat{u}_z = \begin{bmatrix}\dot{\rho}\cos\theta - \rho\dot{\theta}\sin\theta \\ \dot{\rho}\sin\theta + \rho\dot{\theta}\cos\theta \\ \dot{z}\end{bmatrix}\cdot\begin{bmatrix}0 \\ 0 \\ 1\end{bmatrix} = \dot{z} \tag{3.52}$$

To find the acceleration vector $\mathbf{a} = d\mathbf{v}/dt$, we may take a derivative from the Cartesian expression of the velocity vector (3.48),

$$\mathbf{a} = \begin{bmatrix}\left(\ddot{\rho} - \rho\dot{\theta}^2\right)\cos\theta - \left(\rho\ddot{\theta} + 2\dot{\rho}\dot{\theta}\right)\sin\theta \\ \left(\ddot{\rho} - \rho\dot{\theta}^2\right)\sin\theta + \left(\rho\ddot{\theta} + 2\dot{\rho}\dot{\theta}\right)\cos\theta \\ \ddot{z}\end{bmatrix} \tag{3.53}$$

and employ the coordinate orthogonality condition (3.2) to find the acceleration vector in a cylindrical system,

$$\begin{aligned}\mathbf{a} &= (\mathbf{a}\cdot\hat{u}_\rho)\hat{u}_\rho + (\mathbf{a}\cdot\hat{u}_\theta)\hat{u}_\theta + (\mathbf{a}\cdot\hat{k})\hat{k} \\ &= \left(\ddot{\rho} - \rho\dot{\theta}^2\right)\hat{u}_\rho + \left(\rho\ddot{\theta} + 2\dot{\rho}\dot{\theta}\right)\hat{u}_\theta + \ddot{z}\hat{k}\end{aligned} \tag{3.54}$$

where

$$\begin{aligned}\mathbf{a}\cdot\hat{u}_\rho &= \begin{bmatrix}\left(\ddot{\rho} - \rho\dot{\theta}^2\right)\cos\theta - \left(\rho\ddot{\theta} + 2\dot{\rho}\dot{\theta}\right)\sin\theta \\ \left(\ddot{\rho} - \rho\dot{\theta}^2\right)\sin\theta + \left(\rho\ddot{\theta} + 2\dot{\rho}\dot{\theta}\right)\cos\theta \\ \ddot{z}\end{bmatrix}\cdot\begin{bmatrix}\cos\theta \\ \sin\theta \\ 0\end{bmatrix} \\ &= \ddot{\rho} - \rho\dot{\theta}^2\end{aligned} \tag{3.55}$$

$$\begin{aligned}\mathbf{a}\cdot\hat{u}_\theta &= \begin{bmatrix}\left(\ddot{\rho} - \rho\dot{\theta}^2\right)\cos\theta - \left(\rho\ddot{\theta} + 2\dot{\rho}\dot{\theta}\right)\sin\theta \\ \left(\ddot{\rho} - \rho\dot{\theta}^2\right)\sin\theta + \left(\rho\ddot{\theta} + 2\dot{\rho}\dot{\theta}\right)\cos\theta \\ \ddot{z}\end{bmatrix}\cdot\begin{bmatrix}-\sin\theta \\ \cos\theta \\ 0\end{bmatrix} \\ &= \rho\ddot{\theta} + 2\dot{\rho}\dot{\theta}\end{aligned} \tag{3.56}$$

$$\begin{aligned}\mathbf{a}\cdot\hat{k} &= \begin{bmatrix}\left(\ddot{\rho} - \rho\dot{\theta}^2\right)\cos\theta - \left(\rho\ddot{\theta} + 2\dot{\rho}\dot{\theta}\right)\sin\theta \\ \left(\ddot{\rho} - \rho\dot{\theta}^2\right)\sin\theta + \left(\rho\ddot{\theta} + 2\dot{\rho}\dot{\theta}\right)\cos\theta \\ \ddot{z}\end{bmatrix}\cdot\begin{bmatrix}0 \\ 0 \\ 1\end{bmatrix} \\ &= \ddot{z}\end{aligned} \tag{3.57}$$

Taking a derivative from (3.53) and using the coordinate orthogonality condition, we can also find the jerk vector **j** in the cylindrical coordinate system:

$$\begin{aligned}\mathbf{j} &= (\mathbf{j}\cdot\hat{u}_\rho)\hat{u}_\rho + (\mathbf{j}\cdot\hat{u}_\theta)\hat{u}_\theta + (\mathbf{j}\cdot\hat{k})\hat{k} \\ &= \left(\dddot{\rho} - 3\dot{\rho}\dot{\theta}^2 - 3\rho\dot{\theta}\ddot{\theta}\right)\hat{u}_\rho + \left(\rho\dddot{\theta} + 3\ddot{\rho}\dot{\theta} + 3\dot{\rho}\ddot{\theta} - \rho\dot{\theta}^3\right)\hat{u}_\theta + \dddot{z}\hat{k}\end{aligned} \tag{3.58}$$

■

Example 183 Orthogonality of the Cylindrical Coordinate System The cylindrical coordinate system with unit vectors $\hat{u}_\rho$, $\hat{u}_\theta$, and $\hat{k}$ is an orthogonal right-handed coordinate system because

$$\hat{u}_\rho \cdot \hat{u}_\theta = 0 \qquad \hat{u}_\theta \cdot \hat{k} = 0 \qquad \hat{k} \cdot \hat{u}_\rho = 0 \tag{3.59}$$

and

$$\hat{u}_\rho \times \hat{u}_\theta = \hat{k} \qquad \hat{u}_\theta \times \hat{k} = \hat{u}_\rho \qquad \hat{k} \times \hat{u}_\rho = \hat{u}_\theta \tag{3.60}$$

Example 184 Alternative Method for Cylindrical Kinematics An alternative and more practical method to find the kinematic vectors is to find the position vector in a cylindrical coordinate using the orthogonal coordinate condition

$$\mathbf{r} = \rho\hat{u}_\rho + z\hat{k} \tag{3.61}$$

and take the derivative from this equation,

$$\mathbf{v} = \dot{\rho}\hat{u}_\rho + \rho\frac{d}{dt}\hat{u}_\rho + \dot{z}\hat{k} + z\frac{d}{dt}\hat{k} \tag{3.62}$$

$$\mathbf{a} = \ddot{\rho}\hat{u}_\rho + 2\dot{\rho}\frac{d}{dt}\hat{u}_\rho + \rho\frac{d^2}{dt^2}\hat{u}_\rho + \ddot{z}\hat{k} + 2\dot{z}\frac{d}{dt}\hat{k} + z\frac{d^2}{dt^2}\hat{k} \tag{3.63}$$

To simplify these equations, we need to find the derivative of the unit vectors. As derived in Equations (3.41)–(3.43), the Cartesian expressions of the unit vectors $\hat{u}_\rho$, $\hat{u}_\theta$, $\hat{u}_z$ with components in a cylindrical system are

$$\hat{u}_\rho = \cos\theta\hat{\imath} + \sin\theta\hat{\jmath} = \begin{bmatrix}\cos\theta \\ \sin\theta \\ 0\end{bmatrix} \tag{3.64}$$

$$\hat{u}_\theta = -\sin\theta\hat{\imath} + \cos\theta\hat{\jmath} = \begin{bmatrix}-\sin\theta \\ \cos\theta \\ 0\end{bmatrix} \tag{3.65}$$

$$\hat{u}_z = \hat{k} = \begin{bmatrix}0 \\ 0 \\ 1\end{bmatrix} \tag{3.66}$$

Taking derivatives shows that

$$\frac{d}{dt}\hat{u}_\rho = -\dot{\theta}\sin\theta\hat{\imath} + \dot{\theta}\cos\theta\hat{\jmath} = \begin{bmatrix} -\dot{\theta}\sin\theta \\ \dot{\theta}\cos\theta \\ 0 \end{bmatrix} \tag{3.67}$$

$$\frac{d}{dt}\hat{u}_\theta = -\dot{\theta}\cos\theta\hat{\imath} - \dot{\theta}\sin\theta\hat{\jmath} = \begin{bmatrix} -\dot{\theta}\cos\theta \\ -\dot{\theta}\sin\theta \\ 0 \end{bmatrix} \tag{3.68}$$

$$\frac{d}{dt}\hat{u}_z = \begin{bmatrix} 0 \\ 0 \\ 0 \end{bmatrix} \tag{3.69}$$

Now, employing the orthogonality condition (3.2), we find the time derivatives of the unit vectors $\hat{u}_\rho$, $\hat{u}_\theta$, $\hat{u}_z$ in a cylindrical system:

$$\frac{d}{dt}\hat{u}_\rho = \left(\frac{d}{dt}\hat{u}_\rho \cdot \hat{u}_\rho\right)\hat{u}_\rho + \left(\frac{d}{dt}\hat{u}_\rho \cdot \hat{u}_\theta\right)\hat{u}_\theta + \left(\frac{d}{dt}\hat{u}_\rho \cdot \hat{u}_z\right)\hat{u}_z = \dot{\theta}\hat{u}_\theta \tag{3.70}$$

$$\frac{d}{dt}\hat{u}_\theta = \left(\frac{d}{dt}\hat{u}_\theta \cdot \hat{u}_\rho\right)\hat{u}_\rho + \left(\frac{d}{dt}\hat{u}_\theta \cdot \hat{u}_\theta\right)\hat{u}_\theta + \left(\frac{d}{dt}\hat{u}_\theta \cdot \hat{u}_z\right)\hat{u}_z = -\dot{\theta}\hat{u}_\rho \tag{3.71}$$

$$\frac{d}{dt}\hat{u}_z = \left(\frac{d}{dt}\hat{u}_z \cdot \hat{u}_\rho\right)\hat{u}_\rho + \left(\frac{d}{dt}\hat{u}_z \cdot \hat{u}_\theta\right)\hat{u}_\theta + \left(\frac{d}{dt}\hat{u}_z \cdot \hat{u}_z\right)\hat{u}_z = 0 \tag{3.72}$$

The time derivatives of unit vectors make a closed set, which is enough to find the derivatives of the position vector $\mathbf{r}$ in a cylindrical system as many times as we wish. So, the velocity, acceleration, and jerk vectors would be

$$\begin{aligned}
\mathbf{v} &= \frac{d}{dt}\mathbf{r} = \frac{d}{dt}\left(\rho\hat{u}_\rho + z\hat{k}\right) = \dot{\rho}\hat{u}_\rho + \rho\frac{d}{dt}\hat{u}_\rho + \dot{z}\hat{k} + z\frac{d}{dt}\hat{k} \\
&= \dot{\rho}\hat{u}_\rho + \rho\dot{\theta}\hat{u}_\theta + \dot{z}\hat{k}
\end{aligned} \tag{3.73}$$

$$\begin{aligned}
\mathbf{a} &= \frac{d}{dt}\mathbf{v} = \frac{d}{dt}\left(\dot{\rho}\hat{u}_\rho + \rho\dot{\theta}\hat{u}_\theta + \dot{z}\hat{k}\right) \\
&= \ddot{\rho}\hat{u}_\rho + \dot{\rho}\frac{d}{dt}\hat{u}_\rho + \dot{\rho}\dot{\theta}\hat{u}_\theta + \rho\ddot{\theta}\hat{u}_\theta + \rho\dot{\theta}\frac{d}{dt}\hat{u}_\theta + \ddot{z}\hat{k} + \dot{z}\frac{d}{dt}\hat{k} \\
&= \ddot{\rho}\hat{u}_\rho + \dot{\rho}\dot{\theta}\hat{u}_\theta + \dot{\rho}\dot{\theta}\hat{u}_\theta + \rho\ddot{\theta}\hat{u}_\theta - \rho\dot{\theta}^2\hat{u}_\rho + \ddot{z}\hat{k} \\
&= \left(\ddot{\rho} - \rho\dot{\theta}^2\right)\hat{u}_\rho + \left(\rho\ddot{\theta} + 2\dot{\rho}\dot{\theta}\right)\hat{u}_\theta + \ddot{z}\hat{k}
\end{aligned} \tag{3.74}$$

$$\begin{aligned}
\mathbf{j} &= \frac{d}{dt}\mathbf{a} = \frac{d}{dt}\left(\left(\ddot{\rho} - \rho\dot{\theta}^2\right)\hat{u}_\rho + \left(\rho\ddot{\theta} + 2\dot{\rho}\dot{\theta}\right)\hat{u}_\theta + \ddot{z}\hat{k}\right) \\
&= \left(\dddot{\rho} - \dot{\rho}\dot{\theta}^2 - 2\rho\dot{\theta}\ddot{\theta}\right)\hat{u}_\rho + \left(\ddot{\rho} - \rho\dot{\theta}^2\right)\dot{\theta}\hat{u}_\theta \\
&\quad + \left(\dot{\rho}\ddot{\theta} + \rho\dddot{\theta} + 2\ddot{\rho}\dot{\theta} + 2\dot{\rho}\ddot{\theta}\right)\hat{u}_\theta - \left(\rho\ddot{\theta} + 2\dot{\rho}\dot{\theta}\right)\dot{\theta}\hat{u}_\rho + \dddot{z}\hat{k} \\
&= \left(\dddot{\rho} - 3\dot{\rho}\dot{\theta}^2 - 3\rho\dot{\theta}\ddot{\theta}\right)\hat{u}_\rho + \left(\rho\dddot{\theta} + 3\ddot{\rho}\dot{\theta} + 3\dot{\rho}\ddot{\theta} - \rho\dot{\theta}^3\right)\hat{u}_\theta + \dddot{z}\hat{k}
\end{aligned} \tag{3.75}$$

These equations are compatible with Equations (3.49) and (3.54).

Example 185 A Constant-Length Vector and Its Derivative Are Perpendicular If the beginning point of a unit vector is fixed, the only path of motion for its tip point is turning on a unit circle. Therefore, the derivative of the unit vector would be a tangent vector to the unit circle in the direction of motion of the tip point. The length of the derivative is equal to the instantaneous angular velocity of the unit vector. These facts are true for any unit vector and derivative with respect to any parameter.

As an example, consider Figure 3.11, which illustrates the intersection of the cylindrical coordinate system with the (x, y)-plane. In a cylindrical coordinate system, we have the unit vectors $\hat{u}_\rho$, $\hat{u}_\theta$, and $\hat{u}_z = \hat{k}$ as

$$\hat{u}_\rho = \begin{bmatrix} \cos\theta \\ \sin\theta \\ 0 \end{bmatrix} \qquad \hat{u}_\theta = \begin{bmatrix} -\sin\theta \\ \cos\theta \\ 0 \end{bmatrix} \qquad \hat{u}_z = \begin{bmatrix} 0 \\ 0 \\ 1 \end{bmatrix} \tag{3.76}$$

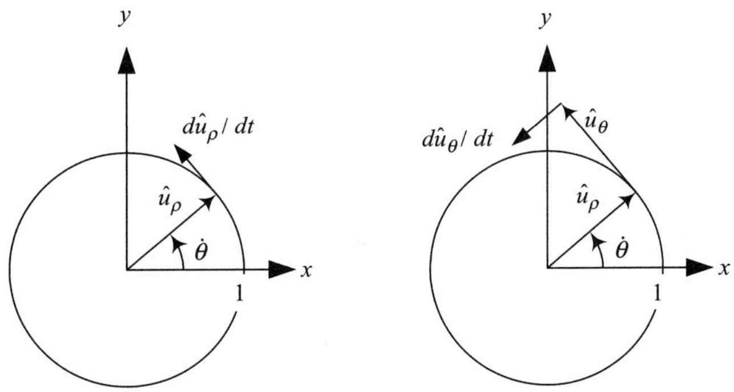

Figure 3.11 Illustration of the unit vectors $\hat{u}_\rho$, $\hat{u}_\theta$ and their time derivatives.

Their derivatives are

$$\frac{d\hat{u}_\rho}{dt} = \begin{bmatrix} -\dot{\theta}\sin\theta \\ \dot{\theta}\cos\theta \\ 0 \end{bmatrix} \qquad \frac{d\hat{u}_\theta}{dt} = \begin{bmatrix} -\dot{\theta}\cos\theta \\ -\dot{\theta}\sin\theta \\ 0 \end{bmatrix} \qquad \frac{d\hat{u}_z}{dt} = \begin{bmatrix} 0 \\ 0 \\ 0 \end{bmatrix} \tag{3.77}$$

and are perpendicular to their associated vectors:

$$\hat{u}_\rho \cdot \frac{d}{dt}\hat{u}_\rho = \begin{bmatrix} \cos\theta \\ \sin\theta \\ 0 \end{bmatrix} \cdot \begin{bmatrix} -\dot{\theta}\sin\theta \\ \dot{\theta}\cos\theta \\ 0 \end{bmatrix} = 0 \tag{3.78}$$

$$\hat{u}_\theta \cdot \frac{d}{dt}\hat{u}_\theta = \begin{bmatrix} -\sin\theta \\ \cos\theta \\ 0 \end{bmatrix} \cdot \begin{bmatrix} -\dot{\theta}\cos\theta \\ -\dot{\theta}\sin\theta \\ 0 \end{bmatrix} = 0 \tag{3.79}$$

$$\hat{u}_z \cdot \frac{d}{dt}\hat{u}_z = \begin{bmatrix} 0 \\ 0 \\ 1 \end{bmatrix} \cdot \begin{bmatrix} 0 \\ 0 \\ 0 \end{bmatrix} = 0 \tag{3.80}$$

Because $\hat{u}_\rho$ and $\hat{u}_\theta$ are turning together with the same angular velocity, the lengths of $(d/dt)\hat{u}_\rho$ and $(d/dt)\hat{u}_\theta$ must be equal and show the rate of rotation of $\hat{u}_\rho$ and $\hat{u}_\theta$:

$$\left|\frac{d}{dt}\hat{u}_\rho\right| = \left|\frac{d}{dt}\hat{u}_\theta\right| = \dot{\theta} \tag{3.81}$$

To show that the perpendicularity of a constant-length vector and its derivative is general, we take the derivative from $r^2 = \mathbf{r} \cdot \mathbf{r}$:

$$\frac{d}{dt}\left(r^2\right) = \frac{d}{dt}\left(\mathbf{r} \cdot \mathbf{r}\right) \tag{3.82}$$

The left-hand side is a scalar and therefore

$$0 = 2\mathbf{r} \cdot \frac{d}{dt}\mathbf{r} \tag{3.83}$$

which shows that $\mathbf{r}$ and $(d/dt)\mathbf{r}$ are perpendicular.

Example 186 Matrix Transformation Using Equations (3.41)–(3.43) we may arrange the unit vector transformation in a matrix equation:

$$\begin{bmatrix} \hat{u}_\rho \\ \hat{u}_\theta \\ \hat{u}_z \end{bmatrix} = \begin{bmatrix} \cos\theta & \sin\theta & 0 \\ -\sin\theta & \cos\theta & 0 \\ 0 & 0 & 1 \end{bmatrix} \begin{bmatrix} \hat{\imath} \\ \hat{\jmath} \\ \hat{\jmath} \end{bmatrix} \tag{3.84}$$

Therefore, we may determine the Cartesian unit vectors in the cylindrical system by a matrix inversion:

$$\begin{aligned} \begin{bmatrix} \hat{\imath} \\ \hat{\jmath} \\ \hat{\jmath} \end{bmatrix} &= \begin{bmatrix} \cos\theta & \sin\theta & 0 \\ -\sin\theta & \cos\theta & 0 \\ 0 & 0 & 1 \end{bmatrix}^{-1} \begin{bmatrix} \hat{u}_\rho \\ \hat{u}_\theta \\ \hat{u}_z \end{bmatrix} \\ &= \begin{bmatrix} \hat{u}_\rho \cos\theta - \hat{u}_\theta \sin\theta \\ \hat{u}_\theta \cos\theta + \hat{u}_\rho \sin\theta \\ \hat{u}_z \end{bmatrix} \end{aligned} \tag{3.85}$$

The position vector in the two systems may be transformed from cylindrical,

$$\mathbf{r} = \rho\hat{u}_\rho + z\hat{k} = \begin{bmatrix} \rho \\ 0 \\ z \end{bmatrix} \tag{3.86}$$

to Cartesian by

$$\begin{bmatrix} \cos\theta & \sin\theta & 0 \\ -\sin\theta & \cos\theta & 0 \\ 0 & 0 & 1 \end{bmatrix}^{-1} \begin{bmatrix} \rho \\ 0 \\ z \end{bmatrix} = \begin{bmatrix} \rho\cos\theta \\ \rho\sin\theta \\ z \end{bmatrix} = \begin{bmatrix} x \\ y \\ z \end{bmatrix} \tag{3.87}$$

Example 187 ★ Gradient in Cylindrical Coordinate System The gradient is an operator to find the directional derivative of a scalar space function $f = f(x, y, z)$:

$$\text{grad} = \nabla = \frac{\partial}{\partial x}\hat{\imath} + \frac{\partial}{\partial y}\hat{\jmath} + \frac{\partial}{\partial z}\hat{k} \tag{3.88}$$

The gradient operator in a cylindrical coordinate system can be found by employing the orthogonality coordinate condition.

Using the chain rule, we find

$$\nabla = \begin{bmatrix} \dfrac{\partial}{\partial x} \\ \dfrac{\partial}{\partial y} \\ \dfrac{\partial}{\partial z} \end{bmatrix} = \begin{bmatrix} \dfrac{\partial}{\partial \rho}\dfrac{\partial \rho}{\partial x} + \dfrac{\partial}{\partial \theta}\dfrac{\partial \theta}{\partial x} + \dfrac{\partial}{\partial z}\dfrac{\partial z}{\partial x} \\ \dfrac{\partial}{\partial \rho}\dfrac{\partial \rho}{\partial y} + \dfrac{\partial}{\partial \theta}\dfrac{\partial \theta}{\partial y} + \dfrac{\partial}{\partial z}\dfrac{\partial z}{\partial y} \\ \dfrac{\partial}{\partial \rho}\dfrac{\partial \rho}{\partial z} + \dfrac{\partial}{\partial \theta}\dfrac{\partial \theta}{\partial z} + \dfrac{\partial}{\partial z}\dfrac{\partial z}{\partial z} \end{bmatrix} \tag{3.89}$$

Using the Cartesian and cylindrical relations

$$\begin{aligned} &\rho = \sqrt{x^2 + y^2} && \theta = \tan^{-1}\frac{y}{x} && z = z \\ &x = \rho\cos\theta && y = \rho\sin\theta && z = z \end{aligned} \tag{3.90}$$

we have

$$\nabla = \begin{bmatrix} \dfrac{\partial}{\partial x} \\ \dfrac{\partial}{\partial y} \\ \dfrac{\partial}{\partial z} \end{bmatrix} = \begin{bmatrix} \cos\theta\dfrac{\partial}{\partial \rho} - \dfrac{1}{\rho}\sin\theta\dfrac{\partial}{\partial \theta} \\ \sin\theta\dfrac{\partial}{\partial \rho} + \dfrac{1}{\rho}\cos\theta\dfrac{\partial}{\partial \theta} \\ \dfrac{\partial}{\partial z} \end{bmatrix} \tag{3.91}$$

Employing the orthogonality condition, we transform ∇ from Cartesian to cylindrical:

$$\begin{aligned} \nabla &= (\nabla \cdot \hat{u}_\rho)\hat{u}_\rho + (\nabla \cdot \hat{u}_\theta)\hat{u}_\theta + (\nabla \cdot \hat{u}_z)\hat{u}_z \\ &= \frac{\partial}{\partial \rho}\hat{u}_\rho + \frac{1}{\rho}\frac{\partial}{\partial \theta}\hat{u}_\theta + \frac{\partial}{\partial z}\hat{u}_z \end{aligned} \tag{3.92}$$

The unit vectors are given in Equations (3.41)–(3.43).

It is also possible to substitute the unit vector expressions

$$\begin{bmatrix} \hat{\imath} \\ \hat{\jmath} \\ \hat{\jmath} \end{bmatrix} = \begin{bmatrix} \hat{u}_\rho\cos\theta - \hat{u}_\theta\sin\theta \\ \hat{u}_\theta\cos\theta + \hat{u}_\rho\sin\theta \\ \hat{u}_z \end{bmatrix} \tag{3.93}$$

into (3.91) and determine the gradient operator in a cylindrical system:

$$\nabla = \frac{\partial}{\partial x}\hat{\imath} + \frac{\partial}{\partial y}\hat{\jmath} + \frac{\partial}{\partial z}\hat{k} = \begin{bmatrix} \frac{\partial}{\partial x} \\ \frac{\partial}{\partial y} \\ \frac{\partial}{\partial z} \end{bmatrix} \cdot \begin{bmatrix} \hat{\imath} \\ \hat{\jmath} \\ \hat{\jmath} \end{bmatrix}$$

$$= \begin{bmatrix} \hat{u}_\rho \cos\theta - \hat{u}_\theta \sin\theta \\ \hat{u}_\theta \cos\theta + \hat{u}_\rho \sin\theta \\ \hat{u}_z \end{bmatrix} \cdot \begin{bmatrix} \cos\theta \frac{\partial}{\partial \rho} - \frac{1}{\rho}\sin\theta \frac{\partial}{\partial \theta} \\ \sin\theta \frac{\partial}{\partial \rho} + \frac{1}{\rho}\cos\theta \frac{\partial}{\partial \theta} \\ \frac{\partial}{\partial z} \end{bmatrix}$$

$$= \frac{\partial}{\partial \rho}\hat{u}_\rho + \frac{1}{\rho}\frac{\partial}{\partial \theta}\hat{u}_\theta + \frac{\partial}{\partial z}\hat{u}_z \tag{3.94}$$

Example 188 Gradient of a Cylindrical Field The gradient of the field

$$V = \rho^3 z \cos\theta \tag{3.95}$$

is

$$\begin{aligned} \nabla V &= \frac{\partial V}{\partial \rho}\hat{u}_\rho + \frac{1}{\rho}\frac{\partial V}{\partial \theta}\hat{u}_\theta + \frac{\partial V}{\partial z}\hat{u}_z \\ &= 3z\rho^2 \cos\theta \hat{u}_\rho - z\rho^2 \sin\theta \hat{u}_\theta + \rho^3 \cos\theta \hat{u}_z \end{aligned} \tag{3.96}$$

Example 189 Motion on a Cardioid Consider a point P that is moving on a 3D cardioid

$$\mathbf{r} = k\,(1 + \cos\theta)\,\hat{u}_\rho + \theta\hat{u}_\theta + z\hat{u}_z \tag{3.97}$$

where k is a constant and Figure 3.12 shows the cardioid for $k = 1$. The velocity and acceleration of P on the cardioid path are

$$\begin{aligned} \mathbf{v} = \frac{d\mathbf{r}}{dt} &= -\dot{\theta}\sin\theta\hat{u}_\rho + (1 + \cos\theta)\,\frac{d}{dt}\hat{u}_\rho + \dot{\theta}\hat{u}_\theta + \theta\frac{d}{dt}\hat{u}_\theta + \dot{z}\hat{u}_z \\ &= -\dot{\theta}\sin\theta\hat{u}_\rho + (1 + \cos\theta)\,\dot{\theta}\hat{u}_\theta + \dot{\theta}\hat{u}_\theta - \theta\dot{\theta}\hat{u}_\rho + \dot{z}\hat{u}_z \\ &= -\,(\sin\theta + \theta)\,\dot{\theta}\hat{u}_\rho + (2 + \cos\theta)\,\dot{\theta}\hat{u}_\theta + \dot{z}\hat{u}_z \end{aligned} \tag{3.98}$$

$$\begin{aligned} \mathbf{a} = \frac{d\mathbf{v}}{dt} &= -\,(\cos\theta + 1)\,\dot{\theta}^2\hat{u}_\rho - (\sin\theta + \theta)\,\ddot{\theta}\hat{u}_\rho - (\sin\theta + \theta)\,\dot{\theta}^2\hat{u}_\theta \\ &\quad - (\sin\theta)\,\dot{\theta}^2\hat{u}_\theta + (2 + \cos\theta)\,\ddot{\theta}\hat{u}_\theta - (2 + \cos\theta)\,\dot{\theta}^2\hat{u}_\rho + \ddot{z}\hat{u}_z \\ &= -((\sin\theta + \theta)\,\ddot{\theta} + (3 + 2\cos\theta)\,\dot{\theta}^2)\hat{u}_\rho \\ &\quad - ((\sin\theta)\,\dot{\theta}^2 - (2 + \cos\theta)\,\ddot{\theta})\hat{u}_\theta + \ddot{z}\hat{u}_z \end{aligned} \tag{3.99}$$

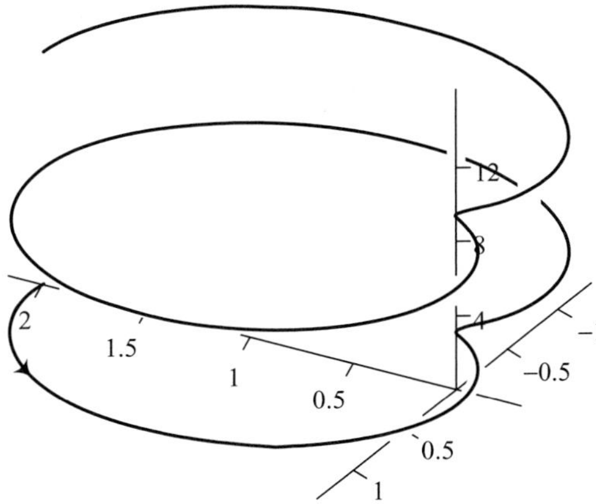

Figure 3.12 A 3D cardioid.

If the point is moving on a planar cardioid with a constant speed v, as is shown in Figure 3.13, then

$$\rho = 1 + \cos\theta \tag{3.100}$$

and its position, velocity, and acceleration would be

$$\mathbf{r} = \rho \hat{u}_\rho = (1 + \cos\theta)\, \hat{u}_\rho \tag{3.101}$$

$$\mathbf{v} = \dot{\rho}\hat{u}_\rho + \rho\dot{\theta}\hat{u}_\theta = -\dot{\theta}\sin\theta\hat{u}_\rho + (1 + \cos\theta)\,\dot{\theta}\hat{u}_\theta \tag{3.102}$$

$$\begin{aligned}\mathbf{a} &= \left(\ddot{\rho} - \rho\dot{\theta}^2\right)\hat{u}_\rho + \left(\rho\ddot{\theta} + 2\dot{\rho}\dot{\theta}\right)\hat{u}_\theta \\ &= -\left(\ddot{\theta}\sin\theta + \dot{\theta}\cos\theta\right)\hat{u}_\rho + \left((1 + \cos\theta)\,\ddot{\theta} - 2\dot{\theta}^2\sin\theta\right)\hat{u}_\theta\end{aligned} \tag{3.103}$$

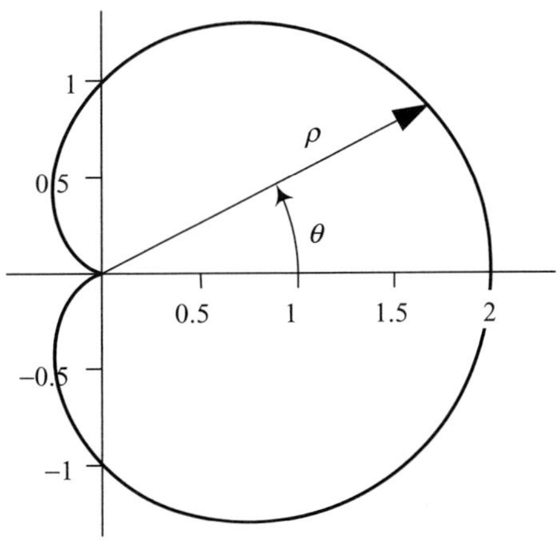

Figure 3.13 A plane cardioid.

However,

$$v = \sqrt{\dot{\rho}^2 + \rho^2\dot{\theta}^2} = \sqrt{2\rho}\dot{\theta} \tag{3.104}$$

and therefore,

$$\dot{\theta} = \frac{v}{\sqrt{2\rho}} \tag{3.105}$$

$$\ddot{\theta} = -\frac{v\dot{\rho}}{2\rho\sqrt{2\rho}} = \frac{v^2 \sin\theta}{4\rho^2} \tag{3.106}$$

which provides

$$\mathbf{v} = -\frac{v}{\sqrt{2\rho}} \sin\theta \hat{u}_\rho + \sqrt{\frac{\rho}{2}} v\hat{u}_\theta \tag{3.107}$$

$$\mathbf{a} = -\frac{3}{4} v\hat{u}_\rho - \frac{3}{4}\frac{\sin\theta}{\rho} v^2 \hat{u}_\theta \tag{3.108}$$

Example 190 Spiral Motions The spiral is a curve that has a periodic rotation with a variable radius. A spiral has an open path and continues to drift outward to infinity or inward to a fixed point as the angle of rotation increases. A spiral is better expressed in cylindrical coordinates as

$$\rho = f(\theta) \tag{3.109}$$

where $f(\theta)$ is a monotonically increasing or decreasing function. The simplest spiral is the *Archimedes spiral*,

$$\rho = a\theta \tag{3.110}$$

which indicates a moving point on a radial line at a constant speed while the line is turning about the origin at a constant angular velocity. If the radial speed of the point is not constant, then the equation of the spiral has the general form

$$\rho = a\theta^n \tag{3.111}$$

where $n = \frac{1}{2}$ indicates the *Fermat spiral*, $n = -1$ indicates the *hyperbolic spiral*, and $n = -\frac{1}{2}$ indicates the *lituus*.

A moving point P on a spiral has the following kinematics:

$$\mathbf{r} = a\theta^n \, \hat{u}_\rho \tag{3.112}$$

$$\mathbf{v} = an\theta^{n-1}\dot{\theta} \, \hat{u}_\rho + a\theta^n\dot{\theta} \, \hat{u}_\theta \tag{3.113}$$

$$\mathbf{a} = a\theta^{n-2} \left\{ n\theta\ddot{\theta} + \left[n(n-1) - \theta^2 \right] \dot{\theta}^2 \right\} \hat{u}_\rho + a\theta^{n-1} \left(\theta\ddot{\theta} + 2n\dot{\theta}^2 \right) \hat{u}_\theta \tag{3.114}$$

When the radial component ρ changes exponentially with θ, the curve is called the *logarithmic spiral*:

$$\rho = c\, e^{a\theta} \tag{3.115}$$

Example 191 ★ Parabolic Cylindrical Coordinate System The parabolic cylindrical coordinate system (u,v,z) relates to the Cartesian system by

$$x = \frac{1}{2}\left(u^2 - v^2\right) \qquad y = uv \qquad z = z \tag{3.116}$$

Figure 3.14 illustrates the intersection of the system with the (x,y)-plane.

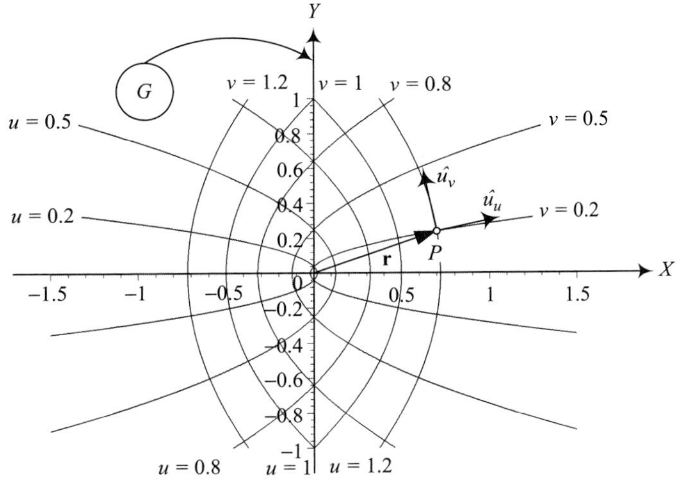

Figure 3.14 Parabolic cylindrical coordinate system.

To find the kinematics of a moving point P in the system, we start with a Cartesian position vector

$$\mathbf{r} = x\hat{\imath} + y\hat{\jmath} + z\hat{k} = \begin{bmatrix} \frac{1}{2}\left(u^2 - v^2\right) \\ uv \\ z \end{bmatrix} \tag{3.117}$$

and employ the definition (1.200) to determine the Cartesian expression of the unit vectors of the parabolic cylindrical system:

$$\hat{u}_u = \frac{\partial \mathbf{r}/\partial u}{|\partial \mathbf{r}/\partial u|} = \frac{u\,\hat{\imath} + v\,\hat{\jmath}}{\sqrt{2}} = \begin{bmatrix} \dfrac{u}{\sqrt{2}} \\ \dfrac{v}{\sqrt{2}} \\ 0 \end{bmatrix} \tag{3.118}$$

$$\hat{u}_v = \frac{\partial \mathbf{r}/\partial v}{|\partial \mathbf{r}/\partial v|} = \frac{-v\,\hat{\imath} + u\,\hat{\jmath}}{\sqrt{2}} = \begin{bmatrix} \dfrac{-v}{\sqrt{2}} \\ \dfrac{u}{\sqrt{2}} \\ 0 \end{bmatrix} \tag{3.119}$$

$$\hat{u}_z = \frac{\partial \mathbf{r}/\partial z}{|\partial \mathbf{r}/\partial z|} = \hat{k} = \begin{bmatrix} 0 \\ 0 \\ 1 \end{bmatrix} \tag{3.120}$$

Using the coordinate orthogonality condition (3.2) we can transform the position vector from a Cartesian to a cylindrical expression:

$$\begin{aligned}\mathbf{r} &= (\mathbf{r}\cdot\hat{u}_u)\hat{u}_u + (\mathbf{r}\cdot\hat{u}_v)\hat{u}_v + (\mathbf{r}\cdot\hat{u}_z)\hat{u}_z \\ &= \frac{\sqrt{2}}{4}u\left(u^2+v^2\right)\hat{u}_u + \frac{\sqrt{2}}{4}v\left(u^2+v^2\right)\hat{u}_v + z\hat{u}_z\end{aligned} \tag{3.121}$$

To find the velocity vector $\mathbf{v} = d\mathbf{r}/dt$, we take a derivative from the Cartesian expression of the position vector (3.117),

$$\mathbf{v} = (u\dot{u} - v\dot{v})\,\hat{\imath} + (u\dot{v} + v\dot{u})\,\hat{\jmath} + \dot{z}\hat{k} = \begin{bmatrix} u\dot{u} - v\dot{v} \\ u\dot{v} + v\dot{u} \\ \dot{z} \end{bmatrix} \tag{3.122}$$

and employ the coordinate orthogonality condition (3.2),

$$\begin{aligned}\mathbf{v} &= (\mathbf{v}\cdot\hat{u}_u)\hat{u}_u + (\mathbf{v}\cdot\hat{u}_v)\hat{u}_v + (\mathbf{v}\cdot\hat{u}_z)\hat{u}_z \\ &= \frac{\sqrt{2}}{2}\dot{u}\left(u^2+v^2\right)\hat{u}_u + \frac{\sqrt{2}}{2}\dot{v}\left(u^2+v^2\right)\hat{u}_v + \dot{z}\hat{u}_z\end{aligned} \tag{3.123}$$

To find the acceleration vector $\mathbf{a} = d\mathbf{v}/dt$, we may take a derivative from the Cartesian expression of the velocity vector (3.122),

$$\mathbf{a} = \begin{bmatrix} u\ddot{u} + \dot{u}^2 - v\ddot{v} - \dot{v}^2 \\ u\ddot{v} + v\ddot{u} + 2\dot{v}\dot{u} \\ \ddot{z} \end{bmatrix} \tag{3.124}$$

and employ the coordinate orthogonality condition (3.2) again,

$$\begin{aligned}\mathbf{a} &= (\mathbf{a}\cdot\hat{u}_u)\hat{u}_u + (\mathbf{a}\cdot\hat{u}_v)\hat{u}_v + (\mathbf{a}\cdot\hat{u}_z)\hat{u}_z \\ &= \frac{\sqrt{2}}{2}\left(\ddot{u}\left(u^2+v^2\right) + u\dot{u}^2 - u\dot{v}^2 + 2v\dot{v}\dot{u}\right)\hat{u}_u \\ &\quad + \frac{\sqrt{2}}{2}\left(\ddot{v}\left(u^2+v^2\right) + 2u\dot{v}\dot{u} + v\left(\dot{v}^2 - \dot{u}^2\right)\right)\hat{u}_v + \ddot{z}\hat{k}\end{aligned} \tag{3.125}$$

Another derivative of (3.124) is given as

$$\mathbf{j} = \begin{bmatrix} 3\dot{u}\ddot{u} - 3\dot{v}\ddot{v} + u\dddot{u} - v\dddot{v} \\ 3\dot{u}\ddot{v} + 3\dot{v}\ddot{u} + u\dddot{v} + v\dddot{u} \\ \dddot{z} \end{bmatrix} \tag{3.126}$$

and using the coordinate orthogonality condition (3.2) provides the jerk vector in a parabolic cylindrical system:

$$\begin{aligned}\mathbf{j} &= (\mathbf{a}\cdot\hat{u}_u)\hat{u}_u + (\mathbf{a}\cdot\hat{u}_v)\hat{u}_v + (\mathbf{a}\cdot\hat{u}_z)\hat{u}_z \\ &= \frac{\sqrt{2}}{2}(\dddot{u}\left(u^2+v^2\right) + 3\ddot{u}\left(u\dot{u}+v\dot{v}\right) - 3\ddot{v}\left(u\dot{v}-v\dot{u}\right))\hat{u}_u \\ &\quad + \frac{\sqrt{2}}{2}(\dddot{v}\left(u^2+v^2\right) + 3\ddot{v}\left(u\dot{u}+v\dot{v}\right) + 3\ddot{u}\left(u\dot{v}-v\dot{u}\right))\hat{u}_v + \dddot{z}\hat{u}_z\end{aligned} \tag{3.127}$$

3.3 SPHERICAL COORDINATE SYSTEM

In a spherical coordinate system, the coordinate of a point is at the intersection of a plane, a cone, and a sphere, as shown in Figure 3.15. The coordinates of the spherical and Cartesian coordinate systems are related by

$$x = r\sin\varphi\cos\theta \qquad y = r\sin\varphi\sin\theta \qquad z = r\cos\varphi \tag{3.128}$$

or inversely by

$$r = \sqrt{x^2+y^2+z^2} \qquad \varphi = \tan^{-1}\frac{\sqrt{x^2+y^2}}{z} \qquad \theta = \tan^{-1}\frac{y}{x} \tag{3.129}$$

The position, velocity, and acceleration of a moving point in a spherical coordinate system are

$$\mathbf{r} = r\hat{u}_r \tag{3.130}$$

$$\mathbf{v} = \dot{r}\,\hat{u}_r + r\dot{\varphi}\,\hat{u}_\varphi + r\dot{\theta}\sin\varphi\,\hat{u}_\theta \tag{3.131}$$

$$\begin{aligned}\mathbf{a} = {} & \left(\ddot{r} - r\dot{\varphi}^2 - r\dot{\theta}^2\sin^2\varphi\right)\hat{u}_r + \left(r\ddot{\varphi} + 2\dot{r}\dot{\varphi} - r\dot{\theta}^2\sin\varphi\cos\varphi\right)\hat{u}_\varphi \\ & + \left[\left(r\ddot{\theta} + 2\dot{r}\dot{\theta}\right)\sin\varphi + 2r\dot{\theta}\dot{\varphi}\cos\varphi\right]\hat{u}_\theta\end{aligned} \tag{3.132}$$

Proof: Starting with a Cartesian position vector

$$\mathbf{r} = x\hat{\imath} + y\hat{\jmath} + z\hat{k} \tag{3.133}$$

and substituting the Cartesian–spherical relations (3.128), we find the Cartesian expression of the position vector with spherical components:

$$\mathbf{r} = \begin{bmatrix} r\sin\varphi\cos\theta \\ r\sin\varphi\sin\theta \\ r\cos\varphi \end{bmatrix} \tag{3.134}$$

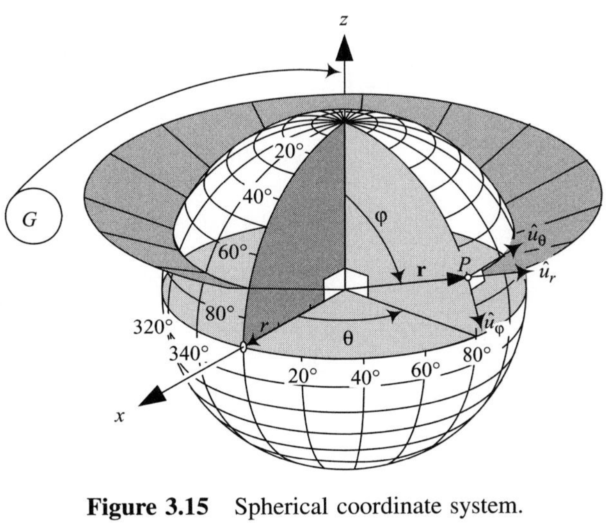

Figure 3.15 Spherical coordinate system.

The unit vectors of the spherical system expressed in a Cartesian system can be calculated by employing the definitions of unit vectors (1.200):

$$\hat{u}_r = \frac{\partial \mathbf{r}/\partial r}{|\partial \mathbf{r}/\partial r|} = \begin{bmatrix} \sin\varphi\cos\theta \\ \sin\varphi\sin\theta \\ \cos\varphi \end{bmatrix} \tag{3.135}$$

$$\hat{u}_\varphi = \frac{\partial \mathbf{r}/\partial \varphi}{|\partial \mathbf{r}/\partial \varphi|} = \begin{bmatrix} \cos\varphi\cos\theta \\ \cos\varphi\sin\theta \\ -\sin\varphi \end{bmatrix} \tag{3.136}$$

$$\hat{u}_\theta = \frac{\partial \mathbf{r}/\partial \theta}{|\partial \mathbf{r}/\partial \theta|} = \begin{bmatrix} -\sin\theta \\ \cos\theta \\ 0 \end{bmatrix} \tag{3.137}$$

To calculate the derivative kinematic vectors, we determine the derivative of the unit vectors (3.133)–(3.135) and take the derivative of the position vector (1.200):

$$\frac{d\hat{u}_r}{dt} = \begin{bmatrix} \dot{\varphi}\cos\theta\cos\varphi - \dot{\theta}\sin\theta\sin\varphi \\ \dot{\theta}\cos\theta\sin\varphi + \dot{\varphi}\cos\varphi\sin\theta \\ -\dot{\varphi}\sin\varphi \end{bmatrix} \tag{3.138}$$

$$\frac{d\hat{u}_\varphi}{dt} = \begin{bmatrix} -\dot{\varphi}\cos\theta\sin\varphi - \dot{\theta}\sin\theta\cos\varphi \\ \dot{\theta}\cos\theta\cos\varphi - \dot{\varphi}\sin\varphi\sin\theta \\ -\dot{\varphi}\cos\varphi \end{bmatrix} \tag{3.139}$$

$$\frac{d\hat{u}_\theta}{dt} = \begin{bmatrix} -\dot{\theta}\cos\theta \\ -\dot{\theta}\sin\theta \\ 0 \end{bmatrix} \tag{3.140}$$

Employing the coordinate orthogonality condition (3.2), we find the derivatives of the unit vectors $\hat{u}_r$, $\hat{u}_\varphi$, and $\hat{u}_\theta$ in a spherical system:

$$\begin{aligned} \frac{d\hat{u}_r}{dt} &= \left(\frac{d\hat{u}_r}{dt}\cdot\hat{u}_r\right)\hat{u}_r + \left(\frac{d\hat{u}_r}{dt}\cdot\hat{u}_\theta\right)\hat{u}_\theta + \left(\frac{d\hat{u}_r}{dt}\cdot\hat{u}_\varphi\right)\hat{u}_\varphi \\ &= \dot{\varphi}\,\hat{u}_\varphi + \dot{\theta}\sin\varphi\,\hat{u}_\theta \end{aligned} \tag{3.141}$$

$$\begin{aligned} \frac{d\hat{u}_\varphi}{dt} &= \left(\frac{d\hat{u}_\varphi}{dt}\cdot\hat{u}_r\right)\hat{u}_r + \left(\frac{d\hat{u}_\varphi}{dt}\cdot\hat{u}_\theta\right)\hat{u}_\theta + \left(\frac{d\hat{u}_\varphi}{dt}\cdot\hat{u}_\varphi\right)\hat{u}_\varphi \\ &= -\dot{\varphi}\,\hat{u}_r + \dot{\theta}\cos\varphi\,\hat{u}_\theta \end{aligned} \tag{3.142}$$

$$\begin{aligned} \frac{d\hat{u}_\theta}{dt} &= \left(\frac{d\hat{u}_\theta}{dt}\cdot\hat{u}_r\right)\hat{u}_r + \left(\frac{d\hat{u}_\theta}{dt}\cdot\hat{u}_\theta\right)\hat{u}_\theta + \left(\frac{d\hat{u}_\theta}{dt}\cdot\hat{u}_\varphi\right)\hat{u}_\varphi \\ &= -\dot{\theta}\sin\varphi\,\hat{u}_r - \dot{\theta}\cos\varphi\,\hat{u}_\varphi \end{aligned} \tag{3.143}$$

The closed set of the time derivatives of the unit vectors is enough to find the multiple derivatives of the position vector $\mathbf{r}$ in a spherical system. The velocity, acceleration, and jerk vectors would be

$$\mathbf{r} = (\mathbf{r} \cdot \hat{u}_r)\hat{u}_r + (\mathbf{r} \cdot \hat{u}_\varphi)\hat{u}_\varphi + (\mathbf{r} \cdot \hat{u}_\theta)\hat{u}_\theta = r\,\hat{u}_r \tag{3.144}$$

$$\begin{aligned} \mathbf{v} &= \frac{d}{dt}\mathbf{r} = \frac{d}{dt}(r\hat{u}_r) = \dot{r}\,\hat{u}_r + r\,\frac{d\hat{u}_r}{dt} \\ &= \dot{r}\,\hat{u}_r + r\dot{\varphi}\,\hat{u}_\varphi + r\dot{\theta}\sin\varphi\,\hat{u}_\theta \end{aligned} \tag{3.145}$$

$$\begin{aligned} \mathbf{a} &= \frac{d}{dt}\mathbf{v} = \frac{d}{dt}\left(\dot{r}\,\hat{u}_r + r\dot{\varphi}\,\hat{u}_\varphi + r\dot{\theta}\sin\varphi\,\hat{u}_\theta\right) \\ &= \ddot{r}\,\hat{u}_r + \dot{r}\,\frac{d\hat{u}_r}{dt} + \dot{r}\dot{\varphi}\,\hat{u}_\varphi + r\ddot{\varphi}\,\hat{u}_\varphi + r\dot{\varphi}\,\frac{d\hat{u}_\varphi}{dt} + \dot{r}\dot{\theta}\sin\varphi\,\hat{u}_\theta \\ &\quad + r\ddot{\theta}\sin\varphi\,\hat{u}_\theta + r\dot{\theta}\dot{\varphi}\cos\varphi\,\hat{u}_\theta + r\dot{\theta}\sin\varphi\,\frac{d\hat{u}_\theta}{dt} \\ &= \left(\ddot{r} - r\dot{\varphi}^2 - r\dot{\theta}^2\sin^2\varphi\right)\hat{u}_r + \left(r\ddot{\varphi} + 2\dot{r}\dot{\varphi} - r\dot{\theta}^2\sin\varphi\cos\varphi\right)\hat{u}_\varphi \\ &\quad + \left[\left(r\ddot{\theta} + 2\dot{r}\dot{\theta}\right)\sin\varphi + 2r\dot{\theta}\dot{\varphi}\cos\varphi\right]\hat{u}_\theta \end{aligned} \tag{3.146}$$

$$\begin{aligned} \mathbf{j} &= \frac{d}{dt}\mathbf{a} \\ &= \left[\dddot{r} - 3\dot{r}\dot{\varphi}^2 - 3r\dot{\varphi}\ddot{\varphi} - \tfrac{3}{2}r\dot{\theta}^2\dot{\varphi}\sin 2\varphi - 3\left(r\dot{\theta}\ddot{\theta} + \dot{r}\dot{\theta}^2\right)\sin^2\varphi\right]\hat{u}_r \\ &\quad + \left[r\left(\dddot{\varphi} - \dot{\varphi}^3\right) + 3\left(\dot{r}\ddot{\varphi} + \ddot{r}\dot{\varphi}\right)\right]\hat{u}_\varphi \\ &\quad + \left[-\tfrac{3}{2}\dot{\theta}\left(r\ddot{\theta} + \dot{r}\dot{\theta}\right)\sin 2\varphi - 3r\dot{\varphi}\dot{\theta}^2\cos^2\varphi\right]\hat{u}_\varphi \\ &\quad + \left\{r\left(\dddot{\theta} - \dot{\theta}^3\right) + 3\left[\dot{r}\ddot{\theta} + \dot{\theta}(\ddot{r} - r\dot{\varphi}^2)\right]\right\}\sin\varphi\,\hat{u}_\theta \\ &\quad + 3\left[r\ddot{\theta}\dot{\varphi} + \dot{\theta}(r\ddot{\varphi} + \dot{r}\dot{\varphi})\right]\cos\varphi\,\hat{u}_\theta \end{aligned} \tag{3.147}$$

■

Example 192 Orthogonality of Spherical Coordinate System A spherical coordinate frame with unit vectors $\hat{e}_r$, $\hat{e}_\theta$, and $\hat{e}_\varphi$ makes an orthogonal right-handed coordinate system because

$$\hat{e}_r \cdot \hat{e}_\varphi = 0 \qquad \hat{e}_\varphi \cdot \hat{e}_\theta = 0 \qquad \hat{e}_\theta \cdot \hat{e}_r = 0 \tag{3.148}$$

$$\hat{e}_r \times \hat{e}_\varphi = \hat{e}_\theta \quad \hat{e}_\varphi \times \hat{e}_\theta = \hat{e}_r \quad \hat{e}_\theta \times \hat{e}_r = \hat{e}_\varphi \tag{3.149}$$

Example 193 Alternative Method for Spherical Kinematics An alternative method to find the kinematic vectors is to use the Cartesian expression of the position and unit vectors with components in a spherical coordinate and employ the orthogonality condition (3.2). Substituting the Cartesian expression of the position vector with spherical components (3.134) and the unit vectors (3.133)–(3.137) in (3.2), we can transform $\mathbf{r}$ from a Cartesian to a spherical expression,

$$\mathbf{r} = (\mathbf{r} \cdot \hat{u}_r)\hat{u}_r + (\mathbf{r} \cdot \hat{u}_\varphi)\hat{u}_\varphi + (\mathbf{r} \cdot \hat{u}_\theta)\hat{u}_\theta = r\hat{u}_r \tag{3.150}$$

where

$$\mathbf{r}\cdot\hat{u}_r = \begin{bmatrix} r\sin\varphi\cos\theta \\ r\sin\varphi\sin\theta \\ r\cos\varphi \end{bmatrix} \cdot \begin{bmatrix} \sin\varphi\cos\theta \\ \sin\varphi\sin\theta \\ \cos\varphi \end{bmatrix} = r \tag{3.151}$$

$$\mathbf{r}\cdot\hat{u}_\theta = \begin{bmatrix} r\sin\varphi\cos\theta \\ r\sin\varphi\sin\theta \\ r\cos\varphi \end{bmatrix} \cdot \begin{bmatrix} -\sin\theta \\ \cos\theta \\ 0 \end{bmatrix} = 0 \tag{3.152}$$

$$\mathbf{r}\cdot\hat{u}_\varphi = \begin{bmatrix} r\sin\varphi\cos\theta \\ r\sin\varphi\sin\theta \\ r\cos\varphi \end{bmatrix} \cdot \begin{bmatrix} \cos\varphi\cos\theta \\ \cos\varphi\sin\theta \\ -\sin\varphi \end{bmatrix} = 0 \tag{3.153}$$

To find the velocity vector $\mathbf{v} = d\mathbf{r}/dt$, we take a derivative from the position vector (3.134),

$$\mathbf{v} = \begin{bmatrix} \dot{r}\cos\theta\sin\varphi + r\dot{\varphi}\cos\theta\cos\varphi - r\dot{\theta}\sin\theta\sin\varphi \\ \dot{r}\sin\theta\sin\varphi + r\dot{\varphi}\cos\theta\sin\varphi + r\dot{\theta}\sin\theta\cos\varphi \\ \dot{r}\cos\varphi - r\dot{\varphi}\sin\varphi \end{bmatrix} \tag{3.154}$$

Employing the coordinate orthogonality condition (3.2), we find the velocity vector in a spherical system:

$$\begin{aligned} \mathbf{v} &= (\mathbf{v}\cdot\hat{u}_r)\hat{u}_r + (\mathbf{v}\cdot\hat{u}_\theta)\hat{u}_\theta + (\mathbf{v}\cdot\hat{u}_\varphi)\hat{u}_\varphi \\ &= \dot{r}\hat{u}_r + r\dot{\theta}\sin\varphi\,\hat{u}_\theta + r\dot{\varphi}\hat{u}_\varphi \end{aligned} \tag{3.155}$$

To find the acceleration vector $\mathbf{a} = d\mathbf{v}/dt$, we may take a derivative from the Cartesian expression of the velocity vector (3.154),

$$\mathbf{a} = \begin{bmatrix} \left(\ddot{r} - r\dot{\theta}^2 - r\dot{\varphi}^2\right)\cos\theta\sin\varphi + (r\ddot{\varphi} + 2\dot{r}\dot{\varphi})\cos\theta\cos\varphi \\ -\left(r\ddot{\theta} + 2\dot{r}\dot{\theta}\right)\sin\theta\sin\varphi - 2r\dot{\theta}\dot{\varphi}\cos\varphi\sin\theta \\ \left(\ddot{r} - r\dot{\theta}^2 - r\dot{\varphi}^2\right)\sin\theta\sin\varphi + (r\ddot{\varphi} + 2\dot{r}\dot{\varphi})\cos\varphi\sin\theta \\ +\left(r\ddot{\theta} + 2\dot{r}\dot{\theta}\right)\cos\theta\sin\varphi + 2r\dot{\theta}\dot{\varphi}\cos\theta\cos\varphi \\ \left(\ddot{r} - r\dot{\varphi}^2\right)\cos\varphi - (r\ddot{\varphi} + 2\dot{r}\dot{\varphi})\sin\varphi \end{bmatrix} \tag{3.156}$$

and use the orthogonality condition (3.2),

$$\begin{aligned} \mathbf{a} &= (\mathbf{a}\cdot\hat{u}_r)\hat{u}_r + (\mathbf{a}\cdot\hat{u}_\theta)\hat{u}_\theta + (\mathbf{a}\cdot\hat{u}_\varphi)\hat{u}_\varphi \\ &= \left(\ddot{r} - r\dot{\varphi}^2 - r\dot{\theta}^2\sin^2\varphi\right)\hat{u}_r + \left[\left(r\ddot{\theta} + 2\dot{r}\dot{\theta}\right)\sin\varphi + 2r\dot{\theta}\dot{\varphi}\cos\varphi\right]\hat{u}_\theta \\ &\quad + \left(r\ddot{\varphi} + 2\dot{r}\dot{\varphi} - r\dot{\theta}^2\sin\varphi\cos\varphi\right)\hat{u}_\varphi \end{aligned} \tag{3.157}$$

Example 194 ★ Gradient in Spherical Coordinate System The orthogonality coordinate condition (3.2) enables us to find the gradient operator

$$\text{grad} = \nabla = \frac{\partial}{\partial x}\hat{\imath} + \frac{\partial}{\partial y}\hat{\jmath} + \frac{\partial}{\partial z}\hat{k} \tag{3.158}$$

in a spherical coordinate system.

Using the chain rule, we find

$$\nabla = \begin{bmatrix} \dfrac{\partial}{\partial x} \\ \dfrac{\partial}{\partial y} \\ \dfrac{\partial}{\partial z} \end{bmatrix} = \begin{bmatrix} \dfrac{\partial}{\partial r}\dfrac{\partial r}{\partial x} + \dfrac{\partial}{\partial \varphi}\dfrac{\partial \varphi}{\partial x} + \dfrac{\partial}{\partial \theta}\dfrac{\partial \theta}{\partial x} \\ \dfrac{\partial}{\partial r}\dfrac{\partial r}{\partial y} + \dfrac{\partial}{\partial \varphi}\dfrac{\partial \varphi}{\partial y} + \dfrac{\partial}{\partial \theta}\dfrac{\partial \theta}{\partial y} \\ \dfrac{\partial}{\partial r}\dfrac{\partial r}{\partial z} + \dfrac{\partial}{\partial \varphi}\dfrac{\partial \varphi}{\partial z} + \dfrac{\partial}{\partial \theta}\dfrac{\partial \theta}{\partial z} \end{bmatrix} \tag{3.159}$$

Having the Cartesian and cylindrical coordinate relations

$$r = \sqrt{x^2 + y^2 + z^2} \qquad \varphi = \tan^{-1}\frac{\sqrt{x^2 + y^2}}{z} \qquad \theta = \tan^{-1}\frac{y}{x}$$

$$x = r \sin\varphi \cos\theta \qquad y = r \sin\varphi \sin\theta \qquad z = r \cos\varphi \tag{3.160}$$

we have

$$\nabla f = \begin{bmatrix} \dfrac{\partial}{\partial x} \\ \dfrac{\partial}{\partial y} \\ \dfrac{\partial}{\partial z} \end{bmatrix} = \begin{bmatrix} \sin\varphi\cos\theta\dfrac{\partial}{\partial r} + \cos\theta\cos\varphi\dfrac{\partial}{\partial \varphi} - \dfrac{1}{r}\dfrac{\sin\theta}{\sin\varphi}\dfrac{\partial}{\partial \theta} \\ \sin\varphi\sin\theta\dfrac{\partial}{\partial r} + \cos\varphi\sin\theta\dfrac{\partial}{\partial \varphi} + \dfrac{1}{r}\dfrac{\cos\theta}{\sin\varphi}\dfrac{\partial}{\partial \theta} \\ \cos\varphi\dfrac{\partial}{\partial r} - \sin\varphi\dfrac{\partial}{\partial \varphi} \end{bmatrix} \tag{3.161}$$

Using the orthogonality condition, we have

$$\begin{aligned} \nabla &= (\nabla \cdot \hat{u}_\rho)\hat{u}_\rho + (\nabla \cdot \hat{u}_\varphi)\hat{u}_\varphi + (\nabla \cdot \hat{u}_\theta)\hat{u}_\theta \\ &= \frac{\partial}{\partial r}\hat{u}_r + \frac{1}{r}\frac{\partial}{\partial \varphi}\hat{u}_\varphi + \frac{1}{r\sin\varphi}\frac{\partial}{\partial \theta}\hat{u}_\theta \end{aligned} \tag{3.162}$$

Example 195 Gradient of a Spherical Field The gradient of the field

$$V = r^3 \frac{\cos\theta}{\sin^2\varphi} \tag{3.163}$$

is

$$\begin{aligned} \nabla V &= \frac{\partial}{\partial r}\hat{u}_r + \frac{1}{r}\frac{\partial}{\partial \varphi}\hat{u}_\varphi + \frac{1}{r\sin\varphi}\frac{\partial}{\partial \theta}\hat{u}_\theta \\ &= 3r^2\frac{\cos\theta}{\sin^2\varphi}\hat{u}_r - 2r^2\cos\theta\frac{\cos\varphi}{\sin^3\varphi}\hat{u}_\varphi - r^2\frac{\sin\theta}{\sin^3\varphi}\hat{u}_\theta \end{aligned} \tag{3.164}$$

Example 196 ★ Elliptic–Hyperbolic Cylindrical Coordinate System The eilliptic–hyperbolic coordinate system (u, v, z) relates to the Cartesian system by

$$x = a\cosh u \cos v \qquad y = a \sinh u \sin v \qquad z = z \tag{3.165}$$

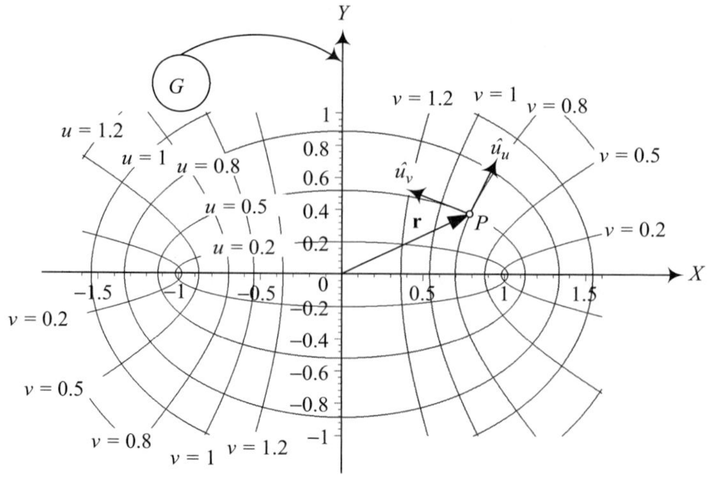

Figure 3.16 Elliptic–hyperbolic cylindrical coordinate system.

Figure 3.16 illustrates the intersection of the system with the (x, y)-plane for $a = 1$. The Cartesian expression of the position vector **r** with eilliptic–hyperbolic components is

$$\mathbf{r} = x\hat{\imath} + y\hat{\jmath} + z\hat{k} = \begin{bmatrix} a \cosh u \cos v \\ a \sinh u \sin v \\ z \end{bmatrix} \tag{3.166}$$

To determine the kinematics of a moving point, we calculate the unit vectors of eilliptic–hyperbolic systems expressed in the Cartesian system:

$$\hat{u}_u = \frac{\partial \mathbf{r}/\partial u}{|\partial \mathbf{r}/\partial u|} = \frac{1}{\sqrt{\cosh^2 u - \cos^2 v}} \begin{bmatrix} \cos v \sinh u \\ \sin v \cosh u \\ 0 \end{bmatrix} \tag{3.167}$$

$$\hat{u}_v = \frac{\partial \mathbf{r}/\partial v}{|\partial \mathbf{r}/\partial v|} = \frac{\sqrt{2}}{\sqrt{\cosh^2 u - \cos^2 v}} \begin{bmatrix} -\sin v \cosh u \\ \cos v \sinh u \\ 0 \end{bmatrix} \tag{3.168}$$

$$\hat{u}_z = \frac{\partial \mathbf{r}/\partial z}{|\partial \mathbf{r}/\partial z|} = \begin{bmatrix} 0 \\ 0 \\ 1 \end{bmatrix} \tag{3.169}$$

Employing the coordinate orthogonality condition (3.2), we find the derivatives of the unit vectors $\hat{u}_u$, $\hat{u}_v$, and $\hat{u}_z$ in the eilliptic–hyperbolic system:

$$\begin{aligned} \frac{d\hat{u}_u}{dt} &= \left(\frac{d\hat{u}_u}{dt} \cdot \hat{u}_u\right)\hat{u}_u + \left(\frac{d\hat{u}_u}{dt} \cdot \hat{u}_v\right)\hat{u}_v + \left(\frac{d\hat{u}_u}{dt} \cdot \hat{u}_z\right)\hat{u}_z \\ &= \frac{\cosh u \sinh u\, \dot{v} - \cos v \sin v\, \dot{u}}{\cosh^2 u - \cos^2 v}\,\hat{u}_v \end{aligned} \tag{3.170}$$

$$\frac{d\hat{u}_v}{dt} = \left(\frac{d\hat{u}_z}{dt} \cdot \hat{u}_u\right)\hat{u}_u + \left(\frac{d\hat{u}_z}{dt} \cdot \hat{u}_v\right)\hat{u}_v + \left(\frac{d\hat{u}_z}{dt} \cdot \hat{u}_z\right)\hat{u}_z$$

$$= -\frac{\cosh u \sinh u\, \dot{v} - \cos v \sin v\, \dot{u}}{\cosh^2 u - \cos^2 v}\hat{u}_u \tag{3.171}$$

$$\frac{d\hat{u}_z}{dt} = \left(\frac{d\hat{u}_v}{dt} \cdot \hat{u}_u\right)\hat{u}_u + \left(\frac{d\hat{u}_v}{dt} \cdot \hat{u}_v\right)\hat{u}_v + \left(\frac{d\hat{u}_v}{dt} \cdot \hat{u}_z\right)\hat{u}_z = 0 \tag{3.172}$$

This closed set of time derivatives of the unit vectors is enough to find the multiple derivatives of the position vector $\mathbf{r}$ in a spherical system. The velocity and acceleration vectors would be

$$\mathbf{r} = (\mathbf{r} \cdot \hat{u}_u)\hat{u}_u + (\mathbf{r} \cdot \hat{u}_v)\hat{u}_v + (\mathbf{r} \cdot \hat{u}_z)\hat{u}_z$$

$$= \frac{a \cosh u \sinh u}{\sqrt{\cosh^2 u - \cos^2 v}}\hat{u}_u - \frac{a \cos v \sin v}{\sqrt{\cosh^2 u - \cos^2 v}}\hat{u}_v + z\hat{u}_z \tag{3.173}$$

$$\mathbf{v} = \frac{d}{dt}\mathbf{r} = a\sqrt{\cosh^2 u - \cos^2 v}\, \dot{u}\, \hat{u}_u$$

$$+ a\sqrt{\cosh^2 u - \cos^2 v}\, \dot{v}\, \hat{u}_v + \dot{z}\, \hat{u}_z \tag{3.174}$$

$$\mathbf{a} = \frac{d}{dt}\mathbf{v} = a\ddot{u}\sqrt{\cosh^2 u - \cos^2 v}\, \hat{u}_u$$

$$+ a\frac{\left(\dot{u}^2 - \dot{v}^2\right) \cosh u \sinh u + 2\dot{u}\dot{v} \cos v \sin v}{\sqrt{\cosh^2 u - \cos^2 v}}\hat{u}_u$$

$$+ a\ddot{v}\sqrt{\cosh^2 u - \cos^2 v}\, \hat{u}_v$$

$$- a\frac{\left(\dot{u}^2 - \dot{v}^2\right) \cos v \sin v + 2\dot{u}\dot{v} \cosh u \sinh u}{\sqrt{\cosh^2 u - \cos^2 v}}\hat{u}_v + \ddot{z}\, \hat{u}_z \tag{3.175}$$

3.4 ★ NONORTHOGONAL COORDINATE FRAMES

There are applications in which we may prefer to interpret a dynamic problem in a nonorthogonal coordinate system. A *nonorthogonal* or *oblique coordinate frame* is made when the three scaled and straight axes *OA, OB, OC* of a triad *OABC* are nonorthogonal and non-coplanar.

3.4.1 ★ Reciprocal Base Vectors

Consider a nonorthogonal triad *OABC* to be used as a coordinate frame. Defining three unit vectors $\hat{b}_1$, $\hat{b}_2$, and $\hat{b}_3$ along the axes *OA, OB, OC*, respectively, we can express any vector $\mathbf{r}$ by a unique linear combination of the three unit vectors $\hat{b}_1$, $\hat{b}_2$, and $\hat{b}_3$ as

$$\mathbf{r} = r_1\hat{b}_1 + r_2\hat{b}_2 + r_3\hat{b}_3 \tag{3.176}$$

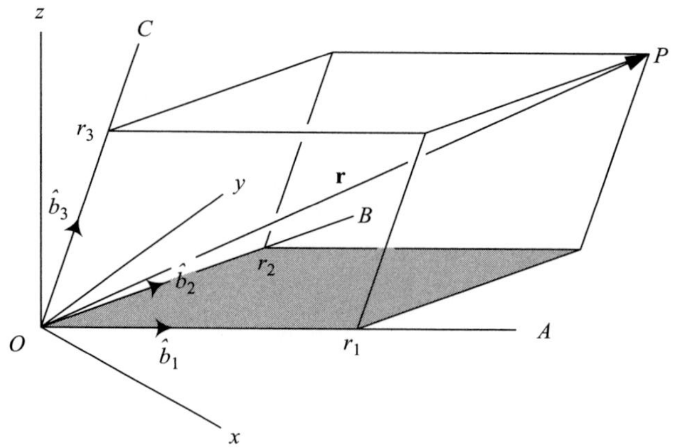

Figure 3.17 Expression of a vector **r** along the three axes of a nonorthogonal triad $OABC$.

where

$$r_1 = \frac{\mathbf{r} \cdot \hat{b}_2 \times \hat{b}_3}{\hat{b}_1 \cdot \hat{b}_2 \times \hat{b}_3} = \frac{\left[\mathbf{r}\hat{b}_2\hat{b}_3\right]}{\left[\hat{b}_1\hat{b}_2\hat{b}_3\right]} \tag{3.177}$$

$$r_2 = \frac{\mathbf{r} \cdot \hat{b}_3 \times \hat{b}_1}{\hat{b}_1 \cdot \hat{b}_2 \times \hat{b}_3} = \frac{\left[\mathbf{r}\hat{b}_3\hat{b}_1\right]}{\left[\hat{b}_1\hat{b}_2\hat{b}_3\right]} \tag{3.178}$$

$$r_3 = \frac{\mathbf{r} \cdot \hat{b}_1 \times \hat{b}_2}{\hat{b}_1 \cdot \hat{b}_2 \times \hat{b}_3} = \frac{\left[\mathbf{r}\hat{b}_1\hat{b}_2\right]}{\left[\hat{b}_1\hat{b}_2\hat{b}_3\right]} \tag{3.179}$$

The scalars r_1, r_2, and r_3 are the components of **r** along OA, OB, and OC, as shown in Figure 3.17. To determine r_1, we draw a line from point P, parallel to $\hat{b}_1$, to intersect the $(\hat{b}_2, \hat{b}_3)$-plane and measure the length of the line segment. The scalars r_2 and r_3 are calculated similarly.

For every nonorthogonal coordinate triad $OABC$ with unit vectors $\hat{b}_1$, $\hat{b}_2$, and $\hat{b}_3$, there exists a set of *reciprocal base vectors* $\mathbf{b}_1^\star$, $\mathbf{b}_2^\star$, $\mathbf{b}_3^\star$ that are perpendicular to the planes $\left(\hat{b}_2, \hat{b}_3\right)$, $\left(\hat{b}_3, \hat{b}_1\right)$, $\left(\hat{b}_1, \hat{b}_2\right)$ and represent the planes respectively:

$$\mathbf{b}_1^\star = \frac{\hat{b}_2 \times \hat{b}_3}{\hat{b}_1 \cdot \hat{b}_2 \times \hat{b}_3} = \frac{\hat{b}_2 \times \hat{b}_3}{\left[\hat{b}_1\hat{b}_2\hat{b}_3\right]} \tag{3.180}$$

$$\mathbf{b}_2^\star = \frac{\hat{b}_3 \times \hat{b}_1}{\hat{b}_1 \cdot \hat{b}_2 \times \hat{b}_3} = \frac{\hat{b}_3 \times \hat{b}_1}{\left[\hat{b}_1\hat{b}_2\hat{b}_3\right]} \tag{3.181}$$

$$\mathbf{b}_3^\star = \frac{\hat{b}_1 \times \hat{b}_2}{\hat{b}_1 \cdot \hat{b}_2 \times \hat{b}_3} = \frac{\hat{b}_1 \times \hat{b}_2}{\left[\hat{b}_1\hat{b}_2\hat{b}_3\right]} \tag{3.182}$$

Using the reciprocal base vectors $\mathbf{b}_1^\star$, $\mathbf{b}_2^\star$, $\mathbf{b}_3^\star$, we may determine the components r_1, r_2, and r_3 as

$$r_1 = \mathbf{r} \cdot \mathbf{b}_1^\star \tag{3.183}$$

$$r_2 = \mathbf{r} \cdot \mathbf{b}_2^\star \tag{3.184}$$

$$r_3 = \mathbf{r} \cdot \mathbf{b}_3^\star \tag{3.185}$$

and show the vector $\mathbf{r}$ as

$$\mathbf{r} = \left(\mathbf{r} \cdot \mathbf{b}_1^\star\right) \hat{b}_1 + \left(\mathbf{r} \cdot \mathbf{b}_2^\star\right) \hat{b}_2 + \left(\mathbf{r} \cdot \mathbf{b}_3^\star\right) \hat{b}_3 \tag{3.186}$$

The inner product of the principal unit vectors $\hat{b}_i$ and reciprocal base vectors $\mathbf{b}_i^\star$ is

$$\hat{b}_i \cdot \mathbf{b}_j^\star = \delta_{ij} \tag{3.187}$$

where δ_{ij} is the Kronecker delta (1.125). Equation (3.187) is called the *reciprocality condition*.

The reciprocal vectors may also be called the *reverse vectors*.

Proof: Let us express the vector $\mathbf{r}$ and unit vectors $\hat{b}_1$, $\hat{b}_2$, $\hat{b}_3$ in a Cartesian coordinate frame as

$$\mathbf{r} = x\hat{\imath} + y\hat{\jmath} + z\hat{k} \tag{3.188}$$

$$\hat{b}_1 = b_{11}\hat{\imath} + b_{12}\hat{\jmath} + b_{13}\hat{k} \tag{3.189}$$

$$\hat{b}_2 = b_{21}\hat{\imath} + b_{22}\hat{\jmath} + b_{23}\hat{k} \tag{3.190}$$

$$\hat{b}_3 = b_{31}\hat{\imath} + b_{32}\hat{\jmath} + b_{33}\hat{k} \tag{3.191}$$

Substituting (3.189)–(3.191) into (3.176) and comparing with (3.188) provide a set of equations

$$\begin{bmatrix} x \\ y \\ z \end{bmatrix} = \begin{bmatrix} b_{11} & b_{12} & b_{13} \\ b_{21} & b_{22} & b_{23} \\ b_{31} & b_{32} & b_{33} \end{bmatrix} \begin{bmatrix} r_1 \\ r_2 \\ r_3 \end{bmatrix} \tag{3.192}$$

that can be solved for the components r_1, r_2, and r_3:

$$\begin{bmatrix} r_1 \\ r_2 \\ r_3 \end{bmatrix} = \begin{bmatrix} b_{11} & b_{12} & b_{13} \\ b_{21} & b_{22} & b_{23} \\ b_{31} & b_{32} & b_{33} \end{bmatrix}^{-1} \begin{bmatrix} x \\ y \\ z \end{bmatrix} \tag{3.193}$$

We may express the solutions by vector–scalar triple products:

$$r_1 = \frac{\mathbf{r} \cdot \hat{b}_2 \times \hat{b}_3}{\hat{b}_1 \cdot \hat{b}_2 \times \hat{b}_3} = \frac{\left[\mathbf{r}\hat{b}_2\hat{b}_3\right]}{\left[\hat{b}_1\hat{b}_2\hat{b}_3\right]} = \mathbf{r} \cdot \frac{\hat{b}_2 \times \hat{b}_3}{\hat{b}_1 \cdot \hat{b}_2 \times \hat{b}_3} \tag{3.194}$$

$$r_2 = \frac{\mathbf{r} \cdot \hat{b}_3 \times \hat{b}_1}{\hat{b}_1 \cdot \hat{b}_2 \times \hat{b}_3} = \frac{\left[\mathbf{r}\hat{b}_3\hat{b}_1\right]}{\left[\hat{b}_1\hat{b}_2\hat{b}_3\right]} = \mathbf{r} \cdot \frac{\hat{b}_3 \times \hat{b}_1}{\hat{b}_1 \cdot \hat{b}_2 \times \hat{b}_3} \tag{3.195}$$

$$r_3 = \frac{\mathbf{r} \cdot \hat{b}_1 \times \hat{b}_2}{\hat{b}_1 \cdot \hat{b}_2 \times \hat{b}_3} = \frac{\left[\mathbf{r}\hat{b}_1\hat{b}_2\right]}{\left[\hat{b}_1\hat{b}_2\hat{b}_3\right]} = \mathbf{r} \cdot \frac{\hat{b}_1 \times \hat{b}_2}{\hat{b}_1 \cdot \hat{b}_2 \times \hat{b}_3} \tag{3.196}$$

The set of equations (3.193) are solvable if

$$\hat{b}_1 \cdot \hat{b}_2 \times \hat{b}_3 \neq 0 \tag{3.197}$$

which means $\hat{b}_1$, $\hat{b}_2$, $\hat{b}_3$ are not coplanar.

Substituting the components r_1, r_2, r_3 in (3.176), we may define three reciprocal vectors $\mathbf{b}_1^\star$, $\mathbf{b}_2^\star$, $\mathbf{b}_3^\star$ to write $\mathbf{r}$ in a decomposed expression similar to the orthogonality condition (3.2):

$$\begin{aligned} \mathbf{r} &= r_1\hat{b}_1 + r_2\hat{b}_2 + r_3\hat{b}_3 \\ &= \left(\mathbf{r} \cdot \mathbf{b}_1^\star\right)\hat{b}_1 + \left(\mathbf{r} \cdot \mathbf{b}_2^\star\right)\hat{b}_2 + \left(\mathbf{r} \cdot \mathbf{b}_3^\star\right)\hat{b}_3 \end{aligned} \tag{3.198}$$

where

$$\begin{aligned} \mathbf{b}_1^\star &= \left(\hat{b}_2 \times \hat{b}_3\right) \div \left(\hat{b}_1 \cdot \hat{b}_2 \times \hat{b}_3\right) \\ \mathbf{b}_2^\star &= \left(\hat{b}_3 \times \hat{b}_1\right) \div \left(\hat{b}_1 \cdot \hat{b}_2 \times \hat{b}_3\right) \\ \mathbf{b}_3^\star &= \left(\hat{b}_1 \times \hat{b}_2\right) \div \left(\hat{b}_1 \cdot \hat{b}_2 \times \hat{b}_3\right) \end{aligned} \tag{3.199}$$

and

$$r_1 = \mathbf{r} \cdot \mathbf{b}_1^\star \qquad r_2 = \mathbf{r} \cdot \mathbf{b}_2^\star \qquad r_3 = \mathbf{r} \cdot \mathbf{b}_3^\star. \tag{3.200}$$

Using Equations (3.199), we may show that the inner product of principal unit vectors $\hat{b}_i$ and reciprocal base vectors $\mathbf{b}_i^\star$ are unity:

$$\begin{aligned} \hat{b}_1 \cdot \mathbf{b}_1^\star &= \hat{b}_1 \cdot \frac{\hat{b}_2 \times \hat{b}_3}{\hat{b}_1 \cdot \hat{b}_2 \times \hat{b}_3} = 1 \\ \hat{b}_2 \cdot \mathbf{b}_2^\star &= \hat{b}_2 \cdot \frac{\hat{b}_3 \times \hat{b}_1}{\hat{b}_1 \cdot \hat{b}_2 \times \hat{b}_3} = 1 \\ \hat{b}_3 \cdot \mathbf{b}_3^\star &= \hat{b}_3 \cdot \frac{\hat{b}_1 \times \hat{b}_2}{\hat{b}_1 \cdot \hat{b}_2 \times \hat{b}_3} = 1 \end{aligned} \tag{3.201}$$

However, because $\hat{b}_i$ and $\mathbf{b}_i^\star$ are not orthogonal, the $\mathbf{b}_i^\star$ are not necessarily unit vectors. Furthermore, because the reciprocal base vectors $\mathbf{b}_1^\star$, $\mathbf{b}_2^\star$, $\mathbf{b}_3^\star$ respectively represent the planes $\left(\hat{b}_2, \hat{b}_3\right)$, $\left(\hat{b}_3, \hat{b}_1\right)$, $\left(\hat{b}_1, \hat{b}_2\right)$, we have

$$\begin{aligned} \hat{b}_1 \cdot \mathbf{b}_2^\bigstar = \hat{b}_2 \cdot \mathbf{b}_3^\bigstar = \hat{b}_3 \cdot \mathbf{b}_1^\bigstar = 0 \\ \hat{b}_1 \cdot \mathbf{b}_3^\bigstar = \hat{b}_2 \cdot \mathbf{b}_1^\bigstar = \hat{b}_3 \cdot \mathbf{b}_2^\bigstar = 0 \end{aligned} \tag{3.202}$$

Using the Kronecker delta δ_{ij}, we can show Equations (3.201) and (3.202) by a single equation:

$$\hat{b}_i \cdot \mathbf{b}_j^\bigstar = \delta_{ij} \tag{3.203}$$

■

Example 197 ★ Position Vector of a Point and Reciprocal Base Vectors Consider a point P at

$$\mathbf{r} = 5\hat{\imath} + 6\hat{\jmath} + 4\hat{k} \tag{3.204}$$

We would like to express $\mathbf{r}$ in a nonorthogonal coordinate frame

$$\mathbf{r} = r_1\hat{b}_1 + r_2\hat{b}_2 + r_3\hat{b}_3 \tag{3.205}$$

with $\hat{b}_1$, $\hat{b}_2$, $\hat{b}_3$ as unit vectors:

$$\hat{b}_1 = \begin{bmatrix} 0.97334 \\ -0.17194 \\ -0.15198 \end{bmatrix} \qquad \hat{b}_2 = \begin{bmatrix} -0.64219 \\ 0.7662 \\ -0.02352 \end{bmatrix} \qquad \hat{b}_3 = \begin{bmatrix} 0 \\ 0 \\ 1 \end{bmatrix} \tag{3.206}$$

To calculate r_1, r_2, and r_3, we use Equation (3.181) to find the reciprocal base vectors of the nonorthogonal frame:

$$\mathbf{b}_1^\bigstar = \frac{\hat{b}_2 \times \hat{b}_3}{\hat{b}_1 \cdot \hat{b}_2 \times \hat{b}_3} = \frac{1}{0.63535}\begin{bmatrix} 0.7662 \\ 0.64219 \\ 0 \end{bmatrix} = \begin{bmatrix} 1.2059 \\ 1.0108 \\ 0 \end{bmatrix} \tag{3.207}$$

$$\mathbf{b}_2^\bigstar = \frac{\hat{b}_3 \times \hat{b}_1}{\hat{b}_1 \cdot \hat{b}_2 \times \hat{b}_3} = \frac{1}{0.63535}\begin{bmatrix} 0.17194 \\ 0.97334 \\ 0 \end{bmatrix} = \begin{bmatrix} 0.27062 \\ 1.532 \\ 0 \end{bmatrix} \tag{3.208}$$

$$\mathbf{b}_3^\bigstar = \frac{\hat{b}_1 \times \hat{b}_2}{\hat{b}_1 \cdot \hat{b}_2 \times \hat{b}_3} = \frac{1}{0.63535}\begin{bmatrix} 0.12049 \\ 0.12049 \\ 0.63535 \end{bmatrix} = \begin{bmatrix} 0.18964 \\ 0.18964 \\ 1.0 \end{bmatrix} \tag{3.209}$$

So, the components r_1, r_2, and r_3 are

$$\begin{aligned} r_1 &= \mathbf{r} \cdot \mathbf{b}_1^\bigstar = 12.094 \\ r_2 &= \mathbf{r} \cdot \mathbf{b}_2^\bigstar = 10.545 \\ r_3 &= \mathbf{r} \cdot \mathbf{b}_3^\bigstar = 6.086 \end{aligned} \tag{3.210}$$

Example 198 ★ A Vector in a Nonorthogonal Coordinate Frame Consider the nonorthogonal coordinate frame $OABC$ and the point P in Figure 3.18. Assume that the Cartesian expression of its position vector $\mathbf{r}$ and the unit vectors of the nonorthogonal triad $OABC$ are

$$\mathbf{r} = 5\hat{\imath} + 6\hat{\jmath} + 4\hat{k} \tag{3.211}$$

$$\begin{aligned} \hat{b}_1 &= 0.766\hat{\imath} + 0.642\hat{\jmath} \\ \hat{b}_2 &= 0.174\hat{\imath} + 0.985\hat{\jmath} \\ \hat{b}_3 &= 0.183\hat{\imath} + 0.183\hat{\jmath} + 0.965\hat{k} \end{aligned} \tag{3.212}$$

To show the expression of $\mathbf{r}$ in the nonorthogonal coordinate frame,

$$\mathbf{r} = r_1\hat{b}_1 + r_2\hat{b}_2 + r_3\hat{b}_3 \tag{3.213}$$

we need to calculate r_1, r_2, and r_3:

$$r_1 = \frac{\left[\mathbf{r}\hat{b}_2\hat{b}_3\right]}{\left[\hat{b}_1\hat{b}_2\hat{b}_3\right]} \qquad r_2 = \frac{\left[\mathbf{r}\hat{b}_3\hat{b}_1\right]}{\left[\hat{b}_1\hat{b}_2\hat{b}_3\right]} \qquad r_3 = \frac{\left[\mathbf{r}\hat{b}_1\hat{b}_2\right]}{\left[\hat{b}_1\hat{b}_2\hat{b}_3\right]} \tag{3.214}$$

$$\mathbf{r}\cdot\hat{b}_2\times\hat{b}_3 = \begin{bmatrix}5\\6\\4\end{bmatrix}\cdot\begin{bmatrix}0.174\\0.985\\0\end{bmatrix}\times\begin{bmatrix}0.183\\0.183\\0.965\end{bmatrix} = 3.1515 \tag{3.215}$$

$$\mathbf{r}\cdot\hat{b}_3\times\hat{b}_1 = \begin{bmatrix}5\\6\\4\end{bmatrix}\cdot\begin{bmatrix}0.183\\0.183\\0.965\end{bmatrix}\times\begin{bmatrix}0.766\\0.642\\0\end{bmatrix} = 1.2467 \tag{3.216}$$

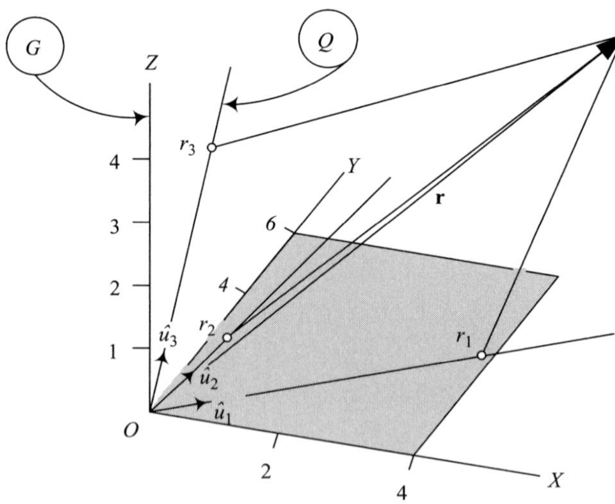

Figure 3.18 A position vector in a nonorthogonal coordinate frame.

$$\mathbf{r} \cdot \hat{b}_1 \times \hat{b}_2 = \begin{bmatrix} 5 \\ 6 \\ 4 \end{bmatrix} \cdot \begin{bmatrix} 0.766 \\ 0.642 \\ 0 \end{bmatrix} \times \begin{bmatrix} 0.174 \\ 0.985 \\ 0 \end{bmatrix} = 2.5712 \tag{3.217}$$

$$\hat{b}_1 \cdot \hat{b}_2 \times \hat{b}_3 = \begin{bmatrix} 0.766 \\ 0.642 \\ 0 \end{bmatrix} \cdot \begin{bmatrix} 0.174 \\ 0.985 \\ 0 \end{bmatrix} \times \begin{bmatrix} 0.183 \\ 0.183 \\ 0.965 \end{bmatrix} = 0.6203 \tag{3.218}$$

$$\begin{aligned} r_1 &= \frac{3.1515}{0.6203} = 5.08 \\ r_2 &= \frac{1.2467}{0.6203} = 2.01 \\ r_3 &= \frac{2.5712}{0.6203} = 4.14 \end{aligned} \tag{3.219}$$

Therefore, the nonorthogonal expression of $\mathbf{r}$ is

$$\mathbf{r} = 5.08\hat{b}_1 + 2.01\hat{b}_2 + 4.14\hat{b}_3 \tag{3.220}$$

We may also use the reciprocal base vectors of the $OABC$ triad,

$$\begin{aligned} \mathbf{b}_1^\star &= \frac{\hat{b}_2 \times \hat{b}_3}{\hat{b}_1 \cdot \hat{b}_2 \times \hat{b}_3} = \begin{bmatrix} 1.5331 \\ -0.27082 \\ -0.23938 \end{bmatrix} \\ \mathbf{b}_2^\star &= \frac{\hat{b}_3 \times \hat{b}_1}{\hat{b}_1 \cdot \hat{b}_2 \times \hat{b}_3} = \begin{bmatrix} -0.99924 \\ 1.1922 \\ -0.0366 \end{bmatrix} \\ \mathbf{b}_3^\star &= \frac{\hat{b}_1 \times \hat{b}_2}{\hat{b}_1 \cdot \hat{b}_2 \times \hat{b}_3} = \begin{bmatrix} 0 \\ 0 \\ 1.0368 \end{bmatrix} \end{aligned} \tag{3.221}$$

to calculate the components r_1, r_2, and r_3:

$$\begin{aligned} r_1 &= \mathbf{r} \cdot \mathbf{b}_1^\star = \begin{bmatrix} 5 \\ 6 \\ 4 \end{bmatrix} \cdot \begin{bmatrix} 1.5331 \\ -0.27082 \\ -0.23938 \end{bmatrix} = 5.08 \\ r_2 &= \mathbf{r} \cdot \mathbf{b}_2^\star = \begin{bmatrix} 5 \\ 6 \\ 4 \end{bmatrix} \cdot \begin{bmatrix} -0.99924 \\ 1.1922 \\ -0.0366 \end{bmatrix} = 2.01 \\ r_3 &= \mathbf{r} \cdot \mathbf{b}_3^\star = \begin{bmatrix} 5 \\ 6 \\ 4 \end{bmatrix} \cdot \begin{bmatrix} 0 \\ 0 \\ 1.0368 \end{bmatrix} = 4.14 \end{aligned} \tag{3.222}$$

***Example 199* ★ *Scalar Triple Product of* $\mathbf{b}_i^\star$** Let us use Equations (3.181) to calculate the scalar triple product of $\mathbf{b}_i^\star$:

$$\begin{aligned}
\mathbf{b}_1^\star \cdot \mathbf{b}_2^\star \times \mathbf{b}_3^\star &= \frac{\hat{b}_2 \times \hat{b}_3}{\hat{b}_1 \cdot \hat{b}_2 \times \hat{b}_3} \cdot \frac{\hat{b}_3 \times \hat{b}_1}{\hat{b}_1 \cdot \hat{b}_2 \times \hat{b}_3} \times \frac{\hat{b}_1 \times \hat{b}_2}{\hat{b}_1 \cdot \hat{b}_2 \times \hat{b}_3} \\
&= \frac{\left(\hat{b}_2 \times \hat{b}_3\right) \cdot \left(\hat{b}_3 \times \hat{b}_1\right) \times \left(\hat{b}_1 \times \hat{b}_2\right)}{\left(\hat{b}_1 \cdot \hat{b}_2 \times \hat{b}_3\right)^3} \\
&= \frac{\left(\hat{b}_2 \times \hat{b}_3\right) \cdot \hat{b}_1 \left(\hat{b}_3 \times \hat{b}_1 \cdot \hat{b}_2\right)}{\left(\hat{b}_1 \cdot \hat{b}_2 \times \hat{b}_3\right)^3} \\
&= \frac{\left(\hat{b}_1 \cdot \hat{b}_2 \times \hat{b}_3\right) \cdot \left(\hat{b}_1 \cdot \hat{b}_2 \times \hat{b}_3\right)}{\left(\hat{b}_1 \cdot \hat{b}_2 \times \hat{b}_3\right)^3} \\
&= \frac{1}{\left(\hat{b}_1 \cdot \hat{b}_2 \times \hat{b}_3\right)}
\end{aligned} \tag{3.223}$$

So, the scalar triple product of $\left[\hat{b}_1\hat{b}_2\hat{b}_3\right]$ and $\left[\mathbf{b}_1^\star\mathbf{b}_2^\star\mathbf{b}_3^\star\right]$ is the inverse of each other. Because $\left[\hat{b}_1\hat{b}_2\hat{b}_3\right]$ is equal to the volume of the parallelepiped made by the vectors $\hat{b}_1, \hat{b}_2, \hat{b}_3$,

$$V = \left[\hat{b}_1\hat{b}_2\hat{b}_3\right] \tag{3.224}$$

and $\left[\mathbf{b}_1^\star\mathbf{b}_2^\star\mathbf{b}_3^\star\right]$ is equal to the volume of the parallelepiped made by the reciprocal base vectors $\mathbf{b}_1^\star, \mathbf{b}_2^\star, \mathbf{b}_3^\star$,

$$V^\star = \left[\mathbf{b}_1^\star\mathbf{b}_2^\star\mathbf{b}_3^\star\right] \tag{3.225}$$

we have

$$VV^\star = 1 \tag{3.226}$$

Example 200* ★ *Short Notation for Principal and Reciprocal Vectors The principal unit vectors and reciprocal base vectors may be shown by using index notation,

$$\hat{b}_i = \left(\mathbf{b}_j^\star \times \mathbf{b}_k^\star\right) V \tag{3.227}$$

$$\mathbf{b}_i^\star = \frac{\hat{b}_j \times \hat{b}_k}{V} \tag{3.228}$$

where the indices *i, j, k* follow a cyclic permutation of *1, 2, 3* .

Example 201 ★ Geometric Interpretation of Reciprocal Base Vectors Consider an oblique coordinate system such that

$$\hat{b}_3 = \hat{k} \tag{3.229}$$

Such a coordinate system is nonorthogonal only in planes perpendicular to $\hat{k}$ and therefore expresses a two-dimensional oblique space because

$$\hat{b}_1 \cdot \hat{k} = 0 \qquad \hat{b}_2 \cdot \hat{k} = 0 \tag{3.230}$$

Figure 3.19 illustrates a two-dimensional oblique coordinate system with unit vectors $\hat{b}_1$ and $\hat{b}_2$. The volume of the parallelepiped made by the vectors $\hat{b}_1, \hat{b}_2, \hat{k}$ is

$$V = \left[\hat{b}_1\hat{b}_2\hat{b}_3\right] = \hat{b}_1 \cdot \hat{b}_2 \times \hat{k} = \hat{b}_1 \times \hat{b}_2 \cdot \hat{k} = \hat{b}_1 \times \hat{b}_2 = \sin\theta \tag{3.231}$$

where θ is the angle between $\hat{b}_1$ and $\hat{b}_2$. The reciprocal base vectors of the system are

$$\mathbf{b}_1^{\bigstar} = \frac{\hat{b}_2 \times \hat{k}}{\hat{b}_1 \cdot \hat{b}_2 \times \hat{k}} = \frac{\hat{b}_2 \times \hat{k}}{V} = \frac{\hat{b}_2 \times \hat{k}}{\sin\theta} \tag{3.232}$$

$$\mathbf{b}_2^{\bigstar} = \frac{\hat{k} \times \hat{b}_1}{\hat{b}_1 \cdot \hat{b}_2 \times \hat{k}} = \frac{\hat{k} \times \hat{b}_1}{V} = \frac{\hat{k} \times \hat{b}_1}{\sin\theta} \tag{3.233}$$

$$\mathbf{b}_3^{\bigstar} = \frac{\hat{b}_1 \times \hat{b}_2}{\hat{b}_1 \cdot \hat{b}_2 \times \hat{k}} = \frac{\hat{b}_1 \times \hat{b}_2}{V} = \hat{k} \tag{3.234}$$

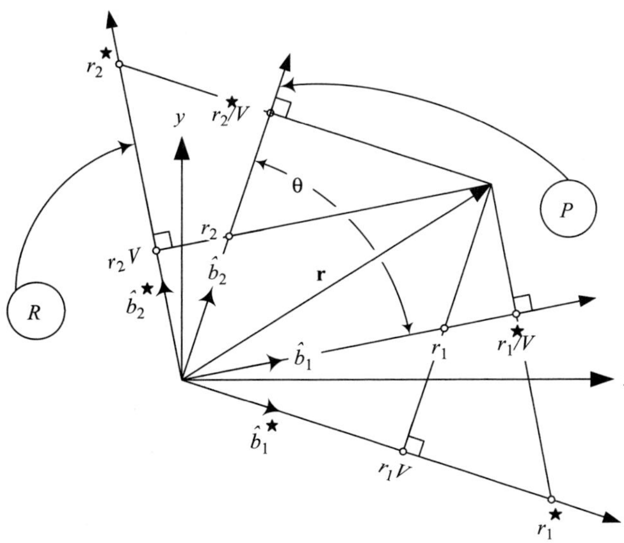

Figure 3.19 Geometric interpretation of reciprocal base vectors in a two-dimensional space.

So, $\mathbf{b}_1^\star$ is a vector in the $\left(\hat{b}_1, \hat{b}_2\right)$-plane and perpendicular to $\hat{b}_2$, and similarly, $\mathbf{b}_2^\star$ is a vector in the $\left(\hat{b}_1, \hat{b}_2\right)$-plane and perpendicular to $\hat{b}_1$. The scalars $r_1 V$ and $r_2 V$ are projections of $\mathbf{r}$ on axes $\mathbf{b}_1^\star$ and $\mathbf{b}_2^\star$, respectively.

3.4.2 ★ Reciprocal Coordinate Frame

When a triad $OABC$ with principal unit vectors $\hat{b}_1, \hat{b}_2, \hat{b}_3$ is nonorthogonal, the reciprocal base vectors $\mathbf{b}_1^\star, \mathbf{b}_2^\star, \mathbf{b}_3^\star$ are also nonorthogonal. The axes on $\mathbf{b}_1^\star, \mathbf{b}_2^\star, \mathbf{b}_3^\star$ define a new coordinate frame $OA^\star B^\star C^\star$ called the *reciprocal coordinate frame* R, while the original triad $OABC$ is called the *principal coordinate frame* P.

The coordinate frames $\hat{b}_1, \hat{b}_2, \hat{b}_3$ and $\mathbf{b}_1^\star, \mathbf{b}_2^\star, \mathbf{b}_3^\star$ are reciprocal to each other. A vector $\mathbf{r}$ in the principal and reciprocal frames P and R are expressed as

$$^P\mathbf{r} = r_1\hat{b}_1 + r_2\hat{b}_2 + r_3\hat{b}_3 \tag{3.235}$$

$$^R\mathbf{r} = r_1^\star\mathbf{b}_1^\star + r_2^\star\mathbf{b}_2^\star + r_3^\star\mathbf{b}_3^\star \tag{3.236}$$

where

$$r_1 = \mathbf{r}\cdot\mathbf{b}_1^\star \qquad r_2 = \mathbf{r}\cdot\mathbf{b}_2^\star \qquad r_3 = \mathbf{r}\cdot\mathbf{b}_3^\star \tag{3.237}$$

$$r_1^\star = \mathbf{r}\cdot\hat{b}_1 \qquad r_2^\star = \mathbf{r}\cdot\hat{b}_2 \qquad r_3^\star = \mathbf{r}\cdot\hat{b}_3 \tag{3.238}$$

$$\mathbf{b}_1^\star = \frac{\hat{b}_2\times\hat{b}_3}{\left[\hat{b}_1\hat{b}_2\hat{b}_3\right]} \qquad \mathbf{b}_2^\star = \frac{\hat{b}_3\times\hat{b}_1}{\left[\hat{b}_1\hat{b}_2\hat{b}_3\right]} \qquad \mathbf{b}_3^\star = \frac{\hat{b}_1\times\hat{b}_2}{\left[\hat{b}_1\hat{b}_2\hat{b}_3\right]} \tag{3.239}$$

$$\hat{b}_1 = \frac{\mathbf{b}_2^\star\times\mathbf{b}_3^\star}{\left[\mathbf{b}_1^\star\mathbf{b}_2^\star\mathbf{b}_3^\star\right]} \qquad \hat{b}_2 = \frac{\mathbf{b}_3^\star\times\mathbf{b}_1^\star}{\left[\mathbf{b}_1^\star\mathbf{b}_2^\star\mathbf{b}_3^\star\right]} \qquad \hat{b}_3 = \frac{\mathbf{b}_1^\star\times\mathbf{b}_2^\star}{\left[\mathbf{b}_1^\star\mathbf{b}_2^\star\mathbf{b}_3^\star\right]} \tag{3.240}$$

The scalars r_1, r_2, r_3 are called the *covariant* components of $\mathbf{r}$, and the scalars $r_1^\star, r_2^\star, r_3^\star$ are called the *contravariant* components of $\mathbf{r}$.

Figure 3.20 illustrates the covariant and contravariant components of a two-dimensional vector $\mathbf{r}$. The expression of a vector $\mathbf{r}$ can be transformed from the principal to reciprocal frames and vice versa by introducing the transformation matrices $^R R_P$ and $^P R_R$:

$$^R\mathbf{r} = {}^R R_P\, {}^P\mathbf{r} = \left[g_{ij}\right]\, {}^P\mathbf{r} \tag{3.241}$$

$$\begin{bmatrix} r_1^\star \\ r_2^\star \\ r_3^\star \end{bmatrix} = \begin{bmatrix} \hat{b}_1\cdot\hat{b}_1 & \hat{b}_1\cdot\hat{b}_2 & \hat{b}_1\cdot\hat{b}_3 \\ \hat{b}_2\cdot\hat{b}_1 & \hat{b}_2\cdot\hat{b}_2 & \hat{b}_2\cdot\hat{b}_3 \\ \hat{b}_3\cdot\hat{b}_1 & \hat{b}_3\cdot\hat{b}_2 & \hat{b}_3\cdot\hat{b}_3 \end{bmatrix} \begin{bmatrix} r_1 \\ r_2 \\ r_3 \end{bmatrix} \tag{3.242}$$

$$^P\mathbf{r} = {}^P R_R\, {}^R\mathbf{r} = \left[g_{ij}^\star\right]\, {}^R\mathbf{r} \tag{3.243}$$

$$\begin{bmatrix} r_1 \\ r_2 \\ r_3 \end{bmatrix} = \begin{bmatrix} \mathbf{b}_1^\star\cdot\mathbf{b}_1^\star & \mathbf{b}_1^\star\cdot\mathbf{b}_2^\star & \mathbf{b}_1^\star\cdot\mathbf{b}_3^\star \\ \mathbf{b}_2^\star\cdot\mathbf{b}_1^\star & \mathbf{b}_2^\star\cdot\mathbf{b}_2^\star & \mathbf{b}_2^\star\cdot\mathbf{b}_3^\star \\ \mathbf{b}_3^\star\cdot\mathbf{b}_1^\star & \mathbf{b}_3^\star\cdot\mathbf{b}_2^\star & \mathbf{b}_3^\star\cdot\mathbf{b}_3^\star \end{bmatrix} \begin{bmatrix} r_1^\star \\ r_2^\star \\ r_3^\star \end{bmatrix} \tag{3.244}$$

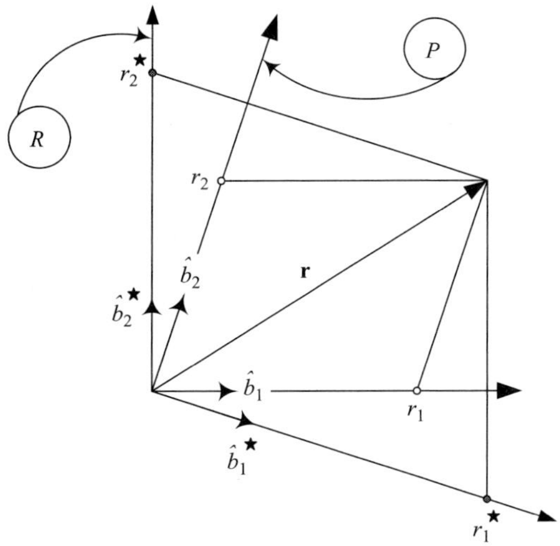

Figure 3.20 Contravariant and covariant components of a two-dimentional vector **r**.

So, the components r_i and $r_i^\star$ are related by

$$r_i = \sum_{j=1}^{3} g_{ij}^\star r_j^\star \tag{3.245}$$

$$r_i^\star = \sum_{j=1}^{3} g_{ij} r_j \tag{3.246}$$

where

$$g_{ij} = g_{ji} = \hat{b}_i \cdot \hat{b}_j \tag{3.247}$$

$$g_{ij}^\star = g_{ji}^\star = \mathbf{b}_i^\star \cdot \mathbf{b}_j^\star \tag{3.248}$$

$$\hat{b}_i \cdot \mathbf{b}_j^\star = \mathbf{b}_i^\star \cdot \hat{b}_j = \delta_{ij} \tag{3.249}$$

The coefficients g_{ij}, $i, j = 1, 2, 3$, are called the *covariant metric quantities* that make a *metric tensor* for the principal coordinate frame. The coefficients $g_{ij}^\star$, $i, j = 1, 2, 3$, are similarly called the *contravariant metric quantities* and make a *metric tensor* for the reciprocal coordinate frame.

Proof: We can use the definition of reciprocal base vectors (3.199) to calculate the reciprocal vectors to $\mathbf{b}_1^\star, \mathbf{b}_2^\star, \mathbf{b}_3^\star$ from $\hat{b}_1, \hat{b}_2, \hat{b}_3$. The reciprocal vector to $\mathbf{b}_1^\star$ is $\hat{b}_1$ because

$$\begin{aligned}\frac{\mathbf{b}_2^\star \times \mathbf{b}_3^\star}{\mathbf{b}_1^\star \cdot \mathbf{b}_2^\star \times \mathbf{b}_3^\star} &= \left(\frac{\hat{b}_3 \times \hat{b}_1}{\hat{b}_1 \cdot \hat{b}_2 \times \hat{b}_3} \times \frac{\hat{b}_1 \times \hat{b}_2}{\hat{b}_1 \cdot \hat{b}_2 \times \hat{b}_3}\right)\left(\hat{b}_1 \cdot \hat{b}_2 \times \hat{b}_3\right) \\ &= \frac{\left(\hat{b}_3 \times \hat{b}_1\right) \times \left(\hat{b}_1 \times \hat{b}_2\right)}{\hat{b}_1 \cdot \hat{b}_2 \times \hat{b}_3} = \hat{b}_1 \end{aligned} \tag{3.250}$$

Similarly, the reciprocal vectors to $\mathbf{b}_2^\star$ and $\mathbf{b}_3^\star$ are $\hat{b}_2$ and $\hat{b}_3$, respectively:

$$\frac{\mathbf{b}_3^\star \times \mathbf{b}_1^\star}{\mathbf{b}_1^\star \cdot \mathbf{b}_2^\star \times \mathbf{b}_3^\star} = \hat{b}_2 \tag{3.251}$$

$$\frac{\mathbf{b}_1^\star \times \mathbf{b}_2^\star}{\mathbf{b}_1^\star \cdot \mathbf{b}_2^\star \times \mathbf{b}_3^\star} = \hat{b}_3 \tag{3.252}$$

Expressing a vector $\mathbf{r}$ and the vectors $\mathbf{b}_1^\star$, $\mathbf{b}_2^\star$, $\mathbf{b}_3^\star$ in a Cartesian coordinate frame,

$$\mathbf{r} = x\hat{\imath} + y\hat{\jmath} + z\hat{k} \tag{3.253}$$

$$\begin{aligned} \mathbf{b}_1^\star &= b_{11}^\star\hat{\imath} + b_{12}^\star\hat{\jmath} + b_{13}^\star\hat{k} \\ \mathbf{b}_2^\star &= b_{21}^\star\hat{\imath} + b_{22}^\star\hat{\jmath} + b_{23}^\star\hat{k} \\ \mathbf{b}_3^\star &= b_{31}^\star\hat{\imath} + b_{32}^\star\hat{\jmath} + b_{33}^\star\hat{k} \end{aligned} \tag{3.254}$$

and using Equation (3.236) provide a set of equations that can be solved for the reciprocal components $r_1^\star$, $r_2^\star$, and $r_3^\star$:

$$\begin{bmatrix} r_1^\star \\ r_2^\star \\ r_3^\star \end{bmatrix} = \begin{bmatrix} b_{11}^\star & b_{12}^\star & b_{13}^\star \\ b_{21}^\star & b_{22}^\star & b_{23}^\star \\ b_{31}^\star & b_{32}^\star & b_{33}^\star \end{bmatrix}^{-1} \begin{bmatrix} x \\ y \\ z \end{bmatrix} \tag{3.255}$$

We may express the solutions by vector–scalar triple products:

$$\begin{aligned} r_1^\star &= \frac{\mathbf{r} \cdot \mathbf{b}_2^\star \times \mathbf{b}_3^\star}{\mathbf{b}_1^\star \cdot \mathbf{b}_2^\star \times \mathbf{b}_3^\star} = \frac{\left[\mathbf{r}\mathbf{b}_2^\star\mathbf{b}_3^\star\right]}{\left[\mathbf{b}_1^\star\mathbf{b}_2^\star\mathbf{b}_3^\star\right]} = \mathbf{r} \cdot \hat{b}_1 \\ r_2^\star &= \frac{\mathbf{r} \cdot \mathbf{b}_3^\star \times \mathbf{b}_1^\star}{\mathbf{b}_1^\star \cdot \mathbf{b}_2^\star \times \mathbf{b}_3^\star} = \frac{\left[\mathbf{r}\mathbf{b}_3^\star\mathbf{b}_1^\star\right]}{\left[\mathbf{b}_1^\star\mathbf{b}_2^\star\mathbf{b}_3^\star\right]} = \mathbf{r} \cdot \hat{b}_2 \\ r_3^\star &= \frac{\mathbf{r} \cdot \mathbf{b}_1^\star \times \mathbf{b}_2^\star}{\mathbf{b}_1^\star \cdot \mathbf{b}_2^\star \times \mathbf{b}_3^\star} = \frac{\left[\mathbf{r}\mathbf{b}_1^\star\mathbf{b}_2^\star\right]}{\left[\mathbf{b}_1^\star\mathbf{b}_2^\star\mathbf{b}_3^\star\right]} = \mathbf{r} \cdot \hat{b}_3 \end{aligned} \tag{3.256}$$

Therefore, the reciprocal expression of a vector $\mathbf{r}$ in the reciprocal frame $OA^\star B^\star C^\star$ is given by (3.236). We may write Equations (3.235) and (3.236) as

$$\begin{aligned} {}^P\mathbf{r} &= r_1\hat{b}_1 + r_2\hat{b}_2 + r_3\hat{b}_3 \\ &= \left(\mathbf{r} \cdot \mathbf{b}_1^\star\right)\hat{b}_1 + \left(\mathbf{r} \cdot \mathbf{b}_2^\star\right)\hat{b}_2 + \left(\mathbf{r} \cdot \mathbf{b}_3^\star\right)\hat{b}_3 \end{aligned} \tag{3.257}$$

$$\begin{aligned} {}^R\mathbf{r} &= r_1^\star\mathbf{b}_1^\star + r_2^\star\mathbf{b}_2^\star + r_3^\star\mathbf{b}_3^\star \\ &= \left(\mathbf{r} \cdot \hat{b}_1\right)\mathbf{b}_1^\star + \left(\mathbf{r} \cdot \hat{b}_2\right)\mathbf{b}_2^\star + \left(\mathbf{r} \cdot \hat{b}_3\right)\mathbf{b}_3^\star \end{aligned} \tag{3.258}$$

Employing Equation (3.257), we may show that

$$r_i^\star = \mathbf{r} \cdot \hat{b}_i = \sum_{j=1}^{3} \left(\mathbf{r} \cdot \mathbf{b}_j^\star\right) \hat{b}_j \cdot \hat{b}_i = \sum_{j=1}^{3} \left(\hat{b}_i \cdot \hat{b}_j\right) r_j \tag{3.259}$$

and determine the transformation matrix ${}^R R_P$ to map the coordinates of a point from a principal frame P to a reciprocal frame R:

$${}^R\mathbf{r} = {}^R R_P \, {}^P\mathbf{r} \tag{3.260}$$

$${}^R R_P = \begin{bmatrix} \hat{b}_1 \cdot \hat{b}_1 & \hat{b}_1 \cdot \hat{b}_2 & \hat{b}_1 \cdot \hat{b}_3 \\ \hat{b}_2 \cdot \hat{b}_1 & \hat{b}_2 \cdot \hat{b}_2 & \hat{b}_2 \cdot \hat{b}_3 \\ \hat{b}_3 \cdot \hat{b}_1 & \hat{b}_3 \cdot \hat{b}_2 & \hat{b}_3 \cdot \hat{b}_3 \end{bmatrix} \tag{3.261}$$

Similarly, we may use Equation (3.258) to find

$$r_i = \mathbf{r} \cdot \mathbf{b}_i^\star = \sum_{j=1}^{3} \left(\mathbf{r} \cdot \hat{b}_j\right) \mathbf{b}_j^\star \cdot \mathbf{b}_i^\star = \sum_{j=1}^{3} \left(\mathbf{b}_i^\star \cdot \mathbf{b}_j^\star\right) r_j^\star \tag{3.262}$$

and determine the transformation matrix ${}^P R_R$ to map the coordinates of a point from the reciprocal frame R to the principal frame P:

$${}^P\mathbf{r} = {}^P R_R \, {}^R\mathbf{r} \tag{3.263}$$

$${}^P R_R = \begin{bmatrix} \mathbf{b}_1^\star \cdot \mathbf{b}_1^\star & \mathbf{b}_1^\star \cdot \mathbf{b}_2^\star & \mathbf{b}_1^\star \cdot \mathbf{b}_3^\star \\ \mathbf{b}_2^\star \cdot \mathbf{b}_1^\star & \mathbf{b}_2^\star \cdot \mathbf{b}_2^\star & \mathbf{b}_2^\star \cdot \mathbf{b}_3^\star \\ \mathbf{b}_3^\star \cdot \mathbf{b}_1^\star & \mathbf{b}_3^\star \cdot \mathbf{b}_2^\star & \mathbf{b}_3^\star \cdot \mathbf{b}_3^\star \end{bmatrix} \tag{3.264}$$

Introducing the notations (3.247) and (3.248) along with (3.203), we realize that the transformation matrices ${}^R R_P$ and ${}^P R_R$ are equal to the covariant and contravariant matrices, respectively:

$${}^R R_P = \left[g_{ij}\right] = \begin{bmatrix} g_{11} & g_{12} & g_{13} \\ g_{21} & g_{22} & g_{23} \\ g_{31} & g_{32} & g_{33} \end{bmatrix} \tag{3.265}$$

$${}^P R_R = \left[g_{ij}^\star\right] = \begin{bmatrix} g_{11}^\star & g_{12}^\star & g_{13}^\star \\ g_{21}^\star & g_{22}^\star & g_{23}^\star \\ g_{31}^\star & g_{32}^\star & g_{33}^\star \end{bmatrix} \tag{3.266}$$

Using the transformation matrices ${}^R R_P$ and ${}^P R_R$, we can relate the principal and reciprocal components:

$$r_i = \sum_{j=1}^{3} g_{ij}^\star r_j^\star \qquad r_i^\star = \sum_{j=1}^{3} g_{ij} r_j \tag{3.267}$$

The transformation matrix ${}^P R_R$ must be the inverse of ${}^R R_P$:

$${}^P R_R^{-1} = {}^R R_P \tag{3.268}$$

Therefore, we have

$$\begin{bmatrix} g_{11}^\star & g_{12}^\star & g_{13}^\star \\ g_{21}^\star & g_{22}^\star & g_{23}^\star \\ g_{31}^\star & g_{32}^\star & g_{33}^\star \end{bmatrix}^{-1} = \begin{bmatrix} g_{11} & g_{12} & g_{13} \\ g_{21} & g_{22} & g_{23} \\ g_{31} & g_{32} & g_{33} \end{bmatrix} \tag{3.269}$$

$$\begin{bmatrix} g_{11} & g_{12} & g_{13} \\ g_{21} & g_{22} & g_{23} \\ g_{31} & g_{32} & g_{33} \end{bmatrix}^{-1} = \begin{bmatrix} g_{11}^\star & g_{12}^\star & g_{13}^\star \\ g_{21}^\star & g_{22}^\star & g_{23}^\star \\ g_{31}^\star & g_{32}^\star & g_{33}^\star \end{bmatrix} \tag{3.270}$$

■

Example 202 Cartesian Coordinate Frame Is Reciprocal to Itself The reciprocal coordinate frame of an orthogonal frame, such as a Cartesian,

$$\hat{b}_1 = \hat{\imath} \qquad \hat{b}_2 = \hat{\jmath} \qquad \hat{b}_3 = \hat{k} \tag{3.271}$$

is an orthogonal coordinate frame that is coaxial to the original one:

$$\begin{aligned} \mathbf{b}_1^\star &= \frac{\hat{b}_2 \times \hat{b}_3}{\hat{b}_1 \cdot \hat{b}_2 \times \hat{b}_3} = \frac{\hat{\jmath} \times \hat{k}}{\hat{\imath} \cdot \hat{\jmath} \times \hat{k}} = \hat{\imath}^\star = \hat{\imath} \\ \mathbf{b}_2^\star &= \frac{\hat{b}_3 \times \hat{b}_1}{\hat{b}_1 \cdot \hat{b}_2 \times \hat{b}_3} = \frac{\hat{k} \times \hat{\imath}}{\hat{\imath} \cdot \hat{\jmath} \times \hat{k}} = \hat{\jmath}^\star = \hat{\jmath} \\ \mathbf{b}_3^\star &= \frac{\hat{b}_1 \times \hat{b}_2}{\hat{b}_1 \cdot \hat{b}_2 \times \hat{b}_3} = \frac{\hat{\imath} \times \hat{\jmath}}{\hat{\imath} \cdot \hat{\jmath} \times \hat{k}} = \hat{k}^\star = \hat{k} \end{aligned} \tag{3.272}$$

Therefore, when the coordinate frame is orthogonal, the covariant and contravariant metric coefficients are equal:

$$\left[g_{ij}\right] = \left[g_{ij}^\star\right] = \begin{bmatrix} \hat{\imath}\cdot\hat{\imath} & \hat{\imath}\cdot\hat{\jmath} & \hat{\imath}\cdot\hat{k} \\ \hat{\jmath}\cdot\hat{\imath} & \hat{\jmath}\cdot\hat{\jmath} & \hat{\jmath}\cdot\hat{k} \\ \hat{k}\cdot\hat{\imath} & \hat{k}\cdot\hat{\jmath} & \hat{k}\cdot\hat{k} \end{bmatrix} = [\mathbf{I}] \tag{3.273}$$

Example 203 ★ Expression of Two Vectors in a Nonorthogonal Frame Consider two vectors $\mathbf{r}_1$, $\mathbf{r}_2$,

$$\mathbf{r}_1 = 2\hat{\imath} + 3\hat{\jmath} + 4\hat{k} \qquad \mathbf{r}_2 = 8\hat{\imath} + 6\hat{\jmath} + 3\hat{k} \tag{3.274}$$

in a nonorthogonal frame with its Cartesian expression of the principal unit vectors:

$$\hat{b}_1 = \begin{bmatrix} 0.97334 \\ -0.17194 \\ -0.15198 \end{bmatrix} \qquad \hat{b}_2 = \begin{bmatrix} -0.64219 \\ 0.7662 \\ -0.02352 \end{bmatrix} \qquad \hat{b}_3 = \begin{bmatrix} 0 \\ 0 \\ 1 \end{bmatrix} \tag{3.275}$$

Using Equation (3.239), we calculate the Cartesian expression of the reciprocal base vectors as

$$\mathbf{b}_1^\star = \begin{bmatrix} 1.2059 \\ 1.0108 \\ 0 \end{bmatrix} \qquad \mathbf{b}_2^\star = \begin{bmatrix} 0.27062 \\ 1.532 \\ 0 \end{bmatrix} \qquad \mathbf{b}_3^\star = \begin{bmatrix} 0.18964 \\ 0.18964 \\ 1 \end{bmatrix} \tag{3.276}$$

to determine the principal and reciprocal components:

$$\begin{aligned} r_{11} &= \mathbf{r}_1 \cdot \mathbf{b}_1^\star = 5.4442 \\ r_{12} &= \mathbf{r}_1 \cdot \mathbf{b}_2^\star = 5.1372 \\ r_{13} &= \mathbf{r}_1 \cdot \mathbf{b}_3^\star = 4.9482 \end{aligned} \tag{3.277}$$

$$\begin{aligned} r_{21} &= \mathbf{r}_2 \cdot \mathbf{b}_1^\star = 15.712 \\ r_{22} &= \mathbf{r}_2 \cdot \mathbf{b}_2^\star = 11.357 \\ r_{23} &= \mathbf{r}_2 \cdot \mathbf{b}_3^\star = 5.655 \end{aligned} \tag{3.278}$$

$$\begin{aligned} r_{11}^\star &= \mathbf{r}_1 \cdot \hat{b}_1 = 0.82294 \\ r_{12}^\star &= \mathbf{r}_1 \cdot \hat{b}_2 = 0.92014 \\ r_{13}^\star &= \mathbf{r}_1 \cdot \hat{b}_3 = 4 \end{aligned} \tag{3.279}$$

$$\begin{aligned} r_{11}^\star &= \mathbf{r}_2 \cdot \hat{b}_1 = 6.299\,1 \\ r_{12}^\star &= \mathbf{r}_2 \cdot \hat{b}_2 = -0.610\,88 \\ r_{13}^\star &= \mathbf{r}_2 \cdot \hat{b}_3 = 3 \end{aligned} \tag{3.280}$$

The expression of vectors $\mathbf{r}_1$, $\mathbf{r}_2$ in the nonorthogonal coordinate frame $\hat{b}_1, \hat{b}_2, \hat{b}_3$ and its reciprocal frame $\mathbf{b}_1^\star, \mathbf{b}_2^\star, \mathbf{b}_3^\star$ would be

$$\begin{aligned} {}^P\mathbf{r}_I &= r_{11}\hat{b}_1 + r_{12}\hat{b}_2 + r_{13}\hat{b}_3 \\ &= 5.4442\hat{b}_1 + 5.1372\hat{b}_2 + 4.9482\hat{b}_3 \end{aligned} \tag{3.281}$$

$$\begin{aligned} {}^R\mathbf{r}_1 &= r_{11}^\star \mathbf{b}_1^\star + r_{12}^\star \mathbf{b}_2^\star + r_{13}^\star \mathbf{b}_3^\star \\ &= 0.82294\mathbf{b}_1^\star + 0.92014\mathbf{b}_2^\star + 4\mathbf{b}_3^\star \end{aligned} \tag{3.282}$$

and

$$\begin{aligned} {}^P\mathbf{r}_2 &= r_{21}\hat{b}_1 + r_{22}\hat{b}_2 + r_{23}\hat{b}_3 \\ &= 15.712\hat{b}_1 + 11.357\hat{b}_2 + 5.655\hat{b}_3 \end{aligned} \tag{3.283}$$

$$\begin{aligned} {}^R\mathbf{r}_2 &= r_{21}^\star \mathbf{b}_1^\star + r_{22}^\star \mathbf{b}_2^\star + r_{23}^\star \mathbf{b}_3^\star \\ &= 6.299\,1\mathbf{b}_1^\star - 0.610\,88\mathbf{b}_2^\star + 3\mathbf{b}_3^\star \end{aligned} \tag{3.284}$$

Example 204 Length of a Vector in Nonorthogonal Coordinate Frame Consider a vector $\mathbf{r}$ in a principal nonorthogonal coordinate frame P with unit vectors $\hat{b}_1, \hat{b}_2, \hat{b}_3$:

$$\hat{b}_1 = \begin{bmatrix} 0.97334 \\ -0.17194 \\ -0.15198 \end{bmatrix} \qquad \hat{b}_2 = \begin{bmatrix} -0.64219 \\ 0.7662 \\ -0.02352 \end{bmatrix} \qquad \hat{b}_3 = \begin{bmatrix} 0 \\ 0 \\ 1 \end{bmatrix} \tag{3.285}$$

The Cartesian and principal expressions of the vector are

$$^G\mathbf{r} = 2\hat{\imath} + 3\hat{\jmath} + 4\hat{k} \tag{3.286}$$

$$^P\mathbf{r} = 5.4442\hat{b}_1 + 5.1372\hat{b}_2 + 4.9482\hat{b}_3 \tag{3.287}$$

The length of $\mathbf{r}$ is

$$r = \sqrt{^G\mathbf{r} \cdot {^G\mathbf{r}}} = \sqrt{\left|^G\mathbf{r}\right| \left|^G\mathbf{r}\right|} = \sqrt{\left|\begin{matrix} 2 \\ 3 \\ 4 \end{matrix}\right| \left|\begin{matrix} 2 \\ 3 \\ 4 \end{matrix}\right|} = 5.3852 \tag{3.288}$$

however, because the unit vectors of the nonorthogonal frame are not orthogonal,

$$r \neq \sqrt{\left|^P\mathbf{r}\right| \left|^P\mathbf{r}\right|} \tag{3.289}$$

Using the nonorthogonal expression of $\mathbf{r}$, the length of the vector must be calculated by

$$\begin{aligned} r^2 &= \sum_{i=1}^{3} \sum_{j=1}^{3} r_i r_j g_{ij} = r_1^2 + r_2^2 + r_3^2 \\ &\quad + 2r_1 r_2 \left(\hat{b}_1 \cdot \hat{b}_2\right) + 2r_2 r_3 \left(\hat{b}_2 \cdot \hat{b}_3\right) + 2r_3 r_1 \left(\hat{b}_3 \cdot \hat{b}_1\right) \end{aligned} \tag{3.290}$$

Therefore, the length r using the nonorthogonal components of $^P\mathbf{r}$ would be

$$r = 5.3852 \tag{3.291}$$

Example 205 Transformation between Principal and Reciprocal Frames Consider a vector

$$\begin{aligned} ^P\mathbf{r} &= r_1\hat{b}_1 + r_2\hat{b}_2 + r_3\hat{b}_3 \\ &= 5.4442\hat{b}_1 + 5.1372\hat{b}_2 + 4.9482\hat{b}_3 \end{aligned} \tag{3.292}$$

in a nonorthogonal frame with the Cartesian expression of the principal unit vectors as

$$\hat{b}_1 = \begin{bmatrix} 0.97334 \\ -0.17194 \\ -0.15198 \end{bmatrix} \qquad \hat{b}_2 = \begin{bmatrix} -0.64219 \\ 0.7662 \\ -0.02352 \end{bmatrix} \qquad \hat{b}_3 = \begin{bmatrix} 0 \\ 0 \\ 1 \end{bmatrix} \tag{3.293}$$

To express $\mathbf{r}$ in the reciprocal frame, we determine the transformation matrix ${}^R R_P$:

$$ {}^R R_P = \left[g_{ij}\right] = \left[\hat{b}_i \cdot \hat{b}_j\right] = \begin{bmatrix} g_{11} & g_{12} & g_{13} \\ g_{21} & g_{22} & g_{23} \\ g_{31} & g_{32} & g_{33} \end{bmatrix} \tag{3.294} $$

$$ = \begin{bmatrix} 1 & -0.75324 & -0.15198 \\ -0.75324 & 1 & -0.02352 \\ -0.15198 & -0.02352 & 1 \end{bmatrix} \tag{3.295} $$

Therefore, the reciprocal expression of $\mathbf{r}$ is

$$ {}^R\mathbf{r} = {}^R R_P \, {}^P\mathbf{r} $$

$$ \begin{bmatrix} r_1^\bigstar \\ r_2^\bigstar \\ r_3^\bigstar \end{bmatrix} = {}^R R_P \begin{bmatrix} 5.4442 \\ 5.1372 \\ 4.9481 \end{bmatrix} = \begin{bmatrix} 0.82263 \\ 0.92003 \\ 4 \end{bmatrix} \tag{3.296} $$

The reverse transformation ${}^P R_R$ can be found by inverting ${}^R R_P$:

$$ {}^P R_R = {}^R R_P^{-1} = \begin{bmatrix} 2.4764 & 1.8752 & 0.42046 \\ 1.8752 & 2.4205 & 0.34192 \\ 0.42046 & 0.34192 & 1.0719 \end{bmatrix} \tag{3.297} $$

Employing ${}^P R_R$, we are able to determine the length of the reciprocal base vectors $\mathbf{b}_i^\bigstar$:

$$ {}^P R_R = \begin{bmatrix} \mathbf{b}_1^\bigstar \cdot \mathbf{b}_1^\bigstar & \mathbf{b}_1^\bigstar \cdot \mathbf{b}_2^\bigstar & \mathbf{b}_1^\bigstar \cdot \mathbf{b}_3^\bigstar \\ \mathbf{b}_2^\bigstar \cdot \mathbf{b}_1^\bigstar & \mathbf{b}_2^\bigstar \cdot \mathbf{b}_2^\bigstar & \mathbf{b}_2^\bigstar \cdot \mathbf{b}_3^\bigstar \\ \mathbf{b}_3^\bigstar \cdot \mathbf{b}_1^\bigstar & \mathbf{b}_3^\bigstar \cdot \mathbf{b}_2^\bigstar & \mathbf{b}_3^\bigstar \cdot \mathbf{b}_3^\bigstar \end{bmatrix} \tag{3.298} $$

$$ \begin{aligned} \left|\mathbf{b}_1^\bigstar\right| &= \sqrt{\mathbf{b}_1^\bigstar \cdot \mathbf{b}_1^\bigstar} = \sqrt{2.4764} = 1.5737 \\ \left|\mathbf{b}_2^\bigstar\right| &= \sqrt{\mathbf{b}_2^\bigstar \cdot \mathbf{b}_2^\bigstar} = \sqrt{2.4205} = 1.5558 \\ \left|\mathbf{b}_3^\bigstar\right| &= \sqrt{\mathbf{b}_3^\bigstar \cdot \mathbf{b}_3^\bigstar} = \sqrt{1.0719} = 1.0353 \end{aligned} \tag{3.299} $$

3.4.3 ★ Inner and Outer Vector Product

In a nonorthogonal coordinate frame $\hat{b}_1, \hat{b}_2, \hat{b}_3$ and its reciprocal frame $\mathbf{b}_1^\bigstar, \mathbf{b}_2^\bigstar, \mathbf{b}_3^\bigstar$, the inner and outer products of a vector $\mathbf{r}$,

$$ {}^P\mathbf{r} = r_1\hat{b}_1 + r_2\hat{b}_2 + r_3\hat{b}_3 \tag{3.300} $$

$$ {}^R\mathbf{r} = r_1^\bigstar \mathbf{b}_1^\bigstar + r_2^\bigstar \mathbf{b}_2^\bigstar + r_3^\bigstar \mathbf{b}_3^\bigstar \tag{3.301} $$

and themselves are

$$ {}^{P}\mathbf{r}\cdot{}^{P}\mathbf{r} = {}^{P}\mathbf{r}^{\mathrm{T}}\left[g_{ij}\right]{}^{P}\mathbf{r} = \sum_{j=1}^{3}\sum_{i=1}^{3} r_i r_j g_{ij} \tag{3.302} $$

$$ {}^{R}\mathbf{r}\cdot{}^{R}\mathbf{r} = {}^{R}\mathbf{r}^{\mathrm{T}}\left[g_{ij}^{\star}\right]{}^{R}\mathbf{r} = \sum_{j=1}^{3}\sum_{i=1}^{3} r_i^{\star} r_j^{\star} g_{ij}^{\star} \tag{3.303} $$

$$ {}^{P}\mathbf{r}\cdot{}^{R}\mathbf{r} = {}^{P}\mathbf{r}\cdot{}^{R}\mathbf{r} = \sum_{i=1}^{3} r_i r_i^{\star} \tag{3.304} $$

$$ {}^{P}\mathbf{r}\times{}^{P}\mathbf{r} = 0 \tag{3.305} $$

$$ {}^{R}\mathbf{r}\times{}^{R}\mathbf{r} = 0 \tag{3.306} $$

$$ {}^{P}\mathbf{r}\times{}^{R}\mathbf{r} = {}^{P}\mathbf{r}\times\left[g_{ij}^{\star}\right]{}^{R}\mathbf{r} \tag{3.307} $$

When two vectors ${}^{P}\mathbf{r}_1$ and ${}^{P}\mathbf{r}_2$ are in the principal frame,

$$ {}^{P}\mathbf{r}_1 = p_1\hat{b}_1 + p_2\hat{b}_2 + p_3\hat{b}_3 \tag{3.308} $$

$$ {}^{P}\mathbf{r}_2 = q_1\hat{b}_1 + q_2\hat{b}_2 + q_3\hat{b}_3 \tag{3.309} $$

the inner product of the vectors is

$$ {}^{P}\mathbf{r}_1\cdot{}^{P}\mathbf{r}_2 = \sum_{j=1}^{3}\sum_{i=1}^{3} p_i q_j g_{ij} \tag{3.310} $$

$$ \begin{aligned} &= p_1q_1g_{11} + p_2q_2g_{22} + p_2q_3g_{23} + \left(p_1q_2 + p_2q_1\right)g_{12} \\ &\quad + \left(p_1q_3 + p_3q_1\right)g_{13} + \left(p_3q_2 + p_3q_3\right)g_{33} \end{aligned} \tag{3.311} $$

and their outer product is

$$ {}^{P}\mathbf{r}_1\times{}^{P}\mathbf{r}_2 = \begin{vmatrix} \hat{b}_2\times\hat{b}_3 & \hat{b}_3\times\hat{b}_1 & \hat{b}_1\times\hat{b}_2 \\ p_1 & p_2 & p_3 \\ q_1 & q_2 & q_3 \end{vmatrix} \tag{3.312} $$

Proof: The position vector of a point in the nonorthogonal coordinate frame $\hat{b}_1, \hat{b}_2, \hat{b}_3$ can be shown in either the principal or reciprocal coordinate frame:

$$ {}^{P}\mathbf{r} = r_1\hat{b}_1 + r_2\hat{b}_2 + r_3\hat{b}_3 \tag{3.313} $$

$$ {}^{R}\mathbf{r} = r_1^{\star}\mathbf{b}_1^{\star} + r_2^{\star}\mathbf{b}_2^{\star} + r_3^{\star}\mathbf{b}_3^{\star} \tag{3.314} $$

Employing $\left|\hat{b}_i\right| = 1$, the inner product of ${}^P\mathbf{r}$ and itself is

$$\begin{aligned}
{}^P\mathbf{r} \cdot {}^P\mathbf{r} &= \left(r_1\hat{b}_1 + r_2\hat{b}_2 + r_3\hat{b}_3\right) \cdot \left(r_1\hat{b}_1 + r_2\hat{b}_2 + r_3\hat{b}_3\right) \\
&= r_1r_1\left(\hat{b}_1 \cdot \hat{b}_1\right) + r_1r_2\left(\hat{b}_1 \cdot \hat{b}_2\right) + r_1r_3\left(\hat{b}_1 \cdot \hat{b}_3\right) \\
&\quad + r_2r_1\left(\hat{b}_2 \cdot \hat{b}_1\right) + r_2r_2\left(\hat{b}_2 \cdot \hat{b}_2\right) + r_2r_3\left(\hat{b}_2 \cdot \hat{b}_3\right) \\
&\quad + r_3r_1\left(\hat{b}_3 \cdot \hat{b}_1\right) + r_3r_2\left(\hat{b}_3 \cdot \hat{b}_2\right) + r_3r_3\left(\hat{b}_3 \cdot \hat{b}_3\right) \\
&= {}^P\mathbf{r}^{\mathrm{T}}\left[g_{ij}\right]{}^P\mathbf{r} = \sum_{j=1}^{3}\sum_{i=1}^{3} r_i r_j g_{ij} = r_1^2 + r_2^2 + r_3^2 \\
&\quad + 2r_1r_2\left(\hat{b}_1 \cdot \hat{b}_2\right) + 2r_2r_3\left(\hat{b}_2 \cdot \hat{b}_3\right) + 2r_3r_1\left(\hat{b}_3 \cdot \hat{b}_1\right)
\end{aligned} \tag{3.315}$$

and with the same method, we have

$$\begin{aligned}
{}^R\mathbf{r} \cdot {}^R\mathbf{r} &= {}^R\mathbf{r}^{\mathrm{T}}\left[g_{ij}^{\star}\right]{}^R\mathbf{r} = \sum_{j=1}^{3}\sum_{i=1}^{3} r_i^{\star} r_j^{\star} g_{ij}^{\star} \\
&= r_1^{\star 2}\left(\mathbf{b}_1^{\star} \cdot \mathbf{b}_1^{\star}\right) + r_2^{\star 2}\left(\mathbf{b}_2^{\star} \cdot \mathbf{b}_2^{\star}\right) + r_3^{\star 2}\left(\mathbf{b}_3^{\star} \cdot \mathbf{b}_3^{\star}\right) \\
&\quad + 2r_1^{\star}r_2^{\star}\left(\mathbf{b}_1^{\star} \cdot \mathbf{b}_2^{\star}\right) + 2r_2^{\star}r_3^{\star}\left(\mathbf{b}_2^{\star} \cdot \mathbf{b}_3^{\star}\right) + 2r_3^{\star}r_1^{\star}\left(\mathbf{b}_3^{\star} \cdot \mathbf{b}_1^{\star}\right)
\end{aligned} \tag{3.316}$$

Using Equations (3.243) and (3.302), the inner product of ${}^P\mathbf{r}$ and ${}^R\mathbf{r}$ is

$$\begin{aligned}
{}^P\mathbf{r} \cdot {}^R\mathbf{r} &= {}^P\mathbf{r} \cdot {}^PR_R\,{}^R\mathbf{r} = {}^P\mathbf{r} \cdot \left[g_{ij}^{\star}\right]{}^R\mathbf{r} = {}^P\mathbf{r}^{\mathrm{T}}\left[g_{ij}\right]\left[g_{ij}^{\star}\right]{}^R\mathbf{r} \\
&= {}^P\mathbf{r}^{\mathrm{T}}\left[\mathbf{I}\right]{}^R\mathbf{r} = r_1r_1^{\star} + r_2r_2^{\star} + r_3r_3^{\star} = {}^R\mathbf{r} \cdot {}^P\mathbf{r}
\end{aligned} \tag{3.317}$$

So, if two vectors ${}^P\mathbf{r}_1$ and ${}^P\mathbf{r}_2$ are both in the principal coordinate frame,

$${}^P\mathbf{r}_1 = p_1\hat{b}_1 + p_2\hat{b}_2 + p_3\hat{b}_3 \tag{3.318}$$

$${}^P\mathbf{r}_2 = q_1\hat{b}_1 + q_2\hat{b}_2 + q_3\hat{b}_3 \tag{3.319}$$

then their inner product would be

$${}^P\mathbf{r}_1 \cdot {}^P\mathbf{r}_2 = {}^P\mathbf{r}_1^{\mathrm{T}}\left[g_{ij}\right]{}^P\mathbf{r}_2 \tag{3.320}$$

which is equal to (3.310).

Because of Equations (3.180)–(3.182), the outer product of ${}^{P}\mathbf{r}$ and itself is zero:

$$
\begin{aligned}
{}^{P}\mathbf{r} \times {}^{P}\mathbf{r} &= \left(r_1\hat{b}_1 + r_2\hat{b}_2 + r_3\hat{b}_3\right) \times \left(r_1\hat{b}_1 + r_2\hat{b}_2 + r_3\hat{b}_3\right) \\
&= r_1r_1\left(\hat{b}_1 \times \hat{b}_1\right) + r_1r_2\left(\hat{b}_1 \times \hat{b}_2\right) + r_1r_3\left(\hat{b}_1 \times \hat{b}_3\right) \\
&\quad + r_2r_1\left(\hat{b}_2 \times \hat{b}_1\right) + r_2r_2\left(\hat{b}_2 \times \hat{b}_2\right) + r_2r_3\left(\hat{b}_2 \times \hat{b}_3\right) \\
&\quad + r_3r_1\left(\hat{b}_3 \times \hat{b}_1\right) + r_3r_2\left(\hat{b}_3 \times \hat{b}_2\right) + r_3r_3\left(\hat{b}_3 \times \hat{b}_3\right) \\
&= r_1r_2\mathbf{b}_3^{\star} - r_1r_3\mathbf{b}_2^{\star} - r_2r_1\mathbf{b}_3^{\star} + r_2r_3\mathbf{b}_1^{\star} + r_3r_1\mathbf{b}_2^{\star} - r_3r_2\mathbf{b}_1^{\star} \\
&= 0
\end{aligned}
\tag{3.321}
$$

Using the same method we can show that

$$
{}^{R}\mathbf{r} \times {}^{R}\mathbf{r} = 0 \tag{3.322}
$$

Therefore,

$$
\hat{b}_1 \times \hat{b}_1 = 0 \qquad \hat{b}_2 \times \hat{b}_2 = 0 \qquad \hat{b}_3 \times \hat{b}_3 = 0 \tag{3.323}
$$

$$
\mathbf{b}_1^{\star} \times \mathbf{b}_1^{\star} = 0 \qquad \mathbf{b}_2^{\star} \times \mathbf{b}_2^{\star} = 0 \qquad \mathbf{b}_3^{\star} \times \mathbf{b}_3^{\star} = 0 \tag{3.324}
$$

Employing these results, we can expand the outer product of two vectors ${}^{P}\mathbf{r}_l$ and ${}^{P}\mathbf{r}_2$ and show Equation (3.312):

$$
\begin{aligned}
{}^{P}\mathbf{r}_l \times {}^{P}\mathbf{r}_2 &= \left(p_1\hat{b}_1 + p_2\hat{b}_2 + p_3\hat{b}_3\right) \times \left(q_1\hat{b}_1 + q_2\hat{b}_2 + q_3\hat{b}_3\right) \\
&= p_1q_1\left(\hat{b}_1 \times \hat{b}_1\right) + p_1q_2\left(\hat{b}_1 \times \hat{b}_2\right) + p_1q_3\left(\hat{b}_1 \times \hat{b}_3\right) \\
&\quad + p_2q_1\left(\hat{b}_2 \times \hat{b}_1\right) + p_2q_2\left(\hat{b}_2 \times \hat{b}_2\right) + p_2q_3\left(\hat{b}_2 \times \hat{b}_3\right) \\
&\quad + p_3q_1\left(\hat{b}_3 \times \hat{b}_1\right) + p_3q_2\left(\hat{b}_3 \times \hat{b}_2\right) + p_3q_3\left(\hat{b}_3 \times \hat{b}_3\right) \\
&= \begin{vmatrix} \hat{b}_2 \times \hat{b}_3 & \hat{b}_3 \times \hat{b}_1 & \hat{b}_1 \times \hat{b}_2 \\ p_1 & p_2 & p_3 \\ q_1 & q_2 & q_3 \end{vmatrix}
\end{aligned}
\tag{3.325}
$$

Substituting $\hat{b}_i \times \hat{b}_j$ from the definition of the reciprocal vectors (3.180)–(3.182), we may also show that

$$
{}^{P}\mathbf{r}_l \times {}^{P}\mathbf{r}_2 = \left[\hat{b}_1\hat{b}_2\hat{b}_3\right] \begin{vmatrix} \mathbf{b}_1^{\star} & \mathbf{b}_2^{\star} & \mathbf{b}_3^{\star} \\ p_1 & p_2 & p_3 \\ q_1 & q_2 & q_3 \end{vmatrix} = V \begin{vmatrix} \mathbf{b}_1^{\star} & \mathbf{b}_2^{\star} & \mathbf{b}_3^{\star} \\ p_1 & p_2 & p_3 \\ q_1 & q_2 & q_3 \end{vmatrix} \tag{3.326}
$$

$$
{}^R\mathbf{r}_1 \times {}^R\mathbf{r}_2 = \left[\mathbf{b}_1^\bigstar \mathbf{b}_2^\bigstar \mathbf{b}_3^\bigstar\right] \begin{vmatrix} \hat{b}_1 & \hat{b}_2 & \hat{b}_3 \\ p_1^\bigstar & p_2^\bigstar & p_3^\bigstar \\ q_1^\bigstar & q_2^\bigstar & q_3^\bigstar \end{vmatrix}
$$

$$
= \frac{1}{V} \begin{vmatrix} \hat{b}_1 & \hat{b}_2 & \hat{b}_3 \\ p_1^\bigstar & p_2^\bigstar & p_3^\bigstar \\ q_1^\bigstar & q_2^\bigstar & q_3^\bigstar \end{vmatrix} \tag{3.327}
$$

To determine the outer product of two vectors in the principal and reciprocal frames, we should transform one vector to the other frame:

$$
{}^P\mathbf{r} \times {}^R\mathbf{r} = {}^P\mathbf{r} \times {}^P R_R \, {}^R\mathbf{r} = {}^P\mathbf{r} \times \left[g_{ij}^\bigstar\right] {}^R\mathbf{r} \tag{3.328}
$$

■

Example 206 ★ Inner Product in the Plane The inner product definition of two vectors $\mathbf{r}_1$ and $\mathbf{r}_2$ in the same coordinate frame is coordinate free:

$$
\mathbf{r}_1 \cdot \mathbf{r}_2 = \mathbf{r}_2 \cdot \mathbf{r}_1 = r_1 r_2 \cos(\mathbf{r}_1, \mathbf{r}_2) \tag{3.329}
$$

If the vectors $\mathbf{r}_1$ and $\mathbf{r}_2$ are orthogonal, then $\cos(\mathbf{r}_1, \mathbf{r}_2) = 0$ and the inner product is zero, $\mathbf{r}_1 \cdot \mathbf{r}_2 = 0$. The inner product may also be defined by the projection principle

$$
\mathbf{r}_1 \cdot \mathbf{r}_2 = r_1 r_2' = r_1' r_2 \tag{3.330}
$$

where r_1' and r_2' are projections of $\mathbf{r}_1$ and $\mathbf{r}_2$ on each other, as are shown in Figure 3.21.

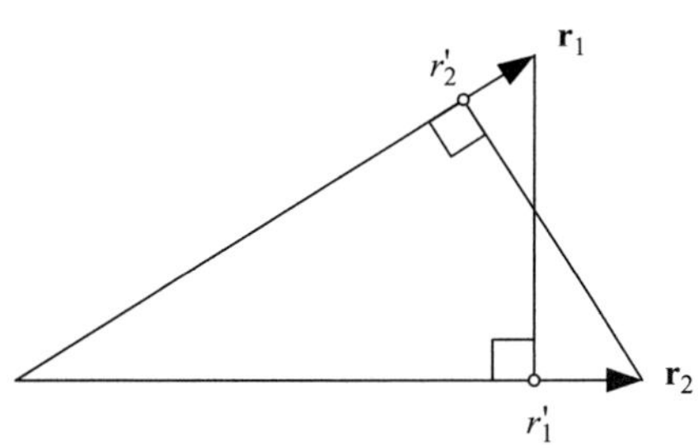

Figure 3.21 Geometric interpretation of the inner product.

Example 207 ★ Inner and Outer Product of Two Vectors Consider two vectors $\mathbf{r}_1$, $\mathbf{r}_2$ in a nonorthogonal frame

$$
\mathbf{r}_1 = 2\hat{\imath} + 3\hat{\jmath} + 4\hat{k} \tag{3.331}
$$

$$
= 5.4442\hat{b}_1 + 5.1372\hat{b}_2 + 4.9482\hat{b}_3 \tag{3.332}
$$

$$
= 0.82294\mathbf{b}_1^\bigstar + 0.92014\mathbf{b}_2^\bigstar + 4\mathbf{b}_3^\bigstar \tag{3.333}
$$

$$\mathbf{r}_2 = 8\hat{\imath} + 6\hat{\jmath} + 3\hat{k} \tag{3.334}$$

$$= 15.712\hat{b}_1 + 11.357\hat{b}_2 + 5.655\hat{b}_3 \tag{3.335}$$

$$= 6.299\,1\mathbf{b}_1^{\star} - 0.610\,88\mathbf{b}_2^{\star} + 3\mathbf{b}_3^{\star} \tag{3.336}$$

with the following unit and reciprocal base vectors:

$$\hat{b}_1 = \begin{bmatrix} 0.97334 \\ -0.17194 \\ -0.15198 \end{bmatrix} \qquad \hat{b}_2 = \begin{bmatrix} -0.64219 \\ 0.7662 \\ -0.02352 \end{bmatrix} \qquad \hat{b}_3 = \begin{bmatrix} 0 \\ 0 \\ 1 \end{bmatrix} \tag{3.337}$$

$$\mathbf{b}_1^{\star} = \begin{bmatrix} 1.2059 \\ 1.0108 \\ 0 \end{bmatrix} \qquad \mathbf{b}_2^{\star} = \begin{bmatrix} 0.27062 \\ 1.532 \\ 0 \end{bmatrix} \qquad \mathbf{b}_3^{\star} = \begin{bmatrix} 0.18964 \\ 0.18964 \\ 1 \end{bmatrix} \tag{3.338}$$

The inner product of $\mathbf{r}_1$and $\mathbf{r}_2$ can be calculated as

$$\mathbf{r}_1 \cdot \mathbf{r}_2 = p_1 q_1^{\star} + p_2 q_2^{\star} + p_3 q_3^{\star}$$
$$= \begin{bmatrix} 5.4442 \\ 5.1372 \\ 4.9482 \end{bmatrix} \cdot \begin{bmatrix} 6.299\,1 \\ -0.610\,88 \\ 3 \end{bmatrix} = 46 \tag{3.339}$$

and the outer product of $\mathbf{r}_1$ and $\mathbf{r}_2$ as

$$\mathbf{r}_1 \times \mathbf{r}_2 = \left[\hat{b}_1 \hat{b}_2 \hat{b}_3\right] \begin{vmatrix} \mathbf{b}_1^{\star} & \mathbf{b}_2^{\star} & \mathbf{b}_3^{\star} \\ p_1 & p_2 & p_3 \\ q_1 & q_2 & q_3 \end{vmatrix}$$
$$= \begin{vmatrix} 0.97334 & -0.17194 & -0.15198 \\ -0.64219 & 0.7662 & -0.02352 \\ 0 & 0 & 1 \end{vmatrix} \begin{vmatrix} \mathbf{b}_1^{\star} & \mathbf{b}_2^{\star} & \mathbf{b}_3^{\star} \\ 5.4442 & 5.1372 & 4.9482 \\ 15.712 & 11.357 & 5.655 \end{vmatrix}$$
$$= -17.247\mathbf{b}_1^{\star} + 29.836\mathbf{b}_2^{\star} - 11.999\mathbf{b}_3^{\star} \tag{3.340}$$

Example 208 ★ Examination of Equation (3.327) To show that Equation (3.327) is applied, we begin with the nonorthogonal expressions of $\mathbf{r}_1$ and $\mathbf{r}_2$:

$$\mathbf{r}_1 = p_1\hat{b}_1 + p_2\hat{b}_2 + p_3\hat{b}_3 = p_1^{\star}\mathbf{b}_1^{\star} + p_2^{\star}\mathbf{b}_2^{\star} + p_3^{\star}\mathbf{b}_3^{\star} \tag{3.341}$$

$$\mathbf{r}_2 = q_1\hat{b}_1 + q_2\hat{b}_2 + q_3\hat{b}_3 = q_1^{\star}\mathbf{b}_1^{\star} + q_2^{\star}\mathbf{b}_2^{\star} + q_3^{\star}\mathbf{b}_3^{\star} \tag{3.342}$$

Using Equations (3.240) will show that the outer product of the vectors is

$$
\begin{aligned}
\mathbf{r}_1 \times \mathbf{r}_2 &= \left(p_1^\star \mathbf{b}_1^\star + p_2^\star \mathbf{b}_2^\star + p_3^\star \mathbf{b}_3^\star\right) \times \left(q_1^\star \mathbf{b}_1^\star + q_2^\star \mathbf{b}_2^\star + q_3^\star \mathbf{b}_3^\star\right) \\
&= p_1^\star q_1^\star \left(\mathbf{b}_1^\star \times \mathbf{b}_1^\star\right) + p_1^\star q_2^\star \left(\mathbf{b}_1^\star \times \mathbf{b}_2^\star\right) + p_1^\star q_3^\star \left(\mathbf{b}_1^\star \times \mathbf{b}_3^\star\right) \\
&\quad + p_2^\star q_1^\star \left(\mathbf{b}_2^\star \times \mathbf{b}_1^\star\right) + p_2^\star q_2^\star \left(\mathbf{b}_2^\star \times \mathbf{b}_2^\star\right) + p_2^\star q_3^\star \left(\mathbf{b}_2^\star \times \mathbf{b}_3^\star\right) \\
&\quad + p_3^\star q_1^\star \left(\mathbf{b}_3^\star \times \mathbf{b}_1^\star\right) + p_3^\star q_2^\star \left(\mathbf{b}_3^\star \times \mathbf{b}_2^\star\right) + p_3^\star q_3^\star \left(\mathbf{b}_3^\star \times \mathbf{b}_3^\star\right) \\
&= \left[\mathbf{b}_1^\star \mathbf{b}_2^\star \mathbf{b}_3^\star\right] \left(p_2^\star q_3^\star - p_3^\star q_2^\star\right) \hat{b}_1 \\
&\quad + \left[\mathbf{b}_1^\star \mathbf{b}_2^\star \mathbf{b}_3^\star\right] \left(p_3^\star q_1^\star - p_1^\star q_3^\star\right) \hat{b}_2 \\
&\quad + \left[\mathbf{b}_1^\star \mathbf{b}_2^\star \mathbf{b}_3^\star\right] \left(p_1^\star q_2^\star - p_2^\star q_1^\star\right) \hat{b}_3 \\
&= \left[\mathbf{b}_1^\star \mathbf{b}_2^\star \mathbf{b}_3^\star\right] \begin{vmatrix} \hat{b}_1 & \hat{b}_2 & \hat{b}_3 \\ p_1^\star & p_2^\star & p_3^\star \\ q_1^\star & q_2^\star & q_3^\star \end{vmatrix} = \frac{1}{V} \begin{vmatrix} \hat{b}_1 & \hat{b}_2 & \hat{b}_3 \\ p_1^\star & p_2^\star & p_3^\star \\ q_1^\star & q_2^\star & q_3^\star \end{vmatrix}
\end{aligned}
\tag{3.343}
$$

Example 209 ★ Inner and Outer Product of Two Vectors Consider the Cartesian position vector of two points P and Q at $\mathbf{p}$ and $\mathbf{q}$,

$$
{}^G\mathbf{p} = 2\hat{\imath} + 2\hat{\jmath} + 2\hat{k} \qquad {}^G\mathbf{q} = -\hat{\imath} - \hat{\jmath} + 4\hat{k} \tag{3.344}
$$

and the unit base vectors of an oblique coordinate frame,

$$
\hat{b}_1 = \hat{\imath} \qquad \hat{b}_2 = \frac{\sqrt{2}}{2}\left(\hat{\imath} + \hat{\jmath}\right) \qquad \hat{b}_3 = \frac{\sqrt{3}}{3}\left(\hat{\imath} + \hat{\jmath} + \hat{k}\right) \tag{3.345}
$$

We can determine the transformation matrix ${}^PR_G$ between the oblique and Cartesian frames:

$$
\begin{bmatrix} \hat{b}_1 \\ \hat{b}_2 \\ \hat{b}_3 \end{bmatrix} = \begin{bmatrix} 1 & 0 & 0 \\ \sqrt{2}/2 & \sqrt{2}/2 & 0 \\ \sqrt{3}/3 & \sqrt{3}/3 & \sqrt{3}/3 \end{bmatrix} \begin{bmatrix} \hat{\imath} \\ \hat{\jmath} \\ \hat{k} \end{bmatrix} \tag{3.346}
$$

$$
{}^PR_G = \begin{bmatrix} 1 & 0 & 0 \\ \sqrt{2}/2 & \sqrt{2}/2 & 0 \\ \sqrt{3}/3 & \sqrt{3}/3 & \sqrt{3}/3 \end{bmatrix} \tag{3.347}
$$

$$
\begin{bmatrix} \hat{\imath} \\ \hat{\jmath} \\ \hat{k} \end{bmatrix} = \begin{bmatrix} 1 & 0 & 0 \\ -1 & \sqrt{2} & 0 \\ 0 & -\sqrt{2} & \sqrt{3} \end{bmatrix} \begin{bmatrix} \hat{b}_1 \\ \hat{b}_2 \\ \hat{b}_3 \end{bmatrix} \tag{3.348}
$$

$$
{}^GR_P = {}^PR_G^{-1} = \begin{bmatrix} 1 & 0 & 0 \\ -1 & \sqrt{2} & 0 \\ 0 & -\sqrt{2} & \sqrt{3} \end{bmatrix} \tag{3.349}
$$

Therefore, **p** and **q** in the principal frame are

$$
\begin{aligned}
{}^{P}\mathbf{p} &= 2\left(\hat{b}_1\right) + 2\left(-\hat{b}_1 + \sqrt{2}\hat{b}_2\right) + 2\left(-\sqrt{2}\hat{b}_2 + \sqrt{3}\hat{b}_3\right) \\
&= 2\sqrt{3}\hat{b}_3 \qquad (3.350) \\
{}^{P}\mathbf{q} &= -\hat{b}_1 - \left(-\hat{b}_1 + \sqrt{2}\hat{b}_2\right) + 4\left(-\sqrt{2}\hat{b}_2 + \sqrt{3}\hat{b}_3\right) \\
&= -5\sqrt{2}\hat{b}_2 + 4\sqrt{3}\hat{b}_3 \qquad (3.351)
\end{aligned}
$$

Using the oblique unit vectors, we are able to determine the metric matrix that is the transformation matrix between the principal and reciprocal frames:

$$
{}^{R}R_P = \left[g_{ij}\right] = \left[\hat{b}_i \cdot \hat{b}_j\right] = \begin{bmatrix} 1 & 0.70711 & 0.57735 \\ 0.70711 & 1 & 0.81650 \\ 0.57735 & 0.81650 & 1 \end{bmatrix} \qquad (3.352)
$$

So, the reciprocal expression of **p** and **q** are

$$
{}^{R}\mathbf{p} = \left[g_{ij}\right]\,{}^{P}\mathbf{p} = \left[g_{ij}\right]\begin{bmatrix} 0 \\ 0 \\ 2\sqrt{3} \end{bmatrix} = 2\mathbf{b}_1^{\star} + 2.8284\mathbf{b}_2^{\star} + 3.4641\mathbf{b}_3^{\star} \qquad (3.353)
$$

$$
{}^{R}\mathbf{q} = \left[g_{ij}\right]\,{}^{P}\mathbf{q} = \left[g_{ij}\right]\begin{bmatrix} 0 \\ -5\sqrt{2} \\ 4\sqrt{3} \end{bmatrix} = -\mathbf{b}_1^{\star} - 1.4142\mathbf{b}_2^{\star} + 1.1547\mathbf{b}_3^{\star} \qquad (3.354)
$$

The inner and outer products of **p** and **q** are

$$
{}^{P}\mathbf{p} \cdot {}^{P}\mathbf{q} = \sum_{j=1}^{3}\sum_{i=1}^{3} p_i q_j g_{ij} = 4 \qquad (3.355)
$$

$$
\begin{aligned}
{}^{P}\mathbf{p} \times {}^{P}\mathbf{q} &= \left[\hat{b}_1\hat{b}_2\hat{b}_3\right]\begin{vmatrix} \mathbf{b}_1^{\star} & \mathbf{b}_2^{\star} & \mathbf{b}_2^{\star} \\ p_1 & p_2 & p_3 \\ q_1 & q_2 & q_3 \end{vmatrix} \\
&= 0.40825\begin{vmatrix} \mathbf{b}_1^{\star} & \mathbf{b}_2^{\star} & \mathbf{b}_2^{\star} \\ 0 & 0 & 2\sqrt{3} \\ 0 & -5\sqrt{2} & 4\sqrt{3} \end{vmatrix} = 10\mathbf{b}_1^{\star} \qquad (3.356)
\end{aligned}
$$

Example 210 ★ Angle between Two Vectors We can use the inner product of two vectors and determine the angle between them. Consider two vectors ${}^{P}\mathbf{r}_1$ and ${}^{P}\mathbf{r}_2$:

$$
{}^{P}\mathbf{r}_1 = p_1\hat{b}_1 + p_2\hat{b}_2 + p_3\hat{b}_3 \qquad (3.357)
$$

$$
{}^{P}\mathbf{r}_2 = q_1\hat{b}_1 + q_2\hat{b}_2 + q_3\hat{b}_3 \qquad (3.358)
$$

The cosine of the angle between ${}^{P}\mathbf{r}_{I}$ and ${}^{P}\mathbf{r}_{2}$ is

$$\cos(\mathbf{r}_1, \mathbf{r}_2) = \frac{\mathbf{r}_1 \cdot \mathbf{r}_2}{|\mathbf{r}_1| \cdot |\mathbf{r}_2|} = \frac{\sum_{j=1}^{3}\sum_{i=1}^{3} p_i q_j g_{ij}}{\sqrt{\sum_{j=1}^{3}\sum_{i=1}^{3} p_i p_j g_{ij}}\sqrt{\sum_{j=1}^{3}\sum_{i=1}^{3} q_i q_j g_{ij}}} \tag{3.359}$$

As an example, let us determine the angle between ${}^{P}\mathbf{r}_{I}$ and ${}^{P}\mathbf{r}_{2}$,

$${}^{P}\mathbf{r}_{I} = 2\sqrt{3}\hat{b}_3 \tag{3.360}$$

$${}^{P}\mathbf{r}_{2} = -5\sqrt{2}\hat{b}_2 + 4\sqrt{3}\hat{b}_3 \tag{3.361}$$

in a curvilinear coordinate system with the following base vectors:

$$\hat{b}_1 = \hat{\imath} \qquad \hat{b}_2 = \frac{\sqrt{2}}{2}(\hat{\imath} + \hat{\jmath}) \qquad \hat{b}_3 = \frac{\sqrt{3}}{3}\left(\hat{\imath} + \hat{\jmath} + \hat{k}\right) \tag{3.362}$$

We should first calculate the $[g_{ij}]$-matrix,

$$[g_{ij}] = {}^{R}R_{P} = \left[\hat{b}_i \cdot \hat{b}_j\right] = \begin{bmatrix} 1 & 0.70711 & 0.57735 \\ 0.70711 & 1 & 0.81650 \\ 0.57735 & 0.81650 & 1 \end{bmatrix} \tag{3.363}$$

and determine the required inner products ${}^{P}\mathbf{r}_{I} \cdot {}^{P}\mathbf{r}_{2}$, ${}^{P}\mathbf{r}_{I} \cdot {}^{P}\mathbf{r}_{I}$, and ${}^{P}\mathbf{r}_{2} \cdot {}^{P}\mathbf{r}_{2}$:

$${}^{P}\mathbf{r}_{I} \cdot {}^{P}\mathbf{r}_{2} = {}^{P}\mathbf{r}_{I}^{\mathrm{T}}\,[g_{ij}]\,{}^{P}\mathbf{r}_{2} = \begin{bmatrix} 0 \\ 0 \\ 2\sqrt{3} \end{bmatrix}^{\mathrm{T}} [g_{ij}] \begin{bmatrix} 0 \\ -5\sqrt{2} \\ 4\sqrt{3} \end{bmatrix} = 4 \tag{3.364}$$

$$|\mathbf{r}_1| = \sqrt{{}^{P}\mathbf{r}_{I} \cdot {}^{P}\mathbf{r}_{1}} = \sqrt{{}^{P}\mathbf{r}_{I}^{\mathrm{T}}\,[g_{ij}]\,{}^{P}\mathbf{r}_{I}} = \sqrt{\begin{bmatrix} 0 \\ 0 \\ 2\sqrt{3} \end{bmatrix}^{\mathrm{T}} [g_{ij}] \begin{bmatrix} 0 \\ 0 \\ 2\sqrt{3} \end{bmatrix}} = \sqrt{12} = 3.4641 \tag{3.365}$$

$$|\mathbf{r}_2| = \sqrt{{}^{P}\mathbf{r}_{2} \cdot {}^{P}\mathbf{r}_{2}} = \sqrt{{}^{P}\mathbf{r}_{2}^{\mathrm{T}}\,[g_{ij}]\,{}^{P}\mathbf{r}_{2}} = \sqrt{\begin{bmatrix} 0 \\ -5\sqrt{2} \\ 4\sqrt{3} \end{bmatrix}^{\mathrm{T}} [g_{ij}] \begin{bmatrix} 0 \\ -5\sqrt{2} \\ 4\sqrt{3} \end{bmatrix}} = \sqrt{18} = 4.2426 \tag{3.366}$$

Therefore, the angle between ${}^{P}\mathbf{r}_{I}$ and ${}^{P}\mathbf{r}_{2}$ is

$$\theta = \arccos(\mathbf{r}_1, \mathbf{r}_2) = \arccos\frac{\mathbf{r}_1 \cdot \mathbf{r}_2}{|\mathbf{r}_1| \cdot |\mathbf{r}_2|} = \arccos\frac{4}{3.4641 \times 4.2426} = \arccos 0.272\,17 = 1.2951\,\mathrm{rad} = 74.204\,\mathrm{deg} \tag{3.367}$$

Example 211 ★ Scalar Triple Product in Oblique Coordinates If **s**, **p**, **q** are three vectors in an oblique coordinate frame P,

$$\mathbf{s} = s_1\hat{b}_1 + s_2\hat{b}_2 + s_3\hat{b}_3 \tag{3.368}$$

$$\mathbf{p} = p_1\hat{b}_1 + p_2\hat{b}_2 + p_3\hat{b}_3 \tag{3.369}$$

$$\mathbf{q} = q_1\hat{b}_1 + q_2\hat{b}_2 + q_3\hat{b}_3 \tag{3.370}$$

then their scalar triple product $\left[\mathbf{spq}\right]$ is

$$\mathbf{s} \cdot \mathbf{p} \times \mathbf{q} = \left(s_1\hat{b}_1 + s_2\hat{b}_2 + s_3\hat{b}_3\right) \cdot \begin{vmatrix} \hat{b}_2 \times \hat{b}_3 & \hat{b}_3 \times \hat{b}_1 & \hat{b}_1 \times \hat{b}_2 \\ p_1 & p_2 & p_3 \\ q_1 & q_2 & q_3 \end{vmatrix} \tag{3.371}$$

The scalar triple product of the principal unit vectors is a scalar:

$$\left[\hat{b}_1\hat{b}_2\hat{b}_3\right] = V \tag{3.372}$$

So, we may show the scalar triple product $\left[\mathbf{spq}\right]$ in a simpler way:

$$\mathbf{s} \cdot \mathbf{p} \times \mathbf{q} = V \begin{vmatrix} s_1 & s_2 & s_3 \\ p_1 & p_2 & p_3 \\ q_1 & q_2 & q_3 \end{vmatrix} \tag{3.373}$$

If the vectors **s**, **p**, **q** are expressed in the reciprocal coordinate frame,

$${}^R\mathbf{s} = s_1^\bigstar \mathbf{b}_1^\bigstar + s_2^\bigstar \mathbf{b}_2^\bigstar + s_3^\bigstar \mathbf{b}_3^\bigstar \tag{3.374}$$

$${}^R\mathbf{p} = p_1^\bigstar \mathbf{b}_1^\bigstar + p_2^\bigstar \mathbf{b}_2^\bigstar + p_3^\bigstar \mathbf{b}_3^\bigstar \tag{3.375}$$

$${}^R\mathbf{q} = q_1^\bigstar \mathbf{b}_1^\bigstar + q_2^\bigstar \mathbf{b}_2^\bigstar + q_3^\bigstar \mathbf{b}_3^\bigstar \tag{3.376}$$

then

$${}^R\mathbf{s} \cdot {}^R\mathbf{p} \times {}^R\mathbf{q} = \frac{1}{V} \begin{vmatrix} s_1^\bigstar & s_2^\bigstar & s_3^\bigstar \\ p_1^\bigstar & p_2^\bigstar & p_3^\bigstar \\ q_1^\bigstar & q_2^\bigstar & q_3^\bigstar \end{vmatrix} \tag{3.377}$$

If the coordinate frame is an orthogonal Cartesian frame, $V = 1$, then

$$V = \hat{\imath} \cdot \hat{\jmath} \times \hat{k} = \begin{vmatrix} 1 & 0 & 0 \\ 0 & 1 & 0 \\ 0 & 0 & 1 \end{vmatrix} = 1 \tag{3.378}$$

and the scalar triple product of three vectors **x**, **y**, **z** in a Cartesian frame reduces to

$$\mathbf{x} \cdot \mathbf{y} \times \mathbf{z} = \begin{vmatrix} x_1 & x_2 & x_3 \\ y_1 & y_2 & y_3 \\ z_1 & z_2 & z_3 \end{vmatrix} \tag{3.379}$$

As an example, consider three vectors **s**, **p**, **q**,

$$\begin{aligned} \mathbf{s} &= 2\sqrt{3}\hat{b}_3 \\ \mathbf{p} &= -5\sqrt{2}\hat{b}_2 + 4\sqrt{3}\hat{b}_3 \\ \mathbf{q} &= \hat{b}_1 - \sqrt{2}\hat{b}_2 + \sqrt{3}\hat{b}_3 \end{aligned} \tag{3.380}$$

in a curvilinear coordinate system with the base vectors

$$\hat{b}_1 = \hat{\imath} \qquad \hat{b}_2 = \frac{\sqrt{2}}{2}\left(\hat{\imath} + \hat{\jmath}\right) \qquad \hat{b}_3 = \frac{\sqrt{3}}{3}\left(\hat{\imath} + \hat{\jmath} + \hat{k}\right) \tag{3.381}$$

The scalar triple product of the base vectors is

$$V = \left[\hat{b}_1\hat{b}_2\hat{b}_3\right] = \begin{vmatrix} 1 & 0 & 0 \\ \frac{\sqrt{2}}{2} & \frac{\sqrt{2}}{2} & 0 \\ \frac{\sqrt{3}}{3} & \frac{\sqrt{3}}{3} & \frac{\sqrt{3}}{3} \end{vmatrix} = 0.40825 \tag{3.382}$$

So, the scalar triple product $\left[\mathbf{spq}\right]$ would be

$$\begin{aligned} \left[\mathbf{spq}\right] &= \mathbf{s} \cdot \mathbf{p} \times \mathbf{q} = V \begin{vmatrix} s_1 & s_2 & s_3 \\ p_1 & p_2 & p_3 \\ q_1 & q_2 & q_3 \end{vmatrix} \\ &= 0.40825 \begin{vmatrix} 0 & 0 & 2\sqrt{3} \\ 0 & -5\sqrt{2} & 4\sqrt{3} \\ 1 & -\sqrt{2} & \sqrt{3} \end{vmatrix} = 10 \end{aligned} \tag{3.383}$$

Example 212 ★ Gram Determinant Consider three vectors **s**, **p**, **q** in an oblique coordinate frame,

$$\mathbf{s} = s_1\hat{b}_1 + s_2\hat{b}_2 + s_3\hat{b}_3 \tag{3.384}$$

$$\mathbf{p} = p_1\hat{b}_1 + p_2\hat{b}_2 + p_3\hat{b}_3 \tag{3.385}$$

$$\mathbf{q} = q_1\hat{b}_1 + q_2\hat{b}_2 + q_3\hat{b}_3 \tag{3.386}$$

The square of their scalar triple product,

$$(\mathbf{s} \cdot \mathbf{p} \times \mathbf{q})^2 = \begin{vmatrix} \mathbf{s}\cdot\mathbf{s} & \mathbf{s}\cdot\mathbf{p} & \mathbf{s}\cdot\mathbf{q} \\ \mathbf{p}\cdot\mathbf{s} & \mathbf{p}\cdot\mathbf{p} & \mathbf{p}\cdot\mathbf{q} \\ \mathbf{q}\cdot\mathbf{s} & \mathbf{q}\cdot\mathbf{p} & \mathbf{q}\cdot\mathbf{q} \end{vmatrix} \tag{3.387}$$

is called the *Gram determinant*. To show this, we may substitute

$$s_i = \mathbf{s} \cdot \mathbf{b}_i^{\star} \qquad p_i = \mathbf{p} \cdot \mathbf{b}_i^{\star} \qquad q_i = \mathbf{q} \cdot \mathbf{b}_i^{\star} \tag{3.388}$$

in their scalar triple product (3.373),

$$\left[\mathbf{spq}\right] = V \begin{vmatrix} s_1 & s_2 & s_3 \\ p_1 & p_2 & p_3 \\ q_1 & q_2 & q_3 \end{vmatrix} = \left[\hat{b}_1\hat{b}_2\hat{b}_3\right] \begin{vmatrix} \mathbf{s}\cdot\mathbf{b}_1^\star & \mathbf{s}\cdot\mathbf{b}_2^\star & \mathbf{s}\cdot\mathbf{b}_3^\star \\ \mathbf{p}\cdot\mathbf{b}_1^\star & \mathbf{p}\cdot\mathbf{b}_2^\star & \mathbf{p}\cdot\mathbf{b}_3^\star \\ \mathbf{q}\cdot\mathbf{b}_1^\star & \mathbf{q}\cdot\mathbf{b}_2^\star & \mathbf{q}\cdot\mathbf{b}_3^\star \end{vmatrix}$$

$$= \frac{1}{\left[\mathbf{b}_1^\star\mathbf{b}_2^\star\mathbf{b}_3^\star\right]} \begin{vmatrix} \mathbf{s}\cdot\mathbf{b}_1^\star & \mathbf{s}\cdot\mathbf{b}_2^\star & \mathbf{s}\cdot\mathbf{b}_3^\star \\ \mathbf{p}\cdot\mathbf{b}_1^\star & \mathbf{p}\cdot\mathbf{b}_2^\star & \mathbf{p}\cdot\mathbf{b}_3^\star \\ \mathbf{q}\cdot\mathbf{b}_1^\star & \mathbf{q}\cdot\mathbf{b}_2^\star & \mathbf{q}\cdot\mathbf{b}_3^\star \end{vmatrix} \tag{3.389}$$

to find

$$\left[\mathbf{spq}\right]\left[\mathbf{b}_1^\star\mathbf{b}_2^\star\mathbf{b}_3^\star\right] = \begin{vmatrix} \mathbf{s}\cdot\mathbf{b}_1^\star & \mathbf{s}\cdot\mathbf{b}_2^\star & \mathbf{s}\cdot\mathbf{b}_3^\star \\ \mathbf{p}\cdot\mathbf{b}_1^\star & \mathbf{p}\cdot\mathbf{b}_2^\star & \mathbf{p}\cdot\mathbf{b}_3^\star \\ \mathbf{q}\cdot\mathbf{b}_1^\star & \mathbf{q}\cdot\mathbf{b}_2^\star & \mathbf{q}\cdot\mathbf{b}_3^\star \end{vmatrix} \tag{3.390}$$

Replacing the reciprocal base vectors $\mathbf{b}_1^\star, \mathbf{b}_2^\star, \mathbf{b}_3^\star$ with arbitrary vectors $\mathbf{u}, \mathbf{v}, \mathbf{w}$ provides

$$\left[\mathbf{spq}\right]\left[\mathbf{uvw}\right] = \begin{vmatrix} \mathbf{s}\cdot\mathbf{u} & \mathbf{s}\cdot\mathbf{v} & \mathbf{s}\cdot\mathbf{w} \\ \mathbf{p}\cdot\mathbf{u} & \mathbf{p}\cdot\mathbf{v} & \mathbf{p}\cdot\mathbf{w} \\ \mathbf{q}\cdot\mathbf{u} & \mathbf{q}\cdot\mathbf{v} & \mathbf{q}\cdot\mathbf{w} \end{vmatrix} \tag{3.391}$$

In the case $\mathbf{u} = \mathbf{s}$, $\mathbf{v} = \mathbf{p}$, and $\mathbf{w} = \mathbf{q}$, we derive the Gram determinant (3.387).

Example 213 ★ Bragg's Condition in an Atomic Lattice Nonorthogonal coordinate frames have applications in dynamic phenomena such as wave propagation in materials with periodic structures. Crystalline or rolled metals may approximately have periodic structures in a nonorthogonal coordinate frame.

Consider an ideal crystal that has the structure shown in Figure 3.22. We may set Cartesian and nonorthogonal coordinate frames at one of the atoms in the lattice. Using the unit vectors $\hat{b}_1$, $\hat{b}_2$, and $\hat{b}_3$, we show the position of each atom by a vector $\mathbf{r}$,

$$\mathbf{r} = r_1\hat{b}_1 + r_2\hat{b}_2 + r_3\hat{b}_3 \tag{3.392}$$

where r_1, r_2, r_3 are integers. The vectors $\hat{b}_1, \hat{b}_2, \hat{b}_3$ are called the crystal primitive translation vectors, and their axes are called the crystal axes. The reciprocal base vectors $\mathbf{b}_1^\star, \mathbf{b}_2^\star, \mathbf{b}_3^\star$ define a reciprocal lattice.

When a very short wavelength electromagnetic radiation, such as an X-ray, collides with a crystal, it is partially absorbed, scattered, and transmitted. Some of the scattered radiation is reflected from the periodically spaced layers of atoms. If the reflected rays satisfy Bragg's condition

$$2d_1 \cos\alpha = m\lambda \qquad m = 1, 2, 3, \ldots \tag{3.393}$$

then they are all in phase and hence make an intense reflected beam. In Bragg's condition (3.393), d_1 is the distance between two adjacent layers, α is the angle between the incoming ray and perpendicular line to the layer, and λ is the wavelength of the electromagnetic radiation.

Using Bragg's condition, the distance d between two neighbor layers of the atomic lattice is

$$d = \frac{r_1}{\sqrt{1/r_1^2 + 1/r_2^2 + 1/r_3^2}} \tag{3.394}$$

To prove Bragg's condition, we can use Figure 3.22 and show that $2d_1 \cos\alpha$ is the extra distance that a reflected ray from a neighbor layer will move. To have two reflected rays from two adjacent layers in phase, the extra distance must be equal to an integer m times the wavelength λ, as shown in Equation (3.393).

In an ideal lattice, there are several sets of planes from which rays may reflect. Two other sets of planes with distances d_2 and d_3 are shown in Figure 3.22. The location of the planes and their distances may be represented by the nonorthogonal coordinate frame unit vectors:

$$\mathbf{r}_1 = r_1\hat{b}_1 \qquad \mathbf{r}_2 = r_2\hat{b}_2 \qquad \mathbf{r}_3 = r_3\hat{b}_3 \tag{3.395}$$

Defining two vectors on the sides of a lattice triangular cell as

$$\mathbf{c}_1 = \mathbf{r}_3 - \mathbf{r}_2 = r_3\hat{b}_3 - r_2\hat{b}_2 \tag{3.396}$$

$$\mathbf{c}_3 = \mathbf{r}_1 - \mathbf{r}_2 = r_1\hat{b}_1 - r_2\hat{b}_2 \tag{3.397}$$

we can determine the unit-normal vector $\hat{n}$ to the layer plane:

$$\begin{aligned}\mathbf{c}_1 \times \mathbf{c}_3 &= (\mathbf{r}_3 - \mathbf{r}_2) \times (\mathbf{r}_1 - \mathbf{r}_2) = \begin{vmatrix} \hat{b}_1 & \hat{b}_2 & \hat{b}_3 \\ 0 & -r_2 & r_3 \\ r_1 & -r_2 & 0 \end{vmatrix} \\ &= r_1 r_2 r_3 \left(\frac{\hat{b}_1}{r_1} + \frac{\hat{b}_2}{r_2} + \frac{\hat{b}_3}{r_3} \right)\end{aligned} \tag{3.398}$$

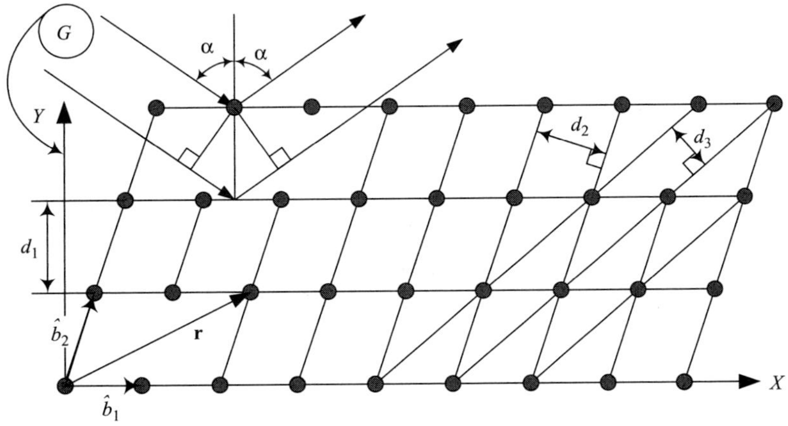

Figure 3.22 An ideal 2D crystal lattice and Bragg diffration.

$$\hat{n} = \frac{\hat{b}_1 + \hat{b}_2 + \hat{b}_3}{\sqrt{1/r_1^2 + 1/r_2^2 + 1/r_3^2}} \tag{3.399}$$

The normal distance between two layers would then be

$$\begin{aligned} d &= \mathbf{r}_1 \cdot \hat{n} = \mathbf{r}_2 \cdot \hat{n} = \mathbf{r}_3 \cdot \hat{n} = r_1\hat{b}_1 \cdot \hat{n} = r_2\hat{b}_2 \cdot \hat{n} = r_3\hat{b}_3 \cdot \hat{n} \\ &= \frac{r_1}{\sqrt{1/r_1^2 + 1/r_2^2 + 1/r_3^2}} \end{aligned} \tag{3.400}$$

The base vectors $\hat{b}_1, \hat{b}_2, \hat{b}_3$ and the normal vector $\hat{n}$ are illustrated in Figure 3.23.

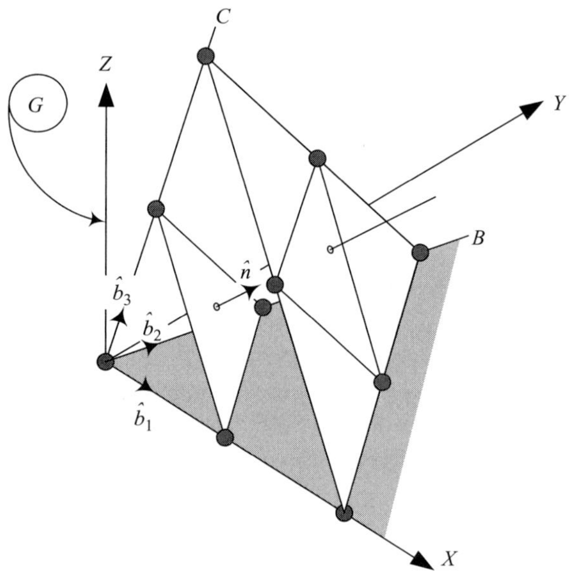

Figure 3.23 The oblique coordinate frame $\hat{b}_1, \hat{b}_2, \hat{b}_3$ and the normal vector $\hat{n}$ to the lattice layers.

3.4.4 ★ Kinematics in Oblique Coordinate Frames

The position vector $\mathbf{r} = \mathbf{r}(t)$ of a moving point in a nonorthogonal coordinate frame with principal unit vectors $\hat{b}_1, \hat{b}_2, \hat{b}_3$ and reciprocal base vectors $\mathbf{b}_1^\star, \mathbf{b}_2^\star, \mathbf{b}_3^\star$ may have variable components

$$\begin{aligned} {}^P\mathbf{r} &= r_1(t)\hat{b}_1 + r_2(t)\hat{b}_2 + r_3(t)\hat{b}_3 \\ {}^R\mathbf{r} &= r_1^\star(t)\mathbf{b}_1^\star + r_2^\star(t)\mathbf{b}_2^\star + r_3^\star(t)\mathbf{b}_3^\star \end{aligned} \tag{3.401}$$

The velocity and acceleration of the point are

$$ {}^{P}\mathbf{v} = \frac{d}{dt}\,{}^{P}\mathbf{r} = \dot{r}_1\hat{b}_1 + \dot{r}_2\hat{b}_2 + \dot{r}_3\hat{b}_3 \tag{3.402} $$

$$ {}^{R}\mathbf{v} = \frac{d}{dt}\,{}^{R}\mathbf{r} = \dot{r}_1^{\star}\mathbf{b}_1^{\star} + \dot{r}_2^{\star}\mathbf{b}_2^{\star} + \dot{r}_3^{\star}\mathbf{b}_3^{\star} \tag{3.403} $$

$$ {}^{P}\mathbf{a} = \frac{d}{dt}\,{}^{P}\mathbf{v} = \ddot{r}_1\hat{b}_1 + \ddot{r}_2\hat{b}_2 + \ddot{r}_3\hat{b}_3 \tag{3.404} $$

$$ {}^{R}\mathbf{a} = \frac{d}{dt}\,{}^{R}\mathbf{v} = \ddot{r}_1^{\star}\mathbf{b}_1^{\star} + \ddot{r}_2^{\star}\mathbf{b}_2^{\star} + \ddot{r}_3^{\star}\mathbf{b}_3^{\star} \tag{3.405} $$

Proof: The principal unit vectors $\hat{b}_1, \hat{b}_2, \hat{b}_3$, as shown in Figure 3.17, have constant length and direction. Furthermore Equation (3.239) indicates that the reciprocal base vectors $\mathbf{b}_1^{\star}, \mathbf{b}_2^{\star}, \mathbf{b}_3^{\star}$ are also constant. So the derivative of a variable vector in the nonorthogonal coordinate frame can be found by taking the derivative of its components regardless of its expression in the principal or reciprocal frames. ■

Example 214 ★ Velocity and Acceleration in an Oblique Frame Consider an oblique coordinate frame with the unit base vectors

$$ \hat{b}_1 = \hat{\imath} \qquad \hat{b}_2 = \frac{\sqrt{2}}{2}\left(\hat{\imath} + \hat{\jmath}\right) \qquad \hat{b}_3 = \frac{\sqrt{3}}{3}\left(\hat{\imath} + \hat{\jmath} + \hat{k}\right) \tag{3.406} $$

A point with position vector $\mathbf{r}$ is moving in the frame,

$$ {}^{G}\mathbf{r} = 2\sin t\,\hat{\imath} + 2e^{t}\,\hat{\jmath} + 2t^2\,\hat{k} \tag{3.407} $$

The principal unit vectors of the oblique frame can be used to determine the transformation matrix between the oblique and Cartesian frames:

$$ \begin{bmatrix} \hat{b}_1 \\ \hat{b}_2 \\ \hat{b}_3 \end{bmatrix} = \begin{bmatrix} 1 & 0 & 0 \\ \sqrt{2}/2 & \sqrt{2}/2 & 0 \\ \sqrt{3}/3 & \sqrt{3}/3 & \sqrt{3}/3 \end{bmatrix} \begin{bmatrix} \hat{\imath} \\ \hat{\jmath} \\ \hat{k} \end{bmatrix} \tag{3.408} $$

$$ \begin{bmatrix} \hat{\imath} \\ \hat{\jmath} \\ \hat{k} \end{bmatrix} = \begin{bmatrix} 1 & 0 & 0 \\ -1 & \sqrt{2} & 0 \\ 0 & -\sqrt{2} & \sqrt{3} \end{bmatrix} \begin{bmatrix} \hat{b}_1 \\ \hat{b}_2 \\ \hat{b}_3 \end{bmatrix} \tag{3.409} $$

Therefore, $\mathbf{r}$ in the principal frame is

$$ \begin{aligned} {}^{P}\mathbf{r} &= 2\sin t\left(\hat{b}_1\right) + 2e^{t}\left(-\hat{b}_1 + \sqrt{2}\hat{b}_2\right) + 2t^2\left(-\sqrt{2}\hat{b}_2 + \sqrt{3}\hat{b}_3\right) \\ &= \left(2\sin t - 2e^{t}\right)\hat{b}_1 + \left(1 - 2t^2\right)\sqrt{2}\hat{b}_2 + 2\sqrt{3}t^2\hat{b}_3 \end{aligned} \tag{3.410} $$

Using the oblique unit vectors, we determine the metric matrix to transform ${}^{P}\mathbf{p}$ to the reciprocal frame:

$$ {}^{R}R_{P} = \left[g_{ij}\right] = \left[\hat{b}_i \cdot \hat{b}_j\right] = \begin{bmatrix} 1 & 0.70711 & 0.57735 \\ 0.70711 & 1 & 0.81650 \\ 0.57735 & 0.81650 & 1 \end{bmatrix} \tag{3.411} $$

$$
\begin{aligned}
{}^{R}\mathbf{r} = \left[g_{ij}\right]\,{}^{P}\mathbf{r} = \left[g_{ij}\right]\begin{bmatrix} 2\sin t - 2e^{t} \\ \left(1-2t^{2}\right)\sqrt{2} \\ 2\sqrt{3}t^{2}\end{bmatrix} \qquad (3.412)\\
= \left(2\sin t - 2e^{t} - 1.0037\times 10^{-5}t^{2} + 1\right)\mathbf{b}_1^{\star} \\
+ \left(1.4142\sin t - 1.4142e^{t} + 1.1844\times 10^{-5}t^{2} + 1.4142\right)\mathbf{b}_2^{\star} \\
+ \left(1.1547\sin t - 1.1547e^{t} + 1.1547t^{2} + 1.1547\right)\mathbf{b}_3^{\star}
\end{aligned}
$$

The velocity and acceleration of the point are

$$
{}^{P}\mathbf{v} = \frac{d}{dt}\,{}^{P}\mathbf{r} = \left(2\cos t - 2e^{t}\right)\hat{b}_1 - 4\sqrt{2}t\hat{b}_2 + 4\sqrt{3}t\hat{b}_3 \qquad (3.413)
$$

$$
\begin{aligned}
{}^{R}\mathbf{v} = \frac{d}{dt}\,{}^{R}\mathbf{r} = \left(2\cos t - 2.0074\times 10^{-5}t - 2e^{t}\right)\mathbf{b}_1^{\star} \\
+ \left(2.3688\times 10^{-5}t + \sqrt{2}\cos t - \sqrt{2}e^{t}\right)\mathbf{b}_2^{\star} \\
+ \left(2.3094t + 1.1547\cos t - 1.1547e^{t}\right)\mathbf{b}_3^{\star} \qquad (3.414)
\end{aligned}
$$

$$
{}^{P}\mathbf{a} = \frac{d}{dt}\,{}^{P}\mathbf{v} = \left(-2e^{t} - 2\sin t\right)\hat{b}_1 - 4\sqrt{2}\hat{b}_2 + 4\sqrt{3}\hat{b}_3
$$

$$
\begin{aligned}
{}^{R}\mathbf{a} = \frac{d}{dt}\,{}^{R}\mathbf{v} = \left(-2e^{t} - 2\sin t - 2.0074\times 10^{-5}\right)\mathbf{b}_1^{\star} \\
+ \left(2.3688\times 10^{-5} - \sqrt{2}\sin t - \sqrt{2}e^{t}\right)\mathbf{b}_2^{\star} \\
+ \left(2.3094 - 1.1547\sin t - 1.1547e^{t}\right)\mathbf{b}_3^{\star} \qquad (3.415)
\end{aligned}
$$

3.5 ★ CURVILINEAR COORDINATE SYSTEM

When a flat coordinate system is not orthogonal, we call it a *nonorthogonal* or *oblique coordinate system*, and when the coordinate system is not flat, we call it *curvilinear*. Figure 3.24 illustrates a nonorthogonal curvilinear coordinate frame $Q\,(q_1, q_2, q_3)$ in which the unit vectors $\hat{u}_i$, $i = 1, 2, 3$, of the coordinate frame are nonorthogonal and are along nonflat coordinate curves q_i, $i = 1, 2, 3$. The orthogonal curvilinear coordinate systems have much more potential for application in dynamics.

Any curvilinear coordinate system $Q\,(q_1, q_2, q_3)$ can be introduced by three equations that relate the new coordinates q_1, q_2, q_3 to Cartesian coordinates x, y, z:

$$
\begin{aligned}
x &= f_1\left(q_1, q_2, q_3\right) = x_1\left(q_1, q_2, q_3\right) \\
y &= f_2\left(q_1, q_2, q_3\right) = x_2\left(q_1, q_2, q_3\right) \qquad (3.416) \\
z &= f_3\left(q_1, q_2, q_3\right) = x_3\left(q_1, q_2, q_3\right)
\end{aligned}
$$

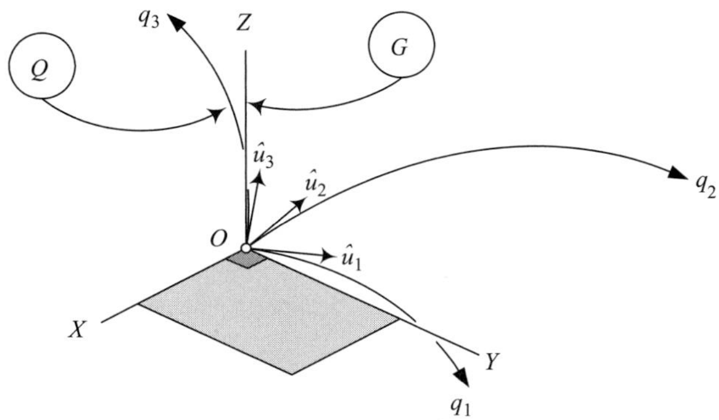

Figure 3.24 A nonorthogonal curvilinear coordinate frame $Q\,(q_1, q_2, q_3)$.

We assume that Equations (3.416) have a unique set of inverse functions to calculate the new Cartesian coordinates:

$$\begin{aligned} q_1 &= q_1\,(x, y, z) = q_1\,(x_1, x_2, x_3 z) \\ q_2 &= q_2\,(x, y, z) = q_2\,(x_1, x_2, x_3 z) \\ q_3 &= q_3\,(x, y, z) = q_3\,(x_1, x_2, x_3 z) \end{aligned} \tag{3.417}$$

Furthermore, to simplify the equation's appearance, we employ the Einstein summation convention introduced in Example 33. Based on the Einstein summation convention, without showing a summation symbol, we sum over every index that is repeated twice in every term of an equation.

3.5.1 ★ Principal and Reciprocal Base Vectors

Having the Cartesian description of a position vector $\mathbf{r}$,

$$\mathbf{r} = x\hat{\imath} + y\hat{\jmath} + z\hat{k} = x_1\hat{\imath} + x_2\hat{\jmath} + x_3\hat{k} \tag{3.418}$$

we determine the *principal unit vectors* $\hat{u}_i$ of the curvilinear coordinate Q-system, by

$$\hat{u}_i = \frac{\Delta\mathbf{r}_{q_i}}{\left|\Delta\mathbf{r}_{q_i}\right|} = \frac{\partial\mathbf{r}/\partial q_i}{\left|\partial\mathbf{r}/\partial q_i\right|} = \frac{1}{b_i}\mathbf{b}_i \tag{3.419}$$

where $\mathbf{b}_i$ is the *principal base vector* of the coordinate system $Q\,(q_1, q_2, q_3)$:

$$\mathbf{b}_i = \frac{\partial\mathbf{r}}{\partial q_i} = b_i\hat{u}_i = \frac{\partial x}{\partial q_i}\hat{\imath} + \frac{\partial y}{\partial q_i}\hat{\jmath} + \frac{\partial z}{\partial q_i}\hat{k} \tag{3.420}$$

$$b_i = |\mathbf{b}_i| = \sqrt{\left(\frac{\partial x}{\partial q_i}\right)^2 + \left(\frac{\partial y}{\partial q_i}\right)^2 + \left(\frac{\partial z}{\partial q_i}\right)^2} \tag{3.421}$$

The scalar b_i is called the *space scale factor* of $Q\,(q_1, q_2, q_3)$. The principal unit vector $\hat{u}_i$ is along the partial derivative $\partial\mathbf{r}/\partial q_i$, which indicates a coordinate curve

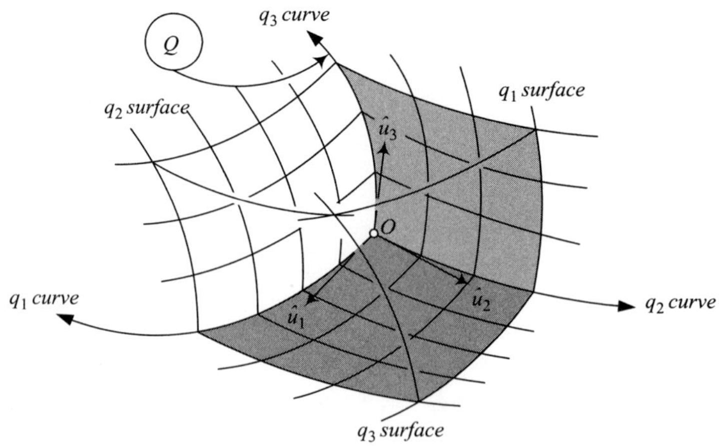

Figure 3.25 The principal unit vector $\hat{u}_i$ indicates the coordinate curve q_i.

q_i, as illustrated in Figure 3.25. The principal unit and base vectors $\hat{u}_i$ and $\mathbf{b}_i$ make an oblique coordinate frame at any point P of the space.

For every curvilinear coordinate system $Q\,(q_1, q_2, q_3)$, we can also define the *reciprocal base vectors* $\mathbf{b}_i^\star$ and *reciprocal unit vectors* $\hat{u}_i^\star$:

$$\mathbf{b}_i^\star = \nabla q_i = \frac{\partial q_i}{\partial x}\hat{\imath} + \frac{\partial q_i}{\partial y}\hat{\jmath} + \frac{\partial q_i}{\partial z}\hat{k} \tag{3.422}$$

$$\hat{u}_i^\star = \frac{\nabla q_i}{|\nabla q_i|} = \frac{1}{b_i^\star}\mathbf{b}_i^\star \tag{3.423}$$

$$b_i^\star = \left|\mathbf{b}_i^\star\right| = \sqrt{\left(\frac{\partial q_i}{\partial x}\right)^2 + \left(\frac{\partial q_i}{\partial y}\right)^2 + \left(\frac{\partial q_i}{\partial z}\right)^2} \tag{3.424}$$

Because ∇q_i is a perpendicular vector to surface $q_i = c$, the reciprocal vectors $\hat{u}_1^\star, \hat{u}_2^\star, \hat{u}_3^\star$ and $\mathbf{b}_1^\star, \mathbf{b}_2^\star, \mathbf{b}_3^\star$ are perpendicular to the coordinate surfaces q_1, q_2, and q_3, respectively, and therefore,

$$\mathbf{b}_i \cdot \mathbf{b}_j^\star = \delta_{ij} \tag{3.425}$$

At any point P of the space, the reciprocal unit and base vectors $\hat{u}_i^\star$ and $\mathbf{b}_i^\star$ make an oblique coordinate frame that is reciprocal to the principal frame of $\hat{u}_i$ and $\mathbf{b}_i$. Figure 3.26 illustrates a principal coordinate frame and its reciprocal coordinate frame.

Any vector $\mathbf{r}$ may be expressed in Cartesian coordinates as well as principal and reciprocal coordinate frames using the unit or base vectors of each coordinate system,

$$\mathbf{r} = x\hat{\imath} + y\hat{\jmath} + z\hat{k} \tag{3.426}$$

$${}^P\mathbf{r} = r_1\mathbf{b}_1 + r_2\mathbf{b}_2 + r_3\mathbf{b}_3 \tag{3.427}$$

$${}^R\mathbf{r} = r_1^\star\mathbf{b}_1^\star + r_2^\star\mathbf{b}_2^\star + r_3^\star\mathbf{b}_3^\star \tag{3.428}$$

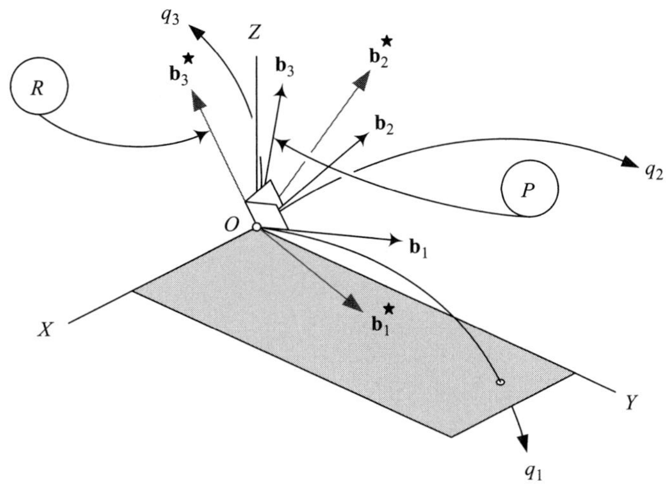

Figure 3.26 A principal coordinate frame and its reciprocal coordinate frame at a point O of space along with the curvilinear coordinate system Q.

where

$$
\begin{aligned}
r_1 &= \mathbf{r} \cdot \frac{\mathbf{b}_2 \times \mathbf{b}_3}{V} \\
r_2 &= \mathbf{r} \cdot \frac{\mathbf{b}_3 \times \mathbf{b}_1}{V} \qquad (3.429) \\
r_3 &= \mathbf{r} \cdot \frac{\mathbf{b}_1 \times \mathbf{b}_2}{V}
\end{aligned}
$$

$$
V = \mathbf{b}_1 \cdot \mathbf{b}_2 \times \mathbf{b}_3 = [\mathbf{b}_1 \mathbf{b}_2 \mathbf{b}_3] \qquad (3.430)
$$

$$
\begin{aligned}
r_1^\star &= \mathbf{r} \cdot \frac{\mathbf{b}_2^\star \times \mathbf{b}_3^\star}{\left[\mathbf{b}_1^\star \mathbf{b}_2^\star \mathbf{b}_3^\star\right]} \\
r_2^\star &= \mathbf{r} \cdot \frac{\mathbf{b}_3^\star \times \mathbf{b}_1^\star}{\left[\mathbf{b}_1^\star \mathbf{b}_2^\star \mathbf{b}_3^\star\right]} \qquad (3.431) \\
r_3^\star &= \mathbf{r} \cdot \frac{\mathbf{b}_1^\star \times \mathbf{b}_2^\star}{\left[\mathbf{b}_1^\star \mathbf{b}_2^\star \mathbf{b}_3^\star\right]}
\end{aligned}
$$

Proof: The transformation (3.417) defines a set of coordinate surfaces and coordinate curves. The coordinate surfaces are defined by the equations

$$
\begin{aligned}
q_1(x, y, z) &= c_1 \\
q_2(x, y, z) &= c_2 \qquad (3.432) \\
q_3(x, y, z) &= c_3
\end{aligned}
$$

where c_1, c_2, c_3 are constants. These surfaces intersect at the coordinate curves q_i with position vector $\mathbf{r}_i$,

$$\begin{aligned}\mathbf{r}_1 &= \mathbf{r}_1(q_1, c_2, c_3)\\ \mathbf{r}_2 &= \mathbf{r}_2(c_1, q_2, c_3)\\ \mathbf{r}_3 &= \mathbf{r}_3(c_1, c_2, q_3)\end{aligned} \tag{3.433}$$

where $\mathbf{r}$ is the Cartesian expression of the position vector with Q-components,

$$\begin{aligned}\mathbf{r} &= x\hat{\imath} + y\hat{\jmath} + z\hat{k}\\ &= x_1(q_1, q_2, q_3)\,\hat{\imath} + x_2(q_1, q_2, q_3)\,\hat{\jmath} + x_3(q_1, q_2, q_3)\,\hat{k}\end{aligned} \tag{3.434}$$

At any point (q_1, q_2, q_3) of the Q-space, we may define two sets of vectors and make two triads to define kinematic vectors. The first set contains tangent vectors to the coordinate curves (3.433) that are found by taking partial derivatives of the position vector (3.420):

$$\mathbf{b}_i = \frac{\partial \mathbf{r}}{\partial q_i} = b_i \hat{u}_i = \frac{\partial x}{\partial q_i}\hat{\imath} + \frac{\partial y}{\partial q_i}\hat{\jmath} + \frac{\partial z}{\partial q_i}\hat{k} \tag{3.435}$$

The triad made up of the principal base vectors $\mathbf{b}_1, \mathbf{b}_2, \mathbf{b}_3$ is called the *principal frame*. The unit vectors $\hat{u}_i$ of the principal frame are aligned with the principal base vectors:

$$\hat{u}_i = \frac{1}{b_i}\mathbf{b}_i \tag{3.436}$$

$$b_i = |\mathbf{b}_i| = \sqrt{\left(\frac{\partial x}{\partial q_i}\right)^2 + \left(\frac{\partial y}{\partial q_i}\right)^2 + \left(\frac{\partial z}{\partial q_i}\right)^2} \tag{3.437}$$

The second set of contains normal vectors to the coordinate surfaces (3.432) that are found by taking the gradient of the surfaces:

$$\mathbf{b}_i^{\star} = \nabla q_i = \frac{\partial q_i}{\partial x}\hat{\imath} + \frac{\partial q_i}{\partial y}\hat{\jmath} + \frac{\partial q_i}{\partial z}\hat{k} \tag{3.438}$$

The triad made up of the *reciprocal base vectors* $\mathbf{b}_1^{\star}, \mathbf{b}_2^{\star}, \mathbf{b}_3^{\star}$ is called the *reciprocal frame*. The reciprocalunit vectors $\hat{u}_i^{\star}$ of the frame are aligned with the reciprocal base vectors:

$$\hat{u}_i^{\star} = \frac{1}{b_i^{\star}}\mathbf{b}_i^{\star} \tag{3.439}$$

$$b_i^{\star} = \left|\mathbf{b}_i^{\star}\right| = \sqrt{\left(\frac{\partial q_i}{\partial x}\right)^2 + \left(\frac{\partial q_i}{\partial y}\right)^2 + \left(\frac{\partial q_i}{\partial z}\right)^2} \tag{3.440}$$

Consider a base vector $\mathbf{b}_1^{\star}$ that is normal to (q_2, q_3)-plane. The base vectors $\mathbf{b}_2$ and $\mathbf{b}_3$ are tangent to the coordinate curves q_2 and q_3 and hence are in the (q_2, q_3)-plane, and therefore,

$$\mathbf{b}_i \cdot \mathbf{b}_j^{\star} = \mathbf{b}_j^{\star} \cdot \mathbf{b}_i = \delta_{ij} \tag{3.441}$$

Substituting the Cartesian expression of $\mathbf{b}_i$,

$$\begin{aligned}\mathbf{b}_1 &= b_{11}\hat{\imath} + b_{12}\hat{\jmath} + b_{13}\hat{k} \\ \mathbf{b}_2 &= b_{21}\hat{\imath} + b_{22}\hat{\jmath} + b_{23}\hat{k} \\ \mathbf{b}_3 &= b_{31}\hat{\imath} + b_{32}\hat{\jmath} + b_{33}\hat{k}\end{aligned} \tag{3.442}$$

in (3.427) and comparing with (3.426) generate

$$\begin{bmatrix} x \\ y \\ z \end{bmatrix} = \begin{bmatrix} b_{11} & b_{12} & b_{13} \\ b_{21} & b_{22} & b_{23} \\ b_{31} & b_{32} & b_{33} \end{bmatrix} \begin{bmatrix} r_1 \\ r_2 \\ r_3 \end{bmatrix} \tag{3.443}$$

and solving for r_i provides

$$\begin{aligned}r_1 &= \frac{\mathbf{r} \cdot \mathbf{b}_2 \times \mathbf{b}_3}{\mathbf{b}_1 \cdot \mathbf{b}_2 \times \mathbf{b}_3} = \mathbf{r} \cdot \frac{\mathbf{b}_2 \times \mathbf{b}_3}{[\mathbf{b}_1\mathbf{b}_2\mathbf{b}_3]} = \mathbf{r} \cdot \frac{\mathbf{b}_2 \times \mathbf{b}_3}{V} \\ r_2 &= \frac{\mathbf{r} \cdot \mathbf{b}_3 \times \mathbf{b}_1}{\mathbf{b}_1 \cdot \mathbf{b}_2 \times \mathbf{b}_3} = \mathbf{r} \cdot \frac{\mathbf{b}_3 \times \mathbf{b}_1}{[\mathbf{b}_1\mathbf{b}_2\mathbf{b}_3]} = \mathbf{r} \cdot \frac{\mathbf{b}_3 \times \mathbf{b}_1}{V} \\ r_3 &= \frac{\mathbf{r} \cdot \mathbf{b}_1 \times \mathbf{b}_2}{\mathbf{b}_1 \cdot \mathbf{b}_2 \times \mathbf{b}_3} = \mathbf{r} \cdot \frac{\mathbf{b}_1 \times \mathbf{b}_2}{[\mathbf{b}_1\mathbf{b}_2\mathbf{b}_3]} = \mathbf{r} \cdot \frac{\mathbf{b}_1 \times \mathbf{b}_2}{V}\end{aligned} \tag{3.444}$$

$$V = \mathbf{b}_1 \cdot \mathbf{b}_2 \times \mathbf{b}_3 = [\mathbf{b}_1\mathbf{b}_2\mathbf{b}_3] \tag{3.445}$$

Therefore,

$${}^P\mathbf{r} = \frac{[\mathbf{r}\mathbf{b}_2\mathbf{b}_3]}{[\mathbf{b}_1\mathbf{b}_2\mathbf{b}_3]}\mathbf{b}_1 + \frac{[\mathbf{r}\mathbf{b}_3\mathbf{b}_1]}{[\mathbf{b}_1\mathbf{b}_2\mathbf{b}_3]}\mathbf{b}_2 + \frac{[\mathbf{r}\mathbf{b}_1\mathbf{b}_2]}{[\mathbf{b}_1\mathbf{b}_2\mathbf{b}_3]}\mathbf{b}_3 \tag{3.446}$$

Substituting the Cartesian expression of $\mathbf{b}_i^\star$,

$$\begin{aligned}\mathbf{b}_1^\star &= b_{11}^\star\hat{\imath} + b_{12}^\star\hat{\jmath} + b_{13}^\star\hat{k} \\ \mathbf{b}_2^\star &= b_{21}^\star\hat{\imath} + b_{22}^\star\hat{\jmath} + b_{23}^\star\hat{k} \\ \mathbf{b}_3^\star &= b_{31}^\star\hat{\imath} + b_{32}^\star\hat{\jmath} + b_{33}^\star\hat{k}\end{aligned} \tag{3.447}$$

in (3.428) and comparing with (3.426) give

$$\begin{bmatrix} x \\ y \\ z \end{bmatrix} = \begin{bmatrix} b_{11}^\star & b_{12}^\star & b_{13}^\star \\ b_{21}^\star & b_{22}^\star & b_{23}^\star \\ b_{31}^\star & b_{32}^\star & b_{33}^\star \end{bmatrix} \begin{bmatrix} r_1^\star \\ r_2^\star \\ r_3^\star \end{bmatrix} \tag{3.448}$$

and solving for $r_i^{\star}$ yields

$$r_1^{\star} = \frac{\mathbf{r} \cdot \mathbf{b}_2^{\star} \times \mathbf{b}_3^{\star}}{\mathbf{b}_1^{\star} \cdot \mathbf{b}_2^{\star} \times \mathbf{b}_3^{\star}} = \mathbf{r} \cdot \frac{\mathbf{b}_2^{\star} \times \mathbf{b}_3^{\star}}{\left[\mathbf{b}_1^{\star}\mathbf{b}_2^{\star}\mathbf{b}_3^{\star}\right]}$$
$$r_2^{\star} = \frac{\mathbf{r} \cdot \mathbf{b}_3^{\star} \times \mathbf{b}_1^{\star}}{\mathbf{b}_1^{\star} \cdot \mathbf{b}_2^{\star} \times \mathbf{b}_3^{\star}} = \mathbf{r} \cdot \frac{\mathbf{b}_3^{\star} \times \mathbf{b}_1^{\star}}{\left[\mathbf{b}_1^{\star}\mathbf{b}_2^{\star}\mathbf{b}_3^{\star}\right]} \tag{3.449}$$
$$r_3^{\star} = \frac{\mathbf{r} \cdot \mathbf{b}_1^{\star} \times \mathbf{b}_2^{\star}}{\mathbf{b}_1^{\star} \cdot \mathbf{b}_2^{\star} \times \mathbf{b}_3^{\star}} = \mathbf{r} \cdot \frac{\mathbf{b}_1^{\star} \times \mathbf{b}_2^{\star}}{\left[\mathbf{b}_1^{\star}\mathbf{b}_2^{\star}\mathbf{b}_3^{\star}\right]}$$

Therefore,

$$^{R}\mathbf{r} = \frac{\left[\mathbf{r}\mathbf{b}_2^{\star}\mathbf{b}_3^{\star}\right]}{\left[\mathbf{b}_1^{\star}\mathbf{b}_2^{\star}\mathbf{b}_3^{\star}\right]}\mathbf{b}_1^{\star} + \frac{\left[\mathbf{r}\mathbf{b}_3^{\star}\mathbf{b}_1^{\star}\right]}{\left[\mathbf{b}_1^{\star}\mathbf{b}_2^{\star}\mathbf{b}_3^{\star}\right]}\mathbf{b}_2^{\star} + \frac{\left[\mathbf{r}\mathbf{b}_1^{\star}\mathbf{b}_2^{\star}\right]}{\left[\mathbf{b}_1^{\star}\mathbf{b}_2^{\star}\mathbf{b}_3^{\star}\right]}\mathbf{b}_3^{\star} \tag{3.450}$$

■

Example 215 ★ Base Vectors of Spherical Coordinate System Consider a Q-coordinate system with the following relations:

$$x = q_1 \sin q_2 \cos q_3 = r \sin\varphi \cos\theta$$
$$y = q_1 \sin q_2 \sin q_3 = r \sin\varphi \sin\theta \tag{3.451}$$
$$z = q_1 \cos q_2 = r \cos\varphi$$

$$q_1 = r = \sqrt{x^2 + y^2 + z^2}$$
$$q_2 = \varphi = \tan^{-1} \frac{\sqrt{x^2 + y^2}}{z} \tag{3.452}$$
$$q_3 = \theta = \tan^{-1} \frac{y}{x}$$

The principal base vectors of the Q-system are

$$\mathbf{b}_1 = \frac{\partial \mathbf{r}}{\partial q_1} = \begin{bmatrix} \sin q_2 \cos q_3 \\ \sin q_2 \sin q_3 \\ \cos q_2 \end{bmatrix} \tag{3.453}$$

$$\mathbf{b}_2 = \frac{\partial \mathbf{r}}{\partial q_2} = \begin{bmatrix} q_1 \cos q_2 \cos q_3 \\ q_1 \cos q_2 \sin q_3 \\ -q_1 \sin q_2 \end{bmatrix} \tag{3.454}$$

$$\mathbf{b}_3 = \frac{\partial \mathbf{r}}{\partial q_3} = \begin{bmatrix} -q_1 \sin q_2 \sin q_3 \\ q_1 \sin q_2 \cos q_3 \\ 0 \end{bmatrix} \tag{3.455}$$

and the reciprocal base vectors of the Q-system are

$$\mathbf{b}_1^{\star} = \nabla q_1 = \frac{1}{\sqrt{x^2+y^2+z^2}} \begin{bmatrix} x \\ y \\ z \end{bmatrix} \tag{3.456}$$

$$\mathbf{b}_2^{\star} = \nabla q_2 = \frac{1}{\left(x^2+y^2+z^2\right)\sqrt{x^2+y^2}} \begin{bmatrix} xz \\ yz \\ -\left(x^2+y^2\right) \end{bmatrix} \tag{3.457}$$

$$\mathbf{b}_3^{\star} = \nabla q_3 = \frac{1}{x^2+y^2} \begin{bmatrix} -y \\ x \\ 0 \end{bmatrix} \tag{3.458}$$

The base vectors $\mathbf{b}_i$ and $\mathbf{b}_i^{\star}$ are in the Cartesian coordinate system. To check Equation (3.441), we may transform either the components of $\mathbf{b}_i$ to the Q-system or the components of $\mathbf{b}_i^{\star}$ to the Cartesian system. In either case, condition (3.441) would be correct. Substituting Equations (3.451) into (3.456)–(3.458) we can determine the reciprocal base vectors $\mathbf{b}_i^{\star}$ with components in the Q-system:

$$\mathbf{b}_1^{\star} = \begin{bmatrix} \sin q_2 \cos q_3 \\ \sin q_2 \sin q_3 \\ \cos q_2 \end{bmatrix} = \mathbf{b}_1 \tag{3.459}$$

$$\mathbf{b}_2^{\star} = \frac{1}{q_1} \begin{bmatrix} \cos q_2 \cos q_3 \\ \cos q_2 \sin q_3 \\ -\sin q_2 \end{bmatrix} = \frac{1}{q_1^2}\mathbf{b}_2 \tag{3.460}$$

$$\mathbf{b}_3^{\star} = \frac{1}{q_1 \sin q_2} \begin{bmatrix} -\sin q_3 \\ \cos q_3 \\ 0 \end{bmatrix} = \frac{1}{q_1^2 \sin^2 q_2}\mathbf{b}_3 \tag{3.461}$$

and therefore,

$$\mathbf{b}_i \cdot \mathbf{b}_j^{\star} = \mathbf{b}_j^{\star} \cdot \mathbf{b}_i = \delta_{ij} \tag{3.462}$$

To express a vector $\mathbf{r}$ in the Q-system, we need to determine the cross products $\mathbf{b}_2 \times \mathbf{b}_3$, $\mathbf{b}_3 \times \mathbf{b}_1$, $\mathbf{b}_1 \times \mathbf{b}_2$ and the scalar triple product $V = [\mathbf{b}_1\mathbf{b}_2\mathbf{b}_3]$:

$$V = \mathbf{b}_1 \cdot \mathbf{b}_2 \times \mathbf{b}_3 = [\mathbf{b}_1\mathbf{b}_2\mathbf{b}_3] = q_1^2 \sin q_2 \tag{3.463}$$

$$\mathbf{b}_2 \times \mathbf{b}_3 = q_1^2 \begin{bmatrix} \cos q_3 \sin^2 q_2 \\ \sin^2 q_2 \sin q_3 \\ \frac{1}{2}\sin 2q_2 \end{bmatrix} \tag{3.464}$$

$$\mathbf{b}_3 \times \mathbf{b}_1 = q_1 \begin{bmatrix} \cos q_2 \cos q_3 \sin q_2 \\ \cos q_2 \sin q_2 \sin q_3 \\ -\sin^2 q_2 \end{bmatrix} \tag{3.465}$$

$$\mathbf{b}_1 \times \mathbf{b}_2 = \begin{bmatrix} -q_1 \sin q_3 \\ q_1 \cos q_3 \\ 0 \end{bmatrix} \tag{3.466}$$

Therefore,

$$\begin{aligned} r_1 &= \mathbf{r} \cdot \frac{\mathbf{b}_2 \times \mathbf{b}_3}{V} = q_1 \\ r_2 &= \mathbf{r} \cdot \frac{\mathbf{b}_3 \times \mathbf{b}_1}{V} = 0 \\ r_3 &= \mathbf{r} \cdot \frac{\mathbf{b}_1 \times \mathbf{b}_2}{V} = 0 \end{aligned} \tag{3.467}$$

and

$${}^P\mathbf{r} = q_1 \mathbf{b}_1 \tag{3.468}$$

To express $\mathbf{r}$ in the R-coordinate frame, we have to determine $r_i^\bigstar$:

$$\left[\mathbf{b}_1^\bigstar \mathbf{b}_2^\bigstar \mathbf{b}_3^\bigstar\right] = \mathbf{b}_1^\bigstar \cdot \mathbf{b}_2^\bigstar \times \mathbf{b}_3^\bigstar = \frac{1}{q_1^2 \sin q_2} \tag{3.469}$$

$$\mathbf{b}_2^\bigstar \times \mathbf{b}_3^\bigstar = \frac{1}{q_1^2} \begin{bmatrix} \cos q_3 \\ \sin q_3 \\ \dfrac{\cos q_2}{\sin q_2} \end{bmatrix} \tag{3.470}$$

$$\mathbf{b}_3^\bigstar \times \mathbf{b}_1^\bigstar = \frac{1}{q_1} \begin{bmatrix} \cos q_3 \cot q_2 \\ \sin q_3 \cot q_2 \\ -1 \end{bmatrix} \tag{3.471}$$

$$\mathbf{b}_1^\bigstar \times \mathbf{b}_2^\bigstar = \frac{1}{q_1} \begin{bmatrix} -\sin q_3 \\ \cos q_3 \\ 0 \end{bmatrix} \tag{3.472}$$

Therefore,

$$\begin{aligned} r_1^\bigstar &= \mathbf{r} \cdot \frac{\mathbf{b}_2^\bigstar \times \mathbf{b}_3^\bigstar}{\left[\mathbf{b}_1^\bigstar \mathbf{b}_2^\bigstar \mathbf{b}_3^\bigstar\right]} = q_1 \\ r_2^\bigstar &= \mathbf{r} \cdot \frac{\mathbf{b}_3^\bigstar \times \mathbf{b}_1^\bigstar}{\left[\mathbf{b}_1^\bigstar \mathbf{b}_2^\bigstar \mathbf{b}_3^\bigstar\right]} = 0 \\ r_3^\bigstar &= \mathbf{r} \cdot \frac{\mathbf{b}_1^\bigstar \times \mathbf{b}_2^\bigstar}{\left[\mathbf{b}_1^\bigstar \mathbf{b}_2^\bigstar \mathbf{b}_3^\bigstar\right]} = 0 \end{aligned} \tag{3.473}$$

and

$$ {}^{R}\mathbf{r} = q_1 \mathbf{b}_1^{\bigstar} \tag{3.474} $$

Example 216 ★ Alternative Proof for $\mathbf{b}_i \cdot \mathbf{b}_j^{\bigstar} = \delta_{ij}$ Substituting (3.416) in $\mathbf{r} = x\hat{\imath} + y\hat{\jmath} + z\hat{k}$ and taking a derivative show that

$$ \begin{aligned} d\mathbf{r} &= dx\hat{\imath} + dy\hat{\jmath} + dz\hat{k} = \frac{\partial \mathbf{r}}{\partial q_1} dq_1 + \frac{\partial \mathbf{r}}{\partial q_2} dq_2 + \frac{\partial \mathbf{r}}{\partial q_3} dq_3 \\ &= \mathbf{b}_1 dq_1 + \mathbf{b}_2 dq_2 + \mathbf{b}_3 dq_3 \end{aligned} \tag{3.475} $$

where

$$ \frac{\partial \mathbf{r}}{\partial q_i} = \frac{\partial x}{\partial q_i}\hat{\imath} + \frac{\partial y}{\partial q_i}\hat{\jmath} + \frac{\partial z}{\partial q_i}\hat{k} \tag{3.476} $$

The infinitesimal vector $d\mathbf{r}$ is the diagonal of the parallelepiped with sides dx, dy, dz or with sides $(\partial \mathbf{r}/\partial q_1)dq_1$, $(\partial \mathbf{r}/\partial q_2)dq_2$, $(\partial \mathbf{r}/\partial q_3)dq_3$. Knowing that

$$ dq_i = \frac{\partial q_i}{\partial x} dx + \frac{\partial q_i}{\partial y} dy + \frac{\partial q_i}{\partial z} dz \tag{3.477} $$

and using (3.475) enable us to write

$$ \begin{aligned} dq_1 &= \nabla q_1 \cdot d\mathbf{r} \\ &= \nabla q_1 \cdot \left(\frac{\partial \mathbf{r}}{\partial q_1} dq_1 + \frac{\partial \mathbf{r}}{\partial q_2} dq_2 + \frac{\partial \mathbf{r}}{\partial q_3} dq_3 \right) \\ &= \left(\nabla q_1 \cdot \frac{\partial \mathbf{r}}{\partial q_1} \right) dq_1 + \left(\nabla q_1 \cdot \frac{\partial \mathbf{r}}{\partial q_2} \right) dq_2 + \left(\nabla q_1 \cdot \frac{\partial \mathbf{r}}{\partial q_3} \right) dq_3 \\ &= \left(\mathbf{b}_1^{\bigstar} \cdot \mathbf{b}_1 \right) dq_1 + \left(\mathbf{b}_1^{\bigstar} \cdot \mathbf{b}_2 \right) dq_2 + \left(\mathbf{b}_1^{\bigstar} \cdot \mathbf{b}_3 \right) dq_3 \end{aligned} \tag{3.478} $$

and show the correctness of (3.425).

Example 217 ★ Curvilinear Coordinate Frames Are Local We define the principal coordinate frame $Q\ (q_1, q_2, q_3)$ and its base vectors by the set of vectors $\mathbf{b}_1, \mathbf{b}_2, \mathbf{b}_3$ that are tangent to and pointing in the positive directions of the coordinate curves q_1, q_2, q_3. Figure 3.27 illustrates a point P along with the principal coordinate frame when it moves on a coordinate curve q_1.

In general, the base vectors $\mathbf{b}_1, \mathbf{b}_2, \mathbf{b}_3$ are neither orthogonal nor of unit length. The base vectors remain constant only in Cartesian coordinates or in flat oblique coordinate systems. The coordinate curves of curvilinear coordinate systems are not flat and hence the base vectors of the associated frame change locally.

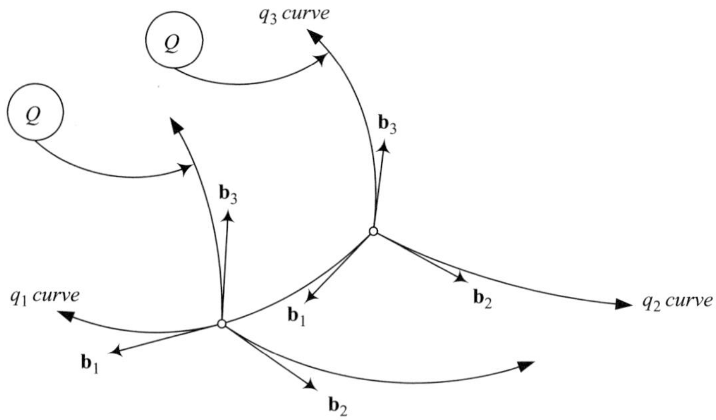

Figure 3.27 The principal coordinate frame when a point P moves on a curve q_1.

Because the direction of the base vectors $\mathbf{b}_1, \mathbf{b}_2, \mathbf{b}_3$ varies from point to point, the coordinate system Q is called local.

Example 218 ★ ★ Reciprocal Coordinates Associated with the system of reciprocal frames $\mathbf{b}_1^\star, \mathbf{b}_2^\star, \mathbf{b}_3^\star$, there is a new set of coordinate curves $q_1^\star, q_2^\star, q_3^\star$ at any point of space such that $q_i^\star$ is tangent to $\mathbf{b}_i^\star$. We may call the new coordinate system the reciprocal system and show it as $Q^\star\left(q_1^\star, q_2^\star, q_3^\star\right)$. There must be a set of reversible relationships between $q_1^\star, q_2^\star, q_3^\star$ and Cartesian coordinates:

$$\begin{aligned} x &= x_1\left(q_1^\star, q_2^\star, q_3^\star\right) \\ y &= x_2\left(q_1^\star, q_2^\star, q_3^\star\right) \\ z &= x_3\left(q_1^\star, q_2^\star, q_3^\star\right) \end{aligned} \tag{3.479}$$

$$\begin{aligned} q_1^\star &= q_1^\star(x, y, z) = q_1^\star(x_1, x_2, x_3) \\ q_2^\star &= q_2^\star(x, y, z) = q_2^\star(x_1, x_2, x_3) \\ q_3^\star &= q_3^\star(x, y, z) = q_3^\star(x_1, x_2, x_3) \end{aligned} \tag{3.480}$$

$$\begin{aligned} x &= f_1(q_1, q_2, q_3) = x_1(q_1, q_2, q_3) \\ y &= f_2(q_1, q_2, q_3) = x_2(q_1, q_2, q_3) \\ z &= f_3(q_1, q_2, q_3) = x_3(q_1, q_2, q_3) \end{aligned} \tag{3.481}$$

We assume that Equations (3.416) have a unique set of inverse functions to calculate the new Cartesian coordinates:

$$\begin{aligned} q_1 &= q_1(x, y, z) = q_1(x_1, x_2, x_3) \\ q_2 &= q_2(x, y, z) = q_2(x_1, x_2, x_3) \\ q_3 &= q_3(x, y, z) = q_3(x_1, x_2, x_3) \end{aligned} \tag{3.482}$$

A position vector $\mathbf{r}$ may be defined in Cartesian, principal, or reciprocal coordinates as

$$\mathbf{r} = x\hat{\imath} + y\hat{\jmath} + z\hat{k} \tag{3.483}$$

$$^{P}\mathbf{r} = r_1\mathbf{b}_1 + r_2\mathbf{b}_2 + r_3\mathbf{b}_3 \tag{3.484}$$

$$^{R}\mathbf{r} = r_1^{\star}\mathbf{b}_1^{\star} + r_2^{\star}\mathbf{b}_2^{\star} + r_3^{\star}\mathbf{b}_3^{\star} \tag{3.485}$$

Similar to

$$\mathbf{b}_i = \frac{\partial \mathbf{r}}{\partial q_i} = \frac{\partial x}{\partial q_i}\hat{\imath} + \frac{\partial y}{\partial q_i}\hat{\jmath} + \frac{\partial z}{\partial q_i}\hat{k} \tag{3.486}$$

$$b_i = |\mathbf{b}_i| = \sqrt{\left(\frac{\partial x}{\partial q_i}\right)^2 + \left(\frac{\partial y}{\partial q_i}\right)^2 + \left(\frac{\partial z}{\partial q_i}\right)^2} \tag{3.487}$$

we must have

$$\mathbf{b}_i^{\star} = \frac{\partial \mathbf{r}}{\partial q_i^{\star}} = \frac{\partial x}{\partial q_i^{\star}}\hat{\imath} + \frac{\partial y}{\partial q_i^{\star}}\hat{\jmath} + \frac{\partial z}{\partial q_i^{\star}}\hat{k} \tag{3.488}$$

$$b_i^{\star} = \left|\mathbf{b}_i^{\star}\right| = \sqrt{\left(\frac{\partial x}{\partial q_i^{\star}}\right)^2 + \left(\frac{\partial y}{\partial q_i^{\star}}\right)^2 + \left(\frac{\partial z}{\partial q_i^{\star}}\right)^2}. \tag{3.489}$$

However, we have

$$\mathbf{b}_i^{\star} = \nabla q_i = \frac{\partial q_i}{\partial x}\hat{\imath} + \frac{\partial q_i}{\partial y}\hat{\jmath} + \frac{\partial q_i}{\partial z}\hat{k} \tag{3.490}$$

$$b_i^{\star} = \left|\mathbf{b}_i^{\star}\right| = \sqrt{\left(\frac{\partial q_i}{\partial x}\right)^2 + \left(\frac{\partial q_i}{\partial y}\right)^2 + \left(\frac{\partial q_i}{\partial z}\right)^2} \tag{3.491}$$

and therefore,

$$\nabla q_i = \frac{\partial \mathbf{r}}{\partial q_i^{\star}} \tag{3.492}$$

$$\frac{\partial q_i}{\partial x} = \frac{\partial x}{\partial q_i^{\star}} \qquad \frac{\partial q_i}{\partial y} = \frac{\partial y}{\partial q_i^{\star}} \qquad \frac{\partial q_i}{\partial z} = \frac{\partial z}{\partial q_i^{\star}} \tag{3.493}$$

Any meaningful physical quantity that is expressible in a principal coordinate Q will have an expression in the reciprocal coordinate $Q^{\star}$.

3.5.2 ★ Principal–Reciprocal Transformation

The principal and reciprocal base vectors make two sets of triads at any point of space and define the principal P- and reciprocal R-coordinate frames. The principal and reciprocal base vectors $\mathbf{b}_i$ and $\mathbf{b}_i^{\star}$ can be expressed as a linear combination of each other:

$$\mathbf{b}_i = \left[g_{ij}^{\star}\right]\mathbf{b}_j^{\star} \tag{3.494}$$

$$\mathbf{b}_i^{\star} = \left[g_{ij}\right]\mathbf{b}_j \tag{3.495}$$

The matrices $[g_{ij}]$ and $\left[g_{ij}^{\star}\right]$ are *metric matrices* of principal and reciprocal coordinate frames where their elements are

$$g_{ij} = g_{ji} = \mathbf{b}_i \cdot \mathbf{b}_j = \frac{\partial \mathbf{r}}{\partial q_i} \cdot \frac{\partial \mathbf{r}}{\partial q_j} = \frac{\partial x_k}{\partial q_i} \frac{\partial x_k}{\partial q_j} \tag{3.496}$$

$$g_{ij}^{\star} = g_{ji}^{\star} = \mathbf{b}_i^{\star} \cdot \mathbf{b}_j^{\star} = \nabla q_i \cdot \nabla q_j = \frac{\partial q_k}{\partial x_i} \frac{\partial q_k}{\partial x_j} \tag{3.497}$$

where

$$g_{ik}\, g_{jk}^{\star} = \delta_{ij} \tag{3.498}$$

The base vectors $\mathbf{b}_i$ and $\mathbf{b}_i^{\star}$ are reciprocal to each other and can be found from

$$\begin{aligned} \mathbf{b}_1^{\star} &= \frac{\mathbf{b}_2 \times \mathbf{b}_3}{V} \\ \mathbf{b}_2^{\star} &= \frac{\mathbf{b}_3 \times \mathbf{b}_1}{V} \\ \mathbf{b}_3^{\star} &= \frac{\mathbf{b}_1 \times \mathbf{b}_2}{V} \end{aligned} \tag{3.499}$$

$$\begin{aligned} \mathbf{b}_1 &= V\left(\mathbf{b}_2^{\star} \times \mathbf{b}_3^{\star}\right) \\ \mathbf{b}_2 &= V\left(\mathbf{b}_3^{\star} \times \mathbf{b}_1^{\star}\right) \\ \mathbf{b}_3 &= V\left(\mathbf{b}_1^{\star} \times \mathbf{b}_2^{\star}\right) \end{aligned} \tag{3.500}$$

$$V = [\mathbf{b}_1 \mathbf{b}_2 \mathbf{b}_3] = \frac{1}{\left[\mathbf{b}_1^{\star} \mathbf{b}_2^{\star} \mathbf{b}_3^{\star}\right]} \tag{3.501}$$

Using these equations, the principal and reciprocal components of a vector $\mathbf{r}$ are

$$r_i = \mathbf{r} \cdot \mathbf{b}_i^{\star} \tag{3.502}$$

$$r_i^{\star} = \mathbf{r} \cdot \mathbf{b}_i \tag{3.503}$$

and hence the principal and reciprocal components of $\mathbf{r}$ would be

$${}^{P}\mathbf{r} = \left(\mathbf{r} \cdot \mathbf{b}_1^{\star}\right) \mathbf{b}_1 + \left(\mathbf{r} \cdot \mathbf{b}_2^{\star}\right) \mathbf{b}_2 + \left(\mathbf{r} \cdot \mathbf{b}_3^{\star}\right) \mathbf{b}_3 \tag{3.504}$$

$${}^{R}\mathbf{r} = (\mathbf{r} \cdot \mathbf{b}_1)\, \mathbf{b}_1^{\star} + (\mathbf{r} \cdot \mathbf{b}_2)\, \mathbf{b}_2^{\star} + (\mathbf{r} \cdot \mathbf{b}_3)\, \mathbf{b}_3^{\star} \tag{3.505}$$

The vectors $\mathbf{b}_i$ and $\mathbf{b}_i^{\star}$ are also called *tangent* and *normal base vectors*, respectively.

Proof: Let us substitute ${}^{P}\mathbf{r}$ from (3.504) in the definition of $r_i^{\star}$, (3.503)

$$r_i^{\star} = \mathbf{r} \cdot \mathbf{b}_i = \left(\mathbf{r} \cdot \mathbf{b}_j^{\star}\right) \mathbf{b}_j \cdot \mathbf{b}_i = \left(\mathbf{b}_i \cdot \mathbf{b}_j\right) r_j \tag{3.506}$$

and determine the transformation matrix ${}^{R}R_{P}$ to map the coordinates of a point from the principal frame P to the reciprocal frame R:

$$ {}^{R}\mathbf{r} = {}^{R}R_{P}\,{}^{P}\mathbf{r} \tag{3.507}$$

$$ {}^{R}R_{P} = \begin{bmatrix} \mathbf{b}_1\cdot\mathbf{b}_1 & \mathbf{b}_1\cdot\mathbf{b}_2 & \mathbf{b}_1\cdot\mathbf{b}_3 \\ \mathbf{b}_2\cdot\mathbf{b}_1 & \mathbf{b}_2\cdot\mathbf{b}_2 & \mathbf{b}_2\cdot\mathbf{b}_3 \\ \mathbf{b}_3\cdot\mathbf{b}_1 & \mathbf{b}_3\cdot\mathbf{b}_2 & \mathbf{b}_3\cdot\mathbf{b}_3 \end{bmatrix} \tag{3.508}$$

Similarly, we may substitute ${}^{R}\mathbf{r}$ from (3.505) in the definition of r_i, (3.502)

$$ r_i = \mathbf{r}\cdot\mathbf{b}_i^\star = \left(\mathbf{r}\cdot\mathbf{b}_j\right)\mathbf{b}_j^\star\cdot\mathbf{b}_i^\star = \left(\mathbf{b}_i^\star\cdot\mathbf{b}_j^\star\right) r_j^\star \tag{3.509}$$

and determine the transformation matrix ${}^{P}R_{R}$ to map the coordinates of a point from the reciprocal frame R to the principal frame P:

$$ {}^{P}\mathbf{r} = {}^{P}R_{R}\,{}^{R}\mathbf{r} \tag{3.510}$$

$$ {}^{P}R_{R} = \begin{bmatrix} \mathbf{b}_1^\star\cdot\mathbf{b}_1^\star & \mathbf{b}_1^\star\cdot\mathbf{b}_2^\star & \mathbf{b}_1^\star\cdot\mathbf{b}_3^\star \\ \mathbf{b}_2^\star\cdot\mathbf{b}_1^\star & \mathbf{b}_2^\star\cdot\mathbf{b}_2^\star & \mathbf{b}_2^\star\cdot\mathbf{b}_3^\star \\ \mathbf{b}_3^\star\cdot\mathbf{b}_1^\star & \mathbf{b}_3^\star\cdot\mathbf{b}_2^\star & \mathbf{b}_3^\star\cdot\mathbf{b}_3^\star \end{bmatrix} \tag{3.511}$$

The base vector $\mathbf{b}_j$ is in the principal frame, and hence

$$ \mathbf{b}_i^\star = {}^{R}R_{P}\,\mathbf{b}_i = \left[g_{ij}\right]\mathbf{b}_j = g_{ij}\mathbf{b}_j \tag{3.512}$$

Similarly, we have

$$ \mathbf{b}_i = {}^{P}R_{R}\,\mathbf{b}_i^\star = \left[g_{ij}^\star\right]\mathbf{b}_j^\star = g_{ij}^\star\mathbf{b}_j^\star \tag{3.513}$$

Using (3.496) and (3.497) indicates that the transformation matrices ${}^{R}R_{P}$ and ${}^{P}R_{R}$ are equal to the *covariant* and *contravariant metric tensors*, respectively:

$$ {}^{R}R_{P} = \left[g_{ij}\right] = \begin{bmatrix} g_{11} & g_{12} & g_{13} \\ g_{21} & g_{22} & g_{23} \\ g_{31} & g_{32} & g_{33} \end{bmatrix} \tag{3.514}$$

$$ {}^{P}R_{R} = \left[g_{ij}^\star\right] = \begin{bmatrix} g_{11}^\star & g_{12}^\star & g_{13}^\star \\ g_{21}^\star & g_{22}^\star & g_{23}^\star \\ g_{31}^\star & g_{32}^\star & g_{33}^\star \end{bmatrix} \tag{3.515}$$

The transformation matrices ${}^{R}R_{P}$ and ${}^{P}R_{R}$ can relate the principal and reciprocal components of a vector $\mathbf{r}$:

$$ r_i = g_{ij}^\star r_j^\star \qquad r_i^\star = g_{ij} r_j \tag{3.516}$$

It shows that the transformation matrix ${}^{P}R_{R}$ is the inverse of ${}^{R}R_{P}$,

$$ {}^{P}R_{R}^{-1} = {}^{R}R_{P} \tag{3.517}$$

and therefore,

$$\begin{bmatrix} g_{11}^{\bigstar} & g_{12}^{\bigstar} & g_{13}^{\bigstar} \\ g_{21}^{\bigstar} & g_{22}^{\bigstar} & g_{23}^{\bigstar} \\ g_{31}^{\bigstar} & g_{32}^{\bigstar} & g_{33}^{\bigstar} \end{bmatrix}^{-1} = \begin{bmatrix} g_{11} & g_{12} & g_{13} \\ g_{21} & g_{22} & g_{23} \\ g_{31} & g_{32} & g_{33} \end{bmatrix} \tag{3.518}$$

$$\begin{bmatrix} g_{11} & g_{12} & g_{13} \\ g_{21} & g_{22} & g_{23} \\ g_{31} & g_{32} & g_{33} \end{bmatrix}^{-1} = \begin{bmatrix} g_{11}^{\bigstar} & g_{12}^{\bigstar} & g_{13}^{\bigstar} \\ g_{21}^{\bigstar} & g_{22}^{\bigstar} & g_{23}^{\bigstar} \\ g_{31}^{\bigstar} & g_{32}^{\bigstar} & g_{33}^{\bigstar} \end{bmatrix} \tag{3.519}$$

Because $\mathbf{b}_k^{\bigstar}$ is perpendicular to the (q_i, q_j)-plane and $\mathbf{b}_i, \mathbf{b}_j$ are tangent to the coordinate curves q_i, q_j, the cross product of $\mathbf{b}_i \times \mathbf{b}_j$ must be proportional to $\mathbf{b}_k^{\bigstar}$,

$$\mathbf{b}_i \times \mathbf{b}_j = A\epsilon_{ijk}\mathbf{b}_k^{\bigstar} \tag{3.520}$$

where A is a constant. Furthermore, $\mathbf{b}_i^{\bigstar}$ and $\mathbf{b}_j^{\bigstar}$ are both perpendicular to $\mathbf{b}_k$, and hence the cross product of $\mathbf{b}_i^{\bigstar} \times \mathbf{b}_j^{\bigstar}$ must be proportional to $\mathbf{b}_k$,

$$\mathbf{b}_i^{\bigstar} \times \mathbf{b}_j^{\bigstar} = B\epsilon_{ijk}\mathbf{b}_k \tag{3.521}$$

where B is another constant. To determine A and B, we form the inner product of (3.520) and (3.521) with $\epsilon_{ijm}\mathbf{b}_m$ and $\epsilon_{ijm}\mathbf{b}_m^{\bigstar}$, respectively,

$$\begin{aligned} \epsilon_{ijm}\mathbf{b}_m \cdot \mathbf{b}_i \times \mathbf{b}_j &= A\epsilon_{ijk}\epsilon_{ijm}\mathbf{b}_m \cdot \mathbf{b}_k^{\bigstar} = A\epsilon_{ijk}\epsilon_{ijm}\delta_{mk} \\ &= A\epsilon_{ijk}\epsilon_{ijk} = 6A \end{aligned} \tag{3.522}$$

$$\begin{aligned} \epsilon_{ijm}\mathbf{b}_m^{\bigstar} \cdot \mathbf{b}_i^{\bigstar} \times \mathbf{b}_j^{\bigstar} &= B\epsilon_{ijk}\epsilon_{ijm}\mathbf{b}_m^{\bigstar} \cdot \mathbf{b}_k = B\epsilon_{ijk}\epsilon_{ijm}\delta_{mk} \\ &= B\epsilon_{ijk}\epsilon_{ijk} = 6B \end{aligned} \tag{3.523}$$

and therefore,

$$A = \tfrac{1}{6}\epsilon_{ijk}\mathbf{b}_k \cdot \mathbf{b}_i \times \mathbf{b}_j \tag{3.524}$$

$$B = \tfrac{1}{6}\epsilon_{ijk}\mathbf{b}_k^{\bigstar} \cdot \mathbf{b}_i^{\bigstar} \times \mathbf{b}_j^{\bigstar} \tag{3.525}$$

Using (3.512) and (3.513), we may substitute the vectors $\mathbf{b}_i$ and $\mathbf{b}_i^{\bigstar}$ in terms of each other and find

$$\begin{aligned} A &= \tfrac{1}{6}\epsilon_{ijk}\, g_{il}^{\bigstar}\, g_{jm}^{\bigstar}\, g_{kn}^{\bigstar}\, \mathbf{b}_n^{\bigstar} \cdot \mathbf{b}_l^{\bigstar} \times \mathbf{b}_m^{\bigstar} = \tfrac{1}{6}\epsilon_{ijk}\,\epsilon_{lmn}\, g_{il}^{\bigstar}\, g_{jm}^{\bigstar}\, g_{kn}^{\bigstar}\, B \\ &= \left|g_{ij}^{\bigstar}\right| B = \det\left[g_{ij}^{\bigstar}\right] B \end{aligned} \tag{3.526}$$

$$\begin{aligned} B &= \tfrac{1}{6}\epsilon_{ijk}\, g_{il}\, g_{jm}\, g_{kn}\, \mathbf{b}_k \cdot \mathbf{b}_i \times \mathbf{b}_j = \tfrac{1}{6}\epsilon_{ijk}\,\epsilon_{lmn}\, g_{il}\, g_{jm}\, g_{kn}\, A \\ &= \left|g_{ij}\right| A = \det\left[g_{ij}\right] A \end{aligned} \tag{3.527}$$

$$\left|g_{ij}^{\bigstar}\right| = \frac{1}{\left|g_{ij}\right|} \tag{3.528}$$

where $|g_{ij}|$ and $\left|g^{\star}_{ij}\right|$ are determinants of the transformation matrices $[g_{ij}]$ and $\left[g^{\star}_{ij}\right]$. The (i, j)-elements of $[g_{ij}]$ and $\left[g^{\star}_{ij}\right]$ are

$$g_{ij} = g_{ji} = \mathbf{b}_i \cdot \mathbf{b}_j = \frac{\partial \mathbf{r}}{\partial q_i} \cdot \frac{\partial \mathbf{r}}{\partial q_j} = \frac{\partial x_k}{\partial q_i}\frac{\partial x_k}{\partial q_j} \tag{3.529}$$

$$g^{\star}_{ij} = g^{\star}_{ji} = \mathbf{b}^{\star}_i \cdot \mathbf{b}^{\star}_j = \nabla q_i \cdot \nabla q_j = \frac{\partial q_k}{\partial x_i}\frac{\partial q_k}{\partial x_j} \tag{3.530}$$

Substituting for $g^{\star}_{ij}$ and g_{ij} in (3.526) and (3.527) yields

$$A = \frac{1}{6}\epsilon_{ijk}\,\epsilon_{lmn}\frac{\partial x_l}{\partial q_i}\frac{\partial x_m}{\partial q_j}\frac{\partial x_n}{\partial q_k} = \det\left[\frac{\partial x_i}{\partial q_j}\right] = \begin{vmatrix} \frac{\partial x_1}{\partial q_1} & \frac{\partial x_1}{\partial q_2} & \frac{\partial x_1}{\partial q_2} \\ \frac{\partial x_2}{\partial q_1} & \frac{\partial x_2}{\partial q_2} & \frac{\partial x_2}{\partial q_3} \\ \frac{\partial x_3}{\partial q_1} & \frac{\partial x_3}{\partial q_2} & \frac{\partial x_3}{\partial q_3} \end{vmatrix}$$

$$= \frac{\partial \mathbf{r}}{\partial q_1} \cdot \frac{\partial \mathbf{r}}{\partial q_2} \times \frac{\partial \mathbf{r}}{\partial q_2} = \mathbf{b}_1 \cdot \mathbf{b}_2 \times \mathbf{b}_3 = [\mathbf{b}_1\mathbf{b}_2\mathbf{b}_3] = V \tag{3.531}$$

$$B = \frac{1}{6}\epsilon_{ijk}\,\epsilon_{lmn}\frac{\partial q_l}{\partial x_i}\frac{\partial q_m}{\partial x_j}\frac{\partial q_n}{\partial x_k} = \det\left[\frac{\partial q_i}{\partial x_j}\right] = \begin{vmatrix} \frac{\partial q_1}{\partial x_1} & \frac{\partial q_1}{\partial x_2} & \frac{\partial q_1}{\partial x_2} \\ \frac{\partial q_2}{\partial x_1} & \frac{\partial q_2}{\partial x_2} & \frac{\partial q_2}{\partial x_3} \\ \frac{\partial q_3}{\partial x_1} & \frac{\partial q_3}{\partial x_2} & \frac{\partial q_3}{\partial x_3} \end{vmatrix}$$

$$= \nabla q_1 \cdot \nabla q_2 \times \nabla q_3 = \mathbf{b}^{\star}_1 \cdot \mathbf{b}^{\star}_2 \times \mathbf{b}^{\star}_3 = \left[\mathbf{b}^{\star}_1\mathbf{b}^{\star}_2\mathbf{b}^{\star}_3\right] = \frac{1}{V} \tag{3.532}$$

and therefore,

$$V = J = [\mathbf{b}_1\mathbf{b}_2\mathbf{b}_3] = \frac{1}{\left[\mathbf{b}^{\star}_1\mathbf{b}^{\star}_2\mathbf{b}^{\star}_3\right]} \tag{3.533}$$

In differential geometry and coordinate transformation, the volume V is called the *Jacobian* and is shown by J. Because of that, from now on, we may use J instead of V.

Substituting A and B in (3.520) and (3.521) yields

$$\frac{\mathbf{b}_i \times \mathbf{b}_j}{J} = \epsilon_{ijk}\mathbf{b}^{\star}_k \tag{3.534}$$

$$J\left(\mathbf{b}^{\star}_i \times \mathbf{b}^{\star}_j\right) = \epsilon_{ijk}\mathbf{b}_k \tag{3.535}$$

which are equivalent to Equations (3.499) and (3.500).

The determinant $|g_{ij}|$ can be simplified based on determinant calculus:

$$\left|g_{ij}\right| = \begin{vmatrix} \sum_{k=1}^{3} \frac{\partial x_k}{\partial q_1}\frac{\partial x_k}{\partial q_1} & \sum_{k=1}^{3} \frac{\partial x_k}{\partial q_1}\frac{\partial x_k}{\partial q_2} & \sum_{k=1}^{3} \frac{\partial x_k}{\partial q_1}\frac{\partial x_k}{\partial q_3} \\ \sum_{k=1}^{3} \frac{\partial x_k}{\partial q_2}\frac{\partial x_k}{\partial q_1} & \sum_{k=1}^{3} \frac{\partial x_k}{\partial q_2}\frac{\partial x_k}{\partial q_2} & \sum_{k=1}^{3} \frac{\partial x_k}{\partial q_2}\frac{\partial x_k}{\partial q_3} \\ \sum_{k=1}^{3} \frac{\partial x_k}{\partial q_3}\frac{\partial x_k}{\partial q_1} & \sum_{k=1}^{3} \frac{\partial x_k}{\partial q_3}\frac{\partial x_k}{\partial q_2} & \sum_{k=1}^{3} \frac{\partial x_k}{\partial q_3}\frac{\partial x_k}{\partial q_3} \end{vmatrix}$$

$$= \begin{vmatrix} \frac{\partial x_1}{\partial q_1} & \frac{\partial x_1}{\partial q_2} & \frac{\partial x_1}{\partial q_2} \\ \frac{\partial x_2}{\partial q_1} & \frac{\partial x_2}{\partial q_2} & \frac{\partial x_2}{\partial q_3} \\ \frac{\partial x_3}{\partial q_1} & \frac{\partial x_3}{\partial q_2} & \frac{\partial x_3}{\partial q_3} \end{vmatrix} \begin{vmatrix} \frac{\partial x_1}{\partial q_1} & \frac{\partial x_1}{\partial q_2} & \frac{\partial x_1}{\partial q_2} \\ \frac{\partial x_2}{\partial q_1} & \frac{\partial x_2}{\partial q_2} & \frac{\partial x_2}{\partial q_3} \\ \frac{\partial x_3}{\partial q_1} & \frac{\partial x_3}{\partial q_2} & \frac{\partial x_3}{\partial q_3} \end{vmatrix} = J^2 \tag{3.536}$$

In shorthand notation we may show $\left|g_{ij}\right|$ as

$$\begin{aligned} \left|g_{ij}\right| &= \left|\frac{\partial x_k}{\partial q_i}\frac{\partial x_k}{\partial q_j}\right| = \left|\frac{\partial x_i}{\partial q_j}\right|\left|\frac{\partial x_i}{\partial q_j}\right| = \left|\frac{\partial x_i}{\partial q_j}\right|^2 \\ &= [\mathbf{b}_1\mathbf{b}_2\mathbf{b}_3]^2 = J^2 \end{aligned} \tag{3.537}$$

and similarly, we find

$$\begin{aligned} \left|g_{ij}^{\star}\right| &= \left|\frac{\partial q_k}{\partial x_i}\frac{\partial q_k}{\partial x_j}\right| = \left|\frac{\partial q_i}{\partial x_j}\right|\left|\frac{\partial q_i}{\partial x_j}\right| = \left|\frac{\partial q_i}{\partial x_j}\right|^2 \\ &= \left[\mathbf{b}_1^{\star}\mathbf{b}_2^{\star}\mathbf{b}_3^{\star}\right]^2 = \frac{1}{J^2} \end{aligned} \tag{3.538}$$

■

Example 219 ★ $\left|g_{ij}^{\star}\right| |g_{ij}| = 1$ Direct multiplication of determinants $\left|g_{ij}^{\star}\right|$ and $|g_{ij}|$ shows that they are inverses of each other:

$$|g_{ij}| \left|g_{ij}^{\star}\right| = \begin{vmatrix} \frac{\partial x_1}{\partial q_1} & \frac{\partial x_1}{\partial q_2} & \frac{\partial x_1}{\partial q_2} \\ \frac{\partial x_2}{\partial q_1} & \frac{\partial x_2}{\partial q_2} & \frac{\partial x_2}{\partial q_3} \\ \frac{\partial x_3}{\partial q_1} & \frac{\partial x_3}{\partial q_2} & \frac{\partial x_3}{\partial q_3} \end{vmatrix} \begin{vmatrix} \frac{\partial q_1}{\partial x_1} & \frac{\partial q_1}{\partial x_2} & \frac{\partial q_1}{\partial x_2} \\ \frac{\partial q_2}{\partial x_1} & \frac{\partial q_2}{\partial x_2} & \frac{\partial q_2}{\partial x_3} \\ \frac{\partial q_3}{\partial x_1} & \frac{\partial q_3}{\partial x_2} & \frac{\partial q_3}{\partial x_3} \end{vmatrix}$$

$$= \begin{vmatrix} \sum_{j=1}^{3} \frac{\partial x_1}{\partial q_j}\frac{\partial q_j}{\partial x_1} & \sum_{j=1}^{3} \frac{\partial x_1}{\partial q_j}\frac{\partial q_j}{\partial x_2} & \sum_{j=1}^{3} \frac{\partial x_1}{\partial q_j}\frac{\partial q_j}{\partial x_3} \\ \sum_{j=1}^{3} \frac{\partial x_2}{\partial q_j}\frac{\partial q_j}{\partial x_1} & \sum_{j=1}^{3} \frac{\partial x_2}{\partial q_j}\frac{\partial q_j}{\partial x_2} & \sum_{j=1}^{3} \frac{\partial x_2}{\partial q_j}\frac{\partial q_j}{\partial x_3} \\ \sum_{j=1}^{3} \frac{\partial x_3}{\partial q_j}\frac{\partial q_j}{\partial x_1} & \sum_{j=1}^{3} \frac{\partial x_3}{\partial q_j}\frac{\partial q_j}{\partial x_2} & \sum_{j=1}^{3} \frac{\partial x_3}{\partial q_j}\frac{\partial q_j}{\partial x_3} \end{vmatrix}$$

$$= \left|g_{ij}^{\star}\right| \left|g_{ij}\right| = \left|g_{ij}g_{ij}^{\star}\right| = \left|\delta_{ij}\right| = 1 \tag{3.539}$$

Example 220 ★ $g_{ii} = b_i^2$ ***and*** $g_{ii}^{\star} = b_i^{\star 2}$ Expansion of the elements g_{ij} and $g_{ij}^{\star}$ for $i = j$ shows that g_{ii} and $g_{ii}^{\star}$ are equal to the length of the base vectors in the principal and reciprocal coordinate frames:

$$g_{ii} = \sum_{k=1}^{3} \frac{\partial x_k}{\partial q_i} \frac{\partial x_k}{\partial q_i} = \left(\frac{\partial x}{\partial q_i}\right)^2 + \left(\frac{\partial y}{\partial q_i}\right)^2 + \left(\frac{\partial z}{\partial q_i}\right)^2 = b_i^2 \tag{3.540}$$

$$g_{ii}^{\star} = \sum_{k=1}^{3} \frac{\partial q_k}{\partial x_i} \frac{\partial q_k}{\partial x_i} = \left(\frac{\partial q_i}{\partial x}\right)^2 + \left(\frac{\partial q_i}{\partial y}\right)^2 + \left(\frac{\partial q_i}{\partial z}\right)^2 = b_i^{\star 2} \tag{3.541}$$

If the coordinate system is orthogonal, then

$$g_{ii} = \frac{1}{g_{ii}^{\star}} \tag{3.542}$$

Example 221 ★ $\mathbf{b}_i^{\star} = \sum_{j=1}^{3} (g_{ij}/b_j)\hat{u}_j$ The reciprocal base vectors $\mathbf{b}_i^{\star}$ can be expressed in the principal frame with a linear combination of principal base vectors. For $\mathbf{b}_1^{\star}$, we have

$$\begin{aligned} \mathbf{b}_1^{\star} &= \left(\mathbf{b}_1^{\star} \cdot \hat{u}_1\right)\hat{u}_1 + \left(\mathbf{b}_1^{\star} \cdot \hat{u}_2\right)\hat{u}_2 + \left(\mathbf{b}_1^{\star} \cdot \hat{u}_3\right)\hat{u}_3 \\ &= \frac{1}{b_1}\left(\mathbf{b}_1^{\star} \cdot \mathbf{b}_1\right)\hat{u}_1 + \frac{1}{b_2}\left(\mathbf{b}_1^{\star} \cdot \mathbf{b}_2\right)\hat{u}_2 + \frac{1}{b_3}\left(\mathbf{b}_1^{\star} \cdot \mathbf{b}_3\right)\hat{u}_3 \\ &= \frac{g_{11}}{b_1}\hat{u}_1 + \frac{g_{12}}{b_2}\hat{u}_2 + \frac{g_{13}}{b_3}\hat{u}_3 \end{aligned} \tag{3.543}$$

and similarly,

$$\mathbf{b}_2^{\star} = \frac{g_{21}}{b_1}\hat{u}_1 + \frac{g_{22}}{b_2}\hat{u}_2 + \frac{g_{23}}{b_3}\hat{u}_3 \tag{3.544}$$

$$\mathbf{b}_3^{\star} = \frac{g_{31}}{b_1}\hat{u}_1 + \frac{g_{32}}{b_2}\hat{u}_2 + \frac{g_{33}}{b_3}\hat{u}_3 \tag{3.545}$$

Example 222 ★ ***Paraboloidal Coordinates*** Consider the paraboloidal coordinate system $Q\,(u, v, w)$ defined by the equations

$$x = uv\cos w \qquad y = uv\sin w \qquad z = \tfrac{1}{2}\left(u^2 - v^2\right) \tag{3.546}$$

or

$$\begin{aligned} u^2 &= \sqrt{x^2 + y^2 + z^2} + z \\ v^2 &= \sqrt{x^2 + y^2 + z^2} - z \\ w &= \tan^{-1}\frac{y}{x} \end{aligned} \tag{3.547}$$

Using Equation (3.420) yields the base vectors $\mathbf{b}_i$:

$$\begin{aligned}\mathbf{b}_1 &= \frac{\partial \mathbf{r}}{\partial u} = v\cos w\hat{\imath} + v\sin w\hat{\jmath} + u\hat{k} \\ \mathbf{b}_2 &= \frac{\partial \mathbf{r}}{\partial v} = u\cos w\hat{\imath} + u\sin w\hat{\jmath} - v\hat{k} \\ \mathbf{b}_3 &= \frac{\partial \mathbf{r}}{\partial w} = -uv\sin w\hat{\imath} + uv\cos w\hat{\jmath}\end{aligned} \tag{3.548}$$

The metric tensor is an identity matrix that shows the principal coordinate frame is orthogonal:

$$\left[g_{ij}\right] = \begin{bmatrix} \mathbf{b}_1\cdot\mathbf{b}_1 & \mathbf{b}_1\cdot\mathbf{b}_2 & \mathbf{b}_1\cdot\mathbf{b}_3 \\ \mathbf{b}_2\cdot\mathbf{b}_1 & \mathbf{b}_2\cdot\mathbf{b}_2 & \mathbf{b}_2\cdot\mathbf{b}_3 \\ \mathbf{b}_3\cdot\mathbf{b}_1 & \mathbf{b}_3\cdot\mathbf{b}_2 & \mathbf{b}_3\cdot\mathbf{b}_3 \end{bmatrix} = \mathbf{I}_3 \tag{3.549}$$

The reciprocal base vectors are calculated from Equation (3.422):

$$\begin{aligned}\mathbf{b}_1^\star &= \nabla q_1 = \frac{v\cos w}{u^2+v^2}\hat{\imath} + \frac{v\sin w}{u^2+v^2}\hat{\jmath} + \frac{u}{u^2+v^2}\hat{k} \\ \mathbf{b}_2^\star &= \nabla q_2 = \frac{u\cos w}{u^2+v^2}\hat{\imath} + \frac{u\sin w}{u^2+v^2}\hat{\jmath} - \frac{v}{u^2+v^2}\hat{k} \\ \mathbf{b}_3^\star &= \nabla q_3 = -\frac{\sin w}{uv}\hat{\imath} + \frac{\cos w}{uv}\hat{\jmath}\end{aligned} \tag{3.550}$$

Because the paraboloidal system $Q\ (u, v, w)$ is orthogonal, the principal and reciprocal coordinate frames are coincident and we have

$$\begin{aligned}\mathbf{b}_1^\star &= \frac{1}{u^2+v^2}\mathbf{b}_1 \\ \mathbf{b}_2^\star &= \frac{1}{u^2+v^2}\mathbf{b}_2 \\ \mathbf{b}_3^\star &= \frac{1}{u^2v^2}\mathbf{b}_3\end{aligned} \tag{3.551}$$

Example 223 ★ Metric of Cylindrical and Spherical Coordinates For cylindrical coordinates, we have

$$\begin{array}{ccc} x = \rho\cos\theta & y = \rho\sin\theta & z = z \\ q_1 = \rho & q_2 = \theta & q_3 = z \end{array} \tag{3.552}$$

Its metric matrix is

$$\left[g_{ij}\right] = \begin{bmatrix} 1 & 0 & 0 \\ 0 & \rho^2 & 0 \\ 0 & 0 & 0 \end{bmatrix} \tag{3.553}$$

For spherical coordinates, we have

$$\begin{array}{ccc} x = r\sin\varphi\cos\theta & y = r\sin\varphi\sin\theta & z = r\cos\varphi \\ q_1 = r & q_2 = \varphi & q_3 = \theta \end{array} \tag{3.554}$$

The metric tensor $[g_{ij}]$ has the following components:

$$[g_{ij}] = \begin{bmatrix} 1 & 0 & 0 \\ 0 & r^2 & 0 \\ 0 & 0 & r^2 \sin^2 \varphi \end{bmatrix} \tag{3.555}$$

Example 224 ★ Vector Products in Curvilinear Coordinate Systems Any vector **r** may be expressed in the principal or reciprocal frames of a curvilinear coordinate:

$$^P\mathbf{r} = r_1\mathbf{b}_1 + r_2\mathbf{b}_2 + r_3\mathbf{b}_3 \tag{3.556}$$

$$^R\mathbf{r} = r_1^\star \mathbf{b}_1^\star + r_2^\star \mathbf{b}_2^\star + r_3^\star \mathbf{b}_3^\star \tag{3.557}$$

The inner products of different expressions of a vector **r** and themselves are

$$^P\mathbf{r} \cdot {}^P\mathbf{r} = {}^P\mathbf{r}^T \left[g_{ij}\right] {}^P\mathbf{r} = r_i r_j g_{ij} \tag{3.558}$$

$$^R\mathbf{r} \cdot {}^R\mathbf{r} = {}^R\mathbf{r}^T \left[g_{ij}^\star\right] {}^R\mathbf{r} = r_i^\star r_j^\star g_{ij}^\star \tag{3.559}$$

$$^P\mathbf{r} \cdot {}^R\mathbf{r} = {}^P\mathbf{r} \cdot {}^R\mathbf{r} = r_i r_i^\star \tag{3.560}$$

and their outer products are

$$^P\mathbf{r} \times {}^P\mathbf{r} = 0 \tag{3.561}$$

$$^R\mathbf{r} \times {}^R\mathbf{r} = 0 \tag{3.562}$$

$$^P\mathbf{r} \times {}^R\mathbf{r} = {}^P\mathbf{r} \times \left[g_{ij}^\star\right] {}^R\mathbf{r} \tag{3.563}$$

Consider three vectors $^P\mathbf{r}_1$, $^P\mathbf{r}_2$, and $^P\mathbf{r}_3$ that are in the principal frame:

$$^P\mathbf{r}_1 = p_1\mathbf{b}_1 + p_2\mathbf{b}_2 + p_3\mathbf{b}_3 \tag{3.564}$$

$$^P\mathbf{r}_2 = q_1\mathbf{b}_1 + q_2\mathbf{b}_2 + q_3\mathbf{b}_3 \tag{3.565}$$

$$^P\mathbf{r}_3 = s_1\mathbf{b}_1 + s_2\mathbf{b}_2 + s_3\mathbf{b}_3 \tag{3.566}$$

Then the inner product of the vectors $^P\mathbf{r}_1$ and $^P\mathbf{r}_2$ is

$$\begin{aligned} ^P\mathbf{r}_1 \cdot {}^P\mathbf{r}_2 &= p_i q_j g_{ij} \\ &= p_1 q_1 g_{11} + p_2 q_2 g_{22} + p_2 q_3 g_{23} + (p_1 q_2 + p_2 q_1)\, g_{12} \\ &\quad + (p_1 q_3 + p_3 q_1)\, g_{13} + (p_3 q_2 + p_3 q_3)\, g_{33} \end{aligned} \tag{3.567}$$

and the outer product of $^P\mathbf{r}_1$ and $^P\mathbf{r}_2$ is

$$^P\mathbf{r}_1 \times {}^P\mathbf{r}_2 = \begin{vmatrix} \mathbf{b}_2 \times \mathbf{b}_3 & \mathbf{b}_3 \times \mathbf{b}_1 & \mathbf{b}_1 \times \mathbf{b}_2 \\ p_1 & p_2 & p_3 \\ q_1 & q_2 & q_3 \end{vmatrix} \tag{3.568}$$

Substituting $\mathbf{b}_i \times \mathbf{b}_j$ from the definition of reciprocal vectors, we may also show that

$$\mathbf{r}_1 \times \mathbf{r}_2 = [\mathbf{b}_1 \mathbf{b}_2 \mathbf{b}_3] \begin{vmatrix} \mathbf{b}_1^\star & \mathbf{b}_2^\star & \mathbf{b}_3^\star \\ p_1 & p_2 & p_3 \\ q_1 & q_2 & q_3 \end{vmatrix} = J \begin{vmatrix} \mathbf{b}_1^\star & \mathbf{b}_2^\star & \mathbf{b}_3^\star \\ p_1 & p_2 & p_3 \\ q_1 & q_2 & q_3 \end{vmatrix} \tag{3.569}$$

$$\mathbf{r}_1 \times \mathbf{r}_2 = \left[\mathbf{b}_1^\star \mathbf{b}_2^\star \mathbf{b}_3^\star\right] \begin{vmatrix} \mathbf{b}_1 & \mathbf{b}_2 & \mathbf{b}_3 \\ p_1^\star & p_2^\star & p_3^\star \\ q_1^\star & q_2^\star & q_3^\star \end{vmatrix} = \frac{1}{J} \begin{vmatrix} \mathbf{b}_1 & \mathbf{b}_2 & \mathbf{b}_3 \\ p_1^\star & p_2^\star & p_3^\star \\ q_1^\star & q_2^\star & q_3^\star \end{vmatrix} \tag{3.570}$$

The outer product of two vectors in the principal and reciprocal frames can be determined by transforming one vector to the other frame:

$${}^P\mathbf{r} \times {}^R\mathbf{r} = {}^P\mathbf{r} \times {}^P R_R \, {}^R\mathbf{r} = {}^P\mathbf{r} \times \left[g_{ij}^\star\right] \, {}^R\mathbf{r} \tag{3.571}$$

The proof of these equations may follow the method of Section 3.4.3.

Furthermore, we can show that

$$\mathbf{r}_1 \cdot \mathbf{r}_2 \times \mathbf{r}_3 = J p_i q_j s_k \epsilon_{ijk} = \frac{1}{J} p_i^\star q_j^\star s_k^\star \epsilon_{ijk} \tag{3.572}$$

3.5.3 ★ Curvilinear Geometry

At every point in space, the principal and reciprocal unit and base vectors of a curvilinear coordinate system define two coordinate frames P and R. A vector $\mathbf{r}$ may be expressed in these frames by their base vectors:

$${}^P\mathbf{r} = r_1 \mathbf{b}_1 + r_2 \mathbf{b}_2 + r_3 \mathbf{b}_3 \tag{3.573}$$

$${}^R\mathbf{r} = r_1^\star \mathbf{b}_1^\star + r_2^\star \mathbf{b}_2^\star + r_3^\star \mathbf{b}_3^\star \tag{3.574}$$

The geometric information in a curvilinear coordinate system $Q(q_1, q_2, q_3)$ depends on the arc element ds, surface element dA, and volume element dV. Figure 3.28 illustrates a principal coordinate frame P and indicates three area elements dA_1, dA_2, dA_3 along with the volume element dV.

The arc element ds can be found by $d\mathbf{r} \cdot d\mathbf{r}$:

$$ds^2 = d\mathbf{r} \cdot d\mathbf{r} = d\mathbf{r}^2 = g_{ij} \, dq_i \, dq_j \tag{3.575}$$

$$g_{ij} = \frac{\partial \mathbf{r}}{\partial q_i} \cdot \frac{\partial \mathbf{r}}{\partial q_j} = \mathbf{b}_i \cdot \mathbf{b}_j \tag{3.576}$$

The surface element dA_1 can be found by a scalar triple product:

$$\begin{aligned} dA_1 &= \hat{u}_1^\star \cdot d\mathbf{r}_2 \times d\mathbf{r}_3 = \hat{u}_1^\star \cdot \mathbf{b}_2 \times \mathbf{b}_3 \, dq_2 \, dq_3 \\ &= J \hat{u}_1^\star \cdot \mathbf{b}_1^\star \, dq_2 \, dq_3 = J b_1^\star \, dq_2 \, dq_3 \end{aligned} \tag{3.577}$$

Similarly, we have

$$dA_2 = J b_2^{\bigstar} \, dq_3 \, dq_1 \tag{3.578}$$

$$dA_3 = J b_3^{\bigstar} \, dq_1 \, dq_2 \tag{3.579}$$

The volume element dV can also be found by a scalar triple product. As is shown in Figure 3.28, dV is equal to the parallelepiped made by $d\mathbf{r}_1, d\mathbf{r}_2, d\mathbf{r}_3$:

$$\begin{aligned} dV &= d\mathbf{r}_1 \cdot d\mathbf{r}_2 \times d\mathbf{r}_3 = \mathbf{b}_1 \cdot \mathbf{b}_2 \times \mathbf{b}_3 \, dq_1 \, dq_2 \, dq_3 \\ &= J \, dq_1 \, dq_2 \, dq_3 \end{aligned} \tag{3.580}$$

$$J = \mathbf{b}_1 \cdot \mathbf{b}_2 \times \mathbf{b}_3 = \sqrt{\det \left[g_{ij} \right]} = \sqrt{\begin{vmatrix} g_{11} & g_{12} & g_{13} \\ g_{21} & g_{22} & g_{23} \\ g_{31} & g_{32} & g_{33} \end{vmatrix}} \tag{3.581}$$

The base vectors are not necessarily orthogonal, but they must be non-coplanar to reach every point of a three-dimensional space. The base vectors are non-coplanar if and only if the *Jacobian* J of the $Q\,(q_1, q_2, q_3)$-system is not zero:

$$J^2 = \det \left[g_{ij} \right] \neq 0 \tag{3.582}$$

Equation (3.582) is called the *non-coplanar condition* and may also be expressed by the triple scalar product of the base vectors:

$$J = \mathbf{b}_1 \cdot \mathbf{b}_2 \times \mathbf{b}_3 = \left| \frac{\partial \,(x, y, z)}{\partial \,(q_1, q_2, q_3)} \right| \neq 0 \tag{3.583}$$

The nonzero Jacobian is also the necessary and sufficient condition that the set of equations (3.417) have a unique inverse solution.

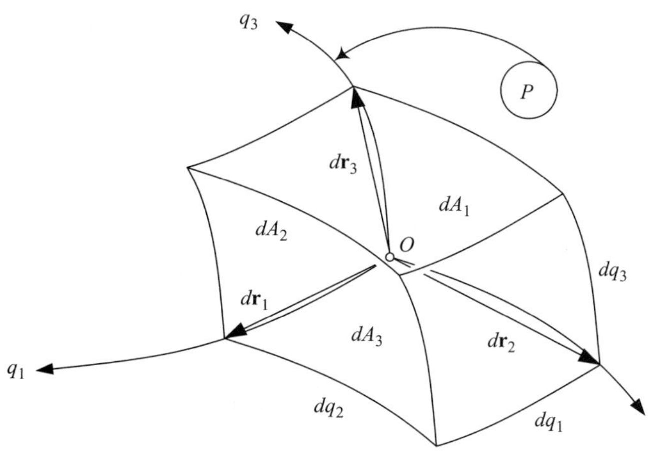

Figure 3.28 A principal coordinate frame P and three area elements dA_1, dA_2, dA_3 along with the volume element dV.

In orthogonal coordinate systems, the equations for arc, area, and volume elements simplify to

$$ds_i = b_i \; dq_i \tag{3.584}$$

$$dA_i = b_j b_k \; dq_j \; dq_k \tag{3.585}$$

$$dV = b_1 b_2 b_3 \; dq_1 \; dq_2 \; dq_3 \tag{3.586}$$

and hence,

$$ds^2 = \left(b_i \; dq_i\right)^2 = \left(b_1 \; dq_1\right)^2 + \left(b_2 \; dq_2\right)^2 + \left(b_3 \; dq_3\right)^2 \tag{3.587}$$

Proof: The Cartesian expression of the position vector of a moving point is

$$\begin{aligned}\mathbf{r} &= x\hat{\imath} + y\hat{\jmath} + z\hat{k} \\ &= f_1\left(q_1, q_2, q_3\right)\hat{\imath} + f_2\left(q_1, q_2, q_3\right)\hat{\jmath} + f_3\left(q_1, q_2, q_3\right)\hat{k}\end{aligned} \tag{3.588}$$

Having $\mathbf{r}$, we may find $d\mathbf{r}$ as

$$\begin{aligned}d\mathbf{r} &= \frac{\partial \mathbf{r}}{\partial q_1} dq_1 + \frac{\partial \mathbf{r}}{\partial q_2} dq_2 + \frac{\partial \mathbf{r}}{\partial q_3} dq_3 = \frac{\partial \mathbf{r}}{\partial q_i} dq_i \\ &= \mathbf{b}_1 \; dq_1 + \mathbf{b}_2 \; dq_2 + \mathbf{b}_3 \; dq_3 = \mathbf{b}_i \; dq_i\end{aligned} \tag{3.589}$$

and calculate the arc element $ds = dr = \sqrt{d\mathbf{r} \cdot d\mathbf{r}}$ by

$$\begin{aligned}ds^2 = d\mathbf{r} \cdot d\mathbf{r} &= \left(\frac{\partial \mathbf{r}}{\partial q_i} \cdot \frac{\partial \mathbf{r}}{\partial q_j}\right) dq_i \; dq_j = \left(\mathbf{b}_i \cdot \mathbf{b}_j\right) dq_i \; dq_j \\ &= g_{ij} \, dq_i \; dq_j\end{aligned} \tag{3.590}$$

$$g_{ij} = \frac{\partial \mathbf{r}}{\partial q_i} \cdot \frac{\partial \mathbf{r}}{\partial q_j} = \mathbf{b}_i \cdot \mathbf{b}_j \tag{3.591}$$

The coefficient g_{ij} represents the *matrices* of the Q-system. If the coordinate system is orthogonal, then

$$g_{ij} = \mathbf{b}_i \cdot \mathbf{b}_j^{\bigstar} = \delta_{ij} \tag{3.592}$$

The area element dA_i is equal to the projection of the cross product of the sides $d\mathbf{r}_j$ and $d\mathbf{r}_k$ on the normal unit vector $\hat{u}_i^{\bigstar}$ on dA_i:

$$\begin{aligned}dA_1 &= \hat{u}_1^{\bigstar} \cdot d\mathbf{r}_2 \times d\mathbf{r}_3 = \hat{u}_1^{\bigstar} \cdot \mathbf{b}_2 \times \mathbf{b}_3 \; dq_2 \, dq_3 \\ &= J\hat{u}_1^{\bigstar} \cdot \mathbf{b}_1^{\bigstar} \; dq_2 \, dq_3 = Jb_1^{\bigstar} dq_2 \, dq_3\end{aligned} \tag{3.593}$$

$$dA_2 = Jb_2^{\bigstar} \; dq_3 \, dq_1 \tag{3.594}$$

$$dA_3 = Jb_3^{\bigstar} \; dq_1 \, dq_2 \tag{3.595}$$

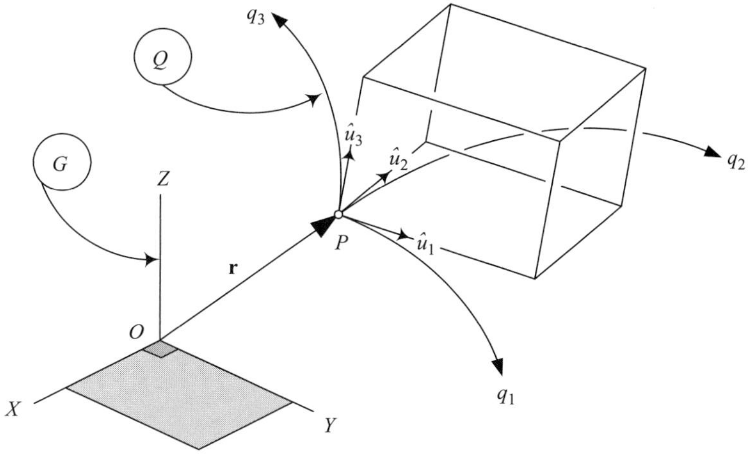

Figure 3.29 The volume element dV in the curvilinear coordinate system (q_1, q_2, q_3).

The volume element dV shown in Figure 3.29 is the volume of the curvilinear parallelepiped by the vectors $\mathbf{b}_1\, dq_1$, $\mathbf{b}_2\, dq_2$, $\mathbf{b}_3\, dq_3$:

$$\begin{aligned} dV &= d\mathbf{r}_1 \cdot d\mathbf{r}_2 \times d\mathbf{r}_3 = \frac{\partial \mathbf{r}}{\partial q_1} dq_1 \cdot \frac{\partial \mathbf{r}}{\partial q_2} dq_2 \times \frac{\partial \mathbf{r}}{\partial q_3} dq_3 \\ &= \mathbf{b}_1 \cdot \mathbf{b}_2 \times \mathbf{b}_3\, dq_1\, dq_2\, dq_3 = [\mathbf{b}_1 \mathbf{b}_2 \mathbf{b}_3]\, dq_1\, dq_2\, dq_3 \\ &= J\, dq_1\, dq_2\, dq_3 \end{aligned} \tag{3.596}$$

■

Example 225 ★ Expression of $d\mathbf{r}$ ***in*** Q ***by the Orthogonality Condition*** Because the arc element ds is a scalar, it is independent of the coordinate system. However, $d\mathbf{r}$ is a vector and therefore is coordinate dependent. The vector $d\mathbf{r}$ in Equation (3.589) is expressed in the Cartesian coordinate system with components in the Q-system. We may express $d\mathbf{r}$ in Q:

$$\begin{aligned} d\mathbf{r} &= (d\mathbf{r} \cdot \hat{u}_1)\hat{u}_1 + (d\mathbf{r} \cdot \hat{u}_2)\hat{u}_2 + (d\mathbf{r} \cdot \hat{u}_3)\hat{u}_3 = (d\mathbf{r} \cdot \hat{u}_j)\hat{u}_j \\ &= \frac{1}{b_j}\left(d\mathbf{r} \cdot \mathbf{b}_j\right)\hat{u}_j = \frac{dq_i}{b_j}\left(\mathbf{b}_i \cdot \mathbf{b}_j\right)\hat{u}_j = \frac{g_{ij}}{b_j} dq_i \hat{u}_j \end{aligned} \tag{3.597}$$

Example 226 ★ Arc Element on a Coordinate Curve The arc length along the coordinate curve q_i is

$$ds = |\mathbf{b}_i|\, dq_i = \sum_{i=1}^{3} \sqrt{\mathbf{b}_i \cdot \mathbf{b}_i}\; dq_i = \sum_{i=1}^{3} \sqrt{g_{ii}}\; dq_i = b_i\, dq_i \tag{3.598}$$

So, the area element in the coordinate surface (q_2, q_3) or $q_1 = \text{const}$ is

$$\begin{aligned} dA_1 &= |\mathbf{b}_2 \times \mathbf{b}_3|\, dq_2\, dq_3 = \sqrt{(\mathbf{b}_2 \times \mathbf{b}_3) \cdot (\mathbf{b}_2 \times \mathbf{b}_3)}\, dq_2\, dq_3 \\ &= \sqrt{(\mathbf{b}_2 \cdot \mathbf{b}_2)(\mathbf{b}_3 \cdot \mathbf{b}_3) - (\mathbf{b}_2 \cdot \mathbf{b}_3)(\mathbf{b}_2 \cdot \mathbf{b}_3)}\, dq_2\, dq_3 \\ &= \sqrt{g_{22}g_{33} - g_{23}^2}\, dq_2\, dq_3 = \sqrt{b_2^2 b_3^2 - g_{23}^2}\, dq_2\, dq_3 \end{aligned} \tag{3.599}$$

Similarly, we have

$$dA_2 = \sqrt{b_3^2 b_1^2 - g_{31}^2}\, dq_3\, dq_1 \tag{3.600}$$

$$dA_3 = \sqrt{b_1^2 b_2^2 - g_{12}^2}\, dq_1\, dq_2 \tag{3.601}$$

which can be shown by index notation as

$$dA_i = \sqrt{b_j^2 b_k^2 - g_{jk}^2}\, dq_j\, dq_k \qquad \text{no summation over } j,\ k \tag{3.602}$$

where i, j, k is a cyclic permutation of $1, 2, 3$.

Example 227 ★ Arc Element of Cartesian, Cylindrical, and Spherical Coordinates In the Cartesian coordinate system, we have

$$b_1 = 1 \qquad b_2 = 1 \qquad b_3 = 1 \tag{3.603}$$

and therefore,

$$ds^2 = dx^2 + dy^2 + dz^2 \tag{3.604}$$

For a cylindrical coordinate system, we have

$$b_1 = 1 \qquad b_2 = \rho \qquad b_3 = 1 \tag{3.605}$$

and therefore,

$$ds^2 = d\rho^2 + (\rho\, d\theta)^2 + dz^2 \tag{3.606}$$

In a spherical coordinate system, we have

$$b_1 = 1 \qquad b_2 = r \sin\theta \qquad b_3 = r \tag{3.607}$$

and therefore,

$$ds^2 = dr^2 + (r \sin\theta\, d\varphi)^2 + (r\, d\theta)^2 . \tag{3.608}$$

Example 228 ★ Surface Element of a Tetrahedron Consider the tetrahedron of Figure 3.30 which is bounded by triangular elements of coordinate surfaces. The areas of these elements are $dA_i/2$. Therefore, the area of the inclined surface dA is

$$dA\, \hat{n} = \tfrac{1}{2} dA_i\, \hat{u}_i^{\star} \tag{3.609}$$

where $\hat{n}$ is a unit-normal vector to dA.

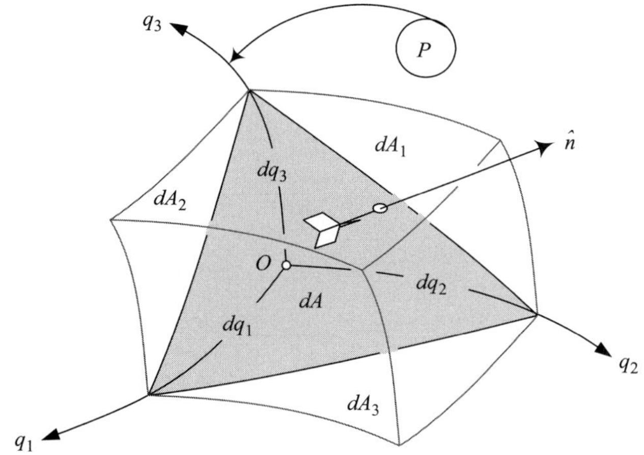

Figure 3.30 A tetrahedron which is bounded by triangular elements of coordinate surfaces.

3.5.4 ★ Curvilinear Kinematics

The geometric and kinematic information in curvilinear coordinate systems $Q(q_1, q_2, q_3)$ depends on the position $\mathbf{r}$, velocity $\mathbf{v}$, and acceleration $\mathbf{a}$ as well as derivative of vectors with respect to the coordinates q_1, q_2, q_3.

The derivative of a reciprocal base vector $\mathbf{b}_i$ with respect to a coordinate q_j is a vector that can be expressed in principal or reciprocal frames as

$$^{P}\frac{\partial \mathbf{b}_i}{\partial q_j} = \Gamma_{ij}^{k}\, \mathbf{b}_k \tag{3.610}$$

$$^{R}\frac{\partial \mathbf{b}_i}{\partial q_j} = \Gamma_{ijk}\, \mathbf{b}_k^{\bigstar} \tag{3.611}$$

where Γ_{ijk} and Γ_{ij}^{k} are called the Christoffel symbols of the first and second kind, respectively:

$$\Gamma_{ijk} = \Gamma_{jik} = \frac{\partial \mathbf{b}_i}{\partial q_j} \cdot \mathbf{b}_k = \frac{1}{2}\left(\frac{\partial g_{jk}}{\partial q_i} + \frac{\partial g_{ik}}{\partial q_j} - \frac{\partial g_{ij}}{\partial q_k}\right) \tag{3.612}$$

$$\Gamma_{ij}^{k} = \Gamma_{ji}^{k} = \frac{\partial \mathbf{b}_i}{\partial q_j} \cdot \mathbf{b}_k^{\bigstar} \tag{3.613}$$

Christoffel symbols of the first and second kinds can be transformed to each other by using the principal–reciprocal transformation matrices $^{R}R_{P} = \left[g_{ij}\right]$ and $^{P}R_{R} = \left[g_{ij}^{\bigstar}\right]$ given in Equations (3.514) and (3.515):

$$\Gamma_{ijk} = g_{km}\Gamma_{ij}^{m} \tag{3.614}$$

$$\Gamma_{ij}^{k} = g_{km}^{\bigstar}\Gamma_{ijm} \tag{3.615}$$

The position vector $\mathbf{r}$, velocity vector $\mathbf{v}$, and acceleration vector $\mathbf{a}$ of a moving point in the orthogonal system $Q\,(q_1, q_2, q_3)$ are given as

$$ {}^{P}\mathbf{r} = \frac{\mathbf{r}\cdot\mathbf{b}_1}{b_1^2}\mathbf{b}_1 + \frac{\mathbf{r}\cdot\mathbf{b}_2}{b_2^2}\mathbf{b}_2 + \frac{\mathbf{r}\cdot\mathbf{b}_3}{b_3^2}\mathbf{b}_3 \tag{3.616} $$

$$ {}^{P}\mathbf{v} = \dot{q}_1\mathbf{b}_1 + \dot{q}_2\mathbf{b}_2 + \dot{q}_3\mathbf{b}_3 = \dot{q}_i\mathbf{b}_i \tag{3.617} $$

$$ v^2 = b_1^2\dot{q}_1^2 + b_2^2\dot{q}_2^2 + b_3^2\dot{q}_3^2 = b_i^2\,\dot{q}_i^2 = g_{ij}\,\dot{q}_i\,\dot{q}_j\delta_{ij} \tag{3.618} $$

$$ {}^{P}\mathbf{a} = a_1\mathbf{b}_1 + a_2\mathbf{b}_2 + a_3\mathbf{b}_3 = a_i\mathbf{b}_i \tag{3.619} $$

$$ a_i = \ddot{q}_i + \Gamma^i_{jk}\dot{q}_j\dot{q}_k \qquad i = 1, 2, 3 \tag{3.620} $$

The applied force on a moving particle may be expressed in a curvilinear coordinate system,

$$ {}^{P}\mathbf{F} = F_1\mathbf{b}_1 + F_2\mathbf{b}_2 + F_3\mathbf{b}_3 \tag{3.621} $$

and therefore, Newton's equation of motion would be

$$ {}^{P}\mathbf{F} = m\mathbf{a} \tag{3.622} $$

$$ F_i = m\left(\ddot{q}_i + \Gamma^i_{jk}\dot{q}_j\dot{q}_k\right) \tag{3.623} $$

Proof: The derivative of a base vector $\mathbf{b}_i$ with respect to a coordinate q_j is a vector:

$$ \frac{\partial\mathbf{b}_i}{\partial q_j} = \frac{\partial^2\mathbf{r}}{\partial q_i\,\partial q_j} = \frac{\partial^2 x_k}{\partial q_i\,\partial q_j}\hat{\imath}_k \tag{3.624} $$

and because the derivatives are assumed to be continuous, we have

$$ \frac{\partial\mathbf{b}_i}{\partial q_j} = \frac{\partial\mathbf{b}_j}{\partial q_i} \tag{3.625} $$

So, the differential of the (i, j)-element of the principal metric g_{ij} to q_k would be

$$ \begin{aligned} \frac{\partial g_{ij}}{\partial q_k} &= \frac{\partial}{\partial q_k}\left(\mathbf{b}_i\cdot\mathbf{b}_j\right) = \frac{\partial\mathbf{b}_i}{\partial q_k}\cdot\mathbf{b}_j + \mathbf{b}_i\cdot\frac{\partial\mathbf{b}_j}{\partial q_k} = \frac{\partial\mathbf{b}_k}{\partial q_i}\cdot\mathbf{b}_j + \mathbf{b}_i\cdot\frac{\partial\mathbf{b}_k}{\partial q_j} \\ &= \frac{\partial}{\partial q_i}\left(\mathbf{b}_j\cdot\mathbf{b}_k\right) - \mathbf{b}_k\cdot\frac{\partial\mathbf{b}_j}{\partial q_i} + \frac{\partial}{\partial q_j}\left(\mathbf{b}_i\cdot\mathbf{b}_k\right) - \mathbf{b}_k\cdot\frac{\partial\mathbf{b}_i}{\partial q_j} \\ &= \frac{\partial g_{jk}}{\partial q_i} + \frac{\partial g_{ik}}{\partial q_j} - 2\mathbf{b}_k\cdot\frac{\partial\mathbf{b}_i}{\partial q_j} \end{aligned} \tag{3.626} $$

and therefore

$$ \mathbf{b}_k\cdot\frac{\partial\mathbf{b}_i}{\partial q_j} = \frac{\partial\mathbf{b}_i}{\partial q_j}\cdot\mathbf{b}_k = \frac{1}{2}\left(\frac{\partial g_{jk}}{\partial q_i} + \frac{\partial g_{ik}}{\partial q_j} - \frac{\partial g_{ij}}{\partial q_k}\right) \tag{3.627} $$

Employing the *Christoffel symbol* of the *first kind*, Γ_{ijk}, we may show this equation in a compact form:

$$\Gamma_{ijk} = \frac{\partial \mathbf{b}_i}{\partial q_j} \cdot \mathbf{b}_k \tag{3.628}$$

$$\Gamma_{ijk} = \frac{1}{2}\left(\frac{\partial g_{jk}}{\partial q_i} + \frac{\partial g_{ik}}{\partial q_j} - \frac{\partial g_{ij}}{\partial q_k}\right) \tag{3.629}$$

Using condition (3.505), we can express the vector $\partial \mathbf{b}_i/\partial q_j$ in the principal coordinate frame as a linear combination of the reciprocal base vectors $\mathbf{b}_i^\bigstar$:

$$\begin{aligned} {}^R\frac{\partial \mathbf{b}_i}{\partial q_j} &= \left(\frac{\partial \mathbf{b}_i}{\partial q_j} \cdot \mathbf{b}_1\right)\mathbf{b}_1^\bigstar + \left(\frac{\partial \mathbf{b}_i}{\partial q_j} \cdot \mathbf{b}_2\right)\mathbf{b}_2^\bigstar + \left(\frac{\partial \mathbf{b}_i}{\partial q_j} \cdot \mathbf{b}_3\right)\mathbf{b}_3^\bigstar \\ &= \Gamma_{ij1}\,\mathbf{b}_1^\bigstar + \Gamma_{ij2}\,\mathbf{b}_2^\bigstar + \Gamma_{ij3}\,\mathbf{b}_3^\bigstar = \Gamma_{ijk}\,\mathbf{b}_k^\bigstar \end{aligned} \tag{3.630}$$

We can also express the vector $\partial \mathbf{b}_i/\partial q_j$ as a linear combination of the principal base vectors $\mathbf{b}_i$,

$$\begin{aligned} {}^P\frac{\partial \mathbf{b}_i}{\partial q_j} &= \left(\frac{\partial \mathbf{b}_i}{\partial q_j} \cdot \mathbf{b}_1^\bigstar\right)\mathbf{b}_1 + \left(\frac{\partial \mathbf{b}_i}{\partial q_j} \cdot \mathbf{b}_2^\bigstar\right)\mathbf{b}_2 + \left(\frac{\partial \mathbf{b}_i}{\partial q_j} \cdot \mathbf{b}_3^\bigstar\right)\mathbf{b}_3 \\ &= \Gamma_{ij}^1\,\mathbf{b}_1 + \Gamma_{ij}^2\,\mathbf{b}_2 + \Gamma_{ij}^3\,\mathbf{b}_3 = \Gamma_{ij}^k\,\mathbf{b}_k \end{aligned} \tag{3.631}$$

and define the *Christoffel symbol* of the *second kind*, Γ_{ij}^k,

$$\Gamma_{ij}^k = \frac{\partial \mathbf{b}_i}{\partial q_j} \cdot \mathbf{b}_k^\bigstar \tag{3.632}$$

Therefore, the principal and reciprocal derivatives of a principal base vector $\mathbf{b}_i$ with respect to a coordinate q_j can be shown by (3.610) or (3.611), in which the coefficients Γ_{ijk} and Γ_{ij}^k are the *Christoffel symbols* of the first and second kinds, respectively.

The definitions of the Christoffel symbols in (3.628) and (3.632) show that

$$\Gamma_{ijk} = \Gamma_{jik} \qquad \Gamma_{ij}^k = \Gamma_{ji}^k \tag{3.633}$$

The inner product of $\partial \mathbf{b}_i/\partial q_j$ from (3.610) by $\mathbf{b}_m$ shows that

$$\Gamma_{ijm} = \frac{\partial \mathbf{b}_i}{\partial q_j} \cdot \mathbf{b}_m = \Gamma_{ij}^k\,\mathbf{b}_k \cdot \mathbf{b}_m = \Gamma_{ij}^k\,g_{km} \tag{3.634}$$

and therefore,

$$\Gamma_{ijm} = g_{km}\,\Gamma_{ij}^k \qquad \Gamma_{ij}^k = g_{km}^\bigstar\,\Gamma_{ijm} \tag{3.635}$$

Using Equation (3.504), we can express a position vector $\mathbf{r}$ in principal and reciprocal frames associated with a coordinate system $Q\,(q_1, q_2, q_3)$:

$${}^P\mathbf{r} = r_i\mathbf{b}_i = \left(\mathbf{r} \cdot \mathbf{b}_i^\bigstar\right)\mathbf{b}_i \tag{3.636}$$

$${}^R\mathbf{r} = r_i^\bigstar\mathbf{b}_i^\bigstar = (\mathbf{r} \cdot \mathbf{b}_i)\,\mathbf{b}_i^\bigstar \tag{3.637}$$

The velocity vector $\mathbf{v}$ in the new coordinate system is found by taking the time derivative of $\mathbf{r}$ and using the chain rule:

$$
\begin{aligned}
{}^{P}\mathbf{v} &= \frac{d\mathbf{r}}{dt} = \frac{\partial \mathbf{r}}{\partial q_1}\dot{q}_1 + \frac{\partial \mathbf{r}}{\partial q_2}\dot{q}_2 + \frac{\partial \mathbf{r}}{\partial q_3}\dot{q}_3 \\
&= \mathbf{b}_1\dot{q}_1 + \mathbf{b}_2\dot{q}_2 + \mathbf{b}_3\dot{q}_3 = \dot{q}_i\mathbf{b}_i = v_i\mathbf{b}_i
\end{aligned}
\tag{3.638}
$$

The reciprocal expression of the velocity vector can be found by a transformation matrix

$$
{}^{R}\mathbf{v} = {}^{R}R_{P}\,{}^{P}\mathbf{v} = \left[g_{ij}\right]\,{}^{P}\mathbf{v} = \left[\mathbf{b}_i\cdot\mathbf{b}_j\right]\,{}^{P}\mathbf{v}
\tag{3.639}
$$

or equivalently by expression (3.637),

$$
{}^{R}\mathbf{v} = (\mathbf{v}\cdot\mathbf{b}_i)\,\mathbf{b}_i^{\star} = v_i^{\star}\mathbf{b}_i^{\star} = g_{ij}v_j\mathbf{b}_i^{\star} = g_{ij}\dot{q}_j\mathbf{b}_i^{\star}
\tag{3.640}
$$

where

$$
v_i = \dot{q}_i
\tag{3.641}
$$

$$
v_i^{\star} = g_{ij}v_j = g_{ij}\dot{q}_j = \dot{q}_i^{\star}
\tag{3.642}
$$

Starting from $\mathbf{v} = \mathbf{b}_i\dot{q}_i$ we find the acceleration vector $\mathbf{a}$ by a time derivative:

$$
\begin{aligned}
{}^{P}\mathbf{a} &= \frac{d}{dt}\left(\mathbf{b}_i\dot{q}_i\right) = \frac{\partial \mathbf{b}_i}{\partial q_j}\dot{q}_j\dot{q}_i + \mathbf{b}_i\ddot{q}_i = \Gamma_{ij}^{k}\,\mathbf{b}_k\dot{q}_i\dot{q}_j + \mathbf{b}_i\ddot{q}_i \\
&= \left(\ddot{q}_i + \Gamma^{i}_{jk}\dot{q}_j\dot{q}_k\right)\mathbf{b}_i
\end{aligned}
\tag{3.643}
$$

The reciprocal expression of the acceleration vector would be

$$
\begin{aligned}
{}^{R}\mathbf{a} &= (\mathbf{a}\cdot\mathbf{b}_i)\,\mathbf{b}_i^{\star} = a_i^{\star}\mathbf{b}_j^{\star} = a_i g_{ij}\mathbf{b}_j^{\star} = \left(\ddot{q}_i + \Gamma^{i}_{jk}\dot{q}_j\dot{q}_k\right)g_{im}\mathbf{b}_m^{\star} \\
&= \left(\ddot{q}_i g_{im} + \Gamma^{i}_{jk}g_{im}\dot{q}_j\dot{q}_k\right)\mathbf{b}_m^{\star} = \left(\ddot{q}_m^{\star} + \Gamma_{jkm}\dot{q}_j\dot{q}_k\right)\mathbf{b}_m^{\star}
\end{aligned}
\tag{3.644}
$$

and therefore,

$$
a_i = \ddot{q}_i + \Gamma^{i}_{jk}\dot{q}_j\dot{q}_k \qquad a_i^{\star} = \left(\ddot{q}_i + \Gamma^{i}_{jk}\dot{q}_j\dot{q}_k\right)g_{im}
\tag{3.645}
$$

The equation of motion $\mathbf{F} = m\mathbf{a}$ for a particle i with mass m_i in a curvilinear coordinate system is

$$
F_i\mathbf{b}_i = m\left(\ddot{q}_i + \Gamma^{i}_{jk}\dot{q}_j\dot{q}_k\right)\mathbf{b}_i
\tag{3.646}
$$

$$
F_i^{\star}\mathbf{b}_i^{\star} = m\left(\ddot{q}_i + \Gamma^{i}_{jk}\dot{q}_j\dot{q}_k\right)g_{in}\mathbf{b}_n^{\star}
\tag{3.647}
$$

in which F_i indicates the force component in direction $\mathbf{b}_i$ on particle i.

Elwin Bruno Christoffel (1829–1900) was a German–French mathematician who improved the theory of differential geometry and mathematical analysis. Christoffel may also be written as Kristoffel. ■

Example 229 ★ Derivative of Reciprocal Base Vector $\mathbf{b}_i^{\star}$ Using Equation (3.441),

$$
\mathbf{b}_i\cdot\mathbf{b}_j^{\star} = \mathbf{b}_j^{\star}\cdot\mathbf{b}_i = \delta_{ij}
\tag{3.648}
$$

and taking a coordinate derivative,

$$\frac{\partial}{\partial q_k}\left(\mathbf{b}_i^\star \cdot \mathbf{b}_j\right) = \frac{\partial \mathbf{b}_i^\star}{\partial q_k} \cdot \mathbf{b}_j + \mathbf{b}_i^\star \cdot \frac{\partial \mathbf{b}_j}{\partial q_k} = \frac{\partial \mathbf{b}_i^\star}{\partial q_k} \cdot \mathbf{b}_j + \Gamma^i_{jk} = 0 \tag{3.649}$$

we find

$$\frac{\partial \mathbf{b}_i^\star}{\partial q_k} \cdot \mathbf{b}_j = -\Gamma^i_{jk} \tag{3.650}$$

and therefore,

$$\frac{\partial \mathbf{b}_i^\star}{\partial q_k} = -\Gamma^i_{jk} \mathbf{b}_j^\star \tag{3.651}$$

It also shows that

$$\Gamma^k_{ij} = -\Gamma^i_{ik} \tag{3.652}$$

and similarly,

$$\Gamma_{ijk} = -\Gamma_{ikj} \tag{3.653}$$

Example 230 ★ Natural Coordinate System Consider the trajectory of a particle in a curvilinear coordinate system that is described by a time-parametric equation:

$$q_i = q_i(t) \qquad i = 1, 2, 3 \tag{3.654}$$

The arc length ds of the trajectory is

$$ds^2 = g_{ij}\, dq_i\, dq_j \tag{3.655}$$

Defining $dq_i/ds = \hat{u}_{t_i}$ as the unit tangent vector to the trajectory q_i provides

$$g_{ij} \frac{dq_i}{ds}\frac{dq_j}{ds} = g_{ij}\, u_{t_i}\, u_{t_j} = \hat{u}_t \cdot \hat{u}_t = 1 \tag{3.656}$$

which is the inner product of $\hat{u}_{t_i} \cdot \hat{u}_{t_i}$ and indicates $\hat{u}_{t_i}$ is a unit vector. Differentiating Equation (3.656) with respect to the arc length s produces

$$g_{ij} \frac{du_{t_i}}{ds} u_{t_j} + g_{ij}\, u_{t_i} \frac{du_{t_j}}{ds} = 0 \tag{3.657}$$

which simplifies to

$$g_{ij}\, u_{t_i} \frac{du_{t_j}}{ds} = \hat{u}_t \cdot \frac{d\hat{u}_t}{ds} = 0 \tag{3.658}$$

So, du_t/ds is a normal vector to u_t. By defining the *curvature* κ and the normal vector $\hat{u}_n$ as

$$\hat{u}_n = \frac{1}{\kappa}\frac{d\hat{u}_t}{ds} \tag{3.659}$$

we may simplify Equation (3.658) to the form

$$g_{ij}\, u_{t_i}\, u_{n_j} = \hat{u}_t \cdot \hat{u}_n = 0 \tag{3.660}$$

An arc derivative of (3.660) provides

$$g_{ij} \frac{du_{t_i}}{ds} u_{n_j} + g_{ij}\, u_{t_i} \frac{du_{n_j}}{ds} = 0 \tag{3.661}$$

or

$$g_{ij}\,\hat{u}_{t_i}\,\frac{du_{n_j}}{ds} = -g_{ij}\,u_{n_j}\,\frac{du_{t_i}}{ds} = -\kappa g_{ij}\,u_{n_i}u_{n_j} = -\kappa \tag{3.662}$$

We may use Equation (3.656) and write (3.662) as

$$g_{ij}\,u_{t_i}\,\frac{du_{n_j}}{ds} = -\kappa g_{ij}\,u_{t_i}\,u_{t_j} = -\kappa \tag{3.663}$$

or

$$g_{ij}\,u_{t_i}\left(\frac{du_{n_j}}{ds} + \kappa u_{t_j}\right) = \hat{u}_t \cdot \left(\frac{d\hat{u}_n}{ds} + \kappa \hat{u}_t\right) = 0 \tag{3.664}$$

So, $d\hat{u}_n/ds + \kappa\hat{u}_t$ is a normal vector to $\hat{u}_t$.

Taking an arc derivative of

$$g_{ij}\,u_{n_i}\,u_{n_j} = 1 \tag{3.665}$$

shows that

$$g_{ij}\,u_{n_i}\,\frac{du_{n_j}}{ds} = u_n \cdot \frac{du_n}{ds} = 0 \tag{3.666}$$

Now, the inner product

$$\hat{u}_n \cdot \left(\frac{d\hat{u}_n}{ds} + \kappa \hat{u}_t\right) = 0 \tag{3.667}$$

indicates that the vector $d\hat{u}_n/ds + \kappa\hat{u}_t$ is orthogonal to both the tangent unit vector $\hat{u}_t$ and the normal unit vector $\hat{u}_n$. Using this vector, we define the binormal unit vector $\hat{u}_b$:

$$\hat{u}_b = \frac{1}{\tau}\left(\frac{d\hat{u}_n}{ds} + \kappa \hat{u}_t\right) \tag{3.668}$$

where τ is the *torsion* of the natural coordinate frame $\hat{u}_t$, $\hat{u}_n$, $\hat{u}_b$. The arc derivative of this equation provides

$$\begin{aligned}
\frac{du_{b_i}}{ds} &= \epsilon_{ijk}\left(u_{t_j}\frac{du_{n_k}}{ds} + \frac{du_{t_j}}{ds}u_{n_k}\right) \\
&= \epsilon_{ijk}\left(u_{t_j}\left(\tau u_{b_k} - \kappa u_{t_k}\right) + \kappa u_{n_j}u_{n_k}\right) \\
&= \tau\epsilon_{ijk}u_{t_j}u_{b_k} = -\tau u_{n_i}
\end{aligned} \tag{3.669}$$

We may summarize Equations (3.659), (3.668), and (3.669) to show the *Frenet–Serret* equations in curvilinear coordinates:

$$\frac{d\hat{u}_t}{ds} = \kappa \hat{u}_n \tag{3.670}$$

$$\frac{d\hat{u}_n}{ds} = \tau \hat{u}_b - \kappa \hat{u}_t \tag{3.671}$$

$$\frac{d\hat{u}_b}{ds} = -\tau \hat{u}_n \tag{3.672}$$

Example 231 ★ Velocity and Acceleration in Natural Coordinate Taking arc derivatives of the position vector provides the velocity and acceleration in the natural coordinate:

$$\mathbf{v} = \frac{d\mathbf{r}}{dt} = \frac{d\mathbf{r}}{ds}\frac{ds}{dt} = \dot{s}\hat{u}_t \tag{3.673}$$

$$\mathbf{a} = \frac{d\mathbf{v}}{dt} = \ddot{s}\hat{u}_t + \dot{s}\frac{d\hat{u}_t}{ds}\frac{ds}{dt} = \ddot{s}\hat{u}_t + \dot{s}^2\kappa\hat{u}_n \tag{3.674}$$

Example 232 ★ Work in Curvilinear Coordinates The work ${}_2W_1$ that is done by a force $\mathbf{F}$ on a moving particle with mass m from point P_1 to point P_2 along a curve $\mathbf{r} = \mathbf{r}(t)$,

$${}^P\mathbf{r}(t) = r_1(t)\,\mathbf{b}_1 + r_2(t)\,\mathbf{b}_2 + r_3(t)\,\mathbf{b}_3 \tag{3.675}$$

is represented by a summation of the tangential components of the force along the path and is defined by the integral

$$\begin{aligned} {}_2W_1 &= \int_{P_1}^{P_2} \mathbf{F}\cdot d\mathbf{r} = \int_{P_1}^{P_2} F_i\,dq_i = \int_{P_1}^{P_2} F_i\frac{dq_i}{dt}dt = \int_{P_1}^{P_2} F_i v_i\,dt \\ &= m\int_{P_1}^{P_2} \left(\ddot{q}_i + \Gamma^i_{jk}\dot{q}_j\dot{q}_k\right)\dot{q}_i\,dt = m\int_{P_1}^{P_2} \left(\ddot{q}_i + \Gamma^i_{jk}\dot{q}_j\dot{q}_k\right)dq \end{aligned} \tag{3.676}$$

If the force $\mathbf{F}$ is conservative, the force is derivable from a scalar potential function

$$V = V(q_i) \qquad i = 1, 2, 3 \tag{3.677}$$

such that

$$\mathbf{F} = -\nabla V \tag{3.678}$$

$$F_i = -\frac{\partial V}{\partial q_i} \qquad i = 1, 2, 3 \tag{3.679}$$

Then

$${}_2W_1 = m\int_{P_1}^{P_2} \left(\ddot{q}_i + \Gamma^i_{jk}\dot{q}_j\dot{q}_k\right)dq = -\Delta V = V(\mathbf{r}_1) - V(\mathbf{r}_2) \tag{3.680}$$

Example 233 ★ Cartesian Velocity Components Substitution of the Cartesian position vector $\mathbf{r}$ in the principal velocity equation (3.638) provides the Cartesian expression of the velocity vector:

$${}^P\mathbf{v} = \frac{d\mathbf{r}}{dt} = \frac{\partial\mathbf{r}}{\partial q_j}\dot{q}_j = \frac{\partial x_i}{\partial q_j}\dot{q}_j\hat{\imath}_i = \dot{x}_i\hat{\imath}_i \tag{3.681}$$

Therefore, the components of velocity in the Cartesian and Q-system are related by the Jacobian matrix:

$$\dot{x}_i = \left[\frac{\partial x_i}{\partial q_j}\right] \dot{q}_j = [J]\,\dot{q}_j \tag{3.682}$$

$$\begin{bmatrix} \dot{x} \\ \dot{y} \\ \dot{z} \end{bmatrix} = \begin{bmatrix} \frac{\partial x}{\partial q_i} & \frac{\partial x}{\partial q_2} & \frac{\partial x}{\partial q_3} \\ \frac{\partial y}{\partial q_i} & \frac{\partial y}{\partial q_2} & \frac{\partial y}{\partial q_3} \\ \frac{\partial z}{\partial q_i} & \frac{\partial z}{\partial q_2} & \frac{\partial z}{\partial q_3} \end{bmatrix} \begin{bmatrix} \dot{q}_1 \\ \dot{q}_2 \\ \dot{q}_3 \end{bmatrix} \tag{3.683}$$

As an example, consider the planar polar coordinate system

$$x = \rho\cos\theta \qquad y = \rho\sin\theta \qquad z = 0 \tag{3.684}$$

for which we have

$$[J] = \left[\frac{\partial x_i}{\partial q_j}\right] = \begin{bmatrix} \frac{\partial x}{\partial \rho} & \frac{\partial x}{\partial \theta} \\ \frac{\partial y}{\partial q_i} & \frac{\partial y}{\partial \theta} \end{bmatrix} = \begin{bmatrix} \cos\theta & -\rho\sin\theta \\ \sin\theta & \rho\cos\theta \end{bmatrix} \tag{3.685}$$

It provides

$$\begin{bmatrix} \dot{x} \\ \dot{y} \end{bmatrix} = \begin{bmatrix} \cos\theta & -\rho\sin\theta \\ \sin\theta & \rho\cos\theta \end{bmatrix} \begin{bmatrix} \dot{\rho} \\ \dot{\theta} \end{bmatrix} = \begin{bmatrix} \dot{\rho}\cos\theta - \rho\dot{\theta}\sin\theta \\ \dot{\rho}\sin\theta + \rho\dot{\theta}\cos\theta \end{bmatrix} \tag{3.686}$$

and therefore,

$$\begin{aligned} \begin{bmatrix} \dot{\rho} \\ \dot{\theta} \end{bmatrix} &= [J]^{-1} \begin{bmatrix} \dot{x} \\ \dot{y} \end{bmatrix} = \begin{bmatrix} \cos\theta & -\rho\sin\theta \\ \sin\theta & \rho\cos\theta \end{bmatrix}^{-1} \begin{bmatrix} \dot{x} \\ \dot{y} \end{bmatrix} \\ &= \begin{bmatrix} \cos\theta & \sin\theta \\ -\frac{1}{\rho}\sin\theta & \frac{1}{\rho}\cos\theta \end{bmatrix} \begin{bmatrix} \dot{x} \\ \dot{y} \end{bmatrix} \\ &= \begin{bmatrix} \dot{x}\cos\theta + \dot{y}\sin\theta \\ -\dot{x}\frac{1}{\rho}\sin\theta + \dot{y}\frac{1}{\rho}\cos\theta \end{bmatrix} \end{aligned} \tag{3.687}$$

Example 234 ★ Base Vectors of Elliptic–Hyperbolic Coordinate System The elliptic–hyperbolic coordinate system (u, v, z) relates to the Cartesian system by

$$\begin{aligned} x &= f_1(q_1, q_2, q_3) = a\cosh u\cos v \\ y &= f_2(q_1, q_2, q_3) = a\sinh u\sin v \\ z &= f_3(q_1, q_2, q_3) = z \end{aligned} \tag{3.688}$$

The principal base vectors of elliptic–hyperbolic systems expressed in the Cartesian system are

$$\mathbf{b}_u = \frac{\partial \mathbf{r}}{\partial u} = \begin{bmatrix} a\cos v \sinh u \\ a \sin v \cosh u \\ 0 \end{bmatrix} \tag{3.689}$$

$$\mathbf{b}_v = \frac{\partial \mathbf{r}}{\partial v} = \begin{bmatrix} a\cos v \sinh u \\ a \sin v \cosh u \\ 0 \end{bmatrix} \tag{3.690}$$

$$\mathbf{b}_z = \frac{\partial \mathbf{r}}{\partial z} = \begin{bmatrix} 0 \\ 0 \\ 1 \end{bmatrix} \tag{3.691}$$

The reciprocal base vectors of elliptic–hyperbolic systems expressed in the Cartesian system are

$$\mathbf{b}_1^\star = \nabla f_1 = \begin{bmatrix} a\cos v \sinh u \\ a \sin v \cosh u \\ 0 \end{bmatrix} \tag{3.692}$$

$$\mathbf{b}_2^\star = \nabla f_2 = \begin{bmatrix} a\sin v \cosh u \\ a \cos v \sinh u \\ 0 \end{bmatrix} \tag{3.693}$$

$$\mathbf{b}_3^\star = \nabla f_3 = \begin{bmatrix} 0 \\ 0 \\ 1 \end{bmatrix} \tag{3.694}$$

Example 235 ★ Christoffel Symbol Taking the derivative of g_{ij} provides

$$\frac{\partial}{\partial q_k} g_{ij} = \frac{\partial \mathbf{b}_i}{\partial q_k} \cdot \mathbf{b}_j + \mathbf{b}_i \cdot \frac{\partial \mathbf{b}_j}{\partial q_k} \qquad i, j, k = 1, 2, 3 \tag{3.695}$$

Interchanging i, j, and k provides two similar equations:

$$\frac{\partial}{\partial q_i} g_{jk} = \frac{\partial \mathbf{b}_j}{\partial q_i} \cdot \mathbf{b}_k + \mathbf{b}_j \cdot \frac{\partial \mathbf{b}_k}{dq_i} \qquad j, k, i = 1, 2, 3 \tag{3.696}$$

$$\frac{\partial}{\partial q_j} g_{ki} = \frac{\partial \mathbf{b}_k}{\partial q_j} \cdot \mathbf{b}_i + \mathbf{b}_k \cdot \frac{\partial \mathbf{b}_i}{dq_j} \qquad k, i, j = 1, 2, 3 \tag{3.697}$$

A combination of these equations using

$$\frac{\partial \mathbf{b}_i}{\partial q_j} = \frac{\partial \mathbf{b}_j}{\partial q_i} \tag{3.698}$$

generates the second kind of Christoffel symbol. The Christoffel symbol may also be called the Christoffel operator:

$$\Gamma^i_{jk} = \frac{1}{2}\left(\frac{\partial g_{ij}}{\partial q_k} + \frac{\partial g_{ik}}{\partial q_j} - \frac{\partial g_{kj}}{\partial q_i}\right) \qquad i, j, k = 1, 2, 3 \tag{3.699}$$

Example 236 ★ Equation of Motion in No Force When there is no force on a particle, its acceleration is zero, and therefore, its equation of motion is

$$\ddot{q}_i = -\Gamma^i_{jk}\dot{q}_j\dot{q}_k \tag{3.700}$$

Example 237 ★ Why Curvilinear Coordinate? Vector and scalar fields, such as a gravitational force field or a double-pole magnetic field, can always be expressed in the orthogonal Cartesian coordinate system; however, sometimes it is much simpler to express them in a proper curvilinear coordinate system. A field is usually better to be expressed by a coordinate system using the lines of flow and potential surfaces. Such a coordinate system may be called a *natural coordinate system*. The behavior of a field at a boundary surface, a potential surface, and singularities and its flow lines are usually defined in the natural coordinate system in a simpler way. The field will have a much simpler expression in a natural coordinate set, whereas in terms of Cartesian coordinates it may have a complicated expression.

Example 238 ★ Principal and Reciprocal Derivatives of a Vector Consider the principal expression of a vector $\mathbf{r}$:

$${}^P\mathbf{r} = r_1\mathbf{b}_1 + r_2\mathbf{b}_2 + r_3\mathbf{b}_3 \tag{3.701}$$

The differential of ${}^P\mathbf{r}$ is

$${}^P d\mathbf{r} = \frac{\partial\mathbf{r}}{\partial q_1}dq_1 + \frac{\partial\mathbf{r}}{\partial q_2}dq_2 + \frac{\partial\mathbf{r}}{\partial q_3}dq_3 \tag{3.702}$$

$${}^R d\mathbf{r} = \frac{\partial\mathbf{r}}{\partial q_1}dq_1 + \frac{\partial\mathbf{r}}{\partial q_2}dq_2 + \frac{\partial\mathbf{r}}{\partial q_3}dq_3 \tag{3.703}$$

where

$$\frac{\partial\mathbf{r}}{\partial q_i} = \sum_{j=1}^{3}\left(\frac{\partial r_j}{\partial q_i}\mathbf{b}_j + r_j\frac{\partial\mathbf{b}_j}{\partial q_i}\right) \tag{3.704}$$

$$\frac{\partial\mathbf{r}}{\partial q_i} = \sum_{j=1}^{3}\left(\frac{\partial r^\star_j}{\partial q_i}\mathbf{b}^\star_j + r^\star_j\frac{\partial\mathbf{b}^\star_j}{\partial q_i}\right) \tag{3.705}$$

Example 239 ★ Christoffel Symbol of the Third Kind, Γ^{ijk} We may determine the differential of the (i, j)-element of the reciprocal metric $g^\star_{ij}$ to q_k and define the

Christoffel symbol of the third kind, Γ^{ijk}:

$$\Gamma^{ijk} = \frac{1}{2}\left(\frac{\partial g^{\star}_{jk}}{\partial q_i} + \frac{\partial g^{\star}_{ik}}{\partial q_j} - \frac{\partial g^{\star}_{ij}}{\partial q_k}\right) \tag{3.706}$$

$$\begin{aligned}
\frac{\partial g_{ij}}{\partial q_k} &= \frac{\partial}{\partial q_k}\left(\mathbf{b}^{\star}_i \cdot \mathbf{b}^{\star}_j\right) \\
&= \frac{\partial \mathbf{b}^{\star}_i}{\partial q_k} \cdot \mathbf{b}^{\star}_j + \mathbf{b}^{\star}_i \cdot \frac{\partial \mathbf{b}^{\star}_j}{\partial q_k} = \frac{\partial \mathbf{b}^{\star}_k}{\partial q_i} \cdot \mathbf{b}^{\star}_j + \mathbf{b}^{\star}_i \cdot \frac{\partial \mathbf{b}^{\star}_k}{\partial q_j} \\
&= \frac{\partial}{\partial q_i}\left(\mathbf{b}^{\star}_j \cdot \mathbf{b}^{\star}_k\right) - \mathbf{b}^{\star}_k \cdot \frac{\partial \mathbf{b}^{\star}_j}{\partial q_i} + \frac{\partial}{\partial q_j}\left(\mathbf{b}^{\star}_i \cdot \mathbf{b}^{\star}_k\right) - \mathbf{b}^{\star}_k \cdot \frac{\partial \mathbf{b}^{\star}_i}{\partial q_j} \\
&= \frac{\partial g^{\star}_{jk}}{\partial q_i} + \frac{\partial g^{\star}_{ik}}{\partial q_j} - 2\mathbf{b}^{\star}_k \cdot \frac{\partial \mathbf{b}^{\star}_i}{\partial q_j}
\end{aligned} \tag{3.707}$$

and therefore,

$$\mathbf{b}^{\star}_k \cdot \frac{\partial \mathbf{b}^{\star}_i}{\partial q_j} = \frac{\partial \mathbf{b}^{\star}_i}{\partial q_j} \cdot \mathbf{b}^{\star}_k = \frac{1}{2}\left(\frac{\partial g^{\star}_{jk}}{\partial q_i} + \frac{\partial g^{\star}_{ik}}{\partial q_j} - \frac{\partial g^{\star}_{ij}}{\partial q_k}\right) \tag{3.708}$$

Employing the Christoffel symbol of the third kind, Γ^{ijk}, we may show this equation in a compact form:

$$\mathbf{b}^{\star}_k \cdot \frac{\partial \mathbf{b}^{\star}_i}{\partial q_j} = \frac{\partial \mathbf{b}^{\star}_i}{\partial q_j} \cdot \mathbf{b}^{\star}_k = \Gamma^{ijk} \tag{3.709}$$

3.5.5 ★ Kinematics in Curvilinear Coordinates

Any position vector $\mathbf{r}$ can be expressed in Cartesian as well as principal ${}^P\mathbf{r}$ or reciprocal ${}^R\mathbf{r}$ coordinate frames using the unit vectors $\hat{u}_i$, $\hat{u}^{\star}_i$ or base vectors $\mathbf{b}_i$, $\mathbf{b}^{\star}_i$ of each coordinate system:

$$\mathbf{r} = x\hat{\imath} + y\hat{\jmath} + z\hat{k} \tag{3.710}$$

$${}^P\mathbf{r} = \sum_{i=1}^{3} r_i \mathbf{b}_i = r_1\mathbf{b}_1 + r_2\mathbf{b}_2 + r_3\mathbf{b}_3 \tag{3.711}$$

$${}^R\mathbf{r} = \sum_{i=1}^{3} r^{\star}_i \mathbf{b}^{\star}_i = r^{\star}_1\mathbf{b}^{\star}_1 + r^{\star}_2\mathbf{b}^{\star}_2 + r^{\star}_3\mathbf{b}^{\star}_3 \tag{3.712}$$

where r_i is the covariant component of $\mathbf{r}$,

$$r_i = \mathbf{r} \cdot \mathbf{b}^{\star}_i = \mathbf{r} \cdot \frac{\mathbf{b}_j \times \mathbf{b}_k}{\left[\mathbf{b}_i\mathbf{b}_j\mathbf{b}_k\right]} \qquad i, j, k \circlearrowright 1, 2, 3 \tag{3.712a}$$

$$\mathbf{b}_i = \frac{\partial \mathbf{r}}{\partial q_i} = b_i\hat{u}_i = \frac{\partial x}{\partial q_i}\hat{\imath} + \frac{\partial y}{\partial q_i}\hat{\jmath} + \frac{\partial z}{\partial q_i}\hat{k} \tag{3.713}$$

$$b_i = |\mathbf{b}_i| = \sqrt{\left(\frac{\partial x}{\partial q_i}\right)^2 + \left(\frac{\partial y}{\partial q_i}\right)^2 + \left(\frac{\partial z}{\partial q_i}\right)^2} \tag{3.714}$$

$$\hat{u}_i = \frac{\Delta \mathbf{r}_{q_i}}{\left|\Delta \mathbf{r}_{q_i}\right|} = \frac{\partial \mathbf{r}/\partial q_i}{|\partial \mathbf{r}/\partial q_i|} = \frac{1}{b_i}\mathbf{b}_i \tag{3.715}$$

and $r_1^\bigstar$ is the contravariant component of $\mathbf{r}$,

$$r_1^\bigstar = \mathbf{r} \cdot \mathbf{b}_i = \mathbf{r} \cdot \frac{\mathbf{b}_j^\bigstar \times \mathbf{b}_k^\bigstar}{\left[\mathbf{b}_i^\bigstar \mathbf{b}_j^\bigstar \mathbf{b}_k^\bigstar\right]} \qquad i, j, k \circlearrowright 1, 2, 3 \tag{3.716}$$

$$\mathbf{b}_i^\bigstar = \nabla q_i = \frac{\partial q_i}{\partial x}\hat{\imath} + \frac{\partial q_i}{\partial y}\hat{\jmath} + \frac{\partial q_i}{\partial z}\hat{k} \tag{3.717}$$

$$b_i^\bigstar = \left|\mathbf{b}_i^\bigstar\right| = \sqrt{\left(\frac{\partial q_i}{\partial x}\right)^2 + \left(\frac{\partial q_i}{\partial y}\right)^2 + \left(\frac{\partial q_i}{\partial z}\right)^2} \tag{3.718}$$

$$\hat{u}_i^\bigstar = \frac{\nabla q_i}{|\nabla q_i|} = \frac{1}{b_i^\bigstar}\mathbf{b}_i^\bigstar \tag{3.719}$$

The velocity vector $\mathbf{v}$ in the Q-system is found by taking a simple time derivative of $\mathbf{r}$ and using the chain rule. The velocity vector $\mathbf{v}$ can be expressed in Cartesian as well as principal ${}^P\mathbf{v}$ or reciprocal ${}^R\mathbf{v}$ coordinate frames:

$$\mathbf{v} = \dot{x}\hat{\imath} + \dot{y}\hat{\jmath} + \dot{z}\hat{k} \tag{3.720}$$

$${}^P\mathbf{v} = \sum_{i=1}^{3} v_i \mathbf{b}_i = v_1\mathbf{b}_1 + v_2\mathbf{b}_2 + v_3\mathbf{b}_3 \tag{3.721}$$

$${}^R\mathbf{v} = \sum_{i=1}^{3} v_i^\bigstar \mathbf{b}_i^\bigstar = v_1^\bigstar\mathbf{b}_1^\bigstar + v_2^\bigstar\mathbf{b}_2^\bigstar + v_3^\bigstar\mathbf{b}_3^\bigstar \tag{3.722}$$

$${}^P\mathbf{v} = \frac{{}^P d}{dt}\,{}^P\mathbf{r} = \sum_{i=1}^{3} \dot{q}_i \frac{\partial \mathbf{r}}{\partial q_i} = \sum_{i=1}^{3} \dot{q}_i b_i \hat{u}_i = \sum_{i=1}^{3} \dot{q}_i \mathbf{b}_i = \sum_{i=1}^{3} v_i \mathbf{b}_i \tag{3.723}$$

$$v_i = \mathbf{v} \cdot \mathbf{b}_i^\bigstar \tag{3.724}$$

$${}^R\mathbf{v} = \frac{{}^R d}{dt}\,{}^R\mathbf{r} = \sum_{j=1}^{3}\sum_{i=1}^{3} g_{ij}\dot{q}_j \mathbf{b}_i^\bigstar = \sum_{j=1}^{3}\sum_{i=1}^{3} g_{ij} v_j \mathbf{b}_i^\bigstar = \sum_{i=1}^{3} v_i^\bigstar \mathbf{b}_i^\bigstar \tag{3.725}$$

$$v_i^\bigstar = \mathbf{v} \cdot \mathbf{b}_i = g_{ij} v_j = g_{ij}\dot{q}_j = \dot{q}_i^\bigstar \tag{3.726}$$

The acceleration vector $\mathbf{a}$ in the Q-system is found by taking a simple time derivative of $\mathbf{v}$ and using the chain rule. The acceleration vector $\mathbf{a}$ can be expressed in Cartesian as well as principal ${}^P\mathbf{a}$ or reciprocal ${}^R\mathbf{a}$ coordinate frames.

We may write the components of **a** as

$$ {}^{P}\mathbf{a} = \frac{{}^{P}d}{dt}\,{}^{P}\mathbf{v} = \sum_{i=1}^{3} \dot{q}_i \frac{\partial \mathbf{v}}{\partial q_i} = \sum_{i=1}^{3} a_i \mathbf{b}_i \tag{3.727} $$

$$ a_i = \mathbf{a} \cdot \mathbf{b}_i^{\bigstar} \tag{3.728} $$

$$ {}^{R}\mathbf{a} = \frac{{}^{R}d}{dt}\,{}^{R}\mathbf{v} = \sum_{i=1}^{3} a_i^{\bigstar} s \mathbf{b}_i^{\bigstar} \tag{3.729} $$

$$ a_i^{\bigstar} = \mathbf{a} \cdot \mathbf{b}_i \tag{3.730} $$

Proof: We may write the components of **a** as

$$ \begin{aligned} a_i &= \mathbf{a} \cdot \hat{u}_i = \frac{d\mathbf{v}}{dt} \cdot \frac{1}{b_i}\frac{\partial \mathbf{r}}{\partial q_i} \\ &= \frac{1}{b_i}\left[\frac{d}{dt}\left(\mathbf{v} \cdot \frac{\partial \mathbf{r}}{\partial q_i}\right) - \mathbf{v} \cdot \frac{d}{dt}\left(\frac{\partial \mathbf{r}}{\partial q_i}\right)\right] \end{aligned} \tag{3.731} $$

Differentiation of (3.738) with respect to $\dot{q}_i$ gives

$$ \frac{\partial \mathbf{v}}{\partial \dot{q}_i} = b_i \hat{u}_i = \frac{\partial \mathbf{r}}{\partial q_i} \tag{3.732} $$

and moreover

$$ \begin{aligned} \frac{d}{dt}\left(\frac{\partial \mathbf{r}}{\partial q_i}\right) &= \sum_{j=1}^{3} \frac{\partial}{\partial q_j}\left(\frac{\partial \mathbf{r}}{\partial q_i}\right)\dot{q}_j = \sum_{j=1}^{3} \frac{\partial}{\partial q_i}\left(\frac{\partial \mathbf{r}}{\partial q_j}\right)\dot{q}_j \\ &= \frac{\partial}{\partial q_i}\sum_{j=1}^{3}\frac{\partial \mathbf{r}}{\partial q_j}\dot{q}_j = \frac{\partial}{\partial q_i}\frac{d\mathbf{r}}{dt} = \frac{\partial \mathbf{v}}{\partial q_i} \end{aligned} \tag{3.733} $$

Therefore,

$$ \begin{aligned} a_i &= \frac{1}{b_i}\left[\frac{d}{dt}\left(\mathbf{v} \cdot \frac{\partial \mathbf{r}}{\partial q_i}\right) - \mathbf{v} \cdot \frac{\partial \mathbf{v}}{\partial q_i}\right] \\ &= \frac{1}{b_i}\left[\frac{d}{dt}\frac{\partial}{\partial \dot{q}_i}\left(\frac{1}{2}v^2\right) - \frac{\partial}{\partial \dot{q}_i}\left(\frac{1}{2}v^2\right)\right] \\ &= \frac{1}{mb_i}\left(\frac{d}{dt}\frac{\partial K}{\partial \dot{q}_i} - \frac{\partial K}{\partial \dot{q}_i}\right) \end{aligned} \tag{3.734} $$

where K is the kinetic energy of the moving particle:

$$ K = \tfrac{1}{2}mv^2 = \tfrac{1}{2}m\left(b_1^2\dot{q}_1^2 + b_2^2\dot{q}_2^2 + b_3^2\dot{q}_3^2\right) = \tfrac{1}{2}mg_{ij}\dot{q}_i\dot{q}_j \tag{3.735} $$

Because the arc element ds is a scalar, it is independent of the coordinate system while $d\mathbf{r}$ is a vector and coordinate dependent. The vector $d\mathbf{r}$ in Equation (3.589) is expressed in the Cartesian coordinate system with components in the Q-system. We may express $d\mathbf{r}$ in Q by using the orthogonality condition (3.2):

$$
\begin{aligned}
d\mathbf{r} &= (d\mathbf{r}\cdot\hat{u}_1)\hat{u}_1 + (d\mathbf{r}\cdot\hat{u}_2)\hat{u}_2 + (d\mathbf{r}\cdot\hat{u}_3)\hat{u}_3 = \sum_{j=1}^{3}(d\mathbf{r}\cdot\hat{u}_j)\hat{u}_j \\
&= \sum_{j=1}^{3}\left(d\mathbf{r}\cdot\frac{\mathbf{b}_j}{b_j}\right)\hat{u}_j = \sum_{j=1}^{3}\sum_{i=1}^{3}\frac{dq_i}{b_j}\left(\mathbf{b}_i\cdot\mathbf{b}_j\right)\hat{u}_j \\
&= \sum_{j=1}^{3}\sum_{i=1}^{3}\frac{dq_i}{b_j}\left(b_i b_j \delta_{ij}\right)\hat{u}_j = \sum_{i=1}^{3} b_i\, dq_i\, \hat{u}_i = \sum_{i=1}^{3} dq_i\, \mathbf{b}_i
\end{aligned}
\tag{3.736}
$$

Using the orthogonality condition (3.2) and the definition of unit vectors (3.419), we transform $\mathbf{r}$ to the coordinate system $Q\,(q_1, q_2, q_3)$:

$$
\begin{aligned}
\mathbf{r} &= (\mathbf{r}\cdot\hat{u}_1)\hat{u}_1 + (\mathbf{r}\cdot\hat{u}_2)\hat{u}_2 + (\mathbf{r}\cdot\hat{u}_3)\hat{u}_3 \\
&= \frac{1}{b_1}\left(\mathbf{r}\cdot\mathbf{b}_1\right)\hat{u}_1 + \frac{1}{b_2}\left(\mathbf{r}\cdot\mathbf{b}_2\right)\hat{u}_2 + \frac{1}{b_3}\left(\mathbf{r}\cdot\mathbf{b}_3\right)\hat{u}_3 \\
&= \frac{1}{b_1^2}\left(\mathbf{r}\cdot\mathbf{b}_1\right)\mathbf{b}_1 + \frac{1}{b_2^2}\left(\mathbf{r}\cdot\mathbf{b}_2\right)\mathbf{b}_2 + \frac{1}{b_3^2}\left(\mathbf{r}\cdot\mathbf{b}_3\right)\mathbf{b}_3 \\
&= \sum_{i=1}^{3}(\mathbf{r}\cdot\hat{u}_i)\hat{u}_i = \sum_{i=1}^{3}\frac{1}{b_i}\left(\mathbf{r}\cdot\mathbf{b}_i\right)\hat{u}_i = \sum_{i=1}^{3}\frac{1}{b_i^2}\left(\mathbf{r}\cdot\mathbf{b}_i\right)\mathbf{b}_i
\end{aligned}
\tag{3.737}
$$

The velocity vector $\mathbf{v}$ in the new coordinate system is found by taking a time derivative of $\mathbf{r}$ and using the chain rule or by dividing $d\mathbf{r}$ in (3.736) by dt:

$$
\begin{aligned}
\mathbf{v} &= \frac{d\mathbf{r}}{dt} = \frac{\partial\mathbf{r}}{\partial q_1}\dot{q}_1 + \frac{\partial\mathbf{r}}{\partial q_2}\dot{q}_2 + \frac{\partial\mathbf{r}}{\partial q_3}\dot{q}_3 \\
&= \dot{q}_1\mathbf{b}_1 + \dot{q}_2\mathbf{b}_2 + \dot{q}_3\mathbf{b}_3 = b_1\dot{q}_1\hat{u}_1 + b_2\dot{q}_2\hat{u}_2 + b_3\dot{q}_3\hat{u}_3 \\
&= \sum_{i=1}^{3} b_i\dot{q}_i\hat{u}_i = \sum_{i=1}^{3}\dot{q}_i\mathbf{b}_i
\end{aligned}
\tag{3.738}
$$

The acceleration vector $\mathbf{a}$ must satisfy the orthogonality condition (3.2):

$$
\mathbf{a} = (\mathbf{a}\cdot\hat{u}_1)\hat{u}_1 + (\mathbf{a}\cdot\hat{u}_2)\hat{u}_2 + (\mathbf{a}\cdot\hat{u}_3)\hat{u}_3 = a_1\hat{u}_1 + a_2\hat{u}_2 + a_3\hat{u}_3 \tag{3.739}
$$

Differentiation of (3.738) with respect to $\dot{q}_i$ gives

$$
\frac{\partial\mathbf{v}}{\partial\dot{q}_i} = b_i\hat{u}_i = \frac{\partial\mathbf{r}}{\partial q_i} \tag{3.740}
$$

Starting from $\mathbf{v} = \sum_{i=1}^{3} \mathbf{b}_i \dot{q}_i$ we have

$$\begin{aligned}
\mathbf{a} &= \frac{d}{dt} \sum_{i=1}^{3} \frac{\partial \mathbf{r}}{\partial q_i} \dot{q}_i = \sum_{i=1}^{3} \frac{\partial^2 \mathbf{r}}{\partial q_i \partial q_j} \dot{q}_i \dot{q}_j + \sum_{i=1}^{3} \frac{\partial \mathbf{r}}{\partial q_i} \ddot{q}_i \\
&= \sum_{i=1}^{3} \frac{\partial \mathbf{b}_i}{\partial q_j} \dot{q}_i \dot{q}_j + \sum_{i=1}^{3} \mathbf{b}_i \ddot{q}_i
\end{aligned} \tag{3.741}$$

$$\mathbf{a} \cdot \mathbf{b}_j = \sum_{i=1}^{3} \frac{1}{b_i} \mathbf{b}_i \cdot \mathbf{b}_j = \sum_{i=1}^{3} b_i$$

$$a_i = \ddot{q}_i + \Gamma^i_{jk} \dot{q}_j \dot{q}_k$$

When the coordinate system is orthogonal, the principal and reciprocal coordinate frames coincide, so the covariant and contravariant components of vectors become equal. Therefore, the kinematic equations become simpler in orthogonal coordinates. The method of Sections 3.1–3.3 is more straight for deriving the velocity and acceleration vectors in an orthogonal coordinate system. ■

Example 240 ★ Differential Operators in Curvilinear Coordinates The gradient, divergence, curl, and Laplacian in general rectangular curvilinear coordinates are

$$\operatorname{grad} f = \frac{1}{b_1} \frac{\partial f}{\partial q_1} \hat{u}_1 + \frac{1}{b_2} \frac{\partial f}{\partial q_2} \hat{u}_2 + \frac{1}{b_3} \frac{\partial f}{\partial q_3} \hat{u}_3 \tag{3.742}$$

$$\begin{aligned}
\operatorname{div} \mathbf{r} &= \nabla \cdot \mathbf{r} \\
&= \frac{1}{b_1 b_2 b_3} \left(\frac{\partial \left(b_2 b_3 r_1\right)}{\partial q_1} + \frac{\partial \left(b_3 b_1 r_2\right)}{\partial q_2} + \frac{\partial \left(b_3 b_1 r_2\right)}{\partial q_3} \right)
\end{aligned} \tag{3.743}$$

$$\operatorname{curl} \mathbf{r} = \nabla \times \mathbf{r} = \frac{1}{b_1 b_2 b_3} \begin{vmatrix} b_1 \hat{u}_1 & b_2 \hat{u}_2 & b_3 \hat{u}_3 \\ \dfrac{\partial}{\partial q_1} & \dfrac{\partial}{\partial q_2} & \dfrac{\partial}{\partial q_3} \\ r_1 b_1 & r_2 b_2 & r_3 b_3 \end{vmatrix} \tag{3.744}$$

$$\begin{aligned}
\nabla^2 f &= \frac{1}{b_1 b_2 b_3} \frac{\partial}{\partial q_1} \left(\frac{b_2 b_3}{b_1} \frac{\partial f}{\partial q_1} \right) + \frac{1}{b_1 b_2 b_3} \frac{\partial}{\partial q_2} \left(\frac{b_3 b_1}{b_2} \frac{\partial f}{\partial q_2} \right) \\
&\quad + \frac{1}{b_1 b_2 b_3} \frac{\partial}{\partial q_3} \left(\frac{b_1 b_2}{b_3} \frac{\partial f}{\partial q_3} \right)
\end{aligned} \tag{3.745}$$

Example 241 Base Vectors of Spherical Coordinate System Using the definition of spherical coordinates (3.128),

$$x = r \sin \varphi \cos \theta \qquad y = r \sin \varphi \sin \theta \qquad z = r \cos \varphi \tag{3.746}$$

we can find the following base vectors by employing (3.713):

$$\mathbf{b}_1 = \hat{u}_r \qquad \mathbf{b}_2 = r \sin \theta \, \hat{u}_\varphi \qquad \mathbf{b}_3 = r \hat{u}_\theta \tag{3.747}$$

where

$$b_1 = \sqrt{\left(\frac{\partial x}{\partial r}\right)^2 + \left(\frac{\partial y}{\partial r}\right)^2 + \left(\frac{\partial z}{\partial r}\right)^2} = 1 \tag{3.748}$$

$$b_2 = \sqrt{\left(\frac{\partial x}{\partial \varphi}\right)^2 + \left(\frac{\partial y}{\partial \varphi}\right)^2 + \left(\frac{\partial z}{\partial \varphi}\right)^2} = r \sin\theta \tag{3.749}$$

$$b_3 = \sqrt{\left(\frac{\partial x}{\partial \theta}\right)^2 + \left(\frac{\partial y}{\partial \theta}\right)^2 + \left(\frac{\partial z}{\partial \theta}\right)^2} = r \tag{3.750}$$

The reciprocal base vectors to $\mathbf{b}_i$ would then be

$$\mathbf{b}_1^\star = \hat{u}_r \qquad \mathbf{b}_2^\star = \frac{1}{r\sin\theta}\hat{u}_\varphi \qquad \mathbf{b}_3^\star = \frac{1}{r}\hat{u}_\theta \tag{3.751}$$

Example 242 ★ Bispherical Coordinate System Consider the bispherical coordinate system φ, θ, ψ with the following relations:

$$\begin{aligned} x &= \frac{\sin\varphi\cos\psi}{\cosh\theta - \cos\varphi} \\ y &= \frac{\sin\varphi\sin\psi}{\cosh\theta - \cos\varphi} \\ z &= \frac{\sinh\theta}{\cosh\theta - \cos\varphi} \end{aligned} \tag{3.752}$$

Defining the position vector $\mathbf{r}$ as

$$\mathbf{r} = \begin{bmatrix} \dfrac{\sin\varphi\cos\psi}{\cosh\theta - \cos\varphi} \\ \dfrac{\sin\varphi\sin\psi}{\cosh\theta - \cos\varphi} \\ \dfrac{\sinh\theta}{\cosh\theta - \cos\varphi} \end{bmatrix} \tag{3.753}$$

we find the unit vectors of the coordinate system as

$$\hat{u}_\varphi = \frac{\dfrac{\partial\mathbf{r}}{\partial\varphi}}{\left|\dfrac{\partial\mathbf{r}}{\partial\varphi}\right|} = -\frac{1}{(\cosh\theta - \cos\varphi)^2}\begin{bmatrix} \cos\psi - \cosh\theta\cos\psi\cos\varphi \\ \sin\psi - \cosh\theta\cos\varphi\sin\psi \\ \sinh\theta\sin\varphi \end{bmatrix} \tag{3.754}$$

$$\hat{u}_\theta = \frac{\dfrac{\partial\mathbf{r}}{\partial\theta}}{\left|\dfrac{\partial\mathbf{r}}{\partial\theta}\right|} = -\frac{1}{(\cosh\theta - \cos\varphi)^2}\begin{bmatrix} \sinh\theta\cos\psi\sin\varphi \\ \sinh\theta\sin\psi\sin\varphi \\ \cosh\theta\cos\varphi - 1 \end{bmatrix} \tag{3.755}$$

$$\hat{u}_\psi = \frac{\dfrac{\partial \mathbf{r}}{\partial \psi}}{\left|\dfrac{\partial \mathbf{r}}{\partial \psi}\right|} = \frac{1}{\cosh\theta - \cos\varphi}\begin{bmatrix} -\sin\psi\sin\varphi \\ \cos\psi\sin\varphi \\ 0 \end{bmatrix} \tag{3.756}$$

The position vector in a bispherical coordinate system is

$$\begin{aligned}\mathbf{r} &= (\mathbf{r}\cdot\hat{u}_\varphi)\hat{u}_\varphi + (\mathbf{r}\cdot\hat{u}_\theta)\hat{u}_\theta + (\mathbf{r}\cdot\hat{u}_\psi)\hat{u}_\psi \\ &= -\frac{\cosh\theta\sin\varphi}{(\cosh\theta - \cos\varphi)^2}\hat{u}_\varphi - \frac{\sinh\theta\cos\varphi}{(\cosh\theta - \cos\varphi)^2}\hat{u}_\theta \end{aligned} \tag{3.757}$$

where

$$\mathbf{r}\cdot\hat{u}_\varphi = -\frac{\cosh\theta\sin\varphi}{(\cosh\theta - \cos\varphi)^2} \tag{3.758}$$

$$\mathbf{r}\cdot\hat{u}_\theta = -\frac{\sinh\theta\cos\varphi}{(\cosh\theta - \cos\varphi)^2} \tag{3.759}$$

$$\mathbf{r}\cdot\hat{u}_\psi = 0 \tag{3.760}$$

Example 243 ★ Some Orthogonal Coordinate Systems The Bipolar coordinate system (u, v, z) relates to the Cartesian system by

$$x = \frac{a\sinh v}{\cosh v - \cos u} \qquad y = \frac{a\sin u}{\cosh v - \cos u} \qquad z = z \tag{3.761}$$

Figure 3.31 illustrates the intersection of the system with the (x, y)-plane for $a = 1$.

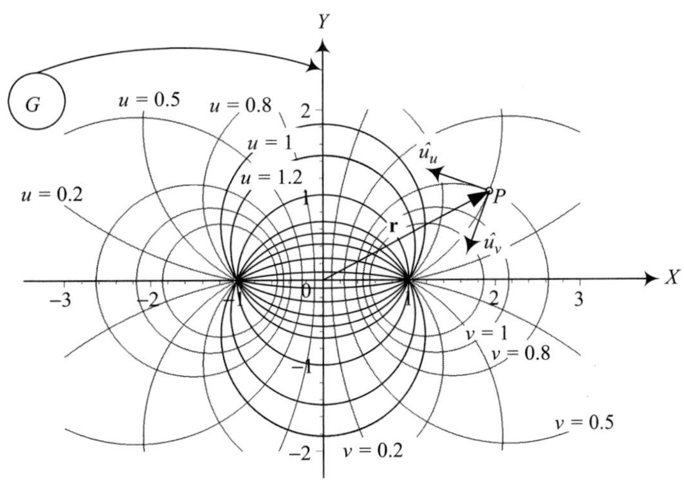

Figure 3.31 Bipolar cylindrical coordinate system.

Bipolarcylindrical (u, v, w):

$$x = a\frac{\sinh v}{\cosh v - \cos u} \qquad y = a\frac{\sin u}{\cosh v - \cos u} \qquad z = w \tag{3.762}$$

Bispherical (u, v, w):

$$x = \frac{\sin u \cos w}{\cosh v - \cos u}$$
$$y = \frac{\sin u \sin w}{\cosh v - \cos u} \tag{3.763}$$
$$z = \frac{\sinh(v)}{\cosh v - \cos u}$$

Cardioidal (u, v, w):

$$x = \frac{uv \cos w}{(u^2 + v^2)^2}$$
$$y = \frac{uv \sin w}{(u^2 + v^2)^2} \tag{3.764}$$
$$z = \frac{1}{2} \frac{u^2 - v^2}{(u^2 + v^2)^2}$$

Cardioidcylindrical (u, v, w):

$$x = \frac{1}{2} \frac{u^2 - v^2}{(u^2 + v^2)^2} \qquad y = \frac{uv}{(u^2 + v^2)^2} \qquad z = w \tag{3.765}$$

Casscylindrical (u, v, w):

$$x = a\frac{\sqrt{2}}{2}\left(\sqrt{e^{2u} + 2e^u \cos v + 1} + \sqrt{e^u \cos v + 1}\right)$$
$$y = a\frac{\sqrt{2}}{2}\left(\sqrt{e^{2u} + 2e^u \cos v + 1} - \sqrt{e^u \cos v + 1}\right) \tag{3.766}$$
$$z = w$$

Confocalellip (u, v, w):

$$x = \sqrt{\frac{a^2 - c^2}{a^2 - b^2}(a^2 - u)(a^2 - v)(a^2 - w)}$$
$$y = \sqrt{\frac{b^2 - a^2}{b^2 - c^2}(a^2 - u)(a^2 - v)(a^2 - w)} \tag{3.767}$$
$$z = \sqrt{\frac{c^2 - a^2}{c^2 - b^2}(a^2 - u)(a^2 - v)(a^2 - w)}$$

Confocalparab (u, v, w):

$$x = \sqrt{\frac{(a^2 - u)(a^2 - v)(a^2 - w)}{b^2 - a^2}}$$

$$y = \sqrt{\frac{(b^2 - u)(b^2 - v)(b^2 - w)}{b^2 - a^2}} \tag{3.768}$$

$$z = \frac{a^2 + b^2 - u - v - w}{2}$$

Conical (u, v, w):

$$x = \frac{uvw}{ab}$$

$$y = \frac{u}{b}\sqrt{\frac{(v^2 - b^2)(b^2 - w^2)}{a^2 - b^2}} \tag{3.769}$$

$$z = \frac{u}{a}\sqrt{\frac{(a^2 - v^2)(a^2 - w^2)}{a^2 - b^2}}$$

Ellcylindrical (u, v, w):

$$x = a \cosh u \cos v \qquad y = a \sinh u \sin v \qquad z = w \tag{3.770}$$

Ellipsoidal (u, v, w):

$$x = \frac{uvw}{ab}$$

$$y = \frac{1}{b}\sqrt{\frac{(u^2 - b^2)(v^2 - b^2)(b^2 - w^2)}{a^2 - b^2}} \tag{3.771}$$

$$z = \frac{1}{a}\sqrt{\frac{(u^2 - a^2)(a^2 - v^2)(a^2 - w^2)}{a^2 - b^2}}$$

Hypercylindrical (u, v, w):

$$x = \sqrt{\sqrt{u^2 + v^2} + u}$$

$$y = \sqrt{\sqrt{u^2 + v^2} - u} \tag{3.772}$$

$$z = w$$

Invcasscylindrical (u, v, w):

$$\begin{aligned} x &= a\frac{\sqrt{2}}{2}\left(1+\sqrt{\frac{e^u \cos v + 1}{e^{2u} + 2e^u \cos v + 1}}\right) \\ y &= a\frac{\sqrt{2}}{2}\left(1-\sqrt{\frac{e^u \cos v + 1}{e^{2u} + 2e^u \cos v + 1}}\right) \\ z &= w \end{aligned} \tag{3.773}$$

Invellcylindrical (u, v, w):

$$\begin{aligned} x &= a\frac{\cosh u \cos v}{\cosh^2 u - \sin^2 v} \\ y &= a\frac{\sinh u \sin v}{\cosh^2 u - \sin^2 v} \\ z &= w \end{aligned} \tag{3.774}$$

Invoblspheroidal (u, v, w):

$$\begin{aligned} x &= a\frac{\cosh u \sin v \cos w}{\cosh^2 u - \cos^2 v} \\ y &= a\frac{\cosh u \sin v \sin w}{\cosh^2 u - \cos^2 v} \\ z &= a\frac{\sinh u \cos v}{\cosh^2 u - \cos^2 v} \end{aligned} \tag{3.775}$$

Logcylindrical (u, v, w):

$$\begin{aligned} x &= \frac{a}{\pi} \ln\left(u^2 + v^2\right) \\ y &= 2\frac{a}{\pi} \arctan\left(\frac{u}{v}\right) \\ z &= w \end{aligned} \tag{3.776}$$

Logcoshcylindrical (u, v, w):

$$\begin{aligned} x &= \frac{a}{\pi} \ln\left(\cosh^2 u - \sin^2 v\right) \\ y &= 2\frac{a}{\pi} \arctan\left(\frac{\tanh u}{\tan v}\right) \\ z &= w \end{aligned} \tag{3.777}$$

Maxwellcylindrical (u, v, w):

$$\begin{aligned} x &= \frac{a}{\pi}\left(u + 1 + e^u \cos v\right) \\ y &= \frac{a}{\pi}\left(u + e^u \sin v\right) \\ z &= w \end{aligned} \tag{3.778}$$

Oblatespheroidal (u, v, w):

$$\begin{aligned} x &= a \cosh u \sin v \cos w \\ y &= a \cosh u \sin v \sin w \\ z &= a \sinh u \cos v \end{aligned} \tag{3.779}$$

Paraboloidal1 (u, v, w):

$$x = uv \cos w \qquad y = uv \sin w \qquad z = \frac{1}{2}\left(u^2 - v^2\right) \tag{3.780}$$

Paraboloidal2 (u, v, w):

$$\begin{aligned} x &= 2\sqrt{\frac{(u-a)(a-v)(a-w)}{a-b}} \\ y &= 2\sqrt{\frac{(u-b)(b-v)(b-w)}{a-b}} \\ z &= u + v + w - a - b \end{aligned} \tag{3.781}$$

Paracylindrical (u, v, w):

$$x = \frac{1}{2}\left(u^2 - v^2\right) \qquad y = uv \qquad z = w \tag{3.782}$$

Rosecylindrical (u, v, w):

$$\begin{aligned} x &= \frac{\sqrt{\sqrt{u^2+v^2}+u}}{\sqrt{u^2+v^2}} \\ y &= \frac{\sqrt{\sqrt{u^2+v^2}-u}}{\sqrt{u^2+v^2}} \\ z &= w \end{aligned} \tag{3.783}$$

Sixsphere (u, v, w):

$$\begin{aligned} x &= \frac{u}{u^2+v^2+w^2} \\ y &= \frac{v}{u^2+v^2+w^2} \\ z &= \frac{w}{u^2+v^2+w^2} \end{aligned} \tag{3.784}$$

Tangentcylindrical (u, v, w):

$$x = \frac{u}{u^2+v^2} \qquad y = \frac{v}{u^2+v^2} \qquad z = w \tag{3.785}$$

Tangentsphere (u, v, w):

$$x = \frac{u}{u^2+v^2}\cos w \qquad y = \frac{u}{u^2+v^2}\sin w \qquad z = \frac{v}{u^2+v^2} \tag{3.786}$$

Toroidal (u, v, w):

$$x = a\frac{\sinh v \cos w}{\cosh v - \cos u}$$
$$y = a\frac{\sinh v \sin w}{\cosh v - \cos u} \tag{3.787}$$
$$z = a\frac{\sin u}{\cosh v - \cos u}$$

KEY SYMBOLS

a, b, c	constant parameters
$\mathbf{a}$	a general vector
$a, \mathbf{a}, \ddot{x}, \dot{\mathbf{v}}$	acceleration
A	surface area
ABC	triad
$\hat{b}_i$	unit vectors of a nonorthogonal oblique triad ABC
$\mathbf{b}_i$	principal base vectors of a nonorthogonal coordinate system
$\mathbf{b}_i^{\star}$	reciprocal base vectors of a nonorthogonal coordinate system
B	body coordinate frame, local coordinate frame
c	cos
C	center
$\mathbf{c}$	relative vector
d	distance, distance between two points
f	function
$\mathbf{F}$	force vector, principal force vector
$\left[g_{ij}\right]$	covariant metric matrix
$\left[g_{ij}^{\star}\right]$	contravariant metric matrix
G	global coordinate frame, fixed coordinate frame
$\mathbf{I} = [\mathbf{I}]$	identity matrix
$\hat{\imath}, \hat{\jmath}, \hat{k}$	local Cartesian coordinate axis unit vectors
$\hat{I}, \hat{J}, \hat{K}$	global Cartesian coordinate axis unit vectors
$\mathbf{j}$	jerk
$J = V$	Jacobian, volume of base parallelogram of $\mathbf{b}_i$, $J = [\mathbf{b}_1\mathbf{b}_2\mathbf{b}_3]$
K	kinetic energy
l	length
m	mass
$\hat{n}$	normal unit vector
$O>$	common origin of B and G
$\mathbf{p}, \mathbf{q}, \mathbf{r}$	vectors
P	principal
P, Q	body points
q_i	curvilinear coordinate
Q	generalized coordinate system, principal coordinate frame
$Q^{\star}$	reciprocal coordinate frame
$r, \varphi, \theta,$	spherical coordinates
$\mathbf{r}$	position vector
r_i	principal components of $\mathbf{r}$ along a principal triad

$r_i^\star$	reciprocal components of $\mathbf{r}$
r_{ij}	the element of row i and column j of a matrix
R	reciprocal, radius,
R	transformation matrix between principal and reciprocal frames
s	sin, arc length
t	time
u, v, w	orthogonal coordinate system
	non-Cartesian coordinates
$\mathbf{u}, \mathbf{v}, \mathbf{w}$	vectors
$\hat{u}$	principal unit vector
$\hat{u}^\star$	reciprocal unit vector
$\mathbf{v}$	velocity vector
V	scalar field
$V = J$	volume of base parallelogram of $\mathbf{b}_i$
$V^\star$	volume of base parallelogram of $\mathbf{b}_i^\star$
W	work
x, y, z	local coordinate axes, Cartesian coordinates
X, Y, Z	global coordinate axes

Greek

α, β, γ	rotation angles about global axes
ϵ_{ijk}	permutation symbol
δ_{ij}	Kronecker delta
φ, θ, ψ	rotation angles, Euler angles
κ	curvature
λ	wavelength
ρ, θ, z	cylindrical coordinates
τ	torsion
Γ	Christoffel symbol
$\omega_x, \omega_y, \omega_z$	angular velocity components
ω	angular velocity vector

Symbol

$[\ \ ]^{-1}$	inverse of the matrix $[\ \ ]$
$[\ \ ]^T$	transpose of the matrix $[\ \ ]$
∇, ˜grad	gradient
↻	cyclic interchange

EXERCISES

1. **Lamé Equation** The direction of scattering X-rays in a crystalline lattice is determined by x, which is the solution of Lamé equation:

$$\mathbf{x} \cdot \mathbf{a} = A \qquad \mathbf{x} \cdot \mathbf{b} = B \qquad \mathbf{x} \cdot \mathbf{c} = C$$

where $\mathbf{a}, \mathbf{b}, \mathbf{c}$ are the basis vectors of the crystal lattice and A, B, C are scalars. Show that the solution of the Lame equation is

$$\mathbf{x} = A\mathbf{a}^\star + B\mathbf{b}^\star + C\mathbf{c}^\star$$

where $\mathbf{a}^\star, \mathbf{b}^\star, \mathbf{c}^\star$ are the reciprocal basis vectors of $\mathbf{a}, \mathbf{b}, \mathbf{c}$.

2. **Base Vectors of Spherical Coordinate System** Show that the base vectors of a spherical coordinate system are

$$\mathbf{b}_1 = \hat{e}_r \qquad \mathbf{b}_2 = r \sin\varphi \hat{e}_\theta \qquad \mathbf{b}_3 = r\hat{e}_\varphi$$

3. **A Length–Length–Angle Coordinate System** A coordinate system is defined by the following coordinate transformations:

$$\begin{aligned} x &= \eta\xi \cos\varphi \\ y &= \eta\xi \sin\varphi \\ z &= \tfrac{1}{2}(\xi^2 - \eta^2) \\ \xi^2 &= \sqrt{x^2 + y^2 + z^2} + z \\ \eta^2 &= \sqrt{x^2 + y^2 + z^2} - z \\ \varphi &= \tan^{-1}\frac{y}{x}. \end{aligned}$$

The base vectors of this coordinate system are

$$\begin{aligned} \mathbf{b}_1 &= \frac{\partial \mathbf{r}}{\partial \xi} = \eta \cos\varphi \hat{\imath} + \eta \sin\varphi \hat{\jmath} + \xi \hat{k} \\ \mathbf{b}_2 &= \frac{\partial \mathbf{r}}{\partial \eta} = \xi \cos\varphi \hat{\imath} + \xi \sin\varphi \hat{\jmath} - \eta \hat{k} \\ \mathbf{b}_3 &= \frac{\partial \mathbf{r}}{\partial \varphi} = -\eta\xi \sin\varphi \hat{\imath} + \eta\xi \cos\varphi \hat{\jmath} \end{aligned}$$

Complete the kinematic information of this coordinate system and find the position, velocity, and acceleration vectors of a moving point in this frame.

4. **Cylindrical Hyperbolic–Elliptic Coordinate System** A cylindrical hyperbolic–elliptic coordinate system is defined by the following coordinate transformations:

$$\begin{aligned} x &= a \cosh\alpha \sin\beta \\ y &= a \sinh\alpha \cos\beta \\ z &= z \end{aligned}$$

where a is a constant. Show that the base vectors of this space are

$$\begin{aligned} \mathbf{b}_1 &= \frac{\partial \mathbf{r}}{\partial \alpha} = a \sinh\alpha \sin\beta \hat{\imath} + a \cosh\alpha \cos\beta \hat{\jmath} \\ \mathbf{b}_2 &= \frac{\partial \mathbf{r}}{\partial \beta} = a \cosh\alpha \cos\beta \hat{\imath} - a \sinh\alpha \sin\beta \hat{\jmath} \\ \mathbf{b}_3 &= \frac{\partial \mathbf{r}}{\partial z} = \hat{k} \end{aligned}$$

and then find the kinematic information of the space.

5. **Jerk on a Space Curve**

$$x = 10\left(1 - e^{-t}\right)\sin\left(2t^2\right)$$

$$y = 10\left(1 - e^{-t}\right)\cos\left(2t^2\right)$$

$$z = xy$$

Determine the velocity, acceleration, and jerk of a particle that moves on the given space curve and express the path of motion.

6. **Constant-Length Vectors** Prove that if $\mathbf{a}$ is a vector with constant length which is dependent on a parameter μ, then

$$\mathbf{a}\cdot\frac{\partial \mathbf{a}}{\partial \mu} = 0$$

and deduce that the dot product of any unit vector and its time derivative is zero, $\hat{e}\cdot d\hat{e}/dt = 0$

7. **Unit Vectors of Orthogonal Coordinate Systems** Prove that if $\hat{e}_i$ and $\hat{e}_i$ are two unit vectors of an orthogonal coordinate system, then

$$\hat{e}_i\cdot\frac{d\hat{e}_i}{dt} = 0$$

$$\hat{e}_i\cdot\frac{d\hat{e}_j}{dt} = -\hat{e}_j\cdot\frac{d\hat{e}_i}{dt}$$

8. **Square of Arc Lengths** Find the square of an element of arc length ds in cylindrical and spherical coordinate systems.

9. **Kinematic Vectors from Jerk** Suppose a particle starts moving with the following jerk:

$$\mathbf{j} = ae^{-t^2}\left[\left(6\omega^2 t - 8t^3 + 12t\right)\sin\omega t - \left(\omega^3 - 12\omega t^2 + 6\omega\right)\cos\omega t\right]\hat{\imath}$$

$$-\,6\frac{10t^{12} - 25t^6 + 1}{\left(t^6+1\right)^3}\hat{\jmath} + \left(3\omega^2\cosh\omega t + \omega^3 t\sinh\omega t\right)\hat{k}$$

Determine the acceleration, velocity, and position vectors of the particle.

10. Curtate Cycloid Kinematics Consider a particle that is moving on the following curtate cycloid as is shown in Figure 3.32:

$$x = bt - c\sin t \qquad y = b - c\cos t$$

Assume $b = 1$ and $c = 1.6$.

(a) Determine the velocity, acceleration, and jerk components of the point.

(b) Determine the angle between velocity vectors when the point crosses the line $y = b$.

(c) Determine the angle between acceleration vectors when the point crosses the line $y = b$.

(d) Determine the angle between jerk vectors when the point crosses the line $y = b$.

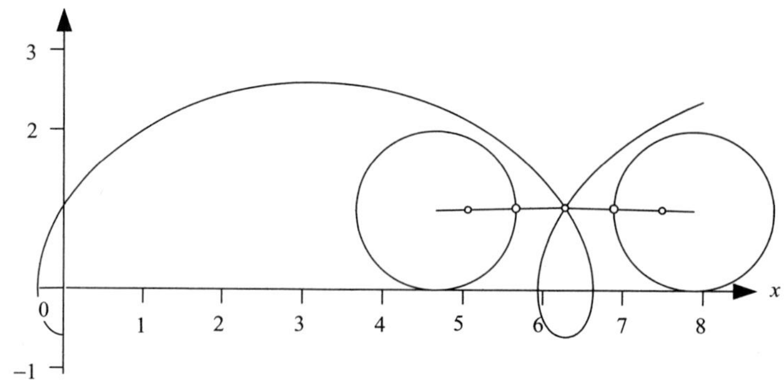

Figure 3.32 A curtate cycloid with $b = 1$ and $c = 1.6$.

11. ★ Modified Curtate Cycloid Consider a particle that is moving on the following path:

$$x = bt - c\sin t \qquad y = b - c\cos t$$

$$c = b\cos\omega t$$

(a) Determine ω such that the point is always between $0 \le y \le 2b$.

(b) Determine the maximum velocity, acceleration, and jerk components of the point for $0 \le t \le 2\pi$.

(c) Determine the angle between velocity vectors when the point crosses the line $y = b$.

(d) Determine the angle between acceleration vectors when the point crosses the line $y = b$.

(e) Determine the angle between jerk vectors when the point crosses the line $y = b$.

12. Comparison of Trochoid with Slider-Crank Mechanisms Figure 3.33 illustrates a trochoid mechanism with the following equations for the path of point A:

$$x = bt - c\sin t \qquad y = b - c\cos t$$

which for the case of $a = 2b = 2c$ produces a pure harmonic motion:

$$x = b\cos t \qquad y = 0$$

The figure also illustrates a slider-crank mechanism.

(a) Determine the position of the slider of the slider-crank mechanism as a function of θ_2 and compare with the position of the trochoid mechanism.

(b) Is it possible to have a harmonic motion by a slider-crank mechanism?

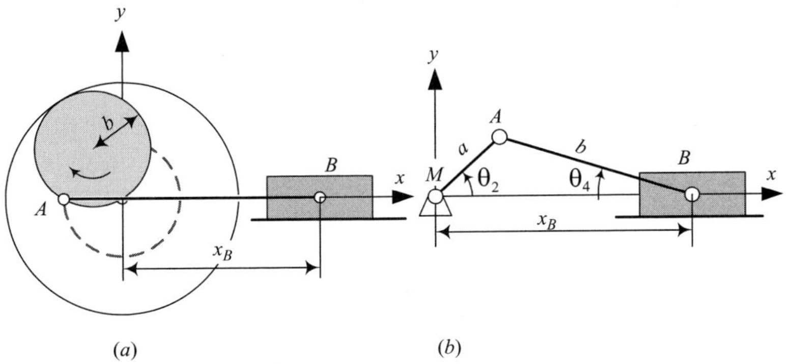

Figure 3.33 A trochoid mechanism and a slider-crank mechanism.

13. **Cylindrical Coordinate System** A point is moving with the following cylindrical characteristics:

$$\rho = 1\,\text{m} \qquad \theta = 30\,\text{deg} \qquad z = 2\,\text{m}$$
$$\dot{\rho} = 0.1\,\text{m/s} \qquad \dot{\theta} = -3\,\text{deg/s} \qquad \dot{z} = 0.2\,\text{m/s}$$
$$\ddot{\rho} = 1\,\text{m/s}^2 \qquad \ddot{\theta} = 2\,\text{deg/s}^2 \qquad \ddot{z} = 0.2\,\text{m/s}^2$$

Determine the position, velocity, and acceleration of the point in a:

(a) Cylindrical coordinate system
(b) Cartesian coordinate system
(c) Spherical coordinate system

14. ★ **Cylindrical–Spherical System Relationship** Assume the position, velocity, and acceleration of a particle are given in a cylindrical coordinate system. Determine the position, velocity, and acceleration of the particle in the spherical system in terms of cylindrical characteristics.

15. ★ **Reciprocal Base Vectors**

(a) Consider a point P at $\mathbf{r}$,

$$\mathbf{r} = \hat{\imath} + 2\hat{\jmath} + 3\hat{k}$$

and a nonorthogonal coordinate frame with

$$\hat{b}_1 = \frac{1}{\sqrt{2}}\begin{bmatrix} 1 \\ 1 \\ 0 \end{bmatrix} \qquad \hat{b}_2 = \frac{1}{1.4526}\begin{bmatrix} 0.9 \\ 0.7 \\ -0.9 \end{bmatrix} \qquad \hat{b}_3 = \frac{1}{1.01}\begin{bmatrix} 0.1 \\ 0.1 \\ 1 \end{bmatrix}$$

(b) Determine the components r_i of $\mathbf{r}$ in the nonorthogonal coordinate frame:

$$\mathbf{r} = r_1\hat{b}_1 + r_2\hat{b}_2 + r_3\hat{b}_3$$

(c) Determine the reciprocal base vectors and the reciprocal components $r_i^\bigstar$ of $\mathbf{r}$ in the reciprocal coordinate frame.

(d) Determine the transformation matrices to transform r_i to $r_i^\bigstar$ and $r_i^\bigstar$ to r_i.

16. Covariant and Contravariant Components

(a) Express the scalar product of two vectors in terms of their covariant and contravariant components.

(b) Find the vector product of two vectors in an oblique coordinate system.

17. Vector Calculation in a Nonorthogonal Coordinate Frame Consider the following two vectors in a principal nonorthogonal coordinate frame:

$$\begin{aligned} {}^P\mathbf{r}_I &= r_{11}\hat{b}_1 + r_{12}\hat{b}_2 + r_{13}\hat{b}_3 \\ &= 5.4442\hat{b}_1 + 5.1372\hat{b}_2 + 4.9482\hat{b}_3 \\ {}^P\mathbf{r}_2 &= r_{21}\hat{b}_1 + r_{22}\hat{b}_2 + r_{23}\hat{b}_3 \\ &= 15.712\hat{b}_1 + 11.357\hat{b}_2 + 5.655\hat{b}_3 \end{aligned}$$

Determine:

(a) ${}^P\mathbf{r}_I \cdot {}^P\mathbf{r}_2$

(b) ${}^P\mathbf{r}_I \times {}^P\mathbf{r}_2$

(c) Angle between ${}^P\mathbf{r}_I$ and ${}^P\mathbf{r}_2$

(d) Length of ${}^P\mathbf{r}_I$ and ${}^P\mathbf{r}_2$

18. Vector Calculation in a Reciprocal Coordinate Frame Consider the following two vectors in a reciprocal nonorthogonal coordinate frame:

$$\begin{aligned} {}^R\mathbf{r}_1 &= r_{11}^{\star}\mathbf{b}_1^{\star} + r_{12}^{\star}\mathbf{b}_2^{\star} + r_{13}^{\star}\mathbf{b}_3^{\star} \\ &= 0.82294\mathbf{b}_1^{\star} + 0.92014\mathbf{b}_2^{\star} + 4\mathbf{b}_3^{\star} \\ {}^R\mathbf{r}_2 &= r_{21}^{\star}\mathbf{b}_1^{\star} + r_{22}^{\star}\mathbf{b}_2^{\star} + r_{23}^{\star}\mathbf{b}_3^{\star} \\ &= 6.299\,1\mathbf{b}_1^{\star} - 0.610\,88\mathbf{b}_2^{\star} + 3\mathbf{b}_3^{\star} \end{aligned}$$

Determine:

(a) ${}^R\mathbf{r}_1 \cdot {}^R\mathbf{r}_2$

(b) ${}^R\mathbf{r}_1 \times {}^R\mathbf{r}_2$

(c) Angle between ${}^R\mathbf{r}_1$ and ${}^R\mathbf{r}_2$

(d) Length of ${}^R\mathbf{r}_1$ and ${}^R\mathbf{r}_2$

19. Base Vectors of Cylindrical Coordinate System Consider the cylindrical coordinate system.

(a) Determine the principal base vectors.

(b) Determine the reciprocal base vectors.

(c) Determine the principal–reciprocal transformation matrix.

★ In Exercises 20–45, determine:

(a) Velocity and acceleration vectors

(b) Principal and reciprocal base vectors

(c) Reciprocal velocity and acceleration vectors

(d) Principal and reciprocal metric matrices

20. ★Bipolar Cylindrical Coordinate System (u, v, z)

$$x = \frac{a \sinh v}{\cosh v - \cos u} \qquad y = \frac{a \sin u}{\cosh v - \cos u} \qquad z = z$$

21. ★ **Bispherical Coordinate System** (u, v, w)

$$x = \frac{\sin u \cos w}{\cosh v - \cos u} \qquad y = \frac{\sin u \sin w}{\cosh v - \cos u} \qquad z = \frac{\sinh(v)}{\cosh v - \cos u}$$

22. ★ **Cardioidal Coordinate System** (u, v, w)

$$x = \frac{uv \cos w}{(u^2 + v^2)^2} \qquad y = \frac{uv \sin w}{(u^2 + v^2)^2} \qquad z = \frac{1}{2} \frac{u^2 - v^2}{(u^2 + v^2)^2}$$

23. ★ **Cardioidcylindrical Coordinate System** (u, v, w)

$$x = \frac{1}{2} \frac{u^2 - v^2}{(u^2 + v^2)^2} \qquad y = \frac{uv}{(u^2 + v^2)^2} \qquad z = w$$

24. ★ **Casscylindrical Coordinate System** (u, v, w)

$$x = a\frac{\sqrt{2}}{2}\left(\sqrt{e^{2u} + 2e^u \cos v + 1} + \sqrt{e^u \cos v + 1}\right)$$

$$y = a\frac{\sqrt{2}}{2}\left(\sqrt{e^{2u} + 2e^u \cos v + 1} - \sqrt{e^u \cos v + 1}\right)$$

$$z = w$$

25. ★ **Confocalellip Coordinate System** (u, v, w)

$$x = \sqrt{\frac{a^2 - c^2}{a^2 - b^2}(a^2 - u)(a^2 - v)(a^2 - w)}$$

$$y = \sqrt{\frac{b^2 - a^2}{b^2 - c^2}(a^2 - u)(a^2 - v)(a^2 - w)}$$

$$z = \sqrt{\frac{c^2 - a^2}{c^2 - b^2}(a^2 - u)(a^2 - v)(a^2 - w)}$$

26. ★ **Confocalparab Coordinate System** (u, v, w)

$$x = \sqrt{\frac{(a^2 - u)(a^2 - v)(a^2 - w)}{b^2 - a^2}}$$

$$y = \sqrt{\frac{(b^2 - u)(b^2 - v)(b^2 - w)}{b^2 - a^2}}$$

$$z = \frac{a^2 + b^2 - u - v - w}{2}$$

27. ★ **Conical Coordinate System** (u, v, w)

$$x = \frac{uvw}{ab}$$

$$y = \frac{u}{b}\sqrt{\frac{(v^2 - b^2)(b^2 - w^2)}{a^2 - b^2}}$$

$$z = \frac{u}{a}\sqrt{\frac{(a^2 - v^2)(a^2 - w^2)}{a^2 - b^2}}$$

28. ★ Ellcylindrical Coordinate System (u, v, w)

$$x = a\cosh u\cos v \qquad y = a\sinh u\sin v \qquad z = w$$

29. ★ Ellipsoidal Coordinate System (u, v, w)

$$x = \frac{uvw}{ab}$$

$$y = \frac{1}{b}\sqrt{\frac{(u^2 - b^2)(v^2 - b^2)(b^2 - w^2)}{a^2 - b^2}}$$

$$z = \frac{1}{a}\sqrt{\frac{(u^2 - a^2)(a^2 - v^2)(a^2 - w^2)}{a^2 - b^2}}$$

30. ★ Hypercylindrical Coordinate System (u, v, w)

$$x = \sqrt{\sqrt{u^2 + v^2} + u} \qquad y = \sqrt{\sqrt{u^2 + v^2} - u} \qquad z = w$$

31. ★ Invcasscylindrical Coordinate System (u, v, w)

$$x = a\frac{\sqrt{2}}{2}\left(1 + \sqrt{\frac{e^u\cos v + 1}{e^{2u} + 2e^u\cos v + 1}}\right)$$

$$y = a\frac{\sqrt{2}}{2}\left(1 - \sqrt{\frac{e^u\cos v + 1}{e^{2u} + 2e^u\cos v + 1}}\right)$$

$$z = w$$

32. ★ Invellcylindrical Coordinate System (u, v, w)

$$x = a\frac{\cosh u\cos v}{\cosh^2 u - \sin^2 v}$$

$$y = a\frac{\sinh u\sin v}{\cosh^2 u - \sin^2 v}$$

$$z = w$$

33. ★ Invoblspheroidal Coordinate System (u, v, w)

$$x = a\frac{\cosh u\sin v\cos w}{\cosh^2 u - \cos^2 v}$$

$$y = a\frac{\cosh u\sin v\sin w}{\cosh^2 u - \cos^2 v}$$

$$z = a\frac{\sinh u\cos v}{\cosh^2 u - \cos^2 v}$$

34. ★ Logcylindrical Coordinate System (u, v, w)

$$x = \frac{a}{\pi}\ln\left(u^2 + v^2\right) \qquad y = 2\frac{a}{\pi}\arctan\left(\frac{u}{v}\right) \qquad z = w$$

35. ★ **Logcoshcylindrical Coordinate System** (u, v, w)

$$x = \frac{a}{\pi} \ln\left(\cosh^2 u - \sin^2 v\right)$$

$$y = 2\frac{a}{\pi} \arctan\left(\frac{\tanh u}{\tan v}\right)$$

$$z = w$$

36. ★ **Maxwellcylindrical Coordinate System** (u, v, w)

$$x = \frac{a}{\pi}\left(u + 1 + e^u \cos v\right) \qquad y = \frac{a}{\pi}\left(u + e^u \sin v\right) \qquad z = w$$

37. ★ **Oblatespheroidal Coordinate System** (u, v, w)

$$x = a \cosh u \sin v \cos w$$

$$y = a \cosh u \sin v \sin w$$

$$z = a \sinh u \cos v$$

38. ★ **Paraboloidal1 Coordinate System** (u, v, w)

$$x = uv \cos w \qquad y = uv \sin w \qquad z = \frac{1}{2}\left(u^2 - v^2\right)$$

39. ★ **Paraboloidal2 Coordinate System** (u, v, w)

$$x = 2\sqrt{\frac{(u-a)(a-v)(a-w)}{a-b}}$$

$$y = 2\sqrt{\frac{(u-b)(b-v)(b-w)}{a-b}}$$

$$z = u + v + w - a - b$$

40. ★ **Paracylindrical Coordinate System** (u, v, w)

$$x = \frac{1}{2}\left(u^2 - v^2\right) \qquad y = uv \qquad z = w$$

41. ★ **Rosecylindrical Coordinate System** (u, v, w)

$$x = \frac{\sqrt{\sqrt{u^2+v^2}+u}}{\sqrt{u^2+v^2}} \qquad y = \frac{\sqrt{\sqrt{u^2+v^2}-u}}{\sqrt{u^2+v^2}} \qquad z = w$$

42. ★ **Sixsphere Coordinate System** (u, v, w)

$$x = \frac{u}{u^2+v^2+w^2}$$

$$y = \frac{v}{u^2+v^2+w^2}$$

$$z = \frac{w}{u^2+v^2+w^2}$$

43. ★ **Tangentcylindrical Coordinate System** (u, v, w)

$$x = \frac{u}{u^2+v^2} \qquad y = \frac{v}{u^2+v^2} \qquad z = w$$

44. ★ Tangentsphere Coordinate System (u, v, w)

$$x = \frac{u}{u^2 + v^2} \cos w \qquad y = \frac{u}{u^2 + v^2} \sin w \qquad z = \frac{v}{u^2 + v^2}$$

45. ★ Toroidal Coordinate System (u, v, w)

$$x = a \frac{\sinh v \cos w}{\cosh v - \cos u}$$

$$y = a \frac{\sinh v \sin w}{\cosh v - \cos u}$$

$$z = a \frac{\sin u}{\cosh v - \cos u}$$

4

Rotation Kinematics

The only possible motion of a rigid body with a fixed point is rotation about the fixed point. We represent the rigid body by a body coordinate frame B that rotates in another coordinate frame G, as shown in Figure 4.1. To determine the orientation of the rigid body, we perform a rotation analysis based on transformation matrices and determine the orientation of B in G.

4.1 ROTATION ABOUT GLOBAL CARTESIAN AXES

Consider a body coordinate frame $B(Oxyz)$ that is representing a rigid body B. The body frame B was originally coincident with a global coordinate frame $G(OXYZ)$. Point O of the body B is fixed on the ground G and is the origin of both coordinate frames.

Assume that the rigid body B rotates α degrees about the Z-axis of G. If the position vector of a body point P is shown by ${}^{B}\mathbf{r}$ and ${}^{G}\mathbf{r}$ in the B- and G-frames, respectively, then the coordinates of P in the local and global coordinate frames are related by

$$
{}^{G}\mathbf{r} = R_{Z,\alpha}\,{}^{B}\mathbf{r} \tag{4.1}
$$

where

$$
{}^{G}\mathbf{r} = \begin{bmatrix} X \\ Y \\ Z \end{bmatrix} \qquad {}^{B}\mathbf{r} = \begin{bmatrix} x \\ y \\ z \end{bmatrix} \tag{4.2}
$$

and $R_{Z,\alpha} = {}^{G}R_{B}$ is the rotation matrix for turning α degrees about the Z-axis:

$$
R_{Z,\alpha} = {}^{G}R_{B} = \begin{bmatrix} \cos\alpha & -\sin\alpha & 0 \\ \sin\alpha & \cos\alpha & 0 \\ 0 & 0 & 1 \end{bmatrix} \tag{4.3}
$$

where $R_{Z,\alpha} = {}^{G}R_{B}$ is the transformation matrix that takes the body coordinates and provides the associated global coordinates. Similarly, rotation β about the Y-axis and γ about the X-axis of the global frame relate the local and global coordinates of the body point P by

$$
{}^{G}\mathbf{r} = R_{Y,\beta}\,{}^{B}\mathbf{r} \tag{4.4}
$$

$$
{}^{G}\mathbf{r} = R_{X,\gamma}\,{}^{B}\mathbf{r} \tag{4.5}
$$

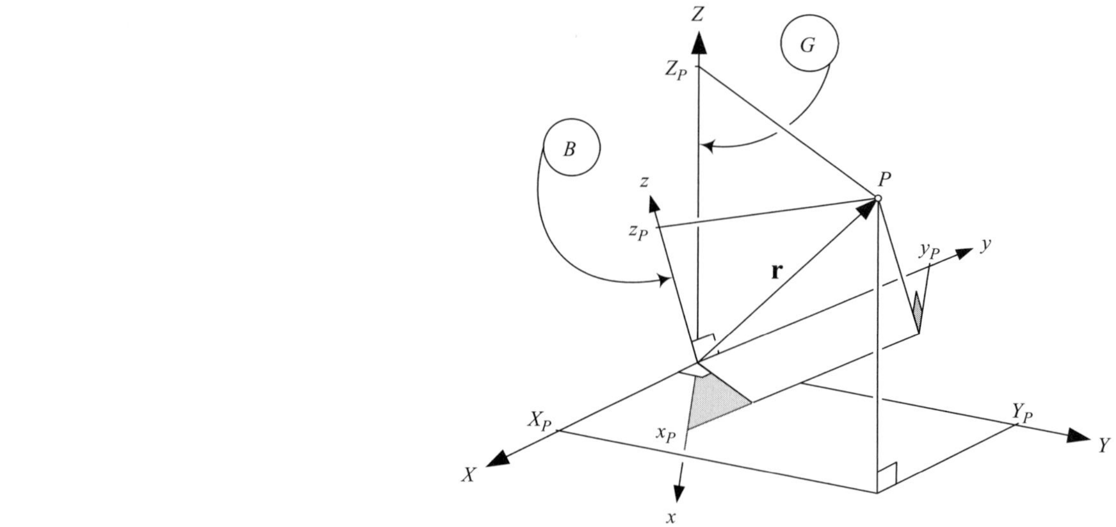

Figure 4.1 A globally fixed G-frame and a body B-frame with a fixed common origin at O.

where

$$R_{Y,\beta} = {}^{G}R_{B} = \begin{bmatrix} \cos\beta & 0 & \sin\beta \\ 0 & 1 & 0 \\ -\sin\beta & 0 & \cos\beta \end{bmatrix} \tag{4.6}$$

$$R_{X,\gamma} = {}^{G}R_{B} = \begin{bmatrix} 1 & 0 & 0 \\ 0 & \cos\gamma & -\sin\gamma \\ 0 & \sin\gamma & \cos\gamma \end{bmatrix} \tag{4.7}$$

Proof: Let $(\hat{\imath}, \hat{\jmath}, \hat{k})$ and $(\hat{I}, \hat{J}, \hat{K})$ be the unit vectors along the coordinate axes of $B(Oxyz)$ and $G(OXYZ)$, respectively. The rigid body has a globally fixed point at O, which is the common origin of the orthogonal triads $Oxyz$ and $OXYZ$. Consider a point P that is a fixed point in B. The B-expression of the position vector of P is a constant vector ${}^{B}\mathbf{r}$:

$${}^{B}\mathbf{r} = x\hat{\imath} + y\hat{\jmath} + z\hat{k} \tag{4.8}$$

Assume that the coordinate frames were initially coincident and then the rigid body turns α degrees about the global Z-axis. After the rotation, the body frame $B\,(Oxyz)$ along with point P and its body position vector ${}^{B}\mathbf{r}$ will move to a new position in G. The G-expression of the position vector of point P after rotation is

$${}^{G}\mathbf{r} = X\hat{I} + Y\hat{J} + Z\hat{K} \tag{4.9}$$

Figure 4.2 illustrates a top view of the initial and final positions of point P and its position vector $\mathbf{r}$ from the view of an observer in $G\,(OXYZ)$.

The orthogonality condition is the connecting link between two orthogonal coordinate frames. Therefore, we may write

$$X = \hat{I} \cdot {}^{B}\mathbf{r} = \hat{I} \cdot x\hat{\imath} + \hat{I} \cdot y\hat{\jmath} + \hat{I} \cdot z\hat{k} \tag{4.10}$$

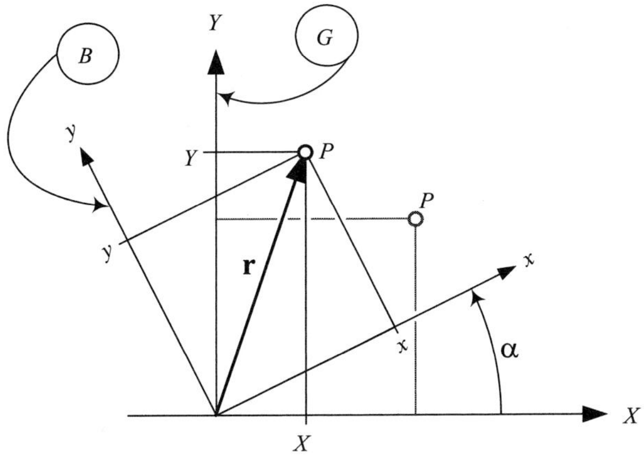

Figure 4.2 Top view of the initial and final positions of a body point P and its position vector $\mathbf{r}$ from the view of an observer in $G(OXYZ)$.

$$Y = \hat{J} \cdot {}^{B}\mathbf{r} = \hat{J} \cdot x\hat{\imath} + \hat{J} \cdot y\hat{\jmath} + \hat{J} \cdot z\hat{k} \tag{4.11}$$

$$Z = \hat{K} \cdot {}^{B}\mathbf{r} = \hat{K} \cdot x\hat{\imath} + \hat{K} \cdot y\hat{\jmath} + \hat{K} \cdot z\hat{k} \tag{4.12}$$

or equivalently

$${}^{G}\mathbf{r} = R_{Z,\alpha}\,{}^{B}\mathbf{r} \tag{4.13}$$

$$\begin{bmatrix} X \\ Y \\ Z \end{bmatrix} = \begin{bmatrix} \hat{I} \cdot \hat{\imath} & \hat{I} \cdot \hat{\jmath} & \hat{I} \cdot \hat{k} \\ \hat{J} \cdot \hat{\imath} & \hat{J} \cdot \hat{\jmath} & \hat{J} \cdot \hat{k} \\ \hat{K} \cdot \hat{\imath} & \hat{K} \cdot \hat{\jmath} & \hat{K} \cdot \hat{k} \end{bmatrix} \begin{bmatrix} x \\ y \\ z \end{bmatrix} \tag{4.14}$$

The elements of the *Z-rotation matrix*, $R_{Z,\alpha}$, are the *direction cosines* of the axes of $B(Oxyz)$ with respect to $G(OXYZ)$. Figure 4.2 indicates that

$$\begin{aligned} &\hat{I} \cdot \hat{\imath} = \cos\alpha && \hat{I} \cdot \hat{\jmath} = -\sin\alpha && \hat{I} \cdot \hat{k} = 0 \\ &\hat{J} \cdot \hat{\imath} = \sin\alpha && \hat{J} \cdot \hat{\jmath} = \cos\alpha && \hat{J} \cdot \hat{k} = 0 \\ &\hat{K} \cdot \hat{\imath} = 0 && \hat{K} \cdot \hat{\jmath} = 0 && \hat{K} \cdot \hat{k} = 1 \end{aligned} \tag{4.15}$$

Combining Equations (4.14) and (4.15) shows that

$$\begin{gathered} {}^{G}\mathbf{r} = R_{Z,\alpha}\,{}^{B}\mathbf{r} \\ \begin{bmatrix} X \\ Y \\ Z \end{bmatrix} = \begin{bmatrix} \cos\alpha & -\sin\alpha & 0 \\ \sin\alpha & \cos\alpha & 0 \\ 0 & 0 & 1 \end{bmatrix} \begin{bmatrix} x \\ y \\ z \end{bmatrix} \end{gathered} \tag{4.16}$$

The vector ${}^{G}\mathbf{r}$ in the global coordinate frame is equal to $R_{Z,\alpha}$ times the position vector in the local coordinate frame ${}^{B}\mathbf{r}$. Hence, we are able to find the global coordinates of a point of a rigid body after rotation about the Z-axis if we have its local coordinates and the angle of rotation α.

Similarly, rotation β about the Y-axis and rotation γ about the X-axis are respectively described by the *Y-rotation matrix* $R_{Y,\beta}$ and the *X-rotation matrix* $R_{X,\gamma}$. The rotation matrices

$$R_{Z,\alpha} = \begin{bmatrix} \cos\alpha & -\sin\alpha & 0 \\ \sin\alpha & \cos\alpha & 0 \\ 0 & 0 & 1 \end{bmatrix} \tag{4.17}$$

$$R_{Y,\beta} = \begin{bmatrix} \cos\beta & 0 & \sin\beta \\ 0 & 1 & 0 \\ -\sin\beta & 0 & \cos\beta \end{bmatrix} \tag{4.18}$$

$$R_{X,\gamma} = \begin{bmatrix} 1 & 0 & 0 \\ 0 & \cos\gamma & -\sin\gamma \\ 0 & \sin\gamma & \cos\gamma \end{bmatrix} \tag{4.19}$$

are called the *basic global rotation matrices*. All of the basic global rotation matrices $R_{Z,\alpha}$, $R_{Y,\beta}$, and $R_{X,\gamma}$ transform a B-expression vector to its G-expression. We show such a rotation matrix by ${}^G R_B$ to indicate that this is a transformation matrix from frame B to G.

We usually refer to the first, second, and third rotations about the axes of the global coordinate frame by α, β, and γ, respectively, regardless of the axis labels. ■

Example 244 Successive Rotation about Global Axes The block shown in Figure 4.3 turns 45 deg about the Z-axis, followed by 45 deg about the X-axis, and then 90 deg about the Y-axis. To find the final global position of the corner $P(1,6,2)$, first we multiply $R_{Z,45}$ by $[1, 6, 2]^T$ to get the new global position after the first rotation:

$$\begin{bmatrix} X_2 \\ Y_2 \\ Z_2 \end{bmatrix} = \begin{bmatrix} \cos\frac{\pi}{4} & -\sin\frac{\pi}{4} & 0 \\ \sin\frac{\pi}{4} & \cos\frac{\pi}{4} & 0 \\ 0 & 0 & 1 \end{bmatrix} \begin{bmatrix} 1 \\ 6 \\ 2 \end{bmatrix} = \begin{bmatrix} -3.5355 \\ 4.9497 \\ 2 \end{bmatrix} \tag{4.20}$$

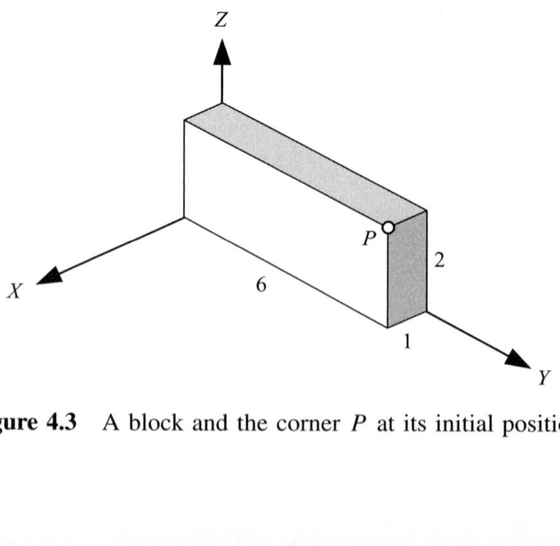

Figure 4.3 A block and the corner P at its initial position.

Then we multiply $R_{X,45}$ by $[-3.5355, 4.9497, 2]^{\mathrm{T}}$ to get the position of P after the second rotation:

$$\begin{bmatrix} X_3 \\ Y_3 \\ Z_3 \end{bmatrix} = \begin{bmatrix} 1 & 0 & 0 \\ 0 & \cos\frac{\pi}{4} & -\sin\frac{\pi}{4} \\ 0 & \sin\frac{\pi}{4} & \cos\frac{\pi}{4} \end{bmatrix} \begin{bmatrix} -3.5355 \\ 4.9497 \\ 2 \end{bmatrix} = \begin{bmatrix} -3.5355 \\ 2.0858 \\ 4.9142 \end{bmatrix} \tag{4.21}$$

Finally we multiply $R_{Y,90}$ by $[-3.5355, 2.0858, 4.9142]^{\mathrm{T}}$ to get the position of P after the third rotation:

$$\begin{bmatrix} X_4 \\ Y_4 \\ Z_4 \end{bmatrix} = \begin{bmatrix} \cos\frac{\pi}{2} & 0 & \sin\frac{\pi}{2} \\ 0 & 1 & 0 \\ -\sin\frac{\pi}{2} & 0 & \cos\frac{\pi}{2} \end{bmatrix} \begin{bmatrix} -3.5355 \\ 2.0858 \\ 4.9142 \end{bmatrix} = \begin{bmatrix} 4.9142 \\ 2.0858 \\ 3.5355 \end{bmatrix} \tag{4.22}$$

The orientations of the block are shown in Figure 4.4.

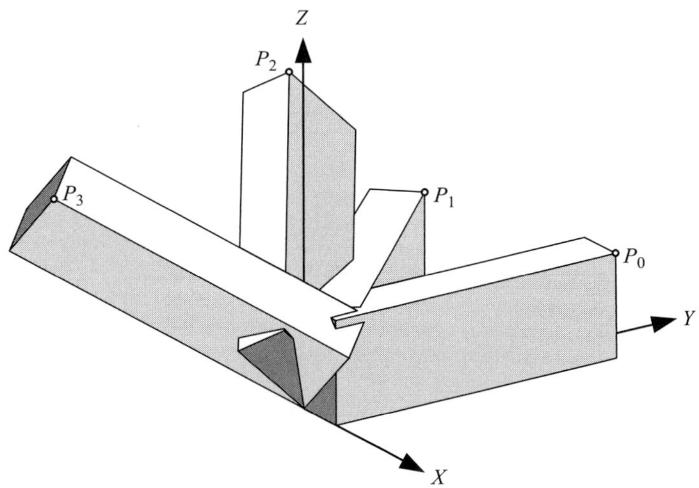

Figure 4.4 The block and body point P at its initial position and after first, second, and third rotations.

Example 245 Invariant Components and Global Rotations Any rotation about the X-axis will not change the value of the x-component of body point P. Similarly, any rotation about the Y- and Z-axis will not change the Y- and Z-components of P, respectively.

Example 246 Global Rotation, Local Position Consider a point P that is moved to ${}^{G}\mathbf{r} = [1, 3, 2]^{\mathrm{T}}$ after a 60 deg rotation about the X-axis. Its position in the local coordinate frame is

$$
{}^{B}\mathbf{r} = R_{X,60}^{-1}\,{}^{G}\mathbf{r}
$$

$$
\begin{bmatrix} x \\ y \\ z \end{bmatrix} = \begin{bmatrix} \cos\frac{\pi}{3} & -\sin\frac{\pi}{3} & 0 \\ \sin\frac{\pi}{3} & \cos\frac{\pi}{3} & 0 \\ 0 & 0 & 1 \end{bmatrix}^{-1} \begin{bmatrix} 1 \\ 3 \\ 2 \end{bmatrix} = \begin{bmatrix} 3.0981 \\ 0.63397 \\ 2 \end{bmatrix} \tag{4.23}
$$

Example 247 Time-Dependent Global Rotation Assume a rigid body B is continuously turning about the Z-axis of G at a rate of 0.2 rad/s. The transformation matrix of the body is

$$
{}^{G}R_{B} = \begin{bmatrix} \cos 0.2t & -\sin 0.2t & 0 \\ \sin 0.2t & \cos 0.2t & 0 \\ 0 & 0 & 1 \end{bmatrix} \tag{4.24}
$$

Applying the rotation on the body B will move any point of B on a circle with radius $R = \sqrt{X^2 + Y^2}$ parallel to the (X, Y)-plane:

$$
\begin{bmatrix} X \\ Y \\ Z \end{bmatrix} = \begin{bmatrix} \cos 0.2t & -\sin 0.2t & 0 \\ \sin 0.2t & \cos 0.2t & 0 \\ 0 & 0 & 1 \end{bmatrix} \begin{bmatrix} x \\ y \\ z \end{bmatrix}
= \begin{bmatrix} x\cos 0.2t - y\sin 0.2t \\ y\cos 0.2t + x\sin 0.2t \\ z \end{bmatrix} \tag{4.25}
$$

$$
\begin{aligned} X^2 + Y^2 &= (x\cos 0.2t - 1.0y\sin 0.2t)^2 + (y\cos 0.2t + x\sin 0.2t)^2 \\ &= x^2 + y^2 = R^2 \end{aligned} \tag{4.26}
$$

Consider a point P at ${}^{B}\mathbf{r} = [1, 0, 0]^{\mathrm{T}}$. The point will be seen at

$$
\begin{bmatrix} X \\ Y \\ Z \end{bmatrix} = \begin{bmatrix} \cos 0.2 & -\sin 0.2 & 0 \\ \sin 0.2 & \cos 0.2 & 0 \\ 0 & 0 & 1 \end{bmatrix} \begin{bmatrix} 1 \\ 0 \\ 0 \end{bmatrix} = \begin{bmatrix} 0.98 \\ 0.198 \\ 0 \end{bmatrix} \tag{4.27}
$$

after $t = 1\,s$ and at

$$
\begin{bmatrix} X \\ Y \\ Z \end{bmatrix} = \begin{bmatrix} \cos 0.6 & -\sin 0.6 & 0 \\ \sin 0.6 & \cos 0.6 & 0 \\ 0 & 0 & 1 \end{bmatrix} \begin{bmatrix} 1 \\ 0 \\ 0 \end{bmatrix} = \begin{bmatrix} 0.825 \\ 0.564 \\ 0 \end{bmatrix} \tag{4.28}
$$

after $t = 3\,s$.

4.2 SUCCESSIVE ROTATIONS ABOUT GLOBAL AXES

After a series of sequential rotations $R_1, R_2, R_3, \ldots, R_n$ about the global axes, the final global position of a body point P can be found by

$$ {}^G\mathbf{r} = {}^GR_B\, {}^B\mathbf{r} \tag{4.29}$$

where

$$ {}^GR_B = R_n \cdots R_3\, R_2\, R_1 \tag{4.30}$$

The vectors ${}^G\mathbf{r}$ and ${}^B\mathbf{r}$ indicate the position vectors of the point P in the global and local coordinate frames, respectively. The matrix ${}^GR_B$, which transforms the local coordinates to their corresponding global coordinates, is called the *global rotation matrix*.

Because matrix multiplications do not commute, the sequence of performing rotations is important and indicates the order of rotations.

Proof: Consider a body frame B that undergoes two sequential rotations R_1 and R_2 about the global axes. Assume that the body coordinate frame B is initially coincident with the global coordinate frame G. The rigid body rotates about a global axis, and the global rotation matrix R_1 gives us the new global coordinate ${}^G\mathbf{r}_1$ of the body point:

$$ {}^G\mathbf{r}_1 = R_1\, {}^B\mathbf{r} \tag{4.31}$$

Before the second rotation the situation is similar to the one before the first rotation. We put the B-frame aside and assume that a new body coordinate frame B_1 is coincident with the global frame. Therefore, the new body coordinate would be ${}^{B_1}\mathbf{r} \equiv {}^G\mathbf{r}_1$. The second global rotation matrix R_2 provides the new global position ${}^G\mathbf{r}_2$ of the body points ${}^{B_1}\mathbf{r}$:

$$ {}^{B_1}\mathbf{r} = R_2\, {}^{B_1}\mathbf{r} \tag{4.32}$$

Substituting (4.31) into (4.32) shows that

$$ {}^G\mathbf{r} = R_2\, R_1\, {}^B\mathbf{r} \tag{4.33}$$

Following the same procedure we can determine the final global position of a body point after a series of sequential rotations $R_1, R_2, R_3, \ldots, R_n$ as (4.30). ■

Example 248 Successive Global Rotations Consider a body point P at $P(1,2,3)$. If the body turns $\gamma = 45$ deg about the X-axis, the global coordinates of the point would be

$$ {}^G\mathbf{r} = R_{X,\gamma}\, {}^B\mathbf{r} $$

$$ \begin{bmatrix} 1 \\ -0.707 \\ 3.535 \end{bmatrix} = \begin{bmatrix} 1 & 0 & 0 \\ 0 & \cos\frac{\pi}{4} & -\sin\frac{\pi}{4} \\ 0 & \sin\frac{\pi}{4} & \cos\frac{\pi}{4} \end{bmatrix} \begin{bmatrix} 1 \\ 2 \\ 3 \end{bmatrix} \tag{4.34}$$

Now the body turns $\alpha = -90$ deg about the Z-axis and moves the point P to

$$^{G}\mathbf{r} = R_{Z,\alpha}\, ^{B_1}\mathbf{r}$$

$$\begin{bmatrix} 0.707 \\ 1 \\ 3.535 \end{bmatrix} = \begin{bmatrix} \cos\frac{\pi}{2} & -\sin\frac{\pi}{2} & 0 \\ \sin\frac{\pi}{2} & \cos\frac{\pi}{2} & 0 \\ 0 & 0 & 1 \end{bmatrix} \begin{bmatrix} 1 \\ -0.707 \\ 3.535 \end{bmatrix} \tag{4.35}$$

The final position of the point P after the two rotations can also be found by applying successive rotations:

$$^{G}\mathbf{r} = R_{Z,\alpha}\, R_{X,\gamma}\, ^{B}\mathbf{r} = R_{Z,\pi/2}\, R_{X,\pi/4} \begin{bmatrix} 1 \\ 2 \\ 3 \end{bmatrix} = \begin{bmatrix} 0.707 \\ 1 \\ 3.535 \end{bmatrix} \tag{4.36}$$

Applying the same rotation matrices, we can find the global coordinates of any other points of the body.

Example 249 Successive Global Rotation Matrix The global rotation matrix after a rotation $R_{Z,\alpha}$ followed by $R_{Y,\beta}$ and then $R_{X,\gamma}$ is

$$\begin{aligned} ^{G}R_{B} &= R_{X,\gamma} R_{Y,\beta} R_{Z,\alpha} \\ &= \begin{bmatrix} c\alpha c\beta & -c\beta s\alpha & s\beta \\ c\gamma s\alpha + c\alpha s\beta s\gamma & c\alpha c\gamma - s\alpha s\beta s\gamma & -c\beta s\gamma \\ s\alpha s\gamma - c\alpha c\gamma s\beta & c\alpha s\gamma + c\gamma s\alpha s\beta & c\beta c\gamma \end{bmatrix} \end{aligned} \tag{4.37}$$

where

$$c \equiv \cos \tag{4.38}$$

$$s \equiv \sin \tag{4.39}$$

Example 250 Successive Global Rotations, Global position Assume that the global position of a body point P after turning $\beta = 45$ deg about the Y-axis and then rotating $\gamma = 60$ deg about the X-axis is located at $^{G}\mathbf{r}$:

$$^{G}\mathbf{r} = \begin{bmatrix} 0.35 \\ 0.5 \\ 1.76 \end{bmatrix} \tag{4.40}$$

To find the coordinates of P in B, we may find the rotation matrix

$$\begin{aligned} ^{G}R_{B} &= R_{X,\gamma}\, R_{Y,\beta} \\ &= \begin{bmatrix} 1 & 0 & 0 \\ 0 & \cos\frac{\pi}{3} & -\sin\frac{\pi}{3} \\ 0 & \sin\frac{\pi}{3} & \cos\frac{\pi}{3} \end{bmatrix} \begin{bmatrix} \cos\frac{\pi}{4} & 0 & \sin\frac{\pi}{4} \\ 0 & 1 & 0 \\ -\sin\frac{\pi}{4} & 0 & \cos\frac{\pi}{4} \end{bmatrix} \\ &= \begin{bmatrix} 0.70711 & 0 & 0.70711 \\ 0.61237 & 0.5 & -0.61237 \\ -0.35355 & 0.86603 & 0.35355 \end{bmatrix} \end{aligned} \tag{4.41}$$

and then find ${}^B\mathbf{r}$ using the matrix inverse:

$$\begin{aligned}
{}^B\mathbf{r} &= {}^G R_B^{-1}\, {}^G\mathbf{r} \\
&= \begin{bmatrix} 0.70711 & 0 & 0.70711 \\ 0.61237 & 0.5 & -0.61237 \\ -0.35355 & 0.86603 & 0.35355 \end{bmatrix}^{-1} \begin{bmatrix} 0.35 \\ 0.5 \\ 1.76 \end{bmatrix} \\
&= \begin{bmatrix} -0.06858 \\ 1.7742 \\ 0.56355 \end{bmatrix}
\end{aligned} \tag{4.42}$$

Example 251 Twelve Independent Triple Global Rotations We may transform a body coordinate frame B from the coincident position with a global frame G to any desired orientation by three rotations about the global axes provided that no two sequential rotations are about the same axis.

There are 12 different independent combinations of triple rotations about the global axes:

$$\begin{aligned}
&1.\ R_{X,\gamma}\, R_{Y,\beta}\, R_{Z,\alpha} \\
&2.\ R_{Y,\gamma}\, R_{Z,\beta}\, R_{X,\alpha} \\
&3.\ R_{Z,\gamma}\, R_{X,\beta}\, R_{Y,\alpha} \\
&4.\ R_{Z,\gamma}\, R_{Y,\beta}\, R_{X,\alpha} \\
&5.\ R_{Y,\gamma}\, R_{X,\beta}\, R_{Z,\alpha} \\
&6.\ R_{X,\gamma}\, R_{Z,\beta}\, R_{Y,\alpha} \\
&7.\ R_{X,\gamma}\, R_{Y,\beta}\, R_{X,\alpha} \\
&8.\ R_{Y,\gamma}\, R_{Z,\beta}\, R_{Y,\alpha} \\
&9.\ R_{Z,\gamma}\, R_{X,\beta}\, R_{Z,\alpha} \\
&10.\ R_{X,\gamma}\, R_{Z,\beta}\, R_{X,\alpha} \\
&11.\ R_{Y,\gamma}\, R_{X,\beta}\, R_{Y,\alpha} \\
&12.\ R_{Z,\gamma}\, R_{Y,\beta}\, R_{Z,\alpha}
\end{aligned} \tag{4.43}$$

The expanded form of the 12 global axis triple rotations are presented in Appendix A.

Example 252 Order of Matrix Multiplication and Order of Rotation Changing the order of rotation matrices is equivalent to changing the order of rotations. Consider a rigid-body point P at ${}^B\mathbf{r}_P = [1, 2, 3]^{\mathrm{T}}$. Its global position after a rotation of 45 deg about the X-axis and then 45 deg about the Y-axis is at

$$\begin{aligned} \left({}^{G}\mathbf{r}_{P}\right)_{1} &= R_{Y,45}\, R_{X,45}\, {}^{B}\mathbf{r}_{P} \\ &= \begin{bmatrix} 0.707 & 0 & 0.707 \\ 0.5 & 0.707 & -0.5 \\ -0.5 & 0.707 & 0.5 \end{bmatrix} \begin{bmatrix} 1 \\ 2 \\ 3 \end{bmatrix} = \begin{bmatrix} 2.828 \\ 0.414 \\ 2.414 \end{bmatrix} \end{aligned} \tag{4.44}$$

If we change the order of rotations and turn the body 45 deg about the Y-axis followed by 45 deg about the X-axis, then the position of P would be at

$$\begin{aligned} \left({}^{G}\mathbf{r}_{P}\right)_{2} &= R_{X,45}\, R_{Y,45}\, {}^{B}\mathbf{r}_{P} \\ &= \begin{bmatrix} 0.707 & 0.5 & 0.5 \\ 0 & 0.707 & -0.707 \\ -0.707 & 0.5 & 0.5 \end{bmatrix} \begin{bmatrix} 1 \\ 2 \\ 3 \end{bmatrix} = \begin{bmatrix} 3.207 \\ -0.707 \\ 1.793 \end{bmatrix} \end{aligned} \tag{4.45}$$

The angle between the two final position vectors of P is

$$\theta = \cos^{-1} \frac{\left({}^{G}\mathbf{r}_{P}\right)_{1} \cdot \left({}^{G}\mathbf{r}_{P}\right)_{2}}{\left|{}^{G}\mathbf{r}_{P}\right|_{1} \left|{}^{G}\mathbf{r}_{P}\right|_{2}} = 0.9362 \text{ rad} \approx 53.64 \text{ deg} \tag{4.46}$$

Example 253 ★ Repeated Rotation about Global Axes Consider a body frame B that turns an angle α about the Z-axis. If

$$\alpha = \frac{2\pi}{n} \qquad n \in \mathbb{N} \tag{4.47}$$

then we need to repeat the rotation n times to get back to the original configuration. It may be checked by multiplying $R_{Z,\alpha}$ by itself until an identity matrix is achieved. In this case any body point will be mapped to the same point in the global frame. To show this, we may find that

$$\begin{aligned} R_{Z,\alpha}^{m} &= \begin{bmatrix} \cos\alpha & -\sin\alpha & 0 \\ \sin\alpha & \cos\alpha & 0 \\ 0 & 0 & 1 \end{bmatrix}^{m} = \begin{bmatrix} \cos\frac{2\pi}{n} & -\sin\frac{2\pi}{n} & 0 \\ \sin\frac{2\pi}{n} & \cos\frac{2\pi}{n} & 0 \\ 0 & 0 & 1 \end{bmatrix}^{m} \\ &= \begin{bmatrix} \cos m\frac{2\pi}{n} & -\sin m\frac{2\pi}{n} & 0 \\ \sin m\frac{2\pi}{n} & \cos m\frac{2\pi}{n} & 0 \\ 0 & 0 & 1 \end{bmatrix} \end{aligned} \tag{4.48}$$

and therefore,

$$R_{Z,\alpha}^{n} = \begin{bmatrix} \cos n\frac{2\pi}{n} & -\sin n\frac{2\pi}{n} & 0 \\ \sin n\frac{2\pi}{n} & \cos n\frac{2\pi}{n} & 0 \\ 0 & 0 & 1 \end{bmatrix} = \begin{bmatrix} 1 & 0 & 0 \\ 0 & 1 & 0 \\ 0 & 0 & 1 \end{bmatrix} \tag{4.49}$$

Repeated rotation about any other global axis provides the same result.

Example 254 ★ Open Problem Consider a body frame B that turns an angle α about the Z-axis followed by a rotation γ about the X-axis such that

$$\alpha = \frac{2\pi}{n_1} \qquad \gamma = \frac{2\pi}{n_2} \qquad \{n_1, n_2\} \in \mathbb{N} \tag{4.50}$$

It seems that we need to repeat the rotations $n = n_1 \times n_2$ times to get back to the original configuration, because we need to multiply ${}^G R_B = R_{X,\gamma}\, R_{Z,\alpha}$ by itself until an identity matrix is achieved. In this case any body point will be mapped to the same point in the global frame. To show this, we may imagine that after n_1 times of combined rotation the configuration with respect to the Z-axis will be the same as the original configuration. However, the body will get the same configuration with respect to the X-axis after n_2 times of combined rotation. So, the body should get back to its original configuration after $n = n_1 \times n_2$ times of combined rotation. However, that is not true for every n_1 and n_2 and for every order of rotation.

As an example, we may check that for $\alpha = 2\pi/4$ and $\gamma = 2\pi/3$ we have

$${}^G R_B = R_{X,\gamma}\, R_{Z,\alpha} = \begin{bmatrix} 0 & -1.0 & 0 \\ -0.5 & 0 & -0.86603 \\ 0.86603 & 0 & -0.5 \end{bmatrix} \tag{4.51}$$

and we need 13 times combined rotation to achieve the original configuration:

$${}^G R_B^{13} = \begin{bmatrix} 0.99979 & -0.01122 & -0.01902 \\ 0.01086 & 0.99979 & -0.019226 \\ 0.019226 & 0.01902 & 0.99967 \end{bmatrix} \approx \mathbf{I} \tag{4.52}$$

If we change the angles to $\alpha = 2\pi/2$ and $\gamma = 2\pi/3$, then we have

$${}^G R_B = R_{X,\gamma}\, R_{Z,\alpha} = \begin{bmatrix} -0.5 & -0.866 & 0 \\ -0.866 & 0.5 & 0 \\ 0 & 0 & -1.0 \end{bmatrix} \tag{4.53}$$

and we need only two combined rotations to achieve the original configuration:

$${}^G R_B^{2} = \begin{bmatrix} 1.0 & 0 & 0 \\ 0 & 1.0 & 0 \\ 0 & 0 & 1.0 \end{bmatrix} = \mathbf{I} \tag{4.54}$$

However, if $n_1 = n_2 = 4$, we only need to apply the combined rotation three times:

$${}^G R_B = R_{X,90}\, R_{Z,90} = \begin{bmatrix} 0 & -1.0 & 0 \\ 0 & 0 & -1.0 \\ 1.0 & 0 & 0 \end{bmatrix} \tag{4.55}$$

$${}^G R_B^{3} = \begin{bmatrix} 1.0 & 0 & 0 \\ 0 & 1.0 & 0 \\ 0 & 0 & 1.0 \end{bmatrix} \tag{4.56}$$

The determination of the required number n to repeat a general combined rotation ${}^{G}R_{B}$ to get back to the original orientation is an open question:

$$ {}^{G}R_{B} = \prod_{j=1}^{m} R_{X_i,\alpha_j} \qquad i = 1, 2, 3 \tag{4.57} $$

$$ \alpha_j = \frac{2\pi}{n_j} \qquad m, n_j \in \mathbb{N} \tag{4.58} $$

$$ {}^{G}R_{B}^{n} = [I] \qquad n = ? \tag{4.59} $$

Example 255 ★ Rotation Matrix and Spherical Trigonometry Any plane passing through the center of a sphere cuts the surface in a circle which is called a *great circle*. Any other plane intersecting the sphere but not passing through the center cuts the surface in a *small circle*. A *spherical triangle* is made up when three great circles intersect. Furthermore, if we are given any three points on the surface of a sphere, we can join them by great-circle arcs to form a spherical triangle.

Consider a vector ${}^{B}\mathbf{r}$ in a body coordinate frame B. The tip point of ${}^{B}\mathbf{r}$ will move on a sphere when the body frame performs multiple rotations. Figure 4.5 depicts a vector at three positions $\mathbf{r}_1$, $\mathbf{r}_2$, and $\mathbf{r}_3$. The vector $\mathbf{r}_1$ indicates the original position of a body point, $\mathbf{r}_2$ indicates the position vector of the body point after a rotation about the Z-axis, and $\mathbf{r}_3$ indicates the position vector of the point after another rotation about the X-axis. However, these rotations may be about any arbitrary axes. The tip point of the vectors $\mathbf{r}_1$, $\mathbf{r}_2$, and $\mathbf{r}_3$ make a spherical triangle that lies on a sphere with radius $R = |\mathbf{r}_1| = |\mathbf{r}_2| = |\mathbf{r}_3|$.

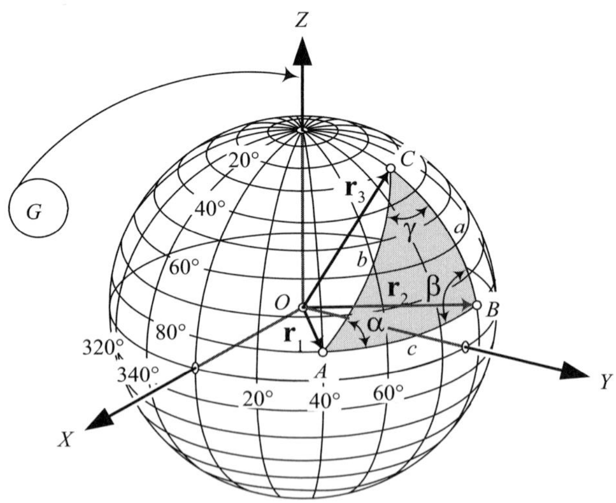

Figure 4.5 Positions of a rotated vector at three positions $\mathbf{r}_1$, $\mathbf{r}_2$, and $\mathbf{r}_3$ and the associated spherical triangle.

Assume $\triangle ABC$ is a triangle on a sphere of unit radius $R = 1$. Let us show the angles of the triangle by α, β, γ and the length of its sides by a,b,c. The arc lengths a,b,c are respectively equal to the plane angles $\angle BOC$, $\angle AOC$, $\angle AOB$ for the unit sphere. The angle α between the planes OAC and OAB is equal to the angle between normal vectors to the planes:

$$\cos\alpha = \frac{(\mathbf{r}_1 \times \mathbf{r}_2) \cdot (\mathbf{r}_1 \times \mathbf{r}_3)}{|\mathbf{r}_1 \times \mathbf{r}_2|\,|\mathbf{r}_1 \times \mathbf{r}_3|} \tag{4.60}$$

However, we have

$$|\mathbf{r}_1 \times \mathbf{r}_2| = \sin c \tag{4.61}$$

$$|\mathbf{r}_1 \times \mathbf{r}_3| = \sin b \tag{4.62}$$

and by using the scalar triple-product equation (1.111) and *bac–cab* rule (1.103), we find

$$\begin{aligned}\cos\alpha &= \frac{\mathbf{r}_1 \cdot [\mathbf{r}_2 \times (\mathbf{r}_1 \times \mathbf{r}_3)]}{\sin b \sin c} = \frac{\mathbf{r}_1 \cdot [\mathbf{r}_1(\mathbf{r}_2 \cdot \mathbf{r}_3) - \mathbf{r}_3(\mathbf{r}_2 \cdot \mathbf{r}_1)]}{\sin b \sin c} \\ &= \frac{\cos a - \cos c \cos b}{\sin b \sin c}\end{aligned} \tag{4.63}$$

Following the same method, we can show the spherical triangles rule:

$$\cos a = \cos b \cos c + \sin b \sin c \cos\alpha \tag{4.64}$$

$$\cos b = \cos c \cos a + \sin c \sin a \cos\beta \tag{4.65}$$

$$\cos c = \cos a \cos b + \sin a \sin b \cos\gamma \tag{4.66}$$

We may also begin with the equation

$$\sin\alpha = \frac{(\mathbf{r}_1 \times \mathbf{r}_2) \times (\mathbf{r}_1 \times \mathbf{r}_3)}{|\mathbf{r}_1 \times \mathbf{r}_2|\,|\mathbf{r}_1 \times \mathbf{r}_3|} \tag{4.67}$$

and show that

$$\frac{\sin\alpha}{\sin a} = \frac{\sin\beta}{\sin b} = \frac{\sin\gamma}{\sin c} \tag{4.68}$$

If the triangle $\triangle ABC$, as is shown in Figure 4.6, is a right triangle on a sphere, then the relationship between the lengths a,b,c and the angles γ and α simplifies to the *Napier rules*:

$$\sin a = \sin b \sin\alpha = \tan c \cot\gamma \tag{4.69}$$

$$\sin c = \sin b \sin\gamma = \tan a \cot\alpha \tag{4.70}$$

$$\cos b = \cos a \cos c = \cos\alpha \cos\gamma \tag{4.71}$$

$$\cos\alpha = \tan c \cot b = \cos a \sin\gamma \tag{4.72}$$

$$\cos\gamma = \cos a \cos c = \cos\alpha \cot\gamma \tag{4.73}$$

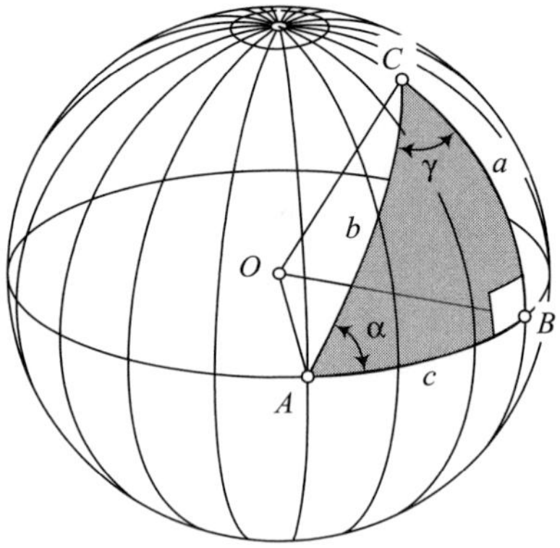

Figure 4.6 A right angle spherical triangle.

4.3 GLOBAL ROLL–PITCH–YAW ANGLES

The rotations about the X-, Y-, and Z-axis of the global coordinate frame are called the *roll*, *pitch*, and *yaw*, respectively. The global *roll–pitch–yaw rotation matrix* is

$$\begin{aligned} {}^{G}R_{B} &= R_{Z,\gamma}\, R_{Y,\beta}\, R_{X,\alpha} \\ &= \begin{bmatrix} c\beta\, c\gamma & -c\alpha\, s\gamma + c\gamma\, s\alpha\, s\beta & s\alpha\, s\gamma + c\alpha\, c\gamma\, s\beta \\ c\beta\, s\gamma & c\alpha\, c\gamma + s\alpha\, s\beta\, s\gamma & -c\gamma\, s\alpha + c\alpha\, s\beta\, s\gamma \\ -s\beta & c\beta\, s\alpha & c\alpha\, c\beta \end{bmatrix} \end{aligned} \tag{4.74}$$

Having the roll, pitch, and yaw angles, we can determine the overall roll–pitch–yaw rotation matrix ${}^{G}R_{B}$ using Equation (4.74). We are also able to find the required roll, pitch, and yaw angles if a rotation matrix is given. Assume that a rotation matrix ${}^{G}R_{B}$ is given and we show its element at row i and column j by r_{ij}. Setting the given matrix equal to (4.74), we may find the roll angle α by

$$\alpha = \tan^{-1}\left(\frac{r_{32}}{r_{33}}\right) \tag{4.75}$$

and the pitch angle β by

$$\beta = -\sin^{-1}(r_{31}) \tag{4.76}$$

and the yaw angle γ by

$$\gamma = \tan^{-1}\left(\frac{r_{21}}{r_{11}}\right) \tag{4.77}$$

provided that $\cos\beta \neq 0$.

Example 256 Determination of Roll–Pitch–Yaw Angles Assume that we would like to determine the required roll–pitch–yaw angles to make the x-axis of the body coordinate B parallel to $\mathbf{u}$ while the y-axis remains in the (X, Y)-plane:

$$\mathbf{u} = \hat{I} + \hat{J} + \hat{K} \tag{4.78}$$

Because the x-axis must be along $\mathbf{u}$, we have

$$^{G}\hat{\imath} = \frac{\mathbf{u}}{|\mathbf{u}|} = \frac{\sqrt{3}}{3}\hat{I} + \frac{\sqrt{3}}{3}\hat{J} + \frac{\sqrt{3}}{3}\hat{K} \tag{4.79}$$

and because the y-axis is in the (X, Y)-plane, we have

$$^{G}\hat{\jmath} = \left(\hat{I}\cdot\hat{\jmath}\right)\hat{I} + \left(\hat{J}\cdot\hat{\jmath}\right)\hat{J} = \cos\theta\hat{I} + \sin\theta\hat{J} \tag{4.80}$$

The dot product of $^{G}\hat{\imath}$ and $^{G}\hat{\jmath}$ must be zero; therefore,

$$\begin{bmatrix}\sqrt{3}/3\\ \sqrt{3}/3\\ \sqrt{3}/3\end{bmatrix}\cdot\begin{bmatrix}\cos\theta\\ \sin\theta\\ 0\end{bmatrix} = 0 \tag{4.81}$$

$$\theta = -45\,\text{deg} \tag{4.82}$$

Then $^{G}\hat{k}$ can be found using the cross product:

$$^{G}\hat{k} = {}^{G}\hat{\imath}\times{}^{G}\hat{\jmath} = \begin{bmatrix}\sqrt{3}/3\\ \sqrt{3}/3\\ \sqrt{3}/3\end{bmatrix}\times\begin{bmatrix}\sqrt{2}/2\\ -\sqrt{2}/2\\ 0\end{bmatrix} = \begin{bmatrix}\sqrt{6}/6\\ \sqrt{6}/6\\ -\sqrt{6}/3\end{bmatrix} \tag{4.83}$$

Hence, the transformation matrix $^{G}R_{B}$ is

$$^{G}R_{B} = \begin{bmatrix}\hat{I}\cdot\hat{\imath} & \hat{I}\cdot\hat{\jmath} & \hat{I}\cdot\hat{k}\\ \hat{J}\cdot\hat{\imath} & \hat{J}\cdot\hat{\jmath} & \hat{J}\cdot\hat{k}\\ \hat{K}\cdot\hat{\imath} & \hat{K}\cdot\hat{\jmath} & \hat{K}\cdot\hat{k}\end{bmatrix} = \begin{bmatrix}\sqrt{3}/3 & \sqrt{2}/2 & \sqrt{6}/6\\ \sqrt{3}/3 & -\sqrt{2}/2 & \sqrt{6}/6\\ \sqrt{3}/3 & 0 & -\sqrt{6}/3\end{bmatrix} \tag{4.84}$$

Now we can determine the required roll–pitch–yaw angles to move the body coordinate frame B from the coincidence orientation with G:

$$\alpha = \tan^{-1}\left(\frac{r_{32}}{r_{33}}\right) = \tan^{-1}\left(\frac{0}{-\sqrt{6}/3}\right) = 0 \tag{4.85}$$

$$\beta = -\sin^{-1}(r_{31}) = -\sin^{-1}\left(\sqrt{3}/3\right) = -0.61548\,\text{rad} \tag{4.86}$$

$$\gamma = \tan^{-1}\left(\frac{r_{21}}{r_{11}}\right) = \tan^{-1}\left(\frac{\sqrt{3}/3}{\sqrt{3}/3}\right) = 0.7854\,\text{rad} \tag{4.87}$$

As a check we may use α, β, γ and determine the matrix (4.74),

$$^{G}R_{B} = R_{Z,\gamma}\,R_{Y,\beta}\,R_{X,\alpha} = \begin{bmatrix}0.57735 & -0.70711 & -0.40825\\ 0.57735 & 0.70711 & -0.40825\\ 0.57735 & 0 & 0.8165\end{bmatrix} \tag{4.88}$$

which is equivalent to (4.84.)

Example 257 Equivalent Roll–Pitch–Yaw Angles Consider a rigid body that turns 45 deg about the Z-axis and then 30 deg about the X-axis. The combined rotation matrix is

$$
\begin{aligned}
{}^{G}R_{B} &= R_{X,30}\ R_{Z,45} \\
&= \begin{bmatrix} 1 & 0 & 0 \\ 0 & \cos\frac{\pi}{6} & -\sin\frac{\pi}{6} \\ 0 & \sin\frac{\pi}{6} & \cos\frac{\pi}{6} \end{bmatrix} \begin{bmatrix} \cos\frac{\pi}{4} & -\sin\frac{\pi}{4} & 0 \\ \sin\frac{\pi}{4} & \cos\frac{\pi}{4} & 0 \\ 0 & 0 & 1 \end{bmatrix} \\
&= \begin{bmatrix} 0.70711 & -0.70711 & 0 \\ 0.61237 & 0.61237 & -0.5 \\ 0.35355 & 0.35355 & 0.86603 \end{bmatrix}
\end{aligned} \tag{4.89}
$$

We may also move the body to the final orientation by a proper roll–pitch–yaw rotation. The required roll angle is

$$
\alpha = \tan^{-1}\frac{r_{32}}{r_{33}} = \tan^{-1}\frac{0.353\,55}{0.866\,03} = 0.387\,59\,\text{rad} \tag{4.90}
$$

and the pitch angle β and yaw angle γ are

$$
\beta = -\sin^{-1} r_{31} = -\sin^{-1} 0.353\,55 = -0.361\,36\,\text{rad} \tag{4.91}
$$

$$
\gamma = \tan^{-1}\frac{r_{21}}{r_{11}} = \tan^{-1}\frac{0.612\,37}{0.707\,11} = 0.713\,72\,\text{rad} \tag{4.92}
$$

We may check that the roll–pitch–yaw combined rotation matrix is the same as (4.89):

$$
\begin{aligned}
{}^{G}R_{B} &= R_{Z,0.713\,72}\ R_{Y,-0.361\,36}\ R_{X,0.387\,59} \\
&= \begin{bmatrix} 0.707\,11 & -0.707\,1 & -2.253\,9\times 10^{-6} \\ 0.612\,37 & 0.612\,38 & -0.499\,99 \\ 0.353\,55 & 0.353\,55 & 0.866\,03 \end{bmatrix}
\end{aligned} \tag{4.93}
$$

Example 258 90 deg Pitch Angle If $\cos\beta = 0$, we have

$$
\beta = 90\,\text{deg} \tag{4.94}
$$

and the roll–pitch–yaw rotation matrix reduces to

$$
\begin{aligned}
{}^{G}R_{B} &= R_{Z,\gamma}\ R_{Y,\beta}\ R_{X,\alpha} \\
&= \begin{bmatrix} 0 & \sin(\alpha-\gamma) & \cos(\alpha-\gamma) \\ 0 & \cos(\alpha-\gamma) & -\sin(\alpha-\gamma) \\ -1 & 0 & 0 \end{bmatrix}
\end{aligned} \tag{4.95}
$$

The angles α and γ are not distinguishable and we have

$$
\alpha - \gamma = \tan^{-1}\frac{r_{12}}{r_{22}} \tag{4.96}
$$

So, the final configuration of the body coordinate frame can be achieved by any combination of α and γ that satisfies Equation (4.96), such as

$$\alpha = 0 \qquad \gamma = -\tan^{-1}\frac{r_{12}}{r_{22}} \tag{4.97}$$

or

$$\gamma = 0 \qquad \alpha = \tan^{-1}\frac{r_{12}}{r_{22}} \tag{4.98}$$

4.4 ROTATION ABOUT LOCAL CARTESIAN AXES

Consider a body coordinate frame *B(Oxyz)* that is coincident with a global coordinate frame *G(OXYZ)*. The origin of both frames is fixed at point O. When B rotates φ about the z-axis of its body coordinate frame, as can be seen in the top view of Figure 4.7, the coordinates of any point P of the rigid body in the local and global coordinate frames are related by

$$^{B}\mathbf{r} = R_{z,\varphi}\,{}^{G}\mathbf{r} \tag{4.99}$$

The vectors $^{G}\mathbf{r}$ and $^{B}\mathbf{r}$ are expressions of the position vectors of the point P in the global and local frames, respectively:

$$^{G}\mathbf{r} = \begin{bmatrix} X \\ Y \\ Z \end{bmatrix} \qquad {}^{B}\mathbf{r} = \begin{bmatrix} x \\ y \\ z \end{bmatrix} \tag{4.100}$$

and $R_{z,\varphi}$ is the *z-rotation matrix*

$$R_{z,\varphi} = \begin{bmatrix} \cos\varphi & \sin\varphi & 0 \\ -\sin\varphi & \cos\varphi & 0 \\ 0 & 0 & 1 \end{bmatrix} \tag{4.101}$$

Similarly, rotation θ about the y-axis and rotation ψ about the x-axis are defined by the *y-rotation matrix* $R_{y,\theta}$ and the *x-rotation matrix* $R_{x,\psi}$, respectively:

$$R_{y,\theta} = \begin{bmatrix} \cos\theta & 0 & -\sin\theta \\ 0 & 1 & 0 \\ \sin\theta & 0 & \cos\theta \end{bmatrix} \tag{4.102}$$

$$R_{x,\psi} = \begin{bmatrix} 1 & 0 & 0 \\ 0 & \cos\psi & \sin\psi \\ 0 & -\sin\psi & \cos\psi \end{bmatrix} \tag{4.103}$$

Proof: Figure 4.7 illustrates the top view of a body coordinate frame *B(Oxyz)* that is rotated φ about the z-axis in a global frame *G(OXYZ)*. Using the unit vectors $(\hat{\imath}, \hat{\jmath}, \hat{k})$ and $(\hat{I}, \hat{J}, \hat{K})$ along the axes of *B(Oxyz)* and *G(OXYZ)*, respectively, the position vector

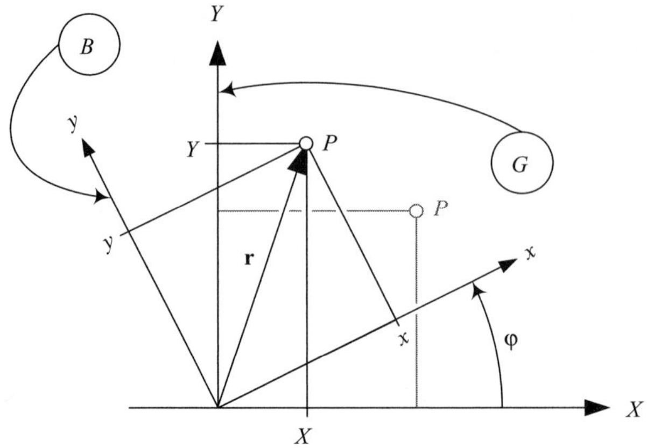

Figure 4.7 Position vectors of point P before and after rotation of the local frame about the z-axis of the local frame.

${}^{B}\mathbf{r}$ of a point P in $B(Oxyz)$ and the final global position vector ${}^{G}\mathbf{r}$ of P after the rotation φ are

$$ {}^{B}\mathbf{r} = x\hat{\imath} + y\hat{\jmath} + z\hat{k} \tag{4.104} $$

$$ {}^{G}\mathbf{r} = X\hat{I} + Y\hat{J} + Z\hat{K} \tag{4.105} $$

Assume that the global vector ${}^{G}\mathbf{r}$ is given; then the components of ${}^{B}\mathbf{r}$ can be found by using the inner product and orthogonality condition:

$$ x = \hat{\imath} \cdot \mathbf{r} = \hat{\imath} \cdot X\hat{I} + \hat{\imath} \cdot Y\hat{J} + \hat{\imath} \cdot Z\hat{K} \tag{4.106} $$

$$ y = \hat{\jmath} \cdot \mathbf{r} = \hat{\jmath} \cdot X\hat{I} + \hat{\jmath} \cdot Y\hat{J} + \hat{\jmath} \cdot Z\hat{K} \tag{4.107} $$

$$ z = \hat{k} \cdot \mathbf{r} = \hat{k} \cdot X\hat{I} + \hat{k} \cdot Y\hat{J} + \hat{k} \cdot Z\hat{K} \tag{4.108} $$

It can be set in a matrix equation:

$$ {}^{B}\mathbf{r} = R_{z,\varphi}\, {}^{G}\mathbf{r} \tag{4.109} $$

$$ \begin{bmatrix} x_2 \\ y_2 \\ z_2 \end{bmatrix} = \begin{bmatrix} \hat{\imath} \cdot \hat{I} & \hat{\imath} \cdot \hat{J} & \hat{\imath} \cdot \hat{K} \\ \hat{\jmath} \cdot \hat{I} & \hat{\jmath} \cdot \hat{J} & \hat{\jmath} \cdot \hat{K} \\ \hat{k} \cdot \hat{I} & \hat{k} \cdot \hat{J} & \hat{k} \cdot \hat{K} \end{bmatrix} \begin{bmatrix} X_2 \\ Y_2 \\ Z_2 \end{bmatrix} \tag{4.110} $$

The elements of the z-rotation matrix $R_{z,\varphi}$ are the *direction cosines* of the axes of $B(Oxyz)$ in $G(OXYZ)$. Figure 4.7 shows that these elements are

$$ \begin{aligned} \hat{\imath} \cdot \hat{I} &= \cos\varphi & \hat{\imath} \cdot \hat{J} &= \sin\varphi & \hat{\imath} \cdot \hat{K} &= 0 \\ \hat{\jmath} \cdot \hat{I} &= -\sin\varphi & \hat{\jmath} \cdot \hat{J} &= \cos\varphi & \hat{\jmath} \cdot \hat{K} &= 0 \\ \hat{k} \cdot \hat{I} &= 0 & \hat{k} \cdot \hat{J} &= 0 & \hat{k} \cdot \hat{K} &= 1 \end{aligned} \tag{4.111} $$

Now Equation (4.110) simplifies to

$$\begin{bmatrix} x \\ y \\ z \end{bmatrix} = \begin{bmatrix} \cos\varphi & \sin\varphi & 0 \\ -\sin\varphi & \cos\varphi & 0 \\ 0 & 0 & 1 \end{bmatrix} \begin{bmatrix} X \\ Y \\ Z \end{bmatrix} \tag{4.112}$$

and we can find the components of ${}^{B}\mathbf{r}$ by multiplying z-rotation matrix $R_{z,\varphi}$ and vector ${}^{G}\mathbf{r}$.

Equation (4.109) states that after rotation about the z-axis of the local coordinate frame the expression of the position vector in the local frame is equal to $R_{z,\varphi}$ times the position vector in the global frame. Therefore, we are able to determine the coordinates of any point of a rigid body in the local coordinate frame if we have its coordinates in the global frame.

Similarly, rotation θ about the y-axis and rotation ψ about the x-axis are described by the y-rotation matrix $R_{y,\theta}$ and the x-rotation matrix $R_{x,\psi}$, respectively. The rotation matrices

$$R_{z,\varphi} = \begin{bmatrix} \cos\varphi & \sin\varphi & 0 \\ -\sin\varphi & \cos\varphi & 0 \\ 0 & 0 & 1 \end{bmatrix} \tag{4.113}$$

$$R_{y,\theta} = \begin{bmatrix} \cos\theta & 0 & -\sin\theta \\ 0 & 1 & 0 \\ \sin\theta & 0 & \cos\theta \end{bmatrix} \tag{4.114}$$

$$R_{x,\psi} = \begin{bmatrix} 1 & 0 & 0 \\ 0 & \cos\psi & \sin\psi \\ 0 & -\sin\psi & \cos\psi \end{bmatrix} \tag{4.115}$$

are called the *basic local rotation matrices* and all transform a G-expression vector to its B-expression. We show such a rotation matrix by ${}^{B}R_{G}$ to indicate that it is a transformation matrix from the G to the B coordinate frame.

We usually indicate the first, second, and third rotations about the local axes by φ, θ, and ψ, respectively. ■

Example 259 Local Rotation, Local Position Assume that a local coordinate frame *B(Oxyz)* has been rotated 60 deg about the z-axis and a point P in the global coordinate frame *G (OXYZ)* is at (3,4,2). Then its coordinates in the local coordinate frame *B (Oxyz)* are

$$\begin{bmatrix} x \\ y \\ z \end{bmatrix} = \begin{bmatrix} \cos\frac{\pi}{3} & \sin\frac{\pi}{3} & 0 \\ -\sin\frac{\pi}{3} & \cos\frac{\pi}{3} & 0 \\ 0 & 0 & 1 \end{bmatrix} \begin{bmatrix} 3 \\ 4 \\ 2 \end{bmatrix} = \begin{bmatrix} 4.9641 \\ -0.59808 \\ 2 \end{bmatrix} \tag{4.116}$$

Example 260 Local Rotation, Global Position A local coordinate frame $B(Oxyz)$ has been rotated 60 deg about the z-axis and a point P in the local coordinate frame is at (3,4,2). Then its position in the global coordinate frame $G(OXYZ)$ is at

$$\begin{bmatrix} X \\ Y \\ Z \end{bmatrix} = \begin{bmatrix} \cos\frac{\pi}{3} & \sin\frac{\pi}{3} & 0 \\ -\sin\frac{\pi}{3} & \cos\frac{\pi}{3} & 0 \\ 0 & 0 & 1 \end{bmatrix}^{\mathrm{T}} \begin{bmatrix} 3 \\ 4 \\ 2 \end{bmatrix} = \begin{bmatrix} -1.9641 \\ 4.5981 \\ 2 \end{bmatrix} \tag{4.117}$$

Example 261 Time-Dependent Local Rotation Assume a rigid body is continuously turning about the X-axis at a rate of 0.2 rad/s. The rotation matrix of the body is

$${}^{B}R_{G} = \begin{bmatrix} 1 & 0 & 0 \\ 0 & \cos 0.2t & \sin 0.2t \\ 0 & -\sin 0.2t & \cos 0.2t \end{bmatrix} \tag{4.118}$$

Consider a point P at ${}^{B}\mathbf{r} = [0, 10, 0]^{\mathrm{T}}$. After $t = 1$ s, the point will globally be seen at

$$\begin{bmatrix} X \\ Y \\ Z \end{bmatrix} = \begin{bmatrix} 1 & 0 & 0 \\ 0 & \cos 0.2 & \sin 0.2 \\ 0 & -\sin 0.2 & \cos 0.2 \end{bmatrix}^{\mathrm{T}} \begin{bmatrix} 0 \\ 10 \\ 0 \end{bmatrix} = \begin{bmatrix} 0 \\ 9.8 \\ 1.986 \end{bmatrix} \tag{4.119}$$

Because B is turning about the x-axis, every point of B moves in a circular fashion on planes perpendicular to the x-axis. To determine the required time t_0 to move P a complete circle, we may solve the following equations for $t = t_0/4$ and determine t_0:

$$\begin{bmatrix} 0 \\ 10 \\ 0 \end{bmatrix} = \begin{bmatrix} 1 & 0 & 0 \\ 0 & \cos 0.2t & \sin 0.2t \\ 0 & -\sin 0.2t & \cos 0.2t \end{bmatrix} \begin{bmatrix} 0 \\ 0 \\ 10 \end{bmatrix} = \begin{bmatrix} 0 \\ 10\sin 0.2t \\ 10\cos 0.2t \end{bmatrix} \tag{4.120}$$

$$10 = 10\sin 0.2t \tag{4.121}$$

$$t = \frac{\pi/2}{0.2} = 7.854 \text{ s} \tag{4.122}$$

$$t_0 = 4t = 31.416 \text{ s} \tag{4.123}$$

4.5 SUCCESSIVE ROTATIONS ABOUT LOCAL AXES

Consider a rigid body B with a local coordinate frame $B(Oxyz)$ that does a series of sequential rotations $R_1, R_2, R_3, \ldots, R_n$ about the local axes. Having the final global position vector ${}^{G}\mathbf{r}$ of a body point P, we can determine its local position vector ${}^{B}\mathbf{r}$ by

$${}^{B}\mathbf{r} = {}^{B}R_{G}\, {}^{G}\mathbf{r} \tag{4.124}$$

where

$$^B R_G = R_n \ldots R_3 R_2 R_1. \tag{4.125}$$

The matrix $^B R_G$ is called the *local rotation matrix* and it maps the global coordinates of body points to their local coordinates.

Proof: Assume that the body coordinate frame B was initially coincident with the global coordinate frame G. The rigid body rotates about a local axis, and a local rotation matrix R_1 relates the global coordinates of a body point to the associated local coordinates:

$$^B \mathbf{r} = R_1 \, ^G \mathbf{r} \tag{4.126}$$

If we introduce an intermediate space-fixed frame G_1 coincident with the new position of the body coordinate frame, then

$$^{G_1} \mathbf{r} \equiv \, ^B \mathbf{r} \tag{4.127}$$

and we may give the rigid body a second rotation about a local coordinate axis. Now another proper local rotation matrix R_2 relates the coordinates in the intermediate fixed frame to the corresponding local coordinates:

$$^B \mathbf{r} = R_2 \, ^{G_1} \mathbf{r} \tag{4.128}$$

Hence, to relate the final coordinates of the point, we must first transform its global coordinates to the intermediate fixed frame and then transform them to the original body frame. Substituting (4.126) in (4.128) shows that

$$^B \mathbf{r} = R_2 \, R_1 \, ^G \mathbf{r} \tag{4.129}$$

Following the same procedure we can determine the final global position of a body point after a series of sequential rotations $R_1, R_2, R_3, \ldots, R_n$ as (4.125).

Rotation about the local coordinate axes is conceptually interesting. This is because in a sequence of rotations each rotation is about one of the axes of the local coordinate frame, which has been moved to its new global position during the last rotation. ■

Example 262 Successive Local Rotation, Local Position Consider a rigid body with a local coordinate frame *B(Oxyz)* that is initially coincident with a global coordinate frame *G(OXYZ)*. The body rotates $\varphi = 30$ deg about the z-axis followed by a rotation $\theta = 30$ deg about the x-axis and then $\psi = 30$ deg about the y-axis. The global coordinates of a point P are $X = 5$, $Y = 3$, $Z = 1$. Its local position vector is

$$\begin{bmatrix} x \\ y \\ z \end{bmatrix} = \, ^B R_G \begin{bmatrix} 5 \\ 3 \\ 1 \end{bmatrix} \tag{4.130}$$

where

$$^B R_G = R_{y,30} R_{x,30} R_{z,30} = \begin{bmatrix} 0.63 & 0.65 & -0.43 \\ -0.43 & 0.75 & 0.50 \\ 0.65 & -0.125 & 0.75 \end{bmatrix} \tag{4.131}$$

and therefore,

$$\begin{bmatrix} x \\ y \\ z \end{bmatrix} = \begin{bmatrix} 0.63 & 0.65 & -0.43 \\ -0.43 & 0.75 & 0.50 \\ 0.65 & -0.125 & 0.75 \end{bmatrix} \begin{bmatrix} 5 \\ 3 \\ 1 \end{bmatrix} = \begin{bmatrix} 4.67 \\ 0.6 \\ 3.625 \end{bmatrix} \tag{4.132}$$

Example 263 Twelve Independent Triple Local Rotations Euler proved a theorem that stated: Any two independent orthogonal coordinate frames can be transformed to each other by a sequence of no more than three rotations about the local coordinate axes provided that no two successive rotations are about the same axis. In general, there are 12 different independent combinations of triple rotation about the local axes:

$$\begin{aligned}
&1.\ R_{x,\psi}\,R_{y,\theta}\,R_{z,\varphi}\\
&2.\ R_{y,\psi}\,R_{z,\theta}\,R_{x,\varphi}\\
&3.\ R_{z,\psi}\,R_{x,\theta}\,R_{y,\varphi}\\
&4.\ R_{z,\psi}\,R_{y,\theta}\,R_{x,\varphi}\\
&5.\ R_{y,\psi}\,R_{x,\theta}\,R_{z,\varphi}\\
&6.\ R_{x,\psi}\,R_{z,\theta}\,R_{y,\varphi}\\
&7.\ R_{x,\psi}\,R_{y,\theta}\,R_{x,\varphi}\\
&8.\ R_{y,\psi}\,R_{z,\theta}\,R_{y,\varphi}\\
&9.\ R_{z,\psi}\,R_{x,\theta}\,R_{z,\varphi}\\
&10.\ R_{x,\psi}\,R_{z,\theta}\,R_{x,\varphi}\\
&11.\ R_{y,\psi}\,R_{x,\theta}\,R_{y,\varphi}\\
&12.\ R_{z,\psi}\,R_{y,\theta}\,R_{z,\varphi}
\end{aligned} \tag{4.133}$$

The expanded forms of the 12 local-axis triple rotations are presented in Appendix B.

Example 264 ★ Repeated Rotation about Body Axes Consider a body frame B that turns φ about the z-axis. If

$$\varphi = \frac{2\pi}{n} \qquad n \in \mathbb{N} \tag{4.134}$$

then we need to repeat the rotation n times to get back to the original configuration. We can check it by multiplying $R_{z,\varphi}$ by itself until an identity matrix is achieved. In this case any global point will be mapped to the same point in the body frame. To show this, we may find that

$$R_{z,\varphi}^{m} = \begin{bmatrix} \cos\varphi & \sin\varphi & 0 \\ -\sin\varphi & \cos\varphi & 0 \\ 0 & 0 & 1 \end{bmatrix}^{m} = \begin{bmatrix} \cos\frac{2\pi}{n} & \sin\frac{2\pi}{n} & 0 \\ -\sin\frac{2\pi}{n} & \cos\frac{2\pi}{n} & 0 \\ 0 & 0 & 1 \end{bmatrix}^{m}$$

$$= \begin{bmatrix} \cos m\frac{2\pi}{n} & \sin m\frac{2\pi}{n} & 0 \\ -\sin m\frac{2\pi}{n} & \cos m\frac{2\pi}{n} & 0 \\ 0 & 0 & 1 \end{bmatrix} \tag{4.135}$$

and therefore

$$R_{z,\varphi}^{n} = \begin{bmatrix} \cos n\frac{2\pi}{n} & \sin n\frac{2\pi}{n} & 0 \\ -\sin n\frac{2\pi}{n} & \cos n\frac{2\pi}{n} & 0 \\ 0 & 0 & 1 \end{bmatrix} = \begin{bmatrix} 1 & 0 & 0 \\ 0 & 1 & 0 \\ 0 & 0 & 1 \end{bmatrix} \tag{4.136}$$

Repeated rotation about any other body axis provides the same result.

Example 265 ★ Open Problem Consider a body frame B that turns φ about the x_i-axis followed by a rotation θ about the x_j-axis and ψ about the x_k-axis such that

$$\varphi = \frac{2\pi}{n_1} \qquad \theta = \frac{2\pi}{n_2} \qquad \psi = \frac{2\pi}{n_3} \qquad \{n_1, n_2, n_3\} \in \mathbb{N} \tag{4.137}$$

Although it seems that we need to repeat the rotations $n = n_1 \times n_2 \times n_3$ times to get back to the original configuration, it is not true in general. The determination of the required number n to repeat a general combined rotation ${}^{B}R_{G}$ to get back to the original orientation is an open question:

$${}^{B}R_{G} = \prod_{j=1}^{m} R_{x_i,\varphi_j} \qquad i = 1, 2, 3 \tag{4.138}$$

$$\varphi_j = \frac{2\pi}{n_j} \qquad m, n_j \in \mathbb{N} \tag{4.139}$$

$${}^{B}R_{G}^{n} = [I] \qquad n = ? \tag{4.140}$$

4.6 EULER ANGLES

The rotation about the Z-axis of the global coordinate frame is called the *precession angle*, the rotation about the x-axis of the local coordinate frame is called the *nutation angle*, and the rotation about the z-axis of the local coordinate frame is called the *spin angle*. The *precession–nutation–spin rotation* angles are also called *Euler angles*. The kinematics and dynamics of rigid bodies have simpler expression based on Euler angles.

The Euler angle rotation matrix ${}^{B}R_{G}$ to transform a position vector from *G(OXYZ)* to *B (Oxyz)*,

$$ {}^{B}\mathbf{r} = {}^{B}R_{G}\,{}^{G}\mathbf{r} \tag{4.141}$$

is

$$\begin{aligned} {}^{B}R_{G} &= R_{z,\psi}\,R_{x,\theta}\,R_{z,\varphi} \\ &= \begin{bmatrix} c\varphi c\psi - c\theta s\varphi s\psi & c\psi s\varphi + c\theta c\varphi s\psi & s\theta s\psi \\ -c\varphi s\psi - c\theta c\psi s\varphi & -s\varphi s\psi + c\theta c\varphi c\psi & s\theta c\psi \\ s\theta s\varphi & -c\varphi s\theta & c\theta \end{bmatrix} \end{aligned} \tag{4.142}$$

So the Euler angle rotation matrix ${}^{G}R_{B}$ to transform a position vector from *B(Oxyz)* to *G(OXYZ)*,

$$ {}^{G}\mathbf{r} = {}^{G}R_{B}\,{}^{B}\mathbf{r} \tag{4.143}$$

is

$$\begin{aligned} {}^{G}R_{B} &= {}^{B}R_{G}^{-1} = {}^{B}R_{G}^{\mathrm{T}} = \left[R_{z,\psi}\,R_{x,\theta}\,R_{z,\varphi}\right]^{\mathrm{T}} \\ &= \begin{bmatrix} c\varphi c\psi - c\theta s\varphi s\psi & -c\varphi s\psi - c\theta c\psi s\varphi & s\theta s\varphi \\ c\psi s\varphi + c\theta c\varphi s\psi & -s\varphi s\psi + c\theta c\varphi c\psi & -c\varphi s\theta \\ s\theta s\psi & s\theta c\psi & c\theta \end{bmatrix} \end{aligned} \tag{4.144}$$

Proof: To find the Euler angle rotation matrix ${}^{B}R_{G}$ for transforming from the global frame *G(OXYZ)* to the final body frame $B(Oxyz)$, we employ a body frame $B'(Ox'y'z')$ as shown in Figure 4.8 that before the first rotation coincides with the global frame. The first rotation φ about the Z-axis relates the coordinate systems by

$$ {}^{G}\mathbf{r} = {}^{G}R_{B'}\,{}^{B'}\mathbf{r} \tag{4.145}$$

where ${}^{G}R_{B'}$ is given in (4.3). Using matrix inversion, we may find the body position for a given global position vector:

$$ {}^{B'}\mathbf{r} = {}^{G}R_{B'}^{-1}\,{}^{G}\mathbf{r} = {}^{B'}R_{G}\,{}^{G}\mathbf{r} \tag{4.146}$$

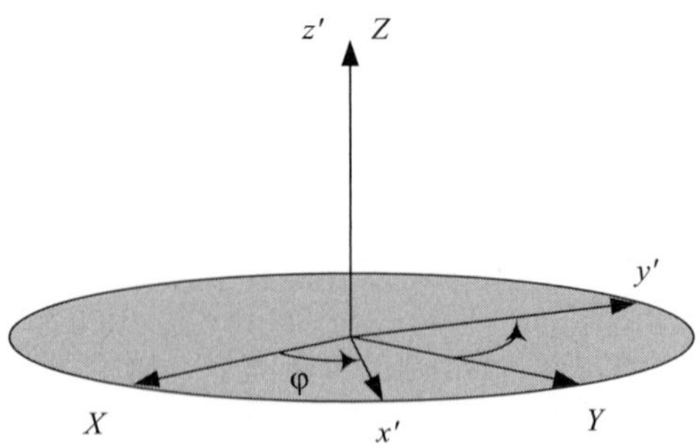

Figure 4.8 First Euler angle.

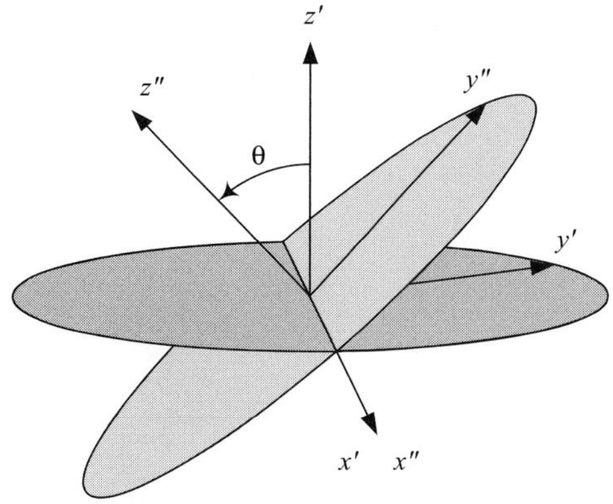

Figure 4.9 Second Euler angle.

Therefore, the first rotation is equivalent to a rotation about the local z-axis by looking for ${}^{B'}\mathbf{r}$:

$$ {}^{B'}\mathbf{r} = {}^{B'}R_G \, {}^G\mathbf{r} \tag{4.147} $$

$$ {}^{B'}R_G = {}^G R_{B'}^{-1} = R_{z,\varphi} = \begin{bmatrix} \cos\varphi & \sin\varphi & 0 \\ -\sin\varphi & \cos\varphi & 0 \\ 0 & 0 & 1 \end{bmatrix} \tag{4.148} $$

Now we consider the $B'(Ox'y'z')$ frame as a new fixed global frame and introduce a new body frame $B''(Ox''y''z'')$ that before the second rotation is coincident with $B'(Ox'y'z')$. Then, we perform a θ rotation about the x''-axis as shown in Figure 4.9. The transformation between $B'(Ox'y'z')$ and $B''(Ox''y''z'')$ is

$$ {}^{B''}\mathbf{r} = {}^{B''}R_{B'} \, {}^{B'}\mathbf{r} \tag{4.149} $$

$$ {}^{B''}R_{B'} = R_{x,\theta} = \begin{bmatrix} 1 & 0 & 0 \\ 0 & \cos\theta & \sin\theta \\ 0 & -\sin\theta & \cos\theta \end{bmatrix} \tag{4.150} $$

Finally we consider $B''(Ox''y''z'')$ as a new fixed global frame that before the third rotation is coincident with a body frame $B(Oxyz)$. The third rotation ψ about the z''-axis is shown in Figure 4.10. The transformation between $B''(Ox''y''z'')$ and $B(Oxyz)$ is

$$ {}^{B}\mathbf{r} = {}^{B}R_{B''} \, {}^{B''}\mathbf{r} \tag{4.151} $$

$$ {}^{B}R_{B''} = R_{z,\psi} = \begin{bmatrix} \cos\psi & \sin\psi & 0 \\ -\sin\psi & \cos\psi & 0 \\ 0 & 0 & 1 \end{bmatrix} \tag{4.152} $$

Having the angles of precession φ, nutation θ, and spin ψ, we can calculate the overall rotation matrix

$$ {}^{B}R_G = R_{z,\psi} R_{x,\theta} R_{z,\varphi} \tag{4.153} $$

as given in Equation (4.142).

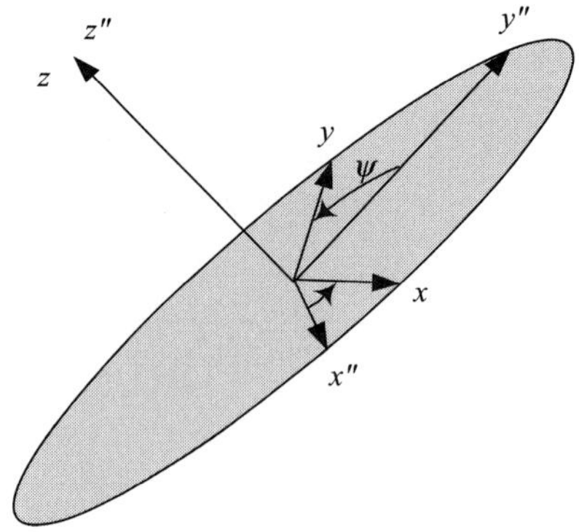

Figure 4.10 Third Euler angle.

When an Euler rotation matrix ${}^{B}R_{G}$ is given, we may calculate the equivalent precession, nutation, and spin angles from

$$\theta = \cos^{-1}(r_{33}) \tag{4.154}$$

$$\varphi = -\tan^{-1}\left(\frac{r_{31}}{r_{32}}\right) \tag{4.155}$$

$$\psi = \tan^{-1}\left(\frac{r_{13}}{r_{23}}\right) \tag{4.156}$$

provided that $\sin\theta \neq 0$. In these equations, r_{ij} indicates the element of row i and column j of the precession–nutation–spin rotation matrix (4.142). ■

Example 266 Euler Angle Rotation Matrix The Euler or precession–nutation–spin rotation matrix for $\varphi = \pi/3$, $\theta = \pi/4$, and $\psi = \pi/6$ would be found by substituting φ, θ, and ψ in Equation (4.142):

$$\begin{aligned} {}^{B}R_{G} &= R_{z,\pi/6}R_{x,\pi/4}R_{z,\pi/3} \\ &= \begin{bmatrix} 0.12683 & 0.92678 & 0.35355 \\ -0.78033 & -0.12683 & 0.61237 \\ 0.61237 & -0.35355 & 0.70711 \end{bmatrix} \end{aligned} \tag{4.157}$$

Example 267 Euler Angles of a Local Rotation Matrix The local rotation matrix after rotating 30 deg about the z-axis, then 30 deg about the x-axis, and then 30 deg about the y-axis is

$${}^{B}R_{G} = R_{y,30}R_{x,30}R_{z,30} = \begin{bmatrix} 0.63 & 0.65 & -0.43 \\ -0.43 & 0.75 & 0.50 \\ 0.65 & -0.125 & 0.75 \end{bmatrix} \tag{4.158}$$

and therefore, the local coordinates of a sample point at $X = 5$, $Y = 30$, $Z = 10$ are

$$\begin{bmatrix} x \\ y \\ z \end{bmatrix} = \begin{bmatrix} 0.63 & 0.65 & -0.43 \\ -0.43 & 0.75 & 0.50 \\ 0.65 & -0.125 & 0.75 \end{bmatrix} \begin{bmatrix} 5 \\ 30 \\ 10 \end{bmatrix} = \begin{bmatrix} 18.35 \\ 25.35 \\ 7 \end{bmatrix} \tag{4.159}$$

The Euler angles of the corresponding precession–nutation–spin rotation matrix are

$$\begin{aligned} \theta &= \cos^{-1}(0.75) = 41.41 \text{ deg} \\ \varphi &= -\tan^{-1}\left(\frac{0.65}{-0.125}\right) = 79.15 \text{ deg} \\ \psi &= \tan^{-1}\left(\frac{-0.43}{0.50}\right) = -40.7 \text{ deg} \end{aligned} \tag{4.160}$$

Hence, $R_{y,30}R_{x,30}R_{z,30} = R_{z,\psi}R_{x,\theta}R_{z,\varphi}$ when $\varphi = 79.15$ deg, $\theta = 41.41$ deg and $\psi = -40.7$ deg. In other words, the rigid body moves to the final configuration by undergoing either three consecutive rotations $\varphi = 79.15$ deg, $\theta = 41.41$ deg, and $\psi = -40.7$ deg about the z, x, and z axes, respectively, or three consecutive rotations 30 deg, 30 deg, and 30 deg about the z, x, and y axes.

Example 268 Relative Rotation Matrix of Two Bodies Consider a body coordinate frame B_1 with a rotation matrix ${}^{B_1}R_G$ made by Euler angles $\varphi = 30$ deg, $\theta = -45$ deg, $\psi = 60$ deg and another body frame B_2 with a rotation matrix ${}^{B_2}R_G$ having $\varphi = 10$ deg, $\theta = 25$ deg, $\psi = -15$ deg.

Using matrix multiplication, we can find a matrix to transform a vector from B_2 to B_1. To find the relative rotation matrix ${}^{B_1}R_{B_2}$ to map the coordinates of the second body frame B_2 to the first body frame B_1, we find the individual rotation matrices first:

$$\begin{aligned} {}^{B_1}R_G &= R_{z,60}R_{x,-45}R_{z,30} \\ &= \begin{bmatrix} 0.127 & 0.78 & -0.612 \\ -0.927 & -0.127 & -0.354 \\ -0.354 & 0.612 & 0.707 \end{bmatrix} \end{aligned} \tag{4.161}$$

$$\begin{aligned} {}^{B_2}R_G &= R_{z,10}R_{x,25}R_{z,-15} \\ &= \begin{bmatrix} 0.992 & -6.33 \times 10^{-2} & -0.109 \\ 0.103 & 0.907 & 0.408 \\ 7.34 \times 10^{-2} & -0.416 & 0.906 \end{bmatrix} \end{aligned} \tag{4.162}$$

The desired rotation matrix ${}^{B_1}R_{B_2}$ can be found by

$${}^{B_1}R_{B_2} = {}^{B_1}R_G\,{}^{G}R_{B_2} \tag{4.163}$$

which is equal to

$$\begin{aligned} {}^{B_1}R_{B_2} &= {}^{B_1}R_G\,{}^{B_2}R_G^{\mathrm{T}} \\ &= \begin{bmatrix} 0.992 & 0.103 & 7.34 \times 10^{-2} \\ -6.33 \times 10^{-2} & 0.907 & -0.416 \\ -0.109 & 0.408 & 0.906 \end{bmatrix} \end{aligned} \tag{4.164}$$

Example 269 Euler Angle Rotation Matrix for Small Angles The Euler rotation matrix ${}^BR_G = R_{z,\psi} R_{x,\theta} R_{z,\varphi}$ for very small Euler angles φ, θ, and ψ is approximated by

$$
{}^BR_G = \begin{bmatrix} 1 & \gamma & 0 \\ -\gamma & 1 & \theta \\ 0 & -\theta & 1 \end{bmatrix} \tag{4.165}
$$

where

$$
\gamma = \varphi + \psi \tag{4.166}
$$

Therefore, for small angles of rotation, the angles φ and ψ are indistinguishable and θ is not important.

When $\theta \to 0$ while φ and ψ are finite, the Euler rotation matrix ${}^BR_G = R_{z,\psi} R_{x,\theta} R_{z,\varphi}$ approaches

$$
\begin{aligned}
{}^BR_G &= \begin{bmatrix} c\varphi c\psi - s\varphi s\psi & c\psi s\varphi + c\varphi s\psi & 0 \\ -c\varphi s\psi - c\psi s\varphi & -s\varphi s\psi + c\varphi c\psi & 0 \\ 0 & 0 & 1 \end{bmatrix} \\
&= \begin{bmatrix} \cos(\varphi + \psi) & \sin(\varphi + \psi) & 0 \\ -\sin(\varphi + \psi) & \cos(\varphi + \psi) & 0 \\ 0 & 0 & 1 \end{bmatrix}
\end{aligned} \tag{4.167}
$$

and therefore, the angles φ and ψ are indistinguishable. Hence, the Euler angles φ, θ, and ψ in rotation matrix (4.142) are not unique when $\theta = 0$. Because of this, the Euler rotation matrix is said to be *singular* at $\theta = 0$.

Example 270 Euler Angle Application in Motion of Rigid Bodies The zxz Euler angles seem to be natural parameters to express the configuration of rotating rigid bodies with a fixed point. Euler angles that show the expression of a top are shown in Figure 4.11 as an example. The rotation of the top about its axis of symmetry is the spin ψ. The angle between the axis of symmetry and the Z-axis is the nutation θ, and the rotation of the axis of symmetry about the Z-axis is the precession φ.

Example 271 Second Type of Euler Rotation Matrix We may call the Euler rotation matrix (4.142) a local 3–1–3 matrix to indicate the order of rotations about $z = x_3$, $x = x_1$, $z = x_3$. Because the indices i, j, and k can be interchanged, the rotation matrices 1–2–1 and 2–3–2 are also mathematically equivalent to the first type of Euler rotation matrix (4.142).

We may change the middle rotation axis from x to y and develop the second type of Euler rotation matrix:

$$
\begin{aligned}
{}^BR_G &= R_{z,\psi} R_{y,\theta} R_{z,\varphi} \\
&= \begin{bmatrix} c\theta c\psi c\varphi - s\psi s\varphi & c\varphi s\psi + c\theta c\psi s\varphi & -c\psi s\theta \\ -c\psi s\varphi - c\theta c\varphi s\psi & c\psi c\varphi - c\theta s\psi s\varphi & s\theta s\psi \\ c\varphi s\theta & s\theta s\varphi & c\theta \end{bmatrix}
\end{aligned} \tag{4.168}
$$

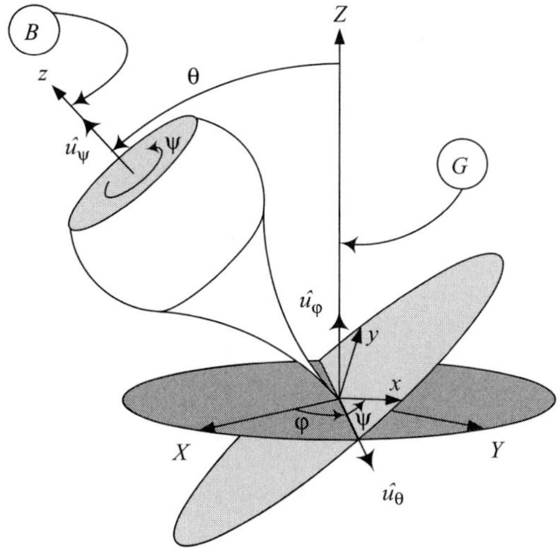

Figure 4.11 Application of Euler angles in describing the configuration of a top.

This is the local 3–2–3 matrix, which is mathematically equivalent to 1–3–1 and 2–1–2 rotations. Therefore, six triple rotations of the 12 combinations in (4.133) are Euler rotation matrices:

$$\begin{array}{lll} 7. & R_{x,\psi}\, R_{y,\theta}\, R_{x,\varphi} & 1\text{–}2\text{–}1 \\ 8. & R_{y,\psi}\, R_{z,\theta}\, R_{y,\varphi} & 2\text{–}3\text{–}2 \\ 9. & R_{z,\psi}\, R_{x,\theta}\, R_{z,\varphi} & 3\text{–}1\text{–}3 \\ 10. & R_{x,\psi}\, R_{z,\theta}\, R_{x,\varphi} & 1\text{–}3\text{–}1 \\ 11. & R_{y,\psi}\, R_{x,\theta}\, R_{y,\varphi} & 2\text{–}1\text{–}2 \\ 12. & R_{z,\psi}\, R_{y,\theta}\, R_{z,\varphi} & 3\text{–}2\text{–}3 \end{array} \tag{4.169}$$

Example 272 ★ Angular Velocity Vector in Terms of Euler Frequencies We may define an Eulerian local frame $E\left(O, \hat{u}_\varphi, \hat{u}_\theta, \hat{u}_\psi\right)$ by introducing the unit vectors $\hat{u}_\varphi$, $\hat{u}_\theta$, and $\hat{u}_\psi$ as are shown in Figure 4.12. The Eulerian frame is not necessarily orthogonal; however, it is very useful in rigid-body kinematic analysis.

The angular velocity vector ${}_G\omega_B$ of the body frame *B(Oxyz)* with respect to the global frame *G(OXYZ)* can be written in the Euler frame *E* as the sum of three Euler angle rate vectors:

$${}_G^E\omega_B = \dot{\varphi}\hat{u}_\varphi + \dot{\theta}\hat{u}_\theta + \dot{\psi}\hat{u}_\psi \tag{4.170}$$

The rates of the Euler angles, $\dot{\varphi}$, $\dot{\theta}$, and $\dot{\psi}$ are called *Euler frequencies*. The angular speeds $\dot{\varphi}$, $\dot{\theta}$, and $\dot{\psi}$ are also called *precession*, *nutation*, and *spin*, respectively.

To find ${}_G\omega_B$, we define the unit vectors $\hat{u}_\varphi$, $\hat{u}_\theta$, and $\hat{u}_\psi$ along the axes of the Euler angles, as shown in Figure 4.12, and express them in the body frame B. The precession unit vector $\hat{u}_\varphi$,

$$\hat{u}_\varphi = \begin{bmatrix} 0 \\ 0 \\ 1 \end{bmatrix} = \hat{K} \tag{4.171}$$

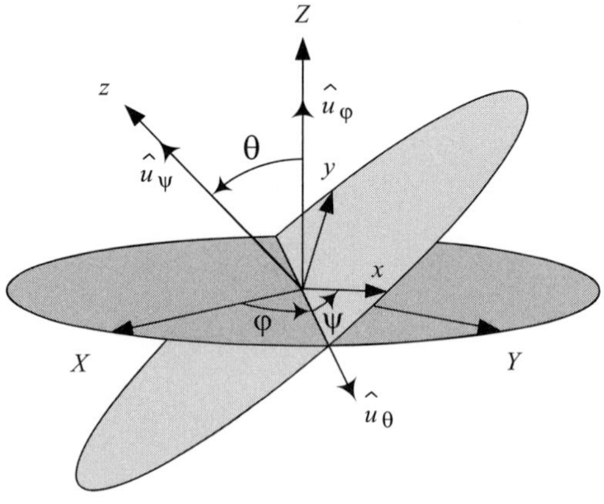

Figure 4.12 Euler angles frame $\hat{u}_\varphi, \hat{u}_\theta, \hat{u}_\psi$.

is a vector in the global frame and can be transformed to the body frame after three rotations:

$$ {}^B\hat{u}_\varphi = {}^BR_G\,\hat{K} = R_{z,\psi}\,R_{x,\theta}\,R_{z,\varphi}\hat{K} = \begin{bmatrix} \sin\theta\ \sin\psi \\ \sin\theta\ \cos\psi \\ \cos\theta \end{bmatrix} \tag{4.172} $$

The nutation unit vector $\hat{u}_\theta$,

$$ \hat{u}_\theta = \begin{bmatrix} 1 \\ 0 \\ 0 \end{bmatrix} = \hat{\imath}' \tag{4.173} $$

is in the intermediate frame $Ox'y'z'$ and needs to get two rotations $R_{x,\theta}$ and $R_{z,\psi}$ to be transformed to the body frame:

$$ {}^B\hat{u}_\theta = {}^BR_{Ox'y'z'}\,\hat{\imath}' = R_{z,\psi}\,R_{x,\theta}\,\hat{\imath}' = \begin{bmatrix} \cos\psi \\ -\sin\psi \\ 0 \end{bmatrix} \tag{4.174} $$

The spin unit vector $\hat{u}_\psi$ is already in the body frame:

$$ \hat{u}_\psi = \begin{bmatrix} 0 \\ 0 \\ 1 \end{bmatrix} = \hat{k} \tag{4.175} $$

Therefore, ${}_G\omega_B$ can be expressed in the body coordinate frame as

$$ \begin{aligned} {}^B_G\omega_B &= \dot{\varphi}\begin{bmatrix} \sin\theta\ \sin\psi \\ \sin\theta\ \cos\psi \\ \cos\theta \end{bmatrix} + \dot{\theta}\begin{bmatrix} \cos\psi \\ -\sin\psi \\ 0 \end{bmatrix} + \dot{\psi}\begin{bmatrix} 0 \\ 0 \\ 1 \end{bmatrix} \\ &= \left(\dot{\varphi}\sin\theta\ \sin\psi + \dot{\theta}\cos\psi\right)\hat{\imath} + \left(\dot{\varphi}\sin\theta\ \cos\psi - \dot{\theta}\sin\psi\right)\hat{\jmath} \\ &\quad + \left(\dot{\varphi}\cos\theta + \dot{\psi}\right)\hat{k} \end{aligned} \tag{4.176} $$

The components of ${}_{G}\omega_{B}$ in the body frame *B(Oxyz)* are related to the Euler frame $E(O\varphi\theta\psi)$ by the relationship

$$
{}_{G}^{B}\omega_{B} = {}^{B}R_{E}\,{}_{G}^{E}\omega_{B} \tag{4.177}
$$

$$
\begin{bmatrix} \omega_x \\ \omega_y \\ \omega_z \end{bmatrix} = \begin{bmatrix} \sin\theta\,\sin\psi & \cos\psi & 0 \\ \sin\theta\,\cos\psi & -\sin\psi & 0 \\ \cos\theta & 0 & 1 \end{bmatrix} \begin{bmatrix} \dot{\varphi} \\ \dot{\theta} \\ \dot{\psi} \end{bmatrix} \tag{4.178}
$$

Then, ${}_{G}\omega_{B}$ can be expressed in the global frame using an inverse transformation of the Euler rotation matrix (4.142),

$$
\begin{aligned}
{}_{G}\omega_{B} &= {}^{B}R_{G}^{-1}\,{}_{G}^{B}\omega_{B} = {}^{B}R_{G}^{-1} \begin{bmatrix} \dot{\varphi}\sin\theta\,\sin\psi + \dot{\theta}\cos\psi \\ \dot{\varphi}\sin\theta\,\cos\psi - \dot{\theta}\sin\psi \\ \dot{\varphi}\cos\theta + \dot{\psi} \end{bmatrix} \\
&= \left(\dot{\theta}\cos\varphi + \dot{\psi}\sin\theta\,\sin\varphi\right)\hat{I} + \left(\dot{\theta}\sin\varphi - \dot{\psi}\cos\varphi\,\sin\theta\right)\hat{J} \\
&\quad + \left(\dot{\varphi} + \dot{\psi}\cos\theta\right)\hat{K}
\end{aligned} \tag{4.179}
$$

and hence components of ${}_{G}\omega_{B}$ in the global coordinate frame *G(OXYZ)* are related to the Euler angle coordinate frame $E(\mathrm{O}\varphi\theta\psi)$ by the relationship

$$
{}_{G}\omega_{B} = {}^{G}R_{E}\,{}_{G}^{E}\omega_{B} \tag{4.180}
$$

$$
\begin{bmatrix} \omega_X \\ \omega_Y \\ \omega_Z \end{bmatrix} = \begin{bmatrix} 0 & \cos\varphi & \sin\theta\,\sin\varphi \\ 0 & \sin\varphi & -\cos\varphi\,\sin\theta \\ 1 & 0 & \cos\theta \end{bmatrix} \begin{bmatrix} \dot{\varphi} \\ \dot{\theta} \\ \dot{\psi} \end{bmatrix} \tag{4.181}
$$

Example 273 ★ Euler Frequencies Based on Cartesian Frequencies The vector ${}_{G}^{B}\omega_{B}$, which indicates the angular velocity of a rigid body B with respect to the global frame G written in frame B, is related to the Euler frequencies by

$$
\begin{aligned}
{}_{G}^{B}\omega_{B} &= {}^{B}R_{E}\,{}_{G}^{E}\omega_{B} \\
&= \begin{bmatrix} \omega_x \\ \omega_y \\ \omega_z \end{bmatrix} = \begin{bmatrix} \sin\theta\sin\psi & \cos\psi & 0 \\ \sin\theta\cos\psi & -\sin\psi & 0 \\ \cos\theta & 0 & 1 \end{bmatrix} \begin{bmatrix} \dot{\varphi} \\ \dot{\theta} \\ \dot{\psi} \end{bmatrix}
\end{aligned} \tag{4.182}
$$

The matrix of coefficients is not an orthogonal matrix because

$$
{}^{B}R_{E}^{\mathrm{T}} = \begin{bmatrix} \sin\theta\sin\psi & \sin\theta\cos\psi & \cos\theta \\ \cos\psi & -\sin\psi & 0 \\ 0 & 0 & 1 \end{bmatrix} \tag{4.183}
$$

$$
{}^{B}R_{E}^{-1} = \frac{1}{\sin\theta} \begin{bmatrix} \sin\psi & \cos\psi & 0 \\ \sin\theta\cos\psi & -\sin\theta\sin\psi & 0 \\ -\cos\theta\sin\psi & -\cos\theta\cos\psi & 1 \end{bmatrix} \tag{4.184}
$$

$$
{}^{B}R_{E}^{\mathrm{T}} \neq {}^{B}R_{E}^{-1} \tag{4.185}
$$

This is because the Euler coordinate frame $E(O\varphi\theta\psi)$ is not an orthogonal frame. For the same reason, the matrix of coefficients that relates the Euler frequencies and the components of ${}_G^G\omega_B$,

$$
\begin{aligned}
{}_G\omega_B &= {}^GR_E\,{}_G^E\omega_B \\
&= \begin{bmatrix} \omega_X \\ \omega_Y \\ \omega_Z \end{bmatrix} = \begin{bmatrix} 0 & \cos\varphi & \sin\theta\sin\varphi \\ 0 & \sin\varphi & -\cos\varphi\sin\theta \\ 1 & 0 & \cos\theta \end{bmatrix} \begin{bmatrix} \dot{\varphi} \\ \dot{\theta} \\ \dot{\psi} \end{bmatrix}
\end{aligned} \tag{4.186}
$$

is not an orthogonal matrix. Therefore, the Euler frequencies based on local and global decomposition of the angular velocity vector ${}_G\omega_B$ must be found by the inverse of the coefficient matrices:

$$
{}_G^E\omega_B = {}^BR_E^{-1}\,{}_G^B\omega_B \tag{4.187}
$$

$$
\begin{bmatrix} \dot{\varphi} \\ \dot{\theta} \\ \dot{\psi} \end{bmatrix} = \frac{1}{\sin\theta} \begin{bmatrix} \sin\psi & \cos\psi & 0 \\ \sin\theta\cos\psi & -\sin\theta\sin\psi & 0 \\ -\cos\theta\sin\psi & -\cos\theta\cos\psi & 1 \end{bmatrix} \begin{bmatrix} \omega_x \\ \omega_y \\ \omega_z \end{bmatrix} \tag{4.188}
$$

$$
{}_G^E\omega_B = {}^GR_E^{-1}\,{}_G\omega_B \tag{4.189}
$$

$$
\begin{bmatrix} \dot{\varphi} \\ \dot{\theta} \\ \dot{\psi} \end{bmatrix} = \frac{1}{\sin\theta} \begin{bmatrix} -\cos\theta\sin\varphi & \cos\theta\cos\varphi & 1 \\ \sin\theta\cos\varphi & \sin\theta\sin\varphi & 0 \\ \sin\varphi & -\cos\varphi & 0 \end{bmatrix} \begin{bmatrix} \omega_X \\ \omega_Y \\ \omega_Z \end{bmatrix} \tag{4.190}
$$

Using (4.187) and (4.189), it can be verified that the transformation matrix ${}^BR_G = {}^BR_E\,{}^GR_E^{-1}$ would be the same as the Euler transformation matrix (4.142). So, the angular velocity vector can be expressed in different frames as

$$
{}_G^B\omega_B = \begin{bmatrix} \hat{\imath} & \hat{\jmath} & \hat{k} \end{bmatrix} \begin{bmatrix} \omega_x \\ \omega_y \\ \omega_z \end{bmatrix} \tag{4.191}
$$

$$
{}_G\omega_B = \begin{bmatrix} \hat{I} & \hat{J} & \hat{K} \end{bmatrix} \begin{bmatrix} \omega_X \\ \omega_Y \\ \omega_Z \end{bmatrix} \tag{4.192}
$$

$$
{}_G^E\omega_B = \begin{bmatrix} \hat{K} & \hat{e}_\theta & \hat{k} \end{bmatrix} \begin{bmatrix} \dot{\varphi} \\ \dot{\theta} \\ \dot{\psi} \end{bmatrix}
$$

Example 274 ★ Integrability of Angular Velocity Components The integrability condition for an arbitrary total differential dz of a function $z = f(x, y)$,

$$
dz = P\,dx + Q\,dy \tag{4.193}
$$

is

$$
\frac{\partial P}{\partial y} = \frac{\partial Q}{\partial x} \tag{4.194}
$$

The angular velocity components ω_x, ω_y, and ω_z along the body coordinate axes x, y, and z cannot be integrated to obtain the associated angles because

$$\omega_x\,dt = \sin\theta\sin\psi\,d\varphi + \cos\psi\,d\theta \tag{4.195}$$

and

$$\frac{\partial(\sin\theta\sin\psi)}{\partial\theta} \neq \frac{\partial\cos\psi}{\partial\varphi} \tag{4.196}$$

However, the integrability condition (4.194) is satisfied by the Euler frequencies. Let us rewrite (4.188) as

$$d\varphi = \frac{\sin\psi}{\sin\theta}\,(\omega_x\,dt) + \frac{\cos\psi}{\sin\theta}\,(\omega_y\,dt) \tag{4.197}$$

$$d\theta = \cos\psi\,(\omega_x\,dt) - \sin\psi\,(\omega_y\,dt) \tag{4.198}$$

$$d\psi = \frac{-\cos\theta\sin\psi}{\sin\theta}\,(\omega_x\,dt) + \frac{-\cos\theta\cos\psi}{\sin\theta}\,(\omega_y\,dt) + \frac{(\omega_z\,dt)}{\sin\theta} \tag{4.199}$$

These equations indicate that

$$\frac{\sin\psi}{\sin\theta} = \frac{\partial\varphi}{\partial\,(\omega_x\,dt)} \qquad \frac{\cos\psi}{\sin\theta} = \frac{\partial\varphi}{\partial\,(\omega_y\,dt)} \tag{4.200a}$$

$$\cos\psi = \frac{\partial\theta}{\partial\,(\omega_x\,dt)} \qquad -\sin\psi = \frac{\partial\theta}{\partial\,(\omega_y\,dt)} \tag{4.200b}$$

$$\frac{-\cos\theta\sin\psi}{\sin\theta} = \frac{\partial\psi}{\partial\,(\omega_x\,dt)} \qquad \frac{-\cos\theta\cos\psi}{\sin\theta} = \frac{\partial\psi}{\partial\,(\omega_y\,dt)} \tag{4.200c}$$

$$\frac{1}{\sin\theta} = \frac{\partial\psi}{\partial\,(\omega_z\,dt)} \tag{4.200d}$$

and therefore,

$$\begin{aligned}
\frac{\partial}{\partial\,(\omega_y\,dt)}\left(\frac{\sin\psi}{\sin\theta}\right) &= \frac{1}{\sin^2\theta}\left(\cos\psi\sin\theta\frac{\partial\psi}{\partial\,(\omega_y\,dt)} - \cos\theta\sin\psi\frac{\partial\theta}{\partial\,(\omega_y\,dt)}\right) \\
&= \frac{1}{\sin^2\theta}\,(-\cos\psi\cos\theta\cos\psi + \cos\theta\sin\psi\sin\psi) \\
&= \frac{-\cos\theta\cos 2\psi}{\sin^2\theta}
\end{aligned} \tag{4.201}$$

$$\begin{aligned}
\frac{\partial}{\partial\,(\omega_x\,dt)}\left(\frac{\cos\psi}{\sin\theta}\right) &= \frac{1}{\sin^2\theta}\left(-\sin\psi\sin\theta\frac{\partial\psi}{\partial\,(\omega_x\,dt)} - \cos\theta\cos\psi\frac{\partial\theta}{\partial\,(\omega_x\,dt)}\right) \\
&= \frac{1}{\sin^2\theta}\,(\sin\psi\cos\theta\sin\psi - \cos\theta\cos\psi\cos\psi) \\
&= \frac{-\cos\theta\cos 2\psi}{\sin^2\theta}
\end{aligned} \tag{4.202}$$

It can be checked that $d\theta$ and $d\psi$ are also integrable.

Example 275 ★ Cardan Angles and Frequencies The system of Euler angles is singular at $\theta = 0$, and as a consequence, φ and ψ become coplanar and indistinguishable. To overcome this problem, we may employ a triple rotation about three different axes. From the 12 angle systems of Appendix B, the rotations about three different axes such as ${}^B R_G = R_{z,\psi} R_{y,\theta} R_{x,\varphi}$ are called *Cardan* or *Bryant angles*. The Cardan angle system is not singular at $\theta = 0$ and has applications in vehicle, aircraft, and attitude dynamics:

$$\begin{aligned} {}^B R_G &= R_{z,\psi} R_{y,\theta} R_{x,\varphi} \\ &= \begin{bmatrix} c\theta c\psi & c\varphi s\psi + s\theta c\psi s\varphi & s\varphi s\psi - c\varphi s\theta c\psi \\ -c\theta s\psi & c\varphi c\psi - s\theta s\varphi s\psi & c\psi s\varphi + c\varphi s\theta s\psi \\ s\theta & -c\theta s\varphi & c\theta c\varphi \end{bmatrix} \end{aligned} \tag{4.203}$$

The angular velocity ω of a rigid body can be expressed in terms of either the components along the axes of *B(Oxyz)* or the *Cardan frequencies* along the axes of the nonorthogonal Cardan frame. The angular velocity in terms of Cardan frequencies is

$${}_G\omega_B = \dot{\varphi} R_{z,\psi} R_{y,\theta} \begin{bmatrix} 1 \\ 0 \\ 0 \end{bmatrix} + \dot{\theta} R_{z,\psi} \begin{bmatrix} 0 \\ 1 \\ 0 \end{bmatrix} + \dot{\psi} \begin{bmatrix} 0 \\ 0 \\ 1 \end{bmatrix} \tag{4.204}$$

Therefore,

$$\begin{bmatrix} \omega_x \\ \omega_y \\ \omega_z \end{bmatrix} = \begin{bmatrix} \cos\theta\cos\psi & \sin\psi & 0 \\ -\cos\theta\sin\psi & \cos\psi & 0 \\ \sin\theta & 0 & 1 \end{bmatrix} \begin{bmatrix} \dot{\varphi} \\ \dot{\theta} \\ \dot{\psi} \end{bmatrix} \tag{4.205}$$

$$\begin{bmatrix} \dot{\varphi} \\ \dot{\theta} \\ \dot{\psi} \end{bmatrix} = \begin{bmatrix} \cos\psi/\cos\theta & -\sin\psi/\cos\theta & 0 \\ \sin\psi & \cos\psi & 0 \\ -\tan\theta\cos\psi & \tan\theta\sin\psi & 1 \end{bmatrix} \begin{bmatrix} \omega_x \\ \omega_y \\ \omega_z \end{bmatrix} \tag{4.206}$$

For small Cardan angles, we have

$${}^B R_G = \begin{bmatrix} 1 & \psi & -\theta \\ -\psi & 1 & \varphi \\ \theta & -\varphi & 1 \end{bmatrix} \tag{4.207}$$

and

$$\begin{bmatrix} \omega_x \\ \omega_y \\ \omega_z \end{bmatrix} = \begin{bmatrix} 1 & \psi & 0 \\ -\psi & 1 & 0 \\ \theta & 0 & 1 \end{bmatrix} \begin{bmatrix} \dot{\varphi} \\ \dot{\theta} \\ \dot{\psi} \end{bmatrix} \tag{4.208}$$

$$\begin{bmatrix} \dot{\varphi} \\ \dot{\theta} \\ \dot{\psi} \end{bmatrix} = \begin{bmatrix} \psi & -\psi & 0 \\ \psi & 1 & 0 \\ -\theta & 0 & 1 \end{bmatrix} \begin{bmatrix} \omega_x \\ \omega_y \\ \omega_z \end{bmatrix} \tag{4.209}$$

4.7 LOCAL ROLL–PITCH–YAW ANGLES

Rotation about the x-axis of the local frame is called *roll* or *bank*, rotation about the y-axis of the local frame is called *pitch* or *attitude*, and rotation about the z-axis of the local frame is called *yaw, spin*, or *heading*. The local roll–pitch–yaw angles are also called the *Cardan or Bryant angles*. The Cardan angles shown in Figure 4.13 represent an applied method to express the orientation of mobile rigid bodies such as vehicles, airplanes, and submarines.

The local *roll–pitch–yaw rotation matrix* is

$$\begin{aligned} {}^{B}R_{G} &= R_{z,\psi}\, R_{y,\theta}\, R_{x,\varphi} \\ &= \begin{bmatrix} c\theta c\psi & c\varphi s\psi + s\theta c\psi s\varphi & s\varphi s\psi - c\varphi s\theta c\psi \\ -c\theta s\psi & c\varphi c\psi - s\theta s\varphi s\psi & c\psi s\varphi + c\varphi s\theta s\psi \\ s\theta & -c\theta s\varphi & c\theta c\varphi \end{bmatrix} \end{aligned} \tag{4.210}$$

or

$$\begin{aligned} {}^{G}R_{B} &= {}^{B}R_{G}^{\mathrm{T}} = [R_{z,\psi}\, R_{y,\theta}\, R_{x,\varphi}]^{\mathrm{T}} = R_{x,\varphi}^{\mathrm{T}}\, R_{y,\theta}^{\mathrm{T}}\, R_{z,\psi}^{\mathrm{T}} = R_{X,\varphi}\, R_{Y,\theta}\, R_{Z,\psi} \\ &= \begin{bmatrix} c\theta c\psi & -c\theta s\psi & s\theta \\ c\varphi s\psi + s\theta c\psi s\varphi & c\varphi c\psi - s\theta s\varphi s\psi & -c\theta s\varphi \\ s\varphi s\psi - c\varphi s\theta c\psi & c\psi s\varphi + c\varphi s\theta s\psi & c\theta c\varphi \end{bmatrix} \end{aligned} \tag{4.211}$$

Having a rotation matrix ${}^{B}R_{G} = \left[r_{ij}\right]$, we are able to determine the equivalent local roll, pitch, and yaw angles by

$$\theta = \sin^{-1} r_{31} \tag{4.212}$$

$$\varphi = -\tan^{-1} \frac{r_{32}}{r_{33}} \tag{4.213}$$

$$\psi = -\tan^{-1} \frac{r_{21}}{r_{11}} \tag{4.214}$$

provided that $\cos\theta \neq 0$.

Example 276 Equivalent Local Roll–Pitch–Yaw for a Set of Euler Angles A body frame B turns 45 deg about the X-axis followed by another 45 deg turn about the Y-axis.

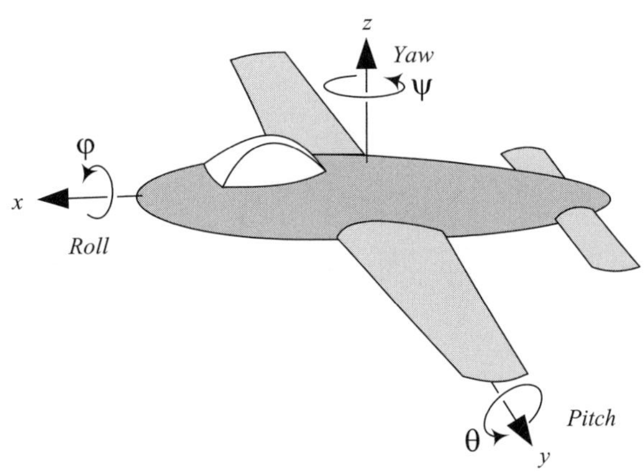

Figure 4.13 Local roll–pitch–yaw angles.

The global coordinates of a body point would be found by a combined rotation:

$$ {}^{G}\mathbf{r} = R_{Y,45}\, R_{X,45}\, {}^{B}\mathbf{r} = {}^{G}R_{B}\, {}^{B}\mathbf{r} \tag{4.215}$$

$$ {}^{G}R_{B} = \begin{bmatrix} \cos\frac{\pi}{4} & 0 & \sin\frac{\pi}{4} \\ 0 & 1 & 0 \\ -\sin\frac{\pi}{4} & 0 & \cos\frac{\pi}{4} \end{bmatrix} \begin{bmatrix} 1 & 0 & 0 \\ 0 & \cos\frac{\pi}{4} & -\sin\frac{\pi}{4} \\ 0 & \sin\frac{\pi}{4} & \cos\frac{\pi}{4} \end{bmatrix} $$

$$ = \begin{bmatrix} 0.70711 & 0.5 & 0.5 \\ 0 & 0.70711 & -0.70711 \\ -0.70711 & 0.5 & 0.5 \end{bmatrix} $$

The same transformation may be done by a local roll–pitch–yaw rotation if the roll, pitch, and yaw angles are found from ${}^{G}R_{B}^{\mathrm{T}}$ according to Equations (4.212)–(4.214):

$$ {}^{B}R_{G} = {}^{G}R_{B}^{\mathrm{T}} = \begin{bmatrix} 0.707 & 0 & -0.707 \\ 0.5 & 0.707 & 0.5 \\ 0.5 & -0.707 & 0.5 \end{bmatrix} \tag{4.216}$$

$$ \theta = \sin^{-1} r_{31} = \sin^{-1} 0.5 = 0.5236\,\mathrm{rad} = 30\,\mathrm{deg} \tag{4.217}$$

$$ \varphi = -\tan^{-1}\frac{r_{32}}{r_{33}} = -\tan^{-1}\frac{-0.707}{0.5} = 0.955\,\mathrm{rad} \approx 54.7\,\mathrm{deg} \tag{4.218}$$

$$ \psi = -\tan^{-1}\frac{r_{21}}{r_{11}} = -\tan^{-1}\frac{0.5}{0.707} = -0.615\,\mathrm{rad} \approx -35.26\,\mathrm{deg} \tag{4.219}$$

As a double check, we may calculate ${}^{B}R_{G}$ from (4.210):

$$ {}^{B}R_{G} = R_{z,-0.615}\, R_{y,0.5236}\, R_{x,0.955} $$

$$ = \begin{bmatrix} 0.70735 & 1.1611\times 10^{-4} & -0.70687 \\ 0.49966 & 0.70727 & 0.50011 \\ 0.5 & -0.70695 & 0.50022 \end{bmatrix} \tag{4.220}$$

The same rotation may also be performed by proper Euler angles from Equations (4.154)–(4.156):

$$ \theta = \cos^{-1} r_{33} = \cos^{-1} 0.5 = 1.0472\,\mathrm{rad} = 60\,\mathrm{deg} \tag{4.221}$$

$$ \varphi = -\tan^{-1}\frac{r_{31}}{r_{32}} = -\tan^{-1}\frac{0.5}{-0.707} = 0.615\,\mathrm{rad} \approx 35.27\,\mathrm{deg} \tag{4.222}$$

$$ \psi = \tan^{-1}\frac{r_{13}}{r_{23}} = \tan^{-1}\frac{-0.707}{0.5} = -0.955\,\mathrm{rad} \approx -54.7\,\mathrm{deg} \tag{4.223}$$

We may calculate ${}^{B}R_{G}$ from (4.142) to check our calculations:

$$ {}^{B}R_{G} = R_{z,\psi}\, R_{x,\theta}\, R_{z,\varphi} $$

$$ = \begin{bmatrix} 0.70714 & 9.8226\times 10^{-5} & -0.70707 \\ 0.4999 & 0.70714 & 0.50005 \\ 0.50005 & -0.70707 & 0.5 \end{bmatrix} \tag{4.224}$$

Example 277 Time-Dependent Local Roll–Pitch–Yaw Rotation A rigid body B is turning about the x-axis with 0.2 rad/s for 2 s and then about the z-axis with 0.3 rad / s for 2 s. The body rotation matrices are

$$R_{x,\psi} = \begin{bmatrix} 1 & 0 & 0 \\ 0 & \cos 0.2t & \sin 0.2t \\ 0 & -\sin 0.2t & \cos 0.2t \end{bmatrix} \tag{4.225}$$

$$R_{z,\varphi} = \begin{bmatrix} \cos 0.3t & \sin 0.3t & 0 \\ -\sin 0.3t & \cos 0.3t & 0 \\ 0 & 0 & 1 \end{bmatrix} \tag{4.226}$$

and therefore, the final rotation matrix between frames B and G is

$$\begin{aligned} {}^{B}R_{G} &= R_{z,0.3\times 2}\, R_{x,0.2\times 2} \\ &= \begin{bmatrix} 0.82534 & 0.52007 & 0.21988 \\ -0.56464 & 0.76018 & 0.3214 \\ 0 & -0.38942 & 0.92106 \end{bmatrix} \end{aligned} \tag{4.227}$$

To check if it is possible to turn about both x- and z-axes at the same time and finish the rotation in just 2 s, we may combine the rotation matrices and find ${}^{B}R_{G}$:

$$\begin{aligned} {}^{B}R_{G} &= R_{z,0.3t}\, R_{x,0.2t} \\ &= \begin{bmatrix} \cos 0.3t & \sin 0.3t \cos 0.2t & \sin 0.3t \sin 0.2t \\ -\sin 0.3t & \cos 0.3t \cos 0.2t & \cos 0.3t \sin 0.2t \\ 0 & -\sin 0.2t & \cos 0.2t \end{bmatrix} \end{aligned} \tag{4.228}$$

The combined rotation matrix ${}^{B}R_{G}$ at $t = 2\,\text{s}$ would be

$${}^{B}R_{G} = \begin{bmatrix} 0.82534 & 0.52007 & 0.21988 \\ -0.56464 & 0.76018 & 0.3214 \\ 0 & -0.38942 & 0.92106 \end{bmatrix} \tag{4.229}$$

which is the same as (4.227). Therefore, the whole rotation maneuver can be done in just 2 s if both rotations are performed simultaneously.

Example 278 ★ Second Type of Cardan Rotation Matrix We may call the Cardan rotation matrix (4.210) a local 1–2–3-matrix to indicate the order of rotations about $x = x_1$, $y = x_2$, $z = x_3$. Interchanging the indices i, j, and k produces the rotation matrices 2–3–1 and 3–1–2, which are mathematically equivalent to the first type of Cardan rotation matrix (4.210).

We may reverse the order of rotation axes from $x-y-z$ to $z-y-x$ and develop the second type of Cardan rotation matrix:

$$\begin{aligned} {}^{B}R_{G} &= R_{x,\psi}\, R_{y,\theta}\, R_{z,\varphi} \\ &= \begin{bmatrix} c\theta c\varphi & c\theta s\varphi & -s\theta \\ c\varphi s\theta s\psi - c\psi s\varphi & c\psi c\varphi + s\theta s\psi s\varphi & c\theta s\psi \\ s\psi s\varphi + c\psi c\varphi s\theta & c\psi s\theta s\varphi - c\varphi s\psi & c\theta c\psi \end{bmatrix} \end{aligned} \tag{4.230}$$

This is the local 3–2–1 matrix that is mathematically equivalent to 1–3–2 and 2–1–3 rotations. Therefore, six triple rotations of the 12 combinations in (4.133) are Cardan rotation matrices:

$$\begin{array}{ll} 1.\ R_{x,\psi}\,R_{y,\theta}\,R_{z,\varphi} & 1\text{–}2\text{–}3 \\ 2.\ R_{y,\psi}\,R_{z,\theta}\,R_{x,\varphi} & 2\text{–}3\text{–}1 \\ 3.\ R_{z,\psi}\,R_{x,\theta}\,R_{y,\varphi} & 3\text{–}1\text{–}2 \\ 4.\ R_{z,\psi}\,R_{y,\theta}\,R_{x,\varphi} & 3\text{–}2\text{–}1 \\ 5.\ R_{y,\psi}\,R_{x,\theta}\,R_{z,\varphi} & 2\text{–}1\text{–}3 \\ 6.\ R_{x,\psi}\,R_{z,\theta}\,R_{y,\varphi} & 1\text{–}3\text{–}2 \end{array} \tag{4.231}$$

Example 279 ★ Angular Velocity and Local Roll–Pitch–Yaw Rate Because the local roll–pitch–yaw rotation matrix (4.210) is the same as the Cardan xyz-matrix (4.203), we show the local roll–pitch–yaw coordinate frame by $C(\varphi, \theta, \psi)$. Using the local roll–pitch–yaw frequencies, the angular velocity of a body B with respect to the global reference frame can be expressed in B or C as

$${}_G^B\omega_B = \omega_x\hat{\imath} + \omega_y\hat{\jmath} + \omega_z\hat{k} \tag{4.232}$$

$${}_G^C\omega_B = \dot{\varphi}\hat{u}_\varphi + \dot{\theta}\hat{u}_\theta + \dot{\psi}\hat{u}_\psi \tag{4.233}$$

Relationships between the components of ${}_G^B\omega_B$ and ${}_G^C\omega_B$ are found when the roll unit vector $\hat{u}_\varphi$, pitch unit vector $\hat{u}_\theta$, and yaw unit vector $\hat{u}_\psi$ are transformed to the body frame. The roll unit vector

$$\hat{u}_\varphi = \begin{bmatrix} 1 \\ 0 \\ 0 \end{bmatrix} \tag{4.234}$$

transforms to the body frame after rotation θ and then rotation ψ:

$${}^B\hat{u}_\varphi = R_{z,\psi}R_{y,\theta}\begin{bmatrix} 1 \\ 0 \\ 0 \end{bmatrix} = \begin{bmatrix} \cos\theta\cos\psi \\ -\cos\theta\sin\psi \\ \sin\theta \end{bmatrix} \tag{4.235}$$

The pitch unit vector

$$\hat{u}_\theta = \begin{bmatrix} 0 \\ 1 \\ 0 \end{bmatrix} \tag{4.236}$$

transforms to the body frame after rotation ψ:

$${}^B\hat{u}_\theta = R_{z,\psi}\begin{bmatrix} 0 \\ 1 \\ 0 \end{bmatrix} = \begin{bmatrix} \sin\psi \\ \cos\psi \\ 0 \end{bmatrix} \tag{4.237}$$

The yaw unit vector

$$\hat{u}_\psi = \begin{bmatrix} 0 \\ 0 \\ 1 \end{bmatrix} \tag{4.238}$$

is already along the local z-axis. Hence, ${}_G^B\omega_B$ can be expressed in body frame $B(Oxyz)$ as

$$
\begin{aligned}
{}_G^B\omega_B &= \begin{bmatrix} \omega_x \\ \omega_y \\ \omega_z \end{bmatrix} = \dot{\varphi} \begin{bmatrix} \cos\theta\cos\psi \\ -\cos\theta\sin\psi \\ \sin\theta \end{bmatrix} + \dot{\theta} \begin{bmatrix} \sin\psi \\ \cos\psi \\ 0 \end{bmatrix} + \dot{\psi} \begin{bmatrix} 0 \\ 0 \\ 1 \end{bmatrix} \\
&= \begin{bmatrix} \cos\theta\cos\psi & \sin\psi & 0 \\ -\cos\theta\sin\psi & \cos\psi & 0 \\ \sin\theta & 0 & 1 \end{bmatrix} \begin{bmatrix} \dot{\varphi} \\ \dot{\theta} \\ \dot{\psi} \end{bmatrix}
\end{aligned}
\tag{4.239}
$$

and therefore, ${}_G^G\omega_B$ in the global frame $G(OXYZ)$ in terms of local roll–pitch–yaw frequencies is

$$
\begin{aligned}
{}_G^G\omega_B &= \begin{bmatrix} \omega_X \\ \omega_Y \\ \omega_Z \end{bmatrix} = {}^BR_G^{-1} \begin{bmatrix} \omega_x \\ \omega_y \\ \omega_z \end{bmatrix} = {}^BR_G^{-1} \begin{bmatrix} \dot{\theta}\sin\psi + \dot{\varphi}\cos\theta\cos\psi \\ \dot{\theta}\cos\psi - \dot{\varphi}\cos\theta\sin\psi \\ \dot{\psi} + \dot{\varphi}\sin\theta \end{bmatrix} \\
&= \begin{bmatrix} \dot{\varphi} + \dot{\psi}\sin\theta \\ \dot{\theta}\cos\varphi - \dot{\psi}\cos\theta\sin\varphi \\ \dot{\theta}\sin\varphi + \dot{\psi}\cos\theta\cos\varphi \end{bmatrix} \\
&= \begin{bmatrix} 1 & 0 & \sin\theta \\ 0 & \cos\varphi & -\cos\theta\sin\varphi \\ 0 & \sin\varphi & \cos\theta\cos\varphi \end{bmatrix} \begin{bmatrix} \dot{\varphi} \\ \dot{\theta} \\ \dot{\psi} \end{bmatrix}.
\end{aligned}
\tag{4.240}
$$

4.8 LOCAL VERSUS GLOBAL ROTATION

The global rotation matrix ${}^GR_B$ is equal to the inverse of the local rotation matrix ${}^BR_G$ and vice versa,

$$
{}^GR_B = {}^BR_G^{-1} \qquad {}^BR_G = {}^GR_B^{-1}
\tag{4.241}
$$

and premultiplication of the global rotation matrix is equal to postmultiplication of the local rotation matrix.

Proof: To distinguish between the two type of rotations, let us show rotation matrices about global axes by $[Q]$ and rotation matrices about local axes by $[A]$. Consider a sequence of global rotations and their resultant global rotation matrix ${}^GQ_B$ to transform a position vector ${}^B\mathbf{r}$ to ${}^G\mathbf{r}$:

$$
{}^G\mathbf{r} = {}^GQ_B\,{}^B\mathbf{r}
\tag{4.242}
$$

The global position vector ${}^G\mathbf{r}$ can also be transformed to ${}^B\mathbf{r}$ using a local rotation matrix ${}^BA_G$:

$$
{}^B\mathbf{r} = {}^BA_G\,{}^G\mathbf{r}
\tag{4.243}
$$

Combining Equations (4.242) and (4.243) leads to

$$^G\mathbf{r} = {}^GQ_B\,{}^BA_G\,{}^G\mathbf{r} \tag{4.244}$$

$$^B\mathbf{r} = {}^BA_G\,{}^GQ_B\,{}^B\mathbf{r} \tag{4.245}$$

and hence,

$$^GQ_B\,{}^BA_G = {}^BA_G\,{}^GQ_B = \mathbf{I} \tag{4.246}$$

Therefore, the global and local rotation matrices are inverses of each other:

$$^GQ_B = {}^BA_G^{-1} \tag{4.247a}$$

$$^GQ_B^{-1} = {}^BA_G \tag{4.247b}$$

Assume that

$$^GQ_B = Q_n \ldots Q_3Q_2Q_1 \tag{4.248}$$

$$^BA_G = A_n \ldots A_3A_2A_1 \tag{4.249}$$

then

$$^GQ_B = {}^BA_G^{-1} = A_1^{-1}A_2^{-1}A_3^{-1}\ldots A_n^{-1} \tag{4.250}$$

$$^BA_G = {}^GQ_B^{-1} = Q_1^{-1}Q_2^{-1}Q_3^{-1}\ldots Q_n^{-1} \tag{4.251}$$

and Equation (4.246) becomes

$$Q_n \cdots Q_2Q_1A_n \ldots A_2A_1 = A_n \cdots A_2A_1Q_n \ldots Q_2Q_1 = \mathbf{I} \tag{4.252}$$

and therefore,

$$Q_n \cdots Q_3Q_2Q_1 = A_1^{-1}A_2^{-1}A_3^{-1}\ldots A_n^{-1} \tag{4.253a}$$

$$A_n \cdots A_3A_2A_1 = Q_1^{-1}Q_2^{-1}Q_3^{-1}\ldots Q_n^{-1} \tag{4.253b}$$

or

$$Q_1^{-1}Q_2^{-1}Q_3^{-1}\cdots Q_n^{-1}Q_n\ldots Q_3Q_2Q_1 = \mathbf{I} \tag{4.254}$$

$$A_1^{-1}A_2^{-1}A_3^{-1}\cdots A_n^{-1}A_n\ldots A_3A_2A_1 = \mathbf{I} \tag{4.255}$$

Hence, the effect of order rotations about the global coordinate axes is equivalent to the effect of the same rotations about the local coordinate axes performed in reverse order:

$$^GQ_B = A_1^{-1}A_2^{-1}A_3^{-1}\cdots A_n^{-1} \tag{4.256}$$

$$^BA_G = Q_1^{-1}Q_2^{-1}Q_3^{-1}\cdots Q_n^{-1} \tag{4.257}$$

■

Example 280 Global Position and Postmultiplication of Rotation Matrix The local position of a point P after a rotation is at $^B\mathbf{r} = [1, 2, 3]^\mathrm{T}$. If the local rotation matrix to transform $^G\mathbf{r}$ to $^B\mathbf{r}$ is given as

$$^BR_{z,\varphi} = \begin{bmatrix} \cos\varphi & \sin\varphi & 0 \\ -\sin\varphi & \cos\varphi & 0 \\ 0 & 0 & 1 \end{bmatrix} = \begin{bmatrix} \cos 30 & \sin 30 & 0 \\ -\sin 30 & \cos 30 & 0 \\ 0 & 0 & 1 \end{bmatrix} \tag{4.258}$$

then we may find the global position vector ${}^G\mathbf{r}$ by postmultiplying ${}^BR_{z,\varphi}$ and the local position vector ${}^B\mathbf{r}^{\mathrm{T}}$,

$$
\begin{aligned}
{}^G\mathbf{r}^{\mathrm{T}} &= {}^B\mathbf{r}^{\mathrm{T}}\ {}^BR_{z,\varphi} = \begin{bmatrix} 1 & 2 & 3 \end{bmatrix} \begin{bmatrix} \cos 30 & \sin 30 & 0 \\ -\sin 30 & \cos 30 & 0 \\ 0 & 0 & 1 \end{bmatrix} \\
&= \begin{bmatrix} -0.13 & 2.23 & 3.0 \end{bmatrix}
\end{aligned} \tag{4.259}
$$

instead of premultiplying of ${}^BR_{z,\varphi}^{-1}$ by ${}^B\mathbf{r}$:

$$
\begin{aligned}
{}^G\mathbf{r} &= {}^BR_{z,\varphi}^{-1}\ {}^B\mathbf{r} \\
&= \begin{bmatrix} \cos 30 & -\sin 30 & 0 \\ \sin 30 & \cos 30 & 0 \\ 0 & 0 & 1 \end{bmatrix} \begin{bmatrix} 1 \\ 2 \\ 3 \end{bmatrix} = \begin{bmatrix} -0.13 \\ 2.23 \\ 3 \end{bmatrix}
\end{aligned} \tag{4.260}
$$

4.9 GENERAL ROTATION

Consider a general rotation of a local coordinate frame *B(Oxyz)* with respect to a global frame $G(OXYZ)$ about their common origin O. The components of any vector $\mathbf{r}$ may be expressed in either frame. There is always a *transformation matrix* ${}^GR_B$ to map the components of $\mathbf{r}$ from the frame *B(Oxyz)* to the other frame *G(OXYZ)*:

$$ {}^G\mathbf{r} = {}^GR_B\ {}^B\mathbf{r} \tag{4.261} $$

In addition, the inverse map ${}^B\mathbf{r} = {}^GR_B^{-1}\ {}^G\mathbf{r}$ can be done by ${}^BR_G$,

$$ {}^B\mathbf{r} = {}^BR_G\ {}^G\mathbf{r} \tag{4.262} $$

where

$$ \left|{}^GR_B\right| = \left|{}^BR_G\right| = 1 \tag{4.263} $$

and

$$ {}^BR_G = {}^GR_B^{-1} = {}^GR_B^{\mathrm{T}} \tag{4.264} $$

When the coordinate frames B and G are orthogonal, the rotation matrix ${}^GR_B$ is called an *orthogonal matrix*. The transpose R^{T} and inverse R^{-1} of an orthogonal matrix $[R]$ are equal:

$$ R^{\mathrm{T}} = R^{-1} \tag{4.265} $$

Because of the matrix orthogonality condition, only three of the nine elements of ${}^GR_B$ are independent.

Proof: Employing the orthogonality condition (3.2) and decomposition of the unit vectors of *G(OXYZ)* along the axes of *B(Oxyz)*,

$$\hat{I} = (\hat{I} \cdot \hat{\imath})\hat{\imath} + (\hat{I} \cdot \hat{\jmath})\hat{\jmath} + (\hat{I} \cdot \hat{k})\hat{k} \tag{4.266}$$

$$\hat{J} = (\hat{J} \cdot \hat{\imath})\hat{\imath} + (\hat{J} \cdot \hat{\jmath})\hat{\jmath} + (\hat{J} \cdot \hat{k})\hat{k} \tag{4.267}$$

$$\hat{K} = (\hat{K} \cdot \hat{\imath})\hat{\imath} + (\hat{K} \cdot \hat{\jmath})\hat{\jmath} + (\hat{K} \cdot \hat{k})\hat{k} \tag{4.268}$$

introduces the transformation matrix ${}^{G}R_{B}$ to map the local axes to the global axes:

$$\begin{bmatrix} \hat{I} \\ \hat{J} \\ \hat{K} \end{bmatrix} = \begin{bmatrix} \hat{I} \cdot \hat{\imath} & \hat{I} \cdot \hat{\jmath} & \hat{I} \cdot \hat{k} \\ \hat{J} \cdot \hat{\imath} & \hat{J} \cdot \hat{\jmath} & \hat{J} \cdot \hat{k} \\ \hat{K} \cdot \hat{\imath} & \hat{K} \cdot \hat{\jmath} & \hat{K} \cdot \hat{k} \end{bmatrix} \begin{bmatrix} \hat{\imath} \\ \hat{\jmath} \\ \hat{k} \end{bmatrix} = {}^{G}R_{B} \begin{bmatrix} \hat{\imath} \\ \hat{\jmath} \\ \hat{k} \end{bmatrix} \tag{4.269}$$

where

$$\begin{aligned} {}^{G}R_{B} &= \begin{bmatrix} \hat{I} \cdot \hat{\imath} & \hat{I} \cdot \hat{\jmath} & \hat{I} \cdot \hat{k} \\ \hat{J} \cdot \hat{\imath} & \hat{J} \cdot \hat{\jmath} & \hat{J} \cdot \hat{k} \\ \hat{K} \cdot \hat{\imath} & \hat{K} \cdot \hat{\jmath} & \hat{K} \cdot \hat{k} \end{bmatrix} \\ &= \begin{bmatrix} \cos(\hat{I}, \hat{\imath}) & \cos(\hat{I}, \hat{\jmath}) & \cos(\hat{I}, \hat{k}) \\ \cos(\hat{J}, \hat{\imath}) & \cos(\hat{J}, \hat{\jmath}) & \cos(\hat{J}, \hat{k}) \\ \cos(\hat{K}, \hat{\imath}) & \cos(\hat{K}, \hat{\jmath}) & \cos(\hat{K}, \hat{k}) \end{bmatrix} \end{aligned} \tag{4.270}$$

Each column of ${}^{G}R_{B}$ is the decomposition of a unit vector of the local frame *B(Oxyz)* in the global frame *G(OXYZ)*:

$${}^{G}R_{B} = \begin{bmatrix} {}^{G}\hat{\imath} & {}^{G}\hat{\jmath} & {}^{G}\hat{k} \end{bmatrix} \tag{4.271}$$

Similarly, each row of ${}^{G}R_{B}$ is the decomposition of a unit vector of the global frame *G(OXYZ)* in the local frame *B(Oxyz)*:

$${}^{G}R_{B} = \begin{bmatrix} {}^{B}\hat{I}^{\mathrm{T}} \\ {}^{B}\hat{J}^{\mathrm{T}} \\ {}^{B}\hat{K}^{\mathrm{T}} \end{bmatrix} \tag{4.272}$$

so the elements of ${}^{G}R_{B}$ are directional cosines of the axes of *G(OXYZ)* in *B(Oxyz)* or B in G. This set of nine directional cosines completely specifies the orientation of *B(Oxyz)* in *G(OXYZ)* and can be used to map the coordinates of any point *(x,y,z)* to its corresponding coordinates *(X,Y,Z)*.

Alternatively, using the method of unit-vector decomposition to develop the matrix ${}^{B}R_{G}$ leads to

$${}^{B}\mathbf{r} = {}^{B}R_{G}\,{}^{G}\mathbf{r} = {}^{G}R_{B}^{-1}\,{}^{G}\mathbf{r} \tag{4.273}$$

$$\begin{aligned} {}^{B}R_{G} &= \begin{bmatrix} \hat{\imath} \cdot \hat{I} & \hat{\imath} \cdot \hat{J} & \hat{\imath} \cdot \hat{K} \\ \hat{\jmath} \cdot \hat{I} & \hat{\jmath} \cdot \hat{J} & \hat{\jmath} \cdot \hat{K} \\ \hat{k} \cdot \hat{I} & \hat{k} \cdot \hat{J} & \hat{k} \cdot \hat{K} \end{bmatrix} \\ &= \begin{bmatrix} \cos(\hat{\imath}, \hat{I}) & \cos(\hat{\imath}, \hat{J}) & \cos(\hat{\imath}, \hat{K}) \\ \cos(\hat{\jmath}, \hat{I}) & \cos(\hat{\jmath}, \hat{J}) & \cos(\hat{\jmath}, \hat{K}) \\ \cos(\hat{k}, \hat{I}) & \cos(\hat{k}, \hat{J}) & \cos(\hat{k}, \hat{K}) \end{bmatrix} \end{aligned} \tag{4.274}$$

It shows that the inverse of a transformation matrix is equal to the transpose of the transformation matrix,

$$ {}^G R_B^{-1} = {}^G R_B^{\mathrm{T}} \tag{4.275} $$

or

$$ {}^G R_B \cdot {}^G R_B^{\mathrm{T}} = \mathbf{I} \tag{4.276} $$

A matrix with condition (4.275) is called an *orthogonal matrix*. Orthogonality of ${}^G R_B$ comes from the fact that it maps an orthogonal coordinate frame to another orthogonal coordinate frame.

An orthogonal transformation matrix ${}^G R_B$ has only three *independent* elements. The constraint equations among the elements of ${}^G R_B$ will be found by applying the matrix orthogonality condition (4.275):

$$ \begin{bmatrix} r_{11} & r_{12} & r_{13} \\ r_{21} & r_{22} & r_{23} \\ r_{31} & r_{32} & r_{33} \end{bmatrix} \begin{bmatrix} r_{11} & r_{21} & r_{31} \\ r_{12} & r_{22} & r_{32} \\ r_{13} & r_{23} & r_{33} \end{bmatrix} = \begin{bmatrix} 1 & 0 & 0 \\ 0 & 1 & 0 \\ 0 & 0 & 1 \end{bmatrix} \tag{4.277} $$

Therefore, the inner product of any two different rows of ${}^G R_B$ is zero, and the inner product of any row of ${}^G R_B$ by itself is unity:

$$ \begin{aligned} r_{11}^2 + r_{12}^2 + r_{13}^2 &= 1 \\ r_{21}^2 + r_{22}^2 + r_{23}^2 &= 1 \\ r_{31}^2 + r_{32}^2 + r_{33}^2 &= 1 \\ r_{11}r_{21} + r_{12}r_{22} + r_{13}r_{23} &= 0 \\ r_{11}r_{31} + r_{12}r_{32} + r_{13}r_{33} &= 0 \\ r_{21}r_{31} + r_{22}r_{32} + r_{23}r_{33} &= 0 \end{aligned} \tag{4.278} $$

These relations are also true for columns of ${}^G R_B$ and evidently for rows and columns of ${}^B R_G$. The orthogonality condition can be summarized by the equation

$$ \sum_{i=1}^{3} r_{ij} r_{ik} = \delta_{jk} \qquad j, k = 1, 2, 3 \tag{4.279} $$

where r_{ij} is the element of row i and column j of the transformation matrix ${}^G R_B$ and δ_{jk} is the Kronecker delta (1.125).

Equation (4.279) provides six independent relations that must be satisfied by the nine directional cosines. Therefore, there are only three independent directional cosines. The independent elements of the matrix ${}^G R_B$ cannot be in the same row or column or any diagonal.

The determinant of a transformation matrix is equal to unity,

$$ \left| {}^G R_B \right| = 1 \tag{4.280} $$

because of Equation (4.276) and noting that

$$ \left| {}^G R_B \cdot {}^G R_B^{\mathrm{T}} \right| = \left| {}^G R_B \right| \cdot \left| {}^G R_B^{\mathrm{T}} \right| = \left| {}^G R_B \right| \cdot \left| {}^G R_B \right| = \left| {}^G R_B \right|^2 = 1 \tag{4.281} $$

Using linear algebra and column vectors ${}^G\hat{\imath}$, ${}^G\hat{\jmath}$, and ${}^G\hat{k}$ of ${}^G R_B$, we know that

$$\left|{}^G R_B\right| = {}^G\hat{\imath} \cdot \left({}^G\hat{\jmath} \times {}^G\hat{k}\right) \tag{4.282}$$

and because the coordinate system is right handed, we have ${}^G\hat{\jmath} \times {}^G\hat{k} = {}^G\hat{\imath}$ and therefore,

$$\left|{}^G R_B\right| = {}^G\hat{\imath}^{\mathrm{T}} \cdot {}^G\hat{\imath} = +1 \tag{4.283}$$

■

Example 281 Elements of Transformation Matrix The coordinate frames of a body $B(Oxyz)$ and a global $G(OXYZ)$ are shown in Figure 4.14, and we are interested in determining the rotation matrix ${}^B R_G$.

The row elements of ${}^B R_G$ are the directional cosines of the axes of $B(Oxyz)$ in the coordinate frame $G(OXYZ)$. The x-axis lies in the (X, Z)-plane at 30 deg from the X-axis, and the angle between y and Y is 45 deg.

Therefore,

$$\begin{aligned}
{}^B R_G &= \begin{bmatrix} \hat{\imath}\cdot\hat{I} & \hat{\imath}\cdot\hat{J} & \hat{\imath}\cdot\hat{K} \\ \hat{\jmath}\cdot\hat{I} & \hat{\jmath}\cdot\hat{J} & \hat{\jmath}\cdot\hat{K} \\ \hat{k}\cdot\hat{I} & \hat{k}\cdot\hat{J} & \hat{k}\cdot\hat{K} \end{bmatrix} = \begin{bmatrix} \cos\frac{\pi}{6} & 0 & \sin\frac{\pi}{6} \\ \hat{\jmath}\cdot\hat{I} & \cos\frac{\pi}{4} & \hat{\jmath}\cdot\hat{K} \\ \hat{k}\cdot\hat{I} & \hat{k}\cdot\hat{J} & \hat{k}\cdot\hat{K} \end{bmatrix} \\
&= \begin{bmatrix} 0.866 & 0 & 0.5 \\ \hat{\jmath}\cdot\hat{I} & 0.707 & \hat{\jmath}\cdot\hat{K} \\ \hat{k}\cdot\hat{I} & \hat{k}\cdot\hat{J} & \hat{k}\cdot\hat{K} \end{bmatrix}
\end{aligned} \tag{4.284}$$

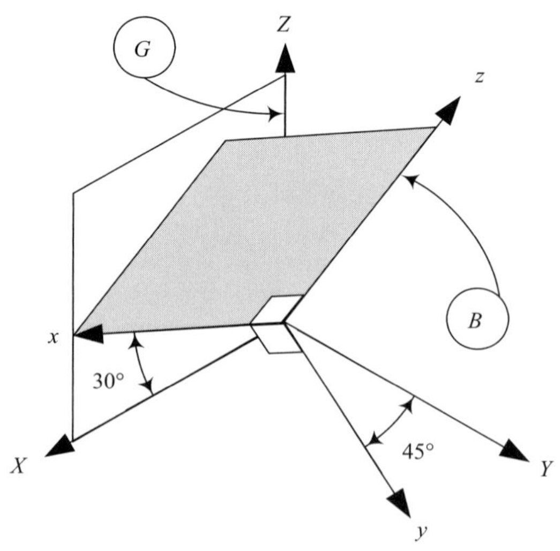

Figure 4.14 Body and global coordinate frames of Example 281.

and using ${}^{B}R_{G}\ {}^{G}R_{B} = {}^{B}R_{G}\ {}^{B}R_{G}^{\mathrm{T}} = I$,

$$\begin{bmatrix} 0.866 & 0 & 0.5 \\ r_{21} & 0.707 & r_{23} \\ r_{31} & r_{32} & r_{33} \end{bmatrix} \begin{bmatrix} 0.866 & r_{21} & r_{31} \\ 0 & 0.707 & r_{32} \\ 0.5 & r_{23} & r_{33} \end{bmatrix} = \begin{bmatrix} 1 & 0 & 0 \\ 0 & 1 & 0 \\ 0 & 0 & 1 \end{bmatrix} \tag{4.285}$$

we obtain a set of equations to find the missing elements:

$$\begin{aligned} 0.866\, r_{21} + 0.5 r_{23} &= 0 \\ 0.866\, r_{31} + 0.5 r_{33} &= 0 \\ r_{21}^2 + r_{23}^2 + 0.5 &= 1 \\ 0.707\, r_{32} + r_{21} r_{31} + r_{23} r_{33} &= 0 \\ r_{31}^2 + r_{32}^2 + r_{33}^2 &= 1 \end{aligned} \tag{4.286}$$

Solving these equations provides the following four possible transformation matrices:

$$\left({}^{B}R_{G}\right)_1 = \begin{bmatrix} 0.86603 & 0 & 0.5 \\ -0.35355 & 0.70711 & 0.61237 \\ 0.35355 & 0.70711 & -0.61237 \end{bmatrix} \tag{4.287}$$

$$\left({}^{B}R_{G}\right)_2 = \begin{bmatrix} 0.86603 & 0 & 0.5 \\ -0.35355 & 0.70711 & 0.61237 \\ -0.35355 & 0.61237 & -0.70711 \end{bmatrix} \tag{4.288}$$

$$\left({}^{B}R_{G}\right)_3 = \begin{bmatrix} 0.86603 & 0 & 0.5 \\ 0.35355 & 0.70711 & -0.61237 \\ 0.35355 & -0.70711 & -0.61237 \end{bmatrix} \tag{4.289}$$

$$\left({}^{B}R_{G}\right)_4 = \begin{bmatrix} 0.86603 & 0 & 0.5 \\ 0.35355 & 0.70711 & -0.61237 \\ -0.35355 & 0.70711 & 0.61237 \end{bmatrix} \tag{4.290}$$

The first, third, and fourth matrices satisfy the matrix orthogonality condition and are possible rotation matrices.

Example 282 Global Position Using ${}^{B}\mathbf{r}$ and ${}^{B}R_{G}$ The position vector $\mathbf{r}$ of a point P may be described in either the $G(OXYZ)$ or the $B(Oxyz)$ frame. If ${}^{B}\mathbf{r} = 10\hat{\imath} - 5\hat{\jmath} + 15\hat{k}$ and the transformation matrix to map ${}^{G}\mathbf{r}$ to ${}^{B}\mathbf{r}$ is

$${}^{B}\mathbf{r} = {}^{B}R_{G}\ {}^{G}\mathbf{r} = \begin{bmatrix} 0.866 & 0 & 0.5 \\ -0.353 & 0.707 & 0.612 \\ 0.353 & 0.707 & -0.612 \end{bmatrix} {}^{G}\mathbf{r} \tag{4.291}$$

then the components of ${}^{G}\mathbf{r}$ in $G(OXYZ)$ would be

$$ {}^{G}\mathbf{r} = {}^{G}R_B\, {}^{B}\mathbf{r} = {}^{B}R_G^{\mathrm{T}}\, {}^{B}\mathbf{r} = \begin{bmatrix} 15.72 \\ 7.07 \\ -7.24 \end{bmatrix} \tag{4.292} $$

Example 283 Two-Point Transformation Matrix The global position vectors of two points P_1 and P_2 of a rigid body B are

$$ {}^{G}\mathbf{r}_{P_1} = \begin{bmatrix} 1.077 \\ 1.365 \\ 2.666 \end{bmatrix} \qquad {}^{G}\mathbf{r}_{P_2} = \begin{bmatrix} -0.473 \\ 2.239 \\ -0.959 \end{bmatrix} \tag{4.293} $$

The origin of the body $B(Oxyz)$ is fixed on the origin of $G(OXYZ)$, and the points P_1 and P_2 are lying on the local x- and y-axis, respectively.

To find ${}^{G}R_B$, we use the local unit vectors ${}^{G}\hat{\imath}$ and ${}^{G}\hat{\jmath}$,

$$ {}^{G}\hat{\imath} = \frac{{}^{G}\mathbf{r}_{P_1}}{\left|{}^{G}\mathbf{r}_{P_1}\right|} = \begin{bmatrix} 0.338 \\ 0.429 \\ 0.838 \end{bmatrix} \qquad {}^{G}\hat{\jmath} = \frac{{}^{G}\mathbf{r}_{P_2}}{\left|{}^{G}\mathbf{r}_{P_2}\right|} = \begin{bmatrix} -0.191 \\ 0.902 \\ -0.387 \end{bmatrix} \tag{4.294} $$

to obtain

$$ {}^{G}\hat{k} = \hat{\imath} \times \hat{\jmath} = \begin{bmatrix} -0.922 \\ -0.029 \\ 0.387 \end{bmatrix} \tag{4.295} $$

Hence, the transformation matrix ${}^{G}R_B$ would be

$$ {}^{G}R_B = \begin{bmatrix} {}^{G}\hat{\imath} & {}^{G}\hat{\jmath} & {}^{G}\hat{k} \end{bmatrix} = \begin{bmatrix} 0.338 & -0.191 & -0.922 \\ 0.429 & 0.902 & -0.029 \\ 0.838 & -0.387 & 0.387 \end{bmatrix} \tag{4.296} $$

Example 284 Length Invariant of a Position Vector Expressing a vector in different frames utilizing rotation matrices does not affect the length and direction properties of the vector. So, the length of a vector is an invariant property:

$$ |\mathbf{r}| = \left|{}^{G}\mathbf{r}\right| = \left|{}^{B}\mathbf{r}\right| \tag{4.297} $$

The length invariant property can be shown as

$$ \begin{aligned} |\mathbf{r}|^2 &= {}^{G}\mathbf{r}^{\mathrm{T}}\, {}^{G}\mathbf{r} = \left[{}^{G}R_B\, {}^{B}\mathbf{r}\right]^{\mathrm{T}}\, {}^{G}R_B\, {}^{B}\mathbf{r} = {}^{B}\mathbf{r}^{\mathrm{T}}\, {}^{G}R_B^{\mathrm{T}}\, {}^{G}R_B\, {}^{B}\mathbf{r} \\ &= {}^{B}\mathbf{r}^{\mathrm{T}}\, {}^{B}\mathbf{r} \end{aligned} \tag{4.298} $$

Example 285 Inverse of Euler Angle Rotation Matrix Precession–nutation–spin or Euler angle rotation matrix (4.142),

$$
\begin{aligned}
{}^{B}R_{G} &= A_{z,\psi}A_{x,\theta}A_{z,\varphi} \\
&= \begin{bmatrix} c\varphi c\psi - c\theta s\varphi s\psi & c\psi s\varphi + c\theta c\varphi s\psi & s\theta s\psi \\ -c\varphi s\psi - c\theta c\psi s\varphi & -s\varphi s\psi + c\theta c\varphi c\psi & s\theta c\psi \\ s\theta s\varphi & -c\varphi s\theta & c\theta \end{bmatrix}
\end{aligned} \tag{4.299}
$$

must be inverted to be a transformation matrix that maps body coordinates to global coordinates:

$$
\begin{aligned}
{}^{G}R_{B} &= {}^{B}R_{G}^{-1} = {}^{B}R_{G}^{\mathrm{T}} = A_{z,\varphi}^{\mathrm{T}}A_{x,\theta}^{\mathrm{T}}A_{z,\psi}^{\mathrm{T}} \\
&= \begin{bmatrix} c\varphi c\psi - c\theta s\varphi s\psi & -c\varphi s\psi - c\theta c\psi s\varphi & s\theta s\varphi \\ c\psi s\varphi + c\theta c\varphi s\psi & -s\varphi s\psi + c\theta c\varphi c\psi & -c\varphi s\theta \\ s\theta s\psi & s\theta c\psi & c\theta \end{bmatrix}
\end{aligned} \tag{4.300}
$$

The transformation matrix (4.299) is called a local Euler rotation matrix, and (4.300) is called a global Euler rotation matrix.

Example 286 ★ Tracking Transformation Consider a tracking radar that is to track a remote object such as an airplane. The radar is at the origin O of a coordinate frame $G(OXYZ)$ that is attached to a tangent plane on the surface of Earth. We also attach a local coordinate frame $B(Oxyz)$ to the radar. The radar frame B is initially coincident with G. In radar-tracking kinematics we may define the X- and Y-axes of G to be lying in the tangent plane and pointing North and East, respectively. Therefore, the Z-axis is perpendicular to the surface of Earth and points toward the center of Earth. Such a coordinate frame with an axis on Earth's radius is called the *geocentric frame*.

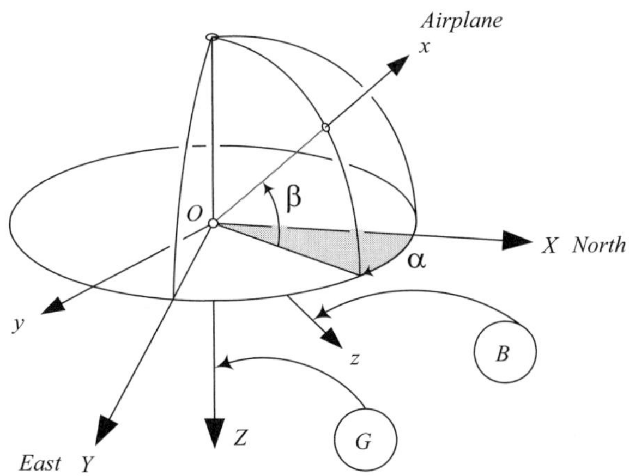

Figure 4.15 The tracking azimuth α and elevation β angles are used to point an object with the x-axis.

To track the airplane, we move the x-axis to point to the airplane. To do so, we turn the radar α degrees about the Z-axis followed by a rotation β about the y-axis. The angle α is called the *azimuth* or *heading angle* and is defined in the tangent plane between North and the projected position vector of the airplane on the (X, Y)-plane. The angle β is called the *elevation angle* and is defined as the angle between the position vector of the airplane and the (X, Y)-plane. The configurations of the coordinate frames are shown in Figure 4.15.

The transformation matrices between B and G are

$$R_1 = {}^G R_B = R_{Z,\alpha} \tag{4.301}$$

$$R_2 = {}^B R_G = R_{y,\beta} \tag{4.302}$$

So, the combined transformation matrix is

$$\begin{aligned} {}^G R_B &= {}^B R_G^{\mathrm{T}} = \left[R_{y,\beta}\, R_{Z,\alpha}^{\mathrm{T}}\right]^{\mathrm{T}} = R_{Z,\alpha}\, R_{y,\beta}^{\mathrm{T}} \\ &= \begin{bmatrix} \cos\alpha & -\sin\alpha & 0 \\ \sin\alpha & \cos\alpha & 0 \\ 0 & 0 & 1 \end{bmatrix} \begin{bmatrix} \cos\beta & 0 & -\sin\beta \\ 0 & 1 & 0 \\ \sin\beta & 0 & \cos\beta \end{bmatrix}^{\mathrm{T}} \\ &= \begin{bmatrix} \cos\alpha\cos\beta & -\sin\alpha & \cos\alpha\sin\beta \\ \cos\beta\sin\alpha & \cos\alpha & \sin\alpha\sin\beta \\ -\sin\beta & 0 & \cos\beta \end{bmatrix} \end{aligned} \tag{4.303}$$

Now assume that at four o'clock and a distance $r = 16\,\mathrm{km}$ an airplane is moving north at height $h = 3\,\mathrm{km}$. Four o'clock means that

$$\alpha = 120\,\mathrm{deg} = \frac{2\pi}{3}\,\mathrm{rad} \tag{4.304}$$

and the distance $r = 16\,\mathrm{km}$ and height $h = 3\,\mathrm{km}$ mean

$${}^B\mathbf{r} = 16\hat{\imath} \tag{4.305a}$$

$$\beta = \arcsin\frac{3}{16} = 0.18862\,\mathrm{rad} \approx 10.807\,\mathrm{deg} \tag{4.305b}$$

Therefore, the global coordinates of the airplane are

$$\begin{bmatrix} X \\ Y \\ Z \end{bmatrix} = {}^G R_B \begin{bmatrix} 16 \\ 0 \\ 0 \end{bmatrix} = \begin{bmatrix} -7.8581 \\ 13.611 \\ -3.0001 \end{bmatrix} \mathrm{km} \tag{4.306}$$

Example 287 ★ Coordinates on the Spherical Earth For many practical problems, we can deal with Earth as a spherical body. The Earth's axis is the diameter that connects the north and south poles. The great circle whose plane is perpendicular to the axis is called the *equator*. Let us set up a coordinate frame at Earth's center such that the Z-axis points to the north pole and the X- and Y-axes lie on the equatorial plane, as shown in Figure 4.16 .

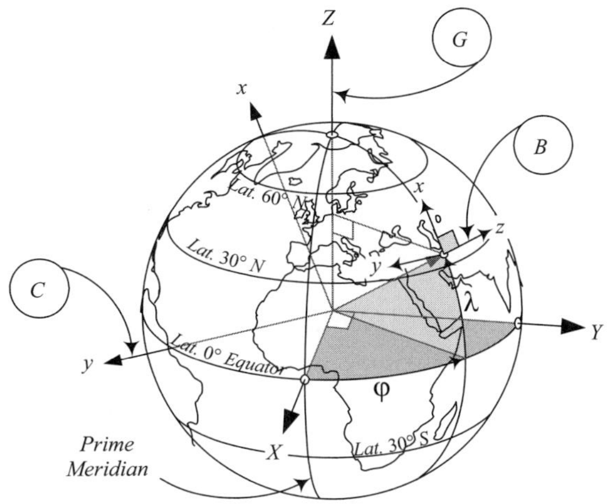

Figure 4.16 Earth's coordinate frame sits at the center such that the Z-axis points to the north pole and the X- and Y-axes lie on the equatorial plane with X pointing to the Greenwich Meridian.

Any location on Earth is expressed by two angles: *longitude* φ and *latitude* λ. The angles of latitude and longitude are coordinates that specify the position on a map, usually measured in degrees, minutes of arc, and seconds of arc (deg, $'$, $''$). The distance between two points subtending an angle of one minute of arc at the center of Earth is known as the *nautical mile*. So, there are $360 \times 60 = 21{,}600$ nautical miles on a great circle around Earth:

$$1 \text{ nmi} = R_0 \frac{1}{60} \frac{\pi}{180} \approx 1.852 \,\text{km} \approx 6076.1 \,\text{ft} \approx 1.1508 \,\text{mi} \tag{4.307}$$

Based on the nautical mile, we may define a knot as the speed equal to one nautical mile in one hour. The knot is used in sea and air navigation:

$$1 \text{ knot} \approx 0.514 \,\text{m/s} \approx 1.852 \,\text{km/h} \approx 1.1508 \,\text{mi/h} \tag{4.308}$$

The latitude λ of a point P on the Earth's surface is the angle between its position vector $\mathbf{r}_P$ and the projection of $\mathbf{r}_P$ on the equatorial plane. So, $\lambda > 0$ north of the equator and $\lambda < 0$ south of the equator. Lines of constant latitude are circles of different sizes. The largest is the equator, whose latitude is zero, and the smallest are two points at the poles where latitude is 90 and -90 deg.

Any semi–great circle terminated by the poles is a *meridian*. So, meridians are lines of constant longitude that extend from pole to pole. The equator is divided into 360 deg and the meridian passing the Royal Astronomical Observatory in Greenwich, located at the eastern edge of London, England, is the one chosen as zero longitude. The longitude φ of a point is the associated angle of its meridian on the equator. Longitudes are measured from 0 deg to 180 deg east of the Greenwich Meridian and from 0 deg to 180 deg west.

The intersection of the Greenwich Meridian and the equator indicates the X-axis of Earth's coordinate frame. Such a coordinate frame is sometimes called the Earth-Centerd Earth-Fixed (*ECEF*) or the Conventional Terrestrial System (CTS).

Figure 4.16 indicates the latitude and longitude of Tehran as a sample point P on Earth:

$$\text{Latitude}: 35^\circ 40' 19'' \text{N} \qquad \text{Longitude}: 51^\circ 25' 28'' \text{ E} \tag{4.309}$$

To have a local coordinate B at P such that its x-axis points North, the y-axis points West, and the z-axis points up, we may define an intermediate coordinate frame C at the center of Earth. The frame G may go to C by a rotation $-(\pi - \varphi)$ about the z-axis followed by a rotation $-(\pi/2 - \lambda)$ about the y-axis. The transformation between G and C is

$$\begin{aligned} {}^C R_G &= R_{y,-(\pi/2-\lambda)} R_{z,-(\pi-\varphi)} \\ &= \begin{bmatrix} -\cos\varphi \sin\lambda & -\sin\lambda \sin\varphi & \cos\lambda \\ \sin\varphi & -\cos\varphi & 0 \\ \cos\lambda \cos\varphi & \cos\lambda \sin\varphi & \sin\lambda \end{bmatrix} \end{aligned} \tag{4.310}$$

Therefore, the global coordinates of the point P at

$$\begin{aligned} \varphi &= 51^\circ 25' 28'' \approx 51.424 \text{ deg} \approx 0.89752 \text{ rad} \\ \lambda &= 35^\circ 40' 19'' \approx 35.672 \text{ deg} \approx 0.62259 \text{ rad} \end{aligned} \tag{4.311}$$

$$r_P \approx R_0 = 6371230 \text{ m} \tag{4.312}$$

would be

$$\begin{aligned} {}^G\mathbf{r} &= {}^C R_G^{\mathrm{T}} \, {}^C\mathbf{r} \\ &= \begin{bmatrix} -0.36362 & -0.45589 & 0.81237 \\ 0.78178 & -0.62355 & 0 \\ 0.50655 & 0.63510 & 0.58314 \end{bmatrix}^{\mathrm{T}} \begin{bmatrix} 0 \\ 0 \\ 6371230 \end{bmatrix} \\ &= \begin{bmatrix} 3.2273 \times 10^6 \\ 4.0464 \times 10^6 \\ 3.7153 \times 10^6 \end{bmatrix} \text{ m} \end{aligned} \tag{4.313}$$

The coordinates of the origin of the B-frame at a point $P(X, Y, Z)$ would be

$$\begin{aligned} {}^G\mathbf{r} &= {}^C R_G^{\mathrm{T}} \, {}^C\mathbf{r} \\ &= \begin{bmatrix} -\cos\varphi \sin\lambda & -\sin\lambda \sin\varphi & \cos\lambda \\ \sin\varphi & -\cos\varphi & 0 \\ \cos\lambda \cos\varphi & \cos\lambda \sin\varphi & \sin\lambda \end{bmatrix}^{\mathrm{T}} \begin{bmatrix} 0 \\ 0 \\ R_0 \end{bmatrix} \\ &= R_0 \begin{bmatrix} \cos\lambda \cos\varphi \\ \cos\lambda \sin\varphi \\ \sin\lambda \end{bmatrix} \end{aligned} \tag{4.314}$$

which can be used to determine the longitude φ and latitude λ for the point P:

$$\tan\varphi = \frac{Y}{X} \tag{4.315}$$

$$\tan\lambda = \frac{Z}{\sqrt{X^2 + Y^2}} \tag{4.316}$$

The body frame B is achieved from C by a translation $\begin{bmatrix} 0 & 0 & R_0 \end{bmatrix}$:

$$^B\mathbf{r} = {}^C\mathbf{r} - R_0\hat{k} \tag{4.317}$$

Therefore, the global coordinates of P at the origin of B are

$$^G\mathbf{r} = {}^C R_G^{\mathrm{T}}\, {}^C\mathbf{r} = {}^C R_G^{\mathrm{T}} \left({}^B\mathbf{r} + R_0\hat{k} \right)$$

$$= {}^C R_G^{\mathrm{T}} \left(\begin{bmatrix} 0 \\ 0 \\ 0 \end{bmatrix} + \begin{bmatrix} 0 \\ 0 \\ 6{,}371{,}230 \end{bmatrix} \right) = \begin{bmatrix} 3.2273 \times 10^6 \\ 4.0464 \times 10^6 \\ 3.7153 \times 10^6 \end{bmatrix} \mathrm{m} \tag{4.318}$$

Example 288 ★ Satellite-Tracking Angles The tracking azimuth and elevation angles are also used to adjust the dish of satellite receivers as shown in Figure 4.17. Let us attach a geocentric coordinate frame $B_1(OXYZ)$ on Earth at the dish point P such that the X-axis points North, the Y-axis points West, and the Z-axis points up. The azimuth angle α is traditionally measured clockwise such that North $= 0$ deg, South $= 180$ deg, East $= 90$ deg, and West $= 270$ deg. Elevation angle β is measured from the tangent surface on Earth to the sky such that $\beta = 90$ deg points straight up.

Let us also attach a local coordinate frame *B(Oxyz)* to the dish such that its x-axis is on the centerline of the dish and B is originally coincident with B_1. To adjust the dish to point to a specific satellite, we need to aim the satellite by the centerline.

To point the x-axis of B to the satellite, we can turn B from a coincident orientation with B_1, $-\alpha$ degrees about the z-axis and then turn it $-\beta$ degrees about the y-axis.

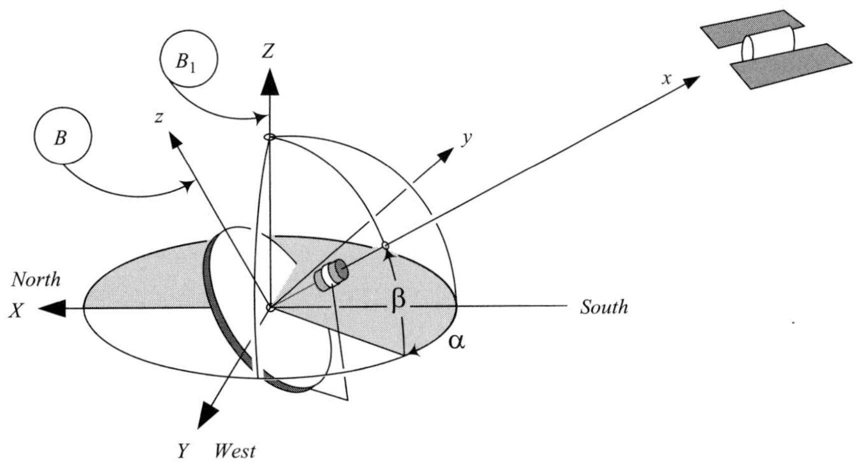

Figure 4.17 The tracking azimuth α and elevation β angles to point a satellite with the x-axis.

So, the transformation matrix between B and B_1 is

$$\begin{aligned} {}^{B}R_1 &= R_{y,-\beta}R_{z,-\alpha} \\ &= \begin{bmatrix} \cos\alpha\cos\beta & -\cos\beta\sin\alpha & \sin\beta \\ \sin\alpha & \cos\alpha & 0 \\ -\cos\alpha\sin\beta & \sin\alpha\sin\beta & \cos\beta \end{bmatrix} \end{aligned} \tag{4.319}$$

The satellite is at ${}^{B}\mathbf{r}_P = [r_P, 0, 0]$ and therefore its position in the tangent frame B_1 is

$$ {}^{1}\mathbf{r} = {}^{B}R_1^{\mathrm{T}}\,{}^{1}\mathbf{r} = {}^{B}R_1^{\mathrm{T}} \begin{bmatrix} r_P \\ 0 \\ 0 \end{bmatrix} = \begin{bmatrix} r_P\cos\beta\cos\alpha \\ -r_P\cos\beta\sin\alpha \\ r_P\sin\beta \end{bmatrix} \tag{4.320}$$

Therefore, if we know the coordinate of the satellite in the tangent plane B_1, we are able to determine the required azimuth α and elevation β to adjust the dish antenna:

$$\tan\alpha = -\frac{Y}{X} \qquad \tan\beta = \frac{Z}{\sqrt{X^2+Y^2}} \tag{4.321}$$

Example 289 ★ Group Property of Transformations A set S together with a binary operation $\otimes$ between the elements of S is called a group $(S, \otimes)$ if it satisfies the following four axioms:

1. *Closure*: If $s_1, s_2 \in S$, then $s_1 \otimes s_2 \in S$.
2. *Identity*: There exists an identity element s_0 such that $s_0 \otimes s = s \otimes s_0 = s \forall s \in S$.
3. *Inverse*: For any $s \in S$ there exists a unique inverse $s^{-1} \in S$ such that $s^{-1} \otimes s = s \otimes s^{-1} = s_0$.
4. *Associativity*: If $s_1, s_2, s_3 \in S$, then $(s_1 \otimes s_2) \otimes s_3 = s_1 \otimes (s_2 \otimes s_3)$.

Three-dimensional coordinate transformations make a group if we define the set of rotation matrices by

$$S = \left\{R \in \mathbb{R}^{3\times 3} : RR^{\mathrm{T}} = R^{\mathrm{T}}R = \mathbf{I}, |R| = 1\right\} \tag{4.322}$$

Therefore, the elements of the set S are transformation matrices $[R]$, the binary operator $\otimes$ means matrix multiplication, the identity matrix is $\mathbf{I}$, and the inverse of element $[R]$ is $[R]^{-1} = [R]^{\mathrm{T}}$.

The set S is also a continuous group because:

1. Binary matrix multiplication is a continuous operation.
2. The inverse of any element in S is a continuous function of that element.

Such a set S is a *differentiable manifold*. A group that is a differentiable manifold is called a *Lie group*.

Example 290 ★ Transformation with Determinant -1 An orthogonal matrix with determinant $+1$ corresponds to a rotation as described in Equation (4.280). In contrast, an orthogonal matrix with determinant -1 describes a reflection. Moreover, it transforms a right-handed coordinate system into a left-handed one, and vice versa. This transformation does not correspond to any physical action on rigid bodies.

4.10 ACTIVE AND PASSIVE ROTATIONS

We call the rotation of a local frame B in a global frame G an *active rotation* if the position vector ${}^B\mathbf{r}$ of a point P is fixed in the local frame and rotates with it. Alternatively, the rotation of a local frame B in a global frame G is called a *passive rotation* if the position vector ${}^G\mathbf{r}$ of a point P is fixed in the global frame and does not rotate with the local frame.

The passive and active transformations are mathematically equivalent. The coordinates of a point P can be transformed from one coordinate to the other by a proper rotation matrix. In an active rotation, the observer is standing in a global frame G and calculating the position of body points in G. However, in a passive rotation, the observer is standing on the body frame B and calculating the position of global points in B.

If $R_1 = {}^GR_B$ is an active rotation matrix about an axis, such as the Z-axis, then

$$ {}^G\mathbf{r} = R_1\,{}^B\mathbf{r} = R_{Z,\alpha}\,{}^B\mathbf{r} \tag{4.323} $$

However, if the same rotation is a passive rotation and $R_2 = {}^BR_G$ is the rotation matrix, then

$$ {}^G\mathbf{r} = R_2^{\mathrm{T}}\,{}^B\mathbf{r} = R_{Z,-\alpha}\,{}^B\mathbf{r} \tag{4.324} $$

In rigid-body kinematics, we usually work with active rotations and examine the rotation of a rigid body in a global frame.

Proof: A local frame $B(Oxyz)$ that is initially coincident with a global frame $G(OXYZ)$ performs a rotation. In an active rotation, a body point P will move with B and its global coordinate will be found by the proper rotation matrix $R_1 = {}^GR_B$. For simplicity, let us assume that the axis of rotation is the Z-axis:

$$ {}^G\mathbf{r} = R_1\,{}^B\mathbf{r} = {}^GR_B\,{}^B\mathbf{r} = R_{Z,\alpha}\,{}^B\mathbf{r} \tag{4.325} $$

In a passive rotation, the point P will keep its global coordinates. We may switch the roll of frames B and G to consider an active rotation of $-\alpha$ for G in B about the z-axis. The coordinate of P in B will be found by the proper rotation matrix $R_1 = {}^BR_G$:

$$ {}^B\mathbf{r} = R_2\,{}^G\mathbf{r} = {}^BR_G\,{}^G\mathbf{r} = R_{z,-\alpha}\,{}^G\mathbf{r} \tag{4.326} $$

So, the global coordinates of P in the passive rotation may be found from (4.326) as

$$ {}^G\mathbf{r} = R_2^{\mathrm{T}}\,{}^B\mathbf{r} = R_{z,-\alpha}^{\mathrm{T}}\,{}^B\mathbf{r} \tag{4.327} $$

However, because of $R_{Z,\alpha} = R_{z,\alpha}^{\mathrm{T}}$, we have

$$ {}^G\mathbf{r} = R_2^{\mathrm{T}}\,{}^B\mathbf{r} = R_{Z,-\alpha}\,{}^B\mathbf{r} \tag{4.328} $$

■

Example 291 Active and Passive Rotation about Z-Axis When the local and global frames B and G are coincident, a body point P is at ${}^{B}\mathbf{r}$:

$$
{}^{B}\mathbf{r} = \begin{bmatrix} 1 \\ 2 \\ 3 \end{bmatrix} \tag{4.329}
$$

A rotation 45 deg about the Z-axis will move the point to ${}^{G}\mathbf{r}$:

$$
\begin{aligned}
{}^{G}\mathbf{r} &= R_{Z,90}\,{}^{B}\mathbf{r} \\
&= \begin{bmatrix} \cos\frac{\pi}{2} & -\sin\frac{\pi}{2} & 0 \\ \sin\frac{\pi}{2} & \cos\frac{\pi}{2} & 0 \\ 0 & 0 & 1 \end{bmatrix} \begin{bmatrix} 1 \\ 2 \\ 3 \end{bmatrix} = \begin{bmatrix} -2 \\ 1 \\ 3 \end{bmatrix}
\end{aligned} \tag{4.330}
$$

Now assume that the point P is a fixed point in G. When B rotates 90 deg about the Z-axis, the coordinates of P in the local frame will change such that

$$
\begin{aligned}
{}^{B}\mathbf{r} &= R_{Z,-90}\,{}^{G}\mathbf{r} \\
&= \begin{bmatrix} \cos\frac{-\pi}{2} & -\sin\frac{-\pi}{2} & 0 \\ \sin\frac{-\pi}{2} & \cos\frac{-\pi}{2} & 0 \\ 0 & 0 & 1 \end{bmatrix} \begin{bmatrix} 1 \\ 2 \\ 3 \end{bmatrix} = \begin{bmatrix} 2 \\ -1 \\ 3 \end{bmatrix}
\end{aligned} \tag{4.331}
$$

Example 292 Multiple Rotation about Global Axes Consider a globally fixed point P at

$$
{}^{G}\mathbf{r} = \begin{bmatrix} 1 \\ 2 \\ 3 \end{bmatrix} \tag{4.332}
$$

The body B will turn 45 deg about the X-axis and then 45 deg about the Y-axis. An observer in B will see P at

$$
\begin{aligned}
{}^{B}\mathbf{r} &= R_{y,-45}\,R_{x,-45}\,{}^{G}\mathbf{r} \\
&= \begin{bmatrix} \cos\frac{-\pi}{4} & 0 & -\sin\frac{-\pi}{4} \\ 0 & 1 & 0 \\ \sin\frac{-\pi}{4} & 0 & \cos\frac{-\pi}{4} \end{bmatrix} \begin{bmatrix} 1 & 0 & 0 \\ 0 & \cos\frac{-\pi}{4} & \sin\frac{-\pi}{4} \\ 0 & -\sin\frac{-\pi}{4} & \cos\frac{-\pi}{4} \end{bmatrix} \begin{bmatrix} 1 \\ 2 \\ 3 \end{bmatrix} \\
&= \begin{bmatrix} 0.707 & 0.5 & 0.5 \\ 0 & 0.707 & -0.707 \\ -0.707 & 0.5 & 0.5 \end{bmatrix} \begin{bmatrix} 1 \\ 2 \\ 3 \end{bmatrix} = \begin{bmatrix} 3.207 \\ -0.707 \\ 1.793 \end{bmatrix}
\end{aligned} \tag{4.333}
$$

To check this result, let us change the role of B and G. So, the body point at

$$
{}^{B}\mathbf{r} = \begin{bmatrix} 1 \\ 2 \\ 3 \end{bmatrix} \tag{4.334}
$$

undergoes an active rotation of 45 deg about the x-axis followed by 45 deg about the y-axis. The global coordinates of the point would be

$$^{B}\mathbf{r} = R_{y,45}\, R_{x,45}\, {}^{G}\mathbf{r} \tag{4.335}$$

so

$$^{G}\mathbf{r} = \left[R_{y,45}\, R_{x,45}\right]^{\mathrm{T}}\, {}^{B}\mathbf{r} = R_{x,45}^{\mathrm{T}}\, R_{y,45}^{\mathrm{T}}\, {}^{B}\mathbf{r} \tag{4.336}$$

Example 293 Multiple Rotations about Body Axes Consider a globally fixed point P at

$$^{G}\mathbf{r} = \begin{bmatrix} 1 \\ 2 \\ 3 \end{bmatrix} \tag{4.337}$$

The body B will turn 45 deg about the x-axis and then 45 deg about the y-axis. An observer in B will see P at

$$\begin{aligned} ^{B}\mathbf{r} &= R_{Y,-45}\, R_{X,-45}\, {}^{G}\mathbf{r} \\ &= \begin{bmatrix} \cos\frac{-\pi}{4} & 0 & \sin\frac{-\pi}{4} \\ 0 & 1 & 0 \\ -\sin\frac{-\pi}{4} & 0 & \cos\frac{-\pi}{4} \end{bmatrix} \begin{bmatrix} 1 & 0 & 0 \\ 0 & \cos\frac{-\pi}{4} & -\sin\frac{-\pi}{4} \\ 0 & \sin\frac{-\pi}{4} & \cos\frac{-\pi}{4} \end{bmatrix} \begin{bmatrix} 1 \\ 2 \\ 3 \end{bmatrix} \\ &= \begin{bmatrix} 0.707 & 0.5 & -0.5 \\ 0 & 0.707 & 0.707 \\ 0.707 & -0.5 & 0.5 \end{bmatrix} \begin{bmatrix} 1 \\ 2 \\ 3 \end{bmatrix} = \begin{bmatrix} 0.20711 \\ 3.5356 \\ 1.2071 \end{bmatrix} \end{aligned} \tag{4.338}$$

4.11 ★ ROTATION OF ROTATED BODY

Consider a rigid body with a fixed point at position B_0 that is not necessarily coincident with the global frame G. The rotation matrix $^{G}R_0$ between B_0 and G is known. If the body rotates from this initial position, then the transformation matrix $^{G}R_B$ between the final position of the body and the global frame is

$$^{G}R_B = {}^{G}R_0\, {}^{0}R_B \tag{4.339}$$

The matrix $^{0}R_B$ is the transformation between initial and final positions of the body with the assumption that B_0 is a fixed position.

Proof: The principal assumption in the theory of rotation kinematics is that the body coordinate frame B always begins to rotate from a coincident configuration with the global frame G. If the body frame is not initially on the global frame, we show it by B_0 and assume that the body is already rotated from the coincident configuration with G. The rotated body B_0 is then going to turn a second rotation.

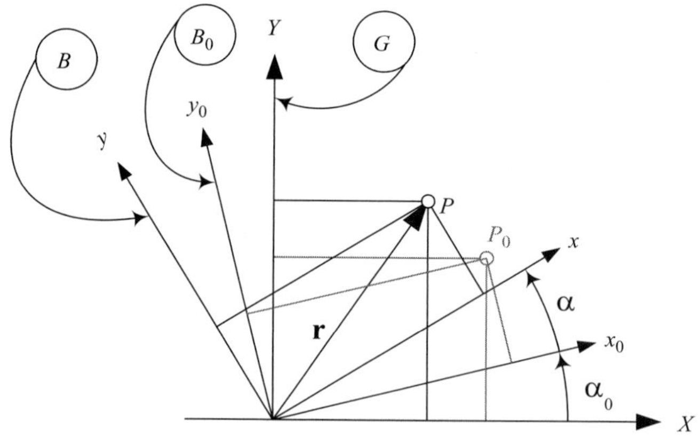

Figure 4.18 Rotation of a rotated rigid body about the Z-axis.

Figure 4.18 illustrates a body frame B at its final position after a rotation of α degrees about the Z-axis from an initial position B_0. The initial position of the body is after a first rotation of α_0 degrees about the Z-axis. In this figure both the first and second rotations are about the global Z-axis.

To prevent confusion and keep the previous definitions, we define three transformation matrices ${}^GR_0$, ${}^0R_B$, ${}^GR_B$. The matrix ${}^GR_0$ indicates the transformation between G and the initial position of the body at B_0. The matrix ${}^0R_B$ is the transformation between the initial position of the body at B_0 and the final position of the body B, and the matrix ${}^GR_B$ is the transformation between G and the final position of the body B.

We assume that ${}^GR_0$ is given as

$$
{}^GR_0 = \begin{bmatrix} \cos\alpha_0 & -\sin\alpha_0 & 0 \\ \sin\alpha_0 & \cos\alpha_0 & 0 \\ 0 & 0 & 1 \end{bmatrix} \tag{4.340}
$$

The matrix ${}^GR_B$ can be found using the regular method. We consider a body frame starts from G and goes to the final position by two rotations α_0 and α:

$$
\begin{aligned}
{}^GR_B &= R_{Z,\alpha_0}\, R_{Z,\alpha} = {}^GR_0\, {}^0R_B \\
&= \begin{bmatrix} \cos(\alpha+\alpha_0) & -\sin(\alpha+\alpha_0) & 0 \\ \sin(\alpha+\alpha_0) & \cos(\alpha+\alpha_0) & 0 \\ 0 & 0 & 1 \end{bmatrix}
\end{aligned} \tag{4.341}
$$

Therefore, the rotation matrix ${}^BR_0$ is given as

$$
{}^BR_0 = {}^GR_B^{\mathrm{T}}\, {}^GR_0 = \begin{bmatrix} \cos\alpha & \sin\alpha & 0 \\ -\sin\alpha & \cos\alpha & 0 \\ 0 & 0 & 1 \end{bmatrix} \tag{4.342}
$$

The method is correct for a general case in which the first and second rotations are about the same axis.

To see the application, let us assume that a point P is at

$$^B\mathbf{r} = x\hat{\imath} + y\hat{\jmath} + z\hat{k} \tag{4.343}$$

Its global position is

$$^G\mathbf{r} = {}^G R_B \, {}^B\mathbf{r} \tag{4.344}$$

and its position in the initial frame B_0 is

$$^0\mathbf{r} = {}^B R_0^{\mathrm{T}} \, {}^B\mathbf{r} \tag{4.345}$$

When the body is at its initial position, the global coordinates of P are

$$^{G_0}\mathbf{r} = {}^G R_0 \, {}^0\mathbf{r} \tag{4.346}$$

■

Example 294 Rotation about X ***of a Rotated Body about*** Z Consider a rigid body that is already rotated 30 deg about the Z-axis and is at B_0:

$$^G R_0 = \begin{bmatrix} \cos\frac{\pi}{6} & -\sin\frac{\pi}{6} & 0 \\ \sin\frac{\pi}{6} & \cos\frac{\pi}{6} & 0 \\ 0 & 0 & 1 \end{bmatrix} \tag{4.347}$$

The body rotates 60 deg about the X-axis from the position B_0. The rotation matrix $^G R_B$ is given as

$$^G R_B = R_{X,\pi/6} \, R_{Z,\pi/3} = \begin{bmatrix} 0.86603 & -0.5 & 0 \\ 0.35355 & 0.61237 & -0.70711 \\ 0.35355 & 0.61237 & 0.70711 \end{bmatrix} \tag{4.348}$$

Therefore,

$$^0 R_B = {}^G R_0^{\mathrm{T}} \, {}^G R_B = \begin{bmatrix} 0.92678 & -0.12683 & -0.35356 \\ -0.12683 & 0.78033 & -0.61238 \\ 0.35355 & 0.61237 & 0.70711 \end{bmatrix} \tag{4.349}$$

If a point P is at

$$^B\mathbf{r} = 1\hat{\imath} + 1\hat{\jmath} + 1\hat{k} \tag{4.350}$$

its global coordinates are

$$^G\mathbf{r} = {}^G R_B \, {}^B\mathbf{r} = {}^G R_B \begin{bmatrix} 1 \\ 1 \\ 1 \end{bmatrix} = \begin{bmatrix} 0.36603 \\ 0.25881 \\ 1.673 \end{bmatrix} \tag{4.351}$$

The coordinates of the current position of P can be found in B_0 as

$$^0\mathbf{r} = {}^0 R_B \, {}^B\mathbf{r} = {}^0 R_B \begin{bmatrix} 1 \\ 1 \\ 1 \end{bmatrix} = \begin{bmatrix} 0.446\,39 \\ 0.041\,12 \\ 1.673 \end{bmatrix} \tag{4.352}$$

Example 295 Rotation about x of a Rotated Body about Z Consider a rigid body that is already rotated 45 deg about the Z-axis and is at B_0:

$$ {}^G R_0 = \begin{bmatrix} \cos\frac{\pi}{4} & -\sin\frac{\pi}{4} & 0 \\ \sin\frac{\pi}{4} & \cos\frac{\pi}{4} & 0 \\ 0 & 0 & 1 \end{bmatrix} \approx \begin{bmatrix} 0.707 & -0.707 & 0 \\ 0.707 & 0.707 & 0 \\ 0 & 0 & 1 \end{bmatrix} \tag{4.353} $$

The body rotates 45 deg about the x-axis from the position B_0. Considering B_0 as a fixed frame, we can determine ${}^0R_B$:

$$ {}^0R_B = {}^B R_0^{\mathrm{T}} = R_{x,\pi/4}^{\mathrm{T}} = \begin{bmatrix} 1 & 0 & 0 \\ 0 & \cos\frac{\pi}{4} & \sin\frac{\pi}{4} \\ 0 & -\sin\frac{\pi}{4} & \cos\frac{\pi}{4} \end{bmatrix}^{\mathrm{T}} \tag{4.354} $$

Therefore,

$$ {}^G R_B = {}^G R_0\, {}^0R_B = \begin{bmatrix} 0.70711 & -0.5 & 0.5 \\ 0.70711 & 0.5 & -0.5 \\ 0 & 0.70711 & 0.70711 \end{bmatrix} \tag{4.355} $$

Let us watch a triangle on the body with corners at O, P_1, and P_2, where

$$ {}^B\mathbf{r}_1 = 1\hat{\imath} \tag{4.356} $$

$$ {}^B\mathbf{r}_2 = 1\hat{\imath} + 1\hat{\jmath} + 1\hat{k} \tag{4.357} $$

The global coordinates of P_1 and P_2 are

$$ {}^G\mathbf{r}_1 = {}^G R_B\, {}^B\mathbf{r}_1 = {}^G R_B \begin{bmatrix} 1 \\ 0 \\ 0 \end{bmatrix} = \begin{bmatrix} 0.707\,11 \\ 0.707\,11 \\ 0 \end{bmatrix} \tag{4.358} $$

$$ {}^G\mathbf{r}_2 = {}^G R_B\, {}^B\mathbf{r}_2 = {}^G R_B \begin{bmatrix} 1 \\ 1 \\ 1 \end{bmatrix} = \begin{bmatrix} 0.707\,11 \\ 0.707\,11 \\ 1.414\,2 \end{bmatrix} \tag{4.359} $$

The coordinates of the current positions of P_1 and P_2 in B_0 are

$$ {}^0\mathbf{r}_1 = {}^0R_B\, {}^B\mathbf{r}_1 = {}^0R_B \begin{bmatrix} 1 \\ 0 \\ 0 \end{bmatrix} = \begin{bmatrix} 1 \\ 0 \\ 0 \end{bmatrix} \tag{4.360} $$

$$ {}^0\mathbf{r}_2 = {}^0R_B\, {}^B\mathbf{r}_2 = {}^0R_B \begin{bmatrix} 1 \\ 1 \\ 1 \end{bmatrix} = \begin{bmatrix} 1 \\ 0 \\ 1.414\,2 \end{bmatrix} \tag{4.361} $$

KEY SYMBOLS

$\mathbf{a}$	general vector
$\tilde{a}$	skew-symmetric matrix of the vector **a**
A	transformation matrix of rotation about a local axis
B	body coordinate frame, local coordinate frame
c	cos
d	distance between two points
$\hat{e}_\varphi, \hat{e}_\theta, \hat{e}_\psi$	coordinate axes of E, local roll–pitch–yaw coordinate axes
E	Eulerian local frame
f, f_1, f_2	function of x and y
G	global coordinate frame, fixed coordinate frame
$\mathbf{I} = [I]$	identity matrix
$\hat{\imath}, \hat{\jmath}, \hat{k}$	local coordinate axis unit vectors
$\tilde{\imath}, \tilde{\jmath}, \tilde{k}$	skew-symmetric matrices of the unit vectors $\hat{\imath}, \hat{\jmath}, \hat{k}$
$\hat{I}, \hat{J}, \hat{K}$	global coordinate axis unit vectors
l	length
m	number of repeating rotation
n	fraction of 2π, number of repeating rotation
O	common origin of B and G
$O\varphi\theta\psi$	Euler angle frame
P	body point, fixed point in B
Q	transformation matrix of rotation about a global axis
$\mathbf{r}$	position vector
r_{ij}	element of row i and column j of a matrix
R	radius of a circle
$\mathbb{R}$	set of real numbers
s	sin, a member of S
S	set
t	time
$\mathbf{u}$	a general axis
$\mathbf{v}$	velocity vector
x, y, z	local coordinate axes
X, Y, Z	global coordinate axes

Greek

α, β, γ	rotation angles about global axes
δ_{ij}	Kronecker's delta
φ, θ, ψ	rotation angles about local axes, Euler angles
$\dot{\varphi}, \dot{\theta}, \dot{\psi}$	Euler frequencies
$\omega_x, \omega_y, \omega_z$	angular velocity components
ω	angular velocity vector

Symbol

$[\ \ ]^{-1}$	inverse of matrix [ ]
$[\ \ ]^{\mathrm{T}}$	transpose of matrix [ ]
$\otimes$	binary operation
$(S, \otimes)$	group

EXERCISES

1. **Body Point and Global Rotations** The point P is at $\mathbf{r}_P = (1, 2, 1)$ in a body coordinate frame $B(Oxyz)$.

   (a) Find the final global position of P after a rotation of 30 deg about the X-axis followed by a 45 deg rotation about the Z-axis.

   (b) Find the final global position of P after a rotation of 45 deg about the X-axis followed by a 30 deg rotation about the Z-axis.

   (c) Find the final global position of P after a rotation of 45 deg about the X-axis followed by a 45 deg rotation about the Y-axis and then a 45 deg rotation about the Z-axis.

   (d) ★ Does it matter if we change the order of rotations in part (c)?

2. **Body Point after Global Rotation**

   (a) Find the position of a point P in the local coordinate frame if it is moved to ${}^G\mathbf{r}_P = [1, 3, 2]^T$ after a 60 deg rotation about the Z-axis.

   (b) Find the position of a point P in the local coordinate frame if it is moved to ${}^G\mathbf{r}_P = [1, 3, 2]^T$ after a 60 deg rotation about the X-axis.

   (c) Find the position of a point P in the local coordinate frame if it is moved to ${}^G\mathbf{r}_P = [1, 3, 2]^T$ after a 60 deg rotation about the Z-axis followed by a 60 deg rotation about the X-axis.

   (d) ★ Is it possible to combine the rotations in part (c) and do only one rotation about the bisector of X- and Z-axes?

3. **Invariant of a Vector** A point was at ${}^B\mathbf{r}_P = [1, 2, z]^T$. After a rotation of 60 deg about the X-axis followed by a 30 deg rotation about the Z-axis, it is at

$$ {}^G\mathbf{r}_P = \begin{bmatrix} X \\ Y \\ 2.933 \end{bmatrix} $$

   Find z, X, and Y.

4. **Constant-Length Vector** Show that the length of a vector will not change by rotation:

$$ \left|{}^G\mathbf{r}\right| = \left|{}^G R_B \, {}^B\mathbf{r}\right| $$

   Show that the distance between two body points will not change by rotation:

$$ \left|{}^B\mathbf{p}_1 - {}^B\mathbf{p}_2\right| = \left|{}^G R_B \, {}^B\mathbf{p}_1 - {}^G R_B \, {}^B\mathbf{p}_2\right| $$

5. **Repeated Global Rotations** Rotate ${}^B\mathbf{r}_P = [2, 2, 3]^T$ 60 deg about the X-axis followed by 30 deg about the Z-axis. Then, repeat the sequence of rotations for 60 deg about the X-axis followed by 30 deg about the Z-axis. After how many rotations will point P be back to its initial global position?

6. ★ **Repeated Global Rotations** How many rotations of $\alpha = \pi/m$ deg about the X-axis followed by $\beta = \pi/n$ deg about the Z-axis are needed to bring a body point to its initial global position if $m, n \in \mathbb{N}$?

7. **Triple Global Rotations** Verify the equations in Appendix A.

8. ★ **Special Triple Rotation** Assume that the first triple rotation in Appendix A brings a body point back to its initial global position. What are the angles $\alpha \neq 0$, $\beta \neq 0$, and $\gamma \neq 0$?

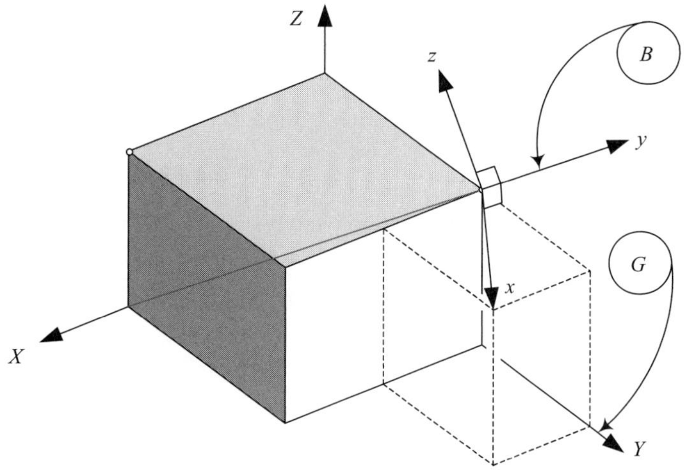

Figure 4.19 A displaced and rotated frame.

9. **★ Combination of Triple Rotations** Any triple rotation in Appendix A can move a body point to its new global position. Assume α_1, β_1, and γ_1 for case 1 are given as $Q_{X,\gamma_1} Q_{Y,\beta_1} Q_{Z,\alpha_1}$. What can α_2, β_2, and γ_2 be (in terms of α_1, β_1, and γ_1) to get the same global position if we use case 2: $Q_{Y,\gamma_2} Q_{Z,\beta_2} Q_{X,\alpha_2}$?

10. **Global Roll–Pitch–Yaw Rotation Matrix** Calculate the global roll–pitch–yaw rotation matrix for $\alpha = 30$, $\beta = 30$, and $\gamma = 30$.

11. **A Displaced and Rotated Frame** Determine the rotation transformation matrix ${}^G R_B$ in Figure 4.19.

12. **Global Roll–Pitch–Yaw Rotation Angles** Calculate the role, pitch, and yaw angles for the following rotation matrix:

$$
{}^B R_G = \begin{bmatrix} 0.53 & -0.84 & 0.13 \\ 0.0 & 0.15 & 0.99 \\ -0.85 & -0.52 & 0.081 \end{bmatrix}
$$

13. **Body Point, Local Rotation** What are the global coordinates of a body point at ${}^B\mathbf{r}_P = [2, 2, 3]^{\mathrm{T}}$ after a rotation of 60 deg about the x-axis?

14. **Two Local Rotations** Find the global coordinates of a body point at ${}^B\mathbf{r}_P = [2, 2, 3]^{\mathrm{T}}$ after a rotation of 60 deg about the x-axis followed by 60 deg about the z-axis.

15. **Triple Local Rotations** Verify the equations in Appendix B.

16. **Combination of Local and Global Rotations** Find the final global position of a body point at ${}^B\mathbf{r}_P = [10, 10, -10]^{\mathrm{T}}$ after a rotation of 45 deg about the x-axis followed by 60 deg about the z-axis.

17. **Combination of Global and Local Rotations** Find the final global position of a body point at ${}^B\mathbf{r}_P = [10, 10, -10]^{\mathrm{T}}$ after a rotation of 45 deg about the x-axis followed by 60 deg about the z-axis.

18. **Repeated Local Rotations** Rotate ${}^B\mathbf{r}_P = [2, 2, 3]^{\mathrm{T}}$ 60 deg about the x-axis followed by 30 deg about the z-axis. Then repeat the sequence of rotations for 60 deg about the x-axis followed by 30 deg about the z-axis. After how many rotations will point P be back to its initial global position?

19. ★ **Repeated Local Rotations** How many rotations of $\alpha = \pi/m$ degrees about the x-axis followed by $\beta = \pi/n$ degrees about the z-axis are needed to bring a body point to its initial global position if $m, n \in \mathbb{N}$?

20. ★ **Remaining Rotation** Find the result of the following sequence of rotations:

$$^{G}R_{B} = A_{y,\theta}^{\mathrm{T}} A_{z,\psi}^{\mathrm{T}} A_{y,-\theta}^{\mathrm{T}}$$

21. **Angles from Rotation Matrix** Find the angles φ, θ, and ψ if the rotation of case 1 in Appendix B is given as $A_{x,\psi} A_{y,\theta} A_{z,\varphi}$.

22. **Euler Angles from Rotation Matrix** Find the Euler angles for the rotation matrix

$$^{B}R_{G} = \begin{bmatrix} 0.53 & -0.84 & 0.13 \\ 0.0 & 0.15 & 0.99 \\ -0.85 & -0.52 & 0.081 \end{bmatrix}$$

23. **Equivalent Euler Angles to Two Rotations** Find the Euler angles corresponding to the rotation matrix $^{B}R_{G} = A_{y,45} A_{x,30}$.

24. **Equivalent Euler Angles to Three Rotations** Find the Euler angles corresponding to the rotation matrix $^{B}R_{G} = A_{z,60} A_{y,45} A_{x,30}$.

25. ★ **Local and Global Positions, Euler Angles** Find the conditions between the Euler angles to transform $^{G}\mathbf{r}_{P} = [1, 1, 0]^{\mathrm{T}}$ to $^{B}\mathbf{r}_{P} = [0, 1, 1]^{\mathrm{T}}$.

26. ★ **Equivalent Euler Angles to Three Rotations** Find the Euler angles for the rotation matrix of case 4 in Appendix B, $A_{z,\psi'} A_{y,\theta'} A_{x,\varphi'}$.

27. ★ **Integrability of Euler Frequencies** Show that $d\theta$ and $d\psi$ are integrable if θ and ψ are second and third Euler angles.

28. ★ **Cardan Angles for Euler Angles** Find the Cardan angles for a given set of Euler angles.

29. ★ **Cardan Frequencies for Euler Frequencies** Find the Euler frequencies in terms of Cardan frequencies.

30. **Elements of Rotation Matrix** The rotation matrix $^{G}R_{B}$ is given as

$$^{G}R_{B} = \begin{bmatrix} \cos(\hat{I}, \hat{\imath}) & \cos(\hat{I}, \hat{\jmath}) & \cos(\hat{I}, \hat{k}) \\ \cos(\hat{J}, \hat{\imath}) & \cos(\hat{J}, \hat{\jmath}) & \cos(\hat{J}, \hat{k}) \\ \cos(\hat{K}, \hat{\imath}) & \cos(\hat{K}, \hat{\jmath}) & \cos(\hat{K}, \hat{k}) \end{bmatrix}$$

Find $^{G}R_{B}$ if $^{G}\mathbf{r}_{P_1} = (0.7071, -1.2247, 1.4142)$ is a point on the x-axis and $^{G}\mathbf{r}_{P_2} = (2.7803, 0.38049, -1.0607)$ is a point on the y-axis.

31. **Completion of Rotation Matrix** Complete the rotation matrix

$$^{G}R_{B} = \begin{bmatrix} \sqrt{3}/3 & \sqrt{6}/6 & \sqrt{2}/2 \\ \sqrt{3}/3 & \sqrt{6}/6 & ? \\ ? & ? & ? \end{bmatrix}$$

32. **Path of Motion in Two Coordinate Frames** Assume a particle is moving on the following path in a global frame G:

$$X = a\cos\omega t \qquad Y = b\sin\omega t$$

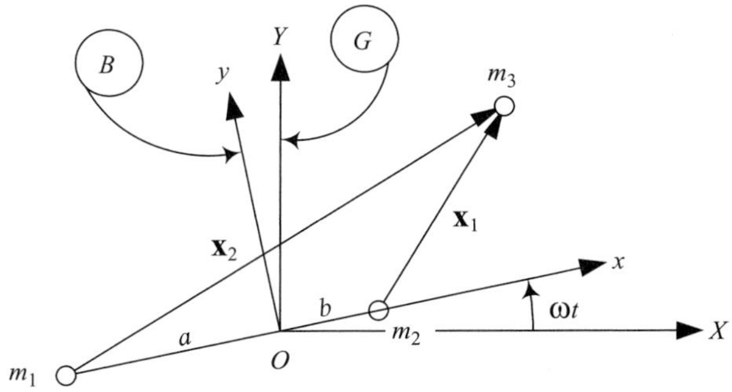

Figure 4.20 The restricted three-body problem in B and G coordinate frames.

A coordinate frame B which is coincident with G at $t = 0$ is turning about the Z-axis with a constant angular velocity ${}_G\omega_B = \Omega\hat{K}$.

(a) Determine the path of motion of the point in G.

(b) Determine the coordinates of the point in B.

(c) Determine the path of motion of the point in B.

(d) Solve parts (b) and (c) for $\omega = \Omega$.

33. ★ **Kinematics of the Restricted Three-Body Problem** Following Victor Szebehely (1921–1997) we define the restricted three-body problem as: Two bodies m_1 and m_2 revolve around their center of mass O in circular orbits under the influence of their mutual gravitational attraction and a third body m_3 (attracted by the previous two but not influencing their motion) moves in the plane defined by the two revolving bodies. The restricted problem of three bodies is to describe the motion of the third body m_3. Figure 4.20 illustrates the problem in B and G coordinate frames:

$$a = \frac{m_1 l}{M} \qquad b = \frac{m_2 l}{M} \qquad a + b = l \qquad M = m_1 + m_2$$

Balance between the gravitational and centrifugal forces requires that

$$G\frac{m_1 m_2}{l^2} = m_2 a\omega^2 = m_1 b\omega^2$$

where G is the gravitational constant and ω is the common angular velocity of m_1 and m_2. From these equations we have

$$Gm_1 = a\omega^2 l^2 \qquad Gm_2 = b\omega^2 l^2 \qquad G\left(m_1 + m_2\right) = \omega^2 l^2$$

where the last equation is Kepler's third law. The equations of motion of m_3 in the global coordinate frame G are

$$\ddot{X} = \frac{\partial f}{\partial X} \qquad \ddot{Y} = \frac{\partial f}{\partial Y}$$

$$f = G\left(\frac{m_1}{r_1} + \frac{m_2}{r_2}\right)$$

$$r_1 = \sqrt{(X - X_1)^2 + (Y - Y_1)^2}$$

$$r_2 = \sqrt{(X - X_2)^2 + (Y - Y_2)^2}$$

where X, Y are the global coordinates of m_3, X_1, Y_1 are the global coordinates of m_1, and X_2, Y_2 are the global coordinates of m_2.

(a) Define X_1, Y_1, X_2, Y_2 as functions of time and show that the differential equations of motion of m_3 are

$$\ddot{X} = -G\left(\frac{m_1\,(X - b\cos\omega t)}{r_1^3} + \frac{m_2\,(X + a\cos\omega t)}{r_2^3}\right)$$

$$\ddot{Y} = -G\left(\frac{m_1\,(Y - b\sin\omega t)}{r_1^3} + \frac{m_2\,(Y + a\sin\omega t)}{r_2^3}\right)$$

(b) Determine the transformation of coordinates of m_3 between B and G to find the equations of motion as

$$\ddot{x} - 2\omega\dot{x} - \omega^2 x = -G\left(m_1\frac{x-b}{r_1^3} + m_2\frac{x+a}{r_2^3}\right)$$

$$\ddot{y} + 2\omega\dot{y} - \omega^2 y = -G\left(m_1\frac{y}{r_1^3} + m_2\frac{y}{r_2^3}\right)$$

34. Linearly Independent Vectors A set of vectors $\mathbf{a}_1, \mathbf{a}_2, \ldots, \mathbf{a}_n$ are considered linearly independent if the equation

$$k_1\mathbf{a}_1 + k_2\mathbf{a}_2 + \cdots + k_n\mathbf{a}_n = 0$$

in which $k_1, k_2, \ldots, k_n$ are unknown coefficients has only one solution:

$$k_1 = k_2 = \cdots = k_n = 0$$

Verify that the unit vectors of a body frame *B(Oxyz)* expressed in the global frame *G(OXYZ)* are linearly independent.

35. Product of Orthogonal Matrices A matrix R is called orthogonal if $R^{-1} = R^{\mathrm{T}}$ where $\left(R^{\mathrm{T}}\right)_{ij} = R_{ji}$. Prove that the product of two orthogonal matrices is also orthogonal.

36. Vector Identity The formula $(a+b)^2 = a^2 + b^2 + 2ab$ for scalars is equivalent to

$$(\mathbf{a}+\mathbf{b})^2 = \mathbf{a}\cdot\mathbf{a} + \mathbf{b}\cdot\mathbf{b} + 2\mathbf{a}\cdot\mathbf{b}$$

for vectors. Show that this formula is equal to

$$(\mathbf{a}+\mathbf{b})^2 = \mathbf{a}\cdot\mathbf{a} + \mathbf{b}\cdot\mathbf{b} + 2\,{}^G R_B\,\mathbf{a}\cdot\mathbf{b}$$

if **a** is a vector in the local frame and **b** is a vector in the global frame.

37. Rotation as a Linear Operation Show that

$$R\,(\mathbf{a}\times\mathbf{b}) = R\mathbf{a}\times R\mathbf{b}$$

where R is a rotation matrix and **a** and **b** are two vectors defined in a coordinate frame.

38. **Scalar Triple Product** Show that for three arbitrary vectors **a**, **b**, and **c** we have

$$\mathbf{a} \cdot (\mathbf{b} \times \mathbf{c}) = (\mathbf{a} \times \mathbf{b}) \cdot \mathbf{c}$$

39. **Decomposition of a Matrix into Symmetric and Skew Symmetric** Any matrix can be decomposed into symmetric and a skew-symmetric matrices. Show the decomposition for the following matrix:

$$\begin{bmatrix} \partial r_x/\partial x & \partial r_x/\partial y & \partial r_x/\partial z \\ \partial r_y/\partial x & \partial r_y/\partial y & \partial r_y/\partial z \\ \partial r_z/\partial x & \partial r_z/\partial y & \partial r_z/\partial z \end{bmatrix}$$

5

Orientation Kinematics

Any rotation ϕ of a rigid body with a fixed point O about a fixed axis $\hat{u}$ can be decomposed into three rotations about three given non-coplanar axes. On the contrary, the finial orientation of the rigid body after a finite number of rotations is equivalent to a unique rotation about a unique axis. Determination of the angle and axis is called the *orientation kinematics* of rigid bodies.

5.1 AXIS–ANGLE ROTATION

Let the body frame $B(Oxyz)$ rotate ϕ about a fixed line in the global frame $G(OXYZ)$ that is indicated by a unit vector $\hat{u}$ with directional cosines u_1, u_2, u_3,

$$\hat{u} = u_1\hat{I} + u_2\hat{J} + u_3\hat{K} \tag{5.1}$$

$$\sqrt{u_1^2 + u_2^2 + u_3^2} = 1 \tag{5.2}$$

This is called the *axis–angle* representation of a rotation. Two parameters are needed to define the axis of rotation that goes through O and one parameter is needed to define the amount of rotation about the axis. So, an angle–axis rotation needs three independent parameters to be defined.

The angle–axis transformation matrix ${}^GR_B$ that transforms the coordinates of the body frame $B(Oxyz)$ to the associated coordinates in the global frame $G(OXYZ)$,

$${}^G\mathbf{r} = {}^GR_B\,{}^B\mathbf{r} \tag{5.3}$$

is

$${}^GR_B = R_{\hat{u},\phi} = \mathbf{I}\cos\phi + \hat{u}\hat{u}^{\mathrm{T}}\ \text{vers}\ \phi + \tilde{u}\sin\phi \tag{5.4}$$

$${}^GR_B = \begin{bmatrix} u_1^2\ \text{vers}\ \phi + c\phi & u_1u_2\ \text{vers}\ \phi - u_3s\phi & u_1u_3\ \text{vers}\ \phi + u_2s\phi \\ u_1u_2\ \text{vers}\ \phi + u_3s\phi & u_2^2\ \text{vers}\ \phi + c\phi & u_2u_3\ \text{vers}\ \phi - u_1s\phi \\ u_1u_3\ \text{vers}\ \phi - u_2s\phi & u_2u_3\ \text{vers}\ \phi + u_1s\phi & u_3^2\ \text{vers}\ \phi + c\phi \end{bmatrix} \tag{5.5}$$

where

$$\text{vers}\ \phi = \text{versin}\ \phi = 1 - \cos\phi = 2\sin^2\frac{\phi}{2} \tag{5.6}$$

and $\tilde{u}$ is the skew-symmetric matrix associated to the vector $\hat{u}$,

$$\tilde{u} = \begin{bmatrix} 0 & -u_3 & u_2 \\ u_3 & 0 & -u_1 \\ -u_2 & u_1 & 0 \end{bmatrix} \tag{5.7}$$

A matrix $\tilde{u}$ is *skew symmetric* if

$$\tilde{u}^{\mathrm{T}} = -\tilde{u} \tag{5.8}$$

For any transformation matrix ${}^{G}R_{B}$, we may obtain the equivalent axis $\hat{u}$ and angle ϕ to provide the same matrix by

$$\tilde{u} = \frac{1}{2\sin\phi}\left({}^{G}R_{B} - {}^{G}R_{B}^{\mathrm{T}}\right) \tag{5.9}$$

$$\cos\phi = \frac{1}{2}\left(\mathrm{tr}\left({}^{G}R_{B}\right) - 1\right) \tag{5.10}$$

Equation (5.5) is called the *angle–axis* or *axis–angle rotation matrix* and is the most general transformation matrix for rotation of a body frame B in a global frame G. If the axis of rotation (5.1) coincides with a global coordinate axis Z, Y, or X, then Equation (5.5) reduces to the principal local rotation matrices (4.101), (4.102), and (4.103), respectively.

Proof: The rotation ϕ about an axis $\hat{u}$ is equivalent to a sequence of rotations about the axes of a body frame B such that the local frame is first rotated to bring one of its axes, say the z-axis, into coincidence with the rotation axis $\hat{u}$ followed by a rotation ϕ about that local axis, then the reverse of the first sequence of rotations.

Figure 5.1 illustrates an axis of rotation $\hat{u} = u_1\hat{I} + u_2\hat{J} + u_3\hat{K}$, the global frame $G(OXYZ)$, and the rotated body frame $B\,(Oxyz)$ when the local z-axis is coincident with $\hat{u}$. Assume that the body and global frames were coincident initially. When we apply a sequence of rotations φ about the z-axis and θ about the y-axis on the body frame $B\,(Oxyz)$, the local z-axis will become coincident with the rotation axis $\hat{u}$. We apply the rotation ϕ about $\hat{u}$ and then perform the sequence of rotations φ and θ backward. Following Equation (4.250), the rotation matrix ${}^{G}R_{B}$ to map the coordinates

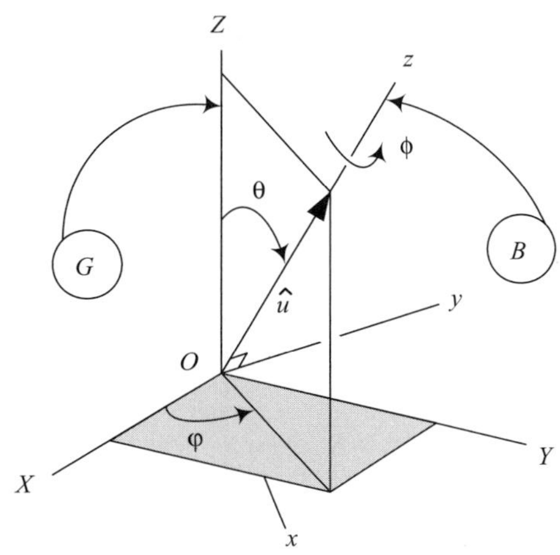

Figure 5.1 Axis of rotation $\hat{u}$ when it is coincident with the local z-axis.

of a point in the body frame to its coordinates in the global frame after rotation ϕ about $\hat{u}$ is

$$\begin{aligned} {}^{G}R_{B} &= {}^{B}R_{G}^{-1} = {}^{B}R_{G}^{\mathrm{T}} = R_{\hat{u},\phi} \\ &= \left[R_{z,-\varphi}\, R_{y,-\theta}\, R_{z,\phi}\, R_{y,\theta}\, R_{z,\varphi}\right]^{\mathrm{T}} \\ &= R_{z,\varphi}^{\mathrm{T}}\, R_{y,\theta}^{\mathrm{T}}\, R_{z,\phi}^{\mathrm{T}}\, R_{y,-\theta}^{\mathrm{T}}\, R_{z,-\varphi}^{\mathrm{T}} \end{aligned} \tag{5.11}$$

Substituting the equations

$$\begin{aligned} \sin\varphi &= \frac{u_2}{\sqrt{u_1^2+u_2^2}} & \cos\varphi &= \frac{u_1}{\sqrt{u_1^2+u_2^2}} \\ \sin\theta &= \sqrt{u_1^2+u_2^2} & \cos\theta &= u_3 \\ \sin\theta\sin\varphi &= u_2 & \sin\theta\cos\varphi &= u_1 \end{aligned} \tag{5.12}$$

in ${}^{G}R_{B}$ will provide the angle–axis rotation matrix

$$\begin{aligned} {}^{G}R_{B} &= R_{\hat{u},\phi} \\ &= \begin{bmatrix} u_1^2 \operatorname{vers}\phi + c\phi & u_1u_2 \operatorname{vers}\phi - u_3 s\phi & u_1u_3 \operatorname{vers}\phi + u_2 s\phi \\ u_1u_2 \operatorname{vers}\phi + u_3 s\phi & u_2^2 \operatorname{vers}\phi + c\phi & u_2u_3 \operatorname{vers}\phi - u_1 s\phi \\ u_1u_3 \operatorname{vers}\phi - u_2 s\phi & u_2u_3 \operatorname{vers}\phi + u_1 s\phi & u_3^2 \operatorname{vers}\phi + c\phi \end{bmatrix} \end{aligned} \tag{5.13}$$

The matrix (5.13) can be decomposed to

$$\begin{aligned} R_{\hat{u},\phi} &= \cos\phi \begin{bmatrix} 1 & 0 & 0 \\ 0 & 1 & 0 \\ 0 & 0 & 1 \end{bmatrix} + (1-\cos\phi) \begin{bmatrix} u_1 \\ u_2 \\ u_3 \end{bmatrix} \begin{bmatrix} u_1 & u_2 & u_3 \end{bmatrix} \\ &\quad + \sin\phi \begin{bmatrix} 0 & -u_3 & u_2 \\ u_3 & 0 & -u_1 \\ -u_2 & u_1 & 0 \end{bmatrix} \end{aligned} \tag{5.14}$$

which can be summarized as (5.4). Showing the rotation matrix by its elements ${}^{G}R_{B} = R_{\hat{u},\phi} = \left[r_{ij}\right]$, we have

$$r_{ij} = \delta_{ij}\cos\phi + u_i u_j\,(1-\cos\phi) - \epsilon_{ijk} u_k \sin\phi \tag{5.15}$$

The *angle–axis rotation equation* (5.4) is also called the *Rodriguez rotation formula* or the *Euler–Lexell–Rodriguez formula*. It is sometimes reported in the literature as the following equivalent forms:

$$R_{\hat{u},\phi} = \begin{cases} \mathbf{I} + \tilde{u}\sin\phi + 2\tilde{u}^2\sin^2\dfrac{\phi}{2} & (5.16) \\ \mathbf{I} + 2\tilde{u}\sin\dfrac{\phi}{2}\left(\mathbf{I}\cos\dfrac{\phi}{2} + \tilde{u}\sin\dfrac{\phi}{2}\right) & (5.17) \\ \mathbf{I} + \tilde{u}\sin\phi + \tilde{u}^2 \operatorname{vers}\phi & (5.18) \\ \left[\mathbf{I} - \hat{u}\hat{u}^{\mathrm{T}}\right]\cos\phi + \tilde{u}\sin\phi + \hat{u}\hat{u}^{\mathrm{T}} & (5.19) \\ \mathbf{I} + \tilde{u}^2 + \tilde{u}\sin\phi - \tilde{u}^2\cos\phi & (5.20) \end{cases}$$

The *inverse of an angle–axis rotation* is

$$ {}^{B}R_{G} = {}^{G}R_{B}^{\mathrm{T}} = R_{\hat{u},-\phi} = \mathbf{I}\cos\phi + \hat{u}\hat{u}^{\mathrm{T}} \operatorname{vers}\phi - \tilde{u}\sin\phi \tag{5.21} $$

which means that the orientation of B in G when B is rotated ϕ about $\hat{u}$ is the same as the orientation of G in B when B is rotated $-\phi$ about $\hat{u}$. The rotation $R_{\hat{u},-\phi}$ is also called the *reverse rotation*.

The 3×3 real orthogonal transformation matrix ${}^{G}R_{B}$ is also called a *rotator* and the skew-symmetric matrix $\tilde{u}$ is called a *spinor*. We can verify that

$$ \tilde{u}\hat{u} = 0 \tag{5.22} $$

$$ \mathbf{I} - \hat{u}\hat{u}^{\mathrm{T}} = \tilde{u}^{2} \tag{5.23} $$

$$ \mathbf{r}^{\mathrm{T}}\tilde{u}\mathbf{r} = 0 \tag{5.24} $$

$$ \hat{u} \times \mathbf{r} = \tilde{u}\mathbf{r} = -\tilde{r}\hat{u} = -\mathbf{r} \times \hat{u} \tag{5.25} $$

We may examine Equations (5.9) and (5.10) by direct substitution and show that

$$ \begin{aligned} {}^{G}R_{B} - {}^{G}R_{B}^{\mathrm{T}} &= \begin{bmatrix} 0 & -2u_3\sin\phi & 2u_2\sin\phi \\ 2u_3\sin\phi & 0 & -2u_1\sin\phi \\ -2u_2\sin\phi & 2u_1\sin\phi & 0 \end{bmatrix} \\ &= 2\sin\phi \begin{bmatrix} 0 & -u_3 & u_2 \\ u_3 & 0 & -u_1 \\ -u_2 & u_1 & 0 \end{bmatrix} = 2\tilde{u}\sin\phi \end{aligned} \tag{5.26} $$

and

$$ \begin{aligned} \operatorname{tr}\left({}^{G}R_{B}\right) &= r_{11} + r_{22} + r_{33} \\ &= 3\cos\phi + u_1^2(1-\cos\phi) + u_2^2(1-\cos\phi) + u_3^2(1-\cos\phi) \\ &= 3\cos\phi + u_1^2 + u_2^2 + u_3^2 - \left(u_1^2 + u_2^2 + u_3^2\right)\cos\phi \\ &= 2\cos\phi + 1 \end{aligned} \tag{5.27} $$

The axis of rotation $\hat{u}$ is also called the Euler axis or the eigenaxis of rotation. ■

Example 296 Skew-Symmetric Matrices for $\hat{\imath}$, $\hat{\jmath}$, and $\hat{k}$ The definition of an skew-symmetric matrix $\tilde{a}$ corresponding to a vector $\mathbf{a}$ is defined by

$$ \tilde{a} = \begin{bmatrix} 0 & -a_3 & a_2 \\ a_3 & 0 & -a_1 \\ -a_2 & a_1 & 0 \end{bmatrix} \tag{5.28} $$

Hence,

$$ \tilde{\imath} = \begin{bmatrix} 0 & 0 & 0 \\ 0 & 0 & -1 \\ 0 & 1 & 0 \end{bmatrix} \tag{5.29} $$

$$\tilde{j} = \begin{bmatrix} 0 & 0 & 1 \\ 0 & 0 & 0 \\ -1 & 0 & 0 \end{bmatrix} \tag{5.30}$$

$$\tilde{k} = \begin{bmatrix} 0 & -1 & 0 \\ 1 & 0 & 0 \\ 0 & 0 & 0 \end{bmatrix} \tag{5.31}$$

Example 297 Angle–Axis of Rotation When $\hat{u} = \hat{K}$ If the local frame $B(Oxyz)$ rotates about the Z-axis, then

$$\hat{u} = \hat{K} \tag{5.32}$$

and the transformation matrix (5.5) reduces to

$$\begin{aligned} {}^G R_B &= \begin{bmatrix} 0 \text{ vers } \phi + \cos\phi & 0 \text{ vers } \phi - 1 \sin\phi & 0 \text{ vers } \phi + 0 \sin\phi \\ 0 \text{ vers } \phi + 1 \sin\phi & 0 \text{ vers } \phi + \cos\phi & 0 \text{ vers } \phi - 0 \sin\phi \\ 0 \text{ vers } \phi - 0 \sin\phi & 0 \text{ vers } \phi + 0 \sin\phi & 1 \text{ vers } \phi + \cos\phi \end{bmatrix} \\ &= \begin{bmatrix} \cos\phi & -\sin\phi & 0 \\ \sin\phi & \cos\phi & 0 \\ 0 & 0 & 1 \end{bmatrix} \end{aligned} \tag{5.33}$$

which is equivalent to the rotation matrix about the Z-axis of the global frame in (4.17).

Example 298 Angle and Axis of a Rotation Matrix A body coordinate frame B undergoes three Euler rotations $(\varphi, \theta, \psi) = (30, 30, 30)$ deg with respect to a global frame G. The rotation matrix to transform the coordinates of B to G is

$$\begin{aligned} {}^G R_B &= {}^B R_G^{\mathrm{T}} = \left[R_{z,\psi} R_{x,\theta} R_{z,\varphi}\right]^{\mathrm{T}} = R_{z,\varphi}^{\mathrm{T}} R_{x,\theta}^{\mathrm{T}} R_{z,\psi}^{\mathrm{T}} \\ &= \begin{bmatrix} 0.875 & -0.216\,51 & 0.433\,01 \\ 0.433\,01 & 0.75 & -0.5 \\ -0.216\,51 & 0.625 & 0.75 \end{bmatrix} \end{aligned} \tag{5.34}$$

The unique angle–axis of rotation for this rotation matrix can then be found by Equations (5.9) and (5.10):

$$\begin{aligned} \phi &= \cos^{-1}\left[\tfrac{1}{2}(\operatorname{tr}({}^G R_B) - 1)\right] = \cos^{-1}(0.6875) \\ &= 0.812\,76 \text{ rad} = 46.568 \text{ deg} \end{aligned} \tag{5.35}$$

$$\tilde{u} = \frac{1}{2\sin\phi}\left({}^G R_B - {}^G R_B^{\mathrm{T}}\right) = \begin{bmatrix} 0 & -0.447 & 0.447 \\ 0.447 & 0 & -0.774 \\ -0.447 & 0.774 & 0 \end{bmatrix} \tag{5.36}$$

$$\hat{u} = \begin{bmatrix} 0.774 \\ 0.447 \\ 0.447 \end{bmatrix} \tag{5.37}$$

As a double check, we may verify the angle–axis of rotation formula and derive the same rotation matrix:

$$\begin{aligned} {}^{G}R_{B} &= R_{\hat{u},\phi} = \mathbf{I}\cos\phi + \hat{u}\hat{u}^{\mathrm{T}} \operatorname{vers}\phi + \tilde{u}\sin\phi \\ &= \begin{bmatrix} 0.875 & -0.21651 & 0.43301 \\ 0.43301 & 0.75 & -0.5 \\ -0.21651 & 0.625 & 0.75 \end{bmatrix} \end{aligned} \tag{5.38}$$

Example 299 Rotation about a Rotated Local Axis When the body coordinate frame $B(Oxyz)$ rotates φ about the global Z-axis, then the x-axis will move to a new orientation in G indicated by $\hat{u}_x$:

$$\hat{u}_x = {}^{G}R_{Z,\varphi}\,\hat{\imath} = \begin{bmatrix} \cos\varphi & -\sin\varphi & 0 \\ \sin\varphi & \cos\varphi & 0 \\ 0 & 0 & 1 \end{bmatrix} \begin{bmatrix} 1 \\ 0 \\ 0 \end{bmatrix} = \begin{bmatrix} \cos\varphi \\ \sin\varphi \\ 0 \end{bmatrix} \tag{5.39}$$

Rotation θ about $\hat{u}_x = (\cos\varphi)\hat{I} + (\sin\varphi)\hat{J}$ can be defined by Rodriguez's formula (5.5):

$$\begin{aligned} &{}^{G}R_{\hat{u}_x,\theta} = \\ &\begin{bmatrix} \cos^2\varphi \operatorname{vers}\theta + \cos\theta & \cos\varphi\sin\varphi \operatorname{vers}\theta & \sin\varphi\sin\theta \\ \cos\varphi\sin\varphi \operatorname{vers}\theta & \sin^2\varphi \operatorname{vers}\theta + \cos\theta & -\cos\varphi\sin\theta \\ -\sin\varphi\sin\theta & \cos\varphi\sin\theta & \cos\theta \end{bmatrix} \end{aligned} \tag{5.40}$$

Therefore, a rotation φ about the global Z-axis followed by a rotation θ about the local x-axis makes a transformation matrix

$$\begin{aligned} {}^{G}R_{B} &= {}^{G}R_{\hat{u}_x,\theta}\,{}^{G}R_{Z,\varphi} \\ &= \begin{bmatrix} \cos\varphi & -\cos\theta\sin\varphi & \sin\theta\sin\varphi \\ \sin\varphi & \cos\theta\cos\varphi & -\cos\varphi\sin\theta \\ 0 & \sin\theta & \cos\theta \end{bmatrix} \end{aligned} \tag{5.41}$$

that must be equal to $[R_{x,\theta}\,R_{z,\varphi}]^{-1} = R_{z,\varphi}^{\mathrm{T}}\,R_{x,\theta}^{\mathrm{T}}$:

$$\begin{aligned} R_{z,\varphi}^{\mathrm{T}}\,R_{x,\theta}^{\mathrm{T}} &= \begin{bmatrix} \cos\varphi & \sin\varphi & 0 \\ -\sin\varphi & \cos\varphi & 0 \\ 0 & 0 & 1 \end{bmatrix}^{\mathrm{T}} \begin{bmatrix} 1 & 0 & 0 \\ 0 & \cos\theta & \sin\theta \\ 0 & -\sin\theta & \cos\theta \end{bmatrix}^{\mathrm{T}} \\ &= \begin{bmatrix} \cos\varphi & -\cos\theta\sin\varphi & \sin\theta\sin\varphi \\ \sin\varphi & \cos\theta\cos\varphi & -\cos\varphi\sin\theta \\ 0 & \sin\theta & \cos\theta \end{bmatrix} \end{aligned} \tag{5.42}$$

Example 300 Orthogonality of Rotation Matrix The orthogonality characteristic of the rotation matrix must be consistent with the Rodriguez formula as well. We show that we can multiply Equation (5.4) and (5.21):

$$\begin{aligned}
{}^{G}R_{B}\,{}^{B}R_{G} &= R_{\hat{u},\phi}R_{\hat{u},-\phi} \\
&= \left(\mathbf{I}\cos\phi + \hat{u}\hat{u}^{\mathrm{T}}\ \mathrm{vers}\ \phi + \tilde{u}\sin\phi\right)\left(\mathbf{I}\cos\phi + \hat{u}\hat{u}^{\mathrm{T}}\ \mathrm{vers}\ \phi - \tilde{u}\sin\phi\right) \\
&= \mathbf{I}\cos^{2}\phi + \hat{u}\hat{u}^{\mathrm{T}}\ \mathrm{vers}\ \phi\cos\phi - \tilde{u}\sin\phi\cos\phi \\
&\qquad + \hat{u}\hat{u}^{\mathrm{T}}\ \mathrm{vers}\ \phi\cos\phi + \hat{u}\hat{u}^{\mathrm{T}}\hat{u}\hat{u}^{\mathrm{T}}\ \mathrm{vers}\ \phi\ \mathrm{vers}\ \phi - \tilde{u}\hat{u}\hat{u}^{\mathrm{T}}\ \mathrm{vers}\ \phi\sin\phi \\
&\qquad + \tilde{u}\sin\phi\cos\phi + \hat{u}\hat{u}^{\mathrm{T}}\tilde{u}\sin\phi\ \mathrm{vers}\ \phi - \tilde{u}\tilde{u}\sin^{2}\phi \\
&= \mathbf{I}
\end{aligned} \tag{5.43}$$

That is because of

$$\hat{u}\hat{u}^{\mathrm{T}} = \begin{bmatrix} u_1^2 & u_1u_2 & u_1u_3 \\ u_1u_2 & u_2^2 & u_2u_3 \\ u_1u_3 & u_2u_3 & u_3^2 \end{bmatrix} \tag{5.44}$$

$$\hat{u}\hat{u}^{\mathrm{T}}\hat{u}\hat{u}^{\mathrm{T}} = \begin{bmatrix} u_1^2 & u_1u_2 & u_1u_3 \\ u_1u_2 & u_2^2 & u_2u_3 \\ u_1u_3 & u_2u_3 & u_3^2 \end{bmatrix} = \hat{u}\hat{u}^{\mathrm{T}} \tag{5.45}$$

$$\tilde{u}\hat{u}\hat{u}^{\mathrm{T}} = \hat{u}\hat{u}^{\mathrm{T}}\tilde{u} = \begin{bmatrix} 0 & 0 & 0 \\ 0 & 0 & 0 \\ 0 & 0 & 0 \end{bmatrix} \tag{5.46}$$

Example 301 Nonuniqueness of Angle–Axis of Rotation The angle–axis of rotation matrices for $(\theta, \hat{u})$, $(-\theta, -\hat{u})$, and $(\theta + 2\pi, \hat{u})$ are the same. So, the equivalent angle and axis for a rotation matrix is not unique. However, the final orientation of B would be the same by applying any of these angle–axis rotations.

Example 302 Elimination of First and Last Local Rotations Consider a body point P at ${}^{B}\mathbf{r} = [2, 3, 1]^{\mathrm{T}}$. The body undergoes three rotations: 90 deg about the z-axis, 90 deg about the x-axis, and −90 deg about the z-axis. Because the first and last rotations are in reverse, they eliminate each other and a single 90 deg rotation about the line $\hat{u}$ that is on the x-axis after the first rotation remains.

To examine this fact, we may find the global position of the corner P:

$$\begin{aligned}
{}^{G}\mathbf{r} &= {}^{G}R_{B}\,{}^{B}\mathbf{r} = \left[R_{z,-\pi}\,R_{x,\pi}\,R_{z,\pi}\right]^{\mathrm{T}}\,{}^{B}\mathbf{r} \\
&= \begin{bmatrix} 1 & 0 & 0 \\ 0 & -1 & 0 \\ 0 & 0 & -1 \end{bmatrix}\begin{bmatrix} 2 \\ 3 \\ 1 \end{bmatrix} = \begin{bmatrix} 2 \\ -3 \\ -1 \end{bmatrix}
\end{aligned} \tag{5.47}$$

We must get the same coordinates if we turn the block about the line $\hat{u}$ where the x-axis points when the block is turned 90 deg about the z-axis, where

$$\hat{u} = R_{z,\pi}\hat{\imath} = \begin{bmatrix} \cos\pi & \sin\pi & 0 \\ -\sin\pi & \cos\pi & 0 \\ 0 & 0 & 1 \end{bmatrix} \begin{bmatrix} 1 \\ 0 \\ 0 \end{bmatrix} = \begin{bmatrix} -1 \\ 0 \\ 0 \end{bmatrix} \tag{5.48}$$

and hence, the rotation matrix ${}^{B}R_{G} = R^{\mathrm{T}}_{\hat{u},\pi}$ is

$$\begin{aligned} {}^{B}R_{G} = {}^{G}R_{B}^{\mathrm{T}} = R_{\hat{u},-\pi} &= \mathbf{I}\cos\pi + \hat{u}\hat{u}^{\mathrm{T}}\ \mathrm{vers}\,\pi - \tilde{u}\sin\pi \\ &= \begin{bmatrix} 1 & 0 & 0 \\ 0 & -1 & 0 \\ 0 & 0 & -1 \end{bmatrix} \end{aligned} \tag{5.49}$$

The global coordinates of P after 90 deg rotation about $\hat{u}$ would be

$${}^{G}\mathbf{r} = R^{\mathrm{T}}_{\hat{u},-\pi}\,{}^{B}\mathbf{r} = \begin{bmatrix} 1 & 0 & 0 \\ 0 & -1 & 0 \\ 0 & 0 & -1 \end{bmatrix}^{\mathrm{T}} \begin{bmatrix} 2 \\ 3 \\ 1 \end{bmatrix} = \begin{bmatrix} 2 \\ -3 \\ -1 \end{bmatrix} \tag{5.50}$$

Example 303 One Rotation, Equivalent to Three Rotations Consider the brick in Figure 5.2 at initial position 1, at position 2 after 90 deg rotation about the z-axis, at position 3 after 90 deg rotation about the x-axis, and at position 4 after −90 deg rotation about the y-axis. As the figure shows, the brick can go to the final position by only one rotation about the x-axis.

To check this, we can find the transformation matrix ${}^{G}R_{B}$ for the two sequence of rotations to go to the final positions and determine the global coordinates of a point P at ${}^{B}\mathbf{r} = [2, 3, 1]^{\mathrm{T}}$. The first transformation matrix is

$${}^{G}R_{B} = {}^{B}R_{G}^{\mathrm{T}} = \left[R_{y,-\pi/2}R_{x,\pi/2}R_{z,\pi/2}\right]^{\mathrm{T}} = \begin{bmatrix} 1 & 0 & 0 \\ 0 & 0 & -1 \\ 0 & 1 & 0 \end{bmatrix} \tag{5.51}$$

and the second transformation matrix is

$${}^{G}R_{B} = R^{\mathrm{T}}_{x,\pi/2} = \begin{bmatrix} 1 & 0 & 0 \\ 0 & \cos\frac{\pi}{2} & \sin\frac{\pi}{2} \\ 0 & -\sin\frac{\pi}{2} & \cos\frac{\pi}{2} \end{bmatrix}^{\mathrm{T}} = \begin{bmatrix} 1 & 0 & 0 \\ 0 & 0 & -1 \\ 0 & 1 & 0 \end{bmatrix} \tag{5.52}$$

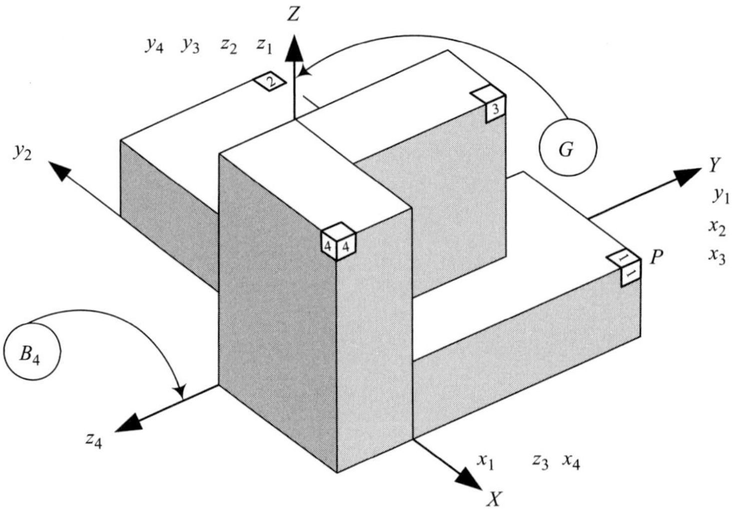

Figure 5.2 A brick at initial position 1 and after rotations of 90 deg about the z-axis, 90 deg about the x-axis, and -90 deg about the y-axis.

It shows that $R_{y,-\pi/2}R_{x,\pi/2}R_{z,\pi/2} \equiv R_{x,\pi/2}$, and therefore the final global position of P is

$$ {}^{G}\mathbf{r} = {}^{B}R_{G}^{\mathrm{T}}\,{}^{B}\mathbf{r} = \begin{bmatrix} 2 \\ -1 \\ 3 \end{bmatrix} \tag{5.53} $$

Example 304 The Reverse Global Rotations Do Not Eliminate Each Other When the first and last rotations are about the same global axis reversely, they do not eliminate each other. Figure 5.2 illustrates a brick at initial position 1, at position 2 after 90 deg rotation about the Z-axis, at position 3 after 90 deg rotation about the Y-axis, and at position 4 after -90 deg rotation about the Z-axis. The transformation matrix of these rotations is

$$ {}^{G}R_{B} = R_{Z,-\pi/2}R_{Y,\pi/2}R_{Z,\pi/2} = \begin{bmatrix} 1 & 0 & 0 \\ 0 & 0 & -1 \\ 0 & 1 & 0 \end{bmatrix} \tag{5.54} $$

It shows that the first rotations, which are reverse rotations about the same axis, do not eliminate each other:

$$ R_{Z,-\pi/2}R_{Y,\pi/2}R_{Z,\pi/2} \neq R_{Y,\pi/2} \tag{5.55} $$

Example 305 ★ Condition of Rigid-Body Rotation Consider some body points $O, P_1, P_2, \ldots$ at $\mathbf{0}, {}^{G}\mathbf{r}_1, {}^{G}\mathbf{r}_2, \ldots$. The body has a fixed point at O that is the origin of the global and body coordinate frames G and B. After a motion of the body, the points will be O, P_1', P_2', ... at $\mathbf{0}, {}^G\mathbf{r}_1', {}^G\mathbf{r}_2', \ldots$. The condition for the motion of the body to be a *rigid body motion* about the fixed point O is that the relative distances of the points remain constant. So, for every pair of points P_i, P_j we have

$$\left(\mathbf{r}_i - \mathbf{r}_j\right) = \left(\mathbf{r}_i' - \mathbf{r}_j'\right) \qquad i, j = 1, 2, 3, \ldots \tag{5.56}$$

Now consider a body point P at $\mathbf{r}$ that moves to P' at $\mathbf{r}'$ after a rigid-body motion about O. As is shown in Figure 5.3, let M be the midpoint of PP' and let $\boldsymbol{\lambda} = \lambda \hat{u}$ indicate a line through O in some direction perpendicular to PP'. Point N is the foot of the perpendicular from M to the axis of $\hat{u}$. Then, PP' is perpendicular to MN, and MN is perpendicular to the axis of $\hat{u}$. In triangle $\triangle PNP'$ we have

$$\angle PNM = \angle P'NM = \tfrac{1}{2}\phi \tag{5.57}$$

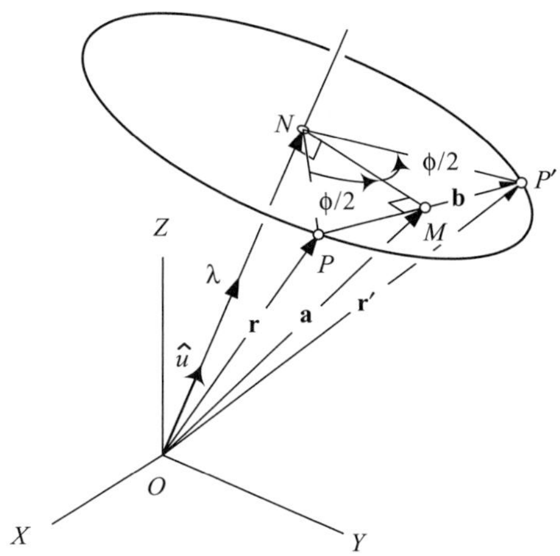

Figure 5.3 Rigid body rotation of a body point P.

The vectors $\overrightarrow{ON}$ and $\overrightarrow{PP'}$ can be expressed by $\mathbf{r}$ and $\mathbf{r}'$:

$$\overrightarrow{OM} = \mathbf{a} = \tfrac{1}{2}(\mathbf{r}' + \mathbf{r}) \tag{5.58}$$

$$\overrightarrow{PP'} = \mathbf{b} = (\mathbf{r}' - \mathbf{r}) \tag{5.59}$$

The vector $\mathbf{b}$ is parallel to $\hat{u} \times \mathbf{a}$, and we have

$$\mathbf{b} = \boldsymbol{\lambda} \times \mathbf{a} = 2(\hat{u} \times \mathbf{a}) \tan \frac{\phi}{2} \tag{5.60}$$

$$\lambda = 2 \tan \frac{\phi}{2} \tag{5.61}$$

Therefore, the point P goes to P' by a rotation ϕ about the axis $\hat{u}$ and its new position vector is determined by Equation (5.60). The condition for a rigid-body rotation about a fixed point is determination of a vector $\boldsymbol{\lambda}$ such that Equation (5.60) is satisfied. Although the rotation is not a vector quantity, $\boldsymbol{\lambda}$ is called the *rotation vector*.

Example 306 ★ Time Derivative of Rotation Matrix Is Skew Symmetric To take the derivative of a matrix, we should take the derivative of every element of the matrix. So, if we show the rotation matrix by its elements, ${}^G R_B = [r_{ij}]$, then ${}^G \dot{R}_B = [\dot{r}_{ij}]$.

The time derivative of the orthogonality condition of rotation matrices (4.276) is given as

$$\frac{d}{dt}\left({}^G R_B^{\mathrm{T}}\,{}^G R_B\right) = {}^G \dot{R}_B^{\mathrm{T}}\,{}^G R_B + {}^G R_B^{\mathrm{T}}\,{}^G \dot{R}_B = 0 \tag{5.62}$$

and leads to

$$\left[{}^G R_B^{\mathrm{T}}\,{}^G \dot{R}_B\right]^{\mathrm{T}} = -{}^G R_B^{\mathrm{T}}\,{}^G \dot{R}_B \tag{5.63}$$

which shows that $\left[{}^G R_B^{\mathrm{T}}\,{}^G \dot{R}_B\right]$ is a skew-symmetric matrix. We may show the skew-symmetric matrix $\left[{}^G R_B^{\mathrm{T}}\,{}^G \dot{R}_B\right]$ by

$$\tilde{\omega} = {}^G R_B^{\mathrm{T}}\,{}^G \dot{R}_B \tag{5.64}$$

and find the following equation for the time derivative of the rotation matrix:

$${}^G \dot{R}_B = {}^G R_B\,\tilde{\omega} \tag{5.65}$$

where $\boldsymbol{\omega}$ is the vector of angular velocity of the frame $B(Oxyz)$ with respect to frame $G(OXYZ)$ and $\tilde{\omega}$ is its skew-symmetric matrix.

Example 307 ★ Parametric Derivative of a Rotation Matrix When a rotation matrix $[R] = R(\tau)$ is a function of a variable τ, we use the orthogonality condition (4.276),

$$[R][R]^{\mathrm{T}} = I \tag{5.66}$$

to determine the derivative of $[R]$ with respect to τ:

$$\frac{d[R]}{d\tau}[R]^{\mathrm{T}} + [R]\frac{d[R]^{\mathrm{T}}}{d\tau} = 0 \tag{5.67}$$

It can be rewritten in the form

$$\frac{d[R]}{d\tau}[R]^{\mathrm{T}} + \left[\frac{d[R]}{d\tau}[R]^{\mathrm{T}}\right]^{\mathrm{T}} = 0 \tag{5.68}$$

which shows that $[(dR/d\tau)R^{\mathrm{T}}]$ is a skew-symmetric matrix.

Example 308 ★ Eigenvalues and Eigenvectors of $^G\boldsymbol{R}_B$ Assume that the rotation matrix for angle ϕ about $\hat{u}$ is ^GR_B. Applying the rotation on the axis of rotation $\hat{u}$ cannot change its direction,

$$^GR_B\,\hat{u} = \lambda\hat{u} \tag{5.69}$$

so the transformation equation implies that

$$\left|^GR_B - \lambda\mathbf{I}\right| = 0 \tag{5.70}$$

The characteristic equation of this determinant is

$$-\lambda^3 + \mathrm{tr}(^GR_B)\lambda^2 - \mathrm{tr}(^GR_B)\lambda + 1 = 0 \tag{5.71}$$

Factoring the left-hand side gives

$$(\lambda - 1)\left[\lambda^2 - \lambda\left(\mathrm{tr}(^GR_B) - 1\right) + 1\right] = 0 \tag{5.72}$$

and shows that $\lambda_1 = 1$ is always an eigenvalue of ^GR_B. Hence, there exist a real vector $\hat{u}$ such that every point on the line indicated by vector $\mathbf{n}_1 = \hat{u}$ remains fixed and invariant under transformation ^GR_B. The remaining eigenvalues are complex conjugates,

$$\lambda_2 = e^{i\phi} = \cos\phi + i\sin\phi \tag{5.73}$$

$$\lambda_3 = e^{-i\phi} = \cos\phi - i\sin\phi \tag{5.74}$$

and their associated eigenvectors are $\mathbf{v}$ and $\overline{\mathbf{v}}$, where $\overline{\mathbf{v}}$ is the complex conjugate of $\mathbf{v}$. Because ^GR_B is orthogonal, its eigenvectors $\mathbf{n}_1$, $\mathbf{v}$, and $\overline{\mathbf{v}}$ are also orthogonal. The eigenvectors $\mathbf{v}$ and $\overline{\mathbf{v}}$ span a plane perpendicular to the axis of rotation $\mathbf{n}_1$. A real basis for this plane can be found by using the following vectors:

$$\mathbf{n}_2 = \tfrac{1}{2}|\mathbf{v} + \overline{\mathbf{v}}| \tag{5.75}$$

$$\mathbf{n}_3 = \tfrac{1}{2}i|\mathbf{v} - \overline{\mathbf{v}}| \tag{5.76}$$

The basis vectors $\mathbf{n}_2$ and $\mathbf{n}_3$ transform to

$$\begin{aligned}
^GR_B\,\mathbf{n}_2 &= \tfrac{1}{2}|\lambda_2\mathbf{v} + \lambda_3\overline{\mathbf{v}}| = \tfrac{1}{2}\left|e^{i\phi}\mathbf{v} + \overline{e^{i\phi}\mathbf{v}}\right| \\
&= \mathbf{v}\cos\phi + \overline{\mathbf{v}}\sin\phi
\end{aligned} \tag{5.77}$$

$$\begin{aligned}
^GR_B\,\mathbf{n}_3 &= \tfrac{1}{2}i|\lambda_2\mathbf{v} - \lambda_3\overline{\mathbf{v}}| = \tfrac{1}{2}\left|e^{i\phi}\mathbf{v} - \overline{e^{i\phi}\mathbf{v}}\right| \\
&= -\mathbf{v}\cos\phi + \overline{\mathbf{v}}\sin\phi
\end{aligned} \tag{5.78}$$

Therefore, the effect of the transformation ^GR_B is to rotate vectors in the planes parallel to the plane that is spanned by $\mathbf{n}_2$ and $\mathbf{n}_3$ through angle ϕ about $\mathbf{n}_1$.

Example 309 ★ Rigid-Body Rotation Theorem We may prove a theorem to show that any displacement of a rigid body with a fixed point is a rotation.

Theorem If O, P_1, and P_2 are three particles of a rigid body, and if the body is displaced about O, then the body undergoes a rotation about O that can be determined by the displacement of P_1 and P_2.

Let us set up global and local coordinate frames G and B at O and show the position vectors of P_1 and P_2 by ${}^G\mathbf{r}_1$ and ${}^G\mathbf{r}_2$, respectively. The position vectors will be at ${}^G\mathbf{r}'_1$ and ${}^G\mathbf{r}'_2$ after displacement of the body. To prove the theorem, we must show that there is a vector $\boldsymbol{\lambda}$ such that if

$$\mathbf{a}_1 = \frac{1}{2}\left(\mathbf{r}'_1 + \mathbf{r}_1\right) \qquad \mathbf{a}_2 = \frac{1}{2}\left(\mathbf{r}'_2 + \mathbf{r}_2\right) \tag{5.79}$$

$$\mathbf{b}_1 = \left(\mathbf{r}'_1 - \mathbf{r}_1\right) \qquad \mathbf{b}_2 = \left(\mathbf{r}'_2 - \mathbf{r}_2\right) \tag{5.80}$$

then

$$\mathbf{b}_1 = \boldsymbol{\lambda} \times \mathbf{a}_1 \qquad \mathbf{b}_2 = \boldsymbol{\lambda} \times \mathbf{a}_2 \tag{5.81}$$

Furthermore, if $\mathbf{r}$ is the position vector of any other body point P, and if

$$\mathbf{a} = \frac{1}{2}\left(\mathbf{r}' + \mathbf{r}\right) \qquad \mathbf{b} = \left(\mathbf{r}' - \mathbf{r}\right) \tag{5.82}$$

then

$$\mathbf{b} = \boldsymbol{\lambda} \times \mathbf{a} \tag{5.83}$$

Because the body is rigid, we must have

$$\mathbf{b}_1 \cdot \mathbf{a}_1 = 0 \qquad \mathbf{b}_2 \cdot \mathbf{a}_2 = 0 \qquad \mathbf{b}_1 \cdot \mathbf{a}_1 + \mathbf{b}_2 \cdot \mathbf{a}_2 = 0 \tag{5.84}$$

Now if there exists a rotation vector $\boldsymbol{\lambda}$, then it must be perpendicular to $\mathbf{b}_1$ and $\mathbf{b}_2$. Therefore,

$$\boldsymbol{\lambda} = c\left(\mathbf{b}_1 \times \mathbf{b}_2\right) \tag{5.85}$$

and conditions (5.81) require that

$$\mathbf{b}_1 = c\left(\mathbf{b}_1 \times \mathbf{b}_2\right) \times \mathbf{a}_1 \qquad \mathbf{b}_2 = c\left(\mathbf{b}_1 \times \mathbf{b}_2\right) \times \mathbf{a}_2 \tag{5.86}$$

or

$$\mathbf{b}_1 = -c\mathbf{b}_1\left(\mathbf{b}_2 \cdot \mathbf{a}_1\right) \qquad \mathbf{b}_2 = c\mathbf{b}_2\left(\mathbf{b}_1 \cdot \mathbf{a}_2\right) \tag{5.87}$$

so

$$c = -\frac{1}{\mathbf{b}_2 \cdot \mathbf{a}_1} = \frac{1}{\mathbf{b}_1 \cdot \mathbf{a}_2} \tag{5.88}$$

If $\mathbf{b}_2 \cdot \mathbf{a}_1 \neq 0$ and $\mathbf{b}_1 \cdot \mathbf{a}_2 \neq 0$, then there exists a rotation vector $\boldsymbol{\lambda}$ such that

$$\mathbf{u} = \frac{\mathbf{b}_1 \times \mathbf{b}_2}{\mathbf{b}_1 \cdot \mathbf{a}_2} = \frac{\mathbf{b}_2 \times \mathbf{b}_1}{\mathbf{b}_2 \cdot \mathbf{a}_1} \tag{5.89}$$

Now if $\mathbf{r}$ is the position vector of an arbitrary body point P and $\mathbf{b} = \boldsymbol{\lambda} \times \mathbf{a}$, $\mathbf{b} \cdot \mathbf{a} = 0$, then

$$\mathbf{b} \cdot \mathbf{a}_1 = (\boldsymbol{\lambda} \times \mathbf{a}) \cdot \mathbf{a}_1 = -(\boldsymbol{\lambda} \times \mathbf{a}_1) \cdot \mathbf{a} = -\mathbf{b}_1 \cdot \mathbf{a} \tag{5.90}$$

$$\mathbf{b} \cdot \mathbf{a}_2 = (\boldsymbol{\lambda} \times \mathbf{a}) \cdot \mathbf{a}_2 = -(\boldsymbol{\lambda} \times \mathbf{a}_2) \cdot \mathbf{a} = -\mathbf{b}_2 \cdot \mathbf{a} \tag{5.91}$$

The point P at $\mathbf{r}$ is therefore rigidly connected to O, P_1, and P_2. The Euler rigid-body rotation theorem may also be expressed as: The displacement of a rigid body with a fixed point at the origin of global and body coordinate frames G and B from an initial to a final orientation is achieved by a rotation ${}^2R_1$ through a certain angle ϕ about an axis $\hat{u}$ which is fixed in both frames. The axis is in the direction of the eigenvector associated with the eigenvalue $\lambda = +1$ of the rotation transformation matrix ${}^2R_1$.

Example 310 ★ Final Rotation Formula We need a formula to determine the rotation ϕ about ${}^G\hat{u}$ of a body coordinate frame B that is not coincident with the global frame G.

Figure 5.4 illustrates a global frame G and a body frame B_0 at an arbitrary configuration. The body frame is supposed to turn ϕ about an axis ${}^G\hat{u}$ from its current position at B_0. For simplicity, we may drop the superscript G from ${}^G\hat{u}$ and remember that $\hat{u}$ indicates the fixed axis in G:

$$\hat{u} = {}^G\hat{u} = u_1\hat{I} + u_2\hat{J} + u_3\hat{K} \tag{5.92}$$

$$\sqrt{u_1^2 + u_2^2 + u_3^2} = 1 \tag{5.93}$$

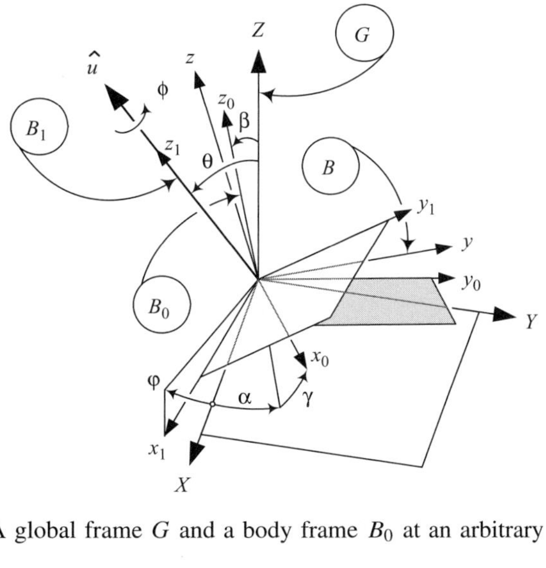

Figure 5.4 A global frame G and a body frame B_0 at an arbitrary configuration.

We can always assume that the body has come to the position B_0 from a coincident position with G by a rotation α about z_0 followed by a rotation β about x_0 and then a rotation γ about z_0.

Consider the body frame B at the coincident position with B_0. When we apply a sequence of rotations φ about the z-axis and θ about the y-axis on the body frame, the local z-axis will coincide with the rotation axis ${}^G\hat{u}$. Let us take a picture of B at this time and indicate it by B_1. Then we apply the rotation ϕ about $z \equiv \hat{u}$ and perform the sequence of rotations $-\theta$ about the y-axis and $-\varphi$ about the z-axis. The resultant of this maneuver would be a rotation ϕ of B about $\hat{u}$ starting from B_0.

The initial relative orientation of the body must be known, and therefore the transformation matrix ${}^GR_0$ between B_0 and G is known:

$$ {}^GR_0 = \left[b_{ij}\right] = \begin{bmatrix} b_{11} & b_{12} & b_{13} \\ b_{21} & b_{22} & b_{23} \\ b_{31} & b_{32} & b_{33} \end{bmatrix} \tag{5.94} $$

Having

$$ \begin{aligned} {}^GR_0 &= R_{z_0,\gamma}\, R_{x_0,\beta}\, R_{z_0,\alpha} \\ &= \begin{bmatrix} c\alpha c\gamma - c\beta s\alpha s\gamma & c\gamma s\alpha + c\alpha c\beta s\gamma & s\beta s\gamma \\ -c\alpha s\gamma - c\beta c\gamma s\alpha & c\alpha c\beta c\gamma - s\alpha s\gamma & c\gamma s\beta \\ s\alpha s\beta & -c\alpha s\beta & c\beta \end{bmatrix} \end{aligned} \tag{5.95} $$

we can determine the angles α, β, and γ:

$$ \alpha = -\arctan \frac{b_{31}}{b_{32}} \tag{5.96} $$

$$ \beta = \arccos b_{33} \tag{5.97} $$

$$ \gamma = \arctan \frac{b_{13}}{b_{23}} \tag{5.98} $$

The transformation matrix between B_0 and B comes from the Rodriguez formula (5.4). However, ${}^G\hat{u}$ must be expressed in B_0 to use Equation (5.4):

$$ {}^0\hat{u} = {}^GR_0^{\mathrm{T}}\, {}^G\hat{u} = {}^GR_0^{\mathrm{T}}\, \hat{u} \tag{5.99} $$

$$ \begin{aligned} {}^0R_B &= \mathbf{I}\cos\phi + {}^0\hat{u}\,{}^0\hat{u}^{\mathrm{T}} \ \operatorname{vers}\phi + {}^0\tilde{u}\sin\phi \\ &= {}^GR_0^{\mathrm{T}}\, \cos\phi + \left({}^GR_0^{\mathrm{T}}\,\hat{u}\right)\left({}^GR_0^{\mathrm{T}}\,\hat{u}\right)^{\mathrm{T}} \ \operatorname{vers}\phi + {}^GR_0^{\mathrm{T}}\,\tilde{u}\sin\phi \\ &= {}^GR_0^{\mathrm{T}}\, \cos\phi + {}^GR_0^{\mathrm{T}}\,\hat{u}\,\hat{u}^{\mathrm{T}}\,{}^GR_0 \ \operatorname{vers}\phi + {}^GR_0^{\mathrm{T}}\,\tilde{u}\sin\phi \end{aligned} \tag{5.100} $$

The transformation matrix ${}^GR_B$ between the final position of the body and the global frame would be

$$ \begin{aligned} {}^GR_B &= {}^GR_0\,{}^0R_B \\ &= {}^GR_0\left[{}^GR_0^{\mathrm{T}}\, \cos\phi + {}^GR_0^{\mathrm{T}}\left[\hat{u}\,\hat{u}^{\mathrm{T}}\right]{}^GR_0 \ \operatorname{vers}\phi + {}^GR_0^{\mathrm{T}}\,\tilde{u}\sin\phi\right] \\ &= \mathbf{I}\cos\phi + \left[\hat{u}\,\hat{u}^{\mathrm{T}}\right]{}^GR_0 \ \operatorname{vers}\phi + \tilde{u}\sin\phi \end{aligned} \tag{5.101} $$

This equation needs a name. Let us call it the *final rotation formula*. It determines the transformation matrix between a body frame B and the global frame G after the rotation ϕ of B about $\hat{u} = {}^G\hat{u}$, starting from a position $B_0 \neq G$.

As an example, consider a body that is rotated 45 deg about the Z-axis and is at B_0:

$$
{}^{G}R_0 = \begin{bmatrix} \cos\frac{\pi}{4} & -\sin\frac{\pi}{4} & 0 \\ \sin\frac{\pi}{4} & \cos\frac{\pi}{4} & 0 \\ 0 & 0 & 1 \end{bmatrix} \approx \begin{bmatrix} 0.707 & -0.707 & 0 \\ 0.707 & 0.707 & 0 \\ 0 & 0 & 1 \end{bmatrix} \tag{5.102}
$$

The body is then supposed to turn 90 deg about ${}^{G}\hat{u}$:

$$
\phi = \tfrac{1}{2}\pi \qquad {}^{G}\hat{u} = \hat{I} \tag{5.103}
$$

Therefore,

$$
\begin{aligned}
{}^{G}R_B &= \begin{bmatrix} 1 & 0 & 0 \\ 0 & 1 & 0 \\ 0 & 0 & 1 \end{bmatrix} \cos\frac{\pi}{2} + \begin{bmatrix} 0 & 0 & 0 \\ 0 & 0 & -1 \\ 0 & 1 & 0 \end{bmatrix} \sin\frac{\pi}{2} \\
&\quad + {}^{G}R_0 \left[\begin{bmatrix} 1 \\ 0 \\ 0 \end{bmatrix} \begin{bmatrix} 1 & 0 & 0 \end{bmatrix} \right] \left(1 - \cos\frac{\pi}{2}\right) \\
&= \begin{bmatrix} 0.70711 & 0 & 0 \\ 0.70711 & 0 & -1 \\ 0 & 1 & 0 \end{bmatrix}
\end{aligned} \tag{5.104}
$$

A body point at ${}^{B}\mathbf{r} = \hat{\imath}$ will be seen at

$$
\begin{aligned}
{}^{G}\mathbf{r} &= {}^{G}R_B\, {}^{B}\mathbf{r} \\
&= \begin{bmatrix} 0.70711 & 0 & 0 \\ 0.70711 & 0 & -1 \\ 0 & 1 & 0 \end{bmatrix} \begin{bmatrix} 1 \\ 0 \\ 0 \end{bmatrix} = \begin{bmatrix} 0.70711 \\ 0.70711 \\ 0 \end{bmatrix}
\end{aligned} \tag{5.105}
$$

***Example 311 ★ Euler Theorem* 1** A rotation about a point is equivalent to a rotation about a line passing through this point.

Proof: Let us suppose that a body has been rotated about the point O. Let us select an arbitrary segment A_1B_1 in the initial position of the body that is not passing through O. Then A_2B_2 would be the corresponding segment at the final position. Let us draw the plane of symmetry π of segment A_1B_1 and A_2B_2 by points O and midpoints of A_1A_2 and B_1B_2. The planes OA_IB_I and OA_2B_2 intersect at a line l which is also in the plane of symmetry. The line l passes through O because l is the locus of points equidistant from A_1 and A_2 as well as B_1 and B_2. While $OA_I = OA_2$ and $OB_I = OB_2$, let us pick a point C on l different than O.

The tetrahedrons OCA_IB_I and OCA_2B_2 are equal and superposable. This is because the vertices O, C, A_1 and O, C, A_2 are placed symmetrically with respect to π. Hence, if the tetrahedrons OCA_1B_1 and OCA_2B_2 were not superposable, the vertices B_1 and B_2 would have to be placed asymmetrically with respect to π.

If we now rotate the rigid body about the axis l so that A_1 falls on A, then the tetrahedron OCA_1B_1 will fall on the tetrahedron OCA_2B_2. After this rotation of the body we will know the position of three points O, A_2, B_2 of the body and therefore we know the positions of every point of the body.

A consequence of the Euler theorem is that: During a rotation of a body about a point, there exists in the body a certain line having the property that its points do not change their position.

Another consequence of the Euler theorem is that: If a rigid body makes two successive rotations about two axes passing through the fixed point O, then the body can be displaced from its initial position to its final position by means of one rotation about an axis passing through O. Therefore, the combination of two rotations about an axis passing through one point is a rotation about an axis passing through the same point. ■

Example 312 ★ Euler–Chasles Theorem A rigid body can be displaced from an initial position to another final position by means of one translation and one rotation.

Proof: Let P_1 be a point of the rigid body B in its initial position B_1 and P_2 the corresponding point in the final position B_2. We first translate the body to an intermediate position B' so that P_1 falls on P_2. If the position B' is identical with B_2, then we have displaced the body by means of one translation, conformably to the requirements of the theorem. Let us assume that the position B' is different than B_2. Because the positions B' and B_2 of the rigid body have a common point P_2, then by Euler theorem 1 of Example 311 we can displace B' to B_2 by a rotation about an axis passing through P_2.

If we choose a different P_1, then in general we will have a different translation and a different rotation about a different axis. ■

Example 313 ★ Euler Theorem 2 A rigid body can be displaced from an initial position to a final position by means of two successive rotations.

Proof: Let P_1 be a point of the rigid body B in its initial position B_1 and P_2 the corresponding point in the final position B_2. We rotate the body 180 deg about the axis l, which is the axis of symmetry of the segment P_1P_2. By this rotation, point P_1 will fall on point P_2. The body will assume the position B' which has the point P_2 in common with position B_2. Consequently, we can go from position B' to B_2 by a rotation about an axis l passing through P_2. ■

5.2 EULER PARAMETERS

Any orientation of a local frame $B(Oxyz)$ relative to a global frame $G(OXYZ)$ can be achieved by a rotation ϕ about an axis ${}^G\hat{u} = u_1\hat{I} + u_2\hat{J} + u_3\hat{K}$. The existence of such an axis of rotation is the analytical representation of Euler's theorem for rigid-body rotation: *The most general displacement of a rigid body with one point fixed is a rotation about some axis.*

An effective way to find the angle ϕ and axis ${}^G\hat{u}$ is the *Euler parameters* e_0, e_1, e_2, e_3 such that e_0 is a scalar and e_1, e_2, e_3 are components of a vector $\mathbf{e}$,

$$e_0 = \cos\frac{\phi}{2} \tag{5.106}$$

$$\mathbf{e} = e_1\hat{I} + e_2\hat{J} + e_3\hat{K} = \hat{u}\sin\frac{\phi}{2} \tag{5.107}$$

so

$$e_1^2 + e_2^2 + e_3^2 + e_0^2 = e_0^2 + \mathbf{e}^{\mathrm{T}}\,\mathbf{e} = 1 \tag{5.108}$$

Using the Euler parameters, the transformation matrix ${}^GR_B$ to satisfy the equation ${}^G\mathbf{r} = {}^GR_B\,{}^B\mathbf{r}$ is

$$\begin{aligned} {}^GR_B = R_{\hat{u},\phi} &= \left(e_0^2 - \mathbf{e}^2\right)\mathbf{I} + 2\mathbf{e}\,\mathbf{e}^{\mathrm{T}} + 2e_0\tilde{e} \\ &= \begin{bmatrix} e_0^2 + e_1^2 - e_2^2 - e_3^2 & 2\left(e_1e_2 - e_0e_3\right) & 2\left(e_0e_2 + e_1e_3\right) \\ 2\left(e_0e_3 + e_1e_2\right) & e_0^2 - e_1^2 + e_2^2 - e_3^2 & 2\left(e_2e_3 - e_0e_1\right) \\ 2\left(e_1e_3 - e_0e_2\right) & 2\left(e_0e_1 + e_2e_3\right) & e_0^2 - e_1^2 - e_2^2 + e_3^2 \end{bmatrix} \end{aligned} \tag{5.109}$$

where $\tilde{e}$ is the *skew-symmetric* matrix associated with $\mathbf{e}$,

$$\tilde{e} = \begin{bmatrix} 0 & -e_3 & e_2 \\ e_3 & 0 & -e_1 \\ -e_2 & e_1 & 0 \end{bmatrix} \tag{5.110}$$

Euler parameters provide a well-suited, redundant, and nonsingular rotation description for arbitrary and large rotations. It is redundant because there are four parameters and a constraint equation to define the required three parameters of angle–axis rotation.

When a transformation matrix ${}^GR_B$ is given as

$${}^GR_B = \left[r_{ij}\right] = \begin{bmatrix} r_{11} & r_{12} & r_{13} \\ r_{21} & r_{22} & r_{23} \\ r_{31} & r_{32} & r_{33} \end{bmatrix} \tag{5.111}$$

where r_{ij} indicates the element of row i and column j of ${}^GR_B$,we may find the Euler parameters e_0, e_1, e_2, e_3 and indicate the axis and angle of rotation from

$$e_0^2 = \tfrac{1}{4}(\mathrm{tr}({}^GR_B) + 1) = \tfrac{1}{4}(r_{11} + r_{22} + r_{33} + 1) \tag{5.112}$$

$$\mathbf{e} = \frac{1}{4e_0}\begin{bmatrix} r_{32} - r_{23} \\ r_{13} - r_{31} \\ r_{21} - r_{12} \end{bmatrix} \tag{5.113}$$

or from one of the following four sets of equations:

$$e_0 = \pm\frac{1}{2}\sqrt{1 + r_{11} + r_{22} + r_{33}}$$

$$e_1 = \frac{1}{4}\frac{r_{32} - r_{23}}{e_0} \qquad e_2 = \frac{1}{4}\frac{r_{13} - r_{31}}{e_0} \qquad e_3 = \frac{1}{4}\frac{r_{21} - r_{12}}{e_0} \tag{5.114}$$

$$e_1 = \pm\frac{1}{2}\sqrt{1 + r_{11} - r_{22} - r_{33}}$$

$$e_2 = \frac{1}{4}\frac{r_{21} + r_{12}}{e_1} \qquad e_3 = \frac{1}{4}\frac{r_{31} + r_{13}}{e_1} \qquad e_0 = \frac{1}{4}\frac{r_{32} + r_{23}}{e_1} \tag{5.115}$$

$$e_2 = \pm\frac{1}{2}\sqrt{1 - r_{11} + r_{22} - r_{33}}$$

$$e_3 = \frac{1}{4}\frac{r_{32} + r_{23}}{e_2} \qquad e_0 = \frac{1}{4}\frac{r_{13} - r_{31}}{e_2} \qquad e_1 = \frac{1}{4}\frac{r_{21} + r_{12}}{e_2} \tag{5.116}$$

$$e_3 = \pm\frac{1}{2}\sqrt{1 - r_{11} - r_{22} + r_{33}}$$

$$e_0 = \frac{1}{4}\frac{r_{21} - r_{12}}{e_3} \qquad e_1 = \frac{1}{4}\frac{r_{31} + r_{13}}{e_3} \qquad e_2 = \frac{1}{4}\frac{r_{32} + r_{23}}{e_3} \tag{5.117}$$

Any set of Equations (5.114)–(5.117) provides the same Euler parameters e_0, e_1, e_2, e_3. To minimize the numerical inaccuracies, it is recommended to use the set with maximum divisor. The plus-and-minus sign indicates that rotation ϕ about $\hat{u}$ is equivalent to rotation $-\phi$ about $-\hat{u}$.

We may obtain the angle of rotation ϕ and the axis of rotation $\hat{u}$ for a given transformation matrix ${}^G R_B$ by

$$\cos\phi = \tfrac{1}{2}(r_{11} + r_{22} + r_{33} - 1) \tag{5.118}$$

$$\hat{u} = \frac{1}{2\sin\phi}\begin{bmatrix} r_{32} - r_{23} \\ r_{13} - r_{31} \\ r_{21} - r_{12} \end{bmatrix} \tag{5.119}$$

Proof: Consider the point P of a rigid body with a fixed point at O, as is shown in Figure 5.5. The position vector of P before rotation is $\mathbf{r}$ and the globally fixed axis of rotation ON is indicated by the unit vector $\hat{u}$. After a rotation ϕ about $\hat{u}$, the point P moves to P' with the position vector $\mathbf{r}'$. To obtain the relationship between $\mathbf{r}$ and $\mathbf{r}'$, we express $\mathbf{r}'$ by the following vector equation:

$$\mathbf{r}' = \overrightarrow{ON} + \overrightarrow{NQ} + \overrightarrow{QP'} \tag{5.120}$$

Employing Figure 5.5 and using $\mathbf{r}$, $\mathbf{r}'$, $\hat{u}$, and ϕ, we may rewrite Equation (5.120):

$$\begin{aligned} \mathbf{r}' &= (\mathbf{r}\cdot\hat{u})\,\hat{u} + \hat{u}\times(\mathbf{r}\times\hat{u})\cos\phi - (\mathbf{r}\times\hat{u})\sin\phi \\ &= (\mathbf{r}\cdot\hat{u})\,\hat{u} + \left[\mathbf{r} - (\mathbf{r}\cdot\hat{u})\,\hat{u}\right]\cos\phi + (\hat{u}\times\mathbf{r})\sin\phi \end{aligned} \tag{5.121}$$

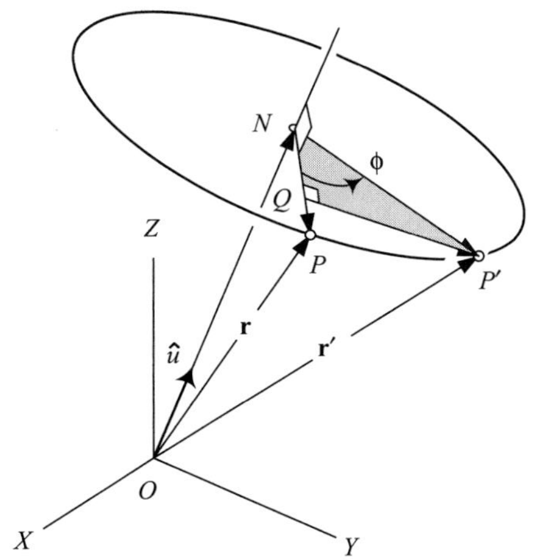

Figure 5.5 Axis and angle of rotation.

Rearranging (5.121) leads to a new form of the Rodriguez rotation formula:

$$\mathbf{r}' = \mathbf{r}\cos\phi + (1 - \cos\phi)(\hat{u}\cdot\mathbf{r})\hat{u} + (\hat{u}\times\mathbf{r})\sin\phi \tag{5.122}$$

Using the Euler parameters in (5.106) and (5.107) along with the trigonometric relations

$$\cos\phi = 2\cos^2\frac{\phi}{2} - 1 \tag{5.123}$$

$$\sin\phi = 2\sin\frac{\phi}{2}\cos\frac{\phi}{2} \tag{5.124}$$

$$1 - \cos\phi = 2\sin^2\frac{\phi}{2} \tag{5.125}$$

we can convert the Rodriguez rotation formula (5.122) to a more useful form:

$$\mathbf{r}' = \mathbf{r}\left(2e_0^2 - 1\right) + 2\mathbf{e}\,(\mathbf{e}\cdot\mathbf{r}) + 2e_0\,(\mathbf{e}\times\mathbf{r}) \tag{5.126}$$

We may define the new position vector of P as $\mathbf{r}' = {}^G\mathbf{r}$ and the initial position vector as $\mathbf{r} = {}^B\mathbf{r}$ to write Equation (5.126) as

$${}^G\mathbf{r} = \left(e_0^2 - \mathbf{e}^2\right){}^B\mathbf{r} + 2\mathbf{e}\left(\mathbf{e}^{\mathrm{T}}\,{}^B\mathbf{r}\right) + 2e_0\left(\tilde{\mathbf{e}}{}^B\mathbf{r}\right). \tag{5.127}$$

When we factor out ${}^B\mathbf{r}$, the transformation matrix ${}^GR_B$ based on the Euler parameters appears as

$${}^G\mathbf{r} = {}^GR_B\,{}^B\mathbf{r} = R_{\hat{u},\phi}\,{}^B\mathbf{r} \tag{5.128}$$

$${}^GR_B = \left(e_0^2 - \mathbf{e}^2\right)\mathbf{I} + 2\mathbf{e}\,\mathbf{e}^{\mathrm{T}} + 2e_0\tilde{\mathbf{e}} \tag{5.129}$$

Direct substitution shows that

$$\mathrm{tr}\left({}^GR_B\right) = r_{11} + r_{22} + r_{33} = 4e_0^2 - 1 = 2\cos\phi + 1 \tag{5.130}$$

and

$$
\begin{aligned}
{}^G R_B - {}^G R_B^{\mathrm{T}} &= \begin{bmatrix} 0 & -r_{21} + r_{12} & r_{13} - r_{31} \\ r_{21} - r_{12} & 0 & -r_{32} + r_{23} \\ -r_{13} + r_{31} & r_{32} - r_{23} & 0 \end{bmatrix} \\
&= \begin{bmatrix} 0 & -2u_3 \sin\phi & 2u_2 \sin\phi \\ 2u_3 \sin\phi & 0 & -2u_1 \sin\phi \\ -2u_2 \sin\phi & 2u_1 \sin\phi & 0 \end{bmatrix} \\
&= \begin{bmatrix} 0 & -4e_0e_3 & 4e_0e_2 \\ 4e_0e_3 & 0 & -4e_0e_1 \\ -4e_0e_2 & 4e_0e_1 & 0 \end{bmatrix}
\end{aligned} \tag{5.131}
$$

Therefore,

$$e_0^2 = \tfrac{1}{4}[\mathrm{tr}({}^G R_B) + 1] \tag{5.132}$$

$$\tilde{e} = \frac{1}{4e_0}({}^G R_B - {}^G R_B^{\mathrm{T}}) \tag{5.133}$$

$$\mathbf{e} = \frac{1}{4e_0} \begin{bmatrix} r_{32} - r_{23} \\ r_{13} - r_{31} \\ r_{21} - r_{12} \end{bmatrix} \tag{5.134}$$

or

$$\cos\phi = \tfrac{1}{2}[\mathrm{tr}({}^G R_B) - 1] \tag{5.135}$$

$$\tilde{u} = \frac{1}{2\sin\phi}({}^G R_B - {}^G R_B^{\mathrm{T}}) \tag{5.136}$$

$$\hat{u} = \frac{1}{2\sin\phi} \begin{bmatrix} r_{32} - r_{23} \\ r_{13} - r_{31} \\ r_{21} - r_{12} \end{bmatrix} \tag{5.137}$$

The first set of Equations (5.114) may be proven by comparing (5.109) and (5.111). The Euler parameter e_0 can be found by summing the diagonal elements r_{11}, r_{22}, and r_{33} to get $\mathrm{tr}\left({}^G R_B\right) = 4e_0^2 - 1$. To find e_1, e_2, and e_3 we need to simplify $r_{32} - r_{23}$, $r_{13} - r_{31}$, and $r_{21} - r_{12}$, respectively. The other sets of solutions (5.115)–(5.117) can also be found by comparison. ■

Example 314 Euler Parameters and Angle–Axis Rotation of ${}^G\mathbf{R}_B$ The Euler parameters for rotation $\phi = 30\,\mathrm{deg}$ about $\hat{u} = (\hat{I} + \hat{J} + \hat{K})/\sqrt{3}$ are

$$e_0 = \cos\frac{\phi}{2} = \cos\frac{\pi}{12} = 0.965\,93 \tag{5.138}$$

$$\mathbf{e} = \hat{u}\sin\frac{\phi}{2} = e_1\hat{I} + e_2\hat{J} + e_3\hat{K} = 0.14943\left(\hat{I} + \hat{J} + \hat{K}\right) \tag{5.139}$$

Therefore, from (5.109) the transformation matrix is

$$
\begin{aligned}
{}^{G}R_B &= \left(e_0^2 - \mathbf{e}^2\right)\mathbf{I} + 2\mathbf{e}\,\mathbf{e}^{\mathrm{T}} + 2e_0\tilde{e} \\
&= \begin{bmatrix} 0.91069 & -0.24402 & 0.33334 \\ 0.33334 & 0.91069 & -0.24402 \\ -0.24402 & 0.33334 & 0.91069 \end{bmatrix}
\end{aligned} \tag{5.140}
$$

As a double check, we may see that any point on the axis of rotation is invariant under the rotation:

$$
\begin{bmatrix} 0.91069 & -0.24402 & 0.33334 \\ 0.33334 & 0.91069 & -0.24402 \\ -0.24402 & 0.33334 & 0.91069 \end{bmatrix} \begin{bmatrix} 1 \\ 1 \\ 1 \end{bmatrix} = \begin{bmatrix} 1 \\ 1 \\ 1 \end{bmatrix} \tag{5.141}
$$

Example 315 Rodriguez Formula in Terms of Euler Parameters Employing the Rodriguez rotation formula (5.17),

$$
R_{\hat{u},\phi} = \mathbf{I} + 2\tilde{u}\sin\frac{\phi}{2}\left(\mathbf{I}\cos\frac{\phi}{2} + \tilde{u}\sin\frac{\phi}{2}\right) \tag{5.142}
$$

we can use the definition of Euler parameters,

$$
e_0 = \cos\frac{\phi}{2} \tag{5.143}
$$

$$
e_1 = u_1\sin\frac{\phi}{2} \qquad e_2 = u_2\sin\frac{\phi}{2} \qquad e_3 = u_3\sin\frac{\phi}{2} \tag{5.144}
$$

and rewrite the Rodriguez formula in terms of Euler parameters:

$$
R_{\hat{u},\phi} = \mathbf{I} + 2\tilde{e}\left(\mathbf{I}e_0 + \tilde{e}\right) \tag{5.145}
$$

Example 316 Euler Parameter and Euler Angle Relationship Comparing the Euler angle rotation matrix (4.142) and the Euler parameter transformation matrix (5.109), we can determine the following relationships between Euler angles and Euler parameters:

$$
e_0 = \cos\frac{\theta}{2}\cos\frac{\psi+\varphi}{2} \tag{5.146}
$$

$$
e_1 = \sin\frac{\theta}{2}\cos\frac{\psi-\varphi}{2} \tag{5.147}
$$

$$
e_2 = \sin\frac{\theta}{2}\sin\frac{\psi-\varphi}{2} \tag{5.148}
$$

$$
e_3 = \cos\frac{\theta}{2}\sin\frac{\psi+\varphi}{2} \tag{5.149}
$$

The reverse relations may be written as

$$\varphi = \cos^{-1} \frac{2\left(e_2 e_3 + e_0 e_1\right)}{\sin\theta} \tag{5.150}$$

$$\theta = \cos^{-1}\left[2\left(e_0^2 + e_3^2\right) - 1\right] \tag{5.151}$$

$$\psi = \cos^{-1} \frac{-2\left(e_2 e_3 - e_0 e_1\right)}{\sin\theta} \tag{5.152}$$

or as

$$\varphi = \tan^{-1}\frac{e_3}{e_0} + \tan^{-1}\frac{e_2}{e_0} \tag{5.153}$$

$$\theta = \cos^{-1}\left(e_0^2 - e_1^2 - e_2^2 + e_3^2\right) \tag{5.154}$$

$$\psi = \tan^{-1}\frac{e_3}{e_0} - \tan^{-1}\frac{e_2}{e_0} \tag{5.155}$$

Example 317 Rotation Matrix for $e_0 = 0$ or $\phi = k\pi$ When the angle of rotation is $\phi = k\pi$, $k = \pm 1, \pm 3, \ldots$, then $e_0 = 0$. In this case, the Euler parameter transformation matrix (5.109) becomes

$$ {}^G R_B = 2\begin{bmatrix} e_1^2 - \frac{1}{2} & e_1 e_2 & e_1 e_3 \\ e_1 e_2 & e_2^2 - \frac{1}{2} & e_2 e_3 \\ e_1 e_3 & e_2 e_3 & e_3^2 - \frac{1}{2} \end{bmatrix} \tag{5.156}$$

which is a symmetric matrix. It indicates that rotations $\phi = k\pi$ and $\phi = -k\pi$ provide equivalent orientations.

Example 318 ★ Vector of Infinitesimal Rotation The Rodriguez rotation formula (5.122) for a differential rotation $d\phi$ becomes

$$\mathbf{r}' = \mathbf{r} + (\hat{u} \times \mathbf{r})\, d\phi \tag{5.157}$$

In this case the difference between $\mathbf{r}'$ and $\mathbf{r}$ is also very small,

$$d\mathbf{r} = \mathbf{r}' - \mathbf{r} = \left(d\phi\, \hat{u}\right) \times \mathbf{r} \tag{5.158}$$

and hence a differential rotation $d\phi$ about the axis $\hat{u}$ is a vector along $\hat{u}$ with magnitude $d\phi$. Dividing both sides by dt leads to

$$\dot{\mathbf{r}} = \frac{d\phi}{dt}\hat{u} \times \mathbf{r} = \dot{\boldsymbol{\phi}} \times \mathbf{r} = \boldsymbol{\omega} \times \mathbf{r} \tag{5.159}$$

which represents the global velocity vector of any point in a rigid body rotating about $\hat{u}$.

Example 319 ★ Exponential Form of Rotation $e^{\phi\tilde{u}}$ Consider a point P in the body frame B with a position vector $\mathbf{r}$. If the rigid body has an angular velocity $\boldsymbol{\omega}$, then the velocity of P in the global coordinate frame is

$$\dot{\mathbf{r}} = \boldsymbol{\omega} \times \mathbf{r} = \tilde{\omega}\mathbf{r} \tag{5.160}$$

This is a first-order linear differential equation that may be integrated to have

$$\mathbf{r}(t) = \mathbf{r}(0)e^{\tilde{\omega}t} \tag{5.161}$$

where $\mathbf{r}(0)$ is the initial position vector of P and $e^{\tilde{\omega}t}$ is a matrix exponential,

$$e^{\tilde{\omega}t} = \mathbf{I} + \tilde{\omega}t + \frac{(\tilde{\omega}t)^2}{2!} + \frac{(\tilde{\omega}t)^3}{3!} + \cdots \tag{5.162}$$

The angular velocity $\boldsymbol{\omega}$ has a magnitude ω and direction indicated by a unit vector $\hat{u}$. Therefore,

$$\boldsymbol{\omega} = \omega\hat{u} \tag{5.163}$$

$$\tilde{\omega} = \omega\tilde{u} \tag{5.164}$$

$$\tilde{\omega}t = \omega t\tilde{u} = \phi\tilde{u} \tag{5.165}$$

and hence

$$\begin{aligned} e^{\tilde{\omega}t} &= e^{\phi\tilde{u}} \\ &= \mathbf{I} + \left(\phi - \frac{\phi^3}{3!} + \frac{\phi^5}{5!} - \cdots\right)\tilde{u} + \left(\frac{\phi^2}{2!} - \frac{\phi^4}{4!} + \frac{\phi^6}{6!}\cdots\right)\tilde{u}^2 \end{aligned} \tag{5.166}$$

or equivalently

$$e^{\phi\tilde{u}} = \mathbf{I} + \tilde{u}\sin\phi + \tilde{u}^2(1 - \cos\phi) \tag{5.167}$$

It is an alternative form of the Rodriguez rotation formula showing that $e^{\phi\tilde{u}}$ is the rotation transformation to map ${}^B\mathbf{r} = \mathbf{r}(0)$ to ${}^G\mathbf{r} = \mathbf{r}(t)$.

We may show that $e^{\phi\tilde{u}} \in S$, where S is the set of rotation matrices

$$S = \{R \in \mathbb{R}^{3\times 3} : RR^{\mathrm{T}} = \mathbf{I}, |R| = 1\} \tag{5.168}$$

by showing that $R = e^{\phi\tilde{u}}$ has the orthogonality property $R^{\mathrm{T}}R = I$ and its determinant is $|R| = 1$.

The orthogonality of $e^{\phi\tilde{u}}$ can be verified by considering

$$\left[e^{\phi\tilde{u}}\right]^{-1} = e^{-\phi\tilde{u}} = e^{\phi\tilde{u}^{\mathrm{T}}} = \left[e^{\phi\tilde{u}}\right]^{\mathrm{T}} \tag{5.169}$$

Therefore, $R^{-1} = R^{\mathrm{T}}$ and consequently $RR^{\mathrm{T}} = \mathbf{I}$. From orthogonality, it follows that $|R| = \pm 1$, and from continuity of the exponential function, it follows that $|e^0| = 1$. Therefore, $|R| = 1$.

Expanding $e^{\phi\tilde{u}}$,

$$e^{\phi\tilde{u}} = \mathbf{I} + \tilde{u}\sin\phi + \tilde{u}^2(1 - \cos\phi) \tag{5.170}$$

gives

$$e^{\phi\tilde{u}} = \begin{bmatrix} u_1^2 \text{ vers } \phi + c\phi & u_1u_2 \text{ vers } \phi - u_3 s\phi & u_1u_3 \text{ vers } \phi + u_2 s\phi \\ u_1u_2 \text{ vers } \phi + u_3 s\phi & u_2^2 \text{ vers } \phi + c\phi & u_2u_3 \text{ vers } \phi - u_1 s\phi \\ u_1u_3 \text{ vers } \phi - u_2 s\phi & u_2u_3 \text{ vers } \phi + u_1 s\phi & u_3^2 \text{ vers } \phi + c\phi \end{bmatrix} \tag{5.171}$$

which is equal to the angle–axis equation (5.5), and therefore

$$e^{\phi\tilde{u}} = R_{\hat{u},\phi} = {}^G R_B = \mathbf{I}\cos\phi + \hat{u}\hat{u}^{\mathrm{T}} \text{ vers } \phi + \tilde{u}\sin\phi \tag{5.172}$$

Example 320 **★ Rodriguez Vector** We may redraw Figure 5.5 as shown in Figure 5.6 and write

$$\cos\frac{\phi}{2}\left|\overrightarrow{MP'}\right| = \sin\frac{\phi}{2}\left|\overrightarrow{NM}\right| \tag{5.173}$$

$$\left|\overrightarrow{NM}\right|\overrightarrow{MP'} = \left|\overrightarrow{MP'}\right|\hat{u}\times\overrightarrow{NM} \tag{5.174}$$

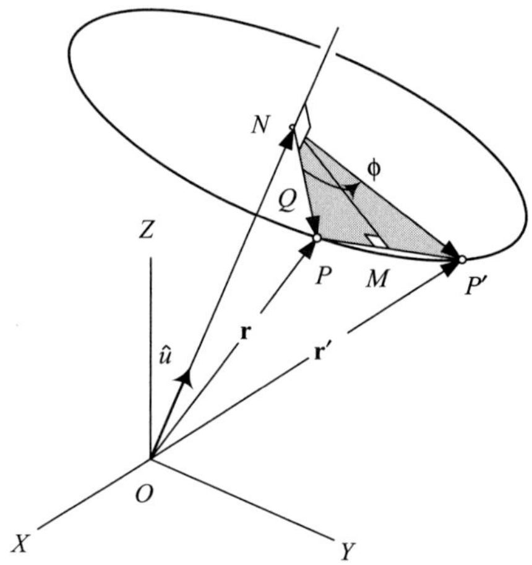

Figure 5.6 Illustration of a rotation of a rigid body to derive a new form of the Rodriguez rotation formula in Example 320.

to find

$$\left(\cos\frac{\phi}{2}\right)\overrightarrow{MP'} = \left(\sin\frac{\phi}{2}\right)\hat{u}\times\overrightarrow{NM} \tag{5.175}$$

Now using the equalities

$$2\overrightarrow{MP'} = \overrightarrow{NP'} - \overrightarrow{NP} \tag{5.176}$$

$$2\overrightarrow{NM} = \overrightarrow{NP'} + \overrightarrow{NP} \tag{5.177}$$

$$\overrightarrow{NP'} - \overrightarrow{NP} = \mathbf{r}' - \mathbf{r} \tag{5.178}$$

$$\hat{u}\times\left(\overrightarrow{NP'} + \overrightarrow{NP}\right) = \hat{u}\times\left(\mathbf{r}' + \mathbf{r}\right) \tag{5.179}$$

we can write an alternative form of the *Rodriguez rotation formula*,

$$\cos\frac{\phi}{2}\left(\mathbf{r}'-\mathbf{r}\right)=\sin\frac{\phi}{2}\hat{u}\times\left(\mathbf{r}'+\mathbf{r}\right) \tag{5.180}$$

or

$$(\mathbf{r}'-\mathbf{r})=\mathbf{q}\times(\mathbf{r}'+\mathbf{r}) \tag{5.181}$$

where

$$\mathbf{q}=\tan\frac{\phi}{2}\hat{u} \tag{5.182}$$

is called the *Rodriguez vector* or the *Gibbs vector*. The Euler parameters are related to the Rodriguez vector according to

$$\mathbf{q}=\frac{\mathbf{e}}{e_0} \tag{5.183}$$

and

$$e_0=\frac{1}{\sqrt{1+\mathbf{q}^{\mathrm{T}}\mathbf{q}}}=\frac{1}{\sqrt{1+\mathbf{q}^2}} \tag{5.184}$$

$$e_i=\frac{q_i}{\sqrt{1+\mathbf{q}^{\mathrm{T}}\mathbf{q}}}=\frac{q_i}{\sqrt{1+\mathbf{q}^2}}\qquad i=1,2,3. \tag{5.185}$$

So, the Rodriguez formula (5.109) can be converted to a new form based on the Rodriguez vector:

$$\begin{aligned}{}^GR_B=R_{\hat{u},\phi}&=\left(e_0^2-\mathbf{e}^2\right)\mathbf{I}+2\mathbf{e}\,\mathbf{e}^{\mathrm{T}}+2e_0\tilde{e}\\&=\frac{1}{1+\mathbf{q}^{\mathrm{T}}\mathbf{q}}[(1-\mathbf{q}^{\mathrm{T}}\mathbf{q})\mathbf{I}+2\mathbf{q}\mathbf{q}^{\mathrm{T}}+2\tilde{q}]\end{aligned} \tag{5.186}$$

Using the Rodriguez vector, we can show that the combination of two rotations $\mathbf{q}'$ and $\mathbf{q}''$ is equivalent to a single rotation $\mathbf{q}$, where

$$\mathbf{q}=\frac{\mathbf{q}''+\mathbf{q}'-\mathbf{q}''\times\mathbf{q}'}{1-\mathbf{q}''\cdot\mathbf{q}'} \tag{5.187}$$

Example 321 Gibbs Rotation Matrix We may rearrange the rotation matrix (5.186) to

$${}^GR_B=R_{\hat{u},\phi}=\left[\mathbf{I}-\tilde{q}\right]\left[\mathbf{I}+\tilde{q}\right]^{-1} \tag{5.188}$$

and call it the *Gibbs rotation formula*. Expressing the Gibbs vector $\mathbf{q}=\tan(\phi/2)\hat{u}$ as

$$\mathbf{q}=\begin{bmatrix}q_1\\q_2\\q_3\end{bmatrix}=\tan\frac{\phi}{2}\begin{bmatrix}u_1\\u_2\\u_3\end{bmatrix} \tag{5.189}$$

and expanding (5.188) or (5.186) provide the Gibbs form of the rotation matrix:

$$
{}^{G}R_{B} = \frac{1}{1+q_1^2+q_2^2+q_3^2} \times \begin{bmatrix} q_1^2-q_2^2-q_3^2+1 & 2q_1q_2-2q_3 & 2q_2+2q_1q_3 \\ 2q_3+2q_1q_2 & -q_1^2+q_2^2-q_3^2+1 & 2q_2q_3-2q_1 \\ 2q_1q_3-2q_2 & 2q_1+2q_2q_3 & -q_1^2-q_2^2+q_3^2+1 \end{bmatrix} \tag{5.190}
$$

Given a rotation matrix ${}^{G}R_{B} = [r_{ij}]$, we can determine the associated Gibbs vector $\mathbf{q} = \tan(\phi/2)\hat{u}$ and the angle of rotations:

$$
\tilde{q} = \frac{1}{4}\frac{{}^{G}R_{B} - {}^{G}R_{B}^{\mathrm{T}}}{1+q_1^2+q_2^2+q_3^2} \tag{5.191}
$$

$$
\mathbf{q} = \frac{1}{4}\frac{1}{1+q_1^2+q_2^2+q_3^2}\begin{bmatrix} r_{32}-r_{23} \\ r_{13}-r_{31} \\ r_{21}-r_{12} \end{bmatrix} \tag{5.192}
$$

$$
\phi = 2\arctan\sqrt{3-\operatorname{tr}\left({}^{G}R_{B}\right)} = \arctan\sqrt{3-(r_{11}+r_{22}+r_{33})} \tag{5.193}
$$

Example 322 ★ Elements of ${}^{G}\boldsymbol{R}_{B}$ Recalling the permutation symbol (1.126) and the Kronecker delta (1.125), we can redefine the elements of the Rodriguez rotation matrix by

$$
r_{ij} = \delta_{ij}\cos\phi + (1-\cos\phi)u_iu_j + \epsilon_{ijk}u_k\sin\phi \tag{5.194}
$$

Example 323 Euler Parameters from ${}^{G}\boldsymbol{R}_{B}$ The following transformation matrix is given:

$$
{}^{G}R_{B} = \begin{bmatrix} 0.5449 & -0.5549 & 0.6285 \\ 0.3111 & 0.8299 & 0.4629 \\ -0.7785 & -0.0567 & 0.6249 \end{bmatrix} \tag{5.195}
$$

To calculate the corresponding Euler parameters, we use Equation (5.114) and find

$$
\begin{aligned} \operatorname{tr}\left({}^{G}R_{B}\right) &= r_{11}+r_{22}+r_{33} \\ &= 0.5449+0.8299+0.6249 = 1.9997 \end{aligned} \tag{5.196}
$$

therefore,

$$
e_0 = \tfrac{1}{2}\sqrt{\operatorname{tr}\left({}^{G}R_{B}\right)+1} = 0.86598 \tag{5.197}
$$

and

$$e_1 = \frac{1}{4}\frac{r_{32} - r_{23}}{e_0} = \frac{1}{4}\frac{-0.0567 - 0.4629}{0.86598} = -0.15$$
$$e_2 = \frac{1}{4}\frac{r_{13} - r_{31}}{e_0} = \frac{1}{4}\frac{0.6285 + 0.7785}{0.86598} = 0.40619 \tag{5.198}$$
$$e_3 = \frac{1}{4}\frac{r_{21} - r_{12}}{e_0} = \frac{1}{4}\frac{0.3111 + 0.5549}{0.86598} = 0.25001$$

Example 324 ★ Euler Parameters When One of Them Is Known Consider the Euler parameter rotation matrix (5.109) corresponding to rotation ϕ about an axis indicated by a unit vector $\hat{u}$. The off-diagonal elements of ${}^{G}R_{B}$,

$$\begin{aligned} e_0e_1 &= \tfrac{1}{4}(r_{32} - r_{23}) \\ e_0e_2 &= \tfrac{1}{4}(r_{13} - r_{31}) \\ e_0e_3 &= \tfrac{1}{4}(r_{21} - r_{12}) \\ e_1e_2 &= \tfrac{1}{4}(r_{12} + r_{21}) \\ e_1e_3 &= \tfrac{1}{4}(r_{13} + r_{31}) \\ e_2e_3 &= \tfrac{1}{4}(r_{23} + r_{32}) \end{aligned} \tag{5.199}$$

can be utilized to find $e_i, i = 0, 1, 2, 3$, if we know one of them.

Example 325 ★ Stanley Method Following an effective method developed by Stanley, we may first find the four e_i^2,

$$\begin{aligned} e_0^2 &= \tfrac{1}{2}[1 + \mathrm{tr}({}^{G}R_{B})] \\ e_1^2 &= \tfrac{1}{4}[1 + 2r_{11} - \mathrm{tr}({}^{G}R_{B})] \\ e_2^2 &= \tfrac{1}{4}[1 + 2r_{22} - \mathrm{tr}({}^{G}R_{B})] \\ e_3^2 &= \tfrac{1}{4}[1 + 2r_{33} - \mathrm{tr}({}^{G}R_{B})] \end{aligned} \tag{5.200}$$

and take the positive square root of the largest e_i^2. Then the other e_i are found by dividing the appropriate three of the six equations (5.199) by the largest e_i.

5.3 ★ QUATERNION

A *quaternion* q is defined as a scalar+vector quantity

$$q = q_0 + \mathbf{q} = q_0 + q_1\hat{I} + q_2\hat{J} + q_3\hat{K} \tag{5.201}$$

where q_0 is a scalar and $\mathbf{q}$ is a vector. A quaternion q can also be shown by a four-element vector,

$$q = \begin{bmatrix} q_0 \\ q_1 \\ q_2 \\ q_3 \end{bmatrix} \tag{5.202}$$

or in a *flag* form,

$$q = q_0 + q_1 i + q_2 j + q_3 k \tag{5.203}$$

where i, j, k are *flags* defined as

$$i^2 = j^2 = k^2 = ijk = -1 \tag{5.204}$$

$$ij = -ji = k \tag{5.205}$$

$$jk = -kj = i \tag{5.206}$$

$$ki = -ik = j \tag{5.207}$$

Addition and multiplication of two quaternions q and p are quaternions:

$$\begin{aligned} q + p &= (q_0 + \mathbf{q}) + (p_0 + \mathbf{p}) \\ &= q_0 + q_1 i + q_2 j + q_3 k + p_0 + p_1 i + p_2 j + p_3 k \\ &= (q_0 + p_0) + (q_1 + p_1)\, i + (q_2 + p_2)\, j + (q_3 + p_3)\, k \end{aligned} \tag{5.208}$$

$$\begin{aligned} qp &= (q_0 + \mathbf{q})\,(p_0 + \mathbf{p}) = q_0 p_0 + q_0 \mathbf{p} + p_0 \mathbf{q} + \mathbf{qp} \\ &= q_0 p_0 - \mathbf{q} \cdot \mathbf{p} + q_0 \mathbf{p} + p_0 \mathbf{q} + \mathbf{q} \times \mathbf{p} \\ &= (q_0 p_0 - q_1 p_1 - q_2 p_2 - q_3 p_3) \\ &\quad + (p_0 q_1 + p_1 q_0 - p_2 q_3 + p_3 q_2)\, i \\ &\quad + (p_0 q_2 + q_0 p_2 + p_1 q_3 - q_1 p_3)\, j \\ &\quad + (p_0 q_3 - p_1 q_2 + q_0 p_3 + p_2 q_1)\, k \end{aligned} \tag{5.209}$$

where $\mathbf{qp}$ is the *quaternion vector product* and is equal to the outer product minus the inner product of $\mathbf{q}$ and $\mathbf{p}$:

$$\mathbf{qp} = \mathbf{q} \times \mathbf{p} - \mathbf{q} \cdot \mathbf{p} \tag{5.210}$$

Quaternion addition is associative and commutative,

$$q + p = p + q \tag{5.211}$$

$$q + (p + r) = (q + p) + r \tag{5.212}$$

while quaternion multiplication is not commutative,

$$qp \neq pq \tag{5.213}$$

However, quaternion multiplication is associative and distributes over addition,

$$(pq)\,r = p\,(qr) \tag{5.214}$$

$$(p+q)\,r = pr + qr \tag{5.215}$$

A quaternion q has a conjugate q^* defined by

$$q^* = q_0 - \mathbf{q} = q_0 - q_1\hat{I} - q_2\hat{J} - q_3\hat{K} \tag{5.216}$$

Multiplication of a quaternion q by its conjugate q^* is given as

$$\begin{aligned} qq^* &= (q_0 + \mathbf{q})\,(q_0 - \mathbf{q}) = q_0q_0 + q_0\mathbf{q} - p_0\mathbf{q} - qq \\ &= q_0q_0 + \mathbf{q}\cdot\mathbf{q} - \mathbf{q}\times\mathbf{q} = q_0^2 + q_1^2 + q_2^2 + q_3^2 = |q|^2 \end{aligned} \tag{5.217}$$

and therefore, we may define the quaternion inverse and quaternion division:

$$|q| = \sqrt{qq^*} = \sqrt{q_0^2 + q_1^2 + q_2^2 + q_3^2} \tag{5.218}$$

$$q^{-1} = \frac{1}{q} = \frac{q^*}{|q|^2} \tag{5.219}$$

A quaternion q is a unit quaternion if $|q| = 1$. When a quaternion q is a unit, we have

$$q^{-1} = q^* \tag{5.220}$$

A quaternion may be variable because of both its scalar and its vector arguments. So, we show a variable quaternion q by $q = q\,(q_0, \mathbf{q})$.

Let us define a unit quaternion $e\left(\phi, \hat{u}\right)$, $\left|e\left(\phi, \hat{u}\right)\right| = 1$ as

$$e\left(\phi, \hat{u}\right) = e_0 + \mathbf{e} = e_0 + e_1\hat{I} + e_2\hat{J} + e_3\hat{K} = \cos\frac{\phi}{2} + \sin\frac{\phi}{2}\hat{u} \tag{5.221}$$

and $\mathbf{r} = 0 + \mathbf{r}$ be a quaternion associated to a pure vector $\mathbf{r}$. The vector $\mathbf{r}$ after a rotation ϕ about ${}^G\hat{u}$ would be

$$\mathbf{r}' = e\left(\phi, \hat{u}\right)\mathbf{r}\,e^*\left(\phi, \hat{u}\right) \tag{5.222}$$

which is equivalent to

$${}^G\mathbf{r} = e\left(\phi, \hat{u}\right)\,{}^B\mathbf{r}\,e^*\left(\phi, \hat{u}\right) \tag{5.223}$$

Therefore, a rotation $R_{\hat{u},\phi}$ can be defined by a quaternion $e\left(\phi, \hat{u}\right) = \cos(\phi/2) + \sin(\phi/2)\hat{u}$, and consequently, two consecutive rotations $R = R_2R_1$ are defined by

$$e\left(\phi, \hat{u}\right) = e_2\left(\phi_2, \hat{u}_2\right)\,e_1\left(\phi_1, \hat{u}_1\right) \tag{5.224}$$

The notations $e_1\left(\phi_1, \hat{u}_1\right)$, $e_2\left(\phi_2, \hat{u}_2\right)$, $\ldots$ are quaternion while e_0, e_1, e_2, e_3 are Euler parameters.

A quaternion may also be represented by a 4×4 matrix,

$$\overleftrightarrow{q} = \begin{bmatrix} q_0 & -q_1 & -q_2 & -q_3 \\ q_1 & q_0 & -q_3 & q_2 \\ q_2 & q_3 & q_0 & -q_1 \\ q_3 & -q_2 & q_1 & q_0 \end{bmatrix} \tag{5.225}$$

which provides the important orthogonality property

$$\overleftrightarrow{q}^{-1} = \overleftrightarrow{q}^{\mathrm{T}} \tag{5.226}$$

The matrix quaternion (5.225) can also be represented by

$$\overleftrightarrow{q} = \begin{bmatrix} q_0 & -\mathbf{q}^{\mathrm{T}} \\ \mathbf{q} & q_0\mathbf{I}_3 - \tilde{q} \end{bmatrix} \tag{5.227}$$

where

$$\tilde{q} = \begin{bmatrix} 0 & -q_3 & q_2 \\ q_3 & 0 & -q_1 \\ -q_2 & q_1 & 0 \end{bmatrix} \tag{5.228}$$

Employing the *matrix quaternion* $\overleftrightarrow{q}$, we can show quaternion multiplication by matrix multiplication:

$$qp = \overleftrightarrow{q}\, p = \begin{bmatrix} q_0 & -q_1 & -q_2 & -q_3 \\ q_1 & q_0 & -q_3 & q_2 \\ q_2 & q_3 & q_0 & -q_1 \\ q_3 & -q_2 & q_1 & q_0 \end{bmatrix} \begin{bmatrix} p_0 \\ p_1 \\ p_2 \\ p_3 \end{bmatrix} \tag{5.229}$$

The matrix description of quaternions relates quaternion manipulations and matrix manipulations. If p, q, and v are three quaternions such that

$$qp = v \tag{5.230}$$

then

$$\overleftrightarrow{q}\,\overleftrightarrow{p} = \overleftrightarrow{v} \tag{5.231}$$

Hence, the quaternion representation of transformation between coordinate frames (5.224) can also be defined by matrix multiplication:

$$\overleftrightarrow{{}^{G}\mathbf{r}} = \overleftrightarrow{e\left(\phi, \hat{u}\right)}\ \overleftrightarrow{{}^{B}\mathbf{r}}\ \overleftrightarrow{e^{*}\left(\phi, \hat{u}\right)} = \overleftrightarrow{e\left(\phi, \hat{u}\right)}\ \overleftrightarrow{{}^{B}\mathbf{r}}\ \overleftrightarrow{e\left(\phi, \hat{u}\right)}^{\mathrm{T}} \tag{5.232}$$

Proof: To show that a unit quaternion $e\left(\phi, \hat{u}\right)$ can work as a rotation matrix ${}^{G}R_{B}$, let us consider a quaternion $e\left(\phi, \hat{u}\right)$ as shown in Equation (5.221). Employing the quaternion multiplication (5.209) we can write

$$\begin{aligned}\mathbf{r}e^* &= e_0\mathbf{r} + \mathbf{r}\times\mathbf{e}^* - \mathbf{r}\cdot\mathbf{e}^* \\ &= e_0\begin{bmatrix} r_1 \\ r_2 \\ r_3\end{bmatrix} + \begin{bmatrix} r_1 \\ r_2 \\ r_3\end{bmatrix} \times \begin{bmatrix} -e_1 \\ -e_2 \\ -e_3\end{bmatrix} + (e_1r_1 + e_2r_2 + e_3r_3) \\ &= (e_1r_1 + e_2r_2 + e_3r_3) + \begin{bmatrix} e_0r_1 + e_2r_3 - e_3r_2 \\ e_0r_2 - e_1r_3 + e_3r_1 \\ e_0r_3 + e_1r_2 - e_2r_1\end{bmatrix}\end{aligned} \tag{5.233}$$

and therefore,

$$\begin{aligned}e\mathbf{r}e^* &= e_0(e_1r_1 + e_2r_2 + e_3r_3) - \begin{bmatrix} e_1 \\ e_2 \\ e_3\end{bmatrix}\cdot\begin{bmatrix} e_0r_1 + e_2r_3 - e_3r_2 \\ e_0r_2 - e_1r_3 + e_3r_1 \\ e_0r_3 + e_1r_2 - e_2r_1\end{bmatrix} \\ &+ e_0\begin{bmatrix} e_0r_1 + e_2r_3 - e_3r_2 \\ e_0r_2 - e_1r_3 + e_3r_1 \\ e_0r_3 + e_1r_2 - e_2r_1\end{bmatrix} + (e_1r_1 + e_2r_2 + e_3r_3)\begin{bmatrix} e_1 \\ e_2 \\ e_3\end{bmatrix} \\ &+ \begin{bmatrix} e_1 \\ e_2 \\ e_3\end{bmatrix}\times\begin{bmatrix} e_0r_1 + e_2r_3 - e_3r_2 \\ e_0r_2 - e_1r_3 + e_3r_1 \\ e_0r_3 + e_1r_2 - e_2r_1\end{bmatrix} = {}^GR_B\begin{bmatrix} r_1 \\ r_2 \\ r_3\end{bmatrix}\end{aligned} \tag{5.234}$$

Here ${}^GR_B$ is equivalent to the Euler parameter transformation matrix (5.109):

$${}^GR_B = \begin{bmatrix} e_0^2 + e_1^2 - e_2^2 - e_3^2 & 2(e_1e_2 - e_0e_3) & 2(e_0e_2 + e_1e_3) \\ 2(e_0e_3 + e_1e_2) & e_0^2 - e_1^2 + e_2^2 - e_3^2 & 2(e_2e_3 - e_0e_1) \\ 2(e_1e_3 - e_0e_2) & 2(e_0e_1 + e_2e_3) & e_0^2 - e_1^2 - e_2^2 + e_3^2\end{bmatrix} \tag{5.235}$$

Employing a similar method, we can also show that

$${}^B\mathbf{r} = e^*(\phi,\hat{u})\,{}^G\mathbf{r}\,e(\phi,\hat{u}) \tag{5.236}$$

which is the inverse transformation of ${}^G\mathbf{r} = e(\phi,\hat{u})\,{}^B\mathbf{r}\,e^*(\phi,\hat{u})$.

Now assume $e_1(\phi_1,\hat{u}_1)$ and $e_2(\phi_2,\hat{u}_2)$ are the quaternions corresponding to the rotation matrices $R_{\hat{u}_1,\phi_1}$ and $R_{\hat{u}_2,\phi_2}$, respectively. The first rotation maps ${}^{B_1}\mathbf{r}$ to ${}^{B_2}\mathbf{r}$ and the second rotation maps ${}^{B_2}\mathbf{r}$ to ${}^{B_3}\mathbf{r}$. Therefore,

$${}^{B_2}\mathbf{r} = e_1(\phi_1,\hat{u}_1)\,{}^{B_1}\mathbf{r}\,e_1^*(\phi_1,\hat{u}_1) \tag{5.237}$$

$${}^{B_3}\mathbf{r} = e_2(\phi_2,\hat{u}_2)\,{}^{B_2}\mathbf{r}\,e_2^*(\phi_2,\hat{u}_2) \tag{5.238}$$

which implies

$$\begin{aligned}{}^{B_3}\mathbf{r} &= e_2(\phi_2,\hat{u}_2)\;e_1(\phi_1,\hat{u}_1)\,{}^{B_1}\mathbf{r}\,e_1^*(\phi_1,\hat{u}_1)\;e_2^*(\phi_2,\hat{u}_2) \\ &= e(\phi,\hat{u})\,{}^{B_1}\mathbf{r}\,e(\phi,\hat{u})\end{aligned} \tag{5.239}$$

showing that

$$e\left(\phi,\hat{u}\right)=e_2\left(\phi_2,\hat{u}_2\right)\,e_1\left(\phi_1,\hat{u}_1\right) \tag{5.240}$$

It is the quaternion equation corresponding to $R=R_2R_1$.

We can use the matrix definition of quaternions and see that

$$\overleftrightarrow{e\left(\phi,\hat{u}\right)}=\begin{bmatrix} e_0 & -e_1 & -e_2 & -e_3 \\ e_1 & e_0 & -e_3 & e_2 \\ e_2 & e_3 & e_0 & -e_1 \\ e_3 & -e_2 & e_1 & e_0 \end{bmatrix} \tag{5.241}$$

$$^{B}\overleftrightarrow{\mathbf{r}}=\begin{bmatrix} 0 & -{}^{B}R_1 & -{}^{B}R_2 & -{}^{B}R_3 \\ {}^{B}R_1 & 0 & -{}^{B}R_3 & {}^{B}R_2 \\ {}^{B}R_2 & {}^{B}R_3 & 0 & -{}^{B}R_1 \\ {}^{B}R_3 & -{}^{B}R_2 & {}^{B}R_1 & 0 \end{bmatrix} \tag{5.242}$$

$$\overleftrightarrow{e\left(\phi,\hat{u}\right)}^{\mathrm{T}}=\begin{bmatrix} e_0 & e_1 & e_2 & e_3 \\ -e_1 & e_0 & e_3 & -e_2 \\ -e_2 & -e_3 & e_0 & e_1 \\ -e_3 & e_2 & -e_1 & e_0 \end{bmatrix} \tag{5.243}$$

Therefore,

$$^{G}\overleftrightarrow{\mathbf{r}}=\overleftrightarrow{e\left(\phi,\hat{u}\right)}\;{}^{B}\overleftrightarrow{\mathbf{r}}\;\overleftrightarrow{e\left(\phi,\hat{u}\right)}^{\mathrm{T}}=\begin{bmatrix} 0 & -{}^{G}R_1 & -{}^{G}R_2 & -{}^{G}R_3 \\ {}^{G}R_1 & 0 & -{}^{G}R_3 & {}^{G}R_2 \\ {}^{G}R_2 & {}^{G}R_3 & 0 & -{}^{G}R_1 \\ {}^{G}R_3 & -{}^{G}R_2 & {}^{G}R_1 & 0 \end{bmatrix} \tag{5.244}$$

where

$$\begin{aligned} {}^{G}R_1 &= {}^{B}R_1\left(e_0^2+e_1^2-e_2^2-e_3^2\right)+{}^{B}R_2\left(2e_1e_2-2e_0e_3\right) \\ &\quad +{}^{B}R_3\left(2e_0e_2+2e_1e_3\right) \end{aligned} \tag{5.245}$$

$$\begin{aligned} {}^{G}R_2 &= {}^{B}R_1\left(2e_0e_3+2e_1e_2\right)+{}^{B}R_2\left(e_0^2-e_1^2+e_2^2-e_3^2\right) \\ &\quad +{}^{B}R_3\left(2e_2e_3-2e_0e_1\right) \end{aligned} \tag{5.246}$$

$$\begin{aligned} {}^{G}R_3 &= {}^{B}R_1\left(2e_1e_3-2e_0e_2\right)+{}^{B}R_2\left(2e_0e_1+2e_2e_3\right) \\ &\quad +{}^{B}R_3\left(e_0^2-e_1^2-e_2^2+e_3^2\right) \end{aligned} \tag{5.247}$$

Equation (5.244) is compatible to Equation (5.234). ■

Example 326 Composition of Rotations Using Quaternions Using quaternions to represent rotations makes it easy to calculate the composition of rotations. If the quaternion $e_1\left(\phi_1,\hat{u}_1\right)$ represents the rotation $R_{\hat{u}_1,\phi_1}$ and $e_2\left(\phi_2,\hat{u}_2\right)$ represents $R_{\hat{u}_2,\phi_2}$, then the product

$$e\left(\phi,\hat{u}\right)=e_2\left(\phi_2,\hat{u}_2\right)\,e_1\left(\phi_1,\hat{u}_1\right) \tag{5.248}$$

represents $R_{\hat{u}_2,\phi_2} R_{\hat{u}_1,\phi_1}$ because

$$\begin{aligned}
e\left(\phi, \hat{u}\right) &= R_{\hat{u}_2,\phi_2} R_{\hat{u}_1,\phi_1} \mathbf{r} \\
&= R_{\hat{u}_2,\phi_2} \left[e_1\left(\phi_1, \hat{u}_1\right) \mathbf{r}\, e_1^*\left(\phi_1, \hat{u}_1\right)\right] \\
&= e_2\left(\phi_2, \hat{u}_2\right) \left[e_1\left(\phi_1, \hat{u}_1\right) \mathbf{r}\, e_1^*\left(\phi_1, \hat{u}_1\right)\right] e_2^*\left(\phi_2, \hat{u}_2\right) \\
&= \left[e_2\left(\phi_2, \hat{u}_2\right) e_1\left(\phi_1, \hat{u}_1\right)\right] \mathbf{r} \left[e_1^*\left(\phi_1, \hat{u}_1\right) e_2^*\left(\phi_2, \hat{u}_2\right)\right] \\
&= \left[e_2\left(\phi_2, \hat{u}_2\right) e_1\left(\phi_1, \hat{u}_1\right)\right] \mathbf{r} \left[e_1\left(\phi_1, \hat{u}_1\right) e_2\left(\phi_2, \hat{u}_2\right)\right]^*
\end{aligned} \tag{5.249}$$

Example 327 Principal Global Rotation Matrices The associated quaternion to the principal global rotation matrices $R_{Z,\alpha}$, $R_{Y,\beta}$, and $R_{X,\gamma}$ are

$$e\left(\alpha, \hat{K}\right) = \left(\cos\frac{\alpha}{2}, \sin\frac{\alpha}{2}\begin{bmatrix}0\\0\\1\end{bmatrix}\right) \tag{5.250}$$

$$e\left(\beta, \hat{J}\right) = \left(\cos\frac{\beta}{2}, \sin\frac{\beta}{2}\begin{bmatrix}0\\1\\0\end{bmatrix}\right) \tag{5.251}$$

$$e\left(\gamma, \hat{I}\right) = \left(\cos\frac{\gamma}{2}, \sin\frac{\gamma}{2}\begin{bmatrix}1\\0\\0\end{bmatrix}\right) \tag{5.252}$$

Therefore, we must be able to derive the same matrices (4.17)–(4.19) by substituting (5.250)–(5.252) in (5.235), respectively. As an example, we check $R_{Z,\alpha}$:

$$\begin{aligned}
R_{Z,\alpha} &= {}^G R_B = \left(e_0^2 - \mathbf{e}^2\right)\mathbf{I} + 2\mathbf{e}\,\mathbf{e}^{\mathrm{T}} + 2e_0\tilde{e} \\
&= \left(\cos^2\frac{\alpha}{2} - \sin^2\frac{\alpha}{2}\right)\mathbf{I} + 2\hat{K}\hat{K}^{\mathrm{T}} + 2\cos\frac{\alpha}{2}\sin\frac{\alpha}{2}\tilde{K} \\
&= \begin{bmatrix}\cos\alpha & -\sin\alpha & 0\\ \sin\alpha & \cos\alpha & 0\\ 0 & 0 & 1\end{bmatrix}
\end{aligned} \tag{5.253}$$

Example 328 Global Roll–Pitch–Yaw Quaternions The three rotations about the X-, Y-, and Z-axes of the global coordinate frame are called the global *roll–pitch–yaw*. The associated rotation matrix is

$$\begin{aligned}
{}^G R_B &= R_{Z,\gamma}\, R_{Y,\beta}\, R_{X,\alpha} \\
&= \begin{bmatrix} c\beta\, c\gamma & -c\alpha\, s\gamma + c\gamma\, s\alpha\, s\beta & s\alpha\, s\gamma + c\alpha\, c\gamma\, s\beta \\ c\beta\, s\gamma & c\alpha\, c\gamma + s\alpha\, s\beta\, s\gamma & -c\gamma\, s\alpha + c\alpha\, s\beta\, s\gamma \\ -s\beta & c\beta\, s\alpha & c\alpha\, c\beta \end{bmatrix}
\end{aligned} \tag{5.254}$$

The global roll–pitch–yaw rotation matrix can be derived by the quaternion as well. The roll quaternion $e\left(\alpha, \hat{I}\right)$, pitch quaternion $e\left(\beta, \hat{J}\right)$, and yaw quaternion $e\left(\gamma, \hat{K}\right)$ are given in (5.250)–(5.252). Multiplying these quaternions creates the global roll–pitch–yaw quaternion $e\left(\phi, \hat{u}\right)$:

$$\begin{aligned} e\left(\phi, \hat{u}\right) &= e\left(\gamma, \hat{K}\right) e\left(\beta, \hat{J}\right) e\left(\alpha, \hat{I}\right) \\ &= \cos\frac{\alpha}{2}\cos\frac{\beta}{2}\cos\frac{\gamma}{2} + \sin\frac{\alpha}{2}\sin\frac{\beta}{2}\sin\frac{\gamma}{2} \\ &\quad + \begin{bmatrix} \cos\frac{\beta}{2}\cos\frac{\gamma}{2}\sin\frac{\alpha}{2} - \cos\frac{\alpha}{2}\sin\frac{\beta}{2}\sin\frac{\gamma}{2} \\ \cos\frac{\alpha}{2}\cos\frac{\gamma}{2}\sin\frac{\beta}{2} + \cos\frac{\beta}{2}\sin\frac{\alpha}{2}\sin\frac{\gamma}{2} \\ \cos\frac{\alpha}{2}\cos\frac{\beta}{2}\sin\frac{\gamma}{2} - \cos\frac{\gamma}{2}\sin\frac{\alpha}{2}\sin\frac{\beta}{2} \end{bmatrix} \end{aligned} \tag{5.255}$$

As a double check, we must get the same matrix (5.254) by substituting (5.255) in (5.235).

Example 329 Expansion of Quaternion Rotation Equation We may substitute for the unit quaternion $e(\phi, \hat{u})$ in the rotation equation

$$^{G}\mathbf{r} = e\left(\phi, \hat{u}\right)\, ^{B}\mathbf{r}\, e^{*}\left(\phi, \hat{u}\right) \tag{5.256}$$

and show that

$$\begin{aligned} ^{G}\mathbf{r} &= \left(e_0 + \mathbf{e}\right)\left(0 + {}^{B}\mathbf{r}\right)\left(e_0 - \mathbf{e}\right) \\ &= \left(2e_0^2 - 1\right) {}^{B}\mathbf{r} + 2\left(\mathbf{e} \cdot {}^{B}\mathbf{r}\right) + 2e_0\left(\mathbf{e} \times {}^{B}\mathbf{r}\right) \end{aligned} \tag{5.257}$$

Now substituting for $\mathbf{e}$ and factoring $^{B}\mathbf{r}$ out generate the same transformation matrix as (5.235). However, it is possible to rearrange the matrix as

$$^{G}R_{B} = \begin{bmatrix} 2e_0^2 + 2e_1^2 - 1 & 2\left(e_1e_2 - e_0e_3\right) & 2\left(e_0e_2 + e_1e_3\right) \\ 2\left(e_0e_3 + e_1e_2\right) & 2e_0^2 + 2e_2^2 - 1 & 2\left(e_2e_3 - e_0e_1\right) \\ 2\left(e_1e_3 - e_0e_2\right) & 2\left(e_0e_1 + e_2e_3\right) & 2e_0^2 + 2e_3^2 - 1 \end{bmatrix} \tag{5.258}$$

Example 330 ★ Rodriguez Rotation Formula Using Quaternion We may simplify Equation (5.234) to have a vectorial form similar to the Rodriguez formula:

$$\begin{aligned} ^{G}\mathbf{r} &= e\left(\phi, \hat{u}\right)\, ^{B}\mathbf{r}\, e^{*}\left(\phi, \hat{u}\right) \\ &= \left(e_0^2 - \mathbf{e} \cdot \mathbf{e}\right)\mathbf{r} + 2e_0\left(\mathbf{e} \times \mathbf{r}\right) + 2\mathbf{e}\left(\mathbf{e} \cdot \mathbf{r}\right) \end{aligned} \tag{5.259}$$

Example 331 ★ Inner Automorphism Property of $\boldsymbol{e(\phi,\hat{u})}$ Because $e(\phi, \hat{u})$ is a unit quaternion,

$$e^{*}\left(\phi, \hat{u}\right) = e^{-1}\left(\phi, \hat{u}\right) \tag{5.260}$$

we may write

$${}^{G}\mathbf{r} = e\left(\phi, \hat{u}\right)\, {}^{B}\mathbf{r}\, e^{-1}\left(\phi, \hat{u}\right) \tag{5.261}$$

In abstract algebra, a mapping of the form $\mathbf{r} = q\,\mathbf{r}\,q^{-1}$, computed by multiplying on the left by an element and on the right by its inverse, is called an *inner automorphism*. Thus, ${}^{G}\mathbf{r}$ is the inner automorphism of ${}^{B}\mathbf{r}$ based on the rotation quaternion $e(\phi, \hat{u})$.

5.4 ★ SPINORS AND ROTATORS

There are two general ways to express the finite rotations: a 3×3 real orthogonal matrix R called a *rotator* and an angle ϕ plus a 3×3 real skew-symmetric matrix $\tilde{u}$ called a *spinor*. Rotator is a short name for *rotation tensor* and spinor is for *spin tensor*.

A rotator is a linear operator that transforms ${}^{B}\mathbf{r}$ to ${}^{G}\mathbf{r}$ when the directional cosines of the axes of the coordinate frames B and G are known:

$${}^{G}\mathbf{r} = {}^{G}R_{B}\, {}^{B}\mathbf{r} \tag{5.262}$$

The spinor $\tilde{u}$ corresponds to the vector $\hat{u}$, which, along with angle ϕ, can be utilized to describe a rotator:

$${}^{G}R_{B} = \left(\mathbf{I}\cos\phi + \hat{u}\hat{u}^{\mathrm{T}}\ \mathrm{vers}\ \phi + \tilde{u}\sin\phi\right) \tag{5.263}$$

The rotator R is a linear function of $\mathbf{I}$, $\tilde{u}$, and $\tilde{u}^{2}$:

$$\begin{aligned} R &= \mathbf{I} + a(\lambda\tilde{u}) + b(\lambda\tilde{u})^{2} \\ &= \begin{bmatrix} -b\lambda^{2}(u_2^2 + u_3^2) + 1 & -a\lambda u_3 + b\lambda^{2}u_1u_2 & a\lambda u_2 + b\lambda^{2}u_1u_3 \\ a\lambda u_3 + b\lambda^{2}u_1u_2 & -b\lambda^{2}(u_1^2 + u_3^2) + 1 & -a\lambda u_1 + b\lambda^{2}u_2u_3 \\ -a\lambda u_2 + b\lambda^{2}u_1u_3 & a\lambda u_1 + b\lambda^{2}u_2u_3 & -b\lambda^{2}(u_1^2 + u_2^2) + 1 \end{bmatrix} \end{aligned} \tag{5.264}$$

where λ is the spinor normalization factor and $a = a(\phi)$ and $b = b(\phi)$ are scalar functions of the rotation angle ϕ.

Table 5.1 presents some representations of rotator R as a function of the coefficients a, b and the spinor $\lambda\tilde{u}$.

Proof: The square of $\tilde{u}$ is

$$\begin{aligned} \tilde{u}^{2} = \tilde{u}\tilde{u} &= \begin{bmatrix} -u_2^2 - u_3^2 & u_1u_2 & u_1u_3 \\ u_1u_2 & -u_1^2 - u_3^2 & u_2u_3 \\ u_1u_3 & u_2u_3 & -u_1^2 - u_2^2 \end{bmatrix} \\ &= -\tilde{u}\,\tilde{u}^{\mathrm{T}} = -\tilde{u}^{\mathrm{T}}\tilde{u} = \hat{u}\hat{u}^{\mathrm{T}} - u^{2}\mathbf{I} \end{aligned} \tag{5.265}$$

Table 5.1 Rotator R as a Function of Spinor $\tilde{u}$

a	b	λ	R
$\sin\phi$	$\sin^2\frac{\phi}{2}$	1	$\mathbf{I}+\sin\phi\tilde{u}+2\sin^2\frac{\phi}{2}\tilde{u}^2$
$2\cos^2\frac{\phi}{2}$	$2\cos^2\frac{\phi}{2}$	$\tan\frac{\phi}{2}$	$\mathbf{I}+2\cos^2\frac{\phi}{2}[\tan\frac{\phi}{2}\tilde{u}+\tan^2\frac{\phi}{2}\tilde{u}^2]$ $=[\mathbf{I}+\tan\frac{\phi}{2}\tilde{u}][\mathbf{I}-\tan\frac{\phi}{2}\tilde{u}]^{-1}$
$2\cos\frac{\phi}{2}$	2	$\sin\frac{\phi}{2}$	$\mathbf{I}+2\cos\frac{\phi}{2}\sin\frac{\phi}{2}\tilde{u}+2\sin^2\frac{\phi}{2}\tilde{u}^2$
$\frac{1}{\phi}\sin\phi$	$\frac{2}{\phi^2}\sin^2\frac{\phi}{2}$	ϕ	$\mathbf{I}+\sin\phi\tilde{u}+2\sin^2\frac{\phi}{2}\tilde{u}^2$

This is a symmetric matrix whose eigenvalues are 0, $-u^2$, $-u^2$ and its trace is

$$\mathrm{tr}[\tilde{u}^2]=-2|\hat{u}|^2=-2u^2=-2(u_1^2+u_2^2+u_3^2) \tag{5.266}$$

This is because $\tilde{u}$ satisfies its own characteristic equation:

$$\tilde{u}^2=-u^2\mathbf{I},\quad \tilde{u}^3=-u^2\tilde{u},\ldots,\tilde{u}^n=-u^2\tilde{u}^{n-2}\qquad n\geq 3 \tag{5.267}$$

So, the odd powers of $\tilde{u}$ are skew symmetric with distinct purely imaginary eigenvalues, while even powers of $\tilde{u}$ are symmetric with repeated real eigenvalues.

A rotator is a function of a spinor, so R can be expanded in a Taylor series of $\tilde{u}$:

$$R=\mathbf{I}+c_1\tilde{u}+c_2\tilde{u}^2+c_3\tilde{u}^3+\cdots \tag{5.268}$$

However, because of (5.267), all powers of order 3 or higher can be eliminated. Therefore, R is a quadratic function of $\tilde{u}$:

$$R=\mathbf{I}+a(\lambda\tilde{u})+b(\lambda\tilde{u})^2 \tag{5.269}$$

For the moment let us forget that $|\hat{u}|=\sqrt{u_1^2+u_2^2+u_3^2}=1$ and develop the theory for nonunit vectors indicating the rotation axis. The parameter λ is the spinor normalization factor, and $a=a(\phi,u)$ and $b=b(\phi,u)$ are scalar functions of rotation angle ϕ and an invariant of $\tilde{u}$.

Assuming $\lambda=1$, we find $\mathrm{tr}R=1+2\cos\phi$, which, because of (5.264), is equal to

$$\mathrm{tr}R=1+2\cos\phi=3-2bu^2 \tag{5.270}$$

and therefore,

$$b=\frac{1-\cos\phi}{u^2}=\frac{2}{u^2}\sin^2\frac{\phi}{2} \tag{5.271}$$

Now the orthogonality condition

$$\begin{aligned}\mathbf{I}=R^{\mathrm{T}}R&=\left(\mathbf{I}-a\tilde{u}+b\tilde{u}^2\right)\left(\mathbf{I}+a\tilde{u}+b\tilde{u}^2\right)\\&=\mathbf{I}+(2b-a^2)\tilde{u}^2+b^2\tilde{u}^4=\mathbf{I}+(2b-a^2-b^2u^2)\tilde{u}^2\end{aligned} \tag{5.272}$$

leads to

$$a = \sqrt{2b - b^2u^2} = \frac{1}{u}\sin\phi \tag{5.273}$$

and therefore,

$$R = \mathbf{I} + \frac{1}{u}\sin\phi\tilde{u} + \frac{2}{u^2}\sin^2\frac{\phi}{2}\tilde{u}^2 = \mathbf{I} + \sin\phi\tilde{u} + \text{ vers }\phi\tilde{u}^2 \tag{5.274}$$

From a numerical viewpoint, the sine-squared form of R is preferred to avoid the cancellation in computing $1 - \cos\phi$ for small ϕ. Replacing a and b in (5.264) provides the explicit rotator in terms of $\tilde{u}$ and ϕ,

$$R = \begin{bmatrix} u_1^2 + \left(u_2^2 + u_3^2\right)c\phi & 2u_1u_2s^2\frac{\phi}{2} - u_3s\phi & 2u_1u_3s^2\frac{\phi}{2} + u_2s\phi \\ 2u_1u_2s^2\frac{\phi}{2} + u_3s\phi & u_2^2 + \left(u_3^2 + u_1^2\right)c\phi & 2u_2u_3s^2\frac{\phi}{2} - u_1s\phi \\ 2u_1u_3s^2\frac{\phi}{2} - u_2s\phi & 2u_2u_3s^2\frac{\phi}{2} + u_1s\phi & u_3^2 + \left(u_1^2 + u_2^2\right)c\phi \end{bmatrix} \tag{5.275}$$

which is equivalent to Equation (5.13).

If $\lambda \neq 1$ and $\lambda \neq 0$, then the answers are $a = (1/\lambda u)\sin\phi$ and $b = \left[2/(\lambda u)^2\right] \sin^2(\phi/2)$ which do not affect R.

Recalling $u = \sqrt{u_1^2 + u_2^2 + u_3^2} = 1$, we may use (5.269) to determine different a, b and λ as are shown in Table 5.1. ■

Example 332 ★ Eigenvalues of a Spinor Consider the axis of rotation indicated by

$$\mathbf{u} = \begin{bmatrix} 6 \\ 2 \\ 3 \end{bmatrix} \qquad u = 7 \tag{5.276}$$

The associated spin matrix and its square are

$$\tilde{u} = \begin{bmatrix} 0 & -3 & 2 \\ 3 & 0 & -6 \\ -2 & 6 & 0 \end{bmatrix} \tag{5.277}$$

$$\tilde{u}^2 = \begin{bmatrix} -13 & 12 & 18 \\ 12 & -45 & 6 \\ 18 & 6 & -40 \end{bmatrix} \tag{5.278}$$

where the eigenvalues of $\tilde{u}$ are $(0, 7i, -7i)$ while those of of $\tilde{u}^2$ are $(0, -49, -49)$.

5.5 ★ PROBLEMS IN REPRESENTING ROTATIONS

The applied rotation analysis of rigid bodies is done by matrix calculus. Therefore, we try to express rotations in matrix form for simpler calculation. There are a number of different methods for representing rigid-body rotations; however, only a few of them are fundamentally distinct. The parameters or coordinates that are required to completely describe the orientation of a rigid body relative to another frame are called *attitude coordinates*. There are two inherent problems in representing rotations, both related to incontrovertible properties of rotations:

1. Rigid-body rotations do not commute.
2. Rigid-body rotations cannot map smoothly in three-dimensional Euclidean space.

The noncommutativity of rotations force us to obey the order of rotations.

The lack of a smooth mapping in three-dimensional Euclidean space means we cannot smoothly represent every kind of rotation by using only one set of three numbers. Any set of three rotational coordinates contains at least one geometric orientation where the coordinates are singular, at which at least two coordinates are undefined or not unique. This problem is similar to defining a coordinate system to locate a point on Earth's surface. Using *longitude* and *latitude* becomes problematic at the north and south poles, where a small displacement can produce a radical change in longitude. We cannot find a superior system because it is not possible to smoothly wrap a sphere with a plane. Similarly, it is not possible to smoothly wrap the space of rotations with three-dimensional Euclidean space. This is the reason why we sometimes describe rotations by using four numbers.

We may only use three-number systems and expect to see singularities or use four numbers and cope with the redundancy. The choice depends on the application and method of calculation. For computer applications, the redundancy is not a problem, so most algorithms use representations with extra numbers. However, engineers prefer to work with the minimum set of numbers.

5.5.1 ★ Rotation Matrix

For many purposes, the rotation matrix representation, based on directional cosines, is the most useful representation method of rigid-body rotations. The two reference frames G and B, having a common origin, are defined through orthogonal right-handed sets of unit vectors $\{G\} = \{\hat{I}, \hat{J}, \hat{K}\}$ and $\{B\} = \{\hat{\imath}, \hat{\jmath}, \hat{k}\}$. The rotation or transformation matrix between the two frames is found by using the orthogonality condition of B and G and describing the unit vectors of one of them in the other:

$$\begin{aligned}\hat{I} &= (\hat{I} \cdot \hat{\imath})\hat{\imath} + (\hat{I} \cdot \hat{\jmath})\hat{\jmath} + (\hat{I} \cdot \hat{k})\hat{k} \\ &= \cos(\hat{I}, \hat{\imath})\hat{\imath} + \cos(\hat{I}, \hat{\jmath})\hat{\jmath} + \cos(\hat{I}, \hat{k})\hat{k}\end{aligned} \tag{5.279}$$

$$\begin{aligned}\hat{J} &= (\hat{J} \cdot \hat{\imath})\hat{\imath} + (\hat{J} \cdot \hat{\jmath})\hat{\jmath} + (\hat{J} \cdot \hat{k})\hat{k} \\ &= \cos(\hat{J}, \hat{\imath})\hat{\imath} + \cos(\hat{J}, \hat{\jmath})\hat{\jmath} + \cos(\hat{J}, \hat{k})\hat{k}\end{aligned} \tag{5.280}$$

$$\begin{aligned}\hat{K} &= (\hat{K} \cdot \hat{\imath})\hat{\imath} + (\hat{K} \cdot \hat{\jmath})\hat{\jmath} + (\hat{K} \cdot \hat{k})\hat{k} \\ &= \cos(\hat{K}, \hat{\imath})\hat{\imath} + \cos(\hat{K}, \hat{\jmath})\hat{\jmath} + \cos(\hat{K}, \hat{k})\hat{k}\end{aligned} \tag{5.281}$$

Therefore, having the rotation matrix

$$
{}^{G}R_{B} = \begin{bmatrix} \hat{I}\cdot\hat{\imath} & \hat{I}\cdot\hat{\jmath} & \hat{I}\cdot\hat{k} \\ \hat{J}\cdot\hat{\imath} & \hat{J}\cdot\hat{\jmath} & \hat{J}\cdot\hat{k} \\ \hat{K}\cdot\hat{\imath} & \hat{K}\cdot\hat{\jmath} & \hat{K}\cdot\hat{k} \end{bmatrix}
= \begin{bmatrix} \cos(\hat{I},\hat{\imath}) & \cos(\hat{I},\hat{\jmath}) & \cos(\hat{I},\hat{k}) \\ \cos(\hat{J},\hat{\imath}) & \cos(\hat{J},\hat{\jmath}) & \cos(\hat{J},\hat{k}) \\ \cos(\hat{K},\hat{\imath}) & \cos(\hat{K},\hat{\jmath}) & \cos(\hat{K},\hat{k}) \end{bmatrix} \tag{5.282}
$$

would be enough to find the coordinates of a point in the coordinate frame G when its coordinates are given in B:

$$
{}^{G}\mathbf{r} = {}^{G}R_{B}\,{}^{B}\mathbf{r} \tag{5.283}
$$

The rotation matrices convert the composition of rotations to matrix multiplication. It provides a simple and convenient method of rotations, especially when rotations are about the global or local principal axes.

Orthogonality is the most important and useful condition of rotation matrices. It shows that the inverse of a rotation matrix is equivalent to its transpose, ${}^{G}R_{B}^{-1} = {}^{G}R_{B}^{\mathrm{T}}$. The null rotation is represented by the identity matrix $\mathbf{I}$.

The primary disadvantage of rotation matrices is that there are so many numbers, which often make rotation matrices hard to interpret. Numerical errors may build up until a normalization is necessary.

5.5.2 ★ Axis–Angle

Axis–angle representation, described by the Rodriguez formula, is a direct result of the Euler rigid-body rotation theorem. In this method, a rotation is expressed by the magnitude of rotation, ϕ, with the positive right-hand direction about the axis of rotation $\hat{u}$:

$$
{}^{G}R_{B} = R_{\hat{u},\phi} = \mathbf{I}\cos\phi + \hat{u}\hat{u}^{\mathrm{T}}\,\mathrm{vers}\,\phi + \tilde{u}\sin\phi \tag{5.284}
$$

Matrix representation of the angle–axis expression is found by expanding it:

$$
{}^{G}R_{B} = \begin{bmatrix} u_1^2\,\mathrm{vers}\,\phi + c\phi & u_1u_2\,\mathrm{vers}\,\phi - u_3s\phi & u_1u_3\,\mathrm{vers}\,\phi + u_2s\phi \\ u_1u_2\,\mathrm{vers}\,\phi + u_3s\phi & u_2^2\,\mathrm{vers}\,\phi + c\phi & u_2u_3\,\mathrm{vers}\,\phi - u_1s\phi \\ u_1u_3\,\mathrm{vers}\,\phi - u_2s\phi & u_2u_3\,\mathrm{vers}\,\phi + u_1s\phi & u_3^2\,\mathrm{vers}\,\phi + c\phi \end{bmatrix} \tag{5.285}
$$

Converting the matrix representation to angle–axis form can be done by matrix manipulation. However, it is usually easier if we convert the matrix to a quaternion and then convert the quaternion to angle–axis form.

Angle–axis representation has also some shortcomings. First, the rotation axis is indeterminate when $\phi = 0$. Second, the angle–axis rotation represents a two-to-one mapping system because

$$
R_{-\hat{u},-\phi} = R_{\hat{u},\phi} \tag{5.286}
$$

Third, it is redundant because for any integer k we have

$$
R_{\hat{u},\phi+2k\pi} = R_{\hat{u},\phi} \tag{5.287}
$$

However, all of these problems can be improved to some extent by restricting ϕ to some suitable range such as $[0, \pi]$ or $[-\pi/2, \pi/2]$. The angle–axis representation is also not an efficient method to find the composition of two rotations and determine the equivalent angle–axis of rotations.

5.5.3 ★ Euler Angles

Euler angles are employed to describe the rotation matrix of rigid bodies by only three numbers:

$$
\begin{aligned}
{}^G R_B &= \left[R_{z,\psi} R_{x,\theta} R_{z,\varphi}\right]^{\mathrm{T}} \\
&= \begin{bmatrix} c\varphi c\psi - c\theta s\varphi s\psi & -c\varphi s\psi - c\theta c\psi s\varphi & s\theta s\varphi \\ c\psi s\varphi + c\theta c\varphi s\psi & -s\varphi s\psi + c\theta c\varphi c\psi & -c\varphi s\theta \\ s\theta s\psi & s\theta c\psi & c\theta \end{bmatrix}
\end{aligned} \tag{5.288}
$$

The first Euler angle φ is about a globally fixed axis, the third Euler angle ψ is about a body-fixed axis, and the second Euler angle θ is between these two axes.

Euler angles and rotation matrices are not generally one to one, and also, they are not convenient for composite rotations. The angles φ and ψ are also not distinguishable when $\theta \to 0$.

The equivalent rotation matrix for a set of Euler angles is directly obtained by matrix multiplication; however, the inverse conversion, from rotation matrix to a set of Euler angles, is not straightforward. It is also not applicable when $\sin\theta = 0$. Using Equation (5.288), we have

$$\theta = \cos^{-1}(r_{33}) \tag{5.289}$$

$$\psi = \tan^{-1}\left(\frac{r_{31}}{r_{32}}\right) \tag{5.290}$$

$$\varphi = -\tan^{-1}\left(\frac{r_{13}}{r_{23}}\right) \tag{5.291}$$

It is possible to use a modified Euler angle method that is more efficient and can handle all cases uniformly. The main idea is to work with the sum and difference of φ and ψ,

$$\sigma = \varphi + \psi \tag{5.292}$$

$$\upsilon = \varphi - \psi \tag{5.293}$$

and then

$$\varphi = \tfrac{1}{2}(\sigma - \upsilon) \tag{5.294}$$

$$\psi = \tfrac{1}{2}(\sigma + \upsilon) \tag{5.295}$$

Therefore,

$$r_{11} + r_{22} = \cos\sigma(1 + \cos\theta) \tag{5.296}$$

$$r_{11} - r_{22} = \cos\upsilon(1 - \cos\theta) \tag{5.297}$$

$$r_{21} - r_{12} = \sin\sigma(1 + \cos\theta) \tag{5.298}$$

$$r_{21} + r_{12} = \sin\upsilon(1 - \cos\theta) \tag{5.299}$$

which leads to

$$\sigma = \tan^{-1}\frac{r_{21} - r_{12}}{r_{11} + r_{22}} \tag{5.300}$$

$$\upsilon = \tan^{-1}\frac{r_{21} + r_{12}}{r_{11} - r_{22}} \tag{5.301}$$

This approach resolves the problem at $\sin\theta = 0$. At $\theta = 0$, we can find σ but υ is undetermined, and at $\theta = \pi$, we can find υ but σ is undetermined. The undetermined values are results of $\tan^{-1}(0/0)$. Besides these singularities, both σ and υ are uniquely determined. The middle rotation angle θ can also be found using the arctan operator:

$$\theta = \tan^{-1}\left(\frac{r_{13}\sin\varphi - r_{23}\cos\varphi}{r_{33}}\right) \tag{5.302}$$

The main advantage of Euler angles is that they use only three numbers. They are integrable, and they provide good visualization of spatial rotation with no redundancy. Every set of Euler angles determines a unique orientation. Euler angles are used in the dynamic analysis of spinning bodies.

The other combinations of Euler angles as well as roll–pitch–yaw angles have the same kinds of problems and similar advantages.

5.5.4 ★ Quaternion and Euler Parameters

Quaternions with special rules for addition and multiplication use four numbers to represent rotations. A rotation quaternion is a unit quaternion that may be expressed by Euler parameters or the angle and axis of rotations:

$$e\left(\phi, \hat{u}\right) = e_0 + \mathbf{e} = e_0 + e_1\hat{\imath} + e_2\hat{\jmath} + e_3\hat{k} = \cos\frac{\phi}{2} + \sin\frac{\phi}{2}\hat{u} \tag{5.303}$$

Euler parameters are the elements of rotation quaternions. There is a direct conversion between rotation quaternions and Euler parameters, which in turn are related to angle–axis parameters. We can obtain the angle ϕ and axis $\hat{u}$ of rotation from Euler parameters or rotation quaternion $e\left(\phi, \hat{u}\right)$ by

$$\phi = 2\tan^{-1}\frac{|\mathbf{e}|}{e_0} \tag{5.304}$$

$$\hat{u} = \frac{\mathbf{e}}{|\mathbf{e}|} \tag{5.305}$$

Unit quaternions provide a suitable base for describing rigid-body rotations, although they need normalization due to the error pile-up problem. In general, quaternions offer superior computational efficiency in most applications.

It is interesting to know that Leonhard Euler (1707–1783) was the first to derive the Rodriguez formula, while Benjamin Rodriguez (1795–1851) was the first to discover the Euler parameters. William Hamilton (1805–1865) introduced quaternions, although Friedrich Gauss (1777–1855) discovered them but never published.

Example 333 ★ Taylor Expansion of Rotation Matrix Assume that the rotation matrix ${}^G R_B$ between coordinate frames B and G is a time-dependent transformation

$$ {}^G R_B = R(t) \tag{5.306} $$

The body frame B is coincident with G at $t = 0$. Therefore, $R(0) = \mathbf{I}$, and we may expand the elements of R in a Taylor series

$$ R(t) = \mathbf{I} + R_1 t + \frac{1}{2!} R_2 t^2 + \frac{1}{3!} R_3 t^3 + \cdots \tag{5.307} $$

in which $R_i (i = 1, 2, 3, \ldots)$ is a constant matrix. The rotation matrix $R(t)$ must be orthogonal for all t; hence,

$$ [R]\,[R]^{\mathrm{T}} = \mathbf{I} \tag{5.308} $$

$$ \left(\mathbf{I} + R_1 t + \frac{1}{2!} R_2 t^2 + \cdots\right)\left(\mathbf{I} + R_1^{\mathrm{T}} t + \frac{1}{2!} R_2^{\mathrm{T}} t^2 + \cdots\right) = \mathbf{I}. \tag{5.309} $$

The coefficient of $t^i (i = 1, 2, 3, \ldots)$ must vanish on the left-hand side. This gives

$$ R_1 + R_1^{\mathrm{T}} = 0 \tag{5.310} $$

$$ R_2 + 2R_1 R_1^{\mathrm{T}} + R_2^{\mathrm{T}} = 0 \tag{5.311} $$

$$ R_3 + 3R_2 R_1^{\mathrm{T}} + 3R_1 R_2^{\mathrm{T}} + R_3^{\mathrm{T}} = 0 \tag{5.312} $$

or in general

$$ \sum_{i=0}^{n} \binom{n}{i} R_{n-i} R_i^{\mathrm{T}} = 0 \tag{5.313} $$

where

$$ R_0 = R_0^{\mathrm{T}} = \mathbf{I} \tag{5.314} $$

Equation (5.310) shows that R_1 is a skew-symmetric matrix, and therefore, $R_1 R_1^{\mathrm{T}} = -R_1^2 = C_1$ is symmetric. Now, from Equation (5.311),

$$ R_2 + R_2^{\mathrm{T}} = -2R_1 R_1^{\mathrm{T}} = -[R_1 R_1^{\mathrm{T}} + [R_1 R_1^{\mathrm{T}}]^{\mathrm{T}}] = 2C_1 \tag{5.315} $$

which leads to

$$ R_2 = C_1 + [C_1 - R_2^{\mathrm{T}}] \tag{5.316} $$

$$ R_2^{\mathrm{T}} = C_1 + [C_1 - R_2] = C_1 + [C_1 - R_2^{\mathrm{T}}]^{\mathrm{T}} \tag{5.317} $$

which shows that $[C_1 - R_2^{\mathrm{T}}]$ is skew symmetric because we must have

$$ [C_1 - R_2^{\mathrm{T}}] + [C_1 - R_2^{\mathrm{T}}]^{\mathrm{T}} = 0 \tag{5.318} $$

Therefore, the matrix product $[C_1 - R_2^{\mathrm{T}}][C_1 - R_2^{\mathrm{T}}]^{\mathrm{T}}$ is symmetric:

$$[C_1 - R_2^{\mathrm{T}}][C_1 - R_2^{\mathrm{T}}]^{\mathrm{T}} = -[C_1 - R_2^{\mathrm{T}}]^2 \tag{5.319}$$

The next step,

$$\begin{aligned} R_3 + R_3^{\mathrm{T}} &= -3[R_1 R_2^{\mathrm{T}} + R_2 R_1^{\mathrm{T}}] = -3[R_1 R_2^{\mathrm{T}} + [R_1 R_2^{\mathrm{T}}]^{\mathrm{T}}] \\ &= 3[R_1[R_1^2 - R_2^{\mathrm{T}}] + [R_1^2 - R_2^{\mathrm{T}}]R_1] = 2C_2 \end{aligned} \tag{5.320}$$

leads to

$$R_3 = C_2 + \left[C_2 - R_3^{\mathrm{T}}\right] \tag{5.321}$$

$$R_3^{\mathrm{T}} = C_2 + [C_2 - R_3] = C_2 + \left[C_2 - R_3^{\mathrm{T}}\right]^{\mathrm{T}} \tag{5.322}$$

which shows that $\left[C_2 - R_3^{\mathrm{T}}\right]$ is skew symmetric because we must have

$$\left[C_2 - R_3^{\mathrm{T}}\right] + \left[C_2 - R_3^{\mathrm{T}}\right]^{\mathrm{T}} = 0 \tag{5.323}$$

Therefore, the matrix product $\left[C_2 - R_3^{\mathrm{T}}\right]\left[C_2 - R_3^{\mathrm{T}}\right]^{\mathrm{T}}$ is also symmetric:

$$\left[C_2 - R_3^{\mathrm{T}}\right]\left[C_2 - R_3^{\mathrm{T}}\right]^{\mathrm{T}} = -\left[C_2 - R_3^{\mathrm{T}}\right]^2 \tag{5.324}$$

Continuing this procedure shows that the expansion of a rotation matrix $R(t)$ around the unit matrix can be written in the form

$$\begin{aligned} R(t) = \mathbf{I} + C_1 t &+ \frac{1}{2!}\left[C_1 + [C_1 - R_2^{\mathrm{T}}]\right] t^2 \\ &+ \frac{1}{3!}\left[C_2 + \left[C_2 - R_3^{\mathrm{T}}\right]\right] t^3 + \cdots \end{aligned} \tag{5.325}$$

where the C_i are symmetric, $[C_i - R_{i+1}^{\mathrm{T}}]$ are skew-symmetric matrices, and

$$C_i = \tfrac{1}{2}\left[R_{i-1} + R_{1-1}^{\mathrm{T}}\right] \tag{5.326}$$

Therefore, the expansion of an inverse rotation matrix can also be written as

$$\begin{aligned} R^{\mathrm{T}}(t) = \mathbf{I} + C_1 t &+ \frac{1}{2!}\left[C_1 + [C_1 - R_2^{\mathrm{T}}]\right] t^2 \\ &+ \frac{1}{3!}\left[C_2 + \left[C_2 - R_3^{\mathrm{T}}\right]\right] t^3 + \cdots \end{aligned} \tag{5.327}$$

5.6 COMPOSITION AND DECOMPOSITION OF ROTATIONS

Determination of a rotation to be equivalent to some given rotations and determination of some rotations to be equivalent to a given rotation matrix represent a challenging problem when the axes of rotations are not orthogonal. This problem is called the *composition and decomposition of rotations*.

5.6.1 Composition of Rotations

Rotation ϕ_1 about $\hat{u}_1$ of a rigid body with a fixed point followed by a rotation ϕ_2 about $\hat{u}_2$ can be composed to a unique rotation ϕ_3 about $\hat{u}_3$. In other words, when a rigid body rotates from an initial position to a middle position, ${}^{B_2}\mathbf{r} = {}^{B_2}R_{B_1}\,{}^{B_1}\mathbf{r}$, and then rotates to a final position, ${}^{B_3}\mathbf{r} = {}^{B_3}R_{B_2}\,{}^{B_2}\mathbf{r}$, the middle position can be skipped to rotate directly to the final position, ${}^{B_3}\mathbf{r} = {}^{B_3}R_{B_1}\,{}^{B_1}\mathbf{r}$.

Proof: To show that two successive rotations of a rigid body with a fixed point is equivalent to a single rotation, we start with the Rodriguez rotation formula (5.181) and rewrite it as

$$\left(\mathbf{r}' - \mathbf{r}\right) = \mathbf{w} \times \left(\mathbf{r}' + \mathbf{r}\right) \tag{5.328}$$

where $\mathbf{w}$ is the *Rodriguez vector*

$$\mathbf{w} = \tan\frac{\phi}{2}\hat{u} \tag{5.329}$$

Rotations $\mathbf{w}_1$ followed by $\mathbf{w}_2$ are given as

$$\mathbf{r}_2 - \mathbf{r}_1 = \mathbf{w}_1 \times (\mathbf{r}_2 + \mathbf{r}_1) \tag{5.330}$$

$$\mathbf{r}_3 - \mathbf{r}_2 = \mathbf{w}_2 \times (\mathbf{r}_3 + \mathbf{r}_2) \tag{5.331}$$

The right-hand side of the first one is perpendicular to $\mathbf{w}_1$ and the second one is perpendicular to $\mathbf{w}_2$. Hence, the dot products of the first one with $\mathbf{w}_1$ and the second one with $\mathbf{w}_2$ show that

$$\mathbf{w}_1 \cdot \mathbf{r}_2 = \mathbf{w}_1 \cdot \mathbf{r}_1 \tag{5.332}$$

$$\mathbf{w}_2 \cdot \mathbf{r}_3 = \mathbf{w}_2 \cdot \mathbf{r}_2 \tag{5.333}$$

Also, the cross products of the first one with $\mathbf{w}_2$ and the second one with $\mathbf{w}_1$ show that

$$\begin{aligned} &\mathbf{w}_2 \times (\mathbf{r}_2 - \mathbf{r}_1) - \mathbf{w}_1 \times (\mathbf{r}_3 - \mathbf{r}_2) \\ &\quad = \mathbf{w}_1 \left[\mathbf{w}_2 \cdot (\mathbf{r}_2 + \mathbf{r}_1)\right] - (\mathbf{w}_1 \cdot \mathbf{w}_2)(\mathbf{r}_2 + \mathbf{r}_1) \\ &\qquad - \mathbf{w}_2 \left[\mathbf{w}_1 \cdot (\mathbf{r}_3 + \mathbf{r}_2)\right] + (\mathbf{w}_1 \cdot \mathbf{w}_2)(\mathbf{r}_3 + \mathbf{r}_2) \end{aligned} \tag{5.334}$$

Rearranging while using Equations (5.332) and (5.333) gives us

$$\begin{aligned} &\mathbf{w}_2 \times (\mathbf{r}_2 - \mathbf{r}_1) - \mathbf{w}_1 \times (\mathbf{r}_3 - \mathbf{r}_2) \\ &\quad = (\mathbf{w}_2 \times \mathbf{w}_1) \times (\mathbf{r}_1 + \mathbf{r}_3) + (\mathbf{w}_1 \cdot \mathbf{w}_2)(\mathbf{r}_3 - \mathbf{r}_1) \end{aligned} \tag{5.335}$$

which can be written as

$$\begin{aligned} (\mathbf{w}_1 + \mathbf{w}_2) \times \mathbf{r}_2 &= \mathbf{w}_2 \times \mathbf{r}_1 + \mathbf{w}_1 \times \mathbf{r}_3 \\ &\quad + (\mathbf{w}_2 \times \mathbf{w}_1) \times (\mathbf{r}_1 + \mathbf{r}_3) + (\mathbf{w}_1 \cdot \mathbf{w}_2)(\mathbf{r}_3 - \mathbf{r}_1) \end{aligned} \tag{5.336}$$

Adding Equations (5.330) and (5.331) to obtain $(\mathbf{w}_1 + \mathbf{w}_2) \times \mathbf{r}_2$ leads to

$$(\mathbf{w}_1 + \mathbf{w}_2) \times \mathbf{r}_2 = \mathbf{r}_3 - \mathbf{r}_1 - \mathbf{w}_1 \times \mathbf{r}_1 - \mathbf{w}_2 \times \mathbf{r}_3 \tag{5.337}$$

Therefore, we obtain the required Rodriguez rotation formula to rotate $\mathbf{r}_1$ to $\mathbf{r}_3$,

$$\mathbf{r}_3 - \mathbf{r}_1 = \mathbf{w}_3 \times (\mathbf{r}_3 + \mathbf{r}_1) \tag{5.338}$$

where

$$\mathbf{w}_3 = \frac{\mathbf{w}_1 + \mathbf{w}_2 + \mathbf{w}_2 \times \mathbf{w}_1}{1 - \mathbf{w}_1 \cdot \mathbf{w}_2} \tag{5.339}$$

■

Example 334 Equivalent Rotation to Two Individual Rotations Consider a rigid body B that undergoes a first rotation $\alpha = 30$ deg about

$$\hat{u} = \frac{1}{\sqrt{3}}\left(\hat{I} + \hat{J} + \hat{K}\right) \tag{5.340}$$

followed by a rotation $\beta = 45$ deg about

$$\hat{v} = \hat{K} \tag{5.341}$$

To determine the equivalent single rotation, we define the vectors $\mathbf{u}$ and $\mathbf{v}$,

$$\mathbf{u} = \tan\frac{\alpha}{2}\hat{u} = \begin{bmatrix} \sqrt{3}/3 \\ \sqrt{3}/3 \\ \sqrt{3}/3 \end{bmatrix} \tan\frac{\pi/6}{2} = \begin{bmatrix} 0.1547 \\ 0.1547 \\ 0.1547 \end{bmatrix} \tag{5.342}$$

$$\mathbf{v} = \tan\frac{\beta}{2}\hat{u} = \begin{bmatrix} 0 \\ 0 \\ 1 \end{bmatrix} \tan\frac{\pi/4}{2} = \begin{bmatrix} 0 \\ 0 \\ 0.41421 \end{bmatrix} \tag{5.343}$$

and calculate the $\mathbf{w}$ from (5.339):

$$\mathbf{w} = \frac{\mathbf{u} + \mathbf{v} + \mathbf{v} \times \mathbf{u}}{1 - \mathbf{u} \cdot \mathbf{v}} = \begin{bmatrix} 0.23376 \\ 9.6827 \times 10^{-2} \\ 0.60786 \end{bmatrix} \tag{5.344}$$

The equivalent Rodriguez vector $\mathbf{w}$ may be decomposed to determine the axis and angle of rotations $\hat{w}$ and ϕ:

$$\mathbf{w} = \begin{bmatrix} 0.23376 \\ 9.68 \times 10^{-2} \\ 0.60786 \end{bmatrix} = 0.658 \begin{bmatrix} 0.355 \\ 0.147 \\ 0.923 \end{bmatrix} = \tan\frac{\phi}{2}\hat{w} \tag{5.345}$$

$$\phi = 2 \arctan 0.658 = 1.1645 \text{ rad} \approx 66.72 \text{ deg} \tag{5.346}$$

5.6.2 ★ Decomposition of Rotations

Consider a rotating rigid body B with a fixed point O in a global frame G. A rotation ϕ_1 about $\hat{u}_1$ can be decomposed into three successive rotations about three arbitrary axes $\hat{a}$, $\hat{b}$, and $\hat{c}$ through unique angles α, β, and γ.

Let ${}^GR_{\hat{a},\alpha}$, ${}^GR_{\hat{b},\beta}$, and ${}^GR_{\hat{c}}, \gamma$ be any three successive rotation matrices about non-coaxis non-coplanar unit vectors $\hat{a}$, $\hat{b}$, and $\hat{c}$ through nonvanishing values α, β, and γ. Then, any other rotation ${}^GR_{\hat{u},\phi}$ can be expressed in terms of ${}^GR_{\hat{a},\alpha}$, ${}^GR_{\hat{b},\beta}$, and ${}^GR_{\hat{c},\gamma}$,

$$ {}^GR_{\hat{u},\phi} = {}^GR_{\hat{c},\gamma}\,{}^GR_{\hat{b},\beta}\,{}^GR_{\hat{a},\alpha} \tag{5.347} $$

if α, β, and γ are properly chosen numbers.

Proof: Using the definition of rotation based on quaternions, we may write

$$ {}^G\mathbf{r} = {}^GR_{\hat{u},\phi}\,{}^B\mathbf{r} = e\left(\phi, \hat{u}\right)\,{}^B\mathbf{r}\,e^*\left(\phi, \hat{u}\right) \tag{5.348} $$

Let us assume that $\mathbf{r}_1$ indicates the position vector $\mathbf{r}$ before rotation and $\mathbf{r}_2$, $\mathbf{r}_3$, and $\mathbf{r}_4$ indicate the position vector $\mathbf{r}$ after rotation $R_{\hat{a},\alpha}$, $R_{\hat{b},\beta}$, and $R_{\hat{c},\gamma}$ respectively. Hence,

$$ \mathbf{r}_2 = a\mathbf{r}_1a^* \tag{5.349} $$

$$ \mathbf{r}_3 = b\mathbf{r}_2b^* \tag{5.350} $$

$$ \mathbf{r}_4 = c\mathbf{r}_3c^* \tag{5.351} $$

$$ \mathbf{r}_4 = e\mathbf{r}_1e^* \tag{5.352} $$

where a, b, c, and e are quaternions corresponding to $\left(\alpha, \hat{a}\right)$, $(\beta, \hat{b})$, $\left(\gamma, \hat{c}\right)$, and $\left(\phi, \hat{u}\right)$, respectively:

$$ a\left(\alpha, \hat{a}\right) = a_0 + \mathbf{a} = \cos\frac{\alpha}{2} + \sin\frac{\alpha}{2}\hat{a} \tag{5.353} $$

$$ b(\beta, \hat{b}) = b_0 + \mathbf{b} = \cos\frac{\beta}{2} + \sin\frac{\beta}{2}\hat{b} \tag{5.354} $$

$$ c\left(\gamma, \hat{c}\right) = c_0 + \mathbf{c} = \cos\frac{\gamma}{2} + \sin\frac{\gamma}{2}\hat{c} \tag{5.355} $$

$$ e\left(\phi, \hat{u}\right) = e_0 + \mathbf{e} = \cos\frac{\phi}{2} + \sin\frac{\phi}{2}\hat{u} \tag{5.356} $$

We define the following scalars to simplify the calculations:

$$ \cos\frac{\alpha}{2} = C_1 \qquad \cos\frac{\beta}{2} = C_2 \qquad \cos\frac{\gamma}{2} = C_3 \qquad \cos\frac{\phi}{2} = C \tag{5.357} $$

$$ \sin\frac{\alpha}{2} = S_1 \qquad \sin\frac{\beta}{2} = S_2 \qquad \sin\frac{\gamma}{2} = S_3 \qquad \sin\frac{\phi}{2} = S \tag{5.358} $$

$$\frac{b_2c_3 - b_3c_2}{S_2S_3} = f_1 \qquad \frac{b_3c_1 - b_1c_3}{S_2S_3} = f_2 \qquad \frac{b_1c_2 - b_2c_1}{S_2S_3} = f_3 \tag{5.359}$$

$$\frac{c_2a_3 - c_3a_2}{S_3S_1} = g_1 \qquad \frac{a_3c_1 - a_1c_3}{S_3S_1} = g_2 \qquad \frac{a_1c_2 - a_2c_1}{S_3S_1} = g_3 \tag{5.360}$$

$$\frac{a_2b_3 - a_3b_2}{S_1S_2} = h_1 \qquad \frac{a_3b_1 - a_1b_3}{S_1S_2} = h_2 \qquad \frac{a_1b_2 - a_2b_1}{S_1S_2} = h_3 \tag{5.361}$$

$$\mathbf{b} \cdot \mathbf{c} = n_1 S_2 S_3 \tag{5.362}$$

$$\mathbf{c} \cdot \mathbf{a} = n_2 S_3 S_1 \tag{5.363}$$

$$\mathbf{a} \cdot \mathbf{b} = n_3 S_1 S_2 \tag{5.364}$$

$$(\mathbf{a} \times \mathbf{b}) \cdot \mathbf{c} = n_4 S_1 S_2 S_3 \tag{5.365}$$

Direct substitution shows that

$$\mathbf{r}_4 = e\mathbf{r}_1 e^* = cba\mathbf{r}_1 a^* b^* c^* \tag{5.366}$$

and therefore,

$$\begin{aligned} e = cba &= c\left(b_0a_0 - \mathbf{b} \cdot \mathbf{a} + b_0\mathbf{a} + a_0\mathbf{b} + \mathbf{b} \times \mathbf{a}\right) \\ &= c_0b_0a_0 - a_0\mathbf{b} \cdot \mathbf{c} - b_0\mathbf{c} \cdot \mathbf{a} - c_0\mathbf{a} \cdot \mathbf{b} + (\mathbf{a} \times \mathbf{b}) \cdot \mathbf{c} \\ &\quad + a_0b_0\mathbf{c} + b_0c_0\mathbf{a} + c_0a_0\mathbf{b} \\ &\quad + a_0\left(\mathbf{b} \times \mathbf{c}\right) + b_0\left(\mathbf{c} \times \mathbf{a}\right) + c_0\left(\mathbf{b} \times \mathbf{a}\right) \\ &\quad - \left(\mathbf{a} \cdot \mathbf{b}\right)\mathbf{c} - \left(\mathbf{b} \cdot \mathbf{c}\right)\mathbf{a} + \left(\mathbf{c} \cdot \mathbf{a}\right)\mathbf{b} \end{aligned} \tag{5.367}$$

Hence,

$$e_0 = c_0b_0a_0 - a_0n_1S_2S_3 - b_0n_2S_3S_1 - c_0n_3S_1S_2 + n_4S_1S_2S_3 \tag{5.368}$$

and

$$\begin{aligned} \mathbf{e} &= a_0b_0\mathbf{c} + b_0c_0\mathbf{a} + c_0a_0\mathbf{b} \\ &\quad + a_0\left(\mathbf{b} \times \mathbf{c}\right) + b_0\left(\mathbf{c} \times \mathbf{a}\right) + c_0\left(\mathbf{b} \times \mathbf{a}\right) \\ &\quad - n_1S_2S_3\mathbf{a} + n_2S_3S_1\mathbf{b} - n_3S_1S_2\mathbf{c} \end{aligned} \tag{5.369}$$

which generate the four equations

$$C_1C_2C_3 - n_1C_1S_2S_3 - n_2S_1C_2S_3 + n_3S_1S_2C_3 - n_4S_1S_2S_3 = C \tag{5.370}$$

$$\begin{aligned} & a_1S_1C_2C_3 + b_1C_1S_2C_3 + c_1C_1C_2S_3 \\ & + f_1C_1S_2S_3 + g_1S_1C_2S_3 + h_1S_1S_2C_3 \\ & - n_1a_1S_1S_2S_3 + n_2b_1S_1S_2S_3 - n_3c_1S_1S_2S_3 = u_1S \end{aligned} \tag{5.371}$$

$$
\begin{aligned}
& a_2 S_1 C_2 C_3 + b_2 C_1 S_2 C_3 + c_2 C_1 C_2 S_3 \\
& + f_2 C_1 S_2 S_3 + g_2 S_1 C_2 S_3 + h_2 S_1 S_2 C_3 \\
& - n_1 a_2 S_1 S_2 S_3 + n_2 b_2 S_1 S_2 S_3 - n_3 c_2 S_1 S_2 S_3 = u_2 S
\end{aligned} \tag{5.372}
$$

$$
\begin{aligned}
& a_3 S_1 C_2 C_3 + b_{13} C_1 S_2 C_3 + c_3 C_1 C_2 S_3 \\
& + f_3 C_1 S_2 S_3 + g_3 S_1 C_2 S_3 + h_3 S_1 S_2 C_3 \\
& - n_1 a_3 S_1 S_2 S_3 + n_2 b_3 S_1 S_2 S_3 - n_3 c_3 S_1 S_2 S_3 = u_3 S
\end{aligned} \tag{5.373}
$$

Because $e_0^2 + e_1^2 + e_2^2 + e_3^2 = 1$, only the first equation and two of the others along with

$$
C_1 = \sqrt{1 - S_1^2} \qquad C_2 = \sqrt{1 - S_2^2} \qquad C_3 = \sqrt{1 - S_3^2} \tag{5.374}
$$

must be utilized to determine C_1, C_2, C_3, S_1, S_2, and S_3. ■

KEY SYMBOLS

A	transformation matrix of rotation about a local axis
B	body coordinate frame, local coordinate frame
C	constant value, cosine of half angle
c	cosine
e	unit quaternion, rotation quaternion, exponential
e_0, e_1, e_2, e_3	Euler parameters
G	global coordinate frame, fixed coordinate frame
i, j, k	flags of a quaternion
$\mathbf{I} = [I]$	identity matrix
$\hat{\imath}, \hat{\jmath}, \hat{k}$	local coordinate axis unit vectors
$\tilde{\imath}, \tilde{\jmath}, \tilde{k}$	skew-symmetric matrices of the unit vectors $\hat{\imath}, \hat{\jmath}, \hat{k}$
$\hat{I}, \hat{J}, \hat{K}$	global coordinate axis unit vectors
$\mathbf{n}$	eigenvectors of R
l	length
O	common origin of B and G
P	a body point, a fixed point in B
p, q, r	general quaternions
$\mathbf{r}$	position vector
r_{ij}	element of row i and column j of a matrix
R	rotation transformation matrix
$\mathbb{R}$	set of real numbers
s	sine
S	sine of half angle
t	time
$\hat{u}$	unit vector on axis of rotation
$\tilde{u}$	skew-symmetric matrix of the vector $\hat{u}$

v	velocity vector, eigenvector of R
w	Rodriguez vector
x,y,z	local coordinate axes
X,Y,Z	global coordinate axes

Greek

α, β, γ	rotation angles about global axes
ϵ_{ijk}	permutation symbol
δ_{ij}	Kronecker's delta
λ	eigenvalues of R
ϕ	angle of rotation about $\hat{u}$
φ, θ, ψ	rotation angles about local axes, Euler angles
$\boldsymbol{\omega}$	angular velocity vector
$\tilde{\omega}$	skew symmetric matrix of the vector $\boldsymbol{\omega}$

Symbol

tr	trace operator
vers	$1 - \cos$
$[\ ]^{-1}$	inverse of the matrix $[\ ]$
$[\ ]^{\mathrm{T}}$	transpose of the matrix $[\ ]$
$\overleftrightarrow{q}$	matrix form of a quaternion q

EXERCISES

1. **Invariant Axis of Rotation** Find out if the axis of rotation $\hat{u}$ is fixed in $B(Oxyz)$ or $G(OXYZ)$.

2. **z-Axis–Angle Rotation Matrix** Expand

$$\begin{aligned} {}^G R_B &= {}^B R_G^{-1} = {}^B R_G^{\mathrm{T}} = R_{\hat{u},\phi} \\ &= \left[A_{z,-\varphi} A_{y,-\theta} A_{z,\phi} A_{y,\theta} A_{z,\varphi}\right]^{\mathrm{T}} \\ &= A_{z,\varphi}^{\mathrm{T}} A_{y,\theta}^{\mathrm{T}} A_{z,\phi}^{\mathrm{T}} A_{y,-\theta}^{\mathrm{T}} A_{z,-\varphi}^{\mathrm{T}} \end{aligned}$$

and verify the axis–angle rotation matrix

$$\begin{aligned} {}^G R_B &= R_{\hat{u},\phi} \\ &= \begin{bmatrix} u_1^2 \operatorname{vers}\phi + c\phi & u_1 u_2 \operatorname{vers}\phi - u_3 s\phi & u_1 u_3 \operatorname{vers}\phi + u_2 s\phi \\ u_1 u_2 \operatorname{vers}\phi + u_3 s\phi & u_2^2 \operatorname{vers}\phi + c\phi & u_2 u_3 \operatorname{vers}\phi - u_1 s\phi \\ u_1 u_3 \operatorname{vers}\phi - u_2 s\phi & u_2 u_3 \operatorname{vers}\phi + u_1 s\phi & u_3^2 \operatorname{vers}\phi + c\phi \end{bmatrix} \end{aligned}$$

3. **x-Axis–Angle Rotation Matrix** Find the axis–angle rotation matrix by transforming the x-axis on the axis of rotation $\hat{u}$.

4. **y-Axis–Angle Rotation Matrix** Find the axis–angle rotation matrix by transforming the y-axis on the axis of rotation $\hat{u}$.

5. **Axis–Angle Rotation and Euler Angles** Find the Euler angles corresponding to the 45 deg rotation about $\mathbf{u} = [1,1,1]^{\mathrm{T}}$.

6. **Euler Angles between Two Local Frames** The Euler angles between the coordinate frame B_1 and G are 20 deg, 35 deg, and -40 deg. The Euler angles between the coordinate frame B_2 and G are 60 deg, -30 deg, and -10 deg. Find the angle and axis of rotation that transforms B_2 to B_1.

7. ★ **Angle and Axis of Rotation Based on Euler Angles** Compare the Euler angle rotation matrix with the angle–axis rotation matrix and find the angle and axis of rotation based on Euler angles.

8. ★ **Euler Angles Based on Angle and Axis of Rotation** Compare the Euler angle rotation matrix with the angle–axis rotation matrix and find the Euler angles based on the angle and axis of rotation.

9. ★ **Repeating Global–Local Rotations** Rotate ${}^{B}\mathbf{r}_P = [6, 2, -3]^{\mathrm{T}}$ 60 deg about the Z-axis followed by 30 deg about the x-axis. Then repeat the sequence of rotations for 60 deg about the Z-axis followed by 30 deg about the x-axis. After how many rotations will point P be back to its initial global position?

10. ★ **Repeating Global–Local Rotations** How many rotations of $\alpha = \pi/m$ degrees about the X-axis followed by $\beta = \pi/k$ degrees about the z-axis are needed to bring a body point to its initial global position if $m, k \in \mathbb{N}$?

11. ★ **Small Rotation Angles** Show that for very small angles of rotation φ, θ, and ψ about the axes of the local coordinate frame the first and third rotations are indistinguishable when they are about the same axis.

12. ★ **Inner Automorphism Properly of $\tilde{\mathbf{a}}$** If R is a rotation matrix and $\mathbf{a}$ is a vector, show that

$$R\tilde{\mathbf{a}}R^{\mathrm{T}} = \widetilde{R\mathbf{a}}$$

13. ★ **Angle–Derivative of Principal Rotation Matrices** Show that

$$\frac{dR_{Z,\alpha}}{d\alpha} = \tilde{K}\,R_{Z,\alpha}$$

$$\frac{dR_{Y,\beta}}{d\beta} = \tilde{J}\,R_{Y,\beta}$$

$$\frac{dR_{X,\gamma}}{d\gamma} = \tilde{I}\,R_{X,\gamma}$$

14. ★ **Euler Angles, Euler Parameters** Compare the Euler angle rotation matrix and Euler parameter transformation matrix and verify the following relationships between Euler angles and Euler parameters:

$$e_0 = \cos\frac{\theta}{2}\cos\frac{\psi+\varphi}{2}$$

$$e_1 = \sin\frac{\theta}{2}\cos\frac{\psi-\varphi}{2}$$

$$e_2 = \sin\frac{\theta}{2}\sin\frac{\psi-\varphi}{2}$$

$$e_3 = \cos\frac{\theta}{2}\sin\frac{\psi+\varphi}{2}$$

and

$$\varphi = \cos^{-1}\frac{2\left(e_2e_3 + e_0e_1\right)}{\sin\theta}$$

$$\theta = \cos^{-1}\left[2\left(e_0^2 + e_3^2\right) - 1\right]$$

$$\psi = \cos^{-1}\frac{-2\left(e_2e_3 - e_0e_1\right)}{\sin\theta}$$

15. **★ Quaternion Definition** Find the unit quaternion $e\left(\phi, \hat{u}\right)$ associated with

$$\hat{u} = \begin{bmatrix} 1/\sqrt{3} \\ 1/\sqrt{3} \\ 1/\sqrt{3} \end{bmatrix}$$

$$\phi = \frac{\pi}{3}$$

and find the result of $e\left(\phi, \hat{u}\right) \hat{\imath}\, e^{*}\left(\phi, \hat{u}\right)$.

16. **★ Quaternion Product** Find pq, qp, $p^{*}q$, qp^{*}, $p^{*}p$, qq^{*}, $p^{*}q^{*}$, and $p^{*}rq^{*}$ if

$$p = 3 + i - 2j + 2k$$
$$q = 2 - i + 2j + 4k$$
$$r = -1 + i + j - 3k$$

17. **★ Quaternion Inverse** Find q^{-1}, p^{-1}, $p^{-1}q^{-1}$, $q^{-1}p^{*}$, $p^{*}p^{-1}$, $q^{-1}q^{*}$, and $p^{*^{-1}}q^{*^{-1}}$ if

$$p = 3 + i - 2j + 2k$$
$$q = 2 - i + 2j + 4k$$

18. **★ Quaternion and Angle–Axis of Rotation** Find the unit quaternion associated with

$$p = 3 + i - 2j + 2k$$
$$q = 2 - i + 2j + 4k$$

and find the angle and axis of rotation for each unit quaternion.

19. **★ Unit Quaternion and Rotation** Use the unit quaternion p,

$$p = \frac{1 + i - j + k}{2}$$

and find the global position of

$${}^{B}\mathbf{r} = \begin{bmatrix} 2 \\ -2 \\ 6 \end{bmatrix}$$

20. **★ Quaternion Matrix** Use the unit quaternion matrices associated with

$$p = 3 + i - 2j + 2k$$
$$q = 2 - i + 2j + 4k$$
$$r = -1 + i + j - 3k$$

and find $\overleftrightarrow{p}\,\overleftrightarrow{r}\,\overleftrightarrow{q}$, $\overleftrightarrow{q}\,\overleftrightarrow{p}$, $\leftrightarrow\mathrm{p}^{*}\overleftrightarrow{q}$, $\overleftrightarrow{q}\,\overleftrightarrow{p^{*}}$, $\overleftrightarrow{p^{*}}\,\overleftrightarrow{p}$, $\overleftrightarrow{q}\,\overleftrightarrow{q^{*}}$, $\overleftrightarrow{p^{*}}\,\overleftrightarrow{q^{*}}$, and $\overleftrightarrow{p^{*}}\,\overleftrightarrow{r}\,\overleftrightarrow{\mathrm{q}^{*}}$.

21. **★ Euler Angles and Quaternion** Find quaternion components in terms of Euler angles and Euler angles in terms of quaternion components.

22. **Angular Velocity Vector** Use the definition ${}^{G}R_{B} = \left[r_{ij}\right]$ and ${}^{G}\dot{R}_{B} = \left[\dot{r}_{ij}\right]$, and find the angular velocity vector $\boldsymbol{\omega}$, where $\tilde{\omega} = {}^{G}\dot{R}_{B}{}^{G}R_{B}^{\mathrm{T}}$.

23. ★ ***bac−cab* Rule** Use the Levi-Civita density ϵ_{ijk} to prove the *bac−cab* rule

$$\mathbf{a} \times (\mathbf{b} \times \mathbf{c}) = \mathbf{b}\,(\mathbf{a} \cdot \mathbf{c}) - \mathbf{c}\,(\mathbf{a} \cdot \mathbf{b})$$

24. ★ ***bac−cab* Rule Application** Use the *bac−cab* rule to show that

$$\mathbf{a} = \hat{n}\left(\mathbf{a} \cdot \hat{\mathbf{n}}\right) + \hat{n} \times \left(\mathbf{a} \times \hat{n}\right)$$

where $\hat{n}$ is any unit vector. What is the geometric significance of this equation?

25. ★ **Two Rotations Are Not Enough** Show that, in general, it is impossible to move a point $P(X, Y, Z)$ from the initial position $P(X_i, Y_i, Z_i)$ to the final position $P(X_f, Y_f, Z_f)$ only by *two* rotations about the global axes.

26. ★ **Three Rotations Are Enough** Show that, in general, it is possible to move a point $P(X, Y, Z)$ from the initial position $P(X_i, Y_i, Z_i)$ to the final position $P(X_f, Y_f, Z_f)$ by *three* rotations about different global axes.

27. ★ **Closure Property** Show the closure property of transformation matrices.

28. **Sum of Two Orthogonal Matrices** Show that the sum of two orthogonal matrices is not, in general, an orthogonal matrix but their product is.

29. **Equivalent Cross Product** Show that if $\mathbf{a} = \left[a_1,\ a_2,\ a_3\right]^{\mathrm{T}}$ and $\mathbf{b} = \left[b_1,\ b_2,\ b_3\right]^{\mathrm{T}}$ are two arbitrary vectors and

$$\tilde{\mathbf{a}} = \begin{bmatrix} 0 & -a_3 & a_2 \\ a_3 & 0 & -a_1 \\ -a_2 & a_1 & 0 \end{bmatrix}$$

is the skew-symmetric matrix corresponding to **a**, then

$$\tilde{\mathbf{a}}\mathbf{b} = \mathbf{a} \times \mathbf{b}$$

30. ★ **Skew-Symmetric Matrices** Use $\mathbf{a} = \left[a_1,\ a_2,\ a_3\right]^{\mathrm{T}}$ and $\mathbf{b} = \left[b_1,\ b_2,\ b_3\right]^{\mathrm{T}}$ to show that

(a) $\tilde{\mathbf{a}}\mathbf{b} = -\tilde{\mathbf{b}}\mathbf{a}$

(b) $\left(\widetilde{\mathbf{a} + \mathbf{b}}\right) = \tilde{\mathbf{a}} + \tilde{\mathbf{b}}$

(c) $\left(\widetilde{\tilde{\mathbf{a}}b}\right) = \mathbf{b}\mathbf{a}^{\mathrm{T}} - \mathbf{a}\mathbf{b}^{\mathrm{T}}$

31. ★ **Rotation Matrix Identity** Show that if A, B, and C are three rotation matrices, then

(a) $(AB)\,C = A\,(BC) = ABC$

(b) $(A + B)^{\mathrm{T}} = A^{\mathrm{T}} + B^{\mathrm{T}}$

(c) $(AB)^{\mathrm{T}} = B^{\mathrm{T}} A^{\mathrm{T}}$

(d) $\left(A^{-1}\right)^{\mathrm{T}} = \left(A^{\mathrm{T}}\right)^{-1}$

32. ★ **Skew-Symmetric Matrix Multiplication** Verify that

(a) $\mathbf{a}^{\mathrm{T}}\,\tilde{\mathbf{a}}^{\mathrm{T}} = -\mathbf{a}^{\mathrm{T}}\,\tilde{\mathbf{a}} = 0$

(b) $\tilde{\mathbf{a}}\tilde{\mathbf{b}} = ba^{\mathrm{T}} - \mathbf{a}^{\mathrm{T}}\mathbf{b}\,\mathbf{I}$

33. ★ **Skew-Symmetric Matrix Derivative** Show that

$$\dot{\tilde{\mathbf{a}}} = \widetilde{\dot{\mathbf{a}}}$$

34. ★ **Time Derivative of** $A = [\mathbf{a}, \tilde{\mathbf{a}}]$ Assume that $\mathbf{a}$ is a time-dependent vector and $A = [\mathbf{a}, \tilde{\mathbf{a}}]$ is a 3×4 matrix. What is the time derivative of $C = AA^{\mathrm{T}}$?

35. ★ **Combined Angle–Axis Rotations** The rotation ϕ_1 about $\hat{u}_1$ followed by rotation ϕ_2 about $\hat{u}_2$ is equivalent to a rotation ϕ about $\hat{u}$. Find the angle ϕ and axis $\hat{u}$ in terms of ϕ_1, $\hat{u}_1$, ϕ_2, and $\hat{u}_2$.

36. ★ **Rodriguez Vector** Using the Rodriguez rotation formula, show that

$$\mathbf{r}' - \mathbf{r} = \tan\frac{\phi}{2}\hat{u} \times (\mathbf{r}' + \mathbf{r})$$

37. ★ **Equivalent Rodriguez Rotation Matrices** Show that the Rodriguez rotation matrix

$${}^{G}R_B = \mathbf{I}\cos\phi + \hat{u}\hat{u}^{\mathrm{T}} \text{ vers } \phi + \tilde{u}\sin\phi$$

can also be written as

$${}^{G}R_B = \mathbf{I} + (\sin\phi)\,\tilde{u} + (\text{ vers } \phi)\,\tilde{u}^2$$

38. ★ **Rotation Matrix and Rodriguez Formula** Knowing the alternative definition of the Rodriguez formula,

$${}^{G}R_B = \mathbf{I} + (\sin\phi)\,\tilde{u} + (\text{ vers } \phi)\,\tilde{u}^2$$

and

$$\tilde{u}^{2n-1} = (-1)^{n-1}\,\tilde{u}$$
$$\tilde{u}^{2n} = (-1)^{n-1}\,\tilde{u}^2$$

examine the equation

$${}^{G}R_B^{\mathrm{T}}\,{}^{G}R_B = {}^{G}R_B\,{}^{G}R_B^{\mathrm{T}}$$

39. ★ **Rodriguez Formula Application** Use the alternative definition of the Rodriguez formula,

$${}^{G}R_B = \mathbf{I} + (\sin\phi)\,\tilde{u} + (\text{ vers } \phi)\,\tilde{u}^2$$

and find the global position of a body point at

$${}^{B}\mathbf{r} = \begin{bmatrix} 1 & 3 & 4 \end{bmatrix}^{\mathrm{T}}$$

after a rotation of 45 deg about the axis indicated by

$$\hat{u} = \begin{bmatrix} \frac{1}{\sqrt{3}} & \frac{1}{\sqrt{3}} & \frac{1}{\sqrt{3}} \end{bmatrix}^{\mathrm{T}}$$

40. Axis and Angle of Rotation Consider the transformation matrix

$$R = \begin{bmatrix} \frac{3}{4} & \frac{\sqrt{6}}{4} & \frac{1}{4} \\ \frac{-\sqrt{6}}{4} & \frac{2}{4} & \frac{\sqrt{6}}{4} \\ \frac{1}{4} & \frac{-\sqrt{6}}{4} & \frac{3}{4} \end{bmatrix}$$

(a) Find the axis and angle of rotation.

(b) Determine the Euler angles.

(c) Determine the Euler parameters.

(d) Determine the eigenvalues and eigenvectors.

41. ★ Axis of Rotation Multiplication Show that

$$\tilde{u}^{2k+1} = (-1)^k \, \tilde{u}$$

and

$$\tilde{u}^{2k} = (-1)^k \left(\mathbf{I} - \hat{u}\hat{u}^{\mathrm{T}}\right)$$

42. ★ Stanley Method Find the Euler parameters of the following rotation matrix based on the Stanley method:

$${}^{G}R_{B} = \begin{bmatrix} 0.5449 & -0.5549 & 0.6285 \\ 0.3111 & 0.8299 & 0.4629 \\ -0.7785 & -0.0567 & 0.6249 \end{bmatrix}$$

6

Motion Kinematics

The general motion of a rigid body in a global frame G is a combination of displacements and rotations. This combination may be defined by a 4×4 matrix to express the rigid-body motion.

6.1 RIGID-BODY MOTION

Consider a rigid body with an attached local coordinate frame $B(oxyz)$ that is moving in a fixed global coordinate frame $G(OXYZ)$. The rigid body can rotate in G, while point o of B can translate relative to the origin O of G, as is shown in Figure 6.1. If the vector ${}^G\mathbf{d}$ indicates the position of the moving origin o relative to the fixed origin O, then the coordinates of a body point P in local and global frames are related by

$$
{}^G\mathbf{r}_P = {}^G R_B \, {}^B\mathbf{r}_P + {}^G\mathbf{d} \tag{6.1}
$$

where

$$
{}^G\mathbf{r} = \begin{bmatrix} X_P \\ Y_P \\ Z_P \end{bmatrix} \qquad {}^B\mathbf{r} = \begin{bmatrix} x_P \\ y_P \\ z_P \end{bmatrix} \qquad {}^G\mathbf{d} = \begin{bmatrix} X_o \\ Y_o \\ Z_o \end{bmatrix} \tag{6.2}
$$

The vector ${}^G\mathbf{d}$ is called the *displacement* or *translation* of B with respect to G, and ${}^G R_B$ is the *rotation matrix* that transforms ${}^B\mathbf{r}$ to ${}^G\mathbf{r}$ when ${}^G\mathbf{d} = 0$. Such a combination of a rotation and translation in Equation (6.1) is called the *rigid-body motion* in which the configuration of the body can be expressed by the position of the origin o of B and the orientation of B.

Decomposition of a rigid motion into a rotation and a translation is the simplest method for representing spatial displacement of rigid bodies. We show the translation by a vector and the rotation by any of the methods described in Chapter 5.

Proof: Consider a body frame B that is initially coincident with a globally fixed frame G. Figure 6.1 illustrates a translated and rotated frame B in a frame G. The most general rotation is represented by the *Rodriguez rotation formula* (5.122), which depends on ${}^B\mathbf{r}_P$, the position vector of a point P measured in the body coordinate frame. All points of the body have the same displacement for the translation ${}^G\mathbf{d}$. Therefore, translation of a rigid body is independent of the local position vector ${}^B\mathbf{r}$. Because of that, we can represent the most general displacement of a rigid body by a rotation and a translation:

$$
\begin{aligned}
{}^G\mathbf{r} &= {}^B\mathbf{r} \cos\phi + (1 - \cos\phi) \left(\hat{u} \cdot {}^B\mathbf{r}\right) \hat{u} + \left(\hat{u} \times {}^B\mathbf{r}\right) \sin\phi + {}^G\mathbf{d} \\
&= {}^G R_B \, {}^B\mathbf{r} + {}^G\mathbf{d}
\end{aligned} \tag{6.3}
$$

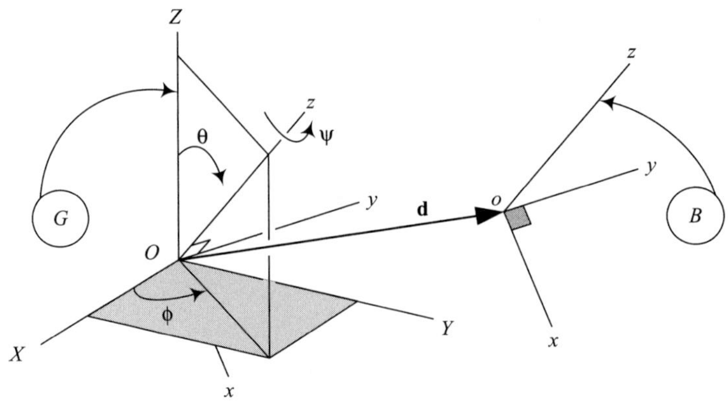

Figure 6.1 Rotation and translation of a body frame.

Equation (6.3) indicates that the most general displacement of a rigid body is a rotation about an axis and a translation along an axis. The choice of the point of reference o is arbitrary; however, when this point is chosen and the body coordinate frame is set up, the rotation and translation are determined.

Based on translation and rotation, the configuration of a body can be uniquely determined by six independent parameters: three translation components X_o, Y_o, Z_o and three rotational components. The rotation parameters may be the Euler angles, Euler parameters, or any set of the triple rotations of Appendices A and B.

If a body moves in such a way that its rotational components remain constant, the motion is called a *pure translation*; if it moves in such a way that X_o, Y_o, and Z_o remain constant, the motion is called a *pure rotation*. Therefore, a rigid body has three translational and three rotational degrees of freedom. ■

Example 335 Rotation and Translation of a Body Coordinate Frame A body coordinate frame $B(oxyz)$ that is originally coincident with global coordinate frame $G(OXYZ)$ rotates 60 deg about the X-axis and translates to $[3,4,5]^{\mathrm{T}}$. The global position of a point at ${}^{B}\mathbf{r} = \left[x, y, z\right]^{\mathrm{T}}$ is

$$
\begin{aligned}
{}^{G}\mathbf{r} &= {}^{G}R_{B}\, {}^{B}\mathbf{r} + {}^{G}\mathbf{d} \\
&= \begin{bmatrix} 1 & 0 & 0 \\ 0 & \cos\dfrac{\pi}{3} & -\sin\dfrac{\pi}{3} \\ 0 & \sin\dfrac{\pi}{3} & \cos\dfrac{\pi}{3} \end{bmatrix} \begin{bmatrix} x \\ y \\ z \end{bmatrix} + \begin{bmatrix} 3 \\ 4 \\ 5 \end{bmatrix} \\
&= (x+3)\hat{I} + \left(\frac{1}{2}y - \frac{\sqrt{3}}{2}z + 4\right)\hat{J} + \left(\frac{1}{2}z + \sqrt{3}y + 5\right)\hat{K} \qquad (6.4)
\end{aligned}
$$

Example 336 Moving-Body Coordinate Frame Figure 6.2 shows a point P at ${}^{B}\mathbf{r}_P = \hat{\imath} + 3\hat{\jmath} + 3\hat{k}$ in a body frame B which is rotated 45 deg about the Z-axis and translated -9 along the X-axis, 5 along the Y-axis, and 2 along the Z-axis.

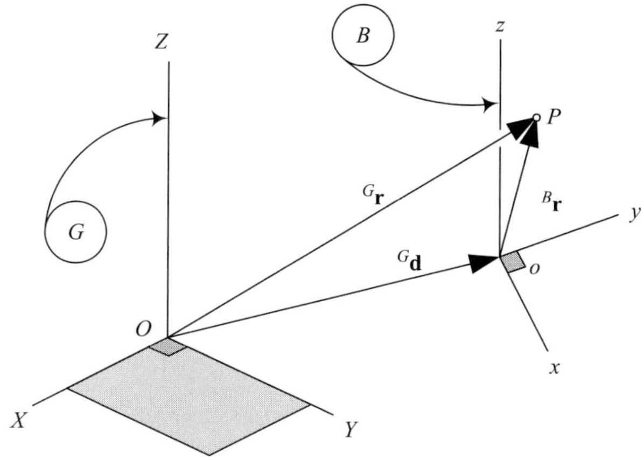

Figure 6.2 A translating and rotating body in a global coordinate frame.

The position of P in a global coordinate frame is

$$^G\mathbf{r} = {}^GR_B\,{}^B\mathbf{r}_P + {}^G\mathbf{d} \tag{6.5}$$

$$= \begin{bmatrix} \cos\frac{\pi}{4} & -\sin\frac{\pi}{4} & 0 \\ \sin\frac{\pi}{4} & \cos\frac{\pi}{4} & 0 \\ 0 & 0 & 1 \end{bmatrix} \begin{bmatrix} 1 \\ 3 \\ 3 \end{bmatrix} + \begin{bmatrix} -9 \\ 5 \\ 2 \end{bmatrix} = \begin{bmatrix} -10.41 \\ 7.83 \\ 5 \end{bmatrix}$$

Example 337 Rotation of a Translated Rigid Body Point P of a rigid body B has the initial position vector

$$^B\mathbf{r}_P = \begin{bmatrix} 1 \\ 2 \\ 3 \end{bmatrix} \tag{6.6}$$

If the body rotates 45 deg about the x-axis and then translates to $^G\mathbf{d} = [3,2,1]^{\mathrm{T}}$, the final position of P would be

$$^G\mathbf{r} = {}^BR_{x,45}^{\mathrm{T}}\;{}^B\mathbf{r}_P + {}^G\mathbf{d} \tag{6.7}$$

$$= \begin{bmatrix} 1 & 0 & 0 \\ 0 & \cos\frac{\pi}{4} & \sin\frac{\pi}{4} \\ 0 & -\sin\frac{\pi}{4} & \cos\frac{\pi}{4} \end{bmatrix}^{\mathrm{T}} \begin{bmatrix} 1 \\ 2 \\ 3 \end{bmatrix} + \begin{bmatrix} 3 \\ 2 \\ 1 \end{bmatrix} = \begin{bmatrix} 4 \\ 1.29 \\ 4.53 \end{bmatrix}$$

Example 338 Robotic Arm Rotation and Elongation The position vector of point P_1 at the tip of a robotic arm, shown in Figure 6.3(a), is

$$ {}^{G}\mathbf{r}_{P_1} = {}^{B}\mathbf{r}_{P_1} = \begin{bmatrix} 1150 \\ 0 \\ 950 \end{bmatrix} \text{mm} \tag{6.8} $$

The arm rotates 60 deg about the global Z-axis and elongates by

$$ {}^{B}\mathbf{d} = \begin{bmatrix} 725.2 \\ 0 \\ 512 \end{bmatrix} \text{mm} \tag{6.9} $$

The final configuration of the arm is shown in Figure 6.3(b).

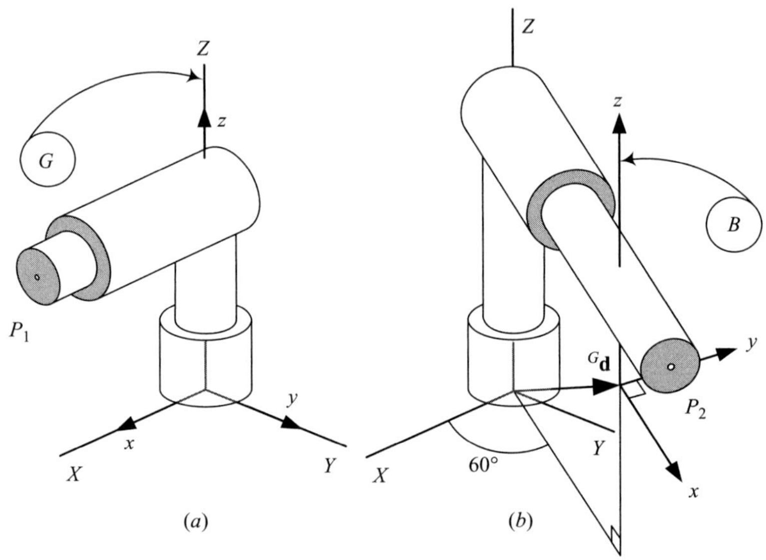

Figure 6.3 A two-arm robot with a revolute and a prismatic joint.

The new position vector of P in G is

$$ {}^{G}\mathbf{r}_{P_2} = {}^{G}R_{B}\,{}^{B}\mathbf{r}_{P_1} + {}^{G}\mathbf{d} = R_{Z,60}\,{}^{B}\mathbf{r}_{P_1} + {}^{G}\mathbf{d} \tag{6.10} $$

where ${}^{G}R_{B} = R_{Z,60}$ is the rotation matrix to transform $\mathbf{r}_{P_2}$ to $\mathbf{r}_{P_1}$ when ${}^{G}\mathbf{d} = 0$. The translation vector ${}^{B}\mathbf{d}$ must be transformed from the body to the global coordinate frame:

$$ {}^{G}\mathbf{d} = {}^{G}R_{B}\,{}^{B}\mathbf{d} \tag{6.11} $$

$$ = \begin{bmatrix} \cos\frac{\pi}{3} & -\sin\frac{\pi}{3} & 0 \\ \sin\frac{\pi}{3} & \cos\frac{\pi}{3} & 0 \\ 0 & 0 & 1 \end{bmatrix} \begin{bmatrix} 725.2 \\ 0 \\ 512 \end{bmatrix} = \begin{bmatrix} 362.6 \\ 628.04 \\ 512 \end{bmatrix} $$

Therefore, the final global position of the tip of the arm is at

$$^{G}\mathbf{r}_{P_2} = {}^{G}R_B\,{}^{B}\mathbf{r}_{P_1} + {}^{G}\mathbf{d} \tag{6.12}$$

$$= \begin{bmatrix} c60 & -s60 & 0 \\ s60 & c60 & 0 \\ 0 & 0 & 1 \end{bmatrix} \begin{bmatrix} 1150 \\ 0 \\ 950 \end{bmatrix} + \begin{bmatrix} 362.6 \\ 628.04 \\ 512 \end{bmatrix} = \begin{bmatrix} 937.6 \\ 1624 \\ 1462 \end{bmatrix}$$

Example 339 Composition of Rigid-Body Motion Consider a rigid motion of body B_1 with respect to body B_2 and then a rigid motion of body B_2 with respect to frame G such that

$$^{2}\mathbf{r} = {}^{2}R_1\,{}^{1}\mathbf{r} + {}^{2}\mathbf{d}_1 \tag{6.13}$$

$$^{G}\mathbf{r} = {}^{G}R_2\,{}^{2}\mathbf{r} + {}^{G}\mathbf{d}_2 \tag{6.14}$$

These two motions may be combined to determine a rigid motion that transforms $^{1}\mathbf{r}$ to $^{G}\mathbf{r}$:

$$^{G}\mathbf{r} = {}^{G}R_2\left({}^{2}R_1\,{}^{1}\mathbf{r} + {}^{2}\mathbf{d}_1\right) + {}^{G}\mathbf{d}_2 = {}^{G}R_2\,{}^{2}R_1\,{}^{1}\mathbf{r} + {}^{G}R_2\,{}^{2}\mathbf{d}_1 + {}^{G}\mathbf{d}_2$$

$$= {}^{G}R_1\,{}^{1}\mathbf{r} + {}^{G}\mathbf{d}_1 \tag{6.15}$$

Therefore,

$$^{G}R_1 = {}^{G}R_2\,{}^{2}R_1 \tag{6.16}$$

$$^{G}\mathbf{d}_1 = {}^{G}R_2\,{}^{2}\mathbf{d}_1 + {}^{G}\mathbf{d}_2 \tag{6.17}$$

which shows that the transformation from frame B_1 to frame G can be done by rotation $^{G}R_1$ and translation $^{G}\mathbf{d}_1$.

6.2 HOMOGENEOUS TRANSFORMATION

Figure 6.4 depicts a body point P where its positions in the local frame B and global frame G are denoted by $^{B}\mathbf{r}_P$ and $^{G}\mathbf{r}_P$. The vector $^{G}\mathbf{d}$ indicates the position of origin

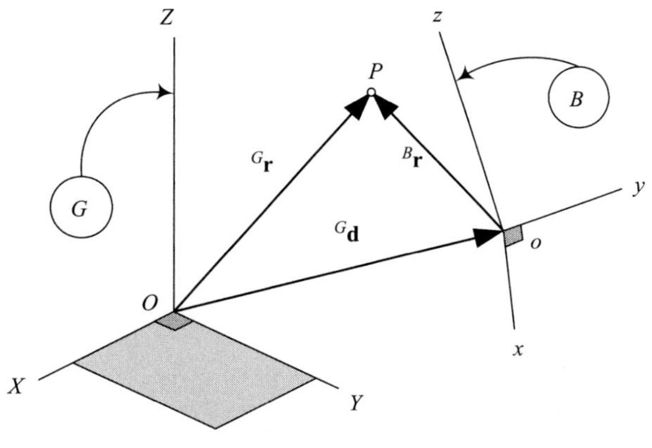

Figure 6.4 Representation of a point P in coordinate frames B and G.

o of B in G. Therefore, a general motion of a rigid body $B(oxyz)$ in the global frame $G(OXYZ)$ is a combination of rotation ${}^GR_B$ and translation ${}^G\mathbf{d}$:

$$^G\mathbf{r} = {}^GR_B\,{}^B\mathbf{r} + {}^G\mathbf{d} \tag{6.18}$$

We may combine a rotation matrix and a translation vector by introducing a 4×4 *homogeneous transformation matrix* ${}^GT_B$ and express a rigid motion by a single matrix transformation

$$^G\mathbf{r} = {}^GT_B\,{}^B\mathbf{r} \tag{6.19}$$

where

$$\begin{aligned} {}^GT_B &= \begin{bmatrix} r_{11} & r_{12} & r_{13} & X_o \\ r_{21} & r_{22} & r_{23} & Y_o \\ r_{31} & r_{32} & r_{33} & Z_o \\ 0 & 0 & 0 & 1 \end{bmatrix} \equiv \left[\begin{array}{c|c} {}^GR_B & {}^G\mathbf{d} \\ \hline 0\ 0\ 0 & 1 \end{array}\right] \\ &\equiv \begin{bmatrix} {}^GR_B & {}^G\mathbf{d} \\ 0 & 1 \end{bmatrix} \end{aligned} \tag{6.20}$$

and

$$^G\mathbf{r} = \begin{bmatrix} X_P \\ Y_P \\ Z_P \\ 1 \end{bmatrix} \qquad {}^B\mathbf{r} = \begin{bmatrix} x_P \\ y_P \\ z_P \\ 1 \end{bmatrix} \qquad {}^G\mathbf{d} = \begin{bmatrix} X_o \\ Y_o \\ Z_o \\ 1 \end{bmatrix} \tag{6.21}$$

The combined matrix ${}^GT_B$ is called the *homogeneous transformation matrix* and is a concise method to represent rigid motions. Introducing the 4×4 matrix ${}^GT_B$ simplifies numerical calculations and transforms the coordinates of a body point between the frames B and G.

Representation of a three-component vector by a four-component vector is called *homogeneous coordinate representation*. The appended element is a *scale factor* w that may be used to separate the magnitude and directional cosines of the vector. Therefore, the *homogeneous expression* of a natural vector is

$$\mathbf{r} = r\hat{u}_r = r\left(u_1\hat{\imath} + u_2\hat{\jmath} + u_3\hat{k}\right) = \begin{bmatrix} u_1 \\ u_2 \\ u_3 \\ r \end{bmatrix} \tag{6.22}$$

and the homogeneous representation of a vector $\mathbf{r} = \left[x, y, z\right]^{\mathrm{T}}$ is

$$\mathbf{r} = \begin{bmatrix} wx \\ wy \\ wz \\ w \end{bmatrix} \tag{6.23}$$

Expressing a vector $\mathbf{r}$ by its homogeneous coordinates shows that the absolute values of the four coordinates are not important. Instead, it is the three ratios x/w, y/w, and

z/w that are important because

$$\mathbf{r} = \begin{bmatrix} x \\ y \\ z \\ w \end{bmatrix} = \begin{bmatrix} x/w \\ y/w \\ z/w \\ 1 \end{bmatrix} \tag{6.24}$$

provided $w \neq 0$ and $w \neq \infty$. The homogeneous vector $w\mathbf{r}$ refers to the same point as $\mathbf{r}$ does.

If $w = 1$, then the homogeneous coordinates of a position vector are the same as the physical coordinates of the vector and the space is a Euclidean space. Hereafter, if no confusion exists and $w = 1$, we will use the regular vectors and their homogeneous representation equivalently.

Proof: We append a fourth element 1 to the coordinates of vectors ${}^G\mathbf{r}_P$, ${}^B\mathbf{r}_P$, ${}^G\mathbf{d}$ and define *homogeneous vectors* as given in (6.21).

Applying the homogeneous transformation (6.19), we find

$${}^G\mathbf{r}_P = {}^GT_B\,{}^B\mathbf{r}_P \tag{6.25}$$

$$\begin{bmatrix} X_P \\ Y_P \\ Z_P \\ 1 \end{bmatrix} = \begin{bmatrix} r_{11} & r_{12} & r_{13} & X_o \\ r_{21} & r_{22} & r_{23} & Y_o \\ r_{31} & r_{32} & r_{33} & Z_o \\ 0 & 0 & 0 & 1 \end{bmatrix} \begin{bmatrix} x_P \\ y_P \\ z_P \\ 1 \end{bmatrix}$$

$$= \begin{bmatrix} X_o + r_{11}x_P + r_{12}y_P + r_{13}z_P \\ Y_o + r_{21}x_P + r_{22}y_P + r_{23}z_P \\ Z_o + r_{31}x_P + r_{32}y_P + r_{33}z_P \\ 1 \end{bmatrix} \tag{6.26}$$

However, the standard method provides

$${}^G\mathbf{r}_P = {}^GR_B\,{}^B\mathbf{r}_p + {}^G\mathbf{d} \tag{6.27}$$

$$\begin{bmatrix} X_P \\ Y_P \\ Z_P \end{bmatrix} = \begin{bmatrix} r_{11} & r_{12} & r_{13} \\ r_{21} & r_{22} & r_{23} \\ r_{31} & r_{32} & r_{33} \end{bmatrix} \begin{bmatrix} x_P \\ y_P \\ z_P \end{bmatrix} + \begin{bmatrix} X_o \\ Y_o \\ Z_o \end{bmatrix}$$

$$= \begin{bmatrix} X_o + r_{11}x_P + r_{12}y_P + r_{13}z_P \\ Y_o + r_{21}x_P + r_{22}y_P + r_{23}z_P \\ Z_o + r_{31}x_P + r_{32}y_P + r_{33}z_P \end{bmatrix} \tag{6.28}$$

which is equivalent to its homogeneous expression (6.26). ■

Example 340 Rotation and Translation of a Body Coordinate Frame A body coordinate frame $B(oxyz)$ that is originally coincident with global coordinate frame $G(OXYZ)$

rotates 60 deg about the X-axis and translates to $[3,4,5,1]^{\mathrm{T}}$. The global position of a point at

$$ {}^{B}\mathbf{r} = \begin{bmatrix} x & y & z & 1 \end{bmatrix}^{\mathrm{T}} \tag{6.29} $$

is

$$ {}^{G}\mathbf{r} = {}^{G}T_{B}\, {}^{B}\mathbf{r} \tag{6.30} $$

$$ = \begin{bmatrix} X \\ Y \\ Z \\ 1 \end{bmatrix} = \begin{bmatrix} 1 & 0 & 0 & 3 \\ 0 & \cos\frac{\pi}{3} & -\sin\frac{\pi}{3} & 4 \\ 0 & \sin\frac{\pi}{3} & \cos\frac{\pi}{3} & 5 \\ 0 & 0 & 0 & 1 \end{bmatrix} \begin{bmatrix} x \\ y \\ z \\ 1 \end{bmatrix} = \begin{bmatrix} x+3 \\ \frac{y}{2} - \frac{\sqrt{3}}{2}z + 4 \\ \frac{z}{2} + \frac{\sqrt{3}}{2}y + 5 \\ 1 \end{bmatrix} $$

Example 341 Rotation and Translation of a Body Coordinate Frame Figure 6.5 illustrates a body coordinate frame $B(oxyz)$ that was originally coincident with global coordinate frame $G(OXYZ)$. The body is rotated 60 deg about the X-axis and then 30 deg about the Z-axis and then translated to $[3,14,5,1]^{\mathrm{T}}$. The homogeneous transformation matrix ${}^{G}T_{B}$ is

$$ {}^{G}T_{B} = \begin{bmatrix} {}^{G}R_{B} & {}^{G}\mathbf{d} \\ 0 & 1 \end{bmatrix} = \begin{bmatrix} 0.86603 & -0.25 & 0.43301 & 3 \\ 0.5 & 0.43301 & -0.75 & 14 \\ 0 & 0.86603 & 0.5 & 5 \\ 0 & 0 & 0 & 1 \end{bmatrix} \tag{6.31} $$

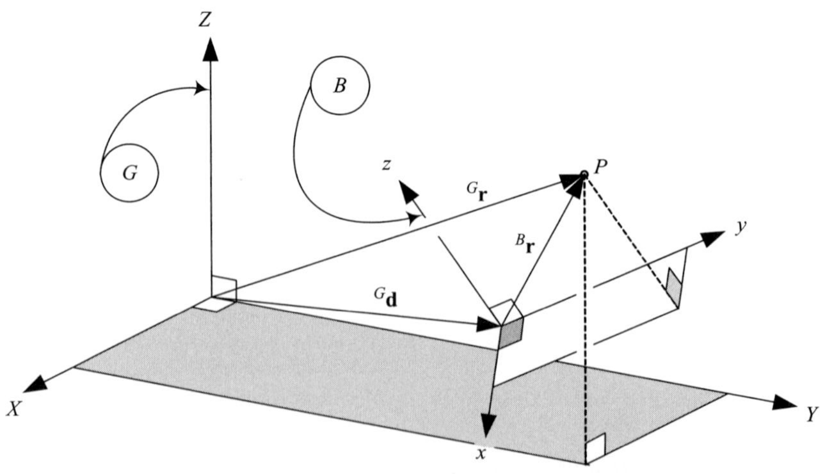

Figure 6.5 A body coordinate frame $B(oxyz)$ after a combined rotation and translation.

because

$$
\begin{aligned}
{}^G R_B &= {}^G R_{Z,30}\,{}^G R_{X,60} \\
&= \begin{bmatrix} \cos\frac{\pi}{6} & -\sin\frac{\pi}{6} & 0 \\ \sin\frac{\pi}{6} & \cos\frac{\pi}{6} & 0 \\ 0 & 0 & 1 \end{bmatrix} \begin{bmatrix} 1 & 0 & 0 \\ 0 & \cos\frac{\pi}{3} & -\sin\frac{\pi}{3} \\ 0 & \sin\frac{\pi}{3} & \cos\frac{\pi}{3} \end{bmatrix} \\
&= \begin{bmatrix} 0.86603 & -0.25 & 0.43301 \\ 0.5 & 0.43301 & -0.75 \\ 0 & 0.86603 & 0.5 \end{bmatrix}
\end{aligned} \tag{6.32}
$$

So, the global position of a point at ${}^B\mathbf{r} = [1, 5, 4, 1]^{\mathrm{T}}$ is

$$
{}^G\mathbf{r} = {}^G T_B\,{}^B\mathbf{r} = {}^G T_B \begin{bmatrix} 1 \\ 5 \\ 1 \\ 1 \end{bmatrix} = \begin{bmatrix} 3.049 \\ 15.915 \\ 9.8302 \\ 1 \end{bmatrix} \tag{6.33}
$$

Example 342 Pure Rotation and Translation Matrices The homogeneous transformation matrix ${}^G T_B$ can be decomposed to a matrix multiplication of a pure rotation matrix ${}^G R_B$ and a pure translation matrix ${}^G D_B$,

$$
\begin{aligned}
{}^G T_B = {}^G D_B\,{}^G R_B &= \begin{bmatrix} 1 & 0 & 0 & X_o \\ 0 & 1 & 0 & Y_o \\ 0 & 0 & 1 & Z_o \\ 0 & 0 & 0 & 1 \end{bmatrix} \begin{bmatrix} r_{11} & r_{12} & r_{13} & 0 \\ r_{21} & r_{22} & r_{23} & 0 \\ r_{31} & r_{32} & r_{33} & 0 \\ 0 & 0 & 0 & 1 \end{bmatrix} \\
&= \begin{bmatrix} r_{11} & r_{12} & r_{13} & X_o \\ r_{21} & r_{22} & r_{23} & Y_o \\ r_{31} & r_{32} & r_{33} & Z_o \\ 0 & 0 & 0 & 1 \end{bmatrix}
\end{aligned} \tag{6.34}
$$

So, the body-to-global homogeneous transformation matrix ${}^G T_B$ of a rigid-body motion can be achieved by a pure rotation first followed by a pure translation (RFDS):

$$
\begin{aligned}
{}^G T_B &= {}^G D_B\,{}^G R_B = \text{RFDS} \\
&= \text{Rotation First, Displacement Second}
\end{aligned} \tag{6.35}
$$

The decomposition of a homogeneous transformation to rotation and translation is not interchangeable:

$$
{}^G T_B = {}^G D_B\,{}^G R_B \neq {}^G R_B\,{}^G D_B \tag{6.36}
$$

However, according to the definition of ${}^G R_B$ and ${}^G D_B$, we have

$$
{}^G T_B = {}^G D_B\,{}^G R_B = {}^G D_B + {}^G R_B - \mathbf{I} = {}^G R_B + {}^G D_B - \mathbf{I} \tag{6.37}
$$

Therefore, if a body coordinate frame $B(oxyz)$ that is originally coincident with global coordinate frame $G(OXYZ)$ translates to

$$^G\mathbf{d} = \begin{bmatrix} d_x \\ d_y \\ d_z \end{bmatrix} \tag{6.38}$$

its motion is a pure translation. The associated homogeneous transformation matrix for such a body point in the global frame is

$$^G\mathbf{r} = {}^GT_B\,{}^B\mathbf{r} = {}^GD_B\,{}^B\mathbf{r} \tag{6.39}$$

$$= \begin{bmatrix} X \\ Y \\ Z \\ 1 \end{bmatrix} = \begin{bmatrix} 1 & 0 & 0 & d_X \\ 0 & 1 & 0 & d_Y \\ 0 & 0 & 1 & d_Z \\ 0 & 0 & 0 & 1 \end{bmatrix} \begin{bmatrix} x \\ y \\ z \\ 1 \end{bmatrix} = \begin{bmatrix} x + d_x \\ y + d_y \\ z + d_z \\ 1 \end{bmatrix}$$

Similarly, the homogeneous representation of rotations about an axis of the global coordinate frame, say rotation α about the Z-axis, is

$$^GT_B = R_{Z,\alpha} = \begin{bmatrix} \cos\alpha & -\sin\alpha & 0 & 0 \\ \sin\alpha & \cos\alpha & 0 & 0 \\ 0 & 0 & 1 & 0 \\ 0 & 0 & 0 & 1 \end{bmatrix} \tag{6.40}$$

Example 343 Rotation about and Translation along a Principal Axis A point P is located at $^B\mathbf{r} = [5,5,5]^{\mathrm{T}}$. If the rigid body rotates 60 deg about the global X-axis and the origin of the body frame translates to $^G\mathbf{d} = [5,5,5]^{\mathrm{T}}$, then the homogeneous transformation can provide the coordinates of the point in the global frame:

$$\begin{bmatrix} 1 & 0 & 0 & 5 \\ 0 & 1 & 0 & 5 \\ 0 & 0 & 1 & 5 \\ 0 & 0 & 0 & 1 \end{bmatrix} \begin{bmatrix} 1 & 0 & 0 & 0 \\ 0 & \cos 60 & -\sin 60 & 0 \\ 0 & \sin 60 & \cos 60 & 0 \\ 0 & 0 & 0 & 1 \end{bmatrix} \begin{bmatrix} 5 \\ 5 \\ 5 \\ 1 \end{bmatrix} = \begin{bmatrix} 10 \\ 3.17 \\ 11.83 \\ 1 \end{bmatrix} \tag{6.41}$$

Now assume that the same point of the rigid body rotates 60 deg about the local x-axis and the origin of the body frame translates to $^G\mathbf{d}$. Then the coordinates of the point in the global frame are

$$\begin{bmatrix} 1 & 0 & 0 & 5 \\ 0 & 1 & 0 & 5 \\ 0 & 0 & 1 & 5 \\ 0 & 0 & 0 & 1 \end{bmatrix} \begin{bmatrix} 1 & 0 & 0 & 0 \\ 0 & \cos 60 & \sin 60 & 0 \\ 0 & -\sin 60 & \cos 60 & 0 \\ 0 & 0 & 0 & 1 \end{bmatrix}^{\mathrm{T}} \begin{bmatrix} 5 \\ 5 \\ 5 \\ 1 \end{bmatrix} = \begin{bmatrix} 10 \\ 3.17 \\ 11.83 \\ 1 \end{bmatrix} \tag{6.42}$$

Example 344 Translation Consider a body point at

$$^B\mathbf{r} = \begin{bmatrix} -1 & 0 & 2 & 1 \end{bmatrix}^{\mathrm{T}} \tag{6.43}$$

that is translated to

$$^G\mathbf{r} = \begin{bmatrix} 0 & 10 & -5 & 1 \end{bmatrix}^{\mathrm{T}} \tag{6.44}$$

The corresponding transformation can be found by

$$\begin{bmatrix} 2 \\ 10 \\ -5 \\ 1 \end{bmatrix} = \begin{bmatrix} 1 & 0 & 0 & d_X \\ 0 & 1 & 0 & d_Y \\ 0 & 0 & 1 & d_Z \\ 0 & 0 & 0 & 1 \end{bmatrix} \begin{bmatrix} 1 \\ 4 \\ 2 \\ 1 \end{bmatrix} \tag{6.45}$$

Therefore,

$$1 + d_X = 2 \qquad 4 + d_Y = 10 \qquad 2 + d_Z = -5 \tag{6.46}$$

and

$$d_X = 1 \qquad d_Y = 6 \qquad d_Z = -7 \tag{6.47}$$

Example 345 Pure Rotation and Translation about Global Axes The basic homogeneous transformations for translation along and rotation about the local x-, y-, and z-axes are given as

$$^BT_G = D_{x,a} = \begin{bmatrix} 1 & 0 & 0 & -a \\ 0 & 1 & 0 & 0 \\ 0 & 0 & 1 & 0 \\ 0 & 0 & 0 & 1 \end{bmatrix} \tag{6.48}$$

$$^BT_G = R_{x,\gamma} = \begin{bmatrix} 1 & 0 & 0 & 0 \\ 0 & \cos\gamma & \sin\gamma & 0 \\ 0 & -\sin\gamma & \cos\gamma & 0 \\ 0 & 0 & 0 & 1 \end{bmatrix} \tag{6.49}$$

$$^BT_G = D_{y,b} = \begin{bmatrix} 1 & 0 & 0 & 0 \\ 0 & 1 & 0 & -b \\ 0 & 0 & 1 & 0 \\ 0 & 0 & 0 & 1 \end{bmatrix} \tag{6.50}$$

$$^BT_G = R_{y,\beta} = \begin{bmatrix} \cos\beta & 0 & -\sin\beta & 0 \\ 0 & 1 & 0 & 0 \\ \sin\beta & 0 & \cos\beta & 0 \\ 0 & 0 & 0 & 1 \end{bmatrix} \tag{6.51}$$

$$^BT_G = D_{z,c} = \begin{bmatrix} 1 & 0 & 0 & 0 \\ 0 & 1 & 0 & 0 \\ 0 & 0 & 1 & -c \\ 0 & 0 & 0 & 1 \end{bmatrix} \tag{6.52}$$

$$
{}^{B}T_{G} = R_{z,\alpha} = \begin{bmatrix} \cos\alpha & \sin\alpha & 0 & 0 \\ -\sin\alpha & \cos\alpha & 0 & 0 \\ 0 & 0 & 1 & 0 \\ 0 & 0 & 0 & 1 \end{bmatrix} \tag{6.53}
$$

The rule (RFDS) of (6.35) for a homogeneous transformation matrix of global-to-body ${}^{B}T_{G}$ is also applicable if we use matrices (6.48)–(6.53):

$$
\begin{aligned} {}^{B}T_{G} &= {}^{B}D_{G}\,{}^{B}R_{G} = \text{RFDS} \\ &= \text{Rotation First, Displacement Second} \end{aligned} \tag{6.54}
$$

Example 346 Pure Rotation and Translation about Local Axes The basic homogeneous transformations for translation along and rotation about the global X-, Y-, and Z-axes are given as

$$
{}^{G}T_{B} = D_{X,a} = \begin{bmatrix} 1 & 0 & 0 & a \\ 0 & 1 & 0 & 0 \\ 0 & 0 & 1 & 0 \\ 0 & 0 & 0 & 1 \end{bmatrix} \tag{6.55}
$$

$$
{}^{G}T_{B} = R_{X,\gamma} = \begin{bmatrix} 1 & 0 & 0 & 0 \\ 0 & \cos\gamma & -\sin\gamma & 0 \\ 0 & \sin\gamma & \cos\gamma & 0 \\ 0 & 0 & 0 & 1 \end{bmatrix} \tag{6.56}
$$

$$
{}^{G}T_{B} = D_{Y,b} = \begin{bmatrix} 1 & 0 & 0 & 0 \\ 0 & 1 & 0 & b \\ 0 & 0 & 1 & 0 \\ 0 & 0 & 0 & 1 \end{bmatrix} \tag{6.57}
$$

$$
{}^{G}T_{B} = R_{Y,\beta} = \begin{bmatrix} \cos\beta & 0 & \sin\beta & 0 \\ 0 & 1 & 0 & 0 \\ -\sin\beta & 0 & \cos\beta & 0 \\ 0 & 0 & 0 & 1 \end{bmatrix} \tag{6.58}
$$

$$
{}^{G}T_{B} = D_{Z,c} = \begin{bmatrix} 1 & 0 & 0 & 0 \\ 0 & 1 & 0 & 0 \\ 0 & 0 & 1 & c \\ 0 & 0 & 0 & 1 \end{bmatrix} \tag{6.59}
$$

$$
{}^{G}T_{B} = R_{Z,\alpha} = \begin{bmatrix} \cos\alpha & -\sin\alpha & 0 & 0 \\ \sin\alpha & \cos\alpha & 0 & 0 \\ 0 & 0 & 1 & 0 \\ 0 & 0 & 0 & 1 \end{bmatrix} \tag{6.60}
$$

Example 347 Homogeneous Transformation as a Vector Addition Figure 6.4 shows that the position of point P can be described by vector addition:

$$^G\mathbf{r}_P = {}^G\mathbf{d} + {}^B\mathbf{r}_P \tag{6.61}$$

Because a vector equation is not meaningful unless all the vectors are expressed in the same coordinate frame, we need to transform either $^B\mathbf{r}_P$ to G or $^G\mathbf{r}_P$ and $^G\mathbf{d}$ to B. Therefore, the correct vector equation is

$$^G\mathbf{r}_P = {}^GR_B\,{}^B\mathbf{r}_P + {}^G\mathbf{d} \tag{6.62}$$

or

$$^BR_G\,{}^G\mathbf{r}_P = {}^BR_G\,{}^G\mathbf{d} + {}^B\mathbf{r}_P \tag{6.63}$$

The first one defines a homogenous transformation from B to G,

$$^G\mathbf{r}_P = {}^GT_B\,{}^B\mathbf{r}_P \tag{6.64}$$

$$^GT_B = \begin{bmatrix} {}^GR_B & {}^G\mathbf{d} \\ 0 & 1 \end{bmatrix} \tag{6.65}$$

and the second one defines a transformation from G to B,

$$^B\mathbf{r}_P = {}^BT_G\,{}^G\mathbf{r}_P \tag{6.66}$$

$$^BT_G = \begin{bmatrix} {}^BR_G & -{}^BR_G\,{}^G\mathbf{d} \\ 0 & 1 \end{bmatrix} = \begin{bmatrix} {}^GR_B^{\mathrm{T}} & -{}^GR_B^{\mathrm{T}}\,{}^G\mathbf{d} \\ 0 & 1 \end{bmatrix} \tag{6.67}$$

Example 348 ★ Point at Infinity A point at infinity indicates a direction. Points at infinity have a convenient representation with homogeneous coordinates. Consider the scale factor w as the fourth coordinate of a point, and hence the homogeneous representation of the point is given by

$$\begin{bmatrix} x \\ y \\ z \\ w \end{bmatrix} = \begin{bmatrix} x/w \\ y/w \\ z/w \\ 1 \end{bmatrix} \tag{6.68}$$

As w tends to zero, the point goes to infinity, and the homogeneous coordinate

$$\begin{bmatrix} x \\ y \\ z \\ 0 \end{bmatrix} \tag{6.69}$$

represents all lines parallel to the vector $\mathbf{r} = \begin{bmatrix} x, y, z \end{bmatrix}^{\mathrm{T}}$ which intersect at a point at infinity. The homogeneous coordinate transformation of points at infinity introduces a proper decomposition of the homogeneous transformation matrices:

$$ {}^{G}T_{B} = \begin{bmatrix} r_{11} & r_{12} & r_{13} & X_o \\ r_{21} & r_{22} & r_{23} & Y_o \\ r_{31} & r_{32} & r_{33} & Z_o \\ 0 & 0 & 0 & 1 \end{bmatrix} \tag{6.70} $$

The first three columns have zero as the fourth coordinate and represent points at infinity, which are the directions corresponding to the three coordinate axes. The fourth column has one as the fourth coordinate and represents the location of the coordinate frame origin.

Example 349 ★ General Homogeneous Transformation The homogeneous transformation (6.20) is a special case of the general transformation. The most general homogeneous transformation, which has been extensively used in the field of computer graphics, is

$$ \begin{aligned} {}^{A}T_{B} &= \left[\begin{array}{c|c} {}^{A}R_{B}(3\times 3) & {}^{A}\mathbf{d}(3\times 1) \\ \hline p(1\times 3) & w(1\times 1) \end{array}\right] \\ &= \left[\begin{array}{c|c} \text{rotation} & \text{translation} \\ \hline \text{perspective} & \text{scale factor} \end{array}\right] \end{aligned} \tag{6.71} $$

In rigid-body dynamics, we always take the last row vector of $[T]$ to be (0,0,0,1). However, a more general form of (6.71) could be useful, for example, when a graphical simulator or a vision system is added to a robotic system or an autonomous vehicle.

The upper left 3×3 submatrix ${}^{A}R_{B}$ denotes the orientation of a moving frame B with respect to another moving frame A. The upper right 3×1 submatrix ${}^{A}\mathbf{d}$ denotes the position of the origin of frame B relative to frame A. The lower left 1×3 submatrix p denotes a perspective transformation, and the lower right element w is a scaling factor.

Example 350 Rigid-Body and Corner Coordinates We may represent a rigid body by an array of homogeneous coordinates of specific points of the body, usually expressed in a body coordinate frame B. The specific points are usually the corners. Figure 6.6(a) illustrates the configuration of a box. The coordinates of the corners of the box in its body frame B are collected in the matrix $[P]$:

$$ \begin{aligned} {}^{B}P &= \begin{bmatrix} \mathbf{r}_1 & \mathbf{r}_2 & \mathbf{r}_3 & \mathbf{r}_4 & \mathbf{r}_5 & \mathbf{r}_6 & \mathbf{r}_7 & \mathbf{r}_8 \end{bmatrix} \\ &= \begin{bmatrix} 1 & 1 & 0 & 0 & 1 & 1 & 0 & 0 \\ 0 & 3 & 3 & 0 & 0 & 3 & 3 & 0 \\ 0 & 0 & 0 & 0 & 5 & 5 & 5 & 5 \\ 1 & 1 & 1 & 1 & 1 & 1 & 1 & 1 \end{bmatrix} \end{aligned} \tag{6.72} $$

The configuration of the box after a rotation of -90 deg about the Z-axis and a translation of three units along the X-axis is shown in Figure 6.6(b). The new

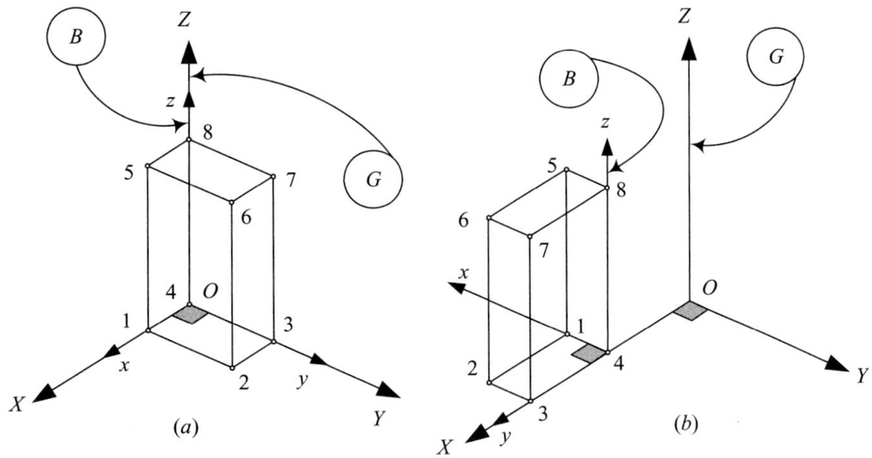

Figure 6.6 Describing the motion of a rigid body in terms of some body points.

coordinates of its corners in the global frame G are found by multiplying the corresponding transformation matrix by $[P]$:

$$
\begin{aligned}
{}^{G}T_B &= D_{X,3}R_{Z,90} \\
&= \begin{bmatrix} 1 & 0 & 0 & 3 \\ 0 & 1 & 0 & 0 \\ 0 & 0 & 1 & 0 \\ 0 & 0 & 0 & 1 \end{bmatrix} \begin{bmatrix} \cos(-90) & -\sin(-90) & 0 & 0 \\ \sin(-90) & \cos(-90) & 0 & 0 \\ 0 & 0 & 1 & 0 \\ 0 & 0 & 0 & 1 \end{bmatrix} \\
&= \begin{bmatrix} 0 & 1 & 0 & 3 \\ -1 & 0 & 0 & 0 \\ 0 & 0 & 1 & 0 \\ 0 & 0 & 0 & 1 \end{bmatrix}
\end{aligned}
\tag{6.73}
$$

Therefore, the global coordinates of corners 1–6 after motion are

$$
\begin{aligned}
{}^{G}P &= {}^{G}T_B \, {}^{B}P \\
&= \begin{bmatrix} 0 & 1 & 0 & 3 \\ -1 & 0 & 0 & 0 \\ 0 & 0 & 1 & 0 \\ 0 & 0 & 0 & 1 \end{bmatrix} \begin{bmatrix} 1 & 1 & 0 & 0 & 1 & 1 & 0 & 0 \\ 0 & 3 & 3 & 0 & 0 & 3 & 3 & 0 \\ 0 & 0 & 0 & 0 & 5 & 5 & 5 & 5 \\ 1 & 1 & 1 & 1 & 1 & 1 & 1 & 1 \end{bmatrix} \\
&= \begin{bmatrix} 3 & 6 & 6 & 3 & 3 & 6 & 6 & 3 \\ -1 & -1 & 0 & 0 & -1 & -1 & 0 & 0 \\ 0 & 0 & 0 & 0 & 5 & 5 & 5 & 5 \\ 1 & 1 & 1 & 1 & 1 & 1 & 1 & 1 \end{bmatrix}
\end{aligned}
\tag{6.74}
$$

Example 351 ★ Rotation and Translation in a Plane Homogeneous transformation (6.20) will be simplified to

$$ {}^{G}T_{B} = \begin{bmatrix} \cos\theta & -\sin\theta & 0 & X_o \\ \sin\theta & \cos\theta & 0 & Y_o \\ 0 & 0 & 1 & 0 \\ 0 & 0 & 0 & 1 \end{bmatrix} \equiv \left[\begin{array}{c|c} {}^{G}R_{Z,\theta} & {}^{G}\mathbf{d} \\ \hline 0\ 0\ 0 & 1 \end{array}\right] \tag{6.75} $$

for a motion parallel to the (X,Y)-plane. Consider a planar body that is displaced from position 1 to position 2 according to Figure 6.7.

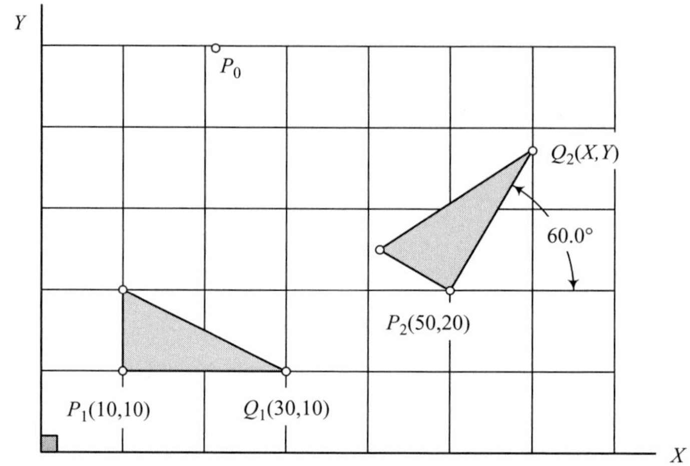

Figure 6.7 Motion in a plane.

New coordinates of Q_2 are

$$ \begin{aligned} \mathbf{r}_{Q_2} &= {}^{2}R_{1}\left(\mathbf{r}_{Q_1} - \mathbf{r}_{P_1}\right) + \mathbf{r}_{P_2} \\ &= \begin{bmatrix} \cos 60 & -\sin 60 & 0 \\ \sin 60 & \cos 60 & 0 \\ 0 & 0 & 1 \end{bmatrix} \left(\begin{bmatrix} 30 \\ 10 \\ 0 \end{bmatrix} - \begin{bmatrix} 10 \\ 10 \\ 0 \end{bmatrix} \right) + \begin{bmatrix} 50 \\ 20 \\ 0 \end{bmatrix} \\ &= \begin{bmatrix} 10 \\ 17.321 \\ 0 \end{bmatrix} + \begin{bmatrix} 50 \\ 20 \\ 0 \end{bmatrix} = \begin{bmatrix} 60 \\ 37.321 \\ 0 \end{bmatrix} \end{aligned} \tag{6.76} $$

or equivalently

$$ \begin{aligned} \mathbf{r}_{Q_2} &= {}^{2}T_{1}\,\mathbf{r}_{Q_1} = \begin{bmatrix} {}^{2}R_{1} & \mathbf{r}_{P_2} - {}^{2}R_{1}\,\mathbf{r}_{P_1} \\ 0 & 1 \end{bmatrix} \mathbf{r}_{Q_1} \\ &= \begin{bmatrix} \cos 60 & -\sin 60 & 0 & 53.66 \\ \sin 60 & \cos 60 & 0 & 6.3397 \\ 0 & 0 & 1 & 0 \\ 0 & 0 & 0 & 1 \end{bmatrix} \begin{bmatrix} 30 \\ 10 \\ 0 \\ 1 \end{bmatrix} = \begin{bmatrix} 60 \\ 37.32 \\ 0 \\ 1 \end{bmatrix} \end{aligned} \tag{6.77} $$

Example 352 ★ Pole of Planar Motion Any planar motion of a rigid body can be substituted with a rotation about a point. Such a point is called the finite rotation pole. Assume that the plane of motion is the (X,Y)-plane. The pole can be determined from the transformation matrix of the planar motion:

$$ {}^{G}T_{B} = \begin{bmatrix} \cos\theta & -\sin\theta & 0 & d_1 \\ \sin\theta & \cos\theta & 0 & d_2 \\ 0 & 0 & 1 & 0 \\ 0 & 0 & 0 & 1 \end{bmatrix} \tag{6.78} $$

The components d_1 and d_2 are the given displacements of a body point. To locate the pole of motion $P_0(X_0, Y_0)$, we search for a point that will not move under the transformation:

$$ \mathbf{r}_{P_0} = {}^{2}T_{1}\,\mathbf{r}_{P_0} \tag{6.79} $$

$$ \begin{bmatrix} X_0 \\ Y_0 \\ 0 \\ 1 \end{bmatrix} = \begin{bmatrix} \cos\theta & -\sin\theta & 0 & d_1 \\ \sin\theta & \cos\theta & 0 & d_2 \\ 0 & 0 & 1 & 0 \\ 0 & 0 & 0 & 1 \end{bmatrix} \begin{bmatrix} X_0 \\ Y_0 \\ 0 \\ 1 \end{bmatrix} $$

Therefore, we will have two equations to determine the coordinates of the pole:

$$ X_0 = d_1 + X_0\cos\theta - Y_0\sin\theta \tag{6.80} $$

$$ Y_0 = d_2 + Y_0\cos\theta + X_0\sin\theta \tag{6.81} $$

The matrix form of the equations,

$$ \begin{bmatrix} 1-\cos\theta & \sin\theta \\ -\sin\theta & 1-\cos\theta \end{bmatrix} \begin{bmatrix} X_0 \\ Y_0 \end{bmatrix} = \begin{bmatrix} d_1 \\ d_2 \end{bmatrix} \tag{6.82} $$

provides the required solutions:

$$ \begin{bmatrix} X_0 \\ Y_0 \end{bmatrix} = \begin{bmatrix} 1-\cos\theta & \sin\theta \\ -\sin\theta & 1-\cos\theta \end{bmatrix}^{-1} \begin{bmatrix} d_1 \\ d_2 \end{bmatrix} = \frac{1}{2}\begin{bmatrix} d_1 + d_2\dfrac{\sin\theta}{\cos\theta - 1} \\ -d_1\dfrac{\sin\theta}{\cos\theta - 1} + d_2 \end{bmatrix} \tag{6.83} $$

As an example, the transformation matrix in Example 351 is

$$ {}^{G}T_{B} = \begin{bmatrix} 0.5 & -0.86603 & 0 & 53.66 \\ 0.86603 & 0.5 & 0 & 6.3397 \\ 0 & 0 & 1.0 & 0 \\ 0 & 0 & 0 & 1.0 \end{bmatrix} $$

which generates two equations to determine the position of the pole of the motion:

$$X_0 = 0.5X_0 - 0.86603Y_0 + 53.66 \tag{6.84}$$

$$Y_0 = 0.86603X_0 + 0.5Y_0 + 6.3397 \tag{6.85}$$

$$\begin{bmatrix} X_0 \\ Y_0 \end{bmatrix} = \begin{bmatrix} 21.339 \\ 49.641 \end{bmatrix} \tag{6.86}$$

6.3 INVERSE AND REVERSE HOMOGENEOUS TRANSFORMATION

If we show a rigid-body motion by the homogeneous transformation

$${}^G T_B = \begin{bmatrix} \mathbf{I} & {}^G\mathbf{d} \\ 0 & 1 \end{bmatrix} \begin{bmatrix} {}^G R_B & 0 \\ 0 & 1 \end{bmatrix} = \begin{bmatrix} {}^G R_B & {}^G\mathbf{d} \\ 0 & 1 \end{bmatrix} \tag{6.87}$$

then the *inverse* of ${}^G T_B$ would be

$${}^B T_G = {}^G T_B^{-1} = \begin{bmatrix} {}^G R_B & {}^G\mathbf{d} \\ 0 & 1 \end{bmatrix}^{-1} = \begin{bmatrix} {}^G R_B^{\mathrm{T}} & -{}^G R_B^{\mathrm{T}}\,{}^G\mathbf{d} \\ 0 & 1 \end{bmatrix} \tag{6.88}$$

and the *reverse* motion of ${}^G T_B$ would be

$$\begin{aligned} {}^G T_{-B} &= \begin{bmatrix} {}^G R_B^{\mathrm{T}} & 0 \\ 0 & 1 \end{bmatrix} \begin{bmatrix} \mathbf{I} & -{}^G\mathbf{d} \\ 0 & 1 \end{bmatrix} \\ &= \begin{bmatrix} {}^G R_B^{\mathrm{T}} & -{}^G R_B^{\mathrm{T}}\,{}^G\mathbf{d} \\ 0 & 1 \end{bmatrix} \end{aligned} \tag{6.89}$$

which yields

$${}^G T_B^{-1}\,{}^G T_B = \mathbf{I}_4 \tag{6.90}$$

$${}^G T_{-B}\,{}^G T_B = \mathbf{I}_4 \tag{6.91}$$

The homogeneous transformation matrix simplifies the rigid-body motion; however, a shortcoming is that they lose the orthogonality property because a transformation matrix is not orthogonal and its inverse is not equal to its transpose:

$${}^G T_B^{-1} \neq {}^G T_B^{\mathrm{T}} \tag{6.92}$$

Although it is traditional to use the inverse matrix notation ${}^G T_B^{-1}$ for ${}^B T_G$,

$${}^G T_B^{-1} = {}^B T_G \tag{6.93}$$

calculating ${}^G T_B^{-1}$ must be done according to Equation (6.88), and not by regular matrix inversion of a 4×4 matrix. The matrix inversion notation makes equations consistent with the multiplication of a matrix $[T]$ by its inverse, $[T]^{-1}$, because

$${}^G T_B^{-1}\,{}^G T_B = {}^B T_G\,{}^G T_B = \mathbf{I}_4 \tag{6.94}$$

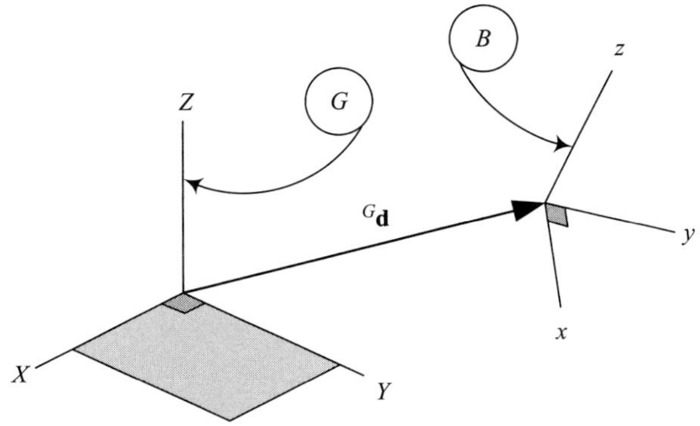

Figure 6.8 Illustration of a rotated and translated body frame $B(oxyz)$ with respect to the global frame $G(OXYZ)$.

Proof: A rotated and translated body frame $B(oxyz)$ with respect to the global frame $G(OXYZ)$ is depicted in Figure 6.8. Transformation of the coordinates of a point P from the global frame to the body frame is ${}^BT_G$, which is the inverse of the transformation ${}^GT_B$.

Starting with the expression of ${}^G\mathbf{r}$ and the definition of ${}^GT_B$ for mapping ${}^B\mathbf{r}$ to ${}^G\mathbf{r}$,

$$ {}^G\mathbf{r} = {}^GR_B\,{}^B\mathbf{r} + {}^G\mathbf{d} = {}^GT_B\,{}^B\mathbf{r} \tag{6.95} $$

$$ \begin{bmatrix} {}^G\mathbf{r} \\ 1 \end{bmatrix} = \begin{bmatrix} {}^GR_B & {}^G\mathbf{d} \\ 0 & 1 \end{bmatrix} \begin{bmatrix} {}^B\mathbf{r} \\ 1 \end{bmatrix} \tag{6.96} $$

we find

$$ {}^B\mathbf{r} = {}^GR_B^{-1}\left({}^G\mathbf{r} - {}^G\mathbf{d}\right) = {}^GR_B^{\mathrm{T}}\,{}^G\mathbf{r} - {}^GR_B^{\mathrm{T}}\,{}^G\mathbf{d} \tag{6.97} $$

to express the transformation matrix ${}^BT_G$ for mapping ${}^G\mathbf{r}$ to ${}^B\mathbf{r}$:

$$ {}^BT_G = {}^GT_B^{-1} = \begin{bmatrix} {}^GR_B^{\mathrm{T}} & -{}^GR_B^{\mathrm{T}}\,{}^G\mathbf{d} \\ 0 & 1 \end{bmatrix} \tag{6.98} $$

Furthermore, a matrix multiplication indicates that

$$ \begin{aligned} {}^GT_B^{-1}\,{}^GT_B &= \begin{bmatrix} {}^GR_B^{\mathrm{T}} & -{}^GR_B^{\mathrm{T}}\,{}^G\mathbf{d} \\ 0 & 1 \end{bmatrix} \begin{bmatrix} {}^GR_B & {}^G\mathbf{d} \\ 0 & 1 \end{bmatrix} \\ &= \begin{bmatrix} {}^GR_B^{\mathrm{T}}\,{}^GR_B & {}^GR_B^{\mathrm{T}}\,{}^G\mathbf{d} - {}^GR_B^{\mathrm{T}}\,{}^G\mathbf{d} \\ 0 & 1 \end{bmatrix} = \begin{bmatrix} \mathbf{I}_3 & 0 \\ 0 & 1 \end{bmatrix} = \mathbf{I}_4 \end{aligned} \tag{6.99} $$

Recalling that a homogeneous motion ${}^GT_B$ is equal to a rotation ${}^GR_B$ plus a translation ${}^GD_B$,

$$ {}^GT_B = {}^GD_B\,{}^GR_B = \begin{bmatrix} \mathbf{I} & {}^G\mathbf{d} \\ 0 & 1 \end{bmatrix} \begin{bmatrix} {}^GR_B & \mathbf{0} \\ 0 & 1 \end{bmatrix} \tag{6.99a} $$

we define the reverse motion by a translation $-{}^GD_B$ followed by a reverse rotation ${}^GR_{-B} = {}^GR_B^{\mathrm{T}}$:

$$
{}^GT_{-B} = \left[{}^GR_{-B}\right]\left[-{}^GD_B\right] = \begin{bmatrix} {}^GR_B^{\mathrm{T}} & \mathbf{0} \\ 0 & 1 \end{bmatrix} \begin{bmatrix} \mathbf{I} & -{}^G\mathbf{d} \\ 0 & 1 \end{bmatrix} \tag{6.99b}
$$

The result of a rigid-body motion ${}^GT_B$ and then the reverse motion ${}^GT_{-B}$ would be an identity matrix:

$$
\begin{aligned}
{}^GT_{-B}\,{}^GT_B &= {}^GT_B\,{}^GT_{-B} \\
&= \begin{bmatrix} {}^GR_B^{\mathrm{T}} & \mathbf{0} \\ 0 & 1 \end{bmatrix} \begin{bmatrix} \mathbf{I} & -{}^G\mathbf{d} \\ 0 & 1 \end{bmatrix} \begin{bmatrix} \mathbf{I} & {}^G\mathbf{d} \\ 0 & 1 \end{bmatrix} \begin{bmatrix} {}^GR_B & \mathbf{0} \\ 0 & 1 \end{bmatrix} \\
&= \begin{bmatrix} {}^GR_B^{\mathrm{T}} & \mathbf{0} \\ 0 & 1 \end{bmatrix} \begin{bmatrix} \mathbf{I}_3 & 0 \\ 0 & 1 \end{bmatrix} \begin{bmatrix} {}^GR_B & \mathbf{0} \\ 0 & 1 \end{bmatrix} = \mathbf{I}_4
\end{aligned} \tag{6.100}
$$

■

Example 353 Alternative Proof for ${}^G\boldsymbol{T}_{\boldsymbol{B}}^{-1}$ ${}^BT_G$ can be found by a geometric expression. Consider the inverse of the rotation matrix ${}^GR_B$,

$$
{}^GR_B^{-1} = {}^GR_B^{\mathrm{T}} = {}^BR_G \tag{6.101}
$$

and the reverse of ${}^G\mathbf{d}$ as the vector ${}^B\mathbf{d}$ to indicate the origin of the global frame with respect to the origin of the body frame,

$$
{}^B\mathbf{d} = {}^BR_G\,{}^G\mathbf{d} = {}^GR_B^{\mathrm{T}}\,{}^G\mathbf{d} \tag{6.102}
$$

They allow us to define the homogeneous transformation ${}^BT_G$:

$$
{}^BT_G = \begin{bmatrix} {}^BR_G & -{}^B\mathbf{d} \\ 0 & 1 \end{bmatrix} = \begin{bmatrix} {}^GR_B^{\mathrm{T}} & -{}^GR_B^{\mathrm{T}}\,{}^G\mathbf{d} \\ 0 & 1 \end{bmatrix} \tag{6.103}
$$

Example 354 Inverse and Reverse of a Homogeneous Matrix Assume that

$$
{}^GT_B = \begin{bmatrix} 1 & 0 & 0 & -1 \\ 0 & \cos 60 & -\sin 60 & 5 \\ 0 & \sin 60 & \cos 60 & 2 \\ 0 & 0 & 0 & 1 \end{bmatrix} = \begin{bmatrix} {}^GR_B & {}^G\mathbf{d} \\ 0 & 1 \end{bmatrix} \tag{6.104}
$$

Then

$$
{}^GR_B = \begin{bmatrix} 1 & 0 & 0 \\ 0 & \cos 60 & -\sin 60 \\ 0 & \sin 60 & \cos 60 \end{bmatrix} \qquad {}^G\mathbf{d} = \begin{bmatrix} -1 \\ 5 \\ 2 \end{bmatrix} \tag{6.105}
$$

The inverse transformation is

$$
{}^{B}T_{G} = {}^{G}T_{B}^{-1} = \begin{bmatrix} {}^{G}R_{B}^{\mathrm{T}} & -{}^{G}R_{B}^{\mathrm{T}}\,{}^{G}\mathbf{d} \\ 0 & 1 \end{bmatrix}
$$

$$
= \begin{bmatrix} 1 & 0 & 0 & 1 \\ 0 & -0.95241 & -0.30481 & 5.3717 \\ 0 & 0.30481 & -0.95241 & 0.38077 \\ 0 & 0 & 0 & 1 \end{bmatrix} \tag{6.106}
$$

and the reverse transformation is

$$
{}^{G}T_{-B} = \begin{bmatrix} {}^{G}R_{-B} & 0 \\ 0 & 1 \end{bmatrix} \begin{bmatrix} \mathbf{I} & -{}^{G}\mathbf{d} \\ 0 & 1 \end{bmatrix}
$$

$$
= \begin{bmatrix} 1 & 0 & 0 & 0 \\ 0 & \cos -60 & -\sin -60 & 0 \\ 0 & \sin -60 & \cos -60 & 0 \\ 0 & 0 & 0 & 1 \end{bmatrix} \begin{bmatrix} 1 & 0 & 0 & 1 \\ 0 & 1 & 0 & -5 \\ 0 & 0 & 1 & -2 \\ 0 & 0 & 0 & 1 \end{bmatrix}
$$

$$
= \begin{bmatrix} 1 & 0 & 0 & 1 \\ 0 & -0.95241 & -0.30481 & 5.3717 \\ 0 & 0.30481 & -0.95241 & 0.38077 \\ 0 & 0 & 0 & 1 \end{bmatrix} \tag{6.107}
$$

Example 355 Transformation Matrix and Coordinate of Points We are able to determine the rigid-body transformation matrix if we have the coordinates of four points of the body in both B and G frames.

Assume A, B, C, and D are four points with the following coordinates in two different frames:

$$
A_1(2,4,1) \qquad B_1(2,6,1) \qquad C_1(1,5,1) \qquad D_1(3,5,2) \tag{6.108}
$$

$$
A_2(5,1,1) \qquad B_2(7,1,1) \qquad C_2(6,2,1) \qquad D_2(6,2,3) \tag{6.109}
$$

The homogeneous transformation matrix $[T]$ maps the coordinates between the two frames,

$$
[T] \begin{bmatrix} 2 & 2 & 1 & 3 \\ 4 & 6 & 5 & 5 \\ 1 & 1 & 2 & 2 \\ 1 & 1 & 1 & 1 \end{bmatrix} = \begin{bmatrix} 5 & 7 & 6 & 6 \\ 1 & 1 & 2 & 2 \\ 1 & 1 & 1 & 3 \\ 1 & 1 & 1 & 1 \end{bmatrix} \tag{6.110}
$$

and hence

$$
[T] = \begin{bmatrix} 5 & 7 & 6 & 6 \\ 1 & 1 & 2 & 2 \\ 1 & 1 & 1 & 3 \\ 1 & 1 & 1 & 1 \end{bmatrix} \begin{bmatrix} 2 & 2 & 1 & 3 \\ 4 & 6 & 5 & 5 \\ 1 & 1 & 2 & 2 \\ 1 & 1 & 1 & 1 \end{bmatrix}^{-1} = \begin{bmatrix} 0 & 1 & 0 & 1 \\ 0 & 0 & 1 & 0 \\ 1 & 0 & 1 & -2 \\ 0 & 0 & 0 & 1 \end{bmatrix} \tag{6.111}
$$

Example 356 Quick Inverse Transformation Decomposition of a transformation matrix into translation and rotation provides a practical numerical method for quick inversion of the matrix.

Consider the transformation matrix

$$[T] = \begin{bmatrix} r_{11} & r_{12} & r_{13} & r_{14} \\ r_{21} & r_{22} & r_{23} & r_{24} \\ r_{31} & r_{32} & r_{33} & r_{34} \\ 0 & 0 & 0 & 1 \end{bmatrix} = [D][R]$$

$$= \begin{bmatrix} 1 & 0 & 0 & r_{14} \\ 0 & 1 & 0 & r_{24} \\ 0 & 0 & 1 & r_{34} \\ 0 & 0 & 0 & 1 \end{bmatrix} \begin{bmatrix} r_{11} & r_{12} & r_{13} & 0 \\ r_{21} & r_{22} & r_{23} & 0 \\ r_{31} & r_{32} & r_{33} & 0 \\ 0 & 0 & 0 & 1 \end{bmatrix} \tag{6.112}$$

Therefore,

$$[T]^{-1} = [DR]^{-1} = R^{-1}D^{-1} = R^{\mathrm{T}}D^{-1}$$

$$= \begin{bmatrix} r_{11} & r_{21} & r_{31} & 0 \\ r_{12} & r_{22} & r_{32} & 0 \\ r_{13} & r_{23} & r_{33} & 0 \\ 0 & 0 & 0 & 1 \end{bmatrix} \begin{bmatrix} 1 & 0 & 0 & -r_{14} \\ 0 & 1 & 0 & -r_{24} \\ 0 & 0 & 1 & -r_{34} \\ 0 & 0 & 0 & 1 \end{bmatrix}$$

$$= \begin{bmatrix} r_{11} & r_{21} & r_{31} & -r_{11}r_{14} - r_{21}r_{24} - r_{31}r_{34} \\ r_{12} & r_{22} & r_{32} & -r_{12}r_{14} - r_{22}r_{24} - r_{32}r_{34} \\ r_{13} & r_{23} & r_{33} & -r_{13}r_{14} - r_{23}r_{24} - r_{33}r_{34} \\ 0 & 0 & 0 & 1 \end{bmatrix} \tag{6.113}$$

Example 357 ★ Inverse of a Combined Matrix Consider a matrix $[T]$ that is a combination of four submatrices $[A]$, $[B]$, $[C]$, and $[D]$ such that

$$[T] = \begin{bmatrix} A & B \\ C & D \end{bmatrix} \tag{6.114}$$

The inverse is given by

$$T^{-1} = \begin{bmatrix} A^{-1} + A^{-1}BE^{-1}CA^{-1} & -A^{-1}BE^{-1} \\ -E^{-1}CA^{-1} & E^{-1} \end{bmatrix} \tag{6.115}$$

where

$$[E] = D - CA^{-1}B \tag{6.116}$$

In the case of the general homogeneous transformation

$$ {}^{A}T_{B} = \left[\begin{array}{c|c} {}^{A}R_{B}(3\times 3) & {}^{A}\mathbf{d}(3\times 1) \\ \hline p(1\times 3) & w(1\times 1) \end{array}\right] \tag{6.117}$$

$$= \left[\begin{array}{c|c} \text{rotation} & \text{translation} \\ \hline \text{perspective} & \text{scale factor} \end{array}\right]$$

we have

$$[T] = {}^A T_B \tag{6.118}$$

$$[A] = {}^A R_B \tag{6.119}$$

$$[B] = {}^A \mathbf{d} \tag{6.120}$$

$$[C] = \begin{bmatrix} p_1 & p_2 & p_3 \end{bmatrix} \tag{6.121}$$

$$[D] = \begin{bmatrix} w_{1\times 1} \end{bmatrix} = w \tag{6.122}$$

and therefore,

$$\begin{aligned} [E] = [E_{1\times 1}] = E &= w - \begin{bmatrix} p_1 & p_2 & p_3 \end{bmatrix} {}^A R_B^{-1}\, {}^A \mathbf{d} \\ &= w - \begin{bmatrix} p_1 & p_2 & p_3 \end{bmatrix} {}^A R_B^{\mathrm{T}}\, {}^A \mathbf{d} \\ &= 1 - p_1(d_1 r_{11} + d_2 r_{21} + d_3 r_{31}) \\ &\quad + p_2(d_1 r_{12} + d_2 r_{22} + d_3 r_{32}) \\ &\quad + p_3(d_1 r_{13} + d_2 r_{23} + d_3 r_{33}) \end{aligned} \tag{6.123}$$

$$-E^{-1} C A^{-1} = \frac{1}{E} \begin{bmatrix} g_1 & g_2 & g_3 \end{bmatrix} \tag{6.124}$$

where

$$g_1 = p_1 r_{11} + p_2 r_{12} + p_3 r_{13} \tag{6.124a}$$

$$g_2 = p_1 r_{21} + p_2 r_{22} + p_3 r_{23} \tag{6.124b}$$

$$g_3 = p_1 r_{31} + p_2 r_{32} + p_3 r_{33} \tag{6.124c}$$

and

$$-A^{-1} B E^{-1} = -\frac{1}{E} \begin{bmatrix} d_1 r_{11} + d_2 r_{21} + d_3 r_{31} \\ d_1 r_{12} + d_2 r_{22} + d_3 r_{32} \\ d_1 r_{13} + d_2 r_{23} + d_3 r_{33} \end{bmatrix} \tag{6.125}$$

$$A^{-1} + A^{-1} B E^{-1} C A^{-1} = [F] = \begin{bmatrix} f_{11} & f_{12} & f_{13} \\ f_{21} & f_{22} & f_{23} \\ f_{31} & f_{32} & r_{33} \end{bmatrix} \tag{6.126}$$

where

$$f_{11} = r_{11} + \frac{1}{E}(p_1 r_{11} + p_2 r_{12} + p_3 r_{13})(d_1 r_{11} + d_2 r_{21} + d_3 r_{31}) \tag{6.126a}$$

$$f_{12} = r_{21} + \frac{1}{E}(p_1 r_{21} + p_2 r_{22} + p_3 r_{23})(d_1 r_{11} + d_1 r_{21} + d_1 r_{31}) \tag{6.126b}$$

$$f_{13} = r_{31} + \frac{1}{E}(p_1 r_{31} + p_2 r_{32} + p_3 r_{33})(d_1 r_{11} + d_2 r_{21} + d_3 r_{31}) \tag{6.126c}$$

$$f_{21} = r_{12} + \frac{1}{E}(p_1 r_{11} + p_2 r_{12} + p_3 r_{13})(d_1 r_{12} + d_2 r_{22} + d_3 r_{32}) \tag{6.126d}$$

$$f_{22} = r_{22} + \frac{1}{E}(p_1 r_{21} + p_2 r_{22} + p_3 r_{23})(d_1 r_{12} + d_2 r_{22} + d_3 r_{32}) \tag{6.126e}$$

$$f_{23} = r_{32} + \frac{1}{E}(p_1 r_{31} + p_2 r_{32} + p_3 r_{33})(d_1 r_{12} + d_2 r_{22} + d_3 r_{32}) \tag{6.126f}$$

$$f_{31} = r_{13} + \frac{1}{E}(p_1 r_{11} + p_2 r_{12} + p_3 r_{13})(d_1 r_{13} + d_2 r_{23} + d_3 r_{33}) \tag{6.126g}$$

$$f_{32} = r_{23} + \frac{1}{E}(p_1 r_{21} + p_2 r_{22} + p_3 r_{23})(d_1 r_{13} + d_2 r_{23} + d_3 r_{33}) \tag{6.126h}$$

$$f_{33} = r_{33} + \frac{1}{E}(p_1 r_{31} + p_2 r_{32} + p_3 r_{33})(d_1 r_{13} + d_2 r_{23} + d_3 r_{33}). \tag{6.126i}$$

This method for a coordinate homogeneous transformation

$$[T] = {}^AT_B \tag{6.127}$$

$$[A] = {}^AR_B \tag{6.128}$$

$$[B] = {}^A\mathbf{d} \tag{6.129}$$

$$[C] = \begin{bmatrix} 0 & 0 & 0 \end{bmatrix} \tag{6.130}$$

$$[D] = [1] \tag{6.131}$$

reduces to

$$[E] = [1] \tag{6.132}$$

$${}^AT_B^{-1} = \begin{bmatrix} {}^AR_B^{\mathrm{T}} & -{}^AR_B^{\mathrm{T}}\,{}^A\mathbf{d} \\ 0 & 1 \end{bmatrix} \tag{6.133}$$

6.4 COMPOUND HOMOGENEOUS TRANSFORMATION

Figure 6.9 shows three reference frames: A, B, and C. The transformation matrices to transform coordinates from frame B to A and from frame C to B are

$${}^AT_B = \begin{bmatrix} {}^AR_B & {}^A\mathbf{d}_1 \\ 0 & 1 \end{bmatrix} \qquad {}^BT_C = \begin{bmatrix} {}^BR_C & {}^B\mathbf{d}_2 \\ 0 & 1 \end{bmatrix} \tag{6.134}$$

The transformation matrix from C to A is

$${}^AT_C = {}^AT_B\,{}^BT_C = \begin{bmatrix} {}^AR_B & {}^A\mathbf{d}_1 \\ 0 & 1 \end{bmatrix}\begin{bmatrix} {}^BR_C & {}^B\mathbf{d}_2 \\ 0 & 1 \end{bmatrix}$$

$$= \begin{bmatrix} {}^AR_B\,{}^BR_C & {}^AR_B\,{}^B\mathbf{d}_2 + {}^A\mathbf{d}_1 \\ 0 & 1 \end{bmatrix} = \begin{bmatrix} {}^AR_C & {}^A\mathbf{d}_2 \\ 0 & 1 \end{bmatrix} \tag{6.135}$$

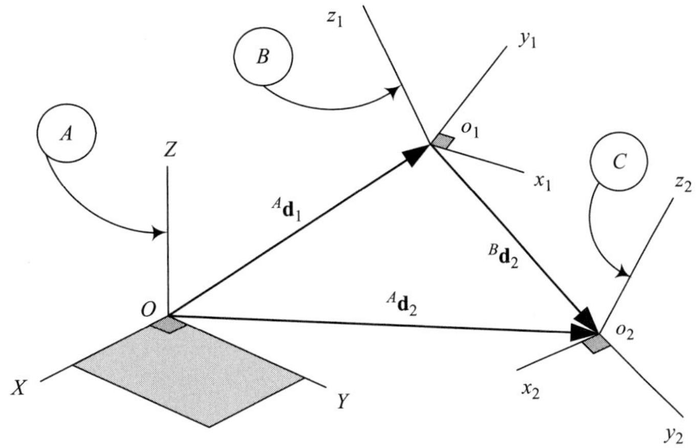

Figure 6.9 Three coordinate frames to analyze compound transformations.

and therefore, the inverse transformation is

$$
\begin{aligned}
{}^C T_A &= \begin{bmatrix} {}^B R_C^{\mathrm{T}} \, {}^A R_B^{\mathrm{T}} & -{}^B R_C^{\mathrm{T}} \, {}^A R_B^{\mathrm{T}} \left[{}^A R_B \, {}^B \mathbf{d}_2 + {}^A \mathbf{d}_1 \right] \\ 0 & 1 \end{bmatrix} \\
&= \begin{bmatrix} {}^B R_C^{\mathrm{T}} \, {}^A R_B^{\mathrm{T}} & -{}^B R_C^{\mathrm{T}} \, {}^B \mathbf{d}_2 \; - \; {}^A R_C^{\mathrm{T}} \, {}^A \mathbf{d}_1 \\ 0 & 1 \end{bmatrix} \\
&= \begin{bmatrix} {}^A R_C^{\mathrm{T}} & -{}^A R_C^{\mathrm{T}} \, {}^A \mathbf{d}_2 \\ 0 & 1 \end{bmatrix}
\end{aligned}
\tag{6.136}
$$

The homogeneous transformations, similar to rotation matrices, follow the rule of in-order transformations. As an example, the combination of four transformations

$$
{}^G T_4 = {}^G T_1 \, {}^1 T_2 \, {}^2 T_3 \, {}^3 T_4 \tag{6.137}
$$

would be equivalent to only one transformation:

$$
\begin{aligned}
{}^G T_4 &= {}^G R_4 \, {}^4 \mathbf{r}_P + {}^G \mathbf{d}_4 \\
&= {}^G R_1 \left({}^1 R_2 \left({}^2 R_3 \left({}^3 R_4 \, {}^4 \mathbf{r}_P + {}^3 \mathbf{d}_4 \right) + {}^2 \mathbf{d}_3 \right) + {}^1 \mathbf{d}_2 \right) + {}^G \mathbf{d}_1
\end{aligned}
\tag{6.138}
$$

Example 358 Homogeneous Transformation for Multiple Frames There are two relatively moving rigid-body frames B_1 and B_2 and a global frame G in Figure 6.10. A point P in the local frame $B_2(x_2 y_2 z_2)$ is at ${}^2\mathbf{r}_P$. The coordinates of P in the global frame $G(OXYZ)$ can be found by using the homogeneous transformation matrices.

The position of P in frame $B_1(x_1 y_1 z_1)$ is

$$
{}^1 \mathbf{r}_P = \begin{bmatrix} x_1 \\ y_1 \\ z_1 \\ 1 \end{bmatrix} = \begin{bmatrix} {}^1 R_2 & {}^1 \mathbf{d}_2 \\ 0 & 1 \end{bmatrix} \begin{bmatrix} x_2 \\ y_2 \\ z_2 \\ 1 \end{bmatrix} \tag{6.139}
$$

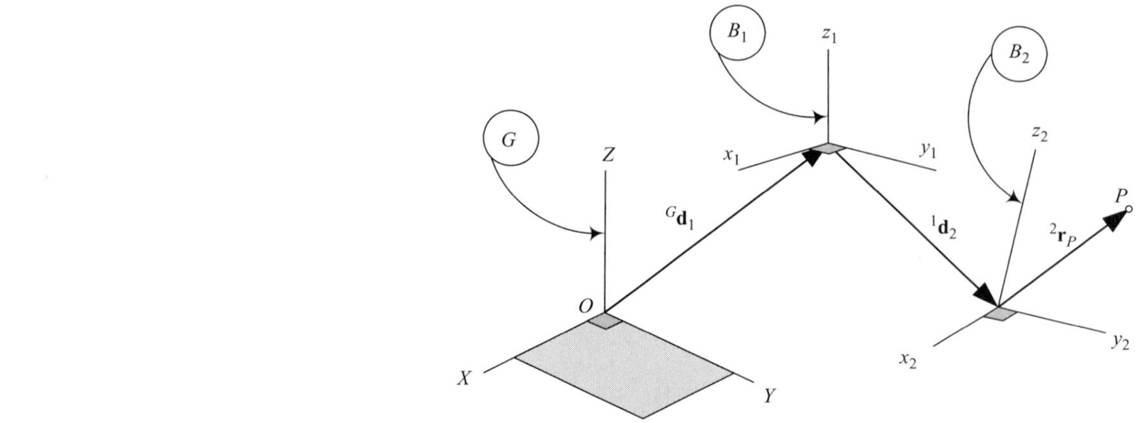

Figure 6.10 Point P in a local frame $B_2\ (x_2y_2z_2)$.

and therefore, its position in the global frame $G(OXYZ)$ would be

$$
\begin{aligned}
{}^G\mathbf{r}_P = \begin{bmatrix} X \\ Y \\ Z \\ 1 \end{bmatrix} &= \begin{bmatrix} {}^GR_1 & {}^G\mathbf{d}_1 \\ 0 & 1 \end{bmatrix} \begin{bmatrix} x_1 \\ y_1 \\ z_1 \\ 1 \end{bmatrix} \\
&= \begin{bmatrix} {}^GR_1 & {}^G\mathbf{d}_1 \\ 0 & 1 \end{bmatrix} \begin{bmatrix} {}^1R_2 & {}^1\mathbf{d}_2 \\ 0 & 1 \end{bmatrix} \begin{bmatrix} x_2 \\ y_2 \\ z_2 \\ 1 \end{bmatrix} \\
&= \begin{bmatrix} {}^GR_1\,{}^1R_2 & {}^GR_1\,{}^1\mathbf{d}_2 + {}^G\mathbf{d}_1 \\ 0 & 1 \end{bmatrix} \begin{bmatrix} x_2 \\ y_2 \\ z_2 \\ 1 \end{bmatrix}
\end{aligned} \tag{6.140}
$$

Example 359 Rotation about an Axis Not Going through Origin The homogeneous transformation matrix can be used to determine the coordinates of body points after rotation about an axis not going through the origin. Such an axis is called an off-center axis. Figure 6.11 indicates an angle of rotation ϕ about the axis $\hat{u}$ passing through a point Q at a position ${}^G\mathbf{d}_Q$.

To find the transformation matrix ${}^GT_B$, we set a local frame B at point Q parallel to the global frame G. Then, a rotation around $\hat{u}$ can be expressed by a translation along $-\mathbf{d}$ to bring the body fame B to the global frame G followed by a homogeneous transformation that is a rotation about $\hat{u}$ and a translation along $\mathbf{d}$,

$$
\begin{aligned}
{}^GT_B = D_{\hat{d},d}\, R_{\hat{u},\phi}\, D_{\hat{d},-d} &= \begin{bmatrix} \mathbf{I} & \mathbf{d} \\ 0 & 1 \end{bmatrix} \begin{bmatrix} R_{\hat{u},\phi} & 0 \\ 0 & 1 \end{bmatrix} \begin{bmatrix} \mathbf{I} & -\mathbf{d} \\ 0 & 1 \end{bmatrix} \\
&= \begin{bmatrix} R_{\hat{u},\phi} & \mathbf{d} - R_{\hat{u},\phi}\mathbf{d} \\ 0 & 1 \end{bmatrix}
\end{aligned} \tag{6.141}
$$

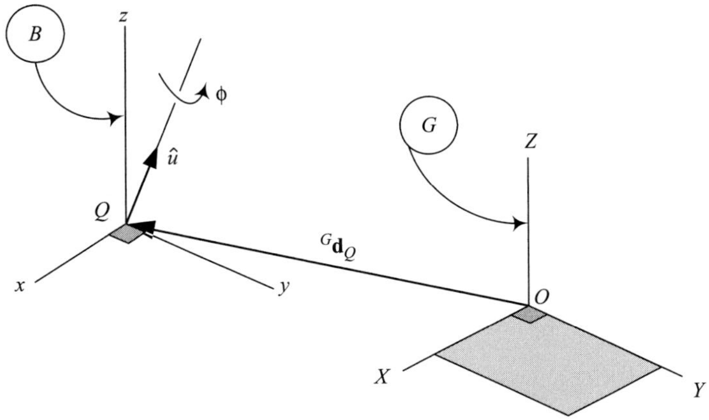

Figure 6.11 Rotation about an axis not going through origin.

where

$$R_{\hat{u},\phi} = \begin{bmatrix} u_1^2 \operatorname{vers}\phi + c\phi & u_1u_2 \operatorname{vers}\phi - u_3 s\phi & u_1u_3 \operatorname{vers}\phi + u_2 s\phi \\ u_1u_2 \operatorname{vers}\phi + u_3 s\phi & u_2^2 \operatorname{vers}\phi + c\phi & u_2u_3 \operatorname{vers}\phi - u_1 s\phi \\ u_1u_3 \operatorname{vers}\phi - u_2 s\phi & u_2u_3 \operatorname{vers}\phi + u_1 s\phi & u_3^2 \operatorname{vers}\phi + c\phi \end{bmatrix} \tag{6.142}$$

and

$$\mathbf{d} - R_{\hat{u},\phi}\mathbf{d} = \begin{bmatrix} d_1(1-u_1^2)\operatorname{vers}\phi - u_1 \operatorname{vers}\phi(d_2u_2 + d_3u_3) + s\phi(d_2u_3 - d_3u_2) \\ d_2(1-u_2^2)\operatorname{vers}\phi - u_2 \operatorname{vers}\phi(d_3u_3 + d_1u_1) + s\phi(d_3u_1 - d_1u_3) \\ d_3(1-u_3^2)\operatorname{vers}\phi - u_3 \operatorname{vers}\phi(d_1u_1 + d_2u_2) + s\phi(d_1u_2 - d_2u_1) \end{bmatrix} \tag{6.143}$$

As an example, let a body frame B with a body point P at

$$^B\mathbf{r}_P = \begin{bmatrix} 1 \\ 0 \\ 0 \end{bmatrix} \tag{6.143a}$$

turn 90 deg about an axis $\hat{u}$ parallel to $\hat{K}$ that is at

$$^G\mathbf{d} = \begin{bmatrix} 1 \\ 1 \\ 0 \end{bmatrix} \tag{6.143b}$$

The associated homogeneous transformation ^GT_B from (6.141) is

$$^GT_B = D_{\hat{d},d}\, R_{\hat{u},\phi}\, D_{\hat{d},-d}$$

$$= \begin{bmatrix} 1 & 0 & 0 & 1 \\ 0 & 1 & 0 & 1 \\ 0 & 0 & 1 & 0 \\ 0 & 0 & 0 & 1 \end{bmatrix} \begin{bmatrix} \cos\frac{\pi}{2} & -\sin\frac{\pi}{2} & 0 & 0 \\ \sin\frac{\pi}{2} & \cos\frac{\pi}{2} & 0 & 0 \\ 0 & 0 & 1 & 0 \\ 0 & 0 & 0 & 1 \end{bmatrix} \begin{bmatrix} 1 & 0 & 0 & -1 \\ 0 & 1 & 0 & -1 \\ 0 & 0 & 1 & 0 \\ 0 & 0 & 0 & 1 \end{bmatrix}$$

$$= \begin{bmatrix} 0 & -1 & 0 & 2 \\ 1 & 0 & 0 & 0 \\ 0 & 0 & 1 & 0 \\ 0 & 0 & 0 & 1 \end{bmatrix} \tag{6.144}$$

So, after the rotation, point P will be seen at ${}^G\mathbf{r}_P$:

$${}^G\mathbf{r}_P = {}^GT_B\,{}^B\mathbf{r}_P = \begin{bmatrix} 0 & -1 & 0 & 2 \\ 1 & 0 & 0 & 0 \\ 0 & 0 & 1 & 0 \\ 0 & 0 & 0 & 1 \end{bmatrix} \begin{bmatrix} 1 \\ 0 \\ 0 \\ 1 \end{bmatrix} = \begin{bmatrix} 2 \\ 1 \\ 0 \\ 1 \end{bmatrix} \tag{6.145}$$

At this moment, the origin o of the body coordinate frame B would be at

$${}^G\mathbf{r}_o = {}^GT_B\,{}^B\mathbf{r}_o = \begin{bmatrix} 0 & -1 & 0 & 2 \\ 1 & 0 & 0 & 0 \\ 0 & 0 & 1 & 0 \\ 0 & 0 & 0 & 1 \end{bmatrix} \begin{bmatrix} 0 \\ 0 \\ 0 \\ 1 \end{bmatrix} = \begin{bmatrix} 2 \\ 0 \\ 0 \\ 1 \end{bmatrix} \tag{6.146}$$

The final positions of point P and frame B are shown in Figure 6.12.

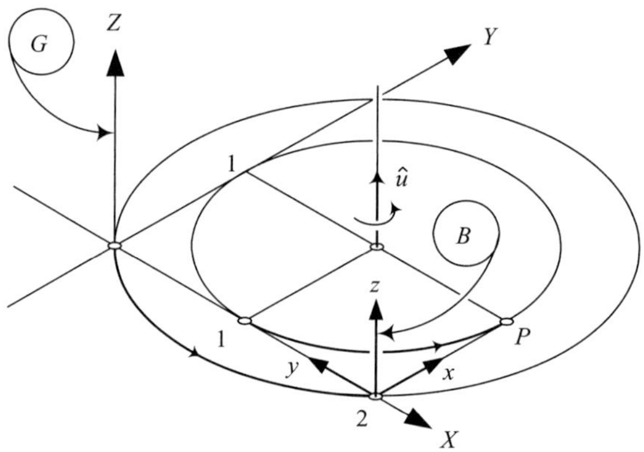

Figure 6.12 Rotation of a body frame B about an off center axis.

Example 360 Tip Point of an RPR Mechanism Point P indicates the tip point of the last arm of the mechanism shown in Figure 6.13. The mechanism has three arms. The first one is the L-shape arm that can turn about the Z-axis with a revolute joint at O. The second arm, which carries the coordinate frame $B_1(x_1y_1z_1)$ at its end, has a prismatic joint with the first arm. The third arm is attached to the second arm by a revolute joint. A coordinate $B_2(x_2y_2z_2)$ is attached to the beginning of the this arm. The position vector of point P in frame $B_2(x_2y_2z_2)$ is ${}^2\mathbf{r}_P$. Frame $B_2(x_2y_2z_2)$ can rotate about z_2 and slide along y_1. Frame $B_1(x_1y_1z_1)$ can rotate about the Z-axis of the global frame $G(OXYZ)$ while its origin is at ${}^G\mathbf{d}_1$. This is neither the best nor the standard method of attaching coordinate frames to connected multibodies.

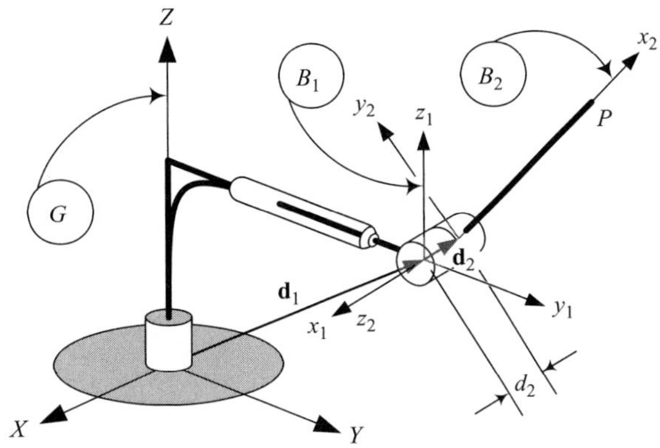

Figure 6.13 An RPR mechanism.

The position of P in $G(OXYZ)$ would be at

$$
\begin{aligned}
{}^G\mathbf{r} &= {}^GR_1\,{}^1R_2\,{}^2\mathbf{r}_P + {}^GR_1\,{}^1\mathbf{d}_2 + {}^G\mathbf{d}_1 = {}^GT_1\,{}^1T_2\,{}^2\mathbf{r}_P \\
&= {}^GT_2\,{}^2\mathbf{r}_P
\end{aligned}
\tag{6.147}
$$

where

$$
{}^1T_2 = \begin{bmatrix} {}^1R_2 & {}^1\mathbf{d}_2 \\ 0 & 1 \end{bmatrix} \tag{6.148}
$$

$$
{}^GT_1 = \begin{bmatrix} {}^GR_1 & {}^G\mathbf{d}_1 \\ 0 & 1 \end{bmatrix} \tag{6.149}
$$

and

$$
{}^GT_2 = \begin{bmatrix} {}^GR_1\,{}^1R_2 & {}^GR_1\,{}^1\mathbf{d}_2 + {}^G\mathbf{d}_1 \\ 0 & 1 \end{bmatrix} \tag{6.150}
$$

Example 361 End Effector of a SCARA Robot Figure 6.14 depicts a model of a SCARA robot. This robot has three arms. A global coordinate frame $G(OXYZ)$ is attached to the base link, while coordinate frames $B_1(o_1x_1y_1z_1)$ and $B_2(o_2x_2y_2z_2)$ are attached to link (1) and the tip of link (3).

To find the coordinates of the tip point in G, we combine the transformations between the coordinate frames. The transformation matrix from B_1 to the base frame G is

$$
{}^GT_{B_1} = \begin{bmatrix} \cos\theta_1 & -\sin\theta_1 & 0 & l_1\cos\theta_1 \\ \sin\theta_1 & \cos\theta_1 & 0 & l_1\sin\theta_1 \\ 0 & 0 & 1 & 0 \\ 0 & 0 & 0 & 1 \end{bmatrix} \tag{6.151}
$$

and the transformation matrix from B_2 to B_1 is

$$
{}^{B_1}T_{B_2} = \begin{bmatrix} \cos\theta_2 & -\sin\theta_2 & 0 & l_2\cos\theta_2 \\ \sin\theta_2 & \cos\theta_2 & 0 & l_2\sin\theta_2 \\ 0 & 0 & 1 & -h \\ 0 & 0 & 0 & 1 \end{bmatrix} \tag{6.152}
$$

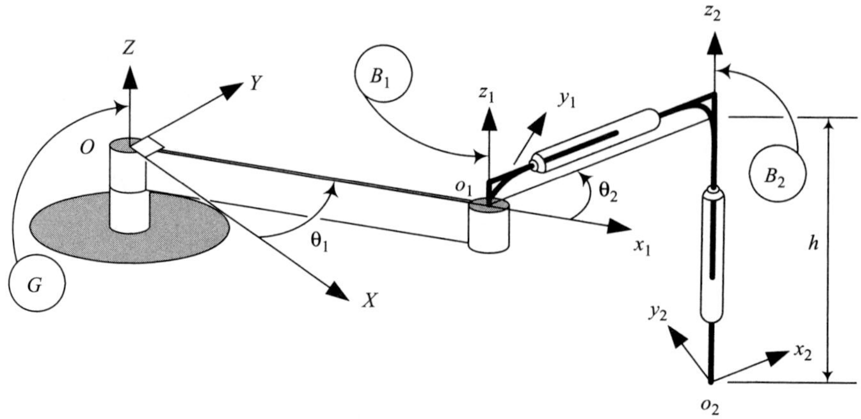

Figure 6.14 The SCARA robot of Example 361.

Therefore, the transformation matrix from B_2 to the base frame G is

$$
\begin{aligned}
{}^{G}T_{B_2} &= {}^{G}T_{B_1}\, {}^{B_1}T_{B_2} \\
&= \begin{bmatrix} c(\theta_1+\theta_2) & -s(\theta_1+\theta_2) & 0 & l_1 c\theta_1 + l_2 c(\theta_1+\theta_2) \\ s(\theta_1+\theta_2) & c(\theta_1+\theta_2) & 0 & l_1 s\theta_1 + l_2 s(\theta_1+\theta_2) \\ 0 & 0 & 1 & -h \\ 0 & 0 & 0 & 1 \end{bmatrix}
\end{aligned}
\tag{6.153}
$$

The origin of the last frame is at ${}^{B_2}\mathbf{r}_{o_2} = [0, 0, 0, 1]^{\mathrm{T}}$. Hence, the position of o_2 in the base coordinate frame is at

$$
{}^{G}\mathbf{r}_2 = {}^{G}T_{B_2}\, {}^{B_2}\mathbf{r}_{o_2} = \begin{bmatrix} l_1 \cos\theta_1 + l_2 \cos(\theta_1+\theta_2) \\ l_1 \sin\theta_1 + l_2 \sin(\theta_1+\theta_2) \\ -h \\ 1 \end{bmatrix}
\tag{6.154}
$$

Example 362 Cylindrical Coordinates To reach a point P in a cylindrical coordinate system, we need a translation r along the X-axis followed by a rotation φ about the Z-axis and finally a translation z along the Z-axis. A set of cylindrical coordinates is shown in Figure 6.15.

The homogeneous transformation matrix for going from cylindrical coordinates $C(Or\varphi z)$ to Cartesian coordinates $G(OXYZ)$ is

$$
\begin{aligned}
{}^{G}T_{C} &= D_{Z,z}\, R_{Z,\varphi}\, D_{X,r} \\
&= \begin{bmatrix} 1 & 0 & 0 & 0 \\ 0 & 1 & 0 & 0 \\ 0 & 0 & 1 & z \\ 0 & 0 & 0 & 1 \end{bmatrix} \begin{bmatrix} \cos\varphi & -\sin\varphi & 0 & 0 \\ \sin\varphi & \cos\varphi & 0 & 0 \\ 0 & 0 & 1 & 0 \\ 0 & 0 & 0 & 1 \end{bmatrix} \begin{bmatrix} 1 & 0 & 0 & r \\ 0 & 1 & 0 & 0 \\ 0 & 0 & 1 & 0 \\ 0 & 0 & 0 & 1 \end{bmatrix}
\end{aligned}
$$

$$= \begin{bmatrix} \cos\varphi & -\sin\varphi & 0 & r\cos\varphi \\ \sin\varphi & \cos\varphi & 0 & r\sin\varphi \\ 0 & 0 & 1 & z \\ 0 & 0 & 0 & 1 \end{bmatrix} \tag{6.155}$$

As an example, consider a point P at $(2, \pi/2, 3)$ in a cylindrical coordinate frame. Then, the Cartesian coordinates of P would be

$$\begin{bmatrix} \cos\frac{\pi}{2} & -\sin\frac{\pi}{2} & 0 & 2\cos\frac{\pi}{2} \\ \sin\frac{\pi}{2} & \cos\frac{\pi}{2} & 0 & 2\sin\frac{\pi}{2} \\ 0 & 0 & 1 & 3 \\ 0 & 0 & 0 & 1 \end{bmatrix} \begin{bmatrix} 0 \\ 0 \\ 0 \\ 1 \end{bmatrix} = \begin{bmatrix} 0 \\ 2 \\ 3 \\ 1 \end{bmatrix} \tag{6.156}$$

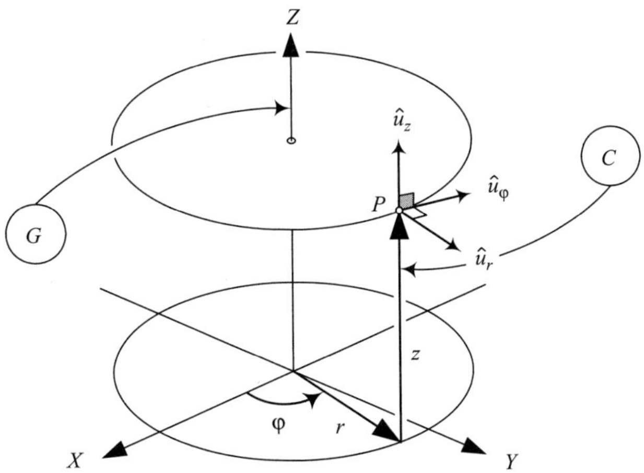

Figure 6.15 Cylindrical coordinates of a point P.

Example 363 ★ Dish Antenna Adjusting to a Satellite, Old Method Figure 6.16 illustrates the relative position of a geostationary satellite S and an observation point P on Earth. Assume the satellite is at a longitude γ and the dish P is at a longitude φ and latitude λ on Earth. To set a dish antenna at P, we should determine the azimuth angle α and the elevation angle β as functions of given parameters γ, φ, λ, R_0, and r_S.

The subsatellite point B on the equator at the longitude γ is the intersection of the geocentric radius vector of the satellite S with Earth's surface. The angle $\angle POB = \eta$ between the radius vectors of points P and B can be obtained from the right spherical triangle $\triangle BPC$ and employing Equation (4.65):

$$\cos\eta = \cos(\varphi - \gamma)\cos\lambda \tag{6.157}$$

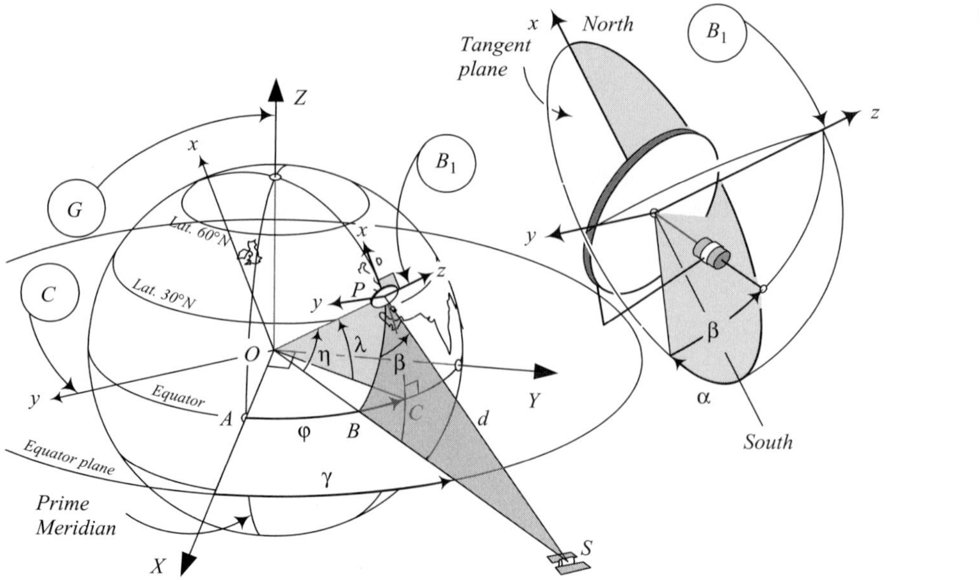

Figure 6.16 The relative position of a geostationary satellite S and an observation point P on Earth.

Using η we can determine the distance d of the satellite S and dish P:

$$d = \sqrt{r_S^2 + r_P^2 - 2r_S r_P \cos\eta} \tag{6.158}$$

$$r_P \approx R_0 \approx 6371230\,\mathrm{m} \tag{6.159}$$

$$r_S = \overline{OS} \approx 42200\,\mathrm{km} \tag{6.160}$$

The distance d is called the *topocentric distance* from the dish to the satellite, R_0 is the radius of Earth, and r_S is the radius of the geostationary orbit.

To determine the elevation angle β, we may look at the plane triangle $\triangle OPS$, as shown in Figure 6.17. Using the sine law

$$\frac{r_S}{\sin\xi} \approx \frac{d}{\sin\eta} \tag{6.161}$$

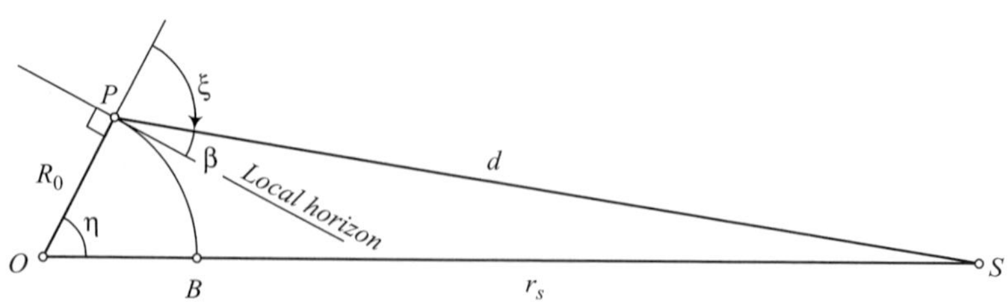

Figure 6.17 The Earth center–dish–satellite triangle $\triangle OPS$.

provides the *zenith angle* $\xi \approx \pi/2 - \beta$ of the satellite at point P:

$$\xi = \arcsin\left(\frac{r_s}{d}\sin\eta\right) \tag{6.162}$$

Then

$$\beta = \frac{\pi}{2} - \xi = \arccos\left(\frac{r_s}{d}\sin\eta\right) \tag{6.163}$$

$$\eta = \arccos\left(\cos\left(\varphi - \gamma\right)\cos\lambda\right) \tag{6.164}$$

To determine the azimuth angle α and set the centerline of the dish to the satellites, we need to determine the angle $\angle BPC = \sigma$ of the spherical right triangle $\triangle BPC$. We can determine σ by employing Equation (4.72):

$$\cos\sigma = \cos(\angle BPC) = \cos\eta\tan\lambda = \cos\left(\varphi - \gamma\right)\sin\lambda \tag{6.165}$$

Therefore, the azimuth angle α would be

$$\alpha = \begin{cases} \pi + \sigma & \lambda > 0, \varphi > \gamma \\ \pi - \sigma & \lambda > 0, \varphi < \gamma \\ \sigma & \lambda < 0, \varphi > \gamma \\ -\sigma & \lambda < 0, \varphi < \gamma \end{cases} \tag{6.166}$$

As an example, consider a point P at

$$\varphi \approx 51.424\,\text{deg} \approx 0.89752\,\text{rad} \tag{6.167}$$

$$\lambda \approx 35.672\,\text{deg} \approx 0.62259\,\text{rad} \tag{6.168}$$

and Hot Bird, a series of communication satellites, at

$$r_S \approx 42200\,\text{km} \tag{6.169}$$

$$\varphi_S \approx 0^\circ 2' \text{ South} \approx 0 \tag{6.170}$$

$$\gamma \approx 12^\circ 59' \text{ East} \approx 12.983\,\text{deg} \approx 0.2266\,\text{rad} \tag{6.171}$$

Using $R_0 \approx 6371230\,\text{m}$, we will have

$$\cos\eta = \cos\left(\gamma - \varphi\right)\cos\lambda = 0.63629 \tag{6.172}$$

$$\eta \approx 0.88112\,\text{rad} \approx 50.484\,\text{deg} \tag{6.173}$$

$$d = \sqrt{r_S^2 + R_0^2 - 2r_S R_0\cos\eta} = 38461\,\text{km} \tag{6.174}$$

and therefore,

$$\beta = \arccos\left(\frac{r_S}{d}\sin\eta\right) = \arccos\left(\frac{42{,}200}{38{,}461}\sin 0.88112\right)$$
$$= 0.56152\,\text{rad} \approx 32.173\,\text{deg} \tag{6.175}$$

$$\cos\sigma = \cos\left(\gamma - \varphi\right)\sin\lambda = 0.45675 \tag{6.176}$$

$$\sigma = 1.0965\,\text{rad} \approx 62.825\,\text{deg} < 90\,\text{deg} \tag{6.177}$$

$$\alpha = \pi + \sigma \approx 180 + 62.825 = 242.83\,\text{deg} \tag{6.178}$$

Example 364 ★ Dish Antenna Adjusting, Transformation Method The relative position of a geostationary satellite S and the dish antenna P on Earth are illustrated in Figure 6.18. The satellite S is always on the (X,Y)-plane of Earth's coordinate frame $G(X, Y, Z)$ at ${}^G\mathbf{r}_S$:

$$ {}^G\mathbf{r}_S = \begin{bmatrix} r_S \cos\gamma \\ r_S \sin\gamma \\ 0 \end{bmatrix} \tag{6.179} $$

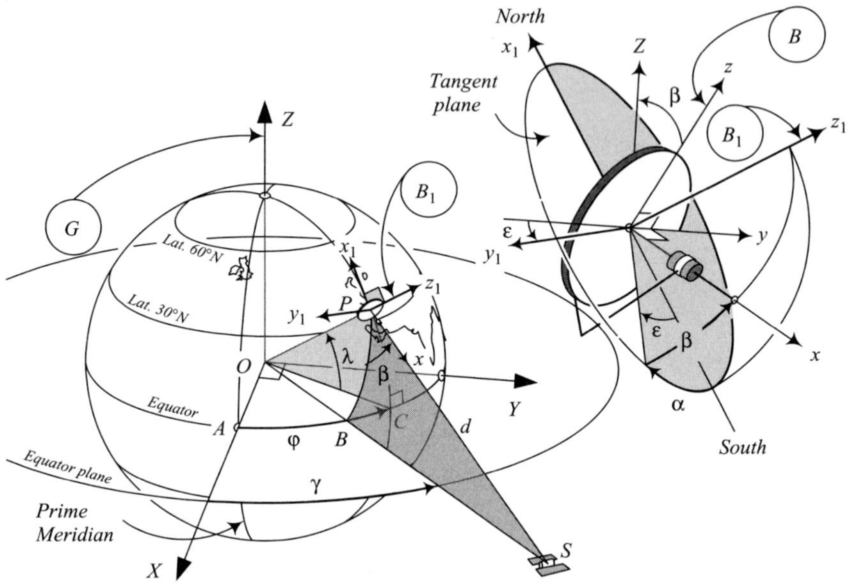

Figure 6.18 The relative position of a geostationary satellite S and the dish antenna P on Earth.

We attach a frame B_1 at point P such that the x_1-axis points North, the y_1-axis West, and the z_1-axis up. So, the (x_1, y_1)-plane indicates the local horizontal plane. We also attach a frame B to the dish at point P such that the x-axis is the normal vector to the plane of the dish and (y,z) is on the plane of the dish. The y-axis is in the local horizontal plane of (x_1, y_1). The point P is at a longitude φ and latitude λ on Earth. To determine the dish angles α and β, let us find the transformation matrices between the frames.

We determine the transformation matrix between B_1 and G by moving B_1 from the coincident configuration with G and rotating φ about the z_1-axis, then a rotation $\pi/2 - \lambda$ about the y_1-axis followed by a rotation π about the z_1-axis, and moving a distance R_0 along the z_1-axis:

$$ \begin{aligned} {}^1R_G &= R_{z,\pi} R_{y,(\pi/2-\lambda)} R_{z,\varphi} \\ &= \begin{bmatrix} -\cos\varphi\sin\lambda & -\sin\lambda\sin\varphi & \cos\lambda \\ \sin\varphi & -\cos\varphi & 0 \\ \cos\lambda\cos\varphi & \cos\lambda\sin\varphi & \sin\lambda \end{bmatrix} \end{aligned} \tag{6.180} $$

$$ {}^1\mathbf{d} = \begin{bmatrix} 0 & 0 & R_0 & 1 \end{bmatrix}^{\mathrm{T}} \tag{6.181} $$

$$ {}^G\mathbf{d} = {}^1R_G^{\mathrm{T}}\,{}^1\mathbf{d} = {}^1R_G^{\mathrm{T}} \begin{bmatrix} 0 \\ 0 \\ R_0 \end{bmatrix} = \begin{bmatrix} R_0 \cos\lambda \cos\varphi \\ R_0 \cos\lambda \sin\varphi \\ R_0 \sin\lambda \end{bmatrix} \tag{6.182} $$

$$ {}^GT_1 = \begin{bmatrix} {}^1R_G^{\mathrm{T}} & {}^G\mathbf{d} \\ 0 & 1 \end{bmatrix} = \begin{bmatrix} {}^GR_1 & {}^G\mathbf{d} \\ 0 & 1 \end{bmatrix} \tag{6.183} $$

$$ = \begin{bmatrix} -\cos\varphi \sin\lambda & \sin\varphi & \cos\lambda \cos\varphi & R_0 \cos\lambda \cos\varphi \\ -\sin\lambda \sin\varphi & -\cos\varphi & \cos\lambda \sin\varphi & R_0 \cos\lambda \sin\varphi \\ \cos\lambda & 0 & \sin\lambda & R_0 \sin\lambda \\ 0 & 0 & 0 & 1 \end{bmatrix} $$

The fourth column of this homogeneous transformation matrix determines the coordinates of any point on Earth provided the longitude φ and latitude λ are given. That means the global coordinates of the point P are

$$ {}^G\mathbf{r}_P = {}^GT_1 \begin{bmatrix} 0 \\ 0 \\ 0 \\ 1 \end{bmatrix} = \begin{bmatrix} R_0 \cos\lambda \cos\varphi \\ R_0 \cos\lambda \sin\varphi \\ R_0 \sin\lambda \\ 1 \end{bmatrix} \tag{6.184} $$

Therefore, the position of the satellite in the B_1-frame is

$$ {}^1\mathbf{r}_S = {}^1T_G\,{}^G\mathbf{r}_S = \begin{bmatrix} {}^GR_1^{\mathrm{T}} & -{}^GR_1^{\mathrm{T}}\,{}^G\mathbf{d} \\ 0 & 1 \end{bmatrix} \begin{bmatrix} r_S \cos\gamma \\ r_S \sin\gamma \\ 0 \\ 1 \end{bmatrix} $$

$$ = \begin{bmatrix} -r_S \sin\lambda \cos(\gamma - \varphi) \\ -r_S \sin(\gamma - \varphi) \\ -R_0 + r_S \cos\lambda \cos(\gamma - \varphi) \\ 1 \end{bmatrix} \tag{6.185} $$

$$ {}^1T_G = \begin{bmatrix} -\cos\varphi \sin\lambda & -\sin\lambda \sin\varphi & \cos\lambda & 0 \\ \sin\varphi & -\cos\varphi & 0 & 0 \\ \cos\lambda \cos\varphi & \cos\lambda \sin\varphi & \sin\lambda & -R_0 \\ 0 & 0 & 0 & 1 \end{bmatrix} \tag{6.186} $$

The coordinate frame B is the dish antenna coordinate frame such that the x-axis of B is on the dish centerline. To point the x-axis of B to the satellite, we may turn B from a coincident orientation with B_1 about the z-axis $-\alpha = -\pi - \epsilon$ degrees and then turn it about the y-axis $-\beta$ degrees. So, the transformation matrix between B and B_1 is

$$ {}^BT_1 = \begin{bmatrix} R_{y,-\beta} R_{z,-\pi-\epsilon} & 0 \\ 0 & 1 \end{bmatrix} = \begin{bmatrix} {}^1R_B & 0 \\ 0 & 1 \end{bmatrix} \tag{6.187} $$

$$ = \begin{bmatrix} -\cos\beta \cos\epsilon & \cos\beta \sin\epsilon & \sin\beta & 0 \\ -\sin\epsilon & -\cos\epsilon & 0 & 0 \\ \cos\epsilon \sin\beta & -\sin\beta \sin\epsilon & \cos\beta & 0 \\ 0 & 0 & 0 & 1 \end{bmatrix} $$

The position of the satellite in dish frame B is

$$
{}^B\mathbf{r}_S = {}^BT_1\,{}^1\mathbf{r}_S = {}^BT_1 \begin{bmatrix} -r_S \sin\lambda \cos(\gamma - \varphi) \\ -r_S \sin(\gamma - \varphi) \\ -R_0 + r_S \cos\lambda \cos(\gamma - \varphi) \\ 1 \end{bmatrix} \tag{6.188}
$$

$$
= \begin{bmatrix} r_S(\cos\lambda\cos\phi\sin\beta - \cos\beta\sin\phi\sin\epsilon + \cos\beta\cos\phi\cos\epsilon\sin\lambda) - R_0\sin\beta \\ r_S\cos\epsilon\sin\phi + r_S\cos\phi\sin\lambda\sin\epsilon \\ r_S(\cos\lambda\cos\phi\cos\beta + \sin\beta\sin\phi\sin\epsilon - \sin\beta\cos\phi\cos\epsilon\sin\lambda) - R_0\cos\beta \\ 1 \end{bmatrix}
$$

$$
\phi = \gamma - \varphi \tag{6.189}
$$

$$
{}^1T_B = \begin{bmatrix} -\cos\beta\cos\epsilon & -\sin\epsilon & \cos\epsilon\sin\beta & 0 \\ \cos\beta\sin\epsilon & -\cos\epsilon & -\sin\beta\sin\epsilon & 0 \\ \sin\beta & 0 & \cos\beta & 0 \\ 0 & 0 & 0 & 1 \end{bmatrix} \tag{6.190}
$$

However, ${}^B\mathbf{r}_S$ must be at $[d,0,0,1]$, where

$$
\begin{aligned} d &= \left|{}^1\mathbf{r}_S\right| = \left|{}^B\mathbf{r}_S\right| \\ &= \sqrt{R_0^2 + r_S^2 - 2R_0 r_S \cos\lambda\cos(\gamma - \varphi)} \end{aligned} \tag{6.191}
$$

Therefore, the second and third components of ${}^B\mathbf{r}_S$ must be zero:

$$
r_S\cos\epsilon\sin\phi + r_S\cos\phi\sin\lambda\sin\epsilon = 0 \tag{6.192a}
$$

$$
r_S(\cos\lambda\cos\phi\cos\beta + \sin\beta\sin\phi\sin\epsilon - \sin\beta\cos\phi\cos\epsilon\sin\lambda) - R_0\cos\beta = 0 \tag{6.192b}
$$

These equations provide the required angles α and β to adjust the antenna. The first equation can be solved for ϵ,

$$
\tan\epsilon = -\frac{\sin\phi}{\cos\phi\sin\lambda} = -\frac{\sin(\lambda - \varphi)}{\cos(\lambda - \varphi)\sin\lambda} \tag{6.193a}
$$

$$
\alpha = \pi + \epsilon \tag{6.193b}
$$

and the first equation for β,

$$
\begin{aligned} \tan\beta &= -\frac{r_S\cos\lambda\cos\phi - R_0}{r_S\sin\phi\sin\epsilon - r_S\cos\phi\cos\epsilon\sin\lambda} \\ &= -\frac{r_S\cos\lambda\cos(\gamma - \varphi) - R_0}{r_S\sin(\lambda - \varphi)\sin\epsilon - r_S\cos(\gamma - \varphi)\cos\epsilon\sin\lambda} \end{aligned} \tag{6.194}
$$

As an example, let us use

$$R_0 \approx 6371230\,\text{m} \tag{6.195}$$

$$r_s \approx 42200\,\text{km} \tag{6.196}$$

$$\varphi \approx 51.424\,\text{deg} \approx 0.89752\,\text{rad} \tag{6.197}$$

$$\lambda \approx 35.672\,\text{deg} \approx 0.62259\,\text{rad} \tag{6.198}$$

$$\gamma \approx 12^{\circ}59'\ East \approx 12.983\,\text{deg} \approx 0.2266\,\text{rad} \tag{6.199}$$

and determine α and β:

$$\beta \approx 0.56154\,\text{rad} \approx 32.174\,\text{deg} \tag{6.200}$$

$$\alpha = 4.0788\,\text{rad} \approx 233.70\,\text{deg} \tag{6.201}$$

Example 365 Foot–Leg Relative Kinematics The foot skeleton is composed of 26 bones in three regions: 7 bones in the tarsal, 5 bones in the metatarsal, and 14 bones in the fingers. Dislocation of any of these bones causes a walking problem that appears as an irregular clinical angle.

To determine the foot–leg clinical angles during the gait and examine their relative kinematics, we attach a sensor to the leg at an arbitrary position and orientation. The sensor reports its position vector ${}^G\mathbf{d}_1$ and a rotation transformation matrix ${}^GR_1$ between its coordinate frame B_1 and the base global frame G. Another sensor on the foot reports the position vector ${}^G\mathbf{d}_2$ and a rotation transformation matrix ${}^GR_2$ between its coordinate frame B_2 and the global frame G. Figure 6.19 illustrates the stationary base frame G along with the leg and foot frames B_1 and B_2.

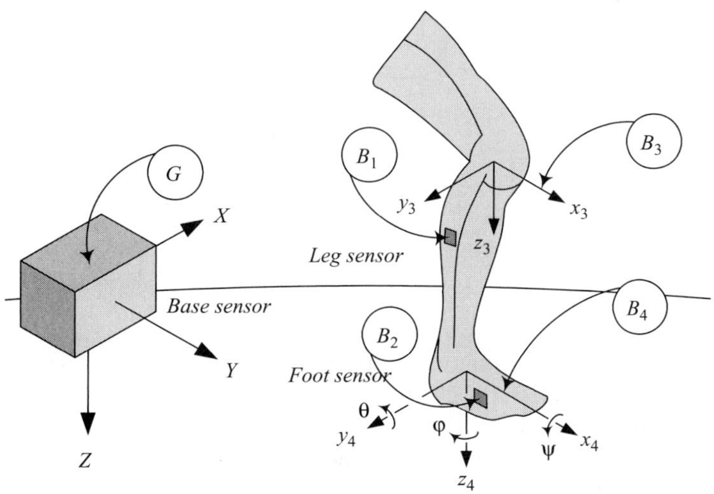

Figure 6.19 A stationary base frame G along with two moving frames B_1 and B_2 attached to the leg and foot.

Let us assume that at the upright standing position the following transformation matrices are initially reported:

$$
{}^{G}T_1 = \begin{bmatrix} {}^{G}R_1 & {}^{G}\mathbf{d}_1 \\ 0 & 1 \end{bmatrix}
= \begin{bmatrix} 0.0360 & -0.1770 & -0.9840 & 21.0150 \\ 0.5060 & 0.8520 & -0.1350 & -5.6340 \\ 0.8620 & -0.4930 & 0.1200 & 3.0810 \\ 0 & 0 & 0 & 1 \end{bmatrix} \quad (6.202)
$$

$$
{}^{G}T_2 = \begin{bmatrix} {}^{G}R_2 & {}^{G}\mathbf{d}_2 \\ 0 & 1 \end{bmatrix}
= \begin{bmatrix} 0.4190 & -0.7290 & 0.5410 & 21.8770 \\ -0.0200 & -0.6030 & -0.7980 & -7.7520 \\ 0.9080 & 0.3230 & -0.2670 & -4.7460 \\ 0 & 0 & 0 & 1 \end{bmatrix} \quad (6.203)
$$

Using the homogeneous transformation calculus, we can determine the position and orientation of the foot with respect to the leg from ${}^{1}T_2$:

$$
{}^{1}T_2 = {}^{G}T_1^{-1}\,{}^{G}T_2 = \begin{bmatrix} {}^{1}R_2 & {}^{1}\mathbf{d}_2 \\ 0 & 1 \end{bmatrix} \quad (6.204)
$$

$$
= \begin{bmatrix} 0.78766 & -0.052936 & -0.61447 & -7.7876 \\ -0.53885 & -0.54396 & -0.64402 & 1.9018 \\ -0.30064 & 0.8375 & -0.45665 & -1.5020 \\ 0 & 0 & 0 & 1 \end{bmatrix}
$$

where

$$
{}^{G}T_1^{-1} = \begin{bmatrix} 0.036 & 0.506 & 0.862 & -0.56156 \\ -0.177 & 0.852 & -0.493 & 10.039 \\ -0.984 & -0.135 & 0.12 & 19.548 \\ 0 & 0 & 0 & 1 \end{bmatrix} \quad (6.205)
$$

$$
{}^{G}T_2^{-1} = \begin{bmatrix} 0.419 & -0.02 & 0.908 & -5.0121 \\ -0.729 & -0.603 & 0.323 & 12.807 \\ 0.541 & -0.798 & -0.267 & -19.289 \\ 0 & 0 & 0 & 1 \end{bmatrix} \quad (6.206)
$$

To determine the clinical angles of the shank–ankle, we prefer to assume that the foot and leg coordinate frames are initially parallel. Then the relative orientation can be decomposed to rotations of the foot about three perpendicular axes of the foot frame. Let us attach frames B_3 and B_4 to the leg and foot, as shown in Figure 6.19. These frames, which are initially parallel, may be attached at the same points as the sensors or at constant distances ${}^{1}\mathbf{d}_3$ and ${}^{2}\mathbf{d}_4$ from them. We assume the transformations between B_1 and B_3 as well as B_2 and B_4 remain constant while the person walks. Therefore, at the initial moment the transformation matrices ${}^{G}T_3$ and ${}^{G}T_4$ are

$$
{}^{G}T_3 = \begin{bmatrix} {}^{G}R_3 & {}^{G}\mathbf{d}_3 \\ 0 & 1 \end{bmatrix} = \begin{bmatrix} {}^{G}R_{Z,90} & {}^{G}\mathbf{d}_1 + {}^{G}R_1 {}^{1}\mathbf{d}_3 \\ 0 & 1 \end{bmatrix} \tag{6.207}
$$

$$
{}^{G}T_4 = \begin{bmatrix} {}^{G}R_4 & {}^{G}\mathbf{d}_4 \\ 0 & 1 \end{bmatrix} = \begin{bmatrix} {}^{G}R_{Z,90} & {}^{G}\mathbf{d}_2 + {}^{G}R_2 {}^{2}\mathbf{d}_4 \\ 0 & 1 \end{bmatrix} \tag{6.208}
$$

The clinical angles are dorsiflexion, plantarflexion, eversion, inversion, inward rotation, and outward rotation. Dorsiflexion is a rotation about the y_4-axis which decreases the angle between the foot and the leg, so that the toes are brought closer to the shin. The movement about the $-y_4$-axis is called plantarflexion. Eversion is the movement of the sole of the foot in the lateral direction about the x_4-axis. Movement in the median direction about the $-x_4$-axis is called inversion. Two more clinical angles which are rotations of the foot about the normal z_3-axis passing through the ankle are called internal and external rotations.

For simplicity, let us assume

$$
{}^{1}\mathbf{d}_3 = 0 \qquad {}^{2}\mathbf{d}_4 = 0 \tag{6.209}
$$

and find the initial matrices ${}^{G}T_3$ and ${}^{G}T_4$ and the constant transformations ${}^{1}T_3$ and ${}^{2}T_4$ as

$$
\left[{}^{G}T_3\right]_0 = \begin{bmatrix} 0 & -1 & 0 & 21.0150 \\ 1 & 0 & 0 & -5.6340 \\ 0 & 0 & 1 & 3.0810 \\ 0 & 0 & 0 & 1 \end{bmatrix} \tag{6.210}
$$

$$
\left[{}^{G}T_4\right]_0 = \begin{bmatrix} 0 & -1 & 0 & 21.8770 \\ 1 & 0 & 0 & -7.7520 \\ 0 & 0 & 1 & -4.7460 \\ 0 & 0 & 0 & 1 \end{bmatrix} \tag{6.211}
$$

$$
{}^{1}T_3 = {}^{G}T_1^{-1} \left[{}^{G}T_3\right]_0 \tag{6.212}
$$

$$
= \begin{bmatrix} 0.506 & -0.036 & 0.862 & 0 \\ 0.852 & 0.177 & -0.493 & 0 \\ -0.135 & 0.984 & 0.12 & 0 \\ 0 & 0 & 0 & 1 \end{bmatrix}
$$

$$
{}^{2}T_4 = {}^{G}T_2^{-1} \left[{}^{G}T_4\right]_0 \tag{6.213}
$$

$$
= \begin{bmatrix} -0.02 & -0.419 & 0.908 & 0 \\ -0.603 & 0.729 & 0.323 & 0 \\ -0.798 & -0.541 & -0.267 & 0 \\ 0 & 0 & 0 & 1 \end{bmatrix}
$$

During walking, the sensors provide the instantaneous matrices ${}^{G}T_1$ and ${}^{G}T_2$. Multiplying these matrices by the constant matrices ${}^{1}T_3$ and ${}^{2}T_4$ determines the instantaneous matrices ${}^{G}T_{3}$ and ${}^{G}T_{4}$:

$$ {}^{G}T_{3} = {}^{G}T_{1}\,{}^{1}T_{3} \tag{6.214}$$

$$ {}^{G}T_{4} = {}^{G}T_{2}\,{}^{2}T_{4} \tag{6.215}$$

Then we can use the relative transformation ${}^{3}T_{4}$ to determine the rotations of the foot about its local axes:

$$ {}^{3}T_{4} = {}^{G}T_{3}^{-1}\,{}^{G}T_{4} \tag{6.216}$$

As an example let us consider the following matrices during the motion:

$$ {}^{G}T_{1} = \begin{bmatrix} 0.0360 & -0.1770 & -0.9840 & 21.0150 \\ 0.5060 & 0.8520 & -0.1350 & -5.6340 \\ 0.8620 & -0.4930 & 0.1200 & 3.0810 \\ 0 & 0 & 0 & 1 \end{bmatrix} \tag{6.217}$$

$$ {}^{G}T_{2} = \begin{bmatrix} 0.4190 & -0.7290 & 0.5410 & 21.8770 \\ -0.0200 & -0.6030 & -0.7980 & -7.7520 \\ 0.9080 & 0.3230 & -0.2670 & -4.7460 \\ 0 & 0 & 0 & 1 \end{bmatrix} \tag{6.218}$$

Therefore,

$$ {}^{G}T_{3} = {}^{G}T_{1}\,{}^{1}T_{3} = \begin{bmatrix} 2.52\times10^{-4} & -1.0009 & 2.13\times10^{-4} & 21.015 \\ 1.0002 & -2.52\times10^{-4} & -6.4\times10^{-5} & -5.634 \\ -6.4\times10^{-5} & -2.13\times10^{-4} & 1.0005 & 3.081 \\ 0 & 0 & 0 & 1 \end{bmatrix} \tag{6.219}$$

$$ {}^{G}T_{4} = {}^{G}T_{2}\,{}^{2}T_{4} = \begin{bmatrix} -5.11\times10^{-4} & -0.99968 & 5.38\times10^{-4} & 21.877 \\ 1.0008 & 5.11\times10^{-4} & 1.37\times10^{-4} & -7.752 \\ 1.37\times10^{-4} & -5.38\times10^{-4} & 1.0001 & -4.746 \\ 0 & 0 & 0 & 1 \end{bmatrix} \tag{6.220}$$

$$ {}^{G}T_{3}^{-1} = \begin{bmatrix} 2.52\times10^{-4} & 1.0002 & -6.4\times10^{-5} & 5.63 \\ -1.0009 & -2.52\times10^{-4} & -2.13\times10^{-4} & 21.033 \\ 2.13\times10^{-4} & -6.4\times10^{-5} & 1.0005 & -3.0874 \\ 0 & 0 & 0 & 1 \end{bmatrix} \tag{6.221}$$

$$ {}^{3}T_{4} = {}^{G}T_{3}^{-1}\,{}^{G}T_{4} = \begin{bmatrix} 1.001 & 2.5922\times10^{-4} & 7.3157\times10^{-5} & -2.1177 \\ 2.5923\times10^{-4} & 1.0006 & -7.5154\times10^{-4} & -0.86072 \\ 7.2908\times10^{-5} & -7.5123\times10^{-4} & 1.0006 & -7.8306 \\ 0 & 0 & 0 & 1 \end{bmatrix} \tag{6.222}$$

Because B_3 is parallel to B_4 initially, we may assign the clinical angles by using the rotations φ about the z_4-axis, θ about the x_4-axis, and ψ about the y_4-axis. The rotation matrix of case 1 in Appendix B provides ${}^4R_3$. Therefore, ${}^4R_3^{\mathrm{T}}$ must be equal to the rotation part of ${}^3T_4$:

$$
\begin{aligned}
{}^4R_3^{\mathrm{T}} &= \left[R_{x,\psi} R_{y,\theta} R_{z,\varphi}\right]^{\mathrm{T}} \\
&= \begin{bmatrix} c\theta c\varphi & c\varphi s\theta s\psi - c\psi s\varphi & s\psi s\varphi + c\psi c\varphi s\theta \\ c\theta s\varphi & c\psi c\varphi + s\theta s\psi s\varphi & c\psi s\theta s\varphi - c\varphi s\psi \\ -s\theta & c\theta s\psi & c\theta c\psi \end{bmatrix}
\end{aligned}
\tag{6.223}
$$

The magnitude of angle φ is

$$
\begin{aligned}
\varphi &= \arctan\frac{r_{21}}{r_{11}} = \arctan\frac{2.5923 \times 10^{-4}}{1.001} \\
&\approx 2.5897 \times 10^{-4}\,\mathrm{rad} \approx 1.4838 \times 10^{-7}\,\mathrm{deg}
\end{aligned}
\tag{6.224}
$$

The magnitude of angle θ is

$$
\begin{aligned}
\theta &= -\arcsin r_{31} = -\arcsin\left(7.2908 \times 10^{-5}\right) \\
&\approx -7.2908 \times 10^{-5}\,\mathrm{rad} \approx -4.1772 \times 10^{-3}\,\mathrm{deg}
\end{aligned}
\tag{6.225}
$$

The magnitude of angle ψ is

$$
\begin{aligned}
\psi &= \arctan\frac{r_{32}}{r_{33}} = \arctan\frac{-7.5123 \times 10^{-4}}{1.0006} \\
&\approx -7.5078 \times 10^{-4}\,\mathrm{rad} \approx -4.3016 \times 10^{-2}\,\mathrm{deg}
\end{aligned}
\tag{6.226}
$$

6.5 ★ SCREW MOTION

A rigid-body motion is called a *screw* if the body translates along an axis while it rotates about that axis. This is called the *Chasles theorem*, which states: *Every rigid-body motion can be reproduced by a screw*. Consider the screw motion illustrated in Figure 6.20. A rigid body rotates about the screw axis $\hat{u}$ and simultaneously translates along the same axis. So, any point on the *screw axis* moves along the axis, while any point off the axis moves on a *helix*.

The angular rotation of a rigid body about the screw axis is called the *twist* ϕ of the screw, and the sliding motion along the screw axis is called the *translation* h. The ratio of translation to twist is called the *pitch* p of the screw:

$$
p = \frac{h}{\phi}
\tag{6.227}
$$

Pitch is the distance through which the rigid body translates parallel to the axis of the screw for a unit rotation. If $p > 0$, then the screw is *right handed*, and if $p < 0$, the screw is *left handed*.

We show a screw by $\check{s}(h, \phi, \hat{u}, \mathbf{s})$ to indicate the unit vector on the screw axis $\hat{u}$, the *location* vector $\mathbf{s}$, the twist angle ϕ, and the translation h. The location vector ${}^G\mathbf{s}$

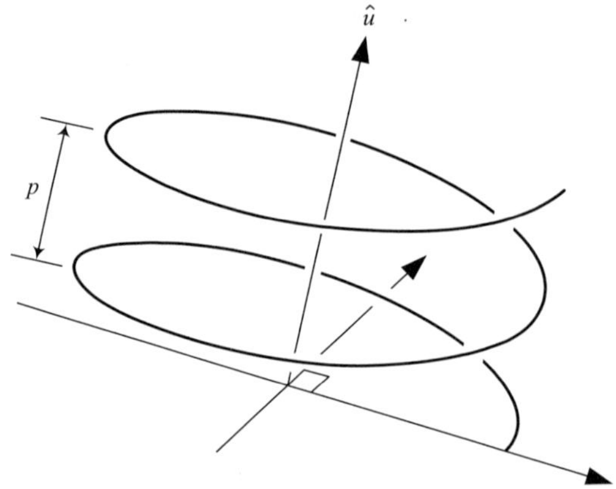

Figure 6.20 A screw motion is the translation along a line combined with a rotation about the line.

indicates the global position of a point on the screw axis. The twist angle ϕ, the twist axis $\hat{u}$, and the translation h are called *screw parameters*. If $\hat{u}$ passes through the origin of the global coordinate frame, then $\mathbf{s} = 0$ and the screw motion is called the *central screw* $\check{s}(h, \phi, \hat{u})$.

The screw $\check{s}(h, \phi, \hat{u}, \mathbf{s})$ is a transformation between the initial and final configurations of a rigid body. The screw transformation can be expressed by a homogeneous matrix that is a combination of rotation and translation. The screw is basically another transformation method to describe the motion of a rigid body. A linear displacement along an axis combined with an angular displacement about the same axis arises repeatedly in multibody dynamics.

For a central screw we have

$$
{}^{G}\check{s}_{B}(h, \phi, \hat{u}) = {}^{G}T_{B} = \begin{bmatrix} {}^{G}R_{B} & h\hat{u} \\ 0 & 1 \end{bmatrix} = D_{\hat{u},h}\; R_{\hat{u},\phi} \tag{6.228}
$$

where

$$
D_{\hat{u},h} = \begin{bmatrix} 1 & 0 & 0 & hu_1 \\ 0 & 1 & 0 & hu_2 \\ 0 & 0 & 1 & hu_3 \\ 0 & 0 & 0 & 1 \end{bmatrix} \tag{6.229}
$$

$$
R_{\hat{u},\phi} = \begin{bmatrix} u_1^2 \operatorname{vers}\phi + c\phi & u_1 u_2 \operatorname{vers}\phi - u_3 s\phi & u_1 u_3 \operatorname{vers}\phi + u_2 s\phi & 0 \\ u_1 u_2 \operatorname{vers}\phi + u_3 s\phi & u_2^2 \operatorname{vers}\phi + c\phi & u_2 u_3 \operatorname{vers}\phi - u_1 s\phi & 0 \\ u_1 u_3 \operatorname{vers}\phi - u_2 s\phi & u_2 u_3 \operatorname{vers}\phi + u_1 s\phi & u_3^2 \operatorname{vers}\phi + c\phi & 0 \\ 0 & 0 & 0 & 1 \end{bmatrix} \tag{6.230}
$$

and hence,

$$
{}^{G}\check{s}_{B}(h, \phi, \hat{u}) = \begin{bmatrix} u_1^2 \operatorname{vers}\phi + c\phi & u_1 u_2 \operatorname{vers}\phi - u_3 s\phi & u_1 u_3 \operatorname{vers}\phi + u_2 s\phi & hu_1 \\ u_1 u_2 \operatorname{vers}\phi + u_3 s\phi & u_2^2 \operatorname{vers}\phi + c\phi & u_2 u_3 \operatorname{vers}\phi - u_1 s\phi & hu_2 \\ u_1 u_3 \operatorname{vers}\phi - u_2 s\phi & u_2 u_3 \operatorname{vers}\phi + u_1 s\phi & u_3^2 \operatorname{vers}\phi + c\phi & hu_3 \\ 0 & 0 & 0 & 1 \end{bmatrix} \tag{6.231}
$$

A central screw transformation matrix reduces to a pure translation for $\phi = 0$ and reduces to a pure rotation for $h = 0$. The negative screw motion is defined by changing the sign of both the translation and twist. Such a motion is called the *reverse central screw* and is denoted by $\check{s}(-h, -\phi, \hat{u})$.

When the screw is not central and $\hat{u}$ is not passing through the origin, a screw motion to move $\mathbf{p}$ to $\mathbf{p}''$ is denoted by

$$\begin{aligned}\mathbf{p}'' &= (\mathbf{p} - \mathbf{s}) \cos\phi + (1 - \cos\phi)\, [\hat{u} \cdot (\mathbf{p} - \mathbf{s})]\hat{u} \\ &\quad + [\hat{u} \times (\mathbf{p} - \mathbf{s})] \sin\phi + \mathbf{s} + h\hat{u}\end{aligned} \tag{6.232}$$

or

$$\mathbf{p}'' = {}^{G}R_{B}\,(\mathbf{p} - \mathbf{s}) + \mathbf{s} + h\hat{u} = {}^{G}R_{B}\,\mathbf{p} + \mathbf{s} - {}^{G}R_{B}\,\mathbf{s} + h\hat{u} \tag{6.233}$$

and therefore,

$$\mathbf{p}'' = \check{s}(h, \phi, \hat{u}, \mathbf{s})\mathbf{p} = [T]\,\mathbf{p} \tag{6.234}$$

where

$$[T] = \begin{bmatrix} {}^{G}R_{B} & {}^{G}\mathbf{s} - {}^{G}R_{B}{}^{G}\mathbf{s} + h\hat{u} \\ 0 & 1 \end{bmatrix} = \begin{bmatrix} {}^{G}R_{B} & {}^{G}\mathbf{d} \\ 0 & 1 \end{bmatrix} \tag{6.235}$$

The location vector ${}^{G}\mathbf{s}$ is the global position of the body frame before the screw motion. The vectors $\mathbf{p}''$ and $\mathbf{p}$ indicate global positions of a point P after and before the screw, as shown in Figure 6.21. The screw axis is indicated by the unit vector $\hat{u}$.

Let us break the screw motion to a twist ϕ about $\hat{u}$ and a translation h along $\hat{u}$. A body point P moves from its first position at $\mathbf{p}$ to its second position P' at $\mathbf{p}'$ by a rotation about $\hat{u}$. Then it moves to P'' at $\mathbf{p}''$ by a translation h parallel to $\hat{u}$.

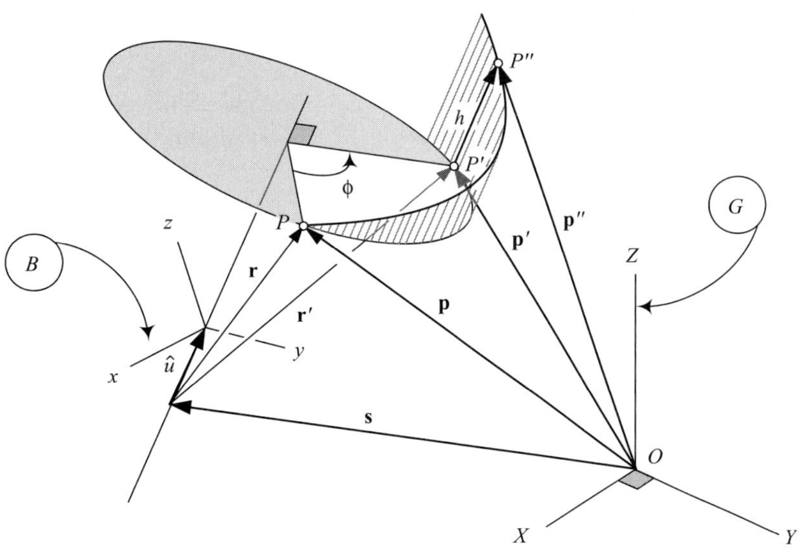

Figure 6.21 Screw motion of a rigid body.

Proof: The axis–angle rotation formula (5.4) relates $\mathbf{r}'$ and $\mathbf{r}$, which are position vectors of P after and before rotation ϕ about $\hat{u}$ when $\mathbf{s} = 0$, $h = 0$:

$$\mathbf{r}' = \mathbf{r}\cos\phi + (1 - \cos\phi)\left(\hat{u}\cdot\mathbf{r}\right)\hat{u} + \left(\hat{u}\times\mathbf{r}\right)\sin\phi \tag{6.236}$$

When the screw axis does not pass through the origin of $G(OXYZ)$, then $\mathbf{r}'$ and $\mathbf{r}$ must accordingly be substituted with the following equations:

$$\mathbf{r} = \mathbf{p} - \mathbf{s} \tag{6.237}$$

$$\mathbf{r}' = \mathbf{p}'' - \mathbf{s} - h\hat{u} \tag{6.238}$$

Substituting (6.237) and (6.238) in (6.236) provides the relationship between the new and old positions of the body point P after a screw motion

$$\begin{aligned}\mathbf{p}'' - (\mathbf{s} + h\hat{u}) &= (\mathbf{p} - \mathbf{s})\cos\phi + [1 - \cos\phi)(\hat{u}\cdot(\mathbf{p} - \mathbf{s})]\hat{u} \\ &\quad + [\hat{u}\times(\mathbf{p} - \mathbf{s})]\sin\phi\end{aligned} \tag{6.239}$$

Equation (6.239) is the *Rodriguez formula* for the most general rigid-body motion.

The vector $\mathbf{r}'$ is a vector after rotation and hence is in the G coordinate frame, and $\mathbf{r}$ is a vector before rotation and hence is in the B coordinate frame:

$${}^{B}\mathbf{r} = {}^{B}\mathbf{p} - {}^{B}\mathbf{s} \tag{6.240}$$

$${}^{G}\mathbf{r} = {}^{G}\mathbf{p} - {}^{G}\mathbf{s} - h{}^{G}\hat{u} \tag{6.241}$$

Using these notations, Equation (6.239) will become

$$\begin{aligned}{}^{G}\mathbf{p} &= \left(\mathbf{I}\cos\phi + \hat{u}\hat{u}^{\mathrm{T}}(1 - \cos\phi) + \tilde{u}\sin\phi\right){}^{B}\mathbf{p} \\ &\quad - \left(\mathbf{I}\cos\phi + \hat{u}\hat{u}^{\mathrm{T}}(1 - \cos\phi) + \tilde{u}\sin\phi\right){}^{B}\mathbf{s} + {}^{G}\mathbf{s} + h{}^{G}\hat{u}\end{aligned} \tag{6.242}$$

The vectors ${}^{B}\mathbf{r}$ and ${}^{G}\mathbf{r}$ are coincident before rotation, and so is the location vector $\mathbf{s}$. Therefore, ${}^{B}\mathbf{s}$, which is the expression of the location vector before rotation, has the same components as ${}^{G}\mathbf{s}$, which remains unchanged in the global frame. Furthermore, the unit vector $\hat{u}$ on the screw axis remains unchanged during the screw motion. We may drop the superscript G from ${}^{G}\hat{u}$ and substitute ${}^{B}\mathbf{s}$ with ${}^{G}\mathbf{s}$ to rewrite Equation (6.242) in a clearer form:

$$\begin{aligned}{}^{G}\mathbf{p} &= \left(\mathbf{I}\cos\phi + \hat{u}\hat{u}^{\mathrm{T}}(1 - \cos\phi) + \tilde{u}\sin\phi\right){}^{B}\mathbf{p} \\ &\quad - \left(\mathbf{I}\cos\phi + \hat{u}\hat{u}^{\mathrm{T}}(1 - \cos\phi) + \tilde{u}\sin\phi\right){}^{G}\mathbf{s} + {}^{G}\mathbf{s} + h\hat{u}\end{aligned} \tag{6.243}$$

Equation (6.243) can be rearranged to show that a screw can be represented by a homogeneous transformation,

$$\begin{aligned}{}^{G}\mathbf{p} &= {}^{G}R_{B}{}^{B}\mathbf{p} + {}^{G}\mathbf{s} - {}^{G}R_{B}{}^{G}\mathbf{s} + h\hat{u} \\ &= {}^{G}R_{B}{}^{B}\mathbf{p} + {}^{G}\mathbf{d} = {}^{G}T_{B}{}^{B}\mathbf{p}\end{aligned} \tag{6.244}$$

$$\begin{aligned}{}^{G}T_{B} &= {}^{G}\check{s}_{B}(h, \phi, \hat{u}, \mathbf{s}) \\ &= \begin{bmatrix}{}^{G}R_{B} & \left(\mathbf{I} - {}^{G}R_{B}\right){}^{G}\mathbf{s} + h\hat{u} \\ 0 & 1\end{bmatrix} = \begin{bmatrix}{}^{G}R_{B} & {}^{G}\mathbf{d} \\ 0 & 1\end{bmatrix}\end{aligned} \tag{6.245}$$

where

$$^G R_B = \mathbf{I}\cos\phi + \hat{u}\hat{u}^{\mathrm{T}}\,(1-\cos\phi) + \tilde{u}\,\sin\phi \tag{6.246}$$

$$^G\mathbf{d} = [(\mathbf{I}-\hat{u}\hat{u}^{\mathrm{T}})(1-\cos\phi) - \tilde{u}\sin\phi]\,^G\mathbf{s} + h\hat{u} \tag{6.247}$$

Direct substitution develops $^G R_B$ and $^G\mathbf{d}$:

$$^G R_B = \begin{bmatrix} u_1^2 \operatorname{vers}\phi + c\phi & u_1u_2 \operatorname{vers}\phi - u_3 s\phi & u_1u_3 \operatorname{vers}\phi + u_2 s\phi \\ u_1u_2 \operatorname{vers}\phi + u_3 s\phi & u_2^2 \operatorname{vers}\phi + c\phi & u_2u_3 \operatorname{vers}\phi - u_1 s\phi \\ u_1u_3 \operatorname{vers}\phi - u_2 s\phi & u_2u_3 \operatorname{vers}\phi + u_1 s\phi & u_3^2 \operatorname{vers}\phi + c\phi \end{bmatrix} \tag{6.248}$$

$$^G\mathbf{d} = \begin{bmatrix} hu_1 + [(1-u_1^2)s_1 - u_1(s_2u_2 + s_3u_3)]\operatorname{vers}\phi + (s_2u_3 - s_3u_2)s\phi \\ hu_2 + [(1-u_2^2)s_2 - u_2(s_3u_3 + s_1u_1)]\operatorname{vers}\phi + (s_3u_1 - s_1u_3)s\phi \\ hu_3 + [(1-u_3^2)s_3 - u_3(s_1u_1 + s_2u_2)]\operatorname{vers}\phi + (s_1u_2 - s_2u_1)s\phi \end{bmatrix} \tag{6.249}$$

This representation of a rigid motion requires six independent parameters, namely one for rotation angle ϕ, one for translation h, two for screw axis $\hat{u}$, and two for location vector $^G\mathbf{s}$. It is because three components of $\hat{u}$ are related to each other according to

$$\hat{u}^{\mathrm{T}}\hat{u} = 1 \tag{6.250}$$

and the location vector $^G\mathbf{s}$ can locate any arbitrary point on the screw axis. It is convenient to choose the point where it has the minimum distance from O and make $^G\mathbf{s}$ perpendicular to $\hat{u}$. Let's indicate the *shortest location vector* by $^G\mathbf{s}_0$; then there is a constraint among the components of the location vector:

$$^G\mathbf{s}_0^{\mathrm{T}}\,\hat{u} = 0 \tag{6.251}$$

If $\mathbf{s} = 0$, then the screw axis passes through the origin of G and (6.245) reduces to the central screw expression (6.231).

The screw parameters ϕ and h, together with the screw axis $\hat{u}$ and location vector $^G\mathbf{s}$, completely define a rigid motion of $B(oxyz)$ in $G(OXYZ)$. So, given the screw parameters and screw axis, we can find the elements of the transformation matrix by Equations (6.248) and (6.249). On the other hand, given the transformation matrix $^G T_B$, we can find the screw angle and axis by

$$\begin{aligned}\cos\phi &= \tfrac{1}{2}\left[\operatorname{tr}\left(^G R_B\right) - 1\right) = \tfrac{1}{2}\left(\operatorname{tr}\left(^G T_B\right) - 2\right] \\ &= \tfrac{1}{2}(r_{11} + r_{22} + r_{33} - 1)\end{aligned} \tag{6.252}$$

$$\tilde{u} = \frac{1}{2\sin\phi}\left(^G R_B - {^G R_B^{\mathrm{T}}}\right) = \frac{1}{2\sin\phi}\begin{bmatrix} r_{32} - r_{23} \\ r_{13} - r_{31} \\ r_{21} - r_{12} \end{bmatrix} \tag{6.253}$$

To find all the required screw parameters, we must also find h and coordinates of one point on the screw axis. Since the points on the screw axis are invariant under the rotation, we must have

$$\begin{bmatrix} r_{11} & r_{12} & r_{13} & r_{14} \\ r_{21} & r_{22} & r_{23} & r_{24} \\ r_{31} & r_{32} & r_{33} & r_{34} \\ 0 & 0 & 0 & 1 \end{bmatrix} \begin{bmatrix} X \\ Y \\ Z \\ 1 \end{bmatrix} = \begin{bmatrix} 1 & 0 & 0 & hu_1 \\ 0 & 1 & 0 & hu_2 \\ 0 & 0 & 1 & hu_3 \\ 0 & 0 & 0 & 1 \end{bmatrix} \begin{bmatrix} X \\ Y \\ Z \\ 1 \end{bmatrix} \tag{6.254}$$

where (X, Y, Z) are coordinates of points on the screw axis.

As a sample point, we may set $X_s = 0$ and find the intersection point of the screw line with the (Y, Z)-plane by searching for $\mathbf{s} = [0, Y_s, Z_s]^{\mathrm{T}}$. Therefore,

$$\begin{bmatrix} r_{11}-1 & r_{12} & r_{13} & r_{14}-hu_1 \\ r_{21} & r_{22}-1 & r_{23} & r_{24}-hu_2 \\ r_{31} & r_{32} & r_{33}-1 & r_{34}-hu_3 \\ 0 & 0 & 0 & 0 \end{bmatrix} \begin{bmatrix} 0 \\ Y_s \\ Z_s \\ 1 \end{bmatrix} = \begin{bmatrix} 0 \\ 0 \\ 0 \\ 0 \end{bmatrix} \tag{6.255}$$

which generates three equations to be solved for Y_s, Z_s, and h:

$$\begin{bmatrix} h \\ Y_s \\ Z_s \end{bmatrix} = \begin{bmatrix} u_1 & -r_{12} & -r_{13} \\ u_2 & 1-r_{22} & -r_{23} \\ u_3 & -r_{32} & 1-r_{33} \end{bmatrix}^{-1} \begin{bmatrix} r_{14} \\ r_{24} \\ r_{34} \end{bmatrix} \tag{6.256}$$

Now we can find the shortest location vector ${}^G\mathbf{s}_0$ by

$${}^G\mathbf{s}_0 = \mathbf{s} - (\mathbf{s} \cdot \hat{u})\hat{u} \tag{6.257}$$

It is interesting to know that the instantaneous screw axis was first used by Giulio Mozzi (1730–1813); however, Michel Chasles (1793–1880) is credited with this discovery. ■

Example 366 ★ Alternative Proof of Screw Transformation Matrix Assume that a body and global frames are parallel initially. To turn the body about an off-center axis $\hat{u}$, we can translate the body frame B along $-\mathbf{s} - h\hat{u}$ to bring the body frame B to be coincident with the global frame G followed by a rigid-body motion that is a rotation about $\hat{u}$ and a translation along $\mathbf{s}$:

$$\begin{aligned} {}^GT_B &= D_{\hat{s},s}\, R_{\hat{u},\phi}\, D_{\hat{s},-s}\, D_{\hat{u},-h} \\ &= \begin{bmatrix} \mathbf{I} & \mathbf{s} \\ 0 & 1 \end{bmatrix} \begin{bmatrix} R_{\hat{u},\phi} & 0 \\ 0 & 1 \end{bmatrix} \begin{bmatrix} \mathbf{I} & -\mathbf{s} \\ 0 & 1 \end{bmatrix} \begin{bmatrix} \mathbf{I} & -h\hat{u} \\ 0 & 1 \end{bmatrix} \\ &= \begin{bmatrix} R_{\hat{u},\phi} & \mathbf{s} - R_{\hat{u},\phi}\mathbf{s} + hR_{\hat{u},\phi}\hat{u} \\ 0 & 1 \end{bmatrix} \\ &= \begin{bmatrix} R_{\hat{u},\phi} & \left(\mathbf{I} - R_{\hat{u},\phi}\right)\mathbf{s} + h\hat{u} \\ 0 & 1 \end{bmatrix} = \begin{bmatrix} {}^GR_B & {}^G\mathbf{d} \\ 0 & 1 \end{bmatrix} \end{aligned} \tag{6.258}$$

The rotation matrix ${}^GR_B$ is given in (6.248) and the displacement vector ${}^G\mathbf{d}$ is given in (6.249). Substituting $h = 0$ reduces ${}^GT_B$ to (6.141) in Example 359, which indicates a rotation about an off-center axis. Rotation of a rigid body about an axis indicated by $\hat{u}$

and passing through a point at ${}^G\mathbf{s}$ such that ${}^G\mathbf{s} \times \hat{u} \neq 0$ is a rotation about an off-center axis. Therefore, the off-center rotation transformation is

$$ {}^G T_B = \begin{bmatrix} {}^G R_B & {}^G\mathbf{s} - {}^G R_B {}^G\mathbf{s} \\ 0 & 1 \end{bmatrix} \tag{6.259} $$

Example 367 ★ Alternative Derivation of Screw Transformation Assume the screw axis does not pass through the origin of G. If ${}^G\mathbf{s}$ is the position vector of some point on the axis $\hat{u}$, then we can derive the matrix representation of the screw $\check{s}(h, \phi, \hat{u}, \mathbf{s})$ by translating the screw axis back to the origin, performing the central screw motion, and translating the line back to its original position:

$$ \begin{aligned} \check{s}(h, \phi, \hat{u}, \mathbf{s}) &= D\left({}^G\mathbf{s}\right) \check{s}\left(h, \phi, \hat{u}\right) D\left(-{}^G\mathbf{s}\right) \\ &= D\left({}^G\mathbf{s}\right) D(h\hat{u})\, R(\hat{u}, \phi)\, D\left(-{}^G\mathbf{s}\right) \\ &= \begin{bmatrix} \mathbf{I} & {}^G\mathbf{s} \\ 0 & 1 \end{bmatrix} \begin{bmatrix} {}^G R_B & h\hat{u} \\ 0 & 1 \end{bmatrix} \begin{bmatrix} \mathbf{I} & -{}^G\mathbf{s} \\ 0 & 1 \end{bmatrix} \\ &= \begin{bmatrix} {}^G R_B & \left(\mathbf{I} - {}^G R_B\right) {}^G\mathbf{s} + h\hat{u} \\ 0 & 1 \end{bmatrix} \end{aligned} \tag{6.260} $$

Example 368 ★ Central Screw Transformation of a Base Unit Vector Consider two initially coincident frames $G(OXYZ)$ and $B(oxyz)$. The body performs a screw motion along the Y-axis for $h = 3$ and $\phi = 90\,\text{deg}$. The position of a body point at ${}^B\mathbf{r} = [1, 1, 0, 1]^{\mathrm{T}}$ can be found by applying the central screw transformation:

$$ \check{s}(h, \phi, \hat{u}) = \check{s}\left(3, \tfrac{1}{2}\pi, \hat{J}\right) = D\left(3\hat{J}\right) R\left(\hat{J}, \tfrac{1}{2}\pi\right) \tag{6.261} $$

$$ = \begin{bmatrix} 1 & 0 & 0 & 0 \\ 0 & 1 & 0 & 3 \\ 0 & 0 & 1 & 0 \\ 0 & 0 & 0 & 1 \end{bmatrix} \begin{bmatrix} 0 & 0 & 1 & 0 \\ 0 & 1 & 0 & 0 \\ -1 & 0 & 0 & 0 \\ 0 & 0 & 0 & 1 \end{bmatrix} = \begin{bmatrix} 0 & 0 & 1 & 0 \\ 0 & 1 & 0 & 3 \\ -1 & 0 & 0 & 0 \\ 0 & 0 & 0 & 1 \end{bmatrix} $$

Therefore,

$$ {}^G\mathbf{r} = \check{s}\left(3, \tfrac{1}{2}\pi, \hat{J}\right) {}^B\mathbf{r} = \begin{bmatrix} 0 & 0 & 1 & 0 \\ 0 & 1 & 0 & 3 \\ -1 & 0 & 0 & 0 \\ 0 & 0 & 0 & 1 \end{bmatrix} \begin{bmatrix} 1 \\ 1 \\ 0 \\ 1 \end{bmatrix} = \begin{bmatrix} 0 \\ 4 \\ -1 \\ 1 \end{bmatrix} \tag{6.262} $$

The pitch of this screw is

$$ p = \frac{h}{\phi} = \frac{3}{\pi/2} = \frac{6}{\pi} = 1.9099\,\text{units/rad} \tag{6.263} $$

Example 369 ★ Screw Transformation of a Point Consider two initially parallel frames $G(OXYZ)$ and $B(oxyz)$. The body performs a screw motion for $h = 5$ and $\phi = 90\,\text{deg}$ along a line that is parallel to the Y-axis and passes through $X = 4$. Therefore, the body coordinate frame is at location $\mathbf{s} = [4, 0, 0]^T$. The position of a body point at ${}^B\mathbf{r} = [3,0,0]^T$ can be found by applying the screw transformation

$$
{}^GT_B = \begin{bmatrix} {}^GR_B & \left(\mathbf{I} - {}^GR_B\right)\mathbf{s} + h\hat{u} \\ 0 & 1 \end{bmatrix} = \begin{bmatrix} 0 & 0 & 1 & 2 \\ 0 & 1 & 0 & 2 \\ -1 & 0 & 0 & 2 \\ 0 & 0 & 0 & 1 \end{bmatrix} \tag{6.264}
$$

because

$$
{}^GR_B = \begin{bmatrix} 0 & 0 & 1 \\ 0 & 1 & 0 \\ -1 & 0 & 0 \end{bmatrix} \qquad \mathbf{s} = \begin{bmatrix} 4 \\ 0 \\ 0 \end{bmatrix} \qquad \hat{u} = \begin{bmatrix} 0 \\ 1 \\ 0 \end{bmatrix} \tag{6.265}
$$

Therefore, the position vector of ${}^G\mathbf{r}$ would be

$$
{}^G\mathbf{r} = {}^GT_B\,{}^B\mathbf{r} = \begin{bmatrix} 0 & 0 & 1 & 4 \\ 0 & 1 & 0 & 5 \\ -1 & 0 & 0 & 4 \\ 0 & 0 & 0 & 1 \end{bmatrix} \begin{bmatrix} 3 \\ 0 \\ 0 \\ 1 \end{bmatrix} = \begin{bmatrix} 4 \\ 5 \\ 1 \\ 1 \end{bmatrix} \tag{6.266}
$$

Any point on the screw axis $\hat{u}$ will only change its coordinate by h on $\hat{u}$. As an example, a point at ${}^B\mathbf{r} = [4, 0, 0]^T$ will change its y-coordinate by h:

$$
\begin{bmatrix} 0 & 0 & 1 & 4 \\ 0 & 1 & 0 & 5 \\ -1 & 0 & 0 & 4 \\ 0 & 0 & 0 & 1 \end{bmatrix} \begin{bmatrix} 4 \\ 0 \\ 0 \\ 1 \end{bmatrix} = \begin{bmatrix} 4 \\ 5 \\ 0 \\ 1 \end{bmatrix} \tag{6.267}
$$

Example 370 ★ Rotation of a Vector Consider a vector ${}^B\mathbf{r}$ that connects a point P_1 at ${}^B\mathbf{r}_1$ to a point P_2 at ${}^B\mathbf{r}_2$. The body frame B is coincident with the global frame G and has a fixed origin. The rotation transformation equation between B and G can be written as

$$
{}^G(\mathbf{r}_2 - \mathbf{r}_1) = {}^GR_B\,{}^B(\mathbf{r}_2 - \mathbf{r}_1) \tag{6.268}
$$

Assume the initial and final positions of point P_1 are along the rotation axis. Equation (6.268) can then be rearranged in a suitable form to calculate the coordinates of the new position of point P_2:

$$
\begin{aligned}
{}^G\mathbf{r}_2 &= {}^GR_B\,{}^B(\mathbf{r}_2 - \mathbf{r}_1) + {}^G\mathbf{r}_1 \\
&= {}^GR_B\,{}^B\mathbf{r}_2 + {}^G\mathbf{r}_1 - {}^GR_B\,{}^B\mathbf{r}_1 = {}^GT_B\,{}^B\mathbf{r}_2
\end{aligned} \tag{6.269}
$$

where

$$
{}^GT_B = \begin{bmatrix} {}^GR_B & {}^G\mathbf{r}_1 - {}^GR_B\,{}^B\mathbf{r}_1 \\ 0 & 1 \end{bmatrix} \tag{6.270}
$$

It is compatible with screw motion (6.245) for $h = 0$.

Example 371 ★ Special Case for Screw Determination When $r_{11} = r_{22} = r_{33} = 1$ and $\phi = 0$, the screw motion is a pure translation h parallel to $\hat{u}$, where

$$\hat{u} = \frac{r_{14} - s_1}{h}\hat{I} + \frac{r_{24} - s_2}{h}\hat{J} + \frac{r_{34} - s_3}{h}\hat{K} \tag{6.271}$$

Example 372 ★ Determination of Screw Parameters We are able to determine the screw parameters when we have the original and final positions of three non-colinear points of a rigid body. Assume $\mathbf{p}_0$, $\mathbf{q}_0$, and $\mathbf{r}_0$ denote the positions of points P, Q, and R before the screw motion and $\mathbf{p}_1$, $\mathbf{q}_1$, and $\mathbf{r}_1$ denote their positions after the screw motion.

To determine the screw parameters ϕ, $\hat{u}$, h, and $\mathbf{s}$, we should solve the following three simultaneous Rodriguez equations:

$$\mathbf{p}_1 - \mathbf{p}_0 = \tan\frac{\phi}{2}\hat{u} \times (\mathbf{p}_1 + \mathbf{p}_0 - 2\mathbf{s}) + h\hat{u} \tag{6.272}$$

$$\mathbf{q}_1 - \mathbf{q}_0 = \tan\frac{\phi}{2}\hat{u} \times (\mathbf{q}_1 + \mathbf{q}_0 - 2\mathbf{s}) + h\hat{u} \tag{6.273}$$

$$\mathbf{r}_1 - \mathbf{r}_0 = \tan\frac{\phi}{2}\hat{u} \times (\mathbf{r}_1 + \mathbf{r}_0 - 2\mathbf{s}) + h\hat{u} \tag{6.274}$$

First, we subtract Equation (6.274) from (6.272) and (6.273):

$$(\mathbf{p}_1 - \mathbf{p}_0) - (\mathbf{r}_1 - \mathbf{r}_0) = \tan\frac{\phi}{2}\hat{u} \times \left[(\mathbf{p}_1 + \mathbf{p}_0) - (\mathbf{r}_1 - \mathbf{r}_0)\right] \tag{6.275}$$

$$(\mathbf{q}_1 - \mathbf{q}_0) - (\mathbf{r}_1 - \mathbf{r}_0) = \tan\frac{\phi}{2}\hat{u} \times \left[(\mathbf{q}_1 + \mathbf{q}_0) - (\mathbf{r}_1 - \mathbf{r}_0)\right] \tag{6.276}$$

Multiplying both sides of (6.275) by $\left[(\mathbf{q}_1 - \mathbf{q}_0) - (\mathbf{r}_1 - \mathbf{r}_0)\right]$, which is perpendicular to $\hat{u}$,

$$\begin{aligned}\left[(\mathbf{q}_1 - \mathbf{q}_0) - (\mathbf{r}_1 - \mathbf{r}_0)\right] \times \left[(\mathbf{p}_1 - \mathbf{p}_0) - (\mathbf{r}_1 - \mathbf{r}_0)\right] &= \tan\frac{\phi}{2}\left[(\mathbf{q}_1 - \mathbf{q}_0) - (\mathbf{r}_1 - \mathbf{r}_0)\right] \\ &\quad \times \left\{\hat{u} \times \left[(\mathbf{p}_1 + \mathbf{p}_0) - (\mathbf{r}_1 - \mathbf{r}_0)\right]\right\}\end{aligned} \tag{6.277}$$

gives us

$$\begin{aligned}&\left[(\mathbf{q}_1 - \mathbf{q}_0) - (\mathbf{r}_1 - \mathbf{r}_0)\right] \times \left[(\mathbf{p}_1 + \mathbf{p}_0) - (\mathbf{r}_1 - \mathbf{r}_0)\right] \\ &\quad = \tan\frac{\phi}{2}\left\{\left[(\mathbf{q}_1 - \mathbf{q}_0) - (\mathbf{r}_1 - \mathbf{r}_0)\right] \cdot \left[(\mathbf{p}_1 + \mathbf{p}_0) - (\mathbf{r}_1 - \mathbf{r}_0)\right]\right\}\hat{u}\end{aligned} \tag{6.278}$$

and therefore, the rotation angle can be found by equating $\tan(\phi/2)$ and the norm of the right-hand side of the following equation:

$$\tan\frac{\phi}{2}\hat{u} = \frac{\left[(\mathbf{q}_1 - \mathbf{q}_0) - (\mathbf{r}_1 - \mathbf{r}_0)\right] \times \left[(\mathbf{p}_1 + \mathbf{p}_0) - (\mathbf{r}_1 - \mathbf{r}_0)\right]}{\left[(\mathbf{q}_1 - \mathbf{q}_0) - (\mathbf{r}_1 - \mathbf{r}_0)\right] \cdot \left[(\mathbf{p}_1 + \mathbf{p}_0) - (\mathbf{r}_1 - \mathbf{r}_0)\right]} \tag{6.279}$$

To find $\mathbf{s}$, we may start with the cross product of $\hat{u}$ with Equation (6.272):

$$\hat{u} \times (\mathbf{p}_1 - \mathbf{p}_0) = \hat{u} \times \left[\tan\frac{\phi}{2}\hat{u} \times (\mathbf{p}_1 + \mathbf{p}_0 - 2\mathbf{s}) + h\hat{u}\right]$$
$$= \tan\frac{\phi}{2}\left\{\left[\hat{u}\cdot(\mathbf{p}_1 + \mathbf{p}_0)\right]\hat{u} - (\mathbf{p}_1 + \mathbf{p}_0) + 2\left[\mathbf{s} - \left(\hat{u}\cdot\mathbf{s}\right)\hat{u}\right]\right\} \quad (6.280)$$

$\mathbf{s}$ is a vector from the origin of the global frame $G(OXYZ)$ to an arbitrary point on the screw axis, and the term $\mathbf{s} - \left(\hat{u}\cdot\mathbf{s}\right)\hat{u}$ is the perpendicular component of $\mathbf{s}$ to $\hat{u}$. This perpendicular component indicates a vector with the shortest distance between O and $\hat{u}$. Let us show the shortest $\mathbf{s}$ by $\mathbf{s}_0$. Therefore,

$$\mathbf{s}_0 = \mathbf{s} - \left(\hat{u}\cdot\mathbf{s}\right)\hat{u}$$
$$= \frac{1}{2}\left[\frac{\hat{u}\times\mathbf{p}_1 - \mathbf{p}_0}{\tan\frac{\phi}{2}} - \left[\hat{u}\cdot(\mathbf{p}_1 + \mathbf{p}_0)\right]\hat{u} + \mathbf{p}_1 + \mathbf{p}_0\right] \quad (6.281)$$

The last parameter of the screw is the pitch h, which can be found from any one of the Equations (6.272), (6.273), and (6.274):

$$h = \hat{u}\cdot(\mathbf{p}_1 - \mathbf{p}_0) = \hat{u}\cdot(\mathbf{q}_1 - \mathbf{q}_0) = \hat{u}\cdot(\mathbf{r}_1 - \mathbf{r}_0) \quad (6.282)$$

Example 373 ★ Principal Central Screw There are three principal central screws, namely the X-screw, Y-screw, and Z-screw, which are given as

$$\check{s}(h_Z, \alpha, \hat{K}) = \begin{bmatrix} \cos\alpha & -\sin\alpha & 0 & 0 \\ \sin\alpha & \cos\alpha & 0 & 0 \\ 0 & 0 & 1 & p_Z\alpha \\ 0 & 0 & 0 & 1 \end{bmatrix} \quad (6.283)$$

$$\check{s}(h_Y, \beta, \hat{J}) = \begin{bmatrix} \cos\beta & 0 & \sin\beta & 0 \\ 0 & 1 & 0 & p_Y\beta \\ -\sin\beta & 0 & \cos\beta & 0 \\ 0 & 0 & 0 & 1 \end{bmatrix} \quad (6.284)$$

$$\check{s}(h_X, \gamma, \hat{I}) = \begin{bmatrix} 1 & 0 & 0 & p_X\gamma \\ 0 & \cos\gamma & -\sin\gamma & 0 \\ 0 & \sin\gamma & \cos\gamma & 0 \\ 0 & 0 & 0 & 1 \end{bmatrix} \quad (6.285)$$

Example 374 ★ Proof of Chasles Theorem Let $[T]$ be an arbitrary spatial displacement and decompose it into a rotation $[R]$ about $\hat{u}$ and a translation $[D]$:

$$[T] = [D][R] \quad (6.286)$$

Furthermore, we may also decompose the translation $[D]$ into parallel and perpendicular components $[D_{\parallel}]$ and $[D_{\perp}]$ to $\hat{u}$:

$$[T] = [D_{\parallel}][D_{\perp}][R] \tag{6.287}$$

Now $[D_{\perp}][R]$ is a planar motion and is therefore equivalent to some rotation $[R'] = [D_{\perp}][R]$ about an axis parallel to the rotation axis $\hat{u}$. So, we have $[T] = [D_{\parallel}][R']$. This decomposition completes the proof because the axis of $[D_{\parallel}]$ can be taken equal to $\hat{u}$.

Example 375 ★ Every Rigid Motion Is a Screw To show that any proper rigid motion can be considered as a screw motion, we must show that a homogeneous transformation matrix

$${}^{G}T_{B} = \begin{bmatrix} {}^{G}R_{B} & {}^{G}\mathbf{d} \\ 0 & 1 \end{bmatrix} \tag{6.288}$$

can be written in the form

$${}^{G}T_{B} = \begin{bmatrix} {}^{G}R_{B} & (\mathbf{I} - {}^{G}R_{B})\,\mathbf{s} + h\hat{u} \\ 0 & 1 \end{bmatrix} \tag{6.289}$$

This problem is then equivalent to the following equation to find h and $\hat{u}$:

$${}^{G}\mathbf{d} = (\mathbf{I} - {}^{G}R_{B})\,\mathbf{s} + h\hat{u} \tag{6.290}$$

The matrix $[\mathbf{I} - {}^{G}R_{B}]$ is singular because ${}^{G}R_{B}$ always has 1 as an eigenvalue. This eigenvalue corresponds to $\hat{u}$ as the eigenvector. Therefore,

$$\left[\mathbf{I} - {}^{G}R_{B}\right]\hat{u} = \left[\mathbf{I} - {}^{G}R_{B}^{\mathrm{T}}\right]\hat{u} = 0 \tag{6.291}$$

and an inner product shows us that

$$\begin{aligned} \hat{u} \cdot {}^{G}\mathbf{d} &= \hat{u} \cdot \left[\mathbf{I} - {}^{G}R_{B}\right]\,\mathbf{s} + \hat{u} \cdot h\hat{u} \\ &= \left[\mathbf{I} - {}^{G}R_{B}\right]\hat{u} \cdot \,\mathbf{s} + \hat{u} \cdot h\hat{u} \end{aligned} \tag{6.292}$$

which leads to

$$h = \hat{u} \cdot {}^{G}\mathbf{d} \tag{6.293}$$

Now we may use h to find $\mathbf{s}$:

$$\mathbf{s} = \left[\mathbf{I} - {}^{G}R_{B}\right]^{-1}\left({}^{G}\mathbf{d} - h\hat{u}\right) \tag{6.294}$$

Example 376 ★ Classification of Rigid-Body Motion Consider a rigid body with a local coordinate frame $B(oxyz)$ that is moving with respect to a global frame $G(OXYZ)$.

The possible motions of $B(oxyz)$ and the transformation between B- and G-frames can be classified as:

1. Rotation ϕ about an axis that is passing through the origin and indicated by the unit vector $\hat{u} = [u_1, u_2, u_3]^{\mathrm{T}}$:

$$
{}^G\mathbf{r} = {}^G R_B \, {}^B\mathbf{r} \tag{6.295}
$$

2. Rotation ${}^G R_B$ plus translation ${}^G\mathbf{d} = [d_1, d_2, d_3]^{\mathrm{T}}$:

$$
{}^G\mathbf{r} = {}^G R_B \, {}^B\mathbf{r} + {}^G\mathbf{d} = {}^G T_B \, {}^B\mathbf{r} \tag{6.296}
$$

3. Rotation ϕ about an axis on the unit vector $\hat{u} = [u_1, u_2, u_3]^{\mathrm{T}}$ that is passing through an arbitrary point indicated by ${}^G\mathbf{s} = [s_1, s_2, s_3]^{\mathrm{T}}$:

$$
{}^G\mathbf{r} = {}^G R_B \, {}^B\mathbf{r} + \left(\mathbf{I} - {}^G R_B\right) {}^G\mathbf{s} \tag{6.297}
$$

4. Screw motion with angle ϕ and displacement h about and along an axis directed by $\hat{u} = [u_1, u_2, u_3]^{\mathrm{T}}$ passing through an arbitrary point indicated by ${}^G\mathbf{s} = [s_1, s_2, s_3]^{\mathrm{T}}$:

$$
{}^G\mathbf{r} = {}^G R_B \, {}^B\mathbf{r} + \left(\mathbf{I} - {}^G R_B\right) {}^G\mathbf{s} + h\hat{u} \tag{6.298}
$$

Besides the above rigid-body motions, the transformation matrices are the main applied tools in computer graphics to calculate motion and reflection. The reflection of a body point can be classified as:

1. Reflection in the xy-plane:

$$
{}^G\mathbf{r} = {}^G R_B \, {}^B\mathbf{r}_{(-z)} \tag{6.299}
$$

where

$$
\mathbf{r}_{(-z)} = \begin{bmatrix} p_1 \\ p_2 \\ -p_3 \end{bmatrix} \tag{6.300}
$$

2. Reflection in the yz-plane:

$$
{}^G\mathbf{r} = {}^G R_B \, {}^B\mathbf{r}_{(-x)} \tag{6.301}
$$

where

$$
\mathbf{r}_{(-z)} = \begin{bmatrix} -p_1 \\ p_2 \\ p_3 \end{bmatrix} \tag{6.302}
$$

3. Reflection in the xz-plane:

$$
{}^G\mathbf{r} = {}^G R_B \, {}^B\mathbf{r}_{(-y)} \tag{6.303}
$$

where

$$
\mathbf{r}_{(-z)} = \begin{bmatrix} p_1 \\ -p_2 \\ p_3 \end{bmatrix} \tag{6.304}
$$

4. Reflection in a plane with equation $u_1 x + u_2 y + u_3 z + h = 0$:

$$ {}^{G}\mathbf{r} = \frac{1}{u_1^2 + u_2^2 + u_3^2} \left({}^{G}R_B \, {}^{B}\mathbf{r} - 2h\hat{u} \right) = {}^{G}R_B \, {}^{B}\mathbf{r} - 2h\hat{u} \tag{6.305} $$

where

$$ {}^{G}R_B = \begin{bmatrix} -u_1^2 + u_2^2 + u_3^2 & -2u_1 u_2 & -2u_3 u_1 \\ -2u_2 u_1 & u_1^2 - u_2^2 + u_3^2 & -2u_2 u_3 \\ -2u_1 u_3 & -2u_3 u_2 & u_1^2 + u_2^2 - u_3^2 \end{bmatrix} \tag{6.306} $$

5. Reflection in a plane going through point $[s_1, s_2, s_3]^{\mathrm{T}}$ and normal to $\hat{u} = [u_1, u_2, u_3]^{\mathrm{T}}$:

$$ {}^{G}\mathbf{r} = {}^{G}R_B \, {}^{B}\mathbf{r} + \left(\mathbf{I} - {}^{G}R_B \right) {}^{G}\mathbf{s} \tag{6.307} $$

where ${}^{G}R_B$ is as in (6.306).

6.6 ★ INVERSE SCREW

The inverse of a screw $\check{s}(h, \phi, \hat{u}, \mathbf{s})$ is defined by

$$ \begin{aligned} {}^{G}\check{s}_B^{-1}(h, \phi, \hat{u}, \mathbf{s}) &= {}^{B}\check{s}_G(h, \phi, \hat{u}, \mathbf{s}) \\ &= \begin{bmatrix} {}^{G}R_B^{\mathrm{T}} & {}^{G}\mathbf{s} - {}^{G}R_B^{\mathrm{T}} \, {}^{G}\mathbf{s} - h\hat{u} \\ 0 & 1 \end{bmatrix} \end{aligned} \tag{6.308} $$

where $\hat{u}$ is a unit vector indicating the axis of the screw, $\mathbf{s}$ is the location vector of a point on the axis of the screw, ϕ is the screw angle, and h is the screw translation. If the screw is central, the axis of the screw passes through the origin and $\mathbf{s} = 0$. Therefore, the inverse of a central screw is

$$ {}^{G}\check{s}_B^{-1}(h, \phi, \hat{u}) = \begin{bmatrix} {}^{G}R_B^{\mathrm{T}} & -h\hat{u} \\ 0 & 1 \end{bmatrix} \tag{6.309} $$

Proof: The homogeneous matrix expression of a screw $\check{s}(h, \phi, \hat{u}, \mathbf{s})$ is

$$ \begin{aligned} {}^{G}T_B &= {}^{G}\check{s}_B(h, \phi, \hat{u}, \mathbf{s}) \\ &= \begin{bmatrix} {}^{G}R_B & {}^{G}\mathbf{s} - {}^{G}R_B \, {}^{G}\mathbf{s} + h\hat{u} \\ 0 & 1 \end{bmatrix} = \begin{bmatrix} {}^{G}R_B & {}^{G}\mathbf{d} \\ 0 & 1 \end{bmatrix} \end{aligned} \tag{6.310} $$

A homogeneous matrix can be inverted according to

$$ {}^{B}T_G = {}^{G}T_B^{-1} = \begin{bmatrix} {}^{G}R_B^{\mathrm{T}} & -{}^{G}R_B^{\mathrm{T}} \, {}^{G}\mathbf{d} \\ 0 & 1 \end{bmatrix} \tag{6.311} $$

To show Equation (6.308), we need to calculate $-{}^G R_B^{\mathrm{T}}\, {}^G\mathbf{d}$:

$$
\begin{aligned}
-{}^G R_B^{\mathrm{T}}\, {}^G\mathbf{d} &= -{}^G R_B^{\mathrm{T}} \left({}^G\mathbf{s} - {}^G R_B\, {}^G\mathbf{s} + h\hat{u}\right) \\
&= -{}^G R_B^{\mathrm{T}}\, {}^G\mathbf{s} + {}^G R_B^{\mathrm{T}}\, {}^G R_B\, {}^G\mathbf{s} - {}^G R_B^{\mathrm{T}}\, h\hat{u} \\
&= -{}^G R_B^{\mathrm{T}}\, {}^G\mathbf{s} + {}^G\mathbf{s} - {}^G R_B^{\mathrm{T}}\, h\hat{u}
\end{aligned}
\tag{6.312}
$$

Since $\hat{u}$ is an invariant vector in both coordinate frames B and G, we have

$$
\hat{u} = {}^G R_B\, \hat{u} = {}^G R_B^{\mathrm{T}}\, \hat{u} \tag{6.313}
$$

and therefore,

$$
-{}^G R_B^{\mathrm{T}}\, {}^G\mathbf{d} = {}^G\mathbf{s} - {}^G R_B^{\mathrm{T}}\, {}^G\mathbf{s} - h\hat{u} \tag{6.314}
$$

This completes the inversion of a general screw:

$$
{}^G\check{s}_B^{-1}(h, \phi, \hat{u}, \mathbf{s}) = \begin{bmatrix} {}^G R_B^{\mathrm{T}} & {}^G\mathbf{s} - {}^G R_B^{\mathrm{T}}\, {}^G\mathbf{s} - h\hat{u} \\ 0 & 1 \end{bmatrix} \tag{6.315}
$$

If the screw is central, the location vector is zero and the inverse of the screw simplifies to

$$
{}^G\check{s}_B^{-1}(h, \phi, \hat{u}) = \begin{bmatrix} {}^G R_B^{\mathrm{T}} & -h\hat{u} \\ 0 & 1 \end{bmatrix} \tag{6.316}
$$

The inversion of a rotation matrix ${}^G R_B = R_{\hat{u},\phi}$ can be found by a rotation $-\phi$ about $\hat{u}$:

$$
{}^G R_B^{-1} = {}^G R_B^{\mathrm{T}} = {}^B R_G = R_{\hat{u},-\phi} \tag{6.317}
$$

So, the inversion of a screw can also be interpreted as a rotation $-\phi$ about $\hat{u}$ plus a translation $-h$ along $\hat{u}$:

$$
{}^G\check{s}_B^{-1}(h, \phi, \hat{u}, \mathbf{s}) = \check{s}(-h, -\phi, \hat{u}, \mathbf{s}) \tag{6.318}
$$

■

Example 377 Checking the Inversion Formula Employing the inversion screw formula, we must have

$$
{}^G\check{s}_B(h, \phi, \hat{u}, \mathbf{s})\, {}^G\check{s}_B^{-1}(h, \phi, \hat{u}, \mathbf{s}) = \mathbf{I}_4 \tag{6.319}
$$

It can be checked by

$$
\begin{aligned}
&\begin{bmatrix} {}^G R_B & {}^G\mathbf{s} - {}^G R_B\, {}^G\mathbf{s} + h\hat{u} \\ 0 & 1 \end{bmatrix} \begin{bmatrix} {}^G R_B^{\mathrm{T}} & {}^G\mathbf{s} - {}^G R_B^{\mathrm{T}}\, {}^G\mathbf{s} - h\hat{u} \\ 0 & 1 \end{bmatrix} \\
&\quad = \begin{bmatrix} \mathbf{I}_3 & {}^G R_B \left({}^G\mathbf{s} - {}^G R_B^{\mathrm{T}}\, {}^G\mathbf{s} - h\hat{u}\right) + \left({}^G\mathbf{s} - {}^G R_B\, {}^G\mathbf{s} + h\hat{u}\right) \\ 0 & 1 \end{bmatrix} \\
&\quad = \begin{bmatrix} \mathbf{I}_3 & {}^G R_B\, {}^G\mathbf{s} - {}^G\mathbf{s} - h\, {}^G R_B\, \hat{u} + {}^G\mathbf{s} - {}^G R_B\, {}^G\mathbf{s} + h\hat{u} \\ 0 & 1 \end{bmatrix} \\
&\quad = \begin{bmatrix} \mathbf{I}_3 & 0 \\ 0 & 1 \end{bmatrix} = \mathbf{I}_4
\end{aligned}
\tag{6.320}
$$

6.7 ★ COMPOUND SCREW TRANSFORMATION

Assume ${}^{1}\check{s}_2(h_1, \phi_1, \hat{u}_1, \mathbf{s}_1)$ is a screw motion to move from coordinate frame B_2 to B_1 and ${}^{G}\check{s}_1(h_0, \phi_0, \hat{u}_0, \mathbf{s}_0)$ is a screw motion to move from coordinate frame B_1 to G. Then, the screw motion to move from B_2 to G is

$$\begin{aligned} {}^{G}\check{s}_2(h, \phi, \hat{u}, \mathbf{s}) &= {}^{G}\check{s}_1(h_0, \phi_0, \hat{u}_0, \mathbf{s}_0)\,{}^{1}\check{s}_2(h_1, \phi_1, \hat{u}_1, \mathbf{s}_1) \\ &= \begin{bmatrix} {}^{G}R_2 & {}^{G}R_1(\mathbf{I} - {}^{1}R_2)\,\mathbf{s}_1 + (\mathbf{I} - {}^{G}R_1)\,\mathbf{s}_0 + h_1\,{}^{G}R_1\hat{u}_1 + h_0\hat{u}_0 \\ 0 & 1 \end{bmatrix} \end{aligned} \tag{6.321}$$

Proof: Direct substitution for ${}^{1}s_2(h_1, \phi_1, \hat{u}_1)$ and ${}^{G}s_1(h_0, \phi_0, \hat{u}_0)$,

$${}^{G}\check{s}_1(h_0, \phi_0, \hat{u}_0, \mathbf{s}_0) = \begin{bmatrix} {}^{G}R_1 & \mathbf{s}_0 - {}^{G}R_1\,\mathbf{s}_0 + h_0\hat{u}_0 \\ 0 & 1 \end{bmatrix} \tag{6.322}$$

$${}^{1}\check{s}_2(h_1, \phi_1, \hat{u}_1, \mathbf{s}_1) = \begin{bmatrix} {}^{1}R_2 & \mathbf{s}_1 - {}^{1}R_2\,\mathbf{s}_1 + h_1\hat{u}_1 \\ 0 & 1 \end{bmatrix} \tag{6.323}$$

shows that

$$\begin{aligned} & {}^{G}\check{s}_2(h, \phi, \hat{u}, \mathbf{s}) \\ &= \begin{bmatrix} {}^{G}R_1 & \mathbf{s}_0 - {}^{G}R_1\,\mathbf{s}_0 + h_0\hat{u}_0 \\ 0 & 1 \end{bmatrix} \begin{bmatrix} {}^{1}R_2 & \mathbf{s}_1 - {}^{1}R_2\,\mathbf{s}_1 + h_1\hat{u}_1 \\ 0 & 1 \end{bmatrix} \\ &= \begin{bmatrix} {}^{G}R_2 & {}^{G}R_1\left(\mathbf{s}_1 - {}^{1}R_2\,\mathbf{s}_1 + h_1\hat{u}_1\right) + \mathbf{s}_0 - {}^{G}R_1\,\mathbf{s}_0 + h_0\hat{u}_0 \\ 0 & 1 \end{bmatrix} \\ &= \begin{bmatrix} {}^{G}R_2 & {}^{G}R_1(\mathbf{I} - {}^{1}R_2)\,\mathbf{s}_1 + (\mathbf{I} - {}^{G}R_1)\,\mathbf{s}_0 + h_1\,{}^{G}R_1\hat{u}_1 + h_0\hat{u}_0 \\ 0 & 1 \end{bmatrix} \end{aligned} \tag{6.324}$$

where

$${}^{G}R_2 = {}^{G}R_1\,{}^{1}R_2 \tag{6.325}$$

To find the screw parameters of the equivalent screw ${}^{G}\check{s}_2(h, \phi, \hat{u}, \mathbf{s})$, we start obtaining $\hat{u}$ and ϕ from ${}^{G}R_2$ based on (6.253) and (6.252). Then, utilizing (6.293) and (6.294), we can find

$$h = \hat{u} \cdot {}^{G}\mathbf{d} \tag{6.326}$$

$$\mathbf{s} = \left[\mathbf{I} - {}^{G}R_2\right]^{-1}({}^{G}\mathbf{d} - h\hat{u}) \tag{6.327}$$

where

$$\begin{aligned} {}^{G}\mathbf{d} &= {}^{G}R_1(\mathbf{I} - {}^{1}R_2)\,\mathbf{s}_1 + (\mathbf{I} - {}^{G}R_1)\,\mathbf{s}_0 + h_1\,{}^{G}R_1\hat{u}_1 + h_0\hat{u}_0 \\ &= \left({}^{G}R_1 - {}^{G}R_2\right)\,\mathbf{s}_1 + {}^{G}R_1\left(h_1\hat{u}_1 - \mathbf{s}_0\right) + \mathbf{s}_0 + h_0\hat{u}_0 \end{aligned} \tag{6.328}$$

■

Example 378 ★ Exponential Representation of a Screw Consider the body point P in the screw motion shown in Figure 6.21. The final position of the point can be given by

$$\mathbf{p}'' = \mathbf{s} + e^{\phi\tilde{u}}\mathbf{r} + h\hat{u} = \mathbf{s} + e^{\phi\tilde{u}}(\mathbf{p} - \mathbf{s}) + h\hat{u} = [T]\mathbf{p} \tag{6.329}$$

where $[T]$ is the exponential representation of screw motion,

$$[T] = \begin{bmatrix} e^{\phi\tilde{u}} & (\mathbf{I} - e^{\phi\tilde{u}})\,\mathbf{s} + h\hat{u} \\ 0 & 1 \end{bmatrix} \tag{6.330}$$

The exponential screw transformation matrix (6.330) is based on the exponential form of the Rodriguez formula (5.167):

$$e^{\phi\tilde{u}} = I + \tilde{u}\sin\phi + \tilde{u}^2(1 - \cos\phi) \tag{6.331}$$

Therefore, the combination of two screws can also be found by

$$\begin{aligned} [T] &= T_1T_2 \\ &= \begin{bmatrix} e^{\phi_1\tilde{u}_1} & (\mathbf{I} - e^{\phi_1\tilde{u}_1})\,\mathbf{s}_1 + h_1\hat{u}_1 \\ 0 & 1 \end{bmatrix} \begin{bmatrix} e^{\phi_2\tilde{u}_2} & (\mathbf{I} - e^{\phi_2\tilde{u}_2})\,\mathbf{s}_2 + h_2\hat{u}_2 \\ 0 & 1 \end{bmatrix} \\ &= \begin{bmatrix} e^{\phi_1\tilde{u}_1 + \phi_2\tilde{u}_2} & {}^G\mathbf{d} \\ 0 & 1 \end{bmatrix} \end{aligned} \tag{6.332}$$

where

$${}^G\mathbf{d} = \left(e^{\phi_1\tilde{u}_1} - e^{\phi_1\tilde{u}_1 + \phi_2\tilde{u}_2}\right)\mathbf{s}_2 + e^{\phi_1\tilde{u}_1}\left(h_2\hat{u}_2 - \mathbf{s}_1\right) + \mathbf{s}_1 + h_1\hat{u}_1 \tag{6.333}$$

Example 379 ★ Combination of Two Principal Central Screws Combination of every two principal central screws can be found by matrix multiplication. As an example, a screw motion about Y followed by another screw motion about X is

$$\begin{aligned} &\check{s}(h_X, \gamma, \hat{I})\,\check{s}(h_Y, \beta, \hat{J}) \\ &= \begin{bmatrix} 1 & 0 & 0 & \gamma\, p_X \\ 0 & c\gamma & -s\gamma & 0 \\ 0 & s\gamma & c\gamma & 0 \\ 0 & 0 & 0 & 1 \end{bmatrix} \begin{bmatrix} c\beta & 0 & s\beta & 0 \\ 0 & 1 & 0 & \beta\, p_Y \\ -s\beta & 0 & c\beta & 0 \\ 0 & 0 & 0 & 1 \end{bmatrix} \\ &= \begin{bmatrix} \cos\beta & 0 & \sin\beta & \gamma p_X \\ \sin\beta\sin\gamma & \cos\gamma & -\cos\beta\sin\gamma & \beta\, p_Y\cos\gamma \\ -\cos\gamma\sin\beta & \sin\gamma & \cos\beta\cos\gamma & \beta\, p_Y\sin\gamma \\ 0 & 0 & 0 & 1 \end{bmatrix} \end{aligned} \tag{6.334}$$

Screw combination is not commutative, and therefore

$$\check{s}(h_X, \gamma, \hat{I})\,\check{s}(h_Y, \beta, \hat{J}) \neq \check{s}(h_Y, \beta, \hat{J})\,\check{s}(h_X, \gamma, \hat{I}) \tag{6.335}$$

Example 380 ★ Decomposition of a Screw Every general screw can be decomposed to three principal central screws:

$$
\begin{aligned}
{}^G\check{s}_B(h, \phi, \hat{u}, \mathbf{s}) &= \begin{bmatrix} {}^GR_B & \mathbf{s} - {}^GR_B\,\mathbf{s} + h\hat{u} \\ 0 & 1 \end{bmatrix} \\
&= \check{s}(h_X, \gamma, \hat{I})\,\check{s}(h_Y, \beta, \hat{J})\,\check{s}(h_Z, \alpha, \hat{K})
\end{aligned} \tag{6.336}
$$

Multiplication of the three screws provides

$$
{}^GR_B = \begin{bmatrix} c\alpha c\beta & -c\beta s\alpha & s\beta \\ c\gamma s\alpha + c\alpha s\beta s\gamma & c\alpha c\gamma - s\alpha s\beta s\gamma & -c\beta s\gamma \\ s\alpha s\gamma - c\alpha c\gamma s\beta & c\alpha s\gamma + c\gamma s\alpha s\beta & c\beta c\gamma \end{bmatrix} \tag{6.337}
$$

and

$$
\mathbf{s} - {}^GR_B\,\mathbf{s} + h\hat{u} = \begin{bmatrix} \gamma p_X + \alpha p_Z \sin\beta \\ \beta p_Y \cos\gamma - \alpha p_Z \cos\beta \sin\gamma \\ \beta p_Y \sin\gamma + \alpha p_Z \cos\beta \cos\gamma \end{bmatrix} = \begin{bmatrix} d_X \\ d_Y \\ d_Z \end{bmatrix} \tag{6.338}
$$

The twist angles α, β, γ can be found from ${}^GR_B$, and the pitches p_X, p_Y, p_Z can be found from (6.338):

$$
p_Z = \frac{d_Z \cos\gamma - d_Y \sin\gamma}{\alpha \cos\beta} \tag{6.339}
$$

$$
p_Y = \frac{d_Z \sin\gamma + d_Y \cos\gamma}{\beta} \tag{6.340}
$$

$$
p_X = \frac{d_X}{\gamma} - \frac{d_Z \cos\gamma - d_Y \sin\gamma}{\gamma \cos\beta} \sin\beta \tag{6.341}
$$

Example 381 ★ Decomposition of a Screw to Principal Central Screws In general, there are six different independent combinations of triple principal central screws, and therefore there are six different methods to decompose a general screw into a combination of principal central screws:

$$
\begin{aligned}
&\textbf{1.}\ \check{s}(h_X, \gamma, \hat{I})\,\check{s}(h_Y, \beta, \hat{J})\,\check{s}(h_Z, \alpha, \hat{K}) \\
&\textbf{2.}\ \check{s}(h_Y, \beta, \hat{J})\,\check{s}(h_Z, \alpha, \hat{K})\,\check{s}(h_X, \gamma, \hat{I}) \\
&\textbf{3.}\ \check{s}(h_Z, \alpha, \hat{K})\,\check{s}(h_X, \gamma, \hat{I})\,\check{s}(h_Y, \beta, \hat{J}) \\
&\textbf{4.}\ \check{s}(h_Z, \alpha, \hat{K})\,\check{s}(h_Y, \beta, \hat{J})\,\check{s}(h_X, \gamma, \hat{I}) \\
&\textbf{5.}\ \check{s}(h_Y, \beta, \hat{J})\,\check{s}(h_X, \gamma, \hat{I})\,\check{s}(h_Z, \alpha, \hat{K}) \\
&\textbf{6.}\ \check{s}(h_X, \gamma, \hat{I})\,\check{s}(h_Z, \alpha, \hat{K})\,\check{s}(h_Y, \beta, \hat{J})
\end{aligned} \tag{6.342}
$$

The expanded forms of the six combinations of principal central screws are presented in Appendix C.

6.8 ★ PLÜCKER LINE COORDINATE

Plücker coordinates are a set of six coordinates used to define and show a directed line in space. Such a line definition is more natural than the traditional methods and simplifies kinematics. Assume that $P_1(X_1, Y_1, Z_1)$ and $P_2(X_2, Y_2, Z_2)$ at $\mathbf{r}_1$ and $\mathbf{r}_2$ are two different points on the line l, as shown in Figure 6.22.

Using the position vectors $\mathbf{r}_1$ and $\mathbf{r}_2$, the equation of line l can be defined by a vector with six elements that are made from the components of two vectors $\hat{u}$ and $\boldsymbol{\rho}$:

$$l = \begin{bmatrix} \hat{u} \\ \boldsymbol{\rho} \end{bmatrix} = \begin{bmatrix} L \\ M \\ N \\ P \\ Q \\ R \end{bmatrix} \tag{6.343}$$

The six elements L, M, N, P, Q, R are called the *Plücker coordinates* of the *directed* line l. The vector $\hat{u}$ is a unit vector along the line l and is referred to as a *direction vector*,

$$\hat{u} = \frac{\mathbf{r}_2 - \mathbf{r}_1}{|\mathbf{r}_2 - \mathbf{r}_1|} = L\hat{I} + M\hat{J} + N\hat{K} \tag{6.344}$$

and vector $\boldsymbol{\rho}$ is the *moment vector* of $\hat{u}$ about the origin,

$$\boldsymbol{\rho} = \mathbf{r}_1 \times \hat{u} = P\hat{I} + Q\hat{J} + R\hat{K} \tag{6.345}$$

Proof: The unit vector $\hat{u}$ along the line l that connects P_1 to P_2 is

$$\begin{aligned} \hat{u} &= \frac{\mathbf{r}_2 - \mathbf{r}_1}{|\mathbf{r}_2 - \mathbf{r}_1|} = \frac{X_2 - X_1}{d}\hat{I} + \frac{Y_2 - Y_1}{d}\hat{J} + \frac{Z_2 - Z_1}{d}\hat{K} \\ &= L\hat{I} + M\hat{J} + N\hat{K} \end{aligned} \tag{6.346}$$

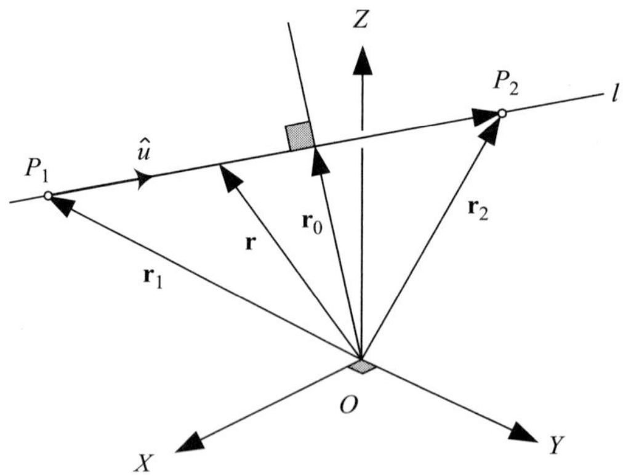

Figure 6.22 A line indicated by two points.

where

$$L^2 + M^2 + N^2 = 1 \tag{6.347}$$

and the distance between P_1 and P_2 is

$$d = \sqrt{(X_2 - X_1)^2 + (Y_2 - Y_1)^2 + (Z_2 - Z_1)^2} \tag{6.348}$$

If $\mathbf{r}$ represents a vector from the origin O to a point on line l, then the vector $\mathbf{r} - \mathbf{r}_1$ is parallel to $\hat{u}$, and therefore the equation of the line l can be written as

$$(\mathbf{r} - \mathbf{r}_1) \times \hat{u} = 0 \tag{6.349}$$

or equivalently as

$$\mathbf{r} \times \hat{u} = \boldsymbol{\rho} \tag{6.350}$$

where $\boldsymbol{\rho}$ is the moment of the direction vector $\hat{u}$ about O:

$$\boldsymbol{\rho} = \mathbf{r}_1 \times \hat{u} \tag{6.351}$$

Furthermore, because vectors $\boldsymbol{\rho}$ and $\hat{u}$ are perpendicular, there is a constraint among their components,

$$\hat{u} \cdot \boldsymbol{\rho} = 0 \tag{6.352}$$

Expanding (6.345) yields

$$\boldsymbol{\rho} = \begin{vmatrix} \hat{I} & \hat{J} & \hat{K} \\ X_1 & Y_1 & Z_1 \\ L & M & N \end{vmatrix} = P\hat{I} + Q\hat{J} + R\hat{K} \tag{6.353}$$

where

$$\begin{aligned} P &= Y_1 N - Z_1 M \\ Q &= Z_1 L - X_1 N \\ R &= X_1 M - Y_1 L \end{aligned} \tag{6.354}$$

and therefore the orthogonality condition (6.352) can be expressed as

$$LP + MQ + NR = 0 \tag{6.355}$$

Because of the constraints (6.347) and (6.355), the Plücker coordinates of the line (6.343) have only four independent components.

Our arrangement of Plücker coordinates in the form of (6.343) is the *line arrangement* and is called *ray coordinates*; in some publication the reverse order in the *axis arrangement* $l = \left[\boldsymbol{\rho}|\hat{u}\right]^{\mathrm{T}}$ is used. In either case, a vertical line as in $\left[\hat{u}|\boldsymbol{\rho}\right]^{\mathrm{T}}$ or a semicolon as in $\left[\hat{u};\boldsymbol{\rho}\right]^{\mathrm{T}}$ may be used to separate the first three elements from the last three. Both arrangements can be used in kinematics efficiently.

The Plücker line coordinates $[\hat{u}, \boldsymbol{\rho}]^{\mathrm{T}}$ are homogeneous because Equation (6.345) shows that the coordinates $[w\hat{u}, w\boldsymbol{\rho}]^{\mathrm{T}}$, where $w \in \mathbb{R}$, define the same line.

Force–moment, angular velocity–translational velocity, and rigid-body motion act like a line vector and can be expressed by Plücker coordinates. The Plücker method is a canonical representation of line definition and therefore is more efficient than other methods such as the parametric form $l(t) = \mathbf{r}_1 + t\hat{u}$, point and direction form $(\mathbf{r}_1, \hat{u})$, or two-point representation $(\mathbf{r}_1, \mathbf{r}_2)$. ■

Example 382 ★ Plücker Coordinates of a Line Connecting Two Points Plücker line coordinates of the line connecting points $P_1(1, 0, 0)$ and $P_2(0, 1, 1)$ are

$$l = \begin{bmatrix} \hat{u} \\ \rho \end{bmatrix} = \begin{bmatrix} -1 & 1 & 1 & 0 & -1 & 1 \end{bmatrix}^{\mathrm{T}} \tag{6.356}$$

because

$$\sqrt{3}\hat{u} = \frac{\mathbf{r}_2 - \mathbf{r}_1}{|\mathbf{r}_2 - \mathbf{r}_1|} = -\hat{I} + \hat{J} + \hat{K} \tag{6.357}$$

and

$$\sqrt{3}\rho = \mathbf{r}_1 \times \sqrt{3}\hat{u} = -\hat{J} + \hat{K} \tag{6.358}$$

Example 383 ★ Plücker Coordinates of Diagonals of a Cube Figure 6.23 depicts a unit cube and two lines on diagonals of two adjacent faces. Line l_1 connecting corners $P_1(1, 0, 1)$ and $P_2(0, 1, 1)$ is given as

$$l_1 = \begin{bmatrix} \hat{u}_1 \\ \rho_1 \end{bmatrix} = \begin{bmatrix} -\frac{\sqrt{2}}{2} & \frac{\sqrt{2}}{2} & 0 & -\frac{\sqrt{2}}{2} & -\frac{\sqrt{2}}{2} & \frac{\sqrt{2}}{2} \end{bmatrix}^{\mathrm{T}} \tag{6.359}$$

because

$$\hat{u}_1 = \frac{\mathbf{p}_2 - \mathbf{p}_1}{|\mathbf{p}_2 - \mathbf{p}_1|} = \frac{-\hat{I} + \hat{J}}{\sqrt{2}} \tag{6.360}$$

and

$$\rho_1 = \mathbf{p}_1 \times \hat{u}_1 = \frac{-\hat{I} - \hat{J} + \hat{K}}{\sqrt{2}} \tag{6.361}$$

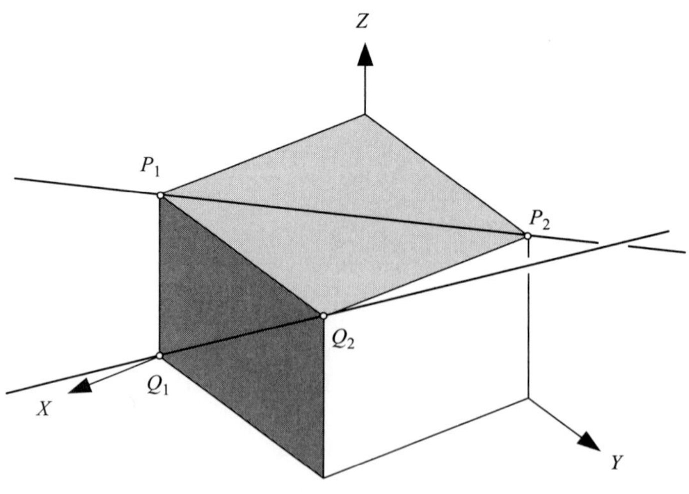

Figure 6.23 A unit cube and two lines along the diagonal of two faces.

Line l_2 connects the corner $Q_1(1,0,0)$ to $Q_2(1,1,1)$ and is expressed by

$$l_2 = \begin{bmatrix} \hat{u}_2 \\ \boldsymbol{\rho}_2 \end{bmatrix} = \begin{bmatrix} 0 & \frac{\sqrt{2}}{2} & \frac{\sqrt{2}}{2} & 0 & -\frac{\sqrt{2}}{2} & \frac{\sqrt{2}}{2} \end{bmatrix}^{\mathrm{T}} \tag{6.362}$$

because

$$\hat{u}_2 = \frac{\mathbf{q}_2 - \mathbf{q}_1}{|\mathbf{q}_2 - \mathbf{q}_1|} = \frac{\hat{J} + \hat{K}}{\sqrt{2}} \tag{6.363}$$

and

$$\boldsymbol{\rho}_2 = \mathbf{q}_1 \times \hat{u}_2 = \frac{-\hat{J} + \hat{K}}{\sqrt{2}} \tag{6.364}$$

Example 384 ★ Grassmanian Matrix to Show Plücker Coordinates The *Grassmanian matrix* is used to show the homogeneous coordinates. The Grassmanian matrix for two points

$$\begin{bmatrix} w_1 & X_1 & Y_1 & Z_1 \\ w_2 & X_2 & Y_2 & Z_2 \end{bmatrix} \tag{6.365}$$

is a short notation for the Plücker coordinates of the line l that connects the two points if we define

$$\begin{aligned} L &= \begin{vmatrix} w_1 & X_1 \\ w_2 & X_2 \end{vmatrix} & P &= \begin{vmatrix} Y_1 & Z_1 \\ Y_2 & Z_2 \end{vmatrix} \\ M &= \begin{vmatrix} w_1 & Y_1 \\ w_2 & Y_2 \end{vmatrix} & Q &= \begin{vmatrix} Z_1 & X_1 \\ Z_2 & X_2 \end{vmatrix} \\ N &= \begin{vmatrix} w_1 & Z_1 \\ w_2 & Z_2 \end{vmatrix} & R &= \begin{vmatrix} X_1 & Y_1 \\ X_2 & Y_2 \end{vmatrix} \end{aligned} \tag{6.366}$$

Example 385 ★ Ray–Axis Arrangement Transformation We may verify that the ray arrangement of Plücker coordinates,

$$l_{\text{ray}} = \begin{bmatrix} \hat{u} \\ \boldsymbol{\rho} \end{bmatrix} \tag{6.367}$$

can be transformed to the axis arrangement,

$$l_{\text{axis}} = \begin{bmatrix} \boldsymbol{\rho} \\ \hat{u} \end{bmatrix} \tag{6.368}$$

and vice versa by using a 6×6 transformation matrix Δ:

$$\begin{bmatrix} \boldsymbol{\rho} \\ \hat{u} \end{bmatrix} = \Delta \begin{bmatrix} \hat{u} \\ \boldsymbol{\rho} \end{bmatrix} \tag{6.369}$$

$$\Delta = \begin{bmatrix} 0 & 0 & 0 & 1 & 0 & 0 \\ 0 & 0 & 0 & 0 & 1 & 0 \\ 0 & 0 & 0 & 0 & 0 & 1 \\ 1 & 0 & 0 & 0 & 0 & 0 \\ 0 & 1 & 0 & 0 & 0 & 0 \\ 0 & 0 & 1 & 0 & 0 & 0 \end{bmatrix} = \begin{bmatrix} 0 & \mathbf{I}_3 \\ \mathbf{I}_3 & 0 \end{bmatrix} \tag{6.370}$$

The transformation matrix Δ is symmetric and satisfies the following equations:

$$\Delta^2 = \Delta\Delta = \mathbf{I} \tag{6.371}$$

$$\Delta^{\mathrm{T}} = \Delta \tag{6.372}$$

Example 386 ★ Classification of Plücker Coordinates There are three cases of Plücker coordinates: *general case*, *line through origin*, and *line at infinity*. These cases are illustrated in Figure 6.24.

In the general case $l = \left[\hat{u}, \boldsymbol{\rho}\right]^{\mathrm{T}}$, illustrated in Figure 6.24(a), both vectors $\hat{u}$ and $\boldsymbol{\rho}$ are nonzero. The vector $\hat{u}$ is parallel to the line l, and $\boldsymbol{\rho}$ is normal to the plane of l and the origin. The magnitude of $|\boldsymbol{\rho}|$ gives the distance of l from the origin.

In the line through origin $l = \left[\hat{u}, \mathbf{0}\right]^{\mathrm{T}}$, as illustrated in Figure 6.24(b), the line passes through the origin and $\boldsymbol{\rho}$ is zero.

In the line at infinity $l = \left[0, \boldsymbol{\rho}\right]^{\mathrm{T}}$, as illustrated in Figure 6.24(c), the distance l from the origin tends to infinity. In this case we may assume $\hat{u}$ is zero. When the line is at infinity, it is better to redefine the Plücker coordinates by normalizing the moment vector

$$l = \begin{bmatrix} \dfrac{\hat{u}}{|\boldsymbol{\rho}|} & \dfrac{\boldsymbol{\rho}}{|\boldsymbol{\rho}|} \end{bmatrix}^{\mathrm{T}} \tag{6.373}$$

Therefore, the direction components of the line tend to zero by increasing the distance while the moment components remain finite.

No line is defined by the zero Plücker coordinates $[0, 0, 0, 0, 0, 0]^{\mathrm{T}}$.

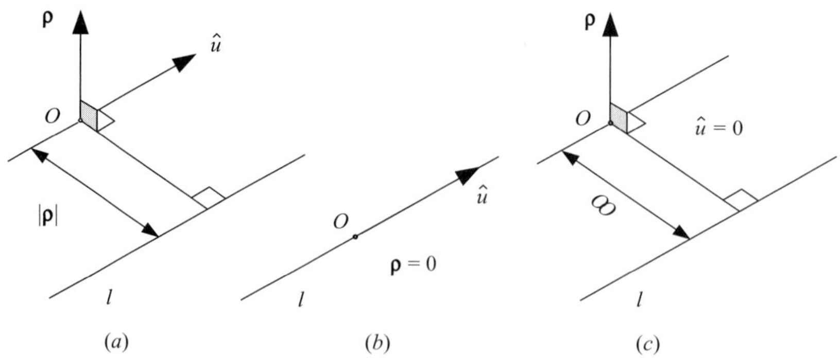

Figure 6.24 Three cases of Plücker coordinates: (a) general case, (b) line through origin, (c) line at infinity.

Example 387 ★ Transformation of a Line Vector Consider the line ${}^{B}l$ in Figure 6.25 defined in a local frame $B(oxyz)$ by

$$ {}^{B}l = \begin{bmatrix} {}^{B}\hat{u} \\ {}^{B}\boldsymbol{\rho} \end{bmatrix} = \begin{bmatrix} {}^{B}\hat{u} \\ {}^{B}\mathbf{r}_P \times {}^{B}\hat{u} \end{bmatrix} \tag{6.374} $$

where $\hat{u}$ is a unit vector parallel to the line l and P is any point on the line. The Plücker coordinates of the line in the global frame $G(OXYZ)$ is expressed by

$$ {}^{G}l = \begin{bmatrix} {}^{G}\hat{u} \\ {}^{G}\boldsymbol{\rho} \end{bmatrix} = \begin{bmatrix} {}^{G}\hat{u} \\ {}^{G}\mathbf{r}_P \times {}^{G}\hat{u} \end{bmatrix} \tag{6.375} $$

where

$$ {}^{G}\hat{u} = {}^{G}R_B \, {}^{B}\hat{u} \tag{6.376} $$

and ${}^{G}\boldsymbol{\rho}$ is the moment of ${}^{G}\hat{u}$ about O:

$$ \begin{aligned} {}^{G}\boldsymbol{\rho} &= {}^{G}\mathbf{r}_P \times {}^{G}\hat{u} = ({}^{G}\mathbf{s}_o + {}^{G}R_B \, {}^{B}\mathbf{r}_P) \times {}^{G}R_B \, {}^{B}\hat{u} \\ &= {}^{G}\mathbf{d}_o \times {}^{G}R_B \, {}^{B}\hat{u} + {}^{G}R_B ({}^{B}\mathbf{r}_P \times {}^{B}\hat{u}) \\ &= {}^{G}\mathbf{d}_o \times {}^{G}R_B \, {}^{B}\hat{u} + {}^{G}R_B \, {}^{B}\boldsymbol{\rho} \\ &= {}^{G}\tilde{\mathbf{d}}_o \, {}^{G}R_B \, {}^{B}\hat{u} + {}^{G}R_B \, {}^{B}\boldsymbol{\rho}. \end{aligned} \tag{6.377} $$

Therefore, the 6×1 Plücker coordinates $\left[\hat{u}, \boldsymbol{\rho}\right]^{\mathrm{T}}$ for a line vector can be transformed from a frame B to another frame G,

$$ {}^{G}l = {}^{G}\Gamma_B \, {}^{B}l \tag{6.378} $$

$$ \begin{bmatrix} {}^{G}\hat{u} \\ {}^{G}\boldsymbol{\rho} \end{bmatrix} = {}^{G}\Gamma_B \begin{bmatrix} {}^{B}\hat{u} \\ {}^{B}\boldsymbol{\rho} \end{bmatrix} \tag{6.379} $$

by a 6×6 transformation matrix ${}^{G}\Gamma_B$ defined as

$$ {}^{G}\Gamma_B = \begin{bmatrix} {}^{G}R_B & 0 \\ {}^{G}\tilde{\mathbf{d}}_o \, {}^{G}R_B & {}^{G}R_B \end{bmatrix} \tag{6.380} $$

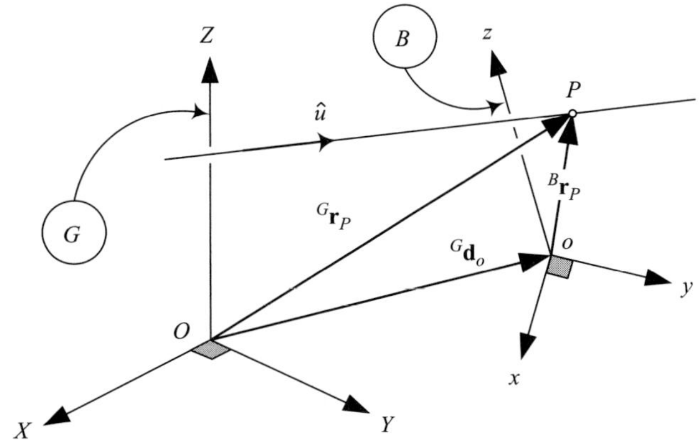

Figure 6.25 A line vector in B- and G-frames.

where

$$ {}^{G}R_{B} = \begin{bmatrix} r_{11} & r_{12} & r_{13} \\ r_{21} & r_{22} & r_{23} \\ r_{31} & r_{32} & r_{33} \end{bmatrix} \tag{6.381} $$

$$ {}^{G}\tilde{\mathbf{d}}_{o} = \begin{bmatrix} 0 & -d_3 & d_2 \\ d_3 & 0 & -d_1 \\ -d_2 & d_1 & 0 \end{bmatrix} \tag{6.382} $$

$$ {}^{G}\tilde{\mathbf{d}}_{o}\,{}^{G}R_{B} = \begin{bmatrix} d_2 r_{31} - d_3 r_{21} & d_2 r_{32} - d_3 r_{22} & d_2 r_{33} - d_3 r_{23} \\ -d_1 r_{31} + d_3 r_{11} & -d_1 r_{32} + d_3 r_{12} & -d_1 r_{33} + d_3 r_{13} \\ d_1 r_{21} - d_2 r_{11} & d_1 r_{22} - d_2 r_{12} & d_1 r_{23} - d_2 r_{13} \end{bmatrix} \tag{6.383} $$

6.9 ★ GEOMETRY OF PLANE AND LINE

Plücker coordinates introduce a suitable method to define the moment between two lines, the shortest distance between two lines, and the angle between two lines.

6.9.1 ★ Moment

Consider two directed lines $l_1 = \left[\hat{u}_1, \boldsymbol{\rho}_1\right]^{\mathrm{T}}$ and $l_2 = \left[\hat{u}_2, \boldsymbol{\rho}_2\right]^{\mathrm{T}}$ as are shown in Figure 6.26. Points P_1 and P_2 on l_1 and l_2 are indicated by vectors $\mathbf{r}_1$ and $\mathbf{r}_2$, respectively. The direction vectors of the lines are $\hat{u}_1$ and $\hat{u}_2$.

The moment of l_2 about P_1 is

$$ (\mathbf{r}_2 - \mathbf{r}_1) \times \hat{u}_2 $$

and we can define the moment of the line l_2 about l_1 by

$$ l_2 \times l_1 = \hat{u}_1 \cdot (\mathbf{r}_2 - \mathbf{r}_1) \times \hat{u}_2 \tag{6.384} $$

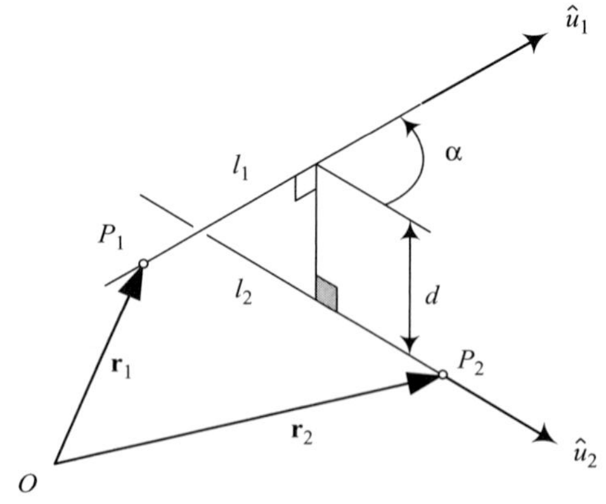

Figure 6.26 Two skew lines and Plücker coordinates.

which, because of $\hat{u}_1 \cdot \mathbf{r}_1 \times \hat{u}_2 = \hat{u}_2 \cdot \hat{u}_1 \times \mathbf{r}_1$, simplifies to

$$l_2 \times l_1 = \hat{u}_1 \cdot \boldsymbol{\rho}_2 + \hat{u}_2 \cdot \boldsymbol{\rho}_1 \tag{6.385}$$

The moment of the lines l_2 about l_1 can be defined by the *reciprocal product* of the Plücker expression of the two lines:

$$l_2 \times l_1 = \begin{bmatrix} \hat{u}_2 \\ \boldsymbol{\rho}_2 \end{bmatrix} \otimes \begin{bmatrix} \hat{u}_1 \\ \boldsymbol{\rho}_1 \end{bmatrix} = \hat{u}_2 \cdot \boldsymbol{\rho}_1 + \hat{u}_1 \cdot \boldsymbol{\rho}_2 \tag{6.386}$$

The reciprocal product, also called a *virtual product*, is commutative and gives the moment between two directed lines.

6.9.2 ★ Angle and Distance

If d is the shortest distance between two lines $l_1 = \left[\hat{u}_1, \boldsymbol{\rho}_1\right]^{\mathrm{T}}$ and $l_2 = \left[\hat{u}_2, \boldsymbol{\rho}_2\right]^{\mathrm{T}}$ and $\alpha \in [0, \pi]$ is the angle between l_1 and l_2, then

$$\sin\alpha = \left|\hat{u}_2 \times \hat{u}_1\right| \tag{6.387}$$

and

$$\begin{aligned} d &= \frac{1}{\sin\alpha}\left|\hat{u}_2 \cdot \boldsymbol{\rho}_1 + \hat{u}_1 \cdot \boldsymbol{\rho}_2\right| = \frac{1}{\sin\alpha}\left\| \begin{bmatrix} \hat{u}_2 \\ \boldsymbol{\rho}_2 \end{bmatrix} \otimes \begin{bmatrix} \hat{u}_1 \\ \boldsymbol{\rho}_1 \end{bmatrix} \right\| \\ &= \frac{1}{\sin\alpha}\left|l_2 \times l_1\right| \end{aligned} \tag{6.388}$$

Therefore, two lines l_1 and l_2 intersect if and only if their reciprocal product is zero. Two parallel lines may be assumed to intersect at infinity. The distance expression does not work for parallel lines.

6.9.3 ★ Plane and Line

The equation of a plane ${}^G\boldsymbol{\pi}$ with a normal unit vector $\hat{n} = n_1\hat{I} + n_2\hat{J} + n_3\hat{K}$ is

$$n_1X + n_2Y + n_3Z = s \tag{6.389}$$

where s is the minimum distance of the plane to the origin O. We may indicate a plane by using a homogeneous representation

$$\boldsymbol{\pi} = \begin{bmatrix} n_1 \\ n_2 \\ n_3 \\ s \end{bmatrix} \tag{6.390}$$

and write the condition $\boldsymbol{\pi}^{\mathrm{T}} \cdot \mathbf{r} = 0$ for a point $\mathbf{r} = [X, Y, Z, w]^{\mathrm{T}}$ to be in the plane by

$$\boldsymbol{\pi}^{\mathrm{T}} \cdot \mathbf{r} = \begin{bmatrix} n_1 & n_2 & n_3 & s \end{bmatrix} \begin{bmatrix} X \\ Y \\ Z \\ w \end{bmatrix} = 0 \tag{6.391}$$

Furthermore, $w = 0$ indicates all points at infinity, and $s = 0$ indicates all planes containing the origin.

The intersection of the π-plane with the X-axis, or the X-intercept, is $X = -s/n_1$, the Y-intercept is $Y = -s/n_2$, and the Z-intercept is $Z = -s/n_3$. The plane is perpendicular to the (X, Y)-plane if $n_3 = 0$. It is perpendicular to the X-axis if $n_2 = n_3 = 0$. There are similar conditions for the other planes and axes. If (X_0, Y_0, Z_0) is a point in the plane (6.390), then

$$n_1(X - X_0) + n_2(Y - Y_0) + n_3(Z - Z_0) = s \tag{6.392}$$

The distance of a point $[X,Y,Z,w]^{\mathrm{T}}$ from the origin is

$$d = \sqrt{\frac{X^2 + Y^2 + Z^2}{w^2}} \tag{6.393}$$

while the distance of a plane $\pi = \begin{bmatrix} m_1 & m_2 & m_3 & s \end{bmatrix}^{\mathrm{T}}$ from the origin is

$$s = \sqrt{\frac{s^2}{m_1^2 + m_2^2 + m_3^2}} \tag{6.394}$$

The equation of a line connecting two points $P_1(X_1, Y_1, Z_1)$ and $P_2(X_2, Y_2, Z_2)$ at $\mathbf{r}_1$ and $\mathbf{r}_2$ can also be expressed by

$$l = \mathbf{r}_1 + m(\mathbf{r}_2 - \mathbf{r}_1) \tag{6.395}$$

and the distance of a point $P(X, Y, Z)$ from any point on l is given by

$$\begin{aligned} s^2 &= (X_1 + m(X_2 - X_1) - X)^2 \\ &\quad + (Y_1 + m(Y_2 - Y_1) - Y)^2 \\ &\quad + (Z_1 + m(Z_2 - Z_1) - Z)^2 \end{aligned} \tag{6.396}$$

which is a minimum for

$$m = -\frac{(X_2 - X_1)(X_1 - X) + (Y_2 - Y_1)(Y_1 - Y) + (Z_2 - Z_1)(Z_1 - Z)}{(X_2 - X_1)^2 + (Y_2 - Y_1)^2 + (Z_2 - Z_1)^2} \tag{6.397}$$

To find the minimum distance of the origin we set $X = Y = Z = 0$.

Example 388 ★ Angle and Distance between Two Diagonals of a Cube The Plücker coordinates of the two diagonals of the unit cube of Figure 6.23 are

$$l_1 = \begin{bmatrix} \hat{u}_1 \\ \rho_1 \end{bmatrix} = \begin{bmatrix} -\frac{\sqrt{2}}{2} & \frac{\sqrt{2}}{2} & 0 & -\frac{\sqrt{2}}{2} & -\frac{\sqrt{2}}{2} & \frac{\sqrt{2}}{2} \end{bmatrix}^{\mathrm{T}} \tag{6.398}$$

$$l_2 = \begin{bmatrix} \hat{u}_2 \\ \rho_2 \end{bmatrix} = \begin{bmatrix} 0 & \frac{\sqrt{2}}{2} & \frac{\sqrt{2}}{2} & 0 & -\frac{\sqrt{2}}{2} & \frac{\sqrt{2}}{2} \end{bmatrix}^{\mathrm{T}} \tag{6.399}$$

The angle between l_1 and l_2 is

$$\alpha = \sin^{-1} \left| \hat{u}_2 \times \hat{u}_1 \right| = \sin^{-1} \frac{\sqrt{3}}{2} = 60 \deg \tag{6.400}$$

and the distance between them is

$$\begin{aligned} d &= \frac{1}{\sin \alpha} \left\| \begin{bmatrix} \hat{u}_1 \\ \boldsymbol{\rho}_1 \end{bmatrix} \otimes \begin{bmatrix} \hat{u}_2 \\ \boldsymbol{\rho}_2 \end{bmatrix} \right\| = \frac{1}{\sin \alpha} \left| \hat{u}_1 \cdot \boldsymbol{\rho}_2 + \hat{u}_2 \cdot \boldsymbol{\rho}_1 \right| \\ &= \frac{2}{\sqrt{3}} \left| -0.5 \right| = \frac{1}{\sqrt{3}} \end{aligned} \tag{6.401}$$

Example 389 Distance of a Point from a Line The equation of a line that connects two points $\mathbf{r}_1 = [-1, 2, 1, 1]^{\mathrm{T}}$ and $\mathbf{r}_2 = [1, -2, -1, 1]^{\mathrm{T}}$ is

$$l = \mathbf{r}_1 + m(\mathbf{r}_2 - \mathbf{r}_1) = \begin{bmatrix} -1 + 2m \\ 2 - 4m \\ 1 - 2m \\ 1 \end{bmatrix} \tag{6.402}$$

So, the distance between point $\mathbf{r}_3 = [1, 1, 0, 1]^{\mathrm{T}}$ and l is given by

$$\begin{aligned} s^2 &= (X_1 + m(X_2 - X_1) - X_3)^2 \\ &\quad + (Y_1 + m(Y_2 - Y_1) - Y_3)^2 \\ &\quad + (Z_1 + m(Z_2 - Z_1) - Z_3)^2 \\ &= 24m^2 - 20m + 6 \end{aligned} \tag{6.403}$$

which is a minimum for

$$m = \tfrac{5}{12} \tag{6.404}$$

Therefore, the point on the line at a minimum distance from $\mathbf{r}_3$ is

$$\mathbf{r} = \begin{bmatrix} -\frac{1}{6} \\ \frac{1}{3} \\ \frac{1}{6} \\ 1 \end{bmatrix} \tag{6.405}$$

Example 390 Distance between Two Lines The line that connects $\mathbf{r}_1 = [-1, 2, 1, 1]^{\mathrm{T}}$ and $\mathbf{r}_2 = [1, -2, -1, 1]^{\mathrm{T}}$ is

$$l = \mathbf{r}_1 + m(\mathbf{r}_2 - \mathbf{r}_1) = \begin{bmatrix} -1 + 2m \\ 2 - 4m \\ 1 - 2m \\ 1 \end{bmatrix} \tag{6.406}$$

and the line connecting $\mathbf{r}_3 = [1, 1, 0, 1]^{\mathrm{T}}$ and $\mathbf{r}_4 = [0, -1, 2, 1]^{\mathrm{T}}$ is

$$l = \mathbf{r}_1 + m(\mathbf{r}_2 - \mathbf{r}_1) = \begin{bmatrix} 1-n \\ 1-2n \\ 2n \\ 1 \end{bmatrix} \tag{6.407}$$

The distance between two arbitrary points on the lines is

$$s^2 = \left(-1 + 2m - 1 + n^2\right) + (2 - 4m - 1 + 2n)^2 + (1 - 2m - 2n)^2 \tag{6.408}$$

The minimum distance is found by minimizing s^2 with respect to m and n:

$$m = 0.443 \qquad n = 0.321 \tag{6.409}$$

So, the two points on the lines that have the minimum distance are at

$$\mathbf{r}_m = \begin{bmatrix} -0.114 \\ 0.228 \\ 0.114 \\ 1 \end{bmatrix} \qquad \mathbf{r}_n = \begin{bmatrix} 0.679 \\ 0.358 \\ 0.642 \\ 1 \end{bmatrix} \tag{6.410}$$

Example 391 ★ Intersection Condition for Two Lines If two lines $l_1 = \left[\hat{u}_1, \boldsymbol{\rho}_1\right]^{\mathrm{T}}$ and $l_2 = \left[\hat{u}_2, \boldsymbol{\rho}_2\right]^{\mathrm{T}}$ intersect and the position of their common point is at $\mathbf{r}$, then

$$\boldsymbol{\rho}_1 = \mathbf{r} \times \hat{u}_1 \tag{6.411}$$

$$\boldsymbol{\rho}_2 = \mathbf{r} \times \hat{u}_2 \tag{6.412}$$

and therefore

$$\boldsymbol{\rho}_1 \cdot \hat{u}_2 = \left(\mathbf{r} \times \hat{u}_1\right) \cdot \hat{u}_2 = \mathbf{r} \cdot \left(\hat{u}_1 \times \hat{u}_2\right) \tag{6.413}$$

$$\boldsymbol{\rho}_2 \cdot \hat{u}_1 = \left(\mathbf{r} \times \hat{u}_2\right) \cdot \hat{u}_1 = \mathbf{r} \cdot \left(\hat{u}_2 \times \hat{u}_1\right) \tag{6.414}$$

which implies

$$\hat{u}_1 \cdot \boldsymbol{\rho}_2 + \hat{u}_2 \cdot \boldsymbol{\rho}_1 = 0 \tag{6.415}$$

or equivalently

$$\begin{bmatrix} \hat{u}_1 \\ \boldsymbol{\rho}_1 \end{bmatrix} \otimes \begin{bmatrix} \hat{u}_2 \\ \boldsymbol{\rho}_2 \end{bmatrix} = 0 \tag{6.416}$$

Example 392 ★ Plücker Coordinates of the Axis of Rotation Consider a homogeneous transformation matrix corresponding to a rotation α about Z along with a translation in the (X, Y)-plane,

$${}^{G}T_{B} = \begin{bmatrix} r_{11} & r_{12} & 0 & X_o \\ r_{21} & r_{22} & 0 & Y_o \\ 0 & 0 & 1 & 0 \\ 0 & 0 & 0 & 1 \end{bmatrix} \tag{6.417}$$

which must be equal to

$$
{}^{G}T_{B} = \begin{bmatrix} \cos\alpha & -\sin\alpha & 0 & X_o \\ \sin\alpha & \cos\alpha & 0 & Y_o \\ 0 & 0 & 1 & 0 \\ 0 & 0 & 0 & 1 \end{bmatrix} \tag{6.418}
$$

The angle of rotation can be obtained by comparison,

$$
\alpha = \tan^{-1}\frac{r_{21}}{r_{11}} \tag{6.419}
$$

The pole of rotation can be found by searching for a point that has the same coordinates in both frames,

$$
\begin{bmatrix} r_{11} & r_{12} & 0 & X_o \\ r_{21} & r_{22} & 0 & Y_o \\ 0 & 0 & 1 & 0 \\ 0 & 0 & 0 & 1 \end{bmatrix} \begin{bmatrix} X_p \\ Y_p \\ 0 \\ 1 \end{bmatrix} = \begin{bmatrix} X_p \\ Y_p \\ 0 \\ 1 \end{bmatrix} \tag{6.420}
$$

which can be written as

$$
\begin{bmatrix} r_{11}-1 & r_{12} & 0 & X_o \\ r_{21} & r_{22}-1 & 0 & Y_o \\ 0 & 0 & 1 & 0 \\ 0 & 0 & 0 & 1 \end{bmatrix} \begin{bmatrix} X_p \\ Y_p \\ 0 \\ 1 \end{bmatrix} = \begin{bmatrix} 0 \\ 0 \\ 0 \\ 1 \end{bmatrix} \tag{6.421}
$$

The solutions of these equations are

$$
X_p = \frac{1}{2}X_o - \frac{1}{2}\frac{r_{21}}{1-r_{11}}Y_o = \frac{1}{2}X_o - \frac{1}{2}\frac{\sin\alpha}{\operatorname{vers}\alpha}Y_o \tag{6.422}
$$

$$
Y_p = \frac{1}{2}Y_o + \frac{1}{2}\frac{r_{21}}{1-r_{11}}X_o = \frac{1}{2}Y_o + \frac{1}{2}\frac{\sin\alpha}{\operatorname{vers}\alpha}X_o \tag{6.423}
$$

The Plücker line $l = \left[\hat{u}, \boldsymbol{\rho}\right]^{\mathrm{T}}$ of the pole axis is then equal to

$$
l = \begin{bmatrix} 0 & 0 & 1 & Y_p & -X_p & 0 \end{bmatrix}^{\mathrm{T}} \tag{6.424}
$$

6.10 ★ SCREW AND PLÜCKER COORDINATE

Consider a screw $\check{s}(h, \phi, \hat{u}, \mathbf{s})$ with a line of action $l = \left[\hat{u}, \boldsymbol{\rho}\right]^{\mathrm{T}}$ and pitch $p = h/\phi$. The screw can be defined by a set of Plücker coordinates

$$
\check{s}(h, \phi, \hat{u}, \mathbf{s}) = \begin{bmatrix} \hat{u} \\ \boldsymbol{\xi} \end{bmatrix} = \begin{bmatrix} \hat{u} \\ \boldsymbol{\rho} + p\hat{u} \end{bmatrix} = \begin{bmatrix} \phi\hat{u} \\ \phi\boldsymbol{\rho} + h\hat{u} \end{bmatrix} \tag{6.425}
$$

If the pitch is infinite, $p = \infty$, then the screw reduces to a pure translation, or equivalently a line at infinity,

$$
\check{s}(h, 0, \hat{u}, \mathbf{r}) = \begin{bmatrix} 0 \\ h\hat{u} \end{bmatrix} \tag{6.426}
$$

A zero pitch screw $p = 0$ corresponds to a pure rotation. Then the screw coordinates are identical to the Plücker coordinates of the screw line:

$$\check{s}(0, \phi, \hat{u}, \mathbf{s}) = \begin{bmatrix} \phi\hat{u} \\ \phi\boldsymbol{\rho} \end{bmatrix} = \begin{bmatrix} \hat{u} \\ \boldsymbol{\rho} \end{bmatrix} \tag{6.427}$$

A central screw is defined by a line through the origin:

$$\check{s}(h, \phi, \hat{u}) = \check{s}(h, \phi, \hat{u}, 0) = \begin{bmatrix} \hat{u} \\ p\hat{u} \end{bmatrix} = \begin{bmatrix} \phi\hat{u} \\ h\hat{u} \end{bmatrix}$$
$$= D(h\hat{u})\, R(\hat{u}, \phi) \tag{6.428}$$

Screw coordinates for differential screw motion is useful in the velocity analysis of connected rigid bodies. Consider a screw axis l, an angular velocity $\boldsymbol{\omega} = \omega\hat{u} = \dot{\phi}\hat{u}$ about l, and a velocity $\mathbf{v}$ along l. If the location vector $\mathbf{s}$ is the position of a point on l, then Plücker coordinates of the line l are

$$l = \begin{bmatrix} \hat{u} \\ \boldsymbol{\rho} \end{bmatrix} = \begin{bmatrix} \hat{u} \\ \mathbf{s} \times \hat{u} \end{bmatrix} \tag{6.429}$$

The pitch of the screw is

$$p = \frac{|\mathbf{v}|}{|\boldsymbol{\omega}|} \tag{6.430}$$

and the direction of the screw is defined by

$$\hat{u} = \frac{\boldsymbol{\omega}}{|\boldsymbol{\omega}|} \tag{6.431}$$

So, the instantaneous screw coordinates $\check{v}(p, \omega, \hat{u}, \mathbf{s})$ are given as

$$\check{v}(p, \omega, \hat{u}, \mathbf{r}) = \begin{bmatrix} \omega\hat{u} & \dfrac{\mathbf{r} \times \boldsymbol{\omega} + |\mathbf{v}|\,\boldsymbol{\omega}}{|\boldsymbol{\omega}|} \end{bmatrix}^{\mathrm{T}} = \begin{bmatrix} \omega\hat{u} \\ \mathbf{s} \times \hat{u} + \mathbf{v} \end{bmatrix}$$
$$= \begin{bmatrix} \boldsymbol{\omega} \\ \mathbf{s} \times \hat{u} + p\boldsymbol{\omega} \end{bmatrix} = \begin{bmatrix} \boldsymbol{\omega} \\ \boldsymbol{\rho} + p\boldsymbol{\omega} \end{bmatrix} \tag{6.432}$$

Example 393 ★ Pitch of an Instantaneous Screw Using Plücker coordinates, we may define the pitch of an instantaneous screw by

$$p = \hat{u} \cdot \boldsymbol{\xi} \tag{6.433}$$
$$\boldsymbol{\xi} = \boldsymbol{\rho} + h\hat{u} \tag{6.434}$$

because $\hat{u} \cdot \boldsymbol{\rho} = 0$, and therefore,

$$\hat{u} \cdot \boldsymbol{\xi} = \hat{u} \cdot \left(\boldsymbol{\rho} + h\hat{u}\right) = \left(\hat{u} \cdot \boldsymbol{\rho} + \phi p\right) = \phi p \tag{6.435}$$

Example 394 ★ Nearest Point on a Screw Axis to the Origin The point on the instantaneous screw axis, nearest to the origin, is indicated by the following position vector:

$$\mathbf{s}_0 = \phi \left(\hat{u} \times \boldsymbol{\xi} \right) \tag{6.436}$$

KEY SYMBOLS

B	body coordinate frame, local coordinate frame
c	cosine
$\mathbf{d}$	translation vector, displacement vector
D	displacement matrix
e	exponential
G	global coordinate frame, fixed coordinate frame
h	translation of a screw
$\mathbf{I} = [I]$	identity matrix
$\hat{\imath}, \hat{\jmath}, \hat{k}$	local coordinate axis unit vectors
$\tilde{\imath}, \tilde{\jmath}, \tilde{k}$	skew-symmetric matrices of the unit vectors $\hat{\imath}, \hat{\jmath}, \hat{k}$
$\hat{I}, \hat{J}, \hat{K}$	global coordinate axis unit vectors
l	line, Plücker coordinates, Plücker line
L, M, N	components of $\hat{u}$
$\hat{n}$	normal unit vector to a plane
p	pitch of a screw
P	body point, fixed point in B, point matrix
P, Q, R	components of $\boldsymbol{\rho}$
$\mathbf{p}, \mathbf{q}, \mathbf{r}$	position vectors, homogeneous position vectors
r_{ij}	element of row i and column j of a matrix
R	rotation transformation matrix
s	sine
$\mathbf{s}$	location vector of a screw
$\check{s}$	screw
T	homogeneous transformation matrix
$\hat{u}$	unit vector on axis of rotation
$\tilde{u}$	skew-symmetric matrix of the vector $\hat{u}$
$\mathbf{v}$	velocity vector
w	weight factor of a homogeneous vector
x, y, z	local coordinate axes
X, Y, Z	global coordinate axes

Greek

α, β, γ	rotation angles about global axes
${}^G\Gamma_B$	transformation matrix for Plücker coordinates
λ	eigenvalues of R
$\boldsymbol{\xi}$	moment vector of a Plücker line
π	homogeneous expression of a plane
$\boldsymbol{\rho}$	moment vector of $\hat{u}$ about origin
ϕ	angle of rotation about $\hat{u}$, rotation of a screw
$\boldsymbol{\omega}$	angular velocity vector
$\tilde{\omega}$	skew symmetric matrix of the vector $\boldsymbol{\omega}$

Symbol	
tr	trace operator
vers	$1-\cos$
$[\ ]^{-1}$	inverse of the matrix [ ]
$[\ ]^{\mathrm{T}}$	transpose of the matrix [ ]
$\triangle$	transformation matrix of ray-to-axis arrangement

EXERCISES

1. **Global Position in Rigid-Body Motion** We move the body coordinate frame B to

$$^{G}\mathbf{d} = \begin{bmatrix} 4 & -3 & 7 \end{bmatrix}^{\mathrm{T}}$$

Find $^{G}\mathbf{r}_P$ if the local position of a point is

$$^{B}\mathbf{r}_P = \begin{bmatrix} 7 & 3 & 2 \end{bmatrix}^{\mathrm{T}}$$

and the orientation of B with respect to the global frame G can be found by a rotation 45 deg about the X-axis and then 45 deg about the Y-axis.

2. **Rotation Matrix Compatibility** It is not possible to find $^{G}R_B$ from the equation of rigid-body motion

$$^{G}\mathbf{r}_P = {}^{G}R_B\, {}^{B}\mathbf{r}_P + {}^{G}\mathbf{d}$$

if $^{G}\mathbf{r}_P$, $^{B}\mathbf{r}_P$, and $^{G}\mathbf{d}$ are given. Explain why and find the required conditions to be able to find $^{G}R_B$.

3. **Global Position with Local Rotation in Rigid-Body Motion** Assume a body coordinate frame B is at

$$^{G}\mathbf{d} = \begin{bmatrix} 4 & -3 & 7 \end{bmatrix}^{\mathrm{T}}$$

Find $^{G}\mathbf{r}_P$ if the local position of a point is

$$^{B}\mathbf{r}_P = \begin{bmatrix} 7 & 3 & 2 \end{bmatrix}^{\mathrm{T}}$$

and the orientation of B with respect to the global frame G can be found by a rotation 45 deg about the x-axis and then 45 deg about the y-axis.

4. **Global and Local Rotation in Rigid-Body Motion** A body coordinate frame B is translated to

$$^{G}\mathbf{d} = \begin{bmatrix} 4 & -3 & 7 \end{bmatrix}^{\mathrm{T}}$$

Find $^{G}\mathbf{r}_P$ if the local position of a point is

$$^{B}\mathbf{r}_P = \begin{bmatrix} 7 & 3 & 2 \end{bmatrix}^{\mathrm{T}}$$

and the orientation of B with respect to the global frame G can be found by a rotation 45 deg about the X-axis, then 45 deg about the y-axis, and finally 45 deg about the z-axis.

5. **Combination of Rigid Motions** The frame B_1 is rotated 35 deg about the z_1-axis and translated to

$$^{B_2}\mathbf{d} = \begin{bmatrix} -40 & 30 & 20 \end{bmatrix}^\mathrm{T}$$

with respect to another frame B_2. The orientation of the frame B_2 in the global frame G can be found by a rotation of 55 deg about

$$\mathbf{u} = \begin{bmatrix} 2 & -3 & 4 \end{bmatrix}$$

Calculate $^G\mathbf{d}_1$, and $^GR_1$.

6. **Possible Transformations** Determine the possible transformations to move $^B\mathbf{r} = [1, 1, 1]^\mathrm{T}$ from $^G\mathbf{r} = [1, 1, 1]^\mathrm{T}$ to $^G\mathbf{r} = [7, 3, 2]^\mathrm{T}$.

7. **Displaced and Rotated Frame** Determine the axis–angle and transformation matrix ^GT_B in Figure 6.27.

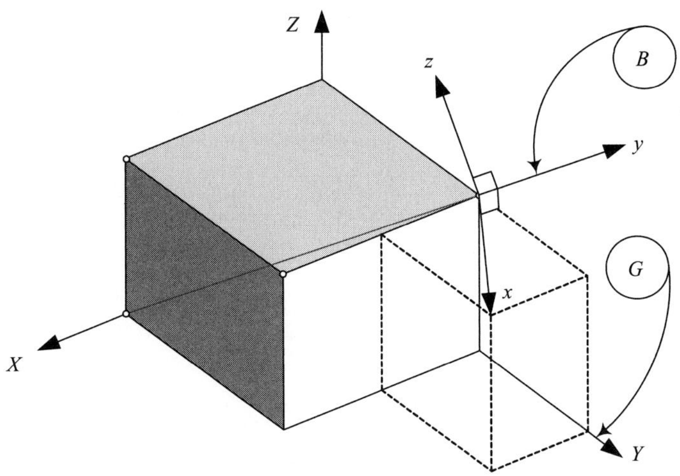

Figure 6.27 A displaced and rotated frame.

8. **Rotation Submatrix in Homogeneous Transformation Matrix** Find the missing elements in the homogeneous transformation matrix

$$[T] = \begin{bmatrix} ? & 0 & ? & 4 \\ 0.707 & ? & ? & 3 \\ ? & ? & 0 & 1 \\ 0 & 0 & 0 & 1 \end{bmatrix}$$

9. **Angle and Axis of Rotation** Find the angle and axis of rotation for $[T]$ and T^{-1}:

$$[T] = \begin{bmatrix} 0.866 & -0.5 & 0 & 4 \\ 0.5 & 0.866 & 0 & 3 \\ 0 & 0 & 1 & 1 \\ 0 & 0 & 0 & 1 \end{bmatrix}$$

10. **Combination of Homogeneous Transformations** Assume that the origin of the frames B_1 and B_2 are at

$$^{2}\mathbf{d}_1 = \begin{bmatrix} -10 & 20 & -20 \end{bmatrix}^{\mathrm{T}}$$

$$^{G}\mathbf{d}_2 = \begin{bmatrix} -15 & -20 & 30 \end{bmatrix}^{\mathrm{T}}$$

The orientation of B_1 in B_2 can be found by a rotation of 60 deg about

$$^{2}\mathbf{u} = \begin{bmatrix} 1 & -2 & 4 \end{bmatrix}$$

and the orientation of B_2 in the global frame G can be found by a rotation of 30 deg about

$$^{G}\mathbf{u} = \begin{bmatrix} 4 & 3 & -4 \end{bmatrix}$$

Calculate the transformation matrices $^{G}T_2$ and $^{G}T_2^{-1}$.

11. **Rotation about an Axis Not Going through Origin** Find the global position of a body point at

$$^{B}\mathbf{r}_P = \begin{bmatrix} 7 & 3 & 2 \end{bmatrix}^{\mathrm{T}}$$

after a rotation of 30 deg about an axis parallel to

$$^{G}\mathbf{u} = \begin{bmatrix} 4 & 3 & -4 \end{bmatrix}$$

and passing through a point at (3, 3, 3).

12. **Rotation about an Off-Center Axis** A body point is at $^{B}\mathbf{r}_P = [7, 3, 2]^{\mathrm{T}}$ when the body and global coordinate frames are coincident. Determine the global coordinate of the point if B turns:

(a) 60 deg about $^{G}\mathbf{u} = [4 \; 3 \; -4]$ that goes through (0, 0, 1)
(b) 45 deg about $^{G}\mathbf{u} = [-4 \; 3 \; 4]$ that goes through (0, 0, 1)
(c) 45 deg about $^{G}\mathbf{u} = [1 \; 1 \; 1]$ that goes through (1, 1, 1)

13. **Inversion of a Square Matrix** The inverse of the 2×2 matrix

$$[A] = \begin{bmatrix} a & b \\ c & d \end{bmatrix}$$

is

$$A^{-1} = \begin{bmatrix} -\dfrac{d}{-ad+bc} & \dfrac{b}{-ad+bc} \\ \dfrac{c}{-ad+bc} & -\dfrac{a}{-ad+bc} \end{bmatrix} \tag{6.437}$$

Use the inverse method of splitting a matrix $[T]$ into

$$[T] = \begin{bmatrix} A & B \\ C & D \end{bmatrix}$$

applying the inverse technique (6.115), verify Equation (6.437), and calculate the inverse of

$$[A] = \begin{bmatrix} 1 & 2 & 3 & 4 \\ 12 & 13 & 14 & 5 \\ 11 & 16 & 15 & 6 \\ 10 & 9 & 8 & 7 \end{bmatrix}$$

14. Combination of Rotations about Noncentral Axes Consider a rotation 30 deg about an axis at the global point (3, 0, 0) followed by another rotation 30 deg about an axis at the global point (0, 3, 0). Find the final global coordinates of

$$^{B}\mathbf{r}_P = \begin{bmatrix} 1 & 1 & 0 \end{bmatrix}^{\mathrm{T}}$$

to

$$^{G}\mathbf{r}_P = \begin{bmatrix} \sqrt{2} & 0 & 3 \end{bmatrix}^{\mathrm{T}}$$

15. Transformation Matrix from Body Points Figure 6.28(a) shows a cube at initial configurations. Label the corners of the cube at the final configuration shown in Figure 6.28(b) and find the associated homogeneous transformation matrix. The length of each side of the cube is 2.

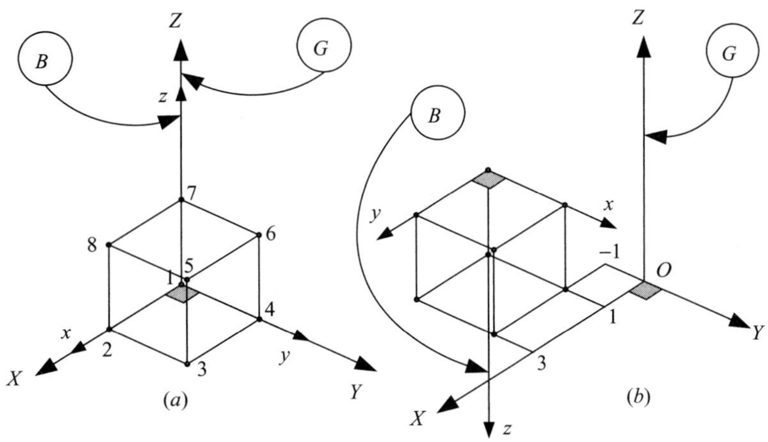

Figure 6.28 A cube before and after a rigid motion.

16. ★ Central Screw Find the central screw that moves the point

$$^{B}\mathbf{r}_P = \begin{bmatrix} 1 & 0 & 0 \end{bmatrix}^{\mathrm{T}}$$

to

$$^{G}\mathbf{r}_P = \begin{bmatrix} 0 & 1 & 4 \end{bmatrix}^{\mathrm{T}}$$

17. ★ Screw Motion Find the global position of

$$^{B}\mathbf{r}_P = \begin{bmatrix} 1 & 0 & 0 & 1 \end{bmatrix}^{\mathrm{T}}$$

after a screw motion $^{G}\check{s}_B(h, \phi, \hat{u}, \mathbf{s}) = {}^{G}\check{s}_B(4{,}60\,\mathrm{deg}, \hat{u}, \mathbf{s})$ where

$$^{G}\mathbf{s} = \begin{bmatrix} 3 & 0 & 0 \end{bmatrix}^{\mathrm{T}} \qquad ^{G}\mathbf{u} = \begin{bmatrix} 1 & 1 & 1 \end{bmatrix}^{\mathrm{T}}$$

18. ★ Pole of Planar Motion Find the pole position of a planar morion if we have the coordinates of two body points before and after the motion as given below:

$$P_1\,(1, 1, 1) \qquad P_2(5, 2, 1)$$
$$Q_1\,(4, 1, 1) \qquad Q_2(7, 2+\sqrt{5}, 1)$$

19. ★ **Screw Parameters** Find the global coordinates of the body points

$$P_1(5, 0, 0) \qquad Q_1(5, 5, 0) \qquad R_1(0, 5, 0)$$

after a rotation of 30 deg about the x-axis followed by a rotation of 40 deg about an axis parallel to the vector

$${}^G\mathbf{u} = \begin{bmatrix} 4 & 3 & -4 \end{bmatrix}$$

and passing through a global point at $(0, 3, 0)$. Use the coordinates of the points and calculate the screw parameters that are equivalent to the two rotations.

20. ★ **Noncentral Rotation** Find the global coordinates of a point at

$${}^B\mathbf{r}_P = \begin{bmatrix} 10 & 20 & -10 \end{bmatrix}^{\mathrm{T}}$$

when the body frame rotates about

$${}^G\mathbf{u} = \begin{bmatrix} 1 & 1 & 1 \end{bmatrix}^{\mathrm{T}}$$

which passes through a point at $(1, -4, 2)$.

21. ★ **Equivalent Screw** Calculate the transformation matrix ${}^GT_B$ for a rotation of 30 deg about the x-axis followed by a translation parallel to

$${}^G\mathbf{s} = \begin{bmatrix} 3 & 2 & -1 \end{bmatrix}^{\mathrm{T}}$$

and then a rotation of 40 deg about an axis parallel to the vector

$${}^G\mathbf{u} = \begin{bmatrix} 2 & -1 & 1 \end{bmatrix}^{\mathrm{T}}$$

Find the screw parameters of ${}^GT_B$.

22. ★ **Central Screw Decomposition** Find a triple central screw for case 1 in Appendix C,

$${}^G\check{s}_B(h, \phi, \hat{u}, \mathbf{s}) = \check{s}(h_X, \gamma, \hat{I})\, \check{s}(h_Y, \beta, \hat{J})\, \check{s}(h_Z, \alpha, \hat{K})$$

to get the same screw as

$${}^G\check{s}_B(h, \phi, \hat{u}, \mathbf{s}) = {}^G\check{s}_B(4, 60, \hat{u}, \mathbf{s})$$

where

$${}^G\mathbf{s} = \begin{bmatrix} 3 & 0 & 0 \end{bmatrix}^{\mathrm{T}} \qquad {}^G\mathbf{u} = \begin{bmatrix} 1 & 1 & 1 \end{bmatrix}^{\mathrm{T}}$$

23. ★ **Central Screw Composition** What is the final position of a point at

$${}^B\mathbf{r}_P = \begin{bmatrix} 10 & 0 & 0 \end{bmatrix}^{\mathrm{T}}$$

after the central screw $\check{s}(4, 30\,\text{deg}, \hat{J})$ followed by $\check{s}(2, 30\,\text{deg}, \hat{I})$ and $\check{s}(-6, 120\,\text{deg}, \hat{K})$?

24. ★ Screw Composition Find the final position of a point at

$$^{B}\mathbf{r}_P = \begin{bmatrix} 10 & 0 & 0 \end{bmatrix}^{\mathrm{T}}$$

after a screw

$$^{1}\check{s}_2(h_1, \phi_1, \hat{u}_1, \mathbf{s}_1) = {}^{1}\check{s}_2\left(1{,}40\,\mathrm{deg}, \begin{bmatrix} 1/\sqrt{3} \\ -1/\sqrt{3} \\ 1/\sqrt{3} \end{bmatrix}, \begin{bmatrix} 2 \\ 3 \\ -1 \end{bmatrix}\right)$$

followed by

$$^{G}\check{s}_1(h_0, \phi_0, \hat{u}_0, \mathbf{s}_0) = {}^{G}\check{s}_1\left(-1{,}45\,\mathrm{deg}, \begin{bmatrix} 1/9 \\ 4/9 \\ 4/9 \end{bmatrix}, \begin{bmatrix} -3 \\ 1 \\ 5 \end{bmatrix}\right)$$

25. ★ Plücker Line Coordinate Find the missing numbers:

$$l = \begin{bmatrix} 1/3 & 1/5 & ? & ? & 2 & -1 \end{bmatrix}^{\mathrm{T}}$$

26. ★ Plücker Lines Find the Plücker lines for *AE*, *BE*, *CE*, *DE* in the local coordinate *B* and calculate the angle between *AE* and the *Z*-axis in the pyramid shown in Figure 6.29. The local coordinate of the edges are

$$A(1, 1, 0) \qquad B(-1, 1, 0) \qquad C(-1, -1, 0)$$
$$D(1, -1, 0) \qquad E(0, 0, 3)$$

Transforms the Plücker lines *AE*, *BE*, *CE*, *DE* to the global coordinate *G*. The global position of *o* is at

$$o(1, 10, 2)$$

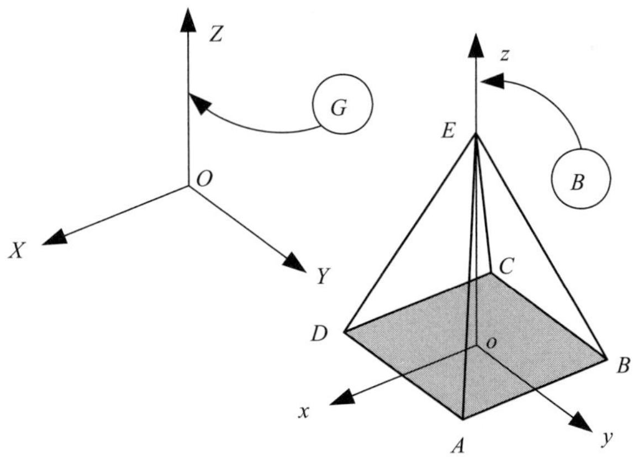

Figure 6.29 A pyramid.

27. ★ **Angle between Two Lines** Find the angle between OE and OD of the pyramid shown in Figure 6.29. The coordinates of the points are

$$D(1, -1, 0) \qquad E(0, 0, 3)$$

28. ★ **Distance from the Origin** The equation of a plane is given as

$$4X - 5Y - 12Z - 1 = 0$$

Determine the perpendicular distance of the plane from the origin.

7

Multibody Kinematics

We can model most multibody systems as connected rigid bodies by revolute or prismatic joints that allow the bodies to move relatively. We apply rigid-body kinematics to the connected bodies and show how to determine their relative motions.

7.1 MULTIBODY CONNECTION

Two rigid bodies that are permanently in contact with a possible relative motion is called a *kinematic pair*. The *multibody* is a mechanical system that is made by connected rigid bodies. Every *body* of the system that can move relative to all other members is called a *link*, *bar*, *arm*, or *object*. Any two or more connected bodies, such that no relative motion can occur among them, are considered a single body. Figure 7.1 illustrates the Shuttle arm as an example of multibody systems.

Every two bodies of a kinematic pair are connected by a *joint*. Most of the joints in multibody dynamics are *revolute* (*rotary*) or *prismatic* (*translatory*). Figure 7.2 illustrates revolute and prismatic joints. A *revolute joint* (R) allows relative rotation and a *prismatic joint* (P) allows a relative translation between the two connected bodies. At every joint, the relative motion of the bodies is qualitatively expressed by a single variable called the *joint coordinate* or *joint variable*. It is an *angle* for a revolute joint and a *distance* for a prismatic joint. The relative rotation or translation between two connected bodies occurs about a line called the *joint axis*.

A joint is called an *active joint* if its coordinate is controlled by an actuator. A *passive joint* does not have any actuator, and its coordinate is a function of the coordinates of active joints and the geometry of the links of the system. Passive joints are also called *inactive* or *free joints*. Prismatic and revolute joints provide one degree of freedom between the two connected bodies.

Although prismatic and revolute joints are the most applied connections, there are other types of joints that are classified according to the number of *degrees of freedom* (*DOF*) they eliminate or they allow. A joint can provide a maximum of five DOF and a minimum of one DOF.

Proof: If there is no contact between two bodies A and B, then B has six DOF with respect to A. Every permanent contact eliminates some rotational or translational DOF. All possible physically realizable kinematic pairs can be classified by the number of rotational, R, or translational, T, degrees of freedom the contact provides.

A joint in class 1 provides one DOF and uses five contact points between A and B. The class 1 joints and their point contact kinematic models are shown in

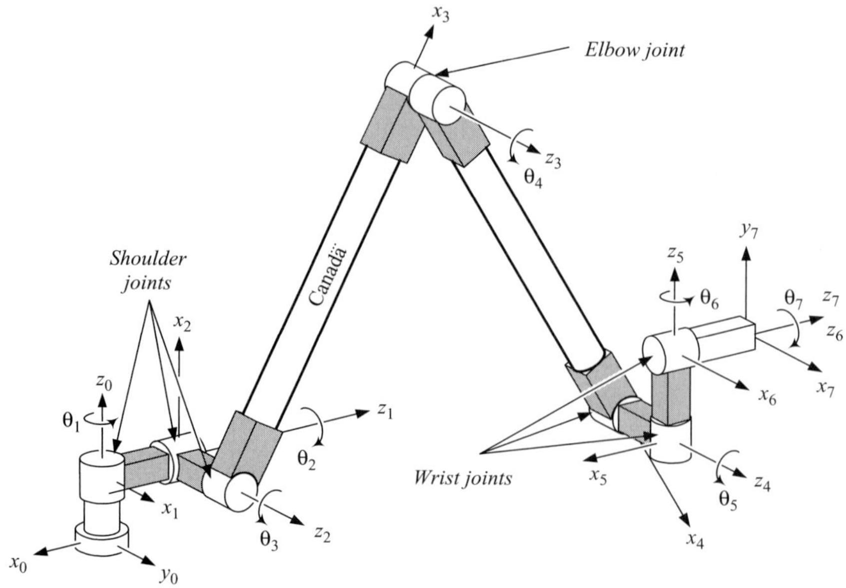

Figure 7.1 Shuttle arm is a multibody system.

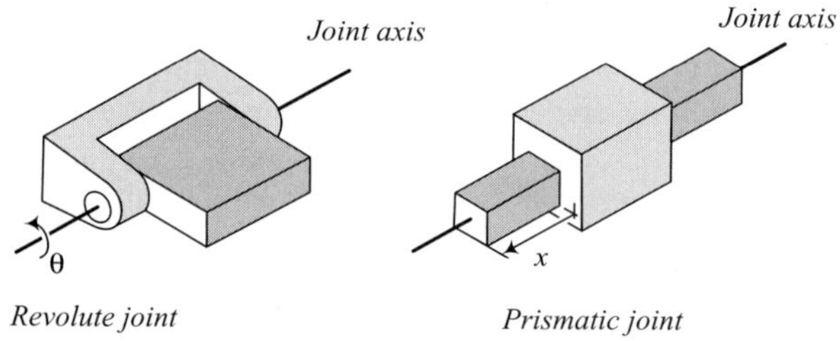

Figure 7.2 A revolute joint and a prismatic joint.

Figure 7.3. There are three types of joints in class 1: revolute, R, with a rotational freedom $R1$; prismatic, P, with a translational freedom $T1$; and helical or screw, S, with a proportional rotational–translational freedom $R(T)$. These joints provide one DOF between two links A and B.

A joint in class 2 provides two DOF and uses four contact points between A and B. The class 2 joints and their point contact kinematic models are shown in Figure 7.4. There are three types of joints in class 2: sphere in slot, with two rotational freedom $R2$; cylindrical, C, with a coaxial rotational and a translational freedom $R1T1$; and disc in slot, with a perpendicular rotational and translational freedom $R1T1$. These joints provide two DOF between two links A and B.

A joint in class 3 provides three DOF and uses three contact points between A and B. The class 3 joints and their point contact kinematic models are shown in Figure 7.5. There are three types of joints in class 3: spherical, S, with three rotational freedom $R3$; sphere in slot, with two rotational and a translational freedom $R2T1$; and disc in

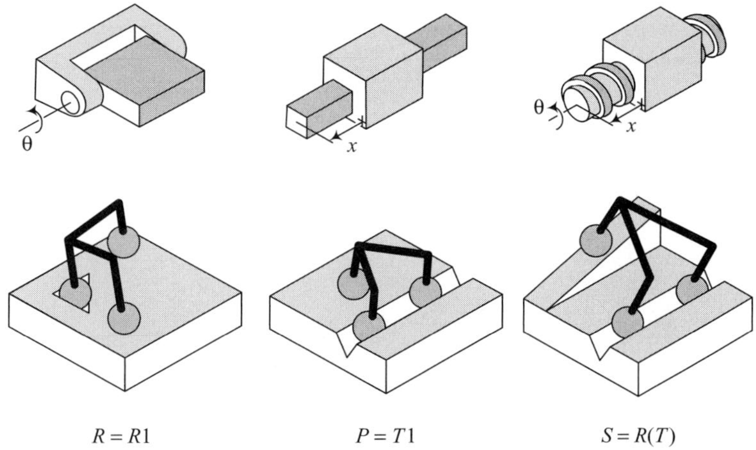

Figure 7.3 Joints of class 1 with one DOF: revolute, R, prismatic, P, and screw, S.

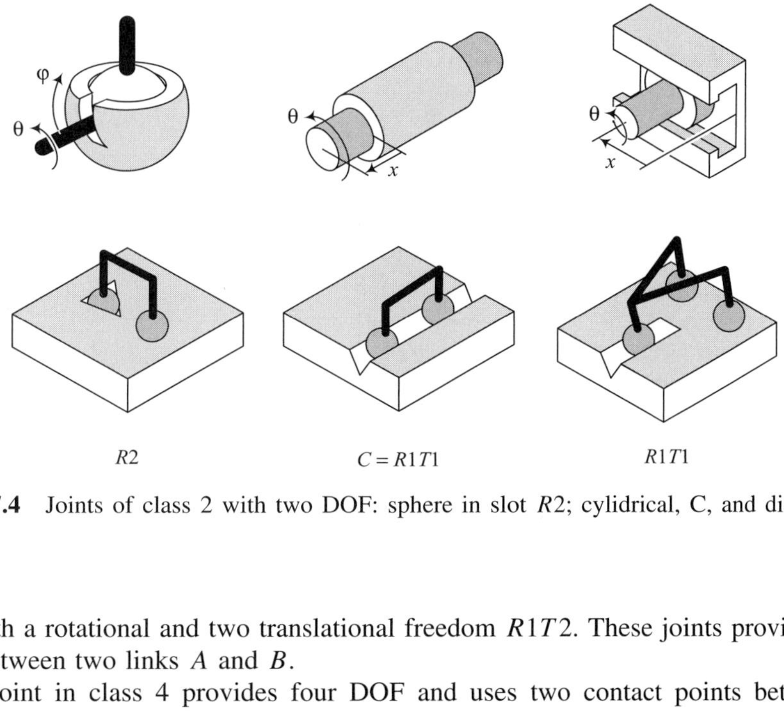

Figure 7.4 Joints of class 2 with two DOF: sphere in slot $R2$; cylidrical, C, and disc in slot $R1T$.

slot, with a rotational and two translational freedom $R1T2$. These joints provide three DOF between two links A and B.

A joint in class 4 provides four DOF and uses two contact points between A and B. The class 4 joints and their point contact kinematic models are shown in Figure 7.6. There are two types of joints in class 4: sphere in slot, with three rotational and one translational freedom $R3T1$, and cylinder in slot, with two rotational and two translational freedom $R2T2$. These joints provide four DOF between two links A and B.

A joint in class 5 provides five DOF and uses one contact point between A and B. The class 5 joint and its point contact kinematic model are shown in Figure 7.7. There is only one type of joints in class 5: sphere on plane, with three rotational and

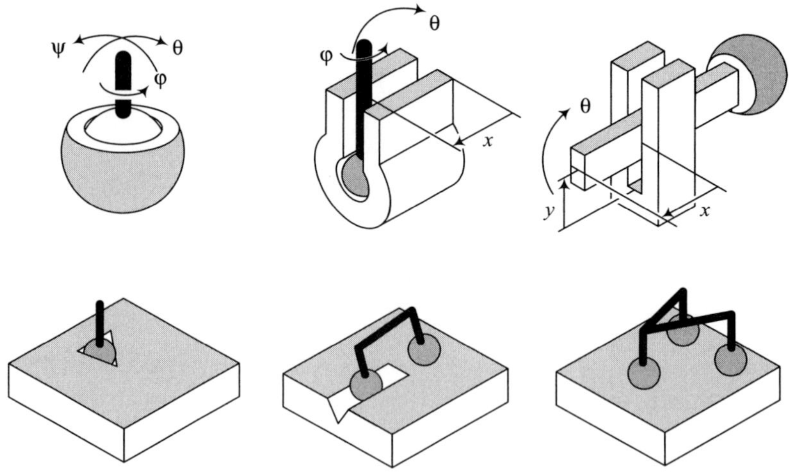

$SC = R3$ $\quad$ $R2T1$ $\quad$ $R1T2$

Figure 7.5 Joints of class 3 with three DOF: spherical, S, sphere in slot, $R2T1$; and disc in slot, $R1T2$.

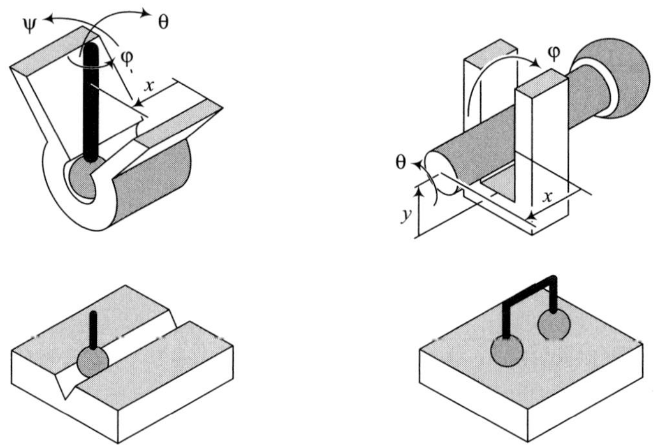

$R3T1$ $\quad$ $R2T2$

Figure 7.6 Joints of class 4 with four DOF: sphere in slot $R3T1$ and cylidrer in slot $R2T2$.

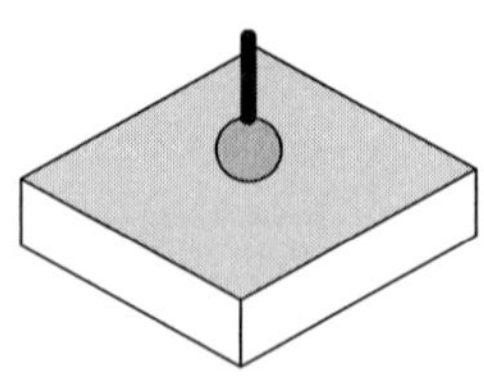

$R3T2$

Figure 7.7 Joint of class 5 with five DOF: sphere on plane $R3T2$.

two translational freedom $R3T2$. This joint provides five DOF between two links A and B.

Therefore, the DOF of a mechanism is equal to

$$f = 6n - j_5 - 2j_4 - 3j_3 - 4j_2 - 5j_1$$

where n is number of links and j_i, $i = 1, 2, 3, 4, 5$, indicates the number of joints of type j_i. ■

Example 395 Number of Links Every triangle that is made by three connected bars counts as one link, because there cannot be any relative motion among the bars. Figure 7.8 illustrates a mechanism with 7 links using 11 bars. There cannot be any relative motion among bars 3, 10, and 11, and they count as one link, say link 3. Bars 6, 8, and 9 also have no relative motion and are counted as one link, say link 6.

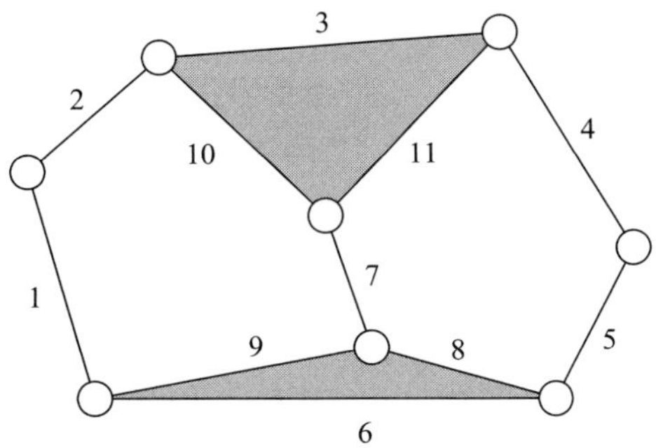

Figure 7.8 A planar linkage with seven links and eight revolute joints.

Example 396 Grübler Formula for DOF Most of the industrial multibodies are planar mechanisms. The links of a planar mechanism move parallel to a fixed plane. If a mechanism is planar, the only possible joints between two links are the revolute R and prismatic P from class 1 and cylindrical C from class 2. Figure 7.9 illustrates the planar joints R, P, and C between two links 1 and 2.

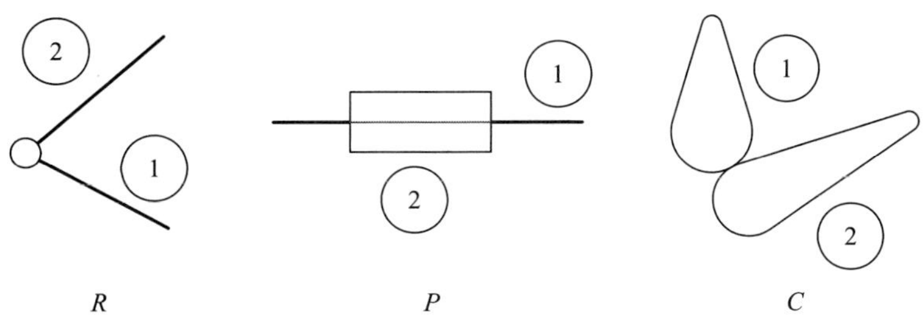

Figure 7.9 Two planar links can be connected by a revolute R, prismatic P, or cylindrical C joint.

Every free link in a planar motion has three DOF: one rotation and two translations. A revolute or prismatic joint from class 1 remains one DOF and eliminates two DOF. A cylindrical joint from class 2 eliminates one DOF and remains two DOF. Let us show the number of links of a mechanism by n, the number of P and R joints by j_1, and the number of C joints by j_2. Then the number of degrees of freedom f of a planar mechanism with one grounded link is

$$f = 3(n-1) - 2j_1 - j_2 \tag{7.1}$$

This is because every free link has three DOF, and if a link is attached to the ground, it loses its three DOF. So, there are $3(n-1)$ DOF for n links of a mechanism. Every planar joint of class 1 eliminates two DOF and each planar joint of class 2 eliminates one DOF. As an example, the planar mechanism in Figure 7.10 has three DOF:

$$n = 7 \qquad j_1 = 7 \qquad j_2 = 1 \tag{7.2}$$

$$f = 3(7-1) - 2 \times 7 - 1 = 3 \tag{7.3}$$

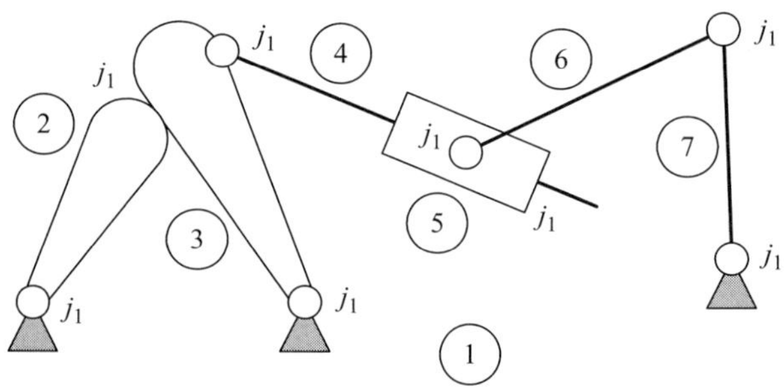

Figure 7.10 A planar linkage with seven links, seven joints j_1, and one joint j_2.

Equation (7.1) is called the *Grübler formula*. The three-dimensional form of the equation, which may also be called the Chebychev–Grübler–Kutzbach formula, is

$$f = 6(n-1) - j_5 - 2j_4 - 3j_3 - 4j_2 - 5j_1 \tag{7.4}$$

These equations may not work, depending on the geometry of the mechanism or size of the links. As an example, Figure 7.11 illustrates a universal joint. There are four links in a universal joint: link (1) is the ground, which has a revolute joint with the input link (2) and the output link (4). The input and the output links are connected with a cross-link (3). The universal joint is a three-dimensional four-bar linkage with one DOF for which the cross-link acts as a coupler link. However, the Kutzbach formula determines a structure with -2 DOF:

$$n = 4 \qquad j_1 = 4 \tag{7.5}$$

$$f = 6(4-1) - 5 \times 4 = -2 \tag{7.6}$$

The universal joint provides one DOF only when the four axes of rotations intersect at a point.

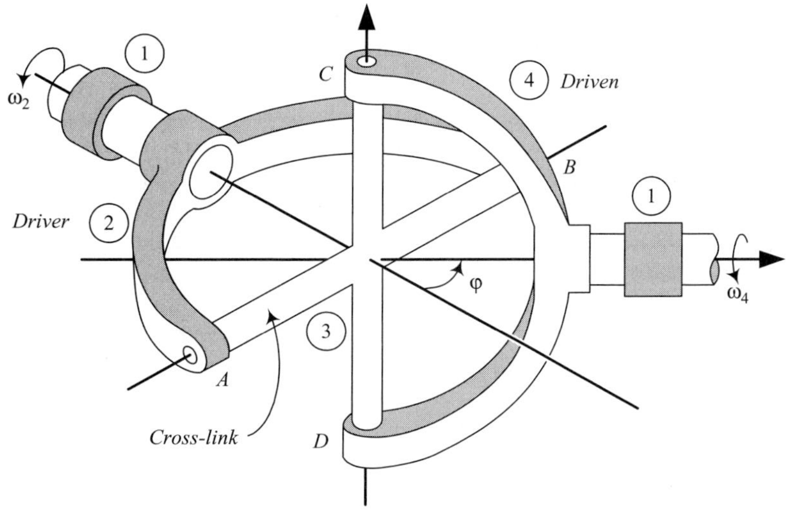

Figure 7.11 A universal joint with four links: link (1) is the ground, link (2) is the input, link (4) is the output, and the cross-link (3) is a coupler link.

Example 397 Compound Links and Multiple Joints When a link is connected to more than two other links, it is called a *compound link*. A compound link needs to be clearly identified because its sides cannot move relatively.

When more than two links are connected at the same point, the contact point is called a *multiple joint*. If n links are connected at a point, there are $n-1$ joints at the connection point.

The mechanism in Figure 7.12 has 16 links where 1, 2, 6, 16, 8, and 9 are compound links. All of the joints except two of them are revolute; however, there are some multiple joints that make $j_1 = 23$, and therefore,

$$f = 3\,(16-1) - 2 \times 23 = -1$$

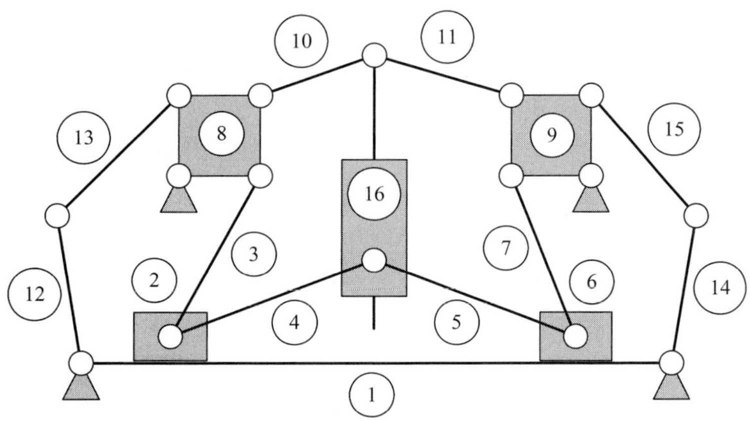

Figure 7.12 A mechanism with 16 links amd 23 joints.

A mechanism with $f = 0$ has no moveability and is equivalent to a rigid body. A mechanism with $f < 0$ is called overrigid or a structure.

Example 398 Equivalent Spherical Joint There are many situations where a link connected to other links must be able to turn about any axis. Such a link should kinematically be connected to the base link by a spherical joint. However, we may use three perpendicular revolute joints to simulate the spherical joint and be able to control the rotations.

Figure 7.13 illustrates a helicopter blade that is kinematically attached to the main rotor using a spherical joint. A practical method of attachment is shown in Figure 7.14 using three revolute joints. Therefore, a spherical joint between two links B_1 and B_4 can be substituted by two intermediate links B_2 and B_3 that are connected to B_1 and B_4 by three mutually perpendicular revolute joints.

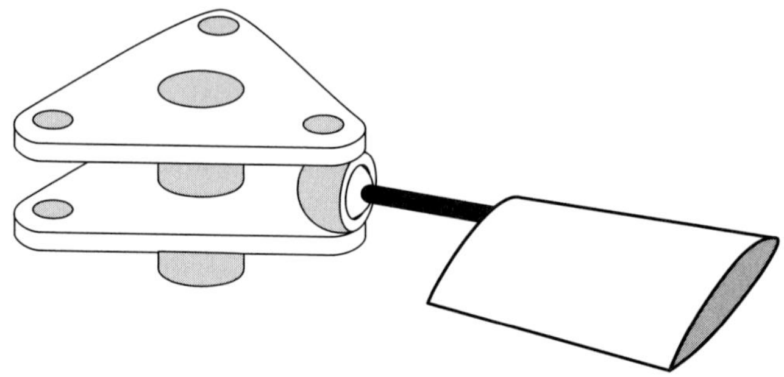

Figure 7.13 A helicopter blade kinematically needs a spherical joint to attach to the main rotor.

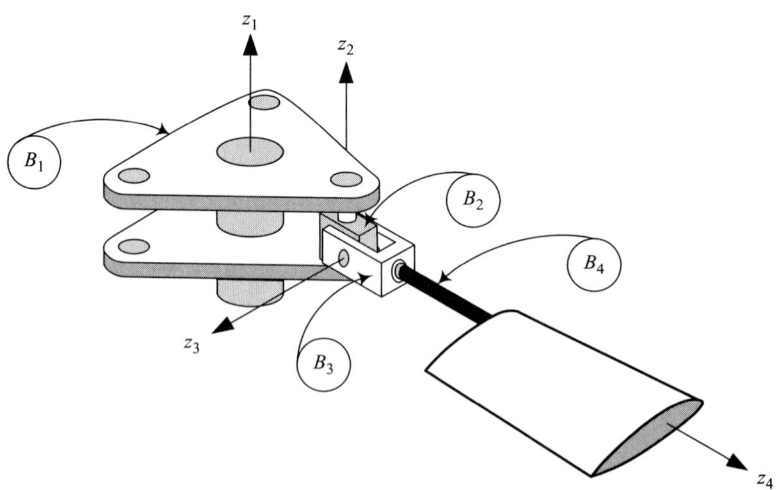

Figure 7.14 Three mutually perpendicular revolute joints are equivalent to a spherical joint.

7.2 DENAVIT–HARTENBERG RULE

Every body of a multibody system needs a coordinate frame to be distinguished from other bodies. We use the coordinate frame transformation calculus to determine the relative orientation of the bodies and also the coordinate of a body point for an observer in any frame. Although we are free to attach a coordinate frame at any point of a rigid body, the mass center or a point on a joint axis provides some advantages.

A serial multibody with n joints will have $n + 1$ links. Numbering of links starts from (0) for the *base link* and increases sequentially up to (n) for the *end-effector* link. Numbering of joints starts from 1 for the joint connecting the first movable link to the base link and increases sequentially up to n. Therefore, link (i) is connected to its *lower link* ($i - 1$) at its *proximal end* by joint i and is connected to its *upper link* ($i + 1$) at its *distal end* by joint $i + 1$. Figure 7.15 illustrates such a link (i).

Figure 7.16 illustrates the links ($i - 1$), (i), and ($i + 1$) of a multibody along with joints $i - 1$, i, and $i + 1$. Each joint is indicated by its z-axis. We rigidly attach a local coordinate frame B_i to each link i at joint $i + 1$ based on the following *standard method*, known as the *Denavit–Hartenberg* (*DH*) method:

1. The z_i-axis is along the $i + 1$ joint axis. Every joint, without exception, is represented by a z-axis. To set up the link coordinate frames, we always begin with identifying the z_i-axes. The positive direction of the z_i-axis is arbitrary. The joint axis for revolute joints is the pin's centerline axis. However, the axis of a prismatic joint may be any axis parallel to the direction of translation. By assigning the z_i-axes, the pairs of links on either side of each joint and also the two joints on either side of each link are identified.

   Generally speaking, the z_i-axes of two joints at the ends of a link may be two skew lines; however, we make the industrial multibody systems such that the z_i-axes are only *parallel*, *perpendicular*, or *orthogonal*. Two parallel joint axes are indicated by a parallel sign (‖). Two joint axes are orthogonal if their axes are intersecting at a right angle. Orthogonal joint axes are indicated by an orthogonal sign (⊢). Two joints are perpendicular if their axes are at

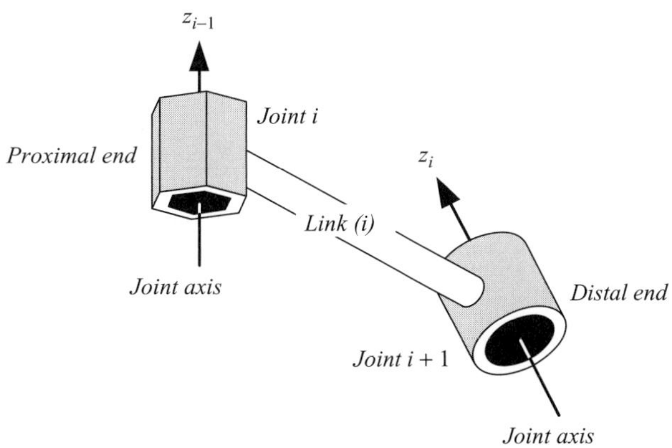

Figure 7.15 A link (i) of multibody that is connected to link ($i - 1$) at joint i and to link ($i + 1$) at joint $i + 1$.

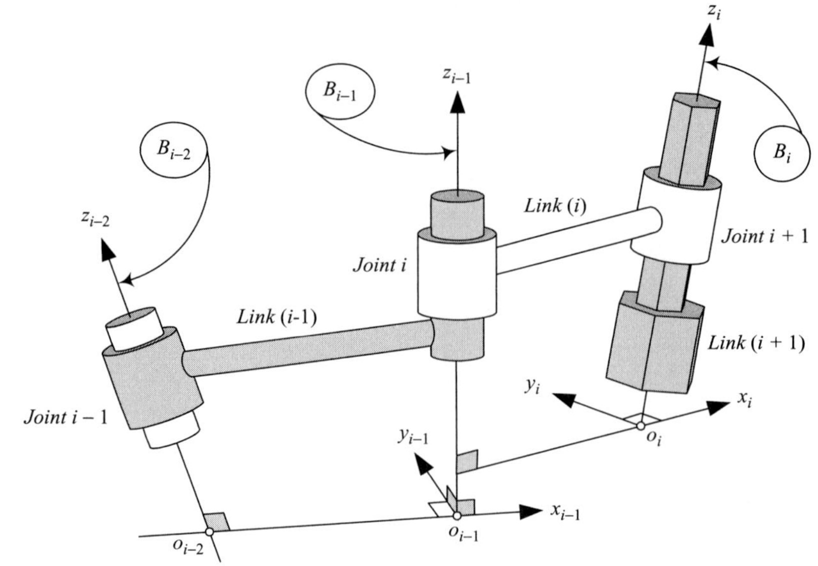

Figure 7.16 Link (i), its previous link $(i-1)$, and the next link $(i+1)$ along with the coordinate frames B_i and B_{i+1}.

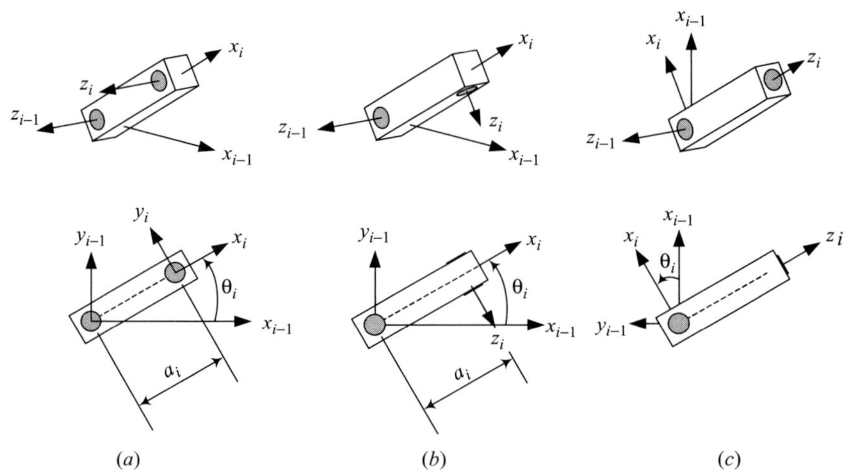

Figure 7.17 A link with two revolute joints at its ends. The joint axes of link (a) are parallel, link (b) are perpendicular, and link (c) are orthogonal.

a right angle with respect to their common normal. Perpendicular joint axes are indicated by a perpendicular sign ($\perp$). Figures 7.17(a)–(c) illustrate two revolute joints at the parallel, perpendicular, and orthogonal configurations.

Every link of a serial multibody, except the base and final links, has two joints at either end, R or P. The axes of the joints can also be parallel ($\|$), perpendicular ($\perp$) or orthogonal ($\vdash$). Therefore, we may indicate a link by the type and configuration of the end joints. As an example P∥R indicates a link where its lower joint is prismatic, its upper joint is revolute, and the joint axes are parallel.

2. The x_i-axis is along the *common normal* between the axes z_{i-1} and z_i, pointing from the z_{i-1}- to the z_i-axis. Even if the z-axes are skew lines, there is always a mutually perpendicular line to the two skew lines. The line is the common normal to the z-axes and has the shortest distance between the two z-axes.

   If the two z-axes are parallel, there are an infinite number of common normals to pick for the x_i-axis. In that case, we may pick the common normal that is collinear with the common normal of the previous joints.

   If the two z-axes are intersecting, there is no common normal between them. In that case, we may assign the x_i-axis perpendicular to the plane formed by the two z-axes in the direction of $z_{i-1} \times z_i$.

   If the two z-axes are collinear, the only nontrivial arrangement of joints is either P∥R or R∥P. So, we assign the x_i-axis such that we have the joint variable $\theta_i = 0$ in the rest position of the multibody.

   The configuration of a multibody at which all the joint variables are zero is called the *home configuration* or *rest position*, which is the reference for all motions of the multibody. The best rest position is where it makes as many axes parallel and coplanar to each other as possible.

3. The y_i-axis is determined by the right-hand rule such that $y_i = z_i \times x_i$.

By applying the DH method, the origin o_i of frame $B_i(o_i, x_i, y_i, z_i)$ attached to link i is placed at the intersection of the axis of joint $i + 1$ with the common normal between the z_{i-1}- and z_i-axes. There are many situations where the DH method may be ignored to set up a more suitable frame for a particular application. However, we always assign reference frames to each link so that one of the three coordinate axes x_i, y_i, or z_i (usually x_i) will be along the axis of the distal joint.

The DH method reduces the required number of parameters to identify two coordinate frames in each other from six parameters to four parameters. The four parameters are indicated by a_i, α_i, θ_i, d_i and are called the DH parameters.

1. The *link length* a_i indicates the distance between the z_{i-1}- and z_i-axes along the x_i-axis, where a_i is the *kinematic length* of link (i).
2. The *link twist* α_i indicates the required rotation of the z_{i-1}-axis about the x_i-axis to become parallel to the z_i-axis.
3. The *joint distance* d_i indicates the distance between the x_{i-1}- and x_i-axes along the z_{i-1}-axis, where d_i is also called the *link offset.*
4. The *joint angle* θ_i indicates the required rotation of the x_{i-1}-axis about the z_{i-1}-axis to become parallel to the x_i-axis.

Figure 7.18 illustrates the DH frames and parameters of the links in Figure 7.16. The parameters θ_i and d_i define the relative positions of two adjacent links at joint i and are called the *joint parameters*. At each joint, either θ_i or d_i is variable and the other is fixed. For a revolute joint (R) at joint i, the value of d_i is fixed, and θ_i is the *joint variable*. For a prismatic joint (P), the value of θ_i is fixed and d_i is the joint variable. The joint parameters θ_i and d_i define a central screw motion $\check{s}(d_i, \theta_i, \hat{k}_{i-1})$ because θ_i is a rotation about the z_{i-1}-axis and d_i is a translation along the z_{i-1}-axis.

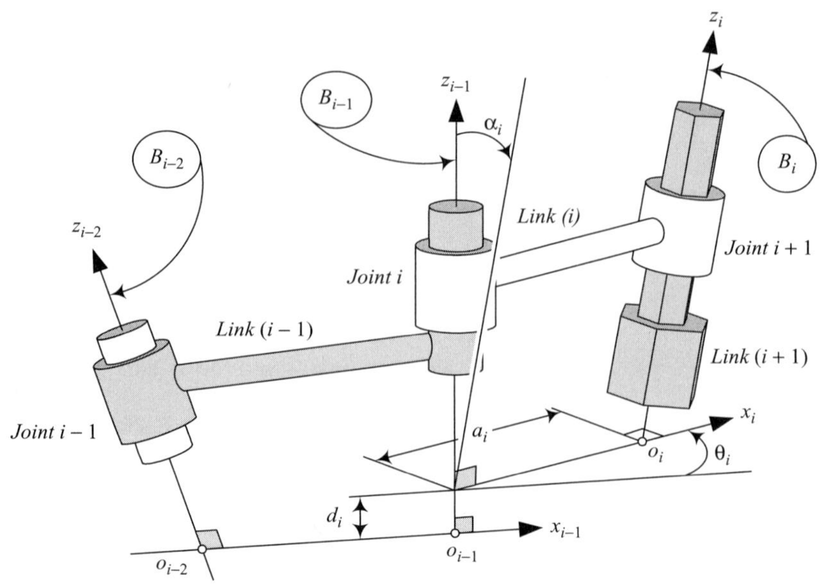

Figure 7.18 The DH parameters $a_i, \alpha_i, d_i, \theta_i$ for the coordinate frame of link (i) at joint i.

The parameters α_i and a_i define the relative positions of joints i and $i+1$ at two ends of link (i) and are called the *link parameters*. The link twist α_i is the angle of rotation of the z_{i-1}-axis about x_i to become parallel to the z_i-axis. The other link parameter, a_i, is the translation along the x_i-axis to bring the z_{i-1}-axis on the z_i-axis. The link parameters α_i and a_i define a central screw motion $\check{s}(a_i, \alpha_i, \hat{\imath})$ because α_i is a rotation about the x_i-axis and a_i is a translation along the x_i-axis.

Combining the two screws, we can move the z_{i-1}-axis to the z_i-axis by a central screw $\check{s}(a_i, \alpha_i, \hat{\imath})$ and move the x_{i-1}-axis to the x_i-axis by a central screw $\check{s}(d_i, \theta_i, \hat{k}_{i-1})$.

Consider the coordinate frame B_i that is fixed to link (i) and the coordinate frame B_{i-1} that is fixed to link $(i-1)$. Applying the Denavit–Hartenberg rules, the homogeneous transformation matrix ${}^{i-1}T_i$ to transform coordinate frame B_i to B_{i-1} can be found using the parameters of link (i) and joint i:

$$ {}^{i-1}T_i = \begin{bmatrix} \cos\theta_i & -\sin\theta_i\cos\alpha_i & \sin\theta_i\sin\alpha_i & a_i\cos\theta_i \\ \sin\theta_i & \cos\theta_i\cos\alpha_i & -\cos\theta_i\sin\alpha_i & a_i\sin\theta_i \\ 0 & \sin\alpha_i & \cos\alpha_i & d_i \\ 0 & 0 & 0 & 1 \end{bmatrix} \tag{7.7} $$

Therefore, we can determine the coordinates of a point in $B_{i-1}(x_{i-1}, y_{i-1}, z_{i-1})$ if we have its coordinates in $B_i(x_i, y_i, z_i)$ using the transformation matrix ${}^{i-1}T_i$:

$$ \begin{bmatrix} x_{i-1} \\ y_{i-1} \\ z_{i-1} \\ 1 \end{bmatrix} = {}^{i-1}T_i \begin{bmatrix} x_i \\ y_i \\ z_i \\ 1 \end{bmatrix} \tag{7.8} $$

The 4×4 matrix ${}^{i-1}T_i$ may be partitioned into two submatrices which represent a unique rotation combined with a unique translation to produce the same rigid motion required to move B_i to B_{i-1},

$$
{}^{i-1}T_i = \begin{bmatrix} {}^{i-1}R_i & {}^{i-1}\mathbf{d}_i \\ 0 & 1 \end{bmatrix} \tag{7.9}
$$

where

$$
{}^{i-1}R_i = \begin{bmatrix} \cos\theta_i & -\sin\theta_i\cos\alpha_i & \sin\theta_i\sin\alpha_i \\ \sin\theta_i & \cos\theta_i\cos\alpha_i & -\cos\theta_i\sin\alpha_i \\ 0 & \sin\alpha_i & \cos\alpha_i \end{bmatrix} \tag{7.10}
$$

and

$$
{}^{i-1}\mathbf{d}_i = \begin{bmatrix} a_i\cos\theta_i \\ a_i\sin\theta_i \\ d_i \end{bmatrix} \tag{7.11}
$$

The inverse of the homogeneous transformation matrix (7.7) is

$$
\begin{aligned}
{}^{i}T_{i-1} &= {}^{i-1}T_i^{-1} \\
&= \begin{bmatrix} \cos\theta_i & \sin\theta_i & 0 & -a_i \\ -\sin\theta_i\cos\alpha_i & \cos\theta_i\cos\alpha_i & \sin\alpha_i & -d_i\sin\alpha_i \\ \sin\theta_i\sin\alpha_i & -\cos\theta_i\sin\alpha_i & \cos\alpha_i & -d_i\cos\alpha_i \\ 0 & 0 & 0 & 1 \end{bmatrix}
\end{aligned} \tag{7.12}
$$

Proof: Consider the coordinate frames $B_i = B_2(x_2, y_2, z_2)$ and $B_{i-1} = B_1(x_1, y_1, z_1)$ in Figure 7.19, which are assumed to be set up based on Denavit–Hartenberg rules.

We may determine the homogeneous transformation matrix ${}^{i}T_{i-1}$ to transform coordinate frame B_{i-1} to B_i by moving B_i to its present position from the coincident

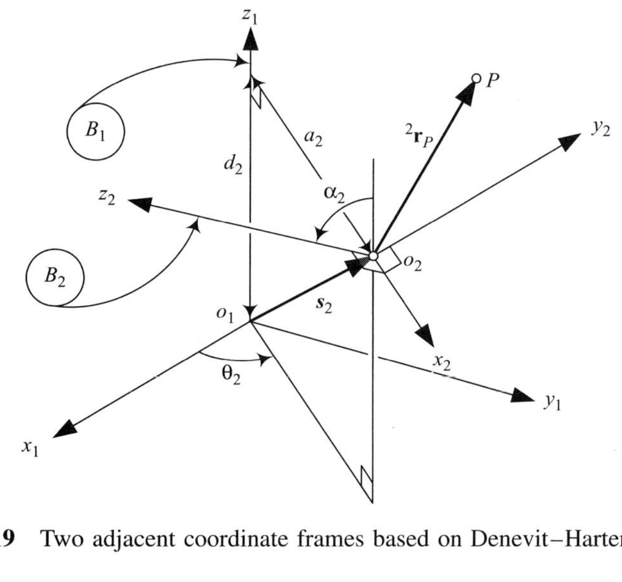

Figure 7.19 Two adjacent coordinate frames based on Denevit–Hartenberg rules.

configuration with B_{i-1}. It can be done by following rotation and translation rules and multiplying four basic transformations:

1. The frame B_i rotates θ_i about the z_i-axis.
2. The rotated frame B_i translates d_i along the z_i-axis.
3. The displaced frame B_i rotates α_i about the local x_i-axis.
4. The displaced frame B_i translates a_i along the local x_i-axis.

Therefore, employing the pure translation and rotation matrices (6.48)–(6.53), the transformation matrix ${}^{i}T_{i-1}$ would be

$$\begin{aligned}{}^{i}T_{i-1} &= D_{x_i,a_i}\ R_{x_i,\alpha_i}\ D_{z_i,d_i}\ R_{z_i,\theta_i} \\ &= \begin{bmatrix} \cos\theta_i & \sin\theta_i & 0 & -a_i \\ -\cos\alpha_i\sin\theta_i & \cos\theta_i\cos\alpha_i & \sin\alpha_i & -d_i\sin\alpha_i \\ \sin\theta_i\sin\alpha_i & -\cos\theta_i\sin\alpha_i & \cos\alpha_i & -d_i\cos\alpha_i \\ 0 & 0 & 0 & 1 \end{bmatrix}\end{aligned} \tag{7.13}$$

where

$$R_{z_i,\theta_i} = {}^{i}T_{i-1} = \begin{bmatrix} \cos\theta_i & \sin\theta_i & 0 & 0 \\ -\sin\theta_i & \cos\theta_i & 0 & 0 \\ 0 & 0 & 1 & 0 \\ 0 & 0 & 0 & 1 \end{bmatrix} \tag{7.14}$$

$$D_{z_i,d_i} = {}^{i}T_{i-1} = \begin{bmatrix} 1 & 0 & 0 & 0 \\ 0 & 1 & 0 & 0 \\ 0 & 0 & 1 & -d_i \\ 0 & 0 & 0 & 1 \end{bmatrix} \tag{7.15}$$

$$R_{x_i,\alpha_i} = {}^{i}T_{i-1} = \begin{bmatrix} 1 & 0 & 0 & 0 \\ 0 & \cos\alpha_i & \sin\alpha_i & 0 \\ 0 & -\sin\alpha_i & \cos\alpha_i & 0 \\ 0 & 0 & 0 & 1 \end{bmatrix} \tag{7.16}$$

$$D_{x_i,a_i} = {}^{i}T_{i-1} = \begin{bmatrix} 1 & 0 & 0 & -a_i \\ 0 & 1 & 0 & 0 \\ 0 & 0 & 1 & 0 \\ 0 & 0 & 0 & 1 \end{bmatrix} \tag{7.17}$$

The inverse of the homogeneous transformation matrix (7.7) is

$$\begin{aligned}{}^{i-1}T_i &= {}^{i}T_{i-1}^{-1} \\ &= \begin{bmatrix} \cos\theta_i & -\cos\alpha_i\sin\theta_i & \sin\theta_i\sin\alpha_i & a_i\cos\theta_i \\ \sin\theta_i & \cos\theta_i\cos\alpha_i & -\cos\theta_i\sin\alpha_i & a_i\sin\theta_i \\ 0 & \sin\alpha_i & \cos\alpha_i & d_i \\ 0 & 0 & 0 & 1 \end{bmatrix}\end{aligned} \tag{7.18}$$

Therefore, we can determine the coordinates of a point in $B_{i-1}(x_{i-1}, y_{i-1}, z_{i-1})$ if we have its coordinates in $B_i(x_i, y_i, z_i)$ using the transformation matrix ${}^{i-1}T_i$:

$$\begin{bmatrix} x_{i-1} \\ y_{i-1} \\ z_{i-1} \\ 1 \end{bmatrix} = {}^{i-1}T_i \begin{bmatrix} x_i \\ y_i \\ z_i \\ 1 \end{bmatrix} \tag{7.19}$$

The 4×4 matrix ${}^{i-1}T_i$ may be partitioned into two submatrices which represent a unique rotation combined with a unique translation to produce the same rigid motion required to move B_i to B_{i-1},

$${}^{i-1}T_i = \begin{bmatrix} {}^{i-1}R_i & {}^{i-1}\mathbf{d}_i \\ 0 & 1 \end{bmatrix} \tag{7.20}$$

where

$${}^{i-1}R_i = \begin{bmatrix} \cos\theta_i & -\sin\theta_i\cos\alpha_i & \sin\theta_i\sin\alpha_i \\ \sin\theta_i & \cos\theta_i\cos\alpha_i & -\cos\theta_i\sin\alpha_i \\ 0 & \sin\alpha_i & \cos\alpha_i \end{bmatrix} \tag{7.21}$$

and

$${}^{i-1}\mathbf{d}_i = \begin{bmatrix} a_i\cos\theta_i \\ a_i\sin\theta_i \\ d_i \end{bmatrix} \tag{7.22}$$

■

Example 399 An Alternative Method to Find ${}^{i}T_{i-1}$ Consider the coordinate frames $B_2(x_2, y_2, z_2)$ and $B_1(x_1, y_1, z_1)$ in Figure 7.19 that are set up based on the Denavit–Hartenberg rules. The point P indicates a body point in B_2. The position vector of the point P in B_2 can be found in frame $B_1(x_1, y_1, z_1)$ using ${}^{2}\mathbf{r}_P$ and ${}^{1}\mathbf{s}_2$,

$${}^{1}\mathbf{r}_P = {}^{1}R_2\,{}^{2}\mathbf{r}_P + {}^{1}\mathbf{s}_2 \tag{7.23}$$

which, in homogeneous coordinate representation, is

$${}^{1}\mathbf{r}_P = \begin{bmatrix} {}^{1}R_2 & {}^{1}\mathbf{s}_2 \\ 0 & 1 \end{bmatrix} {}^{2}\mathbf{r}_P \tag{7.24}$$

$$\begin{bmatrix} x_1 \\ y_1 \\ z_1 \\ 1 \end{bmatrix} = \begin{bmatrix} \cos(\hat{\imath}_2, \hat{\imath}_1) & \cos(\hat{\jmath}_2, \hat{\imath}_1) & \cos(\hat{k}_2, \hat{\imath}_1) & s_{2_x} \\ \cos(\hat{\imath}_2, \hat{\jmath}_1) & \cos(\hat{\jmath}_2, \hat{\jmath}_1) & \cos(\hat{k}_2, \hat{\jmath}_1) & s_{2_y} \\ \cos(\hat{\imath}_2, \hat{k}_1) & \cos(\hat{\jmath}_2, \hat{k}_1) & \cos(\hat{k}_2, \hat{k}_1) & s_{2_z} \\ 0 & 0 & 0 & 1 \end{bmatrix} \begin{bmatrix} x_2 \\ y_2 \\ z_2 \\ 1 \end{bmatrix} \tag{7.25}$$

Using the parameters of Figure 7.19, Equation (7.25) becomes

$$\begin{bmatrix} x_1 \\ y_1 \\ z_1 \\ 1 \end{bmatrix} = \begin{bmatrix} c\theta_2 & -s\theta_2 c\alpha_2 & s\theta_2 s\alpha_2 & a_2 c\theta_2 \\ s\theta_2 & c\theta_2 c\alpha_2 & -c\theta_2 s\alpha_2 & a_2 s\theta_2 \\ 0 & s\alpha_2 & c\alpha_2 & d_2 \\ 0 & 0 & 0 & 1 \end{bmatrix} \begin{bmatrix} x_2 \\ y_2 \\ z_2 \\ 1 \end{bmatrix} \tag{7.26}$$

If we assume the coordinate frame B_1 is $B_{i-1}(x_{i-1}, y_{i-1}, z_{i-1})$, and B_2 is $B_i(x_i, y_i, z_i)$, then we can rewrite the above equation in the required form:

$$\begin{bmatrix} x_{i-1} \\ y_{i-1} \\ z_{i-1} \\ 1 \end{bmatrix} = \begin{bmatrix} c\theta_i & -s\theta_i c\alpha_i & s\theta_i s\alpha_i & a_i c\theta_i \\ s\theta_i & c\theta_i c\alpha_i & -c\theta_i s\alpha_i & a_i s\theta_i \\ 0 & s\alpha_i & c\alpha_i & d_i \\ 0 & 0 & 0 & 1 \end{bmatrix} \begin{bmatrix} x_i \\ y_i \\ z_i \\ 1 \end{bmatrix} \tag{7.27}$$

Following the inversion rule of the homogeneous transformation matrix (6.88), we also find the inverse transformation (7.12):

$$\begin{bmatrix} x_i \\ y_i \\ z_i \\ 1 \end{bmatrix} = \begin{bmatrix} c\theta_i & s\theta_i & 0 & -a_i \\ -s\theta_i c\alpha_i & c\theta_i c\alpha_i & s\alpha_i & -d_i s\alpha_i \\ s\theta_i s\alpha_i & -c\theta_i s\alpha_i & c\alpha_i & -d_i c\alpha_i \\ 0 & 0 & 0 & 1 \end{bmatrix} \begin{bmatrix} x_{i-1} \\ y_{i-1} \\ z_{i-1} \\ 1 \end{bmatrix} \tag{7.28}$$

Example 400 Another Alternative Method to Find ${}^iT_{i-1}$ We can determine ${}^iT_{i-1}$ by following the sequence of translations and rotations that brings the frame B_{i-1} to the present configuration starting from a coincident position with B_i. Considering the frame $B_2 \equiv B_i$ as the body frame and $B_1 \equiv B_{i-1}$ as the global frame, we can assume that all the following rotations and translations are about and along the axes of the local coordinate frame. Figure 7.19 shows that:

1. Frame B_{i-1} is translated $-d_i$ along the local z_i-axis.
2. The displaced frame B_{i-1} is rotated $-\theta_i$ about the local z_i-axis.
3. The displaced frame B_{i-1} is translated $-a_i$ along the local x_i-axis.
4. The displaced frame B_{i-1} is rotated $-\alpha_i$ about the local x_i-axis.

Therefore, the transformation matrix ${}^iT_{i-1}$ would be

$$\begin{aligned} {}^iT_{i-1} &= R_{x_i,-\alpha_i}\, D_{x_i,-a_i}\, R_{z_i,-\theta_i}\, D_{z_i,-d_i} \\ &= \begin{bmatrix} \cos\theta_i & \sin\theta_i & 0 & -a_i \\ -\sin\theta_i\cos\alpha_i & \cos\theta_i\cos\alpha_i & \sin\alpha_i & -d_i\sin\alpha_i \\ \sin\theta_i\sin\alpha_i & -\cos\theta_i\sin\alpha_i & \cos\alpha_i & -d_i\cos\alpha_i \\ 0 & 0 & 0 & 1 \end{bmatrix} \end{aligned} \tag{7.29}$$

where

$$D_{z_i,-d_i} = \begin{bmatrix} 1 & 0 & 0 & 0 \\ 0 & 1 & 0 & 0 \\ 0 & 0 & 1 & -d_i \\ 0 & 0 & 0 & 1 \end{bmatrix} \tag{7.30}$$

$$R_{z_i,-\theta_i} = \begin{bmatrix} \cos\theta_i & -\sin\theta_i & 0 & 0 \\ \sin\theta_i & \cos\theta_i & 0 & 0 \\ 0 & 0 & 1 & 0 \\ 0 & 0 & 0 & 1 \end{bmatrix} \tag{7.31}$$

$$D_{x_i,-a_i} = \begin{bmatrix} 1 & 0 & 0 & -a_i \\ 0 & 1 & 0 & 0 \\ 0 & 0 & 1 & 0 \\ 0 & 0 & 0 & 1 \end{bmatrix} \tag{7.32}$$

$$R_{x_i,-\alpha_i} = \begin{bmatrix} 1 & 0 & 0 & 0 \\ 0 & \cos\alpha_i & -\sin\alpha_i & 0 \\ 0 & \sin\alpha_i & \cos\alpha_i & 0 \\ 0 & 0 & 0 & 1 \end{bmatrix} \tag{7.33}$$

Using ${}^{i-1}T_i = {}^{i}T_{i-1}^{-1}$, we find the inverse transformation:

$$\begin{aligned} {}^{i-1}T_i &= {}^{i}T_{i-1}^{-1} \\ &= \begin{bmatrix} \cos\theta_i & -\sin\theta_i\cos\alpha_i & \sin\theta_i\sin\alpha_i & a_i\cos\theta_i \\ \sin\theta_i & \cos\theta_i\cos\alpha_i & -\cos\theta_i\sin\alpha_i & a_i\sin\theta_i \\ 0 & \sin\alpha_i & \cos\alpha_i & d_i \\ 0 & 0 & 0 & 1 \end{bmatrix} \end{aligned} \tag{7.34}$$

Example 401 Some Tricks in DH Method There are some tricks and comments that may simplify the application of DH rules in multibody kinematics:

1. Showing only z- and x-axes is sufficient to identify a coordinate frame. Drawings are more clear by not showing y-axes.
2. Certain parameters of the frames attached to the first and last links do not need to be defined. If the first and last joints are revolute, R, then

$$a_0 = 0 \qquad a_n = 0 \tag{7.35}$$

$$\alpha_0 = 0 \qquad \alpha_n = 0 \tag{7.36}$$

   In these cases, the zero position for θ_1 and θ_n can be chosen arbitrarily, and link offsets can be set to zero:

$$d_1 = 0 \qquad d_n = 0 \tag{7.37}$$

3. If the first and last joints are prismatic, P, then

$$\theta_1 = 0 \qquad \theta_n = 0 \tag{7.38}$$

   and the zero positions for d_1 and d_n can be chosen arbitrarily.
4. A general applied comment is to set the coordinate frames such that as many parameters as possible are zero at the rest position of the system.

   If the final joint n is R, we may choose x_n to be along with x_{n-1} when $\theta_n = 0$ and the origin of B_n is such that $d_n = 0$.

   If the final joint n is P, we may choose x_n such that $\theta_n = 0$ and the origin of B_n is at the intersection of x_{n-1} and the axis of joint n to make $d_n = 0$.
5. The link parameters a_i and α_i are determined by the geometric design of the robot and are always constant. Because a_i is a length, $a_i \geq 0$. The distance d_i is the offset of the frame B_i with respect to B_{i-1} along the z_{i-1}-axis.

6. The angles α_i and θ_i are directional. The positive direction is determined by the right-hand rule and the directions of x_i and z_{i-1}, respectively.
7. For industrial multibody systems, such as robotic manipulators, the link twist angle α_i is usually a multiple of $\pi/2$ radians.
8. Because the directions of z_i-axes are arbitrary, the DH coordinate frames and hence their transformation matrices are not unique.
9. The base frame $B_0(x_0, y_0, z_0) = G(X, Y, Z)$ is the global frame for an immobile multibody. It is convenient to choose the Z-axis along the axis of joint 1 and set the origin O where the axes of the G-frame are colinear or parallel with the axes of the B_1-frame at the rest position.
10. We can reverse the DH definition for direction of x_i such that it points from z_i to z_{i-1} and still obtains a valid DH coordinate frame. The reverse direction of x_i may be used to set a more convenient reference frame when most of the joint parameters are zero.

Example 402 DH Parameter Table A DH parameter table helps to establish a systematic DH link frame. A DH parameter table is a practical method to set up the link frames in multibodies such as robotic systems. Table 7.1 illustrates the structure of DH tables. It has five columns for frame index and DH parameters. The rows of the DH parameters for each frame will be filled by constant parameters and the joint variable.

As an example, we may examine a 2R planar manipulator. The 2R planar manipulator is an R‖R manipulator with two parallel revolute joints, as is shown in Figure 7.20. The DH parameters of the manipulator are shown in Table 7.2, and the link coordinate frames are set up as in the figure.

To follow the DH rules, we should set the B_2-frame at the center of the second joint such that B_2 and B_1 have coincident origins. However, to determine the tip point of the final link, we needed to instal another frame at the tip point. Such a frame is called the end-effector coordinate frame. The presented coordinate frame eliminates a frame and simplifies the kinematics calculations.

Table 7.1 DH Parameters for Link Frames

Frame No.	a	α	d	θ
1	a_1	α_1	d_1	θ_1
2	a_2	α_2	d_2	θ_2
$\vdots$	$\vdots$	$\vdots$	$\vdots$	$\vdots$
i	a_i	α_i	d_i	θ_i
$\vdots$	$\vdots$	$\vdots$	$\vdots$	$\vdots$
n	a_n	α_n	d_n	θ_n

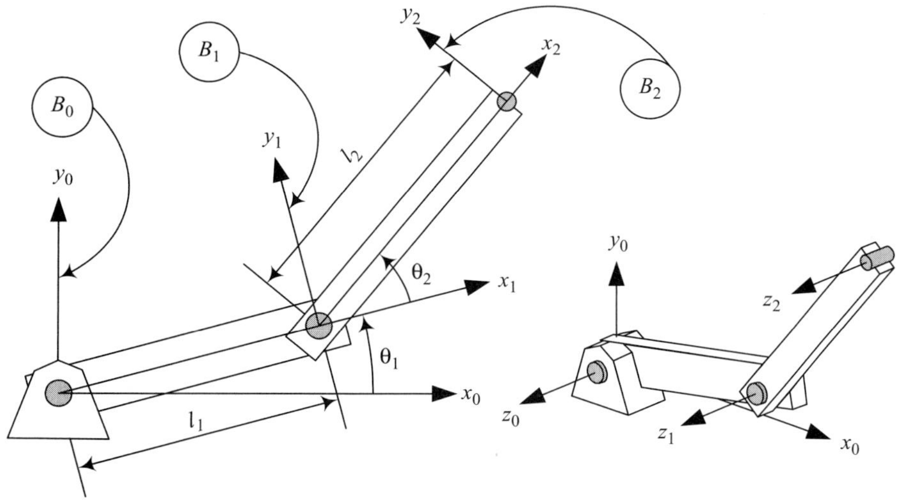

Figure 7.20 An R‖R planar manipulator robot.

Table 7.2 DH Parameters for R‖R Planar Manipulator Robot of Figure 7.20

Frame No.	a	α	d	θ
1	l_1	0	0	θ_1
2	l_2	0	0	θ_2

Example 403 Shortcomings of Denavit–Hartenberg Method The DH method for describing link coordinate frames is neither unique nor the best method. The main drawbacks of the DH method are:

1. The successive coordinate axes are defined in such a way that the origin o_i and axis x_i are defined on the common perpendicular to adjacent link axes. This may be a difficult task depending on the geometry of the links and may produce singularity.
2. The DH notation cannot be extended to ternary and compound links.

Example 404 Classification of Industrial Links A link is identified by its joints at both ends. The relative configuration of the z-axes of the two ends determines the transformation matrix to go from the distal joint coordinate frame B_i to the proximal joint coordinate frame B_{i-1}. There are 12 types of links to make an industrial multibody. The transformation matrix for each type depends solely on the proximal joint and the angle between the z-axes. The 12 types of links are classified in Table 7.3.

Table 7.3 Classification of Industrial Links

1	R∥R(0) or R∥P(0)
2	R∥R(180) or R∥P(180)
3	R⊥R(90) or R⊥P(90)
4	R⊥R(−90) or R⊥P(−90)
5	R⊢R(90) or R⊢P(90)
6	R⊢R(−90) or R⊢P(−90)
7	P∥R(0) or P∥P(0)
8	P∥R(180) or P∥P(180)
9	P⊥R(90) or P⊥P(90)
10	P⊥R(−90) or P⊥P(−90)
11	P⊢R(90) or P⊢P(90)
12	P⊢R(−90) or P⊢P(−90)

The DH transformation matrix for the industrial links are given in Appendix D.

Example 405 DH Coordinate Transformation and Vector Addition The DH transformation from a coordinate frame to the other can also be described by vector addition. Assume P is a point in frame B_2, as shown in Figure 7.21. We may define the position of the point in frame B_1 by a vector equation

$$\overrightarrow{o_1 P} = \overrightarrow{o_2 P} + \overrightarrow{o_1 o_2} \tag{7.39}$$

where

$$^{B_1}\overrightarrow{o_1 o_2} = \begin{bmatrix} s_1 \\ s_2 \\ s_3 \end{bmatrix} \qquad ^{B_1}\overrightarrow{o_1 P} = \begin{bmatrix} x_1 \\ y_1 \\ z_1 \end{bmatrix} \qquad ^{B_2}\overrightarrow{o_2 P} = \begin{bmatrix} x_2 \\ y_2 \\ z_2 \end{bmatrix} \tag{7.40}$$

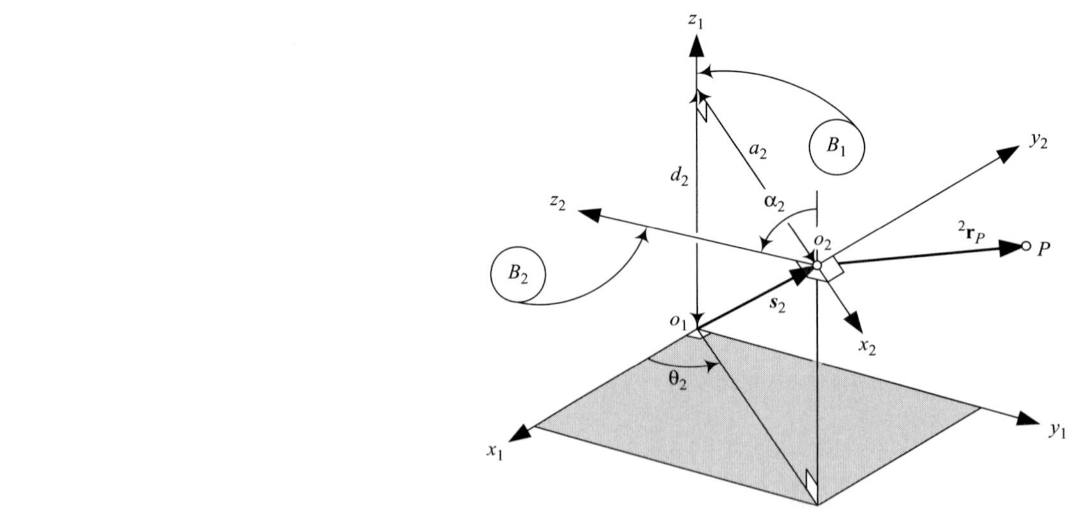

Figure 7.21 Alternative method to derive the Denavit–Hartenberg coordinate transformation.

However, all of the vectors must be expressed in the same coordinate frame to be able to perform the vector calculation. Using the cosines of the angles between axes of the two coordinate frames, we have

$$x_1 = x_2 \cos(x_2, x_1) + y_2 \cos(y_2, x_1) + z_2 \cos(z_2, x_1) + s_1 \tag{7.41a}$$

$$y_1 = x_2 \cos(x_2, y_1) + y_2 \cos(y_2, y_1) + z_2 \cos(z_2, y_1) + s_2 \tag{7.41b}$$

$$z_1 = x_2 \cos(x_2, z_1) + y_2 \cos(y_2, z_1) + z_2 \cos(z_2, z_1) + s_3 \tag{7.41c}$$

$$1 = x_2(0) + y_2(0) + z_2(0) + 1 \tag{7.41d}$$

The transformation (7.41) can be expressed by a homogeneous matrix.

$$\begin{bmatrix} x_1 \\ y_1 \\ z_1 \\ 1 \end{bmatrix} = \begin{bmatrix} \cos(x_2, x_1) & \cos(y_2, x_1) & \cos(z_2, x_1) & s_1 \\ \cos(x_2, y_1) & \cos(y_2, y_1) & \cos(z_2, y_1) & s_2 \\ \cos(x_2, z_1) & \cos(y_2, z_1) & \cos(z_2, z_1) & s_3 \\ 0 & 0 & 0 & 1 \end{bmatrix} \begin{bmatrix} x_2 \\ y_2 \\ z_2 \\ 1 \end{bmatrix} \tag{7.42}$$

The axis x_2 in Figure 7.21 is on the shortest common perpendicular between the z_1- and z_2-axes. The y_2-axis makes B_2 a right-handed coordinate frame. The DH parameters are defined as follows:

1. a_2 is the distance between the z_1- and z_2-axes.
2. α_2 is the twist angle that screws the z_1-axis into the z_2-axis along a.
3. d_2 is the distance from the x_1-axis to the x_2-axis.
4. θ_2 is the angle that screws the x_1-axis into the x_2-axis along d.

Using these definitions, the homogeneous transformation matrix becomes

$$\begin{bmatrix} x_1 \\ y_1 \\ z_1 \\ 1 \end{bmatrix} = \begin{bmatrix} \cos\theta_2 & -\sin\theta_2 \cos\alpha_2 & -\sin\theta_2 \sin\alpha_2 & a_2 \cos\theta_2 \\ \sin\theta_2 & \cos\theta_2 \cos\alpha_2 & \cos\theta_2 \sin\alpha_2 & a_2 \sin\theta_2 \\ 0 & -\sin\alpha_2 & \cos\alpha_2 & d_2 \\ 0 & 0 & 0 & 1 \end{bmatrix} \begin{bmatrix} x_2 \\ y_2 \\ z_2 \\ 1 \end{bmatrix} \tag{7.43}$$

or

$$ {}^1\mathbf{r}_P = {}^1T_2 \, {}^2\mathbf{r}_P \tag{7.44}$$

where

$$ {}^1T_2 = (a, \alpha, d, \theta) \tag{7.45}$$

The DH parameters a, α, θ, d belong to B_2 and define the configuration of B_2 in B_1. In general, the DH parameters $a_i, \alpha_i, \theta_i, d_i$ belong to B_i and define the configuration of B_i with respect to B_{i-1}:

$$ {}^{i-1}T_i = (a_i, \alpha_i, d_i, \theta_i) \tag{7.46}$$

Example 406 Same DH Transformation Matrix In the DH method of setting coordinate frames, because a translation D and a rotation R are along and about the same axis, it is immaterial if we first apply the translation D and then the rotation R or vice versa. Therefore, we can interchange the order of D and R along and about the same axis and obtain the same DH transformation matrix:

$$\begin{aligned} {}^{i-1}T_i &= D_{z_i,d_i}\, R_{z_i,\theta_i}\, D_{x_i,a_i}\, R_{x_i,\alpha_i} = R_{z_i,\theta_i}\, D_{z_i,d_i}\, D_{x_i,a_i}\, R_{x_i,\alpha_i} \\ &= D_{z_i,d_i}\, R_{z_i,\theta_i}\, R_{x_i,\alpha_i}\, D_{x_i,a_i} = R_{z_i,\theta_i}\, D_{z_i,d_i}\, R_{x_i,\alpha_i}\, D_{x_i,a_i} \end{aligned} \tag{7.47}$$

Example 407 ★ DH Application for Spherical Robot Figure 7.22 illustrates a spherical manipulator attached with a spherical wrist. A spherical manipulator is an R⊢R⊢P arm. The associated DH table of the robot is given in Table 7.4.

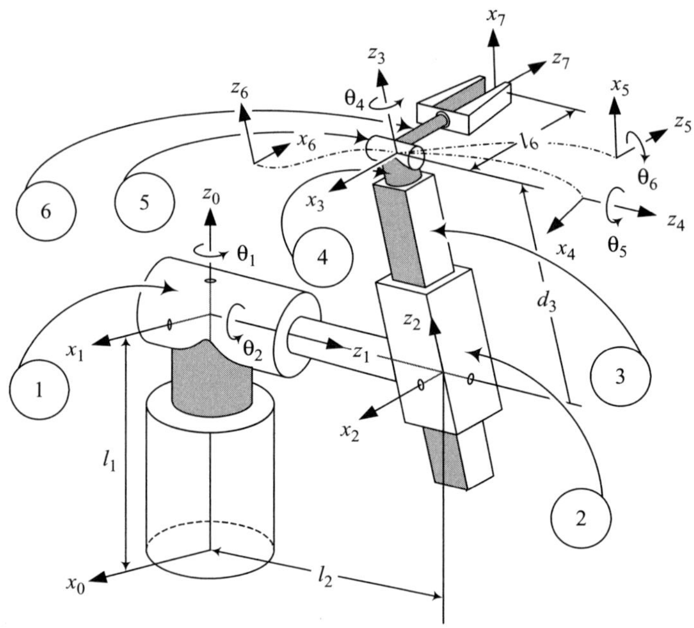

Figure 7.22 A spherical robot made by a spherical manipulator equipped with a spherical wrist.

Table 7.4 DH Parameter Table for Stanford Arm

Frame No.	a_i	α_i	d_i	θ_i
1	0	−90 deg	l_1	θ_1
2	0	90 deg	l_2	θ_2
3	0	0	d_3	0
4	0	−90 deg	0	θ_4
5	0	90 deg	0	θ_5
6	0	0	0	θ_6

The link–joint classifications of the robot are tabulated in Table 7.5.

Table 7.5 Link–Joint Classifications for Stanford Arm

Link No.	Type
1	R⊢R(−90)
2	R⊢P(90)
3	P‖R(0)
4	R⊢R(−90)
5	R⊢R(90)
6	R‖R(0)

Using Appendix D, we can determine the homogeneous transformation matrices for the link–joint classification in Table 7.5 to move from B_i to B_{i-1}:

$$ {}^{0}T_{1} = \begin{bmatrix} \cos\theta_1 & 0 & -\sin\theta_1 & 0 \\ \sin\theta_1 & 0 & \cos\theta_1 & 0 \\ 0 & -1 & 0 & l_1 \\ 0 & 0 & 0 & 1 \end{bmatrix} \tag{7.48} $$

$$ {}^{1}T_{2} = \begin{bmatrix} \cos\theta_2 & 0 & \sin\theta_2 & 0 \\ \sin\theta_2 & 0 & -\cos\theta_2 & 0 \\ 0 & 1 & 0 & l_2 \\ 0 & 0 & 0 & 1 \end{bmatrix} \tag{7.49} $$

$$ {}^{2}T_{3} = \begin{bmatrix} 1 & 0 & 0 & 0 \\ 0 & 1 & 0 & 0 \\ 0 & 0 & 1 & d_3 \\ 0 & 0 & 0 & 1 \end{bmatrix} \tag{7.50} $$

$$ {}^{3}T_{4} = \begin{bmatrix} \cos\theta_4 & 0 & -\sin\theta_4 & 0 \\ \sin\theta_4 & 0 & \cos\theta_4 & 0 \\ 0 & -1 & 0 & 0 \\ 0 & 0 & 0 & 1 \end{bmatrix} \tag{7.51} $$

$$ {}^{4}T_{5} = \begin{bmatrix} \cos\theta_5 & 0 & \sin\theta_5 & 0 \\ \sin\theta_5 & 0 & -\cos\theta_5 & 0 \\ 0 & 1 & 0 & 0 \\ 0 & 0 & 0 & 1 \end{bmatrix} \tag{7.52} $$

$$ {}^{5}T_{6} = \begin{bmatrix} \cos\theta_6 & -\sin\theta_6 & 0 & 0 \\ \sin\theta_6 & \cos\theta_6 & 0 & 0 \\ 0 & 0 & 1 & 0 \\ 0 & 0 & 0 & 1 \end{bmatrix} \tag{7.53} $$

Example 408 ★ DH Application for SSRMS The Space Shuttle Remote Manipulator System (SSRMS) is a robotic arm to act as the hand of the Shuttle or a space station. It may be used for different purposes such as satellite deployment, construction of a space station, transporting a crew member at the end of the arm, surveying and inspecting the outside of the station using a camera.

The SSRMS is shown schematically in Figure 7.23, and the approximate values of its kinematic characteristics are given in Table 7.6.

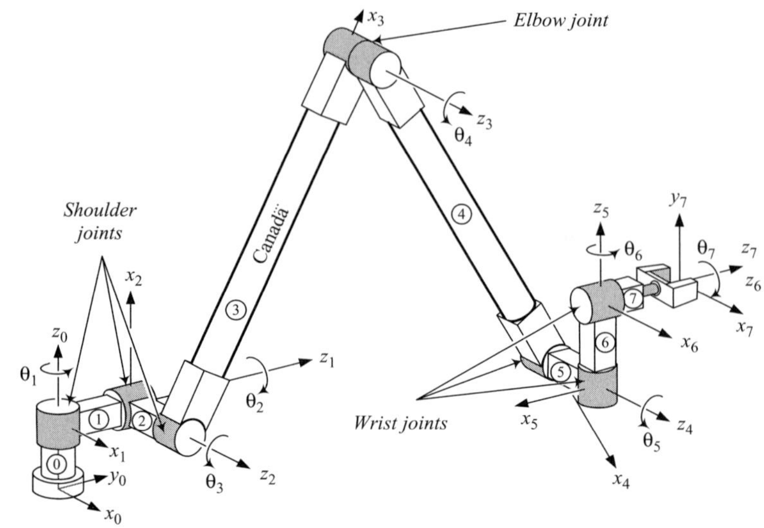

Figure 7.23 Illustration of the space station remote manipulator system.

Table 7.6 Characteristics of SSRMS

Length	14.22 m
Diameter	38.1 cm
Weight	1336 kg
Number of joints	Seven
Handling capacity	116,000 kg (in space)
Maximum velocity of end of arm	Carrying nothing: 37 cm/s Full capacity: 1.2 cm/s
Maximum rotational speed	Approximately: 4 deg/s

Table 7.7 indicates the DH parameters of SSRMS, and Table 7.8 names the link–joint Classifications of SSRMS. Using Appendix D, we can determine the homogeneous transformation matrices for the link–joint classification in Table 7.8 to move from B_i to B_{i-1}.

Links (1) and (2) are R⊢R(−90), and therefore,

$$
{}^{0}T_{1} = \begin{bmatrix} \cos\theta_1 & 0 & -\sin\theta_1 & 0 \\ \sin\theta_1 & 0 & \cos\theta_1 & 0 \\ 0 & -1 & 0 & d_1 \\ 0 & 0 & 0 & 1 \end{bmatrix} \tag{7.54}
$$

Table 7.7 DH Parameters for SSRMS

Frame No.	a_i	α_i	d_i	θ_i
1	0	−90 deg	380 mm	θ_1
2	0	−90 deg	1360 mm	θ_2
3	7110 mm	0	570 mm	θ_3
4	7110 mm	0	475 mm	θ_4
5	0	90 deg	570 mm	θ_5
6	0	−90 deg	635 mm	θ_6
7	0	0	d_7	0

Table 7.8 Link Joint Classifications for SSRMS

Link No.	Type
1	R⊢R(−90)
2	R⊢R(−90)
3	P‖R(0)
4	P‖R(0)
5	R⊢R(90)
6	R⊢R(90)
7	R‖R(0)

$$ {}^{1}T_{2} = \begin{bmatrix} \cos\theta_2 & 0 & -\sin\theta_2 & 0 \\ \sin\theta_2 & 0 & \cos\theta_2 & 0 \\ 0 & -1 & 0 & d_2 \\ 0 & 0 & 0 & 1 \end{bmatrix} \tag{7.55} $$

Links (3) and (4) are R‖R(0), and hence

$$ {}^{2}T_{3} = \begin{bmatrix} \cos\theta_3 & -\sin\theta_3 & 0 & a_3\cos\theta_3 \\ \sin\theta_3 & \cos\theta_3 & 0 & a_3\sin\theta_3 \\ 0 & 0 & 1 & d_3 \\ 0 & 0 & 0 & 1 \end{bmatrix} \tag{7.56} $$

$$ {}^{3}T_{4} = \begin{bmatrix} \cos\theta_4 & -\sin\theta_4 & 0 & a_4\cos\theta_4 \\ \sin\theta_4 & \cos\theta_4 & 0 & a_4\sin\theta_4 \\ 0 & 0 & 1 & d_4 \\ 0 & 0 & 0 & 1 \end{bmatrix} \tag{7.57} $$

Link (5) is R⊢R(90) and link (6) is R⊢R(−90), and therefore

$$ {}^{4}T_{5} = \begin{bmatrix} \cos\theta_5 & 0 & \sin\theta_5 & 0 \\ \sin\theta_5 & 0 & -\cos\theta_5 & 0 \\ 0 & 1 & 0 & d_5 \\ 0 & 0 & 0 & 1 \end{bmatrix} \tag{7.58} $$

$$
{}^5T_6 = \begin{bmatrix} \cos\theta_6 & 0 & -\sin\theta_6 & 0 \\ \sin\theta_6 & 0 & \cos\theta_6 & 0 \\ 0 & -1 & 0 & d_6 \\ 0 & 0 & 0 & 1 \end{bmatrix} \tag{7.59}
$$

Finally link (7) is R⊢R(0) and the coordinate frame attached to the end effector has a translation d_7 with respect to the coordinate frame B_6:

$$
{}^6T_7 = \begin{bmatrix} \cos\theta_7 & -\sin\theta_7 & 0 & 0 \\ \sin\theta_7 & \cos\theta_7 & 0 & 0 \\ 0 & 0 & 1 & d_7 \\ 0 & 0 & 0 & 1 \end{bmatrix} \tag{7.60}
$$

Example 409 Non-DH Transformation Figure 7.24 depicts a polar manipulator with two DOF and a set of coordinate frames that are not set up exactly according to the DH rules.

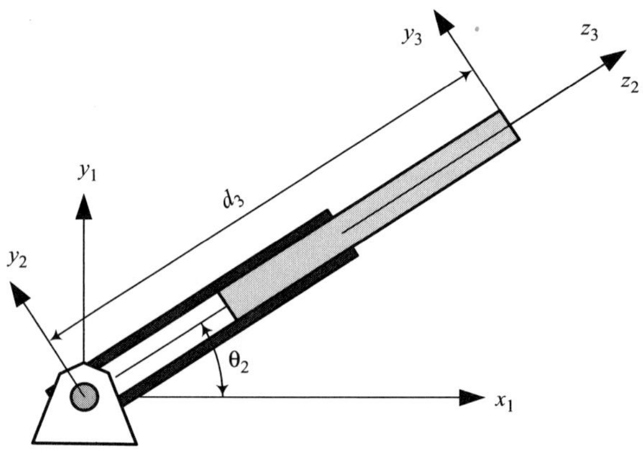

Figure 7.24 A 2 DOF polar manipulator with non-DH coordinate frames.

To determine the link's transformation matrices ${}^1T_2$, ${}^2T_3$, and ${}^1T_3$ in such cases, we may follow the method of homogeneous transformation. The frame B_1 can go to B_2 by a rotation R_{z_2,θ_2} followed by $R_{y_2,90}$. There is no translation between B_1 and B_2. Therefore,

$$
{}^1R_2 = \left[R_{y_2,90}\, R_{z_2,\theta_2}\right]^{\mathrm{T}} = R^{\mathrm{T}}_{z_2,\theta_2}\, R^{\mathrm{T}}_{y_2,90}
$$

$$
= \begin{bmatrix} \cos\theta_2 & \sin\theta_2 & 0 \\ -\sin\theta_2 & \cos\theta_2 & 0 \\ 0 & 0 & 1 \end{bmatrix}^{\mathrm{T}} \begin{bmatrix} \cos\frac{\pi}{2} & 0 & -\sin\frac{\pi}{2} \\ 0 & 1 & 0 \\ \sin\frac{\pi}{2} & 0 & \cos\frac{\pi}{2} \end{bmatrix}^{\mathrm{T}}
$$

$$
= \begin{bmatrix} 0 & -\sin\theta_2 & \cos\theta_2 \\ 0 & \cos\theta_2 & \sin\theta_2 \\ -1 & 0 & 0 \end{bmatrix} \tag{7.61}
$$

and hence,

$$
{}^1T_2 = \begin{bmatrix} 0 & -\sin\theta_2 & \cos\theta_2 & 0 \\ 0 & \cos\theta_2 & \sin\theta_2 & 0 \\ -1 & 0 & 0 & 0 \\ 0 & 0 & 0 & 1 \end{bmatrix} \tag{7.62}
$$

It is also possible to determine ${}^1R_2$ directly from the definition of the directional cosines:

$$
\begin{aligned}
{}^1R_2 &= \begin{bmatrix} \hat{I}\cdot\hat{\imath} & \hat{I}\cdot\hat{\jmath} & \hat{I}\cdot\hat{k} \\ \hat{J}\cdot\hat{\imath} & \hat{J}\cdot\hat{\jmath} & \hat{J}\cdot\hat{k} \\ \hat{K}\cdot\hat{\imath} & \hat{K}\cdot\hat{\jmath} & \hat{K}\cdot\hat{k} \end{bmatrix} = \begin{bmatrix} \hat{\imath}_1\cdot\hat{\imath}_2 & \hat{\imath}_1\cdot\hat{\jmath}_2 & \hat{\imath}_1\cdot\hat{k}_2 \\ \hat{\jmath}_1\cdot\hat{\imath}_2 & \hat{\jmath}_1\cdot\hat{\jmath}_2 & \hat{\jmath}_1\cdot\hat{k}_2 \\ \hat{k}_1\cdot\hat{\imath}_2 & \hat{k}_1\cdot\hat{\jmath}_2 & \hat{k}_1\cdot\hat{k}_2 \end{bmatrix} \\
&= \begin{bmatrix} \cos\frac{\pi}{2} & \cos\left(\frac{\pi}{2}+\theta_2\right) & \cos\theta_2 \\ \cos\frac{\pi}{2} & \cos\theta_2 & \cos\left(\frac{\pi}{2}-\theta_2\right) \\ \cos\pi & \cos\frac{\pi}{2} & \cos\frac{\pi}{2} \end{bmatrix} \\
&= \begin{bmatrix} 0 & -\sin\theta_2 & \cos\theta_2 \\ 0 & \cos\theta_2 & \sin\theta_2 \\ -1 & 0 & 0 \end{bmatrix}
\end{aligned} \tag{7.63}
$$

The final transformation matrix is only a translation along z_3,

$$
{}^2T_3 = \begin{bmatrix} 1 & 0 & 0 & 0 \\ 0 & 1 & 0 & 0 \\ 0 & 0 & 1 & d_3 \\ 0 & 0 & 0 & 1 \end{bmatrix} \tag{7.64}
$$

which provides

$$
{}^1T_3 = {}^1T_2\,{}^2T_3 = \begin{bmatrix} 0 & -\sin\theta_2 & \cos\theta_2 & d_3\cos\theta_2 \\ 0 & \cos\theta_2 & \sin\theta_2 & d_3\sin\theta_2 \\ -1 & 0 & 0 & 0 \\ 0 & 0 & 0 & 1 \end{bmatrix} \tag{7.65}
$$

Example 410 Trebuchet Kinematics as a Multibody The *trebuchet*, shown in Figure 7.25, is a shooting weapon of war powered by a falling massive counterweight m_1. The beam AB is pivoted to the chassis with two unequal sections a and b. The counterweight m_1 is hinged at the shorter arm of the beam at a distance c from the end A. The mass of the projectile is m_2, and it is at the end of a rope with a length l attached to the end B of the longer arm of the beam.

To analyze the trebuchet as a multibody system, we attach a global coordinate frame G at the fixed pivot M and three body frames B, B_1, B_2 to the three moving bodies as shown in the figure. The three independent variable angles α, θ, and γ define the relative positions of the bodies. Let us consider the parameters a, b, c, l, m_1, and m_2 constant and find the transformation matrices to determine the coordinates in the

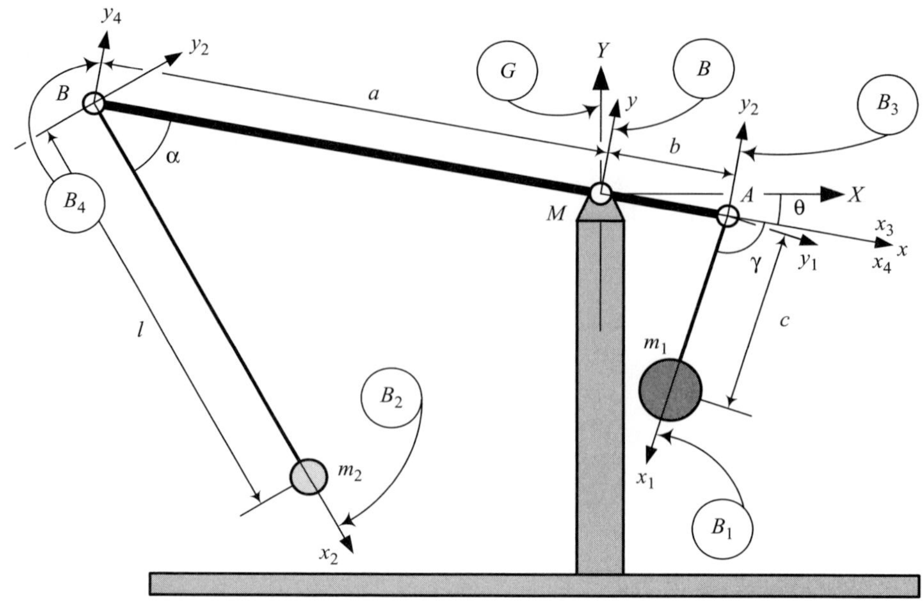

Figure 7.25 The kinematic model of a trebuchet.

relative frames. We also attach two coordinate frames B_3 and B_4 to the beam at joints A and B to simplify the coordinate transformations.

Every link needs at least one coordinate frame, called the *main frame* of the link. However, we may also attach some other frames to the link for better expression of coordinates or a simpler transformation. An extra coordinate frame on a link is called the *Sina frame*. A Sina is a spare frame that can be used instead of the main frame. Usually a Sina frame is related to the main frame of the same link by a constant translation and constant rotation. The Sina frame may also be called the spare, extra, dummy, temporary, intermediate, or jump frame. The frames B_3 and B_4 are the Sina frames for the link AB and B is its main frame.

We may find ${}^{B}T_{G}$ by using the transformation matrix of a zero-length link R‖R(0) or equivalently by turning B a rotation $-\theta$ about the z-axis:

$$
{}^{B}T_{G} = R_{z,-\theta} = \begin{bmatrix} \cos-\theta & \sin-\theta & 0 & 0 \\ -\sin-\theta & \cos-\theta & 0 & 0 \\ 0 & 0 & 1 & 0 \\ 0 & 0 & 0 & 1 \end{bmatrix} \tag{7.66}
$$

The Sina frame B_3 is connected to the main frame B by a translation b along x_3. Similarly, the Sina frame B_4 is connected to B by a translation $-a$ along x_4:

$$
{}^{3}T_{B} = D_{x_3,b} = \begin{bmatrix} 1 & 0 & 0 & -b \\ 0 & 1 & 0 & 0 \\ 0 & 0 & 1 & 0 \\ 0 & 0 & 0 & 1 \end{bmatrix} \tag{7.67}
$$

$$^4T_B = D_{x_4,-a} = \begin{bmatrix} 1 & 0 & 0 & a \\ 0 & 1 & 0 & 0 \\ 0 & 0 & 1 & 0 \\ 0 & 0 & 0 & 1 \end{bmatrix} \tag{7.68}$$

We find $^1T_3$ by turning B_1 a rotation $-\gamma$ about the z_1-axis:

$$^1T_3 = R_{z_1,-\gamma} = \begin{bmatrix} \cos-\gamma & \sin-\gamma & 0 & 0 \\ -\sin-\gamma & \cos-\gamma & 0 & 0 \\ 0 & 0 & 1 & 0 \\ 0 & 0 & 0 & 1 \end{bmatrix} \tag{7.69}$$

Using the same method, we find $^2T_4$ by turning B_2 a rotation $-\alpha$ about the z_2-axis:

$$^2T_4 = R_{z_2,-\alpha} = \begin{bmatrix} \cos-\alpha & \sin-\alpha & 0 & 0 \\ -\sin-\alpha & \cos-\alpha & 0 & 0 \\ 0 & 0 & 1 & 0 \\ 0 & 0 & 0 & 1 \end{bmatrix} \tag{7.70}$$

So,

$$^1T_B = {}^1T_3\,{}^3T_B = \begin{bmatrix} \cos\gamma & -\sin\gamma & 0 & -b\cos\gamma \\ \sin\gamma & \cos\gamma & 0 & -b\sin\gamma \\ 0 & 0 & 1 & 0 \\ 0 & 0 & 0 & 1 \end{bmatrix} \tag{7.71}$$

$$^2T_B = {}^2T_4\,{}^4T_B = \begin{bmatrix} \cos\alpha & -\sin\alpha & 0 & a\cos\alpha \\ \sin\alpha & \cos\alpha & 0 & a\sin\alpha \\ 0 & 0 & 1 & 0 \\ 0 & 0 & 0 & 1 \end{bmatrix} \tag{7.72}$$

By matrix multiplication and a homogeneous matrix inverse calculation, we can determine the required transformation matrices to go to the G-frame:

$$^GT_B = {}^BT_G^{-1} = \begin{bmatrix} \cos\theta & \sin\theta & 0 & 0 \\ -\sin\theta & \cos\theta & 0 & 0 \\ 0 & 0 & 1 & 0 \\ 0 & 0 & 0 & 1 \end{bmatrix} \tag{7.73}$$

$$\begin{aligned} ^GT_1 &= {}^GT_B\,{}^BT_1 = {}^BT_G^{-1}\,{}^1T_B^{-1} = \left[{}^1T_B\,{}^BT_G\right]^{-1} \\ &= \begin{bmatrix} \cos(\theta+\gamma) & \sin(\theta+\gamma) & 0 & b\cos\theta \\ -\sin(\theta+\gamma) & \cos(\theta+\gamma) & 0 & -b\sin\theta \\ 0 & 0 & 1 & 0 \\ 0 & 0 & 0 & 1 \end{bmatrix} \end{aligned} \tag{7.74}$$

$$\begin{aligned} ^GT_2 &= {}^GT_B\,{}^BT_2 = {}^BT_G^{-1}\,{}^2T_B^{-1} = \left[{}^2T_B\,{}^BT_G\right]^{-1} \\ &= \begin{bmatrix} \cos(\theta+\alpha) & \sin(\theta+\alpha) & 0 & -a\cos\theta \\ -\sin(\theta+\alpha) & \cos(\theta+\alpha) & 0 & a\sin\theta \\ 0 & 0 & 1 & 0 \\ 0 & 0 & 0 & 1 \end{bmatrix} \end{aligned} \tag{7.75}$$

Therefore, the global coordinates of m_1 at ${}^1\mathbf{r}_{m_1} = [c, 0, 0]$ and m_2 at ${}^2\mathbf{r}_{m_2} = [l, 0, 0]$ are

$$ {}^G\mathbf{r}_{m_1} = {}^GT_1\,{}^1\mathbf{r}_{m_1} = \begin{bmatrix} b\cos\theta + c\cos(\theta+\gamma) \\ -b\sin\theta - c\sin(\theta+\gamma) \\ 0 \\ 1 \end{bmatrix} \tag{7.76} $$

$$ {}^G\mathbf{r}_{m_2} = {}^GT_2\,{}^2\mathbf{r}_{m_2} = \begin{bmatrix} l\cos(\theta+\alpha) - a\cos\theta \\ a\sin\theta - l\sin(\theta+\alpha) \\ 0 \\ 1 \end{bmatrix} \tag{7.77} $$

Sina (Avicenna) (980–1037) , also known as the "Prince of Physicians," is known as one of the foremost philosophers and scientists of Persia.

Example 411 Transformation Matrix between Two Frames The multibody kinematic problems are always related to a correct expression of the transformation matrix between two coordinate frames. The classical method is to set up the coordinate frames based on DH rules and use the DH transformation matrix (7.7). However, when the DH rules cannot be applied, we recommend determining the homogeneous transformation matrix by a proper combination of principal rotations and translations.

Consider two arbitrary coordinate frames B and G. The homogeneous transformation matrix ${}^BT_G$ can be found by moving B to its present position from the coincident configuration with G and using Equations (6.48)–(6.53). The rotations are positive about the local axes of B and the translations are negative along the local axes of B.

7.3 FORWARD KINEMATICS

The configuration as well as the kinematic information of a multibody may be expressed by joint variables or Cartesian coordinates. *Forward* or *direct kinematics* is the transformation of kinematic information from the joint variables to the Cartesian coordinates. The problem of forward kinematics is the determination of the relative position and orientation of every body of a multibody for a given set of joint variables. This problem can be solved by determining transformation matrices ${}^0T_i$ to express the kinematic information of link (i) in the base coordinate frame B_0.

Although the kinematic information includes configuration, velocity, acceleration, and jerk, forward kinematics generally refers to position and orientation analysis.

The traditional method of forward kinematic analysis for multibodies is to proceed link by link using the Denavit–Hartenberg notations and frames. For an n-DOF multibody, at least n DH transformation matrices, one for each link, are required to determine the global coordinate of any point in any frame. For a given set of joint variables and a set of link coordinate frames, the transformation matrices ${}^jT_i$ are uniquely determined. Therefore, the transformation matrices ${}^jT_i = {}^jT_i(q_k)$ are functions of n joint variables q_k, $k = 1, 2, 3, \ldots, n$. The configuration of the multibody when all the joint variables are zero is called the *rest position*. Determination of the transformation matrices at the rest position is an applied checking procedure.

If the links of the multibody are arranged such that each link (i) has only one coordinate frame B_i and the frames are arranged sequentially, then

$$^{0}T_i = {}^{0}T_1\,{}^{1}T_2\,{}^{2}T_3\,{}^{3}T_4 \ldots {}^{i-1}T_i \qquad i = 1, 2, 3, \ldots, n \tag{7.78}$$

Having $^{0}T_i$, we can determine the coordinates of any point P of link (i) in the base frame when its coordinates are given in frame B_i:

$$^{0}\mathbf{r}_P = {}^{0}T_i\,{}^{i}\mathbf{r}_P \qquad i = 1, 2, 3, \ldots, n \tag{7.79}$$

Generally speaking, the number and labels of coordinate frames do not need to be consequential or increasing. The designer is free to number the frames in any order. However, assigning them sequentially provides a simpler and more meaningful transformation.

Example 412 A 2R Planar Manipulator Figure 7.26 illustrates a 2R or R‖R planar manipulator with two parallel revolute joints. Links (1) and (2) are both R‖R(0) and therefore the transformation matrices $^{0}T_1$, $^{1}T_2$ are

$$^{0}T_1 = \begin{bmatrix} \cos\theta_1 & -\sin\theta_1 & 0 & l_1\cos\theta_1 \\ \sin\theta_1 & \cos\theta_1 & 0 & l_1\sin\theta_1 \\ 0 & 0 & 1 & 0 \\ 0 & 0 & 0 & 1 \end{bmatrix} \tag{7.80}$$

$$^{1}T_2 = \begin{bmatrix} \cos\theta_2 & -\sin\theta_2 & 0 & l_2\cos\theta_2 \\ \sin\theta_2 & \cos\theta_2 & 0 & l_2\sin\theta_2 \\ 0 & 0 & 1 & 0 \\ 0 & 0 & 0 & 1 \end{bmatrix} \tag{7.81}$$

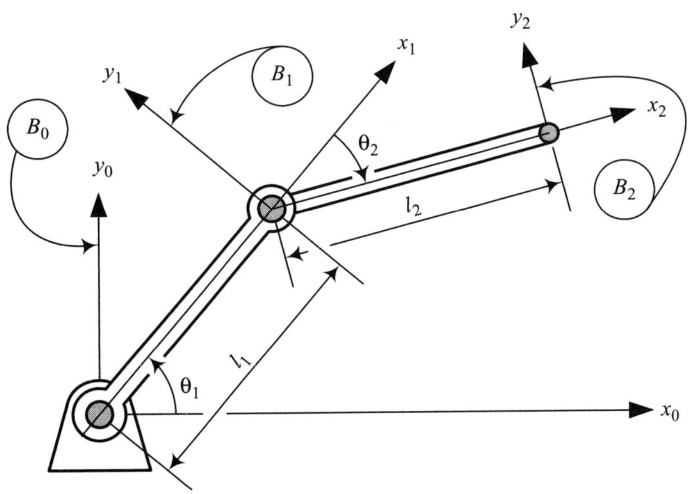

Figure 7.26 A 2R or R‖R planar manipulator.

The forward kinematics of the manipulator is to determine the transformation matrices ${}^{0}T_i$, $i = 1, 2$. The matrix ${}^{0}T_1$ is given in (7.80) and ${}^{0}T_2$ can be found by matrix multiplication:

$$
\begin{aligned}
{}^{0}T_2 &= {}^{0}T_1\,{}^{1}T_2 \\
&= \begin{bmatrix} \cos(\theta_1+\theta_2) & -\sin(\theta_1+\theta_2) & 0 & l_2\cos(\theta_1+\theta_2)+l_1\cos\theta_1 \\ \sin(\theta_1+\theta_2) & \cos(\theta_1+\theta_2) & 0 & l_2\sin(\theta_1+\theta_2)+l_1\sin\theta_1 \\ 0 & 0 & 1 & 0 \\ 0 & 0 & 0 & 1 \end{bmatrix}
\end{aligned}
\tag{7.82}
$$

Example 413 Cycloid Consider the wheel in Figure 7.27 that is turning with angular velocity ω and is slip free on the ground. If the point P of the wheel is on the ground at $t = 0$, then we can find its position in the wheel frame at a time t by employing another coordinate frame M. The frame M is called the rim frame and is stuck to the wheel at its center:

$$
{}^{M}\mathbf{r}_P = \begin{bmatrix} 0 \\ 0 \\ -R_w \end{bmatrix}
\tag{7.83}
$$

Because of spin, the M-frame turns about the y_w-axis, and therefore, the transformation matrix to go from the rim frame to the wheel frame is

$$
{}^{W}R_M = \begin{bmatrix} \cos\omega t & 0 & \sin\omega t \\ 0 & 1 & 0 \\ -\sin\omega t & 0 & \cos\omega t \end{bmatrix}
\tag{7.84}
$$

So the coordinates of P in the wheel frame are

$$
{}^{W}\mathbf{r}_P = {}^{W}R_M\,{}^{M}\mathbf{r}_P = \begin{bmatrix} -R_w\sin t\omega \\ 0 \\ -R_w\cos t\omega \end{bmatrix}
\tag{7.85}
$$

The center of the wheel is moving with speed $v_x = R_w\omega$ and is at ${}^{G}\mathbf{r} = [v_x t, 0, R_w]$ in the global coordinate frame G on the ground. Hence, the coordinates of point P in the global frame G would be

$$
{}^{G}\mathbf{r}_P = {}^{W}\mathbf{r}_P + \begin{bmatrix} v_x t \\ 0 \\ R_w \end{bmatrix} = \begin{bmatrix} R_w(\omega t - \sin t\omega) \\ 0 \\ R_w(1-\cos t\omega) \end{bmatrix}
\tag{7.86}
$$

The path of motion of point P in the (X, Z)-plane can be found by eliminating t between X- and Z-coordinates. However, it is easier to express the path by using ωt as a parameter. Such a path is called a *cycloid*.

In general, point P can be at any distance from the center of the rim frame. If the point is at a distance $d \neq R_w$, then its path of motion is called the *trochoid*. A trochoid is called a *curtate cycloid* if $d < R_w$ and a *prolate cycloid* if $d > R_w$. Figures 7.27(a)–(c) illustrate a cycloid, curtate cycloid, and prolate cycloid, respectively.

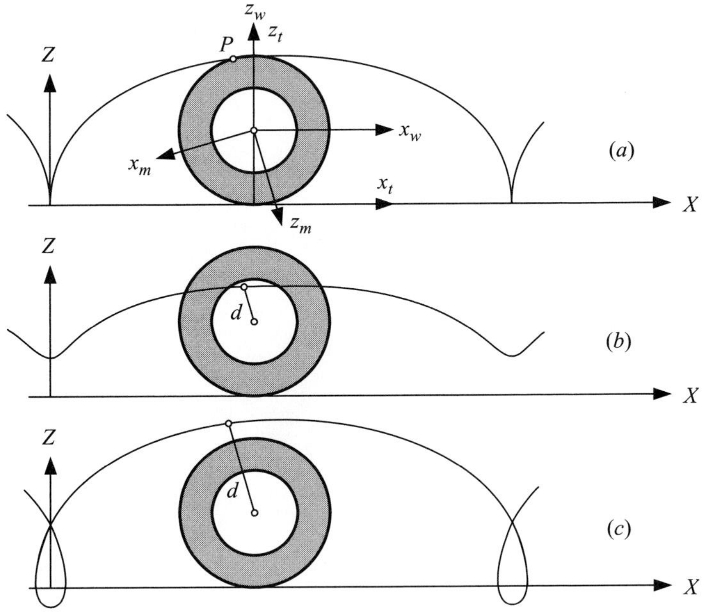

Figure 7.27 A cycloid (a), curtate cycloid (b), and prolate cycloid (c).

Example 414 A Three-Dimensional 2R Planar Manipulator Figure 7.28 illustrates a 3D R‖R planar manipulator with two parallel revolute joints. Links (1) and (2) are both R‖R(0) and move in parallel planes. Their transformation matrices ${}^0T_1$, ${}^1T_2$ are

$$
{}^0T_1 = \begin{bmatrix} \cos\theta_1 & -\sin\theta_1 & 0 & l_1\cos\theta_1 \\ \sin\theta_1 & \cos\theta_1 & 0 & l_1\sin\theta_1 \\ 0 & 0 & 1 & d_1 \\ 0 & 0 & 0 & 1 \end{bmatrix} \tag{7.87}
$$

$$
{}^1T_2 = \begin{bmatrix} \cos\theta_2 & -\sin\theta_2 & 0 & l_2\cos\theta_2 \\ \sin\theta_2 & \cos\theta_2 & 0 & l_2\sin\theta_2 \\ 0 & 0 & 1 & d_2 \\ 0 & 0 & 0 & 1 \end{bmatrix} \tag{7.88}
$$

The forward kinematics of the manipulator is to determine the transformation matrices ${}^0T_i$, $i = 1, 2$. The matrix ${}^0T_1$ is given in (7.80) and ${}^0T_2$ can be found by matrix multiplication:

$$
\begin{aligned}
{}^0T_2 &= {}^0T_1\,{}^1T_2 \\
&= \begin{bmatrix} \cos(\theta_1+\theta_2) & -\sin(\theta_1+\theta_2) & 0 & l_2\cos(\theta_1+\theta_2)+l_1\cos\theta_1 \\ \sin(\theta_1+\theta_2) & \cos(\theta_1+\theta_2) & 0 & l_2\sin(\theta_1+\theta_2)+l_1\sin\theta_1 \\ 0 & 0 & 1 & d_1+d_2 \\ 0 & 0 & 0 & 1 \end{bmatrix}
\end{aligned} \tag{7.89}
$$

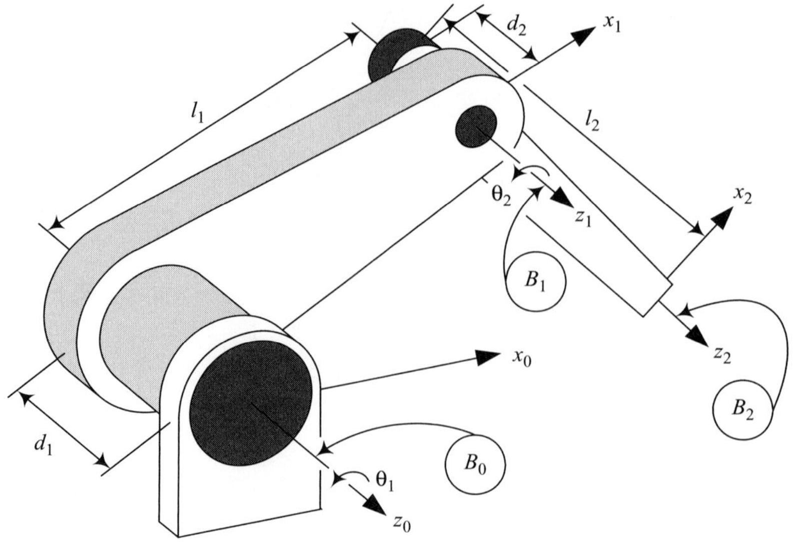

Figure 7.28 A 3D 2R planar manipulator.

Example 415 Space Shuttle Remote Manipulator System The SSRMS is shown in Figure 7.23, with the link–joint classification reported in Table 7.8. The transformation matrices to move from B_i to B_{i-1} are given in Equations (7.54)–(7.60).

The forward kinematics of the SSRMS will be found by direct multiplication of ${}^{i-1}T_i$ $(i = 1, 2, \ldots, 7)$:

$$ {}^{0}T_7 = {}^{0}T_1\, {}^{1}T_2\, {}^{2}T_3\, {}^{3}T_4\, {}^{4}T_5\, {}^{5}T_6\, {}^{6}T_7 \tag{7.90} $$

The forward kinematics of the first three frames are given as

$$ {}^{0}T_1 = \begin{bmatrix} \cos\theta_1 & 0 & -\sin\theta_1 & 0 \\ \sin\theta_1 & 0 & \cos\theta_1 & 0 \\ 0 & -1 & 0 & d_1 \\ 0 & 0 & 0 & 1 \end{bmatrix} \tag{7.91} $$

$$ {}^{0}T_2 = {}^{0}T_1\, {}^{1}T_2 = \begin{bmatrix} c\theta_1 c\theta_2 & s\theta_1 & -c\theta_1 s\theta_2 & -d_2 s\theta_1 \\ c\theta_2 s\theta_1 & -c\theta_1 & -s\theta_1 s\theta_2 & d_2 c\theta_1 \\ -s\theta_2 & 0 & -c\theta_2 & d_1 \\ 0 & 0 & 0 & 1 \end{bmatrix} \tag{7.92} $$

$$ \begin{aligned} {}^{0}T_3 &= {}^{0}T_1\, {}^{1}T_2\, {}^{2}T_3 \\ &= \begin{bmatrix} s\theta_1 s\theta_3 + c\theta_1 c\theta_2 c\theta_3 & c\theta_3 s\theta_1 - c\theta_1 c\theta_2 s\theta_3 & -c\theta_1 s\theta_2 & {}^{0}d_{3x} \\ c\theta_2 c\theta_3 s\theta_1 - c\theta_1 s\theta_3 & -c\theta_1 c\theta_3 - c\theta_2 s\theta_1 s\theta_3 & -s\theta_1 s\theta_2 & {}^{0}d_{3y} \\ -c\theta_3 s\theta_2 & s\theta_2 s\theta_3 & -c\theta_2 & {}^{0}d_{3z} \\ 0 & 0 & 0 & 1 \end{bmatrix} \end{aligned} \tag{7.93} $$

where

$$ {}^{0}\mathbf{d}_{3} = \begin{bmatrix} a_3 s\theta_1 s\theta_3 - d_3 c\theta_1 s\theta_2 - d_2 s\theta_1 + a_3 c\theta_1 c\theta_2 c\theta_3 \\ d_2 c\theta_1 - a_3 c\theta_1 s\theta_3 - d_3 s\theta_1 s\theta_2 + a_3 c\theta_2 c\theta_3 s\theta_1 \\ d_1 - d_3 c\theta_2 - a_3 c\theta_3 s\theta_2 \\ 1 \end{bmatrix} \tag{7.94} $$

and the forward kinematics of the last three frames as

$$ {}^{6}T_{7} = \begin{bmatrix} \cos\theta_7 & -\sin\theta_7 & 0 & 0 \\ \sin\theta_7 & \cos\theta_7 & 0 & 0 \\ 0 & 0 & 1 & d_7 \\ 0 & 0 & 0 & 1 \end{bmatrix} \tag{7.95} $$

$$ {}^{5}T_{7} = {}^{5}T_{6}\,{}^{6}T_{7} = \begin{bmatrix} c\theta_6 c\theta_7 & -c\theta_6 s\theta_7 & -s\theta_6 & -d_7 s\theta_6 \\ c\theta_7 s\theta_6 & -s\theta_6 s\theta_7 & c\theta_6 & d_7 c\theta_6 \\ -s\theta_7 & -c\theta_7 & 0 & d_6 \\ 0 & 0 & 0 & 1 \end{bmatrix} \tag{7.96} $$

$$ \begin{aligned} {}^{4}T_{7} &= {}^{4}T_{5}\,{}^{5}T_{6}\,{}^{6}T_{7} \\ &= \begin{bmatrix} c\theta_5 c\theta_6 c\theta_7 - s\theta_5 s\theta_7 & -c\theta_7 s\theta_5 - c\theta_5 c\theta_6 s\theta_7 & -c\theta_5 s\theta_6 & {}^{4}d_{7x} \\ c\theta_5 s\theta_7 + c\theta_6 c\theta_7 s\theta_5 & c\theta_5 c\theta_7 - c\theta_6 s\theta_5 s\theta_7 & -s\theta_5 s\theta_6 & {}^{4}d_{7y} \\ c\theta_7 s\theta_6 & -s\theta_6 s\theta_7 & c\theta_6 & {}^{4}d_{7z} \\ 0 & 0 & 0 & 1 \end{bmatrix} \end{aligned} \tag{7.97} $$

where

$$ {}^{4}\mathbf{d}_{7} = \begin{bmatrix} d_6 s\theta_5 - d_7 c\theta_5 s\theta_6 \\ -d_6 c\theta_5 - d_7 s\theta_5 s\theta_6 \\ d_5 + d_7 c\theta_6 \\ 1 \end{bmatrix} \tag{7.98} $$

The transformation matrices ${}^{0}T_{4}$, ${}^{0}T_{5}$, ${}^{0}T_{6}$ are determined by matrix multiplication,

$$ {}^{0}T_{4} = {}^{0}T_{3}\,{}^{3}T_{4} \tag{7.99} $$

$$ {}^{0}T_{5} = {}^{0}T_{4}\,{}^{4}T_{5} \tag{7.100} $$

$$ {}^{0}T_{6} = {}^{0}T_{5}\,{}^{5}T_{6} \tag{7.101} $$

and finally, we can find the coordinates of a point in the gripper frame B_7 in the base frame by using ${}^{0}T_{3}\,{}^{3}T_{4}\,{}^{4}T_{7}$:

$$ {}^{0}T_{7} = {}^{0}T_{3} \begin{bmatrix} \cos\theta_4 & -\sin\theta_4 & 0 & a_4\cos\theta_4 \\ \sin\theta_4 & \cos\theta_4 & 0 & a_4\sin\theta_4 \\ 0 & 0 & 1 & d_4 \\ 0 & 0 & 0 & 1 \end{bmatrix} {}^{4}T_{7} \tag{7.102} $$

Example 416 Camera on an Arm of SSRMS The SSRMS is shown in Figure 7.29, with the link transformation matrices given in Equations (7.54)–(7.60). Using the numerical values of Table 7.7 and assuming the values

$$\begin{aligned} \theta_1 &= 0 \\ \theta_2 &= -\tfrac{1}{2}\pi \text{ rad} \\ \theta_3 &= -\tfrac{1}{4}\pi \text{ rad} \\ \theta_4 &= -\tfrac{1}{2}\pi \text{ rad} \\ \theta_5 &= -\tfrac{3}{4}\pi \\ \theta_6 &= \tfrac{1}{2}\pi \text{ rad} \\ \theta_7 &= 0 \\ d_7 &= 500\,\text{mm} \end{aligned} \tag{7.103}$$

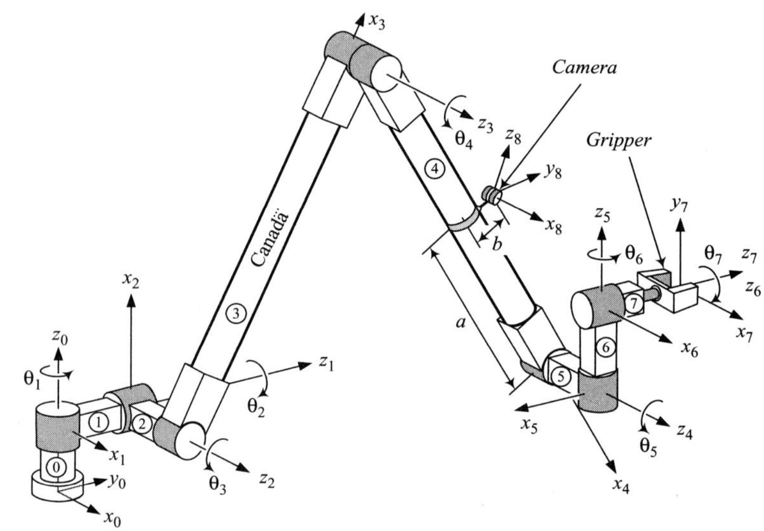

Figure 7.29 The SSRMS with a camera attached to link (4).

for the joint variables and gripper length, we find the following link transformation matrices:

$${}^0T_1 = \begin{bmatrix} 1 & 0 & 0 & 0 \\ 0 & 0 & 1 & 0 \\ 0 & -1 & 0 & 380 \\ 0 & 0 & 0 & 1 \end{bmatrix} \tag{7.104}$$

$${}^1T_2 = \begin{bmatrix} 0 & 0 & 1 & 0 \\ -1 & 0 & 0 & 0 \\ 0 & -1 & 0 & 1360 \\ 0 & 0 & 0 & 1 \end{bmatrix} \tag{7.105}$$

$$ {}^{2}T_{3} = \begin{bmatrix} 0.70711 & 0.70711 & 0 & 5027.5 \\ -0.70711 & 0.70711 & 0 & -5027.5 \\ 0 & 0 & 1 & 570 \\ 0 & 0 & 0 & 1 \end{bmatrix} \tag{7.106} $$

$$ {}^{3}T_{4} = \begin{bmatrix} 0 & 1 & 0 & 0 \\ -1 & 0 & 0 & -7110 \\ 0 & 0 & 1 & 475 \\ 0 & 0 & 0 & 1 \end{bmatrix} \tag{7.107} $$

$$ {}^{4}T_{5} = \begin{bmatrix} -0.70711 & 0 & -0.70711 & 0 \\ -0.70711 & 0 & 0.70711 & 0 \\ 0 & 1 & 0 & 570 \\ 0 & 0 & 0 & 1 \end{bmatrix} \tag{7.108} $$

$$ {}^{5}T_{6} = \begin{bmatrix} 0 & 0 & -1 & 0 \\ 1 & 0 & 0 & 0 \\ 0 & -1 & 0 & 635 \\ 0 & 0 & 0 & 1 \end{bmatrix} \tag{7.109} $$

$$ {}^{6}T_{7} = \begin{bmatrix} 1 & 0 & 0 & 0 \\ 0 & 1 & 0 & 0 \\ 0 & 0 & 1 & 500 \\ 0 & 0 & 0 & 1 \end{bmatrix} \tag{7.110} $$

Assume that there is a camera attached to link (4) to watch the gripper's operation. The camera is at a point P in B_4 with position vector ${}^{4}\mathbf{r}_{P}$ and is aiming the origin of the gripper's frame B_7:

$$ {}^{4}\mathbf{r}_{P} = \begin{bmatrix} -a & b & 0 & 1 \end{bmatrix}^{\mathrm{T}} \tag{7.111} $$

$$ a = 696\,\mathrm{mm} \tag{7.112} $$

$$ b = 154\,\mathrm{mm} \tag{7.113} $$

To determine the position of the camera in the base frame, ${}^{0}\mathbf{r}_{P}$, we need to calculate ${}^{0}T_{4}$,

$$ \begin{aligned} {}^{0}T_{4} &= {}^{0}T_{1}\,{}^{1}T_{2}\,{}^{2}T_{3}\,{}^{3}T_{4} \\ &= \begin{bmatrix} 0 & 0 & 1 & 1045 \\ 0.70711 & 0.70711 & 0 & 11415 \\ -0.70711 & 0.70711 & 0 & 379.95 \\ 0 & 0 & 0 & 1 \end{bmatrix} \end{aligned} \tag{7.114} $$

and find ${}^{0}\mathbf{r}_{P}$,

$$ {}^{0}\mathbf{r}_{P} = {}^{0}T_{4}\,{}^{4}\mathbf{r}_{P} = {}^{0}T_{4} \begin{bmatrix} -696 \\ 154 \\ 0 \\ 1 \end{bmatrix} = \begin{bmatrix} 1045 \\ 11032 \\ 980.99 \\ 1 \end{bmatrix} \tag{7.115} $$

At the configuration (7.103) the transformation matrix ${}^0T_7$ is given as

$$
\begin{aligned}
{}^0T_7 &= {}^0T_1\,{}^1T_2\,{}^2T_3\,{}^3T_4\,{}^4T_5\,{}^5T_6\,{}^6T_7 \\
&= \begin{bmatrix} 1 & 0 & 0 & 1615 \\ 0 & 0 & 1 & 11915 \\ 0 & -1 & 0 & 1015 \\ 0 & 0 & 0 & 1 \end{bmatrix}
\end{aligned} \tag{7.116}
$$

where the position of the gripper point in the base frame is

$$
{}^0\mathbf{d}_7 = \begin{bmatrix} 1615 \\ 11915 \\ 1015 \\ 1 \end{bmatrix} \tag{7.117}
$$

Therefore, the position of the camera in the gripper frame B_7 would be

$$
\begin{aligned}
{}^7\mathbf{r}_P &= {}^7T_4\,{}^4\mathbf{r}_P \\
&= {}^7T_6\,{}^6T_5\,{}^5T_4\,{}^4\mathbf{r}_P = {}^6T_7^{-1}\,{}^5T_6^{-1}\,{}^4T_5^{-1}\,{}^4\mathbf{r}_P = \left[{}^4T_5\,{}^5T_6\,{}^6T_7\right]^{-1}\,{}^4\mathbf{r}_P \\
&= \begin{bmatrix} 0 & 0 & 1 & -570 \\ 0.707 & -0.707 & 0 & 635 \\ 0.707 & 0.707 & 0 & -500 \\ 0 & 0 & 0 & 1 \end{bmatrix} \begin{bmatrix} -696 \\ 154 \\ 0 \\ 1 \end{bmatrix} = \begin{bmatrix} -570 \\ 33.965 \\ -883.25 \\ 1 \end{bmatrix}
\end{aligned} \tag{7.118}
$$

Example 417 Camera Coordinate Frame of SSRMS Consider the SSRMS and its camera in Example 416. To use the camera kinematically, we need to attach a coordinate frame B_8 to the camera at P and determine the configuration of the gripper frame B_7 in the camera frame. The position vectors of the camera in the base and gripper frames are calculated in (7.115) and (7.118) as

$$
{}^0\mathbf{r}_P = \begin{bmatrix} 1045 \\ 11032 \\ 980.99 \\ 1 \end{bmatrix} \qquad {}^7\mathbf{r}_P = \begin{bmatrix} -570 \\ 33.965 \\ -883.25 \\ 1 \end{bmatrix} \tag{7.119}
$$

Assume that the camera is equipped with two electric motors that give it two rotational degrees of freedom. To determine the coordinate frame B_8 at the present configuration, we may begin with calculating the direction of the x_8-axis in the gripper frame B_7,

$$
{}^7\hat{\imath}_8 = \frac{-{}^7\mathbf{r}_P}{\left|{}^7\mathbf{r}_P\right|} = \begin{bmatrix} 0.54193 \\ -3.2292 \times 10^{-2} \\ 0.83975 \end{bmatrix} \tag{7.120}
$$

and determine the x_8-axis in frame B_4,

$$
{}^4\hat{\imath}_8 = {}^4R_7\,{}^7\hat{\imath}_8 = \begin{bmatrix} 0.57096 \\ 0.61663 \\ 0.54193 \\ 0 \end{bmatrix} \tag{7.121}
$$

Let us assume that the first motor of the camera turns the camera about z_8 and the second motor turns it about the displaced y_8. The B_4-frame may act as a global frame to the camera. So, we can find the matrix ${}^8R_4$ by a rotation α about z_8 followed by a rotation β about y_8:

$$\begin{aligned}{}^8R_4 &= R_{y,\beta}\, R_{z,\alpha} \\ &= \begin{bmatrix} \cos\alpha\cos\beta & \cos\beta\sin\alpha & -\sin\beta \\ -\sin\alpha & \cos\alpha & 0 \\ \cos\alpha\sin\beta & \sin\alpha\sin\beta & \cos\beta \end{bmatrix}\end{aligned} \tag{7.122}$$

Using ${}^8\hat{\imath}_8 = [1, 0, 0]$, we have

$$\begin{aligned}{}^4\hat{\imath}_8 = {}^4R_8\, {}^8\hat{\imath}_8 = {}^8R_4^{\mathrm{T}}\, {}^8\hat{\imath}_8 \\ \begin{bmatrix} 0.57096 \\ 0.61663 \\ 0.54193 \end{bmatrix} = {}^8R_4^{\mathrm{T}} \begin{bmatrix} 1 \\ 0 \\ 0 \end{bmatrix} = \begin{bmatrix} \cos\alpha\cos\beta \\ \cos\beta\sin\alpha \\ -\sin\beta \end{bmatrix}\end{aligned} \tag{7.123}$$

where

$${}^4R_8 = {}^8R_4^{\mathrm{T}} = \begin{bmatrix} \cos\alpha\cos\beta & -\sin\alpha & \cos\alpha\sin\beta \\ \cos\beta\sin\alpha & \cos\alpha & \sin\alpha\sin\beta \\ -\sin\beta & 0 & \cos\beta \end{bmatrix} \tag{7.124}$$

The third part of (7.123) provides β,

$$\beta = -\arcsin 0.54193 = -0.57273\,\mathrm{rad} \approx -32.815\,\mathrm{deg} \tag{7.125}$$

and the second part provides α,

$$\alpha = \arcsin\frac{0.61663}{0.84042} = 0.82378\,\mathrm{rad} \approx 47.199\,\mathrm{deg} \tag{7.126}$$

The first part may be used as a double check.

The angles α and β determine the orientation of B_8 at ${}^4\mathbf{d}_8 = [-a, b, 0, 1]^{\mathrm{T}}$. Therefore,

$${}^4T_8 = \begin{bmatrix} 0.57103 & -0.73372 & -0.36821 & -696 \\ 0.61664 & 0.67945 & -0.39762 & 154 \\ 0.54193 & 0 & 0.84042 & 0 \\ 0 & 0 & 0 & 1 \end{bmatrix} \tag{7.127}$$

$${}^8T_4 = \begin{bmatrix} 0.57103 & 0.61664 & 0.54192 & 302.47 \\ -0.73372 & 0.67945 & -6.58\times10^{-8} & -615.31 \\ -0.36822 & -0.39763 & 0.84043 & -195.04 \\ 0 & 0 & 0 & 1 \end{bmatrix} \tag{7.128}$$

Now, we can calculate the position and orientation of the gripper frame in the eye of camera as well as the camera in the base coordinate frame:

$$
\begin{aligned}
{}^{8}T_{7} &= {}^{8}T_{4}\,{}^{4}T_{5}\,{}^{5}T_{6}\,{}^{6}T_{7} \\
&\approx \begin{bmatrix} 0.54192 & -3.225\times10^{-2} & 0.83981 & 1051.8 \\ 0 & -0.99927 & -3.837\times10^{-2} & 0 \\ 0.84043 & 2.08\times10^{-2} & -0.54154 & 0 \\ 0 & 0 & 0 & 1 \end{bmatrix}
\end{aligned} \tag{7.129}
$$

$$
\begin{aligned}
{}^{0}T_{8} &= {}^{0}T_{7}\,{}^{7}T_{8} = {}^{0}T_{7}\,{}^{8}T_{7}^{-1} \\
&= \begin{bmatrix} 0.54192 & -0.03225 & 0.83981 & 2666.8 \\ 0.84043 & 0.0208 & -0.54154 & 11,915 \\ 0 & 0.99927 & 0.03837 & 1015 \\ 0 & 0 & 0 & 1 \end{bmatrix}
\end{aligned} \tag{7.130}
$$

Example 418 ★ Directional Control System The satellite antenna and inspection camera are samples of direction control and detection systems. To have a directional control system, we usually need to control the angles of a flat plate and direct an axis on the plate or its normal vector to the desired direction.

Such a directional control system is usually equipped with two motors that provide two rotational degrees of freedom, as shown in Figure 7.30. The axes of the rotations intersect at a *wrist point*. Let us attach a coordinate frame B_8 to the camera at the wrist point and a Sina coordinate frame B_9 to the base link at the same point. The fixed frame $B_9(x_9, y_9, z_9) \equiv B_9(X, Y, Z)$ acts as the global frame for the local frame $B_8(x_8, y_8, z_8)$.

The first motor turns the control plate about the fixed axis z_9 that is initially coincident with z_8. The second motor turns the plate about the y_8-axis. The transformation

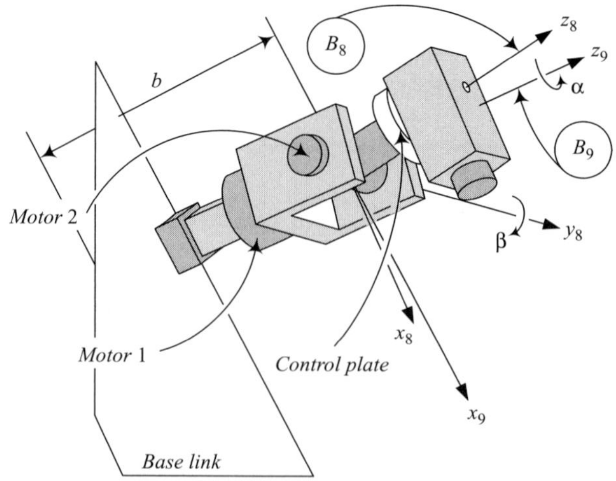

Figure 7.30 A directional control system that is equiped with two motors.

matrix ${}^8R_9$ between B_8 and B_9 from a coincident configuration may be found by a rotation α about z_8 followed by a rotation β about y_8:

$$
{}^8R_9 = R_{y,\beta}\, R_{z,\alpha} = \begin{bmatrix} \cos\alpha\cos\beta & \cos\beta\sin\alpha & -\sin\beta \\ -\sin\alpha & \cos\alpha & 0 \\ \cos\alpha\sin\beta & \sin\alpha\sin\beta & \cos\beta \end{bmatrix} \tag{7.131}
$$

However, because the z_8-axis is not the actual axis of rotation, we must determine ${}^8R_9$ by rotations about the actual rotation axes z_9 and y_8. The first rotation $R_{Z,\alpha}$ is about the global Z-axis,

$$
{}^9R_8 = R_{Z,\alpha} = \begin{bmatrix} \cos\alpha & -\sin\alpha & 0 \\ \sin\alpha & \cos\alpha & 0 \\ 0 & 0 & 1 \end{bmatrix} \tag{7.132}
$$

and the second rotation $R_{y,\beta}$ is about the local y-axis,

$$
{}^8R_9 = R_{y,\beta} = \begin{bmatrix} \cos\beta & 0 & -\sin\beta \\ 0 & 1 & 0 \\ \sin\beta & 0 & \cos\beta \end{bmatrix} \tag{7.133}
$$

To determine the combined rotation matrix, we may find ${}^9R_8 = R^{\mathrm{T}}_{y,\beta} R_{Z,\alpha}$ or ${}^8R_9 = R_{y,\beta} R^{\mathrm{T}}_{Z,\alpha}$:

$$
\begin{aligned}
{}^8R_9 &= R_{y,\beta} R^{\mathrm{T}}_{Z,\alpha} \\
&= \begin{bmatrix} \cos\beta & 0 & -\sin\beta \\ 0 & 1 & 0 \\ \sin\beta & 0 & \cos\beta \end{bmatrix} \begin{bmatrix} \cos\alpha & -\sin\alpha & 0 \\ \sin\alpha & \cos\alpha & 0 \\ 0 & 0 & 1 \end{bmatrix}^{\mathrm{T}} \\
&= \begin{bmatrix} \cos\alpha\cos\beta & \cos\beta\sin\alpha & -\sin\beta \\ -\sin\alpha & \cos\alpha & 0 \\ \cos\alpha\sin\beta & \sin\alpha\sin\beta & \cos\beta \end{bmatrix}
\end{aligned} \tag{7.134}
$$

The transformation matrix (7.134) is the same as (7.131) and works only when the initial configuration of the camera frame B_8 coincides with the Sina frame B_9. Furthermore, the rotations must be in order.

Example 419 ★ Angle–Axis of Rotation of a Directional Control System Figure 7.30 illustrates an inspection camera on a two-DOF directional control system. The two DOF are provided by two motors such that motor 1 turns the control plate of the camera about the z_9-axis and motor 2 turns the plate about the y_8-axis. The axes of rotations intersect at the wrist point that is the origin of the camera coordinate frame B_8 and the Sina frame B_9. The Sina frame is a fixed frame on the base link on which the camera is installed.

Assume that the camera gets the desired direction when motor 1 turns α and motor 2 turns β. Because the two motors can act independently, the order of action of motors is immaterial in determination of the final aiming direction as well as the transformation matrix between B_8 and B_9, according to Equation (7.278). The two motors may also act together and turn their associated angles together. In this case, the motor will perform an axis–angle of rotation. Having the rotation matrix as

$$ {}^{8}R_{9} = R_{y,\beta}\, R_{z,\alpha} = \begin{bmatrix} \cos\alpha\cos\beta & \cos\beta\sin\alpha & -\sin\beta \\ -\sin\alpha & \cos\alpha & 0 \\ \cos\alpha\sin\beta & \sin\alpha\sin\beta & \cos\beta \end{bmatrix} \tag{7.135} $$

we are able to determine the equivalent angle and axis of rotations by Equations (5.9) and (5.10):

$$ \begin{aligned} \cos\phi &= \tfrac{1}{2}[\mathrm{tr}\left({}^{G}R_{B}\right) - 1] \\ &= \tfrac{1}{2}\left(\cos\alpha + \cos\beta + \cos\alpha\cos\beta - 1\right) \end{aligned} \tag{7.136} $$

$$ \begin{aligned} \tilde{u} &= \frac{1}{2\sin\phi}\left({}^{G}R_{B} - {}^{G}R_{B}^{\mathrm{T}}\right) = \frac{1}{2\sin\phi}\left({}^{8}R_{9}^{\mathrm{T}} - {}^{8}R_{9}\right) \\ &= \frac{\begin{bmatrix} 0 & -s\alpha - c\beta s\alpha & s\beta + c\alpha s\beta \\ s\alpha + c\beta s\alpha & 0 & s\alpha s\beta \\ -s\beta - c\alpha s\beta & -s\alpha s\beta & 0 \end{bmatrix}}{2\sqrt{1 - [\frac{1}{2}\left(\cos\alpha + \cos\beta + \cos\alpha\cos\beta - 1\right)]^{2}}} \end{aligned} \tag{7.137} $$

Let us assume that

$$ \alpha = 0.823\,78\,\mathrm{rad} \approx 47.199\,\mathrm{deg} \tag{7.138} $$

$$ \beta = -0.57273\,\mathrm{rad} \approx -32.815\,\mathrm{deg} \tag{7.139} $$

Therefore, the rotation matrix (7.135) is

$$ {}^{8}R_{9} = R_{y,\beta}\, R_{z,\alpha} = \begin{bmatrix} 0.57103 & 0.61664 & 0.54193 \\ -0.73372 & 0.67945 & 0 \\ -0.36821 & -0.39762 & 0.84042 \end{bmatrix} \tag{7.140} $$

and the associated axis and angle are

$$ \cos\phi = \tfrac{1}{2}[\mathrm{tr}\left({}^{8}R_{9}^{\mathrm{T}}\right) - 1] = 0.545\,45\,\mathrm{rad} \approx 31.252\,\mathrm{deg} \tag{7.141} $$

$$ \begin{aligned} \tilde{u} &= \frac{1}{2\sin\phi}\left({}^{8}R_{9}^{\mathrm{T}} - {}^{8}R_{9}\right) \\ &= \begin{bmatrix} 0 & -1.3015 & -0.87715 \\ 1.3015 & 0 & -0.38321 \\ 0.87715 & 0.38321 & 0 \end{bmatrix} \end{aligned} \tag{7.142} $$

To activate both motors together, we may command motors 1 and 2 to act according to the functions

$$ \alpha(t) = 2.4713\frac{t^{2}}{t_{f}^{2}} - 1.6476\frac{t^{3}}{t_{f}^{3}} \tag{7.143} $$

$$ \beta(t) = -1.7182\frac{t^{2}}{t_{f}^{2}} + 1.1455\frac{t^{3}}{t_{f}^{3}} \tag{7.144} $$

where t is time and t_f is the total time it takes for the motors to finish their rotations. These equations guarantee that both motors start from zero angular velocity at $t = 0$ and reach their final angular rotations at $t = t_f$ and stop there. Let us assume

$$t_f = 10\,\text{s} \tag{7.145}$$

to have

$$\alpha(t) = 2.471\,3 \times 10^{-2} t^2 - 1.6476 \times 10^{-3} t^3 \tag{7.146}$$

$$\beta(t) = -1.718\,2 \times 10^{-2} t^2 + 1.1455 \times 10^{-3} t^3 \tag{7.147}$$

Example 420 Closed-Loop Mechanisms When a mechanism makes a closed loop, there would also be a connection between the first and last links. The mathematical condition for a kinematic loop is

$$[T] = {}^0T_1 \, {}^1T_2 \, \ldots \, {}^{i-1}T_i \, \ldots \, {}^{n-1}T_n \, {}^nT_0 = \mathbf{I} \tag{7.148}$$

where ${}^{i-1}T_i$ is the homogeneous transformation matrix to go from the B_i-frame to the B_{i-1}-frame and $\mathbf{I} = [\mathbf{I}]$ indicates a 4×4 identity matrix. In a closed-loop multibody, we may start from the base link and reach any coordinate frame from two different path. Therefore, it is also possible to express the kinematics of a closed-loop mechanism by

$${}^0T_n = {}^0T_1 \, {}^1T_2 \, \ldots \, {}^{i-1}T_i \, \ldots \, {}^{n-1}T_n \tag{7.149}$$

As an example, consider the planar slider–crank mechanism in Figure 7.31. This mechanism can be classified as an R⊥P⊢R‖R‖R. The transformation matrices are functions of a_1, a_2, a_3, d, θ_2, θ_3, and θ_4. The parameters a_2, a_3, and a_4 are constant and d, θ_3, θ_4, and θ_2 are variable. The mechanism has one DOF, and its configuration is controlled by a motor that moves the crank link or a piston that moves the slider block. The loop condition for the mechanism would be

$$[T] = {}^1T_2 \, {}^2T_3 \, {}^3T_4 \, {}^4T_1 = \mathbf{I}_4 \tag{7.150}$$

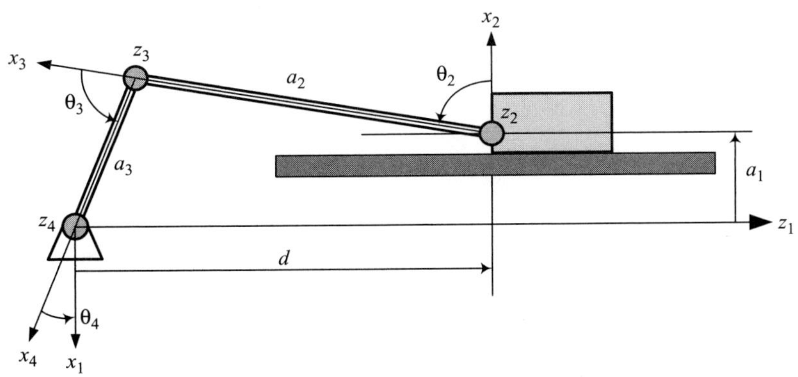

Figure 7.31 A planar slider–crank linkage making a closed loop or parallel mechanism.

Example 421 ★ A 3D Slider–Crank Mechanism Figure 7.32 illustrates a slider that is moving back and forth on the X-axis by the connected bar AB. The bar is attached to a spinning disc that is controlled by a motor and is measured by an input angle θ. The plane of the disc is at an angle 45 deg about the Z-axis. A side view of the mechanism from the Z-axis is shown in Figure 7.33. The global coordinate frame G is set at the fixed center of the disc to measure the distance of the slider, b.

We have only four relatively moving bodies, including the ground. So, we technically need four coordinate frames. However, it might be more clear to set up some extra coordinate frames. Extra frames would simplify the kinematic of the multibody with the penalty of extra matrix multiplications.

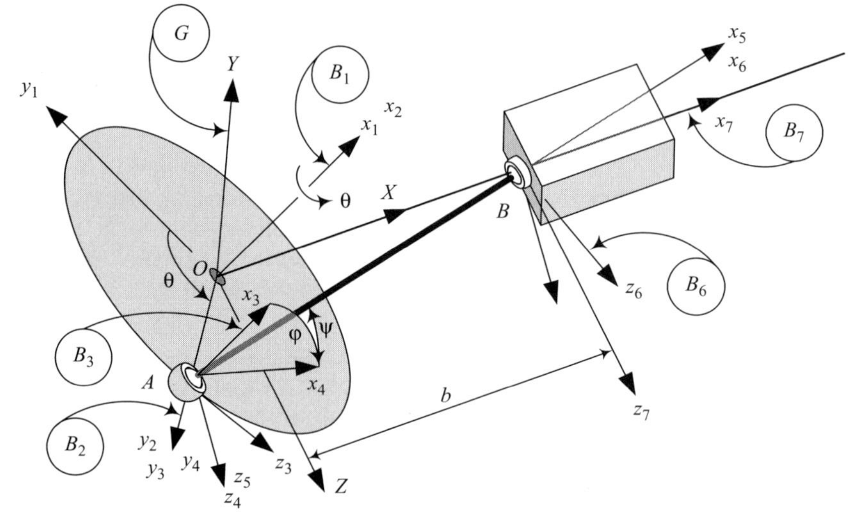

Figure 7.32 A 3D slider–crank mechanism.

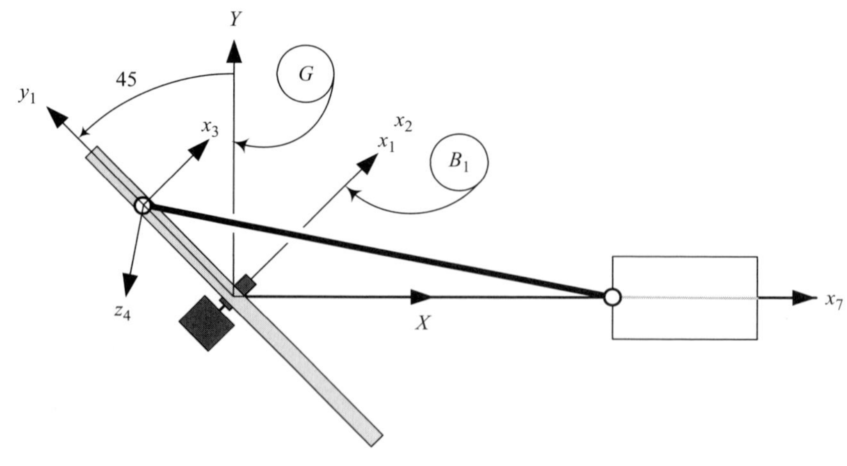

Figure 7.33 Side view of the 3D slider–crank mechanism of Figure 7.32.

The frame B_1 is another globally fixed frame at an angle 45 deg about the Z-axis such that the x_1-axis is perpendicular to the disc and indicates its spin axis. The body frame B_2 is attached to the disc such that x_1 and x_2 are coincident and y_2 points the joint A. The angle between y_1 and y_2 is θ. The connecting bar AB has a constant length l and is attached to the disc at point A at a distance R from the disc center:

$$ {}^2\mathbf{r}_A = \begin{bmatrix} 0 & R & 0 \end{bmatrix}^\mathrm{T} \tag{7.151} $$

A Sina frame B_3 parallel to B_2 is attached to the disc at point A. The frame B_4 is at A such that y_4 is coincident with y_3 and x_4 is seen coincident with AB from the Z viewpoint. The angle between x_3 and x_4 is φ. There is another frame B_5 at A such that z_4 and z_5 are coincident and x_5 is along AB. The angle between x_4 and x_5 is ψ. At point B of the bar AB, we attach a coordinate frame B_6 parallel to B_5. The slider frame B_7, which is parallel to G, is attached to the box at point B:

$$ {}^G\mathbf{r}_B = \begin{bmatrix} b & 0 & 0 \end{bmatrix}^\mathrm{T} \tag{7.152} $$

The transformation between G and B_1 is a constant 45 deg rotation about the Z-axis:

$$ {}^GT_1 = \begin{bmatrix} R_{Z,\pi/4} & 0 \\ 0 & 1 \end{bmatrix} = \begin{bmatrix} \cos(\pi/4) & -\sin(\pi/4) & 0 & 0 \\ \sin(\pi/4) & \cos(\pi/4) & 0 & 0 \\ 0 & 0 & 1 & 0 \\ 0 & 0 & 0 & 1 \end{bmatrix} \tag{7.153} $$

The transformation between B_1 and B_2 is a variable rotation θ about the x_1-axis:

$$ {}^1T_2 = \begin{bmatrix} R_{x_1,\theta} & 0 \\ 0 & 1 \end{bmatrix} = \begin{bmatrix} 1 & 0 & 0 & 0 \\ 0 & \cos\theta & -\sin\theta & 0 \\ 0 & \sin\theta & \cos\theta & 0 \\ 0 & 0 & 0 & 1 \end{bmatrix} \tag{7.154} $$

The transformation between B_2 and B_3 is a constant translation R along the x_2-axis:

$$ {}^2T_3 = \begin{bmatrix} 1 & 0 & 0 & R \\ 0 & 1 & 1 & 0 \\ 0 & 0 & 1 & 0 \\ 0 & 0 & 0 & 1 \end{bmatrix} \tag{7.155} $$

The transformation between B_3 and B_4 is a variable rotation φ about the y_3-axis:

$$ {}^3T_4 = \begin{bmatrix} R_{y_3,\varphi} & 0 \\ 0 & 1 \end{bmatrix} = \begin{bmatrix} \cos\varphi & 0 & \sin\varphi & 0 \\ 0 & 1 & 0 & 0 \\ -\sin\varphi & 0 & \cos\varphi & 0 \\ 0 & 0 & 0 & 1 \end{bmatrix} \tag{7.156} $$

The transformation between B_4 and B_5 is a variable rotation ψ about the z_4-axis:

$$ {}^4T_5 = \begin{bmatrix} R_{z_4,\psi} & 0 \\ 0 & 1 \end{bmatrix} = \begin{bmatrix} \cos\psi & -\sin\psi & 0 & 0 \\ \sin\psi & \cos\psi & 0 & 0 \\ 0 & 0 & 1 & 0 \\ 0 & 0 & 0 & 1 \end{bmatrix} \tag{7.157} $$

The transformation between B_5 and B_6 is a constant translation l along the x_5-axis:

$$ {}^{5}T_{6} = \begin{bmatrix} 1 & 0 & 0 & l \\ 0 & 1 & 1 & 0 \\ 0 & 0 & 1 & 0 \\ 0 & 0 & 0 & 1 \end{bmatrix} \tag{7.158} $$

The transformation between B_7 and G is a variable translation b along the X-axis:

$$ {}^{G}T_{7} = \begin{bmatrix} 1 & 0 & 0 & b \\ 0 & 1 & 1 & 0 \\ 0 & 0 & 1 & 0 \\ 0 & 0 & 0 & 1 \end{bmatrix} \tag{7.159} $$

To find the global position of joint B, we go through $B_1, B_2, \ldots, B_6$ and determine ${}^{G}T_{6}$:

$$ \begin{aligned} {}^{G}T_{6} &= {}^{G}T_{1}\,{}^{1}T_{2}\,{}^{2}T_{3}\,{}^{3}T_{4}\,{}^{4}T_{5}\,{}^{5}T_{6} \\ &= \begin{bmatrix} r_{11} & r_{12} & r_{13} & d_{6X} \\ r_{21} & r_{22} & r_{23} & d_{6Y} \\ r_{31} & r_{32} & r_{33} & d_{6Z} \\ 0 & 0 & 0 & 1 \end{bmatrix} \end{aligned} \tag{7.160} $$

where

$$ \begin{aligned} r_{11} &= \frac{\sqrt{2}}{2}\left(\cos\theta\cos\psi - \cos\psi\sin\theta\right)\sin\varphi \\ &\quad + \frac{\sqrt{2}}{2}\left(\cos\psi\cos\varphi - \cos\theta\sin\psi\right) \end{aligned} \tag{7.161} $$

$$ \begin{aligned} r_{21} &= \frac{\sqrt{2}}{2}\left(-\cos\theta\cos\psi + \cos\psi\sin\theta\right)\sin\varphi \\ &\quad + \frac{\sqrt{2}}{2}\left(\cos\psi\cos\varphi + \cos\theta\sin\psi\right) \end{aligned} \tag{7.162} $$

$$ \begin{aligned} r_{31} &= \tfrac{1}{2}\left(\cos(\theta - \psi) - \cos(\theta + \psi)\right) \\ &\quad - \left(\cos\theta\cos\psi + \cos\psi\sin\theta\right)\sin\varphi \end{aligned} \tag{7.163} $$

$$ \begin{aligned} r_{12} &= \frac{\sqrt{2}}{2}\left(\sin\theta\sin\psi - \cos\theta\sin\psi\right)\sin\varphi \\ &\quad - \frac{\sqrt{2}}{2}\left(\cos\varphi\sin\psi + \cos\theta\cos\psi\right) \end{aligned} \tag{7.164} $$

$$ \begin{aligned} r_{22} &= \frac{\sqrt{2}}{2}\left(\cos\theta\sin\psi - \sin\theta\sin\psi\right)\sin\varphi \\ &\quad + \frac{\sqrt{2}}{2}\left(\cos\theta\cos\psi - \cos\varphi\sin\psi\right) \end{aligned} \tag{7.165} $$

$$ \begin{aligned} r_{32} &= \tfrac{1}{4}\sin(\theta + \psi - \varphi) + \tfrac{1}{4}\sin(\theta - \psi + \varphi) - \tfrac{1}{4}\sin(\theta - \psi - \varphi) \\ &\quad - \tfrac{1}{4}\sin(\theta + \psi + \varphi) + \cos\psi\sin\theta + \cos\theta\sin\psi\sin\varphi \end{aligned} \tag{7.166} $$

$$r_{13} = -\frac{\sqrt{2}}{2}\cos\theta\cos\psi + \frac{\sqrt{2}}{2}(\sin\theta\sin\psi - \cos\theta\sin\psi)\sin\varphi$$
$$+ \frac{\sqrt{2}}{2}(\sin\theta - \cos\theta - \sin\psi)\cos\varphi + \frac{\sqrt{2}}{2}\sin\varphi \tag{7.167}$$

$$r_{23} = \frac{\sqrt{2}}{2}\sin\varphi + \frac{\sqrt{2}}{2}(\cos\theta\sin\psi - \sin\theta\sin\psi)\sin\varphi$$
$$+ \frac{\sqrt{2}}{2}(\cos\psi + \cos\varphi)\cos\theta - \frac{\sqrt{2}}{2}(\sin\theta + \sin\psi)\cos\varphi \tag{7.168}$$

$$r_{33} = \tfrac{1}{4}(\sin(\theta + \psi - \varphi) + \sin(\theta - \psi + \varphi) - \sin(\theta - \psi - \varphi))$$
$$- \tfrac{1}{4}\sin(\theta + \psi + \varphi) + \cos\theta(\cos\varphi + \sin\psi\sin\varphi)$$
$$+ (\cos\psi + \cos\varphi)\sin\theta \tag{7.169}$$

$$d_{6X} = \frac{\sqrt{2}}{2}R + \frac{\sqrt{2}}{2}l\cos\psi\cos\varphi - \frac{\sqrt{2}}{2}l\cos\theta\sin\psi$$
$$+ \frac{\sqrt{2}}{2}l(\cos\theta\cos\psi - \cos\psi\sin\theta)\sin\varphi \tag{7.170}$$

$$d_{6Y} = \frac{\sqrt{2}}{2}R + \frac{\sqrt{2}}{2}l\cos\psi\cos\varphi + \frac{\sqrt{2}}{2}l\cos\theta\sin\psi$$
$$- \frac{\sqrt{2}}{2}l(\cos\theta\cos\psi + \cos\psi\sin\theta)\sin\varphi \tag{7.171}$$

$$d_{6Z} = l\sin\theta\sin\psi - l(\cos\theta\cos\psi + \cos\psi\sin\theta)\sin\varphi \tag{7.172}$$

The X-component of ${}^{G}\mathbf{d}_6$ must be equal to b and the Y- and Z-components of ${}^{G}\mathbf{d}_6$ must be zero:

$$d_{6X} = b \tag{7.173}$$

$$d_{6Y} = 0 \tag{7.174}$$

$$d_{6Z} = 0 \tag{7.175}$$

We should be able to solve these three equations to calculate b, φ, ψ for a given θ.

In a different design, we may assume that the angle between G and B_1 is a controllable angle α. In this case the transformation between G and B_1 is a given angle α rotation about the Z-axis:

$${}^{G}T_1 = \begin{bmatrix} R_{Z,\alpha} & 0 \\ 0 & 1 \end{bmatrix} = \begin{bmatrix} \cos\alpha & -\sin\alpha & 0 & 0 \\ \sin\alpha & \cos\alpha & 0 & 0 \\ 0 & 0 & 1 & 0 \\ 0 & 0 & 0 & 1 \end{bmatrix} \tag{7.176}$$

Reevaluating both sides of Equation (7.160) provides three new equations instead of (7.170)–(7.172) that must be solved for b, φ, ψ, for given θ and α. When $\alpha = 90$ deg, the mechanism reduces to the planar slider–crank mechanism in Figure 7.31, and when $\alpha = 0$, the slider will not move and $b = \text{const}$.

Example 422* ★ *Universal Joint The *universal joint* shown in Figure 7.34 is a common, inexpensive, reliable, and heavy-duty mechanism for connecting two rotating shafts that intersect at an angle φ. The universal joint is also known as Hook's coupling, Hook joint, Cardan joint, U-joint, or yoke joint. The main disadvantage of the joint is fluctuation of the speed ratio between the input and output shafts.

A universal joint has four links: link (1) is the ground, and has a revolute joint with input link (2) and output link (4). The input and the output links are connected with a coupler link (3) that is called the cross-link.

To determine the kinematics of the universal joint, let us separate its links and show them in Figure 7.35. The intersection point of the input and output axes is the common origin of all coordinate frames. Frame B_1 is the ground or chassis frame. Frame B_2 is attached to the input driver link that turns by θ_2 about the x_1-axis. Frame B_3 is attached to the cross-link and turns by α about the y_2-axis. Frame B_4 is attached to the driven output link that turns by β about the z_3-axis. We attach a Sina coordinate frame B_5 to the chassis such that x_5 is coincident with the x_4-axis.

The transformation between B_1 and B_2 is a rotation θ_2 about the x_1-axis:

$$ {}^{1}R_2 = \left[R_{X,\theta_2}\right] = \begin{bmatrix} 1 & 0 & 0 \\ 0 & \cos\theta_2 & -\sin\theta_2 \\ 0 & \sin\theta_2 & \cos\theta_2 \end{bmatrix} \tag{7.177} $$

The transformation between B_2 and B_3 is a rotation α about the y_2-axis:

$$ {}^{2}R_3 = \left[R_{Y,\alpha}\right] = \begin{bmatrix} \cos\alpha & 0 & \sin\alpha \\ 0 & 1 & 0 \\ -\sin\alpha & 0 & \cos\alpha \end{bmatrix} \tag{7.178} $$

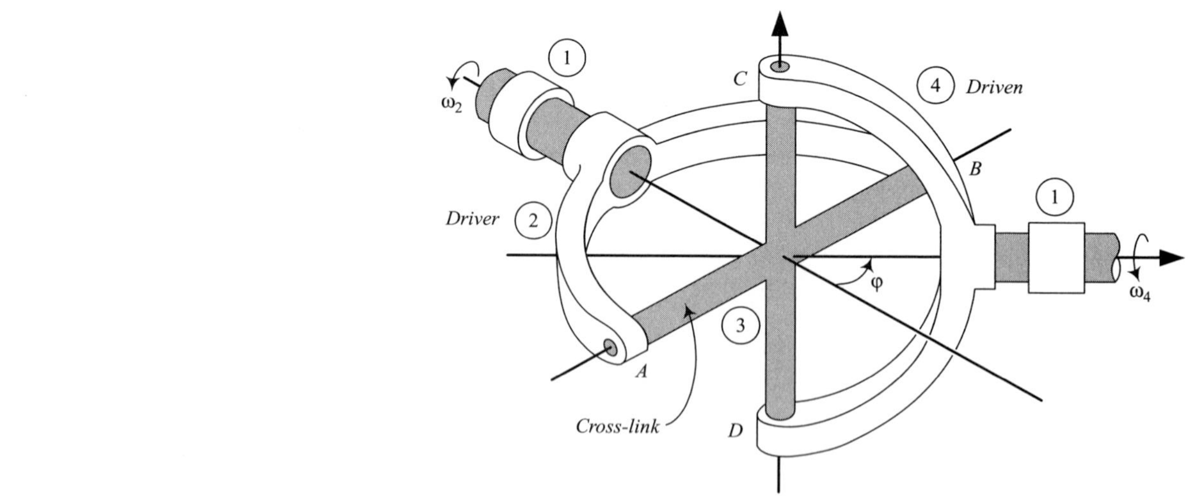

Figure 7.34 A universal joint with four links: link 1 is the ground, link 2 is the input, link 4 is the output, and the cross-link 3 is a coupler link.

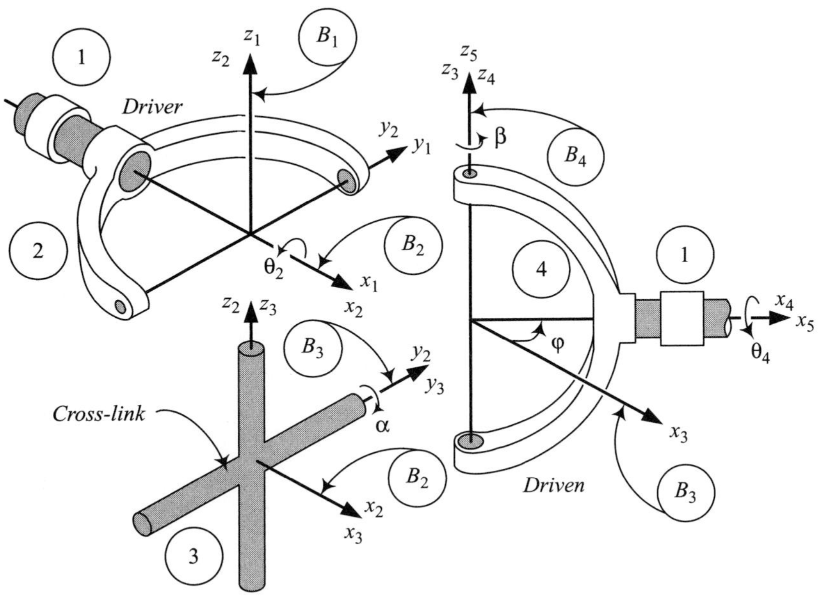

Figure 7.35 A separate illustration of the input, output, and cross-links for a universal joint.

The transformation between B_3 and B_4 is a rotation β about the z_3-axis:

$$
{}^3R_4 = \left[R_{Z,\beta}\right] = \begin{bmatrix} \cos\beta & -\sin\beta & 0 \\ \sin\beta & \cos\beta & 0 \\ 0 & 0 & 1 \end{bmatrix} \tag{7.179}
$$

The transformation between B_4 and B_5 is a rotation $-\theta_4$ about the x_4-axis:

$$
{}^4R_5 = \left[R_{X,-\theta_4}\right] = \begin{bmatrix} 1 & 0 & 0 \\ 0 & \cos\theta_4 & \sin\theta_4 \\ 0 & -\sin\theta_4 & \cos\theta_4 \end{bmatrix} \tag{7.180}
$$

The direct transformation between B_1 and B_5 is a rotation φ about the z_1-axis:

$$
{}^1R_5 = \left[R_{Z,\varphi}\right] = \begin{bmatrix} \cos\varphi & -\sin\varphi & 0 \\ \sin\varphi & \cos\varphi & 0 \\ 0 & 0 & 1 \end{bmatrix} \tag{7.181}
$$

Combining the transformation matrices $B_1, B_2, \ldots, B_5$, we have

$$
{}^1R_5 = {}^1R_2\,{}^2R_3\,{}^3R_4\,{}^4R_5 = \begin{bmatrix} r_{11} & r_{12} & r_{13} \\ r_{21} & r_{22} & r_{23} \\ r_{31} & r_{32} & r_{33} \end{bmatrix} \tag{7.182}
$$

where

$$
r_{11} = \cos\alpha\cos\beta \tag{7.183}
$$

$$
r_{21} = \cos\theta_2\sin\beta + \cos\beta\sin\alpha\sin\theta_2 \tag{7.184}
$$

$$
r_{31} = \sin\beta\sin\theta_2 - \cos\beta\cos\theta_2\sin\alpha \tag{7.185}
$$

$$r_{12} = -\sin\alpha\sin\theta_4 - \cos\alpha\cos\theta_4\sin\beta \tag{7.186}$$

$$r_{22} = (\cos\beta\cos\theta_2 - \sin\alpha\sin\beta\sin\theta_2)\cos\theta_4 + \cos\alpha\sin\theta_2\sin\theta_4 \tag{7.187}$$

$$r_{32} = (\cos\beta\sin\theta_2 + \cos\theta_2\sin\alpha\sin\beta)\cos\theta_4 - \cos\alpha\cos\theta_2\sin\theta_4 \tag{7.188}$$

$$r_{13} = \cos\theta_4\sin\alpha - \cos\alpha\sin\beta\sin\theta_4 \tag{7.189}$$

$$r_{23} = (\cos\beta\cos\theta_2 - \sin\alpha\sin\beta\sin\theta_2)\sin\theta_4 - \cos\alpha\cos\theta_4\sin\theta_2 \tag{7.190}$$

$$r_{33} = (\cos\beta\sin\theta_2 + \cos\theta_2\sin\alpha\sin\beta)\sin\theta_4 + \cos\alpha\cos\theta_2\cos\theta_4 \tag{7.191}$$

However, the transformation ${}^1R_5$ of (7.181) must be equal to ${}^1R_5$ of (7.182). Therefore, we can pick three independent elements, such as r_{11}, r_{31}, r_{13}, to get a set of three independent equations:

$$\cos\alpha\cos\beta = \cos\varphi \tag{7.192}$$

$$\sin\beta\sin\theta_2 - \cos\beta\cos\theta_2\sin\alpha = 0 \tag{7.193}$$

$$\cos\theta_4\sin\alpha - \cos\alpha\sin\beta\sin\theta_4 = 0 \tag{7.194}$$

These equations can theoretically be solved to determine α, β, and θ_4 as functions of θ_2 and φ. Equation (7.193) and (7.194) may be written as

$$\tan\beta\tan\theta_2 = \sin\alpha \tag{7.195}$$

$$\tan\alpha = \sin\beta\tan\theta_4 \tag{7.196}$$

Multiplying these equations yields

$$\tan\alpha\tan\beta\tan\theta_2 = \sin\alpha\sin\beta\tan\theta_4 \tag{7.197}$$

and employing (7.192) provides the equation to calculate θ_4:

$$\tan\theta_4 = \frac{\tan\theta_2}{\cos\varphi} \tag{7.198}$$

Eliminating α between (7.193) and (7.194) and using (7.198) provide the equation to calculate β, and eliminating β between (7.193) and (7.194) and using (7.198) provide the equation to calculate α.

Differentiation of (7.198) for constant φ,

$$\frac{\omega_2}{\csc^2\theta} = \frac{\omega_4}{\csc^2\theta_4}\cos\varphi \tag{7.199}$$

and eliminating θ_4 between (7.199) and (7.198) provide the relationship between the input and output shaft angular velocities:

$$\omega_4 = \frac{\cos\varphi}{\sin^2\theta_2 + \cos^2\theta_2\cos^2\varphi}\omega_2 \tag{7.200}$$

The speed ratio is then given as

$$\Omega = \frac{\omega_4}{\omega_2} = \frac{\cos\varphi}{1 - \sin^2\varphi\cos^2\theta_2} \tag{7.201}$$

Figure 7.36 depicts a three-dimensional plot for Ω. The Ω-surface is plotted for one revolution of the drive shaft and every possible angle between the two shafts:

$$-\pi < \theta < \pi \qquad -\pi < \varphi < \pi \tag{7.202}$$

A two-dimensional view of Ω is depicted in Figure 7.37. When $\varphi \lessapprox 10\,\text{deg}$ there is not much fluctuation in speed ratio; however, when the angle between the two shafts is more than 10 deg, the speed ratio Ω cannot be assumed constant. The universal joint stuck when $\varphi = 90\,\text{deg}$ because theoretically

$$\lim_{\varphi \to 90} \Omega = \text{indefinite} \tag{7.203}$$

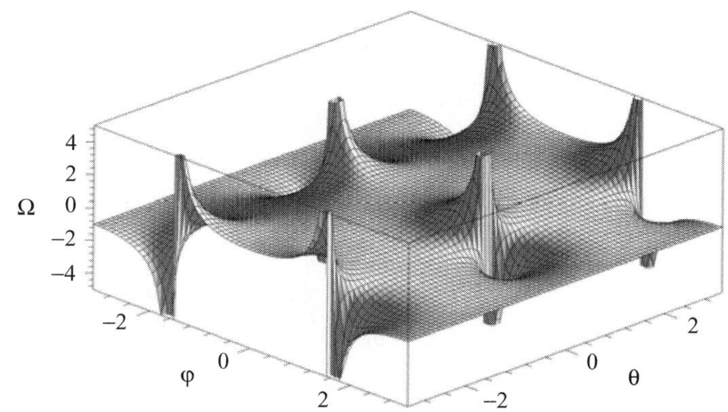

Figure 7.36 A three-dimensional plot for the speed ratio of a universal joint Ω as a function of the input angle θ and the angle between input and output shafts φ.

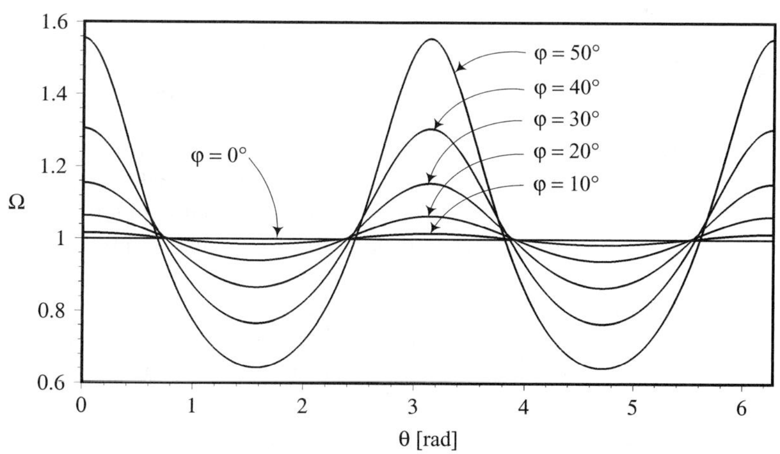

Figure 7.37 A two-dimensional view of Ω as a function of the input angle θ and the angle between input and output shafts φ.

Example 423 ★ Application of Universal Joint To eliminate the inconstant speed ratio of the universal joint between two shafts, we must connect the shafts with two universal joints. Setting the second joint properly, we can compensate for the speed fluctuation and provide a uniform output angular speed. The proper method of coupling, shown in Figure 7.38, consists of two universal joints and a short shaft.

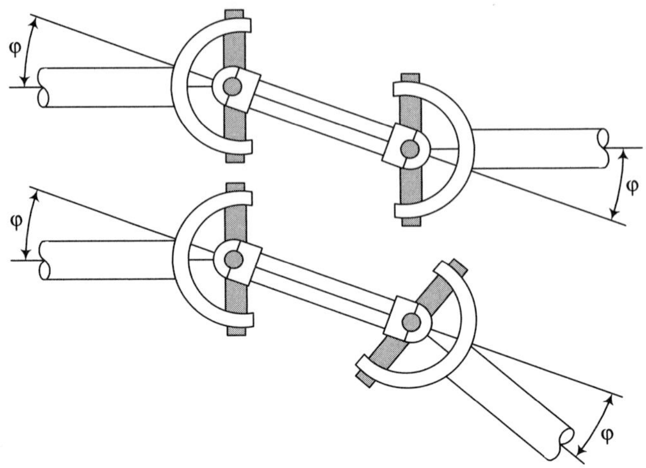

Figure 7.38 The proper methods of coupling to provide a constant speed ratio between the input and output shafts.

The method is used more frequently to keep the input and output shafts parallel. The shaft offset can be varied; however, due to variable length, one shaft must be spline mounted. The velocity transmission is constant between the input and output shafts if the forks are assembled such that they are always in the same plane. Therefore, by coupling two universal joints, we make a constant-velocity joint. The velocity of the central shaft fluctuates during rotation.

Example 424 ★ Tire–Wheel–Vehicle Coordinate Frames The setup of proper coordinate frames is the first step in multibody kinematic analysis. Although each body needs at least one coordinate frame, we may attach a few Sina coordinate frames to a body or to the space to simplify the transformation kinematics. We recommend not leaving it to chance and installing as many frames as needed. You can always eliminate the unused frames or combine relatively fixed frames. Frames with a common origin will more easily become coaxes at the rest position of the multibody.

A suspension vehicle is an example of a multibody with interesting kinematics. We number the tires of a vehicle by starting from the front left as tire number 1. The front right tire takes number 2, and the other tires on the right side of the vehicle sequentially take the numbers 3, 4, ... up to the last tire at the rear right. The numbering then moves to the left side and sequentially increases forward to reach the front left tire.

Figure 7.39 illustrates the first and fourth tires of a four-wheel vehicle. We attach a body coordinate frame $B(x, y, z)$ to the mass center C of the vehicle. We also attach

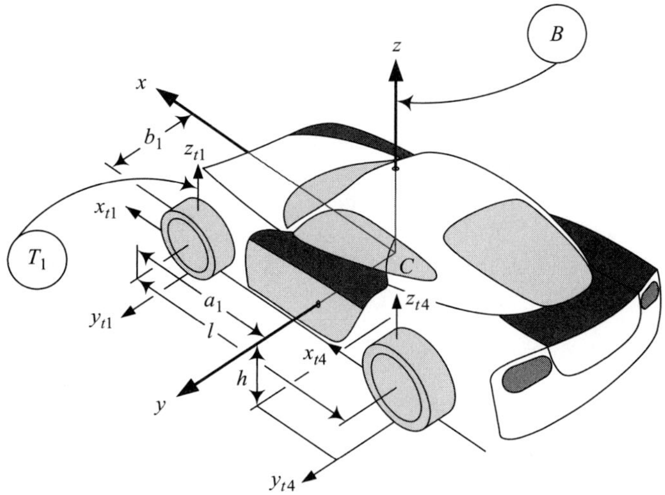

Figure 7.39 The coordinate frames of the first and fourth tires of a four-wheel vehicle and the vehicle body frame.

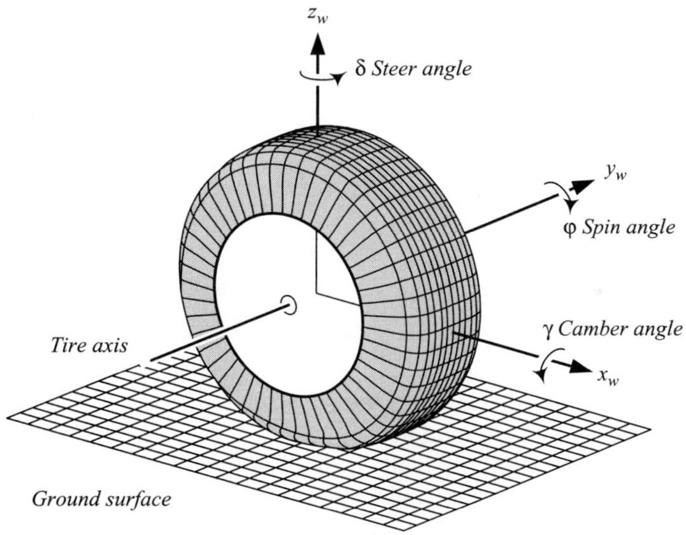

Figure 7.40 Six DOF of a wheel.

tire coordinate frames $T_1(x_{t_1}, y_{t_1}, z_{t_1})$ to tire 1 and $T_4(x_{t_4}, y_{t_4}, z_{t_4})$ to tire 4 at the center of their tireprints. The *tireprint* is the contact area of a tire and the ground.

A wheel, as a rigid body, has six DOF with respect to the vehicle body: three translations and three rotations. To express the motions of a wheel, we attach a wheel coordinate frame $W(ox_w, y_w, z_w)$ to the center of the wheel as is shown in Figure 7.40. The axes x_w, y_w, and z_w indicate the direction of forward, lateral, and vertical translations and rotations. In the position shown in the figure, the rotation about the x_w-axis is the *camber* angle γ, about the y_w-axis is the *spin*, and about the z_w-axis is the *steer* angle δ.

The suspension mechanism eliminates most of the freedoms of a wheel and allows a few. Ideally, translation in the z_w-direction and spin about the y_w-axis are allowed for a nonsteerable wheel, and translation in the z_w-direction, spin about the y_w-axis, and steer about the z_w-axis are allowed for a steerable wheel.

We introduce three coordinate frames to express the orientation of a tire and wheel with respect to the vehicle: the wheel frame W, wheel–body frame C, and tire frame T. The coordinate frame as well as the interested planes of a cambered wheel are illustrated in Figure 7.41.

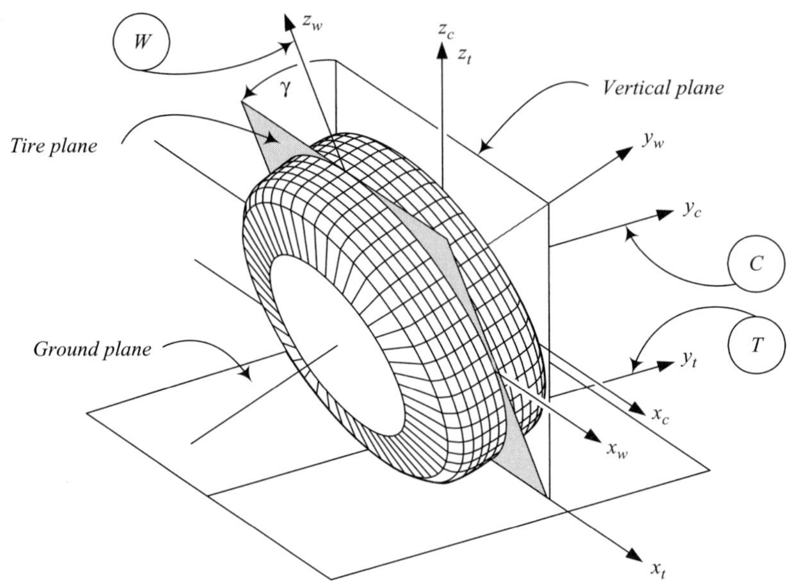

Figure 7.41 Illustration of tire and wheel coordinate frames.

The wheel coordinate frame $W(x_w, y_w, z_w)$ that is attached to the center of a wheel follows every motion of the wheel except the spin. Therefore, the x_w- and z_w-axes are always in the tire plane, while the y_w-axis is always along the spin axis. The *tire plane* is defined by the equivalent disc of a tire.

When the wheel is straight and the W-frame is parallel to the vehicle coordinate frame B, we attach a wheel–body coordinate frame $C(x_c, y_c, z_c)$ at the center of the wheel. The C-frame is attached to the vehicle and parallel to the vehicle coordinate axes. The wheel–body frame C is fixed to the vehicle and does not follow any motion of the wheel.

The tire coordinate frame $T(x_t, y_t, z_t)$ is set up at the center of the tireprint. The z_t-axis is always perpendicular to the ground. The x_t-axis is along the intersection line of the tire plane and the ground. The tire frame does not follows the spin and camber rotations of the tire; however, it follows the steer angle rotation δ about the z_c-axis.

Figure 7.42 illustrates a steered and cambered wheel along with the relative configuration of a wheel–body frame C, a tire frame T, and a wheel frame W. If the

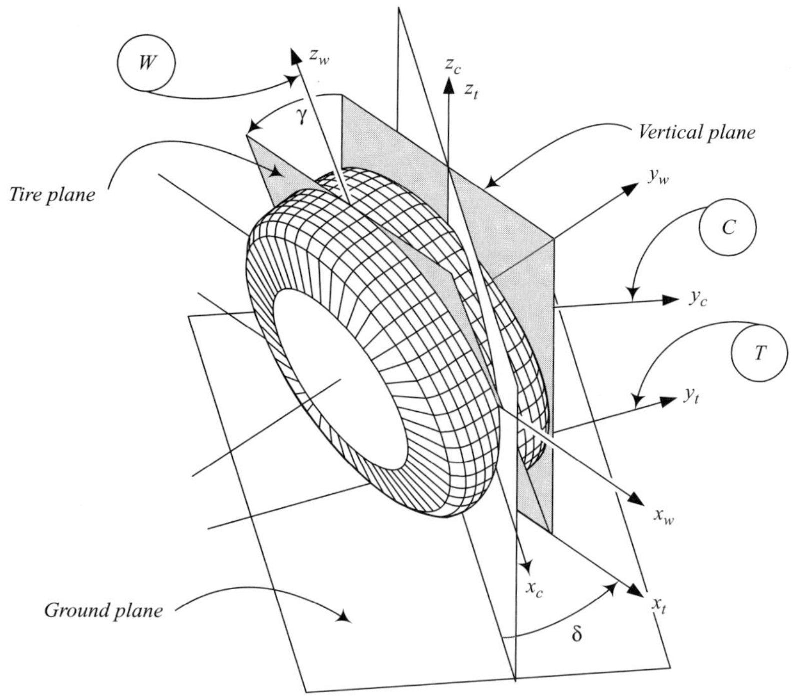

Figure 7.42 Illustration of tire, wheel, and body coordinate frames.

steering axis is along the z_c-axis, then the rotation of the wheel about the z_c-axis is the steer angle δ. However, the steering axis may have any angle and may go through any point of the ground plane.

Example 425 ★ Wheel–Tire Coordinate Frame Transformation Let us use ${}^{T}\mathbf{d}_W$ to indicate the T-expression of the position vector of the origin of the wheel frame relative to the origin of the tire frame. Having the coordinates of a point P in the wheel frame, we can find its coordinates in the tire frame:

$$ {}^{T}\mathbf{r}_P = {}^{T}R_W \, {}^{W}\mathbf{r}_P + {}^{T}\mathbf{d}_W \tag{7.204} $$

If ${}^{W}\mathbf{r}_P$ indicates the position vector of a point P in the wheel frame,

$$ {}^{W}\mathbf{r}_P = \begin{bmatrix} x_P \\ y_P \\ z_P \end{bmatrix} \tag{7.205} $$

then the coordinates of P in the tire frame ${}^{T}\mathbf{r}_P$ are

$$ \begin{aligned} {}^{T}\mathbf{r}_P &= {}^{T}R_W \, {}^{W}\mathbf{r}_P + {}^{T}\mathbf{d} = {}^{T}R_W \, {}^{W}\mathbf{r}_P + {}^{T}R_W \, {}^{W}_{T}\mathbf{d}_W \\ &= \begin{bmatrix} x_P \\ y_P \cos\gamma - R_w \sin\gamma - z_P \sin\gamma \\ R_w \cos\gamma + z_P \cos\gamma + y_P \sin\gamma \end{bmatrix} \end{aligned} \tag{7.206} $$

The vector ${}^W_T\mathbf{d}_W$ is the W-expression of the position vector of the wheel frame in the tire frame, R_w is the radius of the tire, and ${}^T R_W$ is the transformation matrix from the wheel frame W to the tire frame T:

$$ {}^T R_W = \begin{bmatrix} 1 & 0 & 0 \\ 0 & \cos\gamma & -\sin\gamma \\ 0 & \sin\gamma & \cos\gamma \end{bmatrix} \tag{7.207} $$

$$ {}^W_T\mathbf{d}_W = \begin{bmatrix} 0 \\ 0 \\ R_w \end{bmatrix} \tag{7.208} $$

We assume the only motion between W- and T-frames is a camber rotation γ of W about the x_W-axis.

As an example, the center of the wheel ${}^W\mathbf{r}_P = {}^W\mathbf{r}_o = \mathbf{0}$ is the origin of the wheel frame W, that is, at ${}^T\mathbf{r}_o$ in the tire coordinate frame T:

$$ {}^T\mathbf{r}_o = {}^T\mathbf{d}_W = {}^T R_W \, {}^W_T\mathbf{d}_W = \begin{bmatrix} 0 \\ -R_w \sin\gamma \\ R_w \cos\gamma \end{bmatrix} \tag{7.209} $$

The transformation from the wheel frame to the tire frame can also be expressed by a 4×4 homogeneous transformation matrix

$$ \begin{aligned} {}^T\mathbf{r}_P &= {}^T T_W \, {}^W\mathbf{r}_P \\ &= \begin{bmatrix} {}^T R_W & {}^T\mathbf{d}_W \\ 0 & 1 \end{bmatrix} {}^W\mathbf{r}_P \end{aligned} \tag{7.210} $$

where

$$ {}^T T_W = \begin{bmatrix} 1 & 0 & 0 & 0 \\ 0 & \cos\gamma & -\sin\gamma & -R_w \sin\gamma \\ 0 & \sin\gamma & \cos\gamma & R_w \cos\gamma \\ 0 & 0 & 0 & 1 \end{bmatrix} \tag{7.211} $$

The corresponding homogeneous transformation matrix ${}^W T_T$ from the tire frame to the wheel frame would be

$$ {}^W T_T = \begin{bmatrix} {}^W R_T & {}^W\mathbf{d}_T \\ 0 & 1 \end{bmatrix} = \begin{bmatrix} 1 & 0 & 0 & 0 \\ 0 & \cos\gamma & -\sin\gamma & 0 \\ 0 & \sin\gamma & \cos\gamma & -R_w \\ 0 & 0 & 0 & 1 \end{bmatrix} \tag{7.212} $$

It can be checked that ${}^W T_T = {}^T T_W^{-1}$ using the inverse of a homogeneous transformation matrix rule:

$$ \begin{aligned} {}^T T_W^{-1} &= \begin{bmatrix} {}^T R_W & {}^T\mathbf{d}_W \\ 0 & 1 \end{bmatrix}^{-1} = \begin{bmatrix} {}^T R_W^T & -{}^T R_W^T \, {}^T\mathbf{d}_W \\ 0 & 1 \end{bmatrix} \\ &= \begin{bmatrix} {}^W R_T & -{}^W R_T \, {}^T\mathbf{d}_W \\ 0 & 1 \end{bmatrix} \end{aligned} \tag{7.213} $$

Example 426 ★ Tire–Wheel Coordinate Frame Transformation If ${}^{T}\mathbf{r}_P$ indicates the position vector of a point P in the tire coordinate frame,

$$ {}^{T}\mathbf{r}_P = \begin{bmatrix} x_P \\ y_P \\ z_P \end{bmatrix} \tag{7.214} $$

then the position vector ${}^{W}\mathbf{r}_P$ of the point P in the wheel coordinate frame is

$$ {}^{W}\mathbf{r}_P = {}^{W}R_T\, {}^{T}\mathbf{r}_P - {}^{W}_{T}\mathbf{d}_W = \begin{bmatrix} x_P \\ y_P \cos\gamma + z_P \sin\gamma \\ z_P \cos\gamma - R_w - y_P \sin\gamma \end{bmatrix} \tag{7.215} $$

where

$$ {}^{W}R_T = \begin{bmatrix} 1 & 0 & 0 \\ 0 & \cos\gamma & \sin\gamma \\ 0 & -\sin\gamma & \cos\gamma \end{bmatrix} \tag{7.216} $$

$$ {}^{W}\mathbf{d}_T = \begin{bmatrix} 0 \\ 0 \\ R_w \end{bmatrix} \tag{7.217} $$

We may multiply both sides of Equation (7.204) by ${}^{T}R_W^{\mathrm{T}}$ to get

$$ {}^{T}R_W^{\mathrm{T}}\, {}^{T}\mathbf{r}_P = {}^{W}\mathbf{r}_P + {}^{T}R_W^{\mathrm{T}}\, {}^{T}\mathbf{d}_W = {}^{W}\mathbf{r}_P + {}^{W}_{T}\mathbf{d}_W \tag{7.218} $$

$$ {}^{W}\mathbf{r}_P = {}^{W}R_T\, {}^{T}\mathbf{r}_P - {}^{W}_{T}\mathbf{d}_W \tag{7.219} $$

As an example, the center of the tireprint in the wheel frame is at

$$ {}^{W}\mathbf{r}_P = \begin{bmatrix} 1 & 0 & 0 \\ 0 & \cos\gamma & -\sin\gamma \\ 0 & \sin\gamma & \cos\gamma \end{bmatrix}^{\mathrm{T}} \begin{bmatrix} 0 \\ 0 \\ 0 \end{bmatrix} - \begin{bmatrix} 0 \\ 0 \\ R_w \end{bmatrix} = \begin{bmatrix} 0 \\ 0 \\ -R_w \end{bmatrix} \tag{7.220} $$

Example 427 ★ Tire–Wheel Body Frame Transformation The origin of the tire frame is at ${}^{C}\mathbf{d}_T$ in the wheel–body frame:

$$ {}^{C}\mathbf{d}_T = \begin{bmatrix} 0 \\ 0 \\ -R_w \end{bmatrix} \tag{7.221} $$

We assume the tire frame can ideally steer only about the z_c-axis with respect to the wheel–body frame. The associated rotation matrix is

$$ {}^{C}R_T = \begin{bmatrix} \cos\delta & -\sin\delta & 0 \\ \sin\delta & \cos\delta & 0 \\ 0 & 0 & 1 \end{bmatrix} \tag{7.222} $$

Therefore, the transformation between the tire frame and the wheel–body frame can be expressed by

$$ {}^{C}\mathbf{r} = {}^{C}R_T\, {}^{T}\mathbf{r} + {}^{C}\mathbf{d}_T \tag{7.223} $$

or, equivalently, by the homogeneous transformation matrix

$$ {}^{C}T_{T} = \begin{bmatrix} {}^{C}R_{T} & {}^{C}\mathbf{d}_{T} \\ 0 & 1 \end{bmatrix} = \begin{bmatrix} \cos\delta & -\sin\delta & 0 & 0 \\ \sin\delta & \cos\delta & 0 & 0 \\ 0 & 0 & 1 & -R_w \\ 0 & 0 & 0 & 1 \end{bmatrix} \tag{7.224} $$

The homogeneous transformation matrix for the tire frame to the wheel–body frame is

$$ \begin{aligned} {}^{T}T_{C} &= {}^{C}T_{T}^{-1} = \begin{bmatrix} {}^{C}R_{T} & {}^{C}\mathbf{d}_{T} \\ 0 & 1 \end{bmatrix}^{-1} = \begin{bmatrix} {}^{C}R_{T}^{\mathrm{T}} & -{}^{C}R_{T}^{\mathrm{T}}\,{}^{C}\mathbf{d}_{T} \\ 0 & 1 \end{bmatrix} \\ &= \begin{bmatrix} {}^{C}R_{T}^{\mathrm{T}} & -{}^{T}_{C}\mathbf{d}_{T} \\ 0 & 1 \end{bmatrix} = \begin{bmatrix} \cos\delta & \sin\delta & 0 & 0 \\ -\sin\delta & \cos\delta & 0 & 0 \\ 0 & 0 & 1 & R_w \\ 0 & 0 & 0 & 1 \end{bmatrix} \end{aligned} \tag{7.225} $$

Example 428 ★ Wheel Frame to wheel–body Frame Transformation We can find the homogeneous transformation matrix ${}^{C}T_{W}$ between the wheel frame and the wheel–body frame by a combined transformation:

$$ \begin{aligned} {}^{C}T_{W} &= {}^{C}T_{T}\,{}^{T}T_{W} \\ &= \begin{bmatrix} c\delta & -s\delta & 0 & 0 \\ s\delta & c\delta & 0 & 0 \\ 0 & 0 & 1 & -R_w \\ 0 & 0 & 0 & 1 \end{bmatrix} \begin{bmatrix} 1 & 0 & 0 & 0 \\ 0 & c\gamma & -s\gamma & -R_w\sin\gamma \\ 0 & s\gamma & c\gamma & R_w\cos\gamma \\ 0 & 0 & 0 & 1 \end{bmatrix} \\ &= \begin{bmatrix} \cos\delta & -\cos\gamma\sin\delta & \sin\gamma\sin\delta & R_w\sin\gamma\sin\delta \\ \sin\delta & \cos\gamma\cos\delta & -\cos\delta\sin\gamma & -R_w\cos\delta\sin\gamma \\ 0 & \sin\gamma & \cos\gamma & R_w\cos\gamma - R_w \\ 0 & 0 & 0 & 1 \end{bmatrix} \end{aligned} \tag{7.226} $$

If $\mathbf{r}_P$ indicates the position vector of a point P in the wheel coordinate frame,

$$ {}^{W}\mathbf{r}_{P} = \begin{bmatrix} x_P \\ y_P \\ z_P \end{bmatrix} \tag{7.227} $$

then the homogeneous position vector ${}^{C}\mathbf{r}_P$ of the point P in the wheel–body coordinate frame is

$$ \begin{aligned} {}^{C}\mathbf{r}_{P} &= {}^{C}T_{W}\,{}^{W}\mathbf{r}_{P} \\ &= \begin{bmatrix} x_P\cos\delta - y_P\cos\gamma\sin\delta + (R_w + z_P)\sin\gamma\sin\delta \\ x_P\sin\delta + y_P\cos\gamma\cos\delta - (R_w + z_P)\cos\delta\sin\gamma \\ -R_w + (R_w + z_P)\cos\gamma + y_P\sin\gamma \\ 1 \end{bmatrix} \end{aligned} \tag{7.228} $$

The position of the wheel center ${}^{W}\mathbf{r} = \mathbf{0}$ for a cambered and steered wheel is at

$$
{}^{C}\mathbf{r} = {}^{C}T_{W}\,{}^{W}\mathbf{r} = \begin{bmatrix} R_w \sin\gamma \sin\delta \\ -R_w \cos\delta \sin\gamma \\ -R_w(1-\cos\gamma) \\ 1 \end{bmatrix} \tag{7.229}
$$

The equation $x_c = R_w \sin\gamma \sin\delta$ indicates the longitudinal motion of the wheel center, and $y_c = -R_w \cos\delta \sin\gamma$ determines the lateral displacement of the wheel center. The equation $z_c = R_w(\cos\gamma - 1)$ indicates by how much the center of the wheel comes down when the wheel cambers.

If the wheel is not steerable, then $\delta = 0$, and the transformation matrix ${}^{C}T_{W}$ reduces to

$$
{}^{C}T_{W} = \begin{bmatrix} 1 & 0 & 0 & 0 \\ 0 & \cos\gamma & -\sin\gamma & -R_w \sin\gamma \\ 0 & \sin\gamma & \cos\gamma & R_w(\cos\gamma - 1) \\ 0 & 0 & 0 & 1 \end{bmatrix} \tag{7.230}
$$

which yields

$$
\begin{aligned}
{}^{C}\mathbf{r}_P &= {}^{C}T_{W}\,{}^{W}\mathbf{r}_P \\
&= \begin{bmatrix} x_P \\ y_P\cos\gamma - R_w\sin\gamma - z_P\sin\gamma \\ z_P\cos\gamma + y_P\sin\gamma + R_w(\cos\gamma - 1) \\ 1 \end{bmatrix}
\end{aligned} \tag{7.231}
$$

Example 429 ★ Tire-to-Vehicle Coordinate Frame Transformation Assume the origin of the tire coordinate frame T_1 in Figure 7.39 is at ${}^{B}\mathbf{d}_1$,

$$
{}^{B}\mathbf{d}_{T_1} = \begin{bmatrix} a_1 \\ b_1 \\ -h \end{bmatrix} \tag{7.232}
$$

where a_1 is the longitudinal distance between C and the front axle, b_1 is the lateral distance between C and the tireprint of tire 1, and h is the height of C from ground level. If P is a point in the tire frame at

$$
{}^{T_1}\mathbf{r}_P = \begin{bmatrix} x_P \\ y_P \\ z_P \end{bmatrix} \tag{7.233}
$$

then its coordinates in the body frame are

$$
\begin{aligned}
{}^{B}\mathbf{r}_P &= {}^{B}R_{T_1}\,{}^{T_1}\mathbf{r}_P + {}^{B}\mathbf{d}_{T_1} \\
&= \begin{bmatrix} a_1 + x_P\cos\delta_1 - y_P\sin\delta_1 \\ y_P\cos\delta_1 + b_1 + x_P\sin\delta_1 \\ z_P - h \end{bmatrix}
\end{aligned} \tag{7.234}
$$

The rotation matrix ${}^{B}R_{T_1}$ is a result of steering about the z_1-axis,

$$ {}^{B}R_{T_1} = \begin{bmatrix} \cos\delta_1 & -\sin\delta_1 & 0 \\ \sin\delta_1 & \cos\delta_1 & 0 \\ 0 & 0 & 1 \end{bmatrix} \tag{7.235} $$

Employing Equation (7.204), we may examine a wheel point P at

$$ {}^{W}\mathbf{r}_P = \begin{bmatrix} x_P \\ y_P \\ z_P \end{bmatrix} \tag{7.236} $$

and find the body coordinates of the point:

$$ \begin{aligned} {}^{B}\mathbf{r}_P &= {}^{B}R_{T_1}\,{}^{T_1}\mathbf{r}_P + {}^{B}\mathbf{d}_{T_1} \\ &= {}^{B}R_{T_1}\left({}^{T_1}R_W\,{}^{W}\mathbf{r}_P + {}^{T_1}\mathbf{d}_W\right) + {}^{B}\mathbf{d}_{T_1} \\ &= {}^{B}R_{T_1}\,{}^{T_1}R_W\,{}^{W}\mathbf{r}_P + {}^{B}R_{T_1}\,{}^{T_1}\mathbf{d}_W + {}^{B}\mathbf{d}_{T_1} \\ &= {}^{B}R_W\,{}^{W}\mathbf{r}_P + {}^{B}R_{T_1}\,{}^{T_1}\mathbf{d}_W + {}^{B}\mathbf{d}_{T_1} \end{aligned} \tag{7.237} $$

$$ = \begin{bmatrix} a_1 + x_P\cos\delta_1 - y_P\cos\gamma\sin\delta_1 + (R_w + z_P)\sin\gamma\sin\delta_1 \\ x_P\sin\delta_1 + b_1 + y_P\cos\gamma\cos\delta_1 - (R_w + z_P)\cos\delta_1\sin\gamma \\ (R_w + z_P)\cos\gamma + y_P\sin\gamma - h \end{bmatrix} \tag{7.238} $$

where

$$ \begin{aligned} {}^{B}R_W &= {}^{B}R_{T_1}\,{}^{T_1}R_W \\ &= \begin{bmatrix} \cos\delta_1 & -\cos\gamma\sin\delta_1 & \sin\gamma\sin\delta_1 \\ \sin\delta_1 & \cos\gamma\cos\delta_1 & -\cos\delta_1\sin\gamma \\ 0 & \sin\gamma & \cos\gamma \end{bmatrix} \end{aligned} \tag{7.239} $$

$$ {}^{T_1}\mathbf{d}_W = \begin{bmatrix} 0 \\ -R_w\sin\gamma \\ R_w\cos\gamma \end{bmatrix} \tag{7.240} $$

Example 430 ★ Wheel–Body–Vehicle Transformation The wheel–body coordinate frame is always parallel to the vehicle frame. The origin of the wheel–body coordinate frame of wheel 1 is at

$$ {}^{B}\mathbf{d}_{W_1} = \begin{bmatrix} a_1 \\ b_1 \\ -h + R_w \end{bmatrix} \tag{7.241} $$

Hence the transformation between the two frames is only a displacement:

$$ {}^{B}\mathbf{r} = {}^{B}\mathbf{I}_{W_1}\,{}^{W_1}\mathbf{r} + {}^{B}\mathbf{d}_{W_1} \tag{7.242} $$

7.4 ASSEMBLING KINEMATICS

Most modern industrial multibody systems are made modular. They have a main machine and a series of interchangeable tools. The main machine is called the *arm* and is a multibody that holds the main power units and provides a powerful motion of a tip point.

Figure 7.43 illustrates an example of a single-DOF arm as the base for an R‖R planar manipulator. This arm can rotate relative to the global frame by a motor at M_1 and carries another motor at M_2.

The changeable parts are called the *wrist*. They are complex multibodies made to do special jobs. The base of the wrist will be attached to the tip point of the arm. The wrist, which is the actual operator of the multibody, may also be called the *end effector*, *gripper*, *hand*, or *tool*. Figure 7.44 illustrates a sample of a planar wrist that is supposed to be attached to the arm in Figure 7.43.

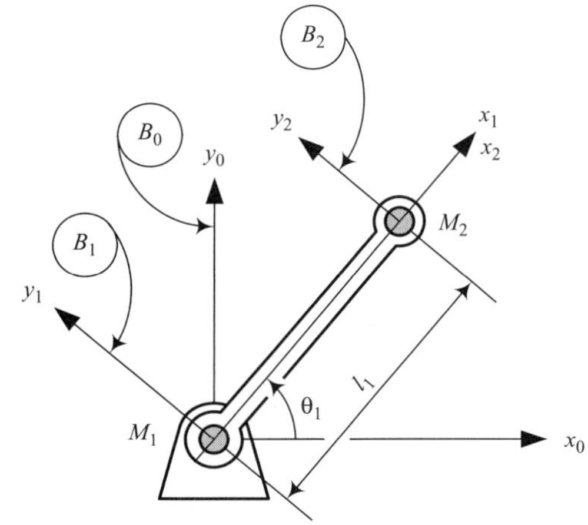

Figure 7.43 A single-DOF arm as the base for a R‖R planar manipulator.

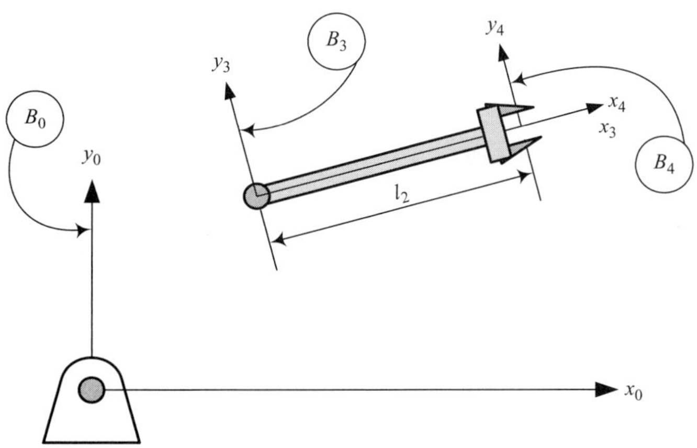

Figure 7.44 A planar wrist.

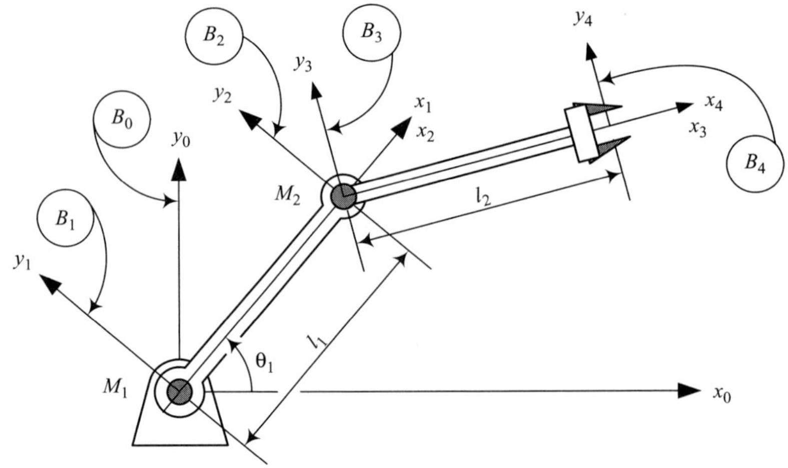

Figure 7.45 The R‖R planar manipulator that is made by assembling a wrist and an arm.

To solve the kinematics of a modular multibody, we consider the arm and the wrist as individual multibodies. However, we attach a Sina coordinate frame at the tip point of the arm and another Sina frame at the base point of the wrist. The coordinate frame at the arm's tip point is called the *takht*, and the coordinate frame at the base of the wrist is called the *neshin* frame. Mating the neshin and takht frames assembles the multibody kinematically. The kinematic mating process of the wrist and arm is called *assembling*.

The coordinate frame B_2 in Figure 7.43 is the takht frame of the arm, and the coordinate frame B_3 in Figure 7.44 is the neshin frame of the wrist. The R‖R planar manipulator that is made by assembling the wrist and arm is shown in Figure 7.45.

The assembled multibody always has some extra coordinate frames. The extra frames require extra transformation matrices that can increase the number of required mathematical calculations. We recommend eliminating the neshin coordinate frame and keep the takht frame to have a Sina frame at the connection point. However, as long as the transformation matrices between the frames are known, having extra coordinate frames is not a significant disadvantage.

The Persian words *takht* and *neshin* mean "chair" and "sit," respectively.

Example 431 Spherical Wrist The spherical wrist is a combined multibody that simulates a spherical joint. Such a combination gives three rotational DOF to a link. The link may be a gripper if the wrist is being used in an assembling and production line. Figure 7.46 illustrates a sample spherical wrist. The axes z_5, z_6, and z_7 are the rotation axes of the wrists.

The coordinate frame B_4 is the neshin frame of the wrist, and B_9 is the tool frame of the wrist. The coordinate frame B_5 is fixed with respect to B_4 and may be considered as the wrist base frame and is called the *wrist dead frame*. B_6 is the frame of the first link in shape ⊏ that rotates about z_5. B_7 is the frame of the middle cylindrical link ◎ that rotates about z_6. B_8 is called the *wrist living frame* and is the fixed frame to the long bar that supports the gripper. B_8 rotates about z_7. Therefore, B_6, B_7, and B_8 are

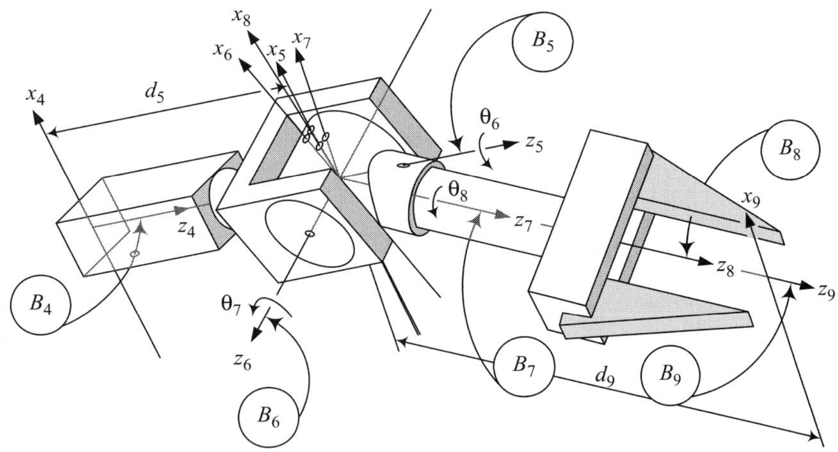

Figure 7.46 A sample of a spherical wrist.

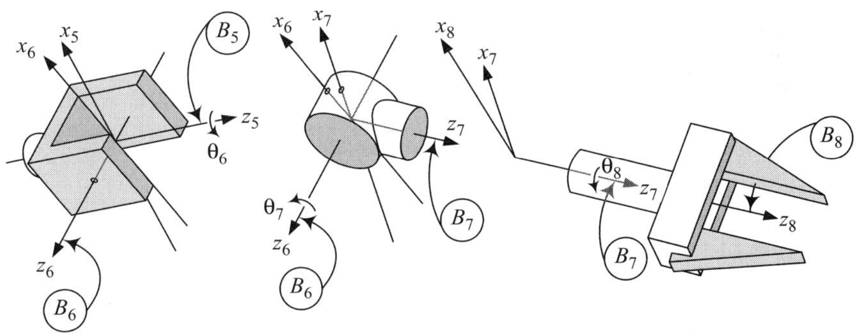

Figure 7.47 The rotating bodies of a spherical wrist.

the frames of the three rotating links. The joint axes of the wrist are intersecting at a common point called the *wrist point*. These three links and their associated frames are shown in Figure 7.47.

This spherical wrist is a common wrist configuration in robotic and multibody applications. The spherical wrist of Figure 7.46 is made of a link R⊢R(−90) attached to another link R⊢R(90) that finally is attached to a spinning gripper link R∥R(0). The final frame B_9 is always parallel to B_8 and is attached to the gripper at a distance d_9 from the wrist point. The final frame B_9 is called the *gripper* or *tool frame*. It is set at a symmetric point between the fingers of an empty gripper.

The wrist will be attached to the final link of a manipulator arm, which usually provides three DOF for positioning the wrist point at a desired coordinate in space.

The kinematic analysis of the spherical wrist begins by deriving the required transformation matrices ${}^5T_6$, ${}^6T_7$, and ${}^7T_8$. The matrix ${}^4T_5$ determines how the wrist dead frame B_5 relates to the neshin frame B_4, and the matrix ${}^8T_9$ determines how the tool frame B_9 connects to the wrist living frame B_8, as are shown in Figure 7.48.

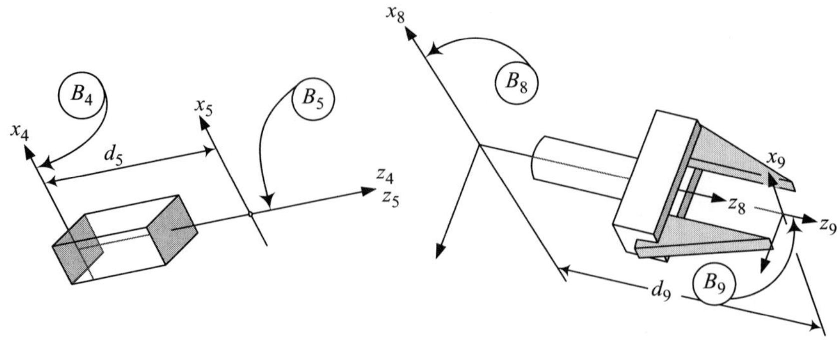

Figure 7.48 The neshin, dead, living, and gripper frames of a spherical wrist.

Link (6) is an R⊢R(−90), link (7) is an R⊢R(90), and link (8) is an R∥R(0); therefore,

$$ {}^5T_6 = \begin{bmatrix} \cos\theta_6 & 0 & -\sin\theta_6 & 0 \\ \sin\theta_6 & 0 & \cos\theta_6 & 0 \\ 0 & -1 & 0 & 0 \\ 0 & 0 & 0 & 1 \end{bmatrix} \tag{7.243} $$

$$ {}^6T_7 = \begin{bmatrix} \cos\theta_7 & 0 & \sin\theta_7 & 0 \\ \sin\theta_7 & 0 & -\cos\theta_7 & 0 \\ 0 & 1 & 0 & 0 \\ 0 & 0 & 0 & 1 \end{bmatrix} \tag{7.244} $$

$$ {}^7T_8 = \begin{bmatrix} \cos\theta_8 & -\sin\theta_8 & 0 & 0 \\ \sin\theta_8 & \cos\theta_8 & 0 & 0 \\ 0 & 0 & 1 & 0 \\ 0 & 0 & 0 & 1 \end{bmatrix} \tag{7.245} $$

The matrix ${}^5T_8 = {}^5T_6\,{}^6T_7\,{}^7T_8$ provides the spherical wrist's transformation. ${}^5T_8$ must reduce to an identity matrix when the wrist is at the rest position and all the angular variables are zero:

$$ \begin{aligned} {}^5T_8 &= {}^5T_6\,{}^6T_7\,{}^7T_8 \\ &= \begin{bmatrix} c\theta_6 c\theta_7 c\theta_8 - s\theta_6 s\theta_8 & -c\theta_8 s\theta_6 - c\theta_6 c\theta_7 s\theta_8 & c\theta_6 s\theta_7 & 0 \\ c\theta_6 s\theta_8 + c\theta_7 c\theta_8 s\theta_6 & c\theta_6 c\theta_8 - c\theta_7 s\theta_6 s\theta_8 & s\theta_6 s\theta_7 & 0 \\ -c\theta_8 s\theta_7 & s\theta_7 s\theta_8 & c\theta_7 & 0 \\ 0 & 0 & 0 & 1 \end{bmatrix} \end{aligned} \tag{7.246} $$

The transformation of the wrist dead frame B_5 to the neshin frame B_4 and the transformation of the wrist living frame B_8 to the tool frame B_9 are just translations

d_5 and d_9, respectively:

$$
{}^4T_5 = \begin{bmatrix} 1 & 0 & 0 & 0 \\ 0 & 1 & 0 & 0 \\ 0 & 0 & 1 & d_5 \\ 0 & 0 & 0 & 1 \end{bmatrix} \tag{7.247}
$$

$$
{}^8T_9 = \begin{bmatrix} 1 & 0 & 0 & 0 \\ 0 & 1 & 0 & 0 \\ 0 & 0 & 1 & d_9 \\ 0 & 0 & 0 & 1 \end{bmatrix} \tag{7.248}
$$

Example 432 Spherical Arm The spherical arm is a multibody that simulates the spherical coordinate for positioning a point in a 3D space. Figure 7.49 illustrates a spherical arm. The coordinate frame B_0 is the global or base frame of the arm. Link (1) can turn about z_0 and link (2) can turn about z_1, which is perpendicular to z_0. These two rotations simulate the two angular motions of the spherical coordinates. The radial coordinate is simulated by link (3), which has a prismatic joint with link (2). There is also a takht coordinate frame at the tip point of the arm at which a wrist can be attached.

Link (1) in Figure 7.49 is an R⊢R(90), link (2) is also an R⊢P(90), and link (3) is an P‖R(0); therefore,

$$
{}^0T_1 = \begin{bmatrix} \cos\theta_1 & 0 & \sin\theta_1 & 0 \\ \sin\theta_1 & 0 & -\cos\theta_1 & 0 \\ 0 & 1 & 0 & d_1 \\ 0 & 0 & 0 & 1 \end{bmatrix} \tag{7.249}
$$

$$
{}^1T_2 = \begin{bmatrix} \cos\theta_2 & 0 & \sin\theta_2 & 0 \\ \sin\theta_2 & 0 & -\cos\theta_2 & 0 \\ 0 & 1 & 0 & 0 \\ 0 & 0 & 0 & 1 \end{bmatrix} \tag{7.250}
$$

$$
{}^2T_3 = \begin{bmatrix} 1 & 0 & 0 & 0 \\ 0 & 1 & 0 & 0 \\ 0 & 0 & 1 & d_3 \\ 0 & 0 & 0 & 1 \end{bmatrix} \tag{7.251}
$$

The transformation matrix from B_3 to the takht frame B_4 is only a translation d_4:

$$
{}^3T_4 = \begin{bmatrix} 1 & 0 & 0 & 0 \\ 0 & 1 & 0 & 0 \\ 0 & 0 & 1 & d_4 \\ 0 & 0 & 0 & 1 \end{bmatrix} \tag{7.252}
$$

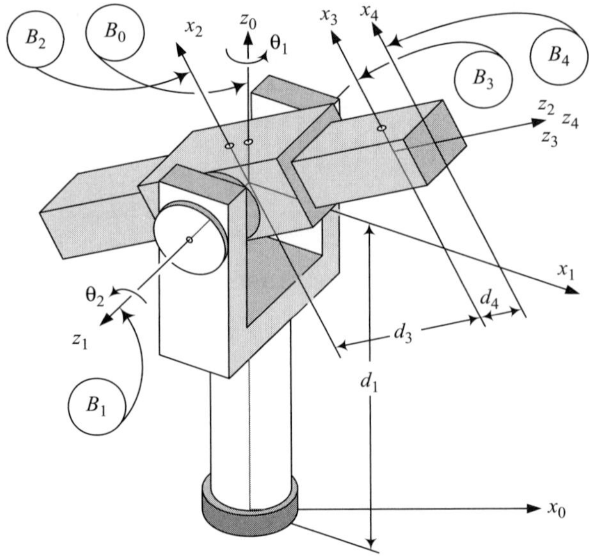

Figure 7.49 A spherical arm.

The transformation matrix of the takht frame B_4 to the base frame B_0 is

$$
\begin{aligned}
{}^0T_4 &= {}^0T_1\,{}^1T_2\,{}^2T_3\,{}^3T_4 \\
&= \begin{bmatrix} c\theta_1 c\theta_2 & s\theta_1 & c\theta_1 s\theta_2 & (d_3+d_4)(c\theta_1 s\theta_2) \\ c\theta_2 s\theta_1 & -c\theta_1 & s\theta_1 s\theta_2 & (d_3+d_4)(s\theta_2 s\theta_1) \\ s\theta_2 & 0 & -c\theta_2 & d_1 - d_3 c\theta_2 - d_4 c\theta_2 \\ 0 & 0 & 0 & 1 \end{bmatrix}
\end{aligned} \tag{7.253}
$$

At the rest position ${}^0T_4$ reduces to

$$
{}^0T_4 = \begin{bmatrix} 1 & 0 & 0 & 0 \\ 0 & -1 & 0 & 0 \\ 0 & 0 & -1 & d_1 - d_4 \\ 0 & 0 & 0 & 1 \end{bmatrix} \tag{7.254}
$$

As a general recommendation, the setup of the DH coordinate frames is better such that the overall transformation matrix at the rest position becomes an identity matrix. If we rearrange the coordinate frame of link (1) to make it an R⊢R(−90), then

$$
{}^0T_1 = \begin{bmatrix} \cos\theta_1 & 0 & -\sin\theta_1 & 0 \\ \sin\theta_1 & 0 & \cos\theta_1 & 0 \\ 0 & -1 & 0 & d_1 \\ 0 & 0 & 0 & 1 \end{bmatrix} \tag{7.255}
$$

and the overall transformation matrix at the rest position becomes an identity matrix.

To make link (1) an R⊢R(−90), we may reverse the direction of the z_1- or x_1-axis. Figure 7.50 illustrates the new arrangement of the coordinate frames. Therefore, the

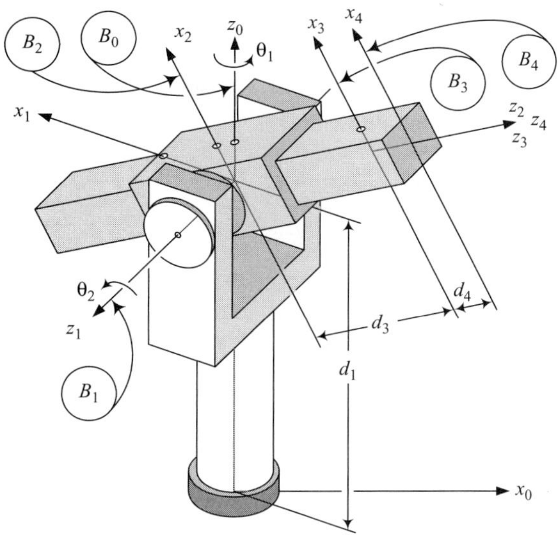

Figure 7.50 A spherical arm with the arrangement of coordinate frames such that the overall transformation matrix reduces to an identity at the rest position.

transformation matrix of the takht frame B_4 to the base frame B_0 at the rest position reduces to

$$
{}^{0}T_{4} = \begin{bmatrix} 1 & 0 & 0 & 0 \\ 0 & 1 & 0 & 0 \\ 0 & 0 & 1 & d_1 + d_4 \\ 0 & 0 & 0 & 1 \end{bmatrix} \tag{7.256}
$$

Example 433 Assembling of a Spherical Wrist and Arm Let us kinematically assemble the spherical wrist of Example 431 to the spherical arm of Example 432. The wrist, arm, and their associated DH coordinate frames are shown in Figures 7.46 and 7.50, respectively.

Assembling of multibodies is a *kinematic surgery* in which during an operation we attach a multibody to the other. In this example we attach a spherical hand to a spherical arm to make an arm–hand manipulator.

The takht coordinate frame B_4 of the arm and the neshin coordinate frame B_4 of the wrist are exactly the same. Therefore, we may assemble the arm and wrist by matching these two frames and make a combined arm–wrist multibody as is shown in Figure 7.51. However, in general, the takht and neshin coordinate frames may have different labels and there may be a constant transformation matrix between them.

The forward kinematics of the tool frame B_9 can be found by matrix multiplication:

$$
{}^{0}T_{9} = {}^{0}T_{1}\,{}^{1}T_{2}\,{}^{2}T_{3}\,{}^{3}T_{4}\,{}^{4}T_{5}\,{}^{5}T_{6}\,{}^{6}T_{7}\,{}^{7}T_{8}\,{}^{8}T_{9} \tag{7.257}
$$

The matrices ${}^{i-1}T_i$ are given in Examples 431 and 432.

Although we can eliminate the coordinate frames B_3 and B_4 to reduce the total number of frames and simplify the matrix calculations, we prefer to keep them and

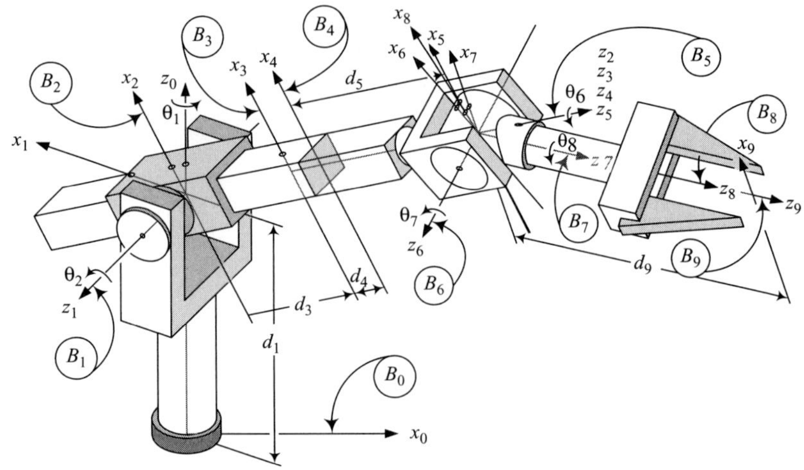

Figure 7.51 Assembling a spherical hand and arm.

simplify the changing process of the wrist. However, if this assembled multibody is supposed to work for a while, we may do the elimination and simplify the multibody to the one in Figure 7.52. We should mathematically substitute the eliminated frames B_3 and B_4 by a transformation matrix ${}^2T_5$:

$$
{}^2T_5 = {}^2T_3 \, {}^3T_4 \, {}^4T_5 = \begin{bmatrix} 1 & 0 & 0 & 0 \\ 0 & 1 & 0 & 0 \\ 0 & 0 & 1 & d_6 \\ 0 & 0 & 0 & 1 \end{bmatrix} \tag{7.258}
$$

$$
d_6 = d_3 + d_4 + d_5 \tag{7.259}
$$

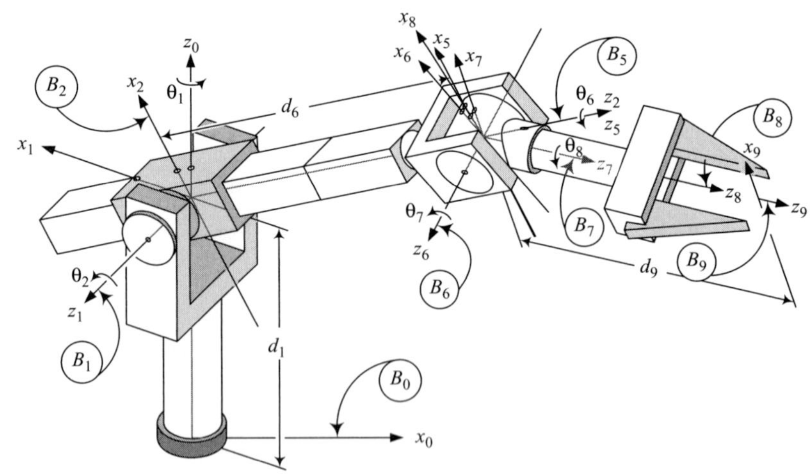

Figure 7.52 Simplification of the coordinate frames for an assembled spherical hand and arm.

Now the forward kinematics of the tool frame B_9 becomes

$$ {}^{0}T_{9} = {}^{0}T_{1}\,{}^{1}T_{2}\,{}^{2}T_{5}\,{}^{5}T_{6}\,{}^{6}T_{7}\,{}^{7}T_{8}\,{}^{8}T_{9} \tag{7.260} $$

Example 434 ★ Best Spherical Wrist A spherical wrist is a combination of links and joints that provides three rotations about three axes intersecting at the wrist point. The spherical wrist is a practical device for an industrial multibody to adjust and control the orientation of a tool.

Figure 7.53 illustrates a wrist with an actual spherical joint at the rest position. The spherical joint is connecting two links: the *forearm* and the *hand*. These two links should be simulated by a wrist with three rotational DOF. The axes of the forearm and hand are assumed to be colinear at the rest position. The axis of the hand is also called the gripper axis.

At the wrist point, we may define two coordinate frames. The first one, the wrist dead frame, is attached to the forearm link, and the second, the wrist living frame, is attached to the hand link. We also introduce a tool or gripper frame. The tool frame of the wrist is denoted by three vectors, $\mathbf{a} \equiv \hat{k}$, $\mathbf{s} \equiv \hat{\imath}$, $\mathbf{n} \equiv \hat{\jmath}$. It is set at a symmetric point between the fingers of an empty hand or at the tip of the tools held by the fingers. The vector $\mathbf{n}$ is called *tilt* and is the normal vector perpendicular to the fingers or jaws. The vector $\mathbf{s}$ is called *twist* and the vector $\mathbf{a}$ is called *turn*. The tool frame is shown in Figure 7.53.

To classify the spherical wrists, let us assume that the rotations of a spherical joint can be decomposed into three rotations about three orthogonal axes. We call the rotations roll, pitch, and yaw, as shown in Figure 7.53. The *roll* is any rotation that turns the gripper about its axis when the wrist is at the rest position. The gripper axis defines a perpendicular plane to the axis called the *gripper wall*. The *pitch* and *yaw* are rotations about two perpendicular axes in the gripper wall at the wrist point. We can consider the first rotation about an axis in the gripper wall as pitch and the rotation about the axis perpendicular to the first as yaw. The roll, pitch, and yaw rotations are defined at the rest position of the wrist.

Practically, we provide the roll, pitch, and yaw rotations by introducing two links and three frames between the dead and living frames. The links will be connected by

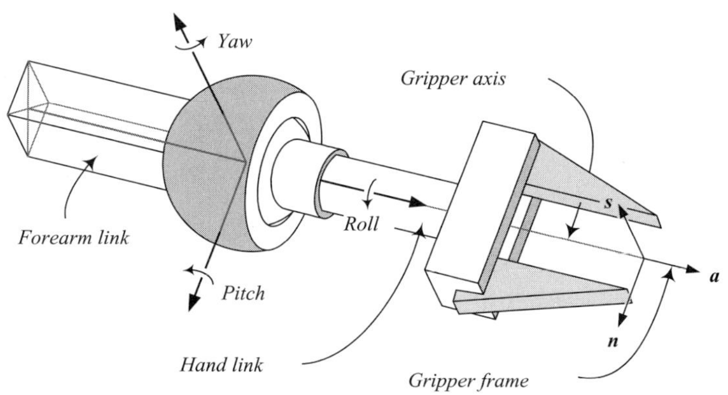

Figure 7.53 The roll, pitch, and yaw rotations of a spherical wrist joint.

three revolute joints. The joint axes must always intersect at the wrist point and be perpendicular when the wrist is at the rest position. The kinematic analysis of such wrists should be conducted by attaching a proper DH coordinate frame to each link. Then the associated DH transformation matrices can be combined to develop the wrist transformation matrices. Figures 7.46–7.48 show an example of simulating a spherical wrist, and Example 431 gives the procedure.

We can also analyze a spherical wrist by defining three coordinate frames at the wrist point and determine their relative transformations. The first orthogonal frame $B_0\,(x_0, y_0, z_0)$ is fixed to the forearm and acts as the wrist dead frame such that z_0 is the joint axis of the forearm and a rotating link. The rotating link is the first wrist link and the joint is the first wrist joint. The directions of the axes x_0 and y_0 are arbitrary.

The second frame $B_1\,(x_1, y_1, z_1)$ is defined such that z_1 is along the gripper axis at the rest position and x_1 is the axis of the second joint. Frame B_1 always turns φ about z_0 and θ about x_1 relative to B_0.

The third frame $B_2\,(x_2, y_2, z_2)$ is the wrist living frame and is defined such that z_2 is always along the gripper axis. If the third joint is a roll, then z_2 is the joint axis; otherwise the third joint is a yaw and x_2 is the joint axis. Therefore, B_2 always turns ψ about z_2 or x_2 relative to B_1.

The coordinate frames B_1 and B_2 indicate virtual rigid bodies and do not necessarily indicate actual bodies. Introducing the coordinate frames B_1 and B_2 simplifies the spherical wrist kinematics by not seeing the interior links of the wrist.

Considering the definition and rotations of B_2 relative to B_1 and B_1 relative to B_0, there are only three types of practical spherical wrists, as classified in Table 7.9. These three wrists are shown in Figures 7.54–7.56.

Table 7.9 Spherical Wrist Classifications

Type	Rotation Order
1	Roll–pitch–roll
2	Roll–pitch–yaw
3	Pitch–yaw–roll

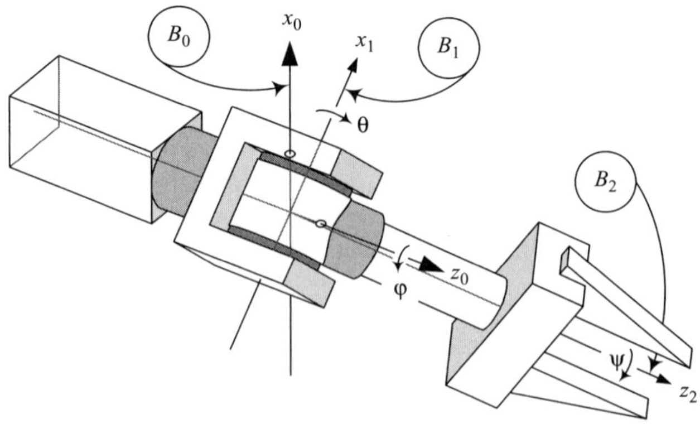

Figure 7.54 Spherical wrist of the roll–pitch–roll type.

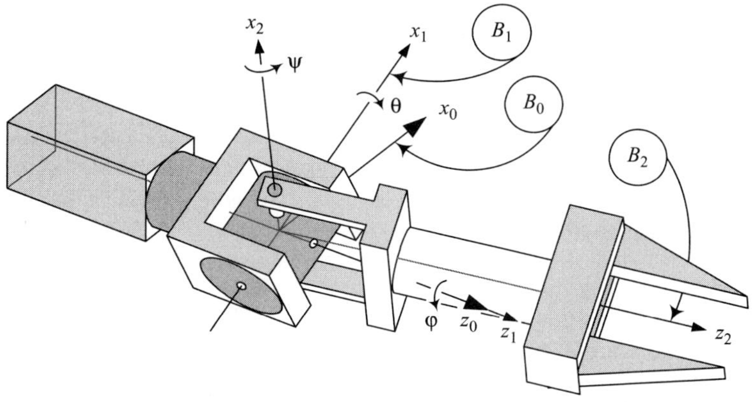

Figure 7.55 Spherical wrist of the roll–pitch–yaw type.

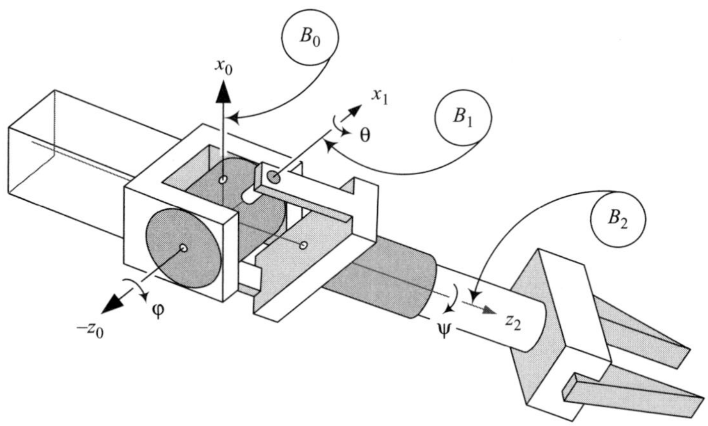

Figure 7.56 Spherical wrist of the pitch–yaw–roll type.

Proof: The first rotation is always about a fixed axis on the forearm link. It is a roll if the joint axis is along the forearm axis, which would also be along the gripper axis at the rest position. If the first axis of rotation is perpendicular to the forearm axis, then we may consider the first rotation as a pitch. So far, the first rotation can be a roll or a pitch.

When the first rotation is a roll, the second rotation is perpendicular to the forearm axis and is considered a pitch. Now there are two possible situations for the third rotation: a roll if it is about the gripper axis and a yaw if it is perpendicular to the axis of the second rotation.

If the first rotation is a pitch, the second rotation can be a roll or a yaw. If it is a yaw, then the third rotation must be a roll to have independent rotations. If it is a roll, then the third rotation must be a yaw. The rotations pitch–yaw–roll and pitch–roll–yaw are not distinguishable, and we may pick pitch–yaw–roll as the only possible spherical wrist with the first rotation as a pitch. ■

Figure 7.54 illustrates a roll–pitch–yaw, Figure 7.55 a roll–pitch–yaw, and Figure 7.56 a pitch–yaw–roll spherical wrist.

From a practical point of view, we prefer a roll as the third rotation to have the best obstacle avoidance and approachability. Therefore, types 1 and 3 are better than type 2. Furthermore, type 1 is the only wrist that can have two joints with 360 deg possible rotation. Therefore type 1, roll–pitch–roll, is the best spherical wrist. The roll–pitch–roll wrist may also be called the Eulerian spherical wrist because it follows the order of Euler angles $Z - x - z$.

Example 435 ★ Type 1, Roll–Pitch–Roll Spherical Wrist Figure 7.54 illustrates a spherical wrist of type 1, roll–pitch–yaw. Frame B_0 indicates its dead coordinate frame and B_2 its living coordinate frame. The transformation matrix ${}^0R_1$ is a rotation φ about the global axis z_0 followed by a rotation θ about the x_1-axis:

$$\begin{aligned}
{}^0R_1 &= {}^1R_0^{\mathrm{T}} = \left[R_{x_1,\theta}\, R_{z_0,\varphi}^{\mathrm{T}}\right]^{\mathrm{T}} = \left[R_{x,\theta}\, R_{Z,\varphi}^{\mathrm{T}}\right]^{\mathrm{T}} \\
&= \left[\begin{bmatrix} 1 & 0 & 0 \\ 0 & \cos\theta & \sin\theta \\ 0 & -\sin\theta & \cos\theta \end{bmatrix} \begin{bmatrix} \cos\varphi & -\sin\varphi & 0 \\ \sin\varphi & \cos\varphi & 0 \\ 0 & 0 & 1 \end{bmatrix}^{\mathrm{T}}\right]^{\mathrm{T}} \\
&= \begin{bmatrix} \cos\varphi & -\cos\theta\sin\varphi & \sin\theta\sin\varphi \\ \sin\varphi & \cos\theta\cos\varphi & -\cos\varphi\sin\theta \\ 0 & \sin\theta & \cos\theta \end{bmatrix}
\end{aligned} \tag{7.261}$$

The transformation matrix ${}^1R_2$ is a rotation ψ about the local axis z_2:

$${}^1R_2 = {}^2R_1^{\mathrm{T}} = R_{z_2,\psi}^{\mathrm{T}} = R_{z,\psi}^{\mathrm{T}} = \begin{bmatrix} \cos\psi & -\sin\psi & 0 \\ \sin\psi & \cos\psi & 0 \\ 0 & 0 & 1 \end{bmatrix} \tag{7.262}$$

Therefore, the transformation matrix between the living and dead wrist frames is

$$\begin{aligned}
{}^0R_2 &= {}^0R_1\, {}^1R_2 = \left[R_{x_1,\theta}\, R_{z_0,\varphi}^{\mathrm{T}}\right]^{\mathrm{T}} R_{z_2,\psi}^{\mathrm{T}} = R_{z_0,\varphi}\, R_{x_1,\theta}^{\mathrm{T}}\, R_{z_2,\psi}^{\mathrm{T}} \\
&= R_{Z,\varphi}\, R_{x,\theta}^{\mathrm{T}}\, R_{z,\psi}^{\mathrm{T}} \\
&= \begin{bmatrix} c\psi c\varphi - c\theta s\psi s\varphi & -c\varphi s\psi - c\theta c\psi s\varphi & s\theta s\varphi \\ c\psi s\varphi + c\theta c\varphi s\psi & c\theta c\psi c\varphi - s\psi s\varphi & -c\varphi s\theta \\ s\theta s\psi & c\psi s\theta & c\theta \end{bmatrix}
\end{aligned} \tag{7.263}$$

Example 436 ★ Type 2, Roll–Pitch–Yaw Spherical Wrist Figure 7.55 illustrates a spherical wrist of type 2, roll–pitch–yaw. Frame B_0 indicates the wrist dead coordinate frame. The main kinematic disdvantage of this type of spherical wrist is that z_1 is not fixed to the gripper. However, we attach a coordinate frame B_2 to the gripper as the wrist living frame such that z_2 is on the gripper axis and x_2 is the third joint axis. The transformation between B_2 and B_1 is only a rotation ψ about the x_2-axis:

$$
{}^1R_2 = R_{x_2,\psi}^{\mathrm{T}} = R_{x,\psi}^{\mathrm{T}} = \begin{bmatrix} 1 & 0 & 0 \\ 0 & \cos\psi & \sin\psi \\ 0 & -\sin\psi & \cos\psi \end{bmatrix}^{\mathrm{T}} \tag{7.264}
$$

To determine the transformation matrix ${}^0R_1$, we turn B_1 first φ deg about the z_0-axis and then θ deg about the x_1-axis:

$$
\begin{aligned}
{}^0R_1 &= {}^1R_0^{\mathrm{T}} = \left[R_{x_1,\theta}\, R_{z_0,\varphi}^{\mathrm{T}}\right]^{\mathrm{T}} = \left[R_{x,\theta}\, R_{Z,\varphi}^{\mathrm{T}}\right]^{\mathrm{T}} \\
&= \left[\begin{bmatrix} 1 & 0 & 0 \\ 0 & c\theta & s\theta \\ 0 & -s\theta & c\theta \end{bmatrix} \begin{bmatrix} c\varphi & -s\varphi & 0 \\ s\varphi & c\varphi & 0 \\ 0 & 0 & 1 \end{bmatrix}^{\mathrm{T}}\right]^{\mathrm{T}} \\
&= \begin{bmatrix} \cos\varphi & -\cos\theta\sin\varphi & \sin\theta\sin\varphi \\ \sin\varphi & \cos\theta\cos\varphi & -\cos\varphi\sin\theta \\ 0 & \sin\theta & \cos\theta \end{bmatrix}
\end{aligned} \tag{7.265}
$$

Therefore, the transformation matrix between the living and dead wrist frames is

$$
\begin{aligned}
{}^0R_2 &= {}^0R_1\, {}^1R_2 = \left[R_{x_1,\theta}\, R_{z_0,\varphi}^{\mathrm{T}}\right]^{\mathrm{T}} R_{x_2,\psi}^{\mathrm{T}} = R_{z_0,\varphi}\, R_{x_1,\theta}^{\mathrm{T}}\, R_{x_2,\psi}^{\mathrm{T}} \\
&= R_{Z,\varphi}\, R_{x,\theta}^{\mathrm{T}}\, R_{x,\psi}^{\mathrm{T}} \\
&= \begin{bmatrix} c\varphi & s\theta s\psi s\varphi - c\theta c\psi s\varphi & c\theta s\psi s\varphi + c\psi s\theta s\varphi \\ s\varphi & c\theta c\psi c\varphi - c\varphi s\theta s\psi & -c\theta c\varphi s\psi - c\psi c\varphi s\theta \\ 0 & c\theta s\psi + c\psi s\theta & c\theta c\psi - s\theta s\psi \end{bmatrix}
\end{aligned} \tag{7.266}
$$

Example 437 ★ Type 3, Pitch–Yaw–Roll Spherical Wrist Figure 7.56 illustrates a spherical wrist of type 3, pitch–yaw–roll. Frame B_0 indicates its dead coordinate frame and B_2 its living coordinate frame. The transformation matrix ${}^1R_2$ is a rotation ψ about the local z_2-axis:

$$
{}^1R_2 = {}^2R_1^{\mathrm{T}} = R_{z_2,\psi}^{\mathrm{T}} = R_{z,\psi}^{\mathrm{T}} = \begin{bmatrix} \cos\psi & -\sin\psi & 0 \\ \sin\psi & \cos\psi & 0 \\ 0 & 0 & 1 \end{bmatrix} \tag{7.267}
$$

To determine the transformation matrix ${}^0R_1$, we turn B_1 first φ deg about the z_0-axis and then θ deg about the x_1-axis:

$$
\begin{aligned}
{}^0R_1 &= {}^1R_0^{\mathrm{T}} = \left[R_{x_1,\theta}\, R_{z_0,\varphi}^{\mathrm{T}}\right]^{\mathrm{T}} = \left[R_{x,\theta}\, R_{Z,\varphi}^{\mathrm{T}}\right]^{\mathrm{T}} \\
&= \left[\begin{bmatrix} 1 & 0 & 0 \\ 0 & c\theta & s\theta \\ 0 & -s\theta & c\theta \end{bmatrix} \begin{bmatrix} c\varphi & -s\varphi & 0 \\ s\varphi & c\varphi & 0 \\ 0 & 0 & 1 \end{bmatrix}^{\mathrm{T}}\right]^{\mathrm{T}} \\
&= \begin{bmatrix} \cos\varphi & -\cos\theta\sin\varphi & \sin\theta\sin\varphi \\ \sin\varphi & \cos\theta\cos\varphi & -\cos\varphi\sin\theta \\ 0 & \sin\theta & \cos\theta \end{bmatrix}
\end{aligned} \tag{7.268}
$$

Therefore, the transformation matrix between the living and dead wrist frames is

$$
\begin{aligned}
{}^{0}R_{2} &= {}^{0}R_{1}\,{}^{1}R_{2} = \left[R_{x_1,\theta}\,R^{\mathrm{T}}_{z_0,\varphi}\right]^{\mathrm{T}} R^{\mathrm{T}}_{z_2,\psi} = R_{z_0,\varphi}\,R^{\mathrm{T}}_{x_1,\theta}\,R^{\mathrm{T}}_{z_2,\psi} \\
&= R_{Z,\varphi}\,R^{\mathrm{T}}_{x,\theta}\,R^{\mathrm{T}}_{z,\psi} \\
&= \begin{bmatrix} c\psi c\varphi - c\theta s\psi s\varphi & -c\varphi s\psi - c\theta c\psi s\varphi & s\theta s\varphi \\ c\psi s\varphi + c\theta c\varphi s\psi & c\theta c\psi c\varphi - s\psi s\varphi & -c\varphi s\theta \\ s\theta s\psi & c\psi s\theta & c\theta \end{bmatrix}
\end{aligned}
\tag{7.269}
$$

7.5 ★ ORDER-FREE ROTATION

We introduce a theorem to generalize and simplify the applied rotation transformation of multibodies. Most industrial multibodies with a common fixed point are similar to spherical wrists with one, two, or three rotary actuators that can work independently. Practically, the order of actuation of the motors is not important. The orientation of the body would be the same no matter which motor acts first. Here is the mathematical explanation.

Theorem *Order-Free Rotations* Consider n bodies $B_1, B_2, \dots, B_n$ with a common fixed point. The body B_1 carries the bodies $B_2, B_3, \dots, B_n$ and turns α_1 about a fixed axis in G. The body B_2 carries the bodies $B_3, B_4, \dots, B_n$ and turns α_2 about a fixed axis in B_1, and so on. The transformation matrix ${}^{G}R_n = {}^{n}R_G^{\mathrm{T}}$ is independent of the order of rotations $\alpha_1, \alpha_2, \dots, \alpha_n$.

Proof: Let us consider a global frame $G\,(OXYZ)$ and two body frames $B_1\,(Ox_1y_1z_1)$ and $B_2\,(Ox_2y_2z_2)$ with a common origin. The body B_1 carries B_2 and rotates with respect to G. The body B_2 rotates in B_1.

The first rotation α of B_1 of such a multibody is always about a globally fixed axis ${}^{G}\hat{u}_1$. The second motion would then be a rotation β of B_2 about a fixed axis ${}^{1}\hat{u}_2$ in B_1. However, we may consider the second rotation β about a local axis ${}^{2}\hat{u}_2$ in B_2. A transformation matrix ${}^{2}R_1$ for rotation β about a local axis ${}^{2}\hat{u}_2$ relates B_2 and B_1. Let us show this rotation matrix by R_β and call it the β-rotation:

$$
{}^{2}R_1 = R_{{}^{2}\hat{u}_2,\beta} = R_\beta \tag{7.270}
$$

To determine a general transformation matrix ${}^{1}R_G$ for the rotation α, we assume B_1 is on G and B_2 is at angle β. We apply a rotation $-\beta$ about the local axis ${}^{2}\hat{u}_2$ to bring B_2 on B_1 and G. Now the axis ${}^{G}\hat{u}_1$ coincides with an axis ${}^{2}\hat{u}_1$ and the rotation α of B_2 is about a local axis. A rotation α of B_2 about ${}^{2}\hat{u}_1$ will equivalently rotate B_1 and B_2 in G. Then, we apply the rotation α about ${}^{2}\hat{u}_1$ and apply a reverse rotation β about the local axis ${}^{2}\hat{u}_2$ to remove the effect of the rotation $-\beta$ about ${}^{2}\hat{u}_2$. Let us show the transformation matrix ${}^{1}R_G$ for rotation α by R_α and call it the α-rotation:

$$
{}^{1}R_G = R_{{}^{2}\hat{u}_2,\beta}\,R_{{}^{2}\hat{u}_1,\alpha}\,R_{{}^{2}\hat{u}_2,-\beta} = R_\alpha \tag{7.271}
$$

The rotations α and β can be interchanged and performed in any order. The final transformation matrix will not be altered by changing the order of rotations α and β.

To examine this fact, let us assume that the frames B_1 and B_2 are on G. Applying the rotation α and then the rotation β provides the transformation matrix

$$\begin{aligned} {}^2R_G &= R_\beta\, R_\alpha = R_{2\hat{u}_2,\beta} \left[R_{2\hat{u}_2,0}\, R_{2\hat{u}_1,\alpha}\, R_{2\hat{u}_2,-0}\right] \\ &= R_{2\hat{u}_2,\beta} \left[\mathbf{I}\; R_{2\hat{u}_1,\alpha}\; \mathbf{I}\right] = R_{2\hat{u}_2,\beta}\, R_{2\hat{u}_1,\alpha} \end{aligned} \tag{7.272}$$

By changing the order of rotation and applying first the rotation β and then α, we find the same transformation matrix:

$$\begin{aligned} {}^2R_G &= R_\alpha\, R_\beta = \left[R_{2\hat{u}_2,\beta}\, R_{2\hat{u}_1,\alpha}\, R_{2\hat{u}_2,-\beta}\right] R_{2\hat{u}_2,\beta} \\ &= R_{2\hat{u}_2,\beta}\, R_{2\hat{u}_1,\alpha} \end{aligned} \tag{7.273}$$

Now consider ${}^3R_2$ as a transformation matrix for rotation of B_3 about ${}^2\hat{u}_3$ in B_2 and ${}^2R_G$ as a the matrix for rotation of B_2 in G. Employing the same procedure, we can find the order-free transformation

$${}^3R_G = {}^3R_2\, {}^2R_G = {}^2R_G\, {}^3R_2 \tag{7.274}$$

We can similarly expand and apply the proof of the theory of order-free rotations to any number of bodies that are similarly connected. ■

Example 438 ★ Different Order of Rotations in Directional Control Systems
Consider the inspection camera with a direction control system of the SSRMS in Figure 7.30. The kinematics of the camera as a part of the Space Shuttle manipulator is analyzed in Examples 416, 417, and 418. To have a directional control system, we need to control the angles of a flat plate and direct an axis on the plate or its normal vector to a desired direction.

Figure 7.30 shows that the two motors of the directional control system are independent. Therefore, we may command motors 1 and 2 to turn in different order or even together. The final configuration of the camera must be the same if we command motor 1 to turn α and motor 2 to turn β in any order. To show this fact, we must define the transformation matrices such that the matrix multiplication is independent of the order of the matrices.

Consider the camera frame B_8 is not coincident with the fixed Sina frame B_9. We may always bring the local z_8-axis of B_8 on the global z_9-axis of B_9 by a rotation $-\beta$ about the y_8-axis. Now rotation α about the globally fixed z_9-axis becomes a local rotation about the z_8-axis. A reverse rotation β about the y_8-axis eliminates the first and third rotations and a rotation α about the fixed z_9-axis remains:

$$\begin{aligned} {}^8R_9 &= R_{Z,\alpha} = R_{y,\beta} R_{z,\alpha} R_{y,-\beta} \\ &= \begin{bmatrix} c\beta & 0 & -s\beta \\ 0 & 1 & 0 \\ s\beta & 0 & c\beta \end{bmatrix} \begin{bmatrix} c\alpha & -s\alpha & 0 \\ s\alpha & c\alpha & 0 \\ 0 & 0 & 1 \end{bmatrix} \begin{bmatrix} c(-\beta) & 0 & -s(-\beta) \\ 0 & 1 & 0 \\ s(-\beta) & 0 & c(-\beta) \end{bmatrix} \\ &= \begin{bmatrix} c\alpha c^2\beta + s^2\beta & -c\beta s\alpha & c\alpha c\beta s\beta - c\beta s\beta \\ c\beta s\alpha & c\alpha & s\alpha s\beta \\ c\alpha c\beta s\beta - c\beta s\beta & -s\alpha s\beta & c^2\beta + c\alpha s^2\beta \end{bmatrix} \end{aligned} \tag{7.275}$$

It reduces to the principal $R_{z,\alpha}$ for $\beta = 0$.

Now we may confirm that the order of rotations is not important. First let us apply a rotation α to the camera frame B_8 about the z_9-axis from a coincident configuration with the Sina frame B_9 followed by a rotation β about the y_8-axis:

$$\begin{aligned}
{}^8R_9 &= R_{y_8,\beta}\, R^{\mathrm{T}}_{z_9,\alpha} \\
&= \begin{bmatrix} \cos\beta & 0 & -\sin\beta \\ 0 & 1 & 0 \\ \sin\beta & 0 & \cos\beta \end{bmatrix} \begin{bmatrix} \cos\alpha & -\sin\alpha & 0 \\ \sin\alpha & \cos\alpha & 0 \\ 0 & 0 & 1 \end{bmatrix}^{\mathrm{T}} \\
&= \begin{bmatrix} \cos\alpha\cos\beta & \cos\beta\sin\alpha & -\sin\beta \\ -\sin\alpha & \cos\alpha & 0 \\ \cos\alpha\sin\beta & \sin\alpha\sin\beta & \cos\beta \end{bmatrix}
\end{aligned} \tag{7.276}$$

Now we turn the camera frame B_8 an angle β from a coincident configuration with the Sina frame B_9 about the y_8-axis followed by a rotation α about the z_9-axis. However, the rotation about the z_9-axis must be considered as (7.275):

$$\begin{aligned}
{}^8R_9 &= R^{\mathrm{T}}_{z_9,\alpha}\, R_{y_8,\beta} \\
&= \begin{bmatrix} c\alpha c^2\beta + s^2\beta & -c\beta s\alpha & c\alpha c\beta s\beta - c\beta s\beta \\ c\beta s\alpha & c\alpha & s\alpha s\beta \\ c\alpha c\beta s\beta - c\beta s\beta & -s\alpha s\beta & c^2\beta + c\alpha s^2\beta \end{bmatrix} \begin{bmatrix} c\beta & 0 & -s\beta \\ 0 & 1 & 0 \\ s\beta & 0 & c\beta \end{bmatrix} \\
&= \begin{bmatrix} \cos\alpha\cos\beta & -\cos\beta\sin\alpha & -\sin\beta \\ \sin\alpha & \cos\alpha & 0 \\ \cos\alpha\sin\beta & -\sin\alpha\sin\beta & \cos\beta \end{bmatrix}
\end{aligned} \tag{7.277}$$

The final rotation matrix (7.277) is the same as (7.276). So, it is immaterial which motor turns first provided that the correct matrices are used.

Therefore, we may define the transformation matrix between the camera frame B_8 and the Sina frame B_9 by matrix multiplication in any order,

$${}^8R_9 = R^{\mathrm{T}}_{z_9,\alpha}\, R_{y_8,\beta} = R_{y_8,\beta}\, R^{\mathrm{T}}_{z_9,\alpha} \tag{7.278}$$

where

$$R^{\mathrm{T}}_{z_9,\alpha} = \begin{bmatrix} c\alpha c^2\beta + s^2\beta & -c\beta s\alpha & c\alpha c\beta s\beta - c\beta s\beta \\ c\beta s\alpha & c\alpha & s\alpha s\beta \\ c\alpha c\beta s\beta - c\beta s\beta & -s\alpha s\beta & c^2\beta + c\alpha s^2\beta \end{bmatrix} \tag{7.279}$$

and

$$R_{y_8,\beta} = \begin{bmatrix} \cos\beta & 0 & -\sin\beta \\ 0 & 1 & 0 \\ \sin\beta & 0 & \cos\beta \end{bmatrix} \tag{7.280}$$

However, to evaluate ${}^8R_9$, we must substitute the current value of β in $R^{\mathrm{T}}_{z_9,\alpha}$.

Example 439 ★ Aiming from a Point to the Other Assume that the camera frame B_8 of the directional control system in Example 418 has already been rotated and is not coincident with the Sina frame B_9. If the rotation α_1 happened before the rotation

β_1, the rotation matrix between B_8 and B_9 is

$$R_1 = {}^8R_9 = \begin{bmatrix} \cos\alpha_1\cos\beta_1 & -\cos\beta_1\sin\alpha_1 & -\sin\beta_1 \\ \sin\alpha_1 & \cos\alpha_1 & 0 \\ \cos\alpha_1\sin\beta_1 & -\sin\alpha_1\sin\beta_1 & \cos\beta_1 \end{bmatrix} \tag{7.281}$$

otherwise, we may use Equation (7.278) to reverse the order of rotations.

To aim the camera to a new direction, we may turn the camera β_2 about y_8 and α_2 about z_9. The rotation matrix between the camera and the base would then be

$$\begin{aligned} {}^8R_9 &= R_1 R_{y_8,\beta_2} R^{\mathrm{T}}_{z_9,\alpha_2} = R_1 R_{y_8,\beta_2}\left[R_{y_8,\beta_1} R_{z_8,\alpha_2} R_{y_8,-\beta_1}\right] \\ &= R_{y_8,\beta_1} R_{z_8,\alpha_1} R_{y_8,\beta_2} R_{y_8,\beta_1} R_{z_8,\alpha_2} R_{y_8,-\beta_1} \\ &= R_{y_8,\beta_1} R_{z_8,\alpha_1} R_{y_8,\beta_1+\beta_2} R_{z_8,\alpha_2} R_{y_8,-\beta_1} \end{aligned} \tag{7.282}$$

or if we change the order of second rotations β_2 and α_2 and use Equation (7.275), then

$$\begin{aligned} {}^8R_9 &= R_1 R^{\mathrm{T}}_{z_9,\alpha_2} R_{y_8,\beta_2} = R_1\left[R_{y_8,\beta_1+\beta_2} R_{z_8,\alpha_2} R_{y_8,-\beta_1-\beta_2}\right] R_{y_8,\beta_2} \\ &= R_{y_8,\beta_1} R_{z_8,\alpha_1} R_{y_8,\beta_1+\beta_2} R_{z_8,\alpha_2} R_{y_8,-\beta_1-\beta_2} R_{y_8,\beta_2} \\ &= R_{y_8,\beta_1} R_{z_8,\alpha_1} R_{y_8,\beta_1+\beta_2} R_{z_8,\alpha_2} R_{y_8,-\beta_1} \end{aligned} \tag{7.283}$$

Example 440 ★ Spherical Wrist and Order Free Rotations All three types of spherical wrists in Figures 7.54–7.56 will act using three independent rotary actuators. The independency of the actuators means we can run the three actuators in any arbitrary order. This freedom requires that the transformation matrix between B_0 and B_2 be order free and independent of the order of rotations.

To determine the order-free transformation matrices, let us consider the Eulerian wrist of Figure 7.57 that is analyzed in Example 431. We assume that none of the angles is zero, and hence, the coordinate frames B_0, B_1, and B_2 are not coincident.

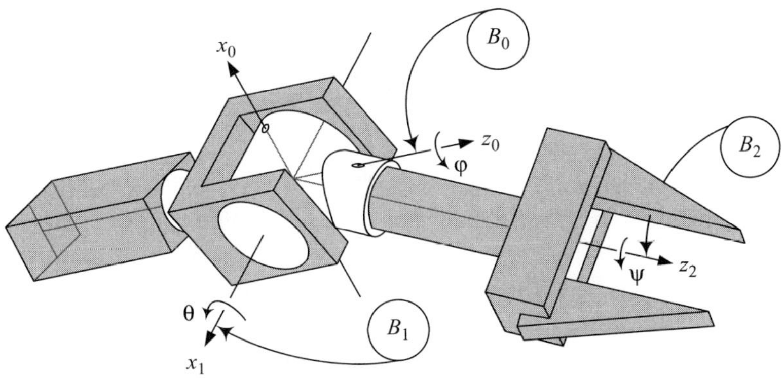

Figure 7.57 Eulerian spherical wrist or roll–pitch–roll type.

The rotation ψ of B_2 is always about the local z_2-axis. This rotation will not move any axis of B_2. Therefore,

$$^2R_1 = R_{z_2,\psi} = R_{z,\psi} = \begin{bmatrix} \cos\psi & \sin\psi & 0 \\ -\sin\psi & \cos\psi & 0 \\ 0 & 0 & 1 \end{bmatrix} \tag{7.284}$$

To find the order-free transformation matrix for rotation θ about x_1, we turn B_2 an angle $-\psi$ about the local z_2 to bring x_2 on x_1. Now rotation θ about x_1 becomes a local rotation about x_2. A reverse rotation ψ about z_2 eliminates the first and third rotations and a rotation θ about x_1 remains:

$$\begin{aligned} ^2R_1 &= R_{x_1,\theta} = R_{z_2,\psi} R_{x_2,\theta} R_{z_2,-\psi} = R_{z,\psi} R_{x,\theta} R_{z,-\psi} \\ &= \begin{bmatrix} c\psi & s\psi & 0 \\ -s\psi & c\psi & 0 \\ 0 & 0 & 1 \end{bmatrix} \begin{bmatrix} 1 & 0 & 0 \\ 0 & c\theta & s\theta \\ 0 & -s\theta & c\theta \end{bmatrix} \begin{bmatrix} c(-\psi) & s(-\psi) & 0 \\ -s(-\psi) & c(-\psi) & 0 \\ 0 & 0 & 1 \end{bmatrix} \\ &= \begin{bmatrix} c^2\psi + c\theta s^2\psi & c\theta c\psi s\psi - c\psi s\psi & s\theta s\psi \\ c\theta c\psi s\psi - c\psi s\psi & c\theta c^2\psi + s^2\psi & c\psi s\theta \\ -s\theta s\psi & -c\psi s\theta & c\theta \end{bmatrix} \end{aligned} \tag{7.285}$$

The matrix (7.285) is the order-free transformation for turning θ about x_1. We can examine and check that

$$^2R_1 = R_{x_1,\theta} R_{z_2,\psi} = R_{z_2,\psi} R_{x_1,\theta} \tag{7.286}$$

provided we use (7.285) for $R_{x_1,\theta}$ and substitute the current value of ψ before multiplication. Assuming a rotation ψ about z_2 and then a rotation θ about x_1, we find

$$\begin{aligned} ^2R_1 &= R_{x_1,\theta} R_{z_2,\psi} \\ &= \begin{bmatrix} c^2\psi + c\theta s^2\psi & c\theta c\psi s\psi - c\psi s\psi & s\theta s\psi \\ c\theta c\psi s\psi - c\psi s\psi & c\theta c^2\psi + s^2\psi & c\psi s\theta \\ -s\theta s\psi & -c\psi s\theta & c\theta \end{bmatrix} \begin{bmatrix} c\psi & s\psi & 0 \\ -s\psi & c\psi & 0 \\ 0 & 0 & 1 \end{bmatrix} \\ &= \begin{bmatrix} \cos\psi & \cos\theta\sin\psi & \sin\theta\sin\psi \\ -\sin\psi & \cos\theta\cos\psi & \cos\psi\sin\theta \\ 0 & -\sin\theta & \cos\theta \end{bmatrix} \end{aligned} \tag{7.287}$$

Now assuming a rotation θ about x_1 and then a rotation ψ about z_2, we find

$$\begin{aligned} ^2R_1 &= R_{z_2,\psi} R_{x_1,\theta} \\ &= \begin{bmatrix} c\psi & s\psi & 0 \\ -s\psi & c\psi & 0 \\ 0 & 0 & 1 \end{bmatrix} \begin{bmatrix} c^2 0 + c\theta s^2 0 & c\theta c0 s0 - c0 s0 & s\theta s0 \\ c\theta c0 s0 - c0 s0 & c\theta c^2 0 + s^2 0 & c0 s\theta \\ -s\theta s0 & -c0 s\theta & c\theta \end{bmatrix} \\ &= \begin{bmatrix} \cos\psi & \cos\theta\sin\psi & \sin\theta\sin\psi \\ -\sin\psi & \cos\theta\cos\psi & \cos\psi\sin\theta \\ 0 & -\sin\theta & \cos\theta \end{bmatrix} \end{aligned} \tag{7.288}$$

because we should substitute $\psi = 0$ in $R_{x_1,\theta}$ to evaluate $R_{z_2,\psi} R_{x_1,\theta}$.

To find the order-free transformation matrix for rotation φ about x_0, we turn B_2 an angle $-\psi$ about the local z_2 to bring x_2 on x_1. Another rotation $-\theta$ about x_2 brings z_2 on z_0. Now rotation φ about x_0 becomes a local rotation about x_2. A reverse rotation θ about x_2 and then ψ about z_2 eliminates the first, second, fourth, and fifth rotations and a rotation φ about x_0 remains:

$$
\begin{aligned}
{}^1R_0 &= R_{x_0,\theta} = R_{z_2,\psi} R_{x_2,\theta} R_{z_2,\varphi} R_{x_2,-\theta} R_{z_2,-\psi} \\
&= R_{z,\psi} R_{x,\theta} R_{z,\varphi} R_{x,-\theta} R_{z,-\psi} = \begin{bmatrix} r_{11} & r_{12} & r_{13} \\ r_{21} & r_{22} & r_{23} \\ r_{31} & r_{32} & r_{33} \end{bmatrix}
\end{aligned} \tag{7.289}
$$

where

$$r_{11} = -(1 - \cos\varphi)\cos^2\psi \sin^2\theta \cos^2\theta + 1 \tag{7.290}$$

$$r_{21} = (1 - \cos\varphi)\sin^2\theta \cos\psi \sin\psi - \cos\theta \sin\varphi \tag{7.291}$$

$$r_{31} = [(1 - \cos\varphi)\cos\theta \sin\psi + \cos\psi \sin\varphi]\sin\theta \tag{7.292}$$

$$r_{12} = (1 - \cos\varphi)\sin^2\theta \cos\psi \sin\psi + \cos\theta \sin\varphi \tag{7.293}$$

$$r_{22} = \cos\varphi + \sin^2\theta \cos^2\psi\,(1 - \cos\varphi) \tag{7.294}$$

$$r_{32} = [(1 - \cos\varphi)\cos\theta \cos\psi - \sin\psi \sin\varphi]\sin\theta \tag{7.295}$$

$$r_{13} = [(1 - \cos\varphi)\cos\theta \sin\psi - \sin\varphi \cos\psi]\sin\theta \tag{7.296}$$

$$r_{23} = [(1 - \cos\varphi)\cos\theta \cos\psi + \sin\psi \sin\varphi]\sin\theta \tag{7.297}$$

$$r_{33} = \cos\varphi + (1 - \cos\varphi)\cos^2\theta \tag{7.298}$$

The matrix (7.289) is the order-free transformation for turning φ about z_0. We can examine and check that the order of rotations φ, θ, and ψ are not important and we have

$$
\begin{aligned}
{}^2R_0 &= R_{x_1,\theta} R_{z_2,\psi} R_{z_0,\varphi} = R_{z_2,\psi} R_{z_0,\varphi} R_{x_1,\theta} \\
&= R_{z_0,\varphi} R_{x_1,\theta} R_{z_2,\psi} = R_{z_0,\varphi} R_{z_2,\psi} R_{x_1,\theta} \\
&= R_{z_2,\psi} R_{x_1,\theta} R_{z_0,\varphi} = R_{x_1,\theta} R_{z_0,\varphi} R_{z_2,\psi}
\end{aligned} \tag{7.299}
$$

We just need to substitute the current values of φ, θ, and ψ before multiplying any transformation matrices.

Example 441 ★ Order-Free Euler Angle Matrix The Euler angle transformation with the order φ, θ, ψ is found in (4.142) as

$$
\begin{aligned}
{}^BR_G &= R_{z,\psi} R_{x,\theta} R_{z,\varphi} \\
&= \begin{bmatrix} c\varphi c\psi - c\theta s\varphi s\psi & c\psi s\varphi + c\theta c\varphi s\psi & s\theta s\psi \\ -c\varphi s\psi - c\theta c\psi s\varphi & -s\varphi s\psi + c\theta c\varphi c\psi & s\theta c\psi \\ s\theta s\varphi & -c\varphi s\theta & c\theta \end{bmatrix}
\end{aligned} \tag{7.300}
$$

where

$$R_{z,\varphi} = \begin{bmatrix} \cos\varphi & \sin\varphi & 0 \\ -\sin\varphi & \cos\varphi & 0 \\ 0 & 0 & 1 \end{bmatrix} \tag{7.301}$$

$$R_{x,\theta} = \begin{bmatrix} 1 & 0 & 0 \\ 0 & \cos\theta & \sin\theta \\ 0 & -\sin\theta & \cos\theta \end{bmatrix} \tag{7.302}$$

$$R_{z,\psi} = \begin{bmatrix} \cos\psi & \sin\psi & 0 \\ -\sin\psi & \cos\psi & 0 \\ 0 & 0 & 1 \end{bmatrix} \tag{7.303}$$

These are in-order rotation matrices associated with φ, θ, ψ. The rotation φ transforms the global coordinate frame to B', rotation θ transforms B' to B'', and rotation ψ transforms B'' to B.

To make them order free, we should define them as a rotation about a local axis of the final coordinate frame. The rotation ψ is already about the local z-axis and is order free. The order-free rotation θ is

$$\begin{aligned} R_{x,\theta} &= R_{z,\psi} R_{x,\theta} R_{z,-\psi} \\ &= \begin{bmatrix} c^2\psi + c\theta s^2\psi & c\theta c\psi s\psi - c\psi s\psi & s\theta s\psi \\ c\theta c\psi s\psi - c\psi s\psi & c\theta c^2\psi + s^2\psi & c\psi s\theta \\ -s\theta s\psi & -c\psi s\theta & c\theta \end{bmatrix} \end{aligned} \tag{7.304}$$

The matrix (7.304) is the order-free transformation for rotation θ about x:

$${}^{B}R_{G} = R_{x,\theta} R_{z,\psi} = R_{z,\psi} R_{x,\theta} \tag{7.305}$$

where

$$\begin{aligned} {}^{B}R_{G} &= R_{x,\theta} R_{z,\psi} \\ &= \begin{bmatrix} c^2\psi + c\theta s^2\psi & (c\theta - 1) & s\theta s\psi \\ (c\theta - 1)\, c\psi s\psi & c\theta c^2\psi + s^2\psi & c\psi s\theta \\ -s\theta s\psi & -c\psi s\theta & c\theta \end{bmatrix} \begin{bmatrix} c\psi & s\psi & 0 \\ -s\psi & c\psi & 0 \\ 0 & 0 & 1 \end{bmatrix} \\ &= \begin{bmatrix} \cos\psi & \cos\theta\sin\psi & \sin\theta\sin\psi \\ -\sin\psi & \cos\theta\cos\psi & \cos\psi\sin\theta \\ 0 & -\sin\theta & \cos\theta \end{bmatrix} \end{aligned} \tag{7.306}$$

and

$$\begin{aligned} {}^{B}R_{G} &= R_{z,\psi} R_{x,\theta} \\ &= \begin{bmatrix} c\psi & s\psi & 0 \\ -s\psi & c\psi & 0 \\ 0 & 0 & 1 \end{bmatrix} \begin{bmatrix} c^2 0 + c\theta s^2 0 & (c\theta - 1)\, c0s0 & s\theta s0 \\ (c\theta - 1)\, c0s0 & c\theta c^2 0 + s^2 0 & c0s\theta \\ -s\theta s0 & -c0s\theta & c\theta \end{bmatrix} \\ &= \begin{bmatrix} \cos\psi & \cos\theta\sin\psi & \sin\theta\sin\psi \\ -\sin\psi & \cos\theta\cos\psi & \cos\psi\sin\theta \\ 0 & -\sin\theta & \cos\theta \end{bmatrix} \end{aligned} \tag{7.307}$$

The order-free transformation matrix for rotation φ is

$$R_{z,\varphi} = R_{z,\psi} R_{x,\theta} R_{z,\varphi} R_{x,-\theta} R_{z,-\psi} = \begin{bmatrix} r_{11} & r_{12} & r_{13} \\ r_{21} & r_{22} & r_{23} \\ r_{31} & r_{32} & r_{33} \end{bmatrix} \tag{7.308}$$

where

$$r_{11} = -(1-\cos\varphi)\cos^2\psi\sin^2\theta\cos^2\theta + 1 \tag{7.309}$$

$$r_{21} = (1-\cos\varphi)\sin^2\theta\cos\psi\sin\psi - \cos\theta\sin\varphi \tag{7.310}$$

$$r_{31} = [(1-\cos\varphi)\cos\theta\sin\psi + \cos\psi\sin\varphi]\sin\theta \tag{7.311}$$

$$r_{12} = (1-\cos\varphi)\sin^2\theta\cos\psi\sin\psi + \cos\theta\sin\varphi \tag{7.312}$$

$$r_{22} = \cos\varphi + \sin^2\theta\cos^2\psi\,(1-\cos\varphi) \tag{7.313}$$

$$r_{32} = [(1-\cos\varphi)\cos\theta\cos\psi - \sin\psi\sin\varphi]\sin\theta \tag{7.314}$$

$$r_{13} = [(1-\cos\varphi)\cos\theta\sin\psi - \sin\varphi\cos\psi]\sin\theta \tag{7.315}$$

$$r_{23} = [(1-\cos\varphi)\cos\theta\cos\psi + \sin\psi\sin\varphi]\sin\theta \tag{7.316}$$

$$r_{33} = \cos\varphi + (1-\cos\varphi)\cos^2\theta \tag{7.317}$$

The matrix (7.308) is the order-free transformation for rotation φ about z. Using the order-free matrices (7.303), (7.304), and (7.308) we can find the Euler angle transformation matrix (4.142) or (7.304) by rotations φ, θ, ψ in any order:

$$\begin{aligned} {}^B R_G &= R_{x,\theta} R_{z,\psi} R_{z,\varphi} = R_{z,\psi} R_{z,\varphi} R_{x,\theta} \\ &= R_{z,\varphi} R_{x,\theta} R_{z,\psi} = R_{z,\varphi} R_{z,\psi} R_{x,\theta} \\ &= R_{z,\psi} R_{x,\theta} R_{z,\varphi} = R_{x,\theta} R_{z,\varphi} R_{z,\psi} \end{aligned} \tag{7.318}$$

Equations (7.309)–(7.317) are the same as (7.290)–(7.298), and it justifies why we call the spherical wrist of Figure 7.57 a Eulerian wrist.

7.6 ★ ORDER-FREE TRANSFORMATION

We introduce a theorem to generalize the applied transformation theory of multibodies. Most of the connected industrial multibodies, such as serial robotic manipulators, are operated by independent actuators. Practically, the order of action of the actuators is not important. The position and orientation of the bodies would be the same no matter which motor acts first. Here is the mathematical explanation.

Theorem *Order-Free Transformations* Consider n connected bodies $B_1, B_2, \ldots, B_n$ such that the body B_1 carries the bodies $B_2, B_3, \ldots, B_n$. The body B_1 can move with respect to G with one DOF that can be a rotation θ_1 about a fixed axis in G or a translation d_1 along a fixed axis in G. The body B_2 carries the bodies $B_3, B_4, \ldots, B_n$ and moves with respect to B_1 and can be a rotation θ_2 about a fixed axis in B_1, a translation d_2 along a fixed axis in B_1, and so on. The homogeneous transformation matrix ${}^G T_n = {}^n T_G^{-1}$ is independent of the order of motions θ_1 or d_1, θ_2 or $d_2, \ldots, \theta_n$ or d_n.

Proof: Let us consider a global frame $G\,(OXYZ)$ and two body frames $B_1\,(Ox_1y_1z_1)$ and $B_2\,(Ox_2y_2z_2)$. The body B_1 carries B_2 and moves with respect to G. The body B_2 moves in B_1.

The transformation ${}^1T_G$ of G to B_1 is a rotation ${}^1R_G$ plus a translation ${}^1D_G$. The only variable of this transformation is either a rotation α or a translation d_1 about or along a globally fixed axis ${}^G\hat{u}_1$:

$$ {}^1T_G = {}^1D_G\,{}^1R_G = {}^1D_{d_1}\,{}^1R_\alpha \tag{7.319} $$

The second transformation ${}^2T_1$ of B_1 to B_2 is a rotation ${}^2R_1$ plus a translation ${}^2D_1$. The only variable of ${}^2T_1$ is either a rotation β or translation d_2 about or along a fixed axis ${}^1\hat{u}_2$ in B_1:

$$ {}^2T_1 = {}^2D_1\,{}^2R_1 = {}^2D_{d_2}\,{}^2R_\beta = T_\beta \tag{7.320} $$

To make ${}^1T_G$ an order-free transformation, we modify the first motion to be about or along a local axis ${}^2\hat{u}_2$ in B_2. To redefine ${}^1T_G$, we move B_2 to B_1 by a translation d_2 plus a rotation β, both along and about their associated local axes in B_2. Now the axes of B_2 are the same as B_1 and ${}^1T_G$ can be performed by a rotation α about a local axis in B_2 plus a translation d_1 along a local axis B_2. Then, we apply the inverse rotation β plus an inverse translation d_2 about and along their axes to move B_2 back to its position. The resultant of these rotations and translations would be an order free ${}^1T_G$:

$$ {}^1T_G = {}^2D_{d_2}\,{}^2R_\beta\;{}^2D_{d_1}\,{}^2R_\alpha\;{}^2R_{-\beta}\,{}^2D_{-d_2} \tag{7.321} $$

$$ = {}^2T_1\,{}^1T_G\,{}^2T_1^{-1} = T_\alpha \tag{7.322} $$

The transformation matrices (7.320) and (7.321) indicate the order-free motions that can be performed in any order to calculate ${}^1T_G$. The final transformation matrix will not be altered by changing the order of motions T_α and T_β. To examine this fact, let us assume that the frames B_1 and B_2 are on G. Applying the motion T_α and then the motion T_β provides the transformation matrix

$$ \begin{aligned} {}^2T_G &= {}^2T_1\,{}^1T_G = T_\beta\,T_\alpha \\ &= {}^2D_{d_2}\,{}^2R_\beta\left[{}^2D_{d_2}\,{}^2R_\beta\;{}^1D_{d_1}\,{}^1R_\alpha\;{}^2R_{-\beta}\,{}^2D_{-d_2}\right] \\ &= {}^2D_{d_2}\,{}^2R_\beta\left[{}^2D_{d_2}\,{}^2R_\beta\;{}^1D_{d_1}\,{}^1R_\alpha\right]{}^2R_{-\beta}\,{}^2D_{-d_2} \\ &= {}^2D_{d_2}\left[{}^2D_{d_2}\,{}^2R_\beta\;{}^1D_{d_1}\,{}^1R_\alpha\right]{}^2D_{-d_2} \\ &= {}^2D_{d_2}\,{}^2R_\beta\;{}^1D_{d_1}\,{}^1R_\alpha \end{aligned} \tag{7.323} $$

By changing the order of motions and applying first the motion T_β and then T_α, we find the same transformation matrix:

$$\begin{aligned}
{}^2T_G &= {}^1T_G\,{}^2T_1 = T_\alpha\,T_\beta \\
&= \left[{}^2D_{d_2}\,{}^2R_\beta\;{}^1D_{d_1}\,{}^1R_\alpha\;{}^2R_{-\beta}\,{}^2D_{-d_2}\right]\,{}^2D_{d_2}\,{}^2R_\beta \\
&= \left[{}^2D_{d_2}\,{}^2R_\beta\;{}^1D_{d_1}\,{}^1R_\alpha\right]\mathbf{I} \\
&= {}^2D_{d_2}\,{}^2R_\beta\;{}^1D_{d_1}\,{}^1R_\alpha
\end{aligned} \tag{7.324}$$

Now consider ${}^3T_2$ a transformation matrix for motion of B_3 about B_2 and ${}^2T_G$ a matrix for motion of B_2 in G. Employing the same procedure, we can find the order-free transformation matrix.

$$ {}^3T_G = {}^3T_2\,{}^2T_G = {}^2T_G\,{}^3T_2 \tag{7.325}$$

We can similarly expand and apply the proof of the theory of order-free transformations to any number of bodies that are similarly connected. ■

Example 442 ★ Order-Free Transformations of a Planar R‖R Manipulator
Figure 7.58 illustrates an R‖R planar manipulator with two parallel revolute joints and variables θ_1 and θ_2. Links (1) and (2) are both R‖R(0) and therefore the transformation matrices ${}^0T_1$, ${}^1T_2$ are

$$ {}^0T_1 = \begin{bmatrix} \cos\theta_1 & -\sin\theta_1 & 0 & l_1\cos\theta_1 \\ \sin\theta_1 & \cos\theta_1 & 0 & l_1\sin\theta_1 \\ 0 & 0 & 1 & 0 \\ 0 & 0 & 0 & 1 \end{bmatrix} \tag{7.326}$$

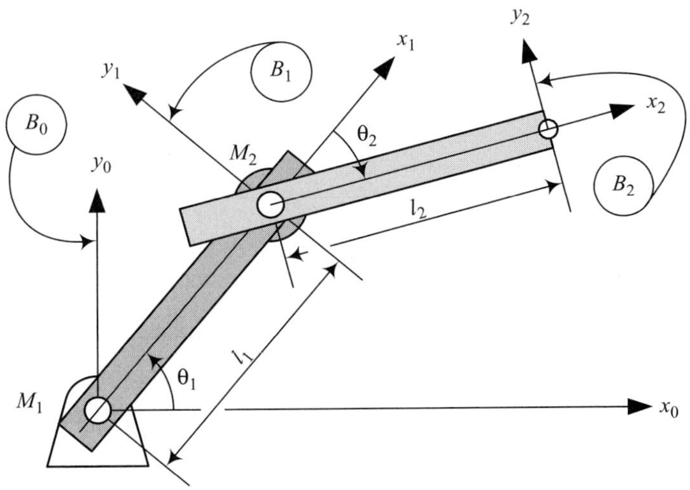

Figure 7.58 A 2R or R‖R planar manipulator.

$$
{}^{1}T_{2} = \begin{bmatrix} \cos\theta_2 & -\sin\theta_2 & 0 & l_2\cos\theta_2 \\ \sin\theta_2 & \cos\theta_2 & 0 & l_2\sin\theta_2 \\ 0 & 0 & 1 & 0 \\ 0 & 0 & 0 & 1 \end{bmatrix} \tag{7.327}
$$

Employing ${}^{0}T_{1}$ and ${}^{1}T_{2}$, we can find the forward kinematics ${}^{0}T_{2}$ only with the assumption that first the grounded actuator M_1 acts to change θ_1 and then the second motor M_2 acts to change θ_2:

$$
\begin{aligned}
{}^{0}T_{2} &= {}^{0}T_{1}\,{}^{1}T_{2} \\
&= \begin{bmatrix} c\,(\theta_1+\theta_2) & -s\,(\theta_1+\theta_2) & 0 & l_2c\,(\theta_1+\theta_2)+l_1c\theta_1 \\ s\,(\theta_1+\theta_2) & c\,(\theta_1+\theta_2) & 0 & l_2s\,(\theta_1+\theta_2)+l_1s\theta_1 \\ 0 & 0 & 1 & 0 \\ 0 & 0 & 0 & 1 \end{bmatrix}
\end{aligned} \tag{7.328}
$$

However, the actuators M_1 and M_2 are independent and may act in any order. To find the order-free transformations, we redefine the transformation matrices ${}^{0}T_{1}$, ${}^{1}T_{2}$ to go from lower body to upper body:

$$
{}^{1}T_{0} = {}^{0}T_{1}^{-1} = \begin{bmatrix} \cos\theta_1 & \sin\theta_1 & 0 & -l_1 \\ -\sin\theta_1 & \cos\theta_1 & 0 & 0 \\ 0 & 0 & 1 & 0 \\ 0 & 0 & 0 & 1 \end{bmatrix} \tag{7.329}
$$

$$
{}^{2}T_{1} = {}^{1}T_{2}^{-1} = \begin{bmatrix} \cos\theta_2 & \sin\theta_2 & 0 & -l_2 \\ -\sin\theta_2 & \cos\theta_2 & 0 & 0 \\ 0 & 0 & 1 & 0 \\ 0 & 0 & 0 & 1 \end{bmatrix} \tag{7.330}
$$

Using ${}^{0}T_{1}$, ${}^{1}T_{2}$ the forward kinematics of the manipulator can be found as

$$
\begin{aligned}
{}^{2}T_{0} &= {}^{2}T_{1}\,{}^{1}T_{0} = {}^{0}T_{2}^{-1} \\
&= \begin{bmatrix} \cos\,(\theta_1+\theta_2) & \sin\,(\theta_1+\theta_2) & 0 & -l_2-l_1\cos\theta_2 \\ -\sin\,(\theta_1+\theta_2) & \cos\,(\theta_1+\theta_2) & 0 & l_1\sin\theta_2 \\ 0 & 0 & 1 & 0 \\ 0 & 0 & 0 & 1 \end{bmatrix}
\end{aligned} \tag{7.331}
$$

We employ the rule (RFDS) of (6.35) to redefine the homogeneous transformation matrices ${}^{0}T_{1}$, and ${}^{1}T_{2}$. The transformation ${}^{2}T_{1} = {}^{2}D_{1}\,{}^{2}R_{1}$ is made by a rotation θ_2 of B_2 about the local axis z_2 plus a translation l_2 of B_2 along the local axis x_2:

$$
\begin{aligned}
{}^{2}T_{1} &= {}^{2}D_{1}\,{}^{2}R_{1} = D_{x,l_2}R_{z,\theta_2} \\
&= R_{z,\alpha} = \begin{bmatrix} 1 & 0 & 0 & -l_2 \\ 0 & 1 & 0 & 0 \\ 0 & 0 & 1 & 0 \\ 0 & 0 & 0 & 1 \end{bmatrix} \begin{bmatrix} \cos\theta_2 & \sin\theta_2 & 0 & 0 \\ -\sin\theta_2 & \cos\theta_2 & 0 & 0 \\ 0 & 0 & 1 & 0 \\ 0 & 0 & 0 & 1 \end{bmatrix}
\end{aligned} \tag{7.332}
$$

The other transformation ${}^1T_0 = {}^1D_0\,{}^1R_0$ is made by a rotation θ_1 of B_1 about the local axis z_1 plus a translation l_1 of B_1 along the local axis x_1. To redefine ${}^1T_0$, we move B_2 to B_1 by a translation l_2 along the local axis x_2 plus a rotation θ_2 about the local axis z_2. Now the axes of B_2 are the same as B_1 and ${}^1T_0$ can be performed by a rotation θ_1 about the local axis z_2 plus a translation l_1 along the local axis x_2. Then, we apply the inverse rotation θ_2 about z_2 plus an inverse translation l_2 along x_2 to move B_2 back to its position. The resultant of these rotations and translations would be ${}^1T_0$:

$$
\begin{aligned}
{}^1T_0 &= D_{x,l_2} R_{z,\theta_2} D_{x,l_1} R_{z_2,\theta_1} R_{z,-\theta_2} D_{x,-l_2} = {}^2T_1\,{}^1T_0\,{}^2T_1^{-1} \\
&= \begin{bmatrix} \cos\theta_1 & \sin\theta_1 & 0 & l_2\cos\theta_1 - l_1\cos\theta_2 - l_2 \\ -\sin\theta_1 & \cos\theta_1 & 0 & l_1\sin\theta_2 - l_2\sin\theta_1 \\ 0 & 0 & 1 & 0 \\ 0 & 0 & 0 & 1 \end{bmatrix}
\end{aligned}
\tag{7.333}
$$

where

$$
D_{x,-l_2} = \begin{bmatrix} 1 & 0 & 0 & l_2 \\ 0 & 1 & 0 & 0 \\ 0 & 0 & 1 & 0 \\ 0 & 0 & 0 & 1 \end{bmatrix} \tag{7.334}
$$

$$
R_{z,-\theta_2} = \begin{bmatrix} \cos-\theta_2 & \sin-\theta_2 & 0 & 0 \\ -\sin-\theta_2 & \cos-\theta_2 & 0 & 0 \\ 0 & 0 & 1 & 0 \\ 0 & 0 & 0 & 1 \end{bmatrix} \tag{7.335}
$$

$$
R_{z_2,\theta_1} = \begin{bmatrix} \cos\theta_1 & \sin\theta_1 & 0 & 0 \\ -\sin\theta_1 & \cos\theta_1 & 0 & 0 \\ 0 & 0 & 1 & 0 \\ 0 & 0 & 0 & 1 \end{bmatrix} \tag{7.336}
$$

$$
D_{x,l_1} = \begin{bmatrix} 1 & 0 & 0 & -l_1 \\ 0 & 1 & 0 & 0 \\ 0 & 0 & 1 & 0 \\ 0 & 0 & 0 & 1 \end{bmatrix} \tag{7.337}
$$

$$
R_{z,\theta_2} = \begin{bmatrix} \cos\theta_2 & \sin\theta_2 & 0 & 0 \\ -\sin\theta_2 & \cos\theta_2 & 0 & 0 \\ 0 & 0 & 1 & 0 \\ 0 & 0 & 0 & 1 \end{bmatrix} \tag{7.338}
$$

$$
D_{x,l_2} = \begin{bmatrix} 1 & 0 & 0 & -l_2 \\ 0 & 1 & 0 & 0 \\ 0 & 0 & 1 & 0 \\ 0 & 0 & 0 & 1 \end{bmatrix} \tag{7.339}
$$

The matrices (7.330) and (7.333) are order free and we can check the result of their product with different orders:

$$
\begin{aligned}
{}^{2}T_{0} &= {}^{1}T_{0}\,{}^{2}T_{1} \\
&= \begin{bmatrix} c\theta_1 & s\theta_1 & 0 & l_2 c\theta_1 - l_1 c\theta_2 - l_2 \\ -s\theta_1 & c\theta_1 & 0 & l_1 s\theta_2 - l_2 s\theta_1 \\ 0 & 0 & 1 & 0 \\ 0 & 0 & 0 & 1 \end{bmatrix} \begin{bmatrix} c\theta_2 & s\theta_2 & 0 & -l_2 \\ -s\theta_2 & c\theta_2 & 0 & 0 \\ 0 & 0 & 1 & 0 \\ 0 & 0 & 0 & 1 \end{bmatrix} \\
&= \begin{bmatrix} c\left(\theta_1+\theta_2\right) & s\left(\theta_1+\theta_2\right) & 0 & -l_2 - l_1 c\theta_2 \\ -s\left(\theta_1+\theta_2\right) & c\left(\theta_1+\theta_2\right) & 0 & l_1 s\theta_2 \\ 0 & 0 & 1 & 0 \\ 0 & 0 & 0 & 1 \end{bmatrix}
\end{aligned} \tag{7.340}
$$

$$
\begin{aligned}
{}^{2}T_{0} &= {}^{1}T_{0}\,{}^{2}T_{1} \\
&= \begin{bmatrix} c\theta_2 & s\theta_2 & 0 & -l_2 \\ -s\theta_2 & c\theta_2 & 0 & 0 \\ 0 & 0 & 1 & 0 \\ 0 & 0 & 0 & 1 \end{bmatrix} \begin{bmatrix} c\theta_1 & s\theta_1 & 0 & 0c\theta_1 - l_1 c0 - 0 \\ -s\theta_1 & c\theta_1 & 0 & l_1 s0 - 0s\theta_1 \\ 0 & 0 & 1 & 0 \\ 0 & 0 & 0 & 1 \end{bmatrix} \\
&= \begin{bmatrix} c\left(\theta_1+\theta_2\right) & s\left(\theta_1+\theta_2\right) & 0 & -l_2 - l_1 c\theta_2 \\ -s\left(\theta_1+\theta_2\right) & c\left(\theta_1+\theta_2\right) & 0 & l_1 s\theta_2 \\ 0 & 0 & 1 & 0 \\ 0 & 0 & 0 & 1 \end{bmatrix}
\end{aligned} \tag{7.341}
$$

Example 443 ★ Order-Free Transformations of an Articulated Manipulator An articulated manipulator is any serial multibody with three links and three revolute joints to reach a point in three dimensions. However, an articulated arm is usually referred to a three-link multibody that is similar to a human hand with a shoulder, arm, and forearm such as the one shown in Figure 7.59.

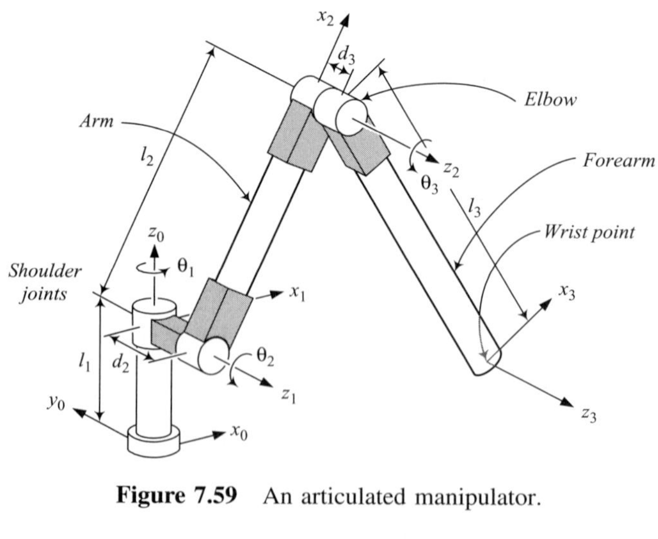

Figure 7.59 An articulated manipulator.

The first link of the arm is an R⊢R(90) with a distance l_1 between x_0 and x_1, and therefore,

$$^0T_1 = \begin{bmatrix} \cos\theta_1 & 0 & \sin\theta_1 & 0 \\ \sin\theta_1 & 0 & -\cos\theta_1 & 0 \\ 0 & 1 & 0 & l_1 \\ 0 & 0 & 0 & 1 \end{bmatrix} \tag{7.342}$$

$$^1T_0 = {}^0T_1^{-1} = \begin{bmatrix} \cos\theta_1 & \sin\theta_1 & 0 & 0 \\ 0 & 0 & 1 & -l_1 \\ \sin\theta_1 & -\cos\theta_1 & 0 & 0 \\ 0 & 0 & 0 & 1 \end{bmatrix} \tag{7.343}$$

The second link of the arm is an R‖R(90) with a distance d_2 between x_1 and x_2 and a distance l_2 between z_1 and z_2:

$$^1T_2 = \begin{bmatrix} \cos\theta_2 & -\sin\theta_2 & 0 & l_2\cos\theta_2 \\ \sin\theta_2 & \cos\theta_2 & 0 & l_2\sin\theta_2 \\ 0 & 0 & 1 & d_2 \\ 0 & 0 & 0 & 1 \end{bmatrix} \tag{7.344}$$

$$^2T_1 = {}^1T_2^{-1} = \begin{bmatrix} \cos\theta_2 & \sin\theta_2 & 0 & -l_2 \\ -\sin\theta_2 & \cos\theta_2 & 0 & 0 \\ 0 & 0 & 1 & -d_2 \\ 0 & 0 & 0 & 1 \end{bmatrix} \tag{7.345}$$

The third link is not connected to any other link at its distal end, and we should set B_3 at its proximal end and attach a takht frame at its tip point. We may also attach only one coordinate frame at the tip point and consider the link as an R⊢R(90) with a distance d_3 between x_2 and x_3 and a distance l_3 between z_2 and z_3:

$$^2T_3 = \begin{bmatrix} \cos\theta_3 & -\sin\theta_3 & 0 & l_3\cos\theta_3 \\ \sin\theta_3 & \cos\theta_3 & 0 & l_3\sin\theta_3 \\ 0 & 0 & 1 & d_3 \\ 0 & 0 & 0 & 1 \end{bmatrix} \tag{7.346}$$

$$^3T_2 = {}^2T_3^{-1} = \begin{bmatrix} \cos\theta_3 & \sin\theta_3 & 0 & -l_3 \\ -\sin\theta_3 & \cos\theta_3 & 0 & 0 \\ 0 & 0 & 1 & -d_3 \\ 0 & 0 & 0 & 1 \end{bmatrix} \tag{7.347}$$

The forward kinematics of the manipulator is

$$\begin{aligned} ^0T_3 &= {}^0T_1\,{}^1T_2\,{}^2T_3 \\ &= \begin{bmatrix} c(\theta_2+\theta_3)\,c\theta_1 & -s(\theta_2+\theta_3)\,c\theta_1 & s\theta_1 & {}^0d_x \\ c(\theta_2+\theta_3)\,s\theta_1 & -s(\theta_2+\theta_3)\,s\theta_1 & -c\theta_1 & {}^0d_y \\ s(\theta_2+\theta_3) & c(\theta_2+\theta_3) & 0 & {}^0d_z \\ 0 & 0 & 0 & 1 \end{bmatrix} \end{aligned} \tag{7.348}$$

$$\begin{bmatrix} {}^0d_x \\ {}^0d_y \\ {}^0d_z \end{bmatrix} = \begin{bmatrix} (l_2c\theta_2 + l_3c(\theta_2+\theta_3))\,c\theta_1 + (d_2+d_3)\,s\theta_1 \\ (l_2c\theta_2 + l_3c(\theta_2+\theta_3))\,s\theta_1 - (d_3+d_2)\,c\theta_1 \\ l_1 + l_2s\theta_2 + l_3s(\theta_2+\theta_3) \end{bmatrix} \tag{7.349}$$

or

$$
{}^{3}T_{0} = {}^{3}T_{2}\,{}^{2}T_{1}\,{}^{1}T_{0}
$$

$$
= \begin{bmatrix} c(\theta_2+\theta_3)\,c\theta_1 & c(\theta_2+\theta_3)\,s\theta_1 & s(\theta_2+\theta_3) & {}^{3}d_x \\ -s(\theta_2+\theta_3)\,c\theta_1 & -s(\theta_2+\theta_3)\,s\theta_1 & c(\theta_2+\theta_3) & {}^{3}d_y \\ s\theta_1 & -c\theta_1 & 0 & {}^{3}d_z \\ 0 & 0 & 0 & 1 \end{bmatrix} \tag{7.350}
$$

$$
\begin{bmatrix} {}^{3}d_x \\ {}^{3}d_y \\ {}^{3}d_z \end{bmatrix} = \begin{bmatrix} -l_1 s(\theta_2+\theta_3) - l_2 c\theta_3 - l_3 \\ -l_1 c(\theta_2+\theta_3) + l_2 s\theta_3 \\ -d_2 - d_3 \end{bmatrix} \tag{7.351}
$$

To determine the order-free transformations ${}^{3}T_{2}$, ${}^{2}T_{1}$, and ${}^{1}T_{0}$, we begin with the final coordinate frame B_3. The transformation ${}^{3}T_{2}$ is already a motion about and along the axes of B_3. The order-free transformation of the second matrix is

$$
{}^{2}T_{1} = {}^{3}T_{2}\,{}^{2}T_{1}\,{}^{3}T_{2}^{-1} = {}^{3}T_{2}\,{}^{2}T_{1}\,{}^{2}T_{3}
$$

$$
= \begin{bmatrix} \cos\theta_2 & \sin\theta_2 & 0 & l_3\cos\theta_2 - l_2\cos\theta_3 - l_3 \\ -\sin\theta_2 & \cos\theta_2 & 0 & l_2\sin\theta_3 - l_3\sin\theta_2 \\ 0 & 0 & 1 & -d_2 \\ 0 & 0 & 0 & 1 \end{bmatrix} \tag{7.352}
$$

We may check and see that

$$
{}^{3}T_{1} = {}^{3}T_{2}\,{}^{2}T_{1} = {}^{2}T_{1}\,{}^{3}T_{2}
$$

$$
= \begin{bmatrix} \cos(\theta_2+\theta_3) & \sin(\theta_2+\theta_3) & 0 & -l_3 - l_2\cos\theta_3 \\ -\sin(\theta_2+\theta_3) & \cos(\theta_2+\theta_3) & 0 & l_2\sin\theta_3 \\ 0 & 0 & 1 & -d_2 - d_3 \\ 0 & 0 & 0 & 1 \end{bmatrix} \tag{7.353}
$$

However, when ${}^{2}T_{1}$ is the first motion, we must substitute the parameters l_3 and θ_3 of the second motion equal to zero.

The order-free transformation of the third matrix is

$$
{}^{1}T_{0} = {}^{3}T_{2}\,{}^{2}T_{1}\,{}^{1}T_{0}\,{}^{2}T_{1}^{-1}\,{}^{3}T_{2}^{-1} = {}^{3}T_{2}\,{}^{2}T_{1}\,{}^{1}T_{0}\,{}^{1}T_{2}\,{}^{2}T_{3} \tag{7.354}
$$

$$
= \frac{1}{2}\begin{bmatrix} c\gamma + c\theta_1 & s\gamma + s\theta_1 & 2s(\theta_2+\theta_3) & {}^{1}d_x \\ s\gamma - s\theta_1 & c\theta_1 - c\gamma & 2c(\theta_2+\theta_3) & {}^{1}d_y \\ 2s(\theta_1-\theta_2-\theta_3) & -2c(\theta_1-\theta_2-\theta_3) & 0 & {}^{1}d_z \\ 0 & 0 & 0 & 1 \end{bmatrix}
$$

$$
\gamma = \theta_1 - 2\theta_2 - 2\theta_3 \tag{7.355}
$$

where

$$
\begin{aligned} {}^{1}d_x &= 2(d_2+d_3-l_1)\sin(\theta_2+\theta_3) + l_3\cos\theta_1 - 2l_2\cos\theta_3 - 2l_3 \\ &\quad + (l_2+l_3)\cos(\theta_1-2\theta_2-2\theta_3) + l_2\cos(\theta_1+\theta_3) \end{aligned} \tag{7.356}
$$

$$
\begin{aligned}
{}^1d_y &= 2\,(d_2 + d_3 - l_1)\cos(\theta_2 + \theta_3) - l_3 \sin\theta_1 + 2l_2 \sin\theta_3 \\
&\quad + (l_2 + l_3)\sin(\theta_1 - 2\theta_2 - 2\theta_3) - l_2 \sin(\theta_1 + \theta_3)
\end{aligned} \tag{7.357}
$$

$$
{}^1d_z = 2[l_3 \sin(\theta_1 - \theta_2 - \theta_3) - d_3 - d_2 + l_2 \sin(\theta_1 - \theta_2)] \tag{7.358}
$$

We may check and see that the transformation ${}^3T_0 = {}^3T_2\,{}^2T_1\,{}^1T_0$ can be found by multiplying ${}^3T_2$, ${}^2T_1$, ${}^1T_0$ in any order:

$$
\begin{aligned}
{}^3T_0 &= {}^3T_2\,{}^2T_1\,{}^1T_0 = {}^2T_1\,{}^3T_2\,{}^1T_0 \\
&= {}^3T_2\,{}^1T_0\,{}^2T_1 = {}^1T_0\,{}^3T_2\,{}^2T_1 \\
&= {}^2T_1\,{}^1T_0\,{}^3T_2 = {}^1T_0\,{}^2T_1\,{}^3T_2
\end{aligned} \tag{7.359}
$$

It indicates that we can run the three motors of the manipulator with any arbitrary order. As long as the associated angles of rotation of the motors remain the same, the final configuration of the manipulator would be the same.

7.7 ★ FORWARD KINEMATICS BY SCREW

It is easier to use screws in forward kinematics and represent a transformation matrix between two adjacent coordinate frames B_i and B_{i-1} that are set up according to DH rules. We can move B_i to B_{i-1} by a central screw $\check{s}(a_i, \alpha_i, \hat{\imath}_{i-1})$ followed by another central screw $\check{s}(d_i, \theta_i, \hat{k}_{i-1})$:

$$
\begin{aligned}
{}^{i-1}T_i &= \check{s}(d_i, \theta_i, \hat{k}_{i-1})\,\check{s}(a_i, \alpha_i, \hat{\imath}_{i-1}) \\
&= \begin{bmatrix}
\cos\theta_i & -\sin\theta_i \cos\alpha_i & \sin\theta_i \sin\alpha_i & a_i \cos\theta_i \\
\sin\theta_i & \cos\theta_i \cos\alpha_i & -\cos\theta_i \sin\alpha_i & a_i \sin\theta_i \\
0 & \sin\alpha_i & \cos\alpha_i & d_i \\
0 & 0 & 0 & 1
\end{bmatrix}
\end{aligned} \tag{7.360}
$$

Proof: The central screw $\check{s}(a_i, \alpha_i, \hat{\imath}_{i-1})$ is given as

$$
\begin{aligned}
\check{s}(a_i, \alpha_i, \hat{\imath}_{i-1}) &= D(a_i, \hat{\imath}_{i-1})R(\hat{\imath}_{i-1}, \alpha_i) = D_{x_{i-1},a_i}\,R_{x_{i-1},\alpha_i} \\
&= \begin{bmatrix} 1 & 0 & 0 & a_i \\ 0 & 1 & 0 & 0 \\ 0 & 0 & 1 & 0 \\ 0 & 0 & 0 & 1 \end{bmatrix}
\begin{bmatrix} 1 & 0 & 0 & 0 \\ 0 & \cos\alpha_i & -\sin\alpha_i & 0 \\ 0 & \sin\alpha_i & \cos\alpha_i & 0 \\ 0 & 0 & 0 & 1 \end{bmatrix} \\
&= \begin{bmatrix} 1 & 0 & 0 & a_i \\ 0 & \cos\alpha_i & -\sin\alpha_i & 0 \\ 0 & \sin\alpha_i & \cos\alpha_i & 0 \\ 0 & 0 & 0 & 1 \end{bmatrix}
\end{aligned} \tag{7.361}
$$

and the central screw $\check{s}(d_i, \theta_i, \hat{k}_{i-1})$ as

$$\check{s}(d_i, \theta_i, \hat{k}_{i-1}) = D(d_i, \hat{k}_{i-1}) R(\hat{k}_{i-1}, \theta_i) = D_{z_{i-1}, d_i} R_{z_{i-1}, \theta_i}$$

$$= \begin{bmatrix} 1 & 0 & 0 & 0 \\ 0 & 1 & 0 & 0 \\ 0 & 0 & 1 & d_i \\ 0 & 0 & 0 & 1 \end{bmatrix} \begin{bmatrix} \cos\theta_i & -\sin\theta_i & 0 & 0 \\ \sin\theta_i & \cos\theta_i & 0 & 0 \\ 0 & 0 & 1 & 0 \\ 0 & 0 & 0 & 1 \end{bmatrix}$$

$$= \begin{bmatrix} \cos\theta_i & -\sin\theta_i & 0 & 0 \\ \sin\theta_i & \cos\theta_i & 0 & 0 \\ 0 & 0 & 1 & d_i \\ 0 & 0 & 0 & 1 \end{bmatrix} \tag{7.362}$$

Therefore, the transformation matrix ${}^{i-1}T_i$ made by the two central screws would be

$${}^{i-1}T_i = \check{s}(d_i, \theta_i, \hat{k}_{i-1})\, \check{s}(a_i, \alpha_i, \hat{\imath}_{i-1})$$

$$= \begin{bmatrix} \cos\theta_i & -\sin\theta_i & 0 & 0 \\ \sin\theta_i & \cos\theta_i & 0 & 0 \\ 0 & 0 & 1 & d_i \\ 0 & 0 & 0 & 1 \end{bmatrix} \begin{bmatrix} 1 & 0 & 0 & a_i \\ 0 & \cos\alpha_i & -\sin\alpha_i & 0 \\ 0 & \sin\alpha_i & \cos\alpha_i & 0 \\ 0 & 0 & 0 & 1 \end{bmatrix}$$

$$= \begin{bmatrix} \cos\theta_i & -\cos\alpha_i \sin\theta_i & \sin\theta_i \sin\alpha_i & a_i \cos\theta_i \\ \sin\theta_i & \cos\theta_i \cos\alpha_i & -\cos\theta_i \sin\alpha_i & a_i \sin\theta_i \\ 0 & \sin\alpha_i & \cos\alpha_i & d_i \\ 0 & 0 & 0 & 1 \end{bmatrix} \tag{7.363}$$

The resultant transformation matrix ${}^{i-1}T_i$ is equivalent to a general screw whose parameters can be found based on Equations (6.248) and (6.249).

The twist of the screw, ϕ, can be computed based on Equation (6.252):

$$\cos\phi = \tfrac{1}{2}[\mathrm{tr}\left({}^{G}R_B\right) - 1]$$
$$= \tfrac{1}{2}(\cos\theta_i + \cos\theta_i \cos\alpha_i + \cos\alpha_i - 1) \tag{7.364}$$

The axis of the screw, $\hat{u}$, can be found by using Equation (6.253):

$$\tilde{u} = \frac{1}{2\sin\phi}\left({}^{G}R_B - {}^{G}R_B^{\mathrm{T}}\right)$$

$$= \frac{1}{2s\phi}\begin{bmatrix} c\theta_i & -c\alpha_i s\theta_i & s\theta_i s\alpha_i \\ s\theta_i & c\theta_i c\alpha_i & -c\theta_i s\alpha_i \\ 0 & s\alpha_i & c\alpha_i \end{bmatrix} - \begin{bmatrix} c\theta_i & s\theta_i & 0 \\ -c\alpha_i s\theta_i & c\theta_i c\alpha_i & s\alpha_i \\ s\theta_i s\alpha_i & -c\theta_i s\alpha_i & c\alpha_i \end{bmatrix}$$

$$= \frac{1}{2s\phi}\begin{bmatrix} 0 & -s\theta_i - c\alpha_i s\theta_i & s\theta_i s\alpha_i \\ s\theta_i + c\alpha_i s\theta_i & 0 & -s\alpha_i - c\theta_i s\alpha_i \\ -s\theta_i s\alpha_i & s\alpha_i + c\theta_i s\alpha_i & 0 \end{bmatrix} \tag{7.365}$$

and therefore,

$$\hat{u} = \frac{1}{2s\phi}\begin{bmatrix} \sin\alpha_i + \cos\theta_i \sin\alpha_i \\ \sin\theta_i \sin\alpha_i \\ \sin\theta_i + \cos\alpha_i \sin\theta_i \end{bmatrix} \tag{7.366}$$

The translation parameter h and the position vector of a point on the screw axis, for instance, $[0, y_{i-1}, z_{i-1}]$, can be found based on Equation (6.256):

$$\begin{aligned}\begin{bmatrix} h \\ y_{i-1} \\ y_{i-1} \end{bmatrix} &= \begin{bmatrix} u_1 & -r_{12} & -r_{13} \\ u_2 & 1-r_{22} & -r_{23} \\ u_3 & -r_{32} & 1-r_{33} \end{bmatrix}^{-1} \begin{bmatrix} r_{14} \\ r_{24} \\ r_{34} \end{bmatrix} \\ &= \frac{1}{2s\phi}\begin{bmatrix} s\alpha_i + c\theta_i s\alpha_i & -s\theta_i\, c\alpha_i & s\theta_i\, s\alpha_i \\ s\theta_i s\alpha_i & 1 - c\theta_i\, c\alpha_i & -c\theta_i\, s\alpha_i \\ s\theta_i + c\alpha_i s\theta_i & s\alpha_i & c\alpha_i \end{bmatrix}^{-1} \begin{bmatrix} a_i\, c\theta_i \\ a_i\, s\theta_i \\ d_i \end{bmatrix}\end{aligned} \tag{7.367}$$

■

Example 444 ★ Classification of Industrial Links by Screw As explained in Example 404 and Appendix D, there are 12 different configurations that are mostly used for industrial links. Each type has its own class of geometric configuration and transformation. Each class is identified by its joints at both ends and has its own transformation matrix to go from the distal joint coordinate frame B_i to the proximal joint coordinate frame B_{i-1}. The transformation matrix of each class depends solely on the proximal joint and the angle between the z-axes. The screw expression for two arbitrary coordinate frames is

$${}^{i-1}T_i = \check{s}(d_i, \theta_i, \hat{k}_{i-1})\, \check{s}(a_i, \alpha_i, \hat{\imath}_{i-1}) \tag{7.368}$$

where

$$\check{s}(d_i, \theta_i, \hat{k}_{i-1}) = D(d_i, \hat{k}_{i-1}) R(\hat{k}_{i-1}, \theta_i) \tag{7.369}$$

$$\check{s}(a_i, \alpha_i, \hat{\imath}_{i-1}) = D(a_i, \hat{\imath}_{i-1}) R(\hat{\imath}_{i-1}, \alpha_i) \tag{7.370}$$

The frame transformations for each class of the industrial links can be expressed by screws as shown in Table 7.10.

As an example, let us examine the first class and find the same result as Equation (D.1) for $d_i = 0$:

$$\begin{aligned}{}^{i-1}T_i &= \check{s}(0, \theta_i, \hat{k}_{i-1})\, \check{s}(a_i, 0, \hat{\imath}_{i-1}) \\ &= D(0, \hat{k}_{i-1}) R(\hat{k}_{i-1}, \theta_i) D(a_i, \hat{\imath}_{i-1}) R(\hat{\imath}_{i-1}, 0) \\ &= \begin{bmatrix} \cos\theta_i & -\sin\theta_i & 0 & 0 \\ \sin\theta_i & \cos\theta_i & 0 & 0 \\ 0 & 0 & 1 & 0 \\ 0 & 0 & 0 & 1 \end{bmatrix} \begin{bmatrix} 1 & 0 & 0 & a_i \\ 0 & 1 & 0 & 0 \\ 0 & 0 & 1 & 0 \\ 0 & 0 & 0 & 1 \end{bmatrix} \\ &= \begin{bmatrix} \cos\theta_i & -\sin\theta_i & 0 & a_i\cos\theta_i \\ \sin\theta_i & \cos\theta_i & 0 & a_i\sin\theta_i \\ 0 & 0 & 1 & 0 \\ 0 & 0 & 0 & 1 \end{bmatrix}\end{aligned} \tag{7.371}$$

Table 7.10 Screw Transformation of Industrial Links

No.	Type of Link	${}^{i-1}T_i =$
1	R‖R(0) or R‖P(0)	$\check{s}(0, \theta_i, \hat{k}_{i-1})\,\check{s}(a_i, 0, \hat{\imath}_{i-1})$
2	R‖R(180) or R‖P(180)	$\check{s}(0, \theta_i, \hat{k}_{i-1})\,\check{s}(a_i, 2\pi, \hat{\imath}_{i-1})$
3	R⊥R(90) or R⊥P(90)	$\check{s}(0, \theta_i, \hat{k}_{i-1})\,\check{s}(a_i, \pi, \hat{\imath}_{i-1})$
4	R⊥R(−90) or R⊥P(−90)	$\check{s}(0, \theta_i, \hat{k}_{i-1})\,\check{s}(a_i, -\pi, \hat{\imath}_{i-1})$
5	R⊢R(90) or R⊢P(90)	$\check{s}(0, \theta_i, \hat{k}_{i-1})\,\check{s}(0, \pi, \hat{\imath}_{i-1})$
6	R⊢R(−90) or R⊢P(−90)	$\check{s}(0, \theta_i, \hat{k}_{i-1})\,\check{s}(0, -\pi, \hat{\imath}_{i-1})$
7	P‖R(0) or P‖P(0)	$\check{s}(d_i, 0, \hat{k}_{i-1})\,\check{s}(a_i, 0, \hat{\imath}_{i-1})$
8	P‖R(180) or P‖P(180)	$\check{s}(d_i, 0, \hat{k}_{i-1})\,\check{s}(a_i, 2\pi, \hat{\imath}_{i-1})$
9	P⊥R(90) or P⊥P(90)	$\check{s}(d_i, 0, \hat{k}_{i-1})\,\check{s}(a_i, \pi, \hat{\imath}_{i-1})$
10	P⊥R(−90) or P⊥P(−90)	$\check{s}(d_i, 0, \hat{k}_{i-1})\,\check{s}(a_i, -\pi, \hat{\imath}_{i-1})$
11	P⊢R(90) or P⊢P(90)	$\check{s}(d_i, 0, \hat{k}_{i-1})\,\check{s}(0, \pi, \hat{\imath}_{i-1})$
12	P⊢R(−90) or P⊢P(−90)	$\check{s}(d_i, 0, \hat{k}_{i-1})\,\check{s}(0, -\pi, \hat{\imath}_{i-1})$

Example 445 ★ Spherical Arm Forward Kinematics Based on Screws Consider the spherical arm in Figure 7.50. The classes of links for the arm are indicated in Table 7.11:

$$ {}^0T_1 = \begin{bmatrix} \cos\theta_1 & 0 & -\sin\theta_1 & 0 \\ \sin\theta_1 & 0 & \cos\theta_1 & 0 \\ 0 & -1 & 0 & d_1 \\ 0 & 0 & 0 & 1 \end{bmatrix} \tag{7.372} $$

$$ {}^1T_2 = \begin{bmatrix} \cos\theta_2 & 0 & \sin\theta_2 & 0 \\ \sin\theta_2 & 0 & -\cos\theta_2 & 0 \\ 0 & 1 & 0 & 0 \\ 0 & 0 & 0 & 1 \end{bmatrix} \tag{7.373} $$

$$ {}^2T_3 = \begin{bmatrix} 1 & 0 & 0 & 0 \\ 0 & 1 & 0 & 0 \\ 0 & 0 & 1 & d_3 \\ 0 & 0 & 0 & 1 \end{bmatrix} \tag{7.374} $$

$$ {}^3T_4 = \begin{bmatrix} 1 & 0 & 0 & 0 \\ 0 & 1 & 0 & 0 \\ 0 & 0 & 1 & d_4 \\ 0 & 0 & 0 & 1 \end{bmatrix} \tag{7.375} $$

Therefore, the configuration of the takht frame B_4 of the spherical arm in the base frame is

$$ \begin{aligned} {}^0T_4 &= {}^0T_1\,{}^1T_2\,{}^2T_3\,{}^3T_4 \\ &= \check{s}(d_1, \theta_1, \hat{k}_0)\,\check{s}(0, -\pi, \hat{\imath}_0)\,\check{s}(0, \theta_2, \hat{k}_1)\,\check{s}(0, \pi, \hat{\imath}_1) \\ &\quad \times \check{s}(d_3, 0, \hat{k}_2)\,\check{s}(0, 0, \hat{\imath}_2)\,\check{s}(d_4, 0, \hat{k}_3)\,\check{s}(0, 0, \hat{\imath}_3) \end{aligned} \tag{7.376} $$

Table 7.11 Screw Transformation for Spherical Arm Shown in Figure 7.50

Link No.	Class	Screw Transformation
1	R⊢R(−90)	${}^{0}T_1 = \check{s}(d_1, \theta_1, \hat{k}_0)\,\check{s}(0, -\pi, \hat{\imath}_0)$
2	R⊢P(90)	${}^{1}T_2 = \check{s}(0, \theta_2, \hat{k}_1)\,\check{s}(0, \pi, \hat{\imath}_1)$
3	P‖R(0)	${}^{2}T_3 = \check{s}(d_3, 0, \hat{k}_2)\,\check{s}(0, 0, \hat{\imath}_2)$
4	R‖R(0)	${}^{3}T_4 = \check{s}(d_4, 0, \hat{k}_3)\,\check{s}(0, 0, \hat{\imath}_3)$

Substituting the associated matrices and multiplying provide the forward kinematics for the takht frame of the arm in the base frame:

$$
\begin{aligned}
{}^{0}T_4 &= {}^{0}T_1\,{}^{1}T_2\,{}^{2}T_3\,{}^{3}T_4 \\
&= \begin{bmatrix} c\theta_1 c\theta_2 & -s\theta_1 & c\theta_1 s\theta_2 & (d_3 + d_4)\, c\theta_1 s\theta_2 \\ c\theta_2 s\theta_1 & c\theta_1 & s\theta_1 s\theta_2 & (d_3 + d_4)\, s\theta_1 s\theta_2 \\ -s\theta_2 & 0 & c\theta_2 & d_1 + d_3 c\theta_2 + d_4 c\theta_2 \\ 0 & 0 & 0 & 1 \end{bmatrix}
\end{aligned}
\tag{7.377}
$$

Example 446 ★ Spherical Wrist Forward Kinematics by Screws Consider the spherical wrist in Figure 7.46. The classes of links for the wrist are indicated in Table 7.12. Therefore, the transformation of the spherical wrist ${}^{7}T_8$ would be

$$
\begin{aligned}
{}^{7}T_8 &= {}^{5}T_6\,{}^{6}T_7\,{}^{7}T_8 \\
&= \check{s}(0, \theta_6, \hat{k}_5)\,\check{s}(0, -\pi, \hat{\imath}_5)\,\check{s}(0, \theta_7, \hat{k}_6)\,\check{s}(0, \pi, \hat{\imath}_6)\,\check{s}(0, \theta_8, \hat{k}_7)\,\check{s}(0, 0, \hat{\imath}_7)
\end{aligned}
\tag{7.378}
$$

Substituting the associated matrices and multiplying provide the same transformation as (7.246).

Table 7.12 Screw Transformation for Spherical Wrist Shown in Figure 7.46

Link No.	Class	Screw Transformation
6	R⊢R(−90)	${}^{5}T_6 = \check{s}(0, \theta_6, \hat{k}_5)\,\check{s}(0, -\pi, \hat{\imath}_5)$
7	R⊢R(−90)	${}^{6}T_7 = \check{s}(0, \theta_7, \hat{k}_6)\,\check{s}(0, \pi, \hat{\imath}_6)$
8	R⊢R(0)	${}^{7}T_8 = \check{s}(0, \theta_8, \hat{k}_7)\,\check{s}(0, 0, \hat{\imath}_7)$

Example 447 ★ Assembled Spherical Arm–Wrist Kinematics by Screws Consider the assembled spherical arm–wrist in Figure 7.52. The classes of links for the arm–wrist assembly are indicated in Table 7.13. Assembling the wrist of Figure 7.46 and arm of Figure 7.50 eliminates the coordinate frames B_3 and B_4. The number of links would then be 1,2,5,6,7,8. Link (1) is the trunk link R⊢R(−90) that holds the manipulator

Table 7.13 Screw Transformation for Spherical Arm–Wrist Assembly Shown in Figure 7.52

Link No.	Class	Screw transformation
1	R⊢R(−90)	${}^0T_1 = \check{s}(d_1, \theta_1, \hat{k}_0)\, \check{s}(0, -\pi, \hat{\imath}_0)$
2	R⊢P(90)	${}^1T_2 = \check{s}(0, \theta_2, \hat{k}_1)\, \check{s}(0, \pi, \hat{\imath}_1)$
5	P‖R(0)	${}^2T_3 = \check{s}(d_6, 0, \hat{k}_2)\, \check{s}(0, 0, \hat{\imath}_2)$
6	R⊢R(−90)	${}^5T_6 = \check{s}(0, \theta_6, \hat{k}_5)\, \check{s}(0, -\pi, \hat{\imath}_5)$
7	R⊢R(−90)	${}^6T_7 = \check{s}(0, \theta_7, \hat{k}_6)\, \check{s}(0, \pi, \hat{\imath}_6)$
8	R⊢R(0)	${}^7T_8 = \check{s}(0, \theta_8, \hat{k}_7)\, \check{s}(0, 0, \hat{\imath}_7)$
9	R‖R(0)	${}^8T_9 = \check{s}(d_9, 0, \hat{k}_8)\, \check{s}(0, 0, \hat{\imath}_8)$

on the ground and turns the arm about the global z_0-axis. Link (2) is the box link R⊢P(90) that holds and turns other links about the z_1-axis. Link (5) is the last link of the arm P‖R(0) and has the wrist attached to its tip point. Links (6), (7), and (8) are the three links of the wrist. Link (8) is the last link of the manipulator that holds the gripper. The coordinate frame B_9 is a Sina frame to indicate the gripper coordinate frame:

$$d_6 = d_3 + d_4 + d_5 \tag{7.379}$$

Example 448 ★ Plücker Coordinate of a Central Screw Using Plücker coordinates we can define a central screw as

$$\check{s}(h, \phi, \hat{u}) = \begin{bmatrix} \phi\hat{u} \\ h\hat{u} \end{bmatrix} \tag{7.380}$$

which is equal to

$$\begin{bmatrix} \phi\hat{u} \\ h\hat{u} \end{bmatrix} = D(h\hat{u})\, R(\hat{u}, \phi) \tag{7.381}$$

Therefore, the central screw $\check{s}(a_i, \alpha_i, \hat{\imath}_{i-1})$ can be expressed by a Plücker coordinate:

$$\check{s}(a_i, \alpha_i, \hat{\imath}_{i-1}) = \begin{bmatrix} \alpha_i\, \hat{\imath}_{i-1} \\ a_i\, \hat{\imath}_{i-1} \end{bmatrix} = D(a_i, \hat{\imath}_{i-1}) R(\hat{\imath}_{i-1}, \alpha_i) \tag{7.382}$$

Similarly, the central screw $\check{s}(d_i, \theta_i, \hat{k}_{i-1})$ can be expressed by a Plücker coordinate:

$$\check{s}(d_i, \theta_i, \hat{k}_{i-1}) = \begin{bmatrix} \theta_i\, \hat{k}_{i-1} \\ d_i\, \hat{k}_{i-1} \end{bmatrix} = D(d_i, \hat{k}_{i-1}) R(\hat{k}_{i-1}, \theta_i) \tag{7.383}$$

Example 449 ★ Intersecting Two Central Screws Two lines (and therefore two screws) are intersecting if their reciprocal product is zero. We can check that the reciprocal product of the intersecting screws $\check{s}(a_i, \alpha_i, \hat{\imath}_{i-1})$ and $\check{s}(d_i, \theta_i, \hat{k}_{i-1})$ is zero:

$$\begin{aligned}\check{s}(d_i, \theta_i, \hat{k}_{i-1}) \times \check{s}(a_i, \alpha_i, \hat{\imath}_{i-1}) &= \begin{bmatrix}\theta_i \hat{k}_{i-1} \\ d_i \hat{k}_{i-1}\end{bmatrix} \otimes \begin{bmatrix}\alpha_i \hat{\imath}_{i-1} \\ a_i \hat{\imath}_{i-1}\end{bmatrix} \\ &= \theta_i \hat{k}_{i-1} \cdot a_i \hat{\imath}_{i-1} + \alpha_i \hat{\imath}_{i-1} \cdot \theta_i \hat{k}_{i-1} \\ &= 0 \end{aligned} \tag{7.384}$$

7.8 ★ CASTER THEORY IN VEHICLES

The forward and assembling kinematics of multibodies can be used to derive the kinematic information of an interested multibody system. The kinematics of a steerable wheel of a vehicle is a good example of an applied mechanical multibody with interesting kinematic behavior. We use the wheel–vehicle system to show how the multibody kinematics can produce new information and ideas. We employ a steerable wheel of the vehicle to apply the multibody kinematics and develop a theory, which we call *caster theory*, to be used in the design of the vehicle's suspension.

The steering axis of a wheel should ideally pass through the center of the tireprint and be perpendicular to the ground plane. However, due to error, physical constraints, or design purposes, the steering axis of a steerable wheel may have any angle and be at any location with respect to the wheel–body coordinate frame $C\,(x_c, y_c, z_c)$. The wheel–body frame C is a fixed frame to the vehicle and parallel to the vehicle coordinate frame B. The frame C is at the center of the wheel when the wheel is at the rest position and does not follow any motion of the wheel.

Figure 7.60 illustrates the front and side views of a wheel and its steering axis. The steering axis has angle φ with the (y_c, z_c)-plane and angle θ with the (x_c, z_c)-plane. The angles φ and θ are measured about the y_c- and x_c-axes, respectively. The angle φ is called the *caster angle*, and the angle θ is called the *lean angle*. The steering axis of the wheel in Figure 7.60 has positive caster and lean angles. The steering

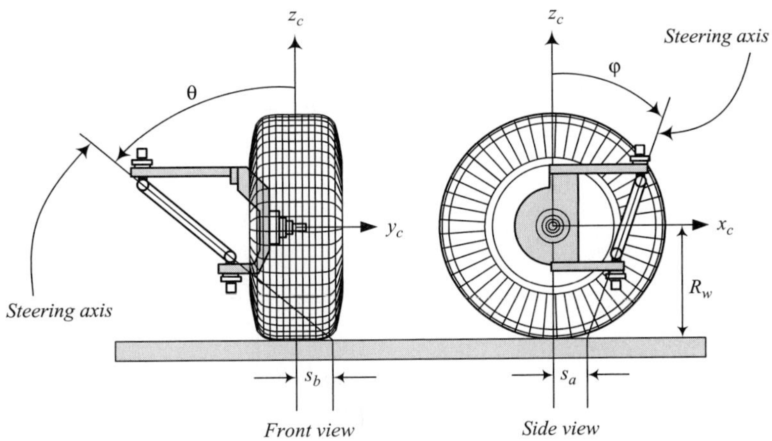

Figure 7.60 The front and side views of a wheel and its steering axis.

axis intersects the ground plane at a point that has coordinates $(s_a, s_b, -R_w)$ in the wheel–body coordinate frame C.

Let us indicate the steering axis by a vector $\hat{u}$. The components of $\hat{u}$ are functions of the caster and lean angles:

$$^C\hat{u} = \begin{bmatrix} u_1 \\ u_2 \\ u_3 \end{bmatrix} = \frac{1}{\sqrt{\cos^2\varphi + \cos^2\theta\sin^2\varphi}} \begin{bmatrix} \cos\theta\sin\varphi \\ -\cos\varphi\sin\theta \\ \cos\theta\cos\varphi \end{bmatrix} \tag{7.385}$$

The *location vector* $^C\mathbf{s}$ indicates the position vector of the intersection point of $\hat{u}$ and the ground plane. The components of $^C\mathbf{s}$ are

$$^C\mathbf{s} = \begin{bmatrix} s_a \\ s_b \\ -R_w \end{bmatrix} \tag{7.386}$$

We can express the rotation of the wheel about the steering axis $\hat{u}$ by a zero-pitch screw motion $\check{s}$:

$$\begin{aligned} ^CT_W &= {}^C\check{s}_W(0, \delta, \hat{u}, \mathbf{s}) \\ &= \begin{bmatrix} ^CR_W & ^C\mathbf{s} - {}^CR_W\,{}^C\mathbf{s} \\ 0 & 1 \end{bmatrix} = \begin{bmatrix} ^CR_W & ^C\mathbf{d}_W \\ 0 & 1 \end{bmatrix} \end{aligned} \tag{7.387}$$

Proof: The steering axis is at the intersection of the *caster plane* π_C and the *lean plane* π_L, both expressed in the wheel–body coordinate frame. The two planes can be indicated by their normal unit vectors $\hat{n}_1$ and $\hat{n}_2$:

$$^C\hat{n}_1 = \begin{bmatrix} 0 \\ \cos\theta \\ \sin\theta \end{bmatrix} \qquad {}^C\hat{n}_2 = \begin{bmatrix} -\cos\varphi \\ 0 \\ \sin\varphi \end{bmatrix} \tag{7.388}$$

The unit vector $\hat{u}$ on the intersection of the caster and lean planes can be found by

$$\hat{u} = \frac{\hat{n}_1 \times \hat{n}_2}{|\hat{n}_1 \times \hat{n}_2|} \tag{7.389}$$

where

$$\hat{n}_1 \times \hat{n}_2 = \begin{bmatrix} \cos\theta\sin\varphi \\ -\cos\varphi\sin\theta \\ \cos\theta\cos\varphi \end{bmatrix} \tag{7.390}$$

$$|\hat{n}_1 \times \hat{n}_2| = \sqrt{\cos^2\varphi + \cos^2\theta\sin^2\varphi} \tag{7.391}$$

and therefore,

$$^C\hat{u} = \begin{bmatrix} u_1 \\ u_2 \\ u_3 \end{bmatrix} = \begin{bmatrix} \dfrac{\cos\theta\sin\varphi}{\sqrt{\cos^2\varphi + \cos^2\theta\sin^2\varphi}} \\ \dfrac{-\cos\varphi\sin\theta}{\sqrt{\cos^2\varphi + \cos^2\theta\sin^2\varphi}} \\ \dfrac{\cos\theta\cos\varphi}{\sqrt{\cos^2\varphi + \cos^2\theta\sin^2\varphi}} \end{bmatrix} \tag{7.392}$$

The intersection point of the steering axis and the ground plane defines the *location vector* **s**:

$$^{C}\mathbf{s} = \begin{bmatrix} s_a \\ s_b \\ -R_w \end{bmatrix} \tag{7.393}$$

The components s_a and s_b are called the *forward* and *lateral steering locations*, respectively.

Using the axis–angle of rotation $(\hat{u}, \delta)$ and the location vector **s**, we can define the steering process as a screw motion $\check{s}$ with zero pitch. Employing Equations (6.245)–(6.249), we find the transformation screw for the wheel frame W to the wheel–body frame C:

$$\begin{aligned}
{}^{C}T_W &= {}^{C}\check{s}_W(0, \delta, \hat{u}, \mathbf{s}) \\
&= \begin{bmatrix} {}^{C}R_W & {}^{C}\mathbf{s} - {}^{C}R_W\,{}^{C}\mathbf{s} \\ 0 & 1 \end{bmatrix} = \begin{bmatrix} {}^{C}R_W & {}^{C}\mathbf{d} \\ 0 & 1 \end{bmatrix} && (7.394) \\
{}^{C}R_W &= \mathbf{I}\cos\delta + \hat{u}\hat{u}^{\mathrm{T}}\operatorname{vers}\delta + \tilde{u}\sin\delta && (7.395) \\
{}^{C}\mathbf{d}_W &= [(\mathbf{I} - \hat{u}\hat{u}^{\mathrm{T}})\operatorname{vers}\delta - \tilde{u}\sin\delta]\,{}^{C}\mathbf{s} && (7.396) \\
\tilde{u} &= \begin{bmatrix} 0 & -u_3 & u_2 \\ u_3 & 0 & -u_1 \\ -u_2 & u_1 & 0 \end{bmatrix} && (7.397) \\
\operatorname{vers}\delta &= 1 - \cos\delta && (7.398)
\end{aligned}$$

Direct substitution shows that

$$^{C}R_W = \begin{bmatrix} u_1^2\operatorname{vers}\delta + c\delta & u_1u_2\operatorname{vers}\delta - u_3s\delta & u_1u_3\operatorname{vers}\delta + u_2s\delta \\ u_1u_2\operatorname{vers}\delta + u_3s\delta & u_2^2\operatorname{vers}\delta + c\delta & u_2u_3\operatorname{vers}\delta - u_1s\delta \\ u_1u_3\operatorname{vers}\delta - u_2s\delta & u_2u_3\operatorname{vers}\delta + u_1s\delta & u_3^2\operatorname{vers}\delta + c\delta \end{bmatrix} \tag{7.399}$$

$$^{C}\mathbf{d}_W = \begin{bmatrix} (s_1 - u_1(s_3u_3 + s_2u_2 + s_1u_1))\operatorname{vers}\delta + (s_2u_3 - s_3u_2)\sin\delta \\ (s_2 - u_2(s_3u_3 + s_2u_2 + s_1u_1))\operatorname{vers}\delta + (s_3u_1 - s_1u_3)\sin\delta \\ (s_3 - u_3(s_3u_3 + s_2u_2 + s_1u_1))\operatorname{vers}\delta + (s_1u_2 - s_2u_1)\sin\delta \end{bmatrix} \tag{7.400}$$

The vector ${}^{C}\mathbf{d}_W$ indicates the position of the wheel center with respect to the wheel–body frame.

The matrix ${}^{C}T_W$ is the homogeneous transformation from the wheel frame W to the wheel–body frame C when the wheel is steered by the angle δ about the steering axis $\hat{u}$. ■

Example 450 ★ Zero Steer Angle To examine the screw transformation, we check the zero steering. Substituting $\delta = 0$ simplifies the steering transformation matrix ${}^{C}R_W$ to [**I**] and the position vector ${}^{C}\mathbf{d}_W$ to **0**:

$$^{C}R_W = \begin{bmatrix} 1 & 0 & 0 \\ 0 & 1 & 0 \\ 0 & 0 & 1 \end{bmatrix} \qquad {}^{C}\mathbf{d}_W = \begin{bmatrix} 0 \\ 0 \\ 0 \end{bmatrix} \tag{7.401}$$

It shows that at zero steering the wheel frame W and wheel–body frame C are coincident.

Example 451* ★ *Zero Lean and Caster Angles Consider an ideal wheel with a steering axis coincident with z_w. Such a wheel has no lean or caster angle. When the wheel is steered by an angle δ, we can find the coordinates of a wheel point P in the wheel–body coordinate frame using the transformation method. Figure 7.61 illustrates a 3D view of such a wheel. The top view of the wheel in Figure 7.62 indicates the relative orientation of the tire frame T, wheel frame W, and wheel–body frame C.

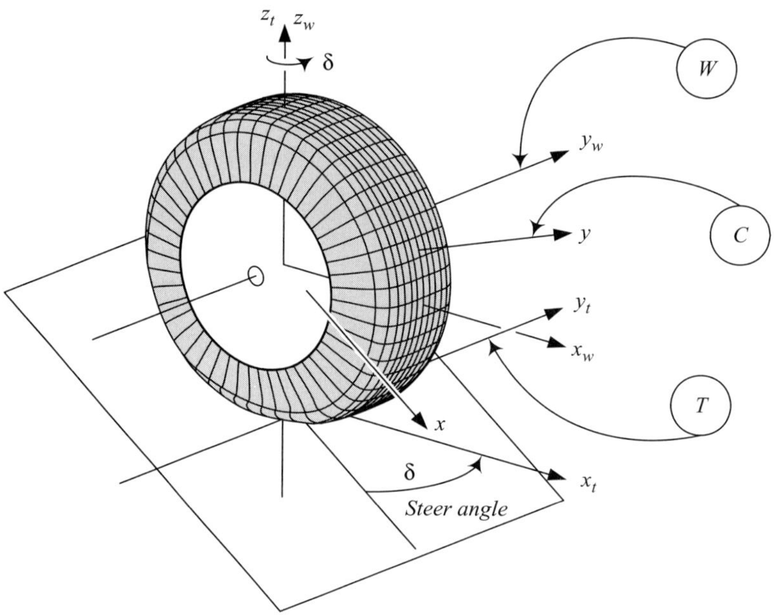

Figure 7.61 A 3D illustration of a steered wheel with a steering axis coincident with z_w.

Consider a point P in the wheel coordinate frame at ${}^W\mathbf{r}_P = \begin{bmatrix} x_w, y_w, z_w \end{bmatrix}^T$. The position vector of P in the wheel–body coordinate frame C is

$$
\begin{aligned}
{}^C\mathbf{r}_P &= {}^C R_W \, {}^W\mathbf{r}_P = R_{z,\delta} \, {}^W\mathbf{r}_P \\
&= \begin{bmatrix} \cos\delta & -\sin\delta & 0 \\ \sin\delta & \cos\delta & 0 \\ 0 & 0 & 1 \end{bmatrix} \begin{bmatrix} x_w \\ y_w \\ z_w \end{bmatrix} = \begin{bmatrix} x_w\cos\delta - y_w\sin\delta \\ y_w\cos\delta + x_w\sin\delta \\ z_w \end{bmatrix}
\end{aligned} \tag{7.402}
$$

We assumed that the wheel–body coordinate frame C is installed at the center of the wheel and is parallel to the vehicle coordinate frame B. Therefore, the transformation from frame W to frame C is a rotation δ about the wheel–body z-axis. There would be no camber angle when the lean and caster angles are zero and steer axis is about the z_w-axis.

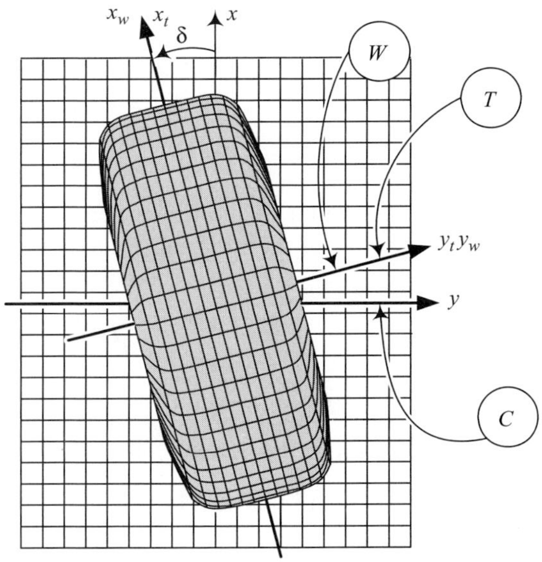

Figure 7.62 Top view of a steered wheel with a steering axis coincident with z_w.

Example 452 ★ Zero Lean, Zero Lateral Location The case of zero lean, $\theta = 0$, and zero lateral location, $s_b = 0$, is important in the caster dynamics of bikes, motorcycles, and bicycle models of vehicles. The screw transformation ${}^C\check{s}_W(0, \delta, \hat{u}, \mathbf{s})$ for this case will be simplified to

$$ {}^C T_W = {}^C\check{s}_W(0, \delta, \hat{u}, \mathbf{s}) = \begin{bmatrix} {}^C R_W & {}^C\mathbf{d} \\ 0 & 1 \end{bmatrix} \tag{7.403} $$

$$ {}^C\hat{u} = \begin{bmatrix} u_1 \\ u_2 \\ u_3 \end{bmatrix} = \begin{bmatrix} \sin\varphi \\ 0 \\ \cos\varphi \end{bmatrix} \tag{7.404} $$

$$ {}^C\mathbf{s} = \begin{bmatrix} s_a \\ 0 \\ -R_w \end{bmatrix} \tag{7.405} $$

$$ {}^C R_W = \begin{bmatrix} \sin^2\varphi \operatorname{vers}\delta + \cos\delta & -\cos\varphi\sin\delta & \sin\varphi\cos\varphi\operatorname{vers}\delta \\ \cos\varphi\sin\delta & \cos\delta & -\sin\varphi\sin\delta \\ \sin\varphi\cos\varphi\operatorname{vers}\delta & \sin\varphi\sin\delta & \cos^2\varphi\operatorname{vers}\delta + \cos\delta \end{bmatrix} \tag{7.406} $$

$$ {}^C\mathbf{d} = \begin{bmatrix} \cos\varphi\,(s_a\cos\varphi + R_w\sin\varphi)\operatorname{vers}\delta \\ -(s_a\cos\varphi + R_w\sin\varphi)\sin\delta \\ -\frac{1}{2}(R_w - R_w\cos 2\varphi + s_a\sin 2\varphi)\operatorname{vers}\delta \end{bmatrix} \tag{7.407} $$

Example 453 ★ Position of Tireprint Center Let us show the position vector of the center of a tireprint in the wheel coordinate frame by $\mathbf{r}_T$:

$$ {}^W\mathbf{r}_T = \begin{bmatrix} 0 \\ 0 \\ -R_w \end{bmatrix} \tag{7.408} $$

If we assume the width of the tire is zero and substitute the tire with the equivalent disc, then the center of the tireprint of the steered wheel would be at

$$ {}^C\mathbf{r}_T = {}^C T_W \, {}^W\mathbf{r}_T = \begin{bmatrix} x_T \\ y_T \\ z_T \end{bmatrix} \tag{7.409} $$

where

$$ x_T = \left(1 - u_1^2\right)(1 - \cos\delta)\, s_a + \left[u_3 \sin\delta - u_1 u_2 (1 - \cos\delta)\right] s_b \tag{7.410} $$

$$ y_T = -\left[u_3 \sin\delta + u_1 u_2 (1 - \cos\delta)\right] s_a + \left(1 - u_2^2\right)(1 - \cos\delta)\, s_b \tag{7.411} $$

$$ \begin{aligned} z_T = {} & \left[u_2 \sin\delta - u_1 u_3 (1 - \cos\delta)\right] s_a \\ & - \left[u_1 \sin\delta + u_2 u_3 (1 - \cos\delta)\right] s_b - R_w \end{aligned} \tag{7.412} $$

or

$$ \begin{aligned} x_T = {} & s_b \left(\frac{\cos\theta \cos\varphi \sin\delta}{\sqrt{\cos^2\theta \sin^2\varphi + \cos^2\varphi}} + \frac{1}{4} \frac{\sin 2\theta \sin 2\varphi \,(1 - \cos\delta)}{\cos^2\theta \sin^2\varphi + \cos^2\varphi} \right) \\ & + s_a \left(1 - \frac{\cos^2\theta \sin^2\varphi}{\cos^2\theta \sin^2\varphi + \cos^2\varphi} \right) (1 - \cos\delta) \end{aligned} \tag{7.413} $$

$$ \begin{aligned} y_T = {} & -s_a \left(\frac{\cos\theta \cos\varphi \sin\delta}{\sqrt{\cos^2\theta \sin^2\varphi + \cos^2\varphi}} - \frac{1}{4} \frac{\sin 2\theta \sin 2\varphi \,(1 - \cos\delta)}{\cos^2\theta \sin^2\varphi + \cos^2\varphi} \right) \\ & + s_b \left(1 - \frac{\cos^2\varphi \sin^2\theta}{\cos^2\theta \sin^2\varphi + \cos^2\varphi} \right) (1 - \cos\delta) \end{aligned} \tag{7.414} $$

$$ \begin{aligned} z_T = {} & -R_w - \frac{s_b \cos\theta \sin\varphi + s_a \cos\varphi \sin\theta}{\sqrt{\cos^2\theta \sin^2\varphi + \cos^2\varphi}} \sin\delta \\ & + \frac{1}{2} \frac{s_b \cos^2\varphi \sin 2\theta - s_a \cos^2\theta \sin 2\varphi}{\cos^2\theta \sin^2\varphi + \cos^2\varphi} (1 - \cos\delta) \end{aligned} \tag{7.415} $$

Example 454 ★ Wheel Center Vertical Displacement We can use the z_T component of ${}^C\mathbf{r}_T$ in (7.412) or (7.415) to determine the height that the center of the tireprint will move in the vertical direction when the wheel is steered. If the steer angle is zero, $\delta = 0$, then

$$ z_T = -R_w \tag{7.416} $$

Because the center of the tireprint is assumed to be on the ground, the height that the center of the wheel will drop during steering is given as

$$\begin{aligned} H &= -R_w - z_T \\ &= \frac{s_b \cos\theta \sin\varphi + s_a \cos\varphi \sin\theta}{\sqrt{\cos^2\theta \sin^2\varphi + \cos^2\varphi}} \sin\delta \\ &\quad - \frac{1}{2} \frac{s_b \cos^2\varphi \sin 2\theta - s_a \cos^2\theta \sin 2\varphi}{\cos^2\theta \sin^2\varphi + \cos^2\varphi} (1 - \cos\delta) \end{aligned} \tag{7.417}$$

The z_T-coordinate of the tireprint may be simplified for different designs:

1. If the lean angle is zero, $\theta = 0$, then

$$z_T = -R_w - \tfrac{1}{2} s_a \sin 2\varphi \,(1 - \cos\delta) - s_b \sin\varphi \sin\delta \tag{7.418}$$

2. If the lean angle and lateral location are zero, $\theta = 0$, $s_b = 0$, then

$$z_T = -R_w - \tfrac{1}{2} s_a \sin 2\varphi \,(1 - \cos\delta) \tag{7.419}$$

In this case, the wheel center drop may be expressed by the dimensionless equation

$$\frac{H}{s_a} = \frac{1}{2} \sin 2\varphi \,(1 - \cos\delta) \tag{7.420}$$

Figure 7.63 illustrates the wheel drop parameter H/s_a at zero lean angle $\theta = 0$ and zero lateral location $s_b = 0$ for various caster angles φ and various steer angles δ.

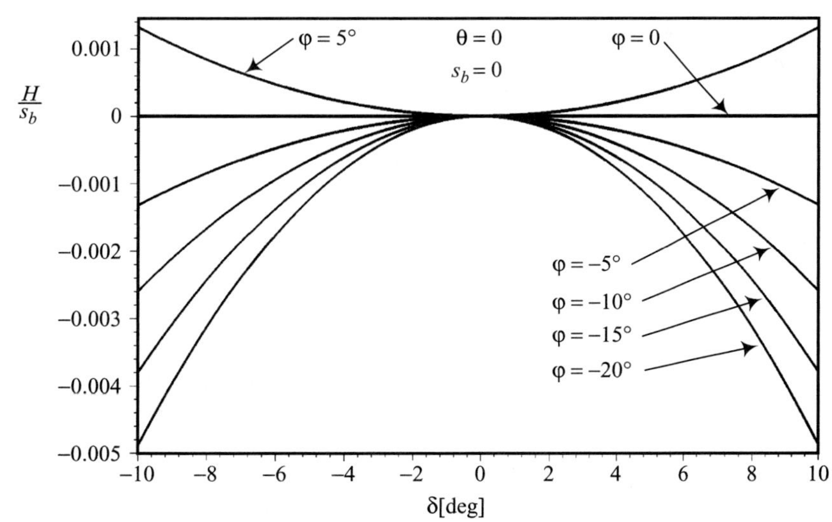

Figure 7.63 H/s_a for caster angles $\varphi = 5$ deg 0, -5 deg, -10 deg, -15 deg, -20 deg and steer angles in the range -10 deg $< \delta < 10$ deg.

3. If the caster angle is zero, $\varphi = 0$, then

$$z_T = -R_w + \tfrac{1}{2} s_b \sin 2\theta \,(1 - \cos\delta) - s_a \sin\theta \sin\delta \tag{7.421}$$

4. If the caster angle and lateral location are zero, $\varphi = 0$, $s_b = 0$, then

$$z_T = -R_w - s_a \sin\theta \sin\delta \tag{7.422}$$

In this case, the wheel center drop may be expressed by the dimensionless equation

$$\frac{H}{s_a} = -\sin\theta \sin\delta \tag{7.423}$$

Figure 7.64 illustrates H/s_a at zero caster angle $\varphi = 0$ and zero lateral location $s_b = 0$ for various lean angles θ and various steer angles δ.

Comparison of Figures 7.63 and 7.64 shows that the lean angle θ has much more effect on the wheel center drop than the caster angle φ.

5. If the lateral location is zero, $s_b = 0$, then

$$\begin{aligned} z_T = {} & -R_w - s_a \frac{\cos\varphi \sin\theta}{\sqrt{\cos^2\theta \sin^2\varphi + \cos^2\varphi}} \sin\delta \\ & - \frac{1}{2} s_a \frac{\cos^2\theta \sin 2\varphi}{\cos^2\theta \sin^2\varphi + \cos^2\varphi} (1 - \cos\delta) \end{aligned} \tag{7.424}$$

and the wheel center drop H may be expressed by the dimensionless equation

$$\frac{H}{s_a} = -\frac{1}{2} \frac{\cos^2\theta \sin^2\varphi \,(1 - \cos\delta)}{\cos^2\theta \sin^2\varphi + \cos^2\varphi} - \frac{\cos\varphi \sin\theta \sin\delta}{\sqrt{\cos^2\theta \sin^2\varphi + \cos^2\varphi}} \tag{7.425}$$

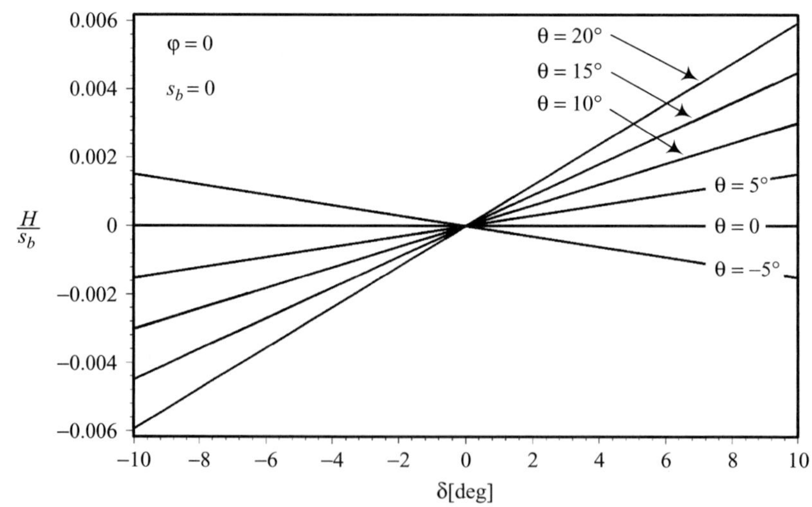

Figure 7.64 H/s_a for lean angles $\theta = 5$ deg 0, -5 deg, -10 deg, -15 deg, -20 deg and steer angles in the range -10 deg $< \delta <$ 10 deg.

Example 455 ★ Position of the Wheel Center As given in Equation (7.400), the wheel center is at ${}^{C}\mathbf{d}_{W}$ with respect to the wheel–body frame C:

$$ {}^{C}\mathbf{d}_{W} = \begin{bmatrix} x_W \\ y_W \\ z_W \end{bmatrix} \tag{7.426}$$

Substituting for $\hat{u}$ and $\mathbf{s}$ from (7.385) and (7.386) in (7.400) provides the coordinates of the wheel center in the wheel–body frame as

$$\begin{aligned} x_W = {} & [s_a - u_1(-R_w u_3 + s_b u_2 + s_a u_1)](1-\cos\delta) \\ & + (s_b u_3 + R_w u_2)\sin\delta \end{aligned} \tag{7.427}$$

$$\begin{aligned} y_W = {} & [s_b - u_2(-R_w u_3 + s_b u_2 + s_a u_1)](1-\cos\delta) \\ & - (R_w u_1 + s_a u_3)\sin\delta \end{aligned} \tag{7.428}$$

$$\begin{aligned} z_W = {} & [-R_w - u_3(-R_w u_3 + s_b u_2 + s_a u_1)](1-\cos\delta) \\ & + (s_a u_2 - s_b u_1)\sin\delta \end{aligned} \tag{7.429}$$

or

$$\begin{aligned} x_W = {} & s_a(1-\cos\delta) \\ & + \frac{\left(\frac{1}{2}R_w\sin 2\varphi - s_a\sin^2\varphi\right)\cos^2\theta + \frac{1}{4}s_b\sin 2\theta\sin 2\varphi}{\cos^2\varphi + \cos^2\theta\sin^2\varphi}(1-\cos\delta) \\ & + \frac{s_b\cos\theta - R_w\sin\theta}{\sqrt{\cos^2\varphi + \cos^2\theta\sin^2\varphi}}\cos\varphi\sin\delta \end{aligned} \tag{7.430}$$

$$\begin{aligned} y_W = {} & s_b(1-\cos\delta) \\ & - \frac{\frac{1}{2}\left(R_w\sin 2\theta + s_b\sin^2\theta\right)\cos^2\varphi - \frac{1}{4}s_a\sin 2\theta\sin 2\varphi}{\cos^2\varphi + \cos^2\theta\sin^2\varphi}(1-\cos\delta) \\ & - \frac{R_w\sin\varphi + s_a\cos\varphi}{\sqrt{\cos^2\varphi + \cos^2\theta\sin^2\varphi}}\cos\theta\sin\delta \end{aligned} \tag{7.431}$$

$$\begin{aligned} z_W = {} & -R_w(1-\cos\delta) \\ & + \frac{\left(R_w\cos^2\theta + \frac{1}{2}s_b\sin 2\theta\right)\cos^2\varphi - \frac{1}{2}s_a\cos^2\theta\sin 2\varphi}{\cos^2\varphi + \cos^2\theta\sin^2\varphi}(1-\cos\delta) \\ & - \frac{s_a\cos\varphi\sin\theta + s_b\cos\theta\sin\varphi}{\sqrt{\cos^2\varphi + \cos^2\theta\sin^2\varphi}}\sin\delta \end{aligned} \tag{7.432}$$

The z_W-coordinate indicates how the center of the wheel will move in the vertical direction with respect to the wheel–body frame when the wheel is steering. It shows that $z_W = 0$ as long as $\delta = 0$.

The z_W-coordinate of the wheel center may be simplified for different designs:

1. If the lean angle is zero, $\theta = 0$, then

$$z_W = -R_w\left(1-\cos^2\varphi\right)(1-\cos\delta) - s_b\sin\varphi\sin\delta - \tfrac{1}{2}s_a\sin 2\varphi\,(1-\cos\delta) \tag{7.433}$$

2. If the lean angle and lateral location are zero, $\theta = 0$, $s_b = 0$, then

$$z_W = -R_w\left(1-\cos^2\varphi\right)(1-\cos\delta) - \tfrac{1}{2}s_a\sin 2\varphi\,(1-\cos\delta) \tag{7.434}$$

3. If the caster angle is zero, $\varphi = 0$, then

$$z_W = -R_w\left(1-\cos^2\theta\right)(1-\cos\delta) - s_a\sin\theta\sin\delta + \tfrac{1}{2}s_b\sin 2\theta\,(1-\cos\delta) \tag{7.435}$$

4. If the caster angle and lateral location are zero, $\varphi = 0$, $s_b = 0$, then

$$z_W = -R_w\left(1-\cos^2\theta\right)(1-\cos\delta) - s_a\sin\theta\sin\delta \tag{7.436}$$

5. If the lateral location is zero, $s_b = 0$, then

$$z_W = -R_w(1-\cos\delta) - \frac{s_a\cos\varphi\sin\theta}{\sqrt{\cos^2\varphi+\cos^2\theta\sin^2\varphi}}\sin\delta + \frac{R_w\cos^2\theta\cos^2\varphi - \dfrac{1}{2}s_a\cos^2\theta\sin 2\varphi}{\cos^2\varphi+\cos^2\theta\sin^2\varphi}(1-\cos\delta) \tag{7.437}$$

In each of the above designs, the height of the wheel center with respect to the ground level can be found by adding H to z_W. The equations for calculating H are found in Example 454.

Example 456 ★ Camber Angle Having nonzero lean and caster angles causes a camber angle γ for a steered wheel. To find the camber angle of a steered wheel, we may determine the angle between the camber line and the vertical direction z_c. The *camber line* is the line connecting the wheel center and the center of the tireprint.

The coordinates of the center of the tireprint (x_T, y_T, z_T) are given in Equations (7.412)–(7.415), and the coordinates of the wheel center (x_W, y_W, z_W) are given in Equations (7.430)–(7.432). Let us indicate the line connecting (x_T, y_T, z_T) to (x_W, y_W, z_W) by a unit vector $\hat{l}_c$:

$$\hat{l}_c = \frac{(x_W-x_T)\,\hat{I} + (y_W-y_T)\,\hat{J} + (z_W-z_T)\,\hat{K}}{\sqrt{(x_W-x_T)^2+(y_W-y_T)^2+(z_W-z_T)^2}} \tag{7.438}$$

in which $\hat{I}$, $\hat{J}$, $\hat{K}$ are the unit vectors of the wheel–body coordinate frame C.

The camber angle is the angle between $\hat{l}_c$ and $\hat{K}$, which can be found by the inner vector product:

$$\gamma = \cos^{-1}\left(\hat{l}_c \cdot \hat{K}\right)$$
$$= \cos^{-1} \frac{(z_W - z_T)}{\sqrt{(x_W - x_T)^2 + (y_W - y_T)^2 + (z_W - z_T)^2}} \tag{7.439}$$

As a special case, let us determine the camber angle when the lean angle and lateral location are zero, $\theta = 0$, $s_b = 0$. In this case, we have

$$x_T = s_a \left(1 - \sin^2 \varphi\right) (\cos \delta - 1) \tag{7.440}$$
$$y_T = -s_a \cos \varphi \sin \delta \tag{7.441}$$
$$z_T = z_T = -R_w - \tfrac{1}{2} s_a \sin 2\varphi \,(1 - \cos \delta) \tag{7.442}$$
$$x_W = [s_a + \tfrac{1}{2} R_w \sin 2\varphi - s_a \sin^2 \varphi] \,(1 - \cos \delta) \tag{7.443}$$
$$y_W = s_b \,(1 - \cos \delta) - R_w \sin \varphi + s_a \cos \varphi \sin \delta \tag{7.444}$$
$$z_W = [R_w \left(\cos^2 \varphi - 1\right) - \tfrac{1}{2} s_a \sin 2\varphi] \,(1 - \cos \delta) \tag{7.445}$$

The lateral force of a tire is a function of sideslip and camber angles. The camber angle can provide a significant effect on the stability of a vehicle to increase the lateral and required force to keep the vehicle on the road. When the suspension and steering mechanisms of a vehicle are designed, Equation (7.439) is the required formula to calculate the camber angle of the tire during steering.

Example 457 ★ Application of Caster Theory The lateral force of a tire is proportional to the sideslip angle α and camber angle γ:

$$F_y = -C_\alpha \,\alpha - C_\gamma \,\gamma \tag{7.446}$$

The definition and a sample of the behavior of the camber and sideslip angles are indicated in Figures 7.65 and 7.66. Both camber and sideslip coefficients C_α and C_γ are proportional to the normal load F_z on the tire. Increasing F_z will increase C_α and C_γ.

By steering a tire, we change the direction of the tire coordinate frame with respect to the wheel–body frame and produce a sideslip angle α. However, the camber angle is produced by the suspension mechanism and the steering axis configuration.

Let us concentrate on the steering axis kinematics. A simple practical steering axis has no lean angle and has a constant lateral location:

$$\theta = 0 \qquad s_b = \text{const} \tag{7.447}$$

We introduce the idea of having a variable and controllable caster angle φ. A variable caster angle mechanism gives us the ability to adjust the camber angle to produce the required lateral force when the sideslip angle cannot be increased or decreased. As long as the left and right wheels are steering together according to a kinematic condition such as Ackerman, the sideslip angle of the inner wheel cannot be increased

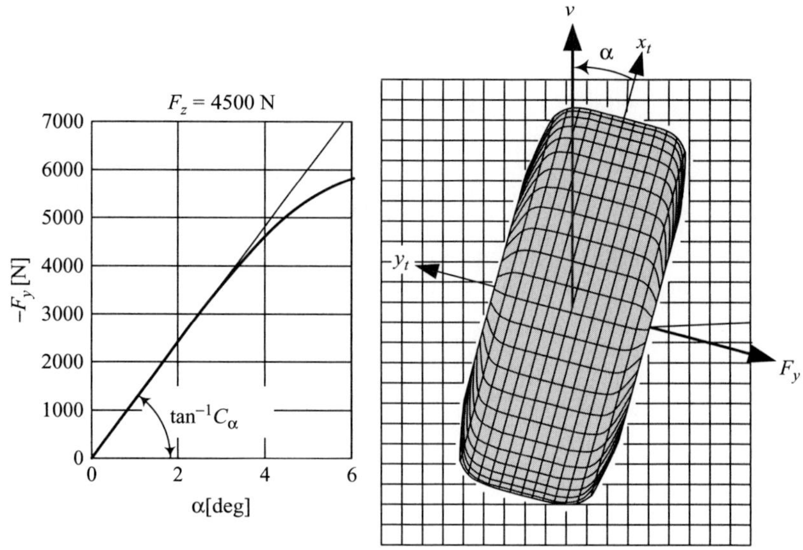

Figure 7.65 Top view of a tire and the generated sideslip force due to sideslip angle α.

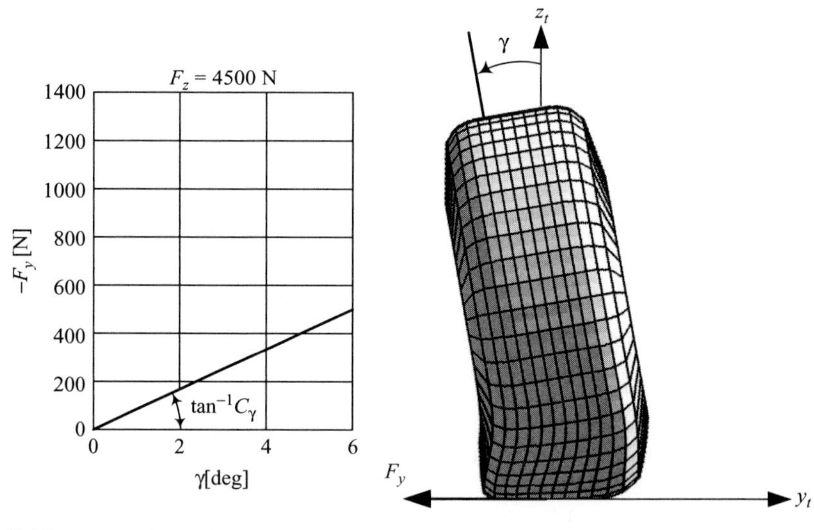

Figure 7.66 Front view of a cambered tire and the generated camber force due to camber angle γ.

independently to alter the reduced lateral force because of weight transfer and lowering the normal load F_z. A variable caster mechanism can adjust the caster angle of the wheels to achieve their top capacity and maximum lateral force, if needed. Such a mechanism will increase the stability and maneuverability of the vehicles.

Figure 7.67 depicts a mechanism to illustrate the variable caster angle idea. We may keep any point of the steering axis stationary with respect to the vehicle body and

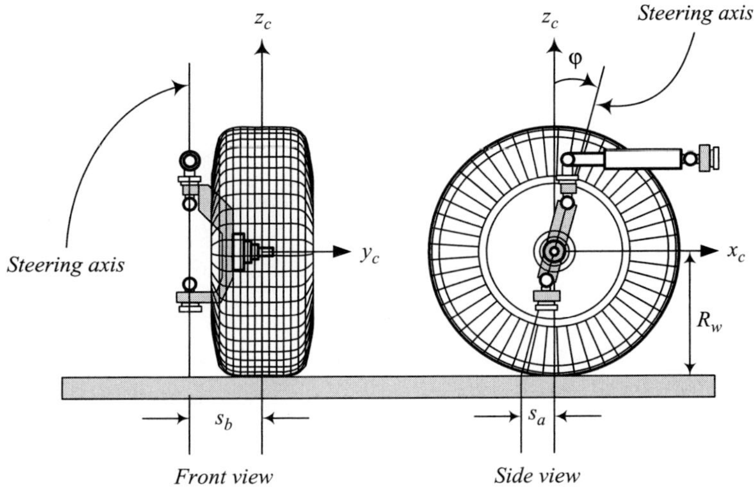

Figure 7.67 A variable caster angle mechanism.

make the steering axis turn about the point. As a sample design we may keep the point on the center of the wheel motionless with respect to the vehicle body. The upper and lower joints of the steering axis of such a design will turn on circles about the wheel center. The longitudinal location s_a of this design would be a function of the caster angle φ:

$$s_a = -R_w \tan\varphi \tag{7.448}$$

The camber–caster relationship of this steerable wheel is

$$\cos\gamma = \frac{z_W - z_T}{\sqrt{(x_W - x_T)^2 + (y_W - y_T)^2 + (z_W - z_T)^2}} \tag{7.449}$$

where

$$x_T = -R_w (1 - \cos\delta) \cos\varphi \sin\varphi + s_b \cos\varphi \sin\delta \tag{7.450}$$

$$y_T = s_b (1 - \cos\delta) + R_w \sin\varphi \sin\delta \tag{7.451}$$

$$z_T = R_w \sin^2\varphi \, (1 - \cos\delta) - R_w - s_b \sin\delta \sin\varphi \tag{7.452}$$

$$x_W = s_b \cos\varphi \sin\delta \tag{7.453}$$

$$y_W = s_b (1 - \cos\delta) \tag{7.454}$$

$$z_W = -s_b \sin\delta \sin\varphi \tag{7.455}$$

Substituting Equations (7.450)–(7.455) in to (7.449) and simplifying provide an applied equation to calculate the camber angle γ as a function of the caster and steer angles φ and δ:

$$\begin{aligned} \cos\gamma &= \cos\delta + (1 - \cos\delta)\cos^2\varphi \\ &= \cos\delta \sin^2\varphi + \cos^2\varphi \end{aligned} \tag{7.456}$$

Figure 7.68 illustrates the behavior of γ as a function of φ and δ.

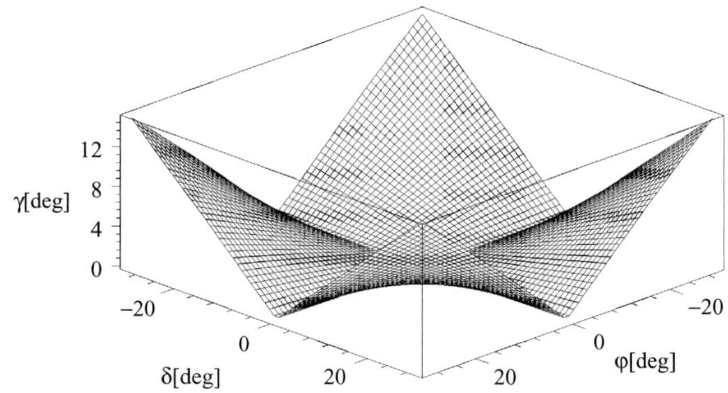

Figure 7.68 Behavior of camber angle γ as a function of caster and steer angles φ and δ.

7.9 INVERSE KINEMATICS

The determination of joint variables of a multibody to be at a specific position and orientation is called *inverse kinematics*. Mathematically, inverse kinematics involves searching for the elements of the joint variable vector $\mathbf{q}$ when transformation matrices are given as functions of the joint variables $q_1, q_2, q_3, \ldots$:

$$^{0}T_n = {}^{0}T_1(q_1)\,{}^{1}T_2(q_2)\,{}^{2}T_3(q_3)\,{}^{3}T_4(q_4)\ \ldots\ {}^{n-1}T_n(q_n) \tag{7.457}$$

Computer-controlled multibodies are usually actuated in the joint variable space; however, positions and orientations are usually expressed in the Cartesian coordinate frame. Therefore, carrying kinematic information back and forth between the joint space and Cartesian space is necessary in multibody applications. To control the configuration of a multibody to reach a position, we must solve the inverse kinematics of the multibody. Therefore, we need to know the required values of the joint variables to reach a desired point in a desired orientation.

Consider a multibody with n prismatic or revolute joints. Such a multibody has n DOF. To have a multibody capable of reaching a particular point at a particular orientation, the multibody needs six DOF, three to position the point and three to adjust the orientation. The result of forward kinematics of such a six-DOF multibody is a 4×4 transformation matrix

$$\begin{aligned} ^{0}T_6 &= {}^{0}T_1\,{}^{1}T_2\,{}^{2}T_3\,{}^{3}T_4\,{}^{4}T_5\,{}^{5}T_6 \\ &= \begin{bmatrix} {}^{0}R_6 & {}^{0}\mathbf{d}_6 \\ 0 & 1 \end{bmatrix} = \begin{bmatrix} r_{11} & r_{12} & r_{13} & r_{14} \\ r_{21} & r_{22} & r_{23} & r_{24} \\ r_{31} & r_{32} & r_{33} & r_{34} \\ 0 & 0 & 0 & 1 \end{bmatrix} \end{aligned} \tag{7.458}$$

The 12 elements of $^{0}T_6$ are trigonometric functions of six unknown joint variables. However, because the upper left 3×3 submatrix of (7.458) is a rotation matrix, only three elements of them are independent. This is because of the orthogonality condition

(4.276). Therefore, only 6 equations of the 12 equations of (7.458) are independent. Theoretically, we should be able to use the 6 independent equations and determine the 6 joint variables.

Trigonometric functions inherently provide multiple solutions. Therefore, we expect multiple inverse kinematic solutions and hence multiple configurations for the multibody.

It is possible to *decouple* the inverse kinematics problem into two subproblems, known as *inverse-position* and *inverse-orientation* kinematics. The practical consequence decoupling is to break the inverse kinematic problem into two independent problems, each with only three unknown parameters. Following the decoupling principle, the overall transformation matrix of a multibody can be decomposed to a translation and a rotation:

$$\begin{aligned} {}^0T_6 &= {}^0D_6\,{}^0R_6 \\ &= \begin{bmatrix} {}^0R_6 & {}^0\mathbf{d}_6 \\ 0 & 1 \end{bmatrix} = \begin{bmatrix} \mathbf{I} & {}^0\mathbf{d}_6 \\ 0 & 1 \end{bmatrix} \begin{bmatrix} {}^0R_6 & \mathbf{0} \\ 0 & 1 \end{bmatrix} \end{aligned} \tag{7.459}$$

The translation matrix ${}^0D_6$ can be solved for the position variables, and the rotation matrix ${}^0R_6$ can be solved for the orientation variables.

Proof: The six-DOF multibodies are similar to robotic manipulators that have a spherical wrist and an arm. The wrist has three revolute joints with intersecting and orthogonal axes at the wrist point. Taking advantage of having a spherical wrist, we can decouple the kinematics of the wrist and arm by decomposing the overall forward kinematics transformation matrix ${}^0T_6$ into the wrist orientation and wrist position:

$${}^0T_6 = {}^0T_3\,{}^3T_6 = \begin{bmatrix} {}^0R_3 & {}^0\mathbf{d}_6 \\ 0 & 1 \end{bmatrix} \begin{bmatrix} {}^3R_6 & \mathbf{0} \\ 0 & 1 \end{bmatrix} \tag{7.460}$$

The wrist orientation matrix is

$${}^3R_6 = {}^0R_3^{\mathrm{T}}\,{}^0R_6 = {}^0R_3^{\mathrm{T}} \begin{bmatrix} r_{11} & r_{12} & r_{13} \\ r_{21} & r_{22} & r_{23} \\ r_{31} & r_{32} & r_{33} \end{bmatrix} \tag{7.461}$$

and the wrist position vector is

$${}^0\mathbf{d}_6 = \begin{bmatrix} r_{14} \\ r_{24} \\ r_{34} \end{bmatrix} \tag{7.462}$$

The wrist position vector only includes the arm joint variables. Therefore, to solve the inverse kinematics of such an arm, we must solve ${}^0T_3$ for the position of the wrist point and then solve ${}^3T_6$ for the orientation of the wrist. The components of the wrist position vector ${}^0\mathbf{d}_6 = \mathbf{d}_w$ provides three equations for the three unknown arm joint variables. By solving the arm's joint variables from $\mathbf{d}_w$, we also know the matrix ${}^3R_6$. Then, the wrist orientation matrix ${}^3R_6$ can be solved for the wrist joint variables.

If we include the gripper coordinate frame in forward kinematics, the decomposition must be done according to the following equation to exclude the effect of gripper distance d_6 from the wrist–arm kinematics:

$$^0T_7 = {}^0T_3\,{}^3T_7 = {}^0T_3\,{}^3T_6\,{}^6T_7$$

$$= \begin{bmatrix} {}^0R_3 & \mathbf{d}_w \\ 0 & 1 \end{bmatrix} \begin{bmatrix} {}^3R_6 & \mathbf{0} \\ 0 & 1 \end{bmatrix} \begin{bmatrix} \mathbf{I} & \begin{matrix} 0 \\ 0 \\ d_6 \end{matrix} \\ 0 & 1 \end{bmatrix} \tag{7.463}$$

In this case, inverse kinematics starts from determination of $^0T_6$, which can be found by

$$^0T_6 = {}^0T_7\,{}^6T_7^{-1}$$

$$= {}^0T_7 \begin{bmatrix} 1 & 0 & 0 & 0 \\ 0 & 1 & 0 & 0 \\ 0 & 0 & 1 & d_6 \\ 0 & 0 & 0 & 1 \end{bmatrix}^{-1} = {}^0T_7 \begin{bmatrix} 1 & 0 & 0 & 0 \\ 0 & 1 & 0 & 0 \\ 0 & 0 & 1 & -d_6 \\ 0 & 0 & 0 & 1 \end{bmatrix} \tag{7.464}$$

■

Example 458 Inverse Kinematics and Multiple Solutions A 2R planar manipulator has two links of R‖R and coordinate frames as shown in Figure 7.69(a).

Using the transformation matrix of two links R‖R(0), we find the forward kinematics of the manipulator as

$$^0T_2 = {}^0T_1\,{}^1T_2$$

$$= \begin{bmatrix} c(\theta_1+\theta_2) & -s(\theta_1+\theta_2) & 0 & l_1 c\theta_1 + l_2 c(\theta_1+\theta_2) \\ s(\theta_1+\theta_2) & c(\theta_1+\theta_2) & 0 & l_1 s\theta_1 + l_2 s(\theta_1+\theta_2) \\ 0 & 0 & 1 & 0 \\ 0 & 0 & 0 & 1 \end{bmatrix} \tag{7.465}$$

The position of the tip point of the manipulator is at

$$\begin{bmatrix} X \\ Y \end{bmatrix} = \begin{bmatrix} l_1 \cos\theta_1 + l_2 \cos(\theta_1+\theta_2) \\ l_1 \sin\theta_1 + l_2 \sin(\theta_1+\theta_2) \end{bmatrix} \tag{7.466}$$

We should be able to use these two equations and determine the required angles θ_1, θ_2 for a given value of X and Y.

To find θ_2, we use

$$X^2 + Y^2 = l_1^2 + l_2^2 - 2l_1 l_2 \cos\theta_2 \tag{7.467}$$

and solve for θ_2:

$$\cos\theta_2 = \frac{X^2 + Y^2 - l_1^2 - l_2^2}{-2l_1 l_2} \tag{7.468}$$

However, as a general recommendation, we should avoid using arcsin and arccos because of inaccuracy. So, we employ the half-angle formula

$$\tan^2\frac{\theta}{2} = \frac{1-\cos\theta}{1+\cos\theta} \tag{7.469}$$

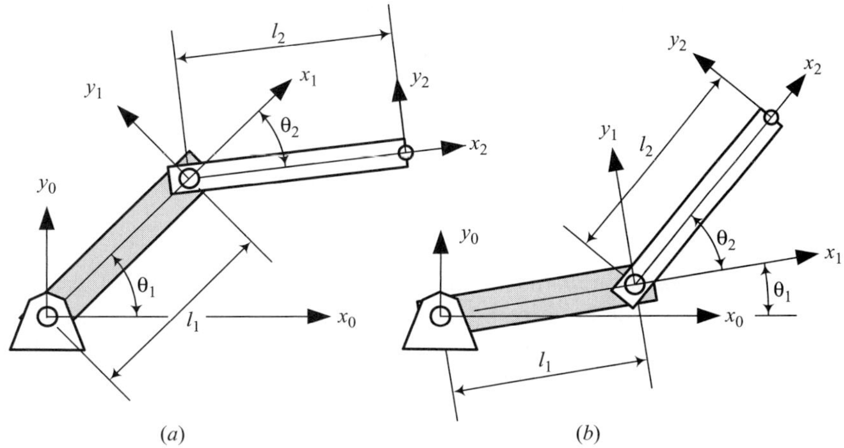

Figure 7.69 Illustration of a 2R planar manipulator in two possible configurations: (*a*) elbow up and (*b*) elbow down.

to find θ_2 using an atan2 function:

$$\theta_2 = \pm 2\text{atan2}\sqrt{\frac{(l_1 + l_2)^2 - (X+Y)^2}{\left(X^2 + Y^2\right) - \left(l_1^2 + l_2^2\right)}} \tag{7.470}$$

The $\pm$ is used because of the square root, and it generates two solutions. These two solutions are called *elbow-up* and *elbow-down* configurations, as shown in Figures 7.69(*a*) and (*b*).

Having θ_2, we can determine the first joint variable θ_1 from

$$\theta_1 = \text{atan2}\frac{X\left(l_1 + l_2\cos\theta_2\right) + Yl_2\sin\theta_2}{Y\left(l_1 + l_2\cos\theta_2\right) - Xl_2\sin\theta_2} \tag{7.471}$$

We will find a θ_1 for every value of θ_2. The associated angles determine the inverse kinematics of the elbow-up and elbow-down configurations.

Example 459 ★ ***Function* $\mathbf{arctan_2(y/x) = atan2(y, x)}$** In the kinematics of multibodies, especially in solving inverse kinematics problems, we need to find an angle based on the sin and cos functions of the angle. The regular arctan cannot show the effect of the individual sign for the numerator and denominator. It always represents an angle in the first or fourth quadrant. To overcome this problem and determine the angle in the correct quadrant, we should employ the atan2 function:

$$\text{atan2}(y, x) = \begin{cases} \text{sgn}y\tan^{-1}\left|\dfrac{y}{x}\right| & \text{if} \quad x > 0,\ y \neq 0 \\ \dfrac{\pi}{2}\text{sgn}y & \text{if} \quad x = 0,\ y \neq 0 \\ \text{sgn}y\left(\pi - \tan^{-1}\left|\dfrac{y}{x}\right|\right) & \text{if} \quad x < 0,\ y \neq 0 \\ \pi - \pi\,\text{sgn}x & \text{if} \quad x \neq 0,\ y = 0 \end{cases} \tag{7.472}$$

where sgn represents the signum function:

$$\operatorname{sgn}(x) = \begin{cases} 1 & \text{if} \quad x > 0 \\ 0 & \text{if} \quad x = 0 \\ -1 & \text{if} \quad x < 0 \end{cases} \tag{7.473}$$

We recommend using $\operatorname{atan2}(y, x)$ wherever $\arctan(y/x)$ is needed.

Example 460 Inverse kinematics of an Articulated Arm An articulated arm is any arm made by three links and three revolute joints to reach a point in three dimensions. However, an articulated arm is usually referred to a three-link multibody with base, shoulder, and elbow joints such as the one shown in Figure 7.70.

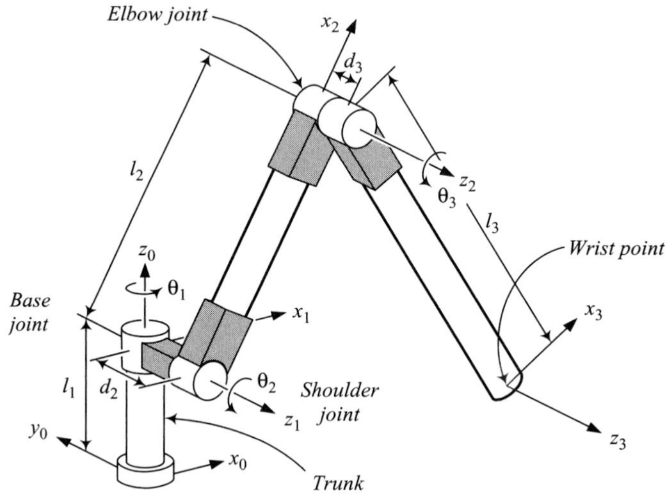

Figure 7.70 An articulated arm.

The first link of the arm is an R⊢R(90) with a distance l_1 between x_0 and x_1. The transformation matrix of the first coordinate frame is

$${}^0T_1 = \begin{bmatrix} \cos\theta_1 & 0 & \sin\theta_1 & 0 \\ \sin\theta_1 & 0 & -\cos\theta_1 & 0 \\ 0 & 1 & 0 & l_1 \\ 0 & 0 & 0 & 1 \end{bmatrix} \tag{7.474}$$

The second link of the arm is an R‖R(90) with a distance d_2 between x_1 and x_2 and a distance l_2 between z_1 and z_2. The transformation matrix of the second coordinate frame is

$${}^1T_2 = \begin{bmatrix} \cos\theta_2 & -\sin\theta_2 & 0 & l_2\cos\theta_2 \\ \sin\theta_2 & \cos\theta_2 & 0 & l_2\sin\theta_2 \\ 0 & 0 & 1 & d_2 \\ 0 & 0 & 0 & 1 \end{bmatrix} \tag{7.475}$$

Because the third link is not connected to the other link at its distal end, we should set B_3 at its proximal end and attach a takht frame at its tip point. For simplicity, we may also attach only one coordinate frame at the tip point and consider the link as an R⊢R(90) with a distance d_3 between x_2 and x_3 and a distance l_3 between z_2 and z_3. The transformation matrix of the third coordinate frame is

$$ {}^2T_3 = \begin{bmatrix} \cos\theta_3 & -\sin\theta_3 & 0 & l_3\cos\theta_3 \\ \sin\theta_3 & \cos\theta_3 & 0 & l_3\sin\theta_3 \\ 0 & 0 & 1 & d_3 \\ 0 & 0 & 0 & 1 \end{bmatrix} \tag{7.476} $$

The forward kinematics of the takht frame B_3 in the base frame is

$$ \begin{aligned} {}^0T_3 &= {}^0T_1\,{}^1T_2\,{}^2T_3 \\ &= \begin{bmatrix} \cos(\theta_2+\theta_3)\cos\theta_1 & -\sin(\theta_2+\theta_3)\cos\theta_1 & \sin\theta_1 & d_x \\ \cos(\theta_2+\theta_3)\sin\theta_1 & -\sin(\theta_2+\theta_3)\sin\theta_1 & -\cos\theta_1 & d_y \\ \sin(\theta_2+\theta_3) & \cos(\theta_2+\theta_3) & 0 & d_z \\ 0 & 0 & 0 & 1 \end{bmatrix} \end{aligned} \tag{7.477} $$

$$ \begin{bmatrix} d_x \\ d_y \\ d_z \end{bmatrix} = \begin{bmatrix} (l_2\cos\theta_2 + l_3\cos(\theta_2+\theta_3))\cos\theta_1 + (d_2+d_3)\sin\theta_1 \\ (l_2\cos\theta_2 + l_3\cos(\theta_2+\theta_3))\sin\theta_1 - (d_3+d_2)\cos\theta_1 \\ l_1 + l_2\sin\theta_2 + l_3\sin(\theta_2+\theta_3) \end{bmatrix} \tag{7.478} $$

The vector ${}^0\mathbf{d}_3 = \begin{bmatrix} d_x & d_y & d_z \end{bmatrix}$ indicates the tip point of the arm in the base coordinate frame. At the rest position the matrix ${}^0T_3$ reduces to

$$ {}^0T_3 = \begin{bmatrix} 1 & 0 & 0 & l_2+l_3 \\ 0 & 0 & -1 & -d_2-d_3 \\ 0 & 1 & 0 & l_1 \\ 0 & 0 & 0 & 1 \end{bmatrix} \tag{7.479} $$

The inverse kinematic problem of the arm is looking for the required angles $\theta_1, \theta_2, \theta_3$ for a given position vector ${}^0\mathbf{d}_3$. There is no standard or unique method of solution. So, the engineer may develop a method based on his or her experience and preference. However, because there is not a very wide variety of industrial arms, most practical arms have already been analyzed and an applied solution of their inverse kinematic problems has been developed.

To find θ_1, we can use

$$ d_x\sin\theta_1 - d_y\cos\theta_1 = d_2 + d_3 \tag{7.480} $$

which provides

$$ \theta_1 = 2\operatorname{atan2}\frac{d_x \pm \sqrt{d_x^2 + d_y^2 - (d_2+d_3)^2}}{d_2+d_3-d_y} \tag{7.481} $$

Equation (7.481) has two solutions for $d_x^2 + d_y^2 > (d_2+d_3)^2$, one solution for $d_x^2 + d_y^2 = (d_2+d_3)^2$, and no real solution for $d_x^2 + d_y^2 < (d_2+d_3)^2$.

To find θ_2, we can combine the first two elements of ${}^0\mathbf{d}_3$,

$$l_3 \cos(\theta_2 + \theta_3) = \pm\sqrt{d_x^2 + d_y^2 - (d_2 + d_3)^2} - l_2 \cos\theta_2 \tag{7.482}$$

and substitute the result in the third element of ${}^0\mathbf{d}_3$ to find

$$l_3^2 = \left(\pm\sqrt{d_x^2 + d_y^2 - (d_2 + d_3)^2} - l_2 \cos\theta_2\right)^2 + (d_z - l_1 - l_2 \sin\theta_2)^2 \tag{7.483}$$

This equation can be rearranged to the form

$$a\cos\theta_2 + b\sin\theta_2 = c \tag{7.484}$$

where

$$a = -2l_2\sqrt{d_x^2 + d_y^2 - (d_2 + d_3)^2} \tag{7.485}$$

$$b = 2l_2\,(l_1 - d_z) \tag{7.486}$$

$$c = d_x^2 + d_y^2 + (l_1 - d_z)^2 - (d_2 + d_3)^2 + l_2^2 - l_3^2 \tag{7.487}$$

with two solutions

$$\theta_2 = \begin{cases} \operatorname{atan2}\left(\dfrac{c}{r}, \pm\sqrt{1 - \dfrac{c^2}{r^2}}\right) - \operatorname{atan2}(a, b) & (7.488) \\ \operatorname{atan2}\left(\dfrac{c}{r}, \pm\sqrt{r^2 - c^2}\right) - \operatorname{atan2}(a, b) & (7.489) \end{cases}$$

where

$$r^2 = a^2 + b^2 \tag{7.490}$$

Having the angle θ_2, we can determine the angle θ_3 from the third element of ${}^0\mathbf{d}_3$:

$$\theta_3 = \arcsin\left(\frac{d_z - l_1 - l_2 \sin\theta_2}{l_3}\right) - 2\theta_2 \tag{7.491}$$

Example 461 ★ Solution of Equation $a\cos\theta + b\sin\theta = c$ The first type of trigonometric equation,

$$a\cos\theta + b\sin\theta = c \tag{7.492}$$

can be solved by introducing two new variables r and ϕ such that

$$a = r\sin\phi \tag{7.493}$$

$$b = r\cos\phi \tag{7.494}$$

and

$$r = \sqrt{a^2 + b^2} \tag{7.495}$$

$$\phi = \operatorname{atan2}(a, b) \tag{7.496}$$

Substituting the new variables shows that

$$\sin(\phi + \theta) = \frac{c}{r} \tag{7.497}$$

$$\cos(\phi+\theta) = \pm\sqrt{1-\frac{c^2}{r^2}} \tag{7.498}$$

Hence, the solutions of the problem are

$$\theta = \text{atan2}\left(\frac{c}{r}, \pm\sqrt{1-\frac{c^2}{r^2}}\right) - \text{atan2}(a,b) \tag{7.499}$$

and

$$\theta = \text{atan2}\left(\frac{c}{r}, \pm\sqrt{r^2-c^2}\right) - \text{atan2}(a,b) \tag{7.500}$$

Therefore, the equation $a\cos\theta + b\sin\theta = c$ has two solutions if $r^2 = a^2+b^2 > c^2$, one solution if $r^2 = c^2$, and no solution if $r^2 < c^2$.

Example 462 Numeric Inverse Kinematics of an Articulated Arm Let us assume the following data for the arm of Example 460:

$$\begin{matrix} l_1 = 0.5\,\text{m} & l_2 = 1.0\,\text{m} & l_3 = 1.2\,\text{m} \\ d_2 = 0.3\,\text{m} & d_3 = 0.15\,\text{m} & \end{matrix} \tag{7.501}$$

The forward kinematics of B_3 in the global frame B_0 would be

$$\begin{aligned} {}^0T_3 &= {}^0T_1\,{}^1T_2\,{}^2T_3 \\ &= \begin{bmatrix} \cos(\theta_2+\theta_3)\cos\theta_1 & -\sin(\theta_2+\theta_3)\cos\theta_1 & \sin\theta_1 & d_x \\ \cos(\theta_2+\theta_3)\sin\theta_1 & -\sin(\theta_2+\theta_3)\sin\theta_1 & -\cos\theta_1 & d_y \\ \sin(\theta_2+\theta_3) & \cos(\theta_2+\theta_3) & 0 & d_z \\ 0 & 0 & 0 & 1 \end{bmatrix} \end{aligned} \tag{7.502}$$

$$\begin{bmatrix} d_x \\ d_y \\ d_z \end{bmatrix} = \begin{bmatrix} (\cos\theta_2 + 1.2\cos(\theta_2+\theta_3))\cos\theta_1 + 0.45\sin\theta_1 \\ (\cos\theta_2 + 1.2\cos(\theta_2+\theta_3))\sin\theta_1 - 0.45\cos\theta_1 \\ 0.5+\sin\theta_2 + 1.2\sin(\theta_2+\theta_3) \end{bmatrix} \tag{7.503}$$

Let us assume the tip point is at

$${}^0\mathbf{d}_3 = \begin{bmatrix} d_x \\ d_y \\ d_z \end{bmatrix} = \begin{bmatrix} 1.8 \\ 0.5 \\ 0.8 \end{bmatrix} \tag{7.504}$$

and the arm is in an elbow-up configuration.

The angle θ_1 would be

$$\begin{aligned} \theta_1 &= 2\text{atan2}\frac{d_x - \sqrt{d_x^2+d_y^2-(d_2+d_3)^2}}{d_2+d_3-d_y} \\ &= 2\text{atan2}\frac{1.8-\sqrt{1.8^2+0.5^2-0.45^2}}{0.45-0.5} \\ &= 0.514\,22\,\text{rad} \approx 29.463\,\text{deg} \end{aligned} \tag{7.505}$$

The angle θ_2 can be found from

$$a\cos\theta_2 + b\sin\theta_2 = c \tag{7.506}$$

where

$$a = -2l_2\sqrt{d_x^2 + d_y^2 - (d_2+d_3)^2} = -3.626\,3 \tag{7.507}$$

$$b = 2l_2\,(l_1 - d_z) = -0.6 \tag{7.508}$$

$$c = d_x^2 + d_y^2 + (l_1 - d_z)^2 - (d_2+d_3)^2 + l_2^2 - l_3^2 = 2.9375 \tag{7.509}$$

Therefore,

$$\theta_2 = \begin{cases} 2.660\,7\,\text{rad} \approx 152.45\,\text{deg} \\ -2.332\,8\,\text{rad} \approx -133.66\,\text{deg} \end{cases} \tag{7.510}$$

The positive value of θ_2 belongs to the desired elbow-up configuration.

The associated angle θ_3 is given as

$$\begin{aligned}\theta_3 &= \arcsin\left(\frac{d_z - l_1 - l_2\sin\theta_2}{l_3}\right) - 2\theta_2 \\ &= \arcsin\left(\frac{0.8 - 0.5 - \sin 2.660\,7}{1.2}\right) - 2\times 2.6607 \\ &= -5.4573\,\text{rad} \approx -312.68\,\text{deg}\end{aligned} \tag{7.511}$$

Example 463 Inverse Kinematics of a Spherical Wrist The forward kinematics of the spherical wrist of Figure 7.46 is

$$\begin{aligned}{}^5T_8 &= {}^5T_6\,{}^6T_7\,{}^7T_8 \\ &= \begin{bmatrix} c\theta_6 c\theta_7 c\theta_8 - s\theta_6 s\theta_8 & -c\theta_8 s\theta_6 - c\theta_6 c\theta_7 s\theta_8 & c\theta_6 s\theta_7 & 0 \\ c\theta_6 s\theta_8 + c\theta_7 c\theta_8 s\theta_6 & c\theta_6 c\theta_8 - c\theta_7 s\theta_6 s\theta_8 & s\theta_6 s\theta_7 & 0 \\ -c\theta_8 s\theta_7 & s\theta_7 s\theta_8 & c\theta_7 & 0 \\ 0 & 0 & 0 & 1 \end{bmatrix}\end{aligned} \tag{7.512}$$

where B_5 is a fixed coordinate frame with respect to the wrist, while the gripper of the wrist is fixed in frame B_8. Numerically,

$${}^5T_8 = \begin{bmatrix} r_{11} & r_{12} & r_{13} & 0 \\ r_{21} & r_{22} & r_{23} & 0 \\ r_{31} & r_{32} & r_{33} & 0 \\ 0 & 0 & 0 & 1 \end{bmatrix} \tag{7.513}$$

and we should be able to determine the angles $\theta_6, \theta_7, \theta_8$:

$$\theta_6 = \arctan\frac{r_{23}}{r_{13}} \tag{7.514}$$

$$\theta_7 = \arccos r_{33} \tag{7.515}$$

$$\theta_8 = -\arctan\frac{r_{32}}{r_{31}} \tag{7.516}$$

Example 464 Inverse Kinematics of an Articulated Manipulator An articulated manipulator is made by assembling a spherical wrist and an articulated arm. Figure 7.71 illustrates an articulated manipulator symbolically along with a new set of DH coordinate frames. Installing the base coordinate frame B_0 such that x_0 and z_1 are colinear simplifies the transformation matrices.

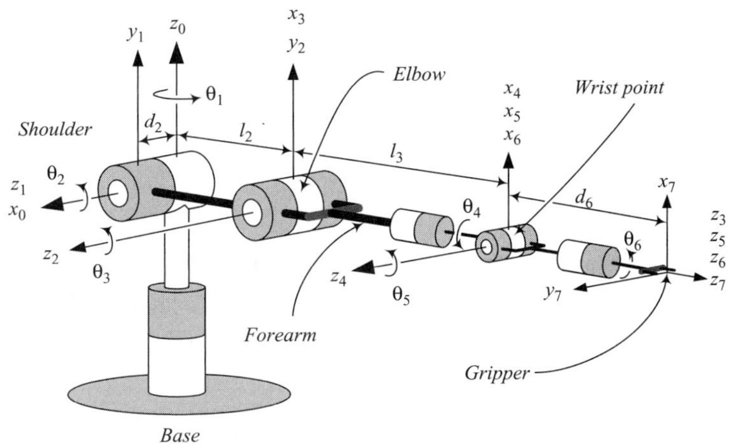

Figure 7.71 An articulated manipulator with 6 DOF.

The class of the links of the manipulator and the transformation matrices ${}^{i-1}T_i$ for this setup of the coordinate frames are

1	R⊢R(90)
2	R‖R(0)
3	R⊢R(90)
4	R⊢R(−90)
5	R⊢R(90)
6	R‖R(0)

$$ {}^{0}T_1 = \begin{bmatrix} \cos\theta_1 & 0 & \sin\theta_1 & 0 \\ \sin\theta_1 & 0 & -\cos\theta_1 & 0 \\ 0 & 1 & 0 & 0 \\ 0 & 0 & 0 & 1 \end{bmatrix} \tag{7.517} $$

$$ {}^{1}T_2 = \begin{bmatrix} \cos\theta_2 & -\sin\theta_2 & 0 & l_2\cos\theta_2 \\ \sin\theta_2 & \cos\theta_2 & 0 & l_2\sin\theta_2 \\ 0 & 0 & 1 & d_2 \\ 0 & 0 & 0 & 1 \end{bmatrix} \tag{7.518} $$

$$ {}^{2}T_3 = \begin{bmatrix} \cos\theta_3 & 0 & \sin\theta_3 & 0 \\ \sin\theta_3 & 0 & -\cos\theta_3 & 0 \\ 0 & 1 & 0 & 0 \\ 0 & 0 & 0 & 1 \end{bmatrix} \tag{7.519} $$

$$
{}^{3}T_{4} = \begin{bmatrix} \cos\theta_4 & 0 & -\sin\theta_4 & 0 \\ \sin\theta_4 & 0 & \cos\theta_4 & 0 \\ 0 & -1 & 0 & l_3 \\ 0 & 0 & 0 & 1 \end{bmatrix} \tag{7.520}
$$

$$
{}^{4}T_{5} = \begin{bmatrix} \cos\theta_5 & 0 & \sin\theta_5 & 0 \\ \sin\theta_5 & 0 & -\cos\theta_5 & 0 \\ 0 & 1 & 0 & 0 \\ 0 & 0 & 0 & 1 \end{bmatrix} \tag{7.521}
$$

$$
{}^{5}T_{6} = \begin{bmatrix} \cos\theta_6 & -\sin\theta_6 & 0 & 0 \\ \sin\theta_6 & \cos\theta_6 & 0 & 0 \\ 0 & 0 & 1 & 0 \\ 0 & 0 & 0 & 1 \end{bmatrix} \tag{7.522}
$$

The gripper is at a distance d_6 from the wrist point and is parallel to B_6:

$$
{}^{6}T_{7} = \begin{bmatrix} 1 & 0 & 0 & 0 \\ 0 & 1 & 0 & 0 \\ 0 & 0 & 1 & d_6 \\ 0 & 0 & 0 & 1 \end{bmatrix} \tag{7.523}
$$

The gripper transformation matrix in the base coordinate frame is

$$
\begin{aligned}
{}^{0}T_{7} &= {}^{0}T_{1}\,{}^{1}T_{2}\,{}^{2}T_{3}\,{}^{3}T_{4}\,{}^{4}T_{5}\,{}^{5}T_{6}\,{}^{6}T_{7} = {}^{0}T_{3}\,{}^{3}T_{6}\,{}^{6}T_{7} \\
&= \begin{bmatrix} t_{11} & t_{12} & t_{13} & t_{14} \\ t_{21} & t_{22} & t_{23} & t_{24} \\ t_{31} & t_{32} & t_{33} & t_{34} \\ 0 & 0 & 0 & 1 \end{bmatrix}
\end{aligned} \tag{7.524}
$$

where

$$
{}^{0}T_{3} = \begin{bmatrix} c\theta_1 c(\theta_2+\theta_3) & s\theta_1 & c\theta_1 s(\theta_2+\theta_3) & l_2 c\theta_1 c\theta_2 + d_2 s\theta_1 \\ s\theta_1 c(\theta_2+\theta_3) & -c\theta_1 & s\theta_1 s(\theta_2+\theta_3) & l_2 c\theta_2 s\theta_1 - d_2 c\theta_1 \\ s(\theta_2+\theta_3) & 0 & -c(\theta_2+\theta_3) & l_2 s\theta_2 \\ 0 & 0 & 0 & 1 \end{bmatrix} \tag{7.525}
$$

$$
{}^{3}T_{6} = \begin{bmatrix} c\theta_4 c\theta_5 c\theta_6 - s\theta_4 s\theta_6 & -c\theta_6 s\theta_4 - c\theta_4 c\theta_5 s\theta_6 & c\theta_4 s\theta_5 & 0 \\ c\theta_5 c\theta_6 s\theta_4 + c\theta_4 s\theta_6 & c\theta_4 c c\theta_6 - c\theta_5 s\theta_4 s\theta_6 & s\theta_4 s\theta_5 & 0 \\ -c\theta_6 s\theta_5 & s\theta_5 s\theta_6 & c\theta_5 & l_3 \\ 0 & 0 & 0 & 1 \end{bmatrix} \tag{7.526}
$$

and

$$
\begin{aligned}
t_{11} = c\theta_1 [c\,(\theta_2+\theta_3)\,(c\theta_4 c\theta_5 c\theta_6 - s\theta_4 s\theta_6) - c\theta_6 s\theta_5 s\,(\theta_2+\theta_3)] \\
+ s\theta_1\,(c\theta_4 s\theta_6 + c\theta_5 c\theta_6 s\theta_4)
\end{aligned} \tag{7.527}
$$

$$
\begin{aligned}
t_{21} = s\theta_1 [c\,(\theta_2+\theta_3)\,(-s\theta_4 s\theta_6 + c\theta_4 c\theta_5 c\theta_6) - c\theta_6 s\theta_5 s\,(\theta_2+\theta_3)] \\
- c\theta_1\,(c\theta_4 s\theta_6 + c\theta_5 c\theta_6 s\theta_4)
\end{aligned} \tag{7.528}
$$

$$
t_{31} = s\,(\theta_2+\theta_3)\,(c\theta_4 c\theta_5 c\theta_6 - s\theta_4 s\theta_6) + c\theta_6 s\theta_5 c\,(\theta_2+\theta_3) \tag{7.529}
$$

$$t_{12} = c\theta_1 [s\theta_5 s\theta_6 s(\theta_2 + \theta_3) - c(\theta_2 + \theta_3)(c\theta_6 s\theta_4 + c\theta_4 c\theta_5 s\theta_6)] + s\theta_1 (c\theta_4 c\theta_6 - c\theta_5 s\theta_4 s\theta_6) \tag{7.530}$$

$$t_{22} = s\theta_1 [s\theta_5 s\theta_6 s(\theta_2 + \theta_3) - c(\theta_2 + \theta_3)(c\theta_6 s\theta_4 + c\theta_4 c\theta_5 s\theta_6)] + c\theta_1 (-c\theta_4 c\theta_6 + c\theta_5 s\theta_4 s\theta_6) \tag{7.531}$$

$$t_{32} = -s\theta_5 s\theta_6 c(\theta_2 + \theta_3) - s(\theta_2 + \theta_3)(c\theta_6 s\theta_4 + c\theta_4 c\theta_5 s\theta_6) \tag{7.532}$$

$$t_{13} = s\theta_1 s\theta_4 s\theta_5 + c\theta_1 [c\theta_5 s(\theta_2 + \theta_3) + c\theta_4 s\theta_5 c(\theta_2 + \theta_3)] \tag{7.533}$$

$$t_{23} = -c\theta_1 s\theta_4 s\theta_5 + s\theta_1 [c\theta_5 s(\theta_2 + \theta_3) + c\theta_4 s\theta_5 c(\theta_2 + \theta_3)] \tag{7.534}$$

$$t_{33} = c\theta_4 s\theta_5 s(\theta_2 + \theta_3) - c\theta_5 c(\theta_2 + \theta_3) \tag{7.535}$$

$$t_{14} = d_6\{s\theta_1 s\theta_4 s\theta_5 + c\theta_1 [c\theta_4 s\theta_5 c(\theta_2 + \theta_3) + c\theta_5 s(\theta_2 + \theta_3)]\} + l_3 c\theta_1 s(\theta_2 + \theta_3) + d_2 s\theta_1 + l_2 c\theta_1 c\theta_2 \tag{7.536}$$

$$t_{24} = d_6\{-c\theta_1 s\theta_4 s\theta_5 + s\theta_1 [c\theta_4 s\theta_5 c(\theta_2 + \theta_3) + c\theta_5 s(\theta_2 + \theta_3)]\} + s\theta_1 s(\theta_2 + \theta_3) l_3 - d_2 c\theta_1 + l_2 c\theta_2 s\theta_1 \tag{7.537}$$

$$t_{34} = d_6[c\theta_4 s\theta_5 s(\theta_2 + \theta_3) - c\theta_5 c(\theta_2 + \theta_3)] + l_2 s\theta_2 + l_3 c(\theta_2 + \theta_3) \tag{7.538}$$

The inverse kinematic problem of the manipulator starts with the wrist position vector ${}^0\mathbf{d}$, which is $\begin{bmatrix} t_{14} & t_{24} & t_{34} \end{bmatrix}^{\mathrm{T}}$ of ${}^0T_7$ for $d_6 = 0$:

$${}^0\mathbf{d} = \begin{bmatrix} c\theta_1 [l_3 s(\theta_2 + \theta_3) + l_2 c\theta_2] + d_2 s\theta_1 \\ s\theta_1 [l_3 s(\theta_2 + \theta_3) + l_2 c\theta_2] - d_2 c\theta_1 \\ l_3 c(\theta_2 + \theta_3) + l_2 s\theta_2 \end{bmatrix} = \begin{bmatrix} d_x \\ d_y \\ d_z \end{bmatrix} \tag{7.539}$$

We must theoretically be able to solve Equation (7.539) for the three joint variables θ_1, θ_2, and θ_3. It can be seen that

$$d_x \sin\theta_1 - d_y \cos\theta_1 = d_2 \tag{7.540}$$

which provides

$$\theta_1 = 2\operatorname{atan2}\left(d_x \pm \sqrt{d_x^2 + d_y^2 - d_2^2}, d_2 - d_y\right) \tag{7.541}$$

Combining the first two elements of ${}^0\mathbf{d}$ gives

$$l_3 \sin(\theta_2 + \theta_3) = \pm\sqrt{d_x^2 + d_y^2 - d_2^2} - l_2 \cos\theta_2 \tag{7.542}$$

then, using the third element of ${}^0\mathbf{d}$ provides

$$l_3^2 = \left(\pm\sqrt{d_x^2 + d_y^2 - d_2^2} - l_2 \cos\theta_2\right)^2 + (d_z - l_2 \sin\theta_2)^2 \tag{7.543}$$

which can be rearranged to following form

$$a \cos\theta_2 + b \sin\theta_2 = c \tag{7.544}$$

where

$$a = 2l_2\sqrt{d_x^2 + d_y^2 - d_2^2} \tag{7.545}$$

$$b = 2l_2 d_z \tag{7.546}$$

$$c = d_x^2 + d_y^2 + d_z^2 - d_2^2 + l_2^2 - l_3^2 \tag{7.547}$$

The solutions for θ_2 are

$$\theta_2 = \begin{cases} \operatorname{atan2}\left(\frac{c}{r}, \pm\sqrt{1 - \frac{c^2}{r^2}}\right) - \operatorname{atan2}(a, b) & (7.548) \\ \operatorname{atan2}\left(\frac{c}{r}, \pm\sqrt{r^2 - c^2}\right) - \operatorname{atan2}(a, b) & (7.549) \end{cases}$$

where

$$r^2 = a^2 + b^2 \tag{7.550}$$

Summing the squares of the elements of ${}^0\mathbf{d}$ gives

$$d_x^2 + d_y^2 + d_z^2 = d_2^2 + l_2^2 + l_3^2 + 2l_2l_3 \sin(2\theta_2 + \theta_3) \tag{7.551}$$

which provides

$$\theta_3 = \arcsin\left(\frac{d_x^2 + d_y^2 + d_z^2 - d_2^2 - l_2^2 - l_3^2}{2l_2l_3}\right) - 2\theta_2 \tag{7.552}$$

Having θ_1, θ_2, and θ_3 means we can find the position of the wrist point in space. However, because the joint variables in ${}^0T_3$ and ${}^3T_6$ are independent, we should find the orientation of the gripper by solving ${}^3T_6$ or ${}^3R_6$ for θ_4, θ_5, and θ_6:

$$\begin{aligned} {}^3R_6 &= \begin{bmatrix} c\theta_4 c\theta_5 c\theta_6 - s\theta_4 s\theta_6 & -c\theta_6 s\theta_4 - c\theta_4 c\theta_5 s\theta_6 & c\theta_4 s\theta_5 \\ c\theta_5 c\theta_6 s\theta_4 + c\theta_4 s\theta_6 & c\theta_4 c\theta_6 - c\theta_5 s\theta_4 s\theta_6 & s\theta_4 s\theta_5 \\ -c\theta_6 s\theta_5 & s\theta_5 s\theta_6 & c\theta_5 \end{bmatrix} \\ &= \begin{bmatrix} s_{11} & s_{12} & s_{13} \\ s_{21} & s_{22} & s_{23} \\ s_{31} & s_{32} & s_{33} \end{bmatrix} \end{aligned} \tag{7.553}$$

The angles θ_4, θ_5, and θ_6 can be found by examining elements of ${}^3R_6$:

$$\theta_4 = \operatorname{atan2}(s_{23}, s_{13}) \tag{7.554}$$

$$\theta_5 = \operatorname{atan2}\left(\sqrt{s_{13}^2 + s_{23}^2}, s_{33}\right) \tag{7.555}$$

$$\theta_6 = \operatorname{atan2}(s_{32}, -s_{31}) \tag{7.556}$$

Example 465 ★ Inverse-Transformation Technique Assume we have the transformation matrix ${}^0T_6$ indicating the global position and the orientation of the gripper of a six-DOF multibody in the base frame B_0. Furthermore, assume the geometry and individual transformation matrices ${}^0T_1(q_1)$, ${}^1T_2(q_2)$, ${}^2T_3(q_3)$, ${}^3T_4(q_4)$, ${}^4T_5(q_5)$, and ${}^5T_6(q_6)$ are given as functions of joint variables q_i, $i = 1, 2, 3, \ldots, 6$.

The forward kinematics of the multibody is

$$
{}^0T_6 = {}^0T_1\,{}^1T_2\,{}^2T_3\,{}^3T_4\,{}^4T_5\,{}^5T_6 = \begin{bmatrix} r_{11} & r_{12} & r_{13} & r_{14} \\ r_{21} & r_{22} & r_{23} & r_{24} \\ r_{31} & r_{32} & r_{33} & r_{34} \\ 0 & 0 & 0 & 1 \end{bmatrix} \tag{7.557}
$$

We can solve the inverse kinematics problem by solving the following equations for the unknown joint variables:

$$
{}^1T_6 = {}^0T_1^{-1}\,{}^0T_6 \tag{7.558}
$$

$$
{}^2T_6 = {}^1T_2^{-1}\,{}^0T_1^{-1}\,{}^0T_6 \tag{7.559}
$$

$$
{}^3T_6 = {}^2T_3^{-1}\,{}^1T_2^{-1}\,{}^0T_1^{-1}\,{}^0T_6 \tag{7.560}
$$

$$
{}^4T_6 = {}^3T_4^{-1}\,{}^2T_3^{-1}\,{}^1T_2^{-1}\,{}^0T_1^{-1}\,{}^0T_6 \tag{7.561}
$$

$$
{}^5T_6 = {}^4T_5^{-1}\,{}^3T_4^{-1}\,{}^2T_3^{-1}\,{}^1T_2^{-1}\,{}^0T_1^{-1}\,{}^0T_6 \tag{7.562}
$$

$$
\mathbf{I} = {}^5T_6^{-1}\,{}^4T_5^{-1}\,{}^3T_4^{-1}\,{}^2T_3^{-1}\,{}^1T_2^{-1}\,{}^0T_1^{-1}\,{}^0T_6 \tag{7.563}
$$

Proof: Let us multiply both sides of the transformation matrix ${}^0T_6$ in Equation (7.557) by ${}^0T_1^{-1}$ to obtain ${}^1T_6$:

$$
{}^0T_1^{-1}\,{}^0T_6 = {}^0T_1^{-1}\left({}^0T_1\,{}^1T_2\,{}^2T_3\,{}^3T_4\,{}^4T_5\,{}^5T_6\right) = {}^1T_6 \tag{7.564}
$$

Remembering that the matrix ${}^0T_1^{-1}$ is the mathematical inverse of the 4×4 matrix ${}^0T_1$, and not an inverse transformation, we must calculate ${}^0T_1^{-1}$ by mathematical matrix inversion.

The left-hand side of Equation (7.564) is a function of the first joint variable q_1. However, the elements of the matrix ${}^1T_6$ on the right-hand side are either zero, constant, or functions of the other joint variables q_2, q_3, q_4, q_5, and q_6. The zero or constant elements on the right-hand side provide an equation for the single unknown variable q_1.

Then, we multiply both sides of (7.564) by ${}^1T_2^{-1}$ to obtain ${}^2T_6$:

$$
\begin{aligned}
{}^1T_2^{-1}\,{}^0T_1^{-1}\,{}^0T_6 &= {}^1T_2^{-1}\,{}^0T_1^{-1}\left({}^0T_1\,{}^1T_2\,{}^2T_3\,{}^3T_4\,{}^4T_5\,{}^5T_6\right) \\
&= {}^2T_6
\end{aligned} \tag{7.565}
$$

The left-hand side of this equation is a function of the second joint variable q_2, while the elements of the matrix ${}^2T_6$ on the right hand side are either zero, constant, or functions of the rest of the joint variables q_3, q_4, q_5, and q_6. Equating the associated element with constant or zero elements on the right-hand side provides the required algebraic equation to solve for q_2.

Following this procedure, we can find the joint variables q_3, q_4, q_5, and q_6 by using the following equalities respectively:

$$\begin{aligned}
&{}^2T_3^{-1}\,{}^1T_2^{-1}\,{}^0T_1^{-1}\,{}^0T_6 \\
&= {}^2T_3^{-1}\,{}^1T_2^{-1}\,{}^0T_1^{-1}\left({}^0T_1\,{}^1T_2\,{}^2T_3\,{}^3T_4\,{}^4T_5\,{}^5T_6\right) \\
&= {}^3T_6
\end{aligned} \tag{7.566}$$

$$\begin{aligned}
&{}^3T_4^{-1}\,{}^2T_3^{-1}\,{}^1T_2^{-1}\,{}^0T_1^{-1}\,{}^0T_6 \\
&= {}^3T_4^{-1}\,{}^2T_3^{-1}\,{}^1T_2^{-1}\,{}^0T_1^{-1}\left({}^0T_1\,{}^1T_2\,{}^2T_3\,{}^3T_4\,{}^4T_5\,{}^5T_6\right) \\
&= {}^4T_6
\end{aligned} \tag{7.567}$$

$$\begin{aligned}
&{}^4T_5^{-1}\,{}^3T_4^{-1}\,{}^2T_3^{-1}\,{}^1T_2^{-1}\,{}^0T_1^{-1}\,{}^0T_6 \\
&= {}^4T_5^{-1}\,{}^3T_4^{-1}\,{}^2T_3^{-1}\,{}^1T_2^{-1}\,{}^0T_1^{-1}\left({}^0T_1\,{}^1T_2\,{}^2T_3\,{}^3T_4\,{}^4T_5\,{}^5T_6\right) \\
&= {}^5T_6
\end{aligned} \tag{7.568}$$

$$\begin{aligned}
&{}^5T_6^{-1}\,{}^4T_5^{-1}\,{}^3T_4^{-1}\,{}^2T_3^{-1}\,{}^1T_2^{-1}\,{}^0T_1^{-1}\,{}^0T_6 \\
&= {}^5T_6^{-1}\,{}^4T_5^{-1}\,{}^3T_4^{-1}\,{}^2T_3^{-1}\,{}^1T_2^{-1}\,{}^0T_1^{-1}\left({}^0T_1\,{}^1T_2\,{}^2T_3\,{}^3T_4\,{}^4T_5\,{}^5T_6\right) \\
&= \mathbf{I}
\end{aligned} \tag{7.569}$$

The *inverse-transformation technique* is sometimes called the *Pieper technique*. ■

Example 466 Inverse Kinematics of a 2R Arm The forward kinematics of a planar 2R manipulator with two links of R‖R that is shown in Figure 7.69 is given as

$${}^0T_1 = \begin{bmatrix} \cos\theta_1 & -\sin\theta_1 & 0 & l_1\cos\theta_1 \\ \sin\theta_1 & \cos\theta_1 & 0 & l_1\sin\theta_1 \\ 0 & 0 & 1 & 0 \\ 0 & 0 & 0 & 1 \end{bmatrix} \tag{7.570}$$

$${}^1T_2 = \begin{bmatrix} \cos\theta_2 & -\sin\theta_2 & 0 & l_2\cos\theta_2 \\ \sin\theta_2 & \cos\theta_2 & 0 & l_2\sin\theta_2 \\ 0 & 0 & 1 & 0 \\ 0 & 0 & 0 & 1 \end{bmatrix} \tag{7.571}$$

$$\begin{aligned}
{}^0T_2 = {}^0T_1\,{}^1T_2 &= \begin{bmatrix} r_{11} & r_{12} & r_{13} & r_{14} \\ r_{21} & r_{22} & r_{23} & r_{24} \\ r_{31} & r_{32} & r_{33} & r_{34} \\ 0 & 0 & 0 & 1 \end{bmatrix} \\
&= \begin{bmatrix} c\left(\theta_1+\theta_2\right) & -s\left(\theta_1+\theta_2\right) & 0 & l_1c\theta_1 + l_2c\left(\theta_1+\theta_2\right) \\ s\left(\theta_1+\theta_2\right) & c\left(\theta_1+\theta_2\right) & 0 & l_1s\theta_1 + l_2s\left(\theta_1+\theta_2\right) \\ 0 & 0 & 1 & 0 \\ 0 & 0 & 0 & 1 \end{bmatrix}
\end{aligned} \tag{7.572}$$

Let us assume that the matrix ${}^0T_2 = [r_{ij}]$ is given. To determine the required angles for a given $[r_{ij}]$, we can multiply ${}^0T_1^{-1}$ to $[r_{ij}]$ and search for constant elements:

$$ {}^0T_1^{-1}\,[r_{ij}] = {}^1T_2 \tag{7.573}$$

$$\begin{bmatrix} r_{11}c\theta_1 + r_{21}s\theta_1 & r_{12}c\theta_1 + r_{22}s\theta_1 & r_{13}c\theta_1 + r_{23}s\theta_1 & r_{14}c\theta_1 - l_1 + r_{24}s\theta_1 \\ r_{21}c\theta_1 - r_{11}s\theta_1 & r_{22}c\theta_1 - r_{12}s\theta_1 & r_{23}c\theta_1 - r_{13}s\theta_1 & r_{24}c\theta_1 - r_{14}s\theta_1 \\ r_{31} & r_{32} & r_{33} & r_{34} \\ 0 & 0 & 0 & 1 \end{bmatrix}$$

$$= \begin{bmatrix} c\theta_2 & -s\theta_2 & 0 & l_2c\theta_2 \\ s\theta_2 & c\theta_2 & 0 & l_2s\theta_2 \\ 0 & 0 & 1 & 0 \\ 0 & 0 & 0 & 1 \end{bmatrix} \tag{7.574}$$

The elements (1, 2) provide an equation

$$ r_{13}\cos\theta_1 + r_{23}\sin\theta_1 = 0 \tag{7.575}$$

that can be solved for θ_1:

$$ \theta_1 = \arctan\frac{r_{13}}{-r_{23}} \tag{7.576}$$

Let us use the elements (2, 1) and determine θ_2:

$$ \theta_2 = \arcsin\left(r_{21}\cos\theta_1 - r_{11}\sin\theta_1\right) \tag{7.577}$$

Example 467 Inverse Kinematics of a Spherical Robot A spherical robot and its link numbers and link coordinate frames are shown in Figure 7.72. The robot as a multibody is made by assembling a spherical wrist and a spherical arm. The transformation matrices of the multibody are

$$ {}^0T_1 = \begin{bmatrix} c\theta_1 & 0 & -s\theta_1 & 0 \\ s\theta_1 & 0 & c\theta_1 & 0 \\ 0 & -1 & 0 & 0 \\ 0 & 0 & 0 & 1 \end{bmatrix} \qquad {}^1T_2 = \begin{bmatrix} c\theta_2 & 0 & s\theta_2 & 0 \\ s\theta_2 & 0 & -c\theta_2 & 0 \\ 0 & 1 & 0 & l_2 \\ 0 & 0 & 0 & 1 \end{bmatrix} \tag{7.578}$$

$$ {}^2T_3 = \begin{bmatrix} 1 & 0 & 0 & 0 \\ 0 & 1 & 0 & 0 \\ 0 & 0 & 1 & d_3 \\ 0 & 0 & 0 & 1 \end{bmatrix} \qquad {}^3T_4 = \begin{bmatrix} c\theta_4 & 0 & -s\theta_4 & 0 \\ s\theta_4 & 0 & c\theta_4 & 0 \\ 0 & -1 & 0 & 0 \\ 0 & 0 & 0 & 1 \end{bmatrix} \tag{7.579}$$

$$ {}^4T_5 = \begin{bmatrix} c\theta_5 & 0 & s\theta_5 & 0 \\ s\theta_5 & 0 & -c\theta_5 & 0 \\ 0 & 1 & 0 & 0 \\ 0 & 0 & 0 & 1 \end{bmatrix} \qquad {}^5T_6 = \begin{bmatrix} c\theta_6 & -s\theta_6 & 0 & 0 \\ s\theta_6 & c\theta_6 & 0 & 0 \\ 0 & 0 & 1 & 0 \\ 0 & 0 & 0 & 1 \end{bmatrix} \tag{7.580}$$

The position and orientation of the wrist for a set of joint variables are the solution of the forward kinematics problem. Therefore, we assume that the matrix ${}^0T_6 = [r_{ij}]$ is

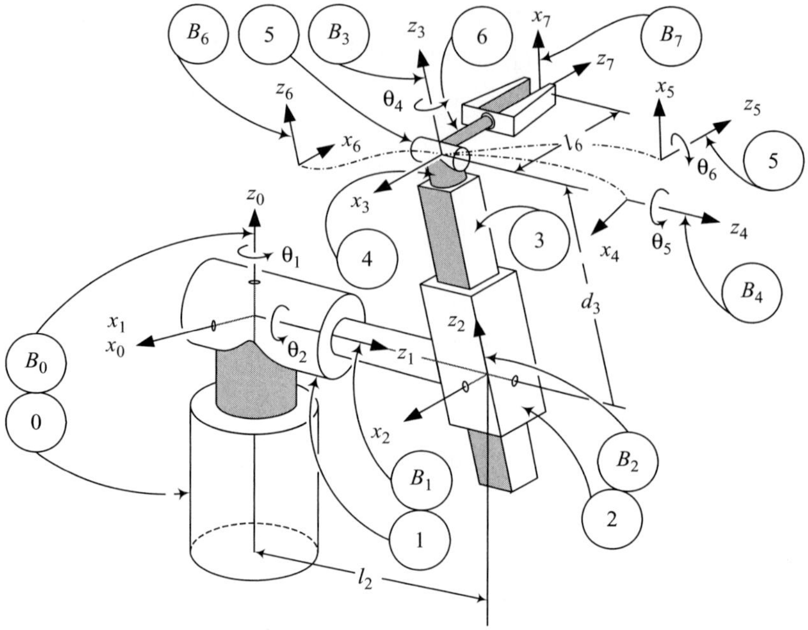

Figure 7.72 A spherical robot made of a spherical manipulator attached to a spherical wrist.

given numerically as

$$
\begin{aligned}
{}^0T_6 &= {}^0T_1\,{}^1T_2\,{}^2T_3\,{}^3T_4\,{}^4T_5\,{}^5T_6 \\
&= \begin{bmatrix} r_{11} & r_{12} & r_{13} & r_{14} \\ r_{21} & r_{22} & r_{23} & r_{24} \\ r_{31} & r_{32} & r_{33} & r_{34} \\ 0 & 0 & 0 & 1 \end{bmatrix}
\end{aligned} \tag{7.581}
$$

Multiplying both sides of (7.581) by ${}^0T_1^{-1}$ provides

$$
\begin{aligned}
{}^0T_1^{-1}\,{}^0T_6 &= \begin{bmatrix} f_{11} & f_{12} & f_{13} & f_{14} \\ f_{21} & f_{22} & f_{23} & f_{24} \\ f_{31} & f_{32} & f_{33} & f_{34} \\ 0 & 0 & 0 & 1 \end{bmatrix} \\
&= \begin{bmatrix} c\theta_1 & s\theta_1 & 0 & 0 \\ 0 & 0 & -1 & 0 \\ -s\theta_1 & c\theta_1 & 0 & 0 \\ 0 & 0 & 0 & 1 \end{bmatrix} \begin{bmatrix} r_{11} & r_{12} & r_{13} & r_{14} \\ r_{21} & r_{22} & r_{23} & r_{24} \\ r_{31} & r_{32} & r_{33} & r_{34} \\ 0 & 0 & 0 & 1 \end{bmatrix}
\end{aligned} \tag{7.582}
$$

where, for $i = 1, 2, 3, 4$,

$$f_{1i} = r_{1i}\cos\theta_1 + r_{2i}\sin\theta_1 \tag{7.583}$$

$$f_{2i} = -r_{3i} \tag{7.584}$$

$$f_{3i} = r_{2i}\cos\theta_1 - r_{1i}\sin\theta_1 \tag{7.585}$$

On the other hand, based on the given transformation matrices,

$$ {}^{1}T_6 = {}^{1}T_2\,{}^{2}T_3\,{}^{3}T_4\,{}^{4}T_5\,{}^{5}T_6 = \begin{bmatrix} f_{11} & f_{12} & f_{13} & f_{14} \\ f_{21} & f_{22} & f_{23} & f_{24} \\ f_{31} & f_{32} & f_{33} & f_{34} \\ 0 & 0 & 0 & 1 \end{bmatrix} \tag{7.586} $$

where

$$ f_{11} = -c\theta_2 s\theta_4 s\theta_6 + c\theta_6 \left(-s\theta_2 s\theta_5 + c\theta_2 c\theta_4 c\theta_5\right) \tag{7.587} $$

$$ f_{21} = -s\theta_2 s\theta_4 s\theta_6 + c\theta_6 \left(c\theta_2 s\theta_5 + c\theta_4 c\theta_5 s\theta_2\right) \tag{7.588} $$

$$ f_{31} = c\theta_4 s\theta_6 + c\theta_5 c\theta_6 s\theta_4 \tag{7.589} $$

$$ f_{12} = -c\theta_2 c\theta_6 s\theta_4 - s\theta_6 \left(-s\theta_2 s\theta_5 + c\theta_2 c\theta_4 c\theta_5\right) \tag{7.590} $$

$$ f_{22} = -c\theta_6 s\theta_2 s\theta_4 - s\theta_6 \left(c\theta_2 s\theta_5 + c\theta_4 c\theta_5 s\theta_2\right) \tag{7.591} $$

$$ f_{32} = c\theta_4 c\theta_6 - c\theta_5 s\theta_4 s\theta_6 \tag{7.592} $$

$$ f_{13} = c\theta_5 s\theta_2 + c\theta_2 c\theta_4 s\theta_5 \tag{7.593} $$

$$ f_{23} = -c\theta_2 c\theta_5 + c\theta_4 s\theta_2 s\theta_5 \tag{7.594} $$

$$ f_{33} = s\theta_4 s\theta_5 \tag{7.595} $$

$$ f_{14} = d_3 s\theta_2 \tag{7.596} $$

$$ f_{24} = -d_3 c\theta_2 \tag{7.597} $$

$$ f_{34} = l_2. \tag{7.598} $$

The only constant element of the matrix (7.586) is $f_{34} = l_2$; therefore,

$$ r_{24}\cos\theta_1 - r_{14}\sin\theta_1 = l_2 \tag{7.599} $$

This is the first kind of trigonometric equation that frequently appears in multibody inverse kinematics. The equation has a standard method of solution. We assume

$$ r_{14} = r\cos\phi \qquad r_{24} = r\sin\phi \tag{7.600} $$

where

$$ r = \sqrt{r_{14}^2 + r_{24}^2} \qquad \phi = \tan^{-1}\frac{r_{24}}{r_{14}} \tag{7.601} $$

and therefore, Equation (7.599) becomes

$$ \frac{l_2}{r} = \sin\phi\cos\theta_1 - \cos\phi\sin\theta_1 = \sin(\phi - \theta_1) \tag{7.602} $$

showing that

$$ \pm\sqrt{1 - (l_2/r)^2} = \cos(\phi - \theta_1) \tag{7.603} $$

Therefore, the solution of Equation (7.599) for θ_1 is

$$\theta_1 = \arctan \frac{r_{24}}{r_{14}} - \arctan \frac{l_2}{\pm\sqrt{r^2 - l_2^2}} \tag{7.604}$$

The minus sign is associated with a left-shoulder configuration as shown in Figure 7.73, and the plus sign is associated with the right-shoulder configuration of Figure 7.72.

The elements f_{14} and f_{24} of matrix (7.586) are functions of θ_1 and θ_2:

$$f_{14} = d_3 \sin\theta_2 = r_{14}\cos\theta_1 + r_{24}\sin\theta_1 \tag{7.605}$$

$$f_{24} = -d_3 \cos\theta_2 = -r_{34} \tag{7.606}$$

Having θ_1, we can determine θ_2 as a function of θ_1:

$$\theta_2 = \tan^{-1}\frac{r_{14}\cos\theta_1 + r_{24}\sin\theta_1}{r_{34}} \tag{7.607}$$

where θ_1 must be substituted from (7.604).

The next step is to multiply ${}^1T_2^{-1}$ by ${}^0T_1^{-1}\,{}^0T_6$ and determine the third joint variable d_3:

$${}^1T_2^{-1}\,{}^0T_1^{-1}\,{}^0T_6 = {}^2T_6 \tag{7.608}$$

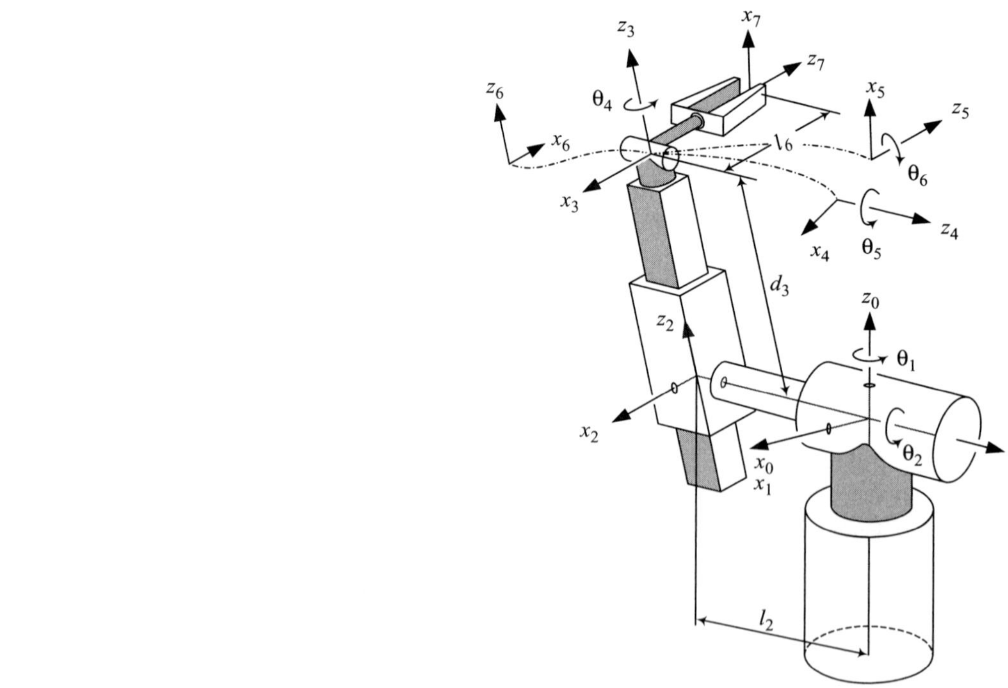

Figure 7.73 Left-shoulder configuration of a spherical robot.

where

$$
{}^{2}T_{6} = \begin{bmatrix} -s\theta_4 s\theta_6 + c\theta_4 c\theta_5 c\theta_6 & -c\theta_6 s\theta_4 - c\theta_4 c\theta_5 s\theta_6 & c\theta_4 s\theta_5 & 0 \\ c\theta_4 s\theta_6 + c\theta_5 c\theta_6 s\theta_4 & c\theta_4 c\theta_6 - c\theta_5 s\theta_4 s\theta_6 & s\theta_4 s\theta_5 & 0 \\ -c\theta_6 s\theta_5 & s\theta_5 s\theta_6 & c\theta_5 & d_3 \\ 0 & 0 & 0 & 1 \end{bmatrix} \tag{7.609}
$$

Employing the elements of the matrices on both sides of Equation (7.608) shows that we can use the element (3,4) to find d_3:

$$
d_3 = r_{34}\cos\theta_2 + r_{14}\cos\theta_1\sin\theta_2 + r_{24}\sin\theta_1\sin\theta_2 \tag{7.610}
$$

There is no other element in Equation (7.608) that is a function of a single variable. Therefore, we move to the next step to find θ_4. However, ${}^{2}T_{3}^{-1}\,{}^{1}T_{2}^{-1}\,{}^{0}T_{1}^{-1}\,{}^{0}T_{6} = {}^{3}T_{6}$ provides no new equation. So, we move another step and find

$$
{}^{3}T_{4}^{-1}\,{}^{2}T_{3}^{-1}\,{}^{1}T_{2}^{-1}\,{}^{0}T_{1}^{-1}\,{}^{0}T_{6} = {}^{4}T_{6} \tag{7.611}
$$

where

$$
{}^{4}T_{6} = \begin{bmatrix} \cos\theta_5\cos\theta_6 & -\cos\theta_5\sin\theta_6 & \sin\theta_5 & 0 \\ \cos\theta_6\sin\theta_5 & -\sin\theta_5\sin\theta_6 & -\cos\theta_5 & 0 \\ \sin\theta_6 & \cos\theta_6 & 0 & 0 \\ 0 & 0 & 0 & 1 \end{bmatrix} \tag{7.612}
$$

The left-hand side of (7.611) shows that

$$
{}^{3}T_{4}^{-1}\,{}^{2}T_{3}^{-1}\,{}^{1}T_{2}^{-1}\,{}^{0}T_{1}^{-1}\,{}^{0}T_{6} = \begin{bmatrix} g_{11} & g_{12} & g_{13} & g_{14} \\ g_{21} & g_{22} & g_{23} & g_{24} \\ g_{31} & g_{32} & g_{33} & g_{34} \\ 0 & 0 & 0 & 1 \end{bmatrix} \tag{7.613}
$$

where, for $i = 1, 2, 3, 4$,

$$
g_{1i} = -r_{3i}c\theta_4 s\theta_2 + r_{2i}\left(c\theta_1 s\theta_4 + c\theta_2 c\theta_4 s\theta_1\right) + r_{1i}\left(-s\theta_1 s\theta_4 + c\theta_1 c\theta_2 c\theta_4\right) \tag{7.614}
$$

$$
g_{2i} = d_3\delta_{4i} - r_{31}c\theta_2 - r_{11}c\theta_1 s\theta_2 - r_{21}s\theta_1 s\theta_2 \tag{7.615}
$$

$$
g_{3i} = r_{31}s\theta_2 s\theta_4 + r_{21}\left(c\theta_1 c\theta_4 - c\theta_2 s\theta_1 s\theta_4\right) + r_{11}\left(-c\theta_4 s\theta_1 - c\theta_1 c\theta_2 s\theta_4\right) \tag{7.616}
$$

The symbol δ_{4i} indicates the Kronecker delta:

$$
\delta_{4i} = \begin{cases} 1 & i = 4 \\ 0 & i \neq 4 \end{cases} \tag{7.617}
$$

Both sides of Equation (7.613) indicate that we can find θ_4 by equating the elements (3,3), θ_5 by equating the elements (1,3) or (2,3), and θ_6 by equating the elements (3,1) or (3,2). The elements (3, 3) provide the equation

$$
r_{13}\left(-c\theta_4 s\theta_1 - c\theta_1 c\theta_2 s\theta_4\right) + r_{23}\left(c\theta_1 c\theta_4 - c\theta_2 s\theta_1 s\theta_4\right) + r_{33}s\theta_2 s\theta_4 = 0 \tag{7.618}
$$

which we can solve for θ_4:

$$
\theta_4 = \arctan\frac{-r_{13}s\theta_1 + r_{23}c\theta_1}{c\theta_2\left(r_{13}c\theta_1 + r_{23}s\theta_1\right) - r_{33}s\theta_2} \tag{7.619}
$$

Based on the plus sign of θ_1 in Equation (7.604), we find the associated θ_4 as

$$\theta_4 = \frac{\pi}{2} + \arctan \frac{-r_{13}s\theta_1 + r_{23}c\theta_1}{c\theta_2 \left(r_{13}c\theta_1 + r_{23}s\theta_1\right) - r_{33}s\theta_2} \tag{7.620}$$

Now we use both of the elements (1, 3) and (2, 3),

$$\sin\theta_5 = r_{23}\left(\cos\theta_1 \sin\theta_4 + \cos\theta_2 \cos\theta_4 \sin\theta_1\right) - r_{33}\cos\theta_4 \sin\theta_2 + r_{13}\left(\cos\theta_1 \cos\theta_2 \cos\theta_4 - \sin\theta_1 \sin\theta_4\right) \tag{7.621}$$

$$-\cos\theta_5 = -r_{33}\cos\theta_2 - r_{13}\cos\theta_1 \sin\theta_2 - r_{23}\sin\theta_1 \sin\theta_2 \tag{7.622}$$

to find θ_5:

$$\theta_5 = \arctan \frac{\sin\theta_5}{\cos\theta_5} \tag{7.623}$$

Finally, θ_6 can be found from the elements (3, 1) and (3, 2):

$$\sin\theta_6 = r_{31}\sin\theta_2 \sin\theta_4 + r_{21}\left(\cos\theta_1 \cos\theta_4 - \cos\theta_2 \sin\theta_1 \sin\theta_4\right) + r_{11}\left(-\cos\theta_4 \sin\theta_1 - \cos\theta_1 \cos\theta_2 \sin\theta_4\right) \tag{7.624}$$

$$\cos\theta_6 = r_{32}\sin\theta_2 \sin\theta_4 + r_{22}\left(\cos\theta_1 \cos\theta_4 - \cos\theta_2 \sin\theta_1 \sin\theta_4\right) + r_{12}\left(-\cos\theta_4 \sin\theta_1 - \cos\theta_1 \cos\theta_2 \sin\theta_4\right) \tag{7.625}$$

$$\theta_6 = \arctan \frac{\sin\theta_6}{\cos\theta_6} \tag{7.626}$$

Example 468 Inverse Kinematics of Euler Angle Transformation Matrix The Euler angle transformation matrix is

$$\begin{aligned} {}^G R_B &= \left[A_{z,\psi}\, A_{x,\theta}\, A_{z,\varphi}\right]^{\mathrm{T}} = R_{Z,\varphi}\, R_{X,\theta}\, R_{Z,\psi} \\ &= \begin{bmatrix} c\varphi c\psi - c\theta s\varphi s\psi & -c\varphi s\psi - c\theta c\psi s\varphi & s\theta s\varphi \\ c\psi s\varphi + c\theta c\varphi s\psi & -s\varphi s\psi + c\theta c\varphi c\psi & -c\varphi s\theta \\ s\theta s\psi & s\theta c\psi & c\theta \end{bmatrix} \\ &= \begin{bmatrix} r_{11} & r_{12} & r_{13} \\ r_{21} & r_{22} & r_{23} \\ r_{31} & r_{32} & r_{33} \end{bmatrix} \end{aligned} \tag{7.627}$$

Having ${}^G R_B$ we should be able to determine the Euler angles φ, θ, ψ. Premultiplying ${}^G R_B$ by $R_{Z,\varphi}^{-1}$ gives

$$\begin{aligned} R_{Z,\varphi}^{-1}\, {}^G R_B &= R_{X,\theta}\, R_{Z,\psi} \\ &= \begin{bmatrix} r_{11}c\varphi + r_{21}s\varphi & r_{12}c\varphi + r_{22}s\varphi & r_{13}c\varphi + r_{23}s\varphi \\ r_{21}c\varphi - r_{11}s\varphi & r_{22}c\varphi - r_{12}s\varphi & r_{23}c\varphi - r_{13}s\varphi \\ r_{31} & r_{32} & r_{33} \end{bmatrix} \end{aligned}$$

$$= \begin{bmatrix} \cos\psi & -\sin\psi & 0 \\ \cos\theta\sin\psi & \cos\theta\cos\psi & -\sin\theta \\ \sin\theta\sin\psi & \sin\theta\cos\psi & \cos\theta \end{bmatrix} \tag{7.628}$$

Equating the elements (1, 3) of both sides,

$$r_{13}\cos\varphi + r_{23}\sin\varphi = 0 \tag{7.629}$$

can be used to determine φ:

$$\varphi = \operatorname{atan2}(r_{13}, -r_{23}) \tag{7.630}$$

Having φ helps us to find ψ by using elements (1, 1) and (1, 2):

$$\cos\psi = r_{11}\cos\varphi + r_{21}\sin\varphi \tag{7.631}$$

$$-\sin\psi = r_{12}\cos\varphi + r_{22}\sin\varphi \tag{7.632}$$

therefore,

$$\psi = \arctan\frac{-r_{12}\cos\varphi - r_{22}\sin\varphi}{r_{11}\cos\varphi + r_{21}\sin\varphi} \tag{7.633}$$

Let us multiply ${}^G R_B$ by $R_{Z,\psi}^{-1}$:

$$\begin{aligned} {}^G R_B\, R_{Z,\psi}^{-1} &= R_{Z,\varphi}\, R_{X,\theta} \\ &= \begin{bmatrix} r_{11}c\psi - r_{12}s\psi & r_{12}c\psi + r_{11}s\psi & r_{13} \\ r_{21}c\psi - r_{22}s\psi & r_{22}c\psi + r_{21}s\psi & r_{23} \\ r_{31}c\psi - r_{32}s\psi & r_{32}c\psi + r_{31}s\psi & r_{33} \end{bmatrix} \\ &= \begin{bmatrix} \cos\varphi & -\cos\theta\sin\varphi & \sin\theta\sin\varphi \\ \sin\varphi & \cos\theta\cos\varphi & -\cos\varphi\sin\theta \\ 0 & \sin\theta & \cos\theta \end{bmatrix} \end{aligned} \tag{7.634}$$

The elements (1, 1) on both sides result in an equation to find ψ:

$$r_{31}\cos\psi - r_{31}\sin\psi = 0 \tag{7.635}$$

$$\psi = \arctan(r_{31}, r_{31}) \tag{7.636}$$

Finally, we can find θ by using elements (3, 2) and (3, 3),

$$r_{32}\cos\psi + r_{31}\sin\psi = \sin\theta \tag{7.637}$$

$$r_{33} = \cos\theta \tag{7.638}$$

which gives

$$\theta = \arctan\frac{r_{32}\cos\psi + r_{31}\sin\psi}{r_{33}} \tag{7.639}$$

KEY SYMBOLS

a	kinematic link length
a, b, c	coefficients of trigonometric equation
$a_i, \alpha_i, d_i, \theta_i$	DH parameters of link i
$\mathbf{a}$	turn vector of end-effector frame
$\tilde{a}$	skew-symmetric matrix of the vector $\mathbf{a}$
A	transformation matrix of rotation about a local axis
B	body coordinate frame, local coordinate frame
c	cosine, constant coefficient
C	wheel–body coordinate frame, cylindrical joint
C_α	sideslip coefficient
C_γ	camber stiffness
d	joint distance, distance between two points
d_x, d_y, d_z	elements of $\mathbf{d}$
$\mathbf{d}$	translation vector, displacement vector
$\mathbf{d}_{\text{wrist}}$	wrist position vector
D	displacement transformation matrix
DH	Denavit–Hartenberg
DOF	degree of freedom
$\hat{e}_\varphi, \hat{e}_\theta, \hat{e}_\psi$	coordinate axes of E, local roll–pitch–yaw coordinate axes
E	Eulerian local frame
f	number of DOF
f_{ij}	element of row i and column j of a matrix
$F, \mathbf{F}$	force
F_x	longitudinal force, forward force
F_y	lateral force
g_{ij}	element of row i and column j of a matrix
$G, 0$	global coordinate frame, fixed coordinate frame
h, H	height
$\mathbf{I} = [I]$	identity matrix
$\hat{\imath}, \hat{\jmath}, \hat{k}$	local coordinate axes unit vectors
$\tilde{\imath}, \tilde{\jmath}, \tilde{k}$	skew symmetric matrices of the unit vector $\hat{\imath}, \hat{\jmath}, \hat{k}$
$\hat{I}, \hat{J}, \hat{K}$	global coordinate axes unit vectors
j	number of joints
J	Jacobian
l	length
n	number of links of a robot, number of joint variables
$\hat{n}$	unit vector
O	common origin of B and G
$O\varphi\theta\psi$	Euler angle frame
p	pitch of a screw
P	prismatic joint, a body point, a fixed point in B
q	joint variable
$\mathbf{q}$	joint variable vector
Q	transformation matrix of rotation about a global axis
$\mathbf{r}$	position vector
r_{ij}	element of row i and column j of a matrix
R	revolute joint, radius of a circle, rotation transformation matrix
R_w	wheel radius

$\mathbb{R}$	set of real numbers
s	sine, a member of S
$\check{s}$	screw
S	screw joint, a set
t	time
T	homogeneous transformation matrix, tool frame, time coordinate frame
T_{arm}	manipulator transformation matrix
T_{wrist}	wrist transformation matrix
$\mathbf{T}$	set of nonlinear algebraic equations of $\mathbf{q}$
$\hat{u}$	unit vector on axis of rotation
$\tilde{u}$	skew-symmetric matrix of the vector $\hat{u}$
$\mathbf{v}$	velocity vector
W	wheel coordinate frame
x, y, z	local coordinate axes
X, Y, Z	global coordinate axes

Greek

α	sideslip angle
α, β, γ	rotation angles about global axes
γ	camber angle
δ	deflection
δ_{ij}	Kronecker's delta
φ, θ, ψ	rotation angles about local axes, Euler angles
$\omega_x, \omega_y, \omega_z$	angular velocity components
ω	angular velocity vector
Ω	speed ratio

Symbol

$[\ \ ]^{-1}$	inverse of the matrix [ ]
$[\ \ ]^{\mathrm{T}}$	transpose of the matrix [ ]
$\otimes$	screw reiprocal product
$\equiv$	equivalent
$\vdash$	orthogonal
(i)	link number i
$\parallel$	parallel sign
$\perp$	perpendicular
$\times$	vector cross product
DH	Denavit-Hartenberg
DOF	degree of freedom
$R1$	1 rotational freedom joint
$R2$	2 rotational freedom joint
$R1T1$	1 rotational and 1 translational freedom joint
$R2T1$	2 rotational and 1 translational freedom joint
$R1T2$	1 rotational and 2 translational freedom joint
$R3T1$	3 rotational and 1 translational freedom joint
$R2T2$	2 rotational and 2 translational freedom joint
$R3T2$	3 rotational and 2 translational freedom joint
$T1$	1 translational freedom joint
sgn	signum function
S	screw joint
SSRMS	space station remote manipulator system

EXERCISES

1. A 4R Planar Manipulator For the 4R planar manipulator, shown in Figure 7.74, find:

(a) DH table

(b) Link-type table

(c) Individual frame transformation matrices ${}^{i-1}T_i$, $i = 1, 2, 3, 4$

(d) Global coordinates of the end effector

(e) Orientation of the end effector φ

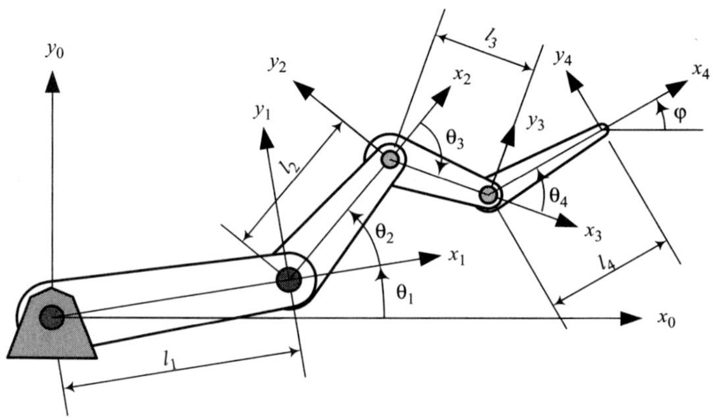

Figure 7.74 A 4R planar manipulator.

2. DH Coordinate Frame for Connected Links

(a) Set up the required link coordinate frames for the manipulators in Figure 7.75(a) and (b) using l_1, l_2, l_3 for the length of the links.

(b) Determine the forward kinematics transformation matrix of the manipulator in Figures 7.75(a) and (b) and find their rest positions.

(c) Determine the global coordinates of the tip point of the manipulator in Figures 7.75(a) and (b) at the position shown.

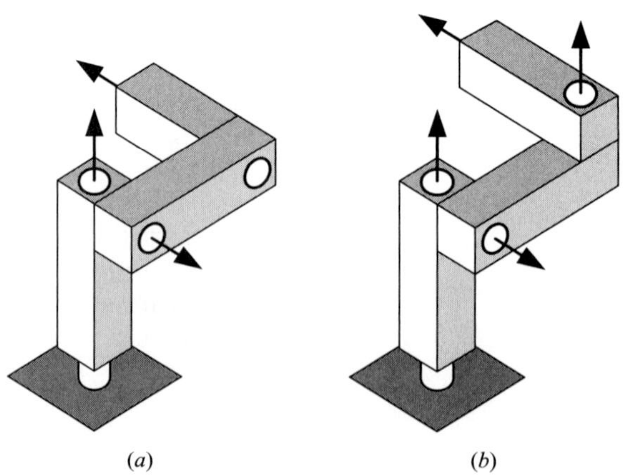

Figure 7.75 Two manipulators that are made by connecting industrial links.

3. **Frame at Center** Attach a link's coordinate frame at the geometric center of the link, $a_i/2$. Using the rigid-motion and homogeneous matrices, and develop the transformation matrices ${}^0T_1$, ${}^1T_2$, and ${}^0T_2$ for the manipulator of Figure 7.76.

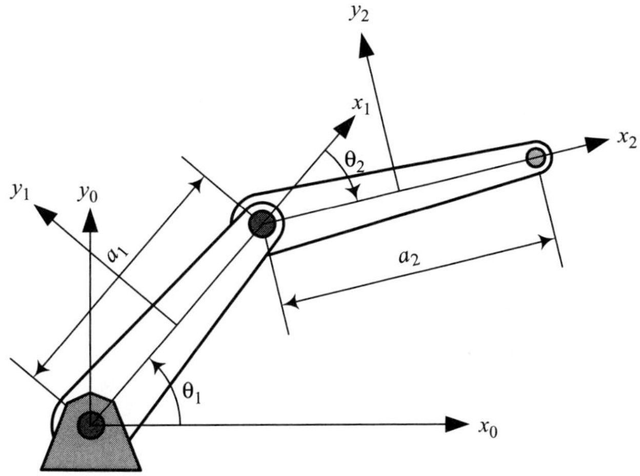

Figure 7.76 A 2R planar manipulator with a coordinate frame at the geometric center of each link.

4. **Free Coordinate Frame** Determine the link's transformation matrices ${}^1T_2$, ${}^2T_3$, and ${}^1T_3$ for the planar Cartesian manipulator shown in Figure 7.77. The coordinate frames are not based on DH rules.

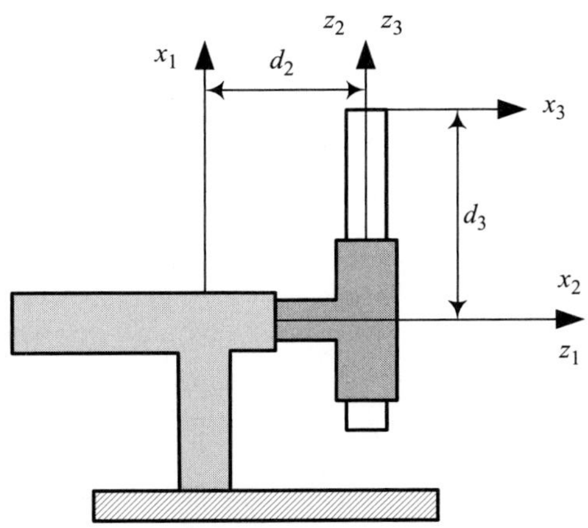

Figure 7.77 A two-DOF Cartesian manipulator.

5. ★ **Manipulator Design** Use industrial links and make a manipulator that:

   (a) Has three prismatic joints and reaches every point in a three-dimensional Cartesian space.

   (b) Has two prismatic joints and one revolute joint and reaches every point in a three-dimensional Cartesian space.

   (c) Has one prismatic joint and two revolute joints and reaches every point in a three-dimensional Cartesian space.

   (d) Has three revolute joints and reaches every point in a three-dimensional Cartesian space.

6. ★ **Special Manipulator Design** Use industrial links and make a manipulator with three DOF such that:

   (a) The tip point of the manipulator traces a circular path about a center point when two joints are locked.

   (b) The tip point of the manipulator traces a circular path about the origin of the global frame when two joints are locked.

   (c) The tip point of the manipulator traces a straight path when two joints are locked.

   (d) The tip point of the manipulator traces a straight path passing through the origin of the global frame.

7. **3R Planar Manipulator Inverse Kinematics** Consider an R$\|$R$\|$R planar manipulator with the following transformation matrices. Solve the inverse kinematics and find θ_1, θ_2, θ_3 for given coordinates x_0, y_0 of the tip point and a given value of φ:

$$
{}^2T_3 = \begin{bmatrix} \cos\theta_3 & -\sin\theta_3 & 0 & l_3\cos\theta_3 \\ \sin\theta_3 & \cos\theta_3 & 0 & l_3\sin\theta_3 \\ 0 & 0 & 1 & 0 \\ 0 & 0 & 0 & 1 \end{bmatrix}
$$

$$
{}^1T_2 = \begin{bmatrix} \cos\theta_2 & -\sin\theta_2 & 0 & l_2\cos\theta_2 \\ \sin\theta_2 & \cos\theta_2 & 0 & l_2\sin\theta_2 \\ 0 & 0 & 1 & 0 \\ 0 & 0 & 0 & 1 \end{bmatrix}
$$

$$
{}^0T_1 = \begin{bmatrix} \cos\theta_1 & -\sin\theta_1 & 0 & l_1\cos\theta_1 \\ \sin\theta_1 & \cos\theta_1 & 0 & l_1\sin\theta_1 \\ 0 & 0 & 1 & 0 \\ 0 & 0 & 0 & 1 \end{bmatrix}
$$

8. **2R Manipulator Tip Point on a Horizontal Path** Consider an elbow-up planar 2R manipulator with $l_1 = l_2 = 1$. The tip point is moving on a straight line from $P_1\ (1, 1.5)$ to $P_2\ (-1, 1.5)$.

   (a) Divide the Cartesian path in 10 equal sections and determine the joint variables at the 11 points.

   (b) ★ Calculate the joint variable θ_1 at P_1 and at P_2. Divide the range of θ_1 into 10 equal sections and determine the coordinates of the tip point at the 11 values of θ_1.

   (c) ★ Calculate the joint variable θ_2 at P_1 and at P_2. Divide the range of θ_1 into 10 equal sections and determine the coordinates of the tip point at the 11 values of θ_2.

9. 2R Manipulator Tip Point on a Tilted Path Consider an elbow-up planar 2R manipulator with $l_1 = l_2 = 1$. The tip point is moving on a straight line from $P_1\,(1, 1.5)$ to $P_2\,(-1, 1)$.

(a) Divide the Cartesian path into 10 equal sections and determine the joint variables at the 11 points.

(b) ★ Calculate the joint variable θ_1 at P_1 and at P_2. Divide the range of θ_1 into 10 equal sections and determine the coordinates of the tip point at the 11 values of θ_1.

(c) ★ Calculate the joint variable θ_2 at P_1 and at P_2. Divide the range of θ_1 into 10 equal sections and determine the coordinates of the tip point at the 11 values of θ_2.

10. 2R Manipulator Motion on a Horizontal Path Consider an elbow-up planar 2R manipulator with $l_1 = l_2 = 1$. The tip point is moving on a straight line from $P_1\,(1, 1.5)$ to $P_2\,(-1, 1.5)$ according to the following functions of time:

$$X = 1 - 6t^2 + 4t^3 \qquad Y = 1.4$$

(a) Calculate and plot θ_1 and θ_2 as functions of time if the time of motion is $0 \le t \le 1$.

(b) ★ Calculate and plot $\dot{\theta}_1$ and $\dot{\theta}_2$ as functions of time.

(c) ★ Calculate and plot $\ddot{\theta}_1$ and $\ddot{\theta}_2$ as functions of time.

(d) ★ Calculate and plot $\dddot{\theta}_1$ and $\dddot{\theta}_2$ as functions of time.

11. 2R Manipulator Motion on a Horizontal Path Consider a planar elbow-up 2R manipulator with $l_1 = l_2 = 1$. The tip point is moving on a straight line from $P_1\,(1, 1.5)$ to $P_2\,(-1, 1.5)$ with a constant speed:

$$X = 1 - vt \qquad Y = 1.5$$

(a) Calculate v and plot θ_1 and θ_2 if the time of motion is $0 \le t \le 1$.

(b) Calculate v and plot θ_1 and θ_2 if the time of motion is $0 \le t \le 5$.

(c) Calculate v and plot θ_1 and θ_2 if the time of motion is $0 \le t \le 10$.

(d) ★ Plot θ_1 and θ_2 as functions of v at point $(0, 1.5)$.

12. 2R Manipulator Motion on a Horizontal Path Consider a planar elbow-up 2R manipulator with $l_1 = l_2 = 1$. The tip point is moving on a straight line from $P_1\,(1, 1.5)$ to $P_2\,(-1, 1.5)$ with a constant acceleration:

$$X = 1 - \tfrac{1}{2}at^2 \qquad Y = 1.5$$

(a) Calculate a and plot θ_1 and θ_2 if the time of motion is $0 \le t \le 1$.

(b) Calculate a and plot θ_1 and θ_2 if the time of motion is $0 \le t \le 5$.

(c) Calculate a and plot θ_1 and θ_2 if the time of motion is $0 \le t \le 10$.

(d) ★ Plot θ_1 and θ_2 as functions of a at point $(0, 1.5)$.

13. ★ 2R Manipulator Kinematics on a Tilted Path Consider a planar elbow-up 2R manipulator with $l_1 = l_2 = 1$. The tip point is moving on a straight line from $P_1\,(1, 1.5)$ to $P_2\,(-1, 1.5)$ with a constant speed:

$$X = 1 - vt \qquad Y = 1.5$$

(a) Calculate and plot θ_1 and θ_2 if the time of motion is $0 \le t \le 1$.

(b) Calculate and plot $\dot{\theta}_1$ and $\dot{\theta}_2$ as functions of time.

(c) Calculate and plot $\ddot{\theta}_1$ and $\ddot{\theta}_2$ as functions of time.

(d) Calculate and plot $\dddot{\theta}_1$ and $\dddot{\theta}_2$ as functions of time.

14. **Acceptable Lengths of a 2R Planar Manipulator** The tip point of a 2R planar manipulator is at (1, 1.1).

(a) Assume $l_1 = 1$. Plot θ_1 and θ_2 versus l_2 and determine the range of possible l_2 for the elbow-up configuration.

(b) Assume $l_2 = 1$. Plot θ_1 and θ_2 versus l_1 and determine the range of possible l_1 for the elbow-up configuration.

(c) Assume $l_1 = 1$. Plot θ_1 and θ_2 versus l_2 and determine the range of possible l_2 for the elbow-down configuration.

(d) Assume $l_2 = 1$. Plot θ_1 and θ_2 versus l_1 and determine the range of possible l_1 for the elbow-down configuration.

15. **SCARA Robot Inverse Kinematics** Consider an R$\|$R$\|$R$\|$P robot with the following transformation matrices. Solve the inverse kinematics and find θ_1, θ_2, θ_3 and d for a given ${}^0T_4$:

$$
{}^0T_1 = \begin{bmatrix} \cos\theta_1 & -\sin\theta_1 & 0 & l_1\cos\theta_1 \\ \sin\theta_1 & \cos\theta_1 & 0 & l_1\sin\theta_1 \\ 0 & 0 & 1 & 0 \\ 0 & 0 & 0 & 1 \end{bmatrix}
$$

$$
{}^1T_2 = \begin{bmatrix} \cos\theta_2 & -\sin\theta_2 & 0 & l_2\cos\theta_2 \\ \sin\theta_2 & \cos\theta_2 & 0 & l_2\sin\theta_2 \\ 0 & 0 & 1 & 0 \\ 0 & 0 & 0 & 1 \end{bmatrix}
$$

$$
{}^2T_3 = \begin{bmatrix} \cos\theta_3 & -\sin\theta_3 & 0 & 0 \\ \sin\theta_3 & \cos\theta_3 & 0 & 0 \\ 0 & 0 & 1 & 0 \\ 0 & 0 & 0 & 1 \end{bmatrix} \qquad {}^3T_4 = \begin{bmatrix} 1 & 0 & 0 & 0 \\ 0 & 1 & 0 & 0 \\ 0 & 0 & 1 & d \\ 0 & 0 & 0 & 1 \end{bmatrix}
$$

$$
\begin{aligned}
{}^0T_4 &= {}^0T_1\,{}^1T_2\,{}^2T_3\,{}^3T_4 \\
&= \begin{bmatrix} c\theta_{123} & -s\theta_{123} & 0 & l_1c\theta_1 + l_2c\theta_{12} \\ s\theta_{123} & c\theta_{123} & 0 & l_1s\theta_1 + l_2s\theta_{12} \\ 0 & 0 & 1 & d \\ 0 & 0 & 0 & 1 \end{bmatrix}
\end{aligned}
$$

$$
\theta_{123} = \theta_1 + \theta_2 + \theta_3 \qquad \theta_{12} = \theta_1 + \theta_2
$$

16. **R⊢R$\|$R Articulated Arm Inverse Kinematics** Consider a three-DOF R⊢R$\|$R manipulator. Use the following transformation matrices and solve the inverse kinematics for θ_1, θ_2, θ_3:

$$
{}^0T_1 = \begin{bmatrix} \cos\theta_1 & 0 & -\sin\theta_1 & 0 \\ \sin\theta_1 & 0 & \cos\theta_1 & 0 \\ 0 & -1 & 0 & d_1 \\ 0 & 0 & 0 & 1 \end{bmatrix}
$$

$$
{}^1T_2 = \begin{bmatrix} \cos\theta_2 & -\sin\theta_2 & 0 & l_2\cos\theta_2 \\ \sin\theta_2 & \cos\theta_2 & 0 & l_2\sin\theta_2 \\ 0 & 0 & 1 & d_2 \\ 0 & 0 & 0 & 1 \end{bmatrix}
$$

$$
{}^{2}T_{3} = \begin{bmatrix} \cos\theta_3 & 0 & \sin\theta_3 & 0 \\ \sin\theta_3 & 0 & -\cos\theta_3 & 0 \\ 0 & 1 & 0 & 0 \\ 0 & 0 & 0 & 1 \end{bmatrix}
$$

17. ★ Space Station Remote Manipulator System Inverse Kinematics The forward kinematics of the SSRMS provides the following transformation matrices. Solve the inverse kinematics for the SSRMS:

$$
{}^{0}T_{1} = \begin{bmatrix} c\theta_1 & 0 & -s\theta_1 & 0 \\ s\theta_1 & 0 & c\theta_1 & 0 \\ 0 & -1 & 0 & d_1 \\ 0 & 0 & 0 & 1 \end{bmatrix} \qquad {}^{1}T_{2} = \begin{bmatrix} c\theta_2 & 0 & -s\theta_2 & 0 \\ s\theta_2 & 0 & c\theta_2 & 0 \\ 0 & -1 & 0 & d_2 \\ 0 & 0 & 0 & 1 \end{bmatrix}
$$

$$
{}^{2}T_{3} = \begin{bmatrix} c\theta_3 & -s\theta_3 & 0 & a_3 c\theta_3 \\ s\theta_3 & c\theta_3 & 0 & a_3 s\theta_3 \\ 0 & 0 & 1 & d_3 \\ 0 & 0 & 0 & 1 \end{bmatrix} \qquad {}^{3}T_{4} = \begin{bmatrix} c\theta_4 & -s\theta_4 & 0 & a_4 c\theta_4 \\ s\theta_4 & c\theta_4 & 0 & a_4 s\theta_4 \\ 0 & 0 & 1 & d_4 \\ 0 & 0 & 0 & 1 \end{bmatrix}
$$

$$
{}^{4}T_{5} = \begin{bmatrix} c\theta_5 & 0 & s\theta_5 & 0 \\ s\theta_5 & 0 & -c\theta_5 & 0 \\ 0 & 1 & 0 & d_5 \\ 0 & 0 & 0 & 1 \end{bmatrix} \qquad {}^{5}T_{6} = \begin{bmatrix} c\theta_6 & 0 & -s\theta_6 & 0 \\ s\theta_6 & 0 & c\theta_6 & 0 \\ 0 & -1 & 0 & d_6 \\ 0 & 0 & 0 & 1 \end{bmatrix}
$$

$$
{}^{6}T_{7} = \begin{bmatrix} c\theta_7 & -s\theta_7 & 0 & 0 \\ s\theta_7 & c\theta_7 & 0 & 0 \\ 0 & 0 & 1 & d_7 \\ 0 & 0 & 0 & 1 \end{bmatrix}
$$

Hint: This robot is a one-DOF redundant robot. It has seven joints, which is one more than the required six DOF to reach a point at a desired orientation. To solve the inverse kinematics of this robot, we need to introduce one extra condition among the joint variables or assign a value to one of the joint variables.

(a) Assume $\theta_1 = 0$ and ${}^{1}T_{7}$ is given. Determine $\theta_2, \theta_3, \theta_4, \theta_5, \theta_6, \theta_7$.

(b) Assume $\theta_2 = 0$ and ${}^{1}T_{7}$ is given. Determine $\theta_1, \theta_3, \theta_4, \theta_5, \theta_6, \theta_7$.

(c) Assume $\theta_3 = 0$ and ${}^{1}T_{7}$ is given. Determine $\theta_1, \theta_2, \theta_4, \theta_5, \theta_6, \theta_7$.

(d) Assume $\theta_5 = 0$ and ${}^{1}T_{7}$ is given. Determine $\theta_1, \theta_2, \theta_3, \theta_4, \theta_6, \theta_7$.

(e) Assume $\theta_6 = 0$ and ${}^{1}T_{7}$ is given. Determine $\theta_1, \theta_2, \theta_3, \theta_4, \theta_5, \theta_7$.

(f) Assume $\theta_7 = 0$ and ${}^{1}T_{7}$ is given. Determine $\theta_1, \theta_2, \theta_3, \theta_4, \theta_5, \theta_6$.

(g) Determine $\theta_1, \theta_2, \theta_3, \theta_4, \theta_5, \theta_6, \theta_7$ such that f is minimized:

$$
f = \theta_1 + \theta_2 + \theta_3 + \theta_4 + \theta_5 + \theta_6 + \theta_7
$$

Part III

Derivative Kinematics

Derivative kinematics explains how the derivatives of vectors are calculated and how they are related to each other.

The simple and mixed forms of the first- and second-derivative transformation formulas for a general vector ${}^{B}\square$ are

$$\frac{{}^{G}d}{dt}\,{}^{B}\square = \frac{{}^{B}d}{dt}\,{}^{B}\square + {}_{G}^{B}\boldsymbol{\omega}_{B} \times {}^{B}\square$$

$${}_{A}^{B}\dot{\square} = {}_{C}^{B}\dot{\square} + \left({}_{A}^{B}\boldsymbol{\omega}_{B} - {}_{C}^{B}\boldsymbol{\omega}_{B}\right) \times {}^{B}\square$$

$$\frac{{}^{G}d}{dt}\frac{{}^{G}d}{dt}\,{}^{B}\square = \frac{{}^{B}d}{dt}\frac{{}^{B}d}{dt}\,{}^{B}\square + {}_{G}^{B}\boldsymbol{\alpha}_{B} \times {}^{B}\square + 2\,{}_{G}^{B}\boldsymbol{\omega}_{B} \times \left(\frac{{}^{B}d}{dt}\,{}^{B}\square + {}_{G}^{B}\boldsymbol{\omega}_{B} \times {}^{B}\square\right)$$

$${}_{A}^{B}\ddot{\square} = {}_{C}^{B}\ddot{\square} + \left({}_{A}^{B}\boldsymbol{\alpha}_{B} - {}_{C}^{B}\boldsymbol{\alpha}_{B}\right) \times {}^{B}\square + 2\left({}_{A}^{B}\boldsymbol{\omega}_{B} - {}_{C}^{B}\boldsymbol{\omega}_{B}\right) \times {}^{B}\dot{\square} + {}_{A}^{B}\boldsymbol{\omega}_{B} \times \left({}_{A}^{B}\boldsymbol{\omega}_{B} \times {}^{B}\square\right) - {}_{C}^{B}\boldsymbol{\omega}_{B} \times \left({}_{C}^{B}\boldsymbol{\omega}_{B} \times {}^{B}\square\right)$$

and are the keys for relating the derivative of vectorial characteristics in different coordinate frames.

The geometric transformation formula

$${}^{G}\mathbf{r} = {}^{G}R_{B}\,{}^{B}\mathbf{r}$$

is not generally correct for nonposition vectors,

$$\frac{d^{n}}{dt^{n}}\,{}^{G}\mathbf{r} \neq {}^{G}R_{B}\,\frac{d^{n}}{dt^{n}}\,{}^{B}\mathbf{r} \qquad \mathbf{n} = 1, 2, 3, \ldots$$

unless the B- and G-frames are fixed relatively.

8

Velocity Kinematics

The angular velocity of a rigid body is the instantaneous rotation of the body with respect to another body. It can be indicated by a vectorial quantity. We review angular velocity calculus to study the time rate of rigid-body motions.

8.1 ANGULAR VELOCITY

Consider a rotating rigid body $B(Oxyz)$ with a fixed point O in a reference frame $G(OXYZ)$, as shown in Figure 8.1. We express the motion of the body by a time-varying rotation transformation matrix between B and G to transform the instantaneous coordinates of body points to their coordinates in the global frame:

$$ {}^{G}\mathbf{r}(t) = {}^{G}R_B(t)\,{}^{B}\mathbf{r} \tag{8.1}$$

The velocity of a body point in the global frame is

$$ {}^{G}\mathbf{v}(t) = {}^{G}\dot{\mathbf{r}}(t) = {}^{G}\dot{R}_B(t)\,{}^{B}\mathbf{r} = {}_{G}\tilde{\omega}_B\,{}^{G}\mathbf{r}(t) = {}_{G}\boldsymbol{\omega}_B \times {}^{G}\mathbf{r}(t) \tag{8.2}$$

where ${}_{G}\boldsymbol{\omega}_B$ is the *angular velocity vector* of B with respect to G. It is equal to a rotation with *angular speed* $\dot{\phi}$ about an *instantaneous axis of rotation* $\hat{u}$:

$$ \boldsymbol{\omega} = \begin{bmatrix} \omega_1 \\ \omega_2 \\ \omega_3 \end{bmatrix} = \dot{\phi}\,\hat{u} \tag{8.3}$$

The angular velocity vector is associated with a skew-symmetric matrix ${}_{G}\tilde{\omega}_B$ called the *angular velocity matrix*,

$$ \tilde{\omega} = \begin{bmatrix} 0 & -\omega_3 & \omega_2 \\ \omega_3 & 0 & -\omega_1 \\ -\omega_2 & \omega_1 & 0 \end{bmatrix} \tag{8.4}$$

where

$$ {}_{G}\tilde{\omega}_B = {}^{G}\dot{R}_B\,{}^{G}R_B^{\mathrm{T}} = \dot{\phi}\,\tilde{u} \tag{8.5}$$

The B-expression of the angular velocity is similarly defined:

$$ {}_{G}^{B}\tilde{\omega}_B = {}^{G}R_B^{\mathrm{T}}\,{}^{G}\dot{R}_B \tag{8.6}$$

Employing the global and body expressions of the angular velocity of the body relative to the global coordinate frame, ${}_{G}\tilde{\omega}_B$ and ${}_{G}^{B}\tilde{\omega}_B$, we determine the global and body

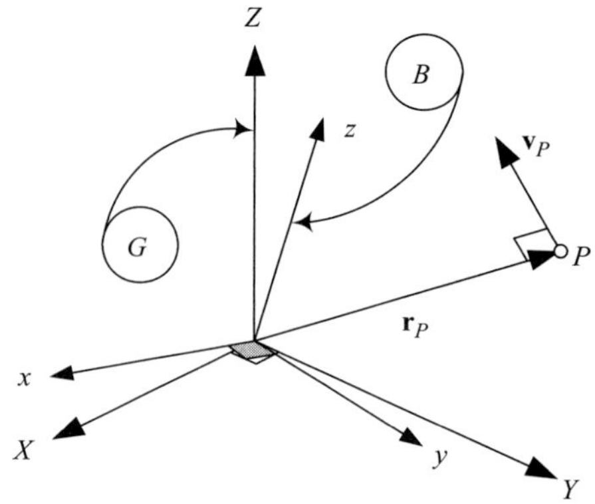

Figure 8.1 A rotating rigid body $B(Oxyz)$ with a fixed point O in a global frame $G(OXYZ)$.

expressions of the velocity of a body point as

$$ {}^G_G\mathbf{v}_P = {}^G_G\boldsymbol{\omega}_B \times {}^G\mathbf{r}_P \tag{8.7}$$

$$ {}^B_G\mathbf{v}_P = {}^B_G\boldsymbol{\omega}_B \times {}^B\mathbf{r}_P \tag{8.8}$$

The G-expression ${}_G\tilde{\omega}_B$ and B-expression ${}^B_G\tilde{\omega}_B$ of the angular velocity matrix can be transformed to each other using the rotation matrix ${}^GR_B$:

$$ {}_G\tilde{\omega}_B = {}^GR_B \, {}^B_G\tilde{\omega}_B \, {}^GR_B^{\mathrm{T}} \tag{8.9}$$

$$ {}^B_G\tilde{\omega}_B = {}^GR_B^{\mathrm{T}} \, {}^G_G\tilde{\omega}_B \, {}^GR_B \tag{8.10}$$

They are also related to each other directly as

$$ {}_G\tilde{\omega}_B \, {}^GR_B = {}^GR_B \, {}^B_G\tilde{\omega}_B^{\mathrm{T}} \tag{8.11}$$

$$ {}^GR_B^{\mathrm{T}} \, {}_G\tilde{\omega}_B = {}^B_G\tilde{\omega}_B \, {}^GR_B^{\mathrm{T}} \tag{8.12}$$

The relative angular velocity vectors of relatively moving rigid bodies can be done only if all the angular velocities are expressed in one coordinate frame:

$$ {}_0\boldsymbol{\omega}_n = {}_0\boldsymbol{\omega}_1 + {}^0_1\boldsymbol{\omega}_2 + {}^0_2\boldsymbol{\omega}_3 + \cdots + {}^0_{n-1}\boldsymbol{\omega}_n = \sum_{i=1}^{n} {}^0_{i-1}\boldsymbol{\omega}_i \tag{8.13}$$

The inverses of the angular velocity matrices ${}_G\tilde{\omega}_B$ and ${}^B_G\tilde{\omega}_B$ are

$$ {}_G\tilde{\omega}_B^{-1} = {}^GR_B \, {}^G\dot{R}_B^{-1} \tag{8.14}$$

$$ {}^B_G\tilde{\omega}_B^{-1} = {}^G\dot{R}_B^{-1} \, {}^GR_B \tag{8.15}$$

Proof: Consider a rigid body with a fixed point O and an attached frame $B(Oxyz)$ as shown in Figure 8.2. The body frame B is initially coincident with the global frame G.

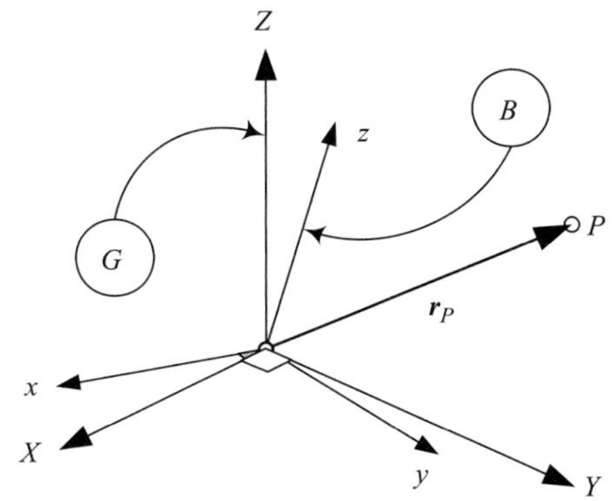

Figure 8.2 A body fixed point P at ${}^B\mathbf{r}$ in the rotating body frame B.

Therefore, the position vector of a body point P at the initial time $t = t_0$ is

$$ {}^G\mathbf{r}(t_0) = {}^B\mathbf{r} \tag{8.16} $$

and at any other time is found by the associated transformation matrix ${}^GR_B(t)$:

$$ {}^G\mathbf{r}(t) = {}^GR_B(t)\,{}^B\mathbf{r} = {}^GR_B(t)\,{}^G\mathbf{r}(t_0) \tag{8.17} $$

The global time derivative of ${}^G\mathbf{r}$ is

$$ \begin{aligned} {}^G\mathbf{v} = {}^G\dot{\mathbf{r}} &= \frac{{}^Gd}{dt}\,{}^G\mathbf{r}(t) = \frac{{}^Gd}{dt}\left[{}^GR_B(t)\,{}^B\mathbf{r}\right] = \frac{{}^Gd}{dt}\left[{}^GR_B(t)\,{}^G\mathbf{r}(t_0)\right] \\ &= {}^G\dot{R}_B(t)\,{}^G\mathbf{r}(t_0) = {}^G\dot{R}_B(t)\,{}^B\mathbf{r} \end{aligned} \tag{8.18} $$

Eliminating ${}^B\mathbf{r}$ between (8.17) and (8.18) determines the velocity of the global point in the global frame:

$$ {}^G\mathbf{v} = {}^G\dot{R}_B(t)\,{}^GR_B^{\mathrm{T}}(t)\,{}^G\mathbf{r}(t) \tag{8.19} $$

We denote the coefficient of ${}^G\mathbf{r}(t)$ by

$$ {}_G\tilde{\omega}_B = {}^G\dot{R}_B\,{}^GR_B^{\mathrm{T}} \tag{8.20} $$

and rewrite Equation (8.19) as

$$ {}^G\mathbf{v} = {}_G\tilde{\omega}_B\,{}^G\mathbf{r}(t) \tag{8.21} $$

or equivalently as

$$ {}^G\mathbf{v} = {}_G\boldsymbol{\omega}_B \times {}^G\mathbf{r}(t) \tag{8.22} $$

where ${}_G\boldsymbol{\omega}_B$ is the *instantaneous angular velocity* of the body B relative to the global frame G as seen from the G-frame.

Transforming ${}^{G}\mathbf{v}$ to the body frame provides the body expression of the velocity vector:

$$\begin{aligned}{}_{G}^{B}\mathbf{v}_{P} &= {}^{G}R_{B}^{\mathrm{T}}\,{}^{G}\mathbf{v} = {}^{G}R_{B}^{\mathrm{T}}\,{}_{G}\tilde{\omega}_{B}\,{}^{G}\mathbf{r} = {}^{G}R_{B}^{\mathrm{T}}\,{}^{G}\dot{R}_{B}\,{}^{G}R_{B}^{\mathrm{T}}\,{}^{G}\mathbf{r} \\ &= {}^{G}R_{B}^{\mathrm{T}}\,{}^{G}\dot{R}_{B}\,{}^{B}\mathbf{r}\end{aligned} \tag{8.23}$$

We denote the coefficient of ${}^{B}\mathbf{r}$ by

$${}_{G}^{B}\tilde{\omega}_{B} = {}^{G}R_{B}^{\mathrm{T}}\,{}^{G}\dot{R}_{B} \tag{8.24}$$

and rewrite Equation (8.23) as

$${}_{G}^{B}\mathbf{v}_{P} = {}_{G}^{B}\tilde{\omega}_{B}\,{}^{B}\mathbf{r}_{P} \tag{8.25}$$

or equivalently as

$${}_{G}^{B}\mathbf{v}_{P} = {}_{G}^{B}\boldsymbol{\omega}_{B} \times {}^{B}\mathbf{r}_{P} \tag{8.26}$$

where ${}_{G}^{B}\boldsymbol{\omega}_{B}$ is the *instantaneous angular velocity* of B relative to the global frame G as seen from the B-frame.

The time derivative of the orthogonality condition, ${}^{G}R_{B}\,{}^{G}R_{B}^{\mathrm{T}} = \mathbf{I}$, introduces an important identity,

$${}^{G}\dot{R}_{B}\,{}^{G}R_{B}^{\mathrm{T}} + {}^{G}R_{B}\,{}^{G}\dot{R}_{B}^{\mathrm{T}} = 0 \tag{8.27}$$

which can be used to show that the angular velocity matrix ${}_{G}\tilde{\omega}_{B} = [{}^{G}\dot{R}_{B}\,{}^{G}R_{B}^{\mathrm{T}}]$ is skew symmetric:

$${}^{G}R_{B}\,{}^{G}\dot{R}_{B}^{\mathrm{T}} = \left[{}^{G}\dot{R}_{B}\,{}^{G}R_{B}^{\mathrm{T}}\right]^{\mathrm{T}} \tag{8.28}$$

Generally speaking, an angular velocity vector is the instantaneous rotation of a coordinate frame A with respect to another frame B that can be expressed in or seen from a third coordinate frame C. We indicate the first coordinate frame A by a right subscript, the second frame B by a left subscript, and the third frame C by a left superscript, ${}_{B}^{C}\boldsymbol{\omega}_{A}$. If the left super- and subscripts are the same, we only show the subscript.

We can transform the G-expression of the global velocity of a body point P, ${}^{G}\mathbf{v}_{P}$, and the B-expression of the global velocity of the point P, ${}_{G}^{B}\mathbf{v}_{P}$, to each other using a rotation matrix:

$$\begin{aligned}{}_{G}^{B}\mathbf{v}_{P} &= {}^{B}R_{G}\,{}^{G}\mathbf{v}_{P} = {}^{B}R_{G}\,{}_{G}\tilde{\omega}_{B}\,{}^{G}\mathbf{r}_{P} = {}^{B}R_{G}\,{}_{G}\tilde{\omega}_{B}\,{}^{G}R_{B}\,{}^{B}\mathbf{r}_{P} \\ &= {}^{B}R_{G}\,{}^{G}\dot{R}_{B}\,{}^{G}R_{B}^{\mathrm{T}}\,{}^{G}R_{B}\,{}^{B}\mathbf{r}_{P} = {}^{B}R_{G}\,{}^{G}\dot{R}_{B}\,{}^{B}\mathbf{r}_{P} \\ &= {}^{G}R_{B}^{\mathrm{T}}\,{}^{G}\dot{R}_{B}\,{}^{B}\mathbf{r}_{P} = {}_{G}^{B}\tilde{\omega}_{B}\,{}^{B}\mathbf{r}_{P} = {}_{G}^{B}\boldsymbol{\omega}_{B} \times {}^{B}\mathbf{r}_{P}\end{aligned} \tag{8.29}$$

$$\begin{aligned}{}^{G}\mathbf{v}_{P} &= {}^{G}R_{B}\,{}_{G}^{B}\mathbf{v}_{P} = {}^{G}R_{B}\,{}_{G}^{B}\tilde{\omega}_{B}\,{}^{B}\mathbf{r}_{P} = {}^{G}R_{B}\,{}_{G}^{B}\tilde{\omega}_{B}\,{}^{G}R_{B}^{\mathrm{T}}\,{}^{G}\mathbf{r}_{P} \\ &= {}^{G}R_{B}\,{}^{G}R_{B}^{\mathrm{T}}\,{}^{G}\dot{R}_{B}\,{}^{G}R_{B}^{\mathrm{T}}\,{}^{G}\mathbf{r}_{P} = {}^{G}\dot{R}_{B}\,{}^{G}R_{B}^{\mathrm{T}}\,{}^{G}\mathbf{r}_{P} \\ &= {}_{G}\tilde{\omega}_{B}\,{}^{G}\mathbf{r}_{P} = {}_{G}\boldsymbol{\omega}_{B} \times {}^{G}\mathbf{r}_{P} = {}^{G}R_{B}\left({}_{G}^{B}\boldsymbol{\omega}_{B} \times {}^{B}\mathbf{r}_{P}\right)\end{aligned} \tag{8.30}$$

From the definitions of ${}_G\tilde{\omega}_B$ and ${}_G^B\tilde{\omega}_B$ in (8.20) and (8.24) and comparing with (8.29) and (8.30), we are able to transform the two angular velocity matrices by

$$
{}_G\tilde{\omega}_B = {}^GR_B \, {}_G^B\tilde{\omega}_B \, {}^GR_B^{\mathrm{T}} \tag{8.31}
$$

$$
{}_G^B\tilde{\omega}_B = {}^GR_B^{\mathrm{T}} \, {}_G\tilde{\omega}_B \, {}^GR_B \tag{8.32}
$$

and derive the following useful equations:

$$
{}^G\dot{R}_B = {}_G\tilde{\omega}_B \, {}^GR_B \tag{8.33}
$$

$$
{}^G\dot{R}_B = {}^GR_B \, {}_G^B\tilde{\omega}_B \tag{8.34}
$$

$$
{}_G\tilde{\omega}_B \, {}^GR_B = {}^GR_B \, {}_G^B\tilde{\omega}_B \tag{8.35}
$$

The angular velocity of B in G is negative of the angular velocity of G in B if both are expressed in the same coordinate frame:

$$
{}_G^G\tilde{\omega}_B = -{}_B^G\tilde{\omega}_G \qquad {}_G^G\boldsymbol{\omega}_B = -{}_B^G\boldsymbol{\omega}_G \tag{8.36}
$$

$$
{}_G^B\tilde{\omega}_B = -{}_B^B\tilde{\omega}_G \qquad {}_G^B\boldsymbol{\omega}_B = -{}_B^B\boldsymbol{\omega}_G \tag{8.37}
$$

The vector ${}_G\boldsymbol{\omega}_B$ can always be expressed in the natural form

$$
{}_G\boldsymbol{\omega}_B = \omega\hat{u} \tag{8.38}
$$

with the magnitude ω and a unit vector $\hat{u}$ parallel to ${}_G\boldsymbol{\omega}_B$ that indicates the *instantaneous axis of rotation*.

Using the Rodriguez rotation formula (5.4) we can show that

$$
\dot{R}_{\hat{u},\phi} = \dot{\phi}\,\tilde{u}\,R_{\hat{u},\phi} \tag{8.39}
$$

and therefore,

$$
\tilde{\omega} = \dot{\phi}\,\tilde{u} \tag{8.40}
$$

or equivalently

$$
\begin{aligned}
{}_G\tilde{\omega}_B &= \lim_{\phi\to 0} \frac{{}^Gd}{dt} R_{\hat{u},\phi} = \lim_{\phi\to 0} \frac{{}^Gd}{dt} \left(-\tilde{u}^2\cos\phi + \tilde{u}\sin\phi + \tilde{u}^2 + \mathbf{I}\right) \\
&= \dot{\phi}\,\tilde{u}
\end{aligned} \tag{8.41}
$$

and therefore,

$$
\boldsymbol{\omega} = \dot{\phi}\,\hat{u} \tag{8.42}
$$

To show the addition of relative angular velocities in Equation (8.13), we start from a combination of rotations

$$
{}^0R_2 = {}^0R_1 \, {}^1R_2 \tag{8.43}
$$

and take a time derivative:

$$
{}^0\dot{R}_2 = {}^0\dot{R}_1 \, {}^1R_2 + {}^0R_1 \, {}^1\dot{R}_2 \tag{8.44}
$$

Substituting the derivative of the rotation matrices with

$$ {}^{0}\dot{R}_{2} = {}_{0}\tilde{\omega}_{2}\, {}^{0}R_{2} \tag{8.45} $$

$$ {}^{0}\dot{R}_{1} = {}_{0}\tilde{\omega}_{1}\, {}^{0}R_{1} \tag{8.46} $$

$$ {}^{1}\dot{R}_{2} = {}_{1}\tilde{\omega}_{2}\, {}^{1}R_{2} \tag{8.47} $$

results in

$$ \begin{aligned} {}_{0}\tilde{\omega}_{2}\, {}^{0}R_{2} &= {}_{0}\tilde{\omega}_{1}\, {}^{0}R_{1}\, {}^{1}R_{2} + {}^{0}R_{1}\, {}_{1}\tilde{\omega}_{2}\, {}^{1}R_{2} \\ &= {}_{0}\tilde{\omega}_{1}\, {}^{0}R_{2} + {}^{0}R_{1}\, {}_{1}\tilde{\omega}_{2}\, {}^{0}R_{1}^{\mathrm{T}}\, {}^{0}R_{1}\, {}^{1}R_{2} \\ &= {}_{0}\tilde{\omega}_{1}\, {}^{0}R_{2} + {}_{1}^{0}\tilde{\omega}_{2}\, {}^{0}R_{2} \end{aligned} \tag{8.48} $$

where

$$ {}^{0}R_{1}\, {}_{1}\tilde{\omega}_{2}\, {}^{0}R_{1}^{\mathrm{T}} = {}_{1}^{0}\tilde{\omega}_{2} \tag{8.49} $$

Therefore, we find

$$ {}_{0}\tilde{\omega}_{2} = {}_{0}\tilde{\omega}_{1} + {}_{1}^{0}\tilde{\omega}_{2} \tag{8.50} $$

which indicates that two angular velocities may be added when they are expressed in the same frame:

$$ {}_{0}\boldsymbol{\omega}_{2} = {}_{0}\boldsymbol{\omega}_{1} + {}_{1}^{0}\boldsymbol{\omega}_{2} \tag{8.51} $$

The expansion of this equation for any number of angular velocities would be Equation (8.13).

Employing the relative angular velocity formula (8.51), we can find the relative velocity formula of a point P in B_2 at ${}^{0}\mathbf{r}_{P}$:

$$ \begin{aligned} {}_{0}\mathbf{v}_{2} &= {}_{0}\boldsymbol{\omega}_{2}\, {}^{0}\mathbf{r}_{P} = \left({}_{0}\boldsymbol{\omega}_{1} + {}_{1}^{0}\boldsymbol{\omega}_{2}\right) {}^{0}\mathbf{r}_{P} = {}_{0}\boldsymbol{\omega}_{1}\, {}^{0}\mathbf{r}_{P} + {}_{1}^{0}\boldsymbol{\omega}_{2}\, {}^{0}\mathbf{r}_{P} \\ &= {}_{0}\mathbf{v}_{1} + {}_{1}^{0}\mathbf{v}_{2} \end{aligned} \tag{8.52} $$

The angular velocity matrices ${}_{G}\tilde{\omega}_{B}$ and ${}_{G}^{B}\tilde{\omega}_{B}$ are skew symmetric and not invertible. However, we can define their inverse by the rules

$$ {}_{G}\tilde{\omega}_{B}^{-1} = {}^{G}R_{B}\, {}^{G}\dot{R}_{B}^{-1} \tag{8.53} $$

$$ {}_{G}^{B}\tilde{\omega}_{B}^{-1} = {}^{G}\dot{R}_{B}^{-1}\, {}^{G}R_{B} \tag{8.54} $$

to get

$$ {}_{G}\tilde{\omega}_{B}^{-1}\, {}_{G}\tilde{\omega}_{B} = {}_{G}\tilde{\omega}_{B}\, {}_{G}\tilde{\omega}_{B}^{-1} = [\mathbf{I}] \tag{8.55} $$

$$ {}_{G}^{B}\tilde{\omega}_{B}^{-1}\, {}_{G}^{B}\tilde{\omega}_{B} = {}_{G}^{B}\tilde{\omega}_{B}\, {}_{G}^{B}\tilde{\omega}_{B}^{-1} = [\mathbf{I}] \tag{8.56} $$

■

Example 469 Rotation of a Body Point about a Global Axis Consider a rigid body is turning about the Z-axis with a constant angular speed $\dot{\alpha} = 10\,\mathrm{deg}/s$. The global velocity of a body point at $P(5,30,10)$ when the body is at $\alpha = 30\,\text{deg}$ is

$$
\begin{aligned}
{}^{G}\mathbf{v}_P &= {}^{G}\dot{R}_B(t)\,{}^{B}\mathbf{r}_P \\
&= \frac{{}^{G}d}{dt}\left(\begin{bmatrix} \cos\alpha & -\sin\alpha & 0 \\ \sin\alpha & \cos\alpha & 0 \\ 0 & 0 & 1 \end{bmatrix}\right)\begin{bmatrix} 5 \\ 30 \\ 10 \end{bmatrix} \\
&= \dot{\alpha}\begin{bmatrix} -\sin\alpha & -\cos\alpha & 0 \\ \cos\alpha & -\sin\alpha & 0 \\ 0 & 0 & 0 \end{bmatrix}\begin{bmatrix} 5 \\ 30 \\ 10 \end{bmatrix} \\
&= \frac{10\pi}{180}\begin{bmatrix} -\sin\frac{1}{6}\pi & -\cos\frac{1}{6}\pi & 0 \\ \cos\frac{1}{6}\pi & -\sin\frac{1}{6}\pi & 0 \\ 0 & 0 & 0 \end{bmatrix}\begin{bmatrix} 5 \\ 30 \\ 10 \end{bmatrix} = \begin{bmatrix} -4.97 \\ -1.86 \\ 0 \end{bmatrix}
\end{aligned} \tag{8.57}
$$

The point P is now at

$$
\begin{aligned}
{}^{G}\mathbf{r}_P &= {}^{G}R_B\,{}^{B}\mathbf{r}_P \\
&= \begin{bmatrix} \cos\frac{1}{6}\pi & -\sin\frac{1}{6}\pi & 0 \\ \sin\frac{1}{6}\pi & \cos\frac{1}{6}\pi & 0 \\ 0 & 0 & 1 \end{bmatrix}\begin{bmatrix} 5 \\ 30 \\ 10 \end{bmatrix} = \begin{bmatrix} -10.67 \\ 28.48 \\ 10 \end{bmatrix}
\end{aligned} \tag{8.58}
$$

Example 470 Rotation of a Global Point about a Global Axis A body point P at ${}^{B}\mathbf{r}_P = [5, 30, 10]^{\mathrm{T}}$ is turned $\alpha = 30\,\text{deg}$ about the Z-axis. The global position of P is at

$$
\begin{aligned}
{}^{G}\mathbf{r}_P &= {}^{G}R_B\,{}^{B}\mathbf{r}_P \\
&= \begin{bmatrix} \cos\frac{1}{6}\pi & -\sin\frac{1}{6}\pi & 0 \\ \sin\frac{1}{6}\pi & \cos\frac{1}{6}\pi & 0 \\ 0 & 0 & 1 \end{bmatrix}\begin{bmatrix} 5 \\ 30 \\ 10 \end{bmatrix} = \begin{bmatrix} -10.67 \\ 28.48 \\ 10 \end{bmatrix}
\end{aligned} \tag{8.59}
$$

If the body is turning with a constant angular speed $\dot{\alpha} = 10\,\text{deg/s}$, the global velocity of the point P would be

$$
\begin{aligned}
{}^{G}\mathbf{v}_P &= {}^{G}\dot{R}_B\,{}^{G}R_B^{\mathrm{T}}\,{}^{G}\mathbf{r}_P \\
&= \frac{10\pi}{180}\begin{bmatrix} -s\frac{1}{6}\pi & -c\frac{1}{6}\pi & 0 \\ c\frac{1}{6}\pi & -s\frac{1}{6}\pi & 0 \\ 0 & 0 & 0 \end{bmatrix}\begin{bmatrix} c\frac{1}{6}\pi & -s\frac{1}{6}\pi & 0 \\ s\frac{1}{6}\pi & c\frac{1}{6}\pi & 0 \\ 0 & 0 & 1 \end{bmatrix}^{\mathrm{T}}\begin{bmatrix} -10.67 \\ 28.48 \\ 10 \end{bmatrix} \\
&= \begin{bmatrix} -4.97 \\ -1.86 \\ 0 \end{bmatrix}
\end{aligned} \tag{8.60}
$$

Example 471 Principal Angular Velocities The principal rotational matrices about the axes X, Y, and Z are

$$R_{X,\gamma} = \begin{bmatrix} 1 & 0 & 0 \\ 0 & \cos\gamma & -\sin\gamma \\ 0 & \sin\gamma & \cos\gamma \end{bmatrix} \tag{8.61}$$

$$R_{Y,\beta} = \begin{bmatrix} \cos\beta & 0 & \sin\beta \\ 0 & 1 & 0 \\ -\sin\beta & 0 & \cos\beta \end{bmatrix} \tag{8.62}$$

$$R_{Z,\alpha} = \begin{bmatrix} \cos\alpha & -\sin\alpha & 0 \\ \sin\alpha & \cos\alpha & 0 \\ 0 & 0 & 1 \end{bmatrix} \tag{8.63}$$

So, their time derivatives are

$$\dot{R}_{X,\gamma} = \dot{\gamma} \begin{bmatrix} 0 & 0 & 0 \\ 0 & -\sin\gamma & -\cos\gamma \\ 0 & \cos\gamma & -\sin\gamma \end{bmatrix} \tag{8.64}$$

$$\dot{R}_{Y,\beta} = \dot{\beta} \begin{bmatrix} -\sin\beta & 0 & \cos\beta \\ 0 & 0 & 0 \\ -\cos\beta & 0 & -\sin\beta \end{bmatrix} \tag{8.65}$$

$$\dot{R}_{Z,\alpha} = \dot{\alpha} \begin{bmatrix} -\sin\alpha & -\cos\alpha & 0 \\ \cos\alpha & -\sin\alpha & 0 \\ 0 & 0 & 0 \end{bmatrix} \tag{8.66}$$

Therefore, the principal angular velocity matrices about axes X, Y, and Z are

$$\tilde{\omega}_X = \dot{R}_{X,\gamma} R_{X,\gamma}^{\mathrm{T}} = \dot{\gamma} \begin{bmatrix} 0 & 0 & 0 \\ 0 & 0 & -1 \\ 0 & 1 & 0 \end{bmatrix} \tag{8.67}$$

$$\tilde{\omega}_Y = \dot{R}_{Y,\beta} R_{Y,\beta}^{\mathrm{T}} = \dot{\beta} \begin{bmatrix} 0 & 0 & 1 \\ 0 & 0 & 0 \\ -1 & 0 & 0 \end{bmatrix} \tag{8.68}$$

$$\tilde{\omega}_Z = \dot{R}_{Z,\alpha} R_{Z,\alpha}^{\mathrm{T}} = \dot{\alpha} \begin{bmatrix} 0 & -1 & 0 \\ 1 & 0 & 0 \\ 0 & 0 & 0 \end{bmatrix} \tag{8.69}$$

which are equivalent to

$$\tilde{\omega}_X = \dot{\gamma}\tilde{I} \tag{8.70}$$

$$\tilde{\omega}_Y = \dot{\beta}\tilde{J} \tag{8.71}$$

$$\tilde{\omega}_Z = \dot{\alpha}\tilde{K} \tag{8.72}$$

and therefore the principal angular velocity vectors are

$$\boldsymbol{\omega}_X = \omega_X \hat{I} = \dot{\gamma}\hat{I} \tag{8.73}$$

$$\boldsymbol{\omega}_Y = \omega_Y \hat{J} = \dot{\beta}\hat{J} \tag{8.74}$$

$$\boldsymbol{\omega}_Z = \omega_Z \hat{K} = \dot{\alpha}\hat{K} \tag{8.75}$$

Using the same technique, we can find the following principal angular velocity matrices about the local axes:

$${}^B_G\tilde{\omega}_x = R^{\mathrm{T}}_{x,\psi}\,\dot{R}_{x,\psi} = \dot{\psi}\begin{bmatrix} 0 & 0 & 0 \\ 0 & 0 & -1 \\ 0 & 1 & 0 \end{bmatrix} = \dot{\psi}\,\tilde{\imath} \tag{8.76}$$

$${}^B_G\tilde{\omega}_y = R^{\mathrm{T}}_{y,\theta}\dot{R}_{y,\theta} = \dot{\theta}\begin{bmatrix} 0 & 0 & 1 \\ 0 & 0 & 0 \\ -1 & 0 & 0 \end{bmatrix} = \dot{\theta}\,\tilde{\jmath} \tag{8.77}$$

$${}^B_G\tilde{\omega}_z = R^{\mathrm{T}}_{z,\varphi}\dot{R}_{z,\varphi} = \dot{\varphi}\begin{bmatrix} 0 & -1 & 0 \\ 1 & 0 & 0 \\ 0 & 0 & 0 \end{bmatrix} = \dot{\varphi}\,\tilde{k} \tag{8.78}$$

Example 472 Time Derivative of Elements of ${}^G R_B$ Expansion of Equation (8.33) yields

$${}^G\dot{R}_B = {}_G\tilde{\omega}_B\,{}^G R_B \tag{8.79}$$

and shows how the time derivative of each element of ${}^G R_B$ is related to components of ${}_G\boldsymbol{\omega}_B$ and elements of ${}^G R_B$:

$$\begin{aligned}
\begin{bmatrix} \dot{r}_{11} & \dot{r}_{12} & \dot{r}_{13} \\ \dot{r}_{21} & \dot{r}_{22} & \dot{r}_{23} \\ \dot{r}_{31} & \dot{r}_{32} & \dot{r}_{33} \end{bmatrix} &= \begin{bmatrix} 0 & -\omega_Z & \omega_Y \\ \omega_Z & 0 & -\omega_X \\ -\omega_Y & \omega_X & 0 \end{bmatrix}\begin{bmatrix} r_{11} & r_{12} & r_{13} \\ r_{21} & r_{22} & r_{23} \\ r_{31} & r_{32} & r_{33} \end{bmatrix} \\
&= \begin{bmatrix} \omega_Y r_{31} - \omega_Z r_{21} & \omega_Y r_{32} - \omega_Z r_{22} & \omega_Y r_{33} - \omega_Z r_{23} \\ \omega_Z r_{11} - \omega_X r_{31} & \omega_Z r_{12} - \omega_X r_{32} & \omega_Z r_{13} - \omega_X r_{33} \\ \omega_X r_{21} - \omega_Y r_{11} & \omega_X r_{22} - \omega_Y r_{12} & \omega_X r_{23} - \omega_Y r_{13} \end{bmatrix}
\end{aligned} \tag{8.80}$$

Expansion of Equation (8.34) yields

$${}^G\dot{R}_B = {}^G R_B\,{}^B_G\tilde{\omega}_B \tag{8.81}$$

and shows how the elements of ${}^G\dot{R}_B$ are related to components of ${}^B_G\boldsymbol{\omega}_B$ and elements of ${}^GR_B$:

$$\begin{bmatrix} \dot{r}_{11} & \dot{r}_{12} & \dot{r}_{13} \\ \dot{r}_{21} & \dot{r}_{22} & \dot{r}_{23} \\ \dot{r}_{31} & \dot{r}_{32} & \dot{r}_{33} \end{bmatrix} = \begin{bmatrix} r_{11} & r_{12} & r_{13} \\ r_{21} & r_{22} & r_{23} \\ r_{31} & r_{32} & r_{33} \end{bmatrix} \begin{bmatrix} 0 & -\omega_z & \omega_y \\ \omega_z & 0 & -\omega_x \\ -\omega_y & \omega_x & 0 \end{bmatrix}$$

$$= \begin{bmatrix} \omega_z r_{12} - \omega_y r_{13} & \omega_x r_{13} - \omega_z r_{11} & \omega_y r_{11} - \omega_x r_{12} \\ \omega_z r_{22} - \omega_y r_{23} & \omega_x r_{23} - \omega_z r_{21} & \omega_y r_{21} - \omega_x r_{22} \\ \omega_z r_{32} - \omega_y r_{33} & \omega_x r_{33} - \omega_z r_{31} & \omega_y r_{31} - \omega_x r_{32} \end{bmatrix} \tag{8.82}$$

Employing these expanded forms, we may determine the angular velocity when a rotation transformation matrix ${}^GR_B$ is given. As an example, assume the following matrix is known:

$${}^GR_B = \begin{bmatrix} \cos\theta\cos\varphi & -\sin\varphi & \cos\varphi\sin\theta \\ \cos\theta\sin\varphi & \cos\varphi & \sin\theta\sin\varphi \\ -\sin\theta & 0 & \cos\theta \end{bmatrix} \tag{8.83}$$

A time derivative yields

$${}^G\dot{R}_B = \begin{bmatrix} -\dot{\theta}c\varphi s\theta - \dot{\varphi}c\theta s\varphi & -\dot{\varphi}c\varphi & \dot{\theta}c\theta c\varphi - \dot{\varphi}s\theta s\varphi \\ \dot{\varphi}c\theta c\varphi - \dot{\theta}s\theta s\varphi & -\dot{\varphi}s\varphi & \dot{\theta}c\theta s\varphi + \dot{\varphi}c\varphi s\theta \\ -\dot{\theta}c\theta & 0 & -\dot{\theta}s\theta \end{bmatrix} \tag{8.84}$$

Let us use any set of three independent elements of ${}^G\dot{R}_B$ to determine components of ${}_G\boldsymbol{\omega}_B$:

$$\dot{r}_{11} = \omega_Y r_{31} - \omega_Z r_{21} = -\dot{\theta}\cos\varphi\sin\theta - \dot{\varphi}\cos\theta\sin\varphi \tag{8.85}$$

$$\dot{r}_{21} = \omega_Z r_{11} - \omega_X r_{31} = \dot{\varphi}\cos\theta\cos\varphi - \dot{\theta}\sin\theta\sin\varphi \tag{8.86}$$

$$\dot{r}_{12} = \omega_Y r_{32} - \omega_Z r_{22} = -\dot{\varphi}\cos\varphi \tag{8.87}$$

Writing these in the matrix form

$$\begin{bmatrix} \dot{r}_{11} \\ \dot{r}_{21} \\ \dot{r}_{12} \end{bmatrix} = \begin{bmatrix} 0 & r_{31} & -r_{21} \\ -r_{31} & 0 & r_{11} \\ 0 & r_{32} & -r_{22} \end{bmatrix} \begin{bmatrix} \omega_X \\ \omega_Y \\ \omega_Z \end{bmatrix} \tag{8.88}$$

provides the solutions

$$\begin{bmatrix} \omega_X \\ \omega_Y \\ \omega_Z \end{bmatrix} = \begin{bmatrix} 0 & r_{31} & -r_{21} \\ -r_{31} & 0 & r_{11} \\ 0 & r_{32} & -r_{22} \end{bmatrix}^{-1} \begin{bmatrix} \dot{r}_{11} \\ \dot{r}_{21} \\ \dot{r}_{12} \end{bmatrix}$$

$$= \begin{bmatrix} -\dot{\theta}\sin\varphi \\ \dot{\theta}\cos\varphi \\ \dot{\varphi} \end{bmatrix} \tag{8.89}$$

Example 473 Decomposition of an Angular Velocity Vector Every angular velocity can be decomposed to three principal angular velocity vectors by employing the orthogonality condition of the coordinate frame (3.1):

$$\begin{aligned} {}_G\boldsymbol{\omega}_B &= \left({}_G\boldsymbol{\omega}_B \cdot \hat{I}\right)\hat{I} + \left({}_G\boldsymbol{\omega}_B \cdot \hat{J}\right)\hat{J} + \left({}_G\boldsymbol{\omega}_B \cdot \hat{K}\right)\hat{K} \\ &= \omega_X\,\hat{I} + \omega_Y\,\hat{J} + \omega_Z\,\hat{K} = \dot{\gamma}\hat{I} + \dot{\beta}\hat{J} + \dot{\alpha}\hat{K} \\ &= \boldsymbol{\omega}_X + \boldsymbol{\omega}_Y + \boldsymbol{\omega}_Z \end{aligned} \tag{8.90}$$

Example 474 ★ Relative Angular Velocity The relative angular velocity formulas (8.51) and (8.13), repeated here as

$${}_0\boldsymbol{\omega}_2 = {}_0\boldsymbol{\omega}_1 + {}_1^0\boldsymbol{\omega}_2 \tag{8.91}$$

$${}_0\boldsymbol{\omega}_n = {}_0\boldsymbol{\omega}_1 + {}_1^0\boldsymbol{\omega}_2 + {}_2^0\boldsymbol{\omega}_3 + \cdots + {}_{n-1}^{\;\;0}\boldsymbol{\omega}_n = \sum_{i=1}^{n} {}_{i-1}^{\;\;0}\boldsymbol{\omega}_i \tag{8.92}$$

are correct if and only if all of the angular velocities are expressed in the B_0-frame. Therefore, any equation of the form

$${}_0\boldsymbol{\omega}_2 \neq {}_0\boldsymbol{\omega}_1 + {}_1\boldsymbol{\omega}_2 \tag{8.93}$$

$$\boldsymbol{\omega}_0 \neq \boldsymbol{\omega}_1 + \boldsymbol{\omega}_2 \tag{8.94}$$

$${}_0\boldsymbol{\omega}_3 \neq {}_0\boldsymbol{\omega}_1 + {}_0\boldsymbol{\omega}_2 \tag{8.95}$$

is wrong or is not completely expressed.

Example 475 A Rotating Cone Figure 8.3 illustrates a cone of mass m that has angular speeds $\dot{\psi}$ about the z-axis and $\dot{\varphi}$ about the Y-axis. It moves such that the z-axis remains in the (X, Z)-plane. The transformation matrix between frames G and B is

$$\begin{aligned} {}^BR_G &= R_{z,\psi}R_{y,\varphi} \\ &= \begin{bmatrix} \cos\psi\cos\varphi & \sin\psi & -\cos\psi\sin\varphi \\ -\cos\varphi\sin\psi & \cos\psi & \sin\psi\sin\varphi \\ \sin\varphi & 0 & \cos\varphi \end{bmatrix} \end{aligned} \tag{8.96}$$

So, the rotation $\dot{\varphi}$ about the Y-axis can be found in the B-frame:

$${}^BR_G \begin{bmatrix} 0 \\ \dot{\varphi} \\ 0 \end{bmatrix} = \begin{bmatrix} \dot{\varphi}\sin\psi \\ \dot{\varphi}\cos\psi \\ 0 \end{bmatrix} \tag{8.97}$$

Therefore, the angular velocity of the cone is

$${}_G^B\boldsymbol{\omega}_B = \dot{\varphi}\sin\psi\,\hat{\imath} + \dot{\varphi}\cos\psi\,\hat{\jmath} + \dot{\psi}\hat{k} \tag{8.98}$$

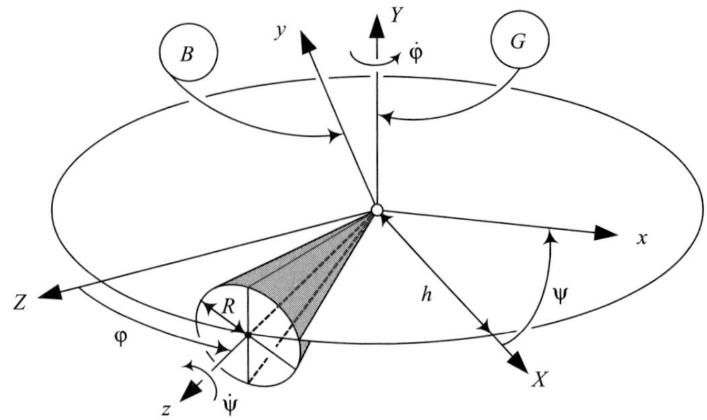

Figure 8.3 A rotating cone with angular speeds $\dot{\psi}$ about the z-axis and $\dot{\varphi}$ about the x-axis.

Example 476 Angular Velocity in Terms of Euler Frequencies The angular velocity vector can be expressed by Euler frequencies as described in Chapter 4:

$$
\begin{aligned}
{}_G^B\boldsymbol{\omega}_B &= \omega_x\hat{\imath} + \omega_y\hat{\jmath} + \omega_z\hat{k} = \dot{\varphi}\hat{e}_\varphi + \dot{\theta}\hat{e}_\theta + \dot{\psi}\hat{e}_\psi \\
&= \dot{\varphi}\begin{bmatrix}\sin\theta\sin\psi \\ \sin\theta\cos\psi \\ \cos\theta\end{bmatrix} + \dot{\theta}\begin{bmatrix}\cos\psi \\ -\sin\psi \\ 0\end{bmatrix} + \dot{\psi}\begin{bmatrix}0 \\ 0 \\ 1\end{bmatrix} \\
&= \begin{bmatrix}\sin\theta\sin\psi & \cos\psi & 0 \\ \sin\theta\cos\psi & -\sin\psi & 0 \\ \cos\theta & 0 & 1\end{bmatrix}\begin{bmatrix}\dot{\varphi} \\ \dot{\theta} \\ \dot{\psi}\end{bmatrix}
\end{aligned}
\tag{8.99}
$$

Therefore, the G-expression of the angular velocity in terms of Euler frequencies is

$$
\begin{aligned}
{}_G\boldsymbol{\omega}_B &= {}^BR_G^{-1}\,{}_G^B\boldsymbol{\omega}_B = {}^BR_G^{-1}\begin{bmatrix}\dot{\varphi}\sin\theta\sin\psi + \dot{\theta}\cos\psi \\ \dot{\varphi}\sin\theta\cos\psi - \dot{\theta}\sin\psi \\ \dot{\varphi}\cos\theta + \dot{\psi}\end{bmatrix} \\
&= \begin{bmatrix}0 & \cos\varphi & \sin\theta\sin\varphi \\ 0 & \sin\varphi & -\cos\varphi\sin\theta \\ 1 & 0 & \cos\theta\end{bmatrix}\begin{bmatrix}\dot{\varphi} \\ \dot{\theta} \\ \dot{\psi}\end{bmatrix}
\end{aligned}
\tag{8.100}
$$

where ${}^BR_G^{-1}$ is the inverse of the Euler transformation matrix:

$$
{}^BR_G^{-1} = \begin{bmatrix}c\varphi c\psi - c\theta s\varphi s\psi & -c\varphi s\psi - c\theta c\psi s\varphi & s\theta s\varphi \\ c\psi s\varphi + c\theta c\varphi s\psi & -s\varphi s\psi + c\theta c\varphi c\psi & -c\varphi s\theta \\ s\theta s\psi & s\theta c\psi & c\theta\end{bmatrix}
\tag{8.101}
$$

Example 477 ★ Angular Velocity and Decomposed Euler Matrix Because ${}^{G}R_{B}$ is a combination of principal rotation matrices in terms of Euler angles, we may use the decomposed form to determine ${}^{G}\dot{R}_{B}$ and ${}_{G}\tilde{\omega}_{B}$:

$$\begin{aligned}
{}^{G}\dot{R}_{B} &= \frac{d}{dt}\left(R_{z,\varphi}^{\mathrm{T}} R_{x,\theta}^{\mathrm{T}} R_{z,\psi}^{\mathrm{T}}\right) = \frac{d}{dt}\left(R_{Z,\varphi} R_{X,\theta} R_{Z,\psi}\right) \\
&= \dot{R}_{Z,\varphi} R_{X,\theta} R_{Z,\psi} + R_{Z,\varphi} \dot{R}_{X,\theta} R_{Z,\psi} + R_{Z,\varphi} R_{X,\theta} \dot{R}_{Z,\psi} \\
&= \dot{\varphi} \hat{K} R_{Z,\varphi} R_{X,\theta} R_{Z,\psi} + \dot{\theta} R_{Z,\varphi} \hat{I} R_{X,\theta} R_{Z,\psi} + \dot{\psi} R_{Z,\varphi} R_{X,\theta} \hat{K} R_{Z,\psi} \qquad (8.102)
\end{aligned}$$

$$\begin{aligned}
{}_{G}\tilde{\omega}_{B} &= {}^{G}\dot{R}_{B}{}^{G}R_{B}^{\mathrm{T}} = {}^{G}\dot{R}_{B}\left(R_{Z,\psi}^{\mathrm{T}} R_{X,\theta}^{\mathrm{T}} R_{Z,\varphi}^{\mathrm{T}}\right) \\
&= \dot{\varphi} \hat{K} + \dot{\theta} R_{Z,\varphi} \hat{I} R_{Z,\varphi}^{\mathrm{T}} + \dot{\psi} R_{Z,\varphi} R_{X,\theta} \hat{K} R_{X,\theta}^{\mathrm{T}} R_{Z,\varphi}^{\mathrm{T}} \qquad (8.103)
\end{aligned}$$

Expanding ${}_{G}\tilde{\omega}_{B}$ provides the following equation, consistent with Equation (8.99):

$$\begin{aligned}
{}_{G}\tilde{\omega}_{B} &= \dot{\varphi} \begin{bmatrix} 0 & -1 & 0 \\ 1 & 0 & 0 \\ 0 & 0 & 0 \end{bmatrix} + \dot{\theta} \begin{bmatrix} 0 & 0 & \sin\varphi \\ 0 & 0 & -\cos\varphi \\ -\sin\varphi & \cos\varphi & 0 \end{bmatrix} \\
&\quad + \dot{\psi} \begin{bmatrix} 0 & -\cos\theta & -\cos\varphi\sin\theta \\ \cos\theta & 0 & -\sin\theta\sin\varphi \\ \cos\varphi\sin\theta & \sin\theta\sin\varphi & 0 \end{bmatrix} \qquad (8.104)
\end{aligned}$$

Example 478 Angular Velocity in Terms of Rotation Frequencies Appendices *A* and *B* show the 12 global and 12 local axis triple rotations. Using those equations, we are able to find the angular velocity of a rigid body in terms of rotation frequencies. As an example, consider the Euler angle transformation matrix in case 9 of Appendix *B*:

$$ {}^{B}R_{G} = R_{z,\psi} R_{x,\theta} R_{z,\varphi} \qquad (8.105)$$

The angular velocity matrix is then equal to

$$\begin{aligned}
{}_{B}\tilde{\omega}_{G} &= {}^{B}\dot{R}_{G}{}^{B}R_{G}^{\mathrm{T}} \\
&= \left(\dot{\varphi}\, R_{z,\psi} R_{x,\theta} \frac{dR_{z,\varphi}}{d\varphi} + \dot{\theta}\, R_{z,\psi} \frac{dR_{x,\theta}}{d\theta} R_{z,\varphi} + \dot{\psi}\, \frac{dR_{z,\psi}}{d\psi} R_{x,\theta} R_{z,\varphi}\right) \\
&\quad \times \left(R_{z,\psi} R_{x,\theta} R_{z,\varphi}\right)^{\mathrm{T}} \\
&= \dot{\varphi}\, R_{z,\psi} R_{x,\theta} \frac{dR_{z,\varphi}}{d\varphi} R_{z,\varphi}^{\mathrm{T}} R_{x,\theta}^{\mathrm{T}} R_{z,\psi}^{\mathrm{T}} + \dot{\theta}\, R_{z,\psi} \frac{dR_{x,\theta}}{d\theta} R_{x,\theta}^{\mathrm{T}} R_{z,\psi}^{\mathrm{T}} \\
&\quad + \dot{\psi}\, \frac{dR_{z,\psi}}{d\psi} R_{z,\psi}^{\mathrm{T}} \qquad (8.106)
\end{aligned}$$

which, in matrix form, is

$$
{}_B\tilde{\omega}_G = \dot{\varphi}\begin{bmatrix} 0 & \cos\theta & -\sin\theta\cos\psi \\ -\cos\theta & 0 & \sin\theta\sin\psi \\ \sin\theta\cos\psi & -\sin\theta\sin\psi & 0 \end{bmatrix}
$$

$$
+\dot{\theta}\begin{bmatrix} 0 & 0 & \sin\psi \\ 0 & 0 & \cos\psi \\ -\sin\psi & -\cos\psi & 0 \end{bmatrix} + \dot{\psi}\begin{bmatrix} 0 & 1 & 0 \\ -1 & 0 & 0 \\ 0 & 0 & 0 \end{bmatrix} \tag{8.107}
$$

or

$$
{}_B\tilde{\omega}_G = \begin{bmatrix} 0 & \dot{\psi}+\dot{\varphi}c\theta & \dot{\theta}s\psi-\dot{\varphi}s\theta c\psi \\ -\dot{\psi}-\dot{\varphi}c\theta & 0 & \dot{\theta}c\psi+\dot{\varphi}s\theta s\psi \\ -\dot{\theta}s\psi+\dot{\varphi}s\theta c\psi & -\dot{\theta}c\psi-\dot{\varphi}s\theta s\psi & 0 \end{bmatrix} \tag{8.108}
$$

The associated angular velocity vector is

$$
{}_B\boldsymbol{\omega}_G = -\begin{bmatrix} \dot{\theta}c\psi+\dot{\varphi}s\theta s\psi \\ -\dot{\theta}s\psi+\dot{\varphi}s\theta c\psi \\ \dot{\psi}+\dot{\varphi}c\theta \end{bmatrix}
$$

$$
= -\begin{bmatrix} \sin\theta\sin\psi & \cos\psi & 0 \\ \sin\theta\cos\psi & -\sin\psi & 0 \\ \cos\theta & 0 & 1 \end{bmatrix}\begin{bmatrix} \dot{\varphi} \\ \dot{\theta} \\ \dot{\psi} \end{bmatrix} \tag{8.109}
$$

However,

$$
{}_B^B\tilde{\omega}_G = -{}_G^B\tilde{\omega}_B \tag{8.110}
$$

$$
{}_B^B\boldsymbol{\omega}_G = -{}_G^B\boldsymbol{\omega}_B \tag{8.111}
$$

and therefore,

$$
{}_G^B\boldsymbol{\omega}_B = \begin{bmatrix} \sin\theta\sin\psi & \cos\psi & 0 \\ \sin\theta\cos\psi & -\sin\psi & 0 \\ \cos\theta & 0 & 1 \end{bmatrix}\begin{bmatrix} \dot{\varphi} \\ \dot{\theta} \\ \dot{\psi} \end{bmatrix} \tag{8.112}
$$

Example 479 ★ ${}_G\tilde{\omega}_B = {}^GR_B\,{}_G^B\tilde{\omega}_B\,{}^GR_B^T$ ***Is Equivalent to*** ${}_G\boldsymbol{\omega}_B = {}^GR_B\,{}_G^B\boldsymbol{\omega}_B$ The vectorial expression of an angular velocity can be expressed in any coordinate frame by employing a coordinate transformation matrix. Therefore, we must have

$$
{}_G\boldsymbol{\omega}_B = {}^GR_B\,{}_G^B\boldsymbol{\omega}_B \tag{8.113}
$$

which is equivalent to

$$ {}_G\tilde{\omega}_B = {}^G R_B \, {}^B_G\tilde{\omega}_B \, {}^G R_B^{\mathrm{T}} \tag{8.114} $$

To show this fact, let us multiply ${}_G\tilde{\omega}_B$ by an arbitrary vector ${}^G\mathbf{r}$ and show it is equal to ${}_G\boldsymbol{\omega}_B \times {}^G\mathbf{r}$:

$$ \begin{aligned} {}^G R_B \, {}^B_G\tilde{\omega}_B \, {}^G R_B^{\mathrm{T}} \, {}^G\mathbf{r} &= {}^G R_B \, {}^B_G\tilde{\omega}_B \, {}^B R_G \, {}^G\mathbf{r} = {}^G R_B \, {}^B_G\tilde{\omega}_B \, {}^B\mathbf{r} \\ &= {}^G R_B \left({}^B_G\boldsymbol{\omega}_B \times {}^B\mathbf{r} \right) = {}_G\boldsymbol{\omega}_B \times {}^G\mathbf{r} \end{aligned} \tag{8.115} $$

Example 480 ★ Coordinate Transformation of Angular Velocity The angular velocity ${}_1\boldsymbol{\omega}_2$ of coordinate frame B_2 with respect to B_1 and expressed in B_1 can be expressed in the base coordinate frame B_0 according to

$$ {}^0 R_1 \, {}_1\tilde{\omega}_2 \, {}^0 R_1^{\mathrm{T}} = {}^0_1\tilde{\omega}_2 \tag{8.116} $$

To show this equation, it is enough to apply both sides on an arbitrary vector ${}^0\mathbf{r}$:

$$ {}^0 R_1 \, {}_1\tilde{\omega}_2 \, {}^0 R_1^{\mathrm{T}} \, {}^0\mathbf{r} = {}^0_1\tilde{\omega}_2 \, {}^0\mathbf{r} \tag{8.117} $$

The left-hand side would be

$$ \begin{aligned} {}^0 R_1 \, {}_1\tilde{\omega}_2 \, {}^0 R_1^{\mathrm{T}} \, {}^0\mathbf{r} &= {}^0 R_1 \, {}_1\tilde{\omega}_2 \, {}^1 R_0 \, {}^0\mathbf{r} = {}^0 R_1 \, {}_1\tilde{\omega}_2 \, {}^1\mathbf{r} \\ &= {}^0 R_1 \left({}_1\boldsymbol{\omega}_2 \times {}^1\mathbf{r} \right) = {}^0_1\boldsymbol{\omega}_2 \times {}^0\mathbf{r} \end{aligned} \tag{8.118} $$

which is equal to the right-hand side:

$$ {}^0_1\tilde{\omega}_2 \, {}^0\mathbf{r} = {}^0_1\boldsymbol{\omega}_2 \times {}^0\mathbf{r} \tag{8.119} $$

Example 481 ★ Time Derivative of Unit Vectors Using Equation (8.7) we can define the time derivative of the unit vectors of a body coordinate frame $B(\hat{\imath}, \hat{\jmath}, \hat{k})$ rotating in the global coordinate frame $G(\hat{I}, \hat{J}, \hat{K})$:

$$ \frac{{}^G d\hat{\imath}}{dt} = {}^B_G\boldsymbol{\omega}_B \times \hat{\imath} \tag{8.120} $$

$$ \frac{{}^G d\hat{\jmath}}{dt} = {}^B_G\boldsymbol{\omega}_B \times \hat{\jmath} \tag{8.121} $$

$$ \frac{{}^G d\hat{k}}{dt} = {}^B_G\boldsymbol{\omega}_B \times \hat{k} \tag{8.122} $$

***Example 482* ★ *Angular Velocity in Natural Coordinate Frame* $\hat{u}_t$, $\hat{u}_n$, $\hat{u}_b$** The time derivatives of the unit vectors $\hat{u}_t$, $\hat{u}_n$, $\hat{u}_b$ of the natural coordinate frame $N\left(\hat{u}_t, \hat{u}_n, \hat{u}_b\right)$ are

$$\frac{d\hat{u}_t}{dt} = \frac{\dot{s}}{\rho}\hat{u}_n \tag{8.123}$$

$$\frac{d\hat{u}_n}{dt} = -\frac{\dot{s}}{\rho}\hat{u}_t + \frac{\dot{s}}{\sigma}\hat{u}_b \tag{8.124}$$

$$\frac{d\hat{u}_b}{dt} = -\frac{\dot{s}}{\sigma}\hat{u}_n \tag{8.125}$$

Using Equation (8.7) we can determine the angular velocity vector in the natural coordinate frame:

$$\frac{d\hat{u}_t}{dt} = {}_G^N\boldsymbol{\omega}_N \times \hat{u}_t \tag{8.126}$$

$$\frac{d\hat{u}_n}{dt} = {}_G^N\boldsymbol{\omega}_N \times \hat{u}_n \tag{8.127}$$

$$\frac{d\hat{u}_b}{dt} = {}_G^N\boldsymbol{\omega}_N \times \hat{u}_b \tag{8.128}$$

$${}_G^N\boldsymbol{\omega}_N = \omega_t\hat{u}_t + \omega_n\hat{u}_n + \omega_b\hat{u}_b = \frac{\dot{s}}{\sigma}\hat{u}_t + \frac{\dot{s}}{\rho}\hat{u}_b \tag{8.129}$$

Example 483* ★ *Derivative of a Quaternion Consider a quaternion q as a function of a scalar parameter s:

$$\begin{aligned} p &= p_0(s) + \mathbf{p}(s) \\ &= p_0(s) + p_1(s)\,i + p_2(s)\,j + p_3(s)\,k \end{aligned} \tag{8.130}$$

The derivative of q with respect to s is

$$\frac{dq}{ds} = \frac{dp_0}{ds} + \frac{dp_1}{ds}i + \frac{dp_2}{ds}j + \frac{dp_3}{ds}k \tag{8.131}$$

If the parameter s is time t, we have

$$\frac{dq}{dt} = \frac{dp_0}{dt} + \frac{dp_1}{dt}i + \frac{dp_2}{dt}j + \frac{dp_3}{dt}k \tag{8.132}$$

Let us show the position vector of a moving particle by a quaternion as

$$\overleftrightarrow{r} = t + \mathbf{r}(t) = t + r_1(t)\,i + r_2(t)\,j + r_3(t)\,k \tag{8.133}$$

The velocity and acceleration vectors of the particle are

$$\overleftrightarrow{v} = \left(\frac{d}{dt} + \mathbf{0}\right)(t + \mathbf{r}) = 1 + \mathbf{v}(t) \tag{8.134}$$

$$\overleftrightarrow{a} = \left(\frac{d}{dt} + \mathbf{0}\right)(1 + \mathbf{v}) = 0 + \mathbf{a}(t) \tag{8.135}$$

Example 484 ★ Angular Velocity, Quaternions, and Euler Parameters We may also express the angular velocity vector by quaternions and Euler parameters. Let us use the orthogonality property of a unit quaternion

$$ee^{-1} = ee^* = 1 \tag{8.136}$$

to determine

$$\dot{e}e^* + e\dot{e}^* = 0 \tag{8.137}$$

Starting from the unit quaternion representation of a finite rotation

$${}^G\mathbf{r} = e(t)\,{}^B\mathbf{r}\,e^*(t) = e(t)\,{}^B\mathbf{r}\,e^{-1}(t) \tag{8.138}$$

where

$$e = e_0 + \mathbf{e} \qquad e^* = e^{-1} = e_0 - \mathbf{e} \tag{8.139}$$

we can find

$$\begin{aligned} {}^G\dot{\mathbf{r}} &= \dot{e}\,{}^B\mathbf{r}\,e^* + e\,{}^B\mathbf{r}\,\dot{e}^* = \dot{e}\,e^{*G}\mathbf{r} + {}^G\mathbf{r}\,e\,\dot{e}^* \\ &= 2\dot{e}\,e^{*G}\mathbf{r} \end{aligned} \tag{8.140}$$

and therefore the angular velocity in terms of the rotation quaternion is

$${}_G\boldsymbol{\omega}_B = 2\dot{e}\,e^* \tag{8.141}$$

We may expand this equation using quaternion products to find the angular velocity components based on Euler parameters and show that ${}_G\boldsymbol{\omega}_B$ is a vector:

$$\begin{aligned} {}_G\boldsymbol{\omega}_B &= 2\dot{e}\,e^* = 2(\dot{e}_0 + \dot{\mathbf{e}})(e_0 - \mathbf{e}) \\ &= 2(\dot{e}_0 e_0 + e_0\dot{\mathbf{e}} - \dot{e}_0\mathbf{e} + \dot{\mathbf{e}}\cdot\mathbf{e} - \dot{\mathbf{e}}\times\mathbf{e}) \\ &= 2\begin{bmatrix} 0 \\ e_0\dot{e}_1 - e_1\dot{e}_0 + e_2\dot{e}_3 - e_3\dot{e}_2 \\ e_0\dot{e}_2 - e_2\dot{e}_0 - e_1\dot{e}_3 + e_3\dot{e}_1 \\ e_0\dot{e}_3 + e_1\dot{e}_2 - e_2\dot{e}_1 - e_3\dot{e}_0 \end{bmatrix} \end{aligned} \tag{8.142}$$

It can be rearranged in matrix form as

$$\begin{bmatrix} 0 \\ \omega_X \\ \omega_Y \\ \omega_Z \end{bmatrix} = 2\begin{bmatrix} \dot{e}_0 & -\dot{e}_1 & -\dot{e}_2 & -\dot{e}_3 \\ \dot{e}_1 & \dot{e}_0 & -\dot{e}_3 & \dot{e}_2 \\ \dot{e}_2 & \dot{e}_3 & \dot{e}_0 & -\dot{e}_1 \\ \dot{e}_3 & -\dot{e}_2 & \dot{e}_1 & \dot{e}_0 \end{bmatrix}\begin{bmatrix} e_0 \\ -e_1 \\ -e_2 \\ -e_3 \end{bmatrix} \tag{8.143}$$

The scalar component of the angular velocity quaternion is zero because

$$\dot{e}_0 e_0 + \dot{\mathbf{e}} \cdot \mathbf{e} = \dot{e}_0 e_0 + e_1 \dot{e}_1 + e_2 \dot{e}_2 + e_3 \dot{e}_3 = 0 \tag{8.144}$$

We can also define an angular velocity quaternion

$$\overleftrightarrow{{}_G\boldsymbol{\omega}_B} = 2 \overleftrightarrow{\dot{e}} \; \overleftrightarrow{e^*} \tag{8.145}$$

to be used for definition of the derivative of a rotation quaternion:

$$\overleftrightarrow{\dot{e}} = \frac{1}{2} \overleftrightarrow{{}_G\boldsymbol{\omega}_B} \; \overleftrightarrow{e} \tag{8.146}$$

$$\begin{bmatrix} \dot{e}_0 & -\dot{e}_1 & -\dot{e}_2 & -\dot{e}_3 \\ \dot{e}_1 & \dot{e}_0 & -\dot{e}_3 & \dot{e}_2 \\ \dot{e}_2 & \dot{e}_3 & \dot{e}_0 & -\dot{e}_1 \\ \dot{e}_3 & -\dot{e}_2 & \dot{e}_1 & \dot{e}_0 \end{bmatrix} = \frac{1}{2} \begin{bmatrix} 0 & -\omega_X & -\omega_Y & -\omega_Z \\ \omega_X & 0 & -\omega_Z & \omega_Y \\ \omega_Y & \omega_Z & 0 & -\omega_X \\ \omega_Z & -\omega_Y & \omega_X & 0 \end{bmatrix} \begin{bmatrix} e_0 & -e_1 & -e_2 & -e_3 \\ e_1 & e_0 & -e_3 & e_2 \\ e_2 & e_3 & e_0 & -e_1 \\ e_3 & -e_2 & e_1 & e_0 \end{bmatrix} \tag{8.147}$$

where

$$\dot{e}_0 = -\omega_X e_1 - \omega_Y e_2 - \omega_Z e_3 \tag{8.148}$$

$$\dot{e}_1 = \omega_X e_0 + \omega_Y e_3 - \omega_Z e_2 \tag{8.149}$$

$$\dot{e}_2 = \omega_Y e_0 - \omega_X e_3 + \omega_Z e_1 \tag{8.150}$$

$$\dot{e}_3 = \omega_X e_2 - \omega_Y e_1 + \omega_Z e_0 \tag{8.151}$$

A coordinate transformation can transform the angular velocity into a body coordinate frame:

$${}^B_G\boldsymbol{\omega}_B = e^* \, {}^G_G\boldsymbol{\omega}_B \, e = 2e^* \, \dot{e} \tag{8.152}$$

$$= 2 \begin{bmatrix} e_0 & e_1 & e_2 & e_3 \\ -e_1 & e_0 & e_3 & -e_2 \\ -e_2 & -e_3 & e_0 & e_1 \\ -e_3 & -e_2 & e_1 & e_0 \end{bmatrix} \begin{bmatrix} \dot{e}_0 \\ \dot{e}_1 \\ \dot{e}_2 \\ \dot{e}_3 \end{bmatrix} \tag{8.153}$$

and therefore,

$$\overleftrightarrow{{}^B_G\boldsymbol{\omega}_B} = 2 \overleftrightarrow{e^*} \; \overleftrightarrow{\dot{e}} \tag{8.154}$$

$$\overleftrightarrow{\dot{e}} = \frac{1}{2} \overleftrightarrow{e} \; \overleftrightarrow{{}^B_G\boldsymbol{\omega}_B} \tag{8.155}$$

$$\begin{bmatrix} \dot{e}_0 & -\dot{e}_1 & -\dot{e}_2 & -\dot{e}_3 \\ \dot{e}_1 & \dot{e}_0 & -\dot{e}_3 & \dot{e}_2 \\ \dot{e}_2 & \dot{e}_3 & \dot{e}_0 & -\dot{e}_1 \\ \dot{e}_3 & -\dot{e}_2 & \dot{e}_1 & \dot{e}_0 \end{bmatrix} = \frac{1}{2} \begin{bmatrix} e_0 & -e_1 & -e_2 & -e_3 \\ e_1 & e_0 & -e_3 & e_2 \\ e_2 & e_3 & e_0 & -e_1 \\ e_3 & -e_2 & e_1 & e_0 \end{bmatrix} \begin{bmatrix} 0 & -\omega_x & -\omega_y & -\omega_z \\ \omega_x & 0 & -\omega_z & \omega_y \\ \omega_y & \omega_z & 0 & -\omega_x \\ \omega_z & -\omega_y & \omega_x & 0 \end{bmatrix} \tag{8.156}$$

where

$$\dot{e}_0 = -\omega_x e_1 - \omega_y e_2 - \omega_z e_3 \tag{8.157}$$

$$\dot{e}_1 = \omega_x e_0 - \omega_y e_3 + \omega_z e_2 \tag{8.158}$$

$$\dot{e}_2 = \omega_y e_0 + \omega_x e_3 - \omega_z e_1 \tag{8.159}$$

$$\dot{e}_3 = \omega_y e_1 - \omega_x e_2 + \omega_z e_0 \tag{8.160}$$

Example 485 ★ Differential of Euler Parameters The rotation matrix ${}^G R_B$ based on Euler parameters is given in Equation (5.109) as

$$\begin{aligned} {}^G R_B &= \begin{bmatrix} e_0^2 + e_1^2 - e_2^2 - e_3^2 & 2(e_1 e_2 - e_0 e_3) & 2(e_0 e_2 + e_1 e_3) \\ 2(e_0 e_3 + e_1 e_2) & e_0^2 - e_1^2 + e_2^2 - e_3^2 & 2(e_2 e_3 - e_0 e_1) \\ 2(e_1 e_3 - e_0 e_2) & 2(e_0 e_1 + e_2 e_3) & e_0^2 - e_1^2 - e_2^2 + e_3^2 \end{bmatrix} \\ &= \begin{bmatrix} r_{11} & r_{12} & r_{13} \\ r_{21} & r_{22} & r_{23} \\ r_{31} & r_{32} & r_{33} \end{bmatrix} \end{aligned} \tag{8.161}$$

The individual parameters can be found from any set of Equations (5.114)–(5.117). The first set indicates that

$$e_0 = \pm\frac{1}{2}\sqrt{1 + r_{11} + r_{22} + r_{33}} \tag{8.162}$$

$$e_1 = \frac{1}{4}\frac{r_{32} - r_{23}}{e_0} \tag{8.163}$$

$$e_2 = \frac{1}{4}\frac{r_{13} - r_{31}}{e_0} \tag{8.164}$$

$$e_3 = \frac{1}{4}\frac{r_{21} - r_{12}}{e_0} \tag{8.165}$$

and therefore,

$$\dot{e}_0 = \frac{\dot{r}_{11} + \dot{r}_{22} + \dot{r}_{33}}{8e_0} \tag{8.166}$$

$$\dot{e}_1 = \frac{1}{4e_0^2}\left[(\dot{r}_{32} - \dot{r}_{23})e_0 - (r_{32} - r_{23})\dot{e}_0\right] \tag{8.167}$$

$$\dot{e}_2 = \frac{1}{4e_0^2}\left[(\dot{r}_{13} - \dot{r}_{31})e_0 - (r_{13} - r_{31})\dot{e}_0\right] \tag{8.168}$$

$$\dot{e}_3 = \frac{1}{4e_0^2}\left[(\dot{r}_{21} - \dot{r}_{12})e_0 - (r_{21} - r_{12})\dot{e}_0\right] \tag{8.169}$$

We may use the differential of the transformation matrix

$${}^G\dot{R}_B = {}_G\tilde{\omega}_B \, {}^G R_B$$

to show that

$$\dot{e}_0 = \frac{1}{2}\left(-e_1\omega_1 - e_2\omega_2 - e_3\omega_3\right) \tag{8.170}$$

$$\dot{e}_1 = \frac{1}{2}\left(e_0\omega_1 + e_2\omega_3 - e_3\omega_2\right) \tag{8.171}$$

$$\dot{e}_2 = \frac{1}{2}\left(e_0\omega_2 - e_1\omega_3 - e_3\omega_1\right) \tag{8.172}$$

$$\dot{e}_3 = \frac{1}{2}\left(e_0\omega_3 + e_1\omega_2 - e_2\omega_1\right) \tag{8.173}$$

Similarly we can find $\dot{e}_1$, $\dot{e}_2$, and $\dot{e}_3$ and set up the final result in matrix form as

$$\begin{bmatrix} \dot{e}_0 \\ \dot{e}_1 \\ \dot{e}_2 \\ \dot{e}_3 \end{bmatrix} = \frac{1}{2}\begin{bmatrix} 0 & -\omega_1 & -\omega_2 & -\omega_3 \\ \omega_1 & 0 & \omega_3 & -\omega_2 \\ \omega_2 & -\omega_3 & 0 & \omega_1 \\ \omega_3 & \omega_2 & -\omega_1 & 0 \end{bmatrix}\begin{bmatrix} e_0 \\ e_1 \\ e_2 \\ e_3 \end{bmatrix} \tag{8.174}$$

or

$$\begin{bmatrix} \dot{e}_0 \\ \dot{e}_1 \\ \dot{e}_2 \\ \dot{e}_3 \end{bmatrix} = \frac{1}{2}\begin{bmatrix} e_0 & -e_3 & -e_2 & -e_1 \\ e_1 & e_0 & -e_3 & e_2 \\ e_2 & e_1 & e_0 & -e_3 \\ e_3 & -e_2 & e_1 & e_0 \end{bmatrix}\begin{bmatrix} 0 \\ \omega_1 \\ \omega_2 \\ \omega_3 \end{bmatrix} \tag{8.175}$$

Example 486 ★ Elements of the Angular Velocity Matrix Employing the permutation symbol introduced in (1.126), we can find the elements of the angular velocity matrix $\tilde{\omega}$ when the angular velocity vector $\boldsymbol{\omega} = [\omega_1, \omega_2, \omega_3]^{\mathrm{T}}$ is given:

$$\tilde{\omega}_{ij} = \epsilon_{ijk}\,\omega_k \tag{8.176}$$

Example 487 ★ Kinematic Differential Equation Consider a variable rotation matrix ${}^{G}R_B(t)$ between frames B and G and a time derivative from the rotation transformation equation:

$${}^{G}\mathbf{r}(t) = {}^{G}R_B(t)\,{}^{G}\mathbf{r}(0) \tag{8.177}$$

$$\frac{{}^{G}d}{dt}\,{}^{G}\mathbf{r}(t) = {}^{G}\dot{R}_B(t)\,{}^{G}\mathbf{r}(0) \tag{8.178}$$

If we define ${}^{G}\dot{R}_B(t)$ as

$${}^{G}\dot{R}_B(t) = {}^{G}\dot{R}_B(0)\,{}^{G}R_B(t) \tag{8.179}$$

then we find the following equation, called a first-order vectorial kinematic differential equation:

$$\frac{{}^{G}d}{dt}\,{}^{G}\mathbf{r}(t) = {}^{G}\dot{R}_B(0)\,{}^{G}R_B(t)\,{}^{G}\mathbf{r}(0) = {}^{G}\dot{R}_B(0)\,{}^{G}\mathbf{r}(t) \tag{8.180}$$

We should be able to determine the solution of this equation in series form:

$$^G\mathbf{r}(t) = {^G\mathbf{r}(0)} + \frac{^Gd\,^G\mathbf{r}(0)}{dt}t + \frac{1}{2!}\frac{^Gd^2\,^G\mathbf{r}(0)}{dt^2}t^2 + \cdots \tag{8.181}$$

Substituting the higher derivatives as

$$\frac{^Gd^2}{dt^2}\,^G\mathbf{r}(t) = {^G\dot{R}_B^2}(0)\,^G\mathbf{r}(t) \tag{8.182}$$

$$\frac{^Gd^3}{dt^3}\,^G\mathbf{r}(t) = {^G\dot{R}_B^3}(0)\,^G\mathbf{r}(t) \tag{8.183}$$

the series solution simplifies to an exponential solution:

$$\begin{aligned} ^G\mathbf{r}(t) &= \left(\mathbf{I} + {^G\dot{R}_B}(0)\,t + \frac{1}{2!}\,{^G\dot{R}_B^2}(0)\,t^2 + \cdots\right)\,^G\mathbf{r}(0) \\ &= e^{^G\dot{R}_B(0)\,t}\,{^G\mathbf{r}(0)} \end{aligned} \tag{8.184}$$

Example 488 ★ Differential Equation for Rotation Matrix Equation (8.33) for defining the angular velocity matrix may be written as a first-order differential equation

$$\frac{d}{dt}\,^GR_B - {_G\tilde{\omega}_B}\,^GR_B = 0 \tag{8.185}$$

The solution of the equation confirms the exponential definition of the rotation matrix as

$$^GR_B\,(t) = {^GR_B}\,(0)\,e^{\tilde{\omega}t} \tag{8.186}$$

or

$$\tilde{\omega}t = \dot{\phi}\,\tilde{u} = \ln {^GR_B}\,(t) - \ln {^GR_B}\,(0) \tag{8.187}$$

Example 489 ★ Alternative Definition of Angular Velocity Vector The angular velocity vector of a rigid body $B(\hat{\imath}, \hat{\jmath}, \hat{k})$ in global frame $G(\hat{I}, \hat{J}, \hat{K})$ can also be defined as

$$^B_G\boldsymbol{\omega}_B = \hat{\imath}\left(\frac{^Gd\hat{\jmath}}{dt}\cdot\hat{k}\right) + \hat{\jmath}\left(\frac{^Gd\hat{k}}{dt}\cdot\hat{\imath}\right) + \hat{k}\left(\frac{^Gd\hat{\imath}}{dt}\cdot\hat{\jmath}\right) \tag{8.188}$$

To prove (8.188), consider a body coordinate frame B moving with a fixed point in the global frame G, as shown in Figure 8.4. In order to describe the motion of the body, it is sufficient to express the motion of the local unit vectors $\hat{\imath}$, $\hat{\jmath}$, $\hat{k}$. Let $\mathbf{r}_P$ be the position vector of a fixed body point P. Then, $^B\mathbf{r}_P$ is a B-vector with constant components:

$$^B\mathbf{r}_P = x\hat{\imath} + y\hat{\jmath} + z\hat{k} \tag{8.189}$$

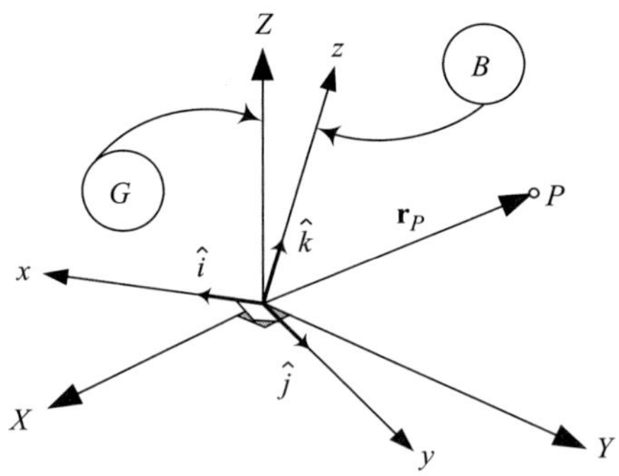

Figure 8.4 A body coordinate frame moving with a fixed point in the global coordinate frame.

When the body moves, the unit vectors $\hat{\imath}$, $\hat{\jmath}$, $\hat{k}$ move relative to the global coordinate frame. Therefore, the vector of differential displacement $d\mathbf{r}$ is

$$d\mathbf{r}_P = x\,d\hat{\imath} + y\,d\hat{\jmath} + z\,d\hat{k} \tag{8.190}$$

The differential $d\mathbf{r}_P$ is the B-expression of the infinitesimal displacement as seen from the G-frame. Using the orthogonality condition (3.1), we can also express $d\mathbf{r}_P$ as

$$d\mathbf{r}_P = \left(d\mathbf{r}_P \cdot \hat{\imath}\right)\hat{\imath} + \left(d\mathbf{r}_P \cdot \hat{\jmath}\right)\hat{\jmath} + \left(d\mathbf{r}_P \cdot \hat{k}\right)\hat{k} \tag{8.191}$$

Substituting (8.190) in the right-hand side of (8.191) results in

$$\begin{aligned} d\mathbf{r}_P = &\left(x\hat{\imath} \cdot d\hat{\imath} + y\hat{\imath} \cdot d\hat{\jmath} + z\hat{\imath} \cdot d\hat{k}\right)\hat{\imath} + \left(x\hat{\jmath} \cdot d\hat{\imath} + y\hat{\jmath} \cdot d\hat{\jmath} + z\hat{\jmath} \cdot d\hat{k}\right)\hat{\jmath} \\ &+ \left(x\hat{k} \cdot d\hat{\imath} + y\hat{k} \cdot d\hat{\jmath} + z\hat{k} \cdot d\hat{k}\right)\hat{k} \end{aligned} \tag{8.192}$$

Employing the unit vector relationships

$$\hat{\jmath} \cdot d\hat{\imath} = -\hat{\imath} \cdot d\hat{\jmath} \qquad \hat{k} \cdot d\hat{\jmath} = -\hat{\jmath} \cdot d\hat{k} \qquad \hat{\imath} \cdot d\hat{k} = -\hat{k} \cdot d\hat{\imath} \tag{8.193}$$

$$\hat{\imath} \cdot d\hat{\imath} = \hat{\jmath} \cdot d\hat{\jmath} = \hat{k} \cdot d\hat{k} = 0 \tag{8.194}$$

$$\hat{\imath} \cdot \hat{\jmath} = \hat{\jmath} \cdot \hat{k} = \hat{k} \cdot \hat{\imath} = 0 \tag{8.195}$$

$$\hat{\imath} \cdot \hat{\imath} = \hat{\jmath} \cdot \hat{\jmath} = \hat{k} \cdot \hat{k} = 1 \tag{8.196}$$

$d\mathbf{r}_P$ reduces to

$$\begin{aligned} d\mathbf{r}_P = &\left(z\hat{\imath} \cdot d\hat{k} - y\hat{\jmath} \cdot d\hat{\imath}\right)\hat{\imath} + \left(x\hat{\jmath} \cdot d\hat{\imath} - z\hat{k} \cdot d\hat{\jmath}\right)\hat{\jmath} \\ &+ \left(y\hat{k} \cdot d\hat{\jmath} - x\hat{\imath} \cdot d\hat{k}\right)\hat{k} \end{aligned} \tag{8.197}$$

This equation can be rearranged to be expressed as a vector product

$$d\mathbf{r}_P = \left[(\hat{k} \cdot d\hat{\jmath})\hat{\imath} + (\hat{\imath} \cdot d\hat{k})\hat{\jmath} + (\hat{\jmath} \cdot d\hat{\imath})\hat{k}\right] \times \left(x\hat{\imath} + y\hat{\jmath} + z\hat{k}\right) \tag{8.198}$$

or

$$
{}_{G}^{B}\dot{\mathbf{r}}_P = \left[\left(\hat{k} \cdot \frac{{}^G d\hat{\jmath}}{dt} \right) \hat{\imath} + \left(\hat{\imath} \cdot \frac{{}^G d\hat{k}}{dt} \right) \hat{\jmath} + \left(\hat{\jmath} \cdot \frac{{}^G d\hat{\imath}}{dt} \right) \hat{k} \right] \times \left(x\hat{\imath} + y\hat{\jmath} + z\hat{k} \right) \tag{8.199}
$$

Comparing this result with

$$
{}_{G}^{B}\dot{\mathbf{r}}_P = {}_{G}^{B}\boldsymbol{\omega}_B \times {}^B\mathbf{r}_P \tag{8.200}
$$

shows that

$$
{}_{G}^{B}\boldsymbol{\omega}_B = \hat{\imath} \left(\frac{{}^G d\hat{\jmath}}{dt} \cdot \hat{k} \right) + \hat{\jmath} \left(\frac{{}^G d\hat{k}}{dt} \cdot \hat{\imath} \right) + \hat{k} \left(\frac{{}^G d\hat{\imath}}{dt} \cdot \hat{\jmath} \right) \tag{8.201}
$$

Example 490 ★ Alternative Proof for Angular Velocity Definition (8.188) The angular velocity definition presented in Equation (8.188) can also be shown by direct substitution for ${}^G R_B$ in the angular velocity matrix ${}_{G}^{B}\tilde{\omega}_B$:

$$
{}_{G}^{B}\tilde{\omega}_B = {}^G R_B^{\mathrm{T}}\, {}^G \dot{R}_B \tag{8.202}
$$

Therefore,

$$
\begin{aligned}
{}_{G}^{B}\tilde{\omega}_B &= \begin{bmatrix} \hat{\imath}\cdot\hat{I} & \hat{\imath}\cdot\hat{J} & \hat{\imath}\cdot\hat{K} \\ \hat{\jmath}\cdot\hat{I} & \hat{\jmath}\cdot\hat{J} & \hat{\jmath}\cdot\hat{K} \\ \hat{k}\cdot\hat{I} & \hat{k}\cdot\hat{J} & \hat{k}\cdot\hat{K} \end{bmatrix} \cdot \frac{{}^G d}{dt} \begin{bmatrix} \hat{I}\cdot\hat{\imath} & \hat{I}\cdot\hat{\jmath} & \hat{I}\cdot\hat{k} \\ \hat{J}\cdot\hat{\imath} & \hat{J}\cdot\hat{\jmath} & \hat{J}\cdot\hat{k} \\ \hat{K}\cdot\hat{\imath} & \hat{K}\cdot\hat{\jmath} & \hat{K}\cdot\hat{k} \end{bmatrix} \\
&= \begin{bmatrix} \hat{\imath}\cdot\frac{{}^G d\hat{\imath}}{dt} & \hat{\imath}\cdot\frac{{}^G d\hat{\jmath}}{dt} & \hat{\imath}\cdot\frac{{}^G d\hat{k}}{dt} \\ \hat{\jmath}\cdot\frac{{}^G d\hat{\imath}}{dt} & \hat{\jmath}\cdot\frac{{}^G d\hat{\jmath}}{dt} & \hat{\jmath}\cdot\frac{{}^G d\hat{k}}{dt} \\ \hat{k}\cdot\frac{{}^G d\hat{\imath}}{dt} & \hat{k}\cdot\frac{{}^G d\hat{\jmath}}{dt} & \hat{k}\cdot\frac{{}^G d\hat{k}}{dt} \end{bmatrix}
\end{aligned} \tag{8.203}
$$

which shows that

$$
{}_{G}^{B}\boldsymbol{\omega}_B = \begin{bmatrix} \frac{{}^G d\hat{\jmath}}{dt}\cdot\hat{k} \\ \frac{{}^G d\hat{k}}{dt}\cdot\hat{\imath} \\ \frac{{}^G d\hat{\imath}}{dt}\cdot\hat{\jmath} \end{bmatrix} \tag{8.204}
$$

Example 491 ★ Another Proof for Angular Velocity Definition (8.188) Let us show the angular velocity ${}_{G}^{B}\boldsymbol{\omega}_B$ by using the orthogonality condition:

$$
{}_{G}^{B}\boldsymbol{\omega}_B = \left({}_{G}^{B}\boldsymbol{\omega}_B \cdot \hat{\imath}\right)\hat{\imath} + \left({}_{G}^{B}\boldsymbol{\omega}_B \cdot \hat{\jmath}\right)\hat{\jmath} + \left({}_{G}^{B}\boldsymbol{\omega}_B \cdot \hat{k}\right)\hat{k} \tag{8.205}
$$

We substitute the unit vectors of the body coordinate frame with a cross product,

$$ {}_G^B\boldsymbol{\omega}_B = \left({}_G^B\boldsymbol{\omega}_B \cdot \hat{\jmath} \times \hat{k}\right)\hat{\imath} + \left({}_G^B\boldsymbol{\omega}_B \cdot \hat{k} \times \hat{\imath}\right)\hat{\jmath} + \left({}_G^B\boldsymbol{\omega}_B \cdot \hat{\imath} \times \hat{\jmath}\right)\hat{k} \tag{8.206} $$

and interchange the dot and cross products,

$$ {}_G^B\boldsymbol{\omega}_B = \left({}_G^B\boldsymbol{\omega}_B \times \hat{\jmath} \cdot \hat{k}\right)\hat{\imath} + \left({}_G^B\boldsymbol{\omega}_B \times \hat{k} \cdot \hat{\imath}\right)\hat{\jmath} + \left({}_G^B\boldsymbol{\omega}_B \times \hat{\imath} \cdot \hat{\jmath}\right)\hat{k} \tag{8.207} $$

Knowing that

$$ {}_G^B\boldsymbol{\omega}_B \times \hat{\jmath} = \frac{{}^G d\hat{\jmath}}{dt} \qquad {}_G^B\boldsymbol{\omega}_B \times \hat{k} = \frac{{}^G d\hat{k}}{dt} \qquad {}_G^B\boldsymbol{\omega}_B \times \hat{\imath} = \frac{{}^G d\hat{\imath}}{dt} \tag{8.208} $$

we can write the angular velocity as defined in Equation (8.188):

$$ {}_G^B\boldsymbol{\omega}_B = \hat{\imath}\left(\frac{{}^G d\hat{\jmath}}{dt} \cdot \hat{k}\right) + \hat{\jmath}\left(\frac{{}^G d\hat{k}}{dt} \cdot \hat{\imath}\right) + \hat{k}\left(\frac{{}^G d\hat{\imath}}{dt} \cdot \hat{\jmath}\right) \tag{8.209} $$

8.2 TIME DERIVATIVE AND COORDINATE FRAMES

The time derivative of a vector depends on the coordinate frame in which we are taking the derivative. Consider an orthogonal global frame $G(OXYZ)$ and an orthogonal body frame $B(Oxyz)$ that is fixed at the point O. When a vector is expressed in the B-frame, we call it a B-vector, and similarly, when a vector is expressed in the G-frame, we call it a G-vector. The first time derivative of a vector $\mathbf{r}$ in G is called the *G-derivative* and is denoted by ${}_G\dot{\mathbf{r}}$,

$$ \frac{{}^G d}{dt}\mathbf{r} = {}_G\dot{\mathbf{r}} \tag{8.210} $$

and the first time derivative of $\mathbf{r}$ in the body frame B is called the *B-derivative* and is denoted by ${}_B\dot{\mathbf{r}}$,

$$ \frac{{}^B d}{dt}\mathbf{r} = {}_B\dot{\mathbf{r}} \tag{8.211} $$

The left superscript on the derivative symbol indicates the coordinate frame in which the derivative is taken. The unit vectors of the derivative frame are considered constant. Therefore, the derivative of the G-vector, ${}^G\mathbf{r} = X\hat{I} + Y\hat{J} + Z\hat{K}$ in G, and the derivative of the B-vector, ${}^B\mathbf{r} = x\hat{\imath} + y\hat{\jmath} + z\hat{k}$ in B, are given as

$$ \frac{{}^G d}{dt}{}^G\mathbf{r} = {}^G\mathbf{v} = {}^G\dot{\mathbf{r}} = \dot{X}\,\hat{I} + \dot{Y}\,\hat{J} + \dot{Z}\,\hat{K} \tag{8.212} $$

$$ \frac{{}^B d}{dt}{}^B\mathbf{r} = {}^B\mathbf{v} = {}^B\dot{\mathbf{r}} = \dot{x}\,\hat{\imath} + \dot{y}\,\hat{\jmath} + \dot{z}\,\hat{k} \tag{8.213} $$

The derivative of a vector in the same frame in which it is expressed is called a *simple derivative*.

It is also possible to find the G-derivative of ${}^B\mathbf{r}$ and the B-derivative of ${}^G\mathbf{r}$. The derivative of a vector in a frame other than the frame in which it is expressed is called a *mixed derivative*. We define the G-derivative of a body vector ${}^B\mathbf{r}$ by

$$\frac{{}^Gd}{dt}{}^B\mathbf{r} = {}^B_G\boldsymbol{\omega}_B \times {}^B\mathbf{r} = {}^B_G\mathbf{v} \tag{8.214}$$

and call ${}^B_G\mathbf{v}$ the B-expression of the G-velocity. Similarly, the B-derivative of a global vector ${}^G\mathbf{r}$ is given as

$$\frac{{}^Bd}{dt}{}^G\mathbf{r} = -{}_G\boldsymbol{\omega}_B \times {}^G\mathbf{r} = {}^G_B\mathbf{v} \tag{8.215}$$

and we call ${}^G_B\mathbf{v}$ the G-expression of the B-velocity.

Whenever it is clear that we are working with velocity vectors, we may use the signs ${}^B_G\mathbf{v}$ and ${}^G_B\mathbf{v}$, which are short notations for $({}^Gd/dt){}^B\mathbf{r}$ and $({}^Bd/dt){}^G\mathbf{r}$, respectively. The left superscript of ${}^B_G\mathbf{v}$ indicates the frame in which $\mathbf{v}$ is expressed and the left subscript indicates the frame in which the derivative is taken. If the left superscript and subscript of a derivative vector are the same, we only keep the superscript. To read ${}^B_G\mathbf{v}$ and ${}^G_B\mathbf{v}$, we may use the expressions *G-velocity of a B-vector* for ${}^B_G\mathbf{v}$ and *B-velocity of a G-vector* for ${}^G_B\mathbf{v}$.

When the interested point P is not a fixed point in B, then, P is moving in frame B and ${}^B\mathbf{r}_P = {}^B\mathbf{r}_P(t)$ is not a constant vector. The G-derivative of ${}^B\mathbf{r}(t)$ is defined by

$$\begin{aligned}\frac{{}^Gd}{dt}{}^B\mathbf{r}(t) &= {}^B_G\mathbf{v} = {}^B_G\dot{\mathbf{r}} = \frac{{}^Bd}{dt}{}^B\mathbf{r} + {}^B_G\boldsymbol{\omega}_B \times {}^B\mathbf{r} \\ &= {}^B\mathbf{v} + {}^B_G\boldsymbol{\omega}_B \times {}^B\mathbf{r}\end{aligned} \tag{8.216}$$

and the B-derivative of ${}^G\mathbf{r}(t)$ is defined by

$$\begin{aligned}\frac{{}^Bd}{dt}{}^G\mathbf{r}(t) &= {}^G_B\mathbf{v} = {}^G_B\dot{\mathbf{r}} = \frac{{}^Gd}{dt}{}^G\mathbf{r}(t) - {}_G\boldsymbol{\omega}_B \times {}^G\mathbf{r} \\ &= {}^G\mathbf{v} - {}_G\boldsymbol{\omega}_B \times {}^G\mathbf{r}\end{aligned} \tag{8.217}$$

The G-derivative of a B-vector is a B-vector and the B-derivative of a G-vector is a G-vector.

Proof: Let $G(OXYZ)$ with unit vectors $\hat{I}$, $\hat{J}$, and $\hat{K}$ be the global coordinate frame, and let $B(Oxyz)$ with unit vectors $\hat{\imath}$, $\hat{\jmath}$, and $\hat{k}$ be a body coordinate frame. The position vector of a moving point P, as shown in Figure 8.5, can be expressed in the body and global frames:

$${}^B\mathbf{r}(t) = x(t)\,\hat{\imath} + y(t)\,\hat{\jmath} + z(t)\,\hat{k} \tag{8.218}$$

$${}^G\mathbf{r}(t) = X(t)\,\hat{I} + Y(t)\,\hat{J} + Z(t)\,\hat{K} \tag{8.219}$$

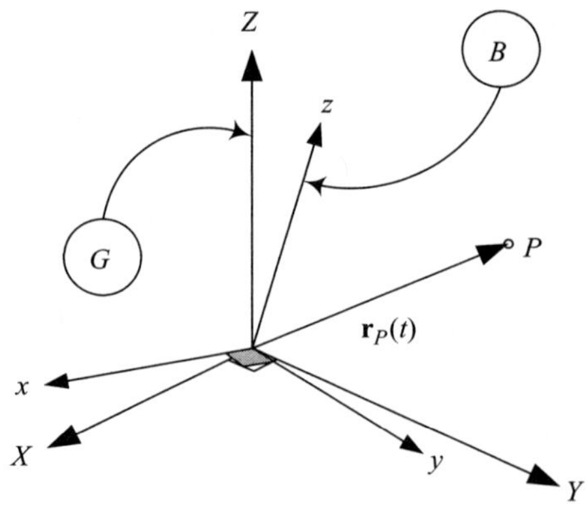

Figure 8.5 A moving body point P at ${}^B\mathbf{r}(t)$ in the rotating body frame B.

The time derivatives of ${}^B\mathbf{r} = {}^B\mathbf{r}_P$ in B and ${}^G\mathbf{r} = {}^G\mathbf{r}_P$ in G are

$$\frac{{}^Bd}{dt}\,{}^B\mathbf{r} = {}^B\mathbf{v} = {}^B\dot{\mathbf{r}} = \dot{x}\,\hat{\imath} + \dot{y}\,\hat{\jmath} + \dot{z}\,\hat{k} \tag{8.220}$$

$$\frac{{}^Gd}{dt}\,{}^G\mathbf{r} = {}^G\mathbf{v} = {}^G\dot{\mathbf{r}} = \dot{X}\,\hat{I} + \dot{Y}\,\hat{J} + \dot{Z}\,\hat{K} \tag{8.221}$$

because the unit vectors of B in Equation (8.218) and the unit vectors of G in Equation (8.219) are considered constant.

To determine the general mixed derivative (8.216), we use Equations (8.120)–(8.122) to find the G-derivative of the B-vector ${}^B\mathbf{r}$ and determine the B-expression of the G-velocity ${}^B_G\mathbf{v}$:

$$\begin{aligned}
\frac{{}^Gd}{dt}\,{}^B\mathbf{r} &= \frac{{}^Gd}{dt}\left(x\hat{\imath} + y\hat{\jmath} + z\hat{k}\right) = \dot{x}\,\hat{\imath} + \dot{y}\,\hat{\jmath} + \dot{z}\,\hat{k} + x\frac{{}^Gd\hat{\imath}}{dt} + y\frac{{}^Gd\hat{\jmath}}{dt} + z\frac{{}^Gd\hat{k}}{dt} \\
&= \dot{x}\,\hat{\imath} + \dot{y}\,\hat{\jmath} + \dot{z}\,\hat{k} + x\left({}^B_G\boldsymbol{\omega}_B \times \hat{\imath}\right) + y\left({}^B_G\boldsymbol{\omega}_B \times \hat{\jmath}\right) + z\left({}^B_G\boldsymbol{\omega}_B \times \hat{k}\right) \\
&= \dot{x}\,\hat{\imath} + \dot{y}\,\hat{\jmath} + \dot{z}\,\hat{k} + {}^B_G\boldsymbol{\omega}_B \times \left(x\hat{\imath} + y\hat{\jmath} + z\hat{k}\right) \\
&= \frac{{}^Bd}{dt}\,{}^B\mathbf{r} + {}^B_G\boldsymbol{\omega}_B \times {}^B\mathbf{r} = {}^B\mathbf{v} + {}^B_G\boldsymbol{\omega}_B \times {}^B\mathbf{r}
\end{aligned} \tag{8.222}$$

We achieved this result because the x-, y-, and z-components of ${}^B\mathbf{r}_P$ are scalar. Scalars are invariant with respect to coordinate frame transformations. Therefore, if x is a scalar, then

$$\frac{{}^Gd}{dt}x = \frac{{}^Bd}{dt}x = \dot{x} \tag{8.223}$$

The B-derivative of ${}^G\mathbf{r}_P$ can be found similarly:

$$\begin{aligned}
\frac{{}^Bd}{dt}{}^G\mathbf{r} &= \frac{{}^Bd}{dt}\left(X\hat{I}+Y\hat{J}+Z\hat{K}\right) \\
&= \dot{X}\,\hat{I}+\dot{Y}\,\hat{J}+\dot{Z}\,\hat{K}+X\frac{{}^Bd\hat{I}}{dt}+Y\frac{{}^Bd\hat{J}}{dt}+Z\frac{{}^Bd\hat{K}}{dt} \\
&= \dot{X}\hat{I}+\dot{Y}\hat{J}+\dot{Z}\hat{K} \\
&\quad + X\left({}_B^G\boldsymbol{\omega}_G\times\hat{I}\right)+Y\left({}_B^G\boldsymbol{\omega}_G\times\hat{J}\right)+Z\left({}_B^G\boldsymbol{\omega}_G\times\hat{K}\right) \\
&= \dot{X}\hat{I}+\dot{Y}\hat{J}+\dot{Z}\hat{K}+{}_B^G\boldsymbol{\omega}_G\times\left(X\hat{I}+Y\hat{J}+Z\hat{K}\right) \\
&= \frac{{}^Gd}{dt}{}^G\mathbf{r}+{}_B^G\boldsymbol{\omega}_G\times{}^G\mathbf{r}={}^G\mathbf{v}+{}_B^G\boldsymbol{\omega}_G\times{}^G\mathbf{r} \\
&= {}^G\mathbf{v}-{}_G\boldsymbol{\omega}_B\times{}^G\mathbf{r}
\end{aligned} \tag{8.224}$$

The angular velocity of B relative to G is a vector quantity and can be expressed in either frames,

$${}_G^G\boldsymbol{\omega}_B=\omega_X\,\hat{I}+\omega_Y\,\hat{J}+\omega_Z\,\hat{K} \tag{8.225}$$

$${}_G^B\boldsymbol{\omega}_B=\omega_x\hat{\imath}+\omega_y\hat{\jmath}+\omega_z\hat{k} \tag{8.226}$$

or be transformed from one frame to the other according to (8.9) and (8.10).

Using Equations of simple and mixed derivatives (8.212)–(8.217), we define the simple integrals as

$${}^G\!\!\int{}^G\mathbf{v}dt={}^G\!\!\int\left(\dot{X}\,\hat{I}+\dot{Y}\,\hat{J}+\dot{Z}\,\hat{K}\right)dt=X\,\hat{\imath}+Y\,\hat{\jmath}+Z\,\hat{k} \tag{8.227}$$

$${}^B\!\!\int{}^B\mathbf{v}dt={}^B\!\!\int\left(\dot{x}\,\hat{\imath}+\dot{y}\,\hat{\jmath}+\dot{z}\,\hat{k}\right)dt=x\,\hat{\imath}+y\,\hat{\jmath}+z\,\hat{k} \tag{8.228}$$

Similarly, the mixed integrals of a body vector are

$${}^G\!\!\int{}_G^B\mathbf{v}\,dt={}^B\mathbf{r}={}_G^B\tilde{\omega}_B^{\mathrm{T}}\,{}_G^B\mathbf{v} \tag{8.229}$$

$${}^B\!\!\int{}_B^G\mathbf{v}dt={}^G\mathbf{r}=-{}_G\tilde{\omega}_B^{\mathrm{T}}\,{}_B^G\mathbf{v} \tag{8.230}$$

If the point P is moving in B, then the mixed integrals would be

$${}^G\!\!\int{}_G^B\mathbf{v}\,dt={}^B\mathbf{r}={}_G^B\tilde{\omega}_B^{\mathrm{T}}\left({}_G^B\mathbf{v}-{}^B\mathbf{v}\right)$$

$$\begin{aligned}
\frac{{}^Gd}{dt}{}^B\mathbf{r}\,(t) &= {}_G^B\mathbf{v}={}_G^B\dot{\mathbf{r}}=\frac{{}^Bd}{dt}{}^B\mathbf{r}+{}_G^B\boldsymbol{\omega}_B\times{}^B\mathbf{r} \\
&= {}^B\mathbf{v}+{}_G^B\boldsymbol{\omega}_B\times{}^B\mathbf{r}
\end{aligned} \tag{8.231}$$

$$\frac{{}^{B}d}{dt}{}^{G}\mathbf{r}(t) = {}^{G}_{B}\mathbf{v} = {}^{G}_{B}\dot{\mathbf{r}} = \frac{{}^{G}d}{dt}{}^{G}\mathbf{r}(t) - {}_{G}\boldsymbol{\omega}_{B} \times {}^{G}\mathbf{r}$$
$$= {}^{G}\mathbf{v} - {}_{G}\boldsymbol{\omega}_{B} \times {}^{G}\mathbf{r} \tag{8.232}$$

■

Example 492 Time Derivative of a Moving Point in B Consider a local frame B rotating in G about the Z-axis with angular velocity $\dot{\alpha}$ and a moving point at ${}^{B}\mathbf{r}_{P}(t) = t\hat{\imath}$. The global position vector of the point is

$$ {}^{G}\mathbf{r}_{P} = {}^{G}R_{B}\,{}^{B}\mathbf{r}_{P} = R_{Z,\alpha}(t)\,{}^{B}\mathbf{r}_{P} = \begin{bmatrix} \cos\alpha & -\sin\alpha & 0 \\ \sin\alpha & \cos\alpha & 0 \\ 0 & 0 & 1 \end{bmatrix} \begin{bmatrix} t \\ 0 \\ 0 \end{bmatrix}$$
$$= t\cos\alpha\,\hat{I} + t\sin\alpha\,\hat{J} \tag{8.233}$$

The angular velocity matrix is

$$ {}_{G}\tilde{\omega}_{B} = {}^{G}\dot{R}_{B}\,{}^{G}R_{B}^{\mathrm{T}} = \dot{\alpha}\tilde{K} \tag{8.234}$$

which gives

$$ {}_{G}\boldsymbol{\omega}_{B} = \dot{\alpha}\hat{K} \tag{8.235}$$

It can also be verified that

$$ {}^{B}_{G}\tilde{\omega}_{B} = {}^{G}R_{B}^{\mathrm{T}}\,{}^{G}_{G}\tilde{\omega}_{B}\,{}^{G}R_{B} = \dot{\alpha}\tilde{k} \tag{8.236}$$

and therefore,

$$ {}^{B}_{G}\boldsymbol{\omega}_{B} = \dot{\alpha}\hat{k} \tag{8.237}$$

Now we can find the following simple derivatives:

$$\frac{{}^{B}d}{dt}{}^{B}\mathbf{r}_{P} = {}^{B}\dot{\mathbf{r}}_{P} = \hat{\imath}$$
$$\frac{{}^{G}d}{dt}{}^{G}\mathbf{r}_{P} = {}^{G}\dot{\mathbf{r}}_{P} \tag{8.238}$$
$$= (\cos\alpha - t\dot{\alpha}\sin\alpha)\,\hat{I} + (\sin\alpha + t\dot{\alpha}\cos\alpha)\,\hat{J} \tag{8.239}$$

For the mixed derivatives we start with

$$\frac{{}^{G}d}{dt}{}^{B}\mathbf{r}_{P} = \frac{{}^{B}d}{dt}{}^{B}\mathbf{r}_{P} + {}^{B}_{G}\boldsymbol{\omega}_{B} \times {}^{B}\mathbf{r}_{P}$$
$$= \begin{bmatrix} 1 \\ 0 \\ 0 \end{bmatrix} + \dot{\alpha}\begin{bmatrix} 0 \\ 0 \\ 1 \end{bmatrix} \times \begin{bmatrix} t \\ 0 \\ 0 \end{bmatrix} = \begin{bmatrix} 1 \\ t\dot{\alpha} \\ 0 \end{bmatrix}$$
$$= \hat{\imath} + t\dot{\alpha}\hat{\jmath} = {}^{B}_{G}\dot{\mathbf{r}}_{P} \tag{8.240}$$

which is the B-expression of the G-velocity of P. We may, however, transform ${}^B_G\dot{\mathbf{r}}_P$ to the global frame and find the global velocity expressed in G as given in Equation (8.239):

$$\begin{aligned}
{}^G\dot{\mathbf{r}}_P &= {}^GR_B\,{}^B_G\dot{\mathbf{r}}_P \\
&= \begin{bmatrix} \cos\alpha & -\sin\alpha & 0 \\ \sin\alpha & \cos\alpha & 0 \\ 0 & 0 & 1 \end{bmatrix} \begin{bmatrix} 1 \\ t\dot{\alpha} \\ 0 \end{bmatrix} = \begin{bmatrix} \cos\alpha - t\dot{\alpha}\sin\alpha \\ \sin\alpha + t\dot{\alpha}\cos\alpha \\ 0 \end{bmatrix} \\
&= (\cos\alpha - t\dot{\alpha}\sin\alpha)\,\hat{I} + (\sin\alpha + t\dot{\alpha}\cos\alpha)\,\hat{J}
\end{aligned} \tag{8.241}$$

The next mixed derivative is

$$\begin{aligned}
\frac{{}^Bd}{dt}{}^G\mathbf{r}_P &= {}^G\dot{\mathbf{r}}_P - {}_G\boldsymbol{\omega}_B \times {}^G\mathbf{r}_P \\
&= \begin{bmatrix} \cos\alpha - t\dot{\alpha}\sin\alpha \\ \sin\alpha + t\dot{\alpha}\cos\alpha \\ 0 \end{bmatrix} - \dot{\alpha}\begin{bmatrix} 0 \\ 0 \\ 1 \end{bmatrix} \times \begin{bmatrix} t\cos\alpha \\ t\sin\alpha \\ 0 \end{bmatrix} \\
&= \begin{bmatrix} \cos\alpha \\ \sin\alpha \\ 0 \end{bmatrix} = (\cos\alpha)\,\hat{I} + (\sin\alpha)\,\hat{J} = {}^G_B\dot{\mathbf{r}}_P
\end{aligned} \tag{8.242}$$

which is the G-expression of the B-velocity of P. To express this velocity in B we apply a frame transformation and find the same result as (8.238):

$$\begin{aligned}
{}^B\dot{\mathbf{r}}_P &= {}^GR_B^{\mathrm{T}}\,{}^G_B\dot{\mathbf{r}}_P \\
&= \begin{bmatrix} \cos\alpha & -\sin\alpha & 0 \\ \sin\alpha & \cos\alpha & 0 \\ 0 & 0 & 1 \end{bmatrix}^{\mathrm{T}} \begin{bmatrix} \cos\alpha \\ \sin\alpha \\ 0 \end{bmatrix} = \begin{bmatrix} 1 \\ 0 \\ 0 \end{bmatrix} = \hat{\imath}
\end{aligned} \tag{8.243}$$

Sometimes it is applied more if we transform the vector to the same frame in which we are taking the derivative and then apply the differential operator. Therefore,

$$\begin{aligned}
\frac{{}^Gd}{dt}{}^B\mathbf{r}_P &= \frac{{}^Gd}{dt}\left({}^GR_B\,{}^B\mathbf{r}_P\right) \\
&= \frac{{}^Gd}{dt}\begin{bmatrix} t\cos\alpha \\ t\sin\alpha \\ 0 \end{bmatrix} = \begin{bmatrix} \cos\alpha - t\dot{\alpha}\sin\alpha \\ \sin\alpha + t\dot{\alpha}\cos\alpha \\ 0 \end{bmatrix}
\end{aligned} \tag{8.244}$$

and

$$\frac{{}^Bd}{dt}{}^G\mathbf{r}_P = \frac{{}^Bd}{dt}\left({}^GR_B^{\mathrm{T}}\,{}^G\mathbf{r}_P\right) = \frac{{}^Bd}{dt}\begin{bmatrix} t \\ 0 \\ 0 \end{bmatrix} = \begin{bmatrix} 1 \\ 0 \\ 0 \end{bmatrix} \tag{8.245}$$

Example 493 Orthogonality of Position and Velocity Vectors The velocity vector of a fixed body point is perpendicular to the vector:

$$\frac{d\mathbf{r}}{dt} \cdot \mathbf{r} = 0 \tag{8.246}$$

To show this property, we may take a derivative from

$$\mathbf{r} \cdot \mathbf{r} = r^2 \tag{8.247}$$

and find

$$\frac{d}{dt}(\mathbf{r} \cdot \mathbf{r}) = \frac{d\mathbf{r}}{dt} \cdot \mathbf{r} + \mathbf{r} \cdot \frac{d\mathbf{r}}{dt} = 2\frac{d\mathbf{r}}{dt} \cdot \mathbf{r} = 0 \tag{8.248}$$

Equation (8.246) is correct in every coordinate frame and for every constant-length vector as long as the vector and the derivative are expressed in the same coordinate frame.

Example 494 ★ Simple Derivative Transformation Formula The B-expression of the G-velocity of a fixed point in the body coordinate frame $B(Oxyz)$ can be found by Equation (8.214). Now consider a point P that can move in $B(Oxyz)$. In this case, the body position vector ${}^B\mathbf{r}_P$ is not constant, and therefore, the B-expression of the G-velocity of such a point is (8.216):

$$\frac{{}^Gd}{dt}{}^B\mathbf{r}_P = {}^B_G\dot{\mathbf{r}}_P = \frac{{}^Bd}{dt}{}^B\mathbf{r}_P + {}^B_G\boldsymbol{\omega}_B \times {}^B\mathbf{r}_P \tag{8.249}$$

Sometimes the result of Equation (8.249) is used to define the transformation of the differential operator on a B-vector ${}^B\square$ from the body to the global coordinate frame:

$$\frac{{}^Gd}{dt}{}^B\square = {}^B_G\dot{\square} = \frac{{}^Bd}{dt}{}^B\square + {}^B_G\boldsymbol{\omega}_B \times {}^B\square \tag{8.250}$$

However, special attention must be paid to the coordinate frame in which the vector ${}^B\square$ and the final result are expressed. The final result is ${}^B_G\dot{\square}$, showing the global (G) time derivative expressed in the body frame (B) or simply the B-expression of the G-derivative of ${}^B\square$. The vector ${}^B\square$ may be any vector quantity such as position, velocity, angular velocity, momentum, angular momentum, a time-varying force vector.

Equation (8.250) is called a *simple derivative transformation formula* and relates the derivative of a B-vector as it would be seen from the G-frame to its derivative as seen from the B-frame. The derivative transformation formula (8.250) is more general and can be applied to every vector for derivative transformation between every two relatively moving coordinate frames.

Example 495 ★ Mixed Derivative Transformation Formula Consider three relatively rotating coordinate frames A, B, and C, as shown in Figure 8.6. The B-expression of the A-velocity of a moving point P in the body coordinate frame $B(Oxyz)$ is

$$\frac{{}^Ad}{dt}{}^B\mathbf{r}_P = {}^B_A\mathbf{v}_P = \frac{{}^Bd}{dt}{}^B\mathbf{r}_P + {}^B_A\boldsymbol{\omega}_B \times {}^B\mathbf{r}_P \tag{8.251}$$

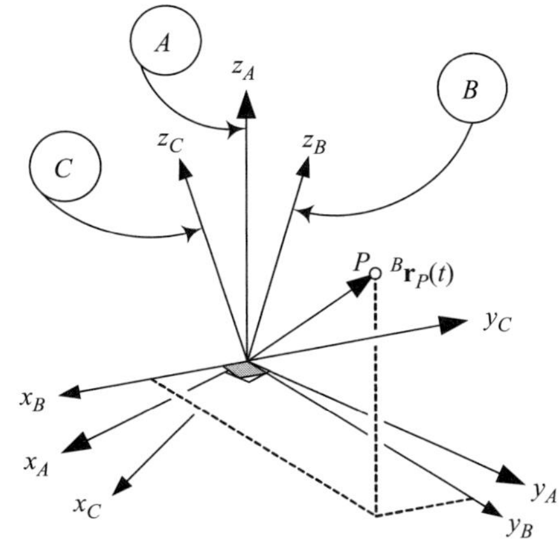

Figure 8.6 Three relatively rotating coordinate frames A, B, and C.

and the B-expression of the C-velocity of a moving point in the body coordinate frame $B(Oxyz)$ is

$$\frac{^C d}{dt}\,^B\mathbf{r}_P = {}^B_C\mathbf{v}_P = \frac{^B d}{dt}\,^B\mathbf{r}_P + {}^B_C\boldsymbol{\omega}_B \times {}^B\mathbf{r}_P \tag{8.252}$$

Combining Equations (8.251) and (8.252), we find

$${}^B_A\mathbf{v}_P - {}^B_A\boldsymbol{\omega}_B \times {}^B\mathbf{r}_P = {}^B_C\mathbf{v}_P - {}^B_C\boldsymbol{\omega}_B \times {}^B\mathbf{r}_P \tag{8.253}$$

Rearranging (8.253) and changing the frame C in which we have taken a derivative of $^B\mathbf{r}_P$ to the frame A in which we need the derivative to be taken yield

$${}^B_A\mathbf{v}_P = {}^B_C\mathbf{v}_P + \left({}^B_A\boldsymbol{\omega}_B - {}^B_C\boldsymbol{\omega}_B\right) \times {}^B\mathbf{r}_P \tag{8.254}$$

We may equivalently show it as

$${}^B_A\dot{\square} = {}^B_C\dot{\square} + \left({}^B_A\boldsymbol{\omega}_B - {}^B_C\boldsymbol{\omega}_B\right) \times {}^B\square \tag{8.255}$$

Equation (8.255) is called the *mixed-derivative transformation formula*. It presents the method used to change the frame in which the derivative of a vector $^B\square$ is taken. Interestingly, mixed-derivative transformation does not involve local derivatives directly.

The mixed-derivative transformation formula (8.255) is more general than the simple-derivative transformation formula (8.250). Equation (8.250) is a special case of (8.255) when $B \equiv C$ or ${}^B_C\boldsymbol{\omega}_B = 0$.

Example 496 ★ Acceleration of a Body Point in the Global Frame The angular acceleration vector of a rigid body $B(Oxyz)$ in the global frame $G(OXYZ)$ is denoted by ${}_G\boldsymbol{\alpha}_B$ and is defined as the global time derivative of ${}_G\boldsymbol{\omega}_B$:

$${}_G\boldsymbol{\alpha}_B = \frac{^G d}{dt}\,{}_G\boldsymbol{\omega}_B \tag{8.256}$$

Using this definition, the acceleration of a fixed body point in the global frame is

$$\begin{aligned} {}^{G}\mathbf{a}_{P} &= \frac{{}^{G}d}{dt}\left({}_{G}\boldsymbol{\omega}_{B} \times {}^{G}\mathbf{r}_{P}\right) = {}_{G}\boldsymbol{\alpha}_{B} \times {}^{G}\mathbf{r}_{P} + {}_{G}\boldsymbol{\omega}_{B} \times \left({}_{G}\boldsymbol{\omega}_{B} \times {}^{G}\mathbf{r}_{P}\right) \\ &= {}_{G}\boldsymbol{\alpha}_{B} \times {}^{G}\mathbf{r}_{P} - {}_{G}\omega_{B}^{2}\,{}^{G}\mathbf{r}_{P} \end{aligned} \tag{8.257}$$

The first term is perpendicular to ${}^{G}\mathbf{r}_{P}$ and may be called *tangential acceleration*. The second term is in the direction of $-{}^{G}\mathbf{r}_{P}$ and may be called *centripetal acceleration*.

Example 497 ★ Second Derivative In general, $d\,\mathbf{r}/dt$ is a variable vector in $G(OXYZ)$ and in any other coordinate frame such as $B(Oxyz)$. Therefore, it can be differentiated in either coordinate frames G or B. However, the order of differentiating is important because

$$\frac{{}^{B}d}{dt}\frac{{}^{G}d\mathbf{r}}{dt} \neq \frac{{}^{G}d}{dt}\frac{{}^{B}d\mathbf{r}}{dt} \tag{8.258}$$

As an example, consider a rotating body coordinate frame about the Z-axis and a variable G-vector as

$${}^{G}\mathbf{r} = t\hat{I} \tag{8.259}$$

The G-derivative of ${}^{G}\mathbf{r}$ is

$$\frac{{}^{G}d\mathbf{r}}{dt} = {}^{G}\dot{\mathbf{r}} = \hat{I} \tag{8.260}$$

We can transform this vector to the B-frame to calculate ${}_{G}^{B}\dot{\mathbf{r}}$:

$$\begin{aligned} {}^{B}\left(\frac{{}^{G}d\mathbf{r}}{dt}\right) &= {}_{G}^{B}\dot{\mathbf{r}} = R_{Z,\varphi}^{\mathrm{T}}\,{}^{G}\dot{\mathbf{r}} = R_{Z,\varphi}^{\mathrm{T}}\left[\hat{I}\right] \\ &= \begin{bmatrix} \cos\varphi & \sin\varphi & 0 \\ -\sin\varphi & \cos\varphi & 0 \\ 0 & 0 & 1 \end{bmatrix}\begin{bmatrix} 1 \\ 0 \\ 0 \end{bmatrix} = \cos\varphi\hat{\imath} - \sin\varphi\hat{\jmath} \end{aligned} \tag{8.261}$$

So far, the first derivative of the G-vector ${}^{G}\mathbf{r}$ is calculated in both frames G and B.

A B-derivative of ${}_{G}^{B}\dot{\mathbf{r}}$ provides

$$\frac{{}^{B}d}{dt}{}_{G}^{B}\dot{\mathbf{r}} = {}_{BG}^{BB}\ddot{\mathbf{r}} = -\dot{\varphi}\sin\varphi\hat{\imath} - \dot{\varphi}\cos\varphi\hat{\jmath} \tag{8.262}$$

where ${}_{BG}^{BB}\ddot{\mathbf{r}}$ is the B-expression of the B-derivative of ${}_{G}^{B}\dot{\mathbf{r}}$, which is the B-expression of the G-derivative of $\mathbf{r}$. Let us transform ${}_{BG}^{BB}\ddot{\mathbf{r}}$ to the G-frame and determine the G-expression of the B-derivative of ${}_{G}^{B}\dot{\mathbf{r}}$:

$${}^{G}\left(\frac{{}^{B}d}{dt}{}_{G}^{B}\dot{\mathbf{r}}\right) = {}_{BG}^{GB}\ddot{\mathbf{r}} = R_{Z,\varphi}\left(\frac{{}^{B}d}{dt}{}_{G}^{B}\dot{\mathbf{r}}\right) = -\dot{\varphi}\hat{J} \tag{8.263}$$

Now, we begin with the B-expression of the vector

$$^{B}\mathbf{r} = R_{Z,\varphi}^{\mathrm{T}}\left[t\hat{I}\right] = t\cos\varphi\hat{\imath} - t\sin\varphi\hat{\jmath} \tag{8.264}$$

and take a B-derivative

$$\frac{^{B}d\mathbf{r}}{dt} = {}^{B}\dot{\mathbf{r}} = (-t\dot{\varphi}\sin\varphi + \cos\varphi)\,\hat{\imath} - (\sin\varphi + t\dot{\varphi}\cos\varphi)\,\hat{\jmath} \tag{8.265}$$

and transform it to the G-frame:

$$\begin{aligned}
{}^{G}\left(\frac{^{B}d\mathbf{r}}{dt}\right) &= {}_{B}^{G}\dot{\mathbf{r}} = R_{Z,\varphi}\,{}^{B}\dot{\mathbf{r}} \\
&= \begin{bmatrix} \cos\varphi & -\sin\varphi & 0 \\ \sin\varphi & \cos\varphi & 0 \\ 0 & 0 & 1 \end{bmatrix} \begin{bmatrix} -t\dot{\varphi}\sin\varphi + \cos\varphi \\ -\sin\varphi - t\dot{\varphi}\cos\varphi \\ 0 \end{bmatrix} \\
&= \hat{I} - t\dot{\varphi}\hat{J}
\end{aligned} \tag{8.266}$$

A G-derivative of ${}_{B}^{G}\dot{\mathbf{r}}$ is the G-expression of the G-derivative of ${}_{B}^{G}\mathbf{r}$, which is the G-expression of the B-derivative of $\mathbf{r}$:

$$\frac{^{G}d}{dt}\,{}_{B}^{G}\dot{\mathbf{r}} = {}_{G\,B}^{G\,G}\ddot{\mathbf{r}} = -\left(\dot{\varphi} + t\ddot{\varphi}\right)\hat{J} \tag{8.267}$$

Equations (8.261) and (8.266) indicate that

$${}_{G}^{B}\dot{\mathbf{r}} \neq {}_{B}^{G}\dot{\mathbf{r}} \tag{8.268}$$

and Equations (8.262) and (8.263) show that

$$\frac{^{B}d}{dt}\,{}_{G}^{B}\dot{\mathbf{r}} \neq \frac{^{G}d}{dt}\,{}_{B}^{G}\dot{\mathbf{r}} \tag{8.269}$$

8.3 MULTIBODY VELOCITY

Consider a rigid body with an attached local coordinate frame B $(oxyz)$ moving freely in a fixed global coordinate frame $G(OXYZ)$, as shown in Figure 8.7. The rigid body can rotate in the global frame, while the origin of the body frame B can translate relative to the origin of G. The coordinates of a body point P in local and global frames are related by the motion kinematic equation

$$^{G}\mathbf{r}_{P} = {}^{G}R_{B}\,{}^{B}\mathbf{r}_{P} + {}^{G}\mathbf{d}_{B} \tag{8.270}$$

where $^{G}\mathbf{d}_{B}$ indicates the position of the moving origin o relative to the fixed origin O.

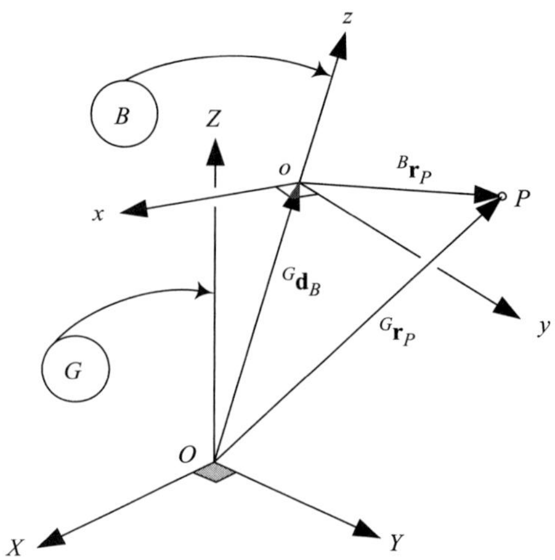

Figure 8.7 A rigid body with an attached coordinate frame $B\ (oxyz)$ moving freely in a global coordinate frame $G(OXYZ)$.

The velocity of the point P in G is

$$\begin{aligned}
{}^G\mathbf{v}_P &= {}^G\dot{\mathbf{r}}_P = {}^G\dot{R}_B\,{}^B\mathbf{r}_P + {}^G\dot{\mathbf{d}}_B = {}_G\tilde{\omega}_B\,{}^G_B\mathbf{r}_P + {}^G\dot{\mathbf{d}}_B \\
&= {}_G\tilde{\omega}_B\left({}^G\mathbf{r}_P - {}^G\mathbf{d}_B\right) + {}^G\dot{\mathbf{d}}_B \\
&= {}_G\boldsymbol{\omega}_B \times \left({}^G\mathbf{r}_P - {}^G\mathbf{d}_B\right) + {}^G\dot{\mathbf{d}}_B
\end{aligned} \tag{8.271}$$

If we have multibodies, we may need to determine the velocity of every body relatively and individually in every coordinate frame. As an example, let us consider a global frame G and two frames B_1 and B_2, as shown in Figure 8.8. Assume that we can determine the kinematics of B_1 relative to G and the kinematics of B_2 relative to B_1. The point P is a fixed point in B_2, so the time derivative of ${}^2\mathbf{r}_P$ in B_2 is zero. Because B_2 is moving relative to B_1, the time derivative of ${}^1\mathbf{r}_P$ in B_1 is a combination of the rotational component due to rotation of B_2 relative to B_1 and the component due to translational velocity of B_2 relative to B_1. The velocity of point P in the base frame G is a combination of the velocity of B_2 relative to B_1 plus the velocity of B_1 relative to $G = B_0$. Such a combination of velocities is called *relative velocity*. The velocity combination is allowed only if all the velocity vectors are expressed in the same frame. The global frame is usually the best choice:

$$\begin{aligned}
{}^0\mathbf{v} &= {}^0\mathbf{v}_1 + {}^0_1\mathbf{v}_2 + {}^0_2\mathbf{v}_P \\
&= {}^0\dot{\mathbf{d}}_1 + {}^0_0\boldsymbol{\omega}_1 \times {}^0_1\mathbf{d}_2 + {}^0R_1\,{}^1\dot{\mathbf{d}}_2 + {}^0_0\boldsymbol{\omega}_2 \times {}^0_2\mathbf{r}_P
\end{aligned} \tag{8.272}$$

Proof: Direct differentiating shows that

$$\begin{aligned}
{}^G\mathbf{v}_P &= \frac{{}^Gd}{dt}\,{}^G\mathbf{r}_P = {}^G\dot{\mathbf{r}}_P = \frac{{}^Gd}{dt}\left({}^GR_B\,{}^B\mathbf{r}_P + {}^G\mathbf{d}_B\right) \\
&= {}^G\dot{R}_B\,{}^B\mathbf{r}_P + {}^G\dot{\mathbf{d}}_B
\end{aligned} \tag{8.273}$$

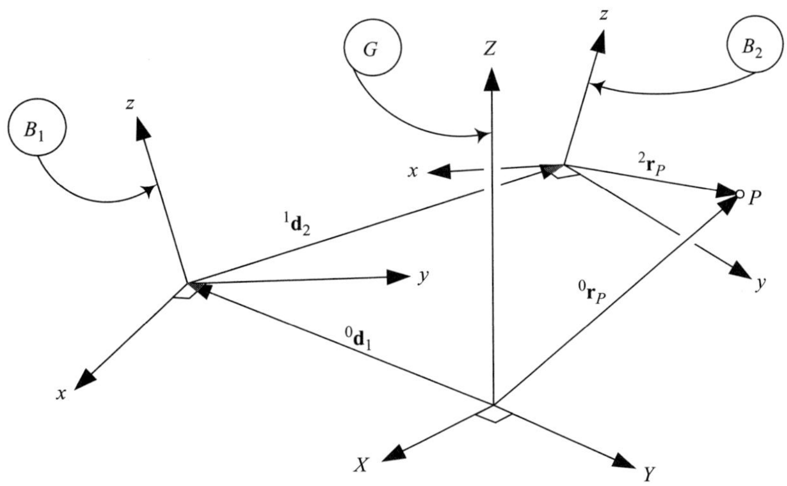

Figure 8.8 A rigid-body coordinate frame B_2 is moving in a frame B_1 that is moving in the base coordinate frame B_0.

where substituting the B-vector ${}^{B}\mathbf{r}_P$ from (8.270) provides

$$\begin{aligned} {}^{G}\mathbf{v}_P &= {}^{G}\dot{R}_B {}^{G}R_B^{\mathrm{T}} \left({}^{G}\mathbf{r}_P - {}^{G}\mathbf{d}_B\right) + {}^{G}\dot{\mathbf{d}}_B = {}_{G}\tilde{\omega}_B \left({}^{G}\mathbf{r}_P - {}^{G}\mathbf{d}_B\right) + {}^{G}\dot{\mathbf{d}}_B \\ &= {}_{G}\boldsymbol{\omega}_B \times \left({}^{G}\mathbf{r}_P - {}^{G}\mathbf{d}_B\right) + {}^{G}\dot{\mathbf{d}}_B \end{aligned} \tag{8.274}$$

Using the relative-position vector, we can write this equation more simply as

$$ {}^{G}\mathbf{v}_P = {}_{G}\boldsymbol{\omega}_B \times {}_{B}^{G}\mathbf{r}_P + {}^{G}\dot{\mathbf{d}}_B \tag{8.275}$$

For the multibodies of Figure 8.8, we first determine the global position vector of the body point P:

$$ {}^{0}\mathbf{r}_P = {}^{0}\mathbf{d}_1 + {}_{1}^{0}\mathbf{d}_2 + {}_{2}^{0}\mathbf{r}_P = {}^{0}\mathbf{d}_1 + {}^{0}R_1 {}^{1}\mathbf{d}_2 + {}^{0}R_2 {}^{2}\mathbf{r}_P \tag{8.276}$$

Then, the G-velocity of point P can be found by taking a G-derivative of this equation. The combination of the relative velocities will be determined by the G-derivative:

$$\begin{aligned} {}^{0}\dot{\mathbf{r}}_P &= {}^{0}\dot{\mathbf{d}}_1 + ({}^{0}\dot{R}_1 {}^{1}\mathbf{d}_2 + {}^{0}R_1 {}^{1}\dot{\mathbf{d}}_2) + {}^{0}\dot{R}_2 {}^{2}\mathbf{r}_P \\ &= {}^{0}\dot{\mathbf{d}}_1 + {}_{0}^{0}\boldsymbol{\omega}_1 \times {}_{1}^{0}\mathbf{d}_2 + {}^{0}R_1 {}^{1}\dot{\mathbf{d}}_2 + {}_{0}^{0}\boldsymbol{\omega}_2 \times {}_{2}^{0}\mathbf{r}_P \end{aligned} \tag{8.277}$$

Most of the time, it is better to use a relative-velocity method and write

$$ {}_{0}^{0}\mathbf{v}_P = {}_{0}^{0}\mathbf{v}_1 + {}_{1}^{0}\mathbf{v}_2 + {}_{2}^{0}\mathbf{v}_P \tag{8.278}$$

because

$$ {}_{0}^{0}\mathbf{v}_1 = {}_{0}^{0}\dot{\mathbf{d}}_1 \tag{8.279}$$

$$ {}_{1}^{0}\mathbf{v}_2 = {}_{0}^{0}\boldsymbol{\omega}_1 \times {}_{1}^{0}\mathbf{d}_2 + {}^{0}R_1 {}^{1}\dot{\mathbf{d}}_2 \tag{8.280}$$

$$ {}_{2}^{0}\mathbf{v}_P = {}_{0}^{0}\boldsymbol{\omega}_2 \times {}_{2}^{0}\mathbf{r}_P \tag{8.281}$$

and therefore,

$$ {}^{0}\mathbf{v}_P = {}^{0}\dot{\mathbf{d}}_1 + {}_{0}^{0}\boldsymbol{\omega}_1 \times {}_{1}^{0}\mathbf{d}_2 + {}^{0}R_1\,{}^{1}\dot{\mathbf{d}}_2 + {}_{0}^{0}\boldsymbol{\omega}_2 \times {}_{2}^{0}\mathbf{r}_P \tag{8.282} $$

■

Example 498 Geometric Interpretation of Motion Velocity Figure 8.9 illustrates a body point P of a moving rigid body. The global velocity of the point P is a vector addition of rotational and translational velocities, both expressed in the global frame:

$$ {}^{G}\mathbf{v}_P = {}_{G}\boldsymbol{\omega}_B \times {}_{B}^{G}\mathbf{r}_P + {}^{G}\dot{\mathbf{d}}_B \tag{8.282a} $$

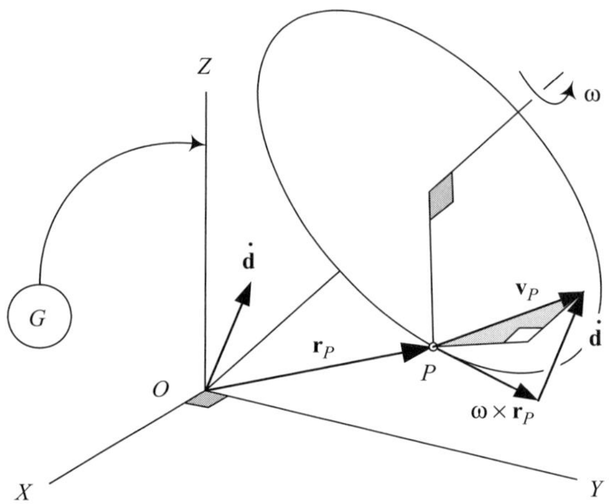

Figure 8.9 Geometric interpretation of rigid-body velocity.

At the moment, the body frame is assumed to be coincident with the global frame, and the body frame has a velocity ${}^{G}\dot{\mathbf{d}}_B$ with respect to the global frame. The translational velocity ${}^{G}\dot{\mathbf{d}}_B$ is a common property of every point of the body, but the rotational velocity ${}_{G}\boldsymbol{\omega}_B \times {}_{B}^{G}\mathbf{r}_P$ differs for different points of the body and depends on ${}^{B}\mathbf{r}_P$.

Example 499 Velocity of a Moving Point in a Moving-Body Frame Assume that point P in Figure 8.7 is moving in the frame B, indicated by a time-varying position vector ${}^{B}\mathbf{r}_P(t)$. The global velocity of P is a composition of the velocity of P in B, rotation of B relative to G, and velocity of B relative to G:

$$ \begin{aligned} \frac{{}^{G}d}{dt}\,{}^{G}\mathbf{r}_P &= \frac{{}^{G}d}{dt}\left({}^{G}\mathbf{d}_B + {}^{G}R_B\,{}^{B}\mathbf{r}_P\right) = \frac{{}^{G}d}{dt}\,{}^{G}\mathbf{d}_B + \frac{{}^{G}d}{dt}\left({}^{G}R_B\,{}^{B}\mathbf{r}_P\right) \\ &= {}^{G}\dot{\mathbf{d}}_B + {}_{B}^{G}\dot{\mathbf{r}}_P + {}_{G}\boldsymbol{\omega}_B \times {}_{B}^{G}\mathbf{r}_P \end{aligned} \tag{8.283} $$

Example 500 Coordinate Frame Velocity Vectors Are Vecfree Velocity vectors of the origin of coordinate frames are free, so to express them in different coordinate frames, we need only premultiply them by a rotation matrix. Hence, considering ${}_j^k\mathbf{v}_i$ as the velocity of the origin of the B_i coordinate frame with respect to the origin of the frame B_j expressed in frame B_k, we can write

$$ {}_j^k\mathbf{v}_i = -{}_i^k\mathbf{v}_j \tag{8.284} $$

and

$$ {}_j^k\mathbf{v}_i = {}^kR_m \, {}_j^m\mathbf{v}_i \tag{8.285} $$

and therefore,

$$ \frac{{}^id}{dt}{}_i^i\mathbf{r}_P = {}^i\mathbf{v}_P = {}_j^i\mathbf{v}_P + {}_i\boldsymbol{\omega}_j \times {}_j^i\mathbf{r}_P \tag{8.286} $$

Example 501 Velocity Analysis of an Inverted Slider–Crank Mechanism The inverted slider–crank mechanism shown in Figure 8.10 is an applied planar four-link mechanism with application in automatic production lines as well as light-vehicle independent suspensions. Link (1) is the ground link, which is the base and reference link. Link (2) $\equiv MA$ is usually the input link, which is controlled by the input angle θ_2. Link (4) is the slider link and is usually considered the output link. The slider link has a revolute joint with the ground and a prismatic joint with coupler link (3) $\equiv AB$. The output variable can be the angle of the slider with the horizon, or the length AB. Link (3) $\equiv AB$ is the coupler link with angular position θ_3.

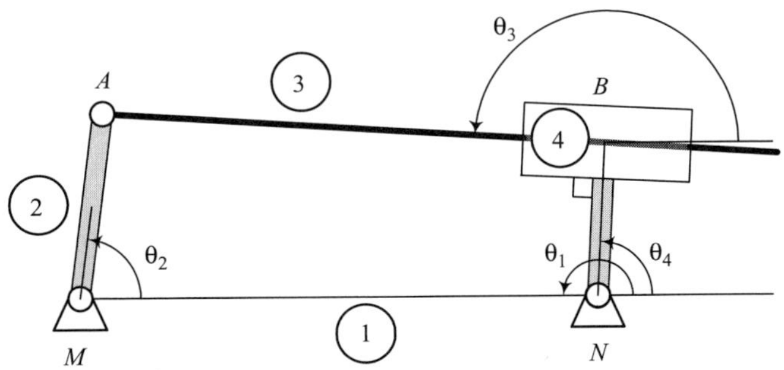

Figure 8.10 An inverted slider–crank mechanism.

We may show the inverted slider–crank mechanism by a vector loop, as shown in Figure 8.11. The direction of each vector is arbitrary; however, the angles should be associated with the vector's direction and be measured with a positive direction of the x-axis. The links and their expression vectors are indicated in Table 8.1.

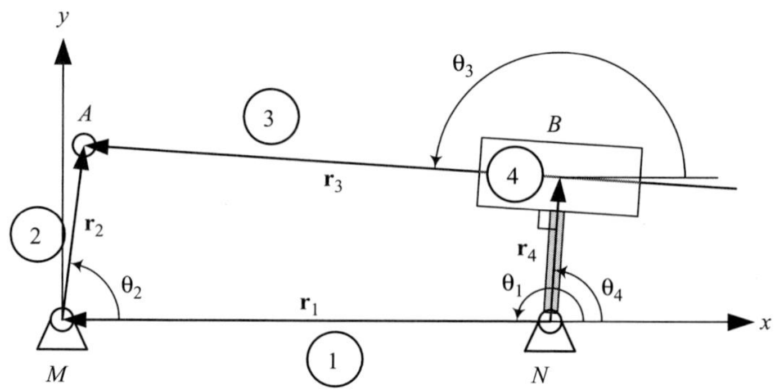

Figure 8.11 Kinematic model of an inverted slider–crank mechanism.

Table 8.1 Vector Representation of Inverted Slider–Crank Mechanism Shown in Figure 8.11

Link	Vector	Length	Angle	Variable
1	${}^G\mathbf{r}_1$	d	$\theta_1 = 180\,\text{deg}$	d
2	${}^G\mathbf{r}_2$	a	θ_2	θ_2
3	${}^G\mathbf{r}_3$	b	θ_3	θ_3 or θ_4
4	${}^G\mathbf{r}_4$	e	$\theta_4 = \theta_3 + 90\,\text{deg}$	–

The angular position of the output slider θ_4 and the length of the coupler link b are functions of the lengths of the links and the value of the input variable θ_2:

$$b = \pm\sqrt{a^2 + d^2 - e^2 - 2ad\cos\theta_2} \tag{8.287}$$

$$\theta_4 = \theta_3 + \frac{\pi}{2} = 2\tan^{-1}\left(\frac{-H \pm \sqrt{H^2 - 4GI}}{2G}\right) \tag{8.288}$$

where

$$G = d - e - a\cos\theta_2 \tag{8.289}$$

$$H = 2a\sin\theta_2 \tag{8.290}$$

$$I = a\cos\theta_2 - d - e \tag{8.291}$$

Assuming θ_2 and $\omega_2 = \dot{\theta}_2$ are given and b, θ_4 are known from Equations (8.287) and (8.288), we can determine the values of $\dot{b}$ and ω_4:

$$\dot{b} = \frac{a}{b}\omega_2 \left[b\cos(\theta_4 - \theta_2) - e\sin(\theta_4 - \theta_2)\right] \tag{8.292}$$

$$\omega_4 = \omega_3 = \frac{a}{b}\omega_2 \sin(\theta_2 - \theta_4) \tag{8.293}$$

Having coordinates θ_2, θ_4, b and velocities ω_2, ω_4, $\dot{b}$ enables us to calculate the absolute and relative velocities of every point of every link of the mechanism. The absolute and relative velocities of points A and B, shown in Figure 8.11, are

$$\begin{aligned} {}^G\mathbf{v}_A &= {}_G\boldsymbol{\omega}_2 \times {}^G\mathbf{r}_2 \\ &= \begin{bmatrix} 0 \\ 0 \\ \omega_2 \end{bmatrix} \times \begin{bmatrix} a\cos\theta_2 \\ a\sin\theta_2 \\ 0 \end{bmatrix} = \begin{bmatrix} -a\omega_2\sin\theta_2 \\ a\omega_2\cos\theta_2 \\ 0 \end{bmatrix} \end{aligned} \tag{8.294}$$

$$\begin{aligned} {}^G\mathbf{v}_{B_4} &= {}_G\boldsymbol{\omega}_4 \times {}^G\mathbf{r}_4 \\ &= \begin{bmatrix} 0 \\ 0 \\ \omega_4 \end{bmatrix} \times \begin{bmatrix} e\cos\theta_4 \\ e\sin\theta_4 \\ 0 \end{bmatrix} = \begin{bmatrix} -e\omega_4\sin\theta_4 \\ e\omega_4\cos\theta_4 \\ 0 \end{bmatrix} \end{aligned} \tag{8.295}$$

$$\begin{aligned} {}^G\mathbf{v}_{B_3/A} &= {}_G\boldsymbol{\omega}_3 \times \left(-{}^G\mathbf{r}_3\right) \\ &= \begin{bmatrix} 0 \\ 0 \\ \omega_4 \end{bmatrix} \times \begin{bmatrix} -b\cos\theta_4 \\ -b\sin\theta_4 \\ 0 \end{bmatrix} = \begin{bmatrix} b\omega_4\sin\theta_4 \\ -b\omega_4\cos\theta_4 \\ 0 \end{bmatrix} \end{aligned} \tag{8.296}$$

$$\begin{aligned} {}^G\mathbf{v}_{B_3} &= {}^G\mathbf{v}_{B_3/A} + {}^G\mathbf{v}_A = \begin{bmatrix} b\omega_4\sin\theta_4 \\ -b\omega_4\cos\theta_4 \\ 0 \end{bmatrix} + \begin{bmatrix} -a\omega_2\sin\theta_2 \\ a\omega_2\cos\theta_2 \\ 0 \end{bmatrix} \\ &= \begin{bmatrix} b\omega_4\sin\theta_4 - a\omega_2\sin\theta_2 \\ a\omega_2\cos\theta_2 - b\omega_4\cos\theta_4 \\ 0 \end{bmatrix} \end{aligned} \tag{8.297}$$

$$\begin{aligned} {}^G\mathbf{v}_{B_3/B_4} &= {}^G\mathbf{v}_{B_3} - {}^G\mathbf{v}_{B_4} \\ &= \begin{bmatrix} b\omega_4\sin\theta_4 - a\omega_2\sin\theta_2 \\ a\omega_2\cos\theta_2 - b\omega_4\cos\theta_4 \\ 0 \end{bmatrix} - \begin{bmatrix} -e\omega_4\sin\theta_4 \\ e\omega_4\cos\theta_4 \\ 0 \end{bmatrix} \\ &= \begin{bmatrix} \omega_4 e\sin\theta_4 - a\omega_2\sin\theta_2 + b\omega_4\sin\theta_4 \\ a\omega_2\cos\theta_2 - \omega_4 e\cos\theta_4 - b\omega_4\cos\theta_4 \\ 0 \end{bmatrix} \end{aligned} \tag{8.298}$$

Example 502 ★ Zero-Velocity Points Consider a moving rigid body with a frame B in a global frame G. If at any time t there exists a point of B that is not moving in G, then such a stationary point is called a *zero-velocity point*.

To examine if there is a point with zero velocity at each time, we can set Equation (8.271) to zero and search for ${}^G\mathbf{r}_0$:

$$ {}_G\tilde{\omega}_B \left({}^G\mathbf{r}_0 - {}^G\mathbf{d}_B\right) + {}^G\dot{\mathbf{d}}_B = 0 \tag{8.299}$$

If there is any, ${}^G\mathbf{r}_0$ indicates the position vector of a zero-velocity point:

$$ {}^G\mathbf{r}_0 = {}^G\mathbf{d}_B - {}_G\tilde{\omega}_B^{-1}\,{}^G\dot{\mathbf{d}}_B \tag{8.300}$$

The skew-symmetric matrix ${}_G\tilde{\omega}_B$ is singular and has no inverse. Therefore, there is no general solution for Equation (8.299).

If we restrict ourselves to planar motions, say the (X, Y)-plane, then ${}_G\boldsymbol{\omega}_B = \omega\hat{K}$ and ${}_G\tilde{\omega}_B^{-1} = 1/\omega$. Equation (8.299) would have a solution at any time t in a $2D$ space, and a zero-velocity point at the position ${}^G\mathbf{r}_0$ can be determined:

$$
{}^G\mathbf{r}_0(t) = {}^G\mathbf{d}_B(t) - \frac{1}{\omega}\,{}^G\dot{\mathbf{d}}_B(t) \tag{8.301}
$$

The zero-velocity point is also called the *pole*, *instant center*, *centro*, or *instantaneous center of rotation*. The position of the pole is generally a function of time and the path of its motion is called a *centroid*.

Example 503 ★ Instantaneous Center of Rotation In a general planar motion of a rigid body at any given instant, the velocities of various points of the body can be expressed as the result of a rotation about an axis perpendicular to the plane. This axis intersects the plane at the instantaneous center of rotation of the body with respect to the ground.

If the directions of the velocities of two different body points A and B of the body are known, the instant center of rotation is at the intersection of the lines perpendicular to the velocity vectors ${}^G\mathbf{v}_A$ and ${}^G\mathbf{v}_B$. If the velocity vectors $\mathbf{v}_A$ and $\mathbf{v}_B$ are perpendicular to the line AB and if their magnitudes are known, the instantaneous center of rotation is at the intersection of AB with the line joining the extremities of the velocity vectors.

There is an instant center of rotation between every two links moving with respect to each other. The instant center is a point common to both bodies that has the same velocity in each body coordinate frame.

Kennedy Theorem: The three instant centers I_{12}, I_{23}, and I_{13} between three links numbered (1), (2), and (3) lie on a straight line.

Consider the three bodies shown in Figure 8.12. The ground is link (1); links (2) and (3) are pivoted to the ground at points M and N and are rotating with angular velocities ω_2 and ω_3. The two links are contacted at point C. The revolute joint at M is the instant center I_{12} and the revolute joint at N is the instant center I_{13}.

The velocity of point C as a point of link (2) is $\mathbf{v}_{C_2}$, perpendicular to the radius MC. Similarly, the velocity of point C as a point of link (3) is $\mathbf{v}_{C_2}$, perpendicular to the radius NC. The instant center of rotation I_{13} must be a common point with the same velocity in both bodies. Let us draw the normal line $n - n$ and tangential line $t - t$ to the curves of links (2) and (3) at the contact point C.

Point C is a common point between the two bodies. The normal components of $\mathbf{v}_{C_2}$ and $\mathbf{v}_{C_3}$ must be equal to keep contact, so the only difference between the velocity of the common point can be in the tangent components. So, the instant center of rotation I_{13} must be at a position where the relative velocity of points C_2 and C_3 with respect to I_{13} are equal and are on the line $t - t$. Hence, it must be on the normal line $n - n$, and the intersection of the normal line $n - n$ with the center line MN is the only possible point for the instant center of rotation I_{13}.

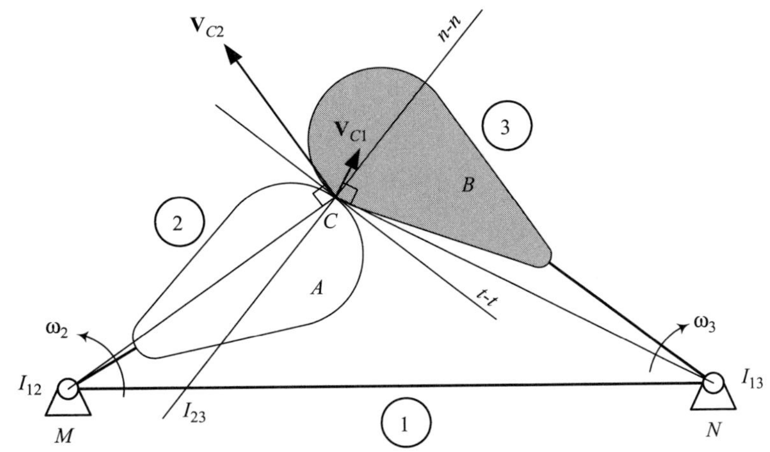

Figure 8.12 A three-link mechanism with the ground as link (1) and two moving links (2) and (3).

We define

$$I_{12}I_{23} = l_2 \tag{8.302}$$

$$I_{13}I_{23} = l_3 \tag{8.303}$$

Then, because the velocities of the two bodies must be equal at the common instant center of rotation, we have

$$l_2\omega_2 = l_3\omega_3 \tag{8.304}$$

or

$$\frac{\omega_2}{\omega_3} = \frac{l_3}{l_2} = \frac{1}{1 + d/l_2} \tag{8.305}$$

where d is the length of the ground link MN.

Example 504 ★ Plane Motion of a Rigid Body The plane motion of a rigid body is such that all points of the body move only in parallel planes. So, to study the motion of the body, it is enough to examine the motion of points in just one plane.

Figure 8.13 illustrates a global coordinate frame G and a rigid body B in a planar motion with the associated body frame B. The position and velocity of a body point P are

$${}^G\mathbf{r}_P = {}^G\mathbf{d}_B + {}^GR_B\, {}^B\mathbf{r}_P = {}^G\mathbf{d}_B + {}^G_B\mathbf{r}_P \tag{8.306}$$

$${}^G\mathbf{v}_P = {}^G\dot{\mathbf{d}}_B + {}_G\boldsymbol{\omega}_B \times \left({}^G\mathbf{r}_P - {}^G\mathbf{d}_B\right) = {}^G\dot{\mathbf{d}}_B + {}_G\boldsymbol{\omega}_B \times {}^G_B\mathbf{r}_P \tag{8.307}$$

The vector ${}^G\mathbf{d}_B$ indicates the position of the moving origin o relative to the fixed origin O. The term ${}^G\dot{\mathbf{d}}_B$ is the velocity of point o and ${}_G\boldsymbol{\omega}_B \times {}^G_B\mathbf{r}_P$ is the velocity of point P relative to o:

$${}^G\mathbf{v}_{P/o} = {}_G\boldsymbol{\omega}_B \times {}^G_B\mathbf{r}_P \tag{8.308}$$

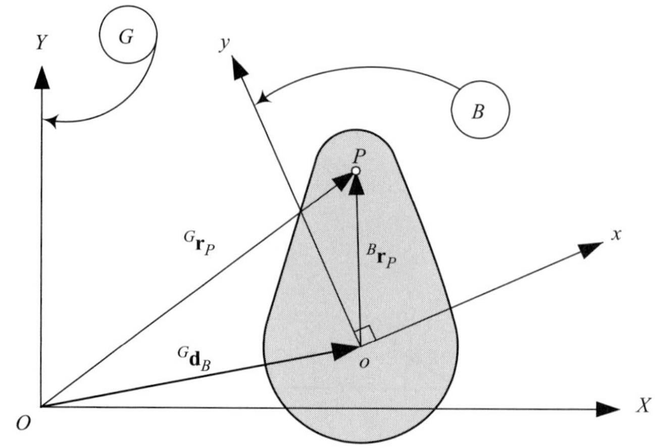

Figure 8.13 A rigid body in a planar motion.

Although it is not a correct view, it might sometimes be helpful if we interpret $^G\dot{\mathbf{d}}_B$ as the translational velocity and $_G\boldsymbol{\omega}_B \times {}^G_B\mathbf{r}_P$ as the rotational velocity components of $^G\mathbf{v}_P$. Then, the velocity of any point P of a rigid body is a superposition of the velocity $^G\dot{\mathbf{d}}_B$ of another arbitrary point o and the angular velocity $_G\boldsymbol{\omega}_B \times {}^G_B\mathbf{r}_P$ of the points P around o.

The relative velocity vector $^G\mathbf{v}_{P/o}$ is perpendicular to the relative position vector ${}^G_B\mathbf{r}_P$. Employing the same concept, we can say that the velocity of points P and o with respect to another point Q are perpendicular to ${}^G_B\mathbf{r}_{P/Q}$ and ${}^G_B\mathbf{r}_{o/Q}$, respectively. We may search for a point Q as the instantaneous center of rotation at which the velocity is zero. Points o, P, and Q are shown in Figure 8.14.

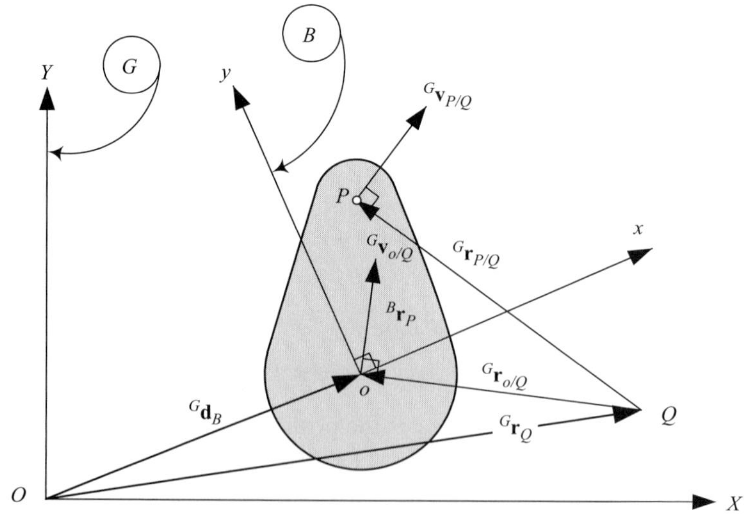

Figure 8.14 Instant center of rotation Q for planar motion of a rigid body.

Assuming a position vector ${}^{G}\mathbf{r}_{o/Q}$ for the instant center point Q, we define

$$ {}^{G}\mathbf{r}_{o/Q} = a_Q {}^{G}\dot{\mathbf{d}}_B + b_Q {}_G\boldsymbol{\omega}_B \times {}^{G}\dot{\mathbf{d}}_B \tag{8.309} $$

Then, following (8.307), the velocity of point Q can be expressed by

$$ \begin{aligned} {}^{G}\mathbf{v}_Q &= {}^{G}\dot{\mathbf{d}}_B + {}_G\boldsymbol{\omega}_B \times {}^{G}\mathbf{r}_{Q/o} = {}^{G}\dot{\mathbf{d}}_B - {}_G\boldsymbol{\omega}_B \times {}^{G}\mathbf{r}_{o/Q} \\ &= {}^{G}\dot{\mathbf{d}}_B - {}_G\boldsymbol{\omega}_B \times \left(a_Q {}^{G}\dot{\mathbf{d}}_B + b_Q {}_G\boldsymbol{\omega}_B \times {}^{G}\dot{\mathbf{d}}_B\right) \\ &= {}^{G}\dot{\mathbf{d}}_B - a_Q {}_G\boldsymbol{\omega}_B \times {}^{G}\dot{\mathbf{d}}_B - b_Q {}_G\boldsymbol{\omega}_B \times \left({}_G\boldsymbol{\omega}_B \times {}^{G}\dot{\mathbf{d}}_B\right) \\ &= 0 \end{aligned} \tag{8.310} $$

Now, using the equations

$$ {}_G\boldsymbol{\omega}_B = \omega \hat{K} \tag{8.311} $$

$$ {}_G\boldsymbol{\omega}_B \times \left({}_G\boldsymbol{\omega}_B \times {}^{G}\dot{\mathbf{d}}_B\right) = \left({}_G\boldsymbol{\omega}_B \cdot {}^{G}\dot{\mathbf{d}}_B\right) {}_G\boldsymbol{\omega}_B - \omega^2 {}^{G}\dot{\mathbf{d}}_B \tag{8.312} $$

$$ {}_G\boldsymbol{\omega}_B \cdot {}^{G}\dot{\mathbf{d}}_B = 0 \tag{8.313} $$

we find

$$ \left(1 + b_Q\, \omega^2\right) {}^{G}\dot{\mathbf{d}}_B - a_Q {}_G\boldsymbol{\omega}_B \times {}^{G}\dot{\mathbf{d}}_B = 0 \tag{8.314} $$

Because ${}^{G}\dot{\mathbf{d}}_B$ and ${}_G\boldsymbol{\omega}_B \times {}^{G}\dot{\mathbf{d}}_B$ must be perpendicular, Equation (8.314) provides

$$ 1 + b_Q\, \omega^2 = 0 \tag{8.315} $$

$$ a_Q = 0 \tag{8.316} $$

and therefore,

$$ {}^{G}\mathbf{r}_{Q/o} = \frac{1}{\omega^2} \left({}_G\boldsymbol{\omega}_B \times {}^{G}\dot{\mathbf{d}}_B\right) \tag{8.317} $$

Example 505 ★ Eulerian and Lagrangian Viewpoints When a variable quantity is measured in a stationary global coordinate frame, it is called the absolute or *Lagrangian* viewpoint. However, if the variable is measured in a moving-body coordinate frame, it is called the relative or *Eulerian viewpoint*.

In $2D$ planar motion of a rigid body, there is always a pole of zero velocity at

$$ {}^{G}\mathbf{r}_0 = {}^{G}\mathbf{d}_B - \frac{1}{\omega} {}^{G}\dot{\mathbf{d}}_B \tag{8.318} $$

The position of the pole in the body coordinate frame can be found by substituting ${}^{G}\mathbf{r}_0$ from (8.270),

$$ {}^{G}R_B {}^{B}\mathbf{r}_0 + {}^{G}\mathbf{d}_B = {}^{G}\mathbf{d}_B - {}_G\tilde{\omega}_B^{-1} {}^{G}\dot{\mathbf{d}}_B \tag{8.319} $$

and solving for the position of the zero-velocity point in the body coordinate frame ${}^{B}\mathbf{r}_0$:

$$ \begin{aligned} {}^{B}\mathbf{r}_0 &= -{}^{G}R_B^{\mathrm{T}} {}_G\tilde{\omega}_B^{-1} {}^{G}\dot{\mathbf{d}}_B = -{}^{G}R_B^{\mathrm{T}} \left[{}^{G}\dot{R}_B {}^{G}R_B^{\mathrm{T}}\right]^{-1} {}^{G}\dot{\mathbf{d}}_B \\ &= -{}^{G}R_B^{\mathrm{T}} \left[{}^{G}R_B {}^{G}\dot{R}_B^{-1}\right] {}^{G}\dot{\mathbf{d}}_B = -{}^{G}\dot{R}_B^{-1} {}^{G}\dot{\mathbf{d}}_B \end{aligned} \tag{8.320} $$

Therefore, ${}^{G}\mathbf{r}_0$ indicates the path of motion of the pole in the global frame, while ${}^{B}\mathbf{r}_0$ indicates the same path in the body frame. The ${}^{G}\mathbf{r}_0$ refers to the Lagrangian centroid and ${}^{B}\mathbf{r}_0$ refers to the Eulerian centroid.

Example 506 ★ Screw Axis and Screw Motion The screw axis may be defined as a line for a moving rigid body B whose points P have velocity parallel to the angular velocity vector ${}_{G}\boldsymbol{\omega}_{B} = \omega\hat{u}$. Such points satisfy

$$ {}^{G}\mathbf{v}_{P} = {}_{G}\boldsymbol{\omega}_{B} \times \left({}^{G}\mathbf{r}_{P} - {}^{G}\mathbf{d}_{B}\right) + {}^{G}\dot{\mathbf{d}}_{B} = p\,{}_{G}\boldsymbol{\omega}_{B} \tag{8.321} $$

where p is a scalar. Since ${}_{G}\boldsymbol{\omega}_{B}$ is perpendicular to ${}_{G}\boldsymbol{\omega}_{B} \times ({}^{G}\mathbf{r} - {}^{G}\mathbf{d})$, a dot product of Equation (8.321) by ${}_{G}\boldsymbol{\omega}_{B}$ yields

$$ p = \frac{1}{\omega^2}\left({}_{G}\boldsymbol{\omega}_{B} \cdot {}^{G}\dot{\mathbf{d}}_{B}\right) \tag{8.322} $$

Introducing a parameter m to indicate different points of the line, the equation of the screw axis would be

$$ {}^{G}\mathbf{r}_{P} = {}^{G}\mathbf{d}_{B} + \frac{1}{\omega^2}\left({}_{G}\boldsymbol{\omega}_{B} \times {}^{G}\dot{\mathbf{d}}_{B}\right) + m\,{}_{G}\boldsymbol{\omega}_{B} \tag{8.323} $$

because if we have $\mathbf{a} \times \mathbf{x} = \mathbf{b}$ and $\mathbf{a} \cdot \mathbf{b} = 0$, then $\mathbf{x} = -\mathbf{a}^{-2}(\mathbf{a} \times \mathbf{b}) + m\mathbf{a}$. In our case, ${}_{G}\boldsymbol{\omega}_{B} \times ({}^{G}\mathbf{r}_{P} - {}^{G}\mathbf{d}_{B}) = p\,{}_{G}\boldsymbol{\omega}_{B} - {}^{G}\dot{\mathbf{d}}_{B}$ and $({}^{G}\mathbf{r}_{P} - {}^{G}\mathbf{d}_{B})$ is perpendicular to ${}_{G}\boldsymbol{\omega}_{B} \times ({}^{G}\mathbf{r}_{P} - {}^{G}\mathbf{d}_{B})$ and hence to $(p\,{}_{G}\boldsymbol{\omega}_{B} - {}^{G}\dot{\mathbf{d}}_{B})$.

Therefore, there exists at any time a line s in space parallel to ${}_{G}\boldsymbol{\omega}_{B}$, which is the locus of points whose velocity is parallel to ${}_{G}\boldsymbol{\omega}_{B}$.

If $\mathbf{s}$ is the position vector of a point on s, then

$$ {}_{G}\boldsymbol{\omega}_{B} \times \left({}^{G}\mathbf{s} - {}^{G}\mathbf{d}_{B}\right) = p\,{}_{G}\boldsymbol{\omega}_{B} - {}^{G}\dot{\mathbf{d}}_{B} \tag{8.324} $$

and the velocity of any point out of $\mathbf{s}$ is

$$ {}^{G}\mathbf{v} = {}_{G}\boldsymbol{\omega}_{B} \times \left({}^{G}\mathbf{r} - {}^{G}\mathbf{s}\right) + p\,{}_{G}\boldsymbol{\omega}_{B} \tag{8.325} $$

It expresses that at any time the velocity of a body point can be decomposed into components perpendicular and parallel to the angular velocity vector ${}_{G}\boldsymbol{\omega}_{B}$. Therefore, the motion of any point of a rigid body is a screw. The parameter p is the pitch that determines the ratio of translation velocity to rotation velocity. In general, $\mathbf{s}$, ${}_{G}\boldsymbol{\omega}_{B}$, and p may be functions of time.

8.4 VELOCITY TRANSFORMATION MATRIX

Consider the motion of a rigid body B in the global coordinate frame G, as shown in Figure 8.7. Assume the body frame $B(Oxyz)$ is coincident at some initial time $t = t_0$ with the global frame $G(OXYZ)$. At any time $t \neq t_0$, B is not necessarily coincident with G, and therefore the homogeneous transformation matrix ${}^G T_B(t)$ between B and G is time varying.

The global position vector ${}^G\mathbf{r}_P(t)$ of a body point P is a function of time, but its local position vector ${}^B\mathbf{r}_P$ is constant and is equal to ${}^G\mathbf{r}_P(t_0)$:

$$ {}^B\mathbf{r}_P \equiv {}^G\mathbf{r}_P(t_0) \tag{8.326} $$

$$ {}^G\mathbf{r}_P(t) = {}^G R_B(t)\, {}^B\mathbf{r}_P + {}^G\mathbf{d}_B(t) \tag{8.327} $$

$$ {}^G\mathbf{r}(t) = {}^G T_B(t)\, {}^B\mathbf{r} = \begin{bmatrix} {}^G R_B(t) & {}^G\mathbf{d}(t) \\ 0 & 1 \end{bmatrix} {}^B\mathbf{r} \tag{8.328} $$

The G-velocity of P is obtained by the G-derivative of the position vector ${}^G\mathbf{r}(t)$:

$$ {}^G\mathbf{v}_P = \frac{d}{dt}\, {}^G\mathbf{r}_P(t) = {}^G\dot{\mathbf{r}}_P \tag{8.329} $$

The G-velocity of a body point may be found by applying a homogeneous *velocity transformation matrix* ${}^G V_B$,

$$ {}^G\mathbf{v}(t) = {}^G V_B\, {}^G\mathbf{r}(t) \tag{8.330} $$

where the velocity matrix is

$$ {}^G V_B = {}^G\dot{T}_B\, {}^G T_B^{-1} \tag{8.331} $$

$$ = \begin{bmatrix} {}^G\dot{R}_B\, {}^G R_B^{\mathrm{T}} & {}^G\dot{\mathbf{d}}_B - {}^G\dot{R}_B\, {}^G R_B^{\mathrm{T}}\, {}^G\mathbf{d}_B \\ 0 & 0 \end{bmatrix} \tag{8.332} $$

$$ = \begin{bmatrix} {}_G\tilde{\omega}_B & {}^G\dot{\mathbf{d}}_B - {}_G\tilde{\omega}_B\, {}^G\mathbf{d}_B \\ 0 & 0 \end{bmatrix} \tag{8.333} $$

The inverse of the velocity transformation matrix is

$$ {}^B V_G = {}^G V_B^{-1} = \begin{bmatrix} {}^G R_B\, {}^G\dot{R}_B^{-1} & -{}^G R_B\, {}^G\dot{R}_B^{-1}\, {}^G\dot{\mathbf{d}} + {}^G\mathbf{d} \\ 0 & 0 \end{bmatrix} \tag{8.334} $$

where

$$ {}^G V_B\, {}^G V_B^{-1} = \mathbf{I} \tag{8.335} $$

$$ {}^G\mathbf{r}_P = {}^G V_B^{-1}\, {}^G\mathbf{v}_P \tag{8.336} $$

The B-expression of the velocity transformation matrix is

$$ {}^B_G V_B = {}^G T_B^{-1}\, {}^G\dot{T}_B = \begin{bmatrix} {}^B_G\tilde{\omega}_B & {}^B\dot{\mathbf{d}} \\ 0 & 0 \end{bmatrix} \tag{8.337} $$

and the transformation of ${}^{G}V_{B}$ to ${}^{B}_{G}V_{B}$ and the expression of ${}^{G}V_{1}$ in another coordinate frame B_2 are

$$
{}^{G}V_{B} = {}^{G}T_{B}\,{}^{B}_{G}V_{B}\,{}^{G}T_{B}^{-1} \tag{8.338}
$$

$$
{}^{B}_{G}V_{B} = {}^{G}T_{B}^{-1}\,{}^{G}V_{B}\,{}^{G}T_{B} = {}^{B}T_{G}\,{}^{G}V_{B}\,{}^{B}T_{G}^{-1} \tag{8.339}
$$

$$
{}^{2}_{G}V_{1} = {}^{2}T_{G}\,{}^{G}V_{1}\,{}^{2}T_{G}^{-1} \tag{8.340}
$$

We can add the velocity transformation matrices when all of the matrices are expressed in the same coordinate frame:

$$
{}^{0}V_{2} = {}^{0}V_{1} + {}^{0}_{1}V_{2} + {}^{0}_{2}V_{3} + \cdots \tag{8.341}
$$

The velocity transformation matrix ${}^{G}V_{B}$ can also be rearranged and shown with a Plücker vector ${}_{G}\mathbf{t}_{B}$:

$$
{}_{G}\mathbf{t}_{B} = \begin{bmatrix} {}^{G}\mathbf{v}_{B} \\ {}_{G}\boldsymbol{\omega}_{B} \end{bmatrix} = \begin{bmatrix} {}^{G}\dot{\mathbf{d}}_{B} - {}_{G}\tilde{\omega}_{B}\,{}^{G}\mathbf{d}_{B} \\ {}_{G}\boldsymbol{\omega}_{B} \end{bmatrix} \tag{8.342}
$$

The Plücker vector ${}_{G}\mathbf{t}_{B}$ is a 6×1 vector called a *velocity transformation vector*, which indicates the translational and rotational velocities of a body point.

Proof: Based on a homogeneous coordinate transformation, we have

$$
{}^{G}\mathbf{r}_{P}(t) = {}^{G}T_{B}(t)\,{}^{B}\mathbf{r}_{P} = {}^{G}T_{B}(t)\,{}^{G}\mathbf{r}_{P}(t_0) \tag{8.343}
$$

and therefore,

$$
\begin{aligned}
{}^{G}\mathbf{v}_{P} &= \frac{{}^{G}d}{dt}\left[{}^{G}T_{B}\,{}^{B}\mathbf{r}_{P}\right] = {}^{G}\dot{T}_{B}\,{}^{B}\mathbf{r}_{P} \\
&= \begin{bmatrix} \frac{{}^{G}d}{dt}{}^{G}R_{B} & \frac{{}^{G}d}{dt}{}^{G}\mathbf{d}_{B} \\ 0 & 0 \end{bmatrix} {}^{B}\mathbf{r}_{P} = \begin{bmatrix} {}^{G}\dot{R}_{B} & {}^{G}\dot{\mathbf{d}}_{B} \\ 0 & 0 \end{bmatrix} {}^{B}\mathbf{r}_{P}
\end{aligned} \tag{8.344}
$$

Substituting for ${}^{B}\mathbf{r}_{P}$ from Equation (8.343) gives

$$
\begin{aligned}
{}^{G}\mathbf{v}_{P} &= {}^{G}\dot{T}_{B}\,{}^{G}T_{B}^{-1}\,{}^{G}\mathbf{r}_{P}(t) \\
&= \begin{bmatrix} {}^{G}\dot{R}_{B} & {}^{G}\dot{\mathbf{d}}_{B} \\ 0 & 0 \end{bmatrix} \begin{bmatrix} {}^{G}R_{B}^{\mathrm{T}} & -{}^{G}R_{B}^{\mathrm{T}}\,{}^{G}\mathbf{d}_{B} \\ 0 & 1 \end{bmatrix} {}^{G}\mathbf{r}_{P}(t) \\
&= \begin{bmatrix} {}^{G}\dot{R}_{B}\,{}^{G}R_{B}^{\mathrm{T}} & {}^{G}\dot{\mathbf{d}}_{B} - {}^{G}\dot{R}_{B}\,{}^{G}R_{B}^{\mathrm{T}}\,{}^{G}\mathbf{d}_{B} \\ 0 & 0 \end{bmatrix} {}^{G}\mathbf{r}_{P}(t) \\
&= \begin{bmatrix} {}_{G}\tilde{\omega}_{B} & {}^{G}\dot{\mathbf{d}}_{B} - {}_{G}\tilde{\omega}_{B}\,{}^{G}\mathbf{d}_{B} \\ 0 & 0 \end{bmatrix} {}^{G}\mathbf{r}_{P}(t)
\end{aligned} \tag{8.345}
$$

Therefore, the G-velocity of any point P of the rigid body B can be obtained by premultiplying the position vector of P in G with the *velocity transformation matrix* ${}^{G}V_{B}$:

$$
{}^{G}\mathbf{v}_{P}(t) = {}^{G}V_{B}\,{}^{G}\mathbf{r}_{P}(t) \tag{8.346}
$$

where

$$
\begin{aligned}
{}^{G}V_{B} &= {}^{G}\dot{T}_{B}\,{}^{G}T_{B}^{-1} \\
&= \begin{bmatrix} {}_{G}\tilde{\omega}_{B} & {}^{G}\dot{\mathbf{d}}_{B} - {}_{G}\tilde{\omega}_{B}\,{}^{G}\mathbf{d}_{B} \\ 0 & 0 \end{bmatrix} = \begin{bmatrix} {}_{G}\tilde{\omega}_{B} & {}^{G}\mathbf{v}_{B} \\ 0 & 0 \end{bmatrix}
\end{aligned}
\tag{8.347}
$$

and

$$
{}_{G}\tilde{\omega}_{B} = {}^{G}\dot{R}_{B}\,{}^{G}R_{B}^{\mathrm{T}} \tag{8.348}
$$

$$
\begin{aligned}
{}^{G}\mathbf{v}_{B} &= {}^{G}\dot{\mathbf{d}}_{B} - {}^{G}\dot{R}_{B}\,{}^{G}R_{B}^{\mathrm{T}}\,{}^{G}\mathbf{d}_{B} = {}^{G}\dot{\mathbf{d}}_{B} - {}_{G}\tilde{\omega}_{B}\,{}^{G}\mathbf{d}_{B} \\
&= {}^{G}\dot{\mathbf{d}}_{B} - {}_{G}\boldsymbol{\omega}_{B} \times {}^{G}\mathbf{d}_{B}
\end{aligned}
\tag{8.349}
$$

The *velocity transformation matrix* ${}^{G}V_{B}$ may be assumed as a matrix operator that provides the global velocity of any point attached to $B(oxyz)$. It consists of the angular velocity matrix ${}_{G}\tilde{\omega}_{B}$ and the frame velocity ${}^{G}\dot{\mathbf{d}}_{B}$, both expressed in $G(OXYZ)$. The velocity matrix ${}^{G}V_{B}$ depends on six parameters: the three components of the angular velocity vector ${}_{G}\boldsymbol{\omega}_{B}$ and the three components of the frame velocity ${}^{G}\dot{\mathbf{d}}_{B}$.

Following the rule of inverse of homogeneous transformation matrices (6.88),

$$
{}^{G}T_{B}^{-1} = \begin{bmatrix} {}^{G}R_{B}^{\mathrm{T}} & -{}^{G}R_{B}^{\mathrm{T}}\,{}^{G}\mathbf{d} \\ 0 & 1 \end{bmatrix} \tag{8.350}
$$

we can introduce the inverse-velocity transformation matrix by

$$
\begin{aligned}
{}^{B}V_{G} &= {}^{G}V_{B}^{-1} \\
&= \begin{bmatrix} \left({}^{G}\dot{R}_{B}\,{}^{G}R_{B}^{\mathrm{T}}\right)^{-1} & -\left({}^{G}\dot{R}_{B}\,{}^{G}R_{B}^{\mathrm{T}}\right)^{-1}\left({}^{G}\dot{\mathbf{d}} - {}^{G}\dot{R}_{B}\,{}^{G}R_{B}^{\mathrm{T}}\,{}^{G}\mathbf{d}\right) \\ 0 & 0 \end{bmatrix} \\
&= \begin{bmatrix} {}^{G}R_{B}\,{}^{G}\dot{R}_{B}^{-1} & -{}^{G}R_{B}\,{}^{G}\dot{R}_{B}^{-1}\left({}^{G}\dot{\mathbf{d}} - {}^{G}\dot{R}_{B}\,{}^{G}R_{B}^{\mathrm{T}}\,{}^{G}\mathbf{d}\right) \\ 0 & 0 \end{bmatrix} \\
&= \begin{bmatrix} {}^{G}R_{B}\,{}^{G}\dot{R}_{B}^{-1} & -{}^{G}R_{B}\,{}^{G}\dot{R}_{B}^{-1}\,{}^{G}\dot{\mathbf{d}} + {}^{G}\mathbf{d} \\ 0 & 0 \end{bmatrix}
\end{aligned}
\tag{8.351}
$$

to get

$$
{}^{G}V_{B}\,{}^{G}V_{B}^{-1} = \mathbf{I} \tag{8.352}
$$

Therefore, having the velocity vector of a body point ${}^{G}\mathbf{v}_{P}$ and the velocity transformation matrix ${}^{G}V_{B}$, we can find the global position of the point by

$$
{}^{G}\mathbf{r}_{P} = {}^{G}V_{B}^{-1}\,{}^{G}\mathbf{v}_{P} \tag{8.353}
$$

The velocity transformation matrix can be expressed in the body coordinate frame B as

$$
\begin{aligned}
{}_{G}^{B}V_{B} &= {}^{G}T_{B}^{-1}\,{}^{G}\dot{T}_{B} \\
&= \begin{bmatrix} {}^{G}R_{B}^{\mathrm{T}} & -{}^{G}R_{B}^{\mathrm{T}}\,{}^{G}\mathbf{d} \\ 0 & 1 \end{bmatrix} \begin{bmatrix} {}^{G}\dot{R}_{B} & {}^{G}\dot{\mathbf{d}} \\ 0 & 0 \end{bmatrix} \\
&= \begin{bmatrix} {}^{G}R_{B}^{\mathrm{T}}\,{}^{G}\dot{R}_{B} & {}^{G}R_{B}^{\mathrm{T}}\,{}^{G}\dot{\mathbf{d}} \\ 0 & 0 \end{bmatrix} = \begin{bmatrix} {}_{G}^{B}\tilde{\omega}_{B} & {}^{B}\dot{\mathbf{d}} \\ 0 & 0 \end{bmatrix}
\end{aligned}
\tag{8.354}
$$

where ${}^{B}_{G}\tilde{\omega}_{B}$ is the angular velocity matrix of B with respect to G expressed in B and ${}^{B}\dot{\mathbf{d}}$ is the velocity of the origin of B in G expressed in B. Using the definition of (8.331) and (8.354), we are able to transform the velocity transformation matrices between the B- and G-frames:

$$
{}^{G}V_{B} = {}^{G}T_{B}\,{}^{B}_{G}V_{B}\,{}^{G}T_{B}^{-1} \tag{8.355}
$$

It can be done backward and also expanded to include a third frame:

$$
{}^{B}_{G}V_{B} = {}^{B}T_{G}\,{}^{G}V_{B}\,{}^{B}T_{G}^{-1} \tag{8.356}
$$

$$
{}^{2}_{G}V_{1} = {}^{2}T_{G}\,{}^{G}V_{1}\,{}^{2}T_{G}^{-1} \tag{8.357}
$$

If we have multibodies, we may need to determine the velocity transformation matrices of every body relatively and individually in every coordinate frame. Let us consider a global frame G and two frames B_1 and B_2. Assume that we can determine the kinematics of B_1 relative to G and the kinematics of B_2 relative to B_1. The point P is a point in B_2. Because B_2 is moving relative to B_1, the velocities of P in B_1 and B_0 are

$$
{}^{1}\mathbf{v}_{P} = {}^{1}V_{2}\,{}^{1}\mathbf{r}_{P} \tag{8.358}
$$

$$
{}^{1}\mathbf{r}_{P} = {}^{1}\mathbf{d}_{2} + {}^{1}_{2}\mathbf{r}_{P} \tag{8.359}
$$

$$
{}^{0}_{1}\mathbf{v}_{P} = {}^{0}_{1}V_{2}\,{}^{0}_{1}\mathbf{r}_{P} \tag{8.360}
$$

$$
{}^{0}_{1}\mathbf{r}_{P} = {}^{0}_{1}\mathbf{d}_{2} + {}^{0}_{2}\mathbf{r}_{P} \tag{8.361}
$$

The velocity of point P in the global frame G is a combination of the velocity of B_2 relative to B_1 plus the velocity of B_1 relative to G. Using the *relative-velocity* principle, the velocity combination is allowed only if all the velocity vectors are expressed in the same frame. The global frame is usually the best choice:

$$
\begin{aligned}
{}^{0}\mathbf{v} &= {}^{0}\mathbf{v}_{1} + {}^{0}_{1}\mathbf{v}_{2} = {}^{0}V_{1}\,{}^{0}\mathbf{d}_{1} + {}^{0}_{1}V_{2}\,({}^{0}_{1}\mathbf{d}_{2} + {}^{0}_{2}\mathbf{r}_{P}) \\
&= ({}^{0}V_{1} + {}^{0}_{1}V_{2})({}^{0}\mathbf{d}_{1} + {}^{0}_{1}\mathbf{d}_{2} + {}^{0}_{2}\mathbf{r}_{P}) = {}^{0}V_{2}\,{}^{0}_{2}\mathbf{r}_{P}
\end{aligned} \tag{8.362a}
$$

$$
{}^{0}V_{2} = {}^{0}V_{1} + {}^{0}_{1}V_{2} \tag{8.362b}
$$

$$
{}^{0}\mathbf{r}_{P} = {}^{0}\mathbf{d}_{1} + {}^{0}_{1}\mathbf{d}_{2} + {}^{0}_{2}\mathbf{r}_{P} \tag{8.362c}
$$

The Plücker coordinates of velocity of a body point is a 6×1 vector called the *velocity transformation vector*:

$$
{}_{G}\mathbf{t}_{B} = \begin{bmatrix} {}^{G}\mathbf{v}_{B} \\ {}_{G}\boldsymbol{\omega}_{B} \end{bmatrix} = \begin{bmatrix} {}^{G}\dot{\mathbf{d}}_{B} - {}_{G}\tilde{\omega}_{B}\,{}^{G}\mathbf{d}_{B} \\ {}_{G}\boldsymbol{\omega}_{B} \end{bmatrix} \tag{8.363}
$$

In analogy to the two representations of the angular velocity, the G-velocity of point of B can be represented either as the velocity transformation matrix ${}^{G}V_{B}$ in (8.347) or as the velocity transformation vector ${}_{G}\mathbf{t}_{B}$ in (8.363). The velocity transformation vector represents a noncommensurate vector because the dimensions of $_G\boldsymbol{\omega}_B$ and ${}^G\mathbf{v}_B$ are not the same. ■

Example 507 Velocity Transformation Matrix Based on Coordinate Transformation Matrix The velocity transformation matrix can be found based on a coordinate transformation matrix. Starting from

$$
{}^G\mathbf{r}(t) = {}^GT_B{}^B\mathbf{r} = \begin{bmatrix} {}^GR_B & {}^G\mathbf{d} \\ 0 & 1 \end{bmatrix} {}^B\mathbf{r} \tag{8.364}
$$

and taking the derivative show that

$$
{}^G\mathbf{v} = \frac{{}^Gd}{dt}\left[{}^GT_B{}^B\mathbf{r}\right] = {}^G\dot{T}_B{}^B\mathbf{r} = \begin{bmatrix} {}^G\dot{R}_B & {}^G\dot{\mathbf{d}} \\ 0 & 0 \end{bmatrix} {}^B\mathbf{r} \tag{8.365}
$$

However,

$$
{}^B\mathbf{r} = {}^GT_B^{-1}{}^G\mathbf{r} \tag{8.366}
$$

and therefore,

$$
\begin{aligned}
{}^G\mathbf{v} &= \begin{bmatrix} {}^G\dot{R}_B & {}^G\dot{\mathbf{d}} \\ 0 & 0 \end{bmatrix} {}^GT_B^{-1}{}^G\mathbf{r} \\
&= \begin{bmatrix} {}^G\dot{R}_B & {}^G\dot{\mathbf{d}} \\ 0 & 0 \end{bmatrix} \begin{bmatrix} {}^GR_B^{\mathrm{T}} & -{}^GR_B^{\mathrm{T}}{}^G\mathbf{d} \\ 0 & 1 \end{bmatrix} {}^G\mathbf{r} \\
&= \begin{bmatrix} {}^G\dot{R}_B{}^GR_B^{\mathrm{T}} & {}^G\dot{\mathbf{d}} - {}^G\dot{R}_B{}^GR_B^{\mathrm{T}}{}^G\mathbf{d} \\ 0 & 0 \end{bmatrix} {}^G\mathbf{r} \\
&= {}^GV_B{}^G\mathbf{r}
\end{aligned} \tag{8.367}
$$

Example 508 Velocity Transformation Matrix in Body Frames The velocity transformation matrix is given by

$$
{}^GV_B = \begin{bmatrix} {}^G\dot{R}_B{}^GR_B^{\mathrm{T}} & {}^G\dot{\mathbf{d}} - {}^G\dot{R}_B{}^GR_B^{\mathrm{T}}{}^G\mathbf{d} \\ 0 & 0 \end{bmatrix} \tag{8.368}
$$

However, the velocity transformation matrix can be expressed in the body coordinate frame B:

$$
{}^B_GV_B = {}^GT_B^{-1}{}^G\dot{T}_B = \begin{bmatrix} {}^B_G\tilde{\omega}_B & {}^B\dot{\mathbf{d}} \\ 0 & 0 \end{bmatrix} \tag{8.369}
$$

where ${}^B_G\tilde{\omega}_B$ is the B-expression of $_G\tilde{\omega}_B$ and ${}^B\dot{\mathbf{d}}$ is the B-velocity of B in G.

It is also possible to use a matrix multiplication to find the velocity transformation matrix in the body coordinate frame:

$$
{}^B_G\mathbf{v}_P = {}^GT_B^{-1}{}^G\mathbf{v}_P = {}^GT_B^{-1}{}^G\dot{T}_B{}^B\mathbf{r}_P = {}^B_GV_B{}^B\mathbf{r}_P \tag{8.370}
$$

We are able to transform the velocity transformation matrices between the B and G frames:

$$ {}^{G}V_{B} = {}^{G}T_{B}\,{}^{B}_{G}V_{B}\,{}^{G}T_{B}^{-1} \tag{8.371} $$

It can also be useful if we define the time derivative of the transformation matrix by

$$ {}^{G}\dot{T}_{B} = {}^{G}V_{B}\,{}^{G}T_{B} \tag{8.372} $$

or

$$ {}^{G}\dot{T}_{B} = {}^{G}T_{B}\,{}^{B}_{G}V_{B} \tag{8.373} $$

Similarly, we may define a velocity transformation matrix from link (i) to link ($i-1$) of a connected multibody by

$$ {}^{i-1}V_{i} = \begin{bmatrix} {}^{i-1}\dot{R}_{i}\,{}^{i-1}R_{i}^{\mathrm{T}} & {}^{i-1}\dot{\mathbf{d}} - {}^{i-1}\dot{R}_{i}\,{}^{i-1}R_{i}^{\mathrm{T}\,i-1}\mathbf{d} \\ 0 & 0 \end{bmatrix} \tag{8.374} $$

and

$$ {}^{i}_{i-1}V_{i} = \begin{bmatrix} {}^{i-1}R_{i}^{\mathrm{T}\,i-1}\dot{R}_{i} & {}^{i-1}R_{i}^{\mathrm{T}\,i-1}\dot{\mathbf{d}} \\ 0 & 0 \end{bmatrix} \tag{8.375} $$

Example 509 Motion with a Fixed Point When a point of a rigid body is fixed to the global frame, it is convenient to set the origins of the moving coordinate frame $B(Oxyz)$ and the global coordinate frame $G(OXYZ)$ on the fixed point. Under these conditions,

$$ {}^{G}\mathbf{d}_{B} = 0 \qquad {}^{G}\dot{\mathbf{d}}_{B} = 0 \tag{8.376} $$

and Equation (8.271) reduces to

$$ {}^{G}\mathbf{v}_{P} = {}_{G}\tilde{\omega}_{B}\,{}^{G}\mathbf{r}_{P}(t) = {}_{G}\boldsymbol{\omega}_{B} \times {}^{G}\mathbf{r}_{P}(t) \tag{8.377} $$

Example 510 Velocity in Spherical Coordinates The homogeneous transformation matrix from spherical coordinates $S(Or\theta\varphi)$ to Cartesian coordinates $G(OXYZ)$ can be found by a translation $D_{Z,r}$ and then a rotation $R_{Y,\theta}$ followed by a rotation $R_{Z,\varphi}$:

$$ {}^{G}T_{S} = R_{Z,\varphi}\,R_{Y,\theta}\,D_{Z,r} = \begin{bmatrix} {}^{G}R_{B} & {}^{G}\mathbf{d} \\ 0 & 1 \end{bmatrix} $$

$$ = \begin{bmatrix} \cos\theta\cos\varphi & -\sin\varphi & \cos\varphi\sin\theta & r\cos\varphi\sin\theta \\ \cos\theta\sin\varphi & \cos\varphi & \sin\theta\sin\varphi & r\sin\theta\sin\varphi \\ -\sin\theta & 0 & \cos\theta & r\cos\theta \\ 0 & 0 & 0 & 1 \end{bmatrix} \tag{8.378} $$

The time derivative of ${}^G T_S$ shows that

$$
\begin{aligned}
{}^G\dot{T}_S &= {}^G V_S {}^G T_S \\
&= \begin{bmatrix} {}^G\tilde{\omega}_S & {}^G\mathbf{v}_S \\ 0 & 0 \end{bmatrix} {}^G T_S \\
&= \begin{bmatrix} 0 & -\dot{\varphi} & \dot{\theta}\cos\varphi & \dot{r}\cos\varphi\sin\theta \\ \dot{\varphi} & 0 & \dot{\theta}\sin\varphi & \dot{r}\sin\theta\sin\varphi \\ -\dot{\theta}\cos\varphi & -\dot{\theta}\sin\varphi & 0 & \dot{r}\cos\theta \\ 0 & 0 & 0 & 0 \end{bmatrix} {}^G T_B
\end{aligned}
\tag{8.379}
$$

Example 511 ★ Velocity of the Gripper of a Planar R‖R Manipulator Figure 8.15 illustrates an R‖R planar manipulator with joint variables θ_1 and θ_2. Links (1) and (2) are both R‖R(0), and therefore the transformation matrices ${}^0T_1$, ${}^1T_2$, and ${}^0T_2$ are

$$
{}^0T_1 = \begin{bmatrix} \cos\theta_1 & -\sin\theta_1 & 0 & l_1\cos\theta_1 \\ \sin\theta_1 & \cos\theta_1 & 0 & l_1\sin\theta_1 \\ 0 & 0 & 1 & 0 \\ 0 & 0 & 0 & 1 \end{bmatrix} \tag{8.380}
$$

$$
{}^1T_2 = \begin{bmatrix} \cos\theta_2 & -\sin\theta_2 & 0 & l_2\cos\theta_2 \\ \sin\theta_2 & \cos\theta_2 & 0 & l_2\sin\theta_2 \\ 0 & 0 & 1 & 0 \\ 0 & 0 & 0 & 1 \end{bmatrix} \tag{8.381}
$$

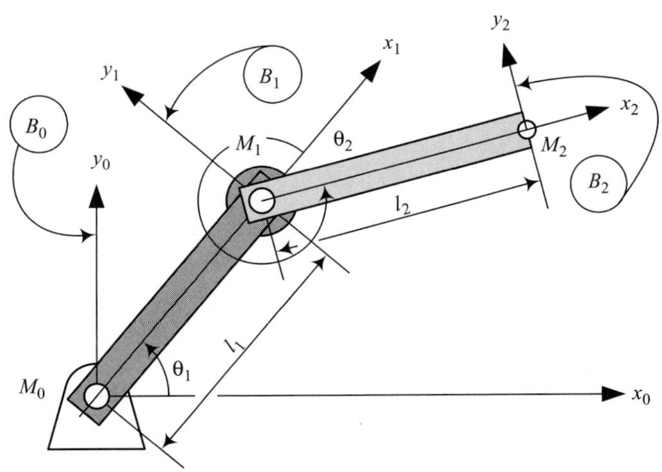

Figure 8.15 A 2R or R‖R planar manipulator.

$$
\begin{aligned}
{}^0T_2 &= {}^0T_1\,{}^1T_2 \\
&= \begin{bmatrix} c(\theta_1+\theta_2) & -s(\theta_1+\theta_2) & 0 & l_2 c(\theta_1+\theta_2) + l_1 c\theta_1 \\ s(\theta_1+\theta_2) & c(\theta_1+\theta_2) & 0 & l_2 s(\theta_1+\theta_2) + l_1 s\theta_1 \\ 0 & 0 & 1 & 0 \\ 0 & 0 & 0 & 1 \end{bmatrix}
\end{aligned}
\tag{8.382}
$$

The points M_1 and M_2 are at

$$
{}^0\mathbf{r}_{M_1} = \begin{bmatrix} l_1\cos\theta_1 \\ l_1\sin\theta_1 \\ 0 \\ 1 \end{bmatrix} \qquad {}^1\mathbf{r}_{M_2} = \begin{bmatrix} l_2\cos\theta_2 \\ l_2\sin\theta_2 \\ 0 \\ 1 \end{bmatrix}
\tag{8.383}
$$

$$
{}^0\mathbf{r}_{M_2} = {}^0T_1\,{}^1\mathbf{r}_{M_2} = \begin{bmatrix} l_2\cos(\theta_1+\theta_2) + l_1\cos\theta_1 \\ l_2\sin(\theta_1+\theta_2) + l_1\sin\theta_1 \\ 0 \\ 1 \end{bmatrix}
\tag{8.384}
$$

To determine the velocity of M_2, we need to calculate ${}^0\dot{T}_2$. However, ${}^0\dot{T}_2$ can be calculated by direct differentiation of ${}^0T_2$,

$$
\begin{aligned}
{}^0\dot{T}_2 &= \frac{d}{dt}\,{}^0T_2 \\
&= \begin{bmatrix} -\dot{\theta}_{12}s\theta_{12} & -\dot{\theta}_{12}c\theta_{12} & 0 & -l_2\dot{\theta}_{12}s\theta_{12} - \dot{\theta}_1 l_1 s\theta_1 \\ \dot{\theta}_{12}c\theta_{12} & -\dot{\theta}_{12}s\theta_{12} & 0 & l_2\dot{\theta}_{12}c\theta_{12} + \dot{\theta}_1 l_1 c\theta_1 \\ 0 & 0 & 0 & 0 \\ 0 & 0 & 0 & 0 \end{bmatrix}
\end{aligned}
\tag{8.385}
$$

$$
\theta_{12} = \theta_1 + \theta_2 \qquad \dot{\theta}_{12} = \dot{\theta}_1 + \dot{\theta}_2
\tag{8.386}
$$

or by the chain rule,

$$
{}^0\dot{T}_2 = \frac{d}{dt}\left({}^0T_1\,{}^1T_2\right) = {}^0\dot{T}_1\,{}^1T_2 + {}^0T_1\,{}^1\dot{T}_2
\tag{8.387}
$$

where

$$
{}^0\dot{T}_1 = \dot{\theta}_1 \begin{bmatrix} -\sin\theta_1 & -\cos\theta_1 & 0 & -l_1\sin\theta_1 \\ \cos\theta_1 & -\sin\theta_1 & 0 & l_1\cos\theta_1 \\ 0 & 0 & 0 & 0 \\ 0 & 0 & 0 & 0 \end{bmatrix}
\tag{8.388}
$$

$$
{}^1\dot{T}_2 = \dot{\theta}_2 \begin{bmatrix} -\sin\theta_2 & -\cos\theta_2 & 0 & -l_2\sin\theta_2 \\ \cos\theta_2 & -\sin\theta_2 & 0 & l_2\cos\theta_2 \\ 0 & 0 & 0 & 0 \\ 0 & 0 & 0 & 0 \end{bmatrix}
\tag{8.389}
$$

Having ${}^{0}\dot{T}_{1}$ and ${}^{1}\dot{T}_{2}$, we can find the velocity transformation matrices ${}^{0}V_{1}$ and ${}^{1}V_{2}$ by using ${}^{0}T_{1}^{-1}$ and ${}^{1}T_{2}^{-1}$:

$$
{}^{0}T_{1}^{-1} = \begin{bmatrix} \cos\theta_1 & \sin\theta_1 & 0 & -l_1 \\ -\sin\theta_1 & \cos\theta_1 & 0 & 0 \\ 0 & 0 & 1 & 0 \\ 0 & 0 & 0 & 1 \end{bmatrix} \tag{8.390}
$$

$$
{}^{1}T_{2}^{-1} = \begin{bmatrix} \cos\theta_2 & \sin\theta_2 & 0 & -l_2 \\ -\sin\theta_2 & \cos\theta_2 & 0 & 0 \\ 0 & 0 & 1 & 0 \\ 0 & 0 & 0 & 1 \end{bmatrix} \tag{8.391}
$$

$$
{}^{0}V_{1} = {}^{0}\dot{T}_{1}\,{}^{0}T_{1}^{-1} = \dot{\theta}_1 \begin{bmatrix} 0 & -1 & 0 & 0 \\ 1 & 0 & 0 & 0 \\ 0 & 0 & 0 & 0 \\ 0 & 0 & 0 & 0 \end{bmatrix} \tag{8.392}
$$

$$
{}^{1}V_{2} = {}^{1}\dot{T}_{2}\,{}^{1}T_{2}^{-1} = \dot{\theta}_2 \begin{bmatrix} 0 & -1 & 0 & 0 \\ 1 & 0 & 0 & 0 \\ 0 & 0 & 0 & 0 \\ 0 & 0 & 0 & 0 \end{bmatrix} \tag{8.393}
$$

Now, we can determine the velocity of points M_1 and M_2 in B_0 and B_1, respectively:

$$
{}^{0}\mathbf{v}_{M_1} = {}^{0}V_{1}\,{}^{0}\mathbf{r}_{M_1} = \dot{\theta}_1 \begin{bmatrix} -l_1 \sin\theta_1 \\ l_1 \cos\theta_1 \\ 0 \\ 0 \end{bmatrix} \tag{8.394}
$$

$$
{}^{1}\mathbf{v}_{M_2} = {}^{1}V_{2}\,{}^{1}\mathbf{r}_{M_2} = \dot{\theta}_2 \begin{bmatrix} -l_2 \sin\theta_2 \\ l_2 \cos\theta_2 \\ 0 \\ 0 \end{bmatrix} \tag{8.395}
$$

To determine the velocity of the tip point M_2 in the base frame, we can use velocity vector addition:

$$
\begin{aligned}
{}^{0}\mathbf{v}_{M_2} &= {}^{0}\mathbf{v}_{M_1} + {}^{0}_{1}\mathbf{v}_{M_2} = {}^{0}\mathbf{v}_{M_1} + {}^{0}T_{1}\,{}^{1}\mathbf{v}_{M_2} \\
&= \begin{bmatrix} -\left(\dot{\theta}_1 + \dot{\theta}_2\right) l_2 \sin(\theta_1 + \theta_2) - \dot{\theta}_1 l_1 \sin\theta_1 \\ \left(\dot{\theta}_1 + \dot{\theta}_2\right) l_2 \cos(\theta_1 + \theta_2) + \dot{\theta}_1 l_1 \cos\theta_1 \\ 0 \\ 0 \end{bmatrix}
\end{aligned} \tag{8.396}
$$

We can also determine ${}^{0}\mathbf{v}_{M_2}$ by using the velocity transformation matrix ${}^{0}V_{2}$,

$$
{}^{0}\mathbf{v}_{M_2} = {}^{0}V_{2}\,{}^{0}\mathbf{r}_{M_2}
$$

where the velocity transformation matrix is

$$ {}^0V_2 = {}^0\dot{T}_2 {}^0T_2^{-1} = \begin{bmatrix} 0 & -\dot{\theta}_1 - \dot{\theta}_2 & 0 & \dot{\theta}_2 l_1 \sin\theta_1 \\ \dot{\theta}_1 + \dot{\theta}_2 & 0 & 0 & -\dot{\theta}_2 l_1 \cos\theta_1 \\ 0 & 0 & 0 & 0 \\ 0 & 0 & 0 & 0 \end{bmatrix} \tag{8.397} $$

where

$$ \begin{aligned} {}^0T_2^{-1} &= {}^2T_1 {}^1T_0 = {}^1T_2^{-1} {}^0T_1^{-1} \\ &= \begin{bmatrix} \cos(\theta_1+\theta_2) & \sin(\theta_1+\theta_2) & 0 & -l_2 - l_1\cos\theta_2 \\ -\sin(\theta_1+\theta_2) & \cos(\theta_1+\theta_2) & 0 & l_1 \sin\theta_2 \\ 0 & 0 & 1 & 0 \\ 0 & 0 & 0 & 1 \end{bmatrix} \end{aligned} \tag{8.398} $$

We can also determine the velocity transformation matrix ${}^0V_2$ using the addition rule:

$$ \begin{aligned} {}^0V_2 &= {}^0V_1 + {}_1^0V_2 \\ &= \begin{bmatrix} 0 & -\dot{\theta}_1 - \dot{\theta}_2 & 0 & \dot{\theta}_2 l_1 \sin\theta_1 \\ \dot{\theta}_1 + \dot{\theta}_2 & 0 & 0 & -\dot{\theta}_2 l_1 \cos\theta_1 \\ 0 & 0 & 0 & 0 \\ 0 & 0 & 0 & 0 \end{bmatrix} \end{aligned} \tag{8.399} $$

where

$$ {}_1^0V_2 = {}^0T_1 {}^1V_2 {}^0T_1^{-1} = \begin{bmatrix} 0 & -\dot{\theta}_2 & 0 & \dot{\theta}_2 l_1 \sin\theta_1 \\ \dot{\theta}_2 & 0 & 0 & -\dot{\theta}_2 l_1 \cos\theta_1 \\ 0 & 0 & 0 & 0 \\ 0 & 0 & 0 & 0 \end{bmatrix} \tag{8.400} $$

Therefore,

$$ \begin{aligned} {}^0\mathbf{v}_{M_2} &= {}^0V_2 {}^0\mathbf{r}_{M_2} \\ &= \begin{bmatrix} -\left(\dot{\theta}_1 + \dot{\theta}_2\right) l_2 \sin(\theta_1+\theta_2) - \dot{\theta}_1 l_1 \sin\theta_1 \\ \left(\dot{\theta}_1 + \dot{\theta}_2\right) l_2 \cos(\theta_1+\theta_2) + \dot{\theta}_1 l_1 \cos\theta_1 \\ 0 \\ 0 \end{bmatrix} \end{aligned} \tag{8.401} $$

8.5 ★ DERIVATIVE OF A HOMOGENEOUS TRANSFORMATION MATRIX

The velocity transformation matrix can be found directly from the homogeneous link transformation matrix. Consider a 4 × 4 homogeneous transformation matrix ${}^GT_B$ to move between two coordinate frames:

$$ {}^{G}T_{B} = \begin{bmatrix} {}^{G}R_{B} & {}^{G}\mathbf{d} \\ 0 & 1 \end{bmatrix} = \begin{bmatrix} r_{11} & r_{12} & r_{13} & r_{14} \\ r_{21} & r_{22} & r_{23} & r_{24} \\ r_{31} & r_{32} & r_{33} & r_{34} \\ 0 & 0 & 0 & 1 \end{bmatrix} \tag{8.402} $$

When the elements of the transformation matrix are time dependent, its derivative is

$$ \frac{{}^{G}dT}{dt} = {}^{G}\dot{T}_{B} = \begin{bmatrix} \frac{dr_{11}}{dt} & \frac{dr_{12}}{dt} & \frac{dr_{13}}{dt} & \frac{dr_{14}}{dt} \\ \frac{dr_{21}}{dt} & \frac{dr_{22}}{dt} & \frac{dr_{23}}{dt} & \frac{dr_{24}}{dt} \\ \frac{dr_{31}}{dt} & \frac{dr_{32}}{dt} & \frac{dr_{33}}{dt} & \frac{dr_{34}}{dt} \\ 0 & 0 & 0 & 0 \end{bmatrix} \tag{8.403} $$

The time derivative of the transformation matrix can be rearranged to be proportional to the transformation matrix,

$$ {}^{G}\dot{T}_{B} = {}^{G}V_{B}\,{}^{G}T_{B} \tag{8.404} $$

where ${}^{G}V_{B}$ is the 4×4 homogeneous velocity transformation matrix or *velocity operator matrix* and is equal to

$$ {}^{G}V_{B} = {}^{G}\dot{T}_{B}\,{}^{G}T_{B}^{-1} = \begin{bmatrix} {}^{G}\dot{R}_{B}\,{}^{G}R_{B}^{\mathrm{T}} & {}^{G}\dot{\mathbf{d}} - {}^{G}\dot{R}_{B}\,{}^{G}R_{B}^{\mathrm{T}}\,{}^{G}\mathbf{d} \\ 0 & 1 \end{bmatrix} \tag{8.405} $$

The homogeneous matrix and its derivative based on the velocity transformation matrix are useful in the forward velocity kinematics of multibody and connected rigid bodies. The matrix ${}^{i-1}\dot{T}_{i}$ for two links connected by a revolute joint is given as

$$ {}^{i-1}\dot{T}_{i} = \dot{\theta}_{i} \begin{bmatrix} -\sin\theta_{i} & -\cos\theta_{i}\cos\alpha_{i} & \cos\theta_{i}\sin\alpha_{i} & -a_{i}\sin\theta_{i} \\ \cos\theta_{i} & -\sin\theta_{i}\cos\alpha_{i} & \sin\theta_{i}\sin\alpha_{i} & a_{i}\cos\theta_{i} \\ 0 & 0 & 0 & 0 \\ 0 & 0 & 0 & 0 \end{bmatrix} \tag{8.406} $$

and for two links connected by a prismatic joint as

$$ {}^{i-1}\dot{T}_{i} = \begin{bmatrix} 0 & 0 & 0 & 0 \\ 0 & 0 & 0 & 0 \\ 0 & 0 & 0 & \dot{d}_{i} \\ 0 & 0 & 0 & 0 \end{bmatrix} \tag{8.407} $$

The associated velocity transformation matrix for a revolute joint R is

$$ {}^{i-1}V_{i} = \dot{\theta}_{i}\,\Delta_{R} = \dot{\theta}_{i} \begin{bmatrix} 0 & -1 & 0 & 0 \\ 1 & 0 & 0 & 0 \\ 0 & 0 & 0 & 0 \\ 0 & 0 & 0 & 0 \end{bmatrix} \tag{8.408} $$

and for a prismatic joint P is

$$
{}^{i-1}V_i = \dot{d}_i \, \Delta_P = \dot{d}_i \begin{bmatrix} 0 & 0 & 0 & 0 \\ 0 & 0 & 0 & 0 \\ 0 & 0 & 0 & 1 \\ 0 & 0 & 0 & 0 \end{bmatrix} \tag{8.409}
$$

Proof: Any transformation matrix can be decomposed into a rotation and a translation:

$$
\begin{aligned}
[T] &= \begin{bmatrix} R_{\hat{u},\phi} & \mathbf{d} \\ 0 & 1 \end{bmatrix} = \begin{bmatrix} \mathbf{I} & \mathbf{d} \\ 0 & 1 \end{bmatrix} \begin{bmatrix} R_{\hat{u},\phi} & 0 \\ 0 & 1 \end{bmatrix} \\
&= [D]\,[R]
\end{aligned} \tag{8.410}
$$

Taking a time derivative, we can find

$$
\begin{aligned}
\dot{T} &= \begin{bmatrix} \dot{R}_{\hat{u},\phi} & \dot{\mathbf{d}} \\ 0 & 0 \end{bmatrix} = \begin{bmatrix} \mathbf{I} & \dot{\mathbf{d}} \\ 0 & 1 \end{bmatrix} \begin{bmatrix} \dot{R}_{\hat{u},\phi} & 0 \\ 0 & 1 \end{bmatrix} - \mathbf{I} \\
&= \left[\mathbf{I} + \dot{D}\right]\left[\mathbf{I} + \dot{R}\right] - \mathbf{I} = [V]\,[T]
\end{aligned} \tag{8.411}
$$

where $[V]$ is the velocity transformation matrix

$$
\begin{aligned}
[V] = \dot{T}\,T^{-1} &= \begin{bmatrix} \dot{R}_{\hat{u},\phi} & \dot{\mathbf{d}} \\ 0 & 0 \end{bmatrix} \begin{bmatrix} R^{\mathrm{T}}_{\hat{u},\phi} & -R^{\mathrm{T}}_{\hat{u},\phi}\,\mathbf{d} \\ 0 & 1 \end{bmatrix} \\
&= \begin{bmatrix} \dot{R}_{\hat{u},\phi}\,R^{\mathrm{T}}_{\hat{u},\phi} & \dot{\mathbf{d}} - \dot{R}_{\hat{u},\phi}\,R^{\mathrm{T}}_{\hat{u},\phi}\,\mathbf{d} \\ 0 & 1 \end{bmatrix} = \begin{bmatrix} \tilde{\omega} & \dot{\mathbf{d}} - \tilde{\omega}\,\mathbf{d} \\ 0 & 1 \end{bmatrix}
\end{aligned} \tag{8.412}
$$

The transformation matrix between two neighbor coordinate frames of a multibody is given in Equation (7.7) based on the DH parameters as

$$
{}^{i-1}T_i = \begin{bmatrix} \cos\theta_i & -\sin\theta_i\cos\alpha_i & \sin\theta_i\sin\alpha_i & a_i\cos\theta_i \\ \sin\theta_i & \cos\theta_i\cos\alpha_i & -\cos\theta_i\sin\alpha_i & a_i\sin\theta_i \\ 0 & \sin\alpha_i & \cos\alpha_i & d_i \\ 0 & 0 & 0 & 1 \end{bmatrix} \tag{8.413}
$$

Direct differentiation shows that if two bodies are connected by a revolute joint, θ_i is the only variable of the DH matrix (8.413), and therefore,

$$
\begin{aligned}
{}^{i-1}\dot{T}_i &= \dot{\theta}_i \begin{bmatrix} -\sin\theta_i & -\cos\theta_i\cos\alpha_i & \cos\theta_i\sin\alpha_i & -a_i\sin\theta_i \\ \cos\theta_i & -\sin\theta_i\cos\alpha_i & \sin\theta_i\sin\alpha_i & a_i\cos\theta_i \\ 0 & 0 & 0 & 0 \\ 0 & 0 & 0 & 0 \end{bmatrix} \\
&= \dot{\theta}_i \begin{bmatrix} 0 & -1 & 0 & 0 \\ 1 & 0 & 0 & 0 \\ 0 & 0 & 0 & 0 \\ 0 & 0 & 0 & 0 \end{bmatrix} {}^{i-1}T_i = \dot{\theta}_i\,\Delta_R\,{}^{i-1}T_i
\end{aligned} \tag{8.414}
$$

which shows the *revolute velocity transformation matrix* ${}^{i-1}V_i$ between coordinate frame B_i and B_{i-1} as

$$
{}^{i-1}V_i = \dot{\theta}_i \, \Delta_R = \dot{\theta}_i \begin{bmatrix} 0 & -1 & 0 & 0 \\ 1 & 0 & 0 & 0 \\ 0 & 0 & 0 & 0 \\ 0 & 0 & 0 & 0 \end{bmatrix} \tag{8.415}
$$

If the two bodies are connected by a prismatic joint, d_i is the only variable of the DH matrix (8.413), and therefore,

$$
{}^{i-1}\dot{T}_i = \dot{d}_i \begin{bmatrix} 0 & 0 & 0 & 0 \\ 0 & 0 & 0 & 0 \\ 0 & 0 & 0 & 1 \\ 0 & 0 & 0 & 0 \end{bmatrix} {}^{i-1}T_i = \dot{d}_i \, \Delta_P \, {}^{i-1}T_i \tag{8.416}
$$

It shows that the *prismatic velocity transformation matrix* is

$$
{}^{i-1}V_i = \dot{d}_i \, \Delta_P = \dot{d}_i \begin{bmatrix} 0 & 0 & 0 & 0 \\ 0 & 0 & 0 & 0 \\ 0 & 0 & 0 & 1 \\ 0 & 0 & 0 & 0 \end{bmatrix} \tag{8.417}
$$

The Δ_R and Δ_P are revolute and prismatic velocity coefficient matrices with application in the velocity analysis of multibodies. ■

***Example 512* ★ *Velocity of Frame* B_i *in* B_0** Consider a multibody with n connected links. The velocity of the frame B_i attached to link (i) with respect to the base coordinate frame B_0 can be found by differentiating ${}^0\mathbf{d}_i$ in the base frame:

$$
\begin{aligned}
{}^0\mathbf{v}_i &= \frac{{}^0d}{dt} {}^0\mathbf{d}_i = \frac{{}^0d}{dt} \left({}^0T_i \, {}^i\mathbf{d}_i \right) \\
&= \left({}^0\dot{T}_1 \, {}^1T_2 \ldots {}^{i-1}T_i + {}^0T_1 \, {}^1\dot{T}_2 \, {}^2T_3 \ldots {}^{i-1}T_i + {}^0T_1 \ldots {}^{i-1}\dot{T}_i \right) {}^i\mathbf{d}_i \\
&= \left[\sum_{j=1}^{i} \frac{\partial \, {}^0T_i}{\partial q_j} \dot{q}_j \right] {}^i\mathbf{d}_i
\end{aligned} \tag{8.418}
$$

However, the partial derivatives $\partial^{i-1}T_i / \partial q_i$ can be found by using the velocity coefficient matrices Δ_i, which is either Δ_R or Δ_P:

$$
\frac{\partial^{i-1}T_i}{\partial q_i} = \Delta_i \, {}^{i-1}T_i \tag{8.419}
$$

Hence,

$$
\frac{\partial \, {}^0T_i}{\partial q_j} = \begin{cases} {}^0T_1 \, {}^1T_2 \ldots {}^{j-2}T_{j-1} \, \Delta_j \, {}^{j-1}T_j \ldots {}^{i-1}T_i & \text{for } j \le i \\ 0 & \text{for } j > i \end{cases} \tag{8.420}
$$

■

Example 513 ★ Differential Rotation and Translation Assume the angle of rotation about the axis $\hat{u}$ is too small and indicated by $d\phi$. Then the differential rotation matrix is

$$\mathbf{I}+dR_{\hat{u},\phi}=\mathbf{I}+R_{\hat{u},d\phi}=\begin{bmatrix} 1 & -u_3d\phi & u_2d\phi & 0 \\ u_3d\phi & 1 & -u_1d\phi & 0 \\ -u_2d\phi & +u_1d\phi & 1 & 0 \\ 0 & 0 & 0 & 1 \end{bmatrix} \tag{8.421}$$

because for very small ϕ we have

$$\sin\phi \simeq d\phi \qquad \cos\phi \simeq 1 \qquad \text{vers}\,\phi \simeq 0 \tag{8.422}$$

Differential translation $d\mathbf{d}$, where

$$d\mathbf{d}=d(d_x\hat{I}+d_y\hat{J}+d_z\hat{K}) \tag{8.423}$$

can be shown by a differential translation matrix

$$\mathbf{I}+dD=\begin{bmatrix} 1 & 0 & 0 & dd_x \\ 0 & 1 & 0 & dd_y \\ 0 & 0 & 1 & dd_z \\ 0 & 0 & 0 & 0 \end{bmatrix} \tag{8.424}$$

and therefore,

$$\begin{aligned} dT &= [\mathbf{I}+dD][\mathbf{I}+dR]-\mathbf{I} \\ &= \begin{bmatrix} 0 & -d\phi u_3 & d\phi u_2 & dd_x \\ d\phi u_3 & 0 & -d\phi u_1 & dd_y \\ -d\phi u_2 & d\phi u_1 & 0 & dd_z \\ 0 & 0 & 0 & 0 \end{bmatrix} \end{aligned} \tag{8.425}$$

Example 514 ★ Combination of Principal Differential Rotations The principal differential rotations about X, Y, Z are

$$R_{X,d\gamma}=\begin{bmatrix} 1 & 0 & 0 & 0 \\ 0 & 1 & -d\gamma & 0 \\ 0 & d\gamma & 1 & 0 \\ 0 & 0 & 0 & 1 \end{bmatrix} \tag{8.426}$$

$$R_{Y,d\beta}=\begin{bmatrix} 1 & 0 & d\beta & 0 \\ 0 & 1 & 0 & 0 \\ -d\beta & 0 & 1 & 0 \\ 0 & 0 & 0 & 1 \end{bmatrix} \tag{8.427}$$

$$R_{Z,d\alpha}=\begin{bmatrix} 1 & -d\alpha & 0 & 0 \\ d\alpha & 1 & 0 & 0 \\ 0 & 0 & 1 & 0 \\ 0 & 0 & 0 & 1 \end{bmatrix} \tag{8.428}$$

Combining the three principal differential rotation matrices about the axes Z, Y, and X yields

$$\begin{aligned}
&\left[\mathbf{I}+R_{X,d\gamma}\right]\left[\mathbf{I}+R_{Y,d\beta}\right]\left[\mathbf{I}+R_{Z,d\alpha}\right] \\
&=\begin{bmatrix} 1 & 0 & 0 & 0 \\ 0 & 1 & -d\gamma & 0 \\ 0 & d\gamma & 1 & 0 \\ 0 & 0 & 0 & 1 \end{bmatrix}\begin{bmatrix} 1 & 0 & d\beta & 0 \\ 0 & 1 & 0 & 0 \\ -d\beta & 0 & 1 & 0 \\ 0 & 0 & 0 & 1 \end{bmatrix}\begin{bmatrix} 1 & -d\alpha & 0 & 0 \\ d\alpha & 1 & 0 & 0 \\ 0 & 0 & 1 & 0 \\ 0 & 0 & 0 & 1 \end{bmatrix} \\
&=\begin{bmatrix} 1 & -d\alpha & d\beta & 0 \\ d\alpha & 1 & -d\gamma & 0 \\ -d\beta & d\gamma & 1 & 0 \\ 0 & 0 & 0 & 1 \end{bmatrix}
\end{aligned} \tag{8.429}$$

The combination of differential rotations is commutative:

$$\begin{aligned}
\left[\mathbf{I}+R_{X,d\gamma}\right]\left[\mathbf{I}+R_{Y,d\beta}\right]\left[\mathbf{I}+R_{Z,d\alpha}\right] &= \left[\mathbf{I}+R_{Y,d\beta}\right]\left[\mathbf{I}+R_{Z,d\alpha}\right]\left[\mathbf{I}+R_{X,d\gamma}\right] \\
&= \left[\mathbf{I}+R_{Z,d\alpha}\right]\left[\mathbf{I}+R_{X,d\gamma}\right]\left[\mathbf{I}+R_{Y,d\beta}\right] \\
&= \left[\mathbf{I}+R_{Z,d\alpha}\right]\left[\mathbf{I}+R_{Y,d\beta}\right]\left[\mathbf{I}+R_{X,d\gamma}\right] \\
&= \left[\mathbf{I}+R_{Y,d\beta}\right]\left[\mathbf{I}+R_{X,d\gamma}\right]\left[\mathbf{I}+R_{Z,d\alpha}\right] \\
&= \left[\mathbf{I}+R_{X,d\gamma}\right]\left[\mathbf{I}+R_{Z,d\alpha}\right]\left[\mathbf{I}+R_{Y,d\beta}\right]
\end{aligned} \tag{8.430}$$

Example 515 ★ Differential of a Transformation Matrix Assume the transformation matrix

$$T=\begin{bmatrix} 0 & 0 & 1 & 4 \\ 1 & 0 & 0 & 4 \\ 0 & 1 & 0 & 4 \\ 0 & 0 & 0 & 1 \end{bmatrix} \tag{8.431}$$

is subjected to a differential rotation and translation given by

$$d\phi\hat{u}=\begin{bmatrix} 0.1 & 0.2 & 0.3 \end{bmatrix} \tag{8.432}$$

$$d\mathbf{d}=\begin{bmatrix} 0.6 & 0.4 & 0.2 \end{bmatrix} \tag{8.433}$$

Then, the differential transformation matrix is

$$\begin{aligned}
dT &= [\mathbf{I}+dD][\mathbf{I}+dR]-\mathbf{I} \\
&=\begin{bmatrix} 0 & -0.3 & 0.2 & 0.6 \\ 0.3 & 0 & -0.1 & 0.4 \\ -0.2 & 0.1 & 0 & 0.2 \\ 0 & 0 & 0 & 0 \end{bmatrix}
\end{aligned} \tag{8.434}$$

Example 516 ★ Derivative of Rotation Matrix Based on the Rodriguez formula, the angle–axis of rotation matrix is

$$R_{\hat{u},\phi} = \mathbf{I}\cos\phi + \hat{u}\hat{u}^{\mathrm{T}} \operatorname{vers}\phi + \tilde{u}\sin\phi \tag{8.435}$$

Therefore, the time rate of the Rodriguez formula is

$$\dot{R}_{\hat{u},\phi} = -\dot{\phi}\sin\phi\,\mathbf{I} + \hat{u}\hat{u}^{\mathrm{T}}\,\dot{\phi}\sin\phi + \tilde{u}\dot{\phi}\cos\phi = \dot{\phi}\tilde{u}R_{\hat{u},\phi} \tag{8.436}$$

8.6 ★ MULTIBODY VELOCITY

Consider a multibody with n connected links. The angular velocity of link (i) in the global coordinate frame B_0 is a summation of global angular velocities of links (j) for $j \le i$,

$${}_0^0\boldsymbol{\omega}_i = \sum_{j=1}^{i} {}_{j-1}^{\;\;0}\boldsymbol{\omega}_j \tag{8.437}$$

where

$${}_{j-1}^{\;\;0}\boldsymbol{\omega}_j = \begin{cases} \dot{\theta}_j\,{}^0\hat{k}_{j-1} & \text{if joint } i \text{ is R} \\ 0 & \text{if joint } i \text{ is P} \end{cases} \tag{8.438}$$

The velocity of the origin of B_i in the base coordinate frame is

$${}_{i-1}^{\;\;0}\dot{\mathbf{d}}_i = \begin{cases} {}_0^0\boldsymbol{\omega}_i \times {}_{i-1}^{\;\;0}\mathbf{d}_i & \text{if joint } i \text{ is R} \\ \dot{d}_i\,{}^0\hat{k}_{i-1} + {}_0^0\boldsymbol{\omega}_i \times {}_{i-1}^{\;\;0}\mathbf{d}_i & \text{if joint } i \text{ is P} \end{cases} \tag{8.439}$$

where θ and d are DH parameters and $\mathbf{d}$ is a frame's origin position vector.

Therefore, if the multibody has n links, the global angular velocity of the final coordinate frame is

$${}_0^0\boldsymbol{\omega}_n = \sum_{i=1}^{n} {}_{i-1}^{\;\;0}\boldsymbol{\omega}_i \tag{8.440}$$

and the global velocity vector of the last link's coordinate frame is

$${}_0^0\dot{\mathbf{d}}_n = \sum_{i=1}^{n} {}_{i-1}^{\;\;0}\dot{\mathbf{d}}_i \tag{8.441}$$

Proof: According to the DH definition, the position vector of the coordinate frame B_i with respect to B_{i-1} is

$${}_{i-1}^{\;\;0}\mathbf{d}_i = d_i\,{}^0\hat{k}_{i-1} + a_i\,{}^0\hat{\imath}_i \tag{8.442}$$

which depends on joint variables q_j for $j \le i$, and therefore, ${}_{i-1}^{\;\;0}\dot{\mathbf{d}}_i$ is a function of joint velocities $\dot{q}_j$ for $j \le i$.

Assume that every joint of a multibody except joint i is locked. Then, the angular velocity of link (i) connected by a revolute joint to link $(i-1)$ is

$$ {}_{i-1}^{0}\boldsymbol{\omega}_i = \dot{\theta}_i\,{}^{0}\hat{k}_{i-1} \quad \text{if the only movable joint } i \text{ is R} \tag{8.443} $$

However, if links (i) and $(i-1)$ are connecting via a prismatic joint, then

$$ {}_{i-1}^{0}\boldsymbol{\omega}_i = 0 \quad \text{if the only movable joint } i \text{ is P} \tag{8.444} $$

The relative position vector (8.442) shows that the velocity of link (i) connected by a revolute joint to link $(i-1)$ is

$$ \begin{aligned} {}_{i-1}^{0}\dot{\mathbf{d}}_i &= \dot{\theta}_i\,{}^{0}\hat{k}_{i-1} \times a_i\,{}^{0}\hat{\imath}_i \\ &= {}_{i-1}^{0}\boldsymbol{\omega}_i \times {}_{i-1}^{0}\mathbf{d}_i \quad \text{if the only movable joint } i \text{ is R} \end{aligned} \tag{8.445} $$

We substitute $a_i{}^{0}\hat{\imath}_i$ by ${}_{i-1}^{0}\mathbf{d}_i$ because the x_i-axis is turning about the z_{i-1}-axis with angular velocity $\dot{\theta}_i$, and therefore, $\dot{\theta}_i{}^{0}\hat{k}_{i-1} \times d_i{}^{0}\hat{k}_{i-1} = 0$. However, if links (i) and $(i-1)$ are connected by a prismatic joint, then

$$ {}_{i-1}^{0}\dot{\mathbf{d}}_i = \dot{d}_i\,{}^{0}\hat{k}_{i-1} \quad \text{if the only movable joint } i \text{ is P} \tag{8.446} $$

Now assume that all lower joints $j \leq i$ are moving. Then, the angular velocity of link (i) in the base coordinate frame is

$$ {}_{0}^{0}\boldsymbol{\omega}_i = \sum_{j=1}^{i} {}_{j-1}^{0}\boldsymbol{\omega}_j \tag{8.447} $$

or

$$ {}_{0}^{0}\boldsymbol{\omega}_i = \begin{cases} \displaystyle\sum_{j=1}^{i} \dot{\theta}_i\,{}^{0}\hat{k}_{i-1} & \text{if joint} j \text{ is R} \\ 0 & \text{if joint } j \text{ is P} \end{cases} \tag{8.448} $$

which can be written in a recursive form

$$ {}_{0}^{0}\boldsymbol{\omega}_i = {}_{0}^{0}\boldsymbol{\omega}_{i-1} + {}_{i-1}^{0}\boldsymbol{\omega}_i \tag{8.449} $$

or

$$ {}_{0}^{0}\boldsymbol{\omega}_i = \begin{cases} {}_{0}^{0}\boldsymbol{\omega}_{i-1} + \dot{\theta}_i\,{}^{0}\hat{k}_{i-1} & \text{if joint } i \text{ is R} \\ {}_{0}^{0}\boldsymbol{\omega}_{i-1} & \text{if joint } i \text{ is P} \end{cases} \tag{8.450} $$

The velocity of the origin of link (i) in the base coordinate frame is

$$ {}_{i-1}^{0}\dot{\mathbf{d}}_i = \begin{cases} {}_{0}^{0}\boldsymbol{\omega}_i \times {}_{i-1}^{0}\mathbf{d}_i & \text{if joint } i \text{ is R} \\ \dot{d}_i{}^{0}\hat{k}_{i-1} + {}_{0}^{0}\boldsymbol{\omega}_i \times {}_{i-1}^{0}\mathbf{d}_i & \text{if joint } i \text{ is P} \end{cases} \tag{8.451} $$

The translation and angular velocities of the last link of an n-link multibody is then a direct application of these results. ■

Example 517 ★ Serial Rigid-Link Angular Velocity Consider a serial multibody with n links and n revolute joints. The global angular velocity of link (i) in terms of the angular velocity of its previous links is

$$ {}^{0}\boldsymbol{\omega}_{i} = {}^{0}\boldsymbol{\omega}_{i-1} + \dot{\theta}_{i}\,{}^{0}\hat{k}_{i-1} \tag{8.452}$$

or in general

$$ {}^{0}\boldsymbol{\omega}_{i} = \sum_{j=1}^{i} \dot{\theta}_{j}\,{}^{0}\hat{k}_{j-1} \tag{8.453}$$

where

$$\begin{aligned} {}_{i-2}^{\;\;0}\boldsymbol{\omega}_{i} &= {}_{i-2}^{\;\;0}\boldsymbol{\omega}_{i-1} + {}_{i-1}^{\;\;0}\boldsymbol{\omega}_{i} = {}_{i-2}^{\;\;0}\boldsymbol{\omega}_{i-1} + \dot{\theta}_{i}\,{}^{0}\hat{k}_{i-1} \\ &= \dot{\theta}_{i-1}\,{}^{0}\hat{k}_{i-2} + \dot{\theta}_{i}\,{}^{0}\hat{k}_{i-1} \end{aligned} \tag{8.454}$$

$$ {}_{i-1}^{\;\;0}\boldsymbol{\omega}_{i} = \dot{\theta}_{i}\,{}^{0}\hat{k}_{i-1} \tag{8.455}$$

$$ {}_{i-2}^{\;\;0}\boldsymbol{\omega}_{i-1} = \dot{\theta}_{i-1}\,{}^{0}\hat{k}_{i-2} \tag{8.456}$$

and therefore,

$$\begin{aligned} {}^{0}\boldsymbol{\omega}_{i} &= \sum_{j=1}^{i-1} {}_{j-1}^{\;\;0}\boldsymbol{\omega}_{j} + \dot{\theta}_{i}\,{}^{0}\hat{k}_{i-1} = \sum_{j=1}^{i-1} \dot{\theta}_{j}\,{}^{0}\hat{k}_{j-1} + \dot{\theta}_{i}\,{}^{0}\hat{k}_{i-1} \\ &= \sum_{j=1}^{i} \dot{\theta}_{j}\,{}^{0}\hat{k}_{j-1} \end{aligned} \tag{8.457}$$

Example 518 ★ Serial Rigid-Link Translational Velocity The global angular velocity of link (i) in a serial manipulator in terms of the angular velocity of its previous links is

$$ {}^{0}\mathbf{v}_{i} = {}^{0}\mathbf{v}_{i-1} + {}_{i-1}^{\;\;0}\boldsymbol{\omega}_{i} \times {}_{i-1}^{\;\;0}\mathbf{d}_{i} \tag{8.458}$$

where

$$ {}^{0}\mathbf{v}_{i} = {}^{0}\dot{\mathbf{d}}_{i} \tag{8.459}$$

or in general

$$ {}^{0}\mathbf{v}_{i} = \sum_{j=1}^{i} \left({}^{0}\hat{k}_{j-1} \times {}_{i-1}^{\;\;0}\mathbf{d}_{i} \right) \dot{\theta}_{j} \tag{8.460}$$

where

$$ {}_{i-1}^{\;\;0}\mathbf{v}_{i} = {}_{0}\boldsymbol{\omega}_{i} \times {}_{i-1}^{\;\;0}\mathbf{d}_{i} \tag{8.461}$$

$$ {}_{i-2}^{\;\;0}\mathbf{v}_{i-1} = {}_{0}\boldsymbol{\omega}_{i-1} \times {}_{i-2}^{\;\;0}\mathbf{d}_{i-1} \tag{8.462}$$

$$\begin{aligned} {}_{i-2}^{\;\;0}\mathbf{v}_{i} &= {}_{i-2}^{\;\;0}\mathbf{v}_{i-1} + {}_{i-1}^{\;\;0}\mathbf{v}_{i} = {}_{i-2}^{\;\;0}\mathbf{v}_{i-1} + {}_{0}\boldsymbol{\omega}_{i} \times {}_{i-1}^{\;\;0}\mathbf{d}_{i} \\ &= {}_{0}\boldsymbol{\omega}_{i-1} \times {}_{i-2}^{\;\;0}\mathbf{d}_{i-1} + {}_{0}\boldsymbol{\omega}_{i} \times {}_{i-1}^{\;\;0}\mathbf{d}_{i} \\ &= \dot{\theta}_{i-1}\,{}^{0}\hat{k}_{i-2} \times {}_{i-2}^{\;\;0}\mathbf{d}_{i-1} + \dot{\theta}_{i}\,{}^{0}\hat{k}_{i-1} \times {}_{i-1}^{\;\;0}\mathbf{d}_{i} \end{aligned} \tag{8.463}$$

and therefore,

$$
{}^0\mathbf{v}_i = \sum_{j=1}^{i-1} {}_{j-1}^{\;\,0}\mathbf{v}_j + {}_{i-1}^{\;\,0}\mathbf{v}_i = \sum_{j=1}^{i} \left({}^0\hat{k}_{j-1} \times {}_{i-1}^{\;\,0}\mathbf{d}_i \right) \dot{\theta}_j
$$

8.7 ★ FORWARD-VELOCITY KINEMATICS

Consider a serial multibody with n connected links. The *forward-velocity kinematics* solves the problem of relating joint speeds $\dot{\mathbf{q}}$ to the end-effector speeds $\dot{\mathbf{X}}$,

$$
\dot{\mathbf{X}} = \mathbf{J}\,\dot{\mathbf{q}} \tag{8.464}
$$

where $\mathbf{q}$ is the *joint variable vector* and $\dot{\mathbf{q}}$ is the *joint speed vector*:

$$
\mathbf{q} = \begin{bmatrix} q_n & q_n & q_n & \cdots & q_n \end{bmatrix}^{\mathrm{T}} \tag{8.465}
$$

$$
\dot{\mathbf{q}} = \begin{bmatrix} \dot{q}_n & \dot{q}_n & \dot{q}_n & \cdots & \dot{q}_n \end{bmatrix}^{\mathrm{T}} \tag{8.466}
$$

Furthermore, $\dot{\mathbf{X}}$ is the *end-effector configuration speed vector*:

$$
\begin{aligned}
\dot{\mathbf{X}} &= \begin{bmatrix} \dot{X}_n & \dot{Y}_n & \dot{Z}_n & \omega_{Xn} & \omega_{Yn} & \omega_{Zn} \end{bmatrix}^{\mathrm{T}} \\
&= \begin{bmatrix} {}^0\dot{\mathbf{d}}_n \\ {}^0\boldsymbol{\omega}_n \end{bmatrix} = \begin{bmatrix} {}^0\mathbf{v}_n \\ {}^0\boldsymbol{\omega}_n \end{bmatrix}
\end{aligned} \tag{8.467}
$$

To solve this problem, we need to determine the $6 \times n$ *Jacobian* transformation matrix $\mathbf{J}(\mathbf{q})$, where n is the number of joint variables.

The global expression of velocity ${}^0\mathbf{v}_n$ of the origin of B_n is proportional to the manipulator joint speeds $\dot{\mathbf{q}}_D$:

$$
{}^0\mathbf{v}_n = \mathbf{J}_D\,\dot{\mathbf{q}}_D \qquad \dot{\mathbf{q}}_D \in \dot{\mathbf{q}} \tag{8.468}
$$

The $3 \times n$ proportionality matrix $\mathbf{J}_D(\mathbf{q})$ is the *displacement Jacobian matrix* of the manipulator:

$$
\mathbf{J}_D = \frac{\partial \mathbf{d}_n\left(\dot{\mathbf{q}}_D\right)}{\partial \dot{\mathbf{q}}_D} = \frac{\partial T\left(\mathbf{q}\right)}{\partial \mathbf{q}} \tag{8.469}
$$

The global expression of angular velocity ${}_0\boldsymbol{\omega}_n$ of B_n is proportional to the rotational components of $\dot{\mathbf{q}}$:

$$
{}_0\boldsymbol{\omega}_n = \mathbf{J}_R\,\dot{\mathbf{q}} \tag{8.470}
$$

The $3 \times n$ proportionality matrix $\mathbf{J}_R(\mathbf{q})$ is the *rotational Jacobian matrix* of the robot:

$$
\mathbf{J}_R = \frac{\partial\, {}_0\boldsymbol{\omega}_n}{\partial \mathbf{q}} \tag{8.471}
$$

It is also possible to break the forward-velocity problem into two subproblems to find the translation and angular velocities of the end frame independently:

$$^0\mathbf{v}_n = \mathbf{J}_D\,\dot{\mathbf{q}} \tag{8.472}$$

$$^0\boldsymbol{\omega}_n = \mathbf{J}_R\,\dot{\mathbf{q}} \tag{8.473}$$

We may combine Equations (8.468) and (8.470) and show the forward-velocity kinematics of a multibody by

$$\dot{\mathbf{X}} = \begin{bmatrix} ^0\mathbf{v}_n \\ ^0\boldsymbol{\omega}_n \end{bmatrix} = \begin{bmatrix} \mathbf{J}_D \\ \mathbf{J}_R \end{bmatrix} \dot{\mathbf{q}} = \mathbf{J}\,\dot{\mathbf{q}} \tag{8.474}$$

Proof: Forward-velocity kinematics is defined as determination of the end-effector translational and angular velocities $^0\mathbf{v}_n$, $^0\boldsymbol{\omega}_n$ for a given set of joint speeds $\dot{q}_i$, $i = 1, 2, \ldots, n$. The components of velocity vectors $^0\mathbf{v}_n$ and $^0\boldsymbol{\omega}_n$ are proportional to the joint speeds $\dot{q}_i$, $i = 1, 2, \ldots, n$:

$$^0\mathbf{v}_n = \mathbf{J}_D\,\dot{\mathbf{q}} \tag{8.475}$$

$$_0\boldsymbol{\omega}_n = \mathbf{J}_R\,\dot{\mathbf{q}} \tag{8.476}$$

The proportionality matrices $\mathbf{J}_D$ and $\mathbf{J}_R$ are called the displacement and rotational Jacobians.

We may combine Equations (8.475) and (8.476) as

$$\dot{\mathbf{X}} = \mathbf{J}\,\dot{\mathbf{q}} \tag{8.477}$$

by defining the Jacobian matrix $\mathbf{J}$ and the vectors $\dot{\mathbf{X}}$ and $\dot{\mathbf{q}}$, known as *end-effector speed vector* and *joint speed vector*, respectively:

$$\mathbf{J} = \begin{bmatrix} \mathbf{J}_D \\ \mathbf{J}_R \end{bmatrix} \tag{8.478}$$

$$\dot{\mathbf{X}} = \begin{bmatrix} ^0\mathbf{v}_n \\ _0\boldsymbol{\omega}_n \end{bmatrix} \tag{8.479}$$

$$\dot{\mathbf{q}} = \begin{bmatrix} \dot{q}_n & \dot{q}_n & \dot{q}_n & \cdots & \dot{q}_n \end{bmatrix}^{\mathrm{T}} \tag{8.480}$$

We may also show $\mathbf{J}_D$ by $\mathbf{J}$ whenever we analyze the velocity kinematics of a manipulator without a wrist.

Consider a robot with six DOF that is made of a three-DOF manipulator to position the wrist point and a spherical wrist with three DOF to orient the end effector. The coordinate transformation of a point in the end-effector coordinate frame B_6 and the base coordinate frame B_0 is

$$\begin{aligned} ^0\mathbf{r} &= {}^0T_6(\mathbf{q})\,{}^6\mathbf{r} = {}^0D_6\,{}^0R_6 = \begin{bmatrix} \mathbf{I} & ^0\mathbf{d}_6 \\ 0 & 1 \end{bmatrix} \begin{bmatrix} ^0R_6 & \mathbf{0} \\ 0 & 1 \end{bmatrix} \\ &= \begin{bmatrix} ^0R_6 & ^0\mathbf{d}_6 \\ 0 & 1 \end{bmatrix} {}^6\mathbf{r} \end{aligned} \tag{8.481}$$

where the transformation matrix ${}^0T_6$ is a function of six joint variables q_i, $i = 1, 2, \ldots, 6$. We can always divide the six joint variables into the end-effector position variables q_1, q_2, q_3 and the end-effector orientation variables q_4, q_5, q_6. The end-effector position variables are the only variables in the position vector ${}^0\mathbf{d}_6$ and the end-effector orientation variables are the only variables in the rotation transformation matrix ${}^0R_6$:

$$ {}^0\mathbf{d}_6 = {}^0\mathbf{d}_6\,(q_1, q_2, q_3) \tag{8.482} $$

$$ {}^0R_6 = {}^0R_6\,(q_4, q_5, q_6) \tag{8.483} $$

The origin of the end-effector frame B_6 is at ${}^6\mathbf{r} = \mathbf{0}$, which is globally at

$$ \begin{aligned} {}^0\mathbf{r} &= \begin{bmatrix} {}^0R_6 & {}^0\mathbf{d}_6 \\ 0 & 1 \end{bmatrix} \begin{bmatrix} \mathbf{0} \\ 1 \end{bmatrix} = \begin{bmatrix} \mathbf{I} & {}^0\mathbf{d}_6 \\ 0 & 1 \end{bmatrix} \begin{bmatrix} \mathbf{0} \\ 1 \end{bmatrix} \\ &= {}^0D_6 \begin{bmatrix} \mathbf{0} \\ 1 \end{bmatrix} = {}^0\mathbf{d}_6 \end{aligned} \tag{8.484} $$

The components of the end-effector displacement vector ${}^0\mathbf{d}_6 = [X, Y, Z]$ are functions of the manipulator joint variables q_1, q_2, q_3:

$$ \begin{bmatrix} X \\ Y \\ Z \end{bmatrix} = \begin{bmatrix} d_1\,(q_1, q_2, q_3) \\ d_2\,(q_1, q_2, q_3) \\ d_3\,(q_1, q_2, q_3) \end{bmatrix} \tag{8.485} $$

Taking a derivative of both sides indicates that each component of ${}^0\mathbf{v}_6 = {}^0\dot{\mathbf{r}}$ is a linear combination of $\dot{q}_1, \dot{q}_2, \dot{q}_3$:

$$ \begin{aligned} \dot{X} &= \frac{\partial d_1}{\partial q_1}\dot{q}_1 + \frac{\partial d_1}{\partial q_2}\dot{q}_2 + \frac{\partial d_1}{\partial q_3}\dot{q}_3 \\ \dot{Y} &= \frac{\partial d_2}{\partial q_1}\dot{q}_1 + \frac{\partial d_2}{\partial q_2}\dot{q}_2 + \frac{\partial d_2}{\partial q_3}\dot{q}_3 \\ \dot{Z} &= \frac{\partial d_3}{\partial q_1}\dot{q}_1 + \frac{\partial d_3}{\partial q_2}\dot{q}_2 + \frac{\partial d_3}{\partial q_3}\dot{q}_3 \end{aligned} \tag{8.486} $$

which indicates that ${}^0\mathbf{v}_6$ is a linear combination of joint speeds q_1, q_2, q_3:

$$ {}^0\mathbf{v}_6 = \mathbf{J}_D\,\dot{\mathbf{q}}_D = \dot{q}_1\frac{\partial\,{}^0\mathbf{d}_6}{\partial q_1} + \dot{q}_2\frac{\partial\,{}^0\mathbf{d}_6}{\partial q_2} + \dot{q}_3\frac{\partial\,{}^0\mathbf{d}_6}{\partial q_3} \tag{8.487} $$

We may show these relations by vector and matrix expressions:

$$ {}^0\mathbf{v}_6 = \frac{\partial \mathbf{d}_6}{\partial \mathbf{q}}\dot{\mathbf{q}}_D = \mathbf{J}_D\,\dot{\mathbf{q}}_D \tag{8.488} $$

$$ \begin{bmatrix} \dot{X} \\ \dot{Y} \\ \dot{Z} \end{bmatrix} = \begin{bmatrix} \frac{\partial d_1}{\partial q_1} & \frac{\partial d_1}{\partial q_2} & \frac{\partial d_1}{\partial q_3} \\ \frac{\partial d_2}{\partial q_1} & \frac{\partial d_2}{\partial q_2} & \frac{\partial d_2}{\partial q_3} \\ \frac{\partial d_3}{\partial q_1} & \frac{\partial d_3}{\partial q_2} & \frac{\partial d_3}{\partial q_3} \end{bmatrix} \begin{bmatrix} \dot{q}_1 \\ \dot{q}_2 \\ \dot{q}_3 \end{bmatrix} \tag{8.489} $$

The displacement Jacobian $\mathbf{J}_D$ is equivalent to the derivative of T with respect to the manipulator joint coordinates:

$$\mathbf{J}_D = \frac{\partial \mathbf{d}_6}{\partial \mathbf{q}} = \frac{\partial\, {}^0D_6}{\partial \mathbf{q}} = \frac{\partial\, {}^0T_6}{\partial \mathbf{q}} = \frac{\partial T\,(\mathbf{q})}{\partial \mathbf{q}} \tag{8.490}$$

The angular velocity of the end effector is

$${}_0\boldsymbol{\omega}_6 = {}^0\dot{R}_6\, {}^0R_6^{\mathrm{T}} \tag{8.491}$$

However, the time derivative of the rotational transformation matrix is

$$\begin{aligned} {}^0\dot{R}_6 &= \frac{d}{dt}\left[{}^0R_1\, {}^1R_2\, {}^2R_3\, {}^3R_4\, {}^4R_5\, {}^5R_6\right] \\ &= \dot{q}_1 \frac{\partial\, {}^0R_1}{\partial q_1}\, {}^1R_2\, {}^2R_3\, {}^3R_4\, {}^4R_5\, {}^5R_6 + \dot{q}_2\, {}^0R_1 \frac{\partial\, {}^1R_2}{\partial q_2}\, {}^2R_3\, {}^3R_4\, {}^4R_5\, {}^5R_6 \\ &\quad + \dot{q}_3\, {}^0R_1\, {}^1R_2 \frac{\partial\, {}^2R_3}{\partial q_3}\, {}^3R_4\, {}^4R_5\, {}^5R_6 + \dot{q}_4\, {}^0R_1\, {}^1R_2\, {}^2R_3 \frac{\partial\, {}^3R_4}{\partial q_4}\, {}^4R_5\, {}^5R_6 \\ &\quad + \dot{q}_5\, {}^0R_1\, {}^1R_2\, {}^2R_3\, {}^3R_4 \frac{\partial\, {}^4R_5}{\partial q_5}\, {}^5R_6 + \dot{q}_6\, {}^0R_1\, {}^1R_2\, {}^2R_3\, {}^3R_4\, {}^4R_5 \frac{\partial\, {}^5R_6}{\partial q_6} \end{aligned} \tag{8.492}$$

and the transpose of ${}^0R_6$ is

$$\begin{aligned} {}^0R_6^{\mathrm{T}} &= \left[{}^0R_1\, {}^1R_2\, {}^2R_3\, {}^3R_4\, {}^4R_5\, {}^5R_6\right]^{\mathrm{T}} \\ &= {}^5R_6^{\mathrm{T}}\, {}^4R_5^{\mathrm{T}}\, {}^3R_4^{\mathrm{T}}\, {}^2R_3^{\mathrm{T}}\, {}^1R_2^{\mathrm{T}}\, {}^0R_1^{\mathrm{T}} \end{aligned} \tag{8.493}$$

Therefore,

$$\begin{aligned} {}_0\boldsymbol{\omega}_6 &= {}^0\dot{R}_6\, {}^0R_6^{\mathrm{T}} \\ &= \dot{q}_1 \frac{\partial\, {}^0R_1}{\partial q_1}\, {}^0R_1^{\mathrm{T}} + \dot{q}_2\, {}^0R_1 \frac{\partial\, {}^1R_2}{\partial q_2}\, {}^0R_2^{\mathrm{T}} \\ &\quad + \dot{q}_3\, {}^0R_2 \frac{\partial\, {}^2R_3}{\partial q_3}\, {}^0R_3^{\mathrm{T}} + \dot{q}_4\, {}^0R_3 \frac{\partial\, {}^3R_4}{\partial q_4}\, {}^0R_4^{\mathrm{T}} \\ &\quad + \dot{q}_5\, {}^0R_4 \frac{\partial\, {}^4R_5}{\partial q_5}\, {}^0R_5^{\mathrm{T}} + \dot{q}_6\, {}^0R_5 \frac{\partial\, {}^5R_6}{\partial q_6}\, {}^0R_6^{\mathrm{T}} \end{aligned} \tag{8.494}$$

$$= {}_0\boldsymbol{\omega}_1 + {}_1^0\boldsymbol{\omega}_2 + {}_2^0\boldsymbol{\omega}_3 + {}_3^0\boldsymbol{\omega}_4 + {}_4^0\boldsymbol{\omega}_5 + {}_5^0\boldsymbol{\omega}_6 \tag{8.495}$$

which indicates that ${}^0\boldsymbol{\omega}_6$ is a linear combination of joint speeds q_i, $i = 1, 2, \ldots, 6$:

$$\begin{aligned} {}_0\boldsymbol{\omega}_6 = \mathbf{J}_R\, \dot{\mathbf{q}} &= \dot{q}_1 \frac{\partial\, {}^0\boldsymbol{\omega}_6}{\partial q_1} + \dot{q}_2 \frac{\partial\, {}^0\boldsymbol{\omega}_6}{\partial q_2} + \dot{q}_3 \frac{\partial\, {}^0\boldsymbol{\omega}_6}{\partial q_3} \\ &\quad + \dot{q}_4 \frac{\partial\, {}^0\boldsymbol{\omega}_6}{\partial q_4} + \dot{q}_5 \frac{\partial\, {}^0\boldsymbol{\omega}_6}{\partial q_5} + \dot{q}_6 \frac{\partial\, {}^0\boldsymbol{\omega}_6}{\partial q_6} \end{aligned} \tag{8.496}$$

where

$$\frac{\partial_0 \boldsymbol{\omega}_6}{\partial q_1} = \frac{\partial^0 R_1}{\partial q_1} {}^0R_1^{\mathrm{T}} \tag{8.497}$$

$$\frac{\partial_0 \boldsymbol{\omega}_6}{\partial q_2} = {}^0R_1 \frac{\partial^1 R_2}{\partial q_2} {}^0R_2^{\mathrm{T}} \tag{8.498}$$

$$\frac{\partial_0 \boldsymbol{\omega}_6}{\partial q_3} = {}^0R_2 \frac{\partial^2 R_3}{\partial q_3} {}^0R_3^{\mathrm{T}} \tag{8.499}$$

$$\frac{\partial_0 \boldsymbol{\omega}_6}{\partial q_4} = {}^0R_3 \frac{\partial^3 R_4}{\partial q_4} {}^0R_4^{\mathrm{T}} \tag{8.500}$$

$$\frac{\partial_0 \boldsymbol{\omega}_6}{\partial q_5} = {}^0R_4 \frac{\partial^4 R_5}{\partial q_5} {}^0R_5^{\mathrm{T}} \tag{8.501}$$

$$\frac{\partial_0 \boldsymbol{\omega}_6}{\partial q_6} = {}^0R_5 \frac{\partial^5 R_6}{\partial q_6} {}^0R_6^{\mathrm{T}} \tag{8.502}$$

Combination of the translational and rotational velocities results in Equation (8.477) for the velocity kinematics of the robot:

$$\dot{\mathbf{X}} = \begin{bmatrix} {}^0\mathbf{v}_n \\ {}_0\boldsymbol{\omega}_n \end{bmatrix} = \begin{bmatrix} \mathbf{J}_D \\ \mathbf{J}_R \end{bmatrix} \dot{\mathbf{q}} = \mathbf{J}\,\dot{\mathbf{q}} \tag{8.503}$$

The Jacobian matrix of the robot is

$$\mathbf{J} = \begin{bmatrix} \frac{\partial^0 \mathbf{d}_6}{\partial q_1} & \frac{\partial^0 \mathbf{d}_6}{\partial q_2} & \frac{\partial^0 \mathbf{d}_6}{\partial q_3} & \mathbf{0} & \mathbf{0} & \mathbf{0} \\ \frac{\partial_0 \boldsymbol{\omega}_6}{\partial q_1} & \frac{\partial_0 \boldsymbol{\omega}_6}{\partial q_2} & \frac{\partial_0 \boldsymbol{\omega}_6}{\partial q_3} & \frac{\partial_0 \boldsymbol{\omega}_6}{\partial q_4} & \frac{\partial_0 \boldsymbol{\omega}_6}{\partial q_5} & \frac{\partial_0 \boldsymbol{\omega}_6}{\partial q_6} \end{bmatrix} \tag{8.504}$$

If the robot has n links and joints, the above equations go from 1 to n instead of 1 to 6. So, in general, the $6 \times n$ Jacobian matrix $\mathbf{J}$ becomes

$$\mathbf{J} = \begin{bmatrix} \frac{\partial^0 \mathbf{d}_n}{\partial q_1} & \frac{\partial^0 \mathbf{d}_n}{\partial q_2} & \frac{\partial^0 \mathbf{d}_n}{\partial q_3} & \cdots\cdots & \frac{\partial^0 \mathbf{d}_n}{\partial q_n} \\ \frac{\partial_0 \boldsymbol{\omega}_n}{\partial q_1} & \frac{\partial_0 \boldsymbol{\omega}_n}{\partial q_2} & \frac{\partial_0 \boldsymbol{\omega}_n}{\partial q_3} & \cdots\cdots & \frac{\partial_0 \boldsymbol{\omega}_n}{\partial q_n} \end{bmatrix} \tag{8.505}$$

■

Example 519 ★ Jacobian Matrix for the 2R Planar Manipulator A 2R planar manipulator with two R‖R links is illustrated in Figure 8.16. The manipulator has been analyzed in Example 412 for forward kinematics and in Example 458 for inverse kinematics.

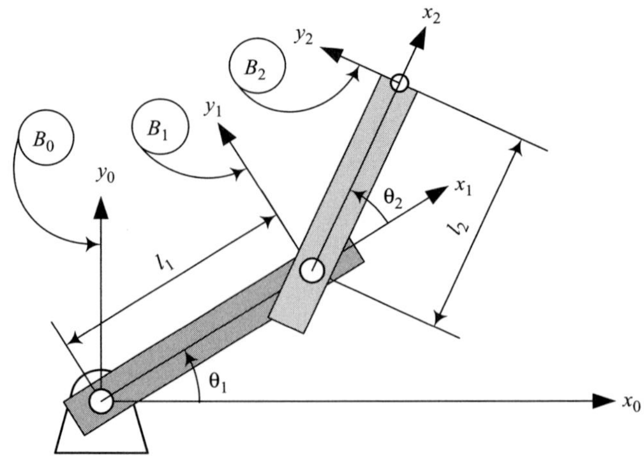

Figure 8.16 A 2R planar manipulator.

The angular velocity of links (1) and (2) are

$$ {}_0\boldsymbol{\omega}_1 = \dot{\theta}_1 {}^0\hat{k}_0 \tag{8.506} $$

$$ {}_1^0\boldsymbol{\omega}_2 = \dot{\theta}_2 {}^0\hat{k}_1 \tag{8.507} $$

$$ {}_0\boldsymbol{\omega}_2 = {}_0\boldsymbol{\omega}_1 + {}_1^0\boldsymbol{\omega}_2 = \left(\dot{\theta}_1 + \dot{\theta}_2\right) {}^0\hat{k}_0 \tag{8.508} $$

and the global velocity of the tip point of the manipulator is

$$ \begin{aligned} {}^0\dot{\mathbf{d}}_2 &= {}^0\dot{\mathbf{d}}_1 + {}_1^0\dot{\mathbf{d}}_2 = {}_0\boldsymbol{\omega}_1 \times {}^0\mathbf{d}_1 + {}_0\boldsymbol{\omega}_2 \times {}_1^0\mathbf{d}_2 \\ &= \dot{\theta}_1 {}^0\hat{k}_0 \times l_1 {}^0\hat{\imath}_1 + \left(\dot{\theta}_1 + \dot{\theta}_2\right) {}^0\hat{k}_0 \times l_2 {}^0\hat{\imath}_2 \\ &= l_1\dot{\theta}_1 {}^0\hat{\jmath}_1 \times l_2\left(\dot{\theta}_1 + \dot{\theta}_2\right) {}^0\hat{\jmath}_2 \end{aligned} \tag{8.509} $$

The unit vectors ${}^0\hat{\jmath}_1$ and ${}^0\hat{\jmath}_2$ can be found by using the coordinate transformation method:

$$ \begin{aligned} {}^0\hat{\jmath}_1 &= R_{z,\theta_1} {}^1\hat{\jmath}_1 \\ &= \begin{bmatrix} \cos\theta_1 & -\sin\theta_1 & 0 \\ \sin\theta_1 & \cos\theta_1 & 0 \\ 0 & 0 & 1 \end{bmatrix} \begin{bmatrix} 0 \\ 1 \\ 0 \end{bmatrix} = \begin{bmatrix} -\sin\theta_1 \\ \cos\theta_1 \\ 0 \end{bmatrix} \end{aligned} \tag{8.510} $$

$$ \begin{aligned} {}^0\hat{\jmath}_2 &= R_{z,\theta_1+\theta_2} {}^2\hat{\jmath}_2 = \begin{bmatrix} c\left(\theta_1+\theta_2\right) & -s\left(\theta_1+\theta_2\right) & 0 \\ s\left(\theta_1+\theta_2\right) & c\left(\theta_1+\theta_2\right) & 0 \\ 0 & 0 & 1 \end{bmatrix} \begin{bmatrix} 0 \\ 1 \\ 0 \end{bmatrix} \\ &= \begin{bmatrix} -s\left(\theta_1+\theta_2\right) \\ c\left(\theta_1+\theta_2\right) \\ 0 \end{bmatrix} \end{aligned} \tag{8.511} $$

Substituting back shows that

$$^0\dot{\mathbf{d}}_2 = l_1\dot{\theta}_1 \begin{bmatrix} -\sin\theta_1 \\ \cos\theta_1 \\ 0 \end{bmatrix} \times l_2\left(\dot{\theta}_1 + \dot{\theta}_2\right) \begin{bmatrix} -\sin(\theta_1+\theta_2) \\ \cos(\theta_1+\theta_2) \\ 0 \end{bmatrix} \tag{8.512}$$

which can be rearranged to

$$\begin{aligned} \begin{bmatrix} \dot{X} \\ \dot{Y} \end{bmatrix} &= \begin{bmatrix} -l_1 s\theta_1 - l_2 s(\theta_1+\theta_2) & -l_2 s(\theta_1+\theta_2) \\ l_1 c\theta_1 + l_2 c(\theta_1+\theta_2) & l_2 c(\theta_1+\theta_2) \end{bmatrix} \begin{bmatrix} \dot{\theta}_1 \\ \dot{\theta}_2 \end{bmatrix} \\ &= \mathbf{J}_D \begin{bmatrix} \dot{\theta}_1 \\ \dot{\theta}_2 \end{bmatrix} \end{aligned} \tag{8.513}$$

Taking advantage of the structural simplicity of the 2R manipulator, we may find its Jacobian simpler. The forward kinematics of the manipulator was found as

$$\begin{aligned} {}^0T_2 &= {}^0T_1\,{}^1T_2 \\ &= \begin{bmatrix} c(\theta_1+\theta_2) & -s(\theta_1+\theta_2) & 0 & l_1 c\theta_1 + l_2 c(\theta_1+\theta_2) \\ s(\theta_1+\theta_2) & c(\theta_1+\theta_2) & 0 & l_1 s\theta_1 + l_2 s(\theta_1+\theta_2) \\ 0 & 0 & 1 & 0 \\ 0 & 0 & 0 & 1 \end{bmatrix} \end{aligned} \tag{8.514}$$

which shows the tip position ${}^0_0\mathbf{d}_2$ of the manipulator is at

$$\begin{bmatrix} X \\ Y \end{bmatrix} = \begin{bmatrix} l_1\cos\theta_1 + l_2\cos(\theta_1+\theta_2) \\ l_1\sin\theta_1 + l_2\sin(\theta_1+\theta_2) \end{bmatrix} \tag{8.515}$$

Direct differentiation gives

$$\begin{bmatrix} \dot{X} \\ \dot{Y} \end{bmatrix} = \begin{bmatrix} -l_1\dot{\theta}_1\sin\theta_1 - l_2\left(\dot{\theta}_1+\dot{\theta}_2\right)\sin(\theta_1+\theta_2) \\ l_1\dot{\theta}_1\cos\theta_1 + l_2\left(\dot{\theta}_1+\dot{\theta}_2\right)\cos(\theta_1+\theta_2) \end{bmatrix} \tag{8.516}$$

which can be rearranged in the matrix form

$$\begin{bmatrix} \dot{X} \\ \dot{Y} \end{bmatrix} = \begin{bmatrix} -l_1 s\theta_1 - l_2 s(\theta_1+\theta_2) & -l_2 s(\theta_1+\theta_2) \\ l_1 c\theta_1 + l_2 c(\theta_1+\theta_2) & l_2 c(\theta_1+\theta_2) \end{bmatrix} \begin{bmatrix} \dot{\theta}_1 \\ \dot{\theta}_2 \end{bmatrix} \tag{8.517}$$

or

$$\dot{\mathbf{X}} = \mathbf{J}_D\,\dot{\boldsymbol{\theta}} \tag{8.518}$$

Example 520 ★ Columns of the Jacobian for the 2R Manipulator We can find the Jacobian of the 2R planar manipulator systematically using the column-by-column method. The global position vectors of the coordinate frames are

$${}^0_1\mathbf{d}_2 = l_2\,{}^0\hat{\imath}_2 \tag{8.519}$$

$${}^0\mathbf{d}_2 = l_1\,{}^0\hat{\imath}_1 + l_2\,{}^0\hat{\imath}_2 \tag{8.520}$$

and therefore,

$$\begin{aligned}{}^{0}\dot{\mathbf{d}}_2 &= {}_0\boldsymbol{\omega}_1 \times {}^{0}\mathbf{d}_2 + {}_1^0\boldsymbol{\omega}_2 \times {}_1^0\mathbf{d}_2 \\ &= \dot{\theta}_1 {}^0\hat{k}_0 \times \left(l_1 {}^0\hat{\imath}_1 + l_2 {}^0\hat{\imath}_2\right) + \dot{\theta}_2 {}^0\hat{k}_1 \times l_2 {}^0\hat{\imath}_2 \\ &= \left[\, {}^0\hat{k}_0 \times \left(l_1 {}^0\hat{\imath}_1 + l_2 {}^0\hat{\imath}_2\right) \;\; {}^0\hat{k}_1 \times l_2 {}^0\hat{\imath}_2 \,\right] \begin{bmatrix} \dot{\theta}_1 \\ \dot{\theta}_2 \end{bmatrix}\end{aligned} \tag{8.521}$$

It can be set in matrix form to show the columns of [**J**]:

$$\begin{bmatrix} {}^0\dot{\mathbf{d}}_2 \\ {}_0\boldsymbol{\omega}_2 \end{bmatrix} = \begin{bmatrix} {}^0\hat{k}_0 \times {}^0\mathbf{d}_2 & {}^0\hat{k}_1 \times {}_1^0\mathbf{d}_2 \\ {}^0\hat{k}_0 & {}^0\hat{k}_1 \end{bmatrix} \begin{bmatrix} \dot{\theta}_1 \\ \dot{\theta}_2 \end{bmatrix} = \mathbf{J}\,\dot{\boldsymbol{\theta}} \tag{8.522}$$

Example 521 ★ Jacobian Matrix of a Planar Polar Manipulator Figure 8.17 illustrates a planar polar manipulator with the following forward kinematics:

$$\begin{aligned}{}^0T_2 = {}^0T_1 {}^1T_2 &= \begin{bmatrix} \cos\theta & -\sin\theta & 0 & 0 \\ \sin\theta & \cos\theta & 0 & 0 \\ 0 & 0 & 1 & 0 \\ 0 & 0 & 0 & 1 \end{bmatrix} \begin{bmatrix} 1 & 0 & 0 & r \\ 0 & 1 & 0 & 0 \\ 0 & 0 & 1 & 0 \\ 0 & 0 & 0 & 1 \end{bmatrix} \\ &= \begin{bmatrix} \cos\theta & -\sin\theta & 0 & r\cos\theta \\ \sin\theta & \cos\theta & 0 & r\sin\theta \\ 0 & 0 & 1 & 0 \\ 0 & 0 & 0 & 1 \end{bmatrix}\end{aligned} \tag{8.523}$$

The tip point of the manipulator is at

$$\begin{bmatrix} X \\ Y \end{bmatrix} = \begin{bmatrix} r\cos\theta \\ r\sin\theta \end{bmatrix} \tag{8.524}$$

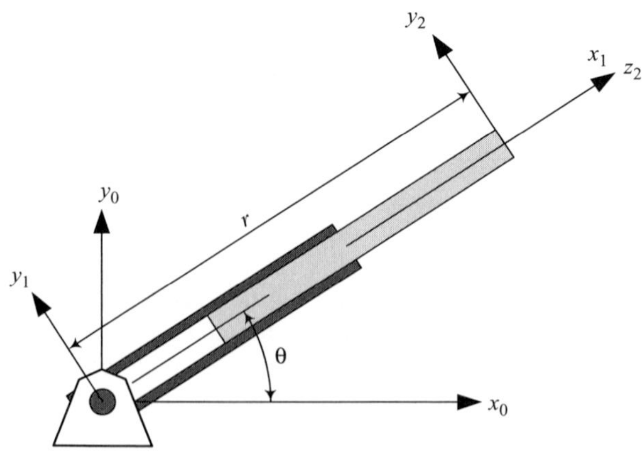

Figure 8.17 A planar polar manipulator.

and therefore its velocity is

$$\begin{bmatrix} \dot{X} \\ \dot{Y} \end{bmatrix} = \begin{bmatrix} \cos\theta & -r\sin\theta \\ \sin\theta & r\cos\theta \end{bmatrix} \begin{bmatrix} \dot{r} \\ \dot{\theta} \end{bmatrix} \tag{8.525}$$

which shows that

$$\mathbf{J}_D = \begin{bmatrix} \cos\theta & -r\sin\theta \\ \sin\theta & r\cos\theta \end{bmatrix} \tag{8.526}$$

8.8 ★ JACOBIAN-GENERATING VECTOR

The Jacobian matrix provides the transformation from joint velocities to Cartesian velocities:

$$\dot{\mathbf{X}} = \mathbf{J}\,\dot{\mathbf{q}} \tag{8.527}$$

$$\mathbf{J} = \begin{bmatrix} {}^0\hat{k}_0 \times {}^0_0\mathbf{d}_n & {}^0\hat{k}_1 \times {}^0_1\mathbf{d}_n & \cdots & {}^0\hat{k}_{n-1} \times {}^{\;\;0}_{n-1}\mathbf{d}_n \\ {}^0\hat{k}_0 & {}^0\hat{k}_1 & \cdots & {}^0\hat{k}_{n-1} \end{bmatrix} \tag{8.528}$$

where $\mathbf{J}$ can be calculated column by column using the *Jacobian-generating vector* $\mathbf{c}_i(\mathbf{q})$, $i = 1, 2, \ldots, n$:

$$\mathbf{c}_i(\mathbf{q}) = \begin{bmatrix} {}^0\hat{k}_{i-1} \times {}^{\;\;0}_{i-1}\mathbf{d}_n \\ {}^0\hat{k}_{i-1} \end{bmatrix} \tag{8.529}$$

To calculate the ith column of the Jacobian matrix, we need to find two vectors ${}^{\;\;0}_{i-1}\mathbf{d}_n$ and ${}^0\hat{k}_{i-1}$. The vector ${}^{\;\;0}_{i-1}\mathbf{d}_n$ indicates the position of the frame B_{i-1} attached to link $i-1$ and the vector ${}^0\hat{k}_{i-1}$ indicates the joint axis unit vector of B_{i-1}, both expressed in the base frame B_0.

Calculating $\mathbf{J}$ based on the Jacobian-generating vectors shows that forward-velocity kinematics is a consequence of the forward kinematics of multibodies.

Proof: Let ${}^0\mathbf{d}_i$ and ${}^0\mathbf{d}_{i-1}$ be the global position vectors of the frames B_i and B_{i-1} and ${}^{i-1}\mathbf{d}_i$ be the position vector of the frame B_i in B_{i-1} as shown in Figure 8.18.

The three position vectors ${}^0\mathbf{d}_i$, ${}^0\mathbf{d}_{i-1}$, and ${}^{i-1}\mathbf{d}_i$ are related by vector addition,

$${}^0\mathbf{d}_i = {}^0\mathbf{d}_{i-1} + {}^0R_{i-1}\,{}^{i-1}\mathbf{d}_i = {}^0\mathbf{d}_{i-1} + d_i\,{}^0\hat{k}_{i-1} + a_i\,{}^0\hat{\imath}_i \tag{8.530}$$

where we have used Equation (8.442). Taking a time derivative,

$$\begin{aligned} {}^0\dot{\mathbf{d}}_i &= {}^0\dot{\mathbf{d}}_{i-1} + {}^0\dot{R}_{i-1}\,{}^{i-1}\mathbf{d}_i + {}^0R_{i-1}\,{}^{i-1}\dot{\mathbf{d}}_i \\ &= {}^0\dot{\mathbf{d}}_{i-1} + {}^0\dot{R}_{i-1}\left(d_i\,{}^{i-1}\hat{k}_{i-1} + a_i\,{}^{i-1}\hat{\imath}_i\right) + {}^0R_{i-1}\,\dot{d}_i\,{}^{i-1}\hat{k}_{i-1} \end{aligned} \tag{8.531}$$

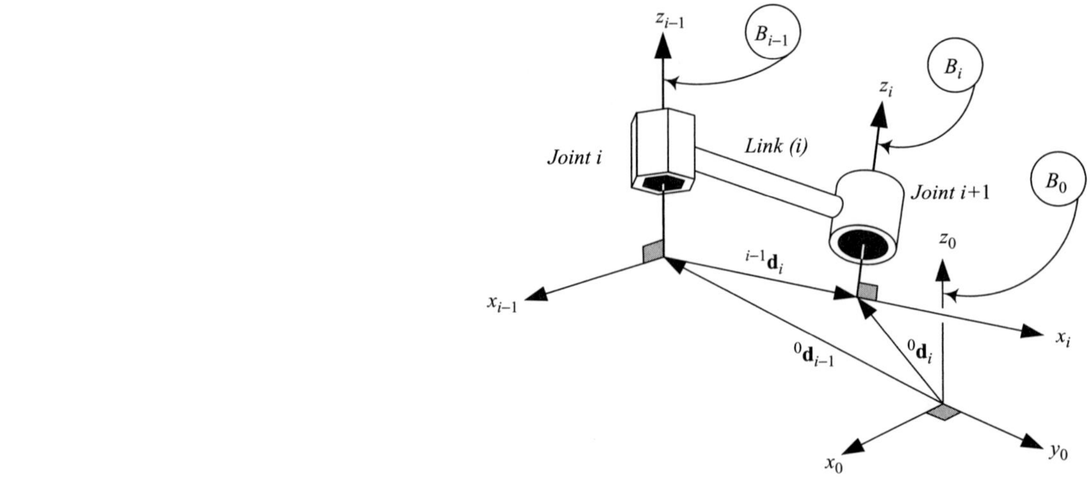

Figure 8.18 Link (i) and associated coordinate frames.

shows that the global velocity of the origin of B_i is a function of the translational and angular velocities of link B_{i-1}. However, we can use the following relations

$$ {}_{i-1}^{\;\;0}\dot{\mathbf{d}}_i = {}^0\dot{\mathbf{d}}_i - {}^0\dot{\mathbf{d}}_{i-1} \tag{8.532}$$

$$\begin{aligned} {}^0\dot{R}_{i-1}{}^{i-1}\mathbf{d}_i &= {}_0\boldsymbol{\omega}_{i-1} \times {}^0R_{i-1}{}^{i-1}\mathbf{d}_i = {}_0\boldsymbol{\omega}_{i-1} \times {}_{i-1}^{\;\;0}\mathbf{d}_i \\ &= \dot{\theta}_i\, {}^0\hat{k}_{i-1} \times {}_{i-1}^{\;\;0}\mathbf{d}_i \end{aligned} \tag{8.533}$$

$$ {}^0R_{i-1}\, \dot{d}_i\, {}^{i-1}\hat{k}_{i-1} = \dot{d}_i\, {}^0R_{i-1}{}^{i-1}\hat{k}_{i-1} = \dot{d}_i\, {}^0\hat{k}_{i-1} \tag{8.534}$$

and conclude that

$$ {}_{i-1}^{\;\;0}\dot{\mathbf{d}}_i = \dot{\theta}_i\, {}^0\hat{k}_{i-1} \times {}_{i-1}^{\;\;0}\mathbf{d}_i + \dot{d}_i\, {}^0\hat{k}_{i-1} \tag{8.535}$$

At each joint, either θ or d is variable. Therefore,

$$ {}_{i-1}^{\;\;0}\dot{\mathbf{d}}_i = {}_0^0\boldsymbol{\omega}_i \times {}_{i-1}^{\;\;0}\mathbf{d}_i \quad \text{if joint } i \text{ is R} \tag{8.536}$$

or

$$ {}_{i-1}^{\;\;0}\dot{\mathbf{d}}_i = \dot{d}_i{}^0\hat{k}_{i-1} + {}_0^0\boldsymbol{\omega}_i \times {}_{i-1}^{\;\;0}\mathbf{d}_i \quad \text{if joint } i \text{ is P} \tag{8.537}$$

The end-effector velocity is then expressed by

$$ {}_0^0\dot{\mathbf{d}}_n = \sum_{i=1}^{n} {}_{i-1}^{\;\;0}\dot{\mathbf{d}}_i = \sum_{i=1}^{n} \dot{\theta}_i{}^0\hat{k}_{i-1} \times {}_{i-1}^{\;\;0}\mathbf{d}_n \tag{8.538}$$

and

$$ {}_0^0\boldsymbol{\omega}_n = \sum_{i=1}^{n} {}_{i-1}^{\;\;0}\boldsymbol{\omega}_i = \sum_{i=1}^{n} \dot{\theta}_i{}^0\hat{k}_{i-1} \tag{8.539}$$

which can be rearranged in matrix form:

$$\begin{bmatrix} {}_0^0\dot{\mathbf{d}}_n \\ {}_0^0\boldsymbol{\omega}_n \end{bmatrix} = \sum_{i=1}^{n} \dot{\theta}_i \begin{bmatrix} {}^0\hat{k}_{i-1} \times {}_{i-1}^{\;\;0}\mathbf{d}_n \\ {}^0\hat{k}_{i-1} \end{bmatrix}$$

$$= \begin{bmatrix} {}^0\hat{k}_0 \times {}_0^0\mathbf{d}_n & {}^0\hat{k}_1 \times {}_1^0\mathbf{d}_n & \dots & {}^0\hat{k}_{n-1} \times {}_{n-1}^{\;\;0}\mathbf{d}_n \\ {}^0\hat{k}_0 & {}^0\hat{k}_1 & \dots & {}^0\hat{k}_{n-1} \end{bmatrix} \begin{bmatrix} \dot{\theta}_1 \\ \dot{\theta}_2 \\ \vdots \\ \dot{\theta}_n \end{bmatrix} \tag{8.540}$$

We usually show this equation by

$$\dot{\mathbf{X}} = \begin{bmatrix} {}_0^0\dot{\mathbf{d}}_n \\ {}_0^0\boldsymbol{\omega}_n \end{bmatrix} = \mathbf{J}\,\dot{\mathbf{q}} \tag{8.541}$$

where the vector $\dot{\mathbf{q}} = \left[\dot{q}_1, \dot{q}_2, \dots, \dot{q}_n\right]^{\mathrm{T}}$ is the *joint velocity vector* and $\mathbf{J}$ is the *Jacobian* matrix,

$$\mathbf{J} = \begin{bmatrix} {}^0\hat{k}_0 \times {}_0^0\mathbf{d}_n & {}^0\hat{k}_1 \times {}_1^0\mathbf{d}_n & \dots & {}^0\hat{k}_{n-1} \times {}_{n-1}^{\;\;0}\mathbf{d}_n \\ {}^0\hat{k}_0 & {}^0\hat{k}_1 & \dots & {}^0\hat{k}_{n-1} \end{bmatrix} \tag{8.542}$$

Employing the *Jacobian-generating vector* $\mathbf{c}_i(\mathbf{q})$ is an applied method to determine $\mathbf{J}$. The vector $\mathbf{c}_i(\mathbf{q})$ is associated with joint i. If joint i is revolute, then

$$\mathbf{c}_i(\mathbf{q}) = \begin{bmatrix} {}^0\hat{k}_{i-1} \times {}_{i-1}^{\;\;0}\mathbf{d}_n \\ {}^0\hat{k}_{i-1} \end{bmatrix} \tag{8.543}$$

and if joint i is prismatic, then $\mathbf{c}_i(\mathbf{q})$ simplifies to

$$\mathbf{c}_i(\mathbf{q}) = \begin{bmatrix} {}^0\hat{k}_{i-1} \\ 0 \end{bmatrix} \tag{8.544}$$

Equation (8.527) provides a set of six equations. The first three relate the translational velocity of the end effector along the x_0-axis. The rest of the equations relate the rotational velocities of the end-effector frame about the base-frame axes. ■

Example 522 ★ Jacobian Matrix of a Spherical Manipulator Figure 8.19 depicts a spherical manipulator.

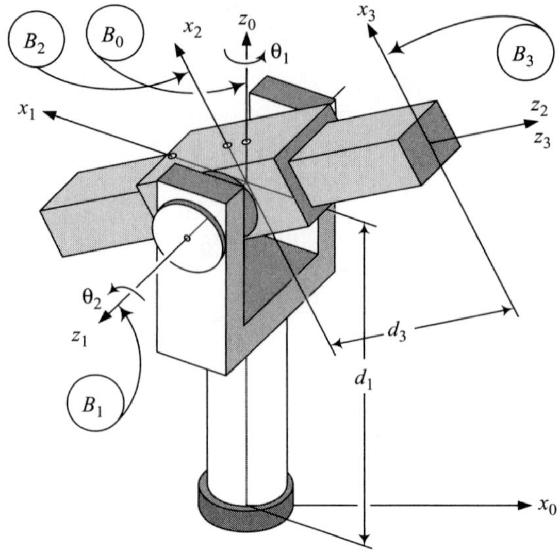

Figure 8.19 A spherical manipulator.

To find its Jacobian matrix, we start with determining the ${}^0\hat{k}_{i-1}$-axes for $i = 1, 2, 3$. It might be easier if we use the homogeneous definitions to find ${}^0\hat{k}_{i-1}$:

$$
{}^0\hat{k}_0 = \begin{bmatrix} 0 \\ 0 \\ 1 \\ 0 \end{bmatrix} \tag{8.545}
$$

$$
\begin{aligned}
{}^0\hat{k}_1 = {}^0T_1\,{}^1\hat{k}_1 &= \begin{bmatrix} \cos\theta_1 & 0 & -\sin\theta_1 & 0 \\ \sin\theta_1 & 0 & \cos\theta_1 & 0 \\ 0 & -1 & 0 & l_0 \\ 0 & 0 & 0 & 1 \end{bmatrix} \begin{bmatrix} 0 \\ 0 \\ 1 \\ 0 \end{bmatrix} \\
&= \begin{bmatrix} -\sin\theta_1 \\ \cos\theta_1 \\ 0 \\ 0 \end{bmatrix}
\end{aligned} \tag{8.546}
$$

$$
\begin{aligned}
{}^0\hat{k}_2 = {}^0T_2\,{}^2\hat{k}_2 &= \begin{bmatrix} c\theta_1 c\theta_2 & -s\theta_1 & c\theta_1 s\theta_2 & 0 \\ c\theta_2 s\theta_1 & c\theta_1 & s\theta_1 s\theta_2 & 0 \\ -s\theta_2 & 0 & c\theta_2 & l_0 \\ 0 & 0 & 0 & 1 \end{bmatrix} \begin{bmatrix} 0 \\ 0 \\ 1 \\ 0 \end{bmatrix} \\
&= \begin{bmatrix} \cos\theta_1 \sin\theta_2 \\ \sin\theta_1 \sin\theta_2 \\ \cos\theta_2 \\ 0 \end{bmatrix}
\end{aligned} \tag{8.547}
$$

Then, we determine the vectors ${}_{i-1}^{\ 0}\mathbf{d}_n$:

$$ {}_0^0\mathbf{d}_3 = l_0{}^0\hat{k}_0 + d_3{}^0\hat{k}_2 = \begin{bmatrix} d_3\cos\theta_1\sin\theta_2 \\ d_3\sin\theta_1\sin\theta_2 \\ l_0 + d_3\cos\theta_2 \\ 0 \end{bmatrix} \tag{8.548} $$

$$ {}_1^0\mathbf{d}_3 = d_3{}^0\hat{k}_2 = \begin{bmatrix} d_3\cos\theta_1\sin\theta_2 \\ d_3\sin\theta_1\sin\theta_2 \\ d_3\cos\theta_2 \\ 0 \end{bmatrix} \tag{8.549} $$

Therefore, the Jacobian of the manipulator is

$$ \begin{aligned} \mathbf{J} &= \begin{bmatrix} {}^0\hat{k}_0 \times {}_0^0\mathbf{d}_3 & {}^0\hat{k}_1 \times {}_1^0\mathbf{d}_3 & {}^0\hat{k}_2 \\ {}^0\hat{k}_0 & {}^0\hat{k}_1 & 0 \end{bmatrix} \\ &= \begin{bmatrix} -d_3\sin\theta_1\sin\theta_2 & d_3\cos\theta_1\cos\theta_2 & \cos\theta_1\sin\theta_2 \\ d_3\cos\theta_1\sin\theta_2 & d_3\cos\theta_2\sin\theta_1 & \sin\theta_1\sin\theta_2 \\ 0 & -d_3\sin\theta_2 & \cos\theta_2 \\ 0 & -\sin\theta_1 & 0 \\ 0 & \cos\theta_1 & 0 \\ 1 & 0 & 0 \end{bmatrix} \end{aligned} \tag{8.550} $$

Example 523 ★ Jacobian Matrix for an Articulated Manipulator A robot is shown in Figure 7.71 and its transformation matrices are calculated in Example 464. Because the articulated manipulator has six *DOF*, its Jacobian would be a 6×6 matrix. The ith column of the Jacobian matrix is

$$ \mathbf{c}_i(\mathbf{q}) = \begin{bmatrix} {}^0\hat{k}_{i-1} \times {}_{i-1}^{\ 0}\mathbf{d}_6 \\ {}^0\hat{k}_{i-1} \end{bmatrix} \tag{8.551} $$

For the first column, we need to find ${}^0\hat{k}_0$ and ${}^0\mathbf{d}_6$. The direction of the z_0-axis in the base coordinate frame is always

$$ {}^0\hat{k}_0 = \begin{bmatrix} 0 \\ 0 \\ 1 \end{bmatrix} \tag{8.552} $$

and the position vector of the end-effector frame B_6 is ${}^0\mathbf{d}_6$. The vector ${}^0\mathbf{d}_6$ can directly be determined from the fourth column of the transformation matrix ${}^0T_6$:

$$\begin{aligned} {}^0T_6 &= {}^0T_1\,{}^1T_2\,{}^2T_3\,{}^3T_4\,{}^4T_5\,{}^5T_6 \\ &= \begin{bmatrix} {}^0R_6 & {}^0\mathbf{d}_6 \\ 0 & 1 \end{bmatrix} = \begin{bmatrix} t_{11} & t_{12} & t_{13} & t_{14} \\ t_{21} & t_{22} & t_{23} & t_{24} \\ t_{31} & t_{32} & t_{33} & t_{34} \\ 0 & 0 & 0 & 1 \end{bmatrix} \end{aligned} \tag{8.553}$$

$${}^0\mathbf{d}_6 = \begin{bmatrix} t_{14} \\ t_{24} \\ t_{34} \end{bmatrix} \tag{8.554}$$

where

$$\begin{aligned} t_{14} = {} & d_6\{s\theta_1 s\theta_4 s\theta_5 + c\theta_1 \left[c\theta_4 s\theta_5 c\left(\theta_2+\theta_3\right]\right\} + c\theta_5 s\left(\theta_2+\theta_3\right)) \\ & + l_3 c\theta_1 s\left(\theta_2+\theta_3\right) + d_2 s\theta_1 + l_2 c\theta_1 c\theta_2 \end{aligned} \tag{8.555}$$

$$\begin{aligned} t_{24} = {} & d_6\{-c\theta_1 s\theta_4 s\theta_5 + s\theta_1 \left[c\theta_4 s\theta_5 c\left(\theta_2+\theta_3\right) + c\theta_5 s\left(\theta_2+\theta_3\right)\right]\} \\ & + s\theta_1 s\left(\theta_2+\theta_3\right) l_3 - d_2 c\theta_1 + l_2 c\theta_2 s\theta_1 \end{aligned} \tag{8.556}$$

$$\begin{aligned} t_{34} = {} & d_6\left[c\theta_4 s\theta_5 s\left(\theta_2+\theta_3\right) - c\theta_5 c\left(\theta_2+\theta_3\right)\right] \\ & + l_2 s\theta_2 + l_3 c\left(\theta_2+\theta_3\right) \end{aligned} \tag{8.557}$$

Therefore,

$${}^0\hat{k}_0 \times {}^0\mathbf{d}_6 = \begin{bmatrix} 0 \\ 0 \\ 1 \end{bmatrix} \times \begin{bmatrix} t_{14} \\ t_{24} \\ t_{34} \end{bmatrix} = \begin{bmatrix} -t_{24} \\ t_{14} \\ 0 \end{bmatrix} \tag{8.558}$$

and the first Jacobian-generating vector is

$$\mathbf{c}_1 = \begin{bmatrix} -t_{24} \\ t_{14} \\ 0 \\ 0 \\ 0 \\ 1 \end{bmatrix} \tag{8.559}$$

For the second column we need to find ${}^0\hat{k}_1$ and ${}^0_1\mathbf{d}_6$. The z_1-axis in the base frame can be found by

$$\begin{aligned} {}^0\hat{k}_1 &= {}^0R_1\,{}^1\hat{k}_1 = {}^0R_1 \begin{bmatrix} 0 \\ 0 \\ 1 \end{bmatrix} = \begin{bmatrix} c\theta_1 & 0 & s\theta_1 \\ s\theta_1 & 0 & -c\theta_1 \\ 0 & 1 & 0 \end{bmatrix} \begin{bmatrix} 0 \\ 0 \\ 1 \end{bmatrix} \\ &= \begin{bmatrix} \sin\theta_1 \\ -\cos\theta_1 \\ 0 \end{bmatrix} \end{aligned} \tag{8.560}$$

The first half of $\mathbf{c}_2$ is ${}^0\hat{k}_1 \times {}^0_1\mathbf{d}_6$. The vector ${}^0_1\mathbf{d}_6$ is the position of the end effector in the coordinate frame B_1. However, it must be expressed in the base frame to be able to perform the cross product. An easier method is to find ${}^1\hat{k}_1 \times {}^1\mathbf{d}_6$ and transform the resultant into the base frame. The vector ${}^1\mathbf{d}_6$ is the fourth column of ${}^1T_6 = {}^1T_2{}^2T_3{}^3T_4{}^4T_5{}^5T_6$, which, from Example 464, is equal to

$$
{}^1\mathbf{d}_6 = \begin{bmatrix} l_2\cos\theta_2 + l_3\sin(\theta_2+\theta_3) \\ l_2\sin\theta_2 - l_3\cos(\theta_2+\theta_3) \\ d_2 \end{bmatrix} \tag{8.561}
$$

Therefore, the first half of $\mathbf{c}_2$ is

$$
\begin{aligned}
{}^0\hat{k}_1 \times {}^0_1\mathbf{d}_6 &= {}^0R_1\left({}^1\hat{k}_1 \times {}^1\mathbf{d}_6\right) \\
&= {}^0R_1\left(\begin{bmatrix} 0 \\ 0 \\ 1 \end{bmatrix} \times \begin{bmatrix} l_2\cos\theta_2 + l_3\sin(\theta_2+\theta_3) \\ l_2\sin\theta_2 - l_3\cos(\theta_2+\theta_3) \\ d_2 \end{bmatrix}\right) \\
&= \begin{bmatrix} \cos\theta_1\left(-l_2\sin\theta_2 + l_3\cos(\theta_2+\theta_3)\right) \\ \sin\theta_1\left(-l_2\sin\theta_2 + l_3\cos(\theta_2+\theta_3)\right) \\ l_2\cos\theta_2 + l_3\sin(\theta_2+\theta_3) \end{bmatrix}
\end{aligned} \tag{8.562}
$$

and

$$
\mathbf{c}_2 = \begin{bmatrix} \cos\theta_1\left(-l_2\sin\theta_2 + l_3\cos(\theta_2+\theta_3)\right) \\ \sin\theta_1\left(-l_2\sin\theta_2 + l_3\cos(\theta_2+\theta_3)\right) \\ l_2\cos\theta_2 + l_3\sin(\theta_2+\theta_3) \\ \sin\theta_1 \\ -\cos\theta_1 \\ 0 \end{bmatrix} \tag{8.563}
$$

The third column is made by ${}^0\hat{k}_2$ and ${}^0_2\mathbf{d}_6$. The vector ${}^0_2\mathbf{d}_6$ is the position of the end effector in the coordinate frame B_2 and is the fourth column of ${}^2T_6 = {}^2T_3{}^3T_4{}^4T_5{}^5T_6$. The z_2-axis in the base frame can be found by

$$
{}^0\hat{k}_2 = {}^0R_2{}^2\hat{k}_2 = {}^0R_1{}^1R_2\begin{bmatrix} 0 \\ 0 \\ 1 \end{bmatrix} = \begin{bmatrix} \sin\theta_1 \\ -\cos\theta_1 \\ 0 \end{bmatrix} \tag{8.564}
$$

and the cross product ${}^0\hat{k}_2 \times {}^0_2\mathbf{d}_6$ can be found by transforming the resultant of ${}^2\hat{k}_2 \times {}^2\mathbf{d}_6$ into the base coordinate frame:

$$
{}^2\hat{k}_2 \times {}^2\mathbf{d}_6 = \begin{bmatrix} l_3\cos\theta_3 \\ l_3\sin\theta_3 \\ 0 \end{bmatrix} \tag{8.565}
$$

$$
{}^0R_2\left({}^2\hat{k}_2 \times {}^2\mathbf{d}_6\right) = \begin{bmatrix} l_3\cos\theta_1\sin(\theta_2+\theta_3) \\ l_3\sin\theta_1\sin(\theta_2+\theta_3) \\ -l_3\cos(\theta_2+\theta_3) \end{bmatrix} \tag{8.566}
$$

Therefore,

$$\mathbf{c}_3 = \begin{bmatrix} l_3 \cos\theta_1 \sin(\theta_2+\theta_3) \\ l_3 \sin\theta_1 \sin(\theta_2+\theta_3) \\ -l_3 \cos(\theta_2+\theta_3) \\ \sin\theta_1 \\ -\cos\theta_1 \\ 0 \end{bmatrix} \tag{8.567}$$

The fourth column needs ${}^0\hat{k}_3$ and ${}^0_3\mathbf{d}_6$. The vector ${}^0\hat{k}_3$ can be found by transforming ${}^3\hat{k}_3$ to the base frame:

$$\begin{aligned} {}^0\hat{k}_3 &= {}^0R_1\,{}^1R_2\,{}^2R_3 \begin{bmatrix} 0 \\ 0 \\ 1 \end{bmatrix} \\ &= \begin{bmatrix} \cos\theta_1 (\cos\theta_2 \sin\theta_3 + \cos\theta_3 \sin\theta_2) \\ \sin\theta_1 (\cos\theta_2 \sin\theta_3 + \sin\theta_2 \cos\theta_3) \\ -\cos(\theta_2+\theta_3) \end{bmatrix} \end{aligned} \tag{8.568}$$

and the first half of $\mathbf{J}_4$ can be found by calculating ${}^3\mathrm{k}_3 \times {}^3\mathbf{d}_6$ and transforming the resultant into the base coordinate frame:

$${}^0R_3 \left({}^3\hat{k}_3 \times {}^3\mathbf{d}_6\right) = {}^0R_3 \left(\begin{bmatrix} 0 \\ 0 \\ 1 \end{bmatrix} \times \begin{bmatrix} 0 \\ 0 \\ l_3 \end{bmatrix} \right) = \begin{bmatrix} 0 \\ 0 \\ 0 \end{bmatrix} \tag{8.569}$$

Therefore,

$$\mathbf{c}_4 = \begin{bmatrix} 0 \\ 0 \\ 0 \\ \cos\theta_1 (\cos\theta_2 \sin\theta_3 + \cos\theta_3 \sin\theta_2) \\ \sin\theta_1 (\cos\theta_2 \sin\theta_3 + \sin\theta_2 \cos\theta_3) \\ -\cos(\theta_2+\theta_3) \end{bmatrix} \tag{8.570}$$

The fifth column needs ${}^0\hat{k}_4$ and ${}^0_4\mathbf{d}_6$. We can find the vector ${}^0\hat{k}_4$ by transforming ${}^4\hat{k}_4$ to the base frame:

$${}^0\hat{k}_4 = {}^0R_4 \begin{bmatrix} 0 \\ 0 \\ 1 \end{bmatrix} = \begin{bmatrix} c\theta_4 s\theta_1 - c\theta_1 s\theta_4 c(\theta_2+\theta_3) \\ -c\theta_1 c\theta_4 - s\theta_1 s\theta_4 c(\theta_2+\theta_3) \\ -s\theta_4 s(\theta_2+\theta_3) \end{bmatrix} \tag{8.571}$$

The first half of $\mathbf{c}_5$ is ${}^4\hat{k}_4 \times {}^4\mathbf{d}_6$ and is expressed in the base coordinate frame:

$${}^0R_4 \left({}^4\hat{k}_4 \times {}^4\mathbf{d}_6\right) = {}^0R_4 \left(\begin{bmatrix} 0 \\ 0 \\ 1 \end{bmatrix} \times \begin{bmatrix} 0 \\ 0 \\ 0 \end{bmatrix} \right) = \begin{bmatrix} 0 \\ 0 \\ 0 \end{bmatrix} \tag{8.572}$$

Therefore,

$$\mathbf{c}_5 = \begin{bmatrix} 0 \\ 0 \\ 0 \\ \cos\theta_4 \sin\theta_1 - \cos\theta_1 \sin\theta_4 \cos(\theta_2+\theta_3) \\ -\cos\theta_1 \cos\theta_4 - \sin\theta_1 \sin\theta_4 \cos(\theta_2+\theta_3) \\ -\sin\theta_4 \sin(\theta_2+\theta_3) \end{bmatrix} \tag{8.573}$$

The sixth column is found by calculating ${}^0\hat{k}_5$ and ${}^0\hat{k}_5 \times {}^0_5\mathbf{d}_6$:

$${}^0\hat{k}_5 = {}^0R_5 \begin{bmatrix} 0 \\ 0 \\ 1 \end{bmatrix} \tag{8.574}$$

$$= \begin{bmatrix} -c\theta_1 c\theta_4 s(\theta_2+\theta_3) - s\theta_4 (s\theta_1 s\theta_4 + c\theta_1 c\theta_4 c(\theta_2+\theta_3)) \\ -s\theta_1 c\theta_4 s(\theta_2+\theta_3) - s\theta_4 (-c\theta_1 s\theta_4 + s\theta_1 c\theta_4 c(\theta_2+\theta_3)) \\ c\theta_4 c(\theta_2+\theta_3) - \frac{1}{2} s(\theta_2+\theta_3)\, s2\theta_4 \end{bmatrix}$$

The first half of $\mathbf{c}_6$ is ${}^5\hat{k}_5 \times {}^5\mathbf{d}_6$ and is expressed in the base coordinate frame:

$${}^0R_5 \left({}^5\hat{k}_5 \times {}^5\mathbf{d}_6\right) = {}^0R_5 \left(\begin{bmatrix} 0 \\ 0 \\ 1 \end{bmatrix} \times \begin{bmatrix} 0 \\ 0 \\ 0 \end{bmatrix} \right) = \begin{bmatrix} 0 \\ 0 \\ 0 \end{bmatrix} \tag{8.575}$$

Therefore,

$$\mathbf{c}_6 = \begin{bmatrix} 0 \\ 0 \\ 0 \\ -c\theta_1 c\theta_4 s(\theta_2+\theta_3) - s\theta_4 (s\theta_1 s\theta_4 + c\theta_1 c\theta_4 c(\theta_2+\theta_3)) \\ -s\theta_1 c\theta_4 s(\theta_2+\theta_3) - s\theta_4 (-c\theta_1 s\theta_4 + s\theta_1 c\theta_4 c(\theta_2+\theta_3)) \\ c\theta_4 c(\theta_2+\theta_3) - \frac{1}{2} s(\theta_2+\theta_3)\, s2\theta_4 \end{bmatrix} \tag{8.576}$$

and the Jacobian matrix for the articulated manipulator would be

$$\mathbf{J} = \begin{bmatrix} \mathbf{c}_1 & \mathbf{c}_2 & \mathbf{c}_3 & \mathbf{c}_4 & \mathbf{c}_5 & \mathbf{c}_6 \end{bmatrix} \tag{8.577}$$

Example 524 ★ Effect of a Spherical Wrist on Jacobian Matrix The Jacobian matrix for a robot having a spherical wrist is always of the form

$$\mathbf{J} = \begin{bmatrix} {}^0\hat{k}_0 \times {}^0_0\mathbf{d}_6 & {}^0\hat{k}_1 \times {}^0_1\mathbf{d}_6 & {}^0\hat{k}_2 \times {}^0_2\mathbf{d}_6 & 0 & 0 & 0 \\ {}^0\hat{k}_0 & {}^0\hat{k}_1 & {}^0\hat{k}_2 & {}^0\hat{k}_3 & {}^0\hat{k}_6 & {}^0\hat{k}_5 \end{bmatrix} \tag{8.578}$$

which shows that the upper 3×3 submatrix is zero. This is because of the spherical wrist structure and having a wrist point as the origin of the wrist coordinate frames B_4, B_5, and B_6.

Example 525 ★ Jacobian Matrix by Direct Differentiation Figure 7.71 illustrates an articulated robot. Its transformation matrices are given in Example 464. Using the result of forward kinematics,

$$ {}^0T_6 = \begin{bmatrix} {}^0R_6 & {}^0\mathbf{d}_6 \\ 0 & 1 \end{bmatrix} \tag{8.579} $$

we know that the position of the end effector is at

$$ {}^0\mathbf{d}_6 = \begin{bmatrix} X_6 \\ Y_6 \\ Z_6 \end{bmatrix} = \begin{bmatrix} t_{14} \\ t_{24} \\ t_{34} \end{bmatrix} \tag{8.580} $$

where

$$ \begin{aligned} t_{14} &= d_6\{s\theta_1 s\theta_4 s\theta_5 + c\theta_1 \left[c\theta_4 s\theta_5 c(\theta_2+\theta_3) + c\theta_5 s(\theta_2+\theta_3)\right]\} \\ &\quad + l_3 c\theta_1 s(\theta_2+\theta_3) + d_2 s\theta_1 + l_2 c\theta_1 c\theta_2 \end{aligned} \tag{8.581} $$

$$ \begin{aligned} t_{24} &= d_6\{-c\theta_1 s\theta_4 s\theta_5 + s\theta_1 \left[c\theta_4 s\theta_5 c(\theta_2+\theta_3) + c\theta_5 s(\theta_2+\theta_3)\right]\} \\ &\quad + s\theta_1 s(\theta_2+\theta_3) l_3 - d_2 c\theta_1 + l_2 c\theta_2 s\theta_1 \end{aligned} \tag{8.582} $$

$$ \begin{aligned} t_{34} &= d_6 \left[c\theta_4 s\theta_5 s(\theta_2+\theta_3) - c\theta_5 c(\theta_2+\theta_3)\right] \\ &\quad + l_2 s\theta_2 + l_3 c(\theta_2+\theta_3) \end{aligned} \tag{8.583} $$

Taking the derivative of X_6 yields

$$ \begin{aligned} \dot{X}_6 &= \frac{\partial X_6}{\partial \theta_1}\dot{\theta}_1 + \frac{\partial X_6}{\partial \theta_2}\dot{\theta}_2 + \cdots + \frac{\partial X_6}{\partial \theta_6}\dot{\theta}_6 \\ &= J_{11}\dot{\theta}_1 + J_{12}\dot{\theta}_2 + \cdots + J_{16}\dot{\theta}_6 \\ &= -t_{24}\dot{\theta}_1 + c\theta_1 \left[-l_2 s\theta_2 + l_3 c(\theta_2+\theta_3)\right]\dot{\theta}_2 \\ &\quad + l_3 c\theta_1 s(\theta_2+\theta_3)\dot{\theta}_3 \end{aligned} \tag{8.584} $$

where

$$ \begin{aligned} J_{11} &= -t_{24} \\ J_{12} &= \cos\theta_1 \left(-l_2 \sin\theta_2 + l_3 \cos\left(\theta_2+\theta_3\right)\right) \\ J_{13} &= l_3 \cos\theta_1 \sin\left(\theta_2+\theta_3\right) \\ J_{14} &= 0 \\ J_{15} &= 0 \\ J_{16} &= 0 \end{aligned} \tag{8.585} $$

Similarly, the derivative of Y_6 is given as

$$\begin{aligned}
\dot{Y}_6 &= \frac{\partial Y_6}{\partial \theta_1}\dot{\theta}_1 + \frac{\partial Y_6}{\partial \theta_2}\dot{\theta}_2 + \cdots + \frac{\partial Y_6}{\partial \theta_6}\dot{\theta}_6 \\
&= J_{21}\dot{\theta}_1 + J_{22}\dot{\theta}_2 + \cdots + J_{26}\dot{\theta}_6 \\
&= t_{14}\dot{\theta}_1 + s\theta_1 \left[-l_2 s\theta_2 + l_3 c(\theta_2 + \theta_3)\right] \dot{\theta}_2 \\
&\quad + l_3 s\theta_1 s\left(\theta_2 + \theta_3\right) \dot{\theta}_3
\end{aligned} \tag{8.586}$$

where

$$\begin{aligned}
J_{21} &= t_{14} \\
J_{22} &= \sin\theta_1 \left(-l_2 \sin\theta_2 + l_3 \cos\left(\theta_2 + \theta_3\right)\right) \\
J_{23} &= l_3 \sin\theta_1 \sin\left(\theta_2 + \theta_3\right) \\
J_{24} &= 0 \\
J_{25} &= 0 \\
J_{26} &= 0
\end{aligned} \tag{8.587}$$

and the derivative of Z_6 is given as

$$\begin{aligned}
\dot{Z}_6 &= \frac{\partial Z_6}{\partial \theta_1}\dot{\theta}_1 + \frac{\partial Z_6}{\partial \theta_2}\dot{\theta}_2 + \cdots + \frac{\partial Z_6}{\partial \theta_6}\dot{\theta}_6 \\
&= J_{31}\dot{\theta}_1 + J_{32}\dot{\theta}_2 + \cdots + J_{36}\dot{\theta}_6 \\
&= \left[l_2 \cos\theta_2 + l_3 \sin(\theta_2 + \theta_3)\right] \dot{\theta}_2 - l_3 \cos(\theta_2 + \theta_3)\dot{\theta}_3
\end{aligned} \tag{8.588}$$

where

$$\begin{aligned}
J_{31} &= 0 \\
J_{32} &= l_2 \cos\theta_2 + l_3 \sin\left(\theta_2 + \theta_3\right) \\
J_{33} &= -l_3 \cos\left(\theta_2 + \theta_3\right) \\
J_{34} &= 0 \\
J_{35} &= 0 \\
J_{36} &= 0.
\end{aligned} \tag{8.589}$$

There is no explicit equation for expressing the rotations of the end effector's frame about the global axes. Therefore, there is no equation to find differential rotations about the axes. This is the reason indirect or more systematic methods for evaluating the Jacobian matrix should be used.

The last three rows of the Jacobian matrix can be found by calculating the angular velocity vector based on the angular velocity matrix,

$$ {}_0\tilde{\omega}_6 = {}^0\dot{R}_6\,{}^0R_6^{\mathrm{T}} = \begin{bmatrix} 0 & -\omega_Z & \omega_Y \\ \omega_Z & 0 & -\omega_X \\ -\omega_Y & \omega_X & 0 \end{bmatrix} \tag{8.590} $$

$$ {}_0\boldsymbol{\omega}_6 = \begin{bmatrix} \omega_X \\ \omega_Y \\ \omega_Z \end{bmatrix} \tag{8.591} $$

and then rearranging the components to show the Jacobian elements:

$$ \omega_X = \frac{\partial \omega_X}{\partial \theta_1}\dot{\theta}_1 + \frac{\partial \omega_X}{\partial \theta_2}\dot{\theta}_2 + \cdots + \frac{\partial \omega_X}{\partial \theta_6}\dot{\theta}_6 \tag{8.592} $$

$$ \omega_Y = \frac{\partial \omega_Y}{\partial \theta_1}\dot{\theta}_1 + \frac{\partial \omega_Y}{\partial \theta_2}\dot{\theta}_2 + \cdots + \frac{\partial \omega_Y}{\partial \theta_6}\dot{\theta}_6 \tag{8.593} $$

$$ \omega_Z = \frac{\partial \omega_Z}{\partial \theta_1}\dot{\theta}_1 + \frac{\partial \omega_Z}{\partial \theta_2}\dot{\theta}_2 + \cdots + \frac{\partial \omega_Z}{\partial \theta_6}\dot{\theta}_6 \tag{8.594} $$

The angular velocity vector of the end-effector frame is given as

$$ \begin{aligned} \omega_X = {} & \sin\theta_1\dot{\theta}_2 + \sin\theta_1\dot{\theta}_3 + \cos\theta_1 \sin\theta_{23}\dot{\theta}_4 \\ & + (\cos\theta_4 \sin\theta_1 - \cos\theta_1 \sin\theta_4 \cos\theta_{23})\,\dot{\theta}_5 \\ & - [c\theta_1 c\theta_4 s\theta_{23} + s\theta_4\,(s\theta_1 s\theta_4 + c\theta_1 c\theta_4 c\theta_{23})]\,\dot{\theta}_6 \end{aligned} \tag{8.595} $$

$$ \begin{aligned} \omega_Y = {} & -\cos\theta_1\dot{\theta}_2 - \cos\theta_1\dot{\theta}_3 + \sin\theta_1 \sin\theta_{23}\dot{\theta}_4 \\ & + (-\cos\theta_1 \cos\theta_4 - \sin\theta_1 \sin\theta_4 \cos\theta_{23})\,\dot{\theta}_5 \\ & + [-s\theta_1 c\theta_4 s\theta_{23} - s\theta_4\,(-c\theta_1 s\theta_4 + s\theta_1 c\theta_4 c\theta_{23})]\,\dot{\theta}_6 \end{aligned} \tag{8.596} $$

$$ \begin{aligned} \omega_Z = {} & \dot{\theta}_1 - \cos(\theta_2+\theta_3)\,\dot{\theta}_4 - \sin\theta_4 \sin(\theta_2+\theta_3)\,\dot{\theta}_5 \\ & + \left(\cos\theta_4 \cos\theta_{23} - \tfrac{1}{2}\sin\theta_{23}\sin 2\theta_4\right)\dot{\theta}_6 \end{aligned} \tag{8.597} $$

and therefore,

$$ \begin{aligned} J_{41} &= 0 \\ J_{42} &= \sin\theta_1 \\ J_{43} &= \sin\theta_1 \\ J_{44} &= \cos\theta_1\,(\cos\theta_2 \sin\theta_3 + \cos\theta_3 \sin\theta_2) \\ J_{45} &= \cos\theta_4 \sin\theta_1 - \cos\theta_1 \sin\theta_4 \cos(\theta_2+\theta_3) \\ J_{46} &= -c\theta_1 c\theta_4 s\,(\theta_2+\theta_3) - s\theta_4\,[s\theta_1 s\theta_4 + c\theta_1 c\theta_4 c(\theta_2+\theta_3)] \end{aligned} \tag{8.598} $$

$$\begin{aligned}
J_{51} &= 0 \\
J_{52} &= -\cos\theta_1 \\
J_{53} &= -\cos\theta_1 \\
J_{54} &= \sin\theta_1 (\cos\theta_2 \sin\theta_3 + \sin\theta_2 \cos\theta_3) \\
J_{55} &= -\cos\theta_1 \cos\theta_4 - \sin\theta_1 \sin\theta_4 \cos(\theta_2 + \theta_3) \\
J_{56} &= -s\theta_1 c\theta_4 s(\theta_2 + \theta_3) - s\theta_4 [-c\theta_1 s\theta_4 + s\theta_1 c\theta_4 c(\theta_2 + \theta_3)]
\end{aligned} \tag{8.599}$$

$$\begin{aligned}
J_{61} &= 1 \\
J_{62} &= 0 \\
J_{63} &= 0 \\
J_{64} &= -\cos(\theta_2 + \theta_3) \\
J_{65} &= -\sin\theta_4 \sin(\theta_2 + \theta_3) \\
J_{66} &= \cos\theta_4 \cos(\theta_2 + \theta_3) - \frac{1}{2}\sin(\theta_2 + \theta_3)\sin 2\theta_4
\end{aligned} \tag{8.600}$$

Example 526 ★ Analytical Jacobian and Geometric Jacobian Assume the global position and orientation of the end-effector frame are specified by a set of six parameters arranged in a vector $\mathbf{X}$,

$$\mathbf{X} = \begin{bmatrix} {}^0\mathbf{r}_n \\ {}^0\phi_n \end{bmatrix} \tag{8.601}$$

where ${}^0\mathbf{r}_n$ is the Cartesian position of the end-effector frame,

$${}^0\mathbf{r}_n = {}^0\mathbf{r}_n(\mathbf{q}) \tag{8.602}$$

and ${}^0\phi_n$ represents three independent rotational parameters,

$${}^0\phi_n = {}^0\phi_n(\mathbf{q}) \tag{8.603}$$

both functions of the joint variable vector $\mathbf{q}$.

The translational velocity of the end-effector frame can be expressed by

$${}^0\mathbf{r}_n = \frac{\partial \mathbf{r}}{\partial \mathbf{q}}\dot{\mathbf{q}} = \mathbf{J}_D(\mathbf{q})\,\dot{\mathbf{q}} \tag{8.604}$$

and the rotational velocity of the end-effector frame can be expressed by

$${}^0\dot{\boldsymbol{\phi}}_n = \frac{\partial \phi}{\partial \mathbf{q}}\dot{\mathbf{q}} = \mathbf{J}_\phi(\mathbf{q})\,\dot{\mathbf{q}} \tag{8.605}$$

The rotational velocity vector $\dot{\boldsymbol{\phi}}$ in general differs from the angular velocity vector $\boldsymbol{\omega}$. The combination of the Jacobian matrices $\mathbf{J}_D$ and $\mathbf{J}_\phi$ in the form

$$\mathbf{J}_A = \begin{bmatrix} \mathbf{J}_D \\ \mathbf{J}_\phi \end{bmatrix} \tag{8.606}$$

is called the *analytical Jacobian* to indicate its difference with the *geometric Jacobian* $\mathbf{J}$.

Having a set of orientation angles, ϕ, it is possible to find the relationship between the angular velocity $\boldsymbol{\omega}$ and the rotational velocity $\dot{\boldsymbol{\phi}}$. As an example, consider the Euler angles $\varphi\theta\psi$ about the zxz-axes. The global angular velocity in terms of Euler frequencies is found in 4.181:

$$\begin{bmatrix} \omega_X \\ \omega_Y \\ \omega_Z \end{bmatrix} = \begin{bmatrix} 0 & \cos\varphi & \sin\theta\sin\varphi \\ 0 & \sin\varphi & -\cos\varphi\sin\theta \\ 1 & 0 & \cos\theta \end{bmatrix} \begin{bmatrix} \dot{\varphi} \\ \dot{\theta} \\ \dot{\psi} \end{bmatrix} \tag{8.607}$$

$$\boldsymbol{\omega} = T(\phi)\,\dot{\boldsymbol{\phi}} \tag{8.608}$$

8.9 ★ INVERSE-VELOCITY KINEMATICS

The *inverse-velocity kinematics problem*, also known as the resolved-rates problem, searches for the joint velocity vector associated with the end-effector velocity vector. For a six-DOF connected multibody, such as a robotic manipulator, six DOF are needed to be able to move the end effector in an arbitrary direction with an arbitrary angular velocity. The speed vector of the end effector $\dot{\mathbf{X}}$ is related to the joint speed vector $\dot{\mathbf{q}}$ by the Jacobian matrix:

$$\dot{\mathbf{X}} = \begin{bmatrix} {}^0\mathbf{v}_n \\ {}^0\boldsymbol{\omega}_n \end{bmatrix} = \begin{bmatrix} \mathbf{J}_D \\ \mathbf{J}_R \end{bmatrix} \dot{\mathbf{q}} = \mathbf{J}\,\dot{\mathbf{q}} \tag{8.609}$$

Consequently, for the inverse-velocity kinematics, we require that the differential change in joint coordinates be expressed in terms of Cartesian translation and the angular velocities of the end effector. If the Jacobian matrix is non-singular at the moment of calculation, the inverse Jacobian $\mathbf{J}^{-1}$ exists and we are able to find the required joint velocity vector by matrix inversion and multiplication:

$$\dot{\mathbf{q}} = \mathbf{J}^{-1}\,\dot{\mathbf{X}} \tag{8.610}$$

A singular configuration is where the determinant of the Jacobian matrix is zero and therefore, $\mathbf{J}^{-1}$ is indeterminate. Equation (8.610) determines the velocity required at the individual joints to produce desired end-effector speeds $\dot{\mathbf{X}}$.

Inverse-velocity kinematics is a consequence of forward-velocity analysis and needs matrix inversion. The inverse-velocity problem is equivalent to the solution of a set of linear algebraic equations. To find $\mathbf{J}^{-1}$, every matrix inversion method may be applied.

Example 527 ★ Inverse Velocity of a 2R Planar Manipulator The forward and inverse kinematics of a 2R planar manipulator have been analyzed in Examples 412 and 458. Its Jacobian and forward-velocity kinematics are also found in Example 519 as

$$\dot{\mathbf{X}} = \mathbf{J}\,\dot{\mathbf{q}} \tag{8.611}$$

$$\begin{bmatrix} \dot{X} \\ \dot{Y} \end{bmatrix} = \begin{bmatrix} -l_1 s\theta_1 - l_2 s\left(\theta_1+\theta_2\right) & -l_2 s\left(\theta_1+\theta_2\right) \\ l_1 c\theta_1 + l_2 c\left(\theta_1+\theta_2\right) & l_2 c\left(\theta_1+\theta_2\right) \end{bmatrix} \begin{bmatrix} \dot{\theta}_1 \\ \dot{\theta}_2 \end{bmatrix} \tag{8.612}$$

For inverse-velocity kinematics, we need to find the inverse of the Jacobian matrix. Therefore,

$$\dot{\mathbf{q}} = \mathbf{J}^{-1}\,\dot{\mathbf{X}} \tag{8.613}$$

$$\begin{bmatrix} \dot{\theta}_1 \\ \dot{\theta}_2 \end{bmatrix} = \begin{bmatrix} -l_1 s\theta_1 - l_2 s\,(\theta_1 + \theta_2) & -l_2 s\,(\theta_1 + \theta_2) \\ l_1 c\theta_1 + l_2 c\,(\theta_1 + \theta_2) & l_2 c\,(\theta_1 + \theta_2) \end{bmatrix}^{-1} \begin{bmatrix} \dot{X} \\ \dot{Y} \end{bmatrix} \tag{8.614}$$

where

$$\mathbf{J}^{-1} = \frac{-1}{l_1 l_2 s\theta_2} \begin{bmatrix} -l_2 c\,(\theta_1 + \theta_2) & -l_2 s\,(\theta_1 + \theta_2) \\ l_1 c\theta_1 + l_2 c\,(\theta_1 + \theta_2) & l_1 s\theta_1 + l_2 s\,(\theta_1 + \theta_2) \end{bmatrix} \tag{8.615}$$

and hence,

$$\dot{\theta}_1 = \frac{\dot{X} c\,(\theta_1 + \theta_2) + \dot{Y} s\,(\theta_1 + \theta_2)}{l_1 s\theta_2} \tag{8.616}$$

$$\dot{\theta}_2 = \frac{\dot{X}[l_1 c\theta_1 + l_2 c\,(\theta_1 + \theta_2)] + \dot{Y}[l_1 s\theta_1 + l_2 s\,(\theta_1 + \theta_2)]}{-l_1 l_2 s\theta_2} \tag{8.617}$$

Singularity occurs when the determinant of the Jacobian is zero:

$$|\mathbf{J}| = l_1 l_2 \sin\theta_2 = 0 \tag{8.618}$$

Therefore, the singular configurations of the manipulator are

$$\theta_2 = 0 \quad \theta_2 = 180\,\text{deg} \tag{8.619}$$

These singular configurations correspond to the fully extended or fully contracted configurations, as shown in Figure 8.20. At the singular configurations, the value of θ_1 is indeterminate and may have any real value. The two columns of the Jacobian matrix become parallel because Equation (8.612) becomes

$$\begin{aligned} \begin{bmatrix} \dot{X} \\ \dot{Y} \end{bmatrix} &= 2l_1 \begin{bmatrix} -s\theta_1 \\ c\theta_1 \end{bmatrix} \dot{\theta}_1 + l_2 \begin{bmatrix} -s\theta_1 \\ c\theta_1 \end{bmatrix} \dot{\theta}_2 \\ &= \left(2l_1\dot{\theta}_1 + l_2\dot{\theta}_2\right) \begin{bmatrix} -s\theta_1 \\ c\theta_1 \end{bmatrix} \end{aligned} \tag{8.620}$$

In this situation, the end point can only move in the direction perpendicular to the arm links.

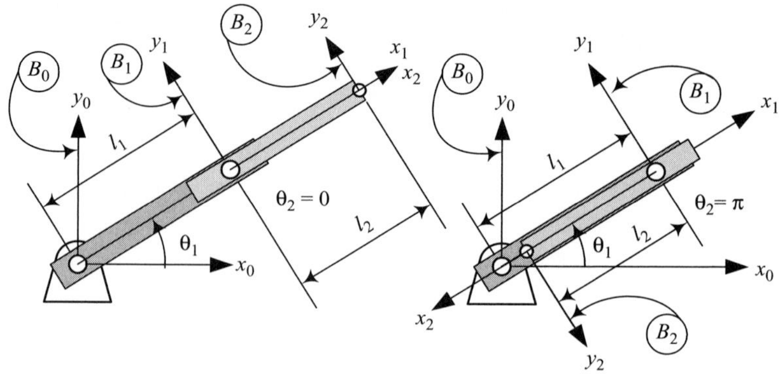

Figure 8.20 Singular configurations of a 2R planar manipulator.

Example 528 ★ Analytic Method for Inverse-Velocity Kinematics Theoretically we must be able to calculate the joint velocities from the forward-velocity equations. However, in general, such a calculation is not easy. As an example, consider the 2R planar manipulator shown in Figure 8.16.

The end-point velocity of the manipulator is expressed in Equation (8.521) as

$$ {}^{0}\dot{\mathbf{d}}_2 = \dot{\theta}_1 {}^{0}\hat{k}_0 \times \left(l_1 {}^{0}\hat{\imath}_1 + l_2 {}^{0}\hat{\imath}_2\right) + \dot{\theta}_2 {}^{0}\hat{k}_1 \times l_2 {}^{0}\hat{\imath}_2 \tag{8.621} $$

Let us find the dot product of this equation with ${}^{0}\hat{\imath}_2$,

$$ \begin{aligned} {}^{0}\dot{\mathbf{d}}_2 \cdot {}^{0}\hat{\imath}_2 &= \dot{\theta}_1 \left({}^{0}\hat{k}_0 \times l_1 {}^{0}\hat{\imath}_1\right) \cdot {}^{0}\hat{\imath}_2 = l_1 \dot{\theta}_1 {}^{0}\hat{k}_0 \cdot \left({}^{0}\hat{\imath}_1 \times {}^{0}\hat{\imath}_2\right) \\ &= l_1 \dot{\theta}_1 {}^{0}\hat{k}_0 \cdot {}^{0}\hat{\imath}_2 \sin\theta_2 = l_1 \dot{\theta}_1 \sin\theta_2 \end{aligned} \tag{8.622} $$

and determine $\dot{\theta}_1$,

$$ \dot{\theta}_1 = \frac{{}^{0}\dot{\mathbf{d}}_2 \cdot {}^{0}\hat{\imath}_2}{l_1 \sin\theta_2} \tag{8.623} $$

Now a dot product of (8.621) with ${}^{0}\hat{\imath}_1$ reduces to

$$ \begin{aligned} {}^{0}\dot{\mathbf{d}}_2 \cdot {}^{0}\hat{\imath}_1 &= \dot{\theta}_1 \left({}^{0}\hat{k}_0 \times l_2 {}^{0}\hat{\imath}_2\right) \cdot {}^{0}\hat{\imath}_1 + \dot{\theta}_2 \left({}^{0}\hat{k}_1 \times l_2 {}^{0}\hat{\imath}_2\right) \cdot {}^{0}\hat{\imath}_1 \\ &= l_2 \left(\dot{\theta}_1 + \dot{\theta}_2\right) {}^{0}\hat{k}_0 \cdot \left({}^{0}\hat{\imath}_2 \times {}^{0}\hat{\imath}_1\right) \\ &= -l_2 \left(\dot{\theta}_1 + \dot{\theta}_2\right) \sin\theta_2 \end{aligned} \tag{8.624} $$

and provides

$$ \dot{\theta}_2 = -\dot{\theta}_1 - \frac{{}^{0}\dot{\mathbf{d}}_2 \cdot {}^{0}\hat{\imath}_1}{l_2 \sin\theta_2} \tag{8.625} $$

Example 529 ★ Inverse Jacobian Matrix of an Articulated Robot The Jacobian matrix **J** of an articulated robot with a spherical wrist is calculated in Example 523. The Jacobian matrix of the manipulator is

$$\mathbf{J} = \begin{bmatrix} {}^0\hat{k}_0 \times {}^0_0\mathbf{d}_6 & {}^0\hat{k}_1 \times {}^0_1\mathbf{d}_6 & {}^0\hat{k}_2 \times {}^0_2\mathbf{d}_6 & 0 & 0 & 0 \\ {}^0\hat{k}_0 & {}^0\hat{k}_1 & {}^0\hat{k}_2 & {}^0\hat{k}_3 & {}^0\hat{k}_4 & {}^0\hat{k}_5 \end{bmatrix} \tag{8.626}$$

The upper right 3×3 submatrix of **J** is zero due to the spherical wrist structure and because the last three position vectors are zero.

Let us split the Jacobian matrix into four 3×3 submatrices and write it as

$$\mathbf{J} = \begin{bmatrix} A & B \\ C & D \end{bmatrix} = \begin{bmatrix} A & 0 \\ C & D \end{bmatrix} \tag{8.627}$$

where

$$[A] = \begin{bmatrix} {}^0\hat{k}_0 \times {}^0_0\mathbf{d}_6 & {}^0\hat{k}_1 \times {}^0_1\mathbf{d}_6 & {}^0\hat{k}_2 \times {}^0_2\mathbf{d}_6 \end{bmatrix} \tag{8.628}$$

$$[C] = \begin{bmatrix} {}^0\hat{k}_0 & {}^0\hat{k}_1 & {}^0\hat{k}_2 \end{bmatrix} \tag{8.629}$$

$$[D] = \begin{bmatrix} {}^0\hat{k}_3 & {}^0\hat{k}_4 & {}^0\hat{k}_5 \end{bmatrix} \tag{8.630}$$

Inversion of such a Jacobian is simpler if we take advantage of $B = 0$. The forward-velocity kinematics of the robot can be written as

$$\dot{\mathbf{X}} = \mathbf{J}\,\dot{\mathbf{q}}$$

$$\begin{bmatrix} {}^0\dot{\mathbf{d}}_2 \\ {}_0\boldsymbol{\omega}_2 \end{bmatrix} = \begin{bmatrix} A & 0 \\ C & D \end{bmatrix} \begin{bmatrix} \dot{\theta}_1 \\ \dot{\theta}_2 \\ \dot{\theta}_3 \\ \dot{\theta}_4 \\ \dot{\theta}_5 \\ \dot{\theta}_6 \end{bmatrix} \tag{8.631}$$

The upper half of the equation is

$${}^0\dot{\mathbf{d}}_2 = [A] \begin{bmatrix} \dot{\theta}_1 \\ \dot{\theta}_2 \\ \dot{\theta}_3 \end{bmatrix} \tag{8.632}$$

which can be inverted to

$$\begin{bmatrix} \dot{\theta}_1 \\ \dot{\theta}_2 \\ \dot{\theta}_3 \end{bmatrix} = A^{-1}\,{}^0\dot{\mathbf{d}}_2 \tag{8.633}$$

The lower half of the equation is

$${}_0\boldsymbol{\omega}_2 = \begin{bmatrix} C & D \end{bmatrix} \begin{bmatrix} \dot{\theta}_1 \\ \dot{\theta}_2 \\ \dot{\theta}_3 \\ \dot{\theta}_4 \\ \dot{\theta}_5 \\ \dot{\theta}_6 \end{bmatrix} = [C] \begin{bmatrix} \dot{\theta}_1 \\ \dot{\theta}_2 \\ \dot{\theta}_3 \end{bmatrix} + [D] \begin{bmatrix} \dot{\theta}_4 \\ \dot{\theta}_5 \\ \dot{\theta}_6 \end{bmatrix} \tag{8.634}$$

and therefore,

$$\begin{bmatrix} \dot{\theta}_4 \\ \dot{\theta}_5 \\ \dot{\theta}_6 \end{bmatrix} = D^{-1} \left({}_0\boldsymbol{\omega}_2 - [C] \begin{bmatrix} \dot{\theta}_1 \\ \dot{\theta}_2 \\ \dot{\theta}_3 \end{bmatrix} \right) \tag{8.635}$$

KEY SYMBOLS

$\mathbf{a}$	turn vector of end-effector frame
B	body coordinate frame
c	cos
d	differential, prismatic joint variable
d_x, d_y, d_z	elements of $\mathbf{d}$
$\mathbf{d}$	translation vector, displacement vector
D	displacement transformation matrix
e	rotation quaternion
e_0, e_1, e_2, e_3	Euler parameters, components of e
G, B_0	global coordinate frame, base coordinate frame
$\hat{\imath}, \hat{\jmath}, \hat{k}$	local coordinate axes unit vectors
$\tilde{\imath}, \tilde{\jmath}, \tilde{k}$	skew-symmetric matrices of the unit vectors $\hat{\imath}, \hat{\jmath}, \hat{k}$
$\hat{I}, \hat{J}, \hat{K}$	global coordinate axes unit vectors
$\mathbf{I} = [I]$	identity matrix
$\mathbf{J}$	Jacobian
l	length
p	pitch of a screw
q	joint coordinate
$\mathbf{q}$	joint coordinate vector
$\mathbf{r}$	position vectors, homogeneous position vector
r_i	element i of $\mathbf{r}$
r_{ij}	element of row i and column j of a matrix
R	rotation transformation matrix
s	sin
$\mathbf{s}$	location vector of a screw
sgn	signum function
SSRMS	space station remote manipulator system
T	homogeneous transformation matrix
T_{arm}	manipulator transformation matrix
T_{wrist}	wrist transformation matrix
$\mathbf{T}$	a set of nonlinear algebraic equations of $\mathbf{q}$
$\mathbf{v}$	velocity vector
V	velocity transformation matrix
$\hat{u}$	unit vector along the axis of $\boldsymbol{\omega}$
$\tilde{u}$	skew symmetric matrix of the vector $\hat{u}$
u_1, u_2, u_3	components of $\hat{u}$
x, y, z	local coordinate axes
X, Y, Z	global coordinate axes
Greek	
α, β, γ	angles of rotation about the axes of global frame
δ	Kronecker function, small increment of a parameter

ϵ	small test number to terminate a procedure
θ	rotary joint angle
θ_{ijk}	$\theta_i + \theta_j + \theta_k$
φ, θ, ψ	angles of rotation about the axes of body frame
ϕ	angle of rotation about $\hat{u}$
$\boldsymbol{\omega}$	angular velocity vector
$\tilde{\omega}$	skew-symmetric matrix of the vector $\boldsymbol{\omega}$
$\omega_1, \omega_2, \omega_3$	components of $\boldsymbol{\omega}$

Symbol	
$[\ \ ]^{-1}$	inverse of the matrix [ ]
$[\ \ ]^{\mathrm{T}}$	transpose of the matrix [ ]
$\vdash$	orthogonal
$\parallel$	parallel
$\perp$	perpendicular
Δ_P	prismatic velocity coefficient matrices
Δ_R	revolute velocity coefficient matrices

EXERCISES

1. **Local Position, Global Velocity** A body is turning about a global principal axis at a constant angular rate. Find the global velocity of a point at ${}^{B}\mathbf{r}$:

$$^{B}\mathbf{r} = \begin{bmatrix} 5 & 20 & 10 \end{bmatrix}^{\mathrm{T}}$$

(a) The axis is the Z-axis and the angular rate $\dot{\alpha} = 2\,\mathrm{rad/s}$ when $\alpha = 30\,\mathrm{deg}$.
(b) The axis is the Y-axis and the angular rate $\dot{\beta} = 2\,\mathrm{rad/s}$ when $\beta = 30\,\mathrm{deg}$.
(c) The axis is the X-axis and the angular rate $\dot{\gamma} = 2\,\mathrm{rad/s}$ when $\gamma = 30\,\mathrm{deg}$.

2. **Parametric Angular Velocity, Global Principal Rotations** A body B is turning in a global frame G. The rotation transformation matrix can be decomposed into principal axes. Determine the angular velocities ${}_{G}\tilde{\omega}_{B}$ and ${}_{G}\boldsymbol{\omega}_{B}$.

(a) ${}^{G}R_{B}$ is the result of a rotation α about the Z-axis followed by β about the Y-axis.
(b) ${}^{G}R_{B}$ is the result of a rotation β about the Y-axis followed by α about the Z-axis.
(c) ${}^{G}R_{B}$ is the result of a rotation α about the Z-axis followed by γ about the X-axis.
(d) ${}^{G}R_{B}$ is the result of a rotation γ about the X-axis followed by α about the Z-axis.
(e) ${}^{G}R_{B}$ is the result of a rotation γ about the X-axis followed by β about the Y-axis.
(f) ${}^{G}R_{B}$ is the result of a rotation β about the Y-axis followed by γ about the X-axis.

3. **Global Position, Constant Angular Velocity** A body is turning about a global principal axis at a constant angular rate. Find the global position of a point at ${}^{B}\mathbf{r}$ after $t = 3\mathrm{s}$ if the body and global coordinate frames were coincident at $t = 0\mathrm{s}$:

$$^{B}\mathbf{r} = \begin{bmatrix} 5 & 20 & 10 \end{bmatrix}^{\mathrm{T}}$$

(a) The axis is the Z-axis and the angular rate $\dot{\alpha} = 2\,\mathrm{rad/s}$.
(b) The axis is the Y-axis and the angular rate $\dot{\beta} = 2\,\mathrm{rad/s}$.
(c) The axis is the X-axis and the angular rate $\dot{\gamma} = 2\,\mathrm{rad/s}$.

4. **Numeric Angular Velocity, Global Principal Rotations** A body B is turning in a global frame G. The rotation transformation matrix can be decomposed into principal axes. Determine the angular velocities $_G\tilde{\omega}_B$ and $_G\boldsymbol{\omega}_B$ for Exercise 2, (a)–(f), using $\dot{\alpha} = 2\,\text{rad/s}$, $\dot{\beta} = 2\,\text{rad/s}$, $\dot{\gamma} = 2\,\text{rad/s}$ and $\alpha = 30\,\text{deg}$, $\beta = 30\,\text{deg}$, $\gamma = 30\,\text{deg}$.

5. **Turning about x-Axis** Find the angular velocity matrix when the body coordinate frame is turning 35 deg/s at 45 deg about the x-axis.

6. **Combined Rotation and Angular Velocity** Find the rotation matrix for a body frame after 30 deg rotation about the Z-axis followed by 30 deg about the X-axis and then 90 deg about the Y-axis. Then calculate the angular velocity of the body if it is turning with $\dot{\alpha} = 20\,\text{deg/s}$, $\dot{\beta} = -40\,\text{deg/s}$, and $\dot{\gamma} = 55\,\text{deg/s}$ about the Z-, Y-, and X-axes, respectively.

7. **Angular Velocity, Expressed in Body Frame** The point P is at $\mathbf{r}_P = (1, 2, 1)$ in a body coordinate $B(Oxyz)$. Find $^B_G\tilde{\omega}_B$ when the body frame is turned 30 deg about the X-axis at a rate $\dot{\gamma} = 75\,\text{deg/s}$ followed by 45 deg about the Z-axis at a rate $\dot{\alpha} = 25\,\text{deg/s}$.

8. **Global Roll–Pitch–Yaw Angular Velocity** Calculate the angular velocity for a global roll–pitch–yaw rotation of $\alpha = 30\,\text{deg}$, $\beta = 30\,\text{deg}$, and $\gamma = 30\,\text{deg}$ with $\dot{\alpha} = 20\,\text{deg/s}$, $\dot{\beta} = -20\,\text{deg/s}$, and $\dot{\gamma} = 20\,\text{deg/s}$.

9. **Roll–Pitch–Yaw Angular Velocity** Find $^B_G\tilde{\omega}_B$ and $_G\tilde{\omega}_B$ for the role, pitch, and yaw rates equal to $\dot{\alpha} = 20\,\text{deg/s}$, $\dot{\beta} = -20\,\text{deg/s}$, and $\dot{\gamma} = 20\,\text{deg/s}$, respectively, and having the rotation matrix

$$^BR_G = \begin{bmatrix} 0.53 & -0.84 & 0.13 \\ 0.0 & 0.15 & 0.99 \\ -0.85 & -0.52 & 0.081 \end{bmatrix}$$

10. **Rolling Disc** A disc of radius R rolls on a horizontal plane and makes a constant angle θ with the Z-axis, and the center of the disc moves on a circle of radius kR with speed v, as shown in Figure 8.21. Determine the angular velocity of the disc.

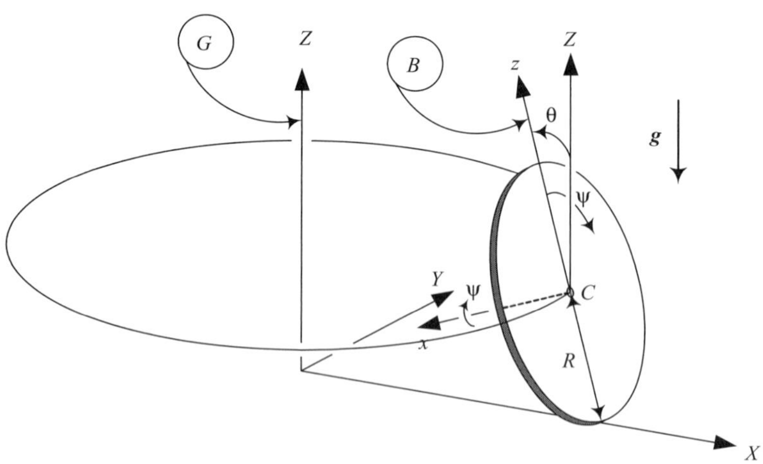

Figure 8.21 A rolling disc at an angle θ with Z-axis.

11. **Angular Velocity from Rodriguez Formula** We may find the time derivative of ${}^G R_B = R_{\hat{u},\phi}$ by

$$ {}^G\dot{R}_B = \frac{d}{dt}{}^G R_B = \dot{\phi}\frac{d}{d\phi}{}^G R_B $$

Use the Rodriguez rotation formula and find ${}_G\tilde{\omega}_B$ and ${}_G^B\tilde{\omega}_B$.

12. **Angular Velocity of End Point of a Stick** Point A of the stick in Figure 8.22 has a constant velocity $\mathbf{v}_A = v\hat{\imath}$ on the x-axis. What is the angular velocity of the stick?

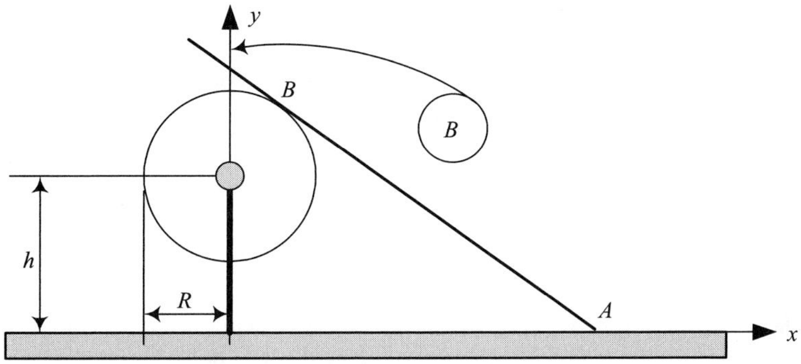

Figure 8.22 A sliding stick that is lined on a circle.

13. **Rotating Slider** Figure 8.23 illustrates a slider link on a rotating arm. Calculate ${}^G d\hat{\imath}/dt, {}^G d\hat{\jmath}/dt, {}^G d\hat{k}/dt$ and ${}^G d^2\hat{\imath}/dt^2, {}^G d^2\hat{\jmath}/dt^2, {}^G d^2\hat{k}/dt^2$ and find ${}^B\mathbf{v}$ and ${}^B\mathbf{a}$ of m at center of mass of the slider by using the rule of mixed derivatives:

$$ \frac{{}^G d}{dt}\left(\frac{{}^B d}{dt}\mathbf{r}\right) = \frac{{}^B d}{dt}\left(\frac{{}^B d}{dt}\mathbf{r}\right) + {}_G\boldsymbol{\omega}_B \times \left(\frac{{}^B d}{dt}\mathbf{r}\right) $$

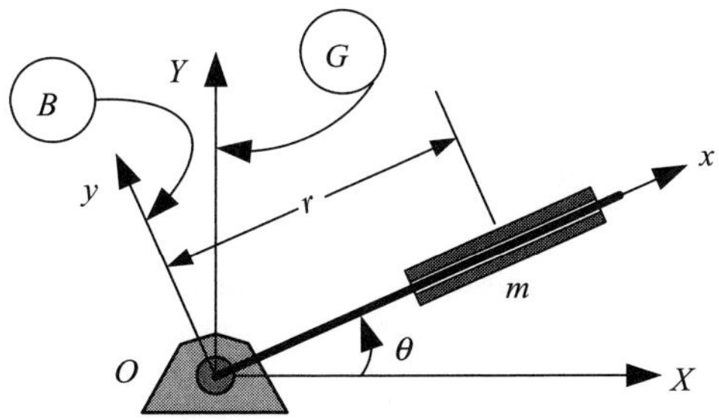

Figure 8.23 A slider on a rotating bar.

14. ★ **Differentiating in Local and Global Frames** Consider a local point at ${}^B\mathbf{r}_P = t\hat{\imath} + \hat{\jmath}$. The local frame B is rotating in G by $\dot{\alpha}$ about the Z-axis. Calculate $({}^B d/dt){}^B\mathbf{r}_P$, $({}^G d/dt){}^G\mathbf{r}_P$, $({}^B d/dt){}^G\mathbf{r}_P$, and $({}^G d/dt){}^B\mathbf{r}_P$.

15. ★ **Skew-Symmetric Identity for Angular Velocity** Show that

$$R\tilde{\omega}R^{\mathrm{T}} = \widetilde{R\boldsymbol{\omega}}$$

16. ★ **Transformation of Angular Velocity Exponents** Show that

$${}_G^B\tilde{\omega}_B^n = {}^GR_B^{\mathrm{T}}\,{}_G\tilde{\omega}_B^n\,{}^GR_B$$

17. ★ **Angular Velocity Matrix Identity** Show that

$$\tilde{\omega}^{2k+1} = (-1)^k\,\omega^{2k}\,\tilde{\omega}$$

and

$$\tilde{\omega}^{2k} = (-1)^k\,\omega^{2(k-1)}\left(\omega^2\,\mathbf{I} - \boldsymbol{\omega}\boldsymbol{\omega}^{\mathrm{T}}\right)$$

18. ★ **Global Triple Angular Velocity Matrix** Determine the angular velocities ${}_G\tilde{\omega}_B$ and ${}_G\boldsymbol{\omega}_B$ for the global triple rotations of Appendix A.

19. ★ **Local Triple Angular Velocity Matrix** Determine the angular velocities ${}_G\tilde{\omega}_B$ and ${}_G\boldsymbol{\omega}_B$ for the local triple rotations of Appendix B.

20. **Angular Velocity, Expressed in Body Frame** A point P is at $\mathbf{r}_P = (1, 2, 1)$ in a body coordinate $B(Oxyz)$.

(a) Find ${}_G^B\tilde{\omega}_B$ when the body frame is turned 30 deg about the X-axis at a rate $\dot{\gamma} = 75$ deg/s, followed by 45 deg about the Z-axis at a rate $\dot{\alpha} = 25$ deg/s.

(b) Find ${}_G^B\tilde{\omega}_B$ when the body frame is turned 45 deg about the Z-axis at a rate $\dot{\alpha} = 25$ deg/s, followed by 30 deg about the X-axis at a rate $\dot{\gamma} = 75$ deg/s.

21. **Global Roll–Pitch–Yaw Angular Velocity** Calculate the angular velocity ${}_G\tilde{\omega}_B$ for a global roll–pitch–yaw rotation of:

(a) $\alpha = 30$ deg, $\beta = 30$ deg, and $\gamma = 30$ deg with $\dot{\alpha} = 20$ deg/s, $\dot{\beta} = -20$ deg/s, and $\dot{\gamma} = 20$ deg/s

(b) $\alpha = 30$ deg, $\beta = 30$ deg, and $\gamma = 30$ deg with $\dot{\alpha} = 0$ deg/s, $\dot{\beta} = -20$ deg/s, and $\dot{\gamma} = 20$ deg/s

(c) $\alpha = 30$ deg, $\beta = 30$ deg, and $\gamma = 30$ deg with $\dot{\alpha} = 20$ deg/s, $\dot{\beta} = 0$ deg/s, and $\dot{\gamma} = 20$ deg/s

(d) $\alpha = 30$ deg, $\beta = 30$ deg, and $\gamma = 30$ deg with $\dot{\alpha} = 20$ deg/s, $\dot{\beta} = -20$ deg/s, and $\dot{\gamma} = 0$ deg/s

(e) $\alpha = 30$ deg, $\beta = 30$ deg, and $\gamma = 30$ deg with $\dot{\alpha} = 0$ deg/s, $\dot{\beta} = 0$ deg/s, and $\dot{\gamma} = 20$ deg/s

22. **Roll–Pitch–Yaw Angular Velocity** Find ${}_G^B\tilde{\omega}_B$ and ${}_G\tilde{\omega}_B$ for the global role, pitch, and yaw rates $\dot{\alpha} = 20$ deg/s, $\dot{\beta} = -20$ deg/s, and $\dot{\gamma} = 20$ deg/s, respectively, and the following rotation matrices:

(a)

$${}^BR_G = \begin{bmatrix} 0.53 & -0.84 & 0.13 \\ 0.0 & 0.15 & 0.99 \\ -0.85 & -0.52 & 0.081 \end{bmatrix}$$

(b)

$${}^GR_B = \begin{bmatrix} 0.53 & -0.84 & 0.13 \\ 0.0 & 0.15 & 0.99 \\ -0.85 & -0.52 & 0.081 \end{bmatrix}$$

23. Angular Velocity from Rodriguez Formula We may find the time derivative of ${}^{G}R_{B} = R_{\hat{u},\phi}$ by

$$ {}^{G}\dot{R}_{B} = \frac{d}{dt}{}^{G}R_{B} = \dot{\phi}\frac{d}{d\phi}{}^{G}R_{B}. $$

Use the Rodriguez rotation formula and find ${}_{G}\tilde{\omega}_{B}$ and ${}^{B}_{G}\tilde{\omega}_{B}$.

24. Skew-Symmetric Matrix Show that any square matrix can be expressed as the sum of a symmetric and skew-symmetric matrix:

$$ A = B + C \qquad B = \tfrac{1}{2}(A + A^{\mathrm{T}}) \qquad C = \tfrac{1}{2}(A - A^{\mathrm{T}}) $$

25. ★ Differentiating in Local and Global Frames Consider a local point at ${}^{B}\mathbf{r}_{P}$. The local frame B is rotating in G by $\dot{\alpha}$ about the Z-axis. Calculate $({}^{B}d/dt)^{B}\mathbf{r}_{P}$, $({}^{G}d/dt)^{G}\mathbf{r}_{P}$, $({}^{B}d/dt)^{G}\mathbf{r}_{P}$, and $({}^{G}d/dt)^{B}\mathbf{r}_{P}$.

(a) ${}^{B}\mathbf{r}_{P} = t\hat{\imath} + \hat{\jmath}$
(b) ${}^{B}\mathbf{r}_{P} = t\hat{\imath} + t\hat{\jmath}$
(c) ${}^{B}\mathbf{r}_{P} = t^{2}\hat{\imath} + \hat{\jmath}$
(d) ${}^{B}\mathbf{r}_{P} = t\hat{\imath} + t^{2}\hat{\jmath}$
(e) ${}^{B}\mathbf{r}_{P} = t\hat{\imath} + t\hat{\jmath} + t\hat{k}$
(f) ${}^{B}\mathbf{r}_{P} = t\hat{\imath} + t^{2}\hat{\jmath} + t\hat{k}$
(g) ${}^{B}\mathbf{r}_{P} = \hat{\imath}\sin t$
(h) ${}^{B}\mathbf{r}_{P} = \hat{\imath}\sin\hat{\imath} + \hat{\jmath}\cos t + \hat{k}$

26. 3R Planar Manipulator Velocity Kinematics Consider an R ∥ R ∥ R planar manipulator with the following transformation matrices:

$$ {}^{2}T_{3} = \begin{bmatrix} c\theta_3 & -s\theta_3 & 0 & l_3 c\theta_3 \\ s\theta_3 & c\theta_3 & 0 & l_3 s\theta_3 \\ 0 & 0 & 1 & 0 \\ 0 & 0 & 0 & 1 \end{bmatrix} \quad {}^{1}T_{2} = \begin{bmatrix} c\theta_2 & -s\theta_2 & 0 & l_2 c\theta_2 \\ s\theta_2 & c\theta_2 & 0 & l_2 s\theta_2 \\ 0 & 0 & 1 & 0 \\ 0 & 0 & 0 & 1 \end{bmatrix} $$

$$ {}^{0}T_{1} = \begin{bmatrix} c\theta_1 & -s\theta_1 & 0 & l_1 c\theta_1 \\ s\theta_1 & c\theta_1 & 0 & l_1 s\theta_1 \\ 0 & 0 & 1 & 0 \\ 0 & 0 & 0 & 1 \end{bmatrix} $$

Calculate the Jacobian matrix $\mathbf{J}$ using direct differentiation and find the Cartesian velocity vector of the end point for:

$$ \begin{aligned} \theta_1 &= 56\,\text{deg} & \theta_2 &= -28\,\text{deg} & \theta_3 &= -10\,\text{deg} \\ l_1 &= 100\,\text{cm} & l_2 &= 55\,\text{cm} & l_3 &= 30\,\text{cm} \\ \dot{\theta}_1 &= 30\,\text{deg/s} & \dot{\theta}_2 &= 10\,\text{deg/s} & \dot{\theta}_3 &= -10\,\text{deg/s} \end{aligned} $$

9

Acceleration Kinematics

Angular acceleration of a rigid body is the time derivative of the instantaneous angular velocity of the body with respect to another body. It is a vectorial quantity in a direction different than the angular velocity. We review acceleration calculus to study the acceleration kinematics of rigid bodies.

9.1 ANGULAR ACCELERATION

Consider a rotating rigid body $B(Oxyz)$ with a fixed point O in a reference frame $G(OXYZ)$ such as shown in Figure 9.1. When the body rotates in G, the global acceleration of a body point P is given as

$$ {}^{G}\mathbf{a} = {}^{G}\dot{\mathbf{v}} = {}^{G}\ddot{\mathbf{r}} = {}_{G}S_{B}\,{}^{G}\mathbf{r} \tag{9.1}$$

$$ = {}_{G}\boldsymbol{\alpha}_{B} \times {}^{G}\mathbf{r} + {}_{G}\boldsymbol{\omega}_{B} \times \left({}_{G}\boldsymbol{\omega}_{B} \times {}^{G}\mathbf{r}\right) \tag{9.2}$$

$$ = \left({}_{G}\tilde{\alpha}_{B} + {}_{G}\tilde{\omega}_{B}^{2}\right) {}^{G}\mathbf{r} = \left({}_{G}\dot{\tilde{\omega}}_{B} + {}_{G}\tilde{\omega}_{B}^{2}\right) {}^{G}\mathbf{r} \tag{9.3}$$

$$ = \left[\ddot{\phi}\tilde{u} + \dot{\phi}\dot{\tilde{u}} + \dot{\phi}^{2}\tilde{u}^{2}\right] {}^{G}\mathbf{r} \tag{9.4}$$

$$ = \left(\ddot{\phi}\hat{u} + \dot{\phi}\dot{\mathbf{u}}\right) \times {}^{G}\mathbf{r} + \dot{\phi}^{2}\hat{u} \times \left(\hat{u} \times {}^{G}\mathbf{r}\right) \tag{9.5}$$

$$ = {}^{G}\ddot{R}_{B}\,{}^{G}R_{B}^{\mathrm{T}}\,{}^{G}\mathbf{r}. \tag{9.6}$$

where ${}_{G}\boldsymbol{\alpha}_{B}$ is the *angular acceleration vector* of B relative to G,

$$ {}_{G}\boldsymbol{\alpha}_{B} = \frac{{}^{G}d}{dt}\,{}_{G}\boldsymbol{\omega}_{B} \tag{9.7}$$

where ${}_{G}\tilde{\alpha}_{B}$ is the *angular acceleration matrix*

$$ {}_{G}\tilde{\alpha}_{B} = {}_{G}\dot{\tilde{\omega}}_{B} = {}^{G}\ddot{R}_{B}\,{}^{G}R_{B}^{\mathrm{T}} + {}^{G}\dot{R}_{B}\,{}^{G}\dot{R}_{B}^{\mathrm{T}} = \ddot{\phi}\tilde{u} + \dot{\phi}\dot{\tilde{u}} \tag{9.8}$$

and ${}_{G}S_{B}$ is the *rotational acceleration transformation*

$$ {}_{G}S_{B} = {}^{G}\ddot{R}_{B}\,{}^{G}R_{B}^{\mathrm{T}} = {}_{G}\tilde{\alpha}_{B} + {}_{G}\tilde{\omega}_{B}^{2} = {}_{G}\tilde{\alpha}_{B} - {}_{G}\tilde{\omega}_{B}\,{}_{G}\tilde{\omega}_{B}^{\mathrm{T}} \tag{9.9}$$

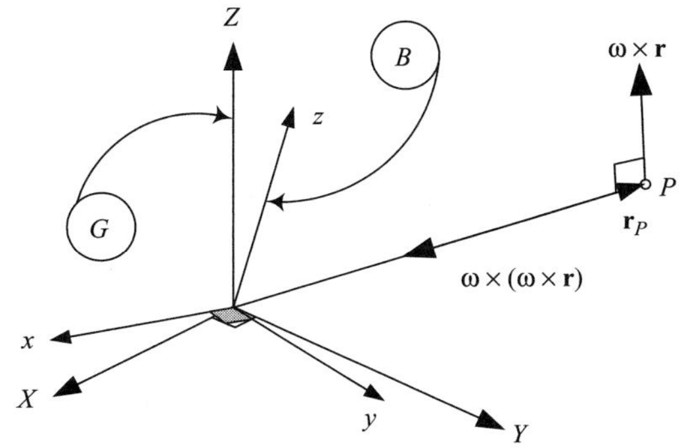

Figure 9.1 A rotating rigid body $B(Oxyz)$ with a fixed point O in a reference frame $G(OXYZ)$.

The angular velocity vector ${}_G\boldsymbol{\omega}_B$ and matrix ${}_G\tilde{\omega}_B$ are

$$ {}_G\tilde{\omega}_B = {}^G\dot{R}_B \, {}^G R_B^{\mathrm{T}} \tag{9.10} $$

$$ {}_G\boldsymbol{\omega}_B = \dot{\phi}\hat{u} = \dot{\phi}\hat{u}_\omega \tag{9.11} $$

The relative angular acceleration of two bodies B_1, B_2 in the global frame G can be combined as

$$ {}_G\boldsymbol{\alpha}_2 = \frac{{}^G d}{dt} {}_G\boldsymbol{\omega}_2 = {}_G\boldsymbol{\alpha}_1 + {}_1^G\boldsymbol{\alpha}_2 \tag{9.12} $$

$$ {}_G S_2 = {}_G S_1 + {}_1^G S_2 + 2 {}_G\tilde{\omega}_1 {}_1^G\tilde{\omega}_2 \tag{9.13} $$

The B-expressions of ${}^G\mathbf{a}$ and ${}_G S_B$ are

$$ {}_G^B\mathbf{a} = {}_G^B\boldsymbol{\alpha}_B \times {}^B\mathbf{r} + {}_G^B\boldsymbol{\omega}_B \times \left({}_G^B\boldsymbol{\omega}_B \times {}^B\mathbf{r} \right) \tag{9.14} $$

$$ {}_G^B S_B = {}^B R_G \, {}^G\ddot{R}_B = {}_G^B\tilde{\alpha}_B + {}_G^B\tilde{\omega}_B^2 \tag{9.15} $$

The global and body expressions of the rotational acceleration transformations ${}_G S_B$ and ${}_G^B S_B$ can be transformed to each other by the following rules:

$$ {}_G S_B = {}^G R_B \, {}_G^B S_B \, {}^G R_B^{\mathrm{T}} \tag{9.16} $$

$$ {}_G^B S_B = {}^G R_B^{\mathrm{T}} \, {}_G S_B \, {}^G R_B \tag{9.17} $$

Proof: The global position and velocity vectors of the body point P are

$$ {}^G\mathbf{r} = {}^G R_B \, {}^B\mathbf{r} \tag{9.18} $$

$$ {}^G\mathbf{v} = {}^G\dot{\mathbf{r}} = {}^G\dot{R}_B \, {}^B\mathbf{r} = {}_G\tilde{\omega}_B \, {}^G\mathbf{r} = {}_G\boldsymbol{\omega}_B \times {}^G\mathbf{r} \tag{9.19} $$

where ${}_G\tilde{\omega}_B$ is also the rotational velocity transformation because it transforms the global position vector of a point, ${}^G\mathbf{r}$, to its velocity vector ${}^G\mathbf{v}$.

Differentiating Equation (9.19) and using the notation ${}_G\boldsymbol{\alpha}_B = ({}^G d/dt) {}_G\boldsymbol{\omega}_B$ provide Equation (9.2):

$$ \begin{aligned} {}^G\mathbf{a} = {}^G\ddot{\mathbf{r}} &= {}_G\dot{\boldsymbol{\omega}}_B \times {}^G\mathbf{r} + {}_G\boldsymbol{\omega}_B \times {}^G\dot{\mathbf{r}} \\ &= {}_G\boldsymbol{\alpha}_B \times {}^G\mathbf{r} + {}_G\boldsymbol{\omega}_B \times \left({}_G\boldsymbol{\omega}_B \times {}^G\mathbf{r} \right) \end{aligned} \tag{9.20} $$

Employing the axis–angle expression of angular velocity,

$$\boldsymbol{\omega} = \dot{\phi}\hat{u} = \dot{\phi}\begin{bmatrix} u_1 \\ u_2 \\ u_3 \end{bmatrix} = \begin{bmatrix} \omega_1 \\ \omega_2 \\ \omega_3 \end{bmatrix} \tag{9.21}$$

$$\tilde{\omega} = \dot{\phi}\tilde{u} = \dot{\phi}\begin{bmatrix} 0 & -u_3 & u_2 \\ u_3 & 0 & -u_1 \\ -u_2 & u_1 & 0 \end{bmatrix} = \begin{bmatrix} 0 & -\omega_3 & \omega_2 \\ \omega_3 & 0 & -\omega_1 \\ -\omega_2 & \omega_1 & 0 \end{bmatrix} \tag{9.22}$$

we can find the angular acceleration vector $\boldsymbol{\alpha}$ and matrix $\tilde{\alpha}$ in terms of the instantaneous axis and angle of rotation:

$$\boldsymbol{\alpha} = \dot{\boldsymbol{\omega}} = \ddot{\phi}\hat{u} + \dot{\phi}\dot{\mathbf{u}} \tag{9.23}$$

$$\tilde{\alpha} = \dot{\tilde{\omega}} = \ddot{\phi}\tilde{u} + \dot{\phi}\dot{\tilde{u}} \tag{9.24}$$

We may substitute the matrix expressions of angular velocity and acceleration in (9.20) to derive Equations (9.3) and (9.4):

$$\begin{aligned} {}^G\ddot{\mathbf{r}} &= {}_G\boldsymbol{\alpha}_B \times {}^G\mathbf{r} + {}_G\boldsymbol{\omega}_B \times \left({}_G\boldsymbol{\omega}_B \times {}^G\mathbf{r}\right) \\ &= {}_G\tilde{\alpha}_B\, {}^G\mathbf{r} + {}_G\tilde{\omega}_B\, {}_G\tilde{\omega}_B\, {}^G\mathbf{r} \\ &= \left({}_G\tilde{\alpha}_B + {}_G\tilde{\omega}_B^2\right) {}^G\mathbf{r} = \left({}_G\dot{\tilde{\omega}}_B + {}_G\tilde{\omega}_B^2\right) {}^G\mathbf{r} \\ &= \left[\ddot{\phi}\tilde{u} + \dot{\phi}\dot{\tilde{u}} + \dot{\phi}^2\tilde{u}^2\right] {}^G\mathbf{r} \end{aligned} \tag{9.25}$$

Substituting the vector expressions of angular velocity (9.11) and acceleration (9.23) in (9.20), we derive Equation (9.5):

$${}^G\ddot{\mathbf{r}} = \left(\ddot{\phi}\hat{u} + \dot{\phi}\dot{\mathbf{u}}\right) \times {}^G\mathbf{r} + \dot{\phi}^2\hat{u} \times \left(\hat{u} \times {}^G\mathbf{r}\right) \tag{9.26}$$

Recalling that

$${}_G\tilde{\omega}_B = {}^G\dot{R}_B\, {}^GR_B^{\mathrm{T}} \tag{9.27}$$

$${}^G\dot{\mathbf{r}}(t) = {}_G\tilde{\omega}_B\, {}^G\mathbf{r}(t) \tag{9.28}$$

we find Equations (9.6) and (9.8):

$$\begin{aligned} {}^G\ddot{\mathbf{r}} &= \frac{{}^Gd}{dt}\left({}^G\dot{R}_B\, {}^GR_B^{\mathrm{T}}\, {}^G\mathbf{r}\right) \\ &= {}^G\ddot{R}_B\, {}^GR_B^{\mathrm{T}}\, {}^G\mathbf{r} + {}^G\dot{R}_B\, {}^G\dot{R}_B^{\mathrm{T}}\, {}^G\mathbf{r} + \left[{}^G\dot{R}_B\, {}^GR_B^{\mathrm{T}}\right]\left[{}^G\dot{R}_B\, {}^GR_B^{\mathrm{T}}\right] {}^G\mathbf{r} \\ &= \left[{}^G\ddot{R}_B\, {}^GR_B^{\mathrm{T}} + {}^G\dot{R}_B\, {}^G\dot{R}_B^{\mathrm{T}} + \left[{}^G\dot{R}_B\, {}^GR_B^{\mathrm{T}}\right]^2\right] {}^G\mathbf{r} \\ &= \left[{}^G\ddot{R}_B\, {}^GR_B^{\mathrm{T}} - \left[{}^G\dot{R}_B\, {}^GR_B^{\mathrm{T}}\right]^2 + \left[{}^G\dot{R}_B\, {}^GR_B^{\mathrm{T}}\right]^2\right] {}^G\mathbf{r} \\ &= {}^G\ddot{R}_B\, {}^GR_B^{\mathrm{T}}\, {}^G\mathbf{r} \end{aligned} \tag{9.29}$$

$$\begin{aligned}
{}_G\tilde{\alpha}_B = {}_G\dot{\tilde{\omega}}_B &= {}^G\ddot{R}_B\, {}^GR_B^{\mathrm{T}} + {}^G\dot{R}_B\, {}^G\dot{R}_B^{\mathrm{T}} \\
&= {}^G\ddot{R}_B\, {}^GR_B^{\mathrm{T}} + {}^G\dot{R}_B\, {}^GR_B^{\mathrm{T}}\, {}^GR_B\, {}^G\dot{R}_B^{\mathrm{T}} \\
&= {}^G\ddot{R}_B\, {}^GR_B^{\mathrm{T}} + \left[{}^G\dot{R}_B\, {}^GR_B^{\mathrm{T}}\right]\left[{}^G\dot{R}_B\, {}^GR_B^{\mathrm{T}}\right]^{\mathrm{T}} \\
&= {}^G\ddot{R}_B\, {}^GR_B^{\mathrm{T}} + {}_G\tilde{\omega}_B\, {}_G\tilde{\omega}_B^{\mathrm{T}} = {}^G\ddot{R}_B\, {}^GR_B^{\mathrm{T}} - {}_G\tilde{\omega}_B^2
\end{aligned} \tag{9.30}$$

which indicates that

$$ {}^G\ddot{R}_B\, {}^GR_B^{\mathrm{T}} = {}_G\tilde{\alpha}_B + {}_G\tilde{\omega}_B^2 = {}_GS_B \tag{9.31}$$

The expanded forms of the angular accelerations ${}_G\boldsymbol{\alpha}_B$, ${}_G\tilde{\alpha}_B$ and *rotational acceleration transformation* ${}_GS_B$ are

$$\begin{aligned}
{}_G\tilde{\alpha}_B = {}_G\dot{\tilde{\omega}}_B &= \ddot{\phi}\tilde{u} + \dot{\phi}\dot{\tilde{u}} = \begin{bmatrix} 0 & -\dot{\omega}_3 & \dot{\omega}_2 \\ \dot{\omega}_3 & 0 & -\dot{\omega}_1 \\ -\dot{\omega}_2 & \dot{\omega}_1 & 0 \end{bmatrix} \\
&= \begin{bmatrix} 0 & -\dot{u}_3\dot{\phi} - u_3\ddot{\phi} & \dot{u}_2\dot{\phi} + u_2\ddot{\phi} \\ \dot{u}_3\dot{\phi} + u_3\ddot{\phi} & 0 & -\dot{u}_1\dot{\phi} - u_1\ddot{\phi} \\ -\dot{u}_2\dot{\phi} - u_2\ddot{\phi} & \dot{u}_1\dot{\phi} + u_1\ddot{\phi} & 0 \end{bmatrix}
\end{aligned} \tag{9.32}$$

$$ {}_G\boldsymbol{\alpha}_B = \begin{bmatrix} \dot{\omega}_1 \\ \dot{\omega}_2 \\ \dot{\omega}_3 \end{bmatrix} = \begin{bmatrix} \dot{u}_1\dot{\phi} + u_1\ddot{\phi} \\ \dot{u}_2\dot{\phi} + u_2\ddot{\phi} \\ \dot{u}_3\dot{\phi} + u_3\ddot{\phi} \end{bmatrix} \tag{9.33}$$

$$\begin{aligned}
{}_GS_B &= {}_G\dot{\tilde{\omega}}_B + {}_G\tilde{\omega}_B^2 = {}_G\tilde{\alpha}_B + {}_G\tilde{\omega}_B^2 \\
&= \begin{bmatrix} -\omega_2^2 - \omega_3^2 & \omega_1\omega_2 - \dot{\omega}_3 & \dot{\omega}_2 + \omega_1\omega_3 \\ \dot{\omega}_3 + \omega_1\omega_2 & -\omega_1^2 - \omega_3^2 & \omega_2\omega_3 - \dot{\omega}_1 \\ \omega_1\omega_3 - \dot{\omega}_2 & \dot{\omega}_1 + \omega_2\omega_3 & -\omega_1^2 - \omega_2^2 \end{bmatrix}
\end{aligned} \tag{9.34}$$

$$\begin{aligned}
{}_GS_B &= \ddot{\phi}\tilde{u} + \dot{\phi}\dot{\tilde{u}} + \dot{\phi}^2\tilde{u}^2 \\
&= \begin{bmatrix} -(1-u_1^2)\dot{\phi}^2 & u_1u_2\dot{\phi}^2 - \dot{u}_3\dot{\phi} - u_3\ddot{\phi} & u_1u_3\dot{\phi}^2 + \dot{u}_2\dot{\phi} + u_2\ddot{\phi} \\ u_1u_2\dot{\phi}^2 + \dot{u}_3\dot{\phi} + u_3\ddot{\phi} & -(1-u_2^2)\dot{\phi}^2 & u_2u_3\dot{\phi}^2 - \dot{u}_1\dot{\phi} - u_1\ddot{\phi} \\ u_1u_3\dot{\phi}^2 - \dot{u}_2\dot{\phi} - u_2\ddot{\phi} & u_2u_3\dot{\phi}^2 + \dot{u}_1\dot{\phi} + u_1\ddot{\phi} & -(1-u_3^2)\dot{\phi}^2 \end{bmatrix}
\end{aligned} \tag{9.35}$$

The angular acceleration vector of B in G can always be expressed in the natural form

$$ {}_G\boldsymbol{\alpha}_B = \alpha\,\hat{u}_\alpha \tag{9.36}$$

where $\hat{u}_\alpha$ is a unit vector parallel to ${}_G\boldsymbol{\alpha}_B$ and α is the angular acceleration magnitude. The angular velocity and angular acceleration vectors are not parallel in general, and therefore,

$$ \hat{u}_\alpha \neq \hat{u}_\omega \tag{9.37}$$

Only if the axis of rotation is fixed in both the G- and B-frames, we have

$$ {}_G\boldsymbol{\alpha}_B = \alpha\,\hat{u} = \dot{\omega}\,\hat{u} = \ddot{\phi}\,\hat{u} \qquad \text{if } \hat{u} = \hat{u}_\alpha = \hat{u}_\omega \text{ is a fixed axis} \tag{9.38}$$

The angular velocity of several bodies rotating relative to each other can be related according to (8.13):

$$ {}_0\boldsymbol{\omega}_n = {}_0\boldsymbol{\omega}_1 + {}_1^0\boldsymbol{\omega}_2 + {}_2^0\boldsymbol{\omega}_3 + \cdots + {}_{n-1}^{\ \ 0}\boldsymbol{\omega}_n \tag{9.39} $$

The angular accelerations of several relatively rotating rigid bodies follow the same rule:

$$ {}_0\boldsymbol{\alpha}_n = {}_0\boldsymbol{\alpha}_1 + {}_1^0\boldsymbol{\alpha}_2 + {}_2^0\boldsymbol{\alpha}_3 + \cdots + {}_{n-1}^{\ \ 0}\boldsymbol{\alpha}_n \tag{9.40} $$

To show this fact and develop the relative acceleration formula, we consider a pair of relatively rotating rigid links in a base coordinate frame B_0 with a fixed point at O. The angular velocities of the links are related as

$$ {}_0\boldsymbol{\omega}_2 = {}_0\boldsymbol{\omega}_1 + {}_1^0\boldsymbol{\omega}_2 \tag{9.41} $$

So, their angular accelerations are

$$ {}_0\boldsymbol{\alpha}_1 = \frac{{}^0d}{dt}\,{}_0\boldsymbol{\omega}_1 \tag{9.42} $$

$$ {}_0\boldsymbol{\alpha}_2 = \frac{{}^0d}{dt}\,{}_0\boldsymbol{\omega}_2 = {}_0\boldsymbol{\alpha}_1 + {}_1^0\boldsymbol{\alpha}_2 \tag{9.43} $$

and therefore,

$$ \begin{aligned} {}_0S_2 &= {}_0\tilde{\alpha}_2 + {}_0\tilde{\omega}_2^2 = {}_0\tilde{\alpha}_1 + {}_1^0\tilde{\alpha}_2 + \left({}_0\tilde{\omega}_1 + {}_1^0\tilde{\omega}_2\right)^2 \\ &= {}_0\tilde{\alpha}_1 + {}_1^0\tilde{\alpha}_2 + {}_0\tilde{\omega}_1^2 + {}_1^0\tilde{\omega}_2^2 + 2\,{}_0\tilde{\omega}_1\,{}_1^0\tilde{\omega}_2 \\ &= {}_0S_1 + {}_1^0S_2 + 2\,{}_0\tilde{\omega}_1\,{}_1^0\tilde{\omega}_2 \end{aligned} \tag{9.44} $$

Equation (9.44) is the required *relative acceleration transformation formula*. It indicates the method of calculation of relative accelerations for a multibody. As a more general case, consider a six-link multibody. The angular acceleration of link (6) in the base frame would be

$$ \begin{aligned} {}_0S_6 = {}&{}_0S_1 + {}_1^0S_2 + {}_2^0S_3 + {}_3^0S_4 + {}_4^0S_5 + {}_5^0S_6 \\ &+ 2\,{}_0\tilde{\omega}_1\left({}_1^0\tilde{\omega}_2 + {}_2^0\tilde{\omega}_3 + {}_3^0\tilde{\omega}_4 + {}_4^0\tilde{\omega}_5 + {}_5^0\tilde{\omega}_6\right) \\ &+ 2\,{}_1^0\tilde{\omega}_2\left({}_2^0\tilde{\omega}_3 + {}_3^0\tilde{\omega}_4 + {}_4^0\tilde{\omega}_5 + {}_5^0\tilde{\omega}_6\right) \\ &\ \ \vdots \\ &+ 2\,{}_4^0\tilde{\omega}_5\left({}_5^0\tilde{\omega}_6\right) \end{aligned} \tag{9.45} $$

We can transform the G- and B-expressions of the global acceleration of a body point P to each other using a rotation matrix:

$$ \begin{aligned} {}_G^B\mathbf{a}_P &= {}^BR_G\ {}^G\mathbf{a}_P = {}^BR_G\ {}_GS_B\ {}^G\mathbf{r}_P = {}^BR_G\ {}_GS_B\ {}^GR_B\ {}^B\mathbf{r}_P \\ &= {}^BR_G\ {}^G\ddot{R}_B\ {}^GR_B^{\mathrm{T}}\ {}^GR_B\ {}^B\mathbf{r}_P = {}^BR_G\ {}^G\ddot{R}_B\ {}^B\mathbf{r}_P \\ &= {}^GR_B^{\mathrm{T}}\ {}^G\ddot{R}_B\ {}^B\mathbf{r}_P = {}_G^BS_B\ {}^B\mathbf{r}_P = \left({}_G^B\tilde{\alpha}_B + {}_G^B\tilde{\omega}_B^2\right)\ {}^B\mathbf{r}_P \\ &= {}_G^B\boldsymbol{\alpha}_B \times {}^B\mathbf{r} + {}_G^B\boldsymbol{\omega}_B \times \left({}_G^B\boldsymbol{\omega}_B \times {}^B\mathbf{r}\right) \end{aligned} \tag{9.46} $$

$$
\begin{aligned}
{}^{G}\mathbf{a}_P &= {}^{G}R_B\, {}^{B}_{G}\mathbf{a}_P = {}^{G}R_B\, {}^{B}_{G}S_B\, {}^{B}\mathbf{r}_P = {}^{G}R_B\, {}^{B}_{G}S_B\, {}^{G}R_B^{\mathrm{T}}\, {}^{G}\mathbf{r}_P \\
&= {}^{G}R_B\, {}^{G}R_B^{\mathrm{T}}\, {}^{G}\ddot{R}_B\, {}^{G}R_B^{\mathrm{T}}\, {}^{G}\mathbf{r}_P = {}^{G}\ddot{R}_B\, {}^{G}R_B^{\mathrm{T}}\, {}^{G}\mathbf{r}_P \\
&= {}_{G}S_B\, {}^{G}\mathbf{r}_P = \left({}_{G}\tilde{\alpha}_B + {}_{G}\tilde{\omega}_B^2\right) {}^{G}\mathbf{r} \\
&= {}_{G}\boldsymbol{\alpha}_B \times {}^{G}\mathbf{r} + {}_{G}\boldsymbol{\omega}_B \times \left({}_{G}\boldsymbol{\omega}_B \times {}^{G}\mathbf{r}\right)
\end{aligned} \tag{9.47}
$$

From the definitions of ${}_{G}S_B$ and ${}^{B}_{G}S_B$ in (9.9) and (9.15) and comparing with (9.46) and (9.47), we are able to transform the two rotational acceleration transformations by

$$
{}_{G}S_B = {}^{G}R_B\, {}^{B}_{G}S_B\, {}^{G}R_B^{\mathrm{T}} \tag{9.48}
$$

$$
{}^{B}_{G}S_B = {}^{G}R_B^{\mathrm{T}}\, {}_{G}S_B\, {}^{G}R_B \tag{9.49}
$$

and derive the useful equations

$$
{}^{G}\ddot{R}_B = {}_{G}S_B\, {}^{G}R_B \tag{9.50}
$$

$$
{}^{G}\ddot{R}_B = {}^{G}R_B\, {}^{B}_{G}S_B \tag{9.51}
$$

$$
{}_{G}S_B\, {}^{G}R_B = {}^{G}R_B\, {}^{B}_{G}S_B \tag{9.52}
$$

The angular acceleration of B in G is negative of the angular acceleration of G in B if both are expressed in the same coordinate frame:

$$
{}_{G}\tilde{\alpha}_B = -\, {}^{G}_{B}\tilde{\alpha}_G \qquad {}_{G}\boldsymbol{\alpha}_B = -\, {}^{G}_{B}\boldsymbol{\alpha}_G \tag{9.53}
$$

$$
{}^{B}_{G}\tilde{\alpha}_B = -\, {}_{B}\tilde{\alpha}_G \qquad {}^{B}_{G}\boldsymbol{\alpha}_B = -\, {}_{B}\boldsymbol{\alpha}_G \tag{9.54}
$$

The term ${}_{G}\boldsymbol{\alpha}_B \times {}^{G}\mathbf{r}$ in (9.20) is called the *tangential acceleration*, which is a function of the angular acceleration of B in G. The term ${}_{G}\boldsymbol{\omega}_B \times \left({}_{G}\boldsymbol{\omega}_B \times {}^{G}\mathbf{r}\right)$ in ${}^{G}\mathbf{a}$ is called *centripetal acceleration* and is a function of the angular velocity of B in G.
■

Example 530 Rotation of a Body Point about a Global Axis Consider a rigid body is turning about the Z-axis with a constant angular acceleration $\ddot{\alpha} = 2\,\mathrm{rad/s^2}$. The global acceleration of a body point at $P(5, 30, 10)$ cm when the body is at $\dot{\alpha} = 10\,\mathrm{rad/s}$ and $\alpha = 30\,\mathrm{deg}$ is

$$
\begin{aligned}
{}^{G}\mathbf{a}_P &= {}^{G}\ddot{R}_B(t)\, {}^{B}\mathbf{r}_P \\
&= \begin{bmatrix} -87.6 & 48.27 & 0 \\ -48.27 & -87.6 & 0 \\ 0 & 0 & 0 \end{bmatrix} \begin{bmatrix} 5 \\ 30 \\ 10 \end{bmatrix} = \begin{bmatrix} 1010 \\ -2869.4 \\ 0 \end{bmatrix} \mathrm{cm/s}
\end{aligned} \tag{9.55}
$$

where

$$
\begin{aligned}
{}^{G}\ddot{R}_B &= \frac{{}^{G}d^2}{dt^2}\, {}^{G}R_B = \dot{\alpha}\frac{{}^{G}d}{d\alpha}\, {}^{G}R_B = \ddot{\alpha}\frac{{}^{G}d}{d\alpha}\, {}^{G}R_B + \dot{\alpha}^2 \frac{{}^{G}d^2}{d\alpha^2}\, {}^{G}R_B \\
&= \ddot{\alpha} \begin{bmatrix} -\sin\alpha & -\cos\alpha & 0 \\ \cos\alpha & -\sin\alpha & 0 \\ 0 & 0 & 0 \end{bmatrix} + \dot{\alpha}^2 \begin{bmatrix} -\cos\alpha & \sin\alpha & 0 \\ -\sin\alpha & -\cos\alpha & 0 \\ 0 & 0 & 0 \end{bmatrix}
\end{aligned} \tag{9.56}
$$

At this moment, the point P is at

$$
\begin{aligned}
{}^G\mathbf{r}_P &= {}^G R_B\, {}^B\mathbf{r}_P \\
&= \begin{bmatrix} \cos\left(\frac{1}{6}\pi\right) & -\sin\left(\frac{1}{6}\pi\right) & 0 \\ \sin\left(\frac{1}{6}\pi\right) & \cos\left(\frac{1}{6}\pi\right) & 0 \\ 0 & 0 & 1 \end{bmatrix} \begin{bmatrix} 5 \\ 30 \\ 10 \end{bmatrix} = \begin{bmatrix} -10.67 \\ 28.48 \\ 10 \end{bmatrix} \text{cm}
\end{aligned} \tag{9.57}
$$

Example 531 Rotation of a Global Point about a Global Axis A body point P at ${}^B\mathbf{r}_P = \begin{bmatrix} 5 & 30 & 10 \end{bmatrix}^{\mathrm{T}}$ cm is turning with a constant angular acceleration $\ddot{\alpha} = 2\,\text{rad/s}^2$ about the Z-axis. When the body frame is at $\alpha = 30\,\text{deg}$, its angular speed $\dot{\alpha} = 10\,\text{deg/s}$.

The transformation matrix ${}^G R_B$ between the B- and G-frames is

$$
{}^G R_B = \begin{bmatrix} \cos\left(\frac{1}{6}\pi\right) & -\sin\left(\frac{1}{6}\pi\right) & 0 \\ \sin\left(\frac{1}{6}\pi\right) & \cos\left(\frac{1}{6}\pi\right) & 0 \\ 0 & 0 & 1 \end{bmatrix} \approx \begin{bmatrix} 0.866 & -0.5 & 0 \\ 0.5 & 0.866 & 0 \\ 0 & 0 & 1 \end{bmatrix} \tag{9.58}
$$

and therefore, the acceleration of point P is

$$
{}^G\mathbf{a}_P = {}^G\ddot{R}_B\, {}^G R_B^{\mathrm{T}}\, {}^G\mathbf{r}_P = \begin{bmatrix} 1010 \\ -2869.4 \\ 0 \end{bmatrix} \text{cm/s}^2 \tag{9.59}
$$

where

$$
\frac{{}^G d^2}{dt^2}\, {}^G R_B = \ddot{\alpha} \frac{{}^G d}{d\alpha}\, {}^G R_B - \dot{\alpha}^2 \frac{{}^G d^2}{d\alpha^2}\, {}^G R_B \tag{9.60}
$$

is the same as (9.56).

Example 532 Principal Angular Accelerations The principal rotational matrices about the axes X, Y, and Z are given in (8.61)–(8.63) for $R_{X,\gamma}$, $R_{Y,\beta}$, and $R_{Z,\alpha}$. Based on their time derivatives in (8.64)–(8.66), the principal angular velocities about the axes X, Y, and Z are given in (8.67)–(8.69) as

$$
\tilde{\omega}_X = \dot{R}_{X,\gamma} R_{X,\gamma}^{\mathrm{T}} = \dot{\gamma}\tilde{I} \qquad \boldsymbol{\omega}_X = \omega_X\, \hat{I} = \dot{\gamma}\hat{I} \tag{9.61}
$$

$$
\tilde{\omega}_Y = \dot{R}_{Y,\beta} R_{Y,\beta}^{\mathrm{T}} = \dot{\beta}\tilde{J} \qquad \boldsymbol{\omega}_Y = \omega_Y\, \hat{J} = \dot{\beta}\hat{J} \tag{9.62}
$$

$$
\tilde{\omega}_Z = \dot{R}_{Z,\alpha} R_{Z,\alpha}^{\mathrm{T}} = \dot{\alpha}\tilde{K} \qquad \boldsymbol{\omega}_Z = \omega_Z\, \hat{K} = \dot{\alpha}\hat{K} \tag{9.63}
$$

Taking another derivative shows that the principal angular accelerations about the axes X, Y, and Z are

$$
\tilde{\alpha}_X = \ddot{R}_{X,\gamma} R_{X,\gamma}^{\mathrm{T}} + \dot{R}_{X,\gamma} \dot{R}_{X,\gamma}^{\mathrm{T}} = \ddot{\gamma}\tilde{I} \qquad \boldsymbol{\alpha}_X = \alpha_X\, \hat{I} = \ddot{\gamma}\hat{I} \tag{9.64}
$$

$$\tilde{\alpha}_Y = \ddot{R}_{Y,\beta} R^{\mathrm{T}}_{Y,\beta} + \dot{R}_{Y,\beta} \dot{R}^{\mathrm{T}}_{Y,\beta} = \ddot{\beta} \tilde{J} \qquad \boldsymbol{\alpha}_Y = \alpha_Y \hat{J} = \ddot{\beta} \hat{J} \tag{9.65}$$

$$\tilde{\alpha}_Z = \ddot{R}_{Z,\alpha} R^{\mathrm{T}}_{Z,\alpha} + \dot{R}_{Z,\alpha} \dot{R}^{\mathrm{T}}_{Z,\alpha} = \ddot{\alpha} \tilde{K} \qquad \boldsymbol{\alpha}_Z = \alpha_Z \hat{K} = \ddot{\alpha} \hat{K} \tag{9.66}$$

and therefore,

$$S_{X,\ddot{\gamma}} = \ddot{R}_{X,\gamma} R^{\mathrm{T}}_{X,\gamma} = \tilde{\alpha}_X + \tilde{\omega}^2_X = \ddot{\gamma} \tilde{I} + \dot{\gamma}^2 \tilde{I} \tilde{I} \tag{9.67}$$

$$S_{Y,\ddot{\beta}} = \ddot{R}_{Y,\beta} R^{\mathrm{T}}_{Y,\beta} = \tilde{\alpha}_Y + \tilde{\omega}^2_Y = \ddot{\beta} \tilde{J} + \dot{\beta}^2 \tilde{J} \tilde{J} \tag{9.68}$$

$$S_{Z,\ddot{\alpha}} = \ddot{R}_{Z,\alpha} R^{\mathrm{T}}_{Z,\alpha} = \tilde{\alpha}_Z + \tilde{\omega}^2_Z = \ddot{\alpha} \tilde{K} + \dot{\alpha}^2 \tilde{K} \tilde{K} \tag{9.69}$$

$$\ddot{R}_{X,\gamma} = \left(\ddot{\gamma} \tilde{I} + \dot{\gamma}^2 \tilde{I} \tilde{I} \right) R_{X,\gamma} \tag{9.70}$$

$$\ddot{R}_{Y,\beta} = \left(\ddot{\beta} \tilde{J} + \dot{\beta}^2 \tilde{J} \tilde{J} \right) R_{Y,\beta} \tag{9.71}$$

$$\ddot{R}_{Z,\alpha} = \left(\ddot{\alpha} \tilde{K} + \dot{\alpha}^2 \tilde{K} \tilde{K} \right) R_{Z,\alpha} \tag{9.72}$$

Example 533 Decomposition of an Angular Acceleration Vector Every angular acceleration can be decomposed to three principal angular acceleration vectors by employing the orthogonality condition (3.1):

$$\begin{aligned} {}_G\boldsymbol{\alpha}_B &= \left({}_G\boldsymbol{\alpha}_B \cdot \hat{I} \right) \hat{I} + \left({}_G\boldsymbol{\alpha}_B \cdot \hat{J} \right) \hat{J} + \left({}_G\boldsymbol{\alpha}_B \cdot \hat{K} \right) \hat{K} \\ &= \alpha_X \hat{I} + \alpha_Y \hat{J} + \alpha_Z \hat{K} = \ddot{\gamma} \hat{I} + \ddot{\beta} \hat{J} + \ddot{\alpha} \hat{K} \\ &= \boldsymbol{\alpha}_X + \boldsymbol{\alpha}_Y + \boldsymbol{\alpha}_Z \end{aligned} \tag{9.73}$$

Example 534 ★ Relative Angular Acceleration The relative angular acceleration formulas (9.43) or (9.40),

$${}_0\alpha_2 = {}_0\alpha_1 + {}_1^0\alpha_2 \tag{9.74}$$

$${}_0\alpha_n = {}_0\alpha_1 + {}_1^0\alpha_2 + {}_2^0\alpha_3 + \cdots + {}_{n-1}^{\;\;0}\alpha_n = \sum_{i=1}^{n} {}_{i-1}^{\;\;0}\alpha_i \tag{9.75}$$

are correct if and only if all of the angular accelerations are expressed in the B_0-frame. Therefore, any equation of the form

$${}_0\alpha_2 \neq {}_0\alpha_1 + {}_1\alpha_2 \tag{9.76}$$

$$\alpha_0 \neq \alpha_1 + \alpha_2 \tag{9.77}$$

$${}_0\alpha_3 \neq {}_0\alpha_1 + {}_0\alpha_2 \tag{9.78}$$

is wrong or is not completely expressed.

Example 535 Angular Acceleration and Euler Angles The angular velocity ${}_G\boldsymbol{\omega}_B$ in terms of Euler angles is given in Equation (4.181) as

$$
\begin{aligned}
{}_G\boldsymbol{\omega}_B &= \begin{bmatrix} \omega_X \\ \omega_Y \\ \omega_Z \end{bmatrix} = \begin{bmatrix} 0 & \cos\varphi & \sin\theta\sin\varphi \\ 0 & \sin\varphi & -\cos\varphi\sin\theta \\ 1 & 0 & \cos\theta \end{bmatrix} \begin{bmatrix} \dot{\varphi} \\ \dot{\theta} \\ \dot{\psi} \end{bmatrix} \\
&= \begin{bmatrix} \dot{\theta}\cos\varphi + \dot{\psi}\sin\theta\sin\varphi \\ \dot{\theta}\sin\varphi - \dot{\psi}\cos\varphi\sin\theta \\ \dot{\varphi} + \dot{\psi}\cos\theta \end{bmatrix}
\end{aligned} \tag{9.79}
$$

The angular acceleration is then equal to

$$
\begin{aligned}
{}_G\boldsymbol{\alpha}_B &= \frac{{}^Gd}{dt}\,{}_G\boldsymbol{\omega}_B \\
&= \begin{bmatrix} \cos\varphi\left(\ddot{\theta} + \dot{\varphi}\dot{\psi}\sin\theta\right) + \sin\varphi\left(\ddot{\psi}\sin\theta + \dot{\theta}\dot{\psi}\cos\theta - \dot{\theta}\dot{\varphi}\right) \\ \sin\varphi\left(\ddot{\theta} + \dot{\varphi}\dot{\psi}\sin\theta\right) + \cos\varphi\left(\dot{\theta}\dot{\varphi} - \ddot{\psi}\sin\theta - \dot{\theta}\dot{\psi}\cos\theta\right) \\ \ddot{\varphi} + \ddot{\psi}\cos\theta - \dot{\theta}\dot{\psi}\sin\theta \end{bmatrix}
\end{aligned} \tag{9.80}
$$

To determine the B-expression of the angular velocity ${}^B_G\boldsymbol{\alpha}_B$, we may take a time derivative of the B-expression of the angular velocity ${}^B_G\boldsymbol{\omega}_B$ given in Equation (4.182):

$$
{}^B_G\boldsymbol{\omega}_B = \begin{bmatrix} \omega_x \\ \omega_y \\ \omega_z \end{bmatrix} = \begin{bmatrix} \sin\theta\sin\psi & \cos\psi & 0 \\ \sin\theta\cos\psi & -\sin\psi & 0 \\ \cos\theta & 0 & 1 \end{bmatrix} \begin{bmatrix} \dot{\varphi} \\ \dot{\theta} \\ \dot{\psi} \end{bmatrix} \tag{9.81}
$$

$$
\begin{aligned}
{}^B_G\boldsymbol{\alpha}_B &= \frac{{}^Bd}{dt}\,{}^B_G\boldsymbol{\omega}_B \\
&= \frac{d}{dt}\begin{bmatrix} \sin\theta\sin\psi & \cos\psi & 0 \\ \sin\theta\cos\psi & -\sin\psi & 0 \\ \cos\theta & 0 & 1 \end{bmatrix} \begin{bmatrix} \dot{\varphi} \\ \dot{\theta} \\ \dot{\psi} \end{bmatrix} \\
&\quad + \begin{bmatrix} \sin\theta\sin\psi & \cos\psi & 0 \\ \sin\theta\cos\psi & -\sin\psi & 0 \\ \cos\theta & 0 & 1 \end{bmatrix} \frac{d}{dt}\begin{bmatrix} \dot{\varphi} \\ \dot{\theta} \\ \dot{\psi} \end{bmatrix} \\
&= \begin{bmatrix} \cos\psi\left(\ddot{\theta} + \dot{\varphi}\dot{\psi}\sin\theta\right) + \sin\psi\left(\ddot{\varphi}\sin\theta + \dot{\theta}\dot{\varphi}\cos\theta - \dot{\theta}\dot{\psi}\right) \\ \cos\psi\left(\ddot{\varphi}\sin\theta + \dot{\theta}\dot{\varphi}\cos\theta - \dot{\theta}\dot{\psi}\right) - \sin\psi\left(\ddot{\theta} + \dot{\varphi}\dot{\psi}\sin\theta\right) \\ \ddot{\varphi}\cos\theta + \ddot{\psi} - \dot{\theta}\dot{\varphi}\sin\theta \end{bmatrix}
\end{aligned} \tag{9.82}
$$

We may also use the transformation matrix ${}^{B}R_{G}$ in terms of Euler angles (4.142) and determine the angular acceleration vector in the body coordinate frame as

$$\begin{aligned}
{}^{B}_{G}\boldsymbol{\alpha}_{B} &= {}^{B}R_{G}\ {}_{G}\boldsymbol{\alpha}_{B} \\
&= \begin{bmatrix} c\varphi c\psi - c\theta s\varphi s\psi & c\psi s\varphi + c\theta c\varphi s\psi & s\theta s\psi \\ -c\varphi s\psi - c\theta c\psi s\varphi & -s\varphi s\psi + c\theta c\varphi c\psi & s\theta c\psi \\ s\theta s\varphi & -c\varphi s\theta & c\theta \end{bmatrix} {}_{G}\boldsymbol{\alpha}_{B} \\
&= \begin{bmatrix} \cos\psi\left(\ddot{\theta} + \dot{\varphi}\dot{\psi}\sin\theta\right) + \sin\psi\left(\ddot{\varphi}\sin\theta + \dot{\theta}\dot{\varphi}\cos\theta - \dot{\theta}\dot{\psi}\right) \\ \cos\psi\left(\ddot{\varphi}\sin\theta + \dot{\theta}\dot{\varphi}\cos\theta - \dot{\theta}\dot{\psi}\right) - \sin\psi\left(\ddot{\theta} + \dot{\varphi}\dot{\psi}\sin\theta\right) \\ \ddot{\varphi}\cos\theta - \ddot{\psi} - \dot{\theta}\dot{\varphi}\sin\theta \end{bmatrix}
\end{aligned} \tag{9.83}$$

Example 536 B-Expression of Angular Acceleration The angular acceleration expressed in the body frame is the body derivative of the angular velocity vector. To show this, we use the derivative transport formula (8.250):

$$\begin{aligned}
{}^{B}_{G}\boldsymbol{\alpha}_{B} &= {}^{B}_{G}\dot{\boldsymbol{\omega}}_{B} = \frac{{}^{G}d}{dt}\,{}^{B}_{G}\boldsymbol{\omega}_{B} \\
&= \frac{{}^{B}d}{dt}\,{}^{B}_{G}\boldsymbol{\omega}_{B} + {}^{B}_{G}\boldsymbol{\omega}_{B} \times {}^{B}_{G}\boldsymbol{\omega}_{B} = \frac{{}^{B}d}{dt}\,{}^{B}_{G}\boldsymbol{\omega}_{B}
\end{aligned} \tag{9.84}$$

Interestingly, the global and body derivatives of ${}^{B}_{G}\boldsymbol{\omega}_{B}$ are equal:

$$\frac{{}^{G}d}{dt}\,{}^{B}_{G}\boldsymbol{\omega}_{B} = \frac{{}^{B}d}{dt}\,{}^{B}_{G}\boldsymbol{\omega}_{B} = {}^{B}_{G}\boldsymbol{\alpha}_{B} \tag{9.85}$$

This is because ${}_{G}\boldsymbol{\omega}_{B}$ is about an axis $\hat{u}$ that is instantaneously fixed in both B and G.

A vector $\boldsymbol{\alpha}$ can generally indicate the angular acceleration of a coordinate frame A with respect to another frame B. It can be expressed in or seen from a third coordinate frame C. We indicate the first coordinate frame A by a right subscript, the second frame B by a left subscript, and the third frame C by a left superscript, ${}^{C}_{B}\boldsymbol{\alpha}_{A}$. If the left super- and subscripts are the same, we only show the subscript. So, the angular acceleration of A with respect to B as seen from C is the C-expression of ${}_{B}\boldsymbol{\alpha}_{A}$:

$${}^{C}_{B}\boldsymbol{\alpha}_{A} = {}^{C}R_{B}\ {}_{B}\boldsymbol{\alpha}_{A} \tag{9.86}$$

Example 537 B-Expression of Acceleration Transforming ${}^{G}\mathbf{a}$ to the body frame provides the body expression of the acceleration vector:

$$\begin{aligned}
{}^{B}_{G}\mathbf{a}_{P} &= {}^{G}R_{B}^{\mathrm{T}}\,{}^{G}\mathbf{a} = {}^{G}R_{B}^{\mathrm{T}}\ {}_{G}S_{B}\ {}^{G}\mathbf{r} = {}^{G}R_{B}^{\mathrm{T}}\ {}^{G}\ddot{R}_{B}\ {}^{G}R_{B}^{\mathrm{T}}\ {}^{G}\mathbf{r} \\
&= {}^{G}R_{B}^{\mathrm{T}}\ {}^{G}\ddot{R}_{B}\ {}^{B}\mathbf{r}
\end{aligned} \tag{9.87}$$

We denote the coefficient of ${}^{B}\mathbf{r}$ by

$${}^{B}_{G}S_{B} = {}^{G}R_{B}^{\mathrm{T}}\ {}^{G}\ddot{R}_{B} \tag{9.88}$$

and rewrite Equation (9.87) as

$$ {}_G^B\mathbf{a}_P = {}_G^B S_B \, {}^B\mathbf{r}_P \tag{9.89} $$

where ${}_G^B S_B$ is the rotational acceleration transformation of the B-frame relative to the G-frame as seen from the B-frame.

Example 538 ★ Technical Point on ${}_G\tilde{\omega}_B$, ${}_G^B S_B$, ${}_G\tilde{\alpha}_B$ The derivative kinematics of a rigid body with a fixed point begins by differentiating the kinematic transformation between the body B and global G coordinate frames:

$$ {}^G\mathbf{r} = {}^G R_B \, {}^B\mathbf{r} \tag{9.90} $$

The kinematic transformation matrix ${}^G R_B$ takes the coordinates of a point in the B-frame and determines the coordinates of the point in the G-frame. The matrix ${}^G R_B$ is called a kinematic or geometric transformation because the dimensions of what it takes and what it provides are the same.

The first derivative of (9.90) indicates the time rate of motion of a body point:

$$ {}^G\mathbf{v} = {}^G\dot{R}_B \, {}^B\mathbf{r} = {}_G\tilde{\omega}_B \, {}^G\mathbf{r} \tag{9.91} $$

where ${}_G\tilde{\omega}_B$ is the angular velocity matrix of B in G. This matrix is skew symmetric and its associated vector is the angular velocity vector. The matrix ${}_G\tilde{\omega}_B$ also acts as an operator and a transformer. It takes the global position vector of a body point, ${}^G\mathbf{r}$, and determines its global velocity vector, ${}^G\mathbf{v}$. So, it is called the rotational velocity transformation.

The second derivative of (9.90) indicates the time rate of velocity:

$$ {}^G\mathbf{a} = {}^G\ddot{R}_B \, {}^B\mathbf{r} = {}_G S_B \, {}^G\mathbf{r} \tag{9.92} $$

where ${}_G S_B$ is not skew symmetric, so it does not indicate a vector. However, ${}_G S_B$ acts as an operator and a transformer. It takes the global position vector of a body point, ${}^G\mathbf{r}$, and determines its global acceleration vector, ${}^G\mathbf{a}$. It is called the rotational acceleration transformation. The matrix ${}_G S_B$ is the sum of two matrices:

$$ {}_G S_B = {}_G\tilde{\alpha}_B + {}_G\tilde{\omega}_B^2 \tag{9.93} $$

The first matrix, ${}_G\tilde{\alpha}_B$, is the time derivative of ${}_G\tilde{\omega}_B$. So, it is a skew-symmetric matrix and indicates the angular acceleration matrix and vector. However, ${}_G\tilde{\alpha}_B$ cannot transform a position vector to its acceleration vector. The second matrix, ${}_G\tilde{\omega}_B^2$, is the square of the angular velocity matrix. It is not skew symmetric and indicates no vector.

We may consider ${}_G\tilde{\alpha}_B$ and ${}_G\tilde{\omega}_B^2$ as transformers because when they operate on ${}^G\mathbf{r}$ they respectively provide the tangential and centripetal components of the acceleration vector ${}^G\mathbf{a}$.

Example 539 Angular Acceleration from Rodriguez Formula Using the Rodriguez rotation formula, we can show that

$$\begin{aligned} {}_G\tilde{\alpha}_B = {}_G\dot{\tilde{\omega}}_B &= \lim_{\phi \to 0} \frac{{}^G d^2}{dt^2} R_{\hat{u},\phi} \\ &= \lim_{\phi \to 0} \frac{{}^G d^2}{dt^2} \left(-\tilde{u}^2 \cos\phi + \tilde{u} \sin\phi + \tilde{u}^2 + \mathbf{I} \right) = \ddot{\phi}\tilde{u} + \dot{\phi}\dot{\tilde{u}} \end{aligned} \tag{9.94}$$

Example 540 ★ Alternative Proof of Relative Acceleration Formula To show addition of the relative angular accelerations in Equations (9.44) and (9.45), we may start from a combination of rotations,

$$ {}^0R_2 = {}^0R_1\, {}^1R_2 \tag{9.95}$$

and take their time derivatives,

$$ {}^0\dot{R}_2 = {}^0\dot{R}_1\, {}^1R_2 + {}^0R_1\, {}^1\dot{R}_2 \tag{9.96}$$

$$ {}^0\ddot{R}_2 = {}^0\ddot{R}_1\, {}^1R_2 + 2\,{}^0\dot{R}_1\, {}^1\dot{R}_2 + {}^0R_1\, {}^1\ddot{R}_2 \tag{9.97}$$

Substituting the derivatives of rotation matrices with

$$ {}^0\ddot{R}_2 = {}_0S_2\, {}^0R_2 \tag{9.98}$$

$$ {}^0\ddot{R}_1 = {}_0S_1\, {}^0R_1 \tag{9.99}$$

$$ {}^1\ddot{R}_2 = {}_1S_2\, {}^1R_2 \tag{9.100}$$

$$ {}^0\dot{R}_2 = {}_0\tilde{\omega}_2\, {}^0R_2 \tag{9.101}$$

$$ {}^0\dot{R}_1 = {}_0\tilde{\omega}_1\, {}^0R_1 \tag{9.102}$$

$$ {}^1\dot{R}_2 = {}_1\tilde{\omega}_2\, {}^1R_2 \tag{9.103}$$

results in

$$\begin{aligned} {}_0S_2\, {}^0R_2 &= {}_0S_1\, {}^0R_1\, {}^1R_2 + 2\, {}_0\tilde{\omega}_1\, {}^0R_1\, {}_1\tilde{\omega}_2\, {}^1R_2 + {}^0R_1\, {}_1S_2\, {}^1R_2 \\ &= {}_0S_1\, {}^0R_2 + 2\, {}_0\tilde{\omega}_1\, {}^0R_1\, {}_1\tilde{\omega}_2\, {}^0R_1^T\, {}^0R_1\, {}^1R_2 + {}^0R_1\, {}_1S_2\, {}^1R_2 \\ &= {}_0S_1\, {}^0R_2 + 2\, {}_0\tilde{\omega}_1\, {}_1^0\tilde{\omega}_2\, {}^0R_2 + {}^0R_1\, {}_1S_2\, {}^0R_1^T\, {}^0R_1\, {}^1R_2 \\ &= {}_0S_1\, {}^0R_2 + 2\, {}_0\tilde{\omega}_1\, {}_1^0\tilde{\omega}_2\, {}^0R_2 + {}_1^0S_2\, {}^0R_2 \end{aligned} \tag{9.104}$$

Therefore, we find

$$ {}_0S_2 = {}_0S_1 + {}_1^0S_2 + 2\, {}_0\tilde{\omega}_1\, {}_1^0\tilde{\omega}_2 \tag{9.105}$$

which is equivalent to

$$ {}_0\tilde{\alpha}_2 + {}_0\tilde{\omega}_2^2 = {}_0\tilde{\alpha}_1 + {}_0\tilde{\omega}_1^2 + {}_1^0\tilde{\alpha}_2 + {}_1^0\tilde{\omega}_2^2 + 2\, {}_0\tilde{\omega}_1\, {}_1^0\tilde{\omega}_2 \tag{9.106}$$

Simplifying this equation shows that

$$
\begin{aligned}
{}_0\tilde{\alpha}_2 &= {}_0\tilde{\alpha}_1 + {}_1^0\tilde{\alpha}_2 + {}_0\tilde{\omega}_1^2 + {}_1^0\tilde{\omega}_2^2 + 2\,{}_0\tilde{\omega}_1\,{}_1^0\tilde{\omega}_2 - {}_0\tilde{\omega}_2^2 \\
&= {}_0\tilde{\alpha}_1 + {}_1^0\tilde{\alpha}_2 + \left({}_0\tilde{\omega}_1 + {}_1^0\tilde{\omega}_2\right)^2 - {}_0\tilde{\omega}_2^2 \\
&= {}_0\tilde{\alpha}_1 + {}_1^0\tilde{\alpha}_2 + {}_0\tilde{\omega}_2^2 - {}_0\tilde{\omega}_2^2 = {}_0\tilde{\alpha}_1 + {}_1^0\tilde{\alpha}_2
\end{aligned}
\tag{9.107}
$$

which indicates that two angular accelerations may be added when they are expressed in the same frame:

$$
{}_0\boldsymbol{\alpha}_2 = {}_0\boldsymbol{\alpha}_1 + {}_1^0\boldsymbol{\alpha}_2 \tag{9.108}
$$

Example 541 Velocity and Acceleration of a Simple Pendulum A point mass attached to a massless rod hanging from a revolute joint is what we call a *simple pendulum*. Figure 9.2 illustrates a simple pendulum. A local coordinate frame B is attached to the pendulum, which rotates in a global frame G about the Z-axis. The kinematic information of the mass is given as

$$
{}^B\mathbf{r} = l\hat{\imath} \tag{9.109}
$$

$$
{}^G\mathbf{r} = {}^GR_B\,{}^B\mathbf{r} = \begin{bmatrix} l\sin\phi \\ -l\cos\phi \\ 0 \end{bmatrix} \tag{9.110}
$$

$$
{}_G^B\boldsymbol{\omega}_B = \dot{\phi}\hat{k} \tag{9.111}
$$

$$
{}_G\boldsymbol{\omega}_B = {}^GR_B^{\mathrm{T}}\,{}_G^B\boldsymbol{\omega}_B = \dot{\phi}\,\hat{K} \tag{9.112}
$$

$$
\begin{aligned}
{}^GR_B &= \begin{bmatrix} \cos\left(\frac{3}{2}\pi+\phi\right) & -\sin\left(\frac{3}{2}\pi+\phi\right) & 0 \\ \sin\left(\frac{3}{2}\pi+\phi\right) & \cos\left(\frac{3}{2}\pi+\phi\right) & 0 \\ 0 & 0 & 1 \end{bmatrix} \\
&= \begin{bmatrix} \sin\phi & \cos\phi & 0 \\ -\cos\phi & \sin\phi & 0 \\ 0 & 0 & 1 \end{bmatrix}
\end{aligned}
\tag{9.113}
$$

Therefore,

$$
{}_G^B\mathbf{v} = {}^B\dot{\mathbf{r}} + {}_G^B\boldsymbol{\omega}_B \times {}_G^B\mathbf{r} = 0 + \dot{\phi}\hat{k} \times l\hat{\imath} = l\,\dot{\phi}\hat{\jmath} \tag{9.114}
$$

$$
{}^G\mathbf{v} = {}^GR_B\,{}^B\mathbf{v} = \begin{bmatrix} l\,\dot{\phi}\cos\phi \\ l\,\dot{\phi}\sin\phi \\ 0 \end{bmatrix} \tag{9.115}
$$

and

$$
{}_G^B\mathbf{a} = {}_G^B\dot{\mathbf{v}} + {}_G^B\boldsymbol{\omega}_B \times {}_G^B\mathbf{v} = l\,\ddot{\phi}\hat{\jmath} + \dot{\phi}\hat{k} \times l\,\dot{\phi}\hat{\jmath} = l\,\ddot{\phi}\hat{\jmath} - l\,\dot{\phi}^2\hat{\imath} \tag{9.116}
$$

$$
{}^G\mathbf{a} = {}^GR_B\,{}^B\mathbf{a} = \begin{bmatrix} l\,\ddot{\phi}\cos\phi - l\,\dot{\phi}^2\sin\phi \\ l\,\ddot{\phi}\sin\phi + l\,\dot{\phi}^2\cos\phi \\ 0 \end{bmatrix} \tag{9.117}
$$

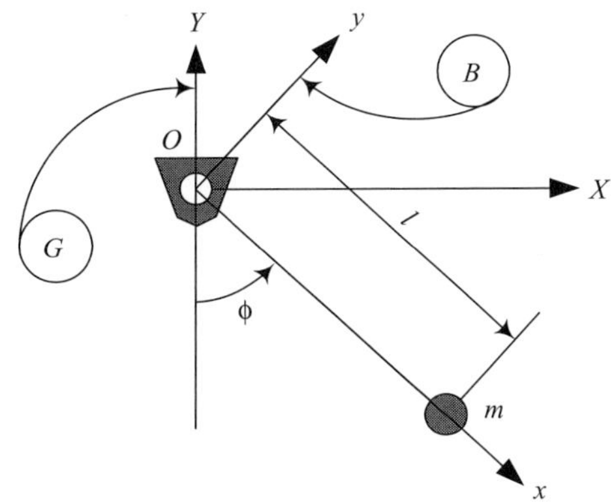

Figure 9.2 Illustration of a simple pendulum.

Example 542 A Moving Vehicle on Earth Consider the motion of a vehicle on Earth at latitude 30 deg that is heading north, as shown in Figure 9.3. The vehicle has a velocity $v = {}_E^B\dot{\mathbf{r}} = 80\hat{\imath}$ km/h $= 22.22\hat{\imath}$ m/s and acceleration $a = {}_E^B\ddot{\mathbf{r}} = 0.1\hat{\imath}$ m/s^2, both with respect to the road and expressed in the vehicle frame B.

To determine the kinematics of the vehicle, we define three coordinate frames: (1) a global frame $G(OXYZ)$ at Earth's center, (2) an Earth fixed frame E at Earth's center, and (3) a vehicle frame B attached to the mass center of the vehicle such that its z-axis is along ${}^B\mathbf{r}$ and its x-axis points to the north pole. The Earth's frame E is turning about the Z-axis with $\omega_E = {}_G\omega_E \approx 2\pi$ rad/d. The frames E and G are assumed coincident at the moment and the vehicle is in the (Z, Y)-plane.

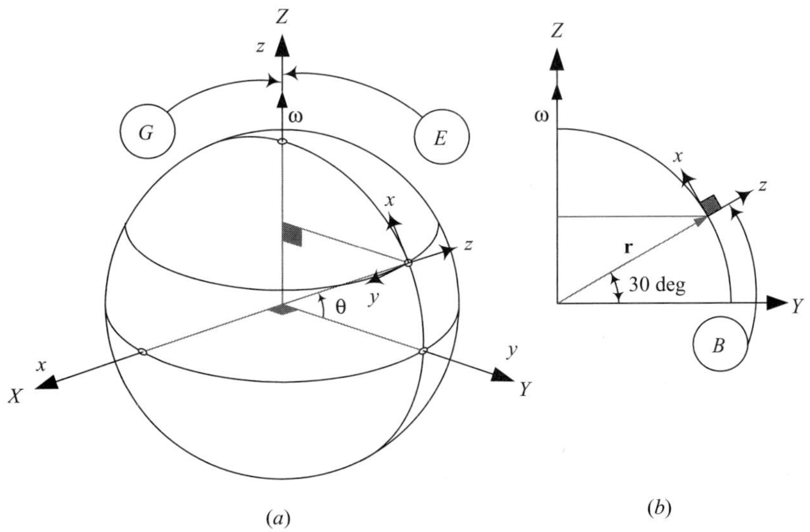

Figure 9.3 A moving vehicle at latitude 30 deg and heading north pole.

The radius of Earth is R, and hence, the vehicle's kinematics are

$$ {}_E^B\mathbf{r} = R\hat{k} \text{ m} \tag{9.118} $$

$$ {}_E^B\dot{\mathbf{r}} = 22.22\hat{\imath} \text{ m/s} \tag{9.119} $$

$$ {}_E^B\ddot{\mathbf{r}} = 0.1\hat{\imath} \text{ m/s}^2 \tag{9.120} $$

$$ \dot{\theta} = \frac{v}{R} \text{ rad/s} \tag{9.121} $$

$$ \ddot{\theta} = \frac{a}{R} \text{ rad/s}^2 \tag{9.122} $$

The angular velocity of B is

$$ \begin{aligned} {}_G^B\boldsymbol{\omega}_B &= {}^B R_G \left({}_G\boldsymbol{\omega}_E + {}_E^G\boldsymbol{\omega}_B \right) = {}^B R_G \left(\omega_E \hat{K} + \dot{\theta}\hat{I} \right) \\ &= (\omega_E \cos\theta)\,\hat{\imath} + (\omega_E \sin\theta)\,\hat{k} + \dot{\theta}\,\hat{\jmath} \\ &= (\omega_E \cos\theta)\,\hat{\imath} + (\omega_E \sin\theta)\,\hat{k} + \frac{v}{R}\hat{\jmath} \end{aligned} \tag{9.123} $$

Therefore, the velocity and acceleration of the vehicle are

$$ {}_G^B\mathbf{v} = {}^B\dot{\mathbf{r}} + {}_G^B\boldsymbol{\omega}_B \times {}_G^B\mathbf{r} = 0 + {}_G^B\boldsymbol{\omega}_B \times R\hat{k} = v\hat{\imath} - (R\omega_E \cos\theta)\,\hat{\jmath} \tag{9.124} $$

$$ \begin{aligned} {}_G^B\mathbf{a} &= {}_G^B\dot{\mathbf{v}} + {}_G^B\boldsymbol{\omega}_B \times {}_G^B\mathbf{v} \\ &= a\hat{\imath} + \left(R\omega_E \dot{\theta} \sin\theta \right) \hat{\jmath} + \begin{bmatrix} \omega_E \cos\theta \\ v/R \\ \omega_E \sin\theta \end{bmatrix} \times \begin{bmatrix} v \\ -R\omega_E \cos\theta \\ 0 \end{bmatrix} \\ &= a\hat{\imath} + \left(R\omega_E \dot{\theta} \sin\theta \right) \hat{\jmath} + \begin{bmatrix} R\omega_E^2 \cos\theta \sin\theta \\ v\omega_E \sin\theta \\ -\frac{1}{R}v^2 - R\omega_E^2 \cos^2\theta \end{bmatrix} \\ &= \begin{bmatrix} a + R\omega_E^2 \cos\theta \sin\theta \\ 2R\omega_E \dot{\theta} \sin\theta \\ -\frac{1}{R}v^2 - R\omega_E^2 \cos^2\theta \end{bmatrix} \end{aligned} \tag{9.125} $$

The term $a\hat{\imath}$ is the acceleration of B relative to Earth, $(2R\omega_E \dot{\theta} \sin\theta)\hat{\jmath}$ is the Coriolis acceleration, $-(v^2/R)\hat{k}$ is the centripetal acceleration due to traveling, and $-R\omega_E^2$ is the centripetal acceleration due to Earth's rotation.

Substituting the numerical values and accepting $R \approx 6.3677 \times 10^6$ m yield

$$ \begin{aligned} {}_G^B\mathbf{v} &= 22.22\hat{\imath} - 6.3677 \times 10^6 \left(\frac{2\pi}{24 \times 3600} \frac{366.25}{365.25} \right) \cos\frac{\pi}{6}\hat{\jmath} \\ &= 22.22\hat{\imath} - 402.13\hat{\jmath} \text{ m/s} \end{aligned} \tag{9.126} $$

$$ {}_G^B\mathbf{a} = 1.5662 \times 10^{-2}\hat{\imath} + 1.6203 \times 10^{-3}\hat{\jmath} - 2.5473 \times 10^{-2}\hat{k} \text{ m/s}^2 \tag{9.127} $$

Example 543 Spherical Pendulum A pendulum free to oscillate in any plane is called a spherical pendulum. This name comes from the codominants that we use to locate the tip mass. Consider a pendulum with a point mass m at the tip point of a long, massless, and straight string with length l. The pendulum is hanging from a point $A(0,0,0)$ in a local coordinate frame $B_1(x_1, y_1, z_1)$.

To indicate the mass m, we attach a coordinate frame $B_2(x_2, y_2, z_2)$ to the pendulum at point A, as shown in Figure 9.4. The pendulum makes an angle β with the vertical z_1-axis. The pendulum swings in the plane (x_2, z_2) and makes an angle γ with the plane (x_1, z_1). Therefore, the transformation matrix between B_2 and B_1 is

$$\begin{aligned} {}^2R_1 &= R_{y_2,-\beta}\, R_{z_2,\gamma} \\ &= \begin{bmatrix} \cos\gamma\cos\beta & \cos\beta\sin\gamma & \sin\beta \\ -\sin\gamma & \cos\gamma & 0 \\ -\cos\gamma\sin\beta & -\sin\gamma\sin\beta & \cos\beta \end{bmatrix} \end{aligned} \tag{9.128}$$

The position vectors of m are

$$ {}^2\mathbf{r} = \begin{bmatrix} 0 \\ 0 \\ -l \end{bmatrix} \qquad {}^1\mathbf{r} = {}^1R_2\,{}^2\mathbf{r} = \begin{bmatrix} l\cos\gamma\sin\beta \\ l\sin\beta\sin\gamma \\ -l\cos\beta \end{bmatrix} \tag{9.129}$$

The equation of motion of m is

$$ {}^1\mathbf{M} = I\ {}_1\boldsymbol{\alpha}_2 \tag{9.130}$$

$$ {}^1\mathbf{r} \times m\,{}^1\mathbf{g} = ml^2\ {}_1\boldsymbol{\alpha}_2 \tag{9.131}$$

$$ \begin{bmatrix} l\cos\gamma\sin\beta \\ l\sin\beta\sin\gamma \\ -l\cos\beta \end{bmatrix} \times m \begin{bmatrix} 0 \\ 0 \\ -g_0 \end{bmatrix} = ml^2\ {}_1\boldsymbol{\alpha}_2 \tag{9.132}$$

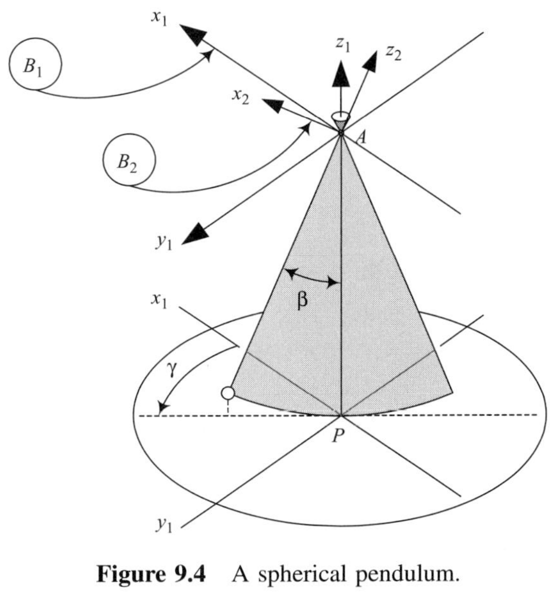

Figure 9.4 A spherical pendulum.

Therefore,

$$
{}_1\boldsymbol{\alpha}_2 = \frac{g_0}{l}\begin{bmatrix} -\sin\beta\sin\gamma \\ \cos\gamma\sin\beta \\ 0 \end{bmatrix} \tag{9.133}
$$

To find the angular acceleration of B_2 in B_1, we use ${}^2R_1$:

$$
\begin{aligned}
{}^1\dot{R}_2 &= \dot{\beta}\frac{d}{d\beta}{}^2R_1 + \dot{\gamma}\frac{d}{d\gamma}{}^2R_1 \\
&= \begin{bmatrix} -\dot{\beta}c\gamma s\beta - \dot{\gamma}c\beta s\gamma & -\dot{\gamma}c\gamma & \dot{\gamma}s\beta s\gamma - \dot{\beta}c\beta c\gamma \\ \dot{\gamma}c\beta c\gamma - \dot{\beta}s\beta s\gamma & -\dot{\gamma}s\gamma & -\dot{\beta}c\beta s\gamma - \dot{\gamma}c\gamma s\beta \\ \dot{\beta}c\beta & 0 & -\dot{\beta}s\beta \end{bmatrix}
\end{aligned} \tag{9.134}
$$

$$
{}_1\tilde{\omega}_2 = {}^1\dot{R}_2\,{}^1R_2^{\mathrm{T}} = \begin{bmatrix} 0 & -\dot{\gamma} & -\dot{\beta}\cos\gamma \\ \dot{\gamma} & 0 & -\dot{\beta}\sin\gamma \\ \dot{\beta}\cos\gamma & \dot{\beta}\sin\gamma & 0 \end{bmatrix} \tag{9.135}
$$

$$
\begin{aligned}
{}^1\ddot{R}_2 &= \ddot{\beta}\frac{d}{d\beta}{}^2R_1 + \dot{\beta}^2\frac{d^2}{d\beta^2}{}^2R_1 + \dot{\beta}\dot{\gamma}\frac{d^2}{d\gamma\, d\beta}{}^2R_1 \\
&\quad + \ddot{\gamma}\frac{d}{d\gamma}{}^2R_1 + \dot{\gamma}\dot{\beta}\frac{d^2}{d\beta\, d\gamma}{}^2R_1 + \dot{\gamma}^2\frac{d^2}{d\gamma^2}{}^2R_1
\end{aligned} \tag{9.136}
$$

$$
\begin{aligned}
{}_1\tilde{\alpha}_2 &= {}^1\ddot{R}_2\,{}^1R_2^{\mathrm{T}} - {}_1\tilde{\omega}_2^2 \\
&= \begin{bmatrix} 0 & -\ddot{\gamma} & -\ddot{\beta}c\gamma + \dot{\beta}\dot{\gamma}s\gamma \\ \ddot{\gamma} & 0 & -\ddot{\beta}s\gamma - \dot{\beta}\dot{\gamma}c\gamma \\ \ddot{\beta}c\gamma - \dot{\beta}\dot{\gamma}s\gamma & \ddot{\beta}s\gamma + \dot{\beta}\dot{\gamma}c\gamma & 0 \end{bmatrix}
\end{aligned} \tag{9.137}
$$

Therefore, the equation of motion of the pendulum would be

$$
\frac{g_0}{l}\begin{bmatrix} -\sin\beta\sin\gamma \\ \cos\gamma\sin\beta \\ 0 \end{bmatrix} = \begin{bmatrix} \ddot{\beta}\sin\gamma + \dot{\beta}\dot{\gamma}\cos\gamma \\ -\ddot{\beta}\cos\gamma + \dot{\beta}\dot{\gamma}\sin\gamma \\ \ddot{\gamma} \end{bmatrix} \tag{9.138}
$$

The third equation indicates that

$$
\dot{\gamma} = \dot{\gamma}_0 \qquad \gamma = \dot{\gamma}_0 t + \gamma_0 \tag{9.139}
$$

The second and third equations can be combined to the form

$$
\ddot{\beta} = -\sqrt{\frac{g_0^2}{l^2}\sin^2\beta + \dot{\beta}^2\dot{\gamma}_0^2} \tag{9.140}
$$

which reduces to the equation of a simple pendulum if $\dot{\gamma}_0 = 0$.

Example 544 ★ Equation of Motion of a Spherical Pendulum Consider a particle P of mass m that is suspended by a string of length l from a point A, as shown in

Figure 9.4. If we show the tension of the string by **T**, then the equation of motion of P is

$$ {}^{1}\mathbf{T} + m\,{}^{1}\mathbf{g} = m\,{}^{1}\ddot{\mathbf{r}} \tag{9.141} $$

or

$$ -T\,{}^{1}\mathbf{r} + m\,{}^{1}\mathbf{g} = m\,{}^{1}\ddot{\mathbf{r}} \tag{9.142} $$

To eliminate ${}^{1}\mathbf{T}$, we multiply the equation by ${}^{1}\mathbf{r}$,

$$ {}^{1}\mathbf{r} \times {}^{1}\mathbf{g} = {}^{1}\mathbf{r} \times {}^{1}\ddot{\mathbf{r}} \tag{9.143} $$

$$ \begin{bmatrix} l\cos\gamma\sin\beta \\ l\sin\beta\sin\gamma \\ -l\cos\beta \end{bmatrix} \times \begin{bmatrix} 0 \\ 0 \\ -g_0 \end{bmatrix} = \begin{bmatrix} l\cos\gamma\sin\beta \\ l\sin\beta\sin\gamma \\ -l\cos\beta \end{bmatrix} \times \begin{bmatrix} \ddot{x} \\ \ddot{y} \\ \ddot{z} \end{bmatrix} $$

and find

$$ \begin{bmatrix} -lg_0\sin\beta\sin\gamma \\ lg_0\cos\gamma\sin\beta \\ 0 \end{bmatrix} = \begin{bmatrix} l\ddot{y}\cos\beta + l\ddot{z}\sin\beta\sin\gamma \\ -l\ddot{x}\cos\beta - l\ddot{z}\cos\gamma\sin\beta \\ l\ddot{y}\cos\gamma - l\ddot{x}\sin\gamma \end{bmatrix} \tag{9.144} $$

These are the equations of motion of m. However, we may express the equations only in terms of γ and β. To do so, we may either take time derivatives of ${}^{1}\mathbf{r}$ or use ${}_{1}\boldsymbol{\alpha}_{2}$ from Example 544 and find ${}^{1}\ddot{\mathbf{r}}$:

$$ {}^{1}\ddot{\mathbf{r}} = {}_{1}\boldsymbol{\alpha}_{2} \times {}^{1}\mathbf{r} \tag{9.145} $$

In either case, Equation (9.138) would be the equation of motion in terms of γ and β.

Example 545 ★ Angular Acceleration Transformation and Euler Angles Using the Euler angle transformation matrix (4.144), we can determine the rotational acceleration transformation ${}_{G}S_{B}$ and the angular acceleration ${}_{G}\boldsymbol{\alpha}_{B}$ in terms of Euler angles and their time rates:

$$ {}_{G}S_{B} = {}^{G}\ddot{R}_{B}\,{}^{G}R_{B}^{\mathrm{T}} = {}_{G}\tilde{\alpha}_{B} + {}_{G}\tilde{\omega}_{B}^{2} \tag{9.146} $$

$$ {}_{G}\tilde{\alpha}_{B} = {}_{G}\dot{\tilde{\omega}}_{B} = {}^{G}\ddot{R}_{B}\,{}^{G}R_{B}^{\mathrm{T}} + {}^{G}\dot{R}_{B}\,{}^{G}\dot{R}_{B}^{\mathrm{T}} \tag{9.147} $$

$$ {}_{G}\tilde{\omega}_{B} = {}^{G}\dot{R}_{B}\,{}^{G}R_{B}^{\mathrm{T}} \tag{9.148} $$

$$ {}_{G}\tilde{\omega}_{B}^{2} = -{}^{G}\dot{R}_{B}\,{}^{G}\dot{R}_{B}^{\mathrm{T}} = \left({}^{G}\dot{R}_{B}\,{}^{G}R_{B}^{\mathrm{T}}\right)^{2} \tag{9.149} $$

The principal decomposed expression of the Euler angle transformation matrix is

$$ \begin{aligned} {}^{G}R_{B} &= {}^{B}R_{G}^{-1} = {}^{B}R_{G}^{\mathrm{T}} = \left[R_{z,\psi}R_{x,\theta}R_{z,\varphi}\right]^{\mathrm{T}} = R_{z,\varphi}^{\mathrm{T}}R_{x,\theta}^{\mathrm{T}}R_{z,\psi}^{\mathrm{T}} \\ &= R_{Z,\varphi}R_{X,\theta}R_{Z,\psi} \end{aligned} \tag{9.150} $$

We may use the decomposed form of ${}^{G}R_{B}$ to determine ${}^{G}\dot{R}_{B}$, ${}^{G}\ddot{R}_{B}$, ${}_{G}\tilde{\omega}_{B}$, and ${}_{G}S_{B}$:

$$ \begin{aligned} {}^{G}\dot{R}_{B} &= \frac{d}{dt}\left(R_{z,\varphi}^{\mathrm{T}}R_{x,\theta}^{\mathrm{T}}R_{z,\psi}^{\mathrm{T}}\right) = \frac{d}{dt}\left(R_{Z,\varphi}R_{X,\theta}R_{Z,\psi}\right) \\ &= \dot{R}_{Z,\varphi}R_{X,\theta}R_{Z,\psi} + R_{Z,\varphi}\dot{R}_{X,\theta}R_{Z,\psi} + R_{Z,\varphi}R_{X,\theta}\dot{R}_{Z,\psi} \end{aligned} \tag{9.151} $$

$$
\begin{aligned}
{}^G\ddot{R}_B = {} & \ddot{R}_{Z,\varphi} R_{X,\theta} R_{Z,\psi} + \dot{R}_{Z,\varphi} \dot{R}_{X,\theta} R_{Z,\psi} + \dot{R}_{Z,\varphi} R_{X,\theta} \dot{R}_{Z,\psi} \\
& + \dot{R}_{Z,\varphi} \dot{R}_{X,\theta} R_{Z,\psi} + R_{Z,\varphi} \ddot{R}_{X,\theta} R_{Z,\psi} + R_{Z,\varphi} \dot{R}_{X,\theta} \dot{R}_{Z,\psi} \\
& + \dot{R}_{Z,\varphi} R_{X,\theta} \dot{R}_{Z,\psi} + R_{Z,\varphi} \dot{R}_{X,\theta} \dot{R}_{Z,\psi} + R_{Z,\varphi} R_{X,\theta} \ddot{R}_{Z,\psi}
\end{aligned} \tag{9.152}
$$

Assume ${}_G\tilde{\omega}_B$ is the same as (8.99):

$$
\begin{aligned}
{}_G\tilde{\omega}_B &= {}^G\dot{R}_B \, {}^G R_B^{\mathrm{T}} = {}^G\dot{R}_B \left(R_{Z,\psi}^{\mathrm{T}} R_{X,\theta}^{\mathrm{T}} R_{Z,\varphi}^{\mathrm{T}} \right) \\
&= \tilde{\omega}_{Z,\dot{\varphi}} + R_{Z,\varphi} \, \tilde{\omega}_{X,\dot{\theta}} R_{Z,\varphi}^{\mathrm{T}} + R_{Z,\varphi} R_{X,\theta} \, \tilde{\omega}_{Z,\dot{\psi}} R_{X,\theta}^{\mathrm{T}} R_{Z,\varphi}^{\mathrm{T}} \\
&= \begin{bmatrix} 0 & -\dot{\varphi} - \dot{\psi} c\theta & \dot{\theta} s\varphi - \dot{\psi} c\varphi s\theta \\ \dot{\varphi} + \dot{\psi} c\theta & 0 & -\dot{\theta} c\varphi - \dot{\psi} s\theta s\varphi \\ \dot{\psi} c\varphi s\theta - \dot{\theta} s\varphi & \dot{\theta} c\varphi + \dot{\psi} s\theta s\varphi & 0 \end{bmatrix}
\end{aligned} \tag{9.153}
$$

Therefore, the angular acceleration matrix is

$$
{}_G\tilde{\alpha}_B = \frac{{}^G d}{dt} \, {}_G\tilde{\omega}_B = \begin{bmatrix} 0 & -a_3 & a_2 \\ a_3 & 0 & -a_1 \\ -a_2 & a_1 & 0 \end{bmatrix} \tag{9.154}
$$

where

$$
\begin{aligned}
a_1 &= \cos\varphi \left(\ddot{\theta} + \dot{\varphi}\dot{\psi} \sin\theta \right) + \sin\varphi \left(\ddot{\psi} \sin\theta + \dot{\theta}\dot{\psi} \cos\theta - \dot{\theta}\dot{\varphi} \right) \\
a_2 &= \sin\varphi \left(\ddot{\theta} + \dot{\varphi}\dot{\psi} \sin\theta \right) + \cos\varphi \left(\dot{\theta}\dot{\varphi} - \ddot{\psi} \sin\theta - \dot{\theta}\dot{\psi} \cos\theta \right) \\
a_3 &= \ddot{\varphi} + \ddot{\psi} \cos\theta - \dot{\theta}\dot{\psi} \sin\theta
\end{aligned} \tag{9.155}
$$

Multiplying ${}_G\tilde{\omega}_B$ by itself yields

$$
{}_G\tilde{\omega}_B^2 = - {}^G\dot{R}_B \, {}^G\dot{R}_B^{\mathrm{T}} = \begin{bmatrix} r_{11} & r_{12} & r_{13} \\ r_{21} = r_{12} & r_{22} & r_{23} \\ r_{31} = r_{13} & r_{32} = r_{23} & r_{33} \end{bmatrix} \tag{9.156}
$$

where

$$
\begin{aligned}
r_{11} &= -\left(\dot{\theta} \sin\varphi - \dot{\psi} \cos\varphi \sin\theta \right)^2 - \left(\dot{\varphi} + \dot{\psi} \cos\theta \right)^2 \\
r_{12} &= \left(\dot{\theta} \cos\varphi + \dot{\psi} \sin\theta \sin\varphi \right) \left(\dot{\theta} \sin\varphi - \dot{\psi} \cos\varphi \sin\theta \right) \\
r_{13} &= \left(\dot{\varphi} + \dot{\psi} \cos\theta \right) \left(\dot{\theta} \cos\varphi + \dot{\psi} \sin\theta \sin\varphi \right)
\end{aligned} \tag{9.157}
$$

$$
\begin{aligned}
r_{22} &= -\left(\dot{\theta} \cos\varphi + \dot{\psi} \sin\theta \sin\varphi \right)^2 - \left(\dot{\varphi} + \dot{\psi} \cos\theta \right)^2 \\
r_{23} &= \left(\dot{\varphi} + \dot{\psi} \cos\theta \right) \left(\dot{\theta} \sin\varphi - \dot{\psi} \cos\varphi \sin\theta \right) \\
r_{33} &= -\left(\dot{\theta} \cos\varphi + \dot{\psi} \sin\theta \sin\varphi \right)^2 - \left(\dot{\theta} \sin\varphi - \dot{\psi} \cos\varphi \sin\theta \right)^2
\end{aligned} \tag{9.158}
$$

We can similarly develop ${}_GS_B$:

$$\begin{aligned}
{}_GS_B &= {}^G\ddot{R}_B\,{}^GR_B^{\mathrm{T}} = {}^G\ddot{R}_B\,\left(R_{Z,\psi}^{\mathrm{T}}R_{X,\theta}^{\mathrm{T}}R_{Z,\varphi}^{\mathrm{T}}\right) \\
&= S_{Z,\ddot{\varphi}} + \dot{R}_{Z,\varphi}\,\tilde{\omega}_{X,\dot{\theta}}\,R_{Z,\varphi}^{\mathrm{T}} + \dot{R}_{Z,\varphi}R_{X,\theta}\,\tilde{\omega}_{Z,\dot{\psi}}\,R_{X,\theta}^{\mathrm{T}}R_{Z,\varphi}^{\mathrm{T}} \\
&\quad + \dot{R}_{Z,\varphi}\,\tilde{\omega}_{X,\dot{\theta}}\,R_{Z,\varphi}^{\mathrm{T}} + R_{Z,\varphi}\,S_{X,\ddot{\theta}}\,R_{Z,\varphi}^{\mathrm{T}} + R_{Z,\varphi}\dot{R}_{X,\theta}\,\tilde{\omega}_{Z,\dot{\psi}}\,R_{X,\theta}^{\mathrm{T}}R_{Z,\varphi}^{\mathrm{T}} \\
&\quad + \dot{R}_{Z,\varphi}R_{X,\theta}\,\tilde{\omega}_{Z,\dot{\psi}}\,R_{X,\theta}^{\mathrm{T}}R_{Z,\varphi}^{\mathrm{T}} + R_{Z,\varphi}\dot{R}_{X,\theta}\,\tilde{\omega}_{Z,\dot{\psi}}\,R_{X,\theta}^{\mathrm{T}}R_{Z,\varphi}^{\mathrm{T}} \\
&\quad + R_{Z,\varphi}R_{X,\theta}\,S_{Z,\ddot{\psi}}\,R_{X,\theta}^{\mathrm{T}}R_{Z,\varphi}^{\mathrm{T}}
\end{aligned} \tag{9.159}$$

Thus,

$${}_GS_B = {}_G\tilde{\alpha}_B + {}_G\tilde{\omega}_B^2 = \begin{bmatrix} r_{11} & r_{12}-a_3 & a_2+r_{13} \\ a_3+r_{12} & r_{22} & r_{23}-a_1 \\ r_{13}-a_2 & a_1+r_{23} & r_{33} \end{bmatrix} \tag{9.160}$$

Example 546 ★ Rotational Jerk Transformation Consider a body coordinate frame B with a fixed point in a global frame G. The B-frame is turning in G with angular velocity ${}_G\boldsymbol{\omega}_B$ and acceleration ${}_G\boldsymbol{\alpha}_B$. The global *jerk* ${}^G\mathbf{j}$ of a body point P at ${}^G\mathbf{r}$ is given as

$$\begin{aligned}
{}^G\mathbf{j} &= {}^G\dddot{\mathbf{r}} = \frac{{}^Gd}{dt}\left({}_GS_B\,{}^G\mathbf{r}\right) = \frac{{}^Gd^2}{dt^2}\left({}_G\tilde{\omega}_B\,{}^G\mathbf{r}\right) = \frac{{}^Gd^3}{dt^3}\,{}^G\mathbf{r} \\
&= \frac{{}^Gd}{dt}\left(\left[\ddot{\phi}\tilde{u} + \dot{\phi}\dot{\tilde{u}} + \dot{\phi}^2\tilde{u}^2\right]{}^G\mathbf{r}\right) \\
&= \left[\dddot{\phi}\tilde{u} + 2\ddot{\phi}\dot{\tilde{u}} + \dot{\phi}\ddot{\tilde{u}} + 3\dot{\phi}\ddot{\phi}\tilde{u}^2 + 2\dot{\phi}^2\dot{\tilde{u}}\tilde{u} + \dot{\phi}^2\tilde{u}\dot{\tilde{u}} + \dot{\phi}^3\tilde{u}^3\right]{}^G\mathbf{r} \\
&= {}_GU_B\,{}^G\mathbf{r}
\end{aligned} \tag{9.161}$$

where ${}_GU_B$ is the *rotational jerk transformation*

$${}_GU_B = \dddot{\phi}\tilde{u} + 2\ddot{\phi}\dot{\tilde{u}} + \dot{\phi}\ddot{\tilde{u}} + 3\dot{\phi}\ddot{\phi}\tilde{u}^2 + 2\dot{\phi}^2\dot{\tilde{u}}\tilde{u} + \dot{\phi}^2\tilde{u}\dot{\tilde{u}} + \dot{\phi}^3\tilde{u}^3 \tag{9.162}$$

Employing ${}^G\mathbf{a} = {}^G\ddot{R}_B\,{}^GR_B^{\mathrm{T}}\,{}^G\mathbf{r}$, we can find the jerk of the body point and jerk transformation based on the rotational transformation matrix ${}^GR_B$:

$$\begin{aligned}
{}^G\mathbf{j} &= \frac{{}^Gd}{dt}\left({}^G\mathbf{a}\right) = \frac{{}^Gd}{dt}\left({}_GS_B\,{}^G\mathbf{r}\right) = \frac{{}^Gd}{dt}\left({}^G\ddot{R}_B\,{}^GR_B^{\mathrm{T}}\,{}^G\mathbf{r}\right) \\
&= \left({}^G\dddot{R}_B\,{}^GR_B^{\mathrm{T}} + {}^G\ddot{R}_B\,{}^G\dot{R}_B^{\mathrm{T}} + {}^G\ddot{R}_B\,{}^GR_B^{\mathrm{T}}\,{}^G\dot{R}_B\,{}^GR_B^{\mathrm{T}}\right){}^G\mathbf{r} \\
&= \left({}^G\dddot{R}_B\,{}^GR_B^{\mathrm{T}} + {}^G\ddot{R}_B\,{}^GR_B^{\mathrm{T}}\,{}^GR_B\,{}^G\dot{R}_B^{\mathrm{T}} + {}^G\ddot{R}_B\,{}^GR_B^{\mathrm{T}}\,{}^G\dot{R}_B\,{}^GR_B^{\mathrm{T}}\right){}^G\mathbf{r} \\
&= \left[{}^G\dddot{R}_B\,{}^GR_B^{\mathrm{T}} + {}^G\ddot{R}_B\,{}^GR_B^{\mathrm{T}}\left({}^GR_B\,{}^G\dot{R}_B^{\mathrm{T}} + \left[{}^GR_B\,{}^G\dot{R}_B^{\mathrm{T}}\right]^{\mathrm{T}}\right)\right]{}^G\mathbf{r} \\
&= {}^G\dddot{R}_B\,{}^GR_B^{\mathrm{T}}\,{}^G\mathbf{r}
\end{aligned} \tag{9.163}$$

$$ {}_GU_B = {}^G\dddot{R}_B\,{}^GR_B^{\mathrm{T}} \tag{9.164}$$

Example 547 ★ Jerk of a Body Point Using ${}^G\mathbf{a} = {}_G\boldsymbol{\alpha}_B \times {}^G\mathbf{r} + {}_G\boldsymbol{\omega}_B \times \left({}_G\boldsymbol{\omega}_B \times {}^G\mathbf{r}\right)$, we can find the vectorial expression formula for the jerk of a body point:

$$\begin{aligned}
{}^G\mathbf{j} &= \frac{{}^Gd}{dt}({}^G\mathbf{a}) = \frac{{}^Gd}{dt}\left[{}_G\boldsymbol{\alpha}_B \times {}^G\mathbf{r} + {}_G\boldsymbol{\omega}_B \times \left({}_G\boldsymbol{\omega}_B \times {}^G\mathbf{r}\right)\right] \\
&= {}_G\boldsymbol{\chi}_B \times {}^G\mathbf{r} + 2\,{}_G\boldsymbol{\alpha}_B \times \left({}_G\boldsymbol{\omega}_B \times {}^G\mathbf{r}\right) + {}_G\boldsymbol{\omega}_B \times \left({}_G\boldsymbol{\alpha}_B \times {}^G\mathbf{r}\right) \\
&\quad + {}_G\boldsymbol{\omega}_B \times \left[{}_G\boldsymbol{\omega}_B \times \left({}_G\boldsymbol{\omega}_B \times {}^G\mathbf{r}\right)\right]
\end{aligned} \tag{9.165}$$

where ${}_G\chi_B$ is the angular jerk vector of B relative to G.

The matrix form of the jerk is

$$\begin{aligned}
{}^G\mathbf{j} &= \frac{{}^Gd}{dt}\left[\left({}_G\tilde{\alpha}_B + {}_G\tilde{\omega}_B^2\right){}^G\mathbf{r}\right] \\
&= \left[{}_G\dot{\tilde{\alpha}}_B + 2\,{}_G\tilde{\omega}_B\,{}_G\tilde{\alpha}_B + \left({}_G\tilde{\alpha}_B + {}_G\tilde{\omega}_B^2\right){}_G\tilde{\omega}_B\right]{}^G\mathbf{r} \\
&= \left({}_G\tilde{\chi}_B + 2\,{}_G\tilde{\omega}_B\,{}_G\tilde{\alpha}_B + {}_G\tilde{\alpha}_B\,{}_G\tilde{\omega}_B + {}_G\tilde{\omega}_B^3\right){}^G\mathbf{r}
\end{aligned} \tag{9.166}$$

and therefore the rotational jerk transformation is

$$ {}_GU_B = {}_G\tilde{\chi}_B + 2\,{}_G\tilde{\omega}_B\,{}_G\tilde{\alpha}_B + {}_G\tilde{\alpha}_B\,{}_G\tilde{\omega}_B + {}_G\tilde{\omega}_B^3 \tag{9.167}$$

Example 548 ★ Angular Jerk The *angular jerk matrix* $\tilde{\chi} = \dot{\tilde{\alpha}} = \ddot{\tilde{\omega}}$ is the time derivative of the angular acceleration matrix (9.8):

$$\begin{aligned}
{}_G\tilde{\chi}_B = {}_G\dot{\tilde{\alpha}}_B = {}_G\ddot{\tilde{\omega}}_B &= \frac{{}^Gd}{dt}\left({}^G\ddot{R}_B\,{}^GR_B^{\mathrm{T}} + {}^G\dot{R}_B\,{}^G\dot{R}_B^{\mathrm{T}}\right) \\
&= {}^G\dddot{R}_B\,{}^GR_B^{\mathrm{T}} + 2\,{}^G\ddot{R}_B\,{}^G\dot{R}_B^{\mathrm{T}} + {}^G\dot{R}_B\,{}^G\ddot{R}_B^{\mathrm{T}} && (9.168) \\
&= {}_GU_B + 2\,{}_GS_B\,{}_G\tilde{\omega}_B^{\mathrm{T}} + {}_G\tilde{\omega}_B\,{}_GS_B^{\mathrm{T}} && (9.169)
\end{aligned}$$

Using the angle–axis expression of the angular acceleration matrix, we find the angle–axis expression of the angular jerk matrix ${}_G\tilde{\chi}_B$:

$$ {}_G\tilde{\chi}_B = {}_G\ddot{\tilde{\omega}}_B = \frac{{}^Gd}{dt}\left(\ddot{\phi}\tilde{u} + \dot{\phi}\dot{\tilde{u}}\right) = \dddot{\phi}\tilde{u} + 2\ddot{\phi}\dot{\tilde{u}} + \dot{\phi}\ddot{\tilde{u}} \tag{9.170}$$

The expanded form is

$$ {}_G\tilde{\chi}_B = \begin{bmatrix} j_{11} & j_{12} & j_{13} \\ j_{21} & j_{22} & j_{23} \\ j_{31} & j_{32} & j_{33} \end{bmatrix} \tag{9.171}$$

where

$$\begin{aligned}
j_{11} &= 3u_1\dot{u}_1\dot{\phi}^2 + 3\left(u_1^2 - 1\right)\dot{\phi}\ddot{\phi} \\
j_{21} &= (2u_2\dot{u}_1 + \dot{u}_2u_1)\,\dot{\phi}^2 + 3u_2u_1\dot{\phi}\ddot{\phi} + \left(\ddot{u}_3\dot{\phi} + 2\dot{u}_3\ddot{\phi} + u_3\dddot{\phi} - u_3\dot{\phi}^3\right) \\
j_{31} &= (2u_3\dot{u}_1 + \dot{u}_3u_1)\,\dot{\phi}^2 + 3u_3u_1\dot{\phi}\ddot{\phi} + \left(\ddot{u}_2\dot{\phi} + 2\dot{u}_2\ddot{\phi} + u_2\dddot{\phi} - u_2\dot{\phi}^3\right) \\
j_{12} &= (2u_1\dot{u}_2 + \dot{u}_1u_2)\,\dot{\phi}^2 + 3u_1u_2\dot{\phi}\ddot{\phi} + \left(\ddot{u}_3\dot{\phi} + 2\dot{u}_3\ddot{\phi} + u_3\dddot{\phi} - u_3\dot{\phi}^3\right) \\
j_{22} &= 3u_2\dot{u}_2\dot{\phi}^2 + 3\left(u_2^2 - 1\right)\dot{\phi}\ddot{\phi} \\
j_{32} &= (2u_3\dot{u}_2 + \dot{u}_3u_2)\,\dot{\phi}^2 + 3u_3u_2\dot{\phi}\ddot{\phi} + \left(\ddot{u}_1\dot{\phi} + 2\dot{u}_1\ddot{\phi} + u_1\dddot{\phi} - u_1\dot{\phi}^3\right) \\
j_{13} &= (2u_1\dot{u}_3 + \dot{u}_1u_3)\,\dot{\phi}^2 + 3u_1u_3\dot{\phi}\ddot{\phi} + \left(\ddot{u}_2\dot{\phi} + 2\dot{u}_2\ddot{\phi} + u_2\dddot{\phi} - u_2\dot{\phi}^3\right) \\
j_{23} &= (2u_2\dot{u}_3 + \dot{u}_2u_3)\,\dot{\phi}^2 + 3u_2u_3\dot{\phi}\ddot{\phi} + \left(\ddot{u}_1\dot{\phi} + 2\dot{u}_1\ddot{\phi} + u_1\dddot{\phi} - u_1\dot{\phi}^3\right) \\
j_{33} &= 3u_3\dot{u}_3\dot{\phi}^2 + 3\left(u_3^2 - 1\right)\dot{\phi}\ddot{\phi}
\end{aligned} \tag{9.172}$$

Example 549 ★ Angular Acceleration in Natural Frame $\hat{u}_t$, $\hat{u}_n$, $\hat{u}_b$ The angular velocity vector in the natural coordinate frame $N(\hat{u}_t, \hat{u}_n, \hat{u}_b)$ is

$${}_G^N\boldsymbol{\omega}_N = \omega_t\hat{u}_t + \omega_n\hat{u}_n + \omega_b\hat{u}_b = \frac{\dot{s}}{\sigma}\hat{u}_t + \frac{\dot{s}}{\rho}\hat{u}_b \tag{9.173}$$

The time derivative of ${}_G^N\boldsymbol{\omega}_N$ provides the angular acceleration vector

$$\begin{aligned}
{}_G^N\boldsymbol{\alpha}_N &= \frac{d}{dt}\,{}_G^N\boldsymbol{\omega}_N = \alpha_t\hat{u}_t + \alpha_b\hat{u}_b + \omega_t\frac{\dot{s}}{\rho}\hat{u}_n - \omega_b\frac{\dot{s}}{\sigma}\hat{u}_n \\
&= \alpha_t\hat{u}_t + \alpha_n\hat{u}_n + \alpha_b\hat{u}_b
\end{aligned} \tag{9.174}$$

where

$$\alpha_t = \frac{\ddot{s}\sigma - \dot{s}\dot{\sigma}}{\sigma^2} \tag{9.175}$$

$$\alpha_n = \frac{\dot{s}^2}{\rho^2} - \frac{\dot{s}^2}{\sigma^2} \tag{9.176}$$

$$\alpha_b = \frac{\ddot{s}\rho - \dot{s}\dot{\rho}}{\rho^2} \tag{9.177}$$

Example 550 ★ Angular Acceleration' Quaternion' and Euler Parameters To express the acceleration of a body point by quaternions and Euler parameters, we consider the velocity equation

$${}^G\dot{\mathbf{r}} = 2\dot{e}\ e^{*\,G}\mathbf{r} \tag{9.178}$$

and take a time derivative

$$\begin{aligned}
{}^G\ddot{\mathbf{r}} &= 2\ddot{e}\ e^{*\,G}\mathbf{r} + 2\dot{e}\ \dot{e}^{*\,G}\mathbf{r} + 4\dot{e}^2\ e^{*2\,G}\mathbf{r} \\
&= 2\left(\ddot{e}\ e^* + \dot{e}\ \dot{e}^* + 2\dot{e}^2\ e^{*2}\right){}^G\mathbf{r}
\end{aligned} \tag{9.179}$$

Therefore, the quaternion expression of the acceleration transformation matrix is

$$_G S_B = 2\left(\ddot{e}\; e^* + \dot{e}\; \dot{e}^* + 2\dot{e}^2\; e^{*2}\right) \tag{9.180}$$

To determine the angular acceleration quaternion, let us use the angular velocity

$$_G\boldsymbol{\omega}_B = 2\dot{e}\, e^* \tag{9.181}$$

and take a derivative

$$_G\boldsymbol{\alpha}_B = 2\ddot{e}\, e^* + 2\dot{e}\, \dot{e}^* \tag{9.182}$$

Using the definition of angular velocity in terms of the rotational quaternion,

$$\overleftrightarrow{_G\boldsymbol{\omega}_B} = 2\,\overleftrightarrow{\dot{e}}\;\overleftrightarrow{e^*} \tag{9.183}$$

$$\overleftrightarrow{{}_G^B\boldsymbol{\omega}_B} = 2\,\overleftrightarrow{e^*}\;\overleftrightarrow{\dot{e}} \tag{9.184}$$

we are able to define the angular acceleration quaternion:

$$\overleftrightarrow{_G\boldsymbol{\alpha}_B} = 2\,\overleftrightarrow{\ddot{e}}\;\overleftrightarrow{e^*} + 2\,\overleftrightarrow{\dot{e}}\;\overleftrightarrow{\dot{e}^*} \tag{9.185}$$

$$\overleftrightarrow{{}_G^B\boldsymbol{\alpha}_B} = 2\,\overleftrightarrow{e^*}\;\overleftrightarrow{\ddot{e}} + 2\,\overleftrightarrow{\dot{e}}\;\overleftrightarrow{\dot{e}^*} \tag{9.186}$$

9.2 SECOND DERIVATIVE AND COORDINATE FRAMES

The time derivative of a vector depends on the coordinate frame in which it is expressed and the frame in which we are taking the derivative. Consider a global frame $G(OXYZ)$ and a body frame $B(Oxyz)$. A vector is called a B-vector if it is expressed in the B-frame, and similarly it is a G-vector if it is expressed in the G-frame. The second derivative of the vector follows the same rule of the first G- and B-derivative of the G- and B-vector as explained in Section 8.2. The derivative of a B-vector ${}^B\mathbf{v} = \dot{x}\hat{\imath} + \dot{y}\hat{\jmath} + \dot{z}\hat{k}$ in B and the derivative of a G-vector ${}^G\mathbf{v} = \dot{X}\hat{I} + \dot{Y}\hat{J} + \dot{Z}\hat{K}$ in G are given as

$${}^B\mathbf{a} = \frac{{}^Bd^2}{dt^2}\,{}^B\mathbf{r} = \frac{{}^Bd}{dt}\,{}^B\mathbf{v} = {}^B\ddot{\mathbf{r}} = \ddot{x}\,\hat{\imath} + \ddot{y}\,\hat{\jmath} + \ddot{z}\,\hat{k} \tag{9.187}$$

$${}^G\mathbf{a} = \frac{{}^Gd^2}{dt^2}\,{}^G\mathbf{r} = \frac{{}^Gd}{dt}\,{}^G\mathbf{v} = {}^G\ddot{\mathbf{r}} = \ddot{X}\,\hat{I} + \ddot{Y}\,\hat{J} + \ddot{Z}\,\hat{K} \tag{9.188}$$

We call ${}^G\mathbf{a}$ and ${}^B\mathbf{a}$ *simple accelerations* because they are *simple derivatives* of the *simple velocities* ${}^G\mathbf{v}$ and ${}^B\mathbf{v}$. We may also calculate the *mixed derivatives* and find the G-derivative of ${}_G^B\mathbf{v}$ and the B-derivative of ${}_B^G\mathbf{v}$.

Consider a body point P at ${}^G\mathbf{r} = {}^GR_B\,{}^B\mathbf{r}$. The G-derivative of ${}_G^B\mathbf{v}$ is

$${}_G^B\mathbf{a} = \frac{{}^Gd}{dt}\,{}_G^B\mathbf{v} = {}_G^B\boldsymbol{\alpha}_B \times {}^B\mathbf{r} + {}_G^B\boldsymbol{\omega}_B \times \left({}_G^B\boldsymbol{\omega}_B \times {}^B\mathbf{r}\right) \tag{9.189}$$

and the B-derivative of ${}_B^G\mathbf{v}$ is

$${}_B^G\mathbf{a} = \frac{{}^Bd}{dt}\,{}_B^G\mathbf{v} = -\,{}_G\boldsymbol{\alpha}_B \times {}^G\mathbf{r} + {}_G\boldsymbol{\omega}_B \times \left({}_G\boldsymbol{\omega}_B \times {}^G\mathbf{r}\right) \tag{9.190}$$

We call ${}_G^B\mathbf{a}$ the B-expression of the G-acceleration and ${}_B^G\mathbf{a}$ the G-expression of the B-acceleration. The left superscript of ${}_G^B\mathbf{a}$ indicates the frame in which $\mathbf{a}$ is expressed, and the left subscript indicates the frame in which that derivative is taken. If the left super- and subscripts of an acceleration vector are the same, it is a simple acceleration vector and we only keep the superscript. To read the mixed accelerations ${}_G^B\mathbf{a}$ and ${}_B^G\mathbf{a}$, we may use the *G-acceleration of a B-velocity* and *B-acceleration of a G-velocity*, respectively.

When the interested point P is not a fixed point in B, then P is moving in frame B with a variable body position ${}^B\mathbf{r}_P = {}^B\mathbf{r}_P(t)$ and velocity ${}^B\mathbf{v}_P = {}^B\mathbf{v}_P(t)$. The mixed derivatives of $\mathbf{v}(t)$ are defined by

$$\begin{aligned} {}_G^B\mathbf{a} &= \frac{{}^Gd}{dt}\,{}_G^B\mathbf{v}(t) \\ &= {}^B\mathbf{a} + {}_G^B\boldsymbol{\alpha}_B \times {}^B\mathbf{r} + 2\,{}_G^B\boldsymbol{\omega}_B \times {}^B\mathbf{v} + {}_G^B\boldsymbol{\omega}_B \times \left({}_G^B\boldsymbol{\omega}_B \times {}^B\mathbf{r}\right) \end{aligned} \tag{9.191}$$

$$\begin{aligned} {}_B^G\mathbf{a} &= \frac{{}^Bd}{dt}\,{}_B^G\mathbf{v}(t) \\ &= {}^G\mathbf{a} - {}_G\boldsymbol{\alpha}_B \times {}^G\mathbf{r} - 2\,{}_G\boldsymbol{\omega}_B \times {}^G\mathbf{v} + {}_G\boldsymbol{\omega}_B \times \left({}_G\boldsymbol{\omega}_B \times {}^G\mathbf{r}\right) \end{aligned} \tag{9.192}$$

Proof: To have more general equations, let us assume that the interested point P is a moving point in the B-frame. If B represents a rigid body, then P is a body point and hence is fixed in B. For a body point, the body position vector ${}^B\mathbf{r}$ is constant in B and its body derivatives would be zero.

Having only one rotating B-frame in a global G-frame, we can define eight different accelerations as the second time derivatives of a position vector $\mathbf{r}$:

1. $$\frac{{}^Gd}{dt}\frac{{}^Gd}{dt}\,{}^G\mathbf{r} = \frac{{}^Gd}{dt}\,{}^G\mathbf{v} = {}^G\mathbf{a} \tag{9.193}$$

2. $$\frac{{}^Gd}{dt}\frac{{}^Gd}{dt}\,{}^B\mathbf{r} = \frac{{}^Gd}{dt}\,{}_G^B\mathbf{v} = {}_{GG}^{BB}\mathbf{a} = {}_G^B\mathbf{a} \tag{9.194}$$

3. $$\frac{{}^Gd}{dt}\frac{{}^Bd}{dt}\,{}^G\mathbf{r} = \frac{{}^Gd}{dt}\,{}_B^G\mathbf{v} = {}_{GB}^{GG}\mathbf{a} \tag{9.195}$$

4. $$\frac{{}^Gd}{dt}\frac{{}^Bd}{dt}\,{}^B\mathbf{r} = \frac{{}^Gd}{dt}\,{}^B\mathbf{v} = {}_{GB}^{BB}\mathbf{a} \tag{9.196}$$

5. $$\frac{{}^Bd}{dt}\frac{{}^Gd}{dt}\,{}^G\mathbf{r} = \frac{{}^Bd}{dt}\,{}^G\mathbf{v} = {}_{BG}^{GG}\mathbf{a} \tag{9.197}$$

6. $$\frac{{}^Bd}{dt}\frac{{}^Gd}{dt}\,{}^B\mathbf{r} = \frac{{}^Bd}{dt}\,{}_G^B\mathbf{v} = {}_{BG}^{BB}\mathbf{a} \tag{9.198}$$

7. $$\frac{{}^Bd}{dt}\frac{{}^Bd}{dt}\,{}^G\mathbf{r} = \frac{{}^Bd}{dt}\,{}_B^G\mathbf{v} = {}_{BB}^{GG}\mathbf{a} = {}_B^G\mathbf{a} \tag{9.199}$$

8. $$\frac{{}^Bd}{dt}\frac{{}^Bd}{dt}\,{}^B\mathbf{r} = \frac{{}^Bd}{dt}\,{}^B\mathbf{v} = {}^B\mathbf{a} \tag{9.200}$$

Only the first and eighth second derivatives are simple accelerations. These two accelerations, given in (9.187) and (9.188), can be found by simple differentiation of a vector in the same frame in which they are expressed. Therefore, the G-derivative of the G-velocity and the B-derivative of the B-velocity provide the G- and B-acceleration, respectively:

$$\begin{aligned} \frac{{}^G d^2}{dt^2}\,{}^G\mathbf{r} &= \frac{{}^G d}{dt}\,{}^G\mathbf{v} = \frac{{}^G d}{dt}\frac{{}^G d}{dt}\,{}^G\mathbf{r} = {}^G\mathbf{a} = {}^G\ddot{\mathbf{r}} \\ &= \ddot{X}\,\hat{I} + \ddot{Y}\,\hat{J} + \ddot{Z}\,\hat{K} \end{aligned} \tag{9.201}$$

$$\begin{aligned} \frac{{}^B d^2}{dt^2}\,{}^B\mathbf{r} &= \frac{{}^B d}{dt}\,{}^B\mathbf{v} = \frac{{}^B d}{dt}\frac{{}^B d}{dt}\,{}^B\mathbf{r} = {}^B\mathbf{a} = {}^B\ddot{\mathbf{r}} \\ &= \ddot{x}\,\hat{\imath} + \ddot{y}\,\hat{\jmath} + \ddot{z}\,\hat{k} \end{aligned} \tag{9.202}$$

Recalling the derivative transfer formula (8.250) and the mixed velocities,

$$\begin{aligned} \frac{{}^G d}{dt}\,{}^B\mathbf{r} &= {}^B_G\mathbf{v} = {}^B_G\dot{\mathbf{r}} = \frac{{}^B d}{dt}\,{}^B\mathbf{r} + {}^B_G\boldsymbol{\omega}_B \times {}^B\mathbf{r} \\ &= {}^B\mathbf{v} + {}^B_G\boldsymbol{\omega}_B \times {}^B\mathbf{r} \end{aligned} \tag{9.203}$$

$$\begin{aligned} \frac{{}^B d}{dt}\,{}^G\mathbf{r} &= {}^G_B\mathbf{v} = {}^G_B\dot{\mathbf{r}} = \frac{{}^G d}{dt}\,{}^G\mathbf{r}(t) - {}_G\boldsymbol{\omega}_B \times {}^G\mathbf{r} \\ &= {}^G\mathbf{v} - {}_G\boldsymbol{\omega}_B \times {}^G\mathbf{r} \end{aligned} \tag{9.204}$$

we can find the mixed accelerations in (9.194)–(9.199) one by one.

The second case, ${}^B_G\mathbf{a} = {}^{BB}_{GG}\mathbf{a}$, is the B-expression of a G-acceleration. It happens when the position vector of a point is given in B while derivatives are taken in G:

$$\begin{aligned} {}^B_G\mathbf{a} = {}^{BB}_{GG}\mathbf{a} &= \frac{{}^G d}{dt}\,{}^B_G\mathbf{v} = \frac{{}^G d}{dt}\left({}^B\mathbf{v} + {}^B_G\boldsymbol{\omega}_B \times {}^B\mathbf{r}\right) \\ &= {}^B\mathbf{a} + {}^B_G\boldsymbol{\omega}_B \times {}^B\mathbf{v} + \left({}^B_G\boldsymbol{\alpha}_B + {}^B_G\boldsymbol{\omega}_B \times {}^B_G\boldsymbol{\omega}_B\right) \times {}^B\mathbf{r} \\ &\quad + {}^B_G\boldsymbol{\omega}_B \times \left({}^B\mathbf{v} + {}^B_G\boldsymbol{\omega}_B \times {}^B\mathbf{r}\right) \\ &= {}^B\mathbf{a} + {}^B_G\boldsymbol{\alpha}_B \times {}^B\mathbf{r} + 2\,{}^B_G\boldsymbol{\omega}_B \times {}^B\mathbf{v} + {}^B_G\boldsymbol{\omega}_B \times \left({}^B_G\boldsymbol{\omega}_B \times {}^B\mathbf{r}\right) \end{aligned} \tag{9.205}$$

The first term, ${}^B\mathbf{a}$, is the body acceleration of the moving point P in B, regardless of the rotation of B in G. So, ${}^B\mathbf{a}$ can be assumed as the acceleration of the moving point in B with respect to a body point that is coincident with P at the moment. The second term, ${}^B_G\boldsymbol{\alpha}_B \times {}^B\mathbf{r}$, is the *tangential acceleration* of a body point that is coincident with P at the moment. The third term, $2\,{}^B_G\boldsymbol{\omega}_B \times {}^B\mathbf{v}$, is called the *Coriolis acceleration* and is the result of a moving point in B while B is rotating in G. The fourth term, ${}^B_G\boldsymbol{\omega}_B \times \left({}^B_G\boldsymbol{\omega}_B \times {}^B\mathbf{r}\right)$, is the *centripetal acceleration* of a body point that is coincident with P at the moment.

The B-expression of the G-acceleration ${}^B_G\mathbf{a}$ is the most applied and practical acceleration in the dynamics of rigid bodies. In a rigid body, the point P is a fixed point in B and the acceleration ${}^B_G\mathbf{a}$ simplifies to

$${}^B_G\mathbf{a} = \frac{{}^G d}{dt}\,{}^B_G\mathbf{v} = {}^B_G\boldsymbol{\alpha}_B \times {}^B\mathbf{r} + {}^B_G\boldsymbol{\omega}_B \times \left({}^B_G\boldsymbol{\omega}_B \times {}^B\mathbf{r}\right) \tag{9.206}$$

A body point will have only a tangential and a centripetal acceleration. Using ${}^B_G\mathbf{a}$ is a practical method of acceleration analysis because we usually prefer to measure the kinematic information of a moving rigid body in the body coordinate frame.

The third case, ${}^{GG}_{GB}\mathbf{a}$, is the G-expression of the G-derivative of ${}^G_B\mathbf{v}$ that is the B-derivative of the G-expression of a position vector:

$$\begin{aligned}{}^{GG}_{GB}\mathbf{a} &= \frac{{}^Gd}{dt}\,{}^G_B\mathbf{v} = \frac{{}^Gd}{dt}\frac{{}^Bd}{dt}\,{}^G\mathbf{r} = \frac{{}^Gd}{dt}\left({}^G\mathbf{v} - {}_G\boldsymbol{\omega}_B \times {}^G\mathbf{r}\right) \\ &= {}^G\mathbf{a} - {}_G\boldsymbol{\alpha}_B \times {}^G\mathbf{r} - {}_G\boldsymbol{\omega}_B \times {}^G\mathbf{v}\end{aligned} \tag{9.207}$$

For a fixed point in B, we have ${}^G_B\mathbf{v} = 0$ and ${}^{GG}_{GB}\mathbf{a} = 0$. Equation (9.207) reduces to the G-expression of (9.206) in this case.

The fourth case, ${}^{BB}_{GB}\mathbf{a}$, is the B-expression of the G-derivative of ${}^B_B\mathbf{v}$ that is the B-derivative of the B-expression of a position vector:

$$\begin{aligned}{}^{BB}_{GB}\mathbf{a} &= \frac{{}^Gd}{dt}\,{}^B\mathbf{v} = \frac{{}^Gd}{dt}\frac{{}^Bd}{dt}\,{}^B\mathbf{r} = \frac{{}^Gd}{dt}\left({}^B\mathbf{v}\right) \\ &= {}^B\mathbf{a} + {}^B_G\boldsymbol{\omega}_B \times {}^B\mathbf{v}\end{aligned} \tag{9.208}$$

The fifth case, ${}^{GG}_{BG}\mathbf{a}$, is the G-expression of the B-derivative of ${}^G_G\mathbf{v}$ that is the G-derivative of the G-expression of a position vector:

$$\begin{aligned}{}^{GG}_{BG}\mathbf{a} &= \frac{{}^Bd}{dt}\,{}^G\mathbf{v} = \frac{{}^Bd}{dt}\frac{{}^Gd}{dt}\,{}^G\mathbf{r} = \frac{{}^Bd}{dt}\left({}^G\mathbf{v}\right) \\ &= {}^G\mathbf{a} - {}_G\boldsymbol{\omega}_B \times {}^G\mathbf{v}\end{aligned} \tag{9.209}$$

The sixth case, ${}^{BB}_{BG}\mathbf{a}$, is the B-expression of the B-derivative of ${}^G_B\mathbf{v}$ that is the B-derivative of the G-expression of a position vector:

$$\begin{aligned}{}^{BB}_{BG}\mathbf{a} &= \frac{{}^Bd}{dt}\,{}^B_G\mathbf{v} = \frac{{}^Bd}{dt}\frac{{}^Gd}{dt}\,{}^B\mathbf{r} = \frac{{}^Bd}{dt}\left({}^B\mathbf{v} + {}^B_G\boldsymbol{\omega}_B \times {}^B\mathbf{r}\right) \\ &= {}^B\mathbf{a} + {}^B_G\boldsymbol{\alpha}_B \times {}^B\mathbf{r} + {}^B_G\boldsymbol{\omega}_B \times {}^B\mathbf{v}\end{aligned} \tag{9.210}$$

The seventh case, ${}^{GG}_{BB}\mathbf{a}$, is the G-expression of the B-derivative of ${}^G_B\mathbf{v}$ that is the B-derivative of the G-expression of a position vector:

$$\begin{aligned}{}^G_B\mathbf{a} = {}^{GG}_{BB}\mathbf{a} &= \frac{{}^Bd}{dt}\,{}^G_B\mathbf{v} = \frac{{}^Bd}{dt}\frac{{}^Bd}{dt}\,{}^G\mathbf{r} = \frac{{}^Bd}{dt}\left({}^G\mathbf{v} - {}_G\boldsymbol{\omega}_B \times {}^G\mathbf{r}\right) \\ &= {}^G\mathbf{a} - {}_G\boldsymbol{\omega}_B \times {}^G\mathbf{v} - \left({}_G\boldsymbol{\alpha}_B - {}_G\boldsymbol{\omega}_B \times {}_G\boldsymbol{\omega}_B\right) \times {}^G\mathbf{r} \\ &\quad - {}_G\boldsymbol{\omega}_B \times \left({}^G\mathbf{v} - {}_G\boldsymbol{\omega}_B \times {}^G\mathbf{r}\right) \\ &= {}^G\mathbf{a} - {}_G\boldsymbol{\alpha}_B \times {}^G\mathbf{r} - 2\,{}_G\boldsymbol{\omega}_B \times {}^G\mathbf{v} + {}_G\boldsymbol{\omega}_B \times \left({}_G\boldsymbol{\omega}_B \times {}^G\mathbf{r}\right)\end{aligned} \tag{9.211}$$

The seventh acceleration, ${}^G_B\mathbf{a}$, is the same as the second acceleration, ${}^B_G\mathbf{a}$, if we switch the name of the coordinate frames B and G. So, ${}^B_G\mathbf{a}$ is the G-acceleration of a moving point in B while the observer is in the B-frame, and ${}^G_B\mathbf{a}$ is the G-acceleration of a moving point in B when the observer is in G.

These accelerations for a point of a rigid body are

$$ {}^{G}\mathbf{a} = \frac{{}^{G}d}{dt}\frac{{}^{G}d}{dt}\,{}^{G}\mathbf{r} = \ddot{X}\,\hat{I} + \ddot{Y}\,\hat{J} + \ddot{Z}\,\hat{K} \tag{9.212} $$

$$ {}^{B}_{G}\mathbf{a} = \frac{{}^{G}d}{dt}\frac{{}^{G}d}{dt}\,{}^{B}\mathbf{r} = {}^{B}_{G}\boldsymbol{\alpha}_{B} \times {}^{B}\mathbf{r} + {}^{B}_{G}\boldsymbol{\omega}_{B} \times \left({}^{B}_{G}\boldsymbol{\omega}_{B} \times {}^{B}\mathbf{r}\right) \tag{9.213} $$

$$ {}^{GG}_{GB}\mathbf{a} = \frac{{}^{G}d}{dt}\frac{{}^{B}d}{dt}\,{}^{G}\mathbf{r} = {}^{G}\mathbf{a} - {}_{G}\boldsymbol{\alpha}_{B} \times {}^{G}\mathbf{r} - {}_{G}\boldsymbol{\omega}_{B} \times {}^{G}\mathbf{v} \tag{9.214} $$

$$ {}^{BB}_{GB}\mathbf{a} = \frac{{}^{G}d}{dt}\frac{{}^{B}d}{dt}\,{}^{B}\mathbf{r} = 0 \tag{9.215} $$

$$ {}^{GG}_{BG}\mathbf{a} = \frac{{}^{B}d}{dt}\frac{{}^{G}d}{dt}\,{}^{G}\mathbf{r} = {}^{G}\mathbf{a} - {}_{G}\boldsymbol{\omega}_{B} \times {}^{G}\mathbf{v} \tag{9.216} $$

$$ {}^{BB}_{BG}\mathbf{a} = \frac{{}^{B}d}{dt}\frac{{}^{G}d}{dt}\,{}^{B}\mathbf{r} = {}^{B}_{G}\boldsymbol{\alpha}_{B} \times {}^{B}\mathbf{r} \tag{9.217} $$

$$ \begin{aligned} {}^{G}_{B}\mathbf{a} &= \frac{{}^{B}d}{dt}\frac{{}^{B}d}{dt}\,{}^{G}\mathbf{r} \\ &= {}^{G}\mathbf{a} - {}_{G}\boldsymbol{\alpha}_{B} \times {}^{G}\mathbf{r} - 2\,{}_{G}\boldsymbol{\omega}_{B} \times {}^{G}\mathbf{v} + {}_{G}\boldsymbol{\omega}_{B} \times \left({}_{G}\boldsymbol{\omega}_{B} \times {}^{G}\mathbf{r}\right) \end{aligned} \tag{9.218} $$

$$ {}^{B}\mathbf{a} = \frac{{}^{B}d}{dt}\frac{{}^{B}d}{dt}\,{}^{B}\mathbf{r} = 0 \tag{9.219} $$

Example 551 Mixed Velocity and Simple Acceleration of a Moving Point in B Consider a local frame $B(Oxyz)$ that is rotating in $G(OXYZ)$ with an angular velocity $\dot{\alpha}$ about the Z-axis and a moving point P in B at

$$ {}^{B}\mathbf{r}_{P}(t) = t^{2}\hat{\imath} \tag{9.220} $$

A top view of the system when B is at an angle α is shown in Figure 9.5. The global position vector of the point is

$$ \begin{aligned} {}^{G}\mathbf{r}_{P} &= {}^{G}R_{B}\,{}^{B}\mathbf{r}_{P} = R_{Z,\alpha}(t)\,{}^{B}\mathbf{r}_{P} = \begin{bmatrix} \cos\alpha & -\sin\alpha & 0 \\ \sin\alpha & \cos\alpha & 0 \\ 0 & 0 & 1 \end{bmatrix} \begin{bmatrix} t^{2} \\ 0 \\ 0 \end{bmatrix} \\ &= t^{2}\cos\alpha\,\hat{I} + t^{2}\sin\alpha\,\hat{J} \end{aligned} \tag{9.221} $$

The angular velocity of B is

$$ {}_{G}\tilde{\omega}_{B} = {}^{G}\dot{R}_{B}\,{}^{G}R_{B}^{\mathrm{T}} = \dot{\alpha}\tilde{K} \qquad {}_{G}\boldsymbol{\omega}_{B} = \dot{\alpha}\hat{K} \tag{9.222} $$

It can also be verified that the local expression of the angular velocity is

$$ {}^{B}_{G}\tilde{\omega}_{B} = {}^{G}R_{B}^{\mathrm{T}}\,{}^{G}_{G}\tilde{\omega}_{B}\,{}^{G}R_{B} = \dot{\alpha}\tilde{k} \qquad {}^{B}_{G}\boldsymbol{\omega}_{B} = \dot{\alpha}\hat{k} \tag{9.223} $$

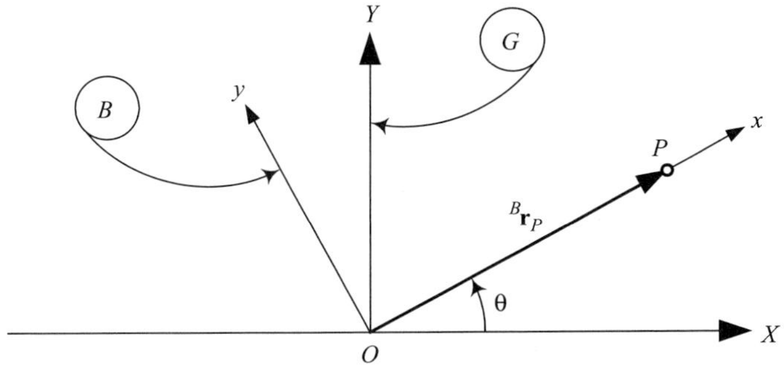

Figure 9.5 Top view of a local frame B that is rotating in G with an angular velocity $\dot{\alpha}$ about the Z-axis and a moving point P in B.

Now we can find the following simple velocities:

$$ {}^B\mathbf{v}_P = \frac{{}^Bd}{dt}\, {}^B\mathbf{r}_P = {}^B\dot{\mathbf{r}}_P = 2t\hat{\imath} \tag{9.224} $$

$$ \begin{aligned} {}^G\mathbf{v}_P &= \frac{{}^Gd}{dt}\, {}^G\mathbf{r}_P = {}^G\dot{\mathbf{r}}_P \\ &= \left(2t\cos\alpha - t^2\dot{\alpha}\sin\alpha\right)\hat{I} + \left(2t\sin\alpha + t^2\dot{\alpha}\cos\alpha\right)\hat{J} \end{aligned} \tag{9.225} $$

For the mixed velocities we start with

$$ \begin{aligned} {}^B_G\mathbf{v}_P &= \frac{{}^Gd}{dt}\, {}^B\mathbf{r}_P = \frac{{}^Bd}{dt}\, {}^B\mathbf{r}_P + {}^B_G\boldsymbol{\omega}_B \times {}^B\mathbf{r}_P \\ &= \begin{bmatrix} 2t \\ 0 \\ 0 \end{bmatrix} + \dot{\alpha}\begin{bmatrix} 0 \\ 0 \\ 1 \end{bmatrix} \times \begin{bmatrix} t^2 \\ 0 \\ 0 \end{bmatrix} = \begin{bmatrix} 2t \\ t^2\dot{\alpha} \\ 0 \end{bmatrix} \\ &= 2t\hat{\imath} + t^2\dot{\alpha}\hat{\jmath} \end{aligned} \tag{9.226} $$

which is the B-expression of the G-velocity of P. It can also be found by a transformation:

$$ {}^B_G\mathbf{v}_P = {}^GR_B^{\mathrm{T}}\, {}^G\mathbf{v}_P = 2t\hat{\imath} + t^2\dot{\alpha}\hat{\jmath} \tag{9.227} $$

The next mixed velocity is

$$ \begin{aligned} {}^G_B\mathbf{v}_P &= \frac{{}^Bd}{dt}\, {}^G\mathbf{r}_P = \frac{{}^Gd}{dt}\, {}^G\mathbf{r}_P - {}_G\boldsymbol{\omega}_B \times {}^G\mathbf{r}_P \\ &= \begin{bmatrix} 2t\cos\alpha - t^2\dot{\alpha}\sin\alpha \\ 2t\sin\alpha + t^2\dot{\alpha}\cos\alpha \\ 0 \end{bmatrix} - \dot{\alpha}\begin{bmatrix} 0 \\ 0 \\ 1 \end{bmatrix} \times \begin{bmatrix} t^2\cos\alpha \\ t^2\sin\alpha \\ 0 \end{bmatrix} \\ &= \begin{bmatrix} 2t\cos\alpha \\ 2t\sin\alpha \\ 0 \end{bmatrix} = (2t\cos\alpha)\hat{I} + (2t\sin\alpha)\hat{J} \end{aligned} \tag{9.228} $$

which is the G-expression of the B-velocity of P. To find this velocity, we can also apply a kinematic transformation ${}^G R_B$ to ${}^B\mathbf{v}_P$:

$$ {}^G_B\mathbf{v}_P = {}^G R_B \, {}^B\mathbf{v}_P = (2t\cos\alpha)\,\hat{I} + (2t\sin\alpha)\,\hat{J} \tag{9.229} $$

The simple accelerations of P are

$$ {}^B\mathbf{a}_P = \frac{{}^Bd}{dt}\,{}^B\mathbf{v}_P = {}^B\ddot{\mathbf{r}}_P = 2\hat{\imath} \tag{9.230} $$

$$ \begin{aligned} {}^G\mathbf{a}_P &= \frac{{}^Gd}{dt}\,{}^G\mathbf{v}_P = {}^G\ddot{\mathbf{r}}_P \\ &= \left(2\cos\alpha - 4t\dot{\alpha}\sin\alpha - t^2\ddot{\alpha}\sin\alpha - t^2\dot{\alpha}^2\cos\alpha\right)\hat{I} \\ &\quad + \left(2\sin\alpha + 4t\dot{\alpha}\cos\alpha + t^2\ddot{\alpha}\cos\alpha - t^2\dot{\alpha}^2\sin\alpha\right)\hat{J} \end{aligned} \tag{9.231} $$

Example 552 Mixed Acceleration of a Moving Point in B The velocities and simple accelerations of the moving point P in B from Example 551 help us to determine the mixed accelerations of P.

The acceleration ${}^B_G\mathbf{a} = {}^{BB}_{GG}\mathbf{a}$ is given as

$$ \begin{aligned} {}^B_G\mathbf{a} &= {}^B\mathbf{a} + {}^B_G\boldsymbol{\alpha}_B \times {}^B\mathbf{r} + 2\,{}^B_G\boldsymbol{\omega}_B \times {}^B\mathbf{v} + {}^B_G\boldsymbol{\omega}_B \times \left({}^B_G\boldsymbol{\omega}_B \times {}^B\mathbf{r}\right) \\ &= \begin{bmatrix} 2 - t^2\dot{\alpha}^2 \\ t\,(4\dot{\alpha} + t\ddot{\alpha}) \\ 0 \end{bmatrix} \end{aligned} \tag{9.232} $$

We must also be able to determine the mixed accelerations ${}^B_G\mathbf{a}$ by a kinematic transformation:

$$ \begin{aligned} {}^B_G\mathbf{a} &= {}^G R_B^{\mathrm{T}}\, {}^G\mathbf{a} \\ &= \begin{bmatrix} \cos\alpha & -\sin\alpha & 0 \\ \sin\alpha & \cos\alpha & 0 \\ 0 & 0 & 1 \end{bmatrix}^{\mathrm{T}} \begin{bmatrix} 2t\cos\alpha - t^2\dot{\alpha}\sin\alpha \\ 2t\sin\alpha + t^2\dot{\alpha}\cos\alpha \\ 0 \end{bmatrix} \\ &= \begin{bmatrix} 2 - t^2\dot{\alpha}^2 \\ t\,(4\dot{\alpha} + t\ddot{\alpha}) \\ 0 \end{bmatrix} \end{aligned} \tag{9.233} $$

The acceleration ${}^{GG}_{GB}\mathbf{a}$ is given as

$$ \begin{aligned} {}^{GG}_{GB}\mathbf{a} &= {}^G\mathbf{a} - {}_G\boldsymbol{\alpha}_B \times {}^G\mathbf{r} - {}_G\boldsymbol{\omega}_B \times {}^G\mathbf{v} \\ &= \begin{bmatrix} 2\cos\alpha - 2t\dot{\alpha}\sin\alpha \\ 2\sin\alpha + 2t\dot{\alpha}\cos\alpha \\ 0 \end{bmatrix} \end{aligned} \tag{9.234} $$

the acceleration ${}^{BB}_{GB}\mathbf{a}$ as

$$ {}^{BB}_{GB}\mathbf{a} = {}^B\mathbf{a} + {}^B_G\boldsymbol{\omega}_B \times {}^B\mathbf{v} = \begin{bmatrix} 2 \\ 2t\dot{\alpha} \\ 0 \end{bmatrix} \tag{9.235} $$

the acceleration ${}_{BG}^{GG}\mathbf{a}$ as

$$ {}_{BG}^{GG}\mathbf{a} = {}^{G}\mathbf{a} - {}_{G}\boldsymbol{\omega}_{B} \times {}^{G}\mathbf{v} = \begin{bmatrix} 2\cos\alpha - (\ddot{\alpha}t + 2\dot{\alpha})\, t\sin\alpha \\ 2\sin\alpha + (\ddot{\alpha}t + 2\dot{\alpha})\, t\cos\alpha \\ 0 \end{bmatrix} \tag{9.236} $$

the acceleration ${}_{BG}^{BB}\mathbf{a}$ as

$$ {}_{BG}^{BB}\mathbf{a} = {}^{B}\mathbf{a} + {}_{G}^{B}\boldsymbol{\alpha}_{B} \times {}^{B}\mathbf{r} + {}_{G}^{B}\boldsymbol{\omega}_{B} \times {}^{B}\mathbf{v} = \begin{bmatrix} 2 \\ (\ddot{\alpha}t + 2\dot{\alpha})\, t \\ 0 \end{bmatrix} \tag{9.237} $$

and the acceleration ${}_{BG}^{BB}\mathbf{a}$ as

$$ \begin{aligned} {}_{B}^{G}\mathbf{a} &= {}_{BB}^{GG}\mathbf{a} \\ &= {}^{G}\mathbf{a} - {}_{G}\boldsymbol{\alpha}_{B} \times {}^{G}\mathbf{r} - 2\,{}_{G}\boldsymbol{\omega}_{B} \times {}^{G}\mathbf{v} + {}_{G}\boldsymbol{\omega}_{B} \times \left({}_{G}\boldsymbol{\omega}_{B} \times {}^{G}\mathbf{r}\right) \\ &= \begin{bmatrix} 2\cos\alpha \\ 2\sin\alpha \\ 0 \end{bmatrix} \end{aligned} \tag{9.238} $$

We can also determine the mixed accelerations ${}_{B}^{G}\mathbf{a}$ by a kinematic transformation:

$$ \begin{aligned} {}_{B}^{G}\mathbf{a} &= {}^{G}R_{B}\, {}^{B}\mathbf{a} \\ &= \begin{bmatrix} \cos\alpha & -\sin\alpha & 0 \\ \sin\alpha & \cos\alpha & 0 \\ 0 & 0 & 1 \end{bmatrix} \begin{bmatrix} 2 \\ 0 \\ 0 \end{bmatrix} = \begin{bmatrix} 2\cos\alpha \\ 2\sin\alpha \\ 0 \end{bmatrix} \end{aligned} \tag{9.239} $$

Example 553 Simplest Way to Find Mixed Derivatives It is simpler and more applied if we transform the position vector to the same frame in which we are taking the derivative and then apply the differential operator. As an example, let us assume that a point is moving in a body coordinate frame with the position vector

$$ {}^{B}\mathbf{r} = \begin{bmatrix} \cos\alpha t \\ \sin\alpha t \\ 0 \end{bmatrix} \qquad \alpha = \text{const} \tag{9.240} $$

while the body coordinate frame is turning about the X-axis with angular velocity and acceleration $\dot{\gamma}$ and $\ddot{\gamma}$:

$$ {}_{G}\boldsymbol{\omega}_{B} = \dot{\gamma}\hat{I} \qquad {}_{G}\boldsymbol{\alpha}_{B} = \ddot{\gamma}\hat{I} \tag{9.241} $$

To determine ${}_{G}^{B}\mathbf{v}$ and ${}_{G}^{B}\mathbf{a}$, we transform ${}^{B}\mathbf{r}$ to the G-frame:

$$ \begin{aligned} {}^{G}\mathbf{r} &= {}^{G}R_{B}\, {}^{B}\mathbf{r} \\ &= \begin{bmatrix} 1 & 0 & 0 \\ 0 & \cos\gamma & -\sin\gamma \\ 0 & \sin\gamma & \cos\gamma \end{bmatrix} \begin{bmatrix} \cos\alpha t \\ \sin\alpha t \\ 0 \end{bmatrix} = \begin{bmatrix} \cos\alpha t \\ \sin\alpha t\cos\gamma \\ \sin\alpha t\sin\gamma \end{bmatrix} \end{aligned} \tag{9.242} $$

Taking derivatives provides the global velocity and acceleration:

$$
{}^{G}\mathbf{v} = \frac{{}^{G}d}{dt}\,{}^{G}\mathbf{r} = \begin{bmatrix} -\alpha \sin \alpha t \\ \alpha \cos \alpha t \cos \gamma - \dot{\gamma} \sin \alpha t \sin \gamma \\ \alpha \cos \alpha t \sin \gamma + \dot{\gamma} \sin \alpha t \cos \gamma \end{bmatrix} \tag{9.243}
$$

$$
\begin{aligned}
{}^{G}\mathbf{a} &= \frac{{}^{G}d}{dt}\,{}^{G}\mathbf{v} \\
&= \begin{bmatrix} -\alpha^2 \cos \alpha t \\ -\left(\alpha^2 + \dot{\gamma}^2\right) \sin \alpha t \cos \gamma - (\ddot{\gamma} \sin \alpha t + 2\alpha\dot{\gamma} \cos \alpha t) \sin \gamma \\ -\left(\alpha^2 + \dot{\gamma}^2\right) \sin \alpha t \sin \gamma + (\ddot{\gamma} \sin \alpha t + 2\alpha\dot{\gamma} \cos \alpha t) \cos \gamma \end{bmatrix}
\end{aligned} \tag{9.244}
$$

Using the kinematic transformation matrix ${}^{G}R_{B}$, we are able to calculate ${}^{B}_{G}\mathbf{v}$ and ${}^{B}_{G}\mathbf{a}$:

$$
{}^{B}_{G}\mathbf{v} = {}^{G}R_{B}^{\mathrm{T}}\,{}^{G}\mathbf{v} = \begin{bmatrix} -\alpha \sin \alpha t \\ \alpha \cos \alpha t \\ \dot{\gamma} \sin \alpha t \end{bmatrix} \tag{9.245}
$$

$$
{}^{B}_{G}\mathbf{a} = {}^{G}R_{B}^{\mathrm{T}}\,{}^{G}\mathbf{a} = \begin{bmatrix} -\alpha^2 \cos \alpha t \\ -\left(\alpha^2 + \dot{\gamma}^2\right) \sin \alpha t \\ \ddot{\gamma} \sin \alpha t + 2\alpha\dot{\gamma} \cos \alpha t \end{bmatrix} \tag{9.246}
$$

Example 554 ★ Transformation of Acceleration Vectors The superscripts of all defined mixed accelerations are always the same. When a transformation matrix ${}^{G}R_{B}$ is given, we may transform the expression of an acceleration from a coordinate frame to the other. Applying a kinematic transformation will change both superscripts simultaneously. Therefore,

$$
{}^{GG}_{BG}\mathbf{a} = {}^{G}R_{B}\,{}^{BB}_{BG}\mathbf{a} \qquad {}^{BB}_{BG}\mathbf{a} = {}^{G}R_{B}^{\mathrm{T}}\,{}^{GG}_{BG}\mathbf{a} \tag{9.247}
$$

$$
{}^{G}_{B}\mathbf{a} = {}^{G}R_{B}\,{}^{B}\mathbf{a} \qquad {}^{B}\mathbf{a} = {}^{G}R_{B}^{\mathrm{T}}\,{}^{G}_{B}\mathbf{a} \tag{9.248}
$$

$$
{}^{GG}_{GB}\mathbf{a} = {}^{G}R_{B}\,{}^{BB}_{GB}\mathbf{a} \qquad {}^{BB}_{GB}\mathbf{a} = {}^{G}R_{B}^{\mathrm{T}}\,{}^{GG}_{GB}\mathbf{a} \tag{9.249}
$$

$$
{}^{G}\mathbf{a} = {}^{G}R_{B}\,{}^{B}_{G}\mathbf{a} \qquad {}^{B}_{G}\mathbf{a} = {}^{G}R_{B}^{\mathrm{T}}\,{}^{G}\mathbf{a} \tag{9.250}
$$

Example 555 ★ Relationship among the Accelerations Comparing the definitions of the eight accelerations (9.193)–(9.200) indicates their relationships. The following equations show some of the relations:

$$
{}^{BB}_{BG}\mathbf{a} = {}^{BB}_{GB}\mathbf{a} + {}^{B}_{G}\boldsymbol{\alpha}_{B} \times {}^{B}\mathbf{r} \tag{9.251}
$$

$$
{}^{GG}_{GB}\mathbf{a} = {}^{GG}_{BG}\mathbf{a} - {}_{G}\boldsymbol{\alpha}_{B} \times {}^{G}\mathbf{r} \tag{9.252}
$$

$$ {}^B_G\mathbf{a} = {}^{BB}_{BG}\mathbf{a} + {}^B_G\boldsymbol{\omega}_B \times {}^B\mathbf{v} + {}^B_G\boldsymbol{\omega}_B \times \left({}^B_G\boldsymbol{\omega}_B \times {}^B\mathbf{r} \right) \tag{9.253} $$

$$ {}^G_B\mathbf{a} = {}^{GG}_{GB}\mathbf{a} - {}_G\boldsymbol{\omega}_B \times {}^G\mathbf{v} + {}_G\boldsymbol{\omega}_B \times \left({}_G\boldsymbol{\omega}_B \times {}^G\mathbf{r} \right) \tag{9.254} $$

Example 556 ★ Second-Derivative Transformation Formula Consider a point P that can move in the body coordinate frame $B(Oxyz)$. The position vector ${}^B\mathbf{r}_P$ is not constant, and therefore, using the derivative transformation formula (8.250), we find the B-expression of the G-velocity as

$$ \frac{{}^Gd}{dt}\, {}^B\mathbf{r}_P = {}^B_G\mathbf{v}_P = \frac{{}^Bd}{dt}\, {}^B\mathbf{r}_P + {}^B_G\boldsymbol{\omega}_B \times {}^B\mathbf{r}_P \tag{9.255} $$

Another G-derivative of this equation provides the B-expression for the global acceleration of P as (9.205):

$$ {}^B_G\mathbf{a} = {}^B\mathbf{a} + {}^B_G\boldsymbol{\alpha}_B \times {}^B\mathbf{r} + 2\,{}^B_G\boldsymbol{\omega}_B \times {}^B\mathbf{v} + {}^B_G\boldsymbol{\omega}_B \times \left({}^B_G\boldsymbol{\omega}_B \times {}^B\mathbf{r} \right) \tag{9.256} $$

Using this result, we can define the *second-derivative transformation formula* of a B-vector ${}^B\square$ from the body to the global coordinate frame:

$$ \begin{aligned} \frac{{}^Gd}{dt}\frac{{}^Gd}{dt}\, {}^B\square &= {}^B_G\ddot{\square} \\ &= \frac{{}^Bd}{dt}\frac{{}^Bd}{dt}\, {}^B\square + {}^B_G\boldsymbol{\alpha}_B \times {}^B\square \\ &\quad + 2\,{}^B_G\boldsymbol{\omega}_B \times \left(\frac{{}^Bd}{dt}\, {}^B\square + {}^B_G\boldsymbol{\omega}_B \times {}^B\square \right) \end{aligned} \tag{9.257} $$

The final result ${}^B_G\ddot{\square}$ shows the second global time derivative expressed in the body frame, or simply the B-expression of the second G-derivative of a vector ${}^B\square$. The vector ${}^B\square$ may be any vector quantity such as position, velocity, angular velocity, momentum, angular momentum, a time-varying force vector.

Example 557 ★ Mixed Second-Derivative Transformation Formula Consider three relatively rotating coordinate frames A, B, and C. The B-expression of the A-acceleration of a moving point P in the body coordinate frame $B(Oxyz)$ is

$$ \begin{aligned} \frac{{}^Ad}{dt}\, {}^B\mathbf{v} = {}^B_A\mathbf{a} &= {}^B\mathbf{a} + {}^B_A\boldsymbol{\alpha}_B \times {}^B\mathbf{r} \\ &\quad + 2\,{}^B_A\boldsymbol{\omega}_B \times {}^B\mathbf{v} + {}^B_A\boldsymbol{\omega}_B \times \left({}^B_A\boldsymbol{\omega}_B \times {}^B\mathbf{r} \right) \end{aligned} \tag{9.258} $$

and the B-expression of the C-acceleration of a moving point in the body coordinate frame $B(Oxyz)$ is

$$ \begin{aligned} \frac{{}^Cd}{dt}\, {}^B\mathbf{v} = {}^B_C\mathbf{a} &= {}^B\mathbf{a} + {}^B_C\boldsymbol{\alpha}_B \times {}^B\mathbf{r} \\ &\quad + 2\,{}^B_C\boldsymbol{\omega}_B \times {}^B\mathbf{v} + {}^B_C\boldsymbol{\omega}_B \times \left({}^B_C\boldsymbol{\omega}_B \times {}^B\mathbf{r} \right) \end{aligned} \tag{9.259} $$

Combining Equations (9.258) and (9.259), we find

$$
\begin{aligned}
&{}_A^B\mathbf{a} - {}_A^B\boldsymbol{\alpha}_B \times {}^B\mathbf{r} - 2\,{}_A^B\boldsymbol{\omega}_B \times {}^B\mathbf{v} - {}_A^B\boldsymbol{\omega}_B \times \left({}_A^B\boldsymbol{\omega}_B \times {}^B\mathbf{r}\right) \\
&\quad = {}_C^B\mathbf{a} - {}_C^B\boldsymbol{\alpha}_B \times {}^B\mathbf{r} - 2\,{}_C^B\boldsymbol{\omega}_B \times {}^B\mathbf{v} - {}_C^B\boldsymbol{\omega}_B \times \left({}_C^B\boldsymbol{\omega}_B \times {}^B\mathbf{r}\right)
\end{aligned} \tag{9.260}
$$

We rearrange it to present the following formula for changing the frame C in which we have taken two derivatives of ${}^B\mathbf{r}_P$ to the frame A in which we need the derivatives to be taken:

$$
\begin{aligned}
{}_A^B\mathbf{a} &= {}_C^B\mathbf{a} + \left({}_A^B\boldsymbol{\alpha}_B - {}_C^B\boldsymbol{\alpha}_B\right) \times {}^B\mathbf{r} + 2\left({}_A^B\boldsymbol{\omega}_B - {}_C^B\boldsymbol{\omega}_B\right) \times {}^B\mathbf{v} \\
&\quad + {}_A^B\boldsymbol{\omega}_B \times \left({}_A^B\boldsymbol{\omega}_B \times {}^B\mathbf{r}\right) - {}_C^B\boldsymbol{\omega}_B \times \left({}_C^B\boldsymbol{\omega}_B \times {}^B\mathbf{r}\right)
\end{aligned} \tag{9.261}
$$

It can equivalently be shown as

$$
\begin{aligned}
{}_A^B\ddot{\Box} &= {}_C^B\ddot{\Box} + \left({}_A^B\boldsymbol{\alpha}_B - {}_C^B\boldsymbol{\alpha}_B\right) \times {}^B\Box + 2\left({}_A^B\boldsymbol{\omega}_B - {}_C^B\boldsymbol{\omega}_B\right) \times {}^B\dot{\Box} \\
&\quad + {}_A^B\boldsymbol{\omega}_B \times \left({}_A^B\boldsymbol{\omega}_B \times {}^B\Box\right) - {}_C^B\boldsymbol{\omega}_B \times \left({}_C^B\boldsymbol{\omega}_B \times {}^B\Box\right)
\end{aligned} \tag{9.262}
$$

We call Equation (9.262) the *mixed second-derivative transformation formula*. It presents the method to change the frame in which the second derivative of a vector ${}^B\Box$ is taken.

The mixed second-derivative transformation formula (9.262) is more general than the simple derivative transformation formula (9.257). Equation (9.257) is a special case of (9.262) when $B \equiv C$ or ${}_C^B\boldsymbol{\omega}_B = 0$ and ${}_C^B\boldsymbol{\alpha}_B = 0$.

Example 558 ★ Alternative Definition of Angular Acceleration Vector Similar to the definition of the angular velocity vector in Equation (8.188),

$$
{}_G^B\boldsymbol{\omega}_B = \hat{\imath}\left(\frac{{}^Gd\hat{\jmath}}{dt}\cdot\hat{k}\right) + \hat{\jmath}\left(\frac{{}^Gd\hat{k}}{dt}\cdot\hat{\imath}\right) + \hat{k}\left(\frac{{}^Gd\hat{\imath}}{dt}\cdot\hat{\jmath}\right) \tag{9.263}
$$

we define the angular acceleration vector of a rigid body $B(\hat{\imath}, \hat{\jmath}, \hat{k})$ in the global frame $G(\hat{I}, \hat{J}, \hat{K})$ as

$$
\begin{aligned}
{}_G^B\boldsymbol{\alpha}_B &= \frac{{}^Gd\hat{\imath}}{dt}\left(\frac{{}^Gd\hat{\jmath}}{dt}\cdot\hat{k}\right) + \frac{1}{2}\left(\frac{{}^Gd^2\hat{\jmath}}{dt^2}\cdot\hat{k} - \frac{{}^Gd^2\hat{k}}{dt^2}\cdot\hat{\jmath}\right)\hat{\imath} \\
&\quad + \frac{{}^Gd\hat{\jmath}}{dt}\left(\frac{{}^Gd\hat{k}}{dt}\cdot\hat{\imath}\right) + \frac{1}{2}\left(\frac{{}^Gd^2\hat{k}}{dt^2}\cdot\hat{\imath} - \frac{{}^Gd^2\hat{\imath}}{dt^2}\cdot\hat{k}\right)\hat{\jmath} \\
&\quad + \frac{{}^Gd\hat{k}}{dt}\left(\frac{{}^Gd\hat{\imath}}{dt}\cdot\hat{\jmath}\right) + \frac{1}{2}\left(\frac{{}^Gd^2\hat{\imath}}{dt^2}\cdot\hat{\jmath} - \frac{{}^Gd^2\hat{\jmath}}{dt^2}\cdot\hat{\imath}\right)\hat{k}
\end{aligned} \tag{9.264}
$$

To prove (9.264), we take a G-derivative from (9.263):

$$
\begin{aligned}
{}_G^B\boldsymbol{\alpha}_B = {} & \frac{{}^Gd\hat{\imath}}{dt}\left(\frac{{}^Gd\hat{\jmath}}{dt}\cdot\hat{k}\right) + \frac{{}^Gd\hat{\jmath}}{dt}\left(\frac{{}^Gd\hat{k}}{dt}\cdot\hat{\imath}\right) + \frac{{}^Gd\hat{k}}{dt}\left(\frac{{}^Gd\hat{\imath}}{dt}\cdot\hat{\jmath}\right) \\
& + \hat{\imath}\left(\frac{{}^Gd^2\hat{\jmath}}{dt^2}\cdot\hat{k} + \frac{{}^Gd\hat{\jmath}}{dt}\cdot\frac{{}^Gd\hat{k}}{dt}\right) \\
& + \hat{\jmath}\left(\frac{{}^Gd^2\hat{k}}{dt^2}\cdot\hat{\imath} + \frac{{}^Gd\hat{k}}{dt}\cdot\frac{{}^Gd\hat{\imath}}{dt}\right) \\
& + \hat{k}\left(\frac{{}^Gd^2\hat{\imath}}{dt^2}\cdot\hat{\jmath} + \frac{{}^Gd\hat{\imath}}{dt}\cdot\frac{{}^Gd\hat{\jmath}}{dt}\right)
\end{aligned} \tag{9.265}
$$

Employing the unit vector relationships

$$\hat{\imath}\cdot\hat{\imath} = \hat{\jmath}\cdot\hat{\jmath} = \hat{k}\cdot\hat{k} = 1 \tag{9.266}$$

$$\hat{\imath}\cdot\hat{\jmath} = \hat{\jmath}\cdot\hat{k} = \hat{k}\cdot\hat{\imath} = 0 \tag{9.267}$$

$$\hat{\imath}\cdot d\hat{\imath} = \hat{\jmath}\cdot d\hat{\jmath} = \hat{k}\cdot d\hat{k} = 0 \tag{9.268}$$

$$\hat{\jmath}\cdot d\hat{\imath} = -\hat{\imath}\cdot d\hat{\jmath} \qquad \hat{k}\cdot d\hat{\jmath} = -\hat{\jmath}\cdot d\hat{k} \qquad \hat{\imath}\cdot d\hat{k} = -\hat{k}\cdot d\hat{\imath} \tag{9.269}$$

$$
\begin{aligned}
\hat{\imath}\cdot d^2\hat{\imath} &= -d\hat{\imath}\cdot d\hat{\imath} \\
\hat{\jmath}\cdot d^2\hat{\jmath} &= -d\hat{\jmath}\cdot d\hat{\jmath} \\
\hat{k}\cdot d^2\hat{k} &= -d\hat{k}\cdot d\hat{k}
\end{aligned} \tag{9.270}
$$

$$
\begin{aligned}
\hat{\imath}\cdot d^2\hat{\jmath} + \hat{\jmath}\cdot d^2\hat{\imath} &= -2d\hat{\imath}\cdot d\hat{\jmath} \\
\hat{\jmath}\cdot d^2\hat{k} + \hat{k}\cdot d^2\hat{\jmath} &= -2d\hat{\jmath}\cdot d\hat{k} \\
\hat{k}\cdot d^2\hat{\imath} + \hat{\imath}\cdot d^2\hat{k} &= -2d\hat{k}\cdot d\hat{\imath}
\end{aligned} \tag{9.271}
$$

we can simplify (9.265) to (9.264).

Example 559 ★ Alternative Definition of Acceleration Vector Consider a body coordinate frame B moving with a fixed point in the global frame G. We can describe the motion of the body by describing the motion of the local unit vectors $\hat{\imath}$, $\hat{\jmath}$, $\hat{k}$. Let ${}^B\mathbf{r}$ be the position vector of a fixed body point P. Then, ${}^B\mathbf{r}$ is a B-vector with constant components:

$${}^B\mathbf{r} = x\hat{\imath} + y\hat{\jmath} + z\hat{k} \tag{9.272}$$

When the body moves, only the unit vectors $\hat{\imath}$, $\hat{\jmath}$, $\hat{k}$ move relative to the global coordinate frame. Therefore, the vectors of differential displacements are

$$d\mathbf{r} = x\,d\hat{\imath} + y\,d\hat{\jmath} + z\,d\hat{k} \tag{9.273}$$

$$d^2\mathbf{r} = x\,d^2\hat{\imath} + y\,d^2\hat{\jmath} + z\,d^2\hat{k} \tag{9.274}$$

The differential $d\mathbf{r}$ is the B-expression of the infinitesimal displacement as seen from the G-frame, and $d^2\mathbf{r}$ is the infinitesimal change of $d\mathbf{r}$ as seen from the G-frame. Using the orthogonality condition (3.1), we can express $d^2\mathbf{r}$ as

$$d^2\mathbf{r} = \left(d^2\mathbf{r} \cdot \hat{\imath}\right)\hat{\imath} + \left(d^2\mathbf{r} \cdot \hat{\jmath}\right)\hat{\jmath} + \left(d^2\mathbf{r} \cdot \hat{k}\right)\hat{k} \tag{9.275}$$

Substituting (9.274) in the right-hand side of (9.275) shows that

$$\begin{aligned} d^2\mathbf{r} = {} & \left(x\hat{\imath} \cdot d^2\hat{\imath} + y\hat{\imath} \cdot d^2\hat{\jmath} + z\hat{\imath} \cdot d^2\hat{k}\right)\hat{\imath} \\ & + \left(x\hat{\jmath} \cdot d^2\hat{\imath} + y\hat{\jmath} \cdot d^2\hat{\jmath} + z\hat{\jmath} \cdot d^2\hat{k}\right)\hat{\jmath} \\ & + \left(x\hat{k} \cdot d^2\hat{\imath} + y\hat{k} \cdot d^2\hat{\jmath} + z\hat{k} \cdot d^2\hat{k}\right)\hat{k} \end{aligned} \tag{9.276}$$

Using the unit vector relationships (9.266)–(9.271), we can modify $d^2\mathbf{r}$ to

$$\begin{aligned} d^2\mathbf{r} = {} & -\left(d\hat{\imath} \cdot d\hat{\imath}\right)x\hat{\imath} - \left(d\hat{\jmath} \cdot d\hat{\jmath}\right)y\hat{\jmath} - \left(d\hat{k} \cdot d\hat{k}\right)z\hat{k} \\ & - 2\left[\left(d\hat{\imath} \cdot d\hat{\jmath}\right)y + \left(d\hat{k} \cdot d\hat{\imath}\right)z\right]\hat{\imath} \\ & - 2\left[\left(d\hat{\jmath} \cdot d\hat{k}\right)z + \left(d\hat{\imath} \cdot d\hat{\jmath}\right)x\right]\hat{\jmath} \\ & - 2\left[\left(d\hat{k} \cdot d\hat{\imath}\right)z + \left(d\hat{\jmath} \cdot d\hat{k}\right)x\right]\hat{k} \\ & - \left(\hat{\jmath} \cdot d^2\hat{\imath}\right)y\hat{\imath} - \left(\hat{k} \cdot d^2\hat{\jmath}\right)z\hat{\jmath} - \left(\hat{\imath} \cdot d^2\hat{k}\right)x\hat{k} \\ & - \left(\hat{k} \cdot d^2\hat{\imath}\right)z\hat{\imath} - \left(\hat{\imath} \cdot d^2\hat{\jmath}\right)x\hat{\jmath} - \left(\hat{\jmath} \cdot d^2\hat{k}\right)y\hat{k} \end{aligned} \tag{9.277}$$

Example 560 Newton Equation of Motion The Newton equation of motion $\mathbf{F} = m\mathbf{a}$ for a particle of mass m is applied in the global coordinate frame G:

$${}^G\mathbf{F} = m\,{}^G\mathbf{a} = m\,\frac{{}^Gd}{dt}\frac{{}^Gd}{dt}\,{}^G\mathbf{r} \tag{9.278}$$

It means that both derivatives of acceleration must be taken in G. If the position vector $\mathbf{r}$ was expressed in the body coordinate frame B, then the Newton equation of motion would be

$${}^B\mathbf{F} = m\,{}^B_G\mathbf{a} = m\,\frac{{}^Gd}{dt}\frac{{}^Gd}{dt}\,{}^B\mathbf{r} \tag{9.279}$$

Example 561 ★ Double Mixed Acceleration Let us take the first and second derivatives of ${}^A\mathbf{r}$ in different coordinate frames B and C. Using the derivative transformation formula, the B-derivative of ${}^A\mathbf{r}$ is

$$\frac{{}^Bd}{dt}\,{}^A\mathbf{r} = {}^A_B\mathbf{v} = \frac{{}^Ad}{dt}\,{}^A\mathbf{r} + {}^A_B\boldsymbol{\omega}_A \times {}^A\mathbf{r} \tag{9.280}$$

A time derivative of this equation in a third frame C would be

$$\begin{aligned}
\frac{^{C}d}{dt}\frac{^{B}d}{dt}\,{}^{A}\mathbf{r} = \frac{^{C}d}{dt}\,{}_{B}^{A}\mathbf{v} = {}_{CB}^{AA}\mathbf{a} &= \frac{^{C}d}{dt}\left(\frac{^{A}d}{dt}\,{}^{A}\mathbf{r}\right) + \frac{^{C}d}{dt}\left({}_{B}^{A}\boldsymbol{\omega}_{A} \times {}^{A}\mathbf{r}\right) \\
&= \frac{^{A}d}{dt}\frac{^{A}d}{dt}\,{}^{A}\mathbf{r} + {}_{C}^{A}\boldsymbol{\omega}_{A} \times \frac{^{A}d}{dt}\,{}^{A}\mathbf{r} \\
&\quad + \left(\frac{^{A}d}{dt}\,{}_{B}^{A}\boldsymbol{\omega}_{A} + {}_{C}^{A}\boldsymbol{\omega}_{A} \times {}_{B}^{A}\boldsymbol{\omega}_{A}\right) \times {}^{A}\mathbf{r} \\
&\quad + {}_{B}^{A}\boldsymbol{\omega}_{A} \times \left(\frac{^{A}d}{dt}\,{}^{A}\mathbf{r} + {}_{C}^{A}\boldsymbol{\omega}_{A} \times {}^{A}\mathbf{r}\right) \\
&= {}^{A}\mathbf{a} + {}_{C}^{A}\boldsymbol{\omega}_{A} \times {}^{A}\mathbf{v} + {}_{B}^{A}\boldsymbol{\alpha}_{A} \times {}^{A}\mathbf{r} + \left({}_{C}^{A}\boldsymbol{\omega}_{A} \times {}_{B}^{A}\boldsymbol{\omega}_{A}\right) \times {}^{A}\mathbf{r} \\
&\quad + {}_{B}^{A}\boldsymbol{\omega}_{A} \times {}^{A}\mathbf{v} + {}_{B}^{A}\boldsymbol{\omega}_{A} \times \left({}_{C}^{A}\boldsymbol{\omega}_{A} \times {}^{A}\mathbf{r}\right)
\end{aligned} \tag{9.281}$$

We call the acceleration ${}_{CB}^{AA}\mathbf{a}$ the *mixed double acceleration*.

9.3 MULTIBODY ACCELERATION

Consider a rigid body with an attached local coordinate frame $B(oxyz)$ moving freely in a fixed global coordinate frame $G(OXYZ)$. The rigid body can rotate in the global frame, while the origin of the body frame B can translate relative to the origin of G. The coordinates of a body point P in local and global frames, as shown in Figure 9.6, are related by the equation

$$^{G}\mathbf{r}_{P} = {}^{G}R_{B}\,{}^{B}\mathbf{r}_{P} + {}^{G}\mathbf{d}_{B} \tag{9.282}$$

where ${}^{G}\mathbf{d}_{B}$ indicates the position of the moving origin o relative to the fixed origin O.

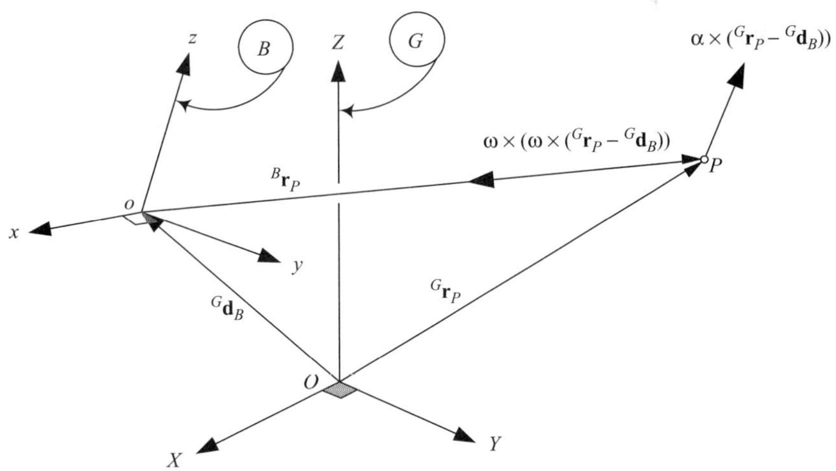

Figure 9.6 A rigid body with coordinate frame $B(oxyz)$ moving freely in a fixed global coordinate frame $G(OXYZ)$.

The acceleration of point P in G is

$$\begin{aligned} {}^{G}\mathbf{a}_P = {}^{G}\dot{\mathbf{v}}_P = {}^{G}\ddot{\mathbf{r}}_P = {}^{G}\ddot{\mathbf{d}}_B + {}_{G}\boldsymbol{\alpha}_B \times \left({}^{G}\mathbf{r}_P - {}^{G}\mathbf{d}_B\right) \\ + {}_{G}\boldsymbol{\omega}_B \times \left[{}_{G}\boldsymbol{\omega}_B \times \left({}^{G}\mathbf{r}_P - {}^{G}\mathbf{d}_B\right)\right] \end{aligned} \tag{9.283}$$

Proof: The acceleration of point P is a consequence of differentiating the velocity equation (8.274) or (8.275):

$$\begin{aligned} {}^{G}\mathbf{a}_P &= \frac{{}^{G}d}{dt}\,{}^{G}\mathbf{v}_P = {}_{G}\boldsymbol{\alpha}_B \times {}^{G}_{B}\mathbf{r}_P + {}_{G}\boldsymbol{\omega}_B \times {}^{G}_{B}\dot{\mathbf{r}}_P + {}^{G}\ddot{\mathbf{d}}_B \\ &= {}_{G}\boldsymbol{\alpha}_B \times {}^{G}_{B}\mathbf{r}_P + {}_{G}\boldsymbol{\omega}_B \times \left({}_{G}\boldsymbol{\omega}_B \times {}^{G}_{B}\mathbf{r}_P\right) + {}^{G}\ddot{\mathbf{d}}_B \\ &= {}^{G}\ddot{\mathbf{d}}_B + {}_{G}\boldsymbol{\alpha}_B \times \left({}^{G}\mathbf{r}_P - {}^{G}\mathbf{d}_B\right) \\ &\quad + {}_{G}\boldsymbol{\omega}_B \times \left[{}_{G}\boldsymbol{\omega}_B \times \left({}^{G}\mathbf{r}_P - {}^{G}\mathbf{d}_B\right)\right] \end{aligned} \tag{9.284}$$

The term ${}_{G}\boldsymbol{\omega}_B \times \left({}_{G}\boldsymbol{\omega}_B \times {}^{G}_{B}\mathbf{r}_P\right)$ is the *centripetal acceleration* and is independent of the angular acceleration. The term ${}_{G}\boldsymbol{\alpha}_B \times {}^{G}_{B}\mathbf{r}_P$ is the *tangential acceleration* and is perpendicular to ${}^{G}_{B}\mathbf{r}_P$. ■

Example 562 Acceleration of a Body Point Consider a rigid body is moving and rotating in a global frame. The acceleration of a body point can be found by taking twice the time derivative of its position vector:

$$ {}^{G}\mathbf{r}_P = {}^{G}R_B\,{}^{B}\mathbf{r}_P + {}^{G}\mathbf{d}_B $$

$$ {}^{G}\dot{\mathbf{r}}_P = {}^{G}\dot{R}_B\,{}^{B}\mathbf{r}_P + {}^{G}\dot{\mathbf{d}}_B \tag{9.285}$$

$$ {}^{G}\ddot{\mathbf{r}}_P = {}^{G}\ddot{R}_B\,{}^{B}\mathbf{r}_P + {}^{G}\ddot{\mathbf{d}}_B \tag{9.286}$$

$$ = {}^{G}\ddot{R}_B\,{}^{G}R_B^{\mathrm{T}}\left({}^{G}\mathbf{r}_P - {}^{G}\mathbf{d}_B\right) + {}^{G}\ddot{\mathbf{d}}_B \tag{9.287}$$

Differentiating the angular velocity matrix

$$ {}_{G}\tilde{\omega}_B = {}^{G}\dot{R}_B\,{}^{G}R_B^{\mathrm{T}} \tag{9.288}$$

shows that

$$\begin{aligned} {}_{G}\dot{\tilde{\omega}}_B &= \frac{{}^{G}d}{dt}\,{}_{G}\tilde{\omega}_B = {}^{G}\ddot{R}_B\,{}^{G}R_B^{\mathrm{T}} + {}^{G}\dot{R}_B\,{}^{G}\dot{R}_B^{\mathrm{T}} \\ &= {}^{G}\ddot{R}_B\,{}^{G}R_B^{\mathrm{T}} + {}_{G}\tilde{\omega}_B\,{}_{G}\tilde{\omega}_B^{\mathrm{T}} \end{aligned} \tag{9.289}$$

and therefore,

$$ {}^{G}\ddot{R}_B\,{}^{G}R_B^{\mathrm{T}} = {}_{G}\dot{\tilde{\omega}}_B - {}_{G}\tilde{\omega}_B\,{}_{G}\tilde{\omega}_B^{\mathrm{T}} \tag{9.290}$$

Hence, the acceleration vector of the body point becomes

$$ {}^{G}\ddot{\mathbf{r}}_P = \left({}_{G}\dot{\tilde{\omega}}_B - {}_{G}\tilde{\omega}_B\,{}_{G}\tilde{\omega}_B^{\mathrm{T}}\right)\left({}^{G}\mathbf{r}_P - {}^{G}\mathbf{d}_B\right) + {}^{G}\ddot{\mathbf{d}}_B \tag{9.291}$$

where

$$
{}_G\dot{\tilde{\omega}}_B = {}_G\tilde{\alpha}_B = \begin{bmatrix} 0 & -\dot{\omega}_3 & \dot{\omega}_2 \\ \dot{\omega}_3 & 0 & -\dot{\omega}_1 \\ -\dot{\omega}_2 & \dot{\omega}_1 & 0 \end{bmatrix} \tag{9.292}
$$

and

$$
{}_G\tilde{\omega}_B \, {}_G\tilde{\omega}_B^{\mathrm{T}} = \begin{bmatrix} \omega_2^2 + \omega_3^2 & -\omega_1\omega_2 & -\omega_1\omega_3 \\ -\omega_1\omega_2 & \omega_1^2 + \omega_3^2 & -\omega_2\omega_3 \\ -\omega_1\omega_3 & -\omega_2\omega_3 & \omega_1^2 + \omega_2^2 \end{bmatrix} \tag{9.293}
$$

Example 563 Turning Panel of a Rotating Satellite The illustrated satellite in Figure 9.7 is rotating about its z_1-axis of symmetry with angular speed $\dot{\alpha}$ and angular acceleration $\ddot{\alpha}$. The panels of the satellite are turning relative to the satellite with angular speed $\dot{\beta}$ and angular acceleration $\ddot{\beta}$ about their longitudinal axis:

$$
\dot{\alpha} = 0.1 \text{ rad/s} \qquad \ddot{\alpha} = 0.05 \text{ rad/s}^2 \tag{9.294}
$$

$$
\dot{\beta} = 0.12 \text{ rad/s} \qquad \ddot{\beta} = -0.1 \text{ rad/s}^2 \tag{9.295}
$$

The angles α and β are now given as

$$
\alpha = 30 \text{ deg} \qquad \beta = 45 \text{ deg} \tag{9.296}
$$

To analyze the satellite as a multibody, we use the intersection point of the satellite axis of symmetry and longitudinal axis of panels as the origin of global $G(OXYZ)$ and body $B_1(Ox_1y_1z_1)$ coordinate frames. A panel $B_2(Ox_2y_2z_2)$ coordinate frame is attached to a panel at its geometric center. The G-frame is assumed to be fixed with

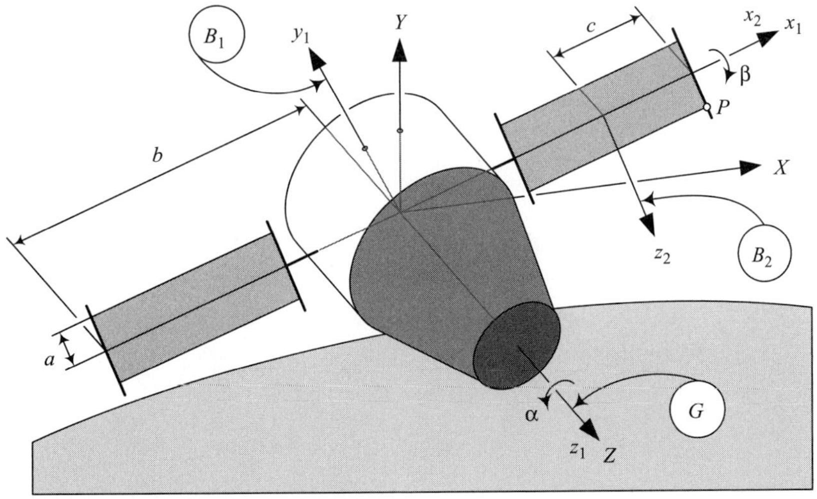

Figure 9.7 A rotating satellite about the Z-axis along with its turning panels about the x_1-axis relative to the satellite.

respect to the satellite. The B_1-frame is attached to the satellite and rotates about the Z-axis. The B_2-frame is attached to the panel and rotates about the x_1-axis.

To locate the satellite and determine its orientation relative to Earth, we usually add another coordinate frame B_3 at O such that its z_3-axis points to the Earth center. There must also be two coordinate frames B_4 and B_5 at the Earth center such that one of them is motionless and the other is attached to Earth and rotates with it. However, we may ignore these coordinate frames when we need to determine the kinematics of any point of the satellite in the satellite G-frame.

To determine the acceleration of a point P at the edge of a panel with size $2a \times 2c$,

$$a = 0.5\,\text{m} \qquad b = 4.2\,\text{m} \qquad c = 1.2\,\text{m} \tag{9.297}$$

we first determine its position vectors and the transformation matrices between the frames:

$$^2\mathbf{r}_P = \begin{bmatrix} c \\ a \\ 0 \end{bmatrix} = \begin{bmatrix} 1.2 \\ 0 \\ 0.5 \end{bmatrix} \text{m} \tag{9.298}$$

$$^1\mathbf{d}_2 = \begin{bmatrix} b-c \\ 0 \\ 0 \end{bmatrix} = \begin{bmatrix} 3 \\ 0 \\ 0 \end{bmatrix} \text{m} \tag{9.299}$$

$$^1R_2 = \begin{bmatrix} 1 & 0 & 0 \\ 0 & \cos\beta & -\sin\beta \\ 0 & \sin\beta & \cos\beta \end{bmatrix} \approx \begin{bmatrix} 1 & 0 & 0 \\ 0 & 0.707 & -0.707 \\ 0 & 0.707 & 0.707 \end{bmatrix} \tag{9.300}$$

$$^GR_1 = \begin{bmatrix} \cos\alpha & -\sin\alpha & 0 \\ \sin\alpha & \cos\alpha & 0 \\ 0 & 0 & 1 \end{bmatrix} \approx \begin{bmatrix} 0.866 & -0.5 & 0 \\ 0.5 & 0.866 & 0 \\ 0 & 0 & 1 \end{bmatrix} \tag{9.301}$$

$$^1\mathbf{r}_P = {}^1\mathbf{d}_2 + {}^1R_2\,{}^2\mathbf{r}_P = \begin{bmatrix} 4.2 \\ -0.35356 \\ 0.35356 \end{bmatrix} \text{m} \tag{9.302}$$

$$^G\mathbf{r}_P = {}^GR_1\,{}^1\mathbf{r}_P = \begin{bmatrix} 3.8141 \\ 1.7938 \\ 0.35356 \end{bmatrix} \text{m} \tag{9.303}$$

Then, we determine the relative angular velocity of the frames and calculate the velocity of the point P:

$${}_1\boldsymbol{\omega}_2 = \begin{bmatrix} \dot\beta \\ 0 \\ 0 \end{bmatrix} = \begin{bmatrix} 0.12 \\ 0 \\ 0 \end{bmatrix} \text{rad/s} \tag{9.304}$$

$${}_G\boldsymbol{\omega}_1 = \begin{bmatrix} 0 \\ 0 \\ \dot\alpha \end{bmatrix} = \begin{bmatrix} 0 \\ 0 \\ 0.1 \end{bmatrix} \text{rad/s} \tag{9.305}$$

$$
{}_G\boldsymbol{\omega}_2 = {}_G\boldsymbol{\omega}_1 + {}_1^G\boldsymbol{\omega}_2 = {}_G\boldsymbol{\omega}_1 + {}^GR_1\,{}_1\boldsymbol{\omega}_2 = \begin{bmatrix} 0.10392 \\ 0.06 \\ 0.1 \end{bmatrix} \text{rad/s} \tag{9.306}
$$

$$
{}^1\mathbf{v}_P = {}^1\dot{\mathbf{d}}_2 + {}_1\boldsymbol{\omega}_2 \times \left({}^1\mathbf{r}_P - {}^1\mathbf{d}_2\right) = \begin{bmatrix} 0 \\ -0.0424 \\ -0.0424 \end{bmatrix} \text{m/s} \tag{9.307}
$$

$$
{}^1\dot{\mathbf{d}}_2 = 0 \tag{9.308}
$$

$$
{}^G\dot{\mathbf{d}}_2 = {}_G\boldsymbol{\omega}_1 \times {}^G\mathbf{d}_2 = {}_G\boldsymbol{\omega}_1 \times {}^GR_1\,{}^1\mathbf{d}_2 = \begin{bmatrix} -0.15 \\ 0.25981 \\ 0 \end{bmatrix} \text{m/s} \tag{9.309}
$$

$$
{}^G\mathbf{v}_P = {}^G\dot{\mathbf{d}}_2 + {}_1^G\mathbf{v}_P = {}^G\dot{\mathbf{d}}_2 + {}^GR_1\,{}^1\mathbf{v}_P = \begin{bmatrix} -0.1288 \\ 0.22309 \\ -0.0424 \end{bmatrix} \text{m/s} \tag{9.310}
$$

Finally, we determine the relative angular acceleration of the frames and calculate the accelerations of the point P:

$$
{}_1\boldsymbol{\alpha}_2 = \begin{bmatrix} \ddot{\beta} \\ 0 \\ 0 \end{bmatrix} = \begin{bmatrix} -0.1 \\ 0 \\ 0 \end{bmatrix} \text{rad/s}^2 \tag{9.311}
$$

$$
{}_G\boldsymbol{\alpha}_1 = \begin{bmatrix} 0 \\ 0 \\ \ddot{\alpha} \end{bmatrix} = \begin{bmatrix} 0 \\ 0 \\ 0.05 \end{bmatrix} \text{rad/s}^2 \tag{9.312}
$$

$$
\begin{aligned}
{}_G\boldsymbol{\alpha}_2 &= {}_G\boldsymbol{\alpha}_1 + {}_1^G\boldsymbol{\alpha}_2 = {}_G\boldsymbol{\alpha}_1 + {}^GR_1\,{}_1\boldsymbol{\alpha}_2 \\
&= \begin{bmatrix} -0.086603 \\ -0.05 \\ 0.05 \end{bmatrix} \text{rad/s}^2
\end{aligned} \tag{9.313}
$$

$$
\begin{aligned}
{}^1\mathbf{a}_P &= {}^1\ddot{\mathbf{d}}_2 + {}_1\boldsymbol{\alpha}_2 \times \left({}^1\mathbf{r}_P - {}^1\mathbf{d}_2\right) + {}_1\boldsymbol{\omega}_2 \times \left({}_1\boldsymbol{\omega}_2 \times \left({}^1\mathbf{r}_P - {}^1\mathbf{d}_2\right)\right) \\
&= \begin{bmatrix} 0 \\ 4.0447 \times 10^{-2} \\ 3.0265 \times 10^{-2} \end{bmatrix} \text{m/s}^2
\end{aligned} \tag{9.314}
$$

$$
{}^1\ddot{\mathbf{d}}_2 = 0 \tag{9.315}
$$

$$
\begin{aligned}
{}^G\ddot{\mathbf{d}}_2 &= {}_G\boldsymbol{\alpha}_1 \times \left({}^G\mathbf{r}_P - {}^G\mathbf{d}_2\right) + {}_G\boldsymbol{\omega}_2 \times \left[{}_G\boldsymbol{\omega}_2 \times \left({}^G\mathbf{r}_P - {}^G\mathbf{d}_2\right)\right] \\
&= {}_G\boldsymbol{\alpha}_1 \times {}^GR_1\left({}^1\mathbf{r}_P - {}^1\mathbf{d}_2\right) + {}_G\boldsymbol{\omega}_2 \times \left[{}_G\boldsymbol{\omega}_2 \times {}^GR_1\left({}^1\mathbf{r}_P - {}^1\mathbf{d}_2\right)\right] \\
&= \begin{bmatrix} -2.5722 \times 10^{-2} \\ 6.4393 \times 10^{-2} \\ 9.3086 \times 10^{-3} \end{bmatrix} \text{m/s}^2
\end{aligned} \tag{9.316}
$$

$$
{}^{G}\mathbf{a}_P = {}^{G}\ddot{\mathbf{d}}_2 + {}^{G}_{1}\mathbf{a}_P = {}^{G}\ddot{\mathbf{d}}_2 + {}^{G}R_1\,{}^{1}\mathbf{a}_P
$$

$$
= \begin{bmatrix} -4.5946 \times 10^{-2} \\ 9.9421 \times 10^{-2} \\ 3.9574 \times 10^{-2} \end{bmatrix} \text{m/s}^2 \qquad (9.317)
$$

Example 564 Acceleration of a Circular Point A moving point A that keeps its distance form another fixed point B is called a circular point. The tip point of the one-link manipulator illustrated in Figure 9.8 is a circular point with respect to the center point at the joint. Knowing that

$$
{}_0\boldsymbol{\omega}_1 = \dot{\theta}_1\,{}^{0}\hat{k}_0 \qquad (9.318)
$$

we can write

$$
{}_0\boldsymbol{\alpha}_1 = {}_0\dot{\boldsymbol{\omega}}_1 = \ddot{\theta}_1\,{}^{0}\hat{k}_0 \qquad (9.319)
$$

$$
{}_0\dot{\boldsymbol{\omega}}_1 \times {}^{0}\mathbf{r}_1 = \ddot{\theta}_1\,{}^{0}\hat{k}_0 \times {}^{0}\mathbf{r}_1 = \ddot{\theta}_1\, R_{Z,\theta+90}\,{}^{0}\mathbf{r}_1 \qquad (9.320)
$$

$$
{}_0\boldsymbol{\omega}_1 \times \left({}_0\boldsymbol{\omega}_1 \times {}^{0}\mathbf{r}_1\right) = -\dot{\theta}_1^{2}\,{}^{0}\mathbf{r}_1 \qquad (9.321)
$$

and calculate the acceleration of the circular point:

$$
{}^{0}\ddot{\mathbf{r}}_1 = \ddot{\theta}_1\, R_{Z,\theta+90}\,{}^{0}\mathbf{r}_1 - \dot{\theta}_1^{2}\,{}^{0}\mathbf{r}_1 \qquad (9.322)
$$

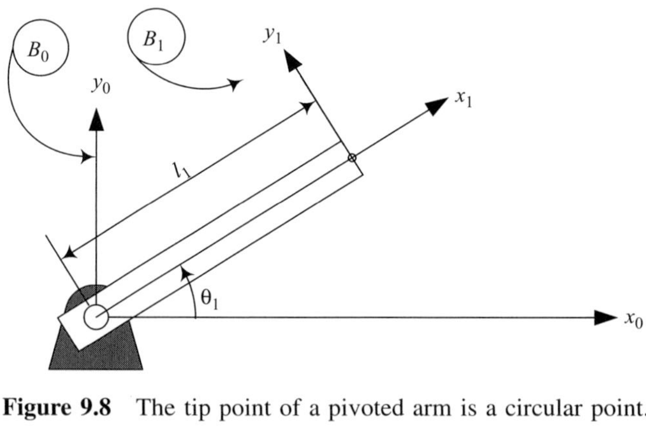

Figure 9.8 The tip point of a pivoted arm is a circular point.

Example 565 Acceleration of a Planar Model of a Vehicle The equations of motion in vehicle dynamics are usually expressed in a set of vehicle coordinate frame $B(Cxyz)$ attached to the vehicle at the mass center C, as shown in Figure 9.9. The x-axis is a longitudinal axis passing through C and directed forward. The y-axis goes laterally to the left from the driver's viewpoint. The z-axis makes the B-frame a right-hand triad.

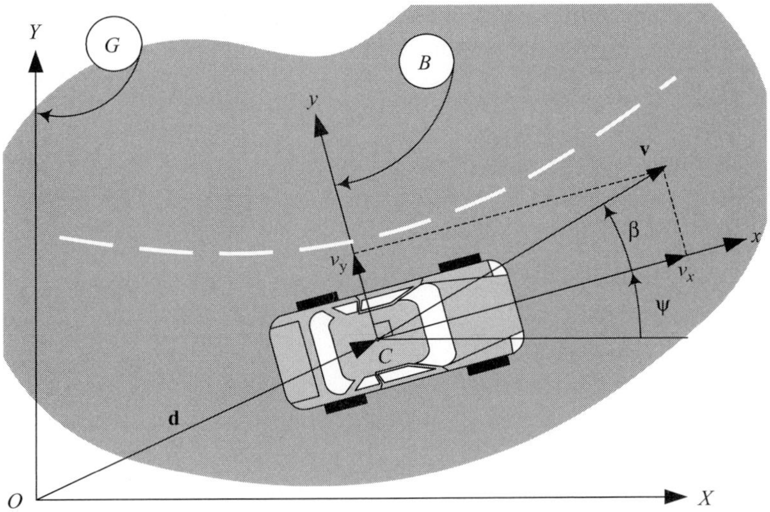

Figure 9.9 A rigid vehicle in planar motion.

The planar model of a car remains flat and parallel to the horizontal road. The z-axis is always perpendicular to the ground, opposite to the gravitational acceleration $\mathbf{g}$.

The position and orientation of the vehicle coordinate frame $B(Cxyz)$ are measured with respect to a grounded fixed coordinate frame $G(OXYZ)$. Analysis of the vehicle motion is equivalent to expressing the position and orientation of $B(Cxyz)$ in $G(OXYZ)$. The angle between the x- and X-axes is the yaw angle ψ and is called the *heading angle*. The vehicle velocity vector $\mathbf{v}$ makes an angle β with the body x-axis called the *sideslip angle* or *attitude angle*. The vehicle's velocity vector $\mathbf{v}$ makes an angle $\beta + \psi$ with the global X-axis called the *cruise angle*.

The B-expressions of the accelerations of the car are

$$ {}^B_G\mathbf{a} = a_x\hat{\imath} + a_y\hat{\jmath} = \left(\dot{v}_x - \omega_z\, v_y\right)\hat{\imath} + \left(\dot{v}_y + \omega_z\, v_x\right)\hat{\jmath} \tag{9.323} $$

$$ {}^B_G\boldsymbol{\alpha}_B = \dot{\omega}_z\hat{k} = \ddot{\psi}\hat{k} \tag{9.324} $$

To show these equations, we decompose the velocity and acceleration of the car at C and express them by their forward and lateral components in B:

$$ {}^B\mathbf{v} = \begin{bmatrix} v_x \\ v_y \\ 0 \end{bmatrix} \qquad {}^B\dot{\mathbf{v}} = \begin{bmatrix} \dot{v}_x \\ \dot{v}_y \\ 0 \end{bmatrix} \tag{9.325} $$

The angular velocity and acceleration vectors of the planar vehicle are

$$ {}^B_G\boldsymbol{\omega}_B = \begin{bmatrix} 0 \\ 0 \\ \omega_z \end{bmatrix} \qquad {}^B_G\dot{\boldsymbol{\omega}}_B = \begin{bmatrix} 0 \\ 0 \\ \dot{\omega}_z \end{bmatrix} \tag{9.326} $$

Substituting the above vectors and matrix in the B-expression of vehicle acceleration, ${}^B_G\mathbf{a}$, provides Equation (9.323):

$$
{}^B_G\mathbf{a} = {}^B\dot{\mathbf{v}}_B + {}^B_G\boldsymbol{\omega}_B \times {}^B\mathbf{v}_B = \begin{bmatrix} \dot{v}_x \\ \dot{v}_y \\ 0 \end{bmatrix} + \begin{bmatrix} 0 \\ 0 \\ \omega_z \end{bmatrix} \times \begin{bmatrix} v_x \\ v_y \\ 0 \end{bmatrix}
$$

$$
= \begin{bmatrix} m\dot{v}_x - m\omega_z v_y \\ m\dot{v}_y + m\omega_z v_x \\ 0 \end{bmatrix} \tag{9.327}
$$

9.4 PARTICLE ACCELERATION

Consider a body coordinate frame B, a global frame G, and a point P in B. There are three applied cases for motion of the point P with respect to B and G.

1. The turning frame B has a fixed point in G and the point P is a fixed point in B. Figure 9.10(a) illustrates a two-dimensional view of this body point situation. The acceleration of the body point P is

$$
{}^G\mathbf{a}_P = {}_G\boldsymbol{\alpha}_B \times {}^G\mathbf{r} + {}_G\boldsymbol{\omega}_B \times ({}_G\boldsymbol{\omega}_B \times {}^G\mathbf{r}) \tag{9.328}
$$

2. The turning frame B has a fixed point in G and the point P is moving in B. Figure 9.10(b) illustrates a two-dimensional view of this particle situation. The acceleration of the point P is

$$
\begin{aligned}
{}^B_G\mathbf{a}_P = {}^B\mathbf{a}_P + {}^B_G\boldsymbol{\alpha}_B \times {}^B\mathbf{r}_P + 2\,{}^B_G\boldsymbol{\omega}_B \times {}^B\mathbf{v}_P \\
+ {}^B_G\boldsymbol{\omega}_B \times \left({}^B_G\boldsymbol{\omega}_B \times {}^B\mathbf{r}_P\right)
\end{aligned} \tag{9.329}
$$

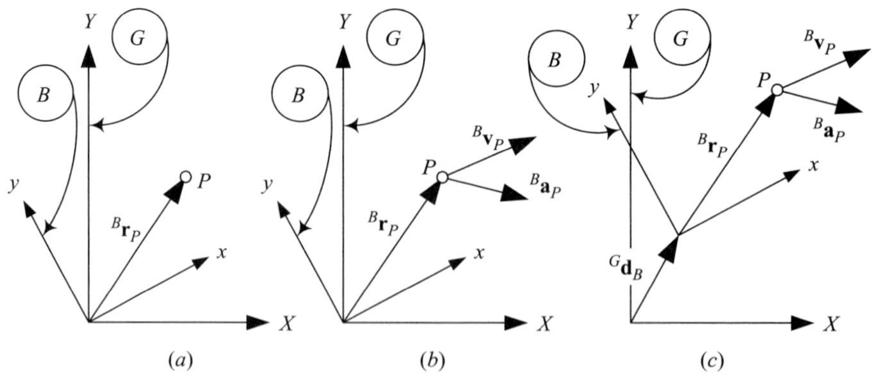

Figure 9.10 A two-dimensional illustration of a moving body B in a global frame G: (a) B has a fixed point O in G and P is a body point; (b) B has a fixed point O in G and P is moving in B; (c) B is moving in G and P is moving in B.

3. The frame B is moving in G and the point P is moving in B. Figure 9.10(c) illustrates a two-dimensional view of this particle situation. The acceleration of the point P is

$$\begin{aligned} {}^G\mathbf{a}_P &= {}^G\ddot{\mathbf{d}}_B + {}^GR_B\, {}^B\mathbf{a} + {}^GR_B \left({}^B_G\boldsymbol{\alpha}_B \times {}^B\mathbf{r}\right) + 2\,{}^GR_B \left({}^B_G\boldsymbol{\omega}_B \times {}^B\mathbf{v}\right) \\ &\quad + {}^GR_B \left[{}^B_G\boldsymbol{\omega}_B \times ({}^B_G\boldsymbol{\omega}_B \times {}^B\mathbf{r})\right] \end{aligned} \tag{9.330}$$

Proof: 1. Consider a body coordinate frame B with a fixed point in G and a fixed point P in B as shown in Figure 9.10(a). The best way to express the position, velocity, and acceleration vectors of a body point P is using the G-frame:

$${}^B\mathbf{r} = {}^B\mathbf{r}_P = x\hat{\imath} + y\hat{\jmath} + z\hat{k} \tag{9.331}$$

$${}^G\mathbf{r} = {}^G\mathbf{r}_P = {}^GR_B\, {}^B\mathbf{r}_P \tag{9.332}$$

$${}^G\mathbf{v}_P = {}^G\dot{\mathbf{r}} = \frac{{}^Gd}{dt}\, {}^G\mathbf{r}_P = {}_G\boldsymbol{\omega}_B \times {}^G\mathbf{r}_P \tag{9.333}$$

$$\begin{aligned} {}^G\mathbf{a}_P &= \frac{{}^Gd^2}{dt^2}\, {}^G\mathbf{r} = {}_G\boldsymbol{\alpha}_B \times {}^G\mathbf{r} + {}_G\boldsymbol{\omega}_B \times {}^G\dot{\mathbf{r}} \\ &= {}_G\boldsymbol{\alpha}_B \times {}^G\mathbf{r} + {}_G\boldsymbol{\omega}_B \times ({}_G\boldsymbol{\omega}_B \times {}^G\mathbf{r}) \end{aligned} \tag{9.334}$$

2. Consider a body coordinate frame B with a fixed point in G and a moving point P in B as shown in Figure 9.10(b). When the point P is moving in B, we can usually measure the velocity and acceleration of P in B as ${}^B\mathbf{v}_P$, ${}^B\mathbf{a}_P$. Therefore, the best way to express the position, velocity, and acceleration vectors of the moving point P is using B-expressions of G-derivatives:

$${}^B\mathbf{r}_P = x\hat{\imath} + y\hat{\jmath} + z\hat{k} \tag{9.335}$$

$${}^G\mathbf{r}_P = {}^GR_B\, {}^B\mathbf{r}_P \tag{9.336}$$

$${}^B_G\mathbf{v}_P = \frac{{}^Gd}{dt}\, {}^B\mathbf{r}_P = {}^B\dot{\mathbf{r}}_P + {}^B_G\boldsymbol{\omega}_B \times {}^B\mathbf{r}_P \tag{9.337}$$

$$\begin{aligned} {}^B_G\mathbf{a}_P &= \frac{{}^Gd^2}{dt^2}\, {}^B\mathbf{r}_P = {}^B\mathbf{a}_P + {}^B_G\boldsymbol{\alpha}_B \times {}^B\mathbf{r}_P \\ &\quad + {}^B_G\boldsymbol{\omega}_B \times {}^B\mathbf{v}_P + {}^B_G\boldsymbol{\omega}_B \times \left({}^B\mathbf{v}_P + {}^B_G\boldsymbol{\omega}_B \times {}^B\mathbf{r}_P\right) \\ &= {}^B\mathbf{a}_P + {}^B_G\boldsymbol{\alpha}_B \times {}^B\mathbf{r}_P \\ &\quad + 2\,{}^B_G\boldsymbol{\omega}_B \times {}^B\mathbf{v}_P + {}^B_G\boldsymbol{\omega}_B \times \left({}^B_G\boldsymbol{\omega}_B \times {}^B\mathbf{r}_P\right) \end{aligned} \tag{9.338}$$

3. Consider a moving-body coordinate frame B in G and a moving point P in B as shown in Figure 9.10(c). In this case, we can usually measure the velocity and acceleration of P in B as ${}^B\mathbf{v}_P$, ${}^B\mathbf{a}_P$ and the velocity of origin of B in G as ${}^G\dot{\mathbf{d}}_B$, ${}^G\ddot{\mathbf{d}}_B$. These vectors are not in the same frame. The best way to express the kinematic of P is to transform the local information to the global frame and express its

position, velocity, and acceleration vectors by G-expressions of G-derivatives:

$$^{B}\mathbf{r} = {}^{B}\mathbf{r}_P = x\hat{\imath} + y\hat{\jmath} + z\hat{k} \tag{9.339}$$

$$^{G}\mathbf{d} = {}^{G}\mathbf{d}_B = d_x\hat{\imath} + d_y\hat{\jmath} + d_z\hat{k} \tag{9.340}$$

$$^{G}\mathbf{r} = {}^{G}\mathbf{r}_P = {}^{G}\mathbf{d}_B + {}^{G}R_B\, {}^{B}\mathbf{r}_P \tag{9.341}$$

$$\begin{aligned} {}^{G}\mathbf{v}_P &= {}^{G}\dot{\mathbf{d}}_B + {}^{G}R_B\, {}^{B}_{G}\mathbf{v} = {}^{G}\dot{\mathbf{d}}_B + {}^{G}R_B \left({}^{B}\dot{\mathbf{r}} + {}^{B}_{G}\boldsymbol{\omega}_B \times {}^{B}\mathbf{r}\right) \\ &= {}^{G}\dot{\mathbf{d}}_B + {}^{G}R_B\, {}^{B}\dot{\mathbf{r}} + {}^{G}R_B \left({}^{B}_{G}\boldsymbol{\omega}_B \times {}^{B}\mathbf{r}\right) \end{aligned} \tag{9.342}$$

$$\begin{aligned} {}^{G}\mathbf{a}_P &= {}^{G}\ddot{\mathbf{d}}_B + {}^{G}R_B\, {}^{B}_{G}\mathbf{a}_P \\ &= {}^{G}\ddot{\mathbf{d}}_B + {}^{G}R_B \left[{}^{B}\ddot{\mathbf{r}} + {}^{B}_{G}\boldsymbol{\omega}_B \times {}^{B}\dot{\mathbf{r}} + {}^{B}_{G}\dot{\boldsymbol{\omega}}_B \times {}^{B}\mathbf{r} + {}^{B}_{G}\boldsymbol{\omega}_B \times \left({}^{B}\dot{\mathbf{r}} + {}^{B}_{G}\boldsymbol{\omega}_B \times {}^{B}\mathbf{r}\right)\right] \\ &= {}^{G}\ddot{\mathbf{d}}_B + {}^{G}R_B\, {}^{B}\mathbf{a} + {}^{G}R_B \left({}^{B}_{G}\boldsymbol{\alpha}_B \times {}^{B}\mathbf{r}\right) \\ &\quad + 2\,{}^{G}R_B \left({}^{B}_{G}\boldsymbol{\omega}_B \times {}^{B}\mathbf{v}\right) + {}^{G}R_B \left[{}^{B}_{G}\boldsymbol{\omega}_B \times \left({}^{B}_{G}\boldsymbol{\omega}_B \times {}^{B}\mathbf{r}\right)\right] \end{aligned} \tag{9.343}$$

■

Example 566 Coriolis Acceleration Consider a rotating frame B in G with angular velocity ${}^{B}_{G}\boldsymbol{\omega}_B$ along with a moving point P in B with velocity ${}^{B}\mathbf{v}$. If we are able to measure the local position, velocity, and acceleration of P, the best way to express the acceleration of P in G will be using Equation (9.338), which is the same as (9.191):

$$^{B}_{G}\mathbf{a} = {}^{B}\mathbf{a} + {}^{B}_{G}\boldsymbol{\alpha}_B \times {}^{B}\mathbf{r} + 2\,{}^{B}_{G}\boldsymbol{\omega}_B \times {}^{B}\mathbf{v} + {}^{B}_{G}\boldsymbol{\omega}_B \times \left({}^{B}_{G}\boldsymbol{\omega}_B \times {}^{B}\mathbf{r}\right) \tag{9.344}$$

where, based on their direction and frame of expression, ${}^{B}\mathbf{a}$ is *local acceleration*, ${}^{B}_{G}\boldsymbol{\alpha}_B \times {}^{B}\mathbf{r}$ is *tangential acceleration*, and ${}^{B}_{G}\boldsymbol{\omega}_B \times ({}^{B}_{G}\boldsymbol{\omega}_B \times {}^{B}\mathbf{r})$ is *centripetal acceleration*. The term $2\,{}^{B}_{G}\boldsymbol{\omega}_B \times {}^{B}\mathbf{v}$ is the *Coriolis acceleration* ${}^{B}_{G}\mathbf{a}_{Co}$:

$$^{B}_{G}\mathbf{a}_{Co} = {}^{B}_{G}\boldsymbol{\omega}_B \times {}^{B}\mathbf{v} + {}^{B}_{G}\boldsymbol{\omega}_B \times {}^{B}\mathbf{v} = 2\,{}^{B}_{G}\boldsymbol{\omega}_B \times {}^{B}\mathbf{v} \tag{9.345}$$

The Coriolis acceleration is perpendicular to both ${}^{B}_{G}\boldsymbol{\omega}_B$ and ${}^{B}\mathbf{v}_P$ and may be interpreted as the required component of acceleration of moving P to follow the rotation of B. The Coriolis acceleration is a combination of two terms, ${}^{B}_{G}\boldsymbol{\omega}_B \times {}^{B}\mathbf{v}$ and ${}^{B}_{G}\boldsymbol{\omega}_B \times {}^{B}\mathbf{v}$. They are mathematically equivalent, so we show the Coriolis acceleration as ${}^{B}_{G}\mathbf{a}_{Co} = 2\,{}^{B}_{G}\boldsymbol{\omega}_B \times {}^{B}\mathbf{v}$. However, the first one comes from

$$^{G}\frac{d}{dt}\,{}^{B}\mathbf{v} = {}^{B}\mathbf{a} + {}^{B}_{G}\boldsymbol{\omega}_B \times {}^{B}\mathbf{v} \tag{9.346}$$

and the second one comes from

$$^{B}_{G}\boldsymbol{\omega}_B \times {}^{G}\frac{d}{dt}\,{}^{B}\mathbf{r} = {}^{B}_{G}\boldsymbol{\omega}_B \times {}^{B}\mathbf{v} + {}^{B}_{G}\boldsymbol{\omega}_B \times \left({}^{B}_{G}\boldsymbol{\omega}_B \times {}^{B}\mathbf{r}\right) \tag{9.347}$$

The term ${}^{B}\mathbf{a}$ in (9.346) is the B-expression of the change of ${}^{B}\mathbf{v}$ in the eyes of an observer in G, and the term ${}^{B}_{G}\boldsymbol{\omega}_B \times {}^{B}\mathbf{v}$ indicates the B-expression of the change in direction of ${}^{B}\mathbf{v}$ due to the rotation of B in the eyes of an observer in G. Equation (9.347) indicates the B-expression of the change in direction of change of ${}^{B}\mathbf{r}$ in the eyes of an observer in G. The first term, ${}^{B}_{G}\boldsymbol{\omega}_B \times {}^{B}\mathbf{v}$, is the direction change of the change of ${}^B\mathbf{r}$ in B, and ${}^B_G\boldsymbol{\omega}_B \times ({}^B_G\boldsymbol{\omega}_B \times {}^B\mathbf{r})$ indicates the change in direction change of ${}^B\mathbf{r}$ due to the direction change of ${}^B\mathbf{r}$ in B.

Practically, we may combine the two terms of ${}^B_G\boldsymbol{\omega}_B \times {}^B\mathbf{v}$ and express them as the Coriolis acceleration (9.345). Such a combination of accelerations appears only in the B-expression of the G-acceleration of a moving point in a rotating frame B, ${}^B_G\mathbf{a} = {}^{BB}_{GG}\mathbf{a} = ({}^G d/dt)\,{}^B_G\mathbf{v} = ({}^G d/dt)({}^G d/dt)\,{}^B\mathbf{r}$, provided the point P follows the rotation of B. An observer in B will not see any motion of P that is caused by rotation of B. Although ${}^B_G\boldsymbol{\omega}_B$ and ${}^B_G\boldsymbol{\alpha}_B$ indicate the angular velocity and acceleration of B relative to G, the acceleration ${}^B_G\mathbf{a}$ of point P would not be (9.344) if P is not following ${}^B_G\boldsymbol{\omega}_B$ and ${}^B_G\boldsymbol{\alpha}_B$.

There is only one ${}^B_G\boldsymbol{\omega}_B \times {}^B\mathbf{v}$ in

$$
{}^{BB}_{GB}\mathbf{a} = ({}^G d/dt)\,{}^B\mathbf{v} = ({}^G d/dt)({}^B d/dt)\,{}^B\mathbf{r}
$$

$$
{}^{BB}_{BG}\mathbf{a} = ({}^B d/dt)\,{}^B_G\mathbf{v} = ({}^B d/dt)({}^G d/dt)\,{}^B\mathbf{r}
$$

We may have a similar discussion for negative Coriolis terms in accelerations ${}^G_B\mathbf{a}$, ${}^{GG}_{GB}\mathbf{a}$, and ${}^{GG}_{BG}\mathbf{a}$.

Example 567 Coriolis Force Proportional to the Coriolis acceleration ${}^B_G\mathbf{a}_{Co}$, we can define a Coriolis force $\mathbf{F}_{Co} = m\,{}^B_G\mathbf{a}_{Co}$, where m is the mass of the point P. We may describe $\mathbf{F}_{Co}$ as the required force on m to make it turn with B while m is moving in B and not in the direction of ${}_G\boldsymbol{\omega}_B$.

To apply $\mathbf{F}_{Co}$ on m, we usually need to provide continued physical contact between B and m. The effect of Coriolis acceleration can be better seen when the Coriolis force is missing and m is not following the rotation of B.

For an applied description, let us consider a rotating disc in a horizontal plane with a constant angular velocity ${}^B_G\boldsymbol{\omega}_B = \omega\hat{k}$. Assume that there is also a moving particle P in a radial groove with a constant velocity ${}^B\mathbf{v} = v\hat{\imath}$. Figures 9.11($a$) and ($b$) illustrate the top view of the system. In Figure 9.11(a), we are standing in the G-frame and watching the rotation of B about $\hat{K}$ and the motion of P in G. In Figure 9.11(b), we are standing in the B-frame and watching the rotation of G about $\hat{k}$ and the motion of P in B. Let us assume that P starts moving in the groove on the x-axis when B and G are coincident.

Because ${}^B\mathbf{a} = 0$ and ${}^B_G\boldsymbol{\alpha}_B = 0$, the acceleration of P is

$$
\begin{aligned}
{}^B_G\mathbf{a} &= 2\,{}^B_G\boldsymbol{\omega}_B \times {}^B\mathbf{v} + {}^B_G\boldsymbol{\omega}_B \times \left({}^B_G\boldsymbol{\omega}_B \times {}^B\mathbf{r}\right) \\
&= 2\begin{bmatrix} 0 \\ 0 \\ \omega \end{bmatrix} \times \begin{bmatrix} v \\ 0 \\ 0 \end{bmatrix} + \begin{bmatrix} 0 \\ 0 \\ \omega \end{bmatrix} \times \left(\begin{bmatrix} 0 \\ 0 \\ \omega \end{bmatrix} \times \begin{bmatrix} r \\ 0 \\ 0 \end{bmatrix}\right) \\
&= \begin{bmatrix} -r\omega^2 \\ 2v\omega \\ 0 \end{bmatrix} = -r\omega^2\hat{\imath} + 2v\omega\hat{\jmath}
\end{aligned}
\tag{9.348}
$$

The component $2v\omega\hat{\jmath}$ is the Coriolis acceleration of P in ${}^B_G\mathbf{a}$. So, there must be a Coriolis force $\mathbf{F}_{Co}$ that the groove applies laterally on the moving mass m to turn it

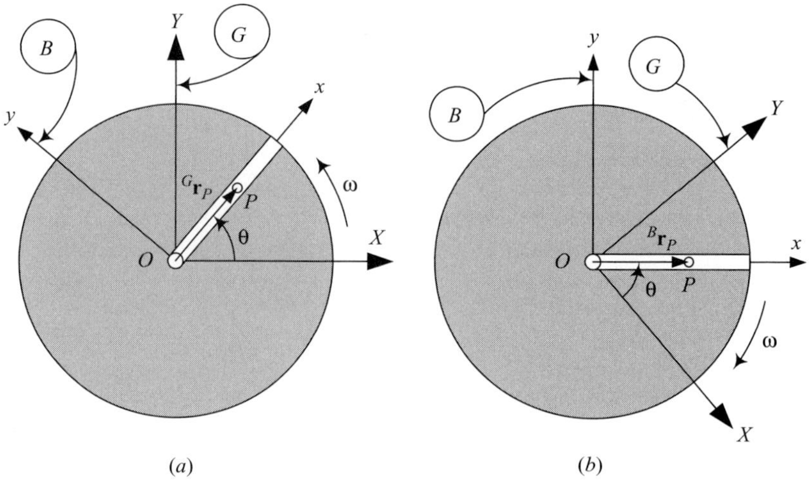

Figure 9.11 A rotating disc in a horizontal plane with a constant angular velocity ${}_G^B\boldsymbol{\omega}_B = \omega\hat{k}$ and a moving particle P in a radial groove with velocity ${}^B\mathbf{v} = v\hat{\imath}$. (a) The observer is in the G-frame. (b) The observer is in the B-frame.

with the disc:

$$ {}^B\mathbf{F}_{Co} = m\,{}_G^B\mathbf{a}_{Co} = 2mv\omega\hat{\jmath} \tag{9.349} $$

The direction of ${}^B\mathbf{F}_{Co}$ is always perpendicular to the radial groove and pointing ${}_G^B\boldsymbol{\omega}_B \times {}^B\mathbf{v}$.

The acceleration ${}_G^B\mathbf{a} = -r\omega^2\hat{\imath} + 2v\omega\hat{\jmath}$ is the B-expression of the global acceleration of P. A transformation to G provides the G-expression of the G-acceleration of P:

$$ {}^G\mathbf{a} = {}^GR_B\,{}_G^B\mathbf{a} = R_{Z,\theta}\,{}_G^B\mathbf{a} = \begin{bmatrix} \cos\theta & -\sin\theta & 0 \\ \sin\theta & \cos\theta & 0 \\ 0 & 0 & 1 \end{bmatrix} \begin{bmatrix} -r\omega^2 \\ 2v\omega \\ 0 \end{bmatrix} $$

$$ = \begin{bmatrix} -r\omega^2\cos\theta - 2v\omega\sin\theta \\ 2v\omega\cos\theta - r\omega^2\sin\theta \\ 0 \end{bmatrix} \tag{9.350} $$

The G-expression of the G-derivative and the B-expression of the B-derivative are the proper equations to integrate. Assuming the initial conditions

$$ r(0) = 0 \qquad \theta(0) = 0 \tag{9.351} $$

we can integrate the acceleration and determine the path of P in the G-frame:

$$ \ddot{X} = -r\omega^2\cos\theta - 2v\omega\sin\theta = -vt\omega^2\cos\omega t - 2v\omega\sin\omega t \tag{9.352} $$

$$ \dot{X} = \int \ddot{X}\,dt = v\cos\omega t - vt\omega\sin\omega t \tag{9.353} $$

$$ X = \int \dot{X}\,dt = vt\cos\omega t \tag{9.354} $$

$$\ddot{Y} = 2v\omega\cos\theta - r\omega^2\sin\theta = 2v\omega\cos\omega t - vt\omega^2\sin\omega t \tag{9.355}$$

$$\dot{Y} = \int \ddot{Y}\,dt = v\sin\omega t + vt\omega\cos\omega t \tag{9.356}$$

$$Y = \int \dot{Y}\,dt = vt\sin\omega t \tag{9.357}$$

A coordinate transformation determines the path of P in the B-frame:

$$^B\mathbf{r} = {^B}R_G\,{^G}\mathbf{r} = {^G}R_B^{\mathrm{T}}\begin{bmatrix} vt\cos\omega t \\ vt\sin\omega t \\ 0 \end{bmatrix} = \begin{bmatrix} vt \\ 0 \\ 0 \end{bmatrix} \tag{9.358}$$

The point P reaches the radius R at the time t_R:

$$t_R = \frac{R}{v} \tag{9.359}$$

Figure 9.12(a) illustrates the path of P in the eyes of an observer in G for different values of v and for

$$\omega = 1\text{ rad/s} \qquad R = 1\text{ m} \tag{9.360}$$

Because $R = 1$ m, $1/v$ indicates the time t_R that it takes for P to reach the edge of the disc. The path for $v = 0.4$ m/s is shown in Figure 9.12(b). It also shows the point P and frame B at two different times. Because $^B\mathbf{r} = vt\hat{\imath}$, the point P is always on the x-axis.

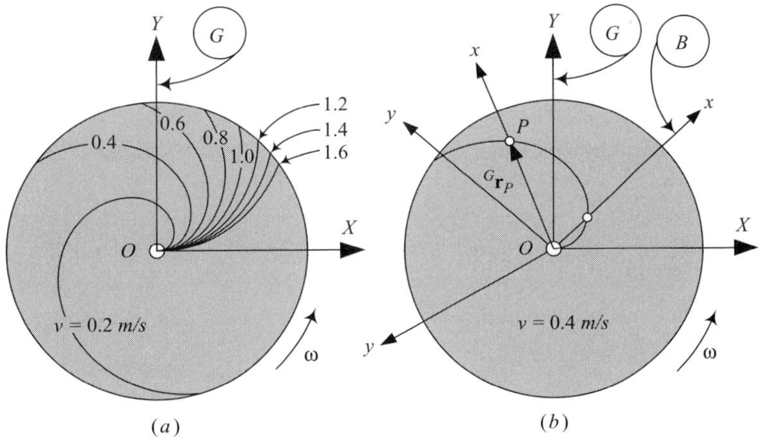

Figure 9.12 The path of P in the eyes of an observer in G. (a) The path for different values of speed v. (b) The position of point P and orientation of frame B at two different times for $v = 0.4$ m/s.

The importance of ${}^B_G\mathbf{a}_{Co}$ was discovered and discussed by the French mathematician *Gaspard Gustave de Coriolis* (1792–1843). Today the terms *work* and *kinetic energy* still retain the meanings introduced by Coriolis.

Example 568 No Coriolis Force Let us consider a frictionless rotating disc in a horizontal plane with a constant angular velocity ${}_G^B\boldsymbol{\omega}_B = \omega\hat{k}$, ${}_G\boldsymbol{\omega}_B = \omega\hat{K}$ and a moving particle P in a radial direction with a constant velocity ${}^G\mathbf{v} = v\hat{I}$. Figures 9.13($a$) and ($b$) illustrate the top view of the system. In Figure 9.13(a), we are standing in the G-frame and watching the rotation of B about $\hat{K}$ and the motion of P in G. In Figure 9.13(b), we are standing in the B-frame and watching the rotation of G about $\hat{k}$ and the motion of P in B. Let us assume that P starts moving on the x-axis from $x = 0$ when B and G are coincident.

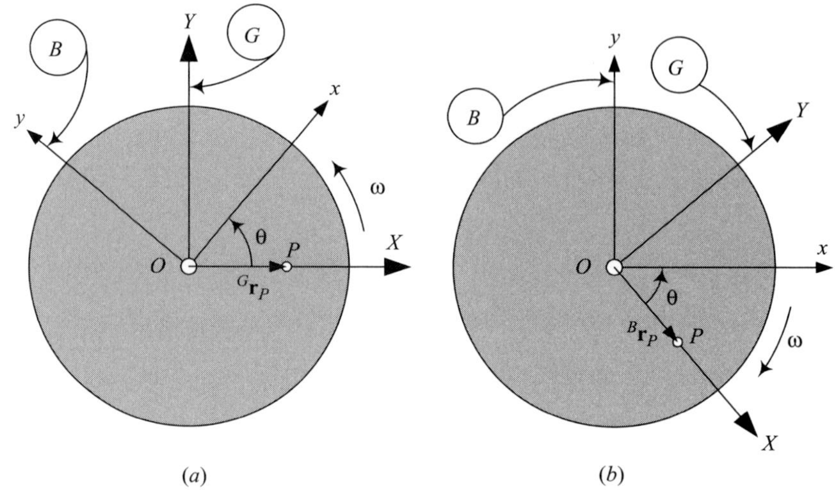

Figure 9.13 A rotating disc in a horizontal plane with a constant angular velocity ${}_G^B\boldsymbol{\omega}_B = \omega\hat{k}$ and a moving particle P with velocity ${}^G\mathbf{v} = v\hat{I}$. ($a$) The observer is in the G-frame. (b) The observer is in the B-frame.

Because P is not following the rotation of B, the best way to analyze the acceleration of P is using ${}_B^G\mathbf{a}$ from (9.192). Knowing that ${}^G\mathbf{a} = 0$ and ${}_G^B\boldsymbol{\alpha}_B = {}_G\boldsymbol{\alpha}_B = 0$, we have

$$
\begin{aligned}
{}_B^G\mathbf{a} &= \frac{{}^Bd}{dt}\,{}_B^G\mathbf{v}\,(t) = -2\,{}_G\boldsymbol{\omega}_B \times {}^G\mathbf{v} + {}_G\boldsymbol{\omega}_B \times \left({}_G\boldsymbol{\omega}_B \times {}^G\mathbf{r}\right) \\
&= -2\begin{bmatrix} 0 \\ 0 \\ \omega \end{bmatrix} \times \begin{bmatrix} v \\ 0 \\ 0 \end{bmatrix} + \begin{bmatrix} 0 \\ 0 \\ \omega \end{bmatrix} \times \left(\begin{bmatrix} 0 \\ 0 \\ \omega \end{bmatrix} \times \begin{bmatrix} r \\ 0 \\ 0 \end{bmatrix}\right) \\
&= \begin{bmatrix} -r\omega^2 \\ -2v\omega \\ 0 \end{bmatrix} = -r\omega^2\hat{I} - 2v\omega\hat{J}
\end{aligned}
\tag{9.361}
$$

The component $-2v\omega\hat{J}$ is the Coriolis acceleration of P in ${}_B^G\mathbf{a}$.

The acceleration ${}_{B}^{G}\mathbf{a} = -r\omega^2\hat{I} - 2v\omega\hat{J}$ is the G-expression of the body acceleration of P. A transformation to B provides the B-expression of the B-acceleration of P:

$$
\begin{aligned}
{}^{B}\mathbf{a} &= {}^{B}R_{G}\,{}_{B}^{G}\mathbf{a} = R_{Z,\theta}^{\mathrm{T}}\,{}_{B}^{G}\mathbf{a} = \begin{bmatrix} \cos\theta & -\sin\theta & 0 \\ \sin\theta & \cos\theta & 0 \\ 0 & 0 & 1 \end{bmatrix}^{\mathrm{T}} \begin{bmatrix} -r\omega^2 \\ -2v\omega \\ 0 \end{bmatrix} \\
&= \begin{bmatrix} -r\omega^2\cos\theta - 2v\omega\sin\theta \\ -2v\omega\cos\theta + r\omega^2\sin\theta \\ 0 \end{bmatrix}
\end{aligned} \tag{9.362}
$$

Assuming the initial conditions

$$r(0) = 0 \qquad \theta(0) = 0 \tag{9.363}$$

we can integrate the acceleration and determine the path of P in the B-frame:

$$\ddot{x} = -r\omega^2\cos\theta - 2v\omega\sin\theta = -vt\omega^2\cos\omega t - 2v\omega\sin\omega t \tag{9.364}$$

$$\dot{x} = \int \ddot{X}dt = v\cos\omega t - vt\omega\sin\omega t \tag{9.365}$$

$$x = \int \dot{X}dt = vt\cos\omega t \tag{9.366}$$

$$\ddot{y} = -2v\omega\cos\theta + r\omega^2\sin\theta = -2v\omega\cos\omega t + vt\omega^2\sin\omega t \tag{9.367}$$

$$\dot{y} = \int \ddot{y}dt = -v\sin\omega t - vt\omega\cos\omega t \tag{9.368}$$

$$y = \int \dot{y}dt = -vt\sin\omega t \tag{9.369}$$

A coordinate transformation determines the path of P in the G-frame:

$${}^{G}\mathbf{r} = {}^{G}R_{B}\,{}^{B}\mathbf{r} = {}^{G}R_{B} \begin{bmatrix} vt\cos\omega t \\ -vt\sin\omega t \\ 0 \end{bmatrix} = \begin{bmatrix} vt \\ 0 \\ 0 \end{bmatrix} \tag{9.370}$$

The point P reaches the radius R at time t_R:

$$t_R = \frac{R}{v} \tag{9.371}$$

Figure 9.14(a) illustrates the path of P in the eyes of an observer in B for different speeds v and for

$$\omega = 1\ \text{rad/s} \qquad R = 1\ \text{m} \tag{9.372}$$

Because $R = 1$ m, $1/v$ also indicates the time t_R that it takes for P to reach the edge of the disc. The path for $v = 0.4$ m/s is shown in Figure 9.14(b). It also shows the point P and frame G at two different times. Because ${}^{G}\mathbf{r} = vt\hat{I}$, the point P is always on the X-axis. Although there exists no applied force on P, an observer in B thinks that

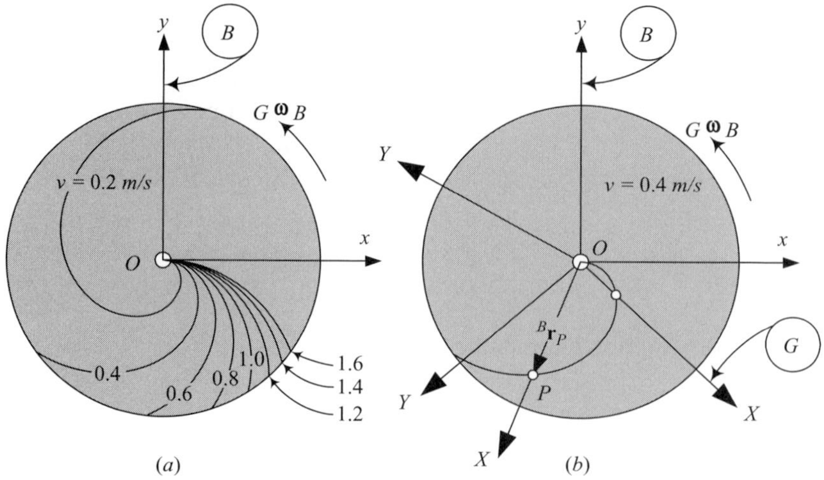

Figure 9.14 The path of P in the eyes of an observer in B. (a) The path for different values of speed v. (b) The position of point P and orientation of frame G at two different times for $v = 0.4\,\text{m/s}$.

P is moving laterally because of a lateral acceleration ${}^G_B\mathbf{a}_{Co} = -2v\omega\hat{J}$. However, an observer in G will see P in a straight radial motion while the ground moves laterally underneath.

This expression of Coriolis acceleration is important in the explanation of the motion of particles close to the ground. The rotation of Earth and the velocity of particles produce a Coriolis acceleration ${}^G_B\mathbf{a}_{Co}$ in the eyes of an observer on Earth.

Example 569 Applied Acceleration When P Is Not Following B Let us summarize the Coriolis effect when a moving point does not follow the rotation of the local frame B. Figure 9.15 illustrates a rotating B-frame in a global G-frame. We, as the observers, are standing in B and watching a moving point P. Practically, we must be able to measure the kinematic information of P in B: ${}^B\mathbf{r}_P$, ${}^B\mathbf{v}_P$, ${}^B\mathbf{a}_P$. Let us also assume that we are aware of the rotation of B in G and have a good estimate of ${}_G\boldsymbol{\omega}_B$ and ${}_G\boldsymbol{\alpha}_B$.

The best way to analyze the motion of P is using the G-expression of the B-acceleration, ${}^G_B\mathbf{a}$, or the B-expression of the G-acceleration, ${}^B_G\mathbf{a}$, from (9.192) and (9.344):

$$ {}^G_B\mathbf{a} = {}^G\mathbf{a} - {}_G\boldsymbol{\alpha}_B \times {}^G\mathbf{r} - 2\,{}_G\boldsymbol{\omega}_B \times {}^G\mathbf{v} + {}_G\boldsymbol{\omega}_B \times \left({}_G\boldsymbol{\omega}_B \times {}^G\mathbf{r}\right) \tag{9.373} $$

$$ {}^B_G\mathbf{a} = {}^B\mathbf{a} + {}^B_G\boldsymbol{\alpha}_B \times {}^B\mathbf{r} + 2\,{}^B_G\boldsymbol{\omega}_B \times {}^B\mathbf{v} + {}^B_G\boldsymbol{\omega}_B \times \left({}^B_G\boldsymbol{\omega}_B \times {}^B\mathbf{r}\right) \tag{9.374} $$

Using either of these equations we assume that P is following the rotation of B in G, so the terms ${}^G_B\mathbf{a}$ and ${}^B_G\mathbf{a}$ are meaningful.

Let us use Equation (9.374) and find ${}^B\mathbf{a}$:

$$ {}^B\mathbf{a} = {}^B_G\mathbf{a} - {}^B_G\boldsymbol{\alpha}_B \times {}^B\mathbf{r} - 2\,{}^B_G\boldsymbol{\omega}_B \times {}^B\mathbf{v} - {}^B_G\boldsymbol{\omega}_B \times \left({}^B_G\boldsymbol{\omega}_B \times {}^B\mathbf{r}\right) \tag{9.375} $$

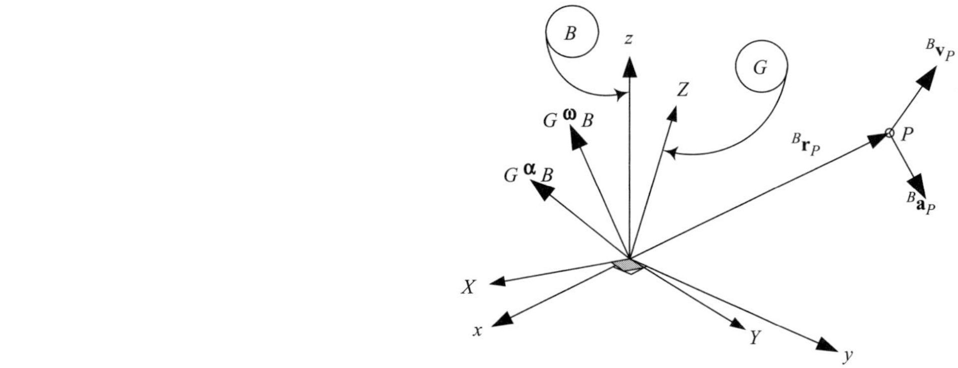

Figure 9.15 Observing a moving point P from the B-frame when the point does not follow the rotation of the frame B.

It shows that the measured local acceleration $^B\mathbf{a}$ is the resultant of the B-expression of the global acceleration $^B_G\mathbf{a}$, local tangential acceleration $-{}^B_G\boldsymbol{\alpha}_B \times {}^B\mathbf{r}$, local Coriolis acceleration $-2\,{}^B_G\boldsymbol{\omega}_B \times {}^B\mathbf{v}$, and local centripetal acceleration $-{}^B_G\boldsymbol{\omega}_B \times \left({}^B_G\boldsymbol{\omega}_B \times {}^B\mathbf{r}\right)$.

An error happens when ω and α are very small and we assume $^B\mathbf{a} = {}^B_G\mathbf{a}$. The simplification of motion of a particle close to the Earth surface is an example of such an error.

Example 570 Newton Equation in a Rotating Frame Consider a spherical rigid body (such as Earth) with a fixed point that is rotating with a constant angular velocity. The equation of motion for a moving point P on the rigid body is found by setting $^G\ddot{\mathbf{d}}_B = {}^B_G\dot{\boldsymbol{\omega}}_B = 0$ in the equation of motion of a moving point in a moving body frame:

$$\begin{aligned} ^G\mathbf{F} &= m\,{}^G\mathbf{a}_P \\ &= m\left[{}^G\ddot{\mathbf{d}}_B + {}^GR_B\left({}^B\mathbf{a}_P + 2\,{}^B_G\boldsymbol{\omega}_B \times {}^B\mathbf{v}_P + {}^B_G\dot{\boldsymbol{\omega}}_B \times {}^B\mathbf{r}_P\right)\right] \\ &\quad + m\,{}^GR_B\left[{}^B_G\boldsymbol{\omega}_B \times \left({}^B_G\boldsymbol{\omega}_B \times {}^B\mathbf{r}_P\right)\right] \end{aligned} \tag{9.376}$$

$$\begin{aligned} ^B\mathbf{F} &= m\,{}^B\mathbf{a}_P + m\,{}^B_G\boldsymbol{\omega}_B \times \left({}^B_G\boldsymbol{\omega}_B \times {}^B\mathbf{r}_P\right) + 2m\,{}^B_G\boldsymbol{\omega}_B \times {}^B\dot{\mathbf{r}}_P \\ &\neq m\,{}^B\mathbf{a}_P \end{aligned} \tag{9.377}$$

It shows that the Newton equation of motion $\mathbf{F} = m\,\mathbf{a}$ is not correct in a rotating frame. The equation of motion of a moving point can be rearranged to

$$^B\mathbf{F} - m\,{}^B_G\boldsymbol{\omega}_B \times \left({}^B_G\boldsymbol{\omega}_B \times {}^B\mathbf{r}_P\right) - 2m\,{}^B_G\boldsymbol{\omega}_B \times {}^B\mathbf{v}_P = m\,{}^B\mathbf{a}_P \tag{9.378}$$

The left-hand side of this equation is called the *effective force*,

$$\mathbf{F}_{\text{eff}} = {}^B\mathbf{F} - m\,{}^B_G\boldsymbol{\omega}_B \times \left({}^B_G\boldsymbol{\omega}_B \times {}^B\mathbf{r}_P\right) - 2m\,{}^B_G\boldsymbol{\omega}_B \times {}^B\mathbf{v}_P \tag{9.379}$$

because it seems that the particle is moving under the influence of this force.

Example 571 Motion Close to Earth Surface An applied example for a free moving point in a rotating frame B is the motion of a particle close to Earth. Such a particle is called free because it is not following the rotation of B. The angular velocity and acceleration of Earth are

$$\omega_E \approx \frac{2\pi}{24 \times 3600} \frac{366.25}{365.25} \text{ rad/s} \tag{9.380}$$

$$\alpha_E \approx 0 \tag{9.381}$$

The measured acceleration ${}^B\mathbf{a}$ of P in Earth's coordinate frame B is

$$\begin{aligned} {}^B\mathbf{a} &= {}^B_G\mathbf{a} - {}^B_G\boldsymbol{\omega}_B \times \left({}^B_G\boldsymbol{\omega}_B \times {}^B\mathbf{r}\right) - 2\,{}^B_G\boldsymbol{\omega}_B \times {}^B\mathbf{v} \\ &= {}^B\mathbf{g} - 2\,{}^B_G\boldsymbol{\omega}_B \times {}^B\mathbf{v} \end{aligned} \tag{9.382}$$

The gravitational acceleration $\mathbf{g}$ measured on the ground is

$${}^B\mathbf{g} = {}^B_G\mathbf{a} - {}^B_G\boldsymbol{\omega}_B \times \left({}^B_G\boldsymbol{\omega}_B \times {}^B\mathbf{r}\right) \tag{9.383}$$

and the Coriolis acceleration of P in the eyes of an observer in B is

$${}^B_G\mathbf{a}_{Co} = -2\,{}^B_G\boldsymbol{\omega}_B \times {}^B\mathbf{v} \tag{9.384}$$

On the surface of Earth, we have

$$\begin{aligned} R &\approx 6.3677 \times 10^6 \text{ m} \\ m_E &\approx 5.9736 \times 10^{24} \text{ kg} \\ G &\approx 6.67259 \times 10^{-11} \text{ N m}^2/\text{kg}^2 \end{aligned} \tag{9.385}$$

The order of magnitude of ${}^B_G\mathbf{a}$ and the maximum value of ${}^B_G\boldsymbol{\omega}_B \times ({}^B_G\boldsymbol{\omega}_B \times {}^B\mathbf{r})$ at the equator are

$${}^B_G\mathbf{a} = -G\frac{m_E}{R^2}\hat{u}_r \approx -9.8303\,\hat{u}_r \text{ m/s}^2 \tag{9.386}$$

$$-{}^B_G\boldsymbol{\omega}_B \times ({}^B_G\boldsymbol{\omega}_B \times {}^B\mathbf{r}) = R\omega^2\hat{u}_r = 0.03386\,\hat{u}_r \text{ m/s}^2 \tag{9.387}$$

These two terms are on the radial axis so it is reasonable if we ignore $-{}^B_G\boldsymbol{\omega}_B \times ({}^B_G\boldsymbol{\omega}_B \times {}^B\mathbf{r})$ and accept $\mathbf{g} \approx {}^B_G\mathbf{a}$. However, the Coriolis term $-2\,{}^B_G\boldsymbol{\omega}_B \times {}^B\mathbf{v}$, which is not necessarily radial, depends on the velocity of the particle. Ignorance of the Coriolis acceleration may cause P to deviate from its predicted path.

Example 572 ★ Free Fall in the Frame of Rotating Earth Consider a particle P with mass m that is falling freely from a height $h > R(\approx 6.3677 \times 10^6\text{m})$ above a point P_1 on the Earth surface. The Earth with radius R, shown in Figure 9.16, is assumed to be spherical with no air. Let us ignore the rotation of Earth about the sun and set a global coordinate frame $G(OXYZ)$ at the Earth center. Another coordinate frame $E(Oxyz)$ is attached to Earth such that its z-axis is coincident with the global Z-axis. The E-frame turns about the Z-axis once every 24 h. We indicate the point P_1 on the Earth surface by the longitude φ and latitude λ. Let us also attach a local coordinate

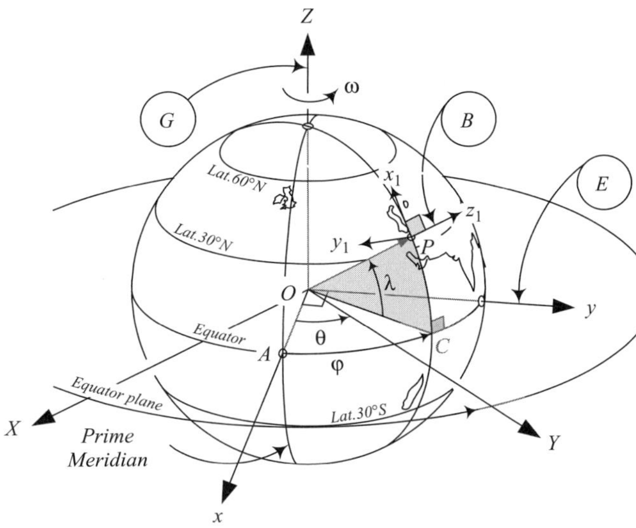

Figure 9.16 The free-fall problem of a particle on Earth needs three coordinate frames: a global frame $G(OXYZ)$ at the Earth center, an Earth frame $E(Oxyz)$ attached to Earth, and a local coordinate frame $B\,(Ox_1y_1z_1)$ on Earth.

frame $B(Ox_1y_1z_1)$ at P_1 such that its z_1-axis points upward to the local frame and its x_1-axis points to the north pole.

The transformation matrices between B and E are

$$
\begin{aligned}
{}^{B}R_{E} &= R_{z_1,\pi}\, R_{y_1,(\pi/2-\lambda)}\, R_{z_1,\varphi} \\
&= \begin{bmatrix} -\cos\varphi\sin\lambda & -\sin\lambda\sin\varphi & \cos\lambda \\ \sin\varphi & -\cos\varphi & 0 \\ \cos\lambda\cos\varphi & \cos\lambda\sin\varphi & \sin\lambda \end{bmatrix}
\end{aligned}
\tag{9.388}
$$

$$
{}^{E}\mathbf{d} = {}^{B}R_{E}^{\mathrm{T}}\, {}^{B}\mathbf{d} = {}^{B}R_{E}^{\mathrm{T}} \begin{bmatrix} 0 \\ 0 \\ R \end{bmatrix} = \begin{bmatrix} R\cos\lambda\cos\varphi \\ R\cos\lambda\sin\varphi \\ R\sin\lambda \end{bmatrix}
\tag{9.389}
$$

$$
\begin{aligned}
{}^{E}T_{B} &= \begin{bmatrix} {}^{E}R_{B} & {}^{E}\mathbf{d} \\ 0 & 1 \end{bmatrix} = \begin{bmatrix} {}^{B}R_{E}^{\mathrm{T}} & {}^{E}\mathbf{d} \\ 0 & 1 \end{bmatrix} \\
&= \begin{bmatrix} -\cos\varphi\sin\lambda & \sin\varphi & \cos\lambda\cos\varphi & R\cos\lambda\cos\varphi \\ -\sin\lambda\sin\varphi & -\cos\varphi & \cos\lambda\sin\varphi & R\cos\lambda\sin\varphi \\ \cos\lambda & 0 & \sin\lambda & R\sin\lambda \\ 0 & 0 & 0 & 1 \end{bmatrix}
\end{aligned}
\tag{9.390}
$$

$$
{}^{B}T_{E} = {}^{E}T_{B}^{-1} = \begin{bmatrix} -\cos\varphi\sin\lambda & -\sin\lambda\sin\varphi & \cos\lambda & 0 \\ \sin\varphi & -\cos\varphi & 0 & 0 \\ \cos\lambda\cos\varphi & \cos\lambda\sin\varphi & \sin\lambda & -R \\ 0 & 0 & 0 & 1 \end{bmatrix}
\tag{9.391}
$$

The acceleration of a moving point close to the Earth surface is found in Equation (9.382):

$$ {}^{E}\mathbf{a} = {}^{E}\mathbf{g} - 2\,{}^{E}_{G}\boldsymbol{\omega}_{E} \times {}^{E}\mathbf{v} \tag{9.392} $$

Using

$$ {}^{E}_{G}\boldsymbol{\omega}_{E} = \omega_{E}\hat{k} \tag{9.393} $$

$$ {}^{E}\mathbf{r} = \begin{bmatrix} x \\ y \\ z \end{bmatrix} \qquad {}^{E}\mathbf{v} = \begin{bmatrix} \dot{x} \\ \dot{y} \\ \dot{z} \end{bmatrix} \qquad {}^{E}\mathbf{a} = \begin{bmatrix} \ddot{x} \\ \ddot{y} \\ \ddot{z} \end{bmatrix} \tag{9.394} $$

$$ {}^{B}\mathbf{g} = -g_0\hat{k} \tag{9.395} $$

we find

$$ \begin{bmatrix} \ddot{x} \\ \ddot{y} \\ \ddot{z} \end{bmatrix} = {}^{E}\mathbf{g} - 2\begin{bmatrix} 0 \\ 0 \\ \omega_E \end{bmatrix} \times \begin{bmatrix} \dot{x} \\ \dot{y} \\ \dot{z} \end{bmatrix} = \begin{bmatrix} 2\dot{y}\omega_E - g_0\cos\lambda\cos\varphi \\ -2\dot{x}\omega_E - g_0\cos\lambda\sin\varphi \\ -g_0\sin\lambda \end{bmatrix} \tag{9.396} $$

where

$$ {}^{E}\mathbf{g} = {}^{E}R_{B}\,{}^{B}\mathbf{g} = {}^{E}R_{B}\begin{bmatrix} 0 \\ 0 \\ -g_0 \end{bmatrix} = \begin{bmatrix} -g_0\cos\lambda\cos\varphi \\ -g_0\cos\lambda\sin\varphi \\ -g_0\sin\lambda \end{bmatrix} \tag{9.397} $$

If the particle is released from rest at a height $h > R$, then its initial conditions are

$$ {}^{B}\mathbf{r}(0) = h\hat{k}_1 = \begin{bmatrix} 0 & 0 & h & 1 \end{bmatrix}^{\mathrm{T}} \tag{9.398} $$

$$ {}^{E}\mathbf{r}\,(0) = {}^{E}T_{B}\,{}^{B}\mathbf{r}\,(0) = \begin{bmatrix} x_0 \\ y_0 \\ z_0 \\ 1 \end{bmatrix} = \begin{bmatrix} (R+h)\cos\varphi\cos\lambda \\ (R+h)\sin\varphi\cos\lambda \\ (R+h)\sin\lambda \\ 1 \end{bmatrix} \tag{9.399} $$

$$ {}^{E}\mathbf{v}\,(0) = \begin{bmatrix} \dot{x}_0 & \dot{y}_0 & \dot{z}_0 \end{bmatrix}^{\mathrm{T}} = \begin{bmatrix} 0 & 0 & 0 \end{bmatrix}^{\mathrm{T}} \tag{9.400} $$

Integrating Equation (9.396) provides

$$ \begin{bmatrix} \dot{x} \\ \dot{y} \\ \dot{z} \end{bmatrix} = \begin{bmatrix} 2y\omega_E - g_0 t\cos\lambda\cos\varphi - 2y_0\omega_E \\ -2x\omega_E - g_0 t\cos\lambda\sin\varphi + 2x_0\omega_E \\ -g_0 t\sin\lambda \end{bmatrix} \tag{9.401} $$

Substituting the first and second equations of (9.401) in the second and first equations of (9.396) separates the variables:

$$ \ddot{x} + 4\omega_E^2 x = -2\omega_E g_0 t\cos\lambda\sin\varphi - g_0\cos\lambda\cos\varphi + 4\omega_E^2\,(R+h)\cos\varphi\cos\lambda \tag{9.402} $$

$$ \ddot{y} + 4\omega_E^2 y = 2\omega_E g_0 t\cos\lambda\cos\varphi - g_0\cos\lambda\sin\varphi + 4\omega_E^2\,(R+h)\sin\varphi\cos\lambda \tag{9.403} $$

The general solutions of (9.402) and (9.403) are

$$x = C_1 \cos 2\omega_E t + C_2 \sin 2\omega_E t - \frac{g_0}{4\omega_E^2}(2\omega_E t \sin\varphi + \cos\varphi)\cos\lambda + (R+h)\cos\varphi\cos\lambda \tag{9.404}$$

$$y = C_3 \cos 2\omega_E t + C_4 \sin 2\omega_E t + \frac{g_0}{4\omega_E^2}(2\omega_E t \cos\varphi - \sin\varphi)\cos\lambda + (R+h)\sin\varphi\cos\lambda \tag{9.405}$$

We determine the coefficients C_1, C_2, C_3, C_4 from initial conditions (9.399) and (9.400), and therefore

$$x = \frac{g_0}{4\omega_E^2}\cos\varphi\cos\lambda\cos 2\omega_E t + \frac{g_0}{4\omega_E^2}\sin\varphi\cos\lambda\sin 2\omega_E t - \frac{g_0}{4\omega_E^2}(2\omega_E t \sin\varphi + \cos\varphi)\cos\lambda + (R+h)\cos\varphi\cos\lambda \tag{9.406}$$

$$y = \frac{g_0}{4\omega_E^2}\sin\varphi\cos\lambda\cos 2\omega_E t - \frac{g_0}{4\omega_E^2}\cos\varphi\cos\lambda\sin 2\omega_E t + \frac{g_0}{4\omega_E^2}(2\omega_E t \cos\varphi - \sin\varphi)\cos\lambda + (R+h)\sin\varphi\cos\lambda \tag{9.407}$$

The third equation of (9.401) is independent of the others and can be integrated independently:

$$z = -\tfrac{1}{2}g_0 t^2 \sin\lambda + (R+h)\sin\lambda \tag{9.408}$$

Equations (9.406)–(9.408) indicate the components of the position vector of the falling particle in the E-frame. We may transform ${}^E\mathbf{r}$ from E to B to examine the components of ${}^B\mathbf{r}$ for an observer in the B-frame:

$${}^B\mathbf{r} = {}^BT_E\,{}^E\mathbf{r} \tag{9.409}$$

$${}^B\mathbf{r} = \begin{bmatrix} -\dfrac{g_0}{8\omega_E^2}\left(\cos 2\omega_E t + 2\omega_E^2 t^2 - 1\right)\sin 2\lambda \\ \dfrac{g_0}{4\omega_E^2}(\sin 2\omega_E t - 2\omega_E t)\cos\lambda \\ \dfrac{g_0}{4\omega_E^2}\left(\cos 2\omega_E t + 2\omega_E^2 t^2 - 1\right)\cos^2\lambda - \dfrac{1}{2}g_0 t^2 + h \end{bmatrix} \tag{9.410}$$

Interestingly, the components of ${}^B\mathbf{r}$ are independent of φ. It indicates that the path of a falling particle at λ are similar for every φ in a local frame B. As a reference, let us recall the free fall of a particle on a flat, level, and stationary ground as

$$x_1 = 0 \qquad y_1 = 0 \qquad z_1 = -\tfrac{1}{2}g_0 t^2 + h \tag{9.411}$$

Recalling that the latitude λ is the principal parameter to cause the different path of a falling particle, we can determine the minimum and maximum deviation from a straight line. The deviation is minimum at the north pole,

$$^{B}\mathbf{r}(t) = \begin{bmatrix} 0 \\ 0 \\ -\frac{1}{2}g_0 t^2 + h \end{bmatrix} \qquad \lambda = 90\,\text{deg} \tag{9.412}$$

and maximum at the equator,

$$^{B}\mathbf{r}(t) = \begin{bmatrix} 0 \\ \dfrac{g_0}{4\omega_E^2}\left(\sin 2\omega_E t - 2\omega_E t\right) \\ \dfrac{g_0}{4\omega_E^2}\left(\cos 2\omega_E t + 2\omega_E^2 t^2\right) - \dfrac{1}{2}g_0 t^2 + h \end{bmatrix} \qquad \lambda = 0 \tag{9.413}$$

Let us examine $^{B}\mathbf{r}(t)$ for the following data:

$$\omega_E \approx 7.2921 \times 10^{-5}\,\text{rad/s}$$
$$g_0 \approx 9.81\,\text{m/s}^2$$
$$h = 100\,\text{m} \tag{9.414}$$

The particle hits the ground at time t_h when $z = 0$. For a nonrotating Earth, the time t_h is given as

$$t_0 = 4.5152\,\text{s} \qquad \omega_E = 0 \tag{9.415}$$

Figure 9.17 illustrates $t_h - t_0$ at different altitudes λ from the south pole to the north pole.

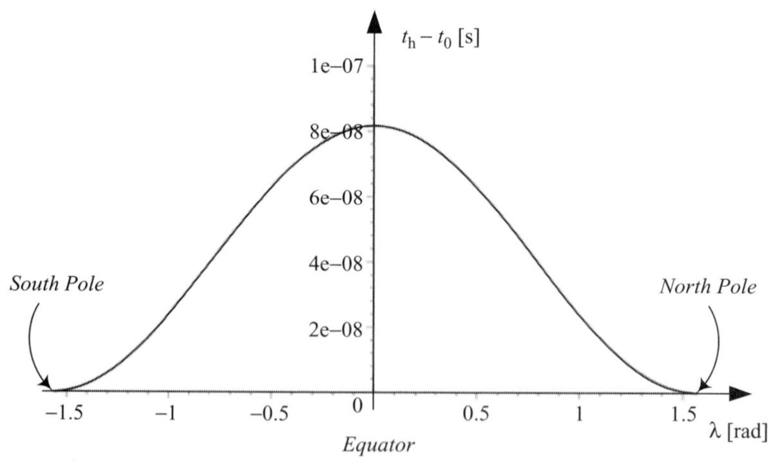

Figure 9.17 Falling time of a particle from $h = 100\,\text{m}$ at different altitudes λ from the south pole to the north pole.

The x-component of ${}^{B}\mathbf{r}(t)$ is depicted in Figure 9.18. The x-axis points North, so a falling body moves south in the northern hemisphere and north in the southern hemisphere. Therefore, the equator acts as an attractive line to the falling objects. The x-displacement is zero at the north pole, south pole, and equator and maximum at $\lambda = \pm 45$ deg.

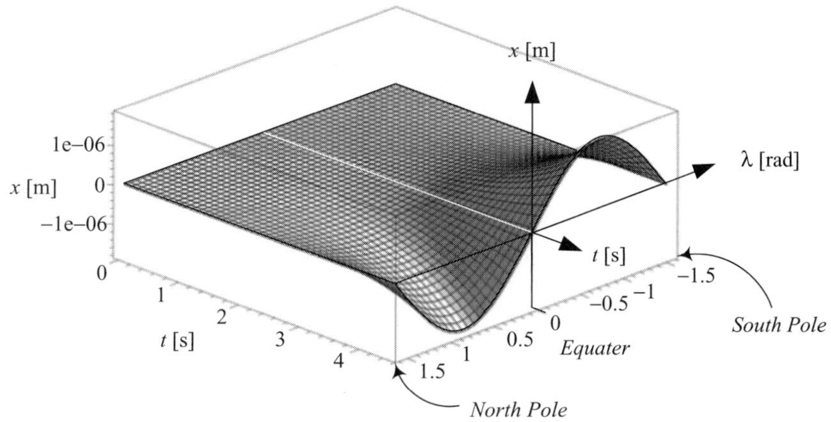

Figure 9.18 The x-component of ${}^{B}\mathbf{r}(t)$ indicates the displacement of a falling body in the north direction.

The y-component of ${}^{B}\mathbf{r}(t)$ is depicted in Figure 9.19. The y-axis points west, so a falling body moves east. The y-displacement is zero at the north and south poles and a maximum at the equator, $\lambda = 0$. The combined x- and y-displacements of the falling body are illustrated in Figure 9.20 at two points in the northern and southern hemispheres.

The z-component of ${}^{B}\mathbf{r}(t)$ is shown in Figures 9.21.

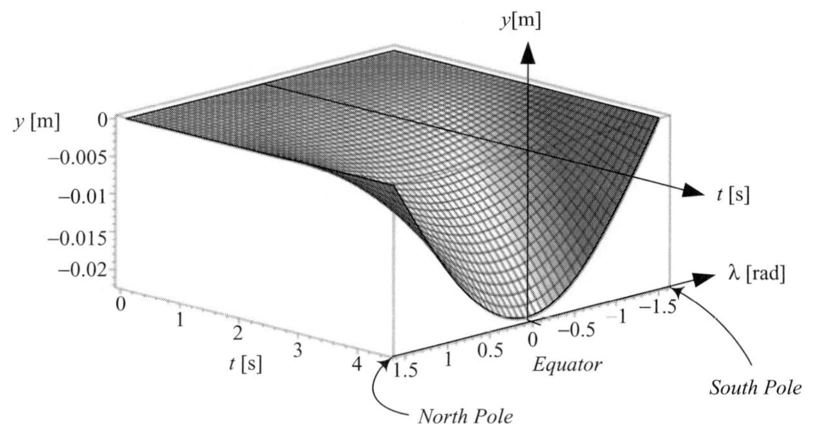

Figure 9.19 The y-component of ${}^{B}\mathbf{r}(t)$ indicates the displacement of a falling body in the east direction.

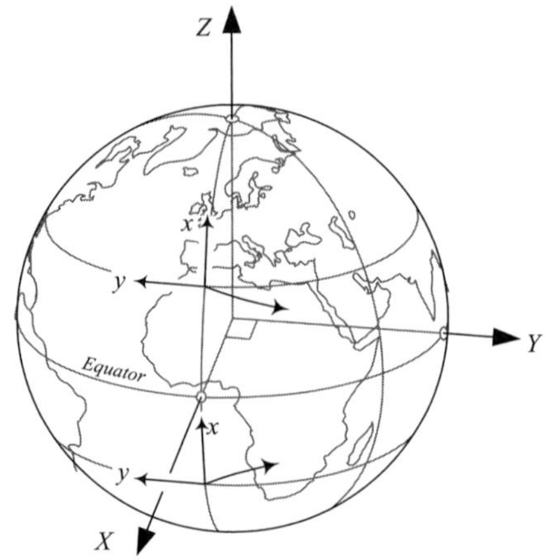

Figure 9.20 The x- and y-displacements of a falling body in the northen and southern hemispheres.

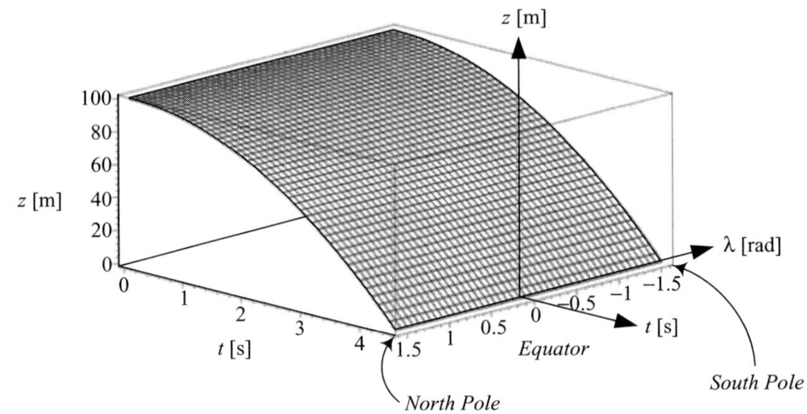

Figure 9.21 The z-component of ${}^{B}\mathbf{r}(t)$ as a function of time t and attitude λ.

Example 573 ★ Global Path of a Free Fall on Rotating Earth The global and Earth coordinate frames $G(OXYZ)$ and $E(Oxyz)$ are shown in Figure 9.16. The Earth frame E is turning with angular velocity ${}_{G}\boldsymbol{\omega}_{E} = \dot{\theta}\hat{K}$ in G. The equations of acceleration of a falling particle are determined in Equation (9.396) and its path of motion in B is derived in Equations (9.406)–(9.408). Employing the transformation matrix between E and G,

$$
{}^{G}R_{E} = \begin{bmatrix} \cos\theta & -\sin\theta & 0 \\ \sin\theta & \cos\theta & 0 \\ 0 & 0 & 1 \end{bmatrix} \tag{9.416}
$$

we can determine the acceleration and path of motion in the G-frame:

$$
{}^{G}\mathbf{r} = {}^{G}R_{E}\ {}^{E}\mathbf{r} \tag{9.417}
$$

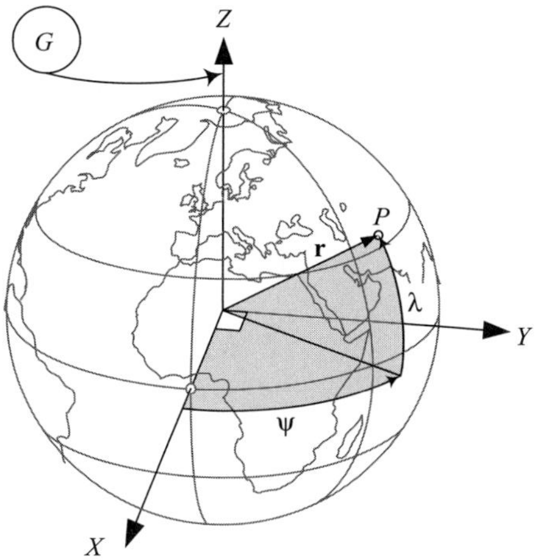

Figure 9.22 Spherical variables ψ, λ, and r used to show a point in the G-frame.

It is simpler if we analyze the free-fall problem of a particle P in the G-frame by integrating its G-acceleration. Let us use a spherical coordinate system and express the Cartesian coordinates of P using the spherical variables ψ, λ, and r as shown in Figure 9.22:

$$ {}^{G}\mathbf{r} = \begin{bmatrix} r\cos\lambda\cos\psi \\ r\cos\lambda\sin\psi \\ r\sin\lambda \end{bmatrix} \tag{9.418} $$

The only acceleration of the free-falling particle in G is the gravitational acceleration ${}^{G}\mathbf{g}$ that is along $-{}^{G}\mathbf{r}$:

$$ {}^{G}\mathbf{a} = \begin{bmatrix} \ddot{X} \\ \ddot{Y} \\ \ddot{Z} \end{bmatrix} = {}^{G}\mathbf{g} = \begin{bmatrix} -g_0\cos\lambda\cos\psi \\ -g_0\cos\lambda\sin\psi \\ -g_0\sin\lambda \end{bmatrix} \tag{9.419} $$

If the particle is initially motionless with respect to E and is released from height h from the surface of Earth, then its initial conditions in G are

$$ {}^{G}\mathbf{r}\,(0) = \begin{bmatrix} X_0 \\ Y_0 \\ Z_0 \end{bmatrix} = \begin{bmatrix} (R+h)\cos\lambda_0\cos\psi_0 \\ (R+h)\cos\lambda_0\sin\psi_0 \\ (R+h)\sin\lambda_0 \end{bmatrix} \tag{9.420} $$

$$ \begin{aligned} {}^{G}\mathbf{v}\,(0) &= {}^{G}_{B}\boldsymbol{\omega}_G \times {}^{G}\mathbf{r}\,(0) = \begin{bmatrix} 0 \\ 0 \\ -\omega_E \end{bmatrix} \times \begin{bmatrix} (R+h)\cos\lambda_0\cos\psi_0 \\ (R+h)\cos\lambda_0\sin\psi_0 \\ (R+h)\sin\lambda_0 \end{bmatrix} \\ &= \begin{bmatrix} \omega_E\,(\cos\lambda_0\sin\psi_0)\,(R+h) \\ -\omega_E\,(\cos\lambda_0\cos\psi_0)\,(R+h) \\ 0 \end{bmatrix} = \begin{bmatrix} \dot{X}_0 \\ \dot{Y}_0 \\ \dot{Z}_0 \end{bmatrix} \end{aligned} \tag{9.421} $$

Integrating Equation (9.419) from λ_0, ψ_0,

$$ {}^{G}\mathbf{v} = \begin{bmatrix} -g_0 t \cos\lambda_0 \cos\psi_0 \\ -g_0 t \cos\lambda_0 \sin\psi_0 \\ -g_0 t \sin\lambda_0 \end{bmatrix} + \begin{bmatrix} \dot{X}_0 \\ \dot{Y}_0 \\ \dot{Z}_0 \end{bmatrix} \tag{9.422} $$

$$ {}^{G}\mathbf{r} = \begin{bmatrix} -\frac{1}{2} g_0 t^2 \cos\lambda_0 \cos\psi_0 \\ -\frac{1}{2} g_0 t^2 \cos\lambda_0 \sin\psi_0 \\ -\frac{1}{2} g_0 t^2 \sin\lambda_0 \end{bmatrix} + \begin{bmatrix} \dot{X}_0 \\ \dot{Y}_0 \\ \dot{Z}_0 \end{bmatrix} t + \begin{bmatrix} X_0 \\ Y_0 \\ Z_0 \end{bmatrix} \tag{9.423} $$

and implementing the initial conditions provide the path of motion in the G-frame:

$$ {}^{G}\mathbf{r} = \begin{bmatrix} \left[-\frac{1}{2} g_0 t^2 \cos\psi_0 + (R+h)\left(\omega_E t \sin\psi_0 + \cos\psi_0\right)\right] \cos\lambda_0 \\ \left[-\frac{1}{2} g_0 t^2 \sin\psi_0 - (R+h)\left(\omega_E t \cos\psi_0 - \sin\psi_0\right)\right] \cos\lambda_0 \\ (R+h)\sin\lambda_0 - \frac{1}{2} t^2 g_0 \sin\lambda \end{bmatrix} \tag{9.424} $$

The particle hits the ground at t_h when we have

$$ \left| {}^{G}\mathbf{r} \right| = R \tag{9.425} $$

This equation results in the following equation to determine t_h:

$$ R^2 = (R+h)^2 \left[\frac{1}{4} g_0^2 t_h^4 - \left(g_0 - \omega_E^2 \cos^2\lambda_0 \right) t_h^2 + 1 \right] \tag{9.426} $$

As expected, t_h is independent of ψ_0.

We can substitute t_h from (9.426) in (9.424) and determine the global coordinates of the particle when it hits the ground. Comparing the hitting coordinate with the projection point P_1 of the initial coordinates of the particle on the ground,

$$ {}^{G}\mathbf{r}_{P_1}(0) = \begin{bmatrix} R\cos\lambda_0 \cos\psi_0 \\ R\cos\lambda_0 \sin\psi_0 \\ R\sin\lambda_0 \end{bmatrix} \tag{9.427} $$

we can determine how much the particle deviates from point P_1 because of the initial velocity. If we calculate the global displacement of P_1 by Earth rotation, we can calculate the hitting coordinate in the E-frame.

Example 574 ★ Better Model of Free Fall on Rotating Earth The gravitational attraction varies with distance from the center of Earth. If ${}^{B}\mathbf{g} = -g_0 \hat{k}_1$ on the surface of Earth, then the gravitational acceleration $\mathbf{g}$ at a point ${}^{E}\mathbf{r}$ is

$$ \begin{aligned} {}^{E}\mathbf{g} &= -g_0 \frac{R^2}{x^2+y^2+z^2} \, {}^{E}R_B \, \hat{k}_1 \\ &= -g_0 \frac{R^2}{x^2+y^2+z^2} \begin{bmatrix} -\cos\lambda\cos\varphi \\ -\cos\lambda\sin\varphi \\ -\sin\lambda \end{bmatrix} \end{aligned} \tag{9.428} $$

where $E(Oxyz)$ is the Earth coordinate frame at the Earth center and $B(Ox_1y_1z_1)$ is a local coordinate frame on the Earth surface, as shown in Figure 9.16. Using

(9.428), (9.388), (9.393), and (9.394), we can modify Equations (9.382) to get a better model describing the acceleration of a moving point close to the Earth surface:

$$^{E}\mathbf{a} = {}^{E}\mathbf{g} - 2\,{}^{E}_{G}\boldsymbol{\omega}_{E} \times {}^{E}\mathbf{v} \tag{9.429}$$

$$\begin{aligned}\begin{bmatrix}\ddot{x}\\ \ddot{y}\\ \ddot{z}\end{bmatrix} &= \frac{-g_0 R^2}{x^2+y^2+z^2}\begin{bmatrix}-\cos\lambda\cos\varphi\\ -\cos\lambda\sin\varphi\\ -\sin\lambda\end{bmatrix} - 2\begin{bmatrix}0\\ 0\\ \omega_E\end{bmatrix}\times\begin{bmatrix}\dot{x}\\ \dot{y}\\ \dot{z}\end{bmatrix}\\ &= \frac{-g_0 R^2}{x^2+y^2+z^2}\begin{bmatrix}2\dot{y}\omega_E - \cos\lambda\cos\varphi\\ -2\dot{x}\omega_E - \cos\lambda\sin\varphi\\ -\sin\lambda\end{bmatrix}\end{aligned} \tag{9.430}$$

Example 575 ★ Free Fall in a Local Frame on Rotating Earth Consider the free fall of a particle P with mass m from a height $h > R(\approx 6.3677 \times 10^6 \text{ m})$ above a point P_1 on the Earth surface. Here P_1 is indicated by the longitude φ and latitude λ. The spherical Earth with no air is shown in Figure 9.16. We set up a global coordinate frame $G(OXYZ)$ at the Earth center. Another frame $E(Oxyz)$ is attached to Earth such that its z-axis is coincident with the Z-axis. We also attach a local coordinate frame $B(Ox_1y_1z_1)$ at P_1 such that its z_1-axis points upward to the local frame and its x_1-axis points to the north pole.

The classical method of analysis of the motion of a free-falling particle on a rotating Earth is expression of the equations of motion in the Earth frame E:

$$^{E}\mathbf{a} = {}^{E}\mathbf{g} - 2\,{}^{E}_{G}\boldsymbol{\omega}_{E} \times {}^{E}\mathbf{v} \tag{9.431}$$

After integrating the equations, we determine the path of the particle in the E-frame. Using coordinate transformation, we may determine the path in any local frame such as $B(Ox_1y_1z_1)$ at P_1. However, because B and E are fixed relatively, we can geometrically transform both sides of Equation (9.431) to B and determine the equations of motion in the B-frame.

Let us show the acceleration, velocity, and position vectors of P in the B-frame by

$$^{B}\mathbf{a} = \begin{bmatrix}\ddot{x}_1\\ \ddot{y}_1\\ \ddot{z}_1\end{bmatrix} \qquad {}^{B}\mathbf{v} = \begin{bmatrix}\dot{x}_1\\ \dot{y}_1\\ \dot{z}_1\end{bmatrix} \qquad {}^{B}\mathbf{r} = \begin{bmatrix}x_1\\ y_1\\ z_1\end{bmatrix} \tag{9.432}$$

and express the equation of motion (9.431) in B:

$$^{B}\mathbf{a} = {}^{B}\mathbf{g} - 2\,{}^{B}_{G}\boldsymbol{\omega}_{E} \times {}^{B}\mathbf{v} \tag{9.433}$$

$$\begin{aligned}\begin{bmatrix}\ddot{x}_1\\ \ddot{y}_1\\ \ddot{z}_1\end{bmatrix} &= \begin{bmatrix}0\\ 0\\ -g_0\end{bmatrix} - 2\begin{bmatrix}\omega_E\cos\lambda\\ 0\\ \omega_E\sin\lambda\end{bmatrix}\times\begin{bmatrix}\dot{x}_1\\ \dot{y}_1\\ \dot{z}_1\end{bmatrix}\\ &= \begin{bmatrix}2\dot{y}_1\omega_E\sin\lambda\\ 2\dot{z}_1\omega_E\cos\lambda - 2\dot{x}_1\omega_E\sin\lambda\\ -2\dot{y}_1\omega_E\cos\lambda - g_0\end{bmatrix}\end{aligned} \tag{9.434}$$

The transformation between B and E is

$$ {}^{B}R_{E} = \begin{bmatrix} -\cos\varphi\sin\lambda & -\sin\lambda\sin\varphi & \cos\lambda \\ \sin\varphi & -\cos\varphi & 0 \\ \cos\lambda\cos\varphi & \cos\lambda\sin\varphi & \sin\lambda \end{bmatrix} \tag{9.435} $$

However, all points at an altitude λ are mathematically equivalent, and we can simplify the problem to $\varphi = 0$ without losing the generality of the solution. The transformation matrix will then be

$$ {}^{B}R_{E} = \begin{bmatrix} -\sin\lambda & 0 & \cos\lambda \\ 0 & -1 & 0 \\ \cos\lambda & 0 & \sin\lambda \end{bmatrix} \tag{9.436} $$

and therefore,

$$ {}_{G}^{B}\boldsymbol{\omega}_{E} = {}^{B}R_{E}\, {}_{G}^{E}\boldsymbol{\omega}_{E} $$

$$ \begin{bmatrix} \omega_E\cos\lambda \\ 0 \\ \omega_E\sin\lambda \end{bmatrix} = \begin{bmatrix} -\sin\lambda & 0 & \cos\lambda \\ 0 & -1 & 0 \\ \cos\lambda & 0 & \sin\lambda \end{bmatrix} \begin{bmatrix} 0 \\ 0 \\ \omega_E \end{bmatrix} \tag{9.437} $$

Assuming the initial conditions

$$ {}^{B}\mathbf{r} = h\hat{k}_1 = \begin{bmatrix} 0 \\ 0 \\ h \end{bmatrix} \qquad {}^{B}\mathbf{v} = \begin{bmatrix} 0 \\ 0 \\ 0 \end{bmatrix} \tag{9.438} $$

we integrate Equation (9.434) to get

$$ \begin{bmatrix} \dot{x}_1 \\ \dot{y}_1 \\ \dot{z}_1 \end{bmatrix} = \begin{bmatrix} 2y_1\omega_E\sin\lambda \\ 2z_1\omega_E\cos\lambda - 2x_1\omega_E\sin\lambda \\ -2y_1\omega_E\cos\lambda - g_0 t \end{bmatrix} \tag{9.439} $$

To separate the variables, we can substitute $\dot{x}_1$ and $\dot{z}_1$ in (9.434) and make its second equation independent:

$$ \ddot{y}_1 + 4\omega_E^2 y_1 = -2\omega_E g_0 t\cos\lambda \tag{9.440} $$

The general solution of (9.440) is

$$ y_1 = C_1\cos 2\omega_E t + C_2\sin 2\omega_E t - \frac{g_0 t\cos\lambda}{2\omega_E} \tag{9.441} $$

where

$$ C_1 = 0 \qquad C_2 = \frac{g_0\cos\lambda}{4\omega_E^2} \tag{9.442} $$

and therefore,

$$ y_1 = \frac{g_0}{4\omega_E^2}\left(\sin 2\omega_E t - 2\omega_E t\right)\cos\lambda \tag{9.443} $$

Now we can integrate the first and third equations of (9.434) to find x_1 and z_1:

$$\begin{aligned} x_1 &= \int 2\left(\frac{g_0}{4\omega_E^2}\left(\sin 2\omega_E t - 2\omega_E t\right)\cos\lambda\right)\omega_E \sin\lambda \, dt \\ &= -\frac{g_0}{8\omega_E^2}\left(\cos 2\omega_E t + 2\omega_E^2 t^2 - 1\right)\sin 2\lambda \end{aligned} \tag{9.444}$$

$$\begin{aligned} z_1 &= \int \left[-2\left(\frac{g_0}{4\omega_E^2}\left(\sin 2\omega_E t - 2\omega_E t\right)\cos\lambda\right)\omega_E \cos\lambda - g_0 t\right] dt \\ &= \frac{g_0}{4\omega_E^2}\left(\cos 2\omega_E t + 2\omega_E^2 t^2 - 1\right)\cos^2\lambda - \tfrac{1}{2} g_0 t^2 + h \end{aligned} \tag{9.445}$$

So, the B-expression of the position vector of the falling particle is

$$ {}^{B}\mathbf{r}(t) = \begin{bmatrix} -\dfrac{g_0}{8\omega_E^2}\left(\cos 2\omega_E t + 2\omega_E^2 t^2 - 1\right)\sin 2\lambda \\ \dfrac{g_0}{4\omega_E^2}\left(\sin 2\omega_E t - 2\omega_E t\right)\cos\lambda \\ \dfrac{g_0}{4\omega_E^2}\left(\cos 2\omega_E t + 2\omega_E^2 t^2 - 1\right)\cos^2\lambda - \dfrac{1}{2} g_0 t^2 + h \end{bmatrix} \tag{9.446}$$

These equations are the same as (9.410).

Example 576 Applied Acceleration When P Is Following B Let us summarize the Coriolis effect when a moving point follows the rotation of the local frame B. In this case there must be continued contact between B and P. So, practically, we must have the B-expression of ${}^{B}\mathbf{r}_P$, ${}^{B}\mathbf{v}_P$, ${}^{B}\mathbf{a}_P$. Figure 9.23 illustrates a moving point P on a given path in a rotating B-frame while B is turning in the global G-frame.

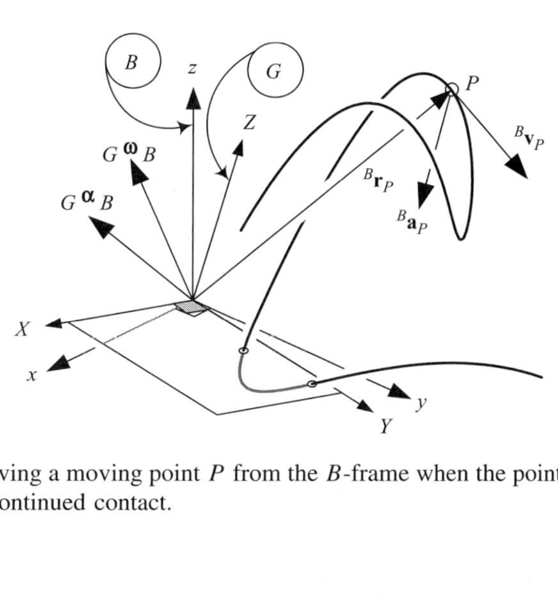

Figure 9.23 Observing a moving point P from the B-frame when the point follows the rotation of the frame B by continued contact.

We, as the observers, are standing in B and watching the motion of P. The best way to analyze the motion of P is using the B-expression of the G-acceleration, ${}^B_G\mathbf{a}$:

$$ {}^B_G\mathbf{a} = {}^B\mathbf{a} + {}^B_G\boldsymbol{\alpha}_B \times {}^B\mathbf{r} + 2\,{}^B_G\boldsymbol{\omega}_B \times {}^B\mathbf{v} + {}^B_G\boldsymbol{\omega}_B \times \left({}^B_G\boldsymbol{\omega}_B \times {}^B\mathbf{r}\right) \tag{9.447} $$

The Coriolis force ${}^B\mathbf{F}_{Co} = m\,{}^B_G\mathbf{a}_{Co} = 2m\,{}^B_G\boldsymbol{\omega}_B \times {}^B\mathbf{v}$ is the required force on m to turn it with B. The force ${}^B\mathbf{F}_{Co}$ must be provided by the path.

Example 577 A Train Heading South Consider a train moving on a meridian with a constant speed $v = R\dot{\lambda}$ from the north pole toward the equator, as shown in Figure 9.24(a). The position of the train on Earth is determined by the latitude φ and altitude λ. The position of Earth frame $E(x, y, z)$ in the global frame $G(X, Y, Z)$ is determined by the rotation θ about the Z-axis.

The position vector of the train and the angular velocity of Earth are

$$ {}^E\mathbf{r} = \begin{bmatrix} R\cos\lambda\cos\varphi \\ R\cos\lambda\sin\varphi \\ R\sin\lambda \end{bmatrix} \qquad {}^E_G\boldsymbol{\omega}_E = \begin{bmatrix} 0 \\ 0 \\ \omega_E \end{bmatrix} \tag{9.448} $$

The velocity vector of the train is

$$ {}^E\mathbf{v} = {}^E\dot{\boldsymbol{\lambda}} \times {}^E\mathbf{r} = -\dot{\lambda}\begin{bmatrix} \sin\varphi \\ -\cos\varphi \\ 0 \end{bmatrix} \times \begin{bmatrix} R\cos\lambda\cos\varphi \\ R\cos\lambda\sin\varphi \\ R\sin\lambda \end{bmatrix} $$

$$ = \dot{\lambda}\begin{bmatrix} R\cos\varphi\sin\lambda \\ R\sin\varphi\sin\lambda \\ -R\cos\lambda \end{bmatrix} = v\begin{bmatrix} \cos\varphi\sin\lambda \\ \sin\varphi\sin\lambda \\ -\cos\lambda \end{bmatrix} \tag{9.449} $$

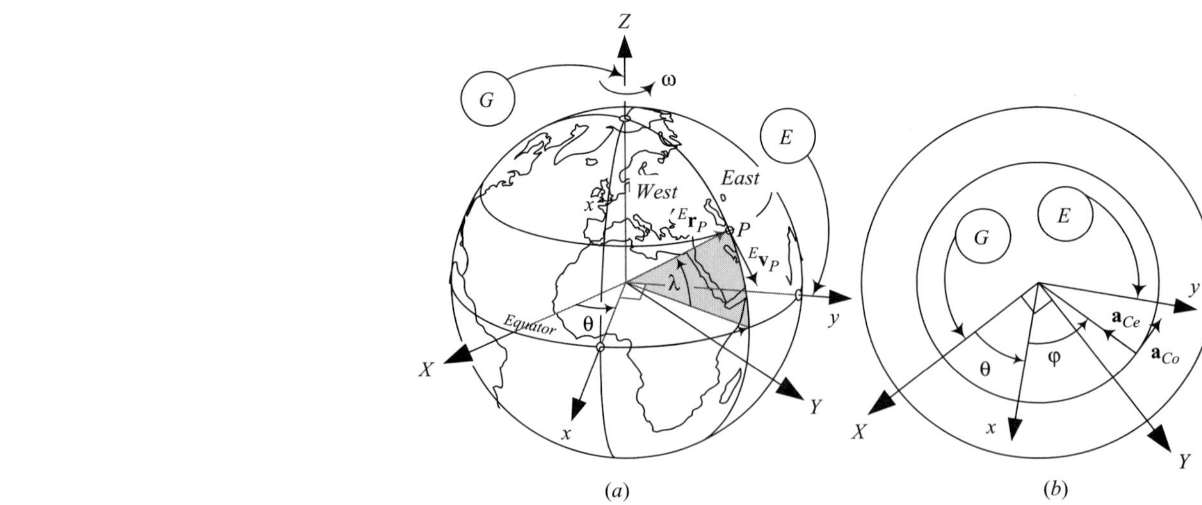

Figure 9.24 A train is moving on a meridian with a constant speed heading south.

Because ${}^{E}_{G}\boldsymbol{\omega}_{E} = const$ and ${}^{E}\dot{\lambda} = const$, the acceleration vector of the train is

$$
\begin{aligned}
{}^{E}_{G}\mathbf{a} &= {}^{B}\mathbf{a} + 2\,{}^{E}_{G}\boldsymbol{\omega}_{E} \times {}^{E}\mathbf{v} + {}^{E}_{G}\boldsymbol{\omega}_{E} \times \left({}^{E}_{G}\boldsymbol{\omega}_{B} \times {}^{E}\mathbf{r}\right) \\
&= {}^{E}\dot{\lambda} \times \left({}^{E}\dot{\lambda} \times {}^{E}\mathbf{r}\right) + 2\,{}^{E}_{G}\boldsymbol{\omega}_{E} \times {}^{E}\mathbf{v} + {}^{E}_{G}\boldsymbol{\omega}_{E} \times \left({}^{E}_{G}\boldsymbol{\omega}_{B} \times {}^{E}\mathbf{r}\right) \\
&= -\frac{v^2}{R}\begin{bmatrix} \cos\lambda\cos\varphi \\ \cos\lambda\sin\varphi \\ \sin\lambda \end{bmatrix} + 2\begin{bmatrix} -v\omega_E \sin\lambda\sin\varphi \\ v\omega_E\cos\varphi\sin\lambda \\ 0 \end{bmatrix} \\
&\quad + \begin{bmatrix} -R\omega_E^2\cos\lambda\cos\varphi \\ -R\omega_E^2\cos\lambda\sin\varphi \\ 0 \end{bmatrix}
\end{aligned}
\tag{9.450}
$$

The first term is the centripetal acceleration ${}^{B}_{G}\mathbf{a}_{C\lambda}$ due to movement in a big circle on Earth and toward the center of Earth. The second term is the Coriolis acceleration ${}^{B}_{G}\mathbf{a}_{Co}$. The third term is the centripetal acceleration ${}^{B}_{G}\mathbf{a}_{Ce}$ due to rotation of Earth. The last two terms are in a plane parallel to the (x, y)-plane and perpendicular to each other. They are shown in a top view of Earth in Figure 9.24(b). The Coriolis acceleration ${}^{B}_{G}\mathbf{a}_{Co}$ is eastward, and the centripetal acceleration ${}^{B}_{G}\mathbf{a}_{Ce}$ is toward the center of the circle of latitude λ.

We may assume that the Coriolis acceleration ${}^{B}_{G}\mathbf{a}_{Co}$ is proportional to a Coriolis force ${}^{B}_{G}\mathbf{F}_{Co} = m\,{}^{B}_{G}\mathbf{a}_{Co}$. The Coriolis force must be provided by the rotating ground. The magnitude of ${}^{B}_{G}\mathbf{F}_{Co}$ is independent of latitude φ:

$$
\begin{aligned}
{}^{B}_{G}\mathbf{F}_{Co} &= m\,{}^{B}_{G}\mathbf{a}_{Co} \\
&= -2mv\omega_E\sin\lambda\sin\varphi\hat{\imath} + 2mv\omega_E\cos\varphi\sin\lambda\hat{\jmath}
\end{aligned}
\tag{9.451}
$$

$$
F_{Co} = 2mv\omega_E\sin\lambda \tag{9.452}
$$

The direction of ${}^{B}_{G}\mathbf{F}_{Co}$ is the same as ${}^{B}_{G}\mathbf{a}_{Co}$. So, the train must lean on the west rail to turn with Earth. Such a weight transfer from the east to the west rail makes the west rail wear faster. The leaning on the west rail will not change if the train goes north.

Everything that moves on Earth leans west. This is the main reason that rivers dig into the west bank and leave dirt on the east bank. If contact between ground and the moving object is missing, then it will move westward with a Coriolis acceleration $-{}^{B}_{G}\mathbf{a}_{Co}$. That is why in the case of a flood rivers will cover the west bank easier than the east bank.

Example 578 ★ Foucault Pendulum, Exact Equations Consider a pendulum with a point mass m at the tip of a long, massless, and straight string with length l. The pendulum is hanging from a point $A(0, 0, l)$ in a local coordinate frame $B_1(x_1, y_1, z_1)$ at a point P on the Earth surface. Point P at longitude φ and latitude λ is indicated by ${}^{E}\mathbf{d}$ in the Earth frame $E(Oxyz)$. The E-frame is turning in a global frame $G(OXYZ)$ about the Z-axis.

To indicate the mass m, we attach a coordinate frame $B_1(x_1, y_1, z_1)$ to the pendulum at point A as shown in Figure 9.25. The pendulum makes an angle β with the vertical

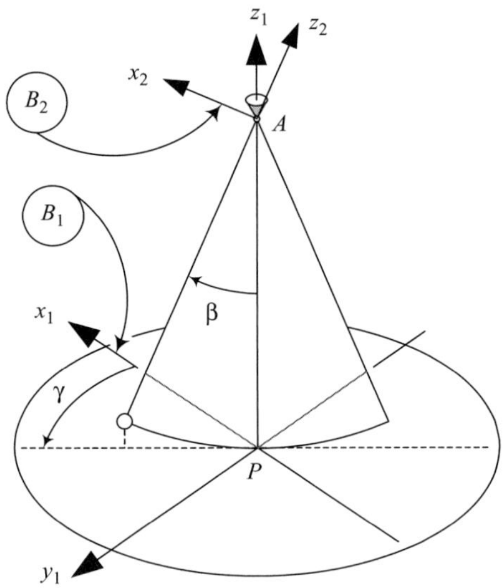

Figure 9.25 Foucault pendulum is a simple pendulum hanging from a point A above a point P on Earth surface.

z_1-axis. The pendulum swings in the plane (x_2, z_2) and makes an angle γ with the plane (x_1, z_1). Therefore, the transformation matrix between B_2 and B_1 is

$$
\begin{aligned}
{}^1T_2 &= {}^1D_2\,{}^1R_2 \\
&= \begin{bmatrix} \cos\gamma\cos\beta & -\sin\gamma & -\cos\gamma\sin\beta & 0 \\ \cos\beta\sin\gamma & \cos\gamma & -\sin\gamma\sin\beta & 0 \\ \sin\beta & 0 & \cos\beta & l \\ 0 & 0 & 0 & 1 \end{bmatrix}
\end{aligned}
\tag{9.453}
$$

The position vector of m is

$$
{}^2\mathbf{r} = \begin{bmatrix} 0 \\ 0 \\ -l \end{bmatrix} \tag{9.454}
$$

$$
{}^1\mathbf{r} = {}^1T_2\,{}^2\mathbf{r} = \begin{bmatrix} x_1 \\ y_1 \\ z_1 \end{bmatrix} = \begin{bmatrix} l\cos\gamma\sin\beta \\ l\sin\beta\sin\gamma \\ l - l\cos\beta \end{bmatrix} \tag{9.455}
$$

Employing the acceleration equation,

$$
{}^1_G\mathbf{a} = {}^1\mathbf{a} + {}^1_G\boldsymbol{\alpha}_1 \times {}^1\mathbf{r} + 2\,{}^1_G\boldsymbol{\omega}_1 \times {}^1\mathbf{v} + {}^1_G\boldsymbol{\omega}_1 \times \left({}^1_G\boldsymbol{\omega}_1 \times {}^1\mathbf{r}\right) \tag{9.456}
$$

we can write the equation of motion of m as

$$
{}^1_G\mathbf{F} - m\,{}^1_G\mathbf{g} = m\,{}^1_G\mathbf{a} \tag{9.457}
$$

where ${}^1\mathbf{F}$ is the applied nongravitational force on m.

Recalling that

$$ {}_G^1\boldsymbol{\alpha}_1 = 0 \tag{9.458} $$

we find the general equation of motion of a particle in frame B_1 as

$$ {}_G^1\mathbf{F} + m\,{}_G^1\mathbf{g} = m\left[{}^1\mathbf{a} + 2\,{}_G^1\boldsymbol{\omega}_1 \times {}^1\mathbf{v} + {}_G^1\boldsymbol{\omega}_1 \times \left({}_G^1\boldsymbol{\omega}_1 \times {}^1\mathbf{r}\right)\right] \tag{9.459} $$

The individual vectors in this equation are

$$ {}^1\mathbf{g} = \begin{bmatrix} 0 \\ 0 \\ -g_0 \end{bmatrix} \qquad {}_G^1\mathbf{F} = \begin{bmatrix} F_x \\ F_y \\ F_z \end{bmatrix} \qquad {}_G^1\boldsymbol{\omega}_1 = \begin{bmatrix} \omega_E \cos\lambda \\ 0 \\ \omega_E \sin\lambda \end{bmatrix} \tag{9.460} $$

$$ {}^1\mathbf{v} = \begin{bmatrix} \dot{x}_1 \\ \dot{y}_1 \\ \dot{z}_1 \end{bmatrix} = \begin{bmatrix} l\dot{\beta}\cos\beta\cos\gamma - l\dot{\gamma}\sin\beta\sin\gamma \\ l\dot{\beta}\cos\beta\sin\gamma + l\dot{\gamma}\cos\gamma\sin\beta \\ l\dot{\beta}\sin\beta \end{bmatrix} \tag{9.461} $$

$$ {}^1\mathbf{a} = \begin{bmatrix} \ddot{x} \\ \ddot{y} \\ \ddot{z} \end{bmatrix} = \begin{bmatrix} l\left(\ddot{\beta}\cos\gamma - \dot{\beta}^2\sin\gamma - \dot{\beta}\dot{\gamma}\sin\gamma\right)\cos\beta \\ \quad -l\left(\ddot{\gamma}\sin\gamma + \dot{\gamma}^2\cos\gamma + \dot{\beta}\dot{\gamma}\cos\gamma\right)\sin\beta \\ l\left(\ddot{\beta}\sin\gamma + \dot{\beta}^2\cos\gamma + \dot{\beta}\dot{\gamma}\cos\gamma\right)\cos\beta \\ \quad +l\left(\ddot{\gamma}\cos\gamma - \dot{\gamma}^2\sin\gamma - \dot{\beta}\dot{\gamma}\sin\gamma\right)\sin\beta \\ l\ddot{\beta}\sin\beta \end{bmatrix} \tag{9.462} $$

In a spherical pendulum, the external force ${}^1\mathbf{F}$ is the tension of the string:

$$ {}_G^1\mathbf{F} = -\frac{F}{l}\,{}^1\mathbf{r} \tag{9.463} $$

Substituting the above vectors in (9.459) provides three coupled ordinary differential equations for two angular variables γ and β. One of the equations is not independent and the others may theoretically be integrated to determine $\gamma = \gamma(t)$ and $\beta = \beta(t)$.

Example 579 ★ Foucault Pendulum, Approximate Solution The equations of motion of a spherical pendulum on the rotating Earth are too complicated to be solved analytically. However, we can simplify the equations and provide an approximate solution to indicate the main dynamic behavior of the pendulum. Substituting the Cartesian expressions of **r**, **v**, **a**, **g** in B_1 along with ${}_G^1\boldsymbol{\omega}_1$ from (9.460) in the equations of motion of the pendulum,

$$ {}_G^1\mathbf{F} + m\,{}_G^1\mathbf{g} = m\left[{}^1\mathbf{a} + 2\,{}_G^1\boldsymbol{\omega}_1 \times {}^1\mathbf{v} + {}_G^1\boldsymbol{\omega}_1 \times \left({}_G^1\boldsymbol{\omega}_1 \times {}^1\mathbf{r}\right)\right] \tag{9.464} $$

and approximating the force ${}_G^1\mathbf{F}$ as

$$ {}_G^1\mathbf{F} = \begin{bmatrix} F_x \\ F_y \\ F_z \end{bmatrix} = \begin{bmatrix} -F\dfrac{x_1}{l} \\ -F\dfrac{y_1}{l} \\ F\dfrac{l - z_1}{l} \end{bmatrix} \tag{9.465} $$

yield

$$\begin{bmatrix} -\dfrac{Fx_1}{ml} \\ -\dfrac{Fy_1}{ml} \\ F\dfrac{l-z_1}{ml} - g_0 \end{bmatrix} = \begin{bmatrix} \ddot{x}_1 - 2\dot{y}_1\omega_E \sin\lambda - \omega_E^2\left(x_1 \sin^2\lambda - z_1 \dfrac{\sin 2\lambda}{2}\right) \\ \ddot{y}_1 + 2\omega_E\left(\dot{x}_1 \sin\lambda - \dot{z}_1 \cos\lambda\right) - \omega_E^2 y_1 \\ \ddot{z}_1 + 2\dot{y}_1\omega_E \cos\lambda + \omega_E^2\left(x_1 \dfrac{\sin 2\lambda}{2} - z_1 \cos^2\lambda\right) \end{bmatrix} \tag{9.466}$$

Simplification starts with ignoring ω_E^2 because of the small value of ω_E for Earth. Furthermore, by the assumption that the rise z is much smaller than l and $\dot{z}$, $\ddot{z}$ are very small, the third equation reduces to

$$\frac{F}{m} = g_0 + 2\dot{y}_1\omega_E \cos\lambda \tag{9.467}$$

Using this equation to eliminate F from the first and second equations and ignoring the very small terms compared to large terms yield

$$\begin{bmatrix} \ddot{x}_1 \\ \ddot{y}_1 \end{bmatrix} = \begin{bmatrix} -\dfrac{x_1}{l} g_0 + 2\dot{y}_1\omega_E \sin\lambda \\ -\dfrac{y_1}{l} g_0 - 2\dot{x}_1\omega_E \sin\lambda \end{bmatrix} \tag{9.468}$$

Introducing the linear natural frequency of a pendulum

$$\omega_n^2 = \frac{g_0}{l} \tag{9.469}$$

reduces the equations to

$$\ddot{x}_1 + \omega_n^2 x_1 - 2\dot{y}_1\omega_E \sin\lambda = 0 \tag{9.470}$$

$$\ddot{y}_1 + \omega_n^2 y_1 + 2\dot{x}_1\omega_E \sin\lambda = 0 \tag{9.471}$$

To solve these equations, we may introduce a complex variable u,

$$u = x_1 + i y_1 \qquad i^2 = -1 \tag{9.472}$$

and combine them to a single differential equation:

$$\ddot{u} + \left(2i\omega_E \sin\lambda\right)\dot{u} + \omega_n^2 u = 0 \tag{9.473}$$

which is a linear ordinary differential equation with constant coefficients. The solution is

$$u = Ae^{st} \tag{9.474}$$

where s must satisfy the characteristic equation

$$s^2 + \left(2i\omega_E \sin\lambda\right)s + \omega_n^2 = 0 \tag{9.475}$$

The characteristic equations provide two possible solutions:

$$s = -i\omega_E \sin\lambda \pm \frac{\sqrt{2}}{2}\sqrt{\omega_E^2 - 2\omega_n^2 - \omega_E^2 \cos 2\lambda} \tag{9.476}$$

which can be approximated as

$$s = \begin{cases} is_1 \\ is_2 \end{cases} \approx -i\omega_E \sin\lambda \pm i\omega_n \tag{9.477}$$

so the general solution of Equation (9.473) is

$$u = A_1 e^{is_1 t} + A_2 e^{is_2 t} \tag{9.478}$$

The complex coefficients A_1 and A_2 are determined by the initial conditions. To calculate A_1 and A_2 let us suppose that the pendulum is released from rest at $t = 0$, $\lambda = 0$ and

$$x(0) = x_0 \qquad y(0) = 0 \tag{9.479}$$

Therefore, the initial conditions of the complex variable u are

$$u(0) = u_0 \qquad \dot{u}(0) = 0 \tag{9.480}$$

Employing these conditions, we determine A_1 and A_2:

$$A_1 = \frac{is_2}{is_2 - is_1} x_0 \qquad A_2 = \frac{-is_1}{is_2 - is_1} x_0 \tag{9.481}$$

It shows that both A_1 and A_2 are real numbers because both characteristic numbers are imaginary.

We may use the Euler identity and rewrite solution (9.478) as

$$u = A_1 \cos s_1 t + A_2 \cos s_2 t + i\,(A_1 \sin s_1 t + A_2 \sin s_2 t) \tag{9.482}$$

which indicates that

$$x = A_1 \cos\left(-\omega_E \sin\lambda + \omega_n\right) t + A_2 \cos\left(-\omega_E \sin\lambda - \omega_n\right) t \tag{9.483}$$

$$y = A_1 \sin\left(-\omega_E \sin\lambda + \omega_n\right) t + A_2 \sin\left(-\omega_E \sin\lambda - \omega_n\right) t \tag{9.484}$$

When $\omega_E \sin\lambda = 0$, the equations reduce to the usual harmonic oscillations of a spherical pendulum. For nonzero $\omega_E \sin\lambda$, the Coriolis effect causes a rotation of the oscillation plane with angular velocity $\omega = -\omega_E \sin\lambda$. So, the plane of oscillation turns $\omega \approx 2\pi$ rad/d at the poles and $\omega = 0$ at the equator.

As an example, let us use

$$\begin{aligned}
\omega_E &\approx 7.2921 \times 10^{-5}\ \text{rad/s} \\
g_0 &\approx 9.81\ \text{m/s}^2 \\
l &= 100\ \text{m} \\
\lambda &= 28^\circ 58' 30''\ N \approx 28.975\ \text{deg}\ N \\
\varphi &= 50^\circ 50' 17''\ E \approx 50.838\ \text{deg}\ E \\
x_0 &= l \cos 10 = 17.365\ \text{m}
\end{aligned} \tag{9.485}$$

and find

$$s = \begin{cases} 0.313\,16i \\ -0.313\,26i \end{cases} \tag{9.486}$$

$$A_1 = 8.6839\,\mathrm{m} \qquad A_2 = 8.6811\,\mathrm{m} \tag{9.487}$$

$$x = 8.6839\cos(0.313\,16t) + 8.6811\cos(-0.313\,26t) \tag{9.488}$$

$$y = 8.6839\sin(0.313\,16t) + 8.6811\sin(-0.313\,26t) \tag{9.489}$$

At the given latitude, which corresponds to Bushehr, Iran, on the Persian Gulf shore, the plane of oscillation turns about the local **g**-axis with an angular speed $\omega = -3.5325 \times 10^{-5}$ rad/s ≈ -87.437 deg/d. These results are independent of longitude. Therefore, the same phenomena will be seen at Orlando, Florida, or New Delhi, India, which are almost at the same latitude. Figure 9.26 depicts the projection of m on the (x, y)-plane for a few oscillations. It takes $T \approx 49.4$ h for the pendulum to turn 2π:

$$T = \frac{2\pi}{3.5325 \times 10^{-5}} = 1.7787 \times 10^5\,\mathrm{s} = 49.408\,\mathrm{h} \tag{9.490}$$

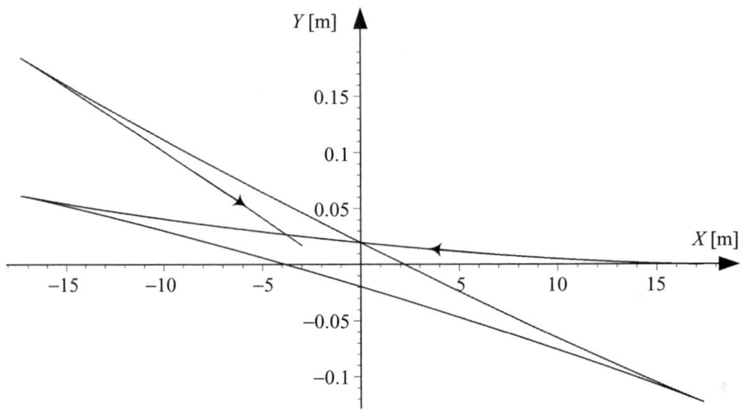

Figure 9.26 The projection of the path of a pendulum with length $l = 100$ m at latitude $\lambda \approx 28.975$ deg N on Earth for a few oscillations (not to scale).

However, the pendulum gets back to the (y, x)-plane after $t = T/2 = 24.704$ h. By that time, the pendulum must have oscillated about $n \approx 4433$ times:

$$n = \frac{\omega_n}{2\pi}\frac{T}{2} = \frac{0.31321}{2\pi}\frac{1.7787 \times 10^5}{2} = 4433.3 \tag{9.491}$$

By shortening the length of the pendulum, say $l = 1$ m, the rotation speed remains the same while the number of oscillations increases to $n \approx 44333$.

9.5 ★ MIXED DOUBLE DERIVATIVE

Consider three relatively rotating frames A, B, and C. The mixed double acceleration ${}^{AA}_{CB}\mathbf{a}$ is the A-expression of the second derivative of a position vector ${}^{A}\mathbf{r}$ when the first derivative is taken in the B-frame and the second derivative is taken in the C-frame:

$$
{}^{AA}_{CB}\mathbf{a} = {}^{A}\mathbf{a} + {}^{A}_{C}\boldsymbol{\omega}_{A} \times {}^{A}\mathbf{v} + {}^{A}_{B}\boldsymbol{\alpha}_{A} \times {}^{A}\mathbf{r} + \left({}^{A}_{C}\boldsymbol{\omega}_{A} \times {}^{A}_{B}\boldsymbol{\omega}_{A}\right) \times {}^{A}\mathbf{r} + {}^{A}_{B}\boldsymbol{\omega}_{A} \times {}^{A}\mathbf{v} + {}^{A}_{B}\boldsymbol{\omega}_{A} \times \left({}^{A}_{C}\boldsymbol{\omega}_{A} \times {}^{A}\mathbf{r}\right) \tag{9.492}
$$

So, the mixed double derivative is produced when we take the first and second derivatives of ${}^{A}\mathbf{r}$ in different coordinate frames B and C.

Proof: Using the derivative transformation formula, the B-derivative of ${}^{A}\mathbf{r}$ is

$$
\frac{{}^{B}d}{dt}\,{}^{A}\mathbf{r} = {}^{A}_{B}\mathbf{v} = \frac{{}^{A}d}{dt}\,{}^{A}\mathbf{r} + {}^{A}_{B}\boldsymbol{\omega}_{A} \times {}^{A}\mathbf{r} \tag{9.493}
$$

A time derivative of this equation in a third frame C would be

$$
\begin{aligned}
\frac{{}^{C}d}{dt}\frac{{}^{B}d}{dt}\,{}^{A}\mathbf{r} &= \frac{{}^{C}d}{dt}\,{}^{A}_{B}\mathbf{v} = {}^{AA}_{CB}\mathbf{a} = \frac{{}^{C}d}{dt}\left(\frac{{}^{A}d}{dt}\,{}^{A}\mathbf{r}\right) + \frac{{}^{C}d}{dt}\left({}^{A}_{B}\boldsymbol{\omega}_{A} \times {}^{A}\mathbf{r}\right) \\
&= \frac{{}^{A}d}{dt}\frac{{}^{A}d}{dt}\,{}^{A}\mathbf{r} + {}^{A}_{C}\boldsymbol{\omega}_{A} \times \frac{{}^{A}d}{dt}\,{}^{A}\mathbf{r} \\
&\quad + \left(\frac{{}^{A}d}{dt}\,{}^{A}_{B}\boldsymbol{\omega}_{A} + {}^{A}_{C}\boldsymbol{\omega}_{A} \times {}^{A}_{B}\boldsymbol{\omega}_{A}\right) \times {}^{A}\mathbf{r} \\
&\quad + {}^{A}_{B}\boldsymbol{\omega}_{A} \times \left(\frac{{}^{A}d}{dt}\,{}^{A}\mathbf{r} + {}^{A}_{C}\boldsymbol{\omega}_{A} \times {}^{A}\mathbf{r}\right) \\
&= {}^{A}\mathbf{a} + {}^{A}_{C}\boldsymbol{\omega}_{A} \times {}^{A}\mathbf{v} + {}^{A}_{B}\boldsymbol{\alpha}_{A} \times {}^{A}\mathbf{r} + \left({}^{A}_{C}\boldsymbol{\omega}_{A} \times {}^{A}_{B}\boldsymbol{\omega}_{A}\right) \times {}^{A}\mathbf{r} \\
&\quad + {}^{A}_{B}\boldsymbol{\omega}_{A} \times {}^{A}\mathbf{v} + {}^{A}_{B}\boldsymbol{\omega}_{A} \times \left({}^{A}_{C}\boldsymbol{\omega}_{A} \times {}^{A}\mathbf{r}\right)
\end{aligned} \tag{9.494}
$$

We call the acceleration ${}^{AA}_{CB}\mathbf{a}$ the *mixed double acceleration*. The first term ${}^{A}\mathbf{a}$ is the local A-acceleration of a moving point P in the A-frame. The combined terms ${}^{A}_{C}\boldsymbol{\omega}_{A} \times {}^{A}\mathbf{v} + {}^{A}_{B}\boldsymbol{\omega}_{A} \times {}^{A}\mathbf{v}$ is the *mixed Coriolis acceleration*:

$$
\mathbf{a}_{\mathrm{Co}} = {}^{AA}_{CB}\mathbf{a}_{\mathrm{Ra}} = {}^{A}_{C}\boldsymbol{\omega}_{A} \times {}^{A}\mathbf{v} + {}^{A}_{B}\boldsymbol{\omega}_{A} \times {}^{A}\mathbf{v} \tag{9.495}
$$

The term ${}^{A}_{B}\boldsymbol{\alpha}_{A} \times {}^{A}\mathbf{r}$ is the *tangential acceleration* of P. The term ${}^{A}_{B}\boldsymbol{\omega}_{A} \times \left({}^{A}_{C}\boldsymbol{\omega}_{A} \times {}^{A}\mathbf{r}\right)$ is the mixed centripetal acceleration. The term $\left({}^{A}_{C}\boldsymbol{\omega}_{A} \times {}^{A}_{B}\boldsymbol{\omega}_{A}\right) \times {}^{A}\mathbf{r}$ is a new term in the acceleration of P that cannot be seen in the simple acceleration ${}^{B}_{G}\mathbf{a}$. We call this term the Rāzī *acceleration* $\mathbf{a}_{Ra}$:

$$
\mathbf{a}_{\mathrm{Ra}} = {}^{AA}_{CB}\mathbf{a}_{\mathrm{Ra}} = \left({}^{A}_{C}\boldsymbol{\omega}_{A} \times {}^{A}_{B}\boldsymbol{\omega}_{A}\right) \times {}^{A}\mathbf{r} \tag{9.496}
$$

■

Example 580 ★ Difference between the Two $\boldsymbol{\omega} \times \mathbf{v}$ of Coriolis Acceleration The difference between the two terms in the Coriolis acceleration is clearer when we consider three coordinate frames B_1, B_2, and G. The frame B_2 is turning in B_1, and B_1 is turning in G. There is also a moving point P in a coordinate frame B_2. The

B_2-expression of the velocity of P in B_1 is

$$ {}_1^2\mathbf{v} = \frac{{}^1d}{dt}\,{}^2\mathbf{r} = \frac{{}^2d}{dt}\,{}^2\mathbf{r} + {}_1^2\boldsymbol{\omega}_2 \times {}^2\mathbf{r} = {}^2\mathbf{v} + {}_1^2\boldsymbol{\omega}_2 \times {}^2\mathbf{r} \tag{9.497} $$

The derivative of ${}_1^2\mathbf{v}$ in G provides a double mixed acceleration ${}_{G1}^{22}\mathbf{a}$,

$$ \begin{aligned} \frac{{}^Gd}{dt}\frac{{}^1d}{dt}\,{}^2\mathbf{r} = {}_{G1}^{22}\mathbf{a} &= \frac{{}^Gd}{dt}\left(\frac{{}^2d}{dt}\,{}^2\mathbf{r}\right) + \frac{{}^Gd}{dt}\left({}_1^2\boldsymbol{\omega}_2 \times {}^2\mathbf{r}\right) \\ &= \frac{{}^2d}{dt}\frac{{}^2d}{dt}\,{}^2\mathbf{r} + {}_G^2\boldsymbol{\omega}_2 \times \frac{{}^2d}{dt}\,{}^2\mathbf{r} + \left(\frac{{}^2d}{dt}\,{}_1^2\boldsymbol{\omega}_2 + {}_G^2\boldsymbol{\omega}_2 \times {}_1^2\boldsymbol{\omega}_2\right) \times {}^2\mathbf{r} \\ &\quad + {}_1^2\boldsymbol{\omega}_2 \times \left(\frac{{}^2d}{dt}\,{}^2\mathbf{r} + {}_G^2\boldsymbol{\omega}_2 \times {}^2\mathbf{r}\right) \end{aligned} \tag{9.498} $$

which can be simplified to

$$ \begin{aligned} \frac{{}^Gd}{dt}\frac{{}^1d}{dt}\,{}^2\mathbf{r} &= {}^2\mathbf{a} + {}_G^2\boldsymbol{\omega}_2 \times {}^2\mathbf{v} + {}_1^2\boldsymbol{\alpha}_2 \times {}^2\mathbf{r} + \left({}_G^2\boldsymbol{\omega}_2 \times {}_1^2\boldsymbol{\omega}_2\right) \times {}^2\mathbf{r} \\ &\quad + {}_1^2\boldsymbol{\omega}_2 \times {}^2\mathbf{v} + {}_1^2\boldsymbol{\omega}_2 \times \left({}_G^2\boldsymbol{\omega}_2 \times {}^2\mathbf{r}\right) \end{aligned} \tag{9.499} $$

which is equivalent to (9.492) and indicates that, if $B_1 \neq G$, then ${}_G^2\boldsymbol{\omega}_2 \times {}^2\mathbf{v} \neq {}_1^2\boldsymbol{\omega}_2 \times {}^2\mathbf{v}$, and there exists a mixed Coriolis acceleration:

$$ \mathbf{a}_{\mathrm{Co}} = {}_{G1}^{22}\mathbf{a}_{\mathrm{Co}} = {}_1^2\boldsymbol{\omega}_2 \times {}^2\mathbf{v} + {}_G^2\boldsymbol{\omega}_2 \times {}^2\mathbf{v} \tag{9.500} $$

Furthermore, when $B_1 \neq G$, there exists a Rāzī acceleration term $\mathbf{a}_{Ra}$ in the acceleration of P that cannot be seen in ${}_G^B\mathbf{a}$:

$$ \mathbf{a}_{\mathrm{Ra}} = {}_{G1}^{22}\mathbf{a}_{\mathrm{Ra}} = \left({}_G^2\boldsymbol{\omega}_2 \times {}_1^2\boldsymbol{\omega}_2\right) \times {}^2\mathbf{r} \tag{9.501} $$

Zakariy Rāzī (865–925) was a Persian mathematician, chemist, and physician and is considered the father of pediatrics.

Example 581 ★ Modified Mixed Double Derivative Recalling the relative angular velocity rule

$$ {}_G^2\boldsymbol{\omega}_2 = {}_G^2\boldsymbol{\omega}_1 + {}_1^2\boldsymbol{\omega}_2 \tag{9.502} $$

we can rewrite (9.499) in a more practical form:

$$ \begin{aligned} \frac{{}^Gd}{dt}\frac{{}^1d}{dt}\,{}^2\mathbf{r} &= {}^2\mathbf{a} + \left({}_G^2\boldsymbol{\omega}_1 + {}_1^2\boldsymbol{\omega}_2\right) \times {}^2\mathbf{v} + {}_1^2\boldsymbol{\alpha}_2 \times {}^2\mathbf{r} \\ &\quad + \left(\left({}_G^2\boldsymbol{\omega}_1 + {}_1^2\boldsymbol{\omega}_2\right) \times {}_1^2\boldsymbol{\omega}_2\right) \times {}^2\mathbf{r} \\ &\quad + {}_1^2\boldsymbol{\omega}_2 \times {}^2\mathbf{v} + {}_1^2\boldsymbol{\omega}_2 \times \left[\left({}_G^2\boldsymbol{\omega}_1 + {}_1^2\boldsymbol{\omega}_2\right) \times {}^2\mathbf{r}\right] \\ &= {}^2\mathbf{a} + {}_G^2\boldsymbol{\omega}_1 \times {}^2\mathbf{v} + {}_1^2\boldsymbol{\omega}_2 \times {}^2\mathbf{v} + {}_1^2\boldsymbol{\alpha}_2 \times {}^2\mathbf{r} \\ &\quad + \left({}_G^2\boldsymbol{\omega}_1 \times {}_1^2\boldsymbol{\omega}_2\right) \times {}^2\mathbf{r} + {}_1^2\boldsymbol{\omega}_2 \times {}^2\mathbf{v} \\ &\quad + {}_1^2\boldsymbol{\omega}_2 \times \left({}_G^2\boldsymbol{\omega}_1 \times {}^2\mathbf{r}\right) + {}_1^2\boldsymbol{\omega}_2 \times \left({}_1^2\boldsymbol{\omega}_2 \times {}^2\mathbf{r}\right) \end{aligned} \tag{9.503} $$

If the point P is not moving in B_2, the double mixed acceleration (9.499) reduces to

$$\frac{{}^{G}d}{dt}\frac{{}^{1}d}{dt}{}^{2}\mathbf{r} = {}_{1}^{2}\boldsymbol{\alpha}_{2} \times {}^{2}\mathbf{r} + \left({}_{G}^{2}\boldsymbol{\omega}_{2} \times {}_{1}^{2}\boldsymbol{\omega}_{2}\right) \times {}^{2}\mathbf{r} + {}_{1}^{2}\boldsymbol{\omega}_{2} \times \left({}_{G}^{2}\boldsymbol{\omega}_{2} \times {}^{2}\mathbf{r}\right) \tag{9.504}$$

or equivalently to

$$\frac{{}^{G}d}{dt}\frac{{}^{1}d}{dt}{}^{2}\mathbf{r} = {}_{1}^{2}\boldsymbol{\alpha}_{2} \times {}^{2}\mathbf{r} + \left({}_{G}^{2}\boldsymbol{\omega}_{1} \times {}_{1}^{2}\boldsymbol{\omega}_{2}\right) \times {}^{2}\mathbf{r} + {}_{1}^{2}\boldsymbol{\omega}_{2} \times \left[\left({}_{G}^{2}\boldsymbol{\omega}_{1} + {}_{1}^{2}\boldsymbol{\omega}_{2}\right) \times {}^{2}\mathbf{r}\right] \tag{9.505}$$

This indicates that when a body B_2 is turning relative to B_1 and G there are three terms in the double mixed acceleration of a body point of B_2: the tangential acceleration ${}_{1}^{2}\boldsymbol{\alpha}_{2} \times {}^{2}\mathbf{r}$, the mixed centripetal acceleration ${}_{1}^{2}\boldsymbol{\omega}_{2} \times \left({}_{G}^{2}\boldsymbol{\omega}_{2} \times {}^{2}\mathbf{r}\right)$, and the Rāzī acceleration $\left({}_{G}^{2}\boldsymbol{\omega}_{2} \times {}_{1}^{2}\boldsymbol{\omega}_{2}\right) \times {}^{2}\mathbf{r}$.

Example 582 ★ Practical Double Mixed Acceleration The Rāzī acceleration appears in the double mixed acceleration when there are a global frame G and two turning body coordinate frames B_1 in B_2 in G. Using the relative motion of the coordinate frames, the B_1-expression of the acceleration of a body point in B_2 is

$$ {}_{G1}^{22}\mathbf{a} = \frac{{}^{G}d}{dt}\frac{{}^{1}d}{dt}{}^{2}\mathbf{r} = \left({}_{G}^{2}\boldsymbol{\omega}_{2} \times {}_{1}^{2}\boldsymbol{\omega}_{2}\right) \times {}^{2}\mathbf{r} + {}_{1}^{2}\boldsymbol{\omega}_{2} \times \left({}_{G}^{2}\boldsymbol{\omega}_{2} \times {}^{2}\mathbf{r}\right) \tag{9.506}$$

Let us consider a point P on a disc with radius R that is mounted on a turning shaft. The disc spins about a tilted axis on the shaft, as shown in Figure 9.27. We set a global coordinate frame G at the disc center O. A coordinate frame B_1 is attached to the disc at O such that y_1 coincides with the Y-axis and (x_2, z_2) is coplanar with (X,Z). The disc is mounted in B_1 at an angle of 45 deg. The disc coordinate frame is B_2. The disc

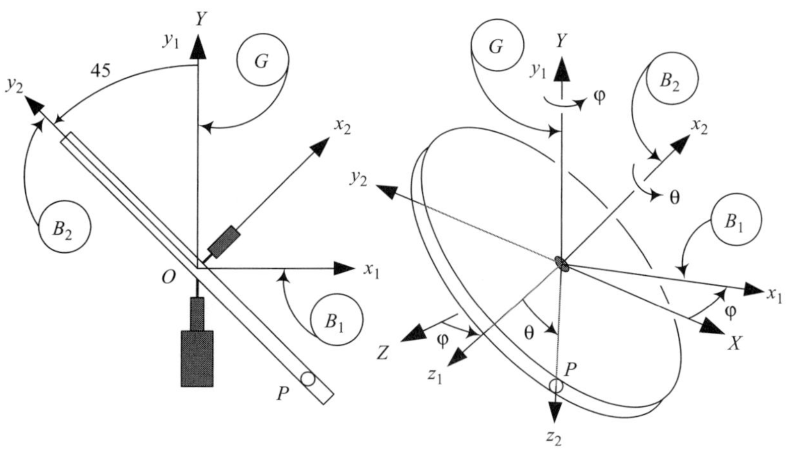

Figure 9.27 A disc spinning about its axis and turning about a globally fixed axis.

spins with constant angular velocity ${}_1^2\boldsymbol{\omega}_2 = \dot{\theta}\hat{\imath}_2$ in B_1, and B_1 turns about the Y-axis with constant angular velocity ${}_G^1\boldsymbol{\omega}_1 = \dot{\varphi}\hat{J}$:

$${}_1^2\boldsymbol{\omega}_2 = \dot{\theta}\hat{\imath}_2 = \begin{bmatrix} \dot{\theta} \\ 0 \\ 0 \end{bmatrix} \qquad {}_G^1\boldsymbol{\omega}_1 = \dot{\varphi}\hat{J} = \begin{bmatrix} 0 \\ \dot{\varphi} \\ 0 \end{bmatrix} \tag{9.507}$$

The point P is at

$${}^2\mathbf{r} = \begin{bmatrix} 0 \\ 0 \\ R \end{bmatrix} \tag{9.508}$$

and the transformation matrices between the frames are

$${}^G R_1 = R_{Y,\varphi} = \begin{bmatrix} \cos\varphi & 0 & \sin\varphi \\ 0 & 1 & 0 \\ -\sin\varphi & 0 & \cos\varphi \end{bmatrix} \tag{9.509}$$

$$\begin{aligned} {}^2R_1 = R_{y,\theta} R_{z,45} &= \begin{bmatrix} \cos\theta & \sin\theta & 0 \\ -\sin\theta & \cos\theta & 0 \\ 0 & 0 & 1 \end{bmatrix} \begin{bmatrix} 1 & 0 & 0 \\ 0 & \cos(\frac{1}{4}\pi) & \sin(\frac{1}{4}\pi) \\ 0 & -\sin(\frac{1}{4}\pi) & \cos(\frac{1}{4}\pi) \end{bmatrix} \\ &= \begin{bmatrix} \cos\theta & 0.70711\sin\theta & 0.70711\sin\theta \\ -\sin\theta & 0.70711\cos\theta & 0.70711\cos\theta \\ 0 & -0.70711 & 0.70711 \end{bmatrix} \end{aligned} \tag{9.510}$$

The velocity ${}_1^2\mathbf{v}$ of P is

$$\begin{aligned} {}_1^2\mathbf{v} &= \frac{{}^1d}{dt}\,{}^2\mathbf{r} = {}_1^2\boldsymbol{\omega}_2 \times {}^2\mathbf{r} \\ &= \begin{bmatrix} \dot{\theta} \\ 0 \\ 0 \end{bmatrix} \times \begin{bmatrix} 0 \\ 0 \\ R \end{bmatrix} = \begin{bmatrix} 0 \\ -R\dot{\theta} \\ 0 \end{bmatrix} \end{aligned} \tag{9.511}$$

If we are standing in B_1, then ${}^1\mathbf{v}$ is the velocity of P that we see:

$${}^1\mathbf{v} = {}^1R_2\,{}_1^2\mathbf{v} = \begin{bmatrix} R\dot{\theta}\sin\theta \\ -0.70711 R\dot{\theta}\cos\theta \\ -0.70711 R\dot{\theta}\cos\theta \end{bmatrix} \tag{9.512}$$

If we are standing in G and watching P, then ${}^G\mathbf{v}$ is the velocity that we see:

$$\begin{aligned} {}^G\mathbf{v} &= {}^GR_2\,{}_G^2\mathbf{v} = {}^GR_2 \frac{{}^Gd}{dt}\,{}^2\mathbf{r} = {}^GR_2\left({}_G^2\boldsymbol{\omega}_2 \times {}^2\mathbf{r}\right) \\ &= {}^GR_2\left[\left({}_G^2\boldsymbol{\omega}_1 + {}_1^2\boldsymbol{\omega}_2\right) \times {}^2\mathbf{r}\right] = {}^GR_2\left[\left({}^2R_1\,{}_G^1\boldsymbol{\omega}_1 + {}_1^2\boldsymbol{\omega}_2\right) \times {}^2\mathbf{r}\right] \\ &= R\begin{bmatrix} 0.146\dot{\theta}\sin(\theta+\varphi) + 0.707\dot{\varphi}\cos\varphi + 0.853\dot{\theta}\sin(\theta-\varphi) \\ -0.707\dot{\theta}\cos\theta \\ 0.146\dot{\theta}\cos(\theta+\varphi) - 0.707\dot{\varphi}\sin\varphi - 0.853\dot{\theta}\cos(\theta-\varphi) \end{bmatrix} \end{aligned} \tag{9.513}$$

where

$$ {}_G^2\boldsymbol{\omega}_1 = {}^2R_1\,{}_G^1\boldsymbol{\omega}_1 = \begin{bmatrix} 0.70711\dot{\varphi}\sin\theta \\ 0.70711\dot{\varphi}\cos\theta \\ -0.70711\dot{\varphi} \end{bmatrix} \tag{9.514} $$

$$ {}_G^2\boldsymbol{\omega}_2 = {}_G^2\boldsymbol{\omega}_1 + {}_1^2\boldsymbol{\omega}_2 = \begin{bmatrix} 0.70711\dot{\varphi}\sin\theta + \dot{\theta} \\ 0.70711\dot{\varphi}\cos\theta \\ -0.70711\dot{\varphi} \end{bmatrix} \tag{9.515} $$

$$ {}_G^2\mathbf{v} = {}_G^2\boldsymbol{\omega}_2 \times {}^2\mathbf{r} = \begin{bmatrix} 0.70711R\dot{\varphi}\cos\theta \\ -R\left(\dot{\theta} + 0.70711\dot{\varphi}\sin\theta\right) \\ 0 \end{bmatrix} \tag{9.516} $$

$$ {}^GR_2 = {}^GR_1\,{}^1R_2 \tag{9.517} $$

The accelerations ${}^1\mathbf{a}$ and ${}^G\mathbf{a}$ are the direct accelerations that we can measure when we are standing in B_1 or G, respectively. However, there are some other second derivatives that can be measured indirectly. The acceleration ${}_1^2\mathbf{a}$ is given as

$$ \begin{aligned} {}_1^2\mathbf{a} &= \frac{{}^1d}{dt}\frac{{}^1d}{dt}{}^2\mathbf{r} = \frac{{}^1d}{dt}{}_1^2\mathbf{v} = {}_1^2\boldsymbol{\alpha}_2 \times {}^2\mathbf{r} + {}_1^2\boldsymbol{\omega}_2 \times \left({}_1^2\boldsymbol{\omega}_2 \times {}^2\mathbf{r}\right) \\ &= \begin{bmatrix} \dot{\theta} \\ 0 \\ 0 \end{bmatrix} \times \left(\begin{bmatrix} \dot{\theta} \\ 0 \\ 0 \end{bmatrix} \times \begin{bmatrix} 0 \\ 0 \\ R \end{bmatrix} \right) = \begin{bmatrix} 0 \\ 0 \\ -R\dot{\theta}^2 \end{bmatrix} \end{aligned} \tag{9.518} $$

and therefore,

$$ {}^1\mathbf{a} = {}^1R_2\,{}_1^2\mathbf{a} = \begin{bmatrix} 0 \\ 0.70711R\dot{\theta}^2 \\ -0.70711R\dot{\theta}^2 \end{bmatrix} \tag{9.519} $$

The acceleration ${}_G^2\mathbf{a}$ is given as

$$ \begin{aligned} {}_G^2\mathbf{a} &= \frac{{}^Gd}{dt}\frac{{}^Gd}{dt}{}^2\mathbf{r} = \frac{{}^Gd}{dt}{}_G^2\mathbf{v} = {}_G^2\boldsymbol{\alpha}_2 \times {}^2\mathbf{r} + {}_G^2\boldsymbol{\omega}_2 \times \left({}_G^2\boldsymbol{\omega}_2 \times {}^2\mathbf{r}\right) \\ &= \begin{bmatrix} 0.707\dot{\varphi}\sin\theta + \dot{\theta} \\ 0.707\dot{\varphi}\cos\theta \\ -0.707\dot{\varphi} \end{bmatrix} \times \left(\begin{bmatrix} 0.707\dot{\varphi}\sin\theta + \dot{\theta} \\ 0.707\dot{\varphi}\cos\theta \\ -0.707\dot{\varphi} \end{bmatrix} \times \begin{bmatrix} 0 \\ 0 \\ R \end{bmatrix} \right) \\ &= \begin{bmatrix} -0.707R\dot{\varphi}\left(\dot{\theta} + 0.707\varphi\sin\theta\right) \\ -0.5R\dot{\varphi}^2\cos\theta \\ -0.5R\dot{\varphi}^2\cos^2\theta - R\left(\dot{\theta} + 0.707\varphi\sin\theta\right)\left(\dot{\theta} + 0.707\dot{\varphi}\sin\theta\right) \end{bmatrix} \end{aligned} \tag{9.520} $$

and therefore,

$$ {}^2\mathbf{a} = {}^GR_2\,{}_G^2\mathbf{a} \tag{9.521} $$

where

$$^GR_2 = {}^GR_1\,{}^1R_2 \tag{9.522}$$
$$= \begin{bmatrix} c\theta c\varphi + 0.707 s\theta s\varphi & 0.707 c\theta s\varphi - c\varphi s\theta & 0.707 s\varphi \\ 0.707 s\theta & 0.707 c\theta & -0.707 \\ 0.707 c\varphi s\theta - c\theta s\varphi & 0.707 c\theta c\varphi + s\theta s\varphi & 0.707 c\varphi \end{bmatrix}$$

The double mixed acceleration ${}^{22}_{G1}\mathbf{a}$ is given as

$$^{22}_{G1}\mathbf{a} = \frac{^Gd}{dt}\frac{^1d}{dt}\,{}^2\mathbf{r} = \left({}^2_G\boldsymbol{\omega}_2 \times {}^2_1\boldsymbol{\omega}_2\right) \times {}^2\mathbf{r} + {}^2_1\boldsymbol{\omega}_2 \times \left({}^2_G\boldsymbol{\omega}_2 \times {}^2\mathbf{r}\right)$$
$$= \begin{bmatrix} -0.707 R\dot{\theta}\dot{\varphi} \\ 0 \\ 0 \end{bmatrix} + \begin{bmatrix} 0 \\ 0 \\ -R\dot{\theta}\left(\dot{\theta} + 0.707\dot{\varphi}\sin\theta\right) \end{bmatrix}$$
$$= \begin{bmatrix} -0.707 R\dot{\theta}\dot{\varphi} \\ 0 \\ -R\dot{\theta}\left(\dot{\theta} + 0.707\dot{\varphi}\sin\theta\right) \end{bmatrix} \tag{9.523}$$

The first term, $-0.707R\dot{\theta}\dot{\varphi}\hat{\imath}_2$, is the Rāzī acceleration and the third term, $-R\dot{\theta}\left(\dot{\theta} + 0.707\dot{\varphi}\sin\theta\right)\hat{k}_2$, is the mixed centripetal acceleration of P.

Employing the mixed-derivative transformation formula (8.255), we can determine ${}^2_1\mathbf{a}$ from ${}^{22}_{G1}\mathbf{a}$ as

$$^2_1\mathbf{a} = {}^{22}_{G1}\mathbf{a} + \left({}^2_1\boldsymbol{\omega}_2 - {}^2_G\boldsymbol{\omega}_2\right) \times {}^2_1\mathbf{v} = \begin{bmatrix} 0 \\ 0 \\ -R\dot{\theta}^2 \end{bmatrix} \tag{9.524}$$

which must be the same as (9.518).

9.6 ★ ACCELERATION TRANSFORMATION MATRIX

Consider the motion of a rigid body B in the global coordinate frame G, as shown in Figure 9.6. Assume the body-fixed frame $B(oxyz)$ is coincident at an initial time t_0 with the global frame $G(OXYZ)$. At any time $t \neq t_0$, B is not necessarily coincident with G, and therefore, the homogeneous kinematic transformation matrix ${}^GT_B(t)$ is time varying.

The acceleration of a body point in the global coordinate frame can be found by applying a homogeneous acceleration transformation matrix ${}^GA_B$,

$$^G\mathbf{a}_P(t) = {}^GA_B\,{}^G\mathbf{r}_P(t) \tag{9.525}$$

where ${}^GA_B$ is the *homogeneous acceleration transformation matrix*

$$^GA_B = \begin{bmatrix} {}_G\tilde{\alpha}_B - {}_G\tilde{\omega}_B\,{}_G\tilde{\omega}_B^{\mathrm{T}} & {}^G\ddot{\mathbf{d}}_B - \left({}_G\tilde{\alpha}_B - {}_G\tilde{\omega}_B\,{}_G\tilde{\omega}_B^{\mathrm{T}}\right){}^G\mathbf{d}_B \\ 0 & 0 \end{bmatrix}$$
$$= \begin{bmatrix} {}_GS_B & {}^G\ddot{\mathbf{d}}_B - {}_GS_B\,{}^G\mathbf{d}_B \\ 0 & 0 \end{bmatrix} \tag{9.526}$$

Proof: Based on the homogeneous coordinate transformation

$$ {}^{G}\mathbf{r}_P(t) = {}^{G}T_B \, {}^{B}\mathbf{r}_P \tag{9.527} $$

$$ {}^{G}T_B = \begin{bmatrix} {}^{G}R_B & {}^{G}\mathbf{d}_B \\ 0 & 1 \end{bmatrix} \tag{9.528} $$

we have

$$ \begin{aligned} {}^{G}\mathbf{v}_P &= {}^{G}\dot{T}_B \, {}^{G}T_B^{-1} \, {}^{G}\mathbf{r}_P(t) \\ &= \begin{bmatrix} {}^{G}\dot{R}_B \, {}^{G}R_B^{\mathrm{T}} & {}^{G}\dot{\mathbf{d}}_B - {}^{G}\dot{R}_B \, {}^{G}R_B^{\mathrm{T}} \, {}^{G}\mathbf{d}_B \\ 0 & 0 \end{bmatrix} {}^{G}\mathbf{r}_P(t) \\ &= \begin{bmatrix} {}_{G}\tilde{\omega}_B & {}^{G}\dot{\mathbf{d}}_B - {}_{G}\tilde{\omega}_B \, {}^{G}\mathbf{d}_B \\ 0 & 0 \end{bmatrix} {}^{G}\mathbf{r}_P(t) \\ &= {}^{G}V_B \, {}^{G}\mathbf{r}_P(t) \end{aligned} \tag{9.529} $$

To find the acceleration of a body point in the global frame, we take the second time derivative from ${}^{G}\mathbf{r}_P(t) = {}^{G}T_B \, {}^{B}\mathbf{r}_P$,

$$ {}^{G}\mathbf{a}_P = \frac{d^2}{dt^2} \, {}^{G}T_B \, {}^{B}\mathbf{r}_P = {}^{G}\ddot{T}_B \, {}^{B}\mathbf{r}_P \tag{9.530} $$

and substitute for ${}^{B}\mathbf{r}_P$:

$$ {}^{G}\mathbf{a}_P = {}^{G}\ddot{T}_B \, {}^{G}T_B^{-1} \, {}^{G}\mathbf{r}_P(t) = {}^{G}A_B \, {}^{G}\mathbf{r}_P(t) \tag{9.531} $$

$$ {}^{G}A_B = {}^{G}\ddot{T}_B \, {}^{G}T_B^{-1} \tag{9.532} $$

Substituting for ${}^{G}\ddot{T}_B$ and ${}^{G}T_B^{-1}$ provides

$$ \begin{aligned} {}^{G}\mathbf{a}_P(t) &= \begin{bmatrix} {}^{G}\ddot{R}_B & {}^{G}\ddot{\mathbf{d}}_B \\ 0 & 0 \end{bmatrix} \begin{bmatrix} {}^{G}R_B^{\mathrm{T}} & -{}^{G}R_B^{\mathrm{T}} \, {}^{G}\mathbf{d}_B \\ 0 & 1 \end{bmatrix} {}^{G}\mathbf{r}_P(t) \\ &= \begin{bmatrix} {}^{G}\ddot{R}_B \, {}^{G}R_B^{\mathrm{T}} & {}^{G}\ddot{\mathbf{d}}_B - {}^{G}\ddot{R}_B \, {}^{G}R_B^{\mathrm{T}} \, {}^{G}\mathbf{d}_B \\ 0 & 0 \end{bmatrix} {}^{G}\mathbf{r}_P(t) \\ &= \begin{bmatrix} {}_{G}\tilde{\alpha}_B - \tilde{\omega}\,\tilde{\omega}^{\mathrm{T}} & {}^{G}\ddot{\mathbf{d}}_B - \left({}_{G}\tilde{\alpha}_B - \tilde{\omega}\,\tilde{\omega}^{\mathrm{T}}\right) {}^{G}\mathbf{d}_B \\ 0 & 0 \end{bmatrix} {}^{G}\mathbf{r}_P(t) \\ &= \begin{bmatrix} {}_{G}S_B & {}^{G}\ddot{\mathbf{d}}_B - {}_{G}S_B \, {}^{G}\mathbf{d}_B \\ 0 & 0 \end{bmatrix} = {}^{G}A_B \, {}^{G}\mathbf{r}_P(t) \end{aligned} \tag{9.533} $$

where

$$ {}^{G}\ddot{R}_B \, {}^{G}R_B^{\mathrm{T}} = {}_{G}\tilde{\alpha}_B - \tilde{\omega}\,\tilde{\omega}^{\mathrm{T}} = {}_{G}S_B \tag{9.534} $$

■

Example 583 Velocity, Acceleration, and Jerk Transformation Matrices The velocity transformation matrix is a matrix to map a position vector to its velocity vector. Assume **p** and **q** denote the position vectors of two body points P and Q:

$$ {}^{G}\mathbf{q} - {}^{G}\mathbf{p} = {}^{G}R_B \left({}^{B}\mathbf{q} - {}^{B}\mathbf{p}\right) \tag{9.535} $$

Assume we have the kinematics of one of the points, say P. Then the position of the other point is

$$ {}^{G}\mathbf{q} = {}^{G}\mathbf{p} + {}^{G}R_{B}\left({}^{B}\mathbf{q} - {}^{B}\mathbf{p}\right) \tag{9.536}$$

This equation is similar to

$$ {}^{G}\mathbf{r} = {}^{G}\mathbf{d} + {}^{G}R_{B}\,{}^{B}\mathbf{r} \tag{9.537}$$

where the origin of the B-frame is at point P and ${}^{B}\mathbf{r}$ indicates the position of Q relative to P. Taking a time derivative shows that

$$ {}^{G}\dot{\mathbf{q}} - {}^{G}\dot{\mathbf{p}} = {}_{G}\tilde{\omega}_{B}\left({}^{G}\mathbf{q} - {}^{G}\mathbf{p}\right) \tag{9.538}$$

which can be converted to

$$ \begin{bmatrix} {}^{G}\dot{\mathbf{q}} \\ 0 \end{bmatrix} = \begin{bmatrix} {}_{G}\tilde{\omega}_{B} & {}^{G}\dot{\mathbf{p}} - {}_{G}\tilde{\omega}_{B}\,{}^{G}\mathbf{p} \\ 0 & 0 \end{bmatrix} \begin{bmatrix} {}^{G}\mathbf{q} \\ 1 \end{bmatrix} = {}^{G}V_{B} \begin{bmatrix} {}^{G}\mathbf{q} \\ 1 \end{bmatrix} \tag{9.539}$$

The matrix $[V]$ is the homogeneous velocity transformation matrix. Similarly, we obtain the acceleration equation

$$ \begin{aligned} {}^{G}\ddot{\mathbf{q}} - {}^{G}\ddot{\mathbf{p}} &= {}_{G}\tilde{\alpha}_{B}\left({}^{G}\mathbf{q} - {}^{G}\mathbf{p}\right) + {}_{G}\tilde{\omega}_{B}\left({}^{G}\dot{\mathbf{q}} - {}^{G}\dot{\mathbf{p}}\right) \\ &= {}_{G}\tilde{\alpha}_{B}\left({}^{G}\mathbf{q} - {}^{G}\mathbf{p}\right) + {}_{G}\tilde{\omega}_{B}\,{}_{G}\tilde{\omega}_{B}\left({}^{G}\mathbf{q} - {}^{G}\mathbf{p}\right) \\ &= \left({}_{G}\tilde{\alpha}_{B} - {}_{G}\tilde{\omega}_{B}\,{}_{G}\tilde{\omega}_{B}^{\mathrm{T}}\right)\left({}^{G}\mathbf{q} - {}^{G}\mathbf{p}\right) \\ &= {}_{G}S_{B}\left({}^{G}\mathbf{q} - {}^{G}\mathbf{p}\right) \end{aligned} \tag{9.540}$$

which can be converted to

$$ \begin{bmatrix} {}^{G}\ddot{\mathbf{q}} \\ 0 \end{bmatrix} = [A] \begin{bmatrix} {}^{G}\mathbf{q} \\ 1 \end{bmatrix} \tag{9.541}$$

where

$$ \begin{aligned} [A] &= \begin{bmatrix} {}_{G}\tilde{\alpha}_{B} - {}_{G}\tilde{\omega}_{B}\,{}_{G}\tilde{\omega}_{B}^{\mathrm{T}} & {}^{G}\ddot{\mathbf{p}} - \left({}_{G}\tilde{\alpha}_{B} - {}_{G}\tilde{\omega}_{B}\,{}_{G}\tilde{\omega}_{B}^{\mathrm{T}}\right){}^{G}\mathbf{p} \\ 0 & 0 \end{bmatrix} \\ &= \begin{bmatrix} {}_{G}S_{B} & {}^{G}\ddot{\mathbf{p}} - {}_{G}S_{B}\,{}^{G}\mathbf{p} \\ 0 & 0 \end{bmatrix} \end{aligned} \tag{9.542}$$

The matrix $[A]$ is the homogeneous acceleration transformation matrix for rigid motion.

The homogeneous jerk transformation matrix can be found by another differentiation:

$$ \begin{aligned} & {}^{G}\dddot{\mathbf{q}} - {}^{G}\dddot{\mathbf{p}} = \\ &= \left[\left({}_{G}\dot{\tilde{\alpha}}_{B} + 2\,{}_{G}\tilde{\omega}_{B}\,{}_{G}\tilde{\alpha}_{B}\right) + \left({}_{G}\tilde{\alpha}_{B} + {}_{G}\tilde{\omega}_{B}^{2}\right){}_{G}\tilde{\omega}_{B}\right]\left({}^{G}\mathbf{q} - {}^{G}\mathbf{p}\right) \\ &= \left({}_{G}\dot{\tilde{\alpha}}_{B} + 2\,{}_{G}\tilde{\omega}_{B}\,{}_{G}\tilde{\alpha}_{B} + \left({}_{G}\tilde{\alpha}_{B} + {}_{G}\tilde{\omega}_{B}^{2}\right){}_{G}\tilde{\omega}_{B}\right)\left({}^{G}\mathbf{q} - {}^{G}\mathbf{p}\right) \\ &= \left({}_{G}\tilde{\chi}_{B} + 2\,{}_{G}\tilde{\omega}_{B}\,{}_{G}\tilde{\alpha}_{B} + {}_{G}\tilde{\alpha}_{B}\,{}_{G}\tilde{\omega}_{B} + {}_{G}\tilde{\omega}_{B}^{3}\right)\left({}^{G}\mathbf{q} - {}^{G}\mathbf{p}\right) \end{aligned} \tag{9.543}$$

which can be converted to

$$\begin{bmatrix} {}^G\dddot{\mathbf{q}} \\ 0 \end{bmatrix} = [J] \begin{bmatrix} \mathbf{q} \\ 1 \end{bmatrix} \tag{9.544}$$

where $[J]$ is the homogeneous jerk transformation matrix

$$[J] = \begin{bmatrix} J_{11} & J_{12} \\ 0 & 0 \end{bmatrix} \tag{9.545}$$

where

$$J_{11} = {}_G\tilde{\chi}_B + 2\,{}_G\tilde{\omega}_B\,{}_G\tilde{\alpha}_B + {}_G\tilde{\alpha}_B\,{}_G\tilde{\omega}_B + {}_G\tilde{\omega}_B^3 \tag{9.546}$$

$$J_{12} = {}^G\ddot{\mathbf{p}} - \left({}_G\tilde{\chi}_B + 2\,{}_G\tilde{\omega}_B\,{}_G\tilde{\alpha}_B + {}_G\tilde{\alpha}_B\,{}_G\tilde{\omega}_B + {}_G\tilde{\omega}_B^3\right) {}^G\mathbf{p} \tag{9.547}$$

Example 584 ★ Homogeneous Jerk Transformation Matrix Following the same pattern we may define a jerk transformation as

$${}^G\mathbf{j}_P(t) = {}_GU_B\,{}^G\mathbf{r}_P(t) \tag{9.548}$$

where

$$\begin{aligned} {}^GJ_B &= \begin{bmatrix} {}^G\dddot{R}_B & {}^G\dddot{\mathbf{d}}_B \\ 0 & 0 \end{bmatrix} \begin{bmatrix} {}^GR_B^{\mathrm{T}} & -{}^GR_B^{\mathrm{T}}\,{}^G\mathbf{d}_B \\ 0 & 1 \end{bmatrix} \\ &= \begin{bmatrix} {}^G\dddot{R}_B\,{}^GR_B^{\mathrm{T}} & {}^G\dddot{\mathbf{d}}_B - {}^G\dddot{R}_B\,{}^GR_B^{\mathrm{T}}\,{}^G\mathbf{d}_B \\ 0 & 0 \end{bmatrix} \end{aligned} \tag{9.549}$$

and ${}_GU_B$ is the jerk transformation matrix:

$$\begin{aligned} {}^G\dddot{R}_B\,{}^GR_B^{\mathrm{T}} &= {}_GU_B \\ &= {}_G\ddot{\tilde{\omega}}_B - 2\left({}_G\dot{\tilde{\omega}}_B - \tilde{\omega}\,\tilde{\omega}^{\mathrm{T}}\right)\tilde{\omega}^{\mathrm{T}} - \tilde{\omega}\left({}_G\dot{\tilde{\omega}}_B - \tilde{\omega}\,\tilde{\omega}^{\mathrm{T}}\right)^{\mathrm{T}} \\ &= {}_G\tilde{\chi}_B - 2\left({}_G\tilde{\alpha}_B - \tilde{\omega}\,\tilde{\omega}^{\mathrm{T}}\right)\tilde{\omega}^{\mathrm{T}} - \tilde{\omega}\left({}_G\tilde{\alpha}_B - \tilde{\omega}\,\tilde{\omega}^{\mathrm{T}}\right)^{\mathrm{T}} \end{aligned} \tag{9.550}$$

Example 585 ★ Kinematics of the Gripper of a Planar R‖R Manipulator Figure 9.28 illustrates an R‖R planar manipulator with joint variables θ_1 and θ_2. Links (1) and (2) are both R‖R(0), and therefore, the transformation matrices are

$${}^0T_1 = \begin{bmatrix} \cos\theta_1 & -\sin\theta_1 & 0 & l_1\cos\theta_1 \\ \sin\theta_1 & \cos\theta_1 & 0 & l_1\sin\theta_1 \\ 0 & 0 & 1 & 0 \\ 0 & 0 & 0 & 1 \end{bmatrix} \tag{9.551}$$

$${}^1T_2 = \begin{bmatrix} \cos\theta_2 & -\sin\theta_2 & 0 & l_2\cos\theta_2 \\ \sin\theta_2 & \cos\theta_2 & 0 & l_2\sin\theta_2 \\ 0 & 0 & 1 & 0 \\ 0 & 0 & 0 & 1 \end{bmatrix} \tag{9.552}$$

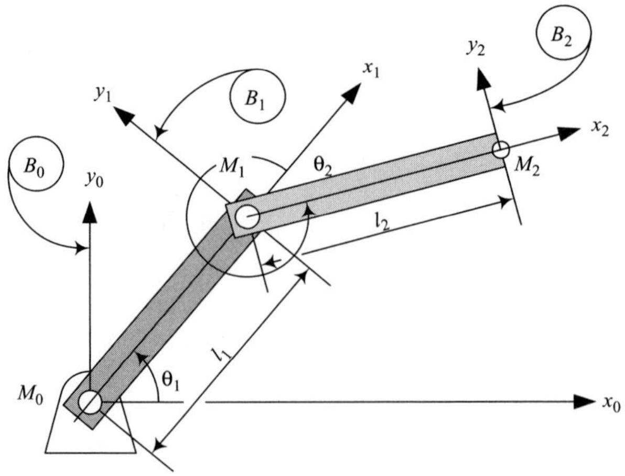

Figure 9.28 An R‖R planar manipulator with joint variables θ_1 and θ_2.

$$
\begin{aligned}
{}^0T_2 &= {}^0T_1\,{}^1T_2 \\
&= \begin{bmatrix} c(\theta_1+\theta_2) & -s(\theta_1+\theta_2) & 0 & l_2 c(\theta_1+\theta_2)+l_1 c\theta_1 \\ s(\theta_1+\theta_2) & c(\theta_1+\theta_2) & 0 & l_2 s(\theta_1+\theta_2)+l_1 s\theta_1 \\ 0 & 0 & 1 & 0 \\ 0 & 0 & 0 & 1 \end{bmatrix}
\end{aligned} \tag{9.553}
$$

The points M_1 and M_2 are at

$$
{}^0\mathbf{r}_{M_1} = \begin{bmatrix} l_1\cos\theta_1 \\ l_1\sin\theta_1 \\ 0 \\ 1 \end{bmatrix} \qquad {}^1\mathbf{r}_{M_2} = \begin{bmatrix} l_2\cos\theta_2 \\ l_2\sin\theta_2 \\ 0 \\ 1 \end{bmatrix} \tag{9.554}
$$

$$
{}^0\mathbf{r}_{M_2} = {}^0T_1\,{}^1\mathbf{r}_{M_2} = \begin{bmatrix} l_2\cos(\theta_1+\theta_2)+l_1\cos\theta_1 \\ l_2\sin(\theta_1+\theta_2)+l_1\sin\theta_1 \\ 0 \\ 1 \end{bmatrix} \tag{9.555}
$$

To determine the velocity and acceleration of M_2, we need to calculate ${}^0\dot{T}_2$, which can be done by direct differentiation of ${}^0T_2$:

$$
\begin{aligned}
{}^0\dot{T}_2 &= \frac{d}{dt}\,{}^0T_2 \\
&= \begin{bmatrix} -\dot\theta_{12}s\theta_{12} & -\dot\theta_{12}c\theta_{12} & 0 & -l_2\dot\theta_{12}s\theta_{12}-\dot\theta_1 l_1 s\theta_1 \\ \dot\theta_{12}c\theta_{12} & -\dot\theta_{12}s\theta_{12} & 0 & l_2\dot\theta_{12}c\theta_{12}+\dot\theta_1 l_1 c\theta_1 \\ 0 & 0 & 0 & 0 \\ 0 & 0 & 0 & 0 \end{bmatrix}
\end{aligned} \tag{9.556}
$$

$$
\theta_{12} = \theta_1 + \theta_2 \tag{9.557}
$$

$$
\dot\theta_{12} = \dot\theta_1 + \dot\theta_2 \tag{9.558}
$$

We can also calculate ${}^0\dot{T}_2$ from ${}^0T_2 = {}^0T_1\,{}^1T_2$ by the chain rule:

$$
{}^0\dot{T}_2 = \frac{d}{dt}\left({}^0T_1\,{}^1T_2\right) = {}^0\dot{T}_1\,{}^1T_2 + {}^0T_1\,{}^1\dot{T}_2 \tag{9.559}
$$

where

$$
{}^0\dot{T}_1 = \dot{\theta}_1 \begin{bmatrix} -\sin\theta_1 & -\cos\theta_1 & 0 & -l_1\sin\theta_1 \\ \cos\theta_1 & -\sin\theta_1 & 0 & l_1\cos\theta_1 \\ 0 & 0 & 0 & 0 \\ 0 & 0 & 0 & 0 \end{bmatrix} \tag{9.560}
$$

$$
{}^1\dot{T}_2 = \dot{\theta}_2 \begin{bmatrix} -\sin\theta_2 & -\cos\theta_2 & 0 & -l_2\sin\theta_2 \\ \cos\theta_2 & -\sin\theta_2 & 0 & l_2\cos\theta_2 \\ 0 & 0 & 0 & 0 \\ 0 & 0 & 0 & 0 \end{bmatrix}. \tag{9.561}
$$

Having ${}^0\dot{T}_1$ and ${}^1\dot{T}_2$, we can find the velocity transformation matrices ${}^0V_1$ and ${}^1V_2$ by using ${}^0T_1^{-1}$ and ${}^1T_2^{-1}$:

$$
{}^0T_1^{-1} = \begin{bmatrix} \cos\theta_1 & \sin\theta_1 & 0 & -l_1 \\ -\sin\theta_1 & \cos\theta_1 & 0 & 0 \\ 0 & 0 & 1 & 0 \\ 0 & 0 & 0 & 1 \end{bmatrix} \tag{9.562}
$$

$$
{}^1T_2^{-1} = \begin{bmatrix} \cos\theta_2 & \sin\theta_2 & 0 & -l_2 \\ -\sin\theta_2 & \cos\theta_2 & 0 & 0 \\ 0 & 0 & 1 & 0 \\ 0 & 0 & 0 & 1 \end{bmatrix} \tag{9.563}
$$

$$
{}^0V_1 = {}^0\dot{T}_1\,{}^0T_1^{-1} = \dot{\theta}_1\,{}_1^0\tilde{k} \tag{9.564}
$$

$$
{}^1V_2 = {}^1\dot{T}_2\,{}^1T_2^{-1} = \dot{\theta}_2\,{}_2^1\tilde{k} \tag{9.565}
$$

Now, we can determine the velocity of points M_1 and M_2 in B_0 and B_1, respectively:

$$
{}^0\mathbf{v}_{M_1} = {}^0V_1\,{}^0\mathbf{r}_{M_1} = \dot{\theta}_1 \begin{bmatrix} -l_1\sin\theta_1 \\ l_1\cos\theta_1 \\ 0 \\ 0 \end{bmatrix} \tag{9.566}
$$

$$
{}^1\mathbf{v}_{M_2} = {}^1V_2\,{}^1\mathbf{r}_{M_2} = \dot{\theta}_2 \begin{bmatrix} -l_2\sin\theta_2 \\ l_2\cos\theta_2 \\ 0 \\ 0 \end{bmatrix} \tag{9.567}
$$

To determine the velocity of the tip point M_2 in the base frame, we can use velocity vector addition:

$$
{}^0\mathbf{v}_{M_2} = {}^0\mathbf{v}_{M_1} + {}_1^0\mathbf{v}_{M_2} = {}^0\mathbf{v}_{M_1} + {}^0T_1\,{}^1\mathbf{v}_{M_2}
$$

$$= \begin{bmatrix} -\left(\dot{\theta}_1 + \dot{\theta}_2\right) l_2 \sin(\theta_1 + \theta_2) - \dot{\theta}_1 l_1 \sin\theta_1 \\ \left(\dot{\theta}_1 + \dot{\theta}_2\right) l_2 \cos(\theta_1 + \theta_2) + \dot{\theta}_1 l_1 \cos\theta_1 \\ 0 \\ 0 \end{bmatrix} \tag{9.568}$$

We can also determine ${}^0\mathbf{v}_{M_2}$ by using the velocity transformation matrix ${}^0V_2$:

$${}^0\mathbf{v}_{M_2} = {}^0V_2\,{}^0\mathbf{r}_{M_2}$$

where the velocity transformation matrix is

$${}^0V_2 = {}^0\dot{T}_2\,{}^0T_2^{-1} = \begin{bmatrix} 0 & -\dot{\theta}_1 - \dot{\theta}_2 & 0 & \dot{\theta}_2 l_1 \sin\theta_1 \\ \dot{\theta}_1 + \dot{\theta}_2 & 0 & 0 & -\dot{\theta}_2 l_1 \cos\theta_1 \\ 0 & 0 & 0 & 0 \\ 0 & 0 & 0 & 0 \end{bmatrix} \tag{9.569}$$

where

$${}^0T_2^{-1} = {}^2T_1\,{}^1T_0 = {}^1T_2^{-1}\,{}^0T_1^{-1} \tag{9.570}$$

$$= \begin{bmatrix} \cos(\theta_1 + \theta_2) & \sin(\theta_1 + \theta_2) & 0 & -l_2 - l_1 \cos\theta_2 \\ -\sin(\theta_1 + \theta_2) & \cos(\theta_1 + \theta_2) & 0 & l_1 \sin\theta_2 \\ 0 & 0 & 1 & 0 \\ 0 & 0 & 0 & 1 \end{bmatrix}$$

Furthermore, we can determine the velocity transformation matrix ${}^0V_2$ using the addition rule:

$${}^0V_2 = {}^0V_1 + {}_1^0V_2 \tag{9.571}$$

where

$${}_1^0V_2 = {}^0T_1\,{}^1V_2\,{}^0T_1^{-1} = \begin{bmatrix} 0 & -\dot{\theta}_2 & 0 & \dot{\theta}_2 l_1 \sin\theta_1 \\ \dot{\theta}_2 & 0 & 0 & -\dot{\theta}_2 l_1 \cos\theta_1 \\ 0 & 0 & 0 & 0 \\ 0 & 0 & 0 & 0 \end{bmatrix} \tag{9.572}$$

Therefore,

$${}^0\mathbf{v}_{M_2} = {}^0V_2\,{}^0\mathbf{r}_{M_2} \tag{9.573}$$

To determine the acceleration of M_2, we need to calculate ${}^0\ddot{T}_2$, which can be done by direct differentiation of ${}^0\dot{T}_2$:

$$\begin{aligned} {}^0\ddot{T}_2 &= \frac{d}{dt}\,{}^0\dot{T}_2 = \frac{d}{dt}\frac{d}{dt}\left({}^0T_1\,{}^1T_2\right) = \frac{d}{dt}\left({}^0\dot{T}_1\,{}^1T_2 + {}^0T_1\,{}^1\dot{T}_2\right) \\ &= {}^0\ddot{T}_1\,{}^1T_2 + 2\,{}^0\dot{T}_1\,{}^1\dot{T}_2 + {}^0T_1\,{}^1\ddot{T}_2 \end{aligned} \tag{9.574}$$

We have

$${}^0\ddot{T}_1 = \frac{d}{dt}\,{}^0\dot{T}_1$$

$$= \begin{bmatrix} -\dot{\theta}_1^2 c\theta_1 - \ddot{\theta}_1 s\theta_1 & \dot{\theta}_1^2 s\theta_1 - \ddot{\theta}_1 c\theta_1 & 0 & -\ddot{\theta}_1 l_1 s\theta_1 - l_1 \dot{\theta}_1^2 c\theta_1 \\ \ddot{\theta}_1 c\theta_1 - \dot{\theta}_1^2 s\theta_1 & -\dot{\theta}_1^2 c\theta_1 - \ddot{\theta}_1 s\theta_1 & 0 & l_1 \ddot{\theta}_1 c\theta_1 - l_1 \dot{\theta}_1^2 s\theta_1 \\ 0 & 0 & 0 & 0 \\ 0 & 0 & 0 & 0 \end{bmatrix} \tag{9.575}$$

$$
{}^{1}\ddot{T}_2 = \frac{d}{dt}\,{}^{1}\dot{T}_2
$$

$$
= \begin{bmatrix} -\dot{\theta}_2^2 c\theta_2 - \ddot{\theta}_2 s\theta_2 & \dot{\theta}_2^2 s\theta_2 - \ddot{\theta}_2 c\theta_2 & 0 & -\ddot{\theta}_2 l_2 s\theta_2 - l_2\dot{\theta}_2^2 c\theta_2 \\ \ddot{\theta}_2 c\theta_2 - \dot{\theta}_2^2 s\theta_2 & -\dot{\theta}_2^2 c\theta_2 - \ddot{\theta}_2 s\theta_2 & 0 & \ddot{\theta}_2 l_2 c\theta_2 - \dot{\theta}_2^2 l_2 s\theta_2 \\ 0 & 0 & 0 & 0 \\ 0 & 0 & 0 & 0 \end{bmatrix} \tag{9.576}
$$

and therefore,

$$
{}^{0}\ddot{T}_2 = {}^{0}\ddot{T}_1\,{}^{1}T_2 + 2\,{}^{0}\dot{T}_1\,{}^{1}\dot{T}_2 + {}^{0}T_1\,{}^{1}\ddot{T}_2
$$

$$
= \begin{bmatrix} r_{11} & r_{12} & 0 & r_{14} \\ r_{21} & r_{22} & 0 & r_{24} \\ 0 & 0 & 0 & 0 \\ 0 & 0 & 0 & 0 \end{bmatrix} \tag{9.577}
$$

where

$$
\begin{aligned}
r_{11} &= -\dot{\theta}_{12}^2 \cos\theta_{12} - \ddot{\theta}_{12} \sin\theta_{12} \\
r_{21} &= -\dot{\theta}_{12}^2 \sin\theta_{12} + \ddot{\theta}_{12} \cos\theta_{12} \\
r_{12} &= \dot{\theta}_{12}^2 \sin\theta_{12} - \ddot{\theta}_{12} \cos\theta_{12} \\
r_{22} &= -\dot{\theta}_{12}^2 \cos\theta_{12} - \ddot{\theta}_{12} \sin\theta_{12} \\
r_{14} &= -\dot{\theta}_{12}^2 l_2 \cos\theta_{12} - \ddot{\theta}_{12} l_2 \sin\theta_{12} - \dot{\theta}_1^2 l_1 \cos\theta_1 - \ddot{\theta}_1 l_1 \sin\theta_1 \\
r_{24} &= \ddot{\theta}_{12} l_2 \cos\theta_{12} - \dot{\theta}_{12}^2 l_2 \sin\theta_{12} - \dot{\theta}_1^2 l_1 \sin\theta_1 + \ddot{\theta}_1 l_1 \cos\theta_1
\end{aligned} \tag{9.578}
$$

Having ${}^{0}\ddot{T}_1$, ${}^{1}\ddot{T}_2$, and ${}^{0}\ddot{T}_2$, we can find the acceleration transformation matrices ${}^{0}A_1$, ${}^{1}A_2$, and ${}^{0}A_2$ by using ${}^{0}T_1^{-1}$, ${}^{1}T_2^{-1}$, and ${}^{0}T_2^{-1}$:

$$
{}^{0}A_1 = {}^{0}\ddot{T}_1\,{}^{0}T_1^{-1} = \begin{bmatrix} -\dot{\theta}_1^2 & -\ddot{\theta}_1 & 0 & 0 \\ \ddot{\theta}_1 & -\dot{\theta}_1^2 & 0 & 0 \\ 0 & 0 & 0 & 0 \\ 0 & 0 & 0 & 0 \end{bmatrix} \tag{9.579}
$$

$$
{}^{1}A_2 = {}^{1}\ddot{T}_2\,{}^{1}T_2^{-1} = \begin{bmatrix} -\dot{\theta}_2^2 & -\ddot{\theta}_2 & 0 & 0 \\ \ddot{\theta}_2 & -\dot{\theta}_2^2 & 0 & 0 \\ 0 & 0 & 0 & 0 \\ 0 & 0 & 0 & 0 \end{bmatrix} \tag{9.580}
$$

$$
{}^{0}A_2 = {}^{0}\ddot{T}_2\,{}^{0}T_2^{-1}
$$

$$
= \begin{bmatrix} -\dot{\theta}_{12}^2 & -\ddot{\theta}_{12} & 0 & l_1\dot{\theta}_2^2 \cos\theta_1 + 2\dot{\theta}_1\dot{\theta}_2 l_1 \cos\theta_1 + \ddot{\theta}_2 l_1 \sin\theta_1 \\ \ddot{\theta}_{12} & -\dot{\theta}_{12}^2 & 0 & l_1\dot{\theta}_2^2 \sin\theta_1 + 2\dot{\theta}_1\dot{\theta}_2 l_1 \sin\theta_1 - \ddot{\theta}_2 l_1 \cos\theta_1 \\ 0 & 0 & 0 & 0 \\ 0 & 0 & 0 & 0 \end{bmatrix} \tag{9.581}
$$

Now, we can determine the velocities of points M_1 and M_2 in B_0 and B_1, respectively:

$$
{}^{0}\mathbf{a}_{M_1} = {}^{0}A_1\,{}^{0}\mathbf{r}_{M_1} = \begin{bmatrix} -l_1\dot{\theta}_1^2\cos\theta_1 - \ddot{\theta}_1 l_1\sin\theta_1 \\ \ddot{\theta}_1 l_1\cos\theta_1 - \dot{\theta}_1^2 l_1\sin\theta_1 \\ 0 \\ 0 \end{bmatrix} \tag{9.582}
$$

$$
{}^{1}\mathbf{a}_{M_2} = {}^{1}A_2\,{}^{1}\mathbf{r}_{M_2} = \begin{bmatrix} -l_2\dot{\theta}_2^2\cos\theta_2 - \ddot{\theta}_2 l_2\sin\theta_2 \\ \ddot{\theta}_2 l_2\cos\theta_2 - \dot{\theta}_2^2 l_2\sin\theta_2 \\ 0 \\ 0 \end{bmatrix} \tag{9.583}
$$

$$
\begin{aligned}
{}^{0}\mathbf{a}_{M_2} &= {}^{0}A_2\,{}^{0}\mathbf{r}_{M_2} \\
&= \begin{bmatrix} -\ddot{\theta}_{12} l_2\left(\cos\theta_{12} + \sin\theta_{12}\right) - \dot{\theta}_1^2 l_1\cos\theta_1 - \ddot{\theta}_1 l_1\sin\theta_1 \\ \ddot{\theta}_{12} l_2\left(\cos\theta_{12} - \sin\theta_{12}\right) - \dot{\theta}_1^2 l_1\sin\theta_1 + \ddot{\theta}_1 l_1\cos\theta_1 \\ 0 \\ 0 \end{bmatrix}
\end{aligned} \tag{9.584}
$$

9.7 ★ FORWARD-ACCELERATION KINEMATICS

The forward-acceleration kinematics problem of multibodies is the method of relating the joint accelerations $\ddot{\mathbf{q}}$ to the link accelerations $\ddot{\mathbf{X}}$,

$$
\ddot{\mathbf{X}} = \mathbf{J}\,\ddot{\mathbf{q}} + \dot{\mathbf{J}}\,\dot{\mathbf{q}} \tag{9.585}
$$

where $\mathbf{J}$ is the Jacobian matrix, $\mathbf{q}$ is the joint variable associate vector, $\dot{\mathbf{q}}$ is the joint velocity associate vector, and $\ddot{\mathbf{q}}$ is the *joint acceleration associate vector:*

$$
\mathbf{q} = \begin{bmatrix} q_1 \\ q_2 \\ \vdots \\ q_n \end{bmatrix} \qquad \dot{\mathbf{q}} = \begin{bmatrix} \dot{q}_1 \\ \dot{q}_2 \\ \vdots \\ \dot{q}_n \end{bmatrix} \qquad \ddot{\mathbf{q}} = \begin{bmatrix} \ddot{q}_1 \\ \ddot{q}_2 \\ \vdots \\ \ddot{q}_n \end{bmatrix} \tag{9.586}
$$

Consider a serial multibody with n connected links. The vector $\mathbf{X}$ is the *final link configuration vector*, $\dot{\mathbf{X}}$ is the *final link configuration velocity vector*, and $\ddot{\mathbf{X}}$ is the *final link configuration acceleration vector*:

$$
\mathbf{X} = \begin{bmatrix} X_n & Y_n & Z_n & \varphi_n & \theta_n & \psi_n \end{bmatrix}^{\mathrm{T}} \tag{9.587}
$$

$$
\dot{\mathbf{X}} = \begin{bmatrix} {}^{0}\mathbf{v}_n \\ {}^{0}\boldsymbol{\omega}_n \end{bmatrix} = \begin{bmatrix} {}^{0}\dot{\mathbf{d}}_n \\ {}^{0}\boldsymbol{\omega}_n \end{bmatrix} \tag{9.588}
$$

$$
= \begin{bmatrix} \dot{X}_n & \dot{Y}_n & \dot{Z}_n & \omega_{Xn} & \omega_{Yn} & \omega_{Zn} \end{bmatrix}^{\mathrm{T}} \tag{9.589}
$$

$$
\begin{aligned}
\ddot{\mathbf{X}} &= \begin{bmatrix} {}^{0}\mathbf{a}_n \\ {}^{0}\boldsymbol{\alpha}_n \end{bmatrix} = \begin{bmatrix} {}^{0}\ddot{\mathbf{d}}_n \\ {}^{0}\dot{\boldsymbol{\omega}}_n \end{bmatrix} \\
&= \begin{bmatrix} \ddot{X}_n & \ddot{Y}_n & \ddot{Z}_n & \dot{\omega}_{Xn} & \dot{\omega}_{Yn} & \dot{\omega}_{Zn} \end{bmatrix}^{\mathrm{T}}
\end{aligned} \tag{9.590}
$$

To calculate the time derivative of the Jacobian matrix $\left[\dot{\mathbf{J}}\right]$, we use Equations (8.536)–(8.539),

$$
{}_{i-1}^{\;\;0}\dot{\mathbf{d}}_i = \begin{cases} {}_0^0\boldsymbol{\omega}_i \times {}_{i-1}^{\;\;0}\mathbf{d}_i & \text{if joint } i \text{ is R} \\ \dot{d}_i\,{}^0\hat{k}_{i-1} + {}_0^0\boldsymbol{\omega}_i \times {}_{i-1}^{\;\;0}\mathbf{d}_i & \text{if joint } i \text{ is P} \end{cases} \tag{9.591}
$$

$$
{}_{i-1}^{\;\;0}\boldsymbol{\omega}_i = \begin{cases} \dot{\theta}_i\,{}^0\hat{k}_{i-1} & \text{if joint } i \text{ is R} \\ 0 & \text{if joint } i \text{ is P} \end{cases} \tag{9.592}
$$

and take a derivative to find the acceleration of link (i) with respect to its previous link $(i-1)$:

$$
{}_{i-1}^{\;\;0}\ddot{\mathbf{d}}_i = \begin{cases} {}_0^0\dot{\boldsymbol{\omega}}_i \times {}_{i-1}^{\;\;0}\mathbf{d}_i + {}_0^0\boldsymbol{\omega}_i \times \left({}_0^0\boldsymbol{\omega}_i \times {}_{i-1}^{\;\;0}\mathbf{d}_i\right) & \text{if joint } i \text{ is R} \\ {}_0^0\dot{\boldsymbol{\omega}}_i \times {}_{i-1}^{\;\;0}\mathbf{d}_i + {}_0^0\boldsymbol{\omega}_i \times \left({}_0^0\boldsymbol{\omega}_i \times {}_{i-1}^{\;\;0}\mathbf{d}_i\right) & \\ \qquad + \ddot{d}_i\,{}^0\hat{k}_{i-1} + 2\dot{d}_i\,{}_0^0\boldsymbol{\omega}_{i-1} \times {}^0\hat{k}_{i-1} & \text{if joint } i \text{ is P} \end{cases} \tag{9.593}
$$

$$
{}_{i-1}^{\;\;0}\dot{\boldsymbol{\omega}}_i = \begin{cases} \ddot{\theta}_i\,{}^0\hat{k}_{i-1} + \dot{\theta}_i\,{}_0^0\boldsymbol{\omega}_{i-1} \times {}^0\hat{k}_{i-1} & \text{if joint } i \text{ is R} \\ 0 & \text{if joint } i \text{ is P} \end{cases} \tag{9.594}
$$

Therefore, the acceleration vectors of the final link are

$$
{}_0^0\ddot{\mathbf{d}}_n = \sum_{i=1}^{n} {}_{i-1}^{\;\;0}\ddot{\mathbf{d}}_i \tag{9.595}
$$

$$
{}_0^0\dot{\boldsymbol{\omega}}_n = \sum_{i=1}^{n} {}_{i-1}^{\;\;0}\boldsymbol{\omega}_i \tag{9.596}
$$

The acceleration relationships can also be rearranged in the recursive form

$$
{}_{i-1}^{\;\;0}\dot{\boldsymbol{\omega}}_i = \begin{cases} \ddot{\theta}_i\,{}^0\hat{k}_{i-1} + \dot{\theta}_i\,{}_0^0\boldsymbol{\omega}_{i-1} \times {}^0\hat{k}_{i-1} & \text{if joint } i \text{ is R} \\ 0 & \text{if joint } i \text{ is P} \end{cases} \tag{9.597}
$$

Example 586 Forward Acceleration of a 2R Planar Manipulator The forward velocity of the $2R$ planar manipulator of Figure 9.8 is

$$
\dot{\mathbf{X}} = \mathbf{J}\,\dot{\mathbf{q}} \tag{9.598}
$$

$$
\begin{bmatrix} \dot{X} \\ \dot{Y} \end{bmatrix} = \begin{bmatrix} -l_1 s\theta_1 - l_2 s\,(\theta_1+\theta_2) & -l_2 s\,(\theta_1+\theta_2) \\ l_1 c\theta_1 + l_2 c\,(\theta_1+\theta_2) & l_2 c\,(\theta_1+\theta_2) \end{bmatrix} \begin{bmatrix} \dot{\theta}_1 \\ \dot{\theta}_2 \end{bmatrix}
$$

The differential of the Jacobian matrix is

$$
\dot{\mathbf{J}} = \begin{bmatrix} \dot{J}_{11} & \dot{J}_{12} \\ \dot{J}_{21} & \dot{J}_{22} \end{bmatrix} \tag{9.599}
$$

where

$$
\begin{aligned}
\dot{J}_{11} &= (-l_1 \cos\theta_1 - l_2 \cos(\theta_1+\theta_2))\,\dot{\theta}_1 - l_2 \cos(\theta_1+\theta_2)\,\dot{\theta}_2 \\
\dot{J}_{12} &= -l_2 \cos(\theta_1+\theta_2)\,\dot{\theta}_1 - l_2 \cos(\theta_1+\theta_2)\,\dot{\theta}_2 \\
\dot{J}_{21} &= (-l_1 \sin\theta_1 - l_2 \sin(\theta_1+\theta_2))\dot{\theta}_1 - l_2 \sin(\theta_1+\theta_2)\,\dot{\theta}_2 \\
\dot{J}_{22} &= -l_2 \sin(\theta_1+\theta_2)\,\dot{\theta}_1 - l_2 \sin(\theta_1+\theta_2)\,\dot{\theta}_2
\end{aligned} \tag{9.600}
$$

Therefore, the forward-acceleration kinematics of the manipulator can be rearranged in the form

$$\ddot{\mathbf{X}} = \mathbf{J}\,\ddot{\mathbf{q}} + \dot{\mathbf{J}}\,\dot{\mathbf{q}} \tag{9.601}$$

The acceleration of the $2R$ manipulator can be arranged in the form

$$\begin{aligned}\begin{bmatrix}\ddot{X}\\ \ddot{Y}\end{bmatrix} &= \begin{bmatrix}-l_1\sin\theta_1 & -l_2\sin(\theta_1+\theta_2)\\ l_1\cos\theta_1 & l_2\cos(\theta_1+\theta_2)\end{bmatrix}\begin{bmatrix}\ddot{\theta}_1\\ \ddot{\theta}_1+\ddot{\theta}_2\end{bmatrix}\\ &\quad - \begin{bmatrix}l_1\cos\theta_1 & l_2\cos(\theta_1+\theta_2)\\ l_1\sin\theta_1 & l_2\sin(\theta_1+\theta_2)\end{bmatrix}\begin{bmatrix}\dot{\theta}_1^2\\ (\dot{\theta}_1+\dot{\theta}_2)^2\end{bmatrix}\end{aligned} \tag{9.602}$$

Example 587 ★ Jacobian Matrix of a Planar Polar Manipulator Figure 8.17 illustrates a planar polar manipulator with the following forward kinematics:

$$\begin{aligned}{}^0T_2 = {}^0T_1\,{}^1T_2 &= \begin{bmatrix}\cos\theta & -\sin\theta & 0 & 0\\ \sin\theta & \cos\theta & 0 & 0\\ 0 & 0 & 1 & 0\\ 0 & 0 & 0 & 1\end{bmatrix}\begin{bmatrix}1 & 0 & 0 & r\\ 0 & 1 & 0 & 0\\ 0 & 0 & 1 & 0\\ 0 & 0 & 0 & 1\end{bmatrix}\\ &= \begin{bmatrix}\cos\theta & -\sin\theta & 0 & r\cos\theta\\ \sin\theta & \cos\theta & 0 & r\sin\theta\\ 0 & 0 & 1 & 0\\ 0 & 0 & 0 & 1\end{bmatrix}\end{aligned} \tag{9.603}$$

The tip point of the manipulator is at

$$\begin{bmatrix}X\\ Y\end{bmatrix} = \begin{bmatrix}r\cos\theta\\ r\sin\theta\end{bmatrix} \tag{9.604}$$

and therefore its velocity is

$$\begin{bmatrix}\dot{X}\\ \dot{Y}\end{bmatrix} = \begin{bmatrix}\cos\theta & -r\sin\theta\\ \sin\theta & r\cos\theta\end{bmatrix}\begin{bmatrix}\dot{r}\\ \dot{\theta}\end{bmatrix} \tag{9.605}$$

which shows that

$$\mathbf{J}_D = \begin{bmatrix}\cos\theta & -r\sin\theta\\ \sin\theta & r\cos\theta\end{bmatrix} \tag{9.606}$$

9.8 ★ INVERSE-ACCELERATION KINEMATICS

Inverse-acceleration kinematics refers to determination of the joint acceleration vector $\ddot{\mathbf{q}}$ from forward-acceleration kinematics:

$$\ddot{\mathbf{X}} = \begin{bmatrix}{}^0\mathbf{a}_n\\ {}^0\boldsymbol{\alpha}_n\end{bmatrix} = \mathbf{J}\,\ddot{\mathbf{q}} + \dot{\mathbf{J}}\,\dot{\mathbf{q}} \tag{9.607}$$

Assuming that the Jacobian matrix $\mathbf{J}$ is square and nonsingular, the joint acceleration vector $\ddot{\mathbf{q}}$ can be found by matrix inversion:

$$\ddot{\mathbf{q}} = \mathbf{J}^{-1}\left(\ddot{\mathbf{X}} - \dot{\mathbf{J}}\,\dot{\mathbf{q}}\right) \tag{9.608}$$

However, calculating $\dot{\mathbf{J}}$ and $\mathbf{J}^{-1}$ becomes more tedious by increasing the DOF of the multibody.

An alternative technique is to write the equation in the new form

$$\ddot{\mathbf{X}} - \dot{\mathbf{J}}\,\dot{\mathbf{q}} = \mathbf{J}\,\ddot{\mathbf{q}} \tag{9.609}$$

$$\begin{bmatrix}\mathbf{m}\\ \mathbf{n}\end{bmatrix} = \begin{bmatrix}A & 0\\ C & D\end{bmatrix}\ddot{\mathbf{q}} \tag{9.610}$$

where

$$\begin{bmatrix}\mathbf{m}\\ \mathbf{n}\end{bmatrix} = \ddot{\mathbf{X}} - \dot{\mathbf{J}}\,\dot{\mathbf{q}} \qquad \dot{\mathbf{q}} = \begin{bmatrix}\ddot{q}_1\\ \ddot{q}_2\\ \ddot{q}_3\\ \ddot{q}_4\\ \ddot{q}_5\\ \ddot{q}_6\end{bmatrix} \tag{9.611}$$

Therefore, the inverse-acceleration kinematics problem can be solved as

$$\begin{bmatrix}\ddot{q}_1\\ \ddot{q}_2\\ \ddot{q}_3\end{bmatrix} = A^{-1}\,[\mathbf{m}] \tag{9.612}$$

and

$$\begin{bmatrix}\ddot{q}_4\\ \ddot{q}_5\\ \ddot{q}_6\end{bmatrix} = D^{-1}\left([\mathbf{n}] - [C]\begin{bmatrix}\ddot{q}_1\\ \ddot{q}_2\\ \ddot{q}_3\end{bmatrix}\right) \tag{9.613}$$

The matrix $\mathbf{J}\dot{q} = \ddot{\mathbf{X}} - \mathbf{J}\,\ddot{\mathbf{q}}$ is called the *acceleration bias vector* and can be calculated by differentiating from

$${}^0_0\dot{\mathbf{d}}_6 = \sum_{i=1}^{3} {}^0_0\boldsymbol{\omega}_i \times {}_{i-1}^{\;\;0}\mathbf{d}_i \tag{9.614}$$

$${}^0_0\boldsymbol{\omega}_6 = \sum_{i=1}^{6} \dot{\theta}_i\, {}^0\hat{k}_{i-1} \tag{9.615}$$

to

$${}^0_0\mathbf{a}_6 = {}^0_0\ddot{\mathbf{d}}_6 = \sum_{i=1}^{3}\left[{}^0_0\dot{\boldsymbol{\omega}}_i \times {}_{i-1}^{\;\;0}\mathbf{d}_i + {}^0_0\boldsymbol{\omega}_i \times \left({}^0_0\boldsymbol{\omega}_i \times {}_{i-1}^{\;\;0}\mathbf{d}_i\right)\right] \tag{9.616}$$

$${}^0_0\boldsymbol{\alpha}_6 = {}^0_0\dot{\boldsymbol{\omega}}_6 = \sum_{i=1}^{6}\left(\ddot{\theta}_i\, {}^0\hat{k}_{i-1} + {}^0_0\boldsymbol{\omega}_i \times \dot{\theta}_i\, {}^0\hat{k}_{i-1}\right) \tag{9.617}$$

The angular acceleration vector ${}_0^0\boldsymbol{\alpha}_6$ is the second half of $\ddot{\mathbf{X}}$. Then, subtracting the second half of $\mathbf{J}\,\ddot{\mathbf{q}}$ from ${}_0^0\boldsymbol{\alpha}_6$ provides the second half of the bias vector:

$$\sum_{i=1}^{6} {}_0^0\boldsymbol{\omega}_i \times \dot{\theta}_i \, {}^0\hat{k}_{i-1} \tag{9.618}$$

We substitute

$${}_0^0\boldsymbol{\omega}_i = \sum_{j=1}^{i} \dot{\theta}_j \, {}^0\hat{k}_{j-1} \tag{9.619}$$

$${}_0^0\dot{\boldsymbol{\omega}}_i = \sum_{j=1}^{i} \left(\ddot{\theta}_j \, {}^0\hat{k}_{j-1} + {}_0^0\boldsymbol{\omega}_{j-1} \times \dot{\theta}_j \, {}^0\hat{k}_{j-1} \right) \tag{9.620}$$

into Equation (9.616),

$$\begin{aligned}
{}_0^0\ddot{\mathbf{d}}_6 &= \sum_{i=1}^{3}\sum_{j=1}^{i} \left(\ddot{\theta}_j \, {}^0\hat{k}_{j-1} + {}_0^0\boldsymbol{\omega}_{j-1} \times \dot{\theta}_j \, {}^0\hat{k}_{j-1} \right) \times {}_{i-1}^{\;\;0}\mathbf{d}_i \\
&\quad + \sum_{i=1}^{3}\sum_{j=1}^{i} \dot{\theta}_j \, {}^0\hat{k}_{j-1} \times \left({}_0^0\boldsymbol{\omega}_i \times {}_{i-1}^{\;\;0}\mathbf{d}_i \right) \\
&= \sum_{i=1}^{3}\sum_{j=1}^{i} \ddot{\theta}_j \, {}^0\hat{k}_{j-1} \times {}_{i-1}^{\;\;0}\mathbf{d}_i + \sum_{i=1}^{3}\sum_{j=1}^{i} \left({}_0^0\boldsymbol{\omega}_{j-1} \times \dot{\theta}_j \, {}^0\hat{k}_{j-1} \right) \times {}_{i-1}^{\;\;0}\mathbf{d}_i \\
&\quad + \sum_{i=1}^{3}\sum_{j=1}^{i} \dot{\theta}_j \, {}^0\hat{k}_{j-1} \times \left({}_0^0\boldsymbol{\omega}_i \times {}_{i-1}^{\;\;0}\mathbf{d}_i \right)
\end{aligned} \tag{9.621}$$

to find the first half of the bias vector:

$$\sum_{i=1}^{3}\sum_{j=1}^{i} \left({}_0^0\boldsymbol{\omega}_{j-1} \times \dot{\theta}_j \, {}^0\hat{k}_{j-1} \right) \times {}_{i-1}^{\;\;0}\mathbf{d}_i + \sum_{i=1}^{3}\sum_{j=1}^{i} \dot{\theta}_j \, {}^0\hat{k}_{j-1} \times \left({}_0^0\boldsymbol{\omega}_i \times {}_{i-1}^{\;\;0}\mathbf{d}_i \right) \tag{9.622}$$

Example 588 Inverse Acceleration of a $2R$ Planar Manipulator The forward velocity and acceleration of the $2R$ planar manipulator are found as

$$\begin{aligned}
\dot{\mathbf{X}} &= \mathbf{J}\,\dot{\mathbf{q}} \\
\begin{bmatrix} \dot{X} \\ \dot{Y} \end{bmatrix} &= \begin{bmatrix} -l_1 s\theta_1 - l_2 s\left(\theta_1 + \theta_2\right) & -l_2 s\left(\theta_1 + \theta_2\right) \\ l_1 c\theta_1 + l_2 c\left(\theta_1 + \theta_2\right) & l_2 c\left(\theta_1 + \theta_2\right) \end{bmatrix} \begin{bmatrix} \dot{\theta}_1 \\ \dot{\theta}_2 \end{bmatrix}
\end{aligned} \tag{9.623}$$

$$\begin{aligned}
\ddot{\mathbf{X}} &= \mathbf{J}\,\ddot{\mathbf{q}} + \dot{\mathbf{J}}\,\dot{\mathbf{q}} \\
\begin{bmatrix} \ddot{X} \\ \ddot{Y} \end{bmatrix} &= \begin{bmatrix} -l_1 \sin\theta_1 & -l_2 \sin\left(\theta_1 + \theta_2\right) \\ l_1 \cos\theta_1 & l_2 \cos\left(\theta_1 + \theta_2\right) \end{bmatrix} \begin{bmatrix} \ddot{\theta}_1 \\ \ddot{\theta}_1 + \ddot{\theta}_2 \end{bmatrix} \\
&\quad - \begin{bmatrix} l_1 \cos\theta_1 & l_2 \cos\left(\theta_1 + \theta_2\right) \\ l_1 \sin\theta_1 & l_2 \sin\left(\theta_1 + \theta_2\right) \end{bmatrix} \begin{bmatrix} \dot{\theta}_1^2 \\ \left(\dot{\theta}_1 + \dot{\theta}_2\right)^2 \end{bmatrix}
\end{aligned} \tag{9.624}$$

The derivative and inverse Jacobian matrices are

$$\dot{\mathbf{J}} = \begin{bmatrix} -l_1\dot{\theta}_1 c\theta_1 - l_2\left(\dot{\theta}_1 + \dot{\theta}_2\right) c\left(\theta_1 + \theta_2\right) & -l_2\left(\dot{\theta}_1 + \dot{\theta}_2\right) c\left(\theta_1 + \theta_2\right) \\ -l_1\dot{\theta}_1 s\theta_1 - l_2\left(\dot{\theta}_1 + \dot{\theta}_2\right) s\left(\theta_1 + \theta_2\right) & -l_2\left(\dot{\theta}_1 + \dot{\theta}_2\right) s\left(\theta_1 + \theta_2\right) \end{bmatrix} \quad (9.625)$$

$$\mathbf{J}^{-1} = \frac{-1}{l_1 l_2 s\theta_2} \begin{bmatrix} -l_2 c\left(\theta_1 + \theta_2\right) & -l_2 s\left(\theta_1 + \theta_2\right) \\ l_1 c\theta_1 + l_2 c\left(\theta_1 + \theta_2\right) & l_1 s\theta_1 + l_2 s\left(\theta_1 + \theta_2\right) \end{bmatrix} \quad (9.626)$$

Therefore, the inverse-acceleration kinematics of the manipulator

$$\ddot{\mathbf{q}} = \mathbf{J}^{-1}\left(\ddot{\mathbf{X}} - \dot{\mathbf{J}}\,\dot{\mathbf{q}}\right) \quad (9.627)$$

can be arranged as

$$\begin{bmatrix} \ddot{\theta}_1 \\ \ddot{\theta}_1 + \ddot{\theta}_2 \end{bmatrix} = \frac{1}{l_1 l_2 s\theta_2} \begin{bmatrix} l_2 \cos\left(\theta_1 + \theta_2\right) & l_2 \sin\left(\theta_1 + \theta_2\right) \\ -l_1 \cos\theta_1 & -l_1 \sin\theta_1 \end{bmatrix} \begin{bmatrix} \ddot{X} \\ \ddot{Y} \end{bmatrix}$$
$$+ \frac{1}{l_1 l_2 s\theta_2} \begin{bmatrix} l_1 l_2 \cos\theta_1 & l_2^2 \\ -l_1^2 & -l_1 l_2 \cos\theta_1 \end{bmatrix} \begin{bmatrix} \dot{\theta}_1^2 \\ \left(\dot{\theta}_1 + \dot{\theta}_2\right)^2 \end{bmatrix} \quad (9.628)$$

KEY SYMBOLS

a	kinematic length of a link
A, B, C, D	submatrices of **J**
B	body coordinate frame
c	cos
$\mathbf{c}$	Jacobian-generating vector
d	differential, prismatic joint variable
d_x, d_y, d_z	elements of **d**
$\mathbf{d}$	translation vector, displacement vector
D	displacement transformation matrix
DOF	degrees of freedom
e	rotation quaternion
e^*	conjugate of e
G, B_0	global coordinate frame, base coordinate frame
$\hat{\imath}, \hat{\jmath}, \hat{k}$	local coordinate axis unit vectors
$\tilde{\imath}, \tilde{\jmath}, \tilde{k}$	skew-symmetric matrices of the unit vectors $\hat{\imath}, \hat{\jmath}, \hat{k}$
$\hat{I}, \hat{J}, \hat{K}$	global coordinate axis unit vectors
$\mathbf{I} = [\mathbf{I}]$	identity matrix
$\mathbf{J}$	Jacobian, geometric Jacobian
$\mathbf{J}_D$	displacement Jacobian
$\mathbf{J}_R$	rotational Jacobian
$\mathbf{J}_\phi$	angular Jacobian
$\mathbf{J}_A$	analytic Jacobian
l	length
P	prismatic joint, point
q	joint coordinate
$\mathbf{q}$	vector joint coordinates
$\mathbf{r}$	position vectors, homogeneous position vector

r_i	element i of $\mathbf{r}$
r_{ij}	element of row i and column j of a matrix
R	rotation transformation matrix, revolute joint
s	sin
S	rotational acceleration transformation
t_{ij}	the element of row i and column j of T
T	homogeneous transformation matrix
T_{arm}	manipulator transformation matrix
T_{wrist}	wrist transformation matrix
$\mathbf{T}$	a set of nonlinear algebraic equations of $\mathbf{q}$
$\mathbf{v}$	velocity vector
V	velocity transformation matrix
$\hat{u}$	unit vector along axis of $\boldsymbol{\omega}$
$\tilde{u}$	skew-symmetric matrix of vector $\hat{u}$
u_1, u_2, u_3	components of $\hat{u}$
x, y, z	local coordinate axes
X, Y, Z	global coordinate axes, coordinates of end-effector

Greek

α, β, γ	angles of rotation about the axes of global frame
δ	Kronecker function, small increment of a parameter
ϵ	small test number to terminate a procedure
θ	rotary joint angle
θ_{ijk}	$\theta_i + \theta_j + \theta_k$
φ, θ, ψ	angles of rotation about the axes of body frame
ϕ	angle of rotation about $\hat{u}$
$\boldsymbol{\omega}$	angular velocity vector
$\hat{u}$	unit vector along axis of $\boldsymbol{\omega}$
$\tilde{\omega}$	skew-symmetric matrix of vector $\boldsymbol{\omega}$
$\omega_1, \omega_2, \omega_3$	components of $\boldsymbol{\omega}$

Symbol

$[\ ]^{-1}$	inverse of matrix [ ]
$[\ ]^{\mathrm{T}}$	transpose of matrix [ ]
$\vdash$	orthogonal
$\parallel$	parallel
$\perp$	perpendicular

EXERCISES

1. **Local Position, Global Acceleration** A body is turning about a global principal axis at a constant angular acceleration of 4 rad/s^2. Find the global velocity and acceleration of a point P at

$$ {}^{B}\mathbf{r} = \begin{bmatrix} 10 \\ 20 \\ 30 \end{bmatrix} $$

   (a) If the axis is the Z-axis, the angular velocity is 3 rad/s, and the angle of rotation is $\pi/3$ rad.

(b) If the axis is the X-axis, the angular velocity is $2\,\mathrm{rad/s}$, and the angle of rotation is $\pi/4\,\mathrm{rad}$.

(c) If the axis is the Y-axis, the angular velocity is $1\,\mathrm{rad}/s$, and the angle of rotation is $\pi/6\,\mathrm{rad}$.

2. **Global Position, Constant Angular Acceleration** A body is turning about the Z-axis at a constant angular acceleration $\ddot{\alpha} = 0.4\,\mathrm{rad/s^2}$. Find the global position of a point at

$$^{B}\mathbf{r} = \begin{bmatrix} 10 \\ 20 \\ 30 \end{bmatrix}$$

after $t = 3\,\mathrm{s}$ when $\dot{\alpha} = 5\mathrm{rad/s}$ if the body and global coordinate frames were coincident at $t = 0\,\mathrm{s}$.

3. **★ Angular Velocity and Acceleration Matrices** A body B is turning in the global frame G. The transformation matrix can be simulated by a rotation α about the Z-axis followed by a rotation β about the X-axis.

(a) Determine the axis and angle of rotation $\hat{u}$ and ϕ as functions of α and β.

(b) Show that if α and β are changing with constant rate, $\hat{u}$ and ϕ also change. Determine the conditions that keep $\hat{u}$ constant and the conditions that keep ϕ constant.

(c) Assume $\hat{u}$ is constant and $\dot{\phi} = 6\,\mathrm{rad/s}$. Determine $\dot{\alpha}$ and $\dot{\beta}$ when $\phi = 30\,\mathrm{deg}$.

(d) Determine $\dot{\phi}$ when $\alpha = 30\,\mathrm{deg}$, $\beta = 45\,\mathrm{deg}$, $\dot{\alpha} = 6\,\mathrm{rad/s}$, and $\dot{\beta} = 6\,\mathrm{rad/s}$.

4. **Turning about x-Axis** Find the angular acceleration matrix when the body coordinate frame is turning $-15\,\mathrm{deg/s^2}$, $30\,\mathrm{deg/s}$ at $45\,\mathrm{deg}$ about the x-axis.

5. **Angular Acceleration and Euler Angles** Calculate the angular velocity and acceleration vectors in body and global coordinate frames if the Euler angles and their rates are:

$$\begin{aligned} &\varphi = 0.5\,\mathrm{rad} && \dot{\varphi} = 2.5\,\mathrm{rad/s} && \ddot{\varphi} = 5\,\mathrm{rad/s^2} \\ &\theta = -0.2\,\mathrm{rad} && \dot{\theta} = -3.5\,\mathrm{rad/s} && \ddot{\theta} = 3.5\,\mathrm{rad/s^2} \\ &\psi = 0.5\,\mathrm{rad} && \dot{\psi} = 3\,\mathrm{rad/s} && \ddot{\psi} = 2.5\,\mathrm{rad/s^2} \end{aligned}$$

6. **Combined Angular Accelerations**

(a) A body B is turning about the z_1-axis of a B_1-frame with angular acceleration $\alpha\,\mathrm{rad/s^2}$, while B_1 is turning about the X-axis of the global G-frame with angular acceleration $\beta\,\mathrm{rad/s^2}$. Determine ${}_B\alpha_1$, ${}_G\alpha_B$, ${}_G\alpha_1$, ${}_BS_1$, ${}_GS_B$, ${}_GS_1$.

(b) The frame B_1 is also turning about the y_2-axis of the B_2-frame with angular acceleration $\gamma\,\mathrm{rad/s^2}$, while B_2 is turning about the X-axis of the global G-frame with angular acceleration $\beta\,\mathrm{rad/s^2}$. Determine ${}_B\alpha_1$, ${}_G\alpha_2$, ${}_G\alpha_1$, ${}_B\alpha_2$, ${}_G\alpha_B$, ${}_2\alpha_1$, ${}_BS_1$, ${}_GS_2$, ${}_GS_1$, ${}_BS_2$, ${}_GS_B$, ${}_2S_1$.

7. **Point on Circumference of a Rolling Disc** Determine the acceleration of point P on the rolling disc of Figure 9.29.

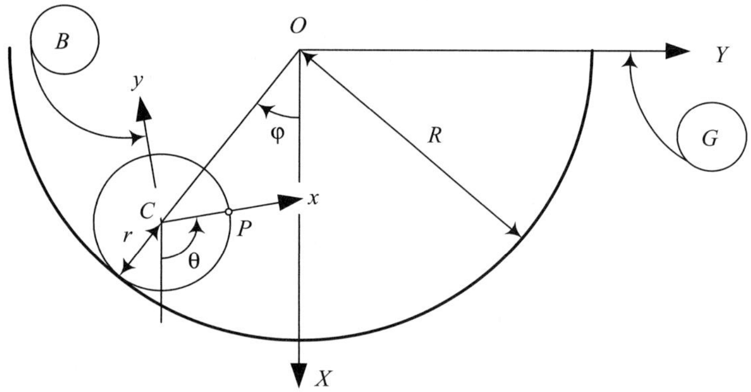

Figure 9.29 A rolling disc in a round ground.

8. **Angular Acceleration by Euler Angles** Employing the Euler angle transformation matrix,

   (a) Determine the relations between the Cartesian angular velocity ${}_G\boldsymbol{\omega}_B$ and $\dot{\varphi}$, $\dot{\theta}$, and $\dot{\psi}$.

   (b) Determine the relations between the Cartesian angular acceleration ${}_G\boldsymbol{\alpha}_B$ and $\ddot{\varphi}$, $\ddot{\theta}$, and $\ddot{\psi}$.

   (c) ★ Determine the relations between the Cartesian angular jerk ${}_G\mathbf{j}_B$ and $\dddot{\varphi}$, $\dddot{\theta}$, and $\dddot{\psi}$

9. **Combined Rotation and Angular Acceleration**

   (a) Find the rotation matrix for a body frame after 45 deg rotation about the Z-axis followed by 45 deg about the X-axis and then 90 deg about the Y-axis.

   (b) Calculate the angular velocity of the body if it is turning with $\dot{\alpha} = 20$ deg/s, $\dot{\beta} = -20$ deg/s, and $\dot{\gamma} = 25$ deg/s about the Z-, Y-, and X-axes, respectively.

   (c) Calculate the angular acceleration of the body if it is turning with $\ddot{\alpha} = 2\,\text{deg/s}^2$, $\ddot{\beta} = 4\,\text{deg}/s^2$, and $\ddot{\gamma} = -6\,\text{deg}/s^2$ about the Z-, Y-, and X-axes.

10. ★ **A Moving Vehicle on Earth** Consider the motion of a vehicle on the Earth surface at latitude λ and longitude φ that is heading north at an angle θ with respect to local north, as shown in Figure 9.30. The vehicle has a velocity ${}^B\mathbf{v}$ with respect to the road and expressed in the vehicle frame B. Determine the acceleration vector of the vehicle in:

    (a) B-frame

    (b) E-frame

    (c) G-frame

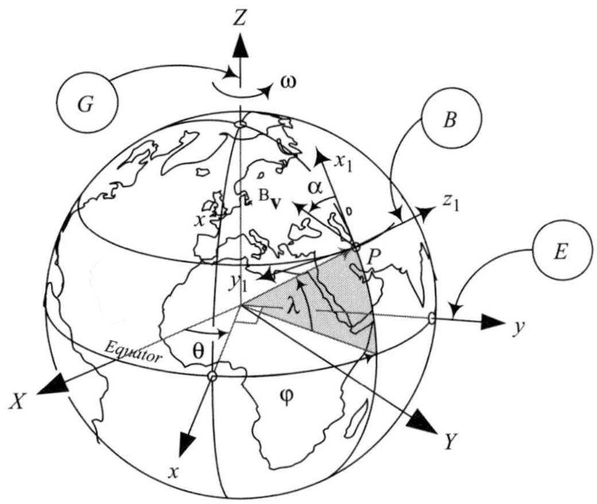

Figure 9.30 A moving vehicle at latitude λ and heading θ with respect to local north.

11. **Deflection of the Path of a Falling Particle** Determine the horizontal deflection of a falling particle in the gravitational field of an airless Earth when the particle is released from a height h above the Earth surface at an altitude λ above the equator plane.

12. **A Ferris Wheel** Determine the acceleration of point P of the Ferris wheel of Figure 9.31.

13. ★ **Differentiation and Coordinate Frame** Define these derivatives:

$$\frac{^Gd}{dt}\frac{^Gd}{dt}\,{}^G\mathbf{r} \qquad \frac{^Gd}{dt}\frac{^Gd}{dt}\,{}^B\mathbf{r} \qquad \frac{^Gd}{dt}\frac{^Bd}{dt}\,{}^G\mathbf{r} \qquad \frac{^Bd}{dt}\frac{^Gd}{dt}\,{}^G\mathbf{r}$$

$$\frac{^Gd}{dt}\frac{^Bd}{dt}\,{}^B\mathbf{r} \qquad \frac{^Bd}{dt}\frac{^Bd}{dt}\,{}^G\mathbf{r} \qquad \frac{^Bd}{dt}\frac{^Gd}{dt}\,{}^B\mathbf{r} \qquad \frac{^Bd}{dt}\frac{^Bd}{dt}\,{}^B\mathbf{r}$$

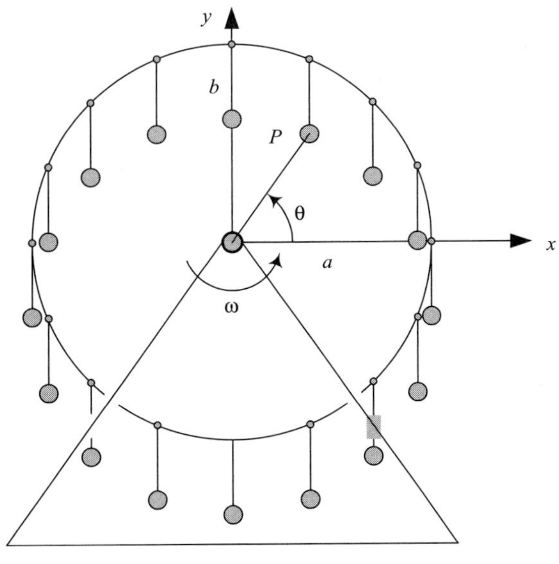

Figure 9.31 A Ferris wheel.

14. ★ **Third Derivative and Coordinate Frames** Consider a global frame $G(OXYZ)$, a body frame $B(Oxyz)$, and a body point P that is moving in the frame B with a variable body position ${}^B\mathbf{r}_P = {}^B\mathbf{r}_P\,(t)$, velocity ${}^B\mathbf{v}_P = {}^B\mathbf{v}_P\,(t)$, and acceleration ${}^B\mathbf{a}_P = {}^B\mathbf{a}_P\,(t)$. Determine how many possible simple and mixed jerks we can define.

15. ★ **Mixed Velocity and Simple Acceleration** Consider a local frame $B(Oxyz)$, that is rotating in $G(OXYZ)$ with an angular velocity $\dot{\alpha} = 10\,\text{rad/s}^2$ about the Z-axis and a moving point P in B at

$$ {}^B\mathbf{r}_P(t) = \sin 2t\,\hat{\imath} $$

Determine ${}_G\tilde{\omega}_B$, ${}^B_G\tilde{\omega}_B$, ${}^B\mathbf{v}$, ${}^B_G\mathbf{v}$, ${}^G_B\mathbf{v}$, ${}^B\mathbf{a}$, ${}^G\mathbf{a}$, ${}^B_G\mathbf{a}$, ${}^G_B\mathbf{a}$, ${}^{GG}_{GB}\mathbf{a}$, ${}^{BB}_{GB}\mathbf{a}$, ${}^{GG}_{BG}\mathbf{a}$, ${}^{BB}_{BG}\mathbf{a}$.

16. ★ **Mixed Second-Derivative Transformation Formula** Consider three relatively rotating coordinate frames A, B, and C. The frame C is turning about the y_B-axis of the B-frame with angular velocity 8 rad/s and angular acceleration $10\,\text{rad/s}^2$, while B is turning about the x_A-axis with angular velocity 5 rad/s and turning about the z_A-axis with angular acceleration $3\,\text{rad/s}^2$. Determine the mixed double acceleration ${}^{AA}_{CB}\mathbf{a}$ if:

   (a) A point P is at ${}^C\mathbf{r} = \begin{bmatrix} 1 & 1 & 1 \end{bmatrix}$ and moving with velocity ${}^C\mathbf{v} = \begin{bmatrix} -10 & 0 & 10 \end{bmatrix}$.

   (b) A point P is at ${}^C\mathbf{r} = \begin{bmatrix} \sin t & 0 & 0 \end{bmatrix}$.

   (c) A point P is at ${}^C\mathbf{r} = \begin{bmatrix} \cot s & \sin t & 0 \end{bmatrix}$.

   (d) A point P is at ${}^C\mathbf{r} = \begin{bmatrix} \cot s & \sin t & t \end{bmatrix}$.

17. **An RPR Manipulator** Determine the velocity and acceleration of point P at the end point of the manipulator shown in Figure 9.32.

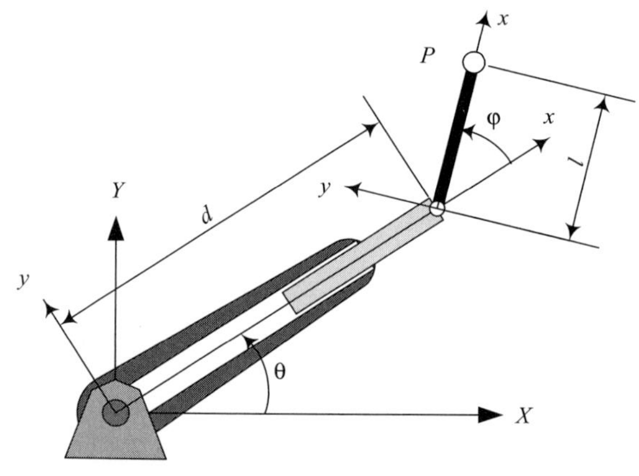

Figure 9.32 A planar RPR manipulator.

18. **A RRP Planar Redundant Manipulator** A three DOF planar manipulator with joint variables θ_1, θ_2, and d_3 is illustrated in Figure 9.33.

   (a) Solve the forward kinematics of the manipulator and calculate the position and orientation of the end-effector X, Y, φ for a given set of joint variables θ_1, θ_2, and d_3, where X, Y are global coordinates of the end-effector frame B_3 and φ is the angular coordinate of B_3.

   (b) Solve the inverse kinematics of the manipulator and determine θ_1, θ_2, and d_3 for given values of X, Y, φ.

(c) Determine the Jacobian matrix of the manipulator and show that the following equation solves the forward-velocity kinematics:

$$\begin{bmatrix} \dot{X} \\ \dot{Y} \\ \dot{\varphi} \end{bmatrix} = \mathbf{J} \begin{bmatrix} \dot{\theta}_1 \\ \dot{\theta}_2 \\ \dot{d}_3 \end{bmatrix}$$

(d) Determine $\mathbf{J}^{-1}$ and solve the inverse-velocity kinematics.

(e) Determine $\dot{\mathbf{J}}$ and solve the forward-acceleration kinematics.

(f) Solve the inverse-acceleration kinematics.

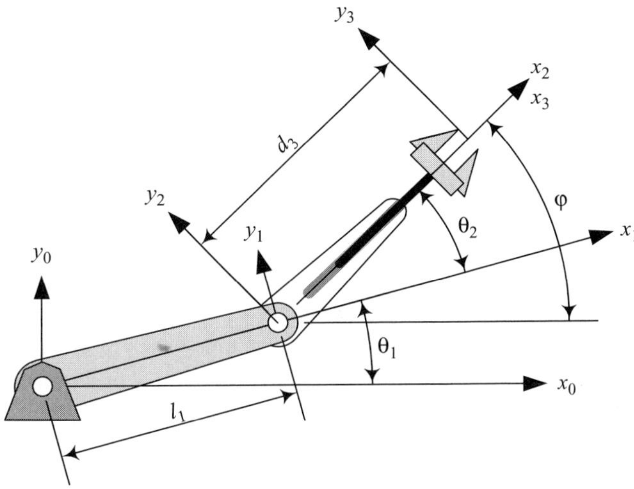

Figure 9.33 A RRP planar redundant manipulator.

19. A RPR Planar Redundant Manipulator

(a) Figure 9.34 illustrates a three-DOF planar manipulator with joint variables θ_1, d_2, and θ_2.

(b) Solve the forward kinematics of the manipulator and calculate the position and orientation of the end-effector X, Y, φ for a given set of joint variables θ_1, θ_2, and d_3, where X, Y are global coordinates of the end-effector frame B_3 and φ is the angular coordinate of B_3.

(c) Solve the inverse kinematics of the manipulator and determine θ_1, θ_2, and d_3 for given values of X, Y, φ.

(d) Determine the Jacobian matrix of the manipulator and show that the following equation solves the forward-velocity kinematics:

$$\begin{bmatrix} \dot{X} \\ \dot{Y} \\ \dot{\varphi} \end{bmatrix} = \mathbf{J} \begin{bmatrix} \dot{\theta}_1 \\ \dot{\theta}_2 \\ \dot{d}_3 \end{bmatrix}$$

(e) Determine $\mathbf{J}^{-1}$ and solve the inverse-velocity kinematics.

(f) Determine $\dot{\mathbf{J}}$ and solve the forward-acceleration kinematics.

(g) Solve the inverse-acceleration kinematics.

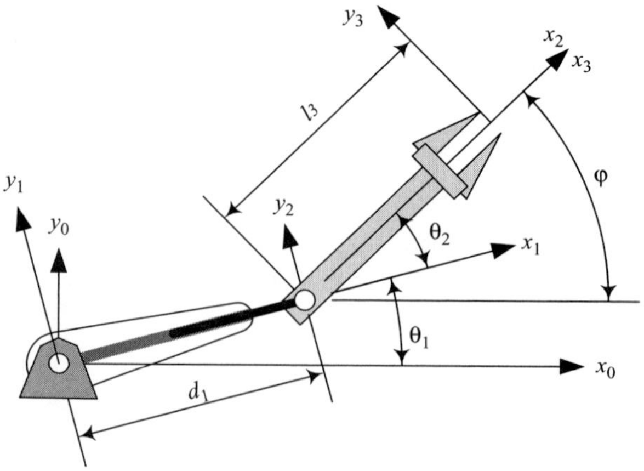

Figure 9.34 A RPR planar redundant manipulator.

20. ★ **An Offset Articulated Manipulator** Figure 9.35 illustrates an offset articulated manipulator.

(a) Solve the forward kinematics of the manipulator.
(b) Solve the inverse kinematics of the manipulator.
(c) Solve the forward-velocity kinematics of the manipulator.
(d) Solve the inverse-velocity kinematics of the manipulator.
(e) Solve the forward-acceleration kinematics of the manipulator.
(f) Solve the inverse-acceleration kinematics of the manipulator.

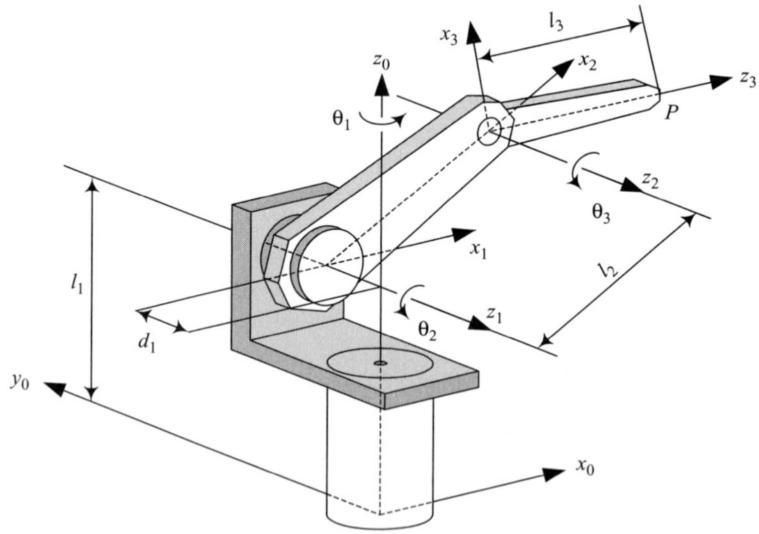

Figure 9.35 An offset articulated manipulator.

21. **Coriolis Acceleration** A disc with radius $R = 1$ m is turning in a horizontal plane with $\boldsymbol{\omega} = \omega\hat{K}$ and $\boldsymbol{\alpha} = \alpha\hat{K}$. We shoot a particle m with speed $\mathbf{v} = 10\hat{I}$ m/s from the center of

the disc which is at the origin of the disc coordinate frame $B(Oxyz)$ and global coordinate frame $G(OXYZ)$. If we ignore any friction between m and the disc, determine the local coordinate of the point where m reaches the periphery of the disc if:

(a) $\omega = 10\,\text{rad/s}$ and $\alpha = 0$
(b) $\omega = 10\,\text{rad/s}$ and $\alpha = 1\,\text{rad/s}^2$
(c) $\omega = 10\,\text{rad/s}$ and $\alpha = -1\,\text{rad/s}^2$
(d) $\omega = \sin t\,\text{rad/s}$

22. Coriolis and Effective Forces A disc with radius $R = 1\,\text{m}$ is turning in a horizontal plane with $\boldsymbol{\omega} = \omega\hat{K}$ and $\boldsymbol{\alpha} = \alpha\hat{K}$. We shoot a particle m in a radial channel with speed $\mathbf{v} = 10\hat{I}\,\text{m/s}$ from the center of the disc which is at the origin of the disc coordinate frame $B(Oxyz)$ and global coordinate frame $G(OXYZ)$. If we ignore any friction between m and the channel, determine the Coriolis and effective forces on m during its motion if:

(a) $\omega = 10\,\text{rad/s}$ and $\alpha = 0$
(b) $\omega = 10\,\text{rad/s}$ and $\alpha = 1\,\text{rad/s}^2$
(c) $\omega = 10\,\text{rad/s}$ and $\alpha = -1\,\text{rad/s}^2$
(d) $\omega = \sin t\,\text{rad/s}$

23. Free Fall from a Long Distance We drop a particle of mass $m = 1\,\text{kg}$ from a height $h = 1000\,\text{m}$ above Shiraz at coordinate 29°36′54”N, 52°32′17”E. Determine the coordinate of impact to the ground relative to the vertical position under the drop point if:

(a) We assume the ground is flat and g is constant.
(b) We asume the ground is spherical and g is constant.
(c) We asume the ground is spherical and g is not constant.
(d) ★ Compare the results of: Shiraz with Fargo, North Dakota, at 46°52′N, 96°48′ W; Toronto, Canada, at 43°40′N, 79°24′ W; New York at 40°48′ N, 73°58′ W; Melbourne, Australia, at 37°47′ S, 144°58′ E; and London GB at 51°32′ N, 0°5′ W.

Shiraz is a city in Iran close to the Persian Gulf and is famous for its wine.

24. Foucault Pendulum, Approximate Solution Use the data

$$\omega_E \approx 7.292\,1 \times 10^{-5}\,\text{rad/s} \qquad g_0 \approx 9.81\,\text{m/s}^2$$
$$\lambda = 29^\circ 36'54''\,\text{N} \qquad \varphi = 52^\circ 32'17''\,\text{E}$$
$$l = 100\,\text{m} \qquad x_0 = l\cos 10 = 17.365\,\text{m}$$

for a Foucault pendulum and determine the approximate angular velocity of the plane of oscillation ω and the path of the projection of m on the (x, y)-plane.

25. ★ Mixed Double or Triple Jerk Consider three relatively rotating frames A, B, and C. Derive an equation for the mixed double or triple jerks ${}^{AAA}_{CBB}\dot{\mathbf{j}}$, ${}^{AAA}_{BCB}\dot{\mathbf{j}}$, ${}^{AAA}_{BBC}\dot{\mathbf{j}}$, ${}^{AAA}_{BCC}\dot{\mathbf{j}}$, ${}^{AAA}_{CBC}\dot{\mathbf{j}}$, ${}^{AAA}_{CCB}\dot{\mathbf{j}}$.

26. ★ Mixed Triple Jerk Consider four relatively rotating frames A, B, C, and D. Derive an equation for the mixed jerks ${}^{AAA}_{DCB}\dot{\mathbf{j}}$.

27. ★ **Rāzī Acceleration** Consider a point P on the disc with radius $R = 1$ m in Figure 9.27 and the following information:

$$ {}_1^2\boldsymbol{\omega}_2 = \dot{\theta}\hat{\imath}_2 = \begin{bmatrix} \dot{\theta} \\ 0 \\ 0 \end{bmatrix} \qquad {}_G^1\boldsymbol{\omega}_1 = \dot{\varphi}\hat{J} = \begin{bmatrix} 0 \\ \dot{\varphi} \\ 0 \end{bmatrix} \qquad {}^2\mathbf{r}_P = \begin{bmatrix} 0 \\ 0 \\ r \end{bmatrix} $$

Determine ${}^G\mathbf{v}_P$, ${}^1\mathbf{v}$, ${}_1^2\mathbf{v}$, ${}_G^2\mathbf{v}$, ${}_1^2\mathbf{a}$, ${}^1\mathbf{a}$, ${}_G^2\mathbf{a}$, ${}_{G1}^{22}\mathbf{a}$, tangential acceleration, mixed centripetal acceleration, mixed Coriolis acceleration, and Rāzī acceleration if:

(a) $r = R$, $\dot{\theta} = 10\,\text{rad/s}$, $\dot{\varphi} = 2\,\text{rad/s}$, and all other variables are zero

(b) $r = R$, $\dot{\theta} = \cos 10t$ rad/s, $\dot{\varphi} = \cos 2t$ rad/s, and all other variables are zero

(c) $r = R$, $\dot{\theta} = 10\,\text{rad/s}$, $\dot{\varphi} = 2\,\text{rad/s}$, $\ddot{\theta} = 1$ rad/s, $\ddot{\varphi} = -0.2\,\text{rad/s}$, and all other variables are zero

(d) $r = t$, $\dot{\theta} = \cos 10t$ rad/s, $\dot{\varphi} = \cos 2t$ rad/s, and all other variables are zero

(e) $r = R \sin t$, $\dot{\theta} = \cos 10t$ rad/s, $\dot{\varphi} = \cos 2t$ rad/s, and all other variables are zero

28. ★ **Acceleration Transformation Matrix** Solve the following exercises by the accceleration transformation matrix method:

(a) 17

(b) 18

(c) 19

(d) 20

29. ★ **Forward- and Inverse-Acceleration Kinematics** Solve the forward- and inverse-acceleration kinematics of the multibody of the following exercises.

(a) 17

(b) 18

(c) 19

(d) 20

10

Constraints

In applied mechanics, there are always some restrictions on the motion of particles and rigid bodies. The mathematical expression of the restrictions is called *constraints*. We classify and review the analytic expressions of constraints to show how they are affecting the actual motions.

10.1 HOMOGENEITY AND ISOTROPY

Consider two points $P_1(x_1, y_1, z_1)$ and $P_2(x_2, y_2, z_2)$ at positions ${}^G\mathbf{r}_1$ and ${}^G\mathbf{r}_2$ expressed in a coordinate frame G attached to a space S. We define their distance as

$$\begin{aligned} d &= \left|{}^G\mathbf{r}_1 - {}^G\mathbf{r}_2\right| \\ &= \sqrt{(x_2 - x_1)^2 + (y_2 - y_1)^2 + (z_2 - z_1)^2} \end{aligned} \tag{10.1}$$

If the distance d is invariant under a coordinate translation, then the space S is called *homogeneous*, and if it is invariant under a coordinate rotation, then the space S is called *isotropic*. A homogeneous and isotropic space is called a *Euclidean* space.

Proof: Assume that we translate the coordinate frame G in space S by a translation vector ${}^G\mathbf{d}$. The new position vectors $\mathbf{r}'_1$, $\mathbf{r}'_2$ of points P_1 and P_2 would be

$${}^G\mathbf{r}'_1 = {}^G\mathbf{r}_1 + {}^G\mathbf{d} \tag{10.2}$$

$${}^G\mathbf{r}'_2 = {}^G\mathbf{r}_2 + {}^G\mathbf{d} \tag{10.3}$$

which shows the distance is invariant under a coordinate translation:

$$d = \left|{}^G\mathbf{r}'_1 - {}^G\mathbf{r}'_2\right| = \left|{}^G\mathbf{r}_1 + {}^G\mathbf{d} - {}^G\mathbf{r}_2 - {}^G\mathbf{d}\right| = \left|{}^G\mathbf{r}_1 - {}^G\mathbf{r}_2\right| \tag{10.4}$$

Now assume that we rotate the coordinate frame G in space S by a rotation matrix R. The new position vectors of points P_1 and P_2 would be

$${}^G\mathbf{r}'_1 = R\,{}^G\mathbf{r}_1 \tag{10.5}$$

$${}^G\mathbf{r}'_2 = R\,{}^G\mathbf{r}_2 \tag{10.6}$$

which shows the distance is invariant under a coordinate rotation:

$$\begin{aligned} d &= \left|{}^G\mathbf{r}'_1 - {}^G\mathbf{r}'_2\right| = \left|R\,{}^G\mathbf{r}_1 - R\,{}^G\mathbf{r}_2\right| = |R|\left|{}^G\mathbf{r}_1 - {}^G\mathbf{r}_2\right| \\ &= \left|{}^G\mathbf{r}_1 - {}^G\mathbf{r}_2\right| \end{aligned} \tag{10.7}$$

A homogeneous and isotropic space is flat, uniform, continuous, compact, and infinite. ■

Example 589 ★ Euclidean Time and Position Spaces The kinematic space in which the physical motion takes place is assumed to be a three-dimensional position Euclidean space E^3. The time t is assumed to be a one-dimensional Euclidean space E^1 and independent of the position space. Motion is shown by the path that a describing point traces when time goes by. A *Euclidean space* is indicated by two characteristics:

1. *Homogeneity*
2. *Isotropy*

The position space is homogeneous and isotropic. The time space is only homogeneous because isotropy is not defined for one-dimensional spaces.

The Euclidean position space, which may also be called Cartesian space or 3-*space*, is the space of triple real numbers (x, y, z). These numbers are scale factors on the axes of a scaled positive orthogonal triad and indicate the coordinates of a point P.

Example 590 ★ Alternative Definition of Euclidean Space In geometry, a space in which the axioms and postulates of Euclidean geometry applies is a Euclidean space. The main Euclidean axioms and postulates are:

1. The sum of angles of a triangle is 180 deg.
2. The distance between two parallel lines is constant.

Example 591 ★ Nonhomogeneous Time Space A time line between two points is defined as the length of a period of time, such as a minute. If a minute at two different times takes the same length, the time space is homogenous; otherwise, the time space is nonhomogeneous. In a nonhomogeneous time space, one minute of today is longer (slower) or shorter (faster) than tomorrow.

Is there any practical method to measure the length of a minute and determine if the time space is nonhomogeneous?

Example 592 ★ Homogeneity of Time Space The homogeneity of time for a dynamic system

$$m\ddot{\mathbf{r}} = \mathbf{F}(\dot{\mathbf{r}}, \mathbf{r}, t) \tag{10.8}$$

is equivalent to have no time in the force function. Such a system is invariant with respect to time translation. The time invariance property of force functions implies that the laws of nature remain constant. A mathematical description of this fact is: If $\mathbf{r} = \mathbf{r}(t)$ is a solution for $m\ddot{\mathbf{r}} = \mathbf{F}$, then $\mathbf{r} = \mathbf{r}(t + \tau)$, $\tau \in \mathbb{R}$, is also a solution. So, there must be no time t in $\mathbf{F}$, explicitly:

$$m\ddot{\mathbf{r}} = \mathbf{F}(\dot{\mathbf{r}}, \mathbf{r}) \tag{10.9}$$

To have an explicitly time-dependent force function, we should have two interacting systems A and B. Then the influence of the subsystem B on A can be expressed by a time-varying force function.

Example 593 ★ Homogeneous and Anisotropic Position Space A uniformly stretched space in a direction is still homogenous but no longer isotropic. Such a position space is similar to a uniformly moving river in which the speed of a boat is different in different directions. However, its speed is the same if the boat is displaced parallel to itself.

Example 594 ★ Particle, Point Mass, and Impenetrability Different points of a three-dimensional Euclidean space E^3 have different position vectors $\mathbf{r}_i$. A *particle* P_i is a point in E^3 that permanently carries a characteristic called the *mass* m_i. Such a particle is a *point mass* and can be shown by m_i or P_i. Every m_i is a positive real constant for all times t. The position vector $\mathbf{r}_i(t)$ of m_i with respect to a fixed point in E^3 is a single-valued, continuous vector function of t. Therefore, every particle occupies one and only one position in E^3 at any given time t. It indicates that, if there exists a single time t_0 at which two particles m_1 and m_2 have different positions, $\mathbf{r}_1(t_0) \neq \mathbf{r}_2(t_0)$, then their positions never coincide, $\mathbf{r}_1(t) \neq \mathbf{r}_2(t)$. Furthermore, if there exists a single time t_0 at which two particles m_1 and m_2 have the same position, $\mathbf{r}_1(t_0) = \mathbf{r}_2(t_0)$, then their positions coincide permanently, $\mathbf{r}_1(t) = \mathbf{r}_2(t)$. This property of point masses is called the *impenetrability*. Figure 10.1 illustrates the impenetrability property.

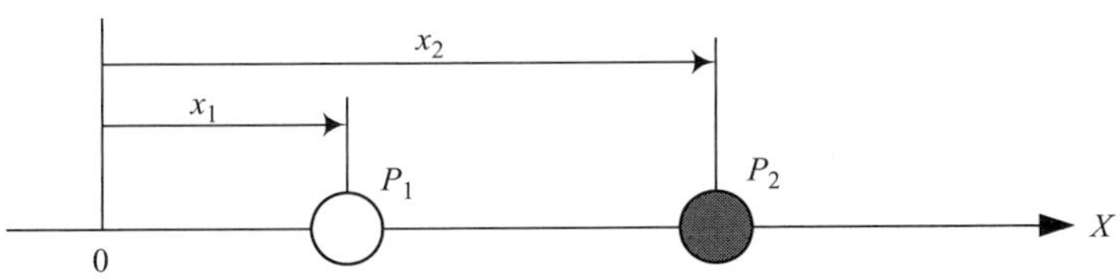

Figure 10.1 Two moving particles in a one-dimensional space will have $x_1(t) \neq x_2(t)$ if $x_1(t_0) \neq x_2(t_0)$.

Example 595 Galileo Relativity Principle To analyze mechanical phenomena, it is necessary to choose a frame of reference. The laws of motion are in general different in form in different frames. So, a simple phenomenon in frame A might be complicated in frame B. A frame in which the position space is homogeneous and isotropic and time is homogeneous is called an *inertial frame*.

Consider a force-free particle in an inertial frame G. Because of homogeneity of position and time, the dynamics of the particle cannot explicitly depend on either the position vector $\mathbf{r}$ of the particle or the time t. So, it must be a function of velocity $\mathbf{v}$ only. Furthermore, because of the isotropy of space, its dynamics must also be independent of the direction of $\mathbf{v}$, and therefore, it is a function only of its magnitude v^2. For such a force-free particle, we have

$$\ddot{\mathbf{r}} = \mathbf{a} = 0 \qquad \dot{\mathbf{r}} = \mathbf{v} = \mathbf{c}_1 \qquad \mathbf{r} = \mathbf{c}_1 t + \mathbf{c}_2 \tag{10.10}$$

In an inertial frame, any free motion takes place with a constant velocity in both magnitude and direction. This is called the *law of inertia*.

If there is another frame moving uniformly in a straight line relative to the inertial frame, then the law of inertia in the other frame will be the same as in the original frame: Free motion takes place with a constant velocity. The two frames are equivalent mechanically, and hence, there is not one but an infinity of inertial frames moving relatively on straight lines with constant velocities. In all of these frames the properties of position and time spaces and the laws of mechanics are the same. This equivalence property in uniformly relative moving frames is called the Galileo relativity principle.

There is no absolute or superior inertial frame which could be preferred to other frames. The position vectors ${}^A\mathbf{r}$ and ${}^B\mathbf{r}$ of a given point P in two different inertial frames A and B, of which the latter moves relative to the former with velocity $\mathbf{v}$, are related by

$${}^A\mathbf{r} = {}^B\mathbf{r} + \mathbf{v}t \tag{10.11}$$

with the assumption that time is the same in the two frames:

$${}^At = {}^Bt \tag{10.12}$$

Equations (10.11) and (10.12) are called a *Galilean transformation* between two inertial frames A and B.

10.2 DESCRIBING SPACE

To describe the motion of a dynamic system, we may use different spaces with different dimensions and application. Configuration, event, state, and state–time are the most applied spaces to express a motion.

10.2.1 Configuration Space

Consider a mechanical system of n particles $P_i (i = 1, 2, \dots, n)$ in a three-dimensional Euclidean space E^3. The positions of the particles are expressed by the vectors ${}^G\mathbf{r}_i = {}^G\mathbf{r}_i(t)$ in a coordinate frame G:

$${}^G\mathbf{r}_i = \begin{bmatrix} x_i(t) \\ y_i(t) \\ z_i(t) \end{bmatrix} \tag{10.13}$$

The motion of the whole system at time t can be expressed by the path of a single describing point in a $3n$-dimensional space made by coordinates x_i, y_i, z_i $(i = 1, 2, \dots, n)$. Such a space is called a *configuration space* (S_C*-space*) and indicated by a set X_C:

$$X_C = \{x_i, y_i, z_i : i = 1, 2, \dots, n\} \tag{10.14}$$

The describing point is called the *configuration point* (S_C*-point*), and the path of an S_C-point is called the *configuration trajectory* (S_C*-trajectory*). The set

$$S_C = \{x_i(t), y_i(t), z_i(t) : i = 1, 2, \ldots, n\} \tag{10.15}$$

indicates the S_C*-trajectory* of the system in the time T-domain:

$$T = \{t : -\infty < t < \infty\} \tag{10.16}$$

Therefore, every configuration of a system is indicated by a point on a S_C-trajectory, and every point of the S_C-trajectory indicates a configuration of the system.

The coordinate axes $x_i(t)$, $y_i(t)$, $z_i(t)$ of an S_C-space can be interchanged, so we may show the S_C-space and S_C-trajectory as

$$X_C = \{u_i : i = 1, 2, \ldots, 3n\} \tag{10.17}$$

$$S_C = \{u_i(t) : i = 1, 2, \ldots, 3n\} \tag{10.18}$$

The degree of freedom of an S_C-point is ideally $3n$, which is equal to the number of dimensions of the S_C-space. However, there might be some regions of an S_C-space that are not allowed to or not reachable by an S_C-point.

The S_C-space is homogeneous and isotropic. The S_C-trajectory is a continuous curve with possible corners and multiple points. A corner is a point of the S_C-trajectory with two tangent vectors.

Example 596 Two Particles on the X-Axis Consider two point masses P_1 and P_2 on the X-axis at positions x_1 and x_2, as shown in Figure 10.1. The configuration space of this system is shown in Figure 10.2.

The impenetrability property splits the two-dimensional S_C-space into two half spaces: possible and forbidden. The two half spaces are divided by the line $x_1 = x_2$.

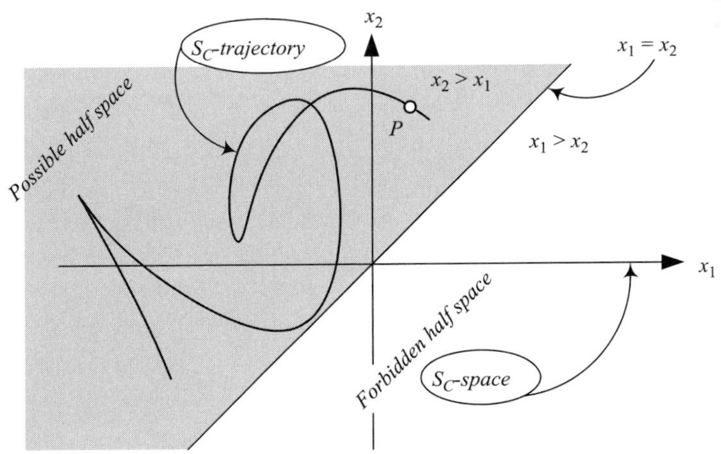

Figure 10.2 The configuration space of two point masses P_1 and P_2 that can move on the X-axis for $x_2 > x_1$.

If for a time t_0 we have $x_2(t_0) > x_1(t_0)$, then the S_C-trajectory of the system can only be in the half space $x_2(t) > x_1(t)$.

Example 597 Rigid-Body Configuration Space Consider a mechanical system of n particles $P_i (i = 1, 2, \ldots, n)$ and m rigid bodies $B_j (j = 1, 2, \ldots, m)$. The positions of the particles are expressed by the vectors ${}^G\mathbf{r}_i = {}^G\mathbf{r}_i(t)$ in a coordinate frame $G(Oxyz)$:

$$ {}^G\mathbf{r}_i = \begin{bmatrix} x_i(t) \\ y_i(t) \\ z_i(t) \end{bmatrix} \tag{10.19} $$

Configuration of the rigid bodies are expressed by the position vectors ${}^G\mathbf{r}_j = {}^G\mathbf{r}_j(t)$ of the origin of their body frames B_j and their orientation angles $\varphi_j, \theta_j, \psi_j$:

$$ {}^G\mathbf{r}_j = \begin{bmatrix} x_j(t) \\ y_j(t) \\ z_j(t) \end{bmatrix} \tag{10.20} $$

$$ {}^G R_{B_j} = R\left(\varphi_j(t),\ \theta_j(t),\ \psi_j(t)\right) \tag{10.21} $$

The motion of the system at time t can be described by an S_C-trajectory in a $(3n + 6m)$-dimensional orthogonal S_C-space made by the coordinates $x_i, y_i, z_i, x_j, y_j, z_j, \varphi_j, \theta_j, \psi_j, i = 1, 2, \ldots, n, j = 1, 2, \ldots, m$.

The configuration of the system is the set

$$ S_C = \left\{ \begin{array}{c} x_i(t), y_i(t), z_i(t), x_j(t), y_j(t), z_j(t), \varphi_j(t), \theta_j(t), \psi_j(t) \\ : i = 1, 2, \ldots, n;\ j = 1, 2, \ldots, m \end{array} \right\} \tag{10.22} $$

which may also be written as

$$ S_C = \{u_i(t) : i = 1, 2, \ldots, 3n + 6m\} \tag{10.23} $$

Example 598 Configuration Space Properties The S_C-trajectories are continuous because every component $u_i(t)$ is a continuous function of time. A discontinuous S_C-trajectory is equivalent to the S_C-point disappearing at some point of the S_C-space and appearing at another point of the S_C-space.

The S_C-trajectories may have multiple points because the S_C-point can be at a point of the S_C-space at different times. An S_C-trajectory without any multiple point indicates a dynamic phenomenon that never repeats itself.

The S_C-trajectory may have corners. A corner is a point at which a direction is not defined.

The position vector $\mathbf{u}$ of an S_C-point is a $3n$-vector is given as

$$ \mathbf{u} = \begin{bmatrix} u_1 & u_2 & u_3 & \ldots & u_{3n} \end{bmatrix}^{\mathrm{T}} \tag{10.24} $$

The velocity and acceleration of an S_C-point are also $3n$-vectors:

$$\dot{\mathbf{u}} = \begin{bmatrix} \dot{u}_1 & \dot{u}_2 & \dot{u}_3 & \dots & \dot{u}_{3n} \end{bmatrix}^{\mathrm{T}} \tag{10.25}$$

$$\ddot{\mathbf{u}} = \begin{bmatrix} \ddot{u}_1 & \ddot{u}_2 & \ddot{u}_3 & \dots & \ddot{u}_{3n} \end{bmatrix}^{\mathrm{T}} \tag{10.26}$$

The direction of an S_C-trajectory is defined by a unit tangent vector $\hat{u}_t$:

$$\hat{u}_t = \frac{\dot{\mathbf{u}}}{\dot{u}} = \frac{\begin{bmatrix} \dot{u}_1 & \dot{u}_2 & \dot{u}_3 & \dots & \dot{u}_{3n} \end{bmatrix}^{\mathrm{T}}}{\left(\sum_{i=1}^{3n} \dot{u}_i^2\right)^{1/2}} \tag{10.27}$$

A corner of an S_C-trajectory is either a *rest point*, at which

$$\sum_{i=1}^{3n} \dot{u}_i^2 = 0 \tag{10.28}$$

or an *impulsive point*, at which the velocity vector is discontinuous:

$$\lim_{\varepsilon \to 0} \dot{\mathbf{u}}(t+\varepsilon) \neq \lim_{\varepsilon \to 0} \dot{\mathbf{u}}(t-\varepsilon) \tag{10.29}$$

At an impulsive point, a sudden force is applied on at least one particle of the system.

Example 599 Two Particles on the x-Axis Consider the position of two particles on the x-axis with the following functions:

$$x_1 = t^2 \qquad x_2 = x_1 + A\cos\omega t \tag{10.30}$$

The S_C-space of this system is

$$X_C = \{x_1, x_2\} \tag{10.31}$$

and its S_C-trajectory is

$$S_C = \left\{t^2, t^2 + A\cos(\omega t)\right\} \tag{10.32}$$

Eliminating t between x_1 and x_2 provides the explicit form of the S_C-trajectory of the system:

$$x_2 = x_1 + A\cos\left(\omega\sqrt{x_1}\right) \tag{10.33}$$

If the system starts moving at time $t_0 = 0$, then

$$x_1(0) = 0 \qquad x_2(0) = A \tag{10.34}$$

If $A > 0$, the impenetrability requires that $x_2(t) > x_1(t)$ at any time t. Therefore, there is a limit in the S_C-space:

$$x_1 < \frac{\pi^2}{4\omega^2} \tag{10.35}$$

This limit is equivalent to having a boundary for time:

$$-\infty < t < \frac{\pi}{2\omega} \tag{10.36}$$

Figure 10.3 illustrates the S_C-space and S_C-trajectory for $A = 1$ and some different ω.

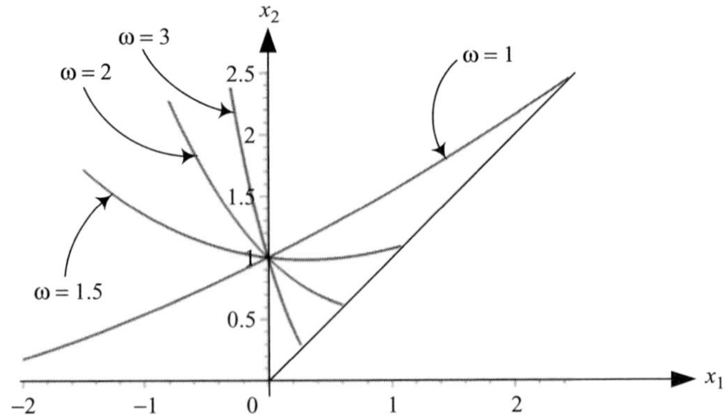

Figure 10.3 S_C-space and S_C-trajectory of a system with two particles at $x_1 = t^2$ and $x_2 = x_1 + A\cos\omega t$ $(x_2 > x_1)$ for $A = 1$ and some different different ω.

Example 600 ★ Configuration Trajectory of a Particle in a Plane Consider a particle that is moving in a two-dimensional configuration space of the (x, y)-plane. The general equations of motion of the particle under a potential force $F = mf_1(x, y)\hat{\imath} + mf_1(x, y)\hat{\jmath}$ are

$$\frac{d^2x}{dt^2} = f_1(x, y) \qquad \frac{d^2y}{dt^2} = f_2(x, y) \tag{10.37}$$

Eliminating t provides a third-order differential equation to determine the trajectory of the particle in the (x, y)-plane:

$$\left(f_2 - y' f_1\right) y''' = \left[\frac{\partial f_2}{\partial x} + \left(\frac{\partial f_2}{\partial y} - \frac{\partial f_1}{\partial x}\right) y' - \frac{\partial f_1}{\partial y} y'^2\right] y'' - 3 f_1 y''^2 \tag{10.38}$$

$$y = y(x) \qquad y' = \frac{dy}{dx} \tag{10.39}$$

The initial conditions of Equations (10.37),

$$x(0) = x_0 \qquad \dot{x}(0) = \dot{x}_0 \qquad y(0) = y_0 \qquad \dot{y}(0) = \dot{y}_0 \tag{10.40}$$

transform to a new set of initial conditions for Equation (10.38):

$$y(0) = y_0 \qquad y'(0) = y'_0 \qquad y''(0) = y''_0 \tag{10.41}$$

As an example, consider a projectile from the origin with velocity v_0 at an angle θ:

$$\mathbf{v} = v_0 \cos\theta\,\hat{\imath} + v_0 \sin\theta\,\hat{\jmath} \tag{10.42}$$

The equations of motion (10.37) are

$$\frac{d^2x}{dt^2} = f_1 = 0 \qquad \frac{d^2y}{dt^2} = f_2 = -g \tag{10.43}$$

so the trajectory equation (10.38) reduces to

$$-gy''' = 0 \tag{10.44}$$

The solution of (10.44) is

$$y'' = c_1 \tag{10.45}$$

$$y' = c_1 x + c_2 \tag{10.46}$$

$$y = \tfrac{1}{2} c_1 x^2 + c_2 x + c_3 \tag{10.47}$$

Employing the initial conditions

$$y(0) = 0 \qquad y'(0) = \frac{\dot{y}_0}{\dot{x}_0} = \tan\theta \qquad \ddot{y}(0) = -g \tag{10.48}$$

$$y''(0) = \frac{dy'}{dt}\frac{dt}{dx} = \frac{\dot{x}_0 \ddot{y}_0 - \ddot{x}_0 \dot{y}_0}{\dot{x}_0^3} = \frac{-g}{v_0^2 \cos^2\theta} \tag{10.49}$$

we find the constants of integration as

$$c_1 = \frac{-g}{v_0^2 \cos^2\theta} \qquad c_2 = \tan\theta \qquad c_3 = 0 \tag{10.50}$$

and determine the configuration trajectory:

$$y = -\frac{1}{2}\frac{g}{v_0^2 \cos^2\theta} x^2 + x \tan\theta \tag{10.51}$$

Example 601 ★ Configuration Trajectory in Space If a particle is moving in a three-dimensional configuration space of (x, y, z) under a potential force such that its equations of motion are

$$\ddot{x} = f_1(x, y, z) \qquad \ddot{y} = f_2(x, y, z) \qquad \ddot{z} = f_3(x, y, z) \tag{10.52}$$

then its configuration trajectory is the solution of a pair of differential equations, one of third order and one of second order:

$$\left(f_2 - y' f_1\right) y''' = \begin{vmatrix} 1 & \dfrac{\partial f_1}{\partial x} + y'\dfrac{\partial f_1}{\partial y} + z'\dfrac{\partial f_1}{\partial z} \\ y' & \dfrac{\partial f_2}{\partial x} + y'\dfrac{\partial f_2}{\partial y} + z'\dfrac{\partial f_2}{\partial z} \end{vmatrix} y'' - 3 f_1 y''^2 \tag{10.53}$$

$$\left(f_2 - y' f_1\right) z'' = \left(f_3 - z' f_1\right) y'' \tag{10.54}$$

These equations come from eliminating t in Equations (10.52).

10.2.2 Event Space

By adding a time axis to a configuration space, we make an *event space*. Consider a mechanical system of n particles P_i $(i = 1, 2, \dots, n)$ in a three-dimensional Euclidean space E^3 with position vectors ${}^G\mathbf{r}_i = {}^G\mathbf{r}_i(t) = x_i(t)\hat{\imath} + y_i(t)\hat{\jmath} + z_i(t)\hat{k}$ expressed in a coordinate frame $G(Oxyz)$. The motion of the system at time t can be expressed by the path of a describing point in $(3n+1)$-dimensional space made up by coordinates x_i, y_i, z_i, t $(i = 1, 2, \dots, n)$. Such a space is called an *event space* or *S_E-space* and is indicated by the set

$$X_E = \{x_i, y_i, z_i, t : i = 1, 2, \dots, n\} \tag{10.55}$$

The describing point is called the *S_E-point*, and the path of an S_E-point is called the *S_E-trajectory*. The S_E-set

$$S_E = \{x_i(t), y_i(t), z_i(t), t : i = 1, 2, \dots, n\} \tag{10.56}$$

indicates the *events* or *S_E-trajectory* of the system in the time T-domain:

$$T = \{t : -\infty < t < \infty\} \tag{10.57}$$

Every event of a dynamic system is indicated by a point on a S_E-trajectory, and every point of the S_E-trajectory indicates an event of the system. When $n = 1$, the S_E-trajectory is called the *time history* of the system.

We may also show the S_E-space and S_E-trajectory as

$$X_E = \{u_i, t : i = 1, 2, \dots, 3n\} \tag{10.58}$$

$$S_E = \{u_i(t), t : i = 1, 2, \dots, 3n\} \tag{10.59}$$

The degree of freedom of an S_E-point is ideally equal to $3n + 1$, where n is the number of particles of the system. However, some regions of an S_E-space might not be allowed or not be reachable by an S_E-point. The S_E-space is homogeneous but is not isotropic.

The S_E-trajectory is a continuous and monotonically increasing curve in the t-direction. The S_E-trajectory might have corners but it never has multiple points.

Proof: To show that the S_E-trajectory is a monotonically increasing curve in the t-direction, we can define the direction $\hat{u}_t$ of the S_E-trajectory and show that $\hat{u}_t$ would never lie on a plane perpendicular to t:

$$\hat{u}_t = \frac{\dot{\mathbf{u}}}{\dot{u}} = \frac{\begin{bmatrix} \dot{u}_1 & \dot{u}_2 & \dot{u}_3 & \dots & \dot{u}_{3n} & 1 \end{bmatrix}^{\mathrm{T}}}{\left(\sum_{i=1}^{3n} \dot{u}_i^2 + 1\right)^{1/2}} \tag{10.60}$$

The direction vector $\hat{u}_t$ is perpendicular to an axis if the associated component to that axis is zero. Any component of $\hat{u}_t$ can be zero except the last one, which is associated to the t-axis. Therefore, $\hat{u}_t$ always has a nonzero component on the t-axis.

In the case that all of the $\dot{u}_i$ $(i = 1, 2, \dots, 3n)$ are zero, $\hat{u}_t$ is parallel to the t-axis and the S_E-trajectory indicates a rest point. ■

Example 602 An Increasing Nonlinear Attraction Force Consider a point mass m on the x-axis that is attracted to the origin by a force F proportional to the nth power of its displacement:

$$F = -kmx^n \qquad n > 0 \tag{10.61}$$

The particle is released from rest at $x(0) = x_0$. The equation of motion of the particle is

$$\ddot{x} = \dot{x}\frac{d\dot{x}}{dx} = -kx^n \tag{10.62}$$

An integral of the equation is

$$\dot{x}^2 = -\frac{kx^{n+1}}{n+1} + C_1 \tag{10.63}$$

$$C_1 = \frac{kx_0^{n+1}}{n+1} \tag{10.64}$$

To determine the event trajectory, we need to find integral of the equation

$$\frac{dx}{\sqrt{\frac{k}{n+1}\left(x_0^{n+1} - x^{n+1}\right)}} = dt \tag{10.65}$$

Example 603 A Decreasing Nonlinear Attraction Force Consider a particle with mass m on the x-axis that is attracted to the origin by a force F:

$$F = -k\frac{m}{x^3} \qquad k = 1 \tag{10.66}$$

The particle is released from rest at $x(0) = x_0$. The equation of motion of the particle is

$$\ddot{x} = \dot{x}\frac{d\dot{x}}{dx} = -\frac{1}{x^3} \tag{10.67}$$

Because of $\lim_{x\to 0} F \to -\infty$, the acceleration of the particle approaches infinity, which indicates the equation of motion is not acceptable at the singular point $x = 0$.

An integral of Equation (10.67) provides

$$\dot{x}^2 = \frac{1}{x^2} - \frac{1}{x_0^2} \tag{10.68}$$

Because $1/x^2 - 1/x_0^2 > 0$, we have $|x| \le x_0$ and the particle that is released at $t = 0$ from $x = x_0$ will never depart farther from the origin than its initial position. The velocity of the particle is directed toward the origin,

$$\dot{x} = -\sqrt{\frac{1}{x^2} - \frac{1}{x_0^2}} \tag{10.69}$$

and is negative during the time interval $0 < t < t_c$, where t_c is the critical time at which the particle reaches the origin. Because of the equation of motion, $\ddot{x} < 0$ for $x > 0$, and therefore, the magnitude of the negative velocity increases while the particle approaches the origin. As $x \to 0$, $|\dot{x}| \to \infty$, and therefore, we can integrate Equation (10.69) from $t = 0$ to $t = t_c - \delta$, where the time $t_c - \delta$ corresponds to $x = \varepsilon > 0$. The limit of the integral for $\epsilon \to 0$ and $\delta \to 0$ provides the S_E-trajectory of the system:

$$x = \frac{1}{x_0}\sqrt{x_0^4 - t^2} \tag{10.70}$$

$$t_c = x_0^2 \tag{10.71}$$

To have a monotonically increasing trajectory with time, t must be limited to

$$0 \le t < x_0^2 \tag{10.72}$$

Figure 10.4 depicts the S_E-space and time history of the particle. The direction of the S_E-trajectory approaches lying on a plane perpendicular to the t-axis. What happens to the particle for $t > t_c$ is not predictable by these results.

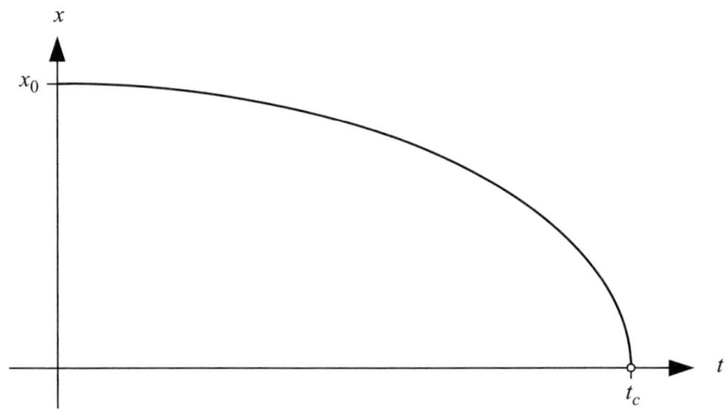

Figure 10.4 S_E-space and S_E-trajectory of a particle under a central force $F = -m/x^3$ released from rest at $x(0) = x_0$.

This problem is similar to the collision problem in central force motions. The mathematical models of attraction force between two particles, $F = -k/x^n$, $n \in N$, is an acceptable model only when the particles are apart. The model cannot approximate the attraction force when the bodies get too close.

Example 604 ★ Regularization Regularization is a general term referring to the techniques that remove singularities. In the nonlinear attraction force problem of Example 603, we need regularization to remove the singularity at $x = 0$ because $\dot{x} \to \infty$.

Let us define new dependent and independent variables u and τ to replace x and t:

$$x = f(u) \qquad \lim_{x \to 0} f(u) = 0 \tag{10.73}$$

$$g(u) = \frac{x'}{\dot{x}} = \frac{dt}{d\tau} \tag{10.74}$$

Employing the new variables, we find

$$u' = \frac{g}{f'}\dot{x} \tag{10.75}$$

and the first integral (10.68) becomes

$$u'^2 = \left(\frac{g}{f'}\right)^2 \dot{x}^2 = \left(\frac{g}{f'}\right)^2 \left(\frac{1}{f^2} - \frac{1}{f_0^2}\right) \tag{10.76}$$

To have a finite velocity u' at collision, we must have $\left(g/f'\right)^2 \to 0$ when $x \to 0$, which is equivalent to having $g/\left(ff'\right)$ finite when $x \to 0$. If the first term for the series expansion of g/f' is Cf,

$$\frac{g}{f'} = Cf + C_1 f^2 + C_2 f^3 + \cdots \tag{10.77}$$

where C is the constant coefficient of the first term. Then $g/\left(ff'\right)$ remains finite at $x \to 0$,

$$\frac{g}{ff'} = C + C_1 f + C_2 f^2 + \cdots \tag{10.78}$$

and the velocity in the system of u and τ is finite at the singularity:

$$u' = C \tag{10.79}$$

As an example, if we try $x = f(u) = u^n$, then we find $g(u) = Cff' = Cnu^{2n-1}u'$.

Introducing a set of new variables, we must be able to reexpress the equation of motion. To transform the equation of motion (10.67), we need to calculate the acceleration $\ddot{x}$. Direct differentiation shows that

$$\dot{x} = \frac{dx}{dt} = \frac{df(u)}{du}\frac{du(\tau)}{d\tau}\frac{d\tau}{dt} = f'u'\frac{1}{g} \tag{10.80}$$

$$\ddot{x} = \frac{d}{dt}\left(\frac{f'}{g}u'\right) = f'u'\frac{d^2\tau}{dt^2} + \left(f'u'' + f''u'^2\right)\left(\frac{d\tau}{dt}\right)^2 \tag{10.81}$$

However,

$$\frac{d^2\tau}{dt^2} = \frac{d}{dt}\frac{1}{g} = -\frac{g'u'}{g^3} \tag{10.82}$$

and therefore,

$$\ddot{x} = -\frac{f'g'}{g^3}u'^2 + \frac{f'u''}{g^2} + \frac{f''}{g^2}u'^2 \tag{10.83}$$

Now, the equation of motion $\ddot{x} = -1/x^3$ becomes

$$u'' = -\frac{g^2}{f'}\left(\frac{1}{f^3} + \frac{1}{g^2}f''u'^2 - \frac{1}{g^3}f'g'u'^2\right) \tag{10.84}$$

Let us choose a simple function for f, such as

$$x = f(u) = u \tag{10.85}$$

$$\frac{dt}{d\tau} = g(u) \tag{10.86}$$

The new velocity u' is

$$u' = \frac{dx}{d\tau} \equiv x' \tag{10.87}$$

and therefore, from (10.76), we have

$$x'^2 = g^2 \left(\frac{1}{x^2} - \frac{1}{x_0^2} \right) \tag{10.88}$$

To have finite x' as $x \to 0$ from (10.77), we must have

$$g = Cx + C_1 x^2 + C_2 x^3 + \cdots \tag{10.89}$$

and therefore, the velocity at collision would be

$$x' = C \tag{10.90}$$

Let us assume $C = 1$ and

$$g = x \tag{10.91}$$

to determine τ:

$$\tau = \int_{t_0}^{t} \frac{dt}{x} \tag{10.92}$$

Some have suggested beginning with a guess function for f and calculating g and τ, including the Finnish scientist Karl Frithiof Sundman (1873–1949), Italian mathematician Tullio Levi-Civita (1873–1941), and Hungarian-American scientist Victor Szebehely (1921–1997).

10.2.3 State Space

The position and velocity of a particle are collectively called the *state* of the particle. By adding new axes for velocity components to a configuration space, we make the *state space* of a system. Consider a mechanical system of n particles P_i $(i = 1, 2, \ldots, n)$ in a three-dimensional Euclidean E^3 with the position and velocity vectors $\mathbf{r}_i(t) = x_i(t)\hat{\imath} + y_i(t)\hat{\jmath} + z_i(t)\hat{k}$ and $\mathbf{v}_i(t) = \dot{x}_i(t)\hat{\imath} + \dot{y}_i(t)\hat{\jmath} + \dot{z}_i(t)\hat{k}$. The motion of the system at time t can be expressed by the path of a describing point in a $6n$-dimensional state space or *S_S-space* made by coordinates $x_i, y_i, z_i, \dot{x}_i, \dot{y}_i, \dot{z}_i$ $(i = 1, 2, \ldots, n)$. The *S_S-space* is indicated by a set X_S:

$$X_S = \{x_i, y_i, z_i, \dot{x}_i, \dot{y}_i, \dot{z}_i : i = 1, 2, \ldots, n\} \tag{10.93}$$

The describing point is called the *S_S-point*, and the path of a S_S-point is called the *S_S-trajectory*. The set

$$S_S = \{x_i(t), y_i(t), z_i(t), \dot{x}_i(t), \dot{y}_i(t), \dot{z}_i(t) : i = 1, 2, \ldots, n\} \tag{10.94}$$

indicates the *state* or *S_S-trajectory* of the system in the time T-domain:

$$T = \{t : -\infty < t < \infty\} \tag{10.95}$$

Every state of a system is indicated by a point on an S_S-trajectory, and every point of the S_S-trajectory indicates a state of the system. When $n = 1$, the S_S-space is called *phase plane* and a set of S_S-trajectories is called the *phase portrait* of the system.

We may also show the S_S-space and S_S-trajectory by the sets

$$X_S = \{u_i, \dot{u}_i, : i = 1, 2, \ldots, 3n\} \tag{10.96}$$

$$S_S = \{u_i(t), \dot{u}_i(t) : i = 1, 2, \ldots, 3n\} \tag{10.97}$$

The degree of freedom of an S_S-point is ideally equal to $6n$, where n is the number of particles of the system. However, there might be some regions of an S_S-space that are not permitted to or not reachable by an S_S-point. The S_S-space is homogeneous and isotropic.

The S_S-trajectory is a continues curve with possible corners and multiple points. The corners of S_S-trajectories may also be called the *equilibria* or *singular points*.

Example 605 A Mass–Spring System Consider a mass m attached to a linear spring with stiffness k as shown in Figure 10.5:

$$m = 1\,\text{kg} \qquad k = 1\,\text{N/m} \tag{10.98}$$

The mass is released to move from the following initial conditions:

$$x(0) = x_0 = a \qquad \dot{x}(0) = \dot{x}_0 = 0 \tag{10.99}$$

The equation of motion of the system is

$$\ddot{x} + x = 0 \tag{10.100}$$

Employing the identity $\ddot{x} = \dot{x}\, dx/dx$, we can integrate the equation of motion and find the energy equation,

$$\tfrac{1}{2}\dot{x}^2 + \tfrac{1}{2}x^2 = E \tag{10.101}$$

where the constant of integration E is the mechanical energy of the system:

$$E = \tfrac{1}{2}\dot{x}_0^2 + \tfrac{1}{2}x_0^2 = \tfrac{1}{2}a^2 \tag{10.102}$$

Equation (10.101) is the S_S-trajectory and $(x, \dot{x})$ is the S_S-space of the system. Figure 10.6 illustrates the phase portrait of the system for a few values of a. Consider the trajectory for $x_0 = a = 0.01$ m. The initial condition provides $E = 0.00005$ J, and the equation of energy indicates a circle. The initial position a determines the level of energy of the system, $E = a^2/2 > 0$. The applied force on m is $F = -kx$, so the

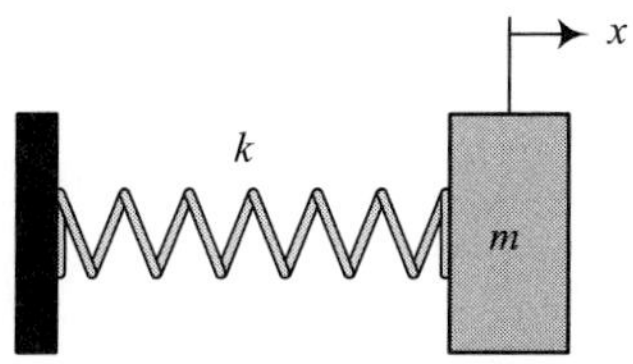

Figure 10.5 A mass–spring vibrating system.

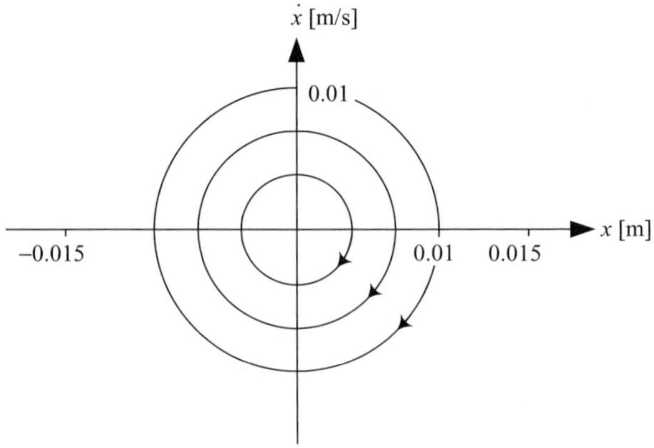

Figure 10.6 The S_S-trajectory and S_S-space of a mass–spring system.

spring attracts m toward the origin by providing a negative acceleration. The negative acceleration generates a negative velocity such that condition (10.101) is satisfied. It shows that the trajectories are traced only in the indicated directions. The direction of an S_S-trajectory at a point can also be determined by $d\dot{x}/dx$ at that point. In this case it is

$$\frac{d\dot{x}}{dx} = \frac{d\dot{x}/dt}{dx/dt} = \frac{-x}{\dot{x}} \tag{10.103}$$

So, when both x and $\dot{x}$ are positive, the slope of $\hat{u}_t$ is negative.

Employing the equation of motion $\ddot{x}/\dot{x} = -x/\dot{x}$ and assigning the x-axis and $\dot{x}$-axis by the unit vectors $\hat{\imath}$ and $\hat{\jmath}$, we can determine the direction of the S_S-trajectory (10.101) as

$$\hat{u}_t = \frac{\dot{x}\hat{\imath} + \ddot{x}\hat{\jmath}}{\sqrt{\dot{x}^2 + \ddot{x}^2}} = \frac{\dot{x}\hat{\imath} - x\hat{\jmath}}{\sqrt{\dot{x}^2 + x^2}} \tag{10.104}$$

The origin $(x, \dot{x}) = (0, 0)$ is the rest point of the system. The *rest positions* are also called *equilibrium* or *singular points*. Liner systems like this mass–spring system have only one equilibrium. Nonlinear systems may have multiple equilibriums with different stability characteristics.

A closed S_S-trajectory indicates a periodic motion. A periodic S_S-trajectory is a level contour of energy and is indicated by the associated value of E.

Example 606 ★ Phase Plane of Linear Systems The equation of motion of a single mass $m\ddot{x} = F(x, \dot{x}, t)$ can always be reduced to two first-order equations in terms of state variables x and $v = \dot{x}$ as

$$\dot{x} = f_2(x, v, t) \qquad \dot{v} = f_2(x, v, t) \tag{10.105}$$

An example is

$$\dot{x} = v \qquad \dot{v} = \frac{1}{m} F(x, v, t) \tag{10.106}$$

The solution of these equations will provide the state variables as functions of time t:

$$x = g_1(t) \qquad v = g_2(t) \tag{10.107}$$

Then the S_S-trajectory will be found by eliminating t between $g_1(t)$ and $g_2(t)$.

A system is called linear if its state equations can be expressed by a set of independent linear equations:

$$\begin{bmatrix} \dot{x} \\ \dot{v} \end{bmatrix} = \begin{bmatrix} a & b \\ c & d \end{bmatrix} \begin{bmatrix} x \\ v \end{bmatrix} \tag{10.108}$$

These equations are equivalent to a single second-order equation:

$$\ddot{x} + p\dot{x} + qx = 0 \tag{10.109}$$

$$p = -(a+d) \qquad q = ad - bc \tag{10.110}$$

To obtain the phase portrait of the linear system, we solve Equations (10.108) for their time history,

$$x = \begin{cases} C_1 e^{s_1 t} + C_2 e^{s_2 t} & s_1 \neq s_2 \qquad (10.111) \\ C_1 e^{s_1 t} + C_2 t e^{s_1 t} & s_1 = s_2 \qquad (10.112) \end{cases}$$

where s_1 and s_2 are the solutions of the characteristic equation:

$$s^2 + ps + q = (s - s_1)(s - s_2) = 0 \tag{10.113}$$

$$s_1 = \frac{1}{2}\left(-q + \sqrt{p^2 - 4q}\right) \qquad s_2 = \frac{1}{2}\left(-q - \sqrt{p^2 - 4q}\right) \tag{10.114}$$

Equations (10.108) are independent provided that the determinant of the coefficient is not zero, so $q = ad - bc \neq 0$. There is only one singular point for the linear systems (10.108) at the origin. However, the shape of the S_S-trajectories in the vicinity of the singular point depends on the characteristic values s_1 and s_2.

There are four possible classes for nonzero s_1 and s_2, as are explained below and shown in Figures 10.7–10.12:

1. *Center*, if s_1 and s_2 are complex conjugates with a zero real part.
2. *Focus*, if s_1 and s_2 are complex conjugates with a nonzero real part.
3. *Node*, if s_1 and s_2 are real and both are positive or both are negative.
4. *Saddle*, if s_1 and s_2 are real and one is positive, the other negative.

The first class indicates steady-state oscillatory motion. When the real part of the characteristic values are zero, there is no divergence or convergence to the singular point. All trajectories are ellipses with a center at the singular point. The phase portrait of an undamped mass–spring system belongs to this class. The phase portrait of a center point is illustrated in Figure 10.7.

The second class indicates an unsteady oscillatory motion. A *stable focus* occurs when the real part of the characteristic values are negative, so x and $\dot{x}$ both converge to zero, as shown in Figure 10.8. An *unstable focus* occurs when the real part of the characteristic values are positive. Then x and $\dot{x}$ both diverge to infinity, as shown in Figure 10.9.

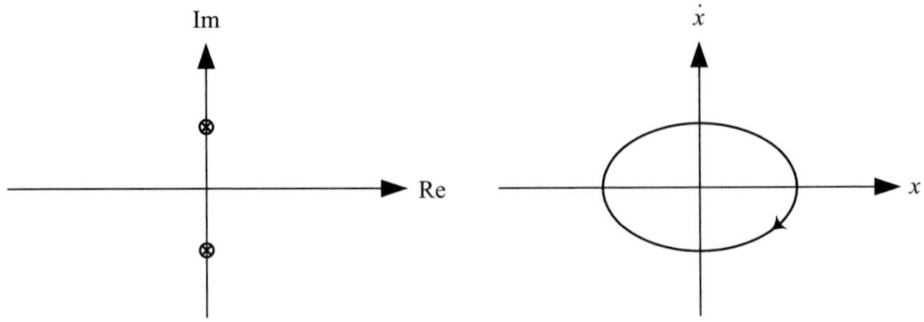

Figure 10.7 When the characteristic values of a linear system are complex conjugates with a zero real part, the singular point is a center.

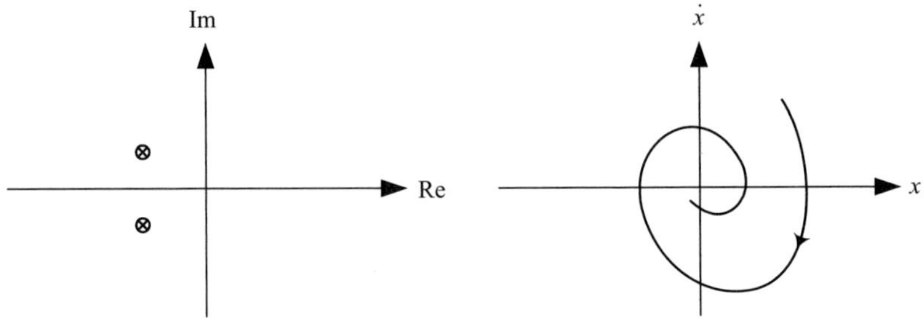

Figure 10.8 When the characteristic values of a linear system are complex conjugates with negative real parts, the singular point is a stable focus.

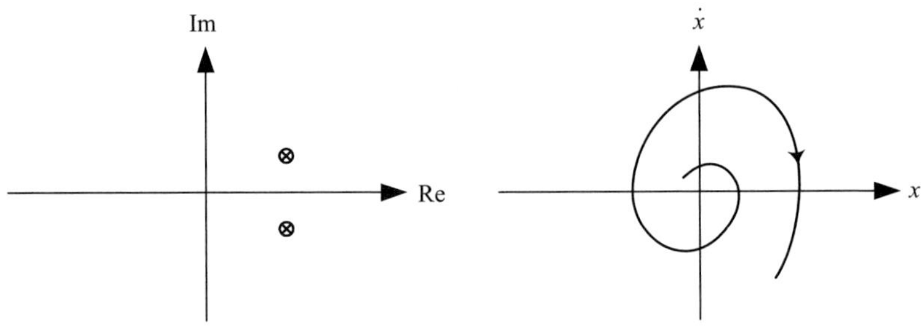

Figure 10.9 When the characteristic values of a linear system are complex conjugates with positive real parts, the singular point is an unstable focus.

The third class indicates an asymmetic attraction or repulsion from the singular point. If the characteristic values s_1 and s_2 are negative, the singular point is called a *stable node* and both x and $\dot{x}$ converge to zero, as shown in Figure 10.10. If both characteristic values are positive, the point is called an *unstable node*, and both x and $\dot{x}$ diverge from zero, as shown in Figure 10.11. Because the characteristic values are real, there is no oscillation in the S_S-trajectories.

The fourth class indicates an unstable system that may approach the singular point before moving asymptotically to infinity. The phase portrait of the system has a saddle

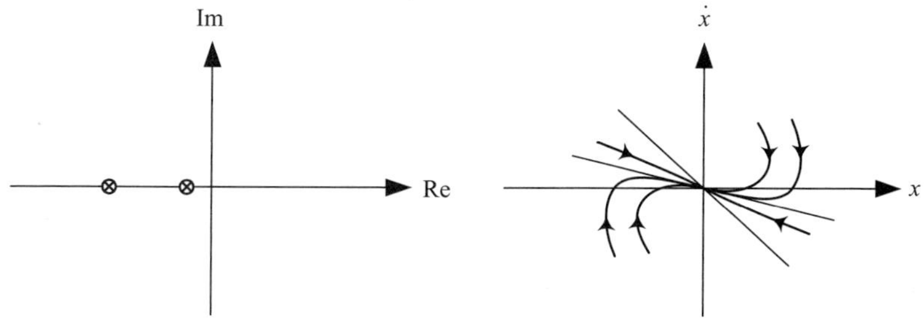

Figure 10.10 When the characteristic values of a linear system are both real and negative, the singular point is a stable node.

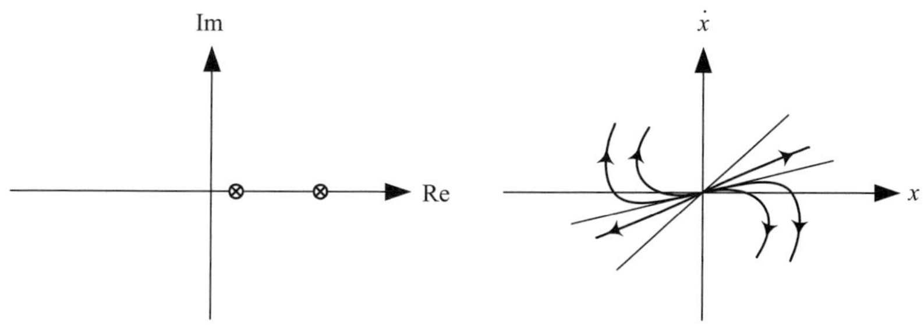

Figure 10.11 When the characteristic values of a linear system are both real and positive, the singular point is a unstable node.

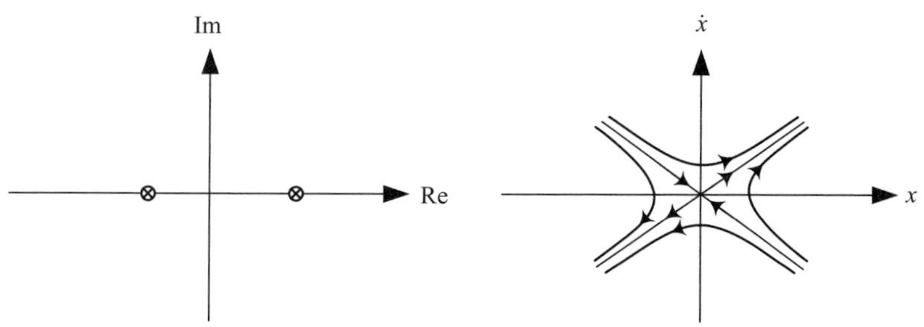

Figure 10.12 When the characteristic values of a linear system are real with one positive and one negative, the singular point is a saddle.

shape as shown in Figure 10.12. There are also two lines passing through the singular point. The diverging line, with arrows pointing to infinity, corresponds to initial conditions which make $s_2 = 0$. The converging line corresponds to initial conditions which make $s_1 = 0$.

Example 607 ★ Rectilinear Motion in a Field Phase plane is a suitable space to express the behavior of dynamic systems, specially when the applied force on a particle is a function of its position and the particle is restricted to move only on an axis:

$$\ddot{x} = F(x) \tag{10.115}$$

The function $F(x)$ defines a force field on the x-axis. Having $F(x)$, we can define a potential function

$$V(x) = -\int_a^x F(x)\,dx \qquad F(x) = -\frac{dV}{dx} \tag{10.116}$$

where a is any convenient value of x in the domain for which $F(x)$ is defined and $V(a) = 0$. The time derivative of V shows that

$$\frac{dV}{dt} = \dot{x}\frac{dV}{dx} = -\dot{x}F(x) \tag{10.117}$$

Therefore, multiplying (10.115) by $\dot{x}$ provides

$$-\frac{d}{dt}(V + K) = 0 \tag{10.118}$$

and hence,

$$V + K = \int_a^x F(x)\,dx + \tfrac{1}{2}\dot{x}^2 = E \tag{10.119}$$

where E is the constant of integration and the mechanical energy of the system. The equation of motion (10.115), which gives $\ddot{x}$ as a function of x, is replaced by the integral of energy (10.119), which gives $\dot{x}^2$ in terms of x. There are two equal values of $\dot{x}$ with opposite signs for every E. Because $K \geq 0$, it follows from the integral of energy that the particle can never leave the region of $V \leq E$.

The integral of energy is the equation of possible S_S-trajectories of the system. The actual trajectory is determined by calculating E from the initial conditions.

Example 608* ★ *Rectilinear Motion with $\dot{x}^2 = g(x)$ The rectilinear motion of a particle under a conservative force field $\ddot{x} = F(x)$ theoretically reduces to an integral of energy of the form $\dot{x}^2 = g(x)$, where $g(x)$ is a continuous and twice-differentiable function $g(x) \in C_2$. The equation $\dot{x}^2 = g(x)$ determines the S_S-trajectories of the system. The dynamic of the particle can be predicted from the graph of $g(x)$.

There are five possible classes for the dynamics of such a particle. To discover the different classes, we may plot the function $y = g(x)$. The ordinate $y = \dot{x}^2$ is proportional to the kinetic energy of the particle, and the slope dy/dx of the graph gives the associated value of $2\ddot{x}$. Motion can only happen on a stretch of the x-axis for which $g(x) \geq 0$.

1. *Rest*: This is an exceptional case in which $g(x)$ and $g'(x) = dg/dx$ both vanish at a point x_0. So, x_0 is the equilibrium point and the graph of $g(x)$ touches the x-axis at x_0 as point c in Figures 10.13(a) and 10.14(a). In the case $x(0) = x_0$, $\dot{x}(0) = 0$, we have $x = x_0$ for all time and the particle rests in equilibrium forever.

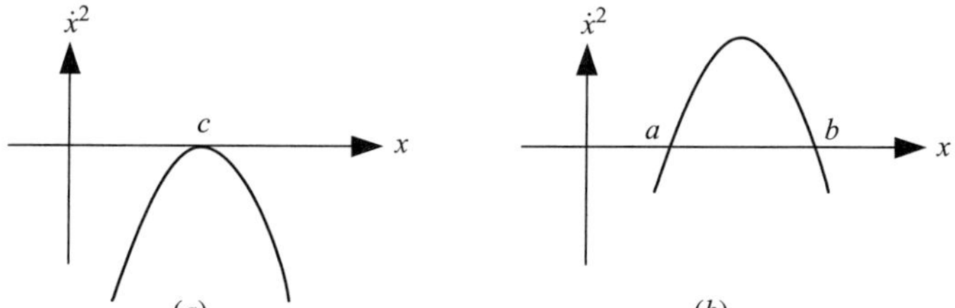

(a) (b)

Figure 10.13 A graph of $\dot{x}^2 = g(x)$ corresponding to a libration.

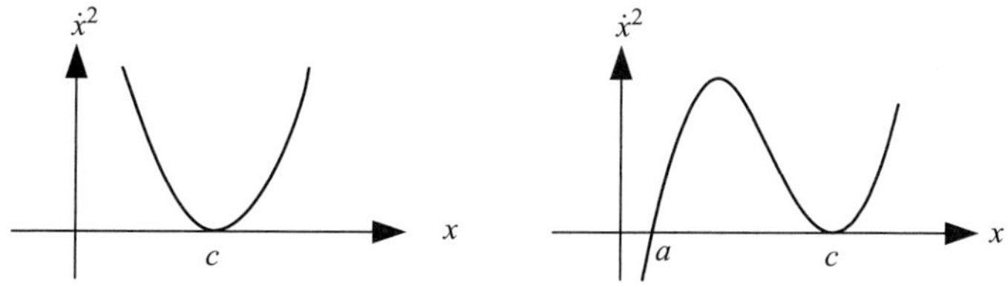

(a) (b)

Figure 10.14 Two possible graphs of $\dot{x}^2 = g(x)$ corresponding to limitation.

2. *Libration*: This is an oscillatory motion in which the particle oscillates continually between two points $x = a$ and $x = b$ on the x-axis. Suppose $x(0) = x_0 \neq 0$ and $g(x_0) > 0$; we have $\dot{x}(0) = \sqrt{g(x_0)} > 0$. The velocity $\dot{x} > 0$ for sufficiently small values of time t. For those values, the relation between x and t is

$$t = \int_{x_0}^{x} \frac{dx}{\sqrt{g(x)}} \tag{10.120}$$

Assume that x_0 is between two zeros a and b $(a < b)$ of $g(x)$ such that $g(x) > 0$ for $a < x < b$, as shown in Figure 10.13(b). Because b is a simple zero of $g(x)$, the integral (10.120) approaches a finite time when $x \to b$. The graph of $g(x)$ crosses the x-axis at a and b with a positive slope at a and a negative slope at b. The particle instantly stops at $x = b$, where the negative acceleration pulls it back. Similarly, the particle reaches $x = a$ in a finite time where it instantly stops and moves to the right again. This back-and-forth motion is called libration. The particle arrives at $x = x_0$ with the same velocity with which it started. The total time of one cycle would be

$$t = \oint \frac{dx}{\sqrt{g(x)}} = 2 \int_{a}^{b} \frac{dx}{\sqrt{g(x)}} \tag{10.121}$$

The particle will show the same behavior if it is released from a or b with $\dot{x} = 0$.

3. *Limitation*: This is an approaching motion in which $x \to c$ as $t \to \infty$. Assume the particle starts from $x(0) = x_0 < c$ with $\dot{x}(0) = \sqrt{g(x_0)} > 0$ and approaches a double zero of $g(x)$ at $x = c$, such as in Figure 10.14(a). Because both the acceleration and velocity of the particle approach zero when $x \to c$, the integral (10.120) diverges and $t \to \infty$ as $x \to c$. In the case $x(0) = x_0 > c$, the particle moves away from $x = c$ such that $x \to \infty$ as $t \to \infty$.

Now consider the situation in which $a < x_0 < c$, where a is a simple zero of $g(x)$ and c is a double zero of $g(x)$, as shown in Figure 10.13(b). If $\dot{x}(0) < 0$, the particle reaches a in a finite time and then returns to $x \to c$ as $t \to \infty$. Point c is the limit of motion that cannot be reached in a finite time. Such a motion is called a limitation.

4. *Lost*: This is a run-away motion in which $x \to \infty$ as $t \to \infty$. Whenever a particle starts from $x(0) = x_0$ with $g(x) > 0$ and $dg/dx > 0$ for $x \geq x_0$, then $x \to \infty$ as $t \to \infty$. Because the position of the particle cannot be determined at a long time, it is called a lost situation. If $x_0 > c$, then Figure 10.14(b) illustrates a lost condition. The mathematical definition of a lost motion is having a divergent integral (10.120) for $x \to \infty$.

5. *Escape*: This is a run-away motion in which $x \to \infty$ in a limited time $t \to t_0$. When a particle starts from $x(0) = x_0$ with $g(x) > 0$ for $x \geq x_0$ and the integral (10.120) converges when $x \to \infty$, the motion is called an escape.

10.2.4 State–Time Space

By adding a time axis to a state space, we make a *state–time space*. Consider a mechanical system of n particles $P_i (i = 1, 2, \ldots, n)$ in a three-dimensional Euclidean space E^3 with position and velocity vectors $\mathbf{r}_i(t) = x_i(t)\hat{\imath} + y_i(t)\hat{\jmath} + z_i(t)\hat{k}$ and $\mathbf{v}_i(t) = \dot{x}_i(t)\hat{\imath} + \dot{y}_i(t)\hat{\jmath} + \dot{z}_i(t)\hat{k}$. The motion of the system at any time t can be expressed by the path of a describing point in a $(6n + 1)$-dimensional space made by coordinates $x_i, y_i, z_i, \dot{x}_i, \dot{y}_i, \dot{z}_i, t$ $(i = 1, 2, \ldots, n)$. A state–time space or S_T-*space* is indicated by the set

$$X_T = \{x_i, y_i, z_i, \dot{x}_i, \dot{y}_i, \dot{z}_i, t : i = 1, 2, \ldots, n\} \tag{10.122}$$

The describing point is called the S_T-*point*, and the path of an S_T-point is called the S_T-*trajectory*. The S_T-set

$$S_T = \left\{ \begin{array}{c} x_i(t), y_i(t), z_i(t), \dot{x}_i(t), \dot{y}_i(t), \dot{z}_i(t), t : \\ i = 1, 2, \ldots, n \end{array} \right\} \tag{10.123}$$

indicates the *state–time* or S_T-*trajectory* of the system in the time T-domain:

$$T = \{t : -\infty < t < \infty\} \tag{10.124}$$

Every state–time of a dynamic system is indicated by a point on an S_T-trajectory, and every point of the S_T-trajectory indicates a state–time of the system.

We may also show the X_T-space and S_T-trajectory as

$$X_T = \{u_i, \dot{u}_i, t : i = 1, 2, \ldots, 3n\} \tag{10.125}$$

$$S_T = \{u_i(t), \dot{u}_i(t), t : i = 1, 2, \ldots, 3n\} \tag{10.126}$$

The degree of freedom of an S_T-point is ideally equal to $6n + 1$, where n is the number of particles of the system. However, there might be some regions of an S_T-space that are not permitted to or not reachable by an S_T-point. The S_T-space is homogeneous and is not isotropic. The S_T-trajectory is a continuous and monotonically increasing curve in the t-direction. The S_T-trajectory might have corners but it never has multiple points.

Example 609 A Damped Vibrating System Consider a linear mass–spring–damper system as shown in Figure 10.15. Let us release the mass from initial conditions $x(0) = x_0 > 0$ and $\dot{x}(0) = 0$. The position is measured from equilibrium, which is the position at which the weight of m and the spring force balance.

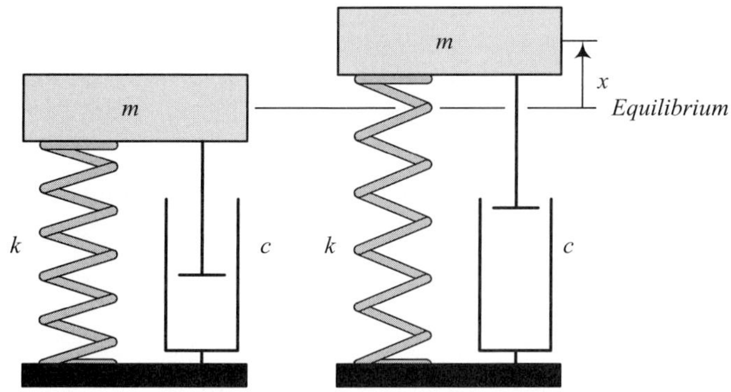

Figure 10.15 A linear mass-spring-damper system.

The equation of motion is

$$m\ddot{x} + c\dot{x} + kx = 0 \qquad c < 2\sqrt{km} \tag{10.127}$$

Using the definitions

$$\xi = \frac{c}{2\sqrt{km}} \qquad \omega_n = \sqrt{\frac{k}{m}} \qquad \omega_d = \omega_n\sqrt{1-\xi^2} \tag{10.128}$$

we rewrite the equation of motion as

$$\ddot{x} + 2\xi\omega_n\,\dot{x} + \omega_n^2 x = 0 \tag{10.129}$$

and determine the solution:

$$x(t) = x_0 e^{-\xi\omega_n t}\cos\omega_d t \tag{10.130}$$

$$\dot{x}(t) = -x_0 e^{-t\xi\omega_n}\left(\omega_d\sin\omega_d t + \xi\omega_n\cos\omega_d t\right) \tag{10.131}$$

This indicates a diminishing oscillatory motion with a constant frequency ω_d and decaying amplitude $x_0 e^{-\xi\omega_n t}$. The S_T- and S_S-trajectories of the motion are shown in Figure 10.16. The S_T-trajectory is a shrinking circular motion around the t-axis and finally lies on the axis. The projection of the S_T-trajectory on the $(x, \dot{x})$-plane is the S_S-trajectory. The S_E-trajectory is the projection of the S_T-trajectory on the (x, t)-plane.

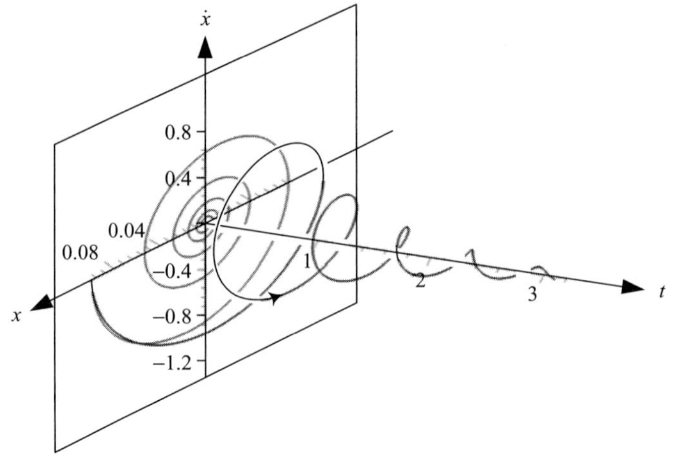

Figure 10.16 The S_T- and S_S-trajectories of a linear mass–spring–damper system released the mass from $x(0) = 0.1 > 0$.

10.2.5 ★ Kinematic Spaces

Consider a mechanical system of n particles P_i $(i = 1, 2, \ldots, n)$ in a three-dimensional Euclidean space E^3 with kinematic vectors

$$
\begin{aligned}
\mathbf{r}_i(t) &= \mathbf{r}_i(t) = x_i(t)\hat{\imath} + y_i(t)\hat{\jmath} + z_i(t)\hat{k} \\
\dot{\mathbf{r}}_i(t) &= \mathbf{v}_i(t) = \dot{x}_i(t)\hat{\imath} + \dot{y}_i(t)\hat{\jmath} + \dot{z}_i(t)\hat{k} \\
\ddot{\mathbf{r}}_i(t) &= \mathbf{a}_i(t) = \ddot{x}_i(t)\hat{\imath} + \ddot{y}_i(t)\hat{\jmath} + \ddot{z}_i(t)\hat{k} \\
&\vdots \\
\mathbf{r}_i^{(m)}(t) &= x_i^{(m)}(t)\hat{\imath} + y_i^{(m)}(t)\hat{\jmath} + z_i^{(m)}(t)\hat{k}
\end{aligned}
\tag{10.132}
$$

We can express the motion of the system at time t by the path of a describing point in different spaces using different kinematic characteristics of the particles.

The combination of acceleration and position is called the *rush space* or S_R*-space*:

$$S_R = \{u_i(t), \ddot{u}_i(t) : i = 1, 2, \ldots, 3n\} \tag{10.133}$$

Adding the time t-axis to a rush space produces the *rush–time space* or S_{RT}*-space*:

$$S_{RT} = \{u_i(t), \ddot{u}_i(t), t : i = 1, 2, \ldots, 3n\} \tag{10.134}$$

The combination of acceleration and velocity is called the *flash space* or S_F*-space*:

$$S_F = \{\dot{u}_i(t), \ddot{u}_i(t) : i = 1, 2, \ldots, 3n\} \tag{10.135}$$

Adding the time t-axis to a flash space produces the *flash–time space* or S_{FT}*-space*:

$$S_{FT} = \{\dot{u}_i(t), \ddot{u}_i(t), t : i = 1, 2, \ldots, 3n\} \tag{10.136}$$

The combination of jerk and position is called the *spark space* or S_P*-space*:

$$S_P = \{u_i(t), \dddot{u}_i(t) : i = 1, 2, \ldots, 3n\} \tag{10.137}$$

Adding the time t-axis to a spark space produces the *spark–time space* or S_{PT}*-space*:

$$S_{PT} = \{u_i(t), \dddot{u}_i(t), t : i = 1, 2, \ldots, 3n\} \tag{10.138}$$

The combination of jerk and velocity is called the *flicker space* or S_K*-space*:

$$S_K = \{\dot{u}_i(t), \dddot{u}_i(t) : i = 1, 2, \ldots, 3n\} \tag{10.139}$$

Adding the time t-axis to a flicker space produces the *flicker–time space* or S_{KT}*-space*:

$$S_{KT} = \{\dot{u}_i(t), \dddot{u}_i(t), t : i = 1, 2, \ldots, 3n\} \tag{10.140}$$

The combination of jerk and acceleration is called the *sparkle space* or S_L*-space*:

$$S_L = \{\ddot{u}_i(t), \dddot{u}_i(t) : i = 1, 2, \ldots, 3n\} \tag{10.141}$$

Adding the time t-axis to a sparkle space produces the *sparkle–time space* or S_{LT}*-space*:

$$S_{LT} = \{\ddot{u}_i(t), \dddot{u}_i(t), t : i = 1, 2, \ldots, 3n\} \tag{10.142}$$

Employing the other kinematic characteristics of the particles of a system, we can define any kinematic space that expresses a motion better. The describing kinematic spaces can also be a combination of more than two kinematic quantities. In any kinematic space, there might be some forbidden parts that the describing point is not allowed to enter or touch. The continuity, multiple point, monotonicity, and corner of the describing trajectory depend on the kinematic characteristics.

Example 610 ★ Harmonic Motion in Flash Space Consider a harmonic motion as

$$x = x_0 \cos \omega t \tag{10.143}$$

where x_0 is the initial position. The acceleration of this motion is

$$\ddot{x} = -x_0 \omega^2 \cos \omega t \tag{10.144}$$

Eliminating t provides the flash trajectory:

$$\ddot{x} + \omega^2 x = 0 \qquad -x_0 \le x \le x_0 \tag{10.145}$$

Equation (10.145) is a segment of a straight line in the $(\ddot{x}, x)$-plane, as shown in Figure 10.17(a) for $x_0 = 1$ and different ω. The slope of the line (10.145) is equivalent to the frequency of the harmonic motion (10.143). The system is faster at higher slopes. The negative slope indicates a stable harmonic system.

Figure 10.17(b) depicts the flash–time trajectory of the motion for $\omega = 1$. The flash trajectory is the projection of the flash–time trajectory on the $(\ddot{x}, x)$-plane.

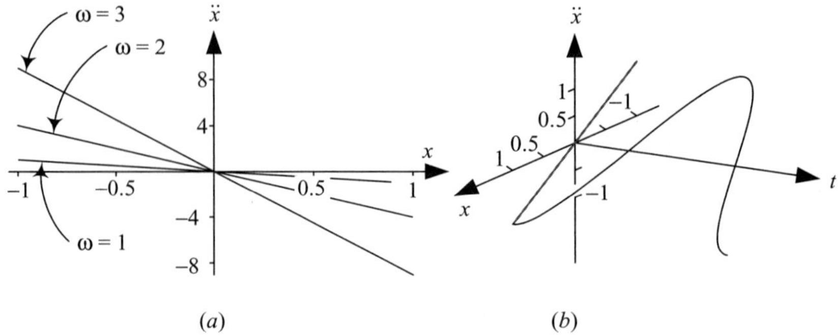

Figure 10.17 The flash and flash–time spaces of the harmonic motion $x = x_0 \cos \omega t$: (*a*) flash trajectories for different ω; (*b*) flash–time trajectory for $\omega = 1$.

Example 611 ★ Comparison of Harmonic and Periodic Motions The flash space is a good environment in which to compare periodic motions with a simple harmonic one. A periodic motion makes a closed curve while a harmonic motion is indicated by a straight line with a negative slope. Comparison of the periodic motion with a straight line can show how close the periodic motion is to a harmonic motion. Such a comparison is informative, for example, in the perturbation approximation of periodic motions.

Let us approximate a periodic motion with a harmonic function using a least-squares analysis. Consider a periodic motion as

$$x_1 = x_0 \cos \omega t + \frac{x_0}{10} \sin 2\omega t \tag{10.146}$$

$$\ddot{x}_1 = -\omega^2 x_0 \cos \omega t - \frac{2}{5}\omega^2 x_0 \sin 2\omega t \tag{10.147}$$

where x_0 is the initial position. Ignoring the second harmonic of the motion, we can approximate the motion with the first harmonic as

$$x_2 = x_0 \cos \omega t \qquad \ddot{x}_2 = -x_0\omega^2 \cos \omega t \tag{10.148}$$

Figure 10.18 illustrates the comparison of the periodic and harmonic motions in a flash space. A harmonic approximation x_3 is derived as

$$x_3 = A \cos \omega t \tag{10.149}$$

We may determine A from the equation

$$\frac{d}{dA} \int_0^{2\pi} \left((x_1 - x_3)^2 + (\ddot{x}_1 - \ddot{x}_3)^2 \right) dt = 0 \tag{10.150}$$

and find

$$A = x_0 \tag{10.151}$$

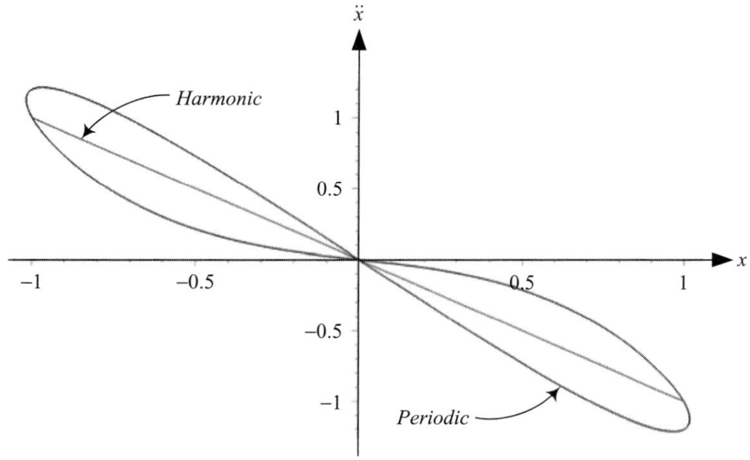

Figure 10.18 A comparison of the periodic motion $x_1 = x_0 \cos \omega t + \dfrac{x_0}{10} \sin 2\omega t$ and the harmonic motion $x_2 = x_0 \cos \omega t$ in flash space.

10.3 HOLONOMIC CONSTRAINT

A *holonomic constraint* on a dynamic system is expressed by an equation of configuration displacement and time:

$$f(\mathbf{u}, t) = 0 \tag{10.152}$$

An equation between the coordinates of the configuration space in the forms

$$f(x_1, y_1, z_1, x_2, y_2, z_2, x_3, y_3, z_3, \ldots, x_{N/3}, y_{N/3}, z_{N/3}) = 0 \tag{10.153}$$

$$f(u_1, u_2, u_3, \ldots, u_N) = f(\mathbf{u}) = 0 \tag{10.154}$$

or reducible to these forms is called a *scleronomic holonomic constraint*. An equation between the coordinates of the event space in the forms

$$f(x_1, y_1, z_1, x_2, y_2, z_2, x_3, y_3, z_3, \ldots, x_{N/3}, y_{N/3}, z_{N/3}, t) = 0 \tag{10.155}$$

$$f(u_1, u_2, u_3, \ldots, u_N, t) = f(\mathbf{u}, t) = 0 \tag{10.156}$$

or reducible to these forms is called a *rheonomic holonomic constraint*. A constraint is *holonomic* only if it can be defined by an equation in the configuration or event space.

Every scleronomic holonomic constraint defines a subspace in the configuration space where the *possible motion* of the system can occur. Similarly, every rheonomic holonomic constraint defines a subspace in the event space where the possible motion of the system can occur. Any motion of a system out of its holonomic subspace is *impossible* as long as the constraint exists.

If a holonomic constraint is imposed on finite displacements u_i of a describing point, there is also a constraint on infinitesimal displacements du_i of the point:

$$df(\mathbf{u}, t) = f_1(d\mathbf{u}, dt) = 0 \tag{10.157}$$

The scleronomic and rheonomic holonomic constraints on infinitesimal displacements are found by differentiating Equations (10.154) and (10.156), respectively:

$$\sum_{i=1}^{N} \frac{\partial f}{\partial u_i} du_i = 0 \tag{10.158}$$

$$\sum_{i=1}^{N} \frac{\partial f}{\partial u_i} du_i + \frac{\partial f}{\partial t} dt = 0 \tag{10.159}$$

Because both of these differential constraints are total differentials and can be integrated to the forms (10.154) and (10.156), they are considered holonomic constraints. The infinitesimal displacement consonants define tangential spaces to the finite displacement constraint surfaces (10.154) and (10.156). If a differential constraint (10.157) is not a total differential and is not integrable, it is not considered holonomic.

Proof: Consider a dynamic system with a three-dimensional configuration space (u_1, u_2, u_3) and a scleronomic holonomic constraint as

$$f(u_1, u_2, u_3) = 0 \tag{10.160}$$

The constraint equation defines a surface in the space (u_1, u_2, u_3). To satisfy the constraint (10.160), the describing S_C-point of the system and its S_C-trajectories must lie in the constraint surface for every possible motion.

When the configuration space has N dimensions, the constraint (10.154) defines an $(N-1)$-dimensional hypersurface in the configuration space in which every S_C-trajectory of the system must lie. Such a constraint surface is rigid and will not change with time. Figure 10.19 illustrates the geometric interpretation of a scleronomic holonomic constraint in a three-dimensional configuration space.

The time-dependent constraint

$$f(u_1, u_2, u_3, t) = 0 \tag{10.161}$$

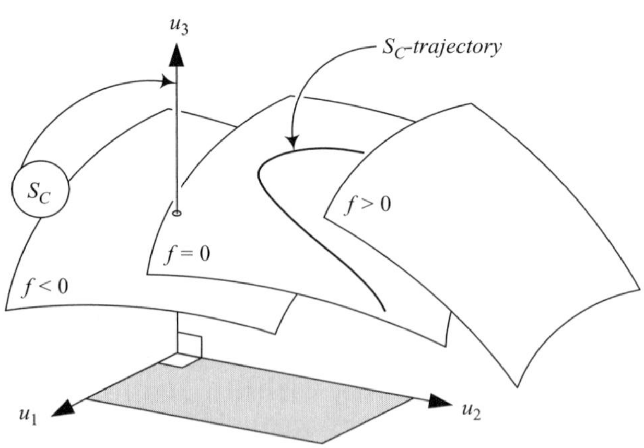

Figure 10.19 A scleronomic constraint in a three-dimensional configuration space and possible S_C-trajectory.

defines a rigid surface in the event space (u_1, u_2, u_3, t). We may also interpret the constraint as a time-variable surface in the configuration space. When the configuration space has N dimensions, a time-dependent constraint defines an $(N-1)$-dimensional deformable hypersurface in the configuration space in which every S_C-trajectory of the system must lie. Such a constraint surface is not rigid and will change with time in a prescribed manner. A scleronomic constraint makes a cylindrical surface in the event space with a rigid projection on the configuration space. Figure 10.20 illustrates the geometric interpretation of a rheonomic constraint and a scleronomic holonomic constraint in a three-dimensional event space.

If we fix t at a time $t = t_0$, then a rheonomic constraint reduces to a scleronomic constraint. Such a scleronomic constraint is called a *frozen constraint*. The fixing-time operation is called a *freezing process* in which the time variable t is eliminated and (10.163) reduces to (10.162) with similar geometric interpretation.

Equation (10.154) imposes a holonomic scleronomic constraint on the *finite displacements* of u_i, $i = 1, 2, 3, \ldots, N$. If the constraint surface is smooth enough such that all first partial derivatives of the function f with respect to all arguments exist and are continuous, we can determine the constraint on *infinitesimal displacements* du_i, $i = 1, 2, 3, \ldots, N$, by differentiation:

$$\sum_{i=1}^{N} \frac{\partial f}{\partial u_i} du_i = \nabla f \cdot d\mathbf{u} = 0 \tag{10.162}$$

Assume the describing point of the system is shown by a point u_i^* on the constraint surface. The term ∇f defines a tangential plane to the constraint surface at u_i^*. Equation (10.162) then indicates that any infinitesimal displacement $d\mathbf{u}$ of the S_C-point must take place in the *constraint plane* ∇f when the partial derivatives are calculated at u_i^*. The infinitesimal displacement constraint for the holonomic rheonomic constraint (10.156) is

$$\sum_{i=1}^{N} \frac{\partial f}{\partial u_i} du_i + \frac{\partial f}{\partial t} dt = \nabla f \cdot d\mathbf{u} + \frac{\partial f}{\partial t} dt = 0 \tag{10.163}$$

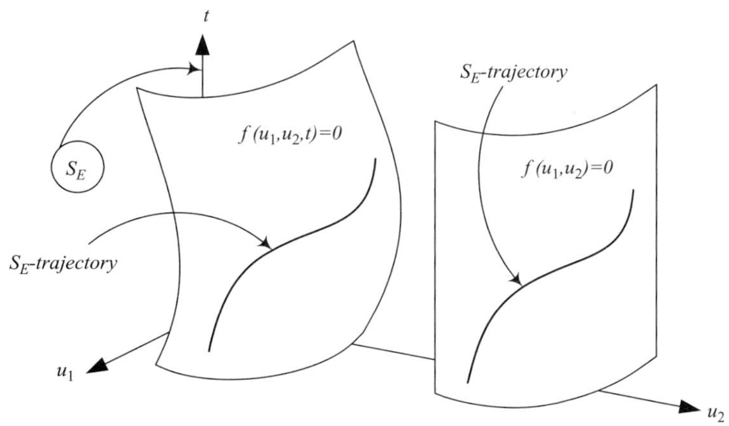

Figure 10.20 A rheonomic and a scleronomic constraint in a three-dimensional event space and possible S_E-trajectories.

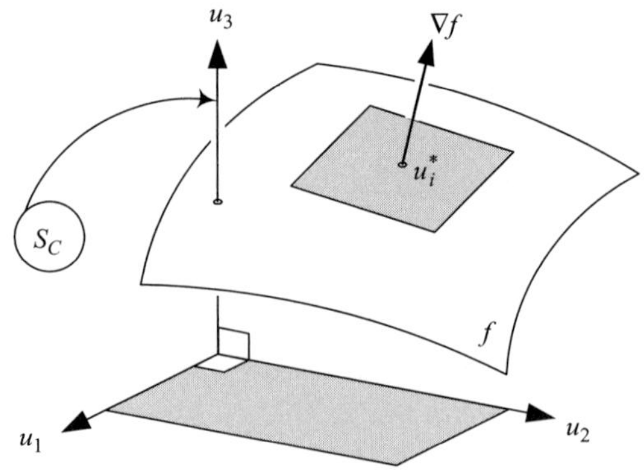

Figure 10.21 A constraint surface in a 3D configuration space and its constraint plane at a point u_i^*.

This equation indicates that any infinitesimal displacement $d\mathbf{u}$ of the S_C-point must be in the *constraint plane* ∇f when the constraint surface is frozen.

The constraint (10.162) is reducible to (10.154) because its integral

$$f(u_1, u_2, u_3, \ldots) - c = 0 \tag{10.164}$$

is of the same form. Similarly, the constraint (10.163) is reducible to (10.156). The constant of integration is not a function of initial conditions $\mathbf{u}_0$ and $\dot{\mathbf{u}}_0$. It can be determined by having the coordinates u_i of any point of the constraint surface including $\mathbf{u}_0$. Figure 10.21 illustrates a scleronomic constraint surface in a 3D configuration space along with the constraint plane at a point u_i^*.

Every trajectory lying in a holonomic constraint surface is called a *possible trajectory*, and any small displacement in the tangent constraint plane is called a *possible displacement*. Out of all possible trajectories of a dynamic system, the one that also satisfies the equations of motion and initial conditions is called the *actual trajectory*. Similarly, the small displacement on the tangent constraint plane that satisfies the equation of motion is called an *actual displacement*. Any displacement that violates the constraint equation (10.154) or (10.162) is called an *impossible displacement*. It is impossible that the describing point leave the constraint plane by an actual displacement.

The equation of a holonomic constraint $f(u_i) = 0$, $i = 1, 2, 3, \ldots, N$, is a regular surface if $N = 3$. It is called a hypersurface if $N > 3$, although we may call it a surface regardless of the value of N.

From the Greek *holonomic* means "lawful," *scleronomic* means "rigid," and *rheonomic* means "flowing". ■

Example 612 A Box in a Channel on a Table Consider the motion of a box on a flat table. We attach a Cartesian coordinate frame G to the table such that the Z-axis is perpendicular to the table. We also attach a body frame B to the box at its geometric center. We count the freedoms of the box by the possible motions of B in G. The box is moving on a fixed plane in G, so it has three freedoms. Therefore there must be three constraints to eliminate one translation and two rotations out of six possible freedoms: X, Y, Z, θ_X, θ_Y, θ_Z. They are imposed by the table, which does not let the box change its Z-component and turn about the X- and Y-axes:

$$Z - Z_0 = 0 \qquad \theta_X = 0 \qquad \theta_Y = 0 \tag{10.165}$$

We may also express the constraint equations in the body frame B:

$$z - z_0 = 0 \qquad \theta_x = 0 \qquad \theta_y = 0 \tag{10.166}$$

Now let us assume that the box is sliding in a channel along the x-axis. A top view of the system is shown in Figure 10.22. The channel also imposes two more constraints,

$$Y - X \tan\alpha = 0 \qquad \theta_Z = 0 \tag{10.167}$$

and reduces its DOF to 1. The value of X of its mass center will clearly identify the position of the box because there is a geometric relation between X and Y that indicates only one of them is independent.

If the channel is mounted on a turntable that is rotating $\omega/2\pi$ times a second about the Z-axis, then the X- and Y-coordinates of the box are related by a rheonomic constraint,

$$X \sin(\omega t) - Y \cos(\omega t) = 0 \qquad \theta_Z = \omega t \tag{10.168}$$

Such a box still has x or X as the only freedom in the configuration space (x, y, z, θ_x, θ_y, θ_z) or (X, Y, Z, θ_X, θ_Y, θ_Z). This example indicates that the equation of a constraint depends on the coordinate frame in which the constraint is expressed. The rheonomic constraints (10.168) in G would be a scleronomic constraints in B:

$$y = 0 \qquad \theta_z = 0 \tag{10.169}$$

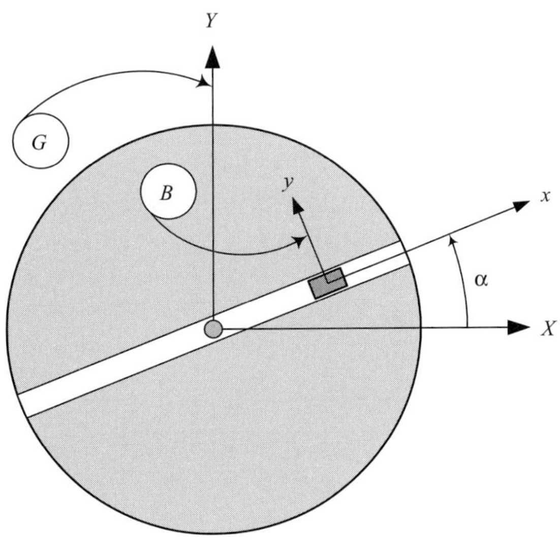

Figure 10.22 A top view of a sliding box in a channel along the x-axis.

Example 613 The Pin of a Tight Joint Two bars with length l are connected between two fixed points with a distance $d = 2l$ apart, as shown in Figure 10.23. The constraints on the planar motion of the joint A are

$$x^2 + y^2 - l^2 = 0 \qquad (x - 2l)^2 + y^2 - l^2 = 0 \tag{10.170}$$

Because of these two holonomic constraints, point A has no DOF. The only possible position of A is the intersection point of the constraint circles of (10.170):

$$x_A = l \qquad y_A = 0 \tag{10.171}$$

Taking a derivative, we can find the constraints on a small displacement of A:

$$\begin{bmatrix} x & y \\ x - 2l & y \end{bmatrix} \begin{bmatrix} dx \\ dy \end{bmatrix} = \begin{bmatrix} 0 \\ 0 \end{bmatrix} \tag{10.172}$$

The determinant of the coefficients of these equations is zero at position (10.171), and therefore, a small displacement of A in dy is possible:

$$\begin{vmatrix} x & y \\ x - 2l & y \end{vmatrix} = 2ly \tag{10.173}$$

If the distance of the fixed joints is $d < 2l$, then the constraints would be

$$x^2 + y^2 - l^2 = 0 \qquad (x - d)^2 + y^2 - l^2 = 0 \tag{10.174}$$

There are two possible positions for joint A; however, it can occupy only one of them:

$$x_A = \tfrac{1}{2}d \qquad y_A = \pm\sqrt{l^2 - \left(\tfrac{1}{2}d\right)^2} \tag{10.175}$$

We must disassemble the bars and break the constraint to be able to assemble them at the other possible position. In this situation, the constraints on a small despoilment of A are

$$\begin{bmatrix} x & y \\ x - d & y \end{bmatrix} \begin{bmatrix} dx \\ dy \end{bmatrix} = \begin{bmatrix} 0 \\ 0 \end{bmatrix} \tag{10.176}$$

The determinant of the coefficients is not zero at both possible positions. Therefore the only possible infinitesimal displacement is at $dx = 0$ and $dy = 0$.

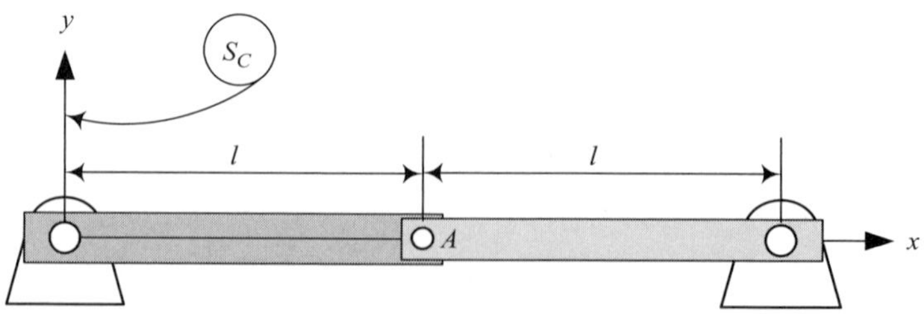

Figure 10.23 Two bars with length l are connected at joint A between two walls with a distance $d = 2l$.

Example 614 ★ Why Constraints? A mechanical system is a combination of relatively moving rigid bodies. We can mathematically define the system by a set of geometric relations between relatively fixed points and differential relations between relatively moving points. The set of relations makes a set of constraint equations that must be satisfied at all time. So, multibody kinematics is nothing but constraint kinematics, and the science of multibody dynamics is not complete without knowing how to analyze and work with constraints. The following dynamic information can be calculated easier by employing constraints:

1. The degree of freedom f_C of a dynamic system is the number of dimensions of its S_C-space, N, minus the number of independent holonomic constraints, L. This number determines the number of independent variables we need to describe the configuration of the system at any time. Furthermore, f_C indicates how many differential equations of motion we must develop to determine the value of the variables as functions of time.
2. Associated to every holonomic constraint, we have a constraint force. All internal forces in the members of a mechanical system as well as contact forces are constraint forces. We need the constraint forces to design the mechanical parts of a system.

Example 615 Multiple Constraints and Independence When there are multiple constraints on the describing point of a dynamic system, they must be linearly independent and consistent. The L holonomic constraints $f_i(u_1, u_2, u_3, \ldots, u_N, t) = 0$, $i = 1, 2, 3, \ldots, L$, in an N-dimensional configuration space are said to be linearly independent if we cannot find L not-all-zero constants a_i such that

$$\sum_{i=1}^{L} a_i f_i \left(u_1, u_2, u_3, \ldots, u_N, t\right) = 0 \tag{10.177}$$

The constraints are consistent if they are linearly independent and all apply within the subspace in which the describing point of the dynamic system can have an actual displacement.

Consider a moving particle in a three-dimensional configuration space (u_1, u_2, u_3) with two holonomic constraints

$$f_1 \left(u_1, u_2, u_3\right) = 0 \qquad f_2 \left(u_1, u_2, u_3\right) = 0 \tag{10.178}$$

Assume that the two constraints are consistent and define two intersecting surfaces. The intersection would be a space curve on which the point must move. Therefore, a moving particle on a prescribed curve in a 3D S_C-space is under two holonomic constraints. Such a particle has only one DOF.

Let us impose a third constraint $f_3(u_1, u_2, u_3) = 0$ on the particle. If the three constraints are consistent and independent, the first two define a curve and the third one intersects the curve at a finite number of points. The particle can occupy only one of the points permanently with no DOF.

As an example, consider the three constraints

$$z - z_0 = 0 \tag{10.179}$$

$$x - x_0 = 0 \tag{10.180}$$

$$x^2 + y^2 + z^2 - R^2 = 0 \tag{10.181}$$

The intersection of these three constraints indicates two points P_1 and P_2 where a particle can actually be in only one of them:

$$P_1 = P_1\left(x_0, \sqrt{R^2 - x_0^2 - z_0^2}, z_0\right) \tag{10.182}$$

$$P_2 = P_2\left(x_0, -\sqrt{R^2 - x_0^2 - z_0^2}, z_0\right) \tag{10.183}$$

If the three constraints are rheonomic, they will define a point that moves in space in a prescribed manner. Consider the three constraints

$$\begin{aligned} z - z_0 \sin \omega t &= 0 \\ x - x_0 &= 0 \\ x^2 + y^2 + z^2 - R^2 &= 0 \end{aligned} \tag{10.184}$$

Their intersections indicate two points P_1 and P_2 with variable coordinates:

$$P_1 = P_1\left(x_0, \sqrt{R^2 - x_0^2 - z_0^2 \sin^2 \omega t}, z_0\right) \tag{10.185}$$

$$P_2 = P_2\left(x_0, -\sqrt{R^2 - x_0^2 - z_0^2 \sin^2 \omega t}, z_0\right) \tag{10.186}$$

Such constraints show that the particle will have one of the following motions:

$$y = \begin{cases} \sqrt{R^2 - x_0^2 - z_0^2 \sin^2 \omega t} & (10.187) \\ -\sqrt{R^2 - x_0^2 - z_0^2 \sin^2 \omega t} & (10.188) \end{cases}$$

Assuming the particle is under the motion (10.187), we can calculate the required force to cause the motion:

$$\begin{aligned} F_y &= m\ddot{y} \\ &= -m\frac{\sqrt{2}z_0^2\omega^2}{4}\frac{z_0^2(3 + \cos 4t\omega) + \left(8R^2 - 8x_0^2 - 4z_0^2\right)\cos 2t\omega}{(2y)^{\frac{3}{2}}} \end{aligned} \tag{10.189}$$

The force F_y is the resultant of the three constraint forces in the direction of the motion.

Now consider a free particle with mass m in a three-dimensional space (x, y, z) under an applied force (10.189). Such a particle will move exactly the same as the particle under the three constraints (10.184).

Example 616 ★ Velocity Vector and Tangent Plane We can take a time derivative from (10.154) and write Equation (10.162) as

$$\sum_{i=1}^{N} \frac{\partial f}{\partial u_i} \frac{du_i}{dt} = \nabla f \cdot \dot{\mathbf{u}} = 0 \tag{10.190}$$

which indicates that the possible velocity vector $\dot{\mathbf{u}}$ is always on the constraint plane ∇f.

Assume that a constraint $f(\mathbf{u})$ is a k-dimensional subspace of an N-dimensional configuration space. At every point $\mathbf{u}^*$ of f we have a k-dimensional tangent space $\nabla f(\mathbf{u}^*)$.

Example 617 Constraint of a Pendulum with Moving Support Consider a planar pendulum in plane (X, Y) with a moving support according to the functions

$$x = x(t) \qquad y = y(t) \tag{10.191}$$

where x and y indicate the position of the support in the configuration plane as shown in Figure 10.24. Because of the constant length l of the pendulum, there exists a constraint between the coordinates X and Y of the pendulum tip point:

$$[X - x(t)]^2 + [Y - y(t)]^2 - l^2 = 0 \tag{10.192}$$

Therefore, the constraint for a small displacement is

$$[X - x(t)][dX - \dot{x}\,dt) + (Y - y(t)](dY - \dot{y}\,dt) = 0 \tag{10.193}$$

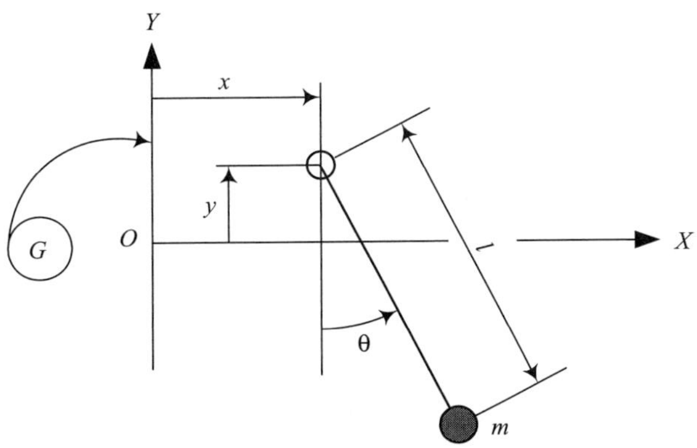

Figure 10.24 A planar pendulum with moving support.

Example 618 ★ Total Differential Constraint Consider a differential constraint

$$A(x,t)dx + B(x,t)dy + C(x,t)dz + D(x,t)dt = 0 \tag{10.194}$$

If there happens to exist a function $f(x, y, z, t)$ such that

$$\frac{\partial f}{\partial x} = A(x, t) \qquad \frac{\partial f}{\partial y} = B(x, t)$$
$$\frac{\partial f}{\partial z} = C(x, t) \qquad \frac{\partial f}{\partial t} = D(x, t) \tag{10.195}$$

then we can write (10.194) in the form

$$df = \frac{\partial f}{\partial x}dx + \frac{\partial f}{\partial y}dy + \frac{\partial f}{\partial z}dz + \frac{\partial f}{\partial t}dt = 0 \tag{10.196}$$

The general solution of such a differential constraint is the holonomic constraint

$$f(x, y, z, t) = c \tag{10.197}$$

called a total differential constraint.

As an example, consider the differential constraint

$$\sin y\, dx + (\sin z + x \cos y)\, dy + (\sin t + y \cos z)\, dz + z \cos t\ dt = 0 \tag{10.198}$$

which is a total differential constraint because there is a function f such that the constraint is df:

$$f(x, y, z, t) = x \sin y + y \sin z + z \sin t = c \tag{10.199}$$

Example 619 ★ Constraint Cannot Be a Function of Acceleration Assume a particle m is under a constraint of position, velocity, acceleration, and time:

$$f(\mathbf{u}, \dot{\mathbf{u}}, \ddot{\mathbf{u}}, t) = 0 \tag{10.200}$$

The constraint equation can theoretically be solved for acceleration as a function of position, velocity, and time:

$$\ddot{\mathbf{u}} = f_1(\mathbf{u}, \dot{\mathbf{u}}, t) \tag{10.201}$$

However, based on Newton's second law, the acceleration of m must be proportional to the applied force $\mathbf{F}$:

$$\ddot{\mathbf{u}} = \frac{1}{m}\mathbf{F}(\mathbf{u}, \dot{\mathbf{u}}, t) \tag{10.202}$$

In general, $f_1 \neq mF$. Therefore, if Newton's second law is applied, there cannot be another constraint on acceleration.

Newton's equation of motion is the only acceptable constraint on accelerations in dynamics. This is why we call this science *Newtonian dynamics*.

As an example assume that a particle with mass m is under the constraint

$$f(x, \dot{x}, \ddot{x}, t) = \ddot{x} + x + t\dot{x} = 0 \tag{10.203}$$

This equation indicates that the particle is under an applied force F,

$$F = -m(x + t\dot{x}) \tag{10.204}$$

Therefore, a constraint equation with a jerk argument such as

$$f(x, \dot{x}, \ddot{x}, \dddot{x}, t) = \dddot{x} + t\ddot{x} + 2\dot{x} = 0 \tag{10.205}$$

must be a time derivative of Newton's equation of motion:

$$\frac{d}{dt}(\ddot{x} + x + t\dot{x}) = \dddot{x} + t\ddot{x} + 2\dot{x} = 0 \tag{10.206}$$

10.4 GENERALIZED COORDINATE

We can define a *configuration degree of freedom* f_C as the number of dimensions of S_C-space, N, minus the number of independent holonomic constraints, L:

$$f_C = N - L \qquad N > L \tag{10.207}$$

The DOF of a dynamic system depends on the space in which we are describing its dynamics. In general, the number of dimensions of its describing space minus the number of independent constraints that can be expressed as a hypersurface is the DOF of the system in that space. We can similarly define a *state degree of freedom* f_S as the number of dimensions of S_S-space minus the number L of independent constraints between u_i and $\dot{u}_i$, $i = 1, 2, 3, \ldots, N$:

$$f_S = 2N - L \tag{10.208}$$

However, the configuration degree of freedom f_C in (10.207) is the most traditional definition of the degree of freedom of a dynamic system.

Because every holonomic constraint defines a subspace in the configuration space in which the S_C-point of the system can move, the dimension f_C of the final subspace indicates the DOF of the system. Therefore, the N-dimensional position vector $d\mathbf{u}$ of the S_C-point in the S_C-space has only f_C-dimensional DOF. So, we can define the f_C number of new variables to determine the N components of $d\mathbf{u}$. The independent variables we choose to show the components of $d\mathbf{u}$ are the *generalized coordinates* q_i, $i = 1, 2, \ldots, f_C$, of the system:

$$q_i = q_i(u_1, u_2, u_3, \ldots, u_N) \qquad i = 1, 2, \ldots, f_C \tag{10.209}$$

Proof: A dynamic system must move such that its kinematics satisfy the constraint equations at all times. If a constraint equation is holonomic, then we can express the constraint as an equation of the coordinates of the configuration space. Theoretically, we can solve a holonomic constraint equation to find one coordinate of the S_C-space in terms of the other coordinates. Having such an equation indicates that the coordinate is not independent and its value is a function of the values of the other coordinates. Therefore, the total number of independent coordinates of S_C-space is equal to the dimensions of the configuration space, N, minus the total number of independent holonomic constraints, L. The number of required independent coordinates is the configuration DOF of the system, f_C.

The f_C independent coordinates that we use to describe the kinematics of a dynamic system are called the *generalized coordinates* of the system. The number of generalized coordinates is equal to the DOF of the S_C-point. Eliminating the dependent coordinates determines the minimum required number of independent coordinates. This number is the number of freedoms. We may use the N-dimensional configuration space along with the L holonomic constraints to express the motion of the S_C-point of a system or define a new f_C-dimensional configuration space with no constraint. We are also free to choose any set of independent variables to express the coordinates of the *constraint-free configuration space*.

Any set of real coordinates $q_i (i = 1, 2, \ldots, f_C)$ which can describe the configuration of a system is called a set of generalized coordinates if and only if the number f_C of its members is necessary and sufficient.

Selection of a set of coordinates to express the configuration space is also optional.

Any set of real coordinates $u_i (i = 1, 2, \ldots, N)$ which can describe the configuration space of a system is a set of configuration coordinates.

Let us introduce a new set of coordinates:

$$v_i = v_i \left(u_1, u_2, u_3, \ldots, u_N\right) \qquad i = 1, 2, \ldots, N \tag{10.210}$$

where the v_i are single-valued functions of their arguments. We may consider the set of functions v_i as mapping functions that transform the coordinates of a point from $(u_1, u_2, u_3, \ldots, u_N)$ to $(v_1, v_2, v_3, \ldots, v_N)$. The set of v_i is another set of coordinates to express the configuration space. To have a one-to-one and reversible mapping, the Jacobian J of the u_i with respect to v_j or vice versa must be nonzero:

$$J = \begin{vmatrix} \frac{\partial v_1}{\partial u_1} & \cdots & \frac{\partial v_1}{\partial u_N} \\ \vdots & \ddots & \vdots \\ \frac{\partial v_N}{\partial u_1} & \cdots & \frac{\partial v_N}{\partial u_N} \end{vmatrix} = \frac{\partial \left(v_1, v_2, v_3, \ldots, v_N\right)}{\partial \left(u_1, u_2, u_3, \ldots, u_N\right)} \tag{10.211}$$

Every one of the constraint equations $f_i\left(u_1, u_2, u_3, \ldots, u_N\right)$, $i = 1, 2, \ldots, L$, will also be transformed to a new constraint equation based on the new coordinates. They express new hypersurfaces in the configuration space:

$$f_i\left(v_1, v_2, v_3, \ldots, v_N\right) \qquad i = 1, 2, \ldots, L \tag{10.212}$$

Hence, the degree of freedom f_C of a dynamic system remains the same regardless of the coordinate system we use to describe the configuration space and constraints:

$$f_C = N - L \tag{10.213}$$

There is no superior option for u_i, v_j, and q_i at this point. Defining a set of configuration or generalized coordinates that simplifies the equations of motion is an art and skill based on experience and preference. However, it seems that canonical coordinates based on Legendre transformation and Hamilton equations are an optimal set. ■

Example 620 A Planar Pendulum Consider a planar pendulum with a mass m attached to a massless rigid bar with length l_0 from the origin of a coordinate frame. The S_C-space is specified by the Cartesian coordinates x and y of m. The mass moves on a circle of radius l_0. The coordinates x and y must satisfy the holonomic constraint

$$x^2 + y^2 - l_0^2 = 0 \tag{10.214}$$

As the two Cartesian coordinates satisfy one holonomic constraint, there is only a single generalized coordinate q_1. We may use the angle θ of the bar as a generalized coordinate, where θ must be a function of x and y:

$$\theta = \arctan \frac{y}{x} \tag{10.215}$$

Employing the constraint equation, x and y must be expressible by θ:

$$x = l_0 \cos\theta \qquad y = l_0 \sin\theta \tag{10.216}$$

We may use l and θ as a new set of configuration coordinates. The constraint (10.214) will then transform to

$$l - l_0 = 0 \tag{10.217}$$

showing that the value of l remains constant and hence the other variable can be used as a generalized coordinate. The Jacobian of new coordinates must be nonzero:

$$J = \begin{vmatrix} \dfrac{\partial x}{\partial l} & \dfrac{\partial x}{\partial \theta} \\ \dfrac{\partial y}{\partial l} & \dfrac{\partial y}{\partial \theta} \end{vmatrix} = l \neq 0 \tag{10.218}$$

Example 621 A Pendulum with Moving Support Consider the planar pendulum in Figure 10.24 that moves in plane (X, Y) such that its support is moving along the x- and y-axes with the given functions

$$x = x(t) \qquad y = y(t) \tag{10.219}$$

The configuration space of the system is the (X, Y)-plane. However, because of the constant length of the pendulum, there exists a constraint between the X- and Y-coordinates of m:

$$[X - x(t)]^2 + [Y - y(t)]^2 - l^2 = 0 \tag{10.220}$$

The constraint equation is of the form (10.156) and therefore is a rheonomic holonomic constraint. Therefore, only one of the coordinates X and Y is independent. Let us accept X as the only independent variable and determine the Y-coordinate of m from (10.220):

$$Y = y(t) \pm \sqrt{l^2 - [X - x(t)]^2} \tag{10.221}$$

For any value of X there exist two possible values for Y where only one of them indicates the actual position of m. The actual value of Y must be determined by the equation of motion and initial conditions of the pendulum. Let us accept the minus sign as the actual position to be consistent with Figure 10.24.

As an another option for a generalized coordinate, we can choose θ instead of X and define X as a function of θ:

$$X = x(t) + l\sin\theta \tag{10.222}$$

Substituting (10.222) in (10.221) provides Y as a function of θ:

$$Y = y(t) - l\cos\theta \tag{10.223}$$

Let us assume that the support is moving with frequency ω on a circle with radius R. Therefore, the functions $x(t)$ and $y(t)$ are

$$x = R\cos\omega t \qquad y = R\sin\omega t \tag{10.224}$$

and the constraint equation (10.220) becomes

$$X^2 + Y^2 + R^2 - 2R\left(X\cos\omega t - Y\sin\omega t\right) - l^2 = 0 \tag{10.225}$$

Example 622 ★ System Configuration Space Figure 10.25 illustrates a $2a \times 2b$ box that is attached to a double pendulum. A point mass m is sliding in the box.

This mechanical system has four parts: two pendulums, a box, and a sliding particle. The configuration of the parts can be uniquely determined by knowing the position of four points: A, B, C, and m. The configuration space of the S_C-point of the system has eight dimensions (X_A, Y_A, X_B, Y_B, X_C, Y_C, X_m, Y_m). The constant distances between points eliminate three DOF and the sliding condition of m eliminates one DOF. Therefore, there must be four holonomic constraints among the coordinates of

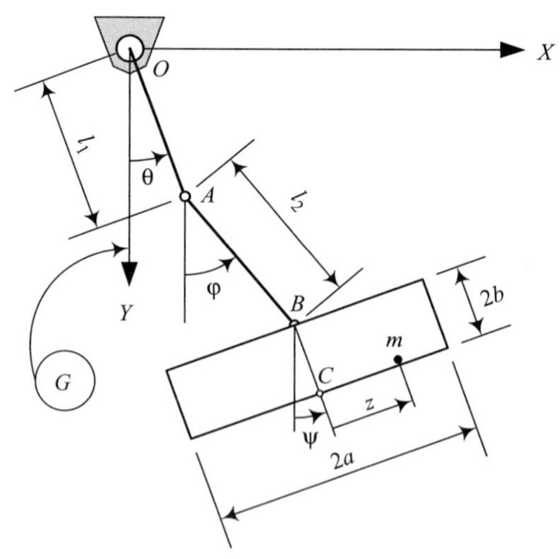

Figure 10.25 A point mass m is sliding in a box that is attached to a double pendulum.

the S_C-space. The constraints are

$$\begin{aligned} X_A^2 + Y_A^2 - l_1^2 &= 0 \\ (X_A - X_B)^2 + (Y_A - Y_B)^2 - l_2^2 &= 0 \\ (X_C - X_B)^2 + (Y_C - Y_B)^2 - 4b^2 &= 0 \\ (X_C - X_B)(X_m - X_C) - (Y_C - Y_B)(Y_m - Y_C) &= 0 \end{aligned} \tag{10.226}$$

Using the above four constraints, we can choose a set of any four coordinates of the configuration space $(X_A, Y_A, X_B, Y_B, X_C, Y_C, X_m, Y_m)$ as the generalized coordinates.

The variables $(\theta, \varphi, \psi, z)$ might be better choices of generalized coordinates. The relations between $(\theta, \varphi, \psi, z)$ and the eight coordinates are

$$\tan\theta = \frac{X_A}{Y_A} \tag{10.227a}$$

$$\tan\varphi = \frac{X_B - X_A}{Y_B - Y_A} \tag{10.227b}$$

$$\tan\psi = \frac{X_C - X_B}{Y_C - Y_B} \tag{10.227c}$$

$$z = \sqrt{(X_C - X_B)^2 + (X_C - X_B)^2} \tag{10.227d}$$

Having the values of the generalized coordinates is enough to determine the configuration of the whole mechanical system as well as every component of the system. If the values of θ, φ, ψ, z are given, we can determine the configuration of the mechanical system of Figure 10.25 by calculating the position of the points A, B, C, and m:

$$\begin{aligned} X_A &= l_1 \sin\theta \qquad & Y_A &= l_1 \cos\theta \\ X_B &= l_1 \sin\theta + l_2 \sin\varphi \qquad & Y_B &= l_1 \cos\theta + l_2 \cos\varphi \\ X_C &= l_1 \sin\theta + l_2 \sin\varphi + 2b \sin\psi \\ Y_C &= l_1 \cos\theta + l_2 \cos\varphi + 2b \cos\psi \\ X_m &= l_1 \sin\theta + l_2 \sin\varphi + 2b \sin\psi + z\cos\psi \\ Y_m &= l_1 \cos\theta + l_2 \cos\varphi + 2b \cos\psi - z\sin\psi \end{aligned} \tag{10.228}$$

Example 623 Advantage of Generalized Coordinates The essential nature or advantage of generalized coordinates is that they provide freedom to be chosen arbitrarily. There is no limit on selecting them as long as they are single valued and C_1 functions of the configuration coordinates. There is also no restriction on the unit dimension of generalized coordinates. While the coordinates of S_C-space usually have no dimension or the dimension of length, the generalized coordinate may have any dimension.

As an example, the motion of a free particle in a plane may be expressed by Cartesian coordinates x and y with dimension of length, or by q_1 and q_2, as

$$x = q_1 \cos q_2 \qquad y = q_1 \sin q_2 \tag{10.229}$$

$$q_2 = \arctan \frac{y}{x} \qquad q_1 = \sqrt{x^2 + y^2} \tag{10.230}$$

where q_1 has the dimension of length and q_2 is dimensionless.

Example 624 ★ Number of DOF, f_C, n, and Equations of Motion We need $n = N - L$ equations of motion for a dynamic system with $N/3$ particles. Such a dynamic system has f_C degrees of freedom. The configuration space of the system is N-dimensional. The number of required equations of motion is equivalent to the configuration degree of freedom $f_C = N - L$, which is equal to the number of generalized coordinates n. Let us summarize the relations of DOF, f_C, n, and number of equations of motion, although all of these are always equal.

1. The N-dimensional coordinate frame that we use to express the position in configuration space is not necessarily Cartesian.
2. It is not true that there is an equation of motion for every generalized coordinate. There are always n independent individual distinguishable generalized coordinates, and there is a set of n coupled differential equations of the n generalized coordinates. Values of the generalized coordinates as functions of time will be the solutions of the equations of motion.
3. Although the number of DOF and generalized coordinates are equal, there is not necessarily a one-to-one relation between a freedom and a generalized coordinate. The n freedoms of massive bodies of a mechanical system are not necessarily indicated by a set of n generalized coordinates. Every freedom of a mechanical system is equivalent to an independent motion of a massive body. We can indicate and measure every motion by a coordinate. The motion coordinates are functions of the generalized coordinates, so the motion coordinates can be determined from the value of the generalized coordinates.
4. The coupling of the n equations of motion depends on the definition of the n generalized coordinates. It is possible to change the type of coupling by changing the coordinates. However, in general, it is not always possible to define a set of coordinates that makes all of the equations decouple.

Example 625 ★ Mapping Generalized Space to Itself There are two approaches to working with holonomic constraints:

1. Solve the L constraint equations for the L number of configuration coordinates as functions of the other $N - L$ coordinates that can be considered independent.
2. Replace the constraint by their associated constraint forces.

Defining $n = N - L$ generalized coordinates q_i $(i = 1, 2, \ldots, n)$, we can define a constraint-free n-dimensional configuration space $(q_1, q_2, \ldots, q_n)$, instead of the N-dimensional configuration space $(u_1, u_2, \ldots, u_N)$ along with the L constraint surfaces $f_i = f_i(u_1, u_2, \ldots, u_N, t)$ $(i = 1, 2, \ldots, L)$. We may call an n-dimensional constraint-free configuration space the *generalized space*.

Assume that we have defined n generalized coordinates as

$$q_1, q_2, \ldots, q_n \tag{10.231}$$

We may also define another set of n generalized coordinates s_j:

$$s_1, s_2, \ldots, s_n \tag{10.232}$$

There must always exist a functional relationship between the two sets of coordinates:

$$s_j = g_j\left(q_1, q_2, \ldots, q_n\right) \qquad j = 1, 2, \ldots, n \tag{10.233}$$

The functions g_j must satisfy the regularity conditions. So, they must be finite, single valued, continuous, and differentiable functions of their arguments. Furthermore, the Jacobian J must be nonzero:

$$J = \begin{vmatrix} \frac{\partial s_1}{\partial q_1} & \cdots & \frac{\partial s_1}{\partial q_n} \\ \vdots & \ddots & \vdots \\ \frac{\partial s_n}{\partial q_1} & \cdots & \frac{\partial s_n}{\partial q_n} \end{vmatrix} \neq 0 \tag{10.234}$$

Differentiation of Equations (10.233) gives

$$ds_j = \frac{\partial g_j}{\partial q_1} dq_1 + \frac{\partial g_j}{\partial q_2} dq_2 + \cdots + \frac{\partial g_j}{\partial q_n} dq_n \qquad j = 1, 2, \ldots, n \tag{10.235}$$

Therefore, the differential of coordinates q_i and s_j are always linearly dependent, regardless of what functional relations exist between the two sets of generalized coordinates.

Expressing the generalized space by two sets q_i and s_j along with the point transformation (10.233) is mapping a closed space to itself. The neighborhood of a point $Q(q_1, q_2, \ldots, q_n)$ in q-space is mapped to the neighborhood of $S(s_1, s_2, \ldots, s_n)$ in s-space. However, a straight line in q-space is no longer a straight line in s-space. This phenomenon becomes less obvious when the size of the line gets smaller. In an infinitesimal region around Q, straight lines are mapped to straight lines and parallel lines remain parallel, although the length and angles are not preserved.

Example 626 ★ Over-, Just, and Underconstraint Systems Consider a dynamic system in an N-dimensional S_C-space along with L holonomic constraint equations. Depending on the relative values of N and L, we may have an over-, just, or under-constraint system.

1. A dynamic system is underconstrained if $N > L$.

The DOF of a dynamic system is the number of dimensions of its S_C-space, N, minus the number of independent holonomic constraints, L:

$$f_C = N - L \qquad N > L \tag{10.236}$$

The constraints are defined by a set of L equations f_i of the coordinates of the configuration space:

$$f_i = f_i\left(u_1, u_2, \ldots, u_N, t\right) \qquad i = 1, 2, \ldots, L \tag{10.237}$$

We must define $n = N - L$ generalized coordinates q_j as algebraic functions of configuration coordinates u_k:

$$q_j = q_j\left(u_1, u_2, \ldots, u_N\right) \qquad j = 1, 2, \ldots, n \tag{10.238}$$

The n functions q_j must be independent from the L constraint equations f_i. Then we must develop n differential equations of motion by employing a proper method such as Newton's second law:

$$F_r = m_r\, \ddot{u}_r \qquad r = 1, 2, \ldots, n \tag{10.239}$$

To determine the N coordinates of the position vector $\mathbf{u} = \mathbf{u}\left(u_1, u_2, \ldots, u_N\right)$ of the S_C-point in S_C-space, we must solve the set of n differential equations of motion. It provides the value of n generalized coordinates $q_j(t)$. Then, we must solve the N algebraic equations (10.237) and (10.238) for $u_1(t), u_2(t), \ldots, u_N(t)$.

2. A dynamic system is just constrained if $N = L$.

When the number of constraints L on an S_C-point is the same as the dimension N of S_C-space, the *DOF* is zero. The S_C-point of the system remains at a point of intersection of all constraint surfaces. Theoretically, we are able to solve the L algebraic equations (10.237) and determine the N components $u_1(t), u_2(t), \ldots, u_N(t)$ of the position vector $\mathbf{u} = \mathbf{u}(u_1, u_2, \ldots, u_N)$ of the S_C-point.

Using $\ddot{u}_1(t), \ddot{u}_2(t), \ldots, \ddot{u}_N(t)$, we can evaluate the required force F_r to provide the acceleration $\ddot{u}_r$ of m_r:

$$F_r = m_r\, \ddot{u}_r \qquad r = 1, 2, \ldots, n \tag{10.240}$$

3. A dynamic system is overconstrained if $N < L$.

If we have more constraints than the dimensions of S_C-space, then there are two situations. First, there are $n = L - N$ redundant and unnecessary constraints. Second, the constraint equations are not consistent and cannot be satisfied all at once.

Example 627 ★ Change of Coordinate Integration of a total differential constraint depends on the ability to determine at least one integrating factor. However, the appearance of the equation of a constraint depends on the coordinates in which the equation is expressed. Change of coordinates and expression of the constraint equations in a separated and integrable form are an alternative method. The following two examples show the idea.

Consider the set of nonlinear and coupled equations

$$\dot{r} = -ar - \tfrac{1}{2}r\sin 2\theta \tag{10.241}$$

$$\dot{\theta} = -\tfrac{1}{2}b - \tfrac{1}{2}\cos 2\theta \tag{10.242}$$

Using polar coordinates

$$x = r\cos\theta \qquad y = r\sin\theta \tag{10.243}$$

we can simplify them to a set of solvable equations:

$$\dot{x} = -ax + \tfrac{1}{2}(b-1)\,y \tag{10.244}$$

$$\dot{y} = -\tfrac{1}{2}(b+1)\,x - ay \tag{10.245}$$

The order one-half Bessel equation

$$t^2\ddot{x} + t\dot{x} + \left(t^2 - \tfrac{1}{4}\right)x = 0 \tag{10.246}$$

can be transformed to the simpler equation

$$\ddot{y} + y = 0 \tag{10.247}$$

if we use a new coordinate y:

$$x = \frac{y}{\sqrt{t}} \tag{10.248}$$

Changing coordinates to get simpler equations dates back to the days of Euler. However, there is no effective and general classical method that can always change the coordinates if the equations are nonlinear.

Example 628 ★ Transforming a Bernoulli Equation to a Linear One There are many methods that can be used to change a variable and make a differential integrable. Transforming a Bernoulli equation to a linear differential form is one that is often applied.

An ordinary differential equation (ODE) of the form

$$\frac{dy}{dx} + p(x)y = q(x) \tag{10.249}$$

is an inhomogeneous linear first-order ODE. This equation is always solvable. Its solution is

$$y = e^{-\int p(x)dx}\left(C_1 + \int q(x)e^{\int p(x)dx}dx\right) \tag{10.250}$$

Consider the constraint equation

$$dx + \left(a(t)x + b(t)x^c\right)dt = 0 \tag{10.251}$$

where c is a constant and $a(t)$, $b(t)$ are arbitrary functions of t. When $c = 0$ the equation is a linear ODE, and when $c = 1$ the equation is separable. For all other values of c, we can introduce a new variable y,

$$y = x^{1-c} \tag{10.252}$$

and transform the equation to

$$\dot{y} + [(1-c)a(t)y + (1-c)b(t)] = 0 \tag{10.253}$$

which is a holonomic constraint because it is linear and integrable.

Equations of the form (10.251) are called Bernoulli equations.

Example 629 ★ Transforming a Non-Bernoulli Equation to a Bernoulli One Consider a constraint of the form

$$\dot{x} - \frac{t}{t^2x^2 + x^5} = 0 \tag{10.254}$$

which is not a Bernoulli equation in x. However, we can exchange the dependent and independent variables,

$$\frac{dt}{dx} - x^2 t - \frac{x^5}{t} = 0 \tag{10.255}$$

and make a Bernoulli equation (10.251) in t for $c = -1$.

10.5 CONSTRAINT FORCE

Every holonomic constraint applies a *constraint force* $\mathbf{F}_C$ on the describing point to keep it in the constraint surface. Here, $\mathbf{F}_C$ is proportional to the gradient of the frozen constraint equation $f(u_1, u_2, u_3, \ldots, u_N) = 0$ and hence is perpendicular to the constraint surface:

$$\mathbf{F}_C = -\lambda\, \nabla f = -\lambda \sum_{i=1}^{N} \frac{\partial f}{\partial u_i} \hat{u}_i. \qquad \lambda \in \mathbb{R} \tag{10.256}$$

$$\frac{F_{C_1}}{\partial f/\partial u_1} = \frac{F_{C_2}}{\partial f/\partial u_2} = \cdots = \frac{F_{C_N}}{\partial f/\partial u_N} \tag{10.257}$$

Figure 10.21 illustrates a constraint surface f and a local constraint plane at a point u^* along with the gradient of f at that point in a 3D configuration space.

Any applied force $\mathbf{F}$ on the S_C-point of a dynamic system can be decomposed into tangential and perpendicular components to the local constraint plane. The normal component of the applied force is a constraint force and the tangential component is called the *given force* $\mathbf{F}_G$:

$$\nabla f \cdot \mathbf{F}_G = 0 \tag{10.258}$$

The decomposition of an applied force into constraint and given components in a 3D configuration space is illustrated in Figure 10.26.

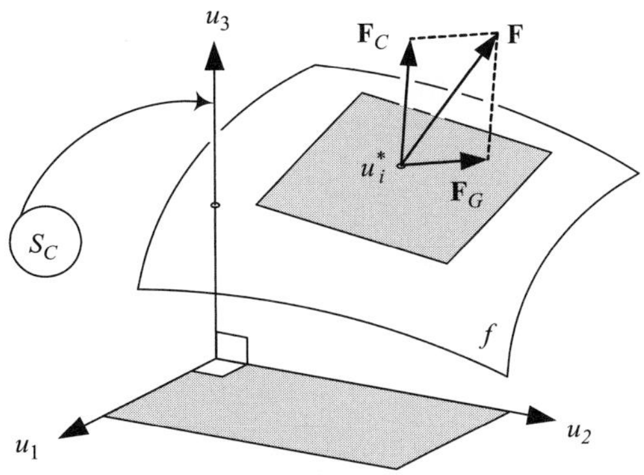

Figure 10.26 Decomposition of an applied force $\mathbf{F}$ into constraint $\mathbf{F}_C$ and given $\mathbf{F}_G$ components in a 3D configuration space.

Proof: The motion of a system can only be changed by applying forces. Every holonomic constraint affects the motion of the describing point of a dynamic system. So a constraint $f(\mathbf{u}, t) = 0$ applies a force on the system, and therefore, we have constraints and forces associated with them or we have no constraints. Such a force is a *constraint force* that can be a function of position and time only.

A holonomic constraint is equivalent to a surface in a proper space. The only possible motion of the describing point is in the constraint surface. Suppose the describing point tends to move out of the surface. Because such a motion is impossible, a force intrinsic to the surface must exist to prevent the motion and make it impossible. The constraint force must be in the opposite direction of the gradient, as given in Equation (10.256). The proportionality coefficient λ of a constraint force is called the *Lagrange multiplier*. If we make ∇f a unit vector, then λ would be the magnitude of the constraint force.

If $\mathbf{F}_C$ is the normal component of the resultant of all applied forces at an S_C-point on the constraint plane, then any other component of the resultant force at the point must be a given force $\mathbf{F}_G$ lying in the constraint plane. The tangential given force $\mathbf{F}_G$ causes the point to move in the tangential plane and produces an actual motion. The force $\mathbf{F}_G$ will not push the point out of the constraint plane. The constraint force $\mathbf{F}_C$ is normal to the plane and will not make the point move. The resultant force of $\mathbf{F}_G$ and $\mathbf{F}_C$ is called the *applied* or *physical force* $\mathbf{F}$.

The physical force and its given and constraint components are position, velocity, and time dependent and may vary at different points of a constraint surface. ■

Example 630 A Spherical Pendulum Consider a particle that is hanging from a point A by a massless rod of length l. The only constraint on the particle is its constant distance from A. Such a constraint defines a sphere on which the finite displacement of the particle may occur:

$$x^2 + y^2 + z^2 - l^2 = 0 \tag{10.259}$$

Taking a derivative determines the constraint on infinitesimal displacement:

$$2x\,dx + 2y\,dy + 2z\,dz = 0 \tag{10.260}$$

Because the configuration space has three dimensions, $N = 3$, and there is only one holonomic constraint, $L = 1$, the pendulum has two degrees of freedom, $f = N - L = 3 - 1 = 2$. Therefore, two independent functions of the Cartesian coordinates can be used to express the two freedoms of the pendulum. Let us choose the angles β and γ as shown in Figure 9.4. The Cartesian coordinates of the pendulum in a frame at A with the z-axis in the direction of $-\mathbf{g}$ are

$$\begin{bmatrix} x \\ y \\ z \end{bmatrix} = \begin{bmatrix} l\cos\gamma\sin\beta \\ l\sin\beta\sin\gamma \\ -l\cos\beta \end{bmatrix} \tag{10.261}$$

These equations show that β and γ are functions of x, y, z:

$$\beta = \arctan\frac{y}{x} \qquad \gamma = -\arccos\frac{z}{l} \tag{10.262}$$

The constraint sphere applies a constraint force $\mathbf{F}_C$ in the direction of the gradient of (10.259):

$$\mathbf{F}_C = \lambda \begin{bmatrix} 2x \\ 2y \\ 2z \end{bmatrix} \tag{10.263}$$

The constraint force is applied on the particle by the connecting rod. So, the force $\mathbf{F}_C$ is proportional to the tension force in the rod.

Example 631 Constraint Force of Two Holonomic Constraints Consider a particle which is constrained to move along a smooth space curve. We consider the curve as the intersection of two smooth constraint surfaces:

$$f_1(u_1, u_2, u_3) = 0 \qquad f_2(u_1, u_2, u_3) = 0 \tag{10.264}$$

The intersecting condition provides that the gradient of these surfaces are not colinear anywhere. The constraint force on the particle must be in a plane perpendicular to the space curve.

Figure 10.27 illustrates two intersecting constraint surfaces $f_1 = 0$ and $f_2 = 0$ on the space curve. The constraint forces of f_1 and f_2 are

$$\mathbf{F}_1 = -\lambda_1 \nabla f_1 \qquad \mathbf{F}_2 = -\lambda_2 \nabla f_2 \tag{10.265}$$

and therefore the constraint force $\mathbf{F}_C$ on the particle is

$$\mathbf{F}_C = -\lambda_1 \nabla f_1 - \lambda_2 \nabla f_2 \tag{10.266}$$

where $\mathbf{F}_C$ lies in a plane normal to the curve, defined by the two gradient vectors. However, the direction of $\mathbf{F}_C$ is not specified unless the ratio λ_1/λ_2 of the Lagrange multipliers is known.

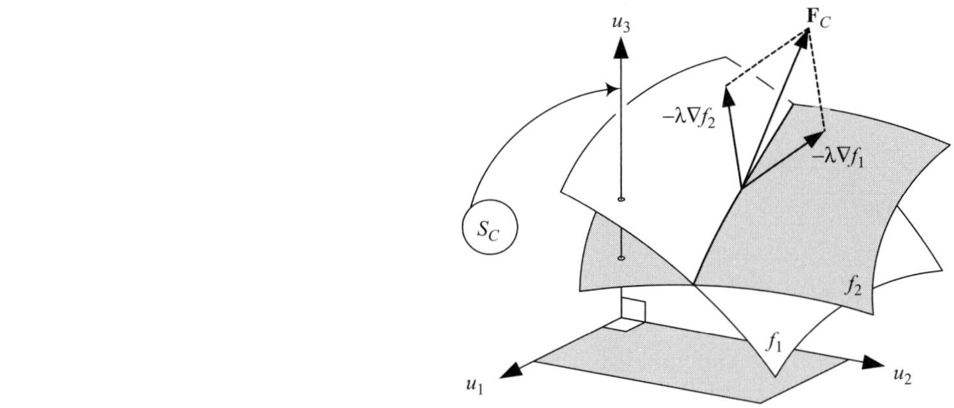

Figure 10.27 Two intersecting constraint surfaces $f_1 = 0$ and $f_2 = 0$ on the space curve.

10.6 VIRTUAL AND ACTUAL WORKS

An arbitrary displacement $\triangle\mathbf{u}$ of a describing S_C-point with respect to a holonomic constraint tangent plane can be decomposed into the tangential and normal components

$$\triangle\mathbf{u} = d\mathbf{u} + \perp\mathbf{u} \tag{10.267}$$

The tangential displacement $d\mathbf{u}$ is the *possible displacement* that satisfies the constraint equations. The normal displacement $\perp\mathbf{u}$ is the *impossible displacement* that can occur only by breaking at least one of the constraints. If the constraint surface is frozen, then any displacement $\delta\mathbf{u}$ in the constraint plane is a *virtual displacement*. A virtual displacement is a possible displacement if the constraint is scleronomic. Figure 10.28 illustrates the tangential and normal components or the constraint and given components of a physical force $\mathbf{F}$ and an arbitrary displacement $\triangle\mathbf{u}$ with respect to a constraint plane at a point $\mathbf{u}^*$ in the constraint surface f.

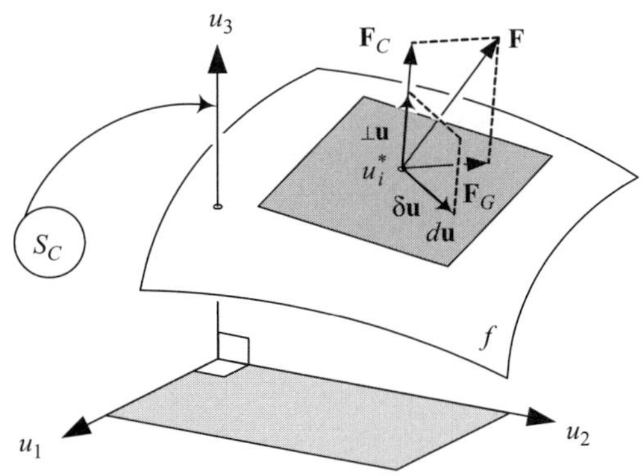

Figure 10.28 The tangential and normal components of a physical force F and an arbiterary displacement $\triangle\mathbf{u}$ on a constraint plane.

The work done by a given force $\mathbf{F}_G$ in an *actual displacement* $d\mathbf{u}$ of a dynamic system is the *actual work* W_A on the system:

$$W_A = \int \mathbf{F}_G \cdot d\mathbf{u} \tag{10.268}$$

The work done by the given force $\mathbf{F}_G$ in a *virtual displacement* $\delta\mathbf{u}$ of a dynamic system is the *virtual work* W_V on the system:

$$W_V = \int \mathbf{F}_G \cdot \delta\mathbf{u} \tag{10.269}$$

Proof: Consider a 3D configuration space with a holonomic constraint $f(\mathbf{u}, t) = 0$. Let us imagine an arbitrary displacement $\Delta\mathbf{u}$ having tangential and perpendicular components $d\mathbf{u}$ and $\perp\mathbf{u}$ with respect to the frozen constraint plane. The tangential component $d\mathbf{u}$ is in the constraint plane and is considered a possible displacement. It is the perpendicular component $\perp\mathbf{u}$ that is impossible and is prevented by the constraint force. Showing the perpendicular direction of a surface at a point by the gradient of the surface at that point, we have

$$\nabla f \cdot d\mathbf{u} = 0 \tag{10.270}$$

$$\perp\mathbf{u} = c\,\nabla f \qquad c \in \mathbb{R} \tag{10.271}$$

Four displacements can be imagined for a describing S_C-point with respect to a frozen holonomic constraint surface and its tangent plane:

1. *Possible displacement*. A displacement $d\mathbf{u}$ lying in the constraint plane is called a *possible displacement*. Possible displacement is consistent with constraint and satisfies the constraint equation.
2. *Impossible displacement*. A displacement $\perp\mathbf{u}$ perpendicular to the constraint plane is called an *impossible displacement*. Impossible displacement is an imaginary motion of the S_C-point normal to the constraint surface by assuming there is no constraint.
3. *Virtual displacement*. The tangential displacement of a frozen constraint is called a *virtual displacement* and is shown by $\delta\mathbf{u}$. Virtual displacement is consistent with frozen constraints and satisfies their frozen equation.
4. *Actual displacement*. Out of all possible displacements, the only one that is caused by the given force $\mathbf{F}_G$ is the *actual displacement* of the dynamic system. So, an actual displacement satisfies both the constraint equation and the equation of motion. The actual displacement is also shown by $d\mathbf{u}$.

The geometric relation of the displacements and the forces implies that the work of the constraint force $\mathbf{F}_C$ is zero in any possible displacement $d\mathbf{u}$ satisfying the scleronomic constraint equation $f(\mathbf{u}) = 0$:

$$\begin{aligned} W &= \int \mathbf{F}_C \cdot d\mathbf{u} = \int \left(F_{C_x}\,dx + F_{C_y}\,dy + F_{C_z}\,dz\right) \\ &= \lambda \int \nabla f \cdot d\mathbf{u} = 0 \end{aligned} \tag{10.272}$$

Assuming a three-dimensional configuration space, the proportionality of $\mathbf{F}_C$ and the gradient of the constraint surface provide

$$\frac{F_{C_x}}{\partial f/\partial x} = \frac{F_{C_y}}{\partial f/\partial y} = \frac{F_{C_z}}{\partial f/\partial z} \tag{10.273}$$

If the constraint is rheonomic, $f(\mathbf{u}, t) = 0$, relation (10.273) is still valid. However, because

$$\begin{aligned} df(\mathbf{u}, t) &= \frac{\partial f}{\partial x}dx + \frac{\partial f}{\partial y}dy + \frac{\partial f}{\partial z}dz + \frac{\partial f}{\partial t}dt \\ &= \nabla f \cdot d\mathbf{u} + \frac{\partial f}{\partial t}dt = 0 \end{aligned} \tag{10.274}$$

the work of the constraint force $\mathbf{F}_C$ is no longer zero:

$$W = \int \mathbf{F}_C \cdot d\mathbf{u} = -\lambda \int \frac{\partial f}{\partial t}dt \tag{10.275}$$

Assuming a frozen constraint, the possible displacements become virtual. Then, we conclude that the *virtual work of a constraint force is always zero*:

$$\begin{aligned} W_V &= \int \mathbf{F}_C \cdot \delta\mathbf{u} = \int \left(F_{C_x}\,\delta x + F_{C_y}\,\delta y + F_{C_z}\,\delta z\right) \\ &= \lambda \int \nabla f \cdot \delta\mathbf{u} = 0 \end{aligned} \tag{10.276}$$

We use this equation as the *principle of virtual work* to define and determine a constraint force. ■

Example 632 Possible and Virtual Displacements of a Particle on a Lift Consider a particle on the floor of an elevator which is rising along the z-axis with speed v_z. The constraint equation for possible displacement is

$$dz - v_z\,dt = 0 \tag{10.277}$$

and the constraint equation for virtual displacement is

$$\delta z = 0 \tag{10.278}$$

which indicate that the possible velocity is

$$\dot{z} = v_z \tag{10.279}$$

and the virtual velocity is

$$\dot{z} = 0 \tag{10.280}$$

Example 633 Friction Force Friction force $\mathbf{F}_f$ is a given force that exists because of a constraint and hence is a function of constraint force F_N. Consider a particle with mass m moving on a rough surface. Then,

$$\mathbf{F}_f = -\mu F_N \frac{\mathbf{v}}{v} \tag{10.281}$$

where μ is the coefficient of friction, F_N is the normal force exerted by the surfaces on the particle, and $\mathbf{v}/v$ is the unit vector in the direction of the particle velocity relative to the surface. This is a simplified Coulomb theory of friction that states: The friction force is proportional to the force pressing two physical surfaces together and is in the plane tangent to the surfaces at the contact point.

Friction force is an example of forces that arise from the presence of holonomic constraints. It is a function of constraint force but its line of action lies in the tangent plane and hence does virtual work.

In this example both the gravitational force mg and the normal force F_N are constraint forces and neither does work in a virtual displacement. However, there is a difference between these two constraint forces. The normal force disappears when the constraint is removed, but the gravitational force becomes a given force when the constraint is removed.

French scientist Charles Augustin de Coulomb (1736–1806) developed a series of two-term equations to model the friction force. The first term is a constant and the second term varies with time, normal force, velocity, or other parameters. Leonardo Da Vinci (1452–1519) was one of the first scholars to systematically study friction.

Example 634 Constraint Force Field Let C be a smooth curve in the (x, y)-plane. If there is a strong attractive force field on C, then a moving point will always be close to C. In the extreme case of an infinite force field, the point must remain on C. In this case we say a constraint is put on the motion of the point.

A constraint force is exerted on a particle of the system from outside the system if it does no work in an arbitrary virtual displacement. Every virtual displacement of a particle is perpendicular to the constraint force. As an example, the tension force of a pendulum is constraint force which is always perpendicular to every motion of the hanging particle.

Example 635 ★ Virtual Velocity Virtual velocity $\delta\dot{u}$ is defined as

$$\delta\dot{u} = \frac{d}{dt}\delta u = \delta\frac{du}{dt} \tag{10.282}$$

Recalling that d/dt is the differential symbol in real time and δ is the differential in frozen time, we show that their order of operation is interchangeable under specified conditions:

$$d\,\delta u = \delta\,du \tag{10.283}$$

Let us interpret du and δu geometrically. Consider a holonomic constraint and two neighboring possible curves as shown in Figure 10.29. Assume that $P_1(t)$ is the current position of the describing point of the dynamics system and C_1 is the actual path of motion. At the same time we can also focus on another point $P_2(t)$ on another possible path C_2 that could be an actual path for another initial condition. The distance between $P_1(t)$ and $P_2(t)$ is a virtual displacement δu at time t. The points $P_1(t)$ and $P_2(t)$ are on a virtual curve Δ_1 that is an isochrone curve indicating different possible positions of the describing point at time t.

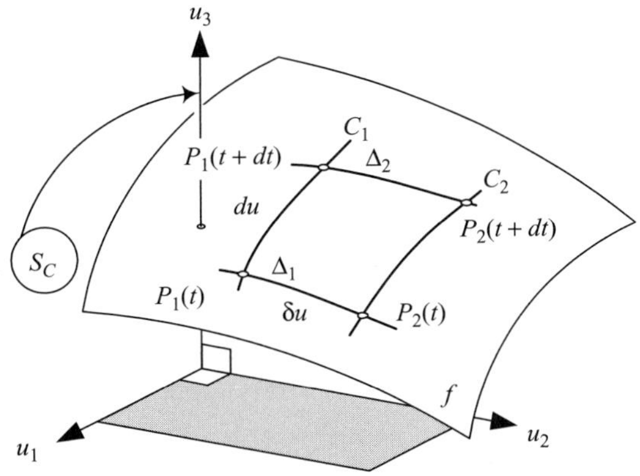

Figure 10.29 A constraint surface and two neighboring actual motion curves C_1 and C_2.

Points P_1 and P_2 would be at $P_1(t+dt)$ and $P_2(t+dt)$ after a time increment dt. The distance between $P_1(dt)$ and $P_1(t+dt)$ is an actual displacement du after dt. Similarly, the distance between $P_1(t+dt)$ and $P_2(t+dt)$ is a virtual displacement at time $t+dt$ and distance between $P_2(t)$ and $P_2(t+dt)$ is an actual displacement after dt.

In general, the virtual displacement at time $t+dt$ is different from time t. Therefore, δu is a function of time, and it is reasonable to consider its rate of change with time. Also, the magnitude of du on two possible curves might be different in the time interval dt. So, we can measure its virtual change.

Now let us generalize the possible and virtual displacements between neighboring curves. Figure 10.30 illustrates a family of possible paths with parameter a and a set

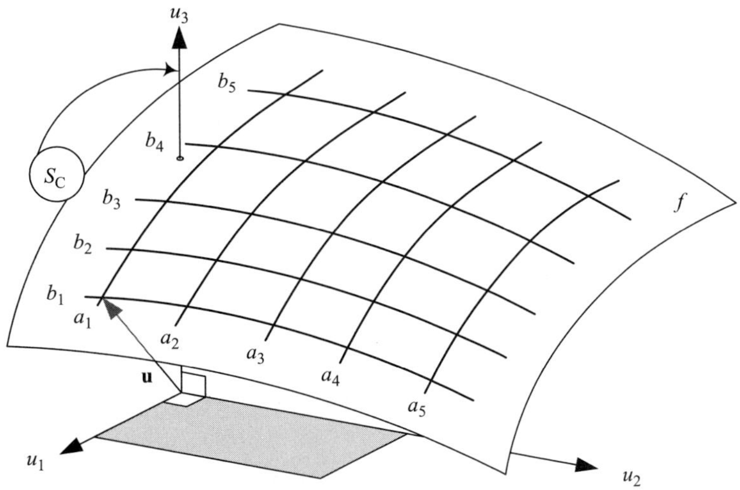

Figure 10.30 A family of possible paths with parameter a and a set of isochrone paths with parameter b.

of isochrone paths with parameter b. The configuration position vector $\mathbf{u}$ of a point in this net is a two-parameter vector:

$$\mathbf{u} = \mathbf{u}(a, b) \tag{10.284}$$

The arc between two neighboring points on an a-constant curve is

$$du = \frac{\partial \mathbf{u}}{\partial b} db \tag{10.285}$$

and the arc between two neighboring points on a b-constant curve is

$$\delta u = \frac{\partial \mathbf{u}}{\partial a} \delta a \tag{10.286}$$

So, we can define the two operators d and δ as

$$d = db \frac{\partial}{\partial b} \qquad \delta = \delta a \frac{\partial}{\partial a} \tag{10.287}$$

Employing (10.285) and (10.286), we can develop the derivatives

$$\delta\, du = \frac{\partial^2 \mathbf{u}}{\partial a\, \partial b} \delta a\, db \qquad d\, \delta u = \frac{\partial^2 \mathbf{u}}{\partial b\, \partial a} db\, \delta a \tag{10.288}$$

The necessary and sufficient condition for (10.283) is that the second derivatives of $\mathbf{u}$ with respect to a and b exist and are continuous:

$$\frac{\partial^2 \mathbf{u}}{\partial a\, \partial b} = \frac{\partial^2 \mathbf{u}}{\partial b\, \partial a} \tag{10.289}$$

This is true when the constraint functions are of class C_2.

Example 636 The Variation Let us call d the derivative and δ the variation operator. To distinguish between d and δ, let us redefine them.

Consider a general scalar function of class C_1 of $n + m + 1$ independent positions, velocities, and time:

$$f = f(u_1, u_2, \ldots, u_n; \dot{u}_1, \dot{u}_2, \ldots, u_m; t) \tag{10.290}$$

The derivative of f is shown by df and is defined as

$$df = \sum_{i=1}^{n} \frac{\partial f}{\partial u_i} du_i + \sum_{j=1}^{m} \frac{\partial f}{\partial \dot{u}_j} d\dot{u}_j + \frac{\partial f}{\partial t} dt \tag{10.291}$$

The variation of f is shown by δf and is defined as

$$\delta f = \sum_{i=1}^{n} \frac{\partial f}{\partial u_i} \delta u_i + \sum_{j=1}^{m} \frac{\partial f}{\partial \dot{u}_j} \delta \dot{u}_j \tag{10.292}$$

Example 637 ★ Virtual Work of a General Force Consider a scalar function W given as

$$W = W(u_1, u_2, \ldots, u_n; \dot{u}_1, \dot{u}_2, \ldots, u_m; t) \tag{10.293}$$

The variation of W is

$$\delta W = \sum_{i=1}^{n} \frac{\partial W}{\partial u_i} \delta u_i + \sum_{j=1}^{m} \frac{\partial W}{\partial \dot{u}_j} \delta \dot{u}_j \tag{10.294}$$

If an applied force $\mathbf{F}$ on a particle, where

$$\mathbf{F} = \mathbf{F}(u_1, u_2, \ldots, u_n; \dot{u}_1, \dot{u}_2, \ldots, u_m; t) \tag{10.295}$$

moves through a displacement $d\mathbf{u}$, then the actual differential work done by the force in this displacement is the inner product of $\mathbf{F}$ and $d\mathbf{u}$,

$$dW = \mathbf{F} \cdot d\mathbf{u} \tag{10.296}$$

and therefore, the total actual work is

$$W_A = \int \mathbf{F} \cdot d\mathbf{u} \tag{10.297}$$

The work W_A is a function of the form (10.290), and therefore its variation would be (10.294). In general, the actual work can be a function of position, velocity, and time.

Similar to (10.296), the work done by $\mathbf{F}$ during a virtual displacement $\delta\mathbf{u}$ can be calculated as

$$\delta W = \mathbf{F} \cdot \delta\mathbf{u} \tag{10.298}$$

Because the differential work δW is calculated by the inner product of the force $\mathbf{F}$ and a virtual displacement, we can also call δW a virtual work increment. Therefore, the total virtual work is

$$W_V = \int \mathbf{F} \cdot \delta\mathbf{u} \tag{10.299}$$

However, the virtual work δW is not in general equal to the variation of W unless $\mathbf{F}$ is not a velocity-dependent force. Therefore, the virtual work cannot generally be found by a variation of work W. The confusion comes from the fact that the sign δW for virtual work (10.298) is just a short notation for $\mathbf{F} \cdot \delta\mathbf{u}$. This confusion is also true for dW as an actual work increment and dW as a differential of W. Actual work in (10.296) is not in general equal to the differential of a work function.

Example 638 ★ Fundamental Equation of Dynamics The general form of an applied force $\mathbf{F}_i$ on a particle m_i of a dynamic system can be decomposed into a constraint force $\mathbf{F}_C$ and a given force $\mathbf{F}_G$:

$$\mathbf{F}_i = \mathbf{F}_{C_i} + \mathbf{F}_{G_i} \tag{10.300}$$

Let us write the Newton equation of motion $m_i \ddot{\mathbf{u}}_i = \mathbf{F}_i$ as

$$m_i \ddot{\mathbf{u}}_i - \mathbf{F}_{G_i} = \mathbf{F}_{C_i} \tag{10.301}$$

When these equations are assumed to be applied on all particles in the dynamic system, we have

$$\sum_{i=1}^{N/3} \left(m_i \ddot{\mathbf{u}}_i - \mathbf{F}_{G_i}\right) = \sum_{i=1}^{N/3} \mathbf{F}_{C_i} \tag{10.302}$$

The virtual work of both sides of Equation (10.302) provides

$$\sum_{i=1}^{N/3} \left(m_i \ddot{\mathbf{u}}_i - \mathbf{F}_{G_i}\right) \cdot \delta \mathbf{u}_i = 0 \tag{10.303}$$

juts because the virtual work of constraint forces is zero. This equation is called the *fundamental equation of dynamics*. It is also referred to as Lagrange's form of the D'Alembert principle. Employing (10.298) and expanding (10.303), we have a new form of the fundamental equation:

$$\delta W - \sum_{i=1}^{N/3} m_i \ddot{\mathbf{u}}_i \cdot \delta \mathbf{u}_i = 0 \tag{10.304}$$

The fundamental equation provides a good environment to show how constraints affect the dynamics of a system quantitatively. Let us write the fundamental equation for all components of the configuration space as

$$\sum_{i=1}^{N} (m_i \ddot{u}_i - F_i) \cdot \delta u_i = 0 \tag{10.305}$$

in which δu_i must satisfy

$$\sum_{i=1}^{N} \frac{\partial f_j}{\partial u_i} \delta u_i = \sum_{i=1}^{N} A_{ji}\, \delta u_i = 0 \qquad j = 1, 2, \ldots, L \tag{10.306}$$

Combining these equations, we have

$$\sum_{i=1}^{N} \left(m_i \ddot{u}_i - F_i + \sum_{j=1}^{L} \lambda_j A_{ji} \right) \cdot \delta u_i = 0 \tag{10.307}$$

Because of the arbitrariness of δu_i, we get

$$m_i \ddot{u}_i - F_i + \sum_{j=1}^{L} \lambda_j A_{ji} = 0 \qquad i = 1, 2, \ldots, N \tag{10.308}$$

British mathematician Leopold Alexander Pars (1896–1985) calls Equation (10.303) the first form of the fundamental equation of dynamics. Jean Le Rond D'Alembert (1717–1783) was a French mathematician who introduced D'Alembert's principle. He found Equation (10.303) based on an expansion of the principle of virtual work in statics,

$$\delta W = \sum_{i=1}^{N/3} (\mathbf{F}_i \cdot \delta \mathbf{u}_i) = 0 \tag{10.309}$$

which was developed by Swiss mathematician Johann Bernoulli (1667–1748).

Example 639 ★ D'Alembert's Principle Let us decompose Equation (10.302) to tangent and normal components with respect to the constraint plane:

$$\sum_{i=1}^{N/3} \left(m_i \ddot{\mathbf{u}}_i - \mathbf{F}_{G_i}\right)_t + \sum_{i=1}^{N/3} \left(m_i \ddot{\mathbf{u}}_i - \mathbf{F}_{G_i}\right)_n = \sum_{i=1}^{N/3} \left(\mathbf{F}_{C_i}\right)_t + \sum_{i=1}^{N/3} \left(\mathbf{F}_{C_i}\right)_n \tag{10.310}$$

The motion happens in the constraint plane, so the normal components of $\ddot{\mathbf{u}}_i$ are zero, and we have

$$\sum_{i=1}^{N/3} \left(m_i \ddot{\mathbf{u}}_i - \mathbf{F}_{G_i}\right)_t = \sum_{i=1}^{N/3} \left(\mathbf{F}_{C_i}\right)_t \tag{10.311}$$

$$\sum_{i=1}^{N/3} \left(\mathbf{F}_{G_i}\right)_n + \sum_{i=1}^{N/3} \left(\mathbf{F}_{C_i}\right)_n = 0 \tag{10.312}$$

D'Alembert's principle states: The totality of the constraint forces may be disregarded in the dynamic analysis of a system of particles:

$$\sum_{i=1}^{N/3} \left(\mathbf{F}_{C_i}\right)_t = 0 \tag{10.313}$$

Based on D'Alembert's principle, the right-hand side of Equation (10.311) is zero and it reduces to the Newton equation for a system of particles:

$$\sum_{i=1}^{N/3} \left(m_i \ddot{\mathbf{u}}_i - \mathbf{F}_{G_i}\right)_t = 0 \tag{10.314}$$

Therefore, the tangent components of the given forces in the constraint plane are the only forces which contribute to particle acceleration. Recalling that the constraint force is normal to the constraint plane, the D'Alembert principle can be justified.

A physical meaning of Equation (10.312) is that the normal component of a physical force is in balance with the constraint forces. The explanation of friction force in Example 633 may clarify this fact.

Although D'Alembert's principle refers to Equation (10.313), some literatures use the D'Alembert principle to refer to the fundamental equation (10.303).

Example 640 Constraint Force Calculation Consider a triangular wedge with inclination θ that is moving with constant velocity v on the X-axis. A particle of mass m is sliding down on the inclined surface of the wedge by gravitational force mg. Figure 10.31 illustrates the dynamic system. The particle m is restricted by a holonomic rheonomic constraint:

$$f(X, Z, t) = X \tan\theta + Z - (vt + l)\tan\theta = 0 \tag{10.315}$$

We show the constraint force by $\mathbf{F}_C$, which is always normal to the moving surface. The virtual displacement $\delta\mathbf{r}$ and actual displacement dr are shown in the figure. The

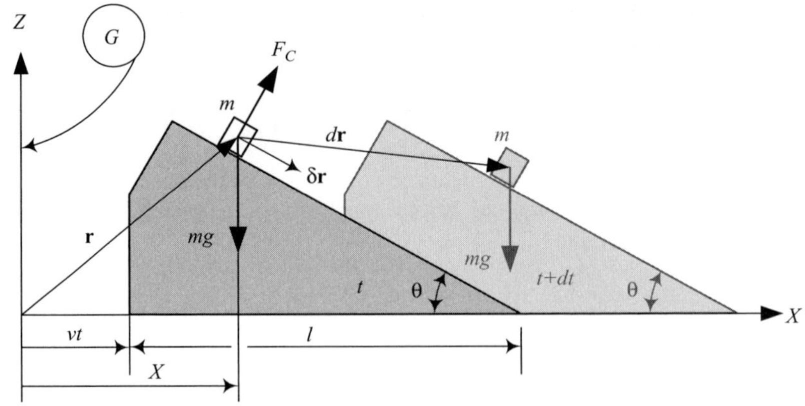

Figure 10.31 A particle of mass m slides down on the inclined surface of a moving wedge.

virtual displacement $\delta\mathbf{r}$ is perpendicular to $\mathbf{F}_C$ at each time. The equations of motion of the particle are

$$m\ddot{x} = \lambda \frac{\partial f}{\partial X} = \lambda \tan\theta \tag{10.316}$$

$$m\ddot{z} = -mg + \lambda \frac{\partial f}{\partial Z} = \lambda - mg \tag{10.317}$$

The differential equations (10.316) and (10.317) along with the constraint equations (10.315) make a set of three equations to be solved for X, Z, and λ:

$$m\ddot{X} = \frac{g\tan\theta}{1+\tan^2\theta} \qquad m\ddot{Z} = -\frac{g\tan\theta}{1+\tan^2\theta} \qquad \lambda = \frac{mg}{1+\tan^2\theta} \tag{10.318}$$

If the initial conditions are

$$X(0) = 0 \qquad Z(0) = h \qquad \dot{X}(0) = v \qquad \dot{Z}(0) = 0 \tag{10.319}$$

we can integrate the equations of motion and determine the motion of m:

$$X = \frac{1}{2}\frac{g\tan\theta}{1+\tan^2\theta}t^2 + vt \tag{10.320}$$

$$Z = h - \frac{1}{2}\frac{g\tan\theta}{1+\tan^2\theta}t^2 \tag{10.321}$$

The constraint force on m is

$$\mathbf{F}_C = \lambda\,\nabla f = \lambda\frac{\partial f}{\partial X}\hat{I} + \lambda\frac{\partial f}{\partial Z}\hat{K} = \lambda\tan\theta\,\hat{I} + \lambda\hat{K}$$

$$= \frac{mg\tan\theta}{1+\tan^2\theta}\hat{I} + \frac{mg}{1+\tan^2\theta}\hat{K} \tag{10.322}$$

$$F_C = \frac{mg}{\sqrt{1+\tan^2\theta}} \tag{10.323}$$

Example 641 Force Balance of a Slider–Crank Mechanism Consider a slider–crank mechanism as shown in Figure 10.32. Using the principle of virtual work in statics (10.309), we can determine the required value of torque **Q** to hold the mechanism when the piston is under a force **F**.

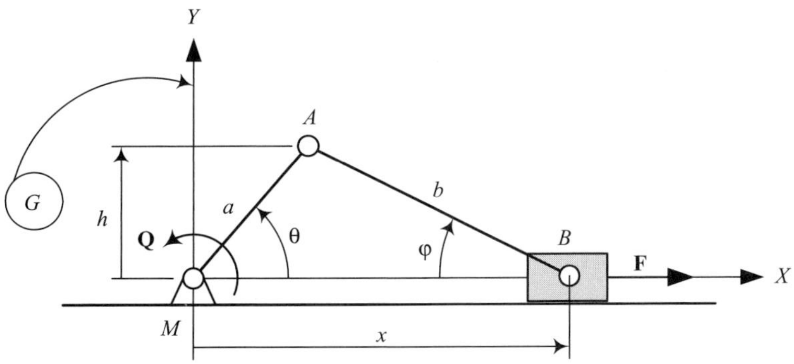

Figure 10.32 A slider–crank mechanism in a force balance situation.

The vectorial expressions of force **F** and torque **Q** are

$$\begin{array}{ll} \mathbf{F} = F\hat{I} & \mathbf{Q} = Q\hat{K} \\ F > 0 & Q > 0 \end{array} \tag{10.324}$$

where **Q** does work by a rotation θ and **F** by a displacement x. Therefore, the virtual work of the force system is

$$\delta W = Q\,\delta\theta - F\delta x = 0 \tag{10.325}$$

Employing the geometry of the mechanism, we have

$$x = a\cos\theta + b\cos\varphi \tag{10.326}$$

$$h = a\sin\theta = b\sin\varphi \tag{10.327}$$

Eliminating φ provides

$$x = a\cos\theta + b\sqrt{1 - \left(\frac{a}{b}\right)^2 \sin^2\theta} \tag{10.328}$$

Therefore, the virtual displacements of δx and $\delta\theta$ are related by the equation

$$\delta x = -a\sin\theta\,\delta\theta - \frac{a^2}{b}\frac{\sin\theta\cos\theta}{\sqrt{1-(a/b)^2\sin^2\theta}}\,\delta\theta \tag{10.329}$$

Substituting (10.329) in (10.325) and simplifying, we find the required relation between F and Q to keep the mechanism at an angle θ:

$$Q = Fa\sin\theta\left(1 + \frac{\cos\theta}{b\sqrt{1-(a/b)^2\sin^2\theta}}\right) \tag{10.330}$$

Example 642 Virtual Work and Equilibrium of Dynamic Systems A double-inverted pendulum is shown in Figure 10.33. Each link has a length l and mass m. Two similar linear springs with stiffness k support the pendulum. If the springs have free length at $\theta_1 = \theta_2 = 0$, we may use the principle of virtual work and determine the positions of the equilibria.

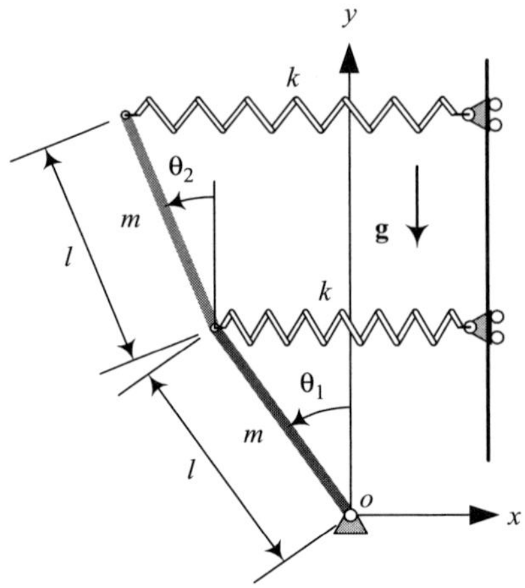

Figure 10.33 A supported double-inverted pendulum.

The forces of the springs are

$$\mathbf{F}_1 = \begin{bmatrix} kl\sin\theta_1 \\ 0 \end{bmatrix} \qquad \mathbf{F}_2 = \begin{bmatrix} kl\left(\sin\theta_1 + \sin\theta_2\right) \\ 0 \end{bmatrix} \tag{10.331}$$

and the gravitational forces are

$$\mathbf{W}_1 = \begin{bmatrix} 0 \\ -mg \end{bmatrix} \qquad \mathbf{W}_2 = \begin{bmatrix} 0 \\ -mg \end{bmatrix} \tag{10.332}$$

The position vectors of the points of application of the above forces are

$$\mathbf{r}_{\mathbf{F}_1} = \begin{bmatrix} -l\sin\theta_1 \\ l\cos\theta_1 \end{bmatrix} \qquad \mathbf{r}_{\mathbf{F}_2} = \begin{bmatrix} -l\left(\sin\theta_1 + \sin\theta_2\right) \\ l\left(\cos\theta_1 + \cos\theta_2\right) \end{bmatrix} \tag{10.333}$$

$$\mathbf{r}_{\mathbf{W}_1} = \begin{bmatrix} -\frac{1}{2}l\sin\theta_1 \\ \frac{1}{2}l\cos\theta_1 \end{bmatrix} \qquad \mathbf{r}_{\mathbf{W}_2} = \begin{bmatrix} -l\left(\sin\theta_1 + \frac{1}{2}\sin\theta_2\right) \\ l\left(\cos\theta_1 + \frac{1}{2}\cos\theta_2\right) \end{bmatrix} \tag{10.334}$$

The virtual displacements of the points are

$$\delta\mathbf{r}_{\mathbf{F}_1} = \begin{bmatrix} -l\,\delta\theta_1\cos\theta_1 \\ -l\,\delta\theta_1\sin\theta_1 \end{bmatrix} \tag{10.335}$$

$$\delta\mathbf{r}_{\mathbf{F}_2} = \begin{bmatrix} -l\left(\delta\theta_1\cos\theta_1 + \delta\theta_2\cos\theta_2\right) \\ -l\left(\delta\theta_1\sin\theta_1 + \delta\theta_2\sin\theta_2\right) \end{bmatrix} \tag{10.336}$$

$$\delta \mathbf{r}_{W_1} = \begin{bmatrix} -\frac{1}{2}l\,\delta\theta_1 \cos\theta_1 \\ -\frac{1}{2}l\,\delta\theta_1 \sin\theta_1 \end{bmatrix} \tag{10.337}$$

$$\delta \mathbf{r}_{W_2} = \begin{bmatrix} -l\left(\delta\theta_1 \cos\theta_1 + \frac{1}{2}\delta\theta_2 \cos\theta_2\right) \\ -l\left(\delta\theta_1 \sin\theta_1 + \frac{1}{2}\delta\theta_2 \sin\theta_2\right) \end{bmatrix} \tag{10.338}$$

So, the virtual work of the system is

$$\begin{aligned} \delta W &= \sum \mathbf{F}_i \cdot \delta \mathbf{r}_i \\ &= \left[\tfrac{3}{2}mg\sin\theta_1 - kl\cos\theta_1\left(2\sin\theta_1 + \sin\theta_2\right)\right] l\delta\theta_1 \\ &\quad + \left[\tfrac{1}{2}mg\sin\theta_2 - kl\cos\theta_2\left(\sin\theta_1 + \sin\theta_2\right)\right] l\delta\theta_2 \end{aligned} \tag{10.339}$$

At equilibrium, we have $\delta W = 0$, and therefore, we find two equations to determine the equilibrium values of (θ_1, θ_2):

$$\tfrac{3}{2}mg\sin\theta_1 - kl\cos\theta_1\left(2\sin\theta_1 + \sin\theta_2\right) = 0 \tag{10.340}$$

$$\tfrac{1}{2}mg\sin\theta_2 - kl\cos\theta_2\left(\sin\theta_1 + \sin\theta_2\right) = 0 \tag{10.341}$$

Four sets of trivial solutions are

$$\theta_1 = \theta_2 = 0 \qquad \theta_1 = \theta_2 = \pm\pi \tag{10.342}$$

as illustrated in Figure 10.34.

To determine the nontrivial equilibria, we set the parameters to the given values, such as

$$m = 1\,\text{kg} \qquad g = 9.81\,\text{m/s}^2 \qquad k = 10\,\text{N/m} \qquad l = 1\,\text{m} \tag{10.343}$$

and plot the functions f_1 and f_2 in Figure 10.35:

$$f_1 = \tfrac{3}{2}mg\sin\theta_1 - kl\cos\theta_1\left(2\sin\theta_1 + \sin\theta_2\right) = 0 \tag{10.344}$$

$$f_2 = \tfrac{1}{2}mg\sin\theta_2 - kl\cos\theta_2\left(\sin\theta_1 + \sin\theta_2\right) = 0 \tag{10.345}$$

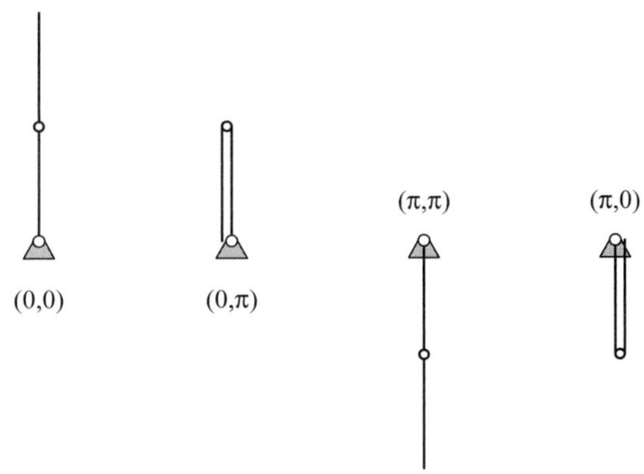

Figure 10.34 Four trivial equilibria of the double-inverted pendulum.

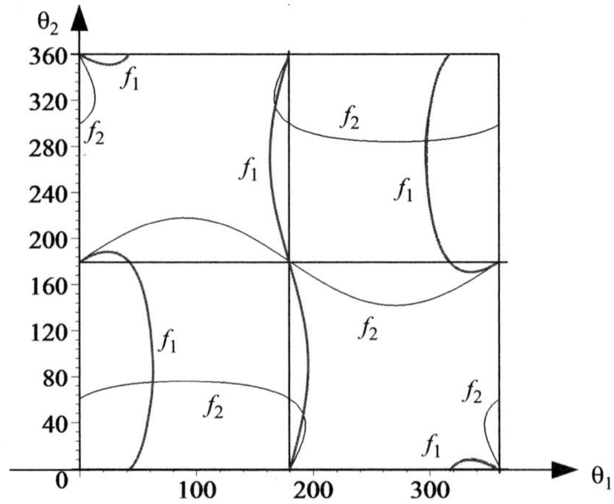

Figure 10.35 Plots of functions f_1 and f_2.

The intersections of f_1 and f_2 indicate the equilibria:

$$\theta_1 \approx 61.64 \,\text{deg} \qquad \theta_2 \approx 75.12 \,\text{deg} \tag{10.346}$$

$$\theta_1 \approx 191.97 \,\text{deg} \qquad \theta_2 \approx 46.65 \,\text{deg} \tag{10.347}$$

$$\theta_1 \approx 168.02 \,\text{deg} \qquad \theta_2 \approx 313.34 \,\text{deg} \tag{10.348}$$

$$\theta_1 \approx 298.35 \,\text{deg} \qquad \theta_2 \approx 284.87 \,\text{deg} \tag{10.349}$$

These equilibrium configurations are illustrated in Figure 10.36.

The number and position of the equilibria depend on the parameters of the system. To investigate the effects of the parameters, let us rewrite the equations f_1 and f_2 as

$$f_1 = \frac{3}{2} c \sin \theta_1 - \cos \theta_1 \left(2 \sin \theta_1 + \sin \theta_2 \right) = 0 \tag{10.350}$$

$$f_2 = \frac{1}{2} c \sin \theta_2 - \cos \theta_2 \left(\sin \theta_1 + \sin \theta_2 \right) = 0 \tag{10.351}$$

where

$$c = \frac{mg}{kl} \tag{10.352}$$

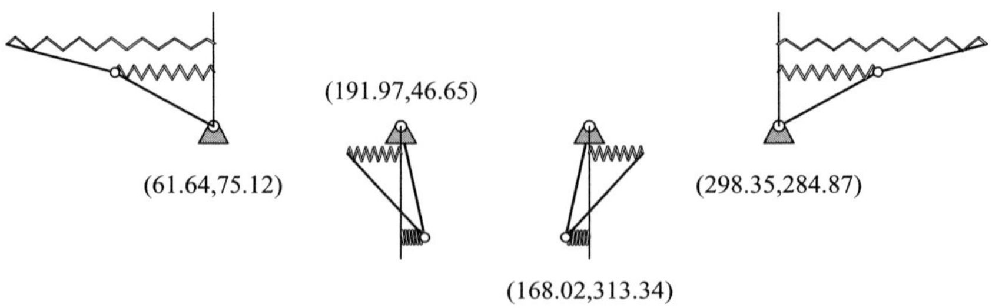

Figure 10.36 Four nontrivial equilibria of the double-inverted pendulum.

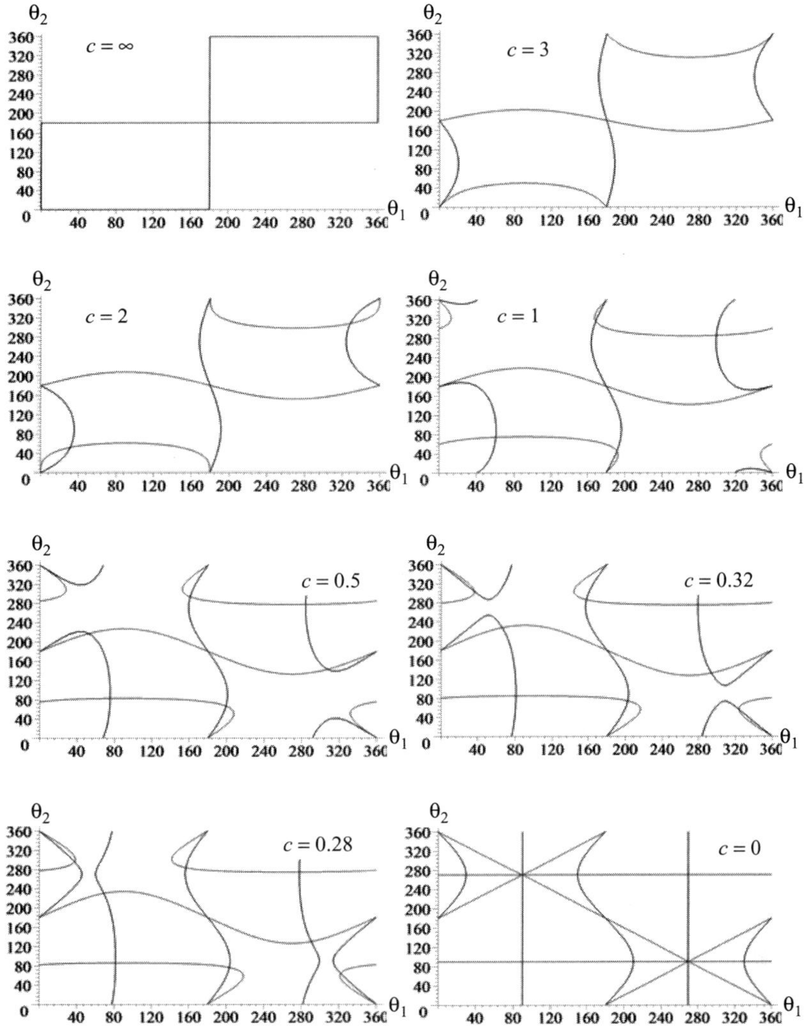

Figure 10.37 Plots of functions f_1 and f_2 for different values of c (not to scale).

So, the common solutions of f_1 and f_2 depend on only one parameter, c. Figure 10.37 illustrates the functions f_1 and f_2 for different values of c. When $c = \infty$, the only equilibria are the trivial positions of (10.342). By decreasing c, up to six more equilibria appear.

Example 643 ★ Energy Equation If all of the constraints on a dynamic system are catastatic, then all of the virtual displacements and velocities would be possible displacements and velocities. Having only catastatic constraint, we have

$$\delta \mathbf{u}_i = d\mathbf{u}_i = \dot{\mathbf{u}}_i dt \tag{10.353}$$

So, we can rewrite the fundamental equation of dynamics (10.303) as

$$\sum_{i=1}^{N/3} m_i \ddot{\mathbf{u}}_i \cdot \dot{\mathbf{u}}_i = \sum_{i=1}^{N/3} \mathbf{F}_{G_i} \cdot \dot{\mathbf{u}}_i \tag{10.354}$$

Let us use the kinetic energy K,

$$K = \sum_{i=1}^{N/3} \frac{1}{2} m_i \dot{\mathbf{u}}_i^2 \tag{10.355}$$

$$\frac{dK}{dt} = \sum_{i=1}^{N/3} m_i \ddot{\mathbf{u}}_i \cdot \dot{\mathbf{u}}_i \tag{10.356}$$

and modify the fundamental equation to

$$\frac{dK}{dt} = \sum_{i=1}^{N/3} \mathbf{F}_{G_i} \cdot \dot{\mathbf{u}}_i \tag{10.357}$$

Equation (10.357) is called the first, or primitive, form of the equation of energy. It is the same as the principle of work energy and expresses that the rate of increase of the kinetic energy of a system is equal to the rate of work of the given forces.

Now let us consider, in addition to catastatic constraint, that the dynamic system is under a conservative force system in which the given forces $\mathbf{F}_{G_i}$ are not functions of velocities $\dot{\mathbf{u}}_i$ and time t. Therefore $\mathbf{F}_{G_i}$ are only functions of displacement $\mathbf{u}_i$, and hence $-\sum_{i=1}^{N/3} \mathbf{F}_{G_i} \cdot \mathbf{u}_i$ is the total differential of a potential energy V:

$$\sum_{i=1}^{N/3} \mathbf{F}_{G_i} \cdot \mathbf{u}_i = -dV \tag{10.358}$$

Substituting this result in Equation (10.357) provides

$$\frac{dK}{dt} = -\frac{dV}{dt} \tag{10.359}$$

and therefore,

$$\frac{d}{dt}(K + V) = 0 \tag{10.360}$$

$$K + V = E \tag{10.361}$$

Equation (10.361) is the second, or classical, form of the equation of energy. So, for a catastatic system under the action of conservative given forces, the sum of the kinetic and potential energies of the system remains constant. The mechanical energy of the system, E, is determined by initial conditions.

Example 644 ★ Second and Third Forms of Fundamental Equation Let us take a derivative of a holonomic constraint and rewrite it as

$$\sum_{i=1}^{N} \frac{\partial f}{\partial u_i} \dot{u}_i + \frac{\partial f}{\partial t} = 0 \tag{10.362}$$

A variation in velocity at the same configuration and the same time gives

$$\sum_{i=1}^{N} \frac{\partial f}{\partial u_i} \left(\dot{u}_i + \delta \dot{u}_i \right) + \frac{\partial f}{\partial t} = 0 \tag{10.363}$$

Therefore,

$$\sum_{i=1}^{N} \frac{\partial f}{\partial u_i} \delta \dot{u}_i = \nabla f \cdot \delta \dot{\mathbf{u}} = 0 \tag{10.364}$$

which shows the virtual velocity and constraint force are orthogonal. So, we may replace virtual displacement in the fundamental equation with virtual velocity and obtain the second form of the fundamental equation of dynamics:

$$\sum_{i=1}^{N/3} \left(m_i \ddot{\mathbf{u}}_i - \mathbf{F}_{G_i} \right) \cdot \delta \dot{\mathbf{u}}_i = 0 \tag{10.365}$$

Instead of a small variation of velocity, we may replace $\delta \dot{u}_i$ with a finite-difference velocity $\Delta \dot{u}_i$ in Equation (10.363) to find the second form:

$$\sum_{i=1}^{N/3} \left(m_i \ddot{\mathbf{u}}_i - \mathbf{F}_{G_i} \right) \cdot \Delta \dot{\mathbf{u}}_i = 0 \tag{10.366}$$

The velocity $\dot{u}_i + \Delta \dot{u}_i$ is another possible velocity at the same configuration and time as $\dot{u}_i$. The second form of the fundamental equation has application in impulsive force problems.

If we differentiate the constraint equation (10.362), then

$$\sum_{i=1}^{N} \left(\frac{\partial f}{\partial u_i} \ddot{u}_i + \dot{u}_i \frac{d}{dt} \frac{\partial f}{\partial u_i} \right) + \frac{d}{dt} \frac{\partial f}{\partial t} = 0 \tag{10.367}$$

where

$$\frac{d}{dt} = \frac{\partial}{\partial t} + \sum_{j=1}^{N} \dot{u}_j \frac{\partial}{\partial u_j} \tag{10.368}$$

Now consider two possible motions with the same configuration and velocity at time t but with different accelerations $\ddot{u}_i$ and $\ddot{u}_i + \delta \ddot{u}_i$. Equation (10.367) becomes

$$\sum_{i=1}^{N} \left(\frac{\partial f}{\partial u_i} \left(\ddot{u}_i + \delta \ddot{u}_i \right) + \dot{u}_i \frac{d}{dt} \frac{\partial f}{\partial u_i} \right) + \frac{d}{dt} \frac{\partial f}{\partial t} = 0 \tag{10.369}$$

and therefore,

$$\sum_{i=1}^{N} \frac{\partial f}{\partial u_i} \delta \ddot{u}_i = \nabla f \cdot \delta \ddot{\mathbf{u}} = 0 \tag{10.370}$$

The variation of acceleration satisfies the same equation as virtual displacement. So, we may replace virtual displacement in the fundamental equation with virtual acceleration and obtain the third form of the fundamental equation of dynamics:

$$\sum_{i=1}^{N/3} \left(m_i \ddot{\mathbf{u}}_i - \mathbf{F}_{G_i} \right) \cdot \delta \ddot{\mathbf{u}}_i = 0 \tag{10.371}$$

Again, instead of a small variation of acceleration, we may replace $\delta \ddot{u}_i$ with a finite-difference velocity $\Delta \ddot{u}_i$ in Equation (10.369) to find the third form:

$$\sum_{i=1}^{N/3} \left(m_i \ddot{\mathbf{u}}_i - \mathbf{F}_{G_i} \right) \cdot \Delta \ddot{\mathbf{u}}_i = 0 \tag{10.372}$$

In the first form of the fundamental equation, we consider an infinitesimal virtual displacement from a given configuration. In the second form, we consider that the configuration is not varied and use the difference between two possible velocities. In the third form, the configuration and velocity are not varied, and we use the difference between two possible accelerations. The second and third forms can be justified by the fact that, if $\delta \mathbf{u}$ is in the constraint surface, then $\delta \dot{\mathbf{u}}$ and $\delta \ddot{\mathbf{u}}$, or $\Delta \dot{\mathbf{u}}_i$ and $\Delta \ddot{\mathbf{u}}_i$ must also be in the constraint surface.

10.7 ★ NONHOLONOMIC CONSTRAINT

The kinematic constraints do not always show up as equality equations between the coordinates of the configuration space. They may also appear as nonintegrable infinitesimal displacement equations and also as inequality relations. The inequality constraints and nonintegrable differential relations are called *nonholonomic constraints*.

Nonholonomic constraints cannot change the DOF of a dynamic system. Therefore, all accessible configurations in the absence of nonholonomic constraints are also accessible in their presence.

10.7.1 ★ Nonintegrable Constraint

A constraint equation that can only be expressed by the differential of the coordinates of the configuration space, du_i, and time, dt, is a *nonholonomic constraint*:

$$\sum_{i=1}^{N} A_i \, du_i = 0 \tag{10.373}$$

$$\sum_{i=1}^{N} A_i \, du_i + A \, dt = 0 \tag{10.374}$$

The coefficients A_i and A are functions of u_i and t:

$$A_i = A_i \left(u_i, t \right) \qquad A = A \left(u_i, t \right) \tag{10.375}$$

The differential forms (10.373) and (10.374) are *total constraints* if there is a function $f(\mathbf{u}, t)$ such that

$$A_i = \frac{\partial f}{\partial u_i} \qquad A = \frac{\partial f}{\partial t} \tag{10.376}$$

Then, the differential constraints (10.373) and (10.374) can be integrated to determine their associated finite displacement holonomic constraints:

$$f(\mathbf{u}) = 0 \tag{10.377}$$

$$f(\mathbf{u}, t) = 0 \tag{10.378}$$

When there is no function $f(\mathbf{u}, t)$ to have (10.376), the differential constraints (10.373) and (10.374) are not integrable and are called *nonholonomic*. The necessary and sufficient conditions for (10.373) and (10.374) to be total differential are

$$\frac{\partial A_i}{\partial u_k} = \frac{\partial A_k}{\partial u_i} \qquad \frac{\partial A}{\partial u_i} = \frac{\partial A_i}{\partial t} \tag{10.379}$$

Proof: Consider a differential constraint of x and t:

$$\sum_{i=1}^{N} A_i \, du_i + A \, dt = 0 \tag{10.380}$$

Suppose the constraint is total and we have a C_2 function $f(\mathbf{u}, t) = 0$ such that

$$A_i = \frac{\partial f}{\partial u_i} \qquad A = \frac{\partial f}{\partial t} \qquad i = 1, 2, \ldots, N \tag{10.381}$$

Because the mixed second partial differentials of f are equal,

$$\frac{\partial^2 f}{\partial u_i \, \partial u_k} = \frac{\partial^2 f}{\partial u_k \, \partial u_i} \qquad \frac{\partial^2 f}{\partial u_i \, \partial t} = \frac{\partial^2 f}{\partial t \, \partial u_i} \tag{10.382}$$

we must have

$$\frac{\partial A_i}{\partial u_k} = \frac{\partial A_k}{\partial u_i} \qquad \frac{\partial A}{\partial u_i} = \frac{\partial A_i}{\partial t} \tag{10.383}$$

which is the necessary condition for (10.380) to be total.

If we show that (10.383) enables us to construct a function f, then it is also a sufficient condition. Let us begin by integrating the first term of (10.380) with respect to u_r, $r \in \{1, 2, \ldots, N\}$:

$$f = \int A_r \, du_r + f_r(u_1, u_2, \ldots, u_{r-1}, u_{r+1}, \ldots, u_N, t) \tag{10.384}$$

The constant of integration must be a function of $u_1, u_2, \ldots, u_{r-1}$, u_{r+1}, u_N, t to disappear under differentiation with respect to u_r. This reduces our problem to that of finding f_r with the property that f, as given by (10.384), satisfies the second condition of (10.383). On differentiating (10.384) with respect to u_s, $s \in \{1, 2, \ldots, N\}$, $s \neq r$, and equating the result to A_s, we get

$$\frac{\partial}{\partial u_s} \int A_r \, du_r + \frac{\partial f_r(u_1, u_2, \ldots, u_{r-1}, u_{r+1}, \ldots, u_N, t)}{\partial u_s} = A_s \tag{10.385}$$

which yields

$$f_r(u_1, u_2, \ldots, u_{r-1}, u_{r+1}, \ldots, u_N, t) = \int \left(A_s - \frac{\partial}{\partial u_s} \int A_r \, du_r \right) du_s \tag{10.386}$$

provided the integrand is a function of u_s. So, a derivative of the integrand with respect to u_r must be zero:

$$\begin{aligned} \frac{\partial}{\partial u_r} \left(A_s - \frac{\partial}{\partial u_s} \int A_r \, du_r \right) &= \frac{\partial A_s}{\partial u_r} - \frac{\partial^2}{\partial u_r \, \partial u_s} \int A_r \, du_r \\ &= \frac{\partial A_s}{\partial u_r} - \frac{\partial A_r}{\partial u_s} = 0 \end{aligned} \tag{10.387}$$

Therefore, (10.380) is an integrable differential constraint if and only if conditions (10.383) are fulfilled. The associated holonomic constraint would be

$$f_r(u_1, u_2, \ldots, u_N, t) = f(\mathbf{u}, t) = c \tag{10.388}$$

The constant c is independent of the initial conditions of the dynamic system and can be determined by having the coordinates of a point $\mathbf{u}$ in the constraint surface at a time t. Substituting t with u_r and A with A_r in (10.384) completes the conditions (10.383).

A nonholonomic constraint in applied mechanics is a system whose state depends on its path of motion. Such a system is described by a set of periodic variables subject to differential constraints. The nonholonomic constraint was introduced by the German physicist Heinrich Rudolf Hertz (1857–1894). ■

Example 645 An Integrable Differential Constraint Consider the differential constraint

$$\left(te^x + 2x\right) dx + e^x dt = 0 \tag{10.389}$$

To check if the constraint is total,

$$df = \frac{\partial f}{\partial x} dx + \frac{\partial f}{\partial t} dt = 0 \tag{10.390}$$

we examine the conditions in (10.376):

$$\frac{\partial^2 f}{\partial t \, \partial x} = e^x \qquad \frac{\partial^2 f}{\partial x \, \partial t} = e^x \tag{10.391}$$

Therefore, the constraint is integrable. To determine the associated holonomic constraint, we integrate $\partial f / \partial t$ to get

$$f = \int e^x \, dt + g(x) = te^x + g(x) \tag{10.392}$$

so

$$\frac{\partial f}{\partial x} = te^x + \frac{dg(x)}{dx} \tag{10.393}$$

Because (10.393) must be equal to $te^x + 2x$, we have $dg(x)/dx = 2x$, which provides $g(x) = x^2$, and therefore the holonomic constraint is

$$te^x + x^2 = c \tag{10.394}$$

If the dynamic system is at $x = x_0$ when $t = 0$, then $c = 0$.

If a differential constraint is in the form

$$A\,dx + B\,dy = 0 \tag{10.395}$$

where

$$\frac{\partial B}{\partial x} = \frac{\partial A}{\partial y} \tag{10.396}$$

then the associated holonomic constraint is

$$A + \int \left(B - \frac{\partial M}{\partial y} \right) dy = c \tag{10.397}$$

Example 646 A Unicycle and Bicycle on a Flat Ground Consider a unicycle that remains vertical on a flat ground as shown in Figure 10.38. The unicycle can move forward and backward and also rotate about a local vertical axis going through the contact point of the wheel and ground. However, it cannot move laterally. This unicycle has a nonintegrable differential constraint because it does not have lateral motion,

$$\dot{X} \sin\theta_z - \dot{Y} \cos\theta_z = 0 \tag{10.398}$$

and it has another constraint because of pure rotation:

$$\dot{X} \cos\theta_z + \dot{Y} \sin\theta_z = R\dot{\varphi} \tag{10.399}$$

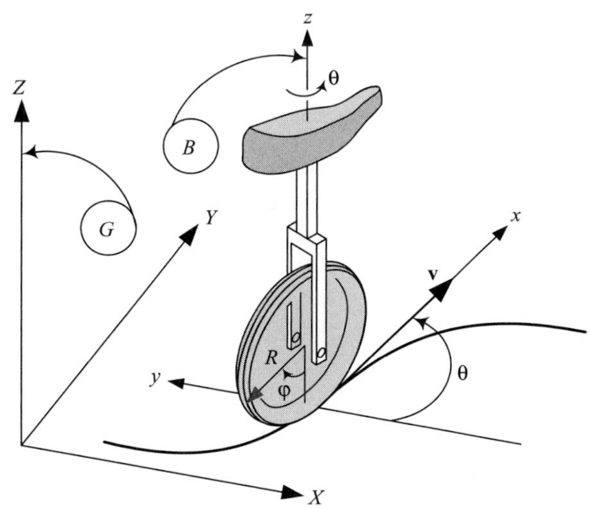

Figure 10.38 A vertical unicycle in a pure rotation.

Nonintegrability of these constraints is equivalent to the fact that we cannot move the unicycle to another point by varying only one coordinate of X, Y, θ, φ while keeping the other three unchanged. Therefore, there cannot be any unique algebraic function among them.

If the unicycle has more than one fixed wheel, such as a nonsteerable bicycle, it cannot turn about θ_z, and we have

$$\dot{\theta}_z = 0 \tag{10.400}$$

Then constraints (10.398) and (10.399) will be integrable and provide three holonomic constraints:

$$\theta_z - \theta_0 = 0 \tag{10.401}$$

$$(X - X_0)\sin\theta_z - (Y - Y_0)\cos\theta_z = 0 \tag{10.402}$$

$$(X - X_0)\cos\theta_z - (Y - Y_0)\sin\theta_z = R(\varphi - \varphi_0) \tag{10.403}$$

Example 647 ★ Breaking an Integrable Constraint Consider the differential constraint

$$dy - g(z)\,dx - xg'(z)\,dz = 0 \tag{10.404}$$

where $g'(z) = dg(z)/dz$. The variables of the differential constraint are separated. So, we can take an integral and determine a holonomic constraint:

$$y - g(z)x - c_1 = 0 \tag{10.405}$$

Now assume we are given a differential form, including the first two terms of (10.404):

$$dy - g(z)\,dx = 0 \tag{10.406}$$

Such a constraint is not integrable and indicates a nonholonomic constraint.

Sometimes a nonholonomic constraint is a broken total differential, although this fact does not help to make a holonomic constraint.

Example 648 Rolling Constraint The constraint imposed on the motion of rolling body A on B is that the velocities of the contact points must be equal. Consider a rolling sphere of radius R on a fixed plane. The velocity of the contact point must be zero.

Let us show the velocity of the center of a rolling sphere by $\mathbf{v}_C$ and its angular velocity by $\boldsymbol{\omega}$. The velocity of any contact point can be found from

$$\mathbf{v}_0 = \mathbf{v}_C + \boldsymbol{\omega} \times \mathbf{r} \tag{10.407}$$

where $\mathbf{r} = -R\hat{u}_n$ and $\hat{u}_n$ is the unit vector normal to the plane. The constraint is to have no sliding at the contact point:

$$\mathbf{v}_C - R\boldsymbol{\omega} \times \hat{u}_n = 0 \tag{10.408}$$

This equation cannot be integrated and indicates a nonholonomic constraint. This is because $\boldsymbol{\omega}$ is not in general the total time derivative of any coordinate, although $\mathbf{v}_C$ is the total time derivative of the position vector of the sphere center.

If we replace the rolling sphere with a rolling cylinder, then the rolling constraint (10.408) becomes holonomic. In this case the fixed axis of the cylinder would be the permanent axis of $\boldsymbol{\omega}$. Then, $\boldsymbol{\omega}$ is a total differential of the angular rotation θ of the cylinder:

$$\omega = \frac{d\theta}{dt} \tag{10.409}$$

Condition (10.408) can therefore be integrated and gives a relation between the angle θ and the coordinates of the cylinder center. In the simple case where the cylinder is rolling on the x-axis, the holonomic rolling constraint is

$$x = x_0 + R\theta \tag{10.410}$$

Practically, if the contact of a rolling body A on B happens at a point, the rolling constraint is nonholonomic, and if it happens on a line, the rolling constraint is holonomic.

Example 649 Motion of a Rolling Sphere Consider a rolling sphere on the (x, y)-plane under an external force $\mathbf{F}$ and torque $\mathbf{M}$. The constraint of motion is (10.408). We can determine the equations of motion of the sphere using the fundamental equation of dynamics (10.303). Denoting the reaction force at the contact point by $\mathbf{N}$, we use the Newton equation $\sum \mathbf{F} = d\mathbf{p}/dt = m\, d\mathbf{v}/dt$ and the Euler equation $\sum \mathbf{r} \times \mathbf{F} = d\mathbf{L}/dt = I\, d\boldsymbol{\omega}/dt$ to begin the analysis:

$$m\frac{d\mathbf{v}}{dt} = \mathbf{F} + \mathbf{N} \tag{10.411}$$

$$I\frac{d\boldsymbol{\omega}}{dt} = \mathbf{M} - R\hat{u}_n \times \mathbf{N} \tag{10.412}$$

Differentiating the constraint equation (10.408) and eliminating $\dot{\boldsymbol{\omega}}$ by (10.412), we obtain an equation that relates $\mathbf{F}$, $\mathbf{N}$, and $\mathbf{M}$:

$$\frac{I}{mR}\left(\mathbf{F} + \mathbf{N}\right) = \mathbf{M} \times \hat{u}_n - R\mathbf{N} + R\hat{u}_n\left(\hat{u}_n \cdot \mathbf{N}\right) \tag{10.413}$$

Writing this equation in components and substituting $I = \frac{2}{5}mR^2$, we have

$$N_x = \frac{5}{7R}M_y - \frac{2}{7}F_x$$

$$N_y = -\frac{5}{7R}M_x - \frac{2}{7}F_y$$

$$N_z = -F_z \tag{10.414}$$

Substituting these equations in (10.411), we obtain the equations of motion involving only the external force system:

$$m\frac{dv_x}{dt} = \frac{5}{7}\left(F_x + \frac{1}{R}M_y\right) \tag{10.415}$$

$$m\frac{dv_y}{dt} = \frac{5}{7}\left(F_y - \frac{1}{R}M_x\right) \tag{10.416}$$

The components ω_x and ω_y of the angular velocity $\boldsymbol{\omega}$ should be calculated from the constraint equation (10.408). The component ω_z is the solution of the z-component of Equation (10.412):

$$\frac{2}{5}mR^2\frac{d\omega_z}{dt} = M_z \tag{10.417}$$

Example 650 A Rolling Disc and Nonholonomic Constraint Figure 10.39 illustrates a thin disc with radius R that is rolling without slipping on a horizontal plane. Let us attach a body coordinate frame B to the disc at the mass center C and a global frame G on the horizontal plane. We use the Euler angles φ, θ, ψ to indicate the orientation of B in G and the coordinates X, Y of the contact point $P(X, Y, Z)$ to indicate the position of the disc. Because of the holonomic constraint $Z = 0$, we can use the five variables X, Y, φ, θ, ψ as the required five generalized coordinates.

The condition of rolling without slipping provides a vectorial nonholonomic constraint:

$${}^G\mathbf{v}_P = {}^G\mathbf{v}_C + {}_G\boldsymbol{\omega}_B \times {}^G_C\mathbf{r}_P = 0 \tag{10.418}$$

To determine scalar equations of the nonholonomic constraint, we need to expand the vectorial equation. Let us introduce two intermediate frames B_1 and B_2, both at C,

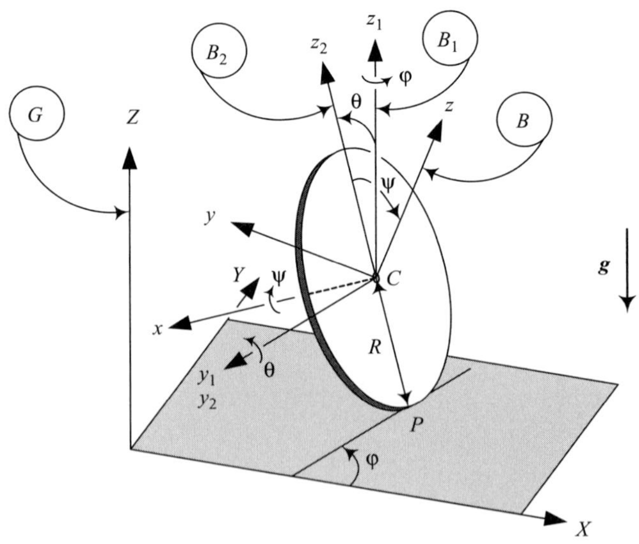

Figure 10.39 A thin disc with radius R is rolling without slipping on a horizontal plane.

to determine the transformation matrix ${}^{G}R_{B}$. The frame B_1 differs with G only by a rotation φ about the z_1-axis:

$$ {}^{1}R_{G} = \begin{bmatrix} \cos\varphi & \sin\varphi & 0 \\ -\sin\varphi & \cos\varphi & 0 \\ 0 & 0 & 1 \end{bmatrix} \tag{10.419} $$

The frame B_2 differs with B_1 only by a rotation θ about the y_2-axis:

$$ {}^{2}R_{1} = \begin{bmatrix} \cos\theta & 0 & -\sin\theta \\ 0 & 1 & 0 \\ \sin\theta & 0 & \cos\theta \end{bmatrix} \tag{10.420} $$

The frame B differs with B_2 only by a rotation ψ about the x-axis:

$$ {}^{B}R_{2} = \begin{bmatrix} 1 & 0 & 0 \\ 0 & \cos\psi & \sin\psi \\ 0 & -\sin\psi & \cos\psi \end{bmatrix} \tag{10.421} $$

Therefore,

$$ \begin{aligned} {}^{B}R_{G} &= {}^{B}R_{2}\, {}^{2}R_{1}\, {}^{1}R_{G} \\ &= \begin{bmatrix} c\theta c\varphi & c\theta s\varphi & -s\theta \\ c\varphi s\theta s\psi - c\psi s\varphi & c\psi c\varphi + s\theta s\psi s\varphi & c\theta s\psi \\ s\psi s\varphi + c\psi c\varphi s\theta & c\psi s\theta s\varphi - c\varphi s\psi & c\theta c\psi \end{bmatrix} \end{aligned} \tag{10.422} $$

$$ \begin{aligned} {}^{G}R_{B} &= {}^{B}R_{G}^{\mathrm{T}} \\ &= \begin{bmatrix} c\theta c\varphi & c\varphi s\theta s\psi - c\psi s\varphi & s\psi s\varphi + c\psi c\varphi s\theta \\ c\theta s\varphi & c\psi c\varphi + s\theta s\psi s\varphi & c\psi s\theta s\varphi - c\varphi s\psi \\ -s\theta & c\theta s\psi & c\theta c\psi \end{bmatrix} \end{aligned} \tag{10.423} $$

Employing the transformation matrix

$$ \begin{aligned} {}^{G}R_{2} &= \left[{}^{2}R_{1}\, {}^{1}R_{G}\right]^{\mathrm{T}} = {}^{1}R_{G}^{\mathrm{T}}\, {}^{2}R_{1}^{\mathrm{T}} \\ &= \begin{bmatrix} \cos\theta\cos\varphi & -\sin\varphi & \cos\varphi\sin\theta \\ \cos\theta\sin\varphi & \cos\varphi & \sin\theta\sin\varphi \\ -\sin\theta & 0 & \cos\theta \end{bmatrix} \end{aligned} \tag{10.424} $$

and the global coordinates of the contact point $P(X, Y)$, we find the global position of C:

$$ \begin{aligned} {}^{G}\mathbf{r}_{C} &= {}^{G}\mathbf{r}_{P} + {}^{G}R_{2}\, {}^{2}_{P}\mathbf{r}_{C} \\ &= \begin{bmatrix} X \\ Y \\ 0 \end{bmatrix} + {}^{G}R_{B} \begin{bmatrix} 0 \\ 0 \\ R \end{bmatrix} = \begin{bmatrix} X + R\cos\varphi\sin\theta \\ Y + R\sin\theta\sin\varphi \\ R\cos\theta \end{bmatrix} \end{aligned} \tag{10.425} $$

Taking a time derivative of ${}^{G}\mathbf{r}_{C}$ provides the velocity of the mass center C:

$$ {}^{G}\mathbf{v}_{C} = \frac{{}^{G}d}{dt}\, {}^{G}\mathbf{r}_{C} = \begin{bmatrix} \dot{X} + R\dot{\theta}\cos\theta\cos\varphi - R\dot{\varphi}\sin\theta\sin\varphi \\ \dot{Y} + R\dot{\theta}\cos\theta\sin\varphi + R\dot{\varphi}\cos\varphi\sin\theta \\ -R\dot{\theta}\sin\theta \end{bmatrix} \tag{10.426} $$

The angular velocity of the disc is given as

$$
\begin{aligned}
{}^{B}_{G}\tilde{\omega}_{B} &= {}^{G}R_{B}^{\mathrm{T}}\,{}^{G}\dot{R}_{B} \qquad (10.427)\\
&= \begin{bmatrix} 0 & \dot{\theta}s\psi - \dot{\varphi}c\theta c\psi & \dot{\theta}c\psi + \dot{\varphi}c\theta s\psi \\ \dot{\varphi}c\theta c\psi - \dot{\theta}s\psi & 0 & \dot{\varphi}s\theta - \dot{\psi} \\ -\dot{\theta}c\psi - \dot{\varphi}c\theta s\psi & \dot{\psi} - \dot{\varphi}s\theta & 0 \end{bmatrix}
\end{aligned}
$$

Employing the above kinematic vectors and transformation matrices, we can write the rolling constraint equation as

$$
\begin{aligned}
{}^{G}\mathbf{v}_{P} &= {}^{G}\mathbf{v}_{C} + {}^{G}R_{B}\left({}^{B}_{G}\boldsymbol{\omega}_{B} \times {}^{G}R_{2}\,{}^{2}_{C}\mathbf{r}_{P}\right) = 0 \\
&= \begin{bmatrix} \dot{X} - R\dot{\psi}\sin\varphi \\ \dot{Y} + R\dot{\psi}\cos\varphi \\ 0 \end{bmatrix} \qquad (10.428)
\end{aligned}
$$

Example 651 ★ Integrating Factor The conditions (10.376) guarantee that a given differential constraint is integrable. However, there exist differential constraints that are integrable while the conditions are not fulfilled. This is because the given constraint is divided by a common factor. The divisor is called integrating factor μ.

We show the existence of μ for a two-dimensional nontotal equation:

$$
A\,dx + B\,dy = 0 \qquad \frac{\partial A}{\partial y} \neq \frac{\partial B}{\partial x} \tag{10.429}
$$

Assume that there exist a holonomic constraint f for this differential:

$$
f(x, y) = c \tag{10.430}
$$

The total differential of f is

$$
df = \frac{\partial f}{\partial x}dx + \frac{\partial f}{\partial y}dy \tag{10.431}
$$

Comparing (10.431) and (10.429) yields

$$
\frac{\partial f/\partial x}{A} = \frac{\partial f/\partial y}{B} = \mu\,(x, y) \tag{10.432}
$$

where we denote the common ratio by $\mu(x, y)$. Therefore, if (10.429) is integrable, then it has at least one integrating factor:

$$
\frac{\partial f}{\partial x} = \mu A \qquad \frac{\partial f}{\partial y} = \mu B \tag{10.433}
$$

$$
\frac{\partial\,(\mu A)}{\partial y} = \frac{\partial\,(\mu B)}{\partial x} \tag{10.434}
$$

We can show that, if $f_1(f)$ is any function of f, then

$$
\mu f_1\,(A\,dx + B\,dy) = f_1\,df = d\left(\int f_1(f)\,df\right) \tag{10.435}
$$

so, $\mu f_1(f)$ is also an integrating factor.

To develop the required conditions to get a total differential constraint, we begin by expanding (10.434) to obtain

$$\frac{1}{\mu}\left(B\frac{\partial \mu}{\partial x} - A\frac{\partial \mu}{\partial y}\right) = \frac{\partial A}{\partial y} - \frac{\partial B}{\partial x} \tag{10.436}$$

It is a partial differential equation for $\mu = \mu(x, y)$. However, any particular solution for $\mu(x, y)$ is enough to make the differential constraint total. Let us examine an integrating factor as a function of x alone. In this case, Equation (10.436) simplifies to

$$\frac{1}{\mu}\frac{d\mu}{dx} = \frac{1}{B}\left(\frac{\partial A}{\partial y} - \frac{\partial B}{\partial x}\right) \tag{10.437}$$

Because the left-hand side of this equation is a function only of x, the right-hand side must also be a function of x:

$$\frac{1}{B}\left(\frac{\partial A}{\partial y} - \frac{\partial B}{\partial x}\right) = g(x) \tag{10.438}$$

Therefore, we have

$$\frac{1}{\mu}\frac{d\mu}{dx} = g(x) \tag{10.439}$$

which can be integrated to

$$\ln \mu = \int g(x)dx \tag{10.440}$$

or

$$\mu = e^{\int g(x)dx} \tag{10.441}$$

We may also search for an integrating factor as a function of y alone and find

$$\mu = e^{\int h(y)dy} \tag{10.442}$$

where

$$h(y) = -\frac{1}{A}\left(\frac{\partial A}{\partial y} - \frac{\partial B}{\partial x}\right) \tag{10.443}$$

Example 652 ★ Making Integrable by an Integrating Factor The differential constraint

$$A\,dx + B\,dy = y\,dx + (x^2 y - x)dy = 0 \tag{10.444}$$

shows that

$$\frac{\partial A}{\partial y} = 1 \qquad \frac{\partial B}{\partial x} = 2xy - 1 \tag{10.445}$$

However, if we multiply the constraint by $1/x^2$, it becomes a total differential constraint:

$$\frac{y}{x^2}dx + \left(y - \frac{1}{x}\right)dy = d\left(\frac{y^2}{2} - \frac{y}{x}\right) = 0 \tag{10.446}$$

Let us use Equation (10.441) to find this integrating factor that is only a function of x:

$$g(x) = \frac{1}{B}\left(\frac{\partial A}{\partial y} - \frac{\partial B}{\partial x}\right) = -\frac{2}{x} \tag{10.447}$$

$$\mu = e^{\int g(x)dx} = x^{-2} \tag{10.448}$$

Example 653 Linear Equation Integrating Factor Consider a constraint equation of the form

$$\dot{x} + p(t)x = q(t) \tag{10.449}$$

where $p(t)$ and $q(t)$ are arbitrary functions of t. This is an inhomogeneous linear first-order ODE. Linear equations have a general integrating factor μ:

$$\mu(t) = e^{\int p(t)dt} \tag{10.450}$$

Multiplying (10.449) by μ yields

$$\frac{d}{dt}(x\,\mu) = q(t)\,\mu \tag{10.451}$$

So, the solution of (10.449) is

$$\begin{aligned} x &= \frac{1}{\mu}C_1 + \frac{1}{\mu}\int q(t)\mu\, dx \\ &= e^{-\int p(t)dx}\left(C_1 + \int q(t)e^{\int p(t)dx}\, dt\right) \end{aligned} \tag{10.452}$$

10.7.2 ★ Inequality Constraint

An *inequality constraint* on a dynamic system is expressed by an equation of displacements and time,

$$f(\mathbf{u}, t) < 0 \tag{10.453}$$

or by their differential:

$$df < 0 \tag{10.454}$$

The first one is called a *limit constraint* and the second one is called a *slip constraint*.

A relation between the coordinates of the configuration space in the form

$$f(u_1, u_2, u_3, \ldots, u_N) = f(\mathbf{u}) < 0 \tag{10.455}$$

or reducible to this form is called a *scleronomic limit constraint*. An equation between the coordinates of the event space of the form

$$f(u_1, u_2, u_3, \ldots, u_N, t) = f(\mathbf{u}, t) < 0 \tag{10.456}$$

or reducible to this form is called a *rheonomic limit constraint*. A limit constraint is always *nonholonomic*.

Each scleronomic limit constraint defines a limit surface in the configuration space that the *possible motion* of the system can only occur on one side of the limit surface. Similarly, each rheonomic limit constraint defines a limit surface in the event space that the possible motion of the system can only occur on one side of the limit surface. Any motion of a system to touch or penetrate the limit surface is not possible as long as the constraint exists.

If a limit constraint is imposed on the finite displacement u_i of a describing point, there is no constraint on infinitesimal displacement du_i of the point.

The differential inequality or slip constraint may exist in dynamic systems regardless of limit constraints.

Proof: Consider a dynamic system with a three-dimensional configuration space (u_1, u_2, u_3) and a scleronomic limit constraint as

$$f(u_1, u_2, u_3) < 0 \tag{10.457}$$

The associated equality constraint equation $f(u_1, u_2, u_3) = 0$ defines a surface in (u_1, u_2, u_3)-space. The local value of the constraint function $f = f(u_1, u_2, u_3)$ is positive on one side and negative on the other. To satisfy the constraint (10.457), the S_C-point of the system and its S_C-trajectories must lie on the side of the constraint surface where $f < 0$. The positive side of a surface f is indicated by ∇f. So, the side $-\nabla f$ is where the S_C-point can move.

A time-dependent limit constraint given as

$$f(u_1, u_2, u_3, t) < 0 \tag{10.458}$$

also separates the configuration space into fields in which $f < 0$ and $f > 0$. The boundary of the fields is defined by the associated constraint equation $f(u_1, u_2, u_3, t) = 0$. This equation defines a rigid surface in the event space (u_1, u_2, u_3, t) or a surface in the configuration space that changes with time in a prescribed manner. At any time t, the S_C-point must be on the side of $-\nabla f$ of the frozen constraint at that time.

Because the describing point of the dynamic system is not supposed to touch the limit surface, the limit surface or any local differential of it has no interaction with the motion of the S_C-point. Therefore, a limit constraint cannot change the DOF of the dynamic system.

Limit constraints cannot be defined by the sign $\leq$ instead of the inequality sign $<$:

$$f(\mathbf{u}, t) \not\leq 0 \tag{10.459}$$

This is because we can always decompose an equal-and-less-than constraint into an equal constraint and a less-than constraint:

$$f(\mathbf{u}, t) \leq 0 \equiv \begin{cases} f(\mathbf{u}, t) < 0 \\ f(\mathbf{u}, t) = 0 \end{cases} \tag{10.460}$$

Assume that the S_C-point of the system is in the field of $f < 0$ where its DOF is f_C. As soon as the S_C-point touches the constraint surface $f = 0$, its DOF drops to $f_C - 1$ and its equations of motion change to include the constraint force of $f = 0$. Therefore, a system with a holonomic constraint $f = 0$ or a limit constraint $f < 0$ has two different dynamics that must be analyzed separately. However, we may use the equal-and-less-than sign ($\leq$) as a short notation to indicate both constraints.

We can transform a greater-than constraint ($f > 0$) to a less-than constraint ($f < 0$) by multiplying by -1. ■

Example 654 A Box on a Turntable with a Wall Consider the motion of a box in a channel on a turntable such as the one explained in Example 612. Assume that we remove one of the channel walls. Some of the kinematic constraints on the motion of the box are not symmetric. The equality constraints on the box are

$$Z_C - Z_0 = 0 \qquad \theta_X = 0 \qquad \theta_Y = 0 \qquad \theta_Z = 2\pi f t \tag{10.461}$$

where (X_C, Y_C, Z_C) are the coordinates of the box center. To express the inequality constraint, we may show the constraints on four corners of the box if we consider the size of the box in the (X, Y)-plane as $2a \times 2b$:

$$Y_C - b\cos\theta_Z - a\sin\theta_Z \geq 0 \tag{10.462}$$

$$Y_C - b\cos\theta_Z + a\sin\theta_Z \geq 0 \tag{10.463}$$

$$Y_C + b\cos\theta_Z + a\sin\theta_Z \geq 0 \tag{10.464}$$

$$Y_C + b\cos\theta_Z - a\sin\theta_Z \geq 0 \tag{10.465}$$

Only two of the above four constraints are needed to make sure that no corner breaks the unilateral wall constraint $Y > 0$.

Example 655 Wiping Blades Figure 10.40 illustrates two sweeping blades A and B that are supposed to weep a rectangular area of $2a \times 4a$ such that $X_A < 0$ and $X_B > 0$. Applying any harmonic motion to the blades with amplitude a, such as the following functions, will do the job:

$$X_B = a[1 + \sin(\omega_B t + \varphi_B)] \tag{10.466}$$

$$X_A = -a[1 + \sin(\omega_A t + \varphi_A)] \tag{10.467}$$

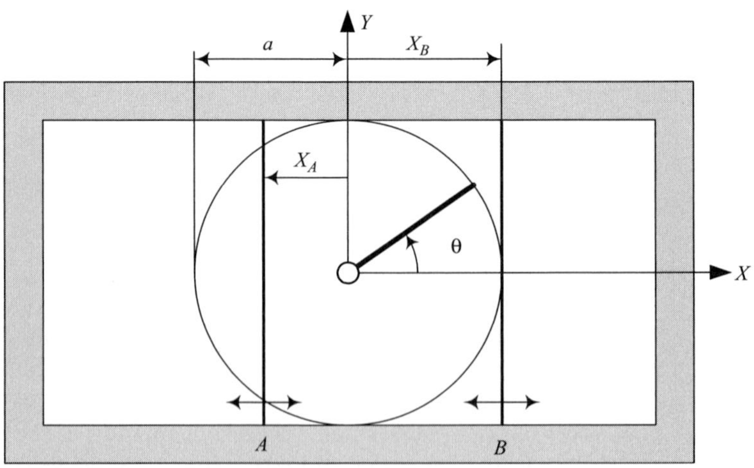

Figure 10.40 Two sweeping blades A and B with $X_A < 0$ and $X_B > 0$.

Let us also add a turning blade with length $l = a$ at the center of the area. If the blade starts turning with a constant angular frequency ω from the horizontal position, then coordinates of the tip point of the blade would be

$$X = a\cos\omega t \qquad Y = a\sin\omega t \tag{10.468}$$

The inequality constraints of the motion of sweeping and turning blades are

$$X_A < X < X_B \tag{10.469}$$

To satisfy the constraints, there must be specific relationships between the frequencies ω, ω_A, ω_B and the initial conditions of the blades:

$$\frac{X_B(0)}{a} = 1 + \sin\varphi_B(0) \tag{10.470}$$

$$\frac{X_A(0)}{a} = -1 - \sin\varphi_A(0) \tag{10.471}$$

Let us assume that

$$\omega = \omega_A = \omega_B \tag{10.472}$$

Then, the constraint $X > X_A$ provides

$$\begin{aligned} e_A &= X - X_A > 0 \\ &= a\cos\omega t + a[1 + \sin(\omega_A t + \varphi_A)] > 0 \end{aligned} \tag{10.473}$$

Figure 10.41 illustrates e_A/a versus φ_A and ωt, both between 0 and 2π. It indicates the regions of the plane $(\varphi_A, \omega t)$ in which $e_A > 0$ and the first constraint is satisfied. We can similarly define an e_B function as

$$\begin{aligned} e_B &= X_B - X > 0 \\ &= a[1 + \sin(\omega_B t + \varphi_B)] - a\cos\omega t > 0 \end{aligned} \tag{10.474}$$

and plot e_B/a versus φ_B and ωt to have an image of the regions of $e_B/a > 0$.

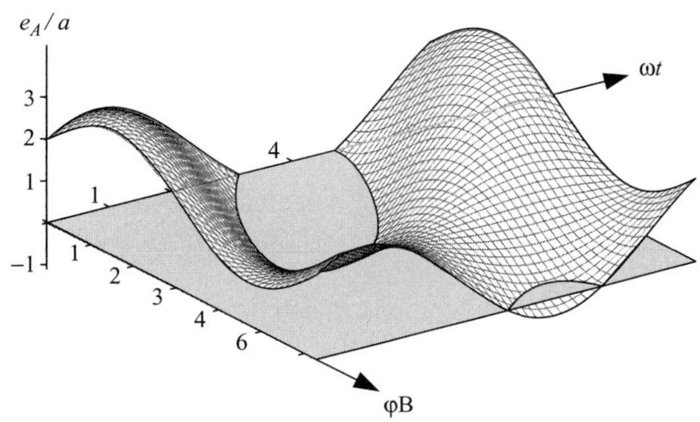

Figure 10.41 Plot of e_A/a versus φ_A and ωt, where $e_A = a\cos\omega t + a[1 + \sin(\omega_A t + \varphi_A)]$.

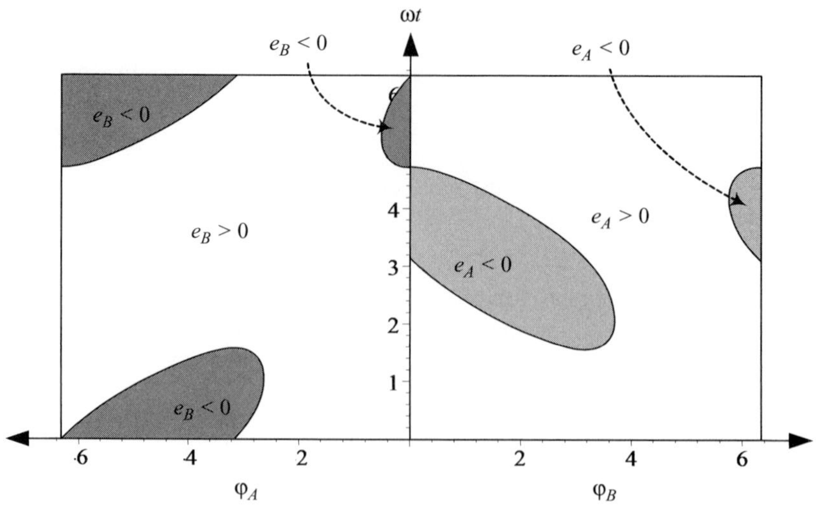

Figure 10.42 Top view of e_B/a and e_B/a, side by side.

To determine the regions of φ_A, φ_B and ωt, in which both constraints are satisfied, let us show the top view of e_B/a and e_B/a side by side, as shown in Figure 10.42. The shaded areas indicate where one of the constraints is not satisfied. Because ωt is periodic, this graph repeats itself vertically. Therefore, having any vertical line on axes φ_A and φ_B that does not hit any shaded area indicates there are safe values that satisfy the constraints.

10.8 ★ DIFFERENTIAL CONSTRAINT

Constraints on infinitesimal displacements are of the forms

$$\sum_{i=1}^{N} A_{ij} du_i = 0 \qquad j = 1, 2, \ldots, L \tag{10.475}$$

$$\sum_{i=1}^{N} A_{ij} du_i + A_j dt = 0 \qquad j = 1, 2, \ldots, L \tag{10.476}$$

where A_{ij} and A_j are C_2 functions of u_i $(i = 1, 2, \ldots, N)$ and t. They are L linear differentials of $N + 1 > L$ variables. The expression of constraints by linear equations of differentials is called the *Pfaffian form*. Holonomic constraints are of the Pfaffian form upon differentiating.

Similar to holonomic constraints, we can call the Pfaffian form (10.475) in which t does not occur explicitly *scleronomic*, and the form (10.476) *rheonomic*. However, to distinguish between finite-displacement holonomic constraints and Pfaffian forms, which many not be holonomic, we call the Pfaffian form (10.475) in which $A_i = 0$ *catastatic* and the Pfaffian form (10.476) in which $A_i \neq 0$ *acatastatic*.

If a Pfaffian form is a total differential, it is integrable and can be reduced to the holonomic form (10.154) or (10.156). If a Pfaffian form is not total and there is no integrating factor, it is a differential constraint.

Consider a single Pfaffian form in a three-dimensional configuration space,

$$P\,dx + Q\,dy + R\,dz = 0 \tag{10.477}$$

where P, Q, and R are functions of x, y, and z. If the conditions (10.383) are not fulfilled, then the necessary and sufficient condition for (10.477) to be integrable is

$$P\left(\frac{\partial Q}{\partial z} - \frac{\partial R}{\partial y}\right) + Q\left(\frac{\partial R}{\partial x} - \frac{\partial P}{\partial z}\right) + R\left(\frac{\partial P}{\partial y} - \frac{\partial Q}{\partial x}\right) = 0 \tag{10.478}$$

Proof: The conditions (10.383) are necessary and sufficient conditions for the differential constraint

$$A\,dx + B\,dy + C\,dz = 0 \tag{10.479}$$

to be total:

$$df = \frac{\partial f}{\partial x}dx + \frac{\partial f}{\partial y}dy + \frac{\partial f}{\partial z}dz = 0 \tag{10.480}$$

If A, B, C have a common factor μ, and if

$$\frac{\partial f}{\partial x} = \mu P \qquad \frac{\partial f}{\partial y} = \mu Q \qquad \frac{\partial f}{\partial z} = \mu R \tag{10.481}$$

then the total differential (10.479) may be written in the form (10.477), which does not necessarily satisfy the conditions (10.477).

To derive the conditions for (10.477) to be integrable, we assume there exist a holonomic function $f = f(x, y, z)$ and an integrating factor $\mu = \mu(x, y, z)$ such that they satisfy conditions (10.481). Then we have

$$\frac{\partial\,(\mu P)}{\partial y} = \frac{\partial^2 f}{\partial y\,\partial x} = \frac{\partial^2 f}{\partial x\,\partial y} = \frac{\partial\,(\mu Q)}{\partial x} \tag{10.482}$$

which can be rearranged as

$$\mu\left(\frac{\partial P}{\partial y} - \frac{\partial Q}{\partial x}\right) = Q\frac{\partial \mu}{\partial x} - P\frac{\partial \mu}{\partial y} \tag{10.483}$$

Similarly we have

$$\mu\left(\frac{\partial Q}{\partial z} - \frac{\partial R}{\partial y}\right) = R\frac{\partial \mu}{\partial y} - Q\frac{\partial \mu}{\partial z} \tag{10.484}$$

$$\mu\left(\frac{\partial R}{\partial x} - \frac{\partial P}{\partial z}\right) = P\frac{\partial \mu}{\partial z} - R\frac{\partial \mu}{\partial x} \tag{10.485}$$

We can respectively multiply these three equations by P, Q, R and sum to eliminate μ:

$$P\left(\frac{\partial Q}{\partial z} - \frac{\partial R}{\partial y}\right) + Q\left(\frac{\partial R}{\partial x} - \frac{\partial P}{\partial z}\right) + R\left(\frac{\partial P}{\partial y} - \frac{\partial Q}{\partial x}\right) = 0 \tag{10.486}$$

This is the necessary condition for differential constraint (10.477) to have an integrating factor and be integrable. It means that, if ϕ is a function of x, y, z and

$$P_1 = \phi P \qquad Q_1 = \phi Q \qquad R_1 = \phi R \tag{10.487}$$

then the condition of integrability (10.478) is satisfied by P_1, Q_1, R_1.

To prove that (10.478) is also a sufficient condition for integrability, we must show that there exists a solution when it is satisfied. Let us assume that one of the variable, say z, is momentarily frozen and is assumed constant. Then Equation (10.477) reduces to

$$P\,dx + Q\,dy = 0 \tag{10.488}$$

where P and Q are functions of variables x, y and parameter z. This equation always has a solution,

$$f_1\,(x, y, z) = c_1 \tag{10.489}$$

where, if $\phi\,(x, y, z)$ is the integrating factor, then

$$\frac{\partial f_1}{\partial x} = \phi P = P_1 \qquad \frac{\partial f_1}{\partial y} = \phi Q = Q_1 \tag{10.490}$$

Although it is not necessarily true that $\partial f_1/\partial z = \phi R = R_1$, we can always write the third condition as

$$R_1 = \phi R = \frac{\partial f_1}{\partial z} + S \tag{10.491}$$

Because P_1, Q_1, R_1 must satisfy (10.486), we have

$$\frac{\partial S}{\partial x}\frac{\partial f_1}{\partial y} - \frac{\partial S}{\partial y}\frac{\partial f_1}{\partial x} = 0 \tag{10.492}$$

which shows that S and f_1 are functionally related:

$$S = S(f_1, z) \tag{10.493}$$

By examination,

$$\begin{aligned}\phi\,(P\,dx + Q\,dy + R\,dz) &= \frac{\partial f_1}{\partial x}dx + \frac{\partial f_1}{\partial y}dy + \frac{\partial f_1}{\partial z}dz + S\,dz \\ &= df_1 + S\,dz\end{aligned} \tag{10.494}$$

we see that the original equation is equivalent to

$$df_1 + S\,dz = 0 \tag{10.495}$$

If $\psi = \psi(f_1, z)$ is an integral factor of this equation, then

$$\psi\phi(P\,dx + Q\,dy + R\,dz) = \psi(df_1 + S\,dz) \tag{10.496}$$

is a total differential. Its original function is

$$f_2\,(f_1, z) = c_2 \tag{10.497}$$

which becomes

$$f(x, y, z) = 0 \tag{10.498}$$

upon replacing f_1 by its expression in x,y, z.

The integrability condition for a general case is

$$A_i\left(\frac{\partial A_j}{\partial u_k}-\frac{\partial A_k}{\partial u_j}\right)+A_j\left(\frac{\partial A_k}{\partial u_i}-\frac{\partial A_i}{\partial u_k}\right)+A_k\left(\frac{\partial A_i}{\partial u_j}-\frac{\partial A_j}{\partial u_i}\right)=0$$

$$i, j, k = 1, 2, \ldots, N \tag{10.499}$$

There are $N(N-1)(N-2)/6$ such equations, of which $(N-1)(N-2)/2$ are independent.

Johann Friedrich Pfaff (1765–1825) was a German mathematician and the first to explain the meaning of nonintegrable differential forms, which are fewer in number than their variable arguments. The condition (10.486) was found by Euler in 1770.

The word *catastatic* comes from the Greek word for "orderly." ■

Example 656 ★ Integrability Analysis Procedure Consider the differential constraint

$$yz(y+z)\,dx + zx(z+x)\,dy + xy(x+y)\,dz = 0 \tag{10.500}$$

This equation satisfies the integrability condition (10.478), and therefore, we can determine a holonomic constraint.

To review the procedure of the analysis in the above Proof, let us freeze z and simplify this equation to

$$yz(y+z)\,dx + zx(z+x)\,dy = 0 \tag{10.501}$$

It has the solution

$$f_1 = \frac{(y+z)(z+x)}{xy} \tag{10.502}$$

The differentials of f_1 are

$$\frac{\partial f_1}{\partial x} = \frac{z(y+z)}{x^2 y} = -\frac{1}{x^2y^2}P \tag{10.503}$$

$$\frac{\partial f_1}{\partial y} = -\frac{z(z+x)}{xy^2} = -\frac{1}{x^2y^2}Q \tag{10.504}$$

so

$$\phi = -\frac{1}{x^2y^2} \tag{10.505}$$

Now we have

$$S = \phi R - \frac{\partial f_1}{\partial z} = -\frac{x+y}{xy} = -2\frac{f_1-1}{z} \tag{10.506}$$

and therefore,

$$\phi(P\,dx + Q\,dy + R\,dz) = df_1 - 2\frac{f_1-1}{z}dz \tag{10.507}$$

An integrating factor of this equation is

$$\psi = \frac{1}{z^2} \tag{10.508}$$

and

$$\psi\phi\left(P\,dx + Q\,dy + R\,dz\right) = \frac{df_1}{z^2} - \frac{2\left(f_1 - 1\right)}{z^2} = d\left(\frac{f_1 - 1}{z^2}\right) \tag{10.509}$$

which yields

$$\frac{f_1 - 1}{z^2} = c_2 \tag{10.510}$$

Replacing f_1 yields the holonomic constraint

$$\frac{x + y + z}{xyz} = c \tag{10.511}$$

This example is adopted form Ince (1926). Edward Lindsay Ince (1891–1941) was a British mathematician who improved the theory of differential equations.

Example 657 ★ Pfaff Problem When the integrability condition (10.478) is not satisfied, the differential constraint (10.477) is not derivable from a single function. Pfaff showed that in this case the differential constraint is equivalent to a pair of algebraic equations known as integral equivalents. For example, the differential constraint of

$$y\,dx + z\,dy + x\,dz = 0 \tag{10.512}$$

can be a result of any set of the coupled algebraic equations

$$\tfrac{1}{2}y^2 + xz = c_1 \qquad x - y = c_2 \tag{10.513}$$

$$\tfrac{1}{2}y^2 + xz = c_1 \qquad y - z = c_2 \tag{10.514}$$

In general, when the integrability condition is not satisfied, then a differential constraint in N or $N + 1$ variables is equivalent to a system of not more than N algebraic equations. This is called the Pfaff problem.

10.9 GENERALIZED MECHANICS

Consider a mechanical system in a constraint-free n-dimensional configuration space. The n coordinates of the space are a set of generalized coordinates q_i, $i = 1, 2, \ldots, n$. The coordinates of the original configuration space u_i, $i = 1, 2, \ldots, N$, and the coordinates of the Cartesian space (x_i, y_i, z_i), $i = 1, 2, \ldots, N/3$, are functions of the generalized coordinates q_i and possibly time t:

$$u_i = u_i\left(q_1, q_2, \ldots, q_n, t\right) \qquad i = 1, 2, \ldots, N \tag{10.515}$$

$$x_i = u_i\left(q_1, q_2, \ldots, q_n, t\right) \qquad i = 1, 2, \ldots, N/3 \tag{10.516}$$

$$y_i = u_i\left(q_1, q_2, \ldots, q_n, t\right) \qquad i = 1, 2, \ldots, N/3 \tag{10.517}$$

$$z_i = u_i\left(q_1, q_2, \ldots, q_n, t\right) \qquad i = 1, 2, \ldots, N/3 \tag{10.518}$$

Therefore, we must be able to express every dynamic characteristic of the system $f(u_i)$ that is expressible by u_i or (x_i, y_i, z_i) by generalized coordinates:

$$f(u_i) = f(q_1, q_2, \ldots, q_n, t) \qquad i = 1, 2, \ldots, N \tag{10.519}$$

Example 658 Generalized Kinetic Energy The kinetic energy of a system with $N/3$ particles is

$$K = \frac{1}{2}\sum_{i=1}^{N/3} m_i \left(\dot{x}_i^2 + \dot{y}_i^2 + \dot{z}_i^2\right) = \frac{1}{2}\sum_{i=1}^{N} m_i \dot{u}_i^2 \tag{10.520}$$

Expressing the configuration coordinate u_i in terms of generalized coordinates q_j, we have

$$\dot{u}_i = \sum_{s=1}^{n} \frac{\partial u_i}{\partial q_s}\dot{q}_s + \frac{\partial u_i}{\partial t} \qquad s = 1, 2, \ldots, N \tag{10.521}$$

Therefore, the kinetic energy in terms of generalized coordinates is

$$\begin{aligned} K &= \frac{1}{2}\sum_{i=1}^{N} m_i \left(\sum_{s=1}^{n} \frac{\partial u_i}{\partial q_s}\dot{q}_s + \frac{\partial u_i}{\partial t}\right)^2 \\ &= \frac{1}{2}\sum_{j=1}^{n}\sum_{k=1}^{n} a_{jk}\dot{q}_j\dot{q}_k + \sum_{j=1}^{n} b_j\dot{q}_j + c \end{aligned} \tag{10.522}$$

where

$$a_{jk} = \sum_{i=1}^{N} m_i \frac{\partial u_i}{\partial q_j}\frac{\partial u_i}{\partial q_k} \tag{10.523}$$

$$b_j = \sum_{i=1}^{N} m_i \frac{\partial u_i}{\partial q_j}\frac{\partial u_i}{\partial t} \tag{10.524}$$

$$c = \frac{1}{2}\sum_{i=1}^{N} m_i \left(\frac{\partial u_i}{\partial t}\right)^2 \tag{10.525}$$

where

$$\begin{aligned} \left(\sum_{s=1}^{n} \frac{\partial u_i}{\partial q_s}\dot{q}_s + \frac{\partial u_i}{\partial t}\right)^2 &= \left(\sum_{j=1}^{n} \frac{\partial u_i}{\partial q_j}\dot{q}_j + \frac{\partial u_i}{\partial t}\right)\left(\sum_{k=1}^{n} \frac{\partial u_i}{\partial q_k}\dot{q}_k + \frac{\partial u_i}{\partial t}\right) \\ &= \sum_{j=1}^{n}\sum_{k=1}^{n}\left(\frac{\partial u_i}{\partial q_j}\frac{\partial u_i}{\partial q_k}\right)\dot{q}_j\dot{q}_k \\ &\quad + 2\sum_{j=1}^{n} \frac{\partial u_i}{\partial q_j}\frac{\partial u_i}{\partial t}\dot{q}_j + \left(\frac{\partial u_i}{\partial t}\right)^2 \end{aligned} \tag{10.526}$$

Using these expressions, we may show the kinetic energy of the dynamic system as

$$K = K_0 + K_1 + K_2 \tag{10.527}$$

where

$$K_0 = \frac{1}{2}\sum_{i=1}^{N} m_i \left(\frac{\partial u_i}{\partial t}\right)^2 \tag{10.528}$$

$$K_1 = \sum_{j=1}^{n}\sum_{i=1}^{N} m_i \frac{\partial u_i}{\partial q_j}\frac{\partial u_i}{\partial t}\dot{q}_j \tag{10.529}$$

$$K_2 = \frac{1}{2}\sum_{j=1}^{n}\sum_{k=1}^{n}\sum_{i=1}^{N} m_i \frac{\partial u_i}{\partial q_j}\frac{\partial u_i}{\partial q_k}\dot{q}_j\dot{q}_k \tag{10.530}$$

Examining a_{kj}, we find that a_{jk} is symmetric:

$$a_{kj} = \sum_{i=1}^{N} m_i \frac{\partial u_i}{\partial q_k}\frac{\partial u_i}{\partial q_j} = a_{jk} \tag{10.531}$$

If the coordinates u_i do not depend explicitly on time t, then $\partial u_i/\partial t = 0$ and we have

$$K = \frac{1}{2}\sum_{j=1}^{n}\sum_{k=1}^{n} a_{jk}\dot{q}_j\dot{q}_k \tag{10.532}$$

Kinetic energy is a scalar quantity and, because of (10.520), must be positive definite. The first term of (10.522) is a positive quadratic form. The third term of (10.522) is also a nonnegative quantity, as indicated by (10.525). The second term of (10.520) can be negative for some $\dot{q}_j$ and t. However, because of (10.520), the sum of all three terms of (10.522) must be positive.

Example 659 Generalized Potential Force If there exists a potential energy function $V = V(u_i)$, $i = 1, 2, \ldots, N$, then the potential force F_j is derivable from the potential energy:

$$F_j = -\frac{\partial V}{\partial u_j} \tag{10.533}$$

Let us assume that we substitute u_i from (10.515) and express V in terms of q_k. Therefore,

$$-\frac{\partial V}{\partial q_r} = -\sum_{j=1}^{N}\frac{\partial V}{\partial u_j}\frac{\partial u_j}{\partial q_r} = \sum_{j=1}^{N} F_j \frac{\partial u_j}{\partial q_r} = F_r \tag{10.534}$$

and the generalized force F_r is related to the potential energy V, given in generalized coordinates by

$$F_r = -\frac{\partial V}{\partial q_r} \tag{10.535}$$

Example 660 ★ Generalized Constraints Consider a mechanical system of $N/3$ particles in the configuration space $u_1, u_2, \ldots, u_N$ with $L\ (< N)$ independent holonomic and nonholonomic constraints:

$$\sum_{i=1}^{N} A_{ji}\, du_i + A_j\, dt = 0 \qquad j = 1, 2, \ldots, L \tag{10.536}$$

Let us assume that L' of the constraints are holonomic,

$$\sum_{i=1}^{N} \frac{\partial f_j}{\partial u_i} du_i + \frac{\partial f_j}{\partial t} dt = 0 \qquad j = 1, 2, \ldots, L' \tag{10.537}$$

and the remaining $L - L'$ constraints are nonholonomic,

$$\sum_{i=1}^{N} A_{ji}\, du_i + A_j\, dt = 0 \qquad j = L' + 1, L' + 2, \ldots, L \tag{10.538}$$

Using the L' holonomic constraints, we reduce the number of configuration coordinates to the minimum required $n = N - L'$ to specify the configuration of a system of $N/3$ particles. Any set of coordinates $q_1, q_2, \ldots, q_n$ is a set of generalized coordinates of a system if and only if the number n of its members is necessary and sufficient to define the configuration of the system uniquely.

Each holonomic constraint in (10.537) is a total differential:

$$df_j\,(u_1, u_2, \ldots, u_N, t) = 0 \qquad j = 1, 2, \ldots, L' \tag{10.539}$$

These equations can be integrated to provide L' equations among the N coordinates u_i, $i = 1, 2, \ldots, N$:

$$f_j(u_1, u_2, \ldots, u_N, t) = c_j \qquad j = 1, 2, \ldots, L' \tag{10.540}$$

The constants c_j are determined from initial conditions $u_1(0)$, $u_2(0)$, ..., $u_N(0)$. Let us define N transformations

$$q_k = q_k\,(u_1, u_2, \ldots, u_N, t) \qquad k = 1, 2, \ldots, N \tag{10.541}$$

where q_k are single-valued functions of $u_1, u_2, \ldots, u_N, t$. We may consider these equations as mapping functions which map the point $(u_1, u_2, \ldots, u_N)$ in u-space to a point $(q_1, q_2, \ldots, q_N)$ in q-space at a fixed time t. The uniqueness of q_k guarantees the uniqueness of the mapping. If there exists a domain U in u-space in which the Jacobian J is not zero for any bounded t,

$$J = \begin{vmatrix} \dfrac{\partial s_1}{\partial u_1} & \cdots & \dfrac{\partial s_1}{\partial u_1} \\ \cdots & \ddots & \cdots \\ \dfrac{\partial s_1}{\partial u_1} & \cdots & \dfrac{\partial s_1}{\partial u_1} \end{vmatrix} \neq 0 \tag{10.542}$$

then, from the implicit function theory, the mapping is one to one. Therefore, the domain U of u-space maps to a domain Q of q-space in which there exist N inverse transformations:

$$u_k = u_k (q_1, q_2, \ldots, q_N, t) \qquad k = 1, 2, \ldots, N \tag{10.543}$$

Using (10.540) and (10.541), we find that the first L' of q_k are constant c_k and fixed by constraint surfaces $f_k = c_k$, $k = 1, 2, \ldots, L'$:

$$q_k = f_k (u_1, u_2, \ldots, u_N, t) = c_k \qquad k = 1, 2, \ldots, L' \tag{10.544}$$

The inverse mapping (10.543) shows that the N configuration coordinates u_k are determined by L' constants c_k, $k = 1, 2, \ldots, L'$, and $N - L'$ variables q_k, $k = L' + 1, L' + 2, \ldots, N$:

$$u_k = u_k (c_1, c_2, \ldots, c_{L'}, q_{L'+1}, q_{L'+2}, \ldots, q_N, t) \qquad k = 1, 2, \ldots, N \tag{10.545}$$

Therefore, only $n = N - L'$ coordinates q_k, $k = 1, 2, \ldots, n$, are required to determine the N configuration coordinates u_k and find the position of the describing S_C-point of the system when the u_k satisfy L' holonomic constraints (10.539) or (10.540). The remaining $n = N - L'$ coordinates q_k, $k = 1, 2, \ldots, n$, are now no longer subject to any holonomic constraints. The n constraint-free coordinates q_i ($i = 1, 2, \ldots, n$) are the generalized coordinates of the system. We may define a constraint-free n-dimensional generalized configuration space $(q_1, q_2, \ldots, q_N)$ in which there is no constraint surface and the describing point of the system may reach any point of the space. However, because of nonholonomic constraints (10.538), the movement of the generalized describing S_Q-point must satisfy Equations (10.538).

From (10.545), the infinitesimal possible and virtual displacements of the S_C-point of the system are

$$du_i = \sum_{j=1}^{n} \frac{\partial u_i}{\partial q_j} dq_j + \frac{\partial u_i}{\partial t} dt \qquad i = 1, 2, \ldots, N \tag{10.546}$$

$$\delta u_i = \sum_{j=1}^{n} \frac{\partial u_i}{\partial q_j} \delta q_j \qquad i = 1, 2, \ldots, N \tag{10.547}$$

The possible displacements du_i satisfy all constraint equations on the system (10.536), which after substituting (10.546),

$$\sum_{i=1}^{N} A_{ji} \left(\sum_{k=1}^{n} \frac{\partial u_i}{\partial q_k} dq_k + \frac{\partial u_i}{\partial t} dt \right) + A_j dt = 0 \tag{10.548}$$
$$j = 1, 2, \ldots, L$$

and simplification become the constraints on the possible displacements of the generalized coordinates dq_k:

$$\sum_{k=1}^{n} \left(\sum_{i=1}^{N} A_{ji} \frac{\partial u_i}{\partial q_k} \right) dq_k + \left(\sum_{i=1}^{N} A_{ji} \frac{\partial u_i}{\partial t} + A_j \right) dt = 0 \tag{10.549}$$
$$j = 1, 2, \ldots, l$$

However, because in the generalized configuration space $(q_1, q_2, \ldots, q_n)$ there is no holonomic constraints, the L' equations of (10.549) are identities and only $l = L - L'$ nonholonomic constraints remain. The nonholonomic constraints (10.549) on the generalized coordinates q_k may be called the generalized constraints . Therefore, any constraint on the generalized coordinates q_k is a generalized constraint and it must be nonholonomic.

Introducing the new notation

$$B_{jk} = \sum_{i=1}^{N} A_{ji} \frac{\partial u_i}{\partial q_k} \tag{10.550}$$

$$B_j = \sum_{i=1}^{N} A_{ji} \frac{\partial u_i}{\partial t} + A_j \tag{10.551}$$

we can write the nonholonomic constraints (10.549) as

$$\sum_{k=1}^{n} B_{jk}\, dq_k + B_j\, dt = 0 \qquad j = 1, 2, \ldots, L - L' \tag{10.552}$$

or as

$$\sum_{k=1}^{n} B_{jk}\, \dot{q}_k + B_j = 0 \qquad j = 1, 2, \ldots, l \qquad l = L - L' \tag{10.553}$$

The possible displacements dq_k must satisfy (10.552) and the possible velocities $\dot{q}_k$ must satisfy (10.553). The nonholonomic constraints on the virtual generalized displacement δq_k are

$$\sum_{k=1}^{n} B_{jk} \delta q_k = 0 \qquad j = 1, 2, \ldots, l \tag{10.554}$$

Example 661 ★ $d\delta$ Operators for Generalized Coordinates It is introduced in Example 635 that d/dt is the differential symbol in real time and δ is the differential in frozen time. It is also shown that the orders of the d and δ operators are interchangeable for configuration position vector $\mathbf{u}$ under specified conditions:

$$d\delta\mathbf{u} = \delta\, d\mathbf{u} \qquad \mathbf{u} = \begin{bmatrix} u_1 & u_2 & u_N \end{bmatrix} \tag{10.555}$$

The same relationship holds for the generalized position vector $\mathbf{q}$:

$$d\delta\mathbf{q} = \delta\, d\mathbf{q} \qquad \mathbf{q} = \begin{bmatrix} q_1 & q_2 & q_n \end{bmatrix} \tag{10.556}$$

To show this, let us use the expression of infinitesimal possible and virtual configuration displacements $\mathbf{u}$ in terms of generalized coordinates $\mathbf{q}$,

$$du_i = \sum_{j=1}^{n} \frac{\partial u_i}{\partial q_j} dq_j + \frac{\partial u_i}{\partial t} dt \qquad i = 1, 2, \ldots, N \tag{10.557}$$

$$\delta u_i = \sum_{j=1}^{n} \frac{\partial u_i}{\partial q_j} \delta q_j \qquad i = 1, 2, \ldots, N \tag{10.558}$$

and substitute them in the ith component of (10.555):

$$\begin{aligned} d\,\delta\mathbf{u} - \delta\,d\mathbf{u} &= d\left(\sum_{j=1}^{n}\frac{\partial u_i}{\partial q_j}\delta q_j\right) - \delta\left(\sum_{j=1}^{n}\frac{\partial u_i}{\partial q_j}dq_j + \frac{\partial u_i}{\partial t}dt\right) \\ &= \sum_{j=1}^{n}\frac{\partial u_i}{\partial q_j}\left(d\,\delta q_j - \delta\,dq_j\right) + \sum_{j=1}^{n} d\left(\frac{\partial u_i}{\partial q_j}\right)\delta q_j \\ &\quad - \sum_{j=1}^{n}\delta\left(\frac{\partial u_i}{\partial q_j}\right)dq_j - \delta\left(\frac{\partial u_i}{\partial t}\right)dt = 0 \end{aligned} \tag{10.559}$$

The sum of the last three terms in this equation is zero:

$$\begin{aligned} &\sum_{j=1}^{n} d\left(\frac{\partial u_i}{\partial q_j}\right)\delta q_j - \sum_{j=1}^{n}\delta\left(\frac{\partial u_i}{\partial q_j}\right)dq_j - \delta\left(\frac{\partial u_i}{\partial t}\right)dt \\ &= \sum_{j,k=1}^{n}\frac{\partial^2 u_i}{\partial q_k\,\partial q_j}dq_k\,\delta q_j + \sum_{j=1}^{n}\frac{\partial^2 u_i}{\partial t\,\partial q_j}dt\,\delta q_j \\ &\quad - \sum_{j,k=1}^{n}\frac{\partial^2 u_i}{\partial q_k\,\partial q_j}\delta q_k\,dq_j - \sum_{j=1}^{n}\frac{\partial^2 u_i}{\partial q_j\,\partial t}\delta q_j\,dt \\ &= \sum_{j,k=1}^{n}\frac{\partial^2 u_i}{\partial q_k\,\partial q_j}dq_k\,\delta q_j - \sum_{k,j=1}^{n}\frac{\partial^2 u_i}{\partial q_j\,\partial q_k}\delta q_j\,dq_k = 0 \end{aligned} \tag{10.560}$$

Therefore, (10.559) reduces to the following equation that must hold for all q and for any transformation $u = u(q)$:

$$\sum_{j=1}^{n}\frac{\partial u_i}{\partial q_j}\left(d\,\delta q_j - \delta\,dq_j\right) = 0 \tag{10.561}$$

which implies that

$$d\,\delta q_j - \delta\,dq_j = 0 \qquad j = 1, 2, \ldots, n \tag{10.562}$$

which shows that

$$d\,\delta\mathbf{q} = \delta\,d\mathbf{q} \tag{10.563}$$

10.10 ★ INTEGRAL OF MOTION

Any equation of the form

$$f(\mathbf{q}, \dot{\mathbf{q}}, t) = c \tag{10.564}$$

where

$$c = f(\mathbf{q}_0, \dot{\mathbf{q}}_0, t_0) \tag{10.565}$$

$$q = \begin{bmatrix} q_1 & q_2 & \cdots & q_n \end{bmatrix} \tag{10.566}$$

with total differential

$$\frac{df}{dt} = \sum_{i=1}^{n} \left(\frac{\partial f}{\partial q_i} \dot{q}_i + \frac{\partial f}{\partial \dot{q}_i} \ddot{q}_i \right) + \frac{\partial f}{\partial t} = 0 \tag{10.567}$$

where the generalized positions $\mathbf{q}$ and velocities $\dot{\mathbf{q}}$ of a dynamic system must satisfy at all times t is called an *integral of motion*. The parameter c whose value depends on initial conditions is called a *constant of motion*. The maximum number of independent integrals of motion for a dynamic system with n degrees of freedom is $2n$. A constant of motion is a quantity whose value remains constant during the motion.

Any integral of motion is a result of a conservation principle or a combination of them. There are only three conservation principles for a dynamic system: energy, momentum, and moment of momentum. Every conservation principle is the result of symmetry in position and time spaces. The conservation of energy indicates the homogeneity of the time space, the conservation of momentum indicates the homogeneity in the position space, and the conservation of the moment of momentum indicates the isotropy in the position space.

Proof: Consider a mechanical system with f_C degrees of freedom. Mathematically the dynamics of the system is expressed by a set of $n = f_C$ second-order differential equations of n generalized coordinates $q_i(t)$, $i = 1, 2, \ldots, n$:

$$\ddot{q}_i = F_i\left(q_i, \dot{q}_i, t\right) \qquad i = 1, 2, \ldots, n \tag{10.568}$$

The general solution of the equations contain $2n$ constants:

$$\dot{q}_i = \dot{q}_i(c_1, c_2, \ldots, c_n, t) \qquad i = 1, 2, \ldots, n \tag{10.569}$$

$$q_i = q_i(c_1, c_2, \ldots, c_{2n}, t) \qquad i = 1, 2, \ldots, n \tag{10.570}$$

To determine these constants and uniquely identify the motion of the system, it is necessary to know the initial conditions $q_i(t_0)$, $\dot{q}_i(t_0)$, which specify the state of the system at some given instant t_0:

$$c_j = c_j\left(\mathbf{q}(t_0), \dot{\mathbf{q}}(t_0), t_0\right) \qquad j = 1, 2, \ldots, 2n \tag{10.571}$$

$$f_j\left(\mathbf{q}(t), \dot{\mathbf{q}}(t), t\right) = c_j\left(\mathbf{q}(t_0), \dot{\mathbf{q}}(t_0), t_0\right) \tag{10.572}$$

Each of the functions f_j is an integral of motion and each c_i is a constant of motion. An integral of motion may also be called a *first integral*, and a constant of motion may also be called a *constant of integral.*

When an integral of motion is given,

$$f_1\left(\mathbf{q}, \dot{\mathbf{q}}, t\right) = c_1 \tag{10.573}$$

we can substitute one of the equations of (10.568) with the first-order equation

$$\dot{q}_1 = f\left(c_1, q_i, \dot{q}_{i+1}, t\right) \qquad i = 1, 2, \ldots, n \tag{10.574}$$

and solve a set of $n - 1$ second-order and one first-order differential equations:

$$\begin{cases} \ddot{q}_{i+1} = F_{i+1}(q_i, \dot{q}_i, t) \\ \quad \dot{q}_1 = f(c_1, q_i, \dot{q}_{i+1}, t) \end{cases} \qquad i = 1, 2, \ldots, n. \tag{10.575}$$

If there exist $2n$ independent first integrals f_j, $j = 1, 2, \ldots, 2n$, then instead of solving n second-order equations of motion (10.568), we can solve a set of $2n$ algebraic equations

$$f_j(\mathbf{q}, \dot{\mathbf{q}}) = c_j(\mathbf{q}(t_0), \dot{\mathbf{q}}(t_0), t_0) \qquad j = 1, 2, \ldots, 2n \tag{10.576}$$

and determine the n generalized coordinates q_i, $i = 1, 2, \ldots, n$:

$$q_i = q_i(c_1, c_2, \ldots, c_{2n}, t) \qquad i = 1, 2, \ldots, n \tag{10.577}$$

Generally speaking, an integral of motion f is a function of generalized coordinates $\mathbf{q}$ and velocities $\dot{\mathbf{q}}$ such that its value remains constant. The value of an integral of motion is the constant of motion c that can be calculated by substituting the given value of variables $\mathbf{q}(t_0)$, $\dot{\mathbf{q}}(t_0)$ at the associated time t_0.

Breaking a constraint increases the degree of freedom and produces a new dynamic system with new equations of motion. However, when a dynamic system and its constraints are set up, the number of degrees of freedom and its generalized coordinates are fixed. If any conservation law is applied on the system, then there exist associated integrals of motion whether we are using or ignoring them. It is not possible to break an integral of motion if it exists. ■

Example 662 A Mass–Spring–Damper Vibrator Consider a mass m attached to a spring with stiffness k and a damper with damping c such as the one shown in Figure 10.15. The equation of motion of the system and its initial conditions are

$$m\ddot{x} + c\dot{x} + kx = 0 \tag{10.578}$$

$$x(0) = x_0 \qquad \dot{x}(0) = \dot{x}_0 \tag{10.579}$$

Its solution is

$$x = c_1 \exp(s_1 t) + c_2 \exp(s_2 t) \tag{10.580}$$

$$s_1 = \frac{c - \sqrt{c^2 - 4km}}{-2m} \qquad s_2 = \frac{c + \sqrt{c^2 - 4km}}{-2m} \tag{10.581}$$

Taking a time derivative, we find

$$\dot{x} = c_1 s_1 \exp(s_1 t) + c_2 s_2 \exp(s_2 t) \tag{10.582}$$

Using x and $\dot{x}$, we determine the integrals of motion f_1 and f_2:

$$f_1 = \frac{\dot{x} - x s_2}{(s_1 - s_2)\exp(s_1 t)} = c_1 \tag{10.583}$$

$$f_2 = \frac{\dot{x} - x s_1}{(s_2 - s_1)\exp(s_2 t)} = c_2 \tag{10.584}$$

Because the constants of the integral remain constant during the motion, we can calculate their value at any particular time, such as $t = 0$:

$$c_1 = \frac{\dot{x}_0 - x_0 s_2}{s_1 - s_2} \qquad c_2 = \frac{\dot{x}_0 - x_0 s_1}{s_2 - s_1} \tag{10.585}$$

Substituting s_1 and s_2 yields

$$c_1 = \frac{\sqrt{c^2 - 4km}\left(cx_0 + x_0\sqrt{c^2 - 4km} + 2m\dot{x}_0\right)}{2\left(c^2 - 4km\right)} \tag{10.586}$$

$$c_2 = \frac{\sqrt{c^2 - 4km}\left(cx_0 - x_0\sqrt{c^2 - 4km} + 2m\dot{x}_0\right)}{2\left(c^2 - 4km\right)} \tag{10.587}$$

Let us take time derivatives of the integrals of motion (10.583) and (10.584) to check if their derivatives are zero:

$$\begin{aligned}\frac{d}{dt}f_1 &= \frac{e^{-ts_1}}{s_1 - s_2}\left(\ddot{x} - \dot{x}\left(s_1 + s_2\right) + xs_1s_2\right)\\ &= \frac{\exp\left(\frac{t}{2m}\left(c - \sqrt{c^2 - 4km}\right)\right)}{\sqrt{c^2 - 4km}}\left(m\ddot{x} + c\dot{x} + kx\right) = 0\end{aligned} \tag{10.588}$$

$$\begin{aligned}\frac{d}{dt}f_2 &= -\frac{e^{-ts_2}}{s_1 - s_2}\left(\ddot{x} - \dot{x}\left(s_1 + s_2\right) + xs_1s_2\right)\\ &= -\frac{\exp\left(\frac{t}{2m}\left(c + \sqrt{c^2 - 4km}\right)\right)}{\sqrt{c^2 - 4km}}\left(m\ddot{x} + c\dot{x} + kx\right) = 0\end{aligned} \tag{10.589}$$

Example 663 ★ Exact Equation An exact equation $D^{(n)}(x)$ is the derivative of an equation of lower order given as

$$D^{(n)}(x) = \frac{d}{dt}\left(D^{(n-1)}(x)\right) \tag{10.590}$$

This equation is simplified by integrating with respect to t:

$$D^{(n-1)}(x) = c_1 \tag{10.591}$$

The nth-order ODE in $x(t)$ is given as

$$D^{(n)}(x) = a_0(t) + a_1(t)x + a_1(t)\dot{x} + \cdots + a_n(t)x^{(n)} \tag{10.592}$$

As an example, consider a particle with mass m that is under a force F:

$$F = -m(x + t\dot{x}) \tag{10.593}$$

The equation of motion of m is

$$\ddot{x} = -x - t\dot{x} \tag{10.594}$$

Example 664 ★ Constraints and Constants of Motion Relationship Consider a dynamic system with $N/3$ particles. The position and velocity of the ith particle are given as $\mathbf{r}_i$ and $\mathbf{v}_i$. Instead of showing the $N/3$ particles by $N/3$ vectors $\mathbf{r}_i$ in three-dimensional space, we may equivalently indicate the motion of the system by a point in an N-dimensional configuration space u_i, $i = 1, 2, \ldots, N$. A holonomic constraint is a function of coordinates u_i and time t, or equivalently a function of positions $\mathbf{r}_i$ and t:

$$f(u_i, t) = 0 \qquad i = 1, 2, \ldots, N \tag{10.595}$$

A differential constraint is a nonintegrable function of du_i and t, or equivalently a nonintegrable function of u_i, $\dot{u}_i$, and t:

$$f(du_i, dt) = 0 \qquad i = 1, 2, \ldots, N \tag{10.596}$$

$$f(u_i, \dot{u}_i, t) = 0 \qquad i = 1, 2, \ldots, N \tag{10.597}$$

A holonomic constraint is an equation among the coordinates u_i of a dynamic system. They must be considered when the dynamic system is being analyzed. If we ignore an existing holonomic constraint among the coordinates u_i, it will not generally be satisfied when the equations of the system are solved.

A nonholonomic constraint is an equation among the velocity components $\dot{u}_i$ of a dynamic system. They must be considered when the dynamic system is being analyzed. If we ignore an existing nonholonomic constraint among the velocity components $\dot{u}_i$, it will not generally be satisfied when the equations of the system are solved. A nonholonomic constraint (10.597) cannot be used to reduce the number of required coordinates and configuration degree of freedom f_C. However, a nonholonomic constraint reduces the state degree of freedom f_S of a dynamic system.

Let us define a $2n$-dimensional generalized state space S_G with coordinates q_i, $\dot{q}_i$, $i = 1, 2, \ldots, n$. This is a holonomic constraint-free space in which the generalized describing point of the system will trace the actual generalized state path of motion. Although there is no holonomic constraint in S_G, a nonholonomic constraint of the form (10.597) will survive and show itself by a new function of generalized coordinates and velocities:

$$f(q_i, \dot{q}_i, t) = 0 \qquad i = 1, 2, \ldots, n \tag{10.598}$$

A nonholonomic constraint among the generalized coordinates and velocities defines a surface in S_G-space in which the S_G-point must move. A nonholonomic constraint will not be satisfied automatically and we must consider its equation when we solve a dynamic system. Therefore, there exists an associated constraint force $\mathbf{F}_C$ in the generalized state space S_G to keep the describing point on the nonholonomic constraint surface:

$$\mathbf{F}_C = -\lambda\,\nabla f = -\lambda \sum_{i=1}^{n} \left(\frac{\partial f}{\partial q_i} \hat{q}_i + \frac{\partial f}{\partial \dot{q}_i} \widehat{\dot{q}}_i \right) \qquad \lambda \in \mathbb{R} \tag{10.599}$$

Although all partial derivatives of $\partial f / \partial q_i$ can be zero, at least one component of $\partial f / \partial \dot{q}_i$ is nonzero.

An integral of motion is a function of generalized coordinates and velocities q_i, $\dot{q}_i$,

$$f(q_i, \dot{q}_i, t) = c \qquad i = 1, 2, \ldots, n \tag{10.600}$$

where n is the number of DOF of the system. We can also interpret an integral of motion as a surface in S_G-space in which the S_G-point will move. An integral of motion will be satisfied automatically and we do not need to consider its equation when we solve a dynamic system. Therefore, there is no associated force in the generalized state space S_G to keep the describing point in the integral surface. The differential and gradient of an integral of motion must be zero:

$$\frac{\partial f}{\partial q_i}\dot{q}_i + \frac{\partial f}{\partial \dot{q}_i}\ddot{q}_i = 0 \qquad i = 1, 2, \ldots, n \tag{10.601}$$

$$\nabla f = \sum_{i=1}^{n}\left(\frac{\partial f}{\partial q_i}\hat{q}_i + \frac{\partial f}{\partial \dot{q}_i}\widehat{\dot{q}}_i\right) = 0 \tag{10.602}$$

We may use an S_G-surface to reduce the order of the system from $2n$ to $2n - 1$ and solve $2n - 1$ first-order differential equations instead of n second-order equations. However, if we ignore an existing integral of motion, it will automatically be satisfied when the equations of the system are solved. An S_G-surface is a function of both q_i and $\dot{q}_i$.

Example 665 ★ Constraint and First Integral of a Pendulum Figure 10.43(a) illustrates a planar pendulum that is analyzed in Example 132. The free-body diagram of Figure 10.43(b) provides two equations of motion:

$$m\ddot{x} = -T\frac{x}{l} \tag{10.603}$$

$$m\ddot{y} = -mg + T\frac{y}{l} \tag{10.604}$$

Eliminating the tension force T, we have one second-order equation of two variables:

$$\ddot{y}x + \ddot{x}y + gx = 0 \tag{10.605}$$

Because of the constant length of the connecting bar, we have the constraint equation between x and y

$$x^2 + y^2 - l^2 = 0 \tag{10.606}$$

Having one constraint in the two-dimensional configuration space (x, y) indicates that we can express the dynamic of the system by only one generalized coordinate. Choosing

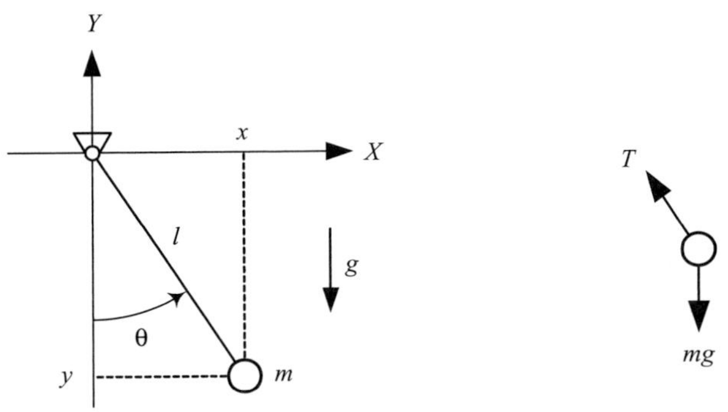

Figure 10.43 A planar pendulum.

θ as the generalized coordinate, we can express x and y by θ to express the equation of motion (10.605) as

$$\ddot{\theta} + \frac{g}{l} \sin\theta = 0 \tag{10.607}$$

Multiplying the equation by $\dot{\theta}$ and integrating yield the integral of energy:

$$f(\theta, \dot{\theta}) = \tfrac{1}{2}\dot{\theta}^2 - \frac{g}{l} \cos\theta = E \tag{10.608}$$

$$E = \tfrac{1}{2}\dot{\theta}_0^2 - \frac{g}{l} \cos\theta_0 \tag{10.609}$$

The integral of motion (10.608) is a first-order differential equation

$$\dot{\theta} = \sqrt{2E + 2\frac{g}{l} \cos\theta} \tag{10.610}$$

This equation expresses the dynamic of the pendulum upon solution.

Let us assume θ is too small such that we can approximate the equation of motion as

$$\ddot{\theta} + \frac{g}{l}\theta = 0 \tag{10.611}$$

The first integral of this equation is

$$f(\theta, \dot{\theta}) = \tfrac{1}{2}\dot{\theta}^2 - \frac{g}{l}\theta = E \tag{10.612}$$

$$E = \tfrac{1}{2}\dot{\theta}_0^2 - \frac{g}{l}\theta_0 \tag{10.613}$$

which provides a separated first-order differential equation

$$\dot{\theta} = \sqrt{2E + 2\frac{g}{l}\theta} \tag{10.614}$$

Its solution is

$$t = \int \frac{d\theta}{\sqrt{2E + 2\frac{g}{l}\theta}} = \sqrt{2}\frac{l}{g}\sqrt{\frac{g}{l}\theta + E} - p \tag{10.615}$$

where p is the second constant of motion,

$$p = \frac{l}{g}\dot{\theta}_0 \tag{10.616}$$

Now, let us ignore the energy integral and solve the second-order equation of motion (10.611):

$$\theta = c_1 \cos\sqrt{\frac{g}{l}}t + c_2 \sin\sqrt{\frac{g}{l}}t \tag{10.617}$$

A time derivative of the solution

$$\sqrt{\frac{l}{g}}\dot{\theta} = -c_1 \sin\sqrt{\frac{g}{l}}t + c_2 \cos\sqrt{\frac{g}{l}}t \tag{10.618}$$

can be used to determine the integrals and constants of motion:

$$f_1 = \theta \cos \sqrt{\frac{g}{l}} t - \sqrt{\frac{l}{g}} \dot{\theta} \sin \sqrt{\frac{g}{l}} t \tag{10.619}$$

$$f_2 = \theta \sin \sqrt{\frac{g}{l}} t + \sqrt{\frac{l}{g}} \dot{\theta} \cos \sqrt{\frac{g}{l}} t \tag{10.620}$$

Using the initial conditions $\theta(0) = \theta_0$, $\dot{\theta}(0) = \dot{\theta}_0$, we have

$$c_1 = \theta_0 \qquad c_2 = \sqrt{\frac{l}{g}} \dot{\theta}_0 \tag{10.621}$$

A second-order equation has only two constants of integrals. Therefore, we should be able to express E and p in terms of c_1 and c_2 or vice versa:

$$E = \frac{1}{2} \dot{\theta}_0^2 - \frac{g}{l} \theta_0 = \frac{1}{2} \frac{g}{l} c_2^2 - \frac{g}{l} c_1 \tag{10.622}$$

$$p = \frac{l}{g} \dot{\theta}_0 = \frac{l}{g} \sqrt{\frac{g}{l}} c_2 \tag{10.623}$$

$$c_2 = \sqrt{\frac{l}{g}} \dot{\theta}_0 = \frac{g}{l} \sqrt{\frac{l}{g}} p \tag{10.624}$$

$$c_1 = \theta_0 = \frac{1}{2} \frac{g}{l} p^2 - \frac{l}{g} E \tag{10.625}$$

where E is the mechanical energy of the pendulum and p is proportional to its moment of momentum.

Example 666 A Hopping Machine Figure 10.44 illustrates a hopping machine in the flight phase. It consists of a body with mass m_1 and mass moment I about its mass center and an extendable leg. There are two actuators on this machine: a motor to turn the leg relative to the body and an actuator to extend and contract the leg. The local configuration of the system is determined by θ, ϕ, and l.

Because there is no external moment on the system while in space, the angular momentum of the system is conserved. Assuming that the mass of the leg is concentrated at its foot, the angular momentum conservation equation is

$$L = I\dot{\theta} + m_2 (d + l)^2 \left(\dot{\theta} + \dot{\varphi}\right) \tag{10.626}$$

We may use this first integral to reduce the order of equations of motion.

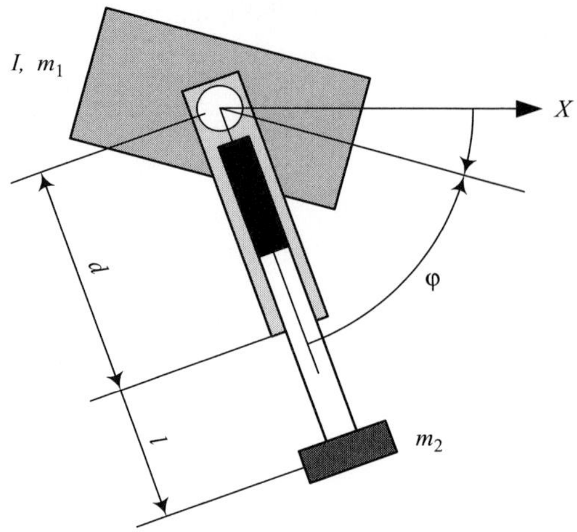

Figure 10.44 A hopping machine.

Example 667 ★ Homogeneity of Time, Integral of Energy Consider a dynamic system with kinetic and potential energies K and V:

$$K = K(\mathbf{q}, \dot{\mathbf{q}}, t) \tag{10.627}$$

$$V = V(\mathbf{q}) \tag{10.628}$$

The difference between these energies is called the *Lagrangian* or *kinetic potential* of the system and is shown by $\mathcal{L}$:

$$\mathcal{L} = K - V = \mathcal{L}(\mathbf{q}, \dot{\mathbf{q}}, t) \tag{10.629}$$

Having the Lagrangian is equivalent to knowing the system. We are able to find the equations of motion of the system from its Lagrangian:

$$\frac{d}{dt}\frac{\partial \mathcal{L}}{\partial \dot{q}_i} - \frac{\partial \mathcal{L}}{\partial q_i} = 0 \qquad i = 1, 2, \ldots, n \tag{10.630}$$

If the Lagrangian of a system does not depend explicitly on time, then its total time derivative is zero. Not having t in $\mathcal{L}$ is equivalent to homogeneity of time, which indicates that the origin of the time axis and the scale of a unit time step are arbitrary:

$$\frac{d\mathcal{L}}{dt} = \sum_{i=1}^{n} \frac{\partial \mathcal{L}}{\partial q_i}\dot{q}_i + \sum_{i=1}^{n} \frac{\partial \mathcal{L}}{\partial \dot{q}_i}\ddot{q}_i = 0 \tag{10.631}$$

Let us substitute $\partial \mathcal{L}/\partial q_i$ from the Lagrange equation (10.630) to obtain

$$\frac{d\mathcal{L}}{dt} = \dot{q}_i \sum_{i=1}^{n} \frac{d}{dt}\left(\frac{\partial \mathcal{L}}{\partial \dot{q}_i}\right) + \sum_{i=1}^{n} \frac{\partial \mathcal{L}}{\partial \dot{q}_i}\ddot{q}_i = \sum_{i=1}^{n} \frac{d}{dt}\left(\dot{q}_i \frac{\partial \mathcal{L}}{\partial \dot{q}_i}\right) \tag{10.632}$$

or

$$\frac{d}{dt}\left(\sum_{i=1}^{n}\dot{q}_i\frac{\partial\mathcal{L}}{\partial\dot{q}_i}-\mathcal{L}\right)=0 \tag{10.633}$$

Therefore,

$$\sum_{i=1}^{n}\dot{q}_i\frac{\partial\mathcal{L}}{\partial\dot{q}_i}-\mathcal{L}=E \tag{10.634}$$

where E is a constant of motion and f_1 is an integral of motion:

$$f_1=\sum_{i=1}^{n}\dot{q}_i\frac{\partial\mathcal{L}}{\partial\dot{q}_i}-\mathcal{L} \tag{10.635}$$

We may use

$$\sum_{i=1}^{n}\dot{q}_i\frac{\partial\mathcal{L}}{\partial\dot{q}_i}=\sum_{i=1}^{n}\dot{q}_i\frac{\partial K}{\partial\dot{q}_i}=2K \tag{10.636}$$

to show that E is the mechanical energy of the system:

$$f_1=2K-(K-V)=K+V=E \tag{10.637}$$

This is equivalent to the principle of conservation of energy in (2.371).

Example 668 ★ Homogeneity of Space, Integral of Momentum Homogeneity of space indicates that we may translate the coordinate system parallel to itself without changing the dynamic properties of the system. This translation may be a constant displacement or a uniform displacement with a constant velocity.

Let us indicate the Lagrangian $\mathcal{L}$ by position and velocity vectors $\mathbf{r}$, $\mathbf{v}$:

$$\mathcal{L}=\mathcal{L}(\mathbf{r}_i,\mathbf{v}_i,t) \tag{10.638}$$

$$\frac{d}{dt}\frac{\partial\mathcal{L}}{\partial\mathbf{v}_i}-\frac{\partial\mathcal{L}}{\partial\mathbf{r}_i}=0 \qquad i=1,2,\ldots,\frac{N}{3} \tag{10.639}$$

A dynamic system is independent of position $\mathbf{r}_i$ if its Lagrangian $\mathcal{L}$ does not depend explicitly on $\mathbf{r}_i$. So, we must have

$$\sum_{i=1}^{N/3}\frac{\partial\mathcal{L}}{\partial\mathbf{r}_i}=0 \tag{10.640}$$

From the Lagrange equation (10.639) we have

$$\sum_{i=1}^{N/3}\frac{d}{dt}\frac{\partial\mathcal{L}}{\partial\mathbf{v}_i}=\frac{d}{dt}\sum_{i=1}^{N/3}\frac{\partial\mathcal{L}}{\partial\mathbf{v}_i}=0 \tag{10.641}$$

Therefore,

$$\sum_{i=1}^{N/3}\frac{\partial\mathcal{L}}{\partial\mathbf{v}_i}=\mathbf{p} \tag{10.642}$$

where $\mathbf{p}$ is a constant of motion and f_2 is an integral of motion:

$$f_2 = \sum_{i=1}^{N/3} \frac{\partial \mathcal{L}}{\partial \mathbf{v}_i} \tag{10.643}$$

Employing the kinetic energy of the system, K, we have

$$K = \frac{1}{2} \sum_{i=1}^{N/3} m_i \mathbf{v}_i^2 \tag{10.644}$$

and knowing that

$$\frac{\partial \mathcal{L}}{\partial \mathbf{v}_i} = \frac{\partial K}{\partial \mathbf{v}_i} \tag{10.645}$$

we can write $\mathbf{p}$ as

$$\mathbf{p} = \sum_{i=1}^{N/3} m_i \mathbf{v}_i \tag{10.646}$$

where $\mathbf{p}$ is the translational momentum of the system.

Not having $\mathbf{r}_i$ in the Lagrangian $\mathcal{L}$ means that the kinetic energy K does not depend on $\mathbf{r}_i$ and we have no variation in V. The potential energy V is generally a function of $\mathbf{r}_i$, where its gradient indicates the force field $\mathbf{F} = -\nabla V(\mathbf{r}_i)$:

$$\frac{\partial \mathcal{L}}{\partial \mathbf{r}_i} = -\frac{\partial V}{\partial \mathbf{r}_i} = \mathbf{F}_i \tag{10.647}$$

Therefore, Equation (10.640) signifies that the sum of the forces on all particles of the system is zero:

$$\sum_{i=1}^{N/3} \mathbf{F}_i = 0 \tag{10.648}$$

The three components of $\mathbf{p}$ may all conserve only in the absence of an external force field in the associated direction.

When the Lagrangian $\mathcal{L}$ is expressed by the generalized coordinates q_i, the derivative of the Lagrangian with respect to generalized velocity $\dot{q}_i$ is called the *generalized momentum* p_i,

$$p_i = \frac{\partial \mathcal{L}}{\partial \dot{q}_i} \tag{10.649}$$

and its derivative with respect to the generalized coordinate q_i is called the *generalized force* Q_i,

$$Q_i = \frac{\partial \mathcal{L}}{\partial q_i} \tag{10.650}$$

Using these notations, the Lagrange equation (10.630) reduces to

$$Q_i = \dot{p}_i \tag{10.651}$$

Example 669 ★ Isotropy of Space, Integral of Moment of Momentum Isotropy of space indicates that we may rotate the coordinate system without changing the dynamic properties of the system.

Let us indicate the Lagrangian $\mathcal{L}$ by position and velocity vectors $\mathbf{r}$, $\mathbf{v}$:

$$\mathcal{L} = \mathcal{L}(\mathbf{r}_i, \mathbf{v}_i, t) \tag{10.652}$$

$$\frac{d}{dt}\frac{\partial \mathcal{L}}{\partial \mathbf{v}_i} - \frac{\partial \mathcal{L}}{\partial \mathbf{r}_i} = 0 \qquad i = 1, 2, \ldots, n \tag{10.653}$$

Virtual rotation $\delta\boldsymbol{\phi}$ provides virtual displacement and velocity as

$$\delta \mathbf{r}_i = \delta\boldsymbol{\phi} \times \mathbf{r}_i \tag{10.654}$$

$$\delta \mathbf{v}_i = \delta\boldsymbol{\phi} \times \mathbf{v}_i \tag{10.655}$$

If the space is isotropic, $\mathcal{L}$ must be independent of the rotation $\delta\boldsymbol{\phi}$:

$$\delta\mathcal{L} = \sum_{i=1}^{N/3} \frac{\partial \mathcal{L}}{\partial \mathbf{r}_i} \cdot \delta \mathbf{r}_i + \sum_{i=1}^{N/3} \frac{\partial \mathcal{L}}{\partial \mathbf{v}_i} \cdot \delta \mathbf{v}_i = 0 \tag{10.656}$$

Substituting $\mathbf{p}_i$ for $\partial\mathcal{L}/\partial\mathbf{v}_i$ and $\dot{\mathbf{p}}_i$ for $\partial\mathcal{L}/\partial\mathbf{r}_i$, we have

$$\sum_{i=1}^{N/3} \left(\dot{\mathbf{p}}_i \cdot \delta\boldsymbol{\phi} \times \mathbf{r}_i + \mathbf{p}_i \cdot \delta\boldsymbol{\phi} \times \mathbf{v}_i\right) = 0 \tag{10.657}$$

Taking advantage of the scalar triple-product equation (1.111) and factoring $\delta\boldsymbol{\phi}$ show that

$$\delta\boldsymbol{\phi} \cdot \sum_{i=1}^{N/3} \left(\mathbf{r}_i \times \dot{\mathbf{p}}_i + \mathbf{v}_i \times \mathbf{p}_i\right) = \delta\boldsymbol{\phi} \cdot \frac{d}{dt} \sum_{i=1}^{N/3} \mathbf{r}_i \times \mathbf{p}_i = 0 \tag{10.658}$$

A nonzero $\delta\boldsymbol{\phi}$ provides

$$\sum_{i=1}^{N/3} \mathbf{r}_i \times \mathbf{p}_i = \mathbf{L} \tag{10.659}$$

where $\mathbf{L}$ is a constant of motion and f_3 is an integral of motion,

$$f_3 = \sum_{i=1}^{N/3} \mathbf{r}_i \times \mathbf{p}_i \tag{10.660}$$

and $\mathbf{L}$ is called the rotational momentum or moment of momentum of the dynamic system.

Example 670 Satellite Orientation Control The planar model of a simplified satellite is illustrated in Figure 10.45. Two similar massless arms with tip mass points m are attached to the main body, which has a mass m_1 and mass moment I. When the satellite is free floating, its angular momentum will be conserved. So, moving the arms causes the main body to rotate.

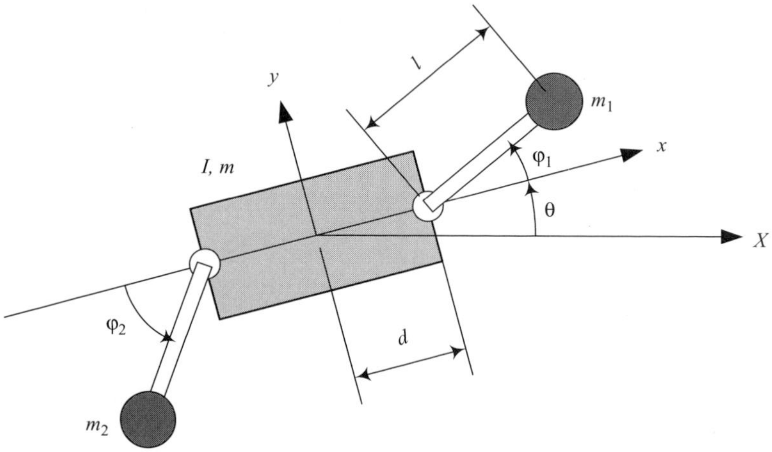

Figure 10.45 A planar satellite with two orientation control arms.

Let us indicate the local coordinates of the masses by (x_1, y_1) and (x_2, y_2) and their angular positions with respect to the body by φ_1 and φ_2. The velocity of the satellite mass center is $\mathbf{v}$ and the global angular position of the satellite is measured by θ. The kinetic energy of the system is

$$\begin{aligned} K &= \tfrac{1}{2}(m + m_1 + m_2)v^2 + \tfrac{1}{2}I\dot{\theta}^2 \\ &\quad + \tfrac{1}{2}m_1(\dot{X}_1^2 + \dot{Y}_1^2) + \tfrac{1}{2}m_1\left(\dot{X}_2^2 + \dot{Y}_2^2\right) \end{aligned} \tag{10.661}$$

where

$$\begin{aligned} X_1 &= d\cos\theta + l\cos(\theta + \varphi_1) \\ Y_1 &= d\sin\theta + l\sin(\theta + \varphi_1) \end{aligned} \tag{10.662}$$

$$\begin{aligned} X_2 &= -d\cos\theta - l\cos(\theta + \varphi_2) \\ Y_2 &= -d\sin\theta - l\sin(\theta + \varphi_2) \end{aligned} \tag{10.663}$$

There is no change in potential energy and therefore $K = \mathcal{L}$. Because there is no θ in the Lagrangian of the system, $L = \partial\mathcal{L}/\partial\dot{\theta}$ is a constant of motion,

$$L = \frac{\partial\mathcal{L}}{\partial\dot{\theta}} = \frac{\partial K}{\partial\dot{\theta}} = \text{const} \tag{10.664}$$

Example 671 ★ Integrals of Motion of Spherical Pendulums A particle of mass m moves under the action of gravity on a smooth sphere of radius R as shown in Figure 10.46.

This dynamic system has two DOF. We take the angles φ and θ as the generalized coordinates. A spherical pendulum has two integrals of motion, namely angular momentum and energy:

$$\tfrac{1}{2}R\left(\dot{\theta}^2 + \dot{\varphi}^2\sin^2\theta\right) - g\cos\theta = E \tag{10.665}$$

$$\dot{\varphi}\sin^2\theta = L \tag{10.666}$$

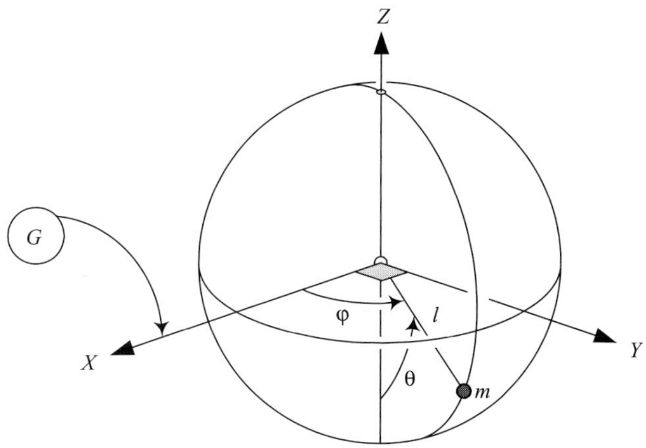

Figure 10.46 A particle moves under the action of gravity on a smooth sphere.

Having two integrals of motion provides two first-order differential equations to determine the time behavior of the generalized coordinates, instead of solving two second-order equations of motion. Eliminating $\dot{\varphi}$ causes the first equations to decouples:

$$\dot{\theta}^2 = 2\frac{E}{R} + 2\frac{g}{R}\cos\theta - \frac{L^2}{\sin^2\theta} \tag{10.667}$$

Employing a new variable

$$z = \cos\theta \tag{10.668}$$

the equation simplifies to

$$\dot{z}^2 = f(z) \tag{10.669}$$

$$f(z) = \frac{2}{R}\left(1 - z^2\right)(E + gz) - L^2 \tag{10.670}$$

This equation is of the form (2.243) and can be solved for $z = z(t)$ by Jacobi elliptic functions.

To determine the path of m on the sphere, we also need to find the time behavior of φ:

$$d\varphi = \frac{L}{1 - z^2}dt = \frac{L}{\left(1 - z^2\right)\sqrt{f(z)}}dz \tag{10.671}$$

Example 672 ★ Integrals of Motion of a Top An axisymmetric rigid body that is rotating in a constant gravitational field about a fixed point on its axis of symmetry is called a top. The fixed point is at a distance l from the body mass center C. Figure 10.47 illustrates a top with a body coordinate frame B and a global frame G at the fixed point O.

We will use Euler angles φ, θ, ψ as the generalized coordinates to describe the motion of B in G. The rotation of the top about its axis of symmetry is shown by ψ. The angle between the axis of symmetry and the Z-axis is shown by θ, and rotation of

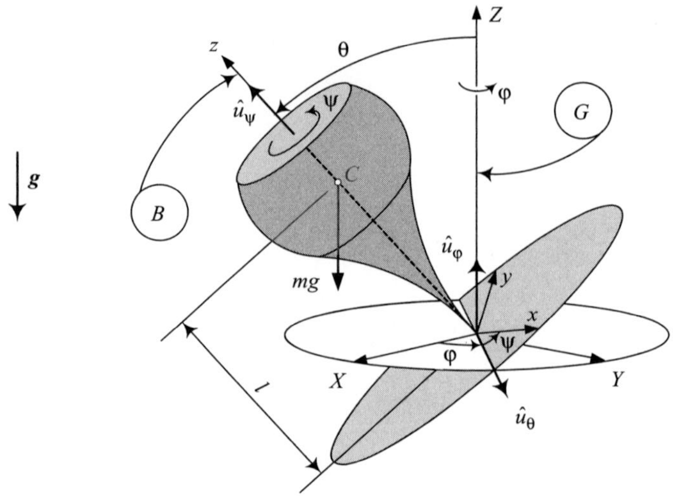

Figure 10.47 A top.

the axis of symmetry about the Z-axis is shown by φ. The moment caused by gravity is

$$
\begin{aligned}
{}^B\mathbf{M} = {}^B\mathbf{r}_C \times m\,{}^B\mathbf{g} &= \begin{bmatrix} 0 \\ 0 \\ l \end{bmatrix} \times mg\,{}^B R_G \begin{bmatrix} 0 \\ 0 \\ -1 \end{bmatrix} \\
&= mgl \begin{bmatrix} \cos\psi \sin\theta \\ -\sin\psi \sin\theta \\ 0 \end{bmatrix}
\end{aligned} \tag{10.672}
$$

where ${}^B R_G$ is given in (4.144).

The rotational equations of motion of a rigid body with a fixed point in a principal body frame are

$$M_x = I_x \dot{\omega}_x - \left(I_y - I_z\right) \omega_y \omega_z \tag{10.673}$$

$$M_y = I_y \dot{\omega}_y - \left(I_z - I_x\right) \omega_z \omega_x \tag{10.674}$$

$$M_z = I_z \dot{\omega}_z - \left(I_x - I_y\right) \omega_x \omega_y \tag{10.675}$$

where I_x, I_y, I_z are the elements of the mass moment matrix $[I]$, which for the axisymmetric top is

$$
{}^B I = \begin{bmatrix} I_x & 0 & 0 \\ 0 & I_x & 0 \\ 0 & 0 & I_z \end{bmatrix} \tag{10.676}
$$

Substituting $[I]$ and ${}^B\mathbf{M}$ provides

$$
mgl \begin{bmatrix} \cos\psi \sin\theta \\ -\sin\psi \sin\theta \\ 0 \end{bmatrix} = \begin{bmatrix} I_x \dot{\omega}_x - \left(I_x - I_z\right) \omega_y \omega_z \\ I_y \dot{\omega}_y - \left(I_z - I_x\right) \omega_z \omega_x \\ I_z \dot{\omega}_z \end{bmatrix} \tag{10.677}
$$

We use Equation (9.81) and replace the components of angular velocity by Euler angles and frequencies. Solving the first two equations for $\dot{\varphi}$ and $\dot{\theta}$ and simplifying, we obtain three differential equations for the Euler angles:

$$I_x\ddot{\theta} + \left(I_z\left(\dot{\psi} + \dot{\varphi}\cos\theta\right) - I_x\dot{\varphi}\cos\theta\right)\dot{\varphi}\sin\theta - mgl\sin\theta = 0 \tag{10.678}$$

$$I_x\ddot{\varphi}\sin\theta + 2I_x\dot{\theta}\dot{\varphi}\cos\theta - I_z\dot{\theta}\left(\dot{\psi} + \dot{\varphi}\cos\theta\right) = 0 \tag{10.679}$$

$$I_z\frac{d}{dt}\left(\dot{\psi} + \dot{\varphi}\cos\theta\right) = 0 \tag{10.680}$$

These three equations provide three integrals of motion. The first integral is the third equation:

$$\dot{\psi} + \dot{\varphi}\cos\theta = \omega_z = \text{const} \tag{10.681}$$

Using this integral of motion, the first and second equations of motion become

$$I_x\ddot{\theta} + \left(I_z\omega_z - I_x\dot{\varphi}\cos\theta\right)\dot{\varphi}\sin\theta - mgl\sin\theta = 0 \tag{10.682}$$

$$I_x\ddot{\varphi}\sin\theta + 2I_x\dot{\theta}\dot{\varphi}\cos\theta - I_z\dot{\theta}\omega_z = 0 \tag{10.683}$$

Multiplying the second equation by $\sin\theta$ provides the second integral of motion:

$$\frac{d}{dt}\left(I_x\dot{\varphi}\sin^2\theta + I_z\omega_z\cos\theta\right) \tag{10.684}$$

The third integral will be obtained by multiplying Equation (12.607) by $\dot{\theta}$ and (12.608) by $\dot{\varphi}\sin\theta$ and adding them:

$$\frac{d}{dt}\left(I_x\left(\dot{\varphi}^2\sin^2\theta + \dot{\theta}^2\right) + 2mgl\cos\theta\right) = 0 \tag{10.685}$$

We must also be able to find the integrals of motion from conservation principles. Gravity does not produce any moment about the z-axis and the Z-axis, and therefore these components of angular momentum conserves:

$$L_z = I_z\omega_z = I_z\left(\dot{\psi} + \dot{\varphi}\cos\theta\right) = \text{const} \tag{10.686}$$

$$\begin{aligned} L_Z &= \omega_z I_z\cos\theta + \omega_y I_x\cos\psi\sin\theta + \omega_x I_x\sin\theta\sin\psi \\ &= \omega_z I_z\cos\theta + \dot{\varphi}I_x\sin^2\theta \end{aligned} \tag{10.687}$$

where L_Z is the third component of ${}^G\mathbf{L}$ after substitution for components of ${}_G^B\boldsymbol{\omega}_B$:

$${}^G\mathbf{L} = {}^BR_G^{\mathrm{T}}\begin{bmatrix} I_x\omega_x \\ I_x\omega_y \\ I_z\omega_z \end{bmatrix} \tag{10.688}$$

The mechanical energy E of the top conserves:

$$E = mgl\cos\theta + \tfrac{1}{2}I_x\left(\omega_x^2 + \omega_y^2\right) + I_z\omega_z^2 \tag{10.689}$$

Therefore,

$$I_x\left(\dot{\varphi}^2\sin^2\theta + \dot{\theta}^2\right) + 2mgl\cos\theta = 2E - I_z\omega_z^2 \tag{10.690}$$

Example 673 ★ The n-Body Problem The problem of how n celestial bodies move under Newtonian gravitational forces kept Newton (1643–1727) busy for almost the last three decades of his life. The first complete mathematical formulation of this problem appeared in *Newton's Principia* (1687). The physical problem may be informally stated as follows: Given only the present positions and velocities of a group of celestial bodies, predict their motion for all future and past times.

More precisely, consider n point masses $m_1, m_2, \ldots, m_n$ in Euclidean three-dimensional space. Suppose that the force of attraction experienced between each pair of particles is Newtonian. If the initial positions and velocities are specified for every particle at some present time t_0, determine the position and velocity of each particle at every future (or past) moment of time. In mathematical terms, this means finding a global solution of the initial-value problem for the differential equations that describe the n-body problem:

$$m_i \ddot{\mathbf{r}}_i = -\frac{\partial V}{\partial \mathbf{r}_i} \qquad i = 1, 2, \ldots, n \tag{10.691}$$

$$V = G \sum_{j=1}^{n} \frac{m_i m_j}{\left|\mathbf{r}_i - \mathbf{r}_j\right|} \qquad i = 1, 2, \ldots, n \tag{10.692}$$

The best way to extract information about the n-body problem is to use the integrals of motion or conservation laws. The number of first integrals depends not on n but on the nature of forces. Therefore, the results obtained using conservation laws for $n = 2$ or $n = 3$ are valid for any n.

Because no external force is acting on the system, the center of mass C will move in a straight line with constant velocity:

$$\sum_{i=1}^{n} m_i \mathbf{v}_i = \mathbf{c}_1 \tag{10.693}$$

$$\sum_{i=1}^{n} m_i \mathbf{r}_i = \mathbf{c}_1 t + \mathbf{c}_2 \tag{10.694}$$

These two vector equations provide six constants of motion that are the components of $\mathbf{c}_1$ and $\mathbf{c}_2$:

$$\mathbf{c}_1 = \sum_{i=1}^{n} m_i \mathbf{v}_i(0) \tag{10.695}$$

$$\mathbf{c}_2 = \sum_{i=1}^{n} m_i \mathbf{r}_i(0) - \frac{1}{t} \sum_{i=1}^{n} m_i \mathbf{v}_i(0) \tag{10.696}$$

Because no external moment is acting on the system, the rotational momentum is conserved:

$$\sum_{i=1}^{n} \mathbf{r}_i \times m_i \mathbf{v}_i = \mathbf{c}_3 \tag{10.697}$$

which provides three more constants of integration:

$$\mathbf{c}_3 = \sum_{i=1}^{n} \mathbf{r}_i(0) \times m_i \mathbf{v}_i(0) \tag{10.698}$$

Because no time t is in the equations of motion, the energy of the system is conserved:

$$\frac{1}{2}\sum_{i=1}^{n} m_i \mathbf{v}_i + V = c_4 \tag{10.699}$$

which provides one more constant of integration. To show (10.699), we may multiply (10.691) by $\dot{\mathbf{r}}_i = \mathbf{v}_i$, add them, and integrate:

$$\begin{aligned}\sum_{i=1}^{n} m_i \int \ddot{\mathbf{r}}_i \cdot \dot{\mathbf{r}}_i \, dt + \sum_{i=1}^{n} \int \frac{\partial V}{\partial \mathbf{r}_i} \cdot \dot{\mathbf{r}}_i \, dt \\ = \sum_{i=1}^{n} m_i \int \ddot{\mathbf{r}}_i \cdot \dot{\mathbf{r}}_i \, dt + \sum_{i=1}^{n} \int \frac{\partial V}{\partial \mathbf{r}_i} \cdot d\mathbf{r} = c_4\end{aligned} \tag{10.700}$$

In the best case, an isolated dynamic system has only 10 contents of motion.

Example 674 The Central-Force Motion Consider the motion of a single particle which is acted upon by a force directed toward the origin of the global coordinate frame $G(O, x, x, z)$. Such a force is called the *central force*. The magnitude of a central force is assumed to be dependent only on the distance between the particle and the origin. Let us assume that the initial velocity of the particle is in the (x, y)-plane. Because the force always acts toward the origin, the force is also in the (x, y)-plane, and this means the particle will always remain in the (x, y)-plane.

The polar or cylindrical coordinate system (ρ, θ) is the best to describe the equations of motion of the particle. The force $\mathbf{F}$ on the particle is given as

$$\mathbf{F} = f(\rho)\hat{u}_\rho \tag{10.701}$$

The acceleration in polar coordinates is

$$\mathbf{a} = \left(\ddot{\rho} - \rho\dot{\theta}^2\right)\hat{u}_\rho + \left(2\dot{\rho}\dot{\theta} + \rho\ddot{\theta}\right)\hat{u}_\theta \tag{10.702}$$

and therefore, the equations of motion of the particle are

$$m\left(\ddot{\rho} - \rho\dot{\theta}^2\right) = f(\rho) \tag{10.703}$$

$$2\dot{\rho}\dot{\theta} + \rho\ddot{\theta} = 0 \tag{10.704}$$

Given

$$\frac{d}{dt}\left(\rho^2\dot{\theta}\right) = 2\rho\dot{\rho}\dot{\theta} + \rho^2\ddot{\theta} = \rho\left(2\dot{\rho}\dot{\theta} + \rho\ddot{\theta}\right) \tag{10.705}$$

Equation (10.704) is an integral of motion and $\rho^2\dot{\theta}$ is a constant of motion. It is actually the moment of momentum L of the system:

$$\mathbf{L} = m\,(\mathbf{r} \times \mathbf{v}) = m\left[\rho\hat{u}_\rho \times \left(\dot{\rho}\hat{u}_\rho + \rho\dot{\theta}\hat{u}_\theta\right)\right] = m\rho^2\dot{\theta}\hat{u}_z \tag{10.706}$$

$$L = |\mathbf{L}| = m\rho^2\dot{\theta} \tag{10.707}$$

The constant of motion L proves that Kepler's second law is true for every central-force motion. The law states: The rate at which the position vector of the particle sweeps out is constant:

$$\dot{A} = \frac{1}{2}\rho^2\dot{\theta} = \frac{L}{2m} \tag{10.708}$$

The path of motion of the particle is called the orbit. The equation of the orbit of the particle is an expression of ρ as a function of θ. To obtain the orbit, we should use the integral of motion (10.707) and eliminate t between (10.707) and (10.703). From (10.707), we have

$$\dot{\theta} = \frac{L}{m\rho^2} \tag{10.709}$$

and therefore, we can develop the following derivatives:

$$\dot{\rho} = \frac{d\rho}{d\theta}\dot{\theta} = \frac{L}{m\rho^2}\frac{d\rho}{d\theta} \tag{10.710}$$

$$\ddot{\rho} = \frac{L}{m\rho^2}\frac{d}{d\theta}\left(\frac{d\rho}{dt}\right) = \frac{L}{m\rho^2}\frac{d}{d\theta}\left(\frac{L}{m\rho^2}\frac{d\rho}{d\theta}\right) \tag{10.711}$$

The equation of motion (10.703) will now become

$$\frac{L^2}{m}\left(\frac{1}{\rho^2}\frac{d}{d\theta}\left(\frac{1}{\rho^2}\frac{d\rho}{d\theta}\right) - \frac{1}{\rho^3}\right) = f(\rho) \tag{10.712}$$

By introducing a new variable $u = 1/\rho$,

$$\frac{1}{\rho^2}\frac{d\rho}{d\theta} = -\frac{du}{d\theta} \tag{10.713}$$

we can simplify the orbit differential equation:

$$-\frac{L^2u^2}{m}\left(\frac{d^2u}{d\theta^2} + u\right) = f(u) \tag{10.714}$$

This equation must be solved to determine the orbit $\rho = \rho(\theta)$ when f is given.

The gravitational force follows the inverse square law,

$$f = -\frac{k}{\rho^2} = -ku^2 \qquad k > 0 \tag{10.715}$$

and reduces Equation (10.714) to

$$\frac{d^2u}{d\theta^2} + u = \frac{mk}{L^2} \tag{10.716}$$

which has the solution

$$u = \frac{mk}{L^2} + C\cos(\theta - \theta_0) \tag{10.717}$$

where C and θ_0 are constants of the integral and may be calculated by initial conditions. The equation of the orbit is then

$$\frac{L^2/mk}{\rho} = 1 + \frac{CL^2}{mk}\cos(\theta - \theta_0) \tag{10.718}$$

This is a conic section with eccentricity $e = CL^2/(mk)$ and semi–latus rectum $L^2/(mk)$. The origin, which is the center of the force, is the focus of the conic section.

Introducing the potential function $V = k/\rho$ to have $\mathbf{F} = -\nabla V$, we can calculate the mechanical energy E of the particle by adding the kinetic energy K and potential energy $V = -k/\rho$:

$$E = K + V = \frac{1}{2}m\left(\dot{\rho}^2 + \rho^2\dot{\theta}^2\right) - \frac{k}{\rho} \tag{10.719}$$

We can eliminate $\dot{\theta}$ by the integral of the moment of momentum (10.707) and eliminate time by (10.710), which yields

$$E = \frac{L^2}{2m}\left[\left(\frac{1}{\rho^2}\frac{d\rho}{d\theta}\right)^2 + \frac{1}{\rho^2}\right] - \frac{k}{\rho} \tag{10.720}$$

or

$$E = \frac{L^2}{2m}\left[\left(\frac{d^2u}{d\theta^2}\right)^2 + u^2\right] - ku \tag{10.721}$$

Substituting (10.717) in (10.721) yields

$$E = \frac{CL^2}{2m} - \frac{mk^2}{2L^2} \tag{10.722}$$

and therefore,

$$C = \frac{1}{L}\sqrt{2mE + \frac{m^2k^2}{L^2}} \tag{10.723}$$

The eccentricity becomes

$$e = \frac{AL^2}{mk} = \sqrt{1 + \frac{2L^2E}{mk^2}} \tag{10.724}$$

It shows that there are four possible orbits:

1. If $E > 0$, then $e < 1$ and the orbit is a hyperbola.
2. If $E = 0$, then $e = 1$ and the orbit is a parabola.
3. If $-mk^2/(2L^2) < E < 0$, then $0 < e < 1$ and the orbit is an ellipse.
4. If $E = -mk^2/(2L^2)$, then $e = 1$ and the orbit is a circle.

10.11 ★ METHODS OF DYNAMICS

Besides the Newton–Euler method to find the equations of motion, there are few other applied methods. The principle of virtual work (10.302), D'Alembert principle (10.303) and (10.313), the fundamental equations of dynamics (10.303), (10.365), and (10.371) are the sources for all other methods to derive the differential equations of motion of dynamic systems. In this section, we review some of the methods of dynamics.

10.11.1 ★ Lagrange Method

The Lagrange equation is a superior method to derive the equations of motion of a dynamic system:

$$\frac{d}{dt}\frac{\partial K}{\partial \dot{q}_i} - \frac{\partial K}{\partial q_i} - Q_i = 0 \qquad i = 1, 2, \ldots, n \tag{10.725}$$

where K is the kinetic energy, Q_i is the generalized force, and q_i are the generalized coordinates of the system. The Lagrange method is based on the fundamental equation of dynamics.

Assuming a potential function V such that

$$\mathbf{F} = -\nabla V \tag{10.726}$$

the Lagrange equation of motion can be written as

$$\frac{d}{dt}\left(\frac{\partial \mathcal{L}}{\partial \dot{q}_r}\right) - \frac{\partial \mathcal{L}}{\partial q_r} = Q_r \qquad r = 1, 2, \ldots, n \tag{10.727}$$

where Q_r is the generalized nonpotential force and

$$\mathcal{L} = K - V \tag{10.728}$$

Proof: The fundamental equation of dynamics (10.303),

$$\sum_{i=1}^{N/3} \left(m_i \ddot{\mathbf{u}}_i - \mathbf{F}_{G_i}\right) \cdot \delta \mathbf{u}_i = 0 \tag{10.729}$$

indicates that there are $N/3$ number of coupled equations between configuration coordinates. It should be expressed by generalized coordinates to have a suitable form to derive the equations of motion of a system.

Recalling that configuration coordinates are functions of generalized coordinates,

$$\mathbf{u}_i = \mathbf{u}_i\left(q_1, q_2, \ldots, q_n, t\right) \tag{10.730}$$

we can express $\mathbf{u}_i$ and $\dot{\mathbf{u}}_i$ as

$$\delta \mathbf{u}_i = \sum_{j=1}^{n} \frac{\partial \mathbf{u}_i}{\partial q_j} \delta q_j \tag{10.731}$$

$$\dot{\mathbf{u}}_i = \frac{d\mathbf{u}_i}{dt} = \sum_{j=1}^{n} \frac{\partial \mathbf{u}_i}{\partial q_j} \delta \dot{q}_j + \frac{\partial \mathbf{u}_i}{\partial t} \tag{10.732}$$

and get

$$\frac{\delta \dot{\mathbf{u}}_i}{\delta \dot{q}_j} = \frac{\partial \mathbf{u}_i}{\partial q_j} \qquad \frac{\delta \dot{u}_k}{\delta \dot{q}_j} = \frac{\partial u_k}{\partial q_j} \tag{10.733}$$

Now, the first term of (10.729) is

$$\begin{aligned}
\sum_{i=1}^{N/3} m_i \ddot{\mathbf{u}}_i \cdot \delta \mathbf{u}_i &= \sum_{j=1}^{n} \sum_{i=1}^{N/3} m_i \ddot{\mathbf{u}}_i \cdot \frac{\partial \mathbf{u}_i}{\partial q_j} \delta q_j = \sum_{j=1}^{n} \delta q_j \sum_{i=1}^{N/3} m_i \ddot{\mathbf{u}}_i \cdot \frac{\partial \mathbf{u}_i}{\partial q_j} \\
&= \sum_{j=1}^{n} \delta q_j \sum_{i=1}^{N/3} \left[\frac{d}{dt} \left(m_i \dot{\mathbf{u}}_i \cdot \frac{\partial \mathbf{u}_i}{\partial q_j} \right) - m_i \dot{\mathbf{u}}_i \cdot \frac{d}{dt} \frac{\partial \mathbf{u}_i}{\partial q_j} \right] \\
&= \sum_{j=1}^{n} \delta q_j \sum_{i=1}^{N/3} \left[\frac{d}{dt} \left(m_i \dot{\mathbf{u}}_i \cdot \frac{\partial \dot{\mathbf{u}}_i}{\partial \dot{q}_j} \right) - m_i \dot{\mathbf{u}}_i \cdot \frac{\partial \dot{\mathbf{u}}_i}{\partial q_j} \right] \\
&= \sum_{j=1}^{n} \delta q_j \left(\frac{d}{dt} \frac{\partial}{\partial \dot{q}_j} \sum_{i=1}^{N/3} \frac{1}{2} m_i \dot{\mathbf{u}}_i^2 - \frac{\partial}{\partial q_j} \sum_{i=1}^{N/3} \frac{1}{2} m_i \dot{\mathbf{u}}_i^2 \right) \\
&= \sum_{j=1}^{n} \delta q_j \left(\frac{d}{dt} \frac{\partial K}{\partial \dot{q}_j} - \frac{\partial K}{\partial q_j} \right)
\end{aligned} \tag{10.734}$$

where

$$K = \sum_{i=1}^{N/3} \frac{1}{2} m_i \dot{\mathbf{u}}_i^2 \tag{10.735}$$

and the second term of (10.729) is:

$$\sum_{i=1}^{N/3} \mathbf{F}_{G_i} \cdot \delta \mathbf{u}_i = \sum_{j=1}^{n} \sum_{i=1}^{N/3} \mathbf{F}_{G_i} \cdot \frac{\partial \mathbf{u}_i}{\partial q_j} \delta q_j = \sum_{j=1}^{n} Q_j \delta q_j \tag{10.736}$$

where

$$Q_j = \sum_{i=1}^{N/3} \mathbf{F}_{G_i} \cdot \frac{\partial \mathbf{u}_i}{\partial q_j} \tag{10.737}$$

K is the kinetic energy of the system, and Q_j is the generalized force associated with coordinate q_j.

Therefore, the expression of the fundamental equation of dynamics in terms of generalized coordinates would be

$$\sum_{j=1}^{n} \left(\frac{d}{dt} \frac{\partial K}{\partial \dot{q}_j} - \frac{\partial K}{\partial q_j} - Q_j \right) \delta q_j = 0 \tag{10.738}$$

Because in general $\delta q_j \neq 0$, we have

$$\frac{d}{dt} \frac{\partial K}{\partial \dot{q}_j} - \frac{\partial K}{\partial q_j} - Q_j = 0 \qquad j = 1, 2, \ldots, n \tag{10.739}$$

This is the *Lagrange equation* and is an applied form to derive the equations of motion of a dynamic system.

If there exists a function $V = V(q_i)$ such that some of the forces are derived from the derivative of V,

$$\mathbf{F}_k = -\nabla V \qquad k = 1, 2, \ldots, n' \qquad n' < n \tag{10.740}$$

then we can write the Lagrange equation as

$$\frac{d}{dt}\frac{\partial K}{\partial \dot{q}_j} - \frac{\partial K}{\partial q_j} + \frac{\partial V}{\partial q_j} - Q_j = 0 \qquad j = 1, 2, \ldots, n \tag{10.741}$$

where Q_j only indicates the nonpotential generalized forces. If we define the Lagrangian as $\mathcal{L} = K - V$, then the Lagrange equation may be written as

$$\frac{d}{dt}\left(\frac{\partial \mathcal{L}}{\partial \dot{q}_j}\right) - \frac{\partial \mathcal{L}}{\partial q_j} = Q_j \qquad j = 1, 2, \ldots, n \tag{10.742}$$

■

Example 675 ★ Unconstrained Particle and Lagrange Method Consider a mass m in a three-dimensional space that is under a force $\mathbf{F}$:

$$\mathbf{F} = F_x\hat{\imath} + F_y\hat{\jmath} + F_z\hat{k} = F_\rho\hat{u}_\rho + F_\theta\hat{u}_\theta + F_z\hat{u}_z \tag{10.743}$$

Let us choose the cylindrical coordinate $(\rho, \theta, z) = (q_1, q_2, q_3)$ to express the position of m:

$$x = \rho\cos\theta \qquad y = \rho\sin\theta \qquad z = z \tag{10.744}$$

$$\mathbf{r} = x\hat{\imath} + y\hat{\jmath} + z\hat{k} = \rho\hat{u}_\rho + \dot{z}\hat{u}_z \tag{10.745}$$

To determine the equations of motion of m, let us calculate its kinetic energy:

$$K = \frac{1}{2}\left(\dot{x}^2 + \dot{y}^2 + \dot{z}^2\right) = \frac{1}{2}\left(\dot{\rho}^2 + \rho\dot{\theta}^2 + \dot{z}^2\right) \tag{10.746}$$

The generalized forces Q_i are

$$Q_1 = Q_\rho = \mathbf{F}\cdot\frac{\partial \mathbf{r}}{\partial \rho} = F_\rho \tag{10.747}$$

$$Q_2 = Q_\theta = \mathbf{F}\cdot\frac{\partial \mathbf{r}}{\partial \theta} = F_\theta \tag{10.748}$$

$$Q_3 = Q_z = \mathbf{F}\cdot\frac{\partial \mathbf{r}}{\partial z} = F_z \tag{10.749}$$

Substituting (10.746)–(10.749) into the Lagrange equation (10.739) provides the equations of motion

$$m\ddot{\rho} - mr\dot{\theta}^2 - F_\rho = 0 \tag{10.750}$$

$$m\rho^2\ddot{\theta} + 2m\rho\dot{\rho}\dot{\theta} - F_\theta = 0 \tag{10.751}$$

$$m\ddot{z} - F_z = 0 \tag{10.752}$$

10.11.2 ★ Gauss Method

Gauss introduced a function Z that is called the *constraint of motion*,

$$Z = \sum_{i=1}^{N/3} \frac{1}{2m_i} \left(m_i \ddot{\mathbf{u}}_i - \mathbf{F}_{G_i} \right)^2 \tag{10.753}$$

and showed that the equations of motion

$$m_i \ddot{\mathbf{u}}_i - \mathbf{F}_{G_i} = 0 \tag{10.754}$$

can be found by minimization of Z with respect to $\ddot{\mathbf{u}}_i$. The Gauss method is called the *principle of least constraint*.

Proof: Because the given forces $\mathbf{F}_{G_i}$ cannot be varied, we can rewrite the third form of the fundamental equation (10.371) as

$$\sum_{i=1}^{N/3} \left(m_i \ddot{\mathbf{u}}_i - \mathbf{F}_{G_i} \right) \cdot \delta \left(\frac{m_i \ddot{\mathbf{u}}_i - \mathbf{F}_{G_i}}{m_i} \right) = 0 \tag{10.755}$$

Simplifying this equation by using the equations of holonomic constraints to calculate $\ddot{\mathbf{u}}_i$ provides

$$\delta \sum_{i=1}^{N/3} \frac{1}{2m_i} \left(m_i \ddot{\mathbf{u}}_i - \mathbf{F}_{G_i} \right)^2 = 0 \tag{10.756}$$

Let us define a quantity Z, called the *constraint of motion*,

$$Z = \sum_{i=1}^{N/3} \frac{1}{2m_i} \left(m_i \ddot{\mathbf{u}}_i - \mathbf{F}_{G_i} \right)^2 \tag{10.757}$$

and call Equation (10.756) the *principle of least constraint*. It states: The actual motion in nature happens in such a way that the constraint of motion becomes minimum. So, the variation of Z with respect to $\ddot{\mathbf{u}}$ is zero for actual motion:

$$\frac{\partial Z}{\partial \ddot{\mathbf{u}}_i} = 0 \qquad i = 1, 2, \ldots, N/3 \tag{10.758}$$

When the particles are free from constraints, then Z becomes zero and we get the Newton equation of motion:

$$m_i \ddot{\mathbf{u}}_i - \mathbf{F}_{G_i} = 0 \tag{10.759}$$

The German mathematician Friedrich Gauss (1777–1855) discovered the principle of least constraints and introduced Z as the constraint of motion. So, we may also call Z a *Gauss function* and Equation (10.756) a *Gauss equation of motion*. ■

Example 676 ★ Application of Principle of Least Constraint Consider a particle under a force $\mathbf{F}$ and the constraint

$$z = f(x, y) \tag{10.760}$$

The constraint indicates that x and y are the only independent variables. An expansion shows that

$$\ddot{z} = \frac{\partial f}{\partial x}\ddot{x} + \frac{\partial f}{\partial y}\ddot{y} + \frac{\partial^2 f}{\partial x^2}\dot{x}^2 + \cdots \tag{10.761}$$

The constraint of motion that should be minimized is

$$Z = (F_x - m\ddot{x})^2 + \left(F_y - m\ddot{y}\right)^2 + (F_z - m\ddot{z})^2 \tag{10.762}$$

Substituting (10.761) into (10.762) yields

$$Z = (F_x - m\ddot{x})^2 + \left(F_y - m\ddot{y}\right)^2 + \left[F_z - m\left(\frac{\partial f}{\partial x}\ddot{x} + \frac{\partial f}{\partial y}\ddot{y}\right)\right]^2 \tag{10.763}$$

and minimizing Z provides the equations of motion

$$(F_x - m\ddot{x}) - (F_z - m\ddot{z})\frac{\partial f}{\partial x} = 0 \tag{10.764}$$

$$\left(F_y - m\ddot{y}\right) - (F_z - m\ddot{z})\frac{\partial f}{\partial y} = 0 \tag{10.765}$$

Example 677 ★ Atwood Machine and Least Constraint Two particles with masses m_1 and $m_2 > m_1$ are attached with a massless constant-length string passing over a massless pulley as shown in Figure 10.48(a). Using the principle of least constraint, we can find the equation of motion of the system.

Let us assume that the acceleration of m_1 is a. The constraint of motion that should be minimized is

$$Z = \tfrac{1}{2}\left[m_1 (a + g)^2 + m_2 (-a + g)^2\right] \tag{10.766}$$

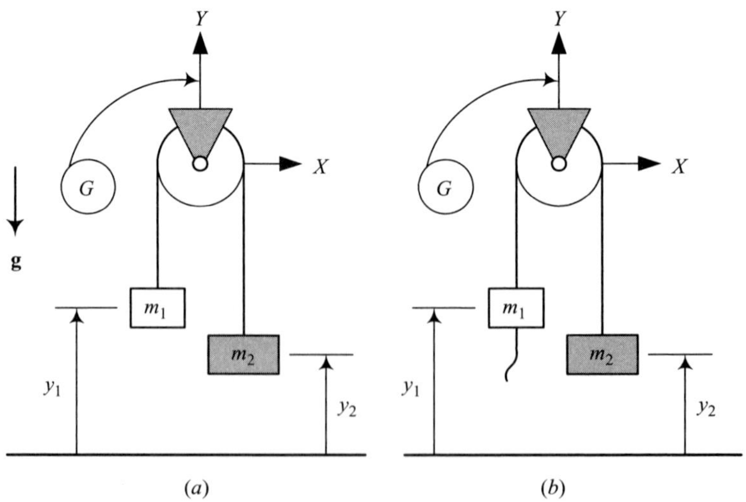

Figure 10.48 Atwood machine.

A derivative of Z gives

$$\frac{dZ}{dA} = am_1 - am_2 + gm_1 + gm_2 = 0 \tag{10.767}$$

and therefore,

$$a = \frac{m_2 - m_1}{m_2 + m_1} g \tag{10.768}$$

Now, let us assume that m_1 is a monkey that climbs a string as illustrated in Figure 10.48(b). The climbing of the monkey with respect to the string is given by a smooth function $f(t)$. If the heights of m_1 and m_2 from a fixed datum are y_1 and y_2, respectively, and the system starts moving from $y_1 = 0$, $y_2 = 0$, $f(0) = 0$, $\dot{f}(0) = 0$, then the constraint of motion is

$$Z = \tfrac{1}{2}\left[m_1(\ddot{y}_1 + g)^2 + m_2(\ddot{y}_2 + g)^2\right] \tag{10.769}$$

Using the conditions $y_1(0) = 0$, $y_2(0) = 0$, $y_1 = f - y_2$, we find

$$Z = \tfrac{1}{2}\left[m_1(\ddot{y}_1 + g)^2 + m_2(\ddot{f} - \ddot{y}_1 + g)^2\right] \tag{10.770}$$

To find $\ddot{y}_1$, we take a derivative to minimize Z:

$$\frac{dZ}{d\ddot{y}_1} = m_1\left(g + \ddot{y}_1\right) - m_2\left(\ddot{f} - \ddot{y}_1 + g\right) = 0 \tag{10.771}$$

Therefore,

$$\ddot{y}_1 = \frac{m_2\left(\ddot{f} + g\right) - gm_1}{m_1 + m_2} \tag{10.772}$$

which provides the position of the monkey upon integration:

$$(m_1 + m_2)y_1 = m_2 f + \tfrac{1}{2}g(m_2 - m_1)t^2 \tag{10.773}$$

If $m_1 = m_2$, we have $y_1 = y_2 = f/2$ and the monkey always remains at the same level as the counterweight.

To demonstrate uniformly accelerated motion due to gravity, the British scientist George Atwood (1745–1807) invented a multiple-pulley machine.

Example 678 ★ A Body Sliding on a Massive Wedge and Least Constraint Consider a body of mass m_1 sliding without friction on the inclined surface of a wedge of mass m_2 that can slide on a horizontal surface as shown in Figure 2.41. If acceleration of m_2 in the x-direction is $\ddot{x}_2$ and acceleration of m_1 with respect to the inclined surface of m_2 is a, then the constraint of motion is

$$Z = \tfrac{1}{2}m_2\ddot{x}_2^2 + \tfrac{1}{2}m_1\left[(-a\cos\theta - \ddot{x}_2)^2 + (-a\sin\theta + g)^2\right] \tag{10.774}$$

Minimizing Z with respect to a and $\ddot{x}_2$ leads to

$$\ddot{x}_2 = -\frac{m_1 \sin\theta\cos\theta}{m_2 + m_1\sin^2\theta} g \tag{10.775}$$

$$a = \frac{(m_1 + m_2)\sin\theta}{m_2 + m_1\sin^2\theta} g \tag{10.776}$$

which shows that the absolute acceleration of m_1 in the y-direction is

$$\ddot{y}_1 = -a \sin\theta = -\frac{(m_1 + m_2)\sin^2\theta}{m_2 + m_1 \sin^2\theta} g \tag{10.777}$$

10.11.3 ★ Hamilton Method

Consider a dynamic system with generalized coordinates q_i, Lagrangian $\mathcal{L} = \mathcal{L}(q_i, \dot{q}_i, t)$, and nonpotential generalized forces Q_i. The equations of motion of the system may be found by the *Hamilton equations*:

$$\frac{\partial H}{\partial q_j} = -\dot{p}_i + Q_i \tag{10.778}$$

$$\frac{\partial H}{\partial p_j} = \dot{q}_i \tag{10.779}$$

where H is the *Hamiltonian function* and p_i is the generalized momentum of the dynamic system:

$$H = H\left(q_j, p_j, t\right) = p_i \dot{q}_i - \mathcal{L} \tag{10.780}$$

$$p_i = \frac{\partial \mathcal{L}}{\partial \dot{q}_i} \tag{10.781}$$

Proof: Consider a dynamic system with the following Lagrangian $\mathcal{L}$, nonpotential generalized force Q_i, and generalized momentum p_i:

$$\mathcal{L} = \mathcal{L}(q_i, \dot{q}_i, t) \qquad i = 1, 2, \ldots, n \tag{10.782}$$

$$Q_i = \frac{d}{dt}\frac{\partial \mathcal{L}}{\partial \dot{q}_i} - \frac{\partial \mathcal{L}}{\partial q_i} \qquad i = 1, 2, \ldots, n \tag{10.783}$$

$$p_i = \frac{\partial \mathcal{L}}{\partial \dot{q}_i} \tag{10.784}$$

The independence of p_i provides the nonzero determinant

$$\left|\frac{\partial^2 \mathcal{L}}{\partial \dot{q}_i \, \partial \dot{q}_j}\right| \neq 0 \tag{10.785}$$

So, we can solve the n equations (10.784) to calculate the generalized velocities $\dot{q}_i$:

$$\dot{q}_i = \dot{q}_i\left(q_j, p_j, t\right) \qquad j = 1, 2, \ldots, n \tag{10.786}$$

Substituting (10.786) into the Hamilton function H,

$$H = p_i \dot{q}_i - \mathcal{L} \tag{10.787}$$

we find H in terms of q_j, p_j, t:

$$H = H\left(q_j, p_j, t\right) \tag{10.788}$$

A variation of H is

$$\delta H = \frac{\partial H}{\partial q_j}\delta q_j + \frac{\partial H}{\partial p_j}\delta p_j + \frac{\partial H}{\partial t}\delta t \tag{10.789}$$

We can also determine the variation of H from (10.787):

$$\begin{aligned}
\delta H &= \dot{q}_i \delta p_j + p_i \delta \dot{q}_i - \left(\frac{\partial \mathcal{L}}{\partial \dot{q}_i}\delta \dot{q}_i + \frac{\partial \mathcal{L}}{\partial q_i}\delta q_i + \frac{\partial \mathcal{L}}{\partial t}\delta t \right) \\
&= \dot{q}_i \delta p_j + p_i \delta \dot{q}_i - \left(p_i \delta \dot{q}_i + (\dot{p}_i - Q_i)\,\delta q_i + \frac{\partial \mathcal{L}}{\partial t}\delta t \right) \\
&= \dot{q}_i \delta p_j - (\dot{p}_i - Q_i)\,\delta q_i - \frac{\partial \mathcal{L}}{\partial t}\delta t
\end{aligned} \tag{10.790}$$

Equating (10.790) and (10.789) indicates that

$$\frac{\partial H}{\partial q_j} = -\dot{p}_i + Q_i \tag{10.791}$$

$$\frac{\partial H}{\partial p_j} = \dot{q}_i \tag{10.792}$$

$$\frac{\partial H}{\partial t} = -\frac{\partial \mathcal{L}}{\partial t} \tag{10.793}$$

The set of equations (10.791) and (10.792) are called the Hamiltonian equations of motion. Hamiltonian equations are also called canonical equations of motion.

If there is no nonpotential force,

$$Q_i = 0 \tag{10.794}$$

then Hamiltonian equations reduce to

$$\frac{\partial H}{\partial q_j} = -\dot{p}_i \tag{10.795}$$

$$\frac{\partial H}{\partial p_j} = \dot{q}_i \tag{10.796}$$

In Hamiltonian mechanics the generalized coordinate q and generalized momentum p are the variables of a dynamic system while in Lagrangian mechanics the generalized coordinate q and generalized velocity $\dot{q}$ are the variables of the dynamic system. ■

Example 679 ★ A Mass–Spring–Damper System Consider a mass–spring system with the following equation of motion:

$$m\ddot{x} + kx = 0 \tag{10.797}$$

The Lagrangian of this system is

$$\mathcal{L} = K - V = \tfrac{1}{2}m\dot{x}^2 - \tfrac{1}{2}kx^2 \tag{10.798}$$

Using the momentum of the system, $p = m\dot{x}$,

$$p = \frac{\partial \mathcal{L}}{\partial \dot{x}} = m\dot{x} \tag{10.799}$$

we can rewrite the Lagrangian as

$$\mathcal{L} = \frac{1}{2}\frac{p^2}{m} - \frac{1}{2}kx^2 \tag{10.800}$$

and define the Hamiltonian as

$$H = p\dot{x} - \mathcal{L} = \frac{p^2}{m} - \frac{1}{2}\frac{p^2}{m} + \frac{1}{2}kx^2 = \frac{1}{2}\frac{p^2}{m} + \frac{1}{2}kx^2 \tag{10.801}$$

Therefore, the Hamiltonian equations of motion are

$$\dot{x} = \frac{\partial H}{\partial p} = \frac{p}{m} \qquad \dot{p} = -\frac{\partial H}{\partial x} = -kx \tag{10.802}$$

Now consider a mass–spring–damper system with the following equation of motion:

$$m\ddot{x} + 2\xi\omega\dot{x} + \omega^2 x = 0 \tag{10.803}$$

where

$$\omega^2 = \frac{k}{m} \qquad \xi = \frac{c}{2m\omega} \tag{10.804}$$

We can define the Lagrangian as

$$\mathcal{L} = \tfrac{1}{2}\left(\dot{x}^2 - \omega^2 x^2\right) e^{2\xi\omega} \tag{10.805}$$

which provides the equation of motion (10.803) by applying the Lagrange equation (10.783).

The generalized momentum of the system

$$p = \frac{\partial \mathcal{L}}{\partial \dot{x}} = \dot{x}e^{2\xi\omega} \tag{10.806}$$

provides the generalized velocity

$$\dot{x} = pe^{-2\xi\omega} \tag{10.807}$$

Therefore, the Hamiltonian of the system is

$$H = p\dot{x} - \mathcal{L} = \tfrac{1}{2}p^2 e^{-2\xi\omega} + \tfrac{1}{2}\omega^2 x^2 e^{2\xi\omega} \tag{10.808}$$

which provides the following canonical equations of motion:

$$\dot{x} = \frac{\partial H}{\partial p} = pe^{-2\xi\omega} \qquad \dot{p} = -\frac{\partial H}{\partial x} = -\omega^2 x e^{2\xi\omega} \tag{10.809}$$

Example 680 ★ Hamiltonian of an Elastic Pendulum Figure 10.49 illustrates an elastic pendulum. If the free length of the pendulum is l and its stretch is u, then the kinetic and potential energies and the Lagrangian of the pendulum are

$$K = \tfrac{1}{2}m\left[\dot{u}^2 + (l+u)^2\dot{\theta}^2\right] \tag{10.810}$$

$$V = \tfrac{1}{2}ku^2 - mb(l+u)(1-\cos\theta) \tag{10.811}$$

$$\mathcal{L} = K - V \tag{10.812}$$

Given

$$p_u = \frac{\partial \mathcal{L}}{\partial \dot{u}} = m\dot{u} \qquad p_\theta = \frac{\partial \mathcal{L}}{\partial \dot{\theta}} = m\,(l+u)^2\,\dot{\theta} \tag{10.813}$$

we have

$$\begin{aligned} H &= \dot{u}\,p_u + \dot{\theta}\,p_\theta - \mathcal{L} \\ &= \frac{1}{2}\left(\frac{p_u^2}{m} + \frac{p_\theta^2}{m\,(l+u)}\right) + mg\,(l+u)\,(1-\cos\theta) + \frac{1}{2}ku^2 \end{aligned} \tag{10.814}$$

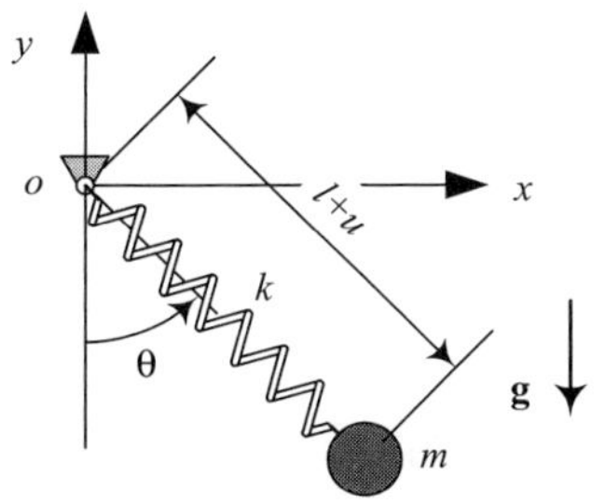

Figure 10.49 An elastic pendulum.

Example 681 ★ Hamiltonian H and Mechanical Energy E If there are no nonpotential forces, $Q_i = 0$, and the dynamic system is scleronomic,

$$\frac{\partial H}{\partial t} = -\frac{\partial \mathcal{L}}{\partial t} = 0 \tag{10.815}$$

then the Hamiltonian H is equal to the mechanical energy of the system and is a constant of motion:

$$H = E = K + V = \text{const} \tag{10.816}$$

To show this, let us take a time derivative of the Lagrangian function (10.782):

$$\begin{aligned} \frac{d\mathcal{L}}{dt} &= \frac{\partial \mathcal{L}}{\partial q_i}\dot{q}_i + \frac{\partial \mathcal{L}}{\partial \dot{q}_i}\ddot{q}_i + \frac{\partial \mathcal{L}}{\partial t} = \frac{d}{dt}\frac{\partial \mathcal{L}}{\partial \dot{q}_i}\dot{q}_i + \frac{\partial \mathcal{L}}{\partial \dot{q}_i}\ddot{q}_i + \frac{\partial \mathcal{L}}{\partial t} \\ &= \frac{d}{dt}\left(\frac{\partial \mathcal{L}}{\partial \dot{q}_i}\dot{q}_i\right) + \frac{\partial \mathcal{L}}{\partial t} \end{aligned} \tag{10.817}$$

We may rearrange this equation as

$$\frac{d}{dt}\left(\frac{\partial \mathcal{L}}{\partial \dot{q}_i}\dot{q}_i - \mathcal{L}\right) = \frac{\partial \mathcal{L}}{\partial t} \tag{10.818}$$

When (10.815) holds, a constant of motion appears:

$$H = \frac{\partial \mathcal{L}}{\partial \dot{q}_i}\dot{q}_i - \mathcal{L} = 2K - \mathcal{L} = E = \text{const} \tag{10.819}$$

It is easy to show that the Hamiltonian of a scleronomic system is constant:

$$\begin{aligned}\frac{d}{dt}H(q_i, p_i) &= \frac{\partial H}{\partial q_i}\dot{q}_i + \frac{\partial H}{\partial p_i}p_i \\ &= \frac{\partial H}{\partial q_i}\left(\frac{\partial H}{\partial p}\right) + \frac{\partial H}{\partial p_i}\left(-\frac{\partial H}{\partial q}\right) = 0\end{aligned} \tag{10.820}$$

A scleronomic system without any nonpotential force is called a *Hamiltonian system*.

Example 682 ★ Hamilton Equations of a Particle in Translation Consider a particle m in a translational motion under a force F:

$$m\ddot{x} = F(x, t) \tag{10.821}$$

Using the momentum p of m,

$$\dot{x} = \frac{p}{m} \tag{10.822}$$

we may write the equation of motion of the particle as

$$\dot{p} = F(x, t) \tag{10.823}$$

The motion of m is expressed by the solution of these two equations. If we use a modified phase plane $(x, p) \equiv (x, m\dot{x})$ to express the motion of m by a moving S_S-point, then the S_S-trajectory and S_S-velocity vectors of m are

$$\mathbf{r} = \begin{bmatrix} x \\ \dot{x} \end{bmatrix} \qquad \mathbf{v} = \begin{bmatrix} p/m \\ F(x, t) \end{bmatrix} \tag{10.824}$$

Because the system has only one DOF, the force F can always be derived from a potential function $V(x, t)$:

$$F(x, t) = -\frac{\partial V(x, t)}{\partial x} \tag{10.825}$$

$$V(x, t) = -\int_{x_0}^{x} F(x, t)\, dx \tag{10.826}$$

We can also define the right-hand side of (10.822) by a derivative:

$$\frac{p}{m} = \frac{d}{dp}\left(\frac{p^2}{2m}\right) \tag{10.827}$$

Let us define the Hamiltonian function of the system as

$$H(x, p) = \frac{p^2}{2m} + V(x, t) \tag{10.828}$$

and rewrite the equations of motion (10.822)–(10.823) and the phase velocity vector **v**:

$$\dot{x} = \frac{\partial H(x, p)}{\partial p} \qquad \dot{p} = -\frac{\partial H(x, p)}{\partial q} \tag{10.829}$$

$$\mathbf{v} = \begin{bmatrix} \dfrac{\partial H}{\partial p} \\ -\dfrac{\partial H}{\partial q} \end{bmatrix} \tag{10.830}$$

The modified phase plane (q, p) has an important area-preserving property, which helps to analyze the motion of dynamic systems.

Example 683 ★ Hamilton Phase Plane and Gradient Consider a Hamiltonian system $H = H(q, p)$. The phase velocity **v** and gradient of H are given as

$$\mathbf{v} = \begin{bmatrix} \dfrac{\partial H}{\partial p} \\ -\dfrac{\partial H}{\partial q} \end{bmatrix} \qquad \nabla H = \begin{bmatrix} \dfrac{\partial H}{\partial q} \\ \dfrac{\partial H}{\partial p} \end{bmatrix} \tag{10.831}$$

The inner product of **v** and ∇H is zero:

$$\mathbf{v} \cdot \nabla H = 0 \tag{10.832}$$

Therefore, the phase velocity of a Hamiltonian system is equal in magnitude and perpendicular in direction to the gradient of the Hamiltonian. Because H is conserved, the S_S-trajectory of motion is along curves of constant H, and the Hamiltonian phase plane is similar to Figure 10.50. The difference between the values of H on neighboring contours remains the same throughout the phase plane. The magnitude of the phase velocity is inversely proportional to the distance between neighboring contours.

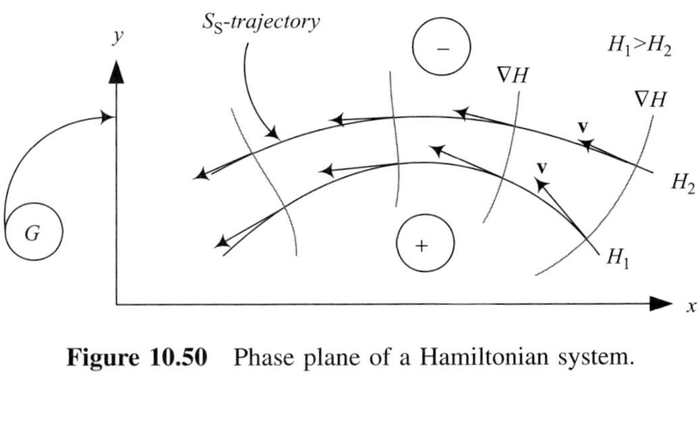

Figure 10.50 Phase plane of a Hamiltonian system.

On each trajectory, the velocity on the right-hand side contours are slower and the velocity on the left-hand side contours are faster.

The contours of the H-plane and the points at which $\nabla H = 0$ are invariant characteristics of a Hamiltonian dynamics system. The invariant points indicate the equilibria of the system.

As an example consider a simple pendulum of length l and mass m such that

$$ml^2 = 1 \tag{10.833}$$

Because the system is conservative, the mechanical energy of the system is the Hamiltonian of the system:

$$H = E = K + V = \tfrac{1}{2}ml^2\dot{\theta}^2 - mgl\cos\theta = \frac{p^2}{2ml^2} - mgl\cos\theta$$

$$= \tfrac{1}{2}p^2 - a^2\cos\theta \tag{10.834}$$

$$a^2 = mgl \tag{10.835}$$

$$p = \frac{\partial \mathcal{L}}{\partial \dot{\theta}} = \frac{\partial (K - V)}{\partial \dot{\theta}} = ml^2\dot{\theta} = \dot{\theta} \tag{10.836}$$

The Hamiltonian equations of motion of the pendulum are

$$\dot{\theta} = p \qquad \dot{p} = -a^2 \sin\theta \tag{10.837}$$

The phase plane of the system is shown in Figure 10.51.

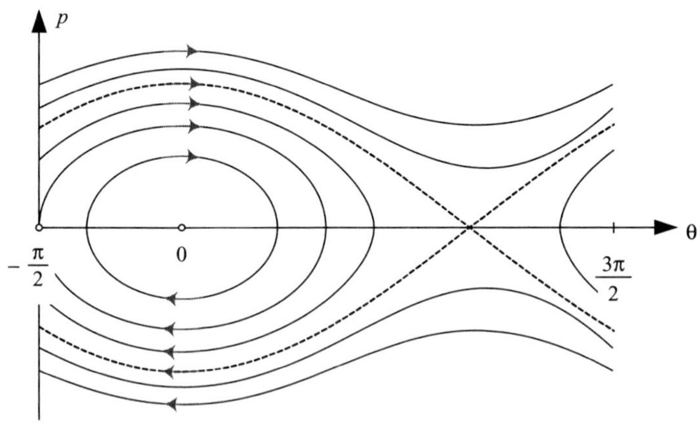

Figure 10.51 Hamiltonian phase plane of a simple pendulum.

Example 684 ★ The Routhian If we want to replace only some of the generalized velocities $\dot{q}$ by momenta p, we may define a Routhian R for the dynamic system,

$$R(q_i, p_i, s_i, \dot{s}_i) = p_i\dot{q}_i - \mathcal{L} \qquad p_i = \frac{\partial \mathcal{L}}{\partial \dot{q}_i} \tag{10.838}$$

in which s and $\dot{s}$ are the generalized coordinate and velocity of the system and q and p are the new generalized coordinate and associated momentum of the system. Let us first suppose that there are only two coordinates q and s and transform the variables from q, s, $\dot{q}$, $\dot{s}$ to q, s, p, $\dot{s}$. The differential of the Lagrangian $\mathcal{L}(q_i, \dot{q}_i, s_i, \dot{s}_i)$ is

$$\begin{aligned} d\mathcal{L} &= \frac{\partial \mathcal{L}}{\partial q_i} dq_i + \frac{\partial \mathcal{L}}{\partial \dot{q}_i} d\dot{q}_i + \frac{\partial \mathcal{L}}{\partial s_i} ds_i + \frac{\partial \mathcal{L}}{\partial \dot{s}_i} d\dot{s}_i \\ &= \dot{p}_i \, dq_i + p_i \, d\dot{q}_i + \frac{\partial \mathcal{L}}{\partial s_i} ds_i + \frac{\partial \mathcal{L}}{\partial \dot{s}_i} d\dot{s}_i \end{aligned} \tag{10.839}$$

Therefore,

$$d(\mathcal{L} - p_i \dot{q}_i) = \dot{p}_i \, dq_i - \dot{q}_i \, dp_i + \frac{\partial \mathcal{L}}{\partial s_i} ds_i + \frac{\partial \mathcal{L}}{\partial \dot{s}_i} d\dot{s}_i \tag{10.840}$$

Now, we define the Routhian as

$$R(q_i, p_i, s_i, \dot{s}_i) = p_i \dot{q}_i - \mathcal{L} \tag{10.841}$$

in which the velocity $\dot{q}$ is expressed in terms of the momentum p using $p_i = \partial \mathcal{L} / \partial \dot{q}_i$. The differential of R is

$$dR = -\dot{p}_i \, dq_i + \dot{q}_i \, dp_i - \frac{\partial \mathcal{L}}{\partial s_i} ds_i - \frac{\partial \mathcal{L}}{\partial \dot{s}_i} d\dot{s}_i \tag{10.842}$$

Therefore, we have

$$\dot{q} = \frac{\partial R}{\partial p} \qquad \dot{p} = -\frac{\partial R}{\partial q} \tag{10.843}$$

$$\frac{\partial \mathcal{L}}{\partial s_i} = -\frac{\partial R}{\partial s_i} \qquad \frac{\partial \mathcal{L}}{\partial \dot{s}_i} = -\frac{\partial R}{\partial \dot{s}_i} \tag{10.844}$$

and substituting these results into the Lagrange equation provides

$$\frac{d}{dt} \frac{\partial R}{\partial \dot{s}_i} - \frac{\partial R}{\partial s_i} = 0 \tag{10.845}$$

Thus the Routhian R is a Hamiltonian with respect to the coordinate q and is a Lagrangian with respect to the coordinate s.

10.11.4 ★ Gibbs–Appell Method

The Gibbs–Appell function of *acceleration energy* is given as

$$\begin{aligned} G &= G\left(q_1, q_2, \ldots, q_n, \dot{q}_1, \dot{q}_2, \ldots, \dot{q}_n, \ddot{q}_1, \ddot{q}_2, \ldots, \ddot{q}_n, t\right) \\ &= \frac{1}{2} \sum_i^N m_i \left(\mathbf{a}_i \cdot \mathbf{a}_i\right) \end{aligned} \tag{10.846}$$

We can determine the equations of motion by applying the Gibbs–Appell equation:

$$\sum_{j=1}^{n} \left(\frac{\partial G}{\partial \ddot{q}_j} - Q_j \right) \delta q_j = 0 \tag{10.847}$$

Proof: Consider a dynamic system of N particles whose positions $\mathbf{r}_i$, $i = 1, 2, \ldots, N$, are expressed by n generalized coordinates q_j, $j = 1, 2, \ldots, n$:

$$\mathbf{r}_i = \mathbf{r}_i\left(q_1, q_2, \ldots, q_n, t\right) \qquad i = 1, 2, \ldots, N \tag{10.848}$$

The velocity and acceleration of each particle are

$$\mathbf{v}_i = \frac{d\mathbf{r}_i}{dt} = \frac{\partial \mathbf{r}_i}{\partial q_j}\dot{q}_j + \frac{\partial \mathbf{r}_i}{\partial t} \tag{10.849}$$

$$\mathbf{a}_i = \frac{d^2\mathbf{r}_i}{dt^2} = \frac{\partial \mathbf{r}_i}{\partial q_j}\ddot{q}_j + \frac{\partial^2 \mathbf{r}_i}{\partial q_k \partial q_j}\dot{q}_j\dot{q}_k + 2\frac{\partial^2 \mathbf{r}_i}{\partial q_j \partial t}\dot{q}_j + \frac{\partial^2 \mathbf{r}_i}{\partial t^2} \tag{10.850}$$

From these equations, we find the relations

$$\frac{\partial \mathbf{r}_i}{\partial q_j} = \frac{\partial \mathbf{v}_i}{\partial \dot{q}_j} = \frac{\partial \mathbf{a}_i}{\partial \ddot{q}_j} \tag{10.851}$$

$$\frac{d}{dt}\frac{\partial \mathbf{v}_i}{\partial \dot{q}_j} = \frac{\partial \mathbf{v}_i}{\partial q_j} \tag{10.852}$$

$$\frac{d}{dt}\frac{\partial \mathbf{a}_i}{\partial \dot{q}_j} = 2\frac{\partial \mathbf{v}_i}{\partial q_j} \tag{10.853}$$

We can combine and expand Equations (10.851) and (10.853) and write them in the general forms

$$\frac{\partial \mathbf{r}_i}{\partial q_j} = \frac{\partial \mathbf{v}_i}{\partial \dot{q}_j} = \frac{\partial \mathbf{a}_i}{\partial \ddot{q}_j} = \cdots = \frac{\partial^{(k)} \mathbf{r}_i}{\partial q_j^{(k)}} \tag{10.854}$$

$$\frac{d}{dt}\frac{\partial^{(k)} \mathbf{r}_i}{\partial q_j^{(k-1)}} = \frac{k}{k-1}\frac{\partial^{(k-1)} \mathbf{r}_i}{\partial q_j^{(k-2)}} \tag{10.855}$$

Employing the kinetic energy

$$K = \frac{1}{2}\sum_i^N m_i\left(\mathbf{v}_i \cdot \mathbf{v}_i\right) \tag{10.856}$$

we can write the Lagrange equation (10.738) as

$$\sum_{j=1}^{n}\left(\frac{d}{dt}\frac{\partial}{\partial \dot{q}_j}\sum_i^N \frac{1}{2}m_i\left(\mathbf{v}_i \cdot \mathbf{v}_i\right) - \frac{\partial}{\partial q_j}\sum_i^N \frac{1}{2}m_i\left(\mathbf{v}_i \cdot \mathbf{v}_i\right) - Q_j\right)\delta q_j = 0 \tag{10.857}$$

Performing the partial and total derivatives, we find

$$\sum_{j=1}^{n}\left(\sum_i^N m_i\mathbf{a}_i \cdot \frac{\partial \mathbf{v}_i}{\partial \dot{q}_j} + m_i\mathbf{v}_i \cdot \frac{d}{dt}\frac{\partial \mathbf{v}_i}{\partial \dot{q}_j} - m_i\mathbf{v}_i \cdot \frac{\partial \mathbf{v}_i}{\partial q_j} - Q_j\right)\delta q_j = 0 \tag{10.858}$$

Based on (10.852), the second and third terms of this equation are equal, and therefore, we have

$$\sum_{j=1}^{n}\left(\sum_i^N m_i\mathbf{a}_i \cdot \frac{\partial \mathbf{a}_i}{\partial \ddot{q}_j} - Q_j\right)\delta q_j = 0 \tag{10.859}$$

which can be transformed to

$$\sum_{j=1}^{n}\left(\frac{\partial}{\partial \ddot{q}_j}\sum_{i}^{N}\frac{1}{2}m_i\left(\mathbf{a}_i\cdot\mathbf{a}_i\right)-Q_j\right)\delta q_j=0 \tag{10.860}$$

Using the *Gibbs–Appell function* of acceleration energy,

$$\begin{aligned} G &= G\left(q_1,q_2,\ldots,q_n,\dot{q}_1,\dot{q}_2,\ldots,\dot{q}_n,\ddot{q}_1,\ddot{q}_2,\ldots,\ddot{q}_n,t\right) \\ &= \frac{1}{2}\sum_{i}^{N}m_i\left(\mathbf{a}_i\cdot\mathbf{a}_i\right) \end{aligned} \tag{10.861}$$

we can write the central equation of dynamics as

$$\sum_{j=1}^{n}\left(\frac{\partial G}{\partial \ddot{q}_j}-Q_j\right)\delta q_j=0 \tag{10.862}$$

Suppose the dynamic system is holonomic so that all the virtual displacements δq_j are independent. In this case, we have the following system of differential equations, known as the *Gibbs–Appell equations of motion*:

$$\frac{\partial G}{\partial \ddot{q}_j}=Q_j \qquad j=1,2,\ldots,n \tag{10.863}$$

The Gibbs–Appell equation (10.863) is fully equivalent to the Lagrange equation (10.739). While the Lagrange equation is based on kinetic energy (10.856), the Gibbs–Appell equation is based on the acceleration energy function (10.861).

When the dynamic system is holonomic and scleronomic, the acceleration (10.850) and the Gibbs–Appell equation (10.861) reduce to

$$\mathbf{a}_i=\frac{\partial \mathbf{r}_i}{\partial q_j}\ddot{q}_j+\frac{\partial^2 \mathbf{r}_i}{\partial q_k \partial q_j}\dot{q}_j\dot{q}_k \qquad j,k=1,2,\ldots,n \tag{10.864}$$

$$G=\tfrac{1}{2}a_{jk}\ddot{q}_j\ddot{q}_k+\Gamma_{jks}\dot{q}_j\dot{q}_k\ddot{q}_s \tag{10.865}$$

where

$$a_{jk}=a_{kj}=\sum_{i}^{N}m_i\frac{\partial \mathbf{r}_i}{\partial q_j}\frac{\partial \mathbf{r}_i}{\partial q_k} \tag{10.866}$$

$$\Gamma_{ijk}=\frac{1}{2}\left(\frac{\partial g_{jk}}{\partial q_i}+\frac{\partial g_{ik}}{\partial q_j}-\frac{\partial g_{ij}}{\partial q_k}\right) \tag{10.867}$$

The Gibbs–Appell equation of motion was introduced by American scientist Josiah Willard Gibbs (1839–1903) and French mathematician Paul Emile Appell (1855–1930). Appell suggested writing Equation (10.863) by introducing the *Appell function A*,

$$A=G-\sum_{i}^{n}Q_i\ddot{q}_i \tag{10.868}$$

and deriving the equations of motion as

$$\frac{\partial A}{\partial \ddot{q}_1} = 0 \qquad \frac{\partial A}{\partial \ddot{q}_2} = 0 \qquad \cdots \qquad \frac{\partial A}{\partial \ddot{q}_n} = 0 \tag{10.869}$$

So, the equations of motion are found by minimization of the Appell function with respect to the generalized accelerations $\ddot{q}_i$. This is similar to the Gauss method, which derives the equations of motion by minimizing the Gauss function Z (10.753) with respect to accelerations $\ddot{\mathbf{u}}_i$. ■

Example 685 ★ Elastic Pendulum and Gibbs–Appell Equation Figure 10.52 illustrates a planar elastic pendulum. If at the equilibrium condition the distance of m and the fulcrum is l_0 and the extra stretch of the length is z, then the Cartesian position of m during the motion is given as

$$x = l \sin\theta \qquad y = -l \cos\theta \tag{10.870}$$

where

$$l = z + l_0 \tag{10.871}$$

where $q_1 = z$ and $q_2 = \theta$ are the generalized coordinates of the system. From (10.870), we have

$$\ddot{x} = \ddot{z} \sin\theta + l\ddot{\theta} \cos\theta - l\dot{\theta}^2 \sin\theta + 2\dot{z}\dot{\theta} \cos\theta \tag{10.872}$$

$$\ddot{y} = -\ddot{z} \cos\theta + l\ddot{\theta} \sin\theta + l\dot{\theta}^2 \cos\theta + 2\dot{z}\dot{\theta} \sin\theta \tag{10.873}$$

The acceleration energy function (10.861) of the elastic pendulum becomes

$$G = \frac{1}{2}\sum_i^N m_i \left(\mathbf{a}_i \cdot \mathbf{a}_i\right) = \frac{1}{2} m \left(\ddot{x}^2 + \ddot{y}^2\right)$$

$$= \frac{1}{2} m \left(l^2\dot{\theta}^4 + l^2\ddot{\theta}^2 - 2l\ddot{z}\dot{\theta}^2 + 4l\dot{z}\dot{\theta}\ddot{\theta} + \ddot{z}^2 + 4\dot{z}^2\dot{\theta}^2\right) \tag{10.874}$$

There are two applied forces on m: the gravitational force $-mg\hat{\jmath}$ and the elastic tension force $\mathbf{F}_k = -kz \sin\theta \hat{\imath} + kz \cos\theta \hat{\jmath}$. The virtual displacement of m can be found from (10.870):

$$\delta x = \delta z \sin\theta + \delta\theta \, l \cos\theta \tag{10.875}$$

$$\delta y = -\delta z \cos\theta + \delta\theta \, l \sin\theta \tag{10.876}$$

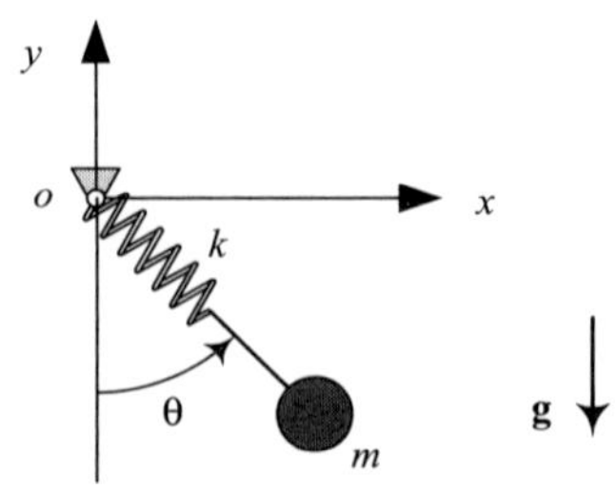

Figure 10.52 An elastic pendulum.

Therefore, the virtual work of the applied forces is

$$\begin{aligned}\delta W &= -kz\sin\theta\,\delta x + (kz\cos\theta - mg)\,\delta y \\ &= -kz\sin\theta\ (\delta z\sin\theta + \delta\theta\, l\cos\theta) \\ &\quad + (kz\cos\theta - mg)\,(-\delta z\cos\theta + \delta\theta\, l\sin\theta) \\ &= (gm\cos\theta - kz)\,\delta z + (-glm\sin\theta)\,\delta\theta \end{aligned} \tag{10.877}$$

and the generalized forces Q_z and Q_θ associated with z and θ are

$$Q_z = gm\cos\theta - kz \qquad Q_\theta = -mgl\sin\theta \tag{10.878}$$

Now the Gibbs–Appell equations (10.862) and (10.863),

$$\frac{\partial G}{\partial \ddot{z}} - Q_z = 0 \qquad \frac{\partial G}{\partial \ddot{\theta}} - Q_\theta = 0 \tag{10.879}$$

would be

$$\ddot{z} - (z + l_0)\,\dot{\theta}^2 - g\cos\theta + \frac{k}{m}z = 0 \tag{10.880}$$

$$(z + l_0)^2\,\ddot{\theta} + 2\,(z + l_0)\,\dot{z}\dot{\theta} + g\,(z + l_0)\sin\theta = 0 \tag{10.881}$$

10.11.5 ★ Kane Method

The Kane equation of motion is

$$Q_i = Q_i^* \tag{10.882}$$

where Q_i is the generalized applied force and Q_i^* the generalized inertia force associated with m_i:

$$Q_j = \sum_{i=1}^{N} \mathbf{F}_i \cdot \frac{\partial \mathbf{v}_i}{\partial \dot{q}_j} \tag{10.883}$$

$$Q_j^* = \frac{d}{dt}\frac{\partial K}{\partial \dot{q}_j} - \frac{\partial K}{\partial q_j} \tag{10.884}$$

where K is the kinetic energy of the system and $\partial \mathbf{v}_i/\partial \dot{q}_j$ is the *partial velocity vector* indicating the change in $\mathbf{v}_i$ for a change in $\dot{q}_j$.

Proof: Consider a dynamic system of N particles whose positions $\mathbf{r}_i$, $i = 1, 2, \ldots, N$, are expressed by n generalized coordinates q_j, $j = 1, 2, \ldots, n$:

$$\mathbf{r}_i = \mathbf{r}_i\,(q_1, q_2, \ldots, q_n, t) \qquad i = 1, 2, \ldots, N \tag{10.885}$$

The velocity of each particle is

$$\mathbf{v}_i = \frac{d\mathbf{r}_i}{dt} = \frac{\partial \mathbf{r}_i}{\partial q_j}\dot{q}_j + \frac{\partial \mathbf{r}_i}{\partial t} \tag{10.886}$$

which provides

$$\mathbf{v}_{ij} = \frac{\partial \mathbf{v}_i}{\partial \dot{q}_j} = \frac{\partial \mathbf{r}_i}{\partial q_j} = \frac{\partial \mathbf{a}_i}{\partial \ddot{q}_j} \tag{10.887}$$

The term $\mathbf{v}_{ij} = \partial \mathbf{v}_i / \partial \dot{q}_j$ is called the *partial velocity vector* and indicates the change in $\mathbf{v}_i$ given a change in $\dot{q}_j$.

Let us begin with the fundamental equation of dynamics (10.303),

$$\sum_{i=1}^{N} (m_i \ddot{\mathbf{r}}_i - \mathbf{F}_i) \cdot \delta \mathbf{r}_i = 0 \tag{10.888}$$

and rewrite it as Equation (10.302):

$$\sum_{i=1}^{N} \left(m_i \ddot{\mathbf{r}}_i - \mathbf{F}_i - \mathbf{F}_{C_i} \right) = 0 \tag{10.889}$$

where $\mathbf{F}_i$ is the given force on m_i and $\mathbf{F}_{C_i}$ is the constraint force, where we always have

$$\sum_{i=1}^{N} \mathbf{F}_{C_i} \cdot \delta \mathbf{r}_i = 0 \tag{10.890}$$

The Kane equations of motion are obtained by projecting the equilibrium equations (10.889) on the partial velocity direction:

$$\sum_{i=1}^{N} m_i \ddot{\mathbf{r}}_i \cdot \frac{\partial \mathbf{v}_i}{\partial \dot{q}_j} = \sum_{i=1}^{N} \left(\mathbf{F}_i \cdot \frac{\partial \mathbf{v}_i}{\partial \dot{q}_j} - \mathbf{F}_{C_i} \cdot \frac{\partial \mathbf{v}_i}{\partial \dot{q}_j} \right) \tag{10.891}$$

The constraint forces $\mathbf{F}_{C_i}$ are assumed to satisfy

$$\sum_{i=1}^{N} \mathbf{F}_{C_i} \cdot \frac{\partial \mathbf{v}_i}{\partial \dot{q}_j} = \sum_{i=1}^{N} \mathbf{F}_{C_i} \cdot \frac{\partial \mathbf{r}_i}{\partial q_j} = 0 \tag{10.892}$$

which is equivalent to D'Alembert's principle (10.890), which states that the virtual work of constraint forces is zero:

$$\sum_{i=1}^{N} \mathbf{F}_{C_i} \cdot \delta \mathbf{r}_i = \sum_{i=1}^{N} \mathbf{F}_{C_i} \cdot \frac{\partial \mathbf{r}_i}{\partial q_j} \delta q_j = 0 \tag{10.893}$$

Substituting (10.892) into (10.891), we write the Kane equations of motion as

$$Q_i = Q_i^* \tag{10.894}$$

where Q_i is the generalized applied force and Q_i^* is the generalized inertia force associated with m_i:

$$Q_j = \sum_{i=1}^{N} \mathbf{F}_i \cdot \frac{\partial \mathbf{v}_i}{\partial \dot{q}_j} \tag{10.895}$$

$$
\begin{aligned}
Q_j^* &= \sum_{i=1}^{N} m_i \ddot{\mathbf{r}}_i \cdot \frac{\partial \mathbf{v}_i}{\partial \dot{q}_j} = \sum_{i=1}^{N} m_i \ddot{\mathbf{r}}_i \cdot \frac{\partial \mathbf{r}_i}{\partial q_j} \\
&= \frac{d}{dt} \sum_{i=1}^{N} m_i \dot{\mathbf{r}}_i \cdot \frac{\partial \mathbf{r}_i}{\partial q_j} - \sum_{i=1}^{N} m_i \dot{\mathbf{r}}_i \cdot \frac{d}{dt} \frac{\partial \mathbf{r}_i}{\partial q_j} \\
&= \frac{d}{dt} \sum_{i=1}^{N} m_i \mathbf{v}_i \cdot \frac{\partial \mathbf{v}_i}{\partial \dot{q}_j} - \sum_{i=1}^{N} m_i \mathbf{v}_i \cdot \frac{\partial \mathbf{v}_i}{\partial q_j} \\
&= \frac{d}{dt} \frac{\partial K}{\partial \dot{q}_j} - \frac{\partial K}{\partial q_j}
\end{aligned}
\tag{10.896}
$$

The kinetic energy of the system is given as

$$
K = \sum_{i=1}^{N} \left(\frac{1}{2} m_i \mathbf{v}_i \cdot \mathbf{v}_i \right) \tag{10.897}
$$

The generalized applied force Q_i and the generalized inertia force Q_i^* in the Kane equation (10.894) can be determined from Equations (10.895) and (10.896).

Consider a rigid body that is made of N rigidly connected particles m_i. The angular velocity ${}_G\boldsymbol{\omega}_B$ and acceleration ${}_G\boldsymbol{\alpha}_B = {}_G\dot{\boldsymbol{\omega}}_B$ of the body are expressed in terms of the generalized coordinates q_j and velocities $\dot{q}_j$. The translational velocity $\mathbf{v}_i$ and acceleration $\mathbf{a}_i$ of the body are calculated at the mass center C. By defining the *partial angular velocity vector* as

$$
\boldsymbol{\omega}_{ij} = \frac{\partial \boldsymbol{\omega}_i}{\partial \dot{q}_j} \tag{10.898}
$$

we can calculate the generalized applied and inertia forces as

$$
Q_i = \mathbf{v}_{ij} \cdot \sum_{i=1}^{N} \mathbf{F}_i + \boldsymbol{\omega}_{ij} \cdot \sum_{i=1}^{N} \mathbf{r}_i \times \mathbf{F}_i \tag{10.899}
$$

$$
\begin{aligned}
Q_i^* &= \mathbf{v}_{ij} \cdot \left(\sum_{i=1}^{N} m_i \right) {}^G\mathbf{a}_i + \boldsymbol{\omega}_{ij} \cdot {}^G I \; {}_G\boldsymbol{\alpha}_B \\
&= \mathbf{v}_{ij} \cdot \left(\sum_{i=1}^{N} m_i \right) \left({}_G^B\mathbf{a}_B + {}_G^B\boldsymbol{\omega}_B \times {}^B\mathbf{v}_B \right) \\
&\quad + \boldsymbol{\omega}_{ij} \cdot \left[{}^B I \; {}_G^B\dot{\boldsymbol{\omega}}_B + {}_G^B\boldsymbol{\omega}_B \times \left({}^B I \; {}_G^B\boldsymbol{\omega}_B \right) \right]
\end{aligned}
\tag{10.900}
$$

In a rigid body with mass $m = \sum_{i=1}^{N} m_i$ under the resultant force system of $\mathbf{F}$ and $\mathbf{M}$, the above generalized applied and inertia forces simplify to

$$
Q_i = \mathbf{v}_{ij} \cdot \mathbf{F} + \boldsymbol{\omega}_{ij} \cdot \mathbf{M} \tag{10.901}
$$

$$
\begin{aligned}
Q_i^* &= \mathbf{v}_{ij} \cdot m \left({}_G^B\mathbf{a}_B + {}_G^B\boldsymbol{\omega}_B \times {}^B\mathbf{v}_B \right) \\
&\quad + \boldsymbol{\omega}_{ij} \cdot \left[{}^B I \; {}_G^B\dot{\omega}_B + {}_G^B\boldsymbol{\omega}_B \times \left({}^B I \; {}_G^B\boldsymbol{\omega}_B \right) \right]
\end{aligned}
\tag{10.902}
$$

■

Example 686 ★ Double Pendulum and Kane Equations Figure 10.53 illustrates a double pendulum with massless rods of lengths a and b and concentrated masses m_1 and m_2. Let us use the angles θ_1 and θ_2 between $\mathbf{g}$ and the rods as the generalized coordinates of the system. The position vectors of m_1 and m_2 are

$$\mathbf{r}_1 = a\sin\theta_1\,\hat{\imath} + a\cos\theta_1\,\hat{\jmath} \tag{10.903}$$

$$\mathbf{r}_2 = (a\sin\theta_1 + b\sin\theta_2)\,\hat{\imath} + (a\cos\theta_1 + b\cos\theta_2)\,\hat{\jmath} \tag{10.904}$$

Differentiation provides the velocity $\mathbf{v}_i$ of m_i and the kinetic energy of the system K:

$$\mathbf{v}_1 = a\dot{\theta}_1\cos\theta_1\,\hat{\imath} - a\dot{\theta}_1\sin\theta_1\,\hat{\jmath} \tag{10.905}$$

$$\mathbf{v}_2 = \left(a\dot{\theta}_1\cos\theta_1 + b\dot{\theta}_2\cos\theta_2\right)\hat{\imath} - \left(a\dot{\theta}_1\sin\theta_1 + b\dot{\theta}_2\sin\theta_2\right)\hat{\jmath} \tag{10.906}$$

$$K = \tfrac{1}{2}m_1 a^2\dot{\theta}_1^2 + \tfrac{1}{2}m_2\left(a^2\dot{\theta}_1^2 + 2ab\dot{\theta}_1\dot{\theta}_2\cos(\theta_1 - \theta_2) + b^2\dot{\theta}_2^2\right) \tag{10.907}$$

To determine the equation of motion using the Kane method, we need to determine the partial velocities $\mathbf{v}_{ij}$. The unit vectors $\hat{u}_i$ in directions θ_i are

$$\hat{u}_i = -\sin\theta_i\,\hat{\imath} + \cos\theta_i\,\hat{\jmath} \tag{10.908}$$

Therefore, the partial velocities $\mathbf{v}_{ij}$ are

$$\mathbf{v}_{11} = \frac{\partial\mathbf{v}_1}{\partial\dot{\theta}_1} = a\cos\theta_1\,\hat{\imath} - a\sin\theta_1\,\hat{\jmath} \tag{10.909}$$

$$\mathbf{v}_{12} = \frac{\partial\mathbf{v}_1}{\partial\dot{\theta}_2} = 0 \tag{10.910}$$

$$\mathbf{v}_{21} = \frac{\partial\mathbf{v}_2}{\partial\dot{\theta}_1} = a\cos\theta_1\,\hat{\imath} - a\sin\theta_1\,\hat{\jmath} \tag{10.911}$$

$$\mathbf{v}_{22} = \frac{\partial\mathbf{v}_2}{\partial\dot{\theta}_2} = b\cos\theta_2\,\hat{\imath} - b\sin\theta_2\,\hat{\jmath} \tag{10.912}$$

The inner products of the active forces $\mathbf{F}_i$,

$$\mathbf{F}_1 = m_1 g\,\hat{\jmath} \qquad \mathbf{F}_2 = m_2 g\,\hat{\jmath} \tag{10.913}$$

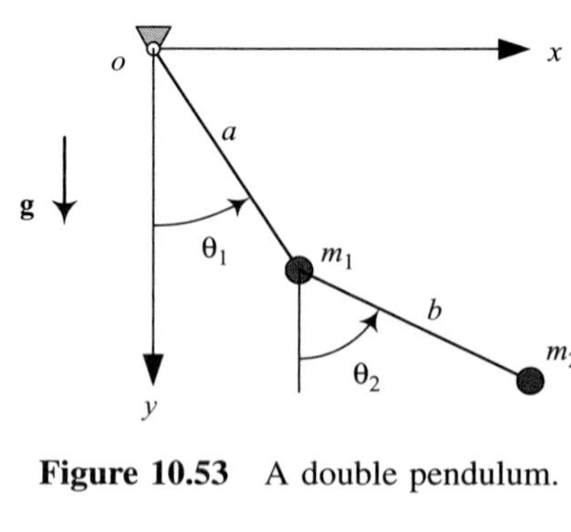

Figure 10.53 A double pendulum.

and the partial velocities (10.909)–(10.912) provide the generalized active forces Q_i:

$$Q_1 = \mathbf{F}_1 \cdot \mathbf{v}_{11} + \mathbf{F}_2 \cdot \mathbf{v}_{21} = -m_1 g a \sin\theta_1 - m_2 g a \sin\theta_1 \tag{10.914}$$

$$Q_2 = \mathbf{F}_1 \cdot \mathbf{v}_{12} + \mathbf{F}_2 \cdot \mathbf{v}_{22} = -m_2 g b \sin\theta_2 \tag{10.915}$$

Using the kinetic energy equation (10.907), we determine the generalized inertia forces:

$$Q_j^* = \frac{d}{dt}\frac{\partial K}{\partial \dot{\theta}_j} - \frac{\partial K}{\partial \theta_j} \tag{10.916}$$

$$\begin{aligned} Q_1^* &= \frac{d}{dt}\left\{m_1 a^2 \dot{\theta}_1 + m_2\left[a^2\dot{\theta}_1 + 2ab\dot{\theta}_2\cos(\theta_1 - \theta_2)\right]\right\} \\ &\quad + m_2[ab\dot{\theta}_1\dot{\theta}_2\sin(\theta_1 - \theta_2)] \end{aligned} \tag{10.917}$$

$$\begin{aligned} Q_2^* &= m_2\frac{d}{dt}\left[ab\dot{\theta}_1\cos(\theta_1 - \theta_2) + b^2\dot{\theta}_2\right] \\ &\quad - m_2\left[ab\dot{\theta}_1\dot{\theta}_2\sin(\theta_1 - \theta_2)\right] \end{aligned} \tag{10.918}$$

Therefore, the Kane equations of motion are

$$\begin{aligned} &\frac{d}{dt}\left[m_1 a^2\dot{\theta}_1 + m_2\left(a^2\dot{\theta}_1 + ab\dot{\theta}_2\cos(\theta_1 - \theta_2)\right)\right] + m_2\left(ab\dot{\theta}_1\dot{\theta}_2\sin(\theta_1 - \theta_2)\right) \\ &\quad = -m_1 g a\sin\theta_1 - m_2 g a\sin\theta_1 \end{aligned} \tag{10.919}$$

$$\begin{aligned} &m_2\frac{d}{dt}\left(ab\dot{\theta}_1\cos(\theta_1 - \theta_2) + b^2\dot{\theta}_2\right) - m_2\left(ab\dot{\theta}_1\dot{\theta}_2\sin(\theta_1 - \theta_2)\right) \\ &\quad = -m_2 g b\sin\theta_2 \end{aligned} \tag{10.920}$$

Taking the derivatives and simplifying, we determine the following equations of motion:

$$\begin{aligned} &(m_1 + m_2)\,a^2\ddot{\theta}_1 + m_2 ab\ddot{\theta}_2\cos(\theta_1 - \theta_2) + m_2 ab\dot{\theta}_2^2\sin(\theta_1 - \theta_2) \\ &+ (m_1 + m_2)\,g a\sin\theta_1 = 0 \end{aligned} \tag{10.921}$$

$$\begin{aligned} &m_2 b^2\ddot{\theta}_2 + m_2 ab\ddot{\theta}_1\cos(\theta_1 - \theta_2) - m_2 ab\dot{\theta}_1^2\sin(\theta_1 - \theta_2) \\ &+ m_2 g b\sin\theta_2 = 0 \end{aligned} \tag{10.922}$$

10.11.6 ★ Nielsen Method

If $\mathcal{L} = \mathcal{L}(q_i, \dot{q}_i, t)$ is the Lagrangian and Q_i is the nonpotential generalized force of a dynamic system, then the equations of motion of the system may be found by the *Nielsen equation*:

$$2\frac{d}{dt}\frac{\partial \dot{\mathcal{L}}}{\partial \ddot{q}_i} - \frac{\partial \dot{\mathcal{L}}}{\partial \dot{q}_i} = Q_i \tag{10.923}$$

Proof: Consider a dynamic system with the Lagrangian

$$\mathcal{L} = \mathcal{L}(q_i, \dot{q}_i, t) \qquad i = 1, 2, \ldots, n \tag{10.924}$$

Suppose the system is also subject to nonpotential generalized forces $Q_1, Q_2, \ldots, Q_n$:

$$Q_i = Q_i\left(q_i, \dot{q}_i, t\right) \tag{10.925}$$

The Lagrange equations are given as

$$\frac{d}{dt}\frac{\partial \mathcal{L}}{\partial \dot{q}_i} - \frac{\partial \mathcal{L}}{\partial q_i} - Q_i = 0 \qquad i = 1, 2, \ldots, n \tag{10.926}$$

The time derivative of $\mathcal{L}$,

$$\dot{\mathcal{L}} = \frac{\partial \mathcal{L}}{\partial \dot{q}_i}\ddot{q}_i + \frac{\partial \mathcal{L}}{\partial q_i}\dot{q}_i + \frac{\partial \mathcal{L}}{\partial t} \tag{10.927}$$

shows that

$$\frac{\partial \dot{\mathcal{L}}}{\partial \ddot{q}_i} = \frac{\partial \mathcal{L}}{\partial \dot{q}_i} \tag{10.928}$$

Using the Lagrange equation (10.926), a time derivative of this equation yields

$$\frac{d}{dt}\frac{\partial \dot{\mathcal{L}}}{\partial \ddot{q}_i} = \frac{d}{dt}\frac{\partial \mathcal{L}}{\partial \dot{q}_i} = \frac{\partial \mathcal{L}}{\partial q_i} + Q_i \tag{10.929}$$

From the equations

$$\frac{d}{dt}\frac{\partial \mathcal{L}}{\partial \dot{q}_i} = \frac{\partial^2 \mathcal{L}}{\partial \dot{q}_i \partial \dot{q}_j}\ddot{q}_j + \frac{\partial^2 \mathcal{L}}{\partial \dot{q}_i \partial q_j}\dot{q}_j + \frac{\partial^2 \mathcal{L}}{\partial \dot{q}_i \partial t} \tag{10.930}$$

$$\frac{\partial \dot{\mathcal{L}}}{\partial \dot{q}_i} = \frac{\partial^2 \mathcal{L}}{\partial \dot{q}_i \partial \dot{q}_j}\ddot{q}_j + \frac{\partial^2 \mathcal{L}}{\partial q_i \partial q_j}\dot{q}_j + \frac{\partial^2 \mathcal{L}}{\partial \dot{q}_i \partial t} + \frac{\partial \mathcal{L}}{\partial q_i} \tag{10.931}$$

we find

$$\frac{\partial \dot{\mathcal{L}}}{\partial \dot{q}_i} = \frac{d}{dt}\frac{\partial \mathcal{L}}{\partial \dot{q}_i} + \frac{\partial \mathcal{L}}{\partial q_i} \tag{10.932}$$

Substituting for $\partial \mathcal{L}/\partial q_i$ from (10.929) generates the Nielsen equation:

$$2\frac{d}{dt}\frac{\partial \dot{\mathcal{L}}}{\partial \ddot{q}_i} - \frac{\partial \dot{\mathcal{L}}}{\partial \dot{q}_i} = Q_i \tag{10.933}$$

The Nielsen method is also called the Mangerone–Deleanu method. ■

Example 687 ★ A Double Pendulum Consider a double pendulum as shown in Figure 10.53. Using θ_1 and θ_2 as the generalized coordinates, the kinetic energy of the system is found in (10.907):

$$K = \tfrac{1}{2}m_1 a^2\dot{\theta}_1^2 + \tfrac{1}{2}m_2\left[a^2\dot{\theta}_1^2 + 2ab\dot{\theta}_1\dot{\theta}_2\cos\left(\theta_1 - \theta_2\right) + b^2\dot{\theta}_2^2\right] \tag{10.934}$$

Calculating the potential energy as

$$V = -m_1 g a\cos\theta_1 - m_2 g\left(a\cos\theta_1 + b\cos\theta_2\right) \tag{10.935}$$

we find

$$\begin{aligned}\mathcal{L} &= K - V \\ &= \tfrac{1}{2}m_1 a^2\dot{\theta}_1^2 + \tfrac{1}{2}m_2\left[a^2\dot{\theta}_1^2 + 2ab\dot{\theta}_1\dot{\theta}_2\cos(\theta_1-\theta_2) + b^2\dot{\theta}_2^2\right] \\ &\quad + m_1 g a\cos\theta_1 + m_2 g\left(a\cos\theta_1 + b\cos\theta_2\right)\end{aligned} \tag{10.936}$$

The time derivative of $\mathcal{L}$ is given as

$$\begin{aligned}\dot{\mathcal{L}} &= (m_1+m_2)\,a^2\dot{\theta}_1\ddot{\theta}_1 + m_2 b^2\dot{\theta}_2\ddot{\theta}_2 \\ &\quad + m_2 ab\ddot{\theta}_1\dot{\theta}_2\cos(\theta_1-\theta_2) + m_2 ab\dot{\theta}_1\ddot{\theta}_2\cos(\theta_1-\theta_2) \\ &\quad - m_2 ab\dot{\theta}_1^2\dot{\theta}_2\sin(\theta_1-\theta_2) + m_2 ab\dot{\theta}_1\dot{\theta}_2^2\sin(\theta_1-\theta_2) \\ &\quad - m_1 g a\dot{\theta}_1\sin\theta_1 - m_2 g\left(a\dot{\theta}_1\sin\theta_1 + b\dot{\theta}_2\sin\theta_2\right)\end{aligned} \tag{10.937}$$

The required partial derivatives of $\dot{\mathcal{L}}$ are

$$\begin{aligned}\frac{\partial\dot{\mathcal{L}}}{\partial\dot{\theta}_1} &= (m_1+m_2)\,a^2\ddot{\theta}_1 + m_2 ab\ddot{\theta}_2\cos(\theta_1-\theta_2) - 2m_2 ab\dot{\theta}_1\dot{\theta}_2\sin(\theta_1-\theta_2) \\ &\quad + m_2 ab\dot{\theta}_2^2\sin(\theta_1-\theta_2) - m_1 g a\sin\theta_1 - m_2 g a\sin\theta_1\end{aligned} \tag{10.938}$$

$$\begin{aligned}\frac{\partial\dot{\mathcal{L}}}{\partial\dot{\theta}_2} &= m_2 b^2\ddot{\theta}_2 + m_2 ab\ddot{\theta}_1\cos(\theta_1-\theta_2) - m_2 ab\dot{\theta}_1^2\sin(\theta_1-\theta_2) \\ &\quad + 2m_2 ab\dot{\theta}_1\dot{\theta}_2\sin(\theta_1-\theta_2) - m_2 g b\sin\theta_2\end{aligned} \tag{10.939}$$

$$\begin{aligned}\frac{\partial\dot{\mathcal{L}}}{\partial\ddot{\theta}_1} &= (m_1+m_2)\,a^2\dot{\theta}_1 + m_2 ab\dot{\theta}_2\cos(\theta_1-\theta_2) \\ \frac{\partial\dot{\mathcal{L}}}{\partial\ddot{\theta}_2} &= m_2 b^2\dot{\theta}_2 + m_2 ab\dot{\theta}_1\cos(\theta_1-\theta_2)\end{aligned} \tag{10.940}$$

$$\begin{aligned}\frac{d}{dt}\frac{\partial\dot{\mathcal{L}}}{\partial\ddot{\theta}_1} &= (m_1+m_2)\,a^2\ddot{\theta}_1 + m_2 ab\ddot{\theta}_2\cos(\theta_1-\theta_2) \\ &\quad - m_2 ab\dot{\theta}_2\left(\dot{\theta}_1-\dot{\theta}_2\right)\sin(\theta_1-\theta_2)\end{aligned} \tag{10.941}$$

$$\begin{aligned}\frac{d}{dt}\frac{\partial\dot{\mathcal{L}}}{\partial\ddot{\theta}_2} &= m_2 b^2\ddot{\theta}_2 + m_2 ab\ddot{\theta}_1\cos(\theta_1-\theta_2) \\ &\quad - m_2 ab\dot{\theta}_1\left(\dot{\theta}_1-\dot{\theta}_2\right)\sin(\theta_1-\theta_2)\end{aligned} \tag{10.942}$$

Therefore, the equations of motion of the system would be

$$\begin{aligned}&(m_1+m_2)\,a^2\ddot{\theta}_1 + m_2 ab\ddot{\theta}_2\cos(\theta_1-\theta_2) + m_2 ab\dot{\theta}_2^2\sin(\theta_1-\theta_2) \\ &\quad + (m_1+m_2)\,g a\sin\theta_1 = 0\end{aligned} \tag{10.943}$$

$$\begin{aligned}&m_2 b^2\ddot{\theta}_2 + m_2 ab\ddot{\theta}_1\cos(\theta_1-\theta_2) - m_2 ab\dot{\theta}_1^2\sin(\theta_1-\theta_2) \\ &\quad + m_2 g b\sin\theta_2 = 0\end{aligned} \tag{10.944}$$

Example 688 ★ Nielsen Equation for Nonpotential Forces If there is no potential force, then the Nielsen equation simplifies to

$$2\frac{d}{dt}\frac{\partial \dot{K}}{\partial \ddot{q}_i} - \frac{\partial \dot{K}}{\partial \dot{q}_i} = Q_i \tag{10.945}$$

We may use Equation (10.929),

$$\frac{d}{dt}\frac{\partial \dot{K}}{\partial \ddot{q}_i} = \frac{\partial K}{\partial q_i} + Q_i \tag{10.946}$$

to derive another form of the Nielsen equation:

$$\frac{\partial \dot{K}}{\partial \dot{q}_i} - 2\frac{\partial K}{\partial q_i} = Q_i \qquad i = 1, 2, \ldots, n \tag{10.947}$$

Example 689 ★ Tzénoff Method and Higher Derivatives of $\mathcal{L}$ Employing higher derivatives of the Lagrangian, $\ddot{\mathcal{L}}$, $\dddot{\mathcal{L}}$, ..., and repeating the procedure used to derive the Nielson equation, we can find new methods of dynamics.

The second derivative of $\mathcal{L} = \mathcal{L}(q_i, \dot{q}_i, t)$ is

$$\begin{aligned}
\ddot{\mathcal{L}} &= \frac{d}{dt}\dot{\mathcal{L}} = \frac{d}{dt}\left(\frac{\partial \mathcal{L}}{\partial \dot{q}_i}\ddot{q}_i + \frac{\partial \mathcal{L}}{\partial q_i}\dot{q}_i + \frac{\partial \mathcal{L}}{\partial t}\right) \\
&= \frac{\partial \mathcal{L}}{\partial \dot{q}_i}\dddot{q}_i + \frac{\partial^2 \mathcal{L}}{\partial \dot{q}_i^2}\ddot{q}_i^2 + \frac{\partial^2 \mathcal{L}}{\partial q_i^2}\dot{q}_i^2 + \frac{\partial^2 \mathcal{L}}{\partial t^2} + \frac{\partial \mathcal{L}}{\partial q_i}\ddot{q}_i \\
&\quad + 2\frac{\partial^2 \mathcal{L}}{\partial q_i \partial \dot{q}_i}\dot{q}_i\ddot{q}_i + 2\frac{\partial^2 \mathcal{L}}{\partial \dot{q}_i \partial t}\ddot{q}_i + 2\frac{\partial^2 \mathcal{L}}{\partial q_i \partial t}\dot{q}_i
\end{aligned} \tag{10.948}$$

which yields

$$\frac{\partial \ddot{\mathcal{L}}}{\partial \dddot{q}_i} = \frac{\partial \mathcal{L}}{\partial \dot{q}_i} \tag{10.949}$$

Let us take a time derivative of this equation and use the Lagrange equation (10.926) to get

$$\frac{d}{dt}\frac{\partial \ddot{\mathcal{L}}}{\partial \dddot{q}_i} = \frac{d}{dt}\frac{\partial \mathcal{L}}{\partial \dot{q}_i} = \frac{\partial \mathcal{L}}{\partial q_i} + Q_i \tag{10.950}$$

From Equation (10.948) and the Lagrange equation, we have

$$\begin{aligned}
\frac{\partial \ddot{\mathcal{L}}}{\partial \ddot{q}_i} &= 2\left(\frac{\partial^2 \mathcal{L}}{\partial \dot{q}_i^2}\ddot{q}_i + \frac{\partial^2 \mathcal{L}}{\partial q_i \partial \dot{q}_i}\dot{q}_i + \frac{\partial^2 \mathcal{L}}{\partial \dot{q}_i \partial t}\right) + \frac{\partial \mathcal{L}}{\partial q_i} \\
&= 2\left(\frac{d}{dt}\frac{\partial \mathcal{L}}{\partial \dot{q}_i}\right) + \frac{\partial \mathcal{L}}{\partial q_i} = 2Q_i + 3\frac{\partial \mathcal{L}}{\partial q_i}
\end{aligned} \tag{10.951}$$

which can be written as

$$\frac{1}{2}\left(\frac{\partial \ddot{\mathcal{L}}}{\partial \ddot{q}_i} - 3\frac{\partial \mathcal{L}}{\partial q_i}\right) = Q_i \tag{10.952}$$

Substituting $\partial\mathcal{L}/\partial q_i$ from (10.950) yields

$$3\frac{d}{dt}\frac{\partial\ddot{\mathcal{L}}}{\partial\dddot{q}_i}-\frac{\partial\ddot{\mathcal{L}}}{\partial\ddot{q}_i}=Q_i \tag{10.953}$$

Equations (10.952) and (10.953) are called Tzénoff equations. If there is no potential force, then these equations simplify to

$$\frac{1}{2}\left(\frac{\partial\ddot{K}}{\partial\ddot{q}_i}-3\frac{\partial K}{\partial q_i}\right)=Q_i \tag{10.954}$$

$$3\frac{d}{dt}\frac{\partial\ddot{K}}{\partial\dddot{q}_i}-\frac{\partial\ddot{K}}{\partial\ddot{q}_i}=Q_i \tag{10.955}$$

Considering the Nielson equation (10.923) and the Tzénoff equation (10.953), we can derive a generalized method of dynamics for a given Lagrangian and its rth-order time derivative:

$$(r-1)\,\frac{d}{dt}\frac{\partial\mathcal{L}^{(r)}}{\partial q_i^{(r+1)}}-\frac{\partial\mathcal{L}^{(r)}}{\partial q_i^{(r)}}=Q_i \tag{10.956}$$

$$\frac{1}{2}\left(\frac{\partial K^{(r)}}{\partial q_i^{(r)}}-(r-1)\,\frac{\partial K}{\partial q_i}\right)=Q_i \tag{10.957}$$

KEY SYMBOLS

a	constant value, acceleration,
a, b	constant parameter curves, length of bars, time functions
$\mathbf{a}=\ddot{\mathbf{r}}$	acceleration vector
A	constant, amplitude, general coordinate frame Appell function
A, B, C, D	coefficients of a differential constraint
B	body coordinate frame, local coordinate frame, general coordinate frame
c	constant value, damping, a constant coefficient
$\mathbf{c}$	constant vector
C	constant value, mass center
d	distance between two points, derivative operator differential of possible displacement operator
$\mathbf{d}$	translation vector
D	derivative operator
e	eccentricity
E	Eulerian space, mechanical energy
f	constraint function, general function, cyclic frequency
f_C	configuration degree of freedom
f_S	state degree of freedom
F	force value

$\mathbf{F}$	force vector, force function
$\mathbf{F}_C$	constraint force
$\mathbf{F}_G$	given force
g	general function, gravitational acceleration
G	global coordinate frame, fixed coordinate frame Gibbs–Appell function
H	Hamiltonian
$I = [I]$	mass moment
$\mathbf{I} = [\mathbf{I}]$	identity matrix
$\hat{\imath}, \hat{\jmath}, \hat{k}$	local coordinate axes unit vectors
$\hat{I}, \hat{J}, \hat{K}$	global coordinate axes unit vectors
J	Jacobian
k	stiffness of spring
K	kinetic energy
l	length
L	number of holonomic constraint
$\mathbf{L}$	rotational momentum
m	mass, point mass, number of rigid bodies in a mechanical system
$\mathbf{M}$	torque, moment
n	number of particles in a dynamic system, number of generalized coordinates
N	dimension of configuration space
$\mathbf{N}$	contact force
p	momentum integral of motion, constant-coefficient generalized momentum
$\mathbf{p}$	momentum
P	point mass, body point, fixed point in B, time function
q	generalized coordinate, time function, constant coefficient
$\dot{q}$	generalized velocity
$\mathbf{q}$	generalized position vector
$\dot{\mathbf{q}}$	generalized velocity vector
Q	transformation matrix of rotation about a global axis, torque
$\mathbf{r}$	position vector
r_{ij}	element of row i and column j of a matrix
R	radius of a circle, rotation transformation matrix
$\mathbb{R}$	set of real numbers
s	sin, a member of S, characteristic value, generalized coordinate alternative generalized coordinates
S	physical space
S_C	configuration trajectory, S_C-trajectory
S_E	event trajectory, S_E-trajectory
S_F	flash trajectory, S_F-trajectory
S_G	generalized state space
S_S	state trajectory, S_S-trajectory
S_T	state–time trajectory, S_T-trajectory
t	time
t_c	critical time
T	time domain set, tension force

u	coordinate of configuration space
	replaced variable for x in regularization
$\hat{u}_t$	unit tangent vector
u_i	ith component of $\mathbf{u}$
$\dot{u}_i$	ith component of $\dot{\mathbf{u}}$
$\mathbf{u}$	position vector in configuration space
$\dot{\mathbf{u}}$	velocity vector in configuration space
$\mathbf{v} = \dot{\mathbf{r}}$	velocity vector
v	velocity, coordinate of configuration space
V	potential energy
W	work
W_A	actual work
W_V	virtual work
x, y, z	local coordinate axes, coordinates in a Cartesian frame
X, Y, Z	global coordinate axes
X_C	configuration space, S_C-*space*
X_E	event space, S_E-*space*
X_F	flash space, S_F-space
X_S	state space, S_S-space
X_T	state–time space, S_T-space
Z	constraint of motion

Greek

α, β, γ	angles
δ	virtual operator, variation operator
δ_{ij}	Kronecker delta
ε	time increment
λ	Lagrange multiplier
μ	friction coefficient, integrating factor
ξ	damping ratio
τ	replaced variable for t in regularization
ϕ	rotation of coordinate frame
φ, θ, ψ	rotation angles about local axes, Euler angles
$\omega, \boldsymbol{\omega}$	angular frequency
ω_N	natural frequency

Symbol

$\mathcal{L}$	Lagrangian
Re	real
Im	imaginary
$\perp$	vertical component
$\triangle$	arbitrary displacement
$[\ \ ]^{-1}$	inverse of the matrix [ ]
$[\ \ ]^{\mathrm{T}}$	transpose of the matrix [ ]
∇	gradient
★	specific point
$\otimes$	binary operation
$(S, \otimes)$	group
∞	infinity
$'$	derivative of a function with respect to its argument

EXERCISES

1. **Possible Configuration Trajectory** Examine the following paths in the (x, y)-plane and determine which one can be a configuration trajectory:

   (a) $x = t,\ y = 1/t$
   (b) $x = t,\ (y^2 + t)^2 - a^2\left(t - y^2\right) = 0$
   (c) $x = a \sin \omega t,\ \ y = e^x$
   (d) $x = a \sin \omega t,\ \ (y^2 + by + x^2)^2 - a^2(x^2 + y^2) = 0$
   (e) $x = \sinh t,\ \ y = \cosh t$
   (f) $v^2 = 2(x \sin x + \cos x)$
   (g) $v^2 = 4x - x^2$

2. **Constraint Particle on a Table** Consider a particle that rests on a smooth horizontal table at point A as shown in Figure 10.54. The particle is attached by a string of length a to a fixed point B on the table at a distance b from point C.

   (a) Determine the constraint of the particle and its degree of freedom and suggest proper generalized coordinates.

   (b) Assume that the points A, B, and C are initially on the same straight line. The table is then made to rotate with a constant angular velocity ω. Express the coordinates of the particle.

   (c) Determine the kinetic energy of the particle:

   $$K = \tfrac{1}{2} m \left(\dot{X}^2 + \dot{Y}^2\right)$$

   (d) Employ the following Lagrange equation and find the equation of motion of the particle:

   $$\frac{d}{dt}\frac{\partial K}{\partial \dot{\theta}} + \frac{\partial K}{\partial \theta} = 0$$

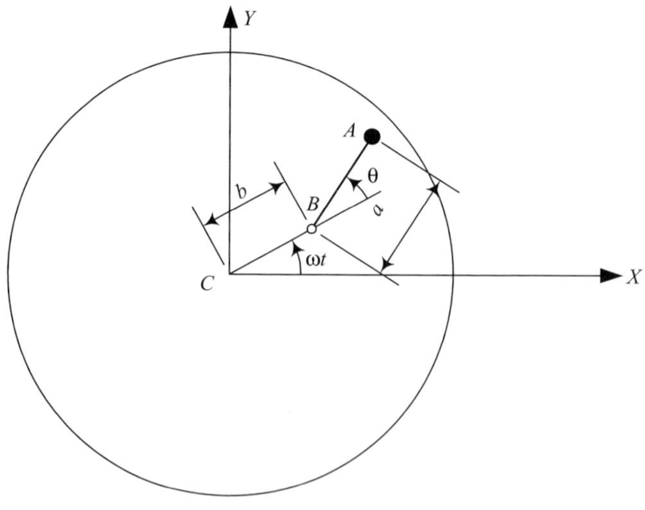

Figure 10.54 A particle attached to a point on a table.

3. **Two Particles with a Given Distance** Particles $P_1(x_1, y_1, z_1)$ and $P_2(x_2, y_2, z_2)$ are moving such that their distance d must follow a given function of time $d = f(t)$. Determine the equations of constraints in finite and infinitesimal displacements when:

   (a) The particles are free to move in 3D Cartesian space.

   (b) P_1 is moving on the x-axis by $x_1 = g(t)$.

   (c) P_1 is moving on the sphere $x^2 + y^2 + z^2 = R$.

4. **Constraints of a Ping Pong Ball** Figure 10.55 illustrates a ping pong table, two rackets, and a ball. Determine the constraints of the ball center if it remains in the (X, Y)-plane and the rockets are able to move up and down only.

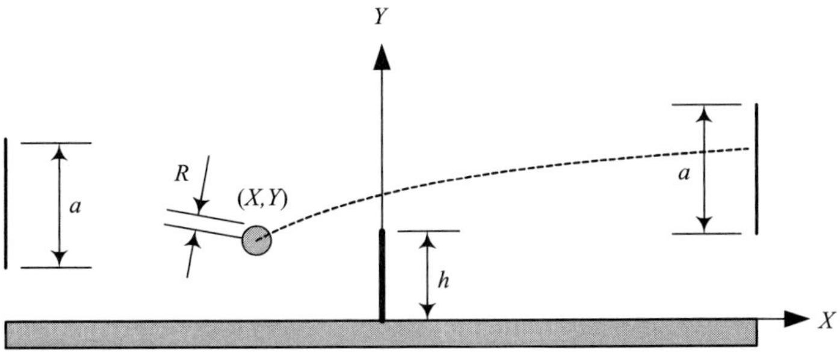

Figure 10.55 A side view of a ping pong table.

5. **Billiard Ball** Consider a billiard table of size $a \times b$.

   (a) Write the constraints of the ball center if there is only one ball of size R on the table.

   (b) Write the constraints of a ball center if there are 10 more balls of size R on the table.

6. ★ **Differential Constraint and Integrability** Determine which of the following deferential constraints is total and determine the holonomic constraint:

   (a)
   $$(t + 2/x)dx + x\,dt = 0$$

   (b)
   $$\left(x - t^3\right)dx + \left(x^3 + t\right)dt = 0$$

   (c)
   $$\left(2tx^4 + \sin x\right)dx + \left(4t^2x^3 + t\cos x\right)dt = 0$$

7. **A Sliding Stick** The sliding stick in Figure 10.56 has one DOF. If we show the position of the center by (x, y) and the angle by θ, write the constraint equations.

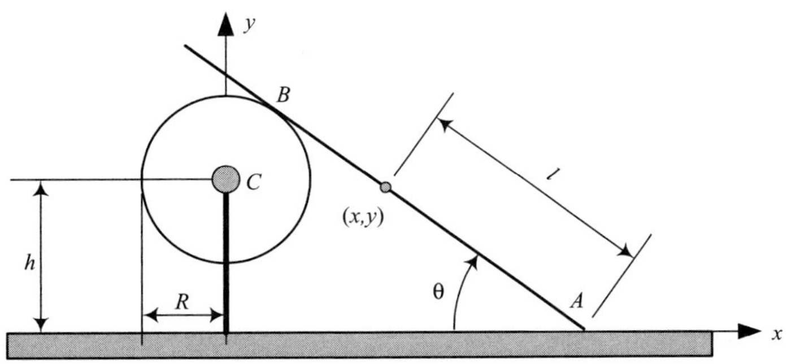

Figure 10.56 A sliding stick.

8. **Velocity of a Constraint Particle** Consider a particle m which is under the following two holonomic constraints:

$$f_i(x, y, z) = 0 \qquad i = 1, 2$$

Show that the velocity vector $\mathbf{v}$ of the particle is subjected to the condition

$$\mathbf{v} = \frac{\dot{f}_i}{\left|\nabla f_i\right|^2} \nabla f_i + \mathbf{c}_i \qquad \mathbf{c}_i \perp \nabla f_i \qquad i = 1, 2$$

and the acceleration vector $\mathbf{a}$ of the particle is subjected to the condition

$$\mathbf{a} = \frac{D^2 f_i}{\left|\nabla f_i\right|^2} \nabla f_i + \mathbf{b}_i \qquad \mathbf{b}_i \perp \nabla f_i \qquad i = 1, 2$$

where $\mathbf{c}$ and $\mathbf{b}$ are arbitrary constant vectors.

9. ★ **A Stick between Two Circular Walls** Determine the constraints of the stick in Figure 10.57 if the ends A and B cannot leave the walls.

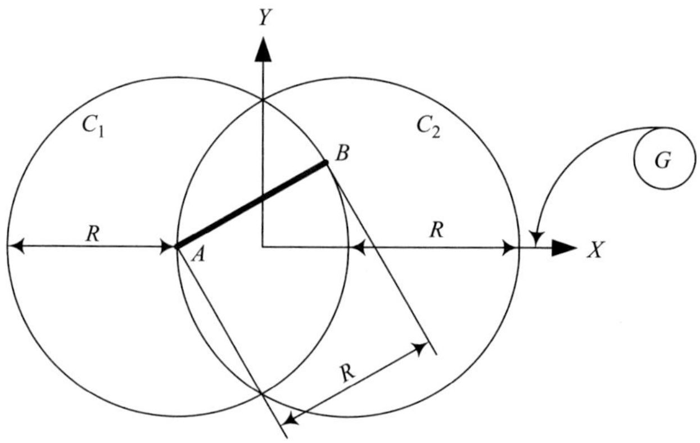

Figure 10.57 A stick between two circular walls.

10. **Constraint of a Rolling Disc on a Curve** Determine the constraints on the motion of a rolling disc on a given path, as shown in Figure 10.58.

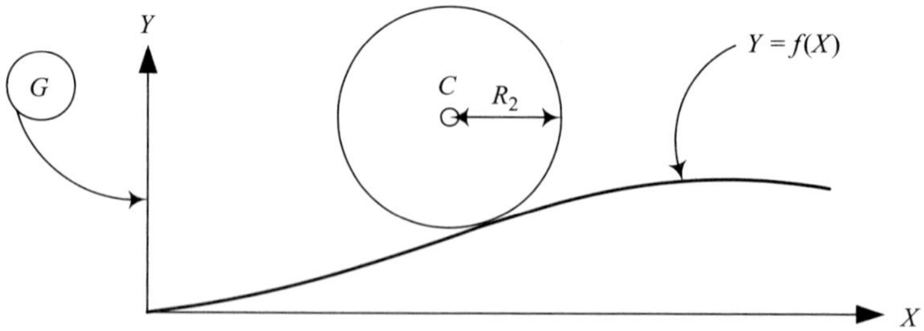

Figure 10.58 A rolling disc on a curve.

11. **Constraint of a Bicycle on a Curve** Figure 10.59 illustrates a bicycle that is moving on a given path. Determine the constraints on the bar AB if the wheels roll without slip.

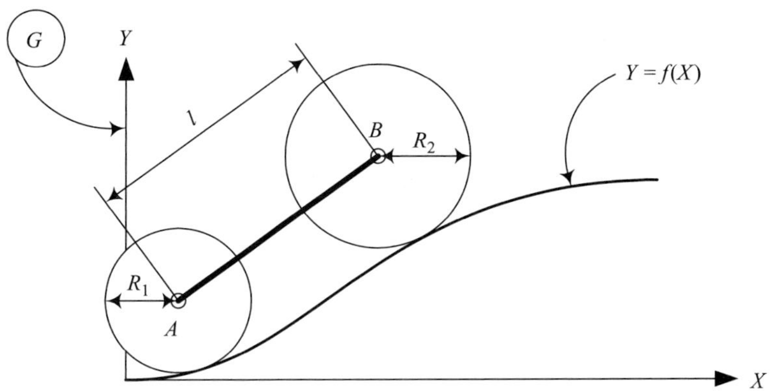

Figure 10.59 A bicycle on a curve.

12. **Differential Constraint** Consider the following differential constraints:

(a)

$$\dot{q}_1 + \dot{q}_2/q_2 - \dot{q}_3/q_2 = 0$$

(b)

$$\dot{q}_1 q_2 q_3 + q_1 \dot{q}_2 q_3 + q_1 q_2 \dot{q}_3 = 0$$

Check if they are exact or integrable.

13. **Chasing Problem** Figure 10.60 shows a particle A that is moving on the X-axis with a given function of time. A chasing point B is going toward A with a velocity $\mathbf{v}$. What is the required constraint if $\mathbf{v}$ is supposed to point A at every time t?

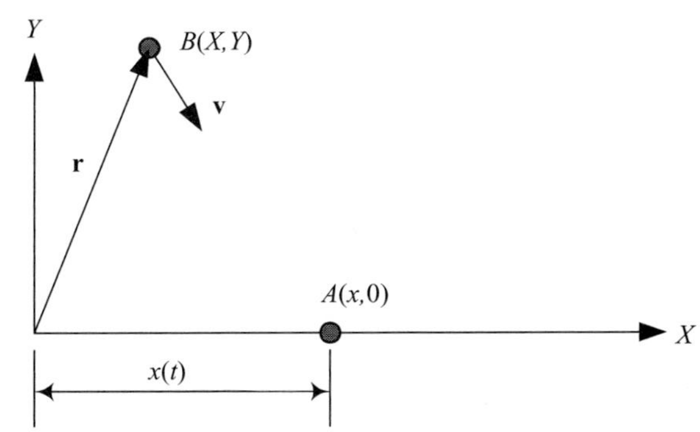

Figure 10.60 A moving point B is pointing another moving point A.

14. ★ **Differential Constraint and Integrability** Determine which of the following deferential constraints is total and determine the holonomic constraint:

(a)

$$(t + 2/x)\, dx + x\, dt = 0$$

(b)

$$\left(x - t^3\right) dx + \left(x^3 + t\right) dt = 0$$

(c)

$$\left(2tx^4 + \sin x\right) dx + \left(4t^2x^3 + t\cos x\right) dt = 0$$

15. **Equilibrium Position of Two Pivoted Massive Bars** Using virtual work:

(a) Determine the equilibrium angle θ for the system of Figure 10.61 if the free length of the spring is at $\theta = 45$ deg.

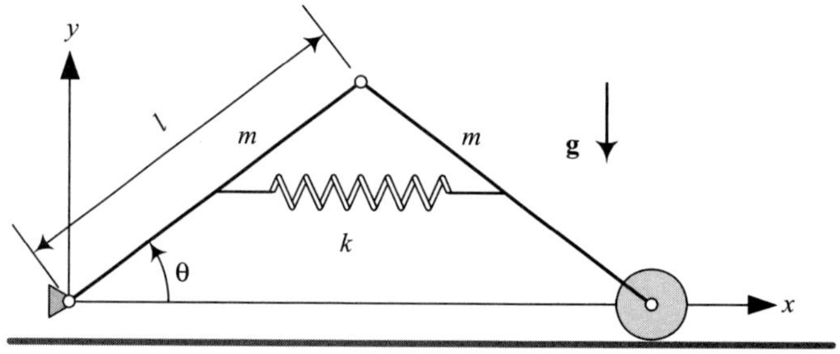

Figure 10.61 Two pivoted massive bars.

(b) What should the stiffness k be to have an equilibrium position at $\theta = 30$ deg?

16. **Equilibrium Position of Two Pivoted Massive Bars** Figure 10.62 illustrates two pivoted massive bars on a frictionless circle of radius R. Employ virtual work and determine the equilibrium angle θ for the system.

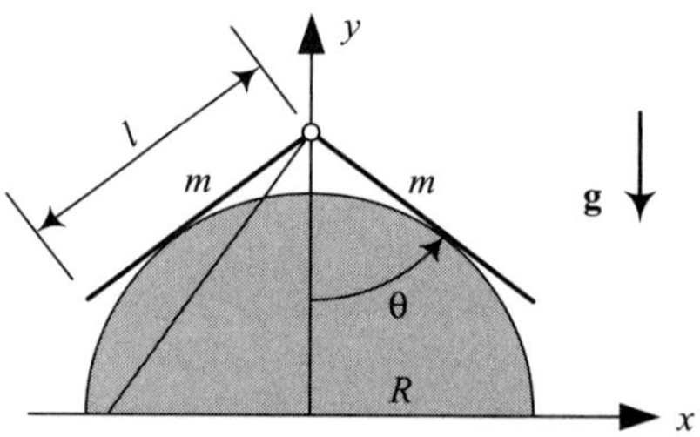

Figure 10.62 Two pivoted massive bars on a circle of radius R.

17. **A Constrained Moving Bar** Point A of the bar in Figure 10.63 is constrained to move on the x-axis according to a given function $x_A = f(t)$. Determine the constraints of the motion of the bar if:

(a) The motion of the bar is such that the midpoint P is moving parallel to the y-axis. Determine the coordinates of B.

(b) The angular motion of θ is based on a given function $\theta = g(t)$. Determine the coordinates of P and B. Determine $f(t)$ and $g(t)$ such that B moves on an ellipse.

(c) $f(t) = a\sin\omega_1 t$ and $g(t) = a\sin\omega_2 t$. Discuss the path of motion of B for different ratios of ω_1/ω_2.

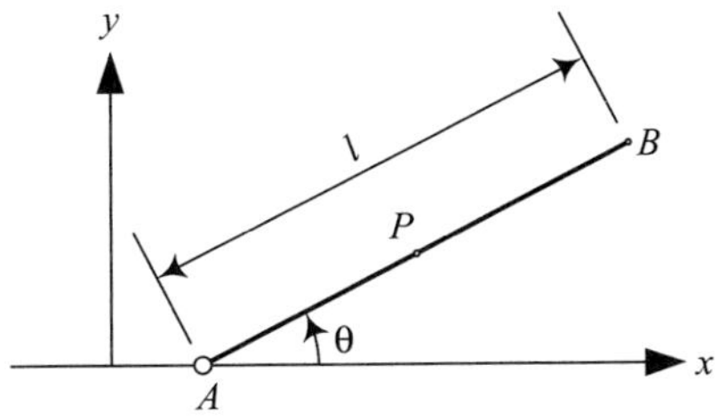

Figure 10.63 A constrained moving bar.

18. **Equilibrium of a Double Pendulum** A double pendulum with massive bars is illustrated in Figure 10.64. If a horizontal force F is applied at the tip point of the pendulum, what would be the equilibrium positions of the pendulum?

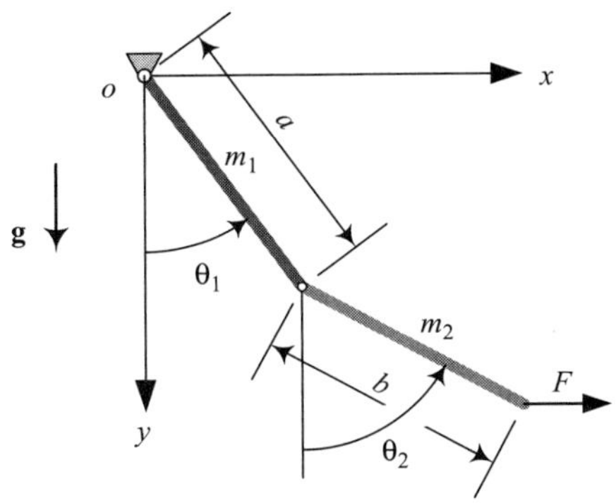

Figure 10.64 Equilibrium positions of a double pendulum with an applied force.

19. ★ **Routhian and Energy** Show that the energy of a dynamic system is given as

$$E = \dot{q}\frac{\partial \mathcal{L}}{\partial \dot{q}} + \dot{s}\frac{\partial \mathcal{L}}{\partial \dot{s}} - \mathcal{L} = \dot{q}p + \dot{s}\frac{\partial \mathcal{L}}{\partial \dot{s}} - \mathcal{L}$$

which in terms of the Routhian would be

$$E = R - \dot{s}\frac{\partial \mathcal{L}}{\partial \dot{s}}$$

20. ★ **Routhian for a Symmetric Top** Determine the Lagrangian and Routhian of a symmetric top in an external gravitational field, in terms of the Eulerian angles (φ, θ, ψ).

21. Two Elastically Connected Particles A particle P_1 moves in a horizontal plane along the following circle:

$$x_1 = R\cos\theta \qquad y_1 = R\sin\theta$$

A second particle P_2 is constrained to move on the following space curve:

$$x_2 = R\sin\varphi \qquad y_2 = R\sin\varphi \qquad z_2 = h\sin\varphi$$

The two particles are connected by a linear spring with stiffness k.

(a) Determine the Lagrangian and equations of motion of the dynamic system.

(b) Determine the Hamiltonian equations of motion.

Part IV

Dynamics

Dynamics is the science of *motion* and describes why and how a motion occurs when a force system is applied on a dynamic system. The motion can be considered as the evolution of the position and orientation, and their time derivatives.

11

Rigid Body and Mass Moment

In analyzing the motion of rigid bodies, two types of integrals arise that belong to the geometry of the body. The first type defines the mass center and appears when the translational motion of the body is considered. The second is the *second moment of mass*, which appears when the rotational motion of the body is considered. The second moment of mass is also called the *mass moment, centrifugal moment*, *moment of inertia*, or *deviation moment*.

11.1 RIGID BODY

A set of point masses with constant relative distances is called a *rigid body*. We attach a fixed coordinate frame B to the rigid body and call it a *body frame.* The body frame may be set at any arbitrary point; however, we usually set B at the mass center of the body. Because the coordinates of the point masses are constant in the body frame, we examine the position and orientation of the body frame with respect to a global coordinate frame G to determine the motion of the rigid body.

A constraint-free rigid body has six degrees of freedom. We express the freedoms of a rigid body as three translations and three rotations of the body frame.

Proof: Consider n point masses that are connected to each other with fixed connections. The system is called rigid if and only if the distance between any two particles remains constant for all time, regardless of the applied force system. To connect n particles to each other, $n\,(n-1)/2$ connections are needed. So, there are $n\,(n-1)/2$ holonomic constraints among the n particles of a rigid body:

$$r_{ij} = \sqrt{\left(x_i - x_j\right)^2 + \left(y_i - y_j\right)^2 + \left(z_i - z_j\right)^2} = \text{const} \qquad (11.1)$$
$$i, j = 1, 2, \ldots, n$$

However, not all of these constraints are independent.

Figure 11.1 illustrates a rigid body that is made by five rigidly connected particles. The first three particles make a planar rigid body by only three possible connections. Connecting a fourth particle to the first three makes a four-particle rigid body with six connections. Considering each particle as a ball joint, we lose the rigidity of m_1, m_2, m_3, or m_4 upon removal of any one of the six connections among the four particles. Such a system of connected particles, in which the number of connections is the minimum required, is called *just rigid*.

Because a free particle has three DOF, we need only three holonomic constraints to take its freedoms. Therefore, we can add a fifth particle m_5 to the first four particles by

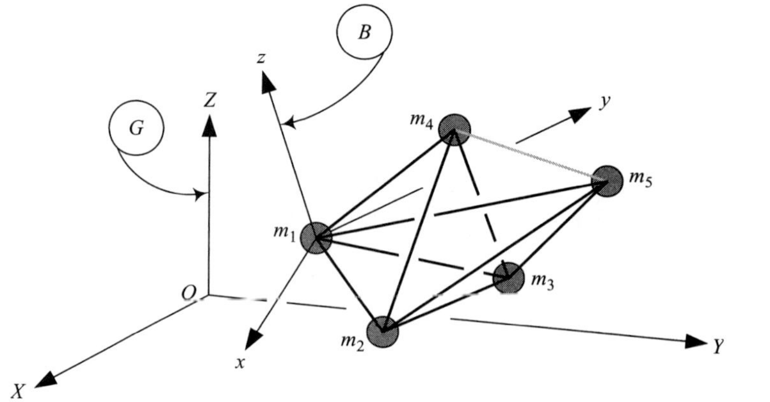

Figure 11.1 A rigid body that is made up of five rigidly connected particles.

connecting m_5 to only three non-colinear other particles. The body would be *overrigid* if we connect m_5 to the fourth particle. Similarly, a sixth particle m_6 may have five connections with only three of them required.

Let us assume that m is the number of particles in excess of 3, so $n = 3 + m$. Each particle which is added to the first three requires three additional non-colinear connections to make a just-rigid body. Therefore, the least number of rigid connections between $n = 3 + m$ particles of a rigid body is $3 + 3m = 3 + 3(n - 3) = 3n - 6$. Because n free particles have $3n$ DOF, a rigid body of n particles will have $f_C = 6$ DOF:

$$f_C = 3n - (3n - 6) = 6 \tag{11.2}$$

Being overrigid will not change the DOF of the rigid body. This is because we take the DOF of a particle by using three connections. Any extra connection is redundant and will not change the DOF of the particle and the rigid body. ■

Example 690 Alternative Explanation of DOF of a Rigid Body A single particle has one DOF if it is restricted to movement on a curve. It has two DOF if it moves on a surface. A particle that moves freely in space has three DOF.

Two particles that are rigidly connected have five DOF because the first particle can freely move in space and the second particle is restricted to movement on a spherical surface around the first one. The radius of the sphere is the length of the connection. Now, consider n rigidly connected particles and single out one of the particles. It has three DOF. A second particle is at a constant distance from the first one and gives two more DOF. A third particle can only move on a circle about the axis connecting the first two particles, so it provides one additional DOF. We can assume that every other particle is connected to the first three and hence provides no more DOF. Therefore, when the motions of the first three particles are specified, the motions of all other particles of the rigid body are uniquely determined. The DOF of the body would be

$$f_C = 3 + 2 + 1 = 6 \tag{11.3}$$

Example 691 ★ Rigid-Body Motion Theorem The rigid-body motion theorem states: The position of all points of a rigid body is determined by the position of three of its points provided the points are not colinear.

Proof: Consider a rigid body at a fixed position in a global coordinate frame G and three non-coplanar points A, B, C of the body occupy the same global coordinates. Consider a fourth point D of the body and suppose there exist two distinct global positions D_1 and D_2 of D in G. Imagine the tetrahedrons $ABCD_1$ and $ABCD_2$ that have a common base ABC and correspondingly equal edges. So, the points D_1 and D_2 are symmetrically placed with respect to the plane ABC, and therefore they cannot be brought into complete coincidence. This is however contrary to the assumption that the first and second positions of the tetrahedron $ABCD$ are $ABCD_1$ and $ABCD_2$, respectively. The unique body point D cannot have different body coordinates, and therefore it cannot have different global coordinates as well. ■

11.2 ELEMENTS OF THE MASS MOMENT MATRIX

In rotational dynamics of rigid bodies, geometric mass distribution integrals appear that are independent of the motion of the body. Because of this independence, we may analyze the integrals beforehand and simplify derivation of the equations of motion.

Every rigid body has a 3×3 mass moment or moment of inertia matrix $[I]$ denoted by

$$[I] = \begin{bmatrix} I_{xx} & I_{xy} & I_{xz} \\ I_{yx} & I_{yy} & I_{yz} \\ I_{zx} & I_{zy} & I_{zz} \end{bmatrix} \tag{11.4}$$

The diagonal elements I_{ij}, $i = j$, are called *polar moments of inertia,*

$$I_{xx} = I_x = \int_B \left(y^2 + z^2\right) dm \tag{11.5}$$

$$I_{yy} = I_y = \int_B \left(z^2 + x^2\right) dm \tag{11.6}$$

$$I_{zz} = I_z = \int_B \left(x^2 + y^2\right) dm \tag{11.7}$$

and the off-diagonal elements I_{ij}, $i \neq j$, are called *products of inertia,*

$$I_{xy} = I_{yx} = -\int_B xy \, dm \tag{11.8}$$

$$I_{yz} = I_{zy} = -\int_B yz \, dm \tag{11.9}$$

$$I_{zx} = I_{xz} = -\int_B zx \, dm \tag{11.10}$$

The elements of $[I]$ are only functions of the mass distribution of the rigid body and may be defined by a single equation

$$I_{ij} = \int_B \left(r^2 \delta_{ij} - x_i x_j\right) dm \qquad i,j = 1,2,3 \tag{11.11}$$

where δ_{ij} is Kronecker's delta (1.125). The elements I_{ij} may also be shown by the following equations:

$$I_{ii} = \int_B \left(x_i^2 + x_j^2\right) dm \quad i \neq j \neq k \tag{11.12}$$

$$I_{ij} = -\int_B x_i x_j \, dm \qquad i \neq j \tag{11.13}$$

The elements of $[I]$ are moments of inertia about a body coordinate frame attached to a point of the body. Therefore, I is a frame-dependent quantity and must be written with a frame indicator, ${}^B I$, to show the frame in which it is computed:

$${}^B I = \int_B \begin{bmatrix} y^2 + z^2 & -xy & -zx \\ -xy & z^2 + x^2 & -yz \\ -zx & -yz & x^2 + y^2 \end{bmatrix} dm \tag{11.14}$$

$$= \int_B \left(r^2 \mathbf{I} - \mathbf{r}\,\mathbf{r}^{\mathrm{T}}\right) dm = \int_B -\tilde{r}\,\tilde{r}\, dm \tag{11.15}$$

Proof: Consider n particles of a rigid body with masses m_i, $i = 1, 2, \ldots, n$, at positions $\mathbf{r}_i$ in a body coordinate frame B. Assume that the rigid body has a fixed point in the global coordinate frame G and turns about an instant axis of rotation with angular velocity ${}_G\boldsymbol{\omega}_B$. The fixed point is the common origin of G and B. The kinetic energy of the rigid body can be found by summation of the kinetic energies of all masses:

$$K = \frac{1}{2}\sum_{i=1}^{n} \mathbf{v}_i^2 m_i = \frac{1}{2}\sum_{i=1}^{n} (\boldsymbol{\omega} \times \mathbf{r}_i) \cdot (\boldsymbol{\omega} \times \mathbf{r}_i)\, m_i$$

$$= \frac{\omega_x^2}{2}\sum_{i=1}^{n}\left(y_i^2 + z_i^2\right) m_i + \frac{\omega_y^2}{2}\sum_{i=1}^{n}\left(z_i^2 + x_i^2\right) m_i + \frac{\omega_z^2}{2}\sum_{i=1}^{n}\left(x_i^2 + y_i^2\right) m_i$$

$$- \omega_x\omega_y \sum_{i=1}^{n} x_i y_i \, m_i - \omega_y\omega_z \sum_{i=1}^{n} y_i z_i \, m_i - \omega_z\omega_x \sum_{i=1}^{n} z_i x_i \, m_i \tag{11.16}$$

The summation over the masses and their coordinates are independent of the components of angular velocity. Let us introduce the local moment of inertia matrix as

$${}^B I = \sum_{i=1}^{n} \begin{bmatrix} y_i^2 + z_i^2 & -x_i y_i & -z_i x_i \\ -x_i y_i & z_i^2 + x_i^2 & -y_i z_i \\ -z_i x_i & -y_i z_i & x_i^2 + y_i^2 \end{bmatrix} m_i \tag{11.17}$$

where the elements are

$$I_{xx} = \sum_{i=1}^{n} \left[m_i \left(y_i^2 + z_i^2\right)\right] \tag{11.18}$$

$$I_{yy} = \sum_{i=1}^{n} \left[m_i \left(z_i^2 + x_i^2 \right) \right] \tag{11.19}$$

$$I_{zz} = \sum_{i=1}^{n} \left[m_i \left(x_i^2 + y_i^2 \right) \right] \tag{11.20}$$

$$I_{xy} = I_{yx} = -\sum_{i=1}^{n} \left(m_i x_i y_i \right) \tag{11.21}$$

$$I_{yz} = I_{zy} = -\sum_{i=1}^{n} \left(m_i y_i z_i \right) \tag{11.22}$$

$$I_{zx} = I_{xz} = -\sum_{i=1}^{n} \left(m_i z_i x_i \right) \tag{11.23}$$

Now, we can rewrite the kinetic energy of the body as

$$\begin{aligned} K &= \tfrac{1}{2} \left(I_{xx} \omega_x^2 + I_{yy} \omega_y^2 + I_{zz} \omega_z^2 \right) \\ &\quad - I_{xy} \omega_x \omega_y - I_{yz} \omega_y \omega_z - I_{zx} \omega_z \omega_x \end{aligned} \tag{11.24}$$

When the number of particles is infinity and the rigid body is a continuous and compact solid, the summations must be replaced by the integral of the kinetic energy of a mass element dm over the whole body:

$$\begin{aligned} K &= \frac{1}{2} \int_B \mathbf{v}^2 dm = \frac{1}{2} \int_B (\boldsymbol{\omega} \times \mathbf{r}) \cdot (\boldsymbol{\omega} \times \mathbf{r})\, dm \\ &= \frac{\omega_x^2}{2} \int_B \left(y^2 + z^2 \right) dm + \frac{\omega_y^2}{2} \int_B \left(z^2 + x^2 \right) dm + \frac{\omega_z^2}{2} \int_B \left(x^2 + y^2 \right) dm \\ &\quad - \omega_x \omega_y \int_B xy\, dm - \omega_y \omega_z \int_B yz\, dm - \omega_z \omega_x \int_B zx\, dm \end{aligned} \tag{11.25}$$

Introducing the mass moment matrix (11.15) with elements (11.5)–(11.10), we are able to write the kinetic energy as (11.24).

The kinetic energy can be rearranged to a matrix multiplication form:

$$K = \tfrac{1}{2}\, {}_B^G\boldsymbol{\omega}_B^{\mathrm{T}}\ {}^B I\ {}_B^G\boldsymbol{\omega}_B = \tfrac{1}{2}\, {}_B^G\boldsymbol{\omega}_B \cdot {}^B\mathbf{L} \tag{11.26}$$

We may utilize the homogeneous position vectors and define a more general moment of inertia $\overline{I}$, called the *pseudo–inertia matrix:*

$$\begin{aligned} {}^B\overline{I} &= \int_B \mathbf{r}\,\mathbf{r}^{\mathrm{T}}\, dm \\ &= \begin{bmatrix} \int_B x^2\, dm & \int_B xy\, dm & \int_B xz\, dm & \int_B x\, dm \\ \int_B xy\, dm & \int_B y^2\, dm & \int_B yz\, dm & \int_B y\, dm \\ \int_B xz\, dm & \int_B yz\, dm & \int_B z^2\, dm & \int_B z\, dm \\ \int_B x\, dm & \int_B y\, dm & \int_B z\, dm & \int_B dm \end{bmatrix} \end{aligned} \tag{11.27}$$

which can be expanded to

$$
{}^{B}\overline{I} = \begin{bmatrix} \dfrac{-I_{xx}+I_{yy}+I_{zz}}{2} & I_{xy} & I_{xz} & mx_C \\ I_{yx} & \dfrac{I_{xx}-I_{yy}+I_{zz}}{2} & I_{yz} & my_C \\ I_{zx} & I_{zy} & \dfrac{I_{xx}+I_{yy}-I_{zz}}{2} & mz_C \\ mx_C & mz_C & my_C & m \end{bmatrix} \tag{11.28}
$$

where

$$
\begin{bmatrix} x_C \\ y_C \\ z_C \end{bmatrix} = \begin{bmatrix} \dfrac{1}{m}\displaystyle\int_B x\,dm \\ \dfrac{1}{m}\displaystyle\int_B y\,dm \\ \dfrac{1}{m}\displaystyle\int_B z\,dm \end{bmatrix} = {}^{B}\mathbf{r}_C \tag{11.29}
$$

and ${}^{B}\mathbf{r}_C$ is the position of the mass center in the body frame. This vector is zero if the body frame is *central*. A body frame is central if it is at the mass center of the body. ■

Example 692 Mass Moment of a Brick and Spherical Rigid Body Consider a homogeneous rectangular brick with mass m, volume V, density $\rho = m/V$, length l, width w, and height h, as shown in Figure 11.2.

The local central coordinate frame is attached to the brick at its mass center. The mass moment matrix of the brick can be found by the integral method. We begin by

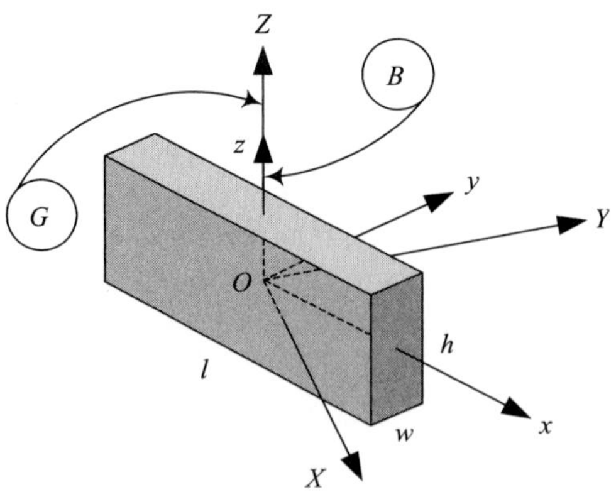

Figure 11.2 A homogeneous rectangular brick.

calculating

$$\begin{aligned} I_{xx} &= \int_B \left(y^2 + z^2\right) dm = \int_V \left(y^2 + z^2\right) \rho dV = \frac{m}{lwh} \int_V \left(y^2 + z^2\right) dV \\ &= \frac{m}{lwh} \int_{-h/2}^{h/2} \int_{-w/2}^{w/2} \int_{-l/2}^{l/2} \left(y^2 + z^2\right) dx\, dy\, dz \\ &= \frac{m}{12} \left(w^2 + h^2\right) \end{aligned} \tag{11.30}$$

which shows that I_{yy} and I_{zz} can be calculated similarly:

$$I_{yy} = \frac{m}{12} \left(h^2 + l^2\right) \tag{11.31}$$

$$I_{zz} = \frac{m}{12} \left(l^2 + w^2\right) \tag{11.32}$$

The coordinate frame B is central and symmetric. The products of inertia in such a frame are zero. To show this, we examine I_{xy}:

$$\begin{aligned} I_{xy} = I_{yx} &= -\int_B xy\, dm = \int_v xy\rho\, dv \\ &= \frac{m}{lwh} \int_{-h/2}^{h/2} \int_{-w/2}^{w/2} \int_{-l/2}^{l/2} xy\, dx\, dy\, dz = 0 \end{aligned} \tag{11.33}$$

Therefore, the mass moment matrix for the rigid rectangular brick in its central frame is

$${}^B I = \begin{bmatrix} \frac{1}{2} m \left(w^2 + h^2\right) & 0 & 0 \\ 0 & \frac{1}{2} m \left(h^2 + l^2\right) & 0 \\ 0 & 0 & \frac{1}{2} m \left(l^2 + w^2\right) \end{bmatrix} \tag{11.34}$$

Assuming that

$$l = w = h = a \tag{11.35}$$

we have

$${}^B I = \begin{bmatrix} \frac{1}{6} ma^2 & 0 & 0 \\ 0 & \frac{1}{6} ma^2 & 0 \\ 0 & 0 & \frac{1}{6} ma^2 \end{bmatrix} \tag{11.36}$$

A rigid body that has a mass moment matrix with equal diagonal elements and no off-diagonal elements, such as (11.36), is called a *spherical rigid body*.

Example 693 Mass Center The mass center of a body B that occupies the volume V with a local density $\rho(x, y, z)$ is a point C at which

$$\int_V \rho\, {}^B\mathbf{r}\, dV = 0 \tag{11.37}$$

Indicating the global position vector of C by ${}^G\mathbf{r}_C$, we have

$$ {}^G\mathbf{r}_C = \frac{\int_V \rho \, {}^G\mathbf{r} \, dV}{\int_V \rho \, dV} \tag{11.38}$$

Example 694 ★ Mass Center Kinematics Let us show the mass center $\mathbf{r}_C$ of a rigid body B by

$$\mathbf{r}_C = \mathbf{r}_0 + \frac{1}{m}\int_V (\mathbf{r} - \mathbf{r}_0)\, dm \tag{11.39}$$

where

$$m = \int_V \rho \, dm \qquad dm = \rho \, dV \tag{11.40}$$

and $\mathbf{r}_0$ is the position of an arbitrary point of B.

1. The expression (11.39) is independent of $\mathbf{r}_0$.

2. $\mathbf{r}_C$ is a unique point in the B-frame such that

$$\int (\mathbf{r} - \mathbf{r}_C)\, dm = 0 \tag{11.41}$$

3. $\mathbf{r}_C$ is in the interior of the convex hull of the B-body.

Proof: To check that the definition of $\mathbf{r}_C$ is independent of $\mathbf{r}_0$, we change $\mathbf{r}_0$ to another point $\mathbf{r}_1$:

$$\begin{aligned}
\mathbf{r}_1 + \frac{1}{m}\int_V (\mathbf{r} - \mathbf{r}_1)\, dm &= \frac{1}{m}\int_V (\mathbf{r} - \mathbf{r}_0)\, dm \\
&\quad + \frac{1}{m}\int_V (\mathbf{r}_0 - \mathbf{r}_1)\, dm + (\mathbf{r}_1 - \mathbf{r}_0) + \mathbf{r}_0 \\
&= \frac{1}{m}\int_V (\mathbf{r} - \mathbf{r}_0)\, dm + (\mathbf{r}_0 - \mathbf{r}_1) + (\mathbf{r}_1 - \mathbf{r}_0) + \mathbf{r}_0 \\
&= \mathbf{r}_0 + \frac{1}{m}\int_V (\mathbf{r} - \mathbf{r}_0)\, dm
\end{aligned} \tag{11.42}$$

Using the definition of $\mathbf{r}_C$ and (11.42), we have

$$\frac{1}{m}\int_V (\mathbf{r} - \mathbf{r}_0)\, dm = m\left(\mathbf{r}_C - \mathbf{r}_C\right) = 0 \tag{11.43}$$

Now assume $\mathbf{r}_1$ is an arbitrary point with the property that

$$\int_V (\mathbf{r} - \mathbf{r}_1)\, dm = 0 \tag{11.44}$$

Then, by (11.42), we have

$$\mathbf{r}_C = \frac{1}{m}\int_V (\mathbf{r} - \mathbf{r}_1)\, dm + \mathbf{r}_1 \tag{11.45}$$

which shows that $\mathbf{r}_C$ is unique:

$$\mathbf{r}_C = \mathbf{r}_1 \tag{11.46}$$

If $\mathbf{r}_C$ is on the boundary of the rigid body B or not in B, then there exists a plane passing through $\mathbf{r}_C$ such that there are points in B which lie on one side of the plane but there are no points in B on the opposite side of the plane. In other words, there exists an $\mathbf{r}$ such that the set S_1 is nonempty,

$$S_1 = \{\mathbf{r} \in \mathbf{B} \mid (\mathbf{r} - \mathbf{r}_C) > 0\} \tag{11.47}$$

but there exists an $\mathbf{r}$ such that the set S_2 is empty:

$$S_2 = \{\mathbf{r} \in \mathbf{B} \mid (\mathbf{r} - \mathbf{r}_C) < 0\} \tag{11.48}$$

It implies that

$$\int_V (\mathbf{r} - \mathbf{r}_C)\, dm > 0 \tag{11.49}$$

which contradicts (11.44). ■

Example 695 ★ Relative Diagonal Moments of Inertia Using the definitions for mass moments (11.5)–(11.7), it is seen that $[I]$ is symmetric:

$$I_{ij} = I_{ji} \qquad i \neq j \tag{11.50}$$

Furthermore, we have

$$\int_B \mathbf{r}^2 dm = \int_B \left(x^2 + y^2 + z^2\right) dm = \tfrac{1}{2}\left(I_{xx} + I_{yy} + I_{zz}\right) \tag{11.51}$$

which is called the mass moment about the origin.

Every diagonal element of $[I]$ is less than the sum of the other two diagonal elements:

$$I_{xx} + I_{yy} \geq I_{zz} \tag{11.52}$$

$$I_{yy} + I_{zz} \geq I_{xx} \tag{11.53}$$

$$I_{zz} + I_{xx} \geq I_{yy} \tag{11.54}$$

Noting that

$$(y - z)^2 \geq 0 \tag{11.55}$$

we have

$$\left(y^2 + z^2\right) \geq 2yz \tag{11.56}$$

and therefore, every diagonal element of $[I]$ is less than the sum of the off-diagonal elements that are not on the same row and column of the element:

$$I_{xx} \geq 2\left|I_{yz}\right| \tag{11.57}$$

$$I_{yy} \geq 2\left|I_{zx}\right| \tag{11.58}$$

$$I_{zz} \geq 2\left|I_{xy}\right| \tag{11.59}$$

We may combine these equations and show them by index notation:

$$I_{ii} + I_{jj} \geq I_{kk} \qquad i \neq j \neq k \tag{11.60}$$

$$I_{ii} \geq 2\left|I_{jk}\right| \qquad i \neq j \neq k \tag{11.61}$$

$$I_{ii} I_{jj} \geq I_{ij}^2 \qquad i \neq j \tag{11.62}$$

Example 696 ★ Short Notation of the Elements of Mass Moment Matrix Taking advantage of the Kronecker delta (1.125) we may write the elements of the mass moment matrix I_{ij} in the short notation forms

$$I_{ij} = \begin{cases} \displaystyle\int_B \left(\left(x_1^2 + x_2^2 + x_3^2\right)\delta_{ij} - x_i x_j\right) dm & (11.63) \\ \displaystyle\int_B \left(r^2 \delta_{ij} - x_i x_j\right) dm & (11.64) \\ \displaystyle\int_B \left(\sum_{k=1}^{3} x_k x_k \delta_{ij} - x_i x_j\right) dm & (11.65) \end{cases}$$

where

$$x_1 = x \qquad x_2 = y \qquad x_3 = z \tag{11.66}$$

Example 697 ★ Mass Moment with Respect to a Plane, a Line, and a Point The mass moment of a system of particles may be defined with respect to a plane, a line, or a point as the sum of the products of the mass of the particles times the square of the perpendicular distances from the particles to the plane, line, or point. For a continuous solid body, the sum would be a definite integral over the volume of the body.

The mass moments with respect to the xy-, yz-, and zx-plane are

$$I_{z^2} = \int_B z^2 \, dm \tag{11.67}$$

$$I_{y^2} = \int_B y^2 \, dm \tag{11.68}$$

$$I_{x^2} = \int_B x^2 \, dm \tag{11.69}$$

The mass moments with respect to the x-, y-, and z-axis are

$$I_x = \int_B \left(y^2 + z^2\right) dm \tag{11.70}$$

$$I_y = \int_B \left(z^2 + x^2\right) dm \tag{11.71}$$

$$I_z = \int_B \left(x^2 + y^2\right) dm \tag{11.72}$$

and therefore,

$$I_x = I_{y^2} + I_{z^2} \tag{11.73}$$

$$I_y = I_{z^2} + I_{x^2} \tag{11.74}$$

$$I_z = I_{x^2} + I_{y^2} \tag{11.75}$$

The moment of inertia with respect to the origin is

$$\begin{aligned} I_o &= \int_B \left(x^2 + y^2 + z^2\right) dm = I_{x^2} + I_{y^2} + I_{z^2} \\ &= \tfrac{1}{2}\left(I_x + I_y + I_z\right) \end{aligned} \tag{11.76}$$

Because the choice of the coordinate frame is arbitrary, we can say that the mass moment with respect to a line is the sum of the mass moments with respect to any two mutually orthogonal planes that pass through the line. The mass moment with respect to a point has a similar meaning for three mutually orthogonal planes intersecting at the point.

Example 698 ★ Geometry-Dependent Integrals Any integral in the form

$$\int_B x^i y^j z^k dm \tag{11.77}$$

that is performed over the volume of a rigid body B, where, i, j, and k are constant integers, is a geometry-dependent integral. It determines the mass center C if

$$i + j + k = 1 \tag{11.78}$$

and the mass moment of inertia I if

$$i + j + k = 2 \tag{11.79}$$

There is no physical definition for $i + j + k > 2$ yet.

Example 699 Radius of Gyration Consider a rigid body with mass m. Assume that the entire mass of the body is concentrated in a particle at a distance k from the point, line, or plane with respect to which the mass moment is I. If k is chosen such that

$$mk^2 = I \tag{11.80}$$

then k is called the *radius of gyration* with respect to the point, line, or plane.

If k is calculated for a line or plane that passes through the mass center, then it is called the *principal radius of gyration* k_C. The principal radius of gyration k_C has the minimum value of k. Assume I is the mass moment of a body with respect to a given axis and k is the radius of gyration about the axis. Let k_C be the principal radius of gyration about a parallel axis that passes through the mass center. Assume x_C, y_C, z_C are the coordinates of the mass center. If the coordinates of a point with respect to C are (a, b, c), then we have

$$x = x_C + a \qquad y = y_C + b \qquad z = z_C + c \tag{11.81}$$

$$\sum m_i a_i = \sum m_i b_i = \sum m_i c_i = 0 \tag{11.82}$$

$$mk_C^2 = \sum m_i \left(a_i^2 + b_i^2 + c_i^2\right) \tag{11.83}$$

Let us calculate the mass moment about an axis, say the z-axis:

$$\begin{aligned} mk^2 &= \sum m_i \left(x_i^2 + y_i^2\right) = \sum m_i \left((x_C + a_i)^2 + (y_C + b)^2\right) \\ &= \sum m_i \left(a_i^2 + b_i^2\right) + m\left(x_C^2 + y_C^2\right) \\ &\quad + 2x_C \sum m_i a_i + 2y_C \sum m_i y_i \end{aligned} \tag{11.84}$$

If r is the perpendicular distance between the two axes, then

$$r^2 = x_C^2 + y_C^2 \tag{11.85}$$

and therefore,

$$mk^2 = mk_C^2 + mr^2 \tag{11.86}$$

Because r^2 is always nonnegative, a principal radius of gyration is smaller than the radius of gyration for any other parallel axis. This is true for any plane, line, or point.

11.3 TRANSFORMATION OF MASS MOMENT MATRIX

Mass moments can be transformed from a coordinate frame B_1 to another coordinate frame B_2, both installed at the mass center of the body, according to the *rotated-axes theorem*:

$${}^2I = {}^2R_1 \; {}^1I \; {}^2R_1^{\mathrm{T}} \tag{11.87}$$

Mass moments of inertia can be transformed from a central frame B_1 to another frame B_2 which is parallel to B_1 according to *parallel-axes theorem*:

$${}^2I = {}^1I + m\,{}^2\tilde{d}_1\,{}^2\tilde{d}_1^{\mathrm{T}} \tag{11.88}$$

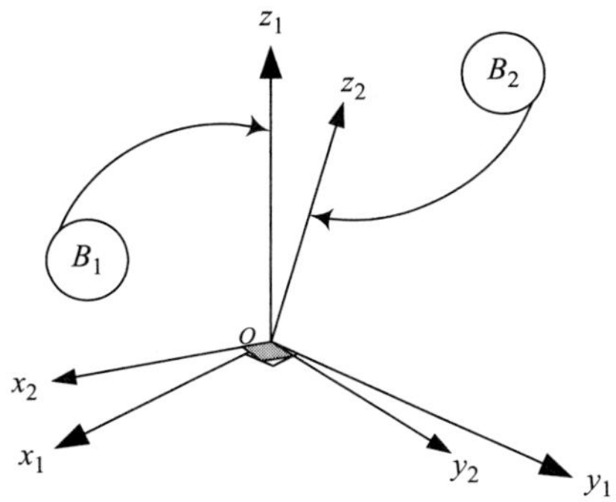

Figure 11.3 Two coordinate frames with a common origin at mass center of a rigid body.

where ${}^2\tilde{d}_1$ is the matrix form of the position vector ${}^2\mathbf{d}_1$ of B_1 in B_2:

$$ {}^2\tilde{d}_1 = \begin{bmatrix} 0 & -{}^2d_z & {}^2d_y \\ {}^2d_z & 0 & -{}^2d_x \\ -{}^2d_y & {}^2d_x & 0 \end{bmatrix} \tag{11.89} $$

Proof: Two coordinate frames with a common origin are shown in Figure 11.3. The angular velocity and angular momentum of a rigid body transform from frame B_1 to B_2 by vector transformation:

$$ {}^2\boldsymbol{\omega} = {}^2R_1\, {}^1\boldsymbol{\omega} \tag{11.90} $$

$$ {}^2\mathbf{L} = {}^2R_1\, {}^1\mathbf{L} \tag{11.91} $$

However, $\mathbf{L}$ and $\boldsymbol{\omega}$ are related according to

$$ {}^1\mathbf{L} = {}^1I\ {}^1\boldsymbol{\omega} \qquad {}^2\mathbf{L} = {}^2I\ {}^2\boldsymbol{\omega} \tag{11.92} $$

and therefore,

$$ {}^2\mathbf{L} = {}^2R_1\, {}^1I\ {}^2R_1^{\mathrm{T}}\, {}^2\boldsymbol{\omega} = {}^2I\ {}^2\boldsymbol{\omega} \tag{11.93} $$

It shows how to transfer the moment of inertia from a coordinate frame B_1 to a rotated frame B_2:

$$ {}^2I\ = {}^2R_1\, {}^1I\ {}^2R_1^{\mathrm{T}} \tag{11.94} $$

Now consider a coordinate frame B_1 at ${}^2\mathbf{d}_1$ as shown in Figure 11.4. Frame B_2 moves in a fixed frame B_1 such that their axes remain parallel. The angular velocities of B_1 and B_2 are equal and the angular momentum of the rigid body transforms from the frame B_1 to B_2 by

$$ {}^2\omega_1 = {}^1\omega_2 = \boldsymbol{\omega} \tag{11.95} $$

$$ {}^2\mathbf{L} = {}^1\mathbf{L} + \left({}^2\mathbf{d}_1 \times m\, {}^2\mathbf{v}_1\right) \tag{11.96} $$

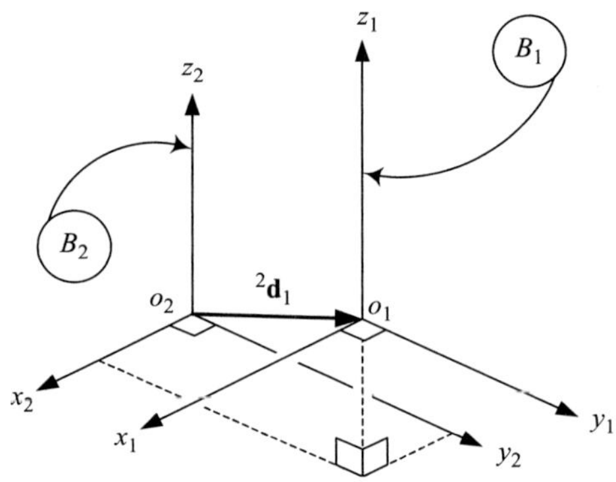

Figure 11.4 A central coordinate frame B_1 and a translated frame B_2.

Therefore,

$$\begin{aligned} {}^2\mathbf{L} &= {}^1\mathbf{L} + m\,{}^2\mathbf{d}_C \times \left(\boldsymbol{\omega} \times {}^2\mathbf{d}_C\right) = {}^1\mathbf{L} + \left(m\,{}^2\tilde{d}_1\,{}^2\tilde{d}_1^{\mathrm{T}}\right)\boldsymbol{\omega} \\ &= \left({}^1I + m\,{}^2\tilde{d}_1\,{}^2\tilde{d}_1^{\mathrm{T}}\right)\boldsymbol{\omega} \end{aligned} \tag{11.97}$$

which shows how to transfer the moment of inertia from frame B_1 to a parallel frame B_2:

$${}^2I = {}^1I + m\,{}^2\tilde{d}_1\,{}^2\tilde{d}_1^{\mathrm{T}} \tag{11.98}$$

The relative translational vectors of two parallel coordinate frames B_1 and B_2 are opposite:

$${}^1\mathbf{d}_2 = -\,{}^2\mathbf{d}_1 \tag{11.99}$$

We can show the relative frames as in Figure 11.5 and rewrite Equation (11.97) as

$${}^2I = {}^1I + m\,{}^1\tilde{d}_2\,{}^1\tilde{d}_2^{\mathrm{T}} \tag{11.100}$$

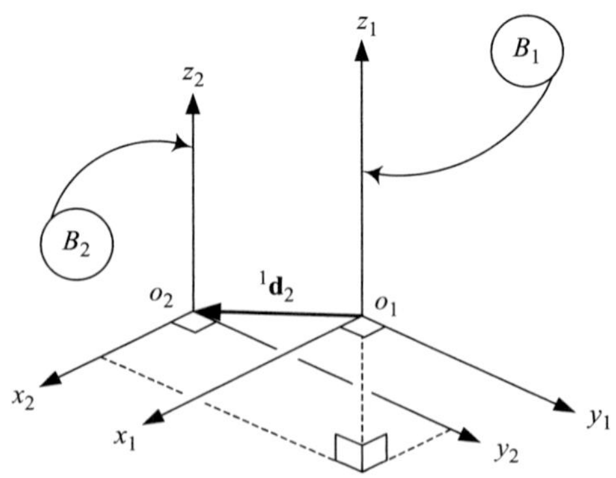

Figure 11.5 A central coordinate frame B_1 and a translated frame B_2.

because for any skew-symmetric matrix $\tilde{r}$ we have

$$\tilde{r}\,\tilde{r}^{\mathrm{T}} = \tilde{r}^{\mathrm{T}}\,\tilde{r} = (-\tilde{r})\,(-\tilde{r})^{\mathrm{T}} \tag{11.101}$$

The parallel-axes theorem is also called the *Huygens–Steiner theorem*. ■

Example 700 Translation of the Inertia Matrix The moment of inertia matrix of the rigid brick shown in Figure 11.6 in the central and symmetric frame $B(oxyz)$ is given in Equation (11.35). The moment of inertia matrix in the parallel frame G can be found by applying the parallel-axes transformation formula (11.97):

$${}^{G}I = {}^{B}I + m\,{}^{G}\tilde{d}_C\,{}^{G}\tilde{d}_C^{\mathrm{T}} \tag{11.102}$$

The position vector of the mass center is

$${}^{G}\mathbf{d}_C = \frac{1}{2}\begin{bmatrix} l \\ w \\ h \end{bmatrix} \tag{11.103}$$

and therefore,

$${}^{G}\tilde{d}_C = \frac{1}{2}\begin{bmatrix} 0 & -h & w \\ h & 0 & -l \\ -w & l & 0 \end{bmatrix} \tag{11.104}$$

which yields

$${}^{G}I = \begin{bmatrix} \frac{1}{3}h^2m + \frac{1}{3}mw^2 & -\frac{1}{4}lmw & -\frac{1}{4}hlm \\ -\frac{1}{4}lmw & \frac{1}{3}h^2m + \frac{1}{3}l^2m & -\frac{1}{4}hmw \\ -\frac{1}{4}hlm & -\frac{1}{4}hmw & \frac{1}{3}l^2m + \frac{1}{3}mw^2 \end{bmatrix} \tag{11.105}$$

Assuming

$$l = 8 \qquad h = 3 \qquad w = 1 \tag{11.106}$$

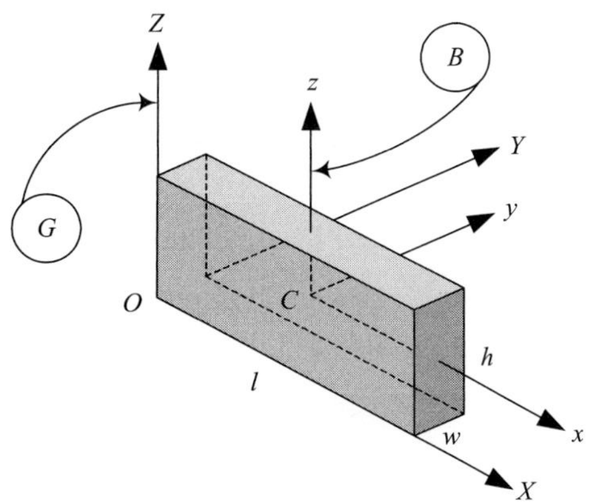

Figure 11.6 A rigid rectangular link in the principal and nonprincipal frames.

from Equation (11.34) we have

$$ {}^{B}I = \begin{bmatrix} \frac{5}{6}m & 0 & 0 \\ 0 & \frac{73}{12}m & 0 \\ 0 & 0 & \frac{65}{12}m \end{bmatrix} \tag{11.107} $$

and

$$ {}^{G}I = {}^{B}I + m\,{}^{G}\tilde{d}_C\,{}^{G}\tilde{d}_C^{\mathrm{T}} \tag{11.108} $$

$$ = \begin{bmatrix} \frac{10}{3}m & -2m & -6m \\ -2m & \frac{73}{3}m & -\frac{3}{4}m \\ -6m & -\frac{3}{4}m & \frac{65}{3}m \end{bmatrix} \tag{11.109} $$

For $l = h = w = a$, we have

$$ {}^{G}I = \frac{ma^2}{6}\begin{bmatrix} 1 & 0 & 0 \\ 0 & 1 & 0 \\ 0 & 0 & 1 \end{bmatrix} + \frac{ma^2}{4}\begin{bmatrix} 0 & -1 & 1 \\ 1 & 0 & -1 \\ -1 & 1 & 0 \end{bmatrix}\begin{bmatrix} 0 & -1 & 1 \\ 1 & 0 & -1 \\ -1 & 1 & 0 \end{bmatrix}^{\mathrm{T}} $$

$$ \times \begin{bmatrix} \frac{2}{3}ma^2 & -\frac{1}{4}ma^2 & -\frac{1}{4}ma^2 \\ -\frac{1}{4}ma^2 & \frac{2}{3}ma^2 & -\frac{1}{4}ma^2 \\ -\frac{1}{4}ma^2 & -\frac{1}{4}ma^2 & \frac{2}{3}ma^2 \end{bmatrix} \tag{11.110} $$

Example 701 Superposition According to the definition of mass moments in Equation (11.17) or (11.11), we are able to split a rigid body to arbitrary sections and add the mass moments of the individual sections to determine the mass moment of the body.

For example, let us determine the mass moment matrix ${}^{B}I$ of the whole cube of Figure 11.7 by splitting it into eight smaller cubes each of mass $m/8$. Using Equation (11.104) for $l = h = w = a/2$, we find ${}^{B}I$ for cube 1:

$$ {}^{B}I_1 = \int_B \begin{bmatrix} y^2+z^2 & -xy & -zx \\ -xy & z^2+x^2 & -yz \\ -zx & -yz & x^2+y^2 \end{bmatrix} dm $$

$$ = \frac{1}{32}\begin{bmatrix} \frac{2}{3}ma^2 & -\frac{1}{4}ma^2 & -\frac{1}{4}ma^2 \\ -\frac{1}{4}ma^2 & \frac{2}{3}ma^2 & -\frac{1}{4}ma^2 \\ -\frac{1}{4}ma^2 & -\frac{1}{4}ma^2 & \frac{2}{3}ma^2 \end{bmatrix} \tag{11.111} $$

By changing the sign of x, we find

$$ {}^{B}I_2 = \frac{1}{32}\begin{bmatrix} \frac{2}{3}ma^2 & \frac{1}{4}ma^2 & \frac{1}{4}ma^2 \\ \frac{1}{4}ma^2 & \frac{2}{3}ma^2 & -\frac{1}{4}ma^2 \\ \frac{1}{4}ma^2 & -\frac{1}{4}ma^2 & \frac{2}{3}ma^2 \end{bmatrix} \tag{11.112} $$

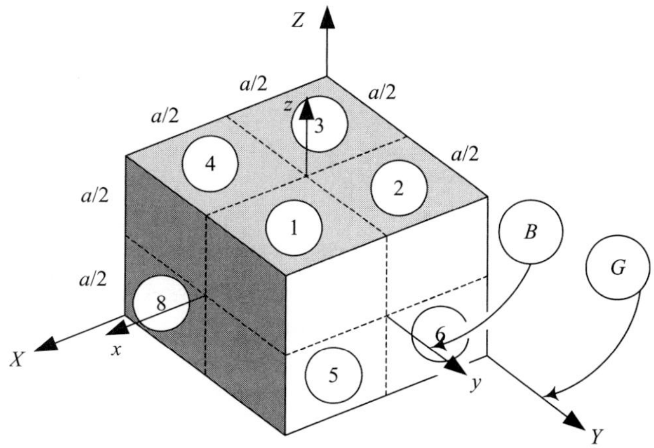

Figure 11.7 A cube as a superposition of eight smaller cubes.

Similarly,

$$
{}^B I_3 = \frac{1}{32} \begin{bmatrix} \frac{2}{3}ma^2 & -\frac{1}{4}ma^2 & \frac{1}{4}ma^2 \\ -\frac{1}{4}ma^2 & \frac{2}{3}ma^2 & \frac{1}{4}ma^2 \\ \frac{1}{4}ma^2 & \frac{1}{4}ma^2 & \frac{2}{3}ma^2 \end{bmatrix} \tag{11.113}
$$

$$
{}^B I_4 = \frac{1}{32} \begin{bmatrix} \frac{2}{3}ma^2 & \frac{1}{4}ma^2 & -\frac{1}{4}ma^2 \\ \frac{1}{4}ma^2 & \frac{2}{3}ma^2 & \frac{1}{4}ma^2 \\ -\frac{1}{4}ma^2 & \frac{1}{4}ma^2 & \frac{2}{3}ma^2 \end{bmatrix} \tag{11.114}
$$

$$
{}^B I_5 = \frac{1}{32} \begin{bmatrix} \frac{2}{3}ma^2 & -\frac{1}{4}ma^2 & \frac{1}{4}ma^2 \\ -\frac{1}{4}ma^2 & \frac{2}{3}ma^2 & \frac{1}{4}ma^2 \\ \frac{1}{4}ma^2 & \frac{1}{4}ma^2 & \frac{2}{3}ma^2 \end{bmatrix} \tag{11.115}
$$

$$
{}^B I_6 = \frac{1}{32} \begin{bmatrix} \frac{2}{3}ma^2 & \frac{1}{4}ma^2 & -\frac{1}{4}ma^2 \\ \frac{1}{4}ma^2 & \frac{2}{3}ma^2 & \frac{1}{4}ma^2 \\ -\frac{1}{4}ma^2 & \frac{1}{4}ma^2 & \frac{2}{3}ma^2 \end{bmatrix} \tag{11.116}
$$

$$
{}^B I_7 = \frac{1}{32} \begin{bmatrix} \frac{2}{3}ma^2 & -\frac{1}{4}ma^2 & -\frac{1}{4}ma^2 \\ -\frac{1}{4}ma^2 & \frac{2}{3}ma^2 & -\frac{1}{4}ma^2 \\ -\frac{1}{4}ma^2 & -\frac{1}{4}ma^2 & \frac{2}{3}ma^2 \end{bmatrix} \tag{11.117}
$$

$$
{}^B I_8 = \frac{1}{32} \begin{bmatrix} \frac{2}{3}ma^2 & \frac{1}{4}ma^2 & \frac{1}{4}ma^2 \\ \frac{1}{4}ma^2 & \frac{2}{3}ma^2 & -\frac{1}{4}ma^2 \\ \frac{1}{4}ma^2 & -\frac{1}{4}ma^2 & \frac{2}{3}ma^2 \end{bmatrix} \tag{11.118}
$$

Combining the eight mass moment matrices I_i, $i = 1, 2, \ldots, 8$, we have

$$
\begin{aligned}
{}^{B}I &= {}^{B}I_1 + {}^{B}I_2 + {}^{B}I_3 + {}^{B}I_4 + {}^{B}I_5 + {}^{B}I_6 + {}^{B}I_7 + {}^{B}I_8 \\
&= \begin{bmatrix} \frac{1}{6}ma^2 & 0 & 0 \\ 0 & \frac{1}{6}ma^2 & 0 \\ 0 & 0 & \frac{1}{6}ma^2 \end{bmatrix}
\end{aligned} \tag{11.119}
$$

Example 702 A Rotating Arm Figure 11.8 shows a rotating arm with a body coordinate frame B_1. Let us assume that ${}^{1}I_1$ is the mass moment matrix of the arm about its mass center C:

$$
{}^{1}I_1 = \begin{bmatrix} I_x & 0 & 0 \\ 0 & I_y & 0 \\ 0 & 0 & I_z \end{bmatrix} \tag{11.120}
$$

From the transformation matrix

$$
{}^{0}R_1 = R_{Z,\theta} = \begin{bmatrix} \cos\theta & -\sin\theta & 0 \\ \sin\theta & \cos\theta & 0 \\ 0 & 0 & 1 \end{bmatrix} \tag{11.121}
$$

we can determine ${}^{0}I_1$:

$$
\begin{aligned}
{}^{0}I_1 &= R_{Z,\theta}\, {}^{1}I_1\, R_{Z,\theta}^{\mathrm{T}} = {}^{0}R_1 \begin{bmatrix} I_x & 0 & 0 \\ 0 & I_y & 0 \\ 0 & 0 & I_z \end{bmatrix} {}^{0}R_1^{\mathrm{T}} \\
&= \begin{bmatrix} I_x\cos^2\theta + I_y\sin^2\theta & \left(I_x - I_y\right)\cos\theta\sin\theta & 0 \\ \left(I_x - I_y\right)\cos\theta\sin\theta & I_y\cos^2\theta + I_x\sin^2\theta & 0 \\ 0 & 0 & I_z \end{bmatrix}
\end{aligned} \tag{11.122}
$$

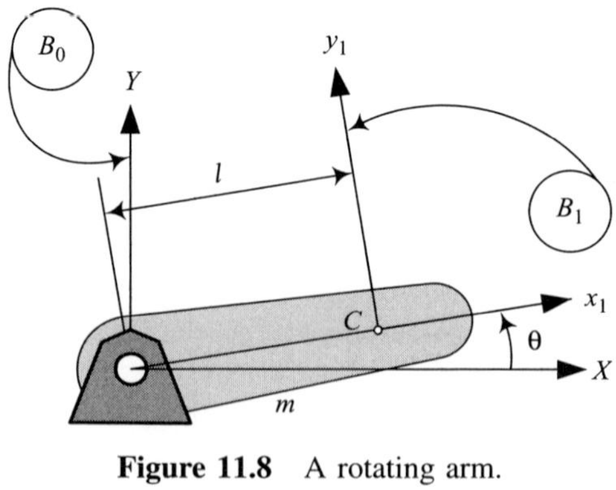

Figure 11.8 A rotating arm.

Example 703 2R Planar Manipulator Mass Moments A $2R$ manipulator that has massive arms is shown in Figure 11.9. The transformation matrices between coordinate frames B_1, B_2, and B_0 are

$$ {}^0R_1 = R_{Z,\theta_1} = \begin{bmatrix} \cos\theta_1 & -\sin\theta_1 & 0 \\ \sin\theta_1 & \cos\theta_1 & 0 \\ 0 & 0 & 1 \end{bmatrix} \tag{11.123} $$

$$ {}^1R_2 = R_{Z,\theta_2} = \begin{bmatrix} \cos\theta_2 & -\sin\theta_2 & 0 \\ \sin\theta_2 & \cos\theta_2 & 0 \\ 0 & 0 & 1 \end{bmatrix} \tag{11.124} $$

$$ {}^0R_2 = R_{z,\theta_1+\theta_2} = \begin{bmatrix} \cos(\theta_1+\theta_2) & -\sin(\theta_1+\theta_2) & 0 \\ \sin(\theta_1+\theta_2) & \cos(\theta_1+\theta_2) & 0 \\ 0 & 0 & 1 \end{bmatrix} \tag{11.125} $$

If we have the mass moment matrices in the local frames B_1 and B_2, we can determine the mass moment matrices in the global coordinate frame B_0:

$$ \begin{aligned} {}^0I_1 &= R_{Z,\theta_1}\, {}^1I_1\, R_{Z,\theta_1}^{\mathrm{T}} = {}^0R_1 \begin{bmatrix} I_{x_1} & 0 & 0 \\ 0 & I_{y_1} & 0 \\ 0 & 0 & I_{z_1} \end{bmatrix} {}^0R_1^{\mathrm{T}} \\ &= \begin{bmatrix} I_{x_1}c^2\theta_1 + I_{y_1}s^2\theta_1 & \left(I_{x_1} - I_{y_1}\right)c\theta_1 s\theta_1 & 0 \\ \left(I_{x_1} - I_{y_1}\right)c\theta_1 s\theta_1 & I_{y_1}c^2\theta_1 + I_{x_1}s^2\theta_1 & 0 \\ 0 & 0 & I_{z_1} \end{bmatrix} \end{aligned} \tag{11.126} $$

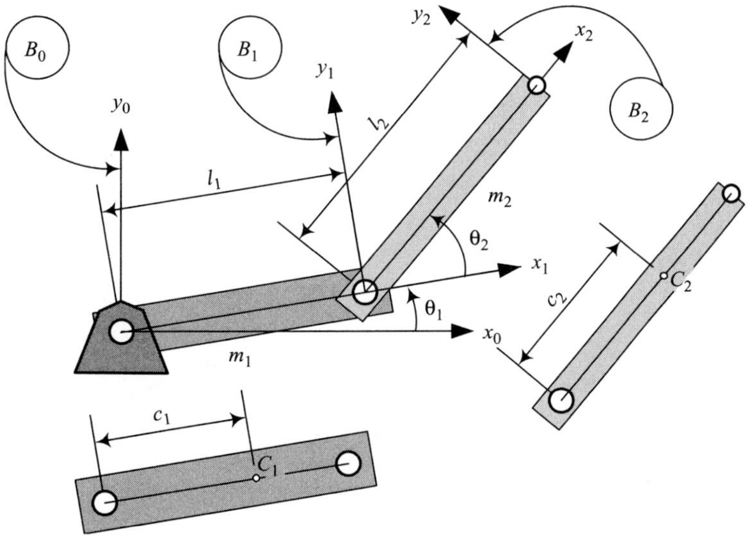

Figure 11.9 A $2R$ manipulator with massive arms.

$$
{}^{0}I_2 = {}^{0}R_2\, {}^{2}I_2\, {}^{0}R_2^{\mathrm{T}} = {}^{0}R_2 \begin{bmatrix} I_{x_2} & 0 & 0 \\ 0 & I_{y_2} & 0 \\ 0 & 0 & I_{z_2} \end{bmatrix} {}^{0}R_2^{\mathrm{T}}
$$

$$
= \begin{bmatrix} I_{x_2}c^2\theta_{12} + I_{y_2}s^2\theta_{12} & \left(I_{x_2} - I_{y_2}\right) c\theta_{12}s\theta_{12} & 0 \\ \left(I_{x_2} - I_{y_2}\right) c\theta_{12}s\theta_{12} & I_{y_2}c^2\theta_{12} + I_{x_2}s^2\theta_{12} & 0 \\ 0 & 0 & I_{z_2} \end{bmatrix} \tag{11.127}
$$

$$
\theta_{12} = \theta_1 + \theta_2 \tag{11.128}
$$

Example 704 ★ General Mass Moment Transformation Consider two body coordinates B and B_1 that are respectively installed at the mass center C of a rigid body and at a point A as shown in Figure 11.10. The coordinate frame B_1 is at a different orientation than B. To determine the relationship between the mass moments of ${}^{1}I$ and ${}^{B}I$, we may set up a new body coordinate B_2 at A to be parallel with B_1:

$$
B_2 \parallel B_1 \tag{11.129}
$$

The relation between ${}^{2}I$ and ${}^{1}I$ is a rotation:

$$
{}^{1}I = {}^{1}R_2\, {}^{2}I\, {}^{1}R_2^{\mathrm{T}} \tag{11.130}
$$

The relation between ${}^{2}I$ and ${}^{B}I$ is a translation. Showing the translation vector by ${}^{B}\mathbf{r}_2 = {}^{B}\mathbf{r}_A$, we have

$$
{}^{2}I = {}^{B}I + m\, {}^{B}\tilde{r}_2\, {}^{B}\tilde{r}_2^{\mathrm{T}} \tag{11.131}
$$

Substituting (11.131) in (11.130) provides the relation of ${}^{1}I$ and ${}^{B}I$:

$$
\begin{aligned}
{}^{1}I &= {}^{1}R_2 \left({}^{B}I + m\, {}^{B}\tilde{r}_2\, {}^{B}\tilde{r}_2^{\mathrm{T}}\right) {}^{1}R_2^{\mathrm{T}} \\
&= {}^{1}R_2\, {}^{B}I\, {}^{1}R_2^{\mathrm{T}} + m\, {}^{1}R_2 \left({}^{B}\tilde{r}_2\, {}^{B}\tilde{r}_2^{\mathrm{T}}\right) {}^{1}R_2^{\mathrm{T}}
\end{aligned} \tag{11.132}
$$

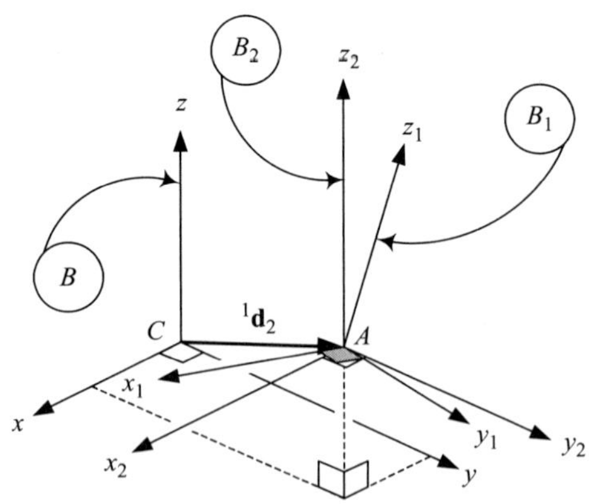

Figure 11.10 Two body coordinates B and B_1 that are respectively installed at points C and A of a rigid body.

Let us write (11.130) and (11.131) as

$$^{2}I = {}^{2}R_1 \, {}^{1}I \, {}^{2}R_1^{\rm T} \tag{11.133}$$

$$^{B}I = {}^{2}I - m \, {}^{B}\tilde{r}_2 \, {}^{B}\tilde{r}_2^{\rm T} \tag{11.134}$$

to determine ^{B}I for a given ^{1}I:

$$\begin{aligned} ^{B}I &= {}^{2}I - m \, {}^{B}\tilde{r}_2 \, {}^{B}\tilde{r}_2^{\rm T} \\ &= {}^{2}R_1 \, {}^{1}I \, {}^{2}R_1^{\rm T} - m \, {}^{B}\tilde{r}_2 \, {}^{B}\tilde{r}_2^{\rm T} \end{aligned} \tag{11.135}$$

Example 705 ★ Mass Moment of a Cube in Two Frames Figure 11.11 illustrates a cube with side a in a global coordinate frame G and the following mass moment matrix:

$$^{G}I = \begin{bmatrix} \frac{2}{3}ma^2 & -\frac{1}{4}ma^2 & -\frac{1}{4}ma^2 \\ -\frac{1}{4}ma^2 & \frac{2}{3}ma^2 & -\frac{1}{4}ma^2 \\ -\frac{1}{4}ma^2 & -\frac{1}{4}ma^2 & \frac{2}{3}ma^2 \end{bmatrix} \tag{11.136}$$

Another coordinate B is set at the corner (a, a, a) such that the y-axis is along the diagonal of the cube and the x-axis is parallel to the (X, Y)-plane. To determine ^{B}I, we need to find $^{G}R_B$ first. The unit vectors of B in G are

$$\hat{\jmath} = \frac{\hat{I} + \hat{J} + \hat{K}}{\sqrt{3}} = \frac{1}{\sqrt{3}} \begin{bmatrix} 1 \\ 1 \\ 1 \end{bmatrix} \tag{11.137}$$

$$\hat{\imath} = r_{11}\hat{I} + r_{12}\hat{J} = \begin{bmatrix} r_{11} \\ r_{12} \\ 0 \end{bmatrix} \tag{11.138}$$

$$\hat{k} = r_{31}\hat{I} + r_{32}\hat{J} + r_{33}\hat{K} = \begin{bmatrix} r_{31} \\ r_{32} \\ r_{33} \end{bmatrix} \tag{11.139}$$

and therefore,

$$^{G}R_B = \begin{bmatrix} r_{11} & \frac{1}{\sqrt{3}} & r_{31} \\ r_{12} & \frac{1}{\sqrt{3}} & r_{32} \\ 0 & \frac{1}{\sqrt{3}} & r_{33} \end{bmatrix} \tag{11.140}$$

Using $^{G}R_B \, {}^{G}R_B^{\rm T} = \mathbf{I}$ yields

$$^{G}R_B = \begin{bmatrix} \frac{\sqrt{2}}{2} & \frac{1}{\sqrt{3}} & -\frac{\sqrt{6}}{6} \\ -\frac{\sqrt{2}}{2} & \frac{1}{\sqrt{3}} & -\frac{\sqrt{6}}{6} \\ 0 & \frac{1}{\sqrt{3}} & \frac{\sqrt{6}}{3} \end{bmatrix} \tag{11.141}$$

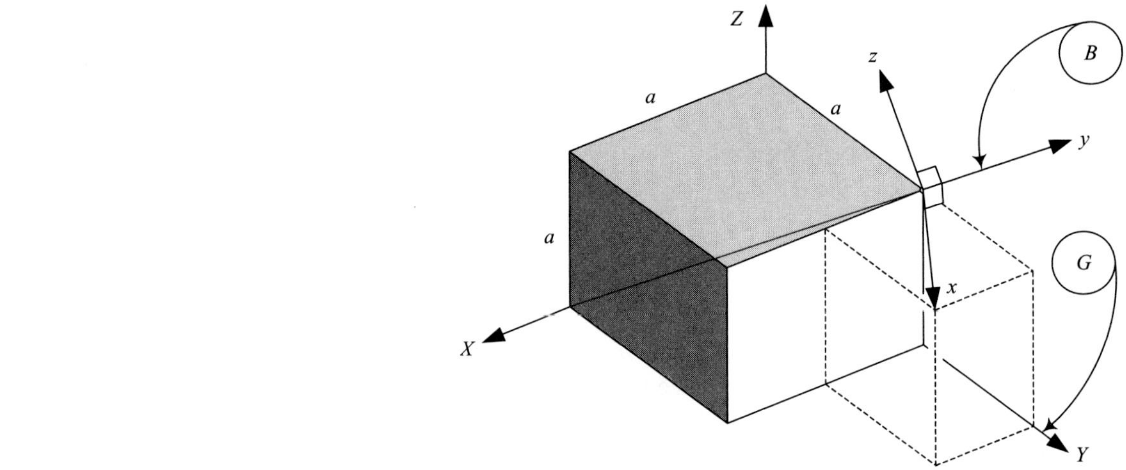

Figure 11.11 Mass moment of a cube in two frames.

Let us rewrite Equation (11.131) using the notation in Figure 11.11:

$$^{B}I = {}^{G}R_B \left({}^{G}I + m \, {}^{G}\tilde{r}_B \, {}^{G}\tilde{r}_B^{\mathrm{T}} \right) {}^{G}R_B^{\mathrm{T}} \tag{11.142}$$

Employing

$$^{G}\tilde{r}_B = \begin{bmatrix} 0 & -a & a \\ a & 0 & -a \\ -a & a & 0 \end{bmatrix} \tag{11.143}$$

we can determine ^{B}I given ^{G}I:

$$\begin{aligned} ^{B}I &= {}^{G}R_B \left({}^{G}I + m \, {}^{G}\tilde{r}_B \, {}^{G}\tilde{r}_B^{\mathrm{T}} \right) {}^{G}R_B^{\mathrm{T}} \\ &= ma^2 \begin{bmatrix} 2.957 & 0.58926 & -1.5266 \\ 0.58926 & 3.5549 & 0.93737 \\ -1.5266 & 0.93737 & 1.4882 \end{bmatrix} \end{aligned} \tag{11.144}$$

Example 706 ★ Mass Moment and Change of Reference Point Consider a rigid body B with a body coordinate frame B at the mass center C and frames B_1 and B_2 at points A and P, respectively, as shown in Figure 11.12. Let us assume that when the body coordinate frame is at A the mass moment matrix is given as ^{1}I. To determine ^{2}I when ^{1}I is given, we establish the relationships of ^{1}I and ^{2}I with the central mass moment ^{B}I:

$$^{B}I = {}^{B}R_1 \, {}^{1}I \, {}^{B}R_1^{\mathrm{T}} - m \, {}^{B}\tilde{r}_1 \, {}^{B}\tilde{r}_1^{\mathrm{T}} \tag{11.145}$$

$$^{2}I = {}^{2}R_B \, {}^{B}I \, {}^{2}R_B^{\mathrm{T}} + m \, {}^{2}R_B \left({}^{B}\tilde{r}_2 \, {}^{B}\tilde{r}_2^{\mathrm{T}} \right) {}^{2}R_B^{\mathrm{T}} \tag{11.146}$$

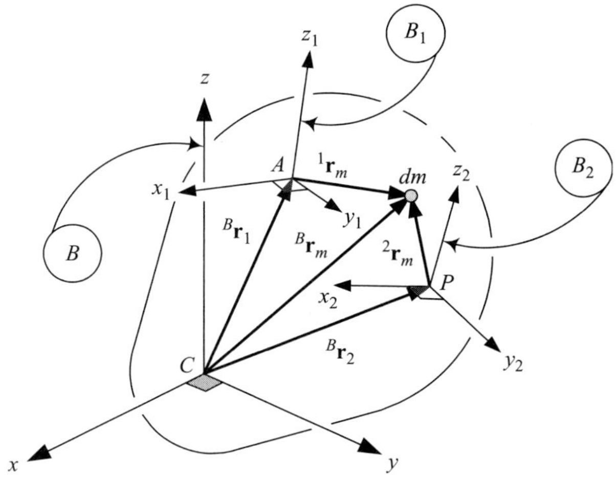

Figure 11.12 A rigid body B with three body coordinates B, B_1, and B_2 at points C, A, and P.

Elimination of ${}^BI$ provides the relationship between ${}^1I$ and ${}^2I$:

$$\begin{aligned}
{}^2I &= {}^2R_B\ \left({}^BR_1\ {}^1I\ {}^BR_1^{\mathrm{T}} - m\ {}^B\tilde{r}_1\ {}^B\tilde{r}_1^{\mathrm{T}}\right)\ {}^2R_B^{\mathrm{T}} \\
&\quad + m\ {}^2R_B\ \left({}^B\tilde{r}_2\ {}^B\tilde{r}_2^{\mathrm{T}}\right)\ {}^2R_B^{\mathrm{T}} \\
&= {}^2R_B\ {}^BR_1\ {}^1I\ {}^BR_1^{\mathrm{T}}\ {}^2R_B^{\mathrm{T}} - m\ {}^2R_B\ \left({}^B\tilde{r}_1\ {}^B\tilde{r}_1^{\mathrm{T}}\right)\ {}^2R_B^{\mathrm{T}} \\
&\quad + m\ {}^2R_B\ \left({}^B\tilde{r}_2\ {}^B\tilde{r}_2^{\mathrm{T}}\right)\ {}^2R_B^{\mathrm{T}} \\
&= {}^2R_1\ {}^1I\ {}^2R_1^{\mathrm{T}} + m\ {}^2R_B\ \left({}^B\tilde{r}_2\ {}^B\tilde{r}_2^{\mathrm{T}} - {}^B\tilde{r}_1\ {}^B\tilde{r}_1^{\mathrm{T}}\right)\ {}^2R_B^{\mathrm{T}}
\end{aligned} \tag{11.147}$$

Example 707 Mass Moment of a Rigid Body about an Axis Let α, β, γ be the direction cosines of a line l which passes through the origin of a coordinate frame B, as shown in Figure 11.13. Consider rigidly connected n point masses m_i. Assume $\mathbf{r}_i = x_i\hat{\imath} + y_i\hat{\jmath} + z_i\hat{k}$ is the position vector of a point mass m_i and d_i is the perpendicular distance of m_i to l. The mass moment of the rigid body about l in B is

$${}^BI_l = \sum_{i=1}^{n} m_i d_i^2 \tag{11.148}$$

Because $\hat{u} = \alpha\hat{\imath} + \beta\hat{\jmath} + \gamma\hat{k}$ is a unit vector on l, we have

$$\begin{aligned}
d_i^2 &= \left(\mathbf{r}_i \times \hat{u}\right)^2 = (\gamma y_i - \beta z_i)^2 + (\alpha z_i - \gamma x_i)^2 + (\beta x_i - \alpha y_i)^2 \\
&= \left(y_i^2 + z_i^2\right)\alpha^2 + \left(x_i^2 + z_i^2\right)\beta^2 + \left(x_i^2 + y_i^2\right)\gamma^2 \\
&\quad - 2x_i y_i \alpha\beta - 2x_i z_i \alpha\gamma - 2y_i z_i \beta\gamma
\end{aligned} \tag{11.149}$$

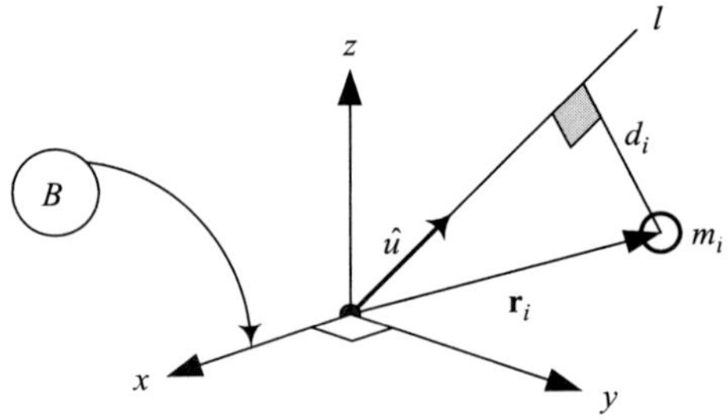

Figure 11.13 Mass moment of a particle about a line l.

and therefore,

$$^{B}I_l = I_{xx}\alpha^2 + I_{yy}\beta^2 + I_{zz}\gamma^2 + 2I_{xy}\alpha\beta + 2I_{yz}\beta\gamma + 2I_{xz}\alpha\gamma \tag{11.150}$$

From Equation (11.149), we may calculate the mass moment about any line through O if the mass moment with respect to the coordinate frame B is known.

Example 708 Mass Moment of a Cone about a Side Line Equation (11.149) calculates the mass moment of a rigid body about a line with direction cosines α, β, γ if the mass moment of the body with respect to the coordinate frame B is known. Let us use the equation to calculate the mass moment of the cone in Figure 11.14 about the side lines l_1 sitting in the (x, y)-plane and about the line $l_2 \perp l_1$, both passing through O.

A principal body coordinate frame B is attached to the cone at its apex with x as its axis of symmetry. The mass moments of the cone in B are

$$I_{xx} = \tfrac{3}{10}ml^2 \tag{11.151}$$

$$I_{yy} = I_{zz} = \tfrac{3}{5}ml^2\left(\cos^2\alpha + \tfrac{1}{4}\sin^2\alpha\right) \tag{11.152}$$

where l is the slant length of the cone and m is its mass. The line l_1 is indicated by

$$\hat{u}_{l_1} = \hat{\imath}\cos\alpha + \hat{k}\sin\alpha \tag{11.153}$$

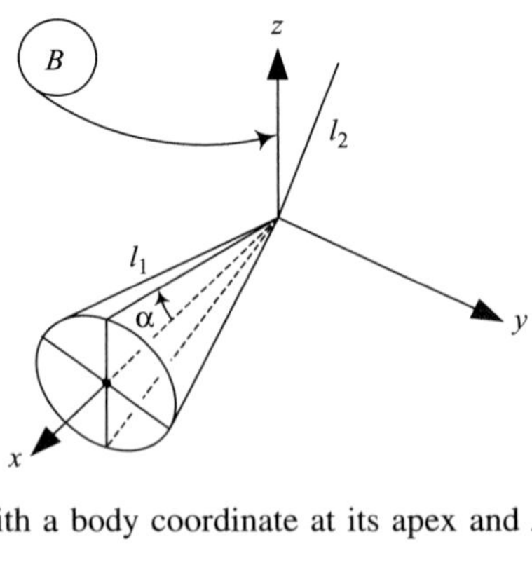

Figure 11.14 A cone with a body coordinate at its apex and x as its axis of symmetry.

and therefore, Equation (11.149) yields

$$ {}^{B}I_{l_1} = \tfrac{3}{4}ml^2 \sin^2\alpha \left(\cos^2\alpha + \tfrac{1}{5}\right) \tag{11.154}$$

Similarly, we can indicate line l_2 by

$$\hat{u}_{l_2} = -\hat{\imath}\sin\alpha + \hat{k}\cos\alpha \tag{11.155}$$

which yields

$$ {}^{B}I_{l_2} = ml^2 \left(\tfrac{3}{4}\sin^4\alpha - \tfrac{21}{20}\sin^2\alpha + \tfrac{3}{5}\right) \tag{11.156}$$

Example 709 ★ Index Notation and Rotation of Frame Consider a mass moment matrix ${}^{1}I$ of a body B in a coordinate frame B_1 attached to B at point O. Another coordinate frame B_2 at O is related to B_1 by a rotation transformation matrix ${}^{2}R_1$:

$${}^{2}r = {}^{2}R_1\, {}^{1}r \tag{11.157}$$

$${}^{2}x_i = \sum_{k=1}^{3} r_{ik}\, {}^{1}x_k \tag{11.158}$$

The elements of ${}^{2}I$ are

$${}^{2}I_{ij} = \int_B \left(r^2\delta_{ij} - {}^{2}x_i\, {}^{2}x_j\right) dm \qquad i, j = 1, 2, 3 \tag{11.159}$$

Substituting for ${}^{2}x_i$ and ${}^{2}x_j$ yields

$${}^{2}I_{ij} = \delta_{ij}\int_B r^2\, dm - \sum_{k=1}^{3}\sum_{l=1}^{3} r_{ik}\, r_{jl} \int_B {}^{2}x_k\, {}^{2}x_l\, dm \tag{11.160}$$

However,

$$\int_B {}^{2}x_k\, {}^{2}x_l\, dm = \delta_{kl}\int_B r^2\, dm - {}^{1}I_{kl} \tag{11.161}$$

and therefore,

$$\begin{aligned}
{}^{2}I_{ij} &= \delta_{ij}\int_B r^2\, dm + \sum_{k=1}^{3}\sum_{l=1}^{3} r_{ik}\, r_{jl} \left({}^{1}I_{kl} - \delta_{kl}\int_B r^2\, dm\right) \\
&= \delta_{ij}\int_B r^2\, dm + \sum_{k=1}^{3}\sum_{l=1}^{3} r_{ik}\, r_{jl}\, {}^{1}I_{kl} - \delta_{ij}\int_B r^2\, dm \\
&= \sum_{k=1}^{3}\sum_{l=1}^{3} r_{ik}\, r_{jl}\, {}^{1}I_{kl}
\end{aligned} \tag{11.162}$$

Similarly, if we have the mass moment ${}^2I$ of a rigid body in a rotated frame B_2, we can determine the mass moment of the body in a nonrotated frame ${}^1I$:

$$ {}^1I_{ij} = \sum_{k=1}^{3}\sum_{l=1}^{3} r_{ki}\, r_{lj}\, {}^2I_{kl} \tag{11.163} $$

11.4 PRINCIPAL MASS MOMENTS

If a local coordinate frame $Oxyz$ is oriented such that the products of inertia vanish, the local coordinate frame is called the *principal coordinate frame* and the associated mass moments are called *principal mass moments*. Principal axes and principal mass moments are eigenvalues and eigenvectors of the mass moment matrix $[I]$. The principal mass moments can be found by solving the following equation for I:

$$ \begin{vmatrix} I_{xx} - I & I_{xy} & I_{xz} \\ I_{yx} & I_{yy} - I & I_{yz} \\ I_{zx} & I_{zy} & I_{zz} - I \end{vmatrix} = 0 \tag{11.164} $$

$$ \det\left(\left[I_{ij}\right] - I\left[\delta_{ij}\right]\right) = 0 \tag{11.165} $$

Equation (11.164) is a cubic equation in I and provides three principal mass moments:

$$ I_1 = I_x \qquad I_2 = I_y \qquad I_3 = I_z \tag{11.166} $$

Associated to each principal mass moment I_i, there exist a principal axis $\mathbf{u}_i$ which can be found from the equation

$$ \left[[I] - I_i\,[\mathbf{I}]\right]\mathbf{u}_i = 0 \tag{11.167} $$

Proof: Referring to Equation (11.93) for the transformation of the mass moment to a rotated frame B_2, let us assume that we can find a frame in which ${}^2I$ is a diagonal matrix:

$$ {}^2I = \begin{bmatrix} I_1 & 0 & 0 \\ 0 & I_2 & 0 \\ 0 & 0 & I_3 \end{bmatrix} \tag{11.168} $$

In such a frame, we have

$$ {}^2R_1\, {}^1I = {}^2I\, {}^2R_1 \tag{11.169} $$

or

$$ \begin{aligned} &\begin{bmatrix} I_{xx} & I_{xy} & I_{xz} \\ I_{yx} & I_{yy} & I_{yz} \\ I_{zx} & I_{zy} & I_{zz} \end{bmatrix} \begin{bmatrix} r_{11} & r_{12} & r_{13} \\ r_{21} & r_{22} & r_{23} \\ r_{31} & r_{32} & r_{33} \end{bmatrix} \\ &\quad = \begin{bmatrix} I_1 & 0 & 0 \\ 0 & I_2 & 0 \\ 0 & 0 & I_3 \end{bmatrix} \begin{bmatrix} r_{11} & r_{12} & r_{13} \\ r_{21} & r_{22} & r_{23} \\ r_{31} & r_{32} & r_{33} \end{bmatrix} \end{aligned} \tag{11.170} $$

which shows that I_1, I_2, and I_3 are eigenvalues of ${}^1I$. These eigenvalues can be found by solving the following equation for λ:

$$\begin{vmatrix} I_{xx}-\lambda & I_{xy} & I_{xz} \\ I_{yx} & I_{yy}-\lambda & I_{yz} \\ I_{zx} & I_{zy} & I_{zz}-\lambda \end{vmatrix} = 0 \tag{11.171}$$

The eigenvalues $\lambda = I_1$, $\lambda = I_2$, and $\lambda = I_3$ are *principal mass moments*, and their associated eigenvectors are called *principal directions*. The coordinate frame made by the eigenvectors is the *principal body coordinate frame*.

According to the determination method of mass moments in Equation (11.15), there exists a mass moment matrix for any coordinate frame at any point of a rigid body. Also, there are principal values and axes for any mass moment matrix.

If a body has three planes of symmetry which are mutually perpendicular, then these planes are the *principal planes*. The triad that is made by intersecting perpendicular planes is the principal coordinate frame of the body. ■

Example 710 Principal Moments of Inertia Consider the mass moment matrix

$$I = \begin{bmatrix} 20 & -2 & 0 \\ -2 & 30 & 0 \\ 0 & 0 & 40 \end{bmatrix} \tag{11.172}$$

To determine the principal moments of inertia, we set up the determinant (11.164),

$$\begin{vmatrix} 20-\lambda & -2 & 0 \\ -2 & 30-\lambda & 0 \\ 0 & 0 & 40-\lambda \end{vmatrix} = 0 \tag{11.173}$$

which leads to the characteristic equation

$$(20-\lambda)(30-\lambda)(40-\lambda) - 4(40-\lambda) = 0 \tag{11.174}$$

Three roots of Equation (11.173) are

$$I_1 = 30.385 \qquad I_2 = 19.615 \qquad I_3 = 40 \tag{11.175}$$

and therefore, the principal moment of inertia matrix is

$$I = \begin{bmatrix} 30.385 & 0 & 0 \\ 0 & 19.615 & 0 \\ 0 & 0 & 40 \end{bmatrix} \tag{11.176}$$

Although in this example we have $I_3 > I_1 > I_2$, it is traditional to sort the principal mass moments such that

$$I_1 \geq I_2 \geq I_3 \tag{11.177}$$

or

$$I_1 \leq I_2 \leq I_3 \tag{11.178}$$

Example 711 Principal Coordinate Frame Consider the mass moment matrix

$$I = \begin{bmatrix} 20 & -2 & 0 \\ -2 & 30 & 0 \\ 0 & 0 & 40 \end{bmatrix} \tag{11.179}$$

The direction of a principal axis x_i with direction cosines $(\alpha_i, \beta_i, \gamma_i)$ is established by solving for direction cosines:

$$\begin{bmatrix} I_{xx} - I_i & I_{xy} & I_{xz} \\ I_{yx} & I_{yy} - I_i & I_{yz} \\ I_{zx} & I_{zy} & I_{zz} - I_i \end{bmatrix} \begin{bmatrix} \alpha_i \\ \beta_i \\ \gamma_i \end{bmatrix} = \begin{bmatrix} 0 \\ 0 \\ 0 \end{bmatrix} \tag{11.180}$$

These three equations are not independent because

$$\alpha_i^2 + \beta_i^2 + \gamma_i^2 = 1 \tag{11.181}$$

Therefore, we should use (11.180) and the two equations of (11.179) to determine $\alpha_i, \beta_i, \gamma_i$.

For the first principal moment of inertia $I_1 = 30.385$, we have

$$\begin{bmatrix} 20 - 30.385 & -2 & 0 \\ -2 & 30 - 30.385 & 0 \\ 0 & 0 & 40 - 30.385 \end{bmatrix} \begin{bmatrix} \cos\alpha_1 \\ \cos\beta_1 \\ \cos\gamma_1 \end{bmatrix} = \begin{bmatrix} 0 \\ 0 \\ 0 \end{bmatrix} \tag{11.182}$$

or

$$-10.385\cos\alpha_1 - 2\cos\beta_1 + 0 = 0 \tag{11.183}$$

$$-2\cos\alpha_1 - 0.385\cos\beta_1 + 0 = 0 \tag{11.184}$$

$$0 + 0 + 9.615\cos\gamma_1 = 0 \tag{11.185}$$

and we obtain

$$\alpha_1 = 79.1 \text{ deg} \tag{11.186}$$

$$\beta_1 = 169.1 \text{ deg} \tag{11.187}$$

$$\gamma_1 = 90.0 \text{ deg} \tag{11.188}$$

Using $I_2 = 19.615$ for the second principal axis,

$$\begin{bmatrix} 20 - 19.62 & -2 & 0 \\ -2 & 30 - 19.62 & 0 \\ 0 & 0 & 40 - 19.62 \end{bmatrix} \begin{bmatrix} \cos\alpha_2 \\ \cos\beta_2 \\ \cos\gamma_2 \end{bmatrix} = \begin{bmatrix} 0 \\ 0 \\ 0 \end{bmatrix} \tag{11.189}$$

we obtain

$$\alpha_2 = 10.9 \text{ deg} \tag{11.190}$$

$$\beta_2 = 79.1 \text{ deg} \tag{11.191}$$

$$\gamma_2 = 90.0 \text{ deg} \tag{11.192}$$

The third principal axis is for $I_3 = 40$,

$$\begin{bmatrix} 20-40 & -2 & 0 \\ -2 & 30-40 & 0 \\ 0 & 0 & 40-40 \end{bmatrix} \begin{bmatrix} \cos\alpha_3 \\ \cos\beta_3 \\ \cos\gamma_3 \end{bmatrix} = \begin{bmatrix} 0 \\ 0 \\ 0 \end{bmatrix} \tag{11.193}$$

which yields

$$\alpha_3 = 90.0\,\text{deg} \tag{11.194}$$

$$\beta_3 = 90.0\,\text{deg} \tag{11.195}$$

$$\gamma_3 = 0.0\,\text{deg} \tag{11.196}$$

Example 712 Principal Rotation Matrix Consider the body inertia matrix

$$I = \begin{bmatrix} \frac{2}{3} & -\frac{1}{2} & -\frac{1}{2} \\ -\frac{1}{2} & \frac{5}{3} & -\frac{1}{4} \\ -\frac{1}{2} & -\frac{1}{4} & \frac{5}{3} \end{bmatrix} \tag{11.197}$$

The eigenvalues and eigenvectors of I are

$$I_1 = 0.2413 \qquad \mathbf{w}_1 = \begin{bmatrix} 2.351 \\ 1 \\ 1 \end{bmatrix} \tag{11.198}$$

$$I_2 = 1.8421 \qquad \mathbf{w}_2 = \begin{bmatrix} -0.851 \\ 1 \\ 1 \end{bmatrix} \tag{11.199}$$

$$I_3 = 1.9167 \qquad \mathbf{w}_3 = \begin{bmatrix} 0 \\ -1 \\ 1 \end{bmatrix} \tag{11.200}$$

The normalized eigenvector matrix W is equal to the transpose of the required transformation matrix to make the inertia matrix diagonal:

$$W = \begin{bmatrix} | & | & | \\ \mathbf{w}_1 & \mathbf{w}_2 & \mathbf{w}_3 \\ | & | & | \end{bmatrix} = {}^2R_1^{\mathrm{T}} \tag{11.201}$$

$$= \begin{bmatrix} 0.8569 & -0.5156 & 0.0 \\ 0.36448 & 0.60588 & -0.70711 \\ 0.36448 & 0.60588 & 0.70711 \end{bmatrix}$$

We may verify that

$$
\begin{aligned}
{}^{2}I &= {}^{2}R_{1}\,{}^{1}I\,{}^{2}R_{1}^{\mathrm{T}} = W^{\mathrm{T}}\,{}^{1}I\,W \\
&= \begin{bmatrix} 0.2413 & -1\times 10^{-4} & 0.0 \\ -1\times 10^{-4} & 1.842\,1 & -1\times 10^{-19} \\ 0.0 & 0.0 & 1.916\,7 \end{bmatrix} \\
&\approx \begin{bmatrix} I_1 & 0 & 0 \\ 0 & I_2 & 0 \\ 0 & 0 & I_3 \end{bmatrix}
\end{aligned}
\tag{11.202}
$$

Example 713 ★ Coefficients of the Characteristic Equation The determinant (11.171) for calculating the principal moments of inertia,

$$
\begin{vmatrix} I_{xx}-\lambda & I_{xy} & I_{xz} \\ I_{yx} & I_{yy}-\lambda & I_{yz} \\ I_{zx} & I_{zy} & I_{zz}-\lambda \end{vmatrix} = 0 \tag{11.203}
$$

leads to a third-degree equation of λ called the *characteristic equation*:

$$
\lambda^3 - a_1\lambda^2 + a_2\lambda - a_3 = 0 \tag{11.204}
$$

The coefficients of the characteristic equation are called the *principal invariants* of $[I]$. The coefficients can directly be found from the equations

$$
a_1 = I_{xx} + I_{yy} + I_{zz} = \mathrm{tr}\,[I] \tag{11.205}
$$

$$
\begin{aligned}
a_2 &= I_{xx}I_{yy} + I_{yy}I_{zz} + I_{zz}I_{xx} - I_{xy}^2 - I_{yz}^2 - I_{zx}^2 \\
&= \begin{vmatrix} I_{xx} & I_{xy} \\ I_{yx} & I_{yy} \end{vmatrix} + \begin{vmatrix} I_{yy} & I_{yz} \\ I_{zy} & I_{zz} \end{vmatrix} + \begin{vmatrix} I_{xx} & I_{xz} \\ I_{zx} & I_{zz} \end{vmatrix} \\
&= \tfrac{1}{2}\left(a_1^2 - \mathrm{tr}\left[I^2\right]\right) = \tfrac{1}{2}\left((\mathrm{tr}\,[I])^2 - \mathrm{tr}\left[I^2\right]\right)
\end{aligned}
\tag{11.206}
$$

$$
\begin{aligned}
a_3 &= I_{xx}I_{yy}I_{zz} + I_{xy}I_{yz}I_{zx} + I_{zy}I_{yx}I_{xz} \\
&\quad - \left(I_{xx}I_{yz}I_{zy} + I_{yy}I_{zx}I_{xz} + I_{zz}I_{xy}I_{yx}\right) \\
&= I_{xx}I_{yy}I_{zz} + 2I_{xy}I_{yz}I_{zx} - \left(I_{xx}I_{yz}^2 + I_{yy}I_{zx}^2 + I_{zz}I_{xy}^2\right) \\
&= \det\,[I]
\end{aligned}
\tag{11.207}
$$

As an example let us find the principal mass moments of $[I]$ in (11.172):

$$
I = \begin{bmatrix} 20 & -2 & 0 \\ -2 & 30 & 0 \\ 0 & 0 & 40 \end{bmatrix} \tag{11.208}
$$

The principal invariants of $[I]$ are

$$a_1 = \operatorname{tr}[I] = 90 \tag{11.209}$$

$$a_2 = \frac{1}{2}\left((\operatorname{tr}[I])^2 - \operatorname{tr}\left[I^2\right]\right) = \frac{1}{2}\left(90^2 - 2908\right) = 2596 \tag{11.210}$$

$$a_3 = \det[I] = 23{,}840 \tag{11.211}$$

and therefore, the characteristic equation of $[I]$ is

$$\lambda^3 - 90\lambda^2 + 2596\lambda - 23{,}840 = 0 \tag{11.212}$$

It has the following roots:

$$\begin{aligned} \lambda_1 &= \sqrt{29} + 25 = 30.385 \\ \lambda_2 &= 25 - \sqrt{29} = 19.615 \\ \lambda_3 &= 40 \end{aligned} \tag{11.213}$$

Example 714 ★ Principal Mass Moments are Coordinate Invariants The roots of the inertia characteristic equation are the principal mass moments and are all real but not necessarily different. The principal mass moments are extreme. This means that the principal mass moments determine the smallest and the largest values of I_{ii} for a rigid body in the considered body frame B. Any translation of B changes the principal mass moments and the principal directions. The principal values of I_{ii} do not depend on the orientation of B. So, the solutions of the characteristic equation are not dependent on the orientation of B. However, the principal directions are functions of the orientation of B. In other words, if I_1, I_2, and I_3 are the principal mass moments for ${}^{B_1}I$, the principal mass moments for ${}^{B_2}I$ are also I_1, I_2, and I_3 when

$${}^{B_2}I = {}^{B_2}R_{B_1}\ {}^{B_1}I\ {}^{B_2}R_{B_1}^{\mathrm{T}} \tag{11.214}$$

We conclude that I_1, I_2, and I_3 are orientation invariants of the matrix $[I]$ and therefore any quantity that depends on I_1, I_2, and I_3 is also orientation invariant. The matrix $[I]$ has only three independent invariants and every other invariant can be expressed in terms of I_1, I_2, and I_3.

Since I_1, I_2, and I_3 are the solutions of the characteristic equation of $[I]$ given in (11.204), we may write the determinant (11.171) in the form

$$(\lambda - I_1)(\lambda - I_2)(\lambda - I_3) = 0 \tag{11.215}$$

The expanded form of this equation is

$$\lambda^3 - (I_1 + I_2 + I_3)\lambda^2 + (I_1 I_2 + I_2 I_3 + I_3 I_1)\, a_2\lambda - I_1 I_2 I_3 = 0 \tag{11.216}$$

By comparing (11.216) and (11.204), we conclude that

$$a_1 = I_{xx} + I_{yy} + I_{zz} = I_1 + I_2 + I_3 \tag{11.217}$$

$$\begin{aligned} a_2 &= I_{xx}I_{yy} + I_{yy}I_{zz} + I_{zz}I_{xx} - I_{xy}^2 - I_{yz}^2 - I_{zx}^2 \\ &= I_1 I_2 + I_2 I_3 + I_3 I_1 \end{aligned} \tag{11.218}$$

$$\begin{aligned} a_3 &= I_{xx}I_{yy}I_{zz} + 2I_{xy}I_{yz}I_{zx} - \left(I_{xx}I_{yz}^2 + I_{yy}I_{zx}^2 + I_{zz}I_{xy}^2\right) \\ &= I_1 I_2 I_3 \end{aligned} \tag{11.219}$$

Being able to express the coefficients a_1, a_2, and a_3 as functions of I_1, I_2, and I_3 determines that the coefficients of the characteristic equation are also orientation invariant.

Example 715 ★ Principal Frame at Two Points of a Rigid Body Figure 11.15 illustrates a cube with side a along with two coordinate frames G and B at two different corners with the following mass moment matrices:

$$ {}^{G}I = ma^2 \begin{bmatrix} 0.66667 & -0.25 & -0.25 \\ -0.25 & 0.66667 & -0.25 \\ -0.25 & -0.25 & 0.66667 \end{bmatrix} \tag{11.220} $$

$$ \begin{aligned} {}^{B}I &= {}^{G}R_B \left({}^{G}I + m\ {}^{G}\tilde{r}_B\ {}^{G}\tilde{r}_B^{\mathrm{T}} \right)\ {}^{G}R_B^{\mathrm{T}} \\ &= ma^2 \begin{bmatrix} 2.957 & 0.58926 & -1.5266 \\ 0.58926 & 3.5549 & 0.93737 \\ -1.5266 & 0.93737 & 1.4882 \end{bmatrix} \end{aligned} \tag{11.221} $$

The coordinate frame B is set at the corner (a, a, a) such that the y-axis is along the diagonal of the cube and the x-axis is parallel to the (X, Y)-plane, and therefore,

$$ {}^{G}R_B = \begin{bmatrix} 0.70711 & 0.57735 & -0.40825 \\ -0.70711 & 0.57735 & -0.40825 \\ 0 & 0.57735 & 0.8165 \end{bmatrix} \tag{11.222} $$

The principal mass moments and axes of ${}^{G}I$ are

$$ {}^{G}I_1 = 0.916\,67ma^2 \qquad {}^{G}\mathbf{u}_1 = \begin{bmatrix} 0 \\ 0.70711 \\ -0.70711 \end{bmatrix} \tag{11.223} $$

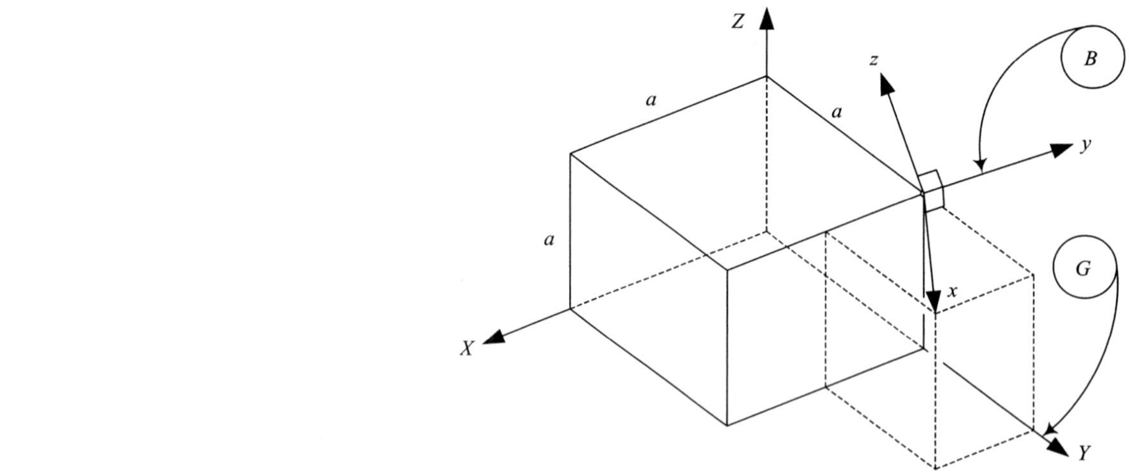

Figure 11.15 Mass moment of a cube at two corner frames.

$$^{G}I_{2}=0.916\,67ma^{2}\qquad {}^{G}\mathbf{u}_{2}=\begin{bmatrix}0.81650\\-0.40825\\-0.40825\end{bmatrix}\tag{11.224}$$

$$^{G}I_{3}=0.166\,67ma^{2}\qquad {}^{G}\mathbf{u}_{3}=\begin{bmatrix}0.57735\\0.57735\\0.57735\end{bmatrix}\tag{11.225}$$

The principal mass moments and axes of ^{B}I are

$$^{B}I_{1}=3.916\,7ma^{2}\qquad {}^{B}\mathbf{u}_{1}=\begin{bmatrix}0.20564\\0.94945\\0.2372\end{bmatrix}\tag{11.226}$$

$$^{B}I_{2}=3.916\,7ma^{2}\qquad {}^{B}\mathbf{u}_{2}=\begin{bmatrix}0.83773\\-0.04.549\\-0.54418\end{bmatrix}\tag{11.227}$$

$$^{B}I_{3}=0.166\,72ma^{2}\qquad {}^{B}\mathbf{u}_{3}=\begin{bmatrix}0.50588\\-0.31062\\0.80474\end{bmatrix}\tag{11.228}$$

KEY SYMBOLS

a	side length of a cube
a, b, c	coordinates with respect to C
a_1, a_2, a_3	principal invariants, coefficients of characteristic equation
B	body coordinate frame, local coordinate frame
c	cosine
C	mass center
d	distance between two points
$\mathbf{d}$	translation vector
$\tilde{d}$	skew-symmetric matrix of the vector $\mathbf{d}$
fc	configuration DOF
G	global coordinate frame, fixed coordinate frame
h	height
$\mathbf{I}$	identity matrix
$I=[I]$	mass moment matrix
I_1, I_2, I_3	principal mass moments
I_{ij}	element of row i and column j of $[I]$
I_o	mass moment about the origin
$\hat{\imath}, \hat{\jmath}, \hat{k}$	local coordinate axes unit vectors
$\hat{I}, \hat{J}, \hat{K}$	global coordinate axes unit vectors
k	radius of gyration
K	kinetic energy
$\mathbf{L}$	moment of momentum
l	length, slant length of a cone

m	mass, number of particles exceeding 3
n	number of particles in a rigid body
O	common origin of B and G
P	body point, fixed point in B
$\mathbf{r}$	position vector
r_i	elements of $\mathbf{r}$
r_{ij}	distance between point P_i and P_j
$\tilde{r}$	skew symmetric matrix of the vector $\mathbf{r}$
R	rotation transformation matrix
s	sin
S	set
$\hat{u}$	unit vector
$\mathbf{u}$	normalized eigenvector
$\mathbf{v}$	velocity vector
V	volume
$\mathbf{w}$	eigenvector
W	normalized eigenvector matrix
x, y, z	local coordinate axes
X, Y, Z	global coordinate axes
Greek	
α, β, γ	direction cosines
δ_{ij}	Kronecker delta
λ	eigenvalue, characteristic value
ρ	density
$\omega, \boldsymbol{\omega}$	angular velocity
Symbol	
$[\ \ ]^{-1}$	inverse of the matrix $[\ \ ]$
$[\ \ ]^{\mathrm{T}}$	transpose of the matrix $[\ \ ]$
DOF	degree of freedom, degrees of freedom
$\parallel$	parallel
$\perp$	perpendicular

EXERCISES

1. **Mass Center of a Hemisphere**

   (a) Determine the position of the mass center of the uniform solid hemisphere in Figure 11.16.

   (b) Using the result of part (a), show that the mass center of a complete uniform sphere with geometric center at the origin of G is at the origin.

2. **Mass Moment of a Hemisphere**

   (a) Determine the mass moment matrix ${}^{G}I$ of the uniform solid hemisphere in Figure 11.16.

   (b) Attach a body coordinate frame B parallel to G at the mass center C of the uniform solid hemisphere in Figure 11.16 and determine the mass moment matrix ${}^{B}I$ of the hemisphere.

   (c) Determine the principal mass moments and frames of the hemisphere in B and G.

   (d) Using the result of (a), determine the mass moment matrix of a complete uniform sphere with geometric center at the origin of G.

   (e) Determine the mass moment matrix of a uniform sphere on a table in a coordinate frame at the contact point.

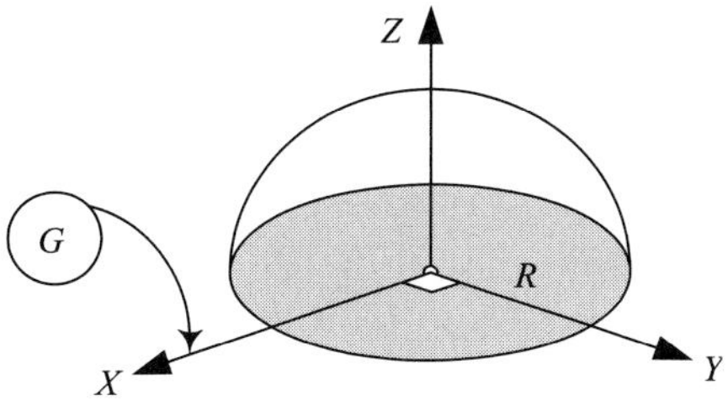

Figure 11.16 A solid hemisphere.

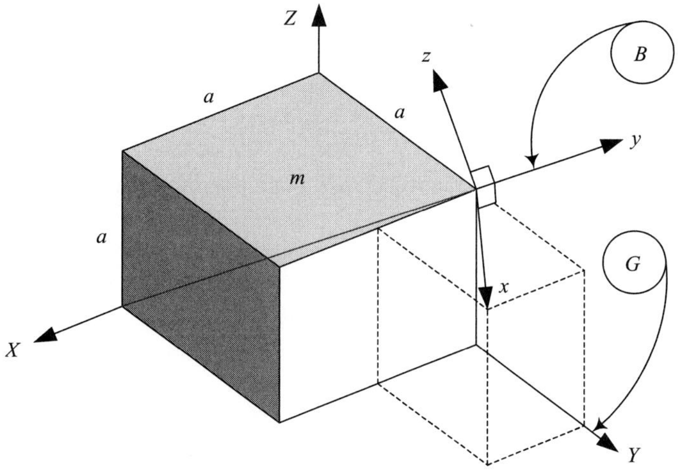

Figure 11.17 A displaced and rotated frame.

3. ★ **A Displaced and Rotated Frame** Figure 11.17 illustrates the same cube and frames G and B as in Example 705.

   (a) Determine the principal mass moment matrices of the cube in both frames G and B.
   (b) Determine the mass moment ${}^{B}I$ by transforming ${}^{G}I$.
   (c) Determine the mass moment ${}^{G}I$ by transforming ${}^{B}I$.
   (d) Determine the principal coordinate frames of the cube in both frames G and B.

4. **Mass Center 1.** Determine the position of the mass center of the shaded area in Figures 11.18 (a) and (b).

5. **Mass Center 2.** Determine the position of the mass center of the shaded area in Figures 11.19 (a) and (b).

6. **Circular Cylinder with an Oblique Plane Face.** Figure 11.20 illustrates a right-circular cylinder with an oblique plane face. Assume h_1 is the greatest height of a side of the cylinder, h_2 is the shortest height of a side of the cylinder, and R is the radius of the cylinder. Then, the area of the curved surface A_S, the total area A, and the volume V of

the cylinder are

$$A_S = \pi R (h_1 + h_2)$$

$$A = \pi R \left(h_1 + h_2 + R + \sqrt{R^2 + \frac{(h_1 + h_2)^2}{2}} \right)$$

$$V = \tfrac{1}{2}\pi R^2 (h_1 + h_2)$$

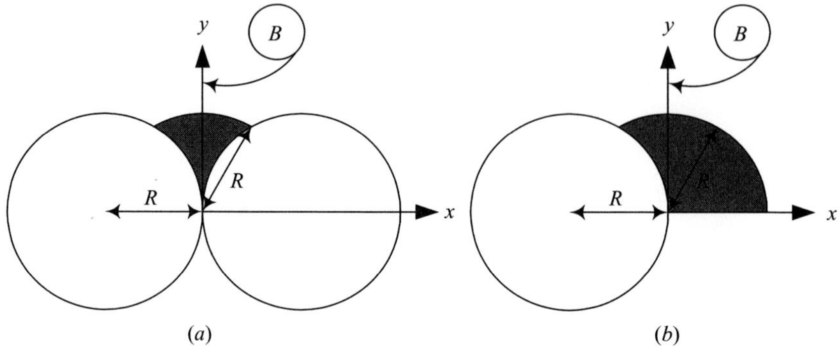

Figure 11.18 A portion of a half circle cut by similar circles.

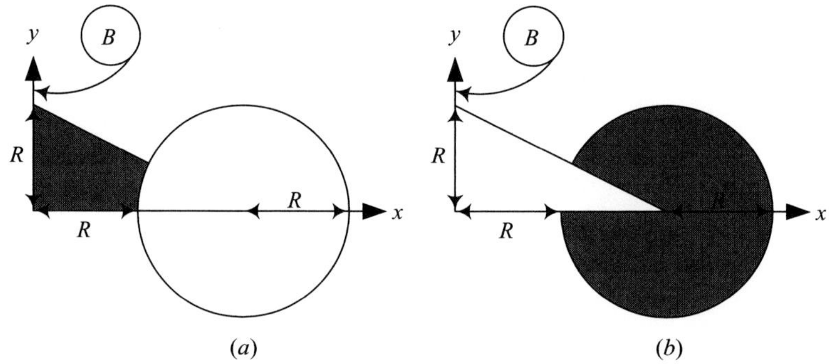

Figure 11.19 Intersection of a triangle and a circle.

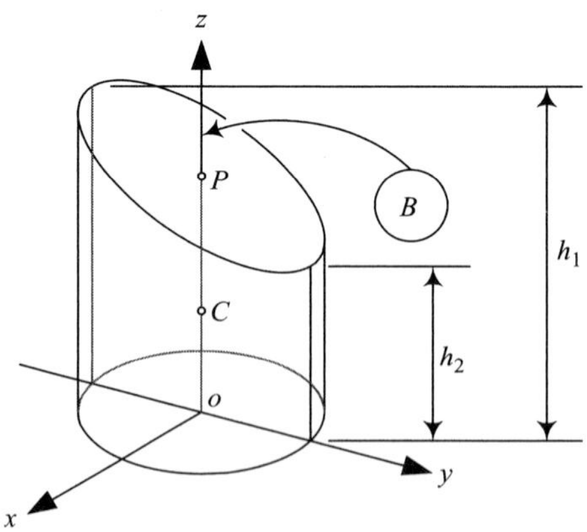

Figure 11.20 A circular cylinder with an oblique plane face.

(a) Show that the coordinates of the mass center C are

$$x_C = 0 \qquad y_C = 0$$
$$z_C = \tfrac{1}{4}(h_1 + h_2) + \frac{1}{16}\frac{(h_1 - h_2)^2}{(h_1 + h_2)}$$

(b) Determine the mass moment matrix in B.

(c) Determine the principal mass moments of ${}^BI$ and the principal coordinate frame at o.

(d) Translate ${}^BI$ to a parallel coordinate frame at C.

(e) Determine the principal mass moments of ${}^CI$ and the principal coordinate frame at C.

(f) Determine the transformation between the principal coordinate frames at o and C.

(g) Translate ${}^BI$ to a parallel coordinate frame at P.

(h) Determine the principal mass moments of ${}^PI$ and the principal coordinate frame at P.

(i) Determine the transformation between the principal coordinate frames at C and P.

7. ★ **Cylindrical Wedge** Figure 11.21 illustrates a cylindrical wedge. Assume R is the radius of the cylinder, h is the height of the wedge, $2a$ is the base chord of the wedge, b is the greatest perpendicular distance from the base chord to the wall of the cylinder measured perpendicular to the axis of the cylinder, and α is the angle subtended at the center o of the normal cross section by the base chord. Then, the area of the curved surface A_S and the volume V of the cylindrical wedge are

$$A_S = \frac{2Rh}{b}\left((b - R)\frac{\alpha}{2} + a\right)$$
$$V = \frac{h}{3b}\left(a\left(3R^2 - a^2\right) + 3R^2(b - R)\frac{\alpha}{2}\right)$$

(a) Determine the coordinates of the mass center C.

(b) Determine the mass moment matrix in B.

(c) Determine the principal mass moments of ${}^BI$ and the principal coordinate frame at o.

(d) Translate ${}^BI$ to a parallel coordinate frame at C.

(e) Determine the principal mass moments of ${}^CI$ and the principal coordinate frame at C.

(f) Determine the transformation between the principal coordinate frames at o and C.

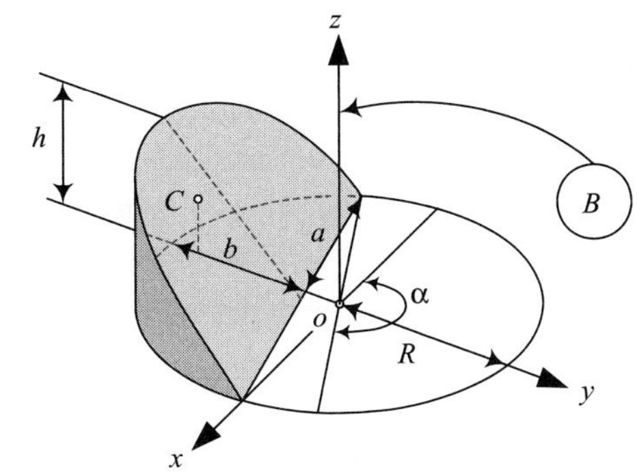

Figure 11.21 A cylindrical wedge.

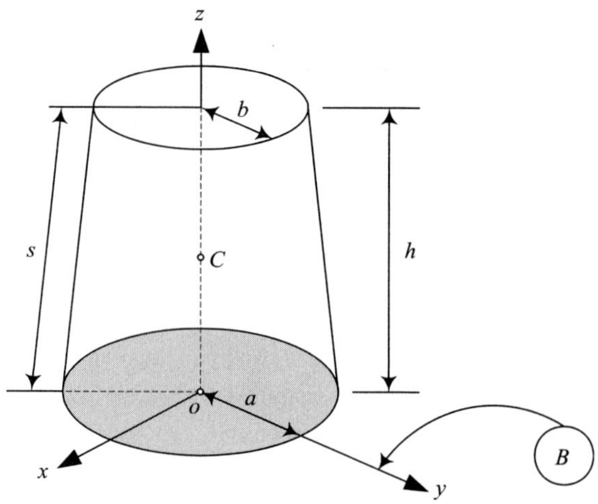

Figure 11.22 A frustrum of a right-circular cone.

8. **Frustrum of a Right-Circular Cone** Figure 11.22 illustrates a frustrum of a right-circular cone. Assume a and b are radii of the base and top circles, h is the height of the cone, and s is the side length of the cone.

$$s = \sqrt{h^2 + (a-b)^2}$$

Given the area of the curved surface A_S, total area A, and volume V of the cylindrical wedge are

$$A_S = \pi s (a+b)$$
$$A = \pi s (a+b) + \pi \left(a^2 + b^2\right)$$
$$V = \tfrac{1}{3}\pi h \left(a^2 + ab + b^2\right)$$

(a) Show that the coordinates of the mass center C are

$$x_C = 0 \qquad y_C = 0$$
$$z_C = \frac{1}{4} h \frac{a^2 + 2ab + 3b^2}{a^2 + ab + b^2}$$

(b) Determine the mass moment matrix in B.

(c) Determine the principal mass moments of ${}^{B}I$.

(d) Translate ${}^{B}I$ to a parallel coordinate frame at C.

(e) Determine the principal mass moments of ${}^{C}I$.

(f) Extend the above results for a complete cone.

(g) Determine the principal mass moments for a parallel coordinate frame at the apex of the complete cone.

9. **Mass Moment of a Cubic Rigid Body** Consider a cubic rigid body B with a coordinate frame $B(Oxyz)$ at the geometric center of the cube $a \times a \times a$. The body is rotating in a global coordinate frame $G(OXYZ)$ with angular velocity ${}_G\boldsymbol{\omega}_B$. Determine the the mass moment matrix of B if the density ρ of the cube is:

(a) $\rho = \dfrac{m}{V} = \dfrac{m}{a \times a \times a} = \text{const}$

(b) $\rho = cr \qquad r = \sqrt{x^2 + y^2 + z^2}$

(c) $\rho = x^2 \, |y| \, |z|$

(d) $\rho = x^2 \, (a/2 - y)$

10. **Cubic Equations and Real Principal Moments** The solution of a cubic equation

$$ax^3 + bx^2 + cx + d = 0$$

where $a \neq 0$ can be found in a systematic way. Transform the equation to a new form with discriminant $4p^3 + q^2$,

$$y^3 + 3py + q = 0$$

using the transformation $x = y - b/3a$, where

$$p = \frac{3ac - b^2}{9a^2} \qquad q = \frac{2b^3 - 9abc + 27a^2 d}{27a^3}.$$

The solutions are then

$$\begin{aligned} y_1 &= \sqrt[3]{\alpha} - \sqrt[3]{\beta} \\ y_2 &= e^{2\pi i/3} \sqrt[3]{\alpha} - e^{4\pi i/3} \sqrt[3]{\beta} \\ y_3 &= e^{4\pi i/3} \sqrt[3]{\alpha} - e^{2\pi i/3} \sqrt[3]{\beta} \end{aligned}$$

where

$$\alpha = \frac{-q + \sqrt{q^2 + 4p^3}}{2} \qquad \beta = \frac{-q + \sqrt{q^2 + 4p^3}}{2}$$

For real values of p and q, if the discriminant is positive, then one root is real and two roots are complex conjugates. If the discriminant is zero, then there are three real roots, of which at least two are equal. If the discriminant is negative, then there are three unequal real roots. Apply this theory for the characteristic equation of the matrix $[I]$ and show that the principal moments of inertia are real.

12

Rigid-Body Dynamics

Employing Newton and Euler equations of motion (2.58) and (2.59) for a particle, we develop the equations of motion of rigid bodies. We usually express the equations of motion of rigid bodies in a local Cartesian coordinate frame attached to their mass center. It is a more practical method.

12.1 RIGID-BODY ROTATIONAL CARTESIAN DYNAMICS

Consider a rigid body B with a fixed point in a global coordinate frame G, as shown in Figure 12.1. The rotational equation of motion of the rigid body is the *Euler equation*

$$ {}^{G}\mathbf{M} = \frac{{}^{G}d}{dt}\ {}^{G}\mathbf{L} \tag{12.1} $$

$$ \begin{aligned} {}^{B}\mathbf{M} &= \frac{{}^{G}d}{dt}\ {}^{B}\mathbf{L} = {}^{B}\dot{\mathbf{L}} + {}_{G}^{B}\boldsymbol{\omega}_{B} \times {}^{B}\mathbf{L} \\ &= {}^{B}I\ {}_{G}^{B}\dot{\boldsymbol{\omega}}_{B} + {}_{G}^{B}\boldsymbol{\omega}_{B} \times \left({}^{B}I\ {}_{G}^{B}\boldsymbol{\omega}_{B}\right) \end{aligned} \tag{12.2} $$

where $\mathbf{L}$ is the *angular momentum* of the body,

$$ {}^{G}\mathbf{L} = {}^{G}I\ {}_{G}\boldsymbol{\omega}_{B} \tag{12.3} $$

$$ {}^{B}\mathbf{L} = {}^{B}I\ {}_{G}^{B}\boldsymbol{\omega}_{B} \tag{12.4} $$

and I is the *mass moment matrix* of the rigid body B,

$$ {}^{B}I = \begin{bmatrix} I_{xx} & I_{xy} & I_{xz} \\ I_{yx} & I_{yy} & I_{yz} \\ I_{zx} & I_{zy} & I_{zz} \end{bmatrix} \tag{12.5} $$

$$ {}^{G}I = \begin{bmatrix} I_{XX} & I_{XY} & I_{XZ} \\ I_{YX} & I_{YY} & I_{YZ} \\ I_{ZX} & I_{ZY} & I_{ZZ} \end{bmatrix} \tag{12.6} $$

The expanded forms of the B-expression of the Euler equation are

$$ \begin{aligned} M_x = {}& I_{xx}\dot{\omega}_x + I_{xy}\dot{\omega}_y + I_{xz}\dot{\omega}_z - \left(I_{yy} - I_{zz}\right)\omega_y\omega_z \\ & - I_{yz}\left(\omega_z^2 - \omega_y^2\right) - \omega_x\left(\omega_z I_{xy} - \omega_y I_{xz}\right) \end{aligned} \tag{12.7} $$

$$ \begin{aligned} M_y = {}& I_{yx}\dot{\omega}_x + I_{yy}\dot{\omega}_y + I_{yz}\dot{\omega}_z - \left(I_{zz} - I_{xx}\right)\omega_z\omega_x \\ & - I_{xz}\left(\omega_x^2 - \omega_z^2\right) - \omega_y\left(\omega_x I_{yz} - \omega_z I_{xy}\right) \end{aligned} \tag{12.8} $$

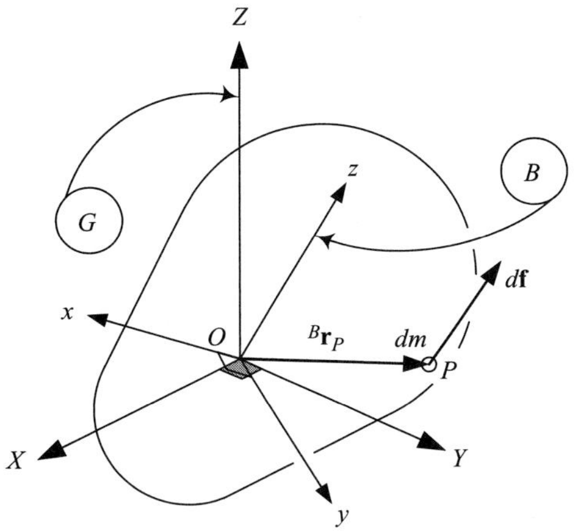

Figure 12.1 A body point mass moving with velocity ${}^G\mathbf{v}_P$ and acted on by force $d\mathbf{f}$.

$$M_z = I_{zx}\dot{\omega}_x + I_{zy}\dot{\omega}_y + I_{zz}\dot{\omega}_z - \left(I_{xx} - I_{yy}\right)\omega_x\omega_y - I_{xy}\left(\omega_y^2 - \omega_x^2\right) - \omega_z\left(\omega_y I_{xz} - \omega_x I_{yz}\right) \tag{12.9}$$

which reduce to the following set of equations in the *principal coordinate frame*:

$$M_1 = I_1\dot{\omega}_1 - (I_2 - I_3)\,\omega_2\omega_3 \tag{12.10}$$

$$M_2 = I_2\dot{\omega}_2 - (I_3 - I_1)\,\omega_3\omega_1 \tag{12.11}$$

$$M_3 = I_3\dot{\omega}_3 - (I_1 - I_2)\,\omega_1\omega_2 \tag{12.12}$$

The principal coordinate frame in which the off-diagonal elements I_{ij}, $i \neq j$, are zero is denoted by numbers 123 to indicate the first, second, and third *principal axes*. The body coordinate frame is assumed to sit at the fixed point of the body.

The kinetic energy of a rotating rigid body is given as

$$K = \tfrac{1}{2}\,{}_G^B\boldsymbol{\omega}_B \cdot {}^B\mathbf{L} = \tfrac{1}{2}\,{}_G^B\boldsymbol{\omega}_B^{\mathrm{T}}\;{}^BI\;{}_G^B\boldsymbol{\omega}_B \tag{12.13}$$

$$= \tfrac{1}{2}\left(I_{xx}\omega_x^2 + I_{yy}\omega_y^2 + I_{zz}\omega_z^2\right) - I_{xy}\omega_x\omega_y - I_{yz}\omega_y\omega_z - I_{zx}\omega_z\omega_x \tag{12.14}$$

which in the principal coordinate frame reduces to

$$K = \tfrac{1}{2}\left(I_1\omega_1^2 + I_2\omega_2^2 + I_3\omega_3^2\right) \tag{12.15}$$

Proof: Let m_i be the mass of the ith particle of a rigid body B made of n particles and let $\mathbf{r}_i = {}^B\mathbf{r}_i$ be the Cartesian position vector of m_i in a central body-fixed coordinate frame $Oxyz$:

$$\mathbf{r}_i = {}^B\mathbf{r}_i = \begin{bmatrix} x_i & y_i & z_i \end{bmatrix}^{\mathrm{T}} \tag{12.16}$$

Assume that $\boldsymbol{\omega} = {}_G^B\boldsymbol{\omega}_B$ is the angular velocity of the rigid body with respect to the G-frame expressed in the body coordinate frame:

$$\boldsymbol{\omega} = {}_G^B\boldsymbol{\omega}_B = \begin{bmatrix} \omega_x & \omega_y & \omega_z \end{bmatrix}^{\mathrm{T}} \tag{12.17}$$

The angular momentum of m_i is

$$\begin{aligned} {}^B\mathbf{L}_i &= \mathbf{r}_i \times m_i \dot{\mathbf{r}}_i = m_i \left[\mathbf{r}_i \times (\boldsymbol{\omega} \times \boldsymbol{r}_i) \right] \\ &= m_i \left[(\mathbf{r}_i \cdot \mathbf{r}_i)\, \boldsymbol{\omega} - (\mathbf{r}_i \cdot \boldsymbol{\omega})\, \mathbf{r}_i \right] \\ &= m_i r_i^2 \boldsymbol{\omega} - m_i \left(\mathbf{r}_i \cdot \boldsymbol{\omega} \right) \mathbf{r}_i \end{aligned} \tag{12.18}$$

Hence, the angular momentum of the rigid body is

$${}^B\mathbf{L} = \boldsymbol{\omega} \sum_{i=1}^{n} m_i r_i^2 - \sum_{i=1}^{n} m_i \left(\mathbf{r}_i \cdot \boldsymbol{\omega} \right) \mathbf{r}_i \tag{12.19}$$

Substitution of $\mathbf{r}_i$ and $\boldsymbol{\omega}$ gives

$$\begin{aligned} {}^B\mathbf{L} &= \left(\omega_x \hat{\imath} + \omega_y \hat{\jmath} + \omega_z \hat{k} \right) \sum_{i=1}^{n} m_i \left(x_i^2 + y_i^2 + z_i^2 \right) \\ &\quad - \sum_{i=1}^{n} m_i \left(x_i \omega_x + y_i \omega_y + z_i \omega_z \right) \cdot \left(x_i \hat{\imath} + y_i \hat{\jmath} + z_i \hat{k} \right) \end{aligned} \tag{12.20}$$

and therefore,

$$\begin{aligned} {}^B\mathbf{L} &= \sum_{i=1}^{n} m_i \left(x_i^2 + y_i^2 + z_i^2 \right) \omega_x \hat{\imath} + \sum_{i=1}^{n} m_i \left(x_i^2 + y_i^2 + z_i^2 \right) \omega_y \hat{\jmath} \\ &\quad + \sum_{i=1}^{n} m_i \left(x_i^2 + y_i^2 + z_i^2 \right) \omega_z \hat{k} - \sum_{i=1}^{n} m_i \left(x_i \omega_x + y_i \omega_y + z_i \omega_z \right) x_i \hat{\imath} \\ &\quad - \sum_{i=1}^{n} m_i \left(x_i \omega_x + y_i \omega_y + z_i \omega_z \right) y_i \hat{\jmath} \\ &\quad - \sum_{i=1}^{n} m_i \left(x_i \omega_x + y_i \omega_y + z_i \omega_z \right) z_i \hat{k} \end{aligned} \tag{12.21}$$

or

$$\begin{aligned} {}^B\mathbf{L} &= \sum_{i=1}^{n} m_i \left[\left(x_i^2 + y_i^2 + z_i^2 \right) \omega_x - \left(x_i \omega_x + y_i \omega_y + z_i \omega_z \right) x_i \right] \hat{\imath} \\ &\quad + \sum_{i=1}^{n} m_i \left[\left(x_i^2 + y_i^2 + z_i^2 \right) \omega_y - \left(x_i \omega_x + y_i \omega_y + z_i \omega_z \right) y_i \right] \hat{\jmath} \\ &\quad + \sum_{i=1}^{n} m_i \left[\left(x_i^2 + y_i^2 + z_i^2 \right) \omega_z - \left(x_i \omega_x + y_i \omega_y + z_i \omega_z \right) z_i \right] \hat{k} \end{aligned} \tag{12.22}$$

which can be rearranged as

$$
\begin{aligned}
{}^{B}\mathbf{L} &= \omega_x \sum_{i=1}^{n} m_i \left(y_i^2 + z_i^2\right) \hat{\imath} - \omega_y \sum_{i=1}^{n} m_i x_i y_i \hat{\imath} - \omega_z \sum_{i=1}^{n} m_i x_i z_i \hat{\imath} \\
&+ \omega_y \sum_{i=1}^{n} m_i \left(z_i^2 + x_i^2\right) \hat{\jmath} - \omega_z \sum_{i=1}^{n} m_i y_i z_i \hat{\jmath} - \omega_x \sum_{i=1}^{n} m_i y_i x_i \\
&+ \omega_z \sum_{i=1}^{n} m_i \left(x_i^2 + y_i^2\right) \hat{k} - \omega_x \sum_{i=1}^{n} m_i z_i x_i \hat{k} - \omega_y \sum_{i=1}^{n} m_i z_i y_i
\end{aligned}
\tag{12.23}
$$

By introducing the mass moment matrix $[I]$ with elements I_{ij} as

$$I_{xx} = \sum_{i=1}^{n} \left[m_i \left(y_i^2 + z_i^2\right)\right] \tag{12.24}$$

$$I_{yy} = \sum_{i=1}^{n} \left[m_i \left(z_i^2 + x_i^2\right)\right] \tag{12.25}$$

$$I_{zz} = \sum_{i=1}^{n} \left[m_i \left(x_i^2 + y_i^2\right)\right] \tag{12.26}$$

$$I_{xy} = I_{yx} = -\sum_{i=1}^{n} \left(m_i x_i y_i\right) \tag{12.27}$$

$$I_{yz} = I_{zy} = -\sum_{i=1}^{n} \left(m_i y_i z_i\right) \tag{12.28}$$

$$I_{zx} = I_{xz} = -\sum_{i=1}^{n} \left(m_i z_i x_i\right) \tag{12.29}$$

we can rewrite the angular momentum in a concise form,

$$L_x = I_{xx}\omega_x + I_{xy}\omega_y + I_{xz}\omega_z \tag{12.30}$$

$$L_y = I_{yx}\omega_x + I_{yy}\omega_y + I_{yz}\omega_z \tag{12.31}$$

$$L_z = I_{zx}\omega_x + I_{zy}\omega_y + I_{zz}\omega_z \tag{12.32}$$

or in a matrix form,

$${}^{B}\mathbf{L} = {}^{B}I \cdot {}_{G}^{B}\boldsymbol{\omega}_B \tag{12.33}$$

$$\begin{bmatrix} L_x \\ L_y \\ L_z \end{bmatrix} = \begin{bmatrix} I_{xx} & I_{xy} & I_{xz} \\ I_{yx} & I_{yy} & I_{yz} \\ I_{zx} & I_{zy} & I_{zz} \end{bmatrix} \begin{bmatrix} \omega_x \\ \omega_y \\ \omega_z \end{bmatrix} \tag{12.34}$$

In a solid rigid body that is a continuous media, the summations will be replaced by integrations over the volume of the body. So the elements of $[I]$ that are only functions of the mass distribution of the rigid body will be defined by

$$I_{ij} = \int_B \left(r^2 \delta_{ij} - x_i x_j\right) dm \qquad i, j = 1, 2, 3 \tag{12.35}$$

where δ_{ij} is Kronecker's delta (1.125).

The B-expression of the Euler equation of motion for a rigid body is

$${}^B\mathbf{M} = \frac{{}^G d}{dt}\, {}^B\mathbf{L} \tag{12.36}$$

where ${}^B\mathbf{M}$ is the B-expression of the resultant external moments applied on the rigid body. Because ${}^B\mathbf{L}$ is a B-vector, its time derivative in the global coordinate frame is

$$\frac{{}^G d\, {}^B\mathbf{L}}{dt} = {}^B\dot{\mathbf{L}} + {}^B_G\boldsymbol{\omega}_B \times {}^B\mathbf{L} \tag{12.37}$$

and therefore the vectorial expression of the Euler equation is

$${}^B\mathbf{M} = {}^B\dot{\mathbf{L}} + {}^B_G\boldsymbol{\omega}_B \times {}^B\mathbf{L} = {}^B I\, {}^B_G\dot{\boldsymbol{\omega}}_B + {}^B_G\boldsymbol{\omega}_B \times \left({}^B I\, {}^B_G\boldsymbol{\omega}_B\right) \tag{12.38}$$

Expansion of this equation provides the most general form of the equation of motion in a Cartesian body frame:

$$\begin{aligned}
{}^B\mathbf{M} = &\left(I_{xx}\dot{\omega}_x + I_{xy}\dot{\omega}_y + I_{xz}\dot{\omega}_z\right)\hat{\imath} \\
&+ \left(I_{yx}\dot{\omega}_x + I_{yy}\dot{\omega}_y + I_{yz}\dot{\omega}_z\right)\hat{\jmath} \\
&+ \left(I_{zx}\dot{\omega}_x + I_{zy}\dot{\omega}_y + I_{zz}\dot{\omega}_z\right)\hat{k} \\
&+ \omega_y\left(I_{xz}\omega_x + I_{yz}\omega_y + I_{zz}\omega_z\right)\hat{\imath} \\
&- \omega_z\left(I_{xy}\omega_x + I_{yy}\omega_y + I_{yz}\omega_z\right)\hat{\imath} \\
&+ \omega_z\left(I_{xx}\omega_x + I_{xy}\omega_y + I_{xz}\omega_z\right)\hat{\jmath} \\
&- \omega_x\left(I_{xz}\omega_x + I_{yz}\omega_y + I_{zz}\omega_z\right)\hat{\jmath} \\
&+ \omega_x\left(I_{xy}\omega_x + I_{yy}\omega_y + I_{yz}\omega_z\right)\hat{k} \\
&- \omega_y\left(I_{xx}\omega_x + I_{xy}\omega_y + I_{xz}\omega_z\right)\hat{k}
\end{aligned} \tag{12.39}$$

We may separate the components of the Euler equation to get the scalar form of the general Euler equations of motion for a rigid body in a body frame:

$$\begin{aligned}
M_x = &\, I_{xx}\dot{\omega}_x + I_{xy}\dot{\omega}_y + I_{xz}\dot{\omega}_z - \left(I_{yy} - I_{zz}\right)\omega_y\omega_z \\
&- I_{yz}\left(\omega_z^2 - \omega_y^2\right) - \omega_x\left(\omega_z I_{xy} - \omega_y I_{xz}\right)
\end{aligned} \tag{12.40}$$

$$\begin{aligned}
M_y = &\, I_{yx}\dot{\omega}_x + I_{yy}\dot{\omega}_y + I_{yz}\dot{\omega}_z - \left(I_{zz} - I_{xx}\right)\omega_z\omega_x \\
&- I_{xz}\left(\omega_x^2 - \omega_z^2\right) - \omega_y\left(\omega_x I_{yz} - \omega_z I_{xy}\right)
\end{aligned} \tag{12.41}$$

$$\begin{aligned}
M_z = &\, I_{zx}\dot{\omega}_x + I_{zy}\dot{\omega}_y + I_{zz}\dot{\omega}_z - \left(I_{xx} - I_{yy}\right)\omega_x\omega_y \\
&- I_{xy}\left(\omega_y^2 - \omega_x^2\right) - \omega_z\left(\omega_y I_{xz} - \omega_x I_{yz}\right)
\end{aligned} \tag{12.42}$$

Assume that we can rotate the body frame about its origin to find an orientation that makes $I_{ij} = 0$ for $i \neq j$. In such a coordinate frame, which is called a *principal frame*, the Euler equations simplify to

$$M_1 = I_1\dot{\omega}_1 - (I_2 - I_3)\,\omega_2\omega_3 \tag{12.43}$$

$$M_2 = I_2\dot{\omega}_2 - (I_3 - I_1)\,\omega_3\omega_1 \tag{12.44}$$

$$M_3 = I_3\dot{\omega}_3 - (I_1 - I_2)\,\omega_1\omega_2 \tag{12.45}$$

The kinetic energy of a rigid body may be found by the integral of the kinetic energy of a mass element dm over the whole body:

$$\begin{aligned} K &= \frac{1}{2}\int_B \mathbf{v}^2 dm = \frac{1}{2}\int_B (\boldsymbol{\omega} \times \boldsymbol{r}) \cdot (\boldsymbol{\omega} \times \boldsymbol{r})\, dm \\ &= \frac{\omega_x^2}{2}\int_B \left(y^2 + z^2\right) dm + \frac{\omega_y^2}{2}\int_B \left(z^2 + x^2\right) dm + \frac{\omega_z^2}{2}\int_B \left(x^2 + y^2\right) dm \\ &\quad - \omega_x\omega_y \int_B xy\, dm - \omega_y\omega_z \int_B yz\, dm - \omega_z\omega_x \int_B zx\, dm \\ &= \frac{1}{2}\left(I_{xx}\omega_x^2 + I_{yy}\omega_y^2 + I_{zz}\omega_z^2\right) \\ &\quad - I_{xy}\omega_x\omega_y - I_{yz}\omega_y\omega_z - I_{zx}\omega_z\omega_x \end{aligned} \tag{12.46}$$

The kinetic energy can be rearranged to a matrix multiplication form:

$$K = \tfrac{1}{2}\,{}^B_G\boldsymbol{\omega}_B^{\mathrm{T}}\;{}^BI\;{}^B_G\boldsymbol{\omega}_B = \tfrac{1}{2}\,{}^B_G\boldsymbol{\omega}_B \cdot {}^B\mathbf{L} \tag{12.47}$$

When the body frame is principal, the kinetic energy will simplify to

$$K = \tfrac{1}{2}\left(I_1\omega_1^2 + I_2\omega_2^2 + I_3\omega_3^2\right) \tag{12.48}$$

Euler equation (12.1), or its B-expression (12.2), is the law of the rotational motion of a rigid body. It is an independent law and cannot be derived from Newton's second law. Similar to the Newton equation of motion, the Euler equation of motion is a simplified model of rotational motions. The limits of applicability of the Euler equation are the same as the Newton equation, expressed in Example 106. ■

Example 716 Why Body Coordinate Frame? Let us derive the global expressions of the Euler equation,

$${}^G\mathbf{M} = \frac{{}^Gd\,{}^G\mathbf{L}}{dt} = {}^G\dot{\mathbf{L}} \tag{12.49}$$

$${}^G\mathbf{L} = {}^GI\;{}_G\boldsymbol{\omega}_B \tag{12.50}$$

by collecting the angular momentum of individual particles of a rigid body. Let m_i be the mass of the ith particle of a rigid body B which is made of n particles and let ${}^G\mathbf{r}_i$ be the Cartesian position vector of m_i in a global coordinate frame $OXYZ$:

$${}^G\mathbf{r}_i = \begin{bmatrix} X & Y & Z \end{bmatrix}^{\mathrm{T}} \tag{12.51}$$

Assume that ${}_G\boldsymbol{\omega}_B$ is the angular velocity of the rigid body with respect to the ground:

$$ {}_G\boldsymbol{\omega}_B = \begin{bmatrix} \omega_X & \omega_Y & \omega_Z \end{bmatrix}^{\mathrm{T}} \tag{12.52} $$

$$ {}^G\mathbf{r}_i = {}^GR_B\,{}^B\mathbf{r}_i \tag{12.53} $$

The angular momentum of m_i in G is

$$ \begin{aligned} {}^G\mathbf{L}_i &= {}^G\mathbf{r}_i \times m_i\,{}^G\dot{\mathbf{r}}_i = m_i\left[{}^G\mathbf{r}_i \times \left({}_G\boldsymbol{\omega}_B \times {}^G\mathbf{r}_i\right)\right] \\ &= m_i\left[\left({}^G\mathbf{r}_i \cdot {}^G\mathbf{r}_i\right){}_G\boldsymbol{\omega}_B - \left({}^G\mathbf{r}_i \cdot {}_G\boldsymbol{\omega}_B\right){}^G\mathbf{r}_i\right] \\ &= m_i r_i^2\,{}_G\boldsymbol{\omega}_B - m_i\left({}^G\mathbf{r}_i \cdot {}_G\boldsymbol{\omega}_B\right){}^G\mathbf{r}_i \end{aligned} \tag{12.54} $$

Therefore, the angular momentum of the rigid body is

$$ {}^G\mathbf{L} = {}_G\boldsymbol{\omega}_B \sum_{i=1}^{n} m_i r_i^2 - \sum_{i=1}^{n} m_i\left({}^G\mathbf{r}_i \cdot {}_G\boldsymbol{\omega}_B\right){}^G\mathbf{r}_i \tag{12.55} $$

Substitution of ${}^G\mathbf{r}_i$ and ${}_G\boldsymbol{\omega}_B$ gives

$$ \begin{aligned} {}^G\mathbf{L} &= \left(\omega_X\hat{\imath} + \omega_Y\hat{\jmath} + \omega_Z\hat{k}\right)\sum_{i=1}^{n} m_i\left(X_i^2 + Y_i^2 + Z_i^2\right) \\ &\quad - \sum_{i=1}^{n} m_i\left(X_i\omega_X + Y_i\omega_Y + Z_i\omega_Z\right)\cdot\left(X_i\hat{\imath} + Y_i\hat{\jmath} + Z_i\hat{k}\right) \end{aligned} \tag{12.56} $$

and therefore,

$$ \begin{aligned} {}^G\mathbf{L} &= \omega_X \sum_{i=1}^{n} m_i\left(Y_i^2 + Z_i^2\right)\hat{\imath} \\ &\quad - \omega_Y \sum_{i=1}^{n} m_i X_i Y_i\hat{\imath} - \omega_Z \sum_{i=1}^{n} m_i X_i Z_i\hat{\imath} \\ &\quad + \omega_Y \sum_{i=1}^{n} m_i\left(Z_i^2 + X_i^2\right)\hat{\jmath} \\ &\quad - \omega_Z \sum_{i=1}^{n} m_i Y_i Z_i\hat{\jmath} - \omega_X \sum_{i=1}^{n} m_i Y_i X_i \\ &\quad + \omega_Z \sum_{i=1}^{n} m_i\left(X_i^2 + Y_i^2\right)\hat{k} \\ &\quad - \omega_X \sum_{i=1}^{n} m_i Z_i X_i\hat{k} - \omega_Y \sum_{i=1}^{n} m_i Z_i Y_i \end{aligned} \tag{12.57} $$

By introducing the mass moment matrix ${}^GI$ with elements ${}^GI_{ij}$ as

$$ {}^GI_{ij} = \int_B\left(r^2\delta_{ij} - X_iX_j\right)dm \qquad i, j = 1, 2, 3 \tag{12.58} $$

$$ X_1 \equiv X \qquad X_2 \equiv Y \qquad X_3 \equiv Z \tag{12.59} $$

we can rewrite the angular momentum in a concise form:

$$ {}^{G}\mathbf{L} = {}^{G}I_{G}\boldsymbol{\omega}_{B} \tag{12.60}$$

$$\begin{bmatrix} L_X \\ L_Y \\ L_Z \end{bmatrix} = \begin{bmatrix} I_{XX} & I_{XY} & I_{XZ} \\ I_{YX} & I_{YY} & I_{YZ} \\ I_{ZX} & I_{ZY} & I_{ZZ} \end{bmatrix} \begin{bmatrix} \omega_X \\ \omega_Y \\ \omega_Z \end{bmatrix} \tag{12.61}$$

The elements of ${}^{G}I$ are related to the elements of ${}^{B}I$ by direction cosines r_{ij}:

$$ {}^{G}I = {}^{G}R_{B}\ {}^{B}I\ {}^{G}R_{B}^{\mathrm{T}} \tag{12.62}$$

$$ {}^{G}I_{ij} = \sum_{k=1}^{3}\sum_{l=1}^{3} r_{ki}\, r_{lj}\ {}^{B}I_{kl} \tag{12.63}$$

When the body B is turning in G, the direction cosines r_{ij} and hence the rotation matrix ${}^{G}R_{B}$ are variable. Therefore, in general, ${}^{G}I$ is a function of orientation of B in G:

$$ {}^{G}I = {}^{G}I\,(\varphi, \theta, \psi) \tag{12.64}$$

So, the global expression of the Euler equation becomes

$$ {}^{G}\mathbf{M} = {}^{G}\frac{d}{dt}\,{}^{G}\mathbf{L} = {}^{G}\dot{I}\ {}_{G}\boldsymbol{\omega}_{B} + {}^{G}I\ {}_{G}\dot{\boldsymbol{\omega}}_{B} \tag{12.65}$$

Having a time-varying ${}^{G}I$ makes the equations of motion very complicated and generally unsolvable. However, the body expression of the Euler equation provides a great advantage by having a constant ${}^{B}I$.

Example 717 ★ Time Derivative of a Variable Mass Moment Matrix The G-derivative of a global mass moment matrix ${}^{G}I$ is

$$ {}^{G}\dot{I} = {}_{G}\tilde{\omega}_{B}\ {}^{G}I - {}^{G}I\ {}_{G}\tilde{\omega}_{B} \tag{12.66}$$

where

$$\begin{aligned} {}^{G}\dot{I} &= {}^{G}\frac{d}{dt}\left({}^{G}R_{B}\ {}^{B}I\ {}^{G}R_{B}^{\mathrm{T}}\right) \\ &= {}^{G}\dot{R}_{B}\ {}^{B}I\ {}^{G}R_{B}^{\mathrm{T}} + {}^{G}R_{B}\ {}^{B}I\ {}^{G}\dot{R}_{B}^{\mathrm{T}} \\ &= {}^{G}\dot{R}_{B}\left({}^{G}R_{B}^{\mathrm{T}}\ {}^{G}I\ {}^{G}R_{B}\right){}^{G}R_{B}^{\mathrm{T}} + {}^{G}R_{B}\left({}^{G}R_{B}^{\mathrm{T}}\ {}^{G}I\ {}^{G}R_{B}\right){}^{G}\dot{R}_{B}^{\mathrm{T}} \\ &= \left({}^{G}\dot{R}_{B}\,{}^{G}R_{B}^{\mathrm{T}}\right){}^{G}I\left({}^{G}R_{B}\,{}^{G}R_{B}^{\mathrm{T}}\right) + \left({}^{G}R_{B}\,{}^{G}R_{B}^{\mathrm{T}}\right){}^{G}I\left({}^{G}R_{B}\,{}^{G}\dot{R}_{B}^{\mathrm{T}}\right) \\ &= {}_{G}\tilde{\omega}_{B}\ {}^{G}I + {}^{G}I\ {}_{G}\tilde{\omega}_{B}^{\mathrm{T}} = {}_{G}\tilde{\omega}_{B}\ {}^{G}I - {}^{G}I\ {}_{G}\tilde{\omega}_{B} \end{aligned} \tag{12.67}$$

Therefore, we can write the G-expression of the Euler equation as

$$\begin{aligned} {}^{G}\mathbf{M} &= {}^{G}\frac{d}{dt}\,{}^{G}\mathbf{L} = {}^{G}\frac{d}{dt}\left({}^{G}I_{G}\boldsymbol{\omega}_{B}\right) = {}^{G}\dot{I}\ {}_{G}\boldsymbol{\omega}_{B} + {}^{G}I\ {}_{G}\dot{\boldsymbol{\omega}}_{B} \\ &= \left({}_{G}\tilde{\omega}_{B}\ {}^{G}I\right){}_{G}\boldsymbol{\omega}_{B} - \left({}^{G}I\ {}_{G}\tilde{\omega}_{B}\right){}_{G}\boldsymbol{\omega}_{B} + {}^{G}I\ {}_{G}\dot{\boldsymbol{\omega}}_{B} \end{aligned} \tag{12.68}$$

Example 718 Solution of Euler Equations Euler equations (12.10)–(12.12) can be integrated in closed form only in a few special cases. It is basically because the right-hand side of the equations are nonlinear and coupled. The general solution of the equations has not developed yet.

We may distinguish two types of problems:

1. The components of the external moment are zero:

$$ {}^{B}\mathbf{M} = \mathbf{0} \tag{12.69} $$

This type of problem is called *torque-free motion*. Having a zero moment, we solve Euler equations for angular accelerations and write them as

$$ \dot{\omega}_1 = \frac{I_2 - I_3}{I_1}\omega_2\omega_3 \tag{12.70} $$

$$ \dot{\omega}_2 = \frac{I_3 - I_1}{I_2}\omega_3\omega_1 \tag{12.71} $$

$$ \dot{\omega}_3 = \frac{I_1 - I_2}{I_3}\omega_1\omega_2 \tag{12.72} $$

Integrability of torque-free equations depends on the relations of the principal mass moments I_1, I_2, I_3. Most classical integrable problems of rigid bodies are of this type.

2. The moment components ${}^{B}M_1$, ${}^{B}M_2$, and ${}^{B}M_3$ are known functions of angular velocities $\omega_1, \omega_2, \omega_3$, orientation parameters φ, θ, ψ, and t:

$$ {}^{B}M_1 = M_1\left(\omega_1, \omega_2, \omega_3, \varphi, \theta, \psi, t\right) \tag{12.73} $$

$$ {}^{B}M_2 = M_2\left(\omega_1, \omega_2, \omega_3, \varphi, \theta, \psi, t\right) \tag{12.74} $$

$$ {}^{B}M_3 = M_3\left(\omega_1, \omega_2, \omega_3, \varphi, \theta, \psi, t\right) \tag{12.75} $$

Although type 2 includes type 1, the application and method of their solutions are different. Being a function of angular velocity

$$ {}^{B}\mathbf{M} = {}^{B}\mathbf{M}\left({}_{G}^{B}\boldsymbol{\omega}_{B}\right) \tag{12.76} $$

means that the source of the moment ${}^{B}\mathbf{M}$ is rotating together with the body. In such cases the rigid body is said to be *self-excited*.

The solution again begins by transforming Euler equations to a set of three first-order ordinary equations for $\dot{\omega}_1, \dot{\omega}_2, \dot{\omega}_3$:

$$ \dot{\omega}_1 = \frac{I_2 - I_3}{I_1}\omega_2\omega_3 - \frac{M_1}{I_1} \tag{12.77} $$

$$ \dot{\omega}_2 = \frac{I_3 - I_1}{I_2}\omega_3\omega_1 - \frac{M_2}{I_2} \tag{12.78} $$

$$ \dot{\omega}_3 = \frac{I_1 - I_2}{I_3}\omega_1\omega_2 - \frac{M_3}{I_3} \tag{12.79} $$

Depending on the mass moments I_1, I_2, I_3 and the simplicity of ${}^{B}\mathbf{M}$, we analytically or numerically solve these equations for $\omega_1 = \omega_1(t)$, $\omega_2 = \omega_2(t)$,

$\omega_3 = \omega_3(t)$. Then the problem reduces to a new set of coupled first-order ordinary differential equations to determine the orientation of B in G.

Depending on the generalized coordinates, the orientation of B in G may be expressed by following methods:

1. Rotation about the Cartesian body frame as given in Equations (8.76)–(8.78):

$$ {}_G^B\boldsymbol{\omega}_B = \omega_1\,\hat{\imath} + \omega_2\,\hat{\jmath} + \omega_3\,\hat{k} = \dot{\alpha}\,\hat{\imath} + \dot{\beta}\,\hat{\jmath} + \dot{\gamma}\,\hat{k} \tag{12.80} $$

$$ \dot{\alpha} = \omega_1(\alpha, \beta, \gamma, t) \tag{12.81} $$

$$ \dot{\beta} = \omega_1(\alpha, \beta, \gamma, t) \tag{12.82} $$

$$ \dot{\gamma} = \omega_1(\alpha, \beta, \gamma, t) \tag{12.83} $$

2. A set of Euler angles as given in Appendix B. Equations (8.99) and (4.188) are the most common sets:

$$ {}_G^B\boldsymbol{\omega}_B = \omega_1\hat{\imath} + \omega_2\hat{\jmath} + \omega_3\hat{k} = \dot{\varphi}\hat{e}_\varphi + \dot{\theta}\hat{e}_\theta + \dot{\psi}\hat{e}_\psi $$

$$ = \begin{bmatrix} \sin\theta\sin\psi & \cos\psi & 0 \\ \sin\theta\cos\psi & -\sin\psi & 0 \\ \cos\theta & 0 & 1 \end{bmatrix} \begin{bmatrix} \dot{\varphi} \\ \dot{\theta} \\ \dot{\psi} \end{bmatrix} \tag{12.84} $$

$$ \begin{bmatrix} \dot{\varphi} \\ \dot{\theta} \\ \dot{\psi} \end{bmatrix} = \frac{1}{\sin\theta} \begin{bmatrix} \sin\psi & \cos\psi & 0 \\ \sin\theta\cos\psi & -\sin\theta\sin\psi & 0 \\ -\cos\theta\sin\psi & -\cos\theta\cos\psi & 1 \end{bmatrix} \begin{bmatrix} \omega_x \\ \omega_y \\ \omega_z \end{bmatrix} \tag{12.85} $$

3. Euler parameters as given in Equation (8.155):

$$ \overset{\leftrightarrow}{\dot{e}} = \frac{1}{2}\overset{\leftrightarrow}{e}\ {}_G^B\overset{\longleftrightarrow}{\boldsymbol{\omega}}_B \tag{12.86} $$

$$ \dot{e}_0 = -\omega_1 e_1 - \omega_2 e_2 - \omega_z e_3 \tag{12.87} $$

$$ \dot{e}_1 = \omega_1 e_0 - \omega_2 e_3 + \omega_z e_2 \tag{12.88} $$

$$ \dot{e}_2 = \omega_2 e_0 + \omega_1 e_3 - \omega_z e_1 \tag{12.89} $$

$$ \dot{e}_3 = \omega_2 e_1 - \omega_1 e_2 + \omega_z e_0 \tag{12.90} $$

Euler parameters also cover the angle–axis and quaternion representations.

Example 719 A Rotating Arm Figure 12.2 shows a rotating arm with a body coordinate frame B. The transformation matrix between B and G and the angular velocity of the arm are given as

$$ {}^G R_B = R_{Z,\theta} = \begin{bmatrix} \cos\theta & -\sin\theta & 0 \\ \sin\theta & \cos\theta & 0 \\ 0 & 0 & 1 \end{bmatrix} \tag{12.91} $$

$$ {}_G^B\omega_B = {}^G R_B^{\mathrm{T}}\ {}^G\dot{R}_B = \dot{\theta}\tilde{k} = \dot{\theta} \begin{bmatrix} 0 & -1 & 0 \\ 1 & 0 & 0 \\ 0 & 0 & 0 \end{bmatrix} \tag{12.92} $$

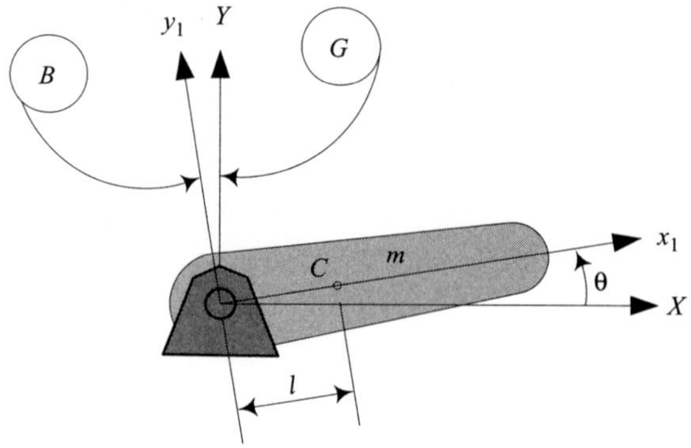

Figure 12.2 A rotating arm.

Assuming a principal mass moment matrix

$$ {}^{B}I = \begin{bmatrix} I_x & 0 & 0 \\ 0 & I_y & 0 \\ 0 & 0 & I_z \end{bmatrix} \tag{12.93}$$

we have

$$ {}^{B}\mathbf{M} = {}^{B}\dot{\mathbf{L}} + {}_{G}^{B}\boldsymbol{\omega}_B \times {}^{B}\mathbf{L} = {}^{B}I\ {}_{G}^{B}\dot{\boldsymbol{\omega}}_B + {}_{G}^{B}\boldsymbol{\omega}_B \times \left({}^{B}I\ {}_{G}^{B}\boldsymbol{\omega}_B\right) \tag{12.94}$$

or

$$\begin{bmatrix} M_x \\ M_y \\ M_z \end{bmatrix} = \begin{bmatrix} 0 \\ 0 \\ I_z\ddot{\theta} \end{bmatrix} \tag{12.95}$$

Using the transformation matrix ${}^{G}R_B$, we can determine ${}^{G}I$:

$$\begin{aligned} {}^{G}I &= {}^{G}R_B\ {}^{B}I\ {}^{G}R_B^{\mathrm{T}} = {}^{G}R_B \begin{bmatrix} I_x & 0 & 0 \\ 0 & I_y & 0 \\ 0 & 0 & I_z \end{bmatrix} {}^{G}R_B^{\mathrm{T}} \\ &= \begin{bmatrix} I_x\cos^2\theta + I_y\sin^2\theta & \left(I_x - I_y\right)\cos\theta\sin\theta & 0 \\ \left(I_x - I_y\right)\cos\theta\sin\theta & I_y\cos^2\theta + I_x\sin^2\theta & 0 \\ 0 & 0 & I_z \end{bmatrix} \end{aligned} \tag{12.96}$$

To determine the G-expression of the Euler equation, we should transform ${}_{G}^{B}\boldsymbol{\omega}_B$ and ${}_{G}^{B}\tilde{\omega}_B$ to G:

$$ {}_{G}\tilde{\omega}_B = {}^{G}R_B\ {}_{G}^{B}\tilde{\omega}_B\ {}^{G}R_B^{\mathrm{T}} = \dot{\theta}\tilde{K} = \dot{\theta}\begin{bmatrix} 0 & -1 & 0 \\ 1 & 0 & 0 \\ 0 & 0 & 0 \end{bmatrix} \tag{12.97}$$

$$ {}_{G}\boldsymbol{\omega}_B = {}^{G}R_B\ {}_{G}^{B}\boldsymbol{\omega}_B = \dot{\theta}\hat{K} = \begin{bmatrix} 0 \\ 0 \\ \dot{\theta} \end{bmatrix} \tag{12.98}$$

Therefore, the G-expression of the Euler equation of the rotating arm is

$$
\begin{aligned}
{}^G\mathbf{M} &= {}^G\frac{d}{dt}\,{}^G\mathbf{L} = {}^G\frac{d}{dt}\left({}^GI_G\boldsymbol{\omega}_B\right) = {}^G\dot{I}\;{}_G\boldsymbol{\omega}_B + {}^GI\;{}_G\dot{\boldsymbol{\omega}}_B \\
&= \left({}_G\tilde{\omega}_B\;{}^GI\right)\;{}_G\boldsymbol{\omega}_B + \left({}^GI\;{}_G\tilde{\omega}_B^{\mathrm{T}}\right)\;{}_G\boldsymbol{\omega}_B + {}^GI\;{}_G\dot{\boldsymbol{\omega}}_B \\
&= I_z\ddot{\theta}
\end{aligned}
\tag{12.99}
$$

or

$$
\begin{bmatrix} M_X \\ M_Y \\ M_Z \end{bmatrix} = \begin{bmatrix} 0 \\ 0 \\ I_z\ddot{\theta} \end{bmatrix}
\tag{12.100}
$$

When a rigid body is turning about a globally fixed principal axis, the B- and G-expressions of the Euler equations look alike.

Example 720 Steady Rotation of a Freely Rotating Rigid Body The Newton–Euler equations of motion for a rigid body are

$$
{}^G\mathbf{F} = m\,{}^G\dot{\mathbf{v}}
\tag{12.101}
$$

$$
{}^B\mathbf{M} = I\;{}^B_G\dot{\boldsymbol{\omega}}_B + {}^B_G\boldsymbol{\omega}_B \times {}^B\mathbf{L}
\tag{12.102}
$$

Consider a situation where the resultant applied force and moment on the body are zero:

$$
\mathbf{F} = 0
\tag{12.103}
$$

$$
\mathbf{M} = 0
\tag{12.104}
$$

Based on the Newton equation, the velocity of the mass center will be constant in the global coordinate frame. However, the Euler equations reduce to

$$
\dot{\omega}_1 = \frac{I_2 - I_3}{I_1}\omega_2\omega_3
\tag{12.105}
$$

$$
\dot{\omega}_2 = \frac{I_3 - I_1}{I_{22}}\omega_3\omega_1
\tag{12.106}
$$

$$
\dot{\omega}_3 = \frac{I_1 - I_2}{I_3}\omega_1\omega_2
\tag{12.107}
$$

showing that the angular velocity of a force-free rigid body can also be constant if

$$
I_1 = I_2 = I_3
\tag{12.108}
$$

angular velocity $\boldsymbol{\omega}$ can also be constant if two principal moments of inertia, say I_1 and I_2, are zero and the third angular velocity, in this case ω_3, is initially zero or if the angular velocity vector is initially parallel to a principal axis.

Example 721 Required Torques to Turn a Disc on a Turntable Consider a uniform disc with mass m, radius R, and mass moment $[I]$ that is mounted on a horizontal shaft as shown in Figure 12.3. The shaft is mounted on a table and is turning with constant angular velocity $\dot{\theta}_2 = \omega$ with respect to the table. If the table is also turning with angular velocity $\dot{\theta}_1 = \Omega$, then we are able to calculate how much force is supported by the bearings of the shaft.

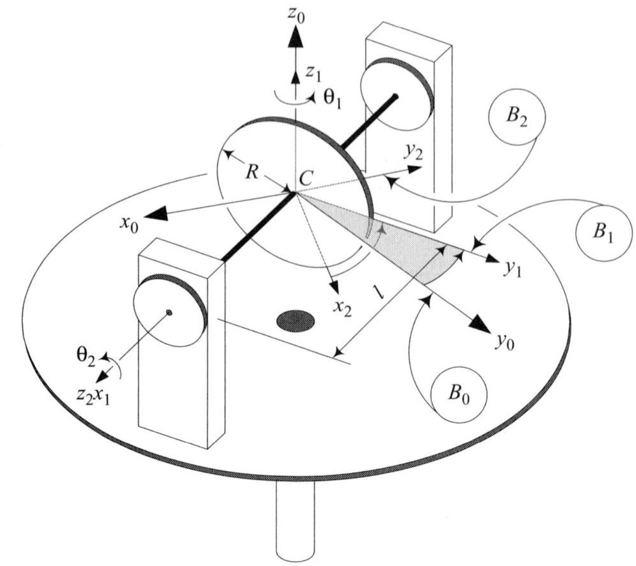

Figure 12.3 A turning disc on a turning table.

The mass center C of the disc is motionless. We attach a global coordinate frame B_0 at C. We also attach a frame B_1 to the table at C and a principal coordinate frame B_2 to the disc. The transformation matrices between the coordinate frames are

$$ {}^0R_1 = R_{Z,\theta_1} = \begin{bmatrix} \cos\theta_1 & -\sin\theta_1 & 0 \\ \sin\theta_1 & \cos\theta_1 & 0 \\ 0 & 0 & 1 \end{bmatrix} \tag{12.109} $$

$$ \begin{aligned} {}^1R_2 &= R_{X,\theta_2}\, R_{Y,90} \\ &= \begin{bmatrix} 1 & 0 & 0 \\ 0 & \cos\theta_2 & -\sin\theta_2 \\ 0 & \sin\theta_2 & \cos\theta_2 \end{bmatrix} \begin{bmatrix} \cos(\frac{1}{2}\pi) & 0 & \sin(\frac{1}{2}\pi) \\ 0 & 1 & 0 \\ -\sin(\frac{1}{2}\pi) & 0 & \cos(\frac{1}{2}\pi) \end{bmatrix} \\ &= \begin{bmatrix} 0 & 0 & 1 \\ \sin\theta_2 & \cos\theta_2 & 0 \\ -\cos\theta_2 & \sin\theta_2 & 0 \end{bmatrix} \end{aligned} \tag{12.110} $$

$$ \begin{aligned} {}^0R_2 &= {}^0R_1\, {}^1R_2 \\ &= \begin{bmatrix} -\sin\theta_1\sin\theta_2 & -\cos\theta_2\sin\theta_1 & \cos\theta_1 \\ \cos\theta_1\sin\theta_2 & \cos\theta_1\cos\theta_2 & \sin\theta_1 \\ -\cos\theta_2 & \sin\theta_2 & 0 \end{bmatrix} \end{aligned} \tag{12.111} $$

The relative angular velocities of B_0, B_1, and B_2 are

$$ {}_0\boldsymbol{\omega}_1 = \begin{bmatrix} 0 \\ 0 \\ \Omega \end{bmatrix} \qquad {}_1\boldsymbol{\omega}_2 = \begin{bmatrix} \omega \\ 0 \\ 0 \end{bmatrix} \tag{12.112} $$

and the angular velocities of the disc in the global frame are

$$ {}_0\boldsymbol{\omega}_2 = \begin{bmatrix} \omega\cos\theta_1 \\ \omega\sin\theta_1 \\ \Omega \end{bmatrix} \qquad {}_0^2\boldsymbol{\omega}_2 = \begin{bmatrix} \omega_1 \\ \omega_2 \\ \omega_3 \end{bmatrix} = \begin{bmatrix} -\Omega\cos\theta_2 \\ \Omega\sin\theta_2 \\ \omega \end{bmatrix} \tag{12.113} $$

where

$$ \begin{aligned} {}_0\tilde{\omega}_2 &= {}_0\tilde{\omega}_1 + {}_1^0\tilde{\omega}_2 = {}_0\tilde{\omega}_1 + {}^0R_1\, {}_1\tilde{\omega}_2\, {}^0R_1^{\mathrm{T}} \\ &= \begin{bmatrix} 0 & -\Omega & 0 \\ \Omega & 0 & 0 \\ 0 & 0 & 0 \end{bmatrix} + {}^0R_1 \begin{bmatrix} 0 & 0 & 0 \\ 0 & 0 & -\omega \\ 0 & \omega & 0 \end{bmatrix} {}^0R_1^{\mathrm{T}} \\ &= \begin{bmatrix} 0 & -\Omega & \omega\sin\theta_1 \\ \Omega & 0 & -\omega\cos\theta_1 \\ -\omega\sin\theta_1 & \omega\cos\theta_1 & 0 \end{bmatrix} \end{aligned} \tag{12.114} $$

$$ \begin{aligned} {}_0^2\tilde{\omega}_2 &= {}^0R_2^{\mathrm{T}}\, {}_0\tilde{\omega}_2\, {}^0R_2 \\ &= \begin{bmatrix} 0 & -\omega & \Omega\sin\theta_2 \\ \omega & 0 & \Omega\cos\theta_2 \\ -\Omega\sin\theta_2 & -\Omega\cos\theta_2 & 0 \end{bmatrix} \end{aligned} \tag{12.115} $$

Taking a time derivative of ${}_0^2\boldsymbol{\omega}_2$, we have

$$ {}_0^2\dot{\boldsymbol{\omega}}_2 = \begin{bmatrix} \dot{\omega}_1 \\ \dot{\omega}_2 \\ \dot{\omega}_3 \end{bmatrix} = \begin{bmatrix} \Omega\omega\sin\theta_2 \\ \Omega\omega\cos\theta_2 \\ 0 \end{bmatrix} \tag{12.116} $$

Substituting (12.113) and (12.116) into (12.10)–(12.12) provides Euler equations of motion:

$$ {}^2M_1 = I_1\Omega\omega\sin\theta_2 - (I_2 - I_3)\,\Omega\omega\sin\theta_2 \tag{12.117} $$

$$ {}^2M_2 = I_2\Omega\omega\cos\theta_2 + (I_3 - I_1)\,\Omega\omega\cos\theta_2 \tag{12.118} $$

$$ {}^2M_3 = (I_1 - I_2)\,\Omega^2\sin\theta_2\cos\theta_2 \tag{12.119} $$

For the uniform disc, we have $I_1 = I_2$, and therefore, these equations simplify to

$$ {}^2M_1 = I_3\Omega\omega\sin\theta_2 \tag{12.120} $$

$$ {}^2M_2 = I_3\Omega\omega\cos\theta_2 \tag{12.121} $$

$$ {}^2M_3 = 0 \tag{12.122} $$

It shows that the third component of the required moment ${}^2\mathbf{M}$ in B_2 is zero and the other two components are periodic functions of θ_2.

To determine the forces in bearings, we should express **M** in B_1 because the bearings are part of the table:

$$
\begin{aligned}
{}^1\mathbf{M} &= {}^1R_2\,{}^2\mathbf{M} \\
&= \begin{bmatrix} 0 & 0 & 1 \\ \sin\theta_2 & \cos\theta_2 & 0 \\ -\cos\theta_2 & \sin\theta_2 & 0 \end{bmatrix} \begin{bmatrix} I_3\Omega\omega\sin\theta_2 \\ I_3\Omega\omega\cos\theta_2 \\ 0 \end{bmatrix} = \begin{bmatrix} 0 \\ \Omega\omega I_3 \\ 0 \end{bmatrix}
\end{aligned}
\tag{12.123}
$$

There is only a y-component for the applied moment of the table to the shaft. Since ${}^1M_{y_1}$ has a constant value, the couple forces F that generate ${}^1M_{y_1}$ are also constant, as shown in Figure 12.4:

$$
F = \frac{\Omega\omega I_3}{2l} \tag{12.124}
$$

The moment in the global frame is periodic function of θ_1:

$$
{}^0\mathbf{M} = {}^0R_2\,{}^2\mathbf{M} = \begin{bmatrix} -\Omega\omega I_3\sin\theta_1 \\ \Omega\omega I_3\cos\theta_1 \\ 0 \end{bmatrix} \tag{12.125}
$$

Let us also examine the required moment to turn the table. The disc makes the mass moment of the table asymmetric:

$$
{}^0I_1 \neq {}^0I_2 \neq {}^0I_3 \tag{12.126}
$$

Substituting the constant angular velocity ${}_0\boldsymbol{\omega}_1$ from (12.112) into Euler equations (12.10)–(12.12), we have

$$
{}^0M_1 = 0 \qquad {}^0M_2 = 0 \qquad {}^0M_3 = 0 \tag{12.127}
$$

So, as long as Ω is constant, no moment is needed to keep the table turning.

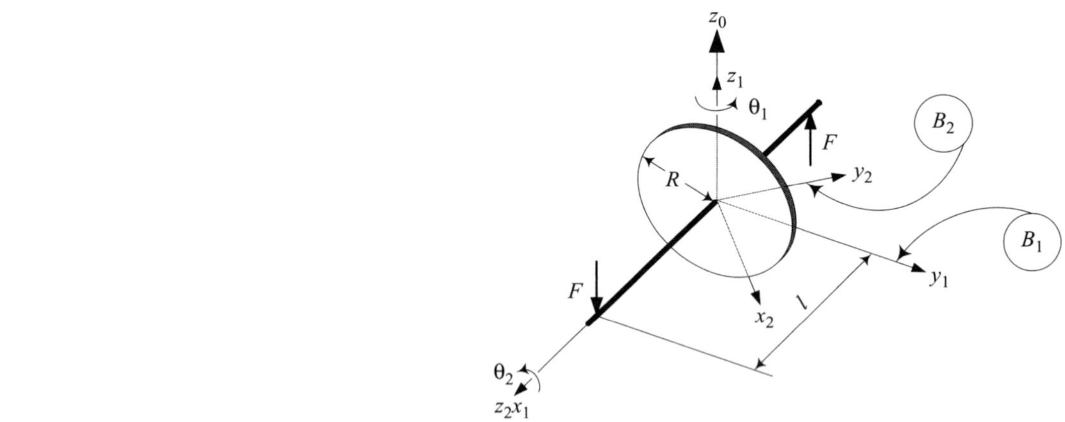

Figure 12.4 Bearing forces F on the shaft.

Example 722 A Turning Disc about a Diagonal on a Turntable Let us change the direction of the disc of Figure 12.3 to be mounted as shown in Figure 12.5. The uniform disc has a mass m, radius R, and mass moment $[I]$. The disc is turning about its diagonal with constant angular velocity $\dot{\theta}_2 = \omega$ in B_1, which is a fixed coordinate frame on the table at the mass center of disc C. The table is also turning with angular velocity $\dot{\theta}_1 = \Omega$ in the global frame B_0. The mass center of the disc is motionless. We attach a principal coordinate frame B_2 to the disc and express Euler equations in B_2.

The transformation matrices between the coordinate frames are given in Equations (12.109)–(12.111). Employing the angular velocity and acceleration of (12.113) and (12.116), we get the same Euler equations of motion as (12.117)–(12.119):

$$M_1 = I_1 \Omega\omega \sin\theta_2 - (I_2 - I_3)\,\Omega\omega \sin\theta_2 \tag{12.128}$$

$$M_2 = I_2 \Omega\omega \cos\theta_2 + (I_3 - I_1)\,\Omega\omega \cos\theta_2 \tag{12.129}$$

$$M_3 = (I_1 - I_2)\,\Omega^2 \sin\theta_2 \cos\theta_2 \tag{12.130}$$

For the diagonal configuration of the disc, we have $I_3 = I_2$, and therefore, these equations simplify to

$$M_1 = I_1 \Omega\omega \sin\theta_2 \tag{12.131}$$

$$M_2 = (2I_2 - I_1)\,\Omega\omega \cos\theta_2 \tag{12.132}$$

$$M_3 = (I_1 - I_2)\,\Omega^2 \sin\theta_2 \cos\theta_2 \tag{12.133}$$

which shows that all component of the required moment ${}^2\mathbf{M} = [M_1, M_2, M_3]^{\mathrm{T}}$ in B_2 are periodic functions of θ_2. Let us transform ${}^2\mathbf{M}$ to B_1 to see what is the applied moment of the table to the shaft. The moment that the table applies to the shaft

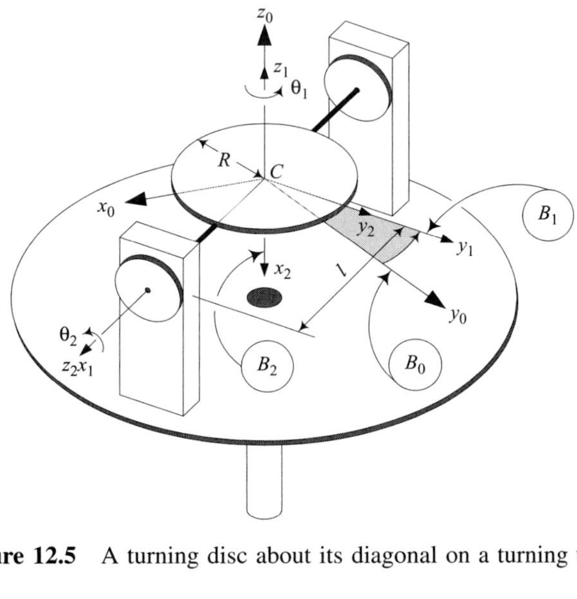

Figure 12.5 A turning disc about its diagonal on a turning table.

through the bearings is given as

$$\begin{aligned} {}^1\mathbf{M} &= {}^1R_2\,{}^2\mathbf{M} \\ &= \begin{bmatrix} 0 & 0 & 1 \\ \sin\theta_2 & \cos\theta_2 & 0 \\ -\cos\theta_2 & \sin\theta_2 & 0 \end{bmatrix} \begin{bmatrix} I_1\Omega\omega\sin\theta_2 \\ (2I_2 - I_1)\,\Omega\omega\cos\theta_2 \\ (I_1 - I_2)\,\Omega^2\sin\theta_2\cos\theta_2 \end{bmatrix} \\ &= \begin{bmatrix} \frac{1}{2}(I_1 - I_2)\,\Omega^2\sin 2\theta_2 \\ \Omega\omega\,(I_2 - (I_1 - I_2)\cos 2\theta_2) \\ -(I_1 - I_2)\,\Omega\omega\sin 2\theta_2 \end{bmatrix} \end{aligned} \tag{12.134}$$

The second and third components of ${}^1\mathbf{M}$ are provided by bearings and are supporting reaction moments. However, having a nonzero first component of ${}^1\mathbf{M}$ in (12.134) compared to (12.123) is interesting. The torque ${}^1M_x$ is the required torque to keep the shaft turning when the table is rotating with angular velocity Ω. So, this shaft needs a motor to apply ${}^1M_x$ to be turned about the x_1-axis.

Example 723 Required Torque to Turn a Cylinder on a Turntable Let us consider a uniform cylinder that is turning by a shaft on a turning table, as shown in Figure 12.6.

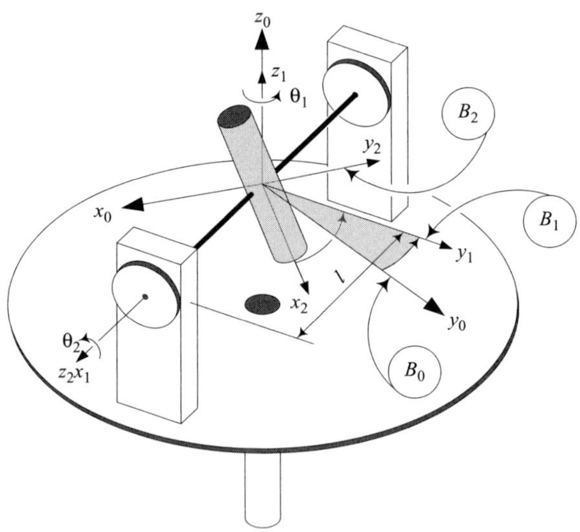

Figure 12.6 A uniform cylinder is turning by a shaft on a turning table.

The cylinder is turning with constant angular velocity $\dot{\theta}_2 = \omega$ in B_1, which is a fixed coordinate frame on the table at the mass center of the cylinder. The table is turning with angular velocity $\dot{\theta}_1 = \Omega$ in the global frame B_0. The mass center of the cylinder is motionless. If B_2 is the principal coordinate frame of the cylinder, then the transformation matrices between the coordinate frames are the same as in Equations

(12.109)–(12.111). Employing the angular velocity and acceleration of (12.113) and (12.116), we get the same Euler equations of motion as (12.117)–(12.119):

$$M_1 = I_1 \Omega\omega \sin\theta_2 - (I_2 - I_3)\,\Omega\omega \sin\theta_2 \tag{12.135}$$

$$M_2 = I_2 \Omega\omega \cos\theta_2 + (I_3 - I_1)\,\Omega\omega \cos\theta_2 \tag{12.136}$$

$$M_3 = (I_1 - I_2)\,\Omega^2 \sin\theta_2 \cos\theta_2 \tag{12.137}$$

For the cylinder configuration shown in Figure 12.6, we have $I_3 = I_2$, and therefore, the equations simplify to

$$M_1 = I_1 \Omega\omega \sin\theta_2 \tag{12.138}$$

$$M_2 = (2I_2 - I_1)\,\Omega\omega \cos\theta_2 \tag{12.139}$$

$$M_3 = (I_1 - I_2)\,\Omega^2 \sin\theta_2 \cos\theta_2 \tag{12.140}$$

These equations are exactly the same as (12.128)–(12.130), which confirms that the dynamic characteristics of rigid bodies in a principal frame are determined by the eigenvalues and eigenvectors of $[I]$, not by the actual geometric shape of the body.

Transforming ${}^2\mathbf{M}$ to B_1, we have

$$\begin{aligned} {}^1\mathbf{M} &= {}^1R_2\, {}^2\mathbf{M} \\ &= \begin{bmatrix} \frac{1}{2}(I_1 - I_2)\,\Omega^2 \sin 2\theta_2 \\ \Omega\omega\,(I_2 - (I_1 - I_2)\cos 2\theta_2) \\ -(I_1 - I_2)\,\Omega\omega \sin 2\theta_2 \end{bmatrix} \end{aligned} \tag{12.141}$$

If Ω is very small, then ${}^1M_x$ is almost zero. However, the required torque to turn the shaft increases rapidly with increasing Ω.

Example 724 Angular Momentum of a Two-Link Manipulator A two-link manipulator is shown in Figure 12.7. Link A rotates with angular velocity $\dot{\varphi}$ about the z-axis of its local coordinate frame. Link B is attached to link A and has angular velocity $\dot{\psi}$ with respect to A about the x_A-axis. Let us attach coordinate frames A and B to links A and B such as shown in the figure. We assume that A and G were coincident at $\varphi = 0$; therefore, the transformation matrix between A and G is

$${}^G R_A = \begin{bmatrix} \cos\varphi(t) & -\sin\varphi(t) & 0 \\ \sin\varphi(t) & \cos\varphi(t) & 0 \\ 0 & 0 & 1 \end{bmatrix} \tag{12.142}$$

The frame B is related to A by Euler angles $\varphi = 90\,\text{deg}$, $\theta = 90\,\text{deg}$, and $\psi = \psi$; hence,

$$\begin{aligned} {}^A R_B &= \begin{bmatrix} c\pi c\psi - c\pi s\pi s\psi & -c\pi s\psi - c\pi c\psi s\pi & s\pi s\pi \\ c\psi s\pi + c\pi c\pi s\psi & -s\pi s\psi + c\pi c\pi c\psi & -c\pi s\pi \\ s\pi s\psi & s\pi c\psi & c\pi \end{bmatrix} \\ &\begin{bmatrix} -\cos\psi & \sin\psi & 0 \\ \sin\psi & \cos\psi & 0 \\ 0 & 0 & -1 \end{bmatrix} \end{aligned} \tag{12.143}$$

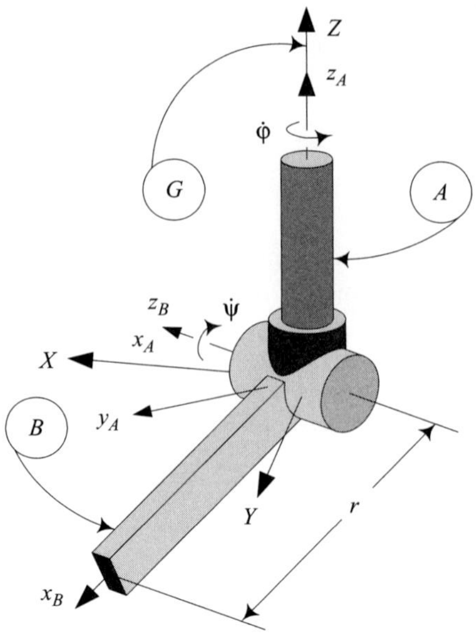

Figure 12.7 A two-link manipulator.

and therefore,

$$
\begin{aligned}
{}^{G}R_B &= {}^{G}R_A \, {}^{A}R_B \\
&= \begin{bmatrix} -\cos\varphi\cos\psi - \sin\varphi\sin\psi & \cos\varphi\sin\psi - \cos\psi\sin\varphi & 0 \\ \cos\varphi\sin\psi - \cos\psi\sin\varphi & \cos\varphi\cos\psi + \sin\varphi\sin\psi & 0 \\ 0 & 0 & -1 \end{bmatrix}
\end{aligned} \tag{12.144}
$$

The angular velocities of A in G and B in A are

$$ {}_{G}\boldsymbol{\omega}_A = \dot{\varphi}\hat{K} \tag{12.145} $$

$$ {}_{A}\boldsymbol{\omega}_B = \dot{\psi}\hat{\imath}_A \tag{12.146} $$

The coordinate frames A and B are assumed to be principal, so the mass moment matrices for the arms A and B can be defined as

$$ {}^{A}I_A = \begin{bmatrix} I_{A_1} & 0 & 0 \\ 0 & I_{A_2} & 0 \\ 0 & 0 & I_{A_3} \end{bmatrix} \tag{12.147} $$

$$ {}^{B}I_B = \begin{bmatrix} I_{B_1} & 0 & 0 \\ 0 & I_{B_2} & 0 \\ 0 & 0 & I_{B_3} \end{bmatrix} \tag{12.148} $$

To analyze multibodies, we should transform all of the kinematic characteristics of the bodies to a common coordinate frame. Most times the global frame is the simplest and the best choice. To determine the total angular momentum of the system, we transform the mass moment matrices to the global frame:

$$
{}^{G}I_{A} = {}^{G}R_{A}\,{}^{A}I_{A}\,{}^{G}R_{A}^{\mathrm{T}}
= \begin{bmatrix} I_{A_1}\cos^2\varphi + I_{A_2}\sin^2\varphi & \left(I_{A_1} - I_{A_2}\right)\cos\varphi\sin\varphi & 0 \\ \left(I_{A_1} - I_{A_2}\right)\cos\varphi\sin\varphi & I_{A_2}\cos^2\varphi + I_{A_1}\sin^2\varphi & 0 \\ 0 & 0 & I_{A_3} \end{bmatrix} \tag{12.149}
$$

$$
{}^{G}I_{B} = {}^{G}R_{B}\,{}^{B}I_{B}\,{}^{G}R_{B}^{\mathrm{T}}
= \begin{bmatrix} I_{B_1}\cos^2\beta - I_{B_2}\sin^2\beta & -\frac{1}{2}\left(I_{B_1} - I_{B_2}\right)\sin 2\beta & 0 \\ -\frac{1}{2}\left(I_{B_1} - I_{B_2}\right)\sin 2\beta & I_{B_2}\cos^2\beta - I_{B_1}\sin^2\beta & 0 \\ 0 & 0 & I_{B_3} \end{bmatrix} \tag{12.150}
$$

$$
\beta = \psi - \varphi \tag{12.151}
$$

The total angular momentum of the manipulator is

$$
{}^{G}\mathbf{L} = {}^{G}\mathbf{L}_{A} + {}^{G}\mathbf{L}_{B} \tag{12.152}
$$

where

$$
{}^{G}\mathbf{L}_{A} = {}^{G}I_{A}\,{}_{G}\boldsymbol{\omega}_{A} = \begin{bmatrix} 0 \\ 0 \\ \dot{\varphi} I_{A_3} \end{bmatrix} \tag{12.153}
$$

$$
\begin{aligned}
{}^{G}\mathbf{L}_{B} &= {}^{G}I_{B}\,{}_{G}\boldsymbol{\omega}_{B} = {}^{G}I_{B}\left({}_{A}^{G}\boldsymbol{\omega}_{B} + {}_{G}\boldsymbol{\omega}_{A}\right) \\
&= {}^{G}I_{B}\left({}^{G}R_{A}\,{}_{A}\boldsymbol{\omega}_{B} + {}_{G}\boldsymbol{\omega}_{A}\right) = {}^{G}I_{B}\begin{bmatrix} \dot{\psi}\cos\varphi \\ \dot{\psi}\sin\varphi \\ \dot{\varphi} \end{bmatrix} \\
&= \begin{bmatrix} \dot{\psi}\left(I_{B_1}\cos^2\beta - I_{B_2}\sin^2\beta\right)\cos\varphi - \frac{1}{2}\dot{\psi}\left(\sin 2\beta\sin\varphi\right)\left(I_{B_1} - I_{B_2}\right) \\ \dot{\psi}\left(I_{B_2}\cos^2\beta - I_{B_1}\sin^2\beta\right)\sin\varphi - \frac{1}{2}\dot{\psi}\left(\sin 2\beta\cos\varphi\right)\left(I_{B_1} - I_{B_2}\right) \\ \dot{\varphi} I_{B_3} \end{bmatrix}
\end{aligned} \tag{12.154}
$$

Example 725 Kinetic Energy of a Planar Rigid Vehicle Consider the planar model of a rigid vehicle as examined in Example 565. The kinetic energy of the vehicle in a planar motion is

$$
\begin{aligned}
K &= \frac{1}{2}\,{}^{G}\mathbf{v}_{B}^{\mathrm{T}}\,m\,{}^{G}\mathbf{v}_{B} + \frac{1}{2}\,{}_{G}\boldsymbol{\omega}_{B}^{\mathrm{T}}\,{}^{G}I\,{}_{G}\boldsymbol{\omega}_{B} \\
&= \frac{1}{2}\begin{bmatrix} v_X \\ v_Y \\ 0 \end{bmatrix}^{\mathrm{T}} m \begin{bmatrix} v_X \\ v_Y \\ 0 \end{bmatrix} + \frac{1}{2}\begin{bmatrix} 0 \\ 0 \\ \omega_Z \end{bmatrix}^{\mathrm{T}} {}^{G}I \begin{bmatrix} 0 \\ 0 \\ \omega_Z \end{bmatrix} \\
&= \frac{1}{2}mv_X^2 + \frac{1}{2}mv_Y^2 + \frac{1}{2}I_3\omega_Z^2 \\
&= \frac{1}{2}m\left(\dot{X}^2 + \dot{Y}^2\right) + \frac{1}{2}I_z\,\dot{\psi}^2
\end{aligned} \tag{12.155}
$$

where

$$
\begin{aligned}
{}^G I &= {}^G R_B \, {}^B I \, {}^G R_B^{\mathrm{T}} \\
&= \begin{bmatrix} \cos\psi & -\sin\psi & 0 \\ \sin\psi & \cos\psi & 0 \\ 0 & 0 & 1 \end{bmatrix} \begin{bmatrix} I_1 & 0 & 0 \\ 0 & I_2 & 0 \\ 0 & 0 & I_3 \end{bmatrix} \begin{bmatrix} \cos\psi & -\sin\psi & 0 \\ \sin\psi & \cos\psi & 0 \\ 0 & 0 & 1 \end{bmatrix}^{\mathrm{T}} \\
&= \begin{bmatrix} I_1\cos^2\psi + I_2\sin^2\psi & (I_1 - I_2)\sin\psi\cos\psi & 0 \\ (I_1 - I_2)\sin\psi\cos\psi & I_2\cos^2\psi + I_1\sin^2\psi & 0 \\ 0 & 0 & I_3 \end{bmatrix}
\end{aligned}
\tag{12.156}
$$

and

$$
{}^G\mathbf{v}_B = \begin{bmatrix} v_X \\ v_Y \\ 0 \end{bmatrix} = \begin{bmatrix} \dot{X} \\ \dot{Y} \\ 0 \end{bmatrix} \tag{12.157}
$$

$$
{}_G\boldsymbol{\omega}_B = \begin{bmatrix} 0 \\ 0 \\ \omega_Z \end{bmatrix} = \begin{bmatrix} 0 \\ 0 \\ r \end{bmatrix} = \begin{bmatrix} 0 \\ 0 \\ \dot{\psi} \end{bmatrix} \tag{12.158}
$$

Example 726 ★ Energy and Momentum Ellipsoids Consider a freely rotating rigid body with an attached principal coordinate frame. Having $\mathbf{M} = 0$ provides a motion under a constant angular momentum and a constant kinetic energy:

$$
\mathbf{L} = I\,\boldsymbol{\omega} = \text{const} \tag{12.159}
$$

$$
K = \tfrac{1}{2}\boldsymbol{\omega}^{\mathrm{T}}\, I\,\boldsymbol{\omega} = \text{const} \tag{12.160}
$$

Because the length of the angular momentum $\mathbf{L}$ is constant, the equation

$$
\begin{aligned}
L^2 &= \mathbf{L}\cdot\mathbf{L} = L_x^2 + L_y^2 + L_z^2 \\
&= I_1^2\omega_1^2 + I_2^2\omega_2^2 + I_3^2\omega_3^2
\end{aligned}
\tag{12.161}
$$

introduces an ellipsoid in the $(\omega_1, \omega_2, \omega_3)$ coordinate frame, called the *momentum ellipsoid*. The tip of all possible angular velocity vectors must lie on the surface of the momentum ellipsoid. The kinetic energy also defines an *energy ellipsoid* in the same coordinate frame so that the tip of the angular velocity vectors must also lie on its surface:

$$
K = \tfrac{1}{2}\left(I_1\omega_1^2 + I_2\omega_2^2 + I_3\omega_3^2\right) \tag{12.162}
$$

Therefore, the dynamics of the moment-free motion of a rigid body requires that the corresponding angular velocity $\boldsymbol{\omega}(t)$ satisfy both Equations (12.161) and (12.162) and therefore lie on the intersection of the momentum and energy ellipsoids.

For a better visualization, let us define the ellipsoids in the (L_x, L_y, L_z) coordinate system as

$$
L_x^2 + L_y^2 + L_z^2 = L^2 \tag{12.163}
$$

$$\frac{L_x^2}{2I_1 K} + \frac{L_y^2}{2I_2 K} + \frac{L_z^2}{2I_3 K} = 1 \tag{12.164}$$

Equation (12.163) is a sphere and Equation (12.164) defines an ellipsoid with $\sqrt{2I_i K}$ as semiaxes. To have a meaningful motion, these two shapes must intersect. The intersection forms a trajectory, such as shown in Figure 12.8.

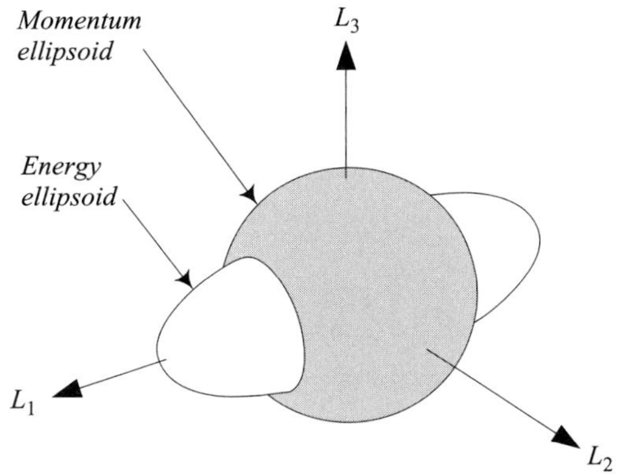

Figure 12.8 Intersection of the momentum and energy ellipsoids.

We conclude that for a given angular momentum there are maximum and minimum limit values for possible kinetic energy. Assuming

$$I_1 > I_3 > I_3 \tag{12.165}$$

the limits of possible kinetic energy are

$$K_{\min} = \frac{L^2}{2I_1} \tag{12.166}$$

$$K_{\max} = \frac{L^2}{2I_3} \tag{12.167}$$

and the corresponding motions are turning about the axes I_1 and I_3, respectively.

Example 727 ★ Alternative Derivation of Euler Equation of Motion Consider a rigid body with a fixed point as shown in Figure 12.9. A small mass dm at ${}^G\mathbf{r}_{dm}$ is under a small force $d\mathbf{f}$. Let us show the moment of the small force $d\mathbf{f}$ by $d\mathbf{m}$:

$${}^G d\mathbf{m} = {}^G\mathbf{r}_{dm} \times {}^G d\mathbf{f} = {}^G\mathbf{r}_{dm} \times {}^G\dot{\mathbf{v}}_{dm}\, dm \tag{12.168}$$

The angular momentum $d\mathbf{l}$ of dm is equal to

$${}^G d\mathbf{l} = {}^G\mathbf{r}_{dm} \times {}^G\mathbf{v}_{dm}\, dm \tag{12.169}$$

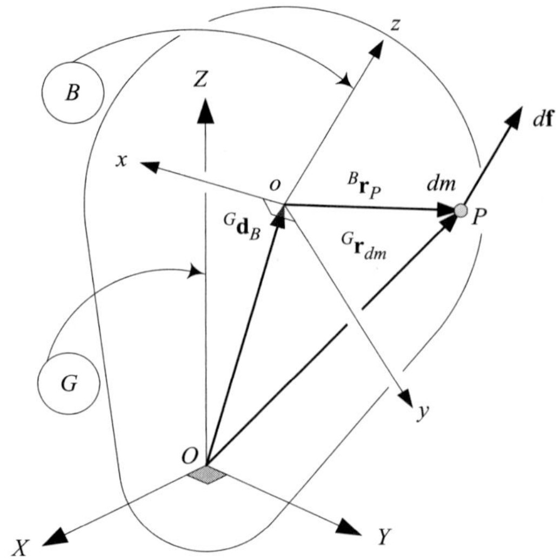

Figure 12.9 A small mass dm at ${}^G\mathbf{r}_{dm}$ of a rigid body with a fixed point at O is under a small force $d\mathbf{f}$.

and according to (12.3), we have

$$d\mathbf{m} = \frac{{}^Gd}{dt} d\mathbf{l} \tag{12.170}$$

$${}^G\mathbf{r}_{dm} \times d\mathbf{f} = \frac{{}^Gd}{dt}\left({}^G\mathbf{r}_{dm} \times {}^G\mathbf{v}_{dm}\, dm\right) \tag{12.171}$$

Integrating over the body gives

$$\begin{aligned} \int_B {}^G\mathbf{r}_{dm} \times d\mathbf{f} &= \int_B \frac{{}^Gd}{dt}\left({}^G\mathbf{r}_{dm} \times {}^G\mathbf{v}_{dm}\, dm\right) \\ &= \frac{{}^Gd}{dt}\int_B \left({}^G\mathbf{r}_{dm} \times {}^G\mathbf{v}_{dm}\, dm\right) \end{aligned} \tag{12.172}$$

However, utilizing

$${}^G\mathbf{r}_{dm} = {}^G\mathbf{d}_B + {}^GR_B\, {}^B\mathbf{r}_{dm} \tag{12.173}$$

where ${}^G\mathbf{d}_B$ is the global position vector of the central body frame, simplifies the left-hand side of the integral to

$$\begin{aligned} \int_B {}^G\mathbf{r}_{dm} \times d\mathbf{f} &= \int_B \left({}^G\mathbf{d}_B + {}^GR_B\, {}^B\mathbf{r}_{dm}\right) \times d\mathbf{f} \\ &= \int_B {}^G\mathbf{d}_B \times d\mathbf{f} + \int_B {}^G_B\mathbf{r}_{dm} \times d\mathbf{f} \\ &= {}^G\mathbf{d}_B \times {}^G\mathbf{F} + {}^G\mathbf{M}_C \end{aligned} \tag{12.174}$$

where $\mathbf{M}_C$ is the resultant of the external moment about the body mass center C. The right-hand side of Equation (12.172) is given as

$$\begin{aligned}
&\frac{^Gd}{dt}\int_B \left({}^G\mathbf{r}_{dm} \times {}^G\mathbf{v}_{dm}\, dm\right) \\
&= \frac{^Gd}{dt}\int_B \left(\left({}^G\mathbf{d}_B + {}^GR_B\, {}^B\mathbf{r}_{dm}\right) \times {}^G\mathbf{v}_{dm}\, dm\right) \\
&= \frac{^Gd}{dt}\int_B \left({}^G\mathbf{d}_B \times {}^G\mathbf{v}_{dm}\right) dm + \frac{^Gd}{dt}\int_B \left({}^G_B\mathbf{r}_{dm} \times {}^G\mathbf{v}_{dm}\right) dm \\
&= \frac{^Gd}{dt}\left({}^G\mathbf{d}_B \times \int_B {}^G\mathbf{v}_{dm} dm\right) + \frac{^Gd}{dt}\mathbf{L}_C \\
&= {}^G\dot{\mathbf{d}}_B \times \int_B {}^G\mathbf{v}_{dm} dm + {}^G\mathbf{d}_B \times \int_B {}^G\dot{\mathbf{v}}_{dm} dm + \frac{d}{dt}\mathbf{L}_C
\end{aligned} \tag{12.175}$$

We use $\mathbf{L}_C$ for angular momentum about the body mass center. Since the body frame is at the center of mass, we have

$$\int_B {}^G\mathbf{r}_{dm}\, dm = m\, {}^G\mathbf{d}_B = m\, {}^G\mathbf{r}_C \tag{12.176}$$

$$\int_B {}^G\mathbf{v}_{dm} dm = m\, {}^G\dot{\mathbf{d}}_B = m\, {}^G\mathbf{v}_C \tag{12.177}$$

$$\int_B {}^G\dot{\mathbf{v}}_{dm} dm = m\, {}^G\ddot{\mathbf{d}}_B = m\, {}^G\mathbf{a}_C \tag{12.178}$$

and therefore,

$$\frac{^Gd}{dt}\int_B \left({}^G\mathbf{r}_{dm} \times {}^G\mathbf{v}_{dm}\, dm\right) = {}^G\mathbf{d}_B \times {}^G\mathbf{F} + \frac{^Gd}{dt}\, {}^G\mathbf{L}_C \tag{12.179}$$

Substituting (12.174) and (12.179) into (12.172) provides the Euler equation of motion in the global frame, indicating that the resultant of the externally applied moments about C is equal to the global derivative of the angular momentum about C:

$${}^G\mathbf{M}_C = \frac{^Gd}{dt}\, {}^G\mathbf{L}_C \tag{12.180}$$

The Euler equation in the body coordinate frame can be found by transforming (12.180) into B:

$$\begin{aligned}
{}^B\mathbf{M}_C &= {}^GR_B^{\mathrm{T}}\, {}^G\mathbf{M}_C = {}^GR_B^{\mathrm{T}} \frac{^Gd}{dt}\mathbf{L}_C = \frac{^Gd}{dt}\, {}^GR_B^{\mathrm{T}}\, \mathbf{L}_C = \frac{^Gd}{dt}\, {}^B\mathbf{L}_C \\
&= {}^B\dot{\mathbf{L}}_C + {}^B_G\boldsymbol{\omega}_B \times {}^B\mathbf{L}_C
\end{aligned} \tag{12.181}$$

12.2 ★ RIGID-BODY ROTATIONAL EULERIAN DYNAMICS

Euler equations for the rotational dynamics of a rigid body may be expressed by using Euler angles and frequencies:

$$\begin{aligned} M_{\varphi}\sin\theta &= \ddot{\varphi}\left(I_1\sin^2\psi + I_2\cos^2\psi\right)\sin\theta - \ddot{\theta}\left(I_2 - I_1\right)\cos\psi\sin\psi \\ &\quad + (I_1 - I_2)\,\dot{\varphi}^2\cos\theta\cos\psi\sin\theta\sin\psi + \dot{\varphi}\dot{\theta}\left(I_1 + I_2 - I_3\right)\cos\theta \\ &\quad + 2\dot{\psi}\dot{\varphi}(I_1 - I_2)\cos\psi\sin\theta\sin\psi + \dot{\theta}\dot{\psi}\left[(I_1 - I_2)\cos 2\psi - I_3\right] \end{aligned} \tag{12.182}$$

$$\begin{aligned} M_{\theta} &= \ddot{\varphi}\left(I_1 - I_2\right)\cos\psi\sin\theta\sin\psi + \ddot{\theta}\left(I_1\sin^2\psi + I_2\cos^2\psi\right) \\ &\quad - \dot{\varphi}^2(I_1\sin^2\psi + I_2\cos^2\psi - I_3)\cos\theta\sin\theta \\ &\quad + \dot{\varphi}\dot{\psi}\left[I_3 + (I_1 - I_2)\cos 2\psi\right]\sin\theta - \dot{\theta}\dot{\psi}(I_1 - I_2)\sin 2\psi \end{aligned} \tag{12.183}$$

$$\begin{aligned} M_{\psi}\sin\theta &= \ddot{\varphi}\left(I_3 - I_1\sin^2\psi - I_2\cos^2\psi\right)\cos\theta\sin\theta + \ddot{\psi}I_3\sin\theta \\ &\quad - \ddot{\theta}\left(I_1 - I_2\right)\cos\theta\sin\psi\cos\psi - \dot{\varphi}^2\left(I_1 - I_2\right)\cos\theta\sin\psi\cos\psi \\ &\quad + \dot{\theta}^2\left(I_1 - I_2\right)\sin\theta\sin\psi\cos\psi - \dot{\theta}\dot{\psi}\left(I_3 + (I_1 - I_2)\cos 2\psi\right)\cos\theta \\ &\quad - \dot{\varphi}\dot{\theta}\left[I_3\cos 2\theta + (I_1 - I_2)(1 - 2\sin^2\theta\cos^2\psi) - 2I_1\cos^2\theta\right] \\ &\quad - 2\dot{\psi}\dot{\varphi}\left(I_1 - I_2\right)\cos\theta\sin\psi\sin\theta\cos\psi \end{aligned} \tag{12.184}$$

Proof: Let us recall the B-expression of the angular velocity and acceleration of B in G as given in Equations (4.182) and (9.82) based on Euler angles and frequencies:

$$\begin{aligned} {}^{B}_{G}\boldsymbol{\omega}_{B} &= \begin{bmatrix} \omega_x \\ \omega_y \\ \omega_z \end{bmatrix} = \begin{bmatrix} \sin\theta\sin\psi & \cos\psi & 0 \\ \sin\theta\cos\psi & -\sin\psi & 0 \\ \cos\theta & 0 & 1 \end{bmatrix} \begin{bmatrix} \dot{\varphi} \\ \dot{\theta} \\ \dot{\psi} \end{bmatrix} \\ &= \begin{bmatrix} \dot{\varphi}\sin\theta\sin\psi + \dot{\theta}\cos\psi \\ \dot{\varphi}\cos\psi\sin\theta - \dot{\theta}\sin\psi \\ \dot{\psi} + \dot{\varphi}\cos\theta \end{bmatrix} \end{aligned} \tag{12.185}$$

$$\begin{aligned} {}^{B}_{G}\boldsymbol{\alpha}_{B} &= \frac{{}^{B}d}{dt}\,{}^{B}_{G}\boldsymbol{\omega}_{B} \\ &= \begin{bmatrix} \left(\ddot{\theta} + \dot{\varphi}\dot{\psi}\sin\theta\right)\cos\psi + \left(\ddot{\varphi}\sin\theta + \dot{\theta}\dot{\varphi}\cos\theta - \dot{\theta}\dot{\psi}\right)\sin\psi \\ \left(\ddot{\varphi}\sin\theta + \dot{\theta}\dot{\varphi}\cos\theta - \dot{\theta}\dot{\psi}\right)\cos\psi - \left(\ddot{\theta} + \dot{\varphi}\dot{\psi}\sin\theta\right)\sin\psi \\ \ddot{\varphi}\cos\theta + \ddot{\psi} - \dot{\theta}\dot{\varphi}\sin\theta \end{bmatrix} \end{aligned} \tag{12.186}$$

Substituting (12.185) and (12.186) in the B-expression of the principal Euler equations (12.10)–(12.12) shows that

$$\begin{aligned} M_1 &= \left[(\ddot{\varphi}\sin\theta - \dot{\theta}\dot{\psi} + \dot{\theta}\dot{\varphi}\cos\theta)\sin\psi + (\ddot{\theta} + \dot{\psi}\dot{\varphi}\sin\theta)\cos\psi\right]I_1 \\ &\quad + \left(\dot{\psi} + \dot{\varphi}\cos\theta\right)\left(\dot{\theta}\sin\psi - \dot{\varphi}\cos\psi\sin\theta\right)(I_2 - I_3) \end{aligned} \tag{12.187}$$

$$\begin{aligned} M_2 &= [(\ddot{\varphi}\sin\theta - \dot{\theta}\dot{\psi} + \dot{\theta}\dot{\varphi}\cos\theta)\cos\psi - (\ddot{\theta} + \dot{\psi}\dot{\varphi}\sin\theta)\sin\psi]I_2 \\ &\quad + \left(\dot{\psi} + \dot{\varphi}\cos\theta\right)\left(\dot{\theta}\cos\psi + \dot{\varphi}\sin\theta\sin\psi\right)(I_1 - I_3) \end{aligned} \tag{12.188}$$

$$
\begin{aligned}
M_3 = {} & \left(\dot{\theta}\cos\psi + \dot{\varphi}\sin\theta\sin\psi\right)\left(\dot{\theta}\sin\psi - \dot{\varphi}\cos\psi\sin\theta\right)(I_1 - I_2) \\
& + \left(\ddot{\varphi}\cos\theta + \ddot{\psi} - \dot{\theta}\dot{\varphi}\sin\theta\right) I_3
\end{aligned} \tag{12.189}
$$

These three coupled ordinary differential equations of $\ddot{\varphi}, \ddot{\theta}, \ddot{\psi}$ are Cartesian expressions of Euler equations in a principal body coordinate frame $B\,(O123)$, using Euler angles φ, θ, ψ of case 9 in Appendix B. The terms M_1, M_2, and M_3 are components of applied moment $\mathbf{M}$ in the orthogonal principal body frame $B\,(O123)$.

Reducing a Cartesian B-vector to its Eulerian components along $\hat{u}_\varphi, \hat{u}_\theta, \hat{u}_\psi$ of the nonorthogonal Euler frame $E(O\varphi\theta\psi)$ is based on Equation (4.182). Therefore, the applied moment $\mathbf{M}(M_1, M_2, M_3)$ is related to its Eulerian components $\mathbf{M}(M_\varphi, M_\theta, M_\psi)$ by

$$
{}^B\mathbf{M} = {}^B R_E \, {}^E\mathbf{M} \tag{12.190}
$$

$$
\begin{aligned}
\begin{bmatrix} M_1 \\ M_2 \\ M_3 \end{bmatrix} &= \begin{bmatrix} \sin\theta\sin\psi & \cos\psi & 0 \\ \sin\theta\cos\psi & -\sin\psi & 0 \\ \cos\theta & 0 & 1 \end{bmatrix} \begin{bmatrix} M_\varphi \\ M_\theta \\ M_\psi \end{bmatrix} \\
&= \begin{bmatrix} M_\theta\cos\psi + M_\varphi\sin\theta\sin\psi \\ M_\varphi\cos\psi\sin\theta - M_\theta\sin\psi \\ M_\psi + M_\varphi\cos\theta \end{bmatrix}
\end{aligned} \tag{12.191}
$$

An inversion determines the Eulerian components based on the Cartesian components:

$$
\begin{aligned}
\begin{bmatrix} M_\varphi \\ M_\theta \\ M_\psi \end{bmatrix} &= \begin{bmatrix} \sin\theta\sin\psi & \cos\psi & 0 \\ \sin\theta\cos\psi & -\sin\psi & 0 \\ \cos\theta & 0 & 1 \end{bmatrix}^{-1} \begin{bmatrix} M_1 \\ M_2 \\ M_3 \end{bmatrix} \\
&= \frac{1}{\sin\theta} \begin{bmatrix} \sin\psi & \cos\psi & 0 \\ \sin\theta\cos\psi & -\sin\theta\sin\psi & 0 \\ -\cos\theta\sin\psi & -\cos\theta\cos\psi & 1 \end{bmatrix} \begin{bmatrix} M_1 \\ M_2 \\ M_3 \end{bmatrix}
\end{aligned} \tag{12.192}
$$

The individual components of ${}^E\mathbf{M}$ are

$$
M_\varphi = \frac{1}{\sin\theta}\left(M_2\cos\psi + M_1\sin\psi\right) \tag{12.193}
$$

$$
M_\theta = M_1\cos\psi - M_2\sin\psi \tag{12.194}
$$

$$
M_\psi = -\frac{1}{\sin\theta}\left(M_2\cos\theta\cos\psi - M_3\sin\theta + M_1\cos\theta\sin\psi\right) \tag{12.195}
$$

Let us substitute the Cartesian components of ${}^B\mathbf{M}$ from (12.187)–(12.189) to find the rotational equations of motion of a rigid body in the Eulerian frame:

$$
\begin{aligned}
M_\varphi\sin\theta = {} & \left(\ddot{\varphi}\sin\theta + 2\dot{\varphi}\dot{\theta}\cos\theta\right) I_1 \\
& + \begin{pmatrix} -[\ddot{\theta} + \left(\dot{\varphi}\cos\theta + 2\dot{\psi}\right)\dot{\varphi}\sin\theta]\cos\psi\sin\psi \\ + \left(\ddot{\varphi}\sin\theta - 2\dot{\psi}\dot{\theta}\right)\cos^2\psi + \dot{\theta}\left(\dot{\varphi}\cos\theta + \dot{\psi}\right) \end{pmatrix} (I_1 - I_2) \\
& - \left(\dot{\psi} + \dot{\varphi}\cos\theta\right)\dot{\theta} I_3
\end{aligned} \tag{12.196}
$$

$$
\begin{aligned}
M_\theta &= \left(\ddot{\theta} - \dot{\varphi}^2 \sin\theta\cos\theta\right) I_1 \\
&+ \begin{pmatrix} \left(\ddot{\varphi}\sin\theta - 2\dot{\psi}\dot{\theta}\right)\cos\psi\sin\psi - \dot{\psi}\dot{\varphi}\sin\theta \\ -\ddot{\theta}\sin^2\psi + \left(2\dot{\psi}\dot{\theta} + \dot{\varphi}^2\cos\theta\right)\sin\theta\cos^2\psi \end{pmatrix} (I_1 - I_2) \\
&+ \left(\dot{\psi} + \dot{\varphi}\cos\theta\right)\dot{\varphi}\sin\theta\, I_3
\end{aligned} \tag{12.197}
$$

$$
\begin{aligned}
M_\psi \sin\theta &= -[(\ddot{\varphi}\sin\theta + 2\dot{\theta}\dot{\varphi}\cos\theta)\cos\theta] I_1 \\
&+ \begin{pmatrix} -\left(\ddot{\theta} + 2\dot{\varphi}\dot{\psi}\sin\theta\right)\cos\theta\cos\psi\sin\psi \\ +\left[(\ddot{\varphi}\sin\theta - 2\dot{\psi}\dot{\theta})\cos\theta - 2\dot{\theta}\dot{\varphi}\sin^2\theta\right]\cos^2\psi \\ -\left(\dot{\varphi}^2 - \dot{\theta}^2\right)\sin\theta\cos\psi\sin\psi + \dot{\theta}\left(\dot{\varphi} + \dot{\psi}\cos\theta\right) \end{pmatrix} (I_1 - I_2) \\
&- \left[(\ddot{\psi} + \ddot{\varphi}\cos\theta)\sin\theta + \dot{\psi}\dot{\theta}\cos\theta + \dot{\varphi}\dot{\theta}(2\cos^2\theta - 1)\right] I_3
\end{aligned} \tag{12.198}
$$

These equations can also be rearranged and are presented as Equations (12.182)–(12.184).

The Euler equations in the Euler frame are generally complicated and cannot be easily solved analytically. ■

Example 728 ★ Axisymmetric Rigid Body and Euler Equations When a rigid body is axisymmetric, the Euler equations in the Euler frame are much simpler. Assume $I_1 = I_2 = I$. Then,

$$
M_\varphi = I\ddot{\varphi} + \dot{\varphi}\dot{\theta}\,(2I - I_3)\cot\theta - \dot{\theta}\dot{\psi} I_3/\sin\theta \tag{12.199}
$$

$$
M_\theta = I\ddot{\theta} - \dot{\varphi}^2 (I - I_3)\cos\theta\sin\theta + \dot{\varphi}\dot{\psi} I_3 \sin\theta \tag{12.200}
$$

$$
\begin{aligned}
M_\psi &= \ddot{\varphi}\,(I_3 - I)\cos\theta + \ddot{\psi} I_3 - \dot{\theta}\dot{\psi} I_3 \cot\theta \\
&- \frac{1}{\sin\theta}\dot{\varphi}\dot{\theta}\left(I_3\cos 2\theta - 2I\cos^2\theta\right)
\end{aligned} \tag{12.201}
$$

Example 729 ★ General Euler Equation in Eulerian Frame The general form of the Euler equations in the Eulerian frame can be determined by substituting ${}^B\mathbf{M}$ from (12.191), ${}^B_G\boldsymbol{\omega}_B$ from (12.185), and ${}^B_G\dot{\boldsymbol{\omega}}_B$ from (12.186) into the equations

$$
\begin{aligned}
M_x &= I_{xx}\dot{\omega}_x + I_{xy}\dot{\omega}_y + I_{xz}\dot{\omega}_z - \left(I_{yy} - I_{zz}\right)\omega_y\omega_z \\
&- I_{yz}\left(\omega_z^2 - \omega_y^2\right) - \omega_x\left(\omega_z I_{xy} - \omega_y I_{xz}\right)
\end{aligned} \tag{12.202}
$$

$$
\begin{aligned}
M_y &= I_{yx}\dot{\omega}_x + I_{yy}\dot{\omega}_y + I_{yz}\dot{\omega}_z - \left(I_{zz} - I_{xx}\right)\omega_z\omega_x \\
&- I_{xz}\left(\omega_x^2 - \omega_z^2\right) - \omega_y\left(\omega_x I_{yz} - \omega_z I_{xy}\right)
\end{aligned} \tag{12.203}
$$

$$
\begin{aligned}
M_z &= I_{zx}\dot{\omega}_x + I_{zy}\dot{\omega}_y + I_{zz}\dot{\omega}_z - \left(I_{xx} - I_{yy}\right)\omega_x\omega_y \\
&- I_{xy}\left(\omega_y^2 - \omega_x^2\right) - \omega_z\left(\omega_y I_{xz} - \omega_x I_{yz}\right)
\end{aligned} \tag{12.204}
$$

Example 730 ★ A Common Mistake The applied moment $\mathbf{M}(M_1, M_2, M_3)$ is related to its Eulerian components $\mathbf{M}\left(M_\varphi, M_\theta, M_\psi\right)$ by Equation (4.182):

$$
{}^B\mathbf{M} = {}^BR_E\ {}^E\mathbf{M} \tag{12.205}
$$

$$
\begin{bmatrix} M_1 \\ M_2 \\ M_3 \end{bmatrix} = \begin{bmatrix} \sin\theta\sin\psi & \cos\psi & 0 \\ \sin\theta\cos\psi & -\sin\psi & 0 \\ \cos\theta & 0 & 1 \end{bmatrix} \begin{bmatrix} M_\varphi \\ M_\theta \\ M_\psi \end{bmatrix}
$$

The matrix ${}^BR_E$ is not a rotational transformation matrix and is not necessarily orthogonal. Therefore, we cannot find ${}^E\mathbf{M}$ by using the transpose of ${}^BR_E$:

$$
{}^E\mathbf{M} \neq {}^BR_E^{\mathrm{T}}\ {}^B\mathbf{M} \tag{12.206}
$$

A common mistake in a few textbooks is to use ${}^BR_E^{\mathrm{T}}$ instead of ${}^BR_E^{-1}$ in the determination of ${}^E\mathbf{M}$.

Example 731 ★ Matrix Form of Eulerian Equations We may rearrange Equations (12.182)–(12.184) and write them in the form

$$
\begin{bmatrix} M_\varphi \\ M_\theta \\ M_\psi \end{bmatrix} = [A] \begin{bmatrix} \ddot{\varphi} \\ \ddot{\theta} \\ \ddot{\psi} \end{bmatrix} + [B] \begin{bmatrix} \dot{\varphi} \\ \dot{\theta} \\ \dot{\psi} \end{bmatrix} \tag{12.207}
$$

where

$$
[A] = \begin{bmatrix} I_1\sin^2\psi + I_2\cos^2\psi & -(I_2 - I_1)\dfrac{\cos\psi\sin\psi}{\sin\theta} & 0 \\ (I_1 - I_2)\cos\psi\sin\theta\sin\psi & I_1\sin^2\psi + I_2\cos^2\psi & 0 \\ \left(I_3 - I_1\sin^2\psi - I_2\cos^2\psi\right)\cos\theta & -(I_1 - I_2)\cot\theta\sin\psi\cos\psi & I_3 \end{bmatrix} \tag{12.208}
$$

$$
[B] = \begin{bmatrix} b_{11} & b_{12} & b_{13} \\ b_{21} & b_{22} & b_{23} \\ b_{31} & b_{32} & b_{33} \end{bmatrix} \tag{12.209}
$$

where

$$
\begin{aligned}
b_{11} &= \tfrac{1}{2}(I_1 - I_2)\,\dot{\varphi}\cos\theta\sin 2\psi \\
b_{21} &= -\tfrac{1}{2}\dot{\varphi}(I_1\sin^2\psi + I_2\cos^2\psi - I_3)\sin 2\theta \\
b_{31} &= -\dot{\varphi}\,(I_1 - I_2)\cot\theta\sin\psi\cos\psi
\end{aligned} \tag{12.210}
$$

$$
\begin{aligned}
b_{12} &= \dot{\varphi}\,(I_1 + I_2 - I_3)\cot\theta + \dot{\psi}\frac{(I_1 - I_2)\cos 2\psi - I_3}{\sin\theta} \\
b_{22} &= 0 \\
b_{32} &= -\dot{\varphi}\frac{I_3\cos 2\theta + (I_1 - I_2)(1 - 2\sin^2\theta\cos^2\psi) - 2I_1\cos^2\theta}{\sin\theta} \\
&\quad + \dot{\theta}\,(I_1 - I_2)\sin\psi\cos\psi
\end{aligned} \tag{12.211}
$$

$$
\begin{aligned}
b_{13} &= \dot{\varphi}\,(I_1 - I_2)\sin 2\psi \\
b_{23} &= \dot{\varphi}\,(I_3 + (I_1 - I_2)\cos 2\psi)\sin\theta - \dot{\theta}\,(I_1 - I_2)\sin 2\psi \\
b_{33} &= -2\dot{\varphi}\,(I_1 - I_2)\cos\theta\sin\psi\cos\psi \\
&\quad - \dot{\theta}\,(I_3 + (I_1 - I_2)\cos 2\psi)\cot\theta
\end{aligned}
\tag{12.212}
$$

Example 732 ★ Angular Acceleration of Eulerian Equations Let us rearrange Equations (12.182)–(12.184) to determine the Eulerian accelerations independently:

$$
\begin{aligned}
I_1 I_2 \ddot{\varphi}\sin^2\theta &= M_\varphi\left[(I_1 - I_2)\cos^2\psi + I_2\right]\sin\theta - M_\theta\,(I_1 - I_2)\sin\psi\cos\psi \\
&+ \dot{\varphi}^2\left[(I_1 - I_2)I_3 + I_2^2 - I_1^2\right]\sin^2\theta\cos\theta\sin\psi\cos\psi \\
&+ \dot{\varphi}\dot{\theta}\left\{\left[(I_1 - I_2)I_3 + I_2^2 - I_1^2\right]\cos^2\psi + I_2(I_3 - I_1 - I_2)\right\}\cos\theta\sin\theta \\
&+ \dot{\theta}\dot{\psi}\left\{\left[(I_1 - I_2)I_3 + I_2^2 - I_1^2\right]\cos^2\psi + I_2(I_3 + I_1 - I_2)\sin\theta\right\} \\
&+ \dot{\psi}\dot{\varphi}\left[(I_1 - I_2)I_3 + I_2^2 - I_1^2\right]\sin^2\theta\sin\psi\cos\psi
\end{aligned}
\tag{12.213}
$$

$$
\begin{aligned}
I_1 I_2 \ddot{\theta}\sin\theta &= M_\theta\left(I_1\sin^2\psi + I_2\cos^2\psi\right) - M_\varphi\,(I_1 - I_2)\sin\theta\sin\psi\cos\psi \\
&+ \dot{\varphi}^2\left\{\left[(I_1 - I_2)I_3 + I_2^2 - I_1^2\right]\cos^2\psi + I_1(I_1 - I_3)\right\}\cos\theta\sin^2\theta \\
&+ \dot{\varphi}\dot{\theta}\left[(I_1 - I_2)I_3 + I_2^2 - I_1^2\right]\cos\psi\sin\psi\cos\theta\sin\theta \\
&+ \dot{\varphi}\dot{\psi}\left\{\left[(I_1 - I_2)I_3 + I_2^2 - I_1^2\right]\cos^2\psi + I_1(I_1 - I_2 - I_3)\right\}\sin^2\theta \\
&+ \dot{\psi}\dot{\theta}\left[(I_1 - I_2)I_3 + I_2^2 - I_1^2\right]\sin\theta\sin\psi\cos\psi
\end{aligned}
\tag{12.214}
$$

$$
\begin{aligned}
I_1 I_2 I_3 \ddot{\psi}\sin^2\theta &= M_\theta I_3\,(I_1 - I_2)\cos\theta\cos\psi\sin\psi + M_\psi I_1 I_2\sin\theta \\
&- M_\varphi I_2\,(I_1 - I_3)\sin\theta\cos\theta\sin^2\psi \\
&- \dot{\varphi}^2\,(2I_1 I_2\,(I_1 - I_2) + I_2 I_3\,(I_2 - I_3) \\
&+ I_3 I_1\,(I_3 - I_1))\cos^2\theta\sin^2\theta\cos\psi\sin\psi \\
&+ \dot{\varphi}^2 I_1 I_2\,(I_1 - I_2)\cos\psi\sin\psi \\
&- \dot{\theta}^2\,(I_1 I_2\,(I_1 - I_2))\sin^2\theta\cos\psi\sin\psi \\
&+ \dot{\varphi}\dot{\theta}[2I_1 I_2\,(I_1 - I_2) + I_2 I_3\,(I_2 - I_3) \\
&+ I_3 I_1\,(I_3 - I_1)]\cos^2\theta\sin\theta\cos^2\psi \\
&+ \dot{\varphi}\dot{\theta}[2I_1 I_2\,(I_1 - I_2)\cos^2\psi - I_1 I_2(I_1 - I_2) - I_1 I_2 I_3]\sin\theta \\
&+ \dot{\varphi}\dot{\theta}[I_1 I_2(I_1 - I_2) + I_2 I_3(I_2 - I_3)]\cos^2\theta\sin\theta \\
&- \dot{\varphi}\dot{\psi}[I_2 I_3(I_2 - I_3) + I_3 I_1(I_3 - I_1)]\cos\theta\sin^2\theta\cos\psi\sin\psi \\
&- \dot{\psi}\dot{\theta}[I_2 I_3(I_2 - I_3) + I_3 I_1(I_3 - I_1)]\cos\theta\sin\theta\cos^2\psi \\
&+ \dot{\psi}\dot{\theta}[I_2 I_3(I_2 - I_3) - I_1 I_2 I_3]\cos\theta\sin\theta
\end{aligned}
\tag{12.215}
$$

If the rigid body is axisymmetric, $I_1 = I_2 = I$, then these equations will simplify to

$$I\ddot{\varphi} = M_\varphi + \dot{\varphi}\dot{\theta}\,(I_3 - 2I)\cot\theta + \dot{\theta}\dot{\psi}I_3/\sin\theta \tag{12.216}$$

$$I\ddot{\theta} = M_\theta + \dot{\varphi}^2(I - I_3)\cos\theta\sin\theta - \dot{\varphi}\dot{\psi}I_3\sin\theta \tag{12.217}$$

$$\ddot{\psi}I_3 = M_\psi + -\ddot{\varphi}\,(I_3 - I)\cos\theta + \dot{\theta}\dot{\psi}I_3\cot\theta + \frac{1}{\sin\theta}\dot{\varphi}\dot{\theta}\left(I_3\cos 2\theta - 2I\cos^2\theta\right) \tag{12.218}$$

12.3 RIGID-BODY TRANSLATIONAL DYNAMICS

Figure 12.10 illustrates a moving body B in a global frame G. Assume that the body coordinate frame B is attached at the mass center C of the rigid body. The Newton equation of motion for the whole body in the global coordinate frame is

$$^G\mathbf{F} = m\,{}^G\mathbf{a}_B \tag{12.219}$$

This equation can be expressed in the body coordinate frame as

$$^B\mathbf{F} = m\,{}^B_G\mathbf{a}_B + m\,{}^B_G\boldsymbol{\omega}_B \times {}^B\mathbf{v}_B \tag{12.220}$$

$$\begin{bmatrix} F_x \\ F_y \\ F_z \end{bmatrix} = \begin{bmatrix} m\left(\dot{v}_x - \left(v_y\omega_z - v_z\omega_y\right)\right) \\ m\left(\dot{v}_y - \left(v_z\omega_x - v_x\omega_z\right)\right) \\ m\left(\dot{v}_z - \left(v_x\omega_y - v_y\omega_x\right)\right) \end{bmatrix} \tag{12.221}$$

In these equations, ${}^G\mathbf{a}_B$ is the global acceleration vector of the body at C, m is the total mass of the body, and $\mathbf{F}$ is the resultant of the external forces acted on the body at C.

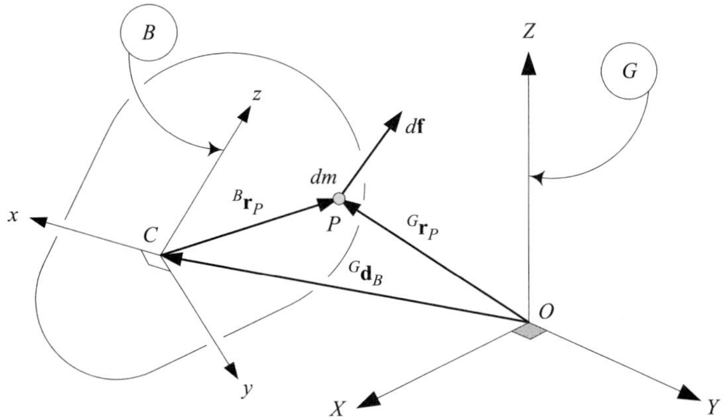

Figure 12.10 A moving body B in a global frame G.

Proof: A body coordinate frame at the center of mass is called a *central frame*. If the frame B is a central frame, then the *center of mass* C is defined such that

$$\int_B {}^B\mathbf{r}_P \, dm = 0 \tag{12.222}$$

Assume that an infinitesimal external force $d\mathbf{f}$ is acted on dm at point P that is at ${}^B\mathbf{r}_P$. The global position vector of dm is related to its local position vector by

$${}^G\mathbf{r}_P = {}^G\mathbf{d}_B + {}^GR_B \, {}^B\mathbf{r}_P \tag{12.223}$$

where ${}^G\mathbf{d}_B$ is the global position vector of the central body frame, and therefore,

$$\begin{aligned}\int_B {}^G\mathbf{r}_P \, dm &= \int_B {}^G\mathbf{d}_B \, dm + {}^GR_B \int_m {}^B\mathbf{r}_P \, dm = \int_B {}^G\mathbf{d}_B \, dm \\ &= {}^G\mathbf{d}_B \int_B dm = m \, {}^G\mathbf{d}_B\end{aligned} \tag{12.224}$$

The time derivative of both sides shows that

$$m \, {}^G\dot{\mathbf{d}}_B = m \, {}^G\mathbf{v}_B = \int_B {}^G\dot{\mathbf{r}}_P \, dm = \int_B {}^G\mathbf{v}_P \, dm \tag{12.225}$$

and another derivative shows that

$$m \, {}^G\dot{\mathbf{v}}_B = m \, {}^G\mathbf{a}_B = \int_B {}^G\dot{\mathbf{v}}_P \, dm \tag{12.226}$$

However, we have $d\mathbf{f} = {}^G\dot{\mathbf{v}}_P \, dm$, and therefore,

$$m \, {}^G\mathbf{a}_B = \int_B d\mathbf{f} \tag{12.227}$$

The integral on the right-hand side includes all the forces acting on particles in the body. The internal forces cancel one another out, so the net result is the vector sum of all the externally applied forces $\mathbf{F}$, and therefore,

$${}^G\mathbf{F} = m \, {}^G\mathbf{a}_B = m \, {}^G\dot{\mathbf{v}}_B \tag{12.228}$$

Transforming this equation to the body coordinate frame, we have

$$\begin{aligned}{}^B\mathbf{F} &= {}^BR_G \, {}^G\mathbf{F} = m \, {}^BR_G \, {}^G\mathbf{a}_B = m \, {}^B_G\mathbf{a}_B \\ &= m \, {}^B\mathbf{a}_B + m \, {}^B_G\boldsymbol{\omega}_B \times {}^B\mathbf{v}_B\end{aligned} \tag{12.229}$$

We use the G-expression of the Newton equation (12.228) when the observer is in the G-frame. In this case, the observer is interested in the motion of the body with respect to herself. We use the B-expression of the Newton equation (12.229) when the observer is in the B-frame and moves with the body. In either case, the result of the analysis would be a set of differential equations. ■

Example 733 A Wound Ribbon Figure 12.11 illustrates a ribbon of negligible weight and thickness that is wound tightly around a uniform massive disc of radius R and mass m. The ribbon is fastened to a rigid support, and the disc is released to roll down

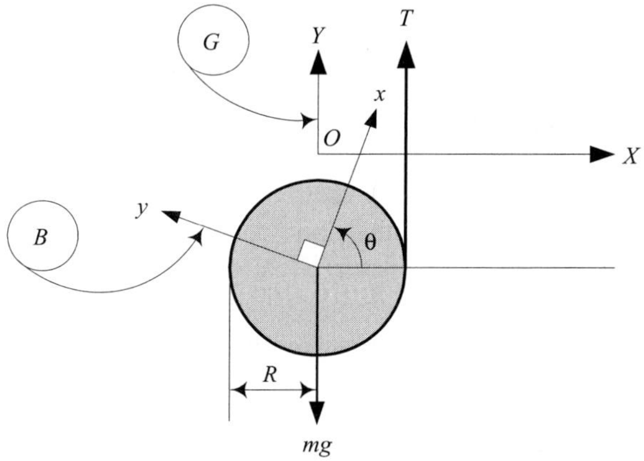

Figure 12.11 A disc is falling by unwinding a ribbon.

vertically. There are two forces acting on the disc during the motion, its weight mg and the tension of the ribbon T. The translational equation of motion of the disc is expressed easier in the global coordinate frame:

$$\sum F_Y = -mg + T = m\ddot{Y} \tag{12.230}$$

The rotational equation of motion is simpler if expressed in the body coordinate frame:

$$\sum M_z = TR = {}^B I \, {}^B_G\dot{\boldsymbol{\omega}}_B + {}^B_G\boldsymbol{\omega}_B \times {}^B I \, {}^B_G\boldsymbol{\omega}_B = I\ddot{\theta} \tag{12.231}$$

There is a constraint between the coordinates Y and θ:

$$Y = Y_0 - R\theta \tag{12.232}$$

To solve the motion, let us eliminate T between (12.230) and (12.231) to obtain

$$m\ddot{Y} = -mg + \frac{I}{R}\ddot{\theta} \tag{12.233}$$

and use the constraint to eliminate $\ddot{Y}$:

$$\ddot{\theta} = \frac{mg}{(I/R + mR)} \tag{12.234}$$

Now, we can find T and $\ddot{Y}$:

$$T = \frac{I}{R}\ddot{\theta} = \frac{I}{mR^2 + I}mg \tag{12.235}$$

$$\ddot{Y} = -\frac{mR^2}{mR^2 + I}g \tag{12.236}$$

For a point mass with $I = 0$, the falling acceleration is the same as the free fall of a particle. However, being a rigid body and having $I \neq 0$, the falling acceleration of the disc is reduced. This is because the kinetic energy of the disc splits between rotation and translation.

Example 734 Rigid-Vehicle Newton–Euler Dynamics Figure 12.12 illustrates a rigid vehicle in planar motion. A global coordinate frame G is attached to the ground and a local coordinate frame B is attached to the vehicle at its mass center C. The Z- and z-axes are parallel, and the orientation of the frame B is indicated by the heading angle ψ between the x- and X-axes. The definitions of kinematics and relative angles are the same as explained in Example 565.

Analysis of ground, air, or space vehicles is more practical when the observer moves with the vehicle. Therefore, in vehicle dynamics we usually employ the B-expression of Newton–Euler equations. We do this especially because applied forces such as traction, braking, and trust are generated by the vehicle and have simple expression in B.

The velocity vector of the vehicle of Figure 12.12, expressed in the body coordinate frame, is

$$ {}^{B}\mathbf{v}_C = \begin{bmatrix} v_x \\ v_y \\ 0 \end{bmatrix} \tag{12.237} $$

where v_x is the forward component and v_y is the lateral component of $\mathbf{v}$. The equations of motion of the vehicle in the body coordinate frame are

$$ \begin{aligned} {}^{B}\mathbf{F} &= {}^{B}R_G\, {}^{G}\mathbf{F} = {}^{B}R_G \left(m\, {}^{G}\mathbf{a}_B\right) = m\, {}^{B}_{G}\mathbf{a}_B \\ &= m\, {}^{B}\dot{\mathbf{v}}_B + m\, {}^{B}_{G}\boldsymbol{\omega}_B \times {}^{B}\mathbf{v}_B \end{aligned} \tag{12.238} $$

$$ \begin{aligned} {}^{B}\mathbf{M} &= \frac{{}^{G}d}{dt}\, {}^{B}\mathbf{L} = {}^{B}_{G}\dot{\mathbf{L}}_B = {}^{B}\dot{\mathbf{L}} + {}^{B}_{G}\boldsymbol{\omega}_B \times {}^{B}\mathbf{L} \\ &= {}^{B}I\, {}^{B}_{G}\dot{\boldsymbol{\omega}}_B + {}^{B}_{G}\boldsymbol{\omega}_B \times \left({}^{B}I\, {}^{B}_{G}\boldsymbol{\omega}_B\right) \end{aligned} \tag{12.239} $$

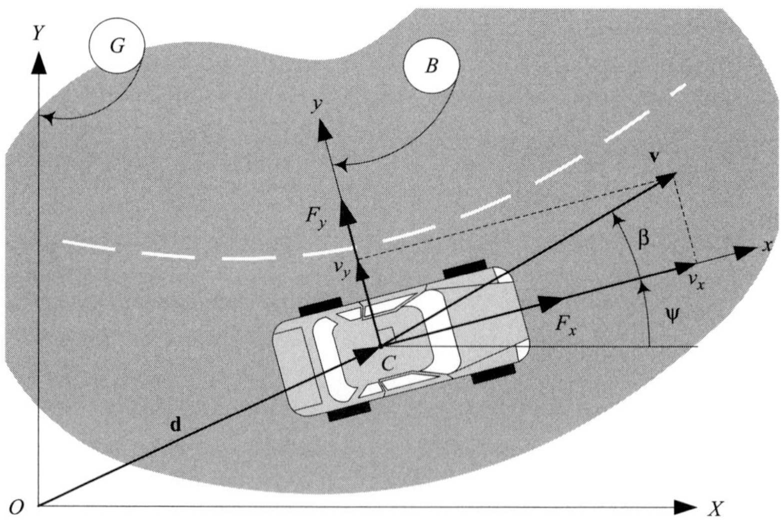

Figure 12.12 A rigid vehicle in planar motion is under a forward traction force F_x and a lateral force F_y.

The force, moment, and kinematic vectors for the rigid vehicle are

$$^B\mathbf{F}_C = \begin{bmatrix} F_x \\ F_y \\ 0 \end{bmatrix} \qquad ^B\mathbf{M}_C = \begin{bmatrix} 0 \\ 0 \\ M_z \end{bmatrix} \tag{12.240}$$

$$^B\dot{\mathbf{v}}_C = \begin{bmatrix} \dot{v}_x \\ \dot{v}_y \\ 0 \end{bmatrix} \qquad {}^B_G\boldsymbol{\omega}_B = \begin{bmatrix} 0 \\ 0 \\ \omega_z \end{bmatrix} \qquad {}^B_G\dot{\boldsymbol{\omega}}_B = \begin{bmatrix} 0 \\ 0 \\ \dot{\omega}_z \end{bmatrix} \tag{12.241}$$

Let us assume that B is the principal coordinate frame of the vehicle to have a diagonal mass moment matrix:

$$^BI = \begin{bmatrix} I_1 & 0 & 0 \\ 0 & I_2 & 0 \\ 0 & 0 & I_3 \end{bmatrix} \tag{12.242}$$

Substituting these vectors and matrices in the equations of motion (12.238)–(12.239) provides the equations

$$\begin{aligned} ^B\mathbf{F} &= m\ ^B\dot{\mathbf{v}}_B + m\ {}^B_G\boldsymbol{\omega}_B \times {}^B\mathbf{v}_B \\ &= m \begin{bmatrix} \dot{v}_x \\ \dot{v}_y \\ 0 \end{bmatrix} + m \begin{bmatrix} 0 \\ 0 \\ \omega_z \end{bmatrix} \times \begin{bmatrix} v_x \\ v_y \\ 0 \end{bmatrix} = \begin{bmatrix} m\dot{v}_x - m\omega_z v_y \\ m\dot{v}_y + m\omega_z v_x \\ 0 \end{bmatrix} \end{aligned} \tag{12.243}$$

$$\begin{aligned} ^B\mathbf{M} &= {}^BI\ {}^B_G\dot{\boldsymbol{\omega}}_B + {}^B_G\boldsymbol{\omega}_B \times \left({}^BI\ {}^B_G\boldsymbol{\omega}_B \right) = \begin{bmatrix} I_1 & 0 & 0 \\ 0 & I_2 & 0 \\ 0 & 0 & I_3 \end{bmatrix} \begin{bmatrix} 0 \\ 0 \\ \dot{\omega}_z \end{bmatrix} \\ &+ \begin{bmatrix} 0 \\ 0 \\ \omega_z \end{bmatrix} \times \left(\begin{bmatrix} I_1 & 0 & 0 \\ 0 & I_2 & 0 \\ 0 & 0 & I_3 \end{bmatrix} \begin{bmatrix} 0 \\ 0 \\ \omega_z \end{bmatrix} \right) = \begin{bmatrix} 0 \\ 0 \\ I_3\,\dot{\omega}_z \end{bmatrix} \end{aligned} \tag{12.244}$$

The first two Newton equations and the third Euler equation are the only nonzero equations that make a set of three equations of motion for the planar rigid vehicle:

$$F_x = m\,\dot{v}_x - m\omega_z\,v_y \tag{12.245}$$

$$F_y = m\,\dot{v}_y + m\omega_z\,v_x \tag{12.246}$$

$$M_z = \dot{\omega}_z\,I_z \tag{12.247}$$

Because the equations of motion are derived in the body frame, the global position vector of the mass center, $^G\mathbf{d}$, does not appear in the equations.

Example 735 Global–Body Transformation of Planar Vehicle Equations Consider a rigid vehicle in planar motion as shown in Figure 12.12. The global coordinate frame is given as G, and B indicates a body coordinate frame that is attached to the vehicle at its mass center C. Let us derive a planar rigid-vehicle equations of motion in the global coordinate frame and transform them to the body frame and rederive Equations

(12.245)–(12.247) in the B-frame. The only angle between B and G is the yaw angle ψ, so the transformation matrix ${}^G R_B$ is

$$ {}^G R_B = \begin{bmatrix} \cos\psi & -\sin\psi & 0 \\ \sin\psi & \cos\psi & 0 \\ 0 & 0 & 1 \end{bmatrix} \tag{12.248} $$

The global expression of the velocity vector is

$$ {}^G\mathbf{v}_C = {}^G R_B \, {}^B\mathbf{v}_C \tag{12.249} $$

$$ \begin{aligned} \begin{bmatrix} v_X \\ v_Y \\ 0 \end{bmatrix} &= \begin{bmatrix} \cos\psi & -\sin\psi & 0 \\ \sin\psi & \cos\psi & 0 \\ 0 & 0 & 1 \end{bmatrix} \begin{bmatrix} v_x \\ v_y \\ 0 \end{bmatrix} \\ &= \begin{bmatrix} v_x \cos\psi - v_y \sin\psi \\ v_y \cos\psi + v_x \sin\psi \\ 0 \end{bmatrix} \end{aligned} \tag{12.250} $$

and therefore, the global acceleration of the vehicle is

$$ \begin{bmatrix} \dot{v}_X \\ \dot{v}_Y \\ 0 \end{bmatrix} = \begin{bmatrix} \left(\dot{v}_x - \dot{\psi}\, v_y\right)\cos\psi - \left(\dot{v}_y + \dot{\psi}\, v_x\right)\sin\psi \\ \left(\dot{v}_y + \dot{\psi}\, v_x\right)\cos\psi + \left(\dot{v}_x - \dot{\psi}\, v_y\right)\sin\psi \\ 0 \end{bmatrix} \tag{12.251} $$

The global Newton equation of motion and the force vector transformation are

$$ {}^G\mathbf{F}_C = m \, {}^G\dot{\mathbf{v}}_C \tag{12.252} $$

$$ {}^G\mathbf{F}_C = {}^G R_B \, {}^B\mathbf{F}_C \tag{12.253} $$

Therefore, the B-expression for the equations of motion is

$$ {}^B\mathbf{F}_C = {}^G R_B^{\mathrm{T}} \, {}^G\mathbf{F}_C = m \, {}^G R_B^{\mathrm{T}} \, {}^G\dot{\mathbf{v}}_C \tag{12.254} $$

Substituting the associated vectors generates the Newton equations of motion in the body coordinate frame:

$$ \begin{aligned} \begin{bmatrix} F_x \\ F_y \\ 0 \end{bmatrix} &= m \, {}^G R_B^{\mathrm{T}} \begin{bmatrix} \left(\dot{v}_x - \dot{\psi}\, v_y\right)\cos\psi - \left(\dot{v}_y + \dot{\psi}\, v_x\right)\sin\psi \\ \left(\dot{v}_y + \dot{\psi}\, v_x\right)\cos\psi + \left(\dot{v}_x - \dot{\psi}\, v_y\right)\sin\psi \\ 0 \end{bmatrix} \\ &= m \begin{bmatrix} \dot{v}_x - \dot{\psi}\, v_y \\ \dot{v}_y + \dot{\psi}\, v_x \\ 0 \end{bmatrix} \end{aligned} \tag{12.255} $$

Applying the same procedure for moment transformation,

$$ \begin{aligned} {}^G\mathbf{M}_C &= {}^G R_B \, {}^B\mathbf{M}_C \\ \begin{bmatrix} 0 \\ 0 \\ M_Z \end{bmatrix} &= \begin{bmatrix} \cos\psi & -\sin\psi & 0 \\ \sin\psi & \cos\psi & 0 \\ 0 & 0 & 1 \end{bmatrix} \begin{bmatrix} 0 \\ 0 \\ M_z \end{bmatrix} = \begin{bmatrix} 0 \\ 0 \\ M_z \end{bmatrix} \end{aligned} \tag{12.256} $$

we find the Euler equation in the body coordinate frame:

$$M_z = \dot{\omega}_z I_z \tag{12.257}$$

Example 736 The Roll Model Vehicle Dynamics Figure 12.13 illustrates a vehicle with a body coordinate frame $B(Cxyz)$ at the mass center C. The x-axis is a longitudinal axis passing through C and directed forward. The y-axis goes laterally to the left from the driver's viewpoint. The z-axis makes the coordinate system a right-hand triad. When the car is parked on a flat horizontal road, the z-axis is perpendicular to the ground, opposite to the gravitational acceleration $\mathbf{g}$. The equations of motion of the vehicle are usually expressed in $B(Cxyz)$.

The angular orientation and angular velocity of the vehicle are expressed by three angles—*roll* φ, *pitch* θ, and *yaw* ψ—and their rates—*roll rate* p, *pitch rate* q, and *yaw rate* r:

$$p = \dot{\varphi} \qquad q = \dot{\theta} \qquad r = \dot{\psi} \tag{12.258}$$

The vehicle force system $(\mathbf{F}, \mathbf{M})$ is the resultant of external forces and moments that the vehicle receives from the ground and environment. The force system may be expressed in the body coordinate frame as

$$^{B}\mathbf{F} = F_x\hat{\imath} + F_y\hat{\jmath} + F_z\hat{k} \tag{12.259}$$

$$^{B}\mathbf{M} = M_x\hat{\imath} + M_y\hat{\jmath} + M_z\hat{k} \tag{12.260}$$

The roll model vehicle dynamics can be expressed by four kinematic variables: the forward motion x, the lateral motion y, the roll angle φ, and the yaw angle ψ. In this model, we do not consider vertical movement z and pitch motion θ. So, there are two holonomic constraints among the six coordinates x, y, z, θ, ψ:

$$z - z_0 = 0 \qquad \theta = 0 \tag{12.261}$$

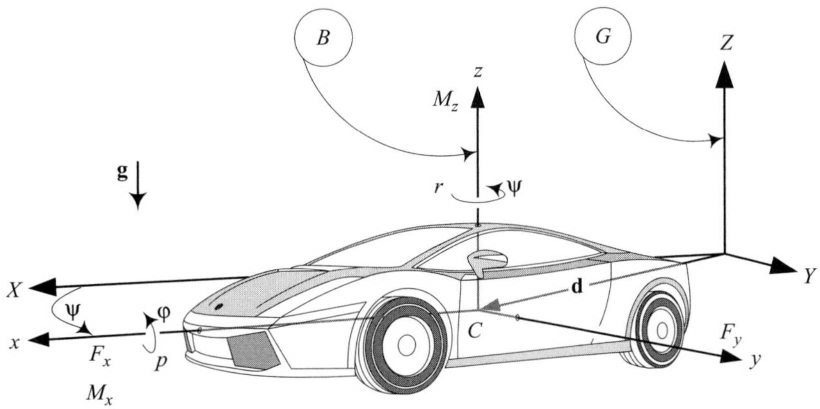

Figure 12.13 A vehicle with roll and yaw rotations.

A global coordinate frame G is fixed on the ground. The orientation of B can be expressed by the heading angle ψ between the x- and X-axes and the roll angle φ between the z- and Z-axes. The global position vector of the mass center is denoted by ${}^G\mathbf{d}$.

The rigid-body equations of motion in the body coordinate frame are

$$\begin{aligned} {}^B\mathbf{F} &= {}^BR_G\,{}^G\mathbf{F} = {}^BR_G\left(m\,{}^G\mathbf{a}_B\right) = m\,{}^B_G\mathbf{a}_B \\ &= m\,{}^B\dot{\mathbf{v}}_B + m\,{}^B_G\boldsymbol{\omega}_B \times {}^B\mathbf{v}_B \end{aligned} \tag{12.262}$$

$$\begin{aligned} {}^B\mathbf{M} &= \frac{{}^Gd}{dt}\,{}^B\mathbf{L} = {}^B_G\dot{\mathbf{L}}_B = {}^B\dot{\mathbf{L}} + {}^B_G\boldsymbol{\omega}_B \times {}^B\mathbf{L} \\ &= {}^BI\,{}^B_G\dot{\boldsymbol{\omega}}_B + {}^B_G\boldsymbol{\omega}_B \times \left({}^BI\,{}^B_G\boldsymbol{\omega}_B\right) \end{aligned} \tag{12.263}$$

The B-expressions of the velocity and acceleration vectors of the vehicle are

$${}^B\mathbf{v}_C = \begin{bmatrix} v_x \\ v_y \\ 0 \end{bmatrix} \qquad {}^B\dot{\mathbf{v}}_C = \begin{bmatrix} \dot{v}_x \\ \dot{v}_y \\ 0 \end{bmatrix} \tag{12.264}$$

where v_x is the forward component and v_y is the lateral component of ${}^B\mathbf{v}$. The angular velocity and acceleration vectors for the rigid vehicle are

$${}^B_G\boldsymbol{\omega}_B = \begin{bmatrix} \omega_x \\ 0 \\ \omega_z \end{bmatrix} = \begin{bmatrix} p \\ 0 \\ r \end{bmatrix} \qquad {}^B_G\dot{\boldsymbol{\omega}}_B = \begin{bmatrix} \dot{\omega}_x \\ 0 \\ \dot{\omega}_z \end{bmatrix} = \begin{bmatrix} \dot{p} \\ 0 \\ \dot{r} \end{bmatrix} \tag{12.265}$$

We may assume that the body coordinate is the principal coordinate frame of the vehicle to have a diagonal mass moment matrix:

$${}^BI = \begin{bmatrix} I_1 & 0 & 0 \\ 0 & I_2 & 0 \\ 0 & 0 & I_3 \end{bmatrix} \tag{12.266}$$

Substituting the above vectors and matrices in the Newton–Euler equations of motion provides the equations

$${}^B\mathbf{F} = m\,{}^B\dot{\mathbf{v}}_B + m\,{}^B_G\boldsymbol{\omega}_B \times {}^B\mathbf{v}_B \tag{12.267}$$

$$\begin{aligned} \begin{bmatrix} F_x \\ F_y \\ 0 \end{bmatrix} &= m\begin{bmatrix} \dot{v}_x \\ \dot{v}_y \\ 0 \end{bmatrix} + m\begin{bmatrix} \omega_x \\ 0 \\ \omega_z \end{bmatrix} \times \begin{bmatrix} v_x \\ v_y \\ 0 \end{bmatrix} \\ &= \begin{bmatrix} m\dot{v}_x - m\omega_z v_y \\ m\dot{v}_y + m\omega_z v_x \\ m\omega_x v_y \end{bmatrix} \end{aligned} \tag{12.268}$$

$${}^B\mathbf{M} = {}^BI\,{}^B_G\dot{\boldsymbol{\omega}}_B + {}^B_G\boldsymbol{\omega}_B \times \left({}^BI\,{}^B_G\boldsymbol{\omega}_B\right) \tag{12.269}$$

$$\begin{bmatrix} M_x \\ 0 \\ M_z \end{bmatrix} = \begin{bmatrix} I_1 & 0 & 0 \\ 0 & I_2 & 0 \\ 0 & 0 & I_3 \end{bmatrix} \begin{bmatrix} \dot{\omega}_x \\ 0 \\ \dot{\omega}_z \end{bmatrix} + \begin{bmatrix} \omega_x \\ 0 \\ \omega_z \end{bmatrix} \times \left(\begin{bmatrix} I_1 & 0 & 0 \\ 0 & I_2 & 0 \\ 0 & 0 & I_3 \end{bmatrix} \begin{bmatrix} \omega_x \\ 0 \\ \omega_z \end{bmatrix} \right) = \begin{bmatrix} I_1 \dot{\omega}_x \\ I_1 \omega_x \omega_z - I_3 \omega_x \omega_z \\ I_3 \dot{\omega}_z \end{bmatrix} \tag{12.270}$$

The first two Newton equations are the equations of motion in the x- and y-directions:

$$\begin{bmatrix} F_x \\ F_y \end{bmatrix} = \begin{bmatrix} m\dot{v}_x - m\omega_z v_y \\ m\dot{v}_y + m\omega_z v_x \end{bmatrix} \tag{12.271}$$

The first and third Euler equations are the equations of motion about the x- and z-axes:

$$\begin{bmatrix} M_x \\ M_z \end{bmatrix} = \begin{bmatrix} I_1 \dot{\omega}_x \\ I_3 \dot{\omega}_z \end{bmatrix} \tag{12.272}$$

Therefore, a rolling rigid vehicle has a motion with four DOF which are translations in the x- and y-directions and rotations about the x- and z-axes. The *Newton–Euler equations of motion* for such a rolling rigid vehicle in the body coordinate frame B are

$$F_x = m\,\dot{v}_x - mr\,v_y \tag{12.273}$$

$$F_y = m\,\dot{v}_y + mr\,v_x \tag{12.274}$$

$$M_z = I_z \dot{\omega}_z = I_z \dot{r} \tag{12.275}$$

$$M_x = I_x \dot{\omega}_x = I_x \dot{p} \tag{12.276}$$

Example 737 Motion of a Six-DOF Vehicle Consider a spacecraft or an airplane such as shown in Figure 12.14 that moves in space. Such a vehicle has six DOF. To develop the equations of motion, we define the kinematic characteristics as

$$^B\mathbf{v}_C = \begin{bmatrix} v_x \\ v_y \\ v_z \end{bmatrix} \qquad ^B\dot{\mathbf{v}}_C = \begin{bmatrix} \dot{v}_x \\ \dot{v}_y \\ \dot{v}_z \end{bmatrix} \tag{12.277}$$

$$^B_G\boldsymbol{\omega}_B = \begin{bmatrix} \omega_x \\ \omega_y \\ \omega_z \end{bmatrix} \qquad ^B_G\dot{\boldsymbol{\omega}}_B = \begin{bmatrix} \dot{\omega}_x \\ \dot{\omega}_y \\ \dot{\omega}_z \end{bmatrix} \tag{12.278}$$

The acceleration vector of the vehicle in the body coordinate frame is

$$^B\mathbf{a} = {}^B\dot{\mathbf{v}}_B + {}^B_G\boldsymbol{\omega}_B \times {}^B\mathbf{v}_B = \begin{bmatrix} \dot{v}_x + \omega_y v_z - \omega_z v_y \\ \dot{v}_y + \omega_z v_x - \omega_x v_z \\ \dot{v}_z + \omega_x v_y - \omega_y v_x \end{bmatrix} \tag{12.279}$$

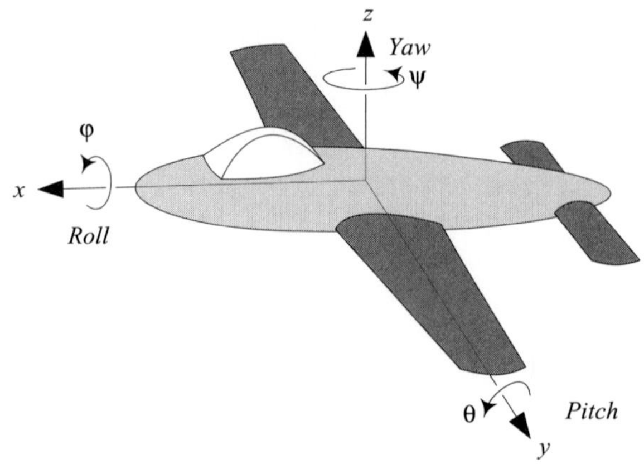

Figure 12.14 An airplane with six-*DOF* motions.

and therefore, Newtons equations of motion for the vehicle are

$$\begin{bmatrix} F_x \\ F_y \\ F_z \end{bmatrix} = m \begin{bmatrix} \dot{v}_x + \omega_y v_z - \omega_z v_y \\ \dot{v}_y + \omega_z v_x - \omega_x v_z \\ \dot{v}_z + \omega_x v_y - \omega_y v_x \end{bmatrix} \tag{12.280}$$

To find the Euler equations of motion,

$$ {}^B\mathbf{M} = {}^BI \, {}_G^B\dot{\boldsymbol{\omega}}_B + {}_G^B\boldsymbol{\omega}_B \times \left({}^BI \, {}_G^B\boldsymbol{\omega}_B \right) \tag{12.281}$$

we need to define the mass moment matrix and perform the required matrix calculations. Assume the body coordinate system is the principal coordinate frame. So,

$$\begin{aligned} & {}^BI \, {}_G^B\dot{\boldsymbol{\omega}}_B + {}_G^B \boldsymbol{\omega}_B \times \left({}^BI \, {}_G^B\boldsymbol{\omega}_B \right) \\ &= \begin{bmatrix} I_1 & 0 & 0 \\ 0 & I_2 & 0 \\ 0 & 0 & I_3 \end{bmatrix} \begin{bmatrix} \dot{\omega}_x \\ \dot{\omega}_y \\ \dot{\omega}_z \end{bmatrix} + \begin{bmatrix} \omega_x \\ \omega_y \\ \omega_z \end{bmatrix} \times \left(\begin{bmatrix} I_1 & 0 & 0 \\ 0 & I_2 & 0 \\ 0 & 0 & I_3 \end{bmatrix} \begin{bmatrix} \omega_x \\ \omega_y \\ \omega_z \end{bmatrix} \right) \\ &= \begin{bmatrix} \dot{\omega}_x I_1 - \omega_y \omega_z I_2 + \omega_y \omega_z I_3 \\ \dot{\omega}_y I_2 + \omega_x \omega_z I_1 - \omega_x \omega_z I_3 \\ \dot{\omega}_z I_3 - \omega_x \omega_y I_1 + \omega_x \omega_y I_2 \end{bmatrix} \end{aligned} \tag{12.282}$$

and therefore, the Euler equations of motion for the vehicle are

$$\begin{bmatrix} M_x \\ M_y \\ M_z \end{bmatrix} = \begin{bmatrix} \dot{\omega}_x I_1 - \omega_y \omega_z I_2 + \omega_y \omega_z I_3 \\ \dot{\omega}_y I_2 + \omega_x \omega_z I_1 - \omega_x \omega_z I_3 \\ \dot{\omega}_z I_3 - \omega_x \omega_y I_1 + \omega_x \omega_y I_2 \end{bmatrix} \tag{12.283}$$

Example 738 Magnus Force and Soccer Ball An expert soccer player is able to kick a ball to move on a curved path. To explain how a ball is able to curve through the air, we need to review the Magnus effect. Figure 12.15 illustrates a simplified side view of

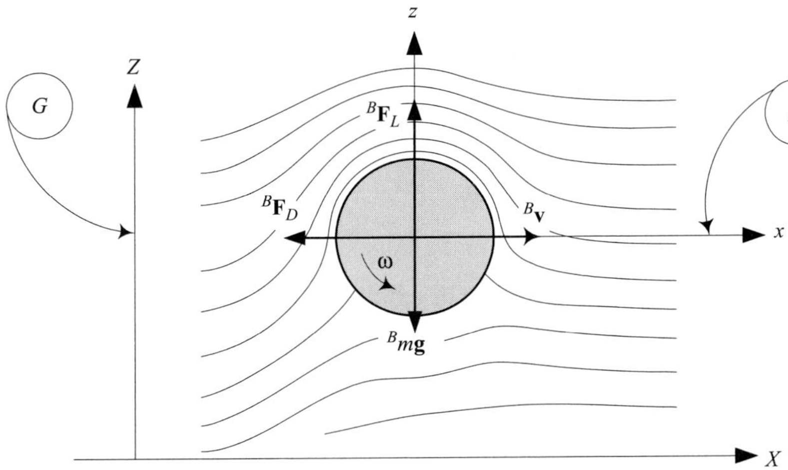

Figure 12.15 Magnus force on a spinning and moving cylinder.

a spinning ball about the y-axis that is moving with velocity $\mathbf{v}$ through air. According to the Magnus equation, the air will apply a lift force $\mathbf{F}_L$ on the ball,

$$^B\mathbf{F}_L = mC_L\, {}_G^B\boldsymbol{\omega}_B \times {}^B\mathbf{v} = mC_L \begin{bmatrix} v_y\omega_z - v_z\omega_y \\ v_z\omega_x - v_x\omega_z \\ v_x\omega_y - v_y\omega_x \end{bmatrix} \tag{12.284}$$

where C_L is a parameter that depends on the speed v and diameter D of the ball, density ρ, and viscosity μ of the air. The body coordinate frame B is attached to the center of the ball, but B is not turning with the spin of the ball. The ball is also under a drag force $\mathbf{F}_D$,

$$^B\mathbf{F}_D = -\tfrac{1}{2}\rho m A C_D \left({}^B\mathbf{v}\cdot{}^B\mathbf{v}\right)\frac{{}^B\mathbf{v}}{\left|{}^B\mathbf{v}\right|} \tag{12.285}$$

$$= -\tfrac{1}{2}\rho m A C_D \sqrt{v_x^2 + v_y^2 + v_z^2}\begin{bmatrix} v_x \\ v_y \\ v_z \end{bmatrix} \tag{12.286}$$

in which A is the frontal area of the ball in contact with the air and C_D is the drag coefficient.

Although the reason for the Magnus force is the viscosity of the air and the boundary layer around the cylinder, let us assume that the resultant moment is zero and therefore the angular velocity of the ball remains constant during the motion. Under this condition, the translational equations of motion of the ball are

$$^B\mathbf{F} = \begin{bmatrix} m\left(\dot{v}_x - \left(v_y\omega_z - v_z\omega_y\right)\right) \\ m\left(\dot{v}_y - \left(v_z\omega_x - v_x\omega_z\right)\right) \\ m\left(\dot{v}_z - \left(v_x\omega_y - v_y\omega_x\right)\right) \end{bmatrix} \tag{12.287}$$

$$
{}^{B}\mathbf{F} = {}^{B}\mathbf{F}_{L} + {}^{B}\mathbf{F}_{D} + m\mathbf{g}
= \begin{bmatrix}
mC_{L}\left(v_{y}\omega_{z} - v_{z}\omega_{y}\right) - \frac{1}{2}Am\rho v_{x}C_{D}\sqrt{v_{x}^{2}+v_{y}^{2}+v_{z}^{2}} \\
mC_{L}\left(v_{z}\omega_{x} - v_{x}\omega_{z}\right) - \frac{1}{2}Am\rho v_{y}C_{D}\sqrt{v_{x}^{2}+v_{y}^{2}+v_{z}^{2}} \\
-mg + mC_{L}\left(v_{x}\omega_{y} - v_{y}\omega_{x}\right) - \frac{1}{2}Am\rho v_{z}C_{D}\sqrt{v_{x}^{2}+v_{y}^{2}+v_{z}^{2}}
\end{bmatrix} \tag{12.288}
$$

In a simple case with the initial conditions

$$
{}^{B}\mathbf{v} = \begin{bmatrix} v_{x} \\ 0 \\ 0 \end{bmatrix} \qquad {}_{G}^{B}\boldsymbol{\omega}_{B} = \begin{bmatrix} 0 \\ 0 \\ \omega_{z} \end{bmatrix} = \text{const} \qquad C_{D} = 0 \tag{12.289}
$$

we have

$$
\dot{v}_{x} = v_{y}\omega_{z} \tag{12.290}
$$

$$
\dot{v}_{y} = -\left(1 + C_{L}\right) v_{x}\omega_{z} \tag{12.291}
$$

$$
\dot{v}_{z} = -g \tag{12.292}
$$

12.4 CLASSICAL PROBLEMS OF RIGID BODIES

There are some classical problems of rigid bodies in which equations of motion can be integrated analytically. We examine these classical problems to review the application of Euler–Newton equations of motion in engineering, physics, and mathematics.

If there is any integral of motion, the best method to develop an analytic solution is employing the integral of motion and reduce the number of equations.

The dynamics of a rigid body depends solely on the mass and principal mass moments of the body. Two rigid bodies are called *equimomental* if they have the same mass and principal mass moments. The equations of motion and dynamics of equimomental rigid bodies are identical subject to equivalent force systems.

12.4.1 Torque-Free Motion

The *torque-free motion* of rigid bodies is an applied classical example with many applications. There are three integrable types of torque-free motions:

1. Spherical torque-free motion, $I_1 = I_2 = I_3$
2. Axisymmetric torque-free motion, $I_1 = I_2 \neq I_3$
3. Asymmetric torque-free motion, $I_1 \neq I_2 \neq I_3$

Torque-free problems of rigid bodies present two integrals of motion: angular momentum conservation L and energy conservation K:

$$
{}_{G}^{B}\boldsymbol{\omega}_{B} \cdot {}^{B}I \; {}_{G}^{B}\boldsymbol{\omega}_{B} = 2K \tag{12.293}
$$

$$
{}^{B}I^{2} \; {}_{G}^{B}\boldsymbol{\omega}_{B}^{2} = L^{2} \tag{12.294}
$$

The integrals of motion in the principal coordinate frame are

$$I_1\omega_1^2 + I_2\omega_2^2 + I_3\omega_3^2 = 2K \tag{12.295}$$

$$I_1^2\omega_1^2 + I_2^2\omega_2^2 + I_3^2\omega_3^2 = L^2 \tag{12.296}$$

Using these two integrals of motion to eliminate two out of three Euler equations is the first step in the analytic treatment of torque-free rigid-body dynamics. Both constants of motion K and L^2 are nonnegative numbers that would be zero only when the rigid body has zero angular velocity.

Proof: Let us determine the inner product of ${}^B_G\boldsymbol{\omega}_B$ and ${}^BI\ {}^B_G\boldsymbol{\omega}_B$ by the Euler equation of motion (12.404):

$$\begin{aligned} & {}^B_G\boldsymbol{\omega}_B \cdot \left({}^BI\ {}^B_G\dot{\boldsymbol{\omega}}_B + {}^B_G\boldsymbol{\omega}_B \times {}^BI\ {}^B_G\boldsymbol{\omega}_B\right) \\ &= {}^B_G\boldsymbol{\omega}_B \cdot {}^BI\ {}^B_G\dot{\boldsymbol{\omega}}_B \\ &= \frac{1}{2}\frac{d}{dt}\left({}^B_G\boldsymbol{\omega}_B \cdot {}^BI\ {}^B_G\boldsymbol{\omega}_B\right) = 0 \end{aligned} \tag{12.297}$$

$$\begin{aligned} & {}^BI\ {}^B_G\boldsymbol{\omega}_B \cdot \left({}^BI\ {}^B_G\dot{\boldsymbol{\omega}}_B + {}^B_G\boldsymbol{\omega}_B \times {}^BI\ {}^B_G\boldsymbol{\omega}_B\right) \\ &= {}^BI\ {}^B_G\boldsymbol{\omega}_B \cdot {}^BI{}^B_G\dot{\boldsymbol{\omega}}_B \\ &= \frac{1}{2}\frac{d}{dt}\left({}^BI{}^B_G\boldsymbol{\omega}_B\right)^2 = 0 \end{aligned} \tag{12.298}$$

These two total differentials lead to two integrals of motion:

$${}^B_G\boldsymbol{\omega}_B \cdot {}^BI\ {}^B_G\boldsymbol{\omega}_B = 2K \tag{12.299}$$

$${}^BI^2\ {}^B_G\boldsymbol{\omega}_B^2 = L^2 \tag{12.300}$$

where K is the kinetic energy and L is the angular momentum of the rigid body. The expanded forms of the integrals of motion are

$$\begin{aligned} L^2 &= \left(I_{xx}\omega_x + I_{xy}\omega_y + I_{xz}\omega_z\right)^2 \\ &\quad + \left(I_{yx}\omega_x + I_{yy}\omega_y + I_{yz}\omega_z\right)^2 \\ &\quad + \left(I_{zx}\omega_x + I_{zy}\omega_y + I_{zz}\omega_z\right)^2 \end{aligned} \tag{12.301}$$

$$\begin{aligned} K &= \tfrac{1}{2}\left(I_{xx}\omega_x^2 + I_{yy}\omega_y^2 + I_{zz}\omega_z^2\right) \\ &\quad - I_{xy}\omega_x\omega_y - I_{yz}\omega_y\omega_z - I_{zx}\omega_z\omega_x \end{aligned} \tag{12.302}$$

These integrals of motion have simpler expressions in the principal frame $B(1, 2, 3)$:

$$\sum_{i=1}^{3} I_i\omega_i^2 = 2K = I_1\omega_1^2 + I_2\omega_2^2 + I_3\omega_3^2 \tag{12.303}$$

$$\sum_{i=1}^{3} I_i^2\omega_i^2 = L^2 = 2DK = I_1^2\omega_1^2 + I_2^2\omega_2^2 + I_3^2\omega_3^2 \tag{12.304}$$

Generally speaking, employing integrals of motion simplifies the analytic solution of Euler dynamic equations. However, if we ignore their existence, they would be automatically satisfied when the differential equations of motion are integrated. ■

Example 739 Kinetic Energy of a Rolling Disc Figure 12.16 illustrates a disc of mass m and radius r that is rolling on a circular path of radius R. To determine the kinetic energy of the disc, we attach body B and global G coordinate frames as shown in the figure. The rotation of B is shown by angle θ and is measured from the X-axis. The position of the disc center C is measured by the angle φ, also from the X-axis.

The kinetic energy of the disc has two parts:

1. The translational kinetic energy K_1 when the disc is considered as a point mass at C:

$$K_1 = \frac{1}{2} m {}^G\mathbf{v}_C \cdot {}^G\mathbf{v}_C \tag{12.305}$$

2. The rotational kinetic energy K_2 when the disc is rotating about a stationary mass center C:

$$K_2 = \frac{1}{2} {}_G^B\boldsymbol{\omega}_B \cdot {}^BI \; {}_G^B\boldsymbol{\omega}_B \tag{12.306}$$

The mass moment matrix of the disc is

$${}^BI = \begin{bmatrix} \frac{1}{4}mR^2 & 0 & 0 \\ 0 & \frac{1}{4}mR^2 & 0 \\ 0 & 0 & \frac{1}{2}mR^2 \end{bmatrix} \tag{12.307}$$

Let us use the instantaneous center which is the contact point with the ground. The angular velocity of the disc would be:

$${}_G^B\boldsymbol{\omega}_B = \frac{R-r}{r}\dot{\varphi}\,\hat{k} \tag{12.308}$$

$${}_G\boldsymbol{\omega}_B = \frac{R-r}{r}\dot{\varphi}\,\hat{K} \tag{12.309}$$

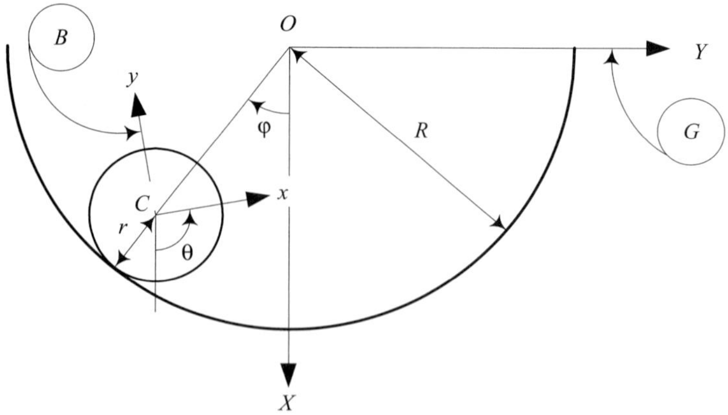

Figure 12.16 A rolling disc on a circular path.

and therefore, the velocity of C is

$$ {}^{G}\mathbf{v}_{C} = {}^{G}\mathbf{r}_{C} \times {}_{G}\boldsymbol{\omega}_{B} = \begin{bmatrix} (R-r)\cos\varphi \\ -(R-r)\sin\varphi \\ 0 \end{bmatrix} \times \begin{bmatrix} 0 \\ 0 \\ \dfrac{R-r}{r}\dot{\varphi} \end{bmatrix} $$

$$ = \begin{bmatrix} -\dfrac{1}{r}\dot{\varphi}(R-r)^2\sin\varphi \\ -\dfrac{1}{r}\dot{\varphi}(R-r)^2\cos\varphi \\ 0 \end{bmatrix} \tag{12.310} $$

So, the kinetic energy of the disc is

$$ \begin{aligned} K &= \frac{1}{2}m\,{}^{G}\mathbf{v}_{C}\cdot{}^{G}\mathbf{v}_{C} + \frac{1}{2}\,{}^{B}_{G}\boldsymbol{\omega}_{B}\cdot{}^{B}I\,{}^{B}_{G}\boldsymbol{\omega}_{B} \\ &= \frac{1}{2}\frac{m}{r^2}\dot{\varphi}^2(R-r)^4 + \frac{1}{r^2}\dot{\varphi}^2 I_3(R-r)^2 \\ &= \frac{1}{2}m\frac{1}{r^2}\left(2R^2-2Rr+r^2\right)(R-r)^2\dot{\varphi}^2 \end{aligned} \tag{12.311} $$

12.4.2 Spherical Torque-Free Rigid Body

A rigid body is called *spherical* or *centrosymmetric* if

$$ I_1 = I_2 = I_3 \tag{12.312} $$

In this case, Euler equations (12.10)–(12.12) will be linear and decoupled:

$$ I_1\dot{\omega}_1 = 0 \tag{12.313} $$

$$ I_1\dot{\omega}_2 = 0 \tag{12.314} $$

$$ I_3\dot{\omega}_3 = 0 \tag{12.315} $$

The solution of these equations indicates that the rigid body will continue its rotation with constant angular velocity:

$$ I_1\omega_1 = \omega_1(0) = \omega_{10} \tag{12.316} $$

$$ I_1\omega_2 = \omega_2(0) = \omega_{20} \tag{12.317} $$

$$ I_3\omega_3 = \omega_3(0) = \omega_{30} \tag{12.318} $$

Proof: Having a rigid body with equal principal mass moments is the simplest situation in rigid-body dynamics. To have equal principal mass moments, a rigid body must have infinity planes of symmetry, such as spheres and cubes. In a spherical rigid body, every line that goes through the mass center is a principal axis. So, the dynamics of the body is independent of its attitude.

Euler equations are dynamically coupled by the variable ω_i that appears in equations for M_j, $j \neq i$. They are also parametrically coupled by the difference of the

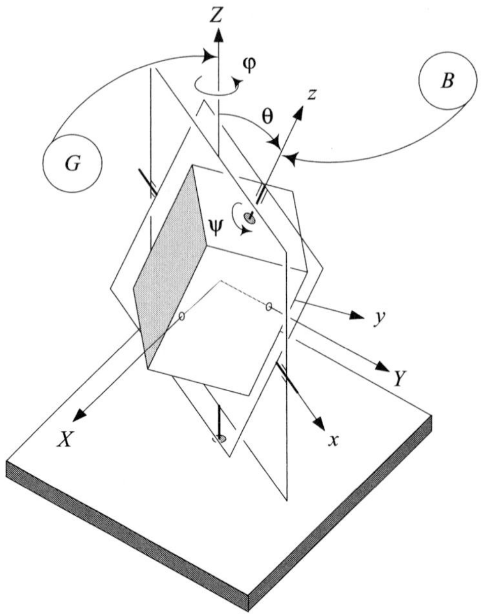

Figure 12.17 A suspended rigid body with a fixed point at geometric center.

principal mass moments. When the principal mass moments $I_i, i = 1, 2, 3$, are equal, the equations uncoupled as Equations (12.313)–(12.315). The uncoupled equations are the simplest differential equations with solutions (12.316)–(12.318). The suspended body in Figure 12.17 is a symmetric rigid body with a fixed point at its mass center if its width, height, and length are equal. ■

12.4.3 Axisymmetric Torque-Free Rigid Body

A rigid body is called *axisymmetric* or *axially symmetric* if

$$I_1 = I_2 \neq I_3 \tag{12.319}$$

The torque-free Euler equations of an axisymmetric rigid body are

$$I_1\dot{\omega}_1 - (I_1 - I_3)\,\omega_2\omega_3 = 0 \tag{12.320}$$

$$I_1\dot{\omega}_2 - (I_3 - I_1)\,\omega_3\omega_1 = 0 \tag{12.321}$$

$$I_3\dot{\omega}_3 = 0 \tag{12.322}$$

It provides a harmonic solution for angular velocity ${}_G^B\boldsymbol{\omega}_B$:

$$\omega_1(t) = \omega_\theta \cos \Omega t \tag{12.323}$$

$$\omega_2(t) = \omega_\theta \sin \Omega t \tag{12.324}$$

$$\omega_3 = \frac{I_1}{I_3 - I_1}\Omega \tag{12.325}$$

The constant parameter Ω is the angular speed of projection of ${}_G^B\boldsymbol{\omega}_B$ on the (x, y)-plane. Angular velocity ${}_G^B\boldsymbol{\omega}_B$ has a constant length and uniformly sweeps a constant cone in the B-frame.

Proof: Assuming a principal body coordinate frame $B(Oxyz) \equiv B(O123)$ and substituting $I_2 = I_1$ in Euler equations (12.10)–(12.12) generate Equations (12.320)–(12.322). The third equation is a total differential, which shows that the third component of ${}_G^B\boldsymbol{\omega}_B$ is always constant:

$$\omega_3 = C_1 \tag{12.326}$$

Let us introduce an axillary constant frequency Ω:

$$\Omega = \frac{I_3 - I_1}{I_1}\omega_3 \tag{12.327}$$

Now we can write the first and second Euler equations as

$$\dot{\omega}_1 + \Omega\omega_2 = 0 \tag{12.328}$$

$$\dot{\omega}_2 - \Omega\omega_1 = 0 \tag{12.329}$$

Multiplying (12.328) by ω_1 and (12.329) by ω_2 and then adding the two equations yield

$$\omega_1\dot{\omega}_1 + \omega_2\dot{\omega}_2 = 0 \tag{12.330}$$

This is a total differential with the integral

$$\sqrt{\omega_1^2 + \omega_2^2} = \omega_\theta = C_2 \tag{12.331}$$

The ω_θ-component of ${}_G^B\boldsymbol{\omega}_B$ lies in the (x, y)-plane and ω_3 is the component of ${}_G^B\boldsymbol{\omega}_B$ on the z-axis. Therefore, the magnitude of the angular velocity remains constant:

$$\left|{}_G^B\boldsymbol{\omega}_B\right| = \sqrt{\omega_1^2 + \omega_2^2 + \omega_3^2} = \sqrt{C_1^2 + C_2^2} = C_3 \tag{12.332}$$

By taking a time derivative of (12.328) and (12.329), we can decouple the equations:

$$\ddot{\omega}_1 + \Omega^2\omega_1 = 0 \tag{12.333}$$

$$\ddot{\omega}_2 + \Omega^2\omega_2 = 0 \tag{12.334}$$

The solutions of these equations are

$$\omega_1(t) = A\sin(\Omega t - \alpha) \tag{12.335}$$

$$\omega_2(t) = B\sin(\Omega t - \beta) \tag{12.336}$$

where α, β, A, and B are related to the initial conditions $\omega_{10} = \omega_1(0)$, $\omega_{20} = \omega_2(0)$, $\dot{\omega}_{10} = \dot{\omega}_1(0)$, and $\dot{\omega}_{20} = \dot{\omega}_2(0)$:

$$\omega_{10} = -A\sin\alpha \qquad \omega_{20} = -B\sin\beta \tag{12.337}$$

$$\dot{\omega}_{10} = A\Omega\cos\alpha \qquad \dot{\omega}_{20} = B\Omega\cos\beta \tag{12.338}$$

Because of (12.331), the tip point of the ω_θ-component of ${}^B_G\boldsymbol{\omega}_B$ in the (x, y)-plane turns on a circle. Therefore, we must have

$$A = B \tag{12.339}$$

$$\alpha = \beta \pm \tfrac{1}{2}\pi \tag{12.340}$$

which yields

$$A = \sqrt{\omega_{10}^2 + \omega_{20}^2} = \omega_\theta = C_2 \tag{12.341}$$

$$\alpha = \arctan \frac{-\omega_{10}\Omega}{\dot{\omega}_{10}} \tag{12.342}$$

With no loss of generality, we may assume $\alpha = 0$ and write the solutions as

$$\omega_1(t) = \omega_\theta \cos \Omega t \tag{12.343}$$

$$\omega_2(t) = \omega_\theta \sin \Omega t \tag{12.344}$$

The component ω_3 is the angular speed of the rigid body about the z-axis and Ω is the angular speed of ${}^B_G\boldsymbol{\omega}_B$ about the z-axis. The tip point of the ω_θ-component of ${}^B_G\boldsymbol{\omega}_B$ in the (x, y)-plane turns about the z-axis with a constant angular velocity Ω. Therefore, the angular velocity vector ${}^B_G\boldsymbol{\omega}_B$ uniformly sweeps a constant cone with the z-axis as the axis of symmetry.

Depending on I_3 and I_1, the body can turn faster or slower than ${}^B_G\boldsymbol{\omega}_B$. For a flat disc with $I_3 > I_1$, we find that the θ-axis rotates faster than the x-axis in the same direction, and for an elongated body with $I_3 < I_1$, the θ-axis rotates slower than the x-axis in the opposite direction. For $I_3 = 2I_1$, we have $\Omega = \omega_3$. The special case $I_3 = I_1$ provides $\Omega = 0$, and therefore, the axis of angular velocity ${}^B_G\boldsymbol{\omega}_B$ remains motionless. ■

Example 740 A Rotating Disc on a Needle Figure 12.18 illustrates a homogeneous disc of mass m and radius R that is suspended at its center on a needle. We attach a body frame B at the center of the disc such that the z-axis is perpendicular to the disc and the (x, y)-plane is the face plane of the disc. We give an initial rotation ${}_G\boldsymbol{\omega}_B = \boldsymbol{\omega}_0$ about an axis that makes an angle α with the z-axis, looking for the consequent motion of the disc.

The mass moment matrix of the disc is

$${}^BI = \begin{bmatrix} \frac{1}{4}mR^2 & 0 & 0 \\ 0 & \frac{1}{4}mR^2 & 0 \\ 0 & 0 & \frac{1}{2}mR^2 \end{bmatrix} \tag{12.345}$$

Let us assume that the initial angular velocity in the (y, z)-plane is

$${}^B_G\boldsymbol{\omega}_B(0) = \omega_0 \sin\alpha\, \hat{\jmath} + \omega_0 \cos\alpha\, \hat{k} \tag{12.346}$$

The motion is torque free; therefore, the third Euler equation provides

$$\omega_3 = \omega_0 \cos\alpha = \text{const} \tag{12.347}$$

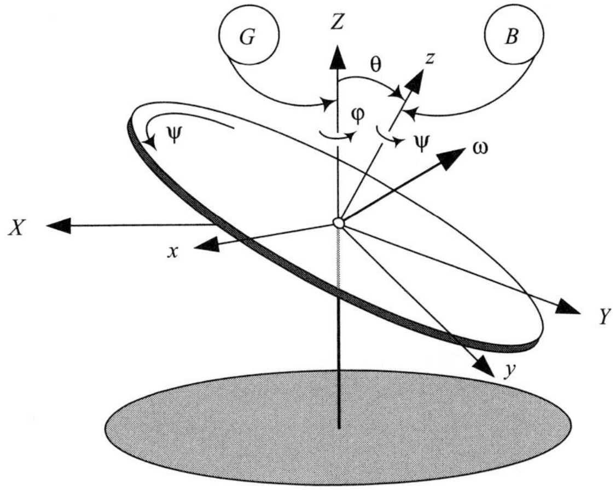

Figure 12.18 A suspended disc on a needle.

and the first and second Euler equations become

$$\dot{\omega}_1 + \omega_2 \omega_0 \cos\alpha = 0 \tag{12.348}$$

$$\dot{\omega}_2 - \omega_1 \omega_0 \cos\alpha = 0 \tag{12.349}$$

These equations are the same as (12.333) and (12.334) with solutions

$$\omega_1(t) = \omega_0 \cos\Omega t \tag{12.350}$$

$$\omega_2(t) = \omega_0 \sin\Omega t \tag{12.351}$$

$$\omega_0 = \omega_\theta = \sqrt{\omega_{10}^2 + \omega_{20}^2} \qquad \Omega = \omega_0 \cos\alpha \tag{12.352}$$

To determine the motion of the disc in G, we recall the relationship of Euler frequencies and the B-expression of angular velocity (8.99):

$$\begin{bmatrix} \omega_x \\ \omega_y \\ \omega_z \end{bmatrix} = \begin{bmatrix} \sin\theta \sin\psi & \cos\psi & 0 \\ \sin\theta \cos\psi & -\sin\psi & 0 \\ \cos\theta & 0 & 1 \end{bmatrix} \begin{bmatrix} \dot{\varphi} \\ \dot{\theta} \\ \dot{\psi} \end{bmatrix} \tag{12.353}$$

It provides three coupled differential equations to be solved for $\varphi(t)$, $\theta(t)$, $\psi(t)$:

$$\begin{aligned} \begin{bmatrix} \dot{\varphi} \\ \dot{\theta} \\ \dot{\psi} \end{bmatrix} &= \begin{bmatrix} \sin\theta \sin\psi & \cos\psi & 0 \\ \sin\theta \cos\psi & -\sin\psi & 0 \\ \cos\theta & 0 & 1 \end{bmatrix}^{-1} \begin{bmatrix} \omega_0 \cos\Omega t \\ \omega_0 \sin\Omega t \\ \omega_0 \cos\alpha \end{bmatrix} \\ &= \begin{bmatrix} \dfrac{\omega_0}{\sin\theta} \sin(\psi + \Omega t) \\ \omega_0 \cos(\psi + \Omega t) \\ \omega_0 \cos\alpha - \omega_0 \cot\theta \sin(\psi + \Omega t) \end{bmatrix} \end{aligned} \tag{12.354}$$

Example 741 ★ Dynamic Discussion and Solution of Example 740 A person working in dynamics is called a dynamician. The duties of a dynamician are modeling, determining the equations of motion, interpreting the solution of equations of motion, and predicting or designing the parameters to obtain a desired solution. In this process, obtaining the solutions is the job of mathematicians. However, mathematicians provide us some tools, algorithms, methods, and sample solved problems to be used for analyzing new problems. As an illustrative example, we examine the results of Example 740.

The set of Equations (12.354) are nonlinear and coupled, with no known general solution. Many dynamic problems of rigid body reach this step or they stop at the starting point where the Euler equations cannot be decoupled. Numerical or approximate solutions are the best methods to evaluate the behavior of a system. Approximate solution methods may be based on linearization, perturbations, or short- or long-term asymptotes. However, none are better that analytic solutions. The best solution to be used to analyze, approximate, predict, examine, and design is the analytic one. Analytic solutions provide a big picture of the behavior of a dynamic system, with clear images of the relative importance of parameters. All approximate solutions have blind spots and none can provide a complete solution.

Let us employ a numerical solution and briefly examine the behavior of the disc of Example 740 by analyzing $\varphi(t)$, $\theta(t)$, $\psi(t)$. Interestingly, the mass and mass moment of the disc are not involved in the differential equations for orientation angles. Substituting (12.352) in (12.354) shows that there are only two parameters involved, ω_0 and α:

$$\begin{bmatrix} \dot{\varphi} \\ \dot{\theta} \\ \dot{\psi} \end{bmatrix} = \begin{bmatrix} \dfrac{\omega_0}{\sin\theta} \sin(\psi + \omega_0 t \cos\alpha) \\ \omega_0 \cos(\psi + \omega_0 t \cos\alpha) \\ \omega_0 \cos\alpha - \omega_0 \cot\theta \sin(\psi + \omega_0 t \cos\alpha) \end{bmatrix} \tag{12.355}$$

For simplicity, we assume

$$\alpha = 0 \tag{12.356}$$

to turn the disc about its axis of symmetry and simplify the equations:

$$\begin{bmatrix} \dot{\varphi} \\ \dot{\theta} \\ \dot{\psi} \end{bmatrix} = \begin{bmatrix} \dfrac{\omega_0}{\sin\theta} \sin(\psi + \omega_0 t) \\ \omega_0 \cos(\psi + \omega_0 t) \\ \omega_0 - \omega_0 \cot\theta \sin(\psi + \omega_0 t) \end{bmatrix} \tag{12.357}$$

Let us set ω_0 to a constant value:

$$\omega_0 = 10 \text{ rad/s} \tag{12.358}$$

Without loss of generality, we may set the initial conditions of φ and ψ to zero:

$$\varphi(0) = 0 \qquad \psi(0) = 0 \tag{12.359}$$

Having $\alpha = 0$ makes the equations singular at $\theta = 0$. The angles φ and ψ are not distinguishable at $\theta = 0$ and the disc can stay in that position regardless of the value of ω_0. So, let us set the initial θ to a small value and examine the behavior of the disc,

$$\theta(0) = 0.1 \text{ rad} \approx 5.7 \text{ deg} \tag{12.360}$$

and find the responses of $\varphi(t)$, $\theta(t)$, $\psi(t)$ as shown in Figures 12.19–12.21, respectively. All of these figures show a periodic behavior. The periodic behavior of $\theta(t)$ indicates that the disc will tilt up to a maximum value and return to $\theta = 0$. Figure 12.21 shows that starting with nonzero $\dot{\psi}(0) = \omega_0 = 10\,\text{rad/s}$ generates a positive periodic behavior for $\dot{\psi}$. To make sure that the repeating behaviors of $\varphi(t)$ and $\theta(t)$ are periodic, we may plot the phase portraits of the variables. To make sure that the periods of the repeating behavior of the variables are equal, we may plot the variables with respect to each other. As an example, we show φ versus θ in Figure 12.22. One closed loop indicates an equal period.

Now we may change the values of the initial condition $\theta(0)$ to see the reaction of the disc to these different values. Figures 12.23–12.26 illustrate the same coordinates as Figures 12.19–12.22 and compare the dynamic behavior for different initial conditions. The time responses of the variables become smoother by increasing $\theta(0)$. Less fluctuation in Figure 12.25 for $\dot{\psi}(t)$ along with decreasing the maximum deviation of $\varphi(t)$ and $\theta(t)$ from zero suggests that we may have a mirror behavior for $\theta(0) > 90\,\text{deg}$. Figure 12.26 shows that by increasing $\theta(0)$ the periodic behaviors of $\varphi(t)$ and $\theta(t)$ become more similar.

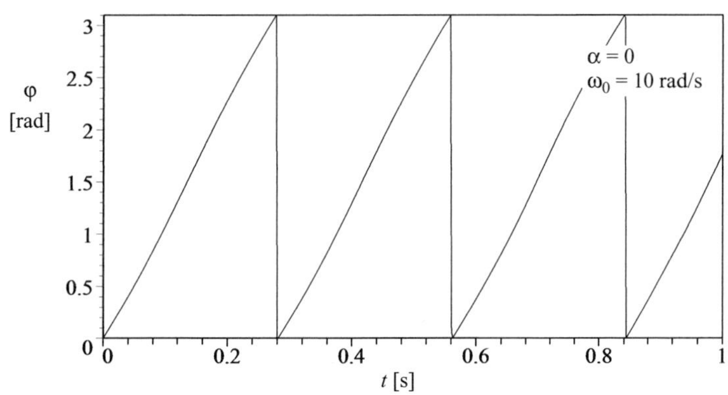

Figure 12.19 Time response of $\varphi(t)$ for $\theta(0) = 0.1\,\text{rad}$.

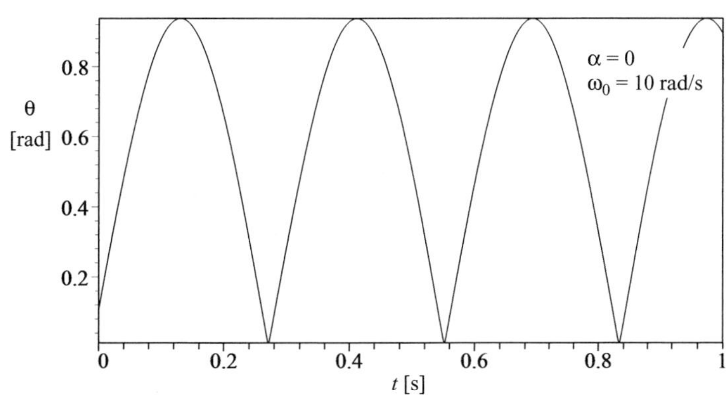

Figure 12.20 Time response of $\theta(t)$ for $\theta(0) = 0.1\,\text{rad}$.

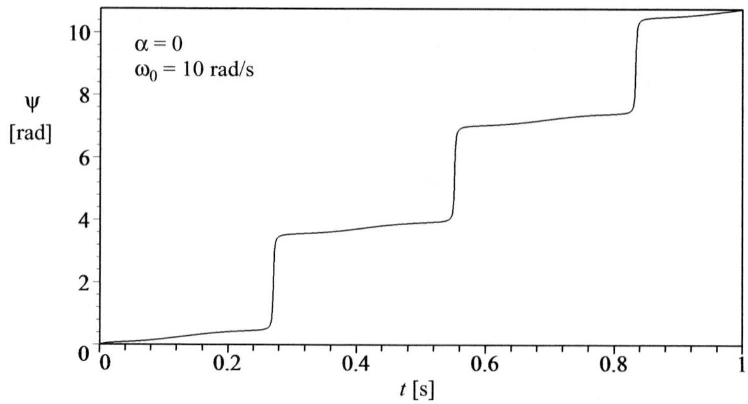

Figure 12.21 Time response of $\psi(t)$ for $\theta(0) = 0.1$ rad.

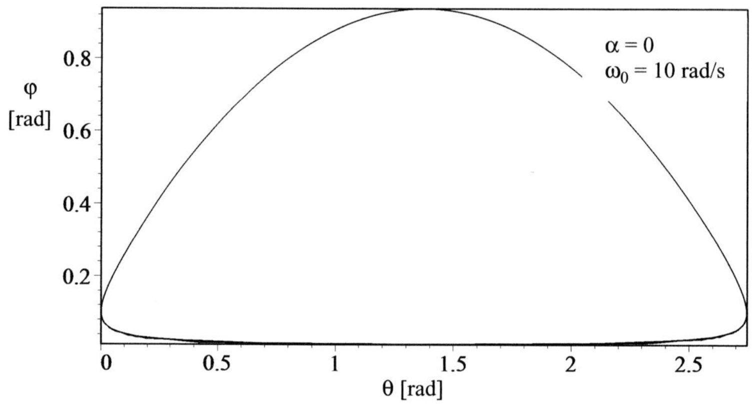

Figure 12.22 Time response of $\varphi(t)$ versus $\theta(t)$ for $\theta(0) = 0.1$ rad.

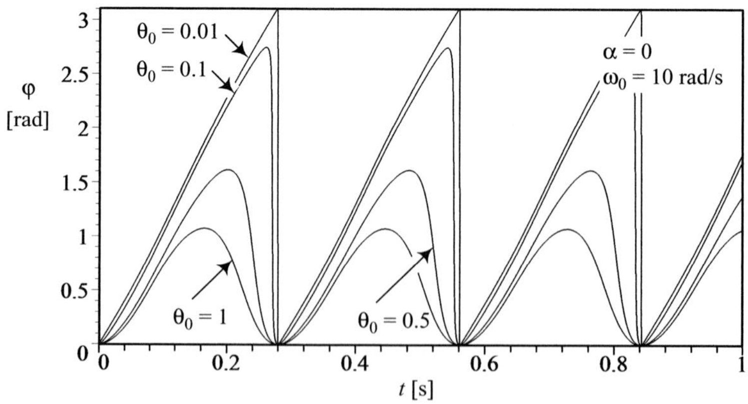

Figure 12.23 Time response of $\varphi(t)$ for different $\theta(0)$.

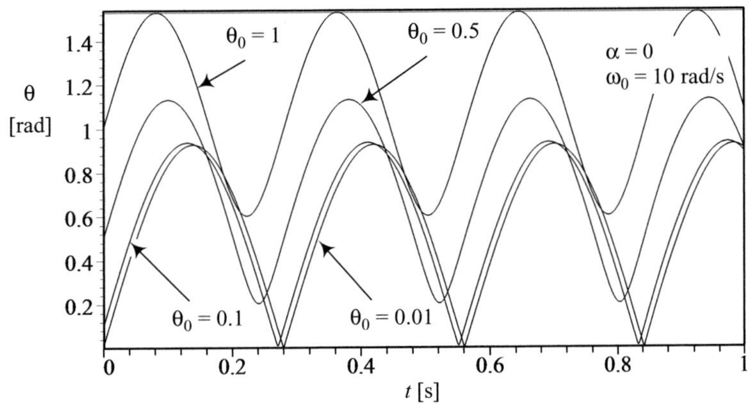

Figure 12.24 Time response of $\theta\,(t)$ for different $\theta\,(0)$.

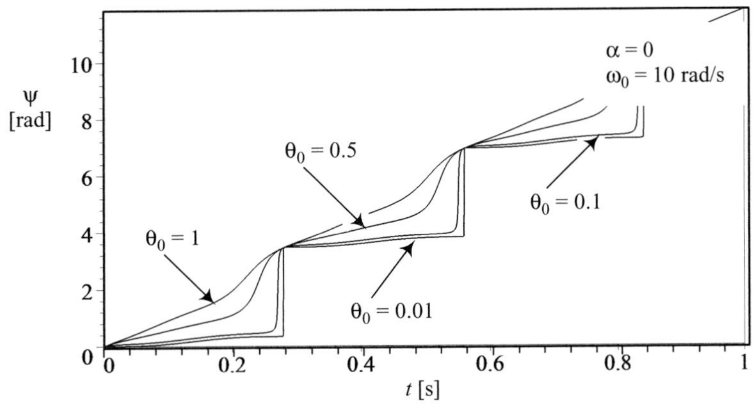

Figure 12.25 Time response of $\psi\,(t)$ versus $\theta\,(0)$.

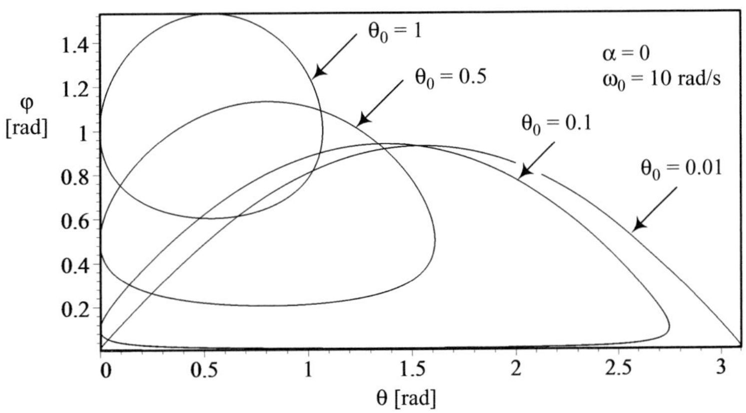

Figure 12.26 Time response of $\varphi\,(t)$ versus $\theta\,(t)$ for different $\theta\,(0)$.

The set of equations (12.357) suggests that increasing the parameter ω_0 will produce the same type of behavior with different amplitude and smaller period. Figures 12.27–12.30 illustrate the same behaviors as Figures 12.19–12.22 for

$$\omega_0 = 20\,\text{rad/s} \tag{12.361}$$

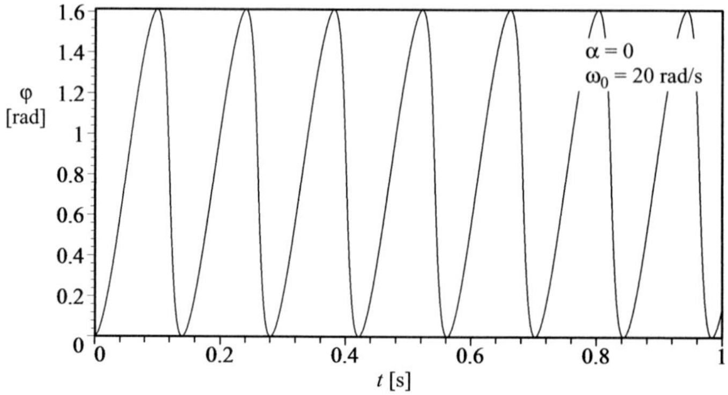

Figure 12.27 Time response of $\varphi\,(t)$ for $\theta\,(0) = 0.1\,\text{rad}$.

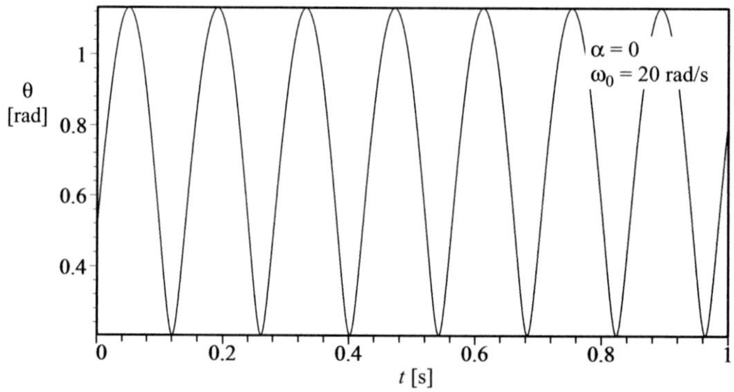

Figure 12.28 Time response of $\theta\,(t)$ for $\theta\,(0) = 0.1\,\text{rad}$.

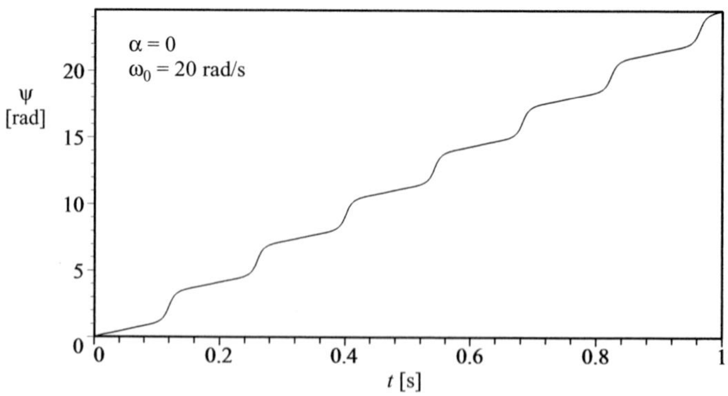

Figure 12.29 Time response of $\psi\,(t)$ for $\theta\,(0) = 0.1\,\text{rad}$.

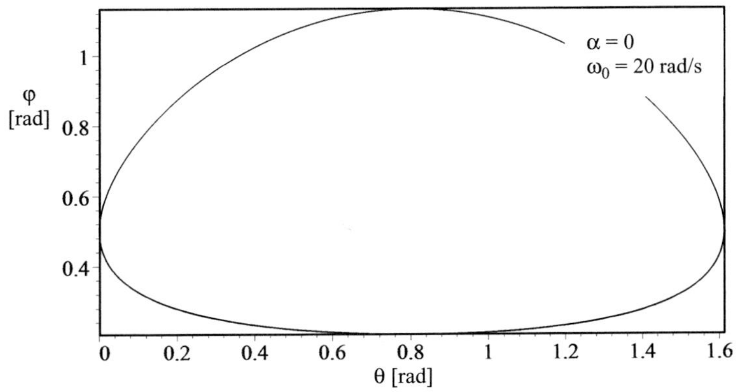

Figure 12.30 Time response of $\varphi(t)$ versus $\theta(t)$ for $\theta(0) = 0.1$ rad.

They show that increasing ω_0 decreases the amplitudes and periods of variables and increases their smoothness.

If the behavior of the system is clear for $\alpha = 0$ and different initial conditions and spins, then it is time to examine the variables $\varphi(t)$, $\theta(t)$, $\psi(t)$ for $\alpha \neq 0$. This is left for the reader to do.

Example 742 ★ Variable Mass Moments Consider a symmetric rigid body with $I_1 = I_2$. Assume that the elements of a principal mass moment matrix are variable such that the product mass moments remain zero:

$$ {}^B\dot{I} = \begin{bmatrix} \dot{I}_1 & 0 & 0 \\ 0 & \dot{I}_1 & 0 \\ 0 & 0 & \dot{I}_3 \end{bmatrix} \tag{12.362}$$

The torque-free Euler equations become

$$ \frac{{}^G d}{dt}\,{}^B\mathbf{L} = {}^BI\,{}^B_G\dot{\boldsymbol{\omega}}_B + {}^B\dot{I}\,{}^B_G\boldsymbol{\omega}_B + {}^B_G\boldsymbol{\omega}_B \times \left({}^BI\,{}^B_G\boldsymbol{\omega}_B\right) = 0 \tag{12.363}$$

or

$$ I_1\dot{\omega}_1 + \dot{I}_1\omega_1 + (I_3 - I_1)\,\omega_2\omega_3 = 0 \tag{12.364}$$

$$ I_1\dot{\omega}_2 + \dot{I}_1\omega_2 - (I_3 - I_1)\,\omega_3\omega_1 = 0 \tag{12.365}$$

$$ I_3\dot{\omega}_3 + \dot{I}_3\omega_3 = 0 \tag{12.366}$$

Equation (12.366) is a total differential and can be integrated:

$$ I_3\omega_3 = C_1 \tag{12.367}$$

Multiplying (12.364) by ω_1 and (12.365) by ω_2 and adding them lead to

$$ I_1\dot{\omega}_\theta + \dot{I}_1\omega_\theta = 0 \tag{12.368}$$

$$ \omega_\theta = \sqrt{\omega_1^2 + \omega_2^2} \tag{12.369}$$

which can be integrated:

$$I_1 \omega_\theta = C_2 \tag{12.370}$$

Equations (12.367) and (12.370) are components of angular momentum ${}^B\mathbf{L}$ and show that the angle of nutation θ remains constant when the mass moments change as in (12.362).

Introducing an angle α such that

$$\frac{d\alpha}{dt} = \frac{I_3 - I_1}{I_1} \omega_3 \tag{12.371}$$

we can combine Equations (12.364) and (12.365) to obtain a second-order differential equation for $I_1 \omega_1$:

$$\frac{d^2 (I_1 \omega_1)}{d\alpha^2} + I_1 \omega_1 = 0 \tag{12.372}$$

The solution provides

$$\omega_1 = \omega_\theta \cos \alpha \tag{12.373}$$

$$\omega_2 = \omega_\theta \sin \alpha \tag{12.374}$$

with

$$\alpha = \int \frac{I_3 - I_1}{I_1} \omega_3 \, dt \tag{12.375}$$

and

$$\omega_\theta = \frac{I_1 (t_0)}{I_1} \omega_\theta (t_0) = \frac{I_{1_0}}{I_1} \omega_{\theta_0} \tag{12.376}$$

Example 743 ★ Application of Variable Mass Moments Satellites usually have long appendages, such as antennas, that should be opened in space. Furthermore, we attach extensible arms to satellites to control their rotation by changing the length of the arms. An extensible member of a satellite is called a boom. Depending on the rigidity of the extended member, we may assume the satellite and its booms are rigid bodies or flexible solid bodies.

Let us consider an axisymmetric satellite with

$${}^B I = \begin{bmatrix} 200 & 0 & 0 \\ 0 & 100 & 0 \\ 0 & 0 & 100 \end{bmatrix} \text{kg m}^2 \tag{12.377}$$

Four particles each of mass $m = 6$ kg uniformly extend radially from the mass center of the satellite. The final length of the extended arms is $l = 50$ m from the mass center. The satellite was initially turning such that θ and its angular momentum are

$$\theta = 6 \text{ deg} \tag{12.378}$$

$$L = 1000 \text{ kg m}^2 \text{ rad}^2/\text{s}^2 \tag{12.379}$$

where θ is the angle between **L** and the z-axis. The rotational kinetic energy of the satellite is

$$K_0 = \frac{1}{2}\left(I_1\omega_{z0}^2 + I_3\omega_{\theta 0}^2\right) = \frac{L^2}{2}\left(\frac{\cos^2\theta}{I_{30}} + \frac{\sin^2\theta}{I_{10}}\right)$$

$$= \frac{1000^2}{2}\left(\frac{\cos^2\theta}{200} + \frac{\sin^2\theta}{100}\right) = 2527.3\,\text{J} \tag{12.380}$$

The new mass moments of the satellite are

$$I_1 = I_{10} + 2\left(\tfrac{1}{3}\right) ml^2 = 10{,}200 \text{ kg m}^2 \tag{12.381}$$

$$I_3 = I_{30} + 4\left(\tfrac{1}{3}\right) ml^2 = 20{,}100 \text{ kg m}^2 \tag{12.382}$$

The components of the new angular velocity are

$$\omega_z = \frac{L\cos\theta}{I_3} = \frac{1000\cos 6}{20100} = 0.049479\,\text{rad/s} \tag{12.383}$$

$$\omega_\theta = \frac{L\sin\theta}{I_1} = \frac{1000\sin 6}{10200} = 0.010248\,\text{rad/s} \tag{12.384}$$

and the new kinetic energy of the satellite is

$$K = \frac{1}{2}\left(I_1\omega_z^2 + I_3\omega_z^2\right) = \frac{L^2}{2}\left(\frac{\cos^2\theta}{I_3} + \frac{\sin^2\theta}{I_1}\right)$$

$$= \frac{1000}{2}\left(\frac{\cos^2\theta}{20{,}100} + \frac{\sin^2\theta}{10{,}200}\right) = 25.139\text{J} \tag{12.385}$$

The difference in kinetic energy is due to the internal friction and braking reaction of the motors.

Example 744 A Spherical Neutron Star Consider a neutron star with radius R whose surface vibrates slowly such that its principal mass moments are harmonic functions of time t:

$$I_1 = I_2 = \tfrac{2}{5} mR^2 \left(1 + \epsilon\cos\Omega t\right) \tag{12.386}$$

$$I_3 = \tfrac{2}{5} mR^2 \left(1 - \tfrac{1}{2}\epsilon\cos\Omega t\right) \tag{12.387}$$

$$\epsilon \ll mR^2 \tag{12.388}$$

Substituting these mass moments in Euler equations for a variable mass moment body,

$$\frac{d}{dt}\left(I_1\omega_1\right) + \left(I_3 - I_1\right)\omega_2\omega_3 = 0 \tag{12.389}$$

$$\frac{d}{dt}\left(I_1\omega_2\right) - \left(I_3 - I_1\right)\omega_3\omega_1 = 0 \tag{12.390}$$

$$\frac{d}{dt}\left(I_3\omega_3\right) = 0 \tag{12.391}$$

provides

$$\frac{d}{dt}(I_1\omega_1) - \tfrac{3}{2}I_0\epsilon\omega_2\omega_3 \cos\Omega t = 0 \tag{12.392}$$

$$\frac{d}{dt}(I_1\omega_2) + \tfrac{3}{2}I_0\epsilon\omega_3\omega_1 \cos\Omega t = 0 \tag{12.393}$$

$$\frac{d}{dt}(I_3\omega_3) = 0 \tag{12.394}$$

$$I_0 = \tfrac{2}{5}mR^2 \tag{12.395}$$

Equation (12.394) has the solution

$$\omega_3 = \frac{\omega_{30}}{1 - \epsilon \sin\Omega t} \tag{12.396}$$

where ω_3 is a weakly time dependent function.

If the frequency of expansion of the star is much less than the third component of angular velocity,

$$\Omega \ll \omega_3 \tag{12.397}$$

then

$$\frac{d}{dt}I_1 = -\tfrac{2}{5}mR^2\Omega\epsilon \sin\Omega t \approx 0 \tag{12.398}$$

and we have

$$I_1\dot{\omega}_1 - \tfrac{3}{2}I_0\epsilon\omega_2\omega_3 \cos\Omega t = 0 \tag{12.399}$$

$$I_1\dot{\omega}_2 + \tfrac{3}{2}I_0\epsilon\omega_3\omega_1 \cos\Omega t = 0 \tag{12.400}$$

Substituting ω_3 from (12.396), we have

$$I_1\dot{\omega}_1 - \frac{3}{2}I_0\epsilon\frac{\omega_{30}\cos\Omega t}{1 - \epsilon\sin\Omega t}\omega_2 = 0 \tag{12.401}$$

$$I_1\dot{\omega}_2 + \frac{3}{2}I_0\epsilon\frac{\omega_{30}\cos\Omega t}{1 - \epsilon\sin\Omega t}\omega_1 = 0 \tag{12.402}$$

12.4.4 ★ Asymmetric Torque-Free Rigid Body

A rigid body is called *asymmetric* if

$$I_1 \neq I_2 \neq I_3 \tag{12.403}$$

When there is no external torque on a rigid body, the B-expression of the Euler equation of rotational motion (12.2) reduces to

$${}^{B}I\ {}_{G}^{B}\dot{\boldsymbol{\omega}}_B + {}_{G}^{B}\boldsymbol{\omega}_B \times {}^{B}I\ {}_{G}^{B}\boldsymbol{\omega}_B = 0 \tag{12.404}$$

which provides three scalar equations in the principal coordinate frame:

$$I_1\dot{\omega}_1 - (I_2 - I_3)\,\omega_2\omega_3 = 0 \tag{12.405}$$

$$I_2\dot{\omega}_2 - (I_3 - I_1)\,\omega_3\omega_1 = 0 \tag{12.406}$$

$$I_3\dot{\omega}_3 - (I_1 - I_2)\,\omega_1\omega_2 = 0 \tag{12.407}$$

Employing the conservation of energy and angular momentum,

$$I_1\omega_1^2 + I_2\omega_2^2 + I_3\omega_3^2 = 2K \tag{12.408}$$

$$I_1^2\omega_1^2 + I_2^2\omega_2^2 + I_3^2\omega_3^2 = L^2 = 2DK \tag{12.409}$$

where

$$D = \frac{I_1^2\omega_1^2 + I_2^2\omega_2^2 + I_3^2\omega_3^2}{I_1\omega_1^2 + I_2\omega_2^2 + I_3\omega_3^2} \tag{12.410}$$

the set of Euler equations reduces to a differential equation to determine ω_2:

$$\dot{\omega}_2 = \sqrt{\frac{(I_1 - I_2)(I_2 - I_3)}{I_1 I_3}\left(a^2 - \omega_2^2\right)\left(b^2 - \omega_2^2\right)} \tag{12.411}$$

$$a^2 = \frac{2K(D - I_3)}{I_2(I_2 - I_3)} \qquad b^2 = \frac{2K(I_1 - D)}{I_2(I_1 - I_2)} \tag{12.412}$$

Assuming $I_1 > I_2 > I_3$, the solution of this equation depends on the relative values of D and I_2.

If $D < I_2$, then

$$\omega_1 = \sqrt{\frac{L^2 - 2KI_3}{I_1(I_1 - I_3)}}\,\mathrm{cn}(\tau, k) \tag{12.413}$$

$$\omega_2 = \sqrt{\frac{L^2 - 2KI_3}{I_2(I_2 - I_3)}}\,\mathrm{sn}(\tau, k) \tag{12.414}$$

$$\omega_3 = -\sqrt{\frac{2KI_1 - L^2}{I_3(I_1 - I_3)}}\,\mathrm{dn}(\tau, k) \tag{12.415}$$

If $D = I_2$, then

$$\omega_1 = \sqrt{\frac{2K(I_2 - I_3)}{I_1(I_1 - I_3)}}\,\frac{1}{\cosh\tau} \tag{12.416}$$

$$\omega_2 = \sqrt{\frac{2K}{I_2}}\,\tanh\tau \tag{12.417}$$

$$\omega_3 = \sqrt{\frac{2K(I_1 - I_2)}{I_3(I_1 - I_3)}}\,\frac{1}{\cosh\tau} \tag{12.418}$$

If $D > I_2$, then

$$\omega_1 = \sqrt{\frac{L^2 - 2KI_3}{I_1 (I_1 - I_3)}} \operatorname{dn}(\tau, k) \tag{12.419}$$

$$\omega_2 = \sqrt{\frac{2KI_1 - L^2}{I_2 (I_1 - I_2)}} \operatorname{sn}(\tau, k) \tag{12.420}$$

$$\omega_3 = -\sqrt{\frac{2KI_1 - L^2}{I_3 (I_1 - I_3)}} \operatorname{cn}(\tau, k) \tag{12.421}$$

where $\operatorname{sn}(\tau, k)$ is the Jacobi elliptic function of τ, k:

$$k = \frac{D - I_3}{D - I_1} \frac{I_1 - I_2}{I_2 - I_3} \tag{12.422}$$

$$\tau = (t - t_0) \sqrt{\frac{(I_1 - I_2)\left(L^2 - 2KI_3\right)}{I_1 I_2 I_3}} \tag{12.423}$$

$$y = \operatorname{sn}(t, k) = \sin(\varphi) \tag{12.424}$$

$$F(y, k) = \int_0^{\varphi} \frac{d\varphi}{\sqrt{1 - k^2 \sin^2 \varphi}} \tag{12.425}$$

Proof: The solution of the Euler equation of rotational motion (12.404) is based on two integrals of motion of kinetic energy K and angular momentum L. The expanded forms of the integrals of motion in the principal frame are

$$\sum_{i=1}^{3} I_i \omega_i^2 = 2K = I_1 \omega_1^2 + I_2 \omega_2^2 + I_3 \omega_3^2 \tag{12.426}$$

$$\sum_{i=1}^{3} I_i^2 \omega_i^2 = L^2 = I_1^2 \omega_1^2 + I_2^2 \omega_2^2 + I_3^2 \omega_3^2 = 2DK \tag{12.427}$$

$$D = \frac{I_1^2 \omega_1^2 + I_2^2 \omega_2^2 + I_3^2 \omega_3^2}{I_1 \omega_1^2 + I_2 \omega_2^2 + I_3 \omega_3^2} \tag{12.428}$$

The parameter D has the same physical dimension as I and is introduced to simplify the equations. Assuming

$$I_1 > I_2 > I_3 \tag{12.429}$$

we can transform the integrals of motion (12.426)–(12.427) to the following forms by multiplying Equation (12.426) with I_1 and with I_3, respectively, and subtracting both equations separately from (12.427):

$$I_1 - D = \frac{I_2 (I_1 - I_2) \omega_2^2 + I_3 (I_1 - I_3) \omega_3^2}{2K} \tag{12.430}$$

$$D - I_3 = \frac{I_1 (I_1 - I_3) \omega_1^2 + I_2 (I_2 - I_3) \omega_2^2}{2K} \tag{12.431}$$

Now, instead of using the three Euler equations (12.405)–(12.407), we use the second Euler equation (12.406) along with the new forms of integrals of motion (12.430) and (12.431). Solving these equations for ω_3 and ω_1 provides

$$\omega_3^2 = \frac{I_2 (I_1 - I_2)}{I_3 (I_1 - I_3)} \left(b^2 - \omega_2^2\right) \tag{12.432}$$

$$b^2 = \frac{2K (I_1 - D)}{I_2 (I_1 - I_2)} \tag{12.433}$$

$$\omega_1^2 = \frac{I_2 (I_2 - I_3)}{I_1 (I_1 - I_3)} \left(a^2 - \omega_2^2\right) \tag{12.434}$$

$$a^2 = \frac{2K (D - I_3)}{I_2 (I_2 - I_3)} \tag{12.435}$$

The parameters a and b are nonnegative and are related to each other:

$$a^2 - b^2 = 2\frac{K}{I_2}\frac{D - I_2}{I_1 - I_2}\frac{I_1 - I_3}{I_2 - I_3} \tag{12.436}$$

Substituting (12.432) and (12.432) in the second Euler equation (12.406) provides a differential equation for ω_2:

$$\dot{\omega}_2 = \sqrt{\frac{(I_2 - I_3)}{I_1}\frac{(I_1 - I_2)}{I_3} \left(a^2 - \omega_2^2\right)\left(b^2 - \omega_2^2\right)} \tag{12.437}$$

The solution of the equation can be expressed by an integral:

$$\int \frac{d\omega_2}{\sqrt{\left(a^2 - \omega_2^2\right)\left(b^2 - \omega_2^2\right)}} = s_2 (t - t_0) \sqrt{\frac{(I_1 - I_2)(I_2 - I_3)}{I_1 I_3}} \tag{12.438}$$

The parameter s_2 indicates the resultant sign of the two square roots. It will be found from initial conditions.

The integral on the left-hand side is the elliptic integral introduced in Section 2.3.3. To reduce it to a normal form, we distinguish three cases:

1. $a^2 < b^2$ or $D < I_2$
2. $a^2 > b^2$ or $D > I_2$
3. $a^2 = b^2$ or $D = I_2$

Case 1. Let us rewrite Equation (12.438) in the form

$$\int \frac{1}{a}\frac{d\omega_2}{\sqrt{\left(1 - \omega_2^2/a^2\right)\left(1 - \omega_2^2/b^2\right)}} = s_2 b (t - t_0) \sqrt{\frac{(I_1 - I_2)(I_2 - I_3)}{I_1 I_3}} \tag{12.439}$$

To simplify the equation and clarify the effect of the integrals of motion, we may use the parameters

$$y = \frac{\omega_2}{a} \qquad k = \frac{a}{b} \tag{12.440}$$

$$\begin{aligned} \tau &= b\,(t - t_0)\sqrt{\frac{(I_1 - I_2)\,(I_2 - I_3)}{I_1 I_3}} \\ &= (t - t_0)\sqrt{\frac{2K\,(I_1 - D)\,(I_2 - I_3)}{I_1 I_2 I_3}} \end{aligned} \tag{12.441}$$

and write the equation in the familiar form of an elliptic integral:

$$\int \frac{dy}{\sqrt{\left(1 - y^2\right)\left(1 - k^2 y^2\right)}} = s_2 \tau \tag{12.442}$$

The solution of (12.442) is the Jacobi elliptic function of τ:

$$y = s_2 \operatorname{sn}(\tau, k) \tag{12.443}$$

or

$$\omega_2 = a y = s_2 \sqrt{\frac{2K\,(D - I_3)}{I_2\,(I_2 - I_3)}} \operatorname{sn}(\tau, k) \tag{12.444}$$

Substituting ω_2 in Equations (12.432) and (12.434),

$$\omega_3 = \frac{I_2\,(I_1 - I_2)}{I_3\,(I_1 - I_3)} \left[b^2 - \operatorname{sn}^2(\tau, k)\right] \tag{12.445}$$

$$\omega_1 = \frac{I_2\,(I_2 - I_3)}{I_1\,(I_1 - I_3)} \left[a^2 - \operatorname{sn}^2(\tau, k)\right] \tag{12.446}$$

and employing the identities

$$\operatorname{sn}^2(\tau, k) + \operatorname{cn}^2(\tau, k) = 1 \tag{12.447}$$

$$\operatorname{dn}^2(\tau, k) + k^2 \operatorname{sn}^2(\tau, k) = 1 \tag{12.448}$$

we obtain the solutions for ω_1 and ω_3:

$$\omega_1 = s_1 \sqrt{\frac{2K\,(D - I_3)}{I_1\,(I_1 - I_3)}} \operatorname{cn}(\tau, k) \tag{12.449}$$

$$\omega_3 = s_3 \sqrt{\frac{2K\,(I_1 - D)}{I_3\,(I_1 - I_3)}} \operatorname{dn}(\tau, k) \tag{12.450}$$

The parameters s_1 and s_3 are also signs of the square roots, which will be determined from initial conditions. Knowing that

$$\frac{d}{d\tau} \operatorname{sn}(\tau, k) = \operatorname{cn}(\tau, k) \operatorname{dn}(\tau, k) \tag{12.451}$$

yields

$$s_1 s_2 s_3 = -1 \tag{12.452}$$

There are four combinations of signs satisfying this relationship.

Case 2. Employing the same method as case 1, we find

$$\omega_1 = s_1 \sqrt{\frac{2K\,(D - I_3)}{I_1\,(I_1 - I_3)}}\,\mathrm{dn}(\tau, k) \tag{12.453}$$

$$\omega_2 = s_2 \sqrt{\frac{2K\,(I_1 - D)}{I_2\,(I_1 - I_2)}}\,\mathrm{sn}(\tau, k) \tag{12.454}$$

$$\omega_3 = s_3 \sqrt{\frac{2K\,(I_1 - D)}{I_3\,(I_1 - I_3)}}\,\mathrm{cn}(\tau, k) \tag{12.455}$$

where

$$k = \frac{a}{b} \tag{12.456}$$

$$\tau = (t - t_0)\sqrt{\frac{2K\,(I_1 - I_2)\,(D - I_3)}{I_1 I_2 I_3}} \tag{12.457}$$

Case 3. The solutions of cases 1 and 2 approach each other when $D \to I_2$, and we have

$$k = 1 \tag{12.458}$$

$$\omega_1 = s_1 \sqrt{\frac{2K\,(I_2 - I_3)}{I_1\,(I_1 - I_3)}}\,\frac{1}{\cosh \tau} \tag{12.459}$$

$$\omega_2 = s_2 \sqrt{\frac{2K}{I_2}}\,\tanh \tau \tag{12.460}$$

$$\omega_3 = s_3 \sqrt{\frac{2K\,(I_1 - I_2)}{I_3\,(I_1 - I_3)}}\,\frac{1}{\cosh \tau} \tag{12.461}$$

If $\tau \to \infty$, then $\omega_2 \to 0$, $\omega_3 \to 0$, and $\omega_2 \to s_2\sqrt{2K/I_2}$, which is a permanent rotation about the second principal axis.

Torque-free rigid body equations are applied to celestial bodies which are almost free of external torques. They also apply on a body which is suspended frictionless such that its mass center is stationary. Figure 12.17 illustrates an example of such a suspended rigid body. ■

***Example 745 ★ Orientation of** B **in** G* We are able to determine the time behavior of the components of ${}_G^B\boldsymbol{\omega}_B$. To determine the orientation of B in G for a given ${}_G^B\boldsymbol{\omega}_B$, we should select a set of generalized orientation coordinates and use the associated method of Example 718. Let us select the Euler angles as the generalized coordinates. Using the associated transformation matrix ${}^B R_G$ of Equation (4.142),

$$
\begin{aligned}
{}^B R_G &= R_{z,\psi} R_{x,\theta} R_{z,\varphi} \\
&= \begin{bmatrix} c\varphi c\psi - c\theta s\varphi s\psi & c\psi s\varphi + c\theta c\varphi s\psi & s\theta s\psi \\ -c\varphi s\psi - c\theta c\psi s\varphi & -s\varphi s\psi + c\theta c\varphi c\psi & s\theta c\psi \\ s\theta s\varphi & -c\varphi s\theta & c\theta \end{bmatrix}
\end{aligned} \tag{12.462}
$$

we can express ${}^B\mathbf{L}$ in global coordinates:

$$
{}^B\mathbf{L} = {}^B R_G \, {}^G\mathbf{L} \tag{12.463}
$$

Because of the conservation of angular momentum, ${}^B\mathbf{L}$ is fixed and hence ${}^G\mathbf{L}$ is also assigned a fixed axis in G:

$$
\left|{}^B\mathbf{L}\right| = \left|{}^G\mathbf{L}\right| = L \tag{12.464}
$$

If we select the Z-axis on ${}^G\mathbf{L}$, then

$$
\begin{bmatrix} I_1\omega_1 \\ I_2\omega_2 \\ I_3\omega_3 \end{bmatrix} = {}^B R_G \begin{bmatrix} 0 \\ 0 \\ L \end{bmatrix} = \begin{bmatrix} L\sin\theta\sin\psi \\ L\sin\theta\cos\psi \\ L\cos\theta \end{bmatrix} \tag{12.465}
$$

Two of the orientation angles can be found by comparison:

$$
\cos\theta = \frac{I_3\omega_3}{L} \tag{12.466}
$$

$$
\tan\psi = \frac{I_1\omega_1}{I_2\omega_2} \tag{12.467}
$$

To find φ, we may use Equation (4.188):

$$
\begin{bmatrix} \dot\varphi \\ \dot\theta \\ \dot\psi \end{bmatrix} = \frac{1}{\sin\theta} \begin{bmatrix} \sin\psi & \cos\psi & 0 \\ \sin\theta\cos\psi & -\sin\theta\sin\psi & 0 \\ -\cos\theta\sin\psi & -\cos\theta\cos\psi & 1 \end{bmatrix} \begin{bmatrix} \omega_1 \\ \omega_2 \\ \omega_3 \end{bmatrix} \tag{12.468}
$$

and determine $\dot\varphi$ as a function of components of ${}^B_G\boldsymbol{\omega}_B$:

$$
\dot\varphi = \omega_2 \frac{\cos\psi}{\sin\theta} + \omega_1 \frac{\sin\psi}{\sin\theta} \tag{12.469}
$$

Substituting ω_1 and ω_2 from (12.465), we obtain an equation to determine φ:

$$
\dot\varphi = \frac{L}{I_2}\cos^2\psi + \frac{L}{I_1}\sin^2\psi \tag{12.470}
$$

***Example 746 ★ Minimum and Maximum Values for** D* Satisfying both Equations (12.426) and (12.427) requires minimum and maximum values for D. Assuming

$I_1 > I_2 > I_3$, we can determine the limits of D by multiplying Equation (12.426) with I_1 and with I_3, respectively, and subtracting both equations separately from (12.427):

$$D = I_1 - \frac{I_2 (I_1 - I_2) \omega_2^2 + I_3 (I_1 - I_3) \omega_3^2}{I_1\omega_1^2 + I_2\omega_2^2 + I_3\omega_3^2} \tag{12.471}$$

$$D = I_3 - \frac{I_1 (I_1 - I_3) \omega_1^2 + I_2 (I_2 - I_3) \omega_2^2}{I_1\omega_1^2 + I_2\omega_2^2 + I_3\omega_3^2} \tag{12.472}$$

The fractions on the right-hand side of these equations are nonnegative. Therefore, the limits of D are

$$I_1 > D > I_3 \tag{12.473}$$

where I_1 and I_3 are the extreme values of D for which the equations of motion have real solution.

Example 747 ★ Polhode The integrals of motion (12.426) and (12.427) introduce two equations that the components of ${}_G^B\boldsymbol{\omega}_B$ must satisfy. These equations geometrically indicate two intersecting ellipsoids in an orthogonal frame $(\omega_1, \omega_2, \omega_3)$. The intersection of the ellipsoids is called a *polhode* and indicates the path of the tip point of $\boldsymbol{\omega}$.

Example 748 ★ Poinsot Geometric Interpretation of Motion The energy integral of motion (12.426) indicates that the inner product of angular velocity ${}_G^B\boldsymbol{\omega}_B$ and angular momentum ${}^B\mathbf{L} = {}^BI\ {}_G^B\boldsymbol{\omega}_B$ is constant:

$${}_G^B\boldsymbol{\omega}_B \cdot {}^B\mathbf{L} = 2K = I_1\omega_1^2 + I_2\omega_2^2 + I_3\omega_3^2 \tag{12.474}$$

The angular momentum integral of motion (12.427) states that the inner product of angular momentum ${}^B\mathbf{L} = {}^BI\ {}_G^B\boldsymbol{\omega}_B$ by itself is constant:

$${}^B\mathbf{L} \cdot {}^B\mathbf{L} = L^2 = I_1^2\omega_1^2 + I_2^2\omega_2^2 + I_3^2\omega_3^2 \tag{12.475}$$

which shows that the magnitude and direction of ${}^B\mathbf{L}$ are both constant. The constant vector ${}^B\mathbf{L}$ indicates an invariant plane in the B-frame. The projection of ${}_G^B\boldsymbol{\omega}_B$ on the invariant direction of ${}^B\mathbf{L}$ is constant. It follows that any increment $\Delta\boldsymbol{\omega}$ between two arbitrary moments of time is perpendicular to ${}^B\mathbf{L}$:

$${}^B\mathbf{L} \cdot \Delta\,{}_G^B\boldsymbol{\omega}_B = 0$$

This equation defines the invariant plane. This geometric interpretation of the integrals of motion and the invariant plane is illustrated in Figure 12.31.

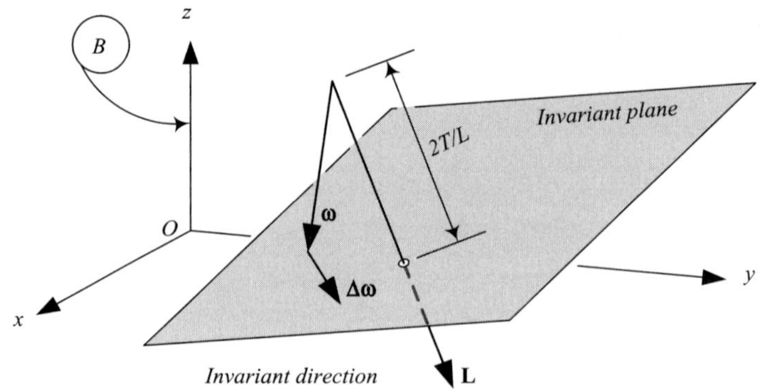

Figure 12.31 Poinsot geometric interpretation of possible angular velocity and momentum.

Example 749 ★ Duffing Equation Representation Let us take a time derivative of torque-free Euler equations (12.405)–(12.407):

$$\ddot{\omega}_1 = b_1 \left(\dot{\omega}_2 \omega_3 + \omega_2 \dot{\omega}_3\right) \tag{12.476}$$

$$\ddot{\omega}_2 = b_2 \left(\dot{\omega}_3 \omega_1 + \omega_3 \dot{\omega}_1\right) \tag{12.477}$$

$$\dot{\omega}_3 = b_3 \left(\dot{\omega}_1 \omega_2 + \omega_1 \dot{\omega}_2\right) \tag{12.478}$$

$$b_1 = \frac{I_2 - I_3}{I_1} \qquad b_2 = \frac{I_3 - I_1}{I_2} \qquad b_3 = \frac{I_1 - I_2}{I_3} \tag{12.479}$$

Substitution of $\dot{\omega}_1, \dot{\omega}_2, \dot{\omega}_3$ from (12.405)–(12.407) provides

$$\ddot{\omega}_1 = b_1 \omega_1 \left(b_2 \omega_3^2 + b_3 \omega_2^2\right) \tag{12.480}$$

$$\ddot{\omega}_2 = b_2 \omega_2 \left(b_3 \omega_1^2 + b_1 \omega_3^2\right) \tag{12.481}$$

$$\dot{\omega}_3 = b_3 \omega_3 \left(b_1 \omega_2^2 + b_2 \omega_1^2\right) \tag{12.482}$$

We can solve (12.474) and (12.475) for ω_2 and ω_3:

$$\omega_2^2 = \frac{2 I_3 K - L^2}{I_2 \left(I_3 - I_2\right)} - \frac{I_1 \left(I_3 - I_1\right)}{I_2 \left(I_3 - I_2\right)} \omega_1^2 \tag{12.483}$$

$$\omega_3^2 = \frac{2 I_2 K - L^2}{I_3 \left(I_2 - I_3\right)} - \frac{I_1 \left(I_2 - I_1\right)}{I_3 \left(I_2 - I_3\right)} \omega_1^2 \tag{12.484}$$

Substituting these equations in (12.480) yields

$$\ddot{\omega}_1 + \frac{\left(I_1 - I_2\right)\left(2 I_3 K - L^2\right) - \left(I_3 - I_1\right)\left(2 I_2 K - L^2\right)}{I_1 I_2 I_3} \omega_1 + \frac{2 \left(I_1 - I_2\right)\left(I_1 - I_3\right)}{I_2 I_3} \omega_1^3 = 0 \tag{12.485}$$

Similarly, we can obtain

$$\ddot{\omega}_2 + \frac{(I_2 - I_3)\left(2I_3K - L^2\right) - (I_1 - I_2)\left(2I_2K - L^2\right)}{I_1I_2I_3}\omega_2$$

$$+ \frac{2\,(I_2 - I_3)\,(I_2 - I_1)}{I_3I_1}\omega_2^3 = 0 \tag{12.486}$$

$$\ddot{\omega}_3 + \frac{(I_3 - I_1)\left(2I_3K - L^2\right) - (I_2 - I_3)\left(2I_2K - L^2\right)}{I_1I_2I_3}\omega_3$$

$$+ \frac{2\,(I_3 - I_1)\,(I_3 - I_2)}{I_1I_2}\omega_3^3 = 0 \tag{12.487}$$

Equations (12.485)–(12.487) are interdependent uncoupled Duffing equations with constant coefficients of the form

$$\ddot{u} + k_1u + k_3u^3 = 0 \tag{12.488}$$

They are interdependent because the parameters I_1, I_2, I_3, K, L must be consistent.

Let us rewrite the equations in the simpler forms

$$\ddot{\omega}_1 + P_1\omega_1 + Q_1\omega_1^3 = 0 \tag{12.489}$$

$$\ddot{\omega}_2 + P_2\omega_2 + Q_2\omega_2^3 = 0 \tag{12.490}$$

$$\ddot{\omega}_3 + P_3\omega_3 + Q_3\omega_3^3 = 0 \tag{12.491}$$

Because there is no term associated with $\dot{\omega}_i$, these equations are of the form (2.165) and have first integrals. Substituting $\ddot{\omega}_i = \dot{\omega}_i\, d\dot{\omega}_i/d\omega$, the first integrals of these equations are

$$\dot{\omega}_1^2 + \tfrac{1}{2}P_1\omega_1^2 + \tfrac{1}{4}Q_1\omega_1^4 = C_x \tag{12.492}$$

$$\dot{\omega}_2^2 + \tfrac{1}{2}P_2\omega_2^2 + \tfrac{1}{4}Q_2\omega_2^4 = C_y \tag{12.493}$$

$$\dot{\omega}_3^2 + \tfrac{1}{2}P_3\omega_3^2 + \tfrac{1}{4}Q_3\omega_3^4 = C_z \tag{12.494}$$

The integral constants C_x, C_y, C_z can be found by setting $\boldsymbol{\omega} = 0$ and using (12.413)–(12.415):

$$C_x = -\frac{\left(L^2 - 2KI_2\right)\left(L^2 - 2KI_3\right)}{I_1^2I_2I_3} \tag{12.495}$$

$$C_y = -\frac{\left(L^2 - 2KI_3\right)\left(L^2 - 2KI_1\right)}{I_1I_2^2I_3} \tag{12.496}$$

$$C_z = -\frac{\left(L^2 - 2KI_1\right)\left(L^2 - 2KI_2\right)}{I_1I_2I_3^2} \tag{12.497}$$

Example 750 ★ Rotational Stability of a Rigid Body about Principal Axes Consider a rigid body in a principal coordinate frame,

$$I = \begin{bmatrix} I_1 & 0 & 0 \\ 0 & I_2 & 0 \\ 0 & 0 & I_3 \end{bmatrix} \tag{12.498}$$

that is turning about the z-axis:

$${}_G^B\boldsymbol{\omega}_B = \begin{bmatrix} 0 \\ 0 \\ \omega_0 \end{bmatrix} \tag{12.499}$$

To examine the stability of this rotation, we perturb ${}_G^B\boldsymbol{\omega}_B$ by adding a small angular velocity $\delta\, {}_G^B\boldsymbol{\omega}_B = \begin{bmatrix} \epsilon_x & \epsilon_y & \epsilon_z \end{bmatrix}^T$:

$${}_G^B\boldsymbol{\omega}_B + \delta\, {}_G^B\boldsymbol{\omega}_B = \begin{bmatrix} \epsilon_x \\ \epsilon_y \\ \omega_0 + \epsilon_z \end{bmatrix} = \begin{bmatrix} \omega_1 \\ \omega_2 \\ \omega_3 \end{bmatrix} \tag{12.500}$$

For torque-free motion, the angular momentum remains constant and Equations (12.405)–(12.407) apply. Substituting the perturbed angular velocity in (12.407) and ignoring the nonlinear terms, we have

$$\dot{\omega}_3 = 0 \tag{12.501}$$

$$\omega_0 + \epsilon_z = \text{const} \tag{12.502}$$

From Equations (12.405) and (12.406), after ignoring the nonlinear terms, we find

$$\dot{\omega}_1 = \frac{I_2 - I_3}{I_1}\omega_0\omega_2 \tag{12.503}$$

$$\dot{\omega}_2 = \frac{I_3 - I_1}{I_2}\omega_0\omega_1 \tag{12.504}$$

Let us eliminate ω_1 between (12.503) and (12.504) to get a second-order differential equation with constant coefficients:

$$\dot{\omega}_2 + \frac{I_3 - I_1}{I_1}\frac{I_3 - I_2}{I_2}\omega_0^2\omega_2 = 0 \tag{12.505}$$

This equation provides a finite and hence stable solution if the coefficient of ω_2 is positive. Therefore, the rotation of a rigid body about a principal z-axis is stable if

$$I_3 > I_1 \qquad \text{and} \qquad I_3 > I_2 \tag{12.506}$$

or if

$$I_3 < I_1 \qquad \text{and} \qquad I_3 < I_2 \tag{12.507}$$

For

$$I_2 < I_3 < I_1 \qquad \text{or} \qquad I_1 > I_3 > I_2 \tag{12.508}$$

the solution of (12.505) is unstable and ω_2 increases with time.

To have a stable rotation of a rigid body about a principal axis, it must be about the axis of the maximum or minimum mass moment.

For example, Figure 12.32 illustrates a prismatic brick with the following mass moment matrix:

$$ {}^{B}I = \frac{m}{12}\begin{bmatrix} (w^2+h^2) & 0 & 0 \\ 0 & (h^2+l^2) & 0 \\ 0 & 0 & (l^2+w^2) \end{bmatrix} \tag{12.509} $$

$$ w < h < l \tag{12.510} $$

Torque-free rotation of the brick is stable only about the axes x_1 and x_2 with the minimum and maximum principal mass moments.

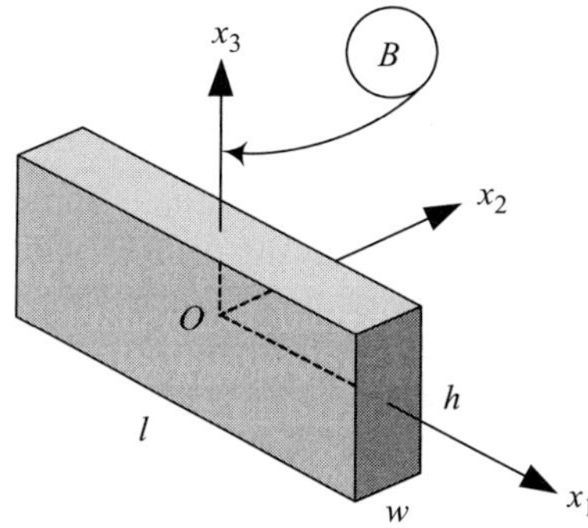

Figure 12.32 A prismatic brick with principal mass moments $I_1 < I_3 < I_2$.

Example 751 ★ Rotational Stability of a Rigid Body about Principal Axes The angular momentum of torque-free rotation of a solid body will remain constant; however, the kinetic energy of the body can decrease by internal damping. The stable attitude is when the kinetic energy is the smallest possible for a given angular momentum.

Consider a solid body that at a time $t = 0$ has the principal mass moment matrix

$$ I = \begin{bmatrix} I_1 & 0 & 0 \\ 0 & I_2 & 0 \\ 0 & 0 & I_3 \end{bmatrix} \tag{12.511} $$

and is turning with an angular velocity ${}^{B}_{G}\boldsymbol{\omega}_B$ and angular momentum ${}^{B}\mathbf{L}$:

$$ {}^{B}_{G}\boldsymbol{\omega}_B = \begin{bmatrix} \omega_{10} \\ \omega_{20} \\ \omega_{30} \end{bmatrix} \tag{12.512} $$

$$ {}^{B}\mathbf{L}_0 = \begin{bmatrix} I_1\omega_{10} \\ I_2\omega_{20} \\ I_3\omega_{30} \end{bmatrix} \qquad L_0^2 = \sqrt{I_1^2\omega_{10}^2 + I_2^2\omega_{20}^2 + I_3^2\omega_{30}^2} \tag{12.513} $$

The total mechanical energy of the solid body is its kinetic energy K, which at time $t = 0$ is

$$K_0 = \tfrac{1}{2}\left(I_1\omega_{10}^2 + I_2\omega_{20}^2 + I_3\omega_{30}^2\right) \tag{12.514}$$

Now we assume that the solid body is not rigid. So, it is made of real materials and its energy is dissipated during small deformations. The mass moment matrix of such a solid body in the principal coordinate frame $B\,(O123)$ of the undeformed solid body would be

$$I = \begin{bmatrix} I_1 & I_{12} & I_{13} \\ I_{21} & I_2 & I_{23} \\ I_{31} & I_{32} & I_3 \end{bmatrix} \tag{12.515}$$

Because the deformation of the solid body is assumed to be very small, the products of inertias are much smaller than the polar terms. So, we may ignore the changes in the principal mass moments I_1, I_2, I_3.

To have continuous dissipation of energy, let us assume that the products of inertias are harmonic functions of time. So deformation of the solid body is a result of small vibrations in the principal coordinate frame. Let us assume that at time $t = t_f$ the kinetic energy reaches the smallest possible value for the given L. At that time, no further deformation and dissipation occur. The mass moment matrix returns to its principal from, and the solid body becomes rigid with constant rotation about the principal axis. Let us assume this axis is the z-axis such that the final angular velocity is

$${}^B_G\boldsymbol{\omega}_B = \begin{bmatrix} 0 \\ 0 \\ \omega_f \end{bmatrix} \tag{12.516}$$

At $t \geq t_f$, the angular momentum must be colinear with ${}^B_G\boldsymbol{\omega}_B$:

$${}^B\mathbf{L} = \begin{bmatrix} 0 \\ 0 \\ I_3\omega_f \end{bmatrix} \tag{12.517}$$

Because during $0 \leq t \leq t_f$ no external torque is applied on the body, its angular momentum must remain constant:

$$\mathbf{L}_f = \mathbf{L}_0 \tag{12.518}$$

The conservation of angular momentum provides

$$\omega_f = \sqrt{\frac{I_1^2}{I_3^2}\omega_{10}^2 + \frac{I_2^2}{I_3^2}\omega_{20}^2 + \omega_{30}^2} \tag{12.519}$$

The final kinetic energy is

$$K_f = \frac{1}{2}I_3\omega_f^2 = \frac{1}{2}\left(\omega_{10}^2\frac{I_1}{I_3}I_1 + \omega_{20}^2\frac{I_2}{I_3}I_2 + \omega_{30}^2 I_3\right) \tag{12.520}$$

The final kinetic energy is assumed to have the minimum possible magnitude. This assumption is correct only if we have

$$I_3 > I_1 \qquad \text{and} \qquad I_3 > I_2 \tag{12.521}$$

From the stability conditions of rigid-body rotation (12.506) and (12.507), we know that the rotation is unstable if I_3 is the intermediate principal mass moment. Therefore, only a rotation about the principal axis of maximum mass moment is stable for a solid body with dissipating energy of small deformation.

The rotation of a torque-free solid body about one of the principal axes is stable only if the rotation is about the axis with the largest mass moment. Rotations about the axes of intermediate and minimum mass moments are unstable. Any other form of rotation is also unstable and decays toward the stable terminal rotation. When angular velocity and angular momentum vectors are colinear, there is no internal energy dissipation. When angular velocity and angular momentum vectors are not colinear, energy drawn from the rotating body and angular velocity change until $\boldsymbol{\omega}$ becomes colinear with $\mathbf{L}$ on the axis with the maximum mass moment.

12.4.5 General Motion

An unconstraint rigid body has six DOF and, hence, it needs six coordinates to specify its configuration. The traditional way to define the motion of a rigid body is to decouple the translational and rotational motions to analyze the motion of the mass center and the motion about the mass center. If a rigid body is under a force system $\left({}^G\mathbf{F},\ {}^G\mathbf{M}\right)$, then the G-expressions of the translational and rotational equations of motion of the body are

$$ {}^G\mathbf{F} = \frac{{}^G d}{dt}\ {}^G\mathbf{p} = m\ {}^G\mathbf{a}_B = m\ {}^G\dot{\mathbf{v}}_B \tag{12.522} $$

$$ {}^G\mathbf{M} = \frac{{}^G d}{dt}\ {}^G\mathbf{L} = {}^GI\ {}_G\boldsymbol{\alpha}_B = {}^GI\ {}_G\dot{\boldsymbol{\omega}}_B \tag{12.523} $$

We refer to the first one as the Newton equation of motion and the second one as the Euler equation of motion. In these equations, ${}^G\mathbf{F}$ is the G-expression of the resultant of all external forces at the mass center C of the body and ${}^G\mathbf{M}$ is the G-expression of the resultant of all external moments on the body. The vector ${}^G\mathbf{a}_B$ is the acceleration of point C, and ${}_G\boldsymbol{\alpha}_B$ is the angular acceleration of B in G. The constant m is the mass of the body and ${}^GI$ is the mass moment matrix of the body as calculated in G.

We attach a body coordinate frame $B\,(oxyz)$ to the rigid body at C and express the equations of motion in B:

$$ {}^B\mathbf{F} = m\ {}^B_G\mathbf{a}_B + m\ {}^B_G\boldsymbol{\omega}_B \times {}^B\mathbf{v}_B \tag{12.524} $$

$$ {}^B\mathbf{M} = {}^BI\ {}^B_G\dot{\boldsymbol{\omega}}_B + {}^B_G\boldsymbol{\omega}_B \times \left({}^BI\ {}^B_G\boldsymbol{\omega}_B\right) \tag{12.525} $$

Although the B- and G-expressions of equations of motion can equivalently describe the motion of a rigid body, the B-expression of the Euler equation has great advantage by having a constant mass moment matrix ${}^GI$.

Proof: Figure 12.33 illustrates a rigid body B that has a general motion in the global coordinate frame G. We attach a body frame B at a point o. Application of the Euler theorem about the rigid-body motion with a fixed point allows us to analyze the general

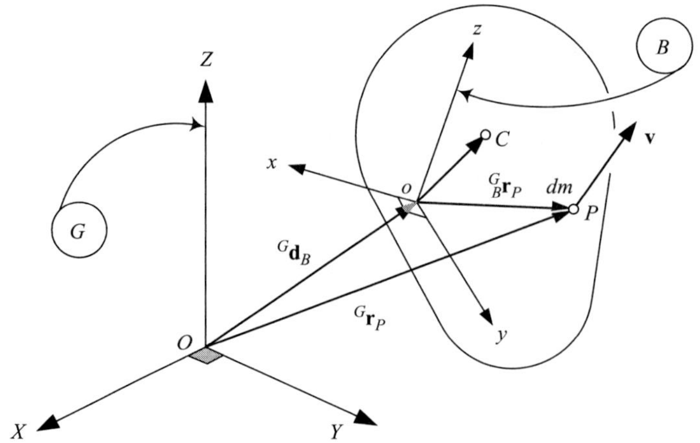

Figure 12.33 A rigid body B in a general motion.

motion of a rigid body by considering the motion of a point o plus the motion of the body about o. To show this, let us determine the kinetic energy of the body.

Point o is an arbitrary fixed point in B, and point C is the mass center of B. The velocity of a mass particle dm at P is

$$\begin{aligned} {}^G\mathbf{v}_P = {}^G\dot{\mathbf{r}}_P &= {}^G\dot{\mathbf{d}}_B + {}_G\boldsymbol{\omega}_B \times \left({}^G\mathbf{r}_P - {}^G\mathbf{d}_B\right) \\ &= {}^G\dot{\mathbf{d}}_B + {}_G\boldsymbol{\omega}_B \times {}^G_B\mathbf{r}_P \end{aligned} \tag{12.526}$$

and hence its kinetic energy is

$$dK = \tfrac{1}{2} dm\, {}^G\mathbf{v}_P \cdot {}^G\mathbf{v}_P \tag{12.527}$$

To determine the total kinetic energy, we integrate dT over the whole body:

$$\begin{aligned} K &= \frac{1}{2}\int_B {}^G\mathbf{v}_P^2\, dm \\ &= \frac{1}{2} m\, {}^G\dot{\mathbf{d}}_B^2 + m\, {}^G\dot{\mathbf{d}}_B \cdot \left({}_G\boldsymbol{\omega}_B \times {}^G_B\mathbf{r}_P\right) + \frac{1}{2}\, {}^G_B\boldsymbol{\omega}_B^{\mathrm{T}}\ {}^BI\ {}^G_B\boldsymbol{\omega}_B \end{aligned} \tag{12.528}$$

Moving the B-frame from o to C provides ${}^G_B\mathbf{r}_P = 0$ and simplifies K to

$$K = \frac{1}{2}\int_B {}^G\mathbf{v}_P^2\, dm = \frac{1}{2} m\, {}^G\dot{\mathbf{d}}_B^2 + \frac{1}{2}\, {}^G_B\boldsymbol{\omega}_B^{\mathrm{T}}\ {}^BI\ {}^G_B\boldsymbol{\omega}_B \tag{12.529}$$

By choosing the mass center C as the interesting point of a rigid body and as the origin of the body coordinate frame B, we decouple the translational and rotational motions of the rigid body. In translational dynamics, we ignore the rotational motion of the body and consider it as a particle at C with mass m. The G-expression of the Newton equation (2.24), which will be written as (12.522) to indicate the global coordinate frame, has the simplest expression. However, when the applied force is generated by the rigid body or moves with it, the B-expression of the Newton equation may be easier to work with. The translational equation of motion of a rigid body in the

B-frame (12.524) has been derived in Section 12.3. The B-expression of the Newton equation provides three second-order scalar differential equations:

$$\begin{bmatrix} F_x \\ F_y \\ F_z \end{bmatrix} = \begin{bmatrix} m[\dot{v}_x - (v_y\omega_z - v_z\omega_y)] \\ m[\dot{v}_y - (v_z\omega_x - v_x\omega_z)] \\ m[\dot{v}_z - (v_x\omega_y - v_y\omega_x)] \end{bmatrix} \tag{12.530}$$

Because in translational motion a rigid body is treated as a particle at its mass center, the orientation of the body coordinate frame will not affect the equations of motion. Therefore, there is no superior orientation for B.

The rotational equation of motion of a rigid body (12.523) has a simple form in the G-frame. However, the mass moment matrix is not necessarily constant in G. To have a constant $[I]$, we should transform the Euler equation to the B-frame, as given in (12.525). Furthermore, when the applied moment is generated by the rigid body or moves with it, the B-expression of the Euler equation is easier to work with. The rotational equation of motion of a rigid body in the B-frame (12.525) has been derived in Section 12.1. The B-expression of the Euler equation provides three second-order scalar differential equations. Because in rotational motion a rigid body is treated as a collection of rigidly connected particles, the orientation of the body coordinate frame will affect the equations of motion by changing $[I]$. Therefore, there is always a superior orientation for B in which the mass moment matrix becomes diagonal. Euler equations get the simplest form in such a principal coordinate frame:

$$\begin{bmatrix} M_1 \\ M_2 \\ M_3 \end{bmatrix} = \begin{bmatrix} I_1\dot{\omega}_1 - (I_2 - I_3)\,\omega_2\omega_3 \\ I_2\dot{\omega}_2 - (I_3 - I_1)\,\omega_3\omega_1 \\ I_3\dot{\omega}_3 - (I_1 - I_2)\,\omega_1\omega_2 \end{bmatrix} \tag{12.531}$$

The B-expression of the Euler equation is usually the best option to analyze the rotational motion of a rigid body.

The decoupling process of translation and rotation conceptually simplifies the analysis of motion of a rigid body. However, the equations of motion for translational and rotational motions are mathematically coupled and should be solved together. ■

Example 752 Rolling Disc on an Inclined Plane A homogeneous disc with mass m and radius R rolls without slipping on an inclined ground plane such that the plane of the disc remains perpendicular to the ground. Figure 12.34 illustrates the rolling disc and the required coordinate frames.

The ground plane is inclined by an angle α with respect to the horizontal plane. The global frame $G(\hat{I}, \hat{J}, \hat{K})$ is defined with gravitational acceleration $\mathbf{g} = -g\hat{K}$, as shown in the figure. To indicate the inclined ground, we attach a temporary coordinate frame G_1 $(OX_1Y_1Z_1)$ to the ground such that the Z_1-axis is set perpendicular to the inclined ground. The ground plane is indicated by axes X_1 and Y_1. Let us attach a principal coordinate frame $B(\hat{\imath}, \hat{\jmath}, \hat{k})$ to the mass center C of the disc such that the y-axis is perpendicular to the disc plane.

We also attach a frame B_1 $(Cx_1y_1z_1)$ to the disc at C such that y_1 and y are colinear and the (x_1, y_1)-plane is coplanar with the (x, y)-plane. The frame B_1 moves with the disc but does not turn about the disc axis. The (x, y)-plane is always parallel to the (X_1, Y_1)-plane.

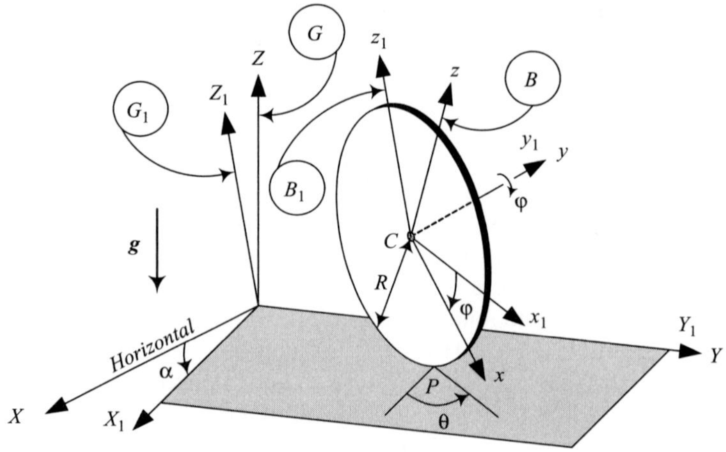

Figure 12.34 A rolling disc on an inclined plane.

The transformation between B and B_1 is a rotation φ about the y-axis:

$$^{B_1}R_B = \begin{bmatrix} \cos\varphi & 0 & \sin\varphi \\ 0 & 1 & 0 \\ -\sin\varphi & 0 & \cos\varphi \end{bmatrix} \tag{12.532}$$

The transformation between B_1 and G_1 is a rotation θ about the Z_1-axis:

$$^{G_1}R_{B_1} = \begin{bmatrix} \cos\theta & -\sin\theta & 0 \\ \sin\theta & \cos\theta & 0 \\ 0 & 0 & 1 \end{bmatrix} \tag{12.533}$$

The transformation between G_1 and G is a rotation α about the Y-axis:

$$^{G}R_{G_1} = \begin{bmatrix} \cos\alpha & 0 & \sin\alpha \\ 0 & 1 & 0 \\ -\sin\alpha & 0 & \cos\alpha \end{bmatrix} \tag{12.534}$$

There are two external forces acting on the disc: the gravitational force mg at C and the ground reaction force at the contact point P. We also assume there is a moment $\mathbf{M}$ that keeps the disc perpendicular to the ground. The moment $\mathbf{M}$ has a magnitude Q and is on the x_1-axis:

$$^{B_1}\mathbf{M} = -Q\,\hat{\imath}_1 \tag{12.535}$$

Let us show the reaction force at P by $\mathbf{F}$. It seems that $\mathbf{F}$ has the simplest expression in B_1. So, we assume that

$$\begin{aligned} ^{B_1}\mathbf{F} &= F_{x_1}\,{}^1\hat{\imath}_1 + F_{y_1}\,{}^1\hat{\jmath}_1 + F_{z_1}\,{}^1\hat{k}_1 \\ &= F_{x_1}\,\hat{\imath}_1 + F_{y_1}\,\hat{\jmath}_1 + F_{z_1}\,\hat{k}_1 \end{aligned} \tag{12.536}$$

We may drop the left superscript 1 from $\hat{\imath}_1, \hat{\jmath}_1, \hat{k}_1$ because the left superscript B_1 on $\mathbf{F}$ indicates the coordinate frame. The G- and B_1-expressions of the gravitational force **W** are

$$^{G}\mathbf{W} = -mg\,\hat{K} \tag{12.537}$$

$$\begin{aligned} ^{B_1}\mathbf{W} &= {}^{B_1}R_{G_1}\,{}^{G_1}R_G\,m\mathbf{g} = -mg\,{}^{G_1}R_{B_1}^{\mathrm{T}}\,{}^{G}R_{G_1}^{\mathrm{T}}\,\hat{K} \\ &= mg\begin{bmatrix} \cos\theta\sin\alpha \\ -\sin\theta\sin\alpha \\ -\cos\alpha \end{bmatrix} \end{aligned} \tag{12.538}$$

Showing the position vector of C by **r**,

$$^{G}\mathbf{r} = X\hat{I} + Y\hat{J} + (R\cos\alpha)\,\hat{K} \tag{12.539}$$

we have the translational equations of motion as

$$\begin{aligned} m\left(\ddot{X}\hat{I} + \ddot{Y}\hat{J} + \ddot{Z}\hat{K}\right) &= {}^{G}\mathbf{F} + {}^{G}\mathbf{W} \\ &= \begin{bmatrix} F_{x_1}\cos\theta\cos\alpha - F_{y_1}\cos\alpha\sin\theta + F_{z_1}\sin\alpha \\ F_{x_1}\sin\theta + F_{y_1}\cos\theta \\ F_{x_1}\cos\theta\sin\alpha + F_{y_1}\sin\theta\sin\alpha - F_{z_1}\cos\alpha - mg \end{bmatrix} \end{aligned} \tag{12.540}$$

because

$$^{G}\mathbf{F} = {}^{G}R_{B_1}\,{}^{B_1}\mathbf{F} \tag{12.541}$$

We may also express the translational equations of motion in the equivalent global frame G_1. Let us show the position vector of C by **r**,

$$^{G_1}\mathbf{r} = X_1\,\hat{I}_1 + Y_1\,\hat{J}_1 + R\,\hat{K}_1 \tag{12.542}$$

We have the translational equations of motion as

$$\begin{aligned} m\left(\ddot{X}_1\,{}^1\hat{I} + \ddot{Y}_1\,{}^1\hat{J}\right) &= {}^{G_1}\mathbf{F} + {}^{G_1}\mathbf{W} \\ &= \begin{bmatrix} F_{x_1}\cos\theta - F_{y_1}\sin\theta + gm\sin\alpha \\ F_{y_1}\cos\theta + F_{x_1}\sin\theta \\ F_{z_1} - gm\cos\alpha \end{bmatrix} \end{aligned} \tag{12.543}$$

where

$$^{G_1}\mathbf{W} = {}^{G_1}R_G\,{}^{G}\mathbf{W} = mg\begin{bmatrix} \sin\alpha \\ 0 \\ -\cos\alpha \end{bmatrix} \tag{12.544}$$

$$^{G_1}\mathbf{F} = {}^{G_1}R_{B_1}\,{}^{B_1}\mathbf{F} = \begin{bmatrix} F_{x_1}\cos\theta - F_{y_1}\sin\theta \\ F_{y_1}\cos\theta + F_{x_1}\sin\theta \\ F_{z_1} \end{bmatrix} \tag{12.545}$$

To obtain the rotational equations of motion, let us find the mass moment matrix of the disc and angular velocity of B:

$$^{B}I = \begin{bmatrix} \frac{1}{4}mR^2 & 0 & 0 \\ 0 & \frac{1}{2}mR^2 & 0 \\ 0 & 0 & \frac{1}{4}mR^2 \end{bmatrix} \tag{12.546}$$

$$
\begin{aligned}
{}^B_G\boldsymbol{\omega}_B &= {}^B_{G_1}\boldsymbol{\omega}_{B_1} + {}^B_{B_1}\boldsymbol{\omega}_B = {}^B R_{G_1} \begin{bmatrix} 0 \\ 0 \\ \dot{\theta} \end{bmatrix} + {}^B R_{B_1} \begin{bmatrix} 0 \\ \dot{\varphi} \\ 0 \end{bmatrix} \\
&= \begin{bmatrix} -\dot{\theta}\sin\varphi \\ \dot{\varphi} \\ \dot{\theta}\cos\varphi \end{bmatrix} = \begin{bmatrix} \omega_x \\ \omega_y \\ \omega_z \end{bmatrix}
\end{aligned} \tag{12.547}
$$

Therefore, the angular momentum of the disc would be

$$
{}^B\mathbf{L} = {}^B I \; {}^B_G\boldsymbol{\omega}_B = \tfrac{1}{4} m R^2 \begin{bmatrix} -\dot{\theta}\sin\varphi \\ 2\dot{\varphi} \\ \dot{\theta}\cos\varphi \end{bmatrix} \tag{12.548}
$$

Substituting these matrices in the Euler equation provides the rotational equations of motion:

$$
\begin{aligned}
{}^B\mathbf{M} &= \frac{{}^G d}{dt}\,{}^B\mathbf{L} = {}^B\dot{\mathbf{L}} + {}^B_G\boldsymbol{\omega}_B \times {}^B\mathbf{L} \\
&= \frac{mR^2}{4} \begin{bmatrix} -\ddot{\theta}\sin\varphi - \dot{\theta}\dot{\varphi}\cos\varphi \\ 2\ddot{\varphi} \\ \ddot{\theta}\cos\varphi - \dot{\theta}\dot{\varphi}\sin\varphi \end{bmatrix} + \frac{mR^2}{4} \begin{bmatrix} -\dot{\theta}\dot{\varphi}\cos\varphi \\ 0 \\ -\dot{\theta}\dot{\varphi}\sin\varphi \end{bmatrix} \\
&= \tfrac{1}{4} m R^2 \begin{bmatrix} -\ddot{\theta}\sin\varphi - 2\dot{\theta}\dot{\varphi}\cos\varphi \\ 2\ddot{\varphi} \\ \ddot{\theta}\cos\varphi - 2\dot{\theta}\dot{\varphi}\sin\varphi \end{bmatrix}
\end{aligned} \tag{12.549}
$$

The external moment ${}^B\mathbf{M}$ comes from the contribution of the reaction force $\mathbf{F}$ at P and the required torque Q that keeps the disc upright:

$$
\begin{aligned}
{}^B\mathbf{M} &= {}^B R_{B_1} \left({}^{B_1}\mathbf{M} + {}^{B_1}\mathbf{r}_P \times {}^{B_1}\mathbf{F} \right) \\
&= {}^B R_{B_1} \left(\begin{bmatrix} -Q \\ 0 \\ 0 \end{bmatrix} + \begin{bmatrix} 0 \\ 0 \\ -R \end{bmatrix} \times \begin{bmatrix} F_{x_1} \\ F_{y_1} \\ F_{z_1} \end{bmatrix} \right) \\
&= \begin{bmatrix} -\left(Q - RF_{y_1}\right)\cos\varphi \\ -RF_{x_1} \\ -\left(Q - RF_{y_1}\right)\sin\varphi \end{bmatrix}
\end{aligned} \tag{12.550}
$$

Therefore, the set of Euler equations of motion is given as

$$
\begin{bmatrix} \left(RF_{y_1} - Q\right)\cos\varphi \\ RF_{x_1} \\ \left(RF_{y_1} - Q\right)\sin\varphi \end{bmatrix} = \frac{1}{4} m R^2 \begin{bmatrix} -\ddot{\theta}\sin\varphi - 2\dot{\theta}\dot{\varphi}\cos\varphi \\ -2\ddot{\varphi} \\ \ddot{\theta}\cos\varphi - 2\dot{\theta}\dot{\varphi}\sin\varphi \end{bmatrix} \tag{12.551}
$$

If we transform the equations to B_1, then they will have a simple expression:

$$
\begin{aligned}
{}^{B_1}\mathbf{M} &= {}^{B_1}R_B \, {}^B\mathbf{M} \\
&= \begin{bmatrix} RF_{y_1} - Q \\ -RF_{x_1} \\ 0 \end{bmatrix} = \begin{bmatrix} -\frac{1}{2}R^2 m\dot{\theta}\dot{\varphi} \\ \frac{1}{2}R^2 m\ddot{\varphi} \\ \frac{1}{4}R^2 m\ddot{\theta} \end{bmatrix}
\end{aligned} \tag{12.552}
$$

Solution of Equations (12.552) is simpler than (12.551).

There exist two holonomic constraints in this system:

$$Z_1 - R = 0 \tag{12.553}$$

$$\omega_{x_1} = 0 \tag{12.554}$$

There is also a nonholonomic constraint due to the pure rolling condition. The contact point of the disc must be at rest. Therefore, the velocity of point P is zero in both coordinate frames G and B:

$$^{B_1}\dot{\mathbf{r}}_P = {}^{G_1}\dot{\mathbf{r}}_P = 0 \tag{12.555}$$

$$\dot{X}_1 \hat{I}_1 + \dot{Y}_1 \hat{J}_1 = {}^{G_1}R_B \left({}^{B}_{G_1}\boldsymbol{\omega}_B \times {}^{B}\mathbf{r}_P \right) = \begin{bmatrix} -R\dot{\varphi}\cos\theta \\ -R\dot{\varphi}\sin\theta \\ 0 \end{bmatrix}$$

$$^{B}\mathbf{r}_P = {}^{B}R_{B_1} \, {}^{B_1}\mathbf{r}_P$$

$$= \begin{bmatrix} \cos\varphi & 0 & \sin\varphi \\ 0 & 1 & 0 \\ -\sin\varphi & 0 & \cos\varphi \end{bmatrix}^{\mathrm{T}} \begin{bmatrix} 0 \\ 0 \\ -R \end{bmatrix} = \begin{bmatrix} R\sin\varphi \\ 0 \\ -R\cos\varphi \end{bmatrix} \tag{12.556}$$

The constraint equation provides two first-order differential relationships that must be satisfied:

$$\dot{X}_1 = -R\dot{\varphi}\cos\theta \qquad \dot{Y}_1 = -R\dot{\varphi}\sin\theta \tag{12.557}$$

The three translational equations of (12.543) and the three rotational equations of (12.552) along with the constraint equations should be solved for the position and orientation of the disc. From the third equation of (12.552), we have

$$\dot{\theta} = C_1 = \text{const} \qquad C_1 = \dot{\theta}(0) \tag{12.558}$$

$$\theta = C_1 t + C_2 \qquad C_2 = \theta(0) \tag{12.559}$$

Then, we can eliminate F_{y_1} between the first two equations of (12.543),

$$m\ddot{X}_1\cos\theta + m\ddot{Y}_1\sin\theta = F_{x_1} + gm\sin\alpha\cos\theta \tag{12.560}$$

and eliminate F_{x_1} by the second equation of (12.552),

$$\ddot{X}_1\cos\theta + \ddot{Y}_1\sin\theta = -\tfrac{1}{2}R\ddot{\varphi} + g\sin\alpha\cos\theta \tag{12.561}$$

and $\ddot{X}_1$, $\ddot{Y}_1$ by (12.557),

$$R\ddot{\varphi} = 2g\sin\alpha\cos\theta \tag{12.562}$$

We have $\theta = \theta(t)$. So,

$$\ddot{\varphi} = 2\frac{g}{R}\sin\alpha\cos(C_1 t + C_2) = C_3\cos(C_1 t + C_2) \tag{12.563}$$

$$C_3 = 2\frac{g}{R}\sin\alpha \tag{12.564}$$

Integration of this equation yields

$$\dot{\varphi} = \frac{C_3}{C_1} \sin (C_2 + tC_1) + \dot{\varphi}(0) \tag{12.565}$$

$$\varphi = -\frac{C_3}{C_1^2} \cos (C_2 + tC_1) + \dot{\varphi}(0)\, t + \varphi(0) \tag{12.566}$$

Substituting $\dot{\varphi}$ in Equations (12.557), we find X and Y as functions of t:

$$\dot{X}_1 = -R \left(\frac{C_3}{C_1} \sin (C_2 + tC_1) + \dot{\varphi}(0) \right) \cos (C_2 + tC_1) \tag{12.567}$$

$$X = \frac{1}{2} \frac{R}{C_1^2} C_3 \cos^2 (C_2 + tC_1) + X(0) \tag{12.568}$$

$$\dot{Y}_1 = -R \left(\frac{C_3}{C_1} \sin (C_2 + tC_1) + \dot{\varphi}(0) \right) \sin (C_2 + tC_1) \tag{12.569}$$

$$Y = \frac{1}{4} \frac{R}{C_1^2} C_3 [\sin 2 (C_2 + tC_1) - 2tC_1] + Y(0) \tag{12.570}$$

Having X, Y, θ, φ as functions of time, we are able to determine F_{x_1}, F_{y_1}, Q from Equations (12.552) and (12.543),

$$F_{x_1} = -\tfrac{1}{2} m R \ddot{\varphi} = -mg \sin\alpha \cos (C_1 t + C_2) \tag{12.571}$$

$$F_{y_1} = \frac{-2RC_3 \cos 2 (C_2 + tC_1) + mg \sin\alpha \sin 2 (C_2 + tC_1)}{2 \cos (C_1 t + C_2)} \tag{12.572}$$

$$\begin{aligned} Q = R &\left(\frac{-2RC_3 \cos 2 (C_2 + tC_1) + mg \sin\alpha \sin 2 (C_2 + tC_1)}{2 \cos (C_1 t + C_2)} \right) \\ &+ \tfrac{1}{2} R^2 m C_1 \left(\frac{C_3}{C_1} \sin (C_2 + tC_1) + \dot{\varphi}(0) \right) \end{aligned} \tag{12.573}$$

and F_{z_1} is already calculated from the third equation of (12.543),

$$F_{z_1} = gm \cos\alpha \tag{12.574}$$

Example 753 ★ Axisymmetric Rigid Body under Planar Moment Assume that an axisymmetric rigid body is under an external moment $\mathbf{M}$,

$${}^B\mathbf{M} = M_1(t)\, \hat{\imath} + M_2(t)\, \hat{\jmath} \tag{12.575}$$

The Euler equation of motion with such a moment becomes

$$\begin{bmatrix} M_1 \\ M_2 \\ 0 \end{bmatrix} = \begin{bmatrix} I_1 \dot{\omega}_1 - (I_1 - I_3)\, \omega_2 \omega_3 \\ I_2 \dot{\omega}_2 - (I_3 - I_1)\, \omega_3 \omega_1 \\ I_3 \dot{\omega}_3 \end{bmatrix} \tag{12.576}$$

The third equation shows that the third component of angular velocity is constant:

$$\omega_3 = C_1 \tag{12.577}$$

Introducing an axillary constant frequency Ω,

$$\Omega = \frac{I_1 - I_1}{I_1}\omega_3 \tag{12.578}$$

reduces the first two Euler equations to

$$\dot{\omega}_1 - \Omega\omega_2 = \frac{M_1}{I_1} \qquad \dot{\omega}_2 + \Omega\omega_1 = \frac{M_2}{I_1} \tag{12.579}$$

By introducing the two complex variables

$$\eta = \omega_1 + i\Omega\omega_2 \tag{12.580}$$

$$Q = \frac{M_1}{I_1} + i\frac{M_2}{I_1} \tag{12.581}$$

we may combine the two equations of (12.579):

$$\dot{\eta} + i\eta = Q \tag{12.582}$$

This is a first-order equation and has the solution

$$\eta = e^{-i\Omega t}\left(\eta_0 + \int_0^t Q\, e^{i\Omega\tau}\, d\tau\right) \tag{12.583}$$

We can separate the real and imaginary parts of the solution to obtain

$$\begin{aligned}\omega_1(t) &= \omega_{10}\cos\Omega t + \omega_{20}\sin\Omega t \\ &\quad + \int_0^t \left(\frac{M_1}{I_1}\cos\Omega(\tau - t) + \frac{M_2}{I_1}\sin\Omega(\tau - t)\right) d\tau\end{aligned} \tag{12.584}$$

$$\begin{aligned}\omega_2(t) &= \omega_{20}\cos\Omega t - \omega_{20}\sin\Omega t \\ &\quad + \int_0^t \left(\frac{M_1}{I_1}\sin\Omega(\tau - t) + \frac{M_2}{I_1}\cos\Omega(\tau - t)\right) d\tau\end{aligned} \tag{12.585}$$

Example 754 ★ Axisymmetric Rigid Body under a Moment Consider an axisymmetric rigid body that is under an external moment $\mathbf{M}$,

$${}^{B}\mathbf{M} = M_1(t)\,\hat{\imath} + M_2(t)\,\hat{\jmath} + M_3(t)\,\hat{k} \tag{12.586}$$

The Euler equation of motion with such a moment becomes

$$\begin{bmatrix} M_1 \\ M_2 \\ M_3 \end{bmatrix} = \begin{bmatrix} I_1\dot{\omega}_1 - (I_1 - I_3)\,\omega_2\omega_3 \\ I_2\dot{\omega}_2 - (I_3 - I_1)\,\omega_3\omega_1 \\ I_3\dot{\omega}_3 \end{bmatrix} \tag{12.587}$$

The third equation can be integrated first:

$$\omega_3(t) = \omega_{30} + \frac{1}{I_3}\int_0^t M_3(\tau)\, d\tau \tag{12.588}$$

We may introduce an axillary variable α that is considered a known function of time:

$$\alpha = \int_0^t \omega_3(\tau)\, d\tau \tag{12.589}$$

Using α, we can redefine the variables in the first and second Euler equations as

$$\dot{\omega}_i = \frac{d\omega_i}{d\alpha}\dot{\alpha} = \omega_i'(\alpha)\,\omega_3(t) \qquad i = 1, 2 \tag{12.590}$$

where the prime indicates the derivative with respect to α. Using these expressions the first two Euler equations become

$$\omega_1' - \Omega\omega_2 = \frac{M_1(t(\alpha))}{I_1\omega_3(t(\alpha))} \qquad \omega_2' + \Omega\omega_1 = \frac{M_2(t(\alpha))}{I_1\omega_3(t(\alpha))} \tag{12.591}$$

$$\Omega = \frac{I_1 - I_1}{I_1}\omega_3 \tag{12.592}$$

Equations (12.591) are mathematically identical to (12.579) in Example 753 with the new independent variable α instead of t. So, following, the same procedure, we find their solutions as

$$\omega_1(\alpha) = \omega_{10}\cos\Omega\alpha + \omega_{20}\sin\Omega\alpha$$
$$+ \int_0^\alpha \left(\frac{M_1}{I_1\omega_3}\cos\Omega(\tau-\alpha) + \frac{M_2}{I_1\omega_3}\sin\Omega(\tau-\alpha)\right) d\tau \tag{12.593}$$

$$\omega_2(\alpha) = \omega_{20}\cos\Omega\alpha - \omega_{20}\sin\Omega\alpha$$
$$+ \int_0^\alpha \left(\frac{M_1}{I_1\omega_3}\sin\Omega(\tau-\alpha) + \frac{M_2}{I_1\omega_3}\cos\Omega(\tau-\alpha)\right) d\tau \tag{12.594}$$

Example 755 Top Dynamics A top is an axisymmetric rigid body that is rotating about a fixed point in a constant gravitational field. The fixed point is on the axis of symmetry at a distance l from the body mass center C. We attach a body coordinate frame B and a global frame G at the fixed point O. Figure 12.35 illustrates a top with mass moment $[I]$:

$${}^BI = \begin{bmatrix} I_1 & 0 & 0 \\ 0 & I_1 & 0 \\ 0 & 0 & I_3 \end{bmatrix} \tag{12.595}$$

Let us choose Euler angles φ, θ, ψ as the generalized coordinates to describe the motion of B in G. The rotation of the top about its axis of symmetry is the spin ψ. The angle

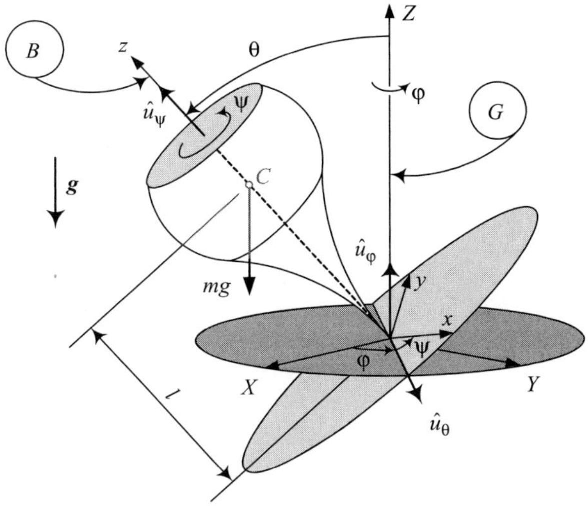

Figure 12.35 A top.

between the axis of symmetry and the Z-axis is the nutation θ, and the rotation of the axis of symmetry about the Z-axis is the precession φ:

$$
{}^{B}R_{G} = \begin{bmatrix} c\varphi c\psi - c\theta s\varphi s\psi & c\psi s\varphi + c\theta c\varphi s\psi & s\theta s\psi \\ -c\varphi s\psi - c\theta c\psi s\varphi & -s\varphi s\psi + c\theta c\varphi c\psi & s\theta c\psi \\ s\theta s\varphi & -c\varphi s\theta & c\theta \end{bmatrix} \tag{12.596}
$$

The moment caused by gravity is

$$
\begin{aligned}
{}^{B}\mathbf{M} &= {}^{B}\mathbf{r}_{C} \times m\, {}^{B}\mathbf{g} = \begin{bmatrix} 0 \\ 0 \\ l \end{bmatrix} \times mg\, {}^{B}R_{G} \begin{bmatrix} 0 \\ 0 \\ -1 \end{bmatrix} \\
&= \begin{bmatrix} 0 \\ 0 \\ l \end{bmatrix} \times mg \begin{bmatrix} -s\theta s\psi \\ -c\psi s\theta \\ -c\theta \end{bmatrix} = mgl \begin{bmatrix} \cos\psi \sin\theta \\ -\sin\psi \sin\theta \\ 0 \end{bmatrix}
\end{aligned} \tag{12.597}
$$

Substituting ${}^{B}\mathbf{M}$ in the B-expression of Euler equations (12.531), we have

$$
mgl \begin{bmatrix} \cos\psi \sin\theta \\ -\sin\psi \sin\theta \\ 0 \end{bmatrix} = \begin{bmatrix} I_1\dot{\omega}_1 - (I_1 - I_3)\,\omega_2\omega_3 \\ I_2\dot{\omega}_2 - (I_3 - I_1)\,\omega_3\omega_1 \\ I_3\dot{\omega}_3 \end{bmatrix} \tag{12.598}
$$

Solving for angular accelerations provides the differential equations

$$
I_1\dot{\omega}_1 = (I_1 - I_3)\,\omega_2\omega_3 + mgl\cos\psi\sin\theta \tag{12.599}
$$

$$
I_2\dot{\omega}_2 = (I_3 - I_1)\,\omega_3\omega_1 - mgl\sin\psi\sin\theta \tag{12.600}
$$

$$
I_3\dot{\omega}_3 = 0 \tag{12.601}
$$

The components of angular velocity can be expressed by Euler angles as given in (9.81):

$$ {}^{B}_{G}\boldsymbol{\omega}_{B} = \begin{bmatrix} \omega_1 \\ \omega_2 \\ \omega_3 \end{bmatrix} = \begin{bmatrix} \dot{\theta}\cos\psi + \dot{\varphi}\sin\theta\sin\psi \\ \dot{\varphi}\cos\psi\sin\theta - \dot{\theta}\sin\psi \\ \dot{\psi} + \dot{\varphi}\cos\theta \end{bmatrix} \tag{12.602} $$

Employing (12.602) and solving the first two equations for $\dot{\varphi}$ and $\dot{\theta}$ provide three differential equations to determine the Euler angles of the top:

$$ I_1\ddot{\theta} + \left(I_3\left(\dot{\psi} + \dot{\varphi}\cos\theta\right) - I_1\dot{\varphi}\cos\theta\right)\dot{\varphi}\sin\theta - mgl\sin\theta = 0 \tag{12.603} $$

$$ I_1\ddot{\varphi}\sin\theta + 2I_1\dot{\theta}\dot{\varphi}\cos\theta - I_3\dot{\theta}\left(\dot{\psi} + \dot{\varphi}\cos\theta\right) = 0 \tag{12.604} $$

$$ I_3\frac{d}{dt}\left(\dot{\psi} + \dot{\varphi}\cos\theta\right) = 0 \tag{12.605} $$

These equations have three integrals of motion. The first one is

$$ \dot{\psi} + \dot{\varphi}\cos\theta = \omega_3 = \text{const} \tag{12.606} $$

Using the first integral of motion, the first and second equations become

$$ I_1\ddot{\theta} + \left(I_3\omega_3 - I_1\dot{\varphi}\cos\theta\right)\dot{\varphi}\sin\theta - mgl\sin\theta = 0 \tag{12.607} $$

$$ I_1\ddot{\varphi}\sin\theta + 2I_1\dot{\theta}\dot{\varphi}\cos\theta - I_3\dot{\theta}\omega_3 = 0 \tag{12.608} $$

Multiplying the second equation by $\sin\theta$ provides the second integral of motion:

$$ \frac{d}{dt}\left(I_1\dot{\varphi}\sin^2\theta + I_3\omega_3\cos\theta\right) = 0 \tag{12.609} $$

$$ I_1\dot{\varphi}\sin^2\theta + I_3\omega_3\cos\theta = L \tag{12.610} $$

The third integral of motion appears when we multiply Equation (12.607) by $\dot{\theta}$ and (12.608) by $\dot{\varphi}\sin\theta$ and add them:

$$ \frac{d}{dt}\left[I_1\left(\dot{\varphi}^2\sin^2\theta + \dot{\theta}^2\right) + 2mgl\cos\theta\right] = 0 \tag{12.611} $$

$$ I_1\left(\dot{\varphi}^2\sin^2\theta + \dot{\theta}^2\right) + 2mgl\cos\theta = 2E - I_3\omega_3^2 \tag{12.612} $$

where

$$ E = mgl\cos\theta + \tfrac{1}{2}I_x\left(\omega_x^2 + \omega_y^2\right) + I_z\omega_z^2 \tag{12.613} $$

These are the same integrals of motion derived in Example 672.

Having three integrals of motion, we should be able to find the three equations for $\dot{\varphi}$, $\dot{\theta}$, $\dot{\psi}$:

$$ \dot{\varphi} = \dot{\varphi}\left(\varphi, \theta, \psi, t\right) \qquad \dot{\theta} = \dot{\theta}\left(\varphi, \theta, \psi, t\right) \qquad \dot{\psi} = \dot{\psi}\left(\varphi, \theta, \psi, t\right) \tag{12.614} $$

Equation (12.609) can be used to find $\dot{\varphi}$ directly:

$$ \dot{\varphi} = \frac{L - I_3\omega_3\cos\theta}{I_1\sin^2\theta} \tag{12.615} $$

Substituting $\dot{\varphi}$ in (12.611) and (12.606) results in the equations for $\dot{\theta}$ and $\dot{\psi}$:

$$I_1\dot{\theta}^2 = 2E - I_3\omega_3^2 - 2mgl\cos\theta - \frac{(L - I_3\omega_3\cos\theta)^2}{I_1\sin^2\theta} \tag{12.616}$$

$$\dot{\psi} = \omega_3 - \frac{L - I_3\omega_3\cos\theta}{I_1\sin^2\theta}\cos\theta \tag{12.617}$$

The decoupled Equation (12.616) is the key to calculate the orientation angles φ, θ, ψ. Upon calculating θ, the angles φ and ψ can be found by integrating (12.615) and (12.617).

To solve (12.616), we employ a new variable,

$$u = \cos\theta \qquad \dot{u} = \dot{\theta}\sin\theta \tag{12.618}$$

to simplify the equation:

$$\dot{u}^2 = \frac{\left(2E - I_3\omega_3^2 - 2mglu\right)\left(1 - u^2\right)}{I_1} - \frac{(L - I_3\omega_3 u)^2}{I_1^2} \tag{12.619}$$

The right-hand side of this equation is a continuous cubic polynomial function of u:

$$\begin{aligned} p(u) &= \frac{2E - I_3\omega_3^2 - 2mglu}{I_1}\left(1 - u^2\right) - \frac{(L - I_3\omega_3 u)^2}{I_1^2} \\ &= \frac{2mgl}{I_1}u^3 - \left(\frac{2E - I_3\omega_3^2}{I_1} + \frac{I_3^2}{I_1^2}\omega_3^2\right)u^2 \\ &\quad + \left(2L\frac{I_3\omega_3}{I_1^2} - \frac{2mgl}{I_1}\right)u + \left(\frac{2E - I_3\omega_3^2}{I_1} - \frac{L^2}{I_1^2}\right) \end{aligned} \tag{12.620}$$

The real roots of $p(u)$ indicate the steady-state motions of the top. For $\dot{u}^2 \geq 0$ and $u = \cos\theta$, the right-hand side accepts only limited values of u:

$$-1 \leq u \leq 1 \tag{12.621}$$

For $u = +1, -1$, the polynomial has negative values. Furthermore, we have

$$\lim_{u\to\pm\infty} p(u) = \pm\infty \tag{12.622}$$

and therefore, $p(u)$ has at least one real root for $u > 1$. Now the nonnegative condition $\dot{u}^2 \geq 0$ dictates that $p(u)$ must have two real roots for $-1 < u < 1$ or a double root. Therefore, $p(u)$ should have a plot similar to Figure 12.36 with three roots u_1, u_2, u_3:

$$u_1 \leq u_2 < u_3 \tag{12.623}$$

The roots u_1 and u_2, which indicate the maximum and minimum values of θ, have either equal or opposite signs. Both cases are physically possible and are considered special cases in the dynamics of tops. These special cases are associated with $\theta = \theta_0 = const$ and $\omega_3 = 0$.

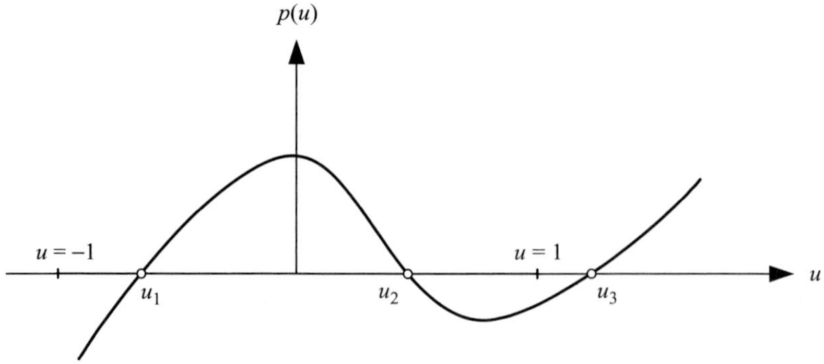

Figure 12.36 Qualitative plot of $p(u)$ versus u.

Let us rewrite Equation (12.619) as

$$\dot{u}^2 = \frac{2mgl}{I_1}(u - u_1)(u - u_2)(u - u_3) \tag{12.624}$$

Calculating the roots u_1, u_2, u_3 and introducing a new variable v,

$$v^2 = \frac{u - u_1}{u_2 - u_1} \tag{12.625}$$

transform Equation (12.624) to

$$\dot{v}^2 = \frac{mgl}{2I_1}(u_3 - u_1)\left(1 - v^2\right)\left(1 - k^2v^2\right) \tag{12.626}$$

$$0 \le k^2 = \frac{u_2 - u_1}{u_3 - u_1} \le 1 \tag{12.627}$$

Separation of variables leads to an elliptic integral of the first kind with modulus k:

$$\int_{v_0}^{v} \frac{dv}{\sqrt{\left(1 - v^2\right)\left(1 - k^2v^2\right)}} = \sqrt{\frac{mgl}{2I_1}(u_3 - u_1)}\,(t - t_0) = \tau \tag{12.628}$$

The solution of the integral is

$$v = sn\tau \tag{12.629}$$

which provides the following solution for θ:

$$\cos\theta = \cos\theta_1 + (\cos\theta_2 - \cos\theta_1)\, sn^2\tau \tag{12.630}$$

$$\cos\theta_1 = u_1 \qquad \cos\theta_2 = u_2 \tag{12.631}$$

Therefore, $\dot{\varphi}$, θ, and $\dot{\psi}$ are elliptic functions of time. The period of these functions is half the period of $sn\tau$. The superposition of periodic changes of $\theta(t)$ onto a precession about the Z-axis with a periodically changing angular velocity $\dot{\varphi}(t)$ generates a wavy motion of the z-axis. It can be visualized on a sphere with the center at the fixed point

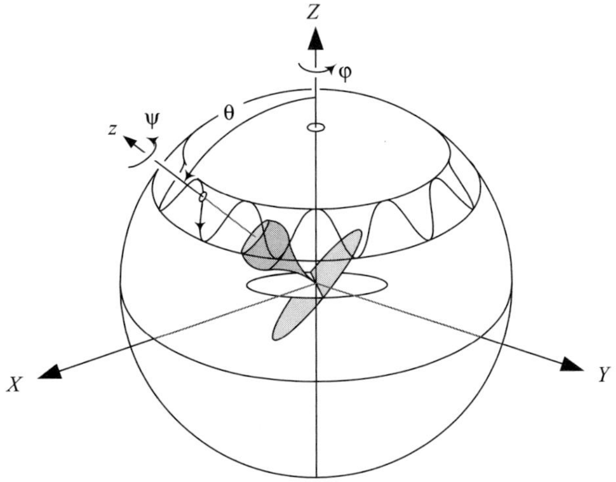

Figure 12.37 Path of the top symmetry axis on a sphere when $\dot{\varphi}(t)$ is changing between two positive extreme values.

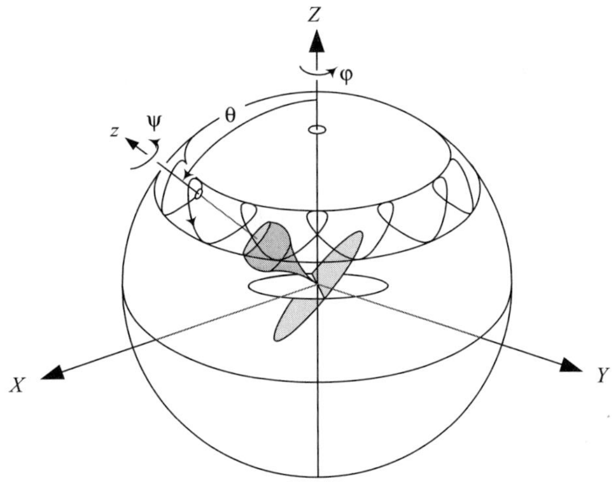

Figure 12.38 Path of the top symmetry axis on a sphere when $\dot{\varphi}(t)$ is changing between negative and positive values.

of the top. The intersection point of the z-axis of the body with this sphere generates paths which indicate the periodic changes of $\theta(t)$ and of $\dot{\varphi}(t)$.

The general paths of motion of the symmetry axis of the top are illustrated in Figures 12.37–12.39. Figure 12.37 depicts the situation in which $\dot{\varphi}(t)$ is changing between two positive extreme values. Figure 12.38 shows the path of the z-axis when $\dot{\varphi}(t)$ is changing between negative and positive values. Figure 12.39 depicts the situation in which $\dot{\varphi}(t)$ is changing between zero and a positive maximum. The periodic dance motion of the top with $\theta(t)$ which is superimposed on the precessional motion is called nutation.

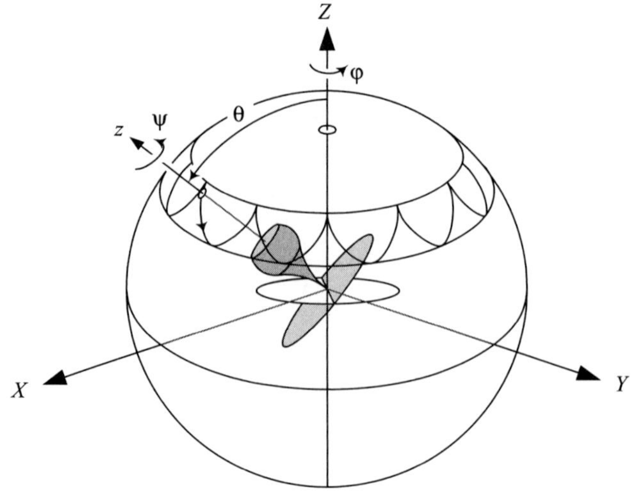

Figure 12.39 Path of the top symmetry axis on a sphere when $\dot{\varphi}(t)$ is changing between negative and positive values.

Example 756 ★ Special Cases of Top Dynamics Some special cases of top dynamics have simpler equations of motion with simpler interpretations. Consider the top kinematics shown in Figure 12.35 and its dynamic equations of motion (12.603)–(12.605):

1. $\omega_3 = 0$. This case is equivalent to a planar pendulum in which θ is the only time-dependent variable. The equations of motion reduce to only one equation:

$$I_1\ddot{\theta} - mgl\sin\theta = 0 \tag{12.632}$$

The planar pendulum case has energy conservation as the only integral of motion:

$$I_1\dot{\theta}^2 + 2mgl\cos\theta = 2E \tag{12.633}$$

2. $\theta = \theta_0 =$ const. This case is dependent on constant precession angular velocity. Equation (12.608) yields $\dot{\varphi} =$ const and Equation (12.605) leads to $\dot{\psi} =$ const. Therefore, the axis of symmetry of the top is moving with a constant precession angular velocity $\dot{\psi}$ around a circular cone with the central Z-axis. These conditions make a quadratic equation of (12.607) to determine $\dot{\varphi}$:

$$-I_1\dot{\varphi}^2\sin\theta\cos\theta + I_3\omega_3\dot{\varphi}\sin\theta - mgl\sin\theta = 0 \tag{12.634}$$

$$\dot{\varphi} = \begin{cases} \dfrac{\omega_3 I_3}{2I_1\cos\theta}\left(1 + \sqrt{1 - \dfrac{4glmI_1}{\omega_3^2 I_3^2}\cos\theta}\right) & \cos\theta \neq 0 \\ gl\dfrac{m}{\omega_3 I_3} & \cos\theta = 0 \end{cases} \tag{12.635}$$

12.5 MULTIBODY DYNAMICS

We consider multibodies as a set of rigid bodies connected to each other by revolute or prismatic joints. Most of the mechanical devices are multibodies. Each body is called a link. We follow the method of assigning numbers and coordinate frames of links as described in Chapter 7.

Figure 12.40 illustrates a link (i) of a multibody along with the velocity and acceleration vectorial characteristics. Figure 12.41 illustrates a free-body diagram of link (i). The force ${}^0\mathbf{F}_{i-1}$ and moment ${}^0\mathbf{M}_{i-1}$ are the global expressions of the resultant force and moment that link $(i-1)$ applies to link (i) at joint i. Similarly, ${}^0\mathbf{F}_i$ and ${}^0\mathbf{M}_i$ are the global expressions of the resultant force and moment that link (i) applies to link $(i+1)$ at joint $i+1$. We measure and show the force systems $({}^0\mathbf{F}_{i-1}, {}^0\mathbf{M}_{i-1})$ and $({}^0\mathbf{F}_i, {}^0\mathbf{M}_i)$ at the origin of the coordinate frames B_{i-1} and B_i, respectively. The sums of the external loads acting on link (i) are shown by $\sum {}^0\mathbf{F}_{e_i}$ and $\sum {}^0\mathbf{M}_{e_i}$.

A multibody with n links will have n sets of Newton–Euler equations. To simplify the solution of the equations, it is better to express all of them in one coordinate frame. The global frame $G \equiv {}^0B$ is the best frame. The G-expressions of Newton–Euler equations (12.219) and (12.1) have the simplest forms:

$$\sum {}^0\mathbf{F} = m\, {}^0\dot{\mathbf{v}} \tag{12.636}$$

$$\sum {}^0\mathbf{M} = {}^0\dot{\mathbf{L}} \tag{12.637}$$

The *Newton–Euler equations of motion* for link (i) in the global coordinate frame are

$${}^0\mathbf{F}_{i-1} - {}^0\mathbf{F}_i + \sum {}^0\mathbf{F}_{e_i} = m_i\, {}^0\mathbf{a}_i \tag{12.638}$$

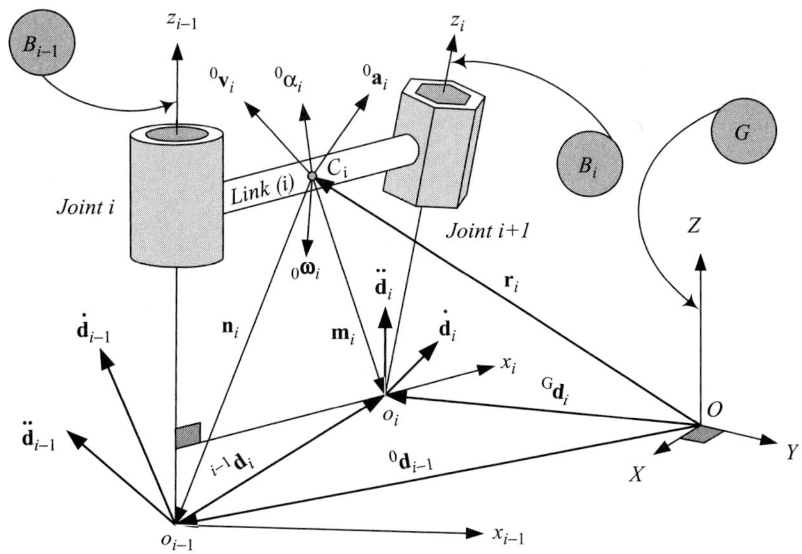

Figure 12.40 A link (i) and its kinematic characteristics.

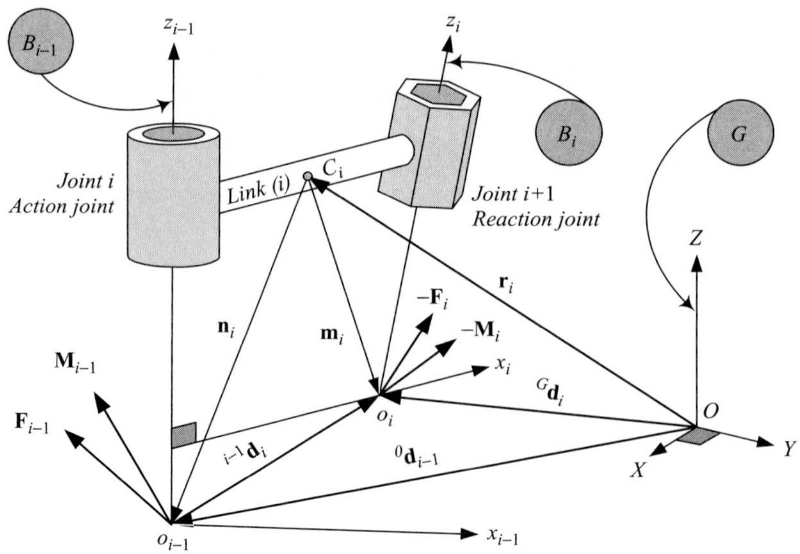

Figure 12.41 A link (i) and its force system.

$$
\begin{aligned}
&{}^0\mathbf{M}_{i-1} - {}^0\mathbf{M}_i + \sum {}^0\mathbf{M}_{e_i} \\
&\quad + \left({}^0\mathbf{d}_{i-1} - {}^0\mathbf{r}_i\right) \times {}^0\mathbf{F}_{i-1} - \left({}^0\mathbf{d}_i - {}^0\mathbf{r}_i\right) \times {}^0\mathbf{F}_i = {}^0I_i\, {}_0\boldsymbol{\alpha}_i \qquad (12.639)
\end{aligned}
$$

Proof: The force system at the distal end of link (i) is made of a force ${}^0\mathbf{F}_i$ and a moment ${}^0\mathbf{M}_i$ measured at the origin of B_i and expressed in B_0. The right subscript on ${}^0\mathbf{F}_i$ and ${}^0\mathbf{M}_i$ is a number indicating link (i) and its coordinate frame B_i.

At joint $i+1$ there is always an *action force* ${}^0\mathbf{F}_i$ where link (i) applies to link $(i+1)$ and a *reaction force* $-{}^0\mathbf{F}_i$ where link $(i+1)$ applies to link (i). Therefore, on link (i) there is always an action force ${}^0\mathbf{F}_{i-1}$ coming from link $(i-1)$, and a reaction force $-{}^0\mathbf{F}_i$ coming from link $(i+1)$. The action force is called the *driving force*, and the reaction force is called the *driven force*. Similarly, at joint $i+1$ there is always an *action moment* ${}^0\mathbf{M}_i$ where link (i) applies to link $(i+1)$, and a *reaction moment* $-{}^0\mathbf{M}_i$ where link $(i+1)$ applies to link (i). So, on link (i) there is always an action moment ${}^0\mathbf{M}_{i-1}$ coming from link $(i-1)$, and a reaction moment $-{}^0\mathbf{M}_i$ coming from link $(i+1)$. The action moment is called the *driving moment*, and the reaction moment is called the *driven moment*.

A link (i) is under two joint force systems: a driving force system $({}^0\mathbf{F}_{i-1}, {}^0\mathbf{M}_{i-1})$ at the origin of the coordinate frame B_{i-1} and a driven force system $({}^0\mathbf{F}_i, {}^0\mathbf{M}_i)$ at the origin of the coordinate frame B_i. The driving force system $({}^0\mathbf{F}_{i-1}, {}^0\mathbf{M}_{i-1})$ gives motion to link (i) and the driven force system $({}^0\mathbf{F}_i, {}^0\mathbf{M}_i)$ gives motion to link $(i+1)$.

In addition to the action and reaction force systems, there might be some external forces acting on link (i) where their resultant makes a force system $(\sum {}^0\mathbf{F}_{e_i}, \sum {}^0\mathbf{M}_{e_i})$ at the mass center C_i. In a multibody application, weight is usually the only external load on the middle links. There are also reactions from the environment that are extra external force systems on the base and end-effector links. The force and moment that the base actuator applies to the first link are ${}^0\mathbf{F}_0$ and ${}^0\mathbf{M}_0$, and the force and moment that the end-effector applies to the environment are ${}^0\mathbf{F}_n$ and ${}^0\mathbf{M}_n$. If weight is the only external load on link (i) and it is in the $-{}^0\hat{k}_0$ direction, then we have

$$\sum {}^0\mathbf{F}_{e_i} = m_i \, {}^0\mathbf{g} = -m_i \, g \, {}^0\hat{k}_0 \tag{12.640}$$

$$\sum {}^0\mathbf{M}_{e_i} = {}^0\mathbf{r}_i \times m_i \, {}^0\mathbf{g} = -\, {}^0\mathbf{r}_i \times m_i \, g \, {}^0\hat{k}_0 \tag{12.641}$$

where $\mathbf{g}$ is the gravitational acceleration vector.

As shown in Figure 12.40, we indicate the global position of the mass center of link (i) by ${}^0\mathbf{r}_i$ and the global positions of the origin of body frames B_i and B_{i-1} by ${}^0\mathbf{d}_i$ and ${}^0\mathbf{d}_{i-1}$, respectively. The link's velocities ${}^0\mathbf{v}_i$, ${}_0\boldsymbol{\omega}_i$ and accelerations ${}^0\mathbf{a}_i$, ${}_0\boldsymbol{\alpha}_i$ are measured and shown at C_i. The physical properties of link (i) are specified by its mass m_i and mass moment ${}^0I_i$ about the link's mass center C_i.

Newton's equation of motion determines that the sum of the forces applied to link (i) is equal to the mass of the link times its acceleration at C_i:

$${}^0\mathbf{F}_{i-1} - {}^0\mathbf{F}_i + \sum {}^0\mathbf{F}_{e_i} = m_i \, {}^0\mathbf{a}_i \tag{12.642}$$

To write the Euler equation, in addition to the action and reaction moments, we must add the moments of the action and reaction forces about C_i. The moments of $-\mathbf{F}_i$ and $\mathbf{F}_{i-1}$ are $-\mathbf{m}_i \times \mathbf{F}_i$ and $\mathbf{n}_i \times \mathbf{F}_{i-1}$, where $\mathbf{m}_i$ is the position vector of o_i from C_i and $\mathbf{n}_i$ is the position vector of o_{i-1} from C_i. Therefore, the Euler equation of motion for link (i) is

$${}^0\mathbf{M}_{i-1} - {}^0\mathbf{M}_i + \sum {}^0\mathbf{M}_{e_i} + {}^0\mathbf{n}_i \times {}^0\mathbf{F}_{i-1} - {}^0\mathbf{m}_i \times {}^0\mathbf{F}_i = {}^0I_i \, {}_0\boldsymbol{\alpha}_i \tag{12.643}$$

If we express the position vectors of $\mathbf{n}_i$ and $\mathbf{m}_i$ by

$${}^0\mathbf{n}_i = {}^0\mathbf{d}_{i-1} - {}^0\mathbf{r}_i \tag{12.644}$$

$${}^0\mathbf{m}_i = {}^0\mathbf{d}_i - {}^0\mathbf{r}_i \tag{12.645}$$

$${}_{i-1}^{\ \ 0}\mathbf{d}_i = {}^0\mathbf{m}_i - {}^0\mathbf{n}_i \tag{12.646}$$

then Equation (12.643) will be the same as Equation (12.639).

There are $2n$ vectorial equations of motion for an n-link multibody. However, there are $2(n+1)$ forces and moments involved. Therefore, one set of force systems (usually $\mathbf{F}_n$ and $\mathbf{M}_n$) must be specified to solve the equations and find the joints' force and moment. ■

Example 757 A Link with Spherical Joint Figure 12.42 illustrates a link attached to the ground by a spherical joint at O. The free-body diagram of the link is made of an external force ${}^G\mathbf{F}_e$ and moment ${}^G\mathbf{M}_e$ at the end point, gravity $m\mathbf{g}$, and driving force ${}^G\mathbf{F}_0$ and moment ${}^G\mathbf{M}_0$ at the joint. The Newton–Euler equations for the link are

$${}^G\mathbf{F}_0 + {}^G\mathbf{F}_e + mg\,\hat{K} = m \, {}^G\mathbf{a}_C \tag{12.647}$$

$${}^G\mathbf{M}_0 + {}^G\mathbf{M}_e + {}^G\mathbf{n} \times {}^G\mathbf{F}_0 + {}^G\mathbf{m} \times {}^G\mathbf{F}_e = {}^GI \, {}_G\boldsymbol{\alpha}_B \tag{12.648}$$

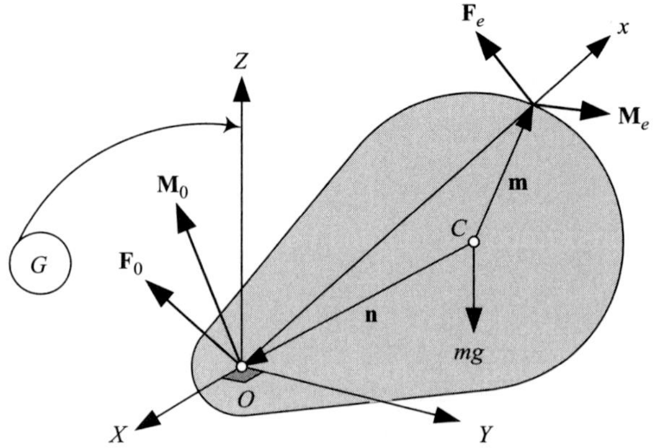

Figure 12.42 A link with a spherical joint.

Example 758 Turning Arms Let us consider the turning uniform arm shown in Figure 12.43(a). Figure 12.43(b) illustrates the free-body diagram of the arm and its relative position vectors $\mathbf{m}$ and $\mathbf{n}$:

$$ {}^0\mathbf{m} = \begin{bmatrix} \frac{l}{2}\cos\theta \\ \frac{l}{2}\sin\theta \\ 0 \end{bmatrix} \qquad {}^0\mathbf{n} = \begin{bmatrix} -\frac{l}{2}\cos\theta \\ -\frac{l}{2}\sin\theta \\ 0 \end{bmatrix} \tag{12.649} $$

The positions of C and B_1 are

$$ {}^0\mathbf{r}_1 = -{}^0\mathbf{n} \tag{12.650} $$

$$ {}^0\mathbf{d}_1 = -{}^0\mathbf{n} + {}^0\mathbf{m} \tag{12.651} $$

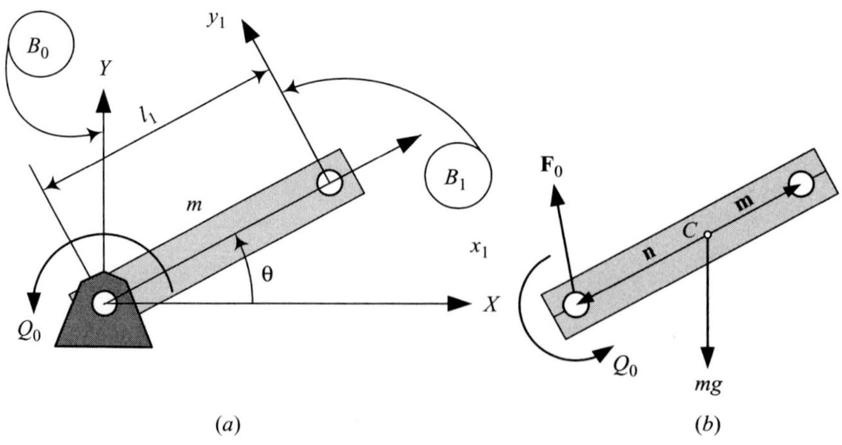

Figure 12.43 A turning uniform arm.

where ${}^0\mathbf{r}_1$ indicates the position of C and ${}^0\mathbf{d}_1$ indicates the position of B_1, both in B_0. Knowing that the arm is turning about the Z-axis, we have

$$
{}_0\boldsymbol{\omega}_1 = \dot{\theta}\,\hat{K} \tag{12.652}
$$

$$
{}_0\boldsymbol{\alpha}_1 = {}_0\dot{\boldsymbol{\omega}}_1 = \ddot{\theta}\,\hat{K} \tag{12.653}
$$

$$
{}^0\mathbf{g} = -g\,\hat{J} \tag{12.654}
$$

$$
\begin{aligned}
{}^0\mathbf{a}_C &= {}_0\boldsymbol{\alpha}_1 \times {}^0\mathbf{r}_1 - {}_0\boldsymbol{\omega}_1 \times \left({}_0\boldsymbol{\omega}_1 \times {}^0\mathbf{r}_1\right) \\
&= \begin{bmatrix} -\frac{l}{2}\ddot{\theta}\sin\theta + \frac{l}{2}\dot{\theta}^2\left(\cos\theta\right) \\ \frac{l}{2}\ddot{\theta}\cos\theta + \frac{l}{2}\dot{\theta}^2\sin\theta \\ 0 \end{bmatrix}
\end{aligned} \tag{12.655}
$$

The forces on the arm are

$$
{}^0\mathbf{F}_0 = \begin{bmatrix} F_X \\ F_Y \\ F_Z \end{bmatrix} \qquad {}^0\mathbf{F}_e = \begin{bmatrix} 0 \\ 0 \\ 0 \end{bmatrix} \tag{12.656}
$$

$$
{}^0\mathbf{M}_0 = \begin{bmatrix} Q_X \\ Q_Y \\ Q_Z \end{bmatrix} \qquad {}^0\mathbf{M}_e = \begin{bmatrix} 0 \\ 0 \\ 0 \end{bmatrix} \tag{12.657}
$$

Let us assume that ${}^1I_1$ is the principal mass moment matrix of the arm about its mass center:

$$
{}^1I_1 = \begin{bmatrix} I_x & 0 & 0 \\ 0 & I_y & 0 \\ 0 & 0 & I_z \end{bmatrix} \tag{12.658}
$$

Employing the rotation transformation matrix

$$
{}^0R_1 = R_{Z,\theta} = \begin{bmatrix} \cos\theta & -\sin\theta & 0 \\ \sin\theta & \cos\theta & 0 \\ 0 & 0 & 1 \end{bmatrix} \tag{12.659}
$$

we determine that

$$
\begin{aligned}
{}^0I_1 &= R_{Z,\theta}\,{}^1I_1\,R_{Z,\theta}^{\mathrm{T}} = {}^0R_1 \begin{bmatrix} I_x & 0 & 0 \\ 0 & I_y & 0 \\ 0 & 0 & I_z \end{bmatrix} {}^0R_1^{\mathrm{T}} \\
&= \begin{bmatrix} I_x\cos^2\theta + I_y\sin^2\theta & \left(I_x - I_y\right)\cos\theta\sin\theta & 0 \\ \left(I_x - I_y\right)\cos\theta\sin\theta & I_y\cos^2\theta + I_x\sin^2\theta & 0 \\ 0 & 0 & I_z \end{bmatrix}
\end{aligned} \tag{12.660}
$$

Substituting the above information in the Newton–Euler equations of the arm,

$$
{}^G\mathbf{F}_0 + {}^G\mathbf{F}_e + mg\,\hat{K} = m\,{}^G\mathbf{a}_C \tag{12.661}
$$

$$
{}^G\mathbf{M}_0 + {}^G\mathbf{M}_e + {}^G\mathbf{n} \times {}^G\mathbf{F}_0 + {}^G\mathbf{m} \times {}^G\mathbf{F}_e = {}^GI\,{}_G\boldsymbol{\alpha}_B \tag{12.662}
$$

provides the equations of motion

$$ {}^0\mathbf{F}_0 + {}^0\mathbf{F}_e + m_1\mathbf{g} = m\,{}^0\mathbf{a}_C \tag{12.663}$$

$$\begin{bmatrix} F_X \\ F_Y \\ F_Z \end{bmatrix} = \begin{bmatrix} -\frac{1}{2}ml\left(\ddot{\theta}\sin\theta - \dot{\theta}^2\cos\theta\right) \\ \frac{1}{2}ml\left(\ddot{\theta}\cos\theta + \dot{\theta}^2\sin\theta\right) + mg \\ 0 \end{bmatrix} \tag{12.664}$$

$$ {}^0\mathbf{M}_0 + {}^0\mathbf{M}_e + {}^0\mathbf{n} \times {}^0\mathbf{F}_0 + {}^0\mathbf{m} \times {}^0\mathbf{F}_e = I\,{}_0\boldsymbol{\alpha}_1 \tag{12.665}$$

$$\begin{bmatrix} Q_X \\ Q_Y \\ Q_Z \end{bmatrix} = \begin{bmatrix} \frac{l}{2}F_Z\sin\theta \\ -\frac{l}{2}F_Z\cos\theta \\ I_z\ddot{\theta} + \frac{l}{2}F_Y\cos\theta - \frac{l}{2}F_X\sin\theta \end{bmatrix} \tag{12.666}$$

Let us substitute the force components from (12.664) to determine the components of the driving moment ${}^0\mathbf{M}_0$:

$$\begin{bmatrix} Q_X \\ Q_Y \\ Q_Z \end{bmatrix} = \begin{bmatrix} 0 \\ 0 \\ \left(I_z + \frac{m_1 l^2}{4}\right)\ddot{\theta} + \frac{1}{2}mgl\cos\theta \end{bmatrix} \tag{12.667}$$

Therefore, $Q_0 = Q_Z$ is the required torque to turn the arm about the Z-axis:

$$Q_0 = \left(I_z + \tfrac{1}{4}m_1 l^2\right)\ddot{\theta} + \tfrac{1}{2}mgl\cos\theta \tag{12.668}$$

Let us consider the motion of the long arm of a clock, as shown in Figure 12.44(a). The arm is supposed to turn with a constant angular velocity ω:

$$\omega = \dot{\theta} = \frac{2\pi}{60}\text{ rad/min} = 6\text{ deg/min} \tag{12.669}$$

Having zero angular acceleration means the motor of the clock should be able to apply a variable torque Q_0,

$$Q_0(\theta) = mga\cos\theta \tag{12.670}$$

where θ is measured from the horizontal axis, as shown in Figure 12.43(a). The torque Q_0 is positive for $90\text{ deg} < \theta < -90\text{ deg}$ or $0 < t < 30$ min and negative for $-90\text{ deg} < \theta < -270\text{ deg}$ or $30\text{ min} < t < 60$ min.

Besides the actuator torque Q_0 and gravitation force mg, there are also a constant frictional torque Q_f and a resistive viscous torque $Q_c = c\dot{\theta}$ that apply on the arm. So, a better model of the dynamics is

$$Q_0 = \left(I_z + m_1 a^2\right)\ddot{\theta} + mga\cos\theta + Q_f + c\dot{\theta} \tag{12.671}$$

which shows that we still need a fluctuating torque to have a constant angular velocity:

$$Q_0(\theta) = mga\cos\theta + Q_f + c\omega \tag{12.672}$$

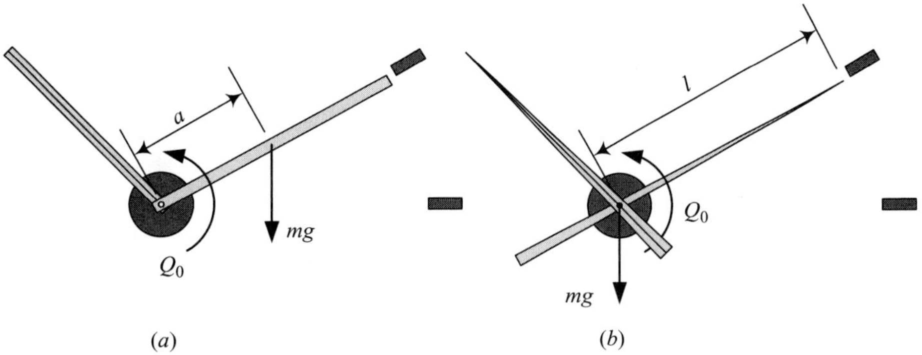

Figure 12.44 A clock with two unbalanced rotating arms in (a) and balanced rotating arms in (b).

The motors of stationary clocks are supposed to apply a constant and uniform torque. The best way to reduce the effect of the variable term of Equation (12.672) is to shorten a. Having a counterbalance and designing the hands of the clock with $a = 0$ as are shown in Figure 12.44(b) can solve the problem.

Example 759 Dynamics of a Four-Bar Linkage Linkages are connected multibodies that make a closed loop. The first and the last links of a linkage are usually connected to the ground link. Figure 12.45(a) illustrates a closed-loop four-bar linkage and Figure 12.45(b) illustrates the free-body diagrams of its links. The positions of the mass centers and the position vectors of ${}^0\mathbf{n}_i$ and ${}^0\mathbf{m}_i$ for each link are known. The Newton–Euler equations for link (i) are

$$ {}^0\mathbf{F}_{i-1} - {}^0\mathbf{F}_i + m_i g\,\hat{J} = m_i\,{}^0\mathbf{a}_i \tag{12.673}$$

$$ {}^0\mathbf{M}_{i-1} - {}^0\mathbf{M}_i + {}^0\mathbf{n}_i \times {}^0\mathbf{F}_{i-1} - {}^0\mathbf{m}_i \times {}^0\mathbf{F}_i = I_i\,{}_0\boldsymbol{\alpha}_i \tag{12.674}$$

The ground link is not moving, and therefore, we have three sets of equations:

$$ {}^0\mathbf{F}_0 - {}^0\mathbf{F}_1 + m_1 g\,\hat{J} = m_1\,{}^0\mathbf{a}_1 \tag{12.675}$$

$$ {}^0\mathbf{M}_0 - {}^0\mathbf{M}_1 + {}^0\mathbf{n}_1 \times {}^0\mathbf{F}_0 - {}^0\mathbf{m}_1 \times {}^0\mathbf{F}_1 = I_1\,{}_0\boldsymbol{\alpha}_1 \tag{12.676}$$

$$ {}^0\mathbf{F}_1 - {}^0\mathbf{F}_2 + m_2 g\,\hat{J} = m_2\,{}^0\mathbf{a}_2 \tag{12.677}$$

$$ {}^0\mathbf{M}_1 - {}^0\mathbf{M}_2 + {}^0\mathbf{n}_2 \times {}^0\mathbf{F}_1 - {}^0\mathbf{m}_2 \times {}^0\mathbf{F}_2 = I_2\,{}_0\boldsymbol{\alpha}_2 \tag{12.678}$$

$$ {}^0\mathbf{F}_2 - {}^0\mathbf{F}_3 + m_3 g\,\hat{J} = m_2\,{}^0\mathbf{a}_2 \tag{12.679}$$

$$ {}^0\mathbf{M}_2 - {}^0\mathbf{M}_3 + {}^0\mathbf{n}_3 \times {}^0\mathbf{F}_2 - {}^0\mathbf{m}_3 \times {}^0\mathbf{F}_3 = I_3\,{}_0\boldsymbol{\alpha}_3 \tag{12.680}$$

Let us assume that there is no friction in the joints and the mechanism is planar. The force vectors are in the (X, Y)-plane, and the moments are parallel to the Z-axis. So,

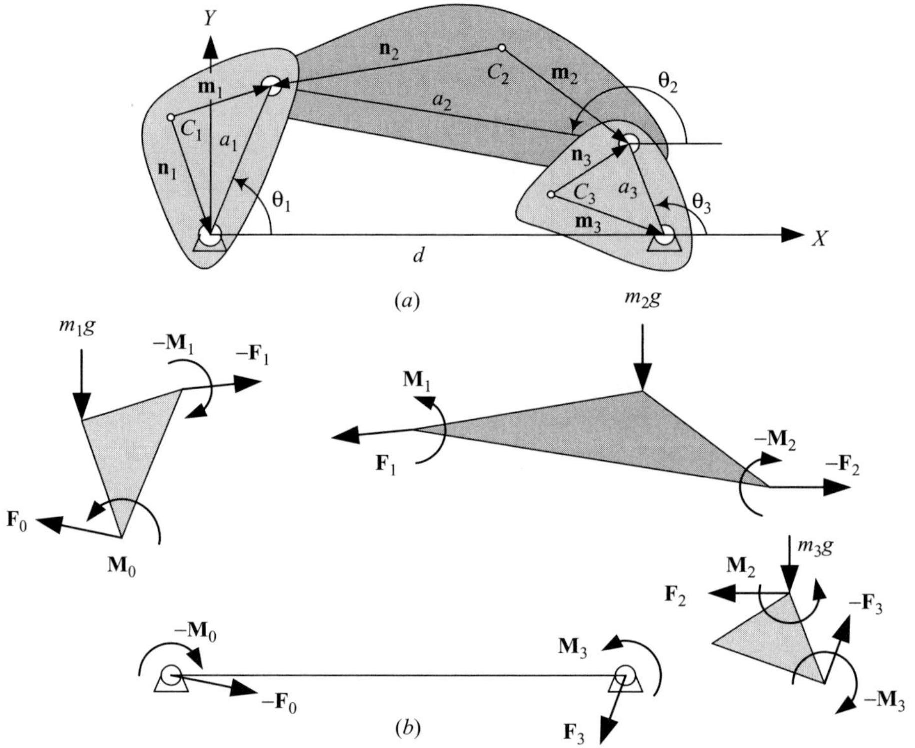

Figure 12.45 A four-bar linkage and free-body diagram of each link.

the equations of motion simplify to

$$ {}^0\mathbf{F}_0 - {}^0\mathbf{F}_1 + m_1 g\,\hat{J} = m_1\,{}^0\mathbf{a}_1 \tag{12.681} $$

$$ {}^0\mathbf{M}_0 + {}^0\mathbf{n}_1 \times {}^0\mathbf{F}_0 - {}^0\mathbf{m}_1 \times {}^0\mathbf{F}_1 = I_1\,{}_0\boldsymbol{\alpha}_1 \tag{12.682} $$

$$ {}^0\mathbf{F}_1 - {}^0\mathbf{F}_2 + m_2 g\,\hat{J} = m_2\,{}^0\mathbf{a}_2 \tag{12.683} $$

$$ {}^0\mathbf{n}_2 \times {}^0\mathbf{F}_1 - {}^0\mathbf{m}_2 \times {}^0\mathbf{F}_2 = I_2\,{}_0\boldsymbol{\alpha}_2 \tag{12.684} $$

$$ {}^0\mathbf{F}_2 - {}^0\mathbf{F}_3 + m_3 g\,\hat{J} = m_2\,{}^0\mathbf{a}_3 \tag{12.685} $$

$$ {}^0\mathbf{n}_3 \times {}^0\mathbf{F}_2 - {}^0\mathbf{m}_3 \times {}^0\mathbf{F}_3 = I_3\,{}_0\boldsymbol{\alpha}_3 \tag{12.686} $$

where ${}^0\mathbf{M}_0$ is the driving torque of the mechanism. These three vectorial equations provide nine scalar equations for the following nine unknowns:

$$ F_{0x},\ F_{0y},\ F_{1x},\ F_{1y},\ F_{2x},\ F_{2y},\ F_{3x},\ F_{3y},\ M_0 \tag{12.687} $$

We can rearrange the set of equations in a matrix form as

$$ [A]\,\mathbf{x} = \mathbf{b} \tag{12.688} $$

where

$$[A] = \begin{bmatrix} 1 & 0 & -1 & 0 & 0 & 0 & 0 & 0 & 0 \\ 0 & 1 & 0 & -1 & 0 & 0 & 0 & 0 & 0 \\ -n_{1y} & n_{1x} & m_{1y} & -m_{1x} & 0 & 0 & 0 & 0 & 1 \\ 0 & 0 & 1 & 0 & -1 & 0 & 0 & 0 & 0 \\ 0 & 0 & 0 & 1 & 0 & -1 & 0 & 0 & 0 \\ 0 & 0 & -n_{2y} & n_{2x} & m_{2y} & -m_{2x} & 0 & 0 & 0 \\ 0 & 0 & 0 & 0 & 1 & 0 & -1 & 0 & 0 \\ 0 & 0 & 0 & 0 & 0 & 1 & 0 & -1 & 0 \\ 0 & 0 & 0 & 0 & -n_{3y} & n_{3x} & m_{3y} & -m_{3x} & 0 \end{bmatrix} \tag{12.689}$$

$$\mathbf{x} = \begin{bmatrix} F_{0x} \\ F_{0y} \\ F_{1x} \\ F_{1y} \\ F_{2x} \\ F_{2y} \\ F_{3x} \\ F_{3y} \\ M_0 \end{bmatrix} \qquad \mathbf{b} = \begin{bmatrix} m_1 a_{1x} \\ m_1 a_{1y} - m_1 g \\ I_1 \alpha_1 \\ m_2 a_{2x} \\ m_2 a_{2y} - m_2 g \\ I_2 \alpha_2 \\ m_3 a_{3x} \\ m_3 a_{3y} - m_3 g \\ I_3 \alpha_3 \end{bmatrix} \tag{12.690}$$

The matrix $[A]$ describes the geometry of the mechanism, the column matrix $\mathbf{x}$ is the unknown forces, and the column matrix $\mathbf{b}$ indicates the dynamic terms. To solve the dynamics of the four-bar mechanism, we must calculate the accelerations ${}^0\mathbf{a}_i$ and ${}_0\boldsymbol{\alpha}_i$ and then find the required driving moment ${}^0\mathbf{M}_0$ and the joint forces.

Example 760 A Turning Arm with a Tip Mass Carrying masses are regular functions of mechanical machinery. The carrying mass will change the position of the mass center as well as the mass moment properties of the machine. To see the effect of such a massive load on the required actuating force systems, let us consider the uniform arm of Figure 12.46(a) with a hanging mass m_2 at the tip point. Figure 12.46(b) illustrates the FBD of the arm. Adding m_2 to the system moves the mass center of the arm to ${}^1\mathbf{r}_1$:

$$\begin{aligned} {}^1\mathbf{r}_1 &= \frac{m_1}{m_1 + m_2} \begin{bmatrix} l/2 \\ 0 \\ 0 \end{bmatrix} + \frac{m_2}{m_1 + m_2} \begin{bmatrix} l \\ 0 \\ 0 \end{bmatrix} \\ &= \begin{bmatrix} \dfrac{m_1 + 2m_2}{2\,(m_1 + m_2)} l \\ 0 \\ 0 \end{bmatrix} = \begin{bmatrix} r_x \\ 0 \\ 0 \end{bmatrix} \end{aligned} \tag{12.691}$$

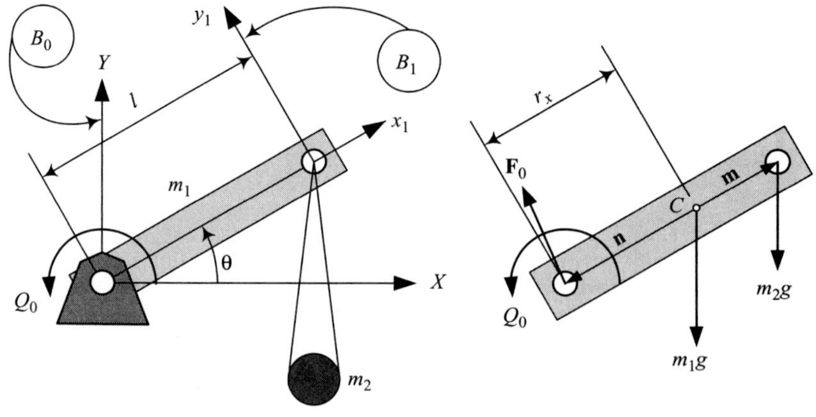

(a) (b)

Figure 12.46 A uniform rotating arm with a hanging weight m_2 at the tip point.

The relative position vectors **m** and **n** of arm (1), which is the only link of the system, are

$$ {}^1\mathbf{n}_1 = -\,{}^1\mathbf{r}_1 = -r_x\hat{\imath} \tag{12.692}$$

$$ {}^1\mathbf{m}_1 = l\hat{\imath} - {}^1\mathbf{r}_1 = (l - r_x)\,\hat{\imath} \tag{12.693}$$

$$ {}^0\mathbf{d}_1 = -\,{}^1\mathbf{n}_1 + {}^1\mathbf{m}_1 = l\hat{\imath} \tag{12.694}$$

$$ {}^0\mathbf{m} = {}^0R_1\,{}^1\mathbf{m}_1 = \begin{bmatrix} (l - r_x)\cos\theta \\ (l - r_x)\sin\theta \\ 0 \end{bmatrix} \tag{12.695}$$

$$ {}^0\mathbf{n} = {}^0R_1\,{}^1\mathbf{n}_1 = \begin{bmatrix} -r_x\cos\theta \\ -r_x\sin\theta \\ 0 \end{bmatrix} \tag{12.696}$$

$$ {}^0\mathbf{r} = -\,{}^0\mathbf{n} = \begin{bmatrix} r_x\cos\theta \\ r_x\sin\theta \\ 0 \end{bmatrix} \tag{12.697}$$

$$ {}^0R_1 = R_{Z,\theta} = \begin{bmatrix} \cos\theta & -\sin\theta & 0 \\ \sin\theta & \cos\theta & 0 \\ 0 & 0 & 1 \end{bmatrix} \tag{12.698}$$

The gravitational acceleration vector **g** and the kinematics of the arm are

$$ {}_0\boldsymbol{\omega}_1 = \dot{\theta}\,\hat{K} \qquad {}_0\boldsymbol{\alpha}_1 = {}_0\dot{\boldsymbol{\omega}}_1 = \ddot{\theta}\,\hat{K} \qquad \mathbf{g} = -g\,\hat{J} \tag{12.699}$$

$$ \begin{aligned} {}^0\mathbf{a}_C &= {}_0\boldsymbol{\alpha}_1 \times {}^0\mathbf{r}_1 + {}_0\boldsymbol{\omega}_1 \times \left({}_0\boldsymbol{\omega}_1 \times {}^0\mathbf{r}_1\right) \\ &= \begin{bmatrix} -r_x\ddot{\theta}\sin\theta - r_x\dot{\theta}^2\cos\theta \\ r_x\ddot{\theta}\cos\theta - r_x\dot{\theta}^2\sin\theta \\ 0 \end{bmatrix} \end{aligned} \tag{12.700}$$

The forces on the arm are

$$
{}^{0}\mathbf{F}_0 = \begin{bmatrix} F_X \\ F_Y \\ F_Z \end{bmatrix} \qquad {}^{0}\mathbf{F}_e = \begin{bmatrix} 0 \\ -(m_1 + m_2)\, g \\ 0 \end{bmatrix} \tag{12.701}
$$

$$
{}^{0}\mathbf{M}_0 = \begin{bmatrix} Q_X \\ Q_Y \\ Q_Z \end{bmatrix} \qquad {}^{0}\mathbf{M}_e = \begin{bmatrix} 0 \\ 0 \\ 0 \end{bmatrix} \tag{12.702}
$$

Let us assume that ${}^{1}I_1$ is the principal mass moment matrix of the arm about its center,

$$
{}^{1}I_1 = \begin{bmatrix} I_x & 0 & 0 \\ 0 & I_y & 0 \\ 0 & 0 & I_z \end{bmatrix} \tag{12.703}
$$

Then the mass moment matrix of the arm about the common mass center at ${}^{1}\mathbf{r}_1$ is

$$
{}^{1}I_1 = \begin{bmatrix} I_x & 0 & 0 \\ 0 & I_y & 0 \\ 0 & 0 & I_3 \end{bmatrix} \tag{12.704}
$$

$$
I_3 = I_z + m_1 \left(r_x - \frac{l}{2} \right)^2 + m_2 \left(l - r_x \right)^2 \tag{12.705}
$$

Because the equations of motion should be written in B_0, we use the transformation matrix ${}^{0}R_1$ to determine ${}^{0}I_1$:

$$
{}^{0}I_1 = R_{Z,\theta}\, {}^{1}I_1\, R_{Z,\theta}^{\mathrm{T}} = {}^{0}R_1 \begin{bmatrix} I_x & 0 & 0 \\ 0 & I_y & 0 \\ 0 & 0 & I_3 \end{bmatrix} {}^{0}R_1^{\mathrm{T}}
$$

$$
= \begin{bmatrix} I_x \cos^2\theta + I_y \sin^2\theta & \left(I_x - I_y\right)\cos\theta \sin\theta & 0 \\ \left(I_x - I_y\right)\cos\theta \sin\theta & I_y \cos^2\theta + I_x \sin^2\theta & 0 \\ 0 & 0 & I_3 \end{bmatrix} \tag{12.706}
$$

Substituting the above information in the equations of motion

$$
{}^{0}\mathbf{F}_0 + {}^{0}\mathbf{F}_e = (m_1 + m_2)\; {}^{0}\mathbf{a}_C \tag{12.707}
$$

$$
{}^{0}\mathbf{M}_0 + {}^{0}\mathbf{M}_e + {}^{0}\mathbf{n} \times {}^{0}\mathbf{F}_0 + {}^{0}\mathbf{m} \times {}^{0}\mathbf{F}_e = {}^{0}I_0\; {}_{0}\boldsymbol{\alpha}_B \tag{12.708}
$$

provides the following set of scalar equations:

$$
{}^{0}\mathbf{F}_0 + {}^{0}\mathbf{F}_e = (m_1 + m_2)\; {}^{0}\mathbf{a}_C \tag{12.709}
$$

$$
\begin{bmatrix} F_X \\ F_Y \\ F_Z \end{bmatrix} = \begin{bmatrix} -(m_1 + m_2)\, r_x \left(\ddot{\theta} \sin\theta + \dot{\theta}^2 \cos\theta\right) \\ (m_1 + m_2)\left(r_x \ddot{\theta} \cos\theta - r_x \dot{\theta}^2 \sin\theta + g\right) \\ 0 \end{bmatrix} \tag{12.710}
$$

$$ {}^0\mathbf{M}_0 + {}^0\mathbf{M}_e + {}^0\mathbf{n} \times {}^0\mathbf{F}_0 + {}^0\mathbf{m} \times {}^0\mathbf{F}_e = {}^0I_1 \, {}_0\boldsymbol{\alpha}_1 \tag{12.711} $$

$$ \begin{bmatrix} Q_X \\ Q_Y \\ Q_Z \end{bmatrix} = \begin{bmatrix} r_x F_Z \sin\theta \\ -r_x F_Z \cos\theta \\ I_3\ddot{\theta} - r_x F_X \, s\theta \\ +[r_x F_Y + (m_1 + m_2)\, g\, (l - r_x)]c\theta \end{bmatrix} \tag{12.712} $$

To calculate the driving moment ${}^0\mathbf{M}_0$, we substitute the force components from the Newton equation (12.710) into the Euler equation (12.712):

$$ \begin{bmatrix} Q_X \\ Q_Y \\ Q_Z \end{bmatrix} = \begin{bmatrix} 0 \\ 0 \\ \left(I_3 + (m_1 + m_2)\, r_x^2\right)\ddot{\theta} + (m_1 + m_2)\, lg \cos\theta \end{bmatrix} \tag{12.713} $$

Substituting r_x from (12.691) provides the required driving torque Q_0:

$$ Q_0 = Q_Z = [I_z + m_1 \left(r_x - \frac{l}{2}\right)^2 + m_2\,(l - r_x)^2 + \frac{(m_1 + 2m_2)^2}{4\,(m_1 + m_2)} l^2] $$
$$ + (m_1 + m_2)\, gl \cos\theta \tag{12.714} $$

Example 761 ***2R Planar Manipulator Newton–Euler Dynamics*** The $2R$ planar manipulators are applied controlled multibodies that can be seen in many robotic designs. An example of a $2R$ manipulator and its free-body diagram are shown in Figure 12.47. Assume $G\left(\hat{I}, \hat{J}, \hat{K}\right) = B_0$ is the global coordinate frame of the manipulator. The driving torques of the actuators are parallel to the Z-axis and are indicated by Q_0 and Q_1. The Newton–Euler equations of motion for the first link are

$$ {}^0\mathbf{F}_0 - {}^0\mathbf{F}_1 + m_1 g\,\hat{J} = m_1\,{}^0\mathbf{a}_1 \tag{12.715} $$

$$ {}^0\mathbf{Q}_0 - {}^0\mathbf{Q}_1 + {}^0\mathbf{n}_1 \times {}^0\mathbf{F}_0 - {}^0\mathbf{m}_1 \times {}^0\mathbf{F}_1 = {}^0I_1\,{}_0\boldsymbol{\alpha}_1 \tag{12.716} $$

and the equations of motion for the second link are

$$ {}^0\mathbf{F}_1 + m_2 g\,\hat{J} = m_2\,{}^0\mathbf{a}_2 \tag{12.717} $$

$$ {}^0\mathbf{Q}_1 + {}^0\mathbf{n}_2 \times {}^0\mathbf{F}_1 = {}^0I_2\,{}_0\boldsymbol{\alpha}_2 \tag{12.718} $$

So, there are four equations for four unknowns $\mathbf{F}_0$, $\mathbf{F}_1$, $\mathbf{Q}_0$, and $\mathbf{Q}_1$. Let us set these equations in matrix form as

$$ [A]\,\mathbf{x} = \mathbf{b} \tag{12.719} $$

$$ [A] = \begin{bmatrix} 1 & 0 & 0 & -1 & 0 & 0 \\ 0 & 1 & 0 & 0 & -1 & 0 \\ n_{1y} & -n_{1x} & 1 & -m_{1y} & m_{1x} & -1 \\ 0 & 0 & 0 & 1 & 0 & 0 \\ 0 & 0 & 0 & 0 & 1 & 0 \\ 0 & 0 & 0 & n_{2y} & -n_{2x} & 1 \end{bmatrix} \tag{12.720} $$

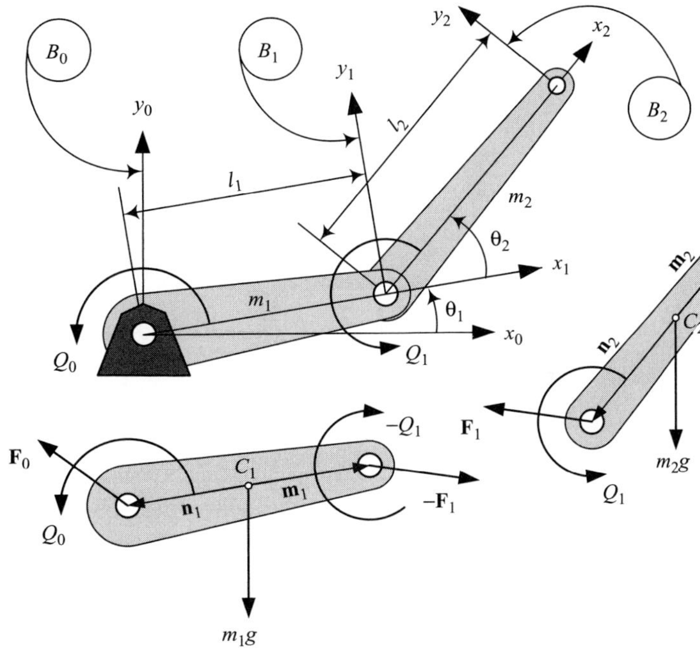

Figure 12.47 Free-body diagram of a 2R planar manipulator.

$$\mathbf{x} = \begin{bmatrix} F_{0x} \\ F_{0y} \\ Q_0 \\ F_{1x} \\ F_{1y} \\ Q_1 \end{bmatrix} \qquad \mathbf{b} = \begin{bmatrix} m_1 a_{1x} \\ m_1 a_{1y} - m_1 g \\ {}^0I_1 \alpha_1 \\ m_2 a_{2x} \\ m_2 a_{2y} - m_2 g \\ {}^0I_2 \alpha_2 \end{bmatrix} \tag{12.721}$$

The matrix A is constant. Having the column matrix $\mathbf{b}$ at any time, we can calculate the column matrix $\mathbf{x}$ to have the joint force and torques.

Example 762 Actuator Torques of a 2R Manipulator In multibody dynamics, we usually do not need to find the joint forces. Actuator commands, in this case the joint torques, are more interesting because we use them to control a multibody. In Example 761 we found four equations for the joint force system of the $2R$ manipulator shown in Figure 12.47:

$${}^0\mathbf{F}_0 - {}^0\mathbf{F}_1 + m_1 g \hat{J} = m_1 {}^0\mathbf{a}_1 \tag{12.722}$$

$${}^0\mathbf{Q}_0 - {}^0\mathbf{Q}_1 + {}^0\mathbf{n}_1 \times {}^0\mathbf{F}_0 - {}^0\mathbf{m}_1 \times {}^0\mathbf{F}_1 = {}^0I_1 \, {}_0\boldsymbol{\alpha}_1 \tag{12.723}$$

$${}^0\mathbf{F}_1 + m_2 g \hat{J} = m_2 {}^0\mathbf{a}_2 \tag{12.724}$$

$${}^0\mathbf{Q}_1 + {}^0\mathbf{n}_2 \times {}^0\mathbf{F}_1 = {}^0I_2 \, {}_0\boldsymbol{\alpha}_2 \tag{12.725}$$

We may eliminate the joint forces $\mathbf{F}_0$ and $\mathbf{F}_1$ from the four equations of motion (12.722)–(12.725) and reduce the number of equations to two for the two torques $\mathbf{Q}_0$ and $\mathbf{Q}_1$. Eliminating $\mathbf{F}_1$ between (12.724) and (12.725) provides

$$ {}^0\mathbf{Q}_1 = {}^0I_2\,{}_0\boldsymbol{\alpha}_2 - {}^0\mathbf{n}_2 \times \left(m_2\,{}^0\mathbf{a}_2 - m_2 g\,\hat{J}\right) \tag{12.726} $$

and eliminating $\mathbf{F}_0$ and $\mathbf{F}_1$ between (12.722) and (12.723) gives

$$ \begin{aligned} {}^0\mathbf{Q}_0 = {}&{}^0\mathbf{Q}_1 + {}^0I_1\,{}_0\boldsymbol{\alpha}_1 + {}^0\mathbf{m}_1 \times \left(m_2\,{}^0\mathbf{a}_2 - m_2 g\,\hat{J}\right) \\ &- {}^0\mathbf{n}_1 \times \left(m_1\,{}^0\mathbf{a}_1 - m_1 g\,\hat{J} + m_2\,{}^0\mathbf{a}_2 - m_2 g\,\hat{J}\right) \end{aligned} \tag{12.727} $$

The forces $\mathbf{F}_1$ and $\mathbf{F}_0$, if we are interested, are

$$ {}^0\mathbf{F}_1 = m_2\,{}^0\mathbf{a}_2 - m_2 g\,\hat{J} \tag{12.728} $$

$$ {}^0\mathbf{F}_0 = m_1\,{}^0\mathbf{a}_1 + m_2\,{}^0\mathbf{a}_2 - (m_1 + m_2)\,g\,\hat{J} \tag{12.729} $$

12.6 ★ RECURSIVE MULTIBODY DYNAMICS

An advantage of the Newton–Euler equations of motion in a multibody application is that we can calculate the joint forces of one link at a time. Therefore, starting from the end-effector link, we can analyze the links one by one and end up at the base link or vice versa. For such an analysis, we need to re-form the Newton–Euler equations of motion to work in the interested link's frame.

The *backward-recursive Newton–Euler equations of motion* for link (i) in its body coordinate frame B_i are

$$ {}^i\mathbf{F}_{i-1} = {}^i\mathbf{F}_i - \sum {}^i\mathbf{F}_{e_i} + m_i\,{}^i_0\mathbf{a}_i \tag{12.730} $$

$$ \begin{aligned} {}^i\mathbf{M}_{i-1} = {}&{}^i\mathbf{M}_i - \sum {}^i\mathbf{M}_{e_i} - \left({}^i\mathbf{d}_{i-1} - {}^i\mathbf{r}_i\right) \times {}^i\mathbf{F}_{i-1} \\ &+ \left({}^i\mathbf{d}_i - {}^i\mathbf{r}_i\right) \times {}^i\mathbf{F}_i + {}^iI_i\,{}^i_0\boldsymbol{\alpha}_i + {}^i_0\boldsymbol{\omega}_i \times {}^iI_i\,{}^i_0\boldsymbol{\omega}_i \end{aligned} \tag{12.731} $$

where

$$ {}^i\mathbf{n}_i = {}^i\mathbf{d}_{i-1} - {}^i\mathbf{r}_i \tag{12.732} $$

$$ {}^i\mathbf{m}_i = {}^i\mathbf{d}_i - {}^i\mathbf{r}_i \tag{12.733} $$

When the driving force system $({}^i\mathbf{F}_{i-1}, {}^i\mathbf{M}_{i-1})$ is found in frame B_i, we can transform them to the frame B_{i-1},

$$ {}^{i-1}\mathbf{F}_{i-1} = {}^{i-1}T_i\,{}^i\mathbf{F}_{i-1} \tag{12.734} $$

$$ {}^{i-1}\mathbf{M}_{i-1} = {}^{i-1}T_i\,{}^i\mathbf{M}_{i-1} \tag{12.735} $$

and apply the Newton–Euler equation for link $(i-1)$. The negative of the converted force system acts as the driven force system $(-{}^{i-1}\mathbf{F}_{i-1}, -{}^{i-1}\mathbf{M}_{i-1})$ for link $(i-1)$.

The *forward-recursive Newton–Euler equations of motion* for the link (i) in its body coordinate frame B_i are

$$
{}^{i}\mathbf{F}_i = {}^{i}\mathbf{F}_{i-1} + \sum {}^{i}\mathbf{F}_{e_i} - m_i\, {}_0^i\mathbf{a}_i \tag{12.736}
$$

$$
\begin{aligned}
{}^{i}\mathbf{M}_i = {}^{i}\mathbf{M}_{i-1} + \sum {}^{i}\mathbf{M}_{e_i} + \left({}^{i}\mathbf{d}_{i-1} - {}^{i}\mathbf{r}_i\right) \times {}^{i}\mathbf{F}_{i-1} \\
- \left({}^{i}\mathbf{d}_{i} - {}^{i}\mathbf{r}_i\right) \times {}^{i}\mathbf{F}_{i} - {}^{i}I_i\, {}_0^i\boldsymbol{\alpha}_i - {}_0^i\boldsymbol{\omega}_i \times {}^{i}I_i\, {}_0^i\boldsymbol{\omega}_i
\end{aligned} \tag{12.737}
$$

where

$$
{}^{i}\mathbf{n}_i = {}^{i}\mathbf{d}_{i-1} - {}^{i}\mathbf{r}_i \tag{12.738}
$$

$$
{}^{i}\mathbf{m}_i = {}^{i}\mathbf{d}_{i} - {}^{i}\mathbf{r}_i. \tag{12.739}
$$

When the reaction force system $({}^{i}\mathbf{F}_i, {}^{i}\mathbf{M}_i)$ is found in frame B_i, we can transform them to frame B_{i+1}:

$$
{}^{i+1}\mathbf{F}_i = {}^{i}T_{i+1}^{-1}\, {}^{i}\mathbf{F}_i \tag{12.740}
$$

$$
{}^{i+1}\mathbf{M}_i = {}^{i}T_{i+1}^{-1}\, {}^{i}\mathbf{M}_i \tag{12.741}
$$

The negative of the converted force system acts as the action force system $(-{}^{i+1}\mathbf{F}_i, -{}^{i+1}\mathbf{M}_i)$ for the link $(i+1)$.

Proof: The Euler equation for a rigid link in the body coordinate frame is

$$
\begin{aligned}
{}^{B}\mathbf{M} &= \frac{{}^{G}d}{dt}\, {}^{B}\mathbf{L} = {}^{B}\dot{\mathbf{L}} + {}_G^B\boldsymbol{\omega}_B \times {}^{B}\mathbf{L} \\
&= {}^{i}I_i\, {}_i\boldsymbol{\alpha}_i + {}_G^B\boldsymbol{\omega}_B \times {}^{i}I_i\, {}_i\boldsymbol{\omega}_i
\end{aligned} \tag{12.742}
$$

where $\mathbf{L}$ is the angular momentum of the link:

$$
{}^{B}\mathbf{L} = {}^{B}I\, {}_G^B\boldsymbol{\omega}_B \tag{12.743}
$$

We may solve the Newton–Euler equations of motion (12.638) and (12.639) for the action force system

$$
{}^{0}\mathbf{F}_{i-1} = {}^{0}\mathbf{F}_i - \sum {}^{0}\mathbf{F}_{e_i} + m_i\, {}^{0}\mathbf{a}_i \tag{12.744}
$$

$$
\begin{aligned}
{}^{0}\mathbf{M}_{i-1} = {}^{0}\mathbf{M}_i - \sum {}^{0}\mathbf{M}_{e_i} - \left({}^{0}\mathbf{d}_{i-1} - {}^{0}\mathbf{r}_i\right) \times {}^{0}\mathbf{F}_{i-1} \\
+ \left({}^{0}\mathbf{d}_{i} - {}^{0}\mathbf{r}_i\right) \times {}^{0}\mathbf{F}_{i} + \frac{{}^{0}d}{dt}\, {}^{0}\mathbf{L}_i
\end{aligned} \tag{12.745}
$$

and then transform the equations to the coordinate frame B_i attached to link (i) to make the backward-recursive form of the Newton–Euler equations of motion:

$$
{}^{i}\mathbf{F}_{i-1} = {}^{0}T_i^{-1}\, {}^{0}\mathbf{F}_{i-1} = {}^{i}\mathbf{F}_i - \sum {}^{i}\mathbf{F}_{e_i} + m_i\, {}_0^i\mathbf{a}_i \tag{12.746}
$$

$$\begin{aligned}
{}^{i}\mathbf{M}_{i-1} &= {}^{0}T_{i}^{-1}\,{}^{0}\mathbf{M}_{i-1} \\
&= {}^{i}\mathbf{M}_{i} - \sum {}^{i}\mathbf{M}_{e_i} - \left({}^{i}\mathbf{d}_{i-1} - {}^{i}\mathbf{r}_{i}\right) \times {}^{i}\mathbf{F}_{i-1} \\
&\quad + \left({}^{i}\mathbf{d}_{i} - {}^{i}\mathbf{r}_{i}\right) \times {}^{i}\mathbf{F}_{i} + \frac{{}^{0}d}{dt}\,{}^{i}\mathbf{L}_{i} \\
&= {}^{i}\mathbf{M}_{i} - \sum {}^{i}\mathbf{M}_{e_i} - \left({}^{i}\mathbf{d}_{i-1} - {}^{i}\mathbf{r}_{i}\right) \times {}^{i}\mathbf{F}_{i-1} \\
&\quad + \left({}^{i}\mathbf{d}_{i} - {}^{i}\mathbf{r}_{i}\right) \times {}^{i}\mathbf{F}_{i} + {}^{i}I_{i}\,{}_{0}^{i}\boldsymbol{\alpha}_{i} + {}_{0}^{i}\boldsymbol{\omega}_{i} \times {}^{i}I_{i}\,{}_{0}^{i}\boldsymbol{\omega}_{i}
\end{aligned} \tag{12.747}$$

The process of starting from link (i) and deriving the equations of motion of the previous link $(i-1)$ is called the *backward Newton–Euler equations of motion*. We may also start from link (i) and derive the equations of motion of the next link $(i+1)$. This method is called the *forward Newton–Euler equations of motion*. Employing the Newton–Euler equations of motion (12.730) and (12.731), we can write them in a *forward-recursive* form in the coordinate frame B_i attached to the link (i):

$$ {}^{i}\mathbf{F}_{i} = {}^{i}\mathbf{F}_{i-1} + \sum {}^{i}\mathbf{F}_{e_i} - m_i\,{}_{0}^{i}\mathbf{a}_{i} \tag{12.748}$$

$$\begin{aligned}
{}^{i}\mathbf{M}_{i} &= {}^{i}\mathbf{M}_{i-1} + \sum {}^{i}\mathbf{M}_{e_i} + \left({}^{i}\mathbf{d}_{i-1} - {}^{i}\mathbf{r}_{i}\right) \times {}^{i}\mathbf{F}_{i-1} \\
&\quad - \left({}^{i}\mathbf{d}_{i} - {}^{i}\mathbf{r}_{i}\right) \times {}^{i}\mathbf{F}_{i} - {}^{i}I_{i}\,{}_{0}^{i}\boldsymbol{\alpha}_{i} - {}_{0}^{i}\boldsymbol{\omega}_{i} \times {}^{i}I_{i}\,{}_{0}^{i}\boldsymbol{\omega}_{i}
\end{aligned} \tag{12.749}$$

$$ {}^{i}\mathbf{n}_{i} = {}^{i}\mathbf{d}_{i-1} - {}^{i}\mathbf{r}_{i} \tag{12.750}$$

$$ {}^{i}\mathbf{m}_{i} = {}^{i}\mathbf{d}_{i} - {}^{i}\mathbf{r}_{i} \tag{12.751}$$

Using the forward Newton–Euler equations of motion (12.748) and (12.749), we can calculate the reaction force system $({}^{i}\mathbf{F}_{i}, {}^{i}\mathbf{M}_{i})$ by having the action force system $({}^{i}\mathbf{F}_{i-1}, {}^{i}\mathbf{M}_{i-1})$. When the reaction force system $({}^{i}\mathbf{F}_{i}, {}^{i}\mathbf{M}_{i})$ is found in frame B_i, we can transform them to frame B_{i+1}:

$$ {}^{i+1}\mathbf{F}_{i} = {}^{i}T_{i+1}^{-1}\,{}^{i}\mathbf{F}_{i} \tag{12.752}$$

$$ {}^{i+1}\mathbf{M}_{i} = {}^{i}T_{i+1}^{-1}\,{}^{i}\mathbf{M}_{i} \tag{12.753}$$

The negative of the converted force system acts as the action force system $(-{}^{i+1}\mathbf{F}_{i}, -{}^{i+1}\mathbf{M}_{i})$ for the link $(i+1)$ and we can apply the Newton–Euler equation to the link $(i+1)$.

The forward Newton–Euler equations of motion allow us to start from a known action force system $({}^{1}\mathbf{F}_{0}, {}^{1}\mathbf{M}_{0})$, where the base link applies to the link (1), and calculate the action force of the next link. Therefore, analyzing the links of a multibody one by one, we end up with the force system where the end effector applies to the environment.

Using the forward- or backward-recursive Newton–Euler equations of motion depends on the measurement and sensory system of the multibody. ■

Example 763 ★ Recursive Dynamics of a 2R Planar Manipulator Consider the $2R$ planar manipulator shown in Figure 12.48. The manipulator is carrying a force system at the end point. We use this manipulator to show how to develop the dynamic equations for a serial multibody.

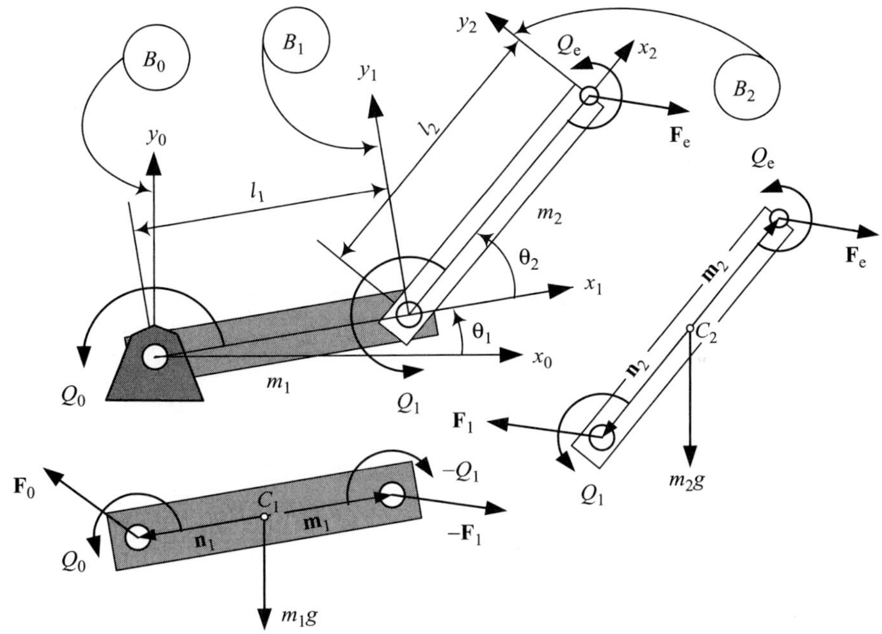

Figure 12.48 A 2R planar manipulator carrying a load at the end point.

The backward-recursive Newton–Euler equations of motion for the first link are

$$\begin{aligned} {}^1\mathbf{F}_0 &= {}^1\mathbf{F}_1 - \sum {}^1\mathbf{F}_{e_1} + m_1\, {}_0^1\mathbf{a}_1 \\ &= {}^1\mathbf{F}_1 - m_1\, {}^1\mathbf{g} + m_1\, {}_0^1\mathbf{a}_1 \end{aligned} \tag{12.754}$$

$$\begin{aligned} {}^1\mathbf{M}_0 &= {}^1\mathbf{M}_1 - \sum {}^1\mathbf{M}_{e_1} - \left({}^1\mathbf{d}_0 - {}^1\mathbf{r}_1\right) \times {}^1\mathbf{F}_0 \\ &\quad + \left({}^1\mathbf{d}_1 - {}^1\mathbf{r}_1\right) \times {}^1\mathbf{F}_1 + {}^1I_1\, {}_0^1\boldsymbol{\alpha}_1 + {}_0^1\boldsymbol{\omega}_1 \times {}^1I_1\, {}_0^1\boldsymbol{\omega}_1 \\ &= {}^1\mathbf{M}_1 - {}^1\mathbf{n}_1 \times {}^1\mathbf{F}_0 + {}^1\mathbf{m}_1 \times {}^1\mathbf{F}_1 \\ &\quad + {}^1I_1\, {}_0^1\boldsymbol{\alpha}_1 + {}_0^1\boldsymbol{\omega}_1 \times {}^1I_1\, {}_0^1\boldsymbol{\omega}_1 \end{aligned} \tag{12.755}$$

and the backward-recursive equations of motion for the second link are

$$\begin{aligned} {}^2\mathbf{F}_1 &= {}^2\mathbf{F}_2 - \sum {}^2\mathbf{F}_{e_2} + m_2\, {}_0^2\mathbf{a}_2 \\ &= -m_2\, {}^2\mathbf{g} - {}^2\mathbf{F}_e + m_2\, {}_0^2\mathbf{a}_2 \end{aligned} \tag{12.756}$$

$$\begin{aligned} {}^2\mathbf{M}_1 &= {}^2\mathbf{M}_2 - \sum {}^2\mathbf{M}_{e_2} - \left({}^2\mathbf{d}_1 - {}^2\mathbf{r}_2\right) \times {}^2\mathbf{F}_1 \\ &\quad + \left({}^2\mathbf{d}_2 - {}^2\mathbf{r}_2\right) \times {}^2\mathbf{F}_2 + {}^2I_2\, {}_0^2\boldsymbol{\alpha}_2 + {}_0^2\boldsymbol{\omega}_2 \times {}^2I_2\, {}_0^2\boldsymbol{\omega}_2 \\ &= -\, {}^2\mathbf{M}_e - {}^2\mathbf{m}_2 \times {}^2\mathbf{F}_e - {}^2\mathbf{n}_2 \times {}^2\mathbf{F}_1 \\ &\quad + {}^2I_2\, {}_0^2\boldsymbol{\alpha}_2 + {}_0^2\boldsymbol{\omega}_2 \times {}^2I_2\, {}_0^2\boldsymbol{\omega}_2 \end{aligned} \tag{12.757}$$

The manipulator consists of two R∥R(0) links; therefore, their transformation matrices ${}^{i-1}T_i$ are of class (D.1). Substituting $d_i = 0$ and $a_i = l_i$ produces the transformation matrices

$$
{}^{0}T_1 = \begin{bmatrix} \cos\theta_1 & -\sin\theta_1 & 0 & l_1\cos\theta_1 \\ \sin\theta_1 & \cos\theta_1 & 0 & l_1\sin\theta_1 \\ 0 & 0 & 1 & 0 \\ 0 & 0 & 0 & 1 \end{bmatrix} \tag{12.758}
$$

$$
{}^{1}T_2 = \begin{bmatrix} \cos\theta_2 & -\sin\theta_2 & 0 & l_2\cos\theta_2 \\ \sin\theta_2 & \cos\theta_2 & 0 & l_2\sin\theta_2 \\ 0 & 0 & 1 & 0 \\ 0 & 0 & 0 & 1 \end{bmatrix} \tag{12.759}
$$

The homogeneous mass moment matrices are

$$
{}^{1}\overline{I}_1 = \frac{m_1 l_1^2}{12}\begin{bmatrix} 0 & 0 & 0 & 0 \\ 0 & 1 & 0 & 0 \\ 0 & 0 & 1 & 0 \\ 0 & 0 & 0 & 0 \end{bmatrix} \qquad {}^{2}\overline{I}_2 = \frac{m_2 l_2^2}{12}\begin{bmatrix} 0 & 0 & 0 & 0 \\ 0 & 1 & 0 & 0 \\ 0 & 0 & 1 & 0 \\ 0 & 0 & 0 & 0 \end{bmatrix} \tag{12.760}
$$

The homogeneous moment of the inertia matrix is obtained by appending a zero row and column to the I-matrix.

The position vectors are

$$
{}^{1}\mathbf{n}_1 = \begin{bmatrix} -l_1/2 \\ 0 \\ 0 \\ 0 \end{bmatrix} \qquad {}^{2}\mathbf{n}_2 = \begin{bmatrix} -l_2/2 \\ 0 \\ 0 \\ 0 \end{bmatrix} \tag{12.761}
$$

$$
{}^{1}\mathbf{m}_1 = \begin{bmatrix} l_1/2 \\ 0 \\ 0 \\ 0 \end{bmatrix} \qquad {}^{2}\mathbf{m}_2 = \begin{bmatrix} l_2/2 \\ 0 \\ 0 \\ 0 \end{bmatrix} \tag{12.762}
$$

$$
{}^{1}\mathbf{r}_1 = -\,{}^{1}\mathbf{n}_1 \qquad {}^{2}\mathbf{r}_2 = -\,{}^{2}\mathbf{n}_1 + {}^{2}\mathbf{m}_2 - {}^{2}\mathbf{n}_2 \tag{12.763}
$$

The angular velocities and accelerations are

$$
{}_{0}^{1}\boldsymbol{\omega}_1 = \begin{bmatrix} 0 \\ 0 \\ \dot{\theta}_1 \\ 0 \end{bmatrix} \qquad {}_{0}^{2}\boldsymbol{\omega}_2 = \begin{bmatrix} 0 \\ 0 \\ \dot{\theta}_1 + \dot{\theta}_2 \\ 0 \end{bmatrix} \tag{12.764}
$$

$$
{}_{0}^{1}\boldsymbol{\alpha}_1 = \begin{bmatrix} 0 \\ 0 \\ \ddot{\theta}_1 \\ 0 \end{bmatrix} \qquad {}_{0}^{2}\boldsymbol{\alpha}_2 = \begin{bmatrix} 0 \\ 0 \\ \ddot{\theta}_1 + \ddot{\theta}_2 \\ 0 \end{bmatrix} \tag{12.765}
$$

The translational acceleration of C_1 is

$$
{}_0^1\mathbf{a}_1 = {}_0^1\boldsymbol{\alpha}_1 \times \left(-{}^1\mathbf{m}_1\right) + {}_0^1\boldsymbol{\omega}_1 \times \left({}_0^1\boldsymbol{\omega}_1 \times \left(-{}^1\mathbf{m}_1\right)\right) + {}_0^1\ddot{\mathbf{d}}_1
= \begin{bmatrix} -\frac{1}{2}l_1\dot{\theta}_1^2 \\ \frac{1}{2}l_1\ddot{\theta}_1 \\ 0 \\ 0 \end{bmatrix} \tag{12.766}
$$

where

$$
{}^1\ddot{\mathbf{d}}_1 = 2\ {}^1\mathbf{a}_1 \tag{12.767}
$$

The translational acceleration of C_2 is

$$
{}_0^2\mathbf{a}_2 = {}_0^2\boldsymbol{\alpha}_2 \times \left(-{}^2\mathbf{m}_2\right) + {}_0^2\boldsymbol{\omega}_2 \times \left[{}_0^2\boldsymbol{\omega}_2 \times \left(-{}^2\mathbf{m}_2\right)\right] + {}_0^2\ddot{\mathbf{d}}_2
= \begin{bmatrix} -\frac{1}{2}l_2\left(\dot{\theta}_1 + \dot{\theta}_2\right)^2 \\ \frac{1}{2}l_2\left(\ddot{\theta}_1 + \ddot{\theta}_2\right) \\ 0 \\ 0 \end{bmatrix} \tag{12.768}
$$

where

$$
{}^2\ddot{\mathbf{d}}_2 = 2\ {}^2\mathbf{a}_2 \tag{12.769}
$$

The gravitational acceleration vectors in the links' frame are

$$
{}^1\mathbf{g} = {}^0T_1^{-1}\ {}^0\mathbf{g} = \begin{bmatrix} -g\sin\theta_2 \\ g\cos\theta_2 \\ 0 \\ 0 \end{bmatrix} \tag{12.770}
$$

$$
{}^2\mathbf{g} = {}^0T_2^{-1}\ {}^0\mathbf{g} = \begin{bmatrix} -g\sin\left(\theta_1 + \theta_2\right) \\ g\cos\left(\theta_1 + \theta_2\right) \\ 0 \\ 0 \end{bmatrix} \tag{12.771}
$$

The external load is usually given in the global coordinate frame. We must transform them to the interested link's frame to apply the recursive equations of motion. Therefore, the external force system expressed in B_2 is

$$
{}^2\mathbf{F}_e = {}^0T_2^{-1}\,{}^0\mathbf{F}_e = \begin{bmatrix} F_{ex}\cos\left(\theta_1 + \theta_2\right) + F_{ey}\sin\left(\theta_1 + \theta_2\right) \\ F_{ey}\cos\left(\theta_1 + \theta_2\right) - F_{ex}\sin\left(\theta_1 + \theta_2\right) \\ 0 \\ 0 \end{bmatrix} \tag{12.772}
$$

$$
{}^2\mathbf{M}_e = {}^0T_2^{-1}\,{}^0\mathbf{M}_e = \begin{bmatrix} 0 \\ 0 \\ M_e \\ 0 \end{bmatrix} \tag{12.773}
$$

Now, we begin from the final link and calculate its action force system. The backward Newton equation for link (2) is

$$ {}^{2}\mathbf{F}_{1} = -m_{2}\,{}^{2}\mathbf{g} - {}^{2}\mathbf{F}_{e} + m_{2}\,{}_{0}^{2}\mathbf{a}_{2} = \begin{bmatrix} {}^{2}F_{1x} \\ {}^{2}F_{1y} \\ 0 \\ 0 \end{bmatrix} \tag{12.774} $$

where

$$ \begin{aligned} {}^{2}F_{1x} &= -\tfrac{1}{2}l_{2}m_{2}\left(\dot{\theta}_{1}+\dot{\theta}_{2}\right)^{2} - F_{ex}\cos\left(\theta_{1}+\theta_{2}\right) \\ &\quad - \left(F_{ey}-gm_{2}\right)\sin\left(\theta_{1}+\theta_{2}\right) \end{aligned} \tag{12.775} $$

$$ \begin{aligned} {}^{2}F_{1y} &= \tfrac{1}{2}l_{2}m_{2}\left(\ddot{\theta}_{1}+\ddot{\theta}_{2}\right) + F_{ex}\sin\left(\theta_{1}+\theta_{2}\right) \\ &\quad - \left(F_{ey}+gm_{2}\right)\cos\left(\theta_{1}+\theta_{2}\right) \end{aligned} \tag{12.776} $$

and the backward Euler equation for link (2) is

$$ \begin{aligned} {}^{2}\mathbf{M}_{1} &= -{}^{2}\mathbf{M}_{e} - {}^{2}\mathbf{m}_{2}\times{}^{2}\mathbf{F}_{e} - {}^{2}\mathbf{n}_{2}\times{}^{2}\mathbf{F}_{1} \\ &\quad + {}^{2}I_{2}\,{}_{0}^{2}\boldsymbol{\alpha}_{2} + {}_{0}^{2}\boldsymbol{\omega}_{2}\times{}^{2}I_{2}\,{}_{0}^{2}\boldsymbol{\omega}_{2} = \begin{bmatrix} 0 \\ 0 \\ {}^{2}M_{1z} \\ 0 \end{bmatrix} \end{aligned} \tag{12.777} $$

where

$$ \begin{aligned} {}^{2}M_{1z} &= -M_{e} + l_{2}F_{ex}\sin\left(\theta_{1}+\theta_{2}\right) - l_{2}F_{ey}\cos\left(\theta_{1}+\theta_{2}\right) \\ &\quad + \tfrac{1}{3}l_{2}^{2}m_{2}\left(\ddot{\theta}_{1}+\ddot{\theta}_{2}\right) - \tfrac{1}{2}gl_{2}m_{2}\cos\left(\theta_{1}+\theta_{2}\right) \end{aligned} \tag{12.778} $$

Finally the action force on link (1) is

$$ \begin{aligned} {}^{1}\mathbf{F}_{0} &= {}^{1}\mathbf{F}_{1} - m_{1}\,{}^{1}\mathbf{g} + m_{1}\,{}_{0}^{1}\mathbf{a}_{1} \\ &= {}^{1}T_{2}\,{}^{2}\mathbf{F}_{1} - m_{1}\,{}^{1}\mathbf{g} + m_{1}\,{}_{0}^{1}\mathbf{a}_{1} = \begin{bmatrix} {}^{1}F_{0x} \\ {}^{1}F_{0y} \\ 0 \\ 0 \end{bmatrix} \end{aligned} \tag{12.779} $$

where

$$ \begin{aligned} {}^{1}F_{0x} &= -F_{ex}\cos\theta_{1} - \left(F_{ey}-gm_{1}\right)\sin\theta_{1} \\ &\quad - \tfrac{1}{2}l_{2}m_{2}\left(\ddot{\theta}_{1}+\ddot{\theta}_{2}\right)\sin\theta_{2} - \tfrac{1}{2}l_{2}m_{2}\left(\dot{\theta}_{1}+\dot{\theta}_{2}\right)^{2}\cos\theta_{2} \\ &\quad + gm_{2}\sin\left(2\theta_{2}+\theta_{1}\right) - \tfrac{1}{2}l_{1}m_{1}\dot{\theta}_{1}^{2} \end{aligned} \tag{12.780} $$

$$ \begin{aligned} {}^{1}F_{0y} &= F_{ex}\sin\theta_{1} - \left(F_{ey}+gm_{1}\right)\cos\theta_{1} \\ &\quad + \tfrac{1}{2}l_{2}m_{2}\left(\ddot{\theta}_{1}+\ddot{\theta}_{2}\right)\cos\theta_{2} - \tfrac{1}{2}l_{2}m_{2}\left(\dot{\theta}_{1}+\dot{\theta}_{2}\right)^{2}\sin\theta_{2} \\ &\quad - gm_{2}\cos\left(2\theta_{2}+\theta_{1}\right) + \tfrac{1}{2}l_{1}m_{1}\ddot{\theta}_{1} \end{aligned} \tag{12.781} $$

and the action moment on link (1) is

$$\begin{aligned}
{}^1\mathbf{M}_0 &= {}^1\mathbf{M}_1 - {}^1\mathbf{n}_1 \times {}^1\mathbf{F}_0 + {}^1\mathbf{m}_1 \times {}^1\mathbf{F}_1 \\
&\quad + {}^1I_1\, {}^1_0\boldsymbol{\alpha}_1 + {}^1_0\boldsymbol{\omega}_1 \times {}^1I_1\, {}^1_0\boldsymbol{\omega}_1 \\
&= {}^1T_2\, {}^2\mathbf{M}_1 - {}^1\mathbf{n}_1 \times {}^1\mathbf{F}_0 + {}^1\mathbf{m}_1 \times {}^1T_2\, {}^2\mathbf{F}_1 \\
&\quad + {}^1I_1\, {}^1_0\boldsymbol{\alpha}_1 + {}^1_0\boldsymbol{\omega}_1 \times {}^1I_1\, {}^1_0\boldsymbol{\omega}_1 = \begin{bmatrix} 0 \\ 0 \\ {}^1M_{0z} \\ 0 \end{bmatrix}
\end{aligned} \tag{12.782}$$

where

$$\begin{aligned}
{}^1M_{0z} &= -M_e + \tfrac{1}{3}l_2^2 m_2 \left(\ddot{\theta}_1 + \ddot{\theta}_2\right) + \tfrac{1}{3}l_1^2 m_1 \ddot{\theta}_1 \\
&\quad - \left(F_{ey} l_2 + \tfrac{1}{2} g l_2 m_2\right) \cos\left(\theta_1 + \theta_2\right) \\
&\quad - \tfrac{1}{2} l_1 m_1 g \cos\theta_1 + F_{ex} l_2 \sin\left(\theta_1 + \theta_2\right)
\end{aligned} \tag{12.783}$$

Example 764 ★ Actuator's Force and Torque Applying a backward-recursive force analysis ends up with a set of known force systems at joints. Each joint is driven by a motor known as an actuator that applies a force in a P joint or a torque in an R joint. When the joint i is prismatic, the force of the driving actuator is along the z_{i-1}-axis,

$$F_m = {}^0\hat{k}_{i-1}^{\mathrm{T}}\, {}^0\mathbf{F}_i \tag{12.784}$$

showing that the $\hat{k}_{i-1}$-component of the joint force $\mathbf{F}_i$ is supported by the actuator. The $\hat{\imath}_{i-1}$- and $\hat{\jmath}_{i-1}$-components of $\mathbf{F}_i$ must be supported by the bearings of the joint. Similarly, when the joint i is revolute, the torque of the driving actuator is along the z_{i-1}-axis,

$$M_m = {}^0\hat{k}_{i-1}^{\mathrm{T}}\, {}^0\mathbf{M}_i \tag{12.785}$$

showing that the $\hat{k}_{i-1}$-component of the joint torque $\mathbf{M}_i$ is supported by the actuator. The $\hat{\imath}_{i-1}$- and $\hat{\jmath}_{i-1}$-components of $\mathbf{M}_i$ must be supported by the bearings of the joint.

KEY SYMBOLS

$\mathbf{0}$	zero vector
a	acceleration, length
$\mathbf{a}$	acceleration vector
A	body coordinate frame
$[A]$	coefficient matrix
$\mathbf{b}$	vector of known values
B	body coordinate frame, local coordinate frame
c	cos, constant coefficient
C	mass center
d	distance between two points

$\mathbf{d}$	translation vector
D	mass moment equivalent
e	Euler parameter
$E\ (O\varphi\theta\psi)$	Eulerian local frame
$F, \mathbf{F}$	force
$\mathbf{g}$	gravitational acceleration vector
G	global coordinate frame, fixed coordinate frame
$I, [I]$	mass moment
$\mathbf{I}, [\mathbf{I}]$	identity matrix
$\hat{\imath}, \hat{\jmath}, \hat{k}$	local coordinate axis unit vectors
$\hat{I}, \hat{J}, \hat{K}$	global coordinate axis unit vectors
k_1, k_2	coefficients of Duffing equation
K	kinetic energy
l	length
$L, \mathbf{L}$	angular momentum
m	mass
$\mathbf{m}, \mathbf{n}$	position vectors of C of a link in its frame
$M, \mathbf{M}$	moment
n	number of links of a multibody, number of particles of a rigid body
p	roll rate
P	a body point, a fixed point in B
q	pitch rate
Q	torque
r	yaw rate
$\mathbf{r}$	position vector
r_{ij}	the element of row i and column j of a matrix
R	rotation transformation matrix, radius
s	sin, sign
t	time
T	homogenous transformation matrix, Tension force
$\mathbf{v}$	velocity vector
$\mathbf{W}$	gravitational force η
x, y, z	local coordinate axes
$\mathbf{x}$	vector of unknown values
X, Y, Z	global coordinate axes

Greek

α	angular acceleration
$\boldsymbol{\alpha}$	angular acceleration vector
α, β, γ	rotation angles about global axes
δ_{ij}	Kronecker delta
η	complex frequency
θ	pitch
ρ	density
τ	time parameter
φ	roll
φ, θ, ψ	rotation angles about local axes, Euler angles
$\dot{\varphi}, \dot{\theta}, \dot{\psi}$	Euler frequencies
ψ	yaw
$\omega_x, \omega_y, \omega_z$	angular velocity components

ω, $\boldsymbol{\omega}$	angular velocity vector
Ω	angular velocity
Symbol	
$\parallel$	parallel
$\perp$	perpendicular
$[\ ]^{-1}$	inverse of the matrix $[\ ]$
$[\ ]^{\mathrm{T}}$	transpose of the matrix $[\ ]$
DOF	degree of freedom, degrees of freedom
FBD	free body diagram
$\mathrm{cn}(u, k)$	Jacobi elliptic function
$\mathrm{sn}(u, k)$	Jacobi elliptic function
$\mathrm{dn}(u, k)$	Jacobi elliptic function

EXERCISES

1. **Kinetic Energy of a Cubic Rigid Body** Consider a cubic rigid body B with a coordinate frame $B\,(Oxyz)$ at the geometric center of the cube $a \times a \times a$. The body is rotating in a global coordinate frame $G\,(OXYZ)$ with angular velocity $_G\boldsymbol{\omega}_B$. Determine the kinetic energy of B if the density ρ of the cube is:

   (a) $$\rho = m/V = m/(a \times a \times a) = \text{const}$$

   (b) $$\rho = cr, \qquad r = \sqrt{x^2 + y^2 + z^2}$$

   (c) $$\rho = x^2\,|y|\,|z|$$

   (d) $$\rho = x^2(a/2 - y)$$

2. **A Tilted Disc on a Turntable** Assume the rotating disc of Figure 12.3 in Example 721 is mounted in an angle α as shown in Figure 12.49.

   (a) Determine the Euler equations of motion of the disc.

   (b) Show that the equations will reduce to (12.117)–(12.119) for $\alpha = 0$ and reduce to (12.128)–(12.130) for $\alpha = 90\,\text{deg}$ with a proper renaming of the principal axes of the disc.

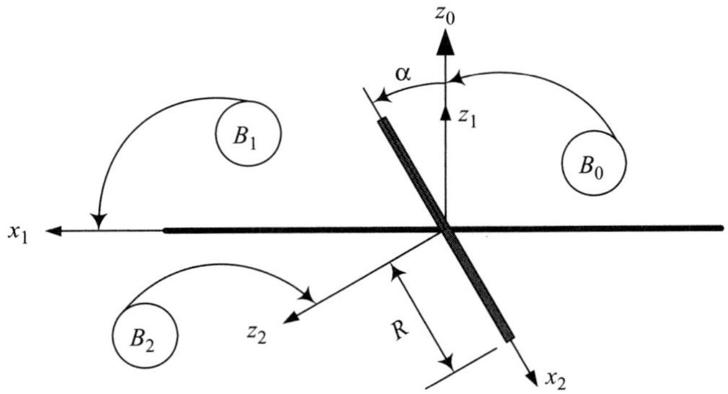

Figure 12.49 A tilted turning disc on a turning table.

3. ★ **Expression for $\mathbf{L}^2$** When $\mathbf{M} = 0$, $d\mathbf{L}/dt = 0$ and hence $\mathbf{L}$ and $\mathbf{L} \cdot \mathbf{L}$ are constant. Determine the expression of ${}^B\mathbf{L} \cdot {}^B\mathbf{L}$ for a general rigid body with a mass moment matrix $[I]$ and simplify the equation for a principal coordinate in which $[I]$ is diagonal.

4. ★ **Expression for $\mathbf{M}^2$** Determine the expression $\mathbf{M}^2 = {}^B\mathbf{M} \cdot {}^B\mathbf{M}$ for a general rigid body. Then simplify $\mathbf{M}^2$ for a principal coordinate frame, an asymmetric rigid body, and a symmetric rigid body.

5. ★ **Expression for K^2** Determine the expression of K^2 for a general rigid body.

6. ★ **Expression for Derivatives of K** Determine $dK/d\boldsymbol{\omega}$, $dK^2/d\boldsymbol{\omega}$, $dK/d\omega_r$, $dK^2/d\omega_r$, $r = 1, 2, 3$.

7. **Dynamic Characteristics of a Torque-Free Rigid Body** Consider a rigid body with the following principal mass moment matrix and initial angular velocity at $t = 0$:

$$
{}^BI = \begin{bmatrix} I_1 & 0 & 0 \\ 0 & I_2 & 0 \\ 0 & 0 & I_3 \end{bmatrix} = \begin{bmatrix} I_1 & 0 & 0 \\ 0 & I_2 & 0 \\ 0 & 0 & I_3 \end{bmatrix} \text{ kg m}^2
$$

$$
{}^B_G\boldsymbol{\omega}_B = \begin{bmatrix} \omega_1 \\ \omega_2 \\ \omega_3 \end{bmatrix} = \begin{bmatrix} 5 \\ 10 \\ 150 \end{bmatrix} \text{ rad/s}
$$

Determine $\mathbf{L}^2$ and K of the rigid body.

8. **A Sliding Ladder** A uniform rod of mass m and length $2a$ is placed like a ladder on a frictionless ground and wall, as shown in Figure 12.50. It is released from $\theta(0) = \theta_0$. Determine the angle θ_m at which the rod loses contact with the ground or wall.

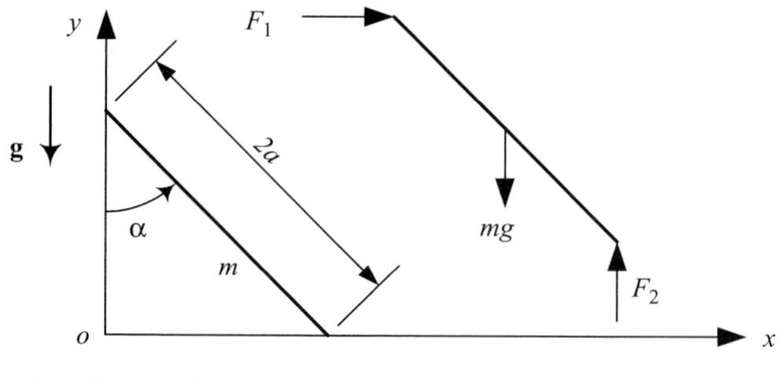

Figure 12.50 A uniform rod of mass m and length $2a$ on a frictionless ground and wall.

9. ★ **Angular Velocity for a Constant K** Consider a symmetric rigid body with a constant kinetic energy K.

   (a) If $\omega_1 = 0$ and $\omega_2 = \omega \sin t$, then what is ω_2?

   (b) If $\omega_1 \neq 0$ and $\omega_2 = \omega \sin t$, then what is ω_2?

10. ★ **Quasi-Kinetic Energy** Determine the following expressions:

(a)
$$N_3 = \frac{1}{3}\int_B \mathbf{v}^3 dm$$

(b)
$$N_4 = \frac{1}{4}\int_B \mathbf{v}^4 dm$$

(c)
$$N_5 = \frac{1}{5}\int_B \mathbf{v}^5 dm$$

(d) Is there a general rule to determine N_n based on $N_2 = K = \frac{1}{2}\int_B \mathbf{v}^2 dm$?

11. **Falling Disc** Assume the disc of Figure 12.51 has an initial velocity v_0 at its center. Show that the velocity of the disc is

$$v_h = \sqrt{v_0^2 + \frac{2mgh}{m + I/R}}$$

if it rolls without slip and drops a height h on a curved path. The disc has a mass m, radius R, and mass moment I.

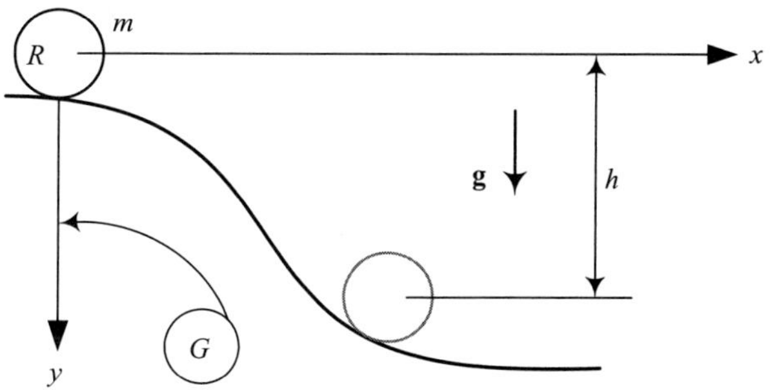

Figure 12.51 A rolling disc on a curved path.

12. **A Turning Pendulum** Figure 12.52 illustrates a pendulum of length l with a hung mass m. The pivot of the pendulum is a circle point that is turning on a circle with radius R and angular speed ω.

(a) Attach a coordinate frame B to the pendulum and determine the transformation matrices.

(b) Determine the angular velocity and acceleration of the pendulum.

(c) Use a free-body diagram and the Newton–Euler method to determine the equations of motion.

(d) Is it possible to decouple the equation for θ from the equations for joint forces?

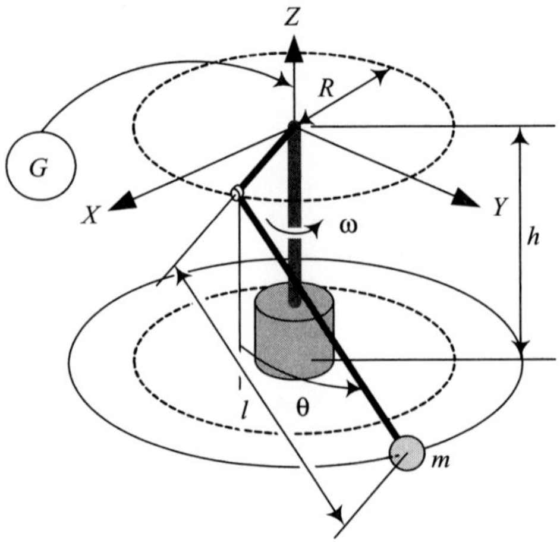

Figure 12.52 A turning pendulum.

13. **★ Euler Equation for a Rigid Body with Nonuniform Density** A rigid body $B\ (Oxyz)$ is turning in a global coordinate frame $G\ (OXYZ)$ with angular velocity ${}_G\boldsymbol{\omega}_B$. Determine the Euler equations of B if the body is a cube with size $a \times a \times a$ and with the following densities:

(a)
$$\rho = x^2(a/2 - y)$$

(b)
$$\rho = x^2(a/2 - y)(a/2 - z)$$

(c)
$$\rho = x^2\left(\frac{a}{2} - y\right)\left(\frac{a^2}{4} - z^2\right)$$

14. **Gyroscopic Effect** A circular disc of mass m and mass moment I about its central axis rotates about its axis of symmetry with angular velocity Ω. At the same time, the axis itself rotates in a horizontal plane with a precessional angular velocity ω as shown in Figure 12.53. A gravity moment mgl acts on the system. Find the value of ω compatible with this steady-state motion.

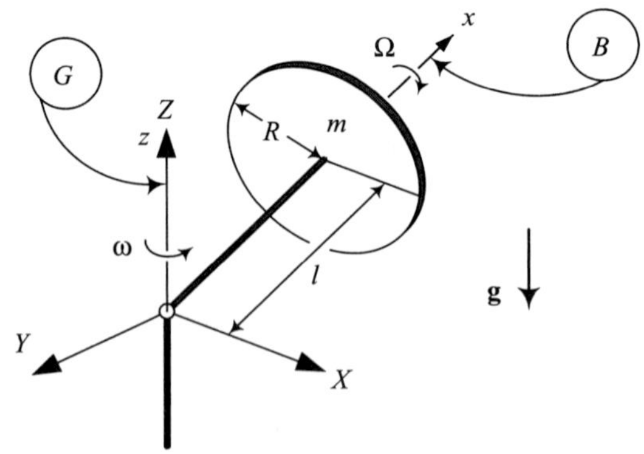

Figure 12.53 A rotating disc about its axis of symmetry and about a varietal axis.

15. A Turning Plate about a Symmetric Line on a Turntable Consider a uniform plate with mass m, side a, and mass moment $[I]$ that is mounted on a horizontal shaft as shown in Figure 12.54. The shaft is mounted on a table and is turning with constant angular velocity $\dot{\theta}_2 = \omega$ with respect to the table. If the table is also turning with angular velocity $\dot{\theta}_1 = \Omega$, then calculate:

(a) How much force is supported by bearings of the shaft.
(b) The moment ${}^0\mathbf{M}$ in the global frame.

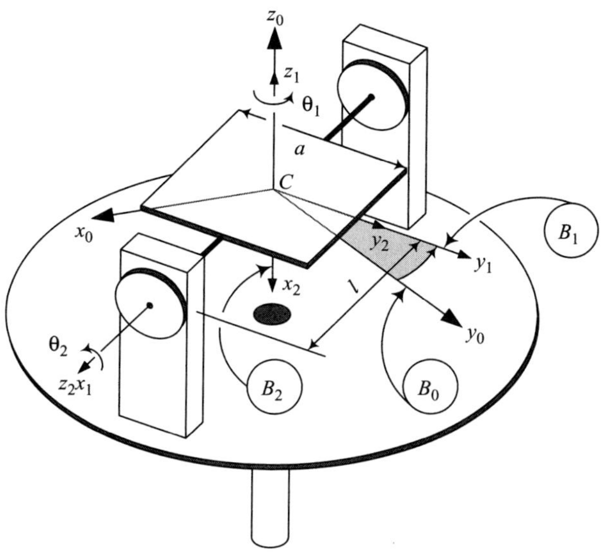

Figure 12.54 A turning square plate about a shaft on a turning table.

16. A Turning Plate about the Centerline on a Turntable Change the orientation of the plate of Problem 15 as shown in Figure 12.55 and solve the problem again.

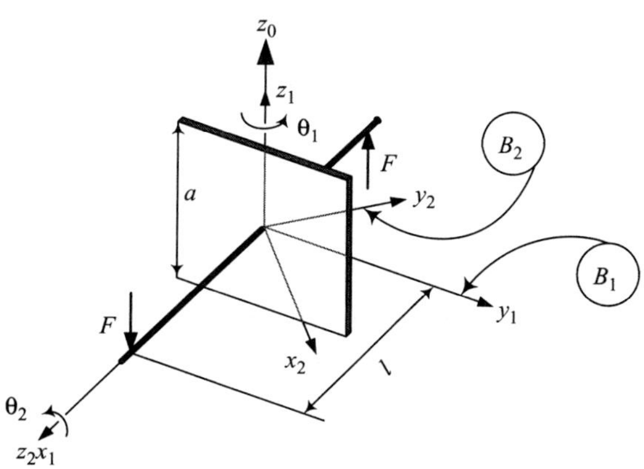

Figure 12.55 A turning plate about the centerline on a turntable.

17. A Turning Plate about Its Diagonal on a Turntable Change the orientation of the plate of Problem 15 as shown in Figure 12.56 and solve the problem again.

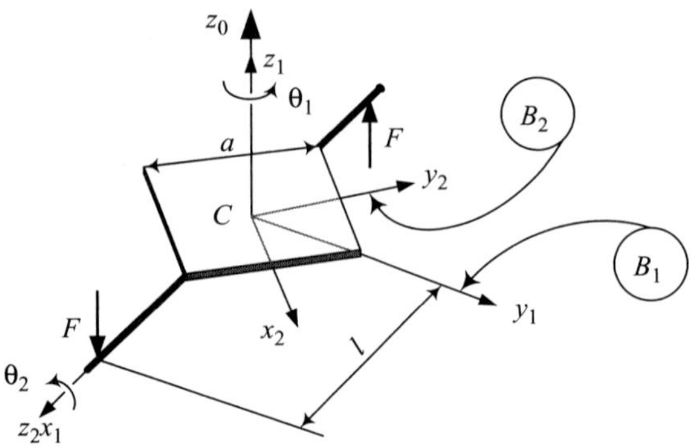

Figure 12.56 A turning plate about its diagonal on a turntable.

18. A Turning Half Square Plate about Its Side on a Turntable Change the plate of Problem 15 to half a square as shown in Figure 12.57 and solve the problem again.

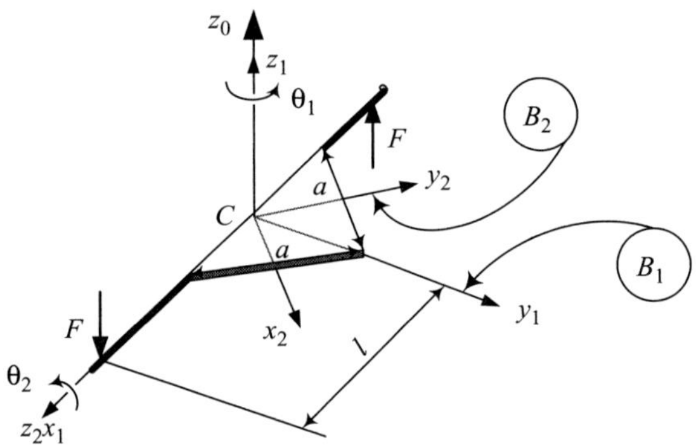

Figure 12.57 A turning half square plate about its diagonal on a turntable.

19. Axisymmetric Rigid-Body Harmonic Equation Combine Equations (12.328) and (12.329) to derive the following equations:

$$\ddot{\omega}_\theta \pm \Omega^2 \omega_\theta = 0$$

$$\omega_1^2 + \omega_2^2 = \omega_\theta^2$$

Then, discuss the solution for (+) and (−) signs.

20. Axisymmetric Rigid-Body Acceleration Ratio Show that for an asymmetric torque-free rigid body with $I_1 = I_2 \neq I_3$ we have

$$\frac{\ddot{\omega}_1}{\ddot{\omega}_2} = \frac{\dot{\omega}_2}{\dot{\omega}_1}$$

21. Asymmetric Torque-Free Rigid Body

(a) Prove Equation (12.411).

(b) Using Equations (12.408) and (12.409), reduce Equations (12.405)–(12.407) to a single equation for ω_1.

22. Orientation of B in G In Example 745 assume the body coordinate frame B is not principal. Discuss if it is possible to determine the orientation of B in G.

23. A Turning Dumbbell Figure 12.58 illustrates a turning dumbbell with a massless rod of length l and two equal masses m. The bar has an angle θ with respect to the Z-axis, which is the axis of rotation. Calculate the moment of momentum of the dumbbell and the required torque to sustain this motion if the angular velocity $\boldsymbol{\omega}$ is constant.

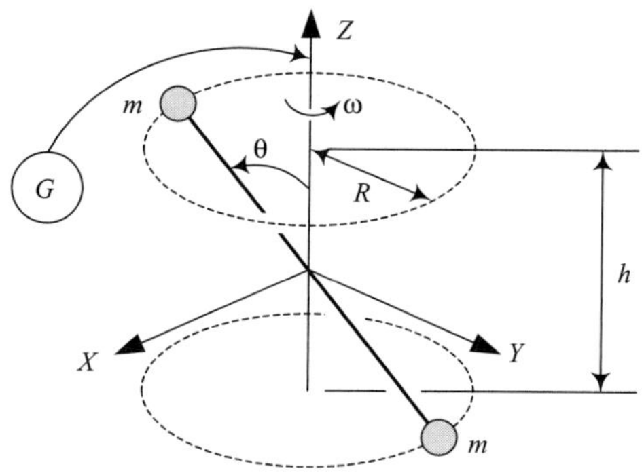

Figure 12.58 A turning dumbbell.

24. A Rotating Cone about Its Tip Point Figure 12.59 illustrates a cone of half angle α and height h at an angle θ with the Z-axis. The cone is turning about the vertical Z-axis with angular velocity $\dot{\varphi}$. At the same time, the cone spins about its axis of symmetry with angular velocity $\dot{\psi}$. For given constant θ and $\dot{\psi}$, determine $\dot{\varphi}$.

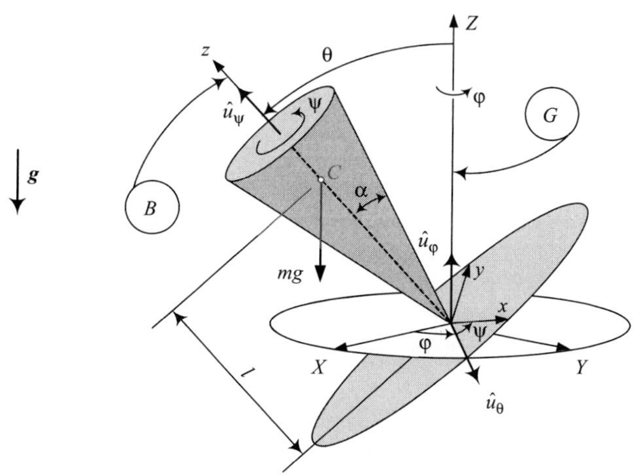

Figure 12.59 A rotating cone about its tip point.

25. Symmetric Rigid Body with Variable Mass Moments The equations of a spherical neutron star is reduced to Equations (12.401) and (12.402):

$$I_1\dot{\omega}_1 - \frac{3}{2}I_0\epsilon\frac{\omega_{30}\cos\Omega t}{1-\epsilon\sin\Omega t}\omega_2 = 0$$

$$I_1\dot{\omega}_2 + \frac{3}{2}I_0\epsilon\frac{\omega_{30}\cos\Omega t}{1-\epsilon\sin\Omega t}\omega_1 = 0$$

Let us rewrite these equations as

$$\dot{\omega}_1 - a\,(t)\,\omega_2 = 0 \qquad \dot{\omega}_2 + a\,(t)\,\omega_1 = 0$$

$$a\,(t) = \frac{3}{2}\frac{I_0}{I_1}\epsilon\frac{\omega_{30}\cos\Omega t}{1-\epsilon\sin\Omega t}$$

Show that these equations can be reduced to

$$\ddot{\omega}_1 + b\,(t)\,\dot{\omega}_1 + a^2\omega_1 = 0 \qquad b\,(t) = -\frac{\dot{a}}{a}$$

or

$$\ddot{\omega}_2 + b\,(t)\,\dot{\omega}_2 + a^2\omega_2 = 0 \qquad b\,(t) = -\frac{\dot{a}}{a}.$$

Reduce the equation to a Hill equation and determine $c\,(t)$,

$$\ddot{u} + c\,(t)\,u = 0$$

by using a new variable u:

$$\omega_2 = ue^{-bt/2}$$

26. ★ Variable Mass Moment Derive the general Euler equations of motion of a rigid body B with variable mass moment matrix $[I] = [I\,(t)]$.

27. A Turning Uniform Link Figure 12.60(a) depicts a triangular link attached to the ground by a revolute joint at O. The free-body diagram of the link shows the gravity and the driving force and moment at the joint, as shown in Figure 12.60(b). Determine the Newton–Euler equations of the triangular link.

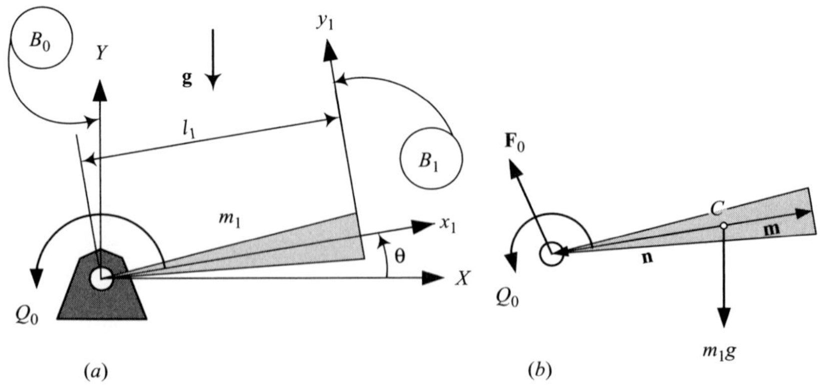

Figure 12.60 A turning uniform triangular link.

28. A Turning Uniform Beam with a Tip Mass Consider the uniform massless beam of Figure 12.61(a) with a hanging mass m_2 at the tip point. Figure 12.61(b) illustrates the free-body diagram of the beam. Determine the equations of motion of the beam.

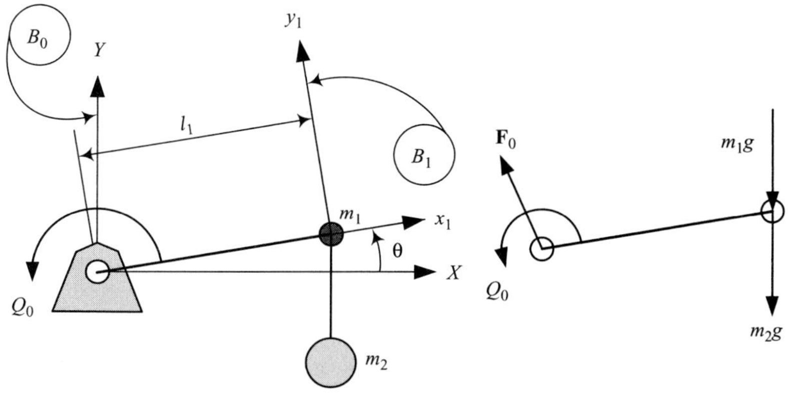

(a) (b)

Figure 12.61 A turning uniform beam with a tip mass.

29. 2R Planar Manipulator Newton–Euler Dynamics A 2R planar manipulator and its free-body diagram are shown in Figure 12.62. The torques of actuators are parallel to the Z-axis and are indicated by Q_0 and Q_1.
Use the multibody Newton–Euler dynamics and determine the equations of motion to find the required torques Q_0 and Q_1 and the joint forces F_0 and F_1.

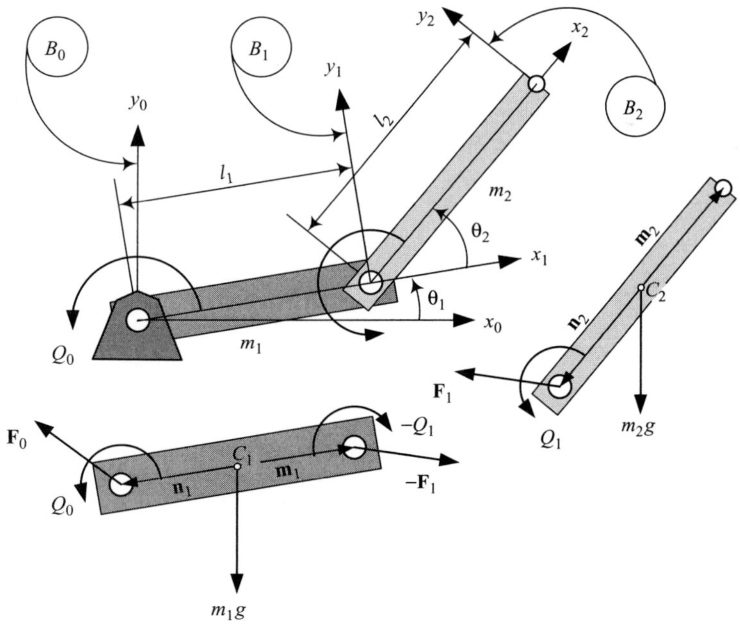

Figure 12.62 Free body diagram of a 2R palanar manipulator.

30. Euler Equation of a Rectangular Rigid Body Determine the Euler equations of motion of a rectangular rigid body $a \times b \times c$ which is turning in a global frame $G\,(OXYZ)$ about its fixed geometric center if:

(a) ${}_G\boldsymbol{\omega}_B = \dot{\alpha}\hat{I}$

(b) ${}_G\boldsymbol{\omega}_B = \dot{\alpha}\hat{I} + \dot{\beta}\hat{J}$

(c) ${}_G\boldsymbol{\omega}_B = \dot{\alpha}\hat{I} + \dot{\beta}\hat{J} + \dot{\gamma}\hat{K}$

31. ★ 2R Planar Manipulator with Massive Arms and Joints A real 2R planar manipulator has a massive motor at joint 0 to turn link (1) and a massive motor at joint 1 to turn the link (2). The manipulator may also carry a massive object by the gripper at the tip point of link (2). The motor at joint 0 is siting on the ground and its weight will not affect the dynamics of the manipulator. Use the free-body diagram of Figure 12.63 and determine the equations of motion of the manipulator.

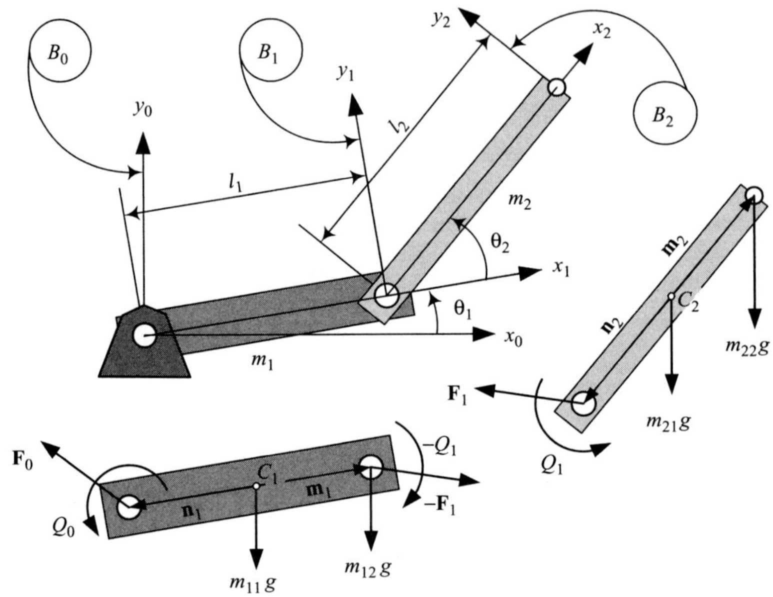

Figure 12.63 A 2R planar manipulator with massive arms and massive joints.

13

Lagrange Dynamics

Lagrangian dynamics seeks the equations of motion of a dynamic system in the generalized configuration space. The Lagrange equation provides n scalar second-order differential equations for an n-DOF dynamic system. Solution of the equations determines the time behavior of the generalized coordinates. We are free to select any set of generalized coordinates so the equations of motion may be simpler to solve in special generalized spaces.

13.1 LAGRANGE FORM OF NEWTON EQUATIONS

Consider a mechanical system of n point masses m_i, $i = 1, 2, \ldots n$. The associated n Newton equations of motion can be transformed to

$$\frac{d}{dt}\left(\frac{\partial K}{\partial \dot{q}_r}\right) - \frac{\partial K}{\partial q_r} = Q_r \qquad r = 1, 2, \ldots n \tag{13.1}$$

where

$$Q_r = \sum_{i=1}^{n}\left(F_{ix}\frac{\partial f_i}{\partial q_1} + F_{iy}\frac{\partial g_i}{\partial q_2} + F_{iz}\frac{\partial h_i}{\partial q_n}\right) \tag{13.2}$$

Equation (13.1) is called the *Lagrange equation of motion*, where K is the kinetic energy of the n-DOF system; q_r, $r = 1, 2, \ldots, n$, are the generalized coordinates of the system; $\mathbf{F} = \left[F_{ix}, F_{iy}, F_{iz}\right]^{\mathrm{T}}$ is the external force acting on the ith particle of the system; and Q_r is the generalized force associated with q_r. The functions f_i, g_i, h_i determine the Cartesian coordinates x_i, y_i, z_i of m_i from the generalized coordinates q_j, $j = 1, 2, \ldots, n$.

Proof: Let (x_i, y_i, z_i) be the Cartesian coordinates of the ith particle in a globally fixed coordinate frame. Assume that the coordinates of every individual particle are functions of a set of generalized coordinates $q_1, q_2, \ldots, q_n$ and possibly time t:

$$x_i = f_i(q_1, q_2, q_3, \ldots, q_n, t) \tag{13.3}$$

$$y_i = g_i(q_1, q_2, q_3, \ldots, q_n, t) \tag{13.4}$$

$$z_i = h_i(q_1, q_2, q_3, \ldots, q_n, t) \tag{13.5}$$

If F_{xi}, F_{yi}, F_{zi} are components of the total force acting on the particle m_i, then the Newton equations of motion for the particle would be

$$F_{xi} = m_i \ddot{x}_i \tag{13.6}$$

$$F_{yi} = m_i \ddot{y}_i \tag{13.7}$$

$$F_{zi} = m_i \ddot{z}_i \tag{13.8}$$

We respectively multiply both sides of these equations by

$$\frac{\partial f_i}{\partial q_r} \qquad \frac{\partial g_i}{\partial q_r} \qquad \frac{\partial h_i}{\partial q_r} \tag{13.9}$$

and add them to get

$$\sum_{i=1}^{n} m_i \left(\ddot{x}_i \frac{\partial f_i}{\partial q_r} + \ddot{y}_i \frac{\partial g_i}{\partial q_r} + \ddot{z}_i \frac{\partial h_i}{\partial q_r} \right) = \sum_{i=1}^{n} \left(F_{xi} \frac{\partial f_i}{\partial q_r} + F_{yi} \frac{\partial g_i}{\partial q_r} + F_{zi} \frac{\partial h_i}{\partial q_r} \right) \tag{13.10}$$

where n is the total number of particles.

Taking a time derivative of Equation (13.3),

$$\dot{x}_i = \frac{\partial f_i}{\partial q_1}\dot{q}_1 + \frac{\partial f_i}{\partial q_2}\dot{q}_2 + \frac{\partial f_i}{\partial q_3}\dot{q}_3 + \cdots + \frac{\partial f_i}{\partial q_n}\dot{q}_n + \frac{\partial f_i}{\partial t} \tag{13.11}$$

we find

$$\frac{\partial \dot{x}_i}{\partial \dot{q}_r} = \frac{\partial}{\partial \dot{q}_r} \left(\frac{\partial f_i}{\partial q_1}\dot{q}_1 + \frac{\partial f_i}{\partial q_2}\dot{q}_2 + \cdots + \frac{\partial f_i}{\partial q_n}\dot{q}_n + \frac{\partial f_i}{\partial t} \right) = \frac{\partial f_i}{\partial q_r} \tag{13.12}$$

and therefore,

$$\ddot{x}_i \frac{\partial f_i}{\partial q_r} = \ddot{x}_i \frac{\partial \dot{x}_i}{\partial \dot{q}_r} = \frac{d}{dt}\left(\dot{x}_i \frac{\partial \dot{x}_i}{\partial \dot{q}_r} \right) - \dot{x}_i \frac{d}{dt}\left(\frac{\partial \dot{x}_i}{\partial \dot{q}_r} \right) \tag{13.13}$$

However,

$$\begin{aligned} \dot{x}_i \frac{d}{dt}\left(\frac{\partial \dot{x}_i}{\partial \dot{q}_r} \right) &= \dot{x}_i \frac{d}{dt}\left(\frac{\partial f_i}{\partial q_r} \right) \\ &= \dot{x}_i \left(\frac{\partial^2 f_i}{\partial q_1 \partial q_r}\dot{q}_1 + \cdots + \frac{\partial^2 f_i}{\partial q_n \partial q_r}\dot{q}_n + \frac{\partial^2 f_i}{\partial t \partial q_r} \right) \\ &= \dot{x}_i \frac{\partial}{\partial q_r}\left(\frac{\partial f_i}{\partial q_1}\dot{q}_1 + \frac{\partial f_i}{\partial q_2}\dot{q}_2 + \cdots + \frac{\partial f_i}{\partial q_n}\dot{q}_n + \frac{\partial f_i}{\partial t} \right) \\ &= \dot{x}_i \frac{\partial \dot{x}_i}{\partial q_r} \end{aligned} \tag{13.14}$$

and we have

$$\ddot{x}_i \frac{\partial \dot{x}_i}{\partial \dot{q}_r} = \frac{d}{dt}\left(\dot{x}_i \frac{\partial \dot{x}_i}{\partial \dot{q}_r} \right) - \dot{x}_i \frac{\partial \dot{x}_i}{\partial q_r} \tag{13.15}$$

which is equal to

$$\ddot{x}_i \frac{\partial f_i}{\partial q_r} = \ddot{x}_i \frac{\dot{x}_i}{\dot{q}_r} = \frac{d}{dt}\left[\frac{\partial}{\partial \dot{q}_r}\left(\frac{1}{2}\dot{x}_i^2\right)\right] - \frac{\partial}{\partial q_r}\left(\frac{1}{2}\dot{x}_i^2\right) \tag{13.16}$$

Following the same procedure, Equations (13.4) and (13.5) provide

$$\ddot{y}_i \frac{\partial g_i}{\partial q_r} = \ddot{y}_i \frac{\dot{y}_i}{\dot{q}_r} = \frac{d}{dt}\left[\frac{\partial}{\partial \dot{q}_r}\left(\frac{1}{2}\dot{y}_i^2\right)\right] - \frac{\partial}{\partial q_r}\left(\frac{1}{2}\dot{y}_i^2\right) \tag{13.17}$$

$$\ddot{z}_i \frac{\partial h_i}{\partial q_r} = \ddot{z}_i \frac{\dot{z}_i}{\dot{q}_r} = \frac{d}{dt}\left[\frac{\partial}{\partial \dot{q}_r}\left(\frac{1}{2}\dot{z}_i^2\right)\right] - \frac{\partial}{\partial q_r}\left(\frac{1}{2}\dot{z}_i^2\right) \tag{13.18}$$

Substituting (13.16), (13.17), and (13.18) on the left-hand side of (13.10) leads to

$$\begin{aligned}
&\sum_{i=1}^{n} m_i \left(\ddot{x}_i \frac{\partial f_i}{\partial q_r} + \ddot{y}_i \frac{\partial g_i}{\partial q_r} + \ddot{z}_i \frac{\partial h_i}{\partial q_r}\right) \\
&\quad = \sum_{i=1}^{n} m_i \frac{d}{dt}\left[\frac{\partial}{\partial \dot{q}_r}\left(\frac{1}{2}\dot{x}_i^2 + \frac{1}{2}\dot{y}_i^2 + \frac{1}{2}\dot{z}_i^2\right)\right] \\
&\quad - \sum_{i=1}^{n} m_i \frac{\partial}{\partial q_r}\left(\frac{1}{2}\dot{x}_i^2 + \frac{1}{2}\dot{y}_i^2 + \frac{1}{2}\dot{z}_i^2\right) \\
&\quad = \frac{1}{2}\sum_{i=1}^{n} m_i \frac{d}{dt}\left[\frac{\partial}{\partial \dot{q}_r}\left(\dot{x}_i^2 + \dot{y}_i^2 + \dot{z}_i^2\right)\right] - \frac{1}{2}\sum_{i=1}^{n} m_i \frac{\partial}{\partial q_r}\left(\dot{x}_i^2 + \dot{y}_i^2 + \dot{z}_i^2\right) \\
&\quad = \frac{d}{dt}\left(\frac{\partial K}{\partial \dot{q}_r}\right) - \frac{\partial K}{\partial q_r}
\end{aligned} \tag{13.19}$$

where K is the *kinetic energy* of the system,

$$\frac{1}{2}\sum_{i=1}^{n} m_i \left(\dot{x}_i^2 + \dot{y}_i^2 + \dot{z}_i^2\right) = K \tag{13.20}$$

Therefore, the Newton equations of motion (13.6)–(13.8) are converted to

$$\frac{d}{dt}\left(\frac{\partial K}{\partial \dot{q}_r}\right) - \frac{\partial K}{\partial q_r} = \sum_{i=1}^{n}\left(F_{xi}\frac{\partial f_i}{\partial q_r} + F_{yi}\frac{\partial g_i}{\partial q_r} + F_{zi}\frac{\partial h_i}{\partial q_r}\right) \tag{13.21}$$

From (13.3)–(13.5), the kinetic energy is a function of the generalized coordinates $q_1, q_2, \ldots, q_n, \dot{q}_1, \dot{q}_2, \ldots, \dot{q}_n$ and time t. The left-hand side of Equation (13.21) includes the kinetic energy of the whole system and the right-hand side is a *generalized force* that shows the effect of changing coordinates from x_i to q_j on the external forces.

Let us consider a virtual displacement such that the coordinate q_r alters to $q_r + \delta q_r$ while the other coordinates $q_1, q_2, q_3, \ldots, q_{r-1}, q_{r+1}, \ldots, q_n$ and time t are unchanged. So, the coordinates x_i, y_i, z_i of m_i are changed to

$$x_i + \frac{\partial f_i}{\partial q_r}\delta q_r \qquad y_i + \frac{\partial g_i}{\partial q_r}\delta q_r \qquad z_i + \frac{\partial h_i}{\partial q_r}\delta q_r \tag{13.22}$$

and the work done in this virtual displacement by all forces acting on the particles of the system is

$$\delta W = \sum_{i=1}^{n} \left(F_{xi} \frac{\partial f_i}{\partial q_r} + F_{yi} \frac{\partial g_i}{\partial q_r} + F_{zi} \frac{\partial h_i}{\partial q_r} \right) \delta q_r \tag{13.23}$$

Because the work done by internal forces appears in opposite pairs, only the work done by external forces remains in Equation (13.23). Let us denote the virtual work by

$$\delta W = Q_r \left(q_1, q_2, q_3, \ldots, q_n, t \right) \delta q_r \tag{13.24}$$

then we have

$$\frac{d}{dt} \left(\frac{\partial K}{\partial \dot{q}_r} \right) - \frac{\partial K}{\partial q_r} = Q_r \tag{13.25}$$

where

$$Q_r = \sum_{i=1}^{n} \left(F_{xi} \frac{\partial f_i}{\partial q_r} + F_{yi} \frac{\partial g_i}{\partial q_r} + F_{zi} \frac{\partial h_i}{\partial q_r} \right) \tag{13.26}$$

Equation (13.25) is the Lagrange form of equations of motion. This equation is true for all values of r from 1 to n. We thus have n second-order scalar ordinary equations in which $q_1, q_2, q_3, \ldots, q_n$ are the dependent variables and t is the independent variable. The generalized coordinates $q_1, q_2, q_3, \ldots, q_n$ can be any measurable parameters to provide the configuration of the system. Because the number of equations and the number of dependent variables are equal, the equations are theoretically sufficient to find all q_i and determine the motion of all m_i.

The term $\partial K / \partial \dot{q}_r$ is called the *generalized momentum* p_r associated with q_r:

$$p_r = \frac{\partial K}{\partial \dot{q}_r} \tag{13.27}$$

So, the Lagrange equation may also be written as

$$\frac{dp_r}{dt} - \frac{\partial K}{\partial q_r} = Q_r \tag{13.28}$$

■

Example 765 Equation of Motion of a Simple Pendulum A simple pendulum is a point mass suspended by a massless string which has a planar motion. A simple pendulum in a constant gravitational field is shown in Figure 13.1. Using x and y for the Cartesian position of m and $\theta = q$ as the generalized coordinate, we have

$$x = f(\theta) = l \sin\theta \tag{13.29}$$

$$y = g(\theta) = l \cos\theta \tag{13.30}$$

$$K = \tfrac{1}{2} m \left(\dot{x}^2 + \dot{y}^2 \right) = \tfrac{1}{2} m l^2 \dot{\theta}^2 \tag{13.31}$$

and therefore,

$$\frac{d}{dt} \left(\frac{\partial K}{\partial \dot{\theta}} \right) - \frac{\partial K}{\partial \theta} = \frac{d}{dt} (m l^2 \dot{\theta}) = m l^2 \ddot{\theta} \tag{13.32}$$

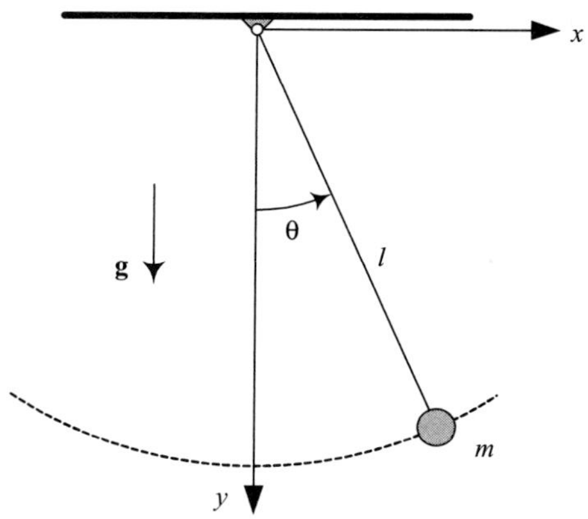

Figure 13.1 A simple pendulum in a constant gravitational field.

The external force components acting on m are

$$F_x = 0 \qquad F_y = mg \tag{13.33}$$

and therefore the generalized force is given as

$$Q_\theta = F_x \frac{\partial f}{\partial \theta} + F_y \frac{\partial g}{\partial \theta} = -mgl \sin\theta \tag{13.34}$$

Hence, the equation of motion of the pendulum is

$$ml^2\ddot{\theta} = -mgl \sin\theta \tag{13.35}$$

Example 766 A Pendulum Attached to an Oscillating Mass Figure 13.2 illustrates a vibrating mass with a hanging pendulum. This mechanical system has two DOF. Let us use x and θ as the generalized coordinates. Such a pendulum can act as a vibration absorber for linearized approximation of the equations if designed properly.

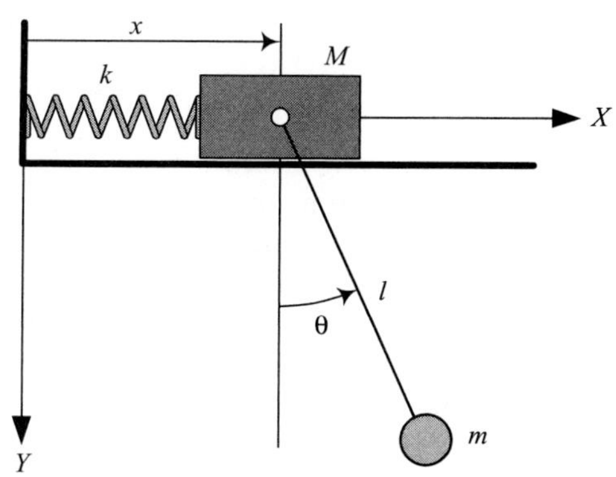

Figure 13.2 A pendulum attached to an oscillating mass.

Beginning with the coordinate relationships

$$X_M = f_M = x \qquad Y_M = g_M = 0 \tag{13.36}$$

$$X_m = f_m = x + l\sin\theta \qquad Y_m = g_m = l\cos\theta \tag{13.37}$$

we may find the kinetic energy in terms of the generalized coordinates x and θ:

$$\begin{aligned} K &= \tfrac{1}{2}M\left(\dot{X}_M^2 + \dot{Y}_M^2\right) + \tfrac{1}{2}m\left(\dot{X}_m^2 + \dot{Y}_m^2\right) \\ &= \tfrac{1}{2}M\dot{x}^2 + \tfrac{1}{2}m\left(\dot{x}^2 + l^2\dot{\theta}^2 + 2l\dot{x}\dot{\theta}\cos\theta\right) \end{aligned} \tag{13.38}$$

Then, the left-hand sides of the Lagrange equations are

$$\frac{d}{dt}\left(\frac{\partial K}{\partial \dot{x}}\right) - \frac{\partial K}{\partial x} = (M+m)\ddot{x} + ml\ddot{\theta}\cos\theta - ml\dot{\theta}^2\sin\theta \tag{13.39}$$

$$\frac{d}{dt}\left(\frac{\partial K}{\partial \dot{\theta}}\right) - \frac{\partial K}{\partial \theta} = ml^2\ddot{\theta} + ml\ddot{x}\cos\theta \tag{13.40}$$

The external forces acting on M and m are

$$F_{X_M} = -kx \qquad F_{Y_M} = 0 \tag{13.41}$$

$$F_{X_m} = 0 \qquad F_{Y_m} = mg \tag{13.42}$$

Therefore, the generalized forces are

$$\begin{aligned} Q_x &= F_{X_M}\frac{\partial f_M}{\partial x} + F_{Y_M}\frac{\partial g_M}{\partial x} + F_{X_m}\frac{\partial f_m}{\partial x} + F_{Y_m}\frac{\partial g_m}{\partial x} \\ &= -kx \end{aligned} \tag{13.43}$$

$$\begin{aligned} Q_\theta &= F_{X_M}\frac{\partial f_M}{\partial \theta} + F_{Y_M}\frac{\partial g_M}{\partial \theta} + F_{X_m}\frac{\partial f_m}{\partial \theta} + F_{Y_m}\frac{\partial g_m}{\partial \theta} \\ &= -mgl\sin\theta \end{aligned} \tag{13.44}$$

and finally the equations of motion are

$$(M+m)\ddot{x} + ml\ddot{\theta}\cos\theta - ml\dot{\theta}^2\sin\theta = -kx \tag{13.45}$$

$$ml^2\ddot{\theta} + ml\ddot{x}\cos\theta = -mgl\sin\theta \tag{13.46}$$

Example 767 A Running Child on a Plank Figure 13.3 illustrates a rough plank of length l and mass m_1 on a smooth inclined surface at an angle θ. A child of mass m_2 runs down the plank and exerts an upward force on the plank. Assume that the child is running fast enough to keep the plank from slipping while applying a backward force F on the plank.

Let us indicate the distance of the upper end of the plank from a fixed point on the slope by x and the distance of the child from the upper end of the plank by y. The kinetic energy of the system is

$$K = \tfrac{1}{2}m_1\dot{x}^2 + \tfrac{1}{2}m_2(\dot{x} + \dot{y})^2 \tag{13.47}$$

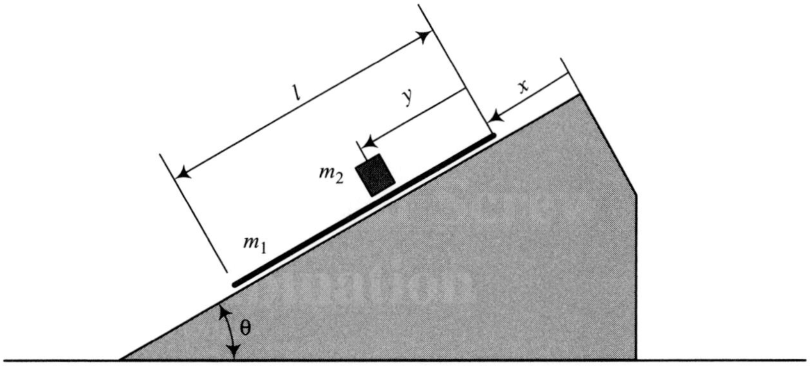

Figure 13.3 A child running on a rough plank that is on a smooth inclined surface.

Taking the required partial derivatives

$$\frac{\partial K}{\partial \dot{x}} = m_1\dot{x} + m_2(\dot{x} + \dot{y}) \tag{13.48}$$

$$\frac{\partial K}{\partial \dot{y}} = m_2(\dot{x} + \dot{y}) \tag{13.49}$$

we use the virtual work form of Lagrange equation,

$$\left[\frac{d}{dt}\left(\frac{\partial K}{\partial \dot{q}_r}\right) - \frac{\partial K}{\partial q_r}\right]\delta q_r = \sum_{i=1}^{n}\left(F_{xi}\frac{\partial f_i}{\partial q_r} + F_{yi}\frac{\partial g_i}{\partial q_r} + F_{zi}\frac{\partial h_i}{\partial q_r}\right)\delta q_r \tag{13.50}$$

$$\left[m_1\ddot{x} + m_2(\ddot{x} + \ddot{y})\right]\delta x = (m_1 + m_2)\, g \sin\theta\, \delta x \tag{13.51}$$

$$\left[m_2(\ddot{x} + \ddot{y})\right]\delta y = (F + (m_1 + m_2)\, g \sin\theta)\,\delta y \tag{13.52}$$

to determine the generalized forces and equations of motion:

$$m_1\ddot{x} + m_2(\ddot{x} + \ddot{y}) = (m_1 + m_2)\, g \sin\theta \tag{13.53}$$

$$m_2(\ddot{x} + \ddot{y}) = F + m_2 g \sin\theta \tag{13.54}$$

If the plank remains stationary, then

$$\ddot{x} = 0 \tag{13.55}$$

and we have

$$m_2\ddot{y} = (m_1 + m_2)\, g \sin\theta \tag{13.56}$$

$$F = m_2\ddot{y} - m_2 g \sin\theta \tag{13.57}$$

Therefore, the required force is given as

$$F = m_1 g \sin\theta \tag{13.58}$$

Let us solve Equation (13.53) and find $\dot{y}$:

$$m_2 \ddot{y}\dot{y} = (m_1 + m_2)\, g\dot{y}\sin\theta \tag{13.59}$$

$$m_2 \dot{y}^2 = (m_1 + m_2)\, gy\sin\theta + C_1 \tag{13.60}$$

$$C_1 = 0 \tag{13.61}$$

The speed of the child when she reaches the tip point of the plank is

$$\dot{y} = \sqrt{\frac{m_1 + m_2}{m_2} gl\sin\theta} \tag{13.62}$$

Example 768 A Sphere on a Table The homogeneous sphere shown in Figure 13.4 is rolling freely on a rough table. Let us examine the possible motions of the sphere.

We attach a fixed coordinate frame G on the table and a body frame B to the center of the sphere. There is one holonomic constraint between B and G,

$$Z_0 - R = 0 \tag{13.63}$$

and therefore the sphere has five DOF. Let (X, Y, R) be the coordinates of the center of the sphere. If φ, θ, and ψ are the Euler angles between B and G, then $X, Y, \varphi, \theta, \psi$ are the generalized coordinates of the sphere. The only force that we have to consider is the friction force F at the contact point with components F_X and F_Y. Using the radius of gyration k, the kinetic energy of the sphere is

$$K = \frac{1}{2} m \left[\dot{X}^2 + \dot{Y}^2 + k^2 \left(\omega_x^2 + \omega_y^2 + \omega_z^2\right)\right] \tag{13.64}$$

Using (4.178), we can define the angular velocity by Euler frequencies,

$$\begin{bmatrix} \omega_x \\ \omega_y \\ \omega_z \end{bmatrix} = \begin{bmatrix} \sin\theta\sin\psi & \cos\psi & 0 \\ \sin\theta\cos\psi & -\sin\psi & 0 \\ \cos\theta & 0 & 1 \end{bmatrix} \begin{bmatrix} \dot{\varphi} \\ \dot{\theta} \\ \dot{\psi} \end{bmatrix} \tag{13.65}$$

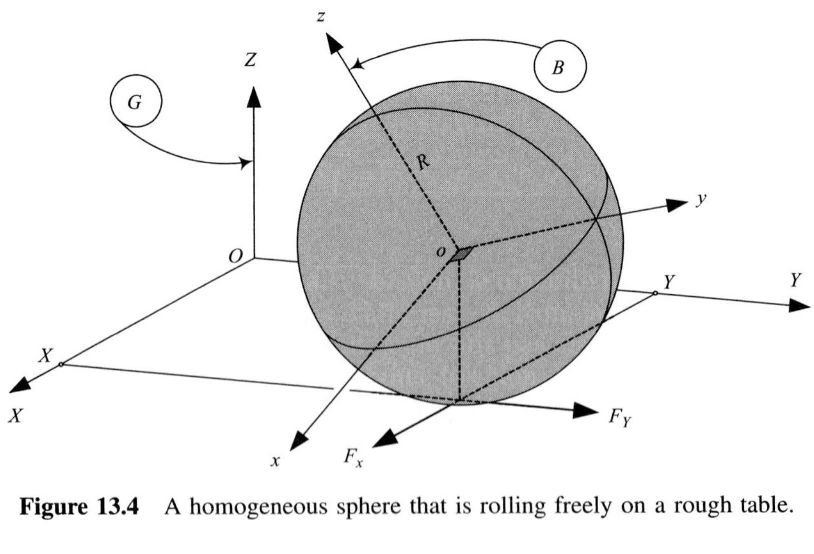

Figure 13.4 A homogeneous sphere that is rolling freely on a rough table.

and therefore,

$$K = \frac{1}{2}m\left[\dot{X}^2 + \dot{Y}^2 + k^2\left(\dot{\varphi}^2 + \dot{\theta}^2 + \dot{\psi}^2 + 2\dot{\psi}\dot{\varphi}\cos\theta\right)\right] \tag{13.66}$$

Employing the Lagrange form (13.25), we find five equations of motion:

$$m\ddot{X} = F_X \tag{13.67}$$

$$m\ddot{Y} = F_Y \tag{13.68}$$

$$mk^2\frac{d}{dt}\left(\dot{\psi} + \dot{\varphi}\cos\theta\right) = 0 \tag{13.69}$$

$$mk^2\frac{d}{dt}\left(\dot{\varphi} + \dot{\psi}\cos\theta\right) = -R\sin\theta\left(F_X\sin\psi - F_Y\cos\psi\right) \tag{13.70}$$

$$mk^2\left(\ddot{\theta} + \dot{\psi}\dot{\varphi}\sin\theta\right) = -R\left(F_X\cos\psi + F_Y\sin\psi\right) \tag{13.71}$$

Because there is no slipping, the rolling nonholonomic constraint applies:

$$\dot{X} - R\omega_y = \dot{X} - R\left(\dot{\varphi}\cos\psi\sin\theta - \dot{\theta}\sin\psi\right) = 0 \tag{13.72}$$

$$\dot{Y} + R\omega_x = \dot{Y} - R\left(\dot{\theta}\cos\psi + \dot{\varphi}\sin\theta\sin\psi\right) = 0 \tag{13.73}$$

We may rewrite Equations (13.70) and (13.71) as

$$\begin{aligned} &mk^2\left[\sin\psi\frac{d}{dt}\left(\dot{\varphi} + \dot{\psi}\cos\theta\right) + \cos\psi\sin\theta\left(\ddot{\theta} + \dot{\psi}\dot{\varphi}\sin\theta\right)\right] \\ &= -RF_X\sin\theta \end{aligned} \tag{13.74}$$

$$\begin{aligned} &mk^2\left[\cos\psi\frac{d}{dt}\left(\dot{\varphi} + \dot{\psi}\cos\theta\right) - \sin\psi\sin\theta\left(\ddot{\theta} + \dot{\psi}\dot{\varphi}\sin\theta\right)\right] \\ &= RF_Y\sin\theta \end{aligned} \tag{13.75}$$

Expanding the derivative terms in (13.74) and (13.75) and eliminating $\ddot{\psi}$ by using (13.69) yield

$$\begin{aligned} mk^2\big[&\ddot{\theta}\cos\psi + \ddot{\varphi}\sin\theta\sin\psi + \dot{\theta}\dot{\varphi}\cos\theta\sin\psi \\ &+\dot{\psi}\dot{\varphi}\left(\cos\psi\sin\theta - \sin\psi\right)\big] = -RF_X \end{aligned} \tag{13.76}$$

$$\begin{aligned} mk^2\big(&-\ddot{\theta}\sin\psi + \ddot{\varphi}\cos\psi\sin\theta + \dot{\theta}\dot{\varphi}\cos\psi\left(\cos\theta - 1\right) \\ &-\dot{\psi}\dot{\varphi}\sin\theta\sin\psi\big) = RF_Y \end{aligned} \tag{13.77}$$

The left-hand sides of (13.76) and (13.77) are $(d/dt)\,\omega_x$ and $(d/dt)\,\omega_y$, respectively. So, we have

$$mk^2\frac{d}{dt}\omega_x = -RF_X \tag{13.78}$$

$$mk^2\frac{d}{dt}\omega_y = RF_Y \tag{13.79}$$

Substituting these equations in the derivatives of the constraint equations (13.72) and (13.73), we get

$$\frac{mk^2}{R}\ddot{X} = RF_Y \tag{13.80}$$

$$\frac{mk^2}{R}\ddot{Y} = RF_X \tag{13.81}$$

Comparing them with Equations (13.67) and (13.68) indicates that

$$\ddot{X} = 0 \qquad \ddot{Y} = 0 \tag{13.82}$$

and from (13.78), (13.79), and (13.69), we have

$$\frac{d}{dt}\omega_x = 0 \qquad \frac{d}{dt}\omega_y = 0 \qquad \frac{d}{dt}\omega_z = 0 \tag{13.83}$$

and

$$F_X = 0 \qquad F_Y = 0 \tag{13.84}$$

Hence, the center of the sphere moves on a straight line with a constant velocity, and the sphere rotates about a locally fixed axis with a constant angular velocity.

Example 769 ★ Rotational Equations of Motion of a Rigid Body Let us use the Lagrange equation and find the rotational equations of motion of a rigid body in the Eulerian frame $E\ (\varphi, \theta, \psi)$.

To simplify the kinetic expression, we usually set a body coordinate at the mass center to decouple the translational and rotational motions. The rotational kinetic energy of a rigid body in a principal Cartesian body coordinate frame is

$$K = \frac{1}{2}\, {}^B_G\omega_B^{\mathrm{T}}\ {}^BI\, {}^B_G\omega_B = \frac{1}{2}\left(I_1\omega_1^2 + I_2\omega_2^2 + I_3\omega_3^2\right) \tag{13.85}$$

Using Euler frequencies (12.185),

$${}^B_G\omega_B = \begin{bmatrix} \dot{\varphi}\sin\theta\sin\psi + \dot{\theta}\cos\psi \\ \dot{\varphi}\cos\psi\sin\theta - \dot{\theta}\sin\psi \\ \dot{\psi} + \dot{\varphi}\cos\theta \end{bmatrix} \tag{13.86}$$

we have

$$\begin{aligned} K &= \frac{1}{2}I_1\left(\dot{\varphi}\sin\theta\sin\psi + \dot{\theta}\cos\psi\right)^2 + \frac{1}{2}I_2\left(\dot{\varphi}\cos\psi\sin\theta - \dot{\theta}\sin\psi\right)^2 \\ &\quad + \frac{1}{2}I_3\left(\dot{\psi} + \dot{\varphi}\cos\theta\right)^2 \\ &= \frac{1}{2}\left(I_1\sin^2\theta\sin^2\psi + I_2\cos^2\psi\sin^2\theta + I_3\cos^2\theta\right)\dot{\varphi}^2 \\ &\quad + \frac{1}{2}\left(I_1\cos^2\psi + I_2\sin^2\theta\right)\dot{\theta}^2 + \frac{1}{2}I_3\dot{\psi}^2 \\ &\quad + \left[(I_1 - I_2)\cos\psi\sin\theta\sin\psi\right]\dot{\varphi}\dot{\theta} + (I_3\cos\theta)\,\dot{\varphi}\dot{\psi} \end{aligned} \tag{13.87}$$

The generalized momenta are

$$p_\varphi = \frac{\partial K}{\partial \dot{\varphi}} = \left(I_1 \sin^2\theta \sin^2\psi + I_2 \cos^2\psi \sin^2\theta + I_3 \cos^2\theta\right)\dot{\varphi}$$
$$+ \left[(I_1 - I_2)\cos\psi \sin\theta \sin\psi\right]\dot{\theta} + (I_3\cos\theta)\,\dot{\psi} \tag{13.88}$$

$$p_\theta = \frac{\partial K}{\partial \dot{\theta}} = \left[(I_1 - I_2)\cos\psi \sin\theta \sin\psi\right]\dot{\varphi}$$
$$+ \left(I_1 \cos^2\psi + I_2 \sin^2\theta\right)\dot{\theta} \tag{13.89}$$

$$p_\psi = \frac{\partial K}{\partial \dot{\psi}} = (I_3\cos\theta)\,\dot{\varphi} + I_3\dot{\psi} \tag{13.90}$$

The generalized momenta are linear functions of Euler frequencies. So, if we define the generalized momenta and coordinates by two column matrices **p** and **q**,

$$\mathbf{p} = \begin{bmatrix} p_\varphi \\ p_\theta \\ p_\psi \end{bmatrix} \qquad \mathbf{q} = \begin{bmatrix} \varphi \\ \theta \\ \psi \end{bmatrix} \tag{13.91}$$

we can write them in matrix form as

$$\begin{bmatrix} p_\varphi \\ p_\theta \\ p_\psi \end{bmatrix} = [m]\begin{bmatrix} \dot{\varphi} \\ \dot{\theta} \\ \dot{\psi} \end{bmatrix} \tag{13.92}$$

where $[m]$ is the inertia matrix of the equation $\mathbf{p} = [m]\,\mathbf{q}$:

$$[m] = \begin{bmatrix} I_1 s^2\theta s^2\psi + I_2 c^2\psi s^2\theta + I_3 c^2\theta & (I_1 - I_2)\,c\psi s\theta s\psi & I_3 c\theta \\ (I_1 - I_2)\,c\psi s\theta s\psi & I_1 c^2\psi + I_2 s^2\theta & 0 \\ I_3 c\theta & 0 & I_3 \end{bmatrix} \tag{13.93}$$

Using the Lagrange equation,

$$\frac{d}{dt}\left(\frac{\partial K}{\partial \dot{q}_r}\right) - \frac{\partial K}{\partial q_r} = Q_r \qquad r = 1, 2, \ldots, n \tag{13.94}$$

the differential equation of rotational motion in the Euler frame are

$$\frac{d}{dt}\begin{pmatrix} \left(I_1 \sin^2\theta \sin^2\psi + I_2 \cos^2\psi \sin^2\theta + I_3 \cos^2\theta\right)\dot{\varphi} \\ + \left[(I_1 - I_2)\cos\psi \sin\theta \sin\psi\right]\dot{\theta} + (I_3\cos\theta)\,\dot{\psi} \end{pmatrix} = Q_\varphi \tag{13.95}$$

$$\frac{d}{dt}\left\{\left[(I_1 - I_2)\cos\psi \sin\theta \sin\psi\right]\dot{\varphi} + \left(I_1\cos^2\psi + I_2\sin^2\theta\right)\dot{\theta}\right\}$$
$$- \begin{pmatrix} \dot{\varphi}^2\left(I_1\sin^2\psi + I_2\cos^2\psi - I_3\right)\sin\theta\cos\theta \\ -(I_1 - I_2)\,\dot{\varphi}\dot{\theta}\cos\psi\sin\psi\cos\theta - I_3\dot{\varphi}\dot{\psi}\sin\theta \end{pmatrix} = Q_\theta \tag{13.96}$$

$$\frac{d}{dt}\left[(I_3\cos\theta)\,\dot{\varphi} + I_3\dot{\psi}\right]$$
$$- \begin{pmatrix} \left[\dot{\varphi}^2(I_1 - I_2)\sin^2\theta - \dot{\theta}^2(I_1 - I_2)\right]\cos\psi\sin\psi \\ -\dot{\varphi}\dot{\theta}\,(I_1 - I_2)\sin\theta\left(1 - 2\cos^2\psi\right) \end{pmatrix} = Q_\psi \tag{13.97}$$

where

$$\frac{\partial K}{\partial \varphi} = 0 \tag{13.98}$$

$$\begin{aligned}\frac{\partial K}{\partial \theta} &= \dot{\varphi}^2 \left(I_1 \sin^2 \psi + I_2 \cos^2 \psi - I_3\right) \sin\theta \cos\theta \\ &\quad - (I_1 - I_2)\, \dot{\varphi}\dot{\theta} \cos\psi \sin\psi \cos\theta - I_3 \dot{\varphi}\dot{\psi} \sin\theta\end{aligned} \tag{13.99}$$

$$\begin{aligned}\frac{\partial K}{\partial \psi} &= \dot{\varphi}^2 (I_1 - I_2) \cos\psi \sin\psi \sin^2\theta - \dot{\theta}^2 (I_1 - I_2) \cos\psi \sin\psi \\ &\quad - \dot{\varphi}\dot{\theta} (I_1 - I_2) \sin\theta \left(1 - 2\cos^2\psi\right)\end{aligned} \tag{13.100}$$

Example 770 ★ Potential Force Field If a system of masses m_i is moving in a potential force field

$$\mathbf{F}_{m_i} = -\nabla_i V \tag{13.101}$$

the Newton equations of motion will be

$$m_i \ddot{\mathbf{r}}_i = -\nabla_i V \qquad i = 1, 2, \ldots, n \tag{13.102}$$

Using the inner product of the equations of motion with $\dot{\mathbf{r}}_i$ and adding the equations,

$$\sum_{i=1}^{n} m_i \dot{\mathbf{r}}_i \cdot \ddot{\mathbf{r}}_i = -\sum_{i=1}^{n} \dot{\mathbf{r}}_i \cdot \nabla_i V \tag{13.103}$$

and then integrating over time,

$$\frac{1}{2}\sum_{i=1}^{n} m_i \dot{\mathbf{r}}_i \cdot \dot{\mathbf{r}}_i = -\int \sum_{i=1}^{n} \mathbf{r}_i \cdot \nabla_i V \tag{13.104}$$

yield

$$K = -\int \sum_{i=1}^{n} \left(\frac{\partial V}{\partial x_i} x_i + \frac{\partial V}{\partial y_i} y_i + \frac{\partial V}{\partial z_i} z_i \right) = -V + E \tag{13.105}$$

where E is the constant of integration and the mechanical energy of the system, equal to the sum of the kinetic and potential energies.

Example 771 Kinetic Energy of Earth The two motions of Earth are called *revolution* about the sun and *rotation* about an axis approximately fixed on Earth. Employing ω_1 and ω_2 to indicate the angular speed of Earth about its axis and the angular speed about the sun, the kinetic energy of Earth due to its rotation would be

$$
\begin{aligned}
K_1 &= \frac{1}{2} I \omega_1^2 \\
&= \frac{1}{2}\frac{2}{5}\left(5.9742 \times 10^{24}\right)\left(\frac{6{,}356{,}912 + 6{,}378{,}388}{2}\right)^2\left(\frac{2\pi}{24 \times 3600}\frac{366.25}{365.25}\right)^2 \\
&= 2.5762 \times 10^{29} J
\end{aligned} \tag{13.106}
$$

and the kinetic energy of Earth due to its revolution is

$$
\begin{aligned}
K_2 &= \frac{1}{2} M r^2 \omega_2^2 \\
&= \frac{1}{2}\left(5.9742 \times 10^{24}\right)\left(1.49475 \times 10^{11}\right)^2\left(\frac{2\pi}{24 \times 3600}\frac{1}{365.25}\right)^2 \\
&= 2.6457 \times 10^{33} \mathrm{J}
\end{aligned} \tag{13.107}
$$

where r is the distance from the sun. The total kinetic energy of Earth is $K = K_1 + K_2$. However, the ratio of the revolutional to rotational kinetic energy is

$$
\frac{K_2}{K_1} = \frac{2.6457 \times 10^{33}}{2.5762 \times 10^{29}} \approx 10{,}000 \tag{13.108}
$$

indicating that the kinetic energy of Earth is mostly due to its revolution about the sun.

Example 772 ★ A Non-Cartesian Coordinate System The parabolic and Cartesian coordinate systems are related by the equations.

$$
x = \eta\xi \cos\varphi \tag{13.109}
$$

$$
y = \eta\xi \sin\varphi \tag{13.110}
$$

$$
z = \frac{\xi^2 - \eta^2}{2} \tag{13.111}
$$

$$
\xi^2 = \sqrt{x^2 + y^2 + z^2} + z \tag{13.112}
$$

$$
\eta^2 = \sqrt{x^2 + y^2 + z^2} - z \tag{13.113}
$$

$$
\varphi = \tan^{-1}\frac{y}{x} \tag{13.114}
$$

Consider an electron in a uniform electric field along the positive z-axis that is under the action of an attractive central force field due to the nuclei of the atom:

$$
\mathbf{F} = -\frac{k}{r^2}\hat{u}_r = -\nabla\left(-\frac{k}{r}\right) \tag{13.115}
$$

The influence of a uniform electric field on the motion of the electrons in atoms is called the *Stark effect*, and it is easier to analyze this motion in a parabolic coordinate system.

The kinetic energy in a parabolic coordinate system is

$$K = \tfrac{1}{2}m\left(\dot{x}^2 + \dot{y}^2 + \dot{z}^2\right) = \tfrac{1}{2}m\left[\left(\eta^2 + \xi^2\right)\left(\dot{\eta}^2 + \dot{\xi}^2\right) + \eta^2\xi^2\dot{\varphi}^2\right] \tag{13.116}$$

and the force acting on the electron is

$$\mathbf{F} = -\nabla\left(-\frac{k}{r} + eEz\right) = -\nabla\left(-\frac{2k}{\xi^2+\eta^2} + \frac{eE}{2}\left(\xi^2 - \eta^2\right)\right) \tag{13.117}$$

which leads to the generalized forces

$$Q_\eta = \mathbf{F}\cdot\mathbf{b}_\eta = -\frac{4k\eta}{\left(\xi^2+\eta^2\right)^2} + eE\eta \tag{13.118}$$

$$Q_\xi = \mathbf{F}\cdot\mathbf{b}_\xi = -\frac{4k\xi}{\left(\xi^2+\eta^2\right)^2} - eE\eta \tag{13.119}$$

$$Q_\varphi = 0 \tag{13.120}$$

where $\mathbf{b}_\xi$, $\mathbf{b}_\eta$, and $\mathbf{b}_\varphi$ are base vectors of the coordinate system:

$$\mathbf{b}_\xi = \frac{\partial\mathbf{r}}{\partial\xi} = \eta\cos\varphi\hat{\imath} + \eta\sin\varphi\hat{\jmath} + \xi\hat{k} \tag{13.121}$$

$$\mathbf{b}_\eta = \frac{\partial\mathbf{r}}{\partial\eta} = \xi\cos\varphi\hat{\imath} + \xi\sin\varphi\hat{\jmath} - \eta\hat{k} \tag{13.122}$$

$$\mathbf{b}_\varphi = \frac{\partial\mathbf{r}}{\partial\varphi} = -\eta\xi\sin\varphi\hat{\imath} + \eta\xi\cos\varphi\hat{\jmath} \tag{13.123}$$

Therefore, following the Lagrange method, the equations of motion of the electron are

$$Q_\eta = \frac{d}{dt}\left[m\dot{\eta}\left(\xi^2+\eta^2\right)\right] - m\eta\left(\dot{\eta}^2 + \dot{\xi}^2\right) - m\eta\xi^2\dot{\varphi}^2 \tag{13.124}$$

$$Q_\xi = \frac{d}{dt}\left[m\dot{\xi}\left(\xi^2+\eta^2\right)\right] - m\eta\left(\dot{\eta}^2 + \dot{\xi}^2\right) - m\xi\eta^2\dot{\varphi}^2 \tag{13.125}$$

$$Q_\varphi = \frac{d}{dt}\left(m\eta^2\xi^2\dot{\varphi}^2\right) \tag{13.126}$$

Example 773 ★ Explicit Form of Lagrange Equations Assume that the coordinates of every particle of a dynamic system is a function of the generalized coordinates $q_1, q_2, q_3, \ldots, q_n$ but not time t. The kinetic energy of the system made of n massive particles can be written as

$$K = \frac{1}{2}\sum_{i=1}^{n} m_i\left(\dot{x}_i^2 + \dot{y}_i^2 + \dot{z}_i^2\right) = \frac{1}{2}\sum_{j=1}^{n}\sum_{k=1}^{n} a_{jk}\dot{q}_j\dot{q}_k \tag{13.127}$$

where the coefficients a_{jk} are functions of $q_1, q_2, q_3, \ldots, q_n$ and

$$a_{jk} = a_{kj} \tag{13.128}$$

The Lagrange equation of motion

$$\frac{d}{dt}\left(\frac{\partial K}{\partial \dot{q}_r}\right) - \frac{\partial K}{\partial q_r} = Q_r \qquad r = 1, 2, \ldots, n \tag{13.129}$$

will then be

$$\frac{d}{dt}\sum_{m=1}^{n} a_{mr}\dot{q}_m - \frac{1}{2}\sum_{j=1}^{n}\sum_{k=1}^{n}\frac{a_{jk}}{\partial q_r}\dot{q}_j\dot{q}_k = Q_r \tag{13.130}$$

or

$$\sum_{m=1}^{n} a_{mr}\ddot{q}_m + \sum_{k=1}^{n}\sum_{n=1}^{n}\Gamma_{rkn}\dot{q}_k\dot{q}_n = Q_r \tag{13.131}$$

where Γ_{ijk} is the Christoffel symbol,

$$\Gamma_{ijk} = \frac{1}{2}\left(\frac{\partial a_{ij}}{\partial q_k} + \frac{\partial a_{ik}}{\partial q_j} - \frac{\partial a_{kj}}{\partial q_i}\right) \tag{13.132}$$

Equation (13.131) indicates that the equations of motion will be a set of n coupled second-order differential equations in the generalized configuration space.

13.2 LAGRANGE EQUATION AND POTENTIAL FORCE

Assume for some forces $\mathbf{F} = \left[F_{ix}, F_{iy}, F_{iz}\right]^{\mathrm{T}}$ there is a function V, called the *potential energy*, such that the force is derivable from V:

$$\mathbf{F} = -\nabla V \tag{13.133}$$

Such a force is called a *potential* or *conservative force*. Then, the Lagrange equation of motion can be written as

$$\frac{d}{dt}\left(\frac{\partial \mathcal{L}}{\partial \dot{q}_r}\right) - \frac{\partial \mathcal{L}}{\partial q_r} = Q_r \qquad r = 1, 2, \ldots, n \tag{13.134}$$

where

$$\mathcal{L} = K - V \tag{13.135}$$

is the *Lagrangian* of the system and Q_r is the nonpotential generalized force.

Proof: Assume the external forces $\mathbf{F} = \left[F_{xi}, F_{yi}, F_{zi}\right]^{\mathrm{T}}$ acting on the system are conservative:

$$\mathbf{F} = -\nabla V \tag{13.136}$$

The work done by these forces in an arbitrary virtual displacement $\delta q_1, \delta q_2, \delta q_3, \ldots, \delta q_n$ is

$$\partial W = -\frac{\partial V}{\partial q_1}\delta q_1 - \frac{\partial V}{\partial q_2}\delta q_2 - \cdots - \frac{\partial V}{\partial q_n}\delta q_n \tag{13.137}$$

Then the Lagrange equation becomes

$$\frac{d}{dt}\left(\frac{\partial K}{\partial \dot{q}_r}\right) - \frac{\partial K}{\partial q_r} = -\frac{\partial V}{\partial q_1} r = 1, 2, \ldots, n \tag{13.138}$$

Introducing the Lagrangian function $\mathcal{L} = K - V$ converts the Lagrange equation of a conservative system to

$$\frac{d}{dt}\left(\frac{\partial \mathcal{L}}{\partial \dot{q}_r}\right) - \frac{\partial \mathcal{L}}{\partial q_r} = 0 \qquad r = 1, 2, \ldots, n \tag{13.139}$$

The Lagrangian function $\mathcal{L}$ is also called the *kinetic potential*.

If there is also a nonconservative force $\mathbf{F}$, then the work done by the force is

$$\begin{aligned} \delta W &= \sum_{i=1}^{n} \left(F_{xi} \frac{\partial f_i}{\partial q_r} + F_{yi} \frac{\partial g_i}{\partial q_r} + F_{zi} \frac{\partial h_i}{\partial q_r} \right) \delta q_r \\ &= Q_r \, \delta q_r \end{aligned} \tag{13.140}$$

and the equation of motion would be

$$\frac{d}{dt}\left(\frac{\partial \mathcal{L}}{\partial \dot{q}_r}\right) - \frac{\partial \mathcal{L}}{\partial q_r} = Q_r \qquad r = 1, 2, \ldots, n \tag{13.141}$$

where Q_r is the nonpotential generalized force doing work in a virtual displacement of the rth generalized coordinate q_r. ■

Example 774 A Spherical Pendulum Figure 13.5 illustrates a spherical pendulum with mass m and length l. The angles φ and θ may be used as generalized describing coordinates of the system.

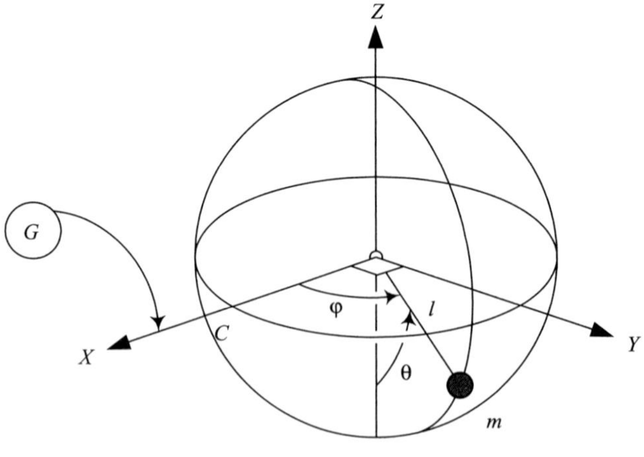

Figure 13.5 A spherical pendulum.

The Cartesian coordinates of m as a function of the generalized coordinates are

$$\begin{bmatrix} X \\ Y \\ Z \end{bmatrix} = \begin{bmatrix} r\cos\varphi\sin\theta \\ r\sin\theta\sin\varphi \\ -r\cos\theta \end{bmatrix} \tag{13.142}$$

and therefore, the kinetic and potential energies of the pendulum are

$$K = \frac{1}{2}m\left(l^2\dot{\theta}^2 + l^2\dot{\varphi}^2\sin^2\theta\right) \tag{13.143}$$

$$V = -mgl\cos\theta \tag{13.144}$$

The kinetic potential function of this system is then equal to

$$\mathcal{L} = \frac{1}{2}m\left(l^2\dot{\theta}^2 + l^2\dot{\varphi}^2\sin^2\theta\right) + mgl\cos\theta \tag{13.145}$$

which leads to the following equations of motion:

$$\ddot{\theta} - \dot{\varphi}^2\sin\theta\cos\theta + \frac{g}{l}\sin\theta = 0 \tag{13.146}$$

$$\ddot{\varphi}\sin^2\theta + 2\dot{\varphi}\dot{\theta}\sin\theta\cos\theta = 0 \tag{13.147}$$

Example 775 A Controlled Arm Figure 13.6 illustrates a controllable arm. Assume that there is a viscous friction in the joint where an ideal motor can apply the torque Q to move the arm. The rotor of an ideal motor has no mass moment by assumption.

The kinetic and potential energies of the arm are

$$K = \frac{1}{2}I\dot{\theta}^2 = \frac{1}{2}\left(I_C + ml^2\right)\dot{\theta}^2 \tag{13.148}$$

$$V = -mg\cos\theta \tag{13.149}$$

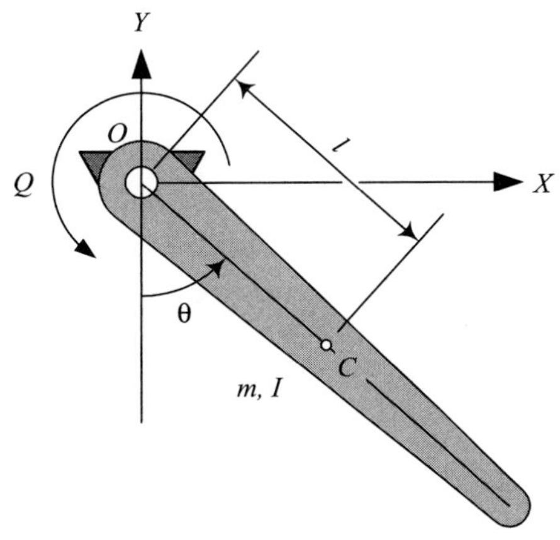

Figure 13.6 A controllable arm.

where m is the mass and I is the moment of inertia of the pendulum about O. The Lagrangian of the manipulator is

$$\mathcal{L} = K - V = \frac{1}{2} I \dot{\theta}^2 + mg \cos\theta \tag{13.150}$$

and therefore, the equation of motion of the manipulator is

$$M = \frac{d}{dt}\left(\frac{\partial \mathcal{L}}{\partial \dot{\theta}}\right) - \frac{\partial \mathcal{L}}{\partial \theta} = I\,\ddot{\theta} + mgl \sin\theta \tag{13.151}$$

The generalized force M is the contribution of the motor torque Q and the viscous friction torque $-c\dot{\theta}$. Hence, the equation of motion of the arm is

$$Q = I\,\ddot{\theta} + c\dot{\theta} + mgl \sin\theta \tag{13.152}$$

Example 776 Elastic Pendulum Figure 13.7 illustrates a planar elastic pendulum. If at the motionless vertical equilibrium condition the distance between m and the fulcrum is l_0 and the extra stretch of the length is z, then the Cartesian positions of m during the motion are

$$x = l \sin\theta \qquad y = -l \cos\theta \tag{13.153}$$

$$l = z + l_0 \tag{13.154}$$

where $q_1 = z$ and $q_2 = \theta$ are the generalized coordinates of the system. The velocity components of m are

$$\dot{x} = \dot{z} \sin\theta + l\dot{\theta} \cos\theta \qquad \dot{y} = -\dot{z} \cos\theta + l\dot{\theta} \sin\theta \tag{13.155}$$

So, the kinetic energy of the pendulum is

$$K = \frac{1}{2} m \left(\dot{x}^2 + \dot{y}^2\right) = \frac{1}{2} m \dot{z}^2 + \frac{1}{2} m l^2 \dot{\theta}^2 \tag{13.156}$$

The gravitational and spring forces are the only external forces on m and are both potential. The potential energy of the pendulum is

$$V = mgy + \frac{1}{2} k z^2 = -mgl \cos\theta + \frac{1}{2} k z^2 \tag{13.157}$$

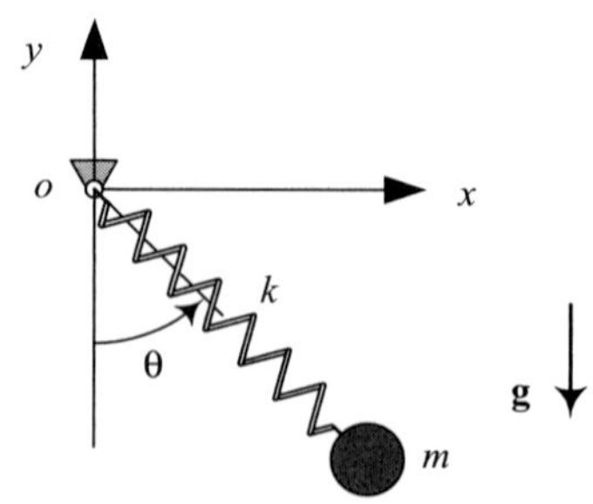

Figure 13.7 An elastic pendulum.

Therefore, the Lagrangian of the system is

$$\mathcal{L} = K - V = \tfrac{1}{2} m \dot{z}^2 + \tfrac{1}{2} m l^2 \dot{\theta}^2 + mgl\cos\theta - \tfrac{1}{2} k z^2 \tag{13.158}$$

Applying the Lagrange equation (13.134),

$$\frac{d}{dt}\left(\frac{\partial \mathcal{L}}{\partial \dot{z}}\right) - \frac{\partial \mathcal{L}}{\partial z} = 0 \qquad \frac{d}{dt}\left(\frac{\partial \mathcal{L}}{\partial \dot{\theta}}\right) - \frac{\partial \mathcal{L}}{\partial \theta} = 0 \tag{13.159}$$

we find the following equations of motion:

$$m\ddot{z} - m\left(z + l_0\right)\dot{\theta}^2 - mg\cos\theta + kz = 0 \tag{13.160}$$

$$m\left(z + l_0\right)^2 \ddot{\theta} + 2m\left(z + l_0\right)\dot{z}\dot{\theta} + mg\left(z + l_0\right)\sin\theta = 0 \tag{13.161}$$

Example 777 An Ideal 2R Planar Manipulator Dynamics An ideal model of a $2R$ planar manipulator is illustrated in Figure 13.8. It is called ideal because we assume that the links are massless and there is no friction. The masses m_1 and m_2 are the masses of the second motor to run the second link and the load at the end point. We take the absolute angle θ_1 and the relative angle θ_2 as the generalized coordinates to express the configuration of the manipulator.

The global positions of m_1 and m_2 are

$$\begin{bmatrix} X_1 \\ Y_2 \end{bmatrix} = \begin{bmatrix} l_1\cos\theta_1 \\ l_1\sin\theta_1 \end{bmatrix} \tag{13.162}$$

$$\begin{bmatrix} X_2 \\ Y_2 \end{bmatrix} = \begin{bmatrix} l_1\cos\theta_1 + l_2\cos\left(\theta_1 + \theta_2\right) \\ l_1\sin\theta_1 + l_2\sin\left(\theta_1 + \theta_2\right) \end{bmatrix} \tag{13.163}$$

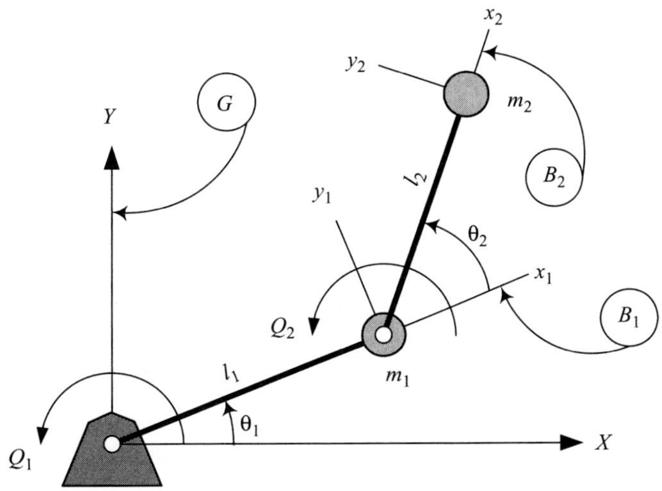

Figure 13.8 An ideal model of a 2R planar manipulator.

and therefore, the global velocities of the masses are

$$\begin{bmatrix} \dot{X}_1 \\ \dot{Y}_1 \end{bmatrix} = \begin{bmatrix} -l_1\dot{\theta}_1 \sin\theta_1 \\ l_1\dot{\theta}_1 \cos\theta_1 \end{bmatrix} \tag{13.164}$$

$$\begin{bmatrix} \dot{X}_2 \\ \dot{Y}_2 \end{bmatrix} = \begin{bmatrix} -l_1\dot{\theta}_1 \sin\theta_1 - l_2\left(\dot{\theta}_1 + \dot{\theta}_2\right)\sin\left(\theta_1 + \theta_2\right) \\ l_1\dot{\theta}_1 \cos\theta_1 + l_2\left(\dot{\theta}_1 + \dot{\theta}_2\right)\cos\left(\theta_1 + \theta_2\right) \end{bmatrix} \tag{13.165}$$

The kinetic energy of this manipulator is the sum of the kinetic energies of the masses and is equal to

$$\begin{aligned} K =& K_1 + K_2 = \tfrac{1}{2}m_1\left(\dot{X}_1^2 + \dot{Y}_1^2\right) + \tfrac{1}{2}m_2\left(\dot{X}_2^2 + \dot{Y}_2^2\right) \\ =& \tfrac{1}{2}m_1 l_1^2\dot{\theta}_1^2 \\ &+ \tfrac{1}{2}m_2\left[l_1^2\dot{\theta}_1^2 + l_2^2\left(\dot{\theta}_1 + \dot{\theta}_2\right)^2 + 2l_1l_2\dot{\theta}_1\left(\dot{\theta}_1 + \dot{\theta}_2\right)\cos\theta_2\right] \end{aligned} \tag{13.166}$$

The potential energy of the manipulator is

$$\begin{aligned} V &= V_1 + V_2 = m_1 g Y_1 + m_2 g Y_2 \\ &= m_1 g l_1 \sin\theta_1 + m_2 g\left[l_1\sin\theta_1 + l_2\sin\left(\theta_1 + \theta_2\right)\right] \end{aligned} \tag{13.167}$$

The Lagrangian is then obtained from Equations (13.166) and (13.167),

$$\begin{aligned} \mathcal{L} = K - V =& \tfrac{1}{2}m_1 l_1^2\dot{\theta}_1^2 \qquad (13.168) \\ &+ \tfrac{1}{2}m_2\left[l_1^2\dot{\theta}_1^2 + l_2^2\left(\dot{\theta}_1 + \dot{\theta}_2\right)^2 + 2l_1l_2\dot{\theta}_1\left(\dot{\theta}_1 + \dot{\theta}_2\right)\cos\theta_2\right] \\ &- \left\{m_1 g l_1 \sin\theta_1 + m_2 g\left[l_1\sin\theta_1 + l_2\sin\left(\theta_1 + \theta_2\right)\right]\right\} \end{aligned}$$

which provides the required partial derivatives as follows:

$$\frac{\partial\mathcal{L}}{\partial\theta_1} = -\left(m_1 + m_2\right) g l_1 \cos\theta_1 - m_2 g l_2 \cos\left(\theta_1 + \theta_2\right) \tag{13.169}$$

$$\begin{aligned} \frac{\partial\mathcal{L}}{\partial\dot{\theta}_1} =& \left(m_1 + m_2\right) l_1^2\dot{\theta}_1 + m_2 l_2^2\left(\dot{\theta}_1 + \dot{\theta}_2\right) \\ &+ m_2 l_1 l_2\left(2\dot{\theta}_1 + \dot{\theta}_2\right)\cos\theta_2 \end{aligned} \tag{13.170}$$

$$\begin{aligned} \frac{d}{dt}\left(\frac{\partial\mathcal{L}}{\partial\dot{\theta}_1}\right) =& \left(m_1 + m_2\right) l_1^2\ddot{\theta}_1 + m_2 l_2^2\left(\ddot{\theta}_1 + \ddot{\theta}_2\right) \\ &+ m_2 l_1 l_2\left(2\ddot{\theta}_1 + \ddot{\theta}_2\right)\cos\theta_2 - m_2 l_1 l_2\dot{\theta}_2\left(2\dot{\theta}_1 + \dot{\theta}_2\right)\sin\theta_2 \end{aligned} \tag{13.171}$$

$$\frac{\partial\mathcal{L}}{\partial\theta_2} = -m_2 l_1 l_2\dot{\theta}_1\left(\dot{\theta}_1 + \dot{\theta}_2\right)\sin\theta_2 - m_2 g l_2 \cos\left(\theta_1 + \theta_2\right) \tag{13.172}$$

$$\frac{\partial\mathcal{L}}{\partial\dot{\theta}_2} = m_2 l_2^2\left(\dot{\theta}_1 + \dot{\theta}_2\right) + m_2 l_1 l_2\dot{\theta}_1\cos\theta_2 \tag{13.173}$$

$$\begin{aligned} \frac{d}{dt}\left(\frac{\partial\mathcal{L}}{\partial\dot{\theta}_2}\right) =& m_2 l_2^2\left(\ddot{\theta}_1 + \ddot{\theta}_2\right) + m_2 l_1 l_2\ddot{\theta}_1\cos\theta_2 \\ &- m_2 l_1 l_2\dot{\theta}_1\dot{\theta}_2\sin\theta_2 \end{aligned} \tag{13.174}$$

Therefore, the equations of motion for the $2R$ manipulator are

$$\begin{aligned} Q_1 &= \frac{d}{dt}\left(\frac{\partial \mathcal{L}}{\partial \dot{\theta}_1}\right) - \frac{\partial \mathcal{L}}{\partial \theta_1} \\ &= (m_1 + m_2)\, l_1^2 \ddot{\theta}_1 + m_2 l_2^2 \left(\ddot{\theta}_1 + \ddot{\theta}_2\right) \\ &\quad + m_2 l_1 l_2 \left(2\ddot{\theta}_1 + \ddot{\theta}_2\right) \cos\theta_2 - m_2 l_1 l_2 \dot{\theta}_2 \left(2\dot{\theta}_1 + \dot{\theta}_2\right) \sin\theta_2 \\ &\quad + (m_1 + m_2)\, g l_1 \cos\theta_1 + m_2 g l_2 \cos(\theta_1 + \theta_2) \end{aligned} \tag{13.175}$$

$$\begin{aligned} Q_2 &= \frac{d}{dt}\left(\frac{\partial \mathcal{L}}{\partial \dot{\theta}_2}\right) - \frac{\partial \mathcal{L}}{\partial \theta_2} \\ &= m_2 l_2^2 \left(\ddot{\theta}_1 + \ddot{\theta}_2\right) + m_2 l_1 l_2 \ddot{\theta}_1 \cos\theta_2 - m_2 l_1 l_2 \dot{\theta}_1 \dot{\theta}_2 \sin\theta_2 \\ &\quad + m_2 l_1 l_2 \dot{\theta}_1 \left(\dot{\theta}_1 + \dot{\theta}_2\right) \sin\theta_2 + m_2 g l_2 \cos(\theta_1 + \theta_2) \end{aligned} \tag{13.176}$$

The generalized forces Q_1 and Q_2 are the required forces to drive the generalized coordinates. In this case, Q_1 is the torque of the base motor and Q_2 is the torque of the motor at m_1.

The equations of motion can be rearranged to a more systematic form:

$$\begin{aligned} Q_1 &= \left[(m_1 + m_2)\, l_1^2 + m_2 l_2 \left(l_2 + 2 l_1 \cos\theta_2\right)\right] \ddot{\theta}_1 \\ &\quad + m_2 l_2 \left(l_2 + l_1 \cos\theta_2\right) \ddot{\theta}_2 \\ &\quad - 2 m_2 l_1 l_2 \sin\theta_2\, \dot{\theta}_1 \dot{\theta}_2 - m_2 l_1 l_2 \sin\theta_2\, \dot{\theta}_2^2 \\ &\quad + (m_1 + m_2)\, g l_1 \cos\theta_1 + m_2 g l_2 \cos(\theta_1 + \theta_2) \end{aligned} \tag{13.177}$$

$$\begin{aligned} Q_2 &= m_2 l_2 \left(l_2 + l_1 \cos\theta_2\right) \ddot{\theta}_1 + m_2 l_2^2 \ddot{\theta}_2 \\ &\quad + m_2 l_1 l_2 \sin\theta_2\, \dot{\theta}_1^2 + m_2 g l_2 \cos(\theta_1 + \theta_2) \end{aligned} \tag{13.178}$$

Example 778 ★ Lagrange Equation from Fundamental Equation Consider a dynamic system on N particles m_i that can be expressed by n generalized coordinates q_i, $i = 1, 2, 3, \ldots, n$. As derived in Example 675, when we substitute the virtual displacement

$$\delta \mathbf{r}_i = \sum_{j=1}^{n} \frac{\partial \mathbf{r}_i}{\partial q_j} \delta q_j \tag{13.179}$$

in the fundamental equation of dynamics (10.303),

$$\sum_{i=1}^{N/3} \left(m_i \ddot{\mathbf{r}}_i - \mathbf{F}_i\right) \cdot \delta \mathbf{r}_i = 0 \tag{13.180}$$

we obtain the Lagrange equation (10.739):

$$\frac{d}{dt}\frac{\partial K}{\partial \dot{q}_j} - \frac{\partial K}{\partial q_j} - Q_j = 0 \qquad j = 1, 2, \ldots, n \tag{13.181}$$

$$Q_j = \sum_{i=1}^{N/3} \mathbf{F}_i \cdot \frac{\partial \mathbf{r}_i}{\partial q_j} \tag{13.182}$$

where K is the kinetic energy of the system, Q_j is the generalized force associated with coordinate q_j, and $\mathbf{F}_i$ is the given or applied force associated with m_i.

If the applied forces $\mathbf{F}_i$ are only functions of positions $\mathbf{r}_i$ and time t and are not dependent on velocities $\dot{\mathbf{r}}_i$, then these types of forces are derivable from a scalar potential function:

$$\mathbf{F}_i = -\nabla V = -\frac{\partial V}{\partial x_i}\hat{\imath} - \frac{\partial V}{\partial y_i}\hat{\jmath} - \frac{\partial V}{\partial z_i}\hat{k} \tag{13.183}$$

$$V = V(x_1, y_1, z_1, x_2, y_2, z_3, \ldots, x_N, y_N, z_N) \tag{13.184}$$

The generalized force Q_j will then be

$$\begin{aligned} Q_j &= -\left(\frac{\partial V}{\partial x_i}\frac{\partial x_i}{\partial q_j} + \frac{\partial V}{\partial y_i}\frac{\partial y_i}{\partial q_j} + \frac{\partial V}{\partial z_i}\frac{\partial z_i}{\partial q_j}\right) \\ &= -\frac{\partial V(q_1, q_2, \ldots, q_n)}{\partial q_j} \\ i &= 1, 2, \ldots, N \qquad j = 1, 2, \ldots, n \end{aligned} \tag{13.185}$$

Substituting for Q_j, the Lagrange equation (13.181) becomes

$$\frac{d}{dt}\left(\frac{\partial K}{\partial \dot{q}_r}\right) - \frac{\partial K}{\partial q_r} + \frac{\partial V}{\partial q_r} = 0 \qquad r = 1, 2, \ldots, n \tag{13.186}$$

Example 779 Lagrange Equations Are Second Order Lagrange equations (13.141) are ordinary differential equations of second order with respect to the generalized coordinates q_i. To show this, we substitute for K,

$$K = K(q_1, q_2, \ldots, q_n, \dot{q}_1, \dot{q}_2, \ldots, \dot{q}_n, t) \tag{13.187}$$

and take the derivatives:

$$\sum_{s=1}^{n} \frac{\partial^2 K}{\partial \dot{q}_r\, \partial \dot{q}_s}\ddot{q}_s + \sum_{s=1}^{n} \frac{\partial^2 K}{\partial \dot{q}_r\, \partial q_s}\dot{q}_s + \frac{\partial^2 K}{\partial \dot{q}_r\, \partial t} - \frac{\partial K}{\partial q_r} + \frac{\partial V}{\partial q_r} = Q_r \tag{13.188}$$

Example 780 ★ Explicit Form of Lagrange Equations In the general case in which the coefficients a_{jk} in the kinetic energy equation (13.127) given as

$$K = \tfrac{1}{2} a_{jk}\dot{q}_j\dot{q}_k \tag{13.189}$$

are functions of $q_1, q_2, q_3, \ldots, q_n, t$, we have

$$\left(\frac{d}{dt}\frac{\partial}{\partial \dot{q}_r} - \frac{\partial}{\partial q_r}\right)\frac{1}{2}a_{jk}\dot{q}_j\dot{q}_k = \frac{d}{dt}\left(a_{js}\dot{q}_j\right) - \frac{1}{2}\frac{\partial a_{jk}}{\partial q_s}\dot{q}_j\dot{q}_k$$

$$= a_{js}\ddot{q}_j + \left(\frac{\partial a_{ks}}{\partial q_j} - \frac{1}{2}\frac{\partial a_{jk}}{\partial q_s}\right)\dot{q}_j\dot{q}_k + \frac{\partial a_{jk}}{\partial t}\dot{q}_j$$

$$= a_{js}\ddot{q}_j + \Gamma_{sjk}\,\dot{q}_j\dot{q}_k + \frac{\partial a_{jk}}{\partial t}\dot{q}_j \tag{13.190}$$

Therefore, the Lagrange equation (13.141) becomes

$$a_{js}\ddot{q}_j + \Gamma_{sjk}\,\dot{q}_j\dot{q}_k = Q_s - \frac{\partial V}{\partial q_s} - \frac{\partial a_{jk}}{\partial t}\dot{q}_j \tag{13.191}$$

where Γ_{ijk} is the Christoffel symbol:

$$\Gamma_{sjk} = \frac{1}{2}\left(\frac{\partial a_{kj}}{\partial q_s} + \frac{\partial a_{sk}}{\partial q_j} - \frac{\partial a_{sj}}{\partial q_k}\right) \tag{13.192}$$

Example 781 ★ Dynamic Coupling and Decoupling The kinetic energy in terms of generalized coordinates is found in Equation (10.522) as

$$K = \frac{1}{2}\sum_{i=1}^{n}\sum_{j=1}^{n} a_{ij}\dot{q}_i\dot{q}_j + \sum_{i=1}^{n} b_i\dot{q}_i + c \tag{13.193}$$

where $a_{ij} = a_{ji}$, b_i, c are functions of the generalized coordinates q_j and time t. To substitute the generalized expression of K in the Lagrange equation, we find

$$\frac{d}{dt}\frac{\partial K}{\partial \dot{q}_s} = \frac{1}{2}\sum_{j=1}^{n}\left(a_{sj} + a_{js}\right)\ddot{q}_j + \frac{1}{2}\sum_{j=1}^{n}\sum_{k=1}^{n}\left(\frac{\partial a_{sj}}{\partial q_k} + \frac{\partial a_{js}}{\partial q_k}\right)\dot{q}_j\dot{q}_k$$

$$+\frac{1}{2}\sum_{j=1}^{n}\left(\frac{\partial a_{sj}}{\partial t} + \frac{\partial a_{js}}{\partial t}\right)\dot{q}_j + \sum_{k=1}^{n}\frac{\partial b_s}{\partial q_k} + \frac{\partial c}{\partial q_s} \tag{13.194}$$

$$\frac{\partial K}{\partial q_s} = \frac{1}{2}\sum_{j=1}^{n}\sum_{k=1}^{n}\frac{\partial a_{ij}}{\partial q_s}\dot{q}_i\dot{q}_j + \sum_{j=1}^{n}\frac{\partial b_i}{\partial q_s}\dot{q}_i + \frac{\partial c}{\partial q_s} \tag{13.195}$$

Let us re-form the second term of (13.194):

$$\frac{1}{2}\sum_{j=1}^{n}\sum_{k=1}^{n}\frac{\partial a_{sj}}{\partial q_k}\dot{q}_j\dot{q}_k + \frac{1}{2}\sum_{j=1}^{n}\sum_{k=1}^{n}\frac{\partial a_{js}}{\partial q_k}\dot{q}_j\dot{q}_k$$

$$= \frac{1}{2}\sum_{j=1}^{n}\sum_{k=1}^{n}\frac{\partial a_{sj}}{\partial q_k}\dot{q}_j\dot{q}_k + \frac{1}{2}\sum_{j=1}^{n}\sum_{k=1}^{n}\frac{\partial a_{sk}}{\partial q_j}\dot{q}_j\dot{q}_k \tag{13.196}$$

It follows that

$$\frac{d}{dt}\frac{\partial K}{\partial \dot{q}_s} - \frac{\partial K}{\partial q_s} = \frac{1}{2}\sum_{j=1}^{n}\left(a_{sj} + a_{js}\right)\ddot{q}_j$$

$$+ \sum_{j=1}^{n}\sum_{k=1}^{n}\Gamma_{kjs}\dot{q}_j\dot{q}_k + \frac{1}{2}\sum_{j=1}^{n}\left(\frac{\partial a_{sj}}{\partial t} + \frac{\partial a_{js}}{\partial t}\right)\dot{q}_j$$

$$+ 2\sum_{k=1}^{n}\Gamma_{ks}\dot{q}_k + \frac{\partial b_s}{\partial t} - \frac{\partial c}{\partial q_s} \tag{13.197}$$

$$\Gamma_{kjs} = \frac{1}{2}\left(\frac{\partial a_{sj}}{\partial q_k} + \frac{\partial a_{ks}}{\partial q_j} - \frac{\partial a_{kj}}{\partial q_s}\right) \tag{13.198}$$

$$\Gamma_{ks} = \frac{1}{2}\left(\frac{\partial b_s}{\partial q_k} - \frac{\partial b_k}{\partial q_s}\right) \tag{13.199}$$

and therefore, the equations of motion of the system are

$$\sum_{j=1}^{n} a_{sj}\ddot{q}_j + \sum_{j=1}^{n}\sum_{k=1}^{n}\Gamma_{kjs}\dot{q}_j\dot{q}_k + \sum_{j=1}^{n}\frac{\partial a_{sj}}{\partial t}\dot{q}_j$$

$$+2\sum_{k=1}^{n}\Gamma_{ks}\dot{q}_k + \frac{\partial b_s}{\partial t} - \frac{\partial c}{\partial q_s} = Q_s \tag{13.200}$$

If the matrix $\left[a_{sj}\right]$ is not diagonal, these equations are coupled by $\ddot{q}_j$. Such a coupling is called dynamic. However, we can always decouple the equations if $\left[a_{sj}\right]$ is nonsingular:

$$\left|a_{sj}\right| \neq 0 \tag{13.201}$$

To show the nonsingularity of $\left[a_{sj}\right]$, let us consider the kinetic energy equation (13.193) that must hold for all bounded values of velocity components $\dot{q}_s$. We may assume that the velocity components are large enough to have

$$K \approx \frac{1}{2}\sum_{i=1}^{n}\sum_{j=1}^{n} a_{ij}\dot{q}_i\dot{q}_j + \cdots \tag{13.202}$$

Now, from

$$K = \frac{1}{2}\sum_{s=1}^{n} m_s\dot{u}_s^2 \geq 0 \tag{13.203}$$

the kinetic energy K in (13.202) is necessarily positive definite, which indicates that (13.201) holds. So, we may multiply Equation (13.200) by $\left[a_{sj}\right]^{-1}$ to remove the dynamic coupling of the equations of motion:

$$\ddot{q}_r + \sum_{j=1}^{n}\sum_{k=1}^{n} \Gamma^r_{kj}\, \dot{q}_j \dot{q}_k + \sum_{j=1}^{n}\left(2\Gamma^r_j + \sum_{s=1}^{n} [a_{rs}]^{-1} \frac{\partial a_{sj}}{\partial t}\right) \dot{q}_j + \sum_{s=1}^{n} [a_{rs}]^{-1} \frac{\partial b_s}{\partial t} - \sum_{s=1}^{n} [a_{rs}]^{-1} \frac{\partial c}{\partial q_s} = Q_r \tag{13.204}$$

where

$$Q_r = \sum_{s=1}^{n} [a_{rs}]^{-1}\, Q_s \tag{13.205}$$

$$\Gamma^r_{kj} = \sum_{s=1}^{n} [a_{rs}]^{-1}\, \Gamma_{kjs} \tag{13.206}$$

$$\Gamma^r_j = \sum_{s=1}^{n} [a_{rs}]^{-1}\, \Gamma_{js} \tag{13.207}$$

when the kinetic energy is not an explicit function of time, the equations of motion simplify to

$$\ddot{q}_r + \sum_{j=1}^{n}\sum_{k=1}^{n} \Gamma^r_{kj}\, \dot{q}_j \dot{q}_k = Q_r \tag{13.208}$$

Example 782 ★ Equivalent Lagrangian When we define the Lagrangian function $\mathcal{L}$ as

$$\mathcal{L} = K - V \tag{13.209}$$

the equations of motion will be obtained from the Lagrange equation (13.134):

$$\frac{d}{dt}\left(\frac{\partial \mathcal{L}}{\partial \dot{q}_r}\right) - \frac{\partial \mathcal{L}}{\partial q_r} = Q_r \qquad r = 1, 2, \ldots, n \tag{13.210}$$

However, we may define other Lagrangian functions $\mathcal{L}^\star$ that provide the same equations of motion as (13.210):

$$\mathcal{L}^\star = c\mathcal{L} + \frac{d}{dt} f\left(q_1, q_2, \ldots, q_n, t\right) \tag{13.211}$$

The Lagrangian $\mathcal{L}^\star$ may differ from $\mathcal{L}$ by a multiplication constant c and addition of the time derivative of a function f of generalized coordinates q_i and time t. In this case, we have

$$\frac{d}{dt}\left(\frac{\partial \mathcal{L}^\star}{\partial \dot{q}_r}\right) - \frac{\partial \mathcal{L}^\star}{\partial q_r} = \frac{d}{dt}\left(\frac{\partial \mathcal{L}}{\partial \dot{q}_r}\right) - \frac{\partial \mathcal{L}}{\partial q_r} \tag{13.212}$$

Because the Lagrangian function in dynamics is not unique, there is the question of finding a proper Lagrangian function $\mathcal{L}$ for a given dynamic system whose equations of motion are given. This question is called the *inverse-Lagrangian problem*.

Example 783 ★ Lagrangian Is Form Invariant Consider a dynamic system and a set of generalized coordinates $(q_1, q_2, \ldots, q_n)$. Suppose the Lagrangian $\mathcal{L}$ of the system is calculated as

$$\mathcal{L} = \mathcal{L}\left(q_1, q_2, \ldots, q_n, \dot{q}_1, \dot{q}_2, \ldots, \dot{q}_n, t\right) \tag{13.213}$$

Now, suppose that we choose another set of generalized coordinates $(s_1, s_2, \ldots, s_n)$ to express the same dynamic system. Independence of the generalized coordinates provides

$$q_i = q_i\left(s_1, s_2, \ldots, s_n, t\right) \tag{13.214}$$

$$s_j = s_j\left(q_1, q_2, \ldots, q_n, t\right) \tag{13.215}$$

$$J = \left|\frac{\partial q_i}{\partial s_i}\right| \neq 0 \tag{13.216}$$

$$\frac{\partial \dot{q}_i}{\partial \dot{s}_i} = \frac{\partial q_i}{\partial s_i} \qquad \frac{\partial \dot{q}_i}{\partial s_i} = \frac{d}{dt}\frac{\partial q_i}{\partial s_i} \tag{13.217}$$

Having s_j, we may recalculate a new Lagrangian function $\mathcal{L}^\star$ based on new K and V:

$$\mathcal{L}^\star = \mathcal{L}^\star\left(s_1, s_2, \ldots, s_n, \dot{s}_1, \dot{s}_2, \ldots, \dot{s}_n, t\right) \tag{13.218}$$

However, we may also substitute $q_i = q_i\left(s_j\right)$ in $\mathcal{L}$ and obtain $\mathcal{L}^\star$ and determine the equations of motion of the system:

$$\mathcal{L}\left(q_1, \ldots, q_n, \dot{q}_1, \ldots, \dot{q}_n, t\right) = \mathcal{L}^\star\left(s_1, \ldots, s_n, \dot{s}_1, \ldots, \dot{s}_n, t\right) \tag{13.219}$$

Let us examine the Lagrange equation for both sets of coordinates:

$$\frac{\partial \mathcal{L}^\star}{\partial \dot{q}_r} = \frac{\partial \mathcal{L}}{\partial \dot{q}_r}\frac{\partial \dot{q}_i}{\partial \dot{s}_i} = \frac{\partial \mathcal{L}}{\partial \dot{q}_r}\frac{\partial q_i}{\partial s_i} \tag{13.220}$$

$$\frac{d}{dt}\left(\frac{\partial \mathcal{L}^\star}{\partial \dot{q}_r}\right) = \left(\frac{d}{dt}\frac{\partial \mathcal{L}}{\partial \dot{q}_r}\right)\frac{\partial q_i}{\partial s_i} + \frac{\partial \mathcal{L}}{\partial \dot{q}_r}\left(\frac{d}{dt}\frac{\partial q_i}{\partial s_i}\right) \tag{13.221}$$

$$\frac{\partial \mathcal{L}^\star}{\partial q_r} = \frac{\partial \mathcal{L}}{\partial q_r}\frac{\partial q_i}{\partial s_i} + \frac{\partial \mathcal{L}}{\partial \dot{q}_r}\frac{\partial \dot{q}_i}{\partial s_i} = \frac{\partial \mathcal{L}}{\partial q_r}\frac{\partial q_i}{\partial s_i} + \frac{\partial \mathcal{L}}{\partial \dot{q}_r}\frac{d}{dt}\frac{\partial q_i}{\partial s_i} \tag{13.222}$$

Therefore, we have

$$\frac{d}{dt}\frac{\partial \mathcal{L}^\star}{\partial \dot{q}_r} - \frac{\partial \mathcal{L}^\star}{\partial q_r} = \left(\frac{d}{dt}\frac{\partial \mathcal{L}}{\partial \dot{q}_r} - \frac{\partial \mathcal{L}}{\partial q_r}\right)\frac{\partial q_i}{\partial s_i} \tag{13.223}$$

which shows that if we change the set of generalized coordinates, the Lagrange equation would have the same form for the Lagrangian in the new coordinates. Such a property is called *form invariant*.

13.3 ★ VARIATIONAL DYNAMICS

Consider a function f of $x(t)$, $\dot{x}(t)$, and t:

$$f = f(x, \dot{x}, t) \tag{13.224}$$

The unknown variable $x(t)$, which is a function of the independent variable t, is called a *path*. Let us assume that the path is connecting the fixed points x_0 and x_f during a given time $t = t_f - t_0$. So, $x = x(t)$ satisfies the boundary conditions

$$x(t_0) = x_0 \qquad x(t_f) = x_f \tag{13.225}$$

The time integral of the function f over $x_0 \leq x \leq x_f$ is $J(x)$ such that its value depends on the path $x(t)$:

$$J(x) = \int_{t_0}^{t_f} f(x, \dot{x}, t)\, dt \tag{13.226}$$

where $J(x)$ is called an *objective function* or a functional.

The particular path $x(t)$ that minimizes $J(x)$ must satisfy the following equation:

$$\frac{\partial f}{\partial x} - \frac{d}{dt}\frac{\partial f}{\partial \dot{x}} = 0 \tag{13.227}$$

This equation is the *Lagrange* or *Euler–Lagrange differential equation* and is in general of second order.

Proof: To show that a path $x = x^{\star}(t)$ is a minimizing path for the functional $J(x) = \int_{t_0}^{t_f} f(x, \dot{x}, t)\, dt$ with boundary conditions (13.225), we need to show that

$$J(x) \geq J(x^{\star}) \tag{13.228}$$

for all continuous paths $x(t)$. Any path $x(t)$ satisfying the boundary conditions (13.225) is called an *admissible path*. To see that $x^{\star}(t)$ is the optimal path, we may examine the integral J for every admissible path. Let us define an admissible path by superposing another admissible path $y(t)$ to $x^{\star}$,

$$x(t) = x^{\star} + \epsilon y(t) \tag{13.229}$$

where

$$y(t_0) = y(t_f) = 0 \tag{13.230}$$

and ϵ is a small parameter,

$$\epsilon \ll 1 \tag{13.231}$$

Substituting $x(t)$ in J of Equation (13.226) and subtracting from $J(x^\star)$ provide

$$\begin{aligned}\Delta J &= J\left[x^\star + \epsilon y(t)\right] - J\left(x^\star\right) \\ &= \int_{t_0}^{t_f} f(x^\star + \epsilon y, \dot{x}^\star + \epsilon \dot{y}, t)\, dt - \int_{t_0}^{t_f} f(x^\star, \dot{x}^\star, t)\, dt\end{aligned} \tag{13.232}$$

Let us expand $f(x^\star + \epsilon y, \dot{x}^\star + \epsilon \dot{y}, t)$ about $(x^\star, \dot{x}^\star)$,

$$\begin{aligned}f(x^\star + \epsilon y, \dot{x}^\star + \epsilon \dot{y}, t) &= f(x^\star, \dot{x}^\star, t) + \epsilon\left(y\frac{\partial f}{\partial x} + \dot{y}\frac{\partial f}{\partial \dot{x}}\right) \\ &\quad + \epsilon^2\left(y^2\frac{\partial^2 f}{\partial x^2} + 2y\dot{y}\frac{\partial^2 f}{\partial x\, \partial \dot{x}} + \dot{y}^2\frac{\partial^2 f}{\partial \dot{x}^2}\right) dt \\ &\quad + O\left(\epsilon^3\right)\end{aligned} \tag{13.233}$$

and find

$$\Delta J = \epsilon V_1 + \epsilon^2 V_2 + O\left(\epsilon^3\right) \tag{13.234}$$

where

$$V_1 = \int_{t_0}^{t_f}\left(y\frac{\partial f}{\partial x} + \dot{y}\frac{\partial f}{\partial \dot{x}}\right) dt \tag{13.235}$$

$$V_2 = \int_{t_0}^{t_f}\left(y^2\frac{\partial^2 f}{\partial x^2} + 2y\dot{y}\frac{\partial^2 f}{\partial x\, \partial \dot{x}} + \dot{y}^2\frac{\partial^2 f}{\partial \dot{x}^2}\right) \tag{13.236}$$

The first integral, V_1, is called the *first variation* of J, and the second integral, V_2, is called the *second variation* of J. All the higher variations are combined and shown as $O\left(\epsilon^3\right)$. If $x^\star$ is the minimizing path, then it is necessary that $\Delta J \geq 0$ for every admissible $y(t)$. If we divide ΔJ by ϵ and make $\epsilon \to 0$, then we find a necessary condition for $x^\star$ to be the optimal path as $V_1 = 0$. This condition is equivalent to

$$\int_{t_0}^{t_f}\left(y\frac{\partial f}{\partial x} + \dot{y}\frac{\partial f}{\partial \dot{x}}\right) dt = 0 \tag{13.237}$$

By integrating by parts, we may write

$$\int_{t_0}^{t_f} \dot{y}\frac{\partial f}{\partial \dot{x}} dt = \left(y\frac{\partial f}{\partial \dot{x}}\right)_{t_0}^{t_f} - \int_{t_0}^{t_f} y\frac{d}{dt}\left(\frac{\partial f}{\partial \dot{x}}\right) dt \tag{13.238}$$

Because of $y(t_0) = y(t_f) = 0$, the first term on the right-hand side is zero. Therefore, the minimization integral condition (13.237) for every admissible $y(t)$ reduces to

$$\int_{t_0}^{t_f} y\left(\frac{\partial f}{\partial x} - \frac{d}{dt}\frac{\partial f}{\partial \dot{x}}\right) dt = 0 \tag{13.239}$$

The terms in the parentheses are continuous functions of t, evaluated on the optimal path $x^\star$, and they do not involve $y(t)$. So, the only way for the bounded integral of the parentheses, $\left(\frac{\partial f}{\partial x} - \frac{d}{dt}\frac{\partial f}{\partial \dot{x}}\right)$, multiplied by a nonzero function $y(t)$ from t_0 and t_f to be zero is

$$\frac{\partial f}{\partial x} - \frac{d}{dt}\frac{\partial f}{\partial \dot{x}} = 0 \tag{13.240}$$

Equation (13.240) is a necessary condition for $x = x^\star(t)$ to be a solution of the minimization problem (13.226). This differential equation is called the *Euler–Lagrange* or *Lagrange* equation. The second necessary condition to have $x = x^\star(t)$ as a minimizing solution is that the second variation, evaluated on $x^\star(t)$, must be negative. ■

Example 784 ★ Basic Lemma Consider two fixed points x_1 and x_2 $(> x_1)$ and $g(x)$ as a continuous function for $x_1 \leq x \leq x_2$. If

$$\int_{x_1}^{x_2} f(x)\, g(x)\, dx = 0 \tag{13.241}$$

for every choice of the continuous and differentiable function $f(x)$ for which

$$f(x_1) = f(x_2) = 0 \tag{13.242}$$

then

$$g(x) = 0 \tag{13.243}$$

identically in $x_1 \leq x \leq x_2$. This result is called the *basic lemma*.

To prove the lemma, let us assume that (13.243) does not hold. So, suppose there is a particular x_0 of x in $x_1 \leq x_0 \leq x_2$ for which $g(x_0) \neq 0$. At the moment, let us assume that $g(x_0) > 0$. Because $g(x)$ is continuous, there must be an interval around x_0 such as $x_{10} \leq x_0 \leq x_{20}$ in which $g(x) > 0$ everywhere. However, (13.241) cannot then hold for every permissible choice of $f(x)$. A similar contradiction is reached if we assume $g(x_0) < 0$. Therefore, the lemma is correct.

Example 785 ★ Lagrange Equation for Extremizing $J = \int_1^2 \dot{x}^2 dt$ The Lagrange equation for extremizing the functional

$$J = \int_1^2 \dot{x}^2\, dt \tag{13.244}$$

is

$$\frac{\partial f}{\partial x} - \frac{d}{dt}\frac{\partial f}{\partial \dot{x}} = -\ddot{x} = 0 \tag{13.245}$$

which shows that optimal path is

$$x = C_1 t + C_2 \tag{13.246}$$

The boundary conditions $x(1)$, $x(2)$ provide C_1 and C_2. For example, assuming boundary conditions $x(1) = 0$, $x(2) = 3$ provides

$$x = 3t - 3 \tag{13.247}$$

Example 786 ★ Geodesics The problem of determining the shortest path between two given points at the same level of a quantitative characteristic is called the geodesic problem.

An example of a geodesic problem is: What is the shortest arc lying on the surface of a sphere and connecting two given points? We can generalize the problem as follows.

Given two points on the surface of

$$g(x, y, z) = 0 \tag{13.248}$$

what is the equation of the shortest arc lying on (13.248) and connecting the points. Let us express the equation of the surface in parametric form using parameters u and v:

$$x = x(u, v) \qquad y = (u, v) \qquad z = z(u, v) \tag{13.249}$$

the differential of the arc length may be written as

$$\begin{aligned}(ds)^2 &= (dx)^2 + (dy)^2 + (dz)^2 \\ &= P(u, v)(du)^2 + 2Q(u, v)\,du\,dv + R(u, v)(dv)^2\end{aligned} \tag{13.250}$$

where

$$P(u, v) = \left(\frac{\partial x}{\partial u}\right)^2 + \left(\frac{\partial y}{\partial u}\right)^2 + \left(\frac{\partial z}{\partial u}\right)^2 \tag{13.251}$$

$$R(u, v) = \left(\frac{\partial x}{\partial v}\right)^2 + \left(\frac{\partial y}{\partial v}\right)^2 + \left(\frac{\partial z}{\partial v}\right)^2 \tag{13.252}$$

$$Q(u, v) = \frac{\partial x}{\partial u}\frac{\partial x}{\partial v} + \frac{\partial y}{\partial u}\frac{\partial y}{\partial v} + \frac{\partial z}{\partial u}\frac{\partial z}{\partial v} \tag{13.253}$$

If the curves $u = \text{const}$ are orthogonal to the curves $v = \text{const}$, the quantity Q is zero. If the given fixed points on the surface are (u_1, v_1) and (u_2, v_2) with $u_2 > u_1$ and we express the arcs and points by

$$v = v(u) \qquad v(u_1) = v_1 \qquad v(u_2) = v_2 \tag{13.254}$$

the length of the arc is given by

$$J = \int_{u_1}^{u_2} \sqrt{P(u, v) + 2Q(u, v)\frac{dv}{du} + R(u, v)\left(\frac{dv}{du}\right)^2} \tag{13.255}$$

Our problem, then, is to find the function $v(u)$ that renders the integral (13.255) a minimum. Employing the Lagrange equation, we find

$$\frac{\dfrac{\partial P}{\partial v} + 2\dfrac{dv}{du}\dfrac{\partial Q}{\partial v} + \left(\dfrac{dv}{du}\right)^2\dfrac{\partial R}{\partial v}}{2\sqrt{P + 2Q\dfrac{dv}{du} + R\left(\dfrac{dv}{du}\right)^2}} - \frac{d}{du}\left(\frac{Q + R\dfrac{dv}{du}}{\sqrt{P + 2Q\dfrac{dv}{du} + R\left(\dfrac{dv}{du}\right)^2}}\right) = 0 \tag{13.256}$$

In the special case where P, Q, and R are explicitly functions of u alone, this last result becomes

$$\frac{Q + R(dv/du)}{\sqrt{P + 2Q(dv/du) + R\,(dv/du)^2}} = C_1 \tag{13.257}$$

If the curves $u =$ const are orthogonal to the curves $v =$ const, we have

$$v = C_1 \int \frac{\sqrt{P}du}{\sqrt{R^2 - C_1^2 R}} \tag{13.258}$$

Still supposing that $Q = 0$ but having P and R as explicit functions of v alone, we have

$$u = C_1 \int \frac{\sqrt{R}dv}{\sqrt{P^2 - C_1^2 P}} \tag{13.259}$$

As a particular case let us consider the geodesic connecting two points on a sphere with radius r. The most convenient parameters u and v for describing position on the sphere surface are the colatitude θ and the longitude φ:

$$x = r\cos\theta\sin\varphi \qquad y = r\sin\theta\sin\varphi \qquad z = r\cos\varphi \tag{13.260}$$

where θ is the angle between the positive z-axis and the line drawn from the sphere center to the designated point and φ is the angle between the (x, z)-plane and the half plane bounded by the z-axis and containing the designated point. Therefore,

$$P = r^2\sin^2\theta \qquad R = r \qquad Q = 0 \tag{13.261}$$

$$\varphi = C_1 \int \frac{dv}{\sqrt{r^2\sin^4\theta - C_1^2\sin^2\theta}} = -\sin^{-1}\frac{\cot\theta}{\sqrt{(r/C_1)^2 - 1}} + C_2 \tag{13.262}$$

from which it follows that

$$r\sin\theta\cos\varphi\sin C_2 - r\sin\theta\sin\varphi\cos C_2 - \frac{z}{\sqrt{(r/C_1)^2 - 1}} = 0 \tag{13.263}$$

Using (13.260) we find that the sphere geodesic lies on the following plane, which passes through the center of the sphere:

$$x\sin C_2 - y\cos C_2 - \frac{z}{\sqrt{(r/C_1)^2 - 1}} = 0 \tag{13.264}$$

Therefore the shortest arc connecting two points on the surface of a sphere is the intersection of the sphere with the plane containing the given points and the center of the sphere. Such an arc is called a *great-circle* arc.

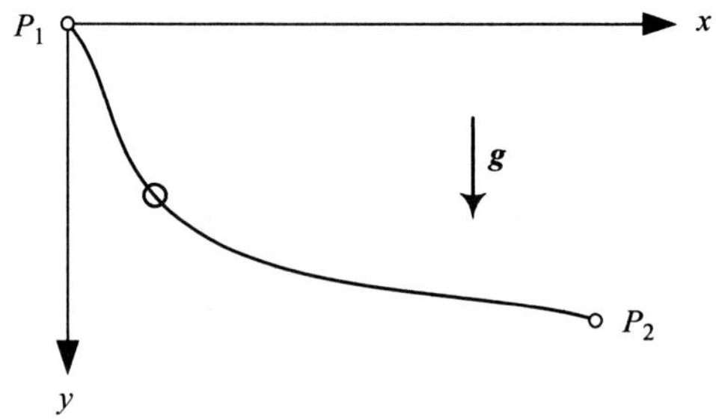

Figure 13.9 A curve joining points P_1 and P_2 and a frictionless sliding point.

Example 787 ★ ***Brachistochrone Problem*** We may use the Lagrange equation and find the frictionless curve joining points P_1 and P_2, as shown in Figure 13.9, along which a particle falling from rest due to gravity travels from the higher to the lower point in minimum time. This is called the brachistochrone problem.

If v is the velocity of the falling point along the curve, then the time required to fall an arc length ds is ds/v. Then, the objective function to find the curve of minimum time is

$$J = \int_{s_1}^{s_2} \frac{ds}{v} \tag{13.265}$$

However,

$$ds = \sqrt{1 + y'^2}dx \qquad y' = \frac{dy}{dx} \tag{13.266}$$

and according to the law of conservation of energy, we have

$$v = \sqrt{2gy} \tag{13.267}$$

Therefore, the objective function simplifies to

$$J = \int_{x_1}^{x_2} \sqrt{\frac{1 + y'^2}{2gy}}dx \tag{13.268}$$

Applying the Lagrange equations, we find

$$y\left(1 + y'^2\right) = 2R \tag{13.269}$$

where R is a constant. The optimal curve starting from $y(0) = 0$ can be expressed by the two parametric equations

$$x = R\left(\theta - \sin\theta\right) \qquad y = R\left(1 - \cos\theta\right) \tag{13.270}$$

The optimal curve is a cycloid. Examples 181 and 413 show more details and application of cycloid curves.

The name of the problem is derived from the Greek words "$\beta\rho\alpha\chi\iota\sigma\tau o\zeta$," meaning "shortest" and "$\chi\rho o\nu o\zeta$," meaning "time." The brachistochrone problem was originally discussed in 1630 by Galileo Galilei (1564–1642) and was solved in 1696 by Johann and Jacob Bernoulli.

Example 788 ★ Lagrange Multiplier Assume $f(x)$ is defined on an open interval $x \in (a, b)$ and has continuous first- and second-order derivatives in some neighborhood of $x_0 \in (a, b)$. The point x_0 is a local extremum of $f(x)$ if

$$\frac{df(x_0)}{dx} = 0 \tag{13.271}$$

Assume $f(\mathbf{x}) = 0$, $\mathbf{x} \in \mathbb{R}^n$, and $g_i(\mathbf{x}) = 0$, $i = 1, 2, \ldots, j$, are functions defined on an open region $\mathbb{R}^n$ and have continuous first- and second-order derivatives in $\mathbb{R}^n$. The necessary condition that $\mathbf{x}_0$ is an extremum of $f(\mathbf{x})$ subject to the constraints $g_i(\mathbf{x}) = 0$ is that there exist j Lagrange multipliers λ_i, $i = 1, 2, \ldots, j$, such that

$$\nabla\left(s + \sum \lambda_i g_i\right) = 0 \tag{13.272}$$

As an example, we can find the minimum of f,

$$f = 1 - x_1^2 - x_2^2 \tag{13.273}$$

subject to the constraint

$$g = x_1^2 + x_2 - 1 = 0 \tag{13.274}$$

by finding the gradient of $f + \lambda g$:

$$\nabla\left[1 - x_1^2 - x_2^2 + \lambda\left(x_1^2 + x_2 - 1\right)\right] = 0 \tag{13.275}$$

which leads to

$$\frac{\partial f}{\partial x_1} = -2x_1 + 2\lambda x_1 = 0 \tag{13.276}$$

$$\frac{\partial f}{\partial x_2} = -2x_2 + \lambda = 0 \tag{13.277}$$

To find the three unknowns x_1, x_2, and λ, we employ Equations (13.276), (13.277), and (13.274). There are two sets of solutions as follows:

$$\begin{array}{lll} x_1 = 0 & x_2 = 1 & \lambda = 2 \\ x_1 = \pm 1/\sqrt{2} & x_2 = 1/2 & \lambda = 1 \end{array} \tag{13.278}$$

Example 789 Dido Problem Consider a planar curve $y(x)$ with a constant length l that connects the points $(-R, 0)$ and $(R, 0)$ as shown in Figure 13.10. The Dido problem is to find the $y(x)$ that maximized the enclosed area. The objective function of the Dido problem is

$$J = \int_{-R}^{R} y\, dx \tag{13.279}$$

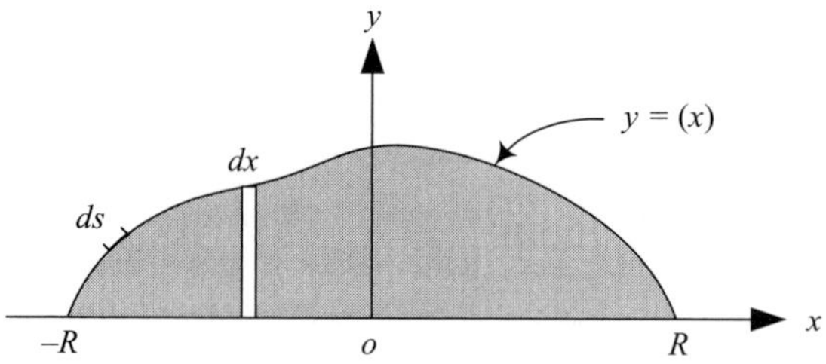

Figure 13.10 Dido problem is to find a planar curve $y(x)$ with a constant length l to maximize the enclosed area.

However, the constant length provides a constraint equation:

$$l = \int_{-R}^{R} ds = \int_{-R}^{R} \sqrt{1 + y'^2}\, dx \tag{13.280}$$

$$ds = \sqrt{1 + y'^2}\, dx \qquad y' = \frac{dy}{dx} \tag{13.281}$$

Therefore, using the Lagrange multiplier λ, the objective function with constraint would be

$$J = \int_{-R}^{R} \left[f\left(y, y', x\right) + \lambda g\left(y, y', x\right) \right] dx \tag{13.282}$$

$$f = y \qquad g = \sqrt{1 + y'^2} \tag{13.283}$$

The Lagrange equation for the constraint objective function (13.282) is

$$\frac{\partial f}{\partial y} - \frac{d}{dx}\frac{\partial f}{\partial y'} + \lambda \left(\frac{\partial g}{\partial y} - \frac{d}{dx}\frac{\partial g}{\partial y'} \right) = 0 \tag{13.284}$$

Equation (13.284) provides

$$\frac{1}{\lambda} = \frac{d}{dx} \frac{y'}{\sqrt{1 + y'^2}} \tag{13.285}$$

This differential equation must be solved to determine the maximizing curve $y\,(x)$. First integration provides

$$\frac{\lambda y'}{\sqrt{1 + y'^2}} = x + C_1 \tag{13.286}$$

Solving this equation for y' yields

$$y' = \frac{\pm\,(x + C_1)}{\sqrt{\lambda^2 - (x + C_1)}} \tag{13.287}$$

and a second integration provides

$$y = \pm\sqrt{\lambda^2 - (x + C_1)} + C_2 \tag{13.288}$$

Satisfying the boundary conditions $(-R, 0)$ and $(R, 0)$, we have

$$C_1 = C_2 = 0 \tag{13.289}$$

$$\lambda = R \tag{13.290}$$

which indicates that the function $y(x)$ is

$$x^2 + y^2 = R^2 \tag{13.291}$$

It is a circle with center at O and radius R.

Example 790 ★ Several Independent Variables We now derive the differential equations that must be satisfied by the twice-differentiable functions $q_1(t), q_2(t), \ldots, q_n(t)$ that extremize the integral J:

$$J = \int_{t_1}^{t_2} f(q_1, q_2, \ldots, q_n, \dot{q}_1, \dot{q}_2, \ldots, \dot{q}_n, t)\, dt \tag{13.292}$$

The functions $q_1(t), q_2(t), \ldots, q_n(t)$ achieve the given values at the fixed limits of integration t_1 and t_2, where $t_1 < t_2$.

Let us show the optimal functions by $q_i^\star(t)$, $i = 1, 2, \ldots, n$. We may examine the integral J for every admissible function. An admissible function may be defined by

$$q_i(t) = q_i^\star + \epsilon y_i(t) \tag{13.293}$$

where

$$y_i(t_1) = y_i(t_2) = 0 \tag{13.294}$$

and ϵ is a small parameter,

$$\epsilon \ll 1 \tag{13.295}$$

Consider a function $f = f(q_i, \dot{q}_i, t)$. The variables $q_i(t)$ satisfy the boundary conditions

$$q_1(t_1) = q_1 \qquad q(t_2) = q_2 \tag{13.296}$$

Substituting $q_i(t)$ in J and subtracting from (13.292) yield.

$$\begin{aligned} \Delta J &= J\left[q_i^\star + \epsilon y_i(t)\right] - J\left(q_i^\star\right) \\ &= \int_{t_0}^{t_f} f(q_i^\star + \epsilon y_i, \dot{q}_i^\star + \epsilon \dot{y}_i, t)\, dt - \int_{t_0}^{t_f} f(q_i^\star, \dot{q}_i^\star, t)\, dt \end{aligned} \tag{13.297}$$

Let us expand $f(x^\star + \epsilon y, \dot{x}^\star + \epsilon \dot{y}, t)$ about $(x^\star, \dot{x}^\star)$:

$$f(q_i^\star + \epsilon y_i, \dot{q}_i^\star + \epsilon \dot{y}_i, t) = f(q_i^\star, \dot{q}_i^\star, t) + \epsilon \left(y_i \frac{\partial f}{\partial q_i} + \dot{y}_i \frac{\partial f}{\partial \dot{q}_i} \right)$$

$$+\epsilon^2\left(y_i^2\frac{\partial^2 f}{\partial q_i^2}+2y_i\dot{y}_j\frac{\partial^2 f}{\partial q_i\,\partial\dot{q}_j}+\dot{y}_i^2\frac{\partial^2 f}{\partial\dot{q}_i^2}\right)dt$$

$$+O\left(\epsilon^3\right) \tag{13.298}$$

and find

$$\Delta J=\epsilon V_1+\epsilon^2 V_2+O\left(\epsilon^3\right) \tag{13.299}$$

where

$$V_1=\int_{t_0}^{t_f}\left(y_i\frac{\partial f}{\partial q_i}+\dot{y}_j\frac{\partial f}{\partial\dot{q}_j}\right)dt \tag{13.300}$$

$$V_2=\int_{t_0}^{t_f}\left(y_i^2\frac{\partial^2 f}{\partial q_i^2}+2y_i\dot{y}_j\frac{\partial^2 f}{\partial q_i\,\partial\dot{q}_j}+\dot{y}_i^2\frac{\partial^2 f}{\partial\dot{q}_i^2}\right). \tag{13.301}$$

If we divide ΔJ by ϵ and make $\epsilon\to 0$, then we find a necessary condition $V_1=0$ for $q_i^\star$ to be the optimal path. By integrating V_1 by parts, we may write

$$\int_{t_0}^{t_f}\dot{y}_1\frac{\partial f}{\partial\dot{q}_1}dt=\left(y_1\frac{\partial f}{\partial\dot{q}_1}\right)_{t_1}^{t_2}-\int_{t_1}^{t_2}y_1\frac{d}{dt}\left(\frac{\partial f}{\partial\dot{q}_1}\right)dt \tag{13.302}$$

Since $y_1(t_1)=y_2(t_2)=0$, the first term on the right-hand side is zero and the integral of V_1 reduces to

$$\int_{t_1}^{t_2}y_1\left(\frac{\partial f}{\partial q_1}-\frac{d}{dt}\frac{\partial f}{\partial\dot{q}_1}\right)dt=0 \tag{13.303}$$

The terms in parentheses are continuous functions of t evaluated on the optimal path $x^\star$, and they do not involve $y_1(t)$. So, the only way for the bounded integral of the parentheses, $\left[\partial f/\partial q_1-(d/dt)(\partial f/\partial\dot{q}_1)\right]$, multiplied by a nonzero function $y_1(t)$ to be zero is if the parentheses are zero. Therefore, the minimization integral condition for every admissible $y_1(t)$ is

$$\frac{\partial f}{\partial q_1}-\frac{d}{dt}\frac{\partial f}{\partial\dot{q}_1}=0 \tag{13.304}$$

Using similar treatment of the successive pairs of terms of (13.302), we derive the following n conditions to minimize (13.292):

$$\frac{\partial f}{\partial q_i}-\frac{d}{dt}\frac{\partial f}{\partial\dot{q}_i}=0\qquad i=1,2,\ldots,n \tag{13.305}$$

Therefore, when a definite integral is given which contains n functions to be determined by the condition that the integral be stationary, we can vary these functions independently. So, the Euler–Lagrange equation can be formed for each function separately. This provides n differential equations.

Example 791 ★ Minimum Time and Bang-Bang Control The most practical function of industrial robots is moving between two points rest to rest. Minimum time control is what we need to increase industrial robot productivity. The objective of time-optimal control is to transfer the end effector of a robot from an initial position to a desired destination in minimum time. Consider a system with the equation of motion

$$\dot{\mathbf{x}} = \mathbf{f}\left(\mathbf{x}(t), \mathbf{Q}(t)\right) \tag{13.306}$$

where $\mathbf{Q}$ is the control input and $\mathbf{x}$ is the state vector of the system:

$$\mathbf{x} = \begin{bmatrix} \mathbf{q} \\ \dot{\mathbf{q}} \end{bmatrix} \tag{13.307}$$

The minimum-time problem is always subject to bounded input, such as

$$|\mathbf{Q}(t)| \le \mathbf{Q}_{\max} \tag{13.308}$$

The solution of the time-optimal control problem subject to bounded input is called bang-bang control. In this solution, at least one of the control variables takes either the maximum or minimum value at each time. The goal of minimum time control is to find the trajectory $\mathbf{x}(t)$ and input $\mathbf{Q}(t)$ starting from an initial state $\mathbf{x}_0(t)$ and arriving at the final state $\mathbf{x}_f(t)$ under the condition that the whole trajectory minimizes the time integral

$$J = \int_{t_0}^{t_f} dt \tag{13.309}$$

Let us define a Hamiltonian function H and a generalized momentum vector $\mathbf{p}$,

$$H(\mathbf{x}, \mathbf{Q}, \mathbf{p}) = \mathbf{p}^{\mathrm{T}} \mathbf{f}\left(\mathbf{x}(t), \mathbf{Q}(t)\right) \tag{13.310}$$

that provide the two equations

$$\dot{\mathbf{x}} = \frac{\partial H}{\partial \mathbf{p}}^{\mathrm{T}} \qquad \dot{\mathbf{p}} = -\frac{\partial H}{\partial \mathbf{x}}^{\mathrm{T}} \tag{13.311}$$

Based on the Pontryagin principle, the optimal input $\mathbf{Q}(t)$ is the one that minimizes the function H. Such an optimal input is to apply the maximum effort $\mathbf{Q}_{\max}$ or $-\mathbf{Q}_{\max}$ over the entire time interval. When the control command takes a value at the boundary of its admissible region, it is said to be saturated. The vector $\mathbf{p}$ is also called a co-state.

To show the application of bang-bang control, let us consider a linear dynamic system given by

$$Q = \ddot{x} \tag{13.312}$$

or

$$\dot{\mathbf{x}} = [A]\,\mathbf{x} + \mathbf{b}Q \tag{13.313}$$

where

$$\mathbf{x} = \begin{bmatrix} x_1 \\ x_2 \end{bmatrix} \qquad [A] = \begin{bmatrix} 0 & 1 \\ 0 & 0 \end{bmatrix} \qquad \mathbf{b} = \begin{bmatrix} 0 \\ 1 \end{bmatrix} \tag{13.314}$$

along with a constraint on the input variable:

$$Q \le 1 \tag{13.315}$$

By defining a co-state vector

$$\mathbf{p} = \begin{bmatrix} p_1 \\ p_2 \end{bmatrix} \tag{13.316}$$

the Hamiltonian (13.310) becomes

$$H(\mathbf{x}, \mathbf{Q}, \mathbf{p}) = \mathbf{p}^{\mathrm{T}} \left([A]\,\mathbf{x} + \mathbf{b}Q \right) \tag{13.317}$$

which provides two first-order differential equations

$$\dot{\mathbf{x}} = \frac{\partial H}{\partial \mathbf{p}}^{\mathrm{T}} = [A]\,\mathbf{x} + \mathbf{b}Q \tag{13.318}$$

$$\dot{\mathbf{p}} = -\frac{\partial H}{\partial \mathbf{x}}^{\mathrm{T}} = -\,[A]\,\mathbf{p} \tag{13.319}$$

Equation (13.319) is

$$\begin{bmatrix} \dot{p}_1 \\ \dot{p}_2 \end{bmatrix} = \begin{bmatrix} 0 \\ -p_2 \end{bmatrix} \tag{13.320}$$

which can be integrated to find **p**:

$$\mathbf{p} = \begin{bmatrix} p_1 \\ p_2 \end{bmatrix} = \begin{bmatrix} C_1 \\ -C_1 t + C_2 \end{bmatrix} \tag{13.321}$$

The Hamiltonian is then equal to

$$H = Qp_2 + p_1 x_2 = \left(-C_1 t + C_2\right) Q + p_1 x_2 \tag{13.322}$$

The control command Q only appears in

$$\mathbf{p}^{\mathrm{T}}\,\mathbf{b}Q = \left(-C_1 t + C_2\right) Q \tag{13.323}$$

which can be maximized by

$$Q(t) = \begin{cases} 1 & \text{if } -C_1 t + C_2 \ge 0 \\ -1 & \text{if } -C_1 t + C_2 < 0 \end{cases} \tag{13.324}$$

This solution implies that $Q(t)$ has a jump point at $t = \frac{C_2}{C_1}$. The jump point at which the control command suddenly changes from maximum to minimum or from minimum to maximum is called the switching point.

Substituting the control input (13.324) into (13.313) gives us two first-order differential equations

$$\begin{bmatrix} \dot{x}_1 \\ \dot{x}_2 \end{bmatrix} = \begin{bmatrix} x_2 \\ Q \end{bmatrix} \tag{13.325}$$

which can be integrated to find the path $\mathbf{x}(t)$:

$$\begin{bmatrix} x_1 \\ x_2 \end{bmatrix} = \begin{cases} \begin{bmatrix} \frac{1}{2}(t + C_3)^2 + C_4 \\ t + C_3 \end{bmatrix} & \text{if } \quad Q = 1 \\ \begin{bmatrix} -\frac{1}{2}(t - C_3)^2 + C_4 \\ -t + C_3 \end{bmatrix} & \text{if } \; Q = -1 \end{cases} \tag{13.326}$$

The constants of integration C_1, C_2, C_3, and C_4 must be calculated based on the boundary conditions

$$\mathbf{x}_0 = \mathbf{x}(t_0) \qquad \mathbf{x}_f = \mathbf{x}(t_f) \tag{13.327}$$

Eliminating t between equations in (13.326) provides the relationship between the state variables x_1 and x_2:

$$x_1 = \begin{cases} \frac{1}{2}x_2^2 + C_4 \text{ if } \quad Q = 1 \\ -\frac{1}{2}x_2^2 + C_4 \text{ if } \; Q = -1 \end{cases} \tag{13.328}$$

These equations show a series of parabolic curves in the (x_1, x_2)-plane with C_4 as a parameter. The parabolas are shown in Figure 13.11 with the arrows indicating the direction of motion on the paths. The (x_1, x_2)-plane is the phase plane.

Considering that there is one switching point in this system, the overall optimal paths are shown in Figure 13.11(*c*). As an example, assume the state of the system at initial and final times are $\mathbf{x}(t_0)$ and $\mathbf{x}(t_f)$, respectively. The motion starts with $Q = 1$, which forces the system to move on the control path $x_1 = \frac{1}{2}x_2^2 + \left(x_{10} - \frac{1}{2}x_{20}^2\right)$ up to the

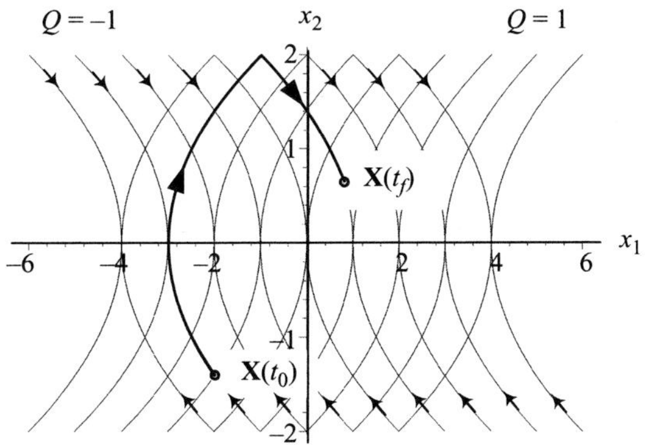

Figure 13.11 Optimal path for $Q = \ddot{x}$ in the phase plane and the mesh of the optimal paths in the phase plane.

intersection point with $x_1 = -\frac{1}{2}x_2^2 + \left(x_{1f} + \frac{1}{2}x_{2f}^2\right)$. The intersection is the switching point at which the control input changes to $Q = -1$. The switching point is at

$$x_1 = \frac{1}{4}\left(2x_{10} + 2x_f - x_{20}^2 + x_{2f}^2\right) \tag{13.329}$$

$$x_2 = \sqrt{\left(x_{1f} + \frac{1}{2}x_{2f}^2\right) - \left(x_{10} - \frac{1}{2}x_{20}^2\right)} \tag{13.330}$$

13.4 ★ HAMILTON PRINCIPLE

The *Hamilton principle* states: *The time integral of the variation of the Lagrangian function $\mathcal{L}$ vanishes along the actual S_T-trajectory connecting two S_T-points $\mathbf{r}_i(t_0)$ and $\mathbf{r}_i(t_1)$ in state–time space*:

$$\int_{t_0}^{t_1} \delta\mathcal{L}\, dt = 0 \tag{13.331}$$

Proof: Beginning from the fundamental equation of dynamics (10.303)

$$\sum_{i=1}^{n} (m_i\ddot{\mathbf{r}}_i - \mathbf{F}_i) \cdot \delta\mathbf{r}_i = 0 \tag{13.332}$$

where $\mathbf{F}_i$ are the given forces, we have

$$\sum_{i=1}^{n} m_i\ddot{\mathbf{r}}_i \cdot \delta\mathbf{r}_i = \sum_{i=1}^{n} \mathbf{F}_i \cdot \delta\mathbf{r}_i = \delta W \tag{13.333}$$

Using

$$\sum_{i=1}^{n} m_i\ddot{\mathbf{r}}_i \cdot \delta\mathbf{r}_i = \frac{d}{dt}\sum_{i=1}^{n} m_i\dot{\mathbf{r}}_i \cdot \delta\dot{\mathbf{r}}_i - \delta K \tag{13.334}$$

we can write Equation (13.333) as

$$\frac{d}{dt}\sum_{i=1}^{n} m_i\dot{\mathbf{r}}_i \cdot \delta\dot{\mathbf{r}}_i = \delta K + \delta W \tag{13.335}$$

On integrating this equation with respect to time over the interval $t_0 \le t \le t_1$, we find

$$\left[\sum_{i=1}^{n} m_i\dot{\mathbf{r}}_i \cdot \delta\dot{\mathbf{r}}_i\right]_{t_0}^{t_1} = \int_{t_0}^{t_1} (\delta K + \delta W)\, dt \tag{13.336}$$

This is the *Hamilton principle in the most general form*. Let us write this equation for virtual displacements $\delta\mathbf{r}_i$ satisfying

$$\delta\mathbf{r}_i(t_0) = \delta\mathbf{r}_i(t_1) = 0 \qquad i = 1, 2, \ldots, n \tag{13.337}$$

to indicate that the actual motion has a S_T-trajectory $\mathbf{r}_i\,(t)$ $(i = 1, 2, \ldots, n)$ with fixed end configurations $\mathbf{r}_i\,(t_0)$ and $\mathbf{r}_i\,(t_1)$, respectively. Consider a neighboring S_T-trajectory $\mathbf{r}'_i\,(t)$,

$$\mathbf{r}'_i\,(t) = \mathbf{r}_i\,(t) + \delta\mathbf{r}_i\,(t) \tag{13.338}$$

satisfying the same boundary conditions (13.337) as the actual S_T-trajectory $\mathbf{r}_i\,(t)$. During the time interval $t_0 \le t \le t_1$, the neighboring S_T-trajectory $\mathbf{r}'_i\,(t)$ are such that the virtual displacements $\delta\mathbf{r}_i$ satisfy the scleronomic holonomic constraint equations. Employing (13.338), the Hamilton principle becomes

$$\int_{t_0}^{t_1} (\delta K + \delta W)\,dt = 0 \tag{13.339}$$

where δK is the virtual change in kinetic energy which results from the virtual displasement and δW is the work done by the given forces in the virtual displasement. Therefore, the Hamilton principle states: The time integral of the sum of the virtual kinetic energy change and virtual work over any time interval vanishes when the virtual displacements of the actual motion and the end configurations are given.

We can express application of the Hamilton principle as this: If we compute the expressions of δK and δW for any arbitrary S_T-trajectory and then set the time integral of their sum equal to zero, we produce a condition that the virtual displacements $\delta\mathbf{r}_i$ were made from the actual S_T-trajectory. Therefore, we have a condition to calculate the actual S_T-trajectory.

When all given forces are potential, we have

$$\delta W = -\delta V \tag{13.340}$$

where δV is a variation of the potential energy V. Therefore, the Hamilton principle becomes

$$\int_{t_0}^{t_1} \delta\,(K - V)\,dt = 0 \tag{13.341}$$

By introducing the Lagrangian function $\mathcal{L}$,

$$\mathcal{L} = K - V \tag{13.342}$$

we write Hamilton principle as Equation (13.331):

$$\int_{t_0}^{t_1} \delta\mathcal{L}\,dt = 0 \tag{13.343}$$

When a system is nonholonomic, we have

$$\int_{t_0}^{t_1} \delta\mathcal{L}\,dt \neq \delta \int_{t_0}^{t_1} \mathcal{L}\,dt = 0 \tag{13.344}$$

However, if a system is holonomic, we may write

$$\int_{t_0}^{t_1} \delta\mathcal{L}\,dt = \delta \int_{t_0}^{t_1} \mathcal{L}\,dt = 0 \tag{13.345}$$

The equation

$$\delta \int_{t_0}^{t_1} \mathcal{L}\, dt = 0 \tag{13.346}$$

defines the problem in the calculus of variation to find stationary values of the integral

$$\int_{t_0}^{t_1} \mathcal{L}\, dt = 0 \tag{13.347}$$

Therefore, only in holonomic systems does the Hamilton principle reduce to a problem in the calculus of variation with the Lagrange equation

$$\frac{\partial \mathcal{L}}{\partial \mathbf{r}_i} - \frac{d}{dt}\frac{\partial \mathcal{L}}{\partial \dot{\mathbf{r}}_i} = 0 \tag{13.348}$$

Equation (13.340) is interesting because δW is virtual work and is not in general a variation of W while δV is the variation of V.

The Hamilton principle is considered as the best known integral principle of mechanics. The Hamilton principle is not, in general, a variational principle. ■

Example 792 ★ The Difference between $\int_{t_0}^{t_1} \delta \mathcal{L}\, dt = 0$ and $\delta \int_{t_0}^{t_1} \mathcal{L}\, dt = 0$ The meaning of

$$\delta \int_{t_0}^{t_1} \mathcal{L}\, dt = 0 \tag{13.349}$$

is different than

$$\int_{t_0}^{t_1} \delta \mathcal{L}\, dt = 0 \tag{13.350}$$

Recalling the definitions of d and δ where $df = 0$ defines the stationary values of f when t is varied and $\delta f = 0$ defines the stationary values of f when t is not varied, we express the meaning of (13.349) as: The time integral of the Lagrangian function is stationary along the actual S_T-trajectory relative to all other possible trajectories having the same end points and differing from the actual trajectory by virtual displacements. However, (13.350) states: The time integral of the variation of the Lagrangian function $\mathcal{L}$ vanishes along the actual S_T-trajectory connecting two S_T-points $r_i(t_0)$ and $r_i(t_1)$ in the state–time space.

Example 793 ★ The Central Principle Expansion of the fundamental equation of dynamics (10.303) is

$$\sum_{i=1}^{n} (m_i \ddot{\mathbf{r}}_i \cdot \delta \mathbf{r}_i - \mathbf{F}_i \cdot \delta \mathbf{r}_i) = 0 \tag{13.351}$$

Let us take a time derivative of $\sum_{i=1}^{n} m_i \dot{\mathbf{r}}_i \cdot \delta \mathbf{r}_i$,

$$\frac{d}{dt}\sum_{i=1}^{n} m_i \dot{\mathbf{r}}_i \cdot \delta \mathbf{r}_i = \sum_{i=1}^{n} m_i \ddot{\mathbf{r}}_i \cdot \delta \mathbf{r}_i + \sum_{i=1}^{n} m_i \dot{\mathbf{r}}_i \cdot \frac{d}{dt}\delta \mathbf{r}_i$$
$$= \sum_{i=1}^{n} m_i \ddot{\mathbf{r}}_i \cdot \delta \mathbf{r}_i + \sum_{i=1}^{n} m_i \dot{\mathbf{r}}_i \cdot \delta \dot{\mathbf{r}}_i \tag{13.352}$$

to derive the identity

$$\sum_{i=1}^{n} m_i \ddot{\mathbf{r}}_i \cdot \delta \mathbf{r}_i = \frac{d}{dt}\sum_{i=1}^{n} m_i \dot{\mathbf{r}}_i \cdot \delta \mathbf{r}_i - \sum_{i=1}^{n} m_i \dot{\mathbf{r}}_i \cdot \delta \dot{\mathbf{r}}_i \tag{13.353}$$

The last term in this identity can be written based on the kinetic energy K:

$$K = \frac{1}{2}\sum_{i=1}^{n} m_i \dot{\mathbf{r}}_i^2 \tag{13.354}$$

where

$$\delta K = \frac{1}{2}\sum_{i=1}^{n} m_i \dot{\mathbf{r}}_i \cdot \delta \dot{\mathbf{r}}_i \tag{13.355}$$

Therefore, we have

$$\sum_{i=1}^{n} m_i \ddot{\mathbf{r}}_i \cdot \delta \mathbf{r}_i = \frac{d}{dt}\sum_{i=1}^{n} m_i \dot{\mathbf{r}}_i \cdot \delta \mathbf{r}_i - \delta K \tag{13.356}$$

This equation is sometimes called the *central principle*. It states: The virtual work done by the inertia forces is equal to the time rate of change of the work done by the momentum minus the virtual change in kinetic energy.

Example 794 ★ Hamilton Principle for a Particle on a Surface Consider a moving particle m on a surface, under its gravitational force. To determine the Hamilton principle, let us write its kinetic and potential energies:

$$K = \tfrac{1}{2}\left(\dot{x}^2 + \dot{y}^2 + \dot{z}^2\right) \qquad V = mgz \tag{13.357}$$

The equation of the surface provides a holonomic constraint between the coordinates of the particle:

$$z = f(x, y) \tag{13.358}$$

Therefore, Equation (13.346) is applicable:

$$\delta \int_{t_0}^{t_1} (K - V)\, dt = \delta \int_{t_0}^{t_1} \frac{m}{2}\left(\dot{x}^2 + \dot{y}^2 + \dot{z}^2 - 2mgz\right) dt$$
$$= \delta \int_{t_0}^{t_1} \frac{m}{2}\left[\dot{x}^2 + \dot{y}^2 + \left(\frac{\partial f}{\partial x}\dot{x} + \frac{\partial f}{\partial y}\dot{y}\right)^2 - 2mgz\right] dt \tag{13.359}$$

Example 795 ★ Nonholonomic Hamilton Principle Consider a moving particle m to be acted on by a potential force of potential energy V:

$$V = V(x, y) \tag{13.360}$$

Assume the particle is constrained such that the slope of its trajectory is proportional to time:

$$\frac{\dot{y}}{\dot{x}} = t \tag{13.361}$$

This is a nonholonomic constraint; therefore, we should use the general form of the Hamilton principle (13.331). The kinetic energy of the particle and its variation are

$$K = \tfrac{1}{2} m \left(\dot{x}^2 + \dot{y}^2\right) \tag{13.362}$$

$$\delta K = m\dot{x}\,\delta\dot{x} + m\dot{y}\,\delta\dot{y} \tag{13.363}$$

The variation of the potential energy is

$$\delta V = \frac{\partial V}{\partial x}\delta x + \frac{\partial V}{\partial y}\delta y \tag{13.364}$$

Using the constraint equation, we have

$$\begin{aligned}\delta \mathcal{L} &= m\dot{x}\,\delta\dot{x} + m\dot{y}\,\delta\dot{y} - \frac{\partial V}{\partial x}\delta x + \frac{\partial V}{\partial y}\delta y \\ &= m\dot{x}\,\delta\dot{x} + (\delta x + t\,\delta\dot{x})\, m\dot{x}t - \left(\frac{\partial V}{\partial x} + t\frac{\partial V}{\partial y}\right)\delta x \\ &= \left[\left(1+t^2\right) m\dot{x}\,\delta\dot{x} + \left(m\dot{x}t - \frac{\partial V}{\partial x} - t\frac{\partial V}{\partial y}\right)\right]\delta x\end{aligned} \tag{13.365}$$

Therefore, the Hamilton principle (13.331) would be

$$\int_{t_0}^{t_1}\left[\left(1+t^2\right) m\dot{x}\,\delta\dot{x} + \left(m\dot{x}t - \frac{\partial V}{\partial x} - t\frac{\partial V}{\partial y}\right)\right]\delta x\, dt = 0 \tag{13.366}$$

13.5 ★ LAGRANGE EQUATION AND CONSTRAINTS

The most general form of the Lagrange equation is

$$\frac{d}{dt}\frac{\partial K}{\partial \dot{q}_i} - \frac{\partial K}{\partial q_i} - Q_i + \sum_{j=1}^{l} \lambda_j \frac{\partial f_j}{\partial \dot{q}_i} = 0 \qquad i = 1, 2, \ldots, n \tag{13.367}$$

where K is the kinetic energy, Q_i the generalized force, q_i the generalized coordinates, $\dot{q}_j$ the generalized velocities, λ_j the Lagrange multipliers, and $f_j (j = 1, 2, \ldots, 1)$ the nonholonomic constraint on q_i:

$$f_j\left(q_1, q_2, \ldots, q_n, \dot{q}_1, \dot{q}_2, \ldots, \dot{q}_n, t\right) = 0 \qquad j = 1, 2, \ldots, l \tag{13.368}$$

So, when a dynamic system is under nonholonomic constraints (13.368), its equations of motion can be found from the general form of the Lagrange equation (13.367).

Proof: The fundamental equation of dynamics in terms of generalized coordinates is found in Equation (10.738) as

$$\sum_{j=1}^{n}\left(\frac{d}{dt}\frac{\partial K}{\partial \dot{q}_j} - \frac{\partial K}{\partial q_j} - Q_j\right)\delta q_j = 0 \tag{13.369}$$

where the kinetic energy K and the generalized force Q are functions of the generalized coordinates q_j, generalized velocities $\dot{q}_j$, and time t. Let us assume that the possible generalized velocities $\dot{q}_j$ must satisfy the nonholonomic generalized constraints (13.368):

$$\sum_{k=1}^{n} B_{jk}\,\dot{q}_k + B_j = 0 \qquad j = 1, 2, \ldots, l \tag{13.370}$$

$$B_{jk} = \frac{\partial f_j}{\partial q_k} \qquad B_j = \frac{\partial f_j}{\partial t} \tag{13.371}$$

Among the possible δq_j, we are looking for those that also satisfy (13.369). The virtual generalized displacements δq_j in (13.369) must satisfy the nonholonomic generalized constraints (10.554):

$$\sum_{k=1}^{n} B_{jk}\,\delta q_k = 0 \qquad j = 1, 2, \ldots, l \tag{13.372}$$

Let us introduce the quantity R_j as

$$R_j = \frac{d}{dt}\frac{\partial K}{\partial \dot{q}_j} - \frac{\partial K}{\partial q_j} - Q_j \tag{13.373}$$

and an n-dimensional vector $\mathbf{R}$ and an l-dimensional vector $\mathbf{B}_j$ as

$$\mathbf{R} = \begin{bmatrix} R_1 & R_2 & \cdots & R_n \end{bmatrix}^{\mathrm{T}} \tag{13.374}$$

$$\mathbf{B}_j = \begin{bmatrix} B_{j1} & B_{j2} & \cdots & B_{jl} \end{bmatrix}^{\mathrm{T}} \tag{13.375}$$

Now, Equations (13.369) and (13.372) become

$$\mathbf{R}\cdot\delta\mathbf{q} = 0 \tag{13.376}$$

$$\mathbf{B}_j\cdot\delta\mathbf{q} = 0 \tag{13.377}$$

showing that $\delta\mathbf{q}$ is orthogonal to both $\mathbf{R}$ and every $\mathbf{B}_j$. Introducing the Lagrange multipliers λ_j, we may combine these two equations:

$$\left(\mathbf{R} + \sum_{j=1}^{l}\lambda_j\mathbf{B}_j\right)\cdot\delta\mathbf{q} = 0 \tag{13.378}$$

which implies that $\mathbf{R}$ is in the same direction as $\sum_{j=1}^{l} \lambda_j \mathbf{B}_j$:

$$\mathbf{R} = -\sum_{j=1}^{l} \lambda_j \mathbf{B}_j \tag{13.379}$$

To show this, let us assume $l = 1$ to have $\mathbf{B}_1$, indicating an n-dimensional vector in Q-space. Then, an arbitrary n-dimensional possible displacement vector $\delta\mathbf{q}$ would lie in the orthogonal plane $\pi_\mathbf{B}$ to $\mathbf{B}_1$. Now, if $\mathbf{R}$ is not on the same axis of $\mathbf{B}_1$, it must have a component $\mathbf{R}_\perp$ on $\pi_\mathbf{B}$ in an orthogonal direction to $\delta\mathbf{q}$. Because $\delta\mathbf{q}$ is an arbitrary virtual displacement vector, it can turn in $\pi_\mathbf{B}$ around the $\mathbf{B}_1$-axis. Such a freedom violates the orthogonality of $\mathbf{R}_\perp$ and $\delta\mathbf{q}$. So, $\mathbf{R}_\perp = 0$, and therefore, $\mathbf{B}_1$ and $\mathbf{R}$ are colinear. In the general case of $\mathbf{B}_j$ $(j = 1, 2, \ldots, l < n)$, the orthogonal plane $\pi_\mathbf{B}$ would be an l-dimensional hyperplane in Q-space on which $\delta\mathbf{q}$ lies. Therefore, (13.379) holds.

Expansion of (13.378) provides

$$\sum_{i=1}^{n} \left(\frac{d}{dt}\frac{\partial K}{\partial \dot{q}_i} - \frac{\partial K}{\partial q_i} - Q_i + \sum_{j=1}^{l} \lambda_j B_{ij} \right) \delta q_i = 0 \qquad i = 1, 2, \ldots, n \tag{13.380}$$

where δq_i are unconstraint virtual displacements. From Equation (13.380), we deduce the fundamental theorem of Lagrangian mechanics:

To satisfy Equation (13.380), it is necessary and sufficient that

$$\frac{d}{dt}\frac{\partial K}{\partial \dot{q}_i} - \frac{\partial K}{\partial q_i} - Q_i + \sum_{j=1}^{l} \lambda_j B_{ij} = 0 \tag{13.381}$$

for every $i = 1, 2, \ldots, n$.

The sufficiency is obvious. To show the necessity, let us assume that one of them is not zero. Then we may choose the associated δq_i in (13.380) to have the same sign as the nonzero parentheses. Such a combination makes the sum (13.380) positive, which contradicts the requirement that it be zero.

Equations (13.367) are the *Lagrange equations of motion*, which are n equations in $n + l$ unknowns q_i, $i = 1, 2, \ldots, n$, and λ_j, $j = 1, 2, \ldots, l$. The n equations (13.381) along with the l nonholonomic constraint equations (13.370) provide the exact number of equations to determine the unknowns. ■

Example 796 A Rolling Disc A thin disc with mass m and radius R is rolling without slipping on a horizontal plane, as is shown in Figure 13.12. To determine its Lagrange equations of motion, we begin with the kinematics of the disc. Let us attach a body coordinate frame B to the disc at the mass center C and a global frame G on the horizontal plane. We use the Euler angles φ, θ, ψ to indicate the orientation of B in G and the coordinates X, Y of the contact point $P(X, Y, Z)$ to indicate the position of the disc. Because of the holonomic constraint $Z = 0$, we can use the variables X, Y, φ, θ, ψ as the required five generalized coordinates.

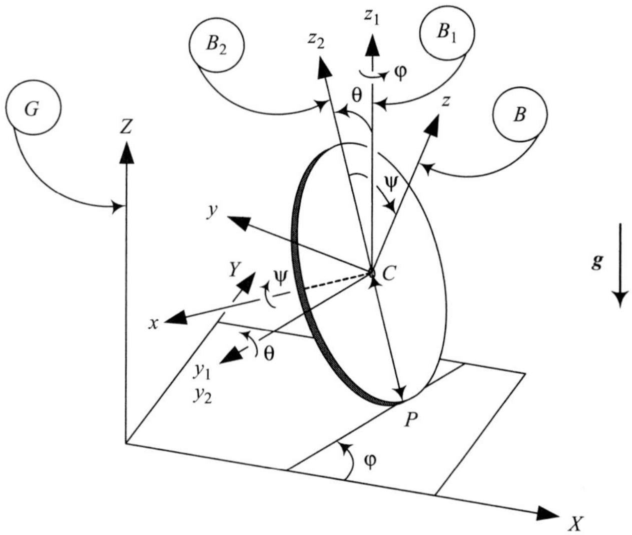

Figure 13.12 A thin disc with mass m and radius R is rolling without slipping on a horizontal plane.

To determine the transformation matrix ${}^G R_B$, we introduce two intermediate frames B_1 and B_2, both at C. The frames B_1 and G are related by a rotation φ about the z_1-axis:

$$
{}^1R_G = \begin{bmatrix} \cos\varphi & \sin\varphi & 0 \\ -\sin\varphi & \cos\varphi & 0 \\ 0 & 0 & 1 \end{bmatrix} \tag{13.382}
$$

The frames B_2 and B_1 are related by a rotation θ about the y_2-axis:

$$
{}^2R_1 = \begin{bmatrix} \cos\theta & 0 & -\sin\theta \\ 0 & 1 & 0 \\ \sin\theta & 0 & \cos\theta \end{bmatrix} \tag{13.383}
$$

The frames B and B_2 are related by a rotation ψ about the x-axis:

$$
{}^BR_2 = \begin{bmatrix} 1 & 0 & 0 \\ 0 & \cos\psi & \sin\psi \\ 0 & -\sin\psi & \cos\psi \end{bmatrix} \tag{13.384}
$$

Therefore,

$$
\begin{aligned}
{}^BR_G &= {}^BR_2\,{}^2R_1\,{}^1R_G \\
&= \begin{bmatrix} c\theta c\varphi & c\theta s\varphi & -s\theta \\ c\varphi s\theta s\psi - c\psi s\varphi & c\psi c\varphi + s\theta s\psi s\varphi & c\theta s\psi \\ s\psi s\varphi + c\psi c\varphi s\theta & c\psi s\theta s\varphi - c\varphi s\psi & c\theta c\psi \end{bmatrix}
\end{aligned} \tag{13.385}
$$

$$
\begin{aligned}
{}^G R_B &= {}^B R_G^{\mathrm{T}} \\
&= \begin{bmatrix} c\theta c\varphi & c\varphi s\theta s\psi - c\psi s\varphi & s\psi s\varphi + c\psi c\varphi s\theta \\ c\theta s\varphi & c\psi c\varphi + s\theta s\psi s\varphi & c\psi s\theta s\varphi - c\varphi s\psi \\ -s\theta & c\theta s\psi & c\theta c\psi \end{bmatrix}
\end{aligned} \tag{13.386}
$$

Employing the transformation matrix

$$
\begin{aligned}
{}^G R_2 &= \left[{}^2 R_1 \, {}^1 R_G\right]^{\mathrm{T}} = {}^1 R_G^{\mathrm{T}} \, {}^2 R_1^{\mathrm{T}} \\
&= \begin{bmatrix} \cos\theta\cos\varphi & -\sin\varphi & \cos\varphi\sin\theta \\ \cos\theta\sin\varphi & \cos\varphi & \sin\theta\sin\varphi \\ -\sin\theta & 0 & \cos\theta \end{bmatrix}
\end{aligned} \tag{13.387}
$$

and the global coordinates of the contact point $P(X, Y)$, we find the global position of C:

$$
\begin{aligned}
{}^G \mathbf{r}_C &= {}^G \mathbf{r}_P + {}^G R_2 \, {}^2_P \mathbf{r}_C \\
&= \begin{bmatrix} X \\ Y \\ 0 \end{bmatrix} + {}^G R_B \begin{bmatrix} 0 \\ 0 \\ R \end{bmatrix} = \begin{bmatrix} X + R\cos\varphi\sin\theta \\ Y + R\sin\theta\sin\varphi \\ R\cos\theta \end{bmatrix}
\end{aligned} \tag{13.388}
$$

Taking a time derivative of ${}^G \mathbf{r}_C$ provides the velocity of the mass center C:

$$
{}^G \mathbf{v}_C = \frac{{}^G d}{dt} {}^G \mathbf{r}_C = \begin{bmatrix} \dot{X} + R\dot{\theta}\cos\theta\cos\varphi - R\dot{\varphi}\sin\theta\sin\varphi \\ \dot{Y} + R\dot{\theta}\cos\theta\sin\varphi + R\dot{\varphi}\cos\varphi\sin\theta \\ -R\dot{\theta}\sin\theta \end{bmatrix} \tag{13.389}
$$

The angular velocity of the disc is

$$
\begin{aligned}
{}^B_G \tilde{\omega}_B &= {}^G R_B^{\mathrm{T}} \, {}^G \dot{R}_B \\
&= \begin{bmatrix} 0 & \dot{\theta} s\psi - \dot{\varphi} c\theta c\psi & \dot{\theta} c\psi + \dot{\varphi} c\theta s\psi \\ \dot{\varphi} c\theta c\psi - \dot{\theta} s\psi & 0 & \dot{\varphi} s\theta - \dot{\psi} \\ -\dot{\theta} c\psi - \dot{\varphi} c\theta s\psi & \dot{\psi} - \dot{\varphi} s\theta & 0 \end{bmatrix}
\end{aligned} \tag{13.390}
$$

Therefore, the kinetic and potential energies of the disc are

$$
\begin{aligned}
K &= \tfrac{1}{2} m \, {}^G \mathbf{v}_C \cdot {}^G \mathbf{v}_C + \tfrac{1}{2} \, {}^B_G \omega_B^{\mathrm{T}} \, {}^B I \, {}^B_G \omega_B \\
&= \tfrac{1}{2} m \left(\dot{X}^2 + \dot{Y}^2\right) + \tfrac{1}{2} m R^2 \dot{\theta}^2 + \tfrac{1}{2} m R^2 \dot{\varphi}^2 \sin^2\theta \\
&\quad + mR\dot{X}\left(\dot{\theta}\cos\varphi\cos\theta - \dot{\varphi}\sin\varphi\sin\theta\right) \\
&\quad + mR\dot{Y}\left(\dot{\theta}\sin\varphi\cos\theta + \dot{\varphi}\cos\varphi\sin\theta\right) \\
&\quad + \tfrac{1}{2} I_1 \dot{\psi}^2 + \tfrac{1}{2} I_2 \dot{\theta}^2 + \tfrac{1}{4}\left(I_1 + I_2\right)\dot{\varphi}^2 \\
&\quad - \tfrac{1}{4}\left(I_1 - I_2\right)\dot{\varphi}^2 \cos 2\theta - I_1 \dot{\psi}\dot{\varphi}\sin\theta
\end{aligned} \tag{13.391}
$$

$$
V = mgR\cos\theta \tag{13.392}
$$

where

$$ {}^{B}I = \begin{bmatrix} I_1 & 0 & 0 \\ 0 & I_2 & 0 \\ 0 & 0 & I_2 \end{bmatrix} \tag{13.393} $$

Rolling without slipping provides a vectorial nonholonomic constraints:

$$ {}^{G}\mathbf{v}_P = {}^{G}\mathbf{v}_C + {}_{G}\omega_B \times {}_{C}^{G}\mathbf{r}_P = 0 \tag{13.394} $$

Employing the above kinematic vectors and transformation matrices, we can write the rolling constraint equation as

$$ \begin{aligned} {}^{G}\mathbf{v}_P &= {}^{G}\mathbf{v}_C + {}^{G}R_B \left({}_{G}^{B}\omega_B \times {}^{G}R_2\, {}_{C}^{2}\mathbf{r}_P\right) \\ &= \begin{bmatrix} \dot{X} - R\dot{\psi}\sin\varphi \\ \dot{Y} + R\dot{\psi}\cos\varphi \\ 0 \end{bmatrix} = 0 \end{aligned} \tag{13.395} $$

The nonholonomic constraints are sceleronomic,

$$ f_1 = \dot{X} - R\dot{\psi}\sin\varphi = 0 \tag{13.396} $$

$$ f_2 = \dot{Y} + R\dot{\psi}\cos\varphi = 0 \tag{13.397} $$

and their Pfaffian forms are

$$ \delta X - R\delta\psi \sin\varphi = 0 \tag{13.398} $$

$$ \delta Y + R\delta\psi \cos\varphi = 0 \tag{13.399} $$

Therefore, the Lagrangian of the disc is

$$ \mathcal{L} = K - V \tag{13.400} $$

and the Lagrange equation is

$$ \sum_{i=1}^{n} \left(\frac{d}{dt}\frac{\partial \mathcal{L}}{\partial \dot{q}_i} - \frac{\partial \mathcal{L}}{\partial q_i} - Q_i + \sum_{j=1}^{l} \lambda_j \frac{\partial f_j}{\partial \dot{q}_i} \right) \delta q_i = 0 \qquad i = 1, 2, \ldots, n \tag{13.401} $$

There is no nonconservative generalized force Q_i. Applying the Lagrange equation to the generalized coordinates $X, Y, \varphi, \theta, \psi$, we find the following equations of motion, respectively:

$$ \frac{d}{dt}\left[m\dot{X} + mR\left(\dot{\theta}\cos\varphi\cos\theta - \dot{\varphi}\sin\varphi\sin\theta\right)\right] + \lambda_1 = 0 \tag{13.402} $$

$$ \frac{d}{dt}\left[m\dot{Y} + mR\left(\dot{\theta}\sin\varphi\cos\theta + \dot{\varphi}\cos\varphi\sin\theta\right)\right] + \lambda_2 = 0 \tag{13.403} $$

$$\frac{d}{dt}\begin{pmatrix}\frac{1}{2}mR^2\dot\varphi\,(1-\cos 2\theta)-mR\dot X\sin\varphi\sin\theta+mR\dot Y\cos\varphi\sin\theta\\ -I_1\left(\dot\psi-\dot\varphi\sin\theta\right)\sin\theta+\frac{1}{2}I_2\dot\varphi\,(1+\cos 2\theta)\end{pmatrix}$$
$$-mR\dot X\left(-\dot\theta\sin\varphi\cos\theta-\dot\varphi\cos\varphi\sin\theta\right)$$
$$-mR\dot Y\left(\dot\theta\cos\varphi\cos\theta-\dot\varphi\sin\varphi\sin\theta\right)=0 \tag{13.404}$$

$$\frac{d}{dt}\left(mR^2\dot\theta+mR\dot X\cos\varphi\cos\theta+mR\dot Y\sin\varphi\cos\theta+I_2\dot\theta\right)$$
$$-\frac{1}{2}mR^2\dot\varphi^2\sin 2\theta-mR\dot X\left(-\dot\theta\cos\varphi\sin\theta-\dot\varphi\sin\varphi\cos\theta\right)$$
$$-mR\dot Y\left(-\dot\theta\sin\varphi\sin\theta+\dot\varphi\cos\varphi\cos\theta\right)+I_2\dot\varphi^2\sin\theta\cos\theta$$
$$+I_1\dot\varphi\left(\dot\psi-\dot\varphi\sin\theta\right)\cos\theta-mgR\sin\theta=0 \tag{13.405}$$

$$I_1\frac{d}{dt}\left(\dot\psi-\dot\varphi\sin\theta\right)-\lambda_1 R\sin\varphi+\lambda_2 R\cos\varphi=0 \tag{13.406}$$

We may introduce two new variables,

$$\omega=\dot\psi-\dot\varphi\sin\theta\qquad\Omega=\dot\psi \tag{13.407}$$

and rewrite the equations of motion simpler.

Equations (13.402)–(13.406), and (13.396)–(13.397) are the seven equations that describe the behavior of the kinematic variables X, Y, φ, θ, ψ along with the Lagrange multipliers λ_1 and λ_2.

Example 797 ★ Generality of Lagrange Equation of Motion The Lagrange equation of motion is more general and can also be employed when the coordinates are not generalized and the constraints are holonomic. To show this, let us assume the coordinates $(s_1, s_2, \dots, s_{n'})$ are not generalized and their number n' exceeds the minimum required number n by l':

$$n'=n+l' \tag{13.408}$$

If there are l constraints

$$f_j\left(s_1, s_2, \dots, s_{n'}, t\right)=0\qquad i=1,2,\dots,l \tag{13.409}$$

then there are l' holonomic constraints

$$f_i\left(s_1, s_2, \dots, s_{n'}, t\right)=0\qquad i=1,2,\dots,l' \tag{13.410}$$

whose Pfaffian form is

$$\sum_{r=1}^{n'}B_{ir}ds_r+B_i dt=0\qquad i=1,2,\dots,l' \tag{13.411}$$

$$B_{ir}=\frac{\partial f_i}{\partial s_r}\qquad B_i=\frac{\partial f_i}{\partial t} \tag{13.412}$$

and there are $l - l'$ nonholonomic constraints

$$f_k(s_1, s_2, \ldots, s_{n'}, t) = 0 \qquad k = l' + 1, l' + 2, \ldots, l \tag{13.413}$$

$$\sum_{r=1}^{n'} B_{kr}\, ds_r + B_k\, dt = 0 \qquad k = l' + 1, l' + 2, \ldots, l \tag{13.414}$$

Introducing the Lagrange multipliers does not depend on the integrability of the constraint equations. Therefore, we can write Lagrange equations of motion as

$$\frac{d}{dt}\frac{\partial K}{\partial \dot{s}_i} - \frac{\partial K}{\partial s_i} - Q_i + \sum_{j=1}^{l'} \lambda_j B_{ji} + \sum_{k=l'+1}^{l} \lambda_k B_{ki} = 0 \qquad i = 1, 2, \ldots, n' \tag{13.415}$$

which can also be written as

$$\frac{d}{dt}\frac{\partial K}{\partial \dot{s}_i} - \frac{\partial K}{\partial s_i} - Q_i + \sum_{r=1}^{l} \lambda_r B_{ri} = 0 \qquad i = 1, 2, \ldots, n' \tag{13.416}$$

Example 798 ★ Lagrange Equation from Central Principle We may begin with the central equation (13.356) of Example 793 to derive the Lagrange equation:

$$\sum_{i=1}^{N/3} m_i \ddot{\mathbf{r}}_i \cdot \delta\mathbf{r}_i = \frac{d}{dt}\sum_{i=1}^{N/3} m_i \dot{\mathbf{r}}_i \cdot \delta\mathbf{r}_i - \delta K \tag{13.417}$$

Let us rewrite the central principle in configuration space as

$$\sum_{i=1}^{N} m_i \ddot{u}_i\, \delta u_i = \frac{d}{dt}\sum_{i=1}^{N} m_i \dot{u}_i\, \delta u_i - \delta K \tag{13.418}$$

Employing

$$\delta u_i = \sum_{k=1}^{n} \frac{\partial u_i}{\partial q_k} \delta q_k \tag{13.419}$$

we have

$$\sum_{i=1}^{N} m_i \dot{u}_i\, \delta u_i = \sum_{i=1}^{N} m_i \dot{u}_i \sum_{k=1}^{n} \frac{\partial u_i}{\partial q_k} \delta q_k = \sum_{k=1}^{n}\sum_{i=1}^{N} \left(m_i \dot{u}_i \frac{\partial u_i}{\partial q_k} \right) \delta q_k$$
$$= \sum_{k=1}^{n}\sum_{i=1}^{N} \left(m_i \dot{u}_i \frac{\partial \dot{u}_i}{\partial \dot{q}_k} \right) \delta q_k \tag{13.420}$$

Also, substituting the equations

$$\sum_{i=1}^{N} m_i \dot{u}_i \, \delta u_i = \sum_{j=1}^{n} \frac{\partial K}{\partial \dot{q}_j} \delta q_j \tag{13.421}$$

$$\delta K = \sum_{j=1}^{n} \frac{\partial K}{\partial q_j} \delta q_j + \sum_{j=1}^{n} \frac{\partial K}{\partial \dot{q}_j} \delta \dot{q}_j \tag{13.422}$$

provides

$$\begin{aligned} \sum_{i=1}^{N} m_i \ddot{u}_i \, \delta u_i &= \frac{d}{dt} \sum_{j=1}^{n} \frac{\partial K}{\partial \dot{q}_j} \delta q_j - \sum_{j=1}^{n} \frac{\partial K}{\partial q_j} \delta q_j - \sum_{j=1}^{n} \frac{\partial K}{\partial \dot{q}_j} \delta \dot{q}_j \\ &= \sum_{j=1}^{n} \frac{d}{dt} \frac{\partial K}{\partial \dot{q}_j} \delta q_j - \sum_{j=1}^{n} \frac{\partial K}{\partial q_j} \delta q_j \end{aligned} \tag{13.423}$$

Using the definition of generalized force Q_j, we have

$$\sum_{i=1}^{N} F_i \, \delta u_i = \sum_{i=1}^{N} F_i \sum_{j=1}^{n} \frac{\partial u_i}{\partial q_j} \delta q_j = \sum_{j=1}^{n} Q_j \, \delta q_j \tag{13.424}$$

Substituting these results in the fundamental equation of dynamics (10.303),

$$\sum_{i=1}^{N} m_i \ddot{u}_i \, \delta u_i - \sum_{i=1}^{N} F_i \, \delta u_i = 0 \tag{13.425}$$

yields

$$\sum_{i=1}^{n} \left(\frac{d}{dt} \frac{\partial K}{\partial \dot{q}_i} - \frac{\partial K}{\partial q_i} - Q_i \right) \delta q_i = 0 \tag{13.426}$$

This is the Lagrange equation for holonomic systems. If the system is nonholonomic, we should add $\sum_{j=1}^{l} \lambda_j B_{ij}$ to the equation of motion.

13.6 CONSERVATION LAWS

Conservation of energy, conservation of momentum, and conservation of the moment of momentum are the only conservation laws in dynamics. Each conservation law is a consequence of a symmetry in time or position spaces and appears in the form of an integral of motion. As reviewed in Section 10.10, if $\mathbf{q}$ and $\dot{\mathbf{q}}$ are the generalized positions and velocities of a dynamic system, then an equation $f = c$ of the form

$$f(\mathbf{q}, \dot{\mathbf{q}}, t) = c \qquad c = f(\mathbf{q}_0, \dot{\mathbf{q}}_0, t_0) \tag{13.427}$$

$$\frac{df}{dt} = \sum_{i=1}^{n} \left(\frac{\partial f}{\partial q_i} \dot{q}_i + \frac{\partial f}{\partial \dot{q}_i} \ddot{q}_i \right) + \frac{\partial f}{\partial t} = 0 \tag{13.428}$$

is an *integral of motion*. The parameter c where its value depends on initial conditions is called a *constant of motion*.

The Lagrangian

$$\mathcal{L} = K - V = \mathcal{L}(\mathbf{q}, \dot{\mathbf{q}}, t) \tag{13.429}$$

and the following equations of motion describe the dynamic behavior of dynamic systems:

$$\frac{d}{dt}\frac{\partial \mathcal{L}}{\partial \dot{q}_i} - \frac{\partial \mathcal{L}}{\partial q_i} = 0 \qquad i = 1, 2, \dots, n \tag{13.430}$$

The Lagrange function is a useful indicator for the existence of a conservation law. If there exists a conservation law in a dynamic system, the Lagrange equation provides the integral of motion as the equation of motion.

13.6.1 Conservation of Energy

If the Lagrangian $\mathcal{L}$ of a system does not depend explicitly on time t,

$$\mathcal{L} = K - V = \mathcal{L}(\mathbf{q}, \dot{\mathbf{q}}) \tag{13.431}$$

then its total time derivative is zero and the mechanical energy E of the system remains constant:

$$\frac{d\mathcal{L}}{dt} = \sum_{i=1}^{n} \frac{\partial \mathcal{L}}{\partial q_i}\dot{q}_i + \sum_{i=1}^{n} \frac{\partial \mathcal{L}}{\partial \dot{q}_i}\ddot{q}_i = 0 \tag{13.432}$$

$$E = \sum_{i=1}^{n} \dot{q}_i \frac{\partial \mathcal{L}}{\partial \dot{q}_i} - \mathcal{L} = K + V \tag{13.433}$$

A Lagrangian of this form is equivalent to the homogeneity of time and indicates that the origin of the time axis and the scale of a unit time step are arbitrary.

Proof: Let us substitute $\partial\mathcal{L}/\partial q_i$ from (13.430) in (13.432),

$$\frac{d\mathcal{L}}{dt} = \dot{q}_i \sum_{i=1}^{n} \frac{d}{dt}\left(\frac{\partial \mathcal{L}}{\partial \dot{q}_i}\right) + \sum_{i=1}^{n} \frac{\partial \mathcal{L}}{\partial \dot{q}_i}\ddot{q}_i = \sum_{i=1}^{n} \frac{d}{dt}\left(\dot{q}_i \frac{\partial \mathcal{L}}{\partial \dot{q}_i}\right) \tag{13.434}$$

to obtain

$$\frac{d}{dt}\left(\sum_{i=1}^{n} \dot{q}_i \frac{\partial \mathcal{L}}{\partial \dot{q}_i} - \mathcal{L}\right) = 0 \tag{13.435}$$

Therefore, we have an integral of motion

$$f_1 = \sum_{i=1}^{n} \dot{q}_i \frac{\partial \mathcal{L}}{\partial \dot{q}_i} - \mathcal{L} = E \tag{13.436}$$

where E is a constant of motion and f_1 is an integral of motion. So, when there is no t in the Lagrangian of a system, the energy of the system is conserved.

To show that E is the mechanical energy of the system, we may use

$$\sum_{i=1}^{n} \dot{q}_i \frac{\partial \mathcal{L}}{\partial \dot{q}_i} = \sum_{i=1}^{n} \dot{q}_i \frac{\partial K}{\partial \dot{q}_i} = 2K \tag{13.437}$$

and find

$$f_1 = 2K - (K - V) = K + V = E \tag{13.438}$$

This is equivalent to the principle of conservation of energy in (2.371). ■

Example 799 Jacobi Integral The Jacobi integral is a generalized form of the energy integral. The necessary conditions for the existence of an energy integral are that the system be catastatic and all given forces be derivable from a potential energy $V = V(q_1, q_2, \dots, q_n)$. Under these conditions, $\mathcal{L}$ would be independent of t.

The Lagrangian form of the fundamental equation of dynamics (10.303),

$$\sum_{i=1}^{n} \left(\frac{d}{dt} \frac{\partial \mathcal{L}}{\partial \dot{q}_i} - \frac{\partial \mathcal{L}}{\partial q_i} \right) \delta q_i = 0 \tag{13.439}$$

becomes

$$\frac{d}{dt} \frac{\partial \mathcal{L}}{\partial \dot{q}_i} = \frac{\partial \mathcal{L}}{\partial q_i} \tag{13.440}$$

$$\frac{d}{dt} \frac{\partial \mathcal{L}}{\partial \dot{q}_i} = \frac{\partial \mathcal{L}}{\partial q_i} - \sum_{j=1}^{l} \lambda_j B_{ij} \tag{13.441}$$

for holonomic and nonholonomic systems, respectively. When the potential energy V is a function of q_i,

$$Q_i = -\frac{\partial V}{\partial q_i} \tag{13.442}$$

and we have

$$\frac{d}{dt} \frac{\partial \mathcal{L}}{\partial \dot{q}_i} - \frac{\partial \mathcal{L}}{\partial q_i} = \frac{d}{dt} \frac{\partial K}{\partial \dot{q}_i} - \frac{\partial K}{\partial q_i} + \frac{\partial V}{\partial q_i} \tag{13.443}$$

If the potential energy V is also a function of $\dot{q}_i$ and t, then

$$Q_i = \frac{d}{dt} \frac{\partial V}{\partial \dot{q}_i} - \frac{\partial V}{\partial q_i} \tag{13.444}$$

and the Lagrange equation may still be written as (13.440) and (13.441).

Let us assume the Lagrangian function does not contain time t explicitly and the system is catastatic. Therefore the fundamental equation may be written as

$$\sum_{i=1}^{n} \left(\frac{d}{dt} \frac{\partial \mathcal{L}}{\partial \dot{q}_i} - \frac{\partial \mathcal{L}}{\partial q_i} \right) \dot{q}_i = 0 \tag{13.445}$$

Employing the following time derivative

$$\frac{d}{dt}\left(\sum_{i=1}^{n}\dot{q}_i\frac{\partial\mathcal{L}}{\partial\dot{q}_i}\right)=\sum_{i=1}^{n}\dot{q}_i\frac{d}{dt}\frac{\partial\mathcal{L}}{\partial\dot{q}_i}+\sum_{i=1}^{n}\ddot{q}_i\frac{\partial\mathcal{L}}{\partial\dot{q}_i}\tag{13.446}$$

and substituting in (13.445) give

$$\frac{d}{dt}\left(\sum_{i=1}^{n}\dot{q}_i\frac{\partial\mathcal{L}}{\partial\dot{q}_i}\right)-\sum_{i=1}^{n}\frac{\partial\mathcal{L}}{\partial\dot{q}_i}\ddot{q}_i-\sum_{i=1}^{n}\frac{\partial\mathcal{L}}{\partial q_i}\dot{q}_i=0\tag{13.447}$$

The last two terms together are time derivatives of the Lagrangian $\mathcal{L}=\mathcal{L}(q_i,\dot{q}_i)$. Therefore, Equation (13.447) is a total derivative

$$\frac{d}{dt}\left(\sum_{i=1}^{n}\dot{q}_i\frac{\partial\mathcal{L}}{\partial\dot{q}_i}-\mathcal{L}\right)=0\tag{13.448}$$

that indicates the Jacobi integral of motion:

$$\sum_{i=1}^{n}\dot{q}_i\frac{\partial\mathcal{L}}{\partial\dot{q}_i}-\mathcal{L}=h=\text{const}\tag{13.449}$$

13.6.2 Conservation of Momentum

If the Lagrangian $\mathcal{L}=\mathcal{L}(q_i,\dot{q}_i,t)$ of an n-DOF system does not depend explicitly on generalized coordinates q_j, $1\le j\le n$,

$$\mathcal{L}=K-V=\mathcal{L}\left(q_1,q_2,\ldots,q_{j-1},q_{j+1},\ldots,q_n,\dot{q}_1,\dot{q}_2,\ldots,q_n,t\right)\tag{13.450}$$

then the total time derivative of the associated generalized momentum p_i is zero, and therefore p_i of the system remains constant:

$$\frac{d}{dt}\frac{\partial\mathcal{L}}{\partial\dot{q}_j}=0\tag{13.451}$$

$$p_j=\frac{\partial\mathcal{L}}{\partial\dot{q}_j}=\text{const}\tag{13.452}$$

The coordinate q_j, which does not appear explicitly in $\mathcal{L}$, is called the *ignorable* or *cyclic coordinate*. Because q_j is a generalized coordinate, the Lagrangian principle of the conservation of momentum includes both translational and rotational momenta.

Proof: It can happen very often that there are generalized coordinates that do not appear in the Lagrangian function $\mathcal{L}$, although their associated generalized velocities appear. If q_j is an ignorable coordinate, then

$$\frac{\partial\mathcal{L}}{\partial q_j}=0\tag{13.453}$$

and from the Lagrange equation we have $\partial\mathcal{L}/\partial\dot{q}_j=p_j=\text{const}$.

We may translate an ignorable coordinate q_j and substitute it with q'_j,

$$q'_j = q_j + C \tag{13.454}$$

where C is a constant. ■

Example 800 Two Particles and a String The particles m_1 and m_2 in Figure 13.13 are connected by a string of length l. The particle m_1 freely moves in a horizontal plane, while m_2 is moving on the z-axis. At $t = 0$, the distance of m_1 from the center hole is r_0, its initial velocity is v_0 perpendicular to r_0, and m_2 is not moving. The system has two DOF with r and φ as the generalized coordinates. The kinetic and potential energies and Lagrangian of the system are

$$K = \frac{1}{2}\left[(m_1 + m_2)\dot{r}^2 + m_1 r^2 \dot{\varphi}^2\right] \tag{13.455}$$

$$V = -m_2 g (l - r) + m_2 g (l - r_0) \tag{13.456}$$

$$\mathcal{L} = \frac{1}{2}\left[(m_1 + m_2)\dot{r}^2 + m_1 r^2 \dot{\varphi}^2\right] - m_2 g r \tag{13.457}$$

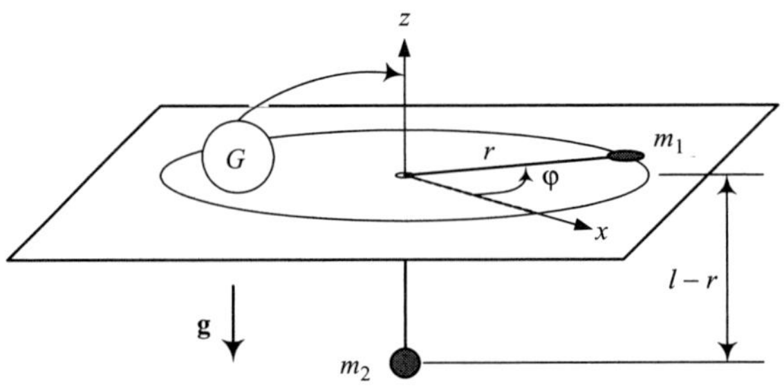

Figure 13.13 The connected particles by a string of length l.

The generalized coordinate φ is ignorable and therefore p_φ is conserved:

$$p_\varphi = \frac{\partial \mathcal{L}}{\partial \dot{\varphi}} = m_1 r^2 \dot{\varphi} = \text{const} \tag{13.458}$$

Furthermore, $\mathcal{L}$ does not depend on time t explicitly. Therefore the energy of the system is also conserved:

$$E = \frac{1}{2}\left((m_1 + m_2)\dot{r}^2 + m_1 r^2 \dot{\varphi}^2\right) + m_2 g r \tag{13.459}$$

These two integrals of motion provide two first-order equations to determine the generalized coordinates r and φ.

13.7 ★ GENERALIZED COORDINATE SYSTEM

The generalized equations of motion of a dynamic system in the generalized curvilinear coordinate space q_i are

$$Q_i = \frac{d}{dt}\frac{\partial K}{\partial \dot{q}_i} - \frac{\partial K}{\partial q_i} \tag{13.460}$$

where K is the kinetic energy of the system,

$$K = \frac{1}{2} m \mathbf{v} \cdot \mathbf{v} \tag{13.461}$$

q_i is the generalized coordinate of the system, and Q_i is the generalized force associated with q_i,

$$Q_i = \mathbf{F} \cdot \mathbf{b}_i = F_x \frac{\partial x}{\partial q_i} + F_y \frac{\partial y}{\partial q_i} + F_z \frac{\partial z}{\partial q_i} \tag{13.462}$$

where $\mathbf{b}_i$ is the principal base vector of the generalized coordinate system q_i.

Proof: Having the Cartesian description of a position vector $\mathbf{r}$,

$$\mathbf{r} = x\hat{\imath} + y\hat{\jmath} + z\hat{k} = x_1\hat{\imath} + x_2\hat{\jmath} + x_3\hat{k} \tag{13.463}$$

we determine the *principal base vectors* $\mathbf{b}_i$ of the generalized coordinate system $Q(q_1, q_2, q_3)$ by

$$\mathbf{b}_i = \frac{\partial \mathbf{r}}{\partial q_i} = b_i \hat{u}_i = \frac{\partial x}{\partial q_i}\hat{\imath} + \frac{\partial y}{\partial q_i}\hat{\jmath} + \frac{\partial z}{\partial q_i}\hat{k} \tag{13.464}$$

$$b_i = |\mathbf{b}_i| = \sqrt{\left(\frac{\partial x}{\partial q_i}\right)^2 + \left(\frac{\partial y}{\partial q_i}\right)^2 + \left(\frac{\partial z}{\partial q_i}\right)^2} \tag{13.465}$$

where the *principal unit vectors* $\hat{u}_i$ of the curvilinear coordinate Q-system is

$$\hat{u}_i = \frac{\Delta \mathbf{r}_{q_i}}{\left|\Delta \mathbf{r}_{q_i}\right|} = \frac{\partial \mathbf{r}/\partial q_i}{|\partial \mathbf{r}/\partial q_i|} = \frac{1}{b_i}\mathbf{b}_i \tag{13.466}$$

The generalized principal base vectors $\mathbf{b}_i$ are assumed to be non-coplanar and hence independent, where $\mathbf{b}_i$ is along the partial derivative $\partial \mathbf{r}/\partial q_i$, which indicates a coordinate curve q_i as illustrated in Figure 3.25. The reciprocal base vectors $\mathbf{b}_i^*$ and *reciprocal unit vectors* $\hat{u}_i^*$ of the curvilinear coordinate Q-system are defined as:

$$\mathbf{b}_i^* = \nabla q_i = \frac{\partial q_i}{\partial x}\hat{\imath} + \frac{\partial q_i}{\partial y}\hat{\jmath} + \frac{\partial q_i}{\partial z}\hat{k} \tag{13.467}$$

$$\hat{u}_i^* = \frac{\nabla q_i}{|\nabla q_i|} = \frac{1}{b_i^*}\mathbf{b}_i^* \tag{13.468}$$

$$b_i^* = \left|\mathbf{b}_i^*\right| = \sqrt{\left(\frac{\partial q_i}{\partial x}\right)^2 + \left(\frac{\partial q_i}{\partial y}\right)^2 + \left(\frac{\partial q_i}{\partial z}\right)^2} \tag{13.469}$$

If the generalized coordinate system $Q(q_1, q_2, q_3)$ is orthogonal, then

$$\mathbf{b}_i^* = \frac{1}{b_i}\hat{u}_i = \frac{1}{b_i^2}\mathbf{b}_i \tag{13.470}$$

We employ the base vectors $\mathbf{b}_i$ to represent a force vector $\mathbf{F} = F_x\hat{\imath} + F_y\hat{\jmath} + F_z\hat{k}$:

$$\mathbf{F} = Q_1\mathbf{b}_1^* + Q_2\mathbf{b}_2^* + Q_3\mathbf{b}_3^* = \sum_{i=1}^{3} Q_i\mathbf{b}_i^* \tag{13.471}$$

$$Q_i = \mathbf{F} \cdot \mathbf{b}_i = F_x \frac{\partial x}{\partial q_i} + F_y \frac{\partial y}{\partial q_i} + F_z \frac{\partial z}{\partial q_i} \tag{13.472}$$

The components Q_i of the generalized expression of the force $\mathbf{F}$ are the components of the *generalized force*.

The generalized equations of motion are the scalar products of the Newton equation of motion $\mathbf{F} = m\mathbf{a}$, with the base vectors $\mathbf{b}_i$ of the generalized coordinate system. The scalar product of $\mathbf{F}$ and $\mathbf{b}_i$ is the generalized force Q_i as given in (13.472). The covariant components of the acceleration vector $\mathbf{a}$ associated with the generalized coordinates q_i are:

$$a_i = \mathbf{a} \cdot \mathbf{b}_i = \ddot{x}\frac{\partial x}{\partial q_i} + \ddot{y}\frac{\partial y}{\partial q_i} + \ddot{z}\frac{\partial z}{\partial q_i} \tag{13.473}$$

To write a_i in the proper form, let us rewrite the first term of (13.473) as

$$\begin{aligned}
\ddot{x}\frac{\partial x}{\partial q_i} &= \frac{d}{dt}\left(\dot{x}\frac{\partial x}{\partial q_i}\right) - \dot{x}\frac{d}{dt}\frac{\partial x}{\partial q_i} = \frac{d}{dt}\left(\dot{x}\frac{\partial x}{\partial q_i}\right) - \dot{x}\frac{\partial \dot{x}}{\partial q_i} \\
&= \frac{d}{dt}\left(\dot{x}\frac{\partial \dot{x}}{\partial \dot{q}_i}\right) - \dot{x}\frac{\partial \dot{x}}{\partial q_i} \\
&= \frac{d}{dt}\frac{\partial}{\partial \dot{q}_i}\left(\frac{1}{2}\dot{x}^2\right) - \frac{\partial}{\partial q_i}\left(\frac{1}{2}\dot{x}^2\right)
\end{aligned} \tag{13.474}$$

Using the same method, we have

$$\begin{aligned}
a_i &= \frac{d}{dt}\frac{\partial}{\partial \dot{q}_i}\left(\frac{\dot{x}^2 + \dot{y}^2 + \dot{z}^2}{2}\right) - \frac{\partial}{\partial \dot{q}_i}\left(\frac{\dot{x}^2 + \dot{y}^2 + \dot{z}^2}{2}\right) \\
&= \left[\frac{d}{dt}\frac{\partial}{\partial \dot{q}_i}\left(\frac{1}{2}v^2\right) - \frac{\partial}{\partial q_i}\left(\frac{1}{2}v^2\right)\right] \\
&= \frac{1}{m}\left(\frac{d}{dt}\frac{\partial K}{\partial \dot{q}_i} - \frac{\partial K}{\partial \dot{q}_i}\right)
\end{aligned} \tag{13.475}$$

Therefore, we can transform the Newton equation of motion $\mathbf{F} = m\mathbf{a}$ to a generalized coordinate system by multiplying with the base vector $\mathbf{b}_i$ of the system:

$$\mathbf{F} \cdot \mathbf{b}_i = m\mathbf{a} \cdot \mathbf{b}_i \tag{13.476}$$

$$Q_i = \frac{d}{dt}\frac{\partial K}{\partial \dot{q}_i} - \frac{\partial K}{\partial \dot{q}_i} \tag{13.477}$$

This is the *Lagrange equation of motion*. It is also applicable to the time-dependent coordinate transformation:

$$x = x\left(q_i, t\right) \qquad y = y\left(q_i, t\right) \qquad z = z\left(q_i, t\right) \tag{13.478}$$

■

Example 801 ★ Generalized Forces in Spherical Coordinate System The base vectors of the spherical coordinate system are

$$\mathbf{b}_1 = \frac{\partial \mathbf{r}}{\partial r} = \hat{u}_r \tag{13.479}$$

$$\mathbf{b}_2 = \frac{\partial \mathbf{r}}{\partial \varphi} = r\hat{u}_\varphi = \hat{u}_\varphi \times \mathbf{r} \tag{13.480}$$

$$\mathbf{b}_3 = \frac{\partial \mathbf{r}}{\partial \theta} = r \sin\varphi\, \hat{u}_\theta = \hat{k} \times \mathbf{r} \tag{13.481}$$

Therefore, the generalized forces of a force $\mathbf{F}$ in the spherical system are

$$Q_1 = \mathbf{F} \cdot \mathbf{b}_1 = \mathbf{F} \cdot \hat{u}_r = F_r \tag{13.482}$$

$$Q_2 = \mathbf{F} \cdot \mathbf{b}_2 = \hat{u}_\varphi \cdot (\mathbf{r} \times \mathbf{F}) \tag{13.483}$$

$$Q_3 = \mathbf{F} \cdot \mathbf{b}_3 = \hat{k} \cdot (\mathbf{r} \times \mathbf{F}) \tag{13.484}$$

where Q_1 is the component of $\mathbf{F}$ in the direction of $\hat{u}_r$ and has a force dimension; however, Q_2 and Q_3 do not have a force dimension and are components of the vector $\mathbf{r} \times \mathbf{F}$ along $\hat{u}_\varphi$ and $\hat{k}$, respectively. The vector $\mathbf{r} \times \mathbf{F}$ is the moment of $\mathbf{F}$ about the origin, so the generalized forces Q_2 and Q_3 have the dimension of moment.

In fact, the generalized force associated with an angular coordinate is always the component of the torque along the axis about which a change in the angular coordinate rotates the position vector. Similarly, the generalized force associated with a translational coordinate is always the component of force along the axis on which a change in the translational coordinate changes the position vector. The dimension of generalized force Q_i is always such that the dimension of $Q_i\, dq_i$ is work.

Example 802 ★ Generalized Work The small work $\triangle W$ done by the force $\mathbf{F}$ when q_i is varied by the small amount Δq_i is

$$\triangle W = \mathbf{F} \cdot \Delta \mathbf{r}_i \tag{13.485}$$

where $\Delta \mathbf{r}_i$ is the change in the position vector $\mathbf{r}$ when q_i is varied:

$$\Delta \mathbf{s}_i = \frac{\partial \mathbf{r}}{\partial q_i} \Delta q_i = \Delta q_i \mathbf{b}_i \tag{13.486}$$

This yields

$$\triangle W = \mathbf{F} \cdot \mathbf{b}_i \Delta q_i = Q_i \Delta q_i \tag{13.487}$$

Therefore, the dimension of the term $Q_i \Delta q_i$ is work. For a general displacement $\Delta \mathbf{r}$,

$$\Delta \mathbf{s} = \Delta q_1\, \mathbf{b}_1 + \Delta q_2\, \mathbf{b}_2 + \Delta q_3\, \mathbf{b}_3 \tag{13.488}$$

we have

$$\triangle W = \mathbf{F} \cdot \Delta \mathbf{r} = \left(\sum_{i=1}^{3} Q_i \mathbf{b}_i^{\star} \right) \cdot \left(\sum_{i=1}^{3} \Delta q_i\, \mathbf{b}_i \right) = \sum_{i=1}^{3} Q_i \Delta q_i \tag{13.489}$$

where

$$\mathbf{b}_i^{\star} \cdot \mathbf{b}_j = \delta_{ij} \tag{13.490}$$

The symbol $\triangle$ may be changed to d for actual displacement and to δ for virtual displacement.

Example 803 ★ Generalized Momentum The components of momentum $\mathbf{p} = m\mathbf{v}$ in the generalized coordinate system $Q\,(q_1, q_2, q_3)$ are

$$p_i = \mathbf{p} \cdot \mathbf{b}_i = m\mathbf{v} \cdot \mathbf{b}_i \tag{13.491}$$

where the three base vectors $\mathbf{b}_i$ are defined at point (q_1, q_2, q_3). The scalar components p_i of the generalized momentum are called the generalized momentum. The covariant components of the velocity $\mathbf{v}$ are expressed as

$$v_i = \mathbf{v} \cdot \mathbf{b}_i = \frac{\partial}{\partial \dot{q}_i}\left(\frac{1}{2}v^2\right) \tag{13.492}$$

and therefore,

$$p_i = \frac{\partial K}{\partial \dot{q}_i} \tag{13.493}$$

$$K = \tfrac{1}{2}mv^2 \tag{13.494}$$

Example 804 ★ Generalized Momentum in Spherical System Employing the base vectors of the spherical coordinate system (13.479)–(13.481), the generalized momenta p_i in the spherical system are

$$p_i = \mathbf{p} \cdot \mathbf{b}_i \tag{13.495}$$

where

$$p_1 = \mathbf{p} \cdot \mathbf{b}_1 = \mathbf{p} \cdot \hat{u}_r = p_r \tag{13.496}$$

$$p_2 = \mathbf{p} \cdot \mathbf{b}_2 = \hat{u}_\varphi \cdot (\mathbf{r} \times \mathbf{p}) \tag{13.497}$$

$$p_3 = \mathbf{p} \cdot \mathbf{b}_3 = \hat{k} \cdot (\mathbf{r} \times \mathbf{p}) \tag{13.498}$$

and $p_1 = p_r$ is the component of momentum $\mathbf{p}$ in the direction of $\hat{u}_r$. The generalized momentum p_2 and p_3 are the components of the moment of momentum $\mathbf{L}$ along $\hat{u}_\varphi$ and $\hat{k}$, respectively. The vector $\mathbf{r} \times \mathbf{F}$ is the moment of $\mathbf{F}$ about the origin. The generalized forces Q_2 and Q_3 thus have the dimension of the moment:

$$\mathbf{L} = \mathbf{r} \times \mathbf{p} \tag{13.499}$$

The components of the generalized momentum may also be found from Equation (13.493):

$$K = \tfrac{1}{2}m\mathbf{v} \cdot \mathbf{v} = \tfrac{1}{2}m\left(\dot{r}^2 + r^2\dot{\varphi}^2 + r^2\dot{\theta}^2\sin^2\varphi\right) \tag{13.500}$$

where

$$p_1 = \frac{\partial K}{\partial \dot{r}} = m\dot{r} \tag{13.501}$$

$$p_2 = \frac{\partial K}{\partial \dot{\varphi}} = mr^2\dot{\varphi} \tag{13.502}$$

$$p_3 = \frac{\partial K}{\partial \dot{\theta}} = mr^2\dot{\theta}\sin^2\varphi \tag{13.503}$$

These results are equivalent to

$$\mathbf{p} = m\mathbf{v} = m\left(\dot{r}\,\hat{u}_r + r\dot{\varphi}\,\hat{u}_\varphi + r\dot{\theta}\sin\varphi\,\hat{u}_\theta\right) \tag{13.504}$$

The contravariant components v_i^* of the velocity vector $\mathbf{v}$ are the generalized velocities $\dot{q}_i$:

$$v_i^* = \mathbf{v}\cdot\mathbf{b}_i^* = \dot{x}\frac{\partial q_i}{\partial x} + \dot{y}\frac{\partial q_i}{\partial y} + \dot{z}\frac{\partial q_i}{\partial z} = \dot{q}_i \tag{13.505}$$

We can define the kinetic energy of a dynamic system using the generalized velocities and the generalized momenta:

$$K = \frac{1}{2}\mathbf{p}\cdot\mathbf{v} = \left(\sum_{i=1}^{3} p_i\mathbf{b}_i^*\right)\cdot\left(\sum_{i=1}^{3}\dot{q}_i\mathbf{b}_i\right) = \frac{1}{2}\sum_{i=1}^{3} p_i\dot{q}_i \tag{13.506}$$

Example 805 ★ Central-Force Motion Let us consider the motion of a particle m in a plane under the action of a central force:

$$\mathbf{F} = -m\omega^2\mathbf{r} = -m\omega^2\left(x\hat{\imath} + x\hat{\jmath}\right) \tag{13.507}$$

The kinetic energy K of m in the Cartesian coordinate frame is

$$K = \tfrac{1}{2}m\left(\dot{x}^2 + \dot{y}^2\right) \tag{13.508}$$

and the base vectors are the Cartesian unit vectors

$$\mathbf{b}_1 = \hat{\imath} \qquad \mathbf{b}_2 = \hat{\jmath} \tag{13.509}$$

Using the kinetic energy (13.508), we find the generalized forces

$$Q_1 = F_x = m\ddot{x} = -m\omega^2 x \tag{13.510}$$

$$Q_2 = F_y = m\ddot{y} = -m\omega^2 y \tag{13.511}$$

The solutions of these two independent equations are

$$x = A\cos(\omega t + \varphi) \qquad y = B\cos(\omega t + \theta) \tag{13.512}$$

where A, B, φ, θ are constant and are to be determined from initial conditions $x_0, \dot{x}_0, y_0, \dot{y}_0$.

We may solve the same problem in the cylindrical coordinate system:

$$K = \tfrac{1}{2}m\left(\dot{r}^2 + r^2\dot{\phi}^2\right) \tag{13.513}$$

$$\mathbf{b}_1 = \hat{u}_r \qquad \mathbf{b}_2 = r\hat{u}_\phi \tag{13.514}$$

$$Q_1 = F_r = -m\omega^2 r = \frac{d}{dt}(m\dot{r}) - mr\dot{\phi}^2 \tag{13.515}$$

$$Q_2 = \hat{k}\cdot(\mathbf{r}\times\mathbf{F}) = \hat{k}\cdot\mathbf{0} = 0 = \frac{d}{dt}\left(mr^2\dot{\phi}\right) \tag{13.516}$$

The second equation in the cylindrical coordinate system indicates that the z-component of the moment of momentum is a constant of motion L:

$$mr^2\dot{\phi} = p_2 = L_z = L \tag{13.517}$$

Because this is a planar problem, the z-component of L is the total magnitude of L. Employing (13.517), we can write Equation (13.515) in a solvable form:

$$m\ddot{r} = -m\omega^2 r + mr\dot{\phi}^2 = -m\omega^2 r + \frac{L^2}{mr^3} \tag{13.518}$$

Using

$$\ddot{r} = \dot{r}\frac{d\dot{r}}{dr} \tag{13.519}$$

the equation of motion (13.518) provides the integral of energy E as the second integral of motion:

$$\tfrac{1}{2}m\dot{r}^2 + \tfrac{1}{2}m\omega^2 r^2 + \frac{L^2}{2mr^2} = E \tag{13.520}$$

So, solution of the problem of central-force motion reduces to the integral

$$t = \int_{r_0}^{r} \frac{dr}{\sqrt{(2/m)E - r^2\omega^2 - L^2/(m^2r^2)}} \tag{13.521}$$

Example 806 ★ A Moving Particle on the Surface of a Sphere Let us examine the motion of a particle on the surface of a sphere as shown in Figure 13.14. The particle m is constrained to move on the surface of radius R and is under the action of gravitational attraction $F_g = mg$ in the direction of $-z$:

$$\mathbf{F}_g = m\mathbf{g} = -mg\hat{K} = -mg\left(\cos\varphi\hat{u}_r - \sin\varphi\hat{u}_\varphi\right) \tag{13.522}$$

The constraint equation $f(r,\theta,\varphi) = 0$ and constraint force $\mathbf{F}_C$ are

$$f(r,\theta,\varphi) = r - R = 0 \tag{13.523}$$

$$\mathbf{F}_C = \lambda\nabla f = \lambda \tag{13.524}$$

The kinetic energy of m in the spherical coordinate system is

$$K = \tfrac{1}{2}m\mathbf{v}\cdot\mathbf{v} = \tfrac{1}{2}m\left(\dot{r}^2 + r^2\dot{\varphi}^2 + r^2\dot{\theta}^2\sin^2\varphi\right) \tag{13.525}$$

Employing the Lagrange equation for a system of n DOF under l constraint equations,

$$\frac{d}{dt}\frac{\partial K}{\partial \dot{q}_i} - \frac{\partial K}{\partial q_i} = Q_i + \sum_{j=1}^{l}\lambda_j\nabla f_j \qquad i = 1, 2, \ldots, n \tag{13.526}$$

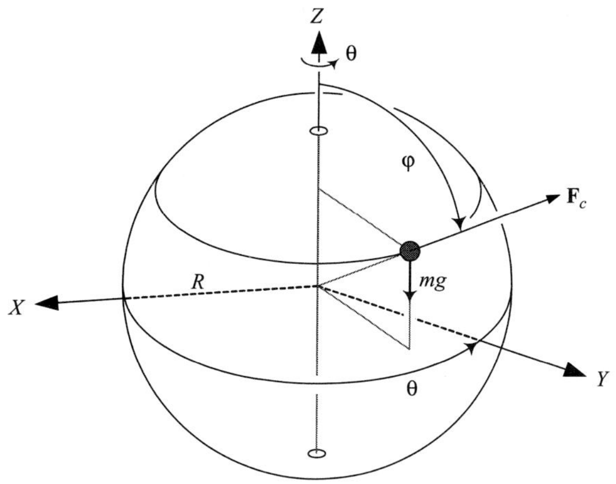

Figure 13.14 A moving particle on the surface of a sphere.

we find the equations of motion

$$m\left(\ddot{r} - r\dot{\varphi}^2 - r\dot{\theta}^2 \sin^2 \varphi\right) = -mg\cos\varphi + \lambda \tag{13.527}$$

$$\frac{d}{dt}\left(mr^2\dot{\varphi}\right) - mr^2\dot{\theta}^2 \sin\varphi\cos\varphi = mgr\sin\varphi \tag{13.528}$$

$$\frac{d}{dt}\left(mr^2\dot{\theta}\sin^2\varphi\right) = 0 \tag{13.529}$$

Equation (13.527) provides a first integral for a constant z-component of the generalized moment of momentum:

$$p_z = mR^2\dot{\theta}\sin^2\varphi = \text{const} \tag{13.530}$$

Substituting p_z in (13.528) provides

$$mR^2\ddot{\varphi} = mgR\sin\varphi + \frac{p_z^2\cos\varphi}{2mR^2\sin^3\varphi} \tag{13.531}$$

This equation is of the form $\ddot{x} = f(x)$ and provides an energy integral of motion:

$$E = \tfrac{1}{2}mR^2\dot{\varphi}^2 + mgR\cos\varphi + \frac{p_z^2}{2mR^2\sin^2\varphi} = \text{const} \tag{13.532}$$

Solution of (13.532) is an elliptic integral.

13.8 ★ MULTIBODY LAGRANGIAN DYNAMICS

The Lagrange method provides a systematic approach to obtain the equations of motion of multibodies. The Lagrangian of a multibody is defined as the difference between the kinetic and potential energies:

$$\mathcal{L} = K - V \tag{13.533}$$

The Lagrange equation of motion for a multibody can be found by employing the Lagrange equation

$$\frac{d}{dt}\left(\frac{\partial \mathcal{L}}{\partial \dot{q}_i}\right) - \frac{\partial \mathcal{L}}{\partial q_i} = Q_i \qquad i = 1, 2, \ldots n \tag{13.534}$$

where q_i is the coordinates by which the energies are expressed and Q_i is the associated generalized nonpotential force that drives q_i.

We can set the equations of motion for an n-link serial multibody in matrix form as

$$\mathbf{D}(\mathbf{q})\,\ddot{\mathbf{q}} + \mathbf{H}(\mathbf{q}, \dot{\mathbf{q}}) + \mathbf{G}(\mathbf{q}) = \mathbf{Q} \tag{13.535}$$

or

$$\mathbf{D}(\mathbf{q})\,\ddot{\mathbf{q}} + \mathbf{C}(\mathbf{q}, \dot{\mathbf{q}})\dot{\mathbf{q}} + \mathbf{G}(\mathbf{q}) = \mathbf{Q} \tag{13.536}$$

or in a summation form as

$$\sum_{j=1}^{n} D_{ij}(q)\,\ddot{q}_j + \sum_{k=1}^{n}\sum_{m=1}^{n} H_{ikm}\dot{q}_k\dot{q}_m + G_i = Q_i \tag{13.537}$$

where D_{ij} is an $n \times n$ inertial-type symmetric matrix,

$$D_{ij} = \sum_{k=1}^{n} \left(\mathbf{J}_{Dk}^{\mathrm{T}}\; m_k\; \mathbf{J}_{Dk} + \tfrac{1}{2}\,\mathbf{J}_{Rk}^{\mathrm{T}}\; {}^{0}I_k\; \mathbf{J}_{Rk}\right) \tag{13.538}$$

The velocity coupling vector is given as

$$H_{ijk} = \sum_{j=1}^{n}\sum_{k=1}^{n} \left(\frac{\partial D_{ij}}{\partial q_k} - \frac{1}{2}\frac{\partial D_{jk}}{\partial q_i}\right) \tag{13.539}$$

and the gravitational vector as

$$G_i = \sum_{j=1}^{n} m_j \mathbf{g}^{\mathrm{T}}\, \mathbf{J}_{Dj}^{(i)} \tag{13.540}$$

Proof: Consider a serial multibody with n links. The kinetic energy of link (i) is

$$K_i = \tfrac{1}{2}\, {}^{0}\mathbf{v}_i^{\mathrm{T}}\, m_i\, {}^{0}\mathbf{v}_i + \tfrac{1}{2}\, {}_0\boldsymbol{\omega}_i^{\mathrm{T}}\, {}^{i}I_i\, {}_0\boldsymbol{\omega}_i \tag{13.541}$$

where m_i is the mass of the link, ${}^{i}I_i$ is the mass moment matrix of the link in the link's frame B_i, ${}^{0}\mathbf{v}_i$ is the global velocity of the link at its mass center C, and ${}_0\boldsymbol{\omega}_i$ is the global angular velocity of the link.

We can express the translational and angular velocity vectors based on the joint coordinate velocities by employing the *Jacobian of the link*, $\mathbf{J}_i$:

$$\dot{\mathbf{X}}_i = \begin{bmatrix} {}^{0}\mathbf{v}_i \\ {}^{0}\boldsymbol{\omega}_i \end{bmatrix} = \begin{bmatrix} \mathbf{J}_{Di} \\ \mathbf{J}_{Ri} \end{bmatrix} \dot{\mathbf{q}} = \mathbf{J}_i\, \dot{\mathbf{q}} \tag{13.542}$$

The link's Jacobian $\mathbf{J}_i$ is a $6 \times n$ matrix that transforms the instantaneous joint coordinate velocities into the instantaneous link's translational and angular velocities. The jth column of $\mathbf{J}_i$ is comprised of $\mathbf{c}_{Di}^{(j)}$ and $\mathbf{c}_{Ri}^{(j)}$, where for $j \leq i$ we have

$$\mathbf{c}_{Di}^{(j)} = \begin{cases} \hat{k}_{j-1} \times {}_{j-1}^{0}\mathbf{r}_i & \text{for an R joint} \\ \hat{k}_{j-1} & \text{for a P joint} \end{cases} \tag{13.543}$$

and

$$\mathbf{c}_{Ri}^{(j)} = \begin{cases} \hat{k}_{j-1} & \text{for an R joint} \\ 0 & \text{for a P joint} \end{cases} \tag{13.544}$$

where ${}_{j-1}^{0}\mathbf{r}_i$ is position of C of the link (i) in the coordinate frame B_{j-1} expressed in the base frame. The columns of $\mathbf{J}_i$ are zero for $j > i$.

The kinetic energy K of the whole multibody is then equal to

$$\begin{aligned} K &= \sum_{i=1}^{n} K_i = \frac{1}{2} \sum_{i=1}^{n} \left({}^0\mathbf{v}_i^{\mathrm{T}}\, m_i \, {}^0\mathbf{v}_i + \frac{1}{2} \, {}_0\boldsymbol{\omega}_i^{\mathrm{T}} \, {}^0I_i \, {}_0\boldsymbol{\omega}_i \right) \\ &= \frac{1}{2} \sum_{i=1}^{n} \left[(\mathbf{J}_{Di}\,\dot{\mathbf{q}}_i)^{\mathrm{T}} \, m_i \, (\mathbf{J}_{Di}\,\dot{\mathbf{q}}_i) + \frac{1}{2} \, (\mathbf{J}_{Ri}\,\dot{\mathbf{q}}_i)^{\mathrm{T}} \, {}^0I_i \, (\mathbf{J}_{Ri}\,\dot{\mathbf{q}}_i) \right] \\ &= \frac{1}{2} \dot{\mathbf{q}}_i^{\mathrm{T}} \left[\sum_{i=1}^{n} \left(\mathbf{J}_{Di}^{\mathrm{T}} \, m_i \, \mathbf{J}_{Di} + \frac{1}{2} \mathbf{J}_{Ri}^{\mathrm{T}} \, {}^0I_i \, \mathbf{J}_{Ri} \right) \right] \dot{\mathbf{q}}_i \end{aligned} \tag{13.545}$$

where ${}^0I_i$ is the mass moment matrix of link (i) about its C and is expressed in the base frame:

$${}^0I_i = {}^0R_i \, {}^iI_i \, {}^0R_i^{\mathrm{T}} \tag{13.546}$$

We may write the kinetic energy of the multibody in the more convenient form

$$K = \tfrac{1}{2} \dot{\mathbf{q}}_i^{\mathrm{T}} \, D \, \dot{\mathbf{q}}_i \tag{13.547}$$

where D is an $n \times n$ matrix called the *multibody inertia matrix:*

$$D = \sum_{i=1}^{n} \left(\mathbf{J}_{Di}^{\mathrm{T}} \, m_i \, \mathbf{J}_{Di} + \tfrac{1}{2} \mathbf{J}_{Ri}^{\mathrm{T}} \, {}^0I_i \, \mathbf{J}_{Ri} \right) \tag{13.548}$$

Suppose the potential energy of link (i) is due to gravity,

$$V_i = -m_i \, {}^0\mathbf{g} \cdot {}^0\mathbf{r}_i \tag{13.549}$$

and therefore, the total potential energy of the multibody is

$$V = \sum_{i=1}^{n} V_i = -\sum_{i=1}^{n} m_i \, {}^0\mathbf{g}^{\mathrm{T}} \, {}^0\mathbf{r}_i \tag{13.550}$$

where ${}^0\mathbf{g}$ is the gravitational acceleration vector expressed in the base frame. Therefore, the Lagrangian of the multibody is

$$\begin{aligned}\mathcal{L} = K - V &= \frac{1}{2}\dot{\mathbf{q}}_i^{\mathrm{T}}\, D\, \dot{\mathbf{q}}_i + \sum_{i=1}^{n} m_i\, {}^0\mathbf{g}^{\mathrm{T}}\, {}^0\mathbf{r}_i \\ &= \frac{1}{2}\sum_{i=1}^{n}\sum_{j=1}^{n} D_{ij}\, \dot{q}_i \dot{q}_j + \sum_{i=1}^{n} m_i\, {}^0\mathbf{g}^{\mathrm{T}}\, {}^0\mathbf{r}_i\end{aligned} \tag{13.551}$$

Employing the Lagrangian $\mathcal{L}$, we can find

$$\begin{aligned}\frac{\partial \mathcal{L}}{\partial q_i} &= \frac{1}{2}\frac{\partial}{\partial q_i}\left(\sum_{j=1}^{n}\sum_{k=1}^{n} D_{jk}\, \dot{q}_j \dot{q}_k\right) + \sum_{j=1}^{n} m_j\, {}^0\mathbf{g}^{\mathrm{T}} \frac{\partial\, {}^0\mathbf{r}_j}{\partial q_i} \\ &= \frac{1}{2}\sum_{j=1}^{n}\sum_{k=1}^{n} \frac{\partial D_{jk}}{\partial q_i}\, \dot{q}_j \dot{q}_k + \sum_{j=1}^{n} m_j\, {}^0\mathbf{g}^{\mathrm{T}}\, \mathbf{J}_{Dj}^{(i)}\end{aligned} \tag{13.552}$$

$$\frac{\partial \mathcal{L}}{\partial \dot{q}_i} = \sum_{j=1}^{n} D_{ij}\, \dot{q}_j \tag{13.553}$$

and

$$\begin{aligned}\frac{d}{dt}\frac{\partial \mathcal{L}}{\partial \dot{q}_i} &= \sum_{j=1}^{n} D_{ij}\, \ddot{q}_j + \sum_{j=1}^{n} \frac{d D_{ij}}{dt}\, \dot{q}_j \\ &= \sum_{j=1}^{n} D_{ij}\, \ddot{q}_j + \sum_{j=1}^{n}\sum_{k=1}^{n} \frac{\partial D_{ij}}{\partial q_k}\, \dot{q}_k\, \dot{q}_j\end{aligned} \tag{13.554}$$

The generalized forces of the Lagrange equations are

$$Q_i = M_i + \mathbf{J}^{\mathrm{T}}\, \mathbf{F}_e \tag{13.555}$$

where M_i is the ith actuator force at joint i and

$$\mathbf{F}_e = \begin{bmatrix} -\mathbf{F}_{en}^{\mathrm{T}} & -\mathbf{M}_{en}^{\mathrm{T}} \end{bmatrix}^{\mathrm{T}} \tag{13.556}$$

is the external force system applied on the end effector.

Finally, the Lagrange equations of motion for an n-link multibody are

$$\sum_{j=1}^{n} D_{ij}(q)\, \ddot{q}_j + H_{ikm}\dot{q}_k\dot{q}_m + G_i = Q_i \tag{13.557}$$

where

$$H_{ijk} = \sum_{j=1}^{n}\sum_{k=1}^{n}\left(\frac{\partial D_{ij}}{\partial q_k} - \frac{1}{2}\frac{\partial D_{jk}}{\partial q_i}\right) \tag{13.558}$$

$$G_i = \sum_{j=1}^{n} m_j \mathbf{g}^{\mathrm{T}}\, \mathbf{J}_{Dj}^{(i)}. \tag{13.559}$$

We can show the equations of motion for a multibody in a more concise matrix form to simplify calculations:

$$\mathbf{D}(\mathbf{q})\,\ddot{\mathbf{q}} + \mathbf{H}(\mathbf{q}, \dot{\mathbf{q}}) + \mathbf{G}(\mathbf{q}) = \mathbf{Q} \tag{13.560}$$

The term $\mathbf{G}(\mathbf{q})$ is called the *gravitational force vector* and the term $\mathbf{H}(\mathbf{q}, \dot{\mathbf{q}})$ is called the *velocity coupling vector*. The velocity coupling vector may sometimes be written in following form

$$\mathbf{H}(\mathbf{q}, \dot{\mathbf{q}}) = \mathbf{C}(\mathbf{q}, \dot{\mathbf{q}})\dot{\mathbf{q}} \tag{13.561}$$

■

Example 807 Lagrange Equation of a One-Link Manipulator Consider the uniform beam of Figure 13.15 with a mass m_2 at the tip point. This is a good example to show the advantage and simplicity of the Lagrange method compared to the Newton–Euler method.

The beam is uniform with a mass center at ${}^0\mathbf{r}_1$ while the tip mass is at ${}^0\mathbf{d}_1$, both in B_0:

$${}^0\mathbf{r}_1 = {}^0R_1\,{}^1\mathbf{r}_1 = \begin{bmatrix} \frac{l}{2}\cos\theta \\ \frac{l}{2}\sin\theta \\ 0 \end{bmatrix} \tag{13.562}$$

$${}^0\mathbf{d}_1 = {}^0R_1\,{}^1\mathbf{d}_1 = \begin{bmatrix} l\cos\theta \\ l\sin\theta \\ 0 \end{bmatrix} \tag{13.563}$$

$${}^0R_1 = R_{Z,\theta} = \begin{bmatrix} \cos\theta & -\sin\theta & 0 \\ \sin\theta & \cos\theta & 0 \\ 0 & 0 & 1 \end{bmatrix} \tag{13.564}$$

The angular velocity of the beam is

$${}_0\boldsymbol{\omega}_1 = \dot{\theta}\,\hat{K} \tag{13.565}$$

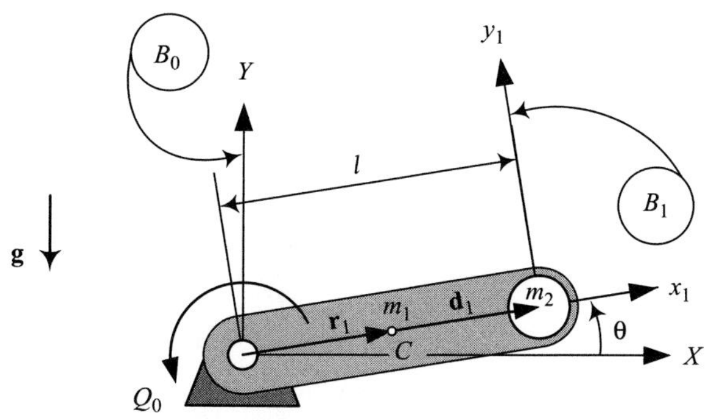

Figure 13.15 A uniform beam with a hanging weight m_2 at the tip point.

and therefore, we find the velocity of C and m_2 as

$$ {}^0\mathbf{v}_1 = {}_0\boldsymbol{\omega}_1 \times {}^0\mathbf{r}_1 = \begin{bmatrix} -\frac{l}{2}\dot{\theta}\sin\theta \\ \frac{l}{2}\dot{\theta}\cos\theta \\ 0 \end{bmatrix} \tag{13.566} $$

$$ {}^0\dot{\mathbf{d}}_1 = {}_0\boldsymbol{\omega}_1 \times {}^0\mathbf{d}_1 = \begin{bmatrix} -l\dot{\theta}\sin\theta \\ l\dot{\theta}\cos\theta \\ 0 \end{bmatrix} \tag{13.567} $$

The kinetic energy of the manipulator is

$$ \begin{aligned} K_2 &= \tfrac{1}{2}m_2\, {}^0\dot{\mathbf{d}}_1 \cdot {}^0\dot{\mathbf{d}}_1 + \tfrac{1}{2}m_1\, {}^0\mathbf{v}_1 \cdot {}^0\mathbf{v}_1 + \tfrac{1}{2}\, {}_0\boldsymbol{\omega}_1^{\mathrm{T}}\, {}^0I_1\, {}_0\boldsymbol{\omega}_1 \\ &= \tfrac{1}{8}l^2\dot{\theta}^2\left(m_1 + 4m_2\right) + \tfrac{1}{2}I_z\dot{\theta}^2 \end{aligned} \tag{13.568} $$

The mass moment matrix of the link in the base frame is

$$ \begin{aligned} {}^0I_1 &= R_{Z,\theta}\, {}^1I_1\, R_{Z,\theta}^{\mathrm{T}} = {}^0R_1 \begin{bmatrix} I_x & 0 & 0 \\ 0 & I_y & 0 \\ 0 & 0 & I_z \end{bmatrix} {}^0R_1^{\mathrm{T}} \\ &= \begin{bmatrix} I_x\cos^2\theta + I_y\sin^2\theta & \left(I_x - I_y\right)\cos\theta\sin\theta & 0 \\ \left(I_x - I_y\right)\cos\theta\sin\theta & I_y\cos^2\theta + I_x\sin^2\theta & 0 \\ 0 & 0 & I_z \end{bmatrix} \end{aligned} \tag{13.569} $$

The potential energy of the manipulator is

$$ \begin{aligned} V &= m_1 g Y_1 + m_2 g Y_2 = m_1 g r_Y + m_2 g d_Y \\ &= m_1 g \frac{l}{2}\sin\theta + m_2 g l \sin\theta \end{aligned} \tag{13.570} $$

and therefore, the Lagrangian of the manipulator is

$$ \begin{aligned} \mathcal{L} = K - V &= \tfrac{1}{8}l^2\dot{\theta}^2\left(m_1 + 4m_2\right) + \tfrac{1}{2}I_z\dot{\theta}^2 \\ &\quad - m_1 g \frac{l}{2}\sin\theta - m_2 g l \sin\theta \end{aligned} \tag{13.571} $$

Applying the Lagrange equation

$$ \frac{d}{dt}\left(\frac{\partial \mathcal{L}}{\partial \dot{\theta}}\right) - \frac{\partial \mathcal{L}}{\partial \theta} = Q_0 \tag{13.572} $$

where

$$ \frac{\partial \mathcal{L}}{\partial \dot{\theta}} = \frac{1}{4}l^2\left(m_1 + 4m_2\right)\dot{\theta} + I_z\dot{\theta} \tag{13.573} $$

$$ \frac{d}{dt}\left(\frac{\partial \mathcal{L}}{\partial \dot{\theta}}\right) = \left(\frac{1}{4}m_1 l^2 + m_2 l^2 + I_z\right)\ddot{\theta} \tag{13.574} $$

$$ \frac{\partial \mathcal{L}}{\partial \theta} = -m_1 g \frac{l}{2}\cos\theta - m_2 g l \cos\theta \tag{13.575} $$

we determine the equation of motion

$$Q_0 = \left(\tfrac{1}{4}m_1 l^2 + m_2 l^2 + I_z\right)\ddot{\theta} + \left(\tfrac{1}{2}m_1 + m_2\right) g l \cos\theta \tag{13.576}$$

Example 808 ★ A Planar Polar One-Link Manipulator Figure 13.16 illustrates a planar polar one-link manipulator with massless link and a massive point m at the tip. The kinetic energy of the manipulator is

$$\begin{aligned} K &= \tfrac{1}{2}m_2\dot{X}_2^2 + \tfrac{1}{2}m_2\dot{Y}_2^2 \\ &= \tfrac{1}{2}m\left(\frac{d}{dt}(q_1\cos q_2)\right)^2 + \tfrac{1}{2}m\left(\frac{d}{dt}(q_1\sin q_2)\right)^2 \\ &= \tfrac{1}{2}m\left(\dot{q}_1^2 + q_1^2\dot{q}_2^2\right) \end{aligned} \tag{13.577}$$

The potential energy of the manipulator is

$$V = mgY_2 = mgq_1\sin q_2 \tag{13.578}$$

and the Lagrangian of the manipulator is

$$\mathcal{L} = K - V = \tfrac{1}{2}m\left(\dot{q}_1^2 + q_1^2\dot{q}_2^2\right) - mgq_1\sin q_2 \tag{13.579}$$

Applying the Lagrange equation

$$\frac{d}{dt}\left(\frac{\partial\mathcal{L}}{\partial\dot{q}_i}\right) - \frac{\partial\mathcal{L}}{\partial q_i} = Q_i$$

yields the equations of motion

$$m\ddot{q}_1 - mq_1\dot{q}_2^2 + mg\sin q_2 = Q_1 \tag{13.580}$$

$$mq_1^2\ddot{q}_2 + 2mq_1\dot{q}_1\dot{q}_2 + mgq_1\cos q_2 = Q_2 \tag{13.581}$$

We may rearrange these equations to the matrix form of (13.536):

$$\mathbf{D}(\mathbf{q})\begin{bmatrix}\ddot{q}_1\\ \ddot{q}_2\end{bmatrix} + \mathbf{C}(\mathbf{q},\dot{\mathbf{q}})\begin{bmatrix}\dot{q}_1\\ \dot{q}_2\end{bmatrix} + \mathbf{G}(\mathbf{q}) = \begin{bmatrix}Q_1\\ Q_2\end{bmatrix} \tag{13.582}$$

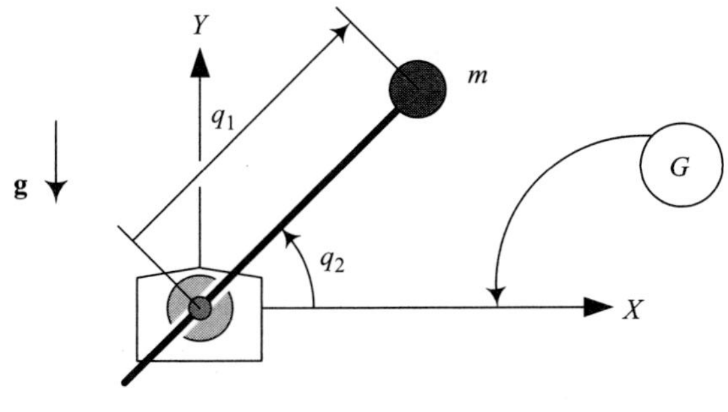

Figure 13.16 A planar polar one-link manipulator.

where

$$\mathbf{D}(\mathbf{q}) = \begin{bmatrix} m & 0 \\ 0 & mq_1^2 \end{bmatrix} \tag{13.583}$$

$$\mathbf{C}(\mathbf{q}, \dot{\mathbf{q}}) = \begin{bmatrix} 0 & -mq_1\dot{q}_2 \\ mq_1\dot{q}_2 & mq_1\dot{q}_1 \end{bmatrix} \tag{13.584}$$

$$\mathbf{G}(\mathbf{q}) = \begin{bmatrix} mg \sin q_2 \\ mgq_1 \cos q_2 \end{bmatrix} \tag{13.585}$$

Example 809 General Equations of 2R Planar Manipulators Consider a general $2R$ manipulator with massive arms and joints while carrying a payload m_0 as shown in Figure 13.17. The first motor drives link (1) and is on the ground. The second motor with mass m_{12} drives link (2) and is mounted on link (1). Assume the mass of the first and second links are m_{11} and m_{21}, respectively.

In a general case, the global position vectors of the link's mass center C_i and massive joints are

$${}^0\mathbf{r}_1 = {}^0R_1\,{}^1\mathbf{r}_1 = R_{Z,\theta_1}\, c_1\,{}^1\hat{\imath}_1 = \begin{bmatrix} c_1 \cos\theta_1 \\ c_1 \sin\theta_1 \\ 0 \end{bmatrix} \tag{13.586}$$

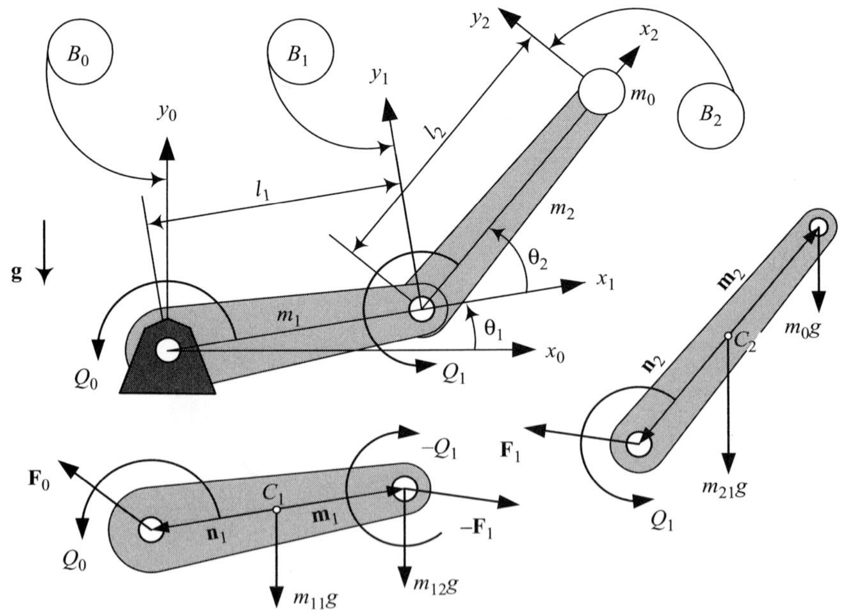

Figure 13.17 A 2R manipulator with massive arms and a carrying payload m_0.

$$
\begin{aligned}
{}^0\mathbf{r}_2 &= {}^0\mathbf{d}_1 + {}^0R_2\,{}^2\mathbf{r}_2 = {}^0\mathbf{d}_1 + R_{Z,\theta_1}\,R_{Z,\theta_2}\,c_2\,{}^2\hat{\imath}_2 \\
&= \begin{bmatrix} l_1\cos\theta_1 + c_2\cos(\theta_1+\theta_2) \\ l_1\sin\theta_1 + c_2\sin(\theta_1+\theta_2) \\ 0 \end{bmatrix}
\end{aligned} \tag{13.587}
$$

$$
{}^0\mathbf{d}_1 = {}^0R_1\,{}^1\mathbf{r}_1 = R_{Z,\theta_1}\,l_1\,{}^1\hat{\imath}_1 = \begin{bmatrix} l_1\cos\theta_1 \\ l_1\sin\theta_1 \\ 0 \end{bmatrix} \tag{13.588}
$$

$$
{}^0\mathbf{d}_2 = {}^0\mathbf{d}_1 + {}^0R_2\,{}^2\mathbf{d}_2 = \begin{bmatrix} l_2\cos(\theta_1+\theta_2) + l_1\cos\theta_1 \\ l_2\sin(\theta_1+\theta_2) + l_1\sin\theta_1 \\ 0 \end{bmatrix} \tag{13.589}
$$

where

$$
{}^0R_1 = R_{Z,\theta_1} = \begin{bmatrix} \cos\theta_1 & -\sin\theta_1 & 0 \\ \sin\theta_1 & \cos\theta_1 & 0 \\ 0 & 0 & 1 \end{bmatrix} \tag{13.590}
$$

$$
{}^1R_2 = R_{Z,\theta_2} = \begin{bmatrix} \cos\theta_2 & -\sin\theta_2 & 0 \\ \sin\theta_2 & \cos\theta_2 & 0 \\ 0 & 0 & 1 \end{bmatrix} \tag{13.591}
$$

$$
{}^0R_2 = {}^0R_1\,{}^1R_2 = \begin{bmatrix} \cos(\theta_1+\theta_2) & -\sin(\theta_1+\theta_2) & 0 \\ \sin(\theta_1+\theta_2) & \cos(\theta_1+\theta_2) & 0 \\ 0 & 0 & 1 \end{bmatrix} \tag{13.592}
$$

The links' angular velocities are

$$
{}_0\boldsymbol{\omega}_1 = \dot{\theta}_1\,\hat{K} \qquad {}_0\boldsymbol{\omega}_2 = \left(\dot{\theta}_1 + \dot{\theta}_2\right)\,\hat{K} \tag{13.593}
$$

The mass moment matrices of the links in the global coordinate frame are

$$
\begin{aligned}
{}^0I_1 &= R_{Z,\theta_1}\,{}^1I_1\,R^{\mathrm{T}}_{Z,\theta_1} = {}^0R_1 \begin{bmatrix} I_{x_1} & 0 & 0 \\ 0 & I_{y_1} & 0 \\ 0 & 0 & I_{z_1} \end{bmatrix} {}^0R_1^{\mathrm{T}} \\
&= \begin{bmatrix} I_{x_1}c^2\theta_1 + I_{y_1}s^2\theta_1 & \left(I_{x_1} - I_{y_1}\right)c\theta_1 s\theta_1 & 0 \\ \left(I_{x_1} - I_{y_1}\right)c\theta_1 s\theta_1 & I_{y_1}c^2\theta_1 + I_{x_1}s^2\theta_1 & 0 \\ 0 & 0 & I_{z_1} \end{bmatrix}
\end{aligned} \tag{13.594}
$$

$$
\begin{aligned}
{}^0I_2 &= {}^0R_2\,{}^2I_2\,{}^0R_2^{\mathrm{T}} = {}^0R_2 \begin{bmatrix} I_{x_2} & 0 & 0 \\ 0 & I_{y_2} & 0 \\ 0 & 0 & I_{z_2} \end{bmatrix} {}^0R_2^{\mathrm{T}} \\
&= \begin{bmatrix} I_{x_2}c^2\theta_{12} + I_{y_2}s^2\theta_{12} & \left(I_{x_2} - I_{y_2}\right)c\theta_{12}s\theta_{12} & 0 \\ \left(I_{x_2} - I_{y_2}\right)c\theta_{12}s\theta_{12} & I_{y_2}c^2\theta_{12} + I_{x_2}s^2\theta_{12} & 0 \\ 0 & 0 & I_{z_2} \end{bmatrix}
\end{aligned} \tag{13.595}
$$

where

$$\theta_{12} = \theta_1 + \theta_2 \tag{13.596}$$

The velocities of C_i and the joints are

$${}^0\mathbf{v}_1 = \frac{{}^0d}{dt}\,{}^0\mathbf{r}_1 = \begin{bmatrix} -c_1\dot{\theta}_1 \sin\theta_1 \\ c_1\dot{\theta}_1 \cos\theta_1 \\ 0 \end{bmatrix} \tag{13.597}$$

$$\begin{aligned} {}^0\mathbf{v}_2 &= \frac{{}^0d}{dt}\,{}^0\mathbf{r}_2 \\ &= \begin{bmatrix} -l_1\dot{\theta}_1 \sin\theta_1 - c_2\left(\dot{\theta}_1 + \dot{\theta}_2\right)\sin\left(\theta_1 + \theta_2\right) \\ l_1\dot{\theta}_1 \cos\theta_1 + c_2\left(\dot{\theta}_1 + \dot{\theta}_2\right)\cos\left(\theta_1 + \theta_2\right) \\ 0 \end{bmatrix} \end{aligned} \tag{13.598}$$

$${}^0\dot{\mathbf{d}}_1 = \begin{bmatrix} -l_1\dot{\theta}_1 \sin\theta_1 \\ l_1\dot{\theta}_1 \cos\theta_1 \\ 0 \end{bmatrix} \tag{13.599}$$

$${}^0\dot{\mathbf{d}}_2 = \begin{bmatrix} -l_1\dot{\theta}_1 \sin\theta_1 - l_2\left(\dot{\theta}_1 + \dot{\theta}_2\right)\sin\left(\theta_1 + \theta_2\right) \\ l_1\dot{\theta}_1 \cos\theta_1 + l_2\left(\dot{\theta}_1 + \dot{\theta}_2\right)\cos\left(\theta_1 + \theta_2\right) \\ 0 \end{bmatrix} \tag{13.600}$$

To calculate the Lagrangian $\mathcal{L} = K - V$, we determine the energies of the manipulator. The kinetic energy of the manipulator is

$$\begin{aligned} K &= \tfrac{1}{2}m_{12}\,{}^0\dot{\mathbf{d}}_1 \cdot {}^0\dot{\mathbf{d}}_1 + \tfrac{1}{2}m_{11}\,{}^0\mathbf{v}_1 \cdot {}^0\mathbf{v}_1 \\ &\quad + \tfrac{1}{2}m_0\,{}^0\dot{\mathbf{d}}_2 \cdot {}^0\dot{\mathbf{d}}_2 + \tfrac{1}{2}m_{21}\,{}^0\mathbf{v}_2 \cdot {}^0\mathbf{v}_2 \\ &\quad + \tfrac{1}{2}\,{}_0\omega_1^{\mathrm{T}}\,{}^0I_1\,{}_0\omega_1 + \tfrac{1}{2}\,{}_0\omega_2^{\mathrm{T}}\,{}^0I_2\,{}_0\omega_2 \end{aligned} \tag{13.601}$$

which, after substituting (13.593) and (13.597)–(13.600), becomes

$$\begin{aligned} K &= \tfrac{1}{2}\left(m_{11}c_1^2 + m_{12}l_1^2 + I_{z_1}\right)\dot{\theta}_1^2 \\ &\quad + \tfrac{1}{2}m_{21}\left[-l_1\dot{\theta}_1 \sin\theta_1 - c_2\left(\dot{\theta}_1 + \dot{\theta}_2\right)\sin\left(\theta_1 + \theta_2\right)\right]^2 \\ &\quad + \tfrac{1}{2}m_{21}\left[l_1\dot{\theta}_1 \cos\theta_1 + c_2\left(\dot{\theta}_1 + \dot{\theta}_2\right)\cos\left(\theta_1 + \theta_2\right)\right]^2 \\ &\quad + \tfrac{1}{2}m_0\left[-l_1\dot{\theta}_1 \sin\theta_1 - l_2\left(\dot{\theta}_1 + \dot{\theta}_2\right)\sin\left(\theta_1 + \theta_2\right)\right]^2 \\ &\quad + \tfrac{1}{2}m_0\left[l_1\dot{\theta}_1 \cos\theta_1 + l_2\left(\dot{\theta}_1 + \dot{\theta}_2\right)\cos\left(\theta_1 + \theta_2\right)\right]^2 \\ &\quad + \tfrac{1}{2}I_{z_2}\left(\dot{\theta}_1 + \dot{\theta}_2\right)^2 \end{aligned} \tag{13.602}$$

The potential energy of the manipulator is

$$\begin{aligned} V &= m_{11} g c_1 \sin\theta_1 + m_{12} g l_1 \sin\theta_1 \\ &\quad + m_{21} g \left[l_1 \sin\theta_1 + c_2 \sin(\theta_1 + \theta_2) \right] \\ &\quad + m_0 g \left[l_1 \sin\theta_1 + l_2 \sin(\theta_1 + \theta_2) \right] \end{aligned} \tag{13.603}$$

Applying the Lagrange equations

$$\frac{d}{dt}\left(\frac{\partial \mathcal{L}}{\partial \dot{\theta}_1}\right) - \frac{\partial \mathcal{L}}{\partial \theta_1} = Q_0 \qquad \frac{d}{dt}\left(\frac{\partial \mathcal{L}}{\partial \dot{\theta}_2}\right) - \frac{\partial \mathcal{L}}{\partial \theta_2} = Q_1 \tag{13.604}$$

determines the general equations of motion:

$$\begin{bmatrix} D_{11} & D_{12} \\ D_{21} & D_{22} \end{bmatrix} \begin{bmatrix} \ddot{\theta}_1 \\ \ddot{\theta}_2 \end{bmatrix} + \begin{bmatrix} C_{11} & C_{12} \\ C_{21} & C_{22} \end{bmatrix} \begin{bmatrix} \dot{\theta}_1 \\ \dot{\theta}_2 \end{bmatrix} + \begin{bmatrix} G_1 \\ G_2 \end{bmatrix} = \begin{bmatrix} Q_o \\ Q_1 \end{bmatrix} \tag{13.605}$$

where

$$\begin{aligned} D_{11} &= 2 l_1 \left(m_{21} c_2 + m_0 l_2 \right) \cos\theta_2 + I_{z_1} + I_{z_2} \\ &\quad + m_{11} c_1^2 + m_{12} l_1^2 + m_{21}\left(c_2^2 + l_1^2\right) + m_0\left(l_1^2 + l_2^2\right) \end{aligned} \tag{13.606}$$

$$D_{12} = l_1 \left(m_{21} c_2 + m_0 l_2 \right) \cos\theta_2 + I_{z_2} + m_0 l_2^2 + m_{21} c_2^2 \tag{13.607}$$

$$D_{21} = l_1 \left(m_{21} c_2 + m_0 l_2 \right) \cos\theta_2 + I_{z_2} + m_{21} c_2^2 + m_0 l_2^2 \tag{13.608}$$

$$D_{22} = I_{z_2} + m_{21} c_2^2 + m_0 l_2^2 \tag{13.609}$$

$$C_{11} = -l_1 \left(m_{21} c_2 + m_0 l_2 \right) \dot{\theta}_2 \sin\theta_2 \tag{13.610}$$

$$C_{12} = -l_1 \left(m_{21} c_2 + m_0 l_2 \right) \left(\dot{\theta}_1 + \dot{\theta}_2 \right) \sin\theta_2 \tag{13.611}$$

$$C_{21} = l_1 \left(m_{21} c_2 + m_0 l_2 \right) \dot{\theta}_1 \sin\theta_2 \tag{13.612}$$

$$C_{22} = 0 \tag{13.613}$$

$$\begin{aligned} G_1 &= \left[\left(m_{21} + m_{12} + m_0 \right) l_1 + m_{11} c_1 \right] \cos\theta_1 \\ &\quad + \left(m_{21} c_2 + m_0 l_2 \right) \cos(\theta_1 + \theta_2) \end{aligned} \tag{13.614}$$

$$G_2 = \left(m_{21} c_2 + m_0 l_2 \right) \cos(\theta_1 + \theta_2) \tag{13.615}$$

KEY SYMBOLS

$\mathbf{a}$	acceleration vector, a general vector
$\tilde{a}$	skew-symmetric matrix of the vector $\mathbf{a}$
A	transformation matrix of rotation about a local axis
$\mathbf{b}$	base vector of a coordinate system
$\mathbf{b}^*$	reciprocal base vector of a coordinate system
B	body coordinate frame, local coordinate frame
B_i, B_{ij}	Pfaffian constraint coefficients
c	cos
C	integration constant
d	distance between two points
$\mathbf{D}(\mathbf{q})$	inertial-type matrix of equation of motion
$\hat{e}_\varphi, \hat{e}_\theta, \hat{e}_\psi$	coordinate axes of E, local roll–pitch–yaw coordinate axes
E	Eulerian local frame, mechanical energy
f	function, constraint equation, final
f, f_1, f_2	functions, functions of x and y, constraint equations
f, g, h	transformation equation from q_i to x, y, z
$F, \mathbf{F}$	force
$\mathbf{F}_C$	constraint force
$g, \mathbf{g}$	gravitational acceleration
G	global coordinate frame, fixed coordinate frame
$\mathbf{G}(\mathbf{q})$	gravitational coefficient matrix of equation of motion
H	Hamiltonian
$\mathbf{H}(\mathbf{q}, \dot{\mathbf{q}})$	velocity coupling vector of equation of motion
$I, [I]$	mass moment
$\mathbf{I}$	identity matrix
$\hat{\imath}, \hat{\jmath}, \hat{k}$	local coordinate axis unit vectors
$\hat{I}, \hat{J}, \hat{K}$	global coordinate axis unit vectors
J	objective function
k	radius of gyration, spring stiffness, constant coefficient
K	kinetic energy
l	length, number of constraints
l'	difference of n' and n
$L, \mathbf{L}$	moment of momentum, angular momentum
$\mathcal{L}$	Lagrangian
m	mass
n	number of DOF, number of generalized coordinates q_i
n'	number of nongeneralized coordinates s_i
N	dimension of configuration space
O	common origin of B and G, order of magnitude
$O\varphi\theta\psi$	Euler angle frame
p	generalized momentum
$\mathbf{p}$	momentum vector, generalized momenta
P	a body point, a fixed point in B
P, Q, R	coefficient functions of ds^2
q_i	generalized coordinate

$\mathbf{q}$	vector of generalized coordinate
Q	transformation matrix of rotation about a global axis, generalized coordinate system $Q\,(q_i)$
Q_i	generalized force associated with q_i
$\mathbf{Q}$	control input
$\mathbf{r}$	position vector
r_{ij}	element of row i and column j of a matrix
R	radius of a circle
$\mathbb{R}$	set of real numbers
s	sin, arc length, a member of S
s_i	nongeneralized coordinates
S	set
t	time
u	coordinate of configuration space
u, v	two-dimensional coordinate system
$\mathbf{u}$	general axis
$\hat{u}$	unit vector, unit base vector
$\hat{u}^*$	reciprocal unit base vector
$\mathbf{v}$	velocity vector
V	potential energy, variation
x, y, z	local coordinates, local coordinate axes
$\mathbf{x}$	state vector
X, Y, Z	global coordinates, global coordinate axes

Greek

α, β, γ	rotation angles about global axes
Γ	Christoffel symbol
δ	virtual displacement, virtual increment
δ_{ij}	Kronecker's delta
ϵ	small parameter
λ	Lagrange multiplier
ξ, η, φ	a curvilinear coordinate system
φ, θ, ψ	rotation angles about local axes, Euler angles
$\dot{\varphi}, \dot{\theta}, \dot{\psi}$	Euler frequencies
$\omega_x, \omega_y, \omega_z$	angular velocity components
$\omega, \boldsymbol{\omega}$	angular velocity vector

Symbol

$[\ \ ]^{-1}$	inverse of the matrix [ ]
$[\ \ ]^{\mathrm{T}}$	transpose of the matrix [ ]
∇	gradient
DOF	degree of freedom
Δ	difference
$\mathcal{L}$	Lagrangian
$\vdash$	orthogonal
(i)	link number i
$\parallel$	parallel
$\perp$	perpendicular

EXERCISES

1. **Equation of Motion in Cylindrical Coordinate** Consider F_r, F_θ, and F_z as the applied forces on a particle in the radial, tangential, and z-directions:

$$x = r\cos\theta \qquad y = r\sin\theta \qquad z = z$$

Using the Lagrange method, determine the equations of motion of the particle in cylindrical coordinates (r, θ, z).

2. **Equation of Motion in Spherical Coordinate** Consider F_r, F_θ, and F_φ as the applied forces on a particle in the r-, θ-, and φ-directions:

$$x = r\cos\theta \qquad y = r\sin\theta\cos\varphi \qquad z = r\sin\theta\sin\varphi$$

Using the Lagrange method, determine the equations of motion of the particle in spherical coordinates (r, θ, φ).

3. **Pendulum in Wind** Figure 13.18 illustrates a simple and an elastic pendulum in wind. The wind exerts a force $\mathbf{F}_w$ on m. Determine the equations of motion of pendulums (a) and (b) for:

(a) $\alpha = 0$, $\mathbf{F}_w = cm\left(\cos\alpha\,\hat{\imath} + \sin\alpha\,\hat{\jmath}\right)$, $c > 0$
(b) $0 < \alpha < 90\,\text{deg}$, $\mathbf{F}_w = cm\left(\cos\alpha\,\hat{\imath} + \sin\alpha\,\hat{\jmath}\right)$, $c > 0$
(c) $\alpha = 90\,\text{deg}$, $\mathbf{F}_w = cm\left(\cos\alpha\,\hat{\imath} + \sin\alpha\,\hat{\jmath}\right)$, $c > 0$
(d) $0 \le \alpha \le 90\,\text{deg}$, $\mathbf{F}_w = cmv\left(\cos\alpha\,\hat{\imath} + \sin\alpha\,\hat{\jmath}\right)$, $c > 0$
(e) $0 \le \alpha \le 90\,\text{deg}$, $\mathbf{F}_w = mv\left(c_1\cos\alpha\,\hat{\imath} + c_2\sin\alpha\,\hat{\jmath}\right)$, $c_1 > 0$, $c_2 > 0$
(f) $0 \le \alpha \le 90\,\text{deg}$, $\mathbf{F}_w = mv^2\left(c_1\cos\alpha\,\hat{\imath} + c_2\sin\alpha\,\hat{\jmath}\right)$, $c_1 > 0$, $c_2 > 0$

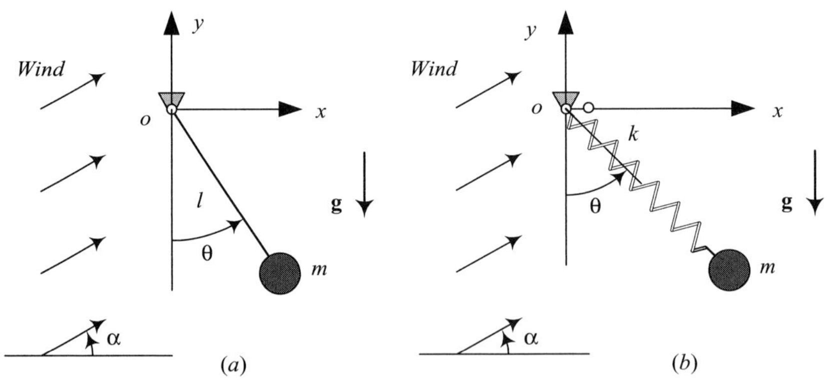

Figure 13.18 Two pendulums in wind.

4. **Inverted Pendulums** Determine the equations of motion of the two controlled inverted pendulums of Figures 13.19 (a) and (b).

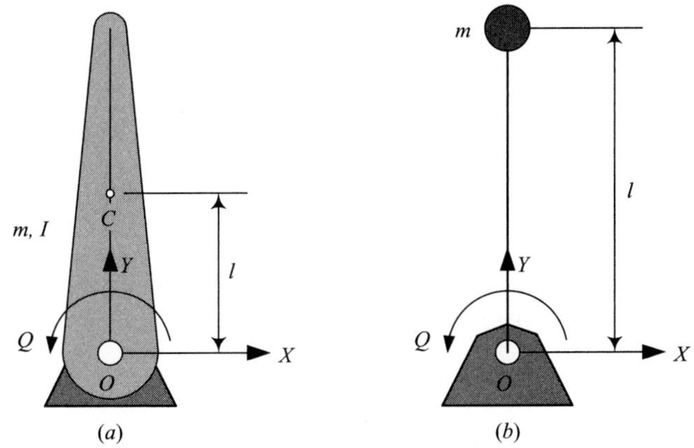

Figure 13.19 Two inverted pendulums.

5. **A Particle in a Viscous Media** A particle of mass m moves on a smooth horizontal circle of radius a with an initial velocity v_0. It is resisted by the air with a force proportional to the square of its velocity v:

$$F = -kv^2$$

   (a) Using the Lagrange method, determine the equation of motion of the particle

   (b) Solve the equation of motion and show that the angular position of the particle as a function of time is

$$\theta = \frac{m}{ka} \ln \left(1 + \frac{k}{m} t\right)$$

   (c) Assume that the particle is forced to move on the circle by a string connected to the center of the circle. Modify the kinetic energy and use the Lagrange method to show that the tension force F_T in the string is

$$F_T = ma\dot{\theta}^2$$

6. **Two Pulleys and Three Hanging Masses** Figure 13.20 shows a mass $4m$ that is attached to a string which passes over a smooth massless stationary pulley. The other end of the string is fastened to a smooth pulley of mass m over which passes a second string attached to masses m and $2m$. The system has two DOF. Use x and y, as are shown, and find the acceleration of the mass $4m$.

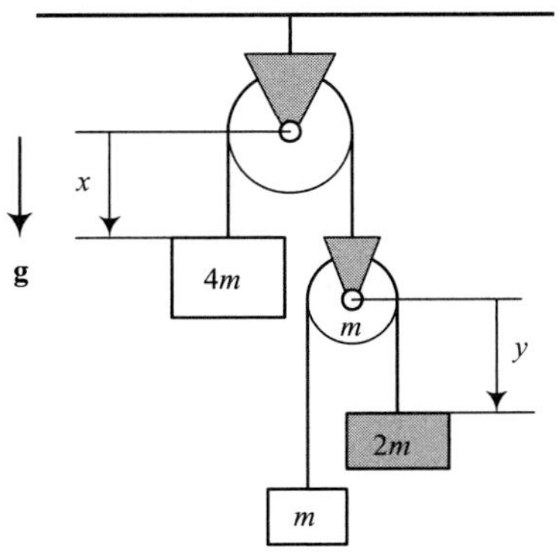

Figure 13.20 Two pulleys and three hanging masses.

7. **★ A Dumbbell on a Smooth Table** Two equal particles, each of mass m, are connected by a massless bar of length l to make a dumbbell. The dumbbell has a force-free motion on a smooth table. Determine the possible motions of the dumbbell.

8. **Two Connected Particles and a Table** Two equal particles of mass m are connected by a string with length l which passes through a hole in a smooth horizontal table. The first particle is set moving on the table at right angles with the string, with a velocity $v_1 = \sqrt{gl}$. The hanging particle is drawn a short distance downward and then released. Let us show the distance of the second particle from its equilibrium position at a time t by x and the angular position of the first particle by θ. Show that the kinetic energy of the system is as given below and determine the subsequent motion of the suspended particle:

$$K = \frac{1}{2}m\left[\dot{x}^2 + \dot{x}^2 + (l - x)^2\dot{\theta}^2\right]$$

9. **★ A Billiard Ball** The homogeneous ball shown in Figure 13.4 is rolling freely on an imperfectly rough table. Assume that the friction coefficient is μ and the ball is supposed to slip. Employing the coordinate frames in Example 768, Equations (13.67)–(13.71) and (13.74)–(13.77) remain the same. Examine the possible motions of the ball.

10. **★ A Gyroscope** Consider a rigid body with a fixed point. The point is the origin of the local principal coordinate frame. The body has two equal principal mass moments.

    (a) Obtain the kinetic energy of the body

    (b) Derive the equations of motion of the body under gravity

11. **A Pendulum with Moving Support** Consider a pendulum in the (x, y)-plane as shown in Figure 13.21. Assume the supporting point O is moving in the y-direction with a given function of time:

$$y_o = f(t)$$

Determine the equation of motion of m.

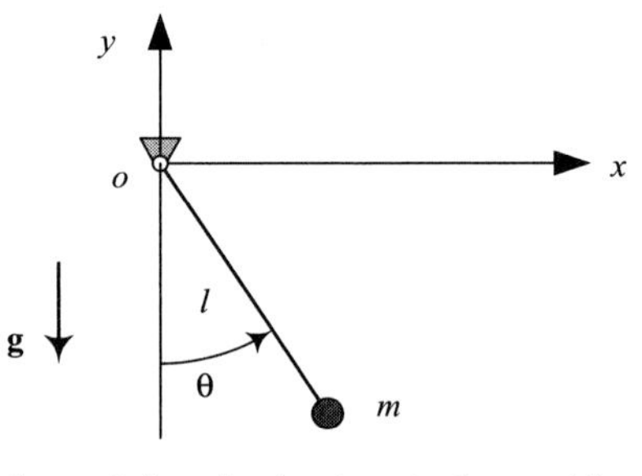

Figure 13.21 A pendulum in the (x, y)-plane with moving support.

12. **★ A Particle in a Two-Dimensional Coordinate Frame** Consider a particle with mass m that is moving freely in a two-dimensional space. Using the coordinates

$$x = \frac{1}{2}(q_1 + q_2) \qquad y = \tan\frac{q_1 - q_2}{2}$$

determine the kinetic energy of the particle,

$$K = \frac{1}{2}m\left(\dot{x}^2 + \dot{y}^2\right)$$

and show that the equations of motion are

$$\ddot{q}_1 + \ddot{q}_2 + (\ddot{q}_1 - \ddot{q}_2)\sec^4\frac{q_1 - q_2}{2} + (\dot{q}_1 - \dot{q}_2)^2\sec^4\frac{q_1 - q_2}{2}\tan\frac{q_1 - q_2}{2}$$

$$= -2g\sec^2\frac{q_1 - q_2}{2}$$

$$\ddot{q}_1 + \ddot{q}_2 + (\ddot{q}_1 - \ddot{q}_2)\sec^4\frac{q_1 - q_2}{2} - (\dot{q}_1 - \dot{q}_2)^2\sec^4\frac{q_1 - q_2}{2}\tan\frac{q_1 - q_2}{2}$$

$$= 2g\sec^2\frac{q_1 - q_2}{2}$$

13. ★ **Geodesic on a General Surface of Revolution** Consider the surface

$$x^2 + y^2 = \left[g(x)\right]^2$$

which is generated by revolving the curve $y = g(x)$, with $g \geq 0$, about the x-axis. Show that using the parametric representation of this surface given as

$$x = u \qquad y = g(x)\cos v \qquad z = g(x)\sin v$$

the geodesic curves on the surface can be found as

$$v = C_1 \int \frac{\sqrt{1 + \left[g'(u)\right]^2}\,du}{g(u)\sqrt{\left[g'(u)\right]^2 - C_1^2}}$$

14. **A Turning Pendulum** Figure 13.22 illustrates a mass m that is attached to a massless rod with length l. The rod is pivoted to a rotating vertical bar that is turning with angular speed ω:

(a) Using conservation of the moment of momentum, draw a graph to show ω versus θ if $m = 2\,\text{kg}$, $l = 1.2\,\text{m}$, and $\omega = 10\,\text{rpm}$ when $\theta = 30\,\text{deg}$.

(b) Assume ω is constant. Determine the equation of motion of the pendulum.

(c) Determine the equilibrium value of θ.

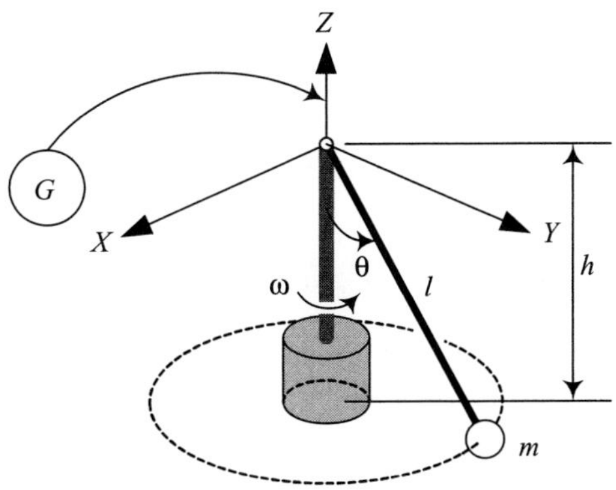

Figure 13.22 A mass m is attached to a massless rod which is turning with angular speed ω.

15. ★ **Minimum Surface of Revolution** Consider two given fixed points (x_1, y_1) and (x_2, y_2), we seek to pass through them the arc $y = y(x)$ whose rotation about the x-axis generates a surface of revolution whose area in $x_1 \le x \le x_2$ is a minimum. Assuming $y_1 > 0$, $y_2 > 0$, and $y(x) \ge 0$ in $x_1 \le x \le x_2$, the problem reduces to the functional

$$J = 2\pi \int_{x_1}^{x_2} y\sqrt{1 + y'^2}dx \qquad y' = \frac{dy}{dx}$$

This is the area of the surface of revolution. Show that the minimum surface of revolution is

$$y = -C_1 \cosh \frac{x - C_2}{-C_1}$$

16. ★ **Generalized Forces in Non-Cartesian Systems**

(a) Determine the generalized forces in a spherical coordinate system.

(b) Determine the generalized momentum in a cylindrical coordinate system.

17. **Generalized Coordinate and Lagrangian** To show that the Lagrangian function and equation are form invariant under a generalized coordinate transformation from q_i to s_j, we must employ Equations (13.217). Show that these two equations are valid.

18. **Pendulum with Flexible Support** Figures 13.23(*a*) and (*b*) illustrate two pendulums with flexible supports in directions of x and y, respectively. Determine the equations of motion for:

(a) A pendulum with a flexible support in the x-direction of Figure 13.23(*a*)

(b) A pendulum with a flexible support in the y-direction of Figure 13.23(*b*)

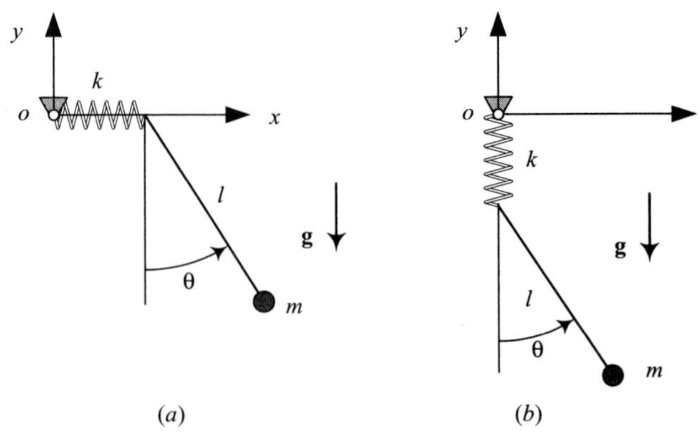

Figure 13.23 Pendulums with flexible support.

19. **Discrete Particles** There are three particles $m_1 = 4\,\text{kg}$, $m_2 = 2\,\text{kg}$, $m_3 = 3\,\text{kg}$ at

$$\mathbf{r}_1 = \begin{bmatrix} 1 \\ -1 \\ 1 \end{bmatrix} \quad \mathbf{r}_2 = \begin{bmatrix} -1 \\ -3 \\ 2 \end{bmatrix} \quad \mathbf{r}_3 = \begin{bmatrix} 2 \\ -1 \\ -3 \end{bmatrix}$$

Their velocities are

$$\mathbf{v}_1 = \begin{bmatrix} 2 \\ 1 \\ 1 \end{bmatrix} \qquad \mathbf{v}_2 = \begin{bmatrix} -1 \\ 0 \\ 2 \end{bmatrix} \qquad \mathbf{v}_3 = \begin{bmatrix} 3 \\ -2 \\ -1 \end{bmatrix}$$

Find the position and velocity of the system at C. Calculate the system's momentum and moment of momentum. Calculate the system's kinetic energy and determine the rotational and translational parts of the kinetic energy.

20. Two Connected Pendulums Assume the free length of the spring in Figure 13.24 is a.

(a) Determine the equations of motions of the connected pendulums of Figure 13.24.

(b) Assume $l_1 = l_2$ and the angles and angular velocity of oscillation are very small. Linearize the system of equations.

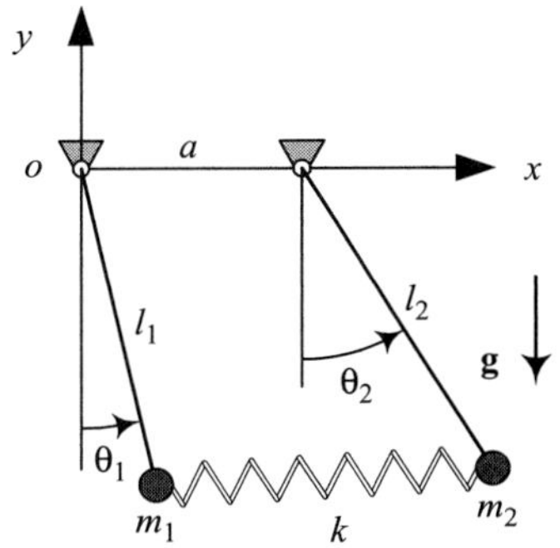

Figure 13.24 Two connected pendulums.

21. A Sliding Plank That Leans against a Wall A uniform plank with mass m and length l is initially leaned against a smooth wall and stands on a friction floor. The plank starts to slide. Derive the equations of motion and the reaction forces on the wall and floor by using (a) the Newton–Euler method and (b) the Lagrange multiplier method.

22. A Pendulum with Moving Support Figure 13.25 illustrates a pendulum with a moving support. Employ the Lagrange equation and determine the equation of motion of the pendulum:

(a) When the support is moving on a circle $x = a\cos\omega t$, $y = a\sin\omega t$ with constant ω as shown in Figure 13.25(a)

(b) ★ When the support is moving on an ellipse $x = a\cos\omega t$, $y = b\sin\omega t$ with constant ω as shown in Figure 13.25(b)

(c) ★ Simplify the equation of part (b) for $b = a$ and recover the equation of part (a).

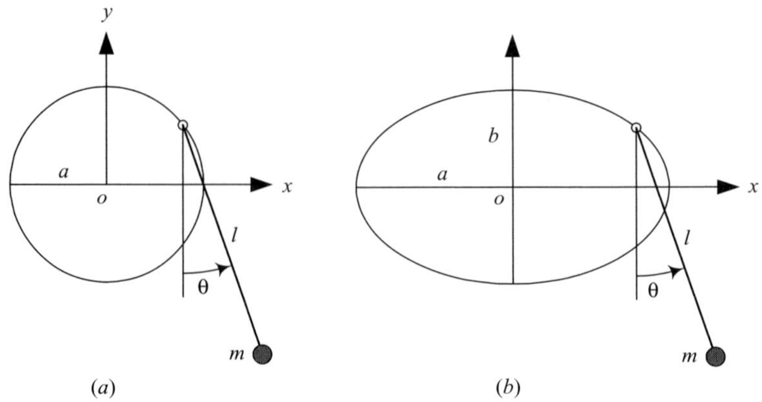

Figure 13.25 Two pendulums with moving support.

23. Equation of Motion for a Given Lagrangian Consider a dynamic system given

$$\mathcal{L} = K - V$$
$$K = \frac{1}{2}\left[m\left(\dot{x}_1^2 + \dot{x}_2^2\right) + I_1\dot{\varphi}^2\sin^2\theta + I_3\left(\dot{\varphi}\cos\theta + \dot{\psi}\right)^2\right]$$
$$V = -mg\left(x_1\sin\alpha - x_3\cos\alpha\right)$$
$$\alpha = \text{const}$$

(a) Derive the equations of motion.

(b) Try to recover the Lagrangian from the equations of motion.

24. A Falling Particle of a Surface A particle with mass m is sliding on the surface of a geometric shape under the gravitational attraction. Determine the constraint force and the angle at which the particle leaves the surface:

(a) When the surface is a circle $x = a\cos\varphi$, $y = a\sin\varphi$ as shown in Figure 13.26(a)

(b) ★ When the surface is an ellipse $x = a\cos\varphi$, $y = b\sin\varphi$ as shown in Figure 13.26(b)

(c) ★ Simplify the equations of part (b) for $b = a$ and recover the equations of part (a).

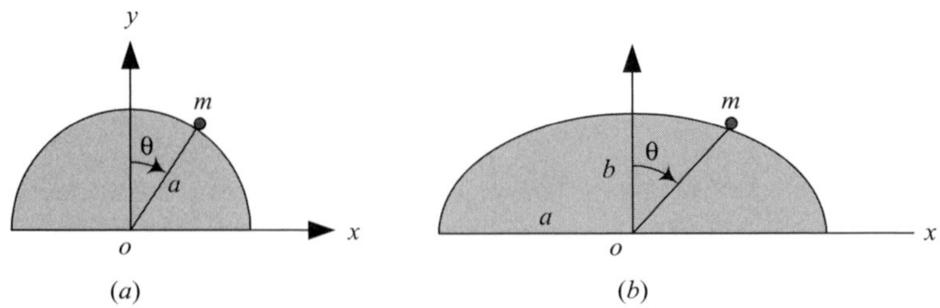

Figure 13.26 A Sliding particle on two surfaces.

25. Application of Conservation Laws Employ the integrals of motion in Example 800 and determine the generalized coordinates as functions of time. Discuss the possible motions of the system.

26. ★ A Variable-Mass Pendulum Let us assume that there are two identical chains with length density of ρ lying on the x-axis as shown in Figure 13.27. A pendulum with mass m will take one of the chains every time it passes $\theta = 0$ and carries the part of the chain which is limited to $x > 0$ or $x < 0$. Let us ignore the size of hooks and m and assume that the weight of the lifted chain, ρx, is concentrated at m.
Determine the equation of motion of this variable-mass pendulum.

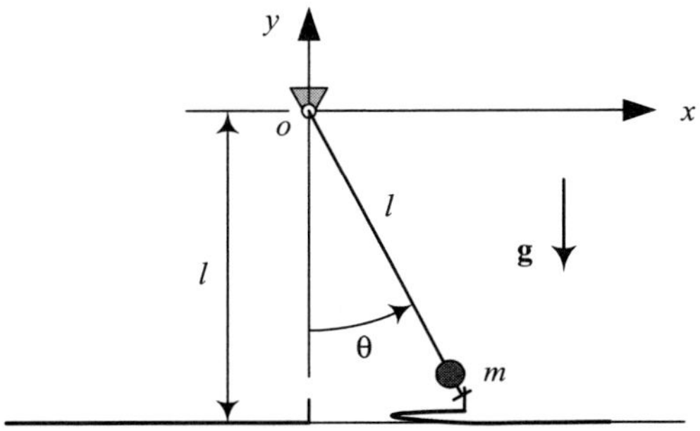

Figure 13.27 A collecting-chain pendulum.

27. Simple and Compound Pendulums Use the Lagrange method and find the equation of motion for the pendulums shown in Figures 13.28(a) and (b). The stiffness of the linear spring is k. Assume the free length of the spring is $a - l$.

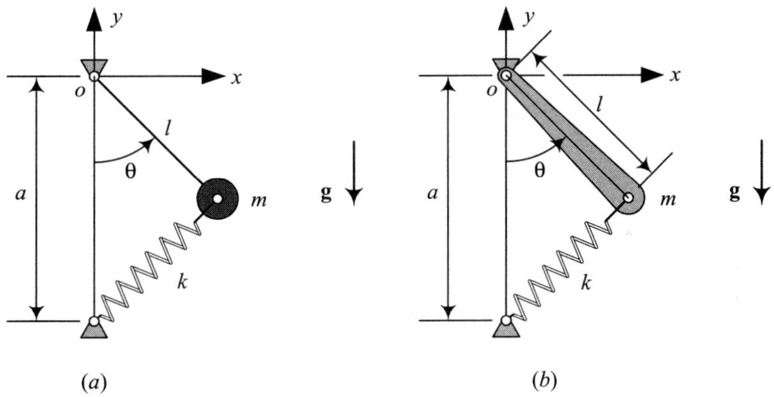

Figure 13.28 A simple and a compound pendulum attached with a linear spring at the tip point.

28. Lagrangian Problem Find the Lagrangian associated with the following equations of motions:

(a) $mr^2\ddot{\theta} + k_1 l_1 \theta + k_2 l_2 \theta + mgl = 0$

(b)

$$\ddot{r} - r\dot{\theta}^2 = 0$$

$$r^2\ddot{\theta} + 2r\dot{r}\dot{\theta} = 0$$

29. ★ **Particle in Electromagnetic Field** Show that equations of motion of a particle with mass m and Lagrangian

$$\mathcal{L} = \frac{1}{2}m\dot{\mathbf{r}}^2 - e\Phi + e\dot{\mathbf{r}} \cdot A$$

are

$$m\ddot{q}_i = e\left(-\frac{\partial \Phi}{\partial q_i} - \frac{\partial A_i}{\partial t}\right) + e\sum_{j=1}^{3}\dot{q}_j\left(\frac{\partial A_j}{\partial q_i} - \frac{\partial A_i}{\partial q_j}\right)$$

where

$$\mathbf{q} = \begin{bmatrix} q_1 \\ q_2 \\ q_3 \end{bmatrix} = \begin{bmatrix} x \\ y \\ z \end{bmatrix}$$

Then convert the equations of motion to a vectorial form

$$m\ddot{\mathbf{r}} = e\mathbf{E}(\mathbf{r}, t) + e\dot{\mathbf{r}} \times \mathbf{B}(\mathbf{r}, t)$$

where $\mathbf{E}$ and $\mathbf{B}$ are electric and magnetic fields:

$$\mathbf{E} = -\nabla\Phi - \frac{\partial \mathbf{A}}{\partial t}$$

$$\mathbf{B} = \nabla \times \mathbf{A}.$$

30. **Connected Springs**

(a) Determine the equations of motion of the systems in Figures 13.29(a) and (b).

(b) Show that by $a \to 0$ the equations of system (b) approach those of (a).

(c) Assume the variations of the coordinates are too small and linearize the equations of motion.

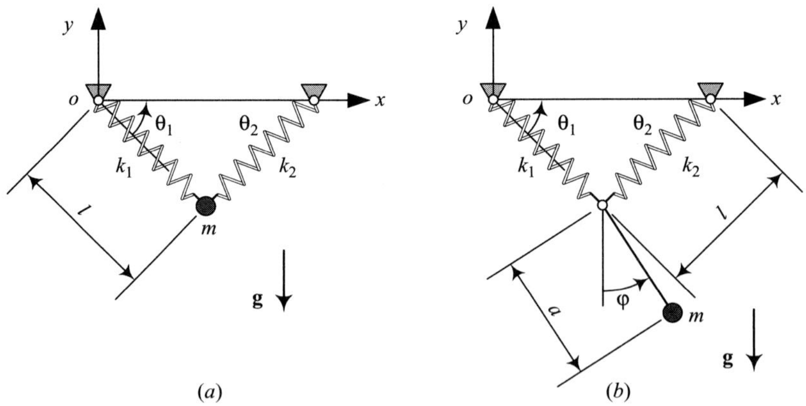

Figure 13.29 A mass and a pendulum suspended by two springs.

31. **Heavy Pendulum** Figure 13.30(a) illustrates a heavy disc with mass m and radius R suspended by a massless rod of length l. Figure 13.30(b) illustrates another heavy disc with mass m and radius R that is attached to a massless rod of length l by a frictionless revolute joint at its center.

(a) Derive the equations of motion for the pendulums in (*a*) and (*b*).

(b) Linearize the equations of motion. Is it possible to compare the period of oscillations?

(c) ★ Assume the disc of Figure 13.30(*b*) has an angular velocity of ω when $\theta = 0$. Determine the equation of motion, linearize the equation, and determine the period of oscillation.

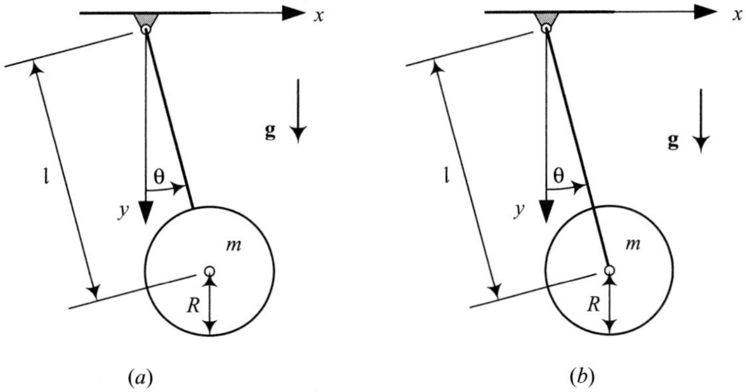

Figure 13.30 Heavy pendulums with and without a revolute joint.

32. Variation and Equilibrium Position Figure 13.31 depicts two discs with masses m_1 and m_2 and radius R that are connected with a bar of mass m and length l. The spring k prevents the system from falling. Assume there is no friction and the spring has a free length of $a + R - l$.

(a) Determine the potential energy V of the system.

(b) Find the equilibrium value of θ by minimizing V.

(c) Derive the equation of motion of the system by the Newton method.

(d) Derive the equation of motion of the system by the Lagrange method.

(e) Derive the tension force in the rod by the Lagrange method.

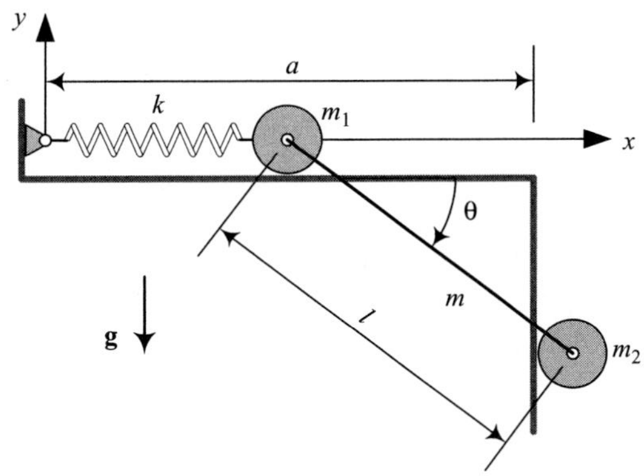

Figure 13.31 Two discs that are connected with a bar and supported by a spring.

33. Strong Nonlinear Vibrating Systems Figures 13.32(a) and (b) illustrate two massive points with a prescribed one-dimensional path of motion while they are attached to a fixed point by a linear spring.

(a) Determine their equations of motion in terms of x.

(b) Determine their equations of motion in terms of θ.

(c) Assume x is too small and simplify the equations of part (a).

(d) Assume θ is too small and simplify the equations of part (b).

(e) Can we linearize the equations of motion for both systems? Are their equations for very small motion equivalent? Does your answer depend on the ratio l/R ?

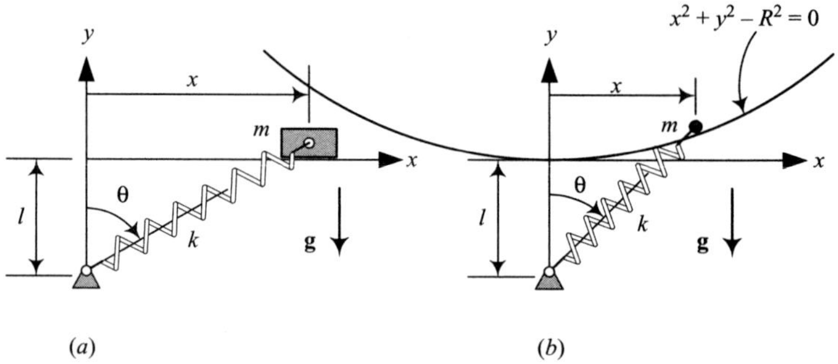

Figure 13.32 Two massive points with prescribed path of motion that are supported by springs.

34. A Partial Circular Wire Figure 13.33 illustrates a circular arc of central angle 2α, arc density ρ, mass m, and radius R. If the minimum distance of the mass center C from the wire is $l = (R \sin\alpha)/\alpha$, then the mass moment of the wire with respect to the center of the circle and the mass center C are $I_o = 2\alpha\rho R^3$ and $I_C = I_o - 2\alpha\rho R l^2$. Determine the Lagrangian and equations of motion of the system.

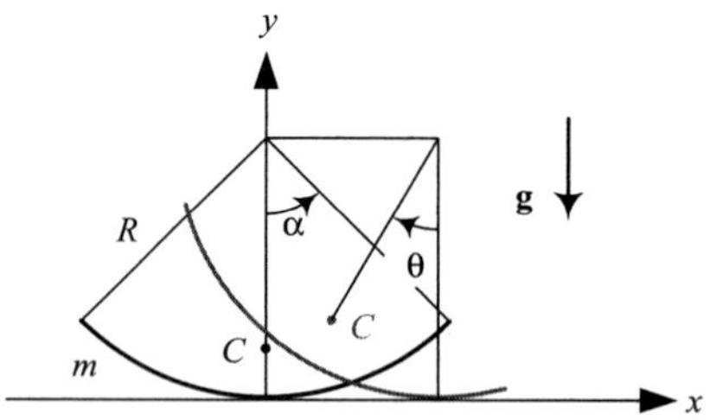

Figure 13.33 A partial circular wire.

35. Lagrangian as a Function of $\ddot{q}$. Assume a Lagrangian $\mathcal{L}$ that is a function of generalized coordinates q_i, velocities $\dot{q}_i$, and accelerations $\ddot{q}_i$:

$$\mathcal{L} = \mathcal{L}(q_i, \dot{q}_i, \ddot{q}_i)$$

Show that the Euler–Lagrange equation of such a Lagrangian is

$$\frac{\partial \mathcal{L}}{\partial q_i} - \frac{d}{dt}\frac{\partial \mathcal{L}}{\partial \dot{q}_i} + \frac{d^2}{dt^2}\frac{\partial \mathcal{L}}{\partial \ddot{q}_i} = 0$$

36. Equivalent Lagrangian Assume $\mathcal{L} = \mathcal{L}(q, \dot{q}, t)$ is the Lagrangian of a one-DOF dynamic system. Let $f(q, t)$ be an arbitrary function and define a new Lagrangian $\mathcal{L}'$ by adding the time derivative of f to $\mathcal{L}$:

$$\mathcal{L}' = \mathcal{L} + \frac{df}{dt} = \mathcal{L} + \frac{\partial f}{\partial q}\dot{q} + \frac{\partial f}{\partial t}$$

Show that the equations of motion for $\mathcal{L}'$ and $\mathcal{L}$ are identical.

37. A Turning Pendulum Figure 13.34 illustrates a pendulum of length l with a suspended mass m. The pivot of the pendulum is a point turning on a circle with radius R and angular speed ω.

(a) Determine the angular velocity and acceleration of the pendulum.
(b) Determine the Lagrangian of the pendulum.
(c) Determine the equations of motion of the pendulum.

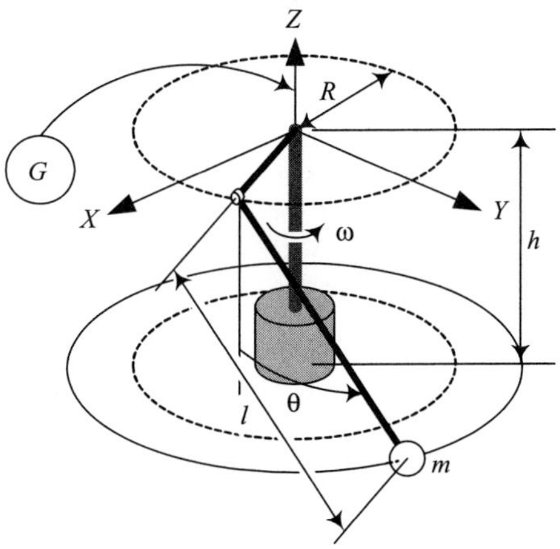

Figure 13.34 A turning pendulum.

38. Kinetic Energy of a Rigid Link Consider a straight and uniform bar as a rigid link of a manipulator. The link has a mass m. Show that the kinetic energy of the bar can be expressed as

$$K = \tfrac{1}{6}m\left(\mathbf{v}_1 \cdot \mathbf{v}_1 + \mathbf{v}_1 \cdot \mathbf{v}_2 + \mathbf{v}_2 \cdot \mathbf{v}_2\right)$$

where $\mathbf{v}_1$ and $\mathbf{v}_2$ are the velocity vectors of the end points of the link.

39. Lagrangian and New Coordinates Consider the Lagrangian

$$\mathcal{L}(x, \dot{x}, y, \dot{y}, t) = \tfrac{1}{2}\left(\dot{x}^2 + \dot{y}^2\right)$$

Determine $\mathcal{L} = \mathcal{L}\left(r, \dot{r}, \theta, \dot{\theta}, t\right)$ and the equations of motion if

$$x = r\cos\theta \qquad y = r\sin\theta$$

40. *n*-Pendulum Equations of Motion Figure 13.35 illustrates an n-pendulum with planar motion in the (x, y)-plane. Determine the kinetic, potential, Lagrangian, and equations of motion of the system.

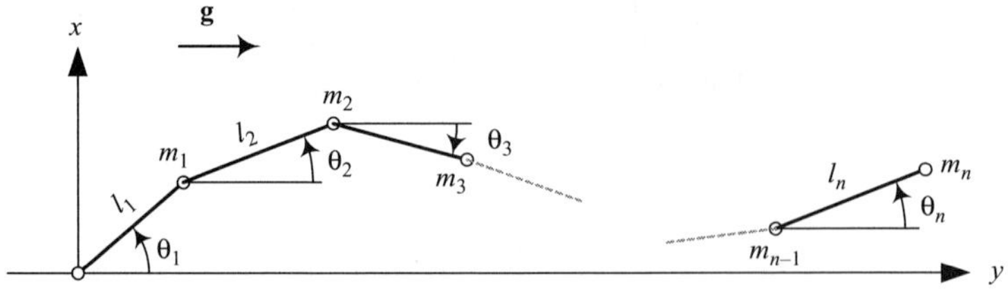

Figure 13.35 An n-pendulum with planar motion in the (x, y)-plane.

41. Velocity-Dependent Potential Show that if a potential function V is a function of q_i and $\dot{q}_i$, then in Newtonian mechanics it must be a linear function of the velocity $\dot{q}_i$; otherwise we will have a force that is a function of acceleration:

$$V = \sum_{i=1}^{n} a_i \dot{q}_i + V_0$$

42. The Ideal 2R Planar Manipulator Dynamics An ideal model of a 2R planar manipulator is illustrated in Figure 13.36. It is called ideal because we assumed that the links are massless and there is no friction. The masses m_1 and m_2 are respectively the mass of the second motor to run the second link and the load at the end point. Use the absolute angle θ_1 and the relative angle θ_2 as the generalized coordinates to express the configuration of the manipulator. Derive the equations of motion.

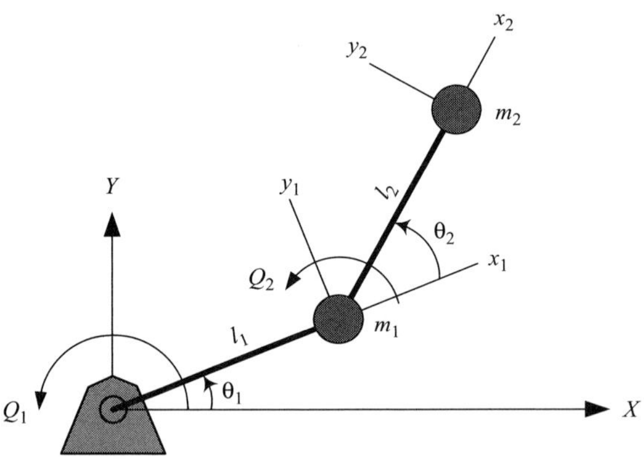

Figure 13.36 An ideal 2R planar manipulator.

43. Ignorable Coordinates and Generalized Momenta Assume the coordinates $q_1, q_2, \ldots, q_m$, $m < n$, of an n-DOF dynamic system are ignorable. Therefore the first m Lagrange equations lead to generalized momentum integrals:

$$p_i = \frac{\partial \mathcal{L}}{\partial \dot{q}_i} = \text{const} \qquad i = 1, 2, \ldots, m \tag{13.616}$$

Why are p_i linear in terms of $\dot{q}_1, \dot{q}_2, \ldots, \dot{q}_m$?

44. An RPR Planar Redundant Manipulator Figure 13.37 illustrates a three-DOF planar manipulator with joint variables θ_1, d_2, and θ_2. Determine the equations of motion of the manipulator if the links are massless and there are two massive points m_1 and m_2.

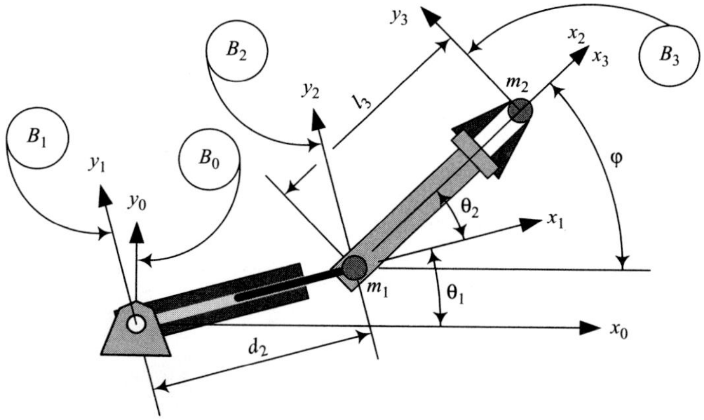

Figure 13.37 An RPR planar redundant manipulator.

45. A Planar Multi-DOF Manipulator Figure 13.38 illustrates a three-DOF planar manipulator. Determine the equations of motion of the manipulator if the links are massless and there are two massive points m_1 and m_2.

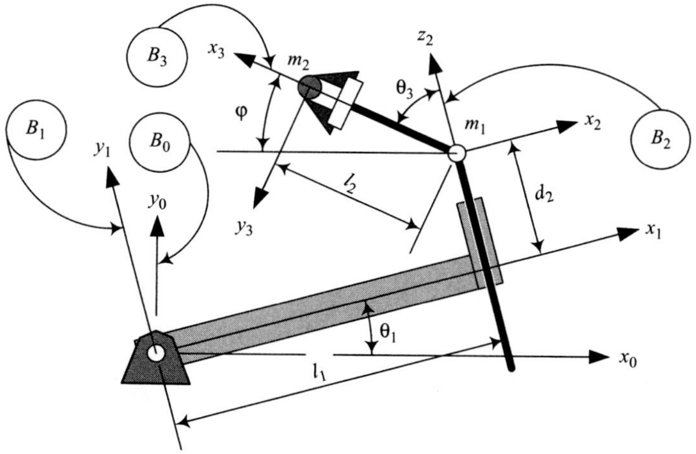

Figure 13.38 A planar multi-DOF manipulator.

46. ★ Articulated Manipulator Figure 13.39 illustrates an articulated manipulator with massive links and a massive load at the tip point. Points C_i, $i = 1, 2, 3$, indicate the mass centers of the links with masses m_i, $i = 1, 2, 3$. The tip point has a mass m_0. A top view of the manipulator is shown in Figure 13.40. Derive the equations of motion of the manipulator. *Hint*: Link (1) of the manipulator is an R⊢R(90) with an extra displacement l_1 along z_1. To determine the transformation matrix ${}^0R_1$, we can begin from a coincident configuration of B_1 and B_0 and move B_1 to its current configuration by a sequence of proper rotations and displacements. The second and third links are respectively R‖R(0) and R⊢R(90).

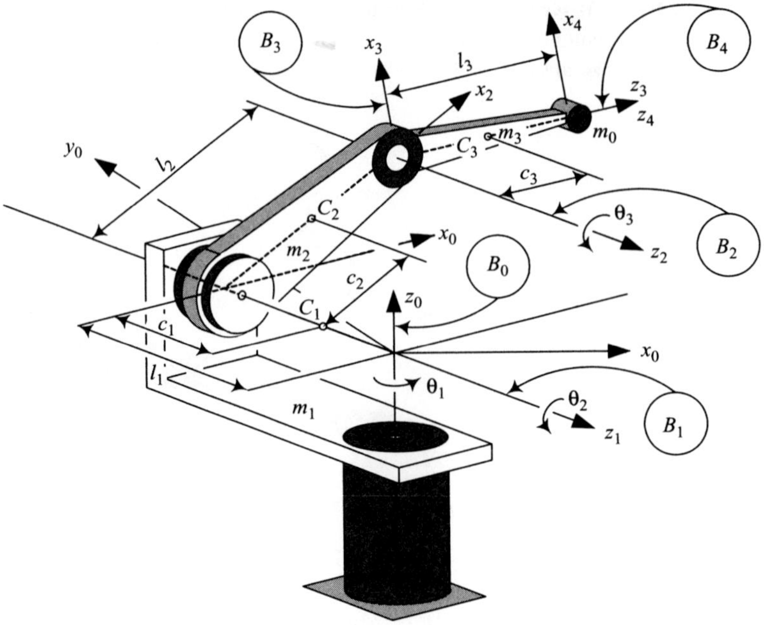

Figure 13.39 An articulated manipulator with massive links and a massive load at the tip point.

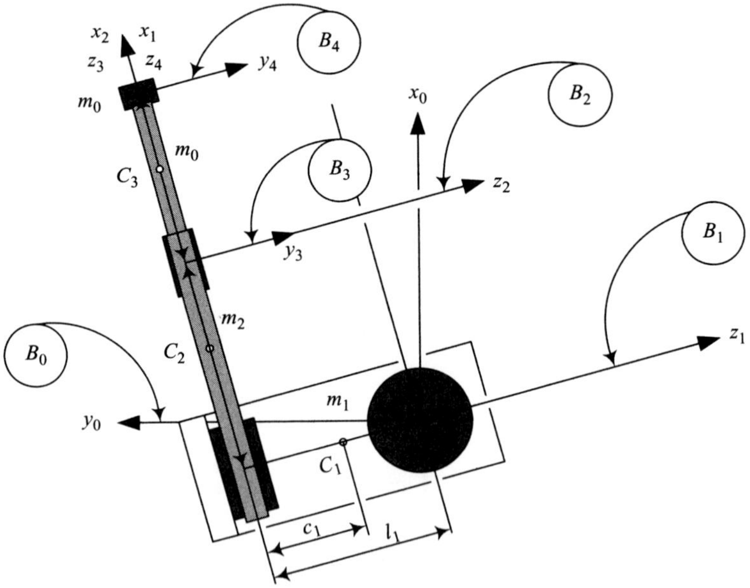

Figure 13.40 A top view of an articulated manipulator with massive links and a massive load at the tip point.

47. **A Hanging and Rotating Bar** Figure 13.41 illustrates a hanging uniform bar that is pivoted to a rotating vertical axis. Determine the differential equation to determine β for a constant ω.

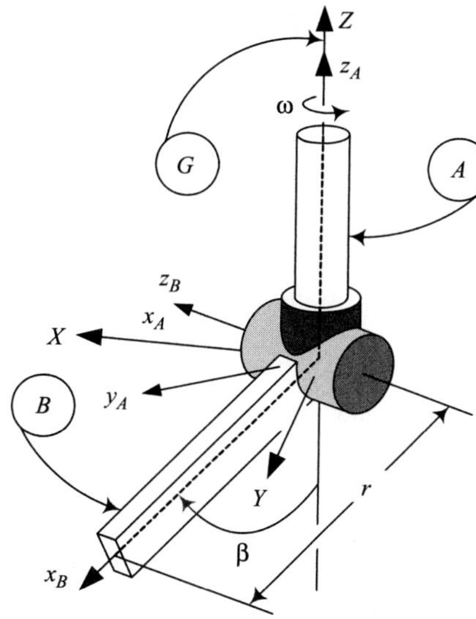

Figure 13.41 A hanging and rotating bar.

48. **Special Cases of 2R Planar Manipulator** Figure 13.17 illustrates a general 2R manipulator with massive arms and joints and a carrying payload m_0. The second motor has a mass m_{12} and is mounted on link (1). The masses of the first and second links are m_{11} and m_{21}, respectively, and their mass centers are at C_1 and C_2. The general equations of motion for the 2R planar manipulator are given in Equations (13.606)–(13.615):

$$\begin{bmatrix} D_{11} & D_{12} \\ D_{21} & D_{22} \end{bmatrix} \begin{bmatrix} \ddot{\theta}_1 \\ \ddot{\theta}_2 \end{bmatrix} + \begin{bmatrix} C_{11} & C_{12} \\ C_{21} & C_{22} \end{bmatrix} \begin{bmatrix} \dot{\theta}_1 \\ \dot{\theta}_2 \end{bmatrix} + \begin{bmatrix} G_1 \\ G_2 \end{bmatrix} = \begin{bmatrix} Q_o \\ Q_1 \end{bmatrix} \tag{13.617}$$

In modeling a special 2R planar manipulator, we may use the equations for simpler models. Simplify the equations and derive the dynamic equations of motion for (a) massless arms and (b) massless joints.

References

Chapter 1

Arnold, V. I., 1978, *Mathematical Methods of Classical Mechanics*, Springer, New York.

Bell, R. J. T., 1963, *An Elementary Treatise on Coordinate Geometry of Three Dimensions*, Macmillan, New York.

Borisenko, A. I., and Tarapov, I. E., 1968, *Vector and Tensor Analysis with Applications* (translated by Silverman, R. A.), Dover, New York.

Coxeter, H. S. M., 1989, *Introduction to Geometry*, Wiley, New York.

Courant, R., and Robbins, H., 1941, *What Is Mathematics?* Oxford University Press, London.

Dugas, R., 1995, *A History of Mechanics*, Switzerland, Editions du Griffon (English translation), Central Book, New York.

Efimov, A. V., Zolotarev, Y. G., and Terpigoreva, V. M., 1985, *Mathematical Analysis: Advanced Topics*, Mir, Moscow.

Hauser, W., 1965, *Introduction to the Principles of Mechanics*, Addison-Wesley, Reading, MA.

Hestenes, D., 1999, *New Foundations for Classical Mechanics*, 2nd ed., Kluwer Academic, New York.

Jazar, R. N., 2010, *Theory of Applied Robotics: Kinematics, Dynamics, and Control*, 2nd ed., Springer, New York.

Kane, T. R., and Levinson, D. A., 1985, *Dynamics: Theory and Applications*, McGraw-Hill, New York.

Milne, E. A., 1948, *Vectorial Mechanics*, Methuen, London.

O'Neill, B., 1966, *Elementary Differential Geometry*, Academic, New York.

Pars, L. A., 1968, *A Treatise on Analytical Dynamics*, Wiley, New York.

Riley, K. F., Hobson, M. P., and Bence, S. J., 2006, *Mathematical Methods for Physics and Engineering*, 3rd ed., Cambridge University Press, New York.

Sharipuv, R. A., 1996, *Course of Differential Geometry*, Bashkir State University Press, Ufa, Russia.

Shaw, J. B., 1922, *Vector Calculus: With Applications to Physics*, Van Nostrand, New York.

Slater, J. C., and Frank, N. H., 1947, *Mechanics*, Mc-Graw-Hill, New York.

Snyder, V., and Sisam, C. H., 1914, *Analytic Geometry of Space*, Henry Holt, New York.

Chapter 2

Arnold, V. I., 1978, *Mathematical Methods of Classical Mechanics*, Springer, New York.

Battin, R. H., 1987, *An Introduction to the Mathematics and Methods of Astrodynamics*, American Institute of Aeronautics and Astronautics, New York.

Boccaletti, D., and Pucacco, G., 1996, *Theory of Orbits*, Vol. 1, Springer, Berlin.

Broucke, R., 1979, On the Isosceles Triangle Configuration in the General Planar Three-Body Problem, *Astronomy and Astrophysics*, *73*, 303–313.

Byerly, W. E., 1916, *An Introduction to the Use of Generalized Coordinates in Mechanics and Physics*, Ginn and Compony, Boston, MA.

Cayley, A., 1895, *Elliptic Functions*, George Bell and Sons, London.

Diacu, F., and Holmes, P., 1996, *Celestial Encounters*, Princeton University Press, Princeton, NJ.

Newton, I., 1687, *Philosophiae Naturalis Principia Mathematica* (*The Mathematical Principles of Natural Phylosiphy*), Royal Society of London, London.

Poincaré, H., 1892–1899, *Les Methodes Nouvelles da la Mecanique Celeste*, Gauthier-Villars, Paris. (English translation, Goroff, D. L., 1993, *New Methods of Celestial Mechanics*, American Institute of Physics, New York.)

Rimrott, F. P. J., 1989, *Introductory Attitude Dynamics*, Springer, New York.

Rosenberg, R., M. 1977, *Analytical Dynamics of Discrete Systems*, Plenum, New York.

Schaub, H., and Junkins, J. L., 2003, *Analytical Mechanics of Space Systems*, AIAA Educational Series, American Institute of Aeronautics and Astronautics, Reston, VA.

Sokolnikoff, I. S., 1951, *Tensor Analysis: Theory and Applications*, Wiley, New York.

Suh, C. H., and Radcliff, C. W., 1978, *Kinematics and Mechanisms Design*, Wiley, New York.

Synder, V., and Sisam, C. H., 1914, *Analytic Geormetry of Space*, Henry Holt, New York.

Szebehely, V., 1967, *Theory of Orbits*; *The Restricted Problem of Three Bodies*, Academic, New York.

Szebehely, V., 1998, *Adventures in Celestial Mechanics*, Wiley, New York.

Whittaker, E. T., 1965, *A Treatise on the Analytical Dynamics of Particles and Rigid Bodies*, Cambridge University Press, London.

Wintner, A., 1941, *The Analytical Foundations of Celestial Mechanics*, Princeton University Press, Princeton, NJ.

Chapter 3

Borisenko, A. I., and Tarapov, I. E., 1968, *Vector and Tensor Analysis with Applications*. (English translation by Silverman, R. A.), Dover, New York.

Buss, S. R., 2003, *3-D Computer Graphics: A Mathematical Introduction with OpenGL*, Cambridge University Press, New York.

Byerly, W. E., 1916, *An Introduction to the Use of Generalized Coordinates in Mechanics and Physics*, Ginn and Company, Boston, MA.

Coxeter, H. S. M., 1989, *Introduction to Geometry*, Wiley, New York.

Lebedov, L. P., and Cloud, M. J., 2003, *Tensor Analysis*, World Scientific, Singapore.

Nikravesh, P., 1988, *Computer-Aided Analysis of Mechanical Systems*, Prentice-Hall, Englewood Cliffs, NJ.

O'Nill, B., 1966, *Elementary Differential Geometry*, Academic, New York.

Rimrott, F. P. J., 1989, *Introductory Attitude Dynamics*, Springer, New York.

Schaub, H., and Junkins, J. L., 2003, *Analytical Mechanics of Space Systems*, AIAA Educational Series, American Institute of Aeronautics and Astronautics, Reston, VA.

Sokolnikoff, I. S., 1951, *Tensor Analysis: Theory and Applications*, Wiley, New York.

Suh, C. H., and Radcliff, C. W., 1978, *Kinematics and Mechanisms Design*, Wiley, New York.

Synder, V., and Sisam, C. H., 1914, *Analytic Geormetry of Space*, Henry Holt, New York.

Chapter 4

Bottema, O., and Roth, B., 1979, *Theoretical Kinematics*, North-Holland, Amsterdam, The Netherlands.

Buss, S. R., 2003, *3-D Computer Graphics: A Mathematical Introduction with OpenGL*, Cambridge University Press, New York.

Cheng, H., and Gupta, K. C., 1989, A Historical Note on Finite Rotations, *Journal of Applied Mechanics*, *56*, 139–145.

Chernousko, F. L., Bolotnik, N. N., and Gradetsky, V. G., 1994, *Manipulation Robots: Dynamics, Control, and Optimization*, CRC Press, Boca Raton, FL.

Coe, C. J., 1934, Displacement of a Rigid Body, *American Mathematical Monthly*, *41*(4), 242–253.

Hunt, K. H., 1978, *Kinematic Geometry of Mechanisms*, Oxford University Press, London.

Jazar, R. N., 2010, *Theory of Applied Robotics: Kinematics, Dynamics, and Control*, 2nd ed., Springer, New York.

Mason, M. T., 2001, *Mechanics of Robotic Manipulation*, MIT Press, Cambridge, MA.

Murray, R. M., Li, Z., and Sastry, S. S. S., 1994, *A Mathematical Introduction to Robotic Manipulation*, CRC Press, Boca Raton, FL.

Paul, B., 1963, On the Composition of Finite Rotations, *American Mathematical Monthly*, *70*(8), 859–862.

Paul, R. P., 1981, *Robot Manipulators: Mathematics, Programming, and Control*, MIT Press, Cambridge, MA.

Schaub, H., and Junkins, J. L., 2003, *Analytical Mechanics of Space Systems*, AIAA Educational Series, American Institute of Aeronautics and Astronautics, Reston, VA.

Selig, J. M., 2005, *Geometric Fundamentals of Robotics*, 2nd ed., Springer, New York.

Spong, M. W., Hutchinson, S., and Vidyasagar, M., 2006, *Robot Modeling and Control*, Wiley, New York.

Suh, C. H., and Radcliff, C. W., 1978, *Kinematics and Mechanisms Design*, Wiley, New York.

Tsai, L. W., 1999, *Robot Analysis*, Wiley, New York.

Chapter 5

Asada, H., and Slotine, J. J. E., 1986, *Robot Analysis and Control*, Wiley, New York.

Ball, R. S., 1900, *A Treatise on the Theory of Screws*, Cambridge University Press, Cambridge, MA.

Bernhardt, R., and Albright, S. L., 2001, *Robot Calibration*, Springer, New York.

Bottema, O., and Roth, B., 1979, *Theoretical Kinematics*, North-Holland, Amsterdam, The Netherlands.

Davidson, J. K., and Hunt, K. H., 2004, *Robots and Screw Theory: Applications of Kinematics and Statics to Robotics*, Oxford University Press, New York.

Hunt, K. H., 1978, *Kinematic Geometry of Mechanisms*, Oxford University Press, London.

Jazar, R. N., 2010, *Theory of Applied Robotics: Kinematics, Dynamics, and Control*, 2nd ed., Springer, New York.

Mason, M. T., 2001, *Mechanics of Robotic Manipulation*, MIT Press, Cambridge, MA.

Nikravesh, P., 1988, *Computer-Aided Analysis of Mechanical Systems*, Prentice-Hall, Englewood Cliffs, NJ.

Paul, R. P., 1981, *Robot Manipulators: Mathematics, Programming, and Control*, MIT Press, Cambridge, MA.

Plücker, J., 1866, Fundamental Views Regarding Mechanics, *Philosophical Transactions*, *156*, 361–380.

Schroer, K., Albright, S. L., and Grethlein, M., 1997, Complete, Minimal and Model-Continuous Kinematic Models for Robot Calibration, *Robotics and Computer-Integrated Manufacturing*, *13*(1), 73–85.

Spong, M. W., Hutchinson, S., and Vidyasagar, M., 2006, *Robot Modeling and Control*, Wiley, New York.

Suh, C. H., and Radcliff, C. W., 1978, *Kinematics and Mechanisms Design*, Wiley, New York.

Tsai, L. W., 1999, *Robot Analysis*, Wiley, New York.

Wang, K., and Lien, T., 1988, Structure, Design & Kinematics of Robot Manipulators, *Robotica*, *6*, 299–306.

Wittenburg, J., and Lilov, L., 2003, Decomposition of a Finite Rotation into Three Rotations About Given Axes, *Multibody System Dynamics*, *9*, 353–375.

Zhuang, H., Roth, Z. S., and Hamano, F., 1992, A Complete, Minimal and Model-Continuous Kinematic Model for Robot Manipulators, *IEEE Transactions on Robotics and Automation*, *8*(4), 451–463.

Chapter 6

Asada, H., and Slotine, J. J. E., 1986, *Robot Analysis and Control*, Wiley, New York.

Jazar, R. N., 2010, *Theory of Applied Robotics: Kinematics, Dynamics, and Control*, 2nd ed., Springer, New York.

Kane, T. R., and Levinson, D. A., 1980, *Dynamics: Theory and Applications*, McGraw-Hill, New York.

Mason, M. T., 2001, *Mechanics of Robotic Manipulation*, MIT Press, Cambridge, MA.

Paul, R. P., 1981, *Robot Manipulators: Mathematics, Programming, and Control*, MIT Press, Cambridge, MA.

Plücker, J., 1866, Fundamental Views Regarding Mechanics, *Philosophical Transactions*, *156*, 361–380.

Spong, M. W., Hutchinson, S., and Vidyasagar, M., 2006, *Robot Modeling and Control*, Wiley, New York.

Suh, C. H., and Radcliff, C. W., 1978, *Kinematics and Mechanisms Design*, Wiley, New York.

Tsai, L. W., 1999, *Robot Analysis*, Wiley, New York.

Wang, K., and Lien, T., 1988, Structure, Design and Kinematics of Robot Manipulators, *Robotica*, *6*, 299–306.

Chapter 7

Bottema, O., and Roth, B., 1979, *Theoretical Kinematics*, North-Holland, Amsterdam, The Netherlands.

Denavit, J., and Hartenberg, R. S., 1955, A Kinematic Notation for Lower-Pair Mechanisms Based on Matrices, *Journal of Applied Mechanics*, *22*(2), 215–221.

Geradin, M., and Cardonna, A., 1989, Kinematics and Dynamics of Rigid and Flexible Mechanisms Using Finite Elements and Quaternion Algebra, *Computational Mechanics*, *4*, 115–135.

Hunt, K. H., 1978, *Kinematic Geometry of Mechanisms*, Oxford University Press, London.

Jazar, R. N., 2008, *Vehicle Dynamics: Theory and Application*, Springer, New York.

Jazar, R. N., 2010, *Theory of Applied Robotics: Kinematics, Dynamics, and Control*, 2nd ed., Springer, New York.

Mason, M. T., 2001, *Mechanics of Robotic Manipulation*, MIT Press, Cambridge, MA.

Nikravesh, P., 1988, *Computer-Aided Analysis of Mechanical Systems*, Prentice-Hall, Englewood Cliffs, NJ.

Schaub, H., and Junkins, J. L., 2003, *Analytical Mechanics of Space Systems*, AIAA Educational Series, American Institute of Aeronautics and Astronautics, Reston, VA.

Spong, M. W., Hutchinson, S., and Vidyasagar, M., 2006, *Robot Modeling and Control*, Wiley, New York.

Suh, C. H., and Radcliff, C. W., 1978, *Kinematics and Mechanisms Design*, Wiley, New York.

Tsai, L. W., 1999, *Robot Analysis*, Wiley, New York.

Chapter 8

Jazar, R. N., 2010, *Theory of Applied Robotics: Kinematics, Dynamics, and Control*, 2nd ed., Springer, New York.

Kane, T. R., and Levinson, D. A., 1980, *Dynamics: Theory and Applications*, McGraw-Hill, New York.

Kane, T. R., Likins, P. W., and Levinson, D. A., 1983, *Spacecraft Dynamics*, McGraw-Hill, New York.

Mason, M. T., 2001, *Mechanics of Robotic Manipulation*, MIT Press, Cambridge, MA.

Rimrott, F. P. J., 1989, *Introductory Attitude Dynamics*, Springer-Verlag, New York.

Schilling, R. J., 1990, *Fundamentals of Robotics: Analysis and Control*, Prentice-Hall, Englewood Cliffs, NJ.

Spong, M. W., Hutchinson, S., and Vidyasagar, M., 2006, *Robot Modeling and Control*, Wiley, New York.

Suh, C. H., and Radcliff, C. W., 1978, *Kinematics and Mechanisms Design*, Wiley, New York.

Talman, R., 2000, *Geometric Mechanics*, Wiley, New York.

Tsai, L. W., 1999, *Robot Analysis*, Wiley, New York.

Chapter 9

Carnahan, B., Luther, H. A., and Wilkes, J. O., 1969, *Applied Numerical Methods*, Wiley, New York.

Eich-Soellner, E., and Führer, C., 1998, *Numerical Methods in Multibody Dynamics*, B.G. Teubner, Stuttgart.

Gerald, C. F., and Wheatley, P. O., 1999, *Applied Numerical Analysis*, 6th ed., Addison-Wesley, New York.

Jazar, R. N., 2010, *Theory of Applied Robotics: Kinematics, Dynamics, and Control*, 2nd ed., Springer, New York.

Kane, T. R., and Levinson, D. A., 1980, *Dynamics: Theory and Applications*, McGraw-Hill, New York.

Nikravesh, P., 1988, *Computer-Aided Analysis of Mechanical Systems*, Prentice-Hall, Englewood Cliffs, NJ.

Suh, C. H., and Radcliff, C. W., 1978, *Kinematics and Mechanisms Design*, Wiley, New York.

Chapter 10

Birkoff, G. D., 1927, *Dynamical Systems*, American Mathematical Society, New York.

Ince, E. L., 1926, Longmans, *Ordinary Differential Equations*, Green and Co., London.

Kansner, A., 1913, *Differential-Geometric Aspects of Dynamics*, American Mathematical Society, New York.

Mason, M. T., 2001, *Mechanics of Robotic Manipulation*, MIT Press, Cambridge, MA.

Nikravesh, P., 1988, *Computer-Aided Analysis of Mechanical Systems*, Prentice-Hall, Englewood Cliffs, NJ.

Pars, L. A., 1965, *Analytical Dynamis*, Wiley, New York.

Rimrott, F. P. J., 1989, *Introductory Attitude Dynamics*, Springer-Verlag, New York.

Rosenberg, R., M. 1977, *Analytical Dynamics of Discrete Systems*, Plenum, New York.

Spong, M. W., Hutchinson, S., and Vidyasagar, M., 2006, *Robot Modeling and Control*, Wiley, New York.

Suh, C. H., and Radcliff, C. W., 1978, *Kinematics and Mechanisms Design*, Wiley, New York.

Szebehely, Victor G., 1967, *Theory of Orbits*, Academic, New York.

Tsai, L. W., 1999, *Robot Analysis*, Wiley, New York.

Chapter 11

Appell, P. E., 1909, *Traté de Mécanique Rationelle*, Vols. I and II, Gauthier-Villars, Paris.

Goldstein, H., Poole, C., and Safko, J., 2002, *Classical Mechanics*, 3rd ed., Addison-Wesley, New York.

MacMillan, W. D., 1936, *Dynamics of Rigid Bodies*, McGraw-Hill, New York.

Meirovitch, L., 1970, *Methods of Analytical Dynamics*, McGraw-Hill, New York.

Nikravesh, P., 1988, *Computer-Aided Analysis of Mechanical Systems*, Prentice-Hall, Englewood Cliffs, NJ.

Pars, L. A., 1965, *Analytical Dynamis*, Wiley, New York.

Rosenberg, R. M., 1977, *Analytical Dynamics of Discrete Systems*, Plenum, New York.

Schaub, H., and Junkins, J. L., 2003, *Analytical Mechanics of Space Systems*, AIAA Educational Series, American Institute of Aeronautics and Astronautics, Reston, VA.

Synge, J. L., and Griffith, B. A., 1942, *Principle of Mechanics*, McGraw-Hill, New York.

Thomson, W. T., 1961, *Introduction to Space Dynamics*, Wiley, New York.

Wittacker, E. T., 1947, *A Treatise on the Analytical Dynamics of Particles and Rigid Bodies*, 4th ed., Cambridge University Press, New York.

Chapter 12

Appell, P. E., 1909, *Traté de Mécanique Rationelle*, Vols I. and II, Gauthier-Villars, Paris.

Brady, M., Hollerbach, J. M., Johnson, T. L., Lozano-Prez, T., and Mason, M. T., 1983, *Robot Motion: Planning and Control*, MIT Press, Cambridge, MA.

Cayley, A., 1895, *Elliptic Functions*, George Bell and Sons, London.

Goldstein, H., Poole, C., and Safko, J., 2002, *Classical Mechanics*, 3rd ed., Addison-Wesley, New York.

Jazar, R. N., 2008, *Vehicle Dynamics: Theory and Application*, Springer, New York.

Jazar, R. N., 2010, *Theory of Applied Robotics: Kinematics, Dynamics, and Control*, 2nd ed., Springer, New York.

Loudon, W. J., 1896, *An Elementary Treatise on Rigid Dynamics*, Macmillan, London.

MacMillan, W. D., 1936, *Dynamics of Rigid Bodies*, McGraw-Hill, New York.

Meirovitch, L., 1970, *Methods of Analytical Dynamics*, McGraw-Hill, New York.

Murray, R. M., Li, Z., and Sastry, S. S. S., 1994, *A Mathematical Introduction to Robotic Manipulation*, CRC Press, Boca Raton, FL.

Nikravesh, P., 1988, *Computer-Aided Analysis of Mechanical Systems*, Prentice-Hall, Englewood Cliffs, NJ.

Pars, L. A., 1965, *Analytical Dynamis*, Wiley, New York.

Paul, R. P., 1981, *Robot Manipulators: Mathematics, Programming, and Control*, MIT Press, Cambridge, MA.

Suh, C. H., and Radcliff, C. W., 1978, *Kinematics and Mechanisms Design*, Wiley, New York.

Synge, J. L., and Griffith, B. A., 1942, *Principle of Mechanics*, McGraw-Hil, New York.

Thomson, W. T., 1961, *Introduction to Space Dynamics*, Wiley, New York.

Tsai, L. W., 1999, *Robot Analysis*, Wiley, New York.

Wittacker, E. T., 1947, *A Treatise on the Analytical Dynamics of Particles and Rigid Bodies*, 4th ed., Cambridge University Press, New York.

Chapter 13

Bricard R., 1927, Leçons de cinématique, t. II: Cinématique Appliquée, Gauthier-Villars, Paris.

Euler L., 1758, Du mouvement de rotation des corps solides autour d'un axe variable. In: Füssli, O. (Ed.) (1964), Leonhard Euler. Opera Omnia: Commentationes Mechanicae ad Theoriam Corporum Rigidorum Pertinentes, pp. 200–236.

Gamkrelidze, R. V., 1958, The Theory of Time Optimal Processes in Linear Systems, *Izvestiya Akademii Nauk SSSR*, *22*, 449–474.

Jazar, R. N., and Naghshinehpour A., 2005, Time Optimal Control Algorithm for Multi-Body Dynamical Systems, *IMechE Part K: Journal of Multi-Body Dynamics*, *219*(3), 225–236.

Krotov, V. F., 1996, *Global Methods in Optimal Control Theory*, Marcel Decker, New York.

Kuo, B. C., and Golnaraghi, F., 2003, *Automatic Control Systems*, Wiley, New York.

Lewis, F. L., and Syrmos, V. L., 1995, *Optimal Control*, Wiley, New York.

Pinch E. R., 1993, *Optimal Control and the Calculus of Variations*, Oxford University Press, New York.

Vincent T. L., and Grantham W. J., 1997, *Nonlinear and Optimal Control Systems*, Wiley, New York.

Vujanovic, B. D., and Atanackovic, T. M., 2004, *An Introduction to Modern Variational Techniques in Mechanics and Engineering*, Birkhäuser, Boston.

A

Global Frame Triple Rotation

In this appendix, the 12 combinations of triple rotation about global fixed axes are presented:

$$Q_{X,\gamma}Q_{Y,\beta}Q_{Z,\alpha} = \begin{bmatrix} c\alpha c\beta & -c\beta s\alpha & s\beta \\ c\gamma s\alpha + c\alpha s\beta s\gamma & c\alpha c\gamma - s\alpha s\beta s\gamma & -c\beta s\gamma \\ s\alpha s\gamma - c\alpha c\gamma s\beta & c\alpha s\gamma + c\gamma s\alpha s\beta & c\beta c\gamma \end{bmatrix} \tag{A.1}$$

$$Q_{Y,\gamma}Q_{Z,\beta}Q_{X,\alpha} = \begin{bmatrix} c\beta c\gamma & s\alpha s\gamma - c\alpha c\gamma s\beta & c\alpha s\gamma + c\gamma s\alpha s\beta \\ s\beta & c\alpha c\beta & -c\beta s\alpha \\ -c\beta s\gamma & c\gamma s\alpha + c\alpha s\beta s\gamma & c\alpha c\gamma - s\alpha s\beta s\gamma \end{bmatrix} \tag{A.2}$$

$$Q_{Z,\gamma}Q_{X,\beta}Q_{Y,\alpha} = \begin{bmatrix} c\alpha c\gamma - s\alpha s\beta s\gamma & -c\beta s\gamma & c\gamma s\alpha + c\alpha s\beta s\gamma \\ c\alpha s\gamma + c\gamma s\alpha s\beta & c\beta c\gamma & s\alpha s\gamma - c\alpha c\gamma s\beta \\ -c\beta s\alpha & s\beta & c\alpha c\beta \end{bmatrix} \tag{A.3}$$

$$Q_{Z,\gamma}Q_{Y,\beta}Q_{X,\alpha} = \begin{bmatrix} c\beta c\gamma & -c\alpha s\gamma + c\gamma s\alpha s\beta & s\alpha s\gamma + c\alpha c\gamma s\beta \\ c\beta s\gamma & c\alpha c\gamma + s\alpha s\beta s\gamma & -c\gamma s\alpha + c\alpha s\beta s\gamma \\ -s\beta & c\beta s\alpha & c\alpha c\beta \end{bmatrix} \tag{A.4}$$

$$Q_{Y,\gamma}Q_{X,\beta}Q_{Z,\alpha} = \begin{bmatrix} c\alpha c\gamma + s\alpha s\beta s\gamma & -c\gamma s\alpha + c\alpha s\beta s\gamma & c\beta s\gamma \\ c\beta s\alpha & c\alpha c\beta & -s\beta \\ -c\alpha s\gamma + c\gamma s\alpha s\beta & s\alpha s\gamma + c\alpha c\gamma s\beta & c\beta c\gamma \end{bmatrix} \tag{A.5}$$

$$Q_{X,\gamma}Q_{Z,\beta}Q_{Y,\alpha} = \begin{bmatrix} c\alpha c\beta & -s\beta & c\beta s\alpha \\ s\alpha s\gamma + c\alpha c\gamma s\beta & c\beta c\gamma & -c\alpha s\gamma + c\gamma s\alpha s\beta \\ -c\gamma s\alpha + c\alpha s\beta s\gamma & c\beta s\gamma & c\alpha c\gamma + s\alpha s\beta s\gamma \end{bmatrix} \tag{A.6}$$

$$Q_{X,\gamma}Q_{Y,\beta}Q_{X,\alpha} = \begin{bmatrix} c\beta & s\alpha s\beta & c\alpha s\beta \\ s\beta s\gamma & c\alpha c\gamma - c\beta s\alpha s\gamma & -c\gamma s\alpha - c\alpha c\beta s\gamma \\ -c\gamma s\beta & c\alpha s\gamma + c\beta c\gamma s\alpha & -s\alpha s\gamma + c\alpha c\beta c\gamma \end{bmatrix} \tag{A.7}$$

$$Q_{Y,\gamma}Q_{Z,\beta}Q_{Y,\alpha} = \begin{bmatrix} -s\alpha s\gamma + c\alpha c\beta c\gamma & -c\gamma s\beta & c\alpha s\gamma + c\beta c\gamma s\alpha \\ c\alpha s\beta & c\beta & s\alpha s\beta \\ -c\gamma s\alpha - c\alpha c\beta s\gamma & s\beta s\gamma & c\alpha c\gamma - c\beta s\alpha s\gamma \end{bmatrix} \tag{A.8}$$

$$Q_{Z,\gamma}Q_{X,\beta}Q_{Z,\alpha}=\begin{bmatrix} c\alpha c\gamma - c\beta s\alpha s\gamma & -c\gamma s\alpha - c\alpha c\beta s\gamma & s\beta s\gamma \\ c\alpha s\gamma + c\beta c\gamma s\alpha & -s\alpha s\gamma + c\alpha c\beta c\gamma & -c\gamma s\beta \\ s\alpha s\beta & c\alpha s\beta & c\beta \end{bmatrix} \tag{A.9}$$

$$Q_{X,\gamma}Q_{Z,\beta}Q_{X,\alpha}=\begin{bmatrix} c\beta & -c\alpha s\beta & s\alpha s\beta \\ c\gamma s\beta & -s\alpha s\gamma + c\alpha c\beta c\gamma & -c\alpha s\gamma - c\beta c\gamma s\alpha \\ s\beta s\gamma & c\gamma s\alpha + c\alpha c\beta s\gamma & c\alpha c\gamma - c\beta s\alpha s\gamma \end{bmatrix} \tag{A.10}$$

$$Q_{Y,\gamma}Q_{X,\beta}Q_{Y,\alpha}=\begin{bmatrix} c\alpha c\gamma - c\beta s\alpha s\gamma & s\beta s\gamma & c\gamma s\alpha + c\alpha c\beta s\gamma \\ s\alpha s\beta & c\beta & -c\alpha s\beta \\ -c\alpha s\gamma - c\beta c\gamma s\alpha & c\gamma s\beta & -s\alpha s\gamma + c\alpha c\beta c\gamma \end{bmatrix} \tag{A.11}$$

$$Q_{Z,\gamma}Q_{Y,\beta}Q_{Z,\alpha}=\begin{bmatrix} -s\alpha s\gamma + c\alpha c\beta c\gamma & -c\alpha s\gamma - c\beta c\gamma s\alpha & c\gamma s\beta \\ c\gamma s\alpha + c\alpha c\beta s\gamma & c\alpha c\gamma - c\beta s\alpha s\gamma & s\beta s\gamma \\ -c\alpha s\beta & s\alpha s\beta & c\beta \end{bmatrix} \tag{A.12}$$

B

Local Frame Triple Rotation

In this appendix, the 12 combinations of triple rotation about local axes are presented:

$$A_{x,\psi}A_{y,\theta}A_{z,\varphi} = \begin{bmatrix} c\theta c\varphi & c\theta s\varphi & -s\theta \\ -c\psi s\varphi + c\varphi s\theta s\psi & c\varphi c\psi + s\theta s\varphi s\psi & c\theta s\psi \\ s\varphi s\psi + c\varphi s\theta c\psi & -c\varphi s\psi + s\theta c\psi s\varphi & c\theta c\psi \end{bmatrix} \tag{B.1}$$

$$A_{y,\psi}A_{z,\theta}A_{x,\varphi} = \begin{bmatrix} c\theta c\psi & s\varphi s\psi + c\varphi s\theta c\psi & -c\varphi s\psi + s\theta c\psi s\varphi \\ -s\theta & c\theta c\varphi & c\theta s\varphi \\ c\theta s\psi & -c\psi s\varphi + c\varphi s\theta s\psi & c\varphi c\psi + s\theta s\varphi s\psi \end{bmatrix} \tag{B.2}$$

$$A_{z,\psi}A_{x,\theta}A_{y,\varphi} = \begin{bmatrix} c\varphi c\psi + s\theta s\varphi s\psi & c\theta s\psi & -c\psi s\varphi + c\varphi s\theta s\psi \\ -c\varphi s\psi + s\theta c\psi s\varphi & c\theta c\psi & s\varphi s\psi + c\varphi s\theta c\psi \\ c\theta s\varphi & -s\theta & c\theta c\varphi \end{bmatrix} \tag{B.3}$$

$$A_{z,\psi}A_{y,\theta}A_{x,\varphi} = \begin{bmatrix} c\theta c\psi & c\varphi s\psi + s\theta c\psi s\varphi & s\varphi s\psi - c\varphi s\theta c\psi \\ -c\theta s\psi & c\varphi c\psi - s\theta s\varphi s\psi & c\psi s\varphi + c\varphi s\theta s\psi \\ s\theta & -c\theta s\varphi & c\theta c\varphi \end{bmatrix} \tag{B.4}$$

$$A_{y,\psi}A_{x,\theta}A_{z,\varphi} = \begin{bmatrix} c\varphi c\psi - s\theta s\varphi s\psi & c\psi s\varphi + c\varphi s\theta s\psi & -c\theta s\psi \\ -c\theta s\varphi & c\theta c\varphi & s\theta \\ c\varphi s\psi + s\theta c\psi s\varphi & s\varphi s\psi - c\varphi s\theta c\psi & c\theta c\psi \end{bmatrix} \tag{B.5}$$

$$A_{x,\psi}A_{z,\theta}A_{y,\varphi} = \begin{bmatrix} c\theta c\varphi & s\theta & -c\theta s\varphi \\ s\varphi s\psi - c\varphi s\theta c\psi & c\theta c\psi & c\varphi s\psi + s\theta c\psi s\varphi \\ c\psi s\varphi + c\varphi s\theta s\psi & -c\theta s\psi & c\varphi c\psi - s\theta s\varphi s\psi \end{bmatrix} \tag{B.6}$$

$$A_{x,\psi}A_{y,\theta}A_{x,\varphi} = \begin{bmatrix} c\theta & s\theta s\varphi & -c\varphi s\theta \\ s\theta s\psi & c\varphi c\psi - c\theta s\varphi s\psi & c\psi s\varphi + c\theta c\varphi s\psi \\ s\theta c\psi & -c\varphi s\psi - c\theta c\psi s\varphi & -s\varphi s\psi + c\theta c\varphi c\psi \end{bmatrix} \tag{B.7}$$

$$A_{y,\psi}A_{z,\theta}A_{y,\varphi} = \begin{bmatrix} -s\varphi s\psi + c\theta c\varphi c\psi & s\theta c\psi & -c\varphi s\psi - c\theta c\psi s\varphi \\ -c\varphi s\theta & c\theta & s\theta s\varphi \\ c\psi s\varphi + c\theta c\varphi s\psi & s\theta s\psi & c\varphi c\psi - c\theta s\varphi s\psi \end{bmatrix} \tag{B.8}$$

$$A_{z,\psi}A_{x,\theta}A_{z,\varphi} = \begin{bmatrix} c\varphi c\psi - c\theta s\varphi s\psi & c\psi s\varphi + c\theta c\varphi s\psi & s\theta s\psi \\ -c\varphi s\psi - c\theta c\psi s\varphi & -s\varphi s\psi + c\theta c\varphi c\psi & s\theta c\psi \\ s\theta s\varphi & -c\varphi s\theta & c\theta \end{bmatrix} \tag{B.9}$$

$$A_{x,\psi}A_{z,\theta}A_{x,\varphi} = \begin{bmatrix} c\theta & c\varphi s\theta & s\theta s\varphi \\ -s\theta c\psi & -s\varphi s\psi + c\theta c\varphi c\psi & c\varphi s\psi + c\theta c\psi s\varphi \\ s\theta s\psi & -c\psi s\varphi - c\theta c\varphi s\psi & c\varphi c\psi - c\theta s\varphi s\psi \end{bmatrix} \tag{B.10}$$

$$A_{y,\psi}A_{x,\theta}A_{y,\varphi} = \begin{bmatrix} c\varphi c\psi - c\theta s\varphi s\psi & s\theta s\psi & -c\psi s\varphi - c\theta c\varphi s\psi \\ s\theta s\varphi & c\theta & c\varphi s\theta \\ c\varphi s\psi + c\theta c\psi s\varphi & -s\theta c\psi & -s\varphi s\psi + c\theta c\varphi c\psi \end{bmatrix} \tag{B.11}$$

$$A_{z,\psi}A_{y,\theta}A_{z,\varphi} = \begin{bmatrix} -s\varphi s\psi + c\theta c\varphi c\psi & c\varphi s\psi + c\theta c\psi s\varphi & -s\theta c\psi \\ -c\psi s\varphi - c\theta c\varphi s\psi & c\varphi c\psi - c\theta s\varphi s\psi & s\theta s\psi \\ c\varphi s\theta & s\theta s\varphi & c\theta \end{bmatrix} \tag{B.12}$$

C

Principal Central Screw Triple Combination

In this appendix, the six combinations of triple principal central screws are presented:

$$\check{s}(h_X, \gamma, \hat{I})\, \check{s}(h_Y, \beta, \hat{J})\, \check{s}(h_Z, \alpha, \hat{K})$$

$$= \begin{bmatrix} c\alpha c\beta & -c\beta s\alpha & s\beta & \gamma p_X + \alpha p_Z s\beta \\ c\gamma s\alpha + c\alpha s\beta s\gamma & c\alpha c\gamma - s\alpha s\beta s\gamma & -c\beta s\gamma & \beta p_Y c\gamma - \alpha p_Z c\beta s\gamma \\ s\alpha s\gamma - c\alpha c\gamma s\beta & c\alpha s\gamma + c\gamma s\alpha s\beta & c\beta c\gamma & \beta p_Y s\gamma + \alpha p_Z c\beta c\gamma \\ 0 & 0 & 0 & 1 \end{bmatrix} \tag{C.1}$$

$$\check{s}(h_Y, \beta, \hat{J})\, \check{s}(h_Z, \alpha, \hat{K})\, \check{s}(h_X, \gamma, \hat{I})$$

$$= \begin{bmatrix} c\alpha c\beta & s\beta s\gamma - c\beta c\gamma s\alpha & c\gamma s\beta + c\beta s\alpha s\gamma & \alpha p_Z s\beta + \gamma p_X c\alpha c\beta \\ s\alpha & c\alpha c\gamma & -c\alpha s\gamma & \beta p_Y + \gamma p_X s\alpha \\ -c\alpha s\beta & c\beta s\gamma + c\gamma s\alpha s\beta & c\beta c\gamma - s\alpha s\beta s\gamma & \alpha p_Z c\beta - \gamma p_X c\alpha s\beta \\ 0 & 0 & 0 & 1 \end{bmatrix} \tag{C.2}$$

$$\check{s}(h_Z, \alpha, \hat{K})\, \check{s}(h_X, \gamma, \hat{I})\, \check{s}(h_Y, \beta, \hat{J})$$

$$= \begin{bmatrix} c\alpha c\beta - s\alpha s\beta s\gamma & -c\gamma s\alpha & c\alpha s\beta + c\beta s\alpha s\gamma & \gamma p_X c\alpha - \beta p_Y c\gamma s\alpha \\ c\beta s\alpha + c\alpha s\beta s\gamma & c\alpha c\gamma & s\alpha s\beta - c\alpha c\beta s\gamma & \gamma p_X s\alpha + \beta p_Y c\alpha c\gamma \\ -c\gamma s\beta & s\gamma & c\beta c\gamma & \alpha p_Z + \beta p_Y s\gamma \\ 0 & 0 & 0 & 1 \end{bmatrix} \tag{C.3}$$

$$\check{s}(h_Z, \alpha, \hat{K})\, \check{s}(h_Y, \beta, \hat{J})\, \check{s}(h_X, \gamma, \hat{I})$$

$$= \begin{bmatrix} c\alpha c\beta & c\alpha s\beta s\gamma - c\gamma s\alpha & s\alpha s\gamma + c\alpha c\gamma s\beta & \gamma p_X c\alpha c\beta - \beta p_Y s\alpha \\ c\beta s\alpha & c\alpha c\gamma + s\alpha s\beta s\gamma & c\gamma s\alpha s\beta - c\alpha s\gamma & \beta p_Y c\alpha + \gamma p_X c\beta s\alpha \\ -s\beta & c\beta s\gamma & c\beta c\gamma & \alpha p_Z - \gamma p_X s\beta \\ 0 & 0 & 0 & 1 \end{bmatrix} \tag{C.4}$$

$$
\check{s}(h_Y, \beta, \hat{J})\, \check{s}(h_X, \gamma, \hat{I})\, \check{s}(h_Z, \alpha, \hat{K})
= \begin{bmatrix} c\alpha c\beta + s\alpha s\beta s\gamma & c\alpha s\beta s\gamma - c\beta s\alpha & c\gamma s\beta & \gamma p_X c\beta + \alpha p_Z c\gamma s\beta \\ c\gamma s\alpha & c\alpha c\gamma & -s\gamma & \beta p_Y - \alpha p_Z s\gamma \\ c\beta s\alpha s\gamma - c\alpha s\beta & s\alpha s\beta + c\alpha c\beta s\gamma & c\beta c\gamma & \alpha p_Z c\beta c\gamma - \gamma p_X s\beta \\ 0 & 0 & 0 & 1 \end{bmatrix} \tag{C.5}
$$

$$
\check{s}(h_X, \gamma, \hat{I})\, \check{s}(h_Z, \alpha, \hat{K})\, \check{s}(h_Y, \beta, \hat{J})
= \begin{bmatrix} c\alpha c\beta & -s\alpha & c\alpha s\beta & \gamma p_X - \beta p_Y s\alpha \\ s\beta s\gamma + c\beta c\gamma s\alpha & c\alpha c\gamma & c\gamma s\alpha s\beta - c\beta s\gamma & \beta p_Y c\alpha c\gamma - \alpha p_Z s\gamma \\ c\beta s\alpha s\gamma - c\gamma s\beta & c\alpha s\gamma & c\beta c\gamma + s\alpha s\beta s\gamma & \alpha p_Z c\gamma + \beta p_Y c\alpha s\gamma \\ 0 & 0 & 0 & 1 \end{bmatrix} \tag{C.6}
$$

D

Industrial Link DH Matrices

Most industrial multibodies are made by connected links that use one of the follwo-ing joint configurations with the associated Denavit–Hartenberg (DH) transformation matrices:

1	R‖R(0)	or	R‖P(0)
2	R‖R(180)	or	R‖P(180)
3	R⊥R(90)	or	R⊥P(90)
4	R⊥R(−90)	or	R⊥P(−90)
5	R⊢R(90)	or	R⊢P(90)
6	R⊢R(−90)	or	R⊢P(−90)
7	P‖R(0)	or	P‖P(0)
8	P‖R(180)	or	P‖P(180)
9	P⊥R(90)	or	P⊥P(90)
10	P⊥R(−90)	or	P⊥P(−90)
11	P⊢R(90)	or	P⊢P(90)
12	P⊢R(−90)	or	P⊢P(−90)

1,2—LINKS WITH R‖R OR R‖P

$$a_i = \text{const} \qquad \alpha_i = 0 \text{ or } 180\,\text{deg} \qquad d_i = 0 \qquad \theta_i = \text{variable}$$

Figure D.1 illustrates a link R‖R(0), and Figure D.2 illustrates a link R‖R(180). If the proximal joint of link (i) is revolute, the distal joint is either revolute or prismatic, and the joint axes at two ends are parallel, then $\alpha_i = 0$ or $\alpha_i = 180\,\text{deg}$, a_i is the distance between the joint axes, and θ_i is the only variable parameter. The joint distance $d_i = \text{const}$ is the distance between the origin of B_i and B_{i-1} along z_i. We usually set (x_i, y_i) and (x_{i-1}, y_{i-1}) coplanar to get $d_i = 0$. The x_i- and x_{i-1}-axes are parallel for a link R‖R at the rest position. Therefore, the transformation matrix ${}^{i-1}T_i$ for such a link with $\alpha_i = 0$ known as R‖R(0) or R‖P(0) is

$$ {}^{i-1}T_i = \begin{bmatrix} \cos\theta_i & -\sin\theta_i & 0 & a_i\cos\theta_i \\ \sin\theta_i & \cos\theta_i & 0 & a_i\sin\theta_i \\ 0 & 0 & 1 & d_i \\ 0 & 0 & 0 & 1 \end{bmatrix} \tag{D.1}$$

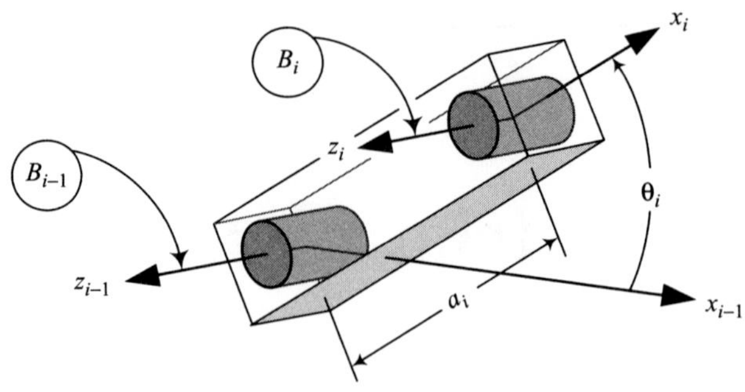

Figure D.1 A link R∥R(0).

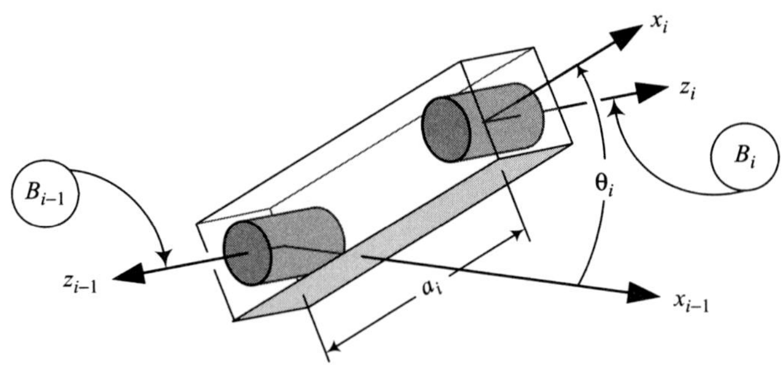

Figure D.2 A link R∥R(180).

while for a link with $\alpha_i = 180\,\text{deg}$ and R∥R(180) or R∥P(180) it is

$$
{}^{i-1}T_i = \begin{bmatrix} \cos\theta_i & \sin\theta_i & 0 & a_i\cos\theta_i \\ \sin\theta_i & -\cos\theta_i & 0 & a_i\sin\theta_i \\ 0 & 0 & -1 & d_i \\ 0 & 0 & 0 & 1 \end{bmatrix} \tag{D.2}
$$

3, 4—LINKS WITH R⊥R OR R⊥P

$$a_i = \text{const} \qquad \alpha_i = 90\,\text{deg or } -90\,\text{deg} \qquad d_i = 0 \qquad \theta_i = \text{variable}$$

Figure D.3 illustrates a link R⊥R(90) and Figure D.4 illustrates a link R⊥R(−90). If the proximal joint of link (i) is revolute, the distal joint is either revolute or prismatic, and the joint axes at two ends are perpendicular, then $\alpha_i = 90\,\text{deg}$ or $\alpha_i = -90\,\text{deg}$, a_i is the distance between the joint axes on x_i, and θ_i is the only variable parameter. The joint distance $d_i = \text{const}$ is the distance between the origin of B_i and B_{i-1} along z_i. We usually set (x_i, y_i) and (x_{i-1}, y_{i-1}) coplanar to get $d_i = 0$.

The R⊥R link is made by twisting the R∥R link 90 deg about its centerline x_{i-1}-axis. The x_i- and x_{i-1}-axes are parallel for a link R⊥R at the rest position. Therefore, the transformation matrix ${}^{i-1}T_i$ for such a link with $\alpha_i = 90\,\text{deg}$ known as R⊥R(90)

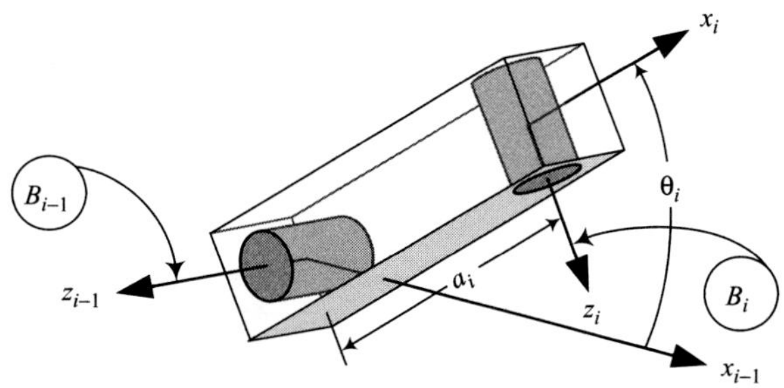

Figure D.3 A R⊥R(90) link.

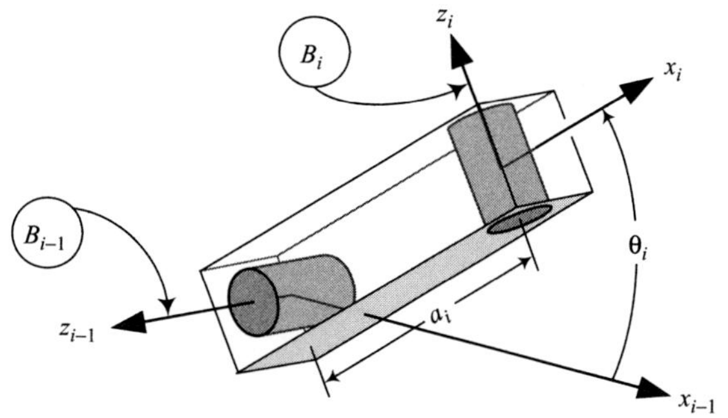

Figure D.4 A R⊥R(−90) link.

or R⊥P(90) is

$$
{}^{i-1}T_i = \begin{bmatrix} \cos\theta_i & 0 & \sin\theta_i & a_i\cos\theta_i \\ \sin\theta_i & 0 & -\cos\theta_i & a_i\sin\theta_i \\ 0 & 1 & 0 & d_i \\ 0 & 0 & 0 & 1 \end{bmatrix} \tag{D.3}
$$

while for a link with $\alpha_i = -90\,\text{deg}$ and R⊥R(−90) or R⊥P(−90) it is

$$
{}^{i-1}T_i = \begin{bmatrix} \cos\theta_i & 0 & -\sin\theta_i & a_i\cos\theta_i \\ \sin\theta_i & 0 & \cos\theta_i & a_i\sin\theta_i \\ 0 & -1 & 0 & d_i \\ 0 & 0 & 0 & 1 \end{bmatrix} \tag{D.4}
$$

5, 6—LINKS WITH R⊢R OR R⊢P

$$
a_i = 0 \qquad \alpha_i = 90\,\text{deg or } -90\,\text{deg} \qquad d_i = 0 \qquad \theta_i = \text{variable}
$$

Figure D.5 illustrates a link R⊢R(90) and Figure D.6 illustrates a link R⊢R(−90). If the proximal joint of link (i) is revolute and the distal joint is either revolute or prismatic and the joint axes at two ends are intersecting orthogonal, then $\alpha_i = 90\,\text{deg}$ or $\alpha_i = -90\,\text{deg}$, $a_i = 0$, $d_i = cte$ is the distance between the coordinate origin on z_i,

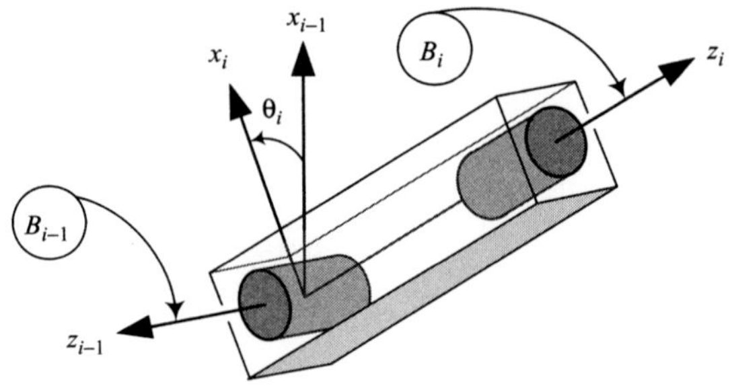

Figure D.5 An R⊢R(90) link.

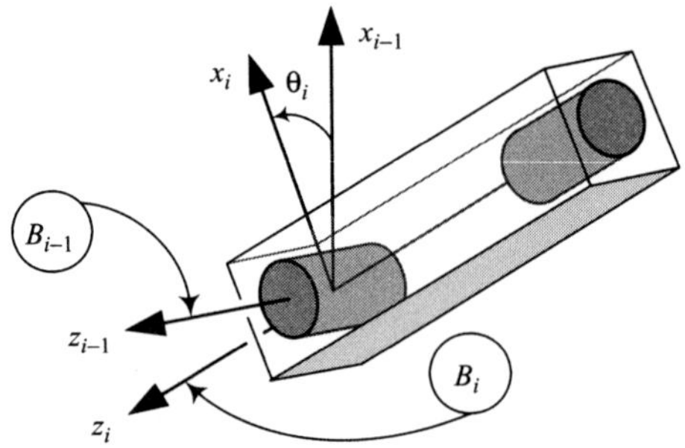

Figure D.6 An R⊢R(−90) link.

and θ_i is the only variable parameter. It is possible to have or assume $d_i = 0$ at the rest position. When $d_i = 0$, the x_i- and x_{i-1}-axes of a link R⊢R are coincident and when $d_i \neq 0$ they are parallel. Therefore, the transformation matrix ${}^{i-1}T_i$ for such a link with $\alpha_i = 90$ deg and R⊢R(90) or R⊢P(90) is

$$
{}^{i-1}T_i = \begin{bmatrix} \cos\theta_i & 0 & \sin\theta_i & 0 \\ \sin\theta_i & 0 & -\cos\theta_i & 0 \\ 0 & 1 & 0 & d_i \\ 0 & 0 & 0 & 1 \end{bmatrix} \tag{D.5}
$$

while for a link with $\alpha_i = -90$ deg and R⊢R(−90) or R⊢P(−90) it is

$$
{}^{i-1}T_i = \begin{bmatrix} \cos\theta_i & 0 & -\sin\theta_i & 0 \\ \sin\theta_i & 0 & \cos\theta_i & 0 \\ 0 & -1 & 0 & d_i \\ 0 & 0 & 0 & 1 \end{bmatrix} \tag{D.6}
$$

7, 8—LINKS WITH P‖R OR P‖P

$$
a_i = \text{const} \qquad \alpha_i = 0 \text{ or } 180\,\text{deg} \qquad d_i = \text{variable} \qquad \theta_i = 0
$$

Figure D.7 illustrates a link P‖R(0) and Figure D.8 illustrates a link P‖R(180). If the proximal joint of link (i) is prismatic, its distal joint is either revolute or prismatic,

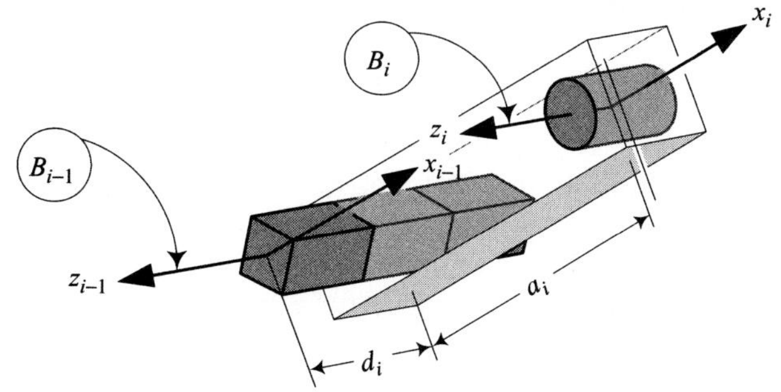

Figure D.7 A P||R(0) link.

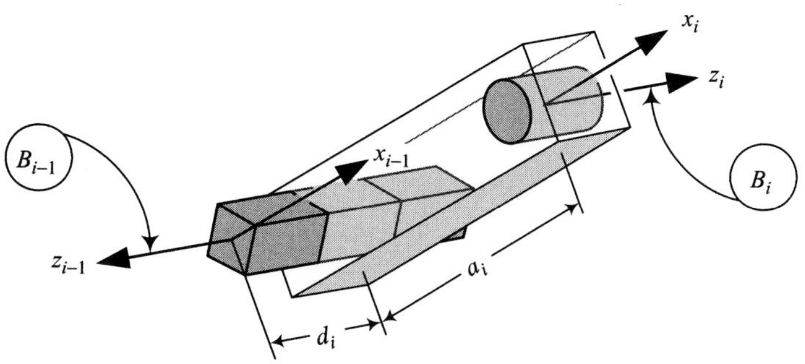

Figure D.8 A P||R(180) link.

and the joint axes at two ends are parallel, then $\alpha_i = 0$ or $\alpha_i = 180\,\text{deg}$, $\theta_i = 0$, $a_i =$ const is the distance between the joint axes on x_i, and d_i is the only variable parameter. It is possible to have $a_i = 0$. The transformation matrix ${}^{i-1}T_i$ for a link with $\alpha_i = 0$ and P||R(0) or P||P(0) is

$$
{}^{i-1}T_i = \begin{bmatrix} 1 & 0 & 0 & a_i \\ 0 & 1 & 0 & 0 \\ 0 & 0 & 1 & d_i \\ 0 & 0 & 0 & 1 \end{bmatrix} \tag{D.7}
$$

while for a link with $\alpha_i = 180\,\text{deg}$ and P||R(180) or P||P(180) it is

$$
{}^{i-1}T_i = \begin{bmatrix} 1 & 0 & 0 & a_i \\ 0 & -1 & 0 & 0 \\ 0 & 0 & -1 & d_i \\ 0 & 0 & 0 & 1 \end{bmatrix} \tag{D.8}
$$

The origin of the B_{i-1}-frame can arbitrarily be chosen at any point on the z_{i-1}-axis or parallel to the z_{i-1}-axis. A simple setup is to locate the origin o_i of a prismatic joint at the previous origin o_{i-1}. This sets $a_i = 0$ and furthermore sets the initial value of the joint variable $d_i = 0$, where d_i will vary when o_i slides up and down parallel to the z_{i-1}-axis.

9, 10—LINKS WITH P⊥R OR P⊥P

$$a_i = \text{const} \qquad \alpha_i = 90 \text{ or } -90\,\text{deg} \qquad d_i = \text{variable} \qquad \theta_i = 0$$

Figure D.9 illustrates a link P⊥R(90) and Figure D.10 illustrates a link P⊥R(−90). If the proximal joint of link (i) is prismatic and its distal joint is either revolute or prismatic with orthogonal, then $\alpha_i = 90\,\text{deg}$ or $\alpha_i = -90\,\text{deg}$, $\theta_i = 0$, $a_i = \text{const}$ is the distance between the joint axes on x_i, and d_i is the only variable parameter. The transformation matrix ${}^{i-1}T_i$ for a link with $\alpha_i = 90\,\text{deg}$ and P⊥R(90) or P⊥P(90) is

$$
{}^{i-1}T_i = \begin{bmatrix} 1 & 0 & 0 & a_i \\ 0 & 0 & -1 & 0 \\ 0 & 1 & 0 & d_i \\ 0 & 0 & 0 & 1 \end{bmatrix} \tag{D.9}
$$

while for a link with $\alpha_i = -90\,\text{deg}$ and P⊥R(−90) or P⊥P(−90) it is

$$
{}^{i-1}T_i = \begin{bmatrix} 1 & 0 & 0 & a_i \\ 0 & 0 & 1 & 0 \\ 0 & -1 & 0 & d_i \\ 0 & 0 & 0 & 1 \end{bmatrix} \tag{D.10}
$$

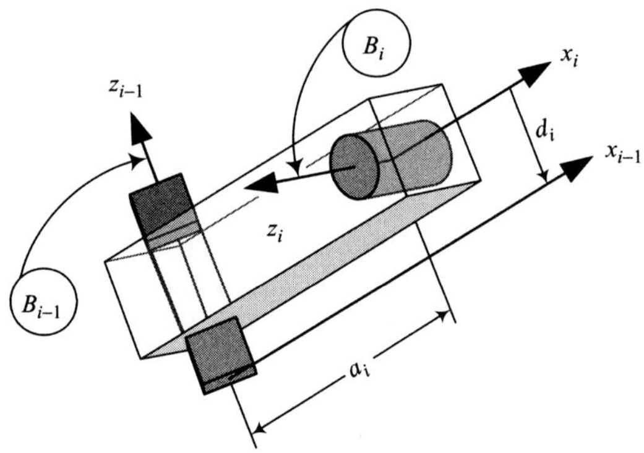

Figure D.9 A P⊥R(90) link.

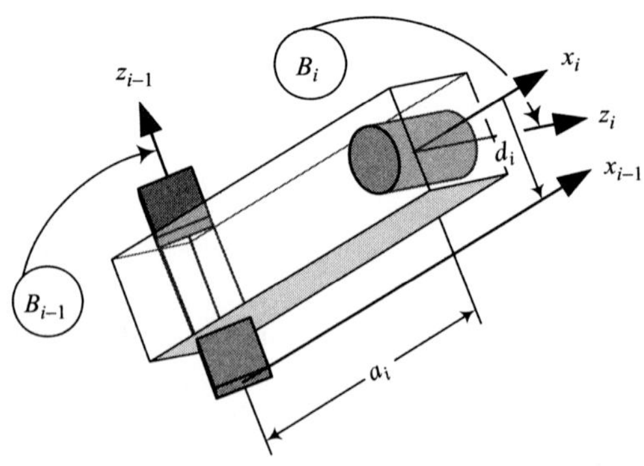

Figure D.10 A P⊥R(−90) link.

11, 12—LINKS WITH P⊢R OR P⊢P

$$a_i = 0 \qquad \alpha_i = 90 \text{ or } -90\,\text{deg} \qquad d_i = \text{variable} \qquad \theta_i = 0$$

Figure D.11 illustrates a link P⊢R(90) and Figure D.12 illustrates a link P⊢R(−90). If the proximal joint of link (i) is prismatic and the distal joint is either revolute or prismatic and the joint axes at two ends are intersecting orthogonal, then $\alpha_i = 90$ deg or $\alpha_i = -90$ deg, $\theta_i = 0$, $a_i = 0$, and d_i is the only variable parameter. The x_i-axis must be perpendicular to the plane of the z_{i-1}- and z_i-axes, and it is possible to have $a_i \neq 0$. Therefore, the transformation matrix ${}^{i-1}T_i$ for a link with $\alpha_i = 90$ deg and P⊢R(90) or P ⊢P(90) is

$$ {}^{i-1}T_i = \begin{bmatrix} 1 & 0 & 0 & 0 \\ 0 & 0 & -1 & 0 \\ 0 & 1 & 0 & d_i \\ 0 & 0 & 0 & 1 \end{bmatrix} \tag{D.11}$$

while for a link with $\alpha_i = -90$ deg and P⊢R(−90) or P⊢P(−90) it is

$$ {}^{i-1}T_i = \begin{bmatrix} 1 & 0 & 0 & 0 \\ 0 & 0 & 1 & 0 \\ 0 & -1 & 0 & d_i \\ 0 & 0 & 0 & 1 \end{bmatrix} \tag{D.12}$$

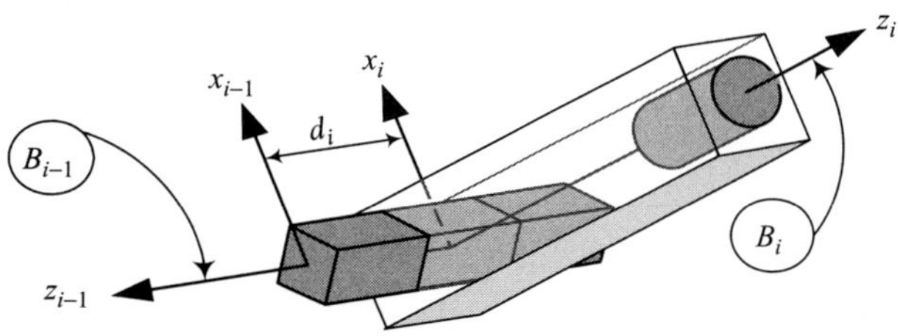

Figure D.11 A P⊢R(90) link.

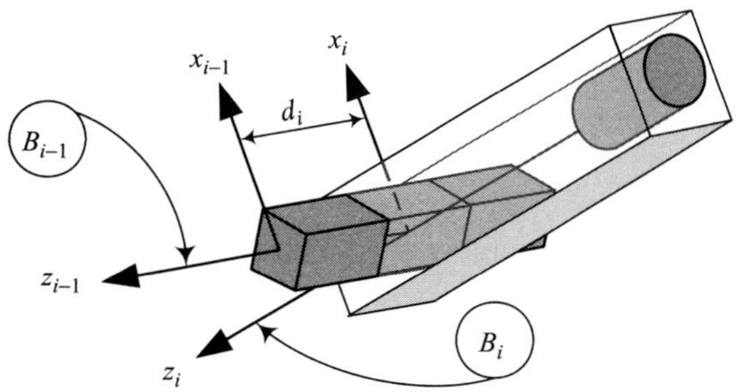

Figure D.12 A P⊢R(−90) link.

E

Trigonometric Formula

Definitions in terms of exponentials:

$$\cos z = \frac{e^{iz} + e^{-iz}}{2} \tag{E.1}$$

$$\sin z = \frac{e^{iz} - e^{-iz}}{2i} \tag{E.2}$$

$$\tan z = \frac{e^{iz} - e^{-iz}}{i\left(e^{iz} + e^{-iz}\right)} \tag{E.3}$$

$$e^{iz} = \cos z + i \sin z \tag{E.4}$$

$$e^{-iz} = \cos z - i \sin z \tag{E.5}$$

Angle sum and difference:

$$\sin(\alpha \pm \beta) = \sin\alpha \cos\beta \pm \cos\alpha \sin\beta \tag{E.6}$$

$$\cos(\alpha \pm \beta) = \cos\alpha \cos\beta \mp \sin\alpha \sin\beta \tag{E.7}$$

$$\tan(\alpha \pm \beta) = \frac{\tan\alpha \pm \tan\beta}{1 \mp \tan\alpha \tan\beta} \tag{E.8}$$

$$\cot(\alpha \pm \beta) = \frac{\cot\alpha \cot\beta \mp 1}{\cot\beta \pm \cot\alpha} \tag{E.9}$$

Symmetry:

$$\sin(-\alpha) = -\sin\alpha \tag{E.10}$$

$$\cos(-\alpha) = \cos\alpha \tag{E.11}$$

$$\tan(-\alpha) = -\tan\alpha \tag{E.12}$$

Multiple angle:

$$\sin(2\alpha) = 2\sin\alpha\cos\alpha = \frac{2\tan\alpha}{1 + \tan^2\alpha} \tag{E.13}$$

$$\cos(2\alpha) = 2\cos^2\alpha - 1 = 1 - 2\sin^2\alpha = \cos^2\alpha - \sin^2\alpha \tag{E.14}$$

$$\tan(2\alpha) = \frac{2\tan\alpha}{1 - \tan^2\alpha} \tag{E.15}$$

$$\cot(2\alpha) = \frac{\cot^2\alpha - 1}{2\cot\alpha} \tag{E.16}$$

$$\sin(3\alpha) = -4\sin^3\alpha + 3\sin\alpha \tag{E.17}$$

$$\cos(3\alpha) = 4\cos^3\alpha - 3\cos\alpha \tag{E.18}$$

$$\tan(3\alpha) = \frac{-\tan^3\alpha + 3\tan\alpha}{-3\tan^2\alpha + 1} \tag{E.19}$$

$$\sin(4\alpha) = -8\sin^3\alpha\cos\alpha + 4\sin\alpha\cos\alpha \tag{E.20}$$

$$\cos(4\alpha) = 8\cos^4\alpha - 8\cos^2\alpha + 1 \tag{E.21}$$

$$\tan(4\alpha) = \frac{-4\tan^3\alpha + 4\tan\alpha}{\tan^4\alpha - 6\tan^2\alpha + 1} \tag{E.22}$$

$$\sin(5\alpha) = 16\sin^5\alpha - 20\sin^3\alpha + 5\sin\alpha \tag{E.23}$$

$$\cos(5\alpha) = 16\cos^5\alpha - 20\cos^3\alpha + 5\cos\alpha \tag{E.24}$$

$$\sin(n\alpha) = 2\sin[(n-1)\alpha)\cos\alpha - \sin((n-2)\alpha] \tag{E.25}$$

$$\cos(n\alpha) = 2\cos[(n-1)\alpha)\cos\alpha - \cos((n-2)\alpha] \tag{E.26}$$

$$\tan(n\alpha) = \frac{\tan[(n-1)\alpha] + \tan\alpha}{1 - \tan((n-1)\alpha)\tan\alpha} \tag{E.27}$$

Half angle:

$$\cos\left(\frac{\alpha}{2}\right) = \pm\sqrt{\frac{1+\cos\alpha}{2}} \tag{E.28}$$

$$\sin\left(\frac{\alpha}{2}\right) = \pm\sqrt{\frac{1-\cos\alpha}{2}} \tag{E.29}$$

$$\tan\left(\frac{\alpha}{2}\right) = \frac{1-\cos\alpha}{\sin\alpha} = \frac{\sin\alpha}{1+\cos\alpha} = \pm\sqrt{\frac{1-\cos\alpha}{1+\cos\alpha}} \tag{E.30}$$

$$\sin\alpha = \frac{2\tan(\alpha/2)}{1+\tan^2(\alpha/2)} \tag{E.31}$$

$$\cos\alpha = \frac{1-\tan^2(\alpha/2)}{1+\tan^2(\alpha/2)} \tag{E.32}$$

Powers of functions:

$$\sin^2\alpha = \tfrac{1}{2}[1 - \cos(2\alpha)] \tag{E.33}$$

$$\sin\alpha\cos\alpha = \tfrac{1}{2}\sin(2\alpha) \tag{E.34}$$

$$\cos^2\alpha = \tfrac{1}{2}[1 + \cos(2\alpha)] \tag{E.35}$$

$$\sin^3\alpha = \tfrac{1}{4}[3\sin(\alpha) - \sin(3\alpha)] \tag{E.36}$$

$$\sin^2\alpha\cos\alpha = \tfrac{1}{4}[\cos\alpha - 3\cos(3\alpha)] \tag{E.37}$$

$$\sin\alpha\cos^2\alpha = \tfrac{1}{4}[\sin\alpha + \sin(3\alpha)] \tag{E.38}$$

$$\cos^3\alpha = \tfrac{1}{4}[\cos(3\alpha) + 3\cos\alpha)] \tag{E.39}$$

$$\sin^4\alpha = \tfrac{1}{8}\,[3 - 4\cos(2\alpha) + \cos(4\alpha)] \tag{E.40}$$

$$\sin^3\alpha\cos\alpha = \tfrac{1}{8}\,[2\sin(2\alpha) - \sin(4\alpha)] \tag{E.41}$$

$$\sin^2\alpha\cos^2\alpha = \tfrac{1}{8}\,[1 - \cos(4\alpha)] \tag{E.42}$$

$$\sin\alpha\cos^3\alpha = \tfrac{1}{8}\,[2\sin(2\alpha) + \sin(4\alpha)] \tag{E.43}$$

$$\cos^4\alpha = \tfrac{1}{8}\,[3 + 4\cos(2\alpha) + \cos(4\alpha)] \tag{E.44}$$

$$\sin^5\alpha = \tfrac{1}{16}\,[10\sin\alpha - 5\sin(3\alpha) + \sin(5\alpha)] \tag{E.45}$$

$$\sin^4\alpha\cos\alpha = \tfrac{1}{16}\,[2\cos\alpha - 3\cos(3\alpha) + \cos(5\alpha)] \tag{E.46}$$

$$\sin^3\alpha\cos^2\alpha = \tfrac{1}{16}\,[2\sin\alpha + \sin(3\alpha) - \sin(5\alpha)] \tag{E.47}$$

$$\sin^2\alpha\cos^3\alpha = \tfrac{1}{16}\,[2\cos\alpha - 3\cos(3\alpha) - 5\cos(5\alpha)] \tag{E.48}$$

$$\sin\alpha\cos^4\alpha = \tfrac{1}{16}\,[2\sin\alpha + 3\sin(3\alpha) + \sin(5\alpha)] \tag{E.49}$$

$$\cos^5\alpha = \tfrac{1}{16}\,[10\cos\alpha + 5\cos(3\alpha) + \cos(5\alpha)] \tag{E.50}$$

$$\tan^2\alpha = \frac{1 - \cos(2\alpha)}{1 + \cos(2\alpha)} \tag{E.51}$$

Products of sin and cos:

$$\cos\alpha\cos\beta = \tfrac{1}{2}\cos(\alpha - \beta) + \tfrac{1}{2}\cos(\alpha + \beta) \tag{E.52}$$

$$\sin\alpha\sin\beta = \tfrac{1}{2}\cos(\alpha - \beta) - \tfrac{1}{2}\cos(\alpha + \beta) \tag{E.53}$$

$$\sin\alpha\cos\beta = \tfrac{1}{2}\sin(\alpha - \beta) + \tfrac{1}{2}\sin(\alpha + \beta) \tag{E.54}$$

$$\cos\alpha\sin\beta = \tfrac{1}{2}\sin(\alpha + \beta) - \tfrac{1}{2}\sin(\alpha - \beta) \tag{E.55}$$

$$\sin(\alpha + \beta)\sin(\alpha - \beta) = \cos^2\beta - \cos^2\alpha = \sin^2\alpha - \sin^2\beta \tag{E.56}$$

$$\cos(\alpha + \beta)\cos(\alpha - \beta) = \cos^2\beta + \sin^2\alpha \tag{E.57}$$

Sum of functions:

$$\sin\alpha \pm \sin\beta = 2\sin\frac{\alpha \pm \beta}{2}\cos\frac{\alpha \pm \beta}{2} \tag{E.58}$$

$$\cos\alpha + \cos\beta = 2\cos\frac{\alpha + \beta}{2}\cos\frac{\alpha - \beta}{2} \tag{E.59}$$

$$\cos\alpha - \cos\beta = -2\sin\frac{\alpha + \beta}{2}\sin\frac{\alpha - \beta}{2} \tag{E.60}$$

$$\tan\alpha \pm \tan\beta = \frac{\sin(\alpha \pm \beta)}{\cos\alpha\cos\beta} \tag{E.61}$$

$$\cot\alpha \pm \cot\beta = \frac{\sin(\beta \pm \alpha)}{\sin\alpha\sin\beta} \tag{E.62}$$

$$\frac{\sin\alpha + \sin\beta}{\sin\alpha - \sin\beta} = \frac{\tan[(\alpha + \beta)/2]}{\tan[(\alpha - \beta)/2]} \tag{E.63}$$

$$\frac{\sin\alpha + \sin\beta}{\cos\alpha - \cos\beta} = \cot\frac{-\alpha + \beta}{2} \tag{E.64}$$

$$\frac{\sin\alpha + \sin\beta}{\cos\alpha + \cos\beta} = \tan\frac{\alpha + \beta}{2} \tag{E.65}$$

$$\frac{\sin\alpha - \sin\beta}{\cos\alpha + \cos\beta} = \tan\frac{\alpha - \beta}{2} \tag{E.66}$$

Trigonometric relations:

$$\sin^2\alpha - \sin^2\beta = \sin(\alpha + \beta)\sin(\alpha - \beta) \tag{E.67}$$

$$\cos^2\alpha - \cos^2\beta = -\sin(\alpha + \beta)\sin(\alpha - \beta) \tag{E.68}$$

$$\begin{aligned} \sin\alpha &= \sqrt{1 - \cos^2\alpha} = \frac{\tan\alpha}{\sqrt{1 + \tan^2\alpha}} \\ &= \frac{1}{\csc\alpha} = \frac{\sqrt{\sec^2\alpha - 1}}{\sec\alpha} = \frac{1}{\sqrt{1 + \cot^2\alpha}} \end{aligned} \tag{E.69}$$

$$\begin{aligned} \cos\alpha &= \sqrt{1 - \sin^2\alpha} = \frac{1}{\sqrt{1 + \tan^2\alpha}} \\ &= \frac{1}{\sec\alpha} = \frac{\sqrt{\csc^2\alpha - 1}}{\csc\alpha} = \frac{\cot\alpha}{\sqrt{1 + \cot^2\alpha}} \end{aligned} \tag{E.70}$$

$$\begin{aligned} \tan\alpha &= \frac{\sin\alpha}{\sqrt{1 - \sin^2\alpha}} = \frac{\sqrt{1 - \cos^2\alpha}}{\cos\alpha} \\ &= \frac{1}{\sqrt{\csc^2\alpha - 1}} = \sqrt{\sec^2\alpha - 1} = \frac{1}{\cot\alpha} \end{aligned} \tag{E.71}$$

Index

F

G

M

Q

R